1990

Mobile DNA

Mobile DNA

Editors

Douglas E. Berg
Department of Microbiology and Immunology
and Department of Genetics
Washington University School of Medicine
St. Louis, Missouri 63110

and

Martha M. Howe
Department of Microbiology and Immunology
The University of Tennessee, Memphis
Memphis, Tennessee 38163

American Society for Microbiology
Washington, D.C.

1913 I Street, N.W.
Washington, DC 20006

Library of Congress Cataloging-in-Publication Data

Mobile DNA.

Includes index.
1. Mobile genetic elements. I. Berg, Douglas E. II. Howe, Martha M.
QH452.3.M63 1989 574.87'3282 88-34982

ISBN 1-55581-005-5

CONTENTS

CONTRIBUTORS†

James W. Ajioka [939] ■ Department of Genetics, Washington University School of Medicine, St. Louis, Missouri 63110-1095

Jorge Almeida [413] ■ Agricultural and Food Research Council, Institute of Plant Science Research, John Innes Institute, Colney Lane, Norwich NR4 7UH, United Kingdom

Frederick W. Alt [693] ■ The Howard Hughes Medical Institute and Departments of Biochemistry and Microbiology, The College of Physicians and Surgeons of Columbia University, 630 W. 168th Street, New York, New York 10032

Alan Barbour [783] ■ Department of Microbiology and Department of Medicine, University of Texas Health Science Center, 7703 Floyd Curl Drive, San Antonio, Texas 78284-7758

Leon Belcour [861] ■ Centre de Génétique Moléculaire du Centre National de la Recherche Scientifique, F-91190 Gif sur Yvette, France

Claire M. Berg [879] ■ Department of Molecular and Cellular Biology, Box U-131, The University of Connecticut, Storrs, Connecticut 06269-2131

Douglas E. Berg [185, 879] ■ Department of Microbiology and Immunology and Department of Genetics, Box 8093, Washington University School of Medicine, 660 S. Euclid, St. Louis, Missouri 63110

Paul M. Bingham [485] ■ Department of Biochemistry, State University of New York, Stony Brook, New York 11794

Ronald K. Blackman [523] ■ Department of Cellular and Developmental Biology, Harvard University, 16 Divinity Avenue, Cambridge, Massachusetts 02138-2097

Jef D. Boeke [335] ■ Department of Molecular Biology and Genetics, Johns Hopkins University School of Medicine, 725 North Wolfe Street, Baltimore, Maryland 21205

Patrick Brown [53] ■ Department of Pediatrics and Department of Biochemistry, Stanford University School of Medicine, Stanford, California 94305

Rosemary Carpenter [413] ■ Agricultural and Food Research Council, Institute of Plant Science Research, John Innes Institute, Colney Lane, Norwich BR4 7UH, United Kingdom

Dana Carroll [567] ■ Department of Biochemistry, University of Utah School of Medicine, Salt Lake City, Utah 84132

Michael Chandler [109] ■ Laboratoire de Génétique du Centre National de la Recherche Scientifique, 118 Route de Narbonne, Toulouse Cedex 31007, France

Robert L. Charlebois [297] ■ Department of Biochemistry, Dalhousie University, Halifax, Nova Scotia, Canada B3H 4H7

Enrico S. Coen [413] ■ Agricultural and Food Research Council, Institute of Plant Science Research, John Innes Institute, Colney Lane, Norwich NR4 7UH, United Kingdom

Stanley N. Cohen [289, 575] ■ Department of Genetics and Department of Medicine, Stanford University School of Medicine, Stanford, California 94305

Michael M. Cox [661] ■ Department of Biochemistry, College of Agricultural and Life Sciences, University of Wisconsin-Madison, Madison, Wisconsin 53706

Nancy L. Craig [211] ■ Department of Microbiology and Immunology, Department of Biochemistry and Biophysics, and the G. W. Hooper Foundation, HSW 1542, University of California, San Francisco, San Francisco, California 94143

Prescott L. Deininger [619] ■ Department of Biochemistry and Molecular Biology, Louisiana State University Medical Center, 1901 Perdido Street, New Orleans, Louisiana 70112

John E. Donelson [763] ■ Department of Biochemistry, University of Iowa, Iowa City, Iowa 52242

† [Brackets indicate beginning page number of respective chapter(s).]

W. Ford Doolittle [297] ■ Department of Biochemistry, Dalhousie University, Halifax, Nova Scotia, Canada B3H 4H7

Bernard Dujon [861] ■ Unité de Génétique Moléculaire des Levures, Département de Biologie Moléculaire, Institut Pasteur, 25 rue du Docteur Roux, F-75724 Paris Cedex 15, France

Marshall H. Edgell [593] ■ Department of Microbiology and Immunology, University of North Carolina at Chapel Hill, Campus Box 7290, FLOB, Chapel Hill, North Carolina 27514

S. Dusko Ehrlich [799] ■ Laboratoire de Génétique Microbienne, Institut des Biotechnologies-INRA, Domaine de Vilvert, 78350 Jouy en Josas, France

William R. Engels [437] ■ Laboratory of Genetics, University of Wisconsin, Madison, Wisconsin 53706

Nina V. Fedoroff [375] ■ Department of Embryology, Carnegie Institute of Washington, 115 West University Parkway, Baltimore, Maryland 21210

D. J. Finnegan [503, 519] ■ Department of Molecular Biology, University of Edinburgh, King's Buildings, Mayfield Road, Edinburgh EH9 3JR, Scotland

Richard A. Firtel [557] ■ Department of Biology, Center for Molecular Genetics, M-034, University of California, San Diego, La Jolla, California 92093

David J. Galas [109] ■ Molecular Biology, University of Southern California, Los Angeles, California 90089-1481

James E. Garrett [567] ■ Institute for Advanced Biomedical Research, Oregon Health Sciences University, Portland, Oregon 97201

William M. Gelbart [523] ■ Department of Cellular and Developmental Biology, Harvard University, 16 Divinity Avenue, Cambridge, Massachusetts 02138-2097

Anna C. Glasgow [637] ■ Department of Biology 147-75, California Institute of Technology, Pasadena, California 91125

Thomas Gridley [927] ■ Whitehead Institute for Biomedical Research, Nine Cambridge Center, Cambridge, Massachusetts 02142, and Department of Biology, Massachusetts Institute of Technology, Cambridge, Massachusetts 02139

Eduardo A. Groisman [879] ■ Department of Molecular Biology, Research Institute of Scripps Clinic, La Jolla, California 92037

Stephen C. Hardies [593] ■ Department of Biochemistry, University of Texas Health Science Center at San Antonio, 7703 Floyd Curl Drive, San Antonio, Texas 78284-7760

Daniel L. Hartl [531, 939] ■ Department of Genetics, Washington University School of Medicine, St. Louis, Missouri 63110-1095

Robert Haselkorn [735] ■ Department of Molecular Genetics and Cell Biology, University of Chicago, 920 East 58th Street, Chicago, Illinois 60637

Barbara Hoffman-Liebermann [575] ■ Department of Biochemistry and Biophysics, University of Pennsylvania School of Medicine, Philadelphia, Pennsylvania 19104

Martha M. Howe ■ Department of Microbiology and Immunology, The University of Tennessee College of Medicine, 858 Madison Avenue, Memphis, Tennessee 38163

Andrew Hudson [413] ■ Agricultural and Food Research Council, Institute of Plant Science Research, John Innes Institute, Colney Lane, Norwich NR4 7UH, United Kingdom

Kelly T. Hughes [637] ■ Department of Biology 147-75, California Institute of Technology, Pasadena, California 91125

Clyde A. Hutchison III [593] ■ Department of Microbiology and Immunology, University of North Carolina at Chapel Hill, Campus Box 7290, FLOB, Chapel Hill, North Carolina 27514

Rudolph Jaenisch [927] ■ Whitehead Institute for Biomedical Research, Nine Cambridge Center, Cambridge, Massachusetts 02142

Amar J. S. Klar [671] ■ Developmental Genetics Laboratory, Building 539, National Cancer Institute Frederick Facility, P.O. Box B, Frederick, Maryland 21701

Nancy Kleckner [227] ■ Department of Biochemistry and Molecular Biology, Harvard University, 7 Divinity Avenue, Cambridge, Massachusetts 02138

Deborah S. Knutzon [567] ■ Department of Biochemistry, University of Utah School of Medicine, Salt Lake City, Utah 84132

J. Michael Koomey [743] ■ Department of Microbiology and Immunology, University of Michigan Medical School, Ann Arbor, Michigan 48109

Arthur Landy [1] ■ Division of Biology and Medicine, Brown University, Providence, Rhode Island 02912

Dan Liebermann [575] ■ Department of Biochemistry and Biophysics, University of Pennsylvania School of Medicine, Philadelphia, Pennsylvania 19104

Daniel D. Loeb [593] ■ Department of Microbiology and Immunology, University of North Carolina at Chapel Hill, Campus Box 7290, FLOB, Chapel Hill, North Carolina 27514

Stuart G. Lutzker [693] ■ Howard Hughes Medical Institute and Departments of Biochemistry and Microbiology, The College of Physicians and Surgeons of Columbia University, 630 W. 168th Street, New York, New York 10032

Mark Meuth [833] ■ Imperial Cancer Research Fund, Clare Hall Laboratories, Blanche Lane, South Mimms, Potters Bar, Hertfordshire EN6 3LD, United Kingdom

D. G. Moerman [537] ■ Department of Zoology, University of British Columbia, Vancouver, British Columbia, Canada V6T 2A9

Ellen Murphy [269] ■ The Public Health Research Institute, 455 First Avenue, New York, New York 10016

Charles A. Omer [289] ■ Central Research and Development Department, Experimental Station, P.O. Box 80228, E. I. DuPont de Nemours and Co., Wilmington, Delaware 19880-0228

Martin L. Pato [23] ■ Department of Molecular and Cellular Biology, National Jewish Center for Immunology and Respiratory Medicine, and Department of Microbiology and Immunology, University of Colorado Health Science Center, Denver, Colorado 80206

Tim P. Robbins [413] ■ Agricultural and Food Research Institute, Institute of Plant Science Research, John Innes Institute, Colney Lane, Norwich NR4 7UH, United Kingdom

W. Ronald Shehee [593] ■ Department of Microbiology and Immunology, University of North Carolina at Chapel Hill, Campus Box 7290, FLOB, Chapel Hill, North Carolina 27514

David Sherratt [163] ■ Institute of Genetics, Glasgow University, Church Street, Glasgow G11 5JS, United Kingdom

Melvin I. Simon [637] ■ Department of Biology 147-75, California Institute of Technology, Pasadena, California 91125

David R. Soll [791] ■ Department of Biology, University of Iowa, Iowa City, Iowa 52242

Philippe Soriano [927] ■ Howard Hughes Medical Institute, Institute for Molecular Genetics and Department of Cell Biology, Baylor College of Medicine, 1 Baylor Plaza, Houston, Texas 77030

John Swanson [743] ■ Laboratory of Microbial Structure and Function, Rocky Mountain Laboratories, National Institute of Allergy and Infectious Diseases, 903 South 4th Street, Hamilton, Montana 59840

John F. Thompson [1] ■ Division of Biology and Medicine, Brown University, Providence, Rhode Island 02912

Harold Varmus [53] ■ Department of Biochemistry and Biophysics and Department of Microbiology and Immunology, HSE 401, University of California, San Francisco, San Francisco, California 94143

R. H. Waterston [537] ■ Department of Genetics, Washington University, St. Louis, Missouri 63110

Meng-Chao Yao [715] ■ Division of Basic Sciences, Fred Hutchinson Cancer Research Center, 1124 Columbia Street, Seattle, Washington 98104

Zuzana Zachar [485] ■ Department of Biochemistry, State University of New York, Stony Brook, New York 11794

Patricia Zambryski [309] ■ Division of Molecular Plant Biology, University of California, Berkeley, California 94720

PREFACE

This book documents the remarkable mobility of DNAs in procaryotic and eucaryotic genomes: the ability of various DNA segments to move to new sites, to invert, and to undergo deletion or amplification, generally without the extensive DNA sequence homology needed for classical recombination. Its chapters describe the variety of mechanisms by which these rearrangements occur, how they are regulated, their biological consequences, and ways in which transposable elements can be exploited as potent research tools. The many examples of mobile DNAs now known provide important exceptions to the general principle that gene and DNA sequence arrangements are stable and transmitted with great fidelity from parents to progeny. Some rearrangements, such as the movement of bacterial antibiotic resistance transposons, variation in surface antigen genes in certain pathogens, and gene amplification in mammalian cells, are typically quite rare, but are brought into prominence by selective pressure in large populations. At the other extreme, antibody gene rearrangements, DNA elimination and amplification in ciliated protozoans, integration of phage λ, and transposition of phage Mu can all be very frequent, but are regulated developmentally or restricted to specific cell types or nuclei.

The reader will be struck by the great diversity of mobile DNAs. *Escherichia coli* K-12 provides a particularly graphic example: the original strain contained one or more copies of at least nine different types of insertion (IS) sequences, a specific inversion system, prophage λ, and several other complete and cryptic prophages. Dozens of additional IS sequences and transposons found in other bacteria can also move in *E. coli* K-12. Spontaneous deletion events, probably involving several mechanisms, can eliminate various bacterial DNA segments including transposable elements. No less striking are the relationships among mobile DNAs in quite disparate organisms. In procaryotes there are many types of elements related to the ampicillin resistance transposon Tn*3*, both in gram-negative and gram-positive bacteria. In eucaryotes, the Ty and *copia* "retrotransposons" of *Saccharomyces cerevisiae* and *Drosophila melanogaster* are closely related to the retroviruses of vertebrates; LINE and SINE elements are found in diverse mammals and, like retrotransposons, use reverse transcriptases to proliferate, although they probably insert by other mechanisms. In addition, the I element of *D. melanogaster* resembles the mammalian LINEs in features of DNA sequence, but is similar to more conventional transposable elements in the regulation of its movement.

The reader will also find great diversity at the mechanistic level. Many events involve DNA breakage and joining, with little or no DNA synthesis, and are well understood, for example: integration and excision of phage λ; the resolution of cointegrates formed during gamma-delta and Tn*3* transposition; and specific DNA inversions that cause flagellar phase variation in *Salmonella typhimurium* and changes in the host range of phage Mu, and which also help drive replication of *S. cerevisiae* 2μm circle DNA. Nonreplicative cut-and-paste mechanisms are apparently involved in the transposition of Tn*5* and Tn*10* in bacteria and *Ac* in maize, and in the insertion of DNA copies of retroviruses and retrotransposons in diverse animal and plant species. Specific DNA breakage underlies macronuclear maturation in protozoa, development of the immune system, and some DNA rearrangements in fungal mitochondria. A paradigm for transposition events intimately linked with DNA replication is provided by the lytic growth of phage Mu. DNA synthesis also figures importantly in many spontaneous deletion and amplification events, mating type interconversion in yeasts, and perhaps changes in surface antigen genes in certain pathogens. The latter two phenomena also involve mechanisms similar to gene conversion seen in classical homologous recombination.

For us and many colleagues, the mobile DNA field is rooted in Barbara McClintock's seminal discovery of transposable controlling elements in maize. Using purely genetic analyses in the 1940s, she demonstrated that there are elements that can transpose to new chromosomal locations, alter the expression of nearby genes, and cause chromosome breakage, all in a developmentally regulated fashion. The prescience of her discoveries is all the more remarkable in that they were made without knowledge of

DNA structure or the current recombinant DNA technologies. They preceded by several decades the discoveries of phage Mu, the bacterial insertion sequences and resistance transposons, and the mobile elements of *D. melanogaster* and other eucaryotes. Much of the initial excitement generated by discoveries of these other elements stemmed from the idea that they might be equivalent to McClintock's intriguing controlling elements. But the current interest in mobile DNA also comes from many other research areas: basic DNA structure, the nature of repetitive DNAs, the causes of mutation, the principles of genome evolution, the control of gene expression, the operation of developmental circuits, the virulence mechanisms of pathogenic microbes, the spread of antibiotic resistance in bacteria, and the emergence of drug resistance in human tumors.

This book is intended to serve scientists with diverse interests and a wide range of experience, from students just discovering the excitement and power of modern molecular genetics to teachers and senior researchers. The chapters detail phenomena from a great range of organisms: humans, their pathogens and parasites, bacteria and their viruses, yeasts and other fungi, protozoans, nematodes, fruit flies, frogs, mice, maize, slime molds, and snapdragons. The chapters differ in length and detail, reflecting variously the sophisticated understanding of intensively studied elements, and the tantalizing early findings of emerging elements or systems. Each chapter provides insights into the current state of our knowledge and points to future research directions.

We are indebted to the authors for their informative, exciting and timely texts; to many colleagues worldwide for helpful critiques of individual chapters; and to members of the ASM Publications Department, Kirk Jensen, Dennis Burke, and Ellie Tupper, for their guidance and superb management of the editing and production process.

Douglas E. Berg
Martha M. Howe

COLOR PLATES

Panel A. Maize kernels exhibiting the phenotypes associated with *Spm* insertion mutations. Kernels shown in the figure carry the wild-type *A* allele of the *a* locus or derivatives of the *a-m1* (b through e) or *a-m2* (f through o) *Spm* insertion alleles of the locus. (a) *A*; (b) *a-m1-5719A1*, no *Spm-s*; (c) *a-m1-5719A1*, with *Spm-s*; (d) *a-m1-6078*, no *Spm-s*; (e) *a-m1-6078*, with *Spm-s*; (f) *a-m2-7995*, no *Spm-s*; (g) *a-m2-7995*, with *Spm-s*; (h) *a-m2-7977B*, with *Spm-s*; (i) *a-m2-8004*, with *Spm-s*; (j) *a-m2-8167B*, with *Spm-s*; (k) *a-m2-8011*, no *Spm-s*; (l) *a-m2-8011*, with *Spm-s*; (m) *a-m2-7991A1*; (n) *a-m2-7991A1 (Spm-i* returning to active state); (o) *a-m2-7995*; preset pattern. (See chapter 14.)

Panel B. *Candida albicans* colonies grown on agar containing phloxine B dye to accentuate white and opaque colonies and sectors. (a) Single opaque and four white colonies; (b) a white colony with two opaque sectors and a narrow intermediate sector; (c) a colony aged on agar after being wrapped with parafilm. (See chapter 37.)

Panel C. Eye color mosaicism in *Drosophila* species, resulting from excision of transposable elements. (a) Eye of a G1 male *Drosophila melanogaster* fly resulting from a P[*ry*(Δ2-3)]♂ × P[*w*(A)]038♀ genetic cross. Red patches correspond to cells retaining the nonautonomous P[*w*(A)] element. Pale yellow-white patches represent clones of cells which sustained early P[*w*(A)] excision events as a result of somatic expression of transposase by the P[*ry*(Δ2-3)] element. Excision and loss of the genetically marked P[*w*(A)] element unmasks the recessive *white* mutant allele on the maternal X chromosome. (See chapter 16; reprinted with permission from F. A. Laski, D. C. Rio, and G. M. Rubin, *Cell* **44**:7–19, 1986 [copyright Cell Press].) (b) A heritable mosaic eye phenotype in *Drosophila simulans* resulting from excision of *mariner* in a *white-peach Mos*/+ fly. (c) Mosaic eye phenotype in *Drosophila simulans*, resulting from excision of *mariner* in a *white-peach* fly lacking the *Mos* factor but showing the maternal-effect excision phenotype that is typical of progeny from a *Mos*/+ female. (See chapter 21; panels b and c reprinted with permission from G. J. Bryan and D. L. Hartl, *Science* **240**:215–217, 1988 [copyright 1988, AAAS].)

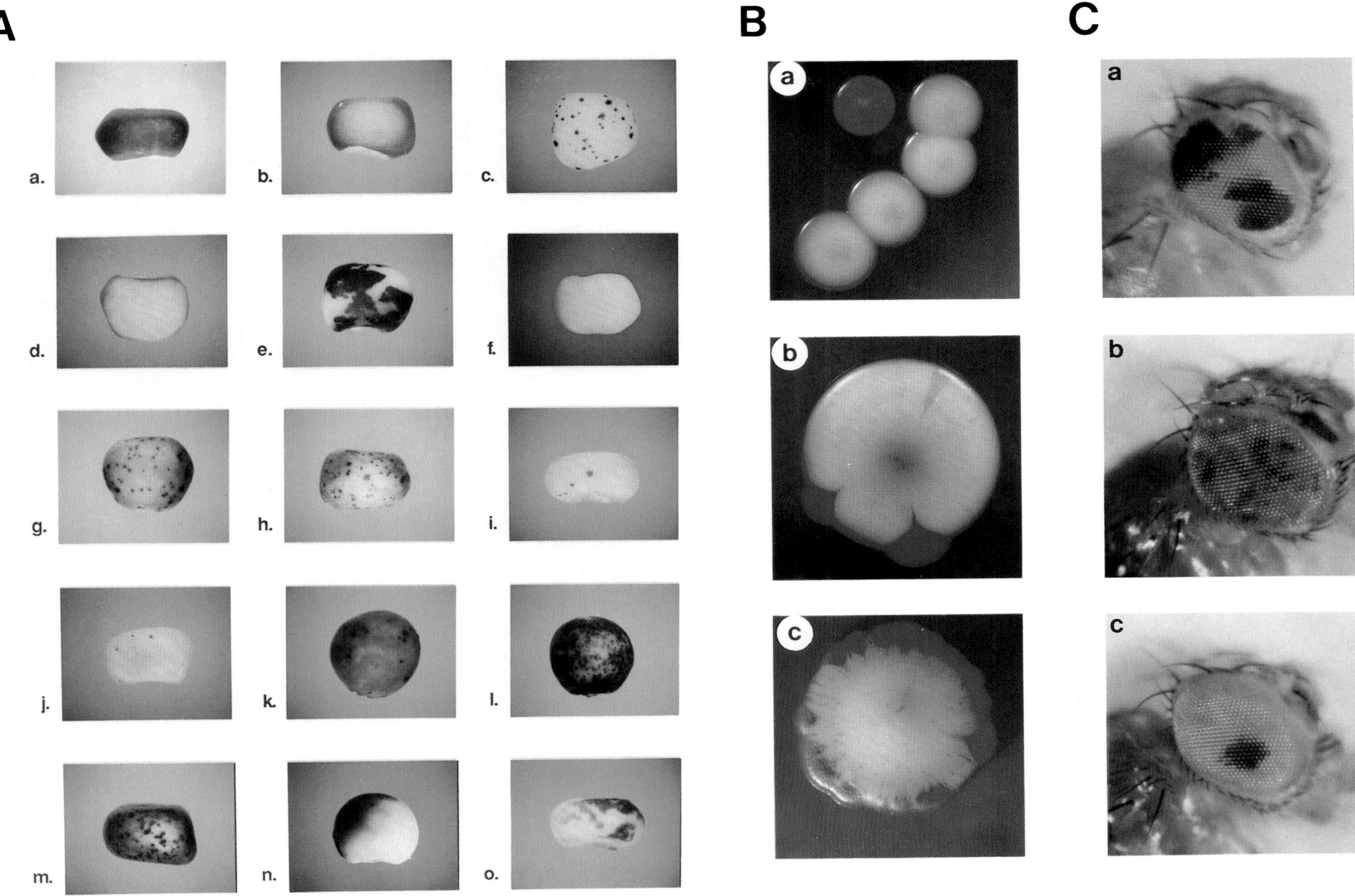
A
a.
b.
c.
d.
e.
f.
g.
h.
i.
j.
k.
l.
m.
n.
o.
B
a
b
c
C
a
b
c

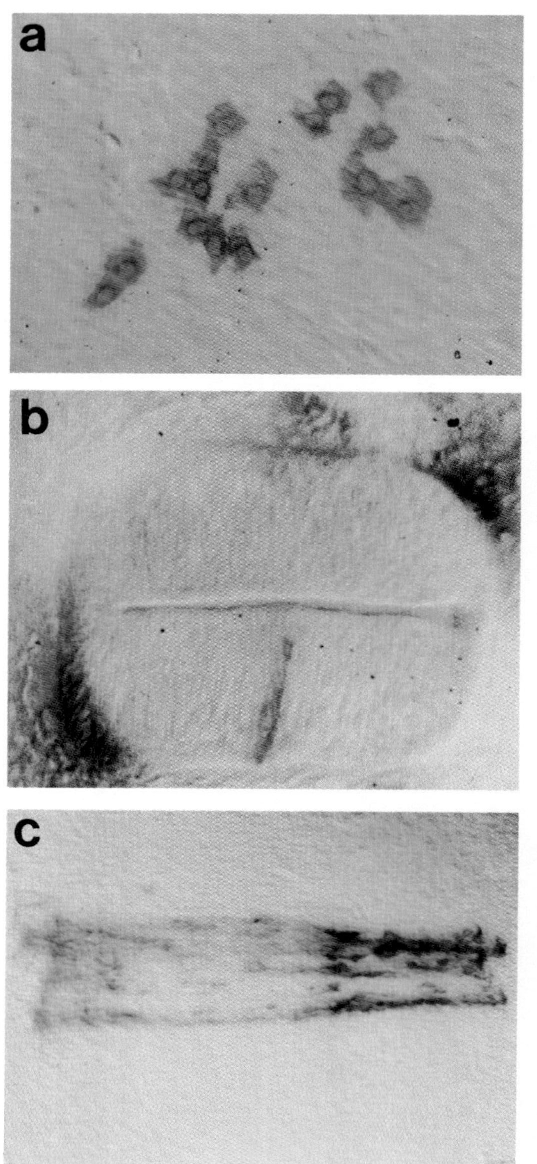
D
a
b
c
d 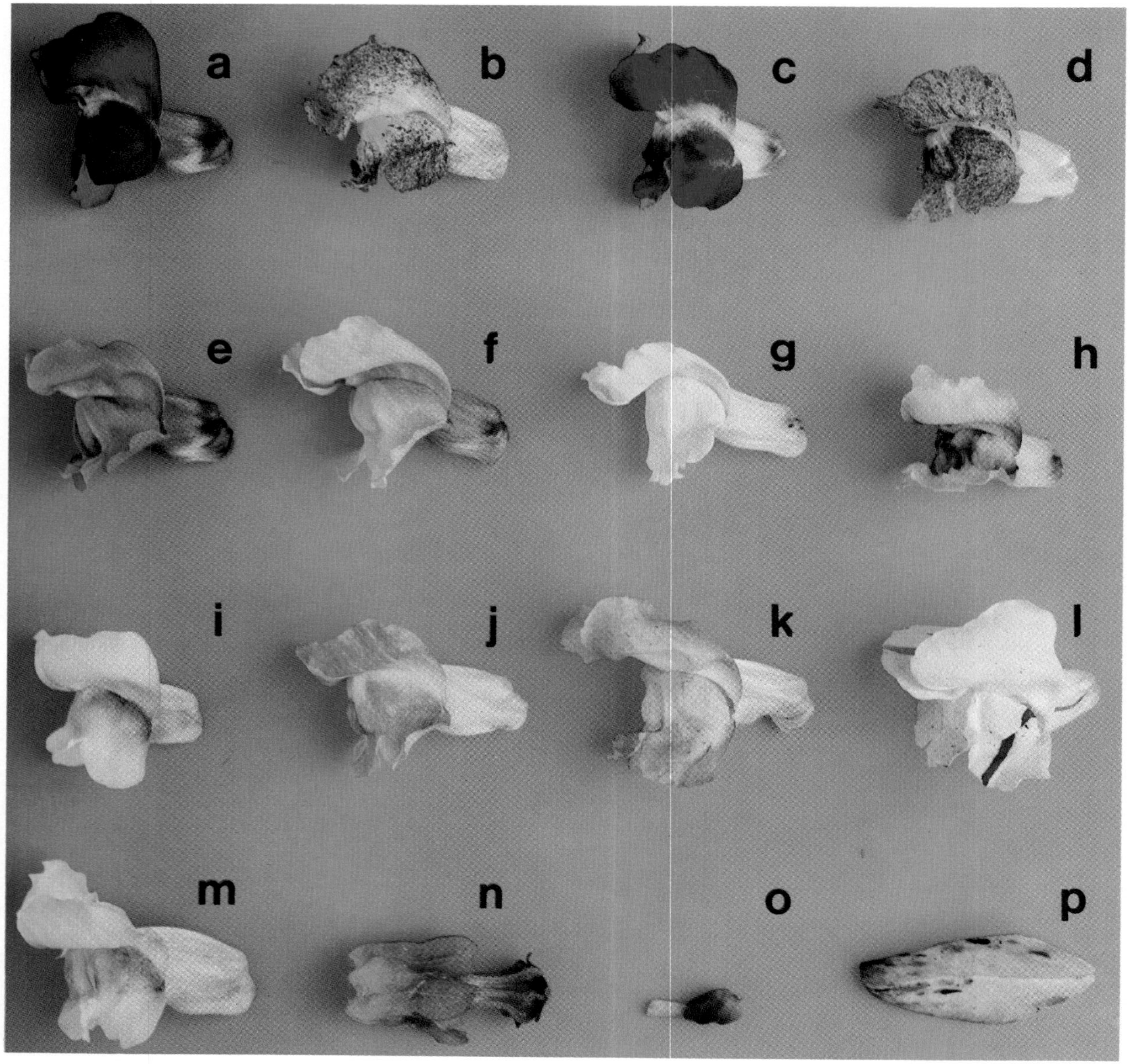
e
f
g
h
i
j
k
l
m
n
o
p
E
a
b
c

Panel D. Examples of the action of transposable elements in *Antirrhinum majus*. Mutations: (a) full red wild type; (b) *pal*^{rec}-2 (the variegated flower studied by Darwin and de Vries); (c) *niv*^{rec}-98; (d) *niv*^{rec}-53, which also carries the *delila* mutation preventing pigmentation of the corolla tube; (e) *pal*-518; (f) *pal*-33; (g) *pal*-15; (h) *pal*-41; (i) *niv*-525; (j) offspring from a cross between a line containing *niv*^{rec}-53 and a line carrying *niv*-44 (both parents carry the *delila* mutation which blocks pigmentation of the corolla tube); (k) *pal*-42; (l) *niv*-540; (m) *niv*-543; (n) *cyloidea*^{radialis}, which also contains the mutations *eosinea* (pink) and *sulfurea* (yellow), giving an overall apricot color; (o) *deficiens*^{globifera}; (p) variegated leaf mutant. (See chapter 15.)

Panel E. Detection of clonally related cells of chicken embryo by using retrovirus-*lacZ*$^+$ vectors injected at different stages during embryonic development. (a) Optical microscope picture of epidermal cells containing a mixture of normal and *lacZ*$^+$ (blue) cells; (b) cross section of spinal cord showing a sector of clonally related *lacZ*$^+$ cells; (c) a radial array of clonally related *lacZ*$^+$ cells in the optic tectum. (See chapter 42; photographs courtesy of G. E. Gray and J. Sanes; panel c reprinted with permission of G. E. Gray, J. C. Glover, J. Majors, and J. R. Sanes [*Proc. Natl. Acad. Sci. USA* **85**:7356–7360, 1988].)

Chapter 1

Regulation of Bacteriophage Lambda Site-Specific Recombination

JOHN F. THOMPSON and ARTHUR LANDY

I. INTRODUCTION

One of the unique features of the bacteriophage lambda site-specific recombination pathway is the complexity and sophistication of its regulation. The regulatory schemes serve to coordinate the recombination reactions with gene expression while allowing cellular input into the developmental decisions. Our focus will be on the manner in which physiological parameters can enter into these decisions, both at the level of the direct intervention of cellular proteins in the reactions as well as at the level of the expression of recombination proteins. We will attempt to incorporate the current understanding of site-specific recombination into that of the overall regulatory circuits of the phage. Other aspects of recombination (40, 42, 129, 153, 182) and the lysis-lysogeny decision (39, 50, 67, 71) have been treated in recent reviews.

After infecting its *Escherichia coli* host, lambda can develop along one of two pathways (shown schematically in Fig. 1). In the lytic pathway, there is an orderly progression of gene expression leading to cell lysis and a phage burst. Alternatively, the lysogenic pathway can be followed: late genes are repressed and the phage DNA is integrated into a specific location on the bacterial chromosome. The lysogenic cell is immune from further phage infection, and the integrated DNA, or prophage, is replicated passively as part of the host chromosome.

John F. Thompson and Arthur Landy ■ Division of Biology and Medicine, Brown University, Providence, Rhode Island 02912.

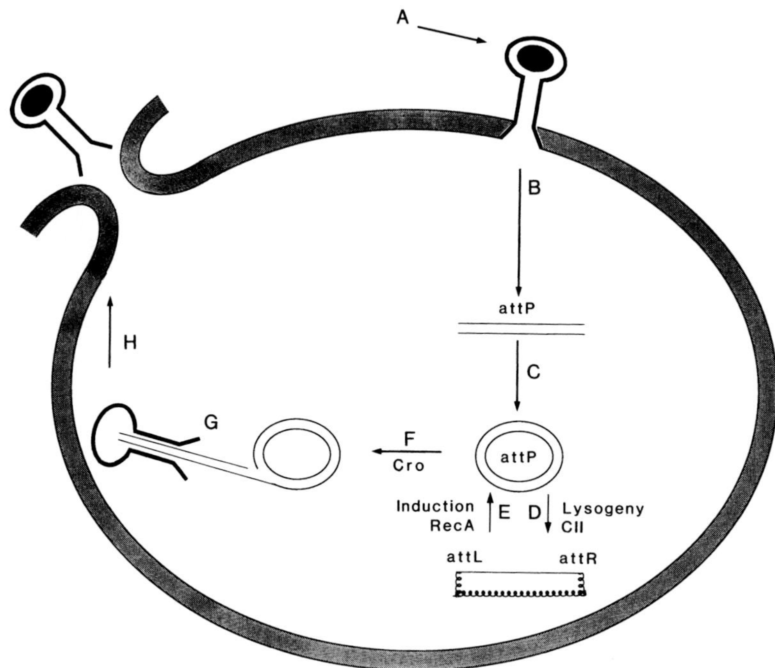

Figure 1. Lambda developmental pathways. A summary of the lambda life cycle and selected proteins is shown. The phage adsorbs to a receptor on the surface of a sensitive cell (A). Linear DNA is injected into the cell (B) and is circularized and supercoiled (C). At this point, either path D or F can be followed. With high concentrations of CII, the lysogenic path (D) is chosen. *attP* is integrated into *attB* and most phage genes are repressed. After many cell generations, the prophage can be induced by RecA (E), regenerating *attP*. The phage DNA is replicated (F) and packaged (G). The host cell is then lysed and progeny phage are released (H).

Upon activation of RecA, lambda repressor (CI) is destroyed, resulting in the expression of lytic and recombination proteins. The prophage is then excised and the lytic pathway resumes.

The ability of lambda and other temperate phage to choose between lysis and lysogeny clearly confers a selective advantage in various circumstances. Lysogeny is favored at high multiplicities of infection, an indication that there are few remaining sensitive cells. By becoming lysogenic, the cell is refractory to further infection yet continues to propagate the phage in its quiescent state. To best exploit this developmental flexibility, it is necessary for the integration and excision reactions to be both highly specific and tightly regulated so that they occur only under the appropriate conditions. This regulation and specificity arises from the direct involvement of multiple host and viral proteins in the recombination reactions as well as the use of many other host and viral genes to control their expression. The role of recombination in regulation of lambda development can occur at two levels: the actual decision-making process or the stabilization of whatever developmental decision is made. We will present arguments that integration stabilizes the lysogenic decision once that pathway is elected based on CII concentration, while, in certain conditions, excision can be one of the factors involved in deciding whether or not lytic development ensues.

Lambda site-specific recombination is a conservative reaction in which DNA replication and high-energy cofactors are not needed. All of the components of the reaction have been purified to homogeneity, and the reactions can be carried out efficiently in vitro. The proteins required for integrative recombination are the virus-encoded integrase (Int) and the bacterially encoded integration host factor (IHF) (128, 131). The excisive recombination requires the same proteins as well as the virus-encoded excisionase (Xis) (4). The bacterially encoded factor for inversion stimulation (FIS) additionally modulates this reaction (174). These components are shown in Fig. 2.

During both recombination reactions, Int cleaves the DNA at specific sites within a 15-base-pair (bp) region of homology (core region) between the recombining partners (33, 123, 132). The energy from the cleavage events is retained by the formation of a transient 3′-phosphotyrosine linkage followed by transfer of the phosphate to a 5′-hydroxyl on another strand (134). This ability to cleave and religate DNA one strand at a time places Int in the class of proteins known as type I topoisomerases (90).

II. RELATION TO OTHER RECOMBINATION SYSTEMS

To classify Int further, sequence comparisons have been performed between Int and other topoisomerases and recombinases (6). The lack of homology between Int and other proteins, even those that are similar in many respects, is striking. Nevertheless, comparisons of a group of recombinases from similar temperate phage revealed a short region near the carboxy terminus in which the amino acids at three positions were completely conserved and those at other positions were moderately conserved (6, 134).

The demonstration of these conserved features with the recombinases from phages lambda, φ80, P1, P2, P4, P22, and 186 (74, 86, 104, 136, 168) has also allowed extension of this paradigm to recombinases from other cellular and plasmid systems (134), including those for the plasmid *oriV*, *E. coli* fimbriae, and the *Saccharomyces cerevisiae* 2μm circle (43, 69, 92, 133). These proteins will be referred to collectively as the "Int family" of recombinases. Not all of these proteins have been fully characterized as to their detailed mode of cutting, but those that have been characterized also proceed via a 3′-phosphotyrosine intermediate (64, 73, 142). Another class of recombinases, including γδ resolvase, Gin, Hin, Cin, and Pin, are apparently unrelated to the Int family and proceed via a 5′-phosphoserine intermediate (70, 145). These proteins are much more highly conserved than the Int family and can even substitute for each other in some instances (137). The classification of recombinases on the basis of their dependence on additional factors, DNA orientation, or topology is not as useful since these characteristics seem to reflect adaptations for the specific regulation and function of a recombination reaction rather than being basic mechanistic properties.

Comparison of the lambda system with other site-specific recombination reactions not only aids in identifying common mechanistic features of the reactions but also allows the regulatory elements unique to lambda development to be discerned. In contrast to the resolvase-related family of DNA invertases, the recombinases related to lambda are diverse as to their function in vivo. Certain phages have recombination systems that are functionally equivalent to those of lambda, allowing them to control their choice between lytic and lysogenic development. Other reactions, such as those from P1 and *oriV*, seem to be important for allowing monomer-size DNA mole-

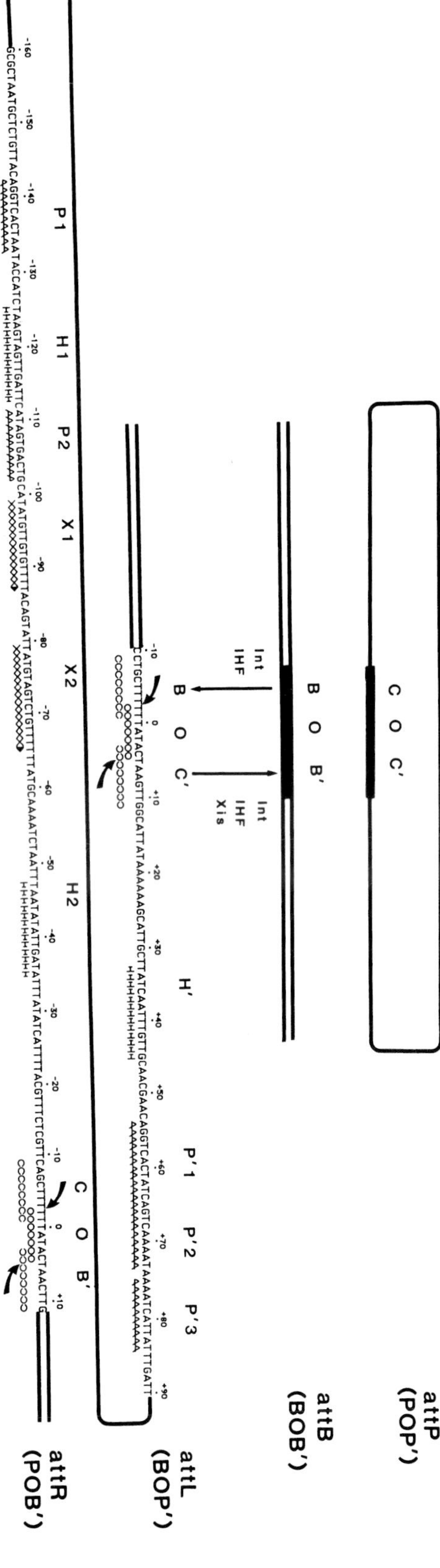

Figure 2. DNA substrates, proteins, and binding sites in lambda site-specific recombination. Supercoiled phage DNA (*attP*; single line) and bacterial DNA (*attB*; double line) undergo integrative recombination to yield *attL* and *attR*, which can undergo excisive recombination to yield *attP* and *attB*, the top-strand DNA sequence is shown (101). The protein-binding sites for IHF (H), Xis (X), arm-type Int (P), and core-type Int (C or B) are numbered from left to right and marked with prime signs if present to the right of the overlap region. The sites of strand exchange in the top strand (down-curved arrow) and omitted bottom strand (up-curved arrow) demark the 7-bp overlap region (O) (123). This figure was adapted from reference 27.

cules to segregate properly during cell division (7). These reactions operate on single-copy DNA circles for which it is critical that newly replicated DNA be separated into monomers. By incorporating a site-specific recombination system, these DNAs can be more readily segregated to daughter cells. Another plasmid system, the *S. cerevisiae* 2μm circle, employs the related recombinase FLP for plasmid amplification (54, 179). The FLP-induced recombination allows the plasmid to overcome the tight regulation on the initiation of DNA replication and to amplify itself to hundreds of copies. Recombination between FLP sites on newly replicated DNA changes the orientation of the replication forks and allows each to replicate the DNA many times (until another recombination event restores the original orientation and permits the forks to meet).

Another related system, the *fimA* inversion reaction, permits the alternate expression of fimbriae in *E. coli* (37, 43, 92). Expression of fimbriae allows *E. coli* to adhere to eucaryotic cells that possess mannose in their outer cell membrane, thereby promoting invasion of epithelial cells. Subsequent inversion of the promoter for this gene turns off fimbria expression and hinders immune system detection of the invading bacteria. The recombinations leading to the on and off states are carried out by the bacterially encoded FimB and FimE proteins, respectively. Both proteins are substantially smaller than the other recombinases in the Int family, but the carboxy termini are clearly homologous with the same region of other proteins in the Int family.

The reactions carried out by Int-family proteins with the simplest substrate requirements and no involvement of accessory factors (P1 and FLP) both are intramolecular and have no apparent need to be coupled with cellular physiology. Thus, the additional complexity of lambda recombination, in terms of both multiple recombinase-binding sites and requirements for accessory factors, is not merely a mechanistic necessity but, rather, is employed to ensure precise reaction between two large DNA molecules and proper regulation with respect to cellular physiology and phage gene expression.

III. COMPONENTS OF THE REACTION

A. DNA

The extent of DNA required for integrative recombination is approximately 240 bp in *attP* and 25 bp in *attB* (79, 124, 125). These lengths are coincident with the outermost protein-binding sites detected by a variety of methods (Fig. 2 and see below). When integration is carried out in vivo or in vitro (with monovalent salt concentrations of 50 to 125 mM and $MgCl_2$ or spermidine), the *attP* DNA must be negatively supercoiled while the *attB* can be linear or supercoiled (122). At higher monovalent salt concentrations in vitro, the reaction will not occur, whereas at lower salt concentrations, the supercoiling requirement is obviated (139). However, the reaction of linear *attP* at low salt concentrations is different from the "normal" reaction in some respects. Exactly how similar these reaction mechanisms are remains to be determined.

The substrates for excisive recombination are *attL* and *attR*. These hybrid *att* sites contain bacterial DNA to the left and right of the overlap region, respectively, with phage DNA on the other side. The DNA required for integration yields products that are fully competent for excision; however, some DNA sequences can be deleted from both *attL* and *attR* without loss of efficiency (27). In contrast to integration, excision does not require supercoiling of either partner (3), but supercoiling of *attL* or *attR* can stimulate the reaction at limiting protein concentrations by enhancing protein binding. As with supercoiling, the excision reaction is less sensitive to salt concentration (139).

The most notable feature of the *attP* sequence is its high AT content (101). The 240-bp minimal *attP* is 73% AT, with one region of 48 bp being 90% AT. Included in these AT-rich regions are several stretches of up to six adenines in a row. This type of sequence has been shown to cause intrinsic curvature when the A tracts are in phase with the helical periodicity (36, 110, 185). Indeed, the *attP* sequence was one of the first shown to have anomalous mobility in polyacrylamide gel electrophoresis (152), one characteristic of curved DNA. The center of curvature has been mapped via the permutation analysis pioneered by Wu and Crothers (185) to −45 to −50 in the P arm (see Fig. 2 for sequence and numbering) (173). This is also the region that was shown to be bromoacetaldehyde modifiable on a supercoiled plasmid (91). Curved DNA has been shown to be easily distorted and to be susceptible to a number of chemical probes when supercoiled (36), so the anomalous mobility and chemical modification are probably manifestations of the same structural feature.

While the intrinsic curvature is readily detectable, it does not appear to be absolutely required for recombination (173). Changing 13 bases near the center of curvature results in an *attP* with the anomalous electrophoretic mobility reduced by half. This *attP* site yields efficient integrative recombination, although the reaction is slower than that of the wild type. Reaction of the analogous *attR* site with *attL* also results in efficient recombination.

B. Structure and Expression of Recombination Proteins

1. Int

The sequence of Int protein has been deduced from the DNA sequence of the *int* gene, leading to a predicted molecular weight of 40,330 (74). This value agrees well with biochemical data obtained via several techniques. Protein sequencing has revealed the posttranslational removal of the amino-terminal methionine (129a), but no other modifications have been detected. Predictions of secondary structure have been attempted by comparison of Int with its related proteins (6). While some features are conserved, secondary structure could be assigned to only a fraction of the protein.

Localization of amino acids responsible for binding to specific DNA sequences has not yet been accomplished. Because the consensus sequence for each of the recombinases varies, one would not expect the residues responsible to be strictly conserved. However, one function is absolutely conserved within this group of proteins: the ability to cleave and ligate DNA. The three absolutely conserved residues, H308, R311, and Y342 (using the numbering of lambda Int), all contain side chains with functional groups that might be useful for such a reaction (6). A combination of chemical and mutagenic studies have established that Y342 is the tyrosine responsible for cleaving the DNA and transiently storing the chemical energy (134). Studies with other proteins in the Int family similarly implicate this tyrosine as being involved in the reaction (73, 142). Models detailing how arginine and histidine might assist in this reaction will have to await additional structural information about the active site.

A second well-characterized mutation in Int involves an E→K change at position 174 (Int-h) (14, 116). Int-h is competent for recombination and has relaxed requirements for IHF and supercoiling of *attP* (102). This mutation is close to the region where homology among the recombinases begins. In the various proteins, however, there are other examples of both E and K at this position (6). A systematic mutational study has not yet been carried out on Int. Other mutations in Int have been identified but not characterized in functional detail (14). Many more mutations are available with the P1 recombinase Cre (183), but it is difficult to extrapolate these results directly onto Int. Experiments to unravel the role of the other 352 amino acids in Int function are under way. Other physical and chemical techniques designed to obtain structural and functional information are being attempted, but complete results have not been obtained.

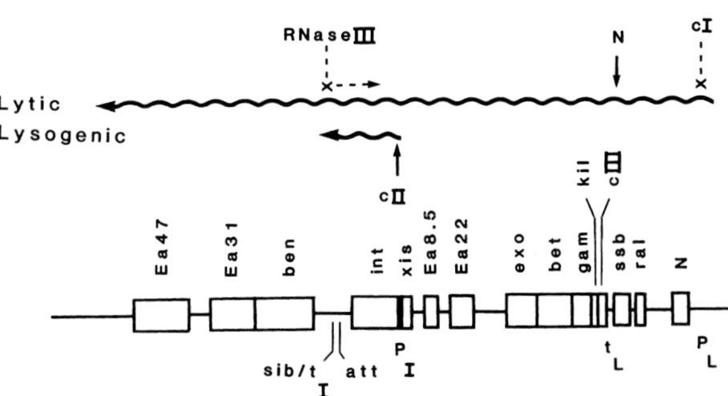

Figure 3. Transcription of *int* and *xis*. The two major transcripts encoding *int* and *xis* are shown. The longer, lytic transcript arises from p_L and is repressed by CI (repressor). When N antiterminates this transcript past t_L and t_I, RNase III acts at the *sib* site and leads to the selective degradation of the *int* gene. The lysogenic transcript starts at p_I, ends at t_I, and codes for *int* but not *xis*.

The *int* gene was located genetically first by using mutant phage that formed abortive lysogens and later by mutations that prevented excision of a prophage (44, 45, 59, 60, 188). Sequence analysis has shown that the *int* gene encompasses bp 28,882 to 27,815 on the lambda chromosome, between the *xis* and *ben* genes (35, 74). It is transcribed in the leftward direction from two promoters, p_I and p_L (157) (Fig. 3). Int protein was first thought to be synthesized only from the p_L transcript. While the *int* gene is transcribed from this promoter in the presence of N protein, Int expression from this RNA is inhibited when the message also contains the *cis*-acting *sib* site downstream of the *att* region. This phenomenon, called retroregulation, occurs when the N-stimulated RNA polymerase reads through an extensive region of dyad symmetry that the p_I-initiated polymerase is unable to traverse (66, 126, 155). When the p_L transcript is released from the DNA, a stable hairpin that is susceptible to RNase III cleavage forms (16, 155). The hairpin at the 3′ end of the p_I transcript is too short to be recognized by RNase III but is long enough to prevent mRNA degradation. The 3′ end of the p_L transcript is unstructured after RNase III cleavage, presumably leading to rapid degradation of the *int* message by cellular nucleases. A truncated transcript that codes for about 85% of the intact protein results (154; G. Plunkett and H. Echols, personal communication). Thus, Int protein is not expressed from p_L upon initial infection but can be expressed from the prophage because the *sib* site is separated from *int* by recombination.

When the host cell is infected, *int* mRNA is

initially transcribed from p_I with positive activation by CII protein. One mutation in p_I increases its match with the consensus promoter and results in constitutive expression from p_I (1, 2, 74, 157) that does not require or respond to CII (20). The transcript initiated from p_I terminates at a simple terminator, t_I, with the resulting transcript coding only for Int.

Because *int* expression from the infecting phage is dependent on positive activation (31, 87), the level of CII plays a major role in determining the level of Int in the cell. *c*II expression is controlled by a complex network of regulatory elements that operates at the levels of transcription, translation, and protein stability. In addition to stimulating *int* expression, CII also activates transcripts from p_{RE} and p_{aQ} that favor the lysogenic response. These transcripts yield repressor and inhibit late gene expression (32, 72, 76, 158, 167). Thus, CII is thought to play the key role in determining whether the lytic or lysogenic pathway is followed (71).

The *c*II gene is transcribed from p_R. This promoter does not require positive activation but can be repressed by both Cro and CI (repressor). *c*II follows *cro* on the p_R mRNA and its expression is stimulated by IHF (78, 109, 115, 135). IHF enhances translation of *c*II and is thought to act by decreasing the pause time for RNA polymerase at t_{R1} (109). t_{R1} is located between *cro* and *c*II and overlaps an IHF-binding site. Pausing of polymerase in this region has been postulated to affect the transcript secondary structure near the *c*II ribosome-binding site. When IHF is present, the ribosome-binding site is exposed and CII protein is translated. This early requirement for IHF allows the lysogenic pathway to be elected only when there is sufficient IHF for the later step of integrative recombination (177). By coupling *c*II expression to the host protein that is later required for successful initiation of the lysogenic pathway, abortive lysogeny is reduced.

The phage-encoded N protein is also required for *c*II expression because it antiterminates transcripts originating from p_R (109). While either N or IHF can stimulate *c*II in vitro, maximal expression is obtained with both proteins (109). In vivo, strains that are mutated in either N or IHF have very low levels of repressor (an indirect indicator of CII levels) (115, 146). While N is required for this step in the lysogenic pathway, it also stimulates transcription of various genes important for the lytic response, so its overall role in the decision between the pathways is not certain (see below).

After translation, CII protein levels are regulated by proteolysis (61, 78), with the chemical half-life of CII varying in different strains (61, 144). Strains that lack a functional Lon protease have accelerated rates of CII degradation, presumably because another protease is present at elevated levels (61), due either to increased synthesis (to compensate for the Lon deficiency) or to increased stability (because of lower Lon levels). Conversely, mutations in the *hflA* and *B* loci lead to stabilization of CII (11, 78, 94). The *hfl* genes (57) are thought to code for a protease involved in CII degradation (11, 78; H. H. Cheng, P. J. Muhlrad, M. A. Hoyt, and H. Echols, personal communication). HflA contains three polypeptides, including subunits of 37,000 and 46,000 molecular weight encoded by *hflA* and a third subunit possibly being a cleavage product of the larger peptide (10; Cheng et al., personal communication). HflB has not been as fully characterized. Several proteins are regulated by *hflA* and *B*, but their identities have not been determined (10, 29). The observation that overproduction of HflA does not affect the lysis-lysogeny decision (10) indicates that HflA-induced degradation is not rate limiting in normal growth conditions. The effect of *hflB* is distinct from that of *hflA* in some respects, so the relation between the two genes is not clear.

Another major determinant of CII (and hence Int) levels is the amount of CIII protein (78, 144, 146). This protein is expressed from the p_L transcript, and its gene is poorly translated due to mRNA secondary structure that blocks the Shine-Dalgarno sequence. Efficient translation is postulated to occur when RNase III binds to but does not cleave the extended secondary structure adjacent to this region (5). CIII protects CII from degradation through an as yet undetermined mechanism (78, 144). CIII similarly protects the unstable heat shock σ^{32} protein (RpoH) from degradation, so the effect is not specific for CII alone (8). σ^{32} is not affected by *hflA* and *B*, so it is not clear whether the same protease is involved in both situations (8). Furthermore, a CIII-like protection can operate with other proteases, as the cellular protein SulB stabilizes SulA from Lon proteolysis (121). Whether CIII (and related proteins) acts by binding directly to one or more proteases or by binding to a conserved region on the susceptible proteins remains to be seen.

CII is metabolically unstable because of its susceptibility to proteolytic cleavage, and this transient nature of expression is accentuated by a phage mechanism that can lead to accelerated degradation of the *c*II message. OOP RNA is transcribed in the antisense orientation and overlaps *c*II (24, 100, 127). When present, OOP RNA hybridizes to the *c*II transcript and leads to its degradation via RNase III-mediated cleavage in the coding region of the RNA (100). Overproduction of OOP results in lambda following the lytic pathway exclusively (171), indicative of lowered CII levels.

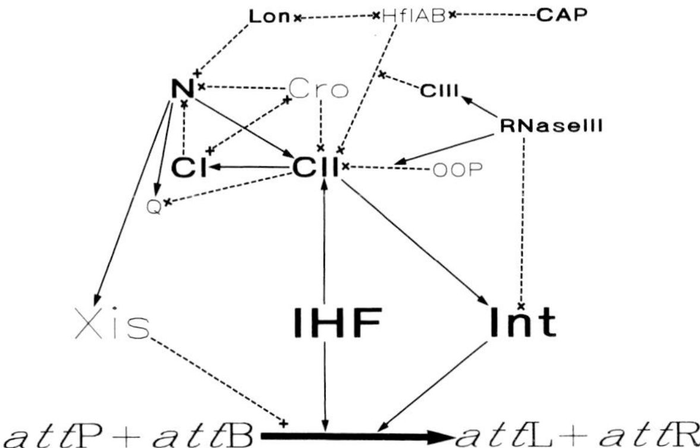

Figure 4. Regulation of proteins that affect integration. The effects of various proteins on gene expression and recombination are diagrammed. Positive effects are shown by solid arrows, and negative effects are indicated by dashed lines. Proteins in boldface type stimulate lysogeny, while those in lightface type stimulate lytic development. This diagram is not all-inclusive (e.g., *c*I and *cro* are autoregulatory and N also stimulates *c*III). Details are described in the text.

In contrast to OOP RNA, most antisense RNAs act by hybridizing to the 5′ end of the target mRNA and blocking the ribosome recognition region (105, 120, 159). The mechanism for OOP inhibition of expression contains aspects similar to the *sib* retroregulation of *int* transcripts, where RNase III cleaves at the 3′ end followed by degradation of the newly exposed RNA. This degradation seems likely to occur most rapidly when there are no translating ribosomes to protect the RNA. Thus, OOP RNA may act by destabilizing inactive transcripts rather than by preventing translation initiation, as do most antisense RNAs. According to this model, OOP RNA plays a role in stabilizing (rather than making) the decision to follow the lytic pathway.

In addition to the first-order effects on *c*II expression described above and summarized in Fig. 4, any cellular or phage protein that affects CI, CIII, RNase III, HflA, HflB, N, IHF, or the myriad other proteins involved will have an indirect effect on *c*II expression. Thus, the regulation of *int* expression is exceedingly complex, but all of this complexity lies at the level of *c*II control. *c*II, in turn, assimilates all of these signals and has a simple effect on the transcription of *int* and the other genes required for execution of the lysogenic response.

2. Xis

As with Int, the primary sequence of Xis has been deduced from its gene (74). The *xis* gene codes for a basic protein of 8,630 daltons, in agreement with the value obtained from sodium dodecyl sulfate-gel electrophoresis measurements (4). Xis has no apparent homology with any other proteins examined (including excisionases from other phage site-specific recombination systems) (103), and there is presently nothing known about its secondary or tertiary structure. Xis is more thermostable than Int and protects Int from thermal inactivation, suggesting a specific interaction between the two proteins even in the absence of DNA (4). Chemical and genetic studies of Xis structure have yet to be carried out. Xis does not seem to have any catalytic activities and appears to exert its effect on recombination by its DNA-binding properties.

The *xis* gene was first identified by selection for mutants that could integrate normally into *attB* but could not be cured (65). Sequence analysis has shown the gene to be located from bp 29,078 to bp 28,863 on the lambda sequence, overlapping the *int* gene and adjacent to an open reading frame of unknown function (35, 74). Like *int*, it is transcribed in the leftward direction (59a). However, since the p_I promoter overlaps the *xis* gene, this *int* transcript is unable to code for *xis* (1, 2, 74). Therefore, *xis* is transcribed only from p_L in the presence of N protein. Xis is regulated posttranslationally by an unknown protease (61). After synthesis, it is rapidly degraded intracellularly (4, 181). This is likely due to a specific cleavage, as Xis is stable once removed from the cell.

3. IHF

IHF consists of two subunits whose genes have been sequenced (47, 113, 117). The predicted molecular weights for the α (11,220) and β (10,580) subunits have been confirmed by sodium dodecyl sulfate-gel electrophoresis (131). IHF is assumed to be a heterodimer, on the basis of genetic evidence, but this has not been shown directly by cross-linking. Sequence comparisons of these subunits with other proteins (47, 113, 117) clearly place it in the type II DNA-binding protein family that includes the major histonelike proteins in *E. coli* (HU) and other bacteria (reviewed in reference 38). The homolog of HU from *Bacillus stearothermophilus* (HBs) has been crystallized, and the structure has been solved to 3-Å (0.3-nm) resolution (172). Because the IHF subunits are so similar in sequence to HBs, the heterodimer can be convincingly superimposed on this structure. The highly conserved, hydrophobic interior of the protein includes all nine phenylalanines present in the two chains. The heterodimer is wedge-shaped, with the narrow end consisting primarily of two alpha helices from each chain. The wider end of the protein has one short alpha helix and three strands of beta sheet from each peptide. Two strands of the beta sheet come out from the protein to form extended arms that are thought to envelop the DNA upon binding. The exact structure of these arms and how the protein conformation might change in the presence of DNA remain to be determined, but this structure is clearly an important framework into which data pertaining to IHF can be integrated.

Two mutations in the β subunit of IHF support the relevance of the HBs structure to IHF (47). A point mutation at residue 49 (*hip-157*) changes a conserved glycine adjacent to a postulated β turn in the interior of the protein to an aspartic acid and completely abolishes activity. In addition, a fusion protein was constructed in which the last four amino acids were altered without any loss of activity. The carboxy terminus of this protein family is variable and exposed to solvent, so the lack of effect on function is not surprising.

More direct evidence for the applicability of the crystal structure of HBs to IHF function has not been obtained, and certain differences between IHF and HBs must be kept in mind. Unlike IHF, HBs contains two identical subunits. This lack of symmetry could be important in IHF binding. More importantly, HBs and most other members of this family bind DNA with little or no sequence specificity. IHF binds with high specificity to its consensus sequence (34), and only at very high concentrations will it interact nonspecifically. While HBs could conceivably interact primarily with the sugar-phosphate backbone, it seems likely that IHF will additionally require base-specific interactions, although structure-specific interactions (bending and twisting of the DNA backbone) may also be important.

The genes for both subunits of IHF were identified and selected for on the basis of their requirement in site-specific recombination (89, 118). Mutations affecting both subunits lead to cells that are viable but have pleiotropic defects in growth and gene expression (see below). The genes have been mapped to min 20 (*hip/himD*) and 37 (*himA*) on the *E. coli* chromosome. Detailed analysis of the transcription of *hip/himD* has not been carried out, but matches to consensus promoter and terminator sequences are present (47). Immediately upstream of *hip/himD* is the gene for ribosomal protein S1 (*rpsA*), followed by a simple terminator. Between *rpsA* and *hip/himD*, there is a good match for a promoter with an IHF-binding site overlapping the −35 region. About half of the *hip/himD* transcripts arise from this promoter, with the remainder presumably coming from the *rpsA* promoter (E. L. Flamm and R. A. Weisberg, personal communication). Following the coding region of *hip/himD* is another simple terminator.

The regulation of *himA* is more complex (Fig. 5).

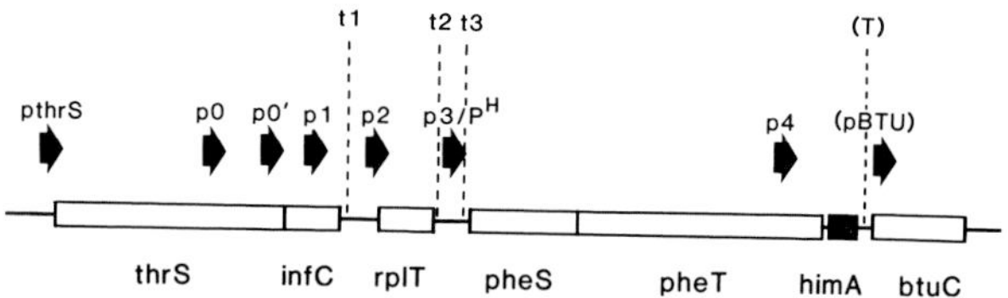

Figure 5. Transcription of *himA*. The genes adjacent to *himA* are shown. Each of the promoters and terminators that have been identified or postulated is indicated (references in text). The putative terminator and promoter downstream of *himA* have been identified only on the basis of homology to consensus sequences.

It is immediately upstream of the *btuCED* operon involved in vitamin B_{12} uptake and is apparently independent of it (51). *himA* is located immediately downstream of the genes encoding the two subunits of phenylalanine tRNA synthetase (*pheST*) and is cotranscribed with them (112, 117). In addition, *pheST* and *himA* are part of an operon (46, 138) that includes genes encoding another tRNA synthetase (*thrS*), an initiation factor (*infC*), and a ribosomal protein (*rplT*). This operon is subject to many regulatory signals (46, 111, 112, 117, 141, 163–165; H. Miller, personal communication). There are promoters located upstream of *thrS* (p_{thrS}) within *thrS* (p_0 and $p_{0'}$), within *infC* (p_1), between *infC* and *rplT* (p_2), between *rplT* and *pheS* (p_3/p_H), and within *pheT* (p_4). The factors affecting initiation at each of these promoters have not been elucidated, but p_{thrS} is apparently under growth rate control (46) and the other promoters may similarly have complex regulatory patterns. In addition, there are terminators downstream of *infC* (t_1), downstream of *rplT* (t_2), and between p_3 and *pheS* (t_3). Each of these terminators is subject to various levels of readthrough, providing potential regulatory sites. There is also an apparent terminator between *himA* and *btuC*.

In vivo mapping of *himA* transcripts indicates that, in the conditions analyzed, most expression arises from the promoters located within *thrS* (p_0 or $p_{0'}$), upstream of *pheS* (p_3/p_H), and within *pheT* (p_4) (111; Miller, personal communication). In order to reach *himA*, the mRNAs from p_0, $p_{0'}$, and p_3/p_H must each traverse terminators. Readthrough of t_2 occurs efficiently both in vivo and in vitro (70%) and has not been observed to be regulated in any fashion (163). t_3 is more efficient and behaves like attenuators found in many amino acid biosynthetic operons (46, 163, 165). A short leader peptide with five phenylalanine residues is located upstream of the terminator. When translated, it allows the formation of an RNA secondary structure that terminates transcription (46). When there is insufficient charged $tRNA^{Phe}$, the leader peptide is not translated and an alternate structure forms that allows readthrough into *pheST* and *himA*. It is not clear whether all transcripts are equally susceptible to this attenuation.

Transcription of *himA* in vitro should be subject to the regulatory effects that are observed in vivo. The observed autoregulation has been proposed to occur by IHF binding to sites overlapping the p_3/p_H and p_4 promoters, leading to repression of transcription, as suggested by the binding and repression seen in vitro (112, 119; Miller, personal communication). Transcription of *himA* (as measured by an operon fusion) is induced in the SOS response, but no LexA-binding site is present upstream of *himA* or *hip/himD*. LexA interacts directly with IHF at p_3/p_H and acts as an antirepressor to stimulate transcription initiation fivefold (Miller, personal communication). The long lag time required for the SOS response to affect IHF transcription (119) suggests that IHF may play a role in the recovery from the SOS response. This is consistent with its apparent role in the recovery from the stationary phase of growth. *himA* mutants grow as well as wild type during exponential phase but stop growing sooner and take longer to recover from poor growth conditions (unpublished results). Other mechanisms for the role of the SOS system in regulating IHF transcription have also been proposed (112). More information in these areas is required before firm conclusions can be drawn.

The physiological role of IHF for *E. coli* is not clear. It both enhances and inhibits the expression of a variety of cellular and viral genes. IHF is required for xylose utilization, flagellar synthesis, isoleucine synthesis from certain precursors, and modulation of *gyrA* (48, 49). Other genes that have pleiotropic effects also have IHF-binding sites in their upstream regulatory regions, but the functional significance of these sites has not been determined. The effect of IHF on gene expression can be greatly magnified by changes in DNA supercoiling. The diversity of genes that IHF affects and the fact that it is not required for growth (and indeed has little or no effect on exponentially growing cells) suggest that IHF may be primarily a modulator of gene expression that "fine tunes" the cell so that energy and resources are used optimally in conditions that are limiting for growth.

4. FIS

FIS is a newly characterized cellular protein that was identified and purified to homogeneity on the basis of its ability to stimulate two site-specific recombination reactions: G inversion in Mu and *hin* inversion in *Salmonella typhimurium* (83–85, 95). It is a homodimer in solution with a subunit molecular weight of 11,240 (82, 95, 95a). Its gene has been sequenced with initial comparisons to other proteins showing homology to a variety of helix-turn-helix-containing DNA-binding proteins (82, 95a). Because the *E. coli* FIS protein can be used to stimulate the *S. typhimurium* Hin reaction (84), the proteins in these two species are assumed to be similar. Like IHF, FIS recovers full activity after heating, but it is not as stable as IHF at extreme pH. FIS has not yet been subjected to biochemical and genetic studies, but with the recent cloning of the gene (82, 95a) these should now be feasible.

The *fis* gene has been mapped near min 72 on the *E. coli* genome, between the *fabE* and *aroE* genes (82, 95a). There are open reading frames immediately adjacent to the *fis* gene, but their identities are unknown. Preliminary results suggest that the upstream gene is cotranscribed with *fis* while the downstream gene is independently transcribed, at least in part (82).

Strains containing a mutation in the *fis* gene are viable, with no gross defects in growth (82, 95a). A *fis* mutant strain with a mutation in *hip/himD* is also viable. FIS has not been shown to have an effect on the expression of any cellular genes, but this situation is likely to change as various systems are examined. FIS undergoes a dramatic drop (at least 70-fold) in intracellular concentration as *E. coli* goes from the exponential to the stationary phase of growth (174). It is not known what controls this degradation, but this observation suggests that FIS may play a role in the regulation of cellular growth.

C. Protein-Binding Sites

Each of the proteins involved in the site-specific recombination reactions binds specifically to one or more locations in the *att* region. These binding sites have been characterized by nuclease protection and chemical modification studies. Nuclease protection experiments allow the region over which the protein resides to be detected, but no information is obtained about which of the protected bases are important for recognition. Dimethyl sulfate (DMS), on the other hand, reacts with only a subset of the nucleotides in a binding region, with only certain positions either protected or enhanced relative to an unbound control DNA. Positions protected by protein binding are thought to be in close contact with the protein and important for recognition. Positions with enhanced reactivity are thought to be more accessible to attack because of deformations in the DNA helix that open one of the grooves (25). Analysis of multiple binding sites with these techniques allows a consensus sequence to be determined for each protein so specific residues can be targeted for mutagenesis and further study.

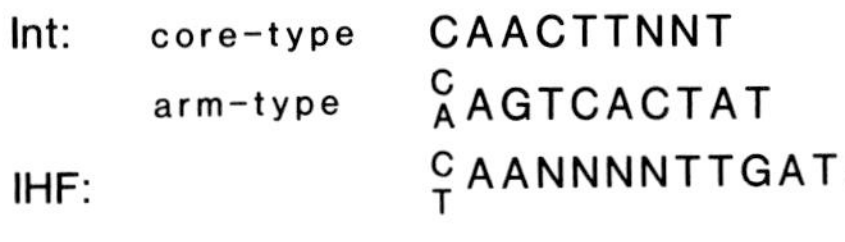

Figure 6. Consensus binding sequences. The consensus recognition sequences for Int (core-type and arm-type sites), IHF, and Xis are shown (see text for references).

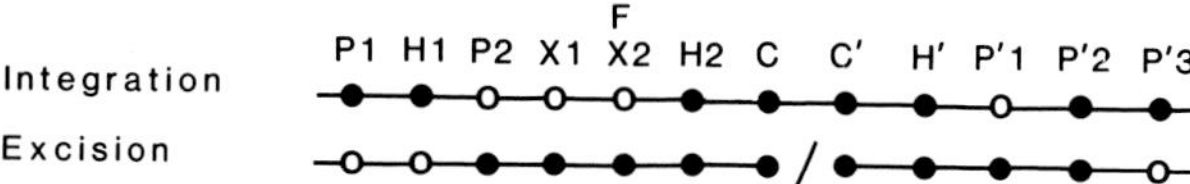

Figure 7. Binding-site requirements for recombination. The protein-binding sites that are required or stimulatory for integration and excision are indicated by filled circles; dispensable sites are indicated by open circles. Uncertainties regarding these assignments are discussed in the text.

1. Int

Nuclease protection studies with Int yielded multiple protected regions, but sequence comparison of these sites resulted in no consistent features among the various regions (151). The uneven size and differential sensitivity of the sites to heparin challenge suggested that all sites were not equivalent. To clarify this, non-*att* DNA was also analyzed for fortuitous binding sites and yielded evidence that there are two distinct sequences involved in binding: "arm-type" sites (distal to the region of strand exchange) and "core-type" sites (adjacent to the points of strand exchange). Several non-*att* sites corresponding to both arm-type and core-type sites have been identified and allow confirmation of the consensus sequences for both types of sites (149, 150) (Fig. 6).

Further evidence for the distinction between arm- and core-type sites can be seen with patterns of DMS modification. Arm-type sites show protection of purines at positions 2–5 and 7–10 in the consensus sequence and enhancement of modification at the completely conserved A6 position (148, 149). Core-type sites show protection of most positions except for enhancements occurring adjacent to the consensus sequences and sometimes at one of the nonspecified positions (150). *N*-Ethylmaleimide modification, which occurs primarily at cysteines, has been shown to affect binding to arm sites more severely than core sites (149).

Binding affinities can be compared among the different arm-type sites with P′1 = P1 > P2 (149, 175, 176). The other P′ sites have not been examined in isolation and are subject to cooperative binding effects when analyzed in the presence of P′1. The P2 site, while having the best match to the consensus sequence, has a lower affinity for Int than do P1 and P′1, reaffirming that the consensus sequence is an average and not necessarily the highest-affinity sequence. Mutagenesis of the P2 site away from the consensus so that it was identical to the P′1 sequence also made it identical in binding affinity (175). However, positions outside the binding sequence can also affect binding affinity. For core-type sites, the order of binding affinity is C′ > C = B > B′ (150). However, this differential affinity is not required for proper recombination, as the relative affinities of these sites can be varied with little or no effect on the reactions (28, 132).

Mutagenesis of Int-binding sites has also been carried out. The only arm-type site mutation found by classical techniques was located in the P′3 site, with the completely conserved C-5 changed to a T. This mutation, referred to as *hen*, resulted in loss of integrative recombination while maintaining excisive recombination (184). Because of the clear change in phenotype generated by this mutation, the same base transition has also been placed in each of the other arm-type sites via in vitro mutagenesis (13). Other, more drastic changes have also been introduced in a subset of the arm-type sites (27, 176). Multiple changes in conserved residues were made because of the highly cooperative nature of protein binding in the *att* site. In some cases, cooperative forces caused by protein-protein interactions can be strong enough to stabilize binding to sites with two changes and a 500-fold loss of binding affinity (56, 177). Thus, interpretation of the phenotype of mutations with less extreme changes must be done carefully when there is no apparent change in function.

In integration, it is clear that the P1, P′2, and P′3 sites are required for reaction while the P2 site is not (13, 27, 176, 184). A point mutation in P′1 does not abolish *attP* function, but Int bound at this site is involved in cooperative interactions with other proteins. Indeed, a low level of Int binding to the mutated site can be detected by nuclease protection (13), so its apparent dispensability should be viewed cautiously. The requirement for the various sites is summarized in Fig. 7.

In excision, none of the point mutations has as large an effect as that seen with integration (13). The largest single effect is with the P′1 mutation, suggesting that this site is required for excision, with some of the defect overcome by cooperative interactions with other sites. *attR* sites with multiple changes in P1 or P2 are functional, indicating that these sites are not absolutely essential (27, 176). However, the P2-mutated *attR* requires more Xis than does wild type for full efficiency, and the reaction between an *attR*-

hen P2 and an *attL-hen* P′2 or *attL-hen* P′3 is defective relative to wild type, suggesting that the P2 site is important for excision, but its effect can be masked by interaction with other sites.

Most mutations in the core region have been useful for analyzing the role of homology and synapsis in recombination rather than the function of core-binding sites (12, 152, 180). Int bound to core participates in strand exchange (33, 132, 134), so these sites are presumed necessary for both recombination reactions.

2. Xis

Nuclease protection studies with Xis revealed a single region of approximately 40 bp that is protected by the protein (186). Xis behaves as a monomer in solution (4), yet its small size relative to the protected area suggests that more than one molecule is involved in the binding. To address this question, protein binding was titrated using electrophoretic gel mobility shifts to assay site occupancy (28). Analysis of these data revealed that cooperative binding between at least two Xis molecules takes place. Mutagenesis of either half of the binding region showed that each could bind Xis independently, albeit with lower affinity (28). Xis concentrations only slightly higher than those necessary to obtain partial protection of single sites resulted in nonspecific protection of DNA, suggesting that it would be difficult to find fortuitous sites in heterologous DNA.

Additional evidence for a two-part binding site was obtained with DMS modification in which a directly repeating pattern of protection was observed (186). No other Xis-binding sites have been found, so the "consensus" binding sequence is of limited value. Mutations in this region include a single-base deletion between the two sites and a series of resected *att* sites in which various extents of the P arm have been replaced with heterologous DNA (27). These mutations result in the loss of excisive recombination.

3. IHF

IHF protects three distinct regions in *attP* from nuclease digestion (34). Other binding sites from a variety of phage, plasmid, and bacterial sources allow the generation of a consensus sequence on the basis of the presence of several highly conserved bases (34, 55, 99, 103). While it is straightforward to align the known IHF-binding sites, the mere presence of the consensus sequence does not guarantee binding. Changes outside the consensus region can decrease the binding affinity by over 100-fold, suggesting that these sequences play a role in binding (L. Moitoso de Vargas and A. Landy, unpublished results). DMS modification of these sites corroborates the importance of the conserved residues with most of these positions protected from attack (34). Enhancement of modification at one conserved position and several nonconserved positions is also observed.

Mutagenesis of each of the *attP* IHF-binding sites has been carried out (56, 177). Two separate 2-bp mutations of highly conserved positions in the H1 site resulted in loss of binding as assayed by nuclease protection, but ambiguous results in functional assays (56, 177). A partial loss of function in integration assays suggested that protein-protein cooperativity was masking the effect of the mutations. To clarify this, changes of 4 and 5 bases in the 9-base consensus were made (56, 177). These constructions demonstrated that the H1 site is required for integration but not excision. The H2 and H′ sites were similarly mutated and found to be required for both recombination reactions (56). However, the nature of these sites may be different in that another host protein (possibly HU) could partially substitute for IHF at these sites (56).

4. FIS

The region protected by FIS is slightly larger than that protected by Xis at the X2 site (174). In addition to the *attP* site, there are also sites in the recombinational enhancer regions of the *gin* family of reactions (68, 83, 95). These sites have been aligned, but show a poor match to the lambda site (26). This contrasts with Int and IHF, in which convincing consensus sequences could be generated when a similar number of sites had been characterized, suggesting that the requirements for FIS binding might be more subtle or might involve specific DNA conformations.

While FIS and Xis both bind to the X2 region, they seem to recognize distinct features, as the DMS modification patterns are completely different (174, 186). Mutagenesis of this region to destroy FIS binding while retaining Xis binding necessitated duplicating the X1 site so the sequence corresponds to X1-X1 rather than X1-X2 (173, 174). This eliminated FIS binding and allowed the effects of FIS and Xis to be separated. FIS has no effect on integration despite its enhancing effect on excision (174).

5. Cooperative interactions

All of the protein-binding sites described above participate in at least one example of cooperative or competitive interactions. Some of these are local in nature, presumably arising from direct contact between neighboring proteins, as with proteins bound at X1 and X2, X1 and F, and X1, X2, and P2 (27, 176, 186). Other contacts result from long-range interactions in higher-order complexes in which proteins bound to widely separated sites on the linear DNA sequence must be folded together in three dimensions. These interactions can be detected by nuclease protection (176) or chemical modification (148) studies as well as by electron microscopy (22, 23). The DNA loops necessary to form these complexes can also be detected by topological analysis of the recombination products (62, 130, 140, 162). These higher-order complexes related to the Int-dependent recombination pathways have been referred to as intasomes (40). The manner in which the linear array of proteins bound to the *att* DNA folds to generate the higher-order structures is not known. Each Int molecule can potentially interact with both a core-type and an arm-type site (149). Unraveling the structure will thus require information about both the protein-protein interactions that occur and how individual Int molecules bind to the DNA.

Long-range interactions have been observed in the *attP* complex in which supercoiling stimulates the binding of Int to the P1 and P′3 sites in a manner that requires other binding sites as well as IHF (148). However, this does not necessarily mean the two sites are adjacent in the complex, as the cooperativity could be mediated by a third protein or through core-binding domains. Higher-order structures requiring Int, IHF, and Xis have been observed with both *attP* and *attR*. With *attP*, binding of IHF to the H1 site is prevented whenever Int and Xis also bind. As the H1 site becomes vacant, the P2 site becomes occupied (176). A similar phenomenon occurs with *attR* with all three proteins bound, but the concentration of Int required to effect the transition in binding from H1 to P2 is much higher (176). The cooperative transition in binding of proteins to the P1, H1, X1, X2, and H2 sites to binding to the P2, X1, X2, and H2 sites occurs over the same range of protein concentrations that is required to overcome IHF inhibition of excision to allow efficient recombination (176, 177).

At one level, the competition between H1 and P2 binding is similar to that observed with Cro and repressor at the o_L and o_R regions: binding of protein to one site blocks another protein from binding to its site. However, with Cro and repressor, competition arises from the recognition sites being on the same region of DNA. In the *att* region, any pairwise combination of proteins can bind, indicating that all the features important for recognition are accessible. The folding of the DNA into a higher-order structure must prevent binding by restricting the space available for each protein or inducing conformational changes in adjacent DNA that make binding unfavorable. This type of three-dimensional competition allows distant sites to interact directly so each protein can play a role in determining recombination efficiency.

IV. REGULATION

Regulation of the integration and excision reactions is critical for the efficient propagation of the phage. The factors important for controlling these events are different for each reaction, with the observed inputs reflecting the varying requirements. In both cases, this necessitates the coordination of information from multiple cellular and phage proteins so that the execution or inhibition of recombination is coupled with the proper expression of a variety of genes. Despite the exposure of *E. coli* to a wide variety of growth conditions in its natural environment, most studies on lambda development have been undertaken in conditions that would not necessarily require the phage's full range of regulatory capabilities. The dominant role of certain regulatory events during exponential growth may be lessened or even supplanted by other processes in conditions not well approximated in normal laboratory study. For this reason, regulatory effects observed only in vitro are discussed in some of the following sections. It is suggested that the subtle nature of these effects on lambda growth explains the lack of detection in vivo in the absence of experiments designed specifically with these models in mind. Indeed, when tested under the appropriate conditions in vivo (C. Ball and R. Johnson, personal communication), FIS was found to be required for excision.

Genetic studies have permitted the placement of most of the regulatory proteins (and RNAs) described into two classes: those that are essential for or stimulate the lysogenic response (CI, CII, CIII, Lon, IHF, CAP, and RNase III) and those that stimulate the lytic response (Cro, HflAB, Q, Xis, and OOP). When proteins within each class interact, the effects are stimulatory, whereas between classes, the interactions are inhibitory. There are two exceptions to this pattern: RNase III, which favors the lytic pathway by inhibiting *c*II and *int* expression but stimulates the lysogenic pathway by allowing *c*III translation, and N, which is required for both pathways. Because mutations in *rnc* (RNase III) lead exclusively to lytic development (5), the enhancing effect on *c*III must predominate at early times during the lysis-lysogeny decision. The RNase III-mediated inhibition of *int* expression would only occur on p_L transcripts after the decision has been made to follow the lytic pathway (or induction). The OOP-mediated RNase III effect on CII probably occurs only when *c*II translation is low and lysogeny is unlikely. RNase III, therefore, serves to regulate the lysis-lysogeny decision via *c*III while stabilizing the lytic response or induction, once those decisions have been made, by facilitating degradation of transcripts for two major components in the lysogenic response. The role of N in the lysis-lysogeny decision is more complex because it affects so many genes. At early times, it probably favors lysogeny by stimulating CIII and CII production relative to Cro. However, it also stimulates expression of *xis* and other genes that favor the lytic response.

A. Integration

For temperate bacteriophage to efficiently take advantage of a lysogenic pathway, there must be a mechanism for ensuring that lysogenic events will not interfere with the lytic pathway when the latter is chosen. Integration of phage DNA during the lytic pathway precludes the formation of phage particles. Lambda avoids this detrimental situation by using Xis to inhibit integration (4, 128). Whenever the lytic pathway is chosen, N protein allows readthrough of the p_L transcript so that Xis is made in abundance and *int* is inhibited by *sib*. On the basis of in vitro results, Xis should inhibit integration so it will not occur once the lytic pathway has been chosen. Additionally, the high Xis and low Int concentrations also prevent the phage DNA from recombining with pseudo-*attB* sites (secondary *att* sites) within the phage or the bacterial chromosome or both. The excised prophage is similarly protected from reintegration once the decision to excise has been made.

For a productive lysogenic cycle to occur, repression of late genes must occur concomitantly with the integration reaction so that late gene expression does not lead to cell lysis without replication of the phage DNA. Conversely, the integration reaction must be stable so that the phage DNA is not lost from the bacterial genome after repression of late genes has been established. Both of these lethal combinations are avoided by intricate genetic control mechanisms centered on the levels of Cro, N, and CII proteins. As discussed earlier, CII levels are subject to many influences from phage and host proteins (71). CII regulates many phage genes so that *int* expression is stimulated concomitant with the synthesis of both CI repressor and an RNA (aQ) that turns off late genes. The involvement of IHF in CII synthesis also ensures that cellular levels of IHF are adequate for integration because the *c*II IHF-binding site is weaker than those in *attP* (177). N levels may also be influenced by physiology as the p_L transcript is processed, and N expression may be affected by translation of a putative leader peptide (166).

While there is the potential for a complex role of cellular physiology in regulating the onset of lysogeny, the control mechanisms that act on *c*II seem designed to be insensitive to the effect of changing growth rate. Variation of the carbon source prior to infection resulted in large changes in growth rate but little change in the frequency of lysogenization (41). Cells growing in a poor carbon source do not divide rapidly, yet those studied were still growing exponentially. When more extreme changes are made in growth conditions, a greater role of physiology is observed. Starvation of cells prior to phage adsorption enhances lysogeny (52, 53, 96, 106). Furthermore, strains with mutations that inactivate the *crp*-cyclic AMP regulatory system (and thus simulate growing cells) have reduced levels of lysogenization (17, 63, 75). Genetic studies suggest this effect is caused by enhanced proteolysis of CII caused by elevated levels of HflA and HflB (11, 29). These results suggest that very poor growth conditions (in which a complete lytic response might not be possible) favor the lysogenic pathway while good growth conditions favor the lytic response.

The other primary modulator of lysogeny found thus far is the multiplicity of infection (96, 106). This allows lambda to conserve its sensitive host population during periods of high phage concentration. The mechanism for multiplicity-of-infection control of lysogeny appears to be mediated by CII and CIII levels (93, 97, 146). The sensitivity of CII to proteolysis is dependent on CIII as well as its initial cellular concentration, with variations in functional half-life of 1.5 to 22 min observed (when overproduced from a plasmid) (144). Without a mechanism to sense the absolute level of CII, the multiplicity of infection would not be effective at regulating lysogeny because the CII/phage ratio would always be constant. Various mechanisms have been proposed for the multiplicity effect (98). Through lysogenization, the phage can infect every sensitive cell yet still preserve a host population that will be able to propagate at later times.

B. Excision

Regulation of prophage excision is vastly different from regulation of integration. During the normal lysogenic state, repressor maintains all phage functions turned off so that the prophage is quiescent. Upon induction of the SOS response and RecA-induced cleavage of repressor, the prophage must rapidly determine the condition of the host cell and determine whether or not induction should proceed. Conceivably, all regulation could occur at the level of active repressor concentration, but the findings discussed below suggest that the regulation is considerably more intricate. In contrast to integrative recombination, where the recombination event is merely a step in lysogeny that occurs after the developmental decision has been made, excisive recombination has the potential to play a more direct role in the induction/quiescence decision.

The first level of regulation in the excision reaction is repressor concentration. Repressor concentration is maintained at a steady state by cooperative interactions at the operator regions of p_L and p_R (81, 156). Binding to o_L, o_{R1}, and o_{R2} represses those two promoters and allows transcription of repressor from p_{RM}. When high concentrations are synthesized, o_{R3} can also be filled, inhibiting further synthesis of repressor until cell division or protein degradation lowers the concentration (80). Because of the cooperative nature of the binding, p_L and p_R can be almost completely repressed (81, 156). Fluctuations in intracellular repressor concentration caused by cell division and nonuniform synthesis must be minimized to prevent spontaneous derepression. This is accomplished through high rates of repressor transcription but relatively low rates of translation (67, 143). In this way, protein is produced continuously at a low rate rather than in bursts. The low frequency of spontaneous phage production attests to the low rate of fluctuation and the tightness of repression (114).

Upon induction of the SOS response by DNA damage, RecA is activated by binding to damaged or single-stranded DNA (reviewed in reference 108). The activated RecA is able to stimulate cleavage of a number of related repressors, including LexA and lambda repressor. This cleavage takes place at a conserved Ala-Gly linkage and functionally inactivates the repressors (77). The cleavage can also occur in the absence of RecA at high pH in vitro (107, 161), suggesting that RecA may assist in the reaction by abstracting a proton from a catalytic amino acid or may allow the repressor to adopt an autocatalytic conformation (58, 161). LexA is cleaved more rapidly than lambda repressor, suggesting that a very mild SOS induction will not result in induction of the prophage (9, 156).

1. Strong SOS response

When a large amount of DNA damage is inflicted on the host cell, the SOS response is rapidly induced and high levels of activated RecA lead to degradation of the lambda repressor. When the concentration of active repressor falls to about 10% of its original value (9), p_R is the first promoter turned on (81). The first gene expressed from this transcript is *cro*. Cro binds to o_{R3} and prevents new synthesis of repressor from p_{RM}. The Cro inhibition of p_{RM} accentuates the derepression and allows *int* and *xis* expression from the p_L transcript. Int and Xis are then translated, with the rate of synthesis dependent on the state of the host cell prior to induction: in growing cells, there will be abundant capacity for transcription and translation and hence large quantities will be synthesized. Cells deprived of nutrients will express these proteins more slowly, but if the SOS response is maintained, sufficient Int and Xis will be synthesized to allow excision of the prophage and continuation of the lytic cycle. Under conditions generating a strong SOS response, the physiology of the cell plays a role only in the timing of the recombination, not in its occurrence. This is shown in Fig. 8 with arbitrary units for repressor concentration and time.

2. Weak SOS response

If the SOS response to mild DNA damage and the resultant drop in repressor concentration are weak, the decision to excise becomes more complicated. Int and Xis cannot be synthesized as rapidly if p_L is only partially derepressed, and the potential for rerepression exists if fluctuations in the operator occupancies allow new synthesis of repressor. Fluctuations of this nature are minimized by cooperative repressor binding and repression by newly synthesized Cro.

It is at this threshold level of repressor cleavage that regulation with respect to cellular physiology becomes important. When there are suboptimal concentrations of Xis for reaction (as would be found in this situation in vivo), FIS stimulates the excision reaction up to 20-fold in vitro (174). Indeed, when Xis concentrations are limiting in vivo, FIS is required for excision (Ball and Johnson, personal communication). These in vivo and in vitro effects occur through interaction of FIS with its binding site on *attR* that allows cooperative binding of Xis to occur simultaneously at the X1 site. Far less Xis is required because only one site must be occupied and binding occurs at a lower concentration than is normally required. Furthermore, FIS levels in vivo change dramatically depending on the growth conditions. In exponentially growing cells, there is sufficient FIS to completely protect the single *attP* site. The levels drop at least 70-fold on going to stationary phase and result in a completely vacant binding site. Thus, FIS

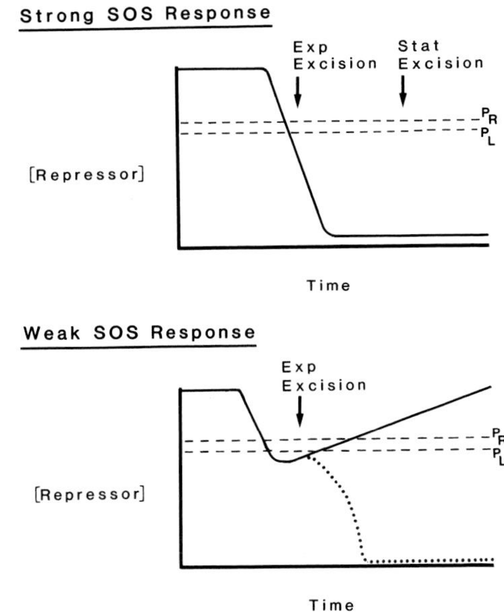

Figure 8. Effect of the SOS response on induction. Two hypothetical curves for intracellular repressor concentration (arbitrary units) are shown as a function of time. When the SOS response is induced, repressor concentration falls below that necessary to repress p_R and p_L (dashed lines), whereupon *int* and *xis* are transcribed. When cells are growing exponentially (Exp), excision takes place rapidly because Int and Xis are translated immediately. In stationary-phase cells (Stat), transcription and translation are less efficient so excision takes longer to occur. With a weak SOS response, repressor concentration could potentially recover, blocking excision (solid line). With exponential-phase cells, this recovery of repressor concentration could be prevented. The DNA degradation initiated by Ben endonuclease would provide a continued stimulation of *recA*, as proposed in the text. In this situation, repressor concentration would continue to fall (dotted line) and allow stable induction to occur.

should be able to stimulate excision in exponentially growing cells but not in stationary-phase cells (174).

IHF also exerts a direct regulatory effect on excision. The H2 and H′ sites must be occupied for excision to occur, while the H1 site must remain vacant or the reaction is inhibited (56, 177). The effect of IHF inhibition is to greatly sensitize the reaction to Int concentration via a series of cooperative and competitive interactions (176, 177). These interactions allow the reaction to go from very low to very high efficiency over a narrow Int range. For example, the efficiency of in vitro excision with a wild-type *attR* goes from 13 to 38% with a twofold increase in Int concentration in one set of conditions (177). Because an *attR* defective in the H1 site (and hence IHF inhibition) requires less Int to begin recombination, an efficiency of 13% is achieved at a far lower Int concentration than with wild type. Going from 13 to 38% recombination requires a 32-fold increase in Int concentration.

The dependence of IHF concentration on growth conditions is less clear than that seen with FIS. An increase in IHF concentration has been detected in stationary-phase cells, but this was found to depend on the conditions used to extract the protein (27, 177). In vivo chemical modification has not shed light on this because all three *attP* sites were found to be fully occupied under the growth conditions tested (174). Variations in concentration above that necessary for complete protection could be taking place but would be undetectable by looking at these sites. Regardless of whether the IHF concentration is changing, however, IHF inhibition is operational in vivo. Because IHF inhibition makes the reaction sensitive to Int concentration, physiological conditions that reduce the rate of Int synthesis will prevent the reaction from occurring (as observed in stationary-phase cells [177]) and potentially allow repressor levels to recover before excision can take place.

During a weak SOS response, FIS stimulation of excision in exponential phase and IHF inhibition of excision in stationary phase complement each other and make recombination much more likely in growing cells. This can be rationalized on the basis of the resources necessary to carry out lytic growth. Healthy cells will have the required protein and DNA synthetic capacity for a proper phage burst, while poorly growing cells might not. Because the weak SOS response is indicative of mild DNA damage that should be readily repairable, it is advantageous for the prophage to remain quiescent if a good phage burst cannot be achieved. This is in contrast to the heavy DNA damage (and resultant strong SOS response) discussed above, where the host cell would be likely to die and the prophage induces regardless of the physiological state of the cell.

While the differential rate of excision in exponential- versus stationary-phase cells has clear potential benefits, the risk of rerepression of the excised phage is especially great with a weak SOS response. Once the decision to excise has been made, it is necessary to ensure that neither recombination nor derepression can be reversed (i.e., the repressor concentration must follow the dotted line rather than the solid line in Fig. 8). To ensure low repressor concentrations, the phage can take advantage of the altered transcription that arises as a result of recombination. Before excision, the p_L transcript extends through the *int* and *xis* genes and into the bacterial DNA beyond *attL*. After recombination, a new set of regulatory signals and genes is introduced into the transcript (Fig. 9).

One such signal transcribed in the excised phage and not in the prophage is the *sib* site (described earlier) that is located just downstream of the *att* region. This region of extensive dyad symmetry is

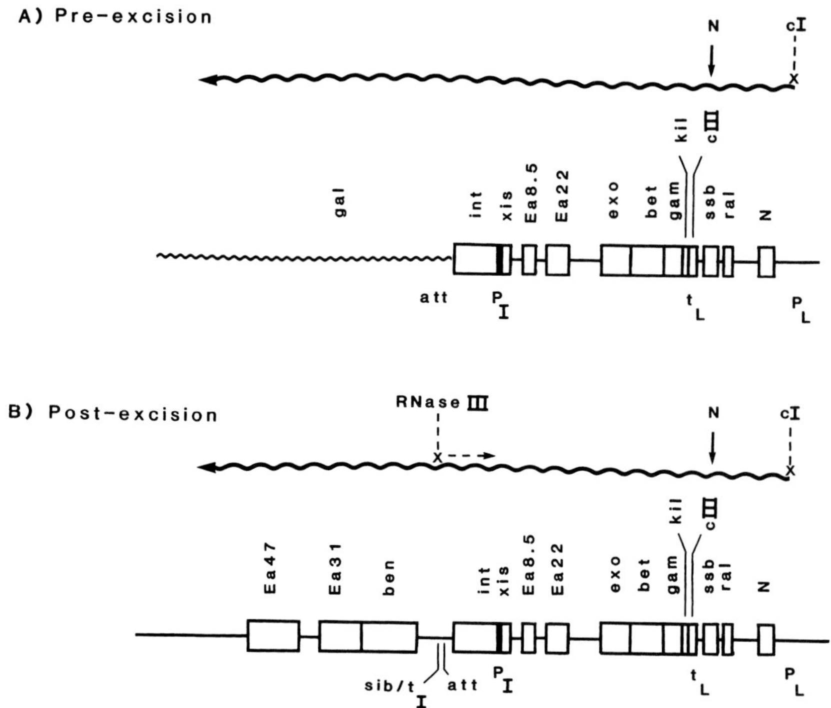

Figure 9. Location of genes and regulatory signals before and after excision. The genes present on the p_L transcripts are shown to scale. In the prophage (A), the bacterial *gal* region is located to the left of *att*. This transcript allows expression of both Int and Xis. After excision (B), phage genes are restored to the left of the *att* site. This places the *sib* site downstream of *int*, leading to its degradation. Also, the *b*-region genes, including *ben*, are expressed after excision.

cleaved by RNase III and causes degradation of the *int* mRNA (16, 66, 126, 155). This reduction in *int* expression immediately after excision, coupled with Xis inhibition of integration (4), makes the excision reaction essentially irreversible.

The excision reaction can be postulated to play a similar role in maintaining derepression. After excision, new genes from the *b* region (further downstream of *sib*) are expressed. The function of these genes has not been determined, but they have no apparent effect on the burst size or growth of the phage (88). Based on the observed enzymatic activity of at least one of the gene products, we are proposing a role for the *b* region in preventing rerepression of lytic genes.

ben is the first gene transcribed downstream of the *att* region and encodes a double-stranded DNA endonuclease (18, 169). Expression of this gene requires N protein and has been observed after prophage induction (18, 169). Ben protein has been purified and studied extensively in vitro (19). It cleaves only supercoiled DNA and requires single-stranded DNA, ATP, and magnesium. Cleavage of supercoiled pBR322 occurs at any of several preferred sites to generate linear products. Other endonucleases encoded in this region have also been identified (15, 30, 147), but these have not been as extensively studied. The differing substrate requirements suggest they are distinct activities.

Any or all of these nucleases could be involved in prolonging the SOS response by cleaving supercoiled domains in the bacterial genome and generating new sites that require repair by the RecA system. After the phage switches to rolling-circle replication, it probably would not require supercoiling and so would be refractory to Ben cleavage. By maintaining the stimulus for activated RecA, lambda repressor would be continuously cleaved and repression of the newly excised phage would be prevented. This phenomenon would not be observed in lysogens with the commonly used temperature-sensitive *c*I857 mutation because it was made from an *ind* mutant stock (170). The repressor made by this allele cannot be cleaved upon RecA induction. In addition, most lysogens with the *S*am7 mutation or *b*-region deletions also contain a mutation or deletion in *ben* (169). There is presently no experimental evidence that directly supports this speculative model, but, if valid, it would not only provide a function for the previously abstruse *b* region but also explain how FIS could enhance excision without a significant loss of phage caused by repression of the excised phage. A number of predictions of this model are experimentally testable.

3. No SOS response

At a frequency of 1 in 10^3 to 10^5 cell generations, the resident prophage spontaneously produces phage in the apparent absence of a signal to trigger the SOS response (although RecA is required for the process) (21, 160). The seemingly random nature and low frequency of this event have limited its study. Some "spontaneous" phage production probably results from DNA damage and a full SOS response in a small fraction of cells. Additionally, a small fraction of cells may produce phage without an SOS response but as a consequence of random fluctuations in repressor concentration. This is one explanation for the observation that overproduction of repressor makes spontaneous phage production undetectable (at least 100-fold lower) (114). In the absence of an SOS response, the derepression could be stabilized by the *ben*-induced RecA cleavage of repressor described above.

Rather than viewing spontaneous phage production as an aberrant reaction, it can be thought of as an important facet of the overall regulation of lambda development. A low rate of spontaneous phage production ensures that all susceptible cells are exposed to free phage while still preserving the bulk of the lysogenic population. In the absence of this process, the phage would be able to escape from the bacterial genome only in the presence of the SOS response when the host cell (and presumably the remainder of the nearby bacterial population) had some degree of DNA damage.

The possible role of FIS in assisting this low-frequency process is not unlike the FIS stimulation of recombination at a rate of 1 in 10^3 to 10^5 cell generations that has also been observed in very different systems (83–85, 95). In bacteriophage Mu, the host range is controlled by a low rate of inversion of tail fibers (178), while in *S. typhimurium*, flagellar antigens are similarly controlled by recombination (187). In both of these cases, as well as spontaneous excision, there are fluctuations in the synthesis of a poorly expressed recombinase followed by an infrequently occurring recombination reaction. The stimulation of all three reactions by FIS suggests that they will occur preferentially in growing cells.

V. CONCLUSIONS

Site-specific recombination has long been appreciated for its role in allowing the phage to be stably integrated in the lysogenic cell. However, this central "housekeeping" function of recombination has overshadowed its potential role in the regulation of the lysogenic pathway. As proposed here, recombination not only helps to regulate the timing of developmental decisions but, just as importantly, can help to ensure that the lysogenic or induction decision is essentially irreversible once it has been made. Furthermore, the cooperative interactions that govern the excision reaction allow it to function as a biological switch that can be coupled with the primary switch mechanism of repressor concentration. These two switches act in concert to allow the prophage to go from stable lysogeny to a fully induced state over a narrow range of conditions. Because of the intricate nature of the control circuits that serve to regulate lambda growth and replication, it is not surprising that the unique aspects of site-specific recombination are exploited to couple the proper signals for gene expression and development.

Acknowledgments. We would like to thank Allan Campbell for discussions that were instrumental in the development of aspects of these regulatory models; Allan Campbell, Hatch Echols, and Bob Weisberg for critical reading of the manuscript; and Hatch Echols, Reid Johnson, Regine Kahmann, Harvey Miller, and Bob Weisberg for communicating results prior to publication. Special thanks are given to Joan Boyles for typing and preparation of figures.

This manuscript was written with support from American Cancer Society grant NP625 (J.F.T.) and Public Health Service grant GM33928 from the National Institutes of Health (A.L.).

LITERATURE CITED

1. **Abraham, J., and H. Echols.** 1981. Regulation of *int* gene transcription by bacteriophage λ: location of the RNA start generated by an *int* constitutive mutation. *J. Mol. Biol.* **146:**157–165.
2. **Abraham, J., D. Mascarenhas, R. Fischer, M. Benedik, A. Campbell, and H. Echols.** 1980. DNA sequence of regulatory region for integration gene of bacteriophage λ. *Proc. Natl. Acad. Sci. USA* 77:2477–2481.
3. **Abremski, K., and S. Gottesman.** 1979. The form of the DNA substrate required for excisive recombination of bacteriophage λ. *J. Mol. Biol.* **131:**637–649.
4. **Abremski, K., and S. Gottesman.** 1982. Purification of the bacteriophage λ *xis* gene product required for λ excisive recombination. *J. Biol. Chem.* **257:**9658–9662.
5. **Altuvia, S., H. Locker-Giladi, S. Koby, O. Ben-Nun, and A. B. Oppenheim.** 1987. RNase III stimulates the translation of the *c*III gene of bacteriophage λ. *Proc. Natl. Acad. Sci. USA* **84:**6511–6515.
6. **Argos, P., A. Landy, K. Abremski, J. B. Egan, E. Haggard-Ljungquist, R. H. Hoess, M. L. Kahn, B. Kalionis, S. V. L. Narayana, L. S. Pierson III, N. Sternberg, and J. Leong.** 1986. The integrase family of site-specific recombinases: regional similarities and global diversity. *EMBO J.* **5:** 433–440.

7. **Austin, S., M. Ziese, and N. Sternberg.** 1981. A novel role for site-specific recombination in maintenance of bacterial replicons. *Cell* **25:**729–736.
8. **Bahl, H., H. Echols, D. B. Straus, D. Court, R. Crowl, and C. P. Georgopoulos.** 1987. Induction of the heat shock response of E. coli through stabilization of σ^{32} by the phage λ CIII protein. *Genes Dev.* **1:**57–64.
9. **Bailone, A., A. Levine, and R. Devoret.** 1979. Inactivation of prophage λ repressor in vivo. *J. Mol. Biol.* **131:**553–572.
10. **Banuett, F., and I. Herskowitz.** 1987. Identification of polypeptides encoded by an *Escherichia coli* locus (*hflA*) that governs the lysis-lysogeny decision of bacteriophage λ. *J. Bacteriol.* **169:**4076–4085.
11. **Banuett, F., M. A. Hoyt, L. McFarlane, H. Echols, and I. Herskowitz.** 1986. *hfl*B, a new *Escherichia coli* locus regulating lysogeny and the level of bacteriophage lambda CII protein. *J. Mol. Biol.* **187:**213–224.
12. **Bauer, C. E., J. F. Gardner, and R. I. Gumport.** 1985. Extent of sequence homology required for bacteriophage lambda site-specific recombination. *J. Mol. Biol.* **181:**187–197.
13. **Bauer, C. E., S. D. Hesse, R. I. Gumport, and J. F. Gardner.** 1986. Mutational analysis of integrase arm-type binding sites of bacteriophage lambda. *J. Mol. Biol.* **192:**513–527.
14. **Bear, S. E., J. B. Clemens, L. W. Enquist, and R. J. Zagursky.** 1987. Mutational analysis of the lambda *int* gene: DNA sequence of dominant mutations. *J. Bacteriol.* **169:** 5880–5883.
15. **Becker, A.** 1970. An endodeoxyribonuclease induced in *E. coli* by infection with λ*pbio* transducing phages. *Biochem. Biophys. Res. Commun.* **41:**63–70.
16. **Belfort, M.** 1980. The *c*II-independent expression of the phage λ *int* gene in RNase III-defective *E. coli. Gene* **11:**149–155.
17. **Belfort, M., and D. Wulff.** 1974. The roles of the lambda *c*III gene and the *Escherichia coli* catabolite gene activation system in the establishment of lysogeny by bacteriophage lambda. *Proc. Natl. Acad. Sci. USA* **71:**779–782.
18. **Benchimol, S., H. Lucko, and A. Becker.** 1982. Bacteriophage λ DNA packaging in vitro. *J. Biol. Chem.* **257:** 5201–5210.
19. **Benchimol, S., H. Lucko, and A. Becker.** 1982. A novel endonuclease specified by bacteriophage λ: purification and properties of the enzyme. *J. Biol. Chem.* **257:**5211–5219.
20. **Benedik, M., D. Mascarenhas, and A. Campbell.** 1982. Probing *c*II and *him*A action at the integrase promoter p_I of bacteriophage lambda. *Gene* **19:**303–311.
21. **Bertani, G.** 1951. The mode of phage liberation by lysogenic *Escherichia coli. J. Bacteriol.* **62:**293–300.
22. **Better, M., C. Lu, R. C. Williams, and H. Echols.** 1982. Site-specific DNA condensation and pairing mediated by the Int protein of bacteriophage λ. *Proc. Natl. Acad. Sci. USA* **79:**5837–5841.
23. **Better, M., S. Wickner, J. Auerbach, and H. Echols.** 1983. Role of the Xis protein of bacteriophage λ in a specific reactive complex at the *att*R prophage attachment site. *Cell* **32:**161–168.
24. **Blattner, F. R., and J. E. Dahlberg.** 1972. RNA synthesis startpoints in bacteriophage λ: are the promoter and operators transcribed? *Nature* (London) *New Biol.* **237:**227–232.
25. **Borowiec, J., and J. D. Gralla.** 1986. High resolution analysis of *lac* transcription complexes inside cells. *Biochemistry* **25:**5051–5057.
26. **Bruist, M. F., A. C. Glasgow, R. C. Johnson, and M. I. Simon.** 1987. Fis binding to the recombinational enhancer of the Hin DNA inversion system. *Genes Dev.* **1:**762–772.
27. **Bushman, W., J. F. Thompson, L. Vargas, and A. Landy.** 1985. Control of directionality in lambda site specific recombination. *Science* **230:**906–911.
28. **Bushman, W., S. Yin, L. L. Thio, and A. Landy.** 1984. Determinants of directionality in lambda site-specific recombination. *Cell* **39:**699–706.
29. **Cheng, H. H., and H. Echols.** 1987. A class of *Escherichia coli* proteins controlled by the *hfl*A locus. *J. Mol. Biol.* **196:**737–740.
30. **Chowdhury, M. R., S. Dunbar, and A. Becker.** 1972. Induction of an endonuclease by some substitution and deletion variants of phage λ. *Virology* **49:**314–318.
31. **Chung, S., and H. Echols.** 1977. Positive regulation of integrative recombination by the *c*II and *c*III genes of bacteriophage λ. *Virology* **79:**312–319.
32. **Court, D., L. Green, and H. Echols.** 1975. Positive and negative regulation by the *c*II and *c*III gene products of bacteriophage lambda. *Virology* **63:**484–491.
33. **Craig, N. L., and H. A. Nash.** 1983. The mechanism of phage λ site-specific recombination: site-specific breakage of DNA by Int topoisomerase. *Cell* **35:**795–803.
34. **Craig, N. L., and H. A. Nash.** 1984. *E. coli* integration host factor binds to specific sites in DNA. *Cell* **39:**707–716.
35. **Daniels, D. L., J. L. Schroder, W. Szybalski, F. Sanger, A. R. Coulson, G. F. Hong, D. F. Hill, G. B. Peterson, and F. R. Blattner.** 1983. Complete annotated lambda sequence, p. 519–676. *In* R. W. Hendrix, J. W. Roberts, F. W. Stahl, and R. A. Weisberg (ed.), *Lambda II.* Cold Spring Harbor Laboratory, Cold Spring Harbor, N.Y.
36. **Diekmann, S., and J. C. Wang.** 1985. On the sequence determinants and flexibility of the kinetoplast DNA fragment with abnormal gel electrophoretic mobilities. *J. Mol. Biol.* **186:**1–11.
37. **Dorman, C. J., and C. F. Higgins.** 1987. Fimbrial phase variation in *Escherichia coli*: dependence on integration host factor and homologies with other site-specific recombinases. *J. Bacteriol.* **169:**3840–3843.
38. **Drlica, K., and J. Rouviere-Yaniv.** 1987. Histone-like proteins of bacteria. *Microbiol. Rev.* **51:**301–319.
39. **Echols, H.** 1986. Bacteriophage λ development: temporal switches and the choice of lysis or lysogeny. *Trends Genet.* **2:**26–30.
40. **Echols, H.** 1986. Multiple DNA-protein interactions governing high-precision DNA transactions. *Science* **233:** 1050–1056.
41. **Echols, H., L. Green, R. Kudrna, and G. Edlin.** 1975. Regulation of phage λ development with the growth rate of host cells: a homeostatic mechanism. *Virology* **65:**344–346.
42. **Echols, H., and G. Guarneros.** 1983. Control of integration and excision, p. 75–92. *In* R. W. Hendrix, J. W. Roberts, F. W. Stahl, and R. A. Weisberg (ed.), *Lambda II.* Cold Spring Harbor Laboratory, Cold Spring Harbor, N.Y.
43. **Eisenstein, B. I., D. S. Sweet, V. Vaughn, and D. I. Friedman.** 1987. Integration host factor is required for the DNA inversion that controls phase variation in *Escherichia coli. Proc. Natl. Acad. Sci. USA* **84:**6506–6510.
44. **Enquist, L. W., and R. A. Weisberg.** 1976. The red plaque test: a rapid method for identification of excision defective variants of bacteriophage λ. *Virology* **72:**147–153.
45. **Enquist, L. W., and R. A. Weisberg.** 1977. A genetic analysis of the *att-int-xis* region of *coli*phage λ. *J. Mol. Biol.* **111:** 97–120.
46. **Fayat, G., J.-F. Mayaux, C. Sacerdot, M. Fromant, M. Springer, M. Grunberg-Manago, and S. Blanquet.** 1983.

Escherichia coli phenylalanyl-tRNA synthetase operon region. *J. Mol. Biol.* **171**:239–261.

47. **Flamm, E. L., and R. A. Weisberg.** 1985. Primary structure of the *hip* gene of *Escherichia coli* and of its product, the β subunit of integration host factor. *J. Mol. Biol.* **183**: 117–128.
48. **Friden, P., K. Voelkel, R. Sternglanz, and M. Freundlich.** 1984. Reduced expression of the isoleucine and valine enzymes in integration host factor mutants of *Escherichia coli. J. Mol. Biol.* **172**:573–579.
49. **Friedman, D. I., E. R. Olson, D. Carver, and M. Gellert.** 1984. Synergistic effects of *him*A and *gyr*B mutations: evidence that *him* functions control expression of *ilv* and *xyl* genes. *J. Bacteriol.* **157**:484–489.
50. **Friedman, D. I., E. R. Olson, C. Georgopoulos, K. Tilly, I. Herskowitz, and F. Banuett.** 1984. Interactions of bacteriophage and host macromolecules in the growth of bacteriophage λ. *Microbiol. Rev.* **48**:299–325.
51. **Friedrich, M. J., L. C. DeVeaux, and R. J. Kadner.** 1986. Nucleotide sequence of the *btuCED* genes involved in vitamin B_{12} transport in *Escherichia coli* and homology with components of periplasmic-binding-protein-dependent transport systems. *J. Bacteriol.* **167**:928–934.
52. **Fry, B. A.** 1959. Conditions of the infection of *Escherichia coli* with lambda phage and the establishment of lysogeny. *J. Gen. Microbiol.* **21**:676–684.
53. **Fry, B. A., and F. Gros.** 1959. The metabolic activities of *Escherichia coli* during the establishment of lysogeny. *J. Gen. Microbiol.* **21**:685–692.
54. **Futcher, A. B.** 1986. Copy number amplification of the 2 μM circle plasmid of *Saccharomyces cerevisiae. J. Theor. Biol.* **119**:197–204.
55. **Gamas, P., M. G. Chandler, P. Prentki, and D. J. Galas.** 1987. *Escherichia coli* integration host factor binds specifically to the ends of the insertion sequence IS1 and to its major insertion hot-spot in pBR322. *J. Mol. Biol.* **195**: 261–272.
56. **Gardner, J. F., and H. A. Nash.** 1986. Role of *Escherichia coli* IHF protein in lambda site-specific recombination. *J. Mol. Biol.* **191**:181–189.
57. **Gautsch, J. W., and D. L. Wulff.** 1974. Fine structure mapping, complementation, and physiology of *Escherichia coli hfl* mutants. *Genetics* **77**:435–448.
58. **Gimble, F. S., and R. T. Sauer.** 1986. λ repressor inactivation: properties of purified *ind*⁻ proteins in the autodigestion and RecA-mediated cleavage reactions. *J. Mol. Biol.* **192**: 39–47.
59. **Gingery, R., and H. Echols.** 1967. Mutants of bacteriophage λ unable to integrate into the host chromosome. *Proc. Natl. Acad. Sci. USA* **58**:1507–1514.

59a. **Gottesman, M. E., and R. A. Weisberg.** 1971. Prophage insertion and excision, p. 113–138. *In* A. D. Hershey (ed.), *The Bacteriophage Lambda.* Cold Spring Harbor Laboratory, Cold Spring Harbor, N.Y.

60. **Gottesman, M. E., and M. Yarmolinsky.** 1968. Integration negative mutants of bacteriophage lambda. *J. Mol. Biol.* **31**:487–505.
61. **Gottesman, S., M. Gottesman, J. E. Shaw, and M. L. Pearson.** 1981. Protein degradation in *E. coli*: the *lon* mutation and bacteriophage lambda N and CII protein stability. *Cell* **24**:225–233.
62. **Griffith, J. D., and H. A. Nash.** 1985. Genetic rearrangement of DNA induces knots with a unique topology: implications for the mechanism of synapsis and crossing-over. *Proc. Natl. Acad. Sci. USA* **82**:3124–3128.
63. **Grodzicker, T., R. R. Arditti, and H. Eisen.** 1972. Establishment of repression by lambdoid phage in catabolite activator protein and adenylate cyclase mutants of *Escherichia coli. Proc. Natl. Acad. Sci. USA* **69**:366–370.
64. **Gronostajski, R. M., and P. D. Sadowski.** 1985. The FLP recombinase of the *Saccharomyces cerevisiae* 2μ plasmid attaches covalently to DNA via a phosphotyrosyl linkage. *Mol. Cell. Biol.* **5**:3274–3279.
65. **Guarneros, G., and H. Echols.** 1970. New mutants of bacteriophage λ. *Virology* **52**:30–38.
66. **Guarneros, G., C. Montanez, T. Hernandez, and D. Court.** 1982. Posttranscriptional control of bacteriophage λ *int* gene expression from a site distal to the gene. *Proc. Natl. Acad. Sci. USA* **79**:238–242.
67. **Gussin, F. N., A. D. Johnson, C. O. Pabo, and R. T. Sauer.** 1983. Repressor and Cro protein: structure, function and role in lysogenization, p. 93–121. *In* R. W. Hendrix, J. W. Roberts, F. W. Stahl, and R. A. Weisberg (ed.), *Lambda II.* Cold Spring Harbor Laboratory, Cold Spring Harbor, N.Y.
68. **Haffter, P., and T. A. Bickle.** 1987. Purification and DNA-binding properties of FIS and Cin, two proteins required for the bacteriophage P1 site-specific recombination system, *cin. J. Mol. Biol.* **198**:579–587.
69. **Hartley, J. L., and J. E. Donelson.** 1980. Nucleotide sequence of the yeast plasmid. *Nature* (London) **286**:860–864.
70. **Hatfull, G. F., and N. D. F. Grindley.** 1986. Analysis of γδ resolvase mutants in vitro: evidence for an interaction between serine-10 of resolvase and site I of *res. Proc. Natl. Acad. Sci. USA* **83**:5429–5433.
71. **Herskowitz, I., and D. Hagen.** 1980. The lysis-lysogeny decision of phage λ: explicit programming and responsiveness. *Annu. Rev. Genet.* **14**:399–445.
72. **Ho, Y. S., and M. Rosenberg.** 1985. Characterization of a third *c*II-dependent, coordinately activated promoter on phage λ involved in lysogenic development. *J. Biol. Chem.* **260**:11838–11844.
73. **Hoess, R. H., and K. Abremski.** 1985. Mechanism of strand cleavage and exchange in the cre-lox site-specific recombination system. *J. Mol. Biol.* **181**:351–362.
74. **Hoess, R. H., C. Foeller, K. Bidwell, and A. Landy.** 1980. Site-specific recombination functions of bacteriophage λ: DNA sequence of regulatory regions and overlapping structural genes for Int and Xis. *Proc. Natl. Acad. Sci. USA* **77**:2482–2486.
75. **Hong, J.-S., G. R. Smith, and B. N. Ames.** 1971. Adenosine 3′:5′-cyclic monophosphate concentration in the bacterial host regulates the viral decision between lysogeny and lysis. *Proc. Natl. Acad. Sci. USA* **68**:2258–2262.
76. **Hoopes, B. C., and W. R. McClure.** 1985. A *c*II-dependent promoter is located within the *Q* gene of bacteriophage λ. *Proc. Natl. Acad. Sci. USA* **82**:3134–3138.
77. **Horii, T., T. Ogawa, T. Nakatani, T. Hase, H. Matsubara, and H. Ogawa.** 1981. Regulation of SOS functions: purification of *E. coli* LexA protein and determination of its specific site cleaved by the RecA protein. *Cell* **27**:515–522.
78. **Hoyt, M., D. Knight, A. Das, H. Miller, and H. Echols.** 1982. Control of phage lambda development by stability and synthesis of CII protein: role of the viral *c*II and host *hfl*A, *him*A and *him*D genes. *Cell* **31**:565–573.
79. **Hsu, P.-L., W. Ross, and A. Landy.** 1980. The λ phage *att* site: functional limits and interaction with Int protein. *Nature* (London) **285**:85–91.
80. **Johnson, A. D., B. J. Meyer, and M. Ptashne.** 1979. Interactions between DNA-bound repressors govern regulation

by the λ phage repressor. *Proc. Natl. Acad. Sci. USA* 76:5061–5065.

81. **Johnson, A. D., A. R. Poteete, G. Lauer, R. T. Sauer, G. K. Ackers, and M. Ptashne.** 1981. λ Repressor and Cro—components of an efficient molecular switch. *Nature* (London) **294**:217–223.
82. **Johnson, R. C., C. A. Ball, D. Pfeffer, and M. I. Simon.** 1988. Isolation of the gene encoding the hin recombinational enhancer binding protein. *Proc. Natl. Acad. Sci. USA* **85**: 3484–3488.
83. **Johnson, R. C., M. F. Bruist, and M. I. Simon.** 1986. Host protein requirements for in vitro site-specific DNA inversion. *Cell* **46**:531–539.
84. **Johnson, R. C., and M. I. Simon.** 1985. Hin-mediated site-specific recombination requires two 26 bp recombination sites and a 60 bp recombinational enhancer. *Cell* **41**:781–791.
85. **Kahmann, R., F. Rudt, C. Koch, and G. Mertens.** 1985. G inversion in bacteriophage Mu DNA is stimulated by a site within the invertase gene and a host factor. *Cell* **41**:771–780.
86. **Kalionis, B., I. F. Dodd, and J. B. Egan.** 1986. Control of gene expression in the P2-related temperate coliphages. III. DNA sequence of the major control region of phage 186. *J. Mol. Biol.* **191**:199–209.
87. **Katzir, N., A. Oppenheim, M. Belfort, and A. B. Oppenheim.** 1976. Activation of the lambda *int* gene by the *c*II and *c*III gene products. *Virology* **74**:324–331.
88. **Kellenberger, G., M. L. Zichichi, and J. Weigle.** 1961. A mutation affecting the DNA content of bacteriophage lambda and its lysogenizing properties. *J. Mol. Biol.* **3**: 399–408.
89. **Kikuchi, A., E. Flamm, and R. A. Weisberg.** 1985. An *Escherichia coli* mutant unable to support site-specific recombination of bacteriophage λ. *J. Mol. Biol.* **183**:129–140.
90. **Kikuchi, Y., and H. A. Nash.** 1979. Nicking-closing activities associated with bacteriophage lambda *int* gene product. *Proc. Natl. Acad. Sci. USA* **76**:3760–3764.
91. **Kitts, P., E. Richet, and H. A. Nash.** 1984. λ Integrative recombination: supercoiling, synapsis, and strand exchange. *Cold Spring Harbor Symp. Quant. Biol.* **49**:735–744.
92. **Klemm, P.** 1980. Two regulatory *fim* genes, *fim*B and *fim*E, control the phase variation of type 1 fimbriae in *Escherichia coli. EMBO J.* **5**:1389–1393.
93. **Knoll, B. J.** 1979. Isolation and characterization of mutations in the *c*III gene of bacteriophage λ which increase the efficiency of lysogenization of Escherichia coli K-12. *Virology* **92**:518–531.
94. **Knoll, B. J.** 1980. Interactions of the λ *c*IIIs1 and the *E. coli hfl*-1 mutations. *Virology* **105**:270–272.
95. **Koch, C., and R. Kahmann.** 1986. Purification and properties of the *Escherichia coli* host factor required for inversion of the G segment in bacteriophage Mu. *J. Biol. Chem.* **261**:15673–15678.
95a. **Koch, C., J. Vandekerckhove, and R. Kahmann.** 1988. *Escherichia coli* host factor for site-specific DNA inversion: cloning and characterization of the *fis* gene. *Proc. Natl. Acad. Sci. USA* **85**:4237–4241.
96. **Kourilsky, P.** 1973. Lysogenization by bacteriophage lambda. *Mol. Gen. Genet.* **122**:183–195.
97. **Kourilsky, P.** 1974. Lysogenization by bacteriophage lambda. II. Identification of genes involved in the multiplicity dependent processes. *Biochimie* **56**:1511–1516.
98. **Kourilsky, P., and A. Knapp.** 1974. Lysogenization by bacteriophage lambda. III. Multiplicity dependent phenomena occurring upon infection by lambda. *Biochimie* **56**: 1517–1523.
99. **Krause, H. M., and N. P. Higgins.** 1986. Positive and negative regulation of the Mu operator by Mu repressor and *Escherichia coli* integration host factor. *J. Biol. Chem.* **261**:3744–3752.
100. **Krinke, L., and D. L. Wulff.** 1987. OOP RNA, produced from multicopy plasmids, inhibits λ *c*II gene expression through an RNase III-dependent mechanism. *Genes Dev.* **1**:1005–1013.
101. **Landy, A., and W. Ross.** 1977. Viral integration and excision: structure of the lambda *att* sites. *Science* **197**: 1147–1160.
102. **Lange-Gustafson, B. J., and H. A. Nash.** 1984. Purification and properties of Int-h, a variant protein involved in site-specific recombination of bacteriophage λ. *J. Biol. Chem.* **259**:12724–12732.
103. **Leong, J. M., S. Nunes-Düby, D. F. Lesser, P. Youderian, M. M. Susskind, and A. Landy.** 1985. The φ80 and P22 attachment sites: primary structure and interaction with *E. coli* integration host factor. *J. Biol. Chem.* **260**:4468–4477.
104. **Leong, J. M., S. E. Nunes-Düby, A. B. Oser, C. F. Lesser, P. Youderian, M. M. Susskind, and A. Landy.** 1986. Structural and regulatory divergence among site-specific recombination genes of lambdoid phage. *J. Mol. Biol.* **189**:603–616.
105. **Liao, S.-M., T.-H. Wu, C. H. Chiang, M. M. Susskind, and W. R. McClure.** 1987. Control of gene expression in bacteriophage P22 by a small antisense RNA. *Genes Dev.* **1**: 197–203.
106. **Lieb, M.** 1953. The establishment of lysogenicity in *Escherichia coli. J. Bacteriol.* **65**:642–651.
107. **Little, J. W.** 1984. Autodigestion of LexA and phage λ repressors. *Proc. Natl. Acad. Sci. USA* **81**:1375–1379.
108. **Little, J. W., and D. W. Mount.** 1982. The SOS regulatory system of *Escherichia coli. Cell* **29**:11–22.
109. **Mahajna, J., A. B. Oppenheim, A. Rattray, and M. Gottesman.** 1986. Translation initiation of bacteriophage lambda gene *c*II requires integration host factor. *J. Bacteriol.* **165**: 167–174.
110. **Marini, J. C., S. D. Levene, D. M. Crothers, and P. T. Englund.** 1982. Bent helical structure in kinetoplast DNA. *Proc. Natl. Acad. Sci. USA* **79**:7664–7668.
111. **Mayaux, J.-F., G. Fayat, M. Fromant, M. Springer, M. Grunberg-Manago, and S. Blanquet.** 1983. Structural and transcriptional evidence for related *thr*S and *inf*C expression. *Proc. Natl. Acad. Sci. USA* **80**:6152–6156.
112. **Mechulam, Y., S. Blanquet, and G. Fayat.** 1987. Dual level control of the *Escherichia coli phe*ST-*him*A operon expression. *J. Mol. Biol.* **197**:453–470.
113. **Mechulam, Y., G. Fayat, and S. Blanquet.** 1985. Sequence of the *Escherichia coli pheST* operon and identification of the *himA* gene. *J. Bacteriol.* **163**:787–791.
114. **Melechen, N. E., G. Go, and H. A. Lozeron.** 1978. Effect of CI repressor level on thymineless and spontaneous induction; specificity of lambda RNA transcription. *Mol. Gen. Genet.* **163**:213–221.
115. **Miller, H.** 1981. Multi-level regulation of bacteriophage λ lysogeny by the *E. coli him*A gene. *Cell* **25**:269–276.
116. **Miller, H., M. A. Mozola, and D. I. Friedman.** 1980. *int*-h: an *int* mutation of phage λ that enhances site-specific recombination. *Cell* **20**:721–729.
117. **Miller, H. I.** 1984. Primary structure of the *himA* gene of *Escherichia coli*: homology with DNA-binding protein HU and association with the phenylalanyl-tRNA synthetase operon. *Cold Spring Harbor Symp. Quant. Biol.* **49**:691–698.

118. **Miller, H. I., and D. I. Friedman.** 1980. An *E. coli* gene product required for λ site-specific recombination. *Cell* **20:**711–719.

119. **Miller, H. I., M. Kirk, and H. Echols.** 1981. SOS induction and autoregulation of the *him*A gene for site-specific recombination in *Escherichia coli. Proc. Natl. Acad. Sci. USA* **78:**6754–6758.

120. **Mizuno, T., M.-Y. Chou, and M. Inouye.** 1984. A unique mechanism regulating gene expression: translational inhibition by a complementary RNA transcript (micRNA). *Proc. Natl. Acad. Sci. USA* **81:**1966–1970.

121. **Mizusawa, S., and S. Gottesman.** 1983. Protein degradation in *Escherichia coli*: the *lon* gene controls the stability of SulA protein. *Proc. Natl. Acad. Sci. USA* **80:**358–362.

122. **Mizuuchi, K., M. Gellert, and H. Nash.** 1978. Involvement of super-twisted DNA in integrative recombination of bacteriophage lambda. *J. Mol. Biol.* **121:**375–392.

123. **Mizuuchi, K., R. Weisberg, L. Enquist, M. Mizuuchi, M. Buraczynska, C. Foeller, P. L. Hsu, W. Ross, and A. Landy.** 1980. Structure and function of the phage lambda *att* site: size, Int-binding sites and location of the crossover point. *Cold Spring Harbor Symp. Quant. Biol.* **45:**429–437.

124. **Mizuuchi, M., and K. Mizuuchi.** 1980. Integrative recombination of bacteriophage λ: extent of the DNA sequence involved in attachment site function. *Proc. Natl. Acad. Sci. USA* **77:**3220–3224.

125. **Mizuuchi, M., and K. Mizuuchi.** 1985. The extent of DNA sequence required for a functional bacterial attachment site of phage lambda. *Nucleic Acids Res.* **13:**1193–1208.

126. **Montanez, C., J. Bueno, U. Schmeissner, D. L. Court, and G. Guarneros.** 1986. Mutations of bacteriophage lambda that define independent but overlapping RNA processing and transcription termination sites. *J. Mol. Biol.* **191:**29–37.

127. **Moore, D., K. Denniston-Thompson, K. Kruger, M. Furth, B. Williams, D. Daniels, and F. Blattner.** 1979. Dissection and comparative anatomy of the origins of replication of lambdoid phages. *Cold Spring Harbor Symp. Quant. Biol.* **43:**155–163.

128. **Nash, H. A.** 1975. Integrative recombination of bacteriophage lambda DNA in vitro. *Proc. Natl. Acad. Sci. USA* **72:**1072–1076.

129. **Nash, H. A.** 1981. Integration and excision of bacteriophage lambda: the mechanisms of conservative site specific recombination. *Annu. Rev. Genet.* **15:**143–167.

129a. **Nash, H. A.** 1981. Site-specific recombination protein of phage lambda, p. 471–480. *In* P. D. Boyer (ed.), *The Enzymes*, vol. 14A. Academic Press, Inc., New York.

130. **Nash, H. A., and T. J. Pollock.** 1983. Site-specific recombination of bacteriophage lambda: the change in topological linking number associated with exchange of DNA strands. *J. Mol. Biol.* **170:**19–38.

131. **Nash, H. A., and C. A. Robertson.** 1981. Purification and properties of the *Escherichia coli* protein factor required for λ integrative recombination. *J. Biol. Chem.* **256:**9246–9253.

132. **Nunes-Düby, S. E., L. Matsumoto, and A. Landy.** 1987. Site-specific recombination intermediates trapped with suicide substrates. *Cell* **50:**779–788.

133. **O'Connor, M. B., J. J. Kilbane, and M. H. Malamy.** 1986. Site-specific and illegitimate recombination in the *oriV1* region of the F factor. *J. Mol. Biol.* **189:**85–102.

134. **Pargellis, C. A., S. E. Nunes-Düby, L. Moitoso de Vargas, and A. Landy.** 1988. Suicide recombination substrates yield covalent λ Int-DNA complexes and lead to identification of the active site tyrosine. *J. Biol. Chem.* **263:**7678–7685.

135. **Peacock, S., H. Weissbach, and H. A. Nash.** 1984. In vitro regulation of phage λ *c*II gene expression by *Escherichia coli* integration host factor. *Proc. Natl. Acad. Sci. USA* **81:**6009–6013.

136. **Pierson, L. S., III, and M. L. Kahn.** 1987. Integration of satellite bacteriophage P4 in *Escherichia coli*: DNA sequences of the phage and host regions involved in site-specific recombination. *J. Mol. Biol.* **196:**487–496.

137. **Plasterk, R. H. A., A. Brinkman, and P. van de Putte.** 1983. DNA inversions in the chromosome of *Escherichia coli* and in bacteriophage Mu: relationship to other site-specific recombination systems. *Proc. Natl. Acad. Sci. USA* **80:**5355–5358.

138. **Plumbridge, J. A., M. Springer, M. Graffe, R. Goursot, and M. Grunberg-Manago.** 1980. Physical localisation and cloning of the structural gene for *E. coli* initiation factor IF3 from a group of genes concerned with translation. *Gene* **11:**33–42.

139. **Pollock, T. J., and K. Abremski.** 1979. DNA without supertwists can be an in vitro substrate for site-specific recombination of bacteriophage λ. *J. Mol. Biol.* **131:**651–654.

140. **Pollock, T. J., and H. A. Nash.** 1983. Knotting of DNA caused by a genetic rearrangement: evidence for a nucleosome-like structure in site-specific recombination of bacteriophage lambda. *J. Mol. Biol.* **170:**1–18.

141. **Pramanik, A., S. J. Wertheimer, J. J. Schwartz, and I. Schwartz.** 1986. Expression of *Escherichia coli infC*: identification of a promoter in an upstream *thrS* coding sequence. *J. Bacteriol.* **168:**746–751.

142. **Prasad, P. V., L.-J. Young, and M. Jayaram.** 1987. Mutations in the two-micrometer circle site specific recombinase that abolish recombination without affecting substrate recognition. *Proc. Natl. Acad. Sci. USA* **84:**2189–2193.

143. **Ptashne, M., K. Backman, M. Z. Humayun, A. Jeffrey, R. Maurer, B. Meyer, and R. T. Sauer.** 1976. Autoregulation and function of a repressor in bacteriophage lambda. *Science* **194:**156–161.

144. **Rattray, A., S. Altuvia, G. Mahajna, A. B. Oppenheim, and M. Gottesman.** 1984. Control of bacteriophage lambda *c*II activity by bacteriophage and host functions. *J. Bacteriol.* **159:**238–242.

145. **Reed, R. R., and N. D. F. Grindley.** 1981. Transposon-mediated site-specific recombination in vitro: DNA cleavage and protein-DNA linkage at the recombination site. *Cell* **25:**721–728.

146. **Reichardt, L. F.** 1975. Control of bacteriophage lambda repressor synthesis after phage infection: the role of the *N*, *c*II, *c*III, and *cro* products. *J. Mol. Biol.* **93:**267–288.

147. **Rhoades, M., and M. Meselson.** 1973. An endonuclease induced by bacteriophage λ. *J. Biol. Chem.* **248:**521–527.

148. **Richet, E., P. Abcarian, and H. A. Nash.** 1986. The interaction of recombination proteins with supercoiled DNA: defining the role of supercoiling in lambda integrative recombination. *Cell* **46:**1011–1021.

149. **Ross, W., and A. Landy.** 1982. Bacteriophage λ Int protein recognizes two classes of sequence in the phage *att* site: characterization of arm-type sites. *Proc. Natl. Acad. Sci. USA* **79:**7724–7728.

150. **Ross, W., and A. Landy.** 1983. Patterns of λ Int recognition in the regions of strand exchange. *Cell* **33:**261–272.

151. **Ross, W., A. Landy, Y. Kikuchi, and H. Nash.** 1979. Interaction of Int protein with specific sites on λ *att* DNA. *Cell* **18:**297–307.

152. **Ross, W., M. Shulman, and A. Landy.** 1982. Biochemical analysis of *att*-defective mutants of the phage lambda site-specific recombination system. *J. Mol. Biol.* **156:**505–529.

153. **Sadowski, P.** 1986. Site-specific recombinases: changing partners and doing the twist. *J. Bacteriol.* **165**:341–347.
154. **Schindler, D., and H. Echols.** 1981. Retroregulation of the *int* gene of bacteriophage λ: control of translation completion. *Proc. Natl. Acad. Sci. USA* **78**:4475–4479.
155. **Schmeissner, U., K. McKenney, M. Rosenberg, and D. Court.** 1984. Removal of a terminator structure by RNA processing regulates *int* gene expression. *J. Mol. Biol.* **176**: 39–53.
156. **Shea, M. A., and G. K. Ackers.** 1985. The O_R control system of bacteriophage lambda: a physical-chemical model for gene regulation. *J. Mol. Biol.* **181**:211–230.
157. **Shimada, K., and A. Campbell.** 1974. *int*-constitutive mutants of bacteriophage lambda. *Proc. Natl. Acad. Sci. USA* **71**:237–241.
158. **Shimatake, H., and M. Rosenberg.** 1981. Purified λ regulatory protein CII positively activates promoters for lysogenic development. *Nature* (London) **292**:128–132.
159. **Simons, R. W., and N. Kleckner.** 1983. Translation control of IS10 transposition. *Cell* **34**:683–691.
160. **Six, E.** 1959. The rate of spontaneous lysis of lysogenic bacteria. *Virology* **7**:328–346.
161. **Slilaty, S. N., J. A. Rupley, and J. W. Little.** 1986. Intramolecular cleavage of LexA and phage λ repressors: dependence of kinetics on repressor concentration, pH, temperature, and solvent. *Biochemistry* **25**:6866–6875.
162. **Spengler, S. J., A. Stasiak, and N. R. Cozzarelli.** 1985. The stereostructure of knots and catenanes produced by phage λ integrative recombination: implications for mechanism and DNA structure. *Cell* **42**:325–334.
163. **Springer, M., J.-F. Mayaux, G. Fayat, J. A. Plumbridge, M. Graffe, S. Blanquet, and M. Grunberg-Manago.** 1985. Attenuation control of the *Escherichia coli* phenylalanyl-tRNA synthetase operon. *J. Mol. Biol.* **181**:467–478.
164. **Springer, M., J. A. Plumbridge, J. S. Butler, M. Graffe, J. Dondon, J. F. Mayaux, G. Fayat, P. Lestienne, S. Blanquet, and M. Grunberg-Manago.** 1985. Autogenous control of *Escherichia coli* threonyl-tRNA synthetase expression in vivo. *J. Mol. Biol.* **185**:93–104.
165. **Springer, M., M. Trudel, M. Graffe, J. Plumbridge, G. Fayat, J. F. Mayaux, C. Sacerdot, S. Blanquet, and M. Grunberg-Manago.** 1983. *Escherichia coli* phenylalanyl-tRNA synthetase operon is controlled by attenuation in vivo. *J. Mol. Biol.* **171**:263–279.
166. **Steege, D. A., K. C. Cone, C. Queen, and M. Rosenberg.** 1987. Bacteriophage λ *N* gene leader RNA. *J. Biol. Chem.* **262**:17651–17658.
167. **Stephenson, F. H.** 1985. A *c*II-responsive promoter within the *Q* gene of bacteriophage lambda. *Gene* **35**:313–320.
168. **Sternberg, N., B. Sauer, R. Hoess, and K. Abremski.** 1986. Bacteriophage P1 *cre* and its regulatory region: evidence for multiple promoters and for regulation by DNA methylation. *J. Mol. Biol.* **187**:197–212.
169. **Sumner-Smith, M., S. Benchimol, H. Murialdo, and A. Becker.** 1982. The *ben* gene of bacteriophage λ: mapping, identification and control of synthesis. *J. Mol. Biol.* **160**: 1–22.
170. **Sussman, R., and F. Jacob.** 1962. Sur un système de repression thermosensible chez le bacteriophage λ d'Escherichia coli. *C.R. Acad. Sci.* **254**:1517–1519.
171. **Takayama, K. M., N. Houba-Herin, and M. Inouye.** 1987. Overproduction of an antisense RNA containing the *oop* RNA sequence of bacteriophage λ induces clear plaque formation. *Mol. Gen. Genet.* **210**:184–186.
172. **Tanaka, I., K. Appelt, J. Dijk, S. W. White, and K. S. Wilson.** 1984. 3-A resolution structure of a protein with histone-like properties in prokaryotes. *Nature* (London) **310**:376–381.
173. **Thompson, J. F., H. F. Mark, B. Franz, and A. Landy.** 1988. Functional and structural characterization of stable DNA curvature in lambda *att*P, p. 119–128. *In* W. K. Olson, M. H. Sarma, R. H. Sarma, and M. Sundaralingam (ed.), *DNA Bending and Curvature.* Adenine Press, Guilderland, N.Y.
174. **Thompson, J. F., L. Moitoso de Vargas, C. Koch, R. Kahmann, and A. Landy.** 1987. Cellular factors couple recombination with growth phase: characterization of a new component in the λ site-specific recombination pathway. *Cell* **50**:901–908.
175. **Thompson, J. F., L. Moitoso de Vargas, S. E. Nunes-Düby, C. Pargellis, S. E. Skinner, and A. Landy.** 1987. Effect of mutations in the P2 Int binding site of bacteriophage lambda, p. 735–744. *In* T. Kelly and R. McMacken (ed.), *Mechanisms of DNA Replication and Recombination.* Alan R. Liss, Inc., New York.
176. **Thompson, J. F., L. Moitoso de Vargas, S. E. Skinner, and A. Landy.** 1987. Protein-protein interactions in a higher-order structure direct lambda site-specific recombination. *J. Mol. Biol.* **195**:481–493.
177. **Thompson, J. F., D. Waechter-Brulla, R. I. Gumport, J. F. Gardner, L. Moitoso de Vargas, and A. Landy.** 1986. Mutations in an integration host factor-binding site: effect on lambda site-specific recombination and regulatory implications. *J. Bacteriol.* **168**:1343–1351.
178. **van de Putte, P., S. Cramer, and M. Giphart-Gassler.** 1980. Invertible DNA determines host specificity of bacteriophage Mu. *Nature* (London) **286**:218–222.
179. **Volkert, F. C., and J. R. Broach.** 1986. Site-specific recombination promotes plasmid amplification in yeast. *Cell* **46**: 541–550.
180. **Weisberg, R. A., L. W. Enquist, C. Foeller, and A. Landy.** 1983. Role for DNA homology in site-specific recombination. The isolation and characterization of a site affinity mutant of coliphage λ. *J. Mol. Biol.* **170**:319–342.
181. **Weisberg, R. A., and M. E. Gottesman.** 1971. The stability of Int and Xis functions, p. 489–500. *In* A. D. Hershey (ed.), *The Bacteriophage Lambda.* Cold Spring Harbor Laboratory, Cold Spring Harbor, N.Y.
182. **Weisberg, R. A., and A. Landy.** 1983. Site-specific recombination in phage lambda, p. 211–250. *In* R. W. Hendrix, J. W. Roberts, F. W. Stahl, and R. A. Weisberg (ed.), *Lambda II.* Cold Spring Harbor Laboratory, Cold Spring Harbor, N.Y.
183. **Wierzbicki, A., M. Kendall, K. Abremski, and R. Hoess.** 1987. A mutational analysis of the bacteriophage P1 recombinase Cre. *J. Mol. Biol.* **195**:785–794.
184. **Winoto, A., S. Chung, J. Abraham, and H. Echols.** 1986. Directional control of site-specific recombination by bacteriophage λ: evidence that a binding site for Int protein far from the crossover point is required for integrative but not excisive recombination. *J. Mol. Biol.* **192**:677–680.
185. **Wu, H.-M., and D. M. Crothers.** 1984. The locus of sequence-directed and protein-induced DNA bending. *Nature* (London) **308**:509–513.
186. **Yin, S., W. Bushman, and A. Landy.** 1985. Interaction of the λ site-specific recombination protein Xis with attachment site DNA. *Proc. Natl. Acad. Sci. USA* **82**:1040–1044.
187. **Zieg, J., M. Hillman, and M. Simon.** 1978. Regulation of gene expression by site-specific inversion. *Cell* **15**:237–244.
188. **Zissler, J.** 1967. Integration negative (int) mutants of phage λ. *Virology* **31**:189.

Chapter 2

Bacteriophage Mu

MARTIN L. PATO

Martin L. Pato ■ Department of Molecular and Cellular Biology, National Jewish Center for Immunology and Respiratory Medicine, and Department of Microbiology and Immunology, University of Colorado Health Science Center, Denver, Colorado 80206.

I. INTRODUCTION

Bacteriophage Mu was accidentally discovered as a contaminant in 1963 by Larry Taylor (187), who recognized that his contaminant was a temperate phage like λ, but one with unusual properties. Among the lysogens formed with this new phage were a significant number of host mutants. Taylor reasoned that this new phage, unlike λ which integrates at a unique site in the host chromosome upon lysogenization, could integrate at numerous sites, if not at random. Insertion into host genes would cause mutations; hence the name Mu, for mutator phage. Subsequent work has validated Taylor's insight and demonstrated that insertion of the Mu genome occurs by a transposition mechanism. Hence, the positioning of this chapter, predictably and appropriately, between a chapter on bacteriophage λ and one on bacterial transposons.

The dual nature of Mu—virus and transposon—has made it extremely useful for elucidating the mechanism of transposition, as transposition can be studied in the entirety of a population by infection of a sensitive host or induction of a lysogen. Biochemical analyses not feasible for systems undergoing only rare transposition events are readily undertaken with Mu, and it is not surprising that the first in vitro transposition system was developed with Mu. This phage also presented us with the first example of gene splicing in procaryotes, the first example of the use of methylation in positive regulation of gene expression, a unique DNA modification system, and a fascinating use of site-specific recombination for altering host range. In addition, it has proved to be an extremely important genetic tool, as discussed by Berg et al. (C. M. Berg, D. E. Berg, and E. A. Groisman, this volume).

I will present a brief overview of the presently accepted main features of the Mu developmental pathway, which includes three important examples of "DNA mobility": conservative or nonreplicative transposition, replicative transposition, and site-specific recombination. In what follows I will present in greater detail the major pieces of evidence which have led to this picture, emphasizing more recent developments, and attempt to point out areas of incompleteness and uncertainty. An extensive monograph on all aspects of Mu biology has recently been published (184), as have reviews emphasizing Mu as a transposon (136, 197) and Mu as a virus (77).

II. OVERVIEW OF Mu DEVELOPMENT

Upon infection of a sensitive host, the viral Mu DNA and associated proteins are injected into the cell, where the linear DNA molecule is converted into a noncovalently closed, circular form. The infecting Mu DNA is integrated into host DNA virtually at random, though with some sequence specificity, by a process of conservative (or nonreplicative) transposition, requiring the Mu A protein, or transposase, and facilitated by the Mu B protein. A few percent of the infected cells persist in a lysogenic state, while the majority of cells continue on into the lytic cycle. During the lytic cycle Mu DNA is amplified about 100-fold by replicative transposition, requiring the Mu A and B proteins and host replication machinery. Mu prophage copies accumulate in the host chromosome until they are packaged by a "headful" mechanism and released as free phage particles. The mature phage particles possess one of two possible sets of tail fibers, determined by the orientation of an internal, invertible region of the Mu genome, which in turn determines the host range for the next infection cycle.

III. INTEGRATION OF INFECTING Mu DNA: CONSERVATIVE TRANSPOSITION

Upon infection, Mu can either lysogenize the infected host or enter the lytic cycle. In either case the

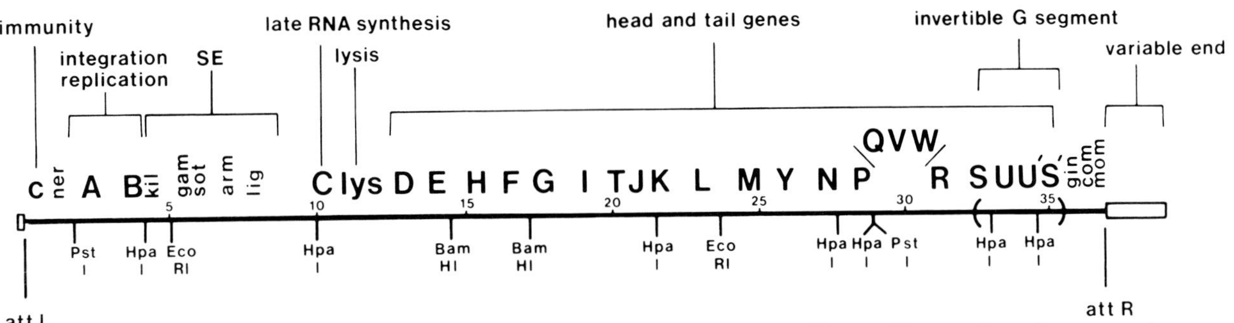

Figure 1. Genetic and physical map of Mu. The solid line represents Mu DNA, and the open boxes at the ends represent the attached host DNA sequences. Vertical markings above the line indicate 5-kb intervals. (Adapted from reference 184.)

initial event is integration of the infecting Mu DNA into the host chromosome. Mu and λ differ in this respect, as λ is integrated into host DNA only during lysogenization. Therefore, when discussing integration of Mu, it is often essential to discriminate between the totality of integration events and those integration events leading to lysogeny. The present view is that all integration occurs by a single mechanism and that a fraction, generally the majority, of integrated Mu genomes then continue into the lytic cycle, while the remainder persist as prophage in a lysogenic state. It is conceivable that the virus particles which yield lysogens are different from the majority of the particles, or that different mechanisms are involved in integration leading to lysogenization or to lytic development; however, there is no evidence presently available to support or refute these alternatives. In what follows, the term **integration** is used specifically for the initial entry of Mu DNA into host DNA following infection.

A. Structure of Mu DNA

The mature virion DNA is unusual in that it contains, in addition to the 37-kilobase (kb) linear Mu genome, host DNA covalently linked to both ends: 50 base pairs (bp) to 150 bp at the left (c or α) end and 0.5 kb to 3 kb at the right (S or β) end (Fig. 1). These heterogeneous host sequences were identified by a combination of electron microscopic analysis of heteroduplex DNA molecules (19, 41, 94) and DNA hybridization (43). The terminal host sequences result from packaging of Mu DNA from sites in the host chromosome, starting with a cleavage upstream of the prophage left end, followed by encapsidation of more than a Mu genome length of DNA and cleavage beyond the right end (19). An additional unusual feature observed in the heteroduplex analysis is a 3-kb invertible segment of the Mu genome, the G region, located about 1.7 kb from the right end.

B. Circularization of Infecting DNA

The genetic maps of the mature virus and prophage are colinear, independent of prophage location (1, 219). This colinearity could be readily explained if the ends of the infecting Mu genome were linked to form an *att* site which was then used for integration into host DNA. However, the heterogeneity of virion DNA ends precludes the use of recombination of permuted ends or reassociation of cohesive ends as a mechanism for circularization, as seen with phages P2 and P22, or phage λ, respectively. Nevertheless, the appeal of a circular integration intermediate led several groups to seek such molecules, and studies with infection of sensitive cells (78, 122) or minicells (158) demonstrated circular, supercoiled molecules observable in the electron microscope. Phenol extraction or treatment with sodium dodecyl sulfate converted the circular DNA to a linear form, suggesting that the circles were not covalently closed. A protein of approximately 64 kilodaltons (kDa) was shown to bind to the physical ends of the infecting DNA in the circular molecules (78). The protein was recently identified as the product of the Mu *N* gene (62); it is required for phage tail assembly and is presumably injected into the cell along with the phage DNA.

Circularization of infecting DNA may have more than one function, one of these being protection of the Mu ends from nucleolytic degradation. Linear Mu DNA is virtually inactive in transfection assays (101), and it is clear that one contributing factor is the nuclease sensitivity of Mu DNA ends. DNA isolated from phage particles by freeze-thaw treatment is complexed with a 65-kDa protein (presumably the N protein) and is more efficient in transfection than DNA isolated by phenol extraction (32), as is a DNA-protein complex isolated upon infection of Mu-immune cells (78); these complexes are relatively resistant in vitro to exonuclease digestion. Transfection frequencies also can be stimulated by the use of *recBC* hosts or by expression of the Mu-encoded *sot* (stimulation of transfection) function (202). *sot* apparently is identical to the previously described Mu *gam* function, which encodes an 18.9-kDa protein that can bind to the ends of DNA and inhibit nuclease digestion (3, 4, 6, 172, 220).

As we shall see, the ends of the Mu genome, rather than the ends of the mature viral DNA, can be brought together by the A protein. Whether circularization of the viral Mu DNA by the N protein is required for integration of infecting DNA, or stimulates integration, or merely protects the DNA from nucleases, is an important and unanswered question. The conditions used to accumulate sizable amounts of the circular forms—infection of Mu lysogens or minicells, or use of phage mutants which cannot integrate—focus necessarily on the nonintegrated molecules; while it is reasonable to suppose that the integrated molecules have gone through the same intermediate, proof is lacking.

C. Physical and Genetic Evidence for Integration

The heterogeneous host DNA sequences at the ends of the mature viral DNA are shed upon integration of Mu into host DNA after infection, as shown by analysis of heteroduplex molecules formed between Mu phage and prophage DNA (95) and by DNA sequencing (7). This is an essential observation allowing us to consider Mu integration to be a legitimate transposition event; i.e., Mu transposes from one site, between the host sequences in the mature viral DNA, to a second site in host DNA. More than semantics is implied here; mechanism is implied, as well.

Measurements of integration have been made with both physical and genetic techniques. As mentioned before, the integration events resulting in lysogeny and detected genetically are assumed to be a subset of all integration events that are measured with physical techniques, though experimental evidence is lacking on this important point.

The physical evidence for integration consists primarily of analysis of the fate of ^{32}P-labeled phage DNA (30, 122). Electrophoretic separation on agarose gels of DNA isolated at various times after infection with labeled phage revealed several labeled species, including nonintegrated phage DNA, as well as fragments of host DNA containing integrated prophage; the latter runs either at the origin of the gel or as a broad band at a position distinguishable from the nonintegrated DNA, apparently depending upon the degree of shearing during DNA isolation. Quantitation of the different species is generally lacking in the published data, but estimates as high as 80% integrated Mu DNA within 15 min after infection at an average multiplicity of infection (MOI) of 3 have been reported (25).

The genetic evidence on the frequencies of lysogenization is complicated by differences in the techniques used in the lysogenization, the MOI, the host used, and the physiological conditions of the host. Most measurements are performed at high MOI to minimize the contribution of uninfected cells and are expressed as lysogens per surviving cell; values between 5 and 40% have been reported (142; M. M. Howe, Ph.D. thesis, Massachusetts Institute of Technology, Cambridge, 1972). A recent study (16a, 16b) used measurements of lysogens per infected cell at low MOI, calculating the number of infected cells from the Poisson distribution. Their values were in the range of 10% (lysogens per infected cell) at low MOI, with the values decreasing with increasing MOI.

D. Integration Is Nonreplicative

Dominated by thought on the replicative nature of transposition during the lytic cycle, to be discussed below, work on integration of infecting DNA initially focused on a replicative mechanism for the initial integration. However, a series of three experiments slowly brought about the acceptance of a nonreplicative, or conservative, mechanism for integration of Mu DNA following infection. The basic distinction is that a conservative integration event results in both strands of the infecting DNA being integrated without being replicated (11, 22).

In the first of these experiments, Liebart et al. (120) showed that infection of isotopically labeled "heavy" cells containing a small plasmid resulted in the formation of some Mu-containing plasmid DNA of a density consistent with integration without replication. Akroyd and Symonds (5) found that a cell transfected with an artificially constructed heteroduplex Mu DNA molecule, containing one wild-type strand and one mutant strand, gave rise to a mixed burst with approximately equal numbers of wild-type and mutant phage, showing that both strands of the infecting DNA were integrated; however, the experiment could not clearly distinguish between a single nonreplicative integration and semiconservative replication followed by integration of both progeny molecules. The third, and most convincing, experiment, that of Harshey (76), showed that infecting phage DNA that was fully methylated (from phage grown on a *dam* methylase-overproducing strain), when injected into a *dam* mutant host, was integrated in a fully methylated form. This was done by showing that the integrated DNA could be cut with the restriction enzyme *Dpn*I, which is specific for fully methylated DNA; replicative integration would have yielded hemimethylated DNA resistant to *Dpn*I cleavage.

E. Integration Targets

Taylor's original observations showed that Mu could integrate at numerous sites in the *Escherichia coli* genome (187). When a collection of about 75 independent insertions into the *lacZ* gene were mapped and each insertion was shown to be in a separate site (20), the notion of random Mu integration was accepted. Later observations, particularly on integration into the *malK-lamB* region of the *E. coli* genome (159, 177), suggested that some site preferences exist.

Sequence analysis beyond the ends of Mu prophage DNA into adjacent host chromosomal DNA revealed that integrated prophage DNA is flanked by 5-bp direct repeats of the DNA at the target site (8, 100). This was a key observation, as such repeats are one of the hallmarks of transposition. One should note that the host sequences adjacent to integrated prophage DNA following infection have only been characterized for two examples and that both are the result of integration leading to lysogeny; that the results can be extended to the entirety of the population of integrated genomes is an assumption, albeit a reasonable one.

One of the ideas initially discussed to explain random choice of target sites for integration involved a role for the host replication fork, because host replication forks are distributed along the chromosome and integration at forks would yield an apparently random pattern. Evidence consistent with this notion came from two sources. First, synchronized cells were infected at different times in the cell cycle and the frequency of mutations in different host genes was measured; peaks of mutation frequency were noted when the time of infection and replication of the gene in question approximately coincided (143, 144). Second, different Hfr strains with aligned chromosome replication were infected at different times and mated with a Mu-sensitive recipient to allow zygotic induction of any integrated copy of Mu in the Hfr chromosome; the observed kinetics of transfer were consistent with integration of Mu in the Hfr organism at host replication forks (53). Note that the former experiment monitored integration leading to lysogeny, while the latter measured total integration. Later experiments by Nakai and Taylor (140) using more convincing physical methods, i.e., hybridization studies of the host DNA attached to the integrated Mu genomes, found no evidence of replication fork involvement in the integration and showed that the results of the Hfr mating experiments were biased by the experimental conditions used. A recent repeat of experiments measuring lysogen formation following infection of synchronized cells failed to show a role for host forks (179a). It is reasonable to conclude that host replication forks do not play any role in selection of target sites for Mu integration. This conclusion is consistent with what we now know about the mechanism of integration.

One factor that may affect the site at which successful integrations occur is the transcriptional activity of DNA into which Mu integrates, indicated by the observation that the frequency of integration of Mu into *lacZ* is reduced when that gene is being actively transcribed (44). In addition, numerous experiments have shown that Mu can integrate in both orientations within a given target gene (16, 219, 222), and lysogens with Mu in *lacZ* are destabilized by isopropyl-β-D-thiogalactopyranoside induction when the left end of the prophage is proximal to the *lac* promoter (L. Taylor, personal communication). It appears that transcriptional activity from host promoters can interfere with successful integration, perhaps by steric interference or by stimulation of induction of the newly integrated prophage.

F. Mu-Encoded Proteins Required for Integration

The Mu A protein, or transposase, is required for integration, and the Mu B protein is required, in addition, for optimal levels of integration. Here we must be careful to distinguish evidence on lysogenization from the physical evidence on integration.

It is clear that the A protein is required for lysogenization; virtually no lysogens were found after infection with several *A*am mutants (79, 142, 192). However, one *A* mutant, *A*am7110, did form lysogens. The *A*am7110 mutation maps at the 3′ end of the gene (79, 192), indicating that the C terminus of the A protein is dispensable for lysogenization, though it is required for entry into the lytic cycle and for Mu DNA replication. Similar arguments apply for a phage containing a small deletion (Δ17) at the 3′ end of the *A* gene (79).

The A protein also is clearly required for integration of the bulk of infecting DNA, as integration of ^{32}P-labeled DNA of nonlysogenizing *A*am mutants was virtually undetectable (25, 123), though a weak signal was detected with *A*am7110 (192).

The role of the B protein in integration is less clear. Most reports show about a 10-fold drop in lysogenization frequency for *B*am phage relative to wild type (142). The studies of Chaconas et al. (25)

on integration of ^{32}P-labeled *B*am phage DNAs showed that most yielded very low levels of integration into the host DNA after infection; the one important exception was *B*am1066, which has a mutation that maps at the 3′ end of the gene and yielded essentially wild-type levels of integration. Precise quantitation in these experiments is difficult, but the level of integration of the *B*am mutants other than *B*am1066 appears to be reduced to less than 5% the level of wild type. (It is of interest that the *B*am1066 allele, which shows high levels of integration, is the allele which was used in most earlier studies of the role of B in Mu biology.) In trying to understand the role(s) of the B protein in integration, it will be necessary to reconcile the severe reduction in integration of *B*am phages other than *B*am1066 with the observation that both classes of *B*am phages yield about the same frequency of lysogenization. Since *B*am1066 integrates at high frequency and cannot proceed on to replicate, it is not clear why the frequency of lysogenization of this mutant is not greatly increased; one suggestion is that the "excess" integration events are reversible (27).

The other Mu-encoded protein that may be required for integration is the N protein, which, as discussed previously, is involved in circularization of incoming DNA.

IV. THE LYTIC CYCLE: REPLICATIVE TRANSPOSITION

Entry into the lytic cycle occurs in the majority of the population following the first integration event, or upon induction of an established lysogen. By a series of replicative transposition events, approximately 100 prophage copies are synthesized which accumulate in the host nucleoid until packaged into viral particles (149). It is assumed that the replicative events following infection and induction are identical, though there is no direct evidence to support the assumption. Induction of Mu development cannot be effected by techniques often used for induction of wild-type λ lysogens, such as UV irradiation; induction of Mu lysogens is accomplished by thermoinduction of mutants with altered repressors (206; Howe, thesis). Our understanding of the replication events during the lytic cycle comes both from biochemical and from genetic approaches; I will focus first on the former. Also, these events have been studied both with prophage initially in the host chromosome or on a second replicon, such as a plasmid; both will be discussed, as the products of the two differ in informative ways.

A. Replication Characteristics

In 1977 Ljungquist and Bukhari (121) reported what turned out to be a pivotal experiment in directing the thinking on Mu replication. They showed that following induction of a lysogen, replication ensued in situ, i.e., without excision of the prophage from its site in the host chromosome. They digested the DNA of a lysogen with a restriction enzyme that cleaves both Mu and host DNA and then separated the resulting fragments on an agarose gel. Two of the fragments contained both Mu and host DNA and corresponded to the junctions between prophage ends and host DNA. These "junction fragments" persisted throughout the lytic cycle, as would be expected if the prophage DNA were being replicated in situ. During the lytic cycle, additional junction fragments of heterogeneous size also accumulated.

The Ljungquist and Bukhari experiment did not address the possibility that the prophage was excised along with adjacent host DNA. Indeed, supercoiled, circular DNA molecules had been observed during the lytic cycle that contained Mu DNA and heterogeneous lengths of host DNA ("HcDNA"), which could have been such excised DNA (173, 206). However, kinetic analysis showed that Mu DNA replication began considerably before the time of appearance of HcDNA (209), suggesting that these extrachromosomal molecules were not the result of early excision events that preceded replication.

Mu DNA replication is initiated as early as 6 to 8 min after thermoinduction of a lysogen, as measured by annealing of labeled DNA to nitrocellulose filters with bound Mu DNA (209, 216). The in situ Mu replication was not associated with amplification of the adjacent host DNA, and replication was presumed to continue up to, but not beyond, the boundaries of the prophage DNA (199, 208).

Experiments involving the annealing of Okazaki fragments to separated strands of Mu DNA indicated that Mu replication is predominantly unidirectional and proceeds from the left to the right end of the genome (66, 218). Similar results were obtained by labeling DNA during a synchronized first round of Mu replication and annealing the labeled DNA to filters carrying either the left or right ends of Mu DNA (161). Both experiments indicated that most, but not all, replication events initiate from the left end of the genome. Electron microscopic observation of replicating molecules deleted for much of the central portion of the genome ("mini-Mu") showed several structures which appeared to be replicating from either end with equal facility (81, 162, 164). In addition, Okazaki fragments isolated during replication of a mini-Mu annealed to either strand of Mu

DNA with approximately equal facility (163). These results indicate that either end of the genome can be used to initiate replication, and it is unclear what factors are responsible for the preferential use of the left end in the complete Mu genomes.

Meselson-Stahl experiments performed during a synchronized round of replication (147) and during in vitro synthesis (89) showed that replication is semiconservative and that, following a round of replication, each prophage copy contains one old and one newly replicated strand of Mu DNA. This latter observation is not trivial, as models in which a nascent strand is segregated, leaving behind both original strands, had been considered by these authors.

B. Products of Replication

The products of a single replication event depend upon the participants, in that intermolecular and intramolecular events lead to different products. The nature of the products was initially explored by genetic analysis, primarily the work of Toussaint and Faelen and their collaborators. These studies demonstrated a variety of Mu-induced chromosomal rearrangements including chromosomal deletions, inversions, and duplications, transposition of chromosomal DNA, and fusion of individual replicons. The findings were instrumental in the formulation of transposition models and, taken with biochemical evidence from in vivo and in vitro analyses, yield a coherent picture for replicative transposition.

1. Intermolecular events

The first studied rearrangement involved formation of a replicon fusion, or cointegrate, in which Mu mediated the integration of a derivative of phage λ that was incapable of normal integration or replication (51, 193). The integrated λ DNA was always sandwiched between two directly repeated copies of Mu; *rec*-mediated recombination between the direct repeats of Mu led to loss of the λ DNA and one copy of the Mu DNA. Similarly, infection of an F′ host by Mu could lead to fusion of the F plasmid and the host chromosome to form an Hfr cell (200). Later physical studies of cells carrying Mu on a plasmid showed that induction resulted in fusion of the plasmid and host chromosome (13, 28), and restriction enzyme analysis showed that the plasmid DNA was flanked by two direct repeats of Mu (29).

An intermolecular replicative transposition leading to cointegrate formation is diagrammed in Fig. 2. Recombination between the Mu copies, mediated either by the host *rec* system or by a specific resolvase

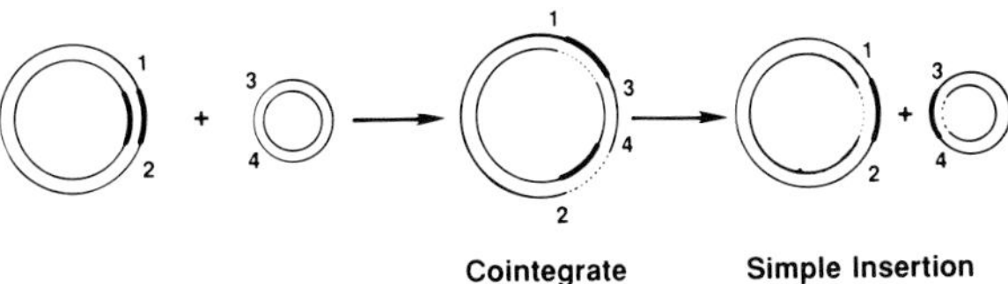

Figure 2. Intermolecular replicative transposition showing formation of a cointegrate, followed by a recombination (resolution) to yield a simple insertion.

equivalent to that encoded by Tn*3*, would lead to resolution of the cointegrate into independent replicons. Though arguments can be made suggesting that Mu does not encode a resolvase, there is no direct evidence on this question.

A transposition end product in which a single copy of Mu DNA is integrated into a DNA molecule is referred to as a simple insertion. Simple insertions can arise from cointegrates by recombination (Fig. 2) or as products of conservative transposition similar to that found after infection. Although most evidence points to cointegrates being the major product of Mu intermolecular replicative transposition, several genetic experiments have yielded both cointegrates and simple insertions as products (34). This is particularly apparent in experiments in which mini-Mu's have been used. M. DuBow and M. Lalumiere (personal communication) found that varying the length of vector sequences outside the ends of a mini-Mu in a small plasmid affected the ratio of transposition products: more than 2 kb of DNA outside the mini-Mu ends led to only cointegrate formation, while mostly simple insertions were found when 406 bp separated the mini-Mu ends. These observations raise the questions of whether a single model can explain formation of both products, and what factors determine the choice of transposition products.

2. Intramolecular events

Intramolecular transpositions most likely use the same mechanism as do intermolecular ones, but the products are different. Potential products—deletions, inversions, and replicative simple insertions—suggested by various models are shown in Fig. 3.

Deletion formation has been observed in genetic analyses, both after infection and after "partial induction" at intermediate temperature, with the deletions extending variable distances from either end of the prophage (48). The HcDNA molecules described earlier are likely to be one of the products of such deletion events. HcDNAs are first observable several minutes after replication is initiated upon induction and reach a level of only about one molecule per cell late in the cycle (209); however, the HcDNA formed

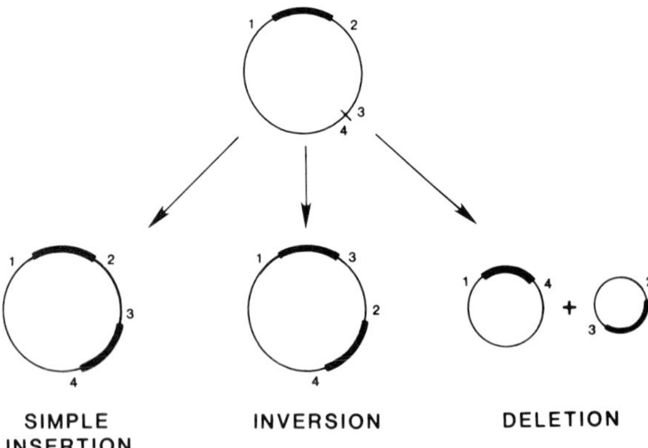

Figure 3. Potential products of intramolecular replicative transposition.

in one round of replicative transposition could be reintegrated into the host chromosome by a subsequent event leading to cointegrate formation. After infection, deletions are often found adjacent to the prophage in isolated lysogens (44, 92; M. M. Howe and D. Zipser, Abstr. Annu. Meet. Am. Soc. Microbiol. 1974, V208, p. 235); these could result from sequential integration and deletion events, but this has not been demonstrated.

Replicative transposition leading to inversions of host DNA (sometimes called insertion-inversions) results in two copies of Mu in inverted orientation flanking the inverted host DNA; e.g., Faelen and Toussaint (49) were able to invert the direction of DNA transfer in an Hfr cell by a replicative transposition event between a prophage on one side of the integrated F factor and a target site on the other side of F. Electron microscopic images of DNA, after denaturation and renaturation, from cells undergoing mini-Mu replication showed stem-loop structures with two inverted Mu strands forming the stems, structures which were most likely the result of inversion events (163).

The replicative simple insertion was addressed in an experiment of Pato and Reich (148). DNA from a lysogen was cleaved with a restriction enzyme that does not cut within Mu DNA, resulting in the appearance, on an agarose gel, of a large junction fragment consisting of Mu DNA with host DNA at both ends. As can be seen from analysis of Fig. 3, the fragment would be expected to persist after formation of a replicative simple insertion following one round of replicative transposition. Deletion or inversion formation would result in loss of this fragment and appearance of a heterogeneous population of fragments. The fragment was observed to disappear after a synchronized round of replicative transposition, showing that replicative simple insertions are not formed frequently, if at all. The experiment of Ljungquist and Bukhari (121) described earlier has often been interpreted as showing that a copy of Mu remains at the "original site" during the lytic cycle. However, during deletion and inversion formation, the "original site" is lost in the sense that there is no longer a copy of Mu flanked by the same host sequences that flanked the original prophage (Fig. 3).

An informative set of experiments by Chaconas et al. (29, 30) illustrated the different products expected from conservative transposition and replicative transposition. F′ plasmids were examined that had gained a copy of Mu DNA after infection or after induction of cells containing a prophage located in the host chromosome. In the former case, the plasmids were of a length equivalent to the sum of the lengths of the F′ and of Mu, as expected for simple insertions resulting from conservative transposition. In the latter case, a heterogeneous population of molecules were found containing deletions and insertions, probably resulting from initial formation of a cointegrate of F and the chromosome followed by a second, intramolecular replicative transposition resulting in deletion formation.

C. Targets for Replicative Transposition

Replicative transposition of Mu or mini-Mu from a plasmid to the chromosome is an **intermolecular** event and probably involves selection of targets throughout the chromosome, though some regions of the chromosome apparently are used as targets more frequently than others (24, 212; B. Wang, L. Liu, and C. Berg, personal communication). A question remains as to whether there is any constraint placed upon target selection during **intramolecular** transposition from one chromosomal site to another; e.g., are proximal sites used more frequently than distal sites? Some evidence indicates that target sites relatively close to the original prophage are used preferentially. For example, HcDNA, which contains Mu DNA and variable lengths of host DNA, is probably the result of intramolecular transpositions leading to deletions; the size of the host DNA in an HcDNA molecule would then be determined by the distance from the transposing copy of Mu to the target site. The average length of the host DNA is roughly 40 kb, or about equal to the length of Mu (209). Also, electron microscopic analysis of replicating mini-Mu DNA (about 10 kb in length) revealed numerous examples of copies of the mini-Mu located close to each other in inverted orientation (163). However, it is difficult to draw firm conclusions from these observations, since use of distant target sites would yield HcDNA too large to be isolated intact with the procedures used, and transposition to distal sites would not be observable with the electron microscopic technique. Preliminary results using pulsed-field gradient electrophoresis to examine the distance between the original prophage site and the target sites used during the first round of replicative transposition following induction of a lysogen indicate that Mu is able to select target sites throughout the *E. coli* genome (M. Pato, unpublished data).

A degree of the nonrandomness of target selection at the level of the nucleotide sequence into which Mu transposes was suggested by analysis of the host sequences immediately adjacent to the Mu ends in a population of mature viral DNA molecules (107). Nucleotide preferences were observed in the first 5 bp of flanking host sequence, but then the sequences were completely random. Specificity is more apparent in the sequences of sites of a large number of insertions into bacteriophage λ, which yielded the consensus sequence 5′-NPy$_{G}^{C}$PuN-3′ beyond the ends of the Mu DNA, with approximately 90% agreement with the consensus sequence (cited in reference 136).

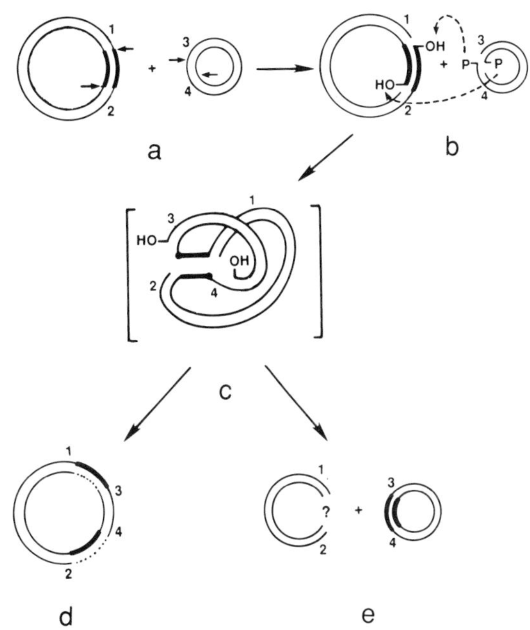

Figure 4. The Shapiro model for intermolecular transposition. (a) The transposon is nicked at the 3′ ends of each DNA strand; staggered nicks are introduced at the target DNA site, producing 5′ overhangs. (b) The 5′ end of each target DNA strand is ligated to a 3′ end of the transposon DNA. (c) The Shapiro transposition intermediate. (d) Replication across the transposon DNA in the intermediate yields a cointegrate. (e) Cleavage at the 5′ ends of the transposon DNA in the intermediate, followed by gap repair, yields a simple insertion.

D. Mu-Encoded Proteins Required for Replication

The Mu-encoded proteins required for replication during the lytic cycle are the A and B proteins. In studies using a DNA-DNA annealing assay, no Mu DNA replication has been detected with any of the *A* or *B* mutants tested (146, 192, 216). However, genetic studies show that B^- mutants can produce deletions and inversions which involve duplication of the Mu genome (47). This is also true for cointegrate formation; an F′ mini-Mu A^+B^- formed a cointegrate with the host chromosome after induction, leading to formation of an Hfr cell, albeit at a much slower rate than a corresponding B^+ plasmid (13). Is the replication associated with these events identical to that in the presence of B protein, only occurring at a reduced rate, or is it different? For example, could a host replication fork, encountering a transposition intermediate which is blocked for the normal replication associated with replicative transposition, proceed through the intermediate, producing the observed rearrangement? The question remains to be answered.

V. A MODEL OF TRANSPOSITION

Several models of transposition have been proposed, and they are discussed at length in the excellent review of Mizuuchi and Craigie (136). Here I will discuss only the model presently accepted for describing Mu transposition, that of Shapiro (176) and Arthur and Sherratt (9), which was based in part on, and most satisfyingly explains, the observations on Mu transposition and which has received its most convincing support from the in vitro analysis of Mu transposition. Future work may show deficiencies in parts of the model, but it would be extremely surprising if its major features proved to be incorrect. The model is referred to as the Shapiro model and is illustrated in Fig. 4 (intermolecular form) and Fig. 5 (intramolecular form). Following the suggestion of Grindley and Sherratt (69), the model proposes that double-stranded, staggered nicks are introduced at the target site (5 bp apart in the case of Mu, to explain the duplication of host DNA at the target), producing 5′ overhangs, and nicks are introduced at the 3′ ends of the transposon DNA in the donor molecule (Fig. 4a). The free 3′ ends of the transposon are ligated to the free 5′ ends of the target DNA (Fig. 4b) (the strand transfer reaction), producing branched structures at both ends of the transposon (Fig. 4c). The structures resemble replication forks, and the 3′ end of the target DNA in the branch structure can be used to prime DNA replication, which results in the formation of a cointegrate in the case of intermolecular replicative transposition (Fig.

and simple insertion products (38). Mu transposition can thus be resolved into two consecutive steps: first, the formation of the intermediate, requiring Mu-encoded proteins; and second, processing of the intermediate into cointegrates or simple insertions by host proteins, independent of Mu proteins.

B. The Purified System: Isolation of Transpososomes

A defined system capable of carrying out the strand transfer reactions involved in formation of the Shapiro intermediate was described by Craigie et al. (36). The substrate requirements are a supercoiled donor molecule and a target molecule which can be supercoiled, relaxed circular, or linear. The protein requirements are the Mu A and B proteins and a host factor, satisfied by purified HU protein. The reaction also requires ATP and magnesium ions.

Having a defined system allowed the further dissection of the strand exchange reaction. Surette et al. (181) found that incubation of a supercoiled donor plasmid molecule with the Mu A protein, HU, and magnesium ions resulted in a DNA protein complex in which the Mu portion was constrained in a supercoiled form, while the remainder of the plasmid was relaxed. The structure was called a "transpososome" and referred to as the "type 1 complex." Treatment of the structure with sodium dodecyl sulfate removed bound proteins and yielded a nicked circular molecule. In a similar study, Craigie and Mizuuchi (40) demonstrated that the nicks introduced in the donor molecule in the presence of the A protein, HU, and magnesium ions were located precisely at the 3′ ends of the Mu DNA.

In a separate reaction with the type 1 complex as donor and with target DNA, B protein, ATP, and magnesium ions, a second complex, type 2, was formed which, upon deproteination, yielded the transposition intermediate. The type 1 complex, purified away from unbound or loosely bound HU and A protein by gel filtration, is fully active in the strand transfer reaction (181); however, if the complex is deproteinized, it cannot efficiently function in the strand exchange reaction even in the presence of additional A protein and the other required components (40). Apparently, the interaction between the A protein and the cleaved end of Mu DNA cannot be disrupted without destroying the ability of the intermediate to function in the strand transfer reaction.

Although the complete reaction requires that the donor be supercoiled, the partial reaction using the type 1 complex with nicks at the 3′ ends of the Mu strands does not. This was most clearly shown by cleaving the type 1 complex between the Mu ends with a restriction enzyme before adding target DNA, B protein, and ATP and finding that the cleaved donor molecule was functional in the strand exchange reaction (181).

Omission of B protein or ATP from the normal in vitro reaction prevented completion of the strand exchange reaction. However, altering the reaction conditions by the addition of glycerol allowed the reaction to proceed, and the products of the reaction were those of intramolecular events (126). One important aspect of this result is the demonstration that ATP is not required for the ligation of Mu DNA and target DNA. Craigie and Mizuuchi (40) pointed out that the energy for the ligation reaction does not come from transfer of energy from cleavage at the Mu ends to a covalent link between the cleaved DNA and A protein in a manner similar to that found with several topoisomerases, since the type 1 complex does not have covalently bound A protein. They suggested that the required energy could come from a direct transfer of the 3′ ends of the Mu strands to 5′ ends of target DNA generated by a nucleophilic attack of the 3′ hydroxyl on the phosphodiester bond at the target site.

G. Gloor and G. Chaconas (personal communication) found that the supercoiled, noncovalently closed virion DNA, held together by N protein bound to the ends of the host DNA flanking the Mu DNA, was proficient in the strand exchange reaction and that the variable host DNA sequences at the ends of the Mu genome were maintained in the type 2 intermediate. These results support the assertion that the conservative transposition occurring during the integration following infection uses a Shapiro intermediate rather than one involving double-strand breaks at the transposon ends, as postulated for Tn*5* and Tn*10* conservative transposition.

The complete strand exchange reaction is diagrammed in Fig. 7. Also shown (Fig. 8) is an electron micrograph taken several years ago (N. Symonds and H. Janzen, personal communication) of a structure isolated after induction from cells containing two plasmids, one of which carried a Mu prophage. The structure is virtually identical to those observed in vitro (181)—a beautiful convergence of in vivo and in vitro studies!

When deproteinized, the type 2 structure is converted into the structure shown in Fig. 6, which is equivalent to the transposition intermediate. As mentioned above, this structure can be processed, by addition to a host cell extract, to yield either a cointegrate or a simple insertion (38).

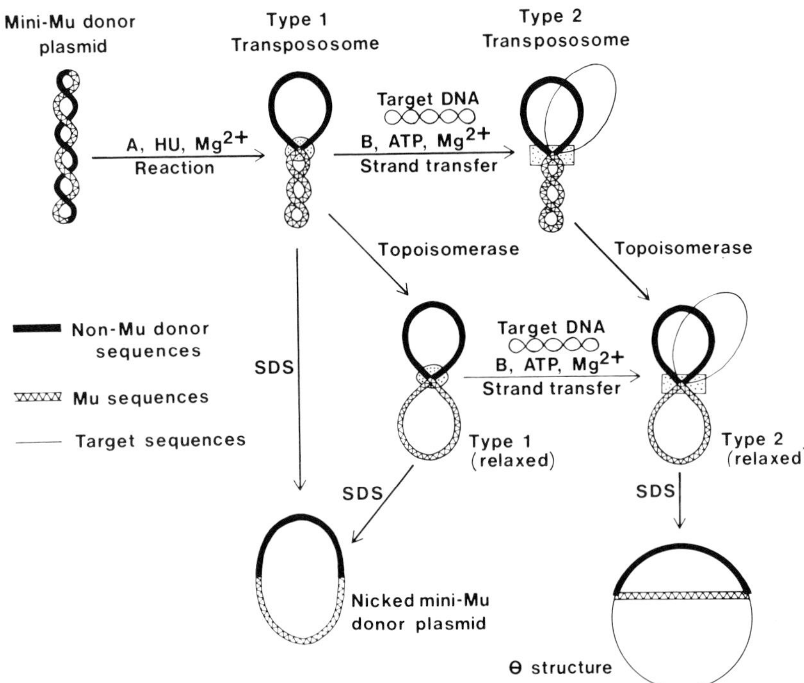

Figure 7. Summary of the structure and reactions of type 1 and type 2 transposition complexes. (Reprinted from reference 181.)

C. Choosing between Cointegrates and Simple Insertions

One of the intriguing, unanswered questions about transposition mechanism involves the choice between formation of cointegrates and simple insertions. The integration occurring after infection produces primarily, if not exclusively, simple insertions. In this case, the donor molecule (the infecting viral DNA) is linear, though it is probably circularized and supercoiled via a noncovalent protein bridge. Replication across the transposition intermediate formed with a noncovalently closed donor would lead to the production of free ends within the target, in this case the host chromosome; the evidence says this does not occur (30). The transposition events occurring during the lytic cycle appear to be mostly, if not exclusively, replicative; a nonreplicative, simple insertion from a prophage would cleave the host chromosome, and the evidence shows that the host nucleoid remains

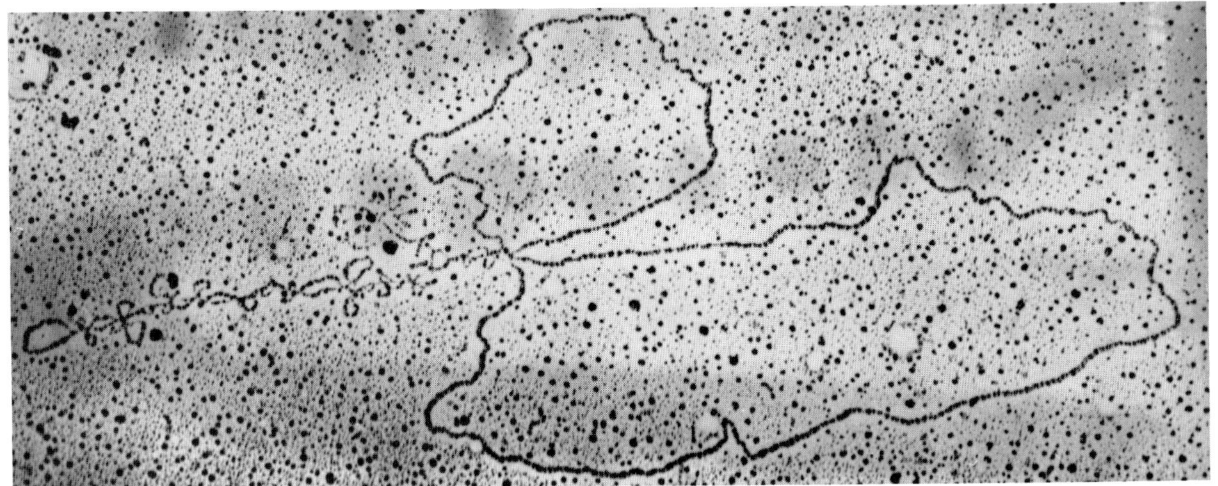

Figure 8. Electron micrograph of the type 2 complex from an in vivo reaction in maxicells of transposition from pML2::Mu into the plasmid R388 as in reference 34 (Symonds and Janzen, personal communication).

intact until late in the lytic cycle (149). As discussed above, both cointegrates and simple insertions have been observed with transposition of mini-Mu's from plasmids (34) or under conditions where the amount of non-Mu DNA outside the ends of Mu is small (DuBow and Lalumiere, personal communication). In vitro, the isolated intermediate devoid of Mu proteins can yield either cointegrates or simple insertions when incubated with an *E. coli* extract (38).

Two factors possibly affecting the choice are the influence of bound proteins and the topology of the DNA. For example, proteins which either enhance or inhibit binding of the intermediate to the host replication machinery could affect the choice, as could proteins which might protect the nicked DNA at the junctions in the intermediate. It is easy to imagine that the topology of the noncovalently closed infecting DNA, or that of the very small circular DNA in the experiments with various amounts of DNA outside the ends of Mu, could be significantly different from that of Mu in the host chromosome.

VII. PROTEINS, SITES, AND TOPOLOGY

A. The A Protein

The A protein is required for all transposition events, whether replicative or conservative, in vivo or in vitro (123, 134, 142, 216). The deduced size of the protein from the nucleotide sequence is 663 amino acids, with a molecular weight of about 75,000 (80); the apparent molecular weight on gels is about 70,000 (60).

The functions of the A protein are multiple and complex. It binds to specific sequences at the ends of Mu DNA (37; see section VII.E below). The ends are probably brought together through the interaction of A proteins bound to the ends (40, 181). Since A^+B^- genomes can lysogenize, and since under appropriate conditions in the in vitro system the strand transfer reaction can be catalyzed by the A protein alone, we can conclude that the A protein can allow binding to target sites (though possibly inefficiently), introduce the critical nicks at the ends of the Mu genome and at the target site, and ligate the Mu and target DNA sequences (40, 181). An additional function is suggested by the observation that the A protein and repressor protein share regions of sequence similarity and can bind to common DNA sites (37, 80); therefore, it is possible that competition for binding of the two proteins at the ends of Mu and at operator sites may be important in regulation.

Partial proteolysis of the purified protein by Nakayama et al. (141) indicated that it is composed of three major domains: an amino-terminal domain of about 30,000 daltons, a core region of about 35,000 daltons, and a carboxy-terminal domain of about 10,000 daltons. The amino-terminal domain can be further subdivided by proteolysis. The role that each of the domains plays in this complex set of reactions is being analyzed (141): the N-terminal domain possesses sequence-dependent DNA-binding properties, binding to a consensus sequence at the ends of Mu DNA and possibly at internal sites on the genome; the central domain has sequence-independent DNA-binding properties; the properties of the C-terminal domain are such that the protein lacking the last 31 amino acids can allow the infecting DNA to lysogenize at near-normal frequency and give a weak signal in the ^{32}P integration assay, but not participate in replication (79, 192). It has been suggested that the C terminus of the A protein may interact with the B protein, and it is a complex of the two proteins that is required for replicative transposition and for high-frequency integration following infection (79, 192). Apparent sequence similarity between a portion of the A protein and the poliovirus genome-linked protein VPg has led to the suggestion that tyrosine 413 of the A protein may be used in a covalent link with DNA (68).

Transposons might be expected to have evolved tight control on the synthesis and activity of their transposases, as high frequencies of transposition could be lethal to the host. In addition to use of transcriptional control (see below) and possible translational control exerted by binding of the A protein to its mRNA (R. Harshey, personal communication), the activity of the protein is unstable in vivo (146); inhibition of synthesis of A during the lytic cycle leads to inhibition of Mu DNA replication with a half-life of about 2 min. Pato and Reich (147) found that the A protein was used stoichiometrically in a system in which the first round of replicative transposition was synchronized; inhibition of synthesis of the A protein, after initiation of the first round, prevented subsequent rounds from occurring. Whether the instability of activity reflects instability of the protein itself, and whether stoichiometric use of the A protein applies during the lytic cycle, is not clear. Isolation of a truncated form of the A protein (12) raises the suggestion that the protein is processed to an inactive form after use. Whatever the mechanism, the functional instability of the A protein requires that it be continually synthesized during the lytic cycle (147).

Several transposases have been shown to function preferentially in *cis* (31, 124, 139, 150). When a dilysogen carrying an A^+ and an A^- prophage was induced, approximately 95% of the progeny phage were A^+, suggesting that the Mu transposase also functions preferentially, but not exclusively, in *cis* (147). It is difficult to assess the degree of preferential use from this result, since it measures the products not of one replication event, but of a large number of events; hence, it is possible that the observed result could reflect the amplification of a small preference for *cis* activity. Inability of the protein to bind efficiently to DNA after being used once could explain the apparent instability of activity.

B. The B Protein

The B protein has an apparent molecular weight of about 33,000 (26, 61), and the size determined from the sequence of the *B* gene is 312 amino acids (131). It is a basic protein which may be associated with the cell inner membrane (14, 174). It has sequence-independent DNA-binding properties and an ATPase activity which is stimulated by the presence of DNA and the A protein (126).

The B protein, by virtue of its sequence-independent DNA-binding properties, may be involved in bringing together the target DNA and the ends of Mu DNA bound with A proteins. Results in vitro (126) and in vivo (35, 75) suggest that transposition in the absence of B protein is shifted towards intramolecular events. Without B protein, random collision may favor intramolecular sites for targets. How then does B protein shift target selection from those sites favored by random collision to intermolecular sites? Perhaps the intramolecular events are normally selected against by a process similar to "transposition immunity" described for several bacterial transposons, in which one copy of a transposon in a plasmid inhibits entry of a second copy. Plasmid molecules containing a Mu end have been shown to be poor targets for transposition, a form of transposition immunity (1a, 44a). This immunity is apparently due to a differential distribution of Mu B protein between plasmid molecules containing or not containing Mu ends; those without Mu ends accumulate more B protein and are preferred targets (1a). The portions of the Mu ends involved in immunity were localized to sequences which include the innermost but not the outermost of the three binding sites for the A protein at each end (44a). A role for the B protein in transposition immunity could explain why the experiments with B^- mutants of Mu in small plasmids, in vivo or in vitro, yield intramolecular events, but this in turn raises the question of why transposition immunity does not prevent intramolecular events in the host chromosome during the lytic cycle.

The role of the B protein clearly involves more than just target selection, as the *Bam*1066 mutation, which removes 18 amino acids from the C terminus of the protein (25, 46), apparently allows normal levels of integration (25), but does not allow replication (216). The C terminus may be required to interact with the host replication machinery in a manner analogous to that observed with the λ P protein.

Partial proteolysis of the B protein yields two fragments (187a). The N-terminal domain binds ATP and DNA, and the C-terminal domain has a hydrophobicity profile that closely resembles that of an HU-like protein.

C. Other Mu-Encoded Proteins

Between the ends of the *B* gene (4.3 kb from the left end of the genome) and the *C* gene (about 10 kb from the left end), which is involved in control of late transcription, there is a region of about 5 to 6 kb of DNA called the SE (semiessential or accessory) region. Phage with a polar insertion slightly beyond the *B* gene are viable, though they make minute or no plaques on most hosts, suggesting that this region, though dispensable, is necessary for optimal levels of phage production (67). Several phenotypes have been attributed to the region, and functions encoded there may play a role in transposition in vivo, though they are clearly not required in vitro. These functions include Kil, a function(s) whose expression results in host lethality (201, 207); Arm, a function(s) which stimulates the rate of Mu DNA replication (67, 210); Gam, a protein which protects linear DNA from exonuclease digestion and is probably identical to Sot (3, 202, 205); and Lig, a function which can complement host ligase mutants (56, 57). This last activity is potentially very interesting, as Lig appears to function by altering the superhelical density of cellular DNA, thus leading to increased synthesis of some host proteins (including ligase) and decreased synthesis of others (P. Ghelardini, J. Liebart, L. Paolozzi, and A. Pedrini, personal communication).

D. Host Proteins

Numerous host proteins are required for Mu development. Those directly involved in Mu transposition include the following.

1. HU

The only host protein required for in vitro DNA strand transfer and formation of the transposition intermediate is HU (36). It is a heterodimer of two closely related subunits, each of about 9,500 daltons (167). HU is one of a class of small, histonelike DNA-binding proteins, able to bind to both single- and double-stranded DNA and to condense DNA into nucleosomelike structures (168). HU is required in addition to the A protein to bring the ends of Mu together (181), though Craigie and Mizuuchi found that HU was no longer required for strand exchange when they used a donor molecule which was constructed with nicks at the 3′ ends of the Mu DNA (40). It therefore appears that HU is required for organizing the Mu DNA into an appropriate form for nicking at the 3′ termini by the A protein.

Single mutations in the genes encoding either of the two subunits of HU do not affect growth of the cells or Mu development, but mutations in both subunits apparently prevent development of Mu (F. Imamoto, personal communication; M. Faelen and O. Huisman, personal communication).

2. IHF

M. Surette and G. Chaconas (personal communication) found that for efficient use in the in vitro reaction, the superhelical density of the donor plasmid had to be greater than the estimated average superhelical density for *E. coli* DNA. Addition of a partially purified host factor allowed inadequately supercoiled plasmid DNA to function; the factor apparently is host IHF (integration host factor). However, in vivo studies showed that IHF is not essential for Mu growth, since Mu will grow in IHF mutants if Mu A and B proteins are supplied from a plasmid (R. K. Yoshida, Ph.D. thesis, University of Wisconsin, Madison, 1984). Perhaps host factors other than IHF can also influence the unusual superhelical density requirements of the donor plasmid.

3. Host replication proteins

Most of the host replication proteins that have been tested are required for replicative transposition of Mu, including the products of *dnaB*, *dnaC*, *dnaE*, *dnaG*, *gyrA*, and *gyrB* (163, 194). Reported results with *dnaA* mutants vary, but seem to show reduced levels of Mu DNA replication (127), and polymerase I mutants yield reduced levels of Mu replication (128).

4. Gyrase

Gyrase is required for Mu development, as shown by studies with gyrase inhibitors, though different mutations in the genes encoding the two subunits, *gyrA* and *gyrB*, lead to widely different yields of phage (165). The demonstrated requirement for a supercoiled donor in the transposition reaction may be sufficient to account for the gyrase requirement; it is not required for transcription of the Mu-encoded replication proteins (S. Shore and M. Howe, personal communication).

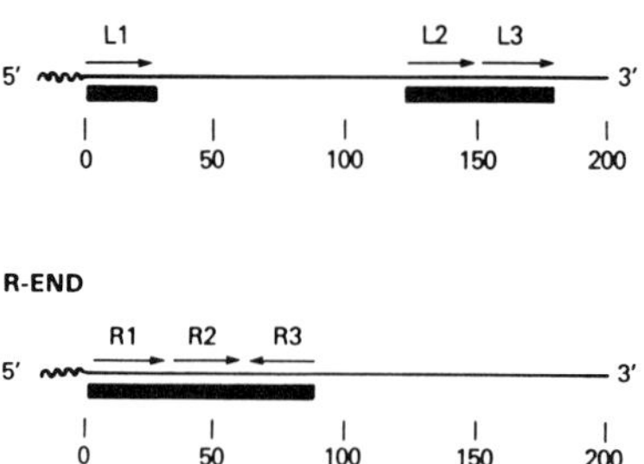

Figure 9. Location of the A protein-binding sites at the ends of Mu DNA. (Reprinted from reference 37.)

E. Sites

The sites on the Mu genome required for normal levels of transposition are the left and right ends in proper orientation. The requirement for intact ends was first suggested by the observation that deletion of the right end inhibits replication (199, 210); since replication proceeds predominantly from the left end of the genome, this result was not trivial. The requirement for correct orientation was shown by Schumm and Howe (175), who constructed mini-Mu's with the ends in various orientations relative to each other and found that only the natural orientation functioned in transposition. Constructions with two right or two left ends can function in transposition, but at a markedly reduced frequency (cited in reference 136), and even single ends can function at very low frequency when using as the second end host sequences with some similarity to the missing end (73).

Two approaches have been used to determine the nature of the required end sequences: footprint analysis with purified proteins, and definition of the minimal required sequences by deletion analysis. Craigie et al. (37) showed that the Mu A protein protects three sites of about 30 nucleotides each at both ends. The spacing and orientation of the sites at the left end (L1, L2, L3) differ from those at the right end (R1, R2, R3) (Fig. 9). A consensus sequence derived from the three stronger binding sites, L1, L3, and R3, TGNTTCAPyTNNAAPuPyPuCGAAAPu, lacks the dyad symmetry often present in protein-binding sites.

Groenen et al. (71) constructed mini-Mu's with different lengths of the right or left ends and measured their transposition potential. The smallest left end with activity was 25 bp, which encompasses the

L1 binding site. Additional DNA up to 147 bp did not increase activity, though it included L2; a left end of 163 bp increased activity 10-fold to full activity, though it added only half of L3. At the right end a fragment with 27 bp containing R1 was active in transposition; fragments with lengths up to 52 bp showed incremental increases in activity, and addition of L3 did not yield any further increase. Mutational analysis showed that the five binding sites other than R3 are important in transposition, as is the terminal nucleotide at each end, and that the spacing of the sites is important, as deletions or insertions between L1 and L2 reduced activity (21, 72).

Electron microscopic analysis of the complexes formed between the A protein and DNA fragments carrying the A-binding sites showed two or three monomers of the protein bound to each end, and the kinetics of binding to DNA fragments containing one, two, or three sites suggest that there is no cooperativity of binding (70).

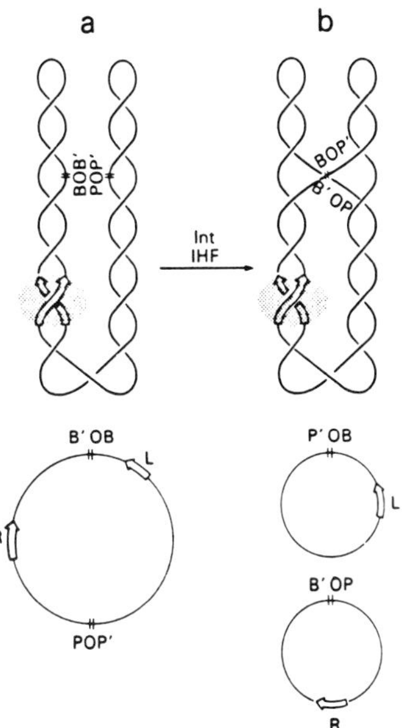

Figure 10. Formation of donor substrates with Mu ends on (a) the same molecule or (b) different molecules. B'OB and POP' are the phage and bacterial *att* sites of λ. R and L are the right and left ends of Mu. Int and IHF are required for site-specific recombination at the λ *att* sites. (Reprinted from reference 39.)

F. DNA Interaction at a Distance and the Role of Topology

Transposition of Mu includes two examples of interaction of distant segments of DNA: first, the ends of Mu DNA must be brought together, presumably via the A proteins bound at the ends, and second, the ends must interact with target DNA, possibly using an interaction between the A and B proteins and DNA.

The question of how distant DNA segments are brought together, particularly when constraints on orientation exist, is much debated. Two models have been proposed: one, the "tracking" model, envisions binding of a protein at one site followed by one-dimensional tracking along DNA until a second site is encountered; the other involves random collision of two sites, probably as protein-DNA complexes.

This question was addressed by Craigie and Mizuuchi (39), who cloned the *att* sites of λ onto a plasmid containing a mini-Mu with one λ *att* inside the mini-Mu and one outside the mini-Mu in the plasmid DNA; recombination between the λ *att* sites in vitro produced either two multiply interlinked, catenated, supercoiled DNAs or a single topologically knotted molecule, depending on the relative orientation of the *att* sites. Starting with a substrate with the Mu ends in the proper orientation, recombination between the λ *att* sites produced either catenated circles with one Mu end in each circle or a knotted molecule with the Mu ends in opposite orientation. Both of these products were active in the in vitro transposition assay! Therefore, any simple tracking model is excluded, as it is impossible to start at one of the Mu ends in either substrate and track linearly around the DNA to encounter the other end in the proper orientation. Rather, the authors conclude, sensing of the orientation of the two ends is determined by the topology of the supercoiled molecules. Negatively supercoiled DNA exists predominantly as a right-handed helix, and in such a helix the appropriate configuration of Mu ends is energetically favored only when they are initially in inverted orientation (as in Fig. 10a); this proper configuration can be maintained even when the geometry is altered (as in the catenane in Fig. 10b).

These experiments rule out any simple tracking model and highlight the importance of supercoiling in determining orientation specificity; however, the actual mechanism of bringing the ends together is still unknown. One of the observations that must be explained by any mechanism is the dependency of transposition frequency on transposon length. Faelen et al. (52) constructed Mu genomes of increasing size by inserting foreign DNA within the ends of a full-length Mu and found a logarithmic decrease of the frequency of transposition of these "maxi-Mu's" with increasing length; doubling of the length reduced transposition by a factor of 10^4. The inhibition was at the level of initiation of replication, suggesting that the defect is in bringing the ends together and not due to progressive aborting of replication along the lengthened transposon, as suggested as one pos-

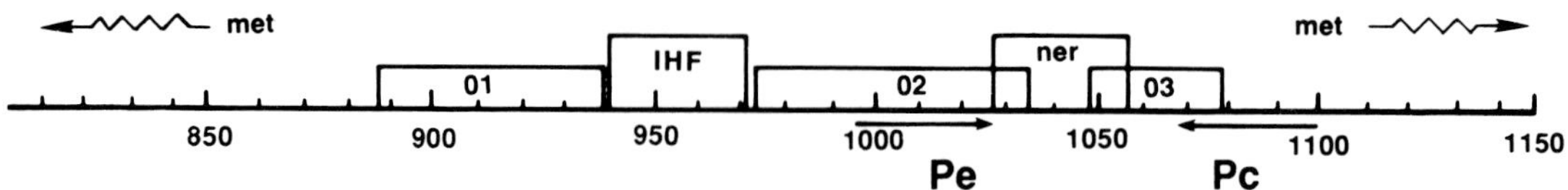

Figure 11. The regulatory region of Mu between the *c* and *ner* genes. Represented are: O1, O2, O3, the operator sequences; *ner* and IHF, the binding sites for the respective proteins; Pe, the promoter for the early transcript; Pc, the major repressor promoter; met, the initiating amino acid for the c and Ner proteins.

sible explanation for length dependency with IS*1* derivatives (31). A potentially important observation with Tn*10*, severe length dependency of transposition from the chromosome (139) but not from a plasmid (213), suggests a role for topology in these effects and raises questions about the topological differences between chromosomal and plasmid DNAs.

The second example of interaction of distant regions of DNA involves the finding of target sites. The question of target selection is interesting in the context of the products of intramolecular transposition. As illustrated in Fig. 5, the choice between inversion and deletion is dictated by which strand of the target DNA is ligated to each Mu end. If there is a bias toward one product, then a mechanism may exist for sensing the target strand to be ligated. For example, deletions are observed much more frequently than inversions with IS*903* (214). Sensing which strand of the target DNA is to be ligated to a Mu end may be accomplished by a tracking transposase or by topological considerations such as those addressed by Craigie and Mizuuchi (39), complicated by the use of target sites located as far as half the length of the *E. coli* chromosome from the original prophage site.

VIII. REGULATION OF Mu DEVELOPMENT

A. Early Transcription of the Mu Genome

Upon infection of a sensitive host, divergent transcription of the repressor gene *c*, proceeding leftwards, and transcription of the early region, proceeding rightwards, is initiated (10, 204, 217). The early region includes the regulatory gene *ner*, the *A* and *B* genes, and genes of the SE region. Entry into lysogeny requires the repression of early region transcription by c protein, while entry into the lytic cycle involves repression of *c* transcription and continuation of transcription of the early region, followed by transcription of C, encoding the positive regulator of late transcription, and then transcription of the late regions.

Figure 11 illustrates the position of the major repressor promoter, Pc (also designated Pcm or Pc-2), and the promoter for the early region transcript, Pe (65, 113, 114). The c repressor inhibits transcription from Pe and autogenously regulates its own synthesis; Ner, the first protein encoded in the early region, inhibits transcription from Pc and reduces transcription from Pe in a manner resembling that of the Cro protein of λ (64, 203). The transcripts from the two promoters overlap such that transcription from one interferes with transcription from the other. Thus the choice between lytic and lysogenic growth is dependent, at least in part, on the balance between use of the two promoters.

1. The c repressor

The c repressor is the only known Mu protein encoded by a transcript synthesized from right to left. In addition to the major Pc promoter, responsible for initiation at 1,066 bp from the left end of the genome, minor promoters have been proposed initiating transcripts at bp 885 (65) and at bp 1088 and 1115 (114). The c protein has a molecular weight of about 19,000 (60, 157). Several mutations resulting in a thermoinducible phenotype were sequenced (J. Vogel and N. P. Higgins, personal communication), and all lie within the *c* gene. The protein binds to three operator regions (O1, O2, O3) located between *c* and *ner* (Fig. 11) (37, 113, 157). Site O2 overlaps the start site of Pe and O3 overlaps the start site of Pc, presumably accounting for regulation of early transcription and autoregulation, respectively. Site O1 overlaps one of the suggested minor promoters. Binding of repressor is apparently cooperative, and O3 is not occupied until O1 and O2 are fully occupied. The O1 and O2 sites have four to five copies of the suggested consensus sequence CTT TPyAPu$^{A}_{T}$AANN$^{A}_{T}$, while O3 has only one copy.

The c repressor and the A protein have considerable sequence similarity (80), and repressor can bind to the A protein-binding sites at the ends of Mu (37), suggesting that high levels of repressor may compete with A protein for binding to these sites. In addition, Goosen and van de Putte (64) argue that transcription from Pc proceeding towards the left end may interfere with A protein binding to the left end of the genome; however, there is no evidence on termination of the Pc transcript and no direct evidence for such an interference.

2. Pe and the Ner repressor

Following an initial burst of early transcription about 4 min after infection or induction, the rate of transcription drops due to expression of *ner*, the first gene of the early transcript; early transcription continues at this reduced rate throughout the lytic cycle (217). In addition to its autoregulatory function, Ner inhibits transcription from Pc (203). Mutation analysis identified a potential Ner-binding site overlapping the start sites of Pe and Pc (63), and footprint analysis with purified Ner protein placed the binding site as shown in Fig. 11 (189). The site contains a dyad symmetry, and binding of Ner to the left or right half may inhibit transcription from Pe or Pc, respectively (63).

Until recently, the early transcript was assumed to extend from the initiation site at 1,028 bp to the C gene at about 10 kb. Two findings have altered this conclusion. The promoter for C has been located about 740 bp upstream of the gene (88, 180), and a promoter for *lig* may be located around 8.2 kb (P. Ghelardini, personal communication). The termination site and the length of the early transcript must now be determined. A potential stem-loop structure at about 9.2 kb serves as a terminator for synthesis of RNA that displays the kinetic and regulatory properties of early RNA, but this does not exclude the possible presence of other, earlier terminators (S. F. Stoddard and M. M. Howe, personal communication).

3. Host IHF

Infection of cells deficient in the HimA or HimD component of the host IHF results in near-normal levels of lysogenization, but no Mu DNA replication or phage production (H. I. Miller and D. I. Friedman, Abstr. Annu. Meet. Am. Soc. Microbiol. 1977, H11, p. 137; Yoshida, thesis). Complementation in *trans* with A and B proteins allows Mu development, suggesting that the inability of Mu DNA to replicate results from a defect in early gene expression in the mutant host strains; indeed, early transcription is reduced, with transcription beyond *A* more severely inhibited than that of the region including *ner* and *A* (Yoshida, thesis).

Genetic analysis and footprinting studies have located an IHF-binding site between O1 and O2 (63, 113; Fig. 11). Binding of IHF to this site markedly alters the balance between Pe and Pc expression, and recent evidence (H. Krause, D. Collier, R. Wells, and N. P. Higgins, personal communication) strongly implicates the topology of the DNA in its response to IHF binding. By comparing the expression of the two promoters on relaxed and supercoiled DNA, in the presence or absence of IHF, Krause et al. found a 40-fold increase in the relative abundance of Pe transcripts on supercoiled DNA in the presence of IHF. In addition, Roulet et al. (166) found that the c repressor binds preferentially to supercoiled DNA. These observations suggest that the circularization and supercoiling of infecting DNA, presumably using the Mu N protein and the host gyrase, respectively, are important in the choice between lysogeny and lysis.

B. *C* and the Middle Transcript

The location of the C promoter revealed that at least two, and possibly four, open reading frames precede C in a small operon (180). The operon apparently functions as a "middle transcript"; transcription is initiated 4 to 8 min after induction, and any interference with DNA replication severely reduces the amount of C transcript (Stoddard and Howe, personal communication). The promoter has a −10 region resembling the consensus sequence, but no normal −35 region (180), and the C gene encodes a 16.5-kDa basic protein (88, 125).

C. Late Transcription

Late transcription is initiated about 20 min into the lytic cycle and requires the C protein (215), though the mechanism of the positive regulation is not known; antitermination (A. van Meeteren, Ph.D. thesis, State University of Leiden, Leiden, The Netherlands, 1980) and sigmalike activity (125) have been suggested. Alternatively, C protein may function as an activator, perhaps like the λ *c*II protein. The positive regulator of late transcription, the C protein, has been shown to be a site-specific DNA-binding protein (139a). Mu DNA replication is apparently also required, as no late transcription is observed with *A* or *B* mutants or with prophage lacking the right end (215; C. F. Marrs and M. M. Howe, personal communication).

Four late promoters have now been identified including ones between genes C and *lys*, upstream of *I*, within *N*, and between *gin* and *mom* (C. F. Marrs, Ph.D. thesis, University of Wisconsin, Madison, 1982; S. F. Stoddard, W. Margolin, G. Rao, and M. M. Howe, personal communication). Conserved sequences were identified in the promoters, having some resemblance to a typical *E. coli* −10 region, but differing in the −35 region (W. Margolin and M. M. Howe, personal communication).

Mu late transcription is inhibited at elevated temperatures in mutant hosts carrying *dnaK*, a member of the *E. coli* heat shock regulon (145). In addition, late transcription is inhibited in a *dnaJ* mutant, another heat shock mutant, when the phage lacks the SE functions. This raises the possibility that there is a function encoded in the SE region that is analogous to the *dnaJ* product. Both *dnaK* and *dnaJ* mutations have no effect on Mu DNA replication, whereas with phage λ they exert their inhibitory effects at initiation of DNA replication (145).

D. *mom*

The *mom* gene, discovered by Toussaint (190), encodes a DNA modification function, which serves to protect modified Mu DNA from host restriction systems. The modification is an unusual one and was shown by Hattman and co-workers (82, 182) to produce α-*N*-(9-β-D-2′-deoxyribofuranosylpurin-6-yl)glycinamide. The adenine residues to be modified form part of the deduced consensus sequence $^{C}_{G}A^{C}_{G}$-NPy (99).

What is most remarkable about the Mom protein is the complexity of the regulation of its expression, apparently designed to delay the expression of Mom until very late in the lytic cycle (105); premature functioning apparently interferes with Mu DNA replication (P. van de Putte, personal communication). Mom is encoded at the extreme right of the genome (191) in an operon with an overlapping gene, *com*, the product of which is involved in regulating the expression of Mom, probably at the level of translation (87, 221); *mom* and *com* encode proteins of 28.3 and 7.4 kDa, respectively (98, 154). Transcription of the *mom* operon is positively regulated by the host Dam methylase (83, 84, 98, 153, 154)—the first demonstration of positive control by a DNA methyltransferase—and by the Mu C protein (86, 88, 125, 154). The role of the Dam system in the regulatory scheme was suggested by the presence of three Dam sites (GATC) located upstream of the *mom* promoter and by the fact that deletion of the sites renders *mom* expression Dam independent. A model by Hattman and Ives (85) postulates that a repressor molecule binds to an operator encompassing the Dam sites when these sites are not methylated, and binding of the repressor prevents binding of the C protein to its appropriate site, in turn preventing *mom* transcription. Deletion of the Dam sites would prevent binding of the repressor and make *mom* expression *dam* independent. The product of the host *mutH* gene, a component of the methyl-directed mismatch repair system, has been proposed to be the repressor, since *mom* expression is *dam* independent in a *mutH* mutant (R. Kahmann, personal communication).

A coherent picture explaining all of the elements of *mom* regulation is slowly forming. The following scenario may describe the regulation of *mom* transcription. Approximately 10 to 20 min into the lytic cycle the positive regulator C is produced, which can activate *mom* expression. However, producing Mom at this time may be premature, as it appears that its early production interferes with Mu replication. To delay the production of Mom, the repressor (MutH?) rapidly binds to the hemimethylated or unmethylated copies of its operator produced by Mu DNA replication, blocking binding of the activator C protein. Eventual titration of the repressor late in the cycle, when a sufficient number of Mu copies have been produced, permits C to bind and leads to initiation of *mom* transcription.

E. Constitutive Transcription: *gin* and *lig*

Inversion of the G region, discussed below, is catalyzed by the *gin* gene product. As this inversion occurs in the lysogenic state, the Gin protein must be produced constitutively and independently of any Mu lytic functions.

Lig was defined as a Mu function which can complement the host *lig* ts7 mutation (57). Though originally observed upon infection, the complementation was also observed in lysogens, suggesting that Lig is produced constitutively. Since *lig* was thought to be part of the SE region encoded on the early transcript, this observation was unexpected. However, *lig* may have its own promoter, which could explain its activity in a lysogen. As Lig is postulated to alter the superhelical density of DNA in host cells, lysogenization by Mu may have greater effects on the host cell than previously imagined.

IX. THE INVERTIBLE G REGION: SITE-SPECIFIC RECOMBINATION

The earliest electron micrographs of heteroduplexes of Mu DNA revealed that some molecules had a 3-kb region of nonrenatured DNA (the "G bubble") about 1.7 kb from the right end of the genome (41, 94). The bubble resulted from inversion of the G region in one of the DNAs supplying a single strand in the heteroduplex. Surprisingly, half the molecules from lysates prepared by induction had G bubbles, while only a couple percent of molecules from lysates prepared by infection had G bubbles (42). This initially baffling observation helped lead to an understanding of the role of the G region and its inversion.

A. G Inversion and Host Range

The dominant orientation of the G region in phage prepared by infection is defined as G+, and the opposite orientation is G−. Kamp et al. (108) found that prophage fixed in the G− orientation could not yield viable phage, defined as phage that could plate on *E. coli* K-12, and Symonds and Coehlo (183) found that only half the cells of a lysogenic population could produce viable phage; those cells had prophages with the G+ orientation, while cells with prophages that had the G− orientation could not produce viable phage. The explanation of these observations is that only G+ phage can infect *E. coli* K-12, and the frequency of inversion of G is sufficiently low that almost all of the phage progeny of the infection are G+; however, inversion occurs in the lysogenic state, eventually resulting in an equilibrium between G+ and G− prophage.

A suggestion of Howe (91) that G− phage might be able to infect cells other than *E. coli* K-12 was proved correct by van de Putte et al. (198), who found that G− phage could infect *Citrobacter freundii*. The host range of G+ phage is now known to include *E. coli* K-12 and various strains of *Salmonella* and *Serratia* spp.; the G− host range includes *E. coli* C and strains of *Citrobacter*, *Shigella*, *Enterobacter*, and *Erwinia* spp. (197).

The inability of G− phage to grow on *E. coli* K-12 results from failure of the phage to adsorb to sensitive cells (17). The adsorption properties of the phage are dependent on the synthesis of proteins encoded in the G region: the G+ orientation allows the synthesis of two proteins, S and U, while inversion of G prevents the synthesis of S and U proteins and allows synthesis of an alternate set of proteins, S′ and U′ (93, 198).

The S and S′ proteins are not encoded entirely within the G region; rather, a remarkable example of gene splicing is responsible for the synthesis of S and S′ proteins with common N-terminal portions and different C-terminal portions. Giphart-Gassler et al. (59) showed that the common 5′ terminus (Sc) is encoded immediately to the left of the G region; transcription initiated in the common portion continues into the variable (Sv or S′v) portion, dependent on the orientation of G. The G+ orientation allows the synthesis of S and U proteins of 55.4 and 20.3 kDa, respectively; the G− orientation yields S′ and U′ proteins of 52.6 and 20.7 kDa, respectively, from the opposite DNA strand. Each pair of genes covers about half the length of the G region with no overlap, and the two halves are separated by a potential stem-loop structure which can function as a transcription terminator (Kahmann, personal communication).

The two sets of genes are responsible for presentation of the tail fibers on the phage particle (74). The S and S′ proteins appear to function like primitive antibodies: the common N-terminal portion presumably anchors the tail fiber to the phage tail, and the variable C-terminal portion serves to recognize alternative sets of phage receptors on the surfaces of sensitive bacteria. The bacterial surface receptor for G+ phage is a terminal glucose in α1-2 linkage within the core lipopolysaccharide; the G− phage receptor of *Erwinia carotovora* is a terminal glucose in a β1-6 linkage within the core lipopolysaccharide (110, 169, 170).

B. Mechanism of G Inversion

Inversion of the G region is accomplished by a site-specific recombination between two 34-bp inverted repeat sequences at the ends of G (102). The inversion is catalyzed by the 21.7-kDa Gin protein, which is encoded by the *gin* gene (118) located to the right of the G region with its promoter overlapping the right inverted repeat sequence (108, 130, 151). Expression of *gin* is constitutive, as shown by the G inversion occurring in the lysogenic state. Kahmann and others (103, 111, 112) demonstrated that efficient recombination requires, in addition to the Gin protein and the inverted repeats, a *cis*-acting sequence called *sis* (sequence for inversion stimulation), located within the *gin* gene, and a host-encoded protein called FIS (factor for inversion stimulation). The *sis* sequence has properties resembling those of eucaryotic enhancers, as its function is independent of orientation and distance relative to the inverted repeats. DNase footprinting studies show that the Gin protein binds to the inverted repeats; FIS protein binds to *sis*, consistent with the observation that FIS is required for *sis* function and vice versa (102). The question of whether the FIS/*sis* system is required for recombination or is stimulatory remains to be answered. The development by Plasterk et al. (152) and Kahmann and co-workers (104, 130) of an in vitro system for G inversion is allowing a detailed analysis of the mechanism of the reaction, as discussed by Glasgow et al. (A. C. Glasgow, K. T. Hughes, and M. I. Simon, this volume).

X. OTHER TRANSPOSING BACTERIOPHAGES

The existence of other transposing phages may eventually allow an understanding of the evolutionary origin of these phages through a comparative analysis. To date, the other transposing phages include a single other coliphage, D108, numerous examples from *Pseudomonas* sp., and two phages from *Vibrio cholerae*.

A. D108

First isolated in 1971 (133), D108 was found to be a transposing phage similar to Mu (96). Heteroduplex analysis between D108 and Mu DNA showed about 90% homology, defined as the ability to form a DNA hybrid visible by electron microscopy (58). The phage DNA contains heterogeneous host DNA sequences at its termini and possesses an invertible region corresponding to the G region, and most antibodies raised against Mu cross-react with D108 (45); hence, it is clear the two phages are closely related. D108 mediates all the chromosomal rearrangements associated with Mu (50), and D108 transposition generates 5-bp repeats of host DNA (185).

The nonhomologous regions between D108 and Mu observed by electron microscope included three regions at the left end of the genomes (estimated to extend from 0.15 to 1.5 kb, 4.9 to 5.2 kb, and 7.9 to 8.5 kb) and a 0.5-kb insertion near the right end of D108 (58). The first region encompasses the *c* and *ner* genes, the intervening control sequences, and the 5′ portion of the *A* genes (137, 195). The repressors of one phage do not bind to the operators of the other; hence, the phages are heteroimmune. Though the nucleotide sequences of the *c* and *ner* genes show little similarity, the amino acid sequences of the proteins do show significant similarity, with the Ner proteins sharing about 50% sequence similarity (137, 188). IHF- and Ner-binding sites have been located between the *c* and *ner* genes (137, 188), but the locations of the divergent promoters have not yet been mapped. Although the overall structure of the regulatory regions of the two phages is similar, significant differences may exist. The D108 c repressor protects a single 77-bp region in footprinting studies, whereas the Mu repressor protects three separate sites, and the D108 repressor-binding site overlaps the IHF-binding site (D. B. Levin and M. S. DuBow, personal communication). Both a full-length D108 c protein (174 amino acids) and a truncated form (142 amino acids) resulting from the use of an alternate, in-frame initiation codon may be expressed (119).

The *A* genes of D108 and Mu show sequence divergence over the 5′ 0.2 kb, but are similar over the remainder of the sequence (195). Complementation experiments using D108 A^- genomes and a source of Mu A protein, as well as the reverse experiment, have given somewhat contradictory results (37, 195); it appears that the two transposases can complement each other, though inefficiently

The second nonhomologous region is located in the 3′ portion of the third open reading frame past the *B* gene (211), and the third region includes the 3′ portion of an open reading frame located 7.6 kb from the left end (M. Banerjee, B. Waggoner, and M. Pato, personal communication).

The extra DNA at the right end of D108 contains an open reading frame for a protein of 104 amino acids, which is transcribed during the D108 lytic cycle (186). The DNA does not resemble an insertion sequence, as it lacks terminal repeats and homology with any known insertion sequence or transposon.

The nucleotide sequences at the termini of the two phages, which include the three A protein-binding sites at each terminus, are very similar (18, 185). The left-end 54 bp encompassing the L1-binding site are identical; the other binding sites contain base changes, perhaps accounting for the different efficiencies with which the heterologous transposases function. Viable hybrid phages carrying the left end of Mu (including *c*, *ner*, and the 5′ end of *A*), as well as the rest of the genome from D108 (called MD phage) and the reverse (called DM phage), have been constructed (195), illustrating that the heterologous termini are capable of functioning together.

B. *Pseudomonas* Transposing Phages

Krylov and co-workers have isolated a large number of transposing phages from clinical isolates of *Pseudomonas* strains (2). Some of the similarities between these phages and the transposing coliphages are striking: the genomes are all about 37 kb in size, with heterogeneous host sequences of about 2.5 kb at the right end of the genomes (115). Indeed, it was this characteristic which was used in their isolation: the DNAs of newly isolated phages were cut with restriction enzymes and run on agarose gels, and the presence or absence of a diffuse band was noted; a diffuse band would correspond to a terminal fragment containing a heterogeneous length of host DNA. A gene responsible for immunity has been mapped to the left end of the genome of the *Pseudomonas* phage D3112, at a site corresponding to the *c* repressor gene of Mu, and three early genes have been mapped to a region between 1.3 and 14.2 kb (15). Despite these resemblances, Rehmat and Shapiro (160) were unable to demonstrate any homology between Mu and the *Pseudomonas* phages by using Southern blot analysis.

Several pieces of evidence suggest that the *Pseudomonas* phages use a mechanism of transposition similar to that used by Mu; e.g., lysogenization results in integration of the phage genomes at numerous sites, leading to auxotrophic mutations in the host, and the integrated genomes are found in both orientations (155). An experiment similar to that of Ljungquist and Bukhari (121) showed that the original prophage in a lysogen is not excised following induction (160).

Though the *Pseudomonas* phages cannot infect *E. coli*, phage D3112 has been introduced into *E. coli* on a transmissible RP4 plasmid. The phage genome can then transpose to the *E. coli* chromosome, yielding auxotrophic host mutants, and spontaneous phage production is observed (116, 156).

C. *V. cholerae* Phages VcA1 and VcA2

The heteroimmune phages VcA1 and VcA2 from *V. cholerae* are thought to be transposing phages similar to Mu. They lysogenize sensitive hosts with integration of phage genomes into numerous host chromosomal sites and produce auxotrophic host mutants (55, 129). A defective form of VcA1 located on a conjugative plasmid can mediate formation of Hfr's in a manner resembling cointegrate formation with F plasmids carrying Mu or mini-Mu genomes (97). Additional information is clearly required to understand the relation of these phages to the known transposing phages.

XI. EVOLUTION OF TRANSPOSING PHAGES

The existence of transposing viruses raises the fascinating question of the evolutionary relationship between viruses and transposable elements. This relationship can be fruitfully analyzed by using the concept of "modules" employed by Campbell and Botstein (23) to describe clusters of genes involved in common functions. The Mu genome can be considered to be composed of several modules: a transposition module, a module for phage functions, the SE region, and the G region (106). These modules can overlap, in the sense that the *A* gene is required for both transposon and phage function.

The best case for modular construction can be made for the G region, which is very closely related to the C region of phage P1 by DNA sequence (33) and function (196). The Gin protein, required for G inversion, and the corresponding Cin protein of P1 can complement each other; indeed, these two proteins, as well as two others involved in site-specific recombination leading to inversion (Hin, which regulates flagellar phase variation in *Salmonella* sp., and the Pin protein of the cryptic e14 prophage), can all complement each other (117, 151, 178). The proteins share about 70% amino acid sequence homology (109, 151). Homology between these proteins and the TnpR resolvase of Tn*3* suggests a relationship between invertases and resolvases (179).

The transposition module has as its minimal requirements the ends of Mu and the *A* gene. Mini-Mu's containing these elements (as well as the *c* and *ner* genes) resemble other transposable elements which transpose at low frequency. Addition of the *B* gene greatly increases the frequency of transposition, and further addition of the SE region again increases the frequency of transposition.

Several points of similarity between Mu and Tn*3* suggest that the Mu transposition module is most closely related to the Tn*3* family of transposons. These include use of a common transposition mechanism (at least for replicative transposition), duplication of 5 bp of host DNA, divergently transcribed transposase and repressor genes, and similarity of nucleotide sequences at the transposon ends. The two transposition modules differ significantly in that Tn*3* apparently does not undergo conservative transposition and Mu encodes a second repressor function, Ner. Also, it is likely that Mu does not encode a resolvase function. A highly sensitive computer analysis of comparisons between the nucleotide sequence of the Mu left end, up to the end of the *B* gene, and the sequences of several other transposable elements showed a significant resemblance between the sequences of Mu and Tn*3* (106).

Comparison of the phage genetic maps of Mu and other bacteriophages suggests a similarity with phage λ (199). Alignment of portions of the maps shows considerable functional agreement. Proceeding rightward from the immunity regions (and eliminating *c*II), one obtains:

Mu:	*c*	*ner*	*A*	*B*	SE region	*C*	lysis genes	late genes
λ:	*c*I	*cro*	*O*	*P*	b1 region	*Q*	lysis genes	late genes

Transcription for both is divergent from promoters located between the repressor genes. The gene pairs *AB* and *OP* are required for replication, but the integration functions associated with the Mu genes have been separated for λ into separate genes located elsewhere on the genome. The Mu SE and λ b1 regions contain numerous small open reading frames that are nonessential for phage development; however, functions localized to the SE region such as *kil* and *gam* have their counterparts in λ, mapping in a different region of the genome. *C* and *Q* are required for positive control of late transcription. The functional alignments are impressive, but no alignments have been observed for corresponding functions at the nucleotide sequence level (106).

Using correspondences such as the above, Kamp (106) suggested a scheme for the evolution of the Mu genome which envisions insertion of a Tn*3*-like transposon into a phage, followed by mobilization of phage modules between two inverted copies of the transposon. One transposon copy at the left of this composite transposon could have retained most of its functions, while the other could have lost its transposase and had its TnpR resolvase changed into the Gin invertase.

Any such scheme is by its nature highly speculative, and there is much to learn before we can meaningfully address the question: which came first, Mu the virus or Mu the transposon?

Acknowledgments. I am indebted to Barbara Waggoner for critical reading of the manuscript and to my colleagues in Mu for supplying unpublished results. This chapter is dedicated to the memory of Ahmad Bukhari, Janet Miller, Harro Priess, and Els van Leerdam.

LITERATURE CITED

1. **Abelson, J., W. Boram, A. I. Bukhari, M. Faelen, M. M. Howe, M. Metlay, A. L. Taylor, A. Toussaint, and C. A. Wijffelman.** 1973. Summary of the genetic mapping of prophage Mu. *Virology* **54:**90–92.

1a. **Adzuma, K., and K. Mizuuchi.** 1988. Target immunity of Mu transposition reflects a differential distribution of Mu B protein. *Cell* **53:**257–266.

2. **Akhverdyan, V. Z., E. A. Khrenova, V. G. Bogush, T. V. Gerasimova, N. B. Kirsanov, and V. N. Krylov.** 1984. Wide distribution of transposable phages in natural populations of Pseudomonas aeruginosa. *Genetika* **20:**1612–1619.

3. **Akroyd, J., B. Barton, P. Lund, S. Maynard-Smith, K. Sultana, and N. Symonds.** 1984. Mapping and properties of the gam and sot genes of phage Mu: their possible roles in recombination. *Cold Spring Harbor Symp. Quant. Biol.* **49:**261–266.

4. **Akroyd, J., E. Clayson, and N. P. Higgins.** 1986. Purification of the gam gene product of bacteriophage Mu, and the determination of the nucleotide sequence of the gam gene. *Nucleic Acids Res.* **14:**6901–6914.

5. **Akroyd, J., and N. Symonds.** 1983. Evidence for a conservative pathway of transposition of bacteriophage Mu. *Nature* (London) **303:**84–86.

6. **Akroyd, J., and N. Symonds.** 1986. Localization of the gam gene of bacteriophage Mu and characterization of the gene product. *Gene* **49:**273–282.

7. **Allet, B.** 1978. Nucleotide sequences at the ends of bacteriophage Mu DNA. *Nature* (London) **274:**553–558.

8. **Allet, B.** 1979. Mu insertion duplicates a 5 base pair sequence at the host inserted site. *Cell* **16:**123–129.

9. **Arthur, A., and D. J. Sherratt.** 1979. Dissection of the transposition process: a transposon-encoded site-specific recombination system. *Mol. Gen. Genet.* **175:**267–274.

10. **Bade, E.** 1972. Asymmetric transcription of bacteriophage Mu-1. *J. Virol.* **10:**1205–1207.

11. **Berg, D. E.** 1977. Insertion and excision of the transposable kanamycin resistence determinant Tn*5*, p. 205–212. *In* A. I. Bukhari, J. A. Shapiro, and S. L. Adhya (ed.), *DNA Insertion Sequences, Plasmids, and Episomes.* Cold Spring Harbor Laboratory, Cold Spring Harbor, N.Y.

12. **Betermier, M., R. Alazard, F. Raguet, E. Roulet, A. Toussaint, and M. Chandler.** 1987. Phage Mu transposase: deletion of the carboxy-terminal end does not abolish DNA binding activity. *Mol. Gen. Genet.* **210:**77–85.

13. **Bialy, H., B. T. Waggoner, and M. L. Pato.** 1980. Fate of plasmids containing Mu DNA: chromosome association and mobilization. *Mol. Gen. Genet.* **180:**377–383.

14. **Boeckh, C., E. G. Bade, H. Delius, and J. N. Reeve.** 1986. Inhibition of bacterial segregation by early functions of phage Mu and association of replication protein B with the inner cell membrane. *Mol. Gen. Genet.* **202:**461–466.

15. **Bogush, V. G., T. G. Plotnikova, and V. N. Krylov.** 1981. Bacteriophages of Pseudomonas aeruginosa with DNA structure similar to that of phage Mu1. Isolation and analysis of hybrid phages D3112 and B39: localization of the immunity region and some genetic factors. *Genetika* **17:**967–976.

16. **Boram, W., and J. Abelson.** 1973. Bacteriophage Mu integration: on the orientation of the prophage. *Virology* **54:**102–108.

16a. **Bourret, R., and M. Fox.** 1988. Lysogenization of *Escherichia coli him*$^+$, *himA*, and *himD* hosts by bacteriophage Mu. *J. Bacteriol.* **170:**1672–1682.

16b. **Bourret, R., and M. Fox.** 1988. Intermediates in bacteriophage Mu lysogenization of *Escherichia coli him* hosts. *J. Bacteriol.* **170:**1683–1690.

17. **Bukhari, A. I., and L. Ambrosio.** 1978. The invertible segment of bacteriophage Mu DNA determines the adsorption properties of Mu particles. *Nature* (London) **271:**575–577.

18. **Bukhari, A. I., J. R. Lupski, P. Svec, and G. N. Godson.** 1985. Comparison of left-end DNA sequences of bacteriophages Mu and D108. *Gene* **33:**235–239.

19. **Bukhari, A. I., and A. L. Taylor.** 1975. Influence of insertions on packaging of host sequences covalently linked to bacteriophage Mu DNA. *Proc. Natl. Acad. Sci. USA* **72:**4399–4403.

20. **Bukhari, A. I., and D. Zipser.** 1972. Random insertion of Mu-1 DNA within a single gene. *Nature* (London) *New Biol.* **236:**240–243.

21. **Burlingame, R. P., M. G. Obukowicz, D. L. Lynn, and M. M. Howe.** 1986. Isolation of point mutations in bacteriophage Mu attachment regions cloned in a λ::mini-Mu phage. *Proc. Natl. Acad. Sci. USA* **83:**6012–6016.

22. **Campbell, A.** 1980. Some general questions about movable elements and their implications. *Cold Spring Harbor Symp. Quant. Biol.* **45:**1–9.

23. **Campbell, A., and D. Botstein.** 1983. Evolution of the lambdoid phages, p. 365–380. *In* R. W. Hendrix, J. W. Roberts, F. W. Stahl, and R. A. Weisberg (ed.), *Lambda II.* Cold Spring Harbor Laboratory, Cold Spring Harbor, N.Y.

24. **Castilho, B. A., P. Olfson, and M. J. Casadaban.** 1984. Plasmid insertion mutagenesis and *lac* gene fusion with

mini-Mu bacteriophage transposons. *J. Bacteriol.* **158:**488–495.

25. **Chaconas, G., E. B. Giddens, J. L. Miller, and G. Gloor.** 1985. A truncated form of the bacteriophage Mu B protein promotes conservative integration, but not replicative transposition, of Mu DNA. *Cell* **41:**857–865.
26. **Chaconas, G., G. Gloor, and J. L. Miller.** 1985. Amplification and purification of the bacteriophage Mu encoded B transposition protein. *J. Biol. Chem.* **260:**2662–2669.
27. **Chaconas, G., G. Gloor, J. L. Miller, D. L. Kennedy, E. B. Giddens, and C. R. Nagainis.** 1984. Transposition of bacteriophage Mu DNA: expression of the A and B proteins from λpL and analysis of infecting Mu DNA. *Cold Spring Harbor Symp. Quant. Biol.* **49:**279–284.
28. **Chaconas, G., R. M. Harshey, and A. I. Bukhari.** 1980. Association of Mu-containing plasmids with the Escherichia coli chromosome upon prophage induction. *Proc. Natl. Acad. Sci. USA* **77:**1778–1782.
29. **Chaconas, G., R. M. Harshey, N. Sarvetnick, and A. I. Bukhari.** 1981. Predominant end-products of prophage Mu DNA transposition during the lytic cycle are replicon fusions. *J. Mol. Biol.* **150:**341–359.
30. **Chaconas, G., D. L. Kennedy, and D. Evans.** 1983. Predominant integration end-products of infecting bacteriophage Mu DNA are simple insertions with no preference for integration of either Mu DNA strand. *Virology* **128:**48–59.
31. **Chandler, M., M. Clerget, and D. Galas.** 1982. The transposition frequency of IS1-flanked transposons is a function of their size. *J. Mol. Biol.* **154:**229–243.
32. **Chase, C. D., and R. H. Benzinger.** 1982. Transfection of *Escherichia coli* spheroplasts with a bacteriophage Mu DNA-protein complex. *J. Virol.* **42:**176–185.
33. **Chow, L. T., and A. I. Bukhari.** 1976. The invertible DNA segments of coliphage Mu and P1 are identical. *Virology* **74:**242–248.
34. **Coelho, A., D. Leach, S. Maynard-Smith, and N. Symond.** 1981. Transposition studies using a ColE1 derivative carrying bacteriophage Mu. *Cold Spring Harbor Symp. Quant. Biol.* **45:**323–328.
35. **Coelho, A., S. Maynard-Smith, and N. Symonds.** 1982. Abnormal cointegrate structures mediated by gene B mutants of phage Mu: their implications with regard to gene function. *Mol. Gen. Genet.* **185:**356–362.
36. **Craigie, R., D. J. Arndt-Jovin, and K. Mizuuchi.** 1985. A defined system for the DNA strand-transfer reaction at the initiation of bacteriophage Mu transposition: protein and DNA substrate requirements. *Proc. Natl. Acad. Sci. USA* **82:**7570–7574.
37. **Craigie, R., M. Mizuuchi, and K. Mizuuchi.** 1984. Site-specific recognition of the bacteriophage Mu ends by the Mu A protein. *Cell* **39:**387–394.
38. **Craigie, R., and K. Mizuuchi.** 1985. Mechanism of transposition of bacteriophage Mu: structure of a transposition intermediate. *Cell* **41:**867–876.
39. **Craigie, R., and K. Mizuuchi.** 1986. Role of DNA topology in Mu transposition: mechanism of sensing the relative orientation of two DNA segments. *Cell* **45:**793–800.
40. **Craigie, R., and K. Mizuuchi.** 1987. Transposition of Mu DNA: joining of Mu to target DNA can be uncoupled from cleavage at the ends of Mu. *Cell* **51:**493–501.
41. **Daniell, E., J. Abelson, J. S. Kim, and N. Davidson.** 1973. Heteroduplex structures of bacteriophage Mu DNA. *Virology* **51:**237–239.
42. **Daniell, E., W. Boram, and J. Abelson.** 1973. Genetic mapping of the inversion loop in bacteriophage Mu DNA. *Proc. Natl. Acad. Sci. USA* **70:**2153–2156.
43. **Daniell, E., D. E. Kohne, and J. Abelson.** 1975. Characterization of the inhomogeneous DNA in virions of bacteriophage Mu by DNA reannealing kinetics. *J. Virol.* **15:**739–743.
44. **Daniell, E., R. Roberts, and J. Abelson.** 1972. Mutations in the lactose operon caused by bacteriophage Mu. *J. Mol. Biol.* **69:**1–8.

44a. **Darzins, A., N. Kent, M. Buckwalter, and M. Casadaban.** 1988. Bacteriophage Mu sites required for transposition immunity. *Proc. Natl. Acad. Sci. USA* **85:**6826–6830.

45. **Dubow, M. S., and A. I. Bukhari.** 1981. The proteins of bacteriophage Mu: composition of the virion and biosynthesis in vivo during lytic growth, p. 47–67. *In* M. S. Dubow (ed.), *Bacteriophage Assembly.* Alan R. Liss, Inc., New York.
46. **Engler, J. A., and M. P. van Bree.** 1981. The nucleotide sequence and protein-coding capability of the transposable element IS5. *Gene* **14:**155–163.
47. **Faelen, M., O. Huisman, and A. Toussaint.** 1978. Involvement of phage Mu-1 early functions in Mu-mediated chromosomal rearrangements. *Nature* (London) **271:**580–582.
48. **Faelen, M., and A. Toussaint.** 1978. Stimulation of deletions in the *Escherichia coli* chromosome by partially induced Mu*cts*62 prophages. *J. Bacteriol.* **136:**477–483.
49. **Faelen, M., and A. Toussaint.** 1980. Inversions induced by temperate bacteriophage Mu-1 in the chromosome of *Escherichia coli* K-12. *J. Bacteriol.* **142:**391–399.
50. **Faelen, M., and A. Toussaint.** 1980. Chromosomal rearrangements induced by temperate bacteriophage D108. *J. Bacteriol.* **143:**1029–1030.
51. **Faelen, M., A. Toussaint, and M. Couturier.** 1971. Mu-1 promoted integration of a λ-gal phage in the chromosome of E. coli. *Mol. Gen. Genet.* **113:**367–370.
52. **Faelen, M., A. Toussaint, B. T. Waggoner, L. Desmet, and M. Pato.** 1986. Transposition and replication of Maxi-Mu derivatives of bacteriophage Mu. *Virology* **153:**70–79.
53. **Fitts, R. A., and A. L. Taylor.** 1980. Integration of bacteriophage Mu at host chromosomal replication forks during lytic development. *Proc. Natl. Acad. Sci. USA* **77:**2801–2805.
54. **Fuller, R. S., J. M. Kaguni, and A. Kornberg.** 1981. Enzymatic replication of the origin of the Escherichia coli chromosome. *Proc. Natl. Acad. Sci. USA* **78:**7370–7374.
55. **Gerdes, J. C., and W. R. Romig.** 1975. Complete and defective bacteriophages of classical *Vibrio cholerae*: relationship to the Kappa type bacteriophage. *J. Virol.* **15:**1231–1238.
56. **Ghelardini, P., J. C. Liebart, C. Marchelli, A. M. Pedrini, and L. Paolozzi.** 1984. *Escherichia coli* K-12 *gyrB* gene product is involved in the lethal effect of the *lig*ts2 mutant of bacteriophage Mu. *J. Bacteriol.* **157:**665–668.
57. **Ghelardini, P., L. Paolozzi, and J. C. Liebart.** 1980. Restoration of ligase activity in E. coli K12 ligts7 strain by bacteriophage Mu and cloning of a DNA fragment harbouring the Mu 'lig' gene. *Nucleic Acids Res.* **8:**3157–3173.
58. **Gill, G. S., R. C. Hull, and R. Curtiss III.** 1981. Mutator bacteriophage D108 and its DNA: an electron microscopic characterization. *J. Virol.* **37:**420–430.
59. **Giphart-Gassler, M., R. Plasterk, and P. van de Putte.** 1982. G inversion in bacteriophage Mu: a novel way of gene splicing. *Nature* (London) **297:**339–342.
60. **Giphart-Gassler, M., J. Reeve, and P. van de Putte.** 1981. Polypeptides encoded by the early region of bacteriophage Mu synthesized in minicells of Escherichia coli. *J. Mol. Biol.* **145:**165–191.

61. **Giphart-Gassler, M., and P. van de Putte.** 1978. Early gene products of bacteriophage Mu: identification of the B gene product. *J. Mol. Biol.* **120:**1–12.
62. **Gloor, G., and G. Chaconas.** 1986. The bacteriophage Mu N gene encodes the 64Kd virion protein which is injected with and circularizes infecting Mu DNA. *J. Biol. Chem.* **261:** 16682–16688.
63. **Goosen, N., and P. van de Putte.** 1984. Regulation of Mu transposition. I. Localization of the presumed recognition sites for HimD and Ner functions controlling bacteriophage Mu transcription. *Gene* **30:**41–46.
64. **Goosen, N., and P. van de Putte.** 1986. Role of Ner protein in bacteriophage Mu transposition. *J. Bacteriol.* **167:** 503–507.
65. **Goosen, N., M. van Heuvel, G. F. Moolenaar, and P. van de Putte.** 1984. Regulation of Mu transposition. II. The Escherichia coli HimD protein positively controls two repressor -promoters and the early promoter of bacteriophage Mu. *Gene* **32:**419–426.
66. **Goosen, T.** 1978. Replication of bacteriophage Mu: direction and possible location of the origin, p. 121–126. *In* I. Molineux and K. Kohiyama (ed.), *DNA Synthesis: Present and Future.* Plenum Publishing Corp., New York.
67. **Goosen, T., M. Giphart-Gassler, and P. van de Putte.** 1982. Bacteriophage Mu DNA replication is stimulated by non-essential early functions. *Mol. Gen. Genet.* **186:**135–139.
68. **Gorbalenya, A. E., A. P. Donchenko, E. V. Koonin, and V. M. Blinov.** 1987. Bacteriophage Mu transposase contains a segment strikingly similar in its amino acid sequence to poliovirus genome-linked protein VPg. *Molek. Genetika* **9:**38–41.
69. **Grindley, N. D. F., and D. J. Sherratt.** 1979. Sequence analysis at IS1 insertion sites: models for transposition. *Cold Spring Harbor Symp. Quant. Biol.* **43:**1257–1261.
70. **Groenen, M., M. Vollering, P. Krijgsman, K. van Drunen, and P. van de Putte.** 1987. Interactions of the transposase with the ends of Mu: formation of specific nucleoprotein structures and non-cooperative binding of the transposase to its binding sites. *Nucleic Acids Res.* **15:**8831–8844.
71. **Groenen, M. A. M., E. Timmers, and P. van de Putte.** 1985. DNA sequences at the ends of the genome of bacteriophage Mu essential for transposition. *Proc. Natl. Acad. Sci. USA* **82:**2087–2091.
72. **Groenen, M. A. M., and P. van de Putte.** 1986. Analysis of the attachment sites of bacteriophage Mu using site-directed mutagenesis. *J. Mol. Biol.* **189:**597–602.
73. **Groenen, M. A. M., and P. van de Putte.** 1986. Transposition of mini-Mu containing only one of the ends of bacteriophage Mu. *EMBO J.* **5:**3687–3690.
74. **Grundy, F. J., and M. M. Howe.** 1984. Involvement of the invertible G segment in bacteriophage Mu tail fiber biosynthesis. *Virology* **134:**296–317.
75. **Harshey, R. M.** 1983. Switch in the transposition products of Mu DNA mediated by proteins: cointegrates versus simple insertions. *Proc. Natl. Acad. Sci. USA* **80:**2012–2016.
76. **Harshey, R. M.** 1984. Transposition without duplication of infecting bacteriophage Mu DNA. *Nature* (London) **311:** 580–581.
77. **Harshey, R. M.** 1988. Phage Mu. *In* P. Calendar (ed.), *The Viruses.* Plenum Publishing Corp., New York, in press.
78. **Harshey, R. M., and A. I. Bukhari.** 1983. Infecting bacteriophage Mu DNA forms a circular DNA-protein complex. *J. Mol. Biol.* **167:**427–441.
79. **Harshey, R. M., and S. D. Cuneo.** 1986. Carboxy-terminal mutants of phage Mu transposase. *J. Genet.* **65:**159–174.
80. **Harshey, R. M., E. D. Getzoff, D. L. Baldwin, J. L. Miller, and G. Chaconas.** 1985. Primary structure of phage Mu transposase: homology to Mu repressor. *Proc. Natl. Acad. Sci. USA* **82:**7676–7680.
81. **Harshey, R. M., R. McKay, and A. I. Bukhari.** 1982. DNA intermediates in transposition of phage Mu. *Cell* **29:**561–571.
82. **Hattman, S.** 1979. Unusual modifications of bacteriophage Mu DNA. *J. Virol.* **32:**468–475.
83. **Hattman, S.** 1982. DNA methyltransferase-dependent transcription of the phage Mu mom gene. *Proc. Natl. Acad. Sci. USA* **79:**5518–5521.
84. **Hattman, S., M. Goradia, C. Monaghan, and A. I. Bukhari.** 1983. Regulation of the DNA-modification function of bacteriophage Mu. *Cold Spring Harbor Symp. Quant. Biol.* **47:**647–653.
85. **Hattman, S., and J. Ives.** 1984. S1 nuclease mapping of the phage Mu mom gene promoter: a model for the regulation of mom expression. *Gene* **29:**185–198.
86. **Hattman, S., J. Ives, W. Margolin, and M. M. Howe.** 1985. Regulation and expression of the bacteriophage Mu mom gene: mapping of the transactivation (Dad) function to the C region. *Gene* **39:**71–76.
87. **Hattman, S., J. Ives, L. Wall, and S. Maric.** 1987. The bacteriophage Mu com gene appears to specify a translation factor required for mom gene expression. *Gene* **55:**345–351.
88. **Heisig, P., and R. Kahmann.** 1986. The sequence and mom transactivation function of the C gene of bacteriophage Mu. *Gene* **43:**59–67.
89. **Higgins, N. P., P. Manlapaz-Romos, R. T. Gandhi, and B. M. Olivera.** 1983. Bacteriophage Mu: a transposing replicon. *Cell* **33:**623–628.
90. **Higgins, N. P., D. Moncecchi, P. Manlapaz-Ramos, and B. M. Olivera.** 1983. Bacteriophage Mu DNA replication in vitro. *J. Biol. Chem.* **258:**4293–4297.
91. **Howe, M. M.** 1978. Invertible DNA in phage Mu. *Nature* (London) **271:**608–610.
92. **Howe, M. M., and J. W. Schumm.** 1981. Transposition of bacteriophage Mu—properties of λ phages containing both ends of Mu. *Cold Spring Harbor Symp. Quant. Biol.* **45:** 337–346.
93. **Howe, M. M., J. W. Schumm, and A. L. Taylor.** 1979. The S and U genes of bacteriophage Mu are located in the invertible G segment of Mu DNA. *Virology* **92:**108–124.
94. **Hsu, M. T., and N. Davidson.** 1972. Structure of inserted bacteriophage Mu-1 DNA and physical mapping of bacterial genes by Mu-1 DNA insertion. *Proc. Natl. Acad. Sci. USA* **69:**2823–2827.
95. **Hsu, M. T., and N. Davidson.** 1974. Electron microscope heteroduplex study of the heterogeneity of Mu phage and prophage DNA. *Virology* **58:**229–239.
96. **Hull, R. A., G. S. Gill, and R. Curtiss III.** 1978. Genetic characterization of Mu-like bacteriophage D108. *J. Virol.* **27:**513–518.
97. **Johnson, S. R., and W. R. Romig.** 1981. *Vibrio cholerae* conjugative plasmid pSJ15 contains transposable prophage dVcA1. *J. Bacteriol.* **146:**632–638.
98. **Kahmann, R.** 1983. Methylation regulates the expression of a DNA-modification function encoded by bacteriophage Mu. *Cold Spring Harbor Symp. Quant. Biol.* **47:**639–646.
99. **Kahmann, R.** 1984. The mom gene of bacteriophage Mu. *Curr. Top. Microbiol. Immunol.* **108:**29–47.
100. **Kahmann, R., and D. Kamp.** 1979. Nucleotide sequences of the attachment sites of bacteriophage Mu DNA. *Nature* (London) **280:**247–250.

101. **Kahmann, R., D. Kamp, and D. Zipser.** 1976. Transfection of Escherichia coli by Mu DNA. *Mol. Gen. Genet.* **149:** 323–328.

102. **Kahmann, R., G. Mertens, A. Klippel, B. Brauer, F. Rudt, and C. Koch.** 1987. The mechanism of G inversion, p. 681–690. *In* T. J. Kelly, Jr., and R. McMacken (ed.), *DNA Replication and Recombination.* Alan R. Liss, Inc., New York.

103. **Kahmann, R., F. Rudt, C. Koch, and G. Mertens.** 1985. G inversion in bacteriophage Mu DNA is stimulated by a site within the invertase gene and a host factor. *Cell* **41:**771–780.

104. **Kahmann, R., F. Rudt, and G. Mertens.** 1984. Substrate and enzyme requirements for in vitro site-specific recombination in bacteriophage Mu. *Cold Spring Harbor Symp. Quant. Biol.* **49:**285–294.

105. **Kahmann, R., A. Seiler, F. G. Wulczyn, and E. Pfaff.** 1985. The mom gene of bacteriophage Mu: a unique regulatory scheme to control a lethal gene. *Gene* **39:**61–70.

106. **Kamp, D.** 1987. Mu evolution, p. 259–269. *In* N. Symonds, A. Toussaint, P. van de Putte, and M. Howe (ed.), *Phage Mu.* Cold Spring Harbor Laboratory, Cold Spring Harbor, N.Y.

107. **Kamp, D., and R. Kahmann.** 1981. Two pathways in bacteriophage Mu transposition? *Cold Spring Harbor Symp. Quant. Biol.* **45:**329–336.

108. **Kamp, D., R. Kahmann, D. Zipser, T. R. Broker, and L. T. Chow.** 1978. Inversion of the G DNA segment of phage Mu controls phage infectivity. *Nature* (London) **271:**577–580.

109. **Kamp, D., E. Kardas, W. Ritthaler, R. Sandulache, R. Schmuker, and B. Stern.** 1984. Comparative analysis of invertible DNA in phage genomes. *Cold Spring Harbor Symp. Quant. Biol.* **49:**301–311.

110. **Kamp, D., and R. Sandulache.** 1983. Recognition of cell surface receptors is controlled by invertible DNA of phage Mu. *FEMS Microbiol. Lett.* **16:**131–135.

111. **Kanaar, R., P. van de Putte, and N. R. Cozzarelli.** 1986. Inversion of the G segment of phage Mu in vitro is stimulated by a host factor. *Biochim. Biophys. Acta* **866:**170–177.

112. **Koch, L., and R. Kahmann.** 1986. Purification and properties of the Escherichia coli host factor required for inversion of the G segment in bacteriophage Mu. *J. Biol. Chem.* **261:**15673–15678.

113. **Krause, H. M., and N. P. Higgins.** 1986. Positive and negative regulation of the Mu operator by Mu repressor and E. coli integration host factor. *J. Biol. Chem.* **261:**3744–3752.

114. **Krause, H. M., M. R. Rothwell, and N. P. Higgins.** 1983. The early promoter of bacteriophage Mu: definition of the site of transcript initiation. *Nucleic Acids Res.* **11:**5483–5495.

115. **Krylov, V. N., V. G. Bogush, and J. Shapiro.** 1979. Bacteriophages of Pseudomonas aeruginosa, the DNA structure of which is similar to the structure of phage Mu1 DNA. I. General description, localization of endonuclease-sensitive sites in DNA, and structure of homoduplexes of phage D3112. *Genetika* **16:**824–832.

116. **Krylov, V. N., T. G. Plotnikova, L. A. Kulakov, T. V. Fedorova, and E. N. Eremenko.** 1982. Integration of Mu-like Pseudomonas aeruginosa bacteriophage D3112 genome into RP4 plasmid and its transfer by the hybrid plasmid into Pseudomonas putida and Escherichia coli C600 bacteria cells. *Genetika* **18:**5–12.

117. **Kutsukake, K., and T. Iino.** 1980. Inversions of specific DNA segments in flagellar phase variation of Salmonella and inversion systems of bacteriophages P1 and Mu. *Proc. Natl. Acad. Sci. USA* **77:**7238–7341.

118. **Kwoh, D. Y., and D. Zipser.** 1981. Identification of the gin protein of bacteriophage Mu. *Virology* **114:**291–296.

119. **Levin, D. B., and M. S. DuBow.** 1987. Cloning and characterization of the repressor gene (c) of the Mu-like transposable phage D108. *FEBS Lett.* **222:**199–203.

120. **Liebart, J. C., P. Ghelardini, and L. Paolozzi.** 1982. Conservative integration of bacteriophage Mu DNA into pBR322 plasmid. *Proc. Natl. Acad. Sci. USA* **79:**4362–4366.

121. **Ljungquist, E., and A. I. Bukhari.** 1977. State of prophage Mu DNA upon induction. *Proc. Natl. Acad. Sci. USA* **74:**3143–3147.

122. **Ljungquist, E., and A. I. Bukhari.** 1979. Behavior of bacteriophage Mu DNA upon infection of E. coli cells. *J. Mol. Biol.* **133:**339–357.

123. **Ljungquist, E., H. Khatoon, M. DuBow, L. Ambrosio, F. deBruijn, and A. I. Bukhari.** 1979. Integration of bacteriophage Mu DNA. *Cold Spring Harbor Symp. Quant. Biol.* **43:**1151–1158.

124. **Machida, Y., C. Machida, H. Ohtsubo, and E. Ohtsubo.** 1982. Factors determining frequency of plasmid cointegration mediated by insertion sequence IS1. *Proc. Natl. Acad. Sci. USA* **79:**277–281.

125. **Margolin, W., and M. M. Howe.** 1986. Localization and DNA sequence analysis of the C gene of bacteriophage Mu, the positive regulator of Mu late transcription. *Nucleic Acids Res.* **14:**4881–4897.

126. **Maxwell, A., R. Craigie, and K. Mizuuchi.** 1987. B protein of bacteriophage Mu is an ATPase that preferentially stimulates intermolecular DNA strand transfer. *Proc. Natl. Acad. Sci. USA* **84:**699–703.

127. **McBeth, D. L., and A. L. Taylor.** 1982. Growth of bacteriophage Mu in *Escherichia coli dnaA* mutants. *J. Virol.* **44:**555–564.

128. **McBeth, D. L., and A. L. Taylor.** 1983. Involvement of *Escherichia coli* K-12 DNA polymerase I in the growth of bacteriophage Mu. *J. Virol.* **48:**149–156.

129. **Mekalanos, J. J., S. L. Moseley, J. R. Murphy, and S. Falkow.** 1982. Isolation of enterotoxin structural gene deletion mutations in Vibrio cholerae induced by two mutagenic vibriophages. *Proc. Natl. Acad. Sci. USA* **79:**151–155.

130. **Mertens, G., A. Hoffmann, H. Blocker, R. Frank, and R. Kahmann.** 1984. Gin-mediated site-specific recombination in bacteriophage Mu DNA: overproduction of the protein and inversion in vitro. *EMBO J.* **3:**2415–2421.

131. **Miller, J. L., S. K. Anderson, D. J. Fujita, G. Chaconas, D. Baldwin, and R. M. Harshey.** 1984. The nucleotide sequence of the B gene of bacteriophage Mu. *Nucleic Acids Res.* **12:**8627–8638.

132. **Miller, J. L., and G. Chaconas.** 1986. Electron microscopic analysis of in vitro transposition intermediates of bacteriophage Mu DNA. *Gene* **48:**101–108.

133. **Mise, K.** 1971. Isolation and characterization of a new generalized transducing bacteriophage different from P1 in *Escherichia coli. J. Virol.* **7:**168–175.

134. **Mizuuchi, K.** 1983. In vitro transposition of bacteriophage Mu: a biochemical approach to a novel replication reaction. *Cell* **35:**785–794.

135. **Mizuuchi, K.** 1984. Mechanism of transposition of bacteriophage Mu: polarity of the strand transfer reaction at the initiation of transposition. *Cell* **39:**395–404.

136. **Mizuuchi, K., and R. Craigie.** 1986. Mechanism of bacteriophage Mu transposition. *Annu. Rev. Genet.* **20:**385–429.

137. **Mizuuchi, M., R. A. Weisberg, and K. Mizuuchi.** 1986. DNA sequence of the control region of phage D108: the N-terminal amino-acid sequence of repressor and trans-

posase are similar both in phage D108 and its relative, phage Mu. *Nucleic Acids Res.* **14:**3813–3825.

138. **Morisato, D., and N. Kleckner.** 1984. Transposase promotes double strand breaks and single strand joints at Tn10 termini in vivo. *Cell* **39:**181–190.

139. **Morisato, D., J. Way, H.-J. Kim, and N. Kleckner.** 1983. Tn-10 transposase acts preferentially on nearby transposon ends in vivo. *Cell* **32:**799–807.

139a. **Nagaraja, V., G. Hecht, and S. Hattman.** 1988. The phage Mu late gene transcription factor, C, is a site-specific DNA binding protein. *Biochem. Pharmacol.* **37:**1809–1810.

140. **Nakai, H., and A. L. Taylor.** 1985. Host DNA replication forks are not preferred targets for bacteriophage Mu transposition. *J. Bacteriol.* **163:**282–290.

141. **Nakayama, C., D. B. Teplow, and R. M. Harshey.** 1987. Structural domains in phage Mu transposase: identification of the site-specific DNA-binding domain. *Proc. Natl. Acad. Sci. USA* **84:**1809–1813.

142. **O'Day, K. J., D. W. Schultz, and M. M. Howe.** 1978. Search for integration-deficient mutants of bacteriophage Mu, p. 48–51. *In* D. Schlessinger (ed.), *Microbiology—1978.* American Society for Microbiology, Washington, D.C.

143. **Paolozzi, L., P. Ghelardini, A. Kepes, and H. Marcovich.** 1979. The mechanism of integration of phage Mu in the chromosome of Escherichia coli. *Biochem. Biophys. Res. Commun.* **88:**111–116.

144. **Paolozzi, L., R. Jucker, and E. Calef.** 1978. Mechanism of phage Mu-1 integration: nalidixic acid treatment causes clustering of Mu-1 induced mutations near replication origin. *Proc. Natl. Acad. Sci. USA* **75:**4940–4943.

145. **Pato, M., M. Banerjee, L. Desmet, and A. Toussaint.** 1987. Involvement of heat shock proteins in bacteriophage Mu development. *J. Bacteriol.* **169:**5505–5509.

146. **Pato, M. L., and C. Reich.** 1982. Instability of transposase activity: evidence from bacteriophage Mu DNA replication. *Cell* **29:**219–225.

147. **Pato, M. L., and C. Reich.** 1984. Stoichiometric use of the transposase of bacteriophage Mu. *Cell* **36:**197–202.

148. **Pato, M. L., and C. Reich.** 1985. Synchronization of bacteriophage Mu replicative transposition: products of the first round after induction, p. 27–35. *In* M. Simon and I. Herskowitz (ed.), *Genome Rearrangement.* Alan R. Liss, Inc., New York.

149. **Pato, M. L., and B. T. Waggoner.** 1981. Cellular location of Mu DNA replicas. *J. Virol.* **38:**249–255.

150. **Phadnis, S. H., C. Sasakawa, and D. E. Berg.** 1986. Localization of the action of the IS50-encoded transposase protein. *Genetics* **112:**421–427.

151. **Plasterk, R. H. A., A. Brinkman, and P. van de Putte.** 1984. DNA inversions in the chromosome of E. coli and in bacteriophage Mu: relationship to other site-specific recombination systems. *Proc. Natl. Acad. Sci. USA* **80:**5355–5358.

152. **Plasterk, R. H. A., R. Kanaar, and P. van de Putte.** 1984. A genetic switch in vitro: DNA inversion by Gin protein of phage Mu. *Proc. Natl. Acad. Sci. USA* **81:**2689–2692.

153. **Plasterk, R. H. A., M. Vollering, A. Brinkman, and P. van de Putte.** 1984. Analysis of the methylation-regulated Mu mom transcript. *Cell* **36:**189–196.

154. **Plasterk, R. H. A., H. Vrieling, and P. van de Putte.** 1983. Transcription initiation of Mu mom depends on methylation of the promoter region and a phage-coded transactivator. *Nature* (London) **301:**344–347.

155. **Plotnikova, T. G., V. Z. Akhverdyan, S. A. Reulets, S. A. Gorbunova, and V. N. Krylov.** 1982. Multiplicity of integration sites of Mu-like bacteriophages in the chromosome and plasmid of Pseudomonas aeruginosa. *Genetika* **19:**1604–1610.

156. **Plotnikova, T. G., A. S. Yanenko, N. B. Kirsanov, and V. N. Krylov.** 1982. Transposition of the D3112 phage genome in Escherichia coli cells. *Genetika* **19:**1611–1615.

157. **Preiss, H., D. Kamp, R. Kahmann, B. Brauer, and H. Delius.** 1982. Nucleotide sequence of the immunity region of bacteriophage Mu. *Mol. Gen. Genet.* **186:**315–321.

158. **Puspurs, A. H., N. J. Trun, and J. N. Reeve.** 1983. Bacteriophage Mu DNA circularizes following infection of Escherichia coli. *EMBO J.* **2:**345–352.

159. **Raibaud, O., M. Roa, C. Braun-Breton, and M. Schwartz.** 1979. Structure of the malB region in E. coli K12. I. Genetic map of the malK-lamB operon. *Mol. Gen. Genet.* **174:** 241–248.

160. **Rehmat, S., and J. A. Shapiro.** 1983. Insertion and replication of the Pseudomonas aeruginosa mutator phage D3112. *Mol. Gen. Genet.* **192:**416–423.

161. **Reich, C., B. T. Waggoner, and M. L. Pato.** 1984. Synchronization of bacteriophage Mu DNA replicative transposition: analysis of the first round after induction. *EMBO J.* **3:**1507–1511.

162. **Resibois, A., M. Colet, and A. Toussaint.** 1982. Localization of mini-Mu in its replication intermediates. *EMBO J.* **1:** 965–969.

163. **Resibois, A., M. Pato, P. Higgins, and A. Toussaint.** 1984. Replication of bacteriophage Mu and its mini-Mu derivatives, p. 69–76. *In* U. Hubscher and S. Spadari (ed.), *Proteins Involved in DNA Replication.* Plenum Publishing Corp., New York.

164. **Resibois, A., A. Toussaint, and M. Colet.** 1982. DNA structures induced by mini-Mu replication. *Virology* **117:** 329–340.

165. **Ross, W., S. H. Shore, and M. M. Howe.** 1986. Mutants of *Escherichia coli* defective for replicative transposition of bacteriophage Mu. *J. Bacteriol.* **167:**905–919.

166. **Roulet, E., B. Allet, and M. Chandler.** 1985. Preferential binding of bacteriophage Mu repressor to supercoiled Mu DNA. *Plasmid* **13:**173–181.

167. **Rouviere-Yaniv, J., and F. Gros.** 1975. Characterization of a novel, low-molecular weight DNA-binding protein from Escherichia coli. *Proc. Natl. Acad. Sci. USA* **72:**3423–3428.

168. **Rouviere-Yaniv, J., M. Yaniv, and J. E. Germond.** 1979. E. coli DNA binding protein HU forms nucleosome-like structures with circular double-stranded DNA. *Cell* **17:**265–274.

169. **Sandulache, R., P. Prehm, D. Expert, A. Toussaint, and D. Kamp.** 1985. The cell wall receptor for bacteriophage MuG(−) in Erwinia and Escherichia coli C. *FEMS Microbiol. Lett.* **28:**307–310.

170. **Sandulache, R., P. Prehm, and D. Kamp.** 1984. Cell wall receptor for bacteriophage Mu G(+). *J. Bacteriol.* **160:** 299–303.

171. **Schaller, H., B. Otto, V. Nusslein, J. Huf, R. Herrmann, and F. Bonhoeffer.** 1972. Deoxyribonucleic acid replication in vitro. *J. Mol. Biol.* **63:**183–200.

172. **Schaus, N. A., and A. Wright.** 1980. Inhibition of Escherichia coli exonuclease V by bacteriophage Mu. *Virology* **102:**214–217.

173. **Schroeder, W., E. G. Bade, and H. Delius.** 1974. Participation of Escherichia coli DNA in the replication of temperate bacteriophage Mu-1. *Virology* **60:**534–542.

174. **Schumann, W., V. Simon, and C. Loegl.** 1984. The bacteriophage Mu gene B product is incorporated into the inner membrane of Escherichia coli. *Gene* **29:**167–173.

175. **Schumm, J. W., and M. M. Howe.** 1981. Mu-specific

properties of λ phages containing both ends of Mu depend on the relative orientation of Mu end DNA fragments. *Virology* **114**:429–450.

176. **Shapiro, J. A.** 1979. Molecular model for the transposition and replication of bacteriophage Mu and other transposable elements. *Proc. Natl. Acad. Sci. USA* **76**:1933–1937.

177. **Silhavy, T. J., E. Brickman, P. J. Bassford, Jr., M. J. Casadaban, H. A. Shuman, V. Schwartz, L. Guarente, M. Schwartz, and J. R. Beckwith.** 1979. Structure of the malB region in E. coli K12. II. Genetic map of the malE,F,G operon. *Mol. Gen. Genet.* **174**:249–259.

178. **Silverman, M., J. Zeig, G. Mandel, and M. Simon.** 1981. Analysis of the functional components of the phase variation system. *Cold Spring Harbor Symp. Quant. Biol.* **45**:17–26.

179. **Simon, M., J. Zieg, M. Silverman, G. Mandel, and R. Doolittle.** 1980. Phase variation: evolution of a controlling element. *Science* **209**:1370–1374.

179a. **Sivan, S., A. Zaritsky, and V. Kagan-Zur.** 1988. Replication forks of *Escherichia coli* are not preferred sites for lysogenic integration of bacteriophage Mu. *J. Bacteriol.* **170**:3089–3093.

180. **Stoddard, S., and M. Howe.** 1987. DNA sequence within the C operon. *Nucleic Acids Res.* **15**:7198.

181. **Surette, M. G., S. J. Buch, and G. Chaconas.** 1987. Transpososomes: stable protein-DNA complexes involved in the in vitro transposition of bacteriophage Mu DNA. *Cell* **49**: 253–262.

182. **Swinton, D., S. Hattman, P. F. Crain, C.-S. Cheng, D. L. Smith, and J. A. McClosky.** 1983. Purification and characterization of the unusual deoxynucleoside, α-N-(9-β-D-2′-deoxyribofuranosylpurin-6-yl)glycinamide, specified by the phage Mu modification function. *Proc. Natl. Acad. Sci. USA* **80**:7400–7404.

183. **Symonds, N., and A. Coelho.** 1978. Role of the G segment in the growth of phage Mu. *Nature* (London) **271**:573–574.

184. **Symonds, N., A. Toussaint, P. van de Putte, and M. Howe (ed.).** 1987. *Phage Mu.* Cold Spring Harbor Laboratory, Cold Spring Harbor, N.Y.

185. **Szatmari, G. B., J. S. Kahn, and M. S. DuBow.** 1986. Orientation and sequence analysis of right ends and target sites of bacteriophage Mu and D108 insertions in the plasmid pSC101. *Gene* **41**:315–319.

186. **Szatmari, G. B., L. Martine, and M. S. Dubow.** 1987. The right end of transposable bacteriophage D108 contains a 520 base pair protein encoding sequence not present in bacteriophage Mu. *Nucleic Acids Res.* **15**:6691–6704.

187. **Taylor, A. L.** 1963. Bacteriophage-induced mutation in E. coli. *Proc. Natl. Acad. Sci. USA* **50**:1043–1051.

187a. **Teplow, D., C. Nakayama, P. Leung, and R. Harshey.** 1988. Structure-function relationships in the transposition protein B of bacteriophage Mu. *J. Biol. Chem* **263**:10851–10857.

188. **Tolias, P. P., and M. S. DuBow.** 1985. The cloning and characterization of the bacteriophage D108 regulatory DNA-binding protein ner. *EMBO J.* **4**:3031–3037.

189. **Tolias, P. P., and M. S. DuBow.** 1986. The overproduction and characterization of the bacteriophage Mu regulatory DNA-binding protein ner. *Virology* **148**:298–311.

190. **Toussaint, A.** 1976. The DNA modification function of temperate phage Mu-1. *Virology* **70**:17–27.

191. **Toussaint, A., L. Desmet, and M. Faelen.** 1980. Mapping of the modification function of temperate phage Mu-1. *Mol. Gen. Genet.* **177**:351–353.

192. **Toussaint, A., L. Desmet, M. Faelen, R. Alazard, M. Chandler, and M. Pato.** 1987. In vivo mutagenesis of bacteriophage Mu transposase. *J. Bacteriol.* **169**:5700–5707.

193. **Toussaint, A., and M. Faelen.** 1973. Connecting two unrelated DNA sequences with a Mu dimer. *Nature* (London) *New Biol.* **242**:1–4.

194. **Toussaint, A., and M. Faelen.** 1974. The dependence of temperate phage Mu-1 upon the replication functions of E. coli K12. *Mol. Gen. Genet.* **131**:209–214.

195. **Toussaint, A., M. Faelen, L. Desmet, and B. Allet.** 1983. The products of gene A of related phages Mu and D108 differ in their specificities. *Mol. Gen. Genet.* **190**:70–79.

196. **Toussaint, A., N. Lefebvre, J. Scott, J. A. Cowan, F. de Bruijn, and A. I. Bukhari.** 1978. Relationships between temperate phages Mu and P1. *Virology* **89**:146–161.

197. **Toussaint, A., and A. Resibois.** 1983. Phage Mu: transposition as a life-style, p. 105–158. *In* J. A. Shapiro (ed.), *Mobile Genetic Elements.* Academic Press, Inc., New York.

198. **van de Putte, P., S. Cramer, and M. Giphart-Gassler.** 1980. Invertible DNA determines host specificity of bacteriophage Mu. *Nature* (London) **286**:218–222.

199. **van de Putte, P., M. Giphart-Gassler, T. Goosen, A. van Meeteren, and C. Wijffelman.** 1978. Is integration essential for Mu development?, p. 33–40. *In* P. Hofschneider and P. Starlinger (ed.), *Integration and Excision of DNA Molecules.* Springer-Verlag, Berlin.

200. **van de Putte, P., and M. Gruijthuijsen.** 1972. Chromosome mobilization and integration of F-factors in the chromosome of recA strains of E. coli under the influence of bacteriophage Mu-1. *Mol. Gen. Genet.* **118**:173–183.

201. **van de Putte, P., G. Westmaas, M. Giphart, and C. Wijffelman.** 1977. On the *kil* gene of bacteriophage Mu, p. 287–294. *In* A. I. Bukhari, J. A. Shapiro, and S. L. Adhya (ed.), *DNA Insertion Elements, Plasmids, and Episomes.* Cold Spring Harbor Laboratory, Cold Spring Harbor, N.Y.

202. **van de Putte, P., G. C. Westmaas, and C. Wijffelman.** 1977. Transfection with Mu-DNA. *Virology* **81**:152–159.

203. **van Leerdam, E., C. Karreman, and P. van de Putte.** 1982. Ner, a cro-like function of bacteriophage Mu. *Virology* **123**:19–28.

204. **van Meeteren, A., and P. van de Putte.** 1980. Transcription of bacteriophage Mu. II. Transcription of the repressor gene. *Mol. Gen. Genet.* **179**:185–189.

205. **van Vliet, F., M. Couturier, J. de Lafonteyne, and E. Jedlick.** 1978. Mu-1 directed inhibition of DNA breakdown in E. coli recA cells. *Mol. Gen. Genet.* **164**:109–112.

206. **Waggoner, B. T., N. S. Gonzalez, and A. L. Taylor.** 1974. Isolation of heterogeneous circular DNA from induced lysogens of bacteriophage Mu-1. *Proc. Natl. Acad. Sci. USA* **71**:1255–1259.

207. **Waggoner, B. T., C. F. Marrs, M. M. Howe, and M. L. Pato.** 1984. Multiple factors and processes involved in host cell killing by bacteriophage Mu: characterization and mapping. *Virology* **136**:168–185.

208. **Waggoner, B. T., and M. L. Pato.** 1978. Early events in the replication of Mu prophage DNA. *J. Virol.* **27**:587–594.

209. **Waggoner, B. T., M. L. Pato, and A. L. Taylor.** 1977. Characterization of covalently closed circular DNA molecules isolated after bacteriophage Mu induction, p. 263–274. *In* A. I. Bukhari, J. A. Shapiro, and S. L. Adhya (ed.), *DNA Insertion Elements, Plasmids, and Episomes.* Cold Spring Harbor Laboratory, Cold Spring Harbor, N.Y.

210. **Waggoner, B. T., M. L. Pato, A. Toussaint, and M. Faelen.** 1981. Replication of mini-Mu prophage DNA. *Virology* **113**:379–387.

211. **Waggoner, B. T., T. Wade, and M. L. Pato.** 1988. Identification of the bacteriophage D108 kil gene and the second

region of sequence non-homology with bacteriophage Mu. *Gene* **62**:111–119.

212. **Wang, B., L. Liu, E. Groisman, M. Casadaban, and C. M. Berg.** 1987. High frequency generalized transduction by miniMu plasmid phage. *Genetics* **116**:201–206.
213. **Way, J., and N. Kleckner.** 1985. Transposition of plasmid-borne Tn10 elements does not exhibit simple length-dependence. *Genetics* **111**:705–713.
214. **Weinert, T. A., K. Derbyshire, F. M. Hughson, and N. D. F. Grindley.** 1984. Replicative and conservative transposition of the insertion sequence IS903. *Cold Spring Harbor Symp. Quant. Biol.* **49**:251–260.
215. **Wijffelman, C., M. Gassler, W. F. Stevens, and P. van de Putte.** 1974. On the control of transcription of bacteriophage Mu. *Mol. Gen. Genet.* **131**:85–96.
216. **Wijffelman, C., and B. Lotterman.** 1977. Kinetics of Mu DNA synthesis. *Mol. Gen. Genet.* **151**:169–174.
217. **Wijffelman, C., and P. van de Putte.** 1974. Transcription of bacteriophage Mu: an analysis of the transcription pattern in the early phase of phage development. *Mol. Gen. Genet.* **135**:327–337.
218. **Wijffelman, C., and P. van de Putte.** 1977. Asymmetric hybridization of Mu strands with short fragments synthesized during Mu DNA replication, p. 329–333. *In* A. I. Bukhari, J. A. Shapiro, and S. L. Adhya (ed.), *DNA Insertion Elements, Plasmids, and Episomes.* Cold Spring Harbor Laboratory, Cold Spring Harbor, N.Y.
219. **Wijffelman, C. A., G. C. Westmass, and P. van de Putte.** 1973. Similarity of vegetative map and prophage map of bacteriophage Mu-1. *Virology* **54**:125–134.
220. **Williams, J. G. K., and C. M. Radding.** 1981. Partial purification and properties of an exonuclease inhibitor induced by bacteriophage Mu-1. *J. Virol.* **39**:548–558.
221. **Wulczyn, G. F., and R. Kahmann.** 1987. Post-transcriptional regulation of the bacteriophage Mu mom gene by the com gene product. *Gene* **51**:139–147.
222. **Zeldis, J. B., A. I. Bukhari, and D. Zipser.** 1973. Orientation of prophage Mu. *Virology* **55**:289–294.

Chapter 3

Retroviruses

HAROLD VARMUS and PATRICK BROWN

Harold Varmus ■ Department of Biochemistry and Biophysics and Department of Microbiology and Immunology, HSE 401, University of California, San Francisco, San Francisco, California 94143. **Patrick Brown** ■ Department of Pediatrics and Department of Biochemistry, Stanford University School of Medicine, Stanford, California 94305.

I. GENERAL INTRODUCTION TO RETROVIROLOGY

Retroviruses that cause tumors in chickens were among the first animal viruses to be described (22, 300), and retroviruses remain vital objects of contemporary scientific interest. There are several reasons for the long-lasting vigor of retrovirology and for the widespread interest in this class of viruses. (i) **Ubiquity:** retroviruses have been found in all vertebrate animals in which they have been sought, including fish, birds, rodents, cats, ungulants, nonhuman primates, and humans (352, 353). (ii) **Biochemistry:** the retrovirus life cycle depends upon unusual biochemical reactions, most obviously the reverse transcription of genomic RNA into DNA and the orderly integration of viral DNA into host chromosomes to form proviruses (363, 368, 369). (iii) **Genetics:** retroviruses perform genetic feats unique among animal viruses, including transduction and insertional mutation of host genes and entry of proviruses into germ lines through which they are vertically transmitted (362). (iv) **Evolution:** studies of the structure of proviruses and of the activity of reverse transcriptases have established evolutionary relationships among retroviruses and a wide variety of transposable elements, viruses, and cellular genes (363), and have implications for early events in the origin of life (378). (v) **Pathology:** retroviruses cause a wide variety of diseases, most commonly cancers (leukemias, sarcomas, and mammary carcinomas), but also immunodeficiencies, anemia, arthritis, and pneumonia (354, 355); two such diseases, the acquired immune deficiency syndrome and adult T-cell leukemia-lymphoma, are important, newly recognized disorders of humans (380). (vi) **Oncogenes:** many retroviruses that cause cancers in experimental animals either carry oncogenes transduced from host genomes or modify the expression of host proto-oncogenes by insertion mutation; the study of these has implicated nearly 50 cellular genes in carcinogenesis (18, 19).

These provocative features have attracted many investigators to the study of retroviruses over the past two decades, and summaries of their accomplishments have been recorded in numerous reviews (17, 42, 356, 363), including a two-volume monograph (381, 382). Our intention in this chapter is not to supersede such encyclopedic exercises, but to offer an updated view of the structure, biological properties, and replication of retroviruses, emphasizing recent findings unavailable to authors of earlier reviews and those aspects of the virus life cycle that are especially germane to the general theme of this book. We have also included enough information about earlier work to create a narrative intelligible to initiates, but it is beyond reason and allotted space to provide more than highly selected references to studies published before the mid-1980s.

A. Retroviruses as Retrotransposons

A provirus introduced into a cell by retroviral infection belongs to a spectrum of insertional elements, now known as retrotransposons, that includes the Ty elements of *Saccharomyces cerevisiae*, the *copia*-like elements of fruit flies, and endogenous proviruses in the germ lines of vertebrates (363; J. D. Boeke, this volume; P. M. Bingham and Z. Zachar, this volume). These mobile elements are united by structure and function: they have long terminal repeats (LTRs), generally a few hundred base pairs in length, and they transpose through RNA intermediates that are copied by an RNA-directed DNA polymerase (reverse transcriptase).

The several types of retrotransposons, and individual examples of each type, may vary in their functional competence. Some lack signals for gene expression or intact reading frames and hence are inactive for transposition unless they can be complemented in *trans*; some can mediate transposition events only within the cells in which they reside because they are incapable of making proteins required for the assembly of extracellular particles; others encode complete virus particles responsible for intercellular movements of genetic information. Regardless of genetic capacity, each of these insertion elements is believed to have originated in a similar fashion, under the influence of enzymes encoded by the element itself or a closely related element. The essential steps, including those by which reverse transcription generates LTRs and those by which integration into the host genome preserves DNA organization, seem to be remarkably similar for all these elements. For this reason, much of the description of retroviral replication provided below is likely to apply as well to elements that are inherently more difficult to study because they lack an infectious extracellular form.

Retroviruses are also related, albeit more distantly, to other viruses (hepadnaviruses and caulimoviruses) and other mobile DNA elements (e.g., F, I, DIRS-1, and long interspersed repeated sequence [LINEs] elements) that transfer their genetic information from an RNA to a DNA form using reverse transcriptases that they may encode. Further details about these viruses and elements are presented below, especially in Section IV.B, and in other chapters of this book.

B. Definition and Taxonomy of Retroviruses

The feature that distinguishes retroviruses from other viruses is the replication of a single-stranded RNA genome through a double-stranded DNA intermediate (368, 369). All viruses that conform to this simple definition also share a multitude of other features, with only minor variations that will be noted in later sections. Thus, retroviruses comprise a relatively homogeneous set of agents, despite their origins in a wide variety of hosts and despite the fact that they exhibit little nucleotide sequence identity between genomes belonging to different retrovirus subgroups (Table 1).

Ideally, the naming and classification of retroviruses proceed synchronously, but, in fact, most retroviruses were named well in advance of modern methods for classification. The standard format for naming retroviruses invokes the host in which the virus is naturally found and the disease prominently associated with the virus itself or with close relatives of the virus. However, many names are also adorned with eponyms honoring discoverers. (See Table 1 for examples.)

Attempts to classify retroviruses have been strongly influenced by available technologies, with morphological distinctions among particles (based upon electron microscopic definition of A, B, C, and D types) gradually giving way to immunological tests for similarities and finally being superseded by molecular genetics (352, 353). Genetic assessments that help classify retroviruses have included tests of function (e.g., are the genomes replication competent? do they carry a viral oncogene derived from a host gene?). The current yardstick for classification is the nucleotide sequence of those elements of the genome required for replication, both regulatory and protein-coding sequences. By these lights, it is possible to organize the most commonly studied retroviruses into at least seven groups: the avian leukosis-sarcoma

prototype of this group, there are at least five genes other than *gag*, *pol*, and *env* (106a): *vif* (previously called *sor*, *P′*, *Q*, or *orfA*), whose product is required for efficient transmission of virus in cell-free medium (96, 343); *vpr* (previously called *R*), an expressed open reading frame without defined function downstream of *vif*; *rev* (previously called *art* or *trs*) and *tat*, both encoded in two exons and involved in regulation of viral gene expression (92, 97, 330, 331) as discussed in Section III; and *nef* (formerly called 3′ *orf*, *E′*, *F*, or *orfB*), encoding a myristylated protein with biochemical and structural similarities to GTP-binding signal transducers (132), but without known biochemical activity, despite a possible negative effect on virus production (230).

The mouse mammary tumor virus (MMTV) LTR contains an extensive open reading frame (*orf*), conserved among many virus strains, to which no function has been ascribed (73, 75).

E. Viral Enzymes

The proteins with catalytic activities encoded in the *gag*-(*prt*)-*pol* region (Fig. 4) have special significance for the major theme of this chapter, the integration of retroviral DNA: PR cleaves polyprotein precursors to produce the instruments for synthesis and integration of viral DNA; RT is responsible for synthesis of viral DNA; and IN is required for the integration reaction itself.

1. Protease (PR)

Retroviral proteases are responsible for the proteolytic scissions in *gag*-containing polyproteins that generate the mature products of the *gag*-(*prt*)-*pol* regions. Though the proteases isolated from virus particles are 13 to 15 kDa in size, the activity is presumably present in the relevant polyprotein precursor(s), permitting the protease to liberate itself. For different retroviruses, the position of the PR-coding sequence varies in interesting ways (see Fig. 4A and 5). In MLV, PR is encoded by the 3′ end of *gag* and the 5′ end of *pol*, and is thus derived from the *gag-pol* polyprotein; in MMTV, PR is entirely encoded by the *prt* frame and released from the *gag-prt* polyprotein; in HIV, PR is encoded only in the *pol* frame, beginning near the end of the region in which *gag* is overlapping with *pol*, and is produced from the *gag-pol* polyprotein; in RSV, PR activity is associated with the carboxy-terminal product of *gag*, though the very low enzymatic activity of this enzyme raises the possibility that another, thus far undetected protein, encoded by both the 3′ end of *gag* and the 5′ portion of *pol*, may be the true protease.

Until recently, little attention has been given to the enzymatic mechanism of cleavage by retroviral proteases. Comparison of cleavage sites of several such proteases reveals some similarities (e.g., Pro is frequently at the amino terminus generated by cleavage, and Phe or Tyr is often at the carboxy terminus [280]), but there is considerable variety among sites attacked by a single protease or among those used by different proteases (359). Nevertheless, all retroviral proteases have related amino acid sequences, including some or all of the short sequence Leu-Leu/Val-Asp-Thr-Gly-Ala-Asp-Lys, which is thought to form the active site for aspartyl (or acidic) proteases (192). Consistent with this observation, pepstatin A, an inhibitor of aspartyl proteases, also inhibits retroviral proteases (192). Major efforts are now being made to identify better inhibitors, in the hope of interfering with the replication of HIV.

Extensive mutagenesis at the 5′ end of the HIV *pol* frame has confirmed the location of PR-coding sequences, using an assay for the enzymatic activity of viral proteins made in *Escherichia coli* (90). Site-directed lesions that change individual amino acids at many positions throughout the 99-residue PR protein reveal the importance of an 11-amino-acid region encompassing the proposed active site (226; D. Loeb and R. Swanstrom, personal communication). Ultimately, it should be possible to design alterations that explore the structure and function of the active site in the manner used for other proteolytic enzymes (59).

Few mutations have been made or found in the presumptive coding regions for PR in viruses other than HIV. When a large, in-frame mutation was made in the PR-coding sequence of MLV *pol*, extracellular virus was produced by cells transfected with the mutant DNA, but *gag* and *gag-pol* polyproteins were not processed, and the particles were not infectious (61).

2. Reverse transcriptase (RT)

The RT proteins differ in size and in enzymatic activities when isolated from different retroviruses, a partial reflection of the proteolytic schemes used to generate them (359) (Fig. 4A). Thus, in RSV, RT is a heterodimer, with an α subunit (63 kDa) that is coextensive with the amino-terminal two-thirds of a β subunit (95 kDa). The α-β complex exhibits several activities: RNA- and DNA-directed DNA polymerase, an RNase specific for RNA-DNA hybrids (RNase H), tRNA-binding activities intrinsic to both subunits, and endonuclease and single- and double-stranded DNA-binding activities attributed to the carboxy terminus of the β subunit. Reverse transcriptase activity can also be elicited from a β-β

complex, the presumptive precursor to the heterodimeric complex, and from isolated α and β subunits. The MLV RT is a monomeric protein of 80 kDa, though polymerase activity can also be detected in *gag-pol* polyproteins in cells or in PR-deficient virions (61). In HIV, RT exists in two forms of 66 and 51 kDa with the same amino terminus (218); functional distinctions between these species have not yet been made. (The larger protein is not analogous to the β subunit of the RSV enzyme complex, since it does not include sequences encoded by the IN domain of *pol*.)

All of the retroviral RT proteins share considerable amino acid sequence similarities with each other and with the reverse transcriptases encoded by hepatitis B viruses and caulimoviruses (357) and by other retrotransposons (39, 254, 358, 397). One particularly well conserved region, centered about the motif . . .Tyr Xaa Asp Asp. . ., is located in the amino-terminal one-third of retroviral RTs, implying that the active site for the polymerase is in this region. Changes in the central motif of HIV RT impair enzymatic activity (194), but a sophisticated genetic analysis of conserved residues throughout RT has not been performed.

There is an enormous literature describing the biochemical properties of reverse transcriptase (109, 368–370). Like other DNA polymerases, it requires a primer for synthesis on the various templates it will accept (hetero- or homopolymeric RNA or DNA). Both RNA and DNA primers will work: although the native primer is tRNA, oligodeoxynucleotides are conventionally used on homopolymeric RNA templates to assay for enzymatic activity. RT will synthesize DNA up to 10 kDa in size from a single priming site on RNA, and it is thought to act in a processive rather than distributive fashion. Unlike most virion polymerases, including the reverse transcriptases of hepatitis B viruses and caulimoviruses, retroviral RT is readily solubilized and purified from virions.

Characteristically, the polymerase activity requires a divalent cation (Mg^{2+} is favored for RSV, MMTV, and HIV, and Mn^{2+} is favored for spleen necrosis virus [SNV] and MLV), exhibits relatively broad optima for pH (usually around pH 8) and temperature (often stable even at values as high as 50°C), and is inhibited by *N*-ethylmaleimide, pyridoxyl phosphate, rifamycin derivatives, and chelators of zinc, but not by aphidicolin (156), an inhibitor of mammalian DNA polymerase α. Actinomycin D inhibits reactions templated by DNA, but not those templated by RNA. Chain-terminating substrates, such as dideoxynucleotides or azidothymidine, also inhibit polymerization and virus replication (249, 267) and have been used to treat HIV infections in humans on this basis (95). The RNase H activity will digest the RNA present in an RNA:DNA hybrid in a 5′-to-3′ or 3′-to-5′ direction; digestion produces oligoribonucleotides, including one that serves as a primer for synthesis of the second DNA strand during infection (see Section IV). Although retroviral RNase H requires a 5′ or 3′ end of RNA to initiate digestion, it appears to be capable of removing tRNA from the 5′ end of minus-strand DNA, removing poly(A) from the 3′ end of viral RNA, and digesting the capped 5′ end of viral RNA after reverse transcription of R-U5 (see Section IV.B). Some inhibitors, especially fluoride ions, have a preferential effect on the RNase H activity of RT proteins.

It has been possible to study early steps in the virus life cycle by using (i) reconstructed reactions with purified enzyme and template-primer, (ii) virions activated by detergents to synthesize DNA from the endogenous template-primer, or (iii) the products of synthesis isolated from infected cells. However, the normal product of reverse transcription in cells, a full-length duplex with LTRs, has never been synthesized in a reconstructed reaction, although, under select conditions, putatively correct products have been synthesized by detergent-disrupted virions of RSV and MLV (12, 21, 111).

Recently, RSV, MLV, and HIV RTs have been produced in *E. coli* (138, 209, 299, 350, 351), offering an abundant source of enzyme for the many current uses for RT (e.g., nucleotide sequencing and cDNA cloning) and opening new avenues for exploring structural and functional properties of RT. For example, numerous linker insertion mutations in MLV *pol* have been tested using products of mutant genes expressed in bacteria (see below). As yet, however, RT has not been crystallized, and no high-resolution structural studies have been reported.

Though many RSV and MLV *pol* mutants with temperature-sensitive or non-conditionally defective reverse transcriptase activity have been isolated by standard virological methods (220, 221), none of these retains normal RNase H activity, and no mutants specifically deficient in RNase H have been isolated in this way. Assignment of the RNase H activity to the carboxy-terminal portion of the RT protein has been recently made on two grounds: protein sequence comparisons with *E. coli* RNase H (180) and linker insertion mutagenesis of MLV *pol*, followed by expression in *E. coli*. MLV RT proteins with lesions in the carboxy-terminal one-third retain normal reverse transcriptase activity but lack RNase H activity; conversely, RT proteins with lesions in the amino-terminal two-thirds retain RNase H activity but lose reverse transcriptase activity (349). Both types of mutation incapacitate virus replication when the mutant sequences are placed within complete proviral units.

3. Integrase (IN)

All examined retroviruses contain a protein of 30 to 46 kDa derived from the carboxy terminus of *gag-pol* (or *gag-prt-pol*) polyproteins. In RSV, the 32-kDa integrase (IN) protein is derived by two proteolytic cleavages, one that also generates the carboxy terminus of the α subunit of the RT complex and another that chips a 37-amino-acid peptide from the end of the *gag-pol* polyprotein (3). (Introduction of stop codons into the coding domain for this small peptide has no apparent influence upon virus growth [194].) In SNV, MLV, and HIV, single cleavages are believed to generate the carboxyl terminus of RT and 34-, 46-, and 34-kDa proteins, respectively, from the end of *gag-pol* fusion proteins.

The strongest evidence implicating IN proteins in proviral integration comes from the study of site-directed mutants of MLV (76, 306), SNV (277), and RSV (150). These mutants are highly defective for virus growth, and integration appears to be the only step in the virus life cycle that is significantly impaired. As described in greater detail below (Section V.C), all tested IN proteins bind to single- and double-stranded DNA, but only the avian sarcoma-leukosis virus (ASLV) IN proteins have demonstrated *att* sequence specificity and endonucleolytic properties.

The predicted sequences of all retroviral IN proteins show substantial similarities (180), including potential sites for coordinated binding of zinc ions (so-called "zinc fingers"), though regions essential for the integration functions have not been defined.

F. Retroviral Genetics

The genetic behavior of retroviruses is influenced by the cardinal biochemical features of these viruses: transcription of genomes from RNA to DNA and from DNA to RNA by enzymes that lack proofreading functions; the diploid nature of the RNA form of the genome; and integration of the DNA form into host chromosomes.

1. Replication errors

When retroviruses are passaged in culture, a variety of genomic changes—deletions, duplications, and point mutations—occur at relatively high frequency (42, 368, 369, 372, 373), but the accurate measurement of error rates is complicated by multiple rounds of infection, recombination, and variable efficiency of virus growth, with or without a helper virus to complement genetic deficiencies. Moreover, the virus life cycle offers several opportunities for mutation: reverse transcription, DNA replication and repair, and RNA synthesis. In this light, estimates of retroviral evolutionary rates based simply on sequence comparisons—in the range of 10^{-3} to 10^{-4} changes per site per year, about 10^5-fold faster than for cellular genes (117)—should be interpreted cautiously with respect to mechanism and error rate per life cycle.

A recent measurement of mutation frequency, performed with virus vectors able to mediate only a single replication cycle, suggested that 0.5% of progeny proviruses are mutant, though the nature of the mutations and the step(s) at which they occur were not determined (79). Sequencing of defined regions of RSV RNA derived from numerous subclones after a limited number of replication cycles indicates a mutation frequency of 1.4×10^{-4} changes per nucleotide per cycle (P. Palese, personal communication). Infidelity of incorporation of nucleotides into homopolymeric and heteropolymeric products by viral reverse transcriptase has been estimated to occur at a frequency of 10^{-4} per nucleotide (124) and is thus much higher than the error rate for host DNA polymerase; however, it is not certain whether this represents the major source of replication errors since the error rate for host RNA polymerase II has not been measured. The mechanisms by which duplications and deletions occur are likewise uncertain; some deletions result from intra- or interchain recombination between homologous sequences, but others show no sign of homology at the deletion endpoints (372).

Replication errors can presumably occur throughout the genome, but evolutionary studies of retroviruses suggest that the rate of divergence is at least twice as great for *env* proteins as for RT and NC (341). The differences in rate are assumed to reflect, among other factors, selective pressures that favor retention of critical functions and alterations in the targets for the immune response.

2. Complementation

As for most other animal viruses, closely related strains of retroviruses can complement the genetic deficiencies of one another. Complementation can be achieved only through coexpression of both viral genomes in a host cell, not by coinfection with replication-deficient particles. Thus, a strain with a mutant *pol* gene can be complemented by a *pol*$^+$

virus if the genomes are coexpressed in the same cell to produce particles in which the mutant genome and wild-type *pol* proteins are packaged together; in other words, once they have been associated with particles, viral proteins cannot diffuse to assist deficient viruses. Studies with retroviruses carrying v-*onc* genes and lacking intact *gag*, *pol*, or *env* genes established long ago that helper viruses could provide all of the essential proteins in *trans*; this principle is now routinely applied in the use of retroviruses as genetic vectors, with the complementing viral DNA often deficient in *cis* components (e.g., the packaging signal), so that replication-competent virus is not present in vector stocks (Soriano et al., this volume).

3. Phenotypic mixing

Complementation for *env* deficiency is a particularly prominent phenomenon, since the resulting viral pseudotypes (with nucleocapsid provided by one genome and envelope proteins by another) have a host range determined mainly by the complementing rather than the packaged genome. Such pseudotypes are common among closely related strains of retroviruses, but they also occur following mixed infection by murine and avian retroviruses and even by retroviruses and rhabdoviruses (380). Pseudotype formation with components from distantly related or unrelated viruses raises questions about the recognition of envelope and core proteins during virus assembly that have yet to be addressed in a satisfactory fashion.

4. Heterozygosity and recombination

Because the retroviral genome is diploid, heterozygotic genomes are presumed to occur regularly, though direct tests for heterozygosis are difficult. Retroviral genomes recombine at high frequency (estimates range as high as 10 to 30% for each cycle of multiplication), and heterodimeric RNAs are thought to be intermediates, with recombination taking place during reverse transcription (42, 220, 221, 363). Recombination appears to be strongly favored by homology, but joining also occurs occasionally between unrelated sequences, e.g., during the latter phase of genetic transduction by retroviruses (Section VI). When viruses are grown in cells that contain related endogenous proviruses, packageable transcripts from those proviruses may participate in recombination reactions with the exogenous virus. This is most dramatically revealed by the repair of deletion mutations in the genome of an exogenous virus in a fashion that superficially resembles gene conversion (47, 50, 307). The mechanism of recombination remains uncertain: copy choice during synthesis of the first DNA strand (42) and exchange of second strands after strand invasion (165, 188) are the dominant hypotheses, but unambiguous tests of these models have yet to be performed.

5. Genetic consequences of integration

Apart from its central role in the virus life cycle, proviral integration can have profound genetic implications for the host (whether a cell in culture or an animal). Since proviruses are inherently quite stable (Section V.J), a provirus inserted into a cell genome is likely to be a permanent acquisition. Proviruses can cause mutations in host genes that contain or are adjacent to sites of integration, causing either inactivating recessive mutations or dominant lesions in which transcription is augmented. Furthermore, activating insertion mutations play a major role in retroviral oncogenesis and appear to be the first step in the transduction of cellular genes, notably to form viral oncogenes. These and related phenomena will be considered in greater detail in Section VI.

III. RETROVIRAL GENE EXPRESSION

The process that ultimately generates new proviruses from an established provirus begins with the expression of viral genes. Viral RNA and protein products are then assembled into particles that later synthesize and integrate new copies of viral DNA during infection of new host cells. In this section, we consider the steps in viral gene expression and the elaboration of progeny virus; in Section IV, we discuss events that accompany the infection of new cells.

A. Transcription of the Provirus into RNA

Integrated into host chromosomes, passively and accurately replicated like adjacent cellular DNA, a provirus serves both as a repository of viral genetic information in an infected cell and as a template for the synthesis of viral RNA. Because proviruses are terminally redundant, with an extra copy of U3 at the 5′ end and an extra copy of U5 at the 3′ end, sites corresponding both to the 5′ end of viral RNA (the initiation site for transcription) and to the 3′ junction with poly(A) lie within viral sequences and, more precisely, within the LTRs (Fig. 3). This has particular importance in the 5′ LTR, where the copy of U3 upstream of the initiation site has a major influence upon transcription of the provirus, even though the polymerase (RNA polymerase II) and cofactors are supplied by the host cell. Thus, the need to provide their own regulatory sequences for transcription partially explains the complex mechanism employed by retroviruses (and retrotransposons) for the synthesis of LTRs during reverse transcription.

The U3 region of proviral DNA includes signals previously defined as components of many eucaryotic promoters. Proviruses invariably contain a TATA-like sequence 25 to 30 nt 5′ of the transcriptional start site to position initiation by RNA polymerase II; most have a CCAAT-like sequence at about position −80 in U3, serving as an upstream activation site; some have GC-rich boxes that are binding sites for the transcription factor Sp1 (83); and other parts of U3 possess enhancer elements that probably act as recognition signals for host DNA-binding proteins that regulate transcriptional efficiency (131). Retroviral enhancers have been mapped by site-directed mutagenesis, shown to function relatively independently of orientation, position, distance, or promoter, and implicated in viral pathogenesis as determinants of cell tropism and as activators of transcription from promoters of proto-oncogenes (68, 368, 369; see below, Section VI). Some retroviruses have recently been cited as harboring enhancers in regions of the viral genome other than U3, such as the *gag* gene (5, 27, 340).

Studies of the transcription factors themselves, however, are still in their infancy. In some cases, binding proteins have been proposed as mediators of transcription: the RSV LTR binds at least two such proteins (310) and the MLV LTR binds at least six (338), but the proteins have not been identified and effects on transcription are uncertain. In other cases, there is strong circumstantial evidence for the regulation of transcription by such factors: a duplicated sequence in the HIV LTR required for optimal expression from the viral promoter binds NF-κB, a phorbol ester-regulated protein that also serves as a positive effector of transcription in B cells (258); the transcription factor Sp1 binds to GC-rich motifs in the HIV LTR (185); and the MMTV LTR binds glucocorticoid hormone-receptor complexes at sites known to be required for hormonal stimulation of transcription (391).

The efficiency of transcription is affected by circumstances beyond the nature of the *cis*-acting sequences in U3, as follows.

1. Host-cell dependence

Although many retroviral LTRs drive efficient expression in many cell types, it is well established that certain LTRs are highly cell type dependent. For example, RSV and some MLV LTRs work poorly in embryos and in embryonal carcinoma lines (248) unless modified by the addition of an embryonal carcinoma cell-specific enhancer from a polyomavirus mutant (223) or altered by mutations in U3 (101). The RSV promoter-enhancer is much more efficient in infected chicken cells than in rat cells (368, 369), and SNV transcription is more efficient in dog cells than in mouse cells (87). Relatively subtle sequence differences in U3 domains can alter growth rates of MLVs in different cell types (69, 152), can change an MLV agent of thymic lymphoma into a cause of erythroleukemia and vice versa (32, 33), or can convert a mammary tumorigenic strain of MMTV into a T-cell leukemia-inducing strain.

2. Influence of proviral integration sites

Examination of levels of viral RNA produced from clones of independently infected cells suggests that some undefined property of an integration site, perhaps determined or reflected by chromatin packaging or methylation status, can strongly influence the efficiency of gene expression (1, 93). Such conclusions must be treated with caution, because the number of proviruses studied is small and because intrinsic genetic variation among proviruses can also explain such findings. For example, when rare clones of infected embryonal carcinoma cells expressing MLV proviruses were selected and studied in detail, at least one was found to contain a provirus with an unexpected lesion in the PBS that relieved the usual block to expression in embryonal carcinoma cells (8). Furthermore, in other clones from this experiment, proviruses were found downstream from active promoters, and these, rather than the viral promoter, were responsible for viral transcripts (8). There is also evidence that rearrangements on the 5′ side of RSV proviruses favor expression in infected rat cells (113, 130, 214).

3. Viral transactivators

Some retroviruses produce viral gene products that affect transcription (Section II.D). The best established of these is the *tax* protein from the X region of HTLVs and bovine leukemia virus, a protein that is essential for virus replication (as judged by site-directed mutagenesis of HTLV-2 DNA [34]) and can also augment transcription of cellular genes. (The possible role of *trans*-regulation of cellular genes during pathogenesis by these viruses is considered in Section VI.) A 21-bp sequence present in three tandem copies in U3 appears to be the major target for the activity of *tax* (25, 104, 296), though sequences in R and U5 may have ancillary roles (296). The

21-bp unit can be manipulated like an enhancer element (i.e., it is active when linked to a heterologous promoter in various arrangements), but the manner in which *tax* protein interacts with it has not been defined. Though the target sequences are protected from nuclease digestion in cells expressing *tax*, the *tax* protein itself does not appear to bind to DNA (176, 266). Moreover, the *rex* protein, encoded by an overlapping reading frame, may have an inhibitory effect upon *tax* activity (298).

The effect of the HIV *tat* protein upon transcription is less clear. The *trans*-activator is a metal-linked dimer (99) essential for virus growth (97), and it can have 100- to 1,000-fold effects upon the expression of viral genes or genes into which the *tat* response element (a sequence encompassing the first 80 nt of the normal HIV transcript) has been incorporated (285, 297, 331). Yet there is disagreement about the steps that are regulated. Some claim a direct effect upon transcription, as measured in nuclear run-on assays (175); others claim an antitermination effect at a site within the *tat* response element (191); others argue for a posttranscriptional effect upon RNA abundance, e.g., by stabilizing transcripts (256, 297); and others claim a dual mechanism, with a major effect upon translational efficiency (63). Further work will be required to judge these contending mechanisms.

One further issue concerning proviral transcription requires brief discussion. At least two sets of measurements support the notion that the 5′ LTR is preferred to the 3′ LTR as a site for transcriptional initiation (64, 147). Why is the 5′ LTR strongly preferred if the LTRs are identical? There is some support for the idea that the activity of the 3′ LTR is suppressed by the arrival of the transcription apparatus from the upstream site ("promoter occlusion"), because placement of a 3′ processing site between the LTRs appears to augment the use of the 3′ LTR as a promoter (64). Similar effects are observed after the deletion of sequences within or adjacent to the 5′ LTR, a phenomenon that is particularly common in transcriptionally activated c-*myc* loci in avian leukosis virus (ALV)-induced bursal lymphomas (105, 122, 279, 385). Since most of the activating ALV proviruses retain their 5′ LTR (122), some sequence downstream from the 5′ LTR (5, 27) may normally account for its favored use as a promoter.

B. Processing of Retroviral RNAs

Primary retroviral transcripts are subjected to at least four kinds of modification that also affect host mRNAs: capping of the 5′ end with an inverted, methylated guanyl nucleotide, methylation at the N^6 position of scattered adenosine residues, 3′ cleavage and polyadenylation, and splicing (43, 45, 106, 368, 369). Though they all exploit host machinery, the latter two modifications are especially interesting in the context of retroviral RNA processing because of the potential insights they offer into signals for these events.

1. 3′ Processing

Transcription of proviral DNA begins within the 5′ LTR, at the U3-R boundary, and extends through the 3′ LTR, stopping somewhere beyond the R-U5 boundary. Though the lengths of primary transcripts of any single provirus have not been accurately determined, it appears from an analysis of RSV RNA in infected chicken fibroblasts that over 20% of transcripts extend beyond the end of U5 into flanking cellular sequences (148). (Indeed, one of the first molecular clones of retroviral DNA was a cDNA clone from such a transcript of RSV, revealing that 2 bp at the 3′ end of U5 was absent from the host-viral junction [392].)

Addition of poly(A) at the correct place on the new transcripts (i.e., at the R-U5 boundary) presumably requires signals present in the RNA in the form of sequence or secondary structure. Virtually all retroviral genomes contain the canonical sequence AAUAAA (or a functional variant such as AGUAAA) about 10 to 20 nt upstream of the polyadenylation site, just as occurs in other eucaryotic genes (16). The importance of this hexameric sequence is further supported by the fact that an RSV mutant exists in which alteration to AAUGAA increases the number of observable "readthrough" transcripts by from 20 to about 80% (148). However, in several retroviral genomes, this sequence is within R and thus it is present twice in viral RNA, near both the 5′ and the 3′ ends (359). Under these circumstances, the AAUAAA sequence could not be sufficient for cleavage and poly(A) addition, or else only short RNAs, the length of R, would then be produced. One possibility is that sequences upstream of R, e.g., from U3 or *env*, would also constitute part of the signal for polyadenylation; because these are only present in the 3′ part of the RNA, wasteful 5′ polyadenylations would not occur. Support for this idea has come from mutagenesis of the SNV U3 region: deletions that remove sequences upstream of AAUAAA markedly impair adenylation and can be rescued by insertion of an intact SV40 polyadenylation signal (80). In addition, analysis of transcripts from RSV proviruses that have been interrupted by insertion of an MLV provirus indicates that an entire MLV LTR provides a poor signal for 3′ processing, suggesting that viral sequences normally upstream of the MLV 3′ LTR may have a role (368, 369).

The life cycle of hepatitis B viruses, similar to that of retroviruses in other ways, presents a very similar problem with respect to polyadenylation: genome-length transcripts are terminally redundant, containing two copies of a sequence of ca. 150 to 200 nt found upstream of the polyadenylation signal in viral DNA (29, 88, 107, 388). It is not clear how stringently or by what mechanism these viruses avoid processing the short transcripts that result from the passage of RNA polymerase through the first copy of the polyadenylation site.

The only evident AAUAAA sequence in the HTLV genomes is over 200 nt 5′ of the polyadenylation site. For such RNA, it has been proposed (but not proven) that secondary structural features of the RNA bring the hexamer sufficiently close to the poly(A) addition site (312).

2. Splicing

All retroviruses use apparently conventional 5′ and 3′ splice sites to produce *env* mRNAs, and some viruses generate other, more complex, spliced RNAs as well (Fig. 3). For example, HTLV mRNAs include a doubly spliced species in which the first two exons are joined at the positions used to make *env* mRNA and the third base of the *env* initiation codon is joined directly to the long open reading frame in the X region (2, 313, 332, 374). HIV and, most likely, related lentiviruses generate even more mRNAs (255), two of which have been well characterized: doubly spliced RNAs for the synthesis of *tat* and *rev* (formerly called *art* or *trs*) (92, 330).

The especially interesting feature of splicing of retroviral RNAs is the requirement for control: the correct proportion of genome-length to subgenomic mRNAs must be preserved in order to optimize production of components needed for virion assembly and (in the cases of HTLV, HIV, and their relatives) to achieve proper levels of viral regulators of gene expression. It is not certain how this is managed, but it appears that both host and viral factors are involved. For example, in an avian cell, about one-half of the RSV RNA is spliced to form *env* and *src* mRNAs in about equal amounts; however, in mammalian cells (nonpermissive for virus production), almost all of viral RNA is spliced to form *src* mRNA (368, 369). Two viral proteins, the *rex* product of HTLV (168) and the *rev* product of HIV (92), may affect the proportion of viral mRNAs, perhaps through an effect upon splicing, but other mechanisms for the action of these proteins have also been proposed.

3. Sorting and stability

All the processed transcripts of any single provirus have the same 5′ and 3′ termini, yet the transcripts have different fates (368, 369): full-length RNA destined for assembly into virus particles enters a pool that is separate from that for similar transcripts that serve as mRNA for *gag*-(*prt*)-*pol*, and the half-life of pregenomic RNA is much shorter than that of mRNA. Furthermore, most spliced RNAs are rarely packaged into virus particles, implying the lack of an assembly determinant or segregation into a cell compartment separated from the assembly process. Little is known about the determinants of sorting and stability of these RNAs.

C. Synthesis and Modification of Retroviral Proteins

1. Major structural proteins

The major structural components of virions, the proteins encoded by *gag* and *env* genes, are synthesized as polyproteins from different mRNAs in different cell compartments (71, 72, 368, 369). Nevertheless, the proteins must ultimately assemble at the plasma membrane for the production of virus particles. Both protein modification and undefined recognition signals probably participate in the latter process.

gag. The primary products of retroviral *gag* genes, proteins of 60 to 80 kDa, are synthesized on free polyribosomes from mRNA believed to be physically indistinguishable from a monomer of virion RNA. (Although in vitro translation tests show that virion RNA is capable of directing the synthesis of *gag* proteins, the structure of *gag* mRNA has not been rigorously examined; for example, subtle alterations in the 3′ half would not have been detected in most studies.)

The initiation of translation at the *gag* AUG is sometimes complicated by the presence of upstream initiation sites; in RSV RNA, for example, there are three AUGs, each followed by a nearby stop codon 5′ of the *gag* AUG, and at least one of these short reading frames is translated (134). Thus, the *gag* mRNA is formally bicistronic. Curiously, the *gag* initiation codon and the preceding initiators are in subgenomic RSV mRNAs as well, since the 5′ splice site for *env* and *src* mRNAs is within the 5′ end of the *gag* gene (94, 135) (Fig. 3). The *gag* AUG is actually used to initiate the synthesis of *env* protein; in *src* mRNA, it is followed by a nearby stop codon, and *src* protein is initiated with a *src* AUG. However, conversion of the stop codon to sense shows that the *gag* AUG is used efficiently (159), implying that *src* mRNA is tricistronic.

The amino termini of *gag* polyproteins are modified by removal of methionine and by acylation. For most or all mammalian viruses, myristic acid is added to Gly-2 (145, 304), but for the ALSVs, the terminus is acetylated (272). Site-directed mutagenesis studies show that myristylation is required to bring *gag* proteins to the plasma membrane, where assembly of mature particles can occur, either before or after formation of a subviral core particle (292; below, Section III.D). Conversion of the *gag* polyprotein to the smaller components found in mature virus particles (Fig. 4A) occurs by PR-mediated proteolysis during or after virus budding. Additional modifications of the *gag* proteins, particularly phosphorylation, have been described (71, 72), but their functions are not known.

MLV has the unusual property of specifying a glycosylated *gag* protein, found on the surface of infected cells and in virus particles, and presumably synthesized on membrane-bound polyribosomes. The protein differs from the normal *gag* product by an extension of ca. 100 amino acids containing a signal peptide at its amino terminus (84). However, the mechanism for making the longer *gag* protein is unknown (initiation at an upstream atypical initiation codon has been proposed), and the longer *gag* protein is not required for efficient virus growth (89, 305).

***env*.** The protein products of *env* are synthesized by the translation of spliced, subgenomic *env* mRNA on membrane-bound polyribosomes (71, 72). Complex N-linked glycosylation occurs at several sites on these proteins, and the proteins are subjected to at least two proteolytic cleavages: a cotranslational cleavage by the host signal peptidase to remove the amino-terminal signal peptide, which varies in length up to 60 amino acids or more; and a posttranslational cleavage by an unidentified (but presumably host) protease to separate the SU protein and the TM protein. A complex between these two proteins is subsequently maintained by disulfide linkages. The complex is held in the plasma membrane by the transmembrane domain of TM, with SU and most of TM in the extracellular compartment, and a usually short carboxy-terminal tail of TM in the cytoplasm.

2. Viral enzymes

Synthesis of the viral enzymes PR, RT, and IN, which appear in virus particles and are encoded in the *gag-(prt)-pol* domain, requires unusual regulatory mechanisms: the *gag*-related polyproteins that serve as precursors to these enzymes are made at 4 to 20% the level of *gag* proteins by occasional stop-codon suppression or ribosomal frameshifting (172, 395). The apparent purposes of such translational control are (i) to make the three catalytic proteins in amounts lower than those of the structural proteins, without having to make additional, specialized mRNAs, and (ii) to assemble all *gag-(prt)-pol* proteins into virus particles through a common signal, presumably provided by the *gag* component.

Though *gag-(prt)-pol* polyproteins have been known for several years, predictions about the strategies for their synthesis were confounded by the differing arrangements of the coding domains in sequenced viral genomes (Fig. 3 and 4A). Thus, the *gag* and *pol* domains of MLV are in the same frame and separated only by a single amber codon, whereas the *gag* and *pol* domains of RSV overlap for 58 nt, with the *pol* reading frame −1 with respect to *gag*. Furthermore, the *gag* and *pol* domains of HTLVs and MMTV are separated by a third open reading frame (*prt*) that overlaps both *gag* and *pol*. Though RNA splicing appeared to reconcile such differences with a single mechanism, no appropriately spliced *gag-pol* mRNAs have been reported, and studies described below have now firmly established that translational mechanisms bypass the termination codons.

Stop-codon suppression. Amino acid sequencing of the MLV PR protein was dramatically revealing: amino acids 1 through 4 are encoded by the last four codons of *gag*; amino acids 6 and higher are encoded by the beginning of the *pol* reading frame; and amino acid 5, corresponding to the *gag* amber termination codon, is Gln (395) (Fig. 5). These results implied that the *gag* amber termination signal is suppressed by a tRNA charged with Gln when read by about 5% of ribosomes. Similar findings were obtained with the closely related feline leukemia virus (394).

Ribosomal frameshifting. Though nonsense suppression could not explain the synthesis of *gag-pol* polyproteins when *gag* and *pol* were in different frames, the results with MLV focused attention upon translational control, in particular ribosomal frameshifting, as a way to solve the problem. Transcription and translation of retroviral genes in vitro demonstrated unequivocally that mammalian ribosomes could shift frames at the expected frequencies to allow synthesis of *gag-pol*, *gag-prt*, and *gag-prt-pol* polyproteins (170–172, 251) (Fig. 4). All the retroviral frameshifts are in the −1 direction, and the efficiency of frameshifting varies from 5% in the *gag-pol* overlap of RSV (172) to 25% in the *gag-prt*

plus strand on a tRNA template that was not removed from the minus strand it had primed. The copied tRNA sequence was shown to be from $tRNA^{Gly}$ rather than the expected $tRNA^{Pro}$ (49), implying that either the genome was a heterodimer (e.g., with a subunit from an endogenous provirus with a PBS for $tRNA^{Gly}$) or the incorrect tRNA was mispaired with the PBS of Mo-MLV. (An unusual MPMV provirus that contains most of a tRNA sequence and most of an extra LTR at its 5′ end [335] can similarly be explained in terms of excessive copying of tRNA.)

(v) Several clones of circles with deletions at the junction site have been reported (363). Clustering of deletion endpoints near a potential recognition site for IN at the 3′ end of the RSV LTR has been interpreted to mean that some of the deletions may be mediated by viral proteins (268).

(vi) Dimeric and trimeric circles have been detected by DNA transfer and hybridization during infection with wild-type virus (125, 161). The single molecularly cloned example differs from those by a large inversion and appears to be the product of a pseudointegration event (326). Most oligomeric circles are probably composed of head-to-tail monomers linked by single complete LTRs, and may arise by eccentric events during DNA synthesis. Evidence for this view comes both from the mapping of unintegrated RSV oligomers (208) and from the analysis of cloned oligomers joined to host chromosomes as a result of infection by integrase-negative mutants of MLV (136).

Some of these circular forms are putative substrates or products of an integration reaction, and they will be considered in more detail in the next section.

V. INTEGRATION OF RETROVIRAL DNA

The final step in the retroviral transposition cycle is the integration of a provirus into the genomic DNA of a new host cell. Genomes of a variety of animal DNA viruses have occasionally been found to integrate into the DNA of their host (22, 36, 197, 240, 386, 390). For most of these viruses integration is a rare accident or, as may be the case for some parvoviruses, an infrequently followed alternative pathway in the life cycle (36). Retroviruses are unique among animal viruses in their dependence on integration for replication. Indeed, this feature of the retroviral life cycle is even more characteristic of retroviruses than is reverse transcription itself, since reverse transcription plays an integral role in the replication not only of retroviruses, but also of hepadnaviruses and caulimoviruses (Section IV.B).

In contrast to the other examples of insertion of animal virus genomes into cellular DNA, retroviral integration produces a provirus and host-viral DNA junctions with invariant structural features, and depends directly on virus-encoded proteins and on specific sequences in the viral DNA.

Pseudogenes and other putative retrotransposed elements that lack LTRs (e.g., LINEs, SINEs) are almost certainly generated via a pathway that parallels the retroviral life cycle (377; Hutchison et al., this volume; Deininger, this volume). Yet the ends of these elements vary from case to case, and they typically lack the short, precise flanking repeats that signify an ordered retroviral-like integration event. Thus, this final step in their transposition is most likely a cell-mediated illegitimate recombination event, quite distinct from retroviral integration.

A. Structural Analysis of Proviral DNA and Integration Sites

Structural analyses of integrated proviruses have provided clues that powerfully constrain the possible models for integration mechanisms and provide a framework for conducting and evaluating further experiments. The principal conclusions from these studies are as follows.

(i) Integrated proviruses are colinear with unintegrated linear viral DNA and are joined to host DNA at the edges of their LTRs. This was initially inferred from restriction analysis (162) and subsequently established by DNA sequencing (70, 160, 235, 322, 323). Sequence analysis of many host-viral DNA junctions from many viral species showed that the viral sequences at the joint are precise and invariant for each virus species (160, 323, 325, 362). These sequences map to sites very close to the ends of the unintegrated linear viral DNA.

The predicted product of viral DNA synthesis is a largely double-stranded linear molecule with blunt ends corresponding to the 5′ termini of plus- and minus-strand strong-stop DNA (Section IV.B). Recent studies have shown that some of the unintegrated viral DNA molecules in MLV-infected cells do indeed have this structure (104a; P. O. Brown, B. Bowerman, J. M. Bishop, and H. E. Varmus, *Proc. Natl. Acad. Sci. USA*, in press). These blunt-ended linear molecules are probably precursors to integrated proviruses, and for the purposes of this discussion we will assume that they are. However, the majority of unintegrated viral DNA molecules found in the infected cells are not blunt ended, but have termini with 3′ ends that are recessed by two bases

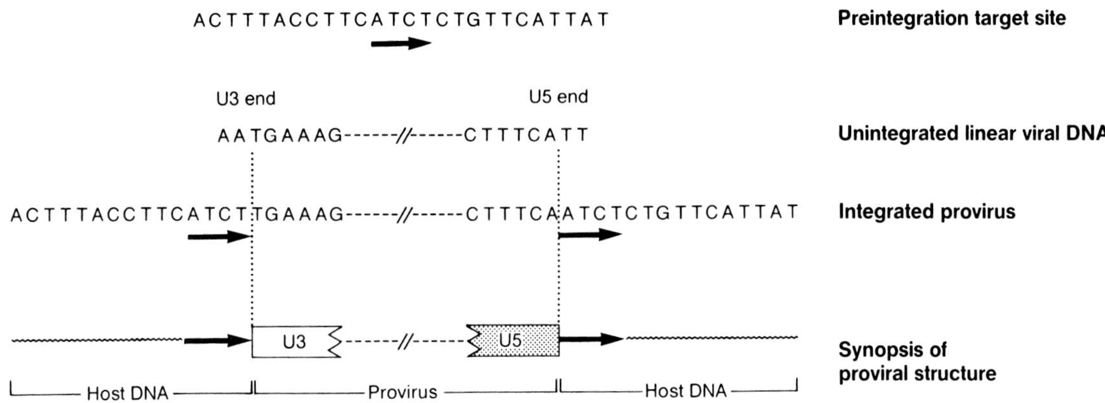

Figure 8. Structure of the integrated provirus and unintegrated precursor. The sequences shown correspond to the plus strand of the MLV DNA and the corresponding target DNA strand. The arrows underline the 4-bp sequence, present in a single copy in the target DNA, that is duplicated upon MLV integration. Note that in the integrated provirus, the terminal 2 bp of the linear precursor has been lost.

relative to the 5′ ends (104a; Brown et al., in press). It is likely that these molecules represent a later intermediate in the integration reaction pathway, generated by removal of the terminal two bases from each 3′ end of a blunt-ended precursors (Section V.B).

Figure 8 illustrates the relationship between the sequences at the ends of an integrated provirus and the corresponding sequences in the blunt-ended linear precursor. The integrated provirus is exactly colinear with the unintegrated linear precursor, except that the terminal 2 bp from each end of the precursor is lost upon integration (160, 323, 325, 362). HIV is a rare exception to this rule in that the ends of the putative linear precursor extend only 1 rather than 2 bp beyond the ends of the integrated HIV provirus (359).

(ii) A small stretch (4 to 6 bp) of host DNA, originally present in a single copy at the target site, is duplicated flanking the integrated provirus (160, 235, 322, 323, 325, 362). The target sequences are otherwise not modified or rearranged. This observation is consistent among many sequenced junctions between proviral and host DNA from several virus-host combinations. With rare, apparently isolated exceptions (64, 250), the length of the duplicated region is uniform for a given virus. The species of the host cell appears not to influence the repeat length, suggesting an intimate role of a virally encoded function in cleaving the target site during the integration reaction (363).

(iii) Many host DNA sites can function as targets for most retroviruses. Comparison of numerous sequenced integration sites for individual host-virus pairs has not revealed any sequence motif that predisposes a host DNA site for integration (28, 323, 325). Even when integration events restricted (by subsequent selection) to a limited DNA region are examined, they are typically scattered at many sites in the region (28, 143, 265, 278) (Fig. 9; see Section V.F). Yet it is clear that not all potential target sites for integration are used with equal frequency. For at least some retroviruses, there are highly preferred host sites for integration, as discussed further in Section V.G.

B. Outline of the Integration Pathway

Based on knowledge of the structure of the integrated provirus, the DNA target, and the putative viral DNA precursors, one can outline a reaction pathway that provides a framework for further discussions (317, 325) (Fig. 10). To facilitate our discussion relating to whether the ultimate integrative precursor is the linear or two-LTR circular form of the viral DNA (Section V.F.), both possibilities are considered here.

In pathway A, integration proceeds via a two-LTR circular intermediate formed by ligation of the ends of the linear form. In pathway B, the linear viral DNA is itself the substrate for integration. In the DNA breakage and joining step, viral and target DNA strands are broken at the sites marked by the arrowheads. The resulting 3′-OH ends of the viral DNA are then joined to the corresponding 5′-P ends of the target DNA. In the intermediate thus formed, the provirus is flanked by short gaps that are the precursors to the flanking repeats in the final product. DNA synthesis primed by the 3′-OH on the target side of the gap can initiate repair of the gap to yield the mature integrated provirus.

Most of the viral DNA molecules in MLV-

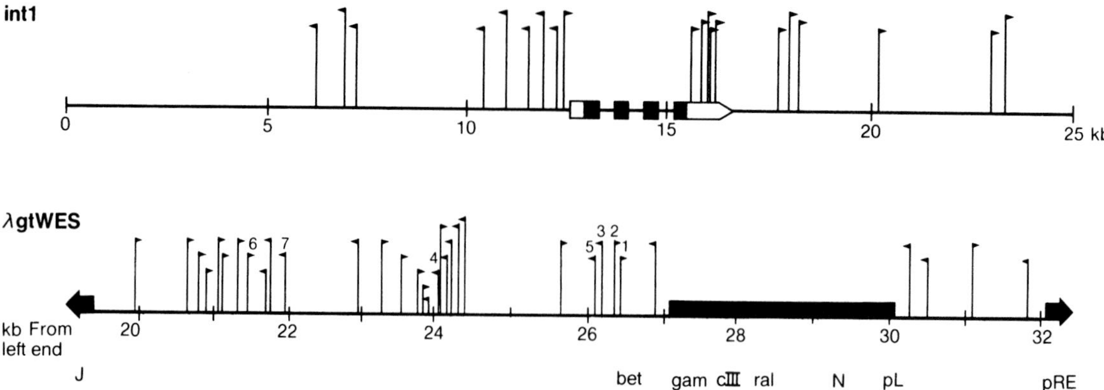

Figure 9. Proviral integration sites in host DNA. The flags indicate the position and orientation of an integrated provirus, and point from 5′ to 3′ relative to the viral (+) strand. (Top) MMTV proviral integration sites near the mouse *int-1* locus in mammary tumors. Each flag corresponds to the site of proviral integration from an individual tumor. The open arrow indicates the approximate extent of the normal *int-1* transcript, the closed boxes represent coding domains, and the intervening lines indicate introns. Note that there are multiple MMTV insertion sites throughout the vicinity of the gene, including some that abut but none that interrupt the coding region (361). (Bottom) In vitro MLV proviral integration sites in λgtWES DNA. Each flag marks the site of the integrated provirus in a single recombinant clone arising from an in vitro integration event. The heavy lines indicate the regions of the lambda genome required for growth on a *recA* host.

infected cells have a recessed 3′-OH end even before they enter the nucleus (104a; Brown et al., in press). Thus, viral DNA cleavage is probably not coupled to target DNA cleavage or to the joining reaction. It is therefore likely that the joining reaction is energetically coupled to cleavage of the target DNA, as discussed in Section V.E.

DNA breakage and joining with a polarity opposite to that shown in Fig. 10, that is, joining an overhanging 3′ end at the target site with a recessed 5′ end produced by cleavage of the viral DNA, is also mechanistically plausible. However, the recent isolation of the gapped intermediate from an in vitro MLV integration reaction has allowed direct verification of the polarity shown in Fig. 10 (104a; Brown et al., in press).

C. Retroviral Integration Requires Viral Proteins

Several lines of evidence indicate an essential role for at least one viral protein in the integration reaction. For example, molecular clones of viral DNA or naked viral DNA isolated from infected cells can be stably introduced with low efficiency into the genome of a target cell by transfection, yielding an active "provirus," but the junctions between the transfected provirus and host DNA occur at many sites in the viral genome and have an erratic structure quite unlike that of a normally integrated provirus (54, 203, 231). Thus, in the absence of the viral proteins normally associated with the viral DNA, proviral insertions occur by the poorly understood DNA transfection pathway (284) rather than by the normal integration process.

1. Genetic evidence

The definitive evidence for a viral function essential for integration comes from the analysis of mutant viruses defective in the 3′ domain of the *pol* coding region. Enzymological studies, described below, had implicated the protein product of this locus (IN) as a likely component of the integration apparatus. Mutations were therefore introduced into this region of the genomes of MLV and SNV and the detailed phenotype of the mutants examined. In each case, the mutant viruses were capable of carrying out the early steps of the transposition cycle, from transcription of the provirus through viral DNA synthesis, nuclear entry, and circularization, but integration was markedly impaired (76, 277, 306). Structural analysis of rare mutant proviruses that managed to insert into the target genome despite the absence of a normal IN protein demonstrated that, as with transfected viral DNA, their junctions with the host sequences were not produced by the retroviral integration mechanism (136). Thus, the IN protein is required specifically for normal integration.

It is difficult to extend this genetic approach to investigate the reasonable possibility that other virus-encoded proteins, such as CA, NC, or RT, might be required for integration. Unlike mutations affecting IN, those in other viral genes typically interfere with earlier steps in the transposition cycle, obscuring

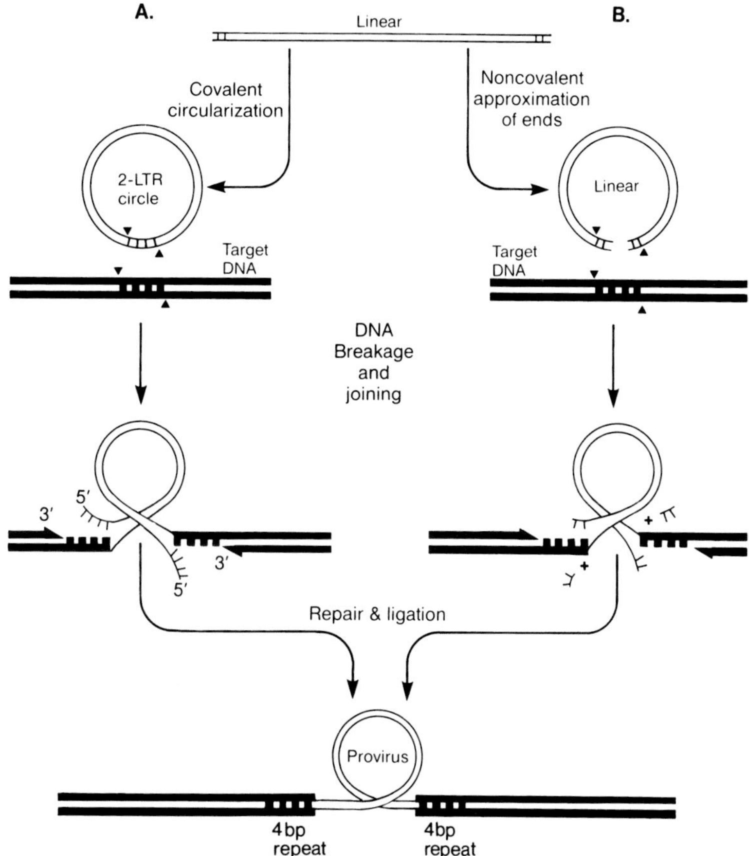

Figure 10. Outline of the integration pathway. The two illustrated pathways are described in the text. Target DNA is indicated by the heavy line, viral DNA by the thin line. The 4 bp that is lost upon integration (2 bp from each end of the linear precursor) and the 4-bp target sequence that is duplicated upon integration are indicated by the line segments perpendicular to the DNA strands.

their possible direct effects on integration (60, 116, 154, 308).

2. Biochemical analysis of the IN protein

Investigations of a possible catalytic function of the IN protein have produced tantalizing results, but their implications are not clear. Studies of p32-*pol* (IN) and the αβ and ββ forms of reverse transcriptase (the C-terminal domain of the β subunit is identical to IN; Section II.F) from ASLV have identified in these proteins an endonuclease activity with intriguing properties (81, 82, 119, 120, 128, 129, 212). The enzyme is more active on single-stranded than on double-stranded DNA, but will cut double-stranded DNA, especially when the DNA is negatively supercoiled. In the presence of a divalent cation, either Mg^{2+} or Mn^{2+}, the enzyme cleaves on the 3′ side of a phosphodiester bond, producing ends terminated by a 3′-OH and a 5′-P (212). Neither end is covalently linked to a protein, as they might be if cleavage resulted from an abortive topoisomerase or recombinase reaction (291, 375; chapters in this volume by J. F. Thompson and A. Landy, M. L. Pato, D. Sherratt, M. M. Cox).

In bacteriophage Mu integration, hydrolytic cleavages in the unintegrated DNA precursor, catalyzed by an endonuclease activity of the Mu gene A product, generate the correct Mu ends for joining to target DNA. To allow correct joining to host DNA, unintegrated retroviral DNA needs to be similarly cleaved (though not necessarily hydrolyzed; see Section V.E) within the terminal inverted repeats at sites two bases proximal to the ends of the linear precursors, or at the corresponding positions in the circle junction of the two-LTR circle. Are these cleavages made by the IN endonuclease activity? The ends of unintegrated linear viral DNA have not been investigated as substrates, but the inverted repeat sequence of a cloned circle junction is indeed a preferred site for cleavage by both the αβ and p32-*pol* forms of the ALV endonuclease (81, 82, 127, 129). However, the preference is only a modest one, and the precise sites at which this sequence is cut in the in vitro reaction

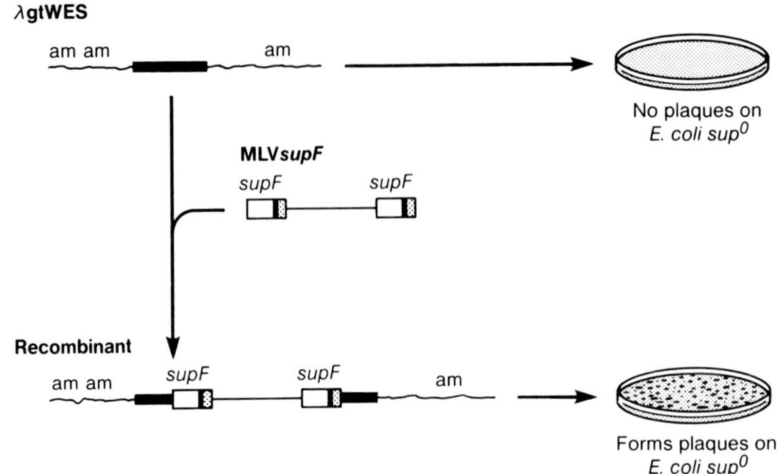

Figure 11. In vitro assay for MLV integration. λgtWES carries amber mutations in three essential genes and is thus unable to make plaques on a sup^0 *E. coli* host. The amber mutations can be suppressed by integration of an MLV provirus containing the *supF* amber suppressor gene, allowing the recombinants to make plaques on the sup^0 lawn. The portion of the λgtWES genome dispensible for lytic growth is indicated by the thick line. See the text for a detailed description of the assay.

are offset under most conditions by one base from the actual sites of integrative recombination. On the other hand, recent work has defined conditions under which the preferred site for cleavage by the purified p32 endonuclease, in a cloned double-stranded RSV circle junction, corresponds exactly to the site of integrative recombination (127, 129). Moreover, there is a rough correlation between the sequence requirements for (inaccurate) in vitro cleavage of a circle junction DNA fragment and the sequences required at the ends of the LTR for normal integration in vivo (41; Section V.F).

The MLV IN protein has not been shown to have intrinsic endonuclease activity (273). Extensive efforts to detect a nuclease activity in the protein produced by the expression of the MLV IN-coding region in *E. coli* have been fruitless, though this protein does have an apparently nonspecific DNA-binding activity (299a). The IN coding domain from SNV has also been expressed in *E. coli* (232). The resulting protein has no discernible endonuclease activity but, like its ASLV and MLV cousins, binds to single- and double-stranded DNA. Ordinary DNA binding by SNV IN protein showed no discernible sequence specificity. However, formation of DNA aggregates held together noncovalently by the SNV IN protein, a mysterious reaction that occurs only at a very high temperature (65°C), seems to require sequences in U5.

In summary, while the significance of the IN endonuclease activity to integration remains to be conclusively demonstrated, the evidence on the whole suggests that, like the gene A protein of bacteriophage Mu, it acts to prepare the viral DNA ends for integration. It is likely that faithful in vitro duplication of the native activity of the IN protein will require that it be assembled in a complex with other viral and perhaps host components.

D. Retroviral Integration In Vitro

A method has recently been developed for accurately reproducing the retroviral integration reaction in a cell-free system (28). This in vitro approach allows the reaction mechanism to be studied at the biochemical level.

1. Strategy

The assay for integration in vitro exploits a powerful genetic selection in *E. coli* to identify recombinants (Fig. 11). Bacteriophage λgtWES carries amber mutations in three genes required for lytic growth and is consequently unable to make plaques on a lawn of wild-type (sup^0) *E. coli*, but it makes plaques with normal efficiency if the *supF* amber suppressor allele is present. MLV*supF* is a replication-competent Mo-MLV derivative that carries the *E. coli supF* gene in its U3 region (224). The principle of the assay is that integration of an MLV*supF* provirus into a nonessential region of λgtWES produces a recombinant lambda genome able to suppress its own amber mutations and thus to make plaques on a nonsuppressor strain of *E. coli*.

Extracts made from cells acutely infected with the MLV*supF* provirus provide unintegrated full-length MLV*supF* DNA precursors and the enzymatic machinery required for their integration. Following incubation of λgtWES DNA with crude or fractionated cellular extracts, the DNA is recovered and packaged into bacteriophage particles in vitro. Overall recovery of the target DNA is determined by plating on a *supF E. coli* strain. To score for integration of the MLV*supF* provirus, the phage are plated on a sup^0 strain of *E. coli* that restricts growth of λgtWES but allows the recombinants (having acquired the *supF* gene) to make plaques (Fig. 11). Between 0.5 and 30% of viral DNA molecules integrate into the target DNA in a typical in vitro reaction (28, 104a; Brown et al., in press). The proviruses integrated in vitro are structurally indistinguishable from normal proviruses, affirming the authenticity of the in vitro reaction. Like integration in vivo, the in vitro reaction displays no evident specificity for target sequence (Fig. 9).

This assay is widely applicable to the study of nonhomologous intermolecular recombination reactions. It has now been used successfully to detect integration of RSV (J. Coffin, personal communication) and the yeast Ty element (84a) in vitro. The more recent development of direct physical methods for detecting the products of integration in vitro has opened up the possibility of analyzing the individual steps in the reaction (104a; Brown et al., in press).

2. Requirements and optimal conditions for integration in vitro

The in vitro integration assay has been used to characterize the conditions that affect MLV integration (28). EDTA reversibly abolishes activity, implying that a divalent cation is essential; Mg^{2+}, Ca^{2+}, or Mn^{2+} is acceptable. The optimal K^+ concentration is about 150 mM, and activity is markedly inhibited by concentrations above 300 mM. The pH range for activity is broad, extending at least from pH 6.6 to pH 8.6. The optimal temperature range for incubation is 15 to 43°C. No detectable integration occurs at 0°C. Pretreatment of the integrative precursors with sodium dodecyl sulfate and proteinase K or with 5 mM *N*-ethylmaleimide abolishes activity, whereas pretreatment with RNase A does not.

Infected cell extracts depleted of nucleotides by gel exclusion chromatography (residual ATP concentration less than 10^{-9} M) were fully active for integration in vitro in the absence of added nucleotides. Thus, neither ATP nor any other ribonucleoside triphosphate or deoxynucleoside triphosphate is required for integration in vitro (28).

E. Energy Conservation and the DNA Breakage and Joining Reaction

Like many previously characterized specialized DNA recombination reactions (290; chapters in this volume by Thompson and Landy, Pato, Sherratt, Cox). Mo-MLV integration can occur in vitro without an extrinsic energy source (28). The DNA breakage and joining reactions are therefore probably coupled. There are many precedents for such energy conservation occurring via a transient high-energy covalent protein-DNA phosphodiester bond (291, 375), and this is a plausible mechanism in the integration reaction (Fig. 12A). There are no recognized instances of the enzyme-DNA bond exchanging directly with a phosphodiester bond in DNA. Instead the joining reaction in all previously defined mechanisms of this kind uses a terminal OH group as the acceptor for the activated phosphoryl end in the intermediate. Thus, if this mechanism is used for integration, cleavage of the second participant in the joining reaction (the viral DNA; Fig. 12) is likely to be necessary in order to generate an "acceptor" OH terminus. A possible alternative mechanism has been proposed for bacteriophage Mu integration (55, 250a; Pato, this volume) (Fig. 12B). In this mechanism, 3′-OH groups are exposed by endonucleolytic hydrolysis of the viral DNA, perhaps catalyzed by the endonuclease activity of the IN protein. An enzyme-catalyzed nucleophilic attack by these 3′-OH groups on the appropriate phosphodiester bonds in the target DNA results in an essentially isoenergetic transesterification, much like that which occurs in RNA splicing (271), producing the recombinant product. The energy kick to make this reaction directional could come from an irreversible conformational change in the integrative complex following DNA bond exchange. Since there is no requirement that the integrative machinery turn over, such an irreversible change is permitted.

A third possible model for the formation of the new bonds in the recombinant is that, rather than conserving the energy released by the cleavage of donor or target DNA, the complex carries an energy charge that can be spent in forming the new bonds (for example, it could carry an adenylated ligase activity).

F. Defining the Viral *att* Site

1. Viral sequences required in *cis* for integration

The existence of normally integrated proviruses with extensive internal deletions or substitutions but intact LTRs focuses attention on the LTRs as the

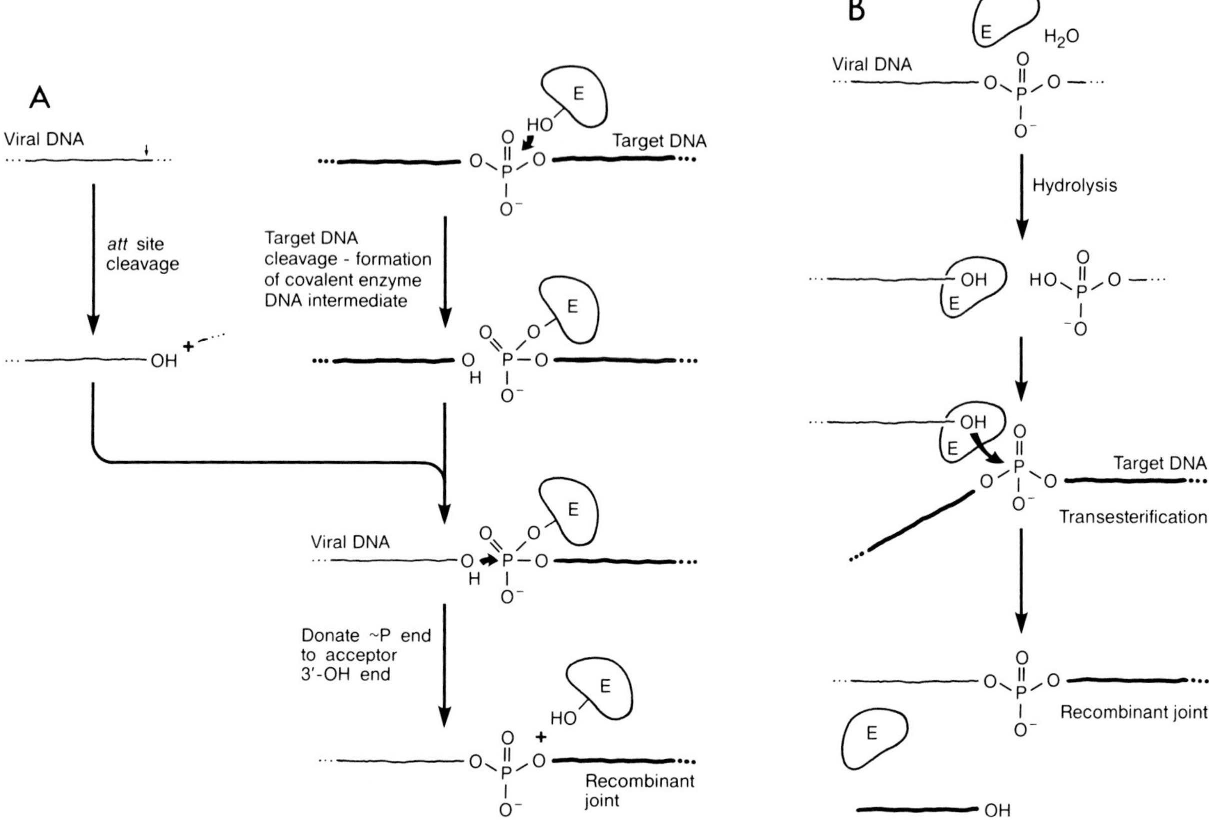

Figure 12. The DNA breakage and joining reaction. (A) Topoisomerase model. In this mechanism, one of the two DNA substrates for recombination (the target DNA in this illustration) is cleaved by exchange of a DNA-DNA phosphodiester bond for a high-energy protein-DNA phosphodiester bond. A hydroxyl group generated by a separate cleavage event in the second DNA substrate (here, the viral DNA) then serves as the acceptor for formation of a new phosphodiester bond, using the energy stored in the protein-DNA bond. The result is a recombinant DNA molecule. In the complete reaction, coordinated bond exchanges involving both strands are required. However, for simplicity, this illustration and the one in panel B depict only the events on one strand. (B) Transesterification. In this mechanism, the terminal hydroxyl group exposed by the hydrolysis of a phosphodiester bond in the first DNA substrate (the viral DNA in this example) is used as the attacking group in a direct enzyme-catalyzed nucleophilic attack on a phosphate group in the second DNA substrate (target DNA). This essentially isoenergetic phosphodiester bond exchange produces the recombinant molecule. Here, the same enzyme is depicted as catalyzing both the first bond cleavage and the transesterification step, though this is not required.

regions of the viral genome required in *cis* for integration. Proviral DNA is joined to host DNA using sites within the short inverted repeats at the outside edges of its long terminal repeats (Fig. 13A), suggesting that the terminal sequences of the LTRs might define the viral *att* sites required in *cis* for integration. To investigate this hypothesis, the effect on integration of mutations in the sequences at the edges of the LTRs (principally in U5) has been investigated for MLV, SNV, and ASLV (41, 48, 51, 52) (Fig. 13).

Because of the overall similarity of their transposition cycles, one would like to generalize conclusions drawn from studies of single viruses and to pool results of experiments using different virus systems, and we have tried to do so judiciously in this review. But it is important to keep in mind the hazards of such an approach. The genetic studies on MLV, SNV, and ASLV support the notion that the sequences necessary to define a viral *att* site are confined to the very distal ends of the LTRs. However, the LTR termini are widely divergent among different retroviruses in the details of their sequences and even in their overall organization. Therefore, we will not attempt to draw general conclusions here regarding the defining characteristics of a viral *att* site but will consider each virus separately.

As is evident from Fig. 13A, the terminal sequences of the retroviral DNA molecule are inverted repeats, ranging from long perfect palindromes with substantial potential for forming alternative secondary structures (e.g., MLV) to hyphenated (e.g., ASLV) or very short (e.g., HTLV-I) repeats with little or no apparent ability to form stable cruciform structures (219). It is therefore unlikely that such

(A)

	U5	U3
RSV	CACCTGCATGAAGCAGAAGGCTTCATT	AATGTAGTCTTATGCAATACTCTTGTAG
MLV	ACTACCCGTCAGCGGGGGTCTTTCATT	AATGAAAGACCCCACCTGTAGGTTTGGC
SNV	GAATCCGTAGTACTTCGGTACAACATT	AATGTGGGAGGGAGCTCTGGGGGGAATA
MMTV	ACCCTCAGGTCAGCCGACTGCGGCATT	AATGCCGCGCCTGCAGCAGAAATGGTTG
HIV	TAGTCAGTGTGGAAAATCTCTAGCAG	CTGGAAGGGCTAATTTGGTCCCAAAGA
HTLV-II	ATTGTCTTCCCGGGGAAGACAAACAAT	GATGACAATGGCGACTAGCCTCCCGAGC

(B)

U5	U3	strain
ACTACCCGTCAGCGGGGGTCTTTCATT	AATGAAAGACCCCACCTGTAGGTTTGGC	MO-MLV
ACTACCCGTCTC-GGGGGTCTTTCATT	AATGAAAGACCCCACCTGTAGGTTTAGC	FR-MLV
ACTACCCGTTAGTGGGGGTCTTTCATT	AATGAAAGACCCCACCTGTAGGTTTGGC	HA-MSV
ACTACCCA-CGACGGGGGTCTTTCATT	AATGAAAGACCCCACCCGTAGG--TGGC	MO-MSV
ACTACCCG-CCTCGGGGGTCTTTCATT	AATGAAAGACCC-ACCATCAGGTTTAGC	CAS-BR-E
ACTGCCCAGCCT-GGGGGTCTTTCATT	AATGAAAGACCCCTTCCTAAGGCTTAGT	FBJ
ACTGCCCAGCCT-GGGGGTCTTTCATT	AATGAAAGACCCCTTCATAAGGCTTAGC	FBR
ACTGCCCA-CCTCGGGGGTCTTTCATT	AATGAAAGACCCCACCAAGTTGCTTAGC	SFFV
ACTACCCGTCAGCGGGGGTCTTTCATT	AATGAAAGACCCCACCAAGTTGCTTAGC	AB-MLV
ACTGCCCAGCCT-GGGGGTCTTTCATT	AATGAAAGACCCCTTCATAAGGCTTAGC	AKV
ACCTACCTAGTTCGGGGGTCTTTCATT	AATGAAAGACCCCACCCCAAAATTTAGC (CT inserted)	FeLV-GA

U5	U3	mutant	integration
ACTACCCGTCAGCGGGGGTCTTTCATT	AATGAAAGACCCCACC	wt	+
ACTACCCGTCAGCGGGGGTCTTTCATA	AATGAAAGACCCCACC	sub594	↓↓
ACTACCCGTCAGCGGGGGTCTTTCAT-	AATGAAAGACCCCACC	dl594	↓
ACTACCCGTCAGCGGGGGTCTTTCA--	AATGAAAGACCCCACC	dl593	0
ACTACCCGTCAGCGGGGGTCTTTCATATA	AATGAAAGACCCCACC	in594-2	↓↓
ACTACCCGTCAGCGGGGGTCTTTCATATT	AATGAAAGACCCCACC	in594-2R	↓
ACTACCCGTCAGCGGGGGTCTTTCAG-	AATGAAAGACCCCACC	in592-R2	↓
ACTACCCGTCAGCGGGGGT--------	AATGAAAGACCCCACC	dl587	0
GCGGGGGTCTTTCACTGGAATTCCAG-	AATGAAAGACCCCACC	in592-12	↓↓↓
GCGGGGGTCTTTCACTGGAATTACAG-	AATGAAAGACCCCACC	in592-12R1	↓
CCAAACCTACAGGTGGGGTCTTTCATT	AATGAAAGACCCCACC	U3-U3	(+)
aCtaccc------gggGGtcTTtCAtt	aATGAAAGACCCcacc	consensus	

(C)

U5	U3	mutant	integration
GAATCCGTAGTACTTCGGTACAACATT	AATGTGGGAGGGAGCTCTGGGGGG	wt	+
GTAGTCAAGCGGAATTCCGACAACATT	AATGTGGGAGGGAGCTCTGGGGGG	183	+
CCTTTTCTACATGACTGAGGATACATT	AATGTGGGAGGGAGCTCTGGGGGG	233	↓
GGGCGCAGGGATCCGGACGACAACATT	AATGTGGGAGGGAGCTCTGGGGGG	211	+
CTGGGTGGGCGCAGGGATCCGGACATT	AATGTGGGAGGGAGCTCTGGGGGG	206	-
CTGGGCGCAGGGATCCGGACTGAATCC	AATGTGGGAGGGAGCTCTGGGGGG	207	-
GAATCCGTAGTACTTCGGTACAACATT	AATGTGGGAGGGGGGAATAGTGC	121	+
GAATCCGTAGTACTTCGGTACAACATT	AATGTGGGAGGGAGCTTTCTGTAC	144	(+)
GAATCCGTAGTACTTCGGTACAACATT	AATGGGGGGAATAGTGCTGGCTC	150	-
GAATCCGTAGTACTTCGGTACAACATT	TCTGTACACATGCT*	147	-
--------------------acaACAT	AATGTGGGAGGG-----------	consensus	

(D)

U5	U3	strain
ACGAGCACCTGCATGAAGCAGAAGGCTTCATT	AATGTAGTCTTATGCAATACTCTTGTAGTC	sra
GCGAACACCTGAATGAAGCTGAAGGCTTCATT	AATGTAGTCTTAATCGTAGGTTAACATGTA	amv
GCGAACACCTGAATGAAGCAGAAGGCTTCATT	AATGTAGTCTTATGCAATACTCCTGTAGTC	prc
ACGAACACCTAAATGAAGCGGAAGGCTTCATT	AATGTAGTCTTATGCAATACTCTTATGTAA	ur5
ACGAGCACCTGAATGAAGCAGAAGGCTTCATT	AATGTAGTCTTATGCAATACTCTTATGTAA	y73
GCGAACACCTGAATGAAGCGGAAGGCTTCATT	GATGTTGCCAGTTAGTCATCATTGCCACTT	fsv
GCGAACACCTGAATGAAGCAGAAGGCTTCATT	AATGTAGTCAAATAGAGCCAGAGGCAACCT	ra0
ACGCGAACCTGAATGAAGCAGAAGGCTTCATT	------------------------------	mc29
--------------------------------	AATGTAGCCTTACACAATAGCCTTACACAA	mh2

U5	U3	mutant	replication
ACGAGCACCTGCATGAAGCAGAAGGCTTCATT	AATGTAGTCTTATGCAATA	sra	+
GATTGCACATGCATGAAGCAGAAGGCTTCATT	AATGTAGTCTTATGCAATA	Δ15	↓↓
ACCGTTGATTGCATGAAGCAGAAGGCTTCATT	AATGTAGTCTTATGCAATA	Δ21	↓
TGGCCGGACCGTTGATTGCAGAAGGCTTCATT	AATGTAGTCTTATGCAATA	Δ28	0
aCgagcaccTgaATGAAGCAGAAGGCTTCATT	aATGtaGtCttataca	CONSENSUS	

Figure 13. Viral *att* site. The plus-strand sequence is shown from 5′ to 3′. The U5 and U3 sequences are shown juxtaposed, as they presumably are during integration. (A) LTR terminal sequences differ markedly among retroviruses (see Table 1 for abbreviations). Vertical arrows indicate the sites at which the integrated proviral DNA is joined to host DNA. Horizontal arrows indicated inverted repeats. (B) The MLV *att* site. The upper panel shows the U5 and U3 ends of several strains of the MLV group. The lower panel shows mutations in the U5 ends that were tested for their effects on integration of Mo-MLV in vivo (48, 51). (+) indicates wild-type levels of integration, 0 indicates a complete block to integration, and the downward arrows indicate reduced integration activity, the number of arrows reflecting the relative reduction. The underlined bases are those that differ from wild-type Mo-MLV. Here and in panels C and D, the consensus sequence reflects bases conserved among the virus strains and mutants depicted. Bases shown in upper case are completely conserved in this group, those in lower case are very well, but not completely conserved. Since the sample size here is small, it is likely that many of the apparently conserved bases reflect sampling error rather than functional significance. (C) The SNV *att* site. Putative *att* site mutants are illustrated, and the effects of the mutations on integration are indicated. The bases that differ from wild-type SNV are underlined. Asterisk signifies that the remainder of the sequence cannot be discerned from the published information (275). (D) The ASLV *att* site. The top panel shows the U5 and U3 ends of several members of the ASLV group of viruses. The bottom panel shows the putative *att* sites of three RSV mutants with deletions near the U5 end (41). The effects of the mutations on replication (integration was not explicitly tested) are indicated. The bases that differ from wild-type Schmidt-Ruppin type A RSV are underlined. Abbreviations not given in Table 1: FR-MLV, Friend murine leukemia virus; CAS-BR-E, Lake Casitas neurotropic wild mouse virus; FBJ, FBJ murine osteogenic sarcoma virus; FBR, FBR murine osteogenic sarcoma virus; AB-MLV, Abelson murine leukemia virus; AKV, AKR murine leukemia virus; FeLV-GA, Gardner-Arnstein strain of feline leukemia virus; sra, Schmidt-Ruppin A strain of RSV; amv, avian myeloblastosis virus; prc, Prague C strain of RSV; UR5, avian sarcoma virus UR5; y73, avian sarcoma virus Y73; fsv, Fujinami sarcoma virus; ra0, Rous-associated virus 0; mc29, avian myelocytomatosis virus MC29; mh2, avian carcinoma virus MH2. Except where noted, sequences shown in this figure were from reference 359.

mouse cells occurs about 100-fold less frequently than would be expected if there were no bias (199).

The most compelling evidence for preferred integration sites comes from a recent study of Rous sarcoma virus (321). Integration "hot spots" were sought by cloning thousands of independent integration sites and probing the entire collection with individual, randomly selected members, looking for sites present in more than one independent clone in the collection. The search yielded clear evidence for such hot spots, indicating that there are on the order of 5,000 such sites in the chicken genome and that about 20% of all RSV integrations occur at a hot spot. In other words, a typical hot spot is used for about 1 in 3,000 integrations, or about 10^6-fold more frequently than expected for a random site. The hot spots appear to be not merely preferred regions for integration, but precise to the nucleotide. No common sequences or potential secondary-structure motifs are discernible within several hundred base pairs on either side of the two such sites that have been sequenced.

The tropism of proviruses for transcriptionally active targets is consistent with the idea that physical accessibility plays a major role in target site selection, but it is unlikely that this is sufficient to account for the acute site specificity of RSV integration into hot spots. Investigating the hot spot sequences as naked DNA targets for integration in vitro (Section V.D), examining their physical state in infected cells, and determining whether their preferential use is RSV specific should provide important insights into the selection of integration target sites.

H. The Machinery for Integration Is in a Stable Nucleoprotein Complex with Viral DNA

A variety of data suggest that viral DNA cannot serve as a substrate for normal retroviral integration unless its presence in the cell is established by infection (38, 54, 203, 231). This apparent confinement of integration activity to authentic viral reverse transcripts is consistent with a model in which subviral structures maintain a high degree of order during replication. Indeed, most of the unintegrated viral DNA produced by reverse transcription in MLV-infected cells can be isolated in stable 160S nucleoprotein complexes that also contain virus-encoded proteins (Bowerman et al., unpublished results). These complexes, when separated from the bulk of cellular constituents by rate-zonal sedimentation or gel exclusion chromatography, retain the full integration activity present in crude cellular extracts in an in vitro integration assay (28; Bowerman et al., unpublished results). The integration activity of the isolated complexes is not appreciably enhanced by supplementation with extracts from infected cells, implying that all the components necessary for integration are intrinsic to the nucleoprotein complex. Moreover, naked viral DNA is not a competent substrate for integration when it is added to extracts from infected cells, implying that the integrative complex cannot readily assemble onto an exogenous DNA molecule (28). Thus, just as the production of mature viral DNA from genomic viral RNA depends on the maintenance of a native subviral structure (Section IV), so it appears that a subviral complex, perhaps a metamorphic derivative of the virion core, is required for normal proviral integration.

The components of the integrative complex remain to be fully defined. The in vivo genetic evidence and enzymatic studies of the IN protein imply that this protein is a critical component of the complex (76, 277, 306). MLV nucleoprotein complexes competent for integration in vitro can be immunoprecipitated efficiently with antisera specific for the CA protein p30*gag* (Bowerman et al., unpublished results). Thus, this protein is also a component of the MLV integrative complex. Further purification and characterization of the integrative complex by using the in vitro integration assay should allow the complete identification of its protein constituents and determination of its structure.

Two recognized types of nucleoprotein complexes, intracellular viral cores and bacteriophage Mu "transpososomes," provide provocative models for thinking about the retroviral integrative complex. Many viruses retain a subviral core structure upon entry into cells. Often, a significant portion of the intracellular life cycle is housed in a semipermeable subviral structure. For example, reovirus virions, upon entry into the cell, are processed by proteolysis to generate an icosahedral subviral core particle that contains the viral genome. Transcription of the minus-strand genome occurs within the core, with export of the plus-strand product.

Reverse transcription of the retroviral genome probably also takes place inside an icosahedral subviral core structure (Section IV). The complexes active in integration are presumably derived from this structure. This idea receives support from the observation that the MLV complexes active in integration contain the CA protein (Bowerman et al., unpublished results), a protein without inherent affinity for DNA and the major constituent of the icosahedral shell of the virion core (Section II.C). However, integration ultimately requires the exposure of the provirus ends to the target DNA substrate. Whether complete or partial exposure of the DNA in the complex occurs concomitantly with DNA synthesis

(A)

	U5	U3
RSV	CACCTGCATGAAGCAGAAGGCTTCATT	AATGTAGTCTTATGCAATACTCTTGTAG
MLV	ACTACCCGTCAGCGGGGGTCTTTCATT	AATGAAAGACCCCACCTGTAGGTTTGGC
SNV	GAATCCGTAGTACTTCGGTACAACATT	AATGTGGGAGGGAGCTCTGGGGGGAATA
MMTV	ACCCTCAGGTCAGCCGACTGCGGCATT	AATGCCGCGCCTGCAGCAGAAATGGTTG
HIV	TAGTCAGTGTGGAAAATCTCTAGCAG	CTGGAAGGGCTAATTTGGTCCCAAAGA
HTLV-II	ATTGTCTTCCCGGGGAAGACAAACAAT	GATGACAATGGCGACTAGCCTCCCGAGC

(B)

U5	U3	strain
ACTACCCGTCAGCGGGGGTCTTTCATT	AATGAAAGACCCCACCTGTAGGTTTGGC	MO-MLV
ACTACCCGTCTC-GGGGGTCTTTCATT	AATGAAAGACCCCACCTGTAGGTTTAGC	FR-MLV
ACTACCCGTTAGTGGGGGTCTTTCATT	AATGAAAGACCCCACCTGTAGGTTTGGC	HA-MSV
ACTACCCA-CGACGGGGGTCTTTCATT	AATGAAAGACCCCACCCGTAGG--TGGC	MO-MSV
ACTACCCG-CCTCGGGGGTCTTTCATT	AATGAAAGACCC-ACCATCAGGTTTAGC	CAS-BR-E
ACTGCCCAGCCT-GGGGGTCTTTCATT	AATGAAAGACCCCTTCCTAAGGCTTAGT	FBJ
ACTGCCCAGCCT-GGGGGTCTTTCATT	AATGAAAGACCCCTTCATAAGGCTTAGC	FBR
ACTGCCCA-CCTCGGGGGTCTTTCATT	AATGAAAGACCCCACCAAGTTGCTTAGC	SFFV
ACTACCCGTCAGCGGGGGTCTTTCATT	AATGAAAGACCCCACCAAGTTGCTTAGC	AB-MLV
ACTGCCCAGCCT-GGGGGTCTTTCATT	AATGAAAGACCCCTTCATAAGGCTTAGC	AKV
ACCTACCTAGTTCGGGGGTCTTTCATT	AATGAAAGACCCCACCCCAAAATTTAGC [CT]	FeLV-GA

		mutant	integration
ACTACCCGTCAGCGGGGGTCTTTCATT	AATGAAAGACCCCACC	wt	+
ACTACCCGTCAGCGGGGGTCTTTCATA	AATGAAAGACCCCACC	sub594	↓↓
ACTACCCGTCAGCGGGGGTCTTTCAT-	AATGAAAGACCCCACC	dl594	↓
ACTACCCGTCAGCGGGGGTCTTTCA--	AATGAAAGACCCCACC	dl593	0
ACTACCCGTCAGCGGGGGTCTTTCATATA	AATGAAAGACCCCACC	in594-2	↓↓
ACTACCCGTCAGCGGGGGTCTTTCATATT	AATGAAAGACCCCACC	in594-2R	↓
ACTACCCGTCAGCGGGGGTCTTTCAG-	AATGAAAGACCCCACC	in592-R2	↓
ACTACCCGTCAGCGGGGGT--------	AATGAAAGACCCCACC	dl587	0
GCGGGGGTCTTTCACTGGAATTCCAG-	AATGAAAGACCCCACC	in592-12	↓↓↓
GCGGGGGTCTTTCACTGGAATTACAG-	AATGAAAGACCCCACC	in592-12R1	↓
CCAAACCTACAGGTGGGGTCTTTCATT	AATGAAAGACCCCACC	U3-U3	(+)
aCtaccc------gggGGtcTTtCAtt	aATGAAAGACCCcacc	consensus	

(C)

U5	U3	mutant	integration
GAATCCGTAGTACTTCGGTACAACATT	AATGTGGGAGGGAGCTCTGGGGGG	wt	+
GTAGTCAAGCGGAATTCCGACAACATT	AATGTGGGAGGGAGCTCTGGGGGG	183	+
CCTTTTCTACATGACTGAGGATACATT	AATGTGGGAGGGAGCTCTGGGGGG	233	↓
GGGCGCAGGGATCCGGACGACAACATT	AATGTGGGAGGGAGCTCTGGGGGG	211	+
CTGGGTGGGCGCAGGGATCCGGACATT	AATGTGGGAGGGAGCTCTGGGGGG	206	-
CTGGGCGCAGGGATCCGGACTGAATCC	AATGTGGGAGGGAGCTCTGGGGGG	207	-
GAATCCGTAGTACTTCGGTACAACATT	AATGTGGGAGGGGGGAATAGTGC	121	+
GAATCCGTAGTACTTCGGTACAACATT	AATGTGGGAGGGAGCTTTCTGTAC	144	(+)
GAATCCGTAGTACTTCGGTACAACATT	AATGGGGGGAATAGTGCTGGCTC	150	-
GAATCCGTAGTACTTCGGTACAACATT	TCTGTACACATGCT*	147	-
--------------------acaACAT	AATGTGGGAGGG------------	consensus	

(D)

U5	U3	strain
ACGAGCACCTGCATGAAGCAGAAGGCTTCATT	AATGTAGTCTTATGCAATACTCTTGTAGTC	sra
GCGAACACCTGAATGAAGCTGAAGGCTTCATT	AATGTAGTCTTAATCGTAGGTTAACATGTA	amv
GCGAACACCTGAATGAAGCAGAAGGCTTCATT	AATGTAGTCTTATGCAATACTCCTGTAGTC	prc
ACGAACACCTAAATGAAGCGGAAGGCTTCATT	AATGTAGTCTTATGCAATACTCTTATGTAA	ur5
ACGAGCACCTGAATGAAGCAGAAGGCTTCATT	AATGTAGTCTTATGCAATACTCTTATGTAA	y73
GCGAACACCTGAATGAAGCGGAAGGCTTCATT	GATGTTGCCAGTTAGTCATCATTGCCACTT	fsv
GCGAACACCTGAATGAAGCAGAAGGCTTCATT	AATGTAGTCAAATAGAGCCAGAGGCAACCT	ra0
ACGCGAACCTGAATGAAGCAGAAGGCTTCATT	------------------------------	mc29
--------------------------------	AATGTAGCCTTACACAATAGCCTTACACAA	mh2

		mutant	replication
ACGAGCACCTGCATGAAGCAGAAGGCTTCATT	AATGTAGTCTTATGCAATA	sra	+
GATTGCACATGCATGAAGCAGAAGGCTTCATT	AATGTAGTCTTATGCAATA	Δ15	↓↓
ACCGTTGATTGCATGAAGCAGAAGGCTTCATT	AATGTAGTCTTATGCAATA	Δ21	↓
TGGCCGGACCGTTGATTGCAGAAGGCTTCATT	AATGTAGTCTTATGCAATA	Δ28	0
aCgagcaccTgaATGAAGCAGAAGGCTTCATT	aATGtaGtCttataca	CONSENSUS	

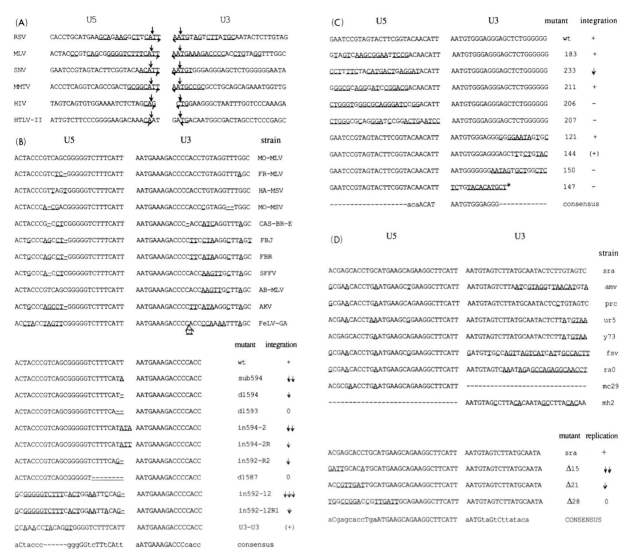

Figure 13. Viral *att* site. The plus-strand sequence is shown from 5′ to 3′. The U5 and U3 sequences are shown juxtaposed, as they presumably are during integration. (A) LTR terminal sequences differ markedly among retroviruses (see Table 1 for abbreviations). Vertical arrows indicate the sites at which the integrated proviral DNA is joined to host DNA. Horizontal arrows indicated inverted repeats. (B) The MLV *att* site. The upper panel shows the U5 and U3 ends of several strains of the MLV group. The lower panel shows mutations in the U5 ends that were tested for their effects on integration of Mo-MLV in vivo (48, 51). (+) indicates wild-type levels of integration, 0 indicates a complete block to integration, and the downward arrows indicate reduced integration activity, the number of arrows reflecting the relative reduction. The underlined bases are those that differ from wild-type Mo-MLV. Here and in panels C and D, the consensus sequence reflects bases conserved among the virus strains and mutants depicted. Bases shown in upper case are completely conserved in this group, those in lower case are very well, but not completely conserved. Since the sample size here is small, it is likely that many of the apparently conserved bases reflect sampling error rather than functional significance. (C) The SNV *att* site. Putative *att* site mutants are illustrated, and the effects of the mutations on integration are indicated. The bases that differ from wild-type SNV are underlined. Asterisk signifies that the remainder of the sequence cannot be discerned from the published information (275). (D) The ASLV *att* site. The top panel shows the U5 and U3 ends of several members of the ASLV group of viruses. The bottom panel shows the putative *att* sites of three RSV mutants with deletions near the U5 end (41). The effects of the mutations on replication (integration was not explicitly tested) are indicated. The bases that differ from wild-type Schmidt-Ruppin type A RSV are underlined. Abbreviations not given in Table 1: FR-MLV, Friend murine leukemia virus; CAS-BR-E, Lake Casitas neurotropic wild mouse virus; FBJ, FBJ murine osteogenic sarcoma virus; FBR, FBR murine osteogenic sarcoma virus; AB-MLV, Abelson murine leukemia virus; AKV, AKR murine leukemia virus; FeLV-GA, Gardner-Arnstein strain of feline leukemia virus; sra, Schmidt-Ruppin A strain of RSV; amv, avian myeloblastosis virus; prc, Prague C strain of RSV; UR5, avian sarcoma virus UR5; y73, avian sarcoma virus Y73; fsv, Fujinami sarcoma virus; ra0, Rous-associated virus 0; mc29, avian myelocytomatosis virus MC29; mh2, avian carcinoma virus MH2. Except where noted, sequences shown in this figure were from reference 359.

alternative secondary structures are of general importance to the integration mechanism. Instead, the presence of terminal inverted repeats points to the likelihood that the juxtaposed ends of the LTRs are recognized by a multimeric protein with a twofold axis of symmetry, perhaps a homodimer.

MLV can provide the "helper" viral functions required in *trans* for the integration of a large group of closely related retroviral genomes (members of the MLV-murine sarcoma virus (MSV) family; Table 1). The 13-bp inverted repeats are the only portions of the LTR termini that are conserved among these viruses (Fig. 13B). Moreover, deletions extending distally toward either end, up to the inverted repeat sequence, fail to prevent integration of MLV (L. Lobel, J. Murphy, and S. Goff, personal communication). Thus, the sequences required for function as a viral attachment site are confined to the inverted repeat at the outer edges of the LTR. The dispensibility of U5 sequences outside the inverted repeat is further emphasized by the identification of an MLV provirus that apparently integrated by using two copies of the U3 terminus in inverted orientation (52). Even sequences within the inverted terminal repeat of MLV can be altered substantially without preventing integration (48, 51). The potential of the inverted repeat sequence to form a cruciform structure by base pairing between the copies at the U3 and U5 ends appears not to be essential for MLV integration, as mutations (e.g., in592-12R1; Fig. 13B) that would destabilize such a structure do not abolish integration. However, mutational analysis of this region is not yet sufficiently complete to allow definition of the sequence features essential for recognition by the integration apparatus.

The actual site on the proviral DNA that is covalently joined to the target DNA is probably always a 5′-. . . CA-3′ (51). This dinucleotide is always conserved at each end of the provirus, even in otherwise divergent retroviruses and retrotransposons. In almost all cases, the CA dinucleotide is situated exactly two bases from each end of the putative linear precursor, and typically the terminal four bases in the precursor are 5′-. . . CATT-3′. However, mutational analysis has shown that at least for MLV the spacing of the CA dinucleotides from the ends of the linear precursor is flexible within a few base pairs; from 1 to at least 4 bp is tolerated distal to the CA at the U5 edge. Moreover, the identity of the bases distal to the joining site is not vitally important for MLV, although the wild-type TT is clearly preferred (51).

The terminal inverted repeat of SNV is substantially shorter than that of MLV, 5 versus 13 bp (359) (Fig. 13A). Deletion analysis with SNV indicates that the sequences required for *att* site function are confined to the terminal 5 to 12 bp of U3 and to the terminal 6 to 8 bp of U5 (275; Fig. 13C).

In ASLV, deletion experiments indicate that viral replication requires specific sequences in the terminal 17 to 23 bp of U5, presumably for integration (41), and except for position 13 from the U5 terminus, these sequences are highly conserved among the viruses in the ASLV family (Fig. 13D). The extent of the U3 terminal sequences required for ASLV *att* site function in vivo has not been investigated by deletion analysis. These sequences are generally much less well conserved among members of the ASLV family than are the sequences at the U5 terminus, but it is worth noting that the terminal 5 to 9 bp is the most highly conserved region. Approximately the same sequences whose importance in replication is implied by sequence conservation and deletion analysis, the terminal 18 to 22 bp of U5, together with the terminal 5 to 8 bp of U3, are required for in vitro circle junction cleavage by the IN protein of ASLV, providing further circumstantial evidence for the role of this activity in integration (41).

Recognition of the viral *att* site cannot simply involve specificity for each of the individual LTR terminal sequences, as these same sequences are present at the internal as well as the external edge of the LTR (Section II.B, Fig. 3). Perhaps the proximity of these sequences to the end of a DNA molecule is important for recognition as an *att* site. Although its physiological significance is not clear, the affinity of Tn*3* transposase for DNA ends per se as well as for the DNA sequence normally present at the ends of Tn*3* may provide an example of this type of specificity (166a). Alternatively, juxtaposition of the sequences at the U3 and U5 ends of the linear precursor may be required for recognition.

2. Topology of the ultimate precursor to the integrated provirus

We noted earlier that the unintegrated viral DNA in the nucleus of an infected cell is a mixture of the linear form and circular forms with either one or two copies of the LTR (Section IV.C). Which of these topological forms is the ultimate integrative precursor?

The first experiment to address this question directly gave results that favored the two-LTR circle as the ultimate precursor (276). A small (57-bp) DNA fragment containing the SNV circle junction was inserted at an internal site between the normal LTRs of a defective SNV proviral clone. The resulting recombinant virus was recovered by cotransfecting this construct with wild-type helper virus DNA into permissive cells, and the progeny virus particles were then used to infect fresh cells. Restriction analysis of the resulting integrated proviral DNA suggested that attachment to host DNA could occur with equal frequency using either the normal LTR edges or the internal circle junction sequence. This result would appear to argue strongly that the circle junction is the viral attachment site for integration, and thus that the two-LTR circle is the integrative precursor. However, efficient use of the internal circle junction site has yet to be confirmed by sequence analysis. Furthermore, the restriction mapping data are consistent only with two-LTR circles and not one-LTR circles integrating at the internal circle junction site. This observation is unexpected and unexplained.

The analogous experiment has since been carried out with MLV (L. Lobel, J. Murphy, and S. Goff, unpublished results). To favor the recovery of proviruses integrated by using the internal site, the genetically marked MLV construct that was used in this experiment carried, in addition to the internal circle junction sequence, a crippling mutation (*dl587*; see Fig. 13B) in the U5 *att* sequence that would normally be used for integration. In spite of this strong selection and a determined search, MLV proviruses joined to host DNA using the transplanted circle junction have not been found.

In vitro, the linear form of unintegrated viral DNA can serve as a precursor to the integrated MLV provirus (28). It does not necessarily follow that the immediate precursor to the integrated provirus is linear, since the linear molecule could be covalently circularized in vitro prior to integration. As shown in Fig. 10, the expected structure of the free 5′ end of the viral DNA in the gapped initial product of the integration reaction differs depending upon whether its immediate precursor is linear or circular. In the intermediate arising directly from a linear DNA precursor (Fig. 10A), the free 5′ end of the viral DNA in the gapped intermediate is identical to the corresponding end of the precursor. In contrast, a reaction involving a circular precursor gives rise to a gapped intermediate in which the viral 5′ end is two bases longer than the corresponding end of the linear precursor (Fig. 10B). We can therefore deduce the topology of the precursor by comparing the free viral DNA ends in the gapped intermediate with the corresponding ends of the unintegrated linear DNA. The intermediate has recently been purified from an in vitro MLV integration reaction (104a; Brown et al., in press). Structural analysis showed that the free 5′ ends of the viral DNA in the intermediate were identical to the corresponding 5′ ends of the unintegrated linear molecule, exactly as expected for a linear precursor. Thus, both the in vivo genetic analysis and direct physical evidence from the in vitro system point compellingly to the linear DNA molecule as the direct precursor to the integrated MLV provirus.

It remains to be seen whether the early evidence for a circular precursor in SNV integration will be corroborated by further investigation, but in view of the great differences among retroviruses in the primary structure of the viral *att* site (Fig. 13), it is not hard to imagine that two viral species might also differ in the preferred topology of the integrative precursor.

G. Characteristics of the DNA Target

Since naked, linear (relaxed) DNA can serve as a target for integration in vitro, the DNA target for integration does not need to be supercoiled or assembled into chromatin (28). It remains possible, however, that a supercoiled or otherwise modified DNA molecule might be a preferred target for integration.

The lack of a requirement for deoxynucleoside triphosphates or ribonucleoside triphosphates implies that neither transcription nor DNA synthesis per se is required for target DNA function. Clearly, generation of the flanking repeat requires DNA synthesis, but formation of a covalent linkage between provirus and target DNA does not (Fig. 10). Thus, the initial covalent linkage at each end of the provirus is on one strand only, and its formation is not inherently coupled with DNA synthesis. Experiments in vivo show that integration occurs at least preferentially in cells that are actively replicating their DNA (164, 366). Perhaps, in the cell, replication physically exposes the DNA to the integration machinery; this would be unnecessary when naked DNA is used as target in the in vitro reaction. Alternatively, these data could reflect a role for cellular functions related to DNA replication in rendering the viral DNA precursor mature for integration (294).

Scrutiny of the integration sites of numerous proviruses integrated in vivo and in vitro has failed to disclose any consistent features that might mark the sites as targets for retroviral integration (28, 326, 363). However, there is considerable evidence that not all sites in the target genome are equally favored for integration in vivo. For example, integration sites are found near transcribed regions and DNase I-hypersensitive sites at a much greater than expected frequency (295, 371), perhaps accounting for the observation that proviruses introduced into cells by natural integration are generally transcribed at a higher level than are proviruses introduced by transfection (166). Moreover, integration of MLV into the hypoxanthine phosphoribosyltransferase gene in F9 mouse cells occurs about 100-fold less frequently than would be expected if there were no bias (199).

The most compelling evidence for preferred integration sites comes from a recent study of Rous sarcoma virus (321). Integration "hot spots" were sought by cloning thousands of independent integration sites and probing the entire collection with individual, randomly selected members, looking for sites present in more than one independent clone in the collection. The search yielded clear evidence for such hot spots, indicating that there are on the order of 5,000 such sites in the chicken genome and that about 20% of all RSV integrations occur at a hot spot. In other words, a typical hot spot is used for about 1 in 3,000 integrations, or about 10^6-fold more frequently than expected for a random site. The hot spots appear to be not merely preferred regions for integration, but precise to the nucleotide. No common sequences or potential secondary-structure motifs are discernible within several hundred base pairs on either side of the two such sites that have been sequenced.

The tropism of proviruses for transcriptionally active targets is consistent with the idea that physical accessibility plays a major role in target site selection, but it is unlikely that this is sufficient to account for the acute site specificity of RSV integration into hot spots. Investigating the hot spot sequences as naked DNA targets for integration in vitro (Section V.D), examining their physical state in infected cells, and determining whether their preferential use is RSV specific should provide important insights into the selection of integration target sites.

H. The Machinery for Integration Is in a Stable Nucleoprotein Complex with Viral DNA

A variety of data suggest that viral DNA cannot serve as a substrate for normal retroviral integration unless its presence in the cell is established by infection (38, 54, 203, 231). This apparent confinement of integration activity to authentic viral reverse transcripts is consistent with a model in which subviral structures maintain a high degree of order during replication. Indeed, most of the unintegrated viral DNA produced by reverse transcription in MLV-infected cells can be isolated in stable 160S nucleoprotein complexes that also contain virus-encoded proteins (Bowerman et al., unpublished results). These complexes, when separated from the bulk of cellular constituents by rate-zonal sedimentation or gel exclusion chromatography, retain the full integration activity present in crude cellular extracts in an in vitro integration assay (28; Bowerman et al., unpublished results). The integration activity of the isolated complexes is not appreciably enhanced by supplementation with extracts from infected cells, implying that all the components necessary for integration are intrinsic to the nucleoprotein complex. Moreover, naked viral DNA is not a competent substrate for integration when it is added to extracts from infected cells, implying that the integrative complex cannot readily assemble onto an exogenous DNA molecule (28). Thus, just as the production of mature viral DNA from genomic viral RNA depends on the maintenance of a native subviral structure (Section IV), so it appears that a subviral complex, perhaps a metamorphic derivative of the virion core, is required for normal proviral integration.

The components of the integrative complex remain to be fully defined. The in vivo genetic evidence and enzymatic studies of the IN protein imply that this protein is a critical component of the complex (76, 277, 306). MLV nucleoprotein complexes competent for integration in vitro can be immunoprecipitated efficiently with antisera specific for the CA protein $p30^{gag}$ (Bowerman et al., unpublished results). Thus, this protein is also a component of the MLV integrative complex. Further purification and characterization of the integrative complex by using the in vitro integration assay should allow the complete identification of its protein constituents and determination of its structure.

Two recognized types of nucleoprotein complexes, intracellular viral cores and bacteriophage Mu "transpososomes," provide provocative models for thinking about the retroviral integrative complex. Many viruses retain a subviral core structure upon entry into cells. Often, a significant portion of the intracellular life cycle is housed in a semipermeable subviral structure. For example, reovirus virions, upon entry into the cell, are processed by proteolysis to generate an icosahedral subviral core particle that contains the viral genome. Transcription of the minus-strand genome occurs within the core, with export of the plus-strand product.

Reverse transcription of the retroviral genome probably also takes place inside an icosahedral subviral core structure (Section IV). The complexes active in integration are presumably derived from this structure. This idea receives support from the observation that the MLV complexes active in integration contain the CA protein (Bowerman et al., unpublished results), a protein without inherent affinity for DNA and the major constituent of the icosahedral shell of the virion core (Section II.C). However, integration ultimately requires the exposure of the provirus ends to the target DNA substrate. Whether complete or partial exposure of the DNA in the complex occurs concomitantly with DNA synthesis or is closely coupled to the integration process (e.g., induced by the recognition of target DNA) requires investigation.

The bacteriophage Mu transposition reaction can be duplicated in vitro with purified DNA containing Mu ends, the Mu gene A and gene B products, and the *E. coli* HU protein (56). An essential early step in the process is the formation of a stable "transpososome" complex of Mu A protein (probably several copies) and HU, with the Mu ends (346). Although the Mu A protein can specifically recognize and bind to isolated sequences found at either Mu end (58), formation of the transpososome complex depends on the presence of both Mu ends in the correct topological relationship (57). Cleavage of the ends of the Mu donor prophage, the initial step in the Mu integration reaction, requires assembly of the transpososome (55, 250a). There are many other well-characterized examples of specialized DNA recombination systems that depend on the assembly of a highly ordered multicomponent nucleoprotein complex for normal activity (see Thompson and Landy, this volume; Pato, this volume). Perhaps, just as Mu DNA must have a specific tertiary structure to allow the formation of the transpososome complex, correct assembly of the retroviral integrative complex may require that the viral genome have a particular structure. For instance, it might need to be encapsidated, or assembly might initiate on an intermediate in viral DNA synthesis.

I. Cellular Functions Influence Retroviral Integration

We have previously noted the evidence that cellular functions may act on the DNA target for integration to influence the efficiency and target site specificity of the reaction. Other cellular components might also influence integration through more direct interactions with the viral integrative complex.

The best-studied candidate for such an interaction is *Fv-1* restriction of MLV replication (182). The mouse *Fv-1* gene serves an unknown cellular function in the absence of infection. The two well-characterized alleles of this gene, *Fv-1^n* and *Fv-1^b*, determine the ability of alternative MLV strains to establish a provirus upon infection of a mouse cell. MLV strains are designated as either N-tropic, B-tropic, or NB-tropic, depending on their ability to replicate in cells of different *Fv-1* genotypes, as follows. Cells carrying the *Fv-1^n* allele do not allow establishment of a provirus by B-tropic MLV strains. Cells carrying the *Fv-1^b* allele restrict N-tropic MLV strains. Cells carrying both alleles restrict both N- and B-tropic viruses. NB-tropic MLV strains are not restricted by either *Fv-1* allele, while phenotypically mixed viruses carrying both determinants are restricted by both alleles. Analysis of recombinants between N- and B-tropic viruses shows that viral N/B tropism is genetically determined (at least in part) by the p30^{gag} (CA) coding region (20, 270). A simple view of the *Fv-1* restriction phenomenon consistent with existing data is that the host *Fv-1* gene product can interact with the cognate CA protein of an infecting subviral core particle so as to interfere with its completion of the transposition cycle. The step(s) blocked by this interaction probably include integration itself, but DNA synthesis, nuclear entry, and circularization of viral DNA may also be impeded by *Fv-1* restriction (183, 184, 393). Therefore, while supporting the idea that CA may be an essential part of the MLV integrative complex, the data do not specifically implicate integration as the target for the action of the *Fv-1* product.

There is evidence suggesting that components of the host apparatus for DNA synthesis might be involved in a poorly defined process of "maturation" of the unintegrated viral DNA precursor (37, 294, 366). Whether this reflects a host cellular role in completing viral DNA synthesis or in some other modification of the DNA is unclear. As in the case of *Fv-1* restriction, distinguishing this kind of apparent effect on integration, i.e., synthesis or modification of the precursor, from a more direct involvement in the reaction itself is difficult in vivo. The in vitro integration system (Section V.D) should make possible a direct examination of the effect of the *Fv-1* gene product and other host cellular components (e.g., those induced by interferons [7]) on this reaction.

J. Proviral Stability

Transposable elements are frequently genetically unstable, and indeed can destabilize the region of the genome into which they are inserted. Typically, this instability depends on interactions between the elements and proteins normally involved in their transposition. Since transposition of a provirus is mediated via an RNA intermediate, the donor provirus is not a direct participant in the integration reaction and thus is not normally a substrate for the viral DNA recombination machinery. Moreover, the proteins involved in proviral integration, although they are encoded by the donor provirus, are not normally processed to their active form until they leave the cell in the budding virion (Section III.D). Thus, quite unlike the elements that transpose through a direct interaction with their target or by excising from their donor site, retroviral proviruses are not likely to be

Table 2. Endogenous proviruses of *Mus musculus*[a]

Name	Number of copies	References
Ecotropic MLV	0–6	341
Non-ecotropic MLV	ca. 10–30[b]	40, 342
MMTV-α	0–10	341
MMTV-β	?	30
IAP	ca. 1,000	206
VL30	ca. 100	196
VL30-GLN	ca. 20–50 ca. 1,000 (LTR only)	169
ETn	ca. 200–1,000	318, 336
MuRRS	?	303

[a] Listed here are the approximate numbers of copies of complete or partial retrotransposons identified in the genomes of laboratory strains of mice. Abbreviations: MLV, murine leukemia virus; MMTV, mouse mammary tumor virus; IAP, intracisternal A particles; VL, viruslike particles; ETn, early transposon; MuRRS, murine retrovirus-related sequence.

[b] There are also several hundred copies of MLV LTRs, some of which have a 190-nt insertion sequence (302).

threatened by element-specific recombination activities.

It is difficult to say with certainty whether proviruses are less stable genetically than ordinary host DNA sequences, but there is no evidence for exceptional genetic instability. Examples of proviral excision (53, 367, 385), deletions and other rearrangements encompassing proviral sequences and adjacent DNA (113, 263, 278), and tandem duplication of a provirus (157) have all been described. The best-characterized examples of excision of proviral sequences result in a residual solo LTR (53, 367). Excisions of this kind could occur by homologous recombination between the repeats, and indeed they happen at the frequency (about 10^{-7} per cell per generation) expected for normal homologous recombination between tandemly repeated sequences of this size (344). Thus, it is unlikely that a specialized mechanism plays a role in these excisions. Precise excision of a provirus has never been demonstrated. A single observed case of a tandem duplication of a provirus during replication of the host cell is likely to have occurred by unequal sister chromatid exchange between the LTRs, presumably mediated by the host cell (157, 344). Many deletions and other rearrangements associated with proviruses have been identified in retrovirally induced tumors (113, 263, 278, 385). It is likely that the rearrangements confer a growth advantage on the host cell by enhancing expression of an adjacent oncogene. The deletions often include the upstream LTR, but this probably reflects selection for this class of deletions rather than any mechanism favoring their occurrence (Section VI). Unlike the characteristic deletions promoted by transposable elements, the proviral breakpoints of these deletions do not appear to be specific. Thus, although our knowledge regarding the frequency of deletions in genomic DNA is insufficient to allow us to dismiss a possible role of the provirus in promoting these events, we consider this unlikely.

In summary, while the genetic consequences of retroviral integration can be profound (Section VI.B), often leading to strong selection for rearrangements that involve the provirus, there is no good evidence that retroviral proviruses are inherently unstable genetic elements, either in the germ line or in somatic cells. Moreover, our current understanding of their transposition mechanism does not lead us to expect such instability.

VI. CONSEQUENCES OF INTEGRATION AND RETROVIRAL PATHOGENESIS

The preceding sections describe the mechanism of integration and the role of the integrated provirus in the virus life cycle. Integration also has consequences for the host, whether the host is an intact animal or a cell in culture. Through the act of integration, proviruses can be added to the germ line, retroviral infection persists indefinitely in cells, and insertion mutations may occur with pathological effects. This section explores the basis for these and related phenomena.

A. Endogenous Proviruses

It is well established that vertebrate genomes contain many copies of DNA elements that resemble retroviral proviruses (44). The inexact numbers are not known for any species because it is difficult to identify elements other than those related to known retroviruses. Judging from studies of *Mus musculus*, for which most information is available, proviruses may account for as much as 1% of the entire genome (Table 2). (Other elements synthesized by reverse transcriptase, such as LINEs, SINEs, and processed pseudogenes, may constitute up to 10% of the mouse genome [Hutchison et al., this volume; Deininger, this volume].)

In general, the proviruses that have been most closely examined—those readily detected with probes derived from infectious retroviruses, such as ALV, MLV, and MMTV—occur in relatively few copies per cell, ranging from 0 to 10 in different chicken flocks and mouse strains. There is strong evidence that these proviruses, each identified by the restriction map of flanking cellular sequences, were acquired long after speciation (6, 44, 46, 161, 179), and, in some cases, during recent breeding of the strains in the laboratory (9, 177, 300). Other proviral elements have been discovered because they are sufficiently related to known retroviruses to cross-hybridize under relaxed conditions for annealing; they make intracellular viruslike particles that can be isolated; they generate RNA that is sometimes packaged in heterologous virus particles; they constitute a significant component of species-specific middle repetitive DNA (387); or they are encountered serendipitously during an analysis of cellular loci in which one happens to reside (44, 234, 341).

Endogenous proviral elements vary greatly in structure and functional competence (44). Among those most closely related to horizontally transmitted viruses and most recently entered into the germ line, it is common to find proviruses that give rise to infectious and even tumorigenic viruses; several examples of fully competent MLV and MMTV proviruses have been described. At the other end of the spectrum are a multitude of solo LTRs, analogs of the "solo deltas" from yeast Ty elements (Boeke, this volume) and thus presumptive remnants of homologous, LTR-LTR recombinations. In between are a wide variety of deficient elements, some with little evidence of intact open reading frames, and others (such as some proviruses for IAPs) apparently able to synthesize *gag* and *pol* products, but no *env* proteins, thereby limiting functional particles to an intracellular role in transposition (245).

The manner in which endogenous proviruses first gained entry to germ lines and the mechanisms by which their numbers reached current levels are still matters of conjecture and debate. In a few instances, endogenous proviruses have been established or increased in number during experimental observation. Infection of preimplantation embryos with the Moloney strain of MLV can establish new endogenous proviruses (173), and MLV present in high titer in members of a mouse colony can infect the germ line repeatedly over the course of a few years, with transmission of the new proviruses to offspring (290, 300). Some endogenous proviruses seem to generate new copies at relatively high frequencies, particularly during crosses between females carrying the proviruses and males of certain genotypes. For example, in crosses between offspring of RF/J females that bear the endogenous ecotropic MLV proviruses *emv*16 and *emv*17 and SWR/J mice, new germ line proviruses appear in the progeny at rates of about 0.35 per animal, only 10-fold lower than those seen for P elements in the progeny of hybrid-dysgenic *Drosophila* crosses (9, 177).

It is less clear how extremely large numbers of some elements have been produced (Table 2). Such elements are not found in large tandem arrays or in arrays of interspersed proviral and host sequences, so gene amplification of the type that expands the copy number for certain proto-oncogenes and drug resistance genes in somatic cell tumors is unlikely to have a role. These elements appear to be unable to make infectious extracellular viruses, but the copy number may increase by intracellular transposition events in which RNA transcripts packaged in cytoplasmic or intracisternal particles are reverse transcribed and integrated at new sites in germ line DNA (44). This possibility is supported by insertions of new IAP proviruses at new sites in tumor cell lines (142, 153, 206). In addition, it has been recently shown that Mo-MLV DNA in which the *env* gene was replaced by a selectable marker can occasionally undergo intracellular transpositions via reverse transcription (144).

Since packaging constraints seem not to be strict, the RNA from endogenous proviruses may form heterozygotic dimers with other retroviral RNAs or viral pseudotypes with proteins provided by horizontally transmitted viruses or other endogenous proviruses. Heterozygotes may lead to the formation of genetic recombinants (Section II.F.4); these have been recognized by the repair of deletion mutations in laboratory strains of virus (47, 53), by the detailed examination of certain viral genomes (e.g., the genomes of Harvey and Kirsten MSV contain portions of VL30 sequence as well as *ras* coding sequences [85]), or by the discovery of endogenous proviruses that are likely to be recombinants between two other classes of endogenous proviruses (40, 44, 302, 342). Pseudotypes or heterozygotes can account for the transposition of proviruses that are deficient in coding potential or of proviruses that encode *env* proteins unable to mediate entry into an appropriate host cell.

The possibility that endogenous proviruses confer some selective advantage upon the host, such as protection against retroviral disease or more efficient evolution via genetic rearrangement, has been often discussed (e.g., 44, 341, 354, 355, 363), but there is scant information on these topics. The occasional insertion mutations caused by endogenous proviruses (see below) show that such proviruses may have disadvantages beyond the simple burden of additional DNA to be replicated. The dearth of fully competent proviruses in the germ line argues either that selective forces work against animals with functional proviruses or that no selective properties foster the retention of such proviruses.

B. Integration and Insertion Mutation

Proviral integration may cause insertion mutations in host genes; insertions that inactivate resident genes and those that augment gene expression have both been observed (362, 363).

1. Recessive insertion mutations

Because inactivating lesions are inherently recessive, they have been sought in cultured cells by using targets for which the cell is hemi- or heterozygous (e.g., a preexisting transforming provirus [367], an X-linked gene such as *hprt* [199], a rearranged immunoglobulin gene [142], or an autosomal gene whose heterologous alleles are distinguishable [100]). If the cultured cell is an embryonal mouse stem cell, it is possible to generate mice carrying the proviral insertion; this feat has been accomplished using *hprt* as the target for an MLV-induced mutation (205). Similar strategies could be employed to obtain insertion mutations of other mammalian genes and thereby study the phenotypic consequences of null alleles. Since heterozygous animals can be bred to generate offspring homozygous for the mutant genes, autosomal loci should also be susceptible to this approach if heterozygous mutants can be identified in cell culture.

Both natural and experimental infections of the germ line by ecotropic strains of MLV have produced recessive mutations that can be phenotypically assessed in this manner (44, 341). In one instance, an embryonic lethal phenotype has been ascribed to MLV proviral DNA inserted in the first intron of the collagen α(I) gene, with impaired initiation of transcription and vascular incompetence due to collagen deficiency (140, 174, 229). In another case, an endogenous MLV provirus on mouse chromosome 9 has been blamed for the *dilute* coat-color phenotype (178), a conclusion strongly supported by the finding that revertants of *dilute* have lost most of the provirus by homologous recombination between the LTRs (53). Interestingly, the same mechanism allows reversion of an MLV insertion mutation in a preexisting RSV provirus in somatic cells and reversion of *hairless*, a recessive mutation on mouse chromosome *14* recently ascribed to an endogenous polytropic MLV provirus (367; J. P. Stoye, W. Frankel, and J. M. Coffin, personal communication). In both germinal and somatic cells, the frequencies of the homologous recombination events are in the vicinity of 10^{-5} to 10^{-7} per generation.

2. Dominant insertion mutations

Insertion mutations that activate transcription have been encountered only in virus-induced tumors affecting presumptive or established proto-oncogenes (18, 19, 364). The target genes for such mutations include several known progenitors of retroviral oncogenes (commonly c-*myc*, c-*myb*, and c-*erbB*; occasionally c-*fms* and c-Ha-*ras* [114, 364, 384]). Other genes were first suspected to be proto-oncogenes when they were found to be repeatedly mutated by proviral insertions; these genes were isolated by the method of transposon tagging, in a fashion analogous to that used to retrieve the *Drosophila white* locus interrupted by a *copia* element (15). The best-established genes on this growing list are those activated by MMTV proviruses (*int-1*, *-2*, *-3*, and *-4* [264]) and some activated by MLV proviruses (e.g., *pim-1*, *fis*, and *pvt* [364]).

Though the regular association of certain retroviral proviruses with certain proto-oncogenes in certain types of tumors remains largely unexplained (e.g., ALV DNA in c-*myc* in bursal lymphomas), there is no evidence for preferred integration at these loci or at specific sites within the loci (Fig. 9). Any suggestion of clustered integration sites in loci for which many mutant alleles have been studied is best explained by selection for cells in which the provirus is well positioned to augment expression without interruption of essential coding sequences. The proviruses are most commonly positioned on the 5′ side of the coding sequence of the cellular gene in the same or opposite transcriptional orientation, or on the 3′ side in the same orientation (364). In some cases, the coding sequence is truncated by the insertions, most often on the 5′ side, with the amino-terminal portion of the resulting protein encoded by a proviral gene.

C. Transduction

An important though uncommon consequence of insertional activation of cellular genes is retroviral transduction, generating the retroviruses that carry host-derived viral oncogenes (Section II.A). Because

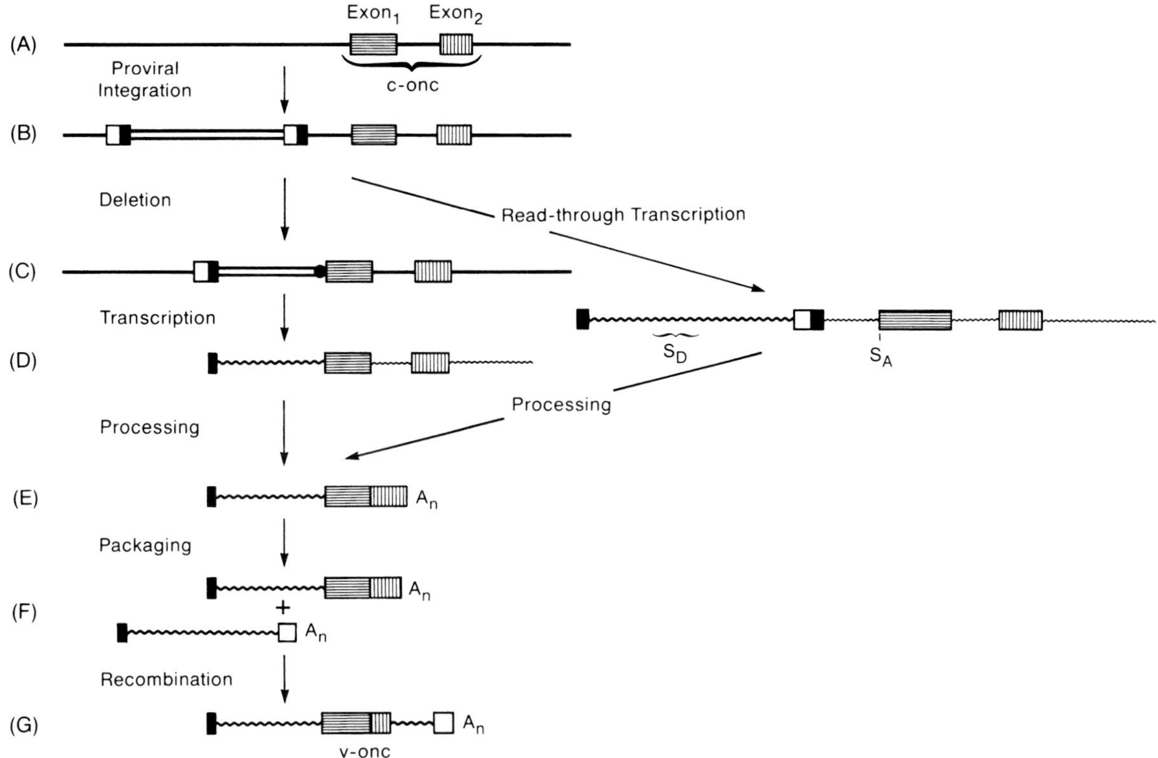

Figure 14. A mechanism for the transduction of a cellular proto-oncogene. (A) The horizontally and vertically striped boxes represent exons in a cellular gene (c-*onc*) destined to be a precursor to a viral transforming gene (v-*onc*, as shown in panel G). (B) Proviral DNA of a retrovirus lacking an *onc* gene integrates on the 5′ side of a c-*onc* gene in the same transcriptional orientation. (Viral DNA is shown by double lines, U3 as open boxes, U5 as closed boxes, and host DNA by single lines.) (C) Two mechanisms may be used to create virus-host transcripts lacking the 3′ end of viral RNA. On the left, a deletion removes a 3′ portion of proviral DNA and adjacent cellular DNA, with the point of rejoining indicated by a closed circle. (The figure is deliberately ambiguous about whether the deletion ends in coding or noncoding sequences of c-*onc*.) Transcriptions from the remaining LTR (in panel D) produce a primary transcript linking sequences from the 5′ end of the viral genome (including at least R, U5, the primer binding site, and the packaging signal) to c-*onc* sequences. (E) The transcript is then processed by splicing and polyadenylation, removing introns between exons of the c-*onc*. On the right, transcription of the undeleted but insertionally mutated locus from the proviral 5′ LTR produces a hybrid transcript that is processed by splicing to yield an RNA similar to that produced by the pathway on the left. (F) The processed RNA is packaged into retrovirus particles (the structural proteins are provided by an intact provirus in the same cell), forming heterozygotic dimers with a subunit of wild-type viral RNA. (G) Recombination during reverse transcription occurs upon infection of neighboring cells, joining sequences from the 3′ end of viral RNA to the 3′ side of the captured *onc* sequences. Further changes (deletions and base substitutions) in the v-*onc*-containing genome are likely to occur during subsequent rounds of infection to produce the conventional strains of highly oncogenic retroviruses. (Reproduced with minor changes from reference 364.)

transduction is rare, it has not been possible to analyze the formation of transducing viruses through direct experiment. Instead, hypotheses about mechanism are simply compatible with available evidence, which consists largely of the detailed structure of the native cellular genes; unfortunately, these genes have been subjected to many rounds of error-prone replication prior to study.

Such exercises in molecular archeology, now applied to a large number of retroviral oncogenes and their progenitors, lead to the following arguments and form the basis for the generally accepted (though hardly inviolate) model shown in Fig. 14 and discussed in references 17–19, 318, 363, and 364.

(i) Recombination must occur between viral and host sequences on both the 5′ and 3′ sides of the transduced sequences.

(ii) The 5′ recombination seems likely to occur at the DNA level, especially in those cases (146, 201, 318) in which part of an intron from the cellular gene is found at the 5′ boundary. In most instances, however, the 5′ boundary joins viral sequences to cellular sequences within an exon. In a few cases, a host 3′ splice site is at the 5′ boundary (167, 190), implying that recombination may have occurred by splicing of a hybrid transcript.

(iii) To produce any of the observed 5′ junctions, a provirus was likely positioned upstream of the sequence to be transduced, in the same transcriptional orientation (Fig. 14). In this arrangement, a deletion could join sequences internal to the provirus with intronic or exonic sequences in the cellular gene, or splicing of a hybrid (proviral-cellular) readthrough transcript could produce the junctions that occur at splice sites. Such spliced readthrough transcripts are sometimes observed in tumors with upstream proviral insertions (123, 262, 319). It is not possible, however, to exclude aberrant *trans*-splicing of RNAs from unlinked templates.

(iv) Proviral insertions adjacent to a proto-oncogene would markedly increase the likelihood of transduction by fostering neoplastic expansion of the clone of cells in which the insertion occurred. This may explain why the transduction of other types of cellular genes has rarely if ever been observed.

(v) All introns that normally reside between complete (or incomplete) exons in the transduced genes have been correctly removed, presumably by RNA splicing.

(vi) Poly(A) is sometimes found at the 3′ host-virus junction (158), but intron sequences are not. For these reasons and others, the 3′ recombination junction is likely to be formed by a mechanism that joins sequences present in RNA form, presumably the same mechanism that mediates recombination between retroviral RNA genomes (Section II.F.5).

(vii) Multiple differences in nucleotide sequence, sometimes including deletions, are often observed when v-*onc* and progenitor c-*onc* sequences are compared (359). These differences are attributable to changes in the transduced sequences during the many cycles of error-prone replication subsequent to transduction, with selection for mutants that encode more potent oncoproteins.

Though retroviral transduction is generally an uncommon event, there are a few situations in which it occurs with surprising regularity: the induction of erythroleukemia by ALV insertion mutations of c-*erbB* (246), the induction of T-cell leukemias by feline leukemia virus insertion mutations of c-*myc* (257, 260), and the recovery of transformation-competent RSV after infection by partial deletion mutants of v-*src* (337). The 3′ recombination can be seen experimentally by transfection with oncogenes linked to a viral LTR on their 5′ but not 3′ side and infection with helper virus (118). Some retroviruses, moreover, carry two genes transduced from host genomes (e.g., one strain of avian erythroblastosis virus contains both v-*erbA* and v-*erbB*, and a strain of avian myeloblastosis virus contains both v-*myc* and v-*mil* [18, 19, 190]).

Despite our limited understanding of low- or high-frequency and single- or multiple-gene transductions, the principle of using retroviruses as vectors to transmit and express foreign genes and the designs of these naturally occurring examples have both strongly influenced experimental work with virus vectors (Soriano et al., this volume).

D. Viral Pathogenesis

Retroviruses first drew scientific attention as causative agents of tumors, and they have come to widespread public attention recently as causative agents of acquired immunodeficiency syndrome. Yet much more is known about the structure and replication of these viruses than about the mechanisms by which they cause disease.

1. Retroviral neoplasia

At least three routes to retroviral oncogenesis have been proposed: (i) expression of retroviral oncogenes, (ii) activation of cellular proto-oncogenes in *cis* by insertion mutation, and (iii) activation of proto-oncogenes in *trans* by viral transcription factors.

(i) Retroviruses that carry their own oncogenes are thought to require no additional factors to induce tumors, thus accounting for uniform neoplastic transformation of cultured cells and rapid, polyclonal tumorigenesis in animals. The focus of attention has thus been shifted to the function of proteins encoded by viral oncogenes, a complex story told in other places (18, 19, 364).

(ii) As discussed in Section VI.B, many retroviruses lacking viral oncogenes cause proviral insertion mutations that activate the expression of proto-oncogenes (364). However, these mutations are unlikely to be sufficient for tumorigenesis: the time course of tumor induction is prolonged, implying a need for multiple events; the activated proto-oncogenic alleles are often insufficient by themselves to cause full transformation of cultured cells or tumors in transgenic animals; and the inciting viruses do not transform cells in culture, even when very large numbers have been infected. Secondary alterations in activated loci are frequently observed: partial deletions of proviral DNA, including recombination between LTRs, and mutations in the coding sequence of the cellular gene. Other host loci are probably affected by genetic (or epigenetic) mechanisms, but the loci and mechanisms are unknown.

(iii) The HTLV-BLV group of retroviruses induce tumors slowly and infrequently, can (inefficiently) transform cultured cells, do not carry transduced cellular genes, and have not been found to activate genes by insertion mutation (35, 380). However, the X gene complex, the overlapping reading frames 3′ of *env*, has been implicated in tumorigenesis by two kinds of experimental results: (i) the *tax* gene can stimulate the production of RNA from certain cellular genes, including the genes for two proteins implicated in T-cell growth control, interleukin-2 and its receptor (62, 239, 327); and (ii) transgenic mice carrying the HTLV-1 X region develop benign tumors, called neurofibromas (149, 261). The prevailing hypothesis is that tumorigenesis by this virus group depends upon *trans*-activation of an unspecified set of cellular genes by the X-encoded proteins.

2. Other retroviral diseases

Much less is known about the mechanisms by which retroviruses induce a wide spectrum of non-neoplastic disease, including immunodeficiency syndromes, anemia, arthritis, pneumonia, peripheral neuropathy, myelopathy, and encephalopathies.

It has been proposed that autoimmune disorders might arise from persistent infection with retroviruses that encode structural proteins sharing antigenic epitopes with host proteins. For example, some mice with an autoimmune disease have antibodies that react both with a component of particles containing small nuclear RNA and with a homologous amino acid sequence in *gag* protein (289).

Cell death is not usually a property of retroviral infection, but it is apparent from the infection of T cells in culture that HIV is capable of killing host cells. The finding of syncytia as a frequent accompaniment of cytotoxicity has stimulated the hypothesis that the machinery that mediates fusion—the viral *env* protein, SU (gp120), and the host receptor, CD4—are the essential factors for cytotoxicity (216, 329). However, fusion is not an inevitable feature of cytotoxicity (334), so multiple or different mechanisms may be at work; in the animal host, these could certainly include the immune response to infection. A few other retroviruses also display cytopathic effects in culture. Cytotoxicity by certain strains of ALV correlates with the use of subgroup B receptors (77), and killing by reticuloendotheliosis viruses has been associated with the presence of unintegrated viral DNA (383).

Regardless of pathogenic mechanism, certain features of the retrovirus replication pathway must affect pathogenic consequences. The tissue tropism of the virus (determined largely by interactions between SU and host receptors) and the tissue-specific efficiency of virus growth (determined in part by interactions between viral transcriptional enhancers and cellular transcription factors) will strongly influence the organ or cells at risk from infection. Proviral integration is probably responsible for virus persistence in most cases (visna virus, since its competence to integrate is still under debate, may be an exception [139]); but the level of viral gene expression can be modulated within a single cell by viral or cellular factors (e.g., the *art* and *tat* genes of HIV, or steroid-receptor complex in MMTV-infected cells) or by the site of integration. In this way, even cytopathic retroviruses such as HIV might remain in a latent state for prolonged periods.

VII. SUMMARY AND PROSPECTS

The proviruses of retroviruses are among the most sophisticated of transposable elements. Encoding the ingredients for infectious, extracellular particles, they have the capacity to spread from cell to cell and from organism to organism. This feature underwrites, at least in part, the ubiquity of retroviruses and their role as agents of disease, characteristics that have brought them to scientific and now to public attention. As recent work reveals, retroviruses also exhibit regulatory phenomena that operate at transcriptional, posttranscriptional, and translational levels, sometimes exploiting and sometimes circumventing host mechanisms for control. Like other retrotransposons, retroviruses use a remarkably precise set of reactions to synthesize DNA from an RNA template and to integrate it into host chromosomes, but retroviruses have the added challenge of gaining entrance to cells, a feat achieved by specific interactions with host cell surface proteins.

In this review, we have emphasized those aspects of retroviruses that account for their complex strategy for replication: the organization of viral genomes, the catalytic functions of virus-coded enzymes, and the mechanisms of gene expression and virion assembly. Despite the wealth of information available on these topics, it is apparent that many crucial aspects of the retroviral life cycle are still poorly understood. Only a single host receptor has been characterized, and little is known about the processes of virus entry and uncoating that trigger reverse transcription. It has been appreciated only recently that the early events in the virus life cycle, DNA synthesis and integration, occur in a nucleoprotein complex, and the composition and organization of that complex are not understood. Most of the major intermediates in reverse transcription have been identified, but the structure and catalytic properties of the enzyme have yet to be studied in a rigorous fashion. Though the products of the integration reaction have been well characterized and the reaction has been performed in a cell-free system, the enzymatic steps that join viral DNA to host DNA remain uncertain. Tantalizing regulatory phenomena affecting retroviruses are now known—cell-dependent transcription, differential efficiency of splicing, RNA sorting, nonsense codon suppression, ribosomal frameshifting, and proteolytic processing—but the viral and cellular signals and factors that govern them are only vaguely comprehended. Moreover, only a crude picture of virus assembly can be drawn, despite the potential for understanding protein-protein and RNA-protein interactions through this process.

The incentives for coming to grips with these problems are now greater than ever because a retrovirus-induced disease, acquired immunodeficiency syndrome, is now a major factor in all our lives. Thus, the study of retroviruses offers a dual prospect, insight into a problem of fundamental interest in biology—retrotransposition—and hope that new knowledge about virus replication and pathogenesis will prompt solutions to one of medicine's most urgent problems.

LITERATURE CITED

1. **Akroyd, J., V. J. Fincham, A. R. Green, P. Levantis, S. Searle, and J. A. Wyke.** 1987. Transcription of Rous sarcoma proviruses in rat cells is determined by chromosomal position effects that fluctuate and can operate over long distances. *Oncogene* **1**:347–355.
2. **Aldovini, A., A. De Rossi, M. B. Feinberg, F. Wong-Staal, and G. Franchini.** 1986. Molecular analysis of a deletion mutant provirus of type I human T-cell lymphotropic virus: evidence for a doubly spliced x-*lor* mRNA. *Proc. Natl. Acad. Sci. USA* **83**:38–43.
3. **Alexander, F., J. Leis, D. A. Soltis, R. M. Crowl, W. Danho, M. S. Poonian, Y.-C. E. Pan, and A. M. Skalka.** 1987. Proteolytic processing of avian sarcoma and leukosis viruses *pol-endo* recombinant proteins reveals another *pol* gene domain. *J. Virol.* **61**:534–542.
4. **Arkhipova, I. R., A. M. Mazo, V. A. Cherkasova, T. V. Gorelova, N. G. Schuppe, and Y. V. Illyin.** 1986. The steps of reverse transcription of *Drosophila* mobile dispersed genetic elements and U3-R-U5 structure of their LTRs. *Cell* **44**: 555–563.
5. **Arrigo, S., M. Yun, and K. Beemon.** 1987. *cis*-acting regulatory elements within *gag* genes of avian retroviruses. *Mol. Cell. Biol.* 7:388–397.
6. **Astrin, S. M., H. L. Robinson, L. B. Crittenden, E. G. Buss, J. Wyban, and W. H. Hayward.** 1980. Ten genetic loci in the chicken that contain structural genes for endogenous avian leukosis viruses. *Cold Spring Harbor Symp. Quant. Biol.* **44**:1105–1109
7. **Avery, R. J., J. D. Norton, J. S. Jones, D. C. Burke, and A. G. Morris.** 1980. Interferon inhibits transformation by murine sarcoma viruses before integration of provirus. *Nature* (London) **288**:93–95.
8. **Barklis, E., R. C. Mulligan, and R. Jaenisch.** 1986. Chromosomal position or virus mutation permits retrovirus expression in embryonal carcinoma cells. *Cell* **47**:391–399.
9. **Bautch, V. L.** 1986. Genetic background affects integration frequency of ecotropic proviral sequences into the mouse germ line. *J. Virol.* **60**:693–701.
10. **Bender, M. A., T. D. Palmer, R. E. Gelinas, and A. D. Miller.** 1987. Evidence that the packaging signal of Moloney murine leukemia virus extends into the *gag* region. *J. Virol.* **61**: 1639–1646.
11. **Bender, W., Y.-H. Chien, S. Chattopadhyay, P. K. Vogt, M. B. Gardner, and N. Davidson.** 1978. High-molecular weight RNAs of AKR, NZB, and wild mouse viruses and avian reticuloendotheliosis virus all have similar dimer structures. *J. Virol.* **25**:888–896.
12. **Benz, E. W., and D. Dina.** 1979. Moloney murine sarcoma virions synthesize full-genome-length double-stranded DNA in vitro. *Proc. Natl. Acad. Sci. USA* **76**:3294–3298.
13. **Berger, E. A., T. R. Fuerst, and B. Moss.** 1988. A soluble recombinant polypeptide comprising the amino-terminal half of the extracellular region of the CD4 molecule contains an active binding site for human immunodeficiency virus. *Proc. Natl. Acad. Sci. USA* **85**:2357–2361.
14. **Bernstein, L. B., S. M. Mount, and A. M. Weiner.** 1983. Pseudogenes for human small nuclear RNA U3 appear to arise by integration of self-primed reverse transcripts of the RNA into new chromosomal sites. *Cell* **32**:461–472.
15. **Bingham, P. M., R. Levis, and G. M. Rubin.** 1981. Cloning of DNA sequences from the *white* locus of D. melanogaster by a novel and general method. *Cell* **25**:693–704.
16. **Birnstiel, M. A.** 1984. Transcription termination and 3′ processing: the end is in site! *Cell* **41**:349–359.
17. **Bishop, J. M.** 1983. Cellular oncogenes and retroviruses. *Annu. Rev. Biochem.* **52**:301–354.
18. **Bishop, J. M., and H. E. Varmus.** 1982. Functions and origins of retroviral transforming genes, p. 999–1108. *In* R. Weiss, N. Teich, H. Varmus, and J. Coffin (ed.), *RNA Tumor Viruses*, vol. 1. Cold Spring Harbor Laboratory, Cold Spring Harbor, N.Y.
19. **Bishop, J. M., and H. E. Varmus.** 1985. Functions and origins of retroviral transforming genes, p. 246–356. *In* R. Weiss, N. Teich, H. Varmus, and J. Coffin (ed.), *RNA Tumor Viruses*, vol. 2. Cold Spring Harbor Laboratory, Cold Spring Harbor, N.Y.
20. **Boone, L. R., F. E. Myer, D. M. Yang, C.-Y. Ou, C. K. Koh, L. E. Roberson, R. W. Tennant, and W. K. Yang.** 1983. Reversal of *Fv-1* host range by in vitro restriction endonuclease fragment exchange between molecular clones of N-tropic and B-tropic murine leukemia virus genomes. *J. Virol.* **48**:110–119.
21. **Boone, L. R., and A. M. Skalka.** 1980. Two species of full-length cDNA are synthesized in high yield by mellitin-treated avian retrovirus particles. *Proc. Natl. Acad. Sci. USA* **77**:847–851.
22. **Botchan, M., W. C. Topp, and J. Sambrook.** 1976. The arrangement of simian virus 40 sequences in the DNA of transformed cells. *Cell* **9**:269–287.
23. **Bova, C. A., J. P. Manfredi, and R. Swanstrom.** 1986. *env* genes of avian retroviruses: nucleotide sequence and molecular recombinants define host range determinants. *Virology* **152**:343–354.
24. **Bova, C. A., J. C. Olsen, and R. Swanstrom.** 1988. The avian retrovirus *env* gene family: molecular analysis of host range and antigenic variants. *J. Virol.* **62**:75–83.
25. **Brady, J., K. T. Jeang, J. Duvall, and G. Khoury.** 1987. Identification of p40x-responsive regulatory sequences

within the human T-cell leukemia virus type I long terminal repeat. *J. Virol.* **61**:2175–2181.

26. **Brierley, I., M. E. G. Boursnell, M. M. Binns, B. Bilimoria, V. C. Blok, T. D. K. Brown, and S. C. Inglis.** 1987. An efficient ribosomal frameshifting signal in the polymerase-encoding region of the coronavirus IBV. *EMBO J.* **6**: 3779–3787.
27. **Broome, S., and W. Gilbert.** 1985. Rous sarcoma virus encodes a transcriptional activator. *Cell* **40**:537–546.
28. **Brown, P. O., B. Bowerman, H. E. Varmus, and J. M. Bishop.** 1987. Correct integration of retroviral DNA in vitro. *Cell* **49**:347–357.
29. **Buscher, M., W. Reiser, H. Will, and H. Schaller.** 1985. Transcripts and the putative RNA pregenome of duck hepatitis B virus: implications for reverse transcriptase. *Cell* **40**:717–724.
30. **Callahan, R., W. Drohan, D. Gallahan, L. D.'Hoostelaere, and M. Potter.** 1982. Novel class of mouse mammary tumor virus-related DNA sequences found in all species of *Mus*, including mice lacking the virus proviral gene. *Proc. Natl. Acad. Sci. USA* **79**:4113–4117.
31. **Cappello, J., K. Handelsman, and H. F. Lodish.** 1985. Sequence of dictyostelium DIRS-1: an apparent retrotransposon with inverted terminal repeats and an internal circle junction sequence. *Cell* **43**:105–115.
32. **Celander, D., and W. A. Haseltine.** 1984. Tissue-specific transcription preference as a determinant of cell tropism and leukaemogenic potential of murine retroviruses. *Nature* (London) **312**:159–163.
33. **Chatis, P. A., C. A. Holland, J. W. Hartley, W. P. Rowe, and N. Hopkins.** 1983. Role for the 3′ end of the genome in determining disease specificity of Friend and Moloney murine leukemia viruses. *Proc. Natl. Acad. Sci. USA* **80**: 4408–4411.
34. **Chen, I. S. Y., D. J. Slamon, J. D. Rosenblatt, N. P. Shah, S. G. Quan, and W. Wachsman.** 1985. The X gene is essential for HTLV replication. *Science* **299**:54–58.
35. **Chen, I. S. Y., W. Wachsman, J. D. Rosenblatt, and A. J. Cann.** 1986. The role of the χ gene in HTLV-associated malignancy. *Cancer Surv.* **5**:329–342.
36. **Cheung, A. M. K., M. D. Hoggan, W. W. Hauswirth, and K. I. Berns.** 1980. Integration of the adeno-associated virus genome into cellular DNA in latently infected Detroit 6 cells. *J. Virol.* **33**:739–748.
37. **Chinsky, J., and R. Soeiro.** 1982. Studies with aphidicolin on the *Fv-1* host restriction of Friend murine leukemia virus. *J. Virol.* **43**:182–190.
38. **Chinsky, J., R. Soeiro, and J. Kopchick.** 1984. *Fv-1* host cell restriction of Friend leukemia virus: microinjection of unintegrated viral DNA. *J. Virol.* **50**:271–274.
39. **Clare, J., and P. Farabaugh.** 1985. Nucleotide sequence of a yeast Ty element: evidence for an unusual mechanism of gene expression. *Proc. Natl. Acad. Sci. USA* **82**:2829–2833.
40. **Cloyd, M. W., and S. K. Chattopadhyay.** 1986. A new class of retrovirus present in many murine leukemia systems. *Virology* **151**:31–40.
41. **Cobrinik, D., R. Katz, R. Terry, A. M. Skalka, and J. Leis.** 1987. Avian sarcoma and leukosis virus *pol*-endonuclease recognition of the tandem long terminal repeat junction: minimum site required for cleavage is also required for viral growth. *J. Virol.* **61**:1999–2008.
42. **Coffin, J. M.** 1979. Structure, replication, and recombination of retrovirus genomes: some unifying hypotheses. *J. Gen. Virol.* **42**:1–26.
43. **Coffin, J. M.** 1982. Structure of the retroviral genome, p. 261–368. *In* R. Weiss, N. Teich, H. Varmus, and J. Coffin (ed.), *RNA Tumor Viruses*, vol. 1. Cold Spring Harbor Laboratory, Cold Spring Harbor, N.Y.
44. **Coffin, J. M.** 1982. Endogenous viruses, p. 1109–1203. *In* R. Weiss, N. Teich, H. Varmus, and J. Coffin (ed.), *RNA Tumor Viruses*, vol. 1. Cold Spring Harbor Laboratory, Cold Spring Harbor, N.Y.
45. **Coffin, J. M.** 1985. Genome structure, p. 17–73. *In* R. Weiss, N. Teich, H. Varmus, and J. Coffin (ed.), *RNA Tumor Viruses*, vol. 2. Cold Spring Harbor Laboratory, Cold Spring Harbor, N.Y.
46. **Cohen, J. C., and H. E. Varmus.** 1979. Endogenous mammary tumor virus DNA varies among wild mice and segregates during inbreeding. *Nature* (London) **278**:418–423.
47. **Colicelli, J., and S. P. Goff.** 1985. Isolation of a recombinant murine leukemia virus utilizing a new primer tRNA. *J. Virol.* **57**:37–45.
48. **Colicelli, J., and S. P. Goff.** 1985. Mutants and pseudo revertants of Moloney murine leukemia virus with alterations at the integration site. *Cell* **42**:573–580.
49. **Colicelli, J., and S. P. Goff.** 1986. Structure of a cloned circular retroviral DNA containing a tRNA sequence between the terminal repeats. *J. Virol.* **57**:674–677.
50. **Colicelli, J., and S. P. Goff.** 1987. Identification of endogenous retroviral sequences as potential donors for recombinational repair of mutant retroviruses: positions of crossover points. *Virology* **160**:518–523.
51. **Colicelli, J., and S. P. Goff.** 1988. Sequence and spacing requirements of a retrovirus integration site. *J. Mol. Biol.* **199**:47–59.
52. **Colicelli, J., and S. P. Goff.** 1988. Isolation of an integrated provirus of Moloney murine leukemia virus with long terminal repeats in inverted orientation: integration utilizing two U3 sequences. *J. Virol.* **62**:633–636.
53. **Copeland, N. G., K. W. Hutchison, and N. A. Jenkins.** 1983. Excision of the DBA ecotropic provirus in dilute coat-color revertants of mice occurs by homologous recombination involving the viral LTRs. *Cell* **33**:379–387.
54. **Copeland, N. G., N. A. Jenkins, and G. M. Cooper.** 1981. Integration of Rous sarcoma virus DNA during transfection. *Cell* **23**:51–60.
55. **Craigie, R., and K. Mizuuchi.** 1987. Transposition of the Mu DNA: joining of Mu to target DNA can be uncoupled from cleavage at the ends of Mu. *Cell* **51**:493–501.
56. **Craigie, R., D. J. Arndt-Jovin, and K. Mizuuchi.** 1985. A defined system for the DNA strand-transfer reaction at the initiation of bacteriophage Mu transposition: protein and DNA substrate requirements. *Proc. Natl. Acad. Sci. USA* **82**:7570–7574.
57. **Craigie, R., and K. Mizuuchi.** 1986. Role of DNA topology in Mu transposition: mechanism of sensing the relative orientation of two DNA segments. *Cell* **45**:793–800.
58. **Craigie, R., M. Mizuuchi, and K. Mizuuchi.** 1984. Site-specific recognition of the bacteriophage Mu ends by the Mu A protein. *Cell* **39**:387–394.
59. **Craik, C. S., C. Largman, T. Fletcher, S. Roczniak, P. J. Barr, R. Fletterick, and W. J. Rutter.** 1985. Redesigning trypsin: alteration of substrate specificity. *Science* **228**:291–297.
60. **Crawford, S., and S. P. Goff.** 1984. Mutations in *gag* proteins P12 and P15 of Moloney murine leukemia virus block early stages of infection. *J. Virol.* **49**:909–917.
61. **Crawford, S., and S. P. Goff.** 1985. A deletion mutation in the 5′ part of the *pol* gene of Moloney murine leukemia virus blocks proteolytic processing of the *gag* and *pol* polyproteins. *J. Virol.* **53**:899–907.

62. **Cross, S. L., M. B. Feinberg, J. B. Wolf, N. J. Holbrook, F. Wong-Staal, and W. J. Leonard.** 1987. Regulation of the human interleukin-2 receptor α chain promoter: activation of a nonfunctional promoter by the transactivator gene of HTLV-1. *Cell* **49:**47–56.

63. **Cullen, B. R.** *Trans*-activation of human immunodeficiency virus occurs via a bimodal mechanism. *Cell* **46:**973–982.

64. **Cullen, B. R., P. T. Lomedico, and G. Ju.** 1984. Transcriptional interference in avian retroviruses: implications for the promoter insertion model of leukemogenesis. *Nature* (London) **307:**241–244.

65. **DeLarco, J., and G. J. Todaro.** 1976. Membrane receptors for murine leukemia viruses: characterization using the purified viral envelope glycoprotein gp71. *Cell* **82:**365–371.

66. **Dalgleish, A. G., P. C. L. Beverly, P. R. Clapham, D. H. Crawford, M. F. Greaves, and R. A. Weiss.** 1984. The CDA (T4) antigen is an essential component of the receptor for the AIDS retrovirus. *Nature* (London) **312:**763–767.

67. **DesGroseillers, L., and P. Jolicoeur.** 1983. Physical mapping of the *Fv-1* tropism host range determinant of BALB/c murine leukemia viruses. *J. Virol.* **48:**685–696.

68. **DesGroseillers, L., and P. Jolicoeur.** 1984. The tandem direct repeats within the long terminal repeat of murine leukemia viruses are the primary determinant of their leukemogenic potential. *J. Virol.* **52:**945–952.

69. **DesGroseillers, L., E. Rassart, and P. Jolicoeur.** 1983. Thymotropism of murine leukemia virus is conferred by its long terminal repeat. *Proc. Natl. Acad. Sci. USA* **80:**4203–4207.

70. **Dhar, R., W. L. McClements, L. W. Enquist, and G. F. Vande Woude.** 1980. Nucleotide sequences of integrated Moloney sarcoma provirus long terminal repeats and their host and viral junctions. *Proc. Natl. Acad. Sci. USA* **77:** 3937–3941.

71. **Dickson, C., R. Eisenman, and H. Fan.** 1985. Protein biosynthesis and assembly, p. 135–145. *In* R. Weiss, N. Teich, H. Varmus, and J. Coffin (ed.), *RNA Tumor Viruses.* Cold Spring Harbor Laboratory, Cold Spring Harbor, N.Y.

72. **Dickson, C., R. Eisenman, H. Fan, E. Hunter, and N. Teich.** 1982. Protein biosynthesis and assembly, p. 513–648. *In* R. Weiss, N. Teich, H. Varmus, and J. Coffin (ed.), *RNA Tumor Viruses.* Cold Spring Harbor Laboratory, Cold Spring Harbor, N.Y.

73. **Dickson, C., and G. Peters.** 1981. Protein-coding potential of mouse mammary tumor virus genome RNA as examined by in vitro translation. *J. Virol.* **37:**36–47.

74. **Di Nocera, P. P., and G. Casari.** 1987. Related polypeptides are encoded by *Drosophila* F elements, I factors, and mammalian L1 sequences. *Proc. Natl. Acad. Sci. USA* **84:** 5843–5847.

75. **Donehower, L. A., A. L. Huang, and G. L. Hager.** 1981. Regulatory and coding potential of the mouse mammary tumor virus long terminal redundancy. *J. Virol.* **37:**226–238.

76. **Donehower, L. A., and H. E. Varmus.** 1984. A mutant murine leukemia virus with a single missense codon in *pol* is defective in a function affecting integration. *Proc. Natl. Acad. Sci. USA* **81:**6461–6465.

77. **Dorner, A. J., and J. M. Coffin.** 1986. Determinants for receptor interaction and cell killing on the avian retrovirus glycoprotein gp85. *Cell* **45:**365–374.

78. **Dorner, A. J., J. P. Stoye, and J. M. Coffin.** 1985. Molecular basis of host range variation in avian retroviruses. *J. Virol.* **53:**32–39.

79. **Dougherty, J. P., and H. M. Temin.** 1986. High mutation rate of a spleen necrosis virus-based retrovirus vector. *Mol. Cell. Biol.* **6:**4387–4395.

80. **Dougherty, J. P., and H. M. Temin.** 1987. A promoterless retroviral vector indicates that there are sequences in U3 required for 3′ RNA processing. *Proc. Natl. Acad. Sci. USA* **84:**1197–1201.

81. **Duyk, G., J. Leis, M. Longiaru, and A. M. Skalka.** 1983. Selective cleavage on the avian retroviral long terminal repeat sequence by the endonuclease associated with the form of avian reverse transcriptase. *Proc. Natl. Acad. Sci. USA* **80:**6745–6749.

82. **Duyk, G., M. Longiaru, D. Cobrinik, R. Kowal, P. deHaseth, A. M. Skalka, and J. Leis.** 1985. Circles with two tandem long terminal repeats are specifically cleaved by *pol* gene-associated endonuclease from avian sarcoma and leukosis viruses: nucleotide sequences required for site-specific cleavage. *J. Virol.* **56:**589–599.

83. **Dynan, W. S., and R. Tjian.** 1985. Control of eukaryotic messenger RNA synthesis by sequence-specific DNA-binding proteins. *Nature* (London) **316:**774–778.

84. **Edwards, S. A., and H. Fan.** 1979. *gag*-related polyproteins of Moloney murine leukemia virus: evidence for independent synthesis of glycosylated and unglycosylated forms. *J. Virol.* **30:**551–563.

84a. **Eichinger, D. J., and J. D. Boeke.** 1988. The DNA intermediate in yeast Ty1 element transposition copurifies with virus-like particles: cell-free Ty1 transposition. *Cell* **54:** 955–966.

85. **Ellis, R. W., D. DeFeo, T. Y. Shih, M. A. Gonda, H. A. Young, N. Tsuchida, D. R. Lowy, and E. M. Scolnick.** 1981. The p21 *src* genes of Harvey and Kirsten sarcoma viruses originate from divergent members of a family of normal vertebrate genes. *Nature* (London) **292:**506–511.

86. **Embretson, J. E., and H. M. Temin.** 1987. Lack of competition results in efficient packaging of heterologous murine retroviral RNAs and reticuloendotheliosis virus encapsidation-minus RNAs by the reticuloendotheliosis virus helper cell line. *J. Virol.* **61:**2675–2683.

87. **Embretson, J. E., and H. M. Temin.** 1987. Transcription from a spleen necrosis virus 5′ long terminal repeat is suppressed in mouse cells. *J. Virol.* **61:**3454–3462.

88. **Enders, G. H., D. Ganem, and H. Varmus.** 1985. Mapping the major transcripts of ground squirrel hepatitis virus: the presumptive template for reverse transcriptase is terminally redundant. *Cell* **42:**297–308.

89. **Fan, H., H. Chute, E. Chao, and M. Feuerman.** 1983. Construction and characterization of Moloney murine leukemia virus mutants unable to synthesize glycosylated *gag* protein. *Proc. Natl. Acad. Sci. USA* **80:**5965–5969.

90. **Farmerie, W. G., D. D. Loeb, N. C. Casavant, C. A. Hutchison III, M. H. Edgell, and R. Swanstrom.** 1987. Expression and processing of the AIDS virus reverse transcriptase in *Escherichia coli. Science* **236:**305–308.

91. **Fawcett, D. H., C. K. Lister, E. Kellett, and D. J. Finnegan.** 1986. Transposable elements controlling I-R hybrid dysgenesis in D. melanogaster are similar to mammalian LINEs. *Cell* **47:**1007–1015.

92. **Feinberg, M. B., R. F. Jarrett, A. Aldovini, R. C. Gallo, and F. Wong-Staal.** 1986. HTLV-III expression and production involve complex regulation at the levels of splicing and translation of viral RNA. *Cell* **46:**807–817.

93. **Feinstein, S. C., S. R. Ross, and K. R. Yamamoto.** 1982. Chromosomal position effects determine transcriptional potential of integrated mammary tumor virus DNA. *J. Mol. Biol.* **156:**549–566.

94. **Ficht, T. A., L.-J. Chang, and C. M. Stoltzfus.** 1984. Avian sarcoma virus *gag* and *env* gene structural protein precursors

contain common amino-terminal sequence. *Proc. Natl. Acad. Sci. USA* **81**:362–366.

95. **Fischl, M. A., D. D. Richman, M. H. Grieco, M. S. Gottlieb, P. A. Volberding, O. L. Laskin, J. M. Leedom, J. E. Groopman, D. Mildvan, R. T. Schooley, G. G. Jackson, D. T. Durack, D. King, and the AZT Collaborative Working Group.** 1987. The efficacy of azidothymidine (AZT) in the treatment of patients with AIDS and AIDS-related complex. *N. Engl. J. Med.* **317**:185–198.
96. **Fisher, A. G., B. Ensoli, L. Ivanoff, M. Chamberlain, S. Peteway, L. Ratner, R. C. Gallo, and F. Wong-Staal.** 1987. The *sor* gene of HIV-1 is required for efficient virus transmission in vitro. *Science* **238**:888–892.
97. **Fisher, A. G., M. B. Feinberg, S. F. Josephs, M. E. Harper, L. M. Marselle, G. Reyes, M. A. Gonda, A. Aldovini, C. Debouk, R. C. Gallo, and F. Wong-Staal.** 1986. The *trans*-activator gene of HTLV-III is essential for virus replication. *Nature* (London) **320**:367–370.
98. **Flavell, A. J., and D. Ish-Horowicz.** 1983. The origin of extrachromosomal circular *copia* elements. *Cell* **34**: 415–419.
99. **Frankel, A. D., D. S. Bredt, and C. O. Pabo.** 1988. Tat protein from human immunodeficiency virus forms a metal-linked dimer. *Science* **240**:70–73.
100. **Frankel, W., T. A. Potter, N. Rosenberg, J. Lenz, and T. V. Rajan.** 1985. Retroviral insertional mutagenesis of a target allele in a heterozygous murine cell line. *Proc. Natl. Acad. Sci. USA* **82**:6600–6604.
101. **Franz, T., F. Holberg, B. Seliger, C. Stocking, and W. Ostertag.** 1986. Retroviral mutants efficiently expressed in embryonal carcinoma cells. *Proc. Natl. Acad. Sci. USA* **83**:3292–3296.
102. **Fritsch, E., and H. M. Temin.** 1977. Inhibition of viral DNA synthesis in stationary chicken embryo fibroblasts infected with avian retroviruses. *J. Virol.* **24**:461–469.
103. **Fu, X., N. Phillips, J. Jentoft, P. T. Tuazon, J. A. Traugh, and J. Leis.** 1985. Site-specific phosphorylation of avian retrovirus nucleocapsid protein pp12 regulates binding to viral RNA: evidence for different protein conformations. *J. Biol. Chem.* **260**:9941–9947.
104. **Fujisawa, J., M. Seiki, M. Sato, and M. Yoshida.** 1986. A transcriptional enhancer sequence of HTLV-I is responsible for *trans*-activation mediated by $p40^x$ of HTLV-1. *EMBO J.* **5**:713–718.

104a. **Fujiwara, T., and K. Mizuuchi.** 1988. Retroviral DNA integration: structure of an integration intermediate. *Cell* **55**:497–504.

105. **Fung, Y.-K. T., A. M. Fadley, L. B. Crittenden, and H.-J. Kung.** 1981. On the mechanism of retrovirus-induced avian lymphoid leukosis: deletion and integration of the proviruses. *Proc. Natl. Acad. Sci. USA* **78**:3418–3422.
106. **Furuichi, Y., A. J. Shatkin, E. Stavnezer, and J. M. Bishop.** 1975. Blocked, methylated 5′-terminal sequence in avian sarcoma virus RNA. *Nature* (London) **257**:618–620.

106a. **Gallo, R., F. Wong-Staal, L. Montagnier, W. Haseltine, and M. Yoshida.** 1988. HIV/HTLV gene nomenclature. *Nature* (London) **333**:504.

107. **Ganem, D., and H. E. Varmus.** 1986. The molecular biology of the hepatitis B viruses. *Annu. Rev. Biochem.* **56**:651–693.
108. **Gelinas, C., and H. M. Temin.** 1986. Nondefective spleen necrosis virus-derived vectors define the upper size limit for packaging reticuloendotheliosis viruses. *Proc. Natl. Acad. Sci. USA* **83**:9211–9215.
109. **Gerard, G. F., and D. P. Grandgenett.** 1980. Retrovirus reverse transcriptase, p. 345–394. *In* J. Stephenson (ed.), *Molecular Biology of RNA Tumor Viruses*. Academic Press, Inc., New York.
110. **Gerlich, W. H., and W. S. Robinson.** 1980. Hepatitis B virus contains protein attached to the 5′ terminus of its complete DNA strand. *Cell* **21**:801–810.
111. **Gilboa, E., S. Goff, A. Shields, F. Yoshimura, S. Mitra, and D. Baltimore.** 1979. In vitro synthesis of a 9-kbp terminally redundant DNA carrying the infectivity of Moloney murine leukemia virus. *Cell* **16**:863–874.
112. **Gilboa, E., S. W. Mitra, S. Goff, and D. Baltimore.** 1979. A detailed model of reverse transcription and tests of crucial aspects. *Cell* **18**:93–100.
113. **Gillespie, D. A. F., K. A. Hart, and J. A. Wyke.** 1985. Rearrangements of viral and cellular DNA are often associated with expression of Rous sarcoma virus in rat cells. *Cell* **41**:279–287.
114. **Gisselbrecht, S., S. Fichelson, B. Sola, D. Bordereaux, A. Hampe, C. Andre, F. Galibert, and P. Tambourin.** 1987. Frequent c-*fms* activation by proviral insertion in mouse myeloblastic leukaemias. *Nature* (London) **329**:259–261.
115. **Gloger, I., and A. Panet.** 1986. Glutamine starvation of murine leukaemia virus-infected cells inhibits the read-through of the *gag-pol* genes and proteolytic processing of the *gag* polyprotein. *J. Gen. Virol.* **67**:2207–2214.
116. **Goff, S. P.** 1984. The genetics of murine leukemia viruses. *Curr. Top. Microbiol. Immunol.* **112**:45–69.
117. **Gojobori, T., and S. Yokoyama.** 1985. Rates of evolution of the retroviral oncogene of Moloney murine sarcoma virus and of its cellular homologues. *Proc. Natl. Acad. Sci. USA* **82**:4198–4201.
118. **Goldfarb, M. P., and R. A. Weinberg.** 1981. Generation of novel, biologically active Harvey sarcoma virus via apparent illegitimate recombination. *J. Virol.* **38**:136–150.
119. **Golomb, M., and D. P. Grandgenett.** 1979. Endonuclease activity of purified RNA directed DNA polymerase from AMV. *J. Biol. Chem.* **254**:1606–1613.
120. **Golomb, M., D. P. Grandgenett, and W. Mason.** 1981. Virus-coded DNA endonuclease from avian retrovirus. *J. Virol.* **38**:548–555.
121. **Gonda, M. A., M. J. Braun, S. G. Carter, T. A. Kost, J. W. Bess, L. O. Arthur, and M. J. Van Der Maaten.** 1987. Characterization and molecular cloning of a bovine lentivirus related to human immunodeficiency virus. *Nature* (London) **330**:388–391.
122. **Goodenow, M. M., and W. S. Hayward.** 1987. 5′ Long terminal repeats of *myc*-associated proviruses appear structurally intact but are functionally impaired in tumors induced by avian leukosis viruses. *J. Virol.* **61**:2489–2498.
123. **Goodwin, R. G., F. M. Rottman, T. Callaghan, H.-J. Kung, P. A. Maroney, and T. W. Nilsen.** 1986. c-*erbB* activation in avian leukosis virus-induced erythroblastosis: multiple epidermal growth factor receptor mRNAs are generated by alternative RNA processing. *Mol. Cell. Biol.* **6**:3128–3133.
124. **Gopinathan, K. P., L. A. Weymouth, T. A. Kunkel, and L. A. Loeb.** 1979. Mutagenesis in vitro by DNA polymerase from an RNA tumor virus. *Nature* (London) **278**:857–859.
125. **Goubin, G., and M. Hill.** 1979. Monomer and multimer covalently closed circular forms of Rous sarcoma virus DNA. *J. Virol.* **29**:799–804.
126. **Grandgenett, D. P., M. Golomb, and A. C. Vora.** 1980. Activation of an Mg^{2+}-dependent DNA endonuclease avian myeloblastosis virus $\alpha\beta$ DNA polymerase by in vitro proteolytic cleavage. *J. Virol.* **33**:264–271.
127. **Grandgenett, D. P., and A. C. Vora.** 1985. Site-specific nicking at the avian retrovirus LTR circle junction by the

viral pp32 DNA endonuclease. *Nucleic Acids Res.* **13**: 6205–6221.

128. **Grandgenett, D. P., A. C. Vora, and R. D. Schiff.** 1978. A 32,000-dalton nucleic acid-binding protein from avian retrovirus cores possesses DNA endonuclease activity. *Virology* **89**:119–132.
129. **Grandgenett, D. P., A. C. Vora, R. Swanstrom, and J. C. Olsen.** 1986. Nuclease mechanism of the avian retrovirus pp32 endonuclease. *J. Virol.* **58**:970–974.
130. **Green, A. R., S. Searle, D. A. F. Gillespie, M. Bissell, and J. A. Wyke.** 1986. Expression of integrated Rous sarcoma viruses: DNA rearrangements 5′ to the provirus are common in transformed rat cells but not seen in infected but untransformed cells. *EMBO J.* **5**:707–712.
131. **Gruss, P., and G. Khoury.** 1983. Enhancer elements. *Cell* **33**:313–314.
132. **Guy, M., M. P. Kieny, Y. Riviere, C. LePeuch, K. Dott, M. Girard, L. Montagnier, and J.-P. Lecocq.** 1987. HIV F/3′ *orf* encodes a phosphorylated GTP-binding protein resembling an oncogene product. *Nature* (London) **330**:266–269.
133. **Guyader, M., M. Emerman, P. Sonigo, F. Clavel, L. Montagnier, and M. Alizon.** 1987. Genome organization and transactivation of the human immunodeficiency virus type 2. *Nature* (London) **326**:662–669.
134. **Hackett, P. B., R. B. Petersen, C. H. Hensel, F. Albericio, S. I. Gunderson, A. C. Palmenberg, and G. Barany.** 1986. Synthesis in vitro of a seven amino acid peptide encoded in the leader RNA of Rous sarcoma virus. *J. Mol. Biol.* **190**:45–58.
135. **Hackett, P. B., R. Swanstrom, H. E. Varmus, and J. M. Bishop.** 1982. The leader sequence of the subgenomic mRNA's of Rous sarcoma virus is approximately 390 nucleotides. *J. Virol.* **41**:527–534.
136. **Hagino-Yamagishi, K., L. A. Donehower, and H. E. Varmus.** 1987. Retroviral DNA integrated during infection by an integration-deficient mutant of murine leukemia virus is oligomeric. *J. Virol.* **61**:1964–1971.
137. **Handelin, B. L., and D. Kabat.** 1985. Cell surface receptors for murine leukemia viruses: two assays and their implications. *Virology* **140**:183–187.
138. **Hansen, J., T. Schulze, and K. Moelling.** 1987. RNase H activity associated with bacterially expressed reverse transcriptase of human T-cell lymphotropic virus III/lymphadenopathy associated virus. *J. Biol. Chem.* **262**:12393–12396.
139. **Harris, J. D., H. Blum, J. Scott, B. Traynor, P. Ventura, and A. Haase.** 1984. Slow virus visna: reproduction in vitro of virus from extrachromosomal DNA. *Proc. Natl. Acad. Sci. USA* **81**:7212–7215.
140. **Hartung, S., R. Jaenisch, and M. Breindl.** 1986. Retrovirus insertion inactivates mouse a1 (I) collagen gene by blocking initiation of transcription. *Nature* (London) **320**:365–366.
141. **Hauber, J., A. Perkins, E. P. Heimer, and B. R. Cullen.** 1987. Trans-activation of human immunodeficiency virus gene expression is mediated by nuclear events. *Proc. Natl. Acad. Sci. USA* **84**:6354–6369.
142. **Hawley, R. G., M. J. Shulman, and N. Hozumi.** 1984. Transposition of two different intracisternal A particle elements into an immunoglobulin kappa-chain gene. *Mol. Cell. Biol.* **4**:2565–2572.
143. **Hayward, W. S., B. G. Neel, and S. M. Astrin.** 1981. Activation of a cellular *onc* gene by promoter insertion in ALV-induced lymphoid leukosis. *Nature* (London) **290**: 475–480.
144. **Heidmann, T., O. Heidmann, and J.-F. Nicolas.** 1988. An indicator gene to demonstrate intracellular transposition of defective retroviruses. *Proc. Natl. Acad. Sci. USA* **85**: 2219–2223.
145. **Henderson, L. E., H. C. Krutzsch, and S. Oroszlan.** 1983. Myristyl amino-terminal acylation of murine retrovirus proteins—an unusual posttranslational protein modification. *Proc. Natl. Acad. Sci. USA* **80**:339–343.
146. **Henry, C., M. Coquillaud, S. Saule, D. Stehelin, and B. Debuire.** 1985. The four C-terminal amino acids of the v-*erbA* polypeptide are encoded by an intronic sequence of the v-*erbB* oncogene. *Virology* **140**:179–182.
147. **Herman, S. A., and J. M. Coffin.** 1986. Differential transcription from the long terminal repeats of integrated avian leukosis virus DNA. *J. Virol.* **60**:497–505.
148. **Herman, S. A., and J. M. Coffin.** 1987. Efficient packaging of readthrough RNA in ALV: implications for oncogene transduction. *Science* **236**:845–847.
149. **Hinrichs, S. H., M. Nerenberg, R. K. Reynolds, G. Khoury, and G. Jay.** 1987. A transgenic mouse model for human neurofibromatosis. *Science* **237**:1340–1347.
150. **Hippenmeyer, P. J., and D. P. Grandgenett.** 1984. Requirement of the avian retrovirus pp32 DNA binding protein domain for replication. *Virology* **137**:358–370.
151. **Hizi, A., L. E. Henderson, T. D. Copeland, R. C. Sowder, C. V. Hixson, and S. Oroszlan.** 1987. Characterization of mouse mammary tumor virus *gag-pro* gene products and the ribosomal frameshift site by protein sequencing. *Proc. Natl. Acad. Sci. USA* **84**:7041–7046.
152. **Holland, C. A., J. Wozney, P. A. Chatis, N. Hopkins, and J. W. Hartley.** 1985. Construction of recombinants between molecular clones of murine retrovirus MCF 247 and Akv: determinant of an in vitro host range property that maps in the long terminal repeat. *J. Virol.* **52**:152–157.
153. **Horowitz, M., S. Luria, G. Rechavi, and D. Givol.** 1984. Mechanism of activation of the mouse c-*mos* oncogene by the LTR of an intracisternal A-particle gene. *EMBO J.* **3**:2937–2943.
154. **Hsu, H. W., P. Schwartzberg, and S. P. Goff.** 1985. Point mutations in the P30 domain of the *gag* gene of Moloney murine leukemia virus. *Virology* **142**:211–214.
155. **Hsu, T. W., J. L. Sabran, E. G. Mark, R. V. Guntaka, and J. M. Taylor.** 1978. Analysis of unintegrated avian RNA tumor virus double-stranded DNA intermediates. *J. Virol.* **28**:810–818.
156. **Hsu, T. W., and J. M. Taylor.** 1982. Effect of aphidicolin on avian sarcoma virus replication. *J. Virol.* **44**:493–498.
157. **Hsu, T. W., J. M. Taylor, C. Aldrich, J. B. Townsend, G. Seal, and W. S. Mason.** 1981. Tandem duplication of the proviral DNA in avian sarcoma virus-transformed quail clone. *J. Virol.* **38**:219–223.
158. **Huang, C.-C., N. Hay, and J. M. Bishop.** 1986. The role of RNA molecules in transduction of the proto-oncogene c-*fps*. *Cell* **44**:935–940.
159. **Hughes, S. H., K. Mellstrom, E. Kosik, T. Tamanoi, and J. Brugge.** 1984. Mutation of a termination codon affects *src* initiation. *Mol. Cell. Biol.* **4**:1738–1746.
160. **Hughes, S. H., A. Mutschler, J. M. Bishop, and H. E. Varmus.** 1981. A Rous sarcoma virus provirus is flanked by short direct repeats of a cellular DNA sequence present in only one copy prior to integration. *Proc. Natl. Acad. Sci. USA* **78**:4299–4303.
161. **Hughes, S. H., F. Payvar, D. Spector, R. T. Schimke, H. L. Robinson, G. S. Payne, J. M. Bishop, and H. E. Varmus.** 1979. Heterogeneity of genetic loci in chickens: analysis of endogenous viral and nonviral genes by cleavage of DNA with restriction endonucleases. *Cell* **18**:347–359.

162. **Hughes, S. H., P. R. Shank, D. H. Spector, H. J. Kung, J. M. Bishop, H. E. Varmus, P. K. Vogt, and M. L. Breitman.** 1978. Proviruses of avian sarcoma virus are terminally redundant, co-extensive with unintegrated linear DNA and integrated at many sites. *Cell* **15**:1397–1410.

163. **Hull, R., and S. N. Covey.** 1983. Does cauliflower mosaic virus replicate by reverse transcription? *Trends Biol. Sci.* **8**:119–126.

164. **Humphries, E. H., C. Glover, and M. E. Reichmann.** 1981. Rous sarcoma virus infection of synchronized cells establishes provirus integration during S phase DNA synthesis prior to cell division. *Proc. Natl. Acad. Sci. USA* **78**: 2601–2605.

165. **Hunter, E.** 1978. The mechanism for genetic recombination in the avian retroviruses. *Curr. Top. Microbiol. Immunol.* **79**:295–309.

166. **Hwang, J. V., and E. Gilboa.** 1984. Expression of genes introduced into cells by retroviral infection is more efficient than that of genes introduced into cells by DNA transfection. *J. Virol.* **50**:417–424.

166a. **Ichikawa, H., K. Ikeda, W. L. Wishart, and E. Ohtsubo.** 1987. Specific binding of transposase to terminal inverted repeats of transposable element Tn*3*. *Proc. Natl. Acad. Sci. USA* **84**:8220–8224.

167. **Ikawa, S., K. Hagino-Yamagishi, S. Kawai, T. Yamamoto, and K. Toyoshima.** 1986. Activation of the cellular *src* gene by transducing retrovirus. *Mol. Cell. Biol.* **6**:2420–2428.

168. **Inoue, J., M. Yoshida, and M. Seiki.** 1987. Transcriptional ($p40^{x}$) and post-transcriptional ($p27^{x\text{-}III}$) regulators are required for the expression and replication of the leukemia virus type I genes. *Proc. Natl. Acad. Sci. USA* **84**: 3653–3657.

169. **Iten, A., and E. Keshet.** 1986. A novel retrovirus like family in mouse DNA. *J. Virol.* **59**:301–307.

169a. **Jacks, T., H. D. Madhani, F. R. Masiarz, and H. E. Varmus.** 1988. Signals for ribosomal frameshifting in the Rouse sarcoma virus *gag-pol* region. *Cell* **55**:447–458.

170. **Jacks, T., M. D. Power, F. R. Masiarz, P. A. Luciw, P. J. Barr, and H. E. Varmus.** 1988. Characterization of ribosomal frameshifting in HIV-1 *gag-pol* expression. *Nature* (London) **231**:280–283.

171. **Jacks, T., K. Townsley, H. E. Varmus, and J. Majors.** 1987. Two efficient ribosomal frameshifting events are required for synthesis of mouse mammary tumor virus *gag*-related polyproteins. *Proc. Natl. Acad. Sci. USA* **84**:4298–4302.

172. **Jacks, T., and H. E. Varmus.** 1985. Expression of Rous sarcoma virus *pol* gene by ribosomal frameshifting. *Science* **230**:1237–1242.

173. **Jaenisch, R.** 1976. Germ line integration and Mendelian transmission of the exogenous Moloney leukemia virus. *Proc. Natl. Acad. Sci. USA* **73**:1260–1264.

174. **Jaenisch, R., K. Harbers, A. Schnicke, J. Lohler, I. Chumakov, D. Jahner, D. Grotkopp, and E. Hoffmann.** 1983. Germ line integration of Moloney murine leukemia virus at the Mov13 locus leads to recessive lethal mutation and early embryonic death. *Cell* **32**:209–216.

175. **Jakobovits, A., D. H. Smith, E. B. Jakobovits, and D. J. Capon.** 1988. A discrete element 3′ of human immunodeficiency virus 1 (HIV-1) and HIV-2 mRNA initiation sites mediates transcriptional activation by an HIV *trans* activator. *Mol. Cell. Biol.* **8**:2555–2561.

176. **Jeang, K.-T., C.-Z. Giam, M. Nerenberg, and G. Khoury.** 1987. Abundant synthesis of functional human T-cell leukemia virus type I $p40^{x}$ protein in eucaryotic cells by using a baculovirus expression vector. *J. Virol.* **61**:708–713.

177. **Jenkins, N. A., and N. G. Copeland.** 1985. High frequency germ-line acquisition of ecotropic MuLV proviruses in SWR/J-RF-J hybrid mice. *Cell* **43**:811–819.

178. **Jenkins, N. A., N. G. Copeland, B. A. Taylor, and B. K. Lee.** 1981. Dilute (*d*) coat colour mutation of DBA/2J mice is associated with the site of integration of an ecotropic MuLV genome. *Nature* (London) **293**:370–374.

179. **Jenkins, N. A., N. G. Copeland, B. A. Taylor, and B. K. Lee.** 1982. Organization, distribution, and stability of endogenous ecotropic murine leukemia virus DNA sequences in chromosomes of *Mus musculus*. *J. Virol.* **43**:26–36.

180. **Johnson, M. S., M. A. McClure, D.-F. Feng, J. Gray, and R. F. Doolittle.** 1986. Computer analysis of retroviral *pol* genes: assignment of enzymatic functions to specific sequences and homologies with nonviral enzymes. *Proc. Natl. Acad. Sci. USA* **83**:7648–7652.

181. **Johnson, P. A., and M. R. Rosner.** 1986. Characterization of murine-specific leukemia virus receptor from L cells. *J. Virol.* **58**:900–908.

182. **Jolicoeur, P.** 1979. The Fv-1 gene of the mouse and its control of murine leukemia virus replication. *Curr. Top. Microbiol. Immunol.* **86**:67–122.

183. **Jolicoeur, P., and E. Rassart.** 1980. Effect of Fv-1 gene product on synthesis of linear and super-coiled viral DNA in cells infected with murine leukemia virus. *J. Virol.* **33**: 183–195.

184. **Jolicoeur, P., and E. Rassart.** 1981. Fate of unintegrated viral DNA in Fv-1 permissive and resistant mouse cells infected with murine leukemia virus. *J. Virol.* **37**:609–619.

185. **Jones, K. A., J. T. Kadonaga, P. A. Luciw, and R. Tjian.** 1986. Activation of the AIDS retrovirus promoter by the cellular transcription factor, Sp1. *Science* **232**:755–759.

186. **Ju, G., L. Boone, and A. M. Skalka.** 1980. Isolation and characterization of recombinant DNA clones of avian retroviruses: size heterogeneity and instability of the direct repeat. *J. Virol.* **33**:1026–1033.

187. **Ju, G., and A. M. Skalka.** 1980. Nucleotide sequence analysis of the long terminal repeat (LTR) of avian retroviruses: structural similarities with transposable elements. *Cell* **22**: 379–386.

188. **Junghans, R. P., L. R. Boone, and A. M. Skalka.** 1982. Retroviral DNA H structures: displacement-assimilation model of recombination. *Cell* **30**:53–62.

189. **Junghans, R. P., L. R. Boone, and A. M. Skalka.** 1982. Products of reverse transcription in avian retrovirus analyzed by electron microscopy. *J. Virol.* **43**:544–554.

190. **Kan, N. C., C. S. Flordellis, G. E. Mark, P. H. Duesberg, and T. S. Papas.** 1984. Nucleotide sequence of avian carcinoma virus MH2: two potential *onc* genes, one related to avian virus MC29 and the other related to murine sarcoma virus 3611. *Proc. Natl. Acad. Sci. USA* **81**:300–3004.

191. **Kao, S. Y., A. F. Calman, P. A. Luciw, and B. M. Peterlin.** 1987. Anti-termination of transcription within the long terminal repeat of HIV-1 by *tat* gene product. *Nature* (London) **330**:489–493.

192. **Katoh, I., T. Yasunaga, Y. Ikawa, and Y. Yoshinaka.** 1987. Inhibition of retroviral protease activity by an aspartyl proteinase inhibitor. *Nature* (London) **329**:654–656.

193. **Katoh, I., Y. Yoshinaka, A. Rein, M. Shibuya, T. Odaka, and S. Oroszlan.** 1985. Murine leukemia virus maturation: protease region required for conversion from "immature" to "mature" core form and for virus infectivity. *Virology* **145**:280–292.

194. **Katz, R. A., and A. M. Skalka.** 1987. A C-terminal domain

in the avian sarcoma-leukosis virus *pol* gene product is not essential for virus replication. *J. Virol.* **62:**528–533.

195. **Katz, R. A., R. W. Terry, and A. M. Skalka.** 1986. A conserved *cis*-acting sequence in the 5′ leader of avian sarcoma virus RNA is required for packaging. *J. Virol.* **59:**163–167.

196. **Keshet, E., Y. Shaul, J. Kaminchik, and H. Aviv.** 1980. Heterogeneity of "virus-like" genes encoding retrovirus-associated 30S RNA and their organization within the mouse genome. *Cell* **20:**431–439.

197. **Ketner, G., and T. J. Kelly.** 1976. Integrated simian virus 40 sequences in transformed cell DNA: analysis using restriction endonucleases. *Proc. Natl. Acad. Sci. USA* **73:** 1102–1106.

198. **Kikuchi, Y., Y. Ando, and T. Shiba.** 1986. Unusual priming mechanism of RNA-directed DNA synthesis in *copia* retrovirus-like particles in *Drosophila*. *Nature* (London) **323:** 824–826.

199. **King, W., M. D. Patel, L. I. Lobel, S. P. Goff, and M. C. Nguyen-Huu.** 1985. Insertion mutagenesis of embryonal carcinoma cells by retroviruses. *Science* **228:**554–558.

200. **Klatzman, D., E. Champagne, S. Chamaret, J. Gruest, D. Guetard, T. Hercend, J.-C. Gluckman, and L. Montagnier.** 1984. T-lymphocyte T4 molecule behaves as the receptor for human retrovirus LAV. *Nature* (London) **312:**767–768.

201. **Klempnauer, K.-H., T. J. Gonda, and J. M. Bishop.** 1982. Nucleotide sequence of the retroviral leukemia gene v-*myb* and its cellular progenitor c-*myb*: the architecture of a transduced oncogene. *Cell* **31:**453–463.

202. **Kowalski, M., J. Potz, L. Basiripour, T. Dorfman, W. C. Goh, E. Terwilliger, A. Dayton, C. Rosen, W. Haseltine, and J. Sodroski.** 1987. Functional regions of the envelope glycoprotein of human immunodeficiency virus type I. *Science* **237:**1351–1356.

203. **Kriegler, M., and M. Botchan.** 1983. Enhanced transformation by a simian virus 40 recombinant virus containing a Harvey murine sarcoma virus long terminal repeat. *Mol. Cell. Biol.* **3:**325–339.

204. **Kuchino, Y., H. Beier, N. Akita, and S. Nishimura.** 1987. Natural UAG suppressor glutamine tRNA is elevated in mouse cells infected with Moloney murine leukemia virus. *Proc. Natl. Acad. Sci. USA* **84:**2668–2672.

205. **Kuehn, M. R., A. Bradley, E. J. Robertson, and M. J. Evans.** 1987. A potential animal model for Lesch-Nyhan syndrome through introduction of HPRT mutations into mice. *Nature* (London) **326:**295–297.

206. **Kuff, E. L., A. Feenstra, K. Lueders, L. Smith, R. Haroley, N. Hozumi, and M. Shulman.** 1983. Intracisternal A-particle genes as movable elements in the mouse genome. *Proc. Natl. Acad. Sci. USA* **80:**1992–1996.

207. **Kung, H. J., Y. K. Fung, J. E. Majors, J. M. Bishop, and H. E. Varmus.** 1981. Synthesis of plus strands of retroviral DNA in cells infected with avian sarcoma virus and mouse mammary tumor virus. *J. Virol.* **37:**127–138.

208. **Kung, H. J., P. R. Shank, J. M. Bishop, and H. E. Varmus.** 1980. Identification and characterization of dimeric and trimeric circular forms of avian sarcoma virus-specific DNA. *Virology* **103:**425–433.

209. **Larder, B., D. Purify, K. Powell, and G. Darby.** 1987. AIDS virus reverse transcriptase defined by high level expression in *Escherichia coli*. *EMBO J.* **6:**3133–3138.

210. **Lasky, L. A., G. Nakamura, D. H. Smith, C. Fennie, C. Shimasaki, E. Patzer, P. Berman, T. Gregory, and D. J. Capon.** 1987. Delineation of a region of the human immunodeficiency virus type I gp120 glycoprotein critical for interaction with the CD4 receptor. *Cell* **50:**975–985.

211. **Leis, J., D. Baltimore, J. M. Bishop, J. Coffin, E. Fleissner, P. Goff, S. Oroszlan, H. Robinson, A. M. Skalka, H. M. Temin, and V. Vogt.** 1988. A standardized and simplified nomenclature for proteins common to all retroviruses. *J. Virol.* **62:**1808–1809.

212. **Leis, J., G. Duyk, S. Johnson, M. Longiaru, and A. Skalka.** 1983. Mechanism of action of the endonuclease associated with *ab* and ββ forms of avian RNA tumor virus reverse transcriptase. *J. Virol.* **45:**727–739.

213. **Leis, J., and J. Jentoft.** 1983. Characteristics and regulation of interaction of avian retrovirus pp12 protein with viral RNA. *J. Virol.* **48:**361–369.

214. **Levantis, P., D. A. F. Gillespie, K. Hart, M. J. Bissell, and J. A. Wyke.** 1986. Control of expression of an integrated Rous sarcoma provirus in rat cells: role of 5′ genomic duplications reveals unexpected patterns of gene transcription and its regulation. *J. Virol.* **57:**907–916.

215. **Lien, J.-M., C. E. Aldrich, and W. S. Mason.** 1986. Evidence that a capped oligoribonucleotide is the primer for duck hepatitis B virus plus-strand DNA synthesis. *J. Virol.* **57:** 229–236.

216. **Lifson, J. D., M. B. Feinberg, G. R. Reyes, L. Rabin, B. Banapour, S. Chakrabarti, B. Moss, F. Wong-Staal, K. S. Steimer, and E. G. Engleman.** 1986. Induction of CD4 dependent cell fusion by the HTLVIII/LAV envelope glycoprotein. *Nature* (London) **323:**725–728.

217. **Lightfoote, M. M., J. E. Coligan, T. M. Folks, A. S. Fauci, M. A. Martin, and S. Venkatesan.** 1986. Structural characterization of reverse transcriptase and endonuclease polypeptides of the acquired immunodeficiency syndrome retrovirus. *J. Virol.* **60:**771–775.

218. **Lilley, D. M. J.** 1983. Dynamic, sequence-dependent DNA structures as exemplified by cruciform extrusion from inverted repeats in negatively supercoiled DNA. *Cold Spring Harbor Symp. Quant. Biol.* **47:**101–112.

219. **Linial, M.** 1987. Creation of a processed pseudogene by retroviral infection. *Cell* **49:**93–102.

220. **Linial, M., and D. Blair.** 1982. Genetics of retroviruses, p. 649–783. *In* R. Weiss, N. Teich, H. Varmus, and J. Coffin (ed.), *RNA Tumor Viruses*, vol. 1. Cold Spring Harbor Laboratory, Cold Spring Harbor, N.Y.

221. **Linial, M., and D. Blair.** 1985. Genetics of retroviruses, p. 147–185. *In* R. Weiss, N. Teich, H. Varmus, and J. Coffin (ed.), *RNA Tumor Viruses*, vol. 2. Cold Spring Harbor Laboratory, Cold Spring Harbor, N.Y.

222. **Linial, M., E. Medeiros, and W. S. Hayward.** 1978. An avian oncovirus mutant (*SE* 21Q1b) deficient in genomic RNA: biological and biochemical characterization. *Cell* **15:** 1371–1381.

223. **Linney, E., B. Davis, J. Overhauser, E. Chao, and H. Fan.** 1984. Nonfunction of a Moloney murine leukemia virus regulatory sequence in F9 embryonal carcinoma cells. *Nature* (London) **308:**470–472.

224. **Lobel, L. I., and S. P. Goff.** 1984. Construction of mutants of Moloney murine leukemia virus by suppressor-linked insertional mutagenesis: positions of viable insertion. *Proc. Natl. Acad. Sci. USA* **81:**4149–4153.

225. **Lobel, L. I., and S. P. Goff.** 1985. Reverse transcription of retroviral genomes: mutations in the terminal repeat sequences. *J. Virol.* **53:**447–455.

226. **Loeb, D. D., C. A. Hutchinson, M. H. Edgell, W. G. Farmerie, and R. Swanstrom.** 1988. Mutational analysis of

the HIV 1 protease suggests functional homology with aspartic proteases. *Science*, in press.

227. **Loeb, D. D., R. W. Padgett, S. C. Hardies, W. R. Shehee, M. B. Comer, M. H. Edgell, and C. A. Hutchison III.** 1986. The sequence of a large L1Md element reveals a tandemly repeated 5′ end and several features found in retrotransposons. *Mol. Cell. Biol.* **6**:168–182.

228. **Loh, T. P., L. L. Sievert, and R. W. Scott.** 1987. Proviral sequences that restrict retroviral expression in mouse embryonal carcinoma cells. *Mol. Cell. Biol.* **7**:3775–3784.

229. **Lohler, J., R. Timpl, and R. Jaenisch.** 1984. Embryonic lethal mutation in mouse collagen I gene causes rupture of blood vessels and is associated with erythropoietic and mesenchymal cell death. *Cell* **38**:597–607.

230. **Luciw, P. A., C. C. Mayer, and J. A. Levy.** 1987. Mutational analysis of the human immunodeficiency virus: the *orf*-B region down-regulates virus replication. *Proc. Natl. Acad. Sci. USA* **84**:1434–1438.

231. **Luciw, P. A., H. Oppermann, J. M. Bishop, and H. E. Varmus.** 1984. Integration and expression of several forms of Rous sarcoma virus DNA used for transfection of mouse cells. *Mol. Cell. Biol.* **4**:1260–1269.

232. **Luk, K. C., T. D. Gilmore, and A. T. Panganiban.** 1987. The spleen necrosis virus *int* gene product expressed in *Escherichia coli* has DNA binding activity and mediates *att* and U5-specific DNA multimer formation in vitro. *Virology* **157**:127–136.

233. **Maddon, P. J., A. G. Dalgleish, J. S. McDougal, P. R. Clapham, R. A. Weiss, and R. Axel.** 1986. The T4 gene encodes the AIDS virus receptor and is expressed in the immune system and the brain. *Cell* **47**:333–348.

234. **Mager, D. L., and J. D. Freeman.** 1987. Human endogenous retrovirus like genome with type C *pol* sequences and *gag* sequences related to human T-cell lymphotropic viruses. *J. Virol.* **61**:4060–4066.

235. **Majors, J. E., and H. E. Varmus.** 1981. Nucleotide sequences at host-proviral junctions for mouse mammary tumour virus. *Nature* (London) **289**:253–258.

236. **Mann, R., and D. Baltimore.** 1985. Varying the position of a retrovirus packaging sequence results in the encapsidation of both unspliced and spliced RNAs. *J. Virol.* **54**:401–407.

237. **Mann, R. S., R. Mulligan, and D. Baltimore.** 1983. Construction of a retrovirus packaging mutant and its use to produce helper-free selective retrovirus. *Cell* **32**:871–879.

238. **Marsh, L. E., and T. J. Guilfoyle.** 1987. Cauliflower mosaic virus replication intermediates are encapsidated into virion-like particles. *Virology* **161**:129–138.

239. **Maruyama, M., H. Shibuya, H. Harada, M. Hatakeyama, M. Seiki, T. Fujita, J. Inoue, M. Yoshida, and T. Taniguchi.** 1987. Evidence for aberrant activation of the interleukin-2 autocrine loop by HTLV-e-encoded p40x and T3/Ti complex triggering. *Cell* **48**:343–350.

240. **Matsuo, T., M. Heller, L. Petti, E. O'Shiro, and E. Kieff.** 1984. Persistence of the entire Epstein-Barr virus genome integrated into human lymphocyte DNA. *Science* **226**: 1322–1324.

241. **McClure, M. A., M. S. Johnson, D.-F. Feng, and R. F. Doolittle.** 1988. Sequence comparisons of retroviral proteins: relative rates of change and general phylogeny. *Proc. Natl. Acad. Sci. USA* **85**:2469–2474.

242. **McClure, M. O., O. J. Sattentau, P. C. L. Beverley, J. P. Hearn, A. K. Fitzgerald, A. J. Zuckerman, and R. A. Weiss.** 1987. HIV infection of primate lymphocytes and conservation of the CD4 receptor. *Nature* (London) **330**:487–489.

243. **McDougal, J. S., M. S. Kennedy, J. M. Sligh, S. P. Cort, A. Mawle, and J. K. A. Nicholson.** 1986. Binding of HTLV-III/LAV to T4$^+$ T cells by a complex of the 110K viral protein and the T4 molecule. *Science* **231**:382–385.

244. **Meric, C., and P.-F. Spahr.** 1986. Rous sarcoma virus nucleic acid-binding protein p12 is necessary for viral 70S RNA dimer formation and packaging. *J. Virol.* **60**:450–459.

245. **Mietz, J. A., Z. Grossman, K. K. Lueders, and E. L. Kuff.** 1987. Nucleotide sequence of a complete mouse intracisternal A-particle genome: relationship to known aspects of particle assembly and function. *J. Virol.* **61**:3020–3029.

246. **Miles, B. D., and H. L. Robinson.** 1985. High-frequency transduction of c-*erb*B in avian leukosis virus-induced erythroblastosis. *J. Virol.* **54**:295–303.

247. **Mims, C. A.** 1986. Virus receptors and cell tropisms. *J. Infect. Dis.* **12**:199–204.

248. **Mitrani, E., J. Coffin, H. Boedtker, and P. Doty.** 1987. Rous sarcoma virus is integrated but not expressed in chicken early embryonic cells. *Proc. Natl. Acad. Sci. USA* **84**:2781–2784.

249. **Mitsuya, H., and S. Broder.** 1986. Strategies for antiviral therapy in AIDS. *Nature* (London) **325**:773–778.

250. **Mizuuchi, K.** 1984. Mechanism of transposition of bacteriophage Mu: polarity of the strand transfer reaction at the initiation of transposition. *Cell* **39**:395–404.

250a. **Mizuuchi, K., and R. Craigie.** 1986. Mechanism of bacteriophage Mu transposition. *Annu. Rev. Genet.* **20**:385–429.

251. **Moore, R., M. Dixon, R. Smith, G. Peters, and C. Dickson.** 1987. Complete nucleotide sequence of a milk-transmitted mouse mammary tumor virus: two frameshift suppression events are required for translation of *gag* and *pol*. *J. Virol.* **61**:480–490.

252. **Morisato, D., and N. Kleckner.** 1987. Tn10 transposition and circle formation in vitro. *Cell* **51**:101–111.

253. **Morris-Vasios, C., J. P. Kochan, and A. M. Skalka.** 1988. ASLV *pol*-endo proteins expressed independently in mammalian cells accumulate in the nucleus but can be directed to other cellular compartments. *J. Virol.* **62**:349–353.

254. **Mount, S. M., and G. M. Rubin.** 1985. Complete nucleotide sequence of the *Drosophila* transposable element copia: homology between copia and retroviral proteins. *Mol. Cell. Biol.* **5**:1630–1638.

255. **Muesing, M. A., D. H. Smith, C. D. Cabradilla, C. V. Benton, L. A. Lasky, and D. J. Capon.** 1985. Nucleic acid structure and expression of the human AIDS/lymphadenopathy retrovirus. *Nature* (London) **313**:450–458.

256. **Muesing, M. A., D. H. Smith, and D. J. Capon.** 1987. Regulation of mRNA accumulation by a human immunodeficiency virus *trans*-activator protein. *Cell* **48**:691–701.

257. **Mullins, J. I., D. S. Brody, R. C. Binari, and S. M. Cotter.** 1984. Viral transduction of the c-*myc* gene in naturally occurring feline leukaemias. *Nature* (London) **308**:856–857.

258. **Nabel, G., and D. Baltimore.** 1987. An inducible transcription factor activates expression of human immunodeficiency virus in T cells. *Nature* (London) **326**:711–714.

259. **Nagashima, K., M. Yoshida, and M. Seiki.** 1986. A single species of *pX* mRNA of human T-cell leukemia virus type I encodes *trans*-activator p40^x and two other phosphoproteins. *J. Virol.* **60**:394–399.

260. **Neil, J. C., D. Hughes, R. McFarlane, N. M. Wilkie, D. E. Onions, G. Lees, and O. Jarrett.** 1984. Transduction and rearrangement of the *myc* gene by feline leukaemia virus in naturally occurring T-cell leukaemias. *Nature* (London) **308**: 814–819.

261. **Nerenberg, M., S. H. Hinrichs, R. K. Reynolds, G. Khoury, and G. Jay.** 1987. The *tat* gene of human T-lymphotropic

virus type 1 induces mesenchymal tumors in transgenic mice. *Science* **227**:1324–1329.

262. **Nilsen, T. W., P. A. Maroney, R. G. Goodwin, F. M. Rottman, L. B. Crittenden, M. A. Raines, and H.-J. Kung.** 1985. c-*erb*B activation in ALV-induced erythroblastosis: novel RNA processing and promoter insertion result in expression of an amino-truncated EGF receptor. *Cell* **41**: 719–726.
263. **Nottenburg, C., E. Stubblefield, and H. E. Varmus.** 1987. An aberrant avian leukosis virus provirus inserted downstream from the chicken c-*myc* coding sequence in a bursal lymphoma results from intrachromosomal recombination between two proviruses and deletion of cellular DNA. *J. Virol.* **61**:1828–1833.
264. **Nusse, R.** 1988. The activation of cellular oncogenes by proviral insertion in murine mammary cancer. *In* M. E. Lippman and R. Dickson (ed.), *Breast Cancer: Cellular and Molecular Biology*, in press.
265. **Nusse, R., A. van Ooyen, D. Cox, Y. K. Fung, and H. E. Varmus.** 1984. Mode of proviral activation of a putative mammary oncogene (*int*-1) on mouse chromosome 15. *Nature* (London) **307**:131–136.
266. **Nyborg, J. K., W. S. Dynan, I. S. Y. Chen, and W. Wachsman.** 1988. Binding of host cell factors to DNA sequences in the HTLV-I LTR: implications for viral gene expression. *Proc. Natl. Acad. Sci. USA* **85**:1457–1461.
267. **Olsen, J. C., P. Furman, J. A. Fyfe, and R. Swanstrom.** 1987. 3′-Azido-3′-deoxythymidine inhibits the replication of avian leukosis virus. *J. Virol.* **61**:2800–2806.
268. **Olsen, J. C., and R. Swanstrom.** 1985. A new pathway in the generation of defective retrovirus DNA. *J. Virol.* **56**: 779–789.
269. **Omer, C. A., R. Resnick, and A. J. Faras.** 1984. Evidence for involvement of an RNA primer in initiation of strong-stop plus DNA synthesis during reverse transcription in vitro. *J. Virol.* **50**:465–470.
270. **Ou, C.-Y., L. R. Boone, C.-K. Koh, R. W. Tennant, and W. K. Yang.** 1983. Nucleotide sequence of *gag-pol* regions that determine the *Fv-1* host range property of BALB/c N-tropic and B-tropic murine leukemia viruses. *J. Virol.* **48**:779–784.
271. **Padgett, R. A., P. J. Grabowski, M. M. Konarska, S. Seiler, and P. A. Sharp.** 1986. Splicing of messenger RNA precursors. *Annu. Rev. Biochem.* **55**:1119–1150.
272. **Palmiter, R. D., J. Gagnon, V. M. Vogt, S. Ripley, and R. N. Eisenman.** 1978. The NH_2-terminal sequence of the avian oncovirus *gag* precursor polyprotein ($Pr76^{gag}$). *Virology* **91**:423–433.
273. **Panet, A., and D. Baltimore.** 1987. Characterization of endonuclease activities in Moloney murine leukemia virus and its replication-defective mutants. *J. Virol.* **61**: 1756–1760.
274. **Panet, A., G. Weil, and R. R. Friis.** 1978. Binding of tryptophanyl-tRNA to the reverse transcriptase of replication-defective avian sarcoma viruses. *J. Virol.* **28**:434–443.
275. **Panganiban, A. T., and H. M. Temin.** 1983. The terminal nucleotides of retrovirus DNA are required for integration but not virus production. *Nature* (London) **306**:155–160.
276. **Panganiban, A. T., and H. M. Temin.** 1984. Circles with two tandem LTRs are precursors to integrated retrovirus DNA. *Cell* **36**:673–679.
277. **Panganiban, A. T., and H. M. Temin.** 1984. The retrovirus *pol* gene encodes a product required for DNA integration: identification of a retrovirus *int* locus. *Proc. Natl. Acad. Sci. USA* **81**:7885–7889.
278. **Payne, G. S., J. M. Bishop, and H. E. Varmus.** 1982. Multiple arrangements of viral DNA and an activated host oncogene in bursal lymphomas. *Nature* (London) **295**: 209–213.
279. **Payne, G. S., S. A. Courtneidge, L. B. Crittenden, A. M. Fadley, J. M. Bishop, and H. E. Varmus.** 1981. Analyses of avian leukosis virus DNA and RNA in bursal tumors suggest a novel mechanism for retroviral oncogenesis. *Cell* **23**: 311–322.
280. **Pearl, L. H., and W. R. Taylor.** 1987. Sequence specificity of retroviral proteases. *Nature* (London) **328**:482–483.
281. **Pepinsky, R. B., R. J. Mattaliano, and V. M. Vogt.** 1986. Structure and processing of the p2 region of avian sarcoma and leukemia virus *gag* precursor polyproteins. *J. Virol.* **58**:50–58.
282. **Pepinsky, R. B., and V. M. Vogt.** 1984. Fine-structure analyses of lipid-protein and protein-protein interactions of *gag* protein p19 of the avian sarcoma and leukemia viruses by cyanogen bromide mapping. *J. Virol.* **52**:145–153.
283. **Perez, L. G., G. L. Davis, and E. Hunter.** 1987. Mutants of the Rous sarcoma virus envelope glycoprotein that lack the transmembrane anchor and cytoplasmic domains: analysis of intracellular transport and assembly into virions. *J. Virol.* **61**:2981–2988.
284. **Perucho, M., D. Hanahan, and M. Wigler.** 1980. Genetic and physical linkage of exogenous sequences in transformed cells. *Cell* **22**:309–317.
285. **Peterlin, B. M., P. A. Luciw, P. J. Barr, and M. D. Walker.** 1987. Elevated levels of mRNA can account for the *trans*-activation of human immunodeficiency virus. *Proc. Natl. Acad. Sci. USA* **83**:9734–9738.
286. **Peters, G. G., and J. Hu.** 1980. Reverse transcriptase as the major determinant for selective packaging of tRNA's into avian sarcoma virus particles. *J. Virol.* **36**:692–700.
287. **Portis, J. L., F. J. Atee, and L. H. Evans.** 1985. Infectious entry of murine retroviruses into mouse cells: evidence of a postadsorption step inhibited by acidic pH. *J. Virol.* **55**: 806–812.
288. **Pugatsch, T., and D. W. Stacey.** 1983. Identification of a sequence likely to be required for avian retroviral packaging. *Virology* **128**:505–511.
289. **Query, C. C., and J. D. Keene.** 1987. A human autoimmune protein associated with U1 RNA contains a region of homology that is cross-reactive with a retroviral $p30^{gag}$ antigen. *Cell* **51**:211–220.
290. **Quint, W., H. van der Putten, F. Janssen, and A. Berns.** 1982. Mobility of endogenous ecotropic murine leukemia viral genomes within mouse chromosomal DNA and integration of a mink cell focus-forming virus-type recombinant provirus in the germ line. *J. Virol.* **41**:901–908.
291. **Reed, R. R., and N. D. F. Grindley.** 1981. Transposon-mediated site-specific recombination *in vitro*: DNA cleavage and protein-DNA linkage at the recombination site. *Cell* **25**:721–728.
292. **Rein, A., M. R. McClure, N. R. Rice, R. B. Luftig, and A. M. Schultz.** 1986. Myristylation site in $Pr65^{gag}$ is essential for virus formation by Moloney murine leukemia virus. *Proc. Natl. Acad. Sci. USA* **83**:7246–7250.
293. **Rhee, S. S., and E. Hunter.** 1987. Myristylation is required for intracellular transport but not for assembly of D-type retrovirus capsids. *J. Virol.* **61**:1045–1053.
294. **Richter, A., H. L. Ozer, L. DesGroseillers, and P. Jolicoeur.** 1984. An X-linked gene affecting mouse cell DNA synthesis also effects production of unintegrated linear and supercoiled DNA of murine leukemia virus. *Mol. Cell. Biol.* **4**:151–159.

295. **Rohdewohld, H., H. Weiher, W. Reik, R. Jaenisch, and M. Breindl.** 1987. Retrovirus integration and chromatin structure: Moloney murine leukemia virus gene expression. *Proc. Natl. Acad. Sci. USA* **84:**4919–4923.

296. **Rosen, C. A., R. Park, J. G. Sodroski, and W. A. Haseltine.** 1987. Multiple sequence elements are required for regulation of human T-cell leukemia virus gene expression. *Proc. Natl. Acad. Sci. USA* **84:**4919–4923.

297. **Rosen, C. A., J. G. Sodroski, W. C. Goh, A. I. Dayton, J. Lipke, and W. A. Haseltine.** 1986. Post-transcriptional regulation accounts for the *trans*-activation of the human T-lymphotropic virus type III. *Nature* (London) **319:** 555–559.

298. **Rosenblatt, J. D., A. J. Cann, D. J. Slamon, I. S. Smalberg, N. P. Shah, J. Fujii, W. Wachsman, and I. S. Y. Chen.** 1988. HTLV-II trans-activation is regulated by two overlapping nonstructural genes. *Science* **240:**916–919.

299. **Roth, M. J., N. Tanese, and S. P. Goff.** 1985. Purification and characterization of murine retroviral reverse transcriptase expressed in *Escherichia coli*. *J. Biol. Chem.* **260:** 9326–9335.

299a. **Roth, M. J., N. Tanese, P. Schwartzberg, and S. P. Goff.** 1988. Gene product of Moloney murine leukemia virus required for proviral integration is a DNA-binding protein. *J. Mol. Biol.* **203:**131–140.

300. **Rowe, W. P., and C. A. Kozak.** 1980. Germ-line reinsertions of AKR murine leukemia virus genomes in *Akv-1* congenic mice. *Proc. Natl. Acad. Sci. USA* **77:**4871–4874.

301. **Saigo, K.** 1986. A potential primer for reverse transcription of mdg3, a Drosophila *copia*-like element, is a leucine tRNA lacking its 3′ terminal 5 bases. *Nucleic Acids Res.* **14:** 4370–4371.

302. **Schmidt, M., K. Gloggler, T. Wirth, and I. Horak.** 1984. Evidence that a major class of mouse endogenous long terminal repeats (LTRs) resulted from recombination between exogenous retroviral LTRs and LTR-like elements (LTR-IS). *Proc. Natl. Acad. Sci. USA* **81:**6696–6700.

303. **Schmidt, M., T. Wirth, B. Kroger, and I. Horak.** 1985. Structure and genomic organization of a new family of murine retrovirus-related DNA sequences (MuRRS). *Nucleic Acids Res.* **13:**3461–3470.

304. **Schultz, A. M., and S. Oroszlan.** 1983. In vivo modification of retroviral *gag* gene-encoded polyproteins by myristic acid. *J. Virol.* **46:**355–361.

305. **Schwartzberg, P., J. Colicelli, and S. P. Goff.** 1983. Deletion mutants of Moloney murine leukemia virus which lack glycosylated *gag* protein are replication competent. *J. Virol.* **46:**538–546.

306. **Schwartzberg, P., J. Colicelli, and S. P. Goff.** 1984. Construction and analysis of deletion mutants in the *pol* gene of Moloney murine leukemia virus: a new viral function required for establishment of the integrated provirus. *Cell* **37:**1043–1052.

307. **Schwartzberg, P., J. Colicelli, and S. P. Goff.** 1985. Recombination between a defective retrovirus and homologous sequences in host DNA: reversion by patch repair. *J. Virol.* **53:**719–726.

308. **Schwartzberg, P., J. Colicelli, M. L. Gordon, and S. P. Goff.** 1984. Mutations in the *gag* gene of Moloney murine leukemia virus: effects on production of virions and reverse transcriptase. *J. Virol.* **49:**918–924.

309. **Scott, M. L., K. McKereghan, H. S. Kaplan, and K. E. Fry.** 1981. Molecular cloning and partial characterization of unintegrated linear DNA from gibbon ape leukemia virus. *Proc. Natl. Acad. Sci. USA* **78:**4213–4217.

310. **Sealey, L., and R. Chalkley.** 1987. At least two nuclear proteins bind specifically to the Rous sarcoma virus long terminal repeat enhancer. *Mol. Cell. Biol.* **7:**787–798.

311. **Seeger, C., D. Ganem, and H. E. Varmus.** 1986. Biochemical and genetic evidence for the hepatitis B virus replication strategy. *Science* **232:**477–484.

312. **Seiki, M., S. Hattori, Y. Hirayama, and M. Yoshida.** 1983. Human adult T-cell leukemia virus: complete nucleotide sequence of the provirus genome integrated in leukemia cell DNA. *Proc. Natl. Acad. Sci. USA* **80:**3618–3622.

313. **Seiki, M., A. Hikikoshi, T. Taniguchi, and M. Yoshida.** 1985. Expression of the *pX* gene of HTLV-1: general splicing mechanism in the HTLV family. *Science* **228:** 1532–1534.

314. **Seiki, M., J. I. Inoue, T. Takeda, and M. Yoshida.** 1986. Direct evidence that p-40^{x} of human T-cell leukemia virus type I is a *trans*-acting transcriptional activator. *EMBO J.* **5:**561–565.

315. **Shank, P. R., J. C. Cohen, H. E. Varmus, K. R. Yamamoto, and G. M. Ringold.** 1978. Mapping of linear and circular forms of mouse mammary tumor virus DNA with restriction endonucleases: evidence for a large specific deletion occurring at high frequency during circularization. *Proc. Natl. Acad. Sci. USA* **75:**2112–2116.

316. **Shank, P. R., S. Hughes, H. J. Kung, J. Majors, N. Quintrell, R. V. Guntaka, J. M. Bishop, and H. E. Varmus.** 1978. Mapping unintegrated avian sarcoma virus DNA: termini of linear DNA bear 300 nucleotides present once or twice in two species of circular DNA. *Cell* **15:**1383–1395.

317. **Shapiro, J. A.** 1979. Molecular model for the transposition and replication of bacteriophage Mu and other transposable elements. *Proc. Natl. Acad. Sci. USA* **76:**1933–1937.

318. **Shell, B., P. Szurek, and W. Dunnick.** 1987. Interruption of two immunoglobulin heavy-chain switch regions in murine plasmacytoma P3.26Bu4 by insertion of retrovirus-like element ETn. *Mol. Cell. Biol.* **7:**1364–1370.

319. **Shen-Ong, G. L., H. C. Morse III, M. Potter, and J. F. Mushinski.** 1986. Two modes of c-*myb* activation in virus-induced mouse myeloid tumors. *Mol. Cell. Biol.* **6:**380–392.

320. **Shields, A., O. N. Witte, E. Rothenberg, and D. Baltimore.** 1978. High frequency of aberrant expression of Moloney murine leukemia virus in clonal infections. *Cell* **14:**601–609.

321. **Shih, C.-C., J. P. Stoye, and J. M. Coffin.** 1988. Highly preferred targets for retrovirus integration. *Cell* **53:**531–537.

322. **Shimotono, K., S. Mizutani, and H. M. Temin.** 1980. Sequence of retrovirus provirus resembles that of bacterial transposable elements. *Nature* (London) **285:**550–554.

323. **Shimotono, K., and H. M. Temin.** 1980. No apparent nucleotide sequence specificity in cellular DNA juxtaposed to retrovirus proviruses. *Proc. Natl. Acad. Sci. USA* **77:** 7357–7361.

324. **Shinnick, T., R. Lerner, and J. G. Sutcliffe.** 1981. Nucleotide sequence of Moloney murine leukemia virus. *Nature* (London) **293:**543–548.

325. **Shoemaker, C., S. Goff, E. Gilboa, M. Pasking, S. W. Mitra, and D. Baltimore.** 1980. Structure of a cloned circular Moloney murine leukemia virus molecule containing an inverted segment: implications for retrovirus integration. *Proc. Natl. Acad. Sci. USA* **77:**3932–3936.

326. **Shoemaker, C., J. Hoffmann, S. P. Goff, and D. Baltimore.** 1981. Intramolecular integration within Moloney murine leukemia virus DNA. *J. Virol.* **40:**164–172.

327. **Siekevitz, M., M. B. Feinberg, N. Holbrook, F. Wong-Staal, and W. C. Greene.** 1987. Activation of interleukin 2 and interleukin 2 receptor (Tac) promoter expression by the

trans-activator (*tat*) gene product of human T-cell leukemia virus, type I. *Proc. Natl. Acad. Sci. USA* **84:**5389–5393.

328. **Smith, J. K., A. Cywinski, and J. M. Taylor.** 1984. Specificity of initiation of plus-strand DNA by Rous sarcoma virus. *J. Virol.* **52:**314–319.

329. **Sodroski, J., W. C. Goh, C. Rosen, K. Campbell, and W. A. Haseltine.** 1986. Role of the HTLV-III/LAV envelope in syncytium formation and cytopathicity. *Nature* (London) **322:**470–474.

330. **Sodroski, J., W. C. Goh, C. Rosen, A. Dayton, E. Terwilliger, and W. Haseltine.** 1986. A second post-transcriptional *trans*-activator gene required for HTLV-III replication. *Nature* (London) **321:**412–416.

331. **Sodroski, J., R. Patarca, C. Rosen, F. Wong-Staal, and W. Haseltine.** 1985. Location of the *trans*-activating region on the genome of human T-cell lymphotropic virus type III. *Science* **229:**74–77.

332. **Sodroski, J., C. Rosen, W. C. Goh, and W. A. Haseltine.** 1985. A transcriptional activator protein encoded by the x-*lor* region of the human T-cell leukemia virus. *Science* **228:**1430–1434.

333. **Sodroski, J., C. A. Rosen, and W. A. Haseltine.** 1984. *Trans*-acting transcriptional activation of the long terminal repeat of human T lymphotropic viruses in infected cells. *Science* **225:**381–385.

334. **Somasundaran, M., and H. L. Robinson.** 1987. A major mechanism of human immunodeficiency virus-induced cell killing does not involve cell fusion. *J. Virol.* **61:**3114–3119.

335. **Sonigo, P., C. Barker, E. Hunter, and S. Wain-Hobson.** 1986. Nucleotide sequence of Mason-Pfizer monkey virus: an immunosuppressive D-type retrovirus. *Cell* **45:**375–385.

336. **Sonigo, P., S. Wain-Hobson, L. Bougueleret, P. Tiollais, F. Jacob, and P. Brulet.** 1987. Nucleotide sequence and evolution of ETn elements. *Proc. Natl. Acad. Sci. USA* **84:** 3768–3771.

337. **Soong, M. M., S. Ijiman, and L. H. Wang.** 1986. Transduction of c-*src* coding and intron sequences by a transformation-defective deletion mutant of Rous sarcoma virus. *J. Virol.* **59:**556–563.

338. **Speck, N. A., and D. Baltimore.** 1987. Six distinct nuclear factors interact with the 75-base-pair repeat of the Moloney murine leukemia virus enhancer. *Mol. Cell. Biol.* **7:** 1101–1110.

339. **Stein, B. S., S. D. Gowda, J. D. Lifson, R. C. Penhallow, K. G. Bensch, and E. G. Engleman.** 1987. pH-independent HIV entry into CD4-positive T cells via virus envelope fusion to the plasma membrane. *Cell* **49:**659–668.

340. **Stoltzfus, C. M., L. J. Chang, T. P. Cripe, and L. P. Turek.** 1987. Efficient transformation by Prague A, Rous sarcoma virus plasmid DNA requires the presence of *cis*-acting regions within the *gag* genes. *J. Virol.* **61:**3401–3409.

341. **Stoye, J. P., and J. M. Coffin.** 1985. Endogenous viruses, p. 357–404. *In* R. Weiss, N. Teich, H. Varmus, and J. Coffin (ed.), *RNA Tumor Viruses*, vol. 2. Cold Spring Harbor Laboratory, Cold Spring Harbor, N.Y.

342. **Stoye, J. P., and J. M. Coffin.** 1987. The four classes of endogenous murine leukemia virus: structural relationships and potential for recombination. *J. Virol.* **61:**2659–2669.

343. **Strebel, K., D. Daugherty, K. Clouse, D. Cohen, T. Folks, and M. A. Martin.** 1987. The HIV 'A' (*sor*) gene product is essential for virus infectivity. *Nature* (London) **328:** 728–730.

344. **Subramani, S., and J. Rubnitz.** 1985. Recombination events after transient infection and stable integration of DNA into mouse cells. *Mol. Cell. Biol.* **5:**659–666.

345. **Summers, J., and W. S. Mason.** 1982. Replication of the genome of a hepatitis B-like virus by reverse transcription of an RNA intermediate. *Cell* **29:**403–415.

346. **Surette, M. G., S. J. Buch, and G. Chaconas.** 1987. Transpososomes: stable protein-DNA complexes involved in the in vitro transposition of bacteriophage Mu DNA. *Cell* **49:** 253–262.

347. **Swanstrom, R., W. J. DeLorbe, J. M. Bishop, and H. E. Varmus.** 1981. Nucleotide sequence of cloned unintegrated avian sarcoma virus DNA: viral DNA contains direct and inverted repeats similar to those in transposable elements. *Proc. Natl. Acad. Sci. USA* **78:**124–128.

348. **Swanstrom, R., R. C. Parker, H. E. Varmus, and J. M. Bishop.** 1983. Transduction of a cellular oncogene: the genesis of Rous sarcoma virus. *Proc. Natl. Acad. Sci. USA* **80:**2519–2523.

349. **Tanese, N., and S. P. Goff.** 1988. Domain structure of the Moloney murine leukemia virus reverse transcriptase: mutational analysis and separate expression of the DNA polymerase and RNase H activities. *Proc. Natl. Acad. Sci. USA* **85:**1777–1781.

350. **Tanese, N., M. Roth, and S. P. Goff.** 1985. Expression of enzymatically active reverse transcriptase in *Escherichia coli*. *Proc. Natl. Acad. Sci. USA* **82:**4944–4948.

351. **Tanese, N., J. Sodroski, W. A. Haseltine, and S. P. Goff.** 1986. Expression of reverse transcriptase activity of human T-lymphotropic virus type III (HTLV-III/LAV) in *Escherichia coli*. *J. Virol.* **59:**743–745.

352. **Teich, N.** 1982. Taxonomy of retroviruses, p. 25–207. *In* R. Weiss, N. Teich, H. Varmus, and J. Coffin (ed.), *RNA Tumor Viruses*, vol. 1. Cold Spring Harbor Laboratory, Cold Spring Harbor, N.Y.

353. **Teich, N.** 1985. Taxonomy of retroviruses, p. 1–16. *In* R. Weiss, N. Teich, H. Varmus, and J. Coffin (ed.), *RNA Tumor Viruses*, vol. 2. Cold Spring Harbor Laboratory, Cold Spring Harbor, N.Y.

354. **Teich, N., J. Wyke, T. Mak, A. Bernstein, and W. Hardy.** 1982. Pathogenesis of retrovirus-induced disease, p. 785–998. *In* R. Weiss, N. Teich, H. Varmus, and J. Coffin (ed.), *RNA Tumor Viruses*, vol. 1. Cold Spring Harbor Laboratory, Cold Spring Harbor, N.Y.

355. **Teich, N., J. Wyke, and P. Kaplan.** 1985. Pathogenesis of retrovirus-induced disease, p. 187–248. *In* R. Weiss, N. Teich, H. Varmus, and J. Coffin (ed.), *RNA Tumor Viruses*, vol. 2. Cold Spring Harbor Laboratory, Cold Spring Harbor, N.Y.

356. **Temin, H. M.** 1985. Reverse transcription in the eukaryotic genome: retroviruses, pararetroviruses, retrotransposons, and retrotranscripts. *Mol. Biol. Evol.* **2:**455–468.

357. **Toh, H., H. Hayashida, and T. Miyata.** 1983. Sequence homology between retroviral reverse transcriptase and putative polymerases of hepatitis B virus and cauliflower mosaic virus. *Nature* (London) **305:**827–829.

358. **Toh, H., R. Kikuno, H. Hayashida, T. Miyata, W. Kugimiya, S. Inouye, S. Yuki, and K. Saigo.** 1985. Close structural resemblance between putative polymerase of a *Drosophila* transposable genetic element 17.6 and *pol* gene product of Moloney murine leukemia virus. *EMBO J.* **4:** 1267–1272.

359. **Van Beveren, C., J. Coffin, and S. Hughes.** 1985. Appendixes, p. 559–1221. *In* R. Weiss, N. Teich, H. Varmus, and J. Coffin (ed.), *RNA Tumor Viruses*, vol. 2. Cold Spring Harbor Laboratory, Cold Spring Harbor, N.Y.

360. **Van Beveren, C., E. Rands, S. K. Chattopadhyay, D. R.**

Lowy, and I. M. Verma. 1982. Long terminal repeat of murine retroviral DNAs: sequence analysis, host-proviral junctions, and preintegration site. *J. Virol.* **41**:542–556.

361. **Van Ooyen, A., and R. Nusse.** 1984. Structure and nucleotide sequence of the putative mammary oncogene *int*-1: proviral insertions leave the protein-coding domain intact. *Cell* **39**:233–240.

362. **Varmus, H. E.** 1982. Form and function of retroviral proviruses. *Science* **216**:812–820.

363. **Varmus, H. E.** 1983. Retroviruses, p. 411–503. *In* J. Shapiro (ed.), *Mobile Genetic Elements.* Academic Press, Inc., New York.

364. **Varmus, H. E.** 1987. Cellular and viral oncogenes, p. 271–346. *In* G. Stamatoyannopoulos, A. W. Nienhuis, P. Leder, and P. W. Majerus (ed.), *Molecular Basis of Blood Diseases.* The W. B. Saunders Co., Philadelphia.

365. **Varmus, H. E.** 1987. Reverse transcription. *Sci. Am.* **257**: 56–66.

366. **Varmus, H. E., T. Padgett, S. Heasley, G. Simon, and J. M. Bishop.** 1977. Cellular functions are required for the synthesis and integration of avian sarcoma virus-specific DNA. *Cell* **11**:307–319.

367. **Varmus, H. E., N. Quintrell, and S. Ortiz.** 1981. Retroviruses as mutagens: insertion and excision of a non-transforming provirus alters expression of a resident transforming provirus. *Cell* **25**:23–36.

368. **Varmus, H. E., and R. Swanstrom.** 1982. Replication of retroviruses, p. 369–512. *In* R. Weiss, N. Teich, H. Varmus, and J. Coffin (ed.), *RNA Tumor Viruses*, vol. 1. Cold Spring Harbor Laboratory, Cold Spring Harbor, N.Y.

369. **Varmus, H. E., and R. Swanstrom.** 1985. Replication of retroviruses, p. 75–134. *In* R. Weiss, N. Teich, H. Varmus, and J. Coffin (ed.), *RNA Tumor Viruses*, vol. 2. Cold Spring Harbor Laboratory, Cold Spring Harbor, N.Y.

370. **Verma, I. M.** 1977. The reverse transcriptase. *Biochim. Biophys. Acta* **473**:1–38.

371. **Vijaya, S., D. L. Steffen, and H. L. Robinson.** 1986. Acceptor sites for retroviral integrations map near DNase I-hypersensitive sites in chromatin. *J. Virol.* **60**:683–692.

372. **Voynow, S. L., and J. M. Coffin.** 1985. Evolutionary variants of Rous sarcoma virus: large deletion mutants do not result from homologous recombination. *J. Virol.* **55**:67–78.

373. **Voynow, S. L., and J. M. Coffin.** 1985. Truncated *gag*-related proteins are produced by large deletion mutants of Rous sarcoma virus and form virus particles. *J. Virol.* **56**:79–85.

374. **Wachsman, W., D. W. Golde, P. A. Temple, E. C. Orr, S. C. Clark, and I. S. Y. Chen.** 1985. HTLV *x*-gene product: requirement for the *env* methionine initiation codon. *Science* **228**:1534–1537.

375. **Wang, J. C.** 1985. DNA topoisomerases. *Annu. Rev. Biochem.* **54**:665–697.

376. **Watanabe, S., and H. M. Temin.** 1982. Encapsidation sequences for spleen necrosis virus, an avian retrovirus, are between the 5′ long terminal repeat and the start of the *gag* gene. *Proc. Natl. Acad. Sci. USA* **79**:5986–5990.

377. **Weiner, A. M., P. L. Deininger, and A. Efstratiadis.** 1987. Nonviral transposons: genes, pseudogenes, and transposable elements generated by the reverse flow of genetic information. *Annu. Rev. Biochem.* **55**:631–662.

378. **Weiner, A. M., and N. Maizels.** 1987. tRNA-like structures tag the 3′ ends of genomic RNA molecules for replication: implications for the origin of protein synthesis. *Proc. Natl. Acad. Sci. USA* **84**:7383–7388.

379. **Weiss, R.** 1982. Experimental biology and assay of RNA tumor viruses, p. 209–260. *In* R. Weiss, N. Teich, H. Varmus, and J. Coffin (ed.), *RNA Tumor Viruses*, vol. 1. Cold Spring Harbor Laboratory, Cold Spring Harbor, N.Y.

380. **Weiss, R.** 1985. Human T-cell retroviruses, p. 405–485. *In* R. Weiss, N. Teich, H. Varmus, and J. Coffin (ed.), *RNA Tumor Viruses*, vol. 2. Cold Spring Harbor Laboratory, Cold Spring Harbor, N.Y.

381. **Weiss, R., N. Teich, H. Varmus, and J. Coffin (ed.).** 1982. *RNA Tumor Viruses*, vol. 1. Cold Spring Harbor Laboratory, Cold Spring Harbor, N.Y.

382. **Weiss, R., N. Teich, H. Varmus, and J. Coffin (ed.).** 1985. *RNA Tumor Viruses*, vol. 2. Cold Spring Harbor Laboratory, Cold Spring Harbor, N.Y.

383. **Weller, S. K., A. E. Joy, and H. M. Temin.** 1980. Correlation between cell killing and massive second-round superinfection by members of some subgroups of avian leukosis virus. *J. Virol.* **33**:494–506.

384. **Westaway, D., J. Papkoff, C. Moscovici, and H. E. Varmus.** 1986. Identification of a provirally activated c-Ha-*ras* oncogene in an avian nephroblastoma via a novel procedure: cDNA cloning of a chimaeric viral-host transcript. *EMBO J.* **5**:301–309.

385. **Westaway, D., G. Payne, and H. E. Varmus.** 1984. Proviral deletions and oncogene base-substitutions in insertionally mutagenized c-*myc* alleles may contribute to the progression of avian bursal tumors. *Proc. Natl. Acad. Sci. USA* **81**: 843–847.

386. **Wettstein, F. O., and J. G. Stevens.** 1982. Variable-sized free episomes of Shope papilloma virus DNA are present in all non-virus-producing neoplasms and integrated episomes are detected in some. *Proc. Natl. Acad. Sci. USA* **79**:790–794.

387. **Wichman, H. A., S. S. Potter, and D. S. Pine.** 1985. *Mys*, a family of mammalian transposable elements isolated by phylogenetic screening. *Nature* (London) **317**:77–80.

388. **Willems, L., A. Gegonne, G. Chen, A. Burny, R. Kettmann, and J. Ghysdael.** 1987. The bovine leukemia virus p34 is a transactivator protein. *EMBO J.* **6**:3385–3389.

389. **Wilson, W., M. H. Malim, J. Mellor, A. J. Kingsman, and S. M. Kingsman.** 1986. Expression strategies of the yeast retrotransposon Ty: a short sequence directs ribosomal frameshifting. *Nucleic Acids Res.* **14**:7001–7016.

390. **Yaginuma, K., M. Kobayashi, E. Yoshida, and K. Koike.** 1985. Hepatitis B virus integration in hepatocellular carcinoma DNA: duplication of cellular flanking sequences at the integration site. *Proc. Natl. Acad. Sci. USA* **82**:4458–4462.

391. **Yamamoto, K. R.** 1985. Steroid receptor regulated transcription of specific genes and gene networks. *Annu. Rev. Genet.* **19**:209–252.

392. **Yamamoto, T., G. Jay, and I. Pastan.** 1980. Unusual features in the nucleotide sequence of a cDNA clone derived from the common region of avian sarcoma virus messenger. *Proc. Natl. Acad. Sci. USA* **77**:176–180.

393. **Yang, W. K., J. O. Kiggans, D.-M. Yang, C.-Y. Ou, R. W. Tennant, A. Brown, and R. H. Bassin.** 1980. Synthesis and circularization of N- and B-tropic retroviral DNA in Fv-1 permissive and restrictive mouse cells. *Proc. Natl. Acad. Sci. USA* **77**:2994–2998.

394. **Yoshinaka, Y., I. Katoh, T. D. Copeland, and S. Oroszlan.** 1985. Translational readthrough of an amber termination codon during synthesis of feline leukemia virus protease. *J. Virol.* **55**:870–873.

395. **Yoshinaka, Y., I. Katoh, T. D. Copeland, and S. Oroszlan.**

1985. Murine leukemia virus protease is encoded by the *gag-pol* gene and is synthesized through suppression of an amber termination codon. *Proc. Natl. Acad. Sci. USA* **82:** 1618–1622.

396. **Yoshinaka, Y., and R. B. Luftig.** 1977. Murine leukemia virus morphogenesis: cleavage of P70 *in vitro* can be accompanied by a shift from a concentrically coiled internal strand ("immature") to a collapsed ("mature") form of the virus core. *Proc. Natl. Acad. Sci. USA* **74:**3446–3450.

397. **Yuki, S., S. Ishimaru, S. Inouye, and K. Saigo.** 1986. Identification of genes for reverse transcriptase-like enzymes in two Drosophila retrotransposons, 412 and gypsy; a rapid detection method of reverse transcriptase genes using YXDD box probes. *Nucleic Acids Res.* **14:**3017–3030.

Chapter 4

Bacterial Insertion Sequences

DAVID J. GALAS and MICHAEL CHANDLER

David J. Galas ■ Molecular Biology, University of Southern California, Los Angeles, California 90089-1481. Michael Chandler ■ Laboratoire de Génetique du Centre National de la Recherche Scientifique, 18 Route de Narbonne, Toulouse Cedex 31007, France.

I. INTRODUCTION

The discovery of bacterial insertion sequences (ISs) and of their capacity for modifying gene expression, sequestering genes, and promoting genome rearrangements has made an important contribution to destroying the concept of the genome as a fundamentally stable entity. The number of distinct bacterial ISs characterized continues to increase, but, unfortunately, only a few are being intensely investigated. Although transposable elements (transposons and ISs) have been studied most thoroughly in bacteria, they have been found in most organisms which have been examined.

Bacterial insertion sequences were discovered during early investigations of the molecular genetics of gene expression in *Escherichia coli* and bacteriophage lambda. Originally isolated as highly polar, somewhat unstable mutations in the galactose (149, 186, 328) and lactose (228, 229) operons of *E. coli* and in the early genes of lambda (37), many of these mutations were soon shown by hybridization and heteroduplex analysis (102, 150) to be insertions of the same few segments of DNA in different positions and orientations. These segments of DNA, isolated repeatedly as insertion mutations, became known as insertion sequences. Their similarity to the genetic elements discovered in maize by McClintock (233) became clear when it was recognized that ISs are natural residents of the *E. coli* genome, and that the observed insertion mutations were examples of their movement to new genetic locations.

They range in length from 800 to 2,500 base pairs (bp) and can be found in the genomes of many different bacteria at multiplicities between a few and a few hundred per genome (Section II.A). They are particularly frequent as components of natural plasmids (Sections II.B and II.C), in which they are often associated with genes responsible for antibiotic resistance: the drug-resistance transposons (Section II.C).

Some of the genetic phenomena associated with IS-induced mutations were at first difficult to understand. For example, most insertions were highly polar, although a few did not reduce the expression of genes downstream of the insertion points. The idea gained currency that ISs could turn genes either off or on (Section III.C). It was proposed that at least some ISs carried "portable promoters" that could activate the transcription of flanking genes. At least some of the initial controversies arose from the unexpectedly complex nature of the changes that could be engendered by IS insertion. In addition to inducing deletions, inversions, and replicon fusions, many elements also appeared to be susceptible to internal rearrangements (Section II.A and III.B). It is now known that the transcription of flanking genes can indeed originate from promoters located within an IS or from "hybrid promoters" created by the insertion event (Section III.C).

In addition to their impact on the study of gene expression, transposable elements continue to provide a rich source of experimental material for studying recombination and repair processes, population dynamics, and the horizontal transmission of genes. They are also widely used to facilitate genetic mapping, as mutagens, as tools for analyzing gene organization, and in gene cloning.

Our purpose here is to review the present state of understanding of the nature, occurrence, and genetic activities of bacterial insertion sequences. We have built on the foundation of several authoritative reviews of a general as well as a more specialized nature. The most comprehensive recent review has been that by Iida et al. (165) in the book edited by Shapiro (331). A particularly incisive review of the mechanistic aspects of transposition has been provided by Grindley and Reed (127). Current knowledge of the mechanisms involved in the transposition of phage Mu, which has contributed much to an understanding of transposition in general, has been recently reviewed by Mizuuchi and Craigie (242). Earlier reviews continue to be useful references for certain aspects of this growing field (46, 52, 342, 343).

Since several chapters of this book are concerned with specific transposable elements, we will try here to provide a more general overview and comparison of their structures, behavior, and effects on the "host" cell. As a focus and a useful fulcrum for this review, however, we will use the insertion sequence IS*1* as a primary example. IS*1* exhibits most of the phenomena we will be concerned with, was one of the first ISs to be characterized, and remains the smallest known element. It is also an element which we have studied extensively in our laboratories over the past few years.

In Section II, we present an overview of the remarkable variety of transposable elements found in bacteria. This survey of the occurrence and variety concerns only ISs found in gram-negative bacteria (the chapter by E. Murphy in this volume deals with ISs from gram-positive bacteria, and the chapter by R. L. Charlebois and W. F. Doolittle deals with archaebacteria) and is further biased toward those found in enteric bacteria, primarily because of the preponderance of published work on these organisms. In Section III, we review the genetic effects of ISs, from the phenomena through which they were first discovered to our present understanding of the wide range of molecular effects they can induce. Later sections deal with the structure and function of the elements themselves, their genes and control mechanisms, and the bacterial host functions with which they interact (Sections IV and V). Our current understanding of the mechanisms of transposition is discussed in Section VI. Finally, we discuss what little is known about the evolutionary implications of transposable elements (Section VII), both as producers of evolutionary change and as evolving genetic elements.

II. OCCURRENCE, VARIETY, AND DISTRIBUTION OF INSERTION SEQUENCES

In addition to the elements found in early investigations (IS*1*, IS*2*, IS*3*, and IS*4*), numerous other sequences have been identified in the genomes, plasmids, and bacteriophages of a wide range of bacterial genera and species. Examples are known in both gram-positive bacteria (Murphy, this volume) and the archaebacteria (Charlebois and Doolittle, this volume), as well as in a number of gram-negative genera such as *Agrobacterium*, *Pseudomonas*, *Rhizobium*, *Salmonella*, and *Shigella*.

A compilation of some insertion sequences in gram-negative bacteria is included in Tables 1 and 2. ISs vary widely in size and, as we discuss later, in genetic organization and behavior. The smallest known, IS*1*, is 768 bp in length (IS*200* is about this size), while most others are between 1,000 and 2,500 bp.

We have included the insertion sequence Tn*1000* (also called gamma delta) even though it is a member of the Tn*3* family and is treated elsewhere (D. Sherratt, this volume) because it plays a particularly important role in plasmid-chromosome interactions in *E. coli* (Section III.B).

A. Distribution in Bacterial Genomes

Early experiments using filter hybridization to determine the copy number of IS*1* and IS*2* in *E. coli* K-12 gave estimates of 6 to 9 and 4 to 6 chromosomal copies, respectively, depending on the strain (312). Subsequent studies using Southern hybridization of total genomic DNA showed that these estimates are remarkably accurate. Extensive data concerning the distribution and genomic copy number are available for IS*1*, *2*, *3*, *4*, *5*, *30*, and *200*. A compilation of the data obtained from well-characterized laboratory strains is presented in Table 3.

Of the sequences listed in Table 3, IS*200* appears to be restricted to *Salmonella* sp. strains, although material which shows some low-level cross hybridization has been observed in two *Shigella* sp. strains (208). More recently, Matsutani et al. (231) isolated a repeated sequence from *Shigella sonnei* that shows significant sequence homology with IS*200* and is estimated to be present in about 10 copies. The others appear in members of the *Enterobacteriaceae* such as *E. coli* and *Shigella* spp., but not in most *Salmonella* spp. In view of the relatedness of *Salmonella typhimurium* and *E. coli* and the documented transfer of genetic information between these species, this apparent "species specificity" is somewhat unexpected.

The copy number of IS*1*, one of the best-documented cases, can vary considerably from strain to strain. Most laboratory strains of *E. coli* K-12 carry from between 6 to 10 copies, while *E. coli* C (from two sources) has 3, and certain strains of *Shigella* spp. have been estimated to carry more than 50 copies, perhaps as many as 200 (259, 265). It is interesting that the high copy number of ISs in *Shigella* spp. is not restricted to IS*1*. Relatively large copy numbers of many previously known and several novel ISs can be detected in a single strain (231; M.-F. Prère and O. Fayet, personal communication). Like most *Salmonella* strains, *E. coli* W is apparently devoid of IS*1*.

Similar variations in copy number have been observed in natural isolates of *E. coli*. In a sample of 71 independent isolates (the ECOR collection), 11 did not carry genomic (as opposed to plasmid-borne) copies of IS*1*, while 16 carried between 10 and 40 copies (317). High genomic copy numbers were also observed in 4 strains in another collection of 10 natural isolates (260).

In general, the copy number of other insertion sequences is lower than that of IS*1*. Although certain individual natural isolates of *E. coli* have been shown to carry up to 17 copies of IS*2*, 14 copies of IS*4*, and 21 copies of IS*5*, the average copy number in the

Table 1. Insertion sequences associated with gram-negative bacteria[a]

Sequence name	Alternate names	Size	Target duplications (bp)	Inverted repeats (bp)	Source	References
Enterics						
IS*1*		768 bp	9 (8–11)	20/23*	*Escherichia coli*	102, 149, 150, 184, 230, 238, 266
IS2		1,327 bp	5	32/41*	*E. coli*	102, 119, 149, 150, 303
IS3		1,258 bp	3	29/40*	*E. coli*	102, 149, 230, 341, 353
IS4		1,426 bp	11–13	16/18	*E. coli*	102, 194
IS5		1,195 bp	4	15/16	*E. coli*	33, 99, 320, 351
IS6*			8	14/14	Tn6	18
	IS*15del*	820 bp			*Salmonella panama* (Tn*1525*)	206, 207, 355, 357
	IS*26*	820 bp			Rts1 (Tn*2680*)	166, 244
	IS*46*				R46	43
	IS*140*				pWP14a	39
	IS*160*				pBP16 (Tn*2440*)	256
	IS*176*	820 bp			NTP16	371
IS*10*		1,329 bp	9	17/22*	R100 (Tn*10*)	54, 135
IS*21*		2,132 bp	4 (5)	10/11	R68.45	367; Reimann et al., in press
	IS*8*				RP4	87
					R68.45	211
	ISP				R68.45	297
IS*30*		1,221 bp	2	23/26	R100 (NR1)-Basel	53, 79
	IS*121*				*E. coli*	261
IS*50*		1,534 bp	9	8/9*	Tn*5*	8, 18
IS*91*		1.8 kb	0	8/9*	pSU233	90, 374
IS*150*		1,443 bp	3	19/24*	*E. coli*	322a
IS*161*		1.7 kb				237
IS*186*		1,338 bp	8–12	23/23	*E. coli*	202, 326
IS*200*		0.7 kb	(2?)	None?	*Salmonella typhimurium*	208, 209
IS*903*		1,057 bp	9	18/18	Tn*903*	124, 258, 268, 269
	IS*102*	1,057 bp			pSC101	22, 226, 263
IS*3411*		1,309 bp	3	23/25*	Tn*3411*	174, 174a
Gamma delta		5,980 bp	5	35/35	F	129, 291
ISSHO*1*	Iso-IS*3*	1,247 bp	3	20/27*	*Shigella dysenteriae*	O. Fayet, M.-F. Prère, and M. Chandler, unpublished data
IS*600*	Iso-IS*3*	1,264 bp	3	19/27*	*Shigella sonnei*	231; Fayet et al., unpublished data
					S. dysenteriae	Fayet et al., unpublished data
Nonenterics						
IS22		7.1 kb			*Pseudomonas aeruginosa* PAO	252
IS*51*		1,311 bp	3	26/26	*P. savastanoi*	372
IS52		1,209 bp	4	9/10	*P. savastanoi*	372
IS222		1.35 kb			*P. aeruginosa*	115
IS401		1.3 kb			*Pseudomonas cepacia*	11
IS402		1 kb			*P. cepacia*	11
IS403		0.8 kb			*P. cepacia*	191, 323
IS404		1.1 kb			*P. cepacia*	191, 323
IS405		1.5 kb			*P. cepacia*	191, 323
IS*411*		1.9 kb			*P. cepacia*	11
IS476		1,225 bp	4	13/13	*Xanthomonas campestris*	191, 323

Continued on following page

Table 1—*Continued*

Sequence name	Alternate names	Size	Target duplications (bp)	Inverted repeats (bp)	Source	References
IS*60*		1.2 kb			*Agrobacterium tumefaciens*	270
IS*66*		2,548 bp	8	18/20	*A. tumefaciens*	31, 226
IS*426*	IS*136*	1,313 bp	9	30/32*	*A. tumefaciens*	358
IS*492*		1,202 bp	5	None	*Pseudomonas atlantica*	12a; D. H. Bartlett and M. Silverman, submitted
IS*4400*	IS*4551*	1.15 kb			*Bacteroides fragilis*	302, 159a, 288a,
	IS*4351*	1,155 bp			(functions in *E. coli*)	319b, 337a
ISR*1*		1.15 kb			*Rhizobium lupini*	283
ISRm*1*		1.4 kb			*Rhizobium meliloti*	310
ISRm2		2.7 kb	8	24/25	*R. meliloti*	94
RSRj-alpha		1,126 bp			*Rhizobium japonicum*	188
RSRj-beta		950 bp			*R. japonicum*	188
IS*701*		1,389 bp	4	22/25*	*Calothrix* sp. PCC7601	D. Mazel, personal communication

[a] We have indicated whether or not sequence data are available by citing the IS length in base pairs (bp) or kilobase pairs (kb), respectively. The original source of the element is also indicated. Target duplications and lengths of the inverted repeats are also indicated. An asterisk indicates nonhomology in the form of the insertion or deletion of bases or limited homology further within the element. In the case of "iso" insertion sequences isolated from *Shigella* spp., we have included only those known to transpose.

collection of 71 strains was determined to be 6.4 for IS*1*, 2.7 for IS2, 1.7 for IS*3*, 2 for IS4, 1.3 for IS*5*, and 0.9 for IS*30* (317). The copy number of IS*5* has also been determined in a set of 25 *E. coli* strains taken from the Murray collection, a collection of *Enterobacteriaceae* isolated in the "preantibiotic era" (between 1917 and 1954 [82]). The average copy number of IS*5* was found to be significantly higher than in the more recently isolated members of the ECOR collection, an observation which may reflect an accumulation of copies during storage of the strains (121). The population dynamics of insertion sequences is discussed in the chapter by J. W. Ajioka and D. L. Hartl in this volume.

Measurement of copy number by the techniques normally used (hybridization) does not imply that the insertion sequences are identical, or even active. Sequence divergence has been observed in the case of copies of IS*1*-like sequences analyzed in *Shigella dysenteriae*, where three major classes of IS*1*-like sequences have been shown to occur having approximately 99, 90, and 60% sequence homology with the original IS*1* (262, 265). In addition to these iso-IS*1* elements, Matsutani et al. (231) have isolated repeated sequences from the *S. sonnei* genome which show sequence homologies with IS2, IS*3* (IS*600*, 40% homologous; IS*629*), and IS*200* (IS*630*). Sequences almost identical to IS*600* have been shown to be proficient in transposition (O. Fayet and M.-F. Prère, unpublished observations). This set of related ISs is discussed further in Section VII. Studies with IS2 in *E. coli* have also shown that this element is subject to significant sequence rearrangements under certain conditions (116–119).

In another case, a degenerate derivative of Tn*1000* has been shown to reside on the plasmid pSC101. This sequence, IS*101*, is much shorter than Tn*1000* (209 rather than 5,980 bp) and requires the transposition functions of Tn*1000* for activity (103, 175, 240, 289). Some evidence of other examples of divergence of Tn*1000* has been obtained for plasmid-associated elements in four *E. coli* strains of the ECOR collection (317), but the sizes of the elements were not determined.

Less information is available concerning the occurrence and distribution of insertion sequences in nonenteric gram-negative bacteria. Three sequences from *Rhizobium* spp., ISR*1*, ISRm*1*, and ISRm2, have been described (94, 283, 310), and three insertion sequences from *Agrobacterium tumefaciens*, IS*60* (270), IS*66* (31, 226), and IS*426* (358), have been characterized. Several sequences having structural features clearly related to those of ISs are present in the genomes of *Rhizobium japonicum* (188) and in *A. tumefaciens* (113, 270, 359). At least 12 different insertion sequences have also been detected in the pseudomonads, another important group of gram-negative bacteria. These are listed in Table 1.

Many of the common insertion sequences resident in the *E. coli* K-12 genome have been located on the genetic map (for review, see reference 86), and

Table 2. Sequences of inverted repeats of known insertion sequences associated with gram-negative bacteria[a]

Sequence name		Inverted repeats
Enterics		
IS*1*	L	GGTgATG CTgCCAACTTAcTGATttAGTGT ATGaTggtgt
	R	GGTaATGaCT CCAACTTAtTGAT AGTGTtttATGtTcaga
IS2	R	TaGAcTgGCCCCcTgA AT cTCCAGACAaCcaaTATCACTT
	L	TgGAtTtGCCCC T AtATt TCCAGACAtCtgtTATCACTTA
IS*3*	L	TGATCtTACCCAgc AATAgTGGACACgcGGC TAAGtGAGt
	R	TGATCcTACCCAcgtAATA TGGACACa GGCcTAAGcGAGg
IS*4*	L	TAATGCCgaTCAGTTAAGGATCagtTGACcGa TCcagT GG
	R	TAATGCCagTCAGTTAAGCAAC TGACtGgcTCtttTtcGG
IS*5*	L	GGAAGGTGCGAAtAAGcggggaaattcttctcggctgact
	R	GGAAGGTGCGAAcAAGtccctgatatgagatcatgtttgt
IS6*	L	GGCACTGTTGCAAA TAGtcggtggtgataaacttatcatc
	R	GGCACTGTTGCAAAgtTAGcgatgaggcagccttttgtct
IS*10*	I	CTGA GAgATCCCCTcATaATTTccccAAAAcgtaaccatg
	O	CTGAtGA ATCCCCTaATgATTTtggtAAAAatcattaagt
IS*21*	L	TGTCAgCGCCAgtgatataagacggtaattcaccatttgg
	R	TGTCAaCGCCAcgatgtttgaccgttatttgccattttca
IS*30*	L	TGTAGATTCAAT TgGTCAAcGCAACAgttatgtgaaaaca
	R	TGTAGATTCAATcT GTCAAtGCAACAcccctttcaattat
IS*50*	I	CTGtCTCTTgatcagatcttgatcccctgcgccatcagat
	O	CTGaCTCTTatacacaagtagcgtcctgaacggaaccttt
IS*91*	L	TCGAgTAGGcAgcctggcggctgcggcttgtcatggtctg
	R	TCGA TAGGaAtttaaatccccaaaagactaaaaaagcacc
IS*150*	L	TGTACTGcACCCatttt GTTGGACgaTcAAAtggaatagc
	R	TGTACTG ACCCcaaaaaGTTGGACagTtAAAcacgaggca
IS*186*	L	CATAAGCGCTAACTTAAGGGTTGtggtATtacgcctgata
	R	CATAAGCGCTAACTTAAGGGTTGaaccATctgaagaatgc
IS*200*	L	TAAATATCCTCCGGCATAGCCGAGCTTTTT CA
	R	GTCTATGGAAACCCCCAGCTAGGCTGGGGG TT
IS*903*	I	GGCTTTGTTGAATAAATCgaacttttgctgagttgaagga
	O	GGCTTTGTTGAATAAATCagatttcgggtaagtctccccc
IS*3411*	L	TGAACCGCCCCGGGaaTCCTGGAGAcTaaacTtcCTGaGA
	R	TGAACCGCCCCGGGttTCCTGGAGAgTgtttTatCTGtGA
Tn*1000*	L	GGGGTTTGAG GGCCAATGGA ACGAAAACGT ACGTTaA GGa
	R	GGGGTTTGAG GGCCAATGGA ACGAAAACGT ACGTTtAtGG
ISSHO*1*	L	TGAAGTGGT CAacaAAaacTGGCCACC GAgtt AGAGtTtt
	R	TGAAGTGGCaCActgAAtt TGGCCACCtGA acAGAGgTga
IS*600*	L	TGAGGTagcCTGagttTAaCGGACACTccttcctgaaata
	R	TGAGGTgtaCTGgcaaTAgCGGACACTaccatttgttctt
Nonenterics		
IS22	NS[b]	
IS*51*	L	TGAACCGCCCCGGGTTTCTCGGAGACcttttg TTTGAGtc
	R	TGAACCGCCCCGGGTTTCTCGGAGACtccaaccTTTGAGa
IS*52*	L	GGAGcCGCTG cAaAAaTAgCcaacagCcCgtaGCCttggc
	R	GGAGaCGCTG aAcAAtTAaCtccgttCgCaccGCCcccat
IS*222*	NS	
IS*401*	NS	
IS*402*	NS	

Continued on following page

Table 2—*Continued*

Sequence name		Inverted repeats
IS*403*	NS	
IS*404*	NS	
IS*405*	NS	
IS*411*	NS	
IS*476*	L	TGACCTGCCCCCActGagCCGTACCAG
	R	TGACCTGCCCCCAtcGt CCGTACCAG
IS*60*	NS	
IS*66*	L	GTAAGCCcaCGGTGAAGGCCgtcttgcgatgacggcctgc
	R	GTAAGCCtcCGGTGAAGGCCacaggtcaggcagcttgagc
IS*426*	L	AcTGCcCCCCATTTCgACCGGACA GTCGGCataagcagaa
	R	ggAgTGCaCCCCATTTC ACCGGACAaGTCGGCtagattga
IS*4400*	NS	
ISR*1*	NS	
ISRm*1*	NS	
ISRm*2*	L	GTAAGCGCTCATTTCCATGCGCgTTgacgggcccgtttga
	R	GTAAGCGCTCATTTCCATGCGCcTTtctcgttgatgatg
IS*701*	L	CCCTgTTTTGTCACTCT GGCAGtagtataataaggtaaaa
	R	CCCTcTTTTGTCACTCTaGGCAGaggaaaagtagaaatca

[a] Further information about these ISs and references to the literature are found in Table 1. See Table 1, footnote *a*, for additional explanations.
[b] NS, Not yet sequenced.

IS*200* has been mapped in *S. typhimurium* using genetic techniques (208). With the development of physical genomic mapping using new gel electrophoretic techniques, however, it has become possible to obtain restriction maps of entire genomes (337). These methods, together with ordered clone libraries of entire bacterial genomes (200), will greatly facilitate the location of insertion sequences in the genomes of many species, and their effects on genome structure and stability should now become amenable to direct analysis.

The presence of large numbers of an insertion sequence might be expected to have a strong influence on the structure and stability of the genome since, in addition to their transposition properties, insertion elements also act as substrates for homologous recombination (Section III.B). Such effects on genome organization have not been investigated in detail, but the distribution of ISs must have been affected by these forces (see Section VII).

B. Distribution in Plasmids and Bacteriophages

At about the same time as it was realized that insertion sequences are an integral part of the *E. coli* genome, it also became clear that they are widely distributed in bacterial plasmids. IS2, IS*3*, and Tn*1000* (gamma delta; D. Sherratt, this volume) were found to reside in the conjugal plasmid F, and IS*1* was found to be a constituent of several conjugal, multiple-antibiotic-resistance plasmids (R factors) and of the prophage P1 (156, 157, 164, 286, 332). These elements were subsequently shown to play an important role in the behavior of their "host" plasmids (Section III.B). Copies of IS*1*, *2*, *3*, *4*, *5*, and *30* are also found relatively frequently in the plasmid population of the ECOR collection of natural *E. coli* isolates (317). IS elements often appear to occur more frequently (per unit length of DNA) in plasmids than in bacterial genomes (e.g., reference 332).

C. As Vectors for Nontransposable Genes

Several plasmid-associated insertion sequences have not yet been demonstrated to have chromosomal equivalents. These include IS*50*, a component of the transposon Tn*5* (D. E. Berg, this volume), originally isolated from an R factor (18); IS*10*, a component of transposon Tn*10* (196; N. Kleckner, this volume), also detected as a single copy in the R factor R1drd19 (68) and as part of the fosfomycin resistance transposon Tn*2921* (253); elements probably identical to the sequences flanking Tn6 (18) (for clarity grouped here under the generic name IS6), IS*15del* (IS*15*), IS26, IS46, IS*140*, and IS*160*, isolated independently as part of several transposons and

Table 3. Distribution of several insertion sequences

Organism	IS*1*	IS2	IS*3*	IS*4*	IS*5*	IS*30*	IS*200*	References
Escherichia coli K-12 wild type	6–8	6	5					155, 259
E. coli K-12 EMG2	6							259
E. coli K-12 Ymel	6							259
E. coli K-12 W1485	6–7	7–8	5		11			155, 215, 259
E. coli K-12 W2252 HfrC	7							259
E. coli K-12 W2367	8							259
E. coli K-12 W3110	7–9					3		53, 259
E. coli K-12 W3350						4		53
E. coli K-12 JE5519	10							259
E. coli K-12 (594)		4		1			0	208, 209
E. coli K-12 (MJ101)							0	208, 209
E. coli K-12 χ101	6	9	5					155
E. coli K-12 C600	8	7	6		10	2		53, 155, 321
E. coli K-12 W6	4	12	5					155
E. coli B	17	1	4	1			0	155, 208, 209
E. coli B/r					0		0	208, 209, 321
E. coli C	3	0	5	0		2	0	53, 155, 208, 209, 259
E. coli W	0							259
E. coli 15T							0	208, 209
Shigella dysenteriae	40						1*[a]	208, 209, 259
Shigella sonnei	30							259
Shigella boydii	2							259
Shigella flexneri	40							259
Shigella sp.		10		10			1*	208, 209
Citrobacter freundii	0	0		0	3		0	208, 209, 259, 321
Edwardsiella tarda	0							259
Enterobacter aerogenes	0	0		0			0	208, 209, 259
Erwinia amylovora	0							259
Klebsiella aerogenes	1	0		0			0	208, 209, 259
Klebsiella pneumoniae		2		0			0	208, 209
Serratia marcescens	2	0		0			0	208, 209, 259
Rhizobium japonicum	0	0	0					38
Proteus mirabilis	0	0		0			0	208, 209, 259
Proteus vulgaris	0	0	0	0			0	38, 208, 209
Proteus morganii	0							259
Pseudomonas aeruginosa	0	0	0					38, 259
Pseudomonas cepacia	0	0	0					38
Acinetobacter sp.	0							259
Caulobacter crescentus	0							259
Myxococcus xanthus	0							259
Bacillus subtilis	0	0	0					38, 259
Clostridium rubrum	0	0	0					38,
Corynebacterium poinsettiae	0	0	0					38
Dactylosporangium aurantiacum	0	0	0					38
Xanthamonas campestris	0	0	0					38
Streptomyces niveus	0	0	0					38
Haemophilus haemolyticus					0			321
Salmonella typhimurium	0	0		0			6–10	208, 209, 259
Salmonella typhi		0		0			18	208, 209
Salmonella pollorum		0		0			1	208, 209
Salmonella gallinarum							1	208, 209
Salmonella heidelberg							7	208, 209
Salmonella agona							0	208, 209
Salmonella dublin							1	208, 209
Salmonella abortus-bovis							7	208, 209
Salmonella typhisuis		0		0			0	208, 209

[a] Asterisk indicates potential homology.

present in a variety of plasmids from a range of bacterial species (39, 43, 244, 255, 355); IS*21* (IS*8*), a component of R68 and other IncP1 plasmids (87, 211, 297, 367); IS*91*, isolated from an alpha-hemolysin plasmid of *E. coli* (90, 375); and IS*903* (IS*602*), associated with transposons Tn*903* (258) and Tn*602* (346) and also present as a single copy, IS*102*, on the plasmid pSC101 (22, 267).

Many (but not all) of the plasmid-associated insertion sequences listed above form part of compound transposons (described more fully in Section IV.A). These are elements composed of two flanking insertion sequences in direct or inverted orientation and a central, non-self-transposable DNA segment. A large variety of interstitial genes have been observed as part of compound transposons, although the majority of known transposons of this type carry antibiotic resistance genes. This bias may not be representative of natural populations, but may rather reflect the widespread use of antibiotics in medicine and animal husbandry over the past 30 years (as well as the ease of manipulation of this type of transposon in the laboratory). While the origins of these resistance genes are outside the scope of this review, it is intriguing that of 84 conjugative plasmids transferred from strains of the preantibiotic era (Murray collection) and established in *E. coli* K-12, not a single example of antibiotic resistance was detected (81, 82, 159). To our knowledge, no information concerning the presence of insertion sequences in this plasmid collection is available.

Some idea of the impact of insertion sequences on the bacterial population can be obtained from the observed number and variety of known transposons. Compound transposons carrying closely related antibiotic genes but flanked by different insertion sequences are not infrequent. The aminoglycoside phosphotransferase (APH-3′ type I) gene, which specifies resistance to aminoglycosides such as kanamycin, has been found flanked by inverted or directly repeated copies of IS*903* (Tn*903* and Tn*602*, respectively [258, 346]), by direct repeats of IS*1* (Tn*2350* [67]), by direct repeats of IS6 (IS26) elements (Tn6 = Tn*2680* [18, 167]), or by direct repeats of IS*15* [IS*15* is an insertion sequence carrying an insertion of IS*15del* (IS6) within itself] (Tn*1525* [205, 206]). In all these cases, significant sequence divergence is observed both upstream and downstream from the aminoglycoside phosphotransferase gene, and in one case, Tn*2680*, an IS*903*-like sequence is also located between the flanking IS26 (IS6) elements (243).

Insertion sequences are also versatile, or promiscuous, in the sense that a given IS is often found flanking different interstitial genes. Sequences similar to IS6 are also found flanking a tetracycline resistance determinant (IS*46* [43]; IS*160* [256]).

Perhaps one of the most promiscuous insertion elements is IS*1*. It forms part of the chloramphenicol resistance transposon Tn*9* (221) (in turn derived from the r determinant of an R-factor IncFII plasmid [136, 160]); the kanamycin resistance transposon Tn*2350* (67); the heat-stable enterotoxin transposon Tn*1681* (340), as part of compound transposons encoding iron uptake (71, 85a, 234, 362); and a large transposon of unknown function apparently resident in the chromosome of *E. coli* HB101 (303).

In summary, insertion sequences represent a large group of "intrachromosomal" genetic elements which have been observed in all bacterial species examined. While they share many common characteristics, they show significant variations in size and nucleotide sequence, in distribution, and, as we discuss below, in the genetic effects they produce.

III. GENETIC EFFECTS OF IS ELEMENTS

A. IS Elements as Mutagens

1. Contribution of insertions to mutation frequency

As indicated in Section I, insertion sequences were first recognized by their ability to induce polar mutations in the *gal* (186, 328), *lac* (229), and phage lambda p_R (37) operons. The body of data accumulated subsequently has amply demonstrated the important role played by insertion sequences in producing spontaneous mutations in bacteria. It is difficult, however, to assess accurately the contribution of insertions to the overall frequency of mutations since many factors can influence transposition frequencies. Factors such as the location of the DNA target (chromosomal or extrachromosomal), the specific affinity of the target for a given insertion sequence (insertion specificity; Section V), the variety and number of insertion sequences carried by the bacterium in question (Section II), the growth conditions of the culture, and the selection or screening procedure are all expected to influence frequency estimates.

Early estimates of the frequency of insertion mutations were made for the *galK* region (186, 328), the *lac* region (229), phage P2 (213), and the *c*II region of phage lambda (27). In each case, it was estimated that between 5 and 15% of spontaneous mutations are IS insertions.

One of the first quantitative assessments of the contribution of insertion mutations to the overall mutation frequency came from a study in which a collection of spontaneous mutants in the *lacI* gene was analyzed at the nucleotide sequence level (50, 101). Of 140 mutations analyzed, 94 were due to frameshift mutations at a single site, 18 were deletions, 20 were point mutations (as judged by the absence of detectable changes in size of the gene), and 2 were insertions of IS*1*. If the contribution of the frameshift hot spot is discounted, IS*1* insertions account for about 4% of spontaneous *lacI* mutations. Other estimates using different systems have indicated much higher contributions. Insertion mutations in the ribosomal protein genes carried by a lambda phage were found to represent 25 to 40% of spontaneous mutations (177), whereas all 10 polar mutations analyzed in the *galT* gene were insertions of IS*1* or IS2 (273), and of 25 spontaneous mutations in the phage lambda *cI* gene (212), 15 were insertions of IS*1* (7 cases), IS*3* (3 cases), and IS*5* (5 cases). A systematic study using the P1 prophage (6) indicated that 27% of non-plaque-forming P1 phage mutants carried insertions. A similar analysis of 17 non-plaque-forming mutants of P1-15, a derivative of P1 which does not carry an IS*1* element, showed that all carried an insertion. These included IS*1*, IS2, IS*5*, IS*30*, and Tn*1000* (5). Moreover, insertion of IS2 occurred at an approximately 10-fold-higher frequency than other insertions. It is important to realize that the specificity of insertion can vary greatly from gene to gene and from element to element (Section V). Indeed, it was subsequently shown that P1-15 carries a highly preferred insertion site for IS2 (324, 326). IS*1* insertion has also been reported to yield temperature-sensitive (128) and cold-sensitive (319a) mutations.

2. Precise and nearly precise excision

Excision, resulting in restored function of a gene inactivated by IS insertion (precise excision) or in relief of polarity on genes located downstream from the insertion (precise or imprecise excision [105, 306]), is another capability of many transposable elements. In the case of Tn*5* (IS*50*) and Tn*10* (IS*10*), this behavior is not dependent on element-specified transposition functions (96, 105; but see reference 192 for phage Mu). It is thought that excision occurs following intrastrand pairing between inverted repeat sequences, leading to the "extrusion" of the intervening DNA segment and its deletion by replication "slippage" across the small directly repeated target sequences (generated by the insertion of an element; see below).

In this model, excision should be influenced by factors affecting the efficiency of formation and repair of the extruded DNA. The necessary intrastrand pairing might be expected to be facilitated within single-stranded regions of DNA, and indeed, excision of elements carried by the F plasmid appears to occur at a much higher frequency than a corresponding chromosomal insertion (96, 154) and is stimulated by conjugation (19, 350), a process involving the transfer of single-stranded DNA. Moreover, Tn*5* insertions in single-strand bacteriophage have been reported to be unstable (146a; H. Schaller, cited in reference 19). Direct proof of the importance of replication slippage has recently been provided (45; see also S. D. Ehrlich, this volume). Activation of a bacteriophage M13 replication origin carried by a plasmid to generate single-stranded DNA and its subsequent conversion to double-stranded DNA greatly stimulates the precise excision of a Tn*10*-based transposon also carried by the plasmid. Using radioactive labeling, it was shown that recombinants are generated during the conversion of single-stranded DNA to double-stranded form.

Factors affecting the stability of the extruded structure would also be expected to influence excision. The excision frequency has been observed to decrease with decreasing length of the inverted sequences (19, 70, 96, 105). This length dependence could explain the differences in frequency of precise excision observed from element to element (154, 218). In addition, extrusion of palindromic sequences in a model plasmid-based system has been shown to depend both on supercoiling and on the local sequence environment (347). Such context effects may explain the observed dependence of excision on the location (nucleotide sequence environment) of the insertion element (105, 154). It has also been suggested that the effects of transcription on the DNA structure can stimulate precise excision in some cases.

As described in Section IV.D, many host mutations which increase excision are in genes implicated in DNA repair. Although their precise role is not known, such (wild-type) functions could intervene by repairing intermediate structures generated by replication slippage. The influence of Dam methylation sites (see Table 5 and Fig. 5) and of minor sequence differences in the ends of many elements (which would generate mismatches when paired in the extruded structures and could therefore stimulate the mismatch repair system) have yet to be investigated.

3. Target duplications

One characteristic of the majority of known insertion elements is that they generate small, directly repeated duplications of the target DNA at the point of insertion (an exception is IS*91* [91]). This is presumably due to the staggered cleavage of target DNA by the transposition enzymes. The length of

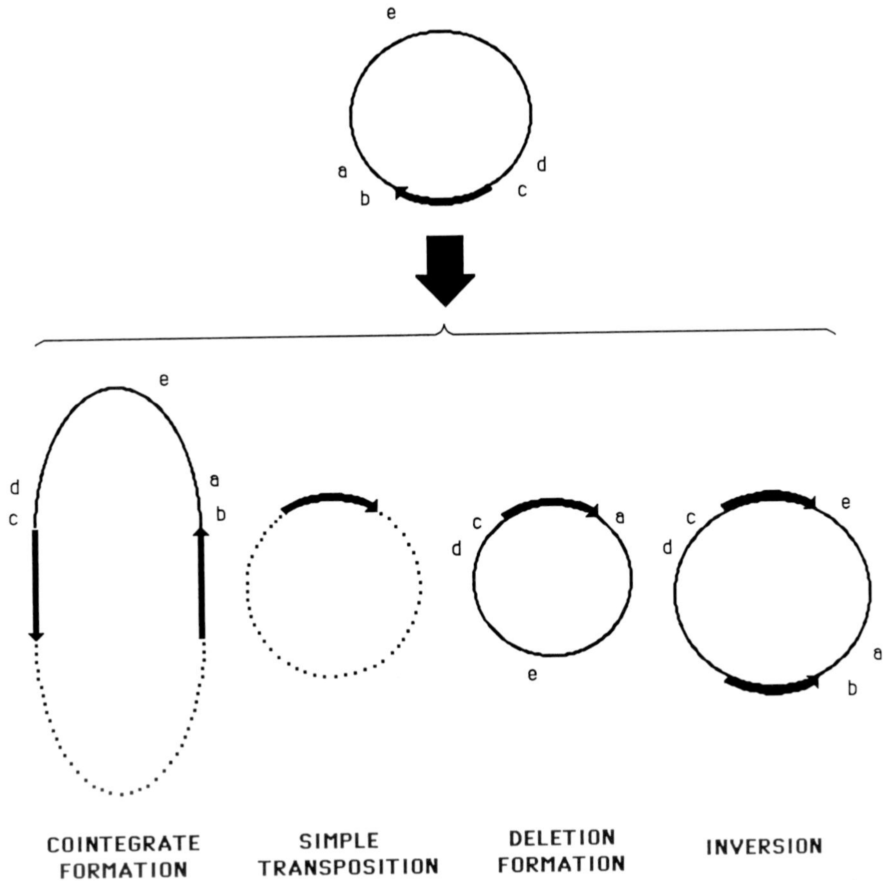

Figure 1. Spectrum of products of transposition events. The products of transpositional recombination diagrammed here represent those that could be produced, according to current models, by a single recombinational event (see Section VI and other chapters in this book). Different models ascribe different pathways to producing these structures. We simply list them here. Note that they all could also be produced by multiple events as well, as noted in the text. (Second replicon indicated by dotted line.)

this duplication is a characteristic proper to each element and ranges from 2 to 13 bp (Table 1).

Although many elements appear to induce a duplication of a fixed number of base pairs, four exceptions have been reported: IS*1*, IS*4*, IS*21*, and IS*186*. Insertions of IS*4* have been shown to generate duplications of 11, 12, and 13 bp (114, 133, 193, 232). Recent experiments with IS*21* have shown that of nine independent insertions, eight resulted in a target duplication of 4 bp, while one resulted in a duplication of 5 bp (C. Reimann, R. Moore, S. Little, A. Savioz, N. Willetts, and D. Haas, *Mol. Gen. Genet.*, in press). Of six insertions of IS*186* analyzed (202, 326), duplications of 8 (one example), 10 (two examples), 11 (two examples), and 12 (one example) bp have been observed.

IS*1* also shows variations in target duplication length. The majority of sequenced IS*1* insertions (52 of 68) generate a duplication of 9 bp (50, 89, 123, 161–163, 185, 190, 222, 263, 282). However, duplications of 7 (1 example), 8 (13 examples), 10 (1 example), and 14 (1 example) bp have been observed, and an example of an 11-bp duplication has also been reported. Two proposals have been advanced to explain the observed variations (162). One possibility is that cleavage of the target also occurs with a 9-bp spacing, but that in some situations, bases are removed from one or the other of the cleaved target ends by nuclease action. An alternative possibility is that duplication of 7, 8, 10, 11, and 14 bp reflects a variation in the spacing of the initial cleavages, which has been attributed to small sequence variations in the different IS*1* elements used. In particular, mutations in two of the IS*1* genes necessary for transposition activity (Section IV) have been associated with this phenomenon (222). These workers report that a 9-bp duplication occurred during all six insertions obtained with one IS*1* derivative, IS*1*R, but occurred in only three of the seven insertions obtained using a derivative, IS*1*K, which carries single-amino-acid changes in InsA and InsB (Section IV). Three were found to have generated an 8-bp and one a 14-bp duplication. Moreover, it was concluded that both gene products must be involved, since the hybrid IS*1* element InsA(K)/InsB(R) generated duplications of 7 (one example), 8 (one example), and 9 (five examples) bp, whereas the reciprocal hybrid, InsA(R)/InsB(K), generated duplications of 8 bp (two examples) and 9 bp (five examples). Mutations in genes whose products are involved in the cleavage reactions may therefore lead to a change in the geometry of cleavage. Factors such as DNA helix geometry (known to be influenced on a local scale by nucleotide sequence [e.g., references 49 and 92] may also be involved. Local variations in parameters such as helix twist or flexibility may well play a subtle role in target cleavage, although no systematic study of this aspect has been reported.

Since variation from the 9-bp IS*1*-induced target duplication is not uncommon, similar variations may also occur with other insertion sequences for which few sequenced insertions are available.

B. Other IS-Mediated Rearrangements

Transposable elements have been shown to mediate a variety of DNA rearrangements in addition to simple insertions. IS-mediated deletions, inversions, and cointegrates (replicon fusions), as well as more complex rearrangements, have been documented. The basic property of transposable elements underlying all these phenomena is their ability to generate a new joint between nonhomologous DNA segments (Fig. 1). However, when carefully examined, not all elements are found to exhibit all these capabilities. The relative frequencies of occurrence of each vary widely among the ISs. It is important to keep in mind that the pathways by which each end product arises may be multiple and include quite distinct processes (Section VI).

In addition to their ability to induce the formation of new DNA joints without the aid of significant homology, insertion sequences can also be used as efficient substrates for the homologous recombination system of the host: they can act as portable segments of homology and thus serve as sites for duplication, inversion, deletion, or replicon fusion via homologous recombination.

1. Deletion formation

The capacity of certain insertion sequences to generate deletions of neighboring DNA was first noted for an IS*1* located in the *gal* operon (293). IS*1* was also shown to be responsible for the generation of miniplasmids from the large (about 90-kilobase [kb]) IncFII plasmid R100.1 (239). The presence of IS*1* in the *gal* operon was found to increase the deletion frequency by 30- to 2,000-fold. This type of deletion event does not result in the loss of the IS*1* element and, at the nucleotide sequence level, the deletions terminate precisely at one end of IS*1* (266). That deletion formation and transposition of IS*1* share common mechanisms was suggested by the observation that IS*1* (in Tn*9*)-mediated deletions into the *lacI* gene terminate preferentially within regions that are preferred for Tn*9* insertion (51), and by the observation that impinging, external transcription reduces IS*1*-mediated transposition, cointegrate formation, and deletion formation (Section IV.A).

The frequency of IS*1*-mediated deletions in the *gal* operon was shown to be exceptionally high (5×10^{-4}) at temperatures below 35°C and to decrease rapidly at higher temperatures to levels similar to those generally observed for transposition and cointegrate formation (72, 293). This behavior is generally interpreted as being due to a temperature-sensitive step in the IS*1* transposition pathway. Host mutations that greatly reduce deletion frequencies have also been reported (254). Subsequent characterization of these mutations indicated that they are located in the *gal* repressor (*galR*) locus (P. Nevers, personal communication), implying that transcription of the *gal* operon plays an important role in determining deletion frequencies in this system. In view of the known inhibitory effect of transcription on transposition activity (Section IV.C), it is possible that the observed temperature sensitivity of deletion formation in this system is due to temperature-dependent changes in impinging transcription. Experiments designed to assess the effect of temperature on intermolecular transposition (IS*1*/Tn*9*-mediated cointegrate formation) have failed to demonstrate such a dramatic temperature dependence (unpublished results).

IS*2* appears to generate deletions at low frequencies relative to transposition frequencies (272, 293). This may indicate that the deletion pathway is only infrequently used by IS*2*. An alternate possibility is that it may simply be due to the rare juxtaposition of sites with relatively high insertion specificity which are characteristic of IS*2* (325).

Of the insertion elements investigated in detail, IS*10* (195, 197, 257, 305), IS*30* (53), and IS*903* (IS*102*) (for example, reference 365 and references therein), all generate similar types of deletion end products. For IS*10* there is also evidence that the deletion endpoints occur in regions of preferred insertion (257). Analysis of a single IS*30*-mediated deletion suggests similar behavior (53). As in the case of IS*1*, this supports the notion that insertion and deletion formation by these elements are directly related to transposition.

In the case of IS*903* and IS*102*, it has been found that deletion results in the appearance of an additional copy of the element (21, 23, 365). Both products of a deletion event (the plasmid carrying the deletion and the deleted fragment) have been isolated and both carry copies of the insertion sequence. This is strong evidence that deletion formation mediated by these elements can involve a replicative transposition pathway (Section VI). Some evidence has been presented which suggests a preference for IS*102* deletion endpoints within transcriptionally active target regions (23a).

Less information is available concerning other insertion sequences. Deletions generated by IS*4* seem to differ in structure from the usual IS-generated deletion in that DNA from both sides of the elements is removed (193). IS*50* has been tested for its ability to induce deletions in plasmid DNA flanking a single copy of the element, in a situation where IS*1* induces deletions at a relatively high frequency in recombination-deficient (*recA* mutant) hosts. No deletions were detected in this system (A. Binda and D. J. Galas, unpublished observations). However, using a similar system, but with an entire Tn*5* rather than a single IS*50*, A. Ahmed (2; personal communication) has observed deletion formation at frequencies approaching 1.5×10^{-7} per cell per generation in recombination-proficient (*recA*$^+$) cells. In *recA* mutant cells, however, such events are reported to occur at frequencies which are 10^2-fold lower. IS*21* has also been reported to mediate deletions of neighboring DNA, but details of the structure of the deletions are less clear (76, 131, 132).

2. Inversion

Another property often considered as characteristic of insertion sequences is their ability to invert a neighboring segment of DNA. Inversion of a DNA segment, accompanied by the appearance of a second copy of the insertion sequence flanking the segment in inverted orientation with respect to the first, has been reported for several elements (6, 23, 72, 305, 364; Ahmed, personal communication), suggesting that these inversions occur by a replicative process. Note, however, that the Tn*10* result cited above is no longer interpreted as replicative in origin (see Section VI).

Care must be taken to distinguish a single-step inversion event from more complex, multistep events. Simple insertion of a second copy of the element in inverse orientation followed by homologous recombination between the two copies, for example, would also lead to the type of end product observed. In at least two of the cases cited above (72, 305), the products were isolated in recombination-proficient hosts. It is therefore possible that the observed inversions occurred via a second, homologous recombination step following simple intramolecular transposition.

3. Cointegrate formation

Cointegrate formation, the fusion of two replicons, is perhaps the best-studied type of genetic rearrangement induced by transposable elements. Because cointegrate assays are relatively easy and do not require an element-associated selectable marker, cointegrate formation is probably the method used most frequently for measuring the transposition activity of newly isolated insertion sequences. The resulting cointegrate structure is composed of donor and recipient replicons separated at each junction by a single, directly repeated copy of the transposable element. The generation of such structures is often considered to be evidence for replicative transposition, since the number of copies of the element increases from one (in the donor molecule) to two (in the cointegrate). As noted later (Section VI), however, cointegrate formation by itself is not necessarily evidence for replicative transposition.

Cointegrate formation was initially observed in the case of members of the Tn*3* family of transposable elements (Sherratt, this volume) and bacteriophage Mu (M. L. Pato, this volume). Numerous examples of replicon fusion induced by other insertion sequences have now been reported. The mechanism leading to replicon fusion is not, however, necessarily the same for each element. In the case of Tn*3*-like elements (which include Tn*1000* and IS*101*), cointegrate formation is an integral step in the transposition process. Cointegrates generated by replicative insertion are unstable. A highly efficient transposon-specified activity (resolvase) catalyzes site-specific recombination between the directly repeated copies of the element (resolution) and leads to the appearance of the donor molecule and a target molecule each carrying a single copy of the element (Sherratt, this volume).

In contrast to the Tn*3* family of elements, fused replicons generated by most insertion sequences are relatively stable in the absence of the host homologous recombination system. The observation that IS*1* (Tn9)-mediated cointegrates are stable, while transposition of Tn9 (without accompanying replicon fusion) is a relatively frequent event, led to the original proposal that cointegrates cannot be obligatory intermediates in the transposition of IS*1* (110).

Although the frequencies of transposition and cointegrate formation by IS*1* (Tn9) can be modulated by the pattern of transcription entering each flanking IS*1* element, the relative frequencies of transposition and cointegrate formation do not vary by more than a factor of 10 (60). Most other elements exhibit much larger differences.

The products of IS6 transposition appear to be exclusively cointegrates (43, 356), whereas cointegrates appear to occur, if at all, as only infrequent products of the transposition of IS*10*, IS*50* (152, 153), IS*91* (90), and IS*3411* (174).

Note that replicon fusions can also be generated by "inverse transposition" (Section IV.A), in which the DNA segment located between repeated insertion sequences carries an entire copy of the plasmid. Such structures can be generated by dimerization of the donor plasmid followed by transposition of one of the plasmid copies flanked by two insertion sequences (17, 29).

An unusual example of transposition-generated replicon fusions is the case of IS*21*. IS*21* exhibits significant activity only when present as a directly repeated tandem dimer (297, 298, 367). For convenience, we refer to the duplicated sequence as IS*21.2*. The sequence IS*21.2*, isolated from the plasmid R68.45, is able to promote replicon fusion at frequencies approaching 10^{-3}. The original duplicated copies of IS*21* in IS*21.2* are separated by 3 bp. Cointegrates generated by this element have only one copy of the IS*21* element at each junction (297) and have lost the intervening 3-bp sequence present in the IS*21.2* element. Replicon fusion by IS*21.2* thus appears to occur by a simple transposition of the entire donor plasmid using two neighboring ends, one from each copy of IS*21* (Reimann et al., in press).

4. Homologous recombination effects: Hfr and F′ formation, amplification, and inversion

In addition to their transposition properties, insertion elements can also act as substrates for homologous recombination. Investigation of the behavior of plasmids such as the sex factor F (IncFI) and multiple-antibiotic-resistance plasmids of the IncFII class in various bacterial hosts provided some of the earliest examples of the importance of insertion sequences in (homologous) recombination. While the structures of these types of plasmids differ in detail, they exhibit a large degree of homology (332) and are structurally related. In particular, both types of plasmid carry a cluster of (different) insertion sequences at a similar relative position on their physical maps (see references 69 and 329 for a comparison of the physical maps of F and IncFII R plasmids). In the case of F, this cluster is composed of IS2, IS*3*, and Tn*1000*; in a similar relative position, IncFII plasmids have a large (approximately 20 kb) region, the *r-det*, which is a series of intercalated transposable elements flanked by directly repeated copies of IS*1* carrying most of the resident antibiotic resistance genes (156, 286).

Homologous recombination between insertion sequences IS2, IS*3*, and Tn*1000* (a member of the Tn*3* family) located on the F plasmid and at various positions on the *E. coli* chromosome appears to be responsible for F integration and the formation of many Hfr strains (F-chromosome cointegrates). Insertion sequences are also involved in the excision of the integrated F plasmid, together with flanking DNA, to produce F′ plasmids. This often involves (but need not) homologous recombination (83a, 133a). There is a strong correlation between the position of IS2, IS*3*, and Tn*1000* elements on the chromosome and the point of origin of several Hfr strains (see reference 86 for review), and Hfr formation occurs at significantly reduced frequencies in *recA* mutant strains (75). In addition, Guyer et al. (130) showed that a chromosomal copy of Tn*1000* acts as a sex factor affinity site (*sfa*) promoting polarized mobilization of chromosomal genes by a resident F^+ plasmid. The process, presumably due to transient integration, depends on the presence of Tn*1000* both on the plasmid and in the chromosome and is RecA dependent. Hfr strains in which Tn*1000* is located at the F-chromosome junctions are unstable. This is presumably because the F-chromosome "cointegrate" is subject to resolution promoted by the Tn*1000* resolvase (Sherratt, this volume). Homologous recombination between transposable elements located on F and at defined sites in the bacterial chromosome (e.g., reference 64) has proved to be a useful method for generating Hfr strains with defined points of origin.

The presence of two directly repeated copies of IS*1* flanking the drug resistance genes of plasmids such as R100.1 (156, 286) can provide an important adaptive function. The resistance genes of this type of plasmid (the r-determinant or *r-det*) undergo spontaneous amplification by tandem duplication in certain species, with a concomitant increase in the level of antibiotic resistance. Such behavior is observed in *Proteus mirabilis* when plasmid-carrying cells are grown in the presence of high levels of chloramphenicol (143, 309); not only do tandem multimers occur, but circular *r-det* molecules can be observed. Other species, however, show quite different behavior. In *S. typhimurium*, a rapid loss of resistance occurs (361), presumably because high levels of recombination between the flanking copies of IS*1* lead to a rapid loss of the interstitial resistance genes. In *E. coli*, the plasmid is stable, but the *r-det* does not undergo amplification. Amplification can be observed, however, using the IS*1*-based transposon Tn*9* (2.6 kb), which is structurally similar to the *r-det* (which is approximately 20 kb) but is much smaller (56, 235). Amplification in this case is RecA dependent and probably occurs by unequal crossing over between the flanking IS*1* elements during replication of the host plasmid (55, 56). It is perhaps worth noting that plasmid-coded functions can also influence this type of recombination. A derivative of R100.1 has been isolated that spontaneously gives rise to circular forms of the *r-det* in *E. coli*. This is probably not due to autonomous replication by the *r-det* by a RecA-dependent pathway (see reference 61 for a discussion of this effect). The presence of a second, related plasmid in the same cell can inhibit *r-det* circle formation. Amplification of DNA flanked by IS*1* elements by tandem duplication in a ColE1-based plasmid has also been observed (58).

As has been shown in the case of Tn*5* (10) and Tn*10* (198), homologous recombination between two insertion sequences in inverted orientation can also give rise to inversion. This type of mechanism has been implicated in chromosomal inversions by IS*3* (312) and IS*5* (215) and has been exploited in generating directed chromosome inversions (106, 318).

Thus, homologous recombination can give rise to many of the products observed following transposition. Among the fully sequenced ISs, IS*186* alone carries a Chi sequence, which is the activation site for *recBCD* endonuclease (338). Although this Chi site has not been shown to be functional, its presence on IS*186* gives this IS an additional capacity in recombination as a portable enhancer of recombination.

C. Effects of IS Elements on Transcription

1. Polarity

In addition to the disruption of gene function, all insertion sequences examined exert strong polar effects on the expression of downstream genes (186, 229, 328). The degree of polarity is generally much higher than that observed with nonsense mutations. The simplest explanation for this behavior is that the inserted elements carry internal transcription terminators. Alternatively, if transcription through the element is not accompanied by concurrent translation, transcription termination may occur at Rho-dependent transcription terminators in the downstream flanking DNA (see reference 278). Both types of mechanism seem likely to occur, in different circumstances.

Early studies demonstrated that IS*1* exerts a polar effect when inserted into a number of operons in either orientation (102, 149, 150), but that the polarity can be partially suppressed by mutations in the *rho* gene or by placing transcription under the control of a lambda p_L promoter in the presence of the lambda antitermination protein N (24, 80, 294). It has been shown more recently that transcription from the *E. coli* rRNA promoter (*rrnX*) is also resistant to IS*1*- (and Tn*9*-) induced termination (42, 333). Although the absence of a detectable Rho-sensitive site in vitro (85) has been taken to indicate that IS*1* exerts its polar effect by introducing nonsense mutations (80), it now seems clear that a transcription terminator responsible for polarity is located in a region between two IS*1* genes (158, 280, 282; S. Hollingshead, personal communication) and that (partially) Rho-dependent termination can occur in either orientation.

Polarity effects, partially suppressible by mutations in *rho*, have also been noted for IS2 (in orientation I [25, 80]) and IS*4* (25). IS2 has been shown to contain a site which is sensitive to Rho in vitro (85). While the polarity effects of IS*1* and IS2 can be suppressed by the phage lambda protein N system (24, 37, 80, 294), this does not appear to be the case for IS*4* (97, 114).

Transcription termination sites have been detected on several other insertion sequences, including IS*5* (204) and IS*30* (78). The role of these sites in polarity (33, 73) remains to be determined, but it might be expected that transcription entering the element from flanking DNA is modulated by such signals.

As described below, certain insertion sequences are able to initiate transcription directed outward into flanking DNA. An absence of accompanying translation would, however, be expected to result in termination at a proximal Rho-dependent termination site, resulting in polarity. Such a situation has been observed in the case of IS*10* (65) inserted into the *hisG* gene of *S. typhimurium*, in which transcription from a promoter, pOUT, located at one end of an IS*10* (Tn*10*) element, terminates at a Rho-dependent site upstream of the next gene in the operon, *hisD*. This polarity can be suppressed by mutation in *rho*, resulting in IS*10*-driven transcription of *hisD*.

2. Transcription mediated by IS elements

Activation of downstream genes has been observed for several insertion elements, including IS*1*, IS2, IS*3*, IS*4*, IS*5*, and IS*10*, and can be due either to the presence of outwardly directed promoters within the element or to the formation of new promoters on insertion. The earliest example of this type of behavior was noted for IS2 insertions in the *gal* operon (313). An F′$_8$ *gal*OP::IS2 plasmid carrying IS2 in one orientation (defined as I) was subjected to selection for the activation of the *gal* operon. It was concluded from the structure of the resulting Gal$^+$ plasmid that fusion of the *gal* operon to an IS2 resident in F in the opposite orientation (defined as II) was responsible for activation, and that IS2 carries a strong promoter able to direct transcription out of the *gal*-proximal end of the element (311). However, in light of sequence rearrangements that have been characterized in IS2 (311), it is probable that IS2 sequence rearrangements were selected and that these are responsible for *gal* activation (148; see also reference 165). Rearrangements involving the deletion of internal transcription terminators would be expected to permit the continuation of transcripts initiated either upstream from the inserted element or from internal IS2 promoters.

Subsequent investigations have provided many examples of the activation of gene expression by the spontaneous insertion of IS2 (12, 32, 36, 41, 44, 120, 178, 248, 277, 360). Studies on IS2 activation of the divergent *argECBH* operon (120) indicated that IS2 in orientation II was not sufficient for gene activation, implying that IS2 does not carry an active outward promoter. It was suggested that a partial promoter sequence may be present within the end of IS2 and that for gene activation, this sequence must be correctly placed with respect to resident signals (120, 147, 148). This hypothesis was confirmed in a study of IS2 activation of the *E. coli ampC* gene (178), where it was shown that IS2 can provide a −35 promoter region which, when correctly spaced from the resident *ampC* −10 region, directs initiation from the normal start point of transcription. In the absence of contradictory data, it is reasonable to assume that activation by the spontaneous insertion of IS2 in the other cases cited above also occurs by this mechanism.

Similar results have been obtained in studies with IS*1*. Insertion in either orientation into the promoter region of the pBR322 beta-lactamase (*bla*) gene can result in IS*1*-directed transcription using a promoter composed of the resident *bla* −10 region together with a −35 sequence located in each end of IS*1* (282). Runoff transcription experiments showed that IS*1*-directed transcription initiates at the normal start point of *bla* transcription.

Recent experiments with IS*21* have indicated the presence of an outward-facing −35 region located close to one end. The tandem duplication of IS*21*,

5′ 20 10 1 3′	Element
CTATCAATAAGTTGGAGTCATTACC	IS1 (R)
AAATCAGTAAGTTGGCAGCATCACC	IS1 (L)
GTCTGGAAATATAGGGGCAAATCCA	IS2 (L)
TCTGGAGATTCAGGGGGCCAGTCTA	IS2 (R)
ACTGATCCTTAACTGATCGGCATTA	IS4 (L)
TCAGTTGCTTAACTGACTGGCATTA	IS4 (R)
TATCAGGGACTTGTTCGCACCTTCC	IS5 (L)
ATTTCCCCGCTTATTCGCACCTTCC	IS5 (R)
ACCACCGACTATTTGCAACAGTGCC	IS6 (R)
TCATCGCTAACTTTGCAACAGTGCC	IS6 (L)
CCGTCTTATATCACTGGCGCTGACA	IS21 (L)
ACGGTCAAACATCGTGGCGTTGACA	IS21 (R)
GTTGCATTGACAGATTGAATCTACA	IS30 (R)
GTTGCGTTGACCAATTGAATCTACA	IS30 (L)
TGTTGGCTATTTTTGCAGCGGCTCC	IS52 (L)
AGATCTGGATTCCACGGGTTTCGCA	IS66 (R)
AAGACGGCCTTCACCGTGGGCTTAC	IS66 (L)
TTCGTACGGTTGGACCTCAAACCCC	IS101 (L)
TTCGTTCCATTGGCCCTCAGACCCC	IS101 (R)
TGTCCGGTGAAATGGGGTGCACTCC	IS136 (R)
CTGTCCGGTCGAAATGGGGGGCAGT	IS136 (L)
ACCACCGACTATTTGCAACAGTGCC	IS140 (R)
CAACCCTTAAGTTAGCGCTTATGGG	IS186 (R)
CAACCCTTAAGTTAGCGCTTATGGG	IS186 (L)
TGTCCGCTATTGCCAGTACACCTCA	IS600 (R)
CTACTGCCAGAGTGACAAAACAGGG	IS701 (L)
CTCTGCCTAGAGTGACAAAAGAGGG	IS701 (R)
GAAATCTGATTTATTCAACAAAGCC	IS903 (L)
AAAGTTCGATTTATTCAACAAAGCC	IS903 (R)
TGGCCAAATTCAGTGTGCCACTTCA	ISSH01 (R)
TGGCCAGTTTTTGTTGACCACTTCA	ISSH01 (L)
TTTAAGACTTTATTGTCCGCCCACA	Tn7 (L)
TTCCCAACTATTTTGTCCGCCCACA	Tn7 (R)
TTCGTTCCATTGGCCCTCAAACCCC	Tn1000 (L&R)

Figure 2. Occurrences of −35 regions in the ends of ISs. Outward-facing −35 regions that are positioned sufficiently close to the ends of ISs to be able to form hybrid promoters on insertion are listed here. This list is extensive, but not necessarily complete, as we have not collected all known IS end sequences. We have included all ISs whose sequences are completely known, as well as IS*140*, which is known to form hybrid promoters (see text). The criterion used for identifying a −35 is that it matches the consensus, TTGACA, in at least four positions (we have not counted a match at the final A), or that it is known to form hybrid promoters (IS*1*, for example).

IS*21.2*, which leads to increased transposition (Section III.B), appears to generate a functional promoter composed of the outward-facing −35 region of one IS*21* and a −10 region located in the end of the adjacent element. It has been proposed that this hybrid promoter directs transcription across the second element and drives the expression of transposition functions (Reimann et al., in press).

Potential outward-facing −35 regions can be seen within the ends of many insertion sequences (Fig. 2). Indeed, it is possible that IS*4* (25, 114, 232), IS*5* (10, 187, 366, 371), IS6, IS*140* (4, 40), and IS*30* (77) are able to direct the transcription of neighboring sequences by generating hybrid promoters. However, since direct evidence concerning the position of transcription initiation is not always available, it remains possible that in several of these cases transcription is initiated from promoters located within the element. This phenomenon has been observed with the transposons Tn*3* (98, 145, 146) and Tn*1000* (P. Linder and G. Churchward, personal communication), where the transcription of the transposase gene can continue into flanking sequences. Observations on the expression of antibiotic resistance genes in naturally occurring transposons have led Piepersberg and collaborators (4, 40) to postulate the existence of IS-created hybrid promoters created by IS*140* and other elements. It seems likely that the formation of hybrid promoters is a relatively common occurrence associated with transposition. In Fig. 2, we show a set of the IS ends and possible −35 regions which might be involved in hybrid promoter formation. We have examined all the sequenced ISs and tabulated here all the possible outward-facing −35 regions with a reasonable match to the consensus or those that have been observed to form hybrid promoters (Fig. 2).

At least two insertion elements carry outward-directed promoters close to one end. Both IS*3* (120) and IS*10* (34, 65) can promote the transcription of flanking genes. IS*150*, a relative of IS*3*, also carries such a promoter (322, 322a). In the case of IS*3*, transcription has been shown to initiate 44 nucleotides within one end (63). For IS*10*, transcription is driven by a promoter (pOUT) that is involved in the control of IS*10* transposition activity and is located at one end of the element (210, 334; Kleckner, this volume).

Recently, several insertion elements able to increase the expression of a plasmid-associated beta-lactamase gene have been isolated in *Pseudomonas cepacia* (323). No information is available concerning the mechanism of this activation.

Another case of IS-directed transcription within a compound transposon occurs in Tn*5*. Transcription of the antibiotic resistance genes internal to the transposon is driven by a promoter generated by a single-base-pair change (relative to IS*50*R) in the end of IS*50*L, which also inactivates the transposition functions of IS*50*L (307, 308).

3. Other effects

In addition to activation resulting from functional or partial promoter sequences furnished by IS elements, IS-induced gene activation can clearly occur by other mechanisms. One example, still poorly understood, involves the *bgl* operon of *E. coli*, which is normally cryptic but can be activated by insertion of either IS*1* or IS*5* (295, 296). Activation does not involve transcription driven from internal or hybrid promoters furnished by these elements. Activating insertions in both orientations have been observed to occur over a 47-bp region (between −77 and −124 bp). Since transcription initiates at the same site in these cases, proximal to the beginning of the first gene, stimulation appears to occur at a distance (295). One possible explanation for this "enhancer-like" phenomenon is that the presence of the element enhances binding of the cyclic AMP-binding protein (CAP), which is also involved in *bgl* expression. Indeed, *bgl* activation can also result from point mutations that increase CAP binding (295). Since mutations in either gyrase subunit are known to activate the *bgl* operon (93, 285), conformational changes induced directly by the insertion or indirectly, by proteins which bind to the element, are implicated. The effect of insertion sequences on local superhelicity and DNA conformation has not been extensively investigated.

IV. STRUCTURAL AND FUNCTIONAL PROPERTIES OF INSERTION SEQUENCES

A. Structural Aspects: Compound Transposons

Many of the IS elements listed in Table 1 are found as components of compound or composite transposons, in which they flank otherwise nontransposable DNA segments and render the entire structure transposable. The central parts of naturally occurring compound transposons include a wide variety of genes, particularly those encoding antibiotic resistance or catabolic functions. Several ISs have only been observed to occur alone (as part of the bacterial chromosome or plasmids), others are found both alone and in naturally occurring transposons, while others, despite subsequent demonstrations that they can transpose autonomously, have only been detected as part of compound transposons (Section II).

This disparity in distribution underscores the important question of how IS elements propagate through, and are maintained in, bacterial populations. In the case of IS*1*, the wide variation in copy number in closely related bacterial species suggests a large random component in its propagation. On the other hand, association with selectable genes in the form of compound transposons probably constitutes an important selective component. In this section, we discuss ways in which compound transposons may be generated and the properties of ISs which affect the activity and stability of such structures.

To form a compound transposon, one copy of an IS must be inserted on either side of the interstitial gene. In general, the flanking IS elements can occur in either relative orientation. This has been shown in the case of natural isolates and laboratory constructs of transposons based on IS*1* (5, 72, 340), for IS*50* (18, 151), for IS*10* (253), and for IS*903* (258, 346). Moreover, a pair of IS elements on a circular plasmid DNA molecule can be considered to flank two interstitial DNA segments and thus constitute two alternative compound transposons. Use of the "alternative" compound transposon (called inverse transposition [56]) was first noted in the case of integration (transposition) of the plasmid R100.1 into the *E. coli* chromosome, mediated by the IS*10* elements flanking a resident Tn*10* transposon and accompanied by the loss of the interstitial Tn*10* tetracycline resistance gene. Other examples of inverse transposition have been reported for Tn*10* (105, 137, 138), Tn*5* (315), and Tn9 (292, 304) and serve to underscore the flexibility of compound transposons.

Several factors determine the probability of formation of new compound transposons and the probability of their subsequent persistence in the population. These factors will be considered using the IS*1* element as a primary example: (i) the specificity of insertion (which will determine the frequency with which a given gene can acquire a nearby IS); (ii) the probability of insertion of a second, adjacent IS, either by inter- or intramolecular events (also expected to be influenced by target specificity); (iii) the coherence and stability of the transposon: activity of the resulting compound structure in transposition compared to the activity of the constituent IS modules.

1. Formation of compound transposons

Although the specificity of insertion is not well understood (see Section V), IS*1* exhibits a pronounced preference for A+T-rich regions and insertion generally occurs over a region of the order of 100 to 200 bp rather than being confined to a precise site. Particularly relevant to our present discussion is the fact that promoters and the neighboring control regions of bacterial genes are often A+T rich (see, e.g., reference 144) and in several instances are known to attract IS*1* insertions (Section V). Transcription terminator regions may also share this A+T richness (see reference 278), and in at least one instance a terminator is known to attract IS*1* insertion. Thus, an initial insertion of IS*1* may favor 5′ or 3′ flanking regions of a particular gene or operon, thereby favoring the mobilization of an entire transcription unit. This is a prerequisite for the formation of a compound transposon carrying a functional genetic determinant. (Note that this may be why IS*1* is so prevalent in compound transposons.) While this is an attractive idea, no systematic study that directly addresses the question of the attractiveness of promoters and terminators for ISs has been undertaken.

The probability of acquisition of a second, flanking IS*1* depends on the specificity of insertion, but may also be influenced by the presence of the initial copy. Studies using a marked IS*1* (distinguishable by the absence of a *Pst*I restriction site) in the *lac* operon (69) have shown that 9 out of 10 mutations which were not due to IS*1*-mediated deletions were due to secondary insertions of the resident IS*1*. Simple insertions, in either orientation, and insertion accompanied by inversion of the intervening DNA (see Section III) occurred in roughly equal numbers. Since the frequency of insertion mutations relative to the overall mutation frequency in the *lac* operon is low in the absence of a neighboring IS*1* (51, 101; see Section II), this observation suggests that the proximity of one copy of IS*1* strongly enhances the probability of a second, nearby insertion. The implications for the formation of compound transposons are clear. They suggest that IS*1*-based compound transposons may form at relatively high frequencies.

The type of mechanism described above is certainly not the only way in which compound transposons can form. An alternate pathway is one in which a single IS element is resident on a plasmid carrying the gene to be incorporated into a compound transposon. If the IS on the plasmid undergoes homologous recombination with a second copy located elsewhere or forms an IS-mediated cointegrate, the entire plasmid will be flanked in the resulting structure by directly repeated copies of the IS. Subsequent deletion could then reduce the length of the internal segment, which may tend to increase the transposition frequency of the entire compound transposon (see Section IV.C for a discussion of the effect of size on transposition frequency). Dimerization of a plasmid carrying a single IS would generate a similar structure (e.g., reference 29). Arber and co-workers (6) have investigated several pathways for such compound transposon formation, particularly for IS*1*.

2. Coherence of compound transposons

We use the term "coherence" here to describe the tendency of a compound transposon to transpose as a unit, as opposed to the independent transposition of the component ISs. The activities of compound transposons, which may be important determinants of their natural occurrence, can be affected by the properties of the sequences surrounding the transposon, as well as the properties of the internal DNA segment. As described elsewhere in this review (Section IV.C), several groups have found that transcription impinging on IS*1* from outside has a strong inhibitory effect on its transposition activity, particularly that involving the proximal end (1, 28, 59, 60, 223, 225). Since we have argued from specificity considerations that a transcription unit is likely to be in the internal DNA segment (the compound transposon is otherwise unlikely to carry useful genetic information), transcription from the internal DNA into one of the IS*1* elements could occur. This would reduce the activity of one of the internal ends (those joined to the internal DNA segment) of the flanking ISs, thereby reducing the independent activity of that IS and increasing the coherence of the compound transposon. It would also be expected to reduce the likelihood of IS-mediated deletions into the internal segment.

Reduction in the activity of the ends of ISs can also occur by mutation of one or both of the ISs and is found in at least two well-studied natural transposons, Tn*5* (171, 307, 308, 315) and Tn*10* (245; Berg, this volume; Kleckner, this volume). In these cases, the ends internal to the transposon have been deactivated by mutations and have also acquired sites for the Dam methylase that reduce the activities of the ends when methylated (Section IV.D). In both these cases, this "asymmetry" is reinforced by the presence of binding sites for host proteins (DnaA in the case of IS*50* and integration host factor [IHF] in the case of IS*10*; Section IV.D) close to the most active end. Such an arrangement would tend to favor transposition of the compound transposon (using the more active outside ends) over solo transposition or "inverse" transposition (using the less active inside ends). In the case of IS*903* (IS*602*), one factor which contributes to this asymmetry appears to be preferential recognition of one end by the element-specified transposase (88; S. Stibitz and W. S. Reznikoff, submitted for publication). An interesting case is that of the transposon Tn*602*, which carries directly repeated IS*602* elements (similar to IS*903*). Inactivation of each of the ISs has shown that transposition of this element is directional. One flanking IS*602* has a tendency to promote the transposition of Tn*602*, while the other promotes "inverse" transposition (Stibitz and Reznikoff, submitted). The most coherent compound structures are therefore expected to be those in which the most active ends of the IS elements occur at the outer extremities of the compound transposon.

The effect of outward-directed transcription from internal promoters of an IS (Section III.C) may also enhance the coherence of the transposon due to the effect that this outward-directed transcription may have on the suppression of the activity of the end. We have discussed the transcriptional effects of ISs in Section III.C and will not reiterate that discussion. However, it is worth noting that IS-induced transcription may have major effects on the selection of compound transposons. Several ISs, including IS*1*, are capable of forming hybrid promoters by insertion (see Section III.C). This possibility may well play a role in the maintenance of compound transposons by making the expression of genetic information in the internal DNA segment dependent on the presence of an IS element located upstream. In Tn*5*, discussed above, the effect of an inactive IS with an internal promoter combines to form a particularly interdependent compound transposon.

Another factor that may influence transposon stability is homologous recombination between the flanking (directly repeated) ISs leading to the deletion of the internal DNA segment. Thus, the large IS*1*-flanked *r-det* carried by the plasmid R100.1 undergoes frequent deletion in *S. typhimurium* by homologous recombination between the directly repeated IS*1* elements (Section III.B), although it is relatively stable in *E. coli*. Some mutant derivatives of R100.1 exhibit a similar instability in *E. coli* (61). Similarly, derivatives of Tn*5* with directly repeated copies of IS*50* (151) and transposons flanked by directly repeated copies of IS6 (IS26 [166]) have also been shown to undergo *recA*-dependent deletion.

3. Pseudo-compound transposons

Most of the examples in the foregoing discussion concern elements whose transposition does not necessarily involve replicon fusion, that is, elements that can transpose directly. For at least one element (IS6) isolated as a component of several compound transposons, however, the exclusive end products of transposition appear to be cointegrates (Section III.B). This observation has interesting implications for the structure of IS6-based compound transposons. Cointegrate formation involves one or the other of the flanking IS6 elements of the IS6-based compound transposon (IS*160* [256]) generating a structure carrying the entire transposon at one junction and a new, directly repeated copy of the IS6 element used to generate the cointegrate at the other. Subsequent transfer of the entire transposon to the recipient molecule must involve recombination between the IS6 copy and the newly generated copy of IS6 carried by the transposon (recombination with the initial copy would regenerate the donor molecule and leave a single copy of IS6 in the recipient molecule). This

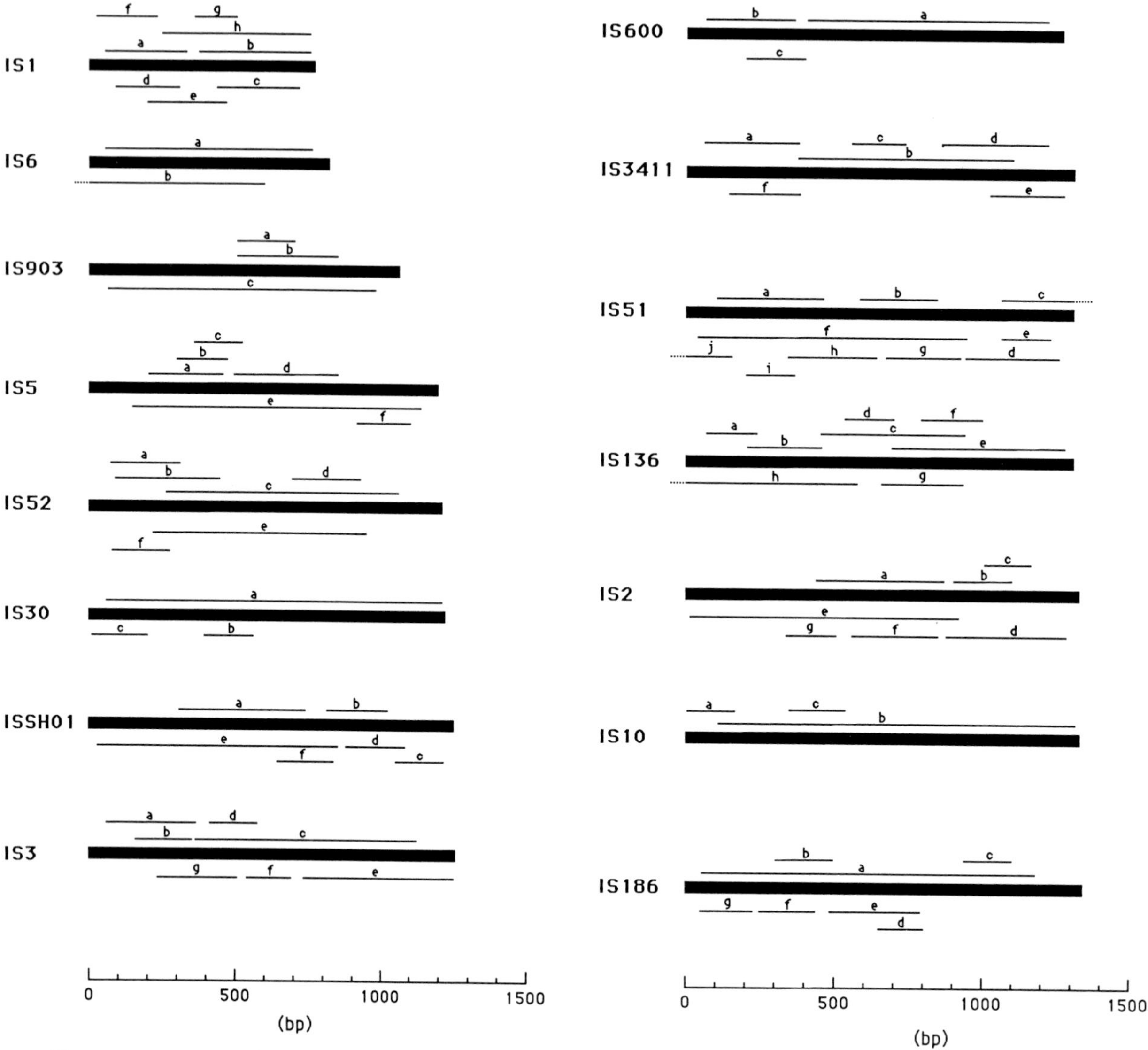

Figure 3. Open reading frames of fully sequenced ISs. All fully sequenced ISs we were aware of (we have not included Tn*3* family elements) were searched for frames with 50 or more codons between an AUG or GUG and a stop. They are indicated on the ISs, listed in order of size, so that those reading from left to right are shown on top, and those reading from right to left on the bottom. The letter designations are completely arbitrary with the exception of IS*1*, for which the lettering corresponds to that used in all published references. A length scale in base pairs is shown at the bottom of the figure.

step is *recA* dependent (43, 356), implying homologous recombination between directly repeated copies of the element. Directly repeated flanking copies of IS*6* must therefore be essential for the "transposition" of IS*6*-based compound transposons. This seems to be the case since, while structures flanked by inverted IS*6* elements are capable of cointegrate formation, no transposition of these structures can be detected (43, 256). From a functional point of view, therefore, this type of transposon does not fall within the strict definition of compound transposons we have used above.

B. Coding Capacities and IS-Encoded Proteins

The internal open reading frames (ORFs) for many IS elements have been derived from their known nucleotide sequences. Several ISs contain one large ORF which extends almost the entire length of the element and one or more short overlapping ORFs on the opposite strand. IS*10*, IS*50*, IS*903*, and Tn*1000* apparently each encodes a single transposase protein (a protein essential for transposition) in the long ORF. Many elements, however, have a set of smaller ORFs, of various lengths and positions, mak-

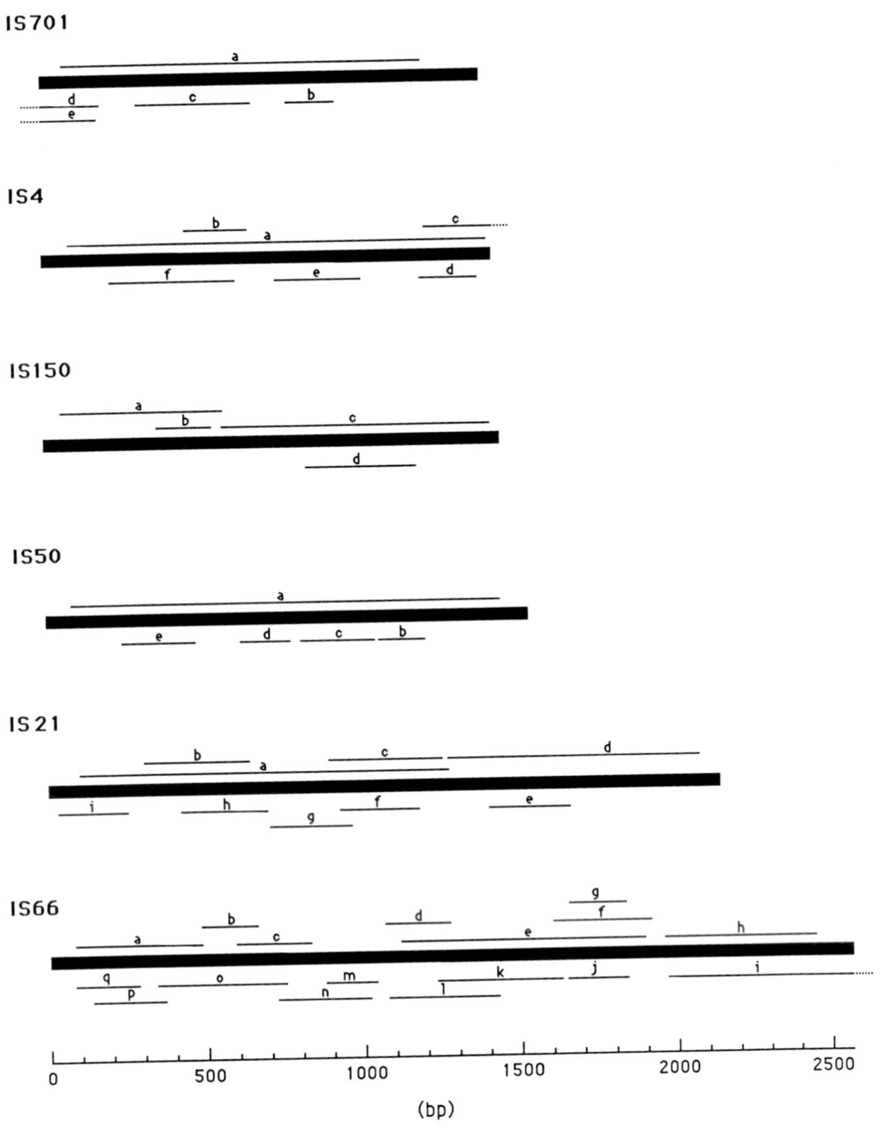

Figure 3. *Continued.*

ing it difficult to ascertain from sequence alone which actually code for proteins.

In Fig. 3, we indicate the ORFs (beginning with either AUG or GUG and having a length of at least 50 codons) for fully sequenced ISs from gram-negative bacteria (excluding Tn*1000* and Tn*3*). Their exact positions are given in Table 4. Only in a few cases has it been demonstrated which ORFs are required for transposition. These include IS*10*, IS*50*, IS*903*, Tn*1000* (gamma delta; and Tn*3*), and IS*1*. We will not discuss here the proteins encoded by IS*10*, IS*50*, and Tn*1000* (Tn*3*), since they are treated in the chapters in this volume by Kleckner, Berg, and Sherratt. There are a large number of small ORFs indicated in the figure. The frequency versus size distribution of the smallest of these is consistent with that expected for random occurrence in DNA, but for longer ORFs (in the range 90 to 140 codons, for example), it is significantly higher than expected. This suggests that several of the ORFs in this size range are probably real. It should be noted here perhaps that the demonstration that a given ORF is essential for transposition does not necessarily mean that it defines its own protein. It is possible that in some cases, where one ORF overlaps or closely follows a second on the same strand, some level of control is exercised by low-frequency frameshifting between the first and the second to generate a fusion protein.

IS*1* is different in several respects from most of the other elements. It does not have a single large ORF, but it has a surprisingly large number of ORFs (eight) for its length, as shown in Fig. 3. Evidence that the two ORFs, called *insA* and *insB*, which together

Table 4. Insertion sequence ORFs[a]

IS	Direct[b]			Complement[b]		
	Coordinates	Reading frame	Length (bp)	Coordinates	Reading frame	Length (bp)
IS*101*						
IS*1*	(a) 27–236	(3)	210	(f) 719–432	(3)	288
	(b) 56–328	(2)	273	(g) 468–202	(1)	267
	(c) 250–750	(1)	501	(h) 304–92	(2)	213
	(d) 353–508	(2)	156			
	(e) 376–750	(1)	375			
IS*6*	(a) 52–765	(1)	714	(b) 591–1	(1)	591+
IS*903*	(a) 499–702	(1)	204	(c) 980–60	(3)	921
	(b) 507–848	(3)	342			
IS*102*	(a) 469–702	(1)	234	(c) 980–60	(3)	921
	(b) 507–848	(3)	342			
IS*5*	(a) 205–459	(1)	255	(e) 1127–150	(3)	978
	(b) 297–473	(3)	177	(f) 1099–914	(2)	186
	(c) 362–517	(2)	156			
	(d) 495–848	(3)	354			
IS*52*	(a) 71–307	(2)	237	(e) 953–216	(3)	738
	(b) 88–438	(1)	351	(f) 274–83	(2)	192
	(c) 264–1061	(3)	798			
	(d) 694–930	(1)	237			
IS*30*	(a) 63–1211	(3)	1,149	(b) 557–391	(1)	168
				(c) 202–11	(2)	192
ISSHO*1*	(a) 307–732	(1)	426	(c) 1213–1064	(2)	150
	(b) 817–1026	(1)	210	(d) 1085–879	(3)	207
				(e) 849–34	(1)	816
				(f) 833–636	(3)	198
IS*3*	(a) 57–362	(3)	306	(e) 1240–968	(2)	273
	(b) 157–357	(1)	201	(f) 691–542	(2)	150
	(c) 362–1225	(2)	864	(g) 505–233	(2)	273
	(d) 415–573	(1)	159			
IS*600*	(a) 65–364	(2)	300	(c) 395–201	(3)	195
	(b) 403–1218	(1)	816			
IS*3411*	(a) 55–378	(1)	324	(e) 1273–1028	(2)	246
	(b) 378–1097	(3)	720	(f) 387–139	(1)	249
	(c) 556–735	(1)	180			
	(d) 863–1222	(2)	360			
IS*51*	(a) 105–461	(3)	357	(d) 1261–941	(2)	321
	(b) 583–840	(1)	258	(e) 1227–1072	(1)	156
	(c) 1065–1311	(3)	247+	(f) 941–42	(3)	900
				(g) 925–674	(2)	252
				(h) 648–343	(1)	306
				(i) 367–209	(2)	159
				(j) 153–1	(1)	153
IS*136*	(a) 71–247	(2)	177	(g) 933–661	(1)	273
	(b) 211–456	(1)	246	(h) 577–1	(2)	577+
	(c) 456–932	(3)	477			
	(d) 532–699	(1)	168			
	(e) 695–1279	(2)	585			
	(f) 793–999	(1)	207			

Continued on following page

Table 4.—*Continued*

IS	Direct[b]			Complement[b]		
	Coordinates	Reading frame	Length (bp)	Coordinates	Reading frame	Length (bp)
IS2R	(a) 441–869	(3)	429	(d) 1285–878	(2)	408
	(b) 905–1102	(2)	198	(e) 917–15	(1)	903
	(c) 1011–1172	(3)	162	(f) 844–554	(2)	291
				(g) 505–332	(2)	174
IS2(Gal)	(a) 441–869	(3)	429	(d) 966–811	(1)	156
	(b) 826–999	(1)	174	(e) 959–15	(3)	945
	(c) 905–1327	(2)	423 +	(f) 844–554	(2)	291
				(g) 505–332	(2)	174
IS*10*	(a) 4–165	(1)	162			
	(b) 108–1313	(3)	1,206			
	(c) 349–534	(1)	186			
IS*186*	(a) 52–1182	(1)	1,131	(d) 797–642	(3)	156
	(b) 305–496	(2)	192	(e) 790–485	(2)	306
	(c) 938–1099	(2)	162	(f) 434–252	(3)	183
				(g) 226–50	(2)	177
IS*701*	(a) 64–1203	(1)	1,140	(b) 929–777	(3)	153
				(c) 665–405	(3)	261
				(d) 188–1	(3)	188 +
				(e) 177–1	(1)	177 +
IS*4*	(a) 85–1410	(1)	1,326	(d) 1382–1197	(3)	186
	(b) 456–653	(3)	198	(e) 1011–739	(2)	273
	(c) 1215–1426	(3)	212 +	(f) 609–217	(2)	393
IS*150*	(a) 48–566	(3)	519	(d) 1180–827	(2)	354
	(b) 358–528	(1)	171			
	(c) 566–1414	(1)	849			
IS*50*	(a) 80–1441	(2)	1,362	(b) 1203–1048	(1)	156
				(c) 1044–805	(1)	240
				(d) 772–614	(2)	159
				(e) 472–242	(2)	231
IS*21*	(a) 102–1271	(3)	1,170	(e) 1651–1400	(2)	252
	(b) 307–633	(1)	327	(f) 1173–922	(1)	252
	(c) 889–1248	(1)	360	(g) 962–600	(3)	363
	(d) 1271–2068	(2)	798	(h) 587–417	(3)	171
				(i) 251–33	(3)	219
IS66	(a) 79–483	(1)	405	(j) 1827–1633	(1)	195
	(b) 483–656	(3)	174	(k) 1626–1228	(1)	399
	(c) 589–825	(1)	237	(l) 1424–1077	(3)	348
	(d) 1067–1267	(2)	201	(m) 1036–875	(1)	162
	(e) 1117–1890	(1)	774	(n) 1017–721	(2)	297
	(f) 1599–1910	(3)	312	(o) 746–339	(3)	408
	(g) 1643–1828	(2)	186	(p) 363–136	(2)	228
	(h) 1955–2443	(2)	489	(q) 281–78	(3)	204
	(i) 1971–2578	(3)	608 +			
Tn*3*	79–240	(1)	162	4680–4417	(1)	264
	331–504	(1)	174	4407–4240	(1)	168
	504–761	(3)	258	3436–3266	(2)	171
	646–1137	(1)	492	3081–37	(1)	3,045
	1196–1459	(2)	264	2599–2441	(2)	159
	2387–2707	(2)	321	1058–831	(3)	228

Continued on following page

Table 4.—*Continued*

IS	Direct[b]			Complement[b]		
	Coordinates	Reading frame	Length (bp)	Coordinates	Reading frame	Length (bp)
	2409–2576	(3)	168			
	2713–3063	(1)	351			
	3140–3298	(2)	159			
	3210–3764	(3)	555			
	3950–4867	(2)	858			
Tn*1000*	1193–1408	(2)	216	5847–5695	(1)	153
	1515–1718	(3)	204	5216–5010	(3)	207
	1726–2043	(1)	318	5100–4951	(1)	150
	3134–3754	(2)	621	3906–3757	(1)	150
	3773–5896	(2)	2,124	3692–3534	(3)	159
				3531–3262	(1)	270
				3042–37	(1)	3,006
				1928–1770	(3)	159
				1766–1380	(3)	387

[a] ISs are listed by size. The coordinates of the open reading frames are listed. The frames are defined by: start with either an AUG or GUG codon, end with any stop codon, and extend for at least 50 codons. The number in parentheses indicates the reading frame relative to the first base pair of the IS (1 indicates in-frame with first base pair). The number following the parentheses is the length in base pairs of the ORF. A plus sign indicates that the ORF does not terminate within the IS. The ORFs of Tn*3* and gamma delta (Tn*1000*) are given for comparison, although these are not shown in Fig. 3.

[b] Direct, Those frames oriented from left to right in the IS as represented in Fig. 3 (the frames indicated on top); Complement, the other direction (indicated on the bottom in Fig. 3).

span most of the length of IS*1*, are required for transposition was reported by Ohtsubo and coworkers (224, 266). Using site-directed mutagenesis to insert amber mutations separately into each of the ORFs in such a way that the changes were neutral in each of the overlapping ORFs (unaltered with respect to amino acid coding), it has now been demonstrated that *insA* and *insB* are the only ORFs encoding proteins essential for IS*1* transposition (176). Two additional reading frames, InsE (176) and InsC (37a, 37b), may also be involved, but are not essential. It is known that there is a relatively weak inward-directed promoter at each end of IS*1* (54, 111, 224). One of these (left) presumably directs transcription of *insA* and *insB*.

Even fewer of the ORFs shown in Fig. 3 have been shown to specify detectable levels of the corresponding protein. All three of the largest ORFs of IS*5* (Fig. 3) have been shown to encode proteins (220, 287, 288). The arrangement of the promoters internal to IS*5* (322) further suggests that the proteins are important components for IS*5* function. Unfortunately, however, no genetic or biochemical evidence clearly attributes a function to any of these genes.

Studies on IS*21* have detected two proteins of 43 and 29 kilodaltons (kDa) (Reimann et al., in press), but, as in the case of IS*5*, it is not clear whether either or both of these proteins are essential. Similarly, a 42-kDa protein product corresponding to the long, open reading frame (presumably the transposase) of IS*30* has been detected (H. Stalder and W. Arber, personal communication).

Proteins coded by both the ORFs *insA* and *insB* of IS*1* have been overproduced in the cell. When a wild-type IS*1* is placed downstream from a strong promoter, a 9.8-kDa protein that corresponds to InsA can be detected in minicells or maxicells (60, 377). No InsB protein, however, could be detected under these conditions (see Section IV.C). Engineered expression constructs can be used to overproduce both InsA and InsB (13-kDa) proteins separately, and the resulting extracts can be used to test for specific DNA-binding activity. Using gel retardation techniques, we have shown that the ends of the element are specifically recognized and bound by the InsA protein (377). There is no evidence, beyond the demonstration that it is required in transposition, that suggests the role played by InsB protein. Although InsA has a sequence-specific binding activity by itself, it may have additional activities when complemented by InsB. It is possible that other, more complicated interactions occur between InsA and InsB. For example, there is no evidence ruling out translational frameshifting, which could result in a single fusion protein from these two ORFs that acts as a transposase.

Among the ISs, there are very few cases in which it has been demonstrated that the ends are recognized and bound by an encoded transposase. The Tn*3* transposase has been shown to recognize and bind to the ends of Tn*3* (159b, 254a, 369), and Grindley and Wiater (122) have found, using footprinting techniques, that the Tn*1000* transposase, which is similar to that of Tn*3*, can recognize and bind to the 38-bp

inverted repeats at the ends of this element. We can assume from its demonstrated activity on IS*10* ends in vitro that the IS*10* transposase also binds to the ends (245).

A notable property of the characterized transposase proteins is that many are quite basic. InsA and InsB, for example, are extremely basic, with estimated pI values of 10.9 and 11.0, respectively. Among the known sequence-specific DNA-binding proteins, only IHF of *E. coli* is as basic (IHF is discussed in Section IV.D). This property of the transposases may play some role in the unusual behavior exhibited by all of those mentioned above. They act efficiently only on IS ends that are present on the DNA molecule encoding the transposase; that is, they act largely in *cis*. Complementation of a transposase-defective element is very inefficient unless the transposase gene is in *cis*, and, in several instances, close linkage is apparently preferred. This *cis* effect is large, in some cases more than 100-fold, and has been demonstrated for IS*10* (247), IS*50* (171), IS*903* (124, 126), and IS*1* (225, 281). The only plausible explanation for this effect appears to be that the proteins bind to the DNA very close to the DNA segment from which they are made, presumably because of the strong nonspecific binding affinity arising from their positive charges. The protein may tend to remain associated with the DNA and diffuse along it to specific binding sites (368).

Some information has been obtained on how the InsA protein of IS*1* interacts with the end of the element (D. Zerbib et al., manuscript in preparation). The minimum essential sequence required for transposition activity of an end of IS*1* is the final 23 bp of either end (they are almost perfect inverted repeats). This sequence, from either end, is sufficient to bind InsA specifically. By testing a set of mutant ends for their abilities to bind InsA and for their transposition activities, it was determined that there are two distinct functional domains in the ends of IS*1*: one that contains the specificity determinants for recognition by InsA, and one (at the very ends) that determines the rate of some step in the transposition process other than InsA binding (Fig. 4). Another protein, IHF, also specifically binds to the ends of IS*1* (Section IV.D) within the same InsA recognition domain. It is clear, however, that the InsA and IHF binding sites are distinct, even though they are almost entirely overlapping. Recent results on the transposition activities of mutant ends of IS*903* (88) have suggested that there may also be two functional domains in these ends. It is not known, however, whether transposase-binding specificity is involved in the same way as for IS*1*.

Since we know that InsA is essential for IS*1*

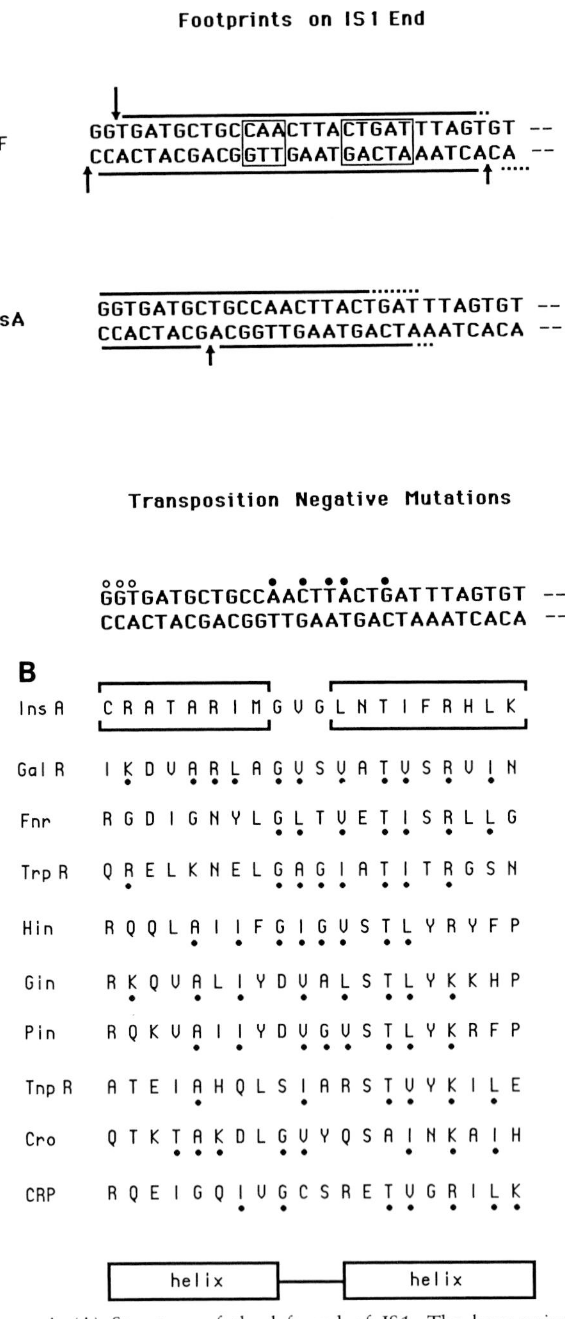

Figure 4. (A) Structure of the left end of IS*1*. The base pairs involved in transposition-negative mutations of the ends are indicated by the circles. Solid circles indicate base pairs that affect protein binding (InsA or IHF, or both). The footprints of IHF and the InsA protein are shown as solid lines for protected regions and arrows indicating bonds that show enhanced sensitivity to DNase I upon protein binding. (InsA footprint is approximate.) (B) Matches of the putative helix-turn-helix region of InsA protein with similar regions of other sequence-specific DNA-binding proteins. The matches with InsA residues (according to equivalence relations whereby K = R and I = V = A). The sources for all the sequences can be found in reference 271.

transposition, it seems likely that it is part of the protein-DNA transposition complex which forms on the ends of IS*1*. Its precise role, however, is unclear. It is possible that in addition to recognizing the ends, InsA itself carries out the cleavage and ligation reactions necessary for transposition (Section VI). Alternatively, InsA may simply provide an anchor via protein-protein interaction for the binding of InsB or possibly other proteins. The specific sequence at the very end is clearly essential for transposition, although it is probably not involved in InsA recognition. It must therefore be involved in another sequence-specific process. Possible candidates for this process are addition of other proteins to the complex, strand cleavage, or some conformational transition of the complex.

The amino acid sequence of InsA is interesting in several respects. The carboxy-terminal region contains a reasonably convincing candidate for a helix-turn-helix motif, present in several specific DNA-binding proteins (Fig. 4B), which may contain the DNA recognition determinants (60). If this is so, InsA is likely to bind DNA in the major groove in a similar way to other DNA-binding proteins like Cro, CRP, and the lambda *c*I repressor (271).

Very few data are available on the stability of the transposase proteins. Both the InsA and InsB proteins of IS*1* have been shown to be stable in vivo (unpublished results). On the other hand, J. Chernak and H. Smith (personal communication; J. Chernak, Ph.D. thesis, Johns Hopkins University, Baltimore, Md., 1987) suggest that one of the small proteins from IS*5* is highly unstable. Very recently, Rak and co-workers have found that the largest IS*5*-encoded protein is rapidly processed (by cleavage of a C-terminal peptide) into a relatively stable 18.5-kDa protein (B. Rak et al., personal communication).

The definition, isolation, and functional characterization of IS-encoded proteins is receiving much attention in various laboratories. These studies are not only a prerequisite to understanding transposition processes at a molecular level, but should provide an important contribution to our understanding of DNA-protein and protein-protein interactions in general.

C. Expression of IS Genes

Transposition at a very high frequency would certainly be detrimental to the host cell, and indeed, all known ISs exhibit relatively low frequencies of transposition. On the other hand, transposition is essential to the propagation and survival of ISs in a bacterial population (Section VII; Ajioka and Hartl, this volume). It would not be surprising, therefore, if transposition activity were regulated not only by the IS element itself, but also by host factors, perhaps those involved in signaling the physiological state of the cell. The few cases in which the control of transposition activity has been studied in some detail have provided examples of some novel regulation mechanisms.

IS*10* is subject to regulation by several systems. It encodes an outward-directed transcript which acts as an antisense RNA repressor of translation (335; Kleckner, this volume). This RNA is apparently responsible for multicopy inhibition of transposition (13, 335) (the reduction in transposition activity of a given IS*10* resulting from the presence of many copies of a distinguishable IS*10* in the same cell) by limiting the translation of the transposase mRNA. A second system makes use of the only two sites in IS*10* for adenine methylation by the host Dam methylase. Both of these sites are implicated in the regulation of transposition by the host cell. One site, in the transposase promoter, reduces the transcription of the transposase gene when methylated (335). This may have the double effect of lowering transposition activity of the end and of timing bursts of transposase synthesis with the passing of a replication fork, which produces transiently hemimethylated DNA. Another site, near the inside end of the element, sharply reduces the activity of this end when methylated (335). This has the effect of increasing the coherence of the compound transposon in situations in which methylation is efficient (see Section III.A).

IS*50* transposition (Berg, this volume) is also subject to several levels of control. Two proteins are encoded by IS*50*. The first is the transposase, and the second is the product of a later translation start within the same ORF (170, 182). IS*50* was found by Biek and Roth (26, 27) to become increasingly inhibited for transposition after it is first introduced into a cell. This negative control acts in *trans*, and some evidence indicates that the second, shorter protein is responsible (170, 182, 183, 216). S. A. McCommas and M. Syvanen (233a) have recently shown that the two different but overlapping transcripts from which the two proteins (the *cis*-acting transposase, P1, and the *trans*-acting repressor, P2) are expressed have different stabilities. When a Tn*5* is introduced into a naive cell (see also reference 181), the initial mRNA ratio of 1:2 drops within 3 h to 1:80.

As for IS*10*, regulation involving host factors also occurs at the ends of IS*50*. The outside end has a binding site for the DnaA protein, which, while not essential for transposition, stimulates the activity of this end (373). Both ends also carry Dam methylation sites (the inside end carries a cluster of these sites; Fig. 5, below). When methylated, the activity of the inside

end is reduced by about 10-fold (372a). In addition, an increase in transposase levels is observed in *dam* mutant cells, which indicates a regulatory effect on expression at the outside end. Furthermore, in *dam* mutant cells the activity of the outside ends also increases by about a factor of 10 (372a).

The several genes encoded by IS*5* have recently been investigated (287, 288; Chernak, Ph.D. thesis). Studies aimed at overproducing the two larger proteins 5A and 5B have led Chernak and Smith (personal communication; Chernak, Ph.D. thesis) to conclude that the genes are inefficiently transcribed and translated and that the proteins are relatively unstable. They also suggest the possibility that the RecA protein may participate in the degradation process, and that the mRNAs may be unstable as well.

For IS*1*, too little is known to suggest the nature of the regulatory circuits. However, several observations are relevant to the question. IS*1* is different from many of the elements mentioned above in that two ORFs are essential for transposition. No multicopy inhibition has been reported for IS*1*, and, moreover, we find that cointegrate formation occurs at almost identical frequencies in *S. typhimurium* LT2, which has no genomic copies of IS*1*, and in *E. coli* (176). (A report of significantly higher frequencies in different strains of *S. typhimurium* has appeared recently [48]. The reason for the difference is unclear.) InsA and InsB proteins are poorly expressed in vivo, as judged by minicell and maxicell assays, and in vitro, as judged by the use of a coupled transcription-translation (Zubay) system (377). Indeed, the putative promoter and ribosome-binding sites appear to be rather poor. Even so, InsA can be overexpressed to a detectable extent by placing a strong promoter upstream, external to IS*1*. In this case, however, InsB is not detectably expressed, although it appears to be contained within the same transcription unit. When the InsA gene is strongly overexpressed by replacing its natural promoter and ribosome-binding site with more active ones, expression of InsB is still undetectable (377). Since the expression of InsA can be stimulated by external, impinging transcription, and external transcription does not seem to stimulate transposition (59; unpublished observations), we suggest that the regulation of transposition probably is accomplished by regulation of the level of InsB. It is possible that the expressions of *insA* and *B* are coupled through the Rho-dependent terminator that has been found at the end of the *insA* gene (158, 282), but the regulatory circuit is not understood. As mentioned in Section IV.C, we have not ruled out at this point the possibility that the products of *insA* and *insB* are fused by translational frameshifting and that regulation occurs at this level.

Several transposable elements code for proteins that act strongly in *cis*. A particularly interesting regulatory situation arises when, as described for IS*50* above, a negative regulatory protein from this element can act in *trans*. This leads to intragenomic competition among copies of the same transposon. Its own transposase proteins act preferentially on itself, while its repressor acts on all copies in the cell (see, e.g., reference 173). In this scenario, the element is indeed "selfish." An intact element preferentially inhibits its coresident ISs. Note that IS*50* and IS*10* both act in this fashion. For IS*10*, the repressor is the antisense RNA, which is known to act in *trans*.

The insertion of an IS element within a transcription unit might be expected a priori to result in the gratuitous expression of IS-encoded genes driven by external transcription. Since the activation of transposition has never been reported, mechanisms must exist to "protect" most ISs from external activation. This might be accomplished simply if transposition were dependent on two (or more) IS proteins encoded on opposite strands of the element. As pointed out in the preceding section, many IS elements have ORFs arranged in this fashion, although in most cases their relevance to transposition remains to be demonstrated (see Reimann et al., in press).

In the case of many elements (IS*10*, IS*50*, IS*903*, Tn*1000*), only one reading frame seems to be essential, or if more than one is involved (e.g., IS*1*), these are encoded on the same strand. The presence of transcription termination signals within an IS could protect downstream genes from overexpression by external transcription. Such terminators have been detected in several ISs (for example, IS*1*, IS2, IS*3*, IS*5*, and IS*30*). External transcription into the transposase gene of IS*10* generates an mRNA with a secondary structure which sequesters the translation initiation signal(s) and prevents translation. The transcript from the resident IS*10* promoter pIN does not include 5′ sequences necessary for the formation of this secondary structure, and hence the initiation signals are available for translation (355). Schwartz et al. (322a) have suggested that a similar arrangement in IS*150* may protect it from impinging transcription.

For IS*1*, external transcription has been shown greatly to increase the expression of the *insA* protein (377). Moreover, such transcription has a large inhibitory effect on transposition activity, as described above. Transcription impinging on either end of the element has a comparable effect, indicating that inhibition is not due to the overproduction of InsA itself (28, 59). IS*50* also exhibits a reduced activity when subject to impinging transcription (28, 315). As in the case of IS*1*, this effect cannot be ascribed to an influence on the expression of IS-encoded genes. It is

possible that in both cases transcription disrupts the structure of the DNA at the ends of the element, inhibiting the formation of a transposition complex (60).

D. Host Factors in Transposition

As relatively simple genetic units, ISs are expected to require some components other than those encoded by the element itself. The role of host factors in the transposition process is in most cases poorly understood. Only in the case of bacteriophage Mu (Pato, this volume), for which a defined in vitro transposition system has been developed (74), are these factors understood to some degree.

The search for bacterial mutants unable to support transposition in vivo has been carried out in several laboratories. Some of these mutants show an effect on transposition as measured by one assay (for example, transposition from an infecting phage into the bacterial chromosome) but not by others. It should be kept in mind that this behavior might reflect effects on the assay system itself rather than on transposition per se. Therefore, only in cases where several transposition assays yield the same qualitative results can a mutation be clearly identified as affecting the transposition process.

1. DNA polymerase I

Syvanen and colleagues (66, 349) have exploited the observation that the activation of the (otherwise cryptic) *bgl* operon of *E. coli* frequently results from insertions of IS*1* or IS*5* (295); see Section III.C) to devise a screening for host mutations. Following ethyl methanesulfonate mutagenesis, mutants were isolated which exhibited a reduced frequency of activation of *bgl*. Several of these exhibited reduced frequencies of transposition of both Tn*5* (IS*50*) and Tn*10* (IS*10*) from bacteriophage lambda to the host chromosome, and one exhibited, in addition, a reduced frequency of Tn*5* and Tn*10* transposition from the bacterial chromosome into a resident F plasmid. This mutation was found to occur in *polA*, the gene encoding DNA polymerase I, and to affect the 5′-3′ exonuclease activity of the enzyme. Other *polA* alleles, however, were found to exhibit only slightly reduced transposition frequencies by both assays. Similar results have been obtained by Sasakawa et al. (316). The role of *polA* in the transposition process remains a matter for speculation.

2. DNA gyrase

Many of the well-studied, site-specific recombination reactions have proved to be sensitive to the degree of supercoiling of donor or recipient DNA. It has been proposed that the energy of supercoiling is used to drive one or more steps of the multistep recombination reactions (251). Sternglanz et al. (345) have reported that strains mutant in DNA topoisomerase I exhibit reduced levels of transposition of Tn*5*, Tn9, and Tn*10* (N. Kleckner and D. Roberts, cited in Sternglanz et al. [345]), but not of Tn*3*. This effect is now thought to be due to secondary mutations in the DNA gyrase subunit genes, *gyrA* and *gyrB*, that result in an overall reduction in the degree of DNA supercoiling in the cell. Indeed, subsequent studies (172) have provided convincing evidence for the involvement of GyrA and GyrB in the transposition of Tn*5* and suggest that the level of supercoiling of the target rather than that of the donor molecule is important in this case.

3. Histonelike proteins

Binding sites for the histonelike protein IHF have been found at both ends of IS*1* within the inverted repeats (111, 112). They occur between the −35 and −10 regions of promoters that direct transcription across the element. IHF induces a strong bend on binding to these sites in vitro (279), protects the ends of IS*1* against DNase I digestion, and increases the accessibility of the terminal phosphate bonds to DNase I cleavage (111). Addition of IHF prior to, or at the same time as, RNA polymerase (RNAP) results in preferential binding to IHF to the left end, although prebound RNAP cannot be displaced by subsequent addition of IHF (111). Moreover, preliminary competition experiments between IHF and the IS*1*-specified protein InsA suggest that both proteins might bind the ends simultaneously (D. Zerbib, unpublished data).

In spite of these suggestive observations, we have been unable to show a convincing effect of IHF in vivo using a number of different assays for transposition activity. Thus, a role for this protein in transposition remains to be demonstrated. It is possible that a closely related host protein may substitute for IHF, masking the effect of mutations in IHF-coding genes on the transposition of IS*1*, or that IHF is involved in IS*1* transposition only under certain yet undetermined physiological conditions. IS*10* is also known to carry an IHF site close to one end, but outside the inverted repeat (N. Kleckner, personal communication). As in the case for IS*1*, a convincing demonstration of the influence of IHF on Tn*10* transposition in vivo has yet to be demonstrated. A recent result (122) demonstrates that an IHF site present near one end of Tn*1000* binds IHF cooperatively with the transposase at an adjacent site. IHF

binding enhances transposase binding. As for IS*1*, however, these results are not reflected in in vivo behavior. IHF-negative mutant strains support apparently normal Tn*1000* transposition. In the next section, we present for comparison a complete list of potential IHF sites in sequenced ISs. The importance of bacterial histonelike proteins in the transposition process is becoming clear from studies using in vitro systems. Hu, a protein related to IHF, has been shown to be essential for the transposition of bacteriophage Mu (Pato, this volume) in a defined in vitro system (74), and both Hu and IHF seem to stimulate Tn*10* circle formation in a cell-free system (246; Kleckner, this volume). In view of their subtle (and pleiotropic) effects on the host cell, an assessment of the role of this type of protein must await further development of efficient in vitro transposition reactions for other transposable elements.

4. Dam methylase

Dam methylation sites are found at the ends of IS*10*, IS*50*, and IS*903*, three elements whose transposition is thought to exhibit a large nonreplicative component (Section VI). Dam sites are also found in several other elements (Fig. 5 and Table 5). The Dam methylation at the ends of IS*10* has been shown not only to influence the expression of IS*10*-encoded transposition functions, but also to determine the intrinsic activity of the ends in transposition (300; Kleckner, this volume). The fact that the ends are more active when these sites are unmethylated or hemimethylated than when fully methylated provides an elegant mechanism by which the transposition of this element could be coupled to the passage of a host replication fork by generating transiently hemimethylated (newly replicated) DNA. A similar picture has emerged from studies of IS*50*, where Dam methylation controls the relative levels of transposase and repressor (203, 233a, 372a) and the activity of the end(s) (227, 372a; Berg, this volume). Dam control of expression of IS*903*-encoded genes has also been reported (D. Roberts et al., cited in reference 300).

5. DnaA protein

Although it is treated in more detail in the chapter by Berg in this volume, it is worth noting here that IS*50* carries a site at one end resembling the consensus sequence for DnaA protein-binding sites (180). This essential protein, involved in the initiation of chromosome replication (see reference 201), has been shown to recognize and bind the IS*50* site in vitro (107) and influences the "activity" of the end in vivo (274, 373). In the next section, we present a complete list of potential DnaA sites in sequenced ISs.

6. Less well-defined mutations

Following mutagenesis, five *E. coli* mutants were identified by their inability to be transduced to chloramphenicol resistance (336). Screening was based on an observation that a P1 *vir* phage previously grown on a strain carrying a particular chromosomal Tn*9* (IS*1*) insertion (Tn*9*A) can transduce chloramphenicol resistance (Cm^r) at high frequency (10^{-4}/PFU). P1 *vir* transduction of Cm^r was 10^4- to 10^5-fold lower in these mutants, and its transduction by a lambda *att80* phage prepared in the same manner was reduced by a factor of 10^6. Since the mutants allow normal phage growth and can be lysogenized by other P1::Tn*9* and lambda::Tn*9* derivatives, it was concluded that the mutations affect the transposition of the apparently highly active Tn*9*A. They map near *metA* and *malB* (168, 169), close to *dnaB*, *ssb*, *uvrA*, and *lexA*, on the *E. coli* linkage map (9). At least two are recessive, and one is thermosensitive. While their effects on the transduction of Tn*9* are by far the largest, tests with Tn*3*, Tn*10*, Tn*5*, and Tn*601* carried by nonreplicating, nonintegrating lambda phage gave various effects: some mutants showed reduced frequencies for all elements, while the others showed a pronounced reduction only in the transduction of Tn*9*A. One of the mutations, *tnm-2*, which affects all the elements tested, appears to be an allele of *dnaB* (C. Georgopoulos and T. S. Ilyina, personal communication).

7. Transcription termination factor Rho

As discussed in previous sections, many insertion sequences have been shown to carry Rho-dependent transcription termination signals that are presumably involved in the control of expression of element-encoded genes. The effects of several *rho* mutations on the transposition of IS*1* or IS*5* (using the *bgl* assay; Section III.C), of Tn*9*, and of Tn*5* (by transposition from the chromosome onto phage lambda) have been investigated (83). Reductions in frequency of 10- to 30-fold were observed. Such an effect on IS*1* may be a reflection of a direct effect of Rho on the expression of the *insB* gene, which lies downstream from the resident IS*1* terminator (Sections IV.B and IV.C), although indirect effects, for example, an increase in transcription impinging on the element (Section IV.C), have not been ruled out.

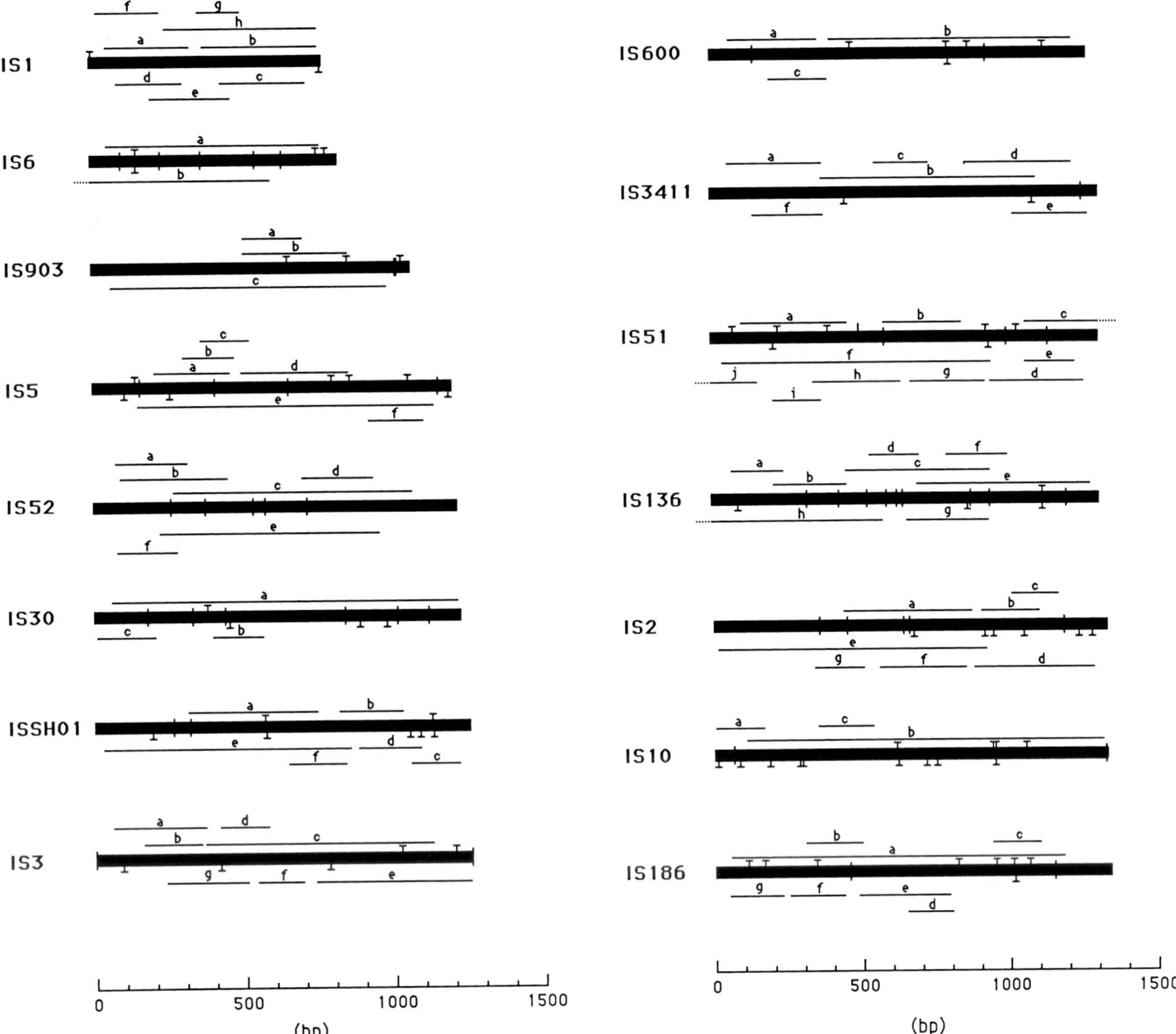

Figure 5. Sequence-specific, host protein-binding sites on ISs. For the same set of ISs shown in Fig. 3, we indicate putative sites (by consensus matching) for IHF (⊤) and Dam methylase (|) with the ORFs indicated for reference. The exact locations of the sites are listed in Table 5. Note that a sequence match does not necessarily indicate a real IHF site (the list is probably inclusive in this case), while for Dam sites it does.

8. Genes affecting precise excision

Although not strictly coupled to the transposition process per se, the excision of transposable elements leading to the restoration of the activity of the target gene represents another aspect of transposon behavior (Section III.A). Several types of host mutation leading to an increase in the frequency of this type of event have been isolated (105, 154, 217, 218, 241).

Studies on host mutants deficient in phage lambda integration have revealed that mutations in the IHF subunit gene *himA* lead to large reductions in the frequency of precise excision of bacteriophage Mu and a moderate (10-fold) reduction in the excision of Tn*10* located in an F′ *pro lac*::Tn*10* plasmid (241). Similar results were obtained for Tn*5*. The effect of IHF on the precise excision of Mu may simply reflect its involvement in the expression of the early operon of this phage (Pato, this volume), whose products are essential for excision (47).

Hopkins et al. (154) have also reported the isolation of host mutations leading to large increases in the excision frequency of several different Tn*5* and a single Tn*10* insertion. These mutations, *uup*, reduce the burst size and frequency of lysogeny of phage Mu. *uup* maps at about 21 min on the *E. coli* genetic map (9), close to the *hip* (*himB*) gene (20.8 min), but

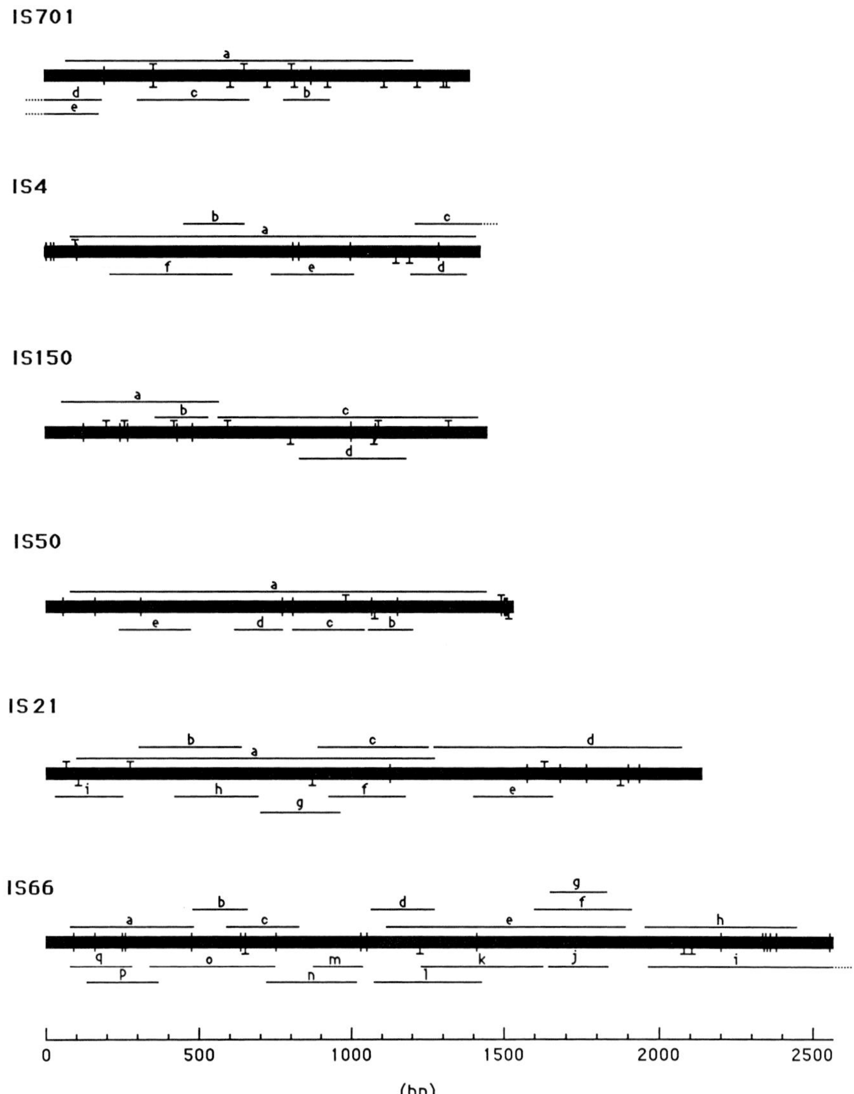

Figure 5. *Continued.*

the mutations have no effect on phage lambda integration.

Another class of mutation, *texA* (105), increases the frequency of precise excision. These mutations are located in the *recBCD* complex (218). Unlike "classic" *recBC* mutations (e.g., *recB21* and *recC22*), which do not increase the frequency of precise excision, *texA* mutations have little effect on homologous recombination. They do, however, exhibit other *recBCD* mutant-associated phenotypes, such as UV sensitivity and the inability to support growth of lambda *red gam* mutant phage. They exhibit an increased frequency of precise excision of Tn*5* and Tn*10* (which are flanked by IS*50* and IS*10*, respectively), but have little, if any, effect on the precise excision of Tn*9* (which carries direct flanking copies of IS*1*). The mutations have been located genetically in the *recB* (*texA344*, *texA345*) and *recC* (*texA343*) genes. They are thought to alter a necessary interaction between the flanking inverted insertion sequences in the excision process. Other *tex* mutations have been localized in the mismatch-repair genes *uvrD* (*texB*), *mutH* (*texC*), *mutS* (*texD*), and *dam* (*texE*) (217). Indeed, previously isolated mutations in these genes also show increased frequencies of precise excision. Like these mutations, *tex* alleles also increase the level of spontaneous mutation. *uvrD* (*texB*) appears to induce the most extreme effect on the excision of Tn*10*, but unlike *texA*, it shows a pronounced effect on Tn*9* excision. Mutations in several other genes involved in recombination and repair pathways also lead to increased excision. These include *mutL*, *mutD*, and *ssb*. The molecular events involved in these processes are unknown.

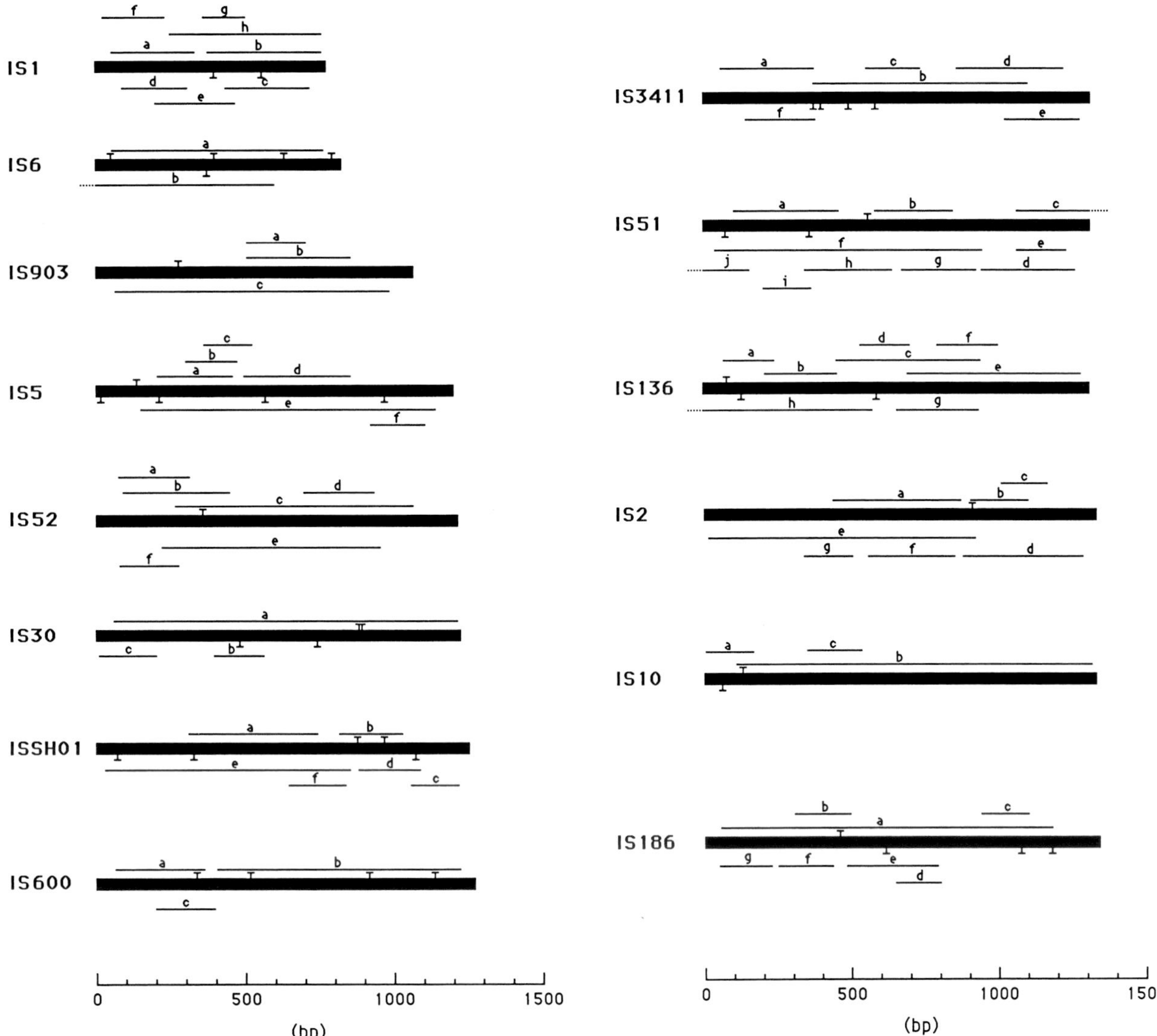

Figure 6. Binding sites for DnaA protein on ISs. The same type of sequence matching procedure as described in Fig. 5 for DnaA-binding sites. The sites were defined as described in Table 5, where the precise coordinates are indicated. The same caveat made for IHF sites in Fig. 5 applies here, that the sites identified may not actually bind DnaA protein.

E. Functional Sites in IS Elements

Several detailed studies of different ISs have demonstrated that the ends are essential components of the transposition apparatus. This has been directly demonstrated for IS*1* (112), IS*10* (246), IS*50* (180, 275, 276), and IS*903* (88) by the analysis of deletions and internal point mutations. In addition to sites concerned with transposase recognition and binding, the ends of ISs may carry an assortment of binding sites for other proteins. There must also be specific sites inside the elements that are involved in gene expression or other functions. Many ISs have been shown, for example, to carry transcription termination signals that presumably modulate the expression of their transposition functions (Section IV.C). We present here a brief survey of the occurrence of a number of sites with well-established consensus sequences within sequenced IS elements. These include: (i) sites for the binding of the host protein IHF, first identified in the ends of IS*1* (111, 112) and subsequently in IS*10* (246) and Tn*1000* (122); (ii) sites for the binding of DnaA protein, involved in the modulation of IS*50* activity; (iii) Dam methylation sites, which play important roles in regulating IS*10* and IS*50* transposition (203, 300, 372a); and (iv) Chi sites, specifying sites of action for RecBCD nuclease. Results of these searches are shown in Fig. 5 and 6 (see also Table 5). The search rules used here were quite simple. For IHF sites, a maximum of one

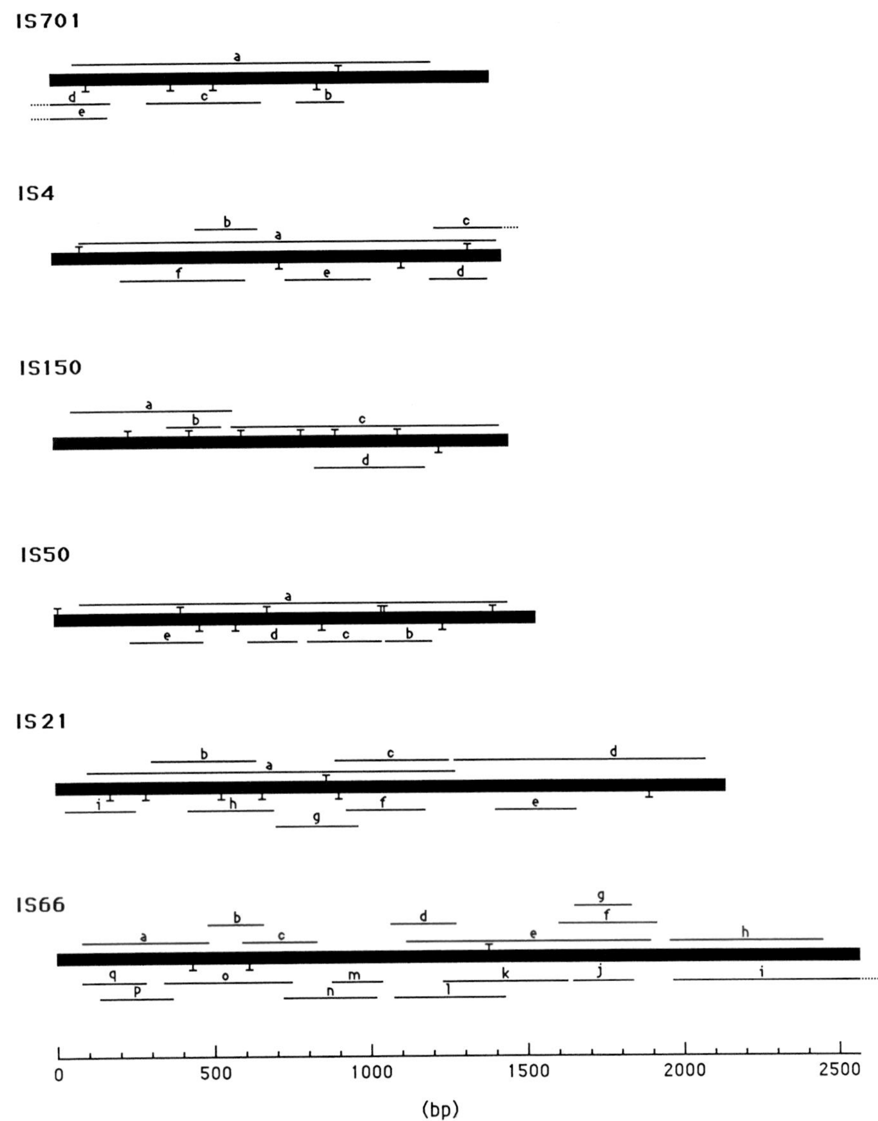

Figure 6. *Continued.*

mismatch from the sequence AAnnnnTTGAT was used. For DnaA sites, up to two mismatches from the sequence TTATCCACA were used. For Dam sites and Chi sites, exact matches with GATC and GCTGGTGG (339), respectively, were used. Using these somewhat oversimplified criteria for the first three categories, we have undoubtedly overlooked several sites and reported sites that are not functional. The listed sites will, however, probably include a large fraction of the existing functional sites, and it is in this spirit that we present this compilation.

The results may be summarized as follows. (i) Several elements carry candidate IHF sites near their ends. These include IS*1*, IS*903*, IS*5*, IS*10*, and IS*50*. These are also several occurrences of neighboring sites, one in each orientation (the sites are nonsymmetric). These occur in IS*6*, ISSHO*1*, IS*136*, IS*10*, and IS*186*. (ii) Dam methylation sites are also found near the ends of several elements; for example, IS*3* has sites at either end, and IS*4* and IS*50* have clusters of sites at one end. In addition, there are several sites near the beginning of ORFs that could be involved in the modulation of expression. (iii) DnaA sites occur near the ends of only two ISs: IS*5* and IS*50*.

Another interesting site, due to its role in stimulating homologous recombination, is Chi (339). Only a single site was found among all 18 ISs. That site is in IS*136* (at position 584), and while its function is unknown, its presence on IS*136* presumably endows this element with the properties of a portable enhancer of recombination. It should also be noted that IS*50* has been shown to carry a sequence which can generate a Chi site by a single-base-pair change.

Table 5. Potential specific binding sites in ISs[a]

IS	Coordinates of binding site: IHF	DnaA	Dam
IS*101*	44/97	None	None
IS*1*	13/756	None/396, 547	None
IS6	153, 747, 776/159, 253	40, 396, 627, 791/357	97, 232, 365, 544, 628
IS*903*	652, 854, 1032/none	274/none	1011, 1016
IS*5*	86, 745, 805, 997/106, 248, 1182	134/15, 212, 562, 966	154, 405, 644, 1164
IS*52*	None/none	355/none	248, 366, 527, 570, 578
IS*30*	376/449, 887, 974	881, 886/475, 742	170, 332, 438, 836, 1008, 1116
ISSHO*1*	565, 1124/189, 572, 955, 1045, 1090, 1131	876, 962/75, 332, 1073	262, 318
IS*3*	1021, 1200/91, 415, 778	None/none	2, 1254
IS*600*	473, 800, 868, 1121/ 806	334, 515, 914, 1133/ none	142, 928
IS*3411*	None/462, 1083	586, 496, 397, 376	1253
IS*51*	74, 223, 393, 931, 1038/211, 938	560/75, 358	585, 998, 1142
IS*136*	1121/94, 864, 1128	82/134, 595	322, 429, 520, 591, 620, 656, 874, 934, 1198
IS2	None/682, 922, 945, 1047, 1242, 1291	910/none	356, 455, 633, 657, 1184
IS*10*	616, 943, 946, 1054/ 41, 78, 184, 289, 298, 623, 715, 747, 953	128/52	67, 1320
IS*186*	113, 164, 338, 820, 956, 1012, 1063/ 1019	462/618, 1075, 1182	463, 1149
IS*701*	356, 655, 805/355, 607, 721, 816, 930, 1112, 1220, 1305, 1309	918/110, 380, 518, 846	193, 871
IS4	98/1158, 1203	83, 1318/713, 1106	8, 19, 31, 106, 810, 832, 997, 1288
IS*150*	201, 264, 421, 597, 1089, 1320/802, 1168	236, 428, 595, 786, 897, 1096/1224	125, 237, 272, 429, 483, 1001, 1076
IS*50*	980, 1490/1074, 1521	8, 397, 676, 1035, 1045, 1395/450, 567, 837, 1320	53, 161, 310, 763, 805, 1063, 1153, 1493, 1510, 1516, 1521
IS*21*	57, 267, 1632/106, 872, 1875	860/169, 283, 526, 655, 898, 1891	1124, 1573, 1681, 1759, 1901, 1928
IS66	None/650, 1208, 2079, 2098	1373/428, 608	102, 186, 258, 272, 485, 639, 648, 741, 1029, 1050, 1419, 1488, 1585, 1689, 1744, 1798, 1892, 1848, 2005, 2038, 2068, 2087

[a] The coordinates of all potential IHF, DnaA, and Dam methylation sites are listed for all of the sequenced ISs (listed in order of size). The asymmetric sites (IHF and DnaA) are listed if they match the consensus with only one mismatch; the Dam site had to be an exact match. The sequences looked for were: IHF, AAnnnnTTGAT; DnaA, TTATCACA; Dam, GATC. The coordinate is listed before the slash if the consensus match is oriented as above, and after it if present in inverse orientation. The positions of the sites are indicated in the figures.

Another protein identified for its effect on site-specific inversion, the Fis (factor for inversion stimulation) protein (179, 199), has at least two binding sites in ISs. The binding of Fis to two sites in IS*5* has been identified by footprinting experiments (B. Rak, personal communication). There has been no general experimental survey of known ISs for binding sites and the consensus for Fis binding is too ill defined to allow scanning known sequences for sites. The function of Fis in IS*5* is not known.

V. INSERTION SPECIFICITY

The rules which govern the choice of target sites are important for understanding the mutagenic effects and evolutionary impact of transposable elements. Since the choice of site must reflect the way in which the transposition complex interacts with the target DNA, these rules may also provide insights into the mechanism of transposition.

Fundamental to our understanding of the biology of transposable elements is the specificity with which they insert themselves into target DNA. Transposable elements in bacteria exhibit a wide range of insertion specificities. Specificity may result from the recognition of a specific target sequence, a structural (e.g., a bend) or functional (e.g., active transcription) feature of the DNA, or a combination of these factors. Sequence specificity will be determined by the size of the site recognized (which determines the expected frequency of the site in the bacterial genome) and the degree of degeneracy of the site tolerated by the IS (which will increase the expected frequency). Certain elements, such as Tn*7*, insert preferentially at a single site in the bacterial genome and, at significantly lower frequencies, at various secondary sites (Tn*7* is discussed in this volume by N. L. Craig). This type of behavior is reminiscent of temperate bacteriophage such as lambda, in which the principal "target" sequence (*attB*) is relatively long and in which variant sites with single-base-pair differences (the secondary sites) are much less attractive. Although a statistically significant sample of sequenced insertions is not available for many ISs, it is clear that specific consensus target sequences of various lengths and degrees of degeneracy are involved in the transposition of several, among them IS*4* (232), IS*5* (99), IS*10* (104, 134), and IS*30* (53). The insertion hot spots for Tn*10* tend to conform to a symmetrical, 6-bp consensus sequence, GCTNAGC, whose degree of match with a target sequence is directly related to the relative insertion frequency at that sequence. It has recently been suggested that specific protein contacts with the consensus base pairs are involved, since the thymine methyl groups are necessary for high-frequency insertion (210). Recent results of Rak and co-workers on IS*150* suggest that this IS may be highly specific in its target selection as well (322, 322a).

Other insertion sequences exhibit little specificity for a given target sequence. For example, for IS*1*, the only notable sequence relation between insertion sites is that the terminal base pairs (positions 1 and 9) of the direct repeats generated at the target site following insertion show a strong bias for G-C (108; P. Prentki and E. Freund, unpublished observations). A G-C bias at one end has also been noted for target repeats generated by IS*50* (20, 250). On the other hand, some of these ISs exhibit pronounced "regional" specificity for segments of DNA of the order of 100 bp in length and insert at many sites within these regions. Among those ISs for which such preference has been observed are IS*1* (108, 236, 376), IS*2* (325, 327), IS*186* (202, 326), and, to a lesser extent, IS*50* (20, 219).

The determinants of regional specificity are more difficult to define and interpret. IS*1* and IS*186* (the latter based on a limited number of insertions) show pronounced preferences for regions rich in A+T and G+C, respectively (108, 202, 236, 325, 326, 376), while IS*2* insertions are found repeatedly in a small region of the phage lambda (314) and P1 (325) genomes that appear to have no obvious sequence features that could account for this property (A. Guidolin and W. Arber, personal communication).

The presence of nearby regions of limited homology with the ends of ISs has often been suggested to play a role in site selection (35, 108, 219, 243, 314), but no convincing evidence of either a direct or statistical nature has been presented.

In the case of IS*1*, at least two properties of the insertion hot spots probably contribute to the observed regional specificity. The initial observations that IS*1* hot spots in the *lac* operon and in bacteriophage P1 were particularly rich in A·T base pairs (108, 236) were subsequently extended to insertions in the plasmid pBR322 (376). This plasmid exhibits two major regions of insertion which have similar (high) A+T densities. However, one region attracts over 90% of all IS*1* insertions into the plasmid, suggesting that A+T density is not the only factor influencing IS*1* site selection. A 60-bp poly(dA) · poly(dT) tract (100% A+T), while attracting some insertions, does not exhibit a "strength" comparable to that of the major site. A striking feature of both the pBR322 hot spots is their aberrant migration on polyacrylamide gels (344), suggesting that these regions are bent. Such structural anomalies might provide the necessary "signal" which was proposed. Interestingly, both pBR322 hot spots carry binding sites for the host protein IHF (111; B. Teter, unpublished results). Although the presence of an IHF-binding site is not necessary for IS*1* insertion (there are many examples of IS*1* insertions in regions of DNA that do not contain obvious sites), IHF has been shown to induce DNA bending (279, 301, 352) and may serve somehow to amplify the preference for this region in vivo. Alternatively, since this protein has also been found to bind to the ends of IS*1* (see Section IV.D), it is possible that the preferential insertion of IS*1* into this region may be related to IHF binding to the DNA and involve direct protein-protein interactions. In view of the possibility that insertion may be influenced by the structure of DNA at the target site, it should be kept in mind that other sequence-dependent differences in DNA structure (for example, bending, abnormally twisted or untwisted stretches, segments with high Z-forming potential) may also influence the insertion of certain ISs.

Transposon-induced deletions are thought to arise by pathways similar to transposition. Thus, we might expect that the selection of deletion endpoints would follow the same preferences as for insertion specificity. Surprisingly few studies have compared these specificities directly (51, 257). The data confirm that the preferences of target sites for both deletions and intermolecular transposition are the same. It has been reported for IS*102* (almost identical to IS*903*) that deletion endpoints occur preferentially in regions undergoing active transcription (21, 22). Although no mechanistic explanation is evident for this phenomenon, it should be noted that transcription can be expected to affect the local conformation of the DNA and may exert its effect indirectly in this way (214, 284). We would expect transposition, therefore, to show the same preference. Lui and Wang (214) have shown that positive supercoils can accumulate in front of a transcription complex. It is tempting to speculate that this may be the source of some of the effects of transcription on transposition. Few data are available concerning the specificity of intermolecular transposition of the related elements IS*102*, IS*903*, and IS*602* at the nucleotide sequence level.

VI. MODELS FOR TRANSPOSITION

Although bacteria carry a wide variety of insertion sequences, apparently unrelated in sequence and in organization, one essential common feature distinguishes the rearrangements they induce from generalized recombination: transposition is entirely independent of sequence homology and leads to the formation of novel DNA joints between the ends of the IS and target DNA. There are, of course, several biochemical events that must form part of the process (a minimum set of DNA strand cleavages and rejoinings) and others, such as replication, that may be involved. There are a number of possible ways of combining these biochemical steps into a temporal pathway to generate a transposition model. Only relatively recently have experimental results from a limited number of systems been obtained that can critically distinguish among some of the possibilities for a particular system, but the diversity of transposable elements leads to serious doubts about the general applicability of any particular model.

In this section, we discuss some of the most relevant experimental results for several elements, describe various models, and attempt to define a framework within which most transposition models can be placed (Fig. 7). Before discussing models in detail, it is perhaps useful to present a brief survey of the history of ideas about how transposition might work.

A. Some Historical Comments

The observation that phage Mu is able to join two unrelated DNA segments provided the basis for the earliest models of transposition which invoke replication of the element (100, 354). Another early hypothesis suggested that the element was transferred from donor to recipient molecule by simple cleavage and reinsertion (15). Transposition would thus be similar to retroviral integration (H. Varmus and P. Brown, this volume). The finding that a short sequence duplication in the target is generated on insertion of IS*1* (50, 123; see Section III.A) suggested that a staggered cut in the target DNA was made during the process, and that repair synthesis produced the repeat. From this starting point, a model for replicative insertion which involves copying the element into the target site was elaborated by Grindley and Sherratt (125). The observation that transposition of the Tn*3* family of elements involved an intermediate in which donor and recipient replicons were fused and separated at each junction by a copy of the element (implying replication of the element; Sherratt, this volume) led Shapiro (330) and Arthur and Sherratt (7) to propose an alternative model which could also explain inversions and deletions in a simple way. A related model was proposed by Kamp and Kahmann (189) to explain probable differences in the lytic and lysogenic pathways of phage Mu. As more information about the behavior of different transposable elements was obtained, evidence accumulated suggesting that not all elements used the same pathways, and several workers attempted to incorporate new data into alternative models. Read et al. (290) suggested that in situ replication of IS*1* could lead to the formation of an active junction which could then recombine with the target (similar ideas were proposed earlier by Faelen et al. [100]). Galas and Chandler (109) and Harshey and Bukhari (140) suggested that a unidirectional replicative model with a choice of strands for ligation in a final step could consolidate several prior ideas and explain how some transposons could make both stable cointegrates and simple insertions. Ohtsubo et al. (264) proposed a modification of the model of Shapiro and of Arthur and Sherratt that could also explain how both cointegrates and simple insertions are generated. The molecular details of these models are presented clearly in the review by Mizuuchi and Craigie (242).

B. Experiments Bearing on Transposition Mechanisms

In the attempt to determine what pathway actually accounts for the transposition of a particular element, most of the evidence derives from examining the structures of transposition products from well-defined donor and target molecules and comparing these with the predictions of models. Great caution, however, must be taken in interpreting the observation of certain products as evidence for a particular mechanism. The assumption usually inherent in such interpretations is that all possible mechanisms have been foreseen, and that the investigator must simply decide among them on the basis of the observed products. This is seldom the case. For example, although cointegrate formation is generally thought to involve replication, the low level of Tn*10*-promoted cointegrate formation, adjacent deletions, and inversions can also be explained in terms of a completely conservative transposition process (Kleckner, this volume). In this case, then, the observation of cointegrates cannot be interpreted as evidence of a replicative process. Another potential source of confusion arises when a transposition product can undergo recombination. The resulting structures can, of course, be quite confounding to attempted inferences about the transposition pathways. With the exception of phage Mu, whose transposition has been analyzed in detail using a defined in vitro system, most of the evidence is rather indirect. Some experiments, however, have provided important insights into transposition mechanisms.

(i) Experiments in which a nonmethylating (*dam*) host was infected with Mu phage whose DNA was fully methylated (139) or in which a suitable host was transfected with heteroduplex phage DNA (3) clearly showed that the initial integration of Mu occurs without replication, although lytic growth of the phage occurs by replicative transposition.

(ii) A similarly elegant demonstration of the conservative nature of the transposition of Tn*10* (IS*10*) was provided by Bender and Kleckner (14). Transposition products of a heteroduplex Tn*10* element containing a *lacZ* gene with a mismatch (along with mismatches at other positions) were examined for the presence of sequence information from both strands. The results demonstrated the presence of information from both strands in a single transposition event, clearly demonstrating that the entire heteroduplex element is transferred during transposition. From this and other evidence, it now seems clear that Tn*10* transposes principally by a conservative (nonreplicative) mechanism (see the chapter by Kleckner in this volume for further discussion of Tn*10*).

(iii) Early studies on Tn*3* demonstrated that transposition proceeds via a cointegrate intermediate and provided strong evidence for the replicative transposition of this family of elements (7, 145). This evidence is reviewed elsewhere (Sherratt, this volume; Pato, this volume).

(iv) There is also convincing evidence that transposition of IS*903* involves replication. Weinert et al. (365) have demonstrated that IS*903* is replicated during the process of deletion formation by recovering and identifying both products of deletion events. Similar results have been obtained for the related IS, IS*102* (23).

(v) Studies of cointegrate formation by IS*1* (29, 59, 110) argue that IS*1* transposition can be replicative.

(vi) Formation of cointegrates by IS*50* is strongly dependent on RecA (153). This effect has the same magnitude (roughly 100-fold) whether Tn*5* is present on a small, high-copy pBR322 derivative or a large, unit-copy F derivative. For this reason, we think it unlikely that the effect can be explained entirely as a result of the direct transposition of an entire plasmid, flanked by Tn*5* elements, from a plasmid dimer, as suggested by Sasakawa and Berg (315). Moreover, a similar RecA dependence has been observed for Tn*5* transposition from the lambda genome (211a) and for Tn*5*-mediated deletion formation (2; Ahmed, personal communication). This supports the idea that either the formation of cointegrates or the stability of an intermediate is enhanced by the presence of RecA. The role of RecA in the transposition process, however, remains unclear (37b, 153, 211a, 378).

(vii) The approximately exponential length dependence of the transposition frequency, first observed for IS*1* compound transposons (57), appears to indicate a rate-limiting, processive step in transposition. One possibility, consistent with replicative models, is that replication itself is involved. The observation of transposition from a lambda::Tn9 derivative (304) of structures vastly different in size (by inverse transposition) at similar frequencies suggests, however, that the processive event may not be

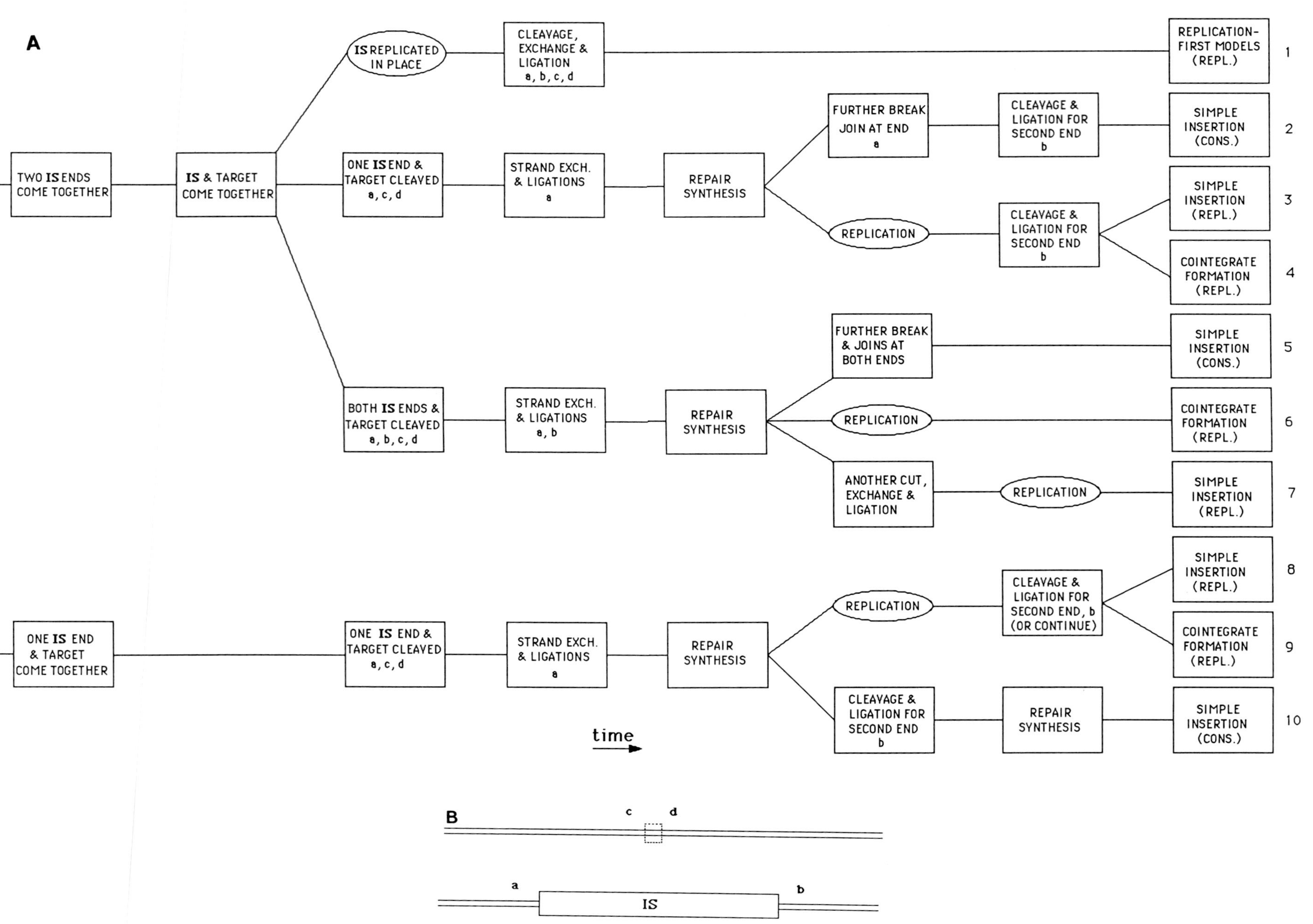

A
TWO IS ENDS COME TOGETHER
ONE IS END & TARGET COME TOGETHER
IS & TARGET COME TOGETHER
IS REPLICATED IN PLACE
CLEAVAGE, EXCHANGE & LIGATION a, b, c, d
REPLICATION-FIRST MODELS (REPL.)
1
ONE IS END & TARGET CLEAVED a, c, d
STRAND EXCH. & LIGATIONS a
REPAIR SYNTHESIS
FURTHER BREAK JOIN AT END a
CLEAVAGE & LIGATION FOR SECOND END b
SIMPLE INSERTION (CONS.)
2
REPLICATION
CLEAVAGE & LIGATION FOR SECOND END b
SIMPLE INSERTION (REPL.)
3
COINTEGRATE FORMATION (REPL.)
4
BOTH IS ENDS & TARGET CLEAVED a, b, c, d
STRAND EXCH. & LIGATIONS a, b
REPAIR SYNTHESIS
FURTHER BREAK & JOINS AT BOTH ENDS
SIMPLE INSERTION (CONS.)
5
REPLICATION
COINTEGRATE FORMATION (REPL.)
6
ANOTHER CUT, EXCHANGE & LIGATION
REPLICATION
SIMPLE INSERTION (REPL.)
7
ONE IS END & TARGET CLEAVED a, c, d
STRAND EXCH. & LIGATIONS a
REPAIR SYNTHESIS
REPLICATION
CLEAVAGE & LIGATION FOR SECOND END, b (OR CONTINUE)
SIMPLE INSERTION (REPL.)
8
COINTEGRATE FORMATION (REPL.)
9
CLEAVAGE & LIGATION FOR SECOND END b
REPAIR SYNTHESIS
SIMPLE INSERTION (CONS.)
10
time
B
c
d
a
b
IS

Figure 7. (A) Flow diagram of events leading to transposition products according to standard hypotheses. The design is described in the text. The letters in some of the boxes indicate the physical location affected according to the labeling of panel B. The final 10 outcomes represent possible models or alternative branches of models. These are described in the text. (B) The sites on the IS donor and target DNA indicated in panel A.

replication. An exponential length dependence of the transposition frequency was later observed for transposition from the bacterial genome of Tn*10*-based transposons (247). It was subsequently shown, however, that the length dependence of transposition from small plasmids was either reduced or nonexistent (363). Since these elements probably do not replicate during transposition, another process is also implicated in the case of Tn*10*. It has recently been demonstrated that transposition of the Mu genome exhibits an exponential decay with increasing length much like that of the IS*1* and IS*10* transposons (100). The observed severe decrease in overall Mu-specific DNA synthesis for longer phage genomes suggests, however, that a step prior to replication is involved. Although the basis of the phenomenon is unknown, the most likely candidate for the processive event to explain all of the above-mentioned effects is the synapsis of the ends and the formation of the transposition complex.

(viii) The defined in vitro system for Mu transposition provides the most direct evidence for the mechanism. These results are reviewed elsewhere in this volume (Pato), so we will simply summarize the conclusions here. First, Mu can transpose by a replicative or a conservative pathway. Both ends of Mu must come together to initiate the event. The polarity of strand transfer is such that the 3′ ends of the element are transferred to the target molecule. The 3′ ends of the target DNA are apparently used as primers in the replicative process. This process as now understood is consistent with the Shapiro model (branch 6 in Fig. 7).

(ix) Finally, a striking example of the diversity of transposition behavior is known, which suggests a diversity of mechanism. IS6 has been shown to form only cointegrates (Sections III.B and IV.A). This may serve to provide a useful insight into models by comparison of the behavior of IS6 with other elements.

C. Classification of Models

It is possible to classify most transposition models (including those cited above) in several distinct ways. One could focus attention, for example, on whether replication of the donor element occurs and call them "replicative" or "conservative," or focus on whether the ends of the element are treated identically throughout the process and call them "symmetric" or "asymmetric." Rather than classify models in this way, however, we have found it useful to separate model processes into their possible constituent steps and show how they may be joined and ordered into hypothetical pathways, that is, into transposition models.

A model, at the level of biochemical detail considered here, consists in knowing which strands of the element are cleaved, which pairs of ends are joined, how replication proceeds, if it does, and in what order each of these events takes place. Figure 7 summarizes the ways in which one can order the constituent events and how the sets of choices are related. In this view, making a model consists in tracing a series of events from the left of the figure (the root of the tree) to the right. Thus, the rightmost nodes of the tree correspond to "transposition models" since they describe an ordered set of events from the initial state to the completed transposition product. Some of the possible paths through the tree can be identified with models that have been previously proposed. As complex as the figure seems, we have simplified a few points to avoid generating excessive numbers of combinations differing in only minor ways.

It is clear from this exercise that some care must be taken about simple classifications of models and, more importantly, in the interpretation of "tests" of models which rely on the examination of transposition products. Note, for example, that there are three distinct endpoints of simple insertion by conservative (nonreplicative) transposition, each of which corresponds to a distinct pathway, and thus a distinct "model." The original proposal of a "cut-and-paste" (conservative model) (15), in which a double-stranded cutting out of the IS is followed by its insertion into the target DNA, could be represented by any of these, since details not proposed in that suggestion distinguish them in the diagram.

In considering possible models, we start with a single IS, resident in a donor molecule, and a target molecule to which it will be joined by a transposition event (Fig. 7). We will consider only intermolecular events here since the description of models is independent of the relationship between the target and donor. The various component steps used in Fig. 7 are (i) synapsis of the ends of the IS, (ii) cleavage and strand exchange, which for the moment are not described in more detail, (iii) a second round of cleavage and joining, and (iv) replication, including either repair synthesis or full replication of the IS.

The first branch of the tree depends on whether or not the ends initially come together. Some subsequent steps depend on this initial synapsis: for example, symmetric models which cleave, join, and replicate both ends in an identical fashion. Work on the in vitro system for Mu strongly implies that this must happen as the initial step for this element (74, 242). Next, after the target and IS have come together, some combination of cleavage, exchange, and joining must occur. Actually, the order of these two steps is somewhat ambiguous since cleavage could precede synapsis, but it seems pointless to make a large expansion of the figure just to take this variation into account. We have therefore tacitly assumed that this variation could occur for any of these branches.

The sense of the staggered cut in the target must be matched to the choice of strand cleavage at the IS ends (Fig. 7). If the staggered cut has a 5′ overhang, then the IS end(s) must be cleaved such that a free 3′ end is available to be joined, and the recessed 3′ end of the target DNA is then available to prime the necessary repair synthesis that creates the short repeat or subsequent replication of the element. If, on the other hand, a 3′ overhang is created by the cleavage of the target, then the IS end must be cleaved to create a 5′ free end, and following joining, the 3′ end of the flanking donor site is available for priming repair synthesis (or replication). A major branching at this point in the tree occurs over the choice of whether one or both ends are joined to the target.

Here again, the work of Mizuuchi and collaborators gives a clear answer about the way these events occur in Mu (242; Pato, this volume). The element is joined at both its 3′ ends to target DNA which has been cleaved with a 5′ overhang.

After the first joints between the target DNA and the IS have been made, it seems likely that some repair synthesis must occur in all cases even if, as seems to be the case for Tn*10*, the entire element is not replicated. It is this repair synthesis that results in the target duplication. From this point in the process, either replication of the entire element can proceed or further strand breakage and joining occurs (with the exception of branch 7, in which both occur; see below).

We have attempted to be reasonably complete in constructing this figure, and have distinguished 10 different pathways, or branches, that can be attributed to eight clearly distinct mechanisms which could be termed models. Of the 10 pathways, 3 lead to conservative transposition and 7 to replicative events. At the top of the figure are the "replication-first" models proposed by Read et al. (290) and Faelen et al. (100). They are represented by branch 1. Several minor variations on the processes could easily have been added but were not because they would multiply the number of models enormously, to little advantage.

The models of Kamp and Kahmann (189), Shapiro (330), and Arthur and Sherratt (7) are grouped under branch 6. The distinction between the polarity of strand cleavage at the IS ends cleaved, and therefore the strands which prime the replication, has not been made in the figure, although this is a clear difference among these proposals. The Kamp and Kahmann model prescribes 5′ ends of the IS being created by the initial cleavage, leading to the priming of replication by DNA flanking the IS. The other models prescribe the opposite polarity of strand transfer, 3′ ends of the IS leading to the priming of replication by the target DNA. In these models, since there are priming sites for replication at both ends, only leading-strand replication is required. There are several options at this point: replication can continue to complement (as in the proposed models, branch 6), breakage and joining to complete the conservative transfer of the IS to the target might occur (as in the Ohtsubo et al. [264] modification, branch 5), or another round of cleavage-exchange-ligation might occur on the new substrates provided by repair synthesis (Fig. 7) following replication. This latter option could also produce simple insertions as well as cointegrates (branch 7) (the details are illustrated in Fig. 8).

The models of Galas and Chandler (109) and Harshey and Bukhari (140), represented by branches 8 and 9, respectively, where the difference between the branches is determined by strand choice at the last step of the process, are both related to the earlier model of Grindley and Sherratt (125). In this case, DNA synthesis is envisioned as proceeding from one end to the other, so that complete replication would require lagging-strand synthesis. In this sense, the models are asymmetric. There is some indication that a mechanism like this may operate for Tn*1721* and Tn*21* (249). Note also that there may be a simple variation on this kind of model that can lead to simple insertion by a conservative process in much the same way as the Ohtsubo variation on the Shapiro model does. This is represented by branch 10. Finally, it is important to note that the Galas-Chandler and Harshey-Bukhari models may also be easily modified to prescribe the coming together of the IS ends at the initial step of transposition. The models resulting from this variation are shown as branches 2, 3, and 4, and correspond in other respects to the "single-end" or "asymmetric" models 10, 9, and 8, respectively. Thus, branches 2, 3, and 4

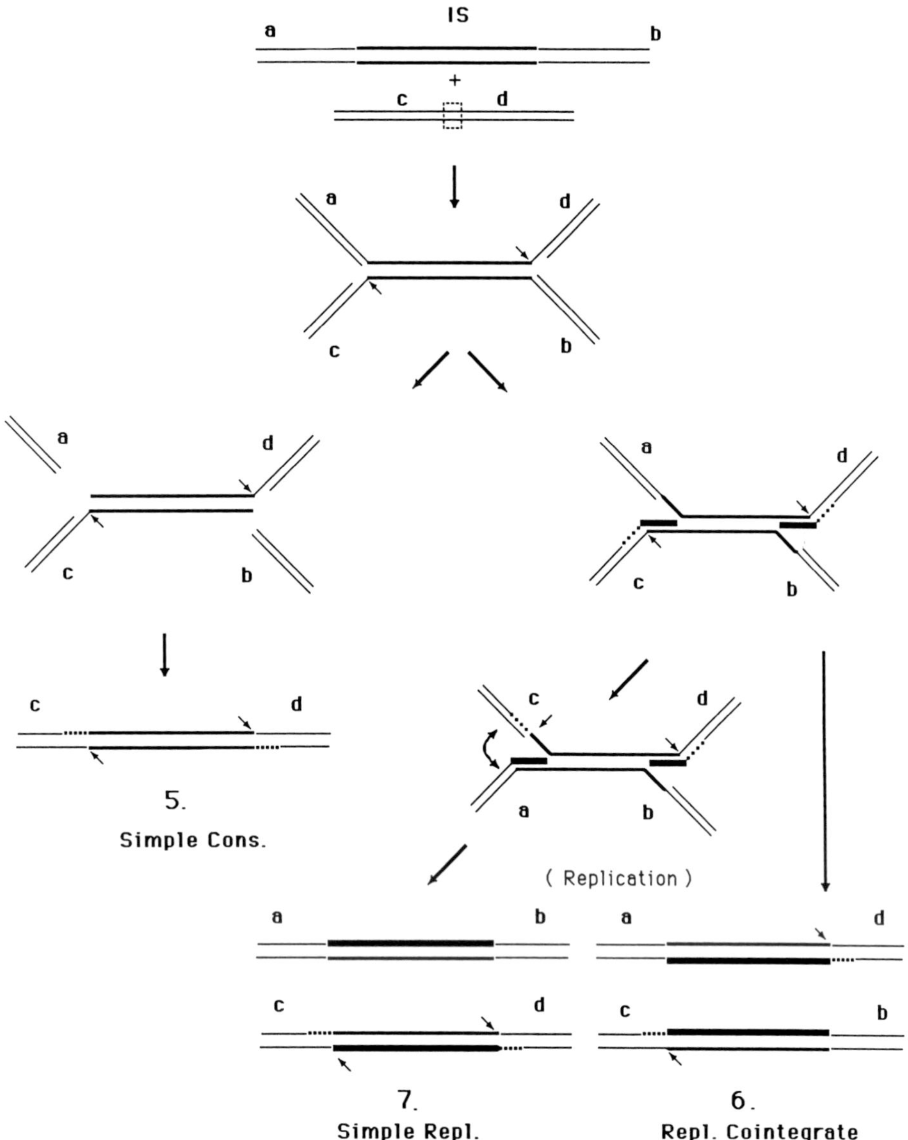

Figure 8. An example of three possible outcomes that could result from variations on the Shapiro model, indicated in Fig. 7 as 5, 6, and 7. The steps are described in the text. Heavy lines, Element; box, potential target site; small arrows, novel joints; dotted lines, newly synthesized target DNA; extra thick lines, replicated element strands.

are "symmetric" model versions of 10, 9, and 8. The major difference dictated by prior choice of the ends to be used in transposition is that the process of choosing an end to terminate transposition must take place before any replication occurs. This means that an idea inherent in the Galas-Chandler and Harshey-Bukhari models, that a "rolling-circle-like" process, terminating with a certain probability at each end encountered, could transpose larger and larger segments at smaller and smaller frequencies, would have to be abandoned.

Much remains to be discovered about transposition mechanisms. In view of the wide variety of known bacterial insertion elements and the diversity already evident, it would be naive to assume that the transposition of all elements occurs in an identical manner. With the sort of framework we have attempted to provide here, it may be possible, however, to foresee many of the possible variations upon the models of transposition that have been proposed.

VII. EVOLUTIONARY ASPECTS OF IS ELEMENTS IN BACTERIA

The discovery that segments of DNA can move within genomes, often duplicating themselves in the process, led naturally to the question of how this phenomenon has affected the evolution of the genome. The question has been evident since McClintock first discussed transposition (see references in reference 233). An obvious question, though one that oversimplifies the problem, is whether transposable genetic elements are for the most part selectively neutral (or negative) in their effects on their hosts (represented by the hypothesis of parasitic or "selfish" DNA), or whether they convey a distinct selective advantage or even play an adaptive role to be discovered. As with the debate on neutrality in molecular evolution, this problem can only be approached by couching the questions in quantitative terms and examining specific situations. The answers to questions about the evolutionary role of IS are unclear, and the topic is too large to deal with comprehensively here. Still, as is evident from the preceding sections, there is some evidence relevant to the problem. The topic has been reviewed relatively recently (348) and is treated in this volume by Ajioka and Hartl from the viewpoint of population dynamics and the distribution of ISs. A brief overview, with a different perspective, however, is in order here. In this section, we will summarize what, in our view, are the important specific questions that are relevant to an understanding of the evolutionary role of ISs and briefly discuss some of the evolutionary relationships among existing transposons.

One fundamental question one can ask about bacterial transposable elements is: what evolutionary models are consistent with the observed distributions of ISs in natural populations? Hartl and Sawyer and co-workers have addressed this question, using some very simple assumptions (95, 142, 317). Their models, when compared with some distribution data from *E. coli* strains, suggest that: (i) some elements (IS*1* and IS*5*) do not autoregulate their numbers, but lead to a strong decrease in the fitness of their host with increasing copy number; (ii) others (IS*3*) exert a weak decrease in fitness and show a strong regulation of number; while (iii) others exert a similarly weak decease in fitness and little regulation in copy number (IS2, IS4, and IS*30*). These results, as emphasized by the authors, must be considered tentative, but they do suggest that experimental, more extensive studies of IS-induced fitness changes and regulation of transposition would be of considerable interest.

Growth competition experiments conducted in a chemostat have demonstrated IS-mediated increases of fitness in two instances. IS*10*-containing strains are selected over IS*10*-free strains. This appears to be due to IS*10*-induced mutation (62). In another instance, IS*50* was shown to increase the growth rate of its host cells without causing specific mutations (30, 141). The cause of this effect is unknown, but depends on the production of IS*50*-encoded proteins. There may well be other, subtle selective effects associated with ISs that remain to be discovered.

In this chapter, we have referred to many IS-induced effects that are likely to have evolutionary significance. Because of the subtlety and universality of natural selection, the table of contents of the chapter can be viewed almost as a list of the properties of ISs important to evolution. In most cases, however, their significance has not been adequately addressed by experiment.

Undoubtedly, one of the most important of these is the variety of effects on gene expression that IS elements can produce. These have been discussed (see Section III.C). Again, the quantitative aspects of these phenomena, which are essential for an understanding of the evolutionary consequences, are poorly understood.

A related set of phenomena concern the propensity of some ISs for the formation of compound transposons, which is discussed in Section IV.A. The increased fitness of a strain containing the transposon-carried gene(s) is associated with the IS, in this case, because of their close genetic linkage. The genes for antibiotic resistance, for example, which are often carried by compound transposons, provide an enormous selective advantage in the presence of the antibiotic. It is difficult to ascertain, however, how often such selective conditions occur in natural populations. Since this sort of selection could be a major factor in evolution, the factors that determine the formation, stability, and potential for the dispersion of compound transposons may strongly influence the evolution of ISs and the genomes they affect. The formation of compound transposons is discussed in Section III.C. Some of the significant factors that should be considered in future studies are listed here. (i) The specificity of insertion of ISs will determine the probability of picking up particular DNA segments (see Section V). For example, the propensity for some ISs to insert into or near promoters will have a major effect. (ii) The specificity and frequency of second insertion events, or of intramolecular events that produce a segment of DNA flanked by two ISs, will determine the probability of formation of a compound transposon from a single insertion. Note that subsequent events, such as deletions, can also alter the structure of the transposon. (iii) The activity of the transposon will be determined by the various factors that affect IS activity, including the flanking DNA sequences, as is discussed in previous sections. (iv) The frequency of insertion of transposons into various conjugal plasmids and phage will be an important determinant of the ability of the transposon to be transmitted into other strains. This,

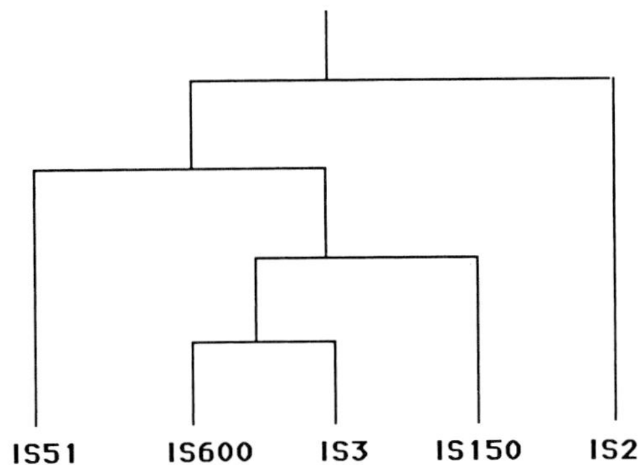

Figure 9. Possible evolutionary relationships of the IS*2*,*3* family of ISs. The tree shown here represents only the topological relationships of the (fully sequenced) ISs in this group. The lengths of the branches do not indicate distance. The relationships were deduced by Schwartz et al. (322a) from the inferred amino acid sequence of the largest ORFs of these elements.

of course, is a complex property, but it is also one that has not been extensively studied. Whether there are major differences among plasmids as targets for transposons and among transposons with respect to their specificity for certain plasmids, whether prior insertion of other ISs affects the frequency of subsequent insertion, and what specific factors determine the attractiveness of a plasmid for transposon insertion are all questions relevant to this problem. Very little is known about any of these questions. It may be, for example, that the correlations found among ISs in their species distribution are causally related to the distributions of ISs on plasmids or phages with the same host ranges.

Finally, there remains the question of paramount interest: what long-term evolutionary impact have ISs had on the organization, structure, and regulation of the bacterial genome? The simple answer to this question is that no one knows. One might hope to find evidence for an evolutionary role by comparing, for example, the genomes of *E. coli* and *S. typhimurium* and noticing whether any of the differences can be attributed to the presence of ISs in either genome. No such evidence (deletions or inversions terminating at ISs, for example) is obvious, but it could be that subsequent IS-related events and other mutations have long since obscured the evidence of early events. However, IS-induced chromosome inversion events have been observed in *E. coli* (215, 316a). It seems that no real conclusion on this question is yet possible. It is curious that there is a marked clustering of known ISs in the *E. coli* genome, however. More than 70% of the known ISs are located in the vicinity of *proAB* and *purE* (86). This kind of clustering could represent molecular fossils from early events that led to changes in this particular region (299), but could also be the result of preferential insertion of ISs near other ISs. In this latter model, the earliest colonization of the genome by an IS might have occurred in this region by chance, and the other insertions followed, attracted by the first one to the same region. Note, however, that copies of IS*200* are found distributed around the map of *S. typhimurium* (208). Now that so many ISs have been characterized and the techniques for the analysis of large segments of DNA make restriction maps of bacterial genomes relatively easy to obtain, future studies should shed some light on these and similar questions.

There is also much to be learned about the evolution of transposons themselves. The largest known family of related transposable elements is the Tn*3* family. The evolution of these can be examined by comparison of their organization and nucleotide sequence, and this has been discussed by several authors, notably Schmitt et al. (see reference 319 and references therein). Recently, it has become apparent that several other independently isolated ISs are probably distinct relatives. When the amino acid sequences of the largest ORFs from several ISs were compared, Schwartz et al. (322a) found some significant matches that suggest homology. The relationships among IS*2*, IS*3*, IS*51*, IS*150*, and IS*600* that are inferred from these comparisons are illustrated in Fig. 9. Another sequence from *S. dysenteriae*, ISSHO*1* (see Table 1), shows significant homology with IS*3*, but the analysis required to place it in the inferred evolutionary tree has not been done.

When comparisons between ISs reveal stronger homologies than those discussed above, this is usually viewed as constituting a set of iso-ISs, though the proper use of the term is poorly defined. The first reported case of such a set was that of the iso-IS*1* elements found in *Shigella* spp. by Ohtsubo and co-workers (259, 260, 262). Three classes of iso-IS*1* elements were detected having different degrees of homology with IS*1* (at the nucleotide sequence level). It appears likely that many iso-ISs are not transpositionally active, though some probably are active. By analogy with the *Drosophila* P elements, we expect that there may be copies that are defective but may transpose by using functions provided by intact, active elements (W. R. Engels, this volume). An example of a defective IS was found some time ago in the plasmid pSC101. IS*101*, only a few hundred base pairs in length, is probably derived from gamma delta (Tn*1000*) (289). It consists of two ends of the element with the resolution site for site-specific recombination between them (all somewhat diverged in sequence from Tn*1000*).

It is probably reasonable to suggest that any process that is capable of producing so many complex changes, physical rearrangements, and regulatory alterations must have had a major impact on evolution if it has been acting for long enough. By this line of reasoning, then, the critical question becomes that of how long they have been around. Finally, the closely related issues of how they originated and have dispersed among bacterial species and in their genetic diversity stand as both fascinating and essentially open questions.

Acknowledgments. We are grateful to our many colleagues who provided us with reprints, preprints, and less formal reports of their current work, and in some cases their critical views on various issues. We appreciate particularly the free exchange of unpublished results that is essential in a review like this and which makes for a vigorous field of research. We thank our co-workers in Los Angeles and Toulouse for stimulating discussions, critical comments on the manuscript, general encouragement, and collaboration in the study of transposition, without which little would have been done. Particular thanks are due Mimi Susskind for carefully reading an early version of the manuscript. Invaluable assistance in the preparation of the manuscript, searching out of references, preparation of figures, analysis of sequences, and compilation of data was provided by Daniel H. Wolf. We are very appreciative of the careful attention and excellent work of our copy editor, Philip Koplin, and our production editor, Ellie Tupper.

Research was supported by grants from the National Institutes of Health and by collaborative grants from Centre National de la Recherche Scientifique/National Science Foundation and NATO.

LITERATURE CITED

1. **Ahmed, A.** 1984. A deletion analysis of transposon Tn9. *J. Mol. Biol.* **173:**523–529.
2. **Ahmed, A.** 1986. Evidence for replicative transposition of Tn5 and Tn9. *J. Mol. Biol.* **191:**75–84.
3. **Akroyd, J. E., and N. Symonds.** 1983. Evidence for a conservative pathway of transposition of bacteriophage Mu. *Nature* (London) **303:**84–86.
4. **Allmansberger, R., B. Brau, and W. Piepersberg.** 1985. Genes for gentamycin-(3)-N-acetyl-transferases III and IV. II. Nucleotide sequences of three AAC(3)-III genes and evolutionary aspects. *Mol. Gen. Genet.* **198:**514–520.
5. **Arber, W., M. Humbelin, P. Caspers, H. J. Reif, S. Iida, and J. Meyer.** 1981. Spontaneous mutations in the Escherichia coli prophage P1 and IS-mediated processes. *Cold Spring Harbor Symp. Quant. Biol.* **45:**38–40.
6. **Arber, W., S. Iida, H. Jutte, P. Caspers, J. Meyer, and C. Hanni.** 1980. Rearrangements of genetic material in Escherichia coli as observed on the bacteriophage P1 plasmid. *Cold Spring Harbor Symp. Quant. Biol.* **43:**1197–1208.
7. **Arthur, A., and D. Sherratt.** 1979. Dissection of the transposition process: a transposon-encoded site-specific recombination system. *Mol. Gen. Genet.* **175:**267–274.
8. **Auerswald, E. A., G. Ludwig, and H. Schaller.** 1981. Structural analysis of Tn5. *Cold Spring Harbor Symp. Quant. Biol.* **45:**107–113.
9. **Bachmann, B. J.** 1987. Linkage map of *Escherichia coli* K-12, edition 7, p. 807–876. *In* F. C. Neidhardt, J. L. Ingraham, K. B. Low, E. Magasanik, M. Schaechter, and H. E. Umbarger (ed.), *Escherichia coli and Salmonella typhimurium: Cellular and Molecular Biology.* American Society for Microbiology, Washington, D.C.
10. **Barany, F., J. D. Boeke, and A. Tomasz.** 1982. Staphylococcal plasmids both replicate and express erythromycin resistance in both Streptococcus pneumoniae and Escherichia coli. *Proc. Natl. Acad. Sci. USA* **79:**2991–2995.
11. **Barsomian, G., and T. G. Lessie.** 1986. Replicon fusions promoted by insertion sequences on Pseudomonas cepacia plasmids pTGL6. *Mol. Gen. Genet.* **204:**273–280.
12. **Barsomian, G., and T. G. Lessie.** 1987. IS2 activates the ilvA gene of Pseudomonas cepacia in Escherichia coli. *J. Bacteriol.* **169:**1777–1779.

12a. **Bartlett, D. H., M. E. Wright, and M. Silverman.** 1988. Variable expression of extracellular polysaccharide in the marine bacterium *Pseudomonas atlantica* is controlled by genome rearrangement. *Proc. Natl. Acad. Sci. USA* **85:** 3923–3928.

13. **Beck, C. F., H. Moyed, and J. L. Ingraham.** 1980. The tetracycline resistance transposon Tn10 inhibits translocation of Tn10. *Mol. Gen. Genet.* **179:**453–455.
14. **Bender, J., and N. Kleckner.** 1986. Genetic evidence that Tn10 transposes by a nonreplicative mechanism. *Cell* **45:** 801–815.
15. **Berg, D. E.** 1977. Insertion and excision of the transposable kanamycin resistance determinants Tn5, p. 205–212. *In* A. I. Bukhari, J. A. Shapiro, and S. L. Adhya (ed.), *DNA Elements, Plasmids, and Episomes.* Cold Spring Harbor Laboratory, Cold Spring Harbor, N.Y.
16. **Berg, D. E.** 1980. Control of gene expression by a mobile recombinational switch. *Proc. Natl. Acad. Sci. USA* **77:** 4880–4884.
17. **Berg, D. E.** 1983. Structural requirement for IS50-mediated gene transposition. *Proc. Natl. Acad. Sci. USA* **80:**792–796.
18. **Berg, D. E., J. Davies, B. Allet, and J. D. Rochaix.** 1975. Transposition of R factor genes to bacteriophage lambda. *Proc. Natl. Acad. Sci. USA* **72:**3628–3632.
19. **Berg, D. E., C. Egner, and J. B. Lowe.** 1983. Mechanism of F factor-enhanced excision of transposon Tn5. *Gene* **22:**1–7.
20. **Berg, D. E., M. A. Schmandt, and J. B. Lowe.** 1983. Specificity of transposon Tn5 insertion. *Genetics* **105:** 813–828.
21. **Bernardi, A., and F. Bernardi.** 1981. Site-specific deletions in the recombinant plasmid pSC101 containing the redB-ori region of phage lambda. *Gene* **13:**103–109.
22. **Bernardi, A., and F. Bernardi.** 1981. Complete sequence of an IS element present in pSC101. *Nucleic Acids Res.* **9:** 2905–2911.
23. **Bernardi, F., and A. Bernardi.** 1986. Intramolecular transposition of IS102. *Gene* **42:**11–19.

23a. **Bernardi, F., and A. Bernardi.** 1988. Transcription of the target is required for IS102-mediated deletion. *Mol. Gen. Genet.* **212:**265–270.

24. **Besemer, J.** 1977. On the polarity of insertion mutations, p. 133–135. *In* A. I. Bukhari, J. A. Shapiro, and S. L. Adhya (ed.), *DNA Insertion Elements, Plasmids, and Episomes.* Cold Spring Harbor Laboratory, Cold Spring Harbor, N.Y.

25. **Besemer, J., and M. Herpers.** 1977. Suppression of polarity of insertion mutations within the gal operon of E. coli. *Mol. Gen. Genet.* **151**:295–304.

26. **Biek, D., and J. R. Roth.** 1980. Regulation of Tn5 transposition in Salmonella typhimurium. *Proc. Natl. Acad. Sci. USA* 77:6047–6051.

27. **Biek, D., and J. R. Roth.** 1981. Regulation of Tn5 transposition. *Cold Spring Harbor Symp. Quant. Biol.* **451**: 189–191.

28. **Biel, S. W., G. Adelt, and D. E. Berg.** 1984. Transcriptional control of IS1 transposition in Escherichia coli. *J. Mol. Biol.* **174**:251–264.

29. **Biel, S. W., and D. E. Berg.** 1984. Mechanism of IS1 transposition in E. coli: choice between simple insertion and cointegration. *Genetics* **108**:319–330.

30. **Biel, S. W., and D. L. Hartl.** 1983. Evolution of transposons: natural selection for Tn5 in Escherichia coli K12. *Genetics* **103**:581–592.

31. **Binns, A. N., D. Sciaky, and H. N. Wood.** 1982. Variation in hormone autonomy and regenerative potential of cells transformed by strain A66 of Agrobacterium tumefaciens. *Cell* **31**:605–612.

32. **Bitoun, R., J. Berman, A. Zilberstein, D. Holland, J. B. Cohen, D. Givol, and A. Zamir.** 1983. Promoter mutations that allow nifA-independent expression of the nitrogen fixation nifHDKY operon. *Proc. Natl. Acad. Sci. USA* **80**: 5812–5816.

33. **Blattner, F. R., M. Fiandt, K. K. Hass, P. A. Twose, and W. Szybalski.** 1974. Deletions and insertions in the immunity region of coliphage lambda: revised measurement of the promoter-startpoint distance. *Virology* **62**:458–471.

34. **Blazey, D. L., and R. O. Burns.** 1982. Transcriptional activity of the transposable element Tn10 in the Salmonella typhimurium ilvGEDA operon. *Proc. Natl. Acad. Sci. USA* 79:5011–5015.

35. **Bossi, L., and M. S. Ciampi.** 1981. DNA sequences at the sites of three insertions of the transposable element Tn5 in the histidine operon of Salmonella. *Mol. Gen. Genet.* **183**: 406–408.

36. **Boyen, A., D. Charlier, M. Crabeel, R. Cunin, S. Palchaudhuri, and N. Glansdorff.** 1978. Studies on the control region of the bipolar argECBH operon of Escherichia coli. I. Effect of regulatory mutations and IS2 insertions. *Mol. Gen. Genet.* **161**:185–196.

37. **Brachet, P., H. Eisen, and A. Rambach.** 1970. Mutations of coliphage lambda affecting the expression of replicative functions O and P. *Mol. Gen. Genet.* **108**:266–276.

37a. **Braedt, G.** 1988. Different reading frames are responsible for IS1-dependent deletions and recombination. *Genetics* **118**: 561–570.

37b. **Braedt, G.** 1985. Recombination in *recA* cells between direct repeats of insertion element IS*1*. *J. Bacteriol.* **162**:529–534.

38. **Brahma, N., A. Schumacher, J. Cullum, and H. Saedler.** 1982. Distribution of the Escherichia coli K12 insertion sequences IS1, IS2 and IS3 among other bacterial species. *J. Gen. Microbiol.* **128**:2229–2234.

39. **Brau, B., and W. Piepersberg.** 1983. Cointegrational transduction and mobilization of gentamycin resistance plasmid pWP14a is mediated by IS140. *Mol. Gen. Genet.* **189**: 298–303.

40. **Brau, B., U. Pilz, and W. Piepersberg.** 1984. Genes for gentamicin-(3)-N-acetyltransferases III and IV. I. Nucleotide sequence of the AAC(3)-IV gene and possible involvement of an IS140 element in its expression. *Mol. Gen. Genet.* **193**: 179–187.

41. **Brennan, M. B., and K. Struhl.** 1980. Mechanisms of increasing expression of a yeast gene in E. coli. *J. Mol. Biol.* **136**:333–338.

42. **Brewster, J. M., and E. A. Morgan.** 1981. Tn9 and IS1 inserts in a ribosomal ribonucleic acid operon of Escherichia coli are incompletely polar. *J. Bacteriol.* **148**:897–903.

43. **Brown, A. M., G. M. Coupland, and N. S. Willetts.** 1984. Characterization of IS46, an insertion sequence found on two IncN plasmids. *J. Bacteriol.* **159**:472–481.

44. **Brunel, F., M. Heusterspreute, M. Merchez, V. Ha Thi, M. F. Pilaete, and J. Davison.** 1983. Gene rearrangements leading to the expression of an insertion-inactivated tetracycline resistance gene in pBR322. *Plasmid* **9**:201–214.

45. **Brunier, D., B. Michel, and S. D. Ehrich.** 1988. Copy choice illegitimate DNA recombination. *Cell* **52**:883–892.

46. **Bukhari, A. I.** 1976. Bacteriophage Mu as a transposition element. *Annu. Rev. Genet.* **10**:389–412.

47. **Bukhari, A. I., and H. Khatoon.** 1982. Low level and high level DNA rearrangements in Escherichia coli. *Basic Life Sci.* **20**:235–244.

48. **Bustos-Martinez, J., and C. Gomez-Eichelmann.** 1987. Frequency of IS1-mediated molecular events in different members of the family Enterobacteriaceae. *J. Bacteriol.* **169**: 4946–4949.

49. **Calladine, C. R.** 1982. Mechanics of sequence dependent stacking of bases in B DNA. *J. Mol. Biol.* **161**:343–352.

50. **Calos, M. P., L. Johnsrud, and J. H. Miller.** 1978. DNA sequence at the integration sites of the insertion element IS1. *Cell* **13**:411–418.

51. **Calos, M. P., and J. H. Miller.** 1980. Molecular consequences of deletion formation mediated by transposon Tn9. *Nature* (London) **285**:38–41.

52. **Calos, M. P., and J. H. Miller.** 1980. Transposable elements. *Cell* **20**:579–595.

53. **Caspers, P., B. Dalrymple, S. Iida, and W. Arber.** 1984. IS30, a new insertion sequence of Escherichia coli K12. *Mol. Gen. Genet.* **196**:68–73.

54. **Chan, P. T., and J. Lebowitz.** 1982. Mapping of RNA polymerase binding sites in R12 derived plasmids carrying the replication-incompatibility region and the insertion element IS1. *Nucleic Acids Res.* **10**:7295–7311.

55. **Chandler, M., B. Allet, E. Gallay, E. Boy de la Tour, and L. Caro.** 1977. Involvement of IS1 in the dissociation of the r-determinant and RTF components of the plasmid R100.1. *Mol. Gen. Genet.* **153**:289–295.

56. **Chandler, M., E. Boy de la Tour, D. Willems, and L. Caro.** 1979. Some properties of the chloramphenicol resistance transposon Tn9. *Mol. Gen. Genet.* **176**:221–231.

57. **Chandler, M., M. Clerget, and D. J. Galas.** 1982. The transposition frequency of IS1-flanked transposons is a function of their size. *J. Mol. Biol.* **154**:229–243.

58. **Chandler, M., and D. J. Galas.** 1983. IS1-mediated tandem duplication of plasmid pBR322. Dependence on recA and on DNA polymerase I. *J. Mol. Biol.* **165**:183–190.

59. **Chandler, M., and D. J. Galas.** 1983. Cointegrate formation mediated by Tn9. II. Activity of IS1 is modulated by external DNA sequences. *J. Mol. Biol.* **170**:61–91.

60. **Chandler, M., and D. J. Galas.** 1985. Studies on the transposition of IS1, p. 53–77. *In* D. R. Helinski, S. N. Cohen, D. B. Clewell, D. A. Jackson, and A. Hollaender (ed.), *Plasmids in Bacteria.* Plenum Publishing Corp., New York.

61. **Chandler, M., J. Sechaud, and L. Caro.** 1982. A mutant of the plasmid R100.1 capable of producing autonomous circular forms of its resistance determinant. *Plasmid* **7**: 251–262.

62. **Chao, L., and S. McBroom.** 1985. Evolution of transposable elements: an IS10 insertion increases fitness in E. coli. *Mol. Biol. Evol.* **2**:359–269.

63. **Charlier, D., J. Piette, and N. Glansdorff.** 1982. IS3 can function as a mobile promoter in E. coli. *Nucleic Acids Res.* **10**:5935–5948.

64. **Chumley, F. G., and J. R. Roth.** 1980. Rearrangement of the bacterial chromosome using Tn10 as a region of homology. *Genetics* **94**:1–14.

65. **Ciampi, M. S., M. B. Schmid, and J. R. Roth.** 1982. Transposon Tn10 provides a promoter for transcription of adjacent sequences. *Proc. Natl. Acad. Sci. USA* **79**: 5016–5020.

66. **Clements, M. B., and M. Syvanen.** 1981. Isolation of polA mutation that affects transposition of insertion sequences and transposons. *Cold Spring Harbor Symp. Quant. Biol.* **451**:201–204.

67. **Clerget, M., M. Chandler, and L. Caro.** 1980. Isolation of an IS1 flanked kanamycin resistance transposon from Rldrd19. *Mol. Gen. Genet.* **180**:123–127.

68. **Clerget, M., M. Chandler, and L. Caro.** 1981. The structure of Rldrd19: a revised physical map of the plasmid. *Mol. Gen. Genet.* **181**:183–191.

69. **Cohen, S. N.** 1977. Special sequences in the structure of cointegrate drug resistance plasmids related to F, p. 672–673. *In* A. I. Bukhari, J. A. Shapiro, and S. L. Adhya (ed.), *DNA Insertion Elements, Plasmids, and Episomes.* Cold Spring Harbor Laboratory, Cold Spring Harbor, N.Y.

70. **Collins, J.** 1980. Instability of palindromic DNA in Escherichia coli. *Cold Spring Harbor Symp. Quant. Biol.* **45**: 409–416.

71. **Colonna, B., M. Nicoletti, P. Visca, M. Casalino, P. Valenti, and F. Maimone.** 1985. Composite IS1 elements encoding hydroxamate-mediated iron uptake in FIme plasmids from epidemic Salmonella spp. *J. Bacteriol.* **162**:307–316.

72. **Cornelis, G., and H. Saedler.** 1980. Deletions and an inversion induced by a resident IS1 of the lactose transposon Tn951. *Mol. Gen. Genet.* **178**:367–374.

73. **Crabeel, M., D. Charlier, R. Cunin, and N. Glansdorff.** 1979. Cloning and endonuclease restriction analysis of argF and of the control region of the argECBH bipolar operon in Escherichia coli. *Gene* **2**:207–231.

74. **Craigie, R., and K. Mizuuchi.** 1985. Mechanism of transposition of bacteriophage Mu: structure of a transposition intermediate. *Cell* **41**:867–876.

75. **Cullum, J., and P. Broda.** 1979. Chromosome transfer and Hfr formation by F in rec+ and RecA strains of Escherichia coli K12. *Plasmid* **2**:358–365.

76. **Currier, T. C., and M. K. Morgan.** 1982. Direct DNA repeat in plasmid R68.45 is associated with deletion formation and concomitant loss of chromosome mobilization ability. *J. Bacteriol.* **150**:251–259.

77. **Dalrymple, B., and W. Arber.** 1985. Promotion of RNA transcription on the insertion element IS30 of E. coli K12. *EMBO J.* **4**:2687–2693.

78. **Dalrymple, B., and W. Arber.** 1986. The characterization of terminators of RNA transcripton on IS30 and an analysis of their role in IS element-mediated polarity. *Gene* **44**:1–10.

79. **Dalrymple, B., P. Caspers, and W. Arber.** 1984. Nucleotide sequence of the prokaryotic mobile genetic element IS30. *EMBO J.* **3**:2145–2149.

80. **Das, A., M. Gottesman, and S. Adhya.** 1977. Polarity of insertion mutations is caused by rho-mediated termination of transcription, p. 93–97. *In* A. I. Bukhari, J. A. Shapiro, and S. L. Adhya (ed.), *DNA Insertion Elements, Plasmids, and Episomes.* Cold Spring Harbor Laboratory, Cold Spring Harbor, N.Y.

81. **Datta, N.** 1985. Plasmids as organisms, p. 3–16. *In* D. R. Helinski, S. N. Cohen, D. B. Clewell, D. A. Jackson, and A. Hollaender (ed.), *Plasmids in Bacteria.* Plenum Publishing Corp., New York.

82. **Datta, N., and V. M. Hughes.** 1983. Plasmids of the same Inc groups in enterobacteria before and after the medical use of antibiotics. *Nature* (London) **306**:616–617.

83. **Datta, A. R., and J. L. Rosner.** 1987. Reduced transposition in rho mutants of Escherichia coli K-12. *J. Bacteriol.* **169**: 888–890.

83a.**Davidson, N., R. C. Deonier, S. Hu, and E. Ohtsubo.** 1975. Electron microscopic heteroduplex studies of sequence relations among plasmids of *Escherichia coli.* X. Deoxyribonucleic acid sequence organization of F and of F-primes, and the sequences involved in Hfr formation, p. 56–65. *In* D. Schlessinger (ed.), *Microbiology—1974.* American Society for Microbiology, Washington, D.C.

84. **Davis, M. A., R. W. Simons, and N. Kleckner.** 1985. Tn10 protects itself at two levels from fortuitous activation by external promoters. *Cell* **43**:379–387.

85. **De Crombugghe, B., S. Adhya, M. Gottesman, and L. Pastan.** 1973. Effect of rho on transcription of bacterial operons. *Nature* (London) **241**:260–264.

85a.**de Lorenzo, Z., M. Herrero, and J. B. Nielands.** 1988. IS1-mediated mobility of the aerobactin system of pColV-K30 in E. coli. *Mol. Gen. Genet.* **213**:487–490.

86. **Deonier, R. C.** 1987. Locations of native insertion sequence elements, p. 982–989. *In* F. C. Neidhardt, J. L. Ingraham, K. B. Low, B. Magasanik, M. Schaechter, and H. E. Umbarger (ed.), *Escherichia coli and Salmonella typhimurium: Cellular and Molecular Biology.* American Society for Microbiology, Washington, D.C.

87. **Depicker, A., M. De Block, D. Inze, M. Van Montagu, and J. Schell.** 1980. IS-like element IS8 in RP4 plasmid and its involvement in cointegration. *Gene* **10**:329–338.

88. **Derbyshire, K. M., L. Hwang, and D. F. Grindley.** 1987. Genetic analysis of the insertion sequence IS903 transposase with its terminal inverted repeats. *Proc. Natl. Acad. Sci. USA* **84**:8049–8053.

89. **Devos, R., R. Contreras, J. van Emmelo, and W. Fiers.** 1979. Identification of the translocatable element IS1 in a molecular chimera constructed with plasmid pBR322 DNA into which a bacteriophage MS2 DNA copy was inserted by the poly(dA) · poly(dT) linker method. *J. Mol. Biol.* **128**: 621–632.

90. **Diaz Aroca, E., F. de la Cruz, J. C. Zabala, and J. M. Ortiz.** 1984. Characterization of the new insertion sequence IS91 from an alpha-hemolysin plasmid of Escherichia coli. *Mol. Gen. Genet.* **193**:493–499.

91. **Diaz Aroca, E., M. V. Mendiola, J. C. Zabala, and F. de la Cruz.** 1987. Transposition of IS91 does not generate a target duplication. *J. Bacteriol.* **169**:442–443.

92. **Dickerson, R. E.** 1983. Base sequence and helix structure variation in B and A DNA. *J. Mol. Biol.* **166**:419–441.

93. **DiNardo, S., K. A. Voelkel, R. Sternglanz, A. E. Reynolds, and A. Wright.** 1982. Escherichia coli DNA topoisomerase mutants have compensatory mutations in gyrase genes. *Cell* **31**:43–51.

94. **Dusha, I., S. Kovalenko, Z. Banfalvi, and A. Kondorosi.** 1987. Rhizobium meliloti insertion element ISRm2 and its use for identification of the fixX gene. *J. Bacteriol.* **169**: 1403–1409.

95. **Dykhuizen, D. E., S. A. Sawyer, L. Green, R. D. Miller, and**

D. L. Hartl. 1985. Joint distribution of insertion elements IS4 and IS5 in natural isolates of Escherichia coli. *Genetics* **111**:219–231.

96. **Egner, C., and D. E. Berg.** 1981. Excision of transposon Tn5 is dependent on the inverted repeats but not on the transposase function of Tn5. *Proc. Natl. Acad. Sci. USA* **78:** 459–463.
97. **Eisen, H., P. Barrand, W. Spiegelman, L. F. Reichart, S. Heinemann, and C. Georgopoulos.** 1982. Mutants in the y region of bacteriophage lambda constitutive for repressor synthesis: their isolation and characterization of the Hyp phenotype. *Gene* **20**:71–81.
98. **Emerick, A. W.** 1982. Read-through transcription from a derepressed Tn3 promoter affects ColE1 functions on a ColE1::Tn3 composite plasmid. *Mol. Gen. Genet.* **185:** 408–417.
99. **Engler, J. A., and M. P. van Bree.** 1981. The nucleotide sequence and protein-coding capability of the transposable element IS5. *Gene* **14**:155–163.
100. **Faelen, M., A. Toussaint, B. Waggoner, L. Desmet, and M. Pato.** 1986. Transposition and replication of maxi-Mu derivatives of bacteriophage Mu. *Virology* **153**:70–79.
101. **Farabaugh, P. J., U. Schmeissner, M. Hofer, and J. H. Miller.** 1978. Genetic studies of the lac repressor. VII. On the molecular nature of spontaneous hotspots in the laci gene of E. coli. *J. Mol. Biol.* **126**:847–863.
102. **Fiandt, M., W. Szybalski, and M. H. Malamy.** 1972. Polar mutations in lac, gal, and phage lambda consist of a few IS-DNA sequences inserted with either orientation. *Mol. Gen. Genet.* **119**:223–231.
103. **Fischhoff, D. A., G. F. Vovis, and N. D. Zinder.** 1980. Organization of chimeras between filamentous bacteriophage f1 and plasmid pSC101. *J. Mol. Biol.* **144**:247–265.
104. **Foster, T. J.** 1977. Insertion of the tetracycline resistance translocation unit Tn10 in the lac operon of Escherichia coli K12. *Mol. Gen. Genet.* **154**:305–309.
105. **Foster, T. J., V. Lundblad, S. Hanley Way, S. M. Halling, and N. Kleckner.** 1981. Three Tn10-associated excision events: relationship to transposition and role of direct and inverted repeats. *Cell* **23**:215–227.
106. **Francois, V., J. Louarn, J. Patte, and J.-M. Louarn.** 1987. A system for in vivo selection of genomic rearrangements with predetermined endpoints in Escherichia coli using modified Tn10 transposons. *Gene* **56**:99–108.
107. **Fuller, R. S., B. E. Funnel, and A. Kornberg.** 1984. The dnaA protein complex with the E. coli chromosome origin (oriC) and other DNA sites. *Cell* **38**:889–900.
108. **Galas, D. J., M. P. Calos, and J. H. Miller.** 1980. Sequence analysis of Tn9 insertions in the lacZ gene. *J. Mol. Biol.* **144**:19–41.
109. **Galas, D. J., and M. Chandler.** 1981. On the molecular mechanisms of transposition. *Proc. Natl. Acad. Sci. USA* **78**:4858–4862.
110. **Galas, D. J., and M. Chandler.** 1982. Structure and stability of Tn9-mediated cointegrates. Evidence for two pathways of transposition. *J. Mol. Biol.* **154**:245–272.
111. **Gamas, P., M. G. Chandler, P. Prentki, and D. J. Galas.** 1987. Escherichia coli integration host factor binds specifically to the ends of the insertion sequence IS1 and to its major insertion hot-spot in pBR322. *J. Mol. Biol.* **195:** 261–272.
112. **Gamas, P., D. Galas, and M. Chandler.** 1985. DNA sequence at the end of IS1 required for transposition. *Nature* (London) **317**:458–460.
113. **Garfinkel, D. J., and E. W. Nester.** 1980. Agrobacterium tumefaciens mutants affected in crown gall tumorigenesis and octopine catabolism. *J. Bacteriol.* **144**:732–743.
114. **Georgopoulos, C., N. McKittrick, G. Herrick, and H. Eisen.** 1982. An IS4 transposition causes a 13-bp duplication of phage lambda DNA and results in the constitutive expression of the cI and cro gene products. *Gene* **20**:83–90.
115. **Gertman, E., B. N. White, D. Berry, and A. M. Kropinski.** 1986. IS222, a new insertion element associated with the genome of Pseudomonas aeruginosa. *J. Bacteriol.* **166**:1134–1136.
116. **Ghosal, D., J. Gross, and H. Saedler.** 1979. DNA sequence of IS2-7 and generation of mini-insertions by replication of IS2 sequences. *Cold Spring Harbor Symp. Quant. Biol.* **43:** 1193–1196.
117. **Ghosal, D., and H. Saedler.** 1978. DNA sequence of the mini-insertion IS2-6 and its relation to the sequence IS2. *Nature* (London) **275**:611–617.
118. **Ghosal, D., and H. Saedler.** 1979. IS2-61 and IS2-611 arise by illegitimate recombination from IS2-6. *Mol. Gen. Genet.* **176**:233–238.
119. **Ghosal, D., H. Sommer, and H. Saedler.** 1979. Nucleotide sequence of the transposable DNA-element IS2. *Nucleic Acids Res.* **6**:1111–1122.
120. **Glansdorff, N., D. Charlier, and M. Zafarullah.** 1981. Activation of gene expression by IS2 and IS3. *Cold Spring Harbor Symp. Quant. Biol.* **45**:153–156.
121. **Green, L., R. D. Miller, D. E. Dykhuizen, and D. L. Hartl.** 1984. Distribution of DNA insertion element IS5 in natural isolates of Escherichia coli. *Proc. Natl. Acad. Sci. USA* **81**:4500–4504.
122. **Grindley, N., and L. A. Wiater.** 1988. γδ transposase and integration host factor bind cooperatively at both ends of γδ. *EMBO J.* **7**:1907–1912.
123. **Grindley, N. D.** 1978. IS1 insertion generates duplication of a nine base pair sequence at its target site. *Cell* **13**:419–426.
124. **Grindley, N. D., and C. M. Joyce.** 1980. Genetic and DNA sequence analysis of the kanamycin resistance transposon Tn903. *Proc. Natl. Acad. Sci. USA* **77**:7176–7180.
125. **Grindley, N. D., and D. J. Sherratt.** 1979. Sequence analysis at IS1 insertion sites: models for transposition. *Cold Spring Harbor Symp. Quant. Biol.* **432**:1257–1261.
126. **Grindley, N. D. F., and C. M. Joyce.** 1981. Analysis of the structure and function of the kanamycin-resistance transposon Tn903. *Cold Spring Harbor Symp. Quant. Biol.* **45:** 125–133.
127. **Grindley, N. D. F., and R. R. Reed.** 1985. Transpositional recombination in prokaryotes. *Annu. Rev. Biochem.* **54:** 863–896.
128. **Gulletta, E., A. Das, and S. Adhya.** 1983. The pleiotropic ts15 mutation of E. coli is an IS1 insertion in the rho structural gene. *Genetics* **105**:265–280.
129. **Guyer, M. S.** 1978. The gamma delta sequence of F is an insertion sequence. *J. Mol. Biol.* **126**:347–365.
130. **Guyer, M. S., R. R. Reed, J. A. Steitz, and K. B. Low.** 1980. Identification of a sex factor affinity site in E. coli as gamma delta. *Cold Spring Harbor Symp. Quant. Biol.* **45**:135–140.
131. **Haas, D., and B. W. Holloway.** 1976. R factor variants with enhanced sex factor activity in Pseudomonas aeruginosa. *Mol. Gen. Genet.* **144**:243–251.
132. **Haas, D., and G. Riess.** 1983. Spontaneous deletions of the chromosome-mobilizing plasmid R68.45 in Pseudomonas aeruginosa. *Plasmid* **9**:42–52.
133. **Habermann, P., R. Klaer, S. Kuhn, and P. Starlinger.** 1979. IS4 is found between eleven or twelve base pair duplications. *Mol. Gen. Genet.* **175**:369–373.

133a.**Hadley, R. G., and R. C. Deonier.** 1980. IS2-IS2 and IS3-IS3 relative recombination frequencies in F integration. *Plasmid* **3:**48–64.

134. **Halling, S. M., and N. Kleckner.** 1982. A symmetrical six-base-pair target site sequence determines Tn10 insertion specificity. *Cell* **28:**155–163.

135. **Halling, S. M., R. W. Simons, J. C. Way, R. B. Walsh, and N. Kleckner.** 1982. DNA sequence organization of IS10-right of Tn10 and comparison with IS10-left. *Proc. Natl. Acad. Sci. USA* 79:2608–2612.

136. **Hanni, C., J. Meyer, S. Iida, and W. Arber.** 1982. Occurrence and properties of the composite transposon Tn2672: evolution of multiple drug resistance transposons. *J. Bacteriol.* **150:**1266–1273.

137. **Harayama, S., T. Oguchi, and T. Iino.** 1984. The E. coli K-12 chromosome flanked by two IS10 sequences transposes. *Mol. Gen. Genet.* **197:**62–66.

138. **Harayama, S., T. Oguchi, and T. Iino.** 1984. Does Tn10 transpose via the cointegrate molecule. *Mol. Gen. Genet.* **194:**444–450.

139. **Harshey, R. M.** 1984. Transposition without duplication of infecting bacteriophage Mu DNA. *Nature* (London) **311:** 580–581.

140. **Harshey, R. M., and A. I. Bukhari.** 1981. A mechanism of DNA transposition. *Proc. Natl. Acad. Sci. USA* **78:** 1090–1094.

141. **Hartl, D. L., D. E. Dykhuizen, R. D. Miller, L. Green, and J. de Framond.** 1983. Transposable element IS50 improves growth rate of E. coli cells without transposition. *Cell* **35:**503–510.

142. **Hartl, D. L., and S. A. Sawyer.** 1988. Why do unrelated insertion sequences occur together in the genome of Escherichia coli? *Genetics* **118:**537–541.

143. **Hashimoto, H., and R. H. Rownd.** 1975. Transition of the R factor NR1 in Proteus mirabilis: level of drug resistance of non-transitioned and transitioned cells. *J. Bacteriol.* **123:** 56–68.

144. **Hawley, D., and W. McClure.** 1983. Prokaryotic promoter sequences S. *Nucleic Acids Res.* **11:**2237–2255.

145. **Heffron, F., B. McCarthy, H. Ohtsubo, and E. Ohtsubo.** 1979. DNA sequence analysis of the transposon Tn3: three genes and three sites involved in transposition of Tn3. *Cell* **18:**1153–1161.

146. **Heffron, F., M. So, and B. J. McCarthy.** 1978. In vitro mutagenesis of a circular DNA molecule by using synthetic restriction sites. *Proc. Natl. Acad. Sci. USA* **75:**6012–6016.

146a.**Herman, R., K. Neugebauer, H. Zentraf, and H. Schaller.** 1978. Transposition of a DNA sequence determining kanamycin resistance into the single-stranded genome of bacteriophage fd. *Mol. Gen. Genet.* **159:**171–178.

147. **Hinton, D. M., and R. E. Musso.** 1982. Transcription initiation sites within an IS2 insertion in a Gal$^-$ constitutive mutant of Escherichia coli. *Nucleic Acids Res.* **10:** 5015–5031.

148. **Hinton, D. M., and R. E. Musso.** 1983. Specific in vitro transcription of the insertion sequence IS2. *J. Mol. Biol.* **169:**53–81.

149. **Hirsch, H. J., H. Saedler, and P. Starlinger.** 1972. Insertion mutations in the control region of the galactose operon of E. coli. II. Physical characterization of the mutations. *Mol. Gen. Genet.* **115:**266–276.

150. **Hirsch, H.-J., P. Starlinger, and P. Brachet.** 1972. Two kinds of insertions in bacterial genes. *Mol. Gen. Genet.* **119:** 191–206.

151. **Hirschel, B. J., and D. Berg.** 1982. *J. Mol. Biol.* **155:** 105–120.

152. **Hirschel, B. J., D. J. Galas, D. E. Berg, and M. Chandler.** 1982. Structure and stability of transposon 5-mediated cointegrates. *J. Mol. Biol.* **159:**557–580.

153. **Hirschel, B. J., D. J. Galas, and M. Chandler.** 1982. Cointegrate formation by Tn5, but not transposition, is dependent on recA. *Proc. Natl. Acad. Sci. USA* **79:**4530–4534.

154. **Hopkins, J. D., M. Clements, and M. Syvanen.** 1983. New class of mutations in Escherichia coli (uup) that affect precise excision of insertion elements and bacteriophage Mu growth. *J. Bacteriol.* **153:**384–389.

155. **Hu, M., and R. C. Deonier.** 1981. Comparison of IS1, IS2, and IS3 copy number in Escherichia coli strains K12, B and C. *Gene* **16:**161–170.

156. **Hu, S., E. Ohtsubo, N. Davidson, and H. Saedler.** 1975. Electron microscopy heteroduplex studies of sequence relations among bacterial plasmids: identification and mapping of the insertion sequences IS1 and IS2 in F and R plasmids. *J. Bacteriol.* **122:**764–775.

157. **Hu, S., K. Ptashne, S. N. Cohen, and N. Davidson.** 1975. alphabeta sequence of F is IS3. *J. Bacteriol.* **123:**687–692.

158. **Huber, P., S. Iida, and W. Arber.** 1987. A transcriptional terminator sequence in the prokaryotic transposable element IS1. *Mol. Gen. Genet.* **206:**485–490.

159. **Hughes, V. M., and N. Datta.** 1983. Conjugal plasmids in bacteria of the 'pre-antibiotic' era. *Nature* (London) **302:** 725–726.

159a.**Hwa, V., N. B. Schoemaker, and A. A. Salyers.** 1988. Direct repeats flanking the *Bacteroides* transposon Tn*4351* are insertion sequence elements. *J. Bacteriol.* **170:**449–451.

159b.**Ichikawa, H., K. Ikeda, W. L. Wishart, and E. Ohtsubo.** 1987. Specific binding of transposase to terminal inverted repeats of transposable element Tn3. *Proc. Natl. Acad. Sci. USA* **84:**8220–8224.

160. **Iida, S.** 1983. On the origin of the chloramphenicol resistance transposon Tn9. *J. Gen. Microbiol.* **129:**1217–1225.

161. **Iida, S., and R. Hiestand Nauer.** 1986. Insertion element IS1 can generate a 10-base pair target duplication. *Gene* **45:** 233–235.

162. **Iida, S., R. Hiestand Nauer, and W. Arber.** 1985. Transposable element IS1 intrinsically generates target duplications of variable length. *Proc. Natl. Acad. Sci. USA* **82:**839–843.

163. **Iida, S., Marcoli, R., and T. A. Bickle.** 1981. Variant insertion element IS1 generates 8-base pair duplications of the target sequence. *Nature* (London) **294:**374–376.

164. **Iida, S., J. Meyer, and W. Arber.** 1978. The insertion element IS1 is a natural constituent of coliphage P1 DNA. *Plasmid* **1:**357–365.

165. **Iida, S., J. Meyer, and W. Arber.** 1983. Prokaryotic IS elements, p. 159–221. *In* J. A. Shapiro (ed.), *Mobile Genetic Elements.* Academic Press, New York.

166. **Iida, S., J. Meyer, P. Linder, N. Goto, R. Nakaya, H. J. Reif, and W. Arber.** 1982. The kanamycin resistance transposon Tn2680 derived from the R plasmid Rts1 and carried by phage p1Km has flanking 0.8-kb-long direct repeats. *Plasmid* **8:**187–198.

167. **Iida, S., B. Mollet, J. Meyer, and W. Arber.** 1984. Functional characterization of the prokaryotic mobile genetic element IS26. *Mol. Gen. Genet.* **198:**84–89.

168. **Ilyina, T. S., E. V. Nechaeva, Y. M. Romanova, and G. B. Smirnov.** 1981. Isolation and mapping of Escherichia coli K12 mutants defective in Tn9 transposition. *Mol. Gen. Genet.* **181:**384–389.

169. **Ilyina, T. S., Y. M. Romanova, and G. B. Smirnov.** 1981.

The effect of tnm mutations of Escherichia coli K12 on transposition of various movable genetic elements. *Mol. Gen. Genet.* **183**:376–379.

170. **Isberg, R. R., A. L. Lazaar, and M. Syvanen.** 1982. Regulation of Tn5 by the right-repeat proteins: control at the level of the transposition reaction. *Cell* **30**:883–892.

171. **Isberg, R. R., and M. Syvanen.** 1981. Replicon fusions promoted by the inverted repeats of Tn5. The right repeat is an insertion sequence. *J. Mol. Biol.* **150**:15–32.

172. **Isberg, R. R., and M. Syvanen.** 1982. DNA gyrase is a host factor required for transposition of Tn5. *Cell* **30**:9–18.

173. **Isberg, R. R., and M. Syvanen.** 1985. Compartmentalization of the proteins encoded by IS50R. *J. Biol. Chem.* **260**: 3645–3651.

174. **Ishiguro, N., and G. Sato.** 1984. Spontaneous deletion of citrate-utilizing ability promoted by insertion sequences. *J. Bacteriol.* **160**:642–650.

174a.**Ishiguro, N., and G. Sato.** 1988. Nucleotide sequence of IS*3411*, which flanks the citrate utilization determinant of transposon Tn*3411*. *J. Bacteriol.* **170**:1902–1906.

175. **Ishizaki, K., and E. Ohtsubo.** 1985. Cointegration and resolution mediated by IS101 present in plasmid p SC101. *Mol. Gen. Genet.* **199**:388–395.

176. **Jakowec, M., P. Prentki, M. Chandler, and D. J. Galas.** 1988. Mutational analysis of the open reading frames in transposable element IS1. *Genetics* **120**:47–55.

177. **Jaskunas, S. R., and M. Nomura.** 1977. Organization of ribosomal protein genes of Escherichia coli as analyzed by polar insertion mutations. *J. Biol. Chem.* **252**:7337–7343.

178. **Jaurin, B., and S. Normark.** 1983. Insertion of IS2 creates a novel ampC promoter in Escherichia coli. *Cell* **32**:809–816.

179. **Johnson, R., M. Bruist, and M. Simon.** 1986. Host protein requirements for in vitro, site-specific DNA inversion. *Cell* **46**:531–539.

180. **Johnson, R. C., and W. S. Reznikoff.** 1983. DNA sequences at the ends of transposon Tn5 required for transposition. *Nature* (London) **304**:280–282.

181. **Johnson, R. C., and W. S. Reznikoff.** 1984. Copy number control of Tn5 transposition. *Genetics* **107**:9–18.

182. **Johnson, R. C., and W. S. Reznikoff.** 1984. Role of the IS50 R proteins in the promotion and control of Tn5 transposition. *J. Mol. Biol.* **177**:645–661.

183. **Johnson, R. C., J. C. Yin, and W. S. Reznikoff.** 1982. Control of Tn5 transposition in Escherichia coli is mediated by protein from the right repeat. *Cell* **30**:873–882.

184. **Johnsrud, L.** 1979. DNA sequence of the transposable element IS1. *Mol. Gen. Genet.* **169**:213–218.

185. **Johnsrud, L., M. P. Calos, and J. H. Miller.** 1978. The transposon Tn9 generates a 9bp repeated sequence during integration. *Cell* **15**:1209–1219.

186. **Jordan, E., H. Saedler, and P. Starlinger.** 1968. 0-zero and strong polar mutations in the gal operon are insertions. *Mol. Gen. Genet.* **102**:353–363.

187. **Jund, R., and G. Loison.** 1982. Activation of transcription of a yeast gene in E. coli by an IS5 element. *Nature* (London) **296**:680–681.

188. **Kaluza, K., M. Hahn, and H. Hennecke.** 1985. Repeated sequences similar to insertion elements clustered around the nif region of the Rhizobium japonicum genome. *J. Bacteriol.* **162**:535–542.

189. **Kamp, D., and R. Kahmann.** 1981. Two pathways in bacteriophage Mu transposition. *Cold Spring Harbor Symp. Quant. Biol.* **451**:329–336.

190. **Kanazawa, H., T. Kiyasu, T. Noumi, M. Futai, and K. Yamaguchi.** 1984. Insertion of transposable elements in the promoter proximal region of the gene cluster for Escherichia coli H+-ATPase: 8 base pair repeat generated by insertion of IS1. *Mol. Gen. Genet.* **194**:179–187.

191. **Kearney, B., P. C. Ronald, D. Dahlbeck, and B. J. Staskawicz.** 1988. Molecular basis for evasion of plant host defence in bacterial spot disease of pepper. *Nature* (London) **332**: 541–543.

192. **Khatoon, H., and A. I. Bukhari.** 1981. DNA rearrangements associated with reversion of bacteriophage Mu-induced mutations. *Genetics* **98**:1–24.

193. **Klaer, R., S. Kuhn, H. J. Fritz, E. Tillmann, I. Saint Girons, P. Habermann, D. Pfeifer, and P. Starlinger.** 1980. Studies on transposition mechanisms and specificity of IS4. *Cold Spring Harbor Symp. Quant. Biol.* **451**:215–224.

194. **Klaer, R., S. Kuhn, E. Tillmann, H. J. Fritz, and P. Starlinger.** 1981. The sequence of IS4. *Mol. Gen. Genet.* **181**:169–175.

195. **Kleckner, N., D. E. Barker, D. G. Ross, D. Botstein, J. A. Swan, and M. Zabeau.** 1978. Properties of the translocatable tetracycline resistance element Tn10 in E. coli and bacteriophage lambda. *Genetics* **90**:427–450.

196. **Kleckner, N., R. K. Chan, B. K. Tye, and D. Botstein.** 1975. Mutagenesis by insertion of a drug resistance element carrying an inverted repetition. *J. Mol. Biol.* **97**:561–575.

197. **Kleckner, N., K. Reichardt, and D. Botstein.** 1979. Inversions and deletions of the Salmonella chromosome generated by the translocatable tetracycline resistance element Tn10. *J. Mol. Biol.* **127**:89–115.

198. **Kleckner, N., and D. G. Ross.** 1980. recA-dependent genetic switch generated by transposon Tn10. *J. Mol. Biol.* **144**: 215–221.

199. **Koch, C., and R. Kahnmann.** 1986. Purification and properties of the E. coli host factor required for inversion of the G segment in bacteriophage Mu. *J. Biol. Chem.* **261**:15673–15678.

200. **Kohara, Y., K. Akiyama, and K. Isono.** 1987. The physical map of the whole E. coli chromosome: application of a new strategy for rapid analysis and sorting of a large genomic library. *Cell* **50**:495–508.

201. **Kornberg, A.** 1980. *DNA Replication*. W. H. Freeman, San Francisco.

202. **Kothary, R. K., D. Jones, and E. P. Candido.** 1985. IS186: an Escherichia coli insertion element isolated from a cDNA library. *J. Bacteriol.* **164**:957–959.

203. **Krebs, M. P., and W. S. Reznikoff.** 1986. Transcriptional and translational initiation sites of IS50: control of transposase and inhibitor expression. *J. Mol. Biol.* **192**:781–791.

204. **Kroger, M., and G. Hobom.** 1982. Structural analysis of insertion sequence IS5. *Nature* (London) **297**:159–162.

205. **Labigne-Roussel, A., S. Briaux-Gerbaud, and P. Courvalin.** 1983. Tn1525, a kanamycin R determinant flanked by two direct copies of IS15. *Mol. Gen. Genet.* **189**:90–101.

206. **Labigne Roussel, A., and P. Courvalin.** 1983. IS15, a new insertion sequence widely spread in R plasmids of gram-negative bacteria. *Mol. Gen. Genet.* **189**:102–112.

207. **Labigne Roussel, A., G. Gerbaud, and P. Courvalin.** 1981. Translocation of sequences encoding antibiotic resistance from the chromosome to a receptor plasmid in Salmonella ordonez. *Mol. Gen. Genet.* **182**:390–408.

208. **Lam, S., and J. R. Roth.** 1983. Genetic mapping of IS200 copies in Salmonella typhimurium strain LT2. *Genetics* **105**:801–811.

209. **Lam, S., and J. R. Roth.** 1983. IS200: a Salmonella-specific insertion sequence. *Cell* **34**:951–960.

210. **Lee, Y., and F. J. Schmidt.** 1985. Characterization of the in

vivo RNA product of the pOUT promoter of IS10R. *J. Bacteriol.* **164**:556–562.

211. **Leemans, J., R. Villarroel, B. Silva, M. van Montagu, and J. Schell.** 1980. Direct repetition of a 1.2Md DNA sequence is involved in site-specific recombination by the P1 plasmid R68. *Gene* **10**:319–328.

211a. **Lichens-Park, A., and M. Syvanen.** 1988. Cointegrate formation by IS50 requires multiple donor molecules. *Mol. Gen. Genet.* **211**:244–251.

212. **Lieb, M.** 1981. A fine structure map of spontaneous and induced mutations in the lambda repressor gene, including insertions of IS elements. *Mol. Gen. Genet.* **184**:364–371.

213. **Lindahl, G., Y. Hirota, and F. Jacob.** 1971. On the process of cellular division in Escherichia coli: replication of the bacterial chromosome under control of prophage P2. *Proc. Natl. Acad. Sci. USA* **68**:2407–2412.

214. **Liu, L. F., and J. C. Wang.** 1987. Supercoiling of the DNA template during transcription. *Proc. Natl. Acad. Sci. USA* **84**:7024–7027.

215. **Louarn, J.-M., J.-P. Bouche, F. Legendre, J. Louarn, and J. Patte.** 1985. Characterization and properties of very large inversions of the E. coli chromosome along the origin-to-terminus axis. *Mol. Gen. Genet.* **201**:467–476.

216. **Lowe, J. B., and D. E. Berg.** 1983. A product of the Tn5 transposase gene inhibits transposition. *Genetics* **103**: 605–615.

217. **Lundblad, V., and N. Kleckner.** 1985. Mismatch repair mutations of Escherichia coli K12 enhance transposon excision. *Genetics* **109**:3–19.

218. **Lundblad, V., A. F. Taylor, G. R. Smith, and N. Kleckner.** 1984. Unusual alleles of recB and recC stimulate excision of inverted repeat transposons Tn10 and Tn5. *Proc. Natl. Acad. Sci. USA* **81**:824–828.

219. **Lupski, J. R., P. Gershon, L. S. Ozaki, and G. N. Godson.** 1984. Specificity of Tn5 insertions into a 36-bp DNA sequence repeated in tandem seven times. *Gene* **30**:99–106.

220. **Lusky, M., M. Kroger, and G. Hobom.** 1981. Detection of replicational inceptor signals in IS5. *Cold Spring Harbor Symp. Quant. Biol.* **45**:173–176.

221. **MacHattie, L. A., and J. B. Jackowski.** 1977. Physical structure and deletion effects of the chloramphenicol resistance element Tn9 in phage lambda, p. 219–228. *In* A. I. Bukhari, J. A. Shapiro, and S. Adhya (ed.), *DNA Insertion Elements, Plasmids, and Episomes.* Cold Spring Harbor Laboratory, Cold Spring Harbor, N.Y.

222. **Machida, C., and Y. Machida.** 1987. Base substitutions in transposable element IS1 cause DNA duplication of variable length at the target site for plasmid cointegration. *EMBO J.* **6**:1799–1803.

223. **Machida, C., Y. Machida, H. C. Wang, K. Ishizaki, and E. Ohtsubo.** 1983. Repression of cointegration ability of insertion element IS1 by transcriptional readthrough from flanking regions. *Cell* **34**:135–142.

224. **Machida, Y., C. Machida, and E. Ohtsubo.** 1984. Insertion element IS1 encodes two structural genes required for its transposition. *J. Mol. Biol.* **177**:229–245.

225. **Machida, Y., C. Machida, H. Ohtsubo, and E. Ohtsubo.** 1982. Factors determining frequency of plasmid cointegration mediated by insertion sequence IS1. *Proc. Natl. Acad. Sci. USA* **79**:277–281.

226. **Machida, Y., M. Sakurai, S. Kiyokawa, A. Ubasawa, Y. Suzuki, and J. E. Ikeda.** 1984. Nucleotide sequence of the insertion sequence found in the T-DNA region of mutant Ti plasmid pTiA66 and distribution of its homologues in octopine Ti plasmid. *Proc. Natl. Acad. Sci. USA* **81**:7495–7499.

227. **Makris, J. C., P. L. Nordman, and W. S. Reznikoff.** 1988. Mutational analysis of insertion sequence 50 (IS50) and transposon 5 (Tn5) ends. *Proc. Natl. Acad. Sci. USA* **85**: 2224–2228.

228. **Malamy, M. H.** 1966. Frameshift mutations in the lactose operon of E. coli. *Cold Spring Harbor Symp. Quant. Biol.* **31**:189.

229. **Malamy, M. H.** 1970. Some properties of insertion mutations in the lac operon, p. 359–373. *In* J. R. Beckwith and D. Zipser (ed.), *The Lactose Operon.* Cold Spring Harbor Laboratory, Cold Spring Harbor, N.Y.

230. **Malamy, M. H., M. Fiandt, and W. Szybalski.** 1972. Electron microscopy of polar insertions in the lac operon of Escherichia coli. *Mol. Gen. Genet.* **119**:207–227.

231. **Matsutani, S., H. Ohtsubo, Y. Maeda, and E. Ohtsubo.** 1987. Isolation and characterization of IS elements repeated in the bacterial chromosome. *J. Mol. Biol.* **196**:445–455.

232. **Mayaux, J. F., M. Springer, M. Graffe, M. Fromant, and G. Fayat.** 1984. IS4 transposition in the attenuator region of the Escherichia coli pheS, T operon. *Gene* **30**:137–146.

233. **McClintock, B.** 1952. Chromosome organization and gene expression. *Cold Spring Harbor Symp. Quant. Biol.* **16**: 13–47.

233a. **McCommas, S. A., and M. Syvanen.** 1988. Temporal control of transposition in Tn*5*. *J. Bacteriol.* **170**:889–894.

234. **McDougall, S., and J. B. Neilands.** 1984. Plasmid- and chromosome-coded aerobactin synthesis in enteric bacteria: insertion sequences flank operon in plasmid-mediated systems. *J. Bacteriol.* **159**:300–305.

235. **Meyer, J., and S. Iida.** 1979. Amplification of chloramphenicol resistance transposons carried by phage P1CM in Escherichia coli. *Mol. Gen. Genet.* **176**:209–219.

236. **Meyer, J., S. Iida, and W. Arber.** 1980. Does the insertion element IS1 transpose preferentially into A+T-rich DNA segments. *Mol. Gen. Genet.* **178**:471–473.

237. **Meyer, J. F., B. A. Nies, J. Kratz, and B. Wiedemann.** 1985. Evolution of Tn21-related transposons: isolation of Tn2425, which harbours IS161. *J. Gen. Microbiol.* **131**:1123–1130.

238. **Michaelis, G., H. Saedler, P. Venkov, and P. Starlinger.** 1969. Two insertions in the galactose operon having different sizes but homologous DNA sequences. *Mol. Gen. Genet.* **104**:371–377.

239. **Mickel, S., E. Ohtsubo, and W. Bauer.** 1977. Heteroduplex mapping of small plasmids derived from R-factor R12: in vivo recombination occurs at IS1 insertion sequences. *Gene* **2**:193–210.

240. **Miller, C. A., and S. N. Cohen.** 1980. F plasmid provides a function that promotes recA-independent site-specific fusions of pSC101 replicon. *Nature* (London) **285**:577–579.

241. **Miller, H. I., and D. I. Friedman.** 1980. An E. coli gene product required for lambda site-specific recombination. *Cell* **20**:711–719.

242. **Mizuuchi, K., and R. Craigie.** 1986. Mechanism of bacteriophage Mu transposition. *Annu. Rev. Genet.* **20**:385–429.

243. **Mollet, B., S. Iida, and W. Arber.** 1985. Gene organization and target specificity of the prokaryotic mobile genetic element IS26. *Mol. Gen. Genet.* **201**:198–203.

244. **Mollet, B., S. Iida, J. Shepherd, and W. Arber.** 1983. Nucleotide sequence of IS26, a new prokaryotic mobile genetic element. *Nucleic Acids Res.* **11**:6319–6330.

245. **Morisato, D., and N. Kleckner.** 1984. Transposase promotes double strand breaks and single strand joints at Tn10 termini. *Cell* **39**:181–190.

246. **Morisato, D., and N. Kleckner.** 1987. Tn10 transposition and circle formation in vitro. *Cell* **51**:101–111.

247. **Morisato, D., J. C. Way, H. J. Kim, and N. Kleckner.** 1983. Tn10 transposase acts preferentially on nearby transposon ends in vivo. *Cell* **32**:799–807.

248. **Mosharrafa, E., W. Pilacinski, J. Zissler, M. Fiandt, and W. Szybalski.** 1976. Insertion sequence IS2 near the gene for prophage lambda excision. *Mol. Gen. Genet.* **147**:103–109.

249. **Motsch, S., and R. Schmitt.** 1984. Replicon fusion mediated by a single-ended derivative of transposon Tn1721. *Mol. Gen. Genet.* **195**:281–287.

250. **Nag, D. K., U. DasGupta, G. Adelt, and D. E. Berg.** 1985. IS50-mediated inverse transposition: specificity and precision. *Gene* **34**:17–26.

251. **Nash, H. A.** 1981. Integration and excision of bacteriophage lambda: the mechanism of conservative site specific recombination. *Annu. Rev. Genet.* **15**:143–167.

252. **Nash, J. H., and V. Krishnapillai.** 1982. Identification of an insertion sequence in the chromosome of Pseudomonas aeruginosa PAO. *J. Bacteriol.* **152**:514–516.

253. **Navas, J., J. M. Garcia Lobo, J. Leon, and J. M. Ortiz.** 1985. Structural and functional analyses of the fosfomycin resistance transposon Tn2921. *J. Bacteriol.* **162**:1061–1067.

254. **Nevers, P., and H. Saedler.** 1978. Mapping and characterization of an E. coli mutant defective in IS1-mediated deletion formation. *Mol. Gen. Genet.* **160**:209–214.

254a. **New, J. H., A. K. Eggleston, and M. Fennewald.** 1988. Binding of Tn3 transposase to the inverted repeats of Tn3. *J. Mol. Biol.* **201**:589–599.

255. **Nies, B. A., J. F. Meyer, J. Kratz, and B. Wiedemann.** 1985. R1767, an example of the evolution of resistance plasmids. *Plasmid* **13**:163–172.

256. **Nies, B. A., J. F. Meyer, and B. Wiedemann.** 1985. Tn2440, a composite tetracycline resistance transposon with direct repeated copies of IS160 at its flanks. *J. Gen. Microbiol.* **131**:2443–2447.

257. **Noel, K. D., and G. Ames.** 1978. Evidence for a common mechanism for the insertion of the Tn10 transposon and for the generation of Tn10-stimulated deletions. *Mol. Gen. Genet.* **166**:217–223.

258. **Nomura, N., H. Yomagishi, and A. Oka.** 1978. Isolation and characterization of transducing coliphage fd carrying a kanamycin resistance determinant. *Gene* **3**:39–51.

259. **Nyman, K., K. Nakamura, H. Ohtsubo, and E. Ohtsubo.** 1981. Distribution of the insertion sequence IS1 in gram-negative bacteria. *Nature* (London) **289**:609–612.

260. **Nyman, K., H. Ohtsubo, D. Davison, and E. Ohtsubo.** 1983. Distribution of insertion element IS1 in natural isolates of Escherichia coli. *Mol. Gen. Genet.* **189**:516–518.

261. **O'Connor, M. B., and M. H. Malamy.** 1983. A new insertion sequence, IS121, is found on the Mu dI1 (Ap lac) bacteriophage and the Escherichia coli K-12 chromosome. *J. Bacteriol.* **156**:669–679.

262. **Ohtsubo, E., H. Ohtsubo, W. Doroszkiewicz, K. Nyman, D. Allen, and D. Davison.** 1984. An evolutionary analysis of iso-IS1 elements from Escherichia coli and Shigella strains. *J. Gen. Appl. Microbiol.* **30**:359–376.

263. **Ohtsubo, E., M. Zenilman, and H. Ohtsubo.** 1980. Plasmids containing insertion elements are potential transposons. *Proc. Natl. Acad. Sci. USA* **77**:750–754.

264. **Ohtsubo, E., M. Zenilman, H. Ohtsubo, M. McCormick, C. Machida, and V. Machida.** 1980. Mechanism of insertion and cointegration by IS*1* and Tn*3*. *Cold Spring Harbor Symp. Quant. Biol.* **45**:283–295.

265. **Ohtsubo, H., K. Nyman, W. Doroszkiewicz, and E. Ohtsubo.** 1981. Multiple copies of iso-insertion sequences of IS1 in Shigella dysenteriae chromosome. *Nature* (London) **292**: 640–643.

266. **Ohtsubo, H., and E. Ohtsubo.** 1978. Nucleotide sequence of an insertion element, IS1. *Proc. Natl. Acad. Sci. USA* **75**: 615–619.

267. **Ohtsubo, H., M. Zenilman, and E. Ohtsubo.** 1980. Insertion element IS102 resides in plasmid pSC101. *J. Bacteriol.* **144**:131–140.

268. **Oka, A., N. Nomura, K. Sugimoto, H. Sugisaki, and M. Takanami.** 1978. Nucleotide sequence at the insertion sites of a kanamycin transposon. *Nature* (London) **276**:845–847.

269. **Oka, A., H. Sugisaki, and M. Takanami.** 1981. Nucleotide sequence of the kanamycin resistance transposon Tn903. *J. Mol. Biol.* **147**:217–226.

270. **Ooms, G., P. J. J. Hooykaas, G. Moolenaar, and R. A. Schilperoort.** 1981. Crown gall plant tumors of abnormal morphology, induced by Agrobacterium tumefaciens carrying mutated octopine Ti plasmid: analysis of T-DNA functions. *Gene* **14**:33–50.

271. **Pabo, C., and R. Sauer.** 1984. Protein-DNA recognition. *Annu. Rev. Biochem.* **53**:293–321.

272. **Peterson, P. A., D. Ghosal, H. Sommer, and H. Saedler.** 1979. Development of a system useful for studying the formation of unstable alleles of IS2. *Mol. Gen. Genet.* **173**:15–21.

273. **Pfeifer, D., P. Habermann, and D. Kubai-Maroni.** 1977. Specific sites for integration of IS elements within the transferase gene of the gal operon of E. coli K12, p. 31–36. *In* A. I. Bukhari, J. A. Shapiro, and S. Adhya (ed.), *DNA Insertion Elements, Plasmids, and Episomes.* Cold Spring Harbor Laboratory, Cold Spring Harbor, N.Y.

274. **Phadnis, S. H., and D. E. Berg.** 1985. recA-independent recombination between repeated IS50 elements is not caused by an IS50-encoded function. *J. Bacteriol.* **161**:928–932.

275. **Phadnis, S. H., and D. E. Berg.** 1987. Identification of base pairs in the IS50 O end needed for IS50 and Tn5 transposition. *Proc. Natl. Acad. Sci. USA* **84**:9118–9122.

276. **Phadnis, S. H., C. Sasakawa, and D. E. Berg.** 1986. Localization of action of the IS50-encoded transposase protein. *Genetics* **112**:421–427.

277. **Pilacinski, W., E. Mosharrafa, R. Edmundson, J. Zissler, M. Fiandt, and W. Szybalski.** 1977. Insertion sequence IS2 associated with int-constitutive mutants of bacteriophage lambda. *Gene* **2**:61–74.

278. **Platt, T.** 1986. Transcription termination and the regulation of gene expression. *Annu. Rev. Biochem.* **55**:339–372.

279. **Prentki, P., M. Chandler, and D. Galas.** 1987. Escherichia coli integration host factor bends the DNA at the ends of IS1 and in an insertion hotspot with multiple IHF binding sites. *EMBO J.* **6**:2479–2487.

280. **Prentki, P., P. Gamas, M. Chandler, and D. Galas.** 1987. Functions of the ends of IS1, p. 719–734. *In* T. Kelley et al. (ed.), *DNA Replication and Recombination.* Alan R. Liss, Inc., New York.

281. **Prentki, P., M.-H. Pham, P. Gamas, M. Chandler, and D. J. Galas.** 1987. Artificial transposable elements in the study of the ends of IS1. *Gene* **61**:91–101.

282. **Prentki, P., B. Teter, M. Chandler, and D. J. Galas.** 1986. Functional promoters created by the insertion of transposable element IS1. *J. Mol. Biol.* **191**:383–393.

283. **Priefer, U. B., H. J. Burkhardt, W. Klipp, and A. Puhler.** 1981. ISR1: an insertion element isolated from the soil bacterium Rhizobium lupini. *Cold Spring Harbor Symp. Quant. Biol.* **45**:87–91.

284. **Pruss, G. J., and K. Drlica.** 1986. Topoisomerase mutants: the gene on pBR322 that encodes resistance to tetracycline affects plasmid supercoiling. *Cold Spring Harbor Symp. Quant. Biol.* **83:**8952–8956.

285. **Pruss, G. J., S. H. Manes, and K. Drlica.** 1982. Escherichia coli DNA topoisomerase I mutants: increased supercoiling is corrected by mutations near gyrase genes. *Cell* **31:**35–42.

286. **Ptashne, K., and S. N. Cohen.** 1975. Occurrence of insertion sequence (IS) regions on plasmid deoxyribonucleic acids as direct and inverted nucleotide sequence duplications. *J. Bacteriol.* **122:**776–781.

287. **Rak, B., M. Lusky, and M. Hable.** 1982. Expression of two proteins from overlapping and oppositely oriented genes on transposable DNA insertion element IS5. *Nature* (London) **297:**124–128.

288. **Rak, B., and M. von Reutern.** 1984. Insertion element IS5 contains a third gene. *EMBO J.* **3:**807–811.

288a.**Rasmussen, J. L., D. A. Odelson, and F. L. Macrina.** 1987. Complete nucleotide sequence of insertion element IS*4351* from *Bacteroides fragilis. J. Bacteriol.* **169:**3573–3580.

289. **Ravetch, J. V., M. Ohsumi, P. Model, G. F. Vovis, D. Fischhoff, and N. D. Zinder.** 1979. Organization of a hybrid between phage f1 and plasmid pSC101. *Proc. Natl. Acad. Sci. USA* **76:**2195–2198.

290. **Read, H. A., S. Das Sarma, and S. R. Jaskunas.** 1980. Fate of donor insertion sequence IS1 during transposition. *Proc. Natl. Acad. Sci. USA* **77:**2514–2518.

291. **Reed, R. R., R. A. Young, J. A. Steitz, N. D. Grindley, and M. S. Guyer.** 1976. Transposition of the Escherichia coli insertion element gamma generates a five-base-pair repeat. *Proc. Natl. Acad. Sci. USA* **76:**4882–4886.

292. **Reif, H. J.** 1980. Genetic evidence for absence of transposition functions from the internal part of Tn981 a relative of Tn9. *Mol. Gen. Genet.* **177:**667–674.

293. **Reif, H. J., and H. Saedler.** 1975. IS1 is involved in deletion formation in the gal region of E. coli K12. *Mol. Gen. Genet.* **137:**17–28.

294. **Reyes, O., M. Gottesman, and S. Adhya.** 1979. Suppression of polarity of insertion mutations in the gal operon and N mutations in bacteriophage lambda. *J. Bacteriol.* **126:** 1108–1112.

295. **Reynolds, A. E., J. Felton, and A. Wright.** 1981. Insertion of DNA activates the cryptic bgl operon in E. coli K12. *Nature* (London) **293:**625–629.

296. **Reynolds, A. E., S. Mahadevan, S. F. LeGrice, and A. Wright.** 1986. Enhancement of bacterial gene expression by insertion elements or by mutation in a CAP-cAMP binding site. *J. Mol. Biol.* **191:**85–95.

297. **Riess, G., B. W. Holloway, and A. Puhler.** 1980. R68.45, a plasmid with chromosome mobilizing ability (Cma) carries a tandem duplication. *Genet. Res.* **36:**99–109.

298. **Riess, G., B. Masepohl, and A. Puehler.** 1983. Analysis of IS21-mediated mobilization of plasmid pACYC184 by R68.45 in Escherichia coli. *Plasmid* **10:**111–118.

299. **Riley, M., and P. Anilionis.** 1978. Evolution of the bacterial genome. *Annu. Rev. Microbiol.* **32:**519–560.

300. **Roberts, D., B. C. Hoopes, W. R. McClure, and N. Kleckner.** 1985. IS10 transposition is regulated by DNA adenine methylation. *Cell* **43:**117–130.

301. **Robertson, C., and H. Nash.** 1988. Bending of bacteriophage lambda attachment site by E. coli integration host factor. *J. Biol. Chem.* **263:**3554–3557.

302. **Robillard, N. J., F. P. Tally, and M. H. Malamy.** 1985. Tn4400, a compound transposon isolated from Bacteroides fragilis, functions in Escherichia coli. *J. Bacteriol.* **164:** 1248–1255.

303. **Romecker, H.-J., and B. Rak.** 1987. Genetic organization of insertion element IS2 based on a revised nucleotide sequence. *Gene* **59:**291–296.

304. **Rosner, J. L., and M. S. Guyer.** 1980. Transposition of IS1-lambdaBI0-IS1 from a bacteriophage lambda derivative carrying the IS1-cat-IS1 transposon (Tn9). *Mol. Gen. Genet.* **178:**111–120.

305. **Ross, D. G., J. Swan, and N. Kleckner.** 1979. Physical structures of Tn10-promoted deletions and inversions: role of 1400bp inverted repetitions. *Cell* **16:**721–731.

306. **Ross, D. G., J. Swan, and N. Kleckner.** 1979. Nearly precise excision: a new type of DNA alteration associated with the translocatable element Tn10. *Cell* **16:**733–738.

307. **Rothstein, S. J., R. A. Jorgensen, K. Postle, and W. S. Reznikoff.** 1980. The inverted repeats of Tn5 are functionally different. *Cell* **19:**795–805.

308. **Rothstein, S. J., and W. S. Reznikoff.** 1981. The functional differences in the inverted repeats of Tn5 are caused by a single base pair nonhomology. *Cell* **23:**191–199.

309. **Rownd, R., and S. Mickel.** 1971. Dissociation and reassociation of RTF and r-determinants of the R factor NR1 in Proteus mirabilis. *Nature* (London) **234:**40–41.

310. **Ruvkun, G. B., S. R. Long, H. M. Meade, R. C. van den Bos, and F. M. Ausubel.** 1982. ISRm1: a rhizobium meliloti insertion sequence that transposes preferentially into nitrogen fixation genes. *J. Mol. Appl. Genet.* **1:**405–418.

311. **Saedler, H.** 1977. IS1 and IS2 in E. coli: implications for the evolution of the chromosome and some plasmids, p. 65–72. *In* A. I. Bukhari, J. A. Shapiro, and S. L. Adhya (ed.), *DNA Insertion Elements, Plasmids, and Episomes.* Cold Spring Harbor Laboratory, Cold Spring Harbor, N.Y.

312. **Saedler, H., and B. Heiss.** 1973. Multiple copies of the insertion-DNA sequences IS1 and IS2 in the chromosome of E. coli K12. *Mol. Gen. Genet.* **122:**267–277.

313. **Saedler, H., H. J. Reif, S. Hu, and N. Davidson.** 1974. IS2, a genetic element for turn-off and turn-on of gene activity in E. coli. *Mol. Gen. Genet.* **132:**265–289.

314. **Saint-Girons, I., H. J. Fritz, C. Shaw, E. Tillmann, and P. Starlinger.** 1981. Integration specificity of an artificial kanamycin transposon constructed by the in vitro insertion of an internal Tn5 fragment into IS2. *Mol. Gen. Genet.* **183:** 45–50.

315. **Sasakawa, C., and D. E. Berg.** 1982. IS50-mediated inverse transposition. Discrimination between the two ends of an IS element. *J. Mol. Biol.* **159:**257–271.

316. **Sasakawa, C., Y. Uno, and M. Yoshikawa.** 1981. The requirement for both DNA polymerase and 5′ to 3′ exonuclease activities of DNA polymerase I during Tn5 transposition. *Mol. Gen. Genet.* **182:**19–24.

316a.**Savic, H.** 1979. Inversion in the lactose operon of *Escherichia coli* K-12. *J. Bacteriol.* **140:**311–319.

317. **Sawyer, S. A., D. E. Dykhuizen, R. F. DuBose, L. Green, T. Mutangadura Mhlanga, D. F. Wolczyk, and D. L. Hartl.** 1987. Distribution and abundance of insertion sequences among natural isolates of Escherichia coli. *Genetics* **115:** 51–63.

318. **Schmidt, M. B., and J. R. Roth.** 1983. Genetic methods for analysis and manipulation of inversion mutations in bacteria. *Genetics* **105:**517–537.

319. **Schmitt, R., P. Rogowsky, S. E. Halford, and J. Grinsted.** 1985. Transposable elements and evolution, p. 91–104. *In* K. H. Schleifer and E. Stackebrandt (ed.), *Evolution of Prokaryotes.* Academic Press, Inc., New York.

319a. **Schnier, J., and K. Isono.** 1984. Insertion of IS1 into the resE gene for ribosomal protein S5 causes cold sensitivity. *Mol. Gen. Genet.* **195:**364–366.

319b. **Schoemaker, N. B., C. Getty, J. F. Gardner, and A. A. Salyers.** 1986. Tn*4351* transposes in *Bacteroides* spp. and mediates the integration of plasmid R751 into the *Bacteroides* chromosome. *J. Bacteriol.* **165:**929–936.

320. **Schoner, B., and M. Kahn.** 1981. The nucleotide sequence of IS5 from Escherichia coli. *Gene* **14:**165–174.

321. **Schoner, B., and R. G. Schoner.** 1981. Distribution of IS5 in bacteria. *Gene* **16:**347–352.

322. **Schwartz, E., C. Herberger, and B. Rak.** 1988. Second-element turn-on of gene expression in an IS1 insertion mutant. *Mol. Gen. Genet.* **211:**282–289.

322a. **Schwartz, E., M. Kroeger, and B. Rak.** 1988. IS50: distribution, nucleotide sequence, and phylogenetic relationship of a new E. coli insertion element. *Nucleic Acids Res.* **16:** 6789–6802.

323. **Scordilis, G. E., H. Ree, and T. G. Lessie.** 1987. Identification of transposable elements which activate gene expression in *Pseudomonas cepacia. J. Bacteriol.* **169:**8–13.

324. **Sengstag, C., and W. Arber.** 1983. IS2 insertion is a major cause of spontaneous mutagenesis of the bacteriophage P1: non-random distribution of target sites. *EMBO J.* **2:**67–71.

325. **Sengstag, C., and W. Arber.** 1987. A cloned DNA fragment from bacteriophage P1 enhances IS2 insertion. *Mol. Gen. Genet.* **206:**344–351.

326. **Sengstag, C., S. Iida, R. Hiestand-Nauer, and W. Arber.** 1986. Terminal inverted repeats of prokaryotic transposable element IS186 which can generate duplications of variable length in an identical target sequence. *Gene* **49:**153–156.

327. **Sengstag, C., J. C. W. Shepherd, and W. Arber.** 1983. The sequence of the bacteriophage P1 genome region serving as hot target for IS2 insertion. *EMBO J.* **2:**1777–1781.

328. **Shapiro, J. A.** 1969. Mutations caused by the insertion of genetic material into the galactose operon of Escherichia coli. *J. Mol. Biol.* **40:**93–105.

329. **Shapiro, J. A.** 1977. F, the E. coli sex factor, p. 671. *In* A. I. Bukhari, J. A. Shapiro, and S. L. Adhya (ed.), *DNA Insertion Elements, Plasmids, and Episomes.* Cold Spring Harbor Laboratory, Cold Spring Harbor, N.Y.

330. **Shapiro, J. A.** 1979. Molecular model for the transposition and replication of bacteriophage Mu and other transposable elements. *Proc. Natl. Acad. Sci. USA* **76:**1933–1937.

331. **Shapiro, J. A. (ed.).** 1983. *Mobile Genetic Elements.* Academic Press, Inc., New York.

332. **Sharp, P. A., S. N. Cohen, and N. Davidson.** 1973. Electron microscope heteroduplex studies of sequence relations among plasmids of Escherichia coli. II. Structure of drug resistance (R) factors and F factors. *J. Mol. Biol.* **75:** 235–255.

333. **Siehnel, R. J., and E. A. Morgan.** 1983. Efficient read-through of Tn9 and IS1 by RNA polymerase molecules that initiate at rRNA promoters. *J. Bacteriol.* **153:**672–684.

334. **Simons, R. W., B. C. Hoopes, W. R. McClure, and N. Kleckner.** 1983. Three promoters near the termini of IS10: pIN, pOUT, and pIII. *Cell* **34:**673–682.

335. **Simons, R. W., and N. Kleckner.** 1983. Translational control of IS10 transposition. *Cell* **34:**683–691.

336. **Smirnov, G. B., T. S. Ilyina, Y. M. Romanova, A. P. Markov, and E. V. Nechaeva.** 1981. Mutants of Escherichia coli affected in the processes of transposition and genomic rearrangements. *Cold Spring Harbor Symp. Quant. Biol.* **45:** 193–200.

337. **Smith, C. L., J. G. Econome, A. Schutt, S. Klico, and C. R. Cantor.** 1987. A physical map of the Escherichia coli K12 genome. *Science* **236:**1448–1453.

337a. **Smith, C. J., and H. Spiegel.** 1987. Transposition of Tn*4551* in *Bacteroides fragilis*: identification and properties of a new transposon from *Bacteroides* spp. *J. Bacteriol.* **169:**3450–3457.

338. **Smith, G.** 1988. Homologous recombination in procaryotes. *Microbiol. Rev.* **52:**1–28.

339. **Smith, G., S. Armundsen, A. Chaudhury, K. Cheng, A. Ponticelli, and A. Taylor.** 1984. Roles of RecBC enzyme and Chi sites in homologous recombination. *Cold Spring Harbor Symp. Quant. Biol.* **49:**485–495.

340. **So, M., F. Heffron, and B. J. McCarthy.** 1979. The E. coli gene encoding heat stable toxin is a bacterial transposon flanked by inverted repeats of IS1. *Nature* (London) **277:** 453–456.

341. **Sommer, H., J. Cullum, and H. Saedler.** 1979. Integration of IS3 into IS2 generates a short sequence duplication. *Mol. Gen. Genet.* **177:**85–89.

342. **Starlinger, P.** 1980. IS elements and transposons. *Plasmid* **3:**241–259.

343. **Starlinger, P., and H. Saedler.** 1976. IS-elements in microorganisms. *Curr. Top. Microbiol. Immunol.* **75:**111–152.

344. **Stellwagen, N.** 1983. Anomalous electrophoresis of DNA restriction fragments on polyacrylamide gels. *Biochemistry* **22:**6186–6193.

345. **Sternglanz, R., S. DiNardo, K. A. Voelkel, Y. Nishimura, Y. Hirota, K. Becherer, L. Zumstein, and J. C. Wang.** 1981. Mutations in the gene coding for Escherichia coli DNA topoisomerase I affect transcription and transposition. *Proc. Natl. Acad. Sci. USA* **78:**2747–2751.

346. **Stibitz, S., and J. E. Davies.** 1987. Tn602: a naturally occurring relative of Tn903 with direct repeats. *Plasmid* **17:**202–209.

347. **Sullivan, K. M., and D. M. J. Lilley.** 1986. A dominant influence of flanking sequences on a local structural transition in DNA. *Cell* **47:**817–827.

348. **Syvanen, M.** 1984. The evolutionary implications of mobile genetic elements. *Annu. Rev. Genet.* **18:**271–293.

349. **Syvanen, M., J. D. Hopkins, and M. Clements.** 1982. A new class of mutants in DNA polymerase I that affects gene transposition. *J. Mol. Biol.* **158:**203–212.

350. **Syvanen, M., J. D. Hopkins, T. J. Griffin, T.-Y. Liang, K. Ippen-Ihler, and R. Kolodner.** 1986. Stimulation of precise excision and recombination by conjugal proficient F′ plasmids. *Mol. Gen. Genet.* **203:**1–7.

351. **Szybalski, W.** 1977. IS elements in Escherichia coli, plasmids, and phage, p. 583–590. *In* A. I. Bukhari, J. A. Shapiro, and S. L. Adhya (ed.), *DNA Insertion Elements, Plasmids, and Episomes.* Cold Spring Harbor Laboratory, Cold Spring Harbor, N.Y.

352. **Thompson, J., L. Moitoso de Vargas, S. Skinner, and A. Landy.** 1987. Protein-protein interactions in a higher order structure direct lambda site specific recombination. *J. Mol. Biol.* **195:**254–263.

353. **Timmerman, K. P., and C.-P. D. Tu.** 1985. Complete sequence of IS3. *Nucleic Acids Res.* **13:**2127–2139.

354. **Toussaint, A., and M. Faelen.** 1973. Connecting two unrelated DNA sequences with a Mu dimer. *Nature* (London) **242:**1–4.

355. **Trieu Cuot, P., and P. Courvalin.** 1984. Nucleotide sequence of the transposable element IS15. *Gene* **30:**113–120.

356. **Trieu Cuot, P., and P. Courvalin.** 1985. Transposition behavior of IS15 and its progenitor IS15-delta: cointegrates are exclusive end products. *Plasmid* **14:**80–89.

357. **Trieu-Cuot, P., A. Labigne-Roussel, and P. Courvalin.** 1983. An IS15 insertion generates an eight-base-pair duplication of the target DNA. *Gene* **24:**125–129.

358. **Vanderleyden, J., J. Desair, C. De Meirsman, K. Michiels, A. P. Van Gool, M. D. Chilton, and G. C. Jen.** 1986. Nucleotide sequence of an insertion sequence (IS) element identified in the T-DNA region of a spontaneous variant of the Ti-plasmid pTI T37. *Nucleic Acids Res.* **1:**6699–6709.

359. **Waldron, C., and A. G. Hepburn.** 1983. Extra DNA in the T region of crown gall Ti plasmid pTiA66. *Plasmid* **10:** 199–203.

360. **Walz, A., B. Ratzkin, and J. Carbon.** 1978. Control of expression of a cloned yeast (Saccharomyces cerevisiae) gene (trp5) by a bacterial insertion element (IS2). *Proc. Natl. Acad. Sci. USA* **75:**6172–6176.

361. **Watanabe, T., and Y. Ogata.** 1970. Genetic stability of various resistance factors in *Escherichia coli* and *Salmonella typhimurium*. *J. Bacteriol.* **102:**363–368.

362. **Waters, V. L., and J. H. Crosa.** 1986. DNA environment of the aerobactin iron uptake system genes in prototypic ColV plasmids. *J. Bacteriol.* **167:**647–654.

363. **Way, J. C., and N. Kleckner.** 1985. Transposition of plasmid-borne Tn10 elements does not exhibit simple length-dependence. *Genetics* **111:**705–713.

364. **Weinert, T. A., K. M. Derbyshire, F. M. Hughson, and N. D. Grindley.** 1984. Replicative and conservative transpositional recombination of insertion sequences. *Cold Spring Harbor Symp. Quant. Biol.* **49:**251–260.

365. **Weinert, T. A., N. A. Schaus, and N. D. Grindley.** 1983. Insertion sequence duplication in transpositional recombination. *Science* **222:**755–765.

366. **Wilhelm, M., and C. P. Hollenberg.** 1984. Selective cloning of Bacillus subtilis xylose isomerase and xylulokinase in Escherichia coli genes by IS5-mediated expression. *EMBO J.* **3:**2555–2560.

367. **Willetts, N. S., C. Crowther, and B. W. Holloway.** 1981. The insertion sequence IS21 of R68.45 and the molecular basis for mobilization of the bacterial chromosome. *Plasmid* **6:** 30–52.

368. **Winter, R., O. Berg, and P. von Hippel.** 1981. *Biochemistry* **20:**6961–6969.

369. **Wishart, W. L., J. R. Broach, and E. Ohtsubo.** 1985. ATP-dependent specific binding of Tn3 transposase to Tn3 inverted repeats. *Nature* (London) **314:**556–558.

370. **Wood, A. G., and J. Konisky.** 1985. Activation of expression of a cloned archaebacterial gene in Escherichia coli by IS2, IS5, or deletions. *Mol. Gen. Genet.* **198:**309–314.

371. **Wrighton, C. J., and P. Strike.** 1987. A pathway for the evolution of the plasmid NTP16 involving the novel kanamycin resistance transposon Tn4352. *Plasmid* **17:**37–45.

372. **Yamada, T., P.-D. Lee, and T. Kosuge.** 1986. Insertion sequence elements of Pseudomonas savastanoi: nucleotide sequence and homology with Agrobacterium tumefaciens transfer DNA. *Proc. Natl. Acad. Sci. USA* **83:**8263–8276.

372a. **Yin, J. C. P., M. P. Krebs, and W. S. Reznikoff.** 1988. The effect of Dam methylation on Tn5 transposition. *J. Mol. Biol.* **199:**35–45.

373. **Yin, J. C. P., and W. S. Reznikoff.** 1987. dnaA, an essential host gene, and Tn5 transposition. *J. Bacteriol.* **169:**3714–3725.

374. **Zabala, J. C., F. de la Cruz, and J. M. Ortiz.** 1982. Several copies of the same insertion sequence are present in alpha-hemolytic plasmids belonging to four different incompatibility groups. *J. Bacteriol.* **151:**472–476.

375. **Zabala, J. C., J. M. Garcia-Lobo, E. Diaz-Aroca, F. de la Cruz, and J. M. Ortiz.** 1984. Escherichia coli alpha-haemolysin synthesis and export genes are flanked by a direct repetition of IS91-like elements. *Mol. Gen. Genet.* **197:** 90–97.

376. **Zerbib, D., P. Gamas, M. Chandler, P. Prentki, S. Bass, and D. Galas.** 1985. Specificity of insertion of IS1. *J. Mol. Biol.* **185:**517–524.

377. **Zerbib, D., M. Jakowec, P. Prentki, D. Galas, and M. Chandler.** 1987. Expression of proteins essential for IS1 transposition: specific binding of InsA to the ends of IS1. *EMBO J.* **6:**3163–3169.

378. **Zupancic, T. J., S. L. Marvo, J. H. Chung, E. G. Peralta, and S. R. Jaskunas.** 1983. RecA independent recombination between direct repeats of IS50. *Cell* **33:**629–637.

Chapter 5

Tn*3* and Related Transposable Elements: Site-Specific Recombination and Transposition

DAVID SHERRATT

I. HISTORICAL BACKGROUND AND GENERAL INTRODUCTION

The ampicillin resistance (Ap^r) transposon Tn*3* was the first transposable element encoding antibiotic resistance to be described by Hedges and Jacob (53), though the movement of penicillin resistance from plasmid to plasmid and from plasmid to chromosome was first described some 3 years earlier (22, 104). The insight of Hedges and Jacobs was to realize that Tn*3*

David Sherratt ■ Institute of Genetics, Glasgow University, Church Street, Glasgow G11 5JS, United Kingdom.

Table 1. Transposons and transposonlike sequences of the Tn*3* family

Elements[a]	Associated determinants[b]	Length (kb)	Terminal repeats	Source and comments	References[c]
Tn*3*	Ap^r	4.957	38/38	R1*drd*19	
Tn*1*, Tn2, Tn*801*, Tn*802*, Tn*901*, Tn*902*, Tn*401*, Tn*1701*, Tn*2601*, Tn*2602*, Tn*2660*	Ap^r	5	38	Plasmids in a range of gram-negative bacteria; these transposons are essentially identical to Tn*3*	Table I of reference 55; 2, 14, 16, 22, 25, 34, 47, 53, 54, 58, 64, 104, 107
Tn*1000* (often called γδ)	Cryptic	5.8	36/37	F plasmid	
IS*101*	None	0.209	35/38	*Salmonella panama*; IS*101* contains only functional termini and *res* site	
Tn*501*	Hg^r	8.2	35/38	*Pseudomonas* plasmid, pUS1	
Tn*1721*, Tn*1771*	Tc^r	11.4	35/38	pSRD1, pFS202	
Tn*2603*	Ox^r, Sm^r, Su^r, Hg^r	22	38	RGN238, *Escherichia coli*	
Tn*21*	Su^r, Hg^r, Ap^r, Sm^r, Su^r	19.6	35/38	R100, *Shigella flexneri*	
Tn*4*	Ap^r, Sm^r, Su^r	23.5	–	R65, *E. coli*	
Tn*2501*	Cryptic	6.3	45/48		81, 82
Tn*551*	Ery^r	5.3	35/35	*Staphylococcus aureus*	55, 67, 89
Tn*917*	Ery^r	5.3	38	Plasmid pAD2 from *Streptococcus faecalis*; Tn*917* is very similar to Tn*551*	113
Tn*4430*	Cryptic	4.1	38	*Bacillus thuringiensis*	77, 79
R46 *res*/resolvase				Plasmid R46 of enterobacter	26, 27
pIP404 *res*/resolvase				*Clostridium perfringens*	38

[a] Elements are divided into the Tn*3* subfamily, the Tn*501* subfamily, gram-positive elements, and transposonlike sequences.
[b] Resistance to ampicillin (Ap^r), mercury (Hg^r), tetracycline (Tc^r), oxacillin (Ox^r), streptomycin (Sm^r), sulfonamide (Su^r), and erythromycin (Ery^r).
[c] Only key references are cited here: see Literature Cited and reference 55 for more extensive lists.

was an element similar to the *Escherichia coli* IS elements already described and characterized (65, 111).

Tn*3* is derived from the IncF plasmid R1*drd*19, while its very similar relatives Tn*1* and Tn2 are derived from the broad-host-range IncP1 plasmid RP4 and the nonconjugative plasmid RSF1030, respectively (55) (Table 1). Tn*3* has been sequenced, is 4.957 kilobases (kb) in size, and is very closely related to a whole group of Ap^r transposons isolated from a range of bacteria (58) (Table 1). It is likely that all of these transposons found in enteric bacteria have interchangeable functions and share more than 90% sequence similarity. In this review, the term Tn*3* is used for both Tn*3* and its close Ap^r relatives. It is our experience that some of these elements have been confused with each other during distribution and subsequent use.

Other members of the Tn*3* family (Table 1) were identified by the similarity to Tn*3* of their inverted termini, which are 35 to 48 bp in length; by their similar mode of transposition; by the 5-bp direct duplication of a sequence adjacent to the insertion site; and by the similarity of their transposition proteins. The organization of genes in the Tn*3* family falls into two classes (Fig. 1). Whether this represents two evolutionary lineages or some functional conservation is not clear. Invariably the 3′ ends of the transposase genes terminate within one of the terminal repeats. Transposons with the gene organization of Tn*3* are described here as the Tn*3* subfamily, and those with the organization of Tn*501* are described as the Tn*501* subfamily. One particularly compact transposon belonging to the Tn*3* subfamily is the 209-bp IS*101*, which consists only of two functional terminal repeats and a *res* site that functions poorly

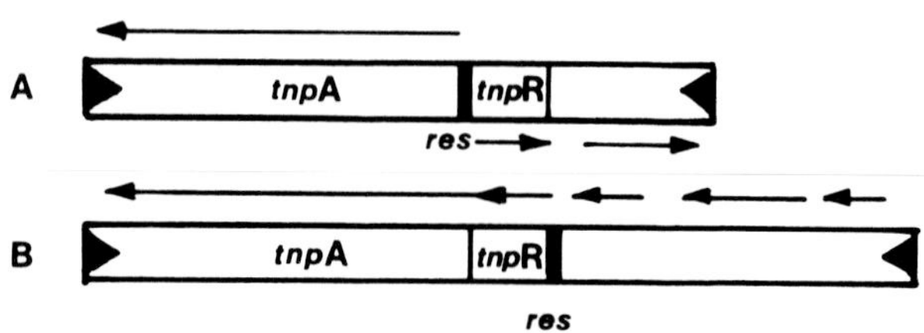

Figure 1. Gene organization of Tn*3* family elements. The two types of gene organization are exemplified. (A) Divergent transcription of *tnpA* and *tnpR* from the central *res* site (Tn*3*, Tn*1000*, Tn*2501*, Tn*4430*). (B) Transcription of *tnpR* and *tnpA* in the same direction; *res* is 5′ of *tnpR* and *tnpA* (Tn*501*, Tn*1721*, Tn*21*, Tn*551*, Tn*917*).

with Tn*3* and Tn*1000* resolvases (34, 64) (Table 1); in this element, fewer than 10 nucleotides are nonfunctional for transposition or resolution. Another cryptic member of the Tn*3* subfamily is Tn*1000* (or γδ). This is a normal constituent of the plasmid F, and its transposition into the *E. coli* chromosome may give rise to transient Hfrs through cointegrate formation. Similarly, homologous recombination between the Tn*1000* of F and any Tn*1000* on the chromosome can give rise to transient Hfrs.

Transposition of Tn*3* family transposons is controlled and normally occurs at frequencies of 10^{-5} to 10^{-7} per cell generation. Under certain conditions transposition frequencies can be up to at least 100 times higher, at least for short periods. Transposition is normally replicative. In nature, Tn*3* family elements have most often been found on plasmids encoding antibiotic resistance in gram-negative bacteria, and their common, widespread occurrence is likely a property of extensive and indiscriminate antibiotic use both in human medicine and in animal husbandry. A number of transposons (and transposon-like sequences) related to the Tn*3* family have also been identified in the gram-positive bacteria *Staphylococcus aureus*, *Streptococcus faecalis*, *Bacillus thuringiensis*, and *Clostridium perfringens* (Table 1).

The most frequent biologically significant effects of transposition are likely to be those that result from intermolecular transposition between two plasmids; the ultimate result is normally the acquisition of a transposon by a plasmid originally lacking a copy of that particular transposon. The transposon donor plasmid is unchanged after the transposition. Such intermolecular events occur in a sequential two-step reaction requiring two Tn*3* gene products, transposase and resolvase, in successive steps (5, 57, 68, 72). The first step involves the true transposition process and results in the ligation of donor transposon sequences to the target and replication of the transposon to give a cointegrate molecule in which the donor and recipient replicons are fused by two directly repeated copies of the transposon (see below, Fig. 7). Such cointegrates are then converted into the normally observed products of transposition by a second, site-specific recombination event, mediated by the transposon-specified enzyme resolvase and the recombination site *res*. This second, site-specific recombination step will be discussed first because we have more detailed knowledge of it than of the action of transposase; useful insight into transposase action can be gained from our understanding of resolvase/*res* interactions.

The properties of Tn*3* family elements have been extensively researched and reviewed (for recent reviews, see references 42, 44, and 55). The recent review by Mizuuchi and Craigie (85) on bacteriophage Mu also makes informative and relevant reading.

II. *res*/RESOLVASE SITE-SPECIFIC RECOMBINATION

A. Introduction

The pioneering work of Heffron and co-workers (41, 56) led to the construction and characterization of Tn*3* mutants with altered transposition properties. Among these were a class that generated fused cointegrates between donor and recipient replicons as transposition end products. Subsequently, it was shown that cointegrates formed by some of these mutants are resolved to the normal end products when resolvase, the product of *tnpR*, is provided in *trans* (5, 57, 68, 70, 72). Such experiments led to the view that cointegrates are normal intermolecular transposition intermediates and that two Tn*3* determinants (*tnpR*, the gene encoding resolvase, and *res*, the recombination site) are required for resolution of cointegrates to normal transposition products. *res* (initially called IRS, internal resolution site [57]) was shown to be a 120-bp sequence between *tnpA* and *tnpR* in the Tn*3* subfamily of transposons. Footprinting experiments (43) showed that resolvase binds to *res* and can protect against DNase I digestion. In vitro recombination experiments showed that recombination occurs at a unique site within *res* (72, 100). *res* also contains the promoters for the *tnpA* and *tnpR* genes (Fig. 2). Therefore, binding of resolvase to *res* can repress transcription of *tnpA* and *tnpR* as well as lead to site-specific recombination. *res* sites and *tnpR* genes have been identified for all of the Tn*3* family elements that specify transposition functions. A feature of the Tn*3* and Tn*1000* *res*/resolvase systems that soon became evident was their directionality and specificity. Resolvase would act efficiently on directly repeated *res* sites within the same molecule, but worked poorly, if at all, on inverted sites and on sites within separate molecules (17, 44, 71, 73, 116). This specificity was observed also in the reaction in vitro. In a similar fashion, "invertases," enzymes showing homology to resolvases and involved in inversion switches in bacteria and phage, were shown also to be directional but preferring to act only on inverted recombination sites (98).

B. The Recombination Site, *res*

Figure 2 shows the sequence of the Tn*3* *res* region and Fig. 3 the consensus sequences of other *res*

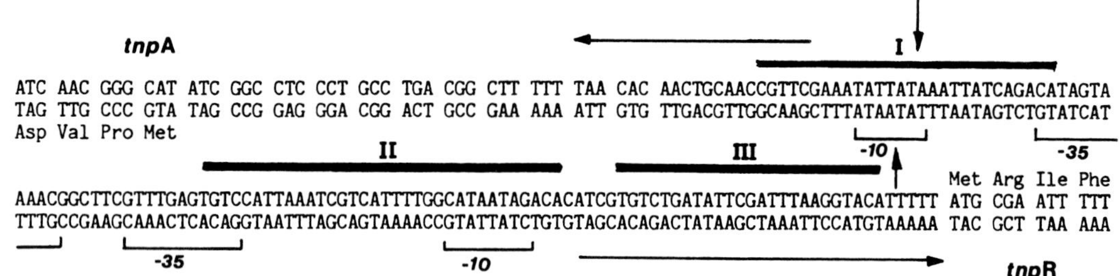

Figure 2. The *res* region of Tn*3*. The three resolvase-binding sites are indicated (thick horizontal bars), as are the *tnpA* and *tnpR* genes and their promoters, and their transcriptional (horizontal arrows) and translational starts. The site of crossover is indicated within site I (vertical arrows).

regions and their organization. Grindley and coworkers (43) originally showed that Tn*1000 res* has three binding sites for resolvase by DNase I footprinting. Those three sites are highly conserved in all *res* regions, and a range of in vitro and in vivo footprinting experiments with various *res* regions, using protection against DNase I cleavage, alkylation, hydroxy radicals, and UV irradiation, as well as alkylation interference experiments in which alkylation of specific bases and phosphates can inhibit resolvase binding (33, 43, 71, 106; J. L. Brown, Ph.D. thesis, Glasgow University, Glasgow, Scotland, 1986), have revealed that the major binding of resolvase is to the DNA major groove at the outer ends of each of the subsites. Each of the three sites in the Tn*3* subfamily is made up of a pair of inverted 9-bp repeats, with the consensus TGTCYRTTA, that span a central spacer; the spacer of site I is 10 bp, that of site II is 16 bp, and that of site III is 7 bp.

Sequence analysis of hybrid *res* regions resulting

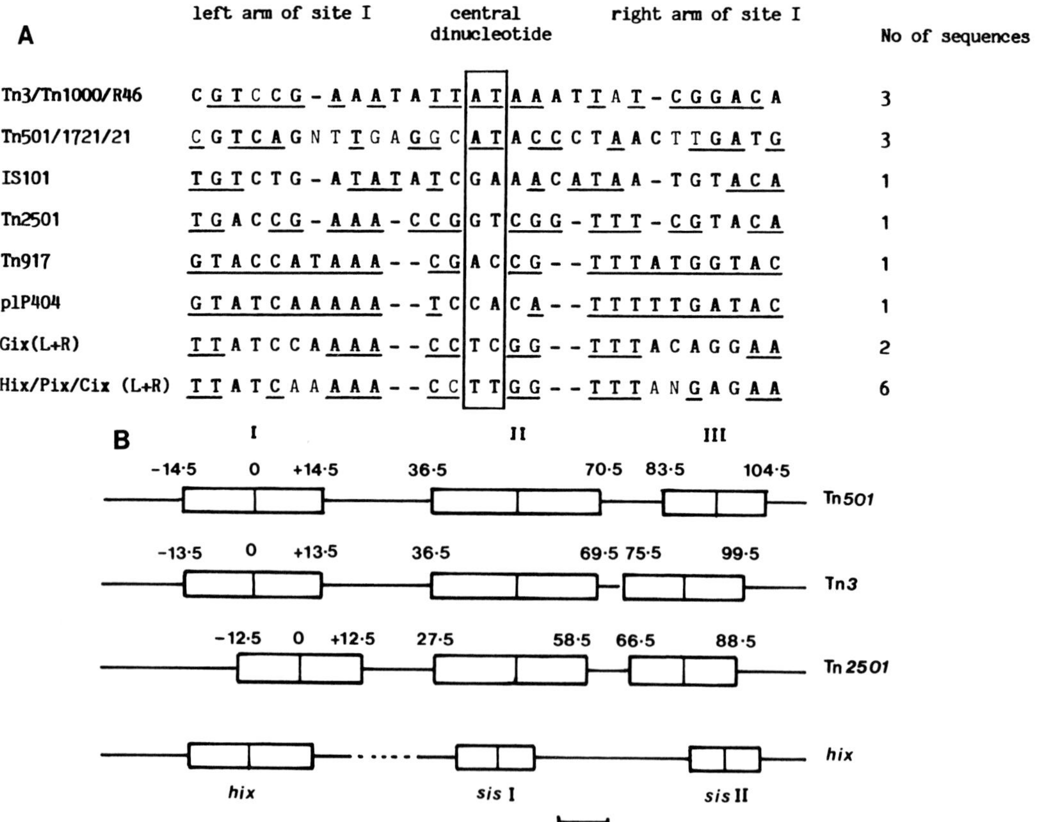

Figure 3. *res* consensus sequences and their spacings. (A) Consensus sequences for site I resolvase-binding sites and their relationship to inversion sites (*gix*, *hix*, *pix*, *cix*). Dyad symmetries are underlined, and the central dinucleotides across which recombination occurs are boxed. Lightface type indicates lack of total conservation within the indicated sequences. (B) The relative sizes and spacings of the resolvase-binding sites. The bar indicates one helical turn (10.5 nucleotides). For comparison, the *hix* site and two FIS-binding sites (*sis* I and II) are indicated.

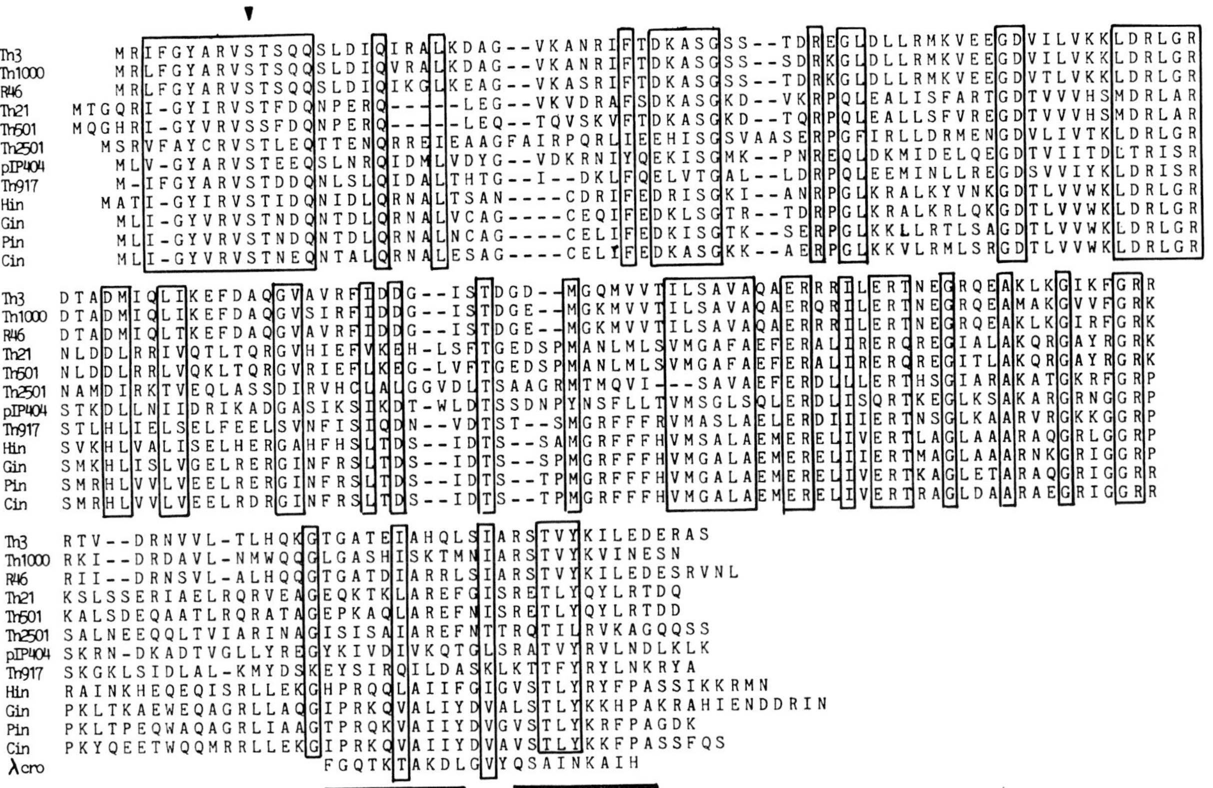

Figure 4. Resolvases of the Tn*3* family elements and invertases. Sequences are aligned to show conserved regions. Highly conserved regions are boxed. The presumptive serine involved in the 5′ phosphoseryl linkage to DNA is indicated by an arrow. The N-terminal catalytic domain (at least of Tn*3*/Tn*1000* resolvase) is depicted in the first two rows of sequence and the C-terminal DNA-binding domain is in the third row. The two thick horizontal bars depict the helix-turn-helix motif (89), and the helix-turn-helix motif of lambda Cro protein is shown for comparison. We have "deleted" an A from the published Tn*917* sequence (113) in order to generate the last 13 amino acids of the Tn*917* resolvase sequence shown.

from resolvase-mediated recombination between Tn*3* and Tn*1000* (72, 100) initially demonstrated that recombination occurs within a 19-bp region of sequence identity between Tn*3* and Tn*1000* within site I. Subsequently, the site of strand cleavage in Tn*1000* site-specific recombination was determined by exploiting the observation that in vitro recombination performed in the absence of Mg^{2+} on substrates containing two directly repeated *res* sites leads to a double-strand 2-bp staggered cleavage at the palindromic sequence at the center of site I (101) (Fig. 2). In this reaction, resolvase became covalently bound to the recessed 5′ ends of cleaved DNA, apparently by a phosphoserine ester linkage through a serine at residue 10 of resolvase (50, 102). It is presumed that this product represents a true intermediate in the recombination reaction, though it is not inconceivable that a conserved tyrosine that is four amino acids nearer to the N terminus is also involved in phosphoester linkage.

Not only are the spacers within sites I, II, and III of variable length, but the distances between sites vary: in the Tn*3* subfamily, 53 bp separates the centers of sites I and II and 34.5 bp separates the centers of II and III. These spacings are not arbitrary, since insertion of a nonintegral number of helical turns between Tn*1000* sites leads to reduced recombination, whereas addition of one to two integral turns between sites I and II has little effect on recombination (52). The model of the synapse structure presented below in Fig. 5 indicates why the spacing between sites is so important.

It seems likely that most if not all *res* regions conform to the sequence organization shown in Fig. 2 and 3, including those of Tn*917*, Tn*551*, and *C. perfringens* plasmid pIP404 (38, 89, 107, 113; unpublished observations).

C. Resolvases

The Tn*3* family resolvases are relatively small proteins of about 185 amino acid residues. All show extensive similarity at the amino acid sequence level, particularly in their N-terminal domains (Fig. 4). As might be expected, sequence conservation is highest within the Tn*3* and Tn*501* subfamilies. Overall, 31% of amino acid residues are identical. Resolvases also show extensive homology with Hin, Gin, Pin, and Cin invertases (97) (Fig. 4): 23 of the amino acid residues conserved within the resolvase family are maintained in the invertase family. Most are in the N-terminal domain, reflecting the likely common recombination mechanism of the resolvases and invertases (see Section IV for more detailed discussion of evolutionary relationships). The proteins have a classical helix-turn-helix motif and show some similarity to other DNA-binding proteins of this class, such as the Cro and CI proteins of phage lambda (95) (Fig. 4).

The Tn*3* and Tn*1000* resolvases are functionally interchangeable and can be separated into two domains after proteolytic cleavage with chymotrypsin (1) (Fig. 4). The 5-kilodalton (kDa) C-terminal fragment can bind specifically to isolated half-sites within *res* as well as to full sites. The 15.5-kDa N-terminal fragment contains the determinants for strand exchange during recombination and for protein-protein interactions. Purified resolvase aggregates and precipitates in low-ionic-strength buffers (<200 mM NaCl) at near neutral pH and in 1 M NaCl solution appear to be mainly in a dimeric state (49). Resolvase is soluble in low-ionic-strength buffers in the presence of DNA presumably because of its binding to DNA. Though it seems likely that each of the three binding sites within each *res* region binds a dimer of resolvase, it is still not clear whether dimeric resolvase molecules in solution are the species that interact with *res* sites in vivo or in vitro. The fact that resolvase binds to sites I, II, and III with comparable efficiency despite the different sizes of spacers within these sites may in part be a consequence of the presence of a flexible hinge between the two resolvase domains.

Many mutations in the Tn*1000* resolvase gene have been isolated (50–52, 92). The mutant resolvases fall into two phenotypic classes: those that bind *res* DNA but have lost recombinational activity, and those that neither recombine nor bind. Mutations in the C-terminal domain eliminate or impair binding to *res*, whereas those in the N-terminal domain can either eliminate binding to *res* or just block recombination. One mutation affecting this domain, a lysine substitution at residue 128, binds to sites I and II but not to site III, unless the spacer within site III is increased from its usual 7 bp to 10 bp. This suggests that this region of resolvase is important in accommodating the different geometries of the *res* subsites (51, 52). Serine at residue 10, which is conserved in all of the resolvases, is likely to become transiently linked to a recessed 5′ phosphate at the crossover site during recombination; its substitution with cysteine or isoleucine destroys recombination activity, though *res* binding still occurs normally (52, 102).

The synthesis of resolvases is invariably controlled. In the case of Tn*3* and Tn*1000*, the autogenous regulation of resolvase synthesis likely results in all *res* sites in a cell being occupied by resolvase. Further discussion of control of resolvase synthesis is given in Section III.E.1.

D. The Recombination Reaction

1. Recombination in vivo

Tn*3* and Tn*1000* *res*/resolvase reactions on directly repeated *res* sites occur efficiently in vivo: during intermolecular transposition, cointegrate intermediates are rarely seen unless special steps are taken to visualize or isolate them. In one set of experiments about 90% of Tn*1* cointegrates were resolved by 40 min of exposure to resolvase in vivo at 42°C, and the equilibrium ratio of Tn*1* cointegrates to resolution products was about 10^5:1 in a *tnpR*$^+$ cell (91). Tn*1* site-specific recombination appears to be more efficient than that of Tn*3* in vivo (unpublished observations). Conjugal transfer has been extensively used to select and transfer cointegrates from cells in which Tn*3* or Tn*1000* transposition is occurring. Indeed, this is the basis of many Tn*1000* and Tn*3* transposon mutagenesis protocols. Experiments with Tn*501* (115) showed that when Tn*501* transposes intermolecularly in cells not exposed to Hg^{2+}, little resolution of Tn*501* cointegrates occurs. Growth of such cointegrates on Hg^{2+}-containing media leads to resolution, suggesting that synthesis of resolvase is Hg^{2+} inducible (70). Though in vivo *res*/resolvase recombination occurs efficiently on directly repeated *res* sites, it occurs inefficiently or not at all on inverted sites or on sites in different molecules (2, 17, 44, 71, 73, 116). This exquisite directionality is discussed further below.

2. Interactions of *res* and resolvase in vitro

Several laboratories (43, 71, 106) reported the use of DNase footprinting (36) on end-labeled linear fragments of the DNA containing the *res* regions of Tn*1000* and Tn*3*. They showed that Tn*1000* resolvase binds to three sites within each of the *res* regions (Fig. 2 and 3) and noted that each site had a twofold axis of symmetry and that a consensus sequence for a half-site binding could be derived. Binding was on one face of the helix at the extremities of each binding site. Subsequent experiments (Brown, Ph.D. thesis), using UV photofootprinting (7), showed that the binding pattern for Tn*3* resolvase was essentially the same on linear and superhelical *res*$^+$ substrates in vitro and on *res*$^+$ plasmid DNA in vivo. Tn*21* and Tn*1721* *res* regions were also shown to bind resolvase at three sites similar to the Tn*3*/Tn*1000* sites (106).

Subsequently, methylation of purines and ethylation of phosphates (117) was used to interfere with resolvase binding (33). This allowed the mapping of contacts between resolvase and the three binding sites within *res*. Resolvase binds specifically at the consensus sequence on the outside of each binding site, regardless of the size of the spacers. Major contacts are in the major groove. Most of these contacts are also made by the isolated C-terminal domain of resolvase (50). Indeed, the only contact detected with intact resolvase and not detected with the C-terminal fragment is a single phosphate 4 bp inside of the conserved 9-bp consensus. Ethylation of six other phosphates (five within the 9-bp consensus) inhibited binding of both the C-terminal domain and intact resolvase with hydroxy radical footprinting. Using the hydroxy radical reagent MPE · Fe(II) (23), a site of enhanced cleavage unique to the center of site I has also been observed (50). This has been interpreted as a localized distortion of DNA structure (a kink) that allows enhanced intercalation at the crossover site.

Binding of resolvase to *res* regions has also been detected by gel retardation assays (35). Binding of resolvase to intact Tn*3*/Tn*1000* *res* regions (52; M. Boocock, A. Bednarz, and D. Sherratt, unpublished results) gives a series of complexes that represent occupation of different resolvase-binding sites. Our own experiments (unpublished results) have detected at least six discrete complexes for DNA fragments containing complete *res* regions. Bending occurs at (or close to) the center of each site. Resolvase also induces bends on each of the individual sites (50; Boocock et al., unpublished results), a major and a minor complex being obtained for each individual site. The molecular basis of these two complexes is not understood. Binding is highly cooperative and indicates an apparent dissociation constant for Tn*3* *res*/resolvase of about 1 nM under the binding conditions most used (10 mM Tris hydrochloride, pH 8.2, 0.1 mM EDTA). A similar dissociation constant in vivo would lead to most cellular resolvase being bound to *res* regions.

3. Recombination in vitro

Reed (99) first defined reaction conditions for the Tn*1000* *res*/resolvase recombination in vitro. Recombination required negatively supercoiled DNA containing two or more directly repeated *res* sites, Mg^{2+}, and purified resolvase. The reaction requires stoichiometric amounts of resolvase and results in a simple −2 catanene (73, 123). The topologically defined state of the product suggests that the two recombining *res* sites must be organized into a topologically unique synapse prior to recombination. Elegant topological analysis (123, 124) suggested that this synapse traps exactly three interdomainal supercoils between the recombination sites (Fig. 5 and 6). This observation and the requirement for directly repeated *res* sites for recombination seemed to support the idea that a processive "tracking" might act to locate and align a pair of directly repeated *res* sites (71, 73, 106). Such a tracking could in principle provide a mechanism to allow recombination between directly repeated sites but not between inverted sites or sites in separate molecules. Subsequently, the notion of tracking has fallen into disfavor as a mechanism for aligning *res* regions (39; see below, Section II.D.4). One set of results that pointed against tracking as a mechanism for bringing *res* sites together came when it was possible to obtain in vitro recombination conditions in which linear and relaxed circular DNA molecules could participate efficiently in resolvase-mediated recombination as well as the normal superhelical substrates (12, 13). Optimal conditions for such recombination simply require the addition of 5 mM spermidine hydrochloride to the normal (stringent) reaction conditions (pH 8.2, 50 mM Tris hydrochloride, 25 to 50 mM NaCl, 5 mM Mg^{2+}, 20% glycerol; 37°C [13]). Under these conditions, intermolecular reactions could occur between *res* sites on a linear molecule and on a circle and inversion can occur between inverted sites on a linear molecule. This indicated that both circularity and negative superhelicity can provide topological constraints that prevent or minimize recombination between other than directly repeated sites.

A type I DNA topoisomerase activity has also been demonstrated for Tn*3*/Tn*1000* resolvases in vitro (73). This activity has the same DNA requirements as does recombination (i.e., two directly repeated *res* sites in a supercoiled molecule). Similarly, a topoisomerase I activity requiring inverted *gix* sites has been demonstrated for Gin invertase (66). The resolvase topoisomerase activity is an intrinsic property of resolvase, and the relative levels of topoisomerization and recombination in an in vitro reaction depend on the reaction conditions and the nature of the DNA substrate (e.g., spacing of *res* sites).

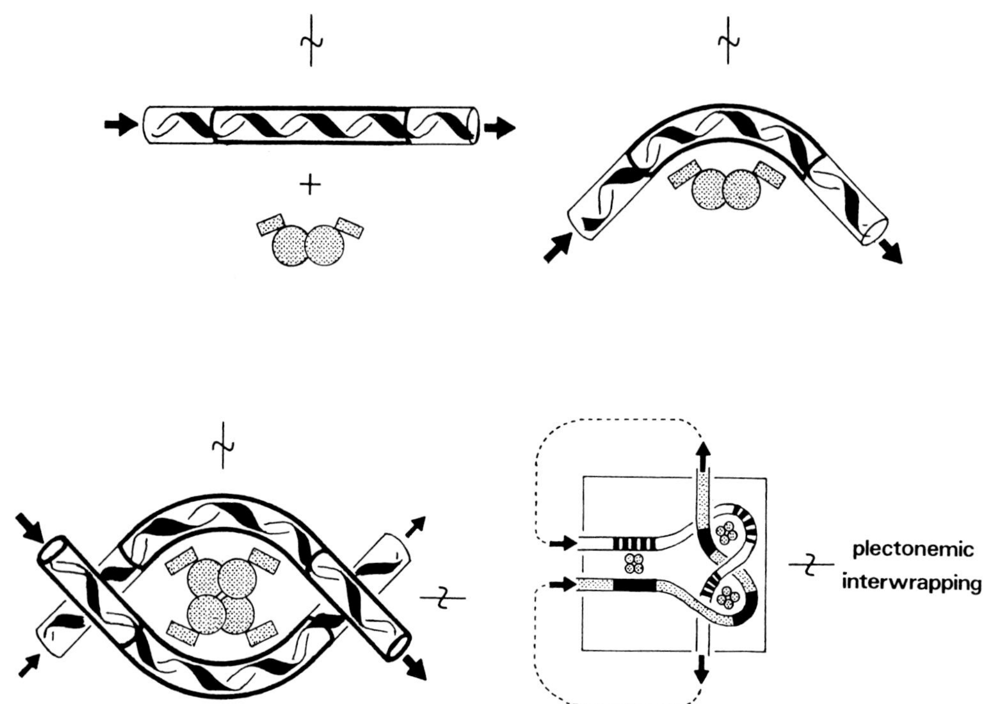

Figure 5. The Tn*3* resolvase synapse. Model for resolvase binding to sites within *res*, through the C-terminal DNA-binding domain of resolvase (shaded areas), inducing a bend within each *res* site, showing how interdomainal supercoils wrap around resolvase tetramers to give a −3 synapse.

4. Structure of the recombinational synapse and the role of DNA topology in directionality and orientation specificity

As indicated above, the resolvases and invertases show strong orientation specificity for their recombination sites, preferably acting intramolecularly on directly repeated and inverted sites, respectively. The first and most intuitive mechanism to be proposed for such directionality was some sort of processive tracking in which an enzyme molecule tightly bound to one site tracks along a DNA molecule (still bound to the initial site) until another site in the correct configuration is encountered (71, 73, 114). Though the inferred unique topological nature of the recombination synapse and the recombination products supported such an idea, as did the observation that adjacent *res* sites are favored for recombination in a substrate containing four directly repeated *res* sites (9), other evidence pointed against a tracking model. For example, in vitro there is no requirement for a high-energy cofactor for recombination, suggesting that any tracking must be a one-dimensional random walk; yet the consequential predicted exponential decrease in recombination frequency as distance between *res* sites increases has never been observed. Also, the failure to partition a pair of "reporter circles," linked by catenation onto the recombination substrate, to just one of the recombinant catenated circles (9) was difficult to reconcile with processive tracking. The ultimate refutation of tracking came from the observation that orientation specificity and intramolecular preference disappear when linear substrates are used for recombination in vitro (12, 13; M. Boocock, unpublished results) and from other experiments in which the in vitro recombination substrates contained multiple *res* sites (Brown, Ph.D. thesis). In these latter experiments, adjacent sites were not necessarily preferred for recombination; for example, in a substrate containing six directly repeated *res* sites, recombination is as efficient between nonadjacent sites that are opposite each other (e.g., sites 1, 4; sites 2, 5) as between adjacent sites. Yet alternate sites (e.g., sites 1, 3) recombine poorly. This observation led to new proposals about how the recombinational synapse is formed and how this provides directionality and specificity.

In the two-step synapsis model (12, 13), resolvase binds to *res* regions (three dimers per *res* region) and then random resolvase-resolvase interactions resulting from three-dimension diffusion bring *res* regions together. It is only when three negative interdomainal supercoils are trapped between the two resolvase/*res* complexes that a stable productive synapse is formed. All other transient synapses, trap-

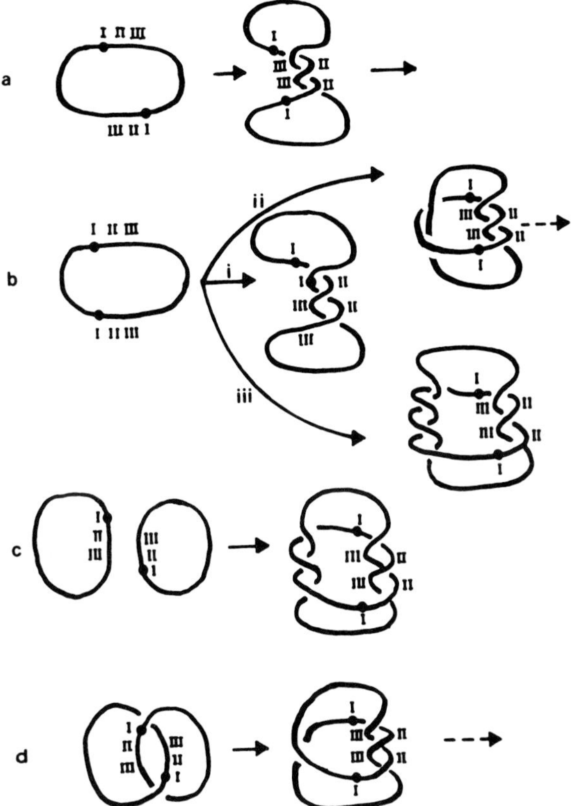

Figure 6. Topological constraints to fusion and inversion. (a) Direct *res* repeats and the −3 synapse. The three resolvase-binding sites in each *res* region are indicated. Recombination occurs between the two type I sites (●). Recombination generates a −2 catenane as a major product and a (+) five-noded catenane (123, 124) as a minor product. (b) Inverted *res* regions and how they might produce a −3 synapse. (i) Aberrant −3 synapse with resolvase-binding sites in wrong relative positions: there are no energetic constraints to formation, but there is no recombination because of incorrect disposition of type I sites. (ii, iii) Potential productive −3 synapses; formation is energetically unfavorable because of the need to trap either two (path ii) or four (path iii) units of positive tangle (positive supercoils). Products (five-noded knots) resulting from pathway (ii) have been observed using a relaxed substrate (13; Boocock, unpublished results). Note: A trefoil (+3 knot) substrate containing inverted *res* sites can produce a productive −3 synapse without energetic constraint and should recombine efficiently (product +3 trefoil). (c) Fusion of unlinked circles. A −3 synapse requires three units of positive tangle (positive supercoils) elsewhere; reaction is energetically unfavorable. (d) Fusion of catenated circles. A −3 synapse requires one unit of positive tangle elsewhere. Products (relaxed circles and four-noded knots) have been observed on a relaxed substrate (13; Boocock, unpublished results).

ping other than three negative interdomainal supercoils, dissociate. The structural requirement for a −3 synapse therefore acts as a topological filter to discard all nonproductive synapses (13, 40). The organization of the three resolvase-binding sites within each *res* region provides the structural basis for a stable productive −3 synapse (13, 123, 124) (Fig. 5). Right-handed plectonemic wrapping of the antiparallel duplexes around the tetramers of resolvase bound to the pairs of interacting sites II and III traps three negative supercoils and aligns type I sites so that recombination between them can occur. The structure and organization of the three resolvase-binding sites provides the means to generate a highly organized and compact synapse. Moreover, the sequence within the *res* regions is such that DNA bending toward the resolvase dimers will be facilitated (Fig. 5). The results outlined below for recombination within multi-*res*-site substrates can easily be accommodated by the two-step model. For example, in a substrate containing six directly repeated *res* regions, the most stable synaptic structure will be one in which all pairs of *res* regions are within isolated −3 synapses, none interfering with any other. This can occur by a 1, 2; 3, 4; 5, 6 type of pairing or by a 1, 2; 3, 6; 4, 5 type of pairing. Results from other multisite substrates are consistent with preferred recombinational products arising from molecules in which the number of independent noninteracting −3 synapses is maximized (Brown, Ph.D. thesis).

In the alternative, "slithering" model of Benjamin and Cozzarelli (8), resolvase bound to a *res* region locates another *res* region within the same molecule (with resolvase bound or possibly unbound) by a slithering process that can continuously "translate" the relative position of one region of a supercoil with respect to the other. Again the productive −3 synapse is favored by the structure of the *res* region.

The need to generate a specific −3 synapse for recombination to occur provides ready explanations of directionality and orientation specificity. Intermolecular reactions are proscribed because the generation of a −3 synapse between unlinked separate molecules would require the addition of three units of "positive tangle" or positive supercoiling (and one unit for molecules in a −2 catenane; Fig. 6c and 6d). Additionally, the slithering model would discriminate between intermolecular reactions by the very nature of the slithering process, which will act only on superhelically interwound DNA segments.

Several factors can act to minimize resolvase-mediated inversion of inverted *res* sites. The simplest and most energetically favorable interaction giving a −3 synapse in a substrate containing inverted *res* sites will have the resolvase-binding sites disposed in the wrong relative position so that the two type I sites are not available to recombine (Fig. 6b). Though it is possible to generate a productive −3 synapse with inverted sites, this again requires the presence of an energetically unfavorable positive tangle elsewhere in the molecule. The ability of linear *res*-containing substrates to recombine intermolecularly with themselves and with circles, as well as intermolecularly (*res* sites inverted or direct), demonstrates the constraints to productive synapsis that topology can introduce. Removal of negative superhelicity can act to reduce the energetic barrier needed to trap positive tangle in order to generate a −3 synapse: recombination can occur in vitro between inverted *res* sites in a relaxed circle and between catenated circles, each containing a single *res* sites, using permissive recombination conditions (13; Boocock, unpublished results).

Similar topological arguments (and experiments) can be used to explain the directionality of inversion systems (we predict a −2 synapse for the Hin, Gin, etc., *hix*, *gix*, etc., systems) and the orientation specificity of transposable element terminal repeats: transposases normally act only on inverted termini within a single duplex molecule. That this is due to topological constraints arising as a consequence of the requirement for a specific "transpososomal" synapse was demonstrated by elegant experiments of Craigie and Mizuuchi using Mu transposition in vitro (21). Separate circular duplexes each containing a single Mu terminus were interwound into multiple interlinked catenanes in which the termini were disposed either in the orientation normally found for a pair of inverted termini in a single molecule or in the opposite orientation. Transposition of the termini from the separate molecules occurred in the former case but not the latter, illustrating the need for a specific topological arrangement rather than an absolute need for inverted termini.

5. The strand exchange mechanism

The structure of a productive −3 synapse was inferred initially from the defined state of the recombination products in vitro: −2 catenanes are the major product and five-noded (+) catenanes are the minor product. These can be generated uniquely from a −3 synapse by one and three 180° right-hand (+) rotations, respectively, during recombination (116, 123, 124). This suggests that recombination can be iterative and that negative superhelicity "drives" the rotation in the right-hand direction. This is supported by the observation that in relaxed circles, rotation can occur in both the right- and left-hand directions (Boocock, unpublished results). Superhelicity is not a prerequisite for strand exchange; it normally determines the direction of duplex rotation and hence the topology of the products.

Just as topological analysis has been a powerful analytical tool for providing insight into how directionality is conferred and into the nature of the resolvase recombinational synapse and the ensuing products, it can be used to dissect the detailed mechanism of strand exchange during recombination. Two simple types of strand exchange mechanism can be formulated. In one (the simple rotation model), a pair of concerted double-strand, staggered cleavages (2 bp; 3′-OH overhang [101, 109]) are made in each of the participating type I resolvase-binding sites, and recombination is normally mediated via a simple right-hand 180° rotation and religation. Assuming a −3 synapse, this mechanism leads to a loss of four superhelical turns from a negatively supercoiled substrate (13).

In the other type (exemplified by the integrase family of recombinases), the reaction occurs by a pair of single-stranded nicks and ligations followed by the second pair of nicks and ligations. With lambda integrase this process is ordered, with a specific pair of nicks always occurring first (69, 94). In principle, this process can occur by either a 90° (anti-Nash) or 270° (Nash) strand rotation and would lead to a loss of either two or six superhelical turns, respectively, from a −3 synapse. If one can then determine the number of negative supercoils lost in a resolvase reaction, it should be possible to infer the reaction mechanism. Using a −17 topoisomer of a substrate containing two Tn*3* *res* sites separated by 219 bp, Boocock et al. determined a superhelicity change of +4, consistent with the double-strand cleavage/simple rotation model (12). The close spacing of the *res* sites was used to demonstrate that all of the superhelicity in the catenated recombination products was in the larger circle, thus unambiguously assigning a superhelicity value to the catenane product. Similar experiments with invertases have shown a +4 change in superhelicity, consistent with an identical mechanism of strand exchange from a −2 synapse (66).

III. TRANSPOSITION

A. Outline of Transposition Mechanisms

Intermolecular transposition of transposon Tn*3* was shown to be a sequential two-step process involving, in the first transposition step, replication of the transposon and an initial fusion of the donor and recipient replicon to create cointegrate transposition intermediates. The second site-specific recombination step involves the resolution of cointegrates to the normal transposition end products. This insight came from experiments in which certain deletion mutants of Tn*3* were shown to produce cointegrates as transposition end products (41, 56). In normal transposi-

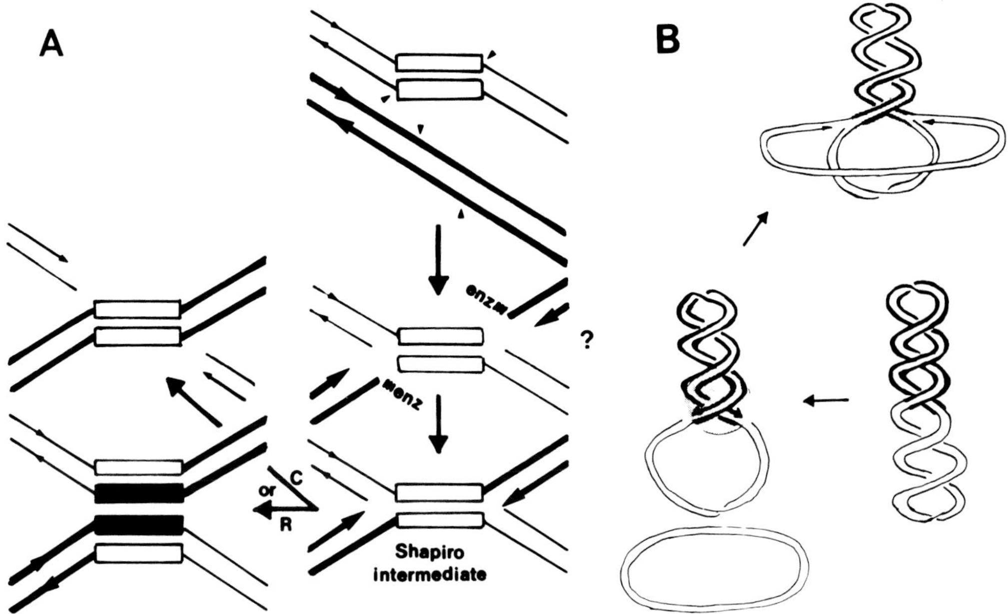

Figure 7. Replicative transposition model. (A) Schematic diagram illustrating the formation of a Shapiro intermediate and its conversion to products via (R) replicative and (C) nonreplicative pathways. In the former cases, the immediate transposition products are cointegrates (intermolecular transposition) or replicative deletions or inversions (intramolecular) (4, 112). (B) Nature of the Mu transpososome (85, 118). Mu A protein is depicted by the shaded areas. Binding of Mu A protein to inverted Mu termini, in the presence of host protein HU, generates a transpososome containing two 3′-OH nicks at the transposon boundaries. Mu DNA remains superhelical and non-Mu DNA is relaxed. This complex then "captures" a potential insertion site, a process promoted by Mu B protein, and transposase-induced double-strand cleavage of the target occurs, followed by ligation of the two transposon strands to the target, generating the Shapiro intermediate.

tion of Tn*3* family elements, cointegrates are rapidly resolved into the observed transposition end products by the action of the Tn*3* site-specific recombination enzyme resolvase on the specific site *res*. Similarly, experiments with bacteriophage Mu showed that its transposition often occurs by a replicative pathway, though Mu specifies no site-specific resolution system (31, 32, 85, 121). Together these observations led to the proposal of similar models for the replicative transposition of Tn*3* transposons and Mu (5, 112) (Fig. 7). In outline, the models explain both the intramolecular and intermolecular transposition of Tn*3* and Mu, and the models as initially presented only differ in the polarity of the polynucleotide chain ends created by transposase nicking at the transposon termini. Subsequent experimental work has supported the proposal (112) that 3′ ends are created at the transposon termini and that transpositional replication is initiated from the recessed 3′-OH group at the staggered breaks in the recipient molecule and proceeds through the element to the donor replicon, generating the fused replicon cointegrate intermediate (Fig. 7).

Other members of the Tn*3* family appear to have the same replicative transposition mechanism and appear to be unique among transposable elements in encoding their own site-specific recombination systems to convert the cointegrate products of intermolecular transposition into the normally observed transposition end products. This suggests that intermolecular transposition is important in the natural history of these elements. Notwithstanding this, in laboratory experiments, intramolecular transposition of Tn*3* to give replicative deletions and inversions occurs at frequencies comparable to those for intermolecular transposition (11). Such deletions and inversions require transposase, but not *res* and resolvase, as predicted by transposition models (Fig. 7). The possibility of retransposition of elements in deleted circular molecules or of homologous recombination between such transposons and homologous sequences in other molecules is always present, and can generate more complex DNA rearrangements. All transposons of the Tn*3* family create 5-bp duplications of DNA at their insertion sites, presumably a consequence of staggered breaks, with 5-bp protrusions being made at the insertion sites. Such duplications are not found bounding a single element in cointegrates, nor in elements that have been involved in the equivalent intramolecular deletions and inversions. Examination of Fig. 7 clearly shows why this is the case: the process of replication during transposition generates two copies of the element within the staggered break.

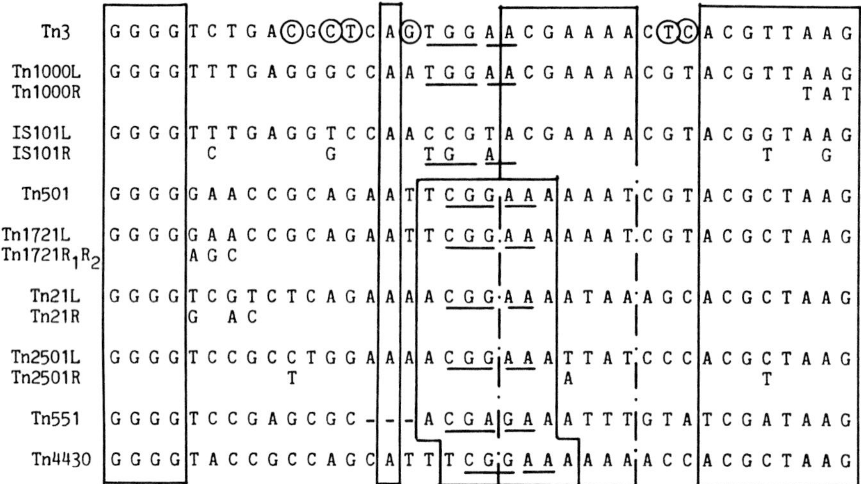

Figure 8. Terminal inverted repeats of the Tn*3* family. The conserved terminal 38 nucleotides are shown for the elements indicated (some have longer [Tn*2501*; Tn*551*] or shorter [Tn*1000*] inverted termini). The conserved GGGG is on the outside of the transposon (with the putative transposon-induced nick on the other strand, immediately 3′ of CCCC), and the conserved ACGPyTAAG is on the inside of the repeat. Most highly conserved motifs are boxed, but for clarity the conserved PyGGAA is underlined. The nucleotides of Tn*3* circled are those which must account for the different specificities of Tn*3* and Tn*1000* transposase. Where the termini of an element are nonidentical, differences are indicated.

Precise excision of Tn*3* family elements by recombination across the 5-bp repeats has not been documented. It is not clear whether this is because it occurs at a very low frequency or because it does not occur at all. There are several precedents of recombination across 9-bp direct repeats flanking a transposon. Such events are stimulated when the DNA is single stranded and when there are substantial inverted repeats immediately inside the direct repeats. It has been suggested that such "recombination" occurs by a copy choice slippage replication mechanism (29). The low frequency of Tn*3* precise excision might reflect the short inverted repeats of Tn*3* or the short 5-bp duplication or both.

B. The Terminal Inverted Repeats and Transposases

Figure 8 shows a comparison of the inverted repeats of members of the Tn*3* family. The repeats are clearly very similar, and there are several motifs and spacings that are especially conserved in all of the sequences. GGGG is invariably found on the outside of the repeats, and ACGPyTAAG is almost always found at the inside of at least one of the two repeats of each pair. An internal heptanucleotide ACGAAAA is highly conserved in the Tn*3* subfamily, and variants of it are also found in the other elements and also in the Mu repeats bound by Mu A protein. In the Tn*3* subfamily, this heptanucleotide is located one helical turn 5′ of the conserved ACGPyTAAG, whereas in the other members of the family, this heptanucleotide is displaced a further three or four nucleotides to the 5′ end of the repeat. Other conserved sequences and motifs are shown in Fig. 8.

Inspection of the repeat sequences also shows that Tn*3*, Tn*1000*, and IS*101* form a closely related group; Tn*501* and Tn*1721* form another very similar group, with Tn*21* a slightly more distant relative. The inverted termini and transposase of the cryptic transposon Tn*2501* appear most closely related to Tn*21*, even though it has the gene organization of the Tn*3* subfamily (81, 82). The terminal sequences of the gram-positive transposons Tn*551* (and the nearly identical Tn*917* [67, 89, 113]) and Tn*4430* (77, 79) are more difficult to place in terms of sequence relationship, but clearly these belong to the Tn*3* family, with Tn*4430* apparently being more similar to the gram-negative elements than is Tn*551*.

Despite the similarity of the different repeats, the interchangeability of transposases in transposition assays is limited. Tn*3* and Tn*1000* transposases are not interchangeable, though it has not been reported if the two transposases bind the termini of their heterologous partners in in vitro assays. The Tn*3* and Tn*1000* termini differ in seven positions, up to nucleotide 35. However, the different specificities for transposase should lie in only one or more of six of these differences, because the 209-bp IS*101* transposes with Tn*1000* but not Tn*3* transposase (55, 64) (Fig. 8). Transposons containing mixed Tn*1000*/Tn*3* termini do not transpose with either transposase (A. Arthur, unpublished results).

Within the Tn*501* family, functional interrelationships are complex: Tn*21* inverted termini only transpose with Tn*21* transposase, while inverted termini of Tn*501* and Tn*1721* transpose at comparable frequencies with either Tn*501* or Tn*1721* transposases (47). This is not surprising, since Tn*501* and Tn*1721* have almost identical termini. Tn*21* transposase can act functionally on Tn*1721* inverted repeats (47), but in Tn*501* it acts preferentially on the *tnpA*-proximal terminal repeat of Tn*501* and a copy of the Tn*21*R repeat that happens to be located 80 bp inside of the *tnpA*-distal Tn*501* terminal repeat (46). All 15 products of "Tn*501* transposition" mediated by Tn*21* transposase analyzed had used one Tn*21* terminus and one Tn*501* terminus. Tn*501* and Tn*1721* transposases normally act on the Tn*501* repeats of Tn*501*, while in a transposon containing one Tn*501* repeat and one Tn*21* repeat, Tn*501*/Tn*1721* transposases function at only about 1% of the level observed for Tn*501* termini. The Tn*21* termini have 30/38 bases in common with Tn*501*/Tn*1721*, about the same difference as the Tn*3*/Tn*1000* termini. The complementation of Tn*1721* by Tn*21* transposase is perhaps therefore surprising, particularly in view of the nonreciprocity of complementation. It appears that either Tn*21* transposase is somewhat promiscuous in its activity or that there are other determinants close to the termini of either Tn*21* or Tn*501*/Tn*1721* (integration host factor [IHF] binding sites?) that affect specificity. Consistent with the apparent promiscuity of Tn*21* transposase is the observation (82) that Tn*21* transposase complements transposition of Tn*2501*, though poorly.

The transposases of these elements are all large (and presumably multifunctional) proteins of about 110,000 kDa. Their sequence relationships are discussed in Section IV. The 3′ ends of the transposase genes all lie within a terminus, in some cases terminating with the UAG termination codon. An inverted repeat involving 5 bp of the terminal repeat sequences just 3′ of the *tnpA* gene and 5 bp just inside the terminal repeat and close to the end of the *tnpA* gene has also been reported (55). The significance of this is not clear: such repeats are absent at the other end of the transposon and are unlikely to act as strong *tnpA* transcriptional terminators, since substantial levels of transcription existing from the *tnpA* end of Tn*3* can be used to express various reporter genes.

Homologous recombination in vivo has been used to generate hybrid transposases between Tn*21* and Tn*501* (30). Not surprisingly, most recombination occurred in the more similar C-terminal regions of the coding sequence, though hybrids involving the 5′ end of transposase were also obtained. Recombinant transposases fell into two sharply defined spec-

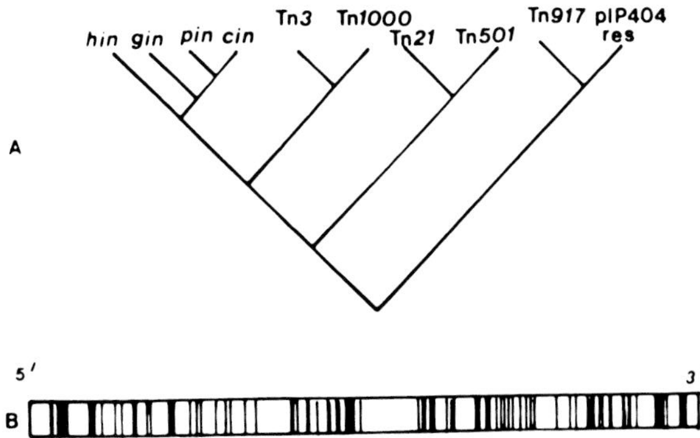

Figure 9. Resolvases/invertases and transposases compared. (A) Phylogenetic tree of resolvases/invertases (38). (B) Depiction of Tn*3* family transposases. Bars indicate positions at which amino acid residues are conserved within the family (Tn*3*, Tn*21*, Tn*501*, Tn*917*, Tn*4430*, part of Tn*2501* [79]).

ificity classes. Their analysis showed that the specificity determinants lie in the region between 28 and 216 amino acids from the N terminus. There is 53% amino acid identity in this region but no clues as to precisely which residues are responsible for specificity. Nevertheless, it is presumed that it is this region that interacts specifically with the transposon terminal repeats. Most workers have suggested, on the basis of the common transposition mechanism, that the transposase catalytic activity resides in the conserved C-terminal region of transposase. A recent publication (79) provides a very clear comparison of the similarities in primary amino acid sequence among the Tn*3* family transposases (Fig. 9). There is little obvious similarity among the Tn*3* family transposases and those of other elements. One might expect to find some similarity between transposases and recombinases/topoisomerases, but no extensive similarity has been reported.

C. In Vitro Transposition Experiments

There have been no reports of successful transposition in vitro for any of the Tn*3* family elements. It is likely that the major mechanism for all of these elements is similar to the replicative mode of Mu transposition that has been demonstrated and characterized in in vitro experiments (85, 118). Three laboratories (63, 86; N. D. F. Grindley, personal communication) have reported progress in the purification of Tn*1000* and Tn*3* transposases in a form that binds specifically to the element inverted repeats. In no case does binding show the ATP dependence reported earlier (126). DNase I footprints have shown transposase binding to the inner 25 bases of the 38-bp inverted repeats (63). Such binding also enhances DNase I cleavage at the outer junction of the transposon termini. Additionally, transposase binds avidly to any DNA termini in a sequence-nonspecific manner. N. D. F. Grindley and colleagues (personal communication) report that Tn*1000* transposase footprints to the whole Tn*1000* inverted repeat. Moreover, they have shown that IHF binds immediately inside the boundaries of each terminus. Binding of IHF and binding of transposase appear to be cooperative. Several laboratories have already reported that Tn*3* transposition occurs in strains mutant for each of the genes that encode for functional IHF (55) synthesis. Given the variety of other IHF-like proteins in *E. coli*, these results are not necessarily inconsistent with models in which the binding of such proteins to the ends is needed along with transposase for transposition.

D. The Transposition Mechanism

1. The bacteriophage Mu transposition mechanism and its applicability to Tn*3* family transposons

Parallels between the mechanisms of Tn*3* and phage Mu transposition were drawn when it became clear that both could transpose replicatively to give cointegrate products. Subsequently, the elegant and painstaking work of Mizuuchi and co-workers on phage Mu transposition in vitro (19–21, 83, 84) has provided a detailed biochemical understanding of its transposition and has confirmed the model for Mu transposition proposed by Shapiro (112) (Fig. 7). Mu transposase, the 70-kDa A protein, binds to the termini of Mu DNA when they are present in inverted orientation on superhelical DNA in the presence of the host protein HU, giving rise to a specific "transpososome" complex (118). Within such complexes, nicking can occur at each of the Mu termini, leading to relaxation of the non-Mu DNA. The Mu DNA remains supercoiled in such complexes because the bound transposase prevents unwinding (118) (Fig. 7). There is no evidence of conservation of phosphodiester bond energy or of protein-DNA intermediates in such complexes. Such transpososomes can now interact with target DNA, which in vitro can be superhelical, circular, or linear. Such interactions are promoted by the 35-kDa Mu B protein and ATP but do not absolutely require them. Mu B protein, especially in the presence of ATP, binds DNA nonspecifically, and would appear to act as a target localization protein for the transpososome complex. Mu B protein hydrolyzes ATP in this interaction. In the absence of Mu B protein or ATP, transposition is less frequent and the events are mainly intramolecular (80, 85).

Interaction of a donor transpososome with a target leads to a complex having the structure of a "Shapiro" intermediate; these have been purified and characterized. The conversion of these to transposition end products via either the replicative or nonreplicative pathway in vitro requires no further participation of Mu proteins, the outcome depending on the availability of a functional replication system (20). Note that in the events mediated by transposase, four phosphodiester bonds are cleaved and two are made, the ligations being to the protruding 5′ extension in the target. It is likely that transposases of the Tn*3* family act in a similar way, also giving rise to a Shapiro intermediate. It is not obvious that Tn*3* family transposons have a counterpart to the Mu B protein. Several workers have noted that the combined size of Mu A and B proteins is similar to the size of Tn*3* family transposases, and have implied that a Mu B-like activity might reside in such transposases. However, note that in *E. coli* infected with phage Mu, the B protein is more abundant than the A protein, consistent with the role of B in target localization; such nonstoichiometry of Mu A and B proteins could not easily be reproduced in Tn*3* family transposases if both activities were encoded by the *tnpA* gene. Alternatively, IHF or a similar protein could be involved in target localization, as seems possible for IS*1* (37).

It is not clear what determines the fate of a Shapiro transposition intermediate in vivo. With the Tn*3* family elements, the intermediate almost always acts as a reactant for the replicative pathway, whereas with Mu, both the replicative and the nonreplicative pathways can proceed. Other elements whose transposition may proceed through Shapiro intermediates (e.g., IS*1*, IS*903*) may use the nonreplicative pathway more frequently than the replicative pathway. Perhaps the precise transpososome structure at the potential replication forks determines the accessibility of replication or repair complexes and hence the pathway used.

2. Apparent nonreplicative transposition of Tn*3* family elements

The majority of transposition events involving Tn*3*-like elements occur replicatively and give rise to cointegrates between donor and recipient replicons in the absence of *res*/resolvase site-specific recombination. However, even in the absence of a functional site-specific recombination system, a few (usually fewer than 2% of total events) products in which a recipient replicon contains a simple transposon insertion are obtained. It has been suggested that these occur with the Tn*3* subfamily nonreplicatively and do not involve cointegrate intermediates (10). These would then be equivalent to nonreplicative Mu insertions that occur both in vivo and in vitro and would most easily arise by the simple insertion pathway from a Shapiro intermediate (Fig. 7). An alternative possibility is that cointegrates are formed as usual but are then resolved by a *recA*-independent, *res*/resolvase-independent pathway. Apparently nonreplicative insertions have also been observed with the Tn*501* subfamily (82).

3. One-ended transposition

Normal transposition requires two transposon termini in inverted orientation. Nevertheless, several groups have observed that a single transposon terminus of a Tn*3*-like element within a plasmid can be used to generate transposase-dependent cointegrations of donor and recipient replicons (4, 6, 59, 87, 88, 110). The frequencies of such intermolecular events are lower than those of normal transposition by factors of 10 (Tn*21*) to 1,000 (Tn*1*/*3*) and give rise to products in which rather more than one copy of the donor replicon containing a transposon terminus is inserted into the recipient plasmid. This has been interpreted as reflecting aberrant replicative transposition in which an initial recombinant joint between one donor transposon terminus and the one end of the cleaved target DNA is made and then followed by rolling circle replication. Subsequently, an aberrant second cleavage is made in a segment of the newly replicated donor molecule (often close to the transposon terminus) and is ligated to the other end of the recipient DNA molecule. Such insertions can be bounded by the normal 5-bp target duplication (88).

4. Insertion site specificity

Members of the Tn*3* family, like most other transposable elements, show some insertion site specificity. They appear to prefer A+T-rich sequences, and also are attracted to sites showing some similarity to transposon ends (45, 120). For example, when both Tn*3* and Tn*501* are present in a cell, Tn*3* often inserts close to Tn*501* ends and vice versa. However, additional factors must also affect Tn*3* target selection: the frequency of Tn*3* insertion close to a Tn*501* terminus was found to vary widely according to the site of the Tn*501* terminus within a plasmid (45). However, productive insertions do not occur into plasmid replicons containing sequences identical to the ends of the functional transposon. This is manifested in the phenomenon termed transposition immunity (60, 105), the process whereby productive Tn*3* transposition into a replicon already containing Tn*3* or a Tn*3* terminal repeat is inhibited (see Section III.E.2). Tn*3* family elements studied in *E. coli* all appear to transpose at higher frequencies into plasmids and phage than into the *E. coli* chromosome (74, 90). It is not clear whether this is a manifestation of transposition immunity or not, though if it is, then the responsible sequences in the chromosome are not likely to be much more than a 38-bp terminal repeat, since DNA hybridization experiments reveal no homology to Tn*3* in the chromosomes of most *E. coli* strains. The gram-positive elements are somewhat different: Tn*917* transposes to both chromosomal and plasmid sites, while Tn*551* prefers to insert into the chromosome (78, 93; E. Murphy, this volume).

Whether there are other preferred insertion sequences for Tn*3* elements remains to be determined. It is interesting to note that IS*1* has a preferred insertion site in pBR322 close to the IHF-binding site of that plasmid, while the IS*1* termini themselves contain IHF-binding sites (37; Grindley, personal communication). This suggests that IHF-IHF interactions may play a role in locating donor and recipient sites. The fact that IHF binds to Tn*3* close to the termini could mean that Tn*3*, too, has some preference for targets close to binding sites for IHF or other "accessory" binding proteins.

5. Host functions and transposition

If, as seems likely, Tn*3* family transposons transpose in a way similar to that followed by phage Mu, their transposases introduce single-stranded nicks at the boundary of transposon and nontransposon DNA at the donor sites and create the double-stranded staggered breaks in the target DNA. In the Mu system, it can be inferred from in vitro experiments that the phosphodiester bonds from the latter target cleavages may be conserved and used during the transposase-mediated ligation of transposon ends into the target: phosphodiester bond energy is not conserved during the nicking at the transposon ends (85; see above, Section III.D.1). For Mu transposition in vitro, the only host function required for this reaction is the host histonelike protein HU. A similar situation is likely to exist for Tn*3* elements.

Host replication functions are probably needed for the replication step and for the final covalent ligation. It is possible that accessory proteins such as IHF or FIS (factor for inversion stimulation [66]) may have some function in transposase gene expression, in stabilizing the transpososome structure, or in target location. However, note that Tn*3* transposition appears to be relatively normal in IHF-defective strains, despite IHF protein binding inside of the Tn*3* termini in vitro (Grindley, personal communication).

Mutations in the *polA* gene can reduce transposition of Tn*5* and Tn*10* 5- to 100-fold, though the effects on Tn*3* transposition were not reported (119). Whether this is a direct consequence of *polA* involvement or is a secondary consequence of the *polA* mutation (e.g., gaps in DNA) is not clear, but the latter seems more likely. Similarly, *dnaA* mutations affect Tn*5* transposition (96), though the possible effect on Tn*3* has not been tested. Mutations that affect supercoiling have also been reported to show differences in transposition frequencies of Tn*3* and other elements (55). This is not surprising in view of the effect of superhelical density on expression of many proteins and of the sensitivity and topology of many protein-DNA interactions.

E. Control of Transposition

1. Expression of transposases and resolvases and its control

The evidence suggests that the transposition of Tn*3* family elements is normally limited by the availability of active transposase; increased levels of synthesis of transposase give rise to increased frequencies of transposition. With Tn*3*, resolvase is a negative regulator of transcription of both the transposase (*tnpA*) and the resolvase (*tnpR*) genes (18, 41, 55, 56) (Fig. 2). This autogenous control of *tnpR* transcription maintains constant levels of resolvase in a cell in which Tn*3* is established. When Tn*3* enters a Tn*3*-free cell, one would expect a burst of transposase synthesis before repression is established. Tn*3* *tnpR* mutants show a 10- to 100-fold increase in transposition and a similar increase in expression from the transposase promoter (18). Examination of the Tn*3* *res* site sequence shows putative promoters for *tnpA* and *tnpR* in sites I and II of *res*, respectively (Fig. 2). Transcripts for the *tnpA* and *tnpR* genes can be initiated from these promoters in vitro (127). In Tn*1000*, a *tnpR* promoter in site II that corresponds to the Tn*3* *tnpR* promoter is known to function in vivo (125). In addition, a weaker *tnpR* promoter with about one-fourth of this activity in vivo lies in site I and overlaps the *tnpA* promoter (125). This latter promoter can initiate Tn*1000* *tnpR* transcription in vitro (103). The significance of these two *tnpR* promoters with Tn*1000* is not clear. When Tn*3* resolvase binds to *res*, it may occupy site II (and III) preferentially before site I (Boocock, personal communication). This could lead to differential repression of the *tnpA* and *tnpR* promoters and to a low level of "leakage" *tnpA* transcription without concomitant *tnpR* expression.

Whereas translation of the *tnpR* gene is highly efficient (R. Slatter and D. Sherratt, unpublished results), translation of the Tn*3* *tnpA* transcript appears rather inefficient, at least partly because of a poor ribosome-binding site preceding the ATG start. Mutations that increase transposase expression have been mapped to this site (15). Examination of the Tn*3* sequence reveals two possible translational initiation sites; the first is a GTG (coordinate 3091), which appears to be rarely if at all used under normal circumstances (55). At a position 33 bp 3′ of this is a more frequently used ATG (24, 55). We estimate that the translation efficiency of the *tnpA* gene is about 5 to 15% that of the *E. coli lacZ* gene, whereas the *tnpA* promoter is comparable in strength to the *lacZ* promoter. In contrast, Tn*3* *tnpR* translation is as efficient as *lacZ* translation, but the *tnpR* promoter is rather weak.

Tn*3* transposition shows startling temperature sensitivity; transposition at 15 to 30°C is about 100 times more frequent than at 37°C, and transposition at 42°C is almost undetectable (74, 91). The molecular basis of this effect is not clear; it appears not to be mediated at the transcriptional level, and several workers have suggested that the level of transposase protein is less at 42 than at 30°C, perhaps as a consequence of increased transposase turnover (e.g., 55; D. J. Sherratt, S. Hettle, and R. Slatter, unpublished results).

Much less is known about the control of transposase synthesis among the Tn*501* subfamily. The gene organization of these elements again provides the potential for repressible coordinate control of resolvase and transposase from a single promoter within *res* (108) (Fig. 1). A Tn*1721* promoter within the *res* region upstream of *tnpR* and *tnpA* that should be controlled by resolvase binding to *res* has been identified (108). In the absence of functional resolvase, the *tnpR* gene (driving *galK*) was translated at a much higher frequency than *tnpA* (3). This could be most simply explained by more efficient translation of *tnpR* than of *tnpA*. Tn*21* and Tn*501* also have putative promoters in their *res* sites, though in these cases there is evidence to suggest that expression of *tnpA* and *tnpR* is not determined solely by such transcription. It has been observed (70; P. Kitts, unpublished results) that both transposition of Tn*501* and resolution of its cointegrates are stimulated by growth of cells in the presence of Hg^{2+}, suggesting either that transcription of *tnpA/R* can initiate in the Hg^{2+}-inducible *mer* operon well upstream of *res* or that the promoter in *res* can respond positively to an Hg^{2+}-induced product. In these experiments, little or no cointegrate resolution was observed in the absence of Hg^{2+}, whereas transposition occurred at a frequency of 4×10^{-5} to 1×10^{-7} per cell generation. Addition of Hg^{2+} resulted in

>90% cointegrate resolution and a >20-fold stimulation in transposition. It is not clear whether these quantitatively different responses to Hg^{2+} suggest different molar protein requirements for resolution and transposition or that expression of resolvase and transposase genes responds differentially to Hg^{2+}.

Another report (62) has proposed that Tn*21* contains a regulatory gene *tnpM* upstream of *res* that encodes a 13-kDa protein which regulates the expression of *tnpR* negatively and *tnpA* positively. The *tnpA* and *tnpR* genes of Tn*501* were also reported to respond to the Tn*21 tnpM* product. The Tn*501* DNA sequence (25) reveals an open reading frame that corresponds to Tn*21 tnpM*. Just how *tnpM* might control *tnpA/R* expression is not clear. Indeed, at least one laboratory (S. Halford, personal communication) has not been able to substantiate the role of *tnpM* in control of transposition functions within the Tn*501* subfamily.

One might imagine that transposition frequencies would be affected by the physiological state of the host, by analogy with how strongly the probability of establishing lysogeny by bacteriophage lambda is affected by the physiological state of the host cells (28). There is no evidence that repressor stability, or indeed the level of repressor and/or transposase activity, responds to cellular signals (other than the Hg^{2+} induction of Tn*501* and erythromycin inducibility of Tn*551* and Tn*917*). Nevertheless, it would be surprising if the transposition of members of the Tn*3* family was not responsive to cellular signals.

2. Transposition immunity

This is the phenomenon by which productive Tn*3* transposition into a plasmid replicon is inhibited when that replicon already contains certain Tn*3* sequences. This trait is probably exhibited by all Tn*3* family transposons but has been most studied with Tn*3*. The original observation (105) was that productive Tn*3* transposition is inhibited into plasmid replicons that already contain Tn*3*. The effect is *cis* specific; i.e., in a cell containing Tn*3*, transposition occurs normally into replicons lacking Tn*3*, but at greatly reduced frequencies into Tn*3*-containing replicons. There is no ultimate barrier to the survival of single replicons containing two or more copies of Tn*3*; these can be readily constructed using in vitro or in vivo techniques. The effect must therefore be exerted at the level of productive transposition into the Tn*3*-containing replicon. Many studies have confirmed the phenomenon and attempted to determine its mechanism (4, 11, 17, 59, 60, 76, 91, 122). Some of the earlier reports contained conflicting conclusions as to whether resolvase is required or not (91, 122). The more recent work has demonstrated, however, that a single 38-bp Tn*3* terminus within a replicon is sufficient for immunity. The only other Tn*3* determinant required is Tn*3* transposase, which of course must be present for the phenomenon to be observed.

Immunity appears to show the same specificity as specific transposase-terminus interactions; e.g., the presence of Tn*1000* or Tn*501* termini in a plasmid does not prevent Tn*3* transposition into it. Moreover, mutations within the 38-bp repeat that reduce or abolish transposition generally abolish or reduce transposition immunity. The level of transposition immunity that has been observed in different experiments has been quite variable. Whether this reflects cellular physiology, different assay conditions, differing levels of transposase, different sequences present in the recipient replicon, or different aspects of recipient replicon biology is not clear. Measured immunity is reduced when high levels of transposase are present (4). Whatever its mechanism, the phenomenon of transposition immunity is important in limiting the level of Tn*3* transposition and the number of Tn*3* transposons within cells.

In replicons containing a single Tn*3* element, intramolecular events occur at frequencies comparable to those for intermolecular events (11). That is, a functional element seems not to inhibit its own transposition into other places in the same DNA molecule. Nevertheless, the insertion of an additional Tn*3* terminus to a Tn*3*-containing replicon blocks intramolecular transposition, presumably as a consequence of immunity (11). Perhaps a requirement for specific topology within the transpososome provides the molecular basis for Tn*3* immunity: the presence of other termini that can interact with the transpososome and interfere with its topology could prevent productive transpososome formation, just as the presence of additional *res* sites within a molecule can interfere with resolvase site-specific recombination between fixed sites (Brown, Ph.D. thesis). The general phenomenon of Tn*3* transposition immunity is not limited to Tn*3* family transposons; equivalent immunity is shown by Tn*7* (M. Rogers and D. Sherratt, unpublished results; N. L. Craig, this volume) and in a slightly different form by phage Mu.

IV. RELATIONSHIPS WITHIN THE Tn*3* FAMILY AND BETWEEN THE Tn*3* FAMILY AND OTHER ELEMENTS

The Tn*3* family of transposable elements constitutes a related yet diverse set of transposable elements that have been highly successful in colonizing a wide range of gram-negative and gram-positive bacteria. As selfish DNA elements they have probably been more successful than any other bacteriophage or procaryotic transposable element. The transposases and terminal repeats of these elements have evolved to use the DNA synthetic machinery of extremely diverse organisms.

The division of the gram-negative elements into two broad classes, the Tn*3* subfamily and the Tn*501* subfamily, seems relatively straightforward. However, even here, Tn*2501*, despite its sequence similarity, occupies a rather uncomfortable place in the Tn*501* subfamily, since its gene organization is more akin to that of the Tn*3* subfamily.

There seems to be no obvious reason behind the different organizations of the two subfamilies. In both cases there is the possibility of control of transposition by resolvase acting to regulate synthesis of both resolvase and transposase. There are many different schemes by which the two gene organizations could have arisen or have been interconverted.

A phylogenetic tree for the relationships among the resolvases/invertases has been derived using partition analysis (38) (Fig. 9). This phylogeny appears similar to what might be derived for the transposon terminal repeats and the transposases, though no formal derivation of a phylogeny for repeats or transposases has been reported. If the phylogenies are truly similar, it suggests that the transposases and resolvases have largely coevolved and have not been subject to frequent reassortment. Nevertheless, it is clear that Tn*4430* has recruited an integraselike resolvase for its resolution (79).

Though the resolvases from the gram-positive elements Tn*917* and *Clostridium* plasmid pIP404 *res* appear to be phylogenetically distinct (Fig. 9), the differences between transposases/terminal repeats of gram-negative and gram-positive transposases do not seem to be greater than those that can be found within just gram-negative elements/bacteria (e.g., compare Tn*4430* terminal repeats with those of the Tn*3* and Tn*501* subfamilies; Fig. 8). This suggests possible horizontal transfer between gram-negative and gram-positive bacteria. Several gram-positive elements have been shown to transpose in *E. coli* (79; E. Murphy, this volume). There are no convincing reports of extensive amino acid similarities between Tn*3* family transposases and other transposases/recombinases/topoisomerases.

The characterized site-specific recombinases clearly fall into two classes (109). First, the resolvases/invertases that are about 20 kDa in size have a C-terminal DNA-binding domain and an N-terminal domain containing the putative serine involved in the 5′-phosphoseryl-ester recombination intermediate. They appear to recombine via a concerted double-strand cleavage ligation, making a 2-bp, 3′ extension stagger (13, 101). They generate a productive synapse of defined topology that gives products of defined topology. The reactions are strongly directional, favoring either direct or inverted repeats; this directionality is determined largely by the synapse topology and the organization of the recombination site (Section II.D.4).

In contrast, the lambda-integrase-like enzymes which constitute the second class vary in size from 20 to 60 kDa and have a tyrosine in the C-terminal segment of the protein that is involved in a phosphotyrosyl protein-DNA ester linkage (48). Neither the productive synapses nor the recombination products have a fixed topology, and there is no fundamental orientation specificity in the reaction mechanism other than by having asymmetric sites (40, 109). Recombination is across a 6- to 8-bp stagger (5′ protrusion) via a pair of ordered single-strand cleavages and ligations (69, 94). Note that with recombination specified both by the resolvases/invertases and integraselike enzymes, the transient protein-DNA phosphoester bond is made to a recessed base (through a 5′-phosphoseryl linkage in the first case and a 3′-phosphotyrosyl linkage in the second), whereas if transposases use a protein-DNA ester link, it must be through an extended 5′-phosphoester (see Fig. 7). *E. coli* DNA gyrase makes such a 5′-phosphotyrosyl ester during topoisomerization (39) and therefore could be mechanistically related to transposases.

The lambda-integrase-like and resolvase/invertase-like enzymes and their respective sites have evolved to function in a variety of processes that are largely plasmid and phage encoded. Their determinants have crossed the gram-positive/gram-negative barrier and into budding *Saccharomyces cerevisiae* (FLP; the determinant involved in inversion switching in the *S. cerevisiae* 2μm plasmid is a lambda-integrase-like protein [3]). Their functions include prophage insertion and excision, resolution of transposition intermediates, maintenance of plasmids in the monomeric state to ensure stable inheritance, inversion switches that control gene expression, and amplification of plasmid copy number (3, 26, 28, 38, 61, 66, 75, 79, 97, 109, 115, 116).

Acknowledgments. I thank Martin Boocock for much inspiration, advice, and discussion. I also thank many colleagues for communication of results prior to publication and Sue Halley for preparing the manuscript.

My own work cited here was funded by the Medical Research Council and Science and Engineering Research Council.

LITERATURE CITED

1. **Abdel-Meguid, S. S., N. D. F. Grindley, N. S. Templeton, and T. A. Steitz.** 1984. Cleavage of the site-specific recombination protein γδ resolvase: the smaller of two fragments binds DNA specifically. *Proc. Natl. Acad. Sci. USA* **81**:2001–2005.
2. **Altenbuchner, J., and R. Schmitt.** 1983. Transposon Tn1721: site-specific recombination generates deletions and inversions. *Mol. Gen. Genet.* **190**:300–308.
3. **Argos, P., A. Landy, K. Abremski, J. B. Egan, E. Haggard-Ljungquist, R. H. Hoess, M. L. Kahn, B. Kalionis, S. V. L. Narayana, L. S. Pierson III, N. Sternberg, and J. M. Leong.** 1986. The integrase family of site-specific recombinases: regional similarities and global diversity. *EMBO J.* **5**:443–450.
4. **Arthur, A., E. R. Nimmo, S. J. H. Hettle, and D. J. Sherratt.** 1984. Transposition and transposition immunity of transposon Tn3 derivatives having different ends. *EMBO J.* **3**: 1723–1729.
5. **Arthur, A., and D. J. Sherratt.** 1979. Dissection of the transposition process: a transposon-encoded site-specific recombination system. *Mol. Gen. Genet.* **175**:267–274.
6. **Avila, P., F. de la Cruz, E. Ward, and J. Grinsted.** 1984. Plasmids containing one end of Tn21 can fuse with other plasmids in the presence of Tn21 transposase. *Mol. Gen. Genet.* **195**:288–293.
7. **Becker, M. M., and J. C. Wang.** 1984. Use of light for footprinting DNA in vivo. *Nature* (London) **309**:682–687.
8. **Benjamin, H. W., and N. R. Cozzarelli.** 1986. DNA-directed synapsis in recombination, slithering and random collison of sites. *Proc. Robert A. Welch Found. Conf. Chem. Res.* **29**:107–126.
9. **Benjamin, H. W., M. M. Matzuk, M. A. Krasnow, and N. R. Cozzarelli.** 1985. Recombination site selection by Tn3 resolvase: topological tests of a tracking mechanism. *Cell* **40**:147–158.
10. **Bennett, P. M., F. de la Cruz, and F. J. Grinsted.** 1983. Cointegrates are not obligatory intermediates in transposition of Tn3 & Tn21. *Nature* (London) **305**:743–744.
11. **Bishop, R., and D. J. Sherratt.** 1984. Transposon Tn1 intra-molecular transposition. *Mol. Gen. Genet.* **196**: 117–122.
12. **Boocock, M. R., J. L. Brown, and D. J. Sherratt.** 1986. Structural and catalytic properties of specific complexes between Tn3 resolvase and the recombination site *res*. *Biochem. Soc. Trans.* **14**:214–216.
13. **Boocock, M. R., J. L. Brown, and D. J. Sherratt.** 1987. Topological specificity in Tn3 resolvase catalysis, p. 703–718. *In* T. J. Kelley and R. McMacken (ed.), *DNA Replication & Recombination*. Alan R. Liss, Inc., New York.
14. **Brown, N. L., J. N. Winnie, D. Fritzinger, and R. D. Pridmore.** 1985. The nucleotide sequence of the *tnpA* gene completes the sequence of the *Pseudomonas* transposon Tn501. *Nucleic Acids Res.* **13**:5657–5669.
15. **Casadaban, M. J., J. Chou, and S. N. Cohen.** 1982. Overproduction of the Tn3 transposition protein and its role in DNA transposition. *Cell* **28**:345–354.
16. **Chen, S.-T., and R. C. Clowes.** 1987. Variations between the nucleotide sequences of Tn*1*, Tn2, and Tn*3* and expression of β-lactamase in *Pseudomonas aeruginosa* and *Escherichia coli*. *J. Bacteriol.* **169**:913–916.
17. **Chiang, S. J., and R. E. Clowes.** 1982. Recombination between two TnA transposon sequences oriented as inverse repeats is found less frequently than between direct repeats. *Mol. Gen. Genet.* **185**:169–175.
18. **Chou, J., M. J. Casadaban, P. G. Lemaux, and S. N. Cohen.** 1979. Identification and characterization of a self-regulated repressor of translocation of the Tn3 element. *Proc. Natl. Acad. Sci. USA* **76**:4020–4024.
19. **Craigie, R., D. J. Arndt-Jovin, and K. Mizuuchi.** 1985. A defined system for the DNA strand-transfer reaction at the initiation of bacteriophage Mu transposition: protein and DNA substrate requirements. *Proc. Natl. Acad. Sci. USA* **82**:7570–7574.
20. **Craigie, R., and K. Mizuuchi.** 1985. Mechanism of transposition of bacteriophage Mu: structure of a transposition intermediate. *Cell* **41**:867–876.
21. **Craigie, R., and K. Mizuuchi.** 1986. Role of DNA topology in Mu transposition: mechanism of sensing the relative orientation of two DNA segments. *Cell* **45**:793–800.
22. **Datta, N., R. W. Hedges, E. J. Shaw, R. Sykes, and M. H. Richmond.** 1971. Properties of an R factor from *Pseudomonas aeruginosa*. *J. Bacteriol.* **108**:1244–1249.
23. **Dervan, P. B.** 1986. Design of sequence-specific DNA-binding molecules. *Science* **232**:464–471.
24. **Ditto, M. D., J. Chou, M. W. Hunkapiller, M. A. Fennewald, S. P. Gerrard, L. E. Hood, S. N. Cohen, and M. J. Casadaban.** 1982. Amino-terminal sequence of Tn*3* transposase protein. *J. Bacteriol.* **149**:407–410.
25. **Diver, W. P., J. Grinsted, D. C. Fritzinger, N. L. Brown, J. Altenbuchner, P. Rogowsky, and R. Schmitt.** 1983. DNA sequences of and complementation by the *tnpA* genes of Tn21, Tn501 and Tn1721. *Mol. Gen. Genet.* **191**:189–193.
26. **Dodd, H. M., and P. M. Bennett.** 1986. Location of the site-specific recombination system of R46: a function necessary for plasmid maintenance. *J. Gen. Microbiol.* **132**: 1009–1020.
27. **Dodd, H. M., and P. M. Bennett.** 1987. The R46 site-specific recombination system is a homologue of the Tn3 and γδ (Tn1000) cointegrate resolution system. *J. Gen. Microbiol.* **133**:2031–2039.
28. **Echols, H.** 1986. Bacteriophage lambda development: temporal switches and the choice of lysis or lysogeny. *Trends Genet.* **2**:26–30.
29. **Egner, C., and D. E. Berg.** 1981. Excision of transposon Tn5 is dependent on the inverted repeats but not on the transposase function of Tn5. *Proc. Natl. Acad. Sci. USA* **78**: 459–463.
30. **Evans, L. R., and N. L. Brown.** 1987. Construction of hybrid Tn501/Tn21 transposases *in vivo*: identification of a region of transposase conferring specificity of recognition of the 38 bp terminal inverted repeats. *EMBO J.* **6**:2849–2853.
31. **Faelen, M., A. Toussaint, and M. Couturier.** 1971. Mu-1 promoted integration of λ-gal phage in the chromosome of *E. coli*. *Mol. Gen. Genet.* **113**:367–370.
32. **Faelen, M., A. Toussaint, and J. De Lafonteyne.** 1975. Model for the enhancement of λ-*gal* integration into partially induced Mu-1 lysogens. *J. Bacteriol.* **121**:873–882.
33. **Falvey, E., and N. D. F. Grindley.** 1987. Contacts between γδ resolvase and the *res* site. *EMBO J.* **6**:815–821.
34. **Fischoff, D. A., G. F. Vovis, and N. F. Zinder.** 1980. Organisation of chimeras between filamentous bacteriophage F1 and plasmid pSC101. *J. Mol. Biol.* **144**:243–265.
35. **Fried, M. G., and D. M. Crothers.** 1983. Kinetics and

mechanism in the reaction of gene regulatory proteins with DNA. *Nucleic Acids Res.* **11**:141–158.
36. **Galas, D. J., and A. Schmitz.** 1978. DN'ase footprinting: a simple method for the detection of protein DNA binding specificity. *Nucleic Acids Res.* **5**:3157–3170.
37. **Gammas, P., M. G. Chandler, P. Prentki, and D. J. Galas.** 1987. *E. coli* integration host factor binds specifically to the ends of the insertion sequence IS1 and to its major insertion hotspot in pBR322. *J. Mol. Biol.* **195**:261–272.
38. **Garnier, T., W. Saurin, and S. T. Cole.** 1987. Molecular characterization of the resolvase gene *res*, carried by a multicopy plasmid from *Clostridium perfringens*: common evolutionary origin for prokaryotic site-specific recombinases. *Mol. Microbiol.* **1**:371–376.
39. **Gellert, M.** 1981. DNA topoisomerases. *Annu. Rev. Biochem.* **50**:879–910.
40. **Gellert, M., and N. Nash.** 1987. Communication between segments of DNA during site-specific recombination. *Nature* (London) **325**:401–404.
41. **Gill, R., E. Heffron, G. Dougan, and S. Falkow.** 1978. Analysis of sequences transposed by complementation of two classes of transposition-deficient mutants of transposition element Tn*3*. *J. Bacteriol.* **136**:742–756.
42. **Grindley, N. D. F.** 1983. Transposition of Tn3 and related transposons. *Cell* **32**:3–5.
43. **Grindley, N. D. F., M. R. Lauth, R. G. Wells, R. J. Wityk, J. J. Salvo, and R. R. Reed.** 1982. Transposon-mediated site-specific recombination: identification of three binding sites for resolvase at the res sites of γδ and Tn3. *Cell* **30**:19–27.
44. **Grindley, N. D. F., and R. R. Reed.** 1985. Transpositional recombination in prokaryotes. *Annu. Rev. Biochem.* **54**: 863–896.
45. **Grinsted, J., P. M. Bennett, S. Higginson, and M. H. Richmond.** 1978. Regional preference of insertion of Tn501 and Tn501 into RP1 and its derivatives. *Mol. Gen. Genet.* **166**:313–320.
46. **Grinsted, J., and N. L. Brown.** 1984. A Tn21 terminal sequence within Tn501: complementation of tnpA gene function and transposon evolution. *Mol. Gen. Genet.* **197**: 497–502.
47. **Grinsted, J., F. de la Cruz, J. Altenbuchner, and R. Schmitt.** 1982. Complementation of *tnp*A mutants of Tn3, Tn21, Tn501 and Tn1721. *Plasmid* **8**:276–286.
48. **Gronostajski, R. M., and P. D. Sadowski.** 1985. The FLP recombinase of the *Saccharomyces cerevisiae* 2μm plasmid attaches covalently to DNA via a phosphotyrosyl linkage. *Mol. Cell. Biol.* **5**:3274–3279.
49. **Halford, S., S. L. Jordan, and E. A. Kirkbride.** 1985. The resolvase protein from the transposon Tn21. *Mol. Gen. Genet.* **200**:169–175.
50. **Hatfull, G. F., and N. D. F. Grindley.** 1986. Analysis of the γδ resolvase mutants *in vitro*. Evidence for an interaction between serine-10 of resolvase and site I of *res*. *Proc. Natl. Acad. Sci. USA* **83**:5429–5433.
51. **Hatfull, G. F., S. M. Noble, and N. D. F. Grindley.** 1987. The γδ resolvase induces an unusual DNA structure at the recombination crossover point. *Cell* **49**:103–110.
52. **Hatfull, G. F., J. J. Salvo, E. E. Falvey, V. Rimphanitchayakit, and N. D. F. Grindley.** 1988. Site-specific recombination of the γδ resolvase. *Symp. Soc. Gen. Microbiol.* **43**:149–181.
53. **Hedges, R. W., and A. F. Jacob.** 1974. Transposition of ampicillin resistance from RP4 to other replicons. *Mol. Gen. Genet.* **132**:31–40.
54. **Hedges, R. W., G. J. Snow, R. Sykes, and M. H. Richmond.** 1971. Properties of an R-factor from *Pseudomonas aeruginosa*. *J. Bacteriol.* **108**:1244–1249.
55. **Heffron, F.** 1983. Tn3 and its relatives, p. 223–260. *In* J. A. Shapiro (ed.), *Mobile Genetic Elements*. Academic Press, Inc., New York.
56. **Heffron, F., P. Bedinger, J. J. Champon, and S. Falkow.** 1977. Deletions affecting the transposition of an antibiotic resistance gene. *Proc. Natl. Acad. Sci. USA* **74**:702–706.
57. **Heffron, F., R. Kostriken, C. Morita, and R. Parker.** 1981. Tn3 encodes a site-specific recombination system: identification of essential sequences, genes and actual site of recombination. *Cold Spring Harbor Symp. Quant. Biol.* **45**: 259–268.
58. **Heffron, F., B. J. McCarthy, H. Ohtsubo, and E. Ohtsubo.** 1979. DNA sequence analysis of the transposon Tn3: three genes and three sites involved in transposition of Tn3. *Cell* **18**:1153–1164.
59. **Heritage, J., and P. M. Bennett.** 1985. Plasmid functions mediated by one end of TnA. *J. Gen. Microbiol.* **131**: 1131–1140.
60. **Huang, C.-J., F. Heffron, Jr., S. Twu, R. H. Schloemer, and C.-H. Lee.** 1986. Analysis of Tn3 sequences required for transposition and immunity. *Gene* **41**:23–31.
61. **Huber, H. E., S. Iida, W. Arber, and T. A. Bickle.** 1985. Site-specific DNA inversion is enhanced by a DNA sequence element in *cis*. *Proc. Natl. Acad. Sci. USA* **82**:3776–3780.
62. **Hyde, D. R., and C.-P. D. Tu.** 1985. *tnp*M: a novel regulatory gene that enhances Tn21 transposition and suppresses cointegrate resolution. *Cell* **42**:629–638.
63. **Ichikawa, H., K. Ikeda, W. L. Wishart, and E. Ohtsubo.** 1987. Specific binding of transposase to terminal inverted repeats of transposable element Tn3. *Proc. Natl. Acad. Sci. USA* **84**:8220–8224.
64. **Ishizaki, K., and E. Ohtsubo.** 1985. Cointegrates and resolution mediated by IS101 present in plasmid pSC101. *Mol. Gen. Genet.* **199**:388–395.
65. **Jordon, E., H. Saedler, and P. Starlinger.** 1968. 0° and strong-polar mutations in the gal operon are insertions. *Mol. Gen. Genet.* **102**:353–365.
66. **Kahmann, R., G. Mertens, A. Klippel, B. Brauer, F. Rudt, and C. Koch.** 1987. The mechanism of G inversion, p. 681–690. *In* T. J. Kelley and R. McMacken (ed.), *DNA Replication & Recombination*. Alan R. Liss, Inc., New York.
67. **Khan, S. A., and R. P. Novick.** 1980. Terminal nucleotide sequence of Tn551, a transposon specifying erythromycin resistance in *S. aureus*: homology with Tn3. *Plasmid* **4**: 148–154.
68. **Kitts, P. A., A. Lamond, and D. J. Sherratt.** 1982. Inter-replicon transposition of Tn1/3 occurs in two sequential genetically separable steps. *Nature* (London) **295**:626–628.
69. **Kitts, P. A., and H. A. Nash.** 1987. Homology-dependent interactions in phage λ site-specific recombination. *Nature* (London) **329**:346–348.
70. **Kitts, P. A., R. Reed, L. Symington, M. Burke, and D. J. Sherratt.** 1982. Transposon-specified site-specific recombination. *Proc. Natl. Acad. Sci. USA* **79**:46–50.
71. **Kitts, P. A., L. S. Symington, P. Dyson, and D. J. Sherratt.** 1983. Transposon-encoded site-specific recombination: nature of the Tn3 sequences which constitute the recombination site *res*. *EMBO J.* **2**:1055–1060.
72. **Kostriken, R., C. Morita, and F. Heffron.** 1981. The transposon Tn3 encodes a site-specific recombination system: identification of essential sequences, genes and actual site of recombination. *Proc. Natl. Acad. Sci. USA* **78**:4041–4045.
73. **Krasnow, M. A., and N. R. Cozzarelli.** 1983. Site-specific

relaxation and recombination by the Tn3 resolvase: recognition of the DNA path between oriented sites. *Cell* **32:** 1313–1324.

74. **Kretschmer, P. J., and S. N. Cohen.** 1979. Effect of temperature on translocation frequency of the Tn*3* element. *J. Bacteriol.* **139:**515–519.

75. **Lane, D., R. de Feyter, M. Kennedy, S.-H. Phua, and D. Semon.** 1986. D protein of miniF acts as a repressor of transcription and as a site-specific resolvase. *Nucleic Acids Res.* **14:**9713–9728.

76. **Lee, C.-H., A. Bhagwat, and F. Heffron.** 1983. Identification of a transposon Tn3 sequence required for transposition immunity. *Proc. Natl. Acad. Sci. USA* **80:**6765–6769.

77. **Lereclus, D., J. Mahillon, G. Menou, and M.-M. Lecadet.** 1986. Identification of Tn4430, a transposon of *Bacillus thuringiensis* functional in *Escherichia coli. Mol. Gen. Genet.* **204:**52–57.

78. **Luchansky, J. B., and P. A. Pattee.** 1984. Isolation of transposon Tn*551* insertions near chromosomal markers of interest in *Staphylococcus aureus. J. Bacteriol.* **159:**894–899.

79. **Mahillon, J., and D. Lereclus.** 1988. Structural and functional analysis of Tn4430: identification of an integrase-like protein involved in the cointegrate resolution process. *EMBO J.* **7:**1515–1526.

80. **Maxwell, A., R. Craigie, and K. Mizuuchi.** 1987. B protein of bacteriophage Mu is an ATPase that preferentially stimulates intermolecular DNA strand transfer. *Proc. Natl. Acad. Sci. USA* **84:**699–703.

81. **Michiels, T., and G. Cornelis.** 1984. Detection and characterization of Tn*2501*, a transposon included within the lactose transposon Tn*951*. *J. Bacteriol.* **158:**866–871.

82. **Michiels, T., G. Cornelis, K. Ellis, and J. Grinsted.** 1987. Tn*2501*, a component of the lactose transposon Tn*951*, is an example of a new category of class II transposable elements. *J. Bacteriol.* **169:**624–631.

83. **Mizuuchi, K.** 1983. *In vitro* transposition of bacteriophage Mu: a biochemical approach to a novel replication reaction. *Cell* **35:**785–794.

84. **Mizuuchi, K.** 1984. Mechanism of transposition in bacteriophage Mu: polarity of the strand transfer reaction at the initiation of transposition. *Cell* **39:**395–404.

85. **Mizuuchi, K., and R. Craigie.** 1986. Mechanism of bacteriophage Mu transposition. *Annu. Rev. Genet.* **20:**385–429.

86. **Morita, M., S. Tsunasawa, and Y. Sugino.** 1987. Overproduction and purification of the Tn3 transposase. *J. Biochem.* **101:**1253–1264.

87. **Mötsch, S., and R. Schmitt.** 1984. Replicon fusion mediated by a single-ended derivative of transposon Tn1721. *Mol. Gen. Genet.* **195:**281–287.

88. **Mötsch, S., R. Schmitt, P. Avila, F. de la Cruz, E. Ward, and J. Grinsted.** 1985. Junction sequences generated by "one-ended transposition." *Nucleic Acids Res.* **13:**3335–3342.

89. **Murphy, E.** 1988. Transposable elements in *Staphylococcus. Symp. Soc. Gen. Microbiol.* **43:**59–89.

90. **Muster, C. J., and J. A. Shapiro.** 1981. Recombination between transposable elements: on replicon fusion. *Cold Spring Harbor Symp. Quant. Biol.* **45:**239–242.

91. **Muster, C. J., L. A. MacHattie, and J. A. Shapiro.** 1983. pλCM system: observations on the roles of transposable elements in the formation and breakdown of plasmids derived from bacteriophage lambda replicons. *J. Bacteriol.* **153:**976–990.

92. **Newman, B. J., and N. D. F. Grindley.** 1984. Mutants of γδ resolvase: a genetic analysis of the recombination function. *Cell* **38:**463–469.

93. **Novick, R. P., I. Edelman, M. D. Schlesinger, A. D. Gruss, E. C. Swanson, and P. A. Pattee.** 1979. Genetic translocation in *Staphylococcus aureus. Proc. Natl. Acad. Sci. USA* **76:** 400–404.

94. **Nunes-Duby, S. E., L. Matsumoto, and A. Laudy.** 1987. Site-specific recombination intermediates trapped with suicide substrates. *Cell* **50:**779–788.

95. **Pabo, C. O., and R. T. Sauer.** 1984. Protein-DNA recognition. *Annu. Rev. Biochem.* **53:**293–321.

96. **Phadnis, S. H., and D. E. Berg.** 1987. Identification of base pairs in the outside end of insertion sequence IS50 that are needed for IS50 and Tn5 transposition. *Proc. Natl. Acad. Sci. USA* **84:**9118–9122.

97. **Plasterk, R. H. A., A. D. Brinkman, and P. van de Putte.** 1983. DNA inversions in the chromosomes of *Escherichia coli* and in bacteriophage Mu: relationship to other site-specific recombination systems. *Proc. Natl. Acad. Sci. USA* **80:**5355–5358.

98. **Plasterk, R. H., T. A. M. Ilmer, and P. van de Putte.** 1983. Site specific recombination of Gin of bacteriophage Mu: inversions and deletions. *Virology* **124:**24–30.

99. **Reed, R. R.** 1981. Transposon-mediated site-specific recombination: a defined *in vitro* system. *Cell* **25:**713–719.

100. **Reed, R. R.** 1981. Resolution of cointegrates between transposons γδ and Tn3 defines the recombination site. *Proc. Natl. Acad. Sci. USA* **78:**3428–3432.

101. **Reed, R. R., and N. D. F. Grindley.** 1981. Transposon-mediated site-specific recombination in vitro: DNA cleavage and protein-DNA linkage at the recombination site. *Cell* **25:**721–728.

102. **Reed, R. R., and C. D. Moser.** 1984. Resolvase-mediated recombination intermediates contain a serine covalently linked to DNA. *Cold Spring Harbor Symp. Quant. Biol.* **49:**245–249.

103. **Reed, R. R., G. I. Shibuya, and J. A. Steitz.** 1982. Nucleotide sequence of the γδ resolvase gene and demonstration that its gene product acts as a repressor of transcription. *Nature* (London) **300:**381–383.

104. **Richmond, M. H., and R. B. Sykes.** 1972. The chromosomal integration of a β-lactamase gene derived from the P-type R-factor RP1 in *E. coli. Genet. Res.* **20:**231–237.

105. **Robinson, M. K., P. M. Bennett, J. Grinsted, and M. H. Richmond.** 1977. Inhibition of TnA translocation by TnA. *J. Bacteriol.* **129:**407–414.

106. **Rogowsky, P., S. E. Halford, and R. Schmitt.** 1985. Definition of three resolvase binding sites at the *res* loci of Tn21 & Tn1721. *EMBO J.* **4:**2135–2141.

107. **Rogowsky, P., and R. Schmitt.** 1984. Resolution of a hybrid cointegrate between transposons Tn501 & Tn1721 defines the recombination site. *Mol. Gen. Genet.* **193:**162–166.

108. **Rogowsky, P., and R. Schmitt.** 1985. Tn1721-encoded resolvase: structure of the *tnp*R gene and its *in vitro* functions. *Mol. Gen. Genet.* **200:**176–181.

109. **Sadowski, P.** 1986. Site-specific recombinases: changing partners and doing the twist. *J. Bacteriol.* **165:**341–347.

110. **Schmitt, R., J. Altenbuchner, and J. Grinsted.** 1981. Complementation of transposition functions encoded by transposons Tn501 (Hg^r) and Tn1721 (Tc^R), p. 359–370. *In* S. B. Levy, B. C. Clowes, and E. L. Konig (ed.), *Molecular Biology, Pathogenicity and Ecology of Bacterial Plasmids.* Plenum Publishing Corp., New York.

111. **Shapiro, J. A.** 1969. Mutations caused by the insertion of genetic material into the galactose operon of *E. coli. J. Mol. Biol.* **40:**93–105.

112. **Shapiro, J. A.** 1979. Molecular model for the transposition

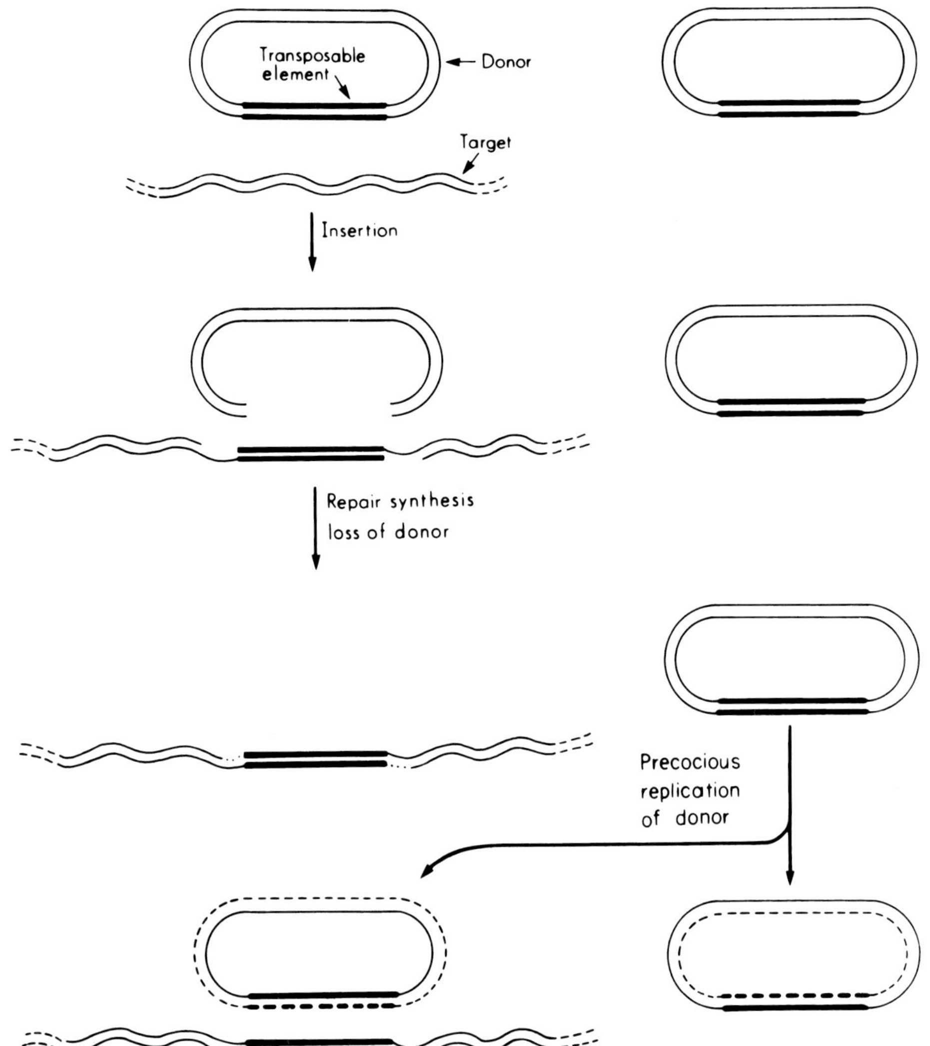

Figure 6. Model for conservative transposition (adapted from references 14 and 16). Thick lines, Transposable element; broken lines, newly synthesized DNA. Both strands at each element-vector junction are cleaved, the target DNA is cleaved (9 bp apart in the case of Tn*5*), the element and protruding ends of target DNAs are joined, and gaps are filled by repair synthesis. Although the DNA molecule that served as the transposition donor is lost, sibling DNAs will persist and replicate precociously to fill the niche created by loss of the donor. In a variant of this model, the gapped donor is repaired by gene conversion from a sibling molecule (Kleckner, this volume); a second variant is shown in Fig. 8, below.

original location or its transposition to new sites. It is thought that the *tnp* or *inh* product makes bacteria more physiologically adaptable to the new (chemostat) environment (30, 67).

III. MECHANISM OF Tn*5* TRANSPOSITION

A. Conservative Transposition

Early studies of the products of transposition from an infecting λ::Tn*5* phage showed that Tn*5* transposition involves the loss of vector DNA and suggested a model of conservative (cut-and-paste) transposition (Fig. 6) (14, 16). In this model, the Tn*5*-encoded transposase and associated host proteins bind element ends and target DNAs, make double-strand breaks at the element-vector junctions, join element and target DNAs together, and fill gaps, thereby generating 9-bp target sequence duplications; the ends of the vector DNA are not joined, and the linear fragment is lost. The surviving transposition product is called a "simple insert."

For Tn*5* transposition between two chromosomal sites, the product is viable if the donor and target sites are in different copies of the chromosome, but not when transposition is between two sites in the same chromosome. In contrast, the products of intrachromosomal **inverse** transposition (Fig. 4) would be viable (see also Fig. 4 of Berg et al., this volume). For transposition from a plasmid to the chromosome, loss of the donating molecule would be compensated by overreplication of its sibling, as in Fig. 6. In both cases the resulting cell lineage will be seen to contain a copy of the element at the donor site as well as one at the new site. Alternatively, in bacteria that degrade linear DNAs less actively than wild-type *E. coli* (*recBC sbcA* [153]), the vector fragment could be repaired using homologous sequences from a sibling DNA molecule (containing the element at the donor site), as in the gene conversion of gapped DNAs (see reference 91).

Cointegrates that consist of complete vector and target sequences joined by direct repeats of the element, such as those made by Tn*3* during its replicative transposition, are not generated by Tn*5*. Efficient site-specific recombination between direct repeats of Tn*3* (resolution) rapidly breaks down the cointegrate to separate donor and target DNAs, each with one Tn*3* element (see Fig. 2 in Berg et al., this volume; Fig. 4 and 5 in M. L. Pato, this volume; and Fig. 7 in D. Sherratt, this volume). The similarity between the final products of conservative transposition and the final product of replicative transposition plus resolution led for a time to the widespread belief that all transposition was replicative (54, 88, 89, 133). It is now clear, however, that there are a variety of transposition mechanisms and that Tn*5* probably employs a conservative one.

B. Absence of IS*50*-Specific Resolvase

A pBR322-related plasmid with direct repeats of IS*50* that exhibited an unusual apparent *recA*-independent recombinational instability has been described. This instability led to the suggestion that Tn*5*, like Tn*3*, encodes a function that can quickly resolve any cointegrates formed during transposition (171). Further tests showed, however, that the instability was due to homologous recombination, which even in *recA* cells occurs sporadically between any duplicate plasmid sequences (45, 50), coupled with inhibition of cell growth by the initial plasmid but not by its recombinant derivative. Thus, the plasmid instability was due to strong selection for rare recombinants, not to frequent recombination (resolution) (126). In addition, other studies showed that various DNAs with direct repeats of IS*50*, such as dimeric pBR322::Tn*5* plasmids and cointegrates such as those diagrammed in Fig. 5, are generally stable in *recA* cells (16, 72, 78). This indicates that IS*50*-containing DNAs are not subject to specific resolution during steady-state growth.

The possibility that Tn*5*-specific resolution might occur only during transposition was tested with the λ::pBR322::Tn*5* cointegrate diagrammed in Fig. 7 (top) by selecting jointly for loss of λ and for retention of the transposon's kanamycin resistance (K in figure). pBR322 with just one copy of IS*50*, the product of a putative resolution reaction, was not found in cells in which Tn*5* had transposed (81). Thus there is no evidence for a Tn*5*-specific resolvase.

C. "Cointegrates" Formed by Conservative Transposition

Cointegrates, consisting of the target, a full set of sequences from the donor, and an extra copy of IS*50* or Tn*5*, constitute up to 2% of the products of transposition involving pBR322::Tn*5* plasmid donors in recombination-deficient (*recA*) cells, and also one-fourth or more of the products formed in *recA*$^+$ cells (16, 73, 74, 78). These cointegrates were sometimes interpreted as products of replicative transposition (2, 74, 78). A study of their origins showed that dimeric and larger forms of pBR322::Tn*5* plasmids are common in *recA*$^+$ cells. When stable monomeric and dimeric pBR322::Tn*5* DNAs in *recA* cells were used as transposition donors, cointegrates arose frequently only from the dimers (16). Although these cointegrates contain a complete set of donor sequences, Fig. 5 illustrates that they do not contain the entire dimeric donor plasmid plus an extra copy of the element, as is the case for products of Tn*3*-type replicative transposition (see Sherratt, this volume). Tn*5*-mediated cointegrates from dimers are thus equivalent to simple insertions from monomers; they are attributable to conservative transposition of just a fragment of the donor DNA (16). The effect of *recA* on the structures of Tn*5* transposition products is thus likely to be indirect, due primarily to its effect on recombinational interactions among donor DNAs, and hence the frequency of dimers in a donor population.

The rare cointegrates found in *recA* cells using pBR322::Tn*5* donors are also consistent with conservative insertion because dimeric and concatemeric derivatives of monomeric pBR322 plasmids are found at low frequency in *recA* cells (45, 50) and may arise by sporadic replication via a rolling-circle mechanism (37). It has also been noted (99) that cointegrates could arise from monomeric DNAs by conservative transposition involving a pair of sibling elements just after passage of a replication fork (Fig. 8).

The origin of the rare cointegrates formed with

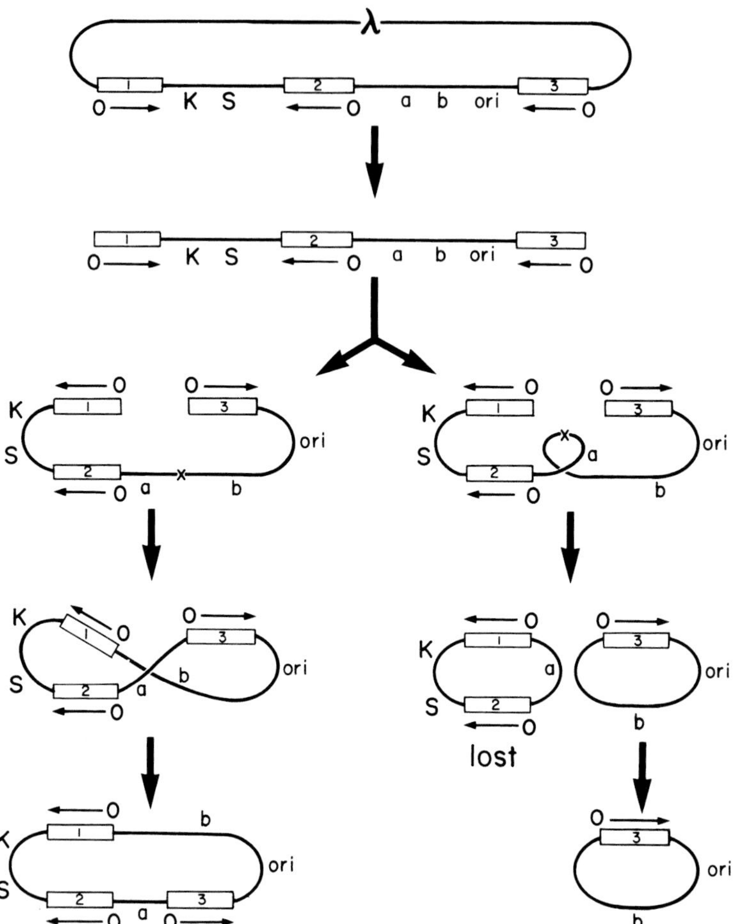

Figure 7. Consequences of conservative transposition from a λ-pBR322::Tn*5* cointegrate to a site within the same molecule. One daughter is lost (right) because it lacks a replication origin (adapted from reference 81 and Z. Koncz-Kalman, O. Olsson, C. Thomas, J. Schell, and C. Koncz, unpublished data).

λ::IS*50* donors (78) was investigated using *recA* cells and a λ::IS*50* phage that was deficient in phage-specific integration and replication and was marked with a resistance gene. Cointegrates were found only in cells infected with two or more phage, and only if the phage were recombination proficient. This indicates that cointegrates formed with λ::IS*50* also arose in two steps: the formation of dimeric DNAs by homologous recombination, and then transposition of just a segment of the dimer, essentially as diagrammed in Fig. 5 and 6 (99).

Products of the breakdown of a λ-pBR322::Tn*5* cointegrate containing three copies of IS*50* (Fig. 7), obtained by a selection for loss of λ genes, provided additional evidence for conservative transposition. One frequent product consisted of the segment with three IS*50* elements, but with an inversion of some sequences between two of the elements. This structure suggested excision of the element and then joining of its ends to an internal site (Fig. 7, left). The other product was a deleted plasmid, probably the result of an equivalent conservative intramolecular transposition, but to a target in the opposite orientation (Fig. 7, right) (81). Comparable rearranged elements have also been obtained, using a Tn*5* element that contains an RK2 plasmid replication origin, by selection for the transposon as an autonomous replicon (Z. Koncz-Kalman and J. Schell, personal communication).

The initial development of a conservative transposition model for Tn*5* (14) was influenced by the precedents of phage λ integration and proposals for the movement of the *Ac* element of maize (62; N. V. Fedoroff, this volume). Additional findings of conservative transposition have been made in the cases of cDNA copies of retroviral and retrotransposon RNAs (H. Varmus and P. Brown, this volume; J. D. Boeke, this volume) and of two other bacterial elements, phage Mu and Tn*10* (Pato, this volume; Kleckner, this volume). It is thus likely that conserv-

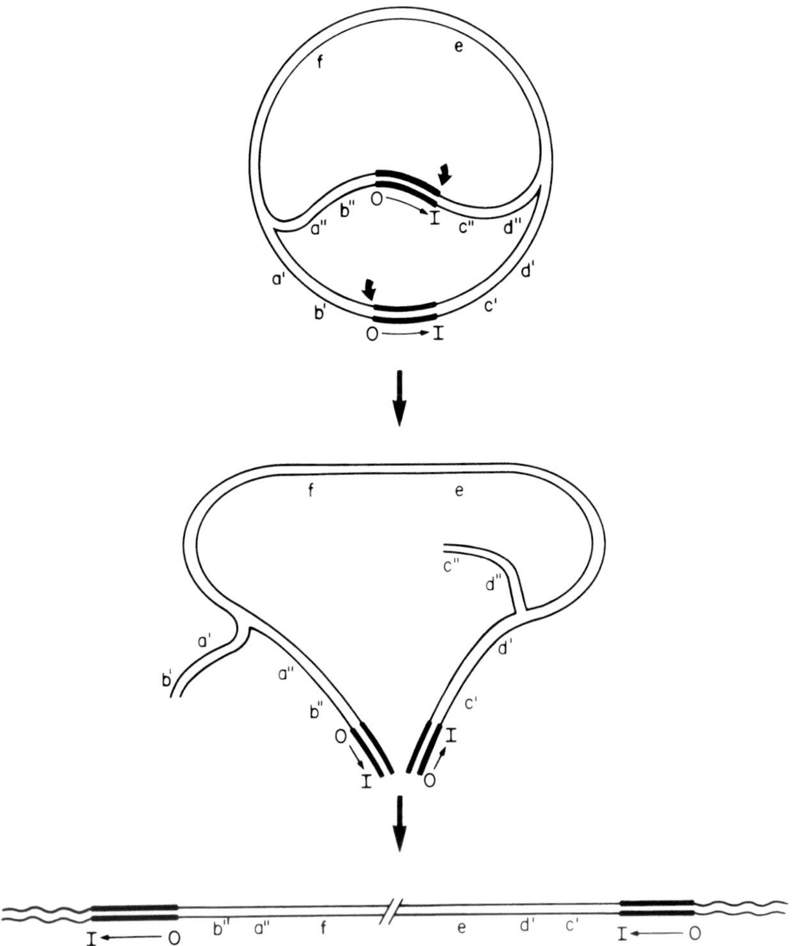

Figure 8. Formation of cointegrates by transposition from a partially replicated monomeric plasmid containing a single transposable element (adapted from reference 99). Thick lines, Transposable element; short curved arrows, positions of double-strand cleavage at element-vector junctions.

ative insertion as envisioned for Tn*5* is commonplace in both procaryotes and eucaryotes.

D. Conservative Transposition and Selfish DNA

Most transposable elements can be thought of as special DNA segments that are selfish in that they tend to increase in copy number each time they move to a new site (139). This idea was first explained in the context of replicative models in which duplication of the element occurs during transposition. Figure 6 illustrates, however, that conservative transposition, as envisioned for Tn*5*, also increases the element copy number, although somewhat less directly: for transposition in cells containing two or more copies of the donor DNA molecule, loss of the DNA from which Tn*5* had transposed would be compensated by an extra replication of a sibling molecule. The resulting cell lineage would thus contain both the transposed copy of the element and a copy at the original donor site (19). Experiments on the control of transposition described below suggest that Tn*5*-related elements transpose frequently just after they have been replicated. Conservative transposition at this time would cause loss of one arm of the replication fork, but not of the other arm, and would thus often give rise to a lineage that also contained an extra copy of the element. Finally, in cells that did not degrade the linear vector DNA, repair of the donor site by a gene conversion-like mechanism would similarly increase the transposable element copy number.

IV. REGULATION OF TRANSPOSITION

The overall frequency, timing, and specificity of Tn*5* and IS*50* transposition are sensitively controlled at several levels: by characteristics of the transposase and inhibitor functions encoded by Tn*5*, by the regulation of the synthesis of these proteins, by the dependence of transposition on several host functions, and by the sequence of the quite divergent O and I ends of IS*50* and the modulation of their substrate activities by methylation and interaction with element- and host-encoded functions.

A. Proteins of IS*50*

The 476-amino-acid transposase (Tnp) (also called protein 1 [p1] or p58 [it is 58 kilodaltons]) (77, 85, 94, 137, 138) and the 421-amino-acid inhibitor (Inh) (also called protein 2 [p2] or p54 [54 kilodaltons]) are encoded in the same large open reading frame in IS*50*R from translational starts that are 55 codons apart (Fig. 3) (77, 85). These proteins were initially found by sodium dodecyl sulfate-polyacrylamide gel electrophoresis of minicell extracts (137, 138) and have also been detected after overproduction from strong promoters in normal cells (80) or as part of *lacZ* fusion proteins after immune precipitation with anti-β-galactosidase antibody (94). The transposase start shown in Fig. 3 was identified rigorously by analyzing *tnp-lacZ* translational fusions and by direct sequencing of the amino terminus of transposase from a transposase–β-galactosidase fusion protein (94).

The large open reading frame in IS*50*L is interrupted by an ochre (TAA) nonsense codon (the only difference between IS*50*L and IS*50*R) 26 codons before its 3′ end (138). The truncated proteins from IS*50*L produced in nonsuppressing strains do not affect the frequency of transposition or the strength of inhibition (77, 85). In ochre-suppressing strains, IS*50*L encodes functional transposase and inhibitor proteins (137, 138). The 2.8 kb of Tn*5* between its IS*50* elements does not encode functions required for transposition (140).

An examination of IS*50*-encoded proteins on sodium dodecyl sulfate gels indicated that in the steady state there is two- to fourfold more inhibitor than transposase and that the amounts of these two proteins are proportional to IS*50*R copy number (77, 85, 94). When Tn*5* is on a multicopy plasmid, about 100 molecules of transposase and 400 molecules of inhibitor are found per cell, or about 5 molecules of transposase and 20 of inhibitor per copy of IS*50*R (83). Cell fractionation experiments using overproducing strains suggested that the transposase is membrane associated and that the inhibitor is cytoplasmic (80).

1. Transposase

A normally expressed *tnp*$^+$ gene does not significantly complement *tnp* mutations in *trans* (77, 78, 85), although some *trans*-complementation of transposase from a *tnp* gene cloned in a high-level expression vector has been reported (169). Even in *cis*, complementation is efficient only when the ends are relatively close to the active *tnp*$^+$ gene (128). The localized action of transposase could reflect additional effects of several factors. (i) If nascent transposase binds DNA, ends adjacent to actively transcribed and translated *tnp* genes would be bound preferentially. (ii) Multiple transposase molecules may be needed for transposition, and the concentrations sufficient for simultaneous binding at distant sites may only rarely be achieved. (iii) Transposase may move only inefficiently on a DNA molecule. (iv) Finally, the protein may be unstable, and, as detailed below, Dam methylation seems to coordinate the synthesis of transposase with transient availability of the I end as a transposition substrate. The net effect, when several copies of an element are present, is that the individual element whose transposase protein is most active will be most prone to transpose and thus to proliferate in the genome.

DNA-binding assays (filter binding and gel retardation) have recently been developed to study transposase interaction with the ends in vitro. Chromatographic fractionation of extracts from strains that overproduce transposase yielded two peaks containing the protein, one consisting of approximately 85% pure transposase and a second consisting of transposase plus two additional transposase-related proteins. Essentially (>95%) pure transposase did not have specific DNA-binding activity, but fractions with the additional proteins showed specific IS*50* O end binding activity (N. de la Cruz and W. S. Reznikoff, personal communication).

2. Inhibitor

A resident IS*50* or Tn*5* element can dramatically decrease the ability of an incoming element to transpose (28). This is due to the *trans*-acting inhibitor protein, which, as noted above, is translated from a second start site in the reading frame for transposase. Studies with *tnp-lacZ* gene fusions showed that the inhibitor does not regulate the synthesis of transposase at either the transcriptional or translational level. It may inhibit transposition by interacting with transposase, with IS*50* ends, or with a host component of the transposition complex (77, 85).

The extent of inhibition was found to range from 2- to more than 100-fold and to depend on the resident IS*50*R copy number. When inhibitor was made under the control of a regulated *lac* promoter, the extent of inhibition was well correlated with the amount of inhibitor proteins synthesized (77, 83, 85,

94). Consistent with this, the total rate of transposition per cell did not depend on element copy number, implying that the chance that a given resident Tn*5* element would transpose was inversely related to the number of copies of IS*50*R present (83, 84). Attempts to study the effects of very low levels of inhibitor or transposase, by making these proteins from IS*50*L in strains carrying ochre suppressors that differ in efficiency (84), were complicated because suppressors insert different amino acids at an ochre codon, and an amino acid substitution at this site has recently been found to affect transposition activity (95). Nevertheless, at normal levels the inhibitor can be an effective regulator of transposition: because inhibitor, unlike transposase, is *trans* acting and its potency increases with concentration (IS*50*R copy number), a resident Tn*5* or IS*50* element can limit its own transposition and thus minimize deleterious insertion mutations. Inhibitor also decreases the chance that an incoming Tn*5*-related element (a potential competitor, considering Tn*5* selfish DNA) will establish itself in the cell lineage by transposition.

B. Transcripts of IS*50*R

The start sites of the two major transcripts from IS*50*R have been identified by inspection of DNA sequences, S1 nuclease mapping of transcripts, and in vitro mutagenesis (Fig. 3). The T1 transcript begins upstream of the translational start site of both transposase and inhibitor, while the T2 transcript begins within the open reading frame for transposase. Measurements using 5′-end-specific probes indicate that in stationary-phase cells the T1 transcript is 1 to 2% as abundant as the T2 transcript and only about half as stable, and that in exponentially growing cells the T1 and T2 transcripts are equally stable (115).

Deletion of the T1 promoter eliminated the synthesis of transposase without affecting the levels of inhibitor. Thus, even though T1 contains the translational start site for the inhibitor, it does not contribute significantly to normal inhibitor levels (94). This could reflect, at least in part, the low abundance of the T1 transcript. Experiments in which a highly active promoter and translational start site were placed at or near the first codon of the *tnp* gene resulted in increased levels of both inhibitor and transposase (80).

Evidence that the T2 transcript specifies transposase was provided by base substitution mutations in the T2 promoter (#1106 and #1110 in Fig. 3). These mutations decreased T2 synthesis and also decreased the level of the inhibitor, measured both as protein and by an inhibition assay. A third transcript, T3, is initiated 12 bp downstream from the start of T2, but is not considered important for inhibitor synthesis in *E. coli* because it is only 1% as abundant as the T2 transcript and because the T2 promoter mutations which sharply reduced inhibitor synthesis did not affect the T3 promoter (94, 170). It is curious that the T2 promoter mutations also decreased the levels of transposase protein, but resulted in a frequency of transposition somewhat higher than that of wild type. The decrease in transposase was postulated to reflect a change in stability or translatability of T1 mRNA, and the enhanced transposition was explained in terms of the effect of a shift in the ratio of transposase to inhibitor (170).

The synthesis of inhibitor was eliminated by changing its ATG translation initiation codon to ATT, which causes a methionine → isoleucine replacement in the transposase protein (mutation #1037, Fig. 3). The element containing this mutation transposed, although only half as efficiently as wild type (despite the absence of inhibitor). This result eliminates the formal possibility that inhibitor protein is essential for transposition; the decreased transposition was attributed to a decrease in the specific activity of the substituted transposase (170). The unexpectedly low transposition activity of this mutant, and the decreased transposase synthesis associated with the mutations affecting the T2 promoter described above, underscore the need for more detailed mutational analyses of these nested genes.

In summary, physical mapping and in vitro mutagenesis have located the likely start points of transcription and translation of the *tnp* and *inh* genes. Difficulties in the quantitative interpretation of the mutational data reflect the nesting of the *inh* gene within *tnp*, so that changes affecting the inhibitor protein often affect the structure of transposase or perhaps the stability or translatability of its mRNA.

C. Dam Methylation

1. Control of transposase synthesis

Tn*5* transposition is some 5- to 20-fold higher in *dam* strains, deficient in adenine methylation at GATC sequences, than in *dam*$^+$ strains (135, 107, 168; K. W. Dodson and D. E. Berg, *Gene*, in press). Two Dam sites over lap the −10 region of the transposase (T1) promoter (Fig. 3), and Dam methylation is known to affect numerous protein-DNA interactions (110, 156). Both S1 nuclease protection assays and protein measurements using a *tnp-lacZ* fusion indicate that Dam methylation decreases transposition by decreasing T1 transcript synthesis (115, 168). In addition, base-substitution mutations that inactivated the *dam* sites without changing conserved nucleotides in the T1 promoter (mutations #1107 and #1108, Fig. 3) resulted in 5-fold more T1 transcript and 10- to 15-fold more transposition in *dam*$^+$ cells; these mutations did not affect the normally high levels of the T1 transcript or of transposition in *dam* cells. This eliminates the possibility that the level of expression of a host gene controlled by Dam methylation actually limits the frequency of Tn*5* transposition (168).

The increased activity of the T1 promoter in *dam* cells was associated with a twofold decrease in levels of the T2 (inhibitor) transcript (168), probably because transcription from one promoter can reduce the activity of a downstream promoter (termed "occlusion") (1). These results further indicate that the level of transposase or the ratio of transposase to inhibitor normally determines the frequency of Tn*5* transposition.

Hemimethylated DNA is produced from fully methylated DNA by normal vegetative replication, bacterial conjugation, or the repair of DNA damage in *dam*$^+$ cells (110, 156). Earlier studies with λ::Tn*5* phage had shown that both Tn*5* transposition and the level of *tnp* gene expression are higher just after infection than at later times (136). These kinetics probably reflect effects of undermethylation of phage DNAs that have been replicated rapidly prior to packaging, compounded by some phage DNA replication in the infected cells (115). The decreased transposition after Tn*5* has become established must additionally reflect the accumulation of the inhibitor protein.

2. Control of I end activity

The I end, unlike the O end, contains Dam methylation (GATC) sites (Fig. 2), and its activity as a substrate appears to be controlled directly by methylation. A specific effect of Dam methylation on I end activity, separate from the effect of changes in transposase levels, was found using a *tnp* gene transcribed from a strong *dam*-insensitive promoter. This engineered *tnp* gene was used to complement an element with one O and one I end in another DNA molecule. Although the transposition frequency was quite low because *trans* complementation is inefficient, the frequency was at least 10-fold higher in *dam* cells than in *dam*$^+$ cells. In a parallel test involving a pair of O ends and the *dam*-insensitive *tnp* promoter, the host *dam* mutation did not affect transposition frequencies (168). These results suggested that methylation of I ends inhibits their use as transposition substrates, although alternative models, e.g., that invoke a Dam-regulated I end-specific host factor, are not excluded.

Assays using *tnp* genes transcribed from normal (*dam*-sensitive) promoters, together with sets of transposons that differ only in their ends, also support the interpretation that I end substrate activity is sensitive to *dam* methylation. Transposition involving pairs of I ends was increased 300- to 1,000-fold by use of *dam* rather than *dam*$^+$ cells, whereas transposition involving pairs of O ends was increased just 4-fold. As a result, in *dam* cells the activity of a pair of I ends is severalfold higher than the activity of a pair of O ends at the same sites (107; Dodson and Berg, in press). In *dam*$^+$ cells, a synthetic transposon in which the relative orientation of the *tnp* gene and O and I ends matches that in IS*50* was fourfold more active than a similar element in which the orientation of the *tnp* gene was reversed. In contrast, in *dam* cells, each element transposed at about the same frequency (Dodson and Berg, in press). This suggests that, as in the case of IS*10* (which is similar to IS*50* in its arrangement of Dam sites; 135), one hemimethylated IS*50* *tnp* gene promoter and IS*50* I end are more active than the fully methylated or the other hemimethylated forms of these sites, and that the most active of the hemimethylated forms are in the same daughter DNA molecule after DNA synthesis. In this model, when the orientation of the *tnp* gene is reversed, transposase, which acts best close to its site of synthesis, would be made from the DNA containing the less active I end sequence, thus accounting for the decreased transposition of rearranged elements in *dam*$^+$ cells (see reference 135 and Kleckner, this volume).

There is likely to be a selective benefit to the link between methylation and transposition. A Tn*5* or IS*50* element will be prone to transpose when brought into the cell by temperate phage infection, which is generally accompanied by limited phage DNA replication, or by bacterial conjugation, which entails the transfer of a single DNA strand (63) and copying in the recipient. This tendency to transpose will increase the chance that the element becomes established in the new bacterial lineage even if the vector DNA is restricted or for some other reason is not maintained. In addition, for an established Tn*5* element, transposition after passage of the replication fork means that a failure to reseal the empty donor site will cause loss of just one arm of the replication fork, not loss of the entire DNA molecule. Finally, the stimulation of transposition following repair of DNA damage may increase the chance that the element will move to the genome of a temperate phage whose lytic growth was induced by the same DNA damage, and thus help the element invade other members of the bacterial population.

D. Genetic Analysis of IS*50* Ends and Effects of Dam Methylation

Analyses of a series of deletion derivatives of IS*50* and Tn*5* indicated that the sites at both the O and I ends needed for transposition are about 19 bp long (82, 141, 143). These segments are noteworthy in being quite divergent in sequence and in containing potential binding sites for several host proteins. Base-substitution mutations have been made throughout the two IS*50* ends to better understand their important features. These studies were carried out in two different laboratories (Berg and Reznikoff), and using different transposition assays, but with generally concordant results.

In the 19-bp O end segment, one or more base-substitution mutations at all but possibly one of the positions reduced transposition activity significantly (Fig. 9). The quantitative effect of the mutations depended on their positions, the nature of the base change, and the conditions of assay, and several mutations reduced transposition to different extents in the two studies (marked with asterisks in Fig. 9). These differences probably reflect the characteristics of the assays used, perhaps the pairing of the mutant O end with an I end in one set of experiments (circled) and with an O end (boxed) in the other.

Even though there are no Dam sites in the O ends, and none were created by the mutations, Dam methylation was found to strongly affect the phenotype of nearly half of the mutant O ends when they were paired with a nonmutant I end (arrows in Fig. 9B) (T. Tomcsanyi and D. E. Berg, unpublished data). At the extreme, the transposition activity of O end mutant 14G (relative to wild type) was 100-fold higher in *dam* than in *dam*$^+$ cells. These results lead to several conclusions. First, the influence of the state of methylation of the I end on how severely specific mutations at the O end affect transposition activity suggests that the set of proteins in the transposition complex is variable; the presence or absence of I end methylation would help determine which proteins participate in the transposition complex. Second, if differences in the results are due to the pairing of the mutant O ends with an I end (127; Tomcsanyi and Berg, unpublished data) versus that with an O end (107), the sequence at the nonmutant end (I versus O) would also seem to affect the composition of the transposition complex. Finally, if most transposition mediated by an I end in Dam$^+$ cells involves hemimethylated (newly replicated) DNA, the results suggest that nonmethylated and hemimethylated DNAs are acted on differently by transposition proteins.

In the 19-bp I end segment, one or more mutations at each position also reduced transposition activity, measured in *dam* cells (to avoid complications due to loss of Dam methylation sites) (Fig. 10). Some mutations caused quite different effects in *dam*$^+$ cells (indicated by ^{M} in Fig. 10). Most of these Dam effects may be due solely to decreased methylation of the I end. However, in one case (mutation 2T, outside of the Dam site) the Dam effect probably reflects the specific sequence change as well as methylation elsewhere in the I end. Finally, although the analysis of nested deletions generated in a linker scanning analysis indicated that 19 bp of wild-type I end sequence is sufficient for normal transposition activity (141), some data obtained using base-substitution mutations suggested that nucleotides beyond position 19 might also contribute to I end transposition activity (see reference 107 and Fig. 10, legend).

In conclusion, the mutational analyses indicate that most or all positions in the O and I ends of IS*50* are important in transposition. Some base pairs will be specifically contacted by transposition proteins, whereas others will only affect local DNA conformation and so influence binding indirectly. Given the distribution of matched and nonmatched sequences in the IS*50* ends, each end may consist of two domains, with the matched domain bound by a common factor such as transposase and the nonmatched domain bound by other factors, at least one specific for the O end and another specific for the I end.

E. Host Functions Needed for Transposition

Searches for host factors involved in the transposition of Tn*5* and its derivatives have implicated DNA gyrase (79), DNA polymerase I (144, 159), DnaA protein (82, 127, 169), integration host factor (IHF) (107; J. C. Makris and W. S. Reznikoff, personal communication), and factors that regulate cell division (145). Host proteins might variously (i) act in a complex with the element-encoded transposase on O or I ends, (ii) bind specific sites in target DNAs, (iii) control donor or target DNA conformation, or (iv) regulate the synthesis of other factors that participate more directly in transposition.

1. DNA gyrase

Negative supercoils, introduced into covalently closed DNAs by the action of DNA gyrase, facilitate transient melting of duplex DNA and can be important in certain protein-DNA interactions (162). Negative supercoils are relaxed by the action of DNA intervening sequences. This flexibility may also contribute to the ability of these elements to transpose and thus to proliferate in diverse bacteria and cellular environments.

Acknowledgments. I am indebted to the late Larry Sandler, whose inspired teaching about McClintock's controlling elements helped set the stage for our discovery of Tn*5*. I am also grateful to the members of my research group for many stimulating discussions, to N. de la Cruz, J. Makris, W. S. Reznikoff, and J. Schell for communication of data in advance of publication, and to C. M. Berg, M. M. Howe, and W. S. Reznikoff for critical readings of the manuscript.

This work was supported by the United States Public Health Service (GM-37138), the National Science Foundation (DMB-8608193), and the Lucille P. Markey Charitable Trust.

LITERATURE CITED

1. **Adhya, S., and M. Gottesman.** 1982. Promoter occlusion: transcription through a promoter may inhibit its activity. *Cell* **29:**939–944.
2. **Ahmed, A.** 1986. Evidence for replicative transposition of Tn*5* and Tn*9*. *J. Mol. Biol.* **191:**75–84.
3. **Albertini, A. M., M. Hofer, M. P. Calos, and J. H. Miller.** 1982. On the formation of spontaneous deletions: the importance of short sequence homologies in the generation of large deletions. *Cell* **29:**319–328.
4. **Andre, D., D. Colau, J. Schell, M. van Montagu, and J.-P. Hernalsteens.** 1986. Gene tagging in plants by a T-DNA insertion mutagen that generates APH(3′)II-plant gene fusions. *Mol. Gen. Genet.* **204:**512–518.
5. **Andrup, L., T. Atlung, N. Ogasawara, H. Yoshikawa, and F. G. Hansen.** 1988. Interaction of the *Bacillus subtilis* DnaA-like protein with the *Escherichia coli* DnaA protein. *J. Bacteriol.* **170:**1333–1338.
6. **Arnau, J., F. J. Murillo, and S. Torres-Martinez.** 1988. Expression of Tn*5*-derived kanamycin resistance in the fungus *Phycomyces blakesleeanus. Mol. Gen. Genet.* **212:** 375–377.
7. **Auerswald, E. A., G. Ludwig, and H. Schaller.** 1981. Structural analysis of Tn*5*. *Cold Spring Harbor Symp. Quant. Biol.* **45:**107–113.
8. **Baker, T. A., K. Sekimuzu, B. E. Funnell, and A. Kornberg.** 1986. Extensive unwinding of the plasmid template during staged enzymatic initiation of DNA replication from the origin of the *Escherichia coli* chromosome. *Cell* **45:**801–806.
9. **Barbeyron, T., K. Kean, and P. Forterre.** 1984. DNA adenine methylation of GATC sequences appeared recently in the *Escherichia coli* lineage. *J. Bacteriol.* **160:**586–590.
10. **Beck, E., G. Ludwig, E. Auerswald, B. Reiss, and H. Schaller.** 1982. Nucleotide sequence and exact localization of the neomycin phosphotransferase gene from transposon Tn*5*. *Gene* **19:**327–336.
11. **Bellofatto, V., L. Shapiro, and D. A. Hodgson.** 1984. Generation of a Tn*5* promoter probe and its use in the study of gene expression in *Caulobacter crescentus. Proc. Natl. Acad. Sci. USA* **81:**1035–1039.
12. **Bender, J., and N. Kleckner.** 1986. Genetic evidence that Tn*10* transposes by a nonreplicative mechanism. *Cell* **45:** 801–815.
13. **Berg, C. M., and D. E. Berg.** 1987. Uses of transposable elements and maps of known insertions, p. 1071–1109. *In* F. C. Neidhardt, J. L. Ingraham, K. B. Low, B. Magasanik, M. Schaechter, and H. E. Umbarger (ed.), *Escherichia coli and Salmonella typhimurium: Cellular and Molecular Biology.* American Society for Microbiology, Washington, D.C.
14. **Berg, D. E.** 1977. Insertion and excision of the transposable kanamycin resistance determinant Tn*5*, p. 205–212. *In* A. I. Bukhari, J. A. Shapiro, and S. L. Adhya (ed.), *Insertion Elements, Plasmids, and Episomes.* Cold Spring Harbor Laboratory, Cold Spring Harbor, N.Y.
15. **Berg, D. E.** 1980. Control of gene expression by a mobile recombinational switch. *Proc. Natl. Acad. Sci. USA* **77:** 4880–4884.
16. **Berg, D. E.** 1983. Structural requirement for IS*50*-mediated gene transposition. *Proc. Natl. Acad. Sci. USA* **80:**792–796.
17. **Berg, D. E.** 1989. Transposable elements in prokaryotes. *In* S. B. Levy and R. Miller (ed.), *Gene Transfer in the Environment.* McGraw-Hill Publishing Co., New York, in press.
18. **Berg, D. E., and C. M. Berg.** 1983. The prokaryotic transposable element Tn*5*. *Bio/Technology* **1:**417–435.
19. **Berg, D. E., C. M. Berg, and C. Sasakawa.** 1984. The bacterial transposon Tn*5*: evolutionary inferences. *Mol. Biol. Evol.* **1:**411–422.
20. **Berg, D. E., J. Davies, B. Allet, and J.-D. Rochaix.** 1975. Transposition of R factor genes to bacteriophage λ. *Proc. Natl. Acad. Sci. USA* **72:**3628–3632.
21. **Berg, D. E., and M. H. Drummond.** 1978. Absence of sequences homologous to Tn*5* (Kan) in the chromosome of *Escherichia coli* K-12. *J. Bacteriol.* **136:**419–422.
22. **Berg, D. E., C. Egner, B. J. Hirschel, J. Howard, L. Johnsrud, R. A. Jorgensen, and T. D. Tlsty.** 1981. Insertion, excision and inversion of Tn*5*. *Cold Spring Harbor Symp. Quant. Biol.* **45:**115–123.
23. **Berg, D. E., C. Egner, and J. B. Lowe.** 1983. Mechanism of F factor-enhanced excision of transposon Tn*5*. *Gene* **22:**1–7.
24. **Berg, D. E., L. Johnsrud, L. McDivitt, R. Ramabhadran, and B. J. Hirschel.** 1982. Inverted repeats of Tn*5* are transposable elements. *Proc. Natl. Acad. Sci. USA* **79:**2632–2635.
25. **Berg, D. E., J. K. Lodge, C. Sasakawa, D. K. Nag, S. H. Phadnis, K. Weston-Hafer, and G. F. Carle.** 1984. Transposon Tn*5*: specific sequence recognition and conservative transposition. *Cold Spring Harbor Symp. Quant. Biol.* **49:** 215–226.
26. **Berg, D. E., M. A. Schmandt, and J. B. Lowe.** 1983. Specificity of transposon Tn*5* insertion. *Genetics* **105:**813–828.
27. **Berg, D. E., A. Weiss, and L. Crossland.** 1980. Polarity of Tn*5* insertion mutations in *Escherichia coli. J. Bacteriol.* **142:**439–446.
28. **Biek, D., and J. R. Roth.** 1980. Regulation of Tn*5* transposition in *Salmonella typhimurium. Proc. Natl. Acad. Sci. USA* **77:**6047–6051.
29. **Biel, S. W., G. Adelt, and D. E. Berg.** 1984. Transcriptional control of IS*1* transposition. *J. Mol. Biol.* **174:**251–264.
30. **Biel, S. W., and D. L. Hartl.** 1983. Evolution of transposons: natural selection for Tn*5* in *Escherichia coli* K-12. *Genetics* **103:**581–592.
31. **Bossi, L., and M. S. Ciampi.** 1981. DNA sequences at the sites of three insertions of the tranposable element Tn*5* in the histidine operon of *Salmonella. Mol. Gen. Genet.* **183:** 106–108.
32. **Bramhill, D., and A. Kornberg.** 1988. Duplex opening by

D. Genetic Analysis of IS*50* Ends and Effects of Dam Methylation

Analyses of a series of deletion derivatives of IS*50* and Tn*5* indicated that the sites at both the O and I ends needed for transposition are about 19 bp long (82, 141, 143). These segments are noteworthy in being quite divergent in sequence and in containing potential binding sites for several host proteins. Base-substitution mutations have been made throughout the two IS*50* ends to better understand their important features. These studies were carried out in two different laboratories (Berg and Reznikoff), and using different transposition assays, but with generally concordant results.

In the 19-bp O end segment, one or more base-substitution mutations at all but possibly one of the positions reduced transposition activity significantly (Fig. 9). The quantitative effect of the mutations depended on their positions, the nature of the base change, and the conditions of assay, and several mutations reduced transposition to different extents in the two studies (marked with asterisks in Fig. 9). These differences probably reflect the characteristics of the assays used, perhaps the pairing of the mutant O end with an I end in one set of experiments (circled) and with an O end (boxed) in the other.

Even though there are no Dam sites in the O ends, and none were created by the mutations, Dam methylation was found to strongly affect the phenotype of nearly half of the mutant O ends when they were paired with a nonmutant I end (arrows in Fig. 9B) (T. Tomcsanyi and D. E. Berg, unpublished data). At the extreme, the transposition activity of O end mutant 14G (relative to wild type) was 100-fold higher in *dam* than in *dam*$^+$ cells. These results lead to several conclusions. First, the influence of the state of methylation of the I end on how severely specific mutations at the O end affect transposition activity suggests that the set of proteins in the transposition complex is variable; the presence or absence of I end methylation would help determine which proteins participate in the transposition complex. Second, if differences in the results are due to the pairing of the mutant O ends with an I end (127; Tomcsanyi and Berg, unpublished data) versus that with an O end (107), the sequence at the nonmutant end (I versus O) would also seem to affect the composition of the transposition complex. Finally, if most transposition mediated by an I end in Dam$^+$ cells involves hemimethylated (newly replicated) DNA, the results suggest that nonmethylated and hemimethylated DNAs are acted on differently by transposition proteins.

In the 19-bp I end segment, one or more mutations at each position also reduced transposition activity, measured in *dam* cells (to avoid complications due to loss of Dam methylation sites) (Fig. 10). Some mutations caused quite different effects in *dam*$^+$ cells (indicated by ^{M} in Fig. 10). Most of these Dam effects may be due solely to decreased methylation of the I end. However, in one case (mutation 2T, outside of the Dam site) the Dam effect probably reflects the specific sequence change as well as methylation elsewhere in the I end. Finally, although the analysis of nested deletions generated in a linker scanning analysis indicated that 19 bp of wild-type I end sequence is sufficient for normal transposition activity (141), some data obtained using base-substitution mutations suggested that nucleotides beyond position 19 might also contribute to I end transposition activity (see reference 107 and Fig. 10, legend).

In conclusion, the mutational analyses indicate that most or all positions in the O and I ends of IS*50* are important in transposition. Some base pairs will be specifically contacted by transposition proteins, whereas others will only affect local DNA conformation and so influence binding indirectly. Given the distribution of matched and nonmatched sequences in the IS*50* ends, each end may consist of two domains, with the matched domain bound by a common factor such as transposase and the nonmatched domain bound by other factors, at least one specific for the O end and another specific for the I end.

E. Host Functions Needed for Transposition

Searches for host factors involved in the transposition of Tn*5* and its derivatives have implicated DNA gyrase (79), DNA polymerase I (144, 159), DnaA protein (82, 127, 169), integration host factor (IHF) (107; J. C. Makris and W. S. Reznikoff, personal communication), and factors that regulate cell division (145). Host proteins might variously (i) act in a complex with the element-encoded transposase on O or I ends, (ii) bind specific sites in target DNAs, (iii) control donor or target DNA conformation, or (iv) regulate the synthesis of other factors that participate more directly in transposition.

1. DNA gyrase

Negative supercoils, introduced into covalently closed DNAs by the action of DNA gyrase, facilitate transient melting of duplex DNA and can be important in certain protein-DNA interactions (162). Negative supercoils are relaxed by the action of DNA

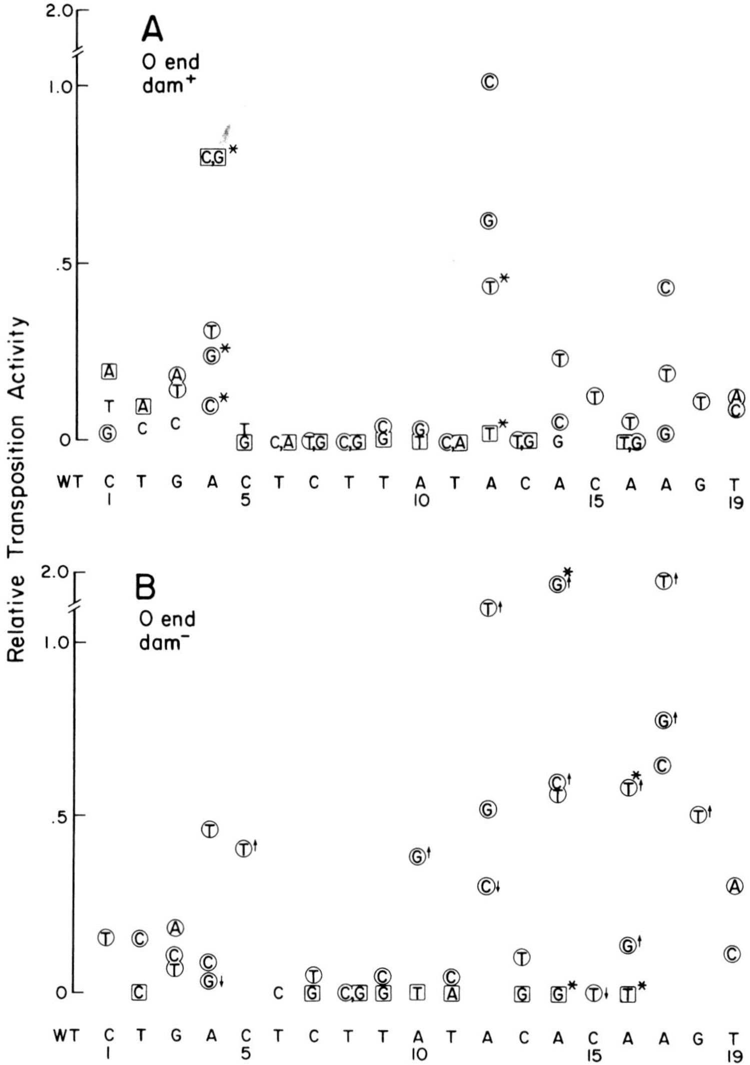

Figure 9. Mutational analysis of the IS*50* O end (from references 107 and 127 and Tomcsanyi and Berg, unpublished data). The wild-type O end sequence is shown at the bottom of each panel. (A) Assays in *dam*$^+$ cells. (B) Assays in isogenic *dam* cells. For circled mutations, mutant O ends were paired with a nonmutant I end, and the frequencies of transposition of the resultant elements to a phage λ target were compared with that obtained with a wild-type O end (about 2×10^{-7} in *dam*$^+$ cells; 2×10^{-4} in *dam* cells [127; Tomcsanyi and Berg, unpublished data]). For boxed mutations, mutant O ends were paired with a nonmutant O end, and the frequencies of transposition of the resultant elements to an F factor target were measured relative to that obtained with a wild-type element at the same site (107). Mutations that are neither circled nor boxed indicate cases in which the same relative transposition frequency was observed in both laboratories. (A) *, Base substitution mutations which behaved significantly differently in the two laboratories. (B) ↑ and ↓, Mutations that were at least severalfold more active or less active, respectively, in *dam* than in *dam*$^+$ cells. None of the mutations creates an adenine methylation (GATC) site in the O end.

topoisomerase I, an enzyme whose action is complementary to that of DNA gyrase. The gyrase inhibitor coumermycin was used to test whether DNA gyrase is needed in Tn*5* transposition (79). Cells carrying a chromosomal Tn*5* element were infected with λ, and the frequencies of new insertions into the intracellular phage DNAs were monitored as a function of when coumermycin was added to the culture. Coumermycin added just before infection blocked both λ DNA supercoiling and transposition to this λ DNA, whereas the drug added just after infection had very little effect on either supercoiling or transposition. In addition, incubation of bacteria with a temperature-sensitive allele of the *gyrA* gene at a restrictive temperature reduced both target DNA supercoiling and transposition; the defects in supercoiling and transposition were partially reversed when the *gyrA* cells were also made defective in the synthesis of topoisomerase I. Additional experiments showed that coumermycin does not affect the synthesis of trans-

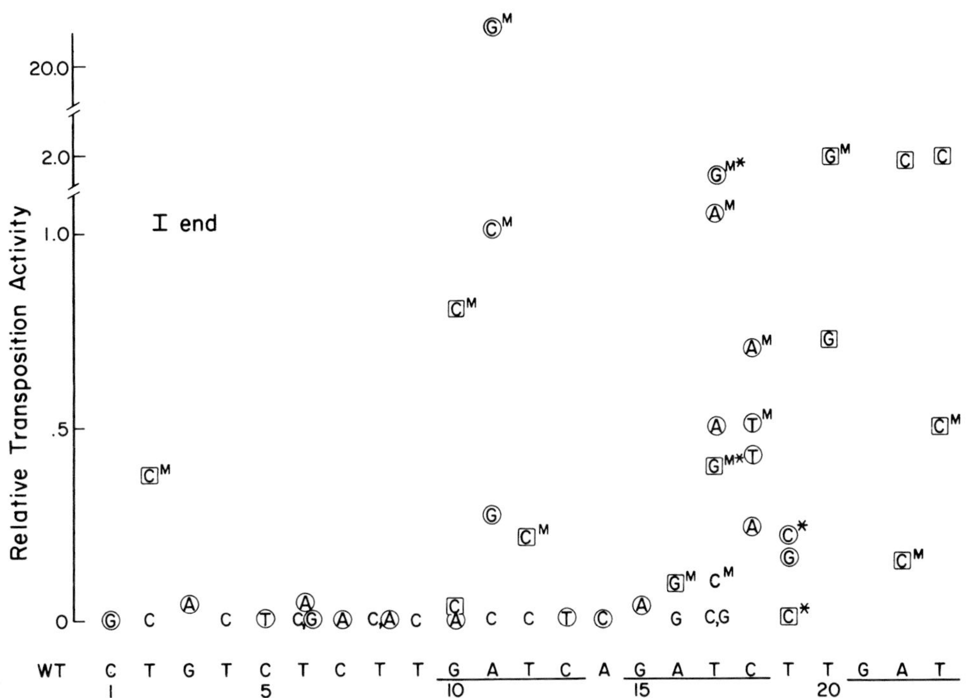

Figure 10. Mutational analysis of the I end. The wild-type I end sequence is shown at the bottom, with the three Dam methylation sites underlined. Mutations in the I ends were generated with synthetic oligonucleotides containing either random inosines (boxed; 107) or random base substitutions (circled; Dodson and Berg, unpublished data). Mutations were paired with a nonmutant O end in each case and were studied in *dam* cells. Some mutations were also studied in dam^+ cells. ^{M} indicates the results of assays in dam^+ cells (where I end sequences would be methylated) in the cases where methylation resulted in a significant change in transposition frequency. The possibility that sequences beyond I end position 19 might contribute to transposition is based on three observations: (i) the presence of a third *dam* site and a possible IHF-binding site (7 of 9 bp match to the consensus) if the I end sequence is extended to position 24 (Fig. 2); (ii) the different phenotypes of the 19C mutation, possibly due to effects of different nearby DNA sequences in the two studies; and (iii) the sevenfold decreased transposition activity of the 22C mutation in the dam^+ strain (the segment mutagenized by Dodson and Berg [unpublished data] extended to I end position 19, whereas the segment mutagenized by Makris et al. [107] extended to I end position 24). Several mutations seem inconsistent with the idea that IHF binding to two sites in the I end is important for transposition, most notably the mutations 20G, 22C, and 23C, which decreased the fit to the consensus, yet had little if any effect on transposition in the *dam* host.

posase (79). Thus, these results indicate that DNA gyrase is needed to supercoil target DNA in preparation for transposition and that supercoiling could be important in facilitating local melting and sequence recognition by the transposition proteins. Since cleavage can relax supercoils, this topological requirement may help ensure that an activated transposition complex does not cleave a target DNA at multiple sites.

The initial studies of *topA* mutations suggested that topoisomerase I might also be important for Tn*5* transposition (157). Subsequent characterizations showed, however, that *topA* mutations severely impair bacterial growth and that compensating mutations that decrease DNA gyrase activity are strongly selected during bacterial growth (44). It is thus not yet clear whether the transposition deficiency was caused by the *top* allele itself or by a compensating mutation.

2. DNA polymerase I

Two groups have shown that DNA polymerase I is important for Tn*5* transposition. In one study (144) the frequency of Tn*5* transposition from an infecting λ::Tn*5* phage was 10-fold lower in a *polA1* mutant strain than in the $polA^+$ control; the transposition of an established Tn*5* element from a chromosomal site was not, however, affected by the *polA1* allele. *polA1* is an amber mutation which eliminates the DNA polymerase but not the 5′→3′ exonuclease activity and is somewhat leaky in strains with low levels of suppression or ribosomal ambiguity. It was therefore postulated that DNA polymerase I is needed for all Tn*5* transposition, that the defect in this *polA1* strain is leaky, and that the residual level of DNA polymerase is only limiting for transposition during λ::Tn*5* infection (probably because transposase is then made in excess). In addition, the *polAex2* mutation (which eliminates the 5′→3′ exonuclease but not the DNA polymerase activity, a defect complementary to that of *polA1*) also strongly reduced the frequency of transposition from an infecting λ::Tn*5* phage (144).

The importance of DNA polymerase I in Tn*5* transposition was also established in a search for bacterial mutations that impaired the transposition of several different elements (159). This analysis exploited findings that most spontaneous mutational activation of the cryptic *bgl* operon is caused by transpositions of IS*1* and IS*5* (132). One mutation, selected for reduced formation of Bgl^+ papillae, impaired Tn*10* transposition and also impaired Tn*5* transposition from an infecting λ::Tn*5* phage as well as from a chromosomal site (in cells containing an established element). This mutation was in the *polA* gene and eliminated both DNA polymerase and 5′→3′ exonuclease activities in in vitro assays (159). A revertant with one-third of wild-type levels of DNA polymerase and 5′→3′ exonuclease activities remained defective for Tn*5* transposition. This revertant had the interesting property that the mutant phenotype was not completely recessive to *polA*$^+$: the introduction of a single *polA*$^+$ allele did not restore transposition proficiency.

DNA polymerase I was originally postulated to help replicate Tn*5*, IS*1*, IS*5*, and Tn*10* during their transposition (144, 159). However, the available evidence now indicates that transpositions of Tn*5* and of Tn*10* are conservative (12, 16, 81, 99). In addition, the existence of *polA* mutant strains with significant DNA polymerase activity (*polAex2* and suppressed *polA34*) that did not support efficient Tn*5* transposition was not consistent with this hypothesis. Although some DNA polymerase I activity may be needed to fill the gaps at element-target junctions or to carry out other DNA repair activities during transposition, it also seems that the DNA polymerase I protein might actually play a structural role as one component of a complex of transposition proteins.

3. DnaA protein and O end-mediated transposition

DnaA protein binds cooperatively to the multiple copies of a 9-bp sequence called the "DnaA box" (Fig. 2) at replication origins in the chromosomes of diverse bacterial species (5, 53, 172; for review, see reference 117). This binding causes local melting of nearby DNA sequences (32), permits the assembly of other proteins of the replication complex, and leads eventually to proper initiation of chromosome replication (8). DnaA protein also binds single DnaA boxes near several genes involved in DNA metabolism, where it helps regulate their transcription. A suggestion that DnaA protein participates in Tn*5* transposition (82) was based on the presence of a DnaA box in the O end of IS*50* (Fig. 2).

Tests of the need for DnaA protein during transposition used strains in which the bacterial DnaA requirement had been suppressed (by replication from an integrated episome whose replication is DnaA independent, or by the absence of RNase III). *dnaA*-null alleles in such backgrounds were found to reduce Tn*5* transposition from an infecting λ::Tn*5* phage to less than 1% of that of *dnaA*$^+$ controls (127, 169). In qualitative agreement, the transposition of resident (established) Tn*5* elements to λ was reduced 7- to 100-fold in *dnaA* strains (169). Thus DnaA protein stimulates but is not absolutely required for Tn*5* transposition.

In the case of an IS*50*-based element with one O and one I end, we found that transposition from an infecting λ phage was reduced in a *dnaA* strain about 50-fold relative to that in an isogenic *dnaA*$^+$ control (Phadnis and Berg, unpublished data). In contrast, the transposition of a resident element with one O and one I end did not seem to be affected by a *dnaA* mutation (less than twofold) (169). However, this second test involved inefficient *trans* complementation for transposase, so that low control values limited quantitative interpretation of the results.

The idea that DnaA protein is important in transposition because it binds to the IS*50* O end was supported by an initial mutational analysis (Fig. 9; 127, 169). For mutant O ends paired with nonmutant I ends and assayed in *dam*$^+$ cells, substitutions at the seven positions corresponding to invariant residues in DnaA boxes (O end positions 8 to 11, 13, 14, and 16) severely reduced transposition, whereas changes at the two variable positions (O end positions 12 and 15) caused no, or only slight, reductions. A more complex picture emerged, however, from assays using *dam* cells: O end mutations 10G, 14C, 14T, and 14G, in conserved positions in the DnaA box, did not drastically reduce transposition (Fig. 9B; Tomcsanyi and Berg, unpublished data), suggesting that DnaA protein is only important in transposition mediated by one O and one I end if the I end is methylated and thus relatively inactive.

How might DnaA protein contribute to transposition involving IS*50* O ends? (i) The possibility that DnaA protein affects transcription of the transposase gene (the DnaA box in the O end is about 13 bp from the transposase promoter) or of nearby regions was ruled out by S1 nuclease mapping of IS*50* RNAs and by a test using transposase made from a promoter that is not close to a DnaA box (169). (ii) The possibility that DnaA protein promotes replication during transposition (82, 169) is at variance with the interpretation that transposition involving O as well as I ends occurs by a cut-and-paste mechanism, without replication (14, 16, 81, 99). (iii) DnaA protein might control the synthesis of a host factor needed directly for O end-mediated transposition. (iv) DnaA protein might cause melting of adjacent sequences in the O end or at the junction with vector DNA (32), rendering them more accessible to recognition or cleavage by other transposition proteins. (v) Finally, DnaA might act structurally, helping to form the transposition complex (analogous to its role in assembling other proteins at the chromosome origin before replication initiates), to stabilize the transposition complex, or to align the element ends (127, 169). Regardless of how DnaA protein operates, if its activity varies during the cell cycle, its involvement in O end-mediated transposition, like the complementary methylation effects on transposase synthesis and I end substrate activity, could help link transposition to periods of rapid DNA replication. As noted above, a tendency to transpose primarily during periods of replication is advantageous when the DNA molecule that serves as the transposition donor is consumed during the transposition process.

4. IHF and I end transposition

Integration host factor (IHF) consists of a heterodimer of two small proteins with homology to eucaryotic histone H1 and the bacterial histonelike protein HU. It binds DNAs containing the 9-bp consensus sequence shown in Fig. 2 and participates in numerous activities, including lambdoid phage integration (55, 98, 161) and morphogenesis (166); controlled expression of several genes (93, 106, 118); and, from in vitro studies, possibly IS*1*, IS*10*, and gamma-delta transposition (121, 130, 165).

Direct evidence that IHF is needed for I end transposition came from the use of a *lac* fusion transposon, Tn*5*ORF*lac*, in an assay in which the formation of Lac$^+$ papillae was an indicator of transposition (95). The level of papillation due to transposition of Tn*5*ORF*lac* with one O and one I end was high in a *dam* strain that contained IHF, but sharply reduced in a *dam* strain that lacked IHF. In contrast, in a *dam*$^+$ strain the lack of IHF caused little if any change in papillation (Makris and Reznikoff, personal communication).

IHF might participate directly in transposition by acting on nonmethylated I ends, or it might function indirectly. There are two potential IHF-binding sites in and overlapping the I end, but their match to the consensus (6 of 9 bp and 7 of 9 bp [107]) (Fig. 2) is weak, and sequences differing by more than 1 bp from the consensus are not efficiently bound by IHF under standard in vitro conditions (55, 161, 166). In addition, two I end mutations that increase the fit to the IHF site cause decreased transposition (3A and 13T), and five I end mutations that decrease the fit to the IHF site (11G, 19C, 20G, 22C, and 23C) cause slight or no reductions in transposition activity (Fig. 2 and 10). Thus, the role of IHF might be indirect, through an effect on expression of a host gene whose product in turn is needed for transposition.

5. *lon*, *sulA*, and cell growth

Tn*5* transposition is reduced in *lon* mutants, but normal in *lon sulA* double mutants or in *lon* mutants grown with DL-pantoyl lactone (145). The *sulA* gene encodes a cell division inhibitor that is highly expressed after DNA damage, and the *lon* gene encodes a protease that cleaves SulA protein. DL-Pantoyl lactone induces septation directly, thereby bypassing the *sulA* inhibition of division in *lon* strains (46, 60). Since the *lon* gene encodes a protease, it might affect Tn*5* transposition by activating a protein in the transposition complex or inactivating an inhibitor.

6. *mutH*

Two findings suggest that MutH protein binds GATC sequences that are not fully methylated: (i) it can mediate repression of transcription of the Mu *mom* gene from a regulatory site containing GATC sequences if the substrate DNA is nonmethylated, but not if it is methylated (148), and (ii) purified MutH protein can nick the nonmethylated strand of hemimethylated DNA at GATC sites and can also nick nonmethylated, but not fully methylated, DNA at these sites (164). Because the I end of IS*50* contains two Dam sites, it will be intriguing to learn whether MutH protein also helps regulate I end-mediated transposition.

F. Position Affects Relative Activity of IS*50* Ends

Initial tests of the transposition activities of the O and I ends involved the use of full-length IS*50* elements bracketing selectable markers, essentially as shown in Fig. 4. In wild-type (*dam*$^+$) cells, pairs of O ends were found to be 100- to 1,000-fold more active than pairs of I ends, as if the I ends were degenerate or not yet fully evolved (24, 64, 78, 140). Further tests (below) showed, however, that the I ends can be highly active and that differences between O and I end activities in full-length IS*50* elements reflect their positions in IS*50*, as well as specific Dam methylation of the I end.

The intrinsic activities of the O and I ends were assessed using a set of transposons that differ only in the sequences at their ends. Tests of these elements in *dam*$^+$ cells indicated that pairs of O ends were only 10-fold more active than pairs of I ends, in contrast to the 100- to 1,000-fold difference found with full-length IS*50* elements (Dodson and Berg, in press). Much of the difference in the activities of O and I ends in IS*50* must thus be due to the organization of the IS element. The transcript for the resistance genes of Tn*5* crosses the I end of IS*50*L, and the *tnp* transcript crosses the I end of IS*50*R (Fig. 1) (137, 138). Transcription is known to interfere directly with the ability of a Tn*5* end to act in transposition (142). This effect of transcription across the I end of IS*50* should help maintain Tn*5* as an intact transposon. The effects of position on O and I end activities might additionally reflect the tendency of transposase to act preferentially near its site of synthesis.

G. Effects of External Promoters on Transposition

The *tnp* gene of IS*50* starts less than 100 bp from flanking host sequences, and consequently the level of transposase synthesis might have been expected to reflect transcription of these host sequences. A direct test showed, however, that little if any transposase protein is made from a transcript initiated at an external promoter (94). Sequences near the beginning of the *tnp* gene constitute a relatively well-matched inverted repeat (nucleotides 54 to 69 and 79 to 97 in Fig. 3). mRNAs made from outside promoters would contain the entire palindromic sequence, whereas the T1 transcript contains only one of the inverted repeats. Intramolecular base pairing in the longer mRNAs should sequester the ribosome binding site and thereby block synthesis of transposase (94).

Other studies showed that transcription across an IS*50* O end also interferes directly with the ability of that end to participate in transposition, possibly by changing the DNA conformation or dislodging proteins bound to the end (142). It is noteworthy that IS*10* transposase levels are also insensitive to transcription from outside promoters (41) and that transposition of IS*10* and IS*1* is similarly inhibited by transcription from external promoters (29, 35, 41, 105). Thus, induced transcription of DNA containing Tn*5* does not lead to its unregulated movement or to sharp increases in the frequency of deleterious insertion mutations.

V. Tn*5* AS A RESEARCH TOOL

Tn*5* has become valuable in molecular genetic analysis and manipulation of procaryotes, primarily because it can transpose efficiently and with relatively little specificity in diverse gram-negative bacteria. Many specialized derivatives of Tn*5* have been made, and some are noted below. These and other derivatives, and the ways in which they have been used, are described in greater detail in references 13 and 18 and elsewhere in this volume (Berg et al.).

A. Transcriptional Regulation

Tn*5* elements with promoterless *lac* (96), *kan* (11), and *lux* (luciferase) (Koncz-Kalman and Schell, personal communication) reporter genes have been constructed to analyze how transcription is regulated. As first demonstrated with promoter probe derivatives of phage Mu (34), the level of expression of the reporter gene accurately reflects the level of transcription of the DNA segment into which the element has been inserted.

B. Translational Control

A Tn*5* element with a *lacZ* reporter gene missing both the promoter and the translational start has been constructed (95). Expression of the reporter gene depends in this case on insertion of the transposon into an expressed target gene in the correct orientation and reading frame. As discussed above, this protein fusion transposon, called Tn*5*ORF*lac*, has been useful in studies of transposition to identify sequence changes in the O and I ends that alter transposition activity (107), to select reversion of the ochre mutant allele of the *tnp* gene in IS*50*L (95), and to test whether IHF is needed for transposition (Makris and Reznikoff, personal communication).

C. Protein Localization and Export

A Tn*5* element has been made that contains a *phoA* (alkaline phosphatase) reporter gene that lacks the sequence encoding the amino-terminal leader for transmembrane transport as well as the promoter and translational start sequences (108, 109). Because alkaline phosphatase is active only when on the cell surface or exported, this transposon is valuable both in global searches for genes encoding surface and exported proteins, and in intensive insertion mutagenesis to identify surface exposed protein domains (108, 109).

D. Portable Promoters and Induction of Conditional Mutations

Tn*5* elements with strong T7, SP6 (123), and *tac* (36) promoters have been constructed. The element with the *tac* promoter is polar on distal gene expression in the absence of the *lac* operon inducer isopropyl-β-D-thiogalactopyranoside and can be used to generate insertion mutations with conditional phenotypes: some exhibit the mutant phenotype only in the absence of inducer, and others exhibit the mutant phenotype only in its presence (36).

E. Transposon-Facilitated DNA Sequencing

Any of the large variety of Tn*5* elements with unique subterminal sequences can be used as a mobile source of primer binding sites for chain-termination DNA sequencing of plasmid and chromosomal DNAs (123). For the sequencing of DNAs cloned in phage λ without extensive subcloning, we have recently constructed a Tn*5* element that is only 264 bp long and contains *supF*, a marker whose insertion into λ is easily selected (S. H. Phadnis, H. V. Huang, and D. E. Berg, unpublished data).

VI. SUMMARY AND OVERVIEW

The Tn*5* transposon is distinguished by its ability to move at relatively high frequency and with low target specificity in diverse bacteria. These properties have encouraged detailed analyses of the mechanism and regulation of its transposition and have made it a favorite tool for molecular genetic research. It can be useful to review some characteristics of Tn*5* in an evolutionary context.

The overall organization of Tn*5* indicates that it could have arisen by insertions of IS*50* elements, perhaps followed by deletions or rearrangements that, in effect, "captured" structural genes from a preexisting resistance operon but without their transcription regulatory sequences. This may have been followed by selection of a base substitution in IS*50*L that created a new promoter, permitting efficient although unregulated expression of the resistance genes. The near identity of IS*50*L and IS*50*R suggests that Tn*5* may be of relatively recent origin. Consistent with this is the apparent rarity of elements closely related to Tn*5* among bacteria isolated from nature, in contrast to the abundance of other Kan^r transposons that contain unrelated IS elements and whose resistance genes are only distantly related to those in Tn*5* (reviewed in reference 17).

The IS*50* element itself seems to be highly evolved, suggesting that it is quite ancient. (i) It transposes by an apparently nonreplicative process that consumes the remainder of the donor molecule and thus tends to increase the element's copy number in the host genome. (ii) The transposase protein operates efficiently only in *cis*, whereas the inhibitor protein operates effectively in *trans*. (iii) IS*50* contributes to bacterial fitness by increasing the rate of adaption to a new environment; this is a physiological effect, separate from induction of favorable mutations, and superimposed on the selection for resistance genes in antibiotic-contaminated medical and agricultural environments. (iv) The overall rate of transposition, and probably its timing in the cell cycle, is regulated by a combination of complementary factors: the interplay of limiting transposase and inhibitor proteins; the control of transposase synthesis and IS*50* I end availability by Dam methylation (in *E. coli*); and a dependence on host factors such as DnaA protein and IHF, whose availability may also change during the cell cycle. Although Dam methylation is relatively rare in other bacterial species (9), DnaA-related proteins are widespread (see reference 5), and histonelike proteins equivalent to IHF are undoubtedly also commonplace. The wide availability of essential host factors is emphasized by the recent finding (R. W. Davies, personal communication) that Tn*5* can transpose in streptomycetes if the *tnp* promoter is replaced by a promoter from gram-positive bacteria. (v) The density of information in IS*50* is extremely high: the coding sequence for inhibitor is nested within that for transposase and in the same reading frame; the *tnp* gene promoter is less than 20 bp from the O end segment upon which transposition proteins act; the *tnp* promoter region contains translational control sequences that preclude inadvertent expression of *tnp* from outside promoters; and the *tnp* and *inh* genes end within the critical I end segment. (vi) The promoters for the *tnp* and *inh* genes are contiguous but differently regulated. The *tnp* promoter starts less than 20 bp from the critical O end recognition signal; nearby is a third promoter that, although not important for *E. coli* growing exponentially in standard laboratory culture, might contribute to inhibitor synthesis in other organisms or environments. (vii) The essential O and I end sites are partially matched in DNA sequence (at 8 of the first 9 bp, but only 4 of the next 10 bp), although most positions in each end are important for efficient transposition. This suggests that both ends are acted upon by transposase and that at least one additional factor is probably specific for the O end (e.g., DnaA protein) and another factor is specific for the I end. It is not known whether the O and I end sequences diverged from a common ancestor or converged from unrelated sequences (for discussion, see references 19 and 141). (viii) Pairs of linked IS*50* elements often transpose in unison, and pairs of O ends, pairs of I ends, and the combination of one O and one I end can each mediate transposition of intervening sequences. This flexibility may also contribute to the ability of these elements to transpose and thus to proliferate in diverse bacteria and cellular environments.

Acknowledgments. I am indebted to the late Larry Sandler, whose inspired teaching about McClintock's controlling elements helped set the stage for our discovery of Tn*5*. I am also grateful to the members of my research group for many stimulating discussions, to N. de la Cruz, J. Makris, W. S. Reznikoff, and J. Schell for communication of data in advance of publication, and to C. M. Berg, M. M. Howe, and W. S. Reznikoff for critical readings of the manuscript.

This work was supported by the United States Public Health Service (GM-37138), the National Science Foundation (DMB-8608193), and the Lucille P. Markey Charitable Trust.

LITERATURE CITED

1. **Adhya, S., and M. Gottesman.** 1982. Promoter occlusion: transcription through a promoter may inhibit its activity. *Cell* **29:**939–944.
2. **Ahmed, A.** 1986. Evidence for replicative transposition of Tn*5* and Tn*9*. *J. Mol. Biol.* **191:**75–84.
3. **Albertini, A. M., M. Hofer, M. P. Calos, and J. H. Miller.** 1982. On the formation of spontaneous deletions: the importance of short sequence homologies in the generation of large deletions. *Cell* **29:**319–328.
4. **Andre, D., D. Colau, J. Schell, M. van Montagu, and J.-P. Hernalsteens.** 1986. Gene tagging in plants by a T-DNA insertion mutagen that generates APH(3′)II-plant gene fusions. *Mol. Gen. Genet.* **204:**512–518.
5. **Andrup, L., T. Atlung, N. Ogasawara, H. Yoshikawa, and F. G. Hansen.** 1988. Interaction of the *Bacillus subtilis* DnaA-like protein with the *Escherichia coli* DnaA protein. *J. Bacteriol.* **170:**1333–1338.
6. **Arnau, J., F. J. Murillo, and S. Torres-Martinez.** 1988. Expression of Tn*5*-derived kanamycin resistance in the fungus *Phycomyces blakesleeanus*. *Mol. Gen. Genet.* **212:** 375–377.
7. **Auerswald, E. A., G. Ludwig, and H. Schaller.** 1981. Structural analysis of Tn*5*. *Cold Spring Harbor Symp. Quant. Biol.* **45:**107–113.
8. **Baker, T. A., K. Sekimuzu, B. E. Funnell, and A. Kornberg.** 1986. Extensive unwinding of the plasmid template during staged enzymatic initiation of DNA replication from the origin of the *Escherichia coli* chromosome. *Cell* **45:**801–806.
9. **Barbeyron, T., K. Kean, and P. Forterre.** 1984. DNA adenine methylation of GATC sequences appeared recently in the *Escherichia coli* lineage. *J. Bacteriol.* **160:**586–590.
10. **Beck, E., G. Ludwig, E. Auerswald, B. Reiss, and H. Schaller.** 1982. Nucleotide sequence and exact localization of the neomycin phosphotransferase gene from transposon Tn*5*. *Gene* **19:**327–336.
11. **Bellofatto, V., L. Shapiro, and D. A. Hodgson.** 1984. Generation of a Tn*5* promoter probe and its use in the study of gene expression in *Caulobacter crescentus*. *Proc. Natl. Acad. Sci. USA* **81:**1035–1039.
12. **Bender, J., and N. Kleckner.** 1986. Genetic evidence that Tn*10* transposes by a nonreplicative mechanism. *Cell* **45:** 801–815.
13. **Berg, C. M., and D. E. Berg.** 1987. Uses of transposable elements and maps of known insertions, p. 1071–1109. *In* F. C. Neidhardt, J. L. Ingraham, K. B. Low, B. Magasanik, M. Schaechter, and H. E. Umbarger (ed.), *Escherichia coli and Salmonella typhimurium: Cellular and Molecular Biology*. American Society for Microbiology, Washington, D.C.
14. **Berg, D. E.** 1977. Insertion and excision of the transposable kanamycin resistance determinant Tn*5*, p. 205–212. *In* A. I. Bukhari, J. A. Shapiro, and S. L. Adhya (ed.), *Insertion Elements, Plasmids, and Episomes*. Cold Spring Harbor Laboratory, Cold Spring Harbor, N.Y.
15. **Berg, D. E.** 1980. Control of gene expression by a mobile recombinational switch. *Proc. Natl. Acad. Sci. USA* **77:** 4880–4884.
16. **Berg, D. E.** 1983. Structural requirement for IS*50*-mediated gene transposition. *Proc. Natl. Acad. Sci. USA* **80:**792–796.
17. **Berg, D. E.** 1989. Transposable elements in prokaryotes. *In* S. B. Levy and R. Miller (ed.), *Gene Transfer in the Environment*. McGraw-Hill Publishing Co., New York, in press.
18. **Berg, D. E., and C. M. Berg.** 1983. The prokaryotic transposable element Tn*5*. *Bio/Technology* **1:**417–435.
19. **Berg, D. E., C. M. Berg, and C. Sasakawa.** 1984. The bacterial transposon Tn*5*: evolutionary inferences. *Mol. Biol. Evol.* **1:**411–422.
20. **Berg, D. E., J. Davies, B. Allet, and J.-D. Rochaix.** 1975. Transposition of R factor genes to bacteriophage λ. *Proc. Natl. Acad. Sci. USA* **72:**3628–3632.
21. **Berg, D. E., and M. H. Drummond.** 1978. Absence of sequences homologous to Tn*5* (Kan) in the chromosome of *Escherichia coli* K-12. *J. Bacteriol.* **136:**419–422.
22. **Berg, D. E., C. Egner, B. J. Hirschel, J. Howard, L. Johnsrud, R. A. Jorgensen, and T. D. Tlsty.** 1981. Insertion, excision and inversion of Tn*5*. *Cold Spring Harbor Symp. Quant. Biol.* **45:**115–123.
23. **Berg, D. E., C. Egner, and J. B. Lowe.** 1983. Mechanism of F factor-enhanced excision of transposon Tn*5*. *Gene* **22:**1–7.
24. **Berg, D. E., L. Johnsrud, L. McDivitt, R. Ramabhadran, and B. J. Hirschel.** 1982. Inverted repeats of Tn*5* are transposable elements. *Proc. Natl. Acad. Sci. USA* **79:**2632–2635.
25. **Berg, D. E., J. K. Lodge, C. Sasakawa, D. K. Nag, S. H. Phadnis, K. Weston-Hafer, and G. F. Carle.** 1984. Transposon Tn*5*: specific sequence recognition and conservative transposition. *Cold Spring Harbor Symp. Quant. Biol.* **49:** 215–226.
26. **Berg, D. E., M. A. Schmandt, and J. B. Lowe.** 1983. Specificity of transposon Tn*5* insertion. *Genetics* **105:**813–828.
27. **Berg, D. E., A. Weiss, and L. Crossland.** 1980. Polarity of Tn*5* insertion mutations in *Escherichia coli*. *J. Bacteriol.* **142:**439–446.
28. **Biek, D., and J. R. Roth.** 1980. Regulation of Tn*5* transposition in *Salmonella typhimurium*. *Proc. Natl. Acad. Sci. USA* **77:**6047–6051.
29. **Biel, S. W., G. Adelt, and D. E. Berg.** 1984. Transcriptional control of IS*1* transposition. *J. Mol. Biol.* **174:**251–264.
30. **Biel, S. W., and D. L. Hartl.** 1983. Evolution of transposons: natural selection for Tn*5* in *Escherichia coli* K-12. *Genetics* **103:**581–592.
31. **Bossi, L., and M. S. Ciampi.** 1981. DNA sequences at the sites of three insertions of the tranposable element Tn*5* in the histidine operon of *Salmonella*. *Mol. Gen. Genet.* **183:** 106–108.
32. **Bramhill, D., and A. Kornberg.** 1988. Duplex opening by

dnaA protein at novel sequences in initiation of replication at the origin of the *E. coli* chromosome. *Cell* **52**:743–755.

33. **Campbell, A., D. E. Berg, D. Botstein, R. Novick, and P. Starlinger.** 1977. Nomenclature of transposable elements in prokaryotes, p. 15–22. *In* A. I. Bukhari, J. A. Shapiro, and S. L. Adhya (ed.), *Insertion Elements, Plasmids, and Episomes.* Cold Spring Harbor Laboratory, Cold Spring Harbor, N.Y.
34. **Casadaban, M. J., and S. N. Cohen.** 1979. Lactose genes fused to exogenous promoters in one step using a Mu-*lac* bacteriophage: in vivo probe for transcriptional control sequences. *Proc. Natl. Acad. Sci. USA* **76**:4530–4533.
35. **Chandler, M., and M. J. Galas.** 1983. Cointegrate formation mediated by Tn*9*. II. Activity of IS*1* is modulated by external DNA sequences. *J. Mol. Biol.* **170**:61–91.
36. **Chow, W.-Y., and D. E. Berg.** 1988. Tn*5*tac1, a derivative of transposon Tn*5* that generates conditional mutations. *Proc. Natl. Acad. Sci. USA* **85**:6468–6472.
37. **Cohen, A., and A. J. Clark.** 1986. Synthesis of linear plasmid multimers in *Escherichia coli* K-12. *J. Bacteriol.* **167**: 327–335.
38. **Collins, J., G. Volckaert, and P. Nevers.** 1982. Precise and nearly precise excision of the symmetrical inverted repeats of Tn*5*: common features of *recA*-independent deletion events in *Escherichia coli. Gene* **19**:139–146.
39. **DasGupta, U., K. Weston-Hafer, and D. E. Berg.** 1987. Local DNA sequence control of deletion formation in plasmid pBR322. *Genetics* **115**:41–49.
40. **Datta, A. R., and J. L. Rosner.** 1987. Reduced transposition in *rho* mutants of *Escherichia coli* K-12. *J. Bacteriol.* **169**: 888–890.
41. **Davis, M. A., R. W. Simons, and N. Kleckner.** 1985. Tn*10* protects itself at two levels from fortuitous activation by external promoters. *Cell* **43**:379–387.
42. **de Bruijn, F. J.** 1987. Transposon Tn*5* mutagenesis to map genes. *Methods Enzymol.* **154**:175–196.
43. **de Vos, G. F., T. M. Finan, E. R. Signer, and G. C. Walker.** 1984. Host-dependent transposon Tn*5*-mediated streptomycin resistance. *J. Bacteriol.* **159**:395–399.
44. **DiNardo, C. J., K. A. Voelkel, and R. Sternglanz.** 1982. *Escherichia coli* DNA topoisomerase mutants have compensatory mutations in DNA gyrase genes. *Cell* **31**:43–51.
45. **Doherty, M. J., P. T. Morrison, and R. Kolodner.** 1983. Genetic recombination of bacterial plasmid DNA: physical and genetic analysis of the products of plasmid recombination in *Escherichia coli. J. Mol. Biol.* **167**:539–560.
46. **Donachie, W. D., and A. C. Robinson.** 1987. Cell division: parameter values and the process, p. 1578–1593. *In* F. C. Neidhardt, J. L. Ingraham, K. B. Low, B. Magasanik, M. Schaechter, and H. E. Umbarger (ed.), *Escherichia coli and Salmonella typhimurium: Cellular and Molecular Biology.* American Society for Microbiology, Washington, D.C.
47. **Egner, C., and D. E. Berg.** 1981. Excision of transposon Tn*5* is dependent on the inverted repeats but not the transposase function of Tn*5*. *Proc. Natl. Acad. Sci. USA* **78**:459–463.
48. **Farabaugh, P. J., U. Schmiessner, M. Hofer, and J. H. Miller.** 1978. Genetic studies of the *lac* repressor. VII. On the molecular nature of spontaneous hotspots in the *lacI* gene of *E. coli. J. Mol. Biol.* **126**:847–863.
49. **Fassler, J. S., G. Ferstandig-Arnold, and I. Tessman.** 1986. Reduced superhelicity of plasmid DNA produced by the *rho*-15 mutation in *Escherichia coli* K-12. *Mol. Gen. Genet.* **204**:424–429.
50. **Fishel, R. A., A. A. James, and R. Kolodner.** 1981. *recA*-independent general genetic recombination of plasmids. *Nature* (London) **294**:184–186.
51. **Foster, T. J., V. Lundblad, S. Hanley-Way, S. Halling, and N. Kleckner.** 1981. Three Tn*10*-associated excision events: relationship to transposition and role of direct and inverted repeats. *Cell* **23**:215–227.
52. **Friedman, D. I., M. J. Imperiale, and S. Adhya.** 1987. RNA 3′ end formation in the control of gene expression. *Annu. Rev. Genet.* **21**:453–488.
53. **Fuller, R. S., B. E. Funnell, and A. Kornberg.** 1984. The *dnaA* protein complex with the *E. coli* chromosomal origin (*oriC*) and other DNA sites. *Cell* **38**:889–900.
54. **Galas, D. J., and M. Chandler.** 1981. On the molecular mechanism of transposition. *Proc. Natl. Acad. Sci. USA* **78**:4858–4862.
55. **Gardner, J. F., and H. A. Nash.** 1986. Role of *Escherichia coli* IHF protein in lambda site specific recombination. A mutational analysis of binding sites. *J. Mol. Biol.* **191**: 181–189.
56. **Gatignol, A., M. Baron, and G. Tiraby.** 1987. Phleomycin resistance encoded by the *ble* gene from transposon Tn*5* as a dominant selectable marker in *Saccharomyces cerevisiae. Mol. Gen. Genet.* **207**:342–348.
57. **Genilloud, O., J. Blazquez, P. Mazodier, and F. Moreno.** 1988. A clinical isolate of transposon Tn*5* expressing streptomycin resistance in *Escherichia coli. J. Bacteriol.* **170**: 1275–1278.
58. **Glickman, B. W., and L. S. Ripley.** 1984. Structural intermediates of deletion mutagenesis: a role for palindromic DNA. *Proc. Natl. Acad. Sci. USA* **81**:512–516.
59. **Gottesman, M. M., and J. L. Rosner.** 1975. Acquisition of a determinant for chloramphenicol resistance by coliphage lambda. *Proc. Natl. Acad. Sci. USA* **72**:5041–5045.
60. **Gottesman, S.** 1987. Regulation by proteolysis, p. 1308–1312. *In* F. C. Neidhardt, J. L. Ingraham, K. B. Low, B. Magasanik, M. Schaechter, and H. E. Umbarger (ed.), *Escherichia coli and Salmonella typhimurium: Cellular and Molecular Biology.* American Society for Microbiology, Washington, D.C.
61. **Gray, G. S., and W. M. Fitch.** 1983. Evolution of antibiotic resistance genes: the DNA sequence of a kanamycin resistance gene from *Staphylococcus aureus. Mol. Biol. Evol.* **1**:57–66.
62. **Greenblatt, I. M., and R. A. Brink.** 1963. Transpositions of Modulator in maize into divided and undivided chromosome segments. *Nature* (London) **197**:412–413.
63. **Gross, J. D., and L. G. Caro.** 1966. DNA transfer in bacterial conjugation. *J. Mol. Biol.* **16**:269–284.
64. **Guarente, L. P., R. R. Isberg, M. Syvanen, and T. J. Silhavy.** 1980. Conferral of transposable properties to a chromosomal gene in *Escherichia coli. J. Mol. Biol.* **141**:235–248.
65. **Hagan, C. E., and G. J. Warren.** 1982. Lethality of palindromic DNA and its use in selection of recombinant plasmids. *Gene* **19**:147–151.
66. **Halling, S. M., and N. Kleckner.** 1982. A symmetrical six-base-pair target site sequence determines Tn*10* insertion specificity. *Cell* **28**:155–163.
67. **Hartl, D. L., D. E. Dykhuizen, R. D. Miller, L. Green, and J. de Framond.** 1983. Transposable element IS*50* improves growth rate of *E. coli* cells without transposition. *Cell* **35**:503–510.
68. **Hedges, R. W., and A. E. Jacob.** 1974. Transposition of ampicillin resistance from RP4 to other replicons. *Mol. Gen. Genet.* **132**:31–40.
69. **Heffron, F., C. Rubens, and S. Falkow.** 1975. The transpo-

sition of a plasmid DNA sequence which mediates ampicillin resistance: molecular nature and specificity of insertion. *Proc. Natl. Acad. Sci. USA* 72:3623–3627.

70. **Heinzel, P., O. Werbitzky, J. Distler, and W. Piepersberg.** 1988. A second streptomycin resistance gene from *Streptomyces griseus* codes for streptomycin-3″-phosphotransferase. Relationships between antibiotic and protein kinases. *Arch. Microbiol.* 150:184–192.

71. **Herrmann, R., K. Neugebauer, H. Zentgraf, and H. Schaller.** 1978. Transposition of a DNA sequence determining kanamycin resistance into the single stranded genome of bacteriophage fd. *Mol. Gen. Genet.* 159:171–178.

72. **Hirschel, B. J., and D. E. Berg.** 1982. A derivative of Tn*5* with direct terminal repeats can transpose. *J. Mol. Biol.* 155:105–120.

73. **Hirschel, B. J., D. J. Galas, D. E. Berg, and M. Chandler.** 1982. Structure and stability of transposon *5*-mediated cointegrates. *J. Mol. Biol.* 159:557–580.

74. **Hirschel, B. J., D. J. Galas, and M. Chandler.** 1982. Cointegrate formation by Tn*5*, but not transposition, is dependent on *recA*. *Proc. Natl. Acad. Sci. USA* 79:4530–4534.

75. **Hopkins, J. D., M. B. Clements, T.-Y. Liang, R. R. Isberg, and M. Syvanen.** 1980. Recombination genes on the *Escherichia coli* sex factor specific for transposable elements. *Proc. Natl. Acad. Sci. USA* 77:2814–2818.

76. **Hopkins, J. D., M. Clements, and M. Syvanen.** 1983. New class of mutations in *Escherichia coli* (*uup*) that affect precise excision of transposable elements and bacteriophage Mu growth. *J. Bacteriol.* 153:384–389.

77. **Isberg, R. R., A. L. Lazaar, and M. Syvanen.** 1982. Regulation of Tn*5* by the right repeat proteins: control at the level of the transposition reactions? *Cell* 30:883–892.

78. **Isberg, R. R., and M. Syvanen.** 1981. Replicon fusion promoted by the inverted repeats of Tn*5*. The right repeat is an insertion sequence. *J. Mol. Biol.* 150:15–32.

79. **Isberg, R. R., and M. Syvanen.** 1982. DNA gyrase is a host factor required for transposition of Tn*5*. *Cell* 30:9–18.

80. **Isberg, R. R., and M. Syvanen.** 1985. Compartmentalization of the proteins encoded by IS*50*R. *J. Biol. Chem.* 260: 3645–3651.

81. **Isberg, R. R., and M. Syvanen.** 1985. Tn*5* transposes independently of cointegrate resolution. Evidence for an alternative model for transposition. *J. Mol. Biol.* 182:69–78.

81a. **Jimenez, A., and J. Davies.** 1980. Expression of a transposable antibiotic resistance element in *Saccharomyces*. *Nature* (London) 287:869–871.

82. **Johnson, R. C., and W. S. Reznikoff.** 1983. DNA sequences at the ends of transposon Tn*5* required for transposition. *Nature* (London) 204:280–282.

83. **Johnson, R. C., and W. S. Reznikoff.** 1984. Copy number control of Tn*5* transposition. *Genetics* 107:9–18.

84. **Johnson, R. C., and W. S. Reznikoff.** 1984. Role of the IS*50*R proteins in the promotion and control of Tn*5* transposition. *J. Mol. Biol.* 177:645–661.

85. **Johnson, R. C., J. C. P. Yin, and W. S. Reznikoff.** 1982. Control of Tn*5* transposition in *Escherichia coli* is mediated by protein from the right repeat. *Cell* 30:873–882.

86. **Jorgensen, R. A., S. J. Rothstein, and W. S. Reznikoff.** 1979. A restriction enzyme cleavage map of Tn*5* and location of a region encoding neomycin resistance. *Mol. Gen. Genet.* 177:65–72.

87. **Kendrick, K. E., and W. S. Reznikoff.** 1988. Transposition of IS*50*L activates downstream genes. *J. Bacteriol.* 170:1965–1988.

88. **Kleckner, N.** 1981. Transposable elements in prokaryotes. *Annu. Rev. Genet.* 15:341–404.

89. **Kleckner, N.** 1983. Transposon Tn*10*, p. 261–298. *In* J. A. Shapiro (ed.), *Mobile Genetic Elements*. Academic Press, Inc., New York.

90. **Kleckner, N., R. K. Chan, B.-K. Tye, and D. Botstein.** 1975. Mutagenesis by insertion of a drug-resistance element carrying an inverted repetition. *J. Mol. Biol.* 97:561–575.

91. **Kobayashi, I., and N. Takahashi.** 1988. Double-strand gap repair by gene conversion in *Escherichia coli*. *Genetics* 119:751–757.

92. **Koncz, C., Z. Koncz-Kalman, and J. Schell.** 1987. Transposon Tn*5* mediated gene transfer into plants. *Mol. Gen. Genet.* 207:99–105.

93. **Krause, H. M., and N. P. Higgins.** 1986. Positive and negative regulation of the Mu operator by Mu repressor and *Escherichia coli* integration host factor. *J. Biol. Chem.* 261:3744–3752.

94. **Krebs, M. P., and W. S. Reznikoff.** 1986. Transcriptional and translational initiation sites of IS*50*. Control of transposase and inhibitor expression. *J. Mol. Biol.* 192:781–792.

95. **Krebs, M. P., and W. S. Reznikoff.** 1988. Use of a Tn*5* derivative that creates *lacZ* translational fusions to obtain a transposition mutant. *Gene* 63:277–285.

96. **Kroos, L., and A. D. Kaiser.** 1984. Construction of Tn*5*-*lac*, a transposon that fuses *lacZ* expression to exogenous promoters and its introduction into *Myxococcus xanthus*. *Proc. Natl. Acad. Sci. USA* 81:5816–5820.

97. **Lech, K. F., C. H. Lee, R. R. Isberg, and M. Syvanen.** 1985. New gene in *Escherichia coli* K-12 (*drpA*): does its product play a role in RNA synthesis? *J. Bacteriol.* 162:117–123.

98. **Leong, J. M., S. E. Nunes-Duby, C. F. Lesser, P. Youderian, M. M. Susskind, and A. Landy.** 1985. The ϕ80 and P22 attachment sites. Primary structure and interaction with *Escherichia coli* integration host factor. *J. Biol. Chem.* 260:4468–4477.

99. **Lichens-Park, A., and M. Syvanen.** 1988. Cointegrate formation by IS*50* requires multiple donor molecules. *Mol. Gen. Genet.* 211:244–251.

100. **Lodge, J. K., K. Weston-Hafer, and D. E. Berg.** 1988. Transposon Tn*5* target specificity: preference for insertion at G/C pairs. *Genetics* 120:645–650.

101. **Lundblad, V., and N. Kleckner.** 1984. Mismatch repair mutations of *Escherichia coli* K12 enhance transposon excision. *Genetics* 109:3–19.

102. **Lundblad, V., A. F. Taylor, G. R. Smith, and N. Kleckner.** 1984. Unusual alleles of *recB* and *recC* stimulate excision of inverted repeat transposons Tn*5* and Tn*10*. *Proc. Natl. Acad. Sci. USA* 81:824–848.

103. **Lupski, J. R., P. Gershon, L. Ozaki, and G. N. Godson.** 1984. Specificity of Tn*5* insertions into a 36-bp DNA sequence repeated in tandem seven times. *Gene* 30:99–106.

104. **Lupski, J. R., S. J. Projan, L. S. Ozaki, and G. N. Godson.** 1986. A temperature dependent pBR322 copy number mutant due to a Tn*5* position effect. *Proc. Natl. Acad. Sci. USA* 83:7381–7385.

105. **Machida, C., Y. Machida, H.-C. Wang, K. Ishizaki, and E. Ohtsubo.** 1983. Repression of cointegration ability of insertion element IS*1* by transcriptional readthrough from flanking regions. *Cell* 34:135–142.

106. **Mahajna, J., A. B. Oppenheim, A. Rattray, and M. Gottesman.** 1986. Translation initiation of bacteriophage lambda *c*II requires integration host factor. *J. Bacteriol.* 165:167–174.

107. **Makris, J. C., P. L. Nordmann, and W. S. Reznikoff.** 1988.

Mutational analysis of insertion sequence *50* (IS*50*) and transposon *5* (Tn*5*) ends. *Proc. Natl. Acad. Sci. USA* **85:** 2224–2228.

108. **Manoil, C., and J. Beckwith.** 1985. Tn*phoA*: a transposon probe for protein export signals. *Proc. Natl. Acad. Sci. USA* **82:**8129–8133.
109. **Manoil, C., and J. Beckwith.** 1986. A genetic approach to analyzing membrane protein topology. *Science* **233:** 1403–1408.
110. **Marinus, M. G.** 1987. Methylation of DNA, p. 697–702. *In* F. C. Neidhardt, J. L. Ingraham, K. B. Low, B. Magasanik, M. Schaechter, and H. E. Umbarger (ed.), *Escherichia coli and Salmonella typhimurium: Cellular and Molecular Biology*. American Society for Microbiology, Washington, D.C.
111. **Mazodier, P., P. Cossart, E. Giraud, and F. Gasser.** 1985. Completion of the nucleotide sequence of the central region of Tn*5* confirms the presence of three resistance genes. *Nucleic Acids Res.* **13:**195–205.
112. **Mazodier, P., O. Genilloud, E. Giraud, and F. Gasser.** 1986. Expression of Tn*5*-encoded streptomycin resistance in *E. coli. Mol. Gen. Genet.* **204:**404–409.
113. **Mazodier, P., E. Giraud, and F. Gasser.** 1982. Tn*5* dependent streptomycin resistance in *Methylobacterium organophilum. FEMS Microbiol. Lett.* **13:**27–30.
114. **Mazodier, P., E. Giraud, and F. Gasser.** 1983. Genetic analysis of the streptomycin resistance encoded by Tn*5*. *Mol. Gen. Genet.* **192:**155–162.
115. **McCommas, S. A., and M. Syvanen.** 1988. Temporal control of Tn*5* transposition. *J. Bacteriol.* **170:**889–894.
116. **McKinnon, R. D., J. S. Waye, D. S. Bautista, and F. L. Graham.** 1985. Nonrandom insertion of Tn*5* into cloned human adenovirus DNA. *Gene* **40:**31–38.
117. **McMacken, R., L. Silver, and C. Georgopoulos.** 1987. DNA replication, p. 564–612. *In* F. C. Neidhardt, J. L. Ingraham, K. B. Low, B. Magasanik, M. Schaechter, and H. E. Umbarger (ed.), *Escherichia coli and Salmonella typhimurium: Cellular and Molecular Biology*. American Society for Microbiology, Washington, D.C.
118. **Mechulam, Y., S. Blanquet, and G. Fayat.** 1987. Dual control of the *Escherichia coli pheSt-himA* operon expression: tRNA(phe)-dependent attenuation and transcriptional operator-repressor control by *himA* and the SOS network. *J. Mol. Biol.* **197:**453–470.
119. **Meyer, R., G. Boch, and J. Shapiro.** 1979. Transposition of DNA inserted into deletions of the Tn*5* kanamycin resistance element. *Mol. Gen. Genet.* **171:**7–13.
120. **Miller, J. H., M. P. Calos, M. Hofer, D. E. Buchel, and B. Muller-Hill.** 1980. Genetic analysis of transpositions in the *lac* region of *Escherichia coli. J. Mol. Biol.* **144:**1–18.
121. **Morisato, D., and N. Kleckner.** 1987. Tn*10* transposition and circle formation *in vitro*. *Cell* **51:**101–111.
122. **Nag, D. K., U. DasGupta, G. Adelt, and D. E. Berg.** 1985. IS*50*-mediated inverse transposition: specificity and precision. *Gene* **34:**17–26.
123. **Nag, D. K., H. V. Huang, and D. E. Berg.** 1988. Bidirectional chain termination DNA sequencing: transposon Tn*5*seq1 as a mobile source of primer sites. *Gene* **64:** 135–145.
124. **Noti, J. D., M. N. Jagadish, and A. Szalay.** 1987. Site-directed Tn*5* and transplacement mutagenesis: methods to identify symbiotic nitrogen fixation genes in slow-growing *Rhizobium. Methods Enzymol.* **154:**197–217.
125. **O'Neill, E. A., G. M. Kiely, and R. A. Bender.** 1984. Transposon Tn*5* encodes streptomycin resistance in nonenteric bacteria. *J. Bacteriol.* **159:**388–389.
126. **Phadnis, S. H., and D. E. Berg.** 1985. *recA*-independent recombination between repeated IS*50* elements is not caused by an IS*50*-encoded function. *J. Bacteriol.* **161:**928–932.
127. **Phadnis, S. H., and D. E. Berg.** 1987. Identification of base pairs in the IS*50* O end needed for IS*50* and Tn*5* transposition. *Proc. Natl. Acad. Sci. USA* **84:**9118–9122.
128. **Phadnis, S. H., G. Sasakawa, and D. E. Berg.** 1986. Localization of action of the IS*50*-encoded transposase protein. *Genetics* **112:**421–427.
129. **Platt, T., and D. G. Bear.** 1983. Role of RNA polymerase, ρ factor, and ribosomes in transcription termination, p. 123–161. *In* J. Beckwith, J. A. Davies, and J. A. Gallant (ed.), *Gene Function in Prokaryotes*. Cold Spring Harbor Laboratory, Cold Spring Harbor, N.Y.
130. **Prentki, P., M. Chandler, and D. J. Galas.** 1987. *Escherichia coli* integration host factor bends DNA at the ends of IS*1* and in an insertion hotspot with multiple IHF binding sites. *EMBO J.* **6:**2479–2487.
131. **Putnoky, D., G. B. Kiss, I. Ott, and A. Kondorosi.** 1983. Tn*5* carries a streptomycin resistance determinant downstream from the kanamycin resistance gene. *Mol. Gen. Genet.* **19:**288–294.
132. **Reynolds, A. E., J. Felton, and A. Wright.** 1981. Insertion of DNA activates the cryptic *bgl* operon in *E. coli* K-12. *Nature* (London) **293:**625–629.
133. **Reznikoff, W. S.** 1983. Some bacterial transposable elements: their organization, mechanics of transposition, and roles in genome evolution, p. 229–252. *In* J. Beckwith, J. A. Davies, and J. A. Gallant (ed.), *Gene Function in Prokaryotes*. Cold Spring Harbor Laboratory, Cold Spring Harbor, N.Y.
134. **Rio, D. C., and G. M. Rubin.** 1985. Transformation of *Drosophila melanogaster* cells with a dominant selectable marker. *Mol. Cell. Biol.* **5:**1833–1838.
135. **Roberts, D., B. C. Hoopes, W. R. McClure, and N. Kleckner.** 1985. IS*10* transposition is regulated by DNA adenine methylation. *Cell* **43:**117–130.
136. **Rosetti, O. L., R. Altman, and R. Young.** 1984. Kinetics of Tn*5* transposition. *Gene* **32:**91–98.
137. **Rothstein, S. J., R. A. Jorgensen, K. Postle, and W. S. Reznikoff.** 1980. The inverted repeats of Tn*5* are functionally different. *Cell* **19:**795–805.
138. **Rothstein, S. J., and W. S. Reznikoff.** 1981. The functional differences in the inverted repeats of Tn*5* are caused by a single base pair nonhomology. *Cell* **23:**191–199.
139. **Sapienza, C., and W. F. Doolittle.** 1981. Genes are things you have whether you want them or not. *Cold Spring Harbor Symp. Quant. Biol.* **45:**177–182.
140. **Sasakawa, C., and D. E. Berg.** 1982. IS*50* mediated inverse transposition: discrimination between the two ends of an IS element. *J. Mol. Biol.* **159:**257–271.
141. **Sasakawa, C., G. F. Carle, and D. E. Berg.** 1983. Sequences essential for transposition at the termini of IS*50*. *Proc. Natl. Acad. Sci. USA* **80:**7293–7297.
142. **Sasakawa, C., J. B. Lowe, L. McDivitt, and D. E. Berg.** 1982. Control of transposon Tn*5* transposition in *Escherichia coli. Proc. Natl. Acad. Sci. USA* **79:**7450–7454.
143. **Sasakawa, C., S. H. Phadnis, G. F. Carle, and D. E. Berg.** 1985. Sequences essential for IS*50* transposition: the first base pair. *J. Mol. Biol.* **182:**487–493.
144. **Sasakawa, C., Y. Uno, and M. Yoshikawa.** 1981. The requirement for both DNA polymerase and 5′ to 3′ exonuclease activities of DNA polymerase I during Tn*5* transposition. *Mol. Gen. Genet.* **182:**19–24.
145. **Sasakawa, C., Y. Uno, and M. Yoshikawa.** 1987. *lon-sulA*

regulatory function affects the efficiency of transposition of Tn*5* from λ *b*221 *c*I857 Pam Oam to the chromosome. *Biochem. Biophys. Res. Commun.* **142:**879–884.

146. **Schaller, H.** 1979. The intergenic region and the origins for filamentous phage DNA replication. *Cold Spring Harbor Symp. Quant. Biol.* **43:**401–408.

147. **Schrenk, W. J., and R. A. Weisberg.** 1975. A simple method for making new transducing lines of coliphage lambda. *Mol. Gen. Genet.* **137:**101–107.

148. **Seiler, A., H. Blocker, R. Frank, and R. Kahmann.** 1986. The *mom* gene of bacteriophage Mu: the mechanism of methylation dependent expression. *EMBO J.* **5:**2719–2728.

149. **Selvaraj, G., and V. N. Iyer.** 1984. Transposon Tn*5* specifies streptomycin resistance in *Rhizobium* sp. *J. Bacteriol.* **158:** 580–589.

150. **Semon, D., N. R. Movva, T. F. Smith, M. El Alama, and J. Davies.** 1987. Plasmid determined bleomycin resistance in *Staphylococcus aureus. Plasmid* **17:**46–53.

151. **Shaw, K. J., and C. M. Berg.** 1979. *Escherichia coli* K-12 auxotrophs induced by insertion of the transposable element Tn*5*. *Genetics* **92:**741–747.

152. **Silverman, M., and M. I. Simon.** 1980. Phase variation: genetic analysis of switching mutants. *Cell* **19:**845–854.

153. **Smith, G. R.** 1988. Homologous recombination in bacteria. *Microbiol. Rev.* **52:**1–28.

154. **Southern, P. J., and P. Berg.** 1982. Transformation of mammalian cells to antibiotic resistance with a bacterial gene under control of the early SV40 promoter. *J. Mol. Appl. Genet.* **1:**327–341.

155. **Starlinger, P., and H. Saedler.** 1972. Insertion mutations in microorganisms. *Biochimie* **54:**177–185.

156. **Sternberg, N.** 1985. Evidence that adenine methylation influences DNA-protein interactions in *Escherichia coli. J. Bacteriol.* **164:**490–493.

157. **Sternglanz, R., S. DiNardo, K. A. Voelker, Y. Nishimura, Y. Hirota, K. Becherer, L. Zumstein, and J. C. Wang.** 1981. Mutations in the gene coding for *Escherichia coli* DNA topoisomerase I affect transcription and transposition. *Proc. Natl. Acad. Sci. USA* **78:**2747–2751.

158. **Stueber, D., and H. Bujard.** 1981. Organization of transcriptional signals in plasmids pBR322 and pACYC184. *Proc. Natl. Acad. Sci. USA* **78:**167–171.

159. **Syvanen, M., J. D. Hopkins, and M. Clements.** 1982. A new class of mutants in DNA polymerase I that affects gene transposition. *J. Mol. Biol.* **158:**203–212.

160. **Syvanen, M., J. D. Hopkins, T. J. Griffith IV, T.-Y. Liang, K. Ippen-Ihler, and R. Kolodner.** 1986. Stimulation of precise excision and recombination by conjugal proficient F′ plasmids. *Mol. Gen. Genet.* **203:**1–7.

161. **Thompson, J. F., D. Waechter-Brulla, R. I. Gumport, J. F. Gardner, L. Moitoso de Vargas, and A. Landy.** 1986. Mutations in an integration host factor-binding site: effect on lambda site-specific recombination and regulatory implications. *J. Bacteriol.* **168:**1343–1351.

162. **Wang, J. C.** 1985. DNA topoisomerases. *Annu. Rev. Biochem.* **54:**665–697.

163. **Weber, P. C., M. Levine, and J. C. Glorioso.** 1987. Rapid identification of nonessential genes of herpes simplex virus type 1 by Tn*5* mutagenesis. *Science* **236:**576–579.

164. **Welsh, K. M., A. L. Lu, S. Clark, and P. Modrich.** 1987. Isolation and characterization of the *Escherichia coli mutH* gene product. *J. Biol. Chem.* **262:**15624–15629.

165. **Wiater, L. A., and N. D. F. Grindley.** 1988. Gamma-delta transposase and integration host factor bind cooperatively at both ends of gamma-delta. *EMBO J.* **7:**1907–1911.

166. **Xin, W., and M. Feiss.** 1988. The interaction of *Escherichia coli* integration host factor with the cohesive end sites of phages λ and 21. *Nucleic Acids Res.* **16:**2015–2030.

167. **Yates, J., N. Murray, D. Reisman, and B. Sugden.** 1984. A *cis*-acting element from the Epstein-Barr viral genome that permits stable replication of recombinant plasmids in latently infected cells. *Proc. Natl. Acad. Sci. USA* **81:** 3806–3810.

168. **Yin, J. C. P., M. P. Krebs, and W. S. Reznikoff.** 1988. The effect of *dam* methylation on Tn*5* transposition. *J. Mol. Biol.* **199:**35–45.

169. **Yin, J. C. P., and W. S. Reznikoff.** 1987. *dnaA*, an essential host gene, and Tn*5* transposition. *J. Bacteriol.* **169:** 4637–4645.

170. **Yin, J. C. P., and W. S. Reznikoff.** 1988. p2 and inhibition of Tn*5* transposition. *J. Bacteriol.* **170:**3008–3015.

171. **Zupancic, T. J., S. L. Marvo, J. H. Chung, E. G. Peralta, and S. R. Jaskunas.** 1983. *RecA*-independent recombination between direct repeats of IS*50*. *Cell* **33:**629–637.

172. **Zyskind, J. W., J. M. Cleary, W. S. A. Brusilow, N. E. Harding, and D. W. Smith.** 1983. Chromosomal replication origin from the marine bacterium *Vibrio harveyi* functions in *Escherichia coli: oriC* consensus sequence. *Proc. Natl. Acad. Sci. USA* **80:**1164–1168.

Chapter 7

Transposon Tn*7*

NANCY L. CRAIG

I. INTRODUCTION

The bacterial transposon Tn7 (7) is a mobile DNA segment that provides resistance to the antifolate agent trimethoprim and to the aminoglycosides streptomycin and spectinomycin. Most transposable elements transpose at low frequency and exhibit little target site specificity upon insertion into large DNA molecules such as bacterial chromosomes (9, 18, 32, 44, 73; chapters by D. J. Galas and M. Chandler, D. Sherratt, D. E. Berg, and N. Kleckner, this volume). Tn7 is distinguished by its capacity to transpose at high frequency to specific target sites in the chromosomes of many different bacteria. In *Escherichia coli*, the organism in which Tn7 has been most extensively studied, this specific insertion ("attachment") site is called *att*Tn7. When *att*Tn7 is unavailable, Tn7 transposes at low frequency to other target sites. There are two distinct classes of Tn7 target sites other than *att*Tn7, one pseudo-*att*Tn7 class and another

Nancy L. Craig ■ Department of Microbiology and Immunology, Department of Biochemistry and Biophysics, and the G. W. Hooper Foundation, HSW 1542, University of California, San Francisco, San Francisco, California 94143.

Table 1. Molecular weights of the *tns* gene products

Gene	Mol wt translated from DNA sequence	References	Mol wt observed	References
tnsA	31,000	Orle and Craig, unpublished observations; Flores et al., personal communication	30,000	68; Orle and Craig, unpublished observations; Rogers and Sherratt, personal communication
tnsB	78,000	Flores et al., personal communication	83,000–85,000	12, 68, 84; Rogers and Sherratt, personal communication; Waddell and Craig, unpublished observations
tnsC	63,000	Flores et al., personal communication	54,000–56,000	12, 84; Waddell and Craig, unpublished observations
			40,000–42,000	68; Rogers and Sherratt, personal communication
tnsD	59,000	Flores et al., personal communication	54,000	68; Rogers and Sherratt, personal communication
			40,000	12, 84; Waddell and Craig, unpublished observations
tnsE	61,000	78	85,000	12
			70,000–75,000	12, 68, 78, 84; Rogers and Sherratt, personal communication; Waddell and Craig, unpublished observations

of Tn7 transposition and are independent of the site from which Tn7 transposes (84).

What are the functions of the Tns proteins in transposition? TnsA, TnsB, and TnsC provide activities essential to all Tn7 transposition events. Biochemical studies have identified a *tnsB*-dependent specific DNA-binding activity that recognizes sequences in Tn7L and Tn7R which participate directly in transposition (51). Thus, TnsB is likely to be involved in the recognition and/or utilization of *cis*-acting transposition sequences at the ends of Tn7. Purification of the *tnsB*-dependent DNA-binding activity has shown that TnsB participates directly in specific DNA binding (R. McKown and N. L. Craig, unpublished observations). TnsB also likely has regulatory roles in *tns* gene expression (see below). The roles of TnsA and TnsC in transposition have not yet been established. Attractive models for the actions of these proteins include additional roles in the recognition or utilization (or both) of *cis*-acting transposition sequences at the ends of Tn7 or in mediating interactions between protein-DNA complexes at the Tn7 ends and the target sites. No *tns* gene encodes only a resolution function like those of the Tn3 family of transposons (9, 18, 32, 44, 73; D. Sherratt, this volume; see below).

Genetic analyses suggest that TnsD and TnsE are the proteins which choose the Tn7 target sites, TnsD directing transposition to *att*Tn7 and pseudo-*att*Tn7 sites (68, 84; Kubo and Craig, unpublished observations) and TnsE directing transposition to sites unrelated to *att*Tn7 (68, 84; Kubo and Craig, unpublished observations). It is attractive to suggest that TnsD and TnsE are themselves DNA-binding proteins which interact directly with their cognate DNA target sites.

2. Host-encoded genes

It has been shown that some other transposable elements employ host-encoded proteins in transposition (19, 56). Host-encoded proteins that participate directly in Tn7 transposition have not yet been identified. One explanation for the surprising number of Tn7-encoded transposition genes is that one (or more) of the *tns* genes encodes a function(s) usually provided by the host to other transposition systems. It has been shown that integration host factor, an *E. coli*-encoded protein involved in at least one other transposition reaction (56) and in conservative site-specific recombination reactions (59), is not essential for Tn7 transposition (Kubo and Craig, unpublished observations; N. Ekaterinaki and D. Sherratt, personal communication).

C. Tn7 Target Sites

1. *tnsABC*+*D* target sites in *E. coli*

The target sites recognized by Tn7 have been most extensively characterized in *E. coli*. Tn7 inserts at high frequency in a specific site called *att*Tn7 at about 84 min of the chromosome (5, 7, 48). The high frequency of Tn7 transposition to *att*Tn7 was ini-

Chapter 7

Transposon Tn*7*

NANCY L. CRAIG

I. INTRODUCTION

The bacterial transposon Tn7 (7) is a mobile DNA segment that provides resistance to the antifolate agent trimethoprim and to the aminoglycosides streptomycin and spectinomycin. Most transposable elements transpose at low frequency and exhibit little target site specificity upon insertion into large DNA molecules such as bacterial chromosomes (9, 18, 32, 44, 73; chapters by D. J. Galas and M. Chandler, D. Sherratt, D. E. Berg, and N. Kleckner, this volume). Tn7 is distinguished by its capacity to transpose at high frequency to specific target sites in the chromosomes of many different bacteria. In *Escherichia coli*, the organism in which Tn7 has been most extensively studied, this specific insertion ("attachment") site is called *att*Tn7. When *att*Tn7 is unavailable, Tn7 transposes at low frequency to other target sites. There are two distinct classes of Tn7 target sites other than *att*Tn7, one pseudo-*att*Tn7 class and another

Nancy L. Craig ■ Department of Microbiology and Immunology, Department of Biochemistry and Biophysics, and the G. W. Hooper Foundation, HSW 1542, University of California, San Francisco, San Francisco, California 94143.

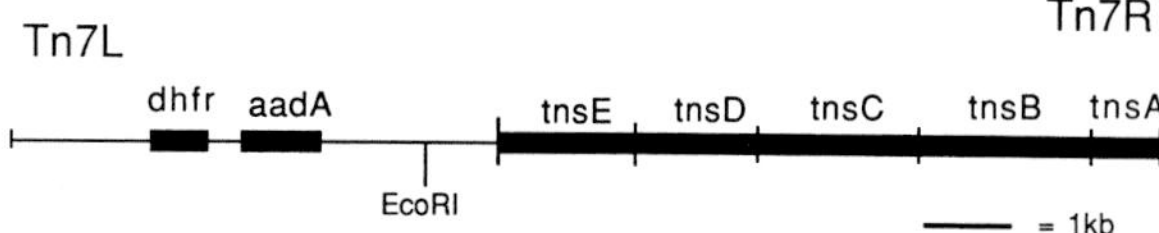

Figure 1. Physical map of Tn7. The locations of the transposition (*tns*) genes of Tn7 as determined by Rogers et al. (68) and Waddell and Craig (84) are shown. The NH_2 termini of all the genes are proximal to Tn7R. Shown are the locations of the Tn7 antibiotic resistance determinants *dhfr*, encoding a trimethoprim-resistant dihydrofolate reductase, and *aadA*, encoding an adenylyltransferase which inactivates streptomycin and spectinomycin, as determined by Fling and Richards (27), Simonsen et al. (76), and Fling et al. (26). The single *Eco*RI site of Tn7 is shown.

class unrelated to *att*Tn7. The same set of transposition genes contained in Tn7 (*tnsABC+D*) directs Tn7 to *att*Tn7 and pseudo-*att*Tn7 sites, whereas a distinct but overlapping set (*tnsABC+E*) directs Tn7 to sites unrelated to *att*Tn7. This capacity for participation in two different transposition pathways also distinguishes Tn7 among transposable elements, as does the large number (five) of Tn7 transposition genes.

Sections II and III of this review summarize our knowledge of the components of the Tn7 transposition machinery and of the mechanism and control of Tn7 transposition. Section IV describes the use of Tn7 transposition as a genetic tool. Section V describes the Tn7-encoded antibiotic resistance determinants.

II. THE Tn*7* TRANSPOSITION MACHINERY

A. The Structure of Tn*7*

Tn7 is a 14-kilobase transposon that contains several antibiotic resistance genes and transposition genes (Fig. 1). Unlike many other large transposons, such as Tn*10* (Kleckner, this volume) and Tn*5* (Berg, this volume), Tn7 is not a composite transposon containing insertion sequence modules. In the *E. coli* chromosome, Tn7 inserts in a single orientation into a specific site which lies about 25 kilobases counterclockwise (leftward) of the origin of replication (48, 49, 85) (Fig. 2). This specific site for Tn7 insertion is termed an "attachment" site (49) or *att*Tn7 (50). Although the nomenclature for specific chromosomal sites of insertion is similar for Tn7 (*att*Tn7) and bacteriophage lambda (*attB*), it cannot be overemphasized that Tn7 inserts via transposition, whereas bacteriophage lambda inserts via conservative site-specific recombination. In *att*Tn7::Tn7, the end of Tn7 containing the transposition genes is proximal to the origin of replication and the end of Tn7 containing the drug resistance genes is distal to the origin (Fig. 2). To reflect this specific chromosomal orientation, the end of Tn7 containing the transposition genes is called the right end of Tn7 (Tn7R) and the end of Tn7 containing the drug resistance genes is called the left end of Tn7 (Tn7L) (Fig. 1 and 2). The reader is cautioned that different groups working on Tn7 have drawn the transposon in different orientations and that a consistent designation for left and right Tn7 ends has not been used in the literature. In the text to follow, the nomenclature given above will be used and instances where a different convention was used in the relevant original literature will be noted.

Transposons very similar (if not identical) to Tn7, such as Tn*73* (20), Tn*1824* (81), and Tn*1527* (29), have been isolated. Tn*1825* (82) is related to but clearly distinct from Tn7. These elements have not been studied in detail.

B. Genes Required for Tn*7* Transposition

1. The *tns* genes

Early studies using deletion mutants revealed that Tn7 encodes several transposition genes (39, 64, 77). More recently, Rogers et al. (68), using complementation between various Tn7 fragments, and Waddell and Craig (84), using insertional mutagenesis, have shown that Tn7 contains five transposition genes: *tnsABCDE* (Fig. 1). The designation *tns* (transposon seven) has been adopted (68, 84) because it does not imply a specific function for the products of these genes. It remains to be determined which of these genes actually encode(s) "transposase" activity, i.e., the capacity to execute the breakage and rejoining events which result in recombination. No other transposable element is known that contains so many transposition genes. DNA sequence analysis of the *tns* region (28, 78; C. Flores, I. Qadri, and C. Lichtenstein, personal communication; E. Nimmo and D. Sherratt, personal communication; K. Orle and N. L. Craig, unpublished observations) has identified candidate open reading frames for each of the five *tns* genes (although exact definition of the translation initiation codon of each gene has not yet been established). Analysis of the predicted protein sequences of the *tns* gene products has not revealed extensive homologies to proteins encoded by other mobile DNA elements, although some very limited homologies have been noted (28; Flores et al., personal communication). The *tns* genes are each oriented in the same direction with the proposed amino terminus of *tnsA* about 100 base pairs (bp) from the

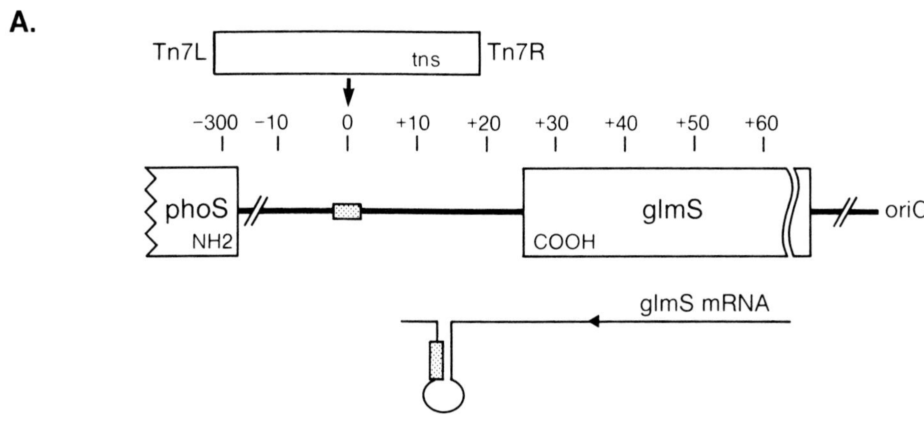

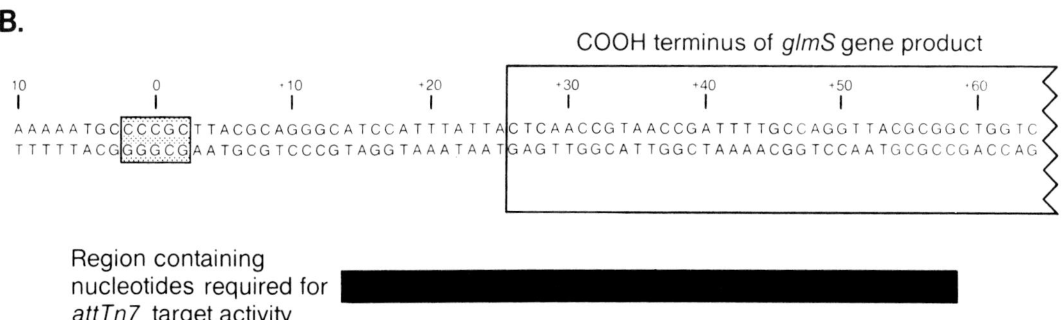

Figure 2. The *E. coli att*Tn*7* region. (A) Physical map of the *att*Tn*7* region. The upper open box is Tn*7* (not to scale) showing its orientation upon insertion into *att*Tn*7* (48). (Lichtenstein and Brenner [48, 49] and Gay et al. [28, 85] use a different nomenclature for the ends of Tn*7* and *att*Tn*7* than is used here.) The next line is the *att*Tn*7* region of the *E. coli* chromosome (28, 85) numbered as originally described by McKown et al. (50). The middle base pair of the 5-bp *E. coli* sequence usually duplicated upon Tn*7* insertion (28, 49, 50) is designated 0, sequences toward *phoS* (leftward) are designated −, and sequences toward *glmS* (rightward) are designated +. The shaded box is the 5-bp sequence usually duplicated upon Tn*7* insertion: *att*Tn*7*(−2 to +2) (28, 49, 50). The next line shows a potential secondary structure at the 3′ end of *glmS* mRNA which provides a transcription terminator (28, 33) (shaded box corresponds to Tn*7*-mediated duplication). (B) Nucleotide sequence of the *att*Tn*7* region (28, 49, 50). The solid bar indicates the region containing the nucleotides essential for *att*Tn*7* target activity, i.e., for high-frequency orientation-specific insertion at the specific insertion point (shaded box). Qadri et al. (personal communication) have shown that essential *att*Tn*7* sequences lie between *att*Tn*7*(+14) and *att*Tn*7*(+58) and that at least some essential sequences are in the interval *att*Tn*7*(+52 to +58). McKown et al. (50) and Gringauz et al. (33) have shown that essential *att*Tn*7* sequences lie between *att*Tn*7*(+14) and *att*Tn*7*(+64) and that some required nucleotides are in the interval *att*Tn*7*(+37 to +64).

terminus of the right end of Tn*7* (Tn7R) (Fig. 1). The proteins encoded by the *tns* genes have been identified by methods such as the analysis of radioactively labeled polypeptides in maxicells (12, 68, 78, 84; M. Rogers and D. Sherratt, personal communication; Orle and Craig, unpublished observations; C. Waddell and N. L. Craig, unpublished observations) (Table 1). In contrast to some other transposons whose transposition proteins are preferentially *cis*-acting (9, 18, 32, 44, 73), the Tns proteins work efficiently in *trans* (75).

The work of Hauer and Shapiro (39) revealed that not all Tn*7* transposition events require the same Tn*7* transposition genes. It is now known that the *tns* genes mediate two distinct but overlapping transposition pathways. *tnsABC* are essential but not sufficient for all Tn*7* transposition events; the presence of either *tnsD* or *tnsE* is additionally required. *tnsABC*+*D* promote high-frequency transposition to *att*Tn*7* (68, 84) and low-frequency transposition to pseudo-*att*Tn*7* sites (K. Kubo and N. L. Craig, unpublished observations), whereas *tnsABC*+*E* promote low-frequency transposition to sites unrelated to *att*Tn*7* (68, 84; Kubo and Craig, unpublished observations). Thus, the large number of *tns* genes is explained in part by the fact that they provide two alternative transposition pathways. The *tns* gene requirements for transposition are defined by the target

Table 1. Molecular weights of the *tns* gene products

Gene	Mol wt translated from DNA sequence	References	Mol wt observed	References
tnsA	31,000	Orle and Craig, unpublished observations; Flores et al., personal communication	30,000	68; Orle and Craig, unpublished observations; Rogers and Sherratt, personal communication
tnsB	78,000	Flores et al., personal communication	83,000–85,000	12, 68, 84; Rogers and Sherratt, personal communication; Waddell and Craig, unpublished observations
tnsC	63,000	Flores et al., personal communication	54,000–56,000	12, 84; Waddell and Craig, unpublished observations
			40,000–42,000	68; Rogers and Sherratt, personal communication
tnsD	59,000	Flores et al., personal communication	54,000	68; Rogers and Sherratt, personal communication
			40,000	12, 84; Waddell and Craig, unpublished observations
tnsE	61,000	78	85,000	12
			70,000–75,000	12, 68, 78, 84; Rogers and Sherratt, personal communication; Waddell and Craig, unpublished observations

of Tn7 transposition and are independent of the site from which Tn7 transposes (84).

What are the functions of the Tns proteins in transposition? TnsA, TnsB, and TnsC provide activities essential to all Tn7 transposition events. Biochemical studies have identified a *tnsB*-dependent specific DNA-binding activity that recognizes sequences in Tn7L and Tn7R which participate directly in transposition (51). Thus, TnsB is likely to be involved in the recognition and/or utilization of *cis*-acting transposition sequences at the ends of Tn7. Purification of the *tnsB*-dependent DNA-binding activity has shown that TnsB participates directly in specific DNA binding (R. McKown and N. L. Craig, unpublished observations). TnsB also likely has regulatory roles in *tns* gene expression (see below). The roles of TnsA and TnsC in transposition have not yet been established. Attractive models for the actions of these proteins include additional roles in the recognition or utilization (or both) of *cis*-acting transposition sequences at the ends of Tn7 or in mediating interactions between protein-DNA complexes at the Tn7 ends and the target sites. No *tns* gene encodes only a resolution function like those of the Tn3 family of transposons (9, 18, 32, 44, 73; D. Sherratt, this volume; see below).

Genetic analyses suggest that TnsD and TnsE are the proteins which choose the Tn7 target sites, TnsD directing transposition to *att*Tn7 and pseudo-*att*Tn7 sites (68, 84; Kubo and Craig, unpublished observations) and TnsE directing transposition to sites unrelated to *att*Tn7 (68, 84; Kubo and Craig, unpublished observations). It is attractive to suggest that TnsD and TnsE are themselves DNA-binding proteins which interact directly with their cognate DNA target sites.

2. Host-encoded genes

It has been shown that some other transposable elements employ host-encoded proteins in transposition (19, 56). Host-encoded proteins that participate directly in Tn7 transposition have not yet been identified. One explanation for the surprising number of Tn7-encoded transposition genes is that one (or more) of the *tns* genes encodes a function(s) usually provided by the host to other transposition systems. It has been shown that integration host factor, an *E. coli*-encoded protein involved in at least one other transposition reaction (56) and in conservative site-specific recombination reactions (59), is not essential for Tn7 transposition (Kubo and Craig, unpublished observations; N. Ekaterinaki and D. Sherratt, personal communication).

C. Tn7 Target Sites

1. *tnsABC+D* target sites in *E. coli*

The target sites recognized by Tn7 have been most extensively characterized in *E. coli*. Tn7 inserts at high frequency in a specific site called *att*Tn7 at about 84 min of the chromosome (5, 7, 48). The high frequency of Tn7 transposition to *att*Tn7 was initially revealed by the observation that after incompatibility exclusion of plasmid RP4::Tn7 in the absence of selection for Tn7, a large (>50%) fraction of the cured bacteria contained Tn7 (5, 7). High-frequency transposition to *att*Tn7 has been also observed with other transposition assays (39, 48, 50, 68, 84). Transposition to *att*Tn7 requires *tnsABC*+*D* (68, 84).

The nucleotide sequence of the *att*Tn7 region has been determined and the point of Tn7 insertion identified (28, 49, 50, 85) (Fig. 2). The point of Tn7 insertion lies between two genes: *phoS*, which encodes a protein involved in phosphate transport, and *glmS*, which encodes a protein involved in cell wall biosynthesis. Tn7 insertion into this site is not obviously deleterious, as protein coding sequences are not disrupted. In *att*Tn7::Tn7, the end of Tn7 encoding the *tns* genes (Tn7R) is adjacent to *glmS*, and the other end of Tn7 (Tn7L) is adjacent to *phoS*; thus, Tn7 insertion into *att*Tn7 is orientation as well as site specific (48). (Note that Lichtenstein and Brenner [48, 49] and Gay et al. [28, 85] use a different designation for the ends of Tn7 than is used here.) Tn7 insertion is accompanied by duplication of a 5-bp chromosomal sequence and most frequently occurs at the position shown in Fig. 2. A small (several-base-pair) variation in the exact position of insertion has occasionally been observed (50); in these cases, Tn7 insertion is also orientation specific and is accompanied by a 5-bp duplication of chromosomal sequence.

In *att*Tn7, the Tn7 insertion point actually lies within the transcriptional terminator of the *glmS* gene (Fig. 2). Gay et al. (28) identified the 3′ ends of *glmS* mRNA, and Gringauz et al. (33) showed that these ends are generated by transcription termination rather than by RNA processing. In *att*Tn7::Tn7, *glmS* transcripts end within Tn7R and do not impinge on the *tns* genes (28; note that Gay et al. use a nomenclature for the Tn7 ends different from the one used in this review). Such protection from external transcription has been found with other transposons (21, 45). Although it is tempting to speculate that transcription could modulate the capacity of *att*Tn7 to act as a target for Tn7 insertion, there is no experimental support for this view. Indeed, the *att*Tn7 target activity of *att*Tn7 segments introduced into plasmids has been shown to be independent of transcription (33, 50).

It is notable that there is no obvious sequence homology between nucleotides surrounding the point of Tn7 insertion in *att*Tn7 (28, 49, 50) and nucleotides at the ends of Tn7 (31, 49). (Although Lichtenstein and Brenner [49] did note some similarity between the Tn7 ends and the insertion point, subsequent work has shown that the insertion point sequence they determined was incorrect [28, 50; C. Lichtenstein, personal communication].) Thus, structural similarities between the ends of the transposon and *att*Tn7 do not direct site-specific insertion of Tn7.

What nucleotides are required to provide *att*Tn7 target activity, i.e., to promote high-frequency site- and orientation-specific Tn7 insertion? Several groups have shown that relatively large segments (several hundred base pairs) containing the specific point of Tn7 insertion have *att*Tn7 target activity when introduced into plasmids (48, 50, 68, 84). Analysis of deletion variants of the *att*Tn7 region (33, 50; I. Qadri, C. Flores, and C. Lichtenstein, personal communication) and of point mutations near the point of insertion (33) have revealed that the nucleotides required for *att*Tn7 target activity do not actually include the specific point of Tn7 insertion and lie entirely to one side (toward *glmS*) of the insertion point (Fig. 2). Thus, the Tn7 transposition machinery resembles certain restriction enzymes (3, 35) whose specific DNA recognition sites are actually displaced from their site of action.

Interestingly, at least some information required for *att*Tn7 target activity is contained within the coding sequence of the *glmS* gene (33, 50; Qadri et al., personal communication) (Fig. 2). Thus, nucleotides in this region have a dual role: they provide protein coding information and recombination signals which direct Tn7 insertion. Why is such a specific and reactive site for Tn7 insertion contained in the *glmS* gene? It is not unreasonable to suggest that a specific insertion site is "advantageous" to Tn7, as it provides a safe haven into which insertion can occur without deleterious impact on the bacterial host. Although it is not clear why the *glmS* gene in particular is used, it is likely that this gene is highly conserved among bacteria because of its critical role in cell wall biosynthesis. Indeed, analysis of site-specific Tn7 insertion in other bacteria has shown that insertion occurs near *glmS*-like genes in these organisms (Qadri et al., personal communication; see below).

The thoughtful reader will note that because the nucleotides required for *att*Tn7 target activity do not include the point of insertion, Tn7 insertion does not physically disrupt the nucleotides required for *att*Tn7 target activity. However, insertion of a second Tn7 element into *att*Tn7::Tn7 has not been observed (39, 48; L. Arciszewska, K. Kubo, and N. L. Craig, unpublished observations). As described below, the ends of Tn7 in a target replicon inhibit subsequent Tn7 insertion into anywhere else within the replicon; i.e., Tn7 has transposition immunity. The inability of *att*Tn7::Tn7 to act as a target for subsequent Tn7 insertion may reflect such transposition immunity.

*att*Tn7 is not the only target site recognized by *tnsABC+D*. When *att*Tn7 is unavailable, *tnsABC+D* promote low-frequency (about 100-fold less than to *att*Tn7) transposition to other sites in the *E. coli* chromosome (Kubo and Craig, unpublished observations). Such *tnsD*-dependent insertions are not random; rather, most insertions occur in one of several preferred sites. DNA sequence analysis of these preferred sites has revealed that they are structurally related to *att*Tn7. In particular, the preferred *tnsD*-dependent sites are similar to *att*Tn7 rightward of the Tn7 insertion point, i.e., in the region of *att*Tn7 that contains the nucleotides essential for *att*Tn7 target activity. Thus, these *tnsD*-dependent sites are pseudo-*att*Tn7 sites. Insertion into pseudo-*att*Tn7 sites is accompanied by a 5-bp duplication of chromosomal sequence; the duplicated sequences are unrelated to those duplicated at *att*Tn7. Rogers et al. (68) have observed low-frequency *tnsD*-dependent transposition to the plasmid R388, but the nucleotide sequence of these target sites is not known. No *tnsD*-dependent transposition has been detected to the IncP plasmids RP4 (39, 64) and pUZ8 (77) or to derivatives of the F plasmid (84).

2. *tnsABC+E* target sites in *E. coli*

tnsABC+E promote low-frequency (about 100-fold less than to *att*Tn7) transposition to plasmids lacking an introduced *att*Tn7 segment (68, 84) and to the *E. coli* chromosome (Kubo and Craig, unpublished observations). *tnsABC+E* do not promote transposition to *att*Tn7 or to the identified pseudo-*att*Tn7 sites (84; Kubo and Craig, unpublished observation). In the *E. coli* chromosome, *tnsE*-dependent insertion occurs at many different sites, and no preferred *tnsE*-dependent sites have been identified (Kubo and Craig, unpublished observations). DNA sequence analysis of several chromosomal *tnsE*-dependent insertion sites has revealed that they are structurally unrelated to *att*Tn7 and to each other. A 5-bp duplication of chromosomal sequence accompanies *tnsE*-dependent insertion. The *tnsE*-dependent insertion sites also lack obvious homology to the ends of Tn7. The fact that *tnsD*- and *tnsE*-dependent target sites are not structurally related suggests that there is no common protein component to recognition of these sites; thus, target site recognition is not mediated, for example, by *tnsC+D* and *tnsC+E*. Analysis of *tnsE*-dependent transposition to the F plasmid (84) and to the plasmid R388 (68) has shown that insertion occurs at many different sites. Analysis of *tnsE*-dependent insertion specificity within relatively small segments, for example, within plasmids such as pBR322, has not yet been carried out.

Transposition to the plasmid RP4 requires *tnsE* (39, 64), suggesting that all target sites recognized by Tn7 in this plasmid are *tnsE*-dependent sites (although it has not been shown directly that *tnsD* is dispensable). A number of Tn7 insertions into RP4 have been generated and characterized (6, 8, 17, 46); some regional specificity in target site selection has been observed. DNA sequence analysis of three independently derived Tn7 RP4 insertions (60) has shown that these insertions occur very close to one another and are accompanied by a 5-bp duplication of plasmid sequences. These insertions were selected on the basis of the insertion phenotype, and thus the apparent insertion specificity may not directly reflect the Tn7 target site selection process. Alternatively, the observed insertion specificity may reflect some target site selectivity in *tnsE*-dependent transposition to this plasmid.

A curious property of Tn7 transposition to IncP1 plasmids such as RP4 (6, 8, 17, 46), the IncP10 plasmid R91-5 (55), and the IncM plasmid R831b (63) is that although insertion occurs at many different sites within the plasmid, virtually all insertions occur in the same orientation with respect to the plasmid backbone. In the RP4 case, these insertions are likely directed by the *tnsE* pathway (39, 64). The basis of this orientation specificity is unknown. Replication of RP4 is unidirectional (54), and it will be interesting to determine if this property of the target replicon directs the orientation of Tn7 insertion.

3. Target site utilization in the presence of *tnsD* and *tnsE*

Several studies have shown that when both *tnsD* and *tnsE* are present and *att*Tn7 is available, transposition occurs exclusively to *att*Tn7 (39, 48, 84; Kubo and Craig, unpublished observations). The much higher reactivity (at least 100-fold) of *att*Tn7 compared to pseudo-*att*Tn7 sites recognized by *tnsD* or non-*att*Tn7 sites recognized by *tnsE* likely precludes detection of insertion at these other sites in the absence of selection. When both *tnsD* and *tnsE* are present and *att*Tn7 is unavailable, transposition to both *tnsD*-dependent and *tnsE*-dependent sites in the *E. coli* chromosome is observed at roughly equal frequency (Kubo and Craig, unpublished observations). Most Tn7 insertions in plasmids appear to result from *tnsE*-dependent transposition (39, 64, 68, 77, 84). The apparent lack of *tnsD*-dependent sites in plasmids and their presence in the *E. coli* chromosome may indicate a paucity of sequences related to *att*Tn7 in plasmids, which are relatively small in size compared to the *E. coli* chromosome.

4. Tn7 target sites in other bacteria

Specific Tn7 insertion sites have been observed in the chromosomes of the following bacteria: *Agrobacterium tumefaciens* (43); *Caulobacter crescentus* (23); *Escherichia coli* (5, 7, 48); *Klebsiella pneumoniae* (Qadri et al., personal communication); *Pseudomonas aeruginosa* (16); *Pseudomonas fluorescens* (4); *Pseudomonas solanacearum* (11); *Rhizobium meliloti* (10); *Rhodopseudomonas capsulata* (88); *Salmonella typhimurium* (A. Mak and N. L. Craig, unpublished observations); *Serratia marcescens* (Qadri et al., personal communication); *Vibrio* spp. (80); and *Xanthomonas campestris* pv. *campestris* (83).

Qadri et al. (personal communication) have determined the nucleotide sequences of the specific Tn7 insertion sites in *K. pneumoniae* and *S. marcescens*. All occur downstream of an open reading frame highly homologous to the *E. coli glmS* gene. It is notable that there is no homology between the point of Tn7 insertion in these other bacteria and the point of Tn7 insertion in *E. coli att*Tn7. Thus, these specific Tn7 insertion sites in other bacteria are structurally related to *E. coli att*Tn7 only within the *glmS* coding sequences, a finding which supports the hypothesis that critical *att*Tn7 sequences lie within the *glmS* gene. In *S. marcesens*, as in *E. coli*, the point of Tn7 insertion is about 30 bp downstream of the *glmS* terminus; in *K. pneumoniae*, the Tn7 insertion point is about 40 bp downstream of this point. The basis of the variation in the exact position of Tn7 insertion among different bacteria is not known. These results indicate that site-specific Tn7 insertion in many bacteria reflects utilization of the *tnsABC*+*D* pathway and substantial conservation of the *glmS* gene. It is also possible, of course, that some specific Tn7 insertion sites in other organisms may be highly preferred *tnsE*-dependent sites (although the lack of homology among characterized *E. coli tnsE* target sites provides no support for this view).

D. The Ends of Tn7

1. DNA sequences that participate directly in transposition

The *cis*-acting Tn7 transposition sequences are located near the ends of Tn7 (4, 39, 50, 64, 68, 77, 84; L. Arciszewska, D. Drake, and N. L. Craig, *J. Mol. Biol.*, in press). Thus, Tn7 transposition does not involve a specialized internal resolution site distinct from its termini as do transposons of the Tn3 family (9, 18, 32, 44, 73; Sherratt, this volume).

As is true of many transposable elements, the extreme termini (about 30 bp) of Tn7 are highly related inverted repeats (31, 49) (Fig. 3). However, the presence of only these terminal inverted repeats in Tn7L (Arciszewska et al., in press) or Tn7R (64, 77; Arciszewska et al., in press) is not sufficient to promote transposition. Analysis of the transposition properties of derivatives containing various extents of the Tn7 ends (Arciszewska et al., in press) has revealed that considerable nucleotide sequence information from each end is required for transposition. In Tn7L about 150 bp is required, and in Tn7R about 75 bp is required (Fig. 3). No difference has been observed in *cis*-acting sequence requirements at the Tn7 termini for *tnsD*-dependent versus *tnsE*-dependent transposition. The essential *cis*-acting transposition sequences in Tn7R contain a GATC methylation site (Fig. 3). No alteration in Tn7 transposition has been observed in *dam* mutant hosts (Ekaterinaki and Sherratt, personal communication), suggesting that in contrast to some other transposons (66, 87), Tn7 transposition is not subject to regulation by DNA adenine methylation.

Because different extents of sequence information are required in Tn7L and Tn7R, the *cis*-acting transposition sequences at the ends of Tn7 are structurally asymmetric. Although asymmetric, the ends of Tn7 are related: each end contains several copies of a closely related 22-bp sequence (49) (Fig. 3). The transposition properties of derivatives containing various numbers of these 22-bp sequences suggest that they are important *cis*-acting transposition sequences (Arciszewska et al., in press). This view is supported by the observation that the 22-bp repeat is specifically recognized by the *tnsB*-dependent binding activity (51).

The ends of Tn7 are also functionally asymmetric. Tn7 end derivatives containing two Tn7L ends do not transpose, whereas those containing two Tn7R ends do transpose (Arciszewska et al., in press). Indeed, the transposition frequency of Tn7R-Tn7R derivatives is equivalent to that of Tn7L-Tn7R derivatives, and Tn7R-Tn7R derivatives insert site specifically into *att*Tn7. Thus, no information is uniquely provided by Tn7L to the transposition machinery. It has also been shown that transposition of a Tn7R-Tn7R derivative requires the same *tns* genes as does that of a Tn7L-Tn7R element (unpublished observations), indicating that no Tns protein acts exclusively within Tn7L.

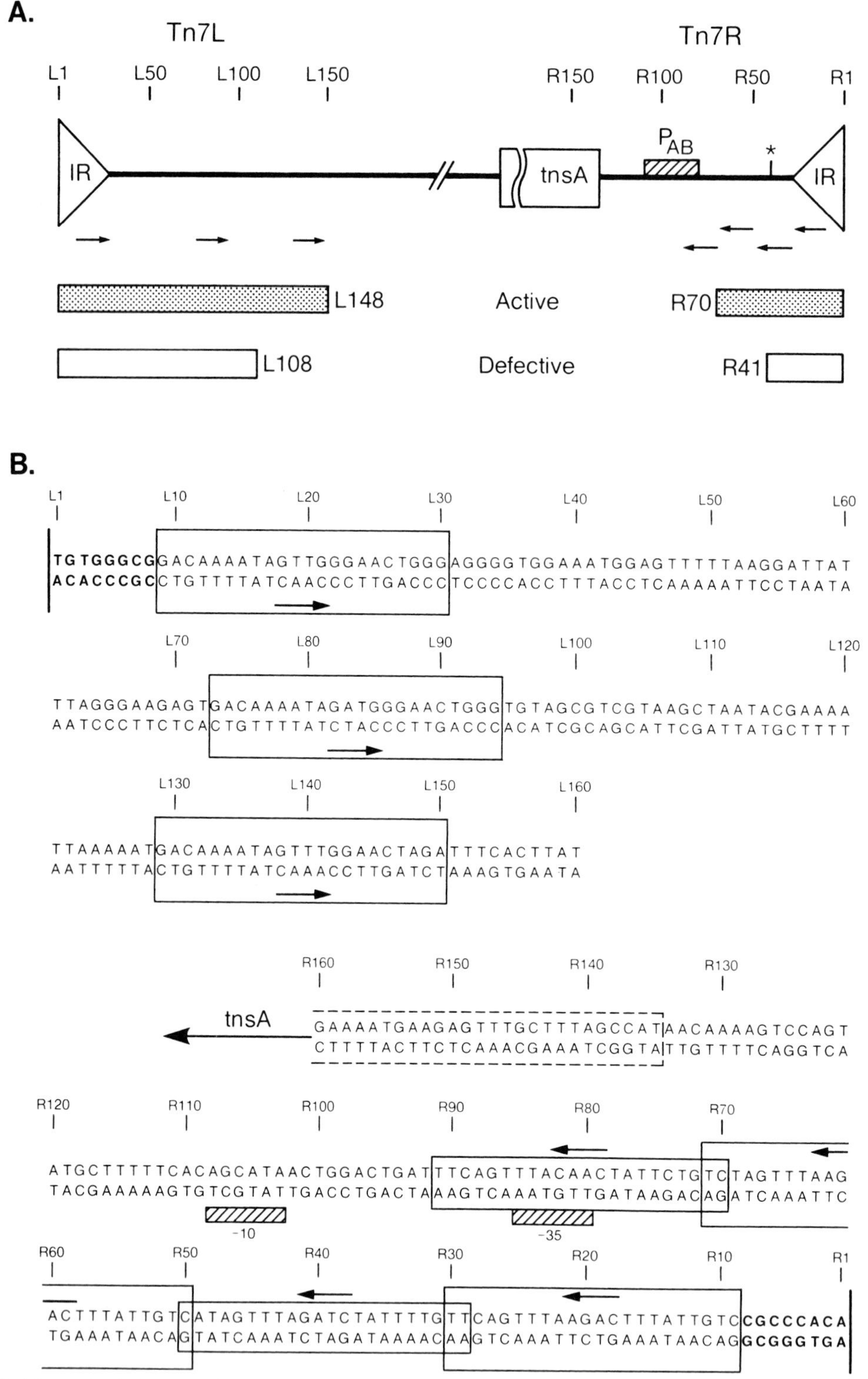

Figure 3. The ends of Tn*7*. (A) Physical map of the ends of Tn*7*. The left and right ends of Tn*7* are shown extending from the termini (designated L1 and R1) inward. The open triangles (IR) are the terminal 30-bp inverted repeats of Tn*7* (31, 49). The amino terminus of *tnsA* and the region of the proposed *tnsAB* promoter (p_{AB}) (28; also see text) are shown. Asterisk indicates the position of a GATC methylation site in Tn*7*R. The arrows indicate the 22-bp repeats (49) thought to be binding sites for the *tnsB*-dependent DNA-binding activity (51, 68). The bottom two lines show the transposition activities of certain Tn*7* end segments, shaded boxes indicating segments which are transposition competent and open boxes indicating segments which are transposition incompetent (Arciszewska et al., in press). (B) Nucleotide sequence of the ends of Tn*7* (31, 49), numbered within the terminal base pair of each end as 1 (L1 and R1) and with the numbers increasing toward the middle of the element. The base pairs at the element termini in boldface type are the segments of the terminal inverted repeats that are not part of the 22-bp repeat sequences, which are indicated with boxes (solid outline and containing arrows). The −10 and −35 regions of the proposed *tnsAB* promoter are shown (cross-hatched boxes). The proposed amino terminus of the *tnsA* gene is shown (box with dashed outline).

2. Tn*7* transposition immunity

Transposition immunity is the capacity of one copy of a transposon located in a target DNA molecule to reduce substantially the frequency of subsequent insertions of another copy of the transposon elsewhere in the target molecule (18, 32, 44, 67, 73; Sherratt, this volume). Hauer and Shapiro (39) were the first to show that Tn7 displays transposition immunity. Hassan and Brevet (38) had previously concluded that Tn7 does not display transposition immunity, but, as Hauer and Shapiro (39) have discussed, some evidence for transposition immunity is provided in the data of Hassan and Brevet (38). The apparent differences in these analyses may reflect the different conditions under which transposition was assayed. With Tn*3*, the degree of transposition immunity can apparently be influenced by the level of transposase (41). It should also be noted that multiple Tn7 insertions within some plasmids have been observed (6, 8, 55), another indication that the degree of Tn7 transposition immunity may be variable.

Arciszewska et al. (in press) have shown that Tn7 displays transposition immunity in transposition to F plasmid derivatives containing Tn7 elements and that immunity is active in both the *tnsABC+D* and *tnsABC+E* pathways. These workers have also shown that immunity is provided to a target replicon by the presence of Tn7 end derivatives containing only the terminal sequences required for efficient transposition, suggesting that the same *cis*-acting sequences may mediate both processes, as has been found with transposon Tn*3* (2, 47). No sequences within Tn7L are required to provide immunity (Arciszewska et al., in press); it remains to be determined if the presence of a single Tn7R end within a target plasmid is sufficient to provide Tn7 transposition immunity. As noted above, transposition immunity provided by the Tn7 ends may be the reason that the *att*Tn7 sequences in *att*Tn7::Tn7 do not provide an active target for subsequent Tn7 insertion.

Transposition immunity is an interesting process, as it represents a profound example of "action at a distance." For example, the presence of several hundred base pairs of Tn7 end sequences within F plasmid derivatives some 50 to 60 kilobase pairs in size substantially reduces (greater than 100-fold) the frequency of subsequent Tn7 insertion elsewhere within the target plasmid. How does transposition immunity operate in molecular terms? Work by Adzuma and Mizuuchi (1a) with bacteriophage Mu has revealed that binding of the transposase Mu A to Mu transposon ends within an immune target molecule may prevent the binding of the Mu B protein to the target DNA molecule which is also required for Mu transposition. Others (47, 74) have suggested that transposition immunity of Tn*3* may reflect a disruption of the transposition apparatus upon encounter with a transposon end within the immune target DNA molecule.

III. THE MECHANISM AND CONTROL OF Tn*7* TRANSPOSITION

A. The Mechanism of Tn*7* Transposition

The products of Tn7 transposition are simple insertions (75). The available evidence, i.e., (i) that plasmids containing two copies of Tn7 are stable (7, 8, 38, 39), (ii) that Tn7 derivatives containing only the ends of Tn7 form simple insertions (39; Arciszewska et al., in press), and (iii) that *tns* gene inactivation does not lead to cointegrate formation (39, 77, 84), suggests that Tn7 does not encode a specialized resolution system as do transposons of the Tn*3* family (18, 32, 44, 73; Sherratt, this volume). There is no information on whether Tn7 produces simple insertions by conservative transposition in which the transposon is cut out from the donor site and pasted into the target site or by replicative transposition in which a copy of Tn7 is generated and translocated (9, 18, 22, 32, 44, 73). Tn7 transposition is not obviously associated with precise (or nearly precise) transposon excision from the donor site (C. Waddell, A. Mak, and N. L. Craig, unpublished observations). Tn7 insertions are highly stable (23, 53; C. Waddell and N. L. Craig, unpublished observations; but see reference 65 for a possible exceptional case).

B. Control of Tn*7* Transposition

Certainly an interesting feature of Tn7 is its capacity for different frequencies of transposition: *tnsABC+D* promote high-frequency transposition to *att*Tn7, whereas *tnsABC+D* and *tnsABC+E* promote low-frequency transposition to sites other than *att*Tn7. There is little information on which protein components of the Tn7 transposition machinery determine transposition frequency. Few differences in transposition have been observed when various *tns* genes have been placed downstream of promoters of different levels of activity (68, 84) (although it should be noted that the levels of the Tns proteins have not been directly measured in these experiments). Several observations hint that the levels of TnsD and TnsE may influence transposition frequency. Waddell and Craig (84) observed a fivefold increase in transposition to *att*Tn7 with *tnsD* in a multicopy plasmid.

Similarly, Smith and Jones (77) observed that the presence of *tnsE* in a multicopy plasmid increased by about 30-fold the frequency of Tn7 transposition from a probable *tnsE* donor site to a *tnsE* target site (although this effect was not observed when Tn7 transposition from *att*Tn7 or from a *tnsE* site to *att*Tn7 was examined [84]).

Schurter and Holloway (72) have observed that Tn7 transposition in *P. aeruginosa* is stimulated by the presence of tandem copies of IS*21* and that IS*21*-mediated rearrangements are also stimulated by the presence of Tn7. The mechanistic basis of this apparently mutual stimulation, which specifically requires tandem IS*21*, is not known. The presence of tandem IS*21* likely increases the expression of an IS*21*-encoded transposition function (71). Perhaps there is some complementation between the *trans*-acting transposition functions encoded by these mobile elements. The relationship between Tn7 and IS*21* at the nucleotide sequence level has not yet been determined.

C. Control of *tns* Gene Expression

Our understanding of the regulatory circuitry of the *tns* genes is rudimentary. The *tns* genes are all oriented in a single direction, with the amino terminus of *tnsA* adjacent to Tn7R (Fig. 1). Gay et al. (28) have identified the 5′ end of the *tnsA* mRNA and thereby mapped the putative *tnsA* promoter to about 100 bp from the terminus of Tn7R (Fig. 3) (note that Gay et al. use a different nomenclature for the Tn7 ends than is used here). There are sequences in this region with considerable homology to the consensus *E. coli* promoter (40). Analysis of the effects of polar insertions in *tnsA* on *tnsB* activity have shown that *tnsA* and *tnsB* form an operon; thus, this promoter is actually a *tnsAB* promoter (84). The *tnsAB* promoter has also been characterized by transcriptional fusion (68). It is likely that expression of the *tnsAB* operon is regulated by *tnsB*. The proposed *tnsAB* promoter lies within sequences recognized by the *tnsB*-dependent binding activity (51). Therefore, TnsB could negatively regulate transcription from the *tnsAB* promoter by excluding RNA polymerase. In support of this hypothesis, Rogers et al. (68) have observed that expression of a transcriptional fusion to the proposed *tnsAB* promoter is decreased threefold in the presence of *tnsB*.

Gay et al. (28) have suggested that transcription from the *tnsAB* promoter may also be affected by transcription from outside Tn7R. These workers have shown that in *att*Tn7::Tn7, *glmS* transcripts actually "run over" the proposed *tnsAB* promoter, although the *glmS* transcripts do not enter the *tnsA* coding sequences. They have proposed that the activity of the *tnsAB* promoter may be inhibited by such external transcription through a "promoter occlusion" (1) mechanism. Gay et al. (28) have also observed different levels of *glmS* transcripts under different host cell growth conditions. If such external transcription from *glmS* does alter *tnsAB* transcription, this could provide a link between cellular metabolism and *tns* gene expression and thus possibly Tn7 transposition frequency.

As the *tns* genes are oriented in a single direction, it might be imagined that they are all contained within a single operon whose transcription is initiated at the *tnsAB* promoter. Some of the potential *tns* reading frames overlap slightly (Flores et al., personal communication; E. Nimmo and D. Sherratt, personal communication) (although it should be emphasized that translation initiation codons have not been directly established). Such overlap could provide translational coupling (61) between *tns* genes, a situation most compatible with the hypothesis that the *tns* genes are contained within a single operon (C. Lichtenstein and D. Sherratt, personal communication). However, the effects of polar insertions within *tnsB*, *tnsC*, and *tnsD* on Tn7 transposition are consistent with the hypothesis that *tnsC*, *tnsD*, and *tnsE* are contained within independent transcription units rather than, for example, a *tnsABCDE* operon (84). Promoters other than the *tnsAB* promoter have been detected within the *tns* region by transcriptional fusion studies (68, 78). A clear understanding of the *tns* gene regulatory circuitry awaits direct analysis of *tns* gene transcription and Tns protein levels. It would not be surprising to find that *tns* gene expression is elaborately controlled at multiple levels. Note that only very low levels of the Tns proteins have been detected in vivo (12; Waddell and Craig, unpublished observations; M. Rogers and D. Sherratt, personal communication).

D. An Inducing Signal for Tn7 Transposition?

Hauer and Shapiro (39) have suggested that Tn7 transposition to *att*Tn7 may be "induced." They and others (5, 7) have observed a very high frequency (approaching 100%) of Tn7 transposition to *att*Tn7 after incompatibility exclusion of RP4::Tn7, whereas relatively little accumulation of Tn7 in *att*Tn7 was observed during vegetative growth of RP4::Tn7-containing cells or when RP4::Tn7 was excluded by another means. Hauer and Shapiro (39) have proposed that the Tn7 transposition machinery may "sense" the state of the donor replicon and be "induced" to transpose when the donor replicon fails to replicate. Lichtenstein and Brenner (48) have also observed a very high-frequency transposition from ColE1::Tn7 to *att*Tn7 following incompatibility exclusion of this plasmid. However, because transposition during vegetative growth was not examined in these experiments, it is unclear whether transposition was actually "induced" by the incompatibility exclusion or occurred at high frequency during vegetative growth. If induction of Tn7 transposition does occur, it may be limited to only some donor replicons. Lower-frequency (albeit substantial, about 10%) transposition of Tn7 from F′*ts*::Tn7 to *att*Tn7 is observed after temperature exclusion of this replicon or after incompatibility exclusion of various F::Tn7 derivatives (Waddell and Craig, unpublished observations).

IV. TRICKS WITH Tn7

Transposons not only provide an interesting arena in which to study the mechanism and control of DNA rearrangements, they also provide useful genetic tools (9; C. M. Berg, D. E. Berg, and E. A. Groisman, this volume). A number of workers have used Tn7 transposition to plasmids to generate insertion mutants (6, 8, 17, 24, 42, 46, 52, 53, 55, 63, 69, 79, 88). With intact Tn7, insertions at many different positions in plasmids have been observed, thus providing useful genetic variants. Tn7 insertions are highly polar (33, 53) and thus are useful for transcriptional analyses as well as for gene disruption. It may now be a more useful stratagem to use Tn7 derivatives containing only the terminal *cis*-acting transposition sequences provided in *trans* with *tnsABC+E* to assure a low insertion specificity. Moreover, if the transposition genes could be removed after insertion, movement of the Tn7 insertions to a new location would be prevented.

The use of Tn7 transposition to bacterial chromosomes for insertional mutagenesis has not been so fruitful, as highly site-specific insertion is often observed (4, 5, 7, 10, 11, 16, 23, 43, 48, 80, 83, 88). Note that all such studies have used intact Tn7 or Tn7 derivatives containing *tnsD*, so that this site-specific chromosomal insertion probably reflects *tnsD*-dependent transposition to *att*Tn7-like sites. The use of Tn7 derivatives lacking *tnsD* would likely assure low-specificity insertion in chromosomes.

Although site-specific insertion of Tn7 may be disadvantageous for mutagenesis, site-specific insertion through Tn7 transposition provides an excellent means by which to easily and stably incorporate exogenous genetic material (and many variants of any particular segment) into a defined chromosomal site, thereby allowing detailed analysis of genes in single copy. Several workers (4, 34) have described bicomponent systems for use in a variety of bacteria in which a Tn7 derivative lacking the *tns* genes is complemented by *tns* genes within another replicon. A bicomponent system is under construction which provides in one plasmid conditional for replication a Tn7 derivative containing only the terminal *cis*-acting transposition sequences, a selectable drug resistance determinant, and a polylinker cloning site, together with another conditional plasmid which provides the *tns* genes (W. Tucker, L. Arciszewska, C. Waddell, N. L. Craig, and N. Gutterson, unpublished data). It is expected that such systems will find wide applicability, particularly in view of the broad host range and high-frequency transposition of Tn7.

V. Tn7-ENCODED DRUG RESISTANCE DETERMINANTS

Tn7 contains two antibiotic resistance determinants (7), trimethoprim resistance due to a type I dihydrofolate reductase (DHFR) encoded by the *dhfr* gene, and streptomycin and spectinomycin resistance due to an aminoglycoside-modifying enzyme encoded by the *aadA* gene (Fig. 1).

A. Trimethoprim Resistance

DHFR catalyzes the reduction of dihydrofolate to tetrahydrofolate in the presence of NADPH; tetrahydrofolate is required in a number of metabolic processes, including DNA synthesis. The DHFR of Tn7 is much less sensitive to inhibition by the folic acid analog trimethoprim than are chromosomal DHFRs (62), and thus bacteria containing Tn7 are resistant to high levels of this compound. Trimethoprim, often in combination with sulfonamides, is widely used in the treatment of urinary tract infections and in veterinary prophylaxis (14, 37). The rapid appearance of trimethoprim-resistant bacteria in response to trimethoprim use is substantially contributed to by both plasmid-borne and chromosomally integrated Tn7 (30).

The nucleotide sequence of the *dhfr* gene of Tn7 has been determined (26, 76), and the polypeptide deduced from the gene sequence has a molecular weight of 17,577, in good agreement with the observed molecular weight of the *dhfr* gene product (25). Tn7 DHFR shows considerable homology (about 30%) with various chromosomal bacterial DHRFs and eucaryotic DHFRs but no detectable homology with other bacterial plasmid-encoded type II DHRFs (26, 76). Tn7 DHFR has been purified and partially characterized (62).

B. Streptomycin and Spectinomycin Resistance

The *aadA* gene of Tn7 encodes a 3″(9)-O-nucleotidyltransferase which inactivates the aminoglycosides streptomycin and spectinomycin through adenylylation of the 3″-hydroxyl of the amino-hexose III ring of streptomycin and the 9-hydroxyl in the actinamine ring of spectinomycin. Thus, bacteria containing Tn7 are resistant to high levels of these antibiotics, which are potent protein synthesis inhibitors (13).

The sequence of the Tn7 *aadA* gene has been determined (26), and the polypeptide deduced from the gene sequence has a molecular weight of 29,207, in good agreement with the observed molecular weight of the *aadA* gene product (24). The Tn7 *aadA* gene product is nearly identical to that of the *aadA* genes of Tn*21*-like transposons and is related (about 35% homology) to the Tn*554 aadA* gene product (26).

C. Evolution of Drug Resistance Determinants

The observation that related transposable elements can contain different arrays of antibiotic resistance genes has led to the hypothesis that the insertion and deletion of DNA segments encoding antibiotic resistance genes may contribute to the evolution of transposable elements (see reference 70 for an example). DNA sequence analysis of the Tn7 *aadA* gene revealed that it is flanked by 59-bp directly repeated sequences, and it has been suggested that this repeat may reflect the insertion of the *aadA* segment into a Tn7 ancestor (26). This 59-bp sequence is also associated with likely "recombination hot spots" in Tn*21*-related transposons (15, 36, 57, 86). A DNA segment containing an open reading frame whose proposed protein product has some modest homology to recombinases which execute conservative site-specific recombination reactions has also been identified in some Tn*21*-like transposons and in Tn7 (upstream of the *dhfr* gene). It has been hypothesized that this proposed protein may be involved with the rearrangement of the antibiotic resistance segments in these transposons (36, 58). It has also been hypothesized that the proposed Tn7 "recombinase" protein may be involved in Tn7 transposition (36, 58). However, this proposed protein is unlikely to be involved in Tn7 transposition, as the *tns* gene region alone can promote Tn7 transposition in cells that lack DNA segments obviously related to the drug resistance determinants associated with the proposed "recombinase" (Waddell and Craig, unpublished observations).

VI. CONCLUDING REMARKS

Tn7 is an elaborate transposable element distinct from other bacterial transposable DNA elements such as insertion sequences and their composite transposons, the Tn*3* family, and bacteriophage Mu. Tn7 is quite distinct from Tn*554*, another mobile element capable of site-specific insertion (E. Murphy, this volume). There is no detectable homology between the *tns* gene products and proteins encoded by Tn*554* (Flores et al., personal communication).

The multitude of Tn7-encoded transposition genes is surprising. Their number is explained in part by the fact that they provide two transposition pathways differing in target site utilization. This capacity to use different classes of target sites is another distinctive feature of Tn7 transposition. The DNA sequences which participate directly in transposition at the ends of Tn7 and at its specific insertion site *att*Tn7 are also elaborate. A most interesting aspect of Tn7 transposition is the capacity of the ends of this element to recognize *att*Tn7 with high specificity although there is no sequence homology between the *cis*-acting recombination signals in the ends of Tn7 and *att*Tn7. It cannot be emphasized too strongly that although Tn7 and bacteriophage lambda both recognize specific "attachment" sites in the *E. coli* chromosome (*att*Tn7 for Tn7 and *attB* for phage lambda), these elements use different recombination pathways for insertion: transposition for Tn7 and conservative site-specific recombination for phage lambda.

Further analysis of Tn7 transposition and its control should provide interesting insights into the mechanism and control of DNA rearrangements and into the control of gene expression.

Acknowledgments. I thank David Sherratt and Conrad Lichtenstein for interesting conversations about Tn7 and for providing me with unpublished information from their laboratories. I also thank Candy Waddell and Susan Michaelis for their comments on the manuscript and Karyl Nakamura for her assistance in preparing the manuscript.

Work in my laboratory has been supported by a grant from the National Institutes of Health and a University of California, San Francisco, Pre-Tenure Award.

LITERATURE CITED

1. **Adhya, S., and M. Gottesman.** 1982. Promoter occlusion: transcription through a promoter may inhibit its activity. *Cell* 29:939–944.

1a. **Adzuma, K., and K. Mizuuchi.** 1988. Target immunity of Mu transposition reflects a differential distribution of Mu B protein. *Cell* **53**:257–266.

2. **Arthur, A., E. Nimmo, S. Hettle, and D. Sherratt.** 1984. Transposition and transposition immunity of transposon Tn3 derivatives having different ends. *EMBO J.* **3**:1723–1729.
3. **Bachi, B., J. Reiser, and V. Pirrotta.** 1979. Methylation and cleavage sequences of the *EcoP1* restriction-modification enzyme. *J. Mol. Biol.* **124**:143–163.
4. **Barry, G.** 1986. Permanent insertion of foreign genes into the chromosomes of soil bacteria. *Bio/Technology* **4**:446–449.
5. **Barth, P., and N. Datta.** 1977. Two naturally occurring transposons indistinguishable from Tn7. *J. Gen. Microbiol.* **102**:129–134.
6. **Barth, P., and N. Grinter.** 1977. Map of plasmid RP4 derived by insertion of transposon C. *J. Mol. Biol.* **113**:455–474.
7. **Barth, P. T., N. Datta, R. W. Hedges, and N. J. Grinter.** 1976. Transposition of a deoxyribonucleic acid sequence encoding trimethoprim and streptomycin resistances from R483 to other replicons. *J. Bacteriol.* **125**:800–810.
8. **Barth, P. T., N. J. Grinter, and D. E. Bradley.** 1978. Conjugal transfer system of plasmid RP4: analysis of transposon 7 insertion. *J. Bacteriol.* **133**:43–52.
9. **Berg, C., and D. Berg.** 1987. Uses of transposable elements and maps of known insertions, p. 1071–1109. *In* F. Neidhardt, J. Ingraham, K. Low, B. Magasanik, M. Schaechter, and H. Umbarger (ed.), *Escherichia coli and Salmonella typhimurium: Cellular and Molecular Biology*. American Society for Microbiology, Washington, D.C.
10. **Bolton, E., P. Glynn, and F. O'Hara.** 1984. Site specific transposition of Tn7 into a *Rhizobium meliloti* megaplasmid. *Mol. Gen. Genet.* **193**:153–157.
11. **Boucher, C., P. Barberis, A. Trigalet, and D. Demery.** 1985. Transposon mutagenesis of *Pseudomonas solanacearum*: isolation of Tn5-induced avirulent mutants. *J. Gen. Microbiol.* **131**:2449–2457.
12. **Brevet, J., F. Faure, and D. Borowski.** 1985. Tn7-encoded proteins. *Mol. Gen. Genet.* **201**:258–264.
13. **Bryan, L.** 1984. Aminoglycoside resistance, p. 241–477. *In* L. Bryan (ed.), *Antimicrobial Drug Resistance*. Academic Press, Inc., Orlando, Fla.
14. **Burchall, J., L. Elwell, and M. Fling.** 1982. Molecular mechanisms of resistance to trimethoprim. *Rev. Infect. Dis.* **4**: 246–254.
15. **Cameron, F., D. Groot Obbink, V. Ackerman, and R. Hall.** 1986. Nucleotide sequence of the AAD (2″) aminoglycoside adenylytransferase determinant *aadB*. Evolutionary relationship of this region with those surrounding *aadA* in R538-1 and *dhfr*II in R388. *Nucleic Acids Res.* **14**:8625–8635.
16. **Caruso, M., and J. Shapiro.** 1982. Interactions in Tn7 and temperate phage F116L of *Pseudomonas aeruginosa*. *Mol. Gen. Genet.* **188**:292–298.
17. **Cowan, P., and V. Krishnapillai.** 1982. Tn7 insertion mutations affecting the host range of the promiscuous IncP-1 plasmid R18. *Plasmid* **8**:164–174.
18. **Craig, N., and N. Kleckner.** 1987. Transposition and site-specific recombination, p. 1054–1070. *In* F. Neidhardt, J. Ingraham, K. Low, B. Magasanik, M. Schaechter, and H. Umbarger (ed.), *Escherichia coli and Salmonella typhimurium: Cellular and Molecular Biology*. American Society for Microbiology, Washington, D.C.
19. **Craigie, R., D. Arndt-Jovin, and K. Mizuuchi.** 1985. A defined system for the DNA strand-transfer reaction at the initiation of bacteriophage Mu transposition: protein and DNA substrate requirements. *Proc. Natl. Acad. Sci. USA* **82**: 7570–7574.
20. **Datta, N., V. Hughes, M. Nugent, and H. Richards.** 1979. Plasmids and transposons and their stability and mutability in bacteria isolated during an outbreak of hospital infection. *Plasmid* **2**:182–191.
21. **Davis, M., R. Simons, and N. Kleckner.** 1985. Tn10 protects itself at two levels from fortuitous activation by external expression. *Cell* **43**:379–387.
22. **Derbyshire, K., and N. Grindley.** 1986. Replicative and conservative transposition in bacteria. *Cell* **47**:325–327.
23. **Ely, B.** 1982. Transposition of Tn7 occurs at a single site on the *Caulobacter crescentus* chromosome. *J. Bacteriol.* **151**: 1056–1058.
24. **Fennewald, M. A., and J. A. Shapiro.** 1979. Transposition of Tn7 in *Pseudomonas aeruginosa* and isolation of *alk*::Tn7 mutations. *J. Bacteriol.* **139**:264–269.
25. **Fling, M., and L. Elwell.** 1980. Protein expression in *Escherichia coli* minicells containing recombinant plasmids specifying trimethoprim-resistant dihydrofolate reductases. *J. Bacteriol.* **141**:779–785.
26. **Fling, M., J. Kopf, and C. Richards.** 1985. Nucleotide sequence of the transposon Tn7 gene encoding an aminoglycoside-modifying enzyme, 3″(9)-O-nucleotidyltransferase. *Nucleic Acids Res.* **13**:7095–7106.
27. **Fling, M., and C. Richards.** 1983. The nucleotide sequence of the trimethoprim-resistant dihydrofolate reductase gene harbored by Tn7. *Nucleic Acids Res.* **11**:5147–5158.
28. **Gay, N., V. Tybulewicz, and J. Walker.** 1986. Insertion of transposon Tn7 into the *Escherichia coli glmS* transcriptional terminator. *Biochem. J.* **234**:111–117.
29. **Goldstein, F., G. Gerbaud, and P. Courvalin.** 1986. Transposable resistance to trimethoprim and O/129 in *Vibrio cholerae*. *Antimicrob. Agents Chemother.* **17**:559–569.
30. **Goldstein, F., B. Papadopoulou, and J. Acor.** 1986. The changing pattern of trimethoprim resistance in Paris, with a review of worldwide experience. *Rev. Infect. Dis.* **8**:725–737.
31. **Gosti-Testu, F., and J. Brevet.** 1982. Determination des sequences terminales du transposon Tn7. *C.R. Acad. Sci.* **294**:193–196.
32. **Grindley, N., and R. Reed.** 1985. Transpositional recombination in prokaryotes. *Annu. Rev. Biochem.* **54**:863–896.
33. **Gringauz, E., K. A. Orle, C. S. Waddell, and N. L. Craig.** 1988. Recognition of *Escherichia coli att*Tn7 by transposon Tn7: lack of specific sequence requirements at the point of Tn7 insertion. *J. Bacteriol.* **170**:2832–2840.
34. **Grinter, N. J.** 1983. A broad host-range cloning vector transposable to various replicons. *Gene* **21**:133–143.
35. **Hadi, S., B. Bachi, J. Shepard, R. Yuan, K. Inelchen, and J. Bickle.** 1979. DNA recognition and cleavage by *EcoP15* restriction endonuclease. *J. Mol. Biol.* **134**:655–666.
36. **Hall, R., and C. Vockler.** 1987. The region of the IncN plasmid R46 coding for resistance to beta-lactam antibiotics, streptomycin/spectinomycin and sulphonamides is closely related to antibiotic resistance segments found in IncW plasmids and in Tn21-like transposons. *Nucleic Acids Res.* **15**: 7491–7501.
37. **Hamilton-Miller, J.** 1984. Resistance to antibacterial agents acting on antifolate metabolism, p. 173–190. *In* L. Bryan (ed.), *Antimicrobial Drug Resistance*. Academic Press, Inc., Orlando, Fla.
38. **Hassan, D., and J. Brevet.** 1983. Absence of *cis*-acting transposition immunity with Tn7. *Plasmid* **10**:31–44.
39. **Hauer, B., and J. Shapiro.** 1984. Control of Tn7 transposition. *Mol. Gen. Genet.* **194**:149–158.
40. **Hawley, D., and W. McClure.** 1983. Compilation and anal-

ysis of *Escherichia coli* promoter DNA sequences. *Nucleic Acids Res.* **8**:2237–2255.

41. **Heritage, J., and P. Bennett.** 1984. The role of TnA transposase in transposition immunity. *Plasmid* **12**:218–221.

42. **Hernalsteens, J., H. De Greve, M. Van Montagu, and J. Schell.** 1978. Mutagenesis by insertion of the drug resistance transposon Tn7 applied to the Ti plasmid of *Agrobacterium tumefaciens*. *Plasmid* **1**:218–225.

43. **Hernalsteens, J., F. Van Vliet, M. De Beuckeleer, A. Picker, G. Engler, M. Lemmers, M. Holsters, M. Van Montagu, and J. Schell.** 1980. The *Agrobacterium tumefaciens* Ti plasmid as a host vector system for introducing foreign DNA into plant cells. *Nature* (London) **287**:654–656.

44. **Kleckner, N.** 1981. Transposable elements in prokaryotes. *Annu. Rev. Genet.* **15**:341–404.

45. **Krebs, M., and W. Reznikoff.** 1986. Transcriptional and translational initiation sites of IS50. Control of transposase and inhibitor expression. *J. Mol. Biol.* **192**:781–791.

46. **Krishnapillai, V., J. Nash, and E. Lanka.** 1984. Insertion mutations in the promiscuous IncP-1 plasmid R18 which affect its host range between *Pseudomonas* species. *Plasmid* **12**:170–180.

47. **Lee, C.-H., A. Bhagwhat, and F. Heffron.** 1983. Identification of a transposon Tn3 sequence required for transposition immunity. *Proc. Natl. Acad. Sci. USA* **80**:6765–6769.

48. **Lichtenstein, C., and S. Brenner.** 1981. Site-specific properties of Tn7 transposition into the *E. coli* chromosome. *Mol. Gen. Genet.* **183**:380–387.

49. **Lichtenstein, C., and S. Brenner.** 1982. Unique insertion site of Tn7 in the *E. coli* chromosome. *Nature* (London) **297**: 601–603.

50. **McKown, R. L., K. A. Orle, T. Chen, and N. L. Craig.** 1988. Sequence requirements of *Escherichia coli att*Tn7, a specific site of transposon Tn7 insertion. *J. Bacteriol.* **170**:352–358.

51. **McKown, R., C. Waddell, L. Arciszewska, and N. L. Craig.** 1987. Identification of a transposon Tn7-dependent DNA binding activity that recognizes the ends of Tn7. *Proc. Natl. Acad. Sci. USA* **84**:7807–7811.

52. **Merrick, M., M. Filser, R. Dixon, C. Elmerich, L. Sibold, and J. Houmard.** 1980. The use of translocatable genetic elements to construct a fine-structure map of the *Klebsiella pneumoniae* nitrogen fixation (*nif*) gene cluster. *J. Gen. Microbiol.* **117**: 509–520.

53. **Merrick, M., M. Filser, C. Kennedy, and R. Dixon.** 1978. Polarity of mutations induced by insertion of transposons Tn5, Tn7 and Tn10 into the *nif* gene cluster of *Klebsiella pneumoniae*. *Mol. Gen. Genet.* **165**:103–111.

54. **Meyers, R., and D. Helinski.** 1977. Unidirectional replication of the P-group plasmid RK2. *Biochim. Biophys. Acta* **47**: 109–113.

55. **Moore, R. J., and V. Krishnapillai.** 1982. Tn7 and Tn*501* insertions into *Pseudomonas aeruginosa* plasmid R91-5: mapping of two transfer regions. *J. Bacteriol.* **149**:276–283.

56. **Morisato, D., and N. Kleckner.** 1987. Tn10 transposition and circle formation *in vitro*. *Cell* **51**:101–111.

57. **Ouellette, M., L. Bissonnette, and P. Roy.** 1987. Precise insertion of antibiotic resistance determinants into Tn21-like transposons: nucleotide sequence of the OXA-1 beta-lactamase gene. *Proc. Natl. Acad. Sci. USA* **84**:7378–7382.

58. **Ouellette, M., and P. Roy.** 1987. Homology of ORFs from Tn2603 and from R46 to site-specific recombinases. *Nucleic Acids Res.* **15**:10055.

59. **Nash, H., and C. Robertson.** 1981. Purification and properties of the *Escherichia coli* protein factor required for lambda integrative recombination. *J. Biol. Chem.* **256**:9246–9253.

60. **Nash, J., and V. Krishnapillai.** 1987. DNA sequence analysis of host range mutants of the promiscuous IncP-1 plasmids R18 and R68 with Tn7 insertions in *oriV*. *Plasmid* **18**:35–45.

61. **Normark, S., S. Bergstrom, T. Edlund, T. Grundstrom, B. Jaurin, F. Lindberg, and O. Olsson.** 1983. Overlapping genes. *Annu. Rev. Genet.* **17**:499–525.

62. **Novak, P., D. Stone, and J. Burchall.** 1983. R plasmid dihydrofolate reductase with a dimeric subunit structure. *J. Biol. Chem.* **258**:10956–10959.

63. **Ogawa, H., C. Tolle, and A. Summers.** 1984. Physical and genetic map of the organomercury resistance (*Omr*) and inorganic mercury resistance (*Hgr*) loci of the IncM plasmid R831b. *Gene* **32**:311–320.

64. **Ouartsi, A., D. Borowski, and J. Brevet.** 1985. Genetic analysis of Tn7 transposition. *Mol. Gen. Genet.* **198**: 221–227.

65. **Owen, D., and A. Ward.** 1985. Transfer of transposable drug-resistance elements Tn5, Tn7 and Tn76 to *Azotobacter beijerinki*: use of plasmid RP4::Tn76 as a suicide vector. *Plasmid* **14**:162–166.

66. **Roberts, D., B. Hoopes, W. McClure, and N. Kleckner.** 1985. IS10 transposition is regulated by DNA adenine methylation. *Cell* **43**:117–130.

67. **Robinson, M., P. Bennett, and M. Richmond.** 1977. Inhibition of TnA translocation by TnA. *J. Bacteriol.* **128**:407–414.

68. **Rogers, M., N. Ekaterinaki, E. Nimmo, and D. Sherratt.** 1986. Analysis of Tn7 transposition. *Mol. Gen. Genet.* **205**: 550–556.

69. **Ronson, C., P. Astwood, and J. Downie.** 1984. Molecular cloning and genetic organization of C4-dicarboxylate transport genes from *Rhizobium leguminosarum*. *J. Bacteriol.* **160**:903–909.

70. **Schmidt, F.** 1984. The role of insertions, deletions, and substitutions in the evolution of R6 related plasmids encoding aminoglycoside transferase ANT-(2″). *Mol. Gen. Genet.* **194**: 248–259.

71. **Schurter, W., and B. Holloway.** 1986. Genetic analysis of promoters on the insertion sequence IS21 of plasmid R68.45. *Plasmid* **15**:8–15.

72. **Schurter, W., and B. Holloway.** 1987. Interactions between the transposable element IS21 on R68.45 and Tn7 in *Pseudomonas aeruginosa* PAO. *Plasmid* **17**:61–64.

73. **Shapiro, J. (ed.).** 1983. *Mobile Genetic Elements*. Academic Press, Inc., Orlando, Fla.

74. **Sherratt, D., A. Arthur, R. Bishop, P. Dyson, P. Kitts, and L. Symington.** 1983. Genetic transposition in bacteria, p. 59–74. *In* K. Chater, C. Cullis, D. Hopwood, A. Johnson, and H. Woolhouse (ed.), *Genetic Re-arrangement. Fifth John Innes Symposium*. Croom Helm, London.

75. **Sherratt, D., A. Arthur, and M. Burke.** 1981. Transposon-specified site specific recombination systems. *Cold Spring Harbor Symp. Quant. Biol.* **45**:275–281.

76. **Simonsen, C., E. Chen, and A. Levionson.** 1983. Identification of the type I trimethoprim-resistant dihydrofolate reductase specified by the *Escherichia coli* R-plasmid 483: comparison with procaryotic and eucaryotic dihydrofolate reductases. *J. Bacteriol.* **155**:1001–1008.

77. **Smith, G., and P. Jones.** 1984. Effects of deletions in transposon Tn7 on its frequency of transposition. *J. Bacteriol.* **157**: 962–964.

78. **Smith, G., and P. Jones.** 1986. Tn7 transposition: a multigene process. Identification of a regulatory gene product. *Nucleic Acids Res.* **14**:7915–7927.

79. **Taylor, D. E.** 1983. Transfer-defective and tetracycline-

sensitive mutants of the incompatibility group HI plasmid R27 generated by insertion of transposon 7. *Plasmid* **9**:227–239.

80. **Thomson, J. A., M. Hendson, and R. M. Magnes.** 1981. Mutagenesis by insertion of drug resistance transposon Tn7 into a Vibrio species. *J. Bacteriol.* **148**:374–378.

81. **Tietze, E., R. Prager, and H. Tschape.** 1982. Characterization of the transposons Tn1822(Tc) and Tn1824(TpSm) and the light they throw on the natural spread of resistance genes. *Plasmid* **8**:253–260.

82. **Tschape, H., E. Tietze, R. Prager, W. Voigt, E. Wolter, and G. Seltman.** 1984. Plasmid-borne streptothricin resistance in gram-negative bacteria. *Plasmid* **12**:189–196.

83. **Turner, P., C. Barber, and M. Daniels.** 1984. Behaviour of the transposons Tn5 and Tn7 in *Xanthomonas campestris* pv. *campestris*. *Mol. Gen. Genet.* **195**:101–107.

84. **Waddell, C., and N. L. Craig.** 1988. Tn7 transposition: two transposition pathways directed by 5 Tn7-encoded genes. *Genes Dev.* **2**:137–149.

85. **Walker, J., N. Gay, M. Saraste, and A. Eberle.** 1984. DNA sequence around the *Escherichia coli unc* operon. *Biochem. J.* **224**:779–815.

86. **Wiedemann, B., J. Meyer, and M. Zuhlsdorf.** 1986. Insertions of resistance genes into Tn21-like transposons. *J. Antimicrob. Chemother.* **18**(Suppl. C):85–92.

87. **Yin, J., M. Krebs, and W. Reznikoff.** 1988. Effect of *dam* methylation on Tn5 transposition. *J. Mol. Biol.* **199**:35–45.

88. **Youvan, D., J. Elder, D. Sandlin, K. Zsebo, P. Alder, N. Panapoulos, B. Marrs, and J. Hearst.** 1982. R-prime site-directed transposon Tn7 mutagenesis of the photosynthetic apparatus in *Rhodopseudomonas capsulata*. *J. Mol. Biol.* **162**:17–41.

Chapter 8

Transposon Tn*10*

NANCY KLECKNER

Nancy Kleckner ■ Department of Biochemistry and Molecular Biology, Harvard University, 7 Divinity Avenue, Cambridge, Massachusetts 02138.

I. INTRODUCTION AND BIOLOGY

A. Structure

Transposon Tn*10* is about 9,300 base pairs (bp) in length (Fig. 1). The ends of the transposon are inverted repeats of the 1,329-bp insertion sequence IS*10* (for insertion sequence *10*)-Right and IS*10*-Left. Between the two IS*10* elements is about 6,700 bp of nonrepeated material, a portion of which specifies resistance to the antibiotic tetracycline.

IS*10*-Right is a fully functional insertion sequence that can transpose as an individual unit. IS*10*-Right specifies a single polypeptide, the "transposase" protein, which acts in conjunction with host proteins to promote Tn*10* and IS*10* transposition and several other types of rearrangements that are mechanistically similar to transposition. The two ends of each IS*10* element are referred to as "outside" and "inside" with respect to their locations in Tn*10*. IS*10*-Left cannot transpose independently; although the two ends of IS*10*-Left are structurally intact, this element specifies a nonfunctional transposase protein (50, 62; R. W. Simons and N. Kleckner, unpublished observations). When IS*10*-Right is being considered as an individual insertion sequence comparable to IS*1*, *2*, *3*, etc., the term IS*10* is used without any additional designation.

Tn*10* confers high-level, inducible resistance to tetracycline in *Escherichia coli* and *Salmonella typhimurium*; the mechanism of resistance involves active efflux of the antibiotic (5, 74, 80, 113). Tetracycline resistance is specified by a 1,900-bp region containing two genes, *tetA* and *tetR*. The *tetA* gene encodes an inner membrane protein of molecular weight 43,000 (apparent molecular weight, 36,000) that is directly responsible for resistance (32, 66, 98, 123). The *tetA* protein appears to consist of two functional domains as revealed by extensive intracistronic complementation among *tetA* mutants (32, 37, 38). The *tetR* gene encodes a repressor protein that negatively regulates transcription of both the *tetA* and *tetR* genes, which are expressed from divergent, overlapping promoters. Tetracycline induces expression from both promoters by binding to the repressor and reducing its affinity for the appropriate operators. For recent work on the regulation of tetracycline resistance and references to earlier work, see references 39, 67, 70, and 162.

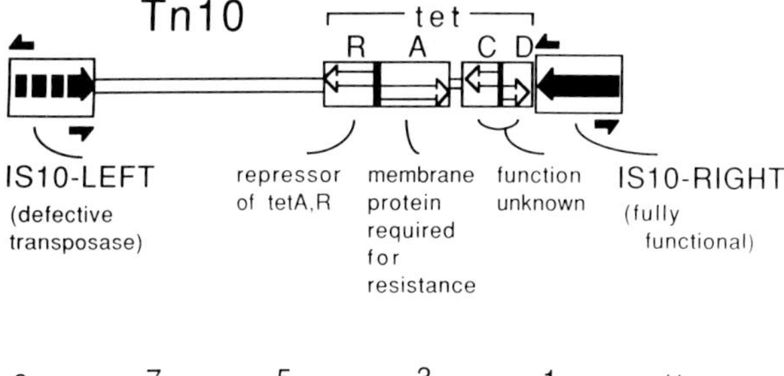

Figure 1. Transposon Tn*10*. IS*10*-Left and IS*10*-Right flank the *tetR*, *tetA*, *tetC*, and *tetD* genes. The short arrows at the termini of the IS*10* elements indicate nearly perfect inverted repeat sequences. Open arrows within the *tet* genes indicate the direction of transcription; the solid arrow within IS*10*-Right indicates transcription of the transposase gene; the broken arrow within IS*10*-Left indicates transcription of the defective transposase gene of this element. The complete DNA sequence for the first 4,800 bp of Tn*10* is known (62, 66, 144); for the rest of the element, some restriction enzyme cleavage sites have been mapped (78, 91).

The *tet* determinant of Tn*10* is not widely used as a selective marker for cloning because strains carrying Tn*10* on a multicopy plasmid exhibit drastically reduced resistance to tetracycline, probably due to overproduction of the *tetA* protein (26, 33, 163; see also reference 119 and references cited therein). However, the *tetA*/*tetR* regulatory region has been used as the basis for a tetracycline-regulated expression vector in which foreign genes are expressed from the *tetA* promoter (41).

Tn*10* contains two additional structural genes, *tetC* and *tetD*, that are also expressed from divergent, overlapping promoters and that fully occupy the 960 bp between the *tetR*/*tetA* region and IS*10*-Right; in fact, the 3 bp of the *tetD* gene stop codon corresponds to base pairs 1328 to 1326 of IS*10*-Right (144). The *tetD* gene product is membrane associated and inducible by tetracycline; little is known about the *tetC* gene product (16). Neither *tetC* nor *tetD* protein is absolutely required for tetracycline resistance, at least when Tn*10* is on a multicopy cloning plasmid (16, 78). However, it is still possible that they play more subtle roles in tetracycline resistance when Tn*10* is present in single copy or under particular conditions in nature.

The remaining 2,700 bp of internal Tn*10* material, adjacent to IS*10*-Left, has not been analyzed. However, this region is presumed to encode some function(s) of importance in nature, since it appears to be preserved in the several Tn*10*-like tetracycline resistance elements examined thus far (76, 147, 156).

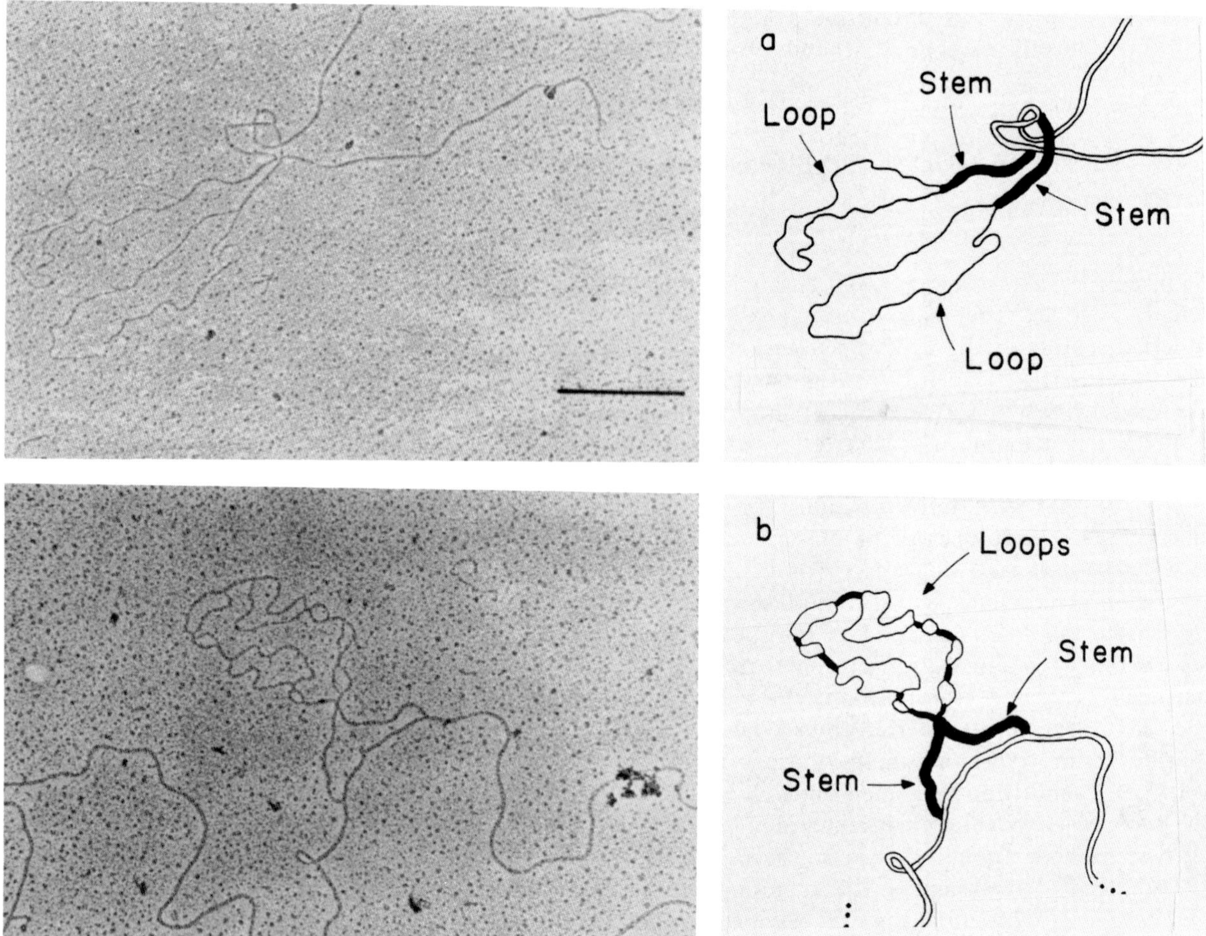

Figure 2. Heteroduplexes between P22Tc-10 and P22Tc-106, P22 derivatives carrying Tn*10* insertions at two different locations; in both phages, the transposon is inserted in the same orientation relative to the P22 genome. (a) A heteroduplex in which the two Tn*10* stem-loop structures are visible and the two loop regions have not interacted. (b) A heteroduplex in which the (complementary) loop regions have partially paired. Complete pairing is not possible because the two interacting DNA single strands are topologically constrained and cannot wind around one another completely. In the interpretive diagrams, the double-stranded P22 DNA is indicated by a pair of parallel thin lines, double-stranded stem segments by a very wide dark line, double-stranded loop segments by a dark line of medium width, and single-stranded loop segments by thin single lines. (Reprinted from reference 87 with permission from Academic Press, Inc.)

B. Discovery

Transposon Tn*10* was originally found on the drug resistance transfer plasmid 222, analyzed by Watanabe et al. (176). The same low-copy conjugative plasmid has also been called NR1 and R100; it is closely related to plasmid R6 (182b). The conjugation regions of these plasmids are closely related to that of the *E. coli* F factor (149). Early genetic analysis revealed translocations of the tetracycline resistance marker from these plasmids to phage P22, the *S. typhimurium* chromosome, and the *E. coli* chromosome (22, 45, 51, 177).

Recognition of Tn*10* as a discrete translocatable element (51, 87) came from physical and genetic analysis of P22 bacteriophages carrying Tn*10*, the prototype for which was P22Tc-10. Heteroduplex mapping of P22Tc-10 revealed a large insertion with an unusual structure, a double-stranded stem and a single-stranded loop (171), interpreted as an inverted duplication separated by nonrepeated sequences. A similar nontandem inverted repeat was also correlated with the tetracycline resistance determinant on plasmid R6 (147). Additional analysis showed that both P22Tc-10 and a second, independently arising P22 *tet* transducing phage carried insertions of identical structure at different positions, and neither phage had suffered any detectable deletion of phage material (87, 171, 177) (Fig. 2). Furthermore, in neither of the two specialized transducing phages was the insertion located adjacent to the phage attachment site as would have been the case had they been generated by aberrant excision of an integrated prophage. Finally, the length of the Tn*10* inverted repeat sequence was similar to that of the bacterial IS elements whose nature as insertion mutations had just been demonstrated, and a tetracycline-sensitive derivative of R6 carried an additional copy of the Tn*10* inverted repeat sequence within its "loop" region as determined by heteroduplex analysis (147). Taken together, these observations suggested that Tn*10* was a transposable element with a composite structure in which genes encoding tetracycline resistance are flanked by inverted repeats of an IS sequence which cooperate to effect transposition of the entire 9.3-kilobase (kb) segment.

Proof of Tn*10* transposition was provided by the demonstration that the Tn*10* element present on a nonreplicating, nonintegrating derivative of P22Tc-10 could transpose out of the phage genome and into the *Salmonella* chromosome (87). This transposition was detected as the stable acquisition of tetracycline resistance genes without concomitant acquisition of phage genes; approximately 1% of such tetracycline-resistant "transductants" contained auxotrophic insertion mutations. That these mutations were caused by Tn*10* insertion was shown genetically by the fact that the tetracycline resistance marker was inextricably linked to the mutation; reversion or generalized transduction to prototrophy resulted in loss of tetracycline resistance, and generalized transduction of tetracycline resistance from such an insertion mutant to a new strain was always accompanied by transduction of the associated auxotrophy.

Tn*10* was subsequently shown to be capable of undergoing multiple cycles of transposition and to retain its physical integrity during these transposition events. Several of the *S. typhimurium* auxotrophic insertions arising by transposition of Tn*10* from P22 were mobilized by conjugation into *E. coli*, and transpositions of the element from those sites into bacteriophage λ were isolated (87). Physical analysis of these λ phages and subsequent physical analysis of Tn*10* insertions into the *S. typhimurium* histidine operon revealed insertions of the same 9.3-kb element originally present in P22Tc-10 (50, 61, 86, 91; S. M. Halling and N. Kleckner, unpublished observations).

One exception to the rule that Tn*10* does not change from one cycle of transposition to another has been observed. The Tn*10* elements on P22::Tn*10* phages undergo transfer of information from IS*10*-Left to IS*10*-Right. The two IS*10* elements differ in DNA sequence at several positions (62). DNA sequence analysis of about 10 independent Tn*10* insertions from P22Tc-10 into the *S. typhimurium* histidine operon revealed that in approximately half, IS*10*-Left had acquired the base changes characteristic of IS*10*-Right, while IS*10*-Right remained unchanged (Halling and Kleckner, unpublished observations). Several lines of evidence, including Southern blot analysis of P22Tc-10 stocks, strongly suggest that this transfer of information is mediated by the extraordinarily active P22 recombination system. Analogous transfer of IS*10*-Left sequences to IS*10*-Right presumably also occurs, but the resulting transposons never give rise to *his*::Tn*10* insertion mutations because their IS*10* sequences are both IS*10*-Left-like and thus defective.

Tn*10* insertions isolated in *E. coli*, from λ or other delivery vehicles, all have the wild-type Tn*10* structure. Only in one special case has transfer of information from one IS*10* to another been observed: a screen for Tn*10* mutants exhibiting higher transposition frequencies yielded one mutant in which IS*10*-Right now contained two base changes from IS*10*-Left (M. A. Davis, Ph.D. thesis, Harvard University, Cambridge, Mass., 1986).

C. Composite Nature

The composite nature of Tn*10* is confirmed by three types of observations.

First, the inside ends of both IS*10*-Right and IS*10*-Left are active in promoting transposition-related events (24, 88, 136).

Second, new composite transposons that contain no Tn*10* sequences other than IS*10*-Right and IS*10*-Left have been isolated experimentally as genetic derivatives of Tn*10*. One such element consists of ampicillin or chloramphenicol resistance genes or both, flanked by inverted repeats of the two IS*10* sequences in the orientation opposite to that found in Tn*10*; it was isolated as an "inside-out" transposition of Tn*10* (50). A second such element consists of *E. coli gal* operon sequences flanked by direct repeats of the two IS*10* sequences; it arose from a Tn*10*-promoted deletion/inversion event (128). Both of these new elements transpose at frequencies comparable to that of Tn*10*. Also, a composite IS*10*-based transposon with internal material different from that of Tn*10* has been isolated from nature (see discussion of Tn*2921*, below).

Third, IS*10*-Right can transpose independently in the absence of any other Tn*10* sequences, and IS*10*-Left can transpose if IS*10*-Right is present to provide transposase. The first example of an IS*10* transposition was that correlated with mutation to tetracycline sensitivity of plasmid R6, discussed above (147). Interestingly, a similar inactivation of tetracycline resistance by IS*10* insertion occurred in an *E. coli* plasmid containing a Tn*10*-like element identified and followed in a human subject during the course of a clinical study (8). Transposition of an individual IS*10*-Right element in the absence of any other Tn*10* sequences has been demonstrated definitively using an IS*10* element marked with a kanamycin resistance determinant (132). Transpositions of individual IS*10* elements can also be inferred from Southern blot analysis of Tn*10*-containing strains (25, 128, 148).

D. Transposition Can Turn Genes Off or On

Insertion of Tn*10* or IS*10* into a structural gene abolishes the expression of that gene. In addition, insertion of either element can either turn off or turn on chromosomal genes, depending upon the exact situation.

In the case of Tn*10*, insertion between a structural gene and its normal promoter, in an operon or an upstream regulatory region, often abolishes gene expression. Many Tn*10* insertions in operons are strongly polar (for early observations, see references 49, 53, 87, and 90).

However, sometimes insertions upstream of a gene place expression of that gene under transposon control. Several cases have been reported in which Tn*10* insertions in operons or upstream of genes have only weak effects on expression of a distal gene(s) (13, 15, 30). In these cases, transcription from the upstream promoter is blocked by Tn*10* but is reinitiated within the element. These "nonpolar" insertions differ from the more frequent polar insertions not with respect to the nature of the Tn*10* element itself, but rather with respect to whether a rho-dependent transcription termination signal is present between the insertion and the downstream gene whose expression is assayed. In the polar cases, transcripts are initiated within the transposon but are then efficiently terminated at the intervening rho-dependent termination site(s); in the nonpolar cases, no such site is present and transcription from within Tn*10* continues unimpeded into the adjacent chromosomal gene(s) (29, 30, 174).

The Tn*10* promoter primarily responsible for transcription of adjacent genes is probably pOUT of IS*10* (see below), which directs transcription toward and across the ends of Tn*10* (149). The level of transcription from pOUT is probably 10 to 25% that of a p*lacUV5* promoter (30, 149).

There may be subtle differences in IS*10*-directed gene turn-on between *S. typhimurium* and *E. coli*. A *hisG*::Tn*10* insertion that turns on the adjacent *hisD* gene in *S. typhimurium* does not give *hisD* expression when transferred to and assayed in *E. coli* (J. Bender and N. Kleckner, unpublished observations); the nature of transcription termination or (less likely) the strength of pOUT may vary somewhat in the two organisms.

Tn*10* appears to be asymmetric with respect to its ability to turn on adjacent genes; it is more effective in turning on adjacent genes when inserted in one of the two possible orientations, the one in which pOUT of IS*10*-Right would be responsible for gene turn-on (13). Other analysis indicates that pOUT of IS*10*-Left is functional, but that sequence differences between IS*10*-Left and IS*10*-Right very near the ends of Tn*10* may influence termination or processing of the pOUT transcripts from IS*10*-Left (R. W. Simons, personal communication). Notable exceptions to this asymmetry are certain Tn*10* insertions which have IS*10*-Right information at both ends (above) and hence can turn on adjacent genes in either orientation; the insertion examined by Ciampi et al. (30) in which IS*10*-Left appears to have turned on *hisD* is of this type (Halling and Kleckner, unpublished observations).

Activation of adjacent genes due to readthrough transcription from other promoters within Tn*10* has not been reported for wild-type Tn*10*. However, mini-Tn*10* derivatives containing the *tetA*/*tetR* region flanked by the outer 70 bp of the Tn*10* ends can give tetracycline-dependent activation of adjacent genes, presumably because of transcription from the *tetA* or *tetR* promoters (181).

IS*10* insertions appear to be less polar, in both orientations, than Tn*10* insertions at the same (polar) site (148; J. Kuo, J. Bender, and N. Kleckner, unpublished observations). The reason for this difference is not known. Appropriately oriented IS*10* insertions can turn on adjacent genes with pOUT as expected given the structure of Tn*10*; IS*10* insertions in the opposite orientation are not effective (174).

E. Tn*10*-Like and Other IS*10*-Based Transposons

The tetracycline resistance determinant of Tn*10*, as identified by DNA hybridization, is widespread in enteric bacteria (109, 110, 114). It has been found in *E. coli* and *Klebsiella*, *Proteus*, *Pseudomonas*, *Salmonella*, *Shigella*, and *Vibrio* spp. and is the only plasmid-transferable tetracycline resistance identified among *Haemophilus* species. In all of the cases examined, this type of tetracycline resistance is present on a DNA segment that is similar or identical to Tn*10*; in some cases, the Tn*10* element appears to be inserted into a second transposon or else secondary and tertiary drug resistances are inserted into Tn*10* (8, 76, 79, 81, 147, 156). All of the Tn*10*-like elements in *Haemophilus* spp. confer constitutive tetracycline resistance; in one case, this is known to be due to a defect in *tetR* gene function (65). It is generally assumed that a particular type of tetracycline resistance became associated with flanking IS*10* sequences and has subsequently been spread by a combination of transposition from plasmid to plasmid plus conjugative transmission of plasmids from one bacterial population to another.

The Tn*10* tetracycline resistance determinant is distinct, by DNA hybridization criteria, from four other classes of tetracycline resistance determinants identified in enteric organisms and from other classes identified in nonenteric organisms (reviewed in reference 97). Among the four other classes, one is the determinant encoded by a different tetracycline resistance transposon, Tn*1721*, and a second one is represented by the multicopy plasmid pSC101 and its laboratory relative pBR322 (108, 114). Tn*10* and at least three of the other classes, including Tn*1721* and pSC101, appear to have evolved from a common ancestor; they all have similarly organized *tetA* and *tetR* genes that encode highly homologous and, in the case of *tetR*, functionally cross-reacting proteins (93, 123, 172, 173, 178).

Interestingly, a second IS*10*-based composite transposon has recently been discovered (120). This 11.8-kb element, Tn*2921*, carries fosfomycin resistance but not tetracycline resistance; it was found on a plasmid present in a clinical isolate of *Serratia marcescens*. In Tn*2921*, the IS*10* elements are oriented as direct repeats. As with Tn*10*, these IS*10* elements can mediate transposition of an intervening segment if placed in other orientations, demonstrating that both ends of both IS*10* elements are functional. Furthermore, none of the internal material in the element is required for transposition, and neither IS*10* element mediates cointegrate formation, as is true for Tn*10*. One of these IS*10* sequences has been completely sequenced and differs from IS*10*-Right by a single nucleotide (J. Navas, personal communication).

Unlike some other IS elements, IS*10* does not naturally occur in the chromosomes of *E. coli* K-12, *S. typhimurium* LT2 (135), or various fosfomycin-sensitive *S. marcescens* strains (120). However, there has been no extensive screen of natural, drug-sensitive strains for homology to IS*10*.

F. Other Transposase-Promoted Events and Their Consequences

IS*10* transposase, in concert with host proteins, can promote a number of different types of DNA rearrangements whose nature depends upon the number and orientations of available IS*10* ends.

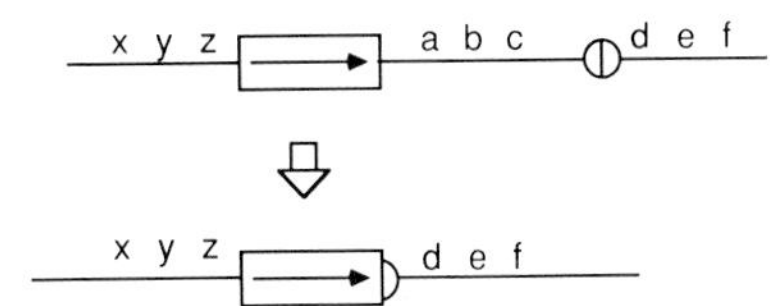

Figure 3. IS*10*-promoted adjacent deletion (rare).

Here, the term IS*10* promoted will be used to refer to events that can be promoted by a single IS*10* element alone, while Tn*10* promoted will be used to refer to events that require two IS*10* elements.

1. IS*10*-promoted events

For a chromosome containing a single IS*10* sequence, IS*10* transposition is the major event (25, 132, 148). A second type of event, an IS*10*-promoted adjacent deletion, has also been observed (D. Roberts, D. Ascherman, and N. Kleckner, manuscript in preparation) (Fig. 3). An IS*10* insertion in the *nadA* locus, adjacent to *gal*, gives rise to IS*10*-promoted *gal* mutant (galactose-resistant) derivatives with the appropriate physical structures. However, this event is quite rare, less than 1/1,000 times as frequent as IS*10* transposition. It is not excluded that IS*10* might promote certain other types of events at similarly low frequencies (see Section II).

2. Tn*10*-promoted events

When a chromosome contains two adjacent IS*10* sequences, additional types of rearrangements are observed (Fig. 4). Such events always involve one IS*10* end from each of the two elements. These events are described below as they occur with wild-type Tn*10* and are referred to as Tn*10*-promoted events; analogous events have been observed with composite IS*10*-based transposons having IS*10* sequences in direct or inside-out inverted orientation. At this level, the inside and outside ends of IS*10* are functionally equivalent.

Three Tn*10*-promoted rearrangements occur efficiently: an **intermolecular** event referred to as inside-out or inverse transposition (Fig. 4A), and two **intramolecular** events, deletion/inversions and (Tn*10*-promoted) adjacent deletions (Fig. 4B) (24, 48, 50, 64, 88, 128, 136). All three events probably occur by the same mechanism as does Tn*10* transposition and differ from transposition only in that they involve an interaction of a target site with the two inside ends of the IS*10* elements of Tn*10* rather than with the two outside ends of Tn*10*. The three events differ from one another simply with respect to the inter- or intramolecular isolation of the target site

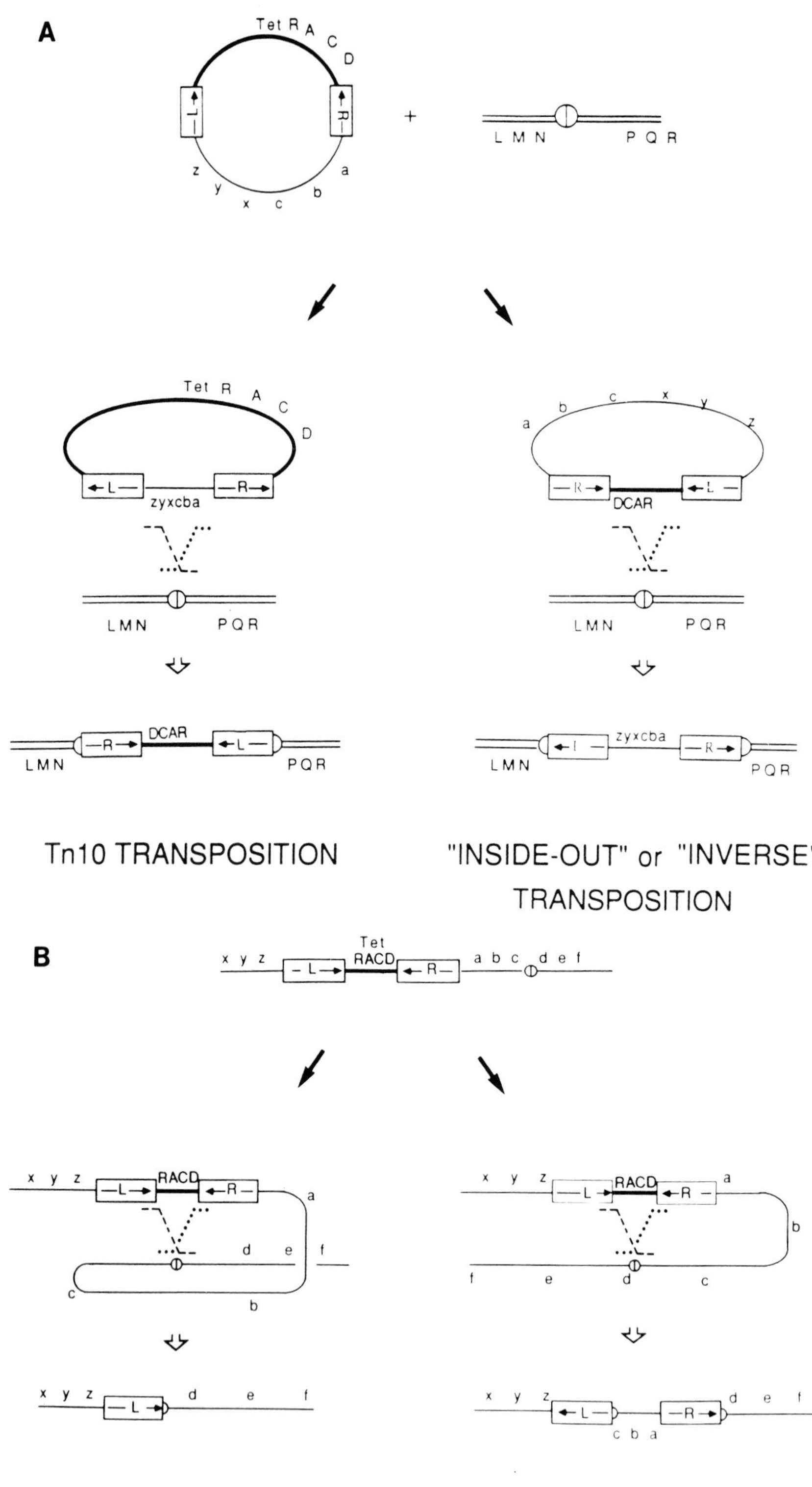

Figure 4. Tn*10*-promoted rearrangements. (A) Intermolecular events: normal Tn*10* transposition (left) and inside-out or inverse transposition (right). (B) Intramolecular events: Tn*10*-promoted adjacent deletion (left) and deletion/inversion (right); only the consequences of the event for the chromosome are drawn. Note that the scale changes for graphic convenience.

and, in the latter case, with respect to the orientation of the adjacent target site relative to the transposon.

Tn*10*-promoted adjacent deletions have the same structure as some of the IS*10*-promoted deletions that could arise from a Tn*10* element (compare Fig. 3 and 4B). That the deletions observed are Tn*10* promoted and not IS*10* promoted is indicated by their high frequency (136, 148).

3. Portable regions of homology

Finally, insertion sequences and transposons serve as "portable regions of homology," DNA segments which are substrates for the host homologous recombination system (28, 89, 90). Homologous recombination between two IS*10* elements at appropriate locations can generate adjacent deletions, cointegrates, or insertion inversions (Fig. 5). Experimentally, transposon-promoted events are distinguished from events that require homologous recombination using $RecA^-$ host strains in which the latter events do not occur.

An example of the ability of IS*10* to serve as a portable homology region has been described in bacteriophage λ (89). In this case, IS*10* transposition has created a genetic switch in which genes are turned on and off by homologous recombination between IS*10* elements. A λ phage carrying Tn*10* in the *rex* gene gave rise, by an IS*10*-promoted event, to a derivative containing a third IS*10* sequence located nearby, just downstream of the major phage promoter p_L. The orientation of the new insertion is such that homologous recombination between it and the nearest IS*10* sequence of the original Tn*10* element inverts the intervening p_L-containing segment, resulting in phenotypic alternation between states in which the nonessential genes distal to p_L are either expressed or not expressed.

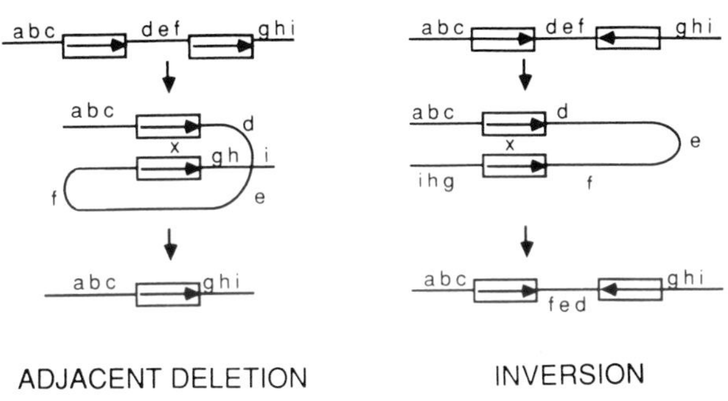

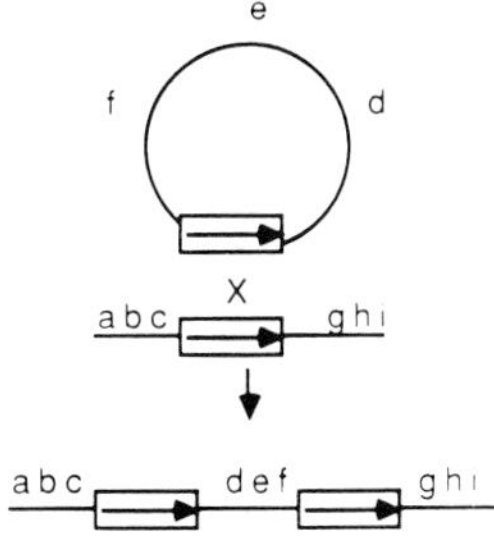

Figure 5. IS*10* as a portable region of homology. Homologous recombination between two IS*10* elements located as direct repeats on the same molecule (top, left), as inverted repeats on the same molecule (top, right), or on two different molecules (bottom) results in an adjacent deletion, an inversion, or a cointegration event, respectively.

4. Approximate rates

The rate of IS*10* transposition is approximately 10^{-4} per element per generation (148), whereas the rate of Tn*10* transposition is approximately 10^{-7} per element per cell generation (118). Tn*10* transposition is rarer primarily because the transposon is longer (118; see below). As discussed below, the rate of Tn*10* (and thus presumably IS*10*) transposition can also vary dramatically among insertions located at different sites. The rates cited above are typical of the majority of chromosomal insertions in the case of Tn*10* (Davis, Ph.D. thesis), and of the several different IS*10* elements examined (132; Roberts et al., unpublished data). The rate of inside-out transposition has not been accurately determined.

The rate of Tn*10*-promoted adjacent deletions and deletion/inversions is intermediate between that of IS*10* and Tn*10* transposition, approximately 10^{-5} per element per generation (88, 128, 148). The rates of these events are relatively constant from one Tn*10* insertion to another (88), presumably because these events involve inside ends which are always present in the same "context."

5. Consequences for genome organization and expression

All of the DNA rearrangements discussed in the sections above can have important biological consequences.

(i) Any DNA segment can be mobilized when it is bracketed by two copies of IS*10* in a composite transposon, regardless of the relative orientations of the two IS*10* elements (see above). New composite transposons can be generated by two types of events. When only a single IS*10* element is present, transposition of that element to an adjacent location without loss of the IS*10* at the original position will create a new transposon containing the segment between the two elements. When two copies of IS*10* are present, deletion/inversion events can generate new transposons containing a formerly adjacent segment (Fig. 4B).

(ii) Tn*10* and IS*10* can also mediate fusion between two replicons. Inside-out transposition will fuse a circular Tn*10*-containing molecule to a target molecule (Fig. 4A). Similarly, intermolecular IS*10* or Tn*10* transposition followed by homologous recombination can integrate a transposon-containing circle into a target molecule (Fig. 5). Such rearrangements can mediate the exchange of information among accessory DNA elements (plasmids and phages) and between such elements and the main bacterial chromosome.

(iii) Tn*10*-promoted deletion/inversions result in the deletion only of unique Tn*10* material and the inversion of one IS*10* sequence plus a segment of chromosomal material adjacent to that element. Transposon-promoted inversions probably make a significant contribution to the total frequency of inversions in bacteria, which are otherwise thought to be rare (130).

(iv) Tn*10*-promoted deletions, which by their nature both eliminate information and fuse previously distant chromosomal segments, occur much more frequently than do most spontaneous deletions not generated by transposons; IS*10*-promoted deletions are as rare as or rarer than spontaneous deletions.

All Tn*10*- and IS*10*-promoted rearrangements have the potential to alter gene expression because they can fuse structural genes to heterologous control sequences. However, an IS*10* sequence is present at every new fusion junction created in these rearrangements. Thus, subsequent excision of all or part of the intervening IS*10* element would usually be required to create a functional new expression unit.

G. Non-Transposase-Promoted Excision

Tn*10* and IS*10* insertions can give rise to a number of different excision events that are not promoted by transposase, and therefore seem to be unrelated to the transposition process itself (Fig. 6). One such event is "precise excision," in which the target DNA sequence is restored to its wild-type, pretransposon form. Precise excision is generally detected genetically as reversion of a Tn*10* or IS*10* insertion mutation (87). Since Tn*10* and IS*10* insertion involves the duplication of a 9-bp target site sequence (see below), precise excision involves the removal of all transposon sequences plus one copy of the 9-bp direct repeat sequence (52). The rate of Tn*10* precise excision from the bacterial chromosome varies over four or five orders of magnitude from one insertion site to another, but is typically about 10^{-9} (92). The frequency of IS*10* precise excision is about 20-fold lower than the frequency of Tn*10* precise excision from the same site (148), presumably due to the absence of extensive inverted repeats (see below).

Tn*10* also undergoes a process called "nearly precise excision" that occurs at a rate of about 10^{-6} per cell generation (52, 103, 137). Nearly precise excisions were originally identified as a distinct class of tetracycline-sensitive deletion derivatives of λ::Tn*10* phages which retained a specific small amount (60 bp) of non-λ material (137). These events correspond to a previously observed genetic class of tetracycline-sensitive derivatives of Tn*10* insertions in the *S. typhimurium hisG* gene; this class of derivatives exhibited relief of the strong polarity of the parental *hisG*::Tn*10* insertions on the downstream *hisD* gene plus an extremely high frequency of subsequent reversion to His^+ (52, 88). Nearly precise excision derivatives of *E. coli lacZ*::Tn*10* insertions have been identified by these same criteria (103).

Nearly precise excision is structurally similar to precise excision. Tn*10* contains a short inverted repeat centered 25 bp from each of its ends. Since the ends of Tn*10* are themselves inverted repeats, these short subterminal sequences are effectively direct repeats, and nearly precise excision is the deletion of all Tn*10* sequences between these direct repeat sequences. IS*10*-Right and IS*10*-Left differ in sequence at four positions in the center of the small inverted repeat sequence, and nearly precise excision derivatives can contain the central information from either end. The high frequency of nearly precise excision may reflect the fact that the directly repeated segments are 24 bp long rather than 9 bp long. All of the Tn*10* insertions examined give nearly precise excisions at about the same frequency (52, 103); perhaps this event is insensitive to chromosomal context, because only sequences internal to Tn*10* are involved.

Nearly precise excision derivatives go on to yield precise excision revertants at a site-dependent frequency that is always about 100 times higher than the frequency of Tn*10* precise excision from the same site (52).

In principle, Tn*10* precise excision might occur in two steps, by nearly precise excision followed by precise excision of the nearly precise excision remnant. However, the relative frequencies of the three events argue that precise excision usually occurs independently in a single step (52).

All three Tn*10*-associated excision events occur independent of the host homologous recombination

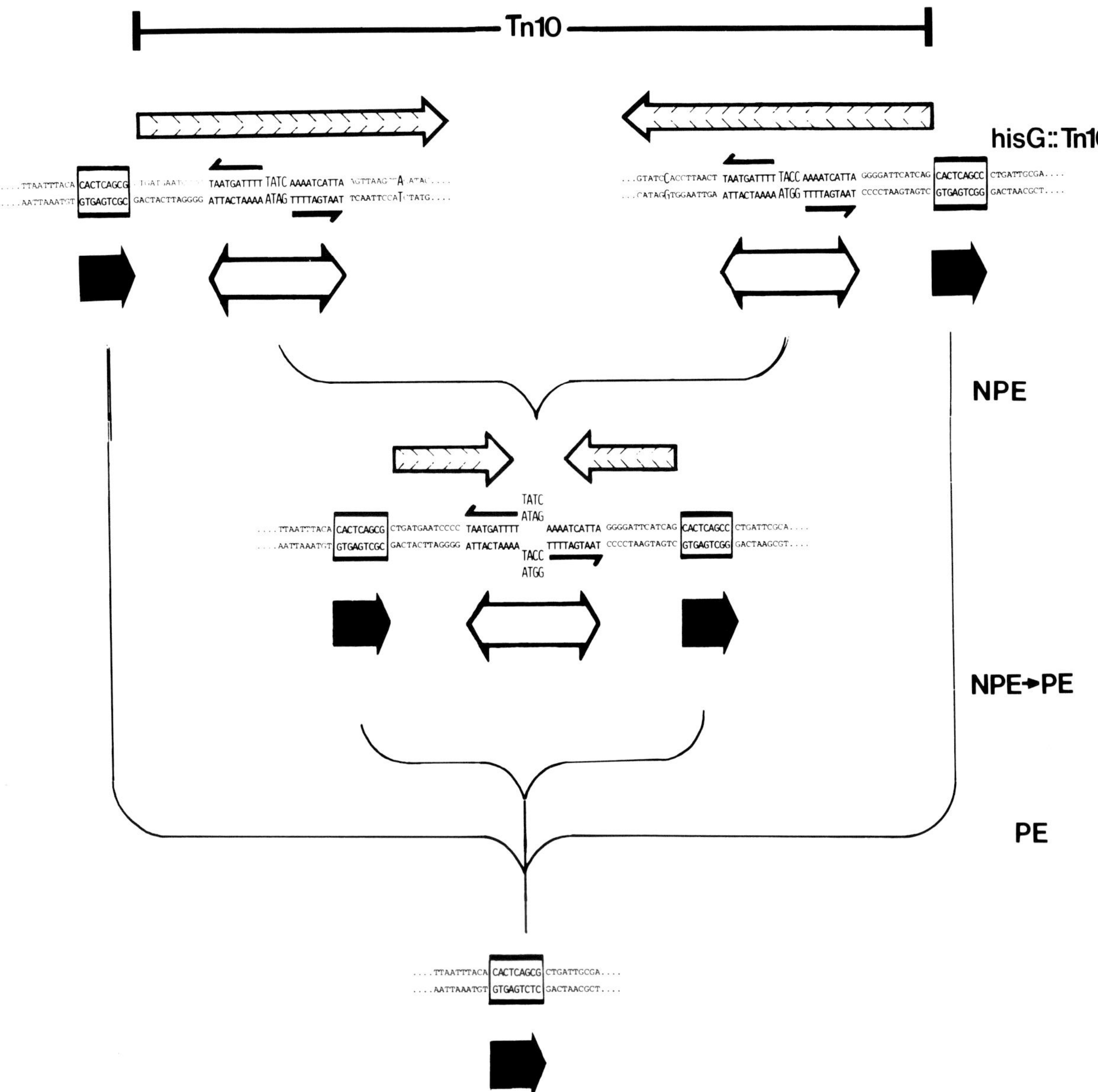

Figure 6. Three non-transposon-promoted Tn*10* excision events. Top line: DNA sequence at the two ends of a Tn*10* insertion in the *S. typhimurium hisG* gene. The transposon sequences are flanked by direct repeats of a 9-bp *hisG* gene sequence, and a short inverted repeat sequence occurs within Tn*10*, very close to each end. Middle line: Nearly precise excision (NPE) involves excision of all Tn*10* sequences internal to the short inverted repeats plus one copy of the inverted repeat sequence itself. Since the ends of Tn*10* are inverted repeats of IS*10*, the two internal inverted repeat sequences near each end can be thought of as direct repeats and nearly precise excision can be described as excision between these direct repeat sequences. Bottom line: Excision of all transposon sequences plus one copy of the 9-bp target site repeat sequence restores the wild-type *hisG* gene sequence. This event, known as precise excision, can occur from an intact Tn*10* element in a single step (PE) or from a nearly precise excision derivative (NPE to PE). In both cases this event can be described as excision between the two 9-bp repeat sequences. (Reprinted from reference 52 with permission from Cell Press.)

function RecA (52, 103). They also occur by a process that does not involve IS*10* transposase: deletion and point mutations in transposase do not reduce the rate of excision (52), and a mutation that increases transposase expression 100-fold does not increase the frequency of excision (unpublished observations).

Several lines of argument support the idea that precise and nearly precise excision involves intermediates in which the two ends of the transposon interact in a single-stranded state (52). Both events are stimulated by the presence of inverted repeats at the ends of Tn*10*: deletion derivatives having shorter inverted repeats exhibit lower frequencies of excision. Intrastrand pairing between lengthy inverted repeat sequences would create a strong secondary structure, and such structures are known to pose a severe block to the progress of DNA polymerase III (57). Finally, recent evidence suggests that excision is strongly stimulated in vivo when mini-Tn*10*-containing molecules are forced to replicate via a single-stranded intermediate on an M13 replicon (19). An alternative class of models for transposon excision, involving interactions between the inverted repeats in the duplex state, seem less likely, since excision occurs in the absence of RecA, the only *E. coli* protein known to recognize homologous DNAs. Transposon Tn*5* also has inverted repeats of an IS element at its ends and exhibits precise excision properties similar to those of Tn*10* (46).

E. coli mutations have been isolated that enhance the frequency of Tn*10* precise and nearly precise excision. The mutations isolated, called *tex* (for transposon excision), fall into two classes: special alleles of the RecB and RecC genes, and null mutations in genes involved in methyl-directed mismatch correction (102, 103). Standard alleles of *ssb* and *mutD* also increase excision (102). The molecular basis for the effects of these mutations is not yet understood. The mutations could either stimulate a normal pathway or create a new one. Some but not all *tex* mutations render transposon excision RecA dependent, suggesting that in these cases new pathways have been created that either are directly dependent on RecA function or else require RecA-dependent induction of SOS functions (103).

Precise excision of the nearly precise excision remnant occurs by a somewhat different pathway from that of precise and nearly precise excision (102). None of the *tex* mutations significantly increase the frequency of this event; in contrast, mutations in the *polA* gene do not alter precise or nearly precise excision, but do stimulate excision of the nearly precise excision remnant. However, there is probably some mechanistic relationship between the excision of the remnant and full precise excision, since the frequencies of the two events vary coordinately from insertion site to insertion site (52). It seems likely that the excision of the remnant almost certainly occurs by a replication slippage mechanism, since the segment excised is a 23-bp inverted repeat that should assume a secondary structure on every discontinuously replicated strand.

Transposon excision events may well be related to spontaneous deletion events that do not involve transposon sequences. Many spontaneous deletions involve excision between short direct repeats of related sequences (47), and it has been suggested that many of these deletions may involve fortuitously located short inverted repeats (2, 59; S. D. Ehrlich, this volume).

Bresler et al. (17) have suggested that precisely excised Tn*10* elements may sometimes undergo subsequent transposition and be recovered at new chromosomal locations. This process might resemble transposition of Tn*10* off of generalized transducing fragments or other apparently linear DNA segments (see Section III.F.2, below).

H. Tn*10* as a Tool for Bacterial Genetics

Tn*10* has been heavily exploited as an experimental tool for bacterial genetics. Applications and specific delivery vehicles were first discussed by Kleckner et al. (90). Since that time, many additional applications have been recognized or developed, new vehicles have been described, and lists of Tn*10* insertions in *E. coli* and *S. typhimurium* have been compiled (4, 9, 27, 28, 31, 69, 138, 142, 179; C. M. Berg, D. E. Berg, and E. A. Groisman, this volume). Tn*10* has also been adapted for convenient analysis of *Saccharomyces cerevisiae* genes that have been cloned in *E. coli* (69, 155).

II. NONREPLICATIVE TN*10* TRANSPOSITION: EVIDENCE AND IMPLICATIONS

A. Evidence

Tn*10* and IS*10* transpose by a mechanism in which the element is excised from the donor site and

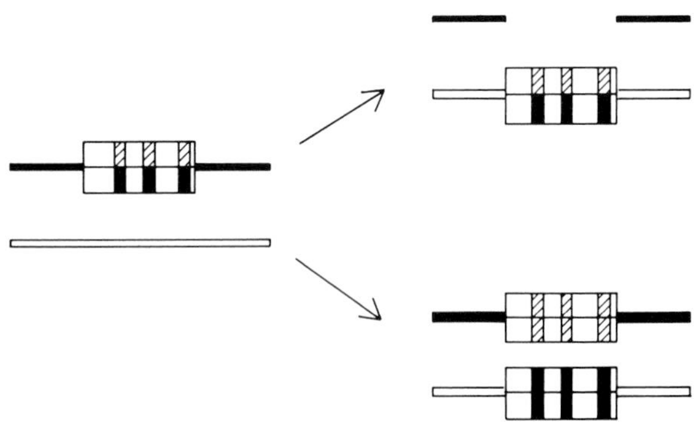

Figure 7. A heteroduplex transposon containing three single-base-pair mismatches is transposing into a target molecule by a nonreplicative mechanism (top) and by a replicative mechanism involving semiconservative replication (bottom). The nonreplicative mechanism integrates the transposon with all three mismatches intact and consequently integrates information from both strands of the original transposon DNA into the new site. In contrast, the replicative mechanism integrates information from only one strand of the original transposon DNA. (Reprinted from reference 6 with permission from Cell Press.)

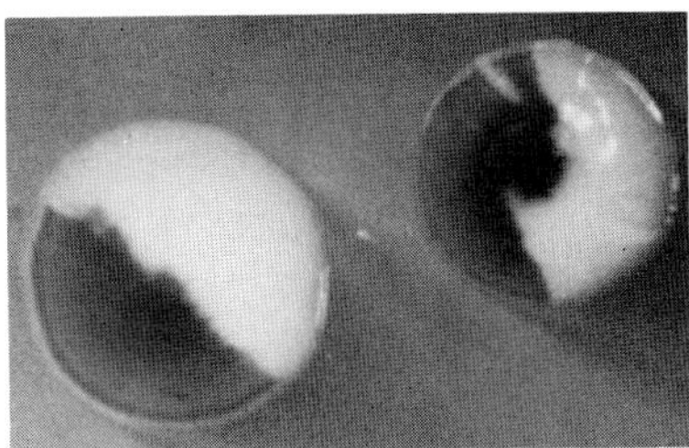

Figure 8. Colonies generated by nonreplicative transposition of a heteroduplex Tn*10* element. The Tn*10* element involved carries both a tetracycline resistance determinant and a *lacZ* gene containing a single-base mismatch such that one strand carries LacZ$^+$ information and one strand carried LacZ$^-$ information. The heteroduplex element is constructed on and delivered into *E. coli* as an insertion in an appropriate bacteriophage λ vector. *E. coli* is infected with phages carrying such elements under conditions where the phage itself cannot replicate, integrate, or kill the cell, and tetracycline-resistant colonies are selected; each such colony results from transposition of an individual Tn*10* element from λ into the bacterial chromosome. When the selective plates also contain the LacZ indicator dye X-Gal, LacZ$^+$ bacteria make white colonies on X-Gal plates, while LacZ$^-$ bacteria make blue colonies. Mixed colonies like those shown here are interpreted to mean that information from both strands, the LacZ$^+$ and LacZ$^-$ strands, of the original Tn*10* element was integrated into the single bacterial chromosome whose progeny are present in cells of the tetracycline-resistant colony (Reprinted from reference 6 with permission from Cell Press.)

inserted into the new target site without significant replication of the transposing segment.

The strongest evidence for nonreplicative Tn*10* transposition is provided by the genetic analysis of Bender and Kleckner (6). These experiments examined the transposition of an artificially constructed heteroduplex Tn*10* element bearing three single-base mismatches (Fig. 7). If transposition occurs without replication of the transposon, information from both strands of the heteroduplex element will be integrated at the new target site; in contrast, if the element is replicated semiconservatively prior to transposition and one replicated copy is inserted into the target while the other remains behind at the donor site, information from only one strand of the original element will be integrated at the target site. Transpositions of heteroduplex Tn*10* elements from a nonreplicating λ genome into the *E. coli* chromosome yielded tetracycline-resistant colonies, each descended from a single transposition event, that contained all of the markers from both strands of the original element. Genetic evidence for heterozygosity of a *lacZ* marker located in the center of the Tn*10* element is presented in Fig. 8. Heterozygosity was also observed by Southern blot analysis for two additional markers that generated restriction fragment polymorphisms located 70 and 1,000 bp from opposite ends of the element; these latter results provide important evidence that the element is not even partially replicated.

The above results are easily explained if Tn*10* uses a nonreplicative transposition mechanism. Certain more complicated models, involving conservative DNA replication, complex cointegrate reduction schemes, or both, are not rigorously excluded, however (6).

A second line of evidence that supports a nonreplicative transposition mechanism is the efficient occurrence of Tn*10*-promoted deletions/inversions (88, 136). These events can be explained very simply as "cut-and-paste" events that are analogous to Tn*10* transposition. In both cases, two IS*10* ends are disconnected from adjacent sequences and joined to a single target site; the two ends used for transposition are the outside ends of Tn*10*, and the two ends used for deletion/inversions are the two inside IS*10* ends of Tn*10*. It is much more complicated to account for these events by replicative models. In support of this argument, Tn*10*-promoted deletion/inversions have been observed in vitro in reactions involving substantially purified transposase and no high-energy cofactors (H. Benjamin and N. Kleckner, unpublished observations).

As expected from a nonreplicative transposition mechanism, Tn*10* and IS*10* do not promote the formation of cointegrates at any significant frequency (6, 63, 120, 182; M. Peifer and N. Kleckner, unpublished observations). Cointegrate formation is one hallmark of all elements known to carry out transposition by a mechanism that involves specific replication of the transposing segment. However, it must be stressed that failure to form cointegrates does not constitute direct evidence against a replicative transposition mechanism; there are perfectly reasonable replicative transposition models that do not result in the formation of cointegrates (reviewed in reference 85).

Finally, nonreplicative transposition is nicely compatible with the possibility that Tn*10* transposition induces a cellular SOS response and with the complex way in which IS*10* transposition is modulated by *dam* methylation (see below).

B. IS*10*-Promoted Adjacent Deletions

Weinert et al. (182) observed that an IS*10* element on a multicopy plasmid gave rise to deletions of the structure expected for IS*10*-promoted adjacent deletions. The possibility that IS*10* might promote such events is interesting; such deletions are known to be efficiently promoted by IS elements that undergo replicative transposition to yield cointegrates,

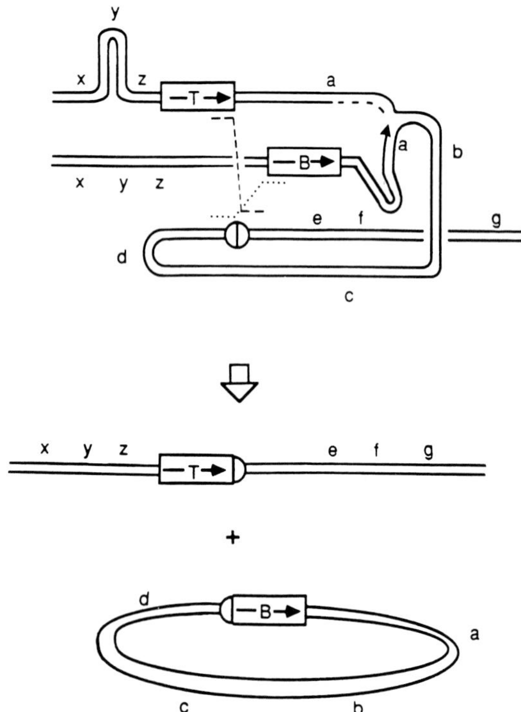

Figure 9. Model for the formation of IS*10*-promoted adjacent deletions by postreplication transposition. The right end of the IS*10* element on the top strand (T) and the left end of the IS*10* element on the bottom strand (B) interact with a target site located in the same DNA molecule. The breaking and joining events are identical to those diagrammed for Tn*10*-promoted adjacent deletions in Fig. 4B. Note: strand polarities at the replication fork are assigned such a that the circular deletion product can ultimately be released from the replicating chromosome by single-strand cleavage of the parental strand that is replicated discontinuously (in region a).

because they are the intramolecular equivalent of cointegrate formation (146).

The occurrence of IS*10*-promoted adjacent deletions was further investigated using a chromosomal insertion of an IS*10* element that transposes at 100 times the wild-type rate. These experiments demonstrate that transposase-dependent, RecA-independent deletions with genetic and physical structures diagnostic of IS*10*-promoted adjacent deletions do occur, but at less than 1/1,000 times the rate of transposition for the same IS*10* element (D. Ascherman, undergraduate thesis, Harvard University, Cambridge, Mass., 1987; Roberts et al., in preparation).

One explanation for these deletions would be that IS*10* promotes replicative events, but at an extremely low frequency. A second possibility would be that such deletions are promoted by an aberrant or abortive nonreplicative event in which only one end of IS*10* is disconnected from the donor site and joined to a new, adjacent target site. A third possibility would be that adjacent deletions arise by a nonreplicative process in which two copies of IS*10* present after passage of a replication fork interact in a manner exactly analogous to transposition, except that one participating IS*10* end is present on one sister chromosome and the second end is present on the other sister (Fig. 9). We call this process postreplicational transposition. It is attractive because it differs from the other known events only in the combination of ends involved, not in the types of strand breakage and joining events required.

Two of the above models, replicative transposition and sister-chromosome events, predict that IS*10* should also generate two other types of structures, cointegrates and insertion/inversions analogous to the final products drawn in Fig. 5. It is not clear whether IS*10* promotes such events. The analysis of Roberts et al. described above would not have detected either event, and most searches for IS*10*-mediated cointegrates have not been sensitive enough to detect events at the low frequency expected for wild-type IS*10*. The one approach that could have been sensitive enough did not detect either type of event (182); however, relatively small decreases in either frequency or recovery might have precluded detection even in this case.

Results complementary to those discussed above for IS*10* have recently been obtained with insertion sequence IS*50*, an element which is similar in many ways to IS*10*. IS*50* has been shown to generate cointegrates, adjacent deletions, and insertions/inversions, all at extremely low frequencies relative to the frequency of transposition (1, 10, 72). Lichens-Park and Syvanen (101) have demonstrated that the formation of IS*50* cointegrates by λ::IS*50* phages is not the consequence of a low-level replicative transposition pathway, because it requires the cooperation of two nonreplicating phage genomes. They present a model similar to that in Fig. 9 to account for their results.

C. Fate of the Transposon Donor Molecule

If Tn*10* transposition is nonreplicative, the double-stranded ends in the donor molecule created by the excision of the transposing segment must be either rejoined as part of the transposition event, dealt with nonspecifically by host DNA-handling enzymes, or lost. Evidence that rigorously distinguishes among these alternatives is not available; however, the evidence argues against transposase-promoted rejoining and in favor of other alternatives.

1. Genetic evidence against resealing

The strongest evidence that the donor molecule is not resealed by any transposase-promoted event is negative: searches for transposition-related excision products have failed to reveal any appropriate candidate(s). These same experiments argue against rejoining of the donor molecule gap by certain types of host-promoted events. The best investigation of this issue is provided by a recent analysis of IS*10* excision products (Kuo et al., unpublished observations). An IS*10*-Right element marked with a Kan^r segment (132) and inserted in the *S. typhimurium hisG* gene is polar on expression of the adjacent downstream gene, *hisD*. Polarity-relief revertants of a strain carrying this insertion occurred at less than 1% the frequency of transposition of the marked element and were not dependent on IS*10* transposase expression. Any excision event that restored the *hisG* translational reading frame should have been recovered in this selection. Thus, these observations rule out efficient transposase- or host-promoted rejoining of the donor chromosome during transpositions, with two exceptions: extensive deletions extending into the *hisD* gene or some specific out-of-frame resealing event. Among the products that should have been recovered is one particularly likely candidate for a transposase-promoted rejoining event, clean excision of the IS*10* base pairs leaving behind both copies of the flanking 9-bp target DNA repeat sequence.

Precise excision, defined above, has also been excluded as a transposition product by the observation that it occurs in the absence of transposition (52) and by the fact that precise excision of Tn*10* and IS*10* occurs at only 1 and 0.01% the respective frequencies of transposition (52, 148; J. Kuo, unpublished observations).

An early analysis of tetracycline-sensitive *his*::Tn*10* derivatives, intended to identify excision products resulting from Tn*10* transposition, was inconclusive because the desired rare events were obscured by more frequent transposon-promoted rearrangements and nearly precise excisions (88).

2. Physical evidence against resealing

Tn*10*-promoted transposon circle formation, discussed in detail below, involves double-strand excision of the element followed by circularization. Both in vivo and in vitro, the other product of circle formation is a linearized donor plasmid backbone segment. Since circle formation is probably closely related to transposition, these observations argue against the existence of any transposase-promoted resealing of the donor site following transposon excision.

3. Transposition induces an SOS response

Additional evidence consistent with the notion that Tn*10* transposition leaves behind a donor molecule gap is provided by the finding that Tn*10* transposition or a closely correlated event induces SOS functions, the set of cellular genes whose expression is turned on by DNA damage (130a). In situations where the rate of Tn*10* or IS*10* transposition is less than but approaching one event per cell per generation, an SOS response dependent upon both transposase and intact transposon ends can be observed. This response is manifested as RecA-dependent induction of λ prophage genes (increased levels of phage production in lysogens and increased expression of marker genes fused to λ promoters) and as induction of expression of a specific cellular SOS gene, *sfiA* (inviability of *lon* mutant strains for which such expression causes irreversible filamentation and death). Data are consistent with the possibility that a single transposition event is sufficient to induce the SOS response and that many or all transposition events do so. Although there is no direct evidence on the nature of the transposon-promoted damage that leads to induction of an SOS response, degradation of sequences around a double-strand gap in the donor molecule is an obvious and attractive possibility. Induction of an SOS response could be important biologically in helping a cell undergoing transposition to recover from or repair damage to the transposon donor chromosome.

4. Genetic evidence for double-strand breaks

A final observation further supports the possibility that Tn*10* promotes the occurrence of double-strand breaks or gaps at the donor site: the presence of an active Tn*10* element on λ stimulates RecA/RecBCD-promoted recombination in the genetically marked interval containing the transposon. At the highest levels of transposase examined, recombination is stimulated twofold (D. Thaler and F. W. Stahl, personal communication). The RecA/RecBCD pathway is known to require double-stranded DNA ends and thus to be stimulated by double-strand breaks (153).

5. Three possible fates

If transposition does indeed leave behind a donor molecule containing a double-stranded gap, three fates for such a molecule can be considered (Fig. 10).

(i) The donor molecule may be destroyed. Since bacteria frequently contain more than a single copy of their chromosome, such a loss could be tolerated.

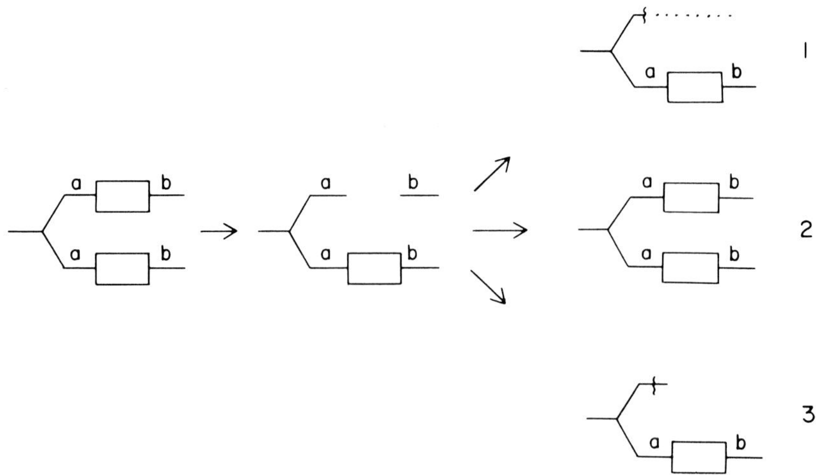

Figure 10. Possible fates of the donor molecule following Tn*10* transposition. (1) Loss of the gapped donor molecule ("donor suicide"). (2) Repair of the donor molecule gap, using homologous sequences on the sister chromosome as template (double-strand gap repair). (3) Resealing of the gapped donor molecule after extensive degradation of material to either side of the gap.

In this situation, all intrachromosomal transposition events are lost, and transposition events are always recovered in chromosomes carrying two copies of the transposon, one at the original donor site and one at the new site.

(ii) Host functions may repair the gap completely using the homologous region of another chromosome in the cell as a template ("double-strand gap repair"). Conceivably, a sister chromosome would be a favored substrate for such repair (see *dam* methylation, below). Such repair would generate one chromosome carrying two copies of the transposon and one chromosome carrying only a single copy at the original donor site. In this situation, transposition is formally replicative; one copy of the transposon is present at the new site and one is still present at the donor site. However, in this case, replication of the transposon is not a mechanistic part of the transposition event itself, but instead is promoted by host DNA repair proteins after the transposon has excised itself from the donor site. Double-strand gap repair is thought to occur in *E. coli* (126, 165).

(iii) Host functions may reseal the donor molecule but only after extensive degradation of sequences on both sides of the original gap. This process would generate "double-ended deletions," deletions extending in both directions from the site vacated by the transposon. Such deletions have been previously seen only with insertion sequence IS*4* (60). They have not been detected with either Tn*10* or *IS10* despite fairly extensive searches, but a definitive analysis has not been performed (88, 128; D. Roberts and D. Ascherman, unpublished observations). Still, it seems likely that nonspecific resealing of the transposon donor is not a common fate for the donor molecule.

Distinguishing experimentally among these possibilities is difficult. The first two processes result in chromosomes that either are indistinguishable from the parental chromosome or contain two copies of the transposon, one at the old site and one at the new site. Pedigree analysis of individual transposition events in individual bacterial clones is probably required. A renewed search for double-ended deletions is also warranted.

D. Evolutionary Implications

The evolution of bacterial transposable elements into stable components of the genome might be favored if they are able to overreplicate their hosts, that is, if the transposon segment is replicated more frequently than the genome as a whole (20). Overreplication is an automatic consequence of replicative transposition modes (43, 125). However, one can argue that it may also be a consequence of certain nonreplicative modes of transposition as well (6).

(i) If the donor molecule is subject to efficient double-strand gap repair, as described above, transposition is effectively replicative. The transposon sequences moved to the new site are replaced by DNA repair.

(ii) Overreplication also occurs if the donor molecule is destroyed during transposition. In this case, one molecule containing two transposons appears and one molecule that originally contained one transposon disappears. The net result is an increase in the average number of transposons per chromosome in the population.

(iii) Finally, even if the donor molecule is resealed without restoration of transposon sequences, overreplication can result in one specific case: if excision of the transposon occurs after the chromosomal replication fork passes the donor site and the element then inserts into a target site ahead of the replication fork. In this situation, donor site resealing without restoration of transposon sequences followed by completion of replication would generate one chromosome containing two transposon copies, at the new and old positions, and one chromosome containing a single transposon copy, this time at the new position. (I thank J. Roth for suggesting this possibility.)

III. MECHANISM OF TRANSPOSITION

A. Simple Description

Tn*10* or *IS10* insertion probably includes the following DNA alterations: (i) excision of a double-stranded transposon fragment from the donor molecule by cleavage of both strands at both ends in one or more steps, (ii) cleavage of the target molecule by a pair of staggered single-strand nicks 9 bp apart, (iii) ligation of the transposon sequences to the overhanging ends of the cleaved target DNA sequences by a pair of symmetrical single-strand joining events, (iv) a small amount of DNA repair synthesis opposite these joints to complete the process.

Double-strand excision of the transposon is a necessary consequence of nonreplicative transposition (6). It is not known whether both strands at each end are broken at the same time or at different times during transposition. Cleavage of the target DNA by a pair of staggered nicks is inferred from the fact that insertions of Tn*10* and *IS10* are flanked by direct repeats of a 9-bp sequence that was present once in the target DNA molecule prior to insertion (84) (Fig. 6). An alternative explanation, that the transposon brings one copy of the sequence with it from the donor site and finds a homologous second copy at the target site, is not correct. A Tn*10* or *IS10* element always chooses the same array of target sites regardless of the particular 9-bp repeat sequence flanking the element at its donor site (84; unpublished observations). Involvement of repair synthesis in transposition is assumed as a necessary consequence of the staggered nicks in the target. Such repair synthesis probably does not usually extend very far into the transposon. In the experiments of Bender and Kleckner (6) described above, a heteroduplex mismatch located 70 bp from one end of Tn*10* is transmitted faithfully to most transposition products; repair synthesis across this position would have destroyed the heterogeneity.

B. Transposase

1. Identity

IS10 transposase is the only **element-encoded** function required for *IS10* or Tn*10* transposition and the other related rearrangements. Protein sequence analysis and site-directed mutagenesis demonstrate that the transposase polypeptide is encoded by a single long open reading frame that begins at base pair 108 of *IS10* and runs nearly the entire length of the element to base pair 1313 (see below, Fig. 13) (62, 145a). The predicted molecular weight of this polypeptide is 46,000; the apparent molecular weight in sodium dodecyl sulfate-polyacrylamide gels is less, about 42,000. There is no information about the functional or physical stability of the protein in vivo.

2. Level is rate determining

The level of transposase protein determines the level of transposition in vivo. When *IS10* is present in a single copy per cell, transposase protein is expressed at a very low level, fewer than 0.15 molecules per cell per generation (129). When transposase expression is inhibited 30-fold by multicopy antisense inhibition (see below), the rate of transposition per *IS10* copy per cell per generation decreases commensurately (151); conversely, when the level of transposase expression is increased, by mutation or by substitution of heterologous transcription and/or translation signals, the level of transposition increases proportionately (118). Increases of five orders of magnitude in transposase expression can bring transposition levels up to one event per cell per generation, or even more, to the point where decreased cell viability makes precise quantitation difficult (M. Septak and N. Kleckner, unpublished observations).

Host factors do not appear to be limiting for transposition under the conditions tested. However, the highest level of transposition examined is about one event per cell per generation.

3. Activities

Transposase protein probably carries out all of the biochemical steps involved in recognition and cleavage of transposon ends on both strands and for ligation of those ends to new sequences. A highly enriched but not completely pure transposase preparation can promote transposon circle formation in vitro with appropriate substrates; circle formation requires both cleavage and ligation at transposon ends (117). However, complete protein purification is required, for example, to exclude the possibility that transposase breaks only one strand at each end of the transposon and a host nuclease breaks the second one. It is generally assumed that transposase also recognizes and cleaves the target DNA, although the only evidence to this effect is the fact that Tn*10*, and every other transposon, has its own characteristic target site specificity (see below).

4. Structure and function

Efforts to correlate specific functions and steps in the transposition reaction with particular sites or domains in transposase protein are in progress. Evidence suggests that the amino-terminal portion of transposase may be involved in recognizing the ends of the transposon. *IS10* transposase will still bind to DNA when its carboxy-terminal one-third is replaced by any of several heterologous polypeptide sequences (J. Leatherwood and N. Kleckner, unpublished observations). Also, certain features of the closely related transposon IS*50* suggest that the amino-terminal end of transposase may see DNA: the full-length IS*50* transposase probably acts preferentially in *cis* (see next section), a property which is presumed to reflect a strong non-sequence-specific interaction with DNA; in contrast, a second IS*50* protein that is identical to transposase except that it is missing 4 kilodaltons of polypeptide chain from the amino terminus acts in *trans* (71, 77).

The carboxy-terminal portion of *IS10* contains a 60-amino-acid region with marked homology to several other insertion sequence transposases; the homology region includes a conserved tyrosine (107). It is speculated that this part of the protein is responsible for recognizing the target DNA, cleaving it, and ligating the broken target strands to the ends of the transposon.

5. Preferential *cis* action

Transposase does not move freely through the cell in three dimensions. Direct evidence for this conclusion comes from in vivo experiments that examine, in the bacterial chromosome, transposition of a pair of Tn*10* ends placed at various distances from an *IS10* transposase gene whose expression is driven from a heterologous (ptac) promoter (118). Under these conditions, complementation can be observed, but its efficiency decreases as the distance between the transposase gene and the transposon ends increases; complementation is less than 1/15 times as efficient at a distance of 50 kb than at a distance of 1 kb. This effect has the important biological implication that each Tn*10* element in the cell is effectively seen only by its own transposase and not by transposase expressed from other *IS10* elements (see additional discussion below of *cis* action in Sections IV and VI).

The observations above imply that transposase makes its first contact with DNA in the immediate vicinity of the gene from which it is expressed. This is easily feasible in a procaryotic organism, since transcription and translation are coupled. If its amino-terminal domain binds DNA, a transposase molecule makes its first contact with DNA before its mRNA is completely transcribed or before it is completely translated and released from the ribosome (see discussion of the *fbi* site in Section VII.A below). Furthermore, after its first contact with DNA, transposase protein does not move freely in three dimensions. Transposase movement might be confined to sliding by one-dimensional diffusion along DNA (e.g., 11, 164) or by other types of restricted protein movement (reviewed in reference 12).

The severity of "*cis* action" in vivo will be determined by the distance of the transposase protein from its gene at the moment that DNA binding becomes possible, by the extent to which it can move after it first contacts DNA, and by the time it has to find its substrate before it is functionally inactivated. The relative contributions of these three factors are not known. Other phenomena that may be indicative of restricted transposase movement during the transposition reactions are discussed below.

Preferential *cis* action due to restricted transposase movement may be a general property of all IS element transposases. Evidence for a severe effect of *cis* action has been provided by Derbyshire et al. (42) in the case of IS*903*. In some cases, however, direct evidence for restricted protein movement is not conclusive. The observation generally cited as evidence for a *cis*-acting transposase is that a defective IS element or transposon is not efficiently complemented by a second, functional transposon present in the same cell (72, 105). This experiment is sufficient to document the phenomenon of preferential *cis* action, but does not constitute evidence for restricted transposase movement, because the same result will also be obtained if, in vivo, action of transposase is somehow coupled to its expression: transposase will not act on elements other than the one encoding it because those elements will not be activated at the time transposase is made. However, exactly such a phenomenon has been documented for the IS*10* insertion module (see below). The experiments described above for Tn*10* avoid this complication by using a transposase gene expressed from a heterologous promoter and examining distance dependence directly.

C. Intermediates in Transposition

1. The three-body problem

Transposition has one unique, and therefore especially interesting, feature in comparison to other recombination reactions: it involves three different DNA sites, the two transposon ends and the target site. In contrast, site-specific recombination events such as λ integration involve reciprocal exchange between two sites.

The special features of Tn*10* and IS*10* transposition that solve this "three-body problem" are not known. One possibility is that the reaction occurs by a succession of two independent two-body interactions. In the first interaction, the two transposon ends come together and the transposon is excised from its donor molecule; in a second interaction, the excised element with its ends held together in a single protein-DNA complex finds and cleaves the target DNA and makes the transposon-target DNA joints. Mu transposition provides another possible solution: the two transposon ends come together, are separated from adjacent flanking DNA by single-strand cleavages at each end without disruption of the DNA-protein complex, and then interact with and become joined to the target DNA molecule, which is processed in subsequent steps to give either nonreplicative or replicative transposition products (36, 159).

2. Transposon circle formation

Both Tn*10* and IS*10* promote the formation in vivo of structures called transposon circles, which have been proposed to represent aberrant transposition products and not transposition intermediates (116). In the presence of high levels of transposase, circles are formed by mini-Tn*10* constructs at levels adequate for physical characterization. Such circles consist of transposon sequences that have been excised from the donor molecule and then circularized by a single-strand joint between the two ends; most or all circles probably also have an additional nick located nonrandomly in the unsealed strand. The termini of the unjoined strands are at or within 1 or 2 bp of the ends of the transposon; the sequence across the new joint has not been analyzed. These circles have been argued not to be transposition intermediates, primarily for reasons of parsimony (116): if transposition involves a circular intermediate in which the two ends are covalently linked, integration at the target site would necessitate cleavage of that covalent linkage.

Weak evidence that both strands at both ends of the transposon are cleaved in a single concerted step is provided by the failure to observe significant levels of substrate molecules that had undergone either double-strand cleavage at only one end or nicking (116). The only other product that has been observed is an excised, linear duplex transposon segment, which might represent double-strand exicision without circularization (D. Morisato, unpublished observations).

D. How Transposon Ends Find Each Other

Analogy with site-specific recombination reactions and Mu transposition suggests that the two ends of a transposing Tn*10* or IS*10* element probably interact in an important way. Two sets of observations provide evidence for an insight into this process.

1. Terminus selection and suppression

Analysis of transposon circle formation with substrates containing different numbers and types of IS*10* ends suggests that the transposase searches among available termini and selects certain ones to the exclusion of others present on the same molecule (116). For example, outside IS*10* ends appear to be more active than inside ends, and one outside end preferentially chooses a second outside end over an otherwise equivalent inside end. Furthermore, if the second outside end contains an appropriate single-base mutation, circle formation by all possible pairs of termini, even those unaffected by the mutation, are suppressed. This latter observation provides evidence for a direct interaction between IS*10* ends, since it is most easily interpreted to mean that the wild-type end is stably sequestered in an inactive complex with the mutant end.

Analogous analysis of substrates containing only one outside IS*10* end also suggest that "nonspecific" sequences are recruited to serve as a second end at a significant frequency, perhaps 1 to 0.1% the frequency at which a normal end would be used if present (116; O. Huisman and L. Signon, personal communication). Occurrence of these secondary products is also suppressed by the presence of a second mutant IS*10* end (116).

Mechanistic explanations for communication among termini involve one-, three-, and intermediate-dimensional searching models. Morisato and Kleckner (116) presented the specific model that "transposase finds one terminus by *cis* action, locates the second terminus by one-dimensional diffusion, and then assumes a configuration appropriate for cleavage at both termini. Selectivity could occur at any step prior to the cleavage reaction, e.g., from higher binding affinities at favored sites and/or from faster or more efficient turnover from the bound state into the cleavage reaction." This type of model can also account for the ability of one wild-type end to recruit a nonspecific partner end.

2. Transposition length dependence

The transposition frequency of Tn*10* and derivative elements decreases exponentially as the length of the transposing segment increases. For Tn*10* elements in the bacterial chromosome or on λ donor phages, the rate of decrease is about 40% per kilobase of transposon length for lengths of 3 to 9.5 kb (118). For mini-Tn*10* elements on multicopy plasmids, the rate of decrease is much less, about 15% per kilobase (181).

Transposition length dependence was first observed for transposon Tn*9* and proposed to reflect the decay of a transposition/replication complex as it moved from one end of the element to the other (23). In light of the nonreplicative transposition mechanism of Tn*10*, a more likely interpretation in this case at least is that the two ends of the transposon must find one another at some point during transposition, with length dependence reflecting the details of that process. Two extreme types of models have been discussed (181; J. C. Way, Ph.D. thesis, Harvard University, Cambridge, Mass., 1984).

One type of model is similar to that discussed above: a protein(s) binds to one end of the transposon and searches in one dimension for the other end. This is particularly attractive, in part because it predicts the observed exponential decay relationship. A second type of model postulates that length dependence reflects the probability that two DNA sites (presumably with proteins already bound) will come together by a three-dimensional collision. In the simplest case, this probability is described by the Jacobson-Stockmayer equation (75, 175), which predicts a different type of decay relationship. The reduced length dependence of plasmid-borne transposons can be accommodated by either model.

The issue of how transposon ends find one another has been discussed for the cases of IS*903* and Tn*5*/IS*50* by Derbyshire et al. (42) and Sasakawa et al. (141).

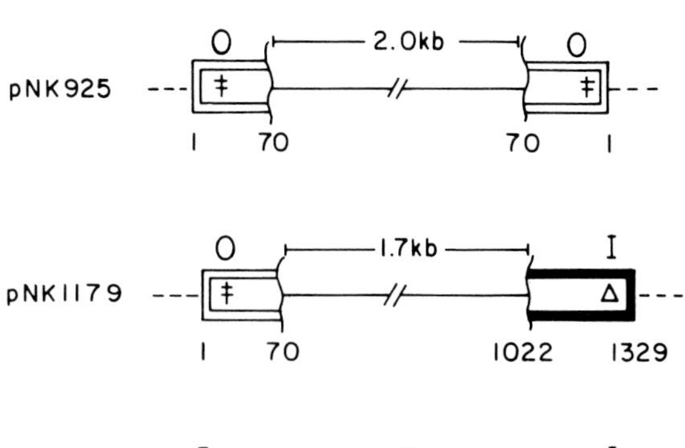

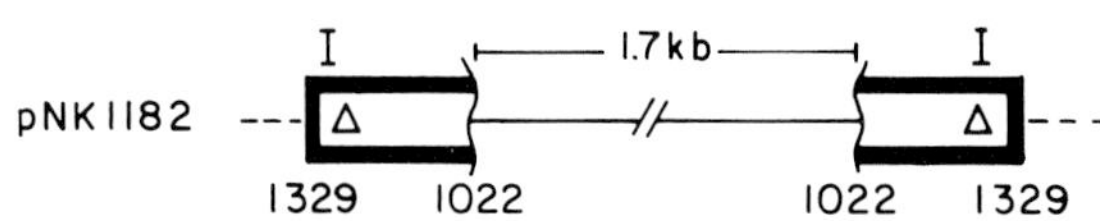

Figure 11. Substrates used for transposon circle formation in vitro. The configurations of IS*10* outside ends (O) and inside ends (I) are indicated. Triangles represent the *dam*-sensitive GATC site present at IS*10* inside ends; double daggers represent the binding site for integration host factor present at IS*10* outside ends. Top: Substrate analogous to Tn*10*, with two outside ends. Middle: Substrate analogous to IS*10*, with one outside and one inside end. Bottom: Substrate with two inside ends, analogous to inside-out transposition and Tn*10*-promoted chromosomal rearrangements. Numbers refer to coordinates in the 1,329-bp IS*10* element. Central 2.0- and 1.7-kb segments contain convenient non-Tn*10* sequences. (Reprinted from reference 117 with permission from Cell Press.)

E. Transposition In Vitro

1. Introduction

Biochemical analysis of Tn*10* transposition has been initiated (117; D. Morisato and N. Kleckner, unpublished observations). This analysis has revealed the involvement of two host proteins, IHF and HU (see below), and has shown that inside and outside ends differ in their host factor requirements.

Tn*10* transposition has been observed upon the addition of appropriate substrates to crude extracts of *E. coli* cells containing transposase; this complete reaction is not yet efficient enough for further analysis. However, insight into the transposition reaction has been obtained by examination in vitro of Tn*10* transposon circle formation. This reaction is convenient because it involves both cleavage and ligation at the transposon termini, it is intramolecular, and its circular product is relatively insensitive to nucleases. Recently, Tn*10*-promoted deletion/inversions have also been observed in in vitro extracts (Benjamin and Kleckner, unpublished observations).

2. Requirements for circle formation

In vitro reactions have been carried out using minitransposon substrates having different combinations of transposon ends: two outside ends, analogous to Tn*10*, two inside ends, analogous to inside-out transpositions of Tn*10* and the other two Tn*10*-promoted rearrangements, and one inside and one outside end, analogous to IS*10* (Fig. 11). All three reactions require IS*10* transposase protein and proceed in a simple buffer which must contain Mg^{2+} but need not contain any high-energy cofactor; presumably, the large amount of energy needed for strand joining comes from one of the cleavage events, as is the case in site-specific recombination, topoisomerase, and Mu transposition reactions. Finally, in all cases, the transposon donor plasmid must be supercoiled; the possible roles of supercoiling are discussed below. The three reactions differ significantly in their host factor requirements, as follows.

Substrates containing two outside ends do not give rise to transposon circles in the presence of transposase alone; host proteins are required. Analysis and fractionation of *E. coli* extracts from cells not containing transposase protein has identified two proteins, each of which can provide "host factor activity." One of these proteins is IHF, the integration host factor originally identified as a host protein required for site-specific integration by bacteriophage λ (82). The second is HU, one of the so-called procaryotic "histonelike" proteins (reviewed in reference 44). Of the total host factor activity in a crude extract, approximately 85% is provided by IHF and 15% is provided by HU. When the two proteins are both present at nonsaturating levels, their effects appear to be roughly additive. The specific activity of IHF as a host factor is much higher than that of HU, since both cells and extracts contain massive amounts of HU and trace amounts of IHF.

Substrates containing two inside ends behave differently. They give rise to transposon circles in the presence of transposase alone without addition of any host proteins. The transposase used in these experiments is about 95% pure; it contains insignificant levels of IHF and HU. Although further purification is needed, transposase is presumed to be acting alone in these experiments. The activity of inside IS*10* ends in vivo is inhibited by *dam* methylation (see below); this inhibition is also observed in vitro for circle formation.

Substrates containing one inside and one outside end exhibit the characteristics of both types of ends: the host factor requirements observed with outside ends and the regulation by *dam* methylation observed with inside ends (see below).

IHF and HU have recently been implicated in transposition by a number of other bacterial elements: HU is required for Mu transposition in vitro (34). IHF is also involved in the transposition of IS*1*, IS*50*, and γδ (55, 56, 182a).

3. Role of IHF and HU

IHF and HU probably play auxiliary roles in the transposition reaction, as they do in the other transposition and site-specific recombination reactions and in DNA replication (reviewed in reference 44). The exact role of these proteins at IS*10* outside ends remains to be elucidated. Both proteins are known to change the structure or conformation of DNA (18, 127, 133, 157; L. Kosturko, Abstr. Cold Spring Harbor Meet. Mol. Genet. Bacteria Phages 1987, p. 232; J. Thompson and A. Landy, *Nucleic Acids Res.*, in press). In Mu transposition, HU is required at an early step (34). Since this HU requirement can be bypassed if donor sequences flanking the Mu ends are cleaved off, HU may act to remove a steric inhibition by those flanking sequences (36). In the case of γδ, IHF and transposase bind cooperatively at both ends of the transposon (182a).

F. Host Factors In Vivo

1. IHF and HU are important in vivo

In vivo, IHF is required for the activity of IS*10* outside ends and exhibits the same discrimination between inside and outside ends as observed in vitro (131; D. E. Roberts, Ph.D. thesis, Harvard University, Cambridge, Mass., 1986). Analogous rearrangements promoted by a Tn*10*-derived element (Tn*Gal*) having direct repeats of IS*10* instead of the usual inverted repeats involve outside and one inside IS*10* end. The frequency of Tn*10*-promoted deletion/inversions and adjacent deletions, which involve two inside IS*10* ends, is the same in isogenic IHF^+ and IHF^- strains. However, analogous rearrangements promoted by a Tn*10*-derived element (Tn*Gal*) having direct repeats involve one outside and one inside end and occur at a sevenfold-lower frequency in IHF^- than in IHF^+ strains. The significant level of rearrangements observed with Tn*Gal* in the total absence of IHF protein is the same as the residual level of circle formation observed in vitro in the absence of IHF and the presence of HU, suggesting that the relative contributions of the two proteins may be the same in both cases.

2. Other host proteins

Transposition in vivo probably depends upon other host functions in addition to IHF and HU. One additional protein likely to be involved, at least indirectly, is DNA gyrase. Transposition of Tn*10* from λ into the *E. coli* chromosome is defective in *gyrA* mutants at temperatures permissive for cell growth (V. Lundblad, Ph.D. thesis, Harvard University, Cambridge, Mass., 1983). It has not been determined in vivo whether these mutations affect the transposition process directly or expression of transposase or both. However, it seems likely that some step(s) in the transposition reaction per se requires supercoiling, because Tn*10* transposon circle formation in vitro requires that the substrate plasmid be supercoiled.

The supercoiling requirement of the transposition reaction merits further attention. Mu transposition requires supercoiling of the donor molecule, and site-specific recombination reactions often require supercoiling of the DNA carrying the recombinase-binding sites (7, 14, 34, 35). However, Tn*10* transposes in vivo off of several DNAs which are generally assumed to be linear and not supercoiled: oversized P22::Tn*10* genomes, P22-packaged transducing particles carrying Tn*10* in chromosomal DNA segments, and conjugally transferred DNA that has been replicated but not integrated into the recipient genome (87, 132, 138). One interesting possibility is that bacteria have a nonspecific mechanism for topologically constraining all or part of a linear DNA, as suggested by analysis of P1-transducing DNAs (139, 140). Another possibility is that Tn*10* transposition requires supercoiling of the target DNA molecule; the supercoiling requirement for circle formation could reflect an aspect of the reaction that is analogous to interaction with a normal target DNA. Interestingly, transposition of Tn*5*, an element very similar to Tn*10*, is reported to require supercoiling of the target molecule in vivo (73).

Tn*10* transposition from λ to the host chromosome is also defective in *polA* and *lig* mutants (Lundblad, Ph.D. thesis). *polA* has also been implicated in the transposition of Tn*5* (141, 160). DNA polymerase I and DNA ligase might be involved directly in transposition, perhaps during the final proposed step involving DNA repair synthesis across the 9-bp repeat sequences. Alternatively, these functions might be involved in the host response to the occurrence of a double-strand break in the donor chromosome or, less likely, in transposase gene expression.

Finally, in vitro analysis of the intermolecular transposition reaction may also reveal other proteins, not required for intramolecular circle formation, that participate directly in target DNA recognition or strand transfer.

Transposition of IS*50* has been shown to require *E. coli dnaA* protein (185). However, IS*10* has no *dnaA* consensus binding sequence, so it probably does not respond to *dnaA*.

3. Why has IS*10* evolved a host factor requirement?

The finding that inside ends give rise to transposon circles in the absence of host factors raises the question of why outside ends have evolved to require these additional proteins. Two (not mutually exclusive) answers are as follows. (i) Host factors may simply make the reaction more efficient. Perfectly isogenic inside and outside end substrates have not been compared, but in the cases examined, outside ends are more active than inside ends (116). (ii) The requirement for host factors may serve a regulatory role, providing an additional means of communication between the transposon and its host. IHF has recently been shown to be involved in regulation of IS*10* activity at another level, antisense inhibition of transposase expression (see below).

G. IS*10* Ends: Structural Analysis and Comparison

The relative structural complexities of the outside and inside ends of IS*10* correspond to their relative functional complexities. The two ends share a 23-bp imperfect terminal inverted repeat that differs at 5 of 23 positions (62) (Fig. 1 and 12; also below, Fig. 13). For full inside-end function, the inverted repeat sequence is both necessary and sufficient (D. Ahmann and D. Morisato, unpublished observations). Thus, transposase probably interacts directly with this terminal inverted repeat sequence both at inside and at outside ends.

Full outside end function requires the presence of a 42-bp sequence (180). The outside end can be divided into three segments (Fig. 12): the inverted repeat sequence (base pairs 1–23), a site required for the binding of IHF in vitro and for efficient transposition in vivo (base pairs 30–42), and the segment in between (base pairs 24–29), whose function is not known (68, 180; Morisato, unpublished observations; O. Huisman, P. Errada, L. Signon, and N. Kleckner, manuscript in preparation).

A large number of mutations in the outer 70 bp of IS*10* have been isolated without respect to phenotype and subsequently characterized. The only mutations with very strong phenotypes occur in a few base pairs in the middle of the inverted repeat sequence (Fig. 12). These and other observations suggest that this may be the region where transposase makes its most intimate contact with DNA, perhaps with the DNA in some type of distorted or "open" form.

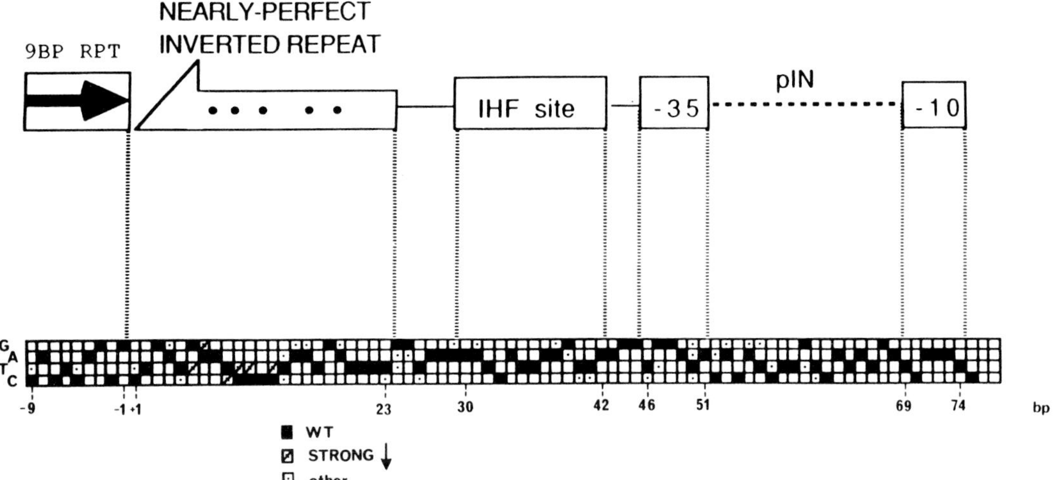

Figure 12. Mutations in the outside end of IS*10*. Single-base mutations have been generated at the outside end of IS*10* by a method that allows identification of mutations without respect to mutant phenotype (115). Mutations have been tested in various combinations of mutant and wild-type IS*10* ends: constructs with two outside ends each carrying the same mutation, constructs with two outside ends, one mutant and one wild type, and constructs with one mutant outside end and one wild-type inside end (in a Dam$^+$ host). Mutations that confer a severe (greater than 50-fold) transposition defect in all three configurations are indicated with a slash in the appropriate box. All other mutations are indicated with a dot in the appropriate box; some of these mutations confer weaker defects in some or all configurations, and others confer no detectable defect in any configuration.

Why do outside ends require host factors for activity while inside ends do not? In one model, base differences in the outside-end inverted repeat sequence weaken its interaction with transposase in a way that can be overcome by host factors. In a second model, the outside end contains inhibitory sequences adjacent to the inverted repeat whose effects are overcome by host factors. Biologically, host factors could serve to make the outside end more reactive; also, the host factor requirement could provide an additional mechanism for host modulation of transposition frequency.

H. Influence of Chromosome Context

The frequency of Tn*10* transposition varies dramatically from one insertion site to another because of local *cis* effects on the activity of its ends. The influence of context has been seen directly: a 75-fold range in transposition frequencies was observed for a set of Tn*10* elements at 18 different sites in a multicopy plasmid when a fixed level of complementing IS*10* transposase protein was provided in *trans* (D. Chin and M. A. Davis, cited in Davis, Ph.D. thesis). Even more dramatic variation was observed in a survey of 81 independent Tn*10* insertions into different sites in the *E. coli* genome. The frequency of transposition of these elements varies over a 1,000-fold range. This diversity is not simply due to variations in transposase level of transposon copy number resulting from differences in distance from the replication origin, although there are minor contributions from these sources. Instead, the diversity is due primarily to *cis*-acting factors or "context effects." For example, the disparities in relative transposition rates remained even when transposase was provided from an overproducer construct at levels sufficient to increase those rates more than 10-fold (Davis, Ph.D. thesis).

An important source of context variability is likely to be the exact nature of the DNA sequences immediately adjacent to the inserted element, within the 9-bp repeat sequence or nearby. One precedent for such effects is provided by restriction enzyme *Eco*RI; variations in base pairs adjacent to the cleavage site resulted in 100-fold variations in cleavage rate (3). Secondarily, high levels of readthrough transcription entering Tn*10* from adjacent chromosomal sequences are known to inhibit the activity of the affected Tn*10* end (see below).

I. Target Site Interactions

1. Insertion specificity

Tn*10* inserts into a large number of different sites in the bacterial genome. Approximately 1 to 2% of all Tn*10* insertions in the *Salmonella* or *E. coli* chromosomes confer an auxotrophic requirement, and a wide range of requirements is represented (86,

87, 90). In addition, Southern blot analysis of 80 independent insertions revealed nearly as many different restriction digest patterns (Davis, Ph.D. thesis). Thus, the *E. coli* chromosome does not contain one or a small number of very highly favored insertion sites that serve as "sinks" for insertions.

However, fine structure genetic mapping of Tn*10* insertions reveals that they do insert preferentially into particular sites, "insertion hot spots" (92). Probably all IS*10*-promoted events prefer the same target sites: major hot spots for Tn*10* insertion have also been shown to be hot spots for Tn*10*-promoted adjacent deletions, for Tn*10*-promoted deletion/inversions, and for IS*10* insertions (124; Bender and Kleckner, unpublished observations).

DNA sequence analysis of many Tn*10* insertion sites has revealed a consensus sequence: a symmetrical 6-bp sequence made up of two 3-bp half-sites separated by 1 bp which occurs in the middle of the 9-bp target site sequence duplicated upon insertion. Major hot spots exhibit near identity to this sequence, and lesser hot spots are more distantly related. The nature and location of the hot spot consensus sequence suggest that it is recognized and cleaved by the symmetrically disposed subunits of a single protein, presumably transposase, in a manner analogous to cleavage by a type II restriction endonuclease.

Detailed analysis of the hot spot consensus sequence suggests the following patterns.

(i) The first two bases of each 3-bp half-site are the most important for insertion. When all sequences are considered together, there is more variation at the third position than at the first two: 90% of sites have G or A at the first position (with G favored) and 98% of sites have C or T at the second position (with C favored); however, at the third position, T, C, A, and G are present at 63, 23, 12, and 2% of sites, respectively.

(ii) The most important single feature that distinguishes a major hot spot from a lesser insertion site is the presence or absence of the thymine 5-methyl group on the consensus T at the third position of the half-site. Two observations support this conclusion. First, at major hot spots, 22/24 half-sites have a T at this position; at lesser hot spots, only 11/28 half-sites contain a T. Second, Tn*10* inserts efficiently into a particular hot spot when the third-position half-sites contain 5-methylcytosine and inefficiently when they contain unmethylated cytosine. The 5-methyl group of 5-methylcytosine is located in the same place in the helix as the 5-methyl group of thymine (95).

(iii) The pattern of preferred bases is compatible with recognition of the consensus sequence in the major groove of B DNA. The aforementioned 5-methyl group of thymine would lie in the major groove of B DNA. Furthermore, the pattern of base preferences at the first two positions in the half-site indicates a strong discrimination between purines and pyrimidines. It has been argued that the major groove provides the best opportunity for such discrimination (111, 145).

The six bases of the hot spot consensus sequence are not the only factors that influence the relative frequency of Tn*10* insertion into different sites (62). The hottest site identified thus far does not have a perfect consensus sequence, and several sites have been identified which have a perfect consensus sequence but are not hot spots. It remains to be determined whether additional influence is exerted only by the nucleotides immediately adjacent to the consensus sequence or whether factors acting over longer distances might also be important.

2. Inter- versus intramolecular target site selection

Several lines of evidence suggest that Tn*10*-promoted deletion/inversions preferentially involve target sites located near the transposon itself; in contrast, IS*10* transposition events seem to occur at target sites located randomly throughout the genome, although the evidence for this conclusion is indirect (148).

The preference of rearrangements for nearby target sites may be another case in which transposase searches in one dimension, this time for a target site rather than for a transposon end. If so, why do transposition events not show the same preference? Two possibilities are the following. (i) If the search for a target molecule occurs subsequent to the complete separation of transposon sequences from flanking chromosomal sequences, simple transposition might involve a free excised transposon segment, and chromosomal target sites could then be found by two- or three-dimensional collision, showing a less strong preference for positions near the donor site. One-dimensional searching by transposase protein bound to the ends of the element would only identify target sites within the transposon itself. (ii) Alternatively, in the case of transposition, intramolecular target sites might be preferentially chosen, but the resulting insertion mutations are usually lost, either due to destruction of the donor molecule or because insertions to sites nearby on the same chromosome are eliminated by subsequent double-strand gap repair of the donor site.

3. Preferential utilization of nontranscribed regions

Tn*10* insertions appear to be sensitive to the transcription of the target DNA. Transposition into an *E. coli* lactose operon assayed in *S. typhimurium* is 10 times more frequent in the absence of inducer than in its presence, and similar effects have been observed with the *S. typhimurium* histidine operon (J. Casadesus and J. Roth, unpublished observations); these investigators speculate that transcription may interfere with transposase molecules or complexes searching for a preferred insertion site by one-dimensional diffusion.

This phenomenon has two important biological consequences. First, it suggests that Tn*10* or *IS10* will tend not to insert into absolutely essential genes or into genes required for growth at the time of insertion, since most such genes are probably actively transcribed. Second, it will help to keep the transposon out of heavily transcribed regions where readthrough transcription might result in unregulated expression of transposase; however, Tn*10* and IS*10* do have specific mechanisms to protect them from being activated by readthrough transcription (see Section IV).

IV. IS*10* GENE EXPRESSION

A. Structure of IS*10*

IS*10* contains a single transposase gene and two important promoters: pIN, the transposase gene promoter, and pOUT, a strong promoter that opposes pIN and directs transcription toward and occasionally across the outside end of IS*10*. The transposase message is also called RNA-IN; the transcript from pOUT is also called RNA-OUT. Two GATC sites involved in regulation by *dam* methylation are located within pIN and within the terminal inverted sequence at the inside end (Fig. 13).

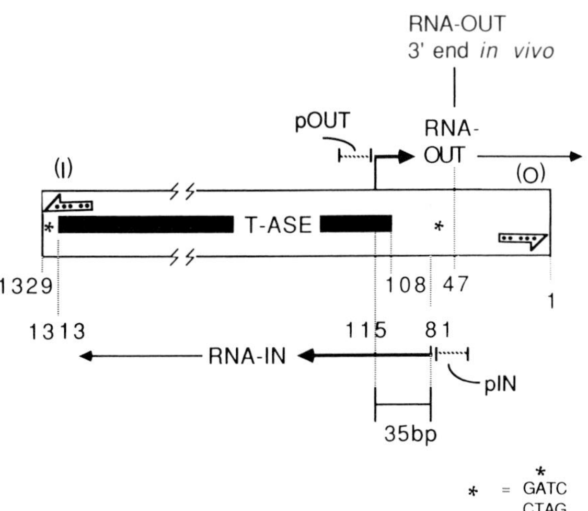

Figure 13. Structure of IS*10*. The following features are indicated: the transposase coding region (base pairs 108–1319, dark bar); the locations of two promoters, pIN and pOUT; the start points of the transposase message (RNA-IN, base pair 81) and the antisense RNA (RNA-OUT, base pair 115); the 35-bp region of complementarity between RNA-IN and RNA-OUT; the 3′ end of the RNA-OUT species observed in vivo; the nearly perfect inverted repeats at the outside (O) and inside (I) ends of IS*10*; and two *dam* methylation sites, one within pIN and one very near the inside end. Note that the orientation of IS*10* is opposite that in Fig. 12.

B. Levels of Transcription and Translation

The normal level of IS*10* transposase expression for a single-copy element is extremely low, fewer than 0.2 molecules of transposase per cell generation as estimated from analysis of a single-copy transposase-*lacZ* protein fusion. So little protein is made that polypeptide chains can only be detected in enzyme assays for β-galactosidase activity if cells are broken open and unassociated transposase–β-galactosidase monomers are allowed to assemble into active tetramers in vitro (129).

IS*10* transcription and translation rates can be estimated by an extension of the fusion analysis involving quantitative comparisons of β-galactosidase activity in whole-cell assays, where only preexisting tetramers give activity, and extract assays, which measures all molecules, preassembled or not. This analysis suggested that the transposase gene is transcribed at the rate of about 0.25 transcripts per element per cell generation and that each transcript has an average probability of being translated once of no more than 60%. Even this latter value is probably an overestimate (129; see Section VII.A, below).

Transcription of the IS*10* transposase gene is infrequent partly because pIN is a relatively weak promoter (149; B. C. Hoopes, Ph.D. thesis, Harvard University, Cambridge, Mass., 1986). Translation is inefficient in part because there is no obvious ribosome-binding-site sequence; single-base mutations that increase transposase translation create sequences that more closely resemble classic Shine-Dalgarno regions (C. Jain and N. Kleckner, unpublished observations). Other factors that modulate transposase expression are discussed below.

C. Sensitive *lacZ* Fusion Vectors

Analysis of IS*10* gene expression has frequently involved fusions between IS*10* sequences and the *E. coli lacZ* gene. Because transcription and translation of IS*10* occur at such low levels, special vectors have been developed in which nonspecific background levels of transcription and translation are low enough that the signals of interest can readily be detected (150). These vectors have two additional important features. First, in order that gene expression and its regulation could be examined both in the chromosome and on multicopy plasmids, the vectors were designed in such a way that constructs could be made on small plasmids and then crossed by in vivo genetic recombination onto appropriate λ phages, which could then be integrated into the chromosome to yield single-copy prophages. Second, features were incorporated that make it possible to generate regulatory mutations in the single-copy prophage fusions and then cross those mutations, again by genetic recombination, back onto multicopy vectors for further characterization.

V. REGULATION BY DNA ADENINE METHYLATION

A. Introduction

IS*10* is regulated in a complex way by *dam* methylation (132). The role of methylation was discovered by analysis of *E. coli* mutants exhibiting increased levels of Tn*10*-promoted rearrangements. Fifty such mutants were identified, and each carries a mutation in or near *dam*, the structural gene for DNA adenine methylase. This enzyme methylates the N-6 position of the adenines on both strands of the symmetrical sequence 5′ GATC. Previously isolated null (or nearly null) *dam* mutations confer the same transposition phenotype as do the new mutations.

The effects of *dam* mutations on IS*10* activity are a direct consequence of the absence of methylation at two specific GATC sites in IS*10*, not an indirect consequence of other changes in cellular DNA metabolism. When the two GATC sites are mutated to a sequence not recognized by *dam* methylase, IS*10* transposition becomes insensitive to the Dam phenotype of the host.

The two GATC sequences of IS*10* are strategically located; each one lies in an important genetic determinant whose activity is modulated by methylation. One site overlaps the −10 region of the transposase gene promoter pIN (Fig. 13; also below, Fig. 15), and absence of methylation at this site increases pIN transcription both in vivo and in vitro. The second site lies at the opposite end of the element, within the transposase binding site at its inside end; absence of methylation at this site activates this end as a substrate for transposition both in vivo and in vitro (116, 117, 132).

The combined effects of methylation at these two GATC sites completely account for the dramatic effects of Dam$^-$ mutations on Tn*10*- and IS*10*-promoted events. The frequency of IS*10* transposition is increased more than 100-fold in Dam$^-$ strains due to the combined effects of nonmethylation at both GATC sites. The frequency of Tn*10*-promoted deletions and deletion/inversions is stimulated even more, at least 500-fold, perhaps because these rearrangements depend not only upon the level of transposase expression, but on the activity of *two* methylation-sensitive inside ends. The frequency of Tn*10* transposition, whose (outside) ends do not respond to methylation, is increased only 10-fold, because it responds only to changes in methylation at the GATC site in pIN.

B. One Hemimethylated IS*10* Is Activated

There is normally no fully unmethylated DNA in wild-type *E. coli* strains. However, hemimethylated DNA is generated transiently upon replication of the transposon. These considerations suggested that IS*10* should be activated when it is hemimethylated, upon passage of the chromosomal replication fork, as well as when it is fully unmethylated. Further analysis revealed this to be the case, and, most importantly, that only one of the two chemically distinguishable hemimethylated IS*10* species is substantially activated for transposition. The most conservative estimate for the approximate relative transposition rates of the fully methylated and the two hemimethylated IS*10* species is 1:12:2,400; a less conservative but reasonable estimate is 1:12:60,000 (Fig. 14).

The conclusions above were reached largely by examination of IS*10* transposition off of hemimethylated DNA (132). IS*10* was transferred from a Dam$^+$ host to a Dam$^-$ host in an Hfr cross, and transposition off of the resulting newly transferred DNA was measured under conditions where integration of the transferred DNA into the recipient chromosome by homologous recombination was blocked by a RecA mutation. Such newly transferred DNA is hemimethylated because a single methylated strand moves from the (Dam$^+$) donor to the recipient and the complementary unmethylated strand is synthesized in the (Dam$^-$) recipient. Appropriate variations in this experiment permitted examination of each of the two chemically different hemimethylated species, assessment of the individual roles of each *dam* site in each species, and estimation of the activity of each hemimethylated species relative to that of a fully methylated element.

C. Three Factors in Activation

Three different factors contribute to the observed activity differences (Fig. 14). (i) Both hemimethylated promoters are activated relative to the

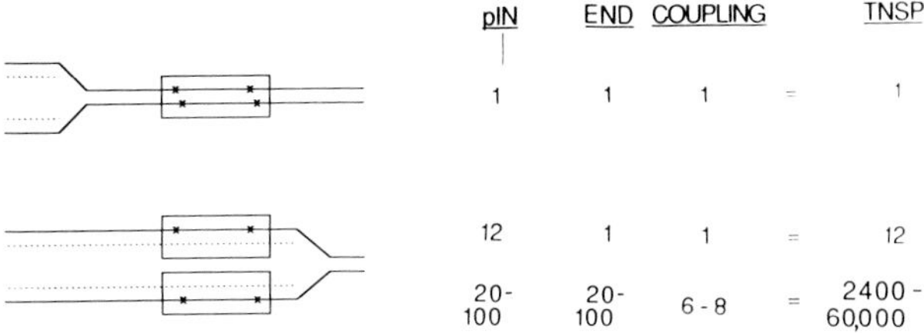

pIN
END
COUPLING
TNSP
1
1
1
=
1
12
1
1
=
12
20-100
20-100
6-8
=
2400-60,000

Figure 14. Effect of hemimethylation on IS*10* transposition. The activities of the two chemically different, hemimethylated IS*10* elements relative to that of fully methylated IS*10* have been estimated from several different lines of evidence (132). Factors that contribute to activation of one hemimethylated IS*10* are indicated. Qualitatively, one hemimethylated pIN element is more active than the other; both are more active than fully methylated pIN. Furthermore, one hemimethylated inside end is much more active than the other, and the less active end has essentially the same activity as a fully methylated inside end. Finally, an additional difference between the transposition frequencies of the two hemimethylated species is provided by the fact that expression and action of transposase are temporarily coupled; the inside end is activated for transposition at precisely the same time that transposase is expressed. Quantitative estimates of these effects can be obtained from in vivo experiments, although each estimate involves certain assumptions. In the most straightforward interpretation of the data (132, Table 4), one hemimethylated promoter is 8 times more active than the other and the less active promoter is a minimum of 12 times more active than the fully methylated promoter; similarly, one hemimethylated end is 90 times more active than the other, and the less active end has virtually the same activity as a fully methylated inside end. A more conservative interpretation, which omits one normalization factor, makes the active hemimethylated promoter 1.6 times more active than the other hemimethylated promoter and the more active hemimethylated end only 20 times more active than the other. The effect of coupling is six- and eightfold (132, Table 5). If the first interpretation is correct, the most active hemimethylated IS*10* element is some 60,000 times more active than a fully methylated element, while the less active element is only 12 times more active than a fully methylated element. In the second interpretation, the more active species is 2,400 times more active than the fully methylated element.

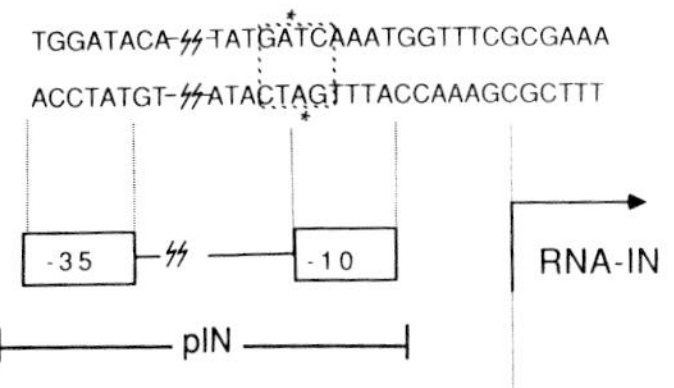

Figure 15. Location of one *dam* methylation site relative to the pIN promoter. Note that the orientation matches that of Fig. 12 and is opposite to that of Fig. 13.

fully methylated promoter, although the promoter of the hemimethylated element with the higher transposition rate is activated severalfold more than the other. (ii) Most importantly, only one of the two hemimethylated inside ends is activated. This asymmetry is the feature which ensures that only one of the two hemimethylated elements transposes. (iii) A third factor, referred to as "coupling," operates to further increase the differential activity of the more active hemimethylated species. The individual, independently assessed contributions of each hemimethylated promoter and terminus do not account for the activity of the more activated species. The missing factor is thought to derive from the fact that transposase is expressed at the same time as the transposon terminus is activated to receive it, that is, because expression and action of transposase protein are coupled. Because of coupling, a given amount of transposase protein results in more transposition events than if the terminus and the promoter were activated independently.

D. Mechanisms of pIN and Terminus Modulation

At the inside end, methylation directly affects the action of IS*10* transposase at that terminus: it inhibits circle formation by inside end substrates in vitro (117). Methylation could affect transposase binding and/or some subsequent step.

At the pIN promoter, the *dam* methylation site overlaps the −10 consensus region known to be important for promoter function (Fig. 15). However, two observations suggest that inhibition of pIN activity may involve some complexities. First, only one of the two methylated adenines lies in the −10 region, and yet **both** hemimethylated promoters are more active than fully methylated pIN; why does the absence of methylation outside of the −10 region improve promoter function? Second, in vitro analysis has revealed the existence of a second methylation-sensitive promoter, pX, that competes with pIN for the formation of open complexes. Open complex formation at pX and pIN respond in opposite ways to methylation: in the fully methylated state, pX is largely occupied while pIN is not; in the unmethylated state, pIN is fully occupied and very little polymerase is found at pX (132; Hoopes, Ph.D. thesis; J. Campbell and N. Kleckner, unpublished observations). When two promoters are coupled in this way, any change in the activity of one promoter changes the activity of the other; thus, methylation may directly affect only pX, only pIN, or both promoters. The location of pX has not been determined.

E. Other Transposons Are Regulated by DNA Adenine Methylation

The activities of transposons Tn*5* and Tn*903* are also regulated by DNA adenine methylation (112, 132, 184).

F. Biological Implications

1. Transposition is coupled to replication

Activation by hemimethylation means that in growing cells IS*10* transposition should be coupled to passage of the chromosomal replication fork. There is no precise estimate of how long the *dam* sites in IS*10* remain hemimethylated. Preliminary experiments suggest that it may be less than 10% of a cell generation time (Campbell and Kleckner, unpublished observations), an estimate consistent with previous estimates of bulk methylation levels (104, 161). If IS*10* remains hemimethylated for 10% of the cell cycle and the hemimethylated element is 1,000 times more active than fully methylated IS*10*, then the ratio of hemimethylated transpositions to fully methylated transpositions should be 100 to 0.9 (1,000 × 10% versus 1 × 90%) and more than 99% of transposition events should occur during the hemimethylated phase.

2. Only one sister chromosome gives rise to an IS*10* transposition

The pattern of the methylation effects ensures that only one of the newly replicated elements will transpose. This type of regulation could be especially important for an element like IS*10* that is thought to leave behind a double-stranded gap in the donor molecule during transposition. It would ensure that transposition only occurs when a second copy of the chromosome or at least a second copy of the critical chromosomal region is present in the cell. In addition, the presence of a nearby sister chromosome could increase the probability that the donor chromosome is successfully repaired by recombinational "double-strand gap repair."

3. Transient induction during single-strand transfer

Activation of hemimethylated IS*10* should also mean that transposition will be transiently induced when IS*10* or Tn*10* enters a new host organism by conjugal transfer, phage infection, or transformation by mechanisms involving uptake of single strands (121, 154). This may be important for the spread of IS*10* among bacteria, because these transfer mechanisms seem to be the primary way in which Tn*10*-encoded tetracycline resistance has spread among different bacterial populations (for a recent summary, see reference 97).

4. A second type of transposase *cis* action

Transposase made by one copy of IS*10* will not act on other IS*10* copies located elsewhere in the cell because the inside ends of those other elements will not be activated at the critical time. Thus, *dam* regulation provides a second mechanism for ensuring that transposase is effectively *cis* acting. This mechanism for *cis* action, which derives from coupling of expression and action, is complementary to the *cis* action that derives from the intrinsic properties of transposase protein (see above). The relative quantitative contributions of the two mechanisms have not been assessed.

VI. MULTICOPY INHIBITION: REGULATION BY ANTISENSE PAIRING AND PREFERENTIAL *cis* ACTION

A. Introduction

1. The phenomenon

The rate of IS*10* or Tn*10* transposition per transposon copy actually decreases as the number of copies per cell increases (151; J. Matsunaya, A. Toyofuku, and R. W. Simons, unpublished observations) (Fig. 16). This effect is observed experimentally as a decrease in the rate of transposition of a single marked copy of Tn*10* in the presence of a plasmid carrying an intact wild-type IS*10* element. This phenomenon has been called multicopy inhibition, and it results from the combination of the following two properties of IS*10* (151).

First, transposase protein made by one copy of IS*10* or Tn*10* almost never reaches a second copy present elsewhere in the cell. For IS*10*, two distinct processes contribute to this effect: limited movement of the protein itself, and coupling of transposase action to its expression by *dam* methylation. For

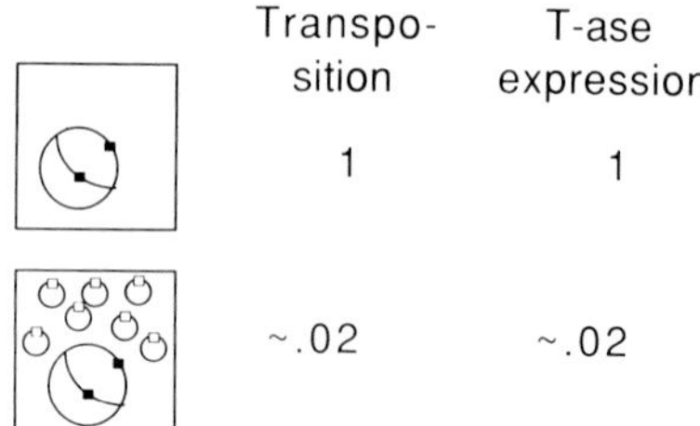

Figure 16. Effects of a multicopy IS*10* plasmid on the activity of a genetically marked single-copy Tn*10* or IS*10* element. Transposition of the single-copy Tn*10* element (solid square) can be monitored specifically because only the complete transposon carries tetracycline resistance. Expression of transposase (T-ase) from a single-copy Tn*10* or IS*10* element can be measured using a single-copy transposase-*lacZ* protein fusion or by a complementation assay; in the latter case, the transposase gene on the multicopy plasmid must be inactivated. (See Fig. 17 for more details about this type of analysis.)

Tn*10*, only the first of these processes contributes (see above). Because transposase acts only in *cis*, the effective transposase concentration per element remains essentially constant as the number of elements per cell increases. Without this feature, the rate of transposition per element would tend to increase as the total cellular transposase concentration increased.

Second, IS*10* encodes a *trans*-acting negative regulator of transposase gene expression whose effectiveness increases with transposon copy number. This negative regulator is an antisense RNA, the transcript from the pOUT promoter, RNA-OUT (Fig. 13). RNA-OUT inhibits transposase gene translation by pairing with the 5′ end of the transposase mRNA, RNA-IN, over a 35-bp region of complementarity provided by the overlap in their DNA template sequences (Fig. 13). As the number of IS*10* copies per cell increases, the rate of transposase expression per copy decreases because having more copies of the element results in a higher concentration of the antisense RNA.

The individual contributions of these two factors are depicted in Fig. 17. The rate of Tn*10* transposition decreases 10-fold in the presence of the IS*10* plasmid even though the total concentration of transposase in the cell, as assayed by complementation of a function-defective element, has increased 40-fold. Furthermore, plasmids that lack transposase but express RNA-OUT inhibit Tn*10* transposition to the same extent as the wild-type IS*10* plasmid; control plasmids lacking RNA-OUT have no effect (151; unpublished observations).

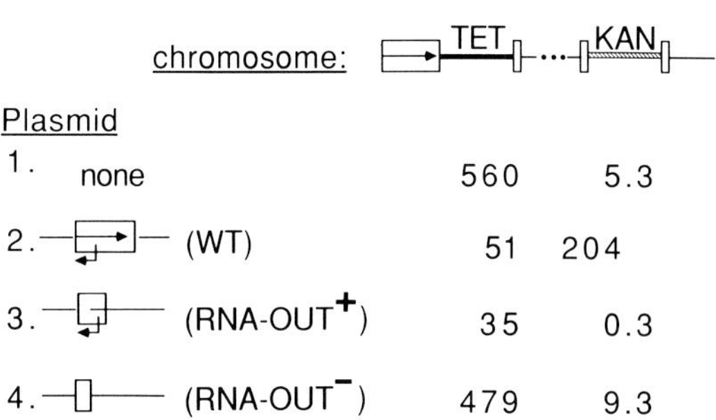

Figure 17. Multicopy inhibition results from the combination of a *cis*-acting transposase protein and a *trans*-acting negative regulator. A strain was constructed carrying chromosomal insertions of both a transposition-proficient Tn*10* element marked with tetracycline resistance genes and a transposase-deleted mini-Tn*10* element consisting of the two outside ends of Tn*10* and a gene for kanamycin resistance. Plasmids were introduced into this strain; such plasmids carried either wild-type IS*10* (line 2), the outer portion of IS*10* (base pairs 1 to 180), which encodes pOUT and RNA-OUT but not transposase (line 3), or a smaller portion of IS*10* (base pairs 1 to 120), which lacks pOUT and thus does not express RNA-OUT (line 4). The resulting strains were then assayed for the frequency of transposition of the two chromosomal elements by a standard "mating-out" assay as described in reference 50. (Line 1) In the absence of any plasmid, transposition of the mini-Tn*10*-Kan^r element is much rarer than transposition of the intact Tn*10* element; this difference reflects the failure of IS*10* transposase to work efficiently in *trans* due to its failure to move freely through the cell. (Line 2) When a wild-type IS*10* plasmid is present, transposition of the mini-Tn*10*-Kan^r element increases dramatically as a consequence of the increased level of transposase present in the cell. In contrast, transposition of the intact Tn*10* element decreases 10-fold. Transposition does not increase, because the increased level of transposase acting inefficiently in *trans* provides less than 10% of the level of transposase protein that the element normally provides for itself in *cis*; transposition decreases because the plasmid-encoded antisense RNA-OUT is present in high concentration and inhibits expression of the transposase made from and destined to act in *cis* upon the intact Tn*10* element itself. (Line 3) The effect of RNA-OUT alone can be seen when the multicopy plasmid does not also make transposase. Now transposition of both the intact Tn*10* element and the mini-Tn*10*-Kan^r element decreases about 10-fold, because transposition of both elements is totally dependent upon transposase expressed from the intact Tn*10*. (Line 4) No inhibition of transposase expression is observed in the presence of a plasmid isogenic to that in line 3 except that the pOUT promoter, responsible for RNA-OUT production, has been deleted.

2. Evidence for antisense RNA pairing

That RNA-OUT is the negative regulator and that it works by RNA-RNA pairing is now well documented (151; J. D. Kittle, R. W. Simons, and N. Kleckner, submitted for publication; R. W. Simons, C. Case, S. Roels, P. Jensen, J. Lee, and N. Kleckner, manuscript in preparation). Mutations that increase or decrease pOUT expression or decrease RNA-OUT stability in vivo have corresponding effects on multicopy inhibition. The ability of RNA-OUT to inhibit transposase expression can be titrated by high levels of appropriate RNA-IN messages provided in *trans*. The paired species is not seen in vivo, but specific cleavage products derived from the paired species are. Finally, pairing between RNA-IN and RNA-OUT has been observed in vitro, and there is a close quantitative correspondence between the effects of mutations on pairing in vitro and multicopy inhibition in vivo. These findings are discussed individually below.

3. Mechanism of antisense inhibition

Pairing between RNA-IN and RNA-OUT prevents expression of the transposase gene at a step subsequent to transcription initiation (151): multicopy inhibition has a 30- to 50-fold effect on *lacZ* expression from a translational pIN-transposase-*lacZ* fusion making a single fusion polypeptide from the transposase translation start signals. In contrast, it reduces expression less than threefold from transcriptional (operon) pIN-transposase-*lacZ* fusions in which *lacZ* is translated from its own ribosome-binding site.

The simplest mechanism by which RNA-OUT could block transposase translation is by preventing binding of ribosomes. The region of RNA-IN that is complementary to RNA-OUT includes both the transposase gene start codon and the upstream region where ribosomes must bind, and ribosomes are generally thought not to bind efficiently to double-stranded RNA (158). An alternative model in which antisense pairing is inhibitory because it leads to degradation of RNA-IN now appears unlikely, as discussed below.

The residual reduction of expression observed with operon fusions could be a secondary consequence either of inhibition of translation or of messenger destabilization following pairing, as discussed below. In favor of the first possibility is the observation that point mutations that alter translation of the transposase gene segment in the operon fusion constructs by altering the AUG start codon or creating chain-terminating frameshift mutations inhibit *lacZ* expression to the same extent as does multicopy inhibition (Jain and Kleckner, unpublished observations).

4. Antisense inhibition is exclusively a *trans* effect

Multicopy inhibition is almost exclusively a diffusible *trans* effect. Mutations that specifically abolish transcription from pOUT or destabilize RNA-OUT in vivo abolish multicopy inhibition but have essentially no effect on transposase expression or transposition of a single-copy Tn*10* element: they increase transposition by twofold or less (Matsunaga et al., unpublished observations). In vitro and in vivo promoter analysis also suggests that no significant *cis* inhibition should be expected; initiation from pIN and pOUT occurs independently and sufficiently infrequently that an IS*10* template is unlikely to initiate both an RNA-OUT and an RNA-IN transcript at the same time.

B. RNA-IN and RNA-OUT: Synthesis and Stability

The primary RNA-OUT species detected in vivo is a 69-base molecule that extends from the pOUT transcription start point at base pair 115 to a 3′ end at base pair 47; minor species beginning at base pairs 116 and 117 or ending at base pair 46 are also detected. RNA-OUT is extremely stable; its in vivo half-life is about 70 min. It is present in about five molecules per copy of IS*10*, which corresponds to

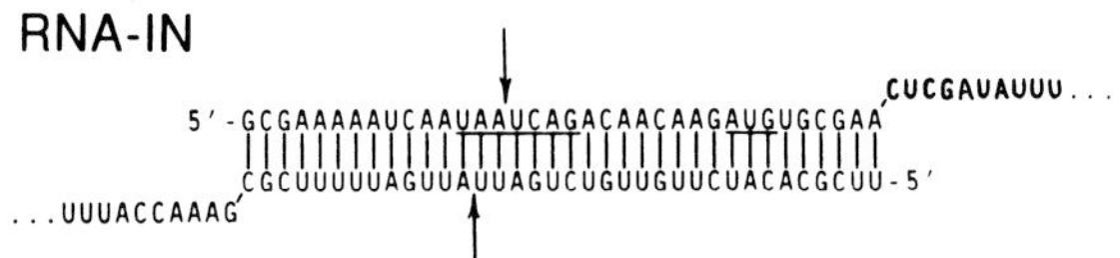

Figure 18. Sites where RNase III cleaves the RNA-IN/RNA-OUT hybrid. Primer extension of RNA-IN and RNA-OUT transcripts from cells expressing both RNAs reveals, for each transcript, one major new 5′ end that is not present in cells expressing only RNA-IN or only RNA-OUT. These two ends reflect the occurrence of a staggered cleavage within the paired region, typical of RNase III cleavage, and these products are not observed in an RNase III$^-$ mutant strain.

about 150 to 500 copies in a cell containing a typical multicopy IS*10* plasmid (21, 96, 143).

In contrast, the level of RNA-IN is less than 1% of the level of RNA-OUT. This paucity of RNA-IN molecules is due in part to infrequent transcription, 0.25 transcripts per IS*10* copy per generation (see above). In addition, RNA-IN is much less stable than RNA-OUT: the half-life of the 5′ end of RNA-IN in vivo is 40 s (C. Case and R. W. Simons, manuscript in preparation). These two factors alone account for at least a 100-fold difference in the relative levels of RNA-OUT and the 5′ end of RNA-IN. RNase III cleavage of paired transcripts further reduces the level of RNA-IN (see below). From these considerations it seems likely that when IS*10* is present on a multicopy plasmid, an RNA-IN molecule is usually newly made in a cell that lacks any other RNA-IN molecules.

C. RNA-IN/RNA-OUT Hybrids Are Cleaved

RNA-IN/RNA-OUT hybrid molecules are cleaved in vivo by RNase III (Case and Simons, in preparation). Primer extension analysis of RNAs extracted from a cell producing both RNA-IN and RNA-OUT reveals species having 5′ ends within the 35-bp region of complementarity. A single major cleavage site is observed on each strand, and the two single-strand cleavages are staggered 2 bp apart in the duplex (Fig. 18). This pattern is typical of the cleavages generated when RNase III acts on double-stranded RNA (134), and no cleavage products are observed in an RNase III$^-$ mutant strain.

The pairing between RNA-IN and RNA-OUT revealed by these cleavage products is essentially stoichiometric. The absolute and relative levels of the RNA-IN and RNA-OUT have been varied over wide ranges. In each case, primer extension analysis only detects a primary 5′ end for the RNA species that is in excess.

RNase III cleavage of RNA-IN/RNA-OUT hybrids is not required for effective antisense inhibition in vivo. Multicopy inhibition is not significantly altered in the RNase III mutant strain where cleavage is not detected. This argues that cleavage is a secondary cellular processing step that plays no role in IS*10* antisense regulation.

D. Structure and Function of RNA-OUT

RNA-OUT has a stem-loop structure both in vitro and in vivo. The two structural domains correspond to two functional domains.

1. RNA-OUT has a stem-loop structure in solution

A thermodynamically stable hairpin secondary structure for RNA-OUT is clear from simple inspection of the sequence (Fig. 19). This structure consists of a strong 21-bp, double-stranded "stem domain" and a "loop domain" which is not precisely defined. The most stable predicted structure of this loop domain consists of two short duplex regions topped by a six-nucleotide loop. The calculated stability of this structure is −22 kcal/mol at 37°C. Complementarity with RNA-IN begins at the top of the RNA-OUT loop and extends down one entire side of the stem.

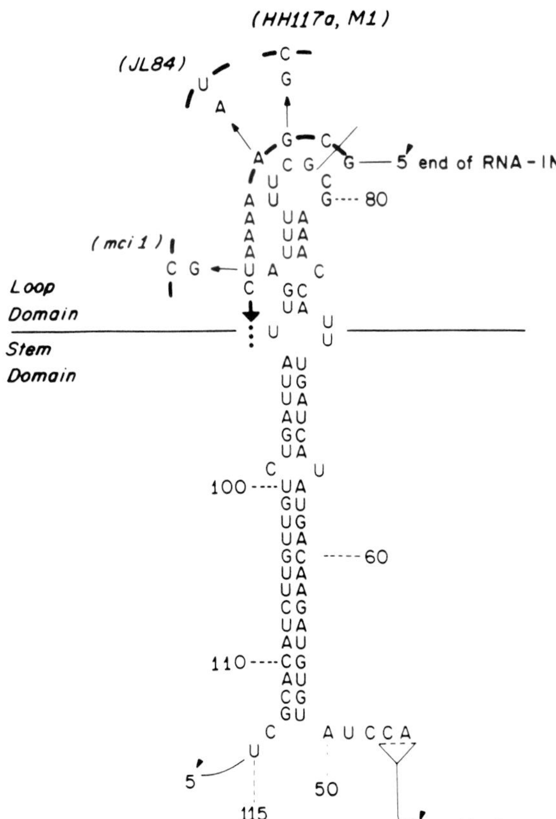

Figure 19. Structure of RNA-OUT, relationship between the sequences of RNA-IN and RNA-OUT, and locations of three important mutations that alter both RNAs. Note that positions correspond to distances from the outside end of IS*10*. RNA-IN starts at position 81.

The proposed structure is supported by the behavior of RNA-OUT in solution. Upon electrophoresis in urea gradient gel, RNA-OUT undergoes a sharp shift to a slower mobility, diagnostic of a major structural transition to a less compact form; furthermore, mutations located throughout the molecule affect RNA-OUT gel mobility under appropriate conditions exactly as predicted from their calculated effects on the thermodynamic stability of the molecule (Kittle et al., submitted).

2. The stem domain confers stability in vivo

Analysis of mutant RNA-OUT species suggests that RNA-OUT has a functionally distinct double-stranded stem domain (Simons et al., in preparation). One important role of this domain in vivo is to ensure the stability of the antisense RNA. Single-base changes that destabilize the RNA-OUT stem domain dramatically reduce the levels of RNA-OUT or its measured half-life in vivo; double-mutant combinations that restore base pairing within the stem domain also restore physical stability. Mutations in the proposed RNA-OUT loop domain do not drastically alter RNA-OUT stability. These observations can be understood according to the current view that the primary mode of single-stranded RNA degradation in *E. coli* is by exonucleolytic digestion from the 3′ end and that such degradation can be blocked by a stable stem-loop structure (summarized in reference 122).

3. The loop domain is critical for RNA pairing

Analysis of pairing between RNA-IN and RNA-OUT suggests that the loop domain of RNA-OUT plays a critical role, described in detail below. Briefly, mutations in the loop domain can have severe effects on the ability of mutant molecules to pair with wild-type or mutant RNA-IN molecules. The same is almost certainly true in vivo, as deduced from in vivo analysis of the mutations characterized in vitro and from the genetic properties of new mutations not yet subjected to in vitro analysis.

The view that the two structural domains of RNA-OUT correspond to two functional domains is underscored by the general finding that most mutations in the loop domain do not affect RNA-OUT stability in vivo, while mutations in the stem domain have no effect on RNA-RNA pairing in vitro.

E. RNA Pairing in Vitro

1. Detection of a second-order reaction

Formation of fully paired complexes between RNA-OUT and RNA-IN in vitro has been observed using gel mobility and RNase digestion assays. The pairing reaction is second order with an apparent rate constant of 3×10^5 M^{-1} s^{-1} at 37°C (Kittle et al., submitted).

2. Mutant analysis and a model for pairing

The rate of formation of fully paired complexes depends critically upon the nature of the RNA-OUT loop domain and the 5′ end of RNA-IN and is insensitive to changes in the RNA-OUT stem domain. From the analysis of mutant RNAs in vitro, it is proposed that pairing between the two RNAs is initiated by an interaction between the 5′ end of RNA-IN and the RNA-OUT loop domain and that this initial interaction is followed by subsequent step(s) in which RNA-IN displaces one strand of the RNA-OUT stem domain to give the full 35 bp of RNA-IN/RNA-OUT duplex pairing (Kittle et al., submitted). Specifically:

(i) Mutations that alter the rate of pairing in vitro can be grouped into three categories whose effects can be accounted for by the proposed model.

(a) Mutations that close up the loop domain of RNA-OUT. Such mutations sequester bases that must normally be free to pair with RNA-IN.

(b) Addition of extra bases at the 5′ end of RNA-IN. The 5′ end of RNA-IN must rotate one time around RNA-OUT for every full turn of RNA-IN/RNA-OUT duplex formed. Normally this is not a problem, because pairing initiates at the 5′ end of RNA-IN and so the nascent duplex can rotate freely around RNA-OUT. When extra bases are present, however, the entire 5′ extension must wind in and out of the RNA-OUT loop domain once for every full turn.

(c) An important class of mutations map to the very 5′ end of RNA-IN and the complementary bases of RNA-OUT. Pairing between RNA-IN and RNA-OUT is critically dependent upon interactions involving these bases. Two such mutations have been analyzed in detail.

(ii) One mutation has been shown to alter the sequence specificity of the RNA pairing reaction in vitro. Mutation *HR117a/M1*, at the third base of RNA-IN and the corresponding base of RNA-OUT, does not interfere with pairing when it is present in **both** RNA-IN and RNA-OUT; in fact, pairing is slightly faster in this homologous mutant than in the wild type. However, the mutation completely abolishes pairing when present in either RNA-IN **or** RNA-OUT; no pairing is observed in either heterologous reaction in which one RNA is mutant and the other is wild type. The behavior of this mutation suggests that hydrogen bonding between bases in the two RNAs at this position is important for pairing.

(iii) A second mutation has more complex effects on the pairing reaction. Mutation *JL84*, at the fourth base of RNA-IN and the corresponding RNA-OUT base, abolishes pairing in one heterologous combination and gives intermediate effects in the other two cases. Factors other than base pairing are presumably significant here.

The *HH117a/M1* and *JL84* mutations have similar effects on RNA pairing in vivo (see below). Additional mutations with in vivo properties similar to those of each of these mutations have been identified; two additional mutations that alter the sequence specificity of the reaction have been identified at the positions corresponding to the second and fifth bases of RNA-IN, and a second change at the fourth base behaves very much like *JL84* (S. Roels, C. Masada, and R. W. Simons, unpublished data).

3. Intermediates

No intermediates in IS*10* antisense RNA pairing have been identified. The genetic analysis demonstrates that formation of a stable complex between RNA-IN and RNA-OUT is more sensitive to single-base changes at the 5′ end of RNA-IN and the loop domain of RNA-OUT than to changes elsewhere in the region of complementarity. However, consideration of data from other systems provides some insight into how the reaction might occur.

RNA-RNA duplexes of about 8 bp in length or longer should be stable under physiological conditions (54; J. D. Kittle, Ph.D. thesis, Harvard University, Cambridge, Mass., 1988). Thus, it seems likely that the formation of a stable complex between RNA-IN and RNA-OUT may require pairing between the 5′ end of RNA-IN and the loop domain of RNA-OUT over just less than a single helical turn.

The antisense pairing reaction analyzed most extensively in vitro, that of ColE1, is initiated by a reversible interaction between the antisense RNA and its target RNA (167–169); evidence for an initial reversible interaction has also recently been obtained for a second antisense reaction, that involving p22 *sar* RNA (99, 100, 183). Furthermore, in the ColE1 case, single-base mutations that alter the sequence specificity of the reaction do so by affecting this particular step.

If IS*10* antisense pairing is analogous, pairing

brids is not required for effective antisense inhibition in vivo. Multicopy inhibition is not significantly altered in the RNase III mutant strain where cleavage is not detected. This argues that cleavage is a secondary cellular processing step that plays no role in IS*10* antisense regulation.

D. Structure and Function of RNA-OUT

RNA-OUT has a stem-loop structure both in vitro and in vivo. The two structural domains correspond to two functional domains.

1. RNA-OUT has a stem-loop structure in solution

A thermodynamically stable hairpin secondary structure for RNA-OUT is clear from simple inspection of the sequence (Fig. 19). This structure consists of a strong 21-bp, double-stranded "stem domain" and a "loop domain" which is not precisely defined. The most stable predicted structure of this loop domain consists of two short duplex regions topped by a six-nucleotide loop. The calculated stability of this structure is −22 kcal/mol at 37°C. Complementarity with RNA-IN begins at the top of the RNA-OUT loop and extends down one entire side of the stem.

The proposed structure is supported by the behavior of RNA-OUT in solution. Upon electrophoresis in urea gradient gel, RNA-OUT undergoes a sharp shift to a slower mobility, diagnostic of a major structural transition to a less compact form; furthermore, mutations located throughout the molecule affect RNA-OUT gel mobility under appropriate conditions exactly as predicted from their calculated effects on the thermodynamic stability of the molecule (Kittle et al., submitted).

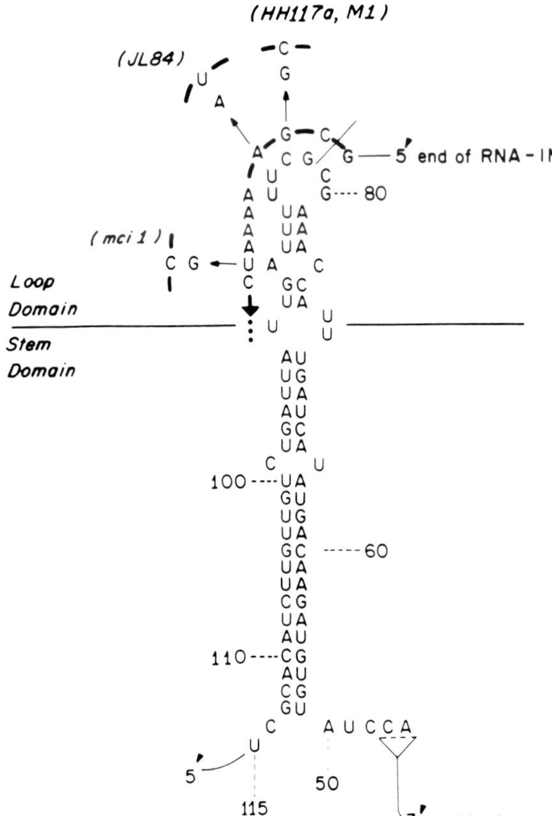

Figure 19. Structure of RNA-OUT, relationship between the sequences of RNA-IN and RNA-OUT, and locations of three important mutations that alter both RNAs. Note that positions correspond to distances from the outside end of IS*10*. RNA-IN starts at position 81.

2. The stem domain confers stability in vivo

Analysis of mutant RNA-OUT species suggests that RNA-OUT has a functionally distinct double-stranded stem domain (Simons et al., in preparation). One important role of this domain in vivo is to ensure the stability of the antisense RNA. Single-base changes that destabilize the RNA-OUT stem domain dramatically reduce the levels of RNA-OUT or its measured half-life in vivo; double-mutant combinations that restore base pairing within the stem domain also restore physical stability. Mutations in the proposed RNA-OUT loop domain do not drastically alter RNA-OUT stability. These observations can be understood according to the current view that the primary mode of single-stranded RNA degradation in *E. coli* is by exonucleolytic digestion from the 3′ end and that such degradation can be blocked by a stable stem-loop structure (summarized in reference 122).

3. The loop domain is critical for RNA pairing

Analysis of pairing between RNA-IN and RNA-OUT suggests that the loop domain of RNA-OUT plays a critical role, described in detail below. Briefly, mutations in the loop domain can have severe effects on the ability of mutant molecules to pair with wild-type or mutant RNA-IN molecules. The same is almost certainly true in vivo, as deduced from in vivo analysis of the mutations characterized in vitro and from the genetic properties of new mutations not yet subjected to in vitro analysis.

The view that the two structural domains of RNA-OUT correspond to two functional domains is underscored by the general finding that most mutations in the loop domain do not affect RNA-OUT stability in vivo, while mutations in the stem domain have no effect on RNA-RNA pairing in vitro.

E. RNA Pairing in Vitro

1. Detection of a second-order reaction

Formation of fully paired complexes between RNA-OUT and RNA-IN in vitro has been observed using gel mobility and RNase digestion assays. The pairing reaction is second order with an apparent rate constant of 3×10^5 M^{-1} s^{-1} at 37°C (Kittle et al., submitted).

2. Mutant analysis and a model for pairing

The rate of formation of fully paired complexes depends critically upon the nature of the RNA-OUT loop domain and the 5′ end of RNA-IN and is insensitive to changes in the RNA-OUT stem domain. From the analysis of mutant RNAs in vitro, it is proposed that pairing between the two RNAs is initiated by an interaction between the 5′ end of RNA-IN and the RNA-OUT loop domain and that this initial interaction is followed by subsequent step(s) in which RNA-IN displaces one strand of the RNA-OUT stem domain to give the full 35 bp of RNA-IN/RNA-OUT duplex pairing (Kittle et al., submitted). Specifically:

(i) Mutations that alter the rate of pairing in vitro can be grouped into three categories whose effects can be accounted for by the proposed model.

(a) Mutations that close up the loop domain of RNA-OUT. Such mutations sequester bases that must normally be free to pair with RNA-IN.

(b) Addition of extra bases at the 5′ end of RNA-IN. The 5′ end of RNA-IN must rotate one time around RNA-OUT for every full turn of RNA-IN/RNA-OUT duplex formed. Normally this is not a problem, because pairing initiates at the 5′ end of RNA-IN and so the nascent duplex can rotate freely around RNA-OUT. When extra bases are present, however, the entire 5′ extension must wind in and out of the RNA-OUT loop domain once for every full turn.

(c) An important class of mutations map to the very 5′ end of RNA-IN and the complementary bases of RNA-OUT. Pairing between RNA-IN and RNA-OUT is critically dependent upon interactions involving these bases. Two such mutations have been analyzed in detail.

(ii) One mutation has been shown to alter the sequence specificity of the RNA pairing reaction in vitro. Mutation *HR117a/M1*, at the third base of RNA-IN and the corresponding base of RNA-OUT, does not interfere with pairing when it is present in **both** RNA-IN and RNA-OUT; in fact, pairing is slightly faster in this homologous mutant than in the wild type. However, the mutation completely abolishes pairing when present in either RNA-IN **or** RNA-OUT; no pairing is observed in either heterologous reaction in which one RNA is mutant and the other is wild type. The behavior of this mutation suggests that hydrogen bonding between bases in the two RNAs at this position is important for pairing.

(iii) A second mutation has more complex effects on the pairing reaction. Mutation *JL84*, at the fourth base of RNA-IN and the corresponding RNA-OUT base, abolishes pairing in one heterologous combination and gives intermediate effects in the other two cases. Factors other than base pairing are presumably significant here.

The *HH117a/M1* and *JL84* mutations have similar effects on RNA pairing in vivo (see below). Additional mutations with in vivo properties similar to those of each of these mutations have been identified; two additional mutations that alter the sequence specificity of the reaction have been identified at the positions corresponding to the second and fifth bases of RNA-IN, and a second change at the fourth base behaves very much like *JL84* (S. Roels, C. Masada, and R. W. Simons, unpublished data).

3. Intermediates

No intermediates in IS*10* antisense RNA pairing have been identified. The genetic analysis demonstrates that formation of a stable complex between RNA-IN and RNA-OUT is more sensitive to single-base changes at the 5′ end of RNA-IN and the loop domain of RNA-OUT than to changes elsewhere in the region of complementarity. However, consideration of data from other systems provides some insight into how the reaction might occur.

RNA-RNA duplexes of about 8 bp in length or longer should be stable under physiological conditions (54; J. D. Kittle, Ph.D. thesis, Harvard University, Cambridge, Mass., 1988). Thus, it seems likely that the formation of a stable complex between RNA-IN and RNA-OUT may require pairing between the 5′ end of RNA-IN and the loop domain of RNA-OUT over just less than a single helical turn.

The antisense pairing reaction analyzed most extensively in vitro, that of ColE1, is initiated by a reversible interaction between the antisense RNA and its target RNA (167–169); evidence for an initial reversible interaction has also recently been obtained for a second antisense reaction, that involving p22 *sar* RNA (99, 100, 183). Furthermore, in the ColE1 case, single-base mutations that alter the sequence specificity of the reaction do so by affecting this particular step.

If IS*10* antisense pairing is analogous, pairing

could initiate by a reversible interaction between the first few bases at the 5′ end of RNA-IN and the complementary RNA-OUT bases. The effects of mutations at the first, third, fourth, and fifth bases of RNA-IN and their RNA-OUT complements have phenotypes analogous to those of the ColE1 mutations that alter the reversible interaction. Interestingly, a single base change near the 5′ end of RNA-IN, *mci1* at the ninth base, has no effect on RNA-IN pairing with wild-type RNA-OUT. This could be an indication that the proposed reversible interaction does not extend as far as the ninth base of RNA-IN.

According to the general pairing model proposed above, pairing between the 5′ end of RNA-IN and the loop domain of RNA-OUT will be followed by propagation of the nascent duplex toward the 3′ ends of the RNAs to form a stable complex. After formation of between 9 and 12 bp, depending upon the exact structure of the RNA-OUT loop domain, additional pairing requires displacement of one RNA-OUT strand in the stem domain by the RNA-IN strand. This aspect of the reaction has not been examined in vitro, except that preliminary observations suggest that most complexes complete this step within 1 or 2 min after formation of a stable complex. Also, it is not known whether full pairing is required in vivo for the inhibition of transposase expression from RNA-IN. One interesting point is that the RNA-OUT stem domain is imperfectly paired while the RNA-IN/RNA-OUT duplex is perfect; the mismatches present in the stem duplex could help to keep strand displacement proceeding in the forward direction, since the backward reaction would necessitate the replacement of perfect base pairs with imperfect ones.

4. Comparison with ColE1 and *sar* antisense pairing reactions

Although the IS*10* reaction has not been characterized in the same elegant detail as the ColE1 reaction (167–170), certain similarities are already clear. Both reactions involve essential interactions between single-stranded regions in both the antisense and target RNAs. Both reactions require the presence of a free 5′ end on one of the RNAs at the point where stable base pairing is initiated, presumably to allow one strand to wind around the other. Both reactions can be disrupted by single-base mutations in critical regions, and in both cases a single-base change is sufficient to change the sequence specificity.

Furthermore, the rate constant for the IS*10* pairing reaction is very similar to that for ColE1 in the absence of its specific facilitating protein, Rom (169). This may be a coincidence. However, the third antisense reaction examined in vitro, P22 *sar*, exhibits exactly the same apparent second-order rate constant as the IS*10* reaction, $3 \times 10^5\ M^{-1}\ s^{-1}$ (99). It is tempting to suppose that the three reactions are fundamentally similar in some way. One possibility is that the rates of all of these reactions are determined by formation of a very small number of RNA-RNA base pairs.

F. Relationship Between Pairing In Vitro and Multicopy Inhibition In Vivo

Steps critical for RNA-RNA pairing in vitro are also important in vivo. Three mutations that affect the loop domain of RNA-OUT or the 5′ end of RNA-IN or both, including one that cleanly alters the sequence specificity of the reaction, have effects on antisense inhibition in vivo that are strikingly similar to those on stable complex formation in vitro. The results for the three mutations discussed above, *HH117a/M1*, *JL84*, and *mci1*, are shown in Fig. 20.

However, despite these close correspondences, the rate in vivo could be significantly faster than it is in vitro. Given reasonable estimates for the level of RNA-OUT, the half-life of RNA-IN, and the role of RNA-IN secondary structure (see Section VII.A, below), the second-order rate constant obtained in vitro does not appear to be sufficient to account for the levels of multicopy inhibition observed (Kittle et al., submitted). It would not be surprising if conditions inside the cell differ significantly from those used in vitro.

G. Host Functions Are Involved

E. coli mutations have been isolated that reduce or eliminate multicopy inhibition as assayed by the expression of a transposase-*lacZ* gene fusion or by transposition (J. Krull and R. W. Simons, manuscript in preparation). These experiments have revealed that *E. coli* integration host factor (IHF) may be involved. One major class of mutations with a strong Mci^- phenotype maps in or near the *himA* gene, the gene encoding one subunit of IHF. Null mutations in either *himA* or the gene encoding the second IHF subunit, *hip*, reduce multicopy inhibition, but to a lesser extent than the newly isolated mutants. One interesting possibility would be that IHF interacts directly with one or both RNAs to stimulate the pairing reaction. Two other observations implicate IHF in RNA interactions: IHF is involved in regulation of λ *cII* gene expression at a posttranscriptional level (106), and HU protein, which is related to IHF, binds RNA and was originally purified as a weakly

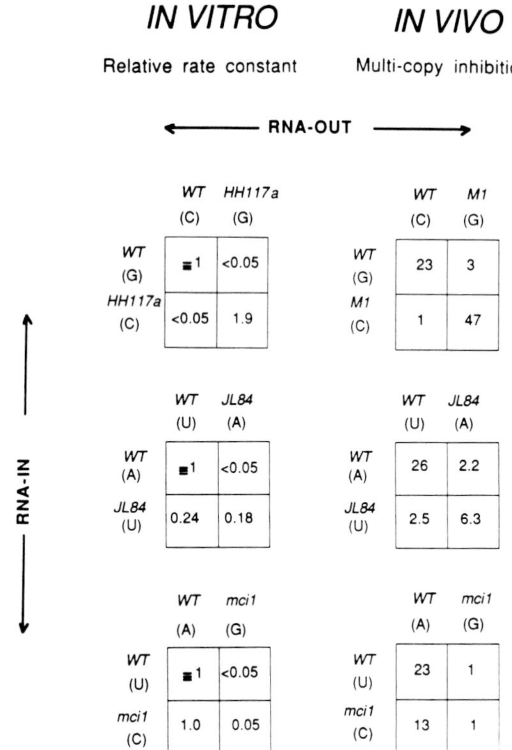

Figure 20. Effects of selected mutations on pairing between RNA-IN and RNA-OUT in vitro and in vivo. The *mci1*, *JL84*, and *HH117a*/*M1* mutations occur in IS*10* base pairs common to the template sequences of both RNA-IN and RNA-OUT (Fig. 19). For each mutation, pairing in vitro and multicopy inhibition in vivo have been examined for the homologous mutant case, in which both RNA-IN and RNA-OUT carry the same mutation, and for the two heterologous cases, in which RNA-IN carries the mutation while RNA-OUT is wild, and vice versa. For each in vitro reaction, the second-order pairing rate constant is presented, normalized to the rate constant for pairing between two wild-type RNAs. For each in vivo experiment, wild-type or mutant RNA-IN was provided by a single-copy Tn*10* element, wild-type or mutant RNA-OUT was provided by a multicopy plasmid not expressing transposase, and the frequency of transposition of the Tn*10* element was measured as in Fig. 17. For each combination, the level of multicopy inhibition is a ratio between the frequency of Tn*10* transposition in the absence of antisense RNA (with an RNA-OUT$^-$ control plasmid) and the frequency of Tn*10* transposition in the presence of a plasmid expressing RNA-OUT of the indicated genotype. The wild-type level of multicopy inhibition is about 25-fold. Mutation *mci1* only affects pairing when present in RNA-OUT because it stabilizes an internal secondary structure within the RNA-OUT loop domain. *HH117a*/*M1* cleanly changes the sequence specificity of the pairing reaction and, in the homologous combination, increases the efficiency of interaction both in vitro and in vivo. Mutation *JL84* has more complex effects. In vitro, mutation *HH117a*/*M1* cleanly changes the specificity of the pairing reaction. The homologous mutant RNAs pair as effectively as, in fact somewhat more effectively than, the two wild-type RNAs; in contrast, the two heterologous combinations do not pair detectably. Mutation *JL84* gives a somewhat more complex pattern of effects, although it significantly decreases the rate of pairing in all cases. Mutation *mci1* is very different from the other two. *mci1* RNA-OUT fails to pair with any RNA-IN, including the homologous *mci1* RNA-IN; this is because it creates additional secondary structure within the RNA-OUT loop domain which sequesters bases that normally are free to interact with the 5′ end of RNA-IN. In contrast, *mci1* RNA-IN pairs normally with wild-type RNA-OUT, suggesting that the pairing reaction is not sensitive to a single-base heterology in this position. The pattern of mutant effects is strikingly similar in vitro and in vivo. *HH117a*/*M1* shows greater than normal levels of antisense inhibition in the homologous mutant combination and little or no inhibition in the other cases. The pattern of *JL84* effects also exhibits the same hierarchy in vivo as in vitro, with no detectable inhibition by *JL84* RNA-OUT of wild-type Tn*10*, significant inhibition in the homologous mutant combination, and an intermediate level of inhibition by wild-type RNA-OUT of *JL84* Tn*10*. Finally, as in vitro, *mci1* RNA-OUT is completely defective in multicopy inhibition of either mutant or wild-type Tn*10* elements, and *mci1* Tn*10* is inhibited by wild-type RNA-OUT to nearly the same level as wild type.

associated component of the ribosome (reviewed in reference 44).

Additional host mutations affecting multicopy inhibition and mapping at several other chromosomal loci have also been isolated and are being characterized further.

H. Biological Significance

Transposase *cis* action and antisense RNA inhibition have the important biological effect of decreasing the rate at which a strain harboring IS*10* or Tn*10* accumulates additional IS*10* copies.

Antisense RNA inhibition is a regulatory mechanism that comes into play when Tn*10* or IS*10* is present in more than a single copy per genome equivalent. It is of negligible importance when IS*10* is present in a single copy; mutations that specifically abolish the multicopy phenomenon without changing other aspects of Tn*10* behavior increase transposition and transposase expression of a chromosomally inserted element less than twofold. However, antisense inhibition may begin to be significant at relatively low IS*10* copy numbers. An increase of only a fewfold in the pOUT activity of a single-copy element is sufficient to inhibit transposase production in *trans* about fivefold (unpublished observations). Furthermore, analysis of transposase expression on otherwise isogenic ColE1 derivatives that maintain different plasmid copy numbers reveals a 20-fold inhibition of transposase gene expression when the plasmid is present in only 10 copies per genome equivalent (Matsunaga et al., unpublished observations).

These observations are biologically reasonable. Tn*10* is often found in nature on conjugative plasmids which are present in only one or two copies per bacterial genome equivalent. If antisense inhibition plays an important role in the biology of IS*10*, it

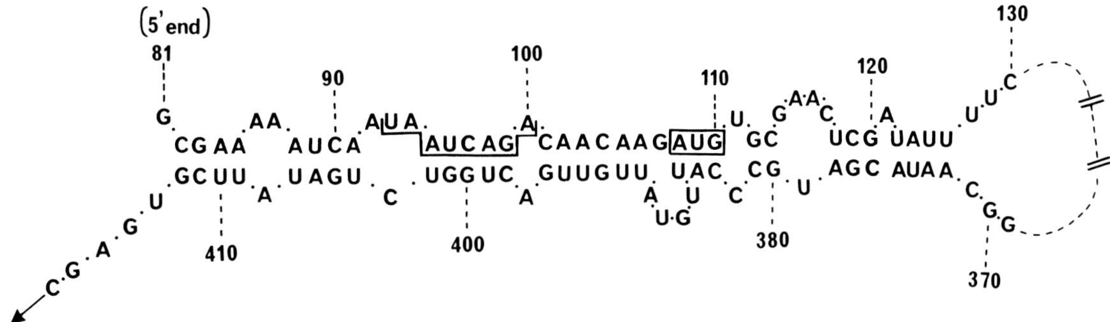

Figure 21. Predicted intramolecular pairing between the two regions of RNA-IN proposed to be responsible for foldback inhibition. IS*10* base pairs 82 to 125 include the probable Shine-Dalgarno region involved in ribosome binding (underlined) and the ATG start codon of the transposase gene (boxed). Distal sequences in the message proposed to interact with the translation start region are encoded by IS*10* base pairs 372 to 412.

would seem important that it do so even as the number of copies per genome just begins to increase.

Multicopy inhibition could assume additional importance in *Haemophilus influenzae*, where tetracycline resistance is usually encoded by Tn*10*-like elements. In this organism, resistance levels are lower than in enteric bacteria, probably because the *tetA* protein does not work as well. To compensate for this defect, some *H. influenzae* plasmids contain two Tn*10* insertions at different locations and others contain tandem amplifications of the entire Tn*10* element, including both IS*10* sequences (76, 156).

VII. ADDITIONAL REGULATORY MECHANISMS

A. RNA-IN Foldback Inhibition

RNA-IN appears to undergo intramolecular secondary structure changes that could be significant for several aspects of IS*10* biology. Short in vitro RNA-IN species synthesized from restriction fragment templates pair efficiently with RNA-OUT, whereas RNA-IN species longer than 315 bases do not pair. At the point of discontinuity, a sequence is present which has the potential to pair with the 5′ end of RNA-IN (Fig. 21) in a way that is expected to prevent that end from interacting with RNA-OUT (Kittle, Ph.D. thesis). This internal site has been named *fbi* for "foldback inhibition" (J. Gonzales and R. W. Simons, unpublished observations).

Characterization of the *fbi* interaction and its consequences in vivo is just beginning. For a single-copy IS*10* element, the *fbi* site might be expected to limit the length of time during which the 5′ end of RNA-IN is freely accessible to ribosomes, and hence to reduce the level of transposase expression. Preliminary analysis of appropriate transposase-*lacZ* fusions suggests that expression is increased severalfold in the absence of the *fbi* interaction (Gonzales and Simons, unpublished observations). The *fbi* interaction should also influence multicopy inhibition by limiting the length of time during which the nascent 5′ end of RNA-IN is available for pairing with RNA-OUT to the time prior to emergence of the *fbi* site from the transcription complex, approximately 5 to 7 s. The *fbi* interaction could mean that RNA-OUT must pair more rapidly than might otherwise be the case; the (physical) half-life of the 5′ end of RNA-IN lacking the *fbi* site is about 40 s (see above). It is known that the *fbi* site can prevent RNA-IN from pairing with RNA-OUT in vivo exactly as predicted from the in vitro analysis: the ability of an RNA-IN species truncated at its 3′ end to "titrate" wild-type RNA-OUT is much greater for RNA-IN species lacking the *fbi* site than for longer species in which the site is present (Gonzalez and Simons, unpublished observations).

The *fbi* site may also play an important role in the *cis*-action phenomenon that depends on restricted movement of transposase. This type of *cis* action requires that translation of the transposase gene occurs while the transcript is still in the vicinity of the gene itself, in the simplest case while it is still connected to its DNA template. Foldback inhibition should prevent expression of transposase from mRNA molecules that have been released from their DNA templates and are free to diffuse through the cell.

B. Protection from Fortuitous Activation

IS*10* contains an elaborate series of controls that minimize the expression of its transposase protein. However, because the nature of the element is to insert randomly in DNA, the element runs the risk of finding itself positioned immediately adjacent to a

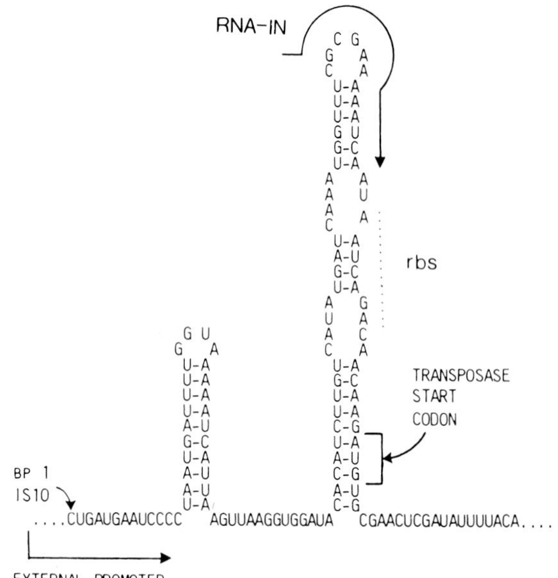

Figure 22. Potential secondary structure of a transcript produced by an external promoter directing transcription across the end of IS*10*. The stability of the larger stem-loop structure is about −17 kcal. The transposase gene start codon, the start point of the normal transposase mRNA, RNA-IN, and the putative ribosome-binding region for the transposase gene are indicated. (Reprinted from reference 40 with permission from Cell Press.)

strong chromosomal promoter which could direct transcription across the end and through the transposase gene of an appropriately oriented IS*10* element. Such readthrough transcription would disrupt both regulation by *dam* methylation, which depends upon activation of pIN, and control by antisense RNA-OUT, which is absolutely specific for target molecules having the precise 5′ end of wild-type RNA-IN.

IS*10* and Tn*10* have several features and mechanisms that protect them from being fortuitously activated by readthrough transcription. In fact, high levels of transcription from an upstream promoter actually decrease the rate of Tn*10* transposition (40). Two levels of protection are involved in this effect.

(i) Externally initiated transcripts are very inefficiently translated; they yield less than 1% as much transposase protein per transcript initiated as do transcripts initiated at pIN (40). Some part of this effect is probably due to premature termination of such transcripts before they traverse the transposase gene. However, inhibition of expression occurs primarily by a posttranscriptional mechanism (40; C. Case and R. W. Simons, unpublished observations). In the model proposed for this inhibition, readthrough transcripts form a strong stem-loop structure, essentially the complement of the RNA-OUT stem-loop structure, with the ribosome-binding site and ATG start codon of the transposase gene sequestered within the stem region. The location of the 5′ end of RNA-IN, within the loop of the readthrough stem-loop structure, is such that no inhibitory secondary structure can form (40, 62) (Fig. 22). In support of this model, point mutations that permit transposase expression from external promoters also disrupt the predicted secondary structure (T. Kolesnikow, J. Rayner, and R. W. Simons, unpublished observations); furthermore, in vivo RNA analysis shows that protection is not mediated by cleavage within the predicted stem-loop structure (Case and Simons, unpublished observations).

These types of protection mechanisms may be quite general. The DNA sequences or several other IS elements contain lengthy inverted repeats that would correspond, in readthrough transcripts, to strong RNA stem-loop structure encompassing transposase gene start signals (58, 83, 94, 152, 166).

(ii) Transcription per se across an IS*10* terminus inhibits its activity, even when transposase protein is provided in *trans* from another element. This direct inhibition is severalfold when transcription is promoted by a fully induced Lac promoter. It is more than sufficient to counteract any small increase in the level of transposase from readthrough transcripts that escapes the other protection mechanisms (40).

Finally, as noted above, Tn*10* and IS*10* also avoid inserting into actively transcribed regions, thus reducing the chances that they will be at risk from external promoters.

VIII. OVERVIEW: MANY FEATURES OF IS*10* LIMIT TRANSPOSITION

Transposition of a single-copy IS*10* element occurs at a low rate in large part because transposase is expressed at a very low level. Not only are transposase gene transcription and translation intrinsically inefficient due to poor initiation signals, transcription is confined to a small fraction of the cell cycle by *dam* methylation and an internal site within the transposase message is postulated to confine translation initiation to incompletely transcribed mRNA molecules.

Transposition is also controlled by *dam* methylation so that it occurs only as the replication fork passes by, a feature that seems particularly convenient for an element that transposes by a nonreplicative mechanism that appears to leave behind a gap at the transposon donor site.

Cells tend to accumulate additional copies of IS*10*. In the absence of special regulatory mechanisms, the more copies of IS*10* present, the faster is the rate of accumulation of additional copies. Three mechanisms act to counteract this tendency. Two mechanisms ensure that transposase made by one copy of IS*10* seldom reaches a second IS*10* copy located elsewhere in the cell: the intrinsic "preferential *cis* action" property of transposase, and the precise nature of regulation by *dam* methylation. In addition, a third regulatory process, antisense RNA control, acts to reduce the level of transposase expression per copy as the number of copies increases. In combination, these three features exert a powerful influence.

Finally, IS*10* has evolved several different mechanisms that protect it from fortuitous activation by readthrough transcription from external promoters. (i) The most important mechanism is the inhibition of translation from readthrough transcripts afforded by a suitably positioned RNA secondary structure. (ii) Readthrough transcripts probably often terminate within IS*10* before they traverse the entire transposase coding region. (iii) Readthrough transcription across the outside end of IS*10* directly inhibits the ability of transposase to work at that site. (iv) Finally, *dam* regulation reduces the chances that the inside end of IS*10* will be active at the time a readthrough transcript passes through. All of these features suggest that IS*10* lives in a highly evolved symbiosis with its bacterial host.

Acknowledgments. This chapter is dedicated to David Botstein and to all of the members of my laboratory who have contributed and are continuing to contribute to our understanding of Tn*10*.

All of the work described in this chapter that was performed in my own laboratory was funded by grants from the National Institutes of Health (GM 25326) and the National Science Foundation (DMB-8518930).

LITERATURE CITED

1. **Ahmed, A.** 1986. Evidence for replicative transposition of Tn5 and Tn9. *J. Mol. Biol.* **191**:75–84.
2. **Albertini, A. M., M. Hofer, M. P. Calos, and J. H. Miller.** 1982. On the formation of spontaneous deletions: the importance of short sequence homologies in the generation of large deletions. *Cell* **29**:319–328.
3. **Alves, J., A. Pingoud, W. Haupt, J. Langowski, F. Peters, G. Maass, and C. Wolff.** 1984. The influence of sequences adjacent to the recognition site on the cleavage of oligodeoxynucleotides by the EcoRI endonuclease. *Eur. J. Biochem.* **140**:83–92.
4. **Anderson, R. P., and J. R. Roth.** 1978. Tandem genetic duplications in *Salmonella typhimurium*: amplification of the histidine operon. *J. Mol. Biol.* **126**:53–71.
5. **Ball, P. R., S. W. Shales, and I. Chopra.** 1980. Plasmid-mediated tetracycline resistance in *Escherichia coli* involves increased efflux of the antibiotic. *Biochem. Biophys. Res. Commun.* **93**:74–81.
6. **Bender, J., and N. Kleckner.** 1986. Genetic evidence that Tn10 transposes by a nonreplicative mechanism. *Cell* **45**: 801–815.
7. **Benjamin, H. W., and N. R. Cozzarelli.** 1988. Isolation and characterization of the Tn3 resolvase synaptic intermediate. *EMBO J.* **7**:1897–1905.
8. **Bennett, P. M., M. H. Richmond, and V. Petrocheilou.** 1980. The inactivation of *tet* genes on a plasmid by the duplication of one inverted repeat of a transposon-like structure which itself mediates tetracycline resistance. *Plasmid* **3**:135–149.
9. **Berg, C. M., and D. E. Berg.** 1987. Uses of transposable elements and maps of known insertions, p. 1071–1109. *In* F. C. Neidhardt, J. L. Ingraham, K. B. Low, B. Magasanik, M. Schaechter, and H. E. Umbarger (ed.), *Escherichia coli and Salmonella typhimurium*. American Society for Microbiology, Washington, D.C.
10. **Berg, D. E., J. Lodge, C. Sasakawa, D. K. Nag, S. H. Phadnis, K. Weston-Hafer, and G. F. Carle.** 1984. Transposon Tn5: specific sequence recognition and conservative transposition. *Cold Spring Harbor Symp. Quant. Biol.* **49**:215–226.
11. **Berg, O. G., R. B. Winter, and P. H. von Hippel.** 1981. Diffusion-driven mechanisms of protein translocation on nucleic acid. 1. Models and theory. *Biochemistry* **20**:6929–6948.
12. **Berg, O. G., R. B. Winter, and P. H. von Hippel.** 1982. How do genome-regulatory proteins locate their DNA target sites? *Trends Biochem. Sci.* **7**:52–55.
13. **Blazey, D. L., and R. P. Burns.** 1982. Transcriptional activity of the transposable element Tn10 in the *Salmonella typhimurium ilg*GEDA operon. *Proc. Natl. Acad. Sci. USA* **79**: 5011–5015.
14. **Boocock, M. R., J. L. Brown, and D. J. Sherratt.** 1985. Structural and catalytic properties of specific complexes between Tn3 resolvase and the recombination site *res*. *Biochem. Soc. Trans.* **14**:214–216.
15. **Brass, J. M., M. D. Manson, and T. J. Larson.** 1984. Transposon Tn10-dependent expression of the *lamB* gene in *Escherichia coli*. *J. Bacteriol.* **159**:93–99.
16. **Braus, G., M. Argast, and C. F. Beck.** 1984. Identification of additional genes on transposon Tn10: *tetC* and *tetD*. *J. Bacteriol.* **160**:504–509.
17. **Bresler, S. E., S. E. Tamm, I. Y. Goryshin, and V. A. Lanzov.** 1983. Postexcision transposition of the transposon Tn10 in *Escherichia coli* K-12. *Mol. Gen. Genet.* **190**:39–142.
18. **Broyles, S., and D. Pettijohn.** 1986. Interaction of the *Escherichia coli* HU protein with DNA. Evidence for formation of nucleosome-like structures with altered DNA helical pitch. *J. Mol. Biol.* **187**:47–60.
19. **Brunier, D., B. Michel, and S. D. Erlich.** 1988. Copy choice illegitimate DNA recombination. *Cell*, in press.
20. **Campbell, A.** 1981. Evolutionary significance of accessory DNA elements in bacteria. *Annu. Rev. Microbiol.* **35**:55–81.
21. **Case, C. C., S. M. Roels, J. E. Gonzales, E. L. Simons, and R. W. Simons.** 1988. Analysis of the promoters and transcripts involved in IS10 anti-sense RNA control. *Gene*, in press.
22. **Chan, R. K., D. Botstein, T. Watanabe, and Y. Ogata.** 1972. Specialized transduction of tetracycline resistance by phage P22 in *Salmonella typhimurium*. II. Properties of a high-frequency-transducing lysate. *Virology* **50**:883–898.
23. **Chandler, M., M. Clerget, and D. J. Galas.** 1982. The transposition frequency of IS11-flanked transposons is a function of their size. *J. Mol. Biol.* **154**:229–243.
24. **Chandler, M., E. Roulet, L. Silver, E. Boy de la Tour, and L. Caro.** 1979. Tn10 mediated integration of the plasmid

R100-1 into the bacterial chromosome: inverse transposition. *Mol. Gen. Genet.* **173:**23–30.

25. **Chao, L., C. Vargas, B. B. Spear, and E. C. Cox.** 1983. Transposable elements as mutator genes in evolution. *Nature* (London) **303:**633–635.
26. **Chopra, I., S. W. Shales, J. M. Ward, and L. J. Wallace.** 1981. Reduced expression of Tn10-mediated tetracycline resistance in *Escherichia coli* containing more than one copy of the transposon. *J. Gen. Microbiol.* **126:**45–54.
27. **Chumley, F. G., R. Menzel, and J. R. Roth.** 1979. Hfr formation directed by Tn10. *Genetics* **91:**639–655.
28. **Chumley, F. G., and J. R. Roth.** 1980. Rearrangement of the bacterial chromosome using Tn10 as a region of homology. *Genetics* **94:**1–14.
29. **Ciampi, M. S., and J. R. Roth.** 1988. Polarity effects in the *hisG* gene of *Salmonella* require a site within the coding sequence. *Genetics* **118:**193–202.
30. **Ciampi, M. S., M. B. Schmid, and J. R. Roth.** 1982. Transposon Tn10 provides a promoter for transcription of adjacent sequences. *Proc. Natl. Acad. Sci. USA* **79:**5016–5020.
31. **Cobbett, C. S., and J. Pittard.** 1980. Formation of a λ(Tn10)*tryR*$^+$ specialized transducing bacteriophage from *Escherichia coli* K-12. *J. Bacteriol.* **144:**877–883.
32. **Coleman, D. C., I. Chopra, S. W. Shales, T. G. B. Howe, and T. J. Foster.** 1983. Analysis of tetracycline resistance encoded by transposon Tn10: deletion mapping of tetracycline-sensitive point mutations and identification of two structural genes. *J. Bacteriol.* **153:**921–929.
33. **Coleman, D. C., and T. J. Foster.** 1981. Analysis of the reduction in expression of tetracycline resistance determined by transposon Tn10 in the multicopy state. *Mol. Gen. Genet.* **182:**171–172.
34. **Craigie, R., D. Arndt-Jovin, and K. Mizuuchi.** 1985. A defined system for the DNA strand-transfer reaction at the initiation of bacteriophage Mu transposition: protein and DNA substrate requirements. *Proc. Natl. Acad. Sci. USA* **82:**7570–7574.
35. **Craigie, R., and K. Mizuuchi.** 1986. Role of DNA topology in Mu transposition: mechanism of sensing the relative orientation of two DNA segments. *Cell* **45:**793–800.
36. **Craigie, R., and K. Mizuuchi.** 1987. Transposition of Mu DNA: joining of Mu to target DNA can be uncoupled from cleavage at the ends of Mu. *Cell* **51:**593–601.
37. **Curiale, M. S., and S. B. Levy.** 1982. Two complementation groups mediate tetracycline resistance determined by Tn10. *J. Bacteriol.* **151:**209–215.
38. **Curiale, M. S., L. M. McMurry, and S. B. Levy.** 1984. Intracistronic complementation of the tetracycline resistance membrane protein of Tn10. *J. Bacteriol.* **157:**211–217.
39. **Daniels, D. W., and K. P. Bertrand.** 1985. Promoter mutations affecting divergent transcription in the Tn10n tetracycline resistance determinant. *J. Mol. Biol.* **184:**599–610.
40. **Davis, M. A., R. W. Simons, and N. Kleckner.** 1985. Tn10 protects itself at two levels against fortuitous activation by external promoters. *Cell* **43:**379–387.
41. **De La Torre, J., J. Ortin, E. Domingo, J. Delamarter, B. Allet, J. Davis, K. P. Bertrand, L. V. Wray, Jr., and W. S. Reznikoff.** 1984. Plasmid vectors based on Tn*10* DNA: gene expression regulated by tetracycline. *Plasmid* **12:**103–110.
42. **Derbyshire, K. M., L. Hwang, and N. D. F. Grindley.** 1987. Genetic analysis of the interaction of the insertion sequence IS903 transposase with its terminal inverted repeats. *Proc. Natl. Acad. Sci. USA* **84:**8049–8053.
43. **Doolittle, W. F., and C. Sapienza.** 1980. Selfish genes, the phenotype paradigm and genome evolution. *Nature* (London) **284:**601–603.
44. **Drlica, K., and J. Rouviere-Yaniv.** 1987. Histone-like proteins of bacteria. *Microbiol. Rev.* **51:**301–319.
45. **Dubnau, E., and B. A. D. Stocker.** 1964. Genetics of plasmids in *Salmonella typhimurium. Nature* (London) **204:** 1112–1113.
46. **Egner, C., and D. E. Berg.** 1981. Excision of transposon Tn5 is dependent on the inverted repeats but not on the transposase function of Tn5. *Proc. Natl. Acad. Sci. USA* **78:** 459–463.
47. **Farabaugh, P. J., U. Schmeissner, M. Hofer, and J. H. Miller.** 1978. Genetic studies on the *lac* repressor. VII. On the molecular nature of spontaneous hotspots in the *lacI* gene of *Escherichia coli. J. Mol. Biol.* **126:**847–863.
48. **Foster, T. J.** 1975. Tetracycline-sensitive mutants of the F0like R factors R100 and R100-1. *Mol. Gen. Genet.* **137:**85–88.
49. **Foster, T. J.** 1977. Insertion of the tetracycline resistance translocation unit Tn10 in the *lac* operon of *Escherichia coli* K12. *Mol. Gen. Genet.* **154:**305–310.
50. **Foster, T., M. A. Davis, K. Takeshita, D. E. Roberts, and N. Kleckner.** 1981. Genetic organization in transposon Tn10. *Cell* **23:**201–213.
51. **Foster, T. J., T. G. B. Howe, and K. M. V. Richmond.** 1975. Translocation of the tetracycline resistance determinant from R100-1 to the *Escherichia coli* K-12 chromosome. *J. Bacteriol.* **124:**1153–1158.
52. **Foster, T., V. Lundblad, S. Hanley-Way, S. Halling, and N. Kleckner.** 1981. Three Tn10-associated excision events: relationship to transposition and role of direct and inverted repeats. *Cell* **23:**215–227.
53. **Foster, T. J., and N. S. Willetts.** 1977. Characterisation of transfer-deficient mutants of the R100-1Tc^s plasmid pDU202, caused by insertion of Tn10. *Mol. Gen. Genet.* **156:**107–114.
54. **Freier, S. M., R. Kierzek, J. A. Jaeger, N. Sugimoto, M. H. Caruthers, T. Neilson, and D. H. Turner.** 1986. Improved free-energy parameters for predictions of RNA duplex stability. *Proc. Natl. Acad. Sci. USA* **83:**9377–9377.
55. **Gamas, P., M. G. Chandler, P. Prentki, and D. J. Galas.** 1987. *Escherichia coli* integration host factor binds specifically to the ends of the insertion sequence IS1 and to its major insertion hot-spot in pBR322. *J. Mol. Biol.* **195:** 261–272.
56. **Gamas, P., D. Galas, and M. Chandler.** 1985. DNA sequence at the end of IS1 required for transposition. *Nature* (London) **317:**458–460.
57. **Gefter, M. L., and L. A. Sherman.** 1977. The role of DNA structure in DNA replication, p. 233–248. *In* E. M. Bradbury and K. Javaherian (ed.), *The Organization and Expression of the Eukaryotic Genome.* Academic Press, Inc., New York.
58. **Ghosal, D., H. Sommer, and H. Saedler.** 1979. Nucleotide sequence of the transposable DNA-element IS2. *Nucleic Acids Res.* **6:**1111–1115.
59. **Glickman, B. W., and L. S. Ripley.** 1984. Structural intermediates of deletion mutagenesis: a role for palindromic DNA. *Proc. Natl. Acad. Sci. USA* **81:**512–516.
60. **Habermann, P., and P. Starlinger.** 1982. Bidirectional deletions associated with IS4. *Mol. Gen. Genet.* **185:**216–222.
61. **Halling, S. M., and N. Kleckner.** 1982. A symmetrical six-basepair target site sequence determines Tn10 insertion specificity. *Cell* **28:**155–163.
62. **Halling, S. M., R. W. Simons, J. C. Way, R. B. Walsh, and N.**

Kleckner. 1982. DNA sequence organization of Tn10's IS10-Right and comparison with IS10-Left. *Proc. Natl. Acad. Sci. USA* **79:**2608–2612.

63. **Harayama, S., T. Oguchi, and T. Iino.** 1984. Does Tn10 transpose via the cointegrate molecule? *Mol. Gen. Genet.* **194:**444–450.

64. **Harayama, S., T. Oguchi, and T. Iino.** 1984. The *E. coli* K-12 chromosome flanked by two IS10 sequences transposases. *Mol. Gen. Genet.* **197:**62–66.

65. **Heuer, C., R. K. Hickman, M. S. Curiale, W. Hillen, and S. B. Levy.** 1987. Constitutive expression of tetracycline resistance mediated by a Tn10-like element in *Haemophilus parainfluenzae* results from a mutation in the repressor gene. *J. Bacteriol.* **169:**990–994.

66. **Hillen, W., and K. Schollmeier.** 1983. Nucleotide sequence of the Tn10 encoded tetracycline resistance gene. *Nucleic Acids Res.* **11:**525–539.

67. **Hillen, W., K. Schollmeier, and C. Gatz.** 1984. Control of expression of the Tn10-encoded tetracycline resistance operon. II. Interaction of RNA polymerase and TET repressor with the *tet* operon regulatory region. *J. Mol. Biol.* **172:** 185–201.

68. **Huisman, O., and N. Kleckner.** 1987. A new generalizable test for detection of mutations affecting Tn10 transposition. *Genetics* **116:**185–189.

69. **Huisman, O., W. Raymond, K. Froehlich, P. Errada, D. Botstein, N. Kleckner, and A. Hoyt.** 1987. A tn*10-lacZ*-KanR-*URA3* gene fusion transposon for insertion mutagenesis and fusion analysis of yeast and bacterial genes. *Genetics* **116:**191–199.

70. **Isackson, P. J., and K. P. Bertrand.** 1985. Dominant negative mutations in the Tn10 *tet* repressor: evidence for use of the conserved helix-turn-helix motif in DNA binding. *Proc. Natl. Acad. Sci. USA* **82:**6226–6230.

71. **Isberg, R. R., A. L. Lazaar, and M. Syvanen.** 1982. Regulation of Tn5 by the right-repeat proteins: control at the level of the transposition reaction? *Cell* **30:**883–892.

72. **Isberg, R. R., and M. Syvanen.** 1981. Replicon fusions promoted by the inverted repeats of Tn5. *J. Mol. Biol.* **150:**15–32.

73. **Isberg, R. R., and M. Syvanen.** 1982. DNA gyrase is a host factor required for transposition of Tn5. *Cell* **30:**9–18.

74. **Izaki, K., K. Kiuchi, and K. Arima.** 1966. Specificity and mechanism of tetracycline resistance in a multiple drug resistant strain of *Escherichia coli. J. Bacteriol.* **91:**628–633.

75. **Jacobson, H., and W. H. Stockmayer.** 1950. Intramolecular reaction in polycondensations. I. The theory of linear systems. *J. Chem. Phys.* **18:**1600–1606.

76. **Jahn, G., R. Laufs, P.-M. Kaulfers, and H. Kolender.** 1979. Molecular nature of two *H. influenzae* R-factors specifying for combined resistances and the multiple integration of drug resistance transposons. *J. Bacteriol.* **138:**584–597.

77. **Johnson, R. C., J. C. P. Yin, and W. S. Reznikoff.** 1982. Control of Tn5 transposition in *Escherichia coli* is mediated by protein from the right repeat. *Cell* **30:**873–882.

78. **Jorgensen, R. A., and W. S. Reznikoff.** 1979. Organization of structural and regulatory genes that mediate tetracycline resistance in transposon Tn10. *J. Bacteriol.* **138:**705–714.

79. **Jorgensen, S. T., B. Oliva, J. Grinsted, and P. M. Bennett.** 1980. New translocation sequence mediating tetracycline resistance found in *Escherichia coli* pathogenic for piglets. *Antimicrob. Agents Chemother.* **18:**200–205.

80. **Kaneko, M., A. Yamaguchi, and T. Sawai.** 1985. Energetics of tetracycline efflux system encoded by Tn10 in *Escherichia coli. FEBS Lett.* **193:**194–198.

81. **Kaulfers, P.-M., R. Laufs, and G. Jahn.** 1978. Molecular properties of transmissible R factors of *Haemophilus influenzae* determining tetracycline resistance. *J. Gen. Microbiol.* **105:**243–252.

82. **Kikuchi, Y., and H. Nash.** 1978. The bacteriophage λ *inbt* gene product. *J. Biol. Chem.* **256:**9246–9253.

83. **Klaer, R., S. Kuhn, E. Tillman, H.-J. Fritz, and P. Starlinger.** 1981. The sequence of IS4. *Mol. Gen. Genet.* **181:**169–175.

84. **Kleckner, N.** 1979. DNA sequence analysis of Tn10 insertions: origin and role of 9-bp flanking repetitions during Tn10 translocation. *Cell* **16:**711–720.

85. **Kleckner, N.** 1981. Transposons and illegitimate recombination in prokaryotes, p. 245–258. *In* J. F. Lemontt and W. M. Generoso (ed.), *Molecular and Cellular Mechanisms of Mutagenesis.* Plenum Publishing Corp., New York.

86. **Kleckner, N., D. F. Barker, D. G. Ross, and D. Botstein.** 1978. Properties of the translocatable tetracycline-resistance element Tn10 in *Escherichia coli* and bacteriophage lambda. *Genetics* **90:**427–450.

87. **Kleckner, N., R. K. Chan, B.-K. Tye, and D. Botstein.** 1975. Mutagenesis by insertion of a drug-resistance element carrying an inverted repetition. *J. Mol. Biol.* **97:**561–575.

88. **Kleckner, N., K. Reichardt, and D. Botstein.** 1979. Inversions and deletions of the *Salmonella* chromosome generated by the translocatable tetracycline-resistance element Tn10. *J. Mol. Biol.* **127:**89–115.

89. **Kleckner, N., and D. G. Ross.** 1980. A *recA*-dependent genetic switch generated by transposon Tn10. *J. Mol. Biol.* **144:**215–221.

90. **Kleckner, N., J. Roth, and D. Botstein.** 1977. Genetic engineering *in vivo* using translocatable drug-resistance elements—new methods in bacterial genetics. *J. Mol. Biol.* **116:**125–159.

91. **Kleckner, N., J. A. Swan, and M. Zabeau.** 1978. Restriction enzyme analysis of Tn10 insertions in the immunity region of bacteriophage lambda. *Genetics* **90:**450–461.

92. **Kleckner, N., D. Steele, K. Reichardt, and D. Botstein.** 1979. Specificity of insertion by the translocatable tetracycline-resistance element Tn10. *Genetics* **92:**1023–1040.

93. **Klock, G., B. Unger, C. Gatz, W. Hillen, J. Altenbuchner, K. Schmid, and R. Schmitt.** 1985. Heterologous repressor-operator recognition among four classes of tetracycline resistance determinants. *J. Bacteriol.* **161:**326–332.

94. **Krebs, M. P., and W. S. Reznikoff.** 1986. Transcriptional and translational initiation sites of IS50. Control of transposase and inhibitor expression. *J. Mol. Biol.* **192:**781–791.

95. **Lee, S., D. Butler, and N. Kleckner.** 1987. Efficient Tn10 transposition into a DNA insertion hot spot *in vivo* requires the 5-methyl groups of symmetrically disposed thymines within the hot-spot consensus sequence. *Proc. Natl. Acad. Sci. USA* **84:**7876–7880.

96. **Lee, Y., and F. J. Schmidt.** 1985. Characterization of the in vivo RNA product of the pOUT promoter of IS*10*R. *J. Bacteriol.* **164:**556–562.

97. **Levy, S. B.** 1984. Resistance to the tetracyclines, p. 191–239. *In* L. F. Bryan (ed.), *Antimicrobial Drug Resistance.* Academic Press, Inc., New York.

98. **Levy, S. B., and L. McMurry.** 1974. Detection of an inducible membrane protein associated with R-factor-mediated tetracycline resistance. *Biochem. Biophys. Res. Commun.* **56:**1060–1068.

99. **Liao, S.-M., and W. R. McClure.** 1988. Structure and some functions of an antisense RNA for bacteriophage P22, p. 17–22. *In* D. Melton (ed.), *Antisense RNA and DNA.* Cold Spring Harbor Laboratory, Cold Spring Harbor, N.Y.

100. **Liao, S.-M., T.-H. Wu, C. H. Chiang, M. M. Susskind, and W. R. McClure.** 1987. Control of gene expression in bacteriophage P22 by a small antisense RNA. I. Characterization *in vitro* of the Psar promoter and the *sar* RNA transcript. *Genes Dev.* **1:**197–203.
101. **Lichens-Park, A., and M. Syvanen.** 1988. Cointegrate formation by IS50 requires multiple donor molecules. *Mol. Gen. Genet.* **211:**244–251.
102. **Lundblad, V., and N. Kleckner.** 1985. Mismatch repair mutations of *Escherichia coli* K12 enhance transposon excision. *Genetics* **109:**3–19.
103. **Lundblad, V., A. Taylor, G. Smith, and N. Kleckner.** 1984. Unusual alleles of RecB and RecC stimulate excision of the inverted repeat transposons Tn10 and Tn5. *Proc. Natl. Acad. Sci. USA* **81:**824–828.
104. **Lyons, S. M., and P. F. Schendel.** 1984. Kinetics of methylation in *Escherichia coli* K-12. *J. Bacteriol.* **159:**421–423.
105. **Machida, Y., C. Machida, H. Ohtsubo, and E. Ohtsubo.** 1982. Factors determining frequency of plasmid cointegration mediated by insertion sequence IS1. *Proc. Natl. Acad. Sci. USA* 79:277–281.
106. **Mahajna, J., A. B. Oppenheim, A. Rattray, and M. Gottesman.** 1986. Translation initiation of bacteriophage λ gene cII requires integration host factor. *J. Bacteriol.* **165:**167–174.
107. **Mahillon, J., J. Seurinck, L. Van Rompuy, J. Delcour, and M. Zabeau.** 1985. Nucleotide sequence and structural organization of an insertion sequence element (IS231) from *Bacillus thuringiensis* strain berlin 1715. *EMBO J.* **4:**3895–3899.
108. **Marshall, B., S. Morrissey, P. Flynn, and S. B. Levy.** 1986. A new tetracycline-resistance determinant, class E, isolated from Enterobacteriaceae. *Gene* **50:**111–117.
109. **Marshall, B., M. Roberts, A. Smith, and S. B. Levy.** 1984. Homogeneity of transferable tetracycline-resistance determinants in *Haemophilus* species. *J. Infect. Dis.* **149:**1028–1029.
110. **Marshall, B., C. Tachibana, and S. B. Levy.** 1983. Frequency of tetracycline resistance determinant classes among lactose-fermenting coliforms. *Antimicrob. Agents Chemother.* **24:** 835–840.
111. **McClarin, J. A., C. A. Frederick, B.-C. Wang, P. Greene, H. W. Boyer, J. Grable, and J. M. Rosenberg.** 1986. Structure of the DNA-Eco RI endonuclease recognition complex at 3 A resolution. *Science* **234:**1526–1540.
112. **McCommas, S. A., and M. Syvanen.** 1988. Temporal control of transposition in Tn5. *J. Bacteriol.* **170:**889–894.
113. **McMurry, L., R. E. Petrucci, Jr., and S. B. Levy.** 1980. Active efflux of tetracycline encoded by four genetically different tetracycline resistance determinants in *Escherichia coli. Proc. Natl. Acad. Sci. USA* 77:3974–3977.
114. **Mendez, B., C. Tachibana, and S. B. Levy.** 1980. Heterogeneity of tetracycline resistance determinants. *Plasmid* **3:** 99–108.
115. **Meyers, R. M., and T. Maniatis.** 1985. A new method for saturated mutagenesis. *Science* 229:242–245.
116. **Morisato, D., and N. Kleckner.** 1984. Transposase promotes double strand breaks and single strand joints at Tn10 termini *in vivo. Cell* 39:181–190.
117. **Morisato, D., and N. Kleckner.** 1987. Tn10 transposition and circle formation *in vitro. Cell* **51:**101–111.
118. **Morisato, D., J. C. Way, H.-J. Kim, and N. Kleckner.** 1983. Tn10 transposase acts preferentially on nearby transposon ends *in vivo. Cell* 32:799–807.
119. **Moyed, H. S., and K. P. Bertrand.** 1983. Mutations in multicopy Tn10 *tet* plasmids that confer resistance to inhibitory effects of inducers of *tet* gene expression. *J. Bacteriol.* 155:557–564.
120. **Navas, J., J. M. Garcia-Lobo, J. Leon, and J. M. Ortiz.** 1985. Structural and functional analyses of the fosfomycin resistance transposon Tn*2921*. *J. Bacteriol.* **162:**1061–1067.
121. **Neidhardt, F. C., J. L. Ingraham, K. B. Low, B. Magasanik, M. Schaechter, and H. E. Umbarger (ed.).** 1987. *Escherichia coli* and *Salmonella typhimurium*. American Society for Microbiology, Washington, D.C.
122. **Newbury, S. F., N. H. Smith, E. C. Robinson, I. D. Hiles, and C. F. Higgins.** 1987. Stabilization of translationally active mRNA by prokaryotic REP sequences. *Cell* **48:**297–310.
123. **Nguyen, T. T., K. Postle, and K. P. Bertrand.** 1983. Sequence homology between the tetracycline-resistance determinants of Tn10 and pBR322. *Gene* **25:**83–92.
124. **Noel, K. D., and G. F. Ames.** 1978. Evidence for a common mechanism for the insertion of the Tn10 transposon and for the generation of Tn10-stimulated deletions. *Mol. Gen. Genet.* **166:**217–223.
125. **Orgel, L. E., and F. H. C. Crick.** 1980. Selfish DNA: the ultimate parasite. *Nature* (London) **284:**604–607.
126. **Picksley, S. M., P. V. Attfield, and R. G. Lloyd.** 1984. Repair of DNA double strand breaks in *Escherichia coli* K12 requires a functional RecN product. *Mol. Gen. Genet.* **195:**267–274.
127. **Prentki, P., M. Chandler, and D. J. Galas.** 1987. *Escherichia coli* integration host factor bends the DNA at the ends of IS1 and as insertion hot spot with multiple IHF sites. *EMBO J.* **6:**2479–2487.
128. **Raleigh, E. A., and N. Kleckner.** 1984. Multiple IS10 rearrangements in *E. coli. J. Mol. Biol.* **173:**437–461.
129. **Raleigh, E. A., and N. Kleckner.** 1986. Quantitation of insertion sequence IS10 *transposase* gene expression by a method generally applicable to any rarely expressed gene. *Proc. Natl. Acad. Sci. USA* **83:**1787–1791.
130. **Riley, M., and S. Krawiec.** 1987. Genome organization, p. 967–981. *In* F. C. Neidhardt, J. L. Ingraham, K. B. Low, B. Magasanik, M. Schaechter, and H. E. Umbarger (ed.), *Escherichia coli* and *Salmonella typhimurium*, vol. 2. American Society for Microbiology, Washington, D.C.
130a. **Roberts, D., and N. Kleckner.** 1988. Tn*10* transposition promotes RecA-dependent induction of a λ prophage. *Proc. Natl. Acad. Sci. USA* **85:**6037–6041.
131. **Roberts, D., D. Morisato, and N. Kleckner.** 1987. The role of the bacterial host in the mechanism and regulation of Tn10 transposition, p. 17–28. *In Eukaryotic Transposable Elements as Mutagenic Agents.* Cold Spring Harbor Laboratory, Cold Spring Harbor, N.Y.
132. **Roberts, D. E., B. C. Hoopes, W. R. McClure, and N. Kleckner.** 1985. IS10 transposition is regulated by DNA adenine methylation. *Cell* **43:**117–130.
133. **Robertson, C. A., and H. Nash.** 1988. Bending of the bacteriophage λ attachment site by *Escherichia coli* integration host factor. *J. Biol. Chem.* **263:**3554–3557.
134. **Robertson, H. D.** 1982. *Escherichia coli* ribonuclease III cleavage sites. *Cell* **30:**669–672.
135. **Ross, D., P. Grisafi, N. Kleckner, and D. Botstein.** 1979. The ends of Tn10 are not IS3. *J. Bacteriol.* **129:**1097–1101.
136. **Ross, D., J. Swan, and N. Kleckner.** 1979. Physical structures of Tn10-promoted deletions and inversions: role of 1400 basepair inverted repetitions. *Cell* **16:**721–731.
137. **Ross, D., J. Swan, and N. Kleckner.** 1979. Nearly precise excision: a new type of DNA alteration associated with the translocatable element Tn10. *Cell* **16:**733–738.

138. **Roth, J. R.** 1981. New genetic techniques using transposable drug resistance elements. *Microbiology* **1981**:402–404.

139. **Sandri, R. M., and H. Berger.** 1980. Bacteriophage P1-mediated generalized transduction in *Escherichia coli*: fate of transduced DNA in Rec^+ and $RecA^-$ recipients. *Virology* **106**:14–29.

140. **Sandri, R. M., and H. Berger.** 1980. Bacteriophage P1-mediated generalized transduction in *Escherichia coli*: structure of abortively transduced DNA. *Virology* **106**:30–40.

141. **Sasakawa, C., Y. Uno, and M. Yoshikawa.** 1981. The requirement for both DNA polymerase and 5′ to 3′ exonuclease activities of DNA polymerase I during Tn5 transposition. *Mol. Gen. Genet.* **182**:19–24.

142. **Schmid, M., and J. R. Roth.** 1980. Circularization of transduced fragments: a mechanism for adding segments to the bacterial chromosome. *Genetics* **94**:15–29.

143. **Schmidt, F. J., R. A. Jorgensen, M. de Wilde, and J. E. Davies.** 1981. A specific tetracycline-induced, low molecular weight RNA encoded by the inverted repeat of Tn10 (IS10). *Plasmid* **6**:148–150.

144. **Schollmeier, K., and W. Hillen.** 1984. Transposon Tn10 contains two structural genes with opposite polarity between *tetA* and IS*10*$_R$. *J. Bacteriol.* **160**:499–503.

145. **Seeman, N. C., J. M. Rosenberg, and A. Rich.** 1976. Sequence-specific recognition of double helical nucleic acids by proteins. *Proc. Natl. Acad. Sci. USA* **73**:804–808.

145a. **Sen, J., M. Septak, C. Jain, and N., Kleckner.** 1988. Translation start of IS10 transposase protein. *Nucleic Acids Res.* **16**:4730.

146. **Shapiro, J. A.** 1979. Molecular model for the transposition and replication of bacteriophage mu and other transposable elements. *Proc. Natl. Acad. Sci. USA* **76**:1933–1937.

147. **Sharp, P. A., S. N. Cohen, and N. Davidson.** 1973. Electron microscope heteroduplex studies of sequence relations among plasmids of *Escherichia coli*. II. Structure of drug resistance (R) factors and F factors. *J. Mol. Biol.* **75**: 235–255.

148. **Shen, M., E. A. Raleigh, and N. Kleckner.** 1987. Physical analysis of IS10-promoted transpositions and rearrangements. *Genetics* **116**:359–369.

149. **Simons, R. W., B. Hoopes, W. McClure, and N. Kleckner.** 1983. Three promoters near the ends of IS10: p-IN, p-OUT and p-III. *Cell* **34**:673–682.

150. **Simons, R. W., F. Houman, and N. Kleckner.** 1987. Improved single and multicopy *lac*-based cloning vectors for protein and operon fusion. *Gene* **53**:85–96.

151. **Simons, R. W., and N. Kleckner.** 1983. Translational control of IS10 transposition. *Cell* **34**:683–691.

152. **Simsek, M., S. DasSarma, U. RajBhandary, and H. G. Khorana.** 1982. A transposable element from *Halobacterium halobium* which inactivates the bacteriorhodopsin gene. *Proc. Natl. Acad. Sci. USA* **79**:7268–7272.

153. **Smith, G. R.** 1987. Mechanism and control of homologous recombination in *Escherichia coli*. *Annu. Rev. Genet.* **21**: 179–201.

154. **Smith, H. O., D. B. Danner, and R. A. Deich.** 1981. Genetic transformation. *Annu. Rev. Biochem.* **50**:41–68.

155. **Snyder, M., S. Elledge, and R. W. Davis.** 1986. Rapid mapping of antigenic coding regions and constructing insertion mutations in yeast by mini-Tn10 "transposon" mutagenesis. *Proc. Natl. Acad. Sci. USA* **83**:730–734.

156. **Spies, T., R. Laufs, and F.-C. Riess.** 1983. Amplification of resistance genes in *Haemophilus influenzae* plasmids. *J. Bacteriol.* **155**:839–846.

157. **Stenzel, T. T., P. Patel, and D. Bastia.** 1985. The integration host factor of *Escherichia coli* binds to bent DNA at the origin of replication of the plasmid pSC101. *Cell* **49**:709–717.

158. **Stormo, G. D.** 1986. Translation initiation, p. 195–224. *In* W. Reznikoff and L. Gold (ed.), *Maximizing Gene Expression*. Butterworths, Boston.

159. **Surette, M. G., S. J. Buch, and G. Chaconas.** 1987. Transposomes: stable protein-DNA complexes involved in the *in vitro* transposition of bacteriophage Mu DNA. *Cell* **49**: 253–262.

160. **Syvanen, M., J. D. Hopkins, and M. Clements.** 1982. A new class of mutants in DNA polymerase I that affects gene transposition. *J. Mol. Biol.* **158**:203–212.

161. **Szyf, M., Y. Gruenbaum, S. Urieli-Shoval, and A. Razin.** 1982. Studies on the biological role of DNA methylation. V. The pattern of *E. coli* DNA methylation. *Nucleic Acids Res.* **10**:7247–7259.

162. **Takahashi, M., L. Altschmied, and W. Hillen.** 1986. Kinetic and equilibrium characterization of the tet repressor-tetracycline complex by fluorescence measurements. Evidence for divalent metal ion requirement and energy transfer. *J. Mol. Biol.* **187**:341–348.

163. **Taylor, D. P., J. Greenberg, and R. H. Rownd.** 1977. Generation of miniplasmids from copy number mutants of the R plasmid NR1. *J. Bacteriol.* **132**:986–995.

164. **Terry, B. J., W. E. Jack, and P. Modrich.** 1985. Facilitated diffusion during catalysis by EcoRI nuclease. Non-specific interactions in EcoRI catalysis. *J. Biol. Chem.* **260**: 13130–13137.

165. **Thaler, D. S., M. M. Stahl, and F. W. Stahl.** 1987. Tests of the double-strand-break repair model for red-mediated recombination of phage λ and plasmid λdv. *Genetics* **116**: 501–511.

166. **Timmerman, K. P., and C.-P. D. Tu.** 1985. Complete sequence of IS3. *Nucleic Acids Res.* **13**:2127–2130.

167. **Tomizawa, J.** 1984. Control of ColE1 plasmid replication: the process of binding of RNAI to the primer transcript. *Cell* **38**:861–870.

168. **Tomizawa, J.** 1985. Control of ColE1 plasmid replication: initial interaction of RNAI and the primer transcript is reversible. *Cell* **40**:527–535.

169. **Tomizawa, J.** 1986. Control of ColE1 plasmid replication: binding of RNAI to RNAII and inhibition of primer formation. *Cell* **47**:89–97.

170. **Tomizawa, J., and T. Som.** 1984. Control of ColE1 plasmid replication: enhancement of binding of RNAI to the primer transcript by Rom protein. *Cell* **38**:871–878.

171. **Tye, B.-K., R. K. Chan, and D. Botstein.** 1974. Packaging of an oversize transducing genome by *Salmonella* phage P22. *J. Mol. Biol.* **85**:485–500.

172. **Unger, B., J. Becker, and W. Hillen.** 1984. Nucleotide sequence of the gene, protein purification and characterization of the pSC101-encoded tetracycline resistance-gene-repressor. *Gene* **31**:103–108.

173. **Unger, B., G. Klock, and W. Hillen.** 1984. Nucleotide sequence of the repressor gene of the RA1 tetracycline resistance determinant: structural and functional comparison with three related Tet repressor genes. *Nucleic Acids Res.* **12**:7693–7703.

174. **Wang, A., and J. R. Roth.** 1988. Activation of silent genes by transposons Tn5 and Tn10. *Genetics*, in press.

175. **Wang, J. C., and N. Davidson.** 1966. On the probability of ring closure of lambda DNA. *J. Mol. Biol.* **19**:469–481.

176. **Watanabe, T., C. Furuse, and S. Sakaizumi.** 1968. Transduc-

tion of various R factors by phage P1 in *Escherichia coli* and by phage P22 in *Salmonella typhimurium*. *J. Bacteriol.* **96**:1791–1795.

177. **Watanabe, T., Y. Ogata, R. K. Chan, and D. Botstein.** 1972. Specialized transduction of tetracycline resistance by phage P22 in *Salmonella typhimurium*. I. Transduction of R factor 222 by phage P22. *Virology* **50**:874–882.

178. **Waters, S. H., P. Rogoswky, J. Grinsted, J. Altenbuchner, and R. Schmitt.** 1983. The tetracycline resistance determinants of RP1 and Tn*1721*: nucleotide sequence analysis. *Nucleic Acids Res.* **11**:6089–6105.

179. **Way, J. C., M. A. Davis, D. Morisato, D. E. Roberts, and N. Kleckner.** 1984. New Tn10 derivatives for transposon mutagenesis and construction of *lacZ* operon fusions by transposition. *Gene* **32**:369–379.

180. **Way, J. C., and N. Kleckner.** 1984. Essential sites at Tn10 termini. *Proc. Natl. Acad. Sci. USA* **81**:3452–3456.

181. **Way, J., and N. Kleckner.** 1985. Transposition of plasmid-borne Tn10 elements does not exhibit simple length-dependence. *Genetics* **111**:705–713.

182. **Weinert, T. A., K. Derbyshire, F. M. Hughson, and N. D. F. Grindley.** 1984. Replicative and conservative transpositional recombination of insertion sequences. *Cold Spring Harbor Symp. Quant. Biol.* **49**:251–260.

182a. **Wiater, L. A., and N. D. F. Grindley.** 1988. γδ transposase and integration host factor bind cooperatively at both ends of γδ. *EMBO J.* **7**:1907–1911.

182b. **Womble, D. D., and R. H. Rownd.** 1988. Genetic and physical map of plasmid NR1: comparison with other IncFII antibiotic resistance plasmids. *Microbiol. Rev.* **52**:433–451.

183. **Wu, T.-H., S.-M. Liao, W. R. McClure, and M. M. Susskind.** 1987. Control of gene expression in bacteriophage P22 by a small antisense RNA. II. Characterization of mutants defective in repression. *Genes Dev.* **1**:204–212.

184. **Yin, J. C. P., M. P. Krebs, and W. S. Reznikoff.** 1988. Effect of *dam* methylation on Tn*5* transposition. *J. Mol. Biol.* **199**:35–45.

185. **Yin, J. C. P., and W. S. Reznikoff.** 1987. dnaA, an essential host function, and Tn*5* transposition. *J. Bacteriol.* **169**: 4637–4645.

Chapter 9

Transposable Elements in Gram-Positive Bacteria

ELLEN MURPHY

I. INTRODUCTION

Gram-positive bacteria possess a variety of transposable elements (Table 1), many of which fall into the three classes already well characterized for transposons of gram-negative bacteria (insertion sequences [IS], composite transposons, and Tn*3*-like elements). Some of the elements, however, are novel. These include site-specific transposons, such as Tn*554* from *Staphylococcus aureus*, which shares some properties with temperate bacteriophages, and self-transmissible elements, such as Tn*916* (*Enterococcus* [*Streptococcus*] *faecalis*) and Tn*1545* (*Streptococcus pneumoniae*), which encode conjugal transfer systems. Among species of *Streptomyces* (see also C. A. Omer and S. N. Cohen, this volume), mobile elements that also exist as circular or linear plasmids have been described. Collectively, these novel elements possess one or more unusual properties, including a lack of terminal inverted repeats, variable termini, and an absence of a target duplication generated during transposition. This suggests that their

Ellen Murphy ■ The Public Health Research Institute, 455 First Avenue, New York, New York 10016.

Table 1. Gram-positive transposable elements and insertion sequences[a]

Element	Resistance markers	Other properties	Size	Terminal repeats	Target duplication (bp)	Insertion specificity	Source	Related elements	References
Conjugative transposons									
Tn*916*	Tc	Conjugation	16.4 kb	20/26-bp IR[b]	0	Moderate	*Enterococcus faecalis*	Tn*1545*	27a, 50
Tn*918*	Tc	Conjugation	16 kb	ND	ND	ND	*E. faecalis*	Tn*916*	27
Tn*919*	Tc	Conjugation	16 kb	ND	ND	ND	*Streptococcus sanguis*	Tn*916*	48
Tn*925*	Tc	Conjugation	16 kb	ND	ND	ND	*E. faecalis*	Tn*916*	25
Tn*1545*	Tc Em Km	Conjugation	25.3 kb	20/26-bp IR	0	Moderate	*Streptococcus pneumoniae*	Tn*916*	19, 36
3951	Cm Em Tc	Conjugation	67 kb	ND	ND	High	*Streptococcus agalactiae*		69
BM6001 element	Cm Tc	Conjugation	65.5 kb	ND	ND	High	*S. pneumoniae*		163
Class II transposons									
Tn*551*	Em		5.3 kb	40-bp IR	5	Low	*Staphylococcus aureus* pI258	Tn*501* (Tn*3*)	75
Tn*917*	Em		5,257 bp	38-bp IR	5	Low	*E. faecalis* pAD2	Tn*501* (Tn*3*)	151, 158
Tn*3871*	Em		5.1 kb	ND	ND	ND	*E. faecalis* pJH1	Tn*917*	8
Tn*4430*	None		4,194 kb	38-bp IR	5	ND	*Bacillus thuringiensis*	Tn*501* (Tn*3*)	96, 104
Tn*4556*	None		6.8 kb	38-bp IR	5	Low	*Streptomyces fradiae*	Tn*3*	26, 130
Tn*4451, 4452*	Cm		6.2 kb	12-bp IR	ND	ND	*Clostridium perfringens*	Tn*3*	1
Other transposons									
Tn*554*	Em Sp		6,691 bp	0	0	High	*S. aureus*	Tn*3853*	122, 123
Tn*3853*	Em Sp		ND	ND	ND	High	*Staphylococcus epidermidis* pWG25, *S. aureus* pWG4	Tn*554*	161
Tn*4001*	Gm Tb Km		4.7 kb	IS*256* (IR)	ND	Low	*S. aureus* pSK1	Tn*3851*	101, 102
Tn*3851*	Gm Tb Km		5.2 kb	ND	ND	ND	*S. aureus*	Tn*4001*	159
Tn*552*	Pc	Inversion	6.7 kb	ND	ND	High	*S. aureus*		7, 124
Tn*4002*	Pc		6.7 kb	80-bp IR	ND	High?	*S. aureus* pSK4	Tn*552*	101
Tn*3852*	Pc		7.3 kb	ND	ND	High	*S. aureus*	Tn*552*	76
Tn*4201*	Pc		6.7 kb	ND	ND	High?	*S. aureus* pCRG1600	Tn*552*	165
Tn*4003*	Tm		3.6 kb	IS*257* (DR)	ND	ND	*S. aureus* pSK1		54

Insertion sequences									
IS*231* (IS*1750*, IR*1*)	None		1,656 bp	20 bp IR	11	ND	*B. thuringiensis*	IS4	105, 106
IS*2150* (IR2)	None		2.1 kb	ND	ND	ND	*B. thuringiensis*		86
IS*256*	None		1.35 kb	ND	ND	ND	Tn*4001*		101, 102
IS*431*-L	None		800 bp	22-bp IR	ND	ND	*S. aureus* pI524	IS26	10
IS*431*-R	None		786 bp	14-bp IR	ND	ND	*S. aureus* pI524	IS26	10
IS*431mec*	None		792 bp	17-bp IR	ND	ND	*S. aureus*	IS26	10; L. Barberis-Maino and B. Berger-Bachi, personal communication
IS*257*	None		0.9 kb	ND	ND	ND	Tn*4003*	IS26	54
ISS*1*	None		820 bp	18-bp IR	8	ND	*Streptococcus lactis*	IS26	138
IS*110*	None		1,550 bp	10/15-bp IR	ND	High	*Streptomyces coelicolor* A3(2)		16, 22
IS*466*	None		~2.5 kb	ND	ND	ND	*S. coelicolor* A3(2)		74
2.6-kb minicircle	None	Circular form found free in cells	2.6 kb	26/33-bp IR	ND	Moderate	*S. coelicolor* A3(2)		23, 99
ISL*1*	None		1,256 bp	21/40-bp IR	3	Low	*Lactobacillus casei*		153

[a] Abbreviations: ND, not done; IR, inverted repeat; DR, direct repeat.
[b] Slash denotes short, imperfect repeats.

transposition mechanisms may be fundamentally different from those of most gram-negative elements. There is also accumulating evidence that genes specifying extracellular toxins and some additional antibiotic resistance markers may be transposable.

The fact that many gram-positive transposons have counterparts among the gram-negative elements strongly suggests that transfer of genetic material between gram-negative and gram-positive species occurs in nature. There are numerous examples of relatedness: Tn*551* and Tn*917*, isolated from *S. aureus* and *E. faecalis*, respectively, closely resemble members of the Tn*3* family of elements (75, 128, 151, 158, 167); ISS*1* from *Streptococcus lactis* and IS*431*/IS*257*, a repeated sequence found in both *S. aureus* and coagulase-negative staphylococcal species, are related to the *Proteus mirabilis* element IS26 (10, 54, 117, 138), and a *Bacillus thuringiensis* insertion sequence, IS*231*/IS*1750*, is distantly related to the *Escherichia coli* element IS4 (79, 106). In the laboratory, transposable elements native to gram-positive organisms, including Tn*916*, Tn*917*, Tn*1545*, and Tn*554*, are often completely or partially functional when transferred to gram-negative hosts (36, 52, 89, 126).

Since almost all of the elements studied carry drug resistance markers, their ability to move among bacterial species assumes considerable medical importance. This is best illustrated by elements studied in *S. aureus* and *Streptococcus* spp. Despite significant advances in the prevention and treatment of infectious diseases, *S. aureus* remains a major cause of nosocomial infection and death in hospitals worldwide, due chiefly to the emergence of multiply antibiotic resistant strains. It is generally accepted that the dissemination of mobile resistance genes has been stimulated by indiscriminate use of antibiotics and that an important factor is the exchange of genetic information between *S. aureus* and ubiquitous but relatively innocuous coagulase-negative species such as *Staphylococcus epidermidis*, *Staphylococcus haemolyticus*, and *Staphylococcus hominis* (5, 111). An example is the recent emergence of both plasmid- and transposon-encoded constitutive erythromycin resistance in *S. aureus* and *S. epidermidis* following the introduction of clindamycin into clinical medicine (93, 157). Prior to this, most erythromycin-resistant strains expressed inducible resistance (90, 115). (Since clindamycin does not induce resistance, such strains remain sensitive to clindamycin.) Another serious clinical problem is the appearance of methicillin-resistant *S. aureus*; such strains commonly carry multiple resistance genes and are responsive only to vancomycin, a costly and toxic drug (59, 133). Methicillin resistance is believed to be trans-

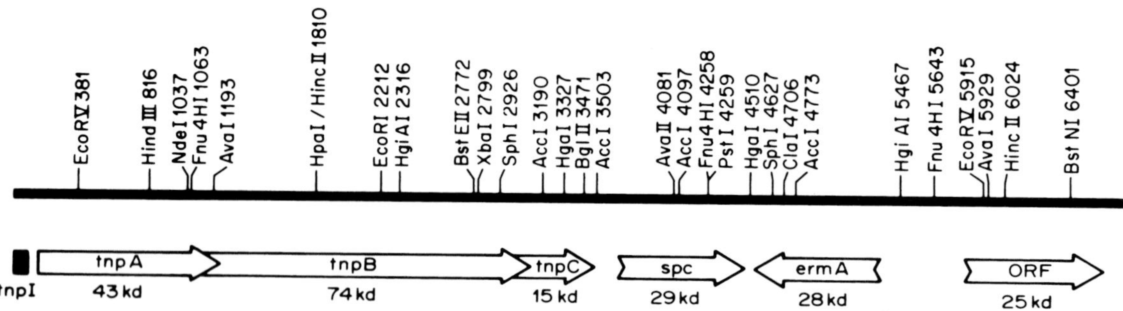

Figure 1. Physical and genetic map of Tn*554* (6,691 bp). For details, see the text. (Reprinted from reference 122 by permission of Cambridge University Press.)

posable, although the element has eluded investigators in the past and is only now beginning to be defined.

A similar situation exists among the streptococci and is best exemplified by the spread of tetracycline resistance among these bacteria. Group A streptococci, like *S. aureus*, are the cause of serious suppurative and systemic infections in humans. Resistance to tetracycline, erythromycin, and chloramphenicol was first reported in the 1950s and has been increasing. (Fortunately, resistance to penicillin is relatively rare and is not found among the group A streptococci.) In gram-positive bacteria, resistance to tetracycline is specified by the nonhomologous genes *tetK*, *tetL*, *tetN*, and *tetM* (17). Resistance is generally due to an active drug efflux mechanism (113); only *tetN* and *tetM*, of all the gram-negative and gram-positive Tc determinants, mediate resistance via a ribosomally associated protein that confers resistance at the level of protein synthesis (18; V. Burdett, personal communication). The *tetL*, *tetN*, and *tetM* genes are found in streptococci; of these, *tetM* is the most commonly isolated and is probably the most ubiquitous antibiotic resistance gene in procaryotes. In addition to its presence in gram-positive bacteria, *tetM* has also been detected in several gram-negative species, including *Neiserria gonorrhoeae* (118), in *Gardnerella vaginalis* (140), and in the mycoplasmas *Mycoplasma hominis* (141) and *Ureaplasma urealyticum* (140). The remarkably broad distribution of *tetM* among procaryotes must be due at least in part to its presence on transposons Tn*916* and Tn*1545*, elements that encode their own conjugal transfer systems in addition to transposition functions.

This chapter focuses on transposable elements unique to gram-positive bacteria, the site-specific elements, and the conjugal transposons. Also included is a discussion of elements analogous to the three classes of transposons found in gram-negative bacteria as well as a brief description of some variable genetic elements of clinical significance. The latter class includes genes specifying enterotoxin B, a cause of staphylococcal food poisoning, and TSST-1, the toxin implicated in toxic shock syndrome.

II. Tn*554*, A SITE-SPECIFIC TRANSPOSABLE ELEMENT

A. General Properties of Tn*554*

Tn*554* (Fig. 1) is a site-specific transposon from *S. aureus* that encodes resistance to the macrolide-lincosamide-streptogramin B (MLS) antibiotics and to spectinomycin (137). Tn*554* has several features that distinguish it from most other procaryotic and eucaryotic transposable elements. Its ends are asymmetric, lacking either inverted or direct terminal repeats; it does not generate a duplication of a target sequence upon transposition; and it is extremely site specific, nearly always inserting in the *S. aureus* chromosome at the same location, a site called *att*Tn*554* (84, 123).

Insertions at *att*Tn*554* occur preferentially in one orientation, designated (+). Tn*554* also exhibits a very high frequency of transposition: with a bacteriophage carrying an insertion of Tn*554*, the transposition frequency per donor molecule is essentially 100% if a target *att*Tn*554* is present (127). Transposition of Tn*554* is markedly reduced if *att*Tn*554* is deleted artificially (123) or naturally as in some *S. epidermidis* strains (161), or if *att*Tn*554* is blocked by a preexisting insertion of Tn*554* (127, 137). Many of these properties resemble those of integrative bacteriophages. However, unlike bacteriophages and the conjugative transposons (see below), Tn*554* does not appear to excise from the donor during transposition (127). In its specificity and its requirement for several transposition proteins (see below), transposition of Tn*554* formally resembles that of the *E. coli* transposon Tn*7* (see N. L. Craig, this volume).

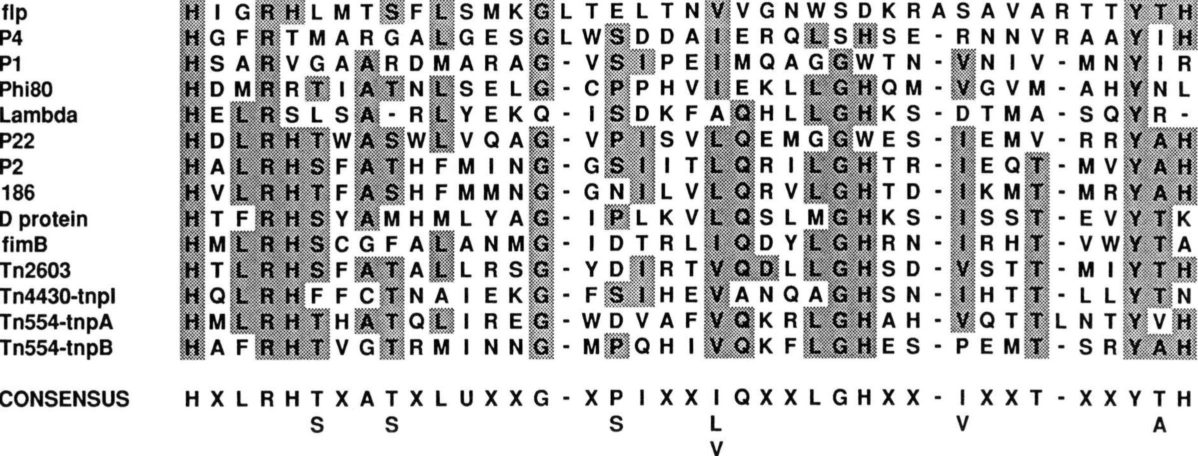

Figure 2. Amino acid homology among the Int family of site-specific recombinases aligned according to Argos et al. (6). Sequences of the seven bacterial integrases (P4, P1, phi80, lambda, P22, P2, and 186) and yeast Flp are from reference 6; also shown are F-factor D protein (94), *fimB* protein (80); Tn*2603* (131); Tn*4430* (104); and Tn*554 tnpA* and *tnpB* proteins (122). (Modified from Fig. 4 in reference 121a.)

B. Genetic Analysis of Tn*554* Transposition

Tn*554* is 6,691 base pairs (bp) in length and contains six open reading frames, five of them reading from left to right (122) (Fig. 1). Those designated *tnpA*, *tnpB*, and *tnpC* are required for transposition; they encode proteins of 74, 43, and 15 kilodaltons, respectively. Two genes correspond to the antibiotic resistance markers *ermA* and *spc*. The former encodes inducible resistance to the MLS antibiotics (121) mediated by an rRNA methylase, the activity and regulation of which have been extensively studied in the closely related *ermC* system (43, 57, 67). Hybridization analysis of clinical isolates indicates that the proportion of erythromycin-resistant staphylococci carrying chromosomal *ermA* has been increasing at the expense of strains containing the plasmid-linked *ermC* gene (158; R. F. Khabbaz, R. C. Cooksey, C. Thornsberry, B. Hill, C. Crowder, W. R. Jarvis, and the NNIS-CNS Study Group, Program Abstr. 27th Intersci. Conf. Antimicrob. Agents Chemother., abstr. no. 1313, 1987), and elements similar or identical to Tn*554* have been isolated in strains of both *S. aureus* and *S. epidermidis* (e.g., Tn*3853*) in both North America and Australia (157, 161). *spc* encodes a spectinomycin adenyltransferase that does not also specify resistance to streptomycin, unlike the product of the related gram-negative *aadA* determinant found on Tn7 and on R plasmids. Despite the difference in substrate specificity, the proteins encoded by *spc* and *aadA* are 36% identical in amino acid sequence (37, 49, 65, 120). The function of the sixth reading frame, designated ORF, is unknown; mutations in ORF affect neither the frequency nor the specificity of transposition (122). An additional site, *tnpI*, located at the left terminus of Tn554, is involved in *trans* inhibition of transposition (119).

The proteins specified by *tnpA* and *tnpB* are members of the Int family of site-specific recombinases, based on amino acid sequence homologies (Fig. 2). This family includes at least two resolvases and an invertase in addition to the bacteriophage integrases. A conserved region of about 40 amino acids, near the carboxy terminus of each protein, includes three completely conserved residues (His-396, Arg-399, and Tyr-433 in the relative alignment of Argos et al. [6]). The conserved tyrosine (corresponding to Tyr-338 and Tyr-500 of TnpA and TnpB, respectively) is likely to be the residue that forms a transient phosphodiester bond with the DNA during recombination. These sequence considerations suggest that both TnpA and TnpB behave as site-specific topoisomerases during transposition and that one or both of them will be found to contain site-specific DNA-binding domains that recognize *att*Tn*554*. The product of *tnpC*, on the other hand, is involved in orientation specificity. *tnpC* is not absolutely required for transposition, since deletion mutants encompassing 80% of *tnpC* transpose at frequencies as high as 1%. Such mutants, although retaining specificity for *att*Tn*554*, display an altered orientation specificity in which the orientation of insertion is dependent upon the orientation of Tn*554* in the donor molecule. In contrast, mutants of *tnpC* in which only the carboxy terminus is affected produce the usual (+) insertions, suggesting that these contain a partially functional *tnpC* product that can support transposition with normal site and orienta-

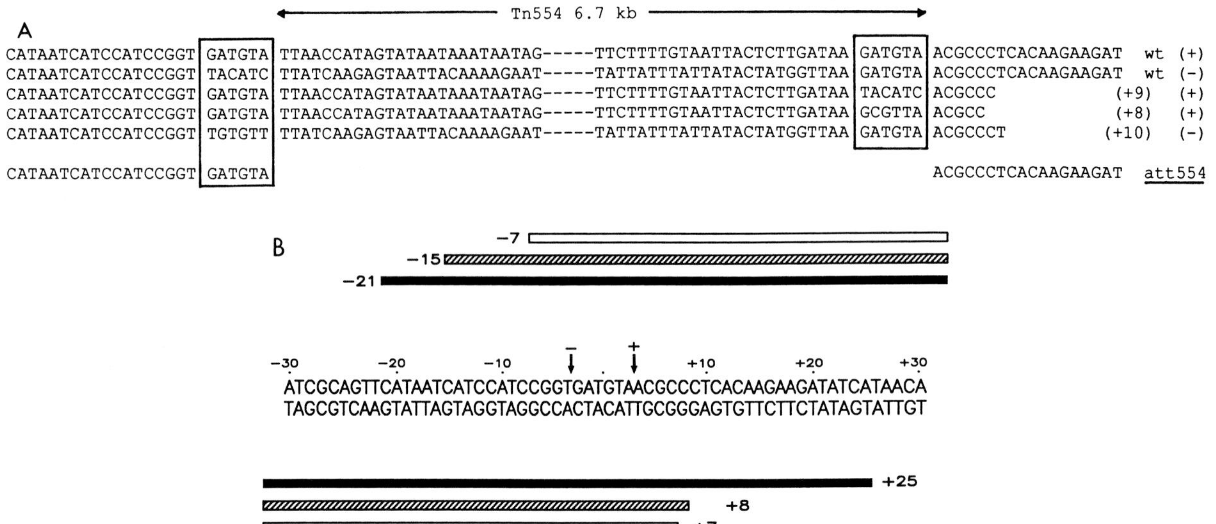

Figure 3. (A) Nucleotide sequence of junctions of (+) and (−) insertions of Tn*554* inserted at *att*Tn*554*, oriented such that the *att*Tn*554* sequences read the same for both insertions (123), and aberrant junctions generated by the insertion of Tn*554* into mutant *att*Tn*554* sites containing deletions on the right side of *att*Tn*554* extending to nucleotides +9, +8, and +10, respectively (unpublished data). The flanking variable sequences are boxed. (B) Deletion analysis of *att*Tn*554*. Bars indicate the extent of sequence still present on each side of the insertion site. Solid bars, Sequences required for full *att*Tn*554* activity; hatched bars, sequences required for partial activity; open bars, sequences present in derivatives with no residual *att*Tn*554* activity. The core 5′-GATGTA-3′ is numbered −3 to +3. (Modified from Fig. 5 in reference 121a.)

tion specificity, although with greatly reduced efficiency (10a).

C. Nucleotide Sequence Analysis of Insertion Sites

There are several noteworthy features of the junction sequences of inserted Tn*554* elements. Tn*554* does not contain inverted or direct repeats at its termini, nor does it generate a duplication of the target DNA (123). A 6-bp "core" sequence, 5′-GATGTA-3′, is present at both insertion junctions and at the insertion site itself (Fig. 3). However, the simplest model for Tn*554* insertion, a Campbell-like crossover between homologous core sequences (20), is insufficient to explain any but the (+) insertions at *att*Tn*554*. In (−) insertions, for example, the left junction carries not 5′-GATGTA-3′ but the complementary sequence 5′-TACATC-3′ (123).

Junction sequences of rare secondary-site insertions into plasmid pI524 suggested that the right-end copy of the GATGTA core was part of Tn*554* and the left-end copy was derived from the insertion site (123). However, recent characterization of additional secondary-site insertions supports the idea that the junction sequences vary from insertion to insertion in a predictable way (unpublished data; D. Dubin, personal communication). Most secondary insertion sites, although lacking the GATGTA core sequence and sharing no significant sequence homology with *att*Tn*554* or with one another, are similar to *att*Tn*554* in that insertions occur in both orientations and with an apparent displacement of 6 to 7 bp. Because the core sequences differ from one another, transposition between these sites yields valuable information concerning the Tn*554* recombination mechanism. We have examined the insertion junctions following serial transposition of Tn*554* between such sites, in some cases for up to five rounds of transposition. For each transposition event, the sequence originally present in the target is found at the left end of Tn*554* in the new insertion, the 6 or 7 bp originally at the left end of Tn*554* become the new right-end junction, and the original right boundary sequence is lost. The insertion junctions of Tn*916* are similarly variable (see below, Section III), except that while the Tn*554* variation is predictably unidirectional, for Tn*916* both of the reciprocal recombinants can be isolated (23, 27a, 146a).

A study of mutant and deleted *att*Tn*554* sites has established that a fully functional *att*Tn*554* site is between 31 and 53 nucleotides in length (Fig. 3B). Deletions extending closer than 10 to 25 bp on either side of the GATGTA core sequence resulted in partial or complete loss of *att*Tn*554* function (unpublished data). Insertion of Tn*554* into mutant derivatives of *att*Tn*554* frequently results in aberrant junction sequences at the right end (Fig. 3A; unpublished observations), indicating the importance of the sequences between +8 and +25 in promoting recombination. Thus, as for bacteriophage λ integrative recombination (143), sequences external to the core are required for Tn*554* transposition. In contrast, for Tn*7* transposition the sequences required for functional *att*Tn*7* lie entirely to one side and do not include the insertion site itself (112; Craig, this volume). However, the actual sequence of the core may be less critical; studies of transposition from mutant and secondary-site inserts to *att*Tn*554* indicate that sequence alterations within the core have a minimal effect on transposition frequency (unpublished data). Thus, a significant difference between Tn*554* insertion and bacteriophage λ integration is the more stringent requirement by the latter for homology between the recombining partners, a feature that is more critical than the specific sequence (39, 64, 166).

D. Regulation of Tn*554* Transposition

Transposition of Tn*554* is highly efficient; however, the choice of insertion sites is very limited, guaranteeing transposition when a site is available but precluding it otherwise. Transposition frequency, as measured by transduction assays, is reduced to 0.1 to 1.0% if the cell already contains Tn*554* inserted at *att*Tn*554*. A burst of transposition, similar to zygotic induction of a prophage, occurs upon transfer to a cell lacking a resident copy of Tn*554* if *att*Tn*554* is present (127, 137).

Tn*554* transposition is also regulated by a site, *tnpI*, which, when present on a high-copy plasmid vector, acts in *trans* to inhibit transposition of Tn*554* to a vacant chromosomal *att*Tn*554* site; *tnpI* activity is thus distinct from the *cis*-acting transposition immunity displayed by Tn*3* (95, 142). *tnpI* was mapped to the leftmost 89 bp of Tn*554*; no similar sequence or regulatory effect is associated with the right end of Tn*554* (119). There are no potential open reading frames in this region (122), nor have any transcripts such as those that regulate Tn*10* transposition (154) been detected. It is likely that inhibition due to *tnpI* is the result of titration of one or more required proteins that bind at the left end of Tn*554*. Since the *tnpA* coding region begins immediately downstream (nucleotide 134) from *tnpI*, it is attractive to speculate that *tnpI* overlaps or is contiguous with the *tnpA* promoter such that binding of TnpA, TnpB, or TnpC to *tnpI* would regulate TnpA production.

III. CONJUGATIVE TRANSPOSABLE ELEMENTS: Tn*916* AND Tn*1545*

A. Description

Tn*916*, its close relatives Tn*918*, Tn*919*, and Tn*925*, and Tn*1545*, a larger and less closely related element, constitute a family of self-transmissible conjugative elements that are ubiquitous among streptococci (for a recent review, see reference 28). Tn*916* is a 16.4-kilobase (kb) transposon which encodes tetracycline resistance (*tetM*) and is found in the chromosome of *E. faecalis* D16. It transposes intracellularly to plasmids such as pAD1 in recombination-deficient strains at a frequency of about 10^{-5} (50). Tn*916* is self-transmissible, promoting conjugation by a mechanism independent of the normal endogenous or plasmid-specified conjugation systems of *E. faecalis*. Tn*916*-mediated conjugation is inefficient (10^{-5} to 10^{-8} per donor) compared to the frequency of intracellular transposition (50, 52), but the range of organisms that can participate in Tn*916*-specified conjugation is broad. Tn*916* and related elements can be transferred among streptococci and to *S. aureus* (27, 71), *Bacillus subtilis* (25), and the mycoplasmas *Acholeplasma laidlawii* and *Mycoplasma pulmonis* (45). Indeed, transformation of *B. subtilis* protoplasts with circular Tn*916* (see below) results in transposition of Tn*916* to the *B. subtilis* chromosome, simultaneously rendering *B. subtilis* competent to conjugate with group A streptococci (146a).

Based on restriction and phenotypic analyses, Tn*916* is identical or very similar to Tn*918* and Tn*925*, both also isolated from *E. faecalis* (25, 27), to Tn*919* from *Streptococcus sanguis*, a major component of dental plaque and often involved in subacute bacterial endocarditis (48), and to an element from *Clostridium difficile* (58).

Tn*1545* is a 25.3-kb element that encodes resistance to tetracycline, erythromycin, and kanamycin. Like Tn*916*, it is self-transmissible (10^{-9} to 10^{-5} per recipient, depending on the donor), it excises precisely at a high (10^{-5}) frequency, and it transposes to different sites within the cell at a frequency of 10^{-6} to 10^{-5} (19, 35). Resistance to tetracycline is specified by *tetM* (35), MLS resistance is specified by a gene homologous to *ermB*/*ermAm* (132) (see Section IV.A, below), and kanamycin resistance is specified by the gene *aphA* encoding aminoglycoside 3′-phosphotransferase type III (34). While related to the type I and type II phosphotransferases specified by Tn*903* and Tn*5*, respectively, aminoglycoside 3′-phosphotransferase type III activity has been found only in gram-positive organisms. The three resistance genes of Tn*1545* are conjugally transferable to recombination-deficient *E. faecalis* (35). Like Tn*916*, Tn*1545* has an exceptionally broad host range. It can be conjugally transferred among *Streptococcus* spp., including the group N lactic streptococci, as well as from *Streptococcus* spp. to *S. aureus*, *B. thuringiensis*, and *Listeria monocytogenes* (35; P. Courvalin, personal communication). In the last-named organism, for which few standard genetic tools are available, Tn*1545* has been useful in genetic analysis (51). In addition, Tn*1545* is transpositionally active in *E. coli* and *B. subtilis*, to which it can be introduced via transformation. Tn*1545* is somewhat more stable than Tn*916* in *E. coli*, although this effect is copy number dependent by an unknown mechanism (36).

B. Transposition, Excision, and Conjugation

Four Tn*916* functions have been distinguished: conjugal transfer with transposition to the recipient chromosome, intracellular transposition (usually measured from the chromosome to a plasmid such as pAD1), zygotically induced transposition following transformation of *E. faecalis*, and excision, generally measured in *E. coli*, where the frequency is much higher than in *Streptococcus* spp. Tn*5* mutagenesis has revealed that over half of Tn*916* specifies transfer functions: insertions within the right-hand ~8 kb of Tn*916* fail to conjugally transfer tetracycline resistance (147). Most *tra* mutants are still capable of intracellular transposition of Tn*916*. A few *tra* mutants are also defective for transposition; these mutants define a function, located within the *tra* region, that is required for intracellular transposition as well as conjugation. Transposition occurring after transformation to a new host bypasses the requirement for all of the genes located on the right half of Tn*916* but requires another function(s) located at the left end (72). Mutants deficient in this latter function also fail to undergo excision in *E. coli*, suggesting that a specific excisionase required for all transposition events may be affected in these mutants (147).

Following transfer by conjugation or transformation of a plasmid carrying a Tn*916* insertion, 50 to 70% of the plasmid-carrying exconjugants are found to lack Tn*916*, while about 10% of the tetracycline-resistant colonies lack the donor plasmid (52, 53). The frequency of excision (i.e., loss of Tn*916*) is much higher (>90%) in *E. coli* than in *Streptococcus* spp. and has been exploited in the analysis of cloned gram-positive genes in *E. coli* (53). Circular, supercoiled DNA molecules corresponding by restriction and hybridization analysis to free, excised Tn*916* have recently been isolated (146a), confirming the hypothesis that Tn*916* transposition proceeds via a free circular intermediate (52). Excision can be precise or imprecise, as shown by reversion of insertions in the pAD1 hemolysin gene to *hly*$^+$ (50) and by direct analysis of the DNA sequences (27a; see below). Both transposition and excision are *rec*-independent functions.

C. Nucleotide Sequence Analysis of Termini and Insertion Sites

Although neither Tn*1545* nor Tn*916* has been described as site specific, a consensus insertion sequence, TTTXTN$_{3-5}$TAAAAA (Fig. 4), is suggested by analysis of several insertion junctions (19). Additionally, there are insertional hot spots; for Tn*916* one of them leads to hyperexpression of the *hly* gene on pAD1 (50). The terminal sequences of Tn*916* reveal an outward-facing, promoterlike sequence; transcription initiated at such a promoter could explain the phenotype of these insertions (72). Insertions of Tn*916* in the chromosomes of other streptococcal species, such as *S. lactis*, exhibit site specificity (63).

An understanding of the recombination events during transposition and excision of Tn*916* and Tn*1545* is beginning to emerge from nucleotide sequence analyses. There is a close relationship between the elements; the ends of Tn*916* and Tn*1545* are identical for at least 100 to 180 bp (19, 27a, 29) and exhibit short, imperfect (20/26-bp) terminal inverted repeats (Fig. 4). Like Tn*554*, Tn*916* and Tn*1545* do not generate a target duplication upon insertion (19, 27a, 29, 72, 123) (there is no homology between Tn*916*/Tn*1545* and Tn*554*, however). All of the DNA molecules participating inTn*916* transposition and excision, including the target site both preinsertion and postexcision, the donor transposon, and the new insertion, can now be isolated, and some of their junction sequences have been determined. Comparison of preinsertion and postexcision target sites (Fig. 4) shows that Tn*916* excision has the same probability of occurring precisely or imprecisely (27a, 29; J. R. Scott, P. A. Kirschman, and M. G. Caparon, personal communication). During imprecise excision, 3 to 5 bp derived from one end of the Tn*916* insertion is substituted for the original target sequence. Although the number of insertions examined is limited and serial transposition has not been studied, the junction sequences of the conjugative elements appear to vary in a manner similar to that observed for Tn*554*. However, Tn*554* variability is predictably unidirectional: inserts always retain the 6-bp sequences previously present at the left end and at the target site, while the previous right-end sequence is lost (see above). With the conjugative elements, the excised transposition intermediates contain with equal probability the 3 to 5 bp derived from the left or right end of the donor insertion. Additional analysis of serial transposition events is needed to follow the fates of these termini.

D. Other Conjugative Transposons

Other, very large (>60 kb) conjugative elements carrying multiple antibiotic resistances have been described in *Streptococcus agalactiae* B109 (Tn*3951*, 67 kb) (69) and *S. pneumoniae* BM6001 (*cat-tet* element, 65.5 kb) (163). Transposition of both elements appears to be site specific in *S. agalactiae*, with both elements showing a preference for the same site (163). The involvement of host recombination systems has not been assessed.

IV. CLASS II TRANSPOSONS RELATED TO Tn*3*

A. Tn*551* and Tn*917*

Tn*551* and Tn*917*, virtually identical 5.3-kb transposons specifying MLS resistance, are members of the Tn*3* family, or class II transposons (75, 128, 151, 158, 167). Tn*551* was discovered in *S. aureus* on plasmid pI258, an IncI penicillinase plasmid isolated in Japan (116). pI258 is unusual in that many closely related plasmids have been found in strains of *S. aureus* from around the world, but only pI258 carries Tn*551*. Tn*917* was isolated from the *E. faecalis* plasmid pAD2 (30). Both transposons specify MLS resistance mediated by an rRNA methylase (92), the product of *ermB* (Tn*551*) or *ermAm* (Tn*917*), and they have been extensively used for insertional mutagenesis in *Streptococcus*, *Staphylococcus*, and *Bacillus* spp. (98, 128, 170). In addition, a Tn*917*-based system analogous to Mu d-*lac* has been developed for analysis of transcription in these bacteria (136, 169).

Like other Tn*3*-related transposons, Tn*551* and Tn*917* contain homologous short terminal inverted repeats of 38 to 40 bp (Fig. 5), and they generate 5-bp duplications of target DNA during transposition (75, 135). The genetic map of Tn*917*, based on its nucleotide sequence (151), places it in the Tn*501* subclass of the Tn*3* family, where the gene order is resistance marker-resolvase (*res*)-transposase (*tnpA*) and where all three genes are transcribed on the same strand (Fig. 6). In contrast, *res* and *tnpA* in the Tn*3* subclass are divergently expressed from overlapping promoters (24, 62; D. Sherratt, this volume).

The published sequence of Tn*917* reveals an open reading frame (ORF) whose predicted amino acid sequence is homologous to that of the resolvase family of proteins (151). Mutations in two additional reading frames, ORFs 5 and 6, downstream from *res*, result in loss of transposition function (135, 151). Although Shaw and Clewell reported no homology

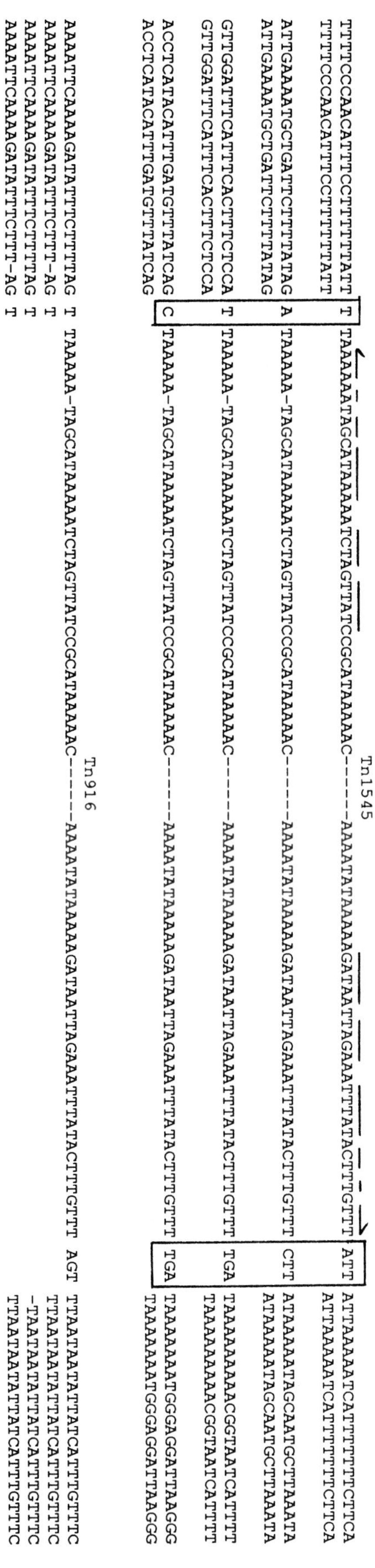

```
                                                                    Tn1545
TTTTCCCAACATTTCCTTTTTTATT T TAAAAAATAGCATAAAAATCTAGTTATCCGCATAAAAAC------AAAATATAAAAGATAATTAGAAATTTATACTTTGTTT ATT ATTAAAAATCATTTTTTTCTTCA
TTTTCCCAACATTTCCTTTTTTATT                                                                                            ATTAAAAATCATTTTTTTCTTCA

ATTGAAAATGCTGATTCTTTTATAG A TAAAAA-TAGCATAAAAATCTAGTTATCCGCATAAAAAC------AAAATATAAAAGATAATTAGAAATTTATACTTTGTTT CTT ATAAAAATAGCAATGCTTAAATA
ATTGAAAATGCTGATTCTTTTATAG                                                                                            ATAAAAATAGCAATGCTTAAATA

GTTGGATTTCATTTCACTTTCTCCA T TAAAAA-TAGCATAAAAATCTAGTTATCCGCATAAAAAC------AAAATATAAAAGATAATTAGAAATTTATACTTTGTTT TGA TAAAAAAAAACGGTAATCATTTT
GTTGGATTTCATTTCACTTTCTCCA                                                                                            TAAAAAAAAACGGTAATCATTTT

ACCTCATACATTTGATGTTTATCAG C TAAAAA-TAGCATAAAAATCTAGTTATCCGCATAAAAAC------AAAATATAAAAGATAATTAGAAATTTATACTTTGTTT TGA TAAAAAAATGGGAGGATTAAGGG
ACCTCATACATTTGATGTTTATCAG                                                                                            TAAAAAAATGGGAGGATTAAGGG

                                                                    Tn916
AAAATTCAAAAGATATTTCTTTTAG T TAAAAA-TAGCATAAAAATCTAGTTATCCGCATAAAAAC------AAAATATAAAAGATAATTAGAAATTTATACTTTGTTT AGT TTAATAATATTATCATTTGTTTC
AAAATTCAAAAGATATTTCTTT-AG T                                                                                          TTAATAATATTATCATTTGTTTC
AAAATTCAAAAGATATTTCTTTTAG T                                                                                          -TAATAATATTATCATTTGTTTC
AAAATTCAAAAGATATTTCTTT-AG T                                                                                          TTAATAATATTATCATTTGTTTC
```

Figure 4. Junction and target sequences of four insertions of Tn*1545* (19) and one of Tn*916*, including two target sequences following excision (29, 72). Target sequences are shown beneath each insertion; for Tn*916* the three target sequences are before insertion, after excision (imprecise), and after excision (precise), respectively. Variable bases at the ends of Tn*1545* are boxed; the imperfect inverted repeats are indicated by arrows. Tn*1545* and Tn*916* are identical in the region shown, except for the variable bases. The Tn*916* sequence shown is the opposite strand from that given in reference 29.

```
Tn3         G G G G T C T G A C G C T C A G T - G G A A C G A A A A C T C A C G T T A A G
Tn21-L      G G G G T C G T C T C A G A A A A C G G A A A A T A A A G C A - C G C T A A G
Tn1721-R    G G G G A G C C C G C A G A A T T C G G A A A A A A A T C G T A - C G C T A A G
Tn501-L     G G G G G A A C C G C A G A A T T C G G A A A A A A A T C G T A - C G C T A A G
Tn917-R     G G G G T C C C G A G C G C T T A - G T G G G A A T T T G T A T C G A T A A G
Tn917-L     G G G G T C C C G A G C G C C T A C G A G G G A A T T T G T A T C G A T A A G
Tn551-R     G G G G T C C - G A G C G C - - A C G A - G A A A T T T G T A T C G A T A A G G G G T A
Tn551-L     G G G G T C C - G A G C G C - - A C G A - G A A A T T T G T A T C G A T A A G A A A T A
Tn4430      G G G G T A C C G C C A G C A T TTC G G A A A A A A A A C C - - A C G C T A A G
Tn4556-R    G G G G G T T G A G G A A C A T C C G A A C G A A A A C C G G - C G C T A A G
Tn4451-R    G G G G T C - G A G T T T G T C A A G A T A C T T T T T G T G GATTTTC T A A
Tn4451-L    G G G C T A - T A C T T T A A T A G G A C A A A A A A A A T T A ACTGTCC T A A
```

Figure 5. Comparison of terminal inverted repeats of Tn*917* (135), Tn*551* (75), Tn*4430* (96), Tn*4556* (130), and Tn*4451* (2) with selected members of the Tn*3* family from gram-negative bacteria: Tn*3* (129), Tn*21* (38), Tn*1721* (146), and Tn*501* (13).

between ORFs 5 and 6 and the *tnpA* proteins of Tn*3*/Tn*501* (15, 62), if one assumes that a sequence determination error was made that changes the reading frame, the two ORFs merge and exhibit 32.5% identity with the deduced amino acid sequence of the Tn*501* transposase.

Transposition of both Tn*501* (*mer*) and Tn*917* (MLS) can be induced by the substrates of the resistance genes (152, 158). It has been proposed that induction of the upstream resistance gene could allow expression of transposase via readthrough of the *tnpA* gene. Such a proposal is supported by the finding of increased levels of a transcript corresponding to the full length of Tn*917* following induction with erythromycin (151). The high basal level of expression of *ermB* in Tn*551*, which results in a constitutive phenotype in spite of the fact that *ermB* expression can actually be induced fourfold with erythromycin (44), might mask such an effect and explain why it has not been demonstrated with this element.

Despite the similarities in sequence, organization, and terminal repeats, members of the Tn*3* family display significant differences in insertion specificity. Tn*3* transposes primarily to plasmids and only rarely to the bacterial chromosome (83). Tn*917* transposes both to chromosomal and plasmid sites, in some cases with a clear preference for the plasmid (164), while Tn*551* transposes preferentially to the chromosome (98, 128). These differences may simply be due to the use of different host systems, or they may be a reflection of subtle mechanistic differences in transposition that are sensitive to factors such as local conformation of the target DNA molecule.

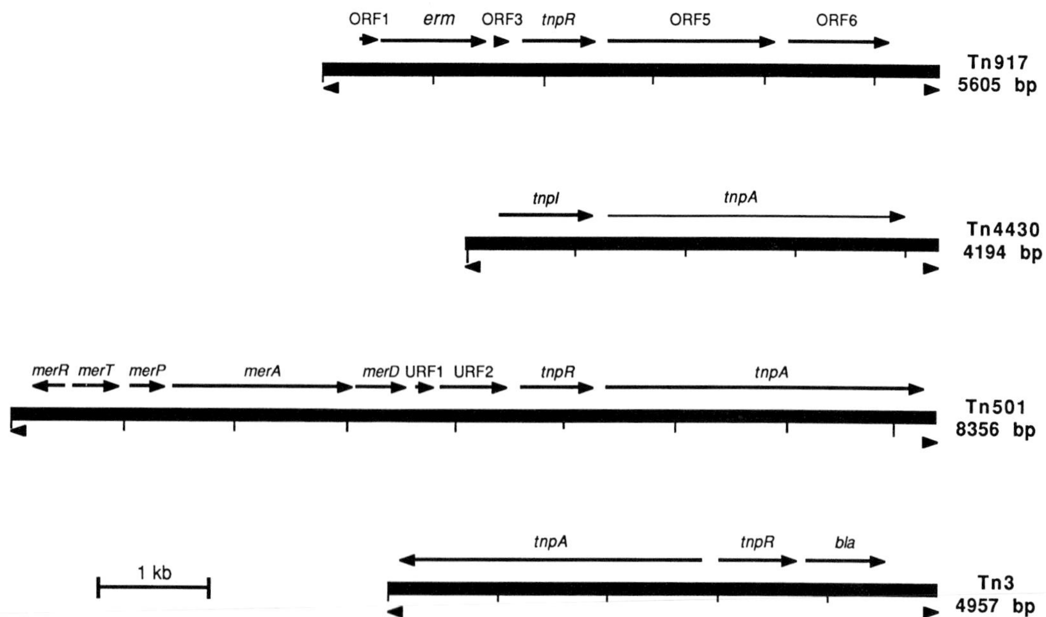

Figure 6. Genetic organization of Tn*917* (151), Tn*4430* (104), Tn*501* (14, 15, 114, 145), and Tn*3* (62). ORF, Open reading frame; *res*, site of cointegrate resolution; *tnpA* and *tnpR* (*tnpI* for Tn*4430*), genes encoding transposase and resolvase, respectively; ⧫, terminal inverted repeats; *bla*, β-lactamase; *erm*, erythromycin resistance; *merR*, *merT*, *merP*, *merA*, *merD*, genes of the mercury resistance operon. In Tn*917*, ORF1 is the *erm* leader peptide, ORF3 is a short ORF of unknown function, and ORFs 5 and 6 together probably encode the transposase (see text). In Tn*501*, URF1 and URF2 are ORFs of unknown function. (Modified from Fig. 2 in reference 121a.)

B. *Bacillus thuringiensis* Element Tn*4430* and Flanking Insertion Sequences

B. thuringiensis strains are characterized by the production of a parasporal inclusion body containing δ-endotoxin or crystal protein, which has insecticidal activity against *Lepidoptera* larvae (for review, see reference 40). Genes encoding δ-endotoxin have been found in a variety of plasmid and chromosomal locations (55, 85). Although this diversity suggests that the gene might be located on a mobile element, transposition of the crystal gene has not been demonstrated. The DNA flanking the δ-endotoxin gene is highly conserved among different subspecies of *B. thuringiensis*, and within that DNA a transposable element, Tn*4430* (previously designated Th sequence), and several insertion sequence (IS)-like elements have been found (86, 87, 96, 97).

Tn*4430*, a member of the Tn*3* family, is 4,194 bp in length (104). It produces a 5-bp duplication of the host DNA and contains 38-bp terminal inverted repeats (96). These repeats seem to act as hot spots for insertion of at least one family of IS elements, IS*231* (105). They are, unexpectedly, more closely related to the termini of the gram-negative elements Tn*21* (30/38 bp identical) and Tn*3* (25/38 bp) than to those of the gram-positive transposons Tn*917* (21/38) and Tn*551* (20/38) (Fig. 5). Since Tn*4430* contains no known genes other than those required for transposition, insertion of a gene specifying kanamycin resistance was necessary to study Tn*4430* transposition (96).

Nucleotide sequence analysis of Tn*4430* revealed two open reading frames, ORF1 and ORF2. Their assignments as resolvase and transposase, respectively, were confirmed by analysis of insertion mutations (105). The two genes are transcribed in the same direction (Fig. 6) and account for virtually all the coding capacity of Tn*4430*. The gene for the transposase, *tnpA*, is similar to *tnpA* genes of other class II transposons (e.g., 41% identity with Tn*917*, 29% with Tn*3*). Unexpectedly, the resolvase protein, coded for by *tnpI*, was found to be unrelated to other resolvase and invertase proteins. Instead, it is a member of the Int family of site-specific recombination proteins (Fig. 2). In this regard, it may be significant that within the region upstream of the *tnpI* gene is a sequence containing a 12/15 match to the core of the lambda attachment site (105).

Two classes of IS-like elements generally flank the crystal gene and Tn*4430* (87, 97, 106). One family of repeated sequence, about 2.1 kb in length, is represented by the elements IS*2150* and IR*2* (86, 97); these are uncharacterized. The second group includes IS*231*, IS*1750*, and IR*1*, which comprise a

```
                * * * * *     * * *     * * * * *
IS4-L       T A A T G C C G A T C A G - T T A A G
IS4-R       T A A T G C C A G T C A G - T T A A G
IS231A-L    C - A T G C C C A T C A A C T T A A G A A
IS231A-R    C - A T G C C C A T C A A C T T A A G A A
IS231B-L    C - A T G C C C A T C A A C T T A A G A A
IS231B-R    T - T A G C A T T T A T A C T T A A G G A
IS231C-L    C - A T G C C C A T C A A C T T A A G A A
IS231C-R    C - A T G C C C A T C A A C T T A A G G A
```

Figure 7. Comparison of the terminal inverted repeats of IS*4* (79) and variants of IS*231* (105, 106). Nucleotides conserved between IS*4* and IS*231* are indicated with asterisks.

family of insertion elements whose sequenced members are highly homologous (85 to 99% nucleotide sequence identity). IS*231* is 1,656 bp in length, contains 20-bp terminal inverted repeats, and produces 11-bp target direct repeats (105, 106). IS*231* contains a long open reading frame specifying a potential protein of 478 amino acids, whose carboxy-terminal portion is distantly related (20% identity) to that of the transposase specified by the *E. coli* insertion element IS*4* (79, 106). Conservation within the ORF is significantly higher than outside the coding region (105), suggesting that the product of the ORF is functional. Limited homology between the IS*4* and IS*231* terminal inverted repeats is also evident (Fig. 7); however it is not known if, like IS*4* (78), the target duplications generated by these elements are of variable length.

Both families of IS elements are present in multiple copies in subspecies of *B. thuringiensis* (86); whether these elements transpose independently of one another or as units including the crystal protein gene or Tn*4430* is not known. The ubiquity of these sequences suggests that these elements play a role, via homologous recombination, in the structural instability of many *B. thuringiensis* plasmids (55), and they might also account for the mobilization of chromosomal markers during plasmid transfer (21).

C. Tn*4556* of *Streptomyces fradiae*

Chung and Olson (26, 130) have described a 6.8-kb element, Tn*4556*, in the actinomycete *S. fradiae*. Tn*4556* has 38-bp inverted repeats related to those of other class II transposons (Fig. 5) and generates a 5-bp target duplication upon transposition (130).

Like Tn*4430*, Tn*4556* is cryptic; derivatives carrying *vph* (viomycin resistance) were constructed in order to study transposition. Tn*4556* transposes to multiple chromosomal locations in *Streptomyces lividans* and to plasmids of *S. fradiae*. Low-copy-number, unstable plasmids of *S. fradiae* appear to be stablized by insertion of Tn*4556*. Efficient resolution of replication-derived plasmid cointegrates, presumed to be mediated by the resolvase function of

Tn*4556*, has been proposed as a mechanism for this stabilization (26).

D. Transposons of *Clostridium perfringens*

More distantly related members of the Tn*3* family have been detected in *C. perfringens* (1). Tn*4451* and Tn*4452*, differing only by inversion of a short segment near one end, are 6.2 kb in length and specify resistance to chloramphenicol. Although transposition of these elements has not been demonstrated in the native host, probably due to the lack of a suitable detection system in *C. perfringens*, these transposons have been shown to transpose to different sites in an *E. coli recA* mutant (1). Tn*4451* has shorter inverted repeat sequences (12 bp, imperfect), which are less well conserved than in the other members of the Tn*3* family (Fig. 5) and appear to generate a 2-bp target duplication (2). The transposon also excises readily (~50% in the absence of selection) and precisely (1, 2), an event more typical of the conjugative elements than of the Tn*3* transposons. Further study of these elements will reveal whether they are truly class II elements.

V. INSERTION SEQUENCES AND COMPOSITE TRANSPOSONS

A. Tn*4001*

Tn*4001* is a recently discovered transposon that encodes resistance to the aminoglycoside antibiotics gentamicin, tobramycin, and kanamycin (Gm^r Tb^r Km^r) in *S. aureus*. Serial transposition in a *rec* host establishes Tn*4001* as a bona fide transposable element (101, 102). Although the reported sizes differ slightly, Tn*4001* is probably identical to Tn*3851*, another Gm^r Tb^r Km^r element also isolated in Australia (159).

Tn*4001* is a composite transposon 4.7 kb in length; a 2-kb unique region specifying the antibiotic resistance is flanked by 1.35-kb inverted repeats designated IS*256* (102). The resistance gene specifies a 56-kilodalton bifunctional modifying enzyme with both aminoglycoside 2″-phosphotransferase (*aphD*, Gm^r) and aminoglycoside 6′-acetyltransferase (*aacA*, Tb^r Km^r) activities (108, 162). The coding region for the modifying enzyme would be large enough to account for most of the unique portion of Tn*4001*. It is likely that, as for other composite transposons, the transposition functions are encoded by one or both of the flanking inverted repeat sequences. Sequences homologous to IS*256* have been detected at multiple sites in some highly resistant clinical isolates (101); whether this represents independent transposition of IS*256* is not known.

Distinct geographical differences exist among gentamicin-resistant strains of *S. aureus*. In Europe, most Gm^r strains carry chromosomal resistance. Although the enzyme with aminoglycoside 2″-phosphotransferase and aminoglycoside 6′-acetyltransferase activities appears similar if not identical to that specified by Tn*4001*, there is no information on their relatedness at the DNA level (47, 73). Plasmid-linked Gm Tb Km resistance is now common in both Australian and North American isolates of *S. aureus*; however, it is carried by unrelated plasmids on the two continents. The plasmids isolated in Australia all closely resemble pSK1, an IncI, nonconjugative plasmid (100, 160). The North American Gm^r plasmids, typified by pGO1, are conjugative and compatible with the IncI plasmids (4). They are widely distributed among *S. aureus* and *S. epidermidis* strains and can be transferred between these species by conjugation (5, 111). Although the North American and Australian plasmids are unrelated to one another, they have in common a 2.5-kb *Hin*dIII fragment which contains the entire unique region of Tn*4001*, including the resistance gene and part of each IS*256* sequence (56, 70, 101, 107). However, the Gm^r Tb^r Km^r marker of the American plasmids is not transposable (5, 101). This is almost certainly due to truncation of the inverted repeat sequences flanking the Gm^r Tb^r Km^r gene on these plasmids (101; G. Archer, personal communication).

B. IS-Like Elements in *Staphylococcus* spp.

In addition to IS*256* associated with Tn*4001*, a second family of closely related IS-like elements, about 800 bp in length, has been detected in multiple plasmid and chromosomal locations in *S. aureus* and in several coagulase-negative staphylococcal species. The sequences known as IS*431*-L and IS*431*-R are found as direct repeats flanking the mercury resistance operon of pI524 and related plasmids (10). IS*431*-L contains a perfect 22-bp terminal inverted repeat; those of IS*431*-R are 14 bp (10) (Fig. 8). Sequences homologous to IS*431*, designated IS*257*, are found as direct repeats flanking the trimethoprim resistance gene of plasmid pSK1. This composite element has been designated Tn*4003* (54); however, as for the ISs themselves, transposition has not yet been directly demonstrated in the laboratory.

Probes specific for IS*431*/IS*257* indicate that homologous sequences are widely distributed among clinical isolates of *S. aureus* and coagulase-negative staphylococci (Archer, personal communication; B. Berger-Bachi, personal communication). Four copies

```
IS431-L     800 bp   AATTCTGGTTCTGTTGCAAAGT----------ACTTTGCAACAGAACCAGAATT
IS431-R     786 bp         ggTTCTGTTGCAAAGT----------ACTTTGCAACAGAAtc
IS431mec    792 bp        CGGTTCTGTTGCAAAGT----------ACTTTGCAACAGAACCG
ISS1        808 bp         GGTTCTGTTGCAAAGTTT------AAACTTTGCAACAGAACC
IS26        820 bp         GGCACTGTTGCAAAgt----------taTTTGCAACAGTGCC
```

Figure 8. Comparison of the ends of IS*431*-L, IS*431*-R, IS*431mec* from *S. aureus* (10; L. Barberis-Maino and B. Berger-Bachi, personal communication), ISS*1* from *S. lactis* (138), and IS26 from *P. mirabilis* (117). (Modified from Fig. 6 in reference 121a.)

of an IS*431*/IS*257*-like sequence are associated with a 25-kb DNA insert unique to methicillin-resistant strains (10, 110). The conjugative plasmid pGO1, found in both *S. aureus* and *S. epidermidis*, contains eight regions that hybridize to IS*431*. These repeats flank the Tp and Gm resistance genes and the *tra* operon. (Plasmid pGO1 also contains sequences hybridizing to IS*256*/Tn*4001* [54], but these are distinct from those hybridizing to IS*431*/IS*257*.) Four of the eight copies on pGO1 are deleted relative to IS*431*, as they fail to hybridize to a 180-bp probe specific for the end of the element (Archer, personal communication). Chromosomal copies of sequences homologous to IS*431* are also widely distributed in multiply antibiotic resistant strains and are more frequently present and more abundant in coagulase-negative isolates than in *S. aureus*. Hybridization analysis reveals that these, too, are often truncated relative to IS*431* (Archer, personal communication).

Nucleotide sequence analysis of IS*431* suggests an evolutionary relationship to other IS elements in both gram-negative and gram-positive organisms. A related IS element found in *S. lactis*, ISS*1*, shares 61% base sequence identity with IS*431* (138). Both sequences contain a long open reading frame whose deduced amino acid sequences are 57% identical with one another and 40 to 44% identical to the transposase specified by a gram-negative insertion sequence, IS*26* (117). IS*26*, discovered as direct repeats flanking the kanamycin resistance transposon Tn*2680* in *Proteus vulgaris*, is 820 bp in length and produces 8-bp target repeats upon transposition (68), as does ISS*1* (138). Although the short terminal inverted repeats for all these elements are very highly conserved (Fig. 8), no directly repeated sequences suggestive of a target duplication were detected adjacent to IS*431*. This suggests that these elements may be vestiges of once active insertion sequences (10, 91), in agreement with the failure by several laboratories to detect transposition of the mercury resistance operon, which is flanked by directly repeated copies of IS*431* on most IncI and IncII penicillinase plasmids. The spontaneous, site-specific deletion of 6.5 kb of DNA associated with loss of mercury resistance (150) is now understood to be a consequence of homologous recombination involving the flanking direct repeats. The finding of apparently identical elements in at least two locations in the *S. aureus* chromosome (168), however, suggests that *mer* may once have been transposable. The constant location of *mer* in the penicillinase plasmids (150) suggests that the loss of transposition function predated the proliferation of these plasmids.

Even if, as seems likely, IS*431*, IS*257*, and related sequences represent remnants of once active IS elements and are incapable of independent transposition, they may play an important role in the evolution of multiply resistant strains via homologous recombination between plasmids or chromosomal markers containing these sequences. On the other hand, these sequences are likely to be more stable than are plasmids, and as such may prove useful as genomic markers in epidemiological studies.

C. IS-Like Elements in *Streptomyces* spp.

Streptomyces spp. contain a number of unusual mobile elements (for a recent review see reference 23). One of these elements, the integrating plasmid SLP1, is treated in detail elsewhere (Omer and Cohen, this volume). Three elements from *Streptomyces coelicolor*, IS*110* (1.5 kb), IS*466* (~2.5 kb), and an element called 2.6-kb minicircle, which all promote chromosomal integration of bacteriophage or plasmid DNA, are discussed briefly here.

IS*110* was first identified as an insertion in the bacteriophage ϕC31 that suppressed a deletion of *attP*; homologous recombination between copies of IS*110* in the phage and in the *S. coelicolor* chromosome allows lysogenation by these mutant bacteriophage (22). IS*110* appears to have a high degree of site specificity, having been reisolated at the same site in ϕC31 on several occasions (16, 23).

IS*466*, which has been found in at least two chromosomal loci in *S. coelicolor* and in the plasmid SCP1 (74), is less well characterized. Recombination between copies of IS*466* leads to the integration of the plasmid SCP1 or to the formation of SCP1′ strains (74), which have enhanced fertility properties (66). SCP1 is a giant (~350 kb) plasmid encoding resistance to and production of the antibiotic methylenomycin A, and is detectable only as a linear molecule (77); thus, the mechanism of its integration may prove to be quite different from that of circular molecules.

The 2.6-kb minicircle is an element found as free

circles at extremely low abundance (<0.1 copy per chromosome). Integration may occurs at different chromosomal sites but occurs commonly at a preferred site, and always involves the same recombination site on the minicircle (99). The recombination sites on the chromosome and on the minicircle, however, do not share homologous sequences (23), which makes this element rather unusual among mobile elements in the actinomycetes. The mechanism by which the free circles arise is unknown, nor is it known whether the circular form is capable of autonomous replication.

VI. PUTATIVE TRANSPOSABLE ELEMENTS

A. Transposition of Antibiotic Resistance Genes

1. Penicillin resistance

In *S. aureus*, resistance to penicillin is mediated by β-lactamase, the product of *blaZ*. This gene is generally carried by the large, nonconjugative penicillinase plasmids belonging to incompatibility groups I and II (150). The prototype penicillinase plasmid, pI524 (31.8 kb), carries a 2.2-kb invertible region flanked by 0.65-kb inverted repeat sequences (124). The *bla* operon is adjacent to this region; based on comparison of plasmids lacking *bla*, the entire *bla* element (designated Tn*552*), including the invertible segment *inv*, is 6.7 kb in length. However, neither the invertible segment nor inversion per se is required for penicillinase expression (124). Within the inverted repeat sequence is a gene, *bin*, that encodes a protein sharing 39% amino acid identity with the Hin family of DNA invertases (144; A. C. Glasgow, K. T. Hughes, and M. I. Simon, this volume). The *bin* product is also more closely related to the resolvases than are the other invertases, raising interesting questions about its evolution and specificity.

Other elements carrying penicillin resistance, Tn*4201*, Tn*4002*, and Tn*3852*, all showing overall similarity in size and structure to Tn*552*, have been isolated (76, 103, 165). Some of these contain at least part of the invertible DNA segment, although not all of them undergo inversion. These elements have been suggested to be transposable, based on the observation of apparent hops from the *S. aureus* chromosome to various plasmids (7, 76, 103, 150, 165). These reactions are usually site specific; similarly, when present as a chromosomal marker, whether in a clinical penicillin-resistant isolate or following transposition from a plasmid, *bla* occupies a unique map position (134, 165). However, neither the *rec* independence of recombination nor the absence of homologous DNA at the target site has been established for these elements. Intermolecular recombination involving the inverted repeat sequences, which can lead to rearrangement, deletion, or duplication of *bla*, is known to be both highly efficient (125) and site specific (150). It is possible that these reactions are also mediated by *bin*; it remains to be determined whether "transposition" of these elements is another manifestation of intermolecular recombination or if a separate transposition apparatus is involved.

2. Transposition of methicillin resistance

Methicillin is a semisynthetic penicillin not susceptible to inactivation by β-lactamase. Although resistant strains of *S. aureus* were first detected in 1961 (9), our understanding of the molecular basis of resistance is still incomplete. Resistance, termed **intrinsic**, is not due to the production of β-lactamase. Recent studies have indicated that all methicillin-resistant strains contain an additional penicillin-binding protein, PBP 2a or PBP 2′, that has very low affinity for methicillin and other β-lactams (60, 61, 139). PBP 2a is believed to be a transpeptidase enzyme (139); substitution of PBP 2a for one (or more?) of the cellular transpeptidases normally required for cell wall synthesis presumably plays a role in resistance.

Both environmental and genetic factors influence the expression of methicillin resistance. Temperature and ionic strength, for example, affect resistance in strains expressing heterogeneous resistance, in which situation only 1 in 10^6 cells in the population is resistant (3, 46). Homogeneous strains, in which all of the cells express full resistance, are unaffected by changes in these variables. Mutations in at least two regulatory loci, unlinked to *mec* but not mapped relative to one another, produce methicillin sensitivity or convert a homogeneously resistant strain to heterogeneous expression (12, 81).

The resistance gene *mec* has been mapped to a specific location on the *S. aureus* chromosome (88). *mec* is associated with a large segment of DNA unique to Mc^r strains (11, 110), as predicted by an earlier finding that methicillin resistance and sensitivity do not behave as allelic markers: sensitivity, but not resistance, is cotransducible with *purA* (156), suggesting that the gene for methicillin resistance is part of an inserted DNA fragment for which there is no homolog in isogenic Mc^s strains. A gene specifying PBP 2a has been cloned independently (109) and found to be identical in nucleotide sequence to a portion of the unique *mec* DNA which confers heterogeneous methicillin resistance when introduced into *E. coli* and *S. aureus* (155). However, PBP 2 expression alone is probably not sufficient for full expression of resistance, since it is present in equivalent amounts in both homogeneous and heterogeneous Mc^r cultures (81).

In addition to the presence of unique DNA in resistant strains, the ability of *mec* to be transferred to a recombination-defective recipient (33, 148) suggests that *mec* might be carried by a transposable element. Although methicillin resistance has often been reported to be transiently associated with markers such as enterotoxin B or with resistance to tetracycline, erythromycin, or cadmium (32, 41, 42, 149), in only one case has a stable physical linkage been demonstrated (110). The cloned Mc^r-specific DNA contains copies of an IS-like element, IS*431* (see above). This may explain the results of a study in which transduction of Mc^r to strain NCTC8325, which lacks IS*431*-like sequences (Archer, personal communication), was found to be dependent upon the presence in the recipient strain of pI524 (31), which carries two copies of IS*431*. Thus, the apparent transposability of methicillin, like that of penicillin resistance, could be explained on the basis of homology-dependent, *recA*-independent recombination between IS*431*-like sequences rather than by transposition. The availability of *mec* probes should help clarify some of these relationships.

B. Transposition of Exoprotein Genes?

Both staphylococci and streptococci specify a wide variety of nonessential extracellular products that play a role in the virulence properties of these bacteria. In *S. aureus* isolates, these genes are highly variable and often have genetic properties suggesting the involvement of mobile elements. Among these genes, those encoding enterotoxin B (*entB*), a cause of staphylococcal food poisoning, and TSST-1, the toxin associated with toxic shock syndrome, are the best candidates for being carried by transposable elements. *entB* can be transferred to a recombination-deficient strain (148), TSST-1 maps at more than one chromosomal location (82), and the DNA for both is generally absent from nonproducing strains (82; S. Khan, personal communication). The availability of cloned DNA for these and a number of additional staphylococcal exoprotein genes is now making possible experiments to determine whether the absence of a particular exoprotein is correlated with the absence of the structural gene and unique flanking sequences. At the same time, manipulation of the cloned genes to insert selectable markers will allow direct tests of their transposability.

Acknowledgments. I thank my colleagues who provided information in advance of publication and K. Drlica for his critical reading of the manuscript.

Work in my laboratory was supported by Public Health Service grant GM27253 from the National Institutes of Health.

LITERATURE CITED

1. **Abraham, L. J., and J. I. Rood.** 1987. Identification of Tn*4451* and Tn*4452*, chloramphenicol resistance transposons from *Clostridium perfingens*. *J. Bacteriol.* **169:** 1579–1584.
2. **Abraham, L. J., and J. I. Rood.** 1988. The *Clostridium perfringens* chloramphenicol resistance transposon Tn*4451* excises precisely in *Escherichia coli*. *Plasmid* **19:**164–168.
3. **Annear, D. I.** 1968. The effect of temperature on resistance of *Staphylococcus aureus* to methicillin and some other antibiotics. *Med. J. Aust.* i:444–446.
4. **Archer, G. L., D. R. Dietrick, and J. L. Johnston.** 1985. Molecular epidemiology of transmissible gentamicin resistance among coagulase-negative staphylococci in a cardiac surgery unit. *J. Infect. Dis.* **151:**243–251.
5. **Archer, G. L., and J. L. Johnston.** 1983. Self-transmissible plasmids in staphylococci that encode resistance to aminoglycosides. *Antimicrob. Agents Chemother.* **24:**70–77.
6. **Argos, P., A. Landy, K. Abremski, J. B. Egan, E. Haggard-Ljungquist, R. H. Hoess, M. L. Kahn, B. Kalionis, S. B. L. Narayana, L. S. Pierson III, N. Sternberg, and J. M. Leong.** 1986. The integrase family of site-specific recombinases: regional similarities and global diversity. *EMBO J.* **5:** 433–440.
7. **Asheshov, E. H.** 1969. The genetics of penicillinase production in *Staphylococcus aureus* strain PS80. *J. Gen. Microbiol.* **59:**289–301.
8. **Banai, M., and D. J. LeBlanc.** 1984. *Streptococcus faecalis* R plasmid pJH1 contains an erythromycin resistance transposon (Tn*3871*) similar to transposon Tn*917*. *J. Bacteriol.* **158:**1172–1174.
9. **Barber, M.** 1961. Methicillin-resistant staphylococci. *J. Clin. Pathol.* **14:**385–393.
10. **Barberis-Maino, L., B. Berger-Bachi, H. Weber, W. D. Beck, and F. H. Kayser.** 1987. IS431, a staphylococcal insertion sequence-like element related to IS26 from *Proteus vulgaris*. *Gene* **59:**107–113.

10a. **Bastos, M. C. F., and E. Murphy.** 1988. Transposon Tn*554* encodes three products required for transposition. *EMBO J.* **7:**2935–2941.

11. **Beck, W. D., B. Berger-Bächi, and F. H. Kayser.** 1986. Additional DNA in methicillin-resistant *Staphylococcus aureus* and molecular cloning of *mec*-specific DNA. *J. Bacteriol.* **165:**373–378.
12. **Berger-Bachi, B.** 1983. Insertional inactivation of staphylococcal methicillin resistance by Tn*551*. *J. Bacteriol.* **154:** 479–487.
13. **Brown, N. L., C. Choi, J. Grinsted, and M. H. Richmond.** 1980. Nucleotide sequences at the ends of the mercury resistance transposon Tn501. *Nucleic Acids Res.* **8:** 1933–1945.
14. **Brown, N. L., T. K. Misra, J. N. Winnie, A. Schmidt, M. Seiff, and S. Silver.** 1986. The nucleotide sequence of the mercuric resistance operons of plasmid R100 and transposon Tn*501*: further evidence for *mer* genes which enhance the

activity of the mercuric ion detoxification system. *Mol. Gen. Genet.* **202**:143–151.
15. **Brown, N. L., J. N. Winnie, D. Fritzinger, and R. D. Pridmore.** 1985. The nucleotide sequence of the *tnpA* gene completes the sequence of the *Pseudomonas* transposon Tn501. *Nucleic Acids Res.* **13**:5657–5669.
16. **Bruton, C. J., and K. F. Chater.** 1987. Nucleotide sequence of IS110, an insertion sequence of *Streptomyces coelicolor* A3(2). *Nucleic Acids Res.* **15**:7053–7065.
17. **Burdett, B., J. Inamine, and S. Rajagopalan.** 1982. Heterogeneity of tetracycline resistance determinants in *Streptococcus. J. Bacteriol.* **149**:995–1004.
18. **Burdett, V.** 1986. Streptococcal tetracycline resistance mediated at the level of protein synthesis. *J. Bacteriol.* **165**: 564–569.
19. **Caillaud, F., and P. Courvalin.** 1987. Nucleotide sequence of the ends of the conjugative shuttle transposon Tn1545. *Mol. Gen. Genet.* **209**:110–115.
20. **Campbell, A. M.** 1962. Episomes. *Adv. Genet.* **11**:101–145.
21. **Carlton, B. C., and M. M. Gonzalez, Jr.** 1985. The genetics and molecular biology of *Bacillus thuringiensis*, p. 211–249. *In* D. A. Dubnau (ed.), *The Molecular Biology of the Bacilli*, vol. 2. Academic Press, Inc., New York.
22. **Chater, K. F., C. J. Bruton, S. G. Foster, and I. Tobek.** 1985. Physical and genetic analysis of IS110, a transposable element of *Streptomyces coelicolor* A3(2). *Mol. Gen. Genet.* **200**:235–239.
23. **Chater, K. F., D. J. Henderson, M. J. Bibb, and D. A. Hopwood.** 1988. Genome flux in *Streptomyces coelicolor* and other streptomyces and its possible relevance to the evolution of mobile antiobiotic resistance determinants. *Symp. Soc. Gen. Microbiol.* **43**:7–42.
24. **Chou, J., M. J. Casadaban, P. G. Lemaux, and S. N. Cohen.** 1979. Identification and characterization of a self-regulated repressor of translocation of the Tn3 element. *Proc. Natl. Acad. Sci. USA* **76**:4020–4024.
25. **Christie, P. J., R. Z. Korman, S. A. Zahler, J. C. Adsit, and G. M. Dunny.** 1987. Two conjugation systems with the *Streptococcus faecalis* plasmid pCF-10: identification of a conjugative transposon that transfers between *S. faecalis* and *Bacillus subtilis. J. Bacteriol.* **169**:2529–2536.
26. **Chung, S. T.** 1987. Tn*4556*, a 6.8-kilobase-pair transposable element of *Streptomyces fradiae. J. Bacteriol.* **169**:4436–4441.
27. **Clewell, D. B., F. Y. An, B. A. White, and C. Gawron-Burke.** 1985. *Streptococcus faecalis* sex pheromone (cAM373) also produced by *Staphylococcus aureus* and identification of a conjugative transposon (Tn*918*). *J. Bacteriol.* **162**:1212–1220.
27a. **Clewell, D. B., S. E. Flannagan, Y. Ike, J. M. Jones, and C. Gawron-Burke.** 1988. Sequence analysis of termini of conjugative transposon Tn*916*. *J. Bacteriol.* **170**:3046–3052.
28. **Clewell, D. B., and C. Gawron-Burke.** 1986. Conjugative transposons and the dissemination of antibiotic resistance in streptococci. *Annu. Rev. Microbiol.* **40**:635–659.
29. **Clewell, D. B., E. Senghas, J. M. Jones, S. E. Flannagan, M. Yamamoto, and C. Gawron-Burke.** 1988. Transposition in *Streptococcus*: structural and genetic properties of the conjugative transposon Tn*916*. *Symp. Soc. Gen. Microbiol.* **43**:43–58.
30. **Clewell, D. B., P. K. Tomich, M. C. Gawron-Burke, A. E. Franke, Y. Yagi, and F. Y. An.** 1982. Mapping of *Streptococcus faecalis* plasmids pAD1 and pAD2 and studies relating to transposition of Tn*917*. *J. Bacteriol.* **152**:1220–1230.
31. **Cohen, S., and H. M. Sweeney.** 1970. Transduction of methicillin resistance in *Staphylococcus aureus* dependent on an unusual specificity of the recipient strain. *J. Bacteriol.* **104**:1158–1167.
32. **Cohen, S., and H. M. Sweeney.** 1973. Effect of the prophage and penicillinase plasmid of the recipient strain upon the transduction and the stability of methicillin resistance in *Staphylococcus aureus. J. Bacteriol.* **116**:803–811.
33. **Cohen, S., H. Sweeney, and S. Basu.** 1977. Mutations in prophage ϕ11 that impair the transducibility of their *Staphylococcus aureus* lysogens for methicillin resistance. *J. Bacteriol.* **129**:237–245.
34. **Collatz, E., C. Carlier, and P. Courvalin.** 1984. Characterization of high-level aminoglycoside resistance in a strain of *Streptococcus pneumoniae. J. Gen. Microbiol.* **130**:1665–1671.
35. **Courvalin, P., and C. Carlier.** 1986. Transposable multiple antibiotic resistance in *Streptococcus pneumoniae. Mol. Gen. Genet.* **205**:291–297.
36. **Courvalin, P., and C. Carlier.** 1987. Tn1545: a conjugative shuttle transposon. *Mol. Gen. Genet.* **206**:259–264.
37. **Davies, J., and D. I. Smith.** 1978. Plasmid-determined resistance to antimicrobial agents. *Annu. Rev. Microbiol.* **32**: 469–518.
38. **De la Cruz, F., and J. Grinsted.** 1981. Genetic and molecular characterization of Tn*21*, a multiple resistance transposon from R100. *J. Bacteriol.* **151**:222–228.
39. **De Massy, B., F. W. Studier, L. Dorgai, E. Appelbaum, and R. A. Weisberg.** 1984. Enzymes and sites of genetic recombination: studies with gene-3 endonuclease of phage T7 and with site-affinity mutants of phage λ. *Cold Spring Harbor Symp. Quant. Biol.* **49**:715–726.
40. **Debabov, B. G.** 1982. The industrial uses of bacilli, p. 331–370. *In* D. A. Dubnau (ed.), *The Molecular Biology of the Bacilli*, vol. 1. Academic Press, Inc., New York.
41. **Dornbusch, K., and H. O. Hallander.** 1973. Transduction of penicillinase production and methicillin resistance-enterotoxin B production in strains of *Staphylococcus aureus. J. Gen. Microbiol.* **76**:1–11.
42. **Dornbusch, K., H. O. Hallander, and F. Lofquist.** 1969. Extrachromosomal control of methicillin resistance and toxin production in *Staphylococcus aureus. J. Bacteriol.* **98**:351–358.
43. **Dubnau, D.** 1985. Induction of *ermC* requires translation of the leader peptide. *EMBO J.* **4**:533–537.
44. **Dubnau, D., and M. Monod.** 1986. The regulation and evolution of MLS resistance, p. 369–385. *In* S. B. Levy and R. P. Novick (ed.), *Antibiotic Resistance Genes: Ecology, Transfer and Expression. Banbury Report*, vol. 24. Cold Spring Harbor Laboratory, Cold Spring Harbor, N.Y.
45. **Dybvig, K., and G. H. Cassell.** 1987. Transposition of gram-positive transposon Tn916 in *Acholeplasma laidlawii* and *Mycoplasma pulmonis. Science* **235**:1392–1394.
46. **Dyke, K. G. H.** 1969. Penicillinase production and intrinsic resistance to penicillins in methicillin-resistant cultures of *Staphylococcus aureus. J. Med. Microbiol.* **2**:261–278.
47. **El Solh, N., N. Moreau, and S. D. Ehrlich.** 1986. Molecular cloning and analysis of *Staphylococcus aureus* chromosomal aminoglycoside resistance genes. *Plasmid* **15**:104–118.
48. **Fitzgerald, G. F., and D. B. Clewell.** 1985. A conjugative transposon (Tn919) in *Streptococcus sanguis. Infect. Immun.* **47**:415–420.
49. **Fling, M. E., J. Knopf, and C. Richards.** 1985. Nucleotide sequence of the transposon Tn7 gene encoding an aminoglycoside-modifying enzyme, 3″(9)-O-nucleotidyltransferase. *Nucleic Acids Res.* **13**:7095–7106.

50. **Franke, A. E., and D. B. Clewell.** 1981. Evidence for a chromosome-borne resistance transposon (Tn*916*) in *Streptococcus faecalis* that is capable of conjugal transfer in the absence of a conjugative plasmid. *J. Bacteriol.* **145**:494–502.

51. **Gaillard, J. L., P. Berche, and P. Sansonetti.** 1986. Transposon mutagenesis as a tool to study the role of hemolysin in the virulence of *Listeria monocytogenes*. *Infect. Immun.* **52**:50–55.

52. **Gawron-Burke, C., and D. B. Clewell.** 1982. A transposon in *Streptococcus faecalis* with fertility properties. *Nature* (London) **300**:281–284.

53. **Gawron-Burke, C., and D. B. Clewell.** 1984. Regeneration of insertionally inactivated streptococcal DNA fragments after excision of transposon Tn*916* in *Escherichia coli*: strategy for targeting and cloning of genes from gram-positive bacteria. *J. Bacteriol.* **159**:214–221.

54. **Gillespie, M. T., B. R. Lyon, L. S. L. Loo, P. R. Matthews, P. R. Stewart, and R. A. Skurray.** 1987. Homologous direct repeat sequences associated with mercury, methicillin, tetracycline and timethoprim resistance determinants in *Staphylococcus aureus*. *FEMS Microbiol. Lett.* **43**:165–171.

55. **Gonzales, Jr., J. M., H. T. Dulmage, and B. C. Carlton.** 1981. Correlation between specific plasmids and delta-endotoxin production in *Bacillus thuringiensis*. *Plasmid* **5**: 351–365.

56. **Gray, G. S., R. T. Huang, and J. Davies.** 1983. Aminocyclitol resistance in *Staphylococcus aureus*: presence of plasmids and aminocyclitol-modifying enzymes. *Plasmid* **9**:147–158.

57. **Gryczan, T. J., G. Grandi, J. Hahn, R. Grandi, and D. Dubnau.** 1980. Conformational alteration of mRNA structure and the posttranscriptional regulation of erythromycin-induced drug resistance. *Nucleic Acids Res.* **8**:6081–6097.

58. **Hachler, H., F. H. Kayser, and B. Berger-Bachi.** 1987. Homology of a transferable tetracycline resistance determinant of *Clostridium difficile* with *Streptococcus* (*Enterococcus*) *faecalis* transposon Tn*916*. *Antimicrob. Agents Chemother.* **31**:1033–1038.

59. **Haley, R. W., A. W. Hightower, R. F. Khabbaz, C. Thornsberry, W. J. Martone, J. R. Allen, and J. M. Hughes.** 1982. The emergence of methicillin-resistant *Staphylococcus aureus* infections in United States hospitals. *Ann. Intern. Med.* **97**:297–303.

60. **Hartman, B. J., and A. Tomasz.** 1984. Low-affinity penicillin-binding protein associated with β-lactam resistance in *Staphylococcus aureus*. *J. Bacteriol.* **158**:513–516.

61. **Hartman, B. J., and A. Tomasz.** 1986. Expression of methicillin resistance in heterogeneous strains of *Staphylococcus aureus*. *Antimicrob. Agents Chemother.* **29**:85–92.

62. **Heffron, F., B. J. McCarthy, H. Ohtsubo, and E. Ohtsubo.** 1979. DNA sequence analysis of the transposon Tn3: three genes and three sites involved in transposition of Tn3. *Cell* **18**:1153–1163.

63. **Hill, C., C. Daly, and G. F. Fitzgerald.** 1985. Conjugative transfer of the transposon Tn919 to lactic acid bacteria. *FEMS Microbiol. Lett.* **30**:115–119.

64. **Hoess, R., K. Abremski, B. Frommer, A. Wierzbicki, and M. Kendall.** 1987. The lox-Cre site-specific recombination system of bacteriophage P1. *UCLA Symp. Mol. Cell. Biol.* **47**:745–756.

65. **Hollingshead, S., and D. Vapnek.** 1985. Nucleotide sequence analysis of a gene encoding a streptomycin/spectinomycin adenyltransferase. *Plasmid* **13**:17–30.

66. **Hopwood, D. A., K. F. Chater, J. E. Dowding, and A. Vivian.** 1973. Recent advances in *Streptomyces coelicolor* genetics. *Bacteriol. Rev.* **37**:371–405.

67. **Horinouchi, S., and B. Weisblum.** 1980. Posttranscriptional modification of mRNA conformation: mechanism that regulates erythromycin-induced resistance. *Proc. Natl. Acad. Sci. USA* **77**:7079–7083.

68. **Iida, S., J. Meyer, P. Linder, N. Goto, R. Nakaya, H. Reif, and W. Arber.** 1982. The kanamycin resistance transposon Tn2680 derived from the R plasmid Rts1 and carried by phage P1Km has flanking 0.8-kb-long direct repeats. *Plasmid* **8**:187–198.

69. **Inamine, J. M., and V. Burdett.** 1985. Structural organization of a 67-kilobase streptococcal conjugative element mediating multiple antibiotic resistance. *J. Bacteriol.* **161**: 620–626.

70. **Jaffe, H. W., H. M. Sweeney, R. A. Weinstein, S. A. Kabins, C. Nathan, and S. Cohen.** 1982. Structural and phenotypic varieties of gentamicin resistance plasmids in hospital strains of *Staphylococcus aureus* and coagulase-negative staphylococci. *Antimicrob. Agents Chemother.* **21**:773–779.

71. **Jones, J., S. Yost, and P. Pattee.** 1987. Transfer of the conjugal tetracycline resistance transposon Tn*916* from *Streptococcus faecalis* to *Staphylococcus aureus* and identification of some insertion sites in the staphylococcal chromosome. *J. Bacteriol.* **169**:2121–2131.

72. **Jones, J. M., C. Gawron-Burke, S. E. Flannagan, M. Yamamoto, E. Senghas, and D. B. Clewell.** 1987. Structural and genetic studies of the conjugative transposon Tn*916*, p. 54–60. *In* J. J. Ferretti and R. Curtiss III (ed.), *Streptococcal Genetics*. American Society for Microbiology, Washington, D.C.

73. **Kayser, F. H., F. Homberger, and M. Devaud.** 1981. Aminocyclitol-modifying enzymes specified by chromosomal genes in *Staphylococcus aureus*. *Antimicrob. Agents Chemother.* **19**:766–772.

74. **Kendall, K., and J. Cullum.** 1986. Identification of a DNA sequence associated with plasmid integration in *Streptomyces coelicolor* A3(2). *Mol. Gen. Genet.* **202**:240–245.

75. **Khan, S. A., and R. P. Novick.** 1980. Terminal nucleotide sequences of Tn551, a transposon specifying erythromycin resistance in *Staphylococcus aureus*: homology with Tn3. *Plasmid* **4**:148–154.

76. **Kigbo, E. P., D. E. Townsend, N. Ashdown, and W. B. Grubb.** 1985. Transposition of penicillinase determinants in methicillin-resistant *Staphylococcus aureus*. *FEMS Microbiol. Lett.* **28**:39–43.

77. **Kinashi, H., M. Shimaji, and A. Sakai.** 1987. Giant linear plasmids in *Streptomyces* which code for antibiotic biosynthesis genes. *Nature* (London) **328**:454–456.

78. **Klaer, R., S. Kuhn, H. J. Fritz, E. Tillmann, I. Saint-Girons, P. Haberman, D. Pfeifer, and P. Starlinger.** 1981. Studies on transposition mechanism and specificity of IS4. *Cold Spring Harbor Symp. Quant. Biol.* **45**:215–224.

79. **Klaer, R., S. Kuhn, E. Tillman, H. Fritz, and P. Starlinger.** 1981. The sequence of IS4. *Mol. Gen. Genet.* **181**:169–175.

80. **Klemm, P.** Two regulatory fim genes, *fimB* and *fimE*, control the phase variation of type 1 fimbriae in *Escherichia coli*. *EMBO J.* **5**:1389–1393.

81. **Kornblum, J., B. J. Hartman, R. P. Novick, and A. Tomasz.** 1986. Conversion of a homogeneously methicillin-resistant strain of *Staphylococcus aureus* to heterogeneous resistance by Tn551-mediated insertional inactivation. *Eur. J. Clin. Microbiol.* **5**:714–718.

82. **Kreiswirth, B. N., J. S. Kornblum, and R. P. Novick.** 1985. Genotypic variability of the toxic shock syndrome exoprotein determinant, p. 105–109. *In* J. Jeljaszewicz (ed.), *The*

Staphylococci. Gustav Fischer Verlag, Stuttgart, Federal Republic of Germany.

83. **Kretschmer, P. J., and S. N. Cohen.** 1977. Selected translocation of plasmid genes: frequency and regional specificity of translocation of the Tn*3* element. *J. Bacteriol.* **130:**888–899.

84. **Krolewski, J. J., E. Murphy, R. P. Novick, and M. G. Rush.** 1981. Site-specificity of the chromosomal insertion of *Staphylococcus aureus* transposon Tn554. *J. Mol. Biol.* **152:** 19–33.

85. **Kronstad, J. W., H. E. Schnept, and H. R. Whiteley.** 1983. Diversity of locations for *Bacillus thuringiensis* crystal protein genes. *J. Bacteriol.* **154:**419–428.

86. **Kronstad, J. W., and H. R. Whiteley.** 1984. Inverted repeat sequences flank a *Bacillus thuringiensis* crystal protein gene. *J. Bacteriol.* **160:**95–102.

87. **Kronstad, J. W., and H. R. Whiteley.** 1986. Three classes of homologous *Bacillus thuringiensis* crystal protein genes. *Gene* **43:**29–40.

88. **Kuhl, S. A., P. A. Pattee, and J. N. Baldwin.** 1978. Chromosomal map location of the methicillin resistance determinant in *Staphylococcus aureus. J. Bacteriol.* **135:**460–465.

89. **Kuramitsul, H. K., and M. J. Casadaban.** 1986. Transposition of the gram-positive transposon Tn*917* in *Escherichia coli. J. Bacteriol.* **167:**711–712.

90. **Lacey, R. W., N. Keyworth, and C. Lincoln.** 1984. Staphylococci in the U.K.: a review. *J. Antimicrob. Chemother.* **14**(Suppl. D):19–25.

91. **Laddaga, R. A., L. Chu, T. K. Misra, and S. Silver.** 1987. Nucleotide sequence and expression of the mercurial-resistance operon from *S. aureus* plasmid pI258. *Proc. Natl. Acad. Sci. USA* **84:**5106–5110.

92. **Lai, C., B. Weisblum, S. R. Fahnestock, and M. Nomura.** 1973. Alteration of 23S ribosomal RNA and erythromycin-induced resistance to lincomycin and spiramycin in *Staphylococcus aureus. J. Mol. Biol.* **74:**67–72.

93. **Lampson, B. C., and J. T. Parisi.** 1986. Naturally occurring *Staphylococcus epidermidis* plasmid expressing constitutive macrolide-lincosamide-streptogramin B resistance contains a deleted attenuator. *J. Bacteriol.* **166:**479–483.

94. **Lane, D., R. de Feyter, M. Kennedy, S. H. Phua, and D. Semon.** 1986. D protein of miniF plasmid acts as a repressor of transcription and as a site-specific resolvase. *Nucleic Acids Res.* **14:**9713–9728.

95. **Lee, C., A. Bhagwat, and F. Heffron.** 1983. Identification of a transposon Tn3 sequence required for transposition immunity. *Proc. Natl. Acad. Sci. USA* **80:**6765–6769.

96. **Lereclus, D., J. Mahillon, G. Menou, and M. Lecadet.** 1986. Identification of Tn4430, a transposon of *Bacillus thuringiensis* functional in *Escherichia coli. Mol. Gen. Genet.* **204:** 52–57.

97. **Lereclus, D., J. Ribier, A. Klier, G. Menou, and M. Lecadet.** 1984. A transposon-like structure related to the δ-endotoxin gene of *Bacillus thuringiensis. EMBO J.* **3:**2561–2567.

98. **Luchansky, J. B., and P. A. Pattee.** 1984. Isolation of transposon Tn*551* insertions near chromosomal markers of interest in *Staphylococcus aureus. J. Bacteriol.* **159:**894–899.

99. **Lydiate, D. J., H. Ikeda, and D. A. Hopwood.** 1986. A 2.6 kb DNA sequence of *Streptomyces coelicolor* A3(2) which functions as a transposable element. *Mol. Gen. Genet.* **203:**79–88.

100. **Lyon, B. R., J. L. Iuorio, J. W. May, and R. A. Skurray.** 1984. Molecular epidemiology of multiresistant *Staphylococcus aureus* in Australian hospitals. *J. Med. Microbiol.* **17:**79–89.

101. **Lyon, B. R., M. T. Gillespie, M. E. Byrne, J. W. May, and R. A. Skurray.** 1987. Plasmid-mediated resistance to gentamicin in *S. aureus*: the involvement of a transposon. *J. Med. Microbiol.* **23:**101–110.

102. **Lyon, B. R., J. W. May, and R. A. Skurray.** 1984. Tn4001: a gentamicin and kanamycin resistance transposon in *Staphylococcus aureus. Mol. Gen. Genet.* **193:**554–556.

103. **Lyon, B. R., and R. Skurray.** 1987. Antimicrobial resistance of *Staphylococcus aureus*: genetic basis. *Microbiol. Rev.* **51:**88–134.

104. **Mahillon, J., and D. Lereclus.** 1988. Structural and functional analysis of Tn4430: identification of an integrase-like protein involved in the cointegrate-resolution process. *EMBO J.* **7:**1515–1526.

105. **Mahillon, J., J. Seurinck, J. Delcour, and M. Zabeau.** 1987. Cloning and nucleotide sequence of different iso-IS*231* elements and their structural association with the Tn*4430* transposon in *Bacillus thuringiensis. Gene* **51:**187–196.

106. **Mahillon, J., J. Seurinck, L. van Rompuy, J. Delcour, and M. Zabeau.** 1985. Nucleotide sequence and structural organization of an insertion sequence element (IS231) from *Bacillus thuringiensis* strain berliner 1715. *EMBO J.* **4:**3895–3899.

107. **Mandel, L. J., E. Murphy, N. H. Steigbigel, and M. H. Miller.** 1984. Gentamicin uptake in *Staphylococcus aureus* possessing plasmid-encoded, aminoglycoside-modifying enzymes. *Antimicrob. Agents Chemother.* **26:**563–569.

108. **Martel, A., M. Masson, N. Moreau, and F. Le Goffic.** 1983. Kinetic studies of aminoglycoside acetyltransferase and phosphotransferase from *Staphylococcus aureus* RPAL. Relationship between the two activities. *Eur. J. Biochem.* **133:**515–521.

109. **Matsuhashi, M., M. D. Song, F. Ishimo, M. Wachi, M. Doi, M. Inoue, K. Ubukata, N. Yamashita, and M. Konno.** 1986. Molecular cloning of the gene of a penicillin-binding protein supposed to cause high resistance to beta-lactam antibiotics in *Staphylococcus aureus. J. Bacteriol.* **167:**975–980.

110. **Matthews, P. R., K. C. Reed, and P. R. Stewart.** 1987. The cloning of chromosomal DNA associated with methicillin and other resistances in *Staphylococcus aureus. J. Gen. Microbiol.* **133:**1919–1929.

111. **McDonnell, R. W., H. M. Sweeney, and S. Cohen.** 1983. Conjugational transfer of gentamicin-resistance plasmids intra- and interspecifically in *Staphylococcus aureus* and *Staphylococcus epidermidis. Antimicrob. Agents Chemother.* **23:**151–160.

112. **McKown, R. L., K. A. Orle, T. Chen, and N. A. Craig.** 1988. Sequence requirements of *Escherichia coli att*Tn7, a specific site of transposon Tn7 insertion. *J. Bacteriol.* **170:**352–358.

113. **McMurry, L., R. Petrucci, and S. B. Levy.** 1980. Active efflux of tetracycline encoded by four genetically different tetracycline resistance determinants in *Escherichia coli. Proc. Natl. Acad. Sci. USA* **77:**3974–3977.

114. **Misra, T. K., N. L. Brown, D. C. Fritzinger, R. D. Pridmore, W. M. Barnes, L. Haberstroh, and S. Silver.** 1984. Mercuric ion-resistance operons of plasmid R100 and transposon Tn501: the beginning of the operon including the regulatory region and the first two structural genes. *Proc. Natl. Acad. Sci. USA* **81:**5975–5979.

115. **Mitsuhashi, S., M. Inoue, K. Kawabe, H. Oshima, and T. Okubo.** 1973. Genetic and biochemical studies of drug resistance in staphylococci, p. 144–165. *In* J. Jeljaszewicz (ed.), *Staphylococci and Staphylococcal Infections. Proceedings of the 2nd International Symposium Held in Warszwa, Poland, September 13–18, 1971.* Polish Medical Publishers, Warsaw.

116. **Mitsuhashi, S., M. Morimura, K. Kono, and H. Oshima.** 1963. Elimination of drug resistance of *Staphylococcus aureus* by treatment with acriflavine. *J. Bacteriol.* **86:** 162–164.
117. **Mollet, B., S. Iida, J. Shepherd, and W. Arber.** 1983. Nucleotide sequence of IS26, a new prokaryotic mobile genetic element. *Nucleic Acids Res.* **11:**6319–6330.
118. **Morse, S. A., S. R. Johnson, J. W. Biddle, and M. C. Roberts.** 1986. High-level tetracycline resistance in *Neisseria gonorrhoeae* is the result of acquisition of streptococcal *tetM* determinant. *Antimicrob. Agents Chemother.* **30:**664–670.
119. **Murphy, E.** 1983. Inhibition of Tn554 transposition: deletion analysis. *Plasmid* **10:**260–269.
120. **Murphy, E.** 1985. Nucleotide sequence of a spectinomycin adenyltransferase AAD(9) determinant from *Staphylococcus aureus* and its relationship to AAD(3″)(9). *Mol. Gen. Genet.* **200:**33–39.
121. **Murphy, E.** 1985. Nucleotide sequence of *ermA*, a macrolide-lincosamide-streptogramin B (MLS) determinant in *Staphylococcus aureus*. *J. Bacteriol.* **162:**633–640.
121a. **Murphy, E.** 1988. Transposable elements in *Staphylococcus*. *Symp. Soc. Gen. Microbiol.* **43:**59–89.
122. **Murphy, E., L. Huwyler, and M. C. F. Bastos.** 1985. Transposon Tn554: complete nucleotide sequence and isolation of transposition-defective and antibiotic-sensitive mutants. *EMBO J.* **4:**3357–3365.
123. **Murphy, E., and S. Löfdahl.** 1984. Transposition of Tn*554* does not generate a target duplication. *Nature* (London) **307:**292–294.
124. **Murphy, E., and R. Novick.** 1979. Physical mapping of *Staphylococcus aureus* penicillinase plasmid pI524: characterization of an invertible region. *Mol. Gen. Genet.* **175:** 19-30.
125. **Murphy, E., and R. Novick.** 1980. Site-specific recombination between plasmids of *Staphylococcus aureus*. *J. Bacteriol.* **141:**316–326.
126. **Murphy, E., and R. P. Novick.** 1984. MLS_B resistance determinants in *Staphylococcus aureus* and their molecular evolution. *J. Antimicrob. Chemother.* **16:**101–110.
127. **Murphy, E., S. Phillips, I. Edelman, and R. P. Novick.** 1981. Tn554: isolation and characterization of plasmid insertions. *Plasmid* **5:**292–305.
128. **Novick, R., I. Edelman, M. D. Schwesinger, A. Gruss, E. Swanson, and P. A. Pattee.** 1979. Genetic translocation in *Staphylococcus aureus*. *Proc. Natl. Acad. Sci. USA* **76:** 400–404.
129. **Ohtsubo, H., H. Ohmori, and E. Ohtsubo.** 1979. Nucleotide sequence analysis of Tn3(Ap): implications for insertion and deletion. *Cold Spring Harbor Symp. Quant. Biol.* **45:** 1269–1278.
130. **Olson, E. R., and S. T. Chung.** 1988. Transposon Tn*4556* of *Streptomyces fradiae*: nucleotide sequence of the ends and the target sites. *J. Bacteriol.* **170:**1955–1957.
131. **Ouellette, M., and P. H. Roy.** 1987. Homology of ORFs from Tn2603 and from R46 to site-specific recombinases. *Nucleic Acids Res.* **15:**10055.
132. **Ounissi, H., and P. Courvalin.** 1982. Heterogeneity of macrolide-lincosamide-streptogramin B-type antibiotic resistance determinants, p. 167–169. *In* D. Schlessinger (ed.), *Microbiology—1982*. American Society for Microbiology, Washington, D.C.
133. **Parker, M. T., and J. H. Hewitt.** 1970. Methicillin resistance in *Staphylococcus aureus*. *Lancet* **i:**800–804.
134. **Pattee, P. A., N. E. Thompson, D. Haubrich, and R. P. Novick.** 1977. Chromosomal map locations of integrated plasmids and related elements in *Staphylococcus aureus*. *Plasmid* **1:**38–51.
135. **Perkins, J. B., and P. J. Youngman.** 1984. A physical and functional analysis of Tn917, a streptococcus transposon in the Tn3 family that functions in *Bacillus*. *Plasmid* **12:** 119–138.
136. **Perkins, J. B., and P. J. Youngman.** 1986. Construction and properties of Tn*917-lac*, a transposon derivative that mediates transcriptional gene fusions in *Bacillus subtilis*. *Proc. Natl. Acad. Sci. USA* **83:**140–144.
137. **Phillips, S., and R. Novick.** 1979. Tn*554*: a repressible site-specific transposon in *Staphylococcus aureus*. *Nature* (London) **278:**476–478.
138. **Polzin, K. M., and M. Shimizu-Kadota.** 1987. Identification of a new insertion element, similar to gram-negative IS26, on the lactose plasmid of *Streptococcus lactis* ML3. *J. Bacteriol.* **169:**5481–5488.
139. **Reynolds, P. E., and C. Fuller.** 1986. Methicillin-resistant strains of *Staphylococcus aureus*; presence of additional penicillin-binding protein in all strains examined. *FEMS Microbiol. Lett.* **33:**251–254.
140. **Roberts, M. C., and G. E. Kenny.** 1986. Dissemination of the *tetM* tetracycline resistance determinant to *Ureaplasma urealyticum*. *Antimicrob. Agents Chemother.* **29:**350–352.
141. **Roberts, M. C., L. A. Koutsky, K. K. Holmes, D. J. LeBlanc, and G. E. Kenny.** 1985. Tetracycline-resistant *Mycoplasma hominis* strains contain streptococcal *tetM* sequences. *Antimicrob. Agents Chemother.* **28:**141–143.
142. **Robinson, M. K., P. M. Bennett, and M. H. Richmond.** 1977. Inhibition of TnA translocation by TnA. *J. Bacteriol.* **129:**407–414.
143. **Ross, W., and A. Landy.** 1982. Bacteriophage λ Int protein recognizes two classes of sequence in the phage *att* site: characterization of arm-type sites. *Proc. Natl. Acad. Sci. USA* **79:**7724–7728.
144. **Rowland, S. J., and K. G. H. Dyke.** 1988. A DNA-invertase from *Staphylococcus aureus* is a member of the Hin family of site-specific recombinases. *FEMS Microbiol. Lett.* **50:** 253–258.
145. **Schmitt, R., J. Altenbuchner, K. Wiebauer, W. Arnold, A. Puhler, and F. Schoffl.** 1981. Basis of transposition and gene amplification by Tn1721 and related tetracycline-resistance transposons. *Cold Spring Harbor Symp. Quant. Biol.* **45:** 59–65.
146. **Schoffl, F., W. Arnold, A. Puhler, J. Altenbuchner, and R. Schmitt.** 1981. The tetracycline resistance transposons Tn1721 and Tn1771 have three 38 base-pair repeats and generate five base-pair direct repeats. *Mol. Gen. Genet.* **181:**87–94.
146a. **Scott, J. R., P. A. Kirchman, and M. G. Caparon.** 1988. An intermediate in transposition of the conjugative transposon Tn*916*. *Proc. Natl. Acad. Sci. USA* **85:**4809–4813.
147. **Senghas, E., J. M. Jones, M. Yamamoto, C. Gawron-Burke, and D. B. Clewell.** 1988. Genetic organization of the bacterial conjugative transposon Tn*916*. *J. Bacteriol.* **170:**245–249.
148. **Shafer, W. M., and J. J. Iandolo.** 1980. Transduction of staphylococcal enterotoxin B synthesis: establishment of the toxin gene in a recombination-deficient mutant. *Infect. Immun.* **27:**280–282.
149. **Shalita, Z., I. Hertman, and S. Sarid.** 1977. Isolation and characterization of a plasmid involved with enterotoxin B production in *Staphylococcus aureus*. *J. Bacteriol.* **129:** 317–325.
150. **Shalita, Z., E. Murphy, and R. P. Novick.** 1980. Penicillinase

plasmids of *Staphylococcus aureus*: structural and evolutionary relationships. *Plasmid* **3**:291–311.

151. **Shaw, J. H., and D. B. Clewell.** 1985. Complete nucleotide sequence of macrolide-lincosamide-streptogramin B-resistance transposon Tn*917* in *Streptococcus faecalis*. *J. Bacteriol.* **164**:782–796.

152. **Sherratt, D., A. Arthur, and M. Burke.** 1980. Transposon-specified site-specific recombination systems. *Cold Spring Harbor Symp. Quant. Biol.* **45**:275–281.

153. **Shimizu-Kadota, M., M. Kiwaki, H. Hirokawa, and N. Tsuchida.** 1985. ISL1: a new transposable element in *Lactobacillus casei*. *Mol. Gen. Genet.* **200**:193–198.

154. **Simons, R. W., and N. Kleckner.** 1983. Translational control of IS10 transposition. *Cell* **34**:683–691.

155. **Song, M. D., M. Wachi, M. Doi, F. Ishino, and M. Matsuhashi.** 1987. Evolution of an inducible penicillin-target protein in methicillin-resistant *Staphylococcus aureus* by gene fusion. *FEBS Lett.* **221**:167–171.

156. **Stewart, G. C., and E. D. Rosenblum.** 1980. Genetic behavior of the methicillin resistance determinant in *Staphylococcus aureus*. *J. Bacteriol.* **144**:1200–1202.

157. **Thakker-Varia, S., W. D. Jenssen, L. Moon-McDermott, M. P. Weinstein, and D. T. Dubin.** 1987. Molecular epidemiology of macrolides-lincosamides-streptogramin B resistance in *Staphylococcus aureus* and coagulase-negative staphylococci. *Antimicrob. Agents Chemother.* **31**:735–743.

158. **Tomich, P. K., F. Y. An, and D. B. Clewell.** 1980. Properties of erythromycin-inducible transposon Tn*917* in *Streptococcus faecalis*. *J. Bacteriol.* **141**:1366–1374.

159. **Townsend, D. E., N. Ashdown, L. C. Greed, and W. B. Grubb.** 1984. Transposition of gentamicin resistance to staphylococcal plasmids encoding resistance to cationic agents. *J. Antimicrob. Chemother.* **14**:115–124.

160. **Townsend, D. E., N. Ashdown, L. C. Greed, and W. B. Grubb.** 1984. Analysis of plasmids mediating gentamicin resistance in methicillin-resistant *Staphylococcus aureus*. *J. Antimicrob. Chemother.* **13**:347–352.

161. **Townsend, D. E., S. Bolton, N. Ashdown, D. I. Annear, and W. B. Grubb.** 1986. Conjugative, staphylococcal plasmids carrying hitch-hiking transposons similar to Tn554: intra- and interspecies dissemination of erythromycin resistance. *Aust. J. Exp. Biol. Med. Sci.* **64**:367–379.

162. **Ubukata, K., N. Yamashita, A. Gotoh, and M. Konno.** 1984. Purification and characterization of aminoglycoside-modifying enzymes from *Staphylococcus aureus* and *Staphylococcus epidermidis*. *Antimicrob. Agents Chemother.* **27**: 754–759.

163. **Vijayakumar, M. N., S. D. Priebe, and W. R. Guild.** 1986. Structure of a conjugative element in *Streptococcus pneumoniae*. *J. Bacteriol.* **166**:978–984.

164. **Weaver, K. E., and D. B. Clewell.** 1987. Transposon Tn*917* delivery vectors for mutagenesis in *Streptococcus faecalis*, p. 17–21. *In* J. J. Ferretti and R. Curtiss III (ed.), *Streptococcal Genetics*. American Society for Microbiology, Washington, D.C.

165. **Weber, D. A., and R. V. Goering.** 1988. Tn*4201*, a β-lactamase transposon in *Staphylococcus aureus*. *Antimicrob. Agents Chemother.* **32**:1164–1169.

166. **Weisberg, R. A., L. W. Enquist, C. Foeller, and A. Landy.** 1983. Role for DNA homology in site-specific recombination. The isolation and characterization of a site affinity mutant of coliphage λ. *J. Mol. Biol.* **170**:319–342.

167. **Weisblum, B., S. B. Holder, and S. M. Halling.** 1979. Deoxyribonucleic acid sequence common to staphylococcal and streptococcal plasmids which specify erythromycin resistance. *J. Bacteriol.* **138**:990–998.

168. **Witte, W., L. Green, T. K. Misra, and S. Silver.** 1986. Resistance to mercury and to cadmium in chromosomally resistant *Staphylococcus aureus*. *Antimicrob. Agents Chemother.* **29**:663–669.

169. **Youngman, P.** 1987. Plasmid vectors for recovering and exploiting Tn917 transpositions in *Bacillus* and other gram-positive bacteria, p. 79–104. *In* K. G. Hardy (ed.), *Plasmids, a Practical Approach*. IRL Press, Oxford.

170. **Youngman, P., J. B. Perkins, and R. Losick.** 1984. Construction of a cloning site near one end of Tn917 into which foreign DNA may be inserted without affecting transposition in *Bacillus subtilis* or expression of the transposon-borne erm gene. *Plasmid* **12**:1–9.

Chapter 10

SLP1: a Paradigm for Plasmids that Site-Specifically Integrate in the Actinomycetes

CHARLES A. OMER and STANLEY N. COHEN

I. INTRODUCTION

Research into the biology of the actinomycetes in recent years has identified a large number (>100) of plasmids (reviewed in reference 20). Traits ascribed to these plasmids and other undefined transmissible elements include production of and resistance to the antibiotics methylenomycin and tylosin and production of the developmentally important substance A-factor (15, 24, 25, 40). Many of the plasmids identified are always extrachromosomal, but other transmissible plasmids capable of site-specific chromosomal integration have also been identified among the actinomycetes (5, 8, 12, 19, 27, 32). The first such element to be discovered, SLP1, originates from the *Streptomyces coelicolor* A3(2) chromosome and is the best studied of the chromosomally integrating, transmissible plasmids (5, 28–31). In this chapter we review the current state of knowledge of these elements, primarily through work that has been done with SLP1.

II. DISCOVERY AND PROPERTIES OF SLP1

SLP1 was identified when spores of *S. coelicolor* were cocultivated on solid media with a lawn of spores of the closely related species *Streptomyces lividans* 66 (5). After germination and growth of the spores, zones of transient inhibition of aerial mycelium formation or "pocks" appeared in the lawn of *S. lividans* cells. These pocks represented 2 to 3% of the number of *S. coelicolor* spores seeded onto the lawn derived from *S. lividans* spores. *S. lividans* cells isolated from these pocks had also gained the ability to form pocks on a lawn of wild-type *S. lividans*. Thus, it appeared that genes for pock formation had been transferred from *S. coelicolor* to *S. lividans*. The pocking phenotype (4) is common to many transmissible-plasmid-bearing strains of *Streptomyces* spp., and it initially was assumed that a plasmid was responsible in this case, too. However, neither the donor *S. coelicolor* strain nor the separately grown *S. lividans* strain contained detectable plasmids. On the other hand, examination of the pock-forming *S. lividans* cells obtained from matings with *S. coelicolor* showed that many, although not all, now contained plasmid DNA. The plasmids varied in length from approximately 10 to 14.5 kilobases (kb) and were designated the SLP1 (*S. lividans* plasmid 1) family (SLP1.1, SLP1.2, SLP1.3, etc.) (5). The smaller SLP1 plasmids contained a subset of the sequences in the larger SLP1 plasmids (Fig. 1).

SLP1 plasmids are capable of transmission on solid media to SLP1$^-$ strains of *S. lividans* as detected by pocking. Pocks resemble temperate bacteriophage plaques; however, while some *Streptomyces* plasmids

Charles A. Omer ■ Central Research and Development Department, Experimental Station, P.O. Box 80228, E. I. DuPont de Nemours and Co., Wilmington, Delaware 19880-0228. **Stanley N. Cohen** ■ Departments of Genetics and Medicine, Stanford University School of Medicine, Stanford, California 94305.

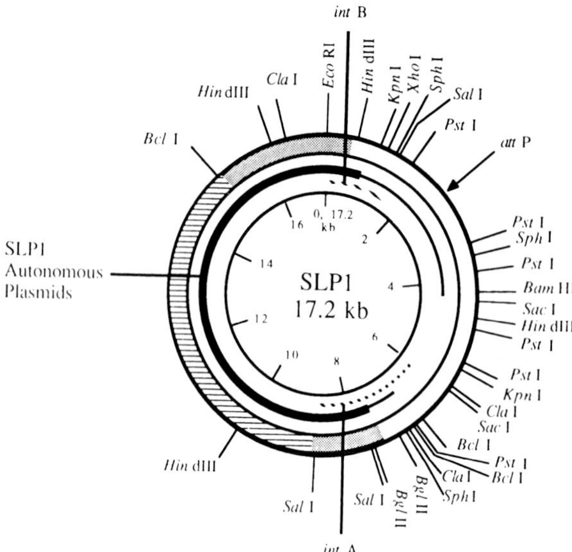

Figure 1. Map of SLP1. The circular form of SLP1 and the locations of restriction endonuclease sites are shown (31). The region of SLP1 in the autonomously replicating, SLP1-derived plasmids is indicated as an arc. The thick section of the arc is present in all of the naturally occurring SLP1-derived plasmids. The thin section at either end of the arc are found only in the larger SLP1 plasmids (5, 29). The locations of the integration site *attP* and the integration essential genes *intA* and *intB* are shown (30, 31). The locations of two regions required for autonomous replication and/or maintenance of SLP1 plasmids are the shaded sections in the outer ring of the figure (reference 17, p. 269). Sequences within the hatched section in the outer ring are required for transfer of SLP1 (reference 17, p. 269).

are prophages (10), no phage or phagelike particles have been found associated with SLP1 pocks. SLP1 plasmids also mediate chromosomal recombination in *S. lividans* (5) (Table 1). This trait appears to be associated with the transfer ability of the plasmids, although the mechanism is not well understood.

The origin of the autonomously replicating SLP1 plasmids and the nature of the *S. lividans* strains that could pock but did not contain detectable plasmids was determined by Southern blot analysis of total chromosomal DNAs. It was found that DNA from the plasmid SLP1.2, the largest of the SLP1 plasmids, did not hybridize to chromosomal DNA of wild-type *S. lividans* 66, but did hybridize to *S. coelicolor* chromosomal DNA. DNA isolated from *S. lividans* strains that had received the pocking phenotype from *S. coelicolor* but contained no detectable plasmids also hybridized to SLP1.2 DNA. Thus, the SLP1 plasmids were derived (and henceforth in this chapter will be referred to as "SLP1-derived") from a DNA sequence (SLP1) that initially was either integrated into the chromosome of *S. coelicolor* or was part of some other, larger, nonisolatable replicon (5). The chromosomal form of SLP1 has been designated SLP1 integrated ($SLP1^{int}$) (28). A chromosomal location for $SLP1^{int}$ near the *strA* locus of *S. coelicolor* has been established genetically by mapping both chromosomal mutations in $SLP1^{int}$ and an antibiotic resistance marker integrated by recombination into $SLP1^{int}$ (5). *S. lividans* strains that can pock but show no SLP1-derived plasmid DNA also have the SLP1 sequence integrated chromosomally at the same physical location as in *S. coelicolor* (28).

SLP1-like elements have been identified in a number of other actinomycetes (Table 1). Hopwood et al. showed that DNA sequences that cross-hybridize to SLP1 were present in 9 of 15 *Streptomyces* strains tested (19). Two of these, *Streptomyces glaucescens* and *Streptomyces parvulus*, were mated with *S. lividans*, and autonomous plasmids presumably originating from the *S. glaucescens* or *S. parvulus* chromosome and homologous to the SLP1 plasmids from *S. coelicolor* were identified. Other integrating genetic elements of the actinomycetes have been shown to exist as plasmids, often of low copy number. Some of these were present naturally, while others appeared after UV irradiation of cells or after regeneration of protoplasts. These plasmids were shown to be present in both chromosomally integrated and autonomous states by hybridization to total DNA from the strain of origin (8, 12, 27, 32).

Some properties of other integrating plasmids of actinomycetes seem similar to SLP1. Most of those tested appear to be transmissible, and several have been shown to mediate chromosome recombination (Table 1). However, whereas autonomously replicating SLP1-derived plasmids are incompatible with $SLP1^{int}$ (5), some other chromosomally integrated plasmids appear to be compatible with their autonomously replicating forms, at least in the same colony (8, 12, 26, 31). One explanation for this apparent difference is that the integrated and excised forms exist in the same colony in different cells. If these elements excise and reintegrate frequently, then in a population of cells there would always be a number of copies in each state. Other explanations may be needed since the pSE101 element of *Saccharopolyspora erythraea* when introduced into *S. lividans* can be found either integrated or autonomous (8). However, once established in *S. lividans*, there is little or no interconversion between the integrated and plasmid forms of pSE101, indicating two alternative fates for the element. Additionally, pSE101 integrates at a unique chromosomal site in *S. erythraea* (*attB*) but at a number of sites in *S. lividans*. Since *S. lividans* is not the natural host for pSE101, there may not be an *attB* site identical to the one in *S. erythraea*. Thus, the multiple sites of integration of pSE101 in *S. lividans* seem analogous to the secondary integration sites for

Table 1. Properties of chromosomally integrating plasmids of the actinomycetes[a]

Name	Species of origin	Length (kb)	Excised form detectable	Homology with other plasmids	Tra$^+$	Incompatibility with other plasmids[b]	Mobilizes host chromosome (frequency)	References
SLP1	*Streptomyces coelicolor* A3(2)	17.2	No	pIJ110, pIJ408[c]	Yes	pIJ110, pIJ408 (vw)	Yes[d] (10^{-4})	5, 21, 28–31
pMEA100	*Nocardia mediterranei* LBG A3136	23.7	Yes		Yes		Yes (10^{-3})	26, 27
pSAM2	*Streptomyces ambofaciens* ATCC 23877, ATCC 15154	11.1	Yes[e]		Yes		ND[f]	32, 38
pSG1	*Streptomyces griseus* NRRL 3851	16.6	Yes[g]		ND		ND	12
pIJ408	*Streptomyces glaucescens* ETH 22794	15.05	Yes[h]	SLP1, pIJ110	Yes	pIJ110 (w), SLP1 (vw)	Yes[d] (10^{-4})	19, 21
pIJ110	*Streptomyces parvulus* ATCC 12434	13.6	Yes[h]	SLP1, pIJ408	Yes	pIJ408 (w), SLP1	Yes[d] (10^{-4})	19, 21
pSE101	*Saccharopolyspora erythraea* NRRL 2338	11.3	Yes		No		No	8
pSE211	*Saccharopolyspora erythraea* NRRL 2338	18	Yes		Yes		Yes	D. Brown and L. Katz, personal communication

[a] Chromosomally integrating plasmids of the actinomycetes are listed along with information about the plasmid, including name of the element, species of origin, length of the plasmid in kilobases, whether or not the excised form can be detected in strains carrying the integrated plasmid, known DNA sequence homology with other integrating plasmids, transmissibility (Tra$^+$), incompatibility of the autonomously replicating form with the autonomously replicating form of other integrating plasmids, ability to mobilize host chromosome, including the frequency of recombinants obtained from a mating (number of recombinants divided by the total number of cells from a mating), and references.

[b] w, Weak incompatibility; approximately threefold increase in plasmid loss and significant ability to resist pock formation by the other plasmid. vw, Very weak incompatibility; at most two- to threefold increase in plasmid loss and little if any ability to resist pock formation by the other plasmid.

[c] DNA homologous to SLP1 was found in total DNA from 7 of 13 other *Streptomyces* strains.

[d] Determined for autonomously replicating form of plasmid.

[e] Excised form found in a strain that had been UV irradiated.

[f] ND, Not determined.

[g] Excised form found in a strain that had been protoplasted and regenerated.

[h] Very low level can be detected by transforming total DNA into *S. lividans* and looking for pocks.

bacteriophage λ in *Escherichia coli* strains that are deleted for λ *attB* (13). DNA sequence analysis of the multiple pSE101 integration sites in *S. lividans* will enable an important comparison to be made with the *attB* site of *S. erythraea*, shedding more light on the specificity of the recombination events.

Elements from bacteria other than the actinomycetes share characteristics with the SLP1-like elements. Examples include the BM6001 element of *Streptococcus pneumoniae* (41) and the Tn*916* family of streptococcal conjugative transposons (11), both of which excise from the chromosome, mobilize themselves to recipient strains, and then integrate into the DNA of the recipient cell. The integration sites used by the Tn*916* family, however, are only moderately specific, unlike the highly site-specific integration of most SLP1-like elements. The *e*14 element found in *E. coli* has some properties that are similar to those of SLP1 (7, 14). *e*14 is integrated chromosomally in most *E. coli* K-12 strains, but not in *E. coli* B/5 or *E. coli* C (14). When *E. coli* K-12 is UV irradiated, *e*14 excises from the chromosome (14). When introduced into *E. coli* C or *E. coli* K-12 strains cured of *e*14, *e*14 integrates into the same chromosomal location at which it normally resides in *E. coli* K-12 (7). It is not clear, however, whether *e*14 is capable of replication and it is unlikely that it can promote its own transfer.

III. STRUCTURE OF THE INTEGRATED FORM OF SLP1 (SLP1^{int}) AND A MODEL FOR ITS EXCISION AND INTEGRATION

The size of SLP1^{int} was found to be 17.2 kb by examination of tandem duplications of SLP1^{int} in *S. lividans*, novel structures that occur in about 10% of intrastrain transfers of SLP1^{int} in *S. lividans* (28, 29). Cleavage at the unique *Bam*HI or *Eco*RI sites within the tandemly duplicated sequence yielded complete copies of the 17.2-kb SLP1^{int}. Ligation of either of these 17.2-kb linear DNAs and introduction into *S. lividans* 66 resulted in transformants containing SLP1^{int} identical in structure and chromosomal location to that in *S. coelicolor* or in *S. lividans* SLP1^{int} strains formed by mating (28). The complete SLP1

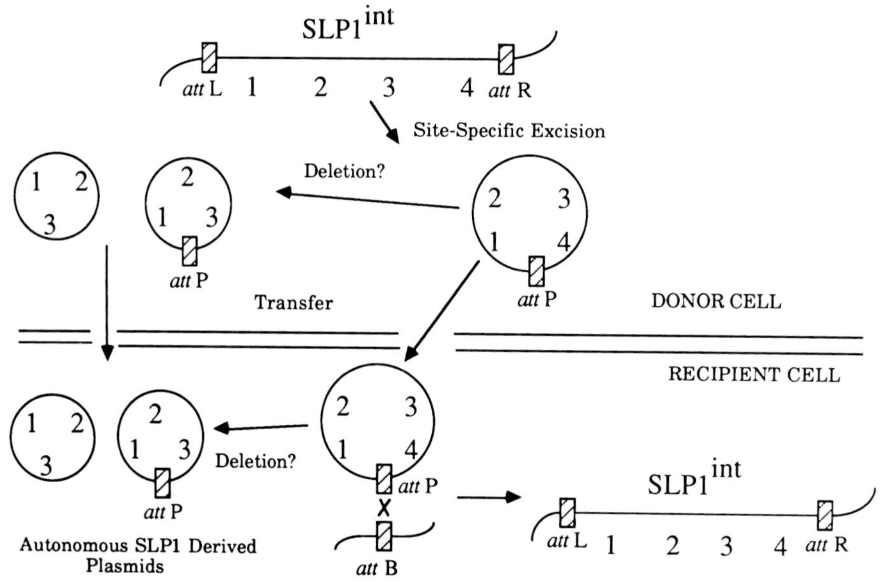

Figure 2. Model for SLP1^{int} excision and integration and SLP1-derived autonomous plasmid formation (28–30). The hatched rectangles indicate the 112-bp *attL*, *attR*, *attB*, and *attP* sites. The numbers 1, 2, 3, and 4 indicate arbitrary sequences within SLP1^{int} and are not intended to indicate specific genes. A double horizontal line separates the donor and recipient cells. The model for SLP1^{int} excision and integration described in the text proposes that in a cell carrying SLP1^{int}, site-specific recombination between the *attL* and *attR* sites occurs, generating a complete, 17.2-kb, circular form of SLP1. This circular molecule can then transfer itself to an SLP1^{-} recipient. After SLP1 entry into the recipient cell, a site-specific recombination between the *attP* site of circular SLP1 and the chromosomal *attB* site of the recipient chromosome integrates SLP1 into the recipient chromosome. Autonomously replicating, SLP1-derived plasmids result from deletions that remove regions required for SLP1 integration. Whether such deletions occur in the donor, recipient, or a fusion derivative of the two is unknown.

sequence has also been cloned in *E. coli* as a 17.2-kb *Bam*HI fragment inserted into pACYC177 (30). The resulting plasmid can transform *S. lividans*, and transformants contain a chromosomally integrated SLP1-pACYC177 hybrid without further rearrangement (31). Additional DNAs inserted into these SLP1-pACYC177 hybrids can also be stably integrated into the *S. lividans* chromosome (31). Thus, such plasmid vectors provide an excellent means of adding specific foreign genes to *S. lividans*.

Analysis of the SLP1-derived plasmids shows that they lack from 2.7 to 7.0 kb of the 17.2-kb SLP1^{int} sequence (28, 29) (Fig. 1). In most cases the missing regions are from the two ends of the SLP1^{int}. In contrast, SLP1.2 contains the sequences from both ends of SLP1^{int}, but lacks a 2.7-kb internal segment (29). Combining this information with the observation that circularized molecules of SLP1^{int} can transform *S. lividans* 66 to SLP1^{int} has led to a model for integration and excision of SLP1^{int} and formation of SLP1-derived plasmids (28) (Fig. 2). Site-specific excision between the two ends of SLP1^{int} (designated *attL* and *attR*) generates a 17.2-kb circular SLP1 molecule containing an *attP* site. This molecule replicates and can transfer to SLP1^{-} strains of *S. lividans*. Once in an SLP1^{-} cell, the *attP* locus of SLP1 inserts specifically into the recipient chromosomal *attB* locus, generating SLP1^{int}.

None of the autonomously replicating SLP1-derived plasmids contain DNA from outside of the 17.2-kb SLP1^{int} sequence (5, 28, 29). SLP1-derived plasmids thus appear to be formed by simple deletions that remove regions of the 17.2-kb circular intermediate required for site-specific recombination with the chromosome but not for replication. Most SLP1-derived plasmids have lost the SLP1 *attP* site. SLP1.2, however, has an intact *attP* site, but is still incapable of chromosomal integration. Such a plasmid can be formed by a deletion of a segment of the complete 17.2-kb SLP1 circle, but not from imprecise excision, since it contains the *attP* region of the 17.2-kb circular intermediate that occurs only upon excision of SLP1^{int} (Fig. 2). The deletion that has occurred in the formation of SLP1.2 has removed a gene(s) required to mediate the site-specific recombination between *attB* and *attP* (see below).

The model for excision and integration of SLP1 is based on the mechanism of excision and integration of the *E. coli* bacteriophage λ (13). It does not involve extensive homology between the SLP1 *attP* site and the chromosomal *attB* site or the SLP1^{int} *attL* and *attR* sites. All of the SLP1 *att* sites have a 112-bp

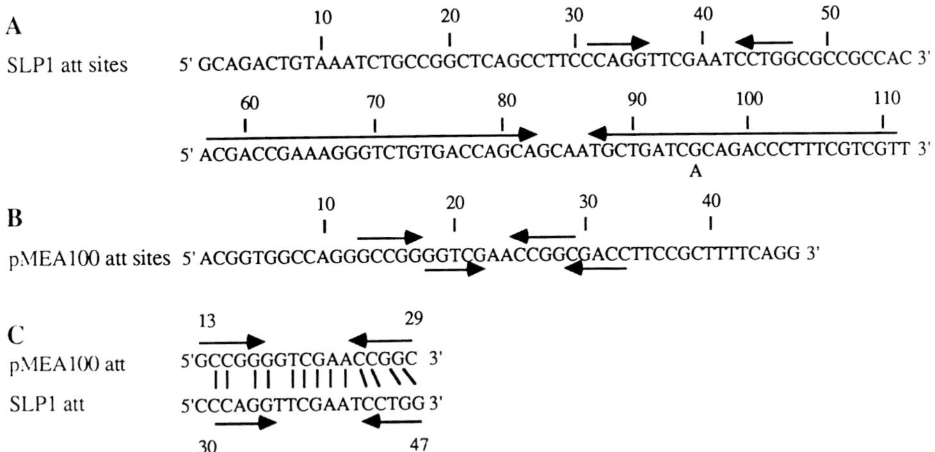

Figure 3. DNA sequences of the *att* sites of SLP1 and pMEA100. (A) The DNA sequences of the 112-bp region common to the SLP1 *att* sites and (B) the 47-bp region common to the pMEA100 *att* sites (26, 30). For SLP1 a G is present at position 95 in *attP* and *attR*, while an A is present at position 95 in *attB* and *attL*. DNA sequences from nucleotides 52 to 112 of the SLP1 *att* site are functionally dispensable (25a). (C) The region of homology between the SLP1 *att* sites and the pMEA100 *att* sites (26). Arrows above and below the sequences indicate the locations of inverted repeat sequences.

region of sequence homology, and it appears that only the left half of it (as shown in Fig. 3) is required for integration and excision (25a, 30). Additionally, as mentioned above, SLP1.2 contains the *attP* site, but has a recessive defect rendering it incapable of chromosomal integration on its own. Thus, something in addition to the 112-bp homology between the *att* sites is required for stable integration. The autonomously replicating, SLP1-derived plasmids resemble λ*dv* plasmids in that they are deletion derivatives of a chromosomally integrating element (3).

Other SLP1-like elements appear to fit the model described for SLP1 excision and integration. In particular, circular forms of several integrating plasmids integrate when introduced into plasmid-free derivatives of the original organism (8, 26, 38). For some of these elements the putative circular intermediate in integration and excision can be detected in the parent strain or can be maintained in other *Streptomyces* species (8, 12, 26, 31) (Table 1). In contrast, the complete 17.2-kb SLP1 element is not readily detectable as an autonomous circle in *S. lividans* or *S. coelicolor*. An additional difference between SLP1 and the other elements is that SLP1-derived plasmids, common in interspecies matings of SLP1 from *S. coelicolor* to *S. lividans*, are rarely formed from other elements.

IV. COMPONENTS OF THE SLP1 SITE-SPECIFIC RECOMBINATION SYSTEM

All but 1 bp of the 112-bp SLP1 *att* sequence is identical in *attB*, *attP*, *attL*, and *attR* (30) (Fig. 3A). These *att* sites contain 25- and 5-bp inverted repeats reminiscent of the inverted repeats that comprise the site-specific recombination sites *loxP* of bacteriophage P1, *FRT* of the 2μm plasmid of *Saccharomyces cerevisiae*, and the sites involved in *hin*-, *gin*-, *cin*-, and *pin*-mediated site-specific inversions (6, 16, 17, 33, 34, 37). A region spanning the 5-bp inverted repeat of SLP1 *att* sites is remarkably similar to a region spanning an inverted repeat in the 47-bp *att* site of the integrating plasmid pMEA100 of the actinomycete *Nocardia mediterranei* (26) (Fig. 3B and C). The 25-bp SLP1 *att*-site inverted repeat, although initially postulated to be important in SLP1 site-specific recombination, now appears to be dispensable (25a, 30). The 5-bp inverted repeat in SLP1 *att* is within the essential region. Thus, the homology between the region spanning the 5-bp repeat of the SLP1 *att* sites and the region spanning the inverted repeat of the pMEA100 *att* sites may have functional significance. Additionally, preliminary sequencing of the *att* sites of the integrating plasmid pSE101 from *Saccharopolyspora erythraea* indicates that each contains an inverted repeat with homology to the 5-bp inverted repeats of the *att* sites of SLP1 and pMEA100 (D. Brown and L. Katz, personal communication).

In the *cre-loxP*, *FRT-FLP*, *hin*, *gin*, *cin*, and *pin* recombination systems only one element-encoded enzyme is necessary to mediate site-specific recombination (1, 2, 9, 22, 23, 34). In bacteriophage λ one phage-encoded protein (Int) is required for λ integration and two phage-encoded proteins (Int and Xis) are required for λ excision (13). For some of these site-specific recombination systems, host-encoded proteins are additionally required. Expression of λ *int* and *xis* genes is tightly regulated (13), and expression of the *FLP* recombinase gene of the *S. cerevisiae* 2μm plasmid has been shown to be regulated at least 100-fold (39). Expression of the site-specific recombination function(s) of SLP1 also seems tightly controlled, because little if any spontaneous excision of SLP1^{int} is detectable in *S. lividans* (25a). Upon mating of an *S. lividans* SLP1^{int} strain with an *S. lividans* SLP1^{-} strain, however, excision, transfer, and integration of SLP1^{int} are efficient and nearly all recipient cells acquire SLP1^{int}.

Two regions of SLP1 in addition to its *attP* site are essential for stable integration (31) (Fig.1). One, designated *intA*, includes a sequence absent from all autonomously replicating, SLP1-derived plasmids. When placed in *trans*, *intA* can mediate integration of a derivative of the SLP1.2 plasmid. The other essential integration gene (*intB*) is found in many SLP1-derived plasmids. It may overlap slightly, but is functionally distinct from, a region required for maintenance and autonomous plasmid replication. It is not known if either of the two genes essential for integration encode the putative site-specific recombinase. It has also not been demonstrated whether either of these genes is required for excision; however, a limited analysis has found no mutations that differentially affect the ability of SLP1 to integrate or excise.

A number of questions remain about the site-specific recombination systems of the SLP1-like plasmids. Although regions of SLP1 required for integration have been identified and mapped, the integrase has yet to be identified for any of these elements. The sites involved in site-specific recombination of SLP1, pMEA100, and pSE101 resemble those in the *loxP-cre* and *FLP-FRT* systems and *gin*-, *hin*-, *cin*-, and *pin*-mediated DNA inversions in that they have an inverted repeat within the DNA target site. If the SLP1, pMEA100, and pSE101 site-specific recombinases follow these examples, then a single element-encoded protein will be necessary for their site-specific recombinations. Given the sequence similarity of SLP1, pMEA100, and pSE101 *att* sites to one another, the *att* sites might be able to substitute for each other, by analogy with the sites acted on by Gin, Hin, Cin, and Pin recombinases.

SLP1^{int} in *S. lividans* does not excise from the chromosome spontaneously at a detectable frequency, but is readily excised and transferred to *S. lividans* SLP1^{-} strains upon cocultivation on solid media. The mechanism regulating this excision process has not been determined. The control of SLP1 replication also merits further study. SLP1-derived plasmids have a copy number of approximately five to seven per chromosome in *S. lividans*. However, SLP1^{int} is replicated as part of the chromosome; its own replication origin appears to be turned off, since there is no SLP1^{int} amplification. The SLP1 replication functions and site-specific recombination functions might be regulated coordinately, since in the complete SLP1 element, the two appear to be expressed only during mating and transfer. Autonomously replicating, SLP1-derived plasmids would have lost the gene(s) responsible for turning off replication as well as those for integration. Consistent with this view are recent observations (S. Grant, S. C. Lee, K. Kendall, and S. N. Cohen, manuscript in preparation) indicating that a gene sequence located in the region of SLP1 absent from SLP1-derived plasmids can prevent maintenance of autonomously replicating, SLP1-derived plasmids.

The widespread occurrence of chromosomally integrating plasmids in the actinomycetes makes it tempting to speculate on roles that they might play in genome organization and gene flux between these organisms. Among the actinomycetes, many *Streptomyces* species and related genera have similar primary metabolic capabilities but vary in the numerous secondary metabolites they make (43). Genes for the biosynthesis of any particular secondary metabolite can be either chromosomal or plasmid borne, but in either case they are usually clustered together (24, 25, 35, 36, 42). We propose that dispensable, chromosomally encoded traits such as secondary metabolite production could be carried on SLP1-like elements. Although the chromosomally integrating elements described here do not code for the production of known secondary metabolites, most are relatively small. As mentioned, however, SLP1 is capable of carrying extra genes, as shown by integration of SLP1-pACYC177. The ability of SLP1-like elements to function as transmissible plasmids as well as to integrate into the chromosome would facilitate transfer of genes between related actinomycetes.

LITERATURE CITED

1. **Abremski, K., and R. Hoess.** 1984. Bacteriophage P1 site-specific recombination: purification and properties of the Cre recombinase protein. *J. Biol. Chem.* **259:**1509–1514.
2. **Andrews, B. J., G. A. Proteau, L. G. Beatty, and P. D. Sadowski.** 1985. The FLP recombinase of the 2μ circle DNA of yeast: interaction with its target sequences. *Cell* **40:** 795–803.
3. **Berg, D. E.** 1974. Genetic evidence for two new types of gene arrangements in new λdv plasmid mutants. *J. Mol. Biol.* **86:**59–68.
4. **Bibb, M. J., R. F. Freeman, and D. A. Hopwood.** 1977. Physical and genetic characterization of a second sex factor,

SCP2, from *Streptomyces coelicolor* A3(2). *Mol. Gen. Genet.* **154:**155–166.

5. **Bibb, M. J., J. M. Ward, T. Kieser, S. N. Cohen, and D. A. Hopwood.** 1981. Excision of chromosomal DNA sequences from *Streptomyces coelicolor* forms a novel family of plasmids detectable in *Streptomyces lividans*. *Mol. Gen. Genet.* **184:** 230–240.
6. **Broach, J. R., V. R. Guarascio, and M. Jayaram.** 1982. Recombination within the yeast plasmid 2μ circle is site-specific. *Cell* **29:**227–234.
7. **Brody, H., A. Greener, and C. W. Hill.** 1985. Excision and reintegration of the *Escherichia coli* K12 chromosomal element e14. *J. Bacteriol.* **161:**112–117.
8. **Brown, D. P., S.-J. D. Chiang, J. S. Tuan, and L. Katz.** 1988. Site-specific integration in *Saccharopolyspora erythraea* and multisite integration in *Streptomyces lividans* of actinomycete plasmid pSE101. *J. Bacteriol.* **170:**2287–2295.
9. **Bruist, M. F., and M. I. Simon.** 1984. Phase variation and the Hin protein: in vivo activity measurements, protein overproduction and purification. *J. Bacteriol.* **159:**71–79.
10. **Chung, S. T.** 1982. Isolation and characterization of *S. fradiae* plasmids which are the prophage of the actinophage ϕSF1. *Gene* **17:**239–246.
11. **Clewell, D. B., and C. Gawron-Burke.** 1986. Conjugative transposons and the dissemination of antibiotic resistance in streptococci. *Annu. Rev. Microbiol.* **40:**635–659.
12. **Cohen, A., D. Bar-Nir, M. E. Goedeke, and Y. Parag.** 1985. The integrated and free states of *Streptomyces griseus* plasmid pSG1. *Plasmid* **13:**41–50.
13. **Gottesman, M. E., and R. A. Weisberg.** 1983. Prophage insertion and excision, p. 113–138. *In* R. W. Hendrix, J. Roberts, F. W. Stahl, and R. A. Weisberg (ed.), *Lambda II*. Cold Spring Harbor Laboratory, Cold Spring Harbor, N.Y.
14. **Greener, A., and C. W. Hill.** 1980. Identification of a novel genetic element in *Escherichia coli* K12. *J. Bacteriol.* **144:** 312–321.
15. **Hara, O., S. Horinouchi, T. Uozumi, and T. Beppu.** 1983. Genetic analysis of A-factor synthesis in *Streptomyces coelicolor* A3(2) and *Streptomyces griseus*. *J. Gen. Microbiol.* **129:**2939–2944.
16. **Hiestand-Nauer, R., and S. Iida.** 1983. Sequence of the site-specific recombinase gene *cin* and its substrates serving in the inversion of the C segment of bacteriophage P1. *EMBO J.* **2:**1733–1740.
17. **Hoess, R. H., M. Ziese, and N. Sternberg.** 1982. P1 site-specific recombination; nucleotide sequence of the recombining sites. *Proc. Natl. Acad. Sci. USA* **79:**3398–3402.
18. **Hopwood, D. A., M. J. Bibb, K. F. Chater, T. Kieser, C. J. Bruton, H. M. Kieser, D. J. Lydiate, C. P. Smith, J. M. Ward, and H. Schrempf.** 1985. *Genetic Manipulation of Streptomyces, a Laboratory Manual.* John Innes Foundation, Norwich, U.K.
19. **Hopwood, D. A., G. Hinterman, T. Kieser, and H. M. Wright.** 1984. Integrated DNA sequences in three streptomycetes form autonomous plasmids after transfer to *Streptomyces lividans*. *Plasmid* **11:**1–16.
20. **Hopwood, D. A., T. Kieser, D. Lydiate, and M. J. Bibb.** 1986. *Streptomyces* plasmids: their biology and use as cloning vectors, p. 159–229. *In* S. W. Queener and L. E. Day (ed.), *The Bacteria. A Treatise on Structure and Function*, vol. IX. *Antibiotic Producing Streptomyces.* Academic Press, Inc., Orlando, Fla.
21. **Hopwood, D. A., T. Kieser, H. M. Wright, and M. J. Bibb.** 1983. Plasmids, recombination and chromosome mapping in *Streptomyces lividans* 66. *J. Gen. Microbiol.* **129:**2257–2269.
22. **Iida, S., J. Meyer, K. Kennedy, and W. Arber.** 1982. A site-specific, conservative recombination system carried by bacteriophage P1. *EMBO J.* **1:**1445–1453.
23. **Kamp, D., L. T. Chow, T. R. Broker, D. Kwoh, D. Zipser, and R. Kahmann.** 1979. Site-specific recombination in phage Mu. *Cold Spring Harbor Symp. Quant. Biol.* **43:**1159–1167.
24. **Kinashi, H., M. Shimaji, and A. Sakarai.** 1987. Giant linear plasmids in *Streptomyces* which code for antibiotic biosynthesis genes. *Nature* (London) **328:**454–456.
25. **Kirby, R., L. F. Wright, and D. A. Hopwood.** 1975. Plasmid determined antibiotic synthesis and resistance in *Streptomyces*. *Nature* (London) **254:**265–267.

25a. **Lee, S. C., C. A. Omer, M. Brasch, and S. N. Cohen.** 1988. Analysis of recombination occurring at SLP1 *att* sites. *J. Bacteriol.* **170:**5806–5813.

26. **Madon, J., P. Moretti, and R. Hutter.** 1987. Site-specific integration and excision of pMEA100 in *Nocardia mediterranei*. *Mol. Gen. Genet.* **209:**257–264.
27. **Moretti, P., G. Hinterman, and R. Hutter.** 1985. Isolation and characterization of an extrachromosomal element from *Nocardia mediterranei*. *Plasmid* **14:**126–133.
28. **Omer, C. A., and S. N. Cohen.** 1984. Plasmid formation in *Streptomyces*: excision and integration of the SLP1 replicon at a specific chromosomal site. *Mol. Gen. Genet.* **196:**429–438.
29. **Omer, C. A., and S. N. Cohen.** 1985. SLP1: transmissible *Streptomyces* chromosomal element capable of site-specific integration, excision, and autonomous replication, p. 449–453. *In* D. Schlessinger (ed.), *Microbiology—1985*. American Society for Microbiology, Washington, D.C.
30. **Omer, C. A., and S. N. Cohen.** 1986. Structural analysis of plasmid and chromosomal loci involved in site-specific excision and integration of the SLP1 element of *Streptomyces coelicolor*. *J. Bacteriol.* **166:**999–1006.
31. **Omer, C. A., D. Stein, and S. N. Cohen.** 1988. Site-specific insertion of biologically functional adventitious genes into the *Streptomyces lividans* chromosome. *J. Bacteriol.* **170:** 2174–2184.
32. **Pernodet, J. L., J. M. Simonet, and M. Guerineau.** 1984. Plasmids in different strains of *Streptomyces ambofaciens*: free and integrated form of plasmid pSAM2. *Mol. Gen. Genet.* **198:**35–41.
33. **Plasterk, R. H. A., A. Brinkman, and P. van de Putte.** 1983. DNA inversion in the chromosome of *E. coli* and the bacteriophage Mu: relationship to other site-specific recombination systems. *Proc. Natl. Acad. Sci. USA* **80:**5355–5358.
34. **Plasterk, R. H. A., and P. van de Putte.** 1985. The invertible P-DNA segment in the chromosome of *E. coli*. *EMBO J.* **4:**237–242.
35. **Rhodes, P. M., N. Winskill, E. J. Friend, and M. Warren.** 1981. Biochemical and genetic characterization of *Streptomyces rimosus* mutants impaired in oxytetracycline biosynthesis. *J. Gen. Microbiol.* **124:**329–338.
36. **Rudd, B. A. M., and D. A. Hopwood.** 1980. A pigmented mycelial antibiotic in *Streptomyces coelicolor*: control by a chromosomal gene cluster. *J. Gen. Microbiol.* **119:**333–340.
37. **Simon, M. I., and M. Silverman.** 1983. Recombinational regulation of gene expression in bacteria, p. 211–227. *In* J. Beckwith, J. Davies, and J. A. Gallant (ed.), *Gene Expression in Prokaryotes*. Cold Spring Harbor Laboratory, Cold Spring Harbor, N.Y.
38. **Simonet, J. M., F. Boccard, J. L. Pernodet, J. Gagnat, and M. Guerineau.** 1987. Excision and integration of a self-transmissible replicon of *Streptomyces ambofaciens*. *Gene* **59:** 137–188.

39. **Som, T., K. A. Armstrong, F. C. Volkert, and J. R. Broach.** 1988. Autoregulation of 2μ circle gene expression provides a model for maintenance of stable plasmid copy levels. *Cell* **52**:27–37.
40. **Stonesifer, J., P. Matsushima, and R. H. Baltz.** 1986. High frequency conjugal transfer of tylosin genes and amplifiable DNA in *Streptomyces fradiae. Mol. Gen. Genet.* **202**: 348–355.
41. **Vijayakumar, M. N., S. D. Priebe, and W. R. Guild.** 1986. Structure of a conjugative element in *Streptococcus pneumoniae. J. Bacteriol.* **166**:978–984.
42. **Weber, M. J., C. K. Wierman, and C. R. Hutchinson.** 1985. Genetic analysis of erythromycin production in *Streptomyces erythreus. J. Bacteriol.* **164**:425–433.
43. **Williams, S. T., M. Goodfellow, G. Alderson, E. M. H. Wellington, P. H. A. Sneath, and M. J. Sackin.** 1983. Numerical classification of *Streptomyces* and related genera. *J. Gen. Microbiol.* **129**:1743–1813.

Chapter 11

Transposable Elements and Genome Structure in Halobacteria

ROBERT L. CHARLEBOIS and W. FORD DOOLITTLE

I. INTRODUCTION: THE ARCHAEBACTERIA

In the early 1970s, Carl Woese and his collaborators began a phylogenetic survey of the procaryotes, using the then recently developed methodology of Sanger to generate "T1-oligonucleotide catalogs" (lists of the sequences of ribo-oligonucleotides released from ^{32}P-labeled 16S rRNA by digestion with RNase T1), and analyzing these catalogs by simple numerical taxonomic procedures (17). Their efforts bore fruit. By the end of the decade, the outlines of a general evolutionary scheme showing relationships between most major bacterial groups and between these and eucaryotic lineages were apparent (18, 63).

This scheme required a radical revision of the way in which we look at cellular evolution. Most well-known bacterial species (eubacteria) belong to one of ten "phyla"; these bear only partial resemblance to higher-order bacterial taxa derived from physiological and structural considerations (63). Eucaryotic cells are chimeras, made up of the descendants of endosymbiotic cyanobacteria and purple bacteria which became modern plastids and mitochondria and a third, "urkaryotic" lineage now em-

Robert L. Charlebois and W. Ford Doolittle ▪ Department of Biochemistry, Dalhousie University, Halifax, Nova Scotia, Canada B3H 4H7.

bodied in the eucaryotic nucleus. This third lineage is not derived from any extant bacterial lineage, as Margulis (31), for instance, would have had us believe, and more recent work, particularly that of Sogin et al. (56), suggests that "urkaryotes" are not only ancient but deeply divergent, having internal branches as low as any in bacterial lineages.

Woese's work also led to the recognition that procaryotes are themselves divided into two assemblages of very ancient divergence: the eubacteria, which include most species in *Bergey's Manual*, and the archaebacteria, a few unusual organisms whose disparate phenotypes mask an underlying biochemical and molecular biological unity. Among this group are obligately anaerobic methanogens, aerobic halophiles, and a collection of thermophilic and usually sulfur-dependent species often isolated from acidic environments. Features they share, and which for the most part distinguish them from eubacteria and eucaryotes, have been reviewed in many places (10, 21, 23, 24, 62, 63). In addition to those aspects of 16S rRNA primary and secondary structure which led to their initial recognition, these features include: (i) unique ribosomal particle morphologies; (ii) unusual 5S rRNA structure; (iii) characteristic tRNA modification patterns; (iv) insensitivity to most antibiotic inhibitors of translation effective against eubacteria and eucaryotes; (v) initiation of translation with methionyl- rather than formylmethionyl-tRNA; (vi) sensitivity to diphtheria toxin; (vii) RNA polymerases with distinct, complex subunit structure and immunological affinities to eucaryotic RNA polymerases; (viii) promoters which, although as yet poorly characterized, differ from either eubacterial or eucaryotic promoters; (ix) introns in some tRNA and at least one rRNA gene; and (x) cell envelopes which lack peptidoglycan and contain ether-linked, rather than ester-linked, lipids.

Recent cladistic analyses (the construction of phylogenetic trees), using the accumulating bank of complete 16S and 23S rDNA (gene) sequences in place of rRNA-derived catalogs, provide a detailed picture of archaebacterial phylogeny (63). The root of the archaebacterial tree was likely thermophilic. The earliest branching separated the last common ancestor of the thermophiles *Thermoproteus*, *Pyrodictium*, *Sulfolobus*, and *Desulfurococcus* spp. from the last common ancestor of the thermophilic genera *Thermoplasma* and *Thermococcus*. The many diverse methanogens and the halophilic archaebacteria (*Halobacterium* and *Halococcus* spp.) derive from this second major branch.

Most sequence-treeing methods support such a branching pattern and show the archaebacteria to be a monophyletic assemblage distinct from either eubacteria or eucaryotes. Lake has proposed, initially on the basis of ribosome morphology, but more recently from sequence analyses, that eucaryotic (nuclear) lineages arose from within the thermophilic archaebacteria, while eubacteria and halobacteria share an ancestor more recent than either does with thermophiles or methanogens (28, 29). What unites the archaebacteria, he argues, is their slower evolutionary rate and thus greater retention of ancestral character states when compared with fast-evolving eubacteria and eucaryotes. Although the balance of evidence supports Woese's initial tripartite scheme (37, 63), it is in any event unnecessary here to judge the merits of Lake's claim. Archaebacteria differ from other organisms by virtue of either separate descent or common retained primitiveness or both. They command the interest of molecular biologists not only because they are there and are different, but because in appreciating their differences, we may come to understand the properties of the last common cellular ancestors of all modern organisms.

II. HALOBACTERIA: GENERAL FEATURES

A. Species of Halobacteria

The family *Halobacteriaceae* has two genera recognized in *Bergey's Manual* (30): *Halobacterium* (gram-negative rods or disks) and *Halococcus* (cocci). Tindall et al. (58) have in addition proposed *Natronobacterium* and *Natronococcus* as genus names for halobacteria which grow at high pH (pH optima, 9 to 10). Most experimental molecular work has used *Halobacterium halobium* and its close relatives *Halobacterium cutirubrum* and *Halobacterium salinarium*. Although these are now all considered strains of a single species, *H. salinarium*, pedigrees of laboratory stocks have not been well kept, and we will use authors' designations. Because of the development of systems for genetic exchange and its relative genetic stability, *Halobacterium volcanii*, which is quite distantly related to the *H. halobium* complex (63), is likely to become the workhorse of halobacterial molecular genetics.

B. Ecology and Physiology

Halobacteria grow best at salt concentrations between 1.5 and 5.2 M (depending on species) and lyse in low salt. They have most commonly been isolated from natural or artificial salt evaporation lagoons and hypersaline lakes (e.g., the Dead Sea or Great Salt Lake) or from salt-preserved foods and hides. They are not the only organisms growing in such environments; certain eubacteria and eucaryotes are quite halotolerant. Halobacteria, specifically, maintain high internal ionic strength, although the principal internal cation is potassium, not sodium. Most halobacterial enzymes require high potassium or sodium levels for activity and often also for stability (27).

Halobacteria are in general mesophilic chemoorganotrophic aerobes. Some species, under appropriate conditions (low oxygen tension), elaborate a "purple membrane" whose sole protein is the 26-kilodalton, retinal-containing pigment bacteriorhodopsin. Bacteriorhodopsin catalyzes light-driven ATP synthesis by causing an ejection of protons from the cell. This simple photosynthetic system has been extensively used as a model for the study of light-dependent proton pumping since the early 1970s (57), long before the discovery of the special phylogenetic position of the halobacteria. That this work also provides an entree into archaebacterial molecular genetics (see below) is thus a happy accident.

C. Genome Structure

Early renaturation kinetic studies by Moore and McCarthy (34) showed that several halobacterial species had genome sizes similar to that of *Escherichia coli* (now known to be 4.7×10^3 kilobase pairs (kbp) [26]). Independent genome size determinations using pulsed-field gel electrophoresis of fragments produced by digestion of chromosomal DNA with certain infrequently cutting restriction endonucleases indicate a genome size of about 4×10^3 kbp for *H. volcanii* (L. C. Schalkwyk, unpublished observation).

Halobacterial species in general exhibit two bands of DNA upon cesium chloride density-gradient centrifugation (22, 33). The main band, comprising 70 to 90% of the total DNA, is of high G+C content (62 to 71 mol%). A satellite band containing relatively AT-rich DNA (55 to 58 mol% G+C) is also almost always present, making up 10 to 30% of the total DNA (*Halobacterium trapanicum* may lack such a satellite [43]).

The distribution of satellite DNAs within genomes of a species and between species has been studied using malachite green-bisacrylamide column chromatography to separate main-band, GC-rich DNA (fraction I or FI) from satellite, relatively AT-rich (fraction II or FII) DNA (13, 15, 38–40, 45). In *H. halobium*, FII DNA comprises three components: (i) a heterogeneous collection of covalently closed circular DNAs present in very low copy number and varying in composition between strains; (ii) a prominent (six to eight copies), large (150 kbp) covalently closed circular (plasmid) DNA, pHH1, found, with various deletions, insertions, and rearrangements, in different strains of this species and *H. cutirubrum* and also found in part integrated into chromosomal DNA in *H. salinarium*; and (iii) "islands" of relatively AT-rich DNA flanked by GC-rich main-band DNA, which are embedded within the halobacterial chromosome. Heterologous hybridization experiments and comparisons of restriction endonuclease digestion patterns reveal that in general FI sequences are more highly conserved among halobacterial species than are FII sequences, although some insertion sequence elements (which are most often embedded within FII DNA) are quite broadly distributed (13–15, 38, 39, 43). Cloned genes important to bacterial growth (rRNA, tRNA, bacterio-opsin, bacterio-opsin-related protein, halo-opsin, several ribosomal proteins) hybridize to FI DNA. Most insertion sequence copies and a gas-vesicle protein gene (*gvpA* [7]) hybridize to FII DNA.

D. Genetics

The development of the molecular biology of the archaebacteria has until recently been impeded by the absence of known processes of genetic exchange. In 1985, Mevarech and Werczberger (32) described a "natural" genetic transfer system. Cultures of auxotrophic mutants of *H. volcanii* (strain DS2) obtained by ethyl methanesulfonate mutagenesis were mixed and filtered together onto nitrocellulose. After suitable incubation, prototrophic recombinants were recovered at frequencies vastly in excess of measured reversion frequencies. This "mating" process requires cell-cell contact, is insensitive to DNase, and appears to be bidirectional. In 1987, Cline and Doolittle (6) described transfection of *H. halobium*, using purified DNA from the *H. halobium*-specific bacteriophage ΦH. *H. volcanii*, which is not sensitive to infection by intact ΦH, could also be transfected, producing plaques on *H. halobium* lawns. Cellular DNA can also be taken up by halobacteria under such conditions, and Mevarech's auxotrophs can be transformed to prototrophy with fragments of purified wild-type DNA (S. W. Cline, L. C. Schalkwyk, and W. F. Doolittle, manuscript in preparation). Most recently, Charlebois et al. (5) described transformation of a pHV2-cured derivative of *H. volcanii* DS2 with pHV2, an endogenous 6,354-base-pair (bp) plasmid found in that species. Efforts to produce a selectable shuttle vector have begun, and it should be possible soon to apply to the study of halobacterial genes, including insertion sequences, the full range of molecular genetic techniques that have led to our current detailed understanding of eubacterial transposable element structure and function.

II. DEMONSTRATION OF THE PRESENCE AND ACTIVITIES OF REPETITIVE AND TRANSPOSABLE ELEMENTS

A. Genetic Instability and Plasmid DNA

H. halobium sports spontaneous mutants which can be detected easily by visual inspection of plates. Gas vacuole-deficient (Vac) strains, found at a frequency of 10^{-2}, appear translucent. Rub mutants, lacking the carotenoid bacterioruberin, look pale and comprise 1 in 10^4 colonies. Pum mutants, which fail to produce the bacteriorhodopsin-containing purple membrane, are equally frequent (46). Sectored colonies, in which such mutations occur at even higher frequencies, are also found.

It is simple to imagine that such high-frequency mutations reflects loss of plasmids. In 1978, Simon (54) indeed showed loss of one of three endogenous plasmids in Vac mutants of *H. salinarium*. With *H. halobium*, Weidinger et al. (60) found that Vac mutants had suffered insertions into, but not loss of, the major plasmid (pHH1) of this species. In a more extensive analysis, Pfeifer et al. (46) reported that all Vac, Rub, and Pum mutants showed alterations in plasmid restriction patterns, in part interpretable as the result of one or more insertions of from 0.5 to 3.0 kbp of DNA. The observed rearrangements were surprisingly complex, as were those in pHH1 isolated from revertants of strains with these phenotypic defects. Pfeifer and collaborators suggested that *vac* genes might be plasmid borne, but that the extraordinarily high overall frequency of rearrangements in plasmid DNA, together with the multiplicity of apparent insertion sites, made it impossible to localize genes affected in Rub and Pum mutants. Subsequent cloning of the gas vesicle protein gene (*gvpA*) confirms that it is on pHH1 (7), while the bacterio-opsin gene (the site of many Pum mutations) is chromosomal (2, 3). Our current appreciation for the extraordinary number and mobility of insertion sequences in *H. halobium*, together with information on their locations (see below), makes the failure of these early attempts to correlate phenotypic change with specific plasmid rearrangements understandable.

B. Repetitive Sequences in the *H. halobium* Genome

Reasoning that genetic instability and frequent plasmid rearrangements might both reflect the presence in *H. halobium* of unusually many or unusually active transposable elements, and that these might be detectable as repetitive sequence, Sapienza and Doolittle (47) looked among random small *Eco*RI (or *Eco*RI-*Bam*HI) clones of *H. halobium* (strain R1) for those which might hybridize to multiple bands of genomic DNA when used as probes in Southern hybridization experiments. In fact, more than 90% of such clones hybridized to several bands. Some clones appeared to contain more than a single type of repetitive DNA, and of the order of 50 different patterns of hybridization to genomic DNA digests were seen, indicating the presence of as many repetitive DNA sequence families, each with from 2 to 20 members. Although many repeats clearly do reside on the rearranged pHH1 borne by strain R1 (see below), many also are "chromosomal" (47). Sapienza et al. (48) also observed frequent rearrangements affecting (and probably involving) these repeated sequences. When a single colony of strain R1 is streaked on a plate and colonies from that plate are used (after inoculation and growth in 1-ml volumes) to prepare DNA for Southern blotting with cloned repeat-sequence probes, rearrangements are already obvious. The frequency of such events (altered number or size of genomic restriction fragments hybridizing to a cloned repetitive-sequence family member) was very high ($>4\times10^{-3}$ events per repeat sequence family). Sapienza et al. (48) calculated that, because of the number and instability (mobility?) of such repeats, two daughter cells produced by the division of a single parental cell of *H. halobium* strain R1 have only an 80% chance of having identically organized genomes. These experiments also suggested that rearrangements occur in "bursts"; clones in which one repeat sequence had suffered rearrangement more often showed genomic alterations involving other repeats. Something of this sort seems also to be involved in the results of Pfeifer et al. (46) cited above. Isolates selected by virtue of having incurred a visibly detectable mutation, which we now know to be due most often to insertion of a transposable element, appeared also to have suffered rearrangements, often complex, of pHH1 DNA, even though the disruption of pHH1 sequences was irrelevant to the alteration in phenotype.

C. Bacterio-Opsin Gene Mutations

The first solid evidence linking these early observations (high frequency of mutation, frequent physical rearrangement, presence of repeated sequences) with the most obvious conclusion—that there are an extraordinary number of active insertion sequence-like elements in the *H. halobium* genome—came from careful studies on the nature of bacteriorhodopsin (Pum$^-$ phenotype) mutants.

The amino acid sequence of bacterio-opsin (Bop)

was known by 1979 (25) and used to design oligonucleotides which allowed cloning of the bacteriorhodopsin apoprotein (*bop*) gene of *H. halobium* in two laboratories (3, 12). Simsek et al. (55) first showed that in two spontaneous Pum mutants deficient in bacteriorhodopsin synthesis, insertions of 1.1 kbp had occurred at identical positions near the 5′ terminus of the bacterio-opsin coding sequence. The inserted DNAs appeared identical, but oppositely oriented, by restriction endonuclease analysis of the two strains. Sequencing of one of the inserts showed it to be 1,118 bp long, to contain imperfect (one mismatch) inverted 8-bp terminal repeats, to contain an 810-bp open reading frame on one strand and a 402-bp open reading frame on the other, and to be flanked by duplications of an 8-bp sequence initially present only once in the wild-type *bop* gene (target) sequence. In short, this insert looks very much like a eubacterial insertion sequence, and Simsek et al. (55) named it ISH*1* (insertion sequence, *Halobacterium*). ISH*1* insertions at this same site in either orientation make up a major class of Bop mutants. ISH*1* is present in one copy in the FI DNA and one copy in FII DNA of the wild-type *H. halobium* strain NRC817 and in from one to four copies in other strains of *H. halobium*, *H. cutirubrum*, and *H. salinarium* (38, 39). In the *bop* gene region, it inserts only at a single position (3, 38, 39), but other insertions, for instance, into pHH1, have been documented (41). Stable 900-nucleotide transcripts of the longer ISH*1* open reading frame are detectable in Northern hybridizations using *H. halobium* cellular RNA (55).

Mutations affecting *bop* expression also result from the insertions of other insertion sequence elements within or near the gene, and in fact all but one or two of the several dozen mutants analyzed so far (38, 39) were so created. ISH2, a frequent disruptor of *bop* gene function, has been found at many sites both within and 5′ to *bop*, as have, in lesser numbers, elements designated ISH*23*, ISH*24*, ISH*26*, ISH*27*, and ISH*28*. Several *bop* mutations at distances up to 1.4 kbp 5′ to the initiation site for *bop* mRNA synthesis indicate the presence there of a regulatory genetic determinant. In sequencing this region, Betlach et al. (1) found an open reading frame beginning 526 bp 5′ to the *bop* initiation site and extending 1,077 bp in the direction opposite to *bop*. They called this locus *brp* (bacterio-opsin-related protein).

D. Bacteriophage ΦH and ISH*1.8*

Bacteriophage ΦH is a lytic virus infecting *H. halobium* and related strains but not, for instance, *H. volcanii*. Its genome is a 59-kbp, linear, partially circularly permuted, terminally redundant, double-stranded DNA, packaged by "headfuls" from tandem concatemeric precursors (53). This DNA suffers frequent insertions and deletions, at least some due to the activity of ISH elements resident in the genome of its host. Schnabel et al. (52) described six (of what must be many more possible) variants of ΦH isolated after limited rounds of replication in *H. halobium* of the "wild-type" (most common) form ΦH1 and one of its variants, ΦH2. Variants differed from ΦH1 by deletions of 150 bp or insertions of 0.05, 0.2, or 1.8 kbp. The 1.8-kbp DNA represented a second copy of an element which they called ISH*1.8*, already present in one copy on ΦH1 and two copies in the FII DNA of the host. The two phage copies in variant ΦH2 are in inverted orientation and further rearrangments involving them are frequent (49, 50, 52). Among these rearrangements are (i) inversion of the 11-kbp region (the L segment) between them, (ii) what appears to be imprecise excision of one of the ISH*1.8* copies from the phage genome, and (iii) circularization of the L segment to produce a 12-kbp, autonomously replicating plasmid, pΦHL, which confers partial immunity to ΦH infection. Virulent variants of ΦH which can lyse cells bearing pΦHL can also be found; these have experienced an insertion of ISH*23* (or ISH*50*; see below) in their L segments.

The extent to which these rearrangements of ΦH DNA involve the activities of a transposase which might be encoded by ISH*1.8* remains unclear. Complex legitimate recombination events are also involved. Circularization and excision of the L segment to produce pΦHL, however, does not involve simple crossing over between ISH*1.8* elements, since these are in inverted orientation and recombination between them inverts the L segment. Instead, the elements themselves contain several 9- or 10-bp inverted repeats, which represent regions of direct repetition on either side of the L segment. Schnabel et al. (51) deduce that in at least one instance, pΦHL excision involves recombination within the sequence TC CCGCCCT, which is present in inverted repeats at one end of each ISH*1.8* element.

IV. STRUCTURES OF ISH ELEMENTS: A MOLECULAR BESTIARY

Given the otherwise depauperate state of archaebacterial molecular biology, a remarkable number of insertion sequence-like elements have been physically described and/or sequenced. We summarize many of their properties below and in Table 1; Pfeifer (39) has recently provided another detailed summary of these insertion sequence features. Except where noted, there is no strong evidence for common evolutionary origins.

A. ISH*1*

The basic features of this *H. halobium* element were discussed above. Heterologous hybridization experiments (38, 41) show ISH*1* to be present also in *H. cutirubrum*, *H. salinarium*, *Halobacterium tunesiensis*, and certain other purple-membrane-producing halobacteria of uncertain species designation. It is not found in *H. volcanii*.

B. ISH*2*

ISH2 was sequenced by DasSarma et al. (8). It is a mere 520 bp long, and its longest open reading frame is only 240 bp. Its termini are 19-bp inverted repeats. Several independent insertions of this element into the *bop* gene have been looked at (1, 42, 44). The target site duplications are variable in length (10, 11, and 20 bp). ISH2 seems to be very active; it is as frequent as ISH*1* in Bop mutants and is often found involved in rearrangements of pHH1 (38, 39).

C. ISH*23* (ISH*50*)

ISH*50* was described by Xu and Doolittle (64). It was identified as an insertion into one of two otherwise identical fragments produced by *Hin*dIII digestion of the rearranged version of pHH1 carried by the Vac mutant R1 of *H. halobium*. The element is 996 bp long and has 29-bp imperfect terminal repeats and oppositely oriented open reading frames of 819 and 366 bp. The sequenced copy is flanked by 8-bp direct repeats of a sequence present once in the element-free *Hin*dIII fragment. ISH*23* is found as an insertion in a Bop mutant (44). Its terminal repeats are identical to those of ISH*50*, and its restriction maps are highly similar, but not identical, to those of ISH*50*. The *bop* gene insertion is flanked by 9-bp direct repeats.

D. ISH*24*

This is the largest characterized transposable element in *H. halobium* and, like several others, it was found in a spontaneous Bop mutant (44). ISH*24* is about 3 kbp long, has 14-bp imperfect terminal repeats, and generates 7-bp target site duplications at its point of insertion, 1.4 kbp 5′ to the *bop* gene initiation site (within the *brp* gene). In wild-type *H. halobium* strain NRC817, its two copies are confined to pHH1 (38, 39).

E. "ISH*25*"

Pfeifer and Betlach put the name of this element in quotation marks since its behavior is anomalous and it may not in fact be a transposable element. ISH*25* was found in a Bop$^+$ revertant of the Bop mutant in which ISH*24* was first detected (42, 44). The restoration of phenotype resulted in this case from what looked like an additional, 588-bp insertion (ISH*25*) within the leftmost (*bop* distal) of the flanking repeats generated by ISH*24* during its insertion. However, an identical arrangement, with "ISH*25*" present in the flanking repeat of a copy of ISH*24*, already exists on pHH1, and ISH*25* lacks the usual insertion sequence features (terminal inverted repeats, generation of target site duplications). Pfeifer et al. (44) conclude that some recombinational event using the homology between the ISH*24* copy near *bop* (which is a chromosomal, FI, sequence) and the ISH*24*+"ISH*25*" copy on the plasmid was involved in phenotypic reversion. Normal levels of *bop* mRNA are restored in the revertant (42), but it is not possible to say how the second event overcame the transcriptional consequences of the first.

F. ISH*26*

ISH*26*, too, was first described as an insertion into *bop*. At least seven copies are present in *H. halobium* strain NRC817. Two size variants have been described (16), of 1,384 and 1,705 bp (the latter containing a duplication of part of the former), and restriction enzyme digestion and hybridization data show that there may also be heterogeneities in sequence between copies. The 16-bp terminal repeats of the sequenced ISH26 show strong sequence similarity to those of ISH2, but the elements are not otherwise closely related. On one strand, ISH*26* (both versions) sports two slightly overlapping open reading frames (441 and 741 bp). ISH*26* is reasonably cosmopolitan in its distribution, being found in the genomes of many *H. halobium*-type strains and the "square" halobacterial strains RS1 and GN (38).

G. ISH*51*

ISH*51* is the only well-characterized insertion sequence-like element from a halobacterium other than *H. halobium* (20). It was identified by restriction map comparisons and sequencing of a variety of clones of *H. volcanii* which all hybridized to a common set of 20 to 30 bands in Southern hybridization experiments against genomic DNA from *H. volcanii*. Two copies in inverted orientation were found flanking a 1.4-kbp stretch of unique-sequence

Table 1. Properties of halobacterial insertion sequence elements

Element	Size	Terminal inverted repeats	Target duplication (bp)	Insertion specificity	ORFs (bp)	Homology	Cloned from:	Reference
ISH*1*	1,118 bp	8/9 TGCCT-GTT	8	Specific, but either orientation in Bop mutants; partial homology with TIR[a]	810		Bop mutants	55
ISH*1.8*	1,895 bp (or 1,887 bp)	None	One site: 5; second site: none	Possible 8-bp sequence in common in both targets examined	672		Bop mutants	51
ISH2	520 bp	19 CATTCGTCTTTAGTTAAGA	10, 11, or 20	Apparently nonspecific	240, 192, 177	TIR of ISH26	Phage ΦH	8
ISH*23*	1.0 kbp	23/29 CGCTCTTG-G--G-- GATTGTTGGT-AGT	9	None apparent	?	ISH*50*	Bop mutant	44
ISH*24*	3 kbp	10/14 ----TGTTGCGAAG	7	?	?		Bop mutant	44
"ISH*25*"	588 bp	None	None	Believed to have moved by a recombination event	?		Bop revertant	41, 42, 44
ISH*26*	1,384 bp	19 CATCCGTCTTTAGTTAAGC	11	CCTC motif implied	441, 741	TIR of ISH2	Bop mutant, pHH1, chromosome	16
ISH*26-1*[b]	1,705 bp	16 CATCCGTCTTTAGTTA						
ISH*27*[c]	1.4 kbp	?	?	?	?		Bop mutant	42
ISH*28*	1.0 kbp	?	?	?	?		Bop mutant	42
ISH*50*	996 bp	23/29 CGCTCTTG-G--G-- GATTGTTGGT-AGT	8	None apparent	810; 366?	ISH*23*	50-kbp plasmid of strain R1	64
ISHS*1*	1.7 kbp	26/27 GGTCG- GTCGCAAAGCGGTCTAGAGTT	8	?	?	TIR of ISH*51*	Bop mutant	36
ISH*51*	1,371–1,379 bp	15/16 TCAGTAC-TCACAAAGC	3	None apparent	None found	TIR of ISHS*1*	*H. volcanii*, by hybridization	20

[a] TIR, Terminal inverted repeat.
[b] Variant of ISH26.
[c] Also known as ISH6*1*.

DNA. Comparisons of these with each other and with another full-length element and a truncated copy allowed several general conclusions. ISH*51* elements are 1,370 to 1,380 bp long and bear 16-bp imperfect terminal inverted repeats, which contain a region similar (8/9 match) to the terminal inverted repeats of ISHS*1* (Table 1). The three and one-half elements sequenced show on average only 85% primary sequence identity and differ from each other by many small insertions or deletions as well as base substitutions. None has an extensive open reading frame, and there can clearly be no conserved protein-coding function. The *H. volcanii* genome contains both "intact" (1.3-kbp) ISH*51* elements and truncated copies containing about half the full sequence. Hofman et al. (20) suggest that transposition of these heterogeneous and seemingly degenerate elements may be under the control of either a few "good" copies or of a different ISH family. One (fortuitous) transposition of ISH*51* has been observed in a laboratory culture; W. L. Lam and R. L. Charlebois (unpublished results) isolated a spontaneous variant of the endogenous 6-kbp *H. volcanii* plasmid pHV2 which had suffered a 1.3-kbp insertion. Hybridization analysis revealed this to be a copy of ISH*51*.

ISH*51* elements are found in both FI and FII DNA of *H. volcanii*. Sequences homologous to it can be identified by Southern hybridization in many halobacterial species other than *H. volcanii*; *H. halobium*, *H. cutirubrum*, and *Halobacterium marismortui* all produce signals, and some hybridization was detectable with *Halobacterium saccharovorum* and *Halococcus morrhuae*.

H. ISH*1.8*

The structure of sequenced ISH*1.8* elements is not typically insertion sequence-like (51). Although its 1,895 bp does contain a substantial (672 bp) open reading frame, ISH*1.8* has no terminal inverted repeats, and only one of the two copies on ΦH2 shows direct flanking target site duplications. The observations that there are two copies of ISH*1.8* in the FII DNA of the *H. halobium* genome, that insertions of ISH*1.8* into the ΦH genome do occur at high frequency, and that elements undergo near-precise and imprecise excision, seem in sum enough to justify its consideration as a transposable element.

V. INSERTION SEQUENCES AND GENOME ORGANIZATION

A restriction on the location of repetitive sequences in *H. halobium* was apparent in the initial random cloning experiments of Sapienza and Doolittle (47). Most small *Eco*RI fragments of *H. halobium* DNA contained repeated sequences, while most *Pst*I fragments did not. Since the recognition sequence of *Eco*RI is richer in A+T than is that of *Pst*I, these enzymes (when used for the production of small fragments) sample different compartments of the genome. Pfeifer (38, 39) has summarized the results of probing fractionated FI (GC-rich) and FII (relatively AT-rich) DNAs from *H. halobium* strain NRC817 with cloned repetitive elements. These experiments also show a nonrandom distribution of elements. One of two copies of ISH*1*, all but one of 8 to 10 copies of ISH2, one of two copies of ISH*23* (ISH*50*), both of two copies of ISH*24*, both of two copies of ISH*1.8*, and all of several copies of ISH*26* and of the preliminarily characterized elements ISH*27* and ISH*28* are found in FII DNA. All cloned genes for *H. halobium* cellular proteins or RNAs for which comparable analyses have been made, with the exception of the gas-vesicle protein (7), are found in FI DNA.

The 150-kbp plasmid pHH1 is a major constituent of *H. halobium* FII DNA. It has been mapped with restriction endonucleases (61) and the known insertion sequence elements it contains have been roughly localized (38, 39). There are four copies of ISH2, one of ISH*23* (ISH*50*), two of ISH*24*, two of ISH26, two of ISH*27*, and one of ISH*28*. These are only the characterized elements; the results of Sapienza and Doolittle (47) indicate that there may well be many more.

Not all of FII DNA is plasmid (covalently closed circles); some reasonable fraction is made up of relatively AT-rich DNA which is linear in the usual DNA preparations. This presumably represents stretches of AT richness embedded in the background of GC-rich chromosomal sequences. Recently, Pfeifer and Betlach (40) have provided graphic demonstration of this kind of organization. They used overlapping cosmid clones to "walk" in either direction from the FII DNA copy of ISH*1* and in this way mapped a 160-kbp region of *H. halobium* chromosomal DNA. The AT-rich (FII) portion of this region, identified by hybridization of clones to fractionated FI and FII DNA, is in its center and is 70 kbp long. Hybridization with specific probes revealed that there are embedded within the 70-kbp FII portion copies of ISH*1*, ISH2, ISH*26*, and at least 10 other (as yet unnamed) repetitive sequences. Pfeifer and Betlach (40) called the 70-kbp, AT-rich segment an island and calculated that there could be 10 more islands of this size embedded within the *H. halobium* chromosome. The 70-kbp island is present, by hybridization, in several strains related to *H. halobium*, but not in more distant purple-membrane-producing isolates, which also lack pHH1 sequences. The FI DNA which flanks the 70-kbp island is, however, found in many halobacterial species. The island, therefore, might represent a recent bloc insertion into the *H. halobium* genome (40).

From all of this, a consistent picture emerges. *H. halobium* ISH elements (which are in general more AT-rich than are genes for cellular functions) preferentially reside in AT-rich portions of the genome. This probably reflects more than just selection against cells in which elements have disrupted genes in GC-rich FI DNA. ISH elements appear as well to prefer to insert into AT-rich targets or into regions of the genome in which insertions have previously occurred. Many such regions are found on pHH1 and presumably on other plasmids in some other strains. Other such regions are found as islands scattered within the chromosome, at least within that of *H. halobium*. Although the majority of ISH transposition events might thus be within or between islands, there is some leakage. ISH elements do occasionally invade GC-rich genomic regions where there are "real" genes. There are so many elements, and they are so active, that spontaneous mutation rates in *H. halobium* are as a result very high, but probably not as high as they would be if movement were not largely confined to islands (48).

VI. GENERAL EVOLUTIONARY CONSIDERATIONS

When the large number of repetitive sequences in *H. halobium*, of the order of 500 or more, was first described, Sapienza and Doolittle (47) suggested that this might be a general feature of archaebacteria, one they share with eucaryotes in general, but not in general with eubacteria. Although insertion sequences have been described in methanogenic archaebacteria (19), large numbers of such elements have not been seen outside the extreme halophiles. Even within the halobacteria, *H. halobium* may be unusually laden with transposable elements. Spontaneous mutations are reportedly less frequent, for instance, in several purple-membrane-producing strains of *Halobacterium* spp. which lack sequences homologous to pHH1 and lack detectable copies of the more common *H. halobium* ISH elements (14, 38).

An appreciation of the prevalence of transposable elements in the genome of *H. halobium* seems often to lead to the assumption that they have some function there. In 1982, Weber and Leighton (59) made explicit a suggestion which underlay many early discussions of genetic instability in halobacteria, that "genomic plasticity is of adaptive value for halobacterial populations by guaranteeing the simultaneous presence of subpopulations," and went on to argue that "some of these may respond better to various environmental stimuli (e.g., seasonal changes of water salinity) and recycle the available biomass thereby ensuring the survival of the species in a highly restrictive ecological niche." This is in a sense a special case of the general notion that transposable elements are good for organisms because they are good for species. Such arguments are subject to strong theoretical objections, and in any event become unnecessary when one recognizes that transposable elements in organisms which can at least sometimes exchange genetic material can be viewed as genetic parasites (11, 35). If one assumes that halobacteria at least sometimes exchange DNA in nature (as the results of Mevarech and Werczberger [32] suggest they might), then it may simply be the case that *H. halobium*, through accidents of history, is unusually afflicted with such parasites.

Acknowledgment. We are grateful to Felicitas Pfeifer for a copy of her related review on halobacterial molecular genetics (39).

LITERATURE CITED

1. **Betlach, M., J. Friedman, H. W. Boyer, and F. Pfeifer.** 1984. Characterization of a halobacterial gene affecting bacterio-opsin gene expression. *Nucleic Acids Res.* **12:**7949–7959.
2. **Betlach, M. C., D. Leong, and H. W. Boyer.** 1986. Bacterio-opsin gene expression in *Halobacterium halobium*. *Syst. Appl. Microbiol.* 7:83–89.
3. **Betlach, M., F. Pfeifer, J. Friedman, and H. W. Boyer.** 1983. Bacterio-opsin mutants of *Halobacterium halobium*. *Proc. Natl. Acad. Sci. USA* **80:**1416–1420.
4. **Blanck, A., and D. Oesterhelt.** 1987. The halo-opsin gene. II. Sequence, primary structure of halorhodopsin and comparison with bacteriorhodopsin. *EMBO J.* **6:**265–273.
5. **Charlebois, R. L., W. L. Lam, S. W. Cline, and W. F. Doolittle.** 1987. Characterization of pHV2 from *Halobacterium volcanii* and its use in demonstrating transformation of an archaebacterium. *Proc. Natl. Acad. Sci. USA* **84:**8530–8534.
6. **Cline, S. W., and W. F. Doolittle.** 1987. Efficient transfection of the archaebacterium *Halobacterium halobium*. *J. Bacteriol.* **169:**1341–1344.
7. **DasSarma, S., T. Damerval, J. G. Jones, and N. Tandeau de Marsac.** 1987. A plasmid-encoded gas vesicle protein gene in a halophilic archaebacterium. *Mol. Microbiol.* **1:**365–370.
8. **DasSarma, S., U. L. RajBhandary, and H. G. Khorana.** 1983. High-frequency spontaneous mutation in the bacterio-opsin gene in *Halobacterium halobium* is mediated by transposable elements. *Proc. Natl. Acad. Sci. USA* **80:**2201–2205.
9. **DasSarma, S., U. L. RajBhandary, and H. G. Khorana.** 1984. Bacterio-opsin mRNA in wild-type and bacterio-opsin-deficient *Halobacterium halobium* strains. *Proc. Natl. Acad. Sci. USA* **81:**125–129.

10. **Dennis, P. P.** 1986. Molecular biology of archaebacteria. *J. Bacteriol.* **168:**471–478.
11. **Doolittle, W. F., and C. Sapienza.** 1980. Selfish genes, the phenotype paradigm and genome evolution. *Nature* (London) **284:**601–603.
12. **Dunn, R., J. McCoy, M. Simsek, A. Majumdar, S. H. Chang, U. L. RajBhandary, and H. G. Khorana.** 1981. The bacteriorhodopsin gene. *Proc. Natl. Acad. Sci. USA* **78:**6744–6748.
13. **Ebert, K., and W. Goebel.** 1985. Conserved and variable regions in the chromosomal and extrachromosomal DNA of halobacteria. *Mol. Gen. Genet.* **200:**96–102.
14. **Ebert, K., W. Goebel, A. Moritz, U. Rdest, and B. Surek.** 1986. Genome and gene structures in halobacteria. *Syst. Appl. Microbiol.* **7:**30–35.
15. **Ebert, K., W. Goebel, and F. Pfeifer.** 1984. Homologies between heterogeneous extrachromosomal DNA populations of *Halobacterium halobium* and four new halobacterial isolates. *Mol. Gen. Genet.* **194:**91–97.
16. **Ebert, K., C. Hanke, H. Delius, W. Goebel, and F. Pfeifer.** 1987. A new insertion element, ISH26, from *Halobacterium halobium*. *Mol. Gen. Genet.* **206:**81–87.
17. **Fox, G. E., K. R. Pechman, and C. R. Woese.** 1977. Comparative cataloging of 16S ribosomal ribonucleic acid: molecular approach to procaryotic systematics. *Int. J. Syst. Bacteriol.* **27:**44–57.
18. **Fox, G. E., E. Stackebrandt, R. B. Hespell, J. Gibson, J. Maniloff, T. A. Dyer, R. S. Wolfe, W. E. Balch, R. S. Tanner, L. J. Magrum, L. B. Zablen, R. Blakemore, R. Gupta, L. Bonen, B. J. Lewis, D. A. Stahl, K. R. Luehrsen, K. N. Chen, and C. R. Woese.** 1980. The phylogeny of prokaryotes. *Science* **209:**457–463.
19. **Hamilton, P. T., and J. N. Reeve.** 1985. Structure of genes and an insertion element in the methane producing archaebacterium *Methanobrevibacter smithii*. *Mol. Gen. Genet.* **200:** 47–59.
20. **Hofman, J. D., L. C. Schalkwyk, and W. F. Doolittle.** 1986. ISH51: a large, degenerate family of insertion sequence-like elements in the genome of the archaebacterium, *Halobacterium volcanii*. *Nucleic Acids Res.* **14:**6983–7000.
21. **Jones, W. J., D. P. Nagle, Jr., and W. B. Whitman.** 1987. Methanogens and the diversity of archaebacteria. *Microbiol. Rev.* **51:**135–177.
22. **Joshi, J. G., W. R. Guild, and P. Handler.** 1963. The presence of two species of DNA in some halobacteria. *J. Mol. Biol.* **6:**34–38.
23. **Kandler, O. (ed.).** 1982. *Archaebacteria*. Gustav Fischer Verlag, Stuttgart, Federal Republic of Germany.
24. **Kandler, O., and W. Zillig (ed.).** 1986. *Archaebacteria '85*. Gustav Fischer Verlag, Stuttgart, Federal Republic of Germany.
25. **Khorana, H. G., G. E. Gerber, W. C. Herlihy, C. P. Gray, R. J. Anderegg, K. Nihei, and K. Biemann.** 1979. Amino acid sequence of bacteriorhodopsin. *Proc. Natl. Acad. Sci. USA* **76:**5046–5050.
26. **Kohara, Y., K. Akiyama, and K. Isono.** 1987. The physical map of the whole *E. coli* chromosome: application of a new strategy for rapid analysis and sorting of a large genomic library. *Cell* **50:**495–508.
27. **Kushner, D. J.** 1985. The Halobacteriaceae, p. 171–214. *In* C. R. Woese and R. S. Wolfe (ed.), *The Bacteria*, vol. VIII. *The Archaebacteria*. Academic Press, Inc., Orlando, Fla.
28. **Lake, J. A.** 1988. Origin of the eukaryotic nucleus determined by rate-invariant analysis of rRNA sequences. *Nature* (London) **331:**184–186.
29. **Lake, J. A., M. W. Clark, E. Henderson, S. P. Fay, M. Oakes, A. Scheinman, J. P. Thornber, and R. A. Mah.** 1985. Eubacteria, halobacteria, and the origin of photosynthesis: the photocytes. *Proc. Natl. Acad. Sci. USA* **82:**3716–3720.
30. **Larsen, H.** 1984. Family V. Halobacteriaceae Gibbons 1974, 269, p. 261–267. *In* N. R. Krieg and J. R. Holt (ed.), *Bergey's Manual of Systematic Bacteriology*, 9th ed., vol. 1. Williams and Wilkins, Baltimore.
31. **Margulis, L.** 1970. *Origin of Eukaryotic Cells*. Yale University Press, New Haven, Conn.
32. **Mevarech, M., and R. Werczberger.** 1985. Genetic transfer in *Halobacterium volcanii*. *J. Bacteriol.* **162:**461–462.
33. **Moore, R. L., and B. J. McCarthy.** 1969. Characterization of the deoxyribonucleic acid of various strains of halophilic bacteria. *J. Bacteriol.* **99:**248–254.
34. **Moore, R. L., and B. J. McCarthy.** 1969. Base sequence homology and renaturation studies of the deoxyribonucleic acid of extremely halophilic bacteria. *J. Bacteriol.* **99:** 255–262.
35. **Orgel, L. E., and F. H. C. Crick.** 1980. Selfish DNA: the ultimate parasite. *Nature* (London) **284:**604–607.
36. **Ovchinnikov, Y. A., S. A. Zozulya, E. M. Zaitseva, S. O. Guriev, E. D. Sverdlov, M. A. Krupenko, and A. A. Aleksandrov.** 1984. The new IS element of *Halobacterium halobium* localized within the bacterioopsin gene. *Bioorgan. Chem.* **10:**560–563.
37. **Pace, N. R., G. J. Olsen, and C. R. Woese.** 1986. Ribosomal RNA phylogeny and the primary lines of evolutionary descent. *Cell* **45:**325–326.
38. **Pfeifer, F.** 1986. Insertion elements and genome organization of *Halobacterium halobium*. *Syst. Appl. Microbiol.* **7:**36–40.
39. **Pfeifer, F.** 1988. Genetics of halobacteria, p. 105–133. *In* F. Rodriguez-Valera (ed.), *Halophilic Bacteria*, vol. 2. CRC Press, Boca Raton, Fla.
40. **Pfeifer, F., and M. Betlach.** 1985. Genome organization in *Halobacterium halobium*: a 70kb island of more (AT) rich DNA in the chromosome. *Mol. Gen. Genet.* **198:**449–455.
41. **Pfeifer, F., M. Betlach, R. Martienssen, J. Friedman, and H. W. Boyer.** 1983. Transposable elements of *Halobacterium halobium*. *Mol. Gen. Genet.* **191:**182–188.
42. **Pfeifer, F., H. W. Boyer, and M. C. Betlach.** 1985. Restoration of bacterioopsin gene expression in a revertant of *Halobacterium halobium*. *J. Bacteriol.* **164:**414–420.
43. **Pfeifer, F., K. Ebert, G. Weidinger, and W. Goebel.** 1982. Structure and functions of chromosomal and extrachromosomal DNA in halobacteria. *Zentralbl. Bakteriol. Hyg. Abt. 1 Orig. Reihe C* **3:**110–119.
44. **Pfeifer, F., J. Friedman, H. W. Boyer, and M. Betlach.** 1984. Characterization of insertions affecting the expression of the bacterio-opsin gene in *Halobacterium halobium*. *Nucleic Acids Res.* **12:**2489–2497.
45. **Pfeifer, F., G. Weidinger, and W. Goebel.** 1981. Characterization of plasmids in halobacteria. *J. Bacteriol.* **145:**369–374.
46. **Pfeifer, F., G. Weidinger, and W. Goebel.** 1981. Genetic variability in *Halobacterium halobium*. *J. Bacteriol.* **145:** 375–381.
47. **Sapienza, C., and W. F. Doolittle.** 1982. Unusual physical organization of the *Halobacterium* genome. *Nature* (London) **295:**384–389.
48. **Sapienza, C., M. R. Rose, and W. F. Doolittle.** 1982. High-frequency genomic rearrangements involving archaebacterial repeat sequence elements. *Nature* (London) **299:**182–185.
49. **Schnabel, H.** 1984. An immune strain of *Halobacterium halobium* carries the invertible L segment of phage ΦH as a plasmid. *Proc. Natl. Acad. Sci. USA* **81:**1017–1020.
50. **Schnabel, H.** 1984. Integration of plasmid pΦHL into phage

genomes during infection of *Halobacterium halobium* R_1-L with phage ΦHL1. *Mol. Gen. Genet.* **197**:19–23.

51. **Schnabel, H., P. Palm, K. Dick, and B. Grampp.** 1984. Sequence analysis of the insertion element ISH1.8 and of associated structural changes in the genome of phage ΦH of the archaebacterium *Halobacterium halobium. EMBO J.* **3**: 1717–1722.
52. **Schnabel, H., E. Schramm, R. Schnabel, and W. Zillig.** 1982. Structural variability in the genome of phage ΦH of *Halobacterium halobium. Mol. Gen. Genet.* **188**:370–377.
53. **Schnabel, H., W. Zillig, M. Pfäffle, R. Schnabel, H. Michel, and H. Delius.** 1982. *Halobacterium halobium* phage ΦH. *EMBO J.* **1**:87–92.
54. **Simon, R. D.** 1978. *Halobacterium* strain 5 contains a plasmid which is correlated with the presence of gas vacuoles. *Nature* (London) **273**:314–317.
55. **Simsek, M., S. DasSarma, U. L. RajBhandary, and H. G. Khorana.** 1982. A transposable element from *Halobacterium halobium* which inactivates the bacteriorhodopsin gene. *Proc. Natl. Acad. Sci. USA* **79**:7268–7272.
56. **Sogin, M. L., H. J. Elwood, and J. H. Gunderson.** 1986. Evolutionary diversity of eukaryotic small-subunit rRNA genes. *Proc. Natl. Acad. Sci. USA* **83**:1383–1387.
57. **Stoeckenius, W., and R. A. Bogomolni.** 1982. Bacteriorhodopsin and related pigments of halobacteria. *Annu. Rev. Biochem.* **51**:587–616.
58. **Tindall, B. J., H. N. M. Ross, and W. D. Grant.** 1984. *Natronobacterium* gen. nov. and *Natronococcus* gen. nov., two new genera of haloalkaliphilic archaebacteria. *Syst. Appl. Microbiol.* **5**:41–57.
59. **Weber, H. J., and T. J. Leighton.** 1982. Genetic instability of *Halobacterium halobium*, p. 350. *In* O. Kandler (ed.), *Archaebacteria.* Gustav Fischer Verlag, Stuttgart, Federal Republic of Germany.
60. **Weidinger, G., G. Klotz, and W. Goebel.** 1979. A large plasmid from *Halobacterium halobium* carrying genetic information for gas vacuole formation. *Plasmid* **2**:377–386.
61. **Weidinger, G., F. Pfeifer, and W. Goebel.** 1982. Plasmids in halobacteria: restriction maps. *Methods Enzymol.* **88**: 374–379.
62. **Woese, C. R.** 1981. Archaebacteria. *Sci. Am.* **244**(6):98–122.
63. **Woese, C. R.** 1987. Bacterial evolution. *Microbiol. Rev.* **51**:221–271.
64. **Xu, W.-L., and W. F. Doolittle.** 1983. Structure of the archaebacterial transposable element ISH50. *Nucleic Acids Res.* **11**:4195–4199.

Chapter 12

Agrobacterium-Plant Cell DNA Transfer

PATRICIA ZAMBRYSKI

Patricia Zambryski ■ Division of Molecular Plant Biology, University of California, Berkeley, California 94720.

Agrobacterium tumefaciens genetically transforms plant cells. In nature this transformation results in crown gall tumors, an agronomically important disease which affects most dicotyledonous plants. This disease forms swellings on fruit trees and vines which should be familiar to anyone who has taken a casual stroll in any field or forest. Initial interest in the disease was aimed at understanding plant neoplasia, with the added hope that some of the fundamental mechanisms involved would provide insight into animal tumor biology. Within the last 10 years, the disease has been shown to be due to the transfer of a particular DNA segment, the T-DNA (transferred DNA), from a large tumor-inducing (Ti) plasmid within the bacterium to the plant cell where the T-DNA element becomes integrated into the plant nuclear genome; its subsequent expression results in the crown gall phenotype (reviewed recently in references 39, 61, and 69). This DNA transfer property promoted *A. tumefaciens* from "plant pathogen" to potential "vector" for introducing genes of interest into plant cells (34, 42, 86, 132). Although *A. tumefaciens* had been studied for nearly 100 years, it was the advent of recombinant DNA technology which led to the analysis of the biological principles underlying the *Agrobacterium*-plant cell interaction and to the modification of this system to genetically engineer plant cells. These most recent findings will be summarized here.

While the study of the interaction between *A. tumefaciens* and the plant cell presents a number of biologically interesting topics, only those relating to the DNA transfer process will be discussed. With regard to the more physiological aspects, such as how the expression of the T-DNA results in the perturbation of plant cell growth and differentiation, several recent papers conclusively show that the T-DNA specifies novel enzymatic pathways resulting in the overproduction of plant growth hormones (1, 50, 90, 108; reviewed in references 51 and 74).

The *Agrobacterium*-plant cell interaction is a unique and highly regulated process constituting the only known example of interkingdom DNA transfer. However, it is becoming more and more evident that the underlying processes are not original, but instead represent novel adaptations of two general procaryotic processes, activation of gene expression in response to external stimuli and conjugative transfer of DNA from a donor to a recipient cell (104).

I. GENERAL CONCEPTS

There are three (known) genetic components carried by *A. tumefaciens* required for plant cell transformation. Two of these components are located on the large, roughly 200-kilobase (kb) Ti plasmid (Fig. 1). The first component, the T-DNA, is the DNA segment which is transferred from *A. tumefaciens* to the plant cell (15). In contrast to transposable elements which can move repeatedly, the T-DNA once transferred is stable; this reflects the fact that T-DNA, unlike transposons, does not itself encode the products which mediate its transfer (58, 66, 134). Only the ends of the T-DNA element, the so-called T-DNA borders, are recognized during the transfer process (e.g., 80, 118). The *trans*-acting functions which mediate T-DNA transit are largely encoded by the Ti plasmid virulence (*vir*) region. This region was first defined genetically as a 35-kb segment, separate from the T-DNA segment, required for crown gall tumor formation (25, 36, 43, 48, 77). Recently, it has been established that the *vir* region is the master switch for the transformation process, since the expression of most of the *vir* loci is regulated to occur only when *A. tumefaciens* is in the presence of susceptible plant cells (98–100). The *vir* region has been extensively mutagenized and shown to consist of six complementation groups, *virA*, *virB*, *virC*, *virD*, *virE*, and *virG*; the genetic and transcriptional organization of the *vir* region is shown in Fig. 1. These loci are either absolutely essential for (*virA*, *virB*, *virD*, *virG*) or enhance the efficiency of (*virC*, *virE*) plant cell transformation (99).

Besides the T-DNA and the *vir* region, there are three chromosomal virulence loci essential to the T-DNA transfer process, *chvA* and *chvB* (28, 29) and *pscA* (106). The products of these loci dramatically affect the surface composition of bacterial cells, and they are essential for the binding of *A. tumefaciens* to plant cells during the infection process. The *chvA* and *chvB* loci are linked on a 15.5-kilobase-pair (kbp) segment of the *A. tumefaciens* chromosome. The 8.5-kbp *chvB* portion encodes a membrane protein of roughly 235 kilodaltons (kDa), which acts as intermediate in the synthesis of cyclic β-1,2-glucan (84; A. Zorreguieta, R. Geremia, S. Cavaignac, G. A. Cangelosi, E. W. Nester, and R. A. Ugalde, manuscript in preparation). The function of the *chvA* locus is unknown. The *pscA* locus is approximately 3.0 kbp in size, and its expression is required for the synthesis of major neutral and acidic polysaccharides (106). In contrast to *vir* gene expression, which is tightly controlled, chromosomal virulence gene expression is constitutive. In addition, there are several chromosomal genes encoding virulence-related proteins (VRPs) (32) whose expression is under *vir* control (see Section II.D below); most of these genes have not been genetically defined, and their functions are unknown.

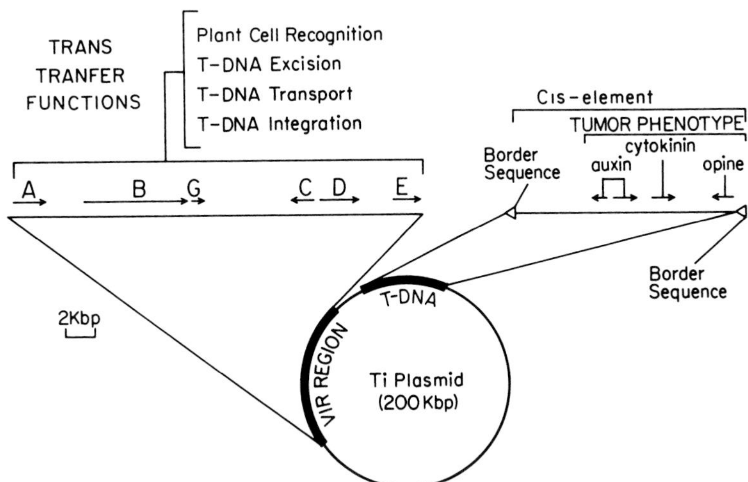

Figure 1. *A. tumefaciens* Ti plasmid and the genetic components of virulence. The transcriptional and functional organization of the virulence (*vir*) region is taken from Stachel and Nester (99). The T-DNA region is that of the nopaline Ti plasmid, which contains a single T-DNA segment (see Fig. 5 for the octopine T-DNA region). The polar (see text) 25-bp T-DNA border repeats are indicated by the open arrows at the end of the element.

Besides bacterial components, there are also plant functions which are essential to the T-DNA transfer process; i.e., the plant cell is not simply a passive recipient of *A. tumefaciens* infection. While it is clear that some basic plant cell machinery such as enzymes for recombination and repair of DNA will be involved in the integration of the T-DNA segment into the plant cell genome, there are additional functions for the synthesis of plant secondary metabolites described below that are absolutely required and have only very recently been uncovered. Since the discovery of crown gall, it had been shown that *A. tumefaciens* will only cause infection on wounded plants; this fact was readily explained by assuming that wounded plant cells present less of a physical barrier to penetration and infection by *A. tumefaciens* than do unwounded cells, which have thick cell walls. However, it has now been shown that wounded cells excrete low-molecular-weight phenolic compounds which are recognized specifically by *A. tumefaciens* as signal molecules that induce *vir* gene expression and thereby activate T-DNA transfer (98, 100).

II. BACTERIAL-PLANT CELL RECOGNITION

In nature *A. tumefaciens* primarily transforms plants at the soil-air junction, the so-called crown of the plant, and hence *A. tumefaciens* is designated as inducing "crown gall" tumors. How does *A. tumefaciens* recognize plant cells which it can specifically infect in the complex biological and chemical environment of the soil? Several years ago, S. Stachel reasoned that the bacterial genes that mediate the T-DNA transfer process might be expressed only when *A. tumefaciens* was in the presence of susceptible plant cells. To test this hypothesis, *Escherichia coli* β-galactosidase (β-Gal) was used as a reporter to monitor *A. tumefaciens*-specific gene expression (97). The coding sequences for β-Gal were incorporated into a transposon and then inserted randomly throughout the Ti plasmid: in T-DNA, *vir*, and other control regions of the *Agrobacterium* Ti plasmid not required for plant cell transformation (97, 99). The fusions showed that most genes were not expressed (as monitored by β-Gal) during vegetative growth of *A. tumefaciens*. However, when *A. tumefaciens* was grown in the presence of plant cells, the β-Gal fusions in the *vir* region specifically were induced to high levels of expression (97, 99, 100). RNA studies confirmed that the induction was at the level of transcription (21, 53, 103). These data demonstrated for the first time that plant cells can induce the expression of (*vir*) genes known to be essential for the plant cell transformation process.

A. Purification and Identification of Plant-Specific Inducers of *vir* Gene Expression

The *vir*::β-Gal fusions were subsequently used as a bioassay to purify and identify the plant cell factor(s) responsible for the activation of *vir* gene expression. The molecules isolated from wounded tobacco cells were shown to be low-molecular-weight phenolic compounds, acetosyringone (AS) and hydroxy-acetosyringone (OH-AS) (98). These mole-

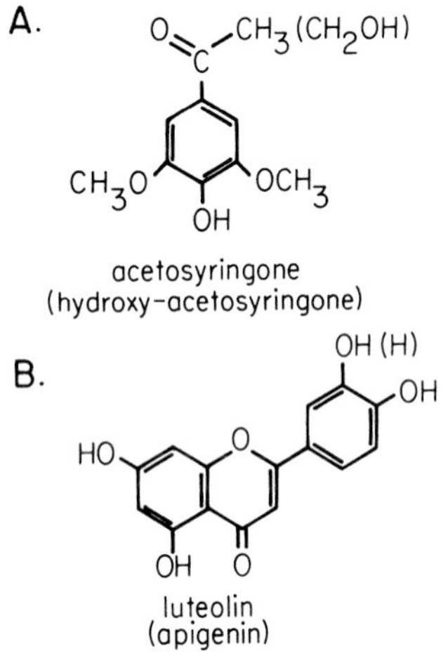

A.

acetosyringone
(hydroxy-acetosyringone)

B.

luteolin
(apigenin)

Figure 2. Plant signal molecules which act as inducers of bacterial gene expression. (A) Phenolic compounds excreted by wounded tobacco cells which act as inducers of *Agrobacterium vir* gene expression (98). (B) Flavonoid molecules which act as inducers of *Rhizobium* nodulation genes (33, 81, 85).

cules resemble products of phenylpropanoid metabolism, one of the pathways for production of the major types of plant secondary metabolites, lignins and flavonoids. These classes of compounds are important to plants during stress or injury. For example, lignin is a major component of the plant cell wall and provides a physical barrier to invading pathogens, while the flavonoids include a range of compounds which produce color, aroma, and taste, as well as specific antimicrobial molecules such as the phytoalexins (19). Figure 2A presents the structures of the two *vir*-inducing plant phenolics AS and OH-AS.

AS and OH-AS are not themselves precursors or end products of the pathway to lignins and flavonoids; since they are single phenolic rings rather than heterocyclic molecules, they are more likely related to lignin-type than to flavonoid-type plant metabolites. Both AS and OH-AS are produced specifically upon wounding of plant tissues (98). This suggests that they might be secondary products released during lignin repair of damaged plant cell walls. Presumably there is a continual excretion of wound-related phenolics during plant growth, caused by abrasion of plant tissues in the soil. *A. tumefaciens* may be attracted to the neighborhood of plants following recognition of these compounds. However, significant levels of *vir* induction with the concomitant events of T-DNA transfer will only occur if the plant wound-signal molecules are present at sufficient levels. As the highest concentrations will be at wound sites, *A. tumefaciens* will effectively infect these sites. This model is supported by recent data which show that AS can act as a chemical attractant to *A. tumefaciens* grown in vitro on AS-containing media (6).

It is interesting to note that the results with *A. tumefaciens* stimulated researchers studying other bacterial-plant cell systems to search for plant compounds which might act as recognition signals to initiate their specific interaction. Thus, it was shown that the expression of a set of bacterial genes required for the formation of nitrogen-fixing nodules by *Rhizobium* sp. are induced by particular flavonoid metabolites of their specific plant hosts (33, 81, 85); two of these *Rhizobium* sp.-inducing molecules are shown in Fig. 2B. No doubt other plant metabolites will be identified as regulators of specific bacterial genes essential for their interaction with plant cells. We have only begun to uncover the ongoing communication in the soil microenvironment.

B. Activation of *vir* Gene Expression in Response to Plant Phenolics

The induction of a particular set of genes within *A. tumefaciens* by plant-wound-specific phenolics implies the existence of a bacterial system to detect the plant signal and to transmit this information from outside to inside the bacterial cell. This recognition system can be direct, resulting in the uptake of the signal molecule through a transport function at the cell membrane, or indirect, activating a secondary messenger at the bacterial surface. Besides events on the outside of the bacterial cell, the signal (either directly or indirectly) must result in the transcriptional activation of the separate *vir* loci. For example, the signal may modify transcriptional activator proteins or the RNA polymerase to interact specifically with *vir* promoter sequences and control their transcription. Thus, the bacterial response to plant factors involves a minimum of two distinct steps, external recognition and signal transmission, followed by internal changes in gene expression.

The *vir*::β-Gal fusions mentioned above provide an assay to identify the genes involved in activation of *vir* gene expression by the plant signal molecule. These fusions can be cloned into small plasmids capable of autonomous replication in *A. tumefaciens*, allowing the quantification of the constitutive and AS-induced β-Gal activities of the fusions in different genetic backgrounds. Both the *virA* and the *virG* loci were found to be required for AS induction of the other *vir* loci (103). Moreover, *virA* and *virG* are the only *vir* loci which are expressed constitutively (at a low but detectable level) in vegetatively growing *A. tumefaciens* (99, 103); this fact is consistent with their role as *vir* regulators, i.e., they must be present to be able to respond to plant factors.

C. How Do *virA* and *virG* Activate the Expression of the Inducible *vir* Loci?

DNA sequencing reveals that *virA* encodes a polypeptide of 829 amino acids (68) and *virG* encodes a polypeptide of 267 amino acids (73, 82, 125). Each polypeptide shows significant amino acid homology to other pairs of bacterial proteins which act as sensors and regulators of gene expression in response to environmental signals (76, 87). The information relating to how these other bacterial systems function should provide insight into how *virA* and *virG* act in *A. tumefaciens*. For example, the *envZ/ompR*, *ntrB/ntrC*, and *phoR/phoB* pairs of genes allow *Escherichia coli* to respond to changes in osmolarity and nitrogen or phosphate concentration, respectively. The first gene in each pair is a membrane protein which directly senses the external environment. The second gene in each pair acts as a positive activator of transcription of the gene(s) that responds to the stimulus. How the presence of the signal is transduced from the sensor to the regulator protein is best known in the *ntrB/ntrC* system; *ntrB* converts *ntrC* between active and inactive forms by phosphorylation and dephosphorylation (64, 75).

By analogy to the *E. coli* systems, the VirA protein most likely functions as a chemoreceptor which senses the presence of plant wound factors such as AS and transmits this information to the inside of the bacterium (potentially by modification of the *virG* product). Like its homologous *E. coli* counterparts, VirA contains a transmembrane domain, and cell fractionation experiments have localized VirA to the inner membrane (68). How *virG* functions to activate *vir* gene expression is unknown. The inducible *vir* loci lack −35 consensus sequences in their promoter regions, yet they possess other hexanucleotide sequences in common which might function as *cis*-regulatory sequences for *vir*-specific transcription (21). Thus, the product of *virG* might interact with these promoter-proximal sequences to enhance the binding of RNA polymerase holoenzyme, or it might interact directly with the holoenzyme to alter its affinity for these regions.

The regulation of *virA* is simple: it is constitutively expressed. The levels of induction of *virA* gene fusions as well as the *virA* transcript itself are identical whether *A. tumefaciens* is grown vegetatively or in the presence of plant cells (103). In contrast, the regulation of *virG* is complex (103). There are two distinct messages that differ at their 5′ termini. A "constitutive" message is produced under both vegetative and plant-induced conditions, and an induced message is produced only during plant induction. The latter message is 50 bases longer at its 5′ terminus, and it is present at 10-fold-higher levels than the constitutive transcript. The significance and functions of the two *virG* messages are not understood, and *virG* expression is further complicated by the fact that *virG* induction by plant cells is regulated at two distinct levels. One level is independent of the presence of a wild-type copy of *virG*; i.e., *virG* mutants (caused by transposon insertion) still give roughly 25% the level of wild-type *virG* transcription. The second level of *virG* induction depends on intact *virG* protein; thus, *virG* positively autoregulates its own expression (103).

Further mysteries must be solved before the induction of *vir* gene expression is understood. For example, *virG*::β-Gal fusions are induced to high levels, indicating that the *virG* product is abundant (99, 103), an unexpected property for a regulatory protein. Potentially *virG* is not very efficient in its role, so that large quantities are needed to produce maximal *vir* gene expression. In support of this notion, a so-called supervirulent strain of *A. tumefaciens* produces at least fivefold-higher levels of *virG* transcript, resulting in three- to fourfold-higher levels of induction of *virG*::β-Gal fusions (56).

D. Protein Products Synthesized when *vir* Gene Expression Is Activated

An ultimate understanding of the detailed reaction steps involved in the transformation of plant cells by *A. tumefaciens* requires examination of the protein products which mediate this event. To this end, there have been numerous studies to clone, sequence, and express *vir*-specific genes. The majority of these analyses have used *E. coli* as a host for the expression and detection of *vir* proteins. In addition, our own work has focused on the identification of *vir*-specific proteins in their natural cellular environment in *A. tumefaciens* following incubation in the presence of the plant signal molecule AS. Together these studies provide an estimate of the complexity of the *vir* response.

1. *vir* gene products

Table 1 summarizes what is known physically, functionally, and phenotypically about *vir* gene products. The complete nucleotide sequence is available for each of the loci for the octopine Ti plasmid and for *virG* (82) and *virE* (45) from the nopaline Ti plasmid. The predicted sizes of the proteins specified by each of the loci, the effect of mutations in each locus on plant transformation, and the assigned and probable functions of *vir* proteins are indicated. *virA* and *virG* are the only monocistronic *vir* loci, speci-

Table 1. *vir*-Specific protein products[a]

Locus	Size (kb)	General role	Protein products: Size (kDa)	Amount	Location	Function	References
virA	2.0	Absolutely essential; regulatory system	70	+	M	AS sensor (kinase?)	68, 99, 103
virG	1.0		30	+ +	C?	*vir* transcriptional regulator	73, 82, 99, 103, 125
virC	2.0	Attenuated virulence; enhance host range	26, 23	+	?	?	99, 128, 129
virE	2.0		7, 60.5	+ + +	C/M?	Single-strand-DNA-binding protein	16, 17a, 20, 32, 41, 44, 45, 99, 124
virD	4.5	Absolutely essential; DNA transfer machinery	16, 47, 21, 75	+	C?	T-DNA border endonuclease (16-, 47-kDa polypeptides); helicase? pilot protein?	54, 83, 99, 102, 115, 130
virB	9.5		16(?), 33, 80, 25, etc.	+ + +	M	Transfer structure?, channel?	32, 99, 120

[a] Abbreviations: M, membrane; C, cytoplasm; AS, acetosyringone.

fying polypeptides of 70 and 30 kDa, respectively. *virC* is 2.0 kbp in size, and its nucleotide sequence predicts two polypeptides of 26 and 23 kDa. The sequence of the 2.0-kbp *virE* locus predicts two polypeptides of 7.0 and 60.5 kDa. The sequence of *virD* predicts four polypeptides of 16, 47, 21, and 75 kDa. *virB* is the largest locus, roughly 9.5 kbp in size. The nucleotide sequence of the *virB* region predicts 11 open reading frames specifying polypeptides of 25.9, 12, 11.7, 21.6, 65.7, 23.5, 31.7, 5.9, 26.1, 72.7, and 38 kDa (120). However, these data do not agree with genetic studies, which identified polypeptides of 33, 80, and 25 kDa encoded by the N-terminal half of the locus in extracts from AS-induced *A. tumefaciens* (32). This apparent conflict remains to be resolved. The sizes of the *virB* proteins in the table are taken from the in vivo data. Only the *virA* protein, the three *virB* polypeptides, and the larger *virE* polypeptide have been observed directly in *A. tumefaciens*. All the other protein assignments are from sequencing or the expression of cloned *vir*-specific sequences in *E. coli*.

The relative amounts and intracellular locations of some *vir*-specific proteins have also been determined; these assignments either confirm or stimulate predictions for *vir* protein function. For example, the VirA protein is localized to the membrane (68), as would be predicted for a sensor of plant signal molecules. *virD*-specific polypeptides are not produced in large amounts, consistent with a role in enzymatic T-DNA processing (see Section III.C.2 below). The most abundant *vir* proteins produced in AS-induced *A. tumefaciens* are *virE* and *virB* polypeptides (32). That *virE* encodes a protein that binds single-stranded (ss) DNA (16, 17a, 20; also see Section III.C.2) which could stoichiometrically cover ss T-DNA molecules fits with its relative abundance in AS-induced cells. While the function(s) of the *virB* products is unknown, their fractionation to the cell envelope (32) suggests that they participate in T-DNA transfer events at the bacterial cell surface. That these identified *virB* proteins are made in large amounts suggests that they are structural rather than enzymatic in function, and as mentioned (Section III.D.2 below), they may form a channel for transfer of the T-DNA molecule to the plant cell.

The only *vir* locus without an assigned or probable function is *virC*. *virC* function is required for transformation of some but not all plant species (99, 128); thus, *virC* has been proposed to be involved in specifying the "host range" for *Agrobacterium* infection. However, the concept of host range is poorly defined, since variation in infectivity occurs not only among different plant genera (*Kalanchoe*, *Vitis*, *Nicotiana*), but also among different varieties of plants within the same genus (*Nicotiana tabacum*, *Nicotiana rustica*, *Nicotiana glauca*) (128). Since *virC* proteins are not abundant in AS-induced *A. tumefaciens* cells (32), they are unlikely to play a major structural role in T-DNA transfer.

2. Non-*vir* gene products

A recent and unexpected finding is that there are at least five proteins, designated virulence-related proteins (VRPs), induced by growth in the presence of AS and which are encoded by genes outside the *vir* region (32). The production of the VRPs is dependent on intact *virA* and *virG*; i.e., they are under the control of the *vir* regulatory system. The genes for two of these VRPs map to the Ti plasmid outside the *vir* region; the other VRPs are most likely chromosomally encoded since they are produced in strains that lack the Ti plasmid if *virA* and *virG* function is provided (for example, by cloning this portion of the *vir* region onto a plasmid replicon). Five VRPs have been identified in [^{35}S]methionine-labeled extracts on one-dimensional protein gels; potentially different labeling protocols combined with higher-resolution analysis of *vir*-induced proteins will reveal additional VRPs.

The existence of the VRPs indicates another level of complexity to the T-DNA transfer process. Somehow other bacterial genes are called into action following recognition of plant wound cell factors. Whether or not VRPs provide a function essential to the T-DNA transfer process is largely unknown; only one VRP, encoded by the *pinF* (plant-inducible locus F) locus, which maps immediately adjacent to the *vir* region, has been shown to be nonessential (99). However, several roles may be envisioned for VRPs. Some VRPs may directly function in T-DNA transfer by providing either a component of the regulatory system (such as an RNA polymerase sigma factor) or a component of the transfer apparatus (such as a DNA- or membrane-associated protein). VRPs may have an indirect role in T-DNA transfer per se; for example, it is likely that the general metabolism of the bacterial cell is altered during the plant cell transformation process. Basic biosynthetic or energy transfer pathways might be redirected from providing substrates for vegetative growth to providing substrates more specific for the interaction of *A. tumefaciens* with plant cells. In this case some VRPs might have a negative regulatory role, for example, to repress genes normally expressed in the free-living, not plant-associated, bacterium.

III. THE T-DNA TRANSFER PROCESS

The activation of *vir* gene expression sets in motion a complex series of events to transfer the T-DNA element from *A. tumefaciens* to the plant cell. In the T-DNA transfer process, a specific DNA segment is recognized and mobilized. This transferable T-DNA copy must be transported through the bacterial inner and outer cell membranes and wall and then through the plant cell and nuclear membranes. Finally, the T-DNA element is integrated as a linear nonpermuted fragment into the plant nuclear genome. The general usefulness of and fundamental mechanisms involved in this DNA transfer system are reviewed in the following sections.

A. General Properties of the T-DNA Transfer System

1. Definition of the T-DNA element

The T-DNA region is defined as that segment of the Ti plasmid that is homologous to sequences present in transformed plant cells. The sizes of T-DNA elements vary in different naturally occurring Ti plasmids. Ti plasmids are classified according to the type of opine (usually amino acid-keto acid or sugar derivative) they induce in crown gall tumor cells. The synthesis of opines in crown gall tumor cells is an intriguing evolutionary strategy, since the opines can be specifically catabolized by the bacteria responsible for inciting the tumor (reviewed in reference 105). Two commonly studied Ti plasmids, nopaline and octopine types, result in nopaline- and octopine-producing tumor tissues, respectively. The nopaline T-DNA is one large, continuous segment roughly 22 kbp in size, and the octopine Ti plasmid contains three adjacent T-DNAs (Fig. 5, below), a left T-DNA (TL) element of 13 kbp, a central T-DNA (TC) element of 1.5 kbp, and a right T-DNA (TR) element of 7.8 kbp. The TL element contains oncogenic functions for tumor initiation and maintenance, and the TR element contains several opine-synthetic genes. TC does not specify a phenotype in transformed plant cells.

The structural limits of different T-DNAs were defined by comparing the nucleotide sequences at the ends of the T-DNA element following integration into the plant genome versus the nucleotide sequence of the corresponding region of the Ti plasmid. These analyses revealed that in all cases the homology between sequences present in the Ti plasmid and those in transformed plant cell DNA ends within or proximal to a 25-base-pair (bp) sequence that flanks the T-DNA region of the Ti plasmids as direct (albeit imperfect) repeats (49, 65, 93, 95, 126, 131, 133). These repeats delimit all T-DNAs analyzed to date. Figure 3 compares six T-DNA terminal repeats, two from the nopaline Ti plasmid and four from the octopine Ti plasmid. Only these 25-bp direct repeats at the ends of the T-DNA are required for its mobilization to the plant cells since its transfer is unaffected by (i) deletions of the internal portion of native T-DNAs (66, 134) or (ii) placement of T-DNA border repeats, either on a separate plasmid (24, 46)

TGGCAGGATATATTGTGGTGTAAAC - nopaline left

TGACAGGATATATTGGCGGGTAAAC - nopaline right

CGGCAGGATATATTCAATTGTAAAT - octopine TL left

TGGCAGGATATATACCGTTGTAATT - octopine TL right

TGGCAGGATATATCGAGGTGTAAAA - octopine TR left

TGGCAGGATATATGCGGTTGTAATT - octopine TR right

Figure 3. Summary of T-DNA border repeat sequences from the nopaline and octopine Ti plasmids (8, 37, 95, 126, 131). See Fig. 5 and text for description of TL, TC, and TR.

or on the *Agrobacterium* chromosome (47), in *trans* to the Ti-plasmid *vir* region. These findings form the basis of the design of modified and simplified T-DNA molecules which are useful as vectors to transform plant cells with cloned DNA fragments of interest.

2. Vectors for the transfer of DNA of interest to plant cells

An *A. tumefaciens* strain useful as a vector should carry the following genetic components: (i) the entire *vir* region, (ii) T-DNA borders, (iii) a marker gene internal to the T-DNA which is expressed in transformed plant cells to allow their detection, and (iv) a sequence internal (see below) to the T-DNA borders which allows the insertion of DNA of interest. No internal T-DNA sequences encoding the traits responsible for the tumorous crown gall phenotype should be present. Two representative types of Ti-plasmid-derived vectors are briefly summarized below. Several recent reviews detail the construction and use of vectors for plant cell transformation (34, 42, 69, 86, 132).

Knowing the location of the T-DNA borders and the functions controlling plant cell dedifferentiation allowed the construction of one of the first Ti plasmid vectors; this plasmid, pGV3850, is a deletion/substitution derivative wherein most of the internal sequences of the T-DNA were deleted and replaced by sequences from the *E. coli* plasmid pBR322 (134). The pGV3850 nononcogenic Ti plasmid is an example of one type of T-DNA vector, designated *cis* vector (see also references 23 and 113). The DNA of interest that will be transferred to plant cells is cloned in a pBR322-based plasmid and then cointegrated (i.e., inserted into the pBR sequences between the T-DNA borders) in *cis* to the *vir* region of the pGV3850 Ti plasmid by homologous recombination (134).

The second type of T-DNA vector, designated *trans* vector, contains modified T-DNA regions on small, autonomously replicating plasmids which exist in *trans* to the Ti-plasmid *vir* region (e.g., 5, 10, 92, 94). *trans* T-DNA vectors should contain (i) the 25-bp T-DNA border sequences without the internal oncogenic T-DNA genes, (ii) a marker for the detection of transformed plant cells, (iii) a bacterial drug resistance marker to select for its maintenance in *A. tumefaciens*, (iv) a wide-host-range origin of replication functional in both *E. coli* (for cloning and manipulation of DNA inserts) and *A. tumefaciens* (for stable maintenance), and (v) a set of restriction enzyme sites ("polylinker") located between the T-DNA borders to facilitate insertion of DNA of interest.

In both *cis* and *trans* vectors, DNA of interest for plant genetic engineering is prepared in *E. coli* and then transferred to *A. tumefaciens* by using helper plasmids capable of mobilizing and transferring compatible plasmid DNA. The two systems can be equally efficient in transferring DNA to plant cells. Each system has advantages and disadvantages: the *cis* system is less dependent on appropriately located restriction sites but is dependent on a recombination step in *A. tumefaciens*; in contrast, the *trans* system is independent of recombination but is dependent on compatible restriction sites between vector and insert DNA. Either type of vector is relatively easy to construct, and a wide variety is now available. One cautionary note is that vectors should contain restriction fragments extending to both sides of the T-DNA borders by at least a few hundred nucleotides. Vectors carrying only the 25-bp border sequences without their normal flanking sequences have a reduced efficiency of T-DNA transfer since the natural sequence context of these borders promotes their most efficient utilization (see Section III.B.2 below).

The use of T-DNA vectors for plant cell transformation requires that the transferred DNA segment also carry a marker gene which, when expressed, allows the detection of transformed plant cells. Many of the common markers (antibiotic resistance, color or light elicitors) used in bacterial and animal systems are applicable to plants (34, 42, 69, 86). The only requirement is that the transcriptional signals to promote their expression must be obtained from sequences known to promote transcription in plant cells.

Note that the vector potential of *A. tumefaciens* can be realized because recipient plant cells regenerate from single cells in culture. This totipotency makes plant cells optimal for genetic engineering: the introduced DNA can be studied in all plant tissues. When the regenerated plants are fertile, inheritance of this DNA in progeny plants can be analyzed.

B. Activity of the T-DNA Border Sequences

1. T-DNA border sequences direct transfer in a polar fashion

Early genetic studies showed that deletion of the region overlapping the left-border 25-bp repeat had little effect on *Agrobacterium* pathogenicity and that deletion of the right-border region abolished crown gall tumor formation (58, 91). When restriction fragments containing the native right border or a clone carrying a synthetic 25-bp repeat sequence were reintroduced at the right-border deletion site, tumor-forming activity was restored (80, 118). Accompanying studies showed that if the orientation of the right-border fragment (or 25-bp sequence) was reversed with respect to its natural orientation, the efficiency of T-DNA transfer was greatly attenuated (80, 118). These results suggested that the T-DNA element might be transferred from right to left, determined by the orientation of the border repeats. Thus, when the right border is present in its natural orientation, transfer will include the T-DNA tumor-specifying genes internal to the T-DNA element; when only the left border is present, transfer would only occur away from the T-DNA element and no tumor-forming genes would be transferred to the plant cell. Recent molecular studies provided a confirmation and explanation for the polar function of the 25-bp T-DNA border repeats (see Section III.C.1 below).

2. Sequence context of the T-DNA border repeats modulates their activity during T-DNA transfer to plant cells

Figure 3 shows that all border sequences identified to date are highly related to each other, suggesting that each might be capable of directing T-DNA transfer. However, nonselective use of all T-DNA borders would lead to nonproductive T-DNA transfer events. Recent data suggest that sequences adjacent to the 25-bp border repeats influence their efficiency in promoting T-DNA transfer. In the case of the nopaline Ti plasmid, cloned DNA fragments containing chemically synthesized native right- or left-border sequences alone are equally active (at about 30% of wild type) in T-DNA transfer (117); however, fragments containing several kilobase pairs of DNA sequences bracketing the native right or left border are very (95%) or less (10 to 40%) active, respectively (117; see also reference 55). Further, a defined 24-bp sequence (5′-TAAPuTPyNCTGTPuT-NTGTTTGTTTG-3′), designated *overdrive* (79), situated to the right and within 60 bp of the right copies of the native 25-bp repeats of the TL and TR elements (see Fig. 5 below) of the octopine Ti plasmid has been shown to enhance the transfer efficiency of constructs carrying only synthetic 25-bp repeats.

Models to explain how the activity of the T-DNA borders is controlled must take into account that the sequences immediately surrounding octopine and nopaline borders are highly dissimilar (e.g., 8, 37) and no sequences with good homology to the octopine 24-bp *overdrive* sequence are near the right nopaline border (117). Recent observations offer a partial explanation; the *overdrive* sequence stimulates T-DNA transmission from the octopine Ti plasmid when placed up to 6 kbp from the borders in either orientation and either upstream or downstream of them (79, 112; W. Ream, unpublished results). A search through 16 kbp of sequences internal to the T-DNA or 1 kbp immediately external to the right or left T-DNA borders of the nopaline Ti plasmid has not revealed any sequences with significant homology to the octopine *overdrive* (117). Thus, any potential *overdrive* sequences in the nopaline Ti plasmid must be either divergent or farther from the borders than in the octopine Ti plasmid. Another possibility is that each Ti plasmid has evolved slightly modified mechanisms to accommodate and use different T-DNA border regions. For example, the *vir*-specific proteins which recognize and act on the T-DNA borders may differ in their specificities for the nucleotide sequence context of the borders, or the nopaline *vir* proteins which react with the border sequences may be inherently more active than their octopine counterparts. This latter hypothesis is supported by the finding that *overdrive* is not required for efficient T-DNA transfer from the octopine Ti plasmid when the concentration of *vir*-specific products is elevated by increasing the copy number of the *vir* region (Ream, unpublished results).

Sequences which act to enhance DNA movement have been best studied in bacterial systems where a DNA segment undergoes a site-specific recombinational inversion, such as the alteration of phage host range by inversion of the G segment of phage Mu and the production of different flagellar antigens by variation of the H segment of *Salmonella typhimurium* (see A. C. Glasgow, K. T. Hughes, and M. I. Simon, this volume, for review). The sizes of Gin (G inversion) and Hin (H inversion) recombination enhancers are 60 and 170 bp, respectively, and they function in *cis* generally in a distance- and orientation-independent manner. The enhancer site, located within 100 to 200 bp of the ends of the invertible DNA segment, specifically binds a host factor Fis (factor for inversion), resulting in up to 150-fold stimulation in recombination. Detailed studies revealed that the spacing between the enhancer and the ends of the invertible segment are critical and reflect the requirement for integral turns of the DNA helix to facilitate Fis binding.

It remains to be shown whether the octopine-specific *overdrive*, or its as yet unknown nopaline counterpart, function similarly to these better-characterized recombinational enhancers. The *overdrive* function has been assayed primarily by measuring the size of tumors formed at the wound site on whole plant tissues (79, 80, 112, 118); these studies are only reliable over approximately a 5- to 10-fold range. One study provided a quantitative and statistical evaluation of the activity of T-DNA border tester constructs to promote between 0 and 100 tumors on potato slices (117). However, there has been no systematic study in which large populations of single plant cell protoplasts are infected with different constructs. Further, tumor formation is not directly related to the initial steps of T-DNA transfer which occurs within the bacterial cell. Direct biochemical studies using purified proteins and wild-type or mutagenized T-DNA templates are required to understand how "enhancer" sequences function during T-DNA border recognition.

C. Generation of Free, Transferable T-DNA Copies in *A. tumefaciens*

1. *vir*-Induced, T-DNA-associated molecular reactions

While the studies described above conclusively demonstrate that at least one T-DNA border is essential for normal T-DNA transfer (and potentially integration), they do not address how the borders are used to transfer the T-DNA copy to the plant cell. Somehow the T-DNA must be removed or copied from the Ti plasmid, and the T-DNA borders might be expected to play a role in such excision or replication.

Several T-DNA-associated molecular reactions were found to occur soon after the induction of *vir* gene expression by AS. Total DNA from AS-induced bacteria was electrophoresed directly, i.e., in the absence of restriction digestion, to assay for the presence of potential T-DNA homologous molecules which had a higher mobility than total uncut DNA. The two-border nopaline Ti plasmid was used to detect a linear, ss copy of the T-DNA region (101). This novel molecule, designated the T-strand, is produced at about one copy per bacterium and corresponds to the "bottom strand" of the nopaline T-DNA region such that its 5′ and 3′ ends map to the right and left T-DNA border repeats, respectively.

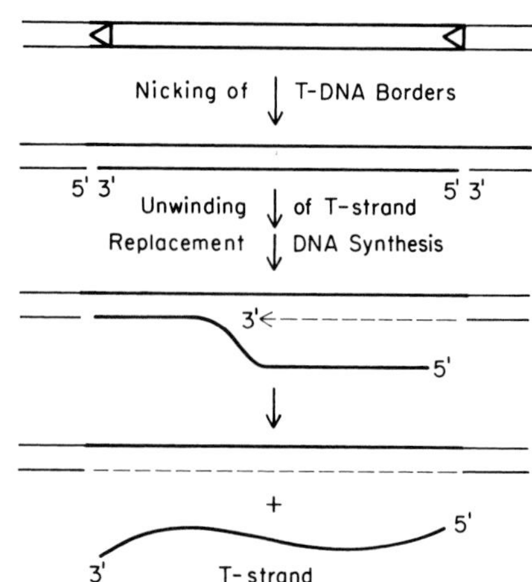

Figure 4. *vir*-induced T-DNA associated molecular reactions. The arrows indicate the 25-bp T-DNA border sequences which act as sites for recognition for the border endonuclease. These sequences are cleaved on their bottom strands. The nicked right and left T-DNA borders serve as initiation and termination sites for the generation of a single-stranded copy of the lower strand of the T-DNA region (T-strand). The 3′ end of the nicked right border serves as a priming site for replacement-strand synthesis to regenerate the bottom strand of the T-DNA region following unwinding of the T-strand from the duplex T-DNA region (101).

The T-strand is a unique and interesting molecule, and its properties both explain genetic studies for polar T-DNA transfer and suggest a possible mechanism for the entire T-DNA transfer process (see below).

Whereas the T-strand is found separate from the Ti plasmid, it seemed that there should be AS- (i.e., *vir*-) induced changes in the T-DNA region within the Ti plasmid and that these changes in Ti-plasmid DNA should be seen only following restriction digestion. Two types of structures were found (101): (i) ss endonucleolytic cleavages within the bottom of the T-DNA border repeats, designated border nicks; and (ii) Ti plasmid molecules carrying T-DNA regions with partial ss or triple-stranded character, designated T-region structures. Detailed studies located the border nicks to between the third and fourth bases of the bottom strand of the 25-bp border repeat sequence (2, 119). It was proposed that border nicks represent initiation and termination sites of T-strand synthesis, respectively, and that the T-region structures correspond to intermediates in this T-strand synthesis (101) (Fig. 4).

The formation of single T-strand molecules from Ti plasmids carrying two border sequences is easy to envisage, and potentially it will be straightforward to define precisely how each border participates in T-strand synthesis. The situation in Ti plasmids such as octopine with four borders is more complex; nevertheless, its analysis reveals some of the functional

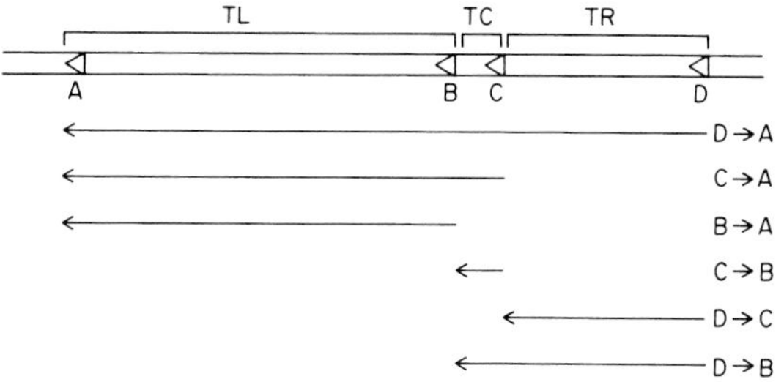

Figure 5. Generation of multiple T-strands from the four-border octopine Ti plasmid. The three adjacent T-DNAs of the octopine Ti plasmid, TL (T-DNA left), TC (T-DNA center), and TR (T-DNA right), are shown delimited by four T-DNA borders, A through D. The six possible border-to-border combinations of T-strands are shown below (102).

properties of the T-DNA borders. Figure 5 diagrams the T-strands which are produced from the octopine T-DNA regions; the four borders bracketing the three T-DNAs TL, TC, and TR lead to six distinct T-strands that correspond exactly to the molecules expected if each possible border-to-border combination is used for T-strand production (102, 115). Thus, left and right borders can be used to both initiate and terminate T-strand production, and borders can be "skipped" to produce composite molecules. There are two possible models for the production of multiple T-strands: processing of a single large progenitor or independent synthesis. Since each of the four borders is nicked independently of the others, the T-strand made in a particular induced cell is probably simply a function of which pairs of borders are cleaved. Thus, T-strand synthesis would initiate at one border nick and proceed 5′ to 3′ until it encountered a second nick, and the skipped borders of composite T-strands would not be cleaved on the parent Ti plasmid. That the copy number of T-strands is less than one per AS-induced bacterium (101, 102) supports the model in which the distribution and placement of border nicks in the Ti plasmid population determine which T-strands are produced.

The fact that T-strands (TC and TC/TL) are produced from a border which lacks an adjacent *overdrive* sequence is in agreement with findings that the stimulatory *overdrive* sequence need not be adjacent to a border site to act. *overdrive* may be an entry site, rather than a reaction site, for a protein necessary for T-strand production.

2. Protein products which function to generate transferable T-DNA copies

Several *vir*-specific products are required for border nicking and T-strand production. These include the products of *virA* and *virG*, the genes that regulate the expression of *vir* genes (102, 115). More specifically, of the other four *vir* loci, only mutations in the first half of the 4.5-kbp *virD* locus block the production of both border nicks (102, 114, 130) and T-strands (102, 115). DNA sequencing suggests that this 2.0-kbp region of the *virD* locus encodes two polypeptides, VirD1 and VirD2, of molecular weight 16.2 and 47.4 kDa, respectively (54, 130). It is assumed that VirD1 and VirD2 proteins act as site-specific endonucleases which recognize and cleave the lower strand of the 25-bp T-DNA border repeat sequences. Thus, these proteins directly affect the first step of T-strand formation; however, it is unknown whether they have any additional role in later steps for unwinding or copying of the T-strand from the Ti plasmid.

Several activities in addition to the border endonuclease, including helicase(s), polymerase(s), repair enzymes, and ss-DNA-binding protein(s) (SSB), are expected to be necessary for the production of T-strand molecules. Since no other *vir* or chromosomal virulence mutants are deficient for T-strand production alone, and since it seems unlikely that the VirD1 and VirD2 polypeptides could specify all these activities (in addition to the border endonuclease), the functions necessary to complete the generation of the T-strand molecule may be essential bacterial functions encoded by the *Agrobacterium* chromosome. These chromosomal functions are likely to be well conserved, since border nicking as well as T-strands are produced in *E. coli* cells expressing VirD1 and VirD2 polypeptides (54; G. DeVos and P. Zambryski, manuscript in preparation). It is also possible that the VirD1/VirD2 endonuclease possesses one additional activity, such as helicase. By analogy, the nicking enzyme responsible for initiating F plasmid mobilization also has a helicase activity which is essential for conjugal DNA transfer (see below).

The largest open reading frame of the *virE* locus, *virE2*, has recently been shown to specify an SSB (16, 17a, 20). Radioactivity labeled ss DNA, but not double-stranded (ds) DNA or RNA, probes were retarded on polyacrylamide gels specifically following incubation with extracts prepared from *vir*-induced cells. A variety of different ss DNAs were all effective substrates for the retardation assay, suggesting that the binding occurred independent of sequence composition. That *A. tumefaciens* has evolved an ss-DNA-binding protein specifically under *vir* control strengthens the argument that an ss DNA, presumably the T-strand, is indeed the DNA intermediate molecule (see below) that *A. tumefaciens* transfers to plant cells. Furthermore, the VirE2 polypeptide is the most abundant AS-induced protein observed (17a, 32); this relative concentration suggests it has a structural rather than an enzymatic role in the T-DNA transfer process. For example, to cover completely an ss 20-kilobase T-strand molecule would require 700 to 350 protein molecules, assuming a 30- to 60-nucleotide substrate. This number of molecules would represent about 0.1% of the total cell protein, consistent with the amount of VirE2 detected. The substrate binding site is by analogy to the size of the binding sites for the smooth-contoured or beaded structure of the *E. coli* SSB (13), and synthetic oligonucleotides at least 36 nucleotides long were required as substrates for the gel retardation assay with *vir*-induced extracts (41).

The major ss-DNA-binding proteins studied are the *E. coli* SSB and the T4 bacteriophage gene 32 protein. These proteins have five distinguishing characteristics: (i) ss DNA binding; (ii) requirement in stoichiometric rather than catalytic amounts; (iii) cooperative DNA binding; (iv) stimulation of DNA polymerase; and (v) absence of intrinsic enzymatic activity (reviewed in reference 13). The VirE2 protein fulfills the first two criteria as already discussed. In addition, VirE2 protein binds tightly and cooperatively to ss DNA (V. Citovsky, M. L. Wong, and P. Zambryski, manuscript in preparation). This cooperativity can be most readily visualized under the electron microscope, where VirE2 can be seen to coat linear ss DNAs, resulting in rodlike DNA-protein complexes. Further, under low VirE2:ss DNA ratios, few linear ss DNAs are coated to completion, while the majority of the DNA is condensed and unassociated with protein. Thus, the VirE2 protein fulfills three characteristics of a true SSB. VirE2 probably does not have a role in replication per se since it does not complement *E. coli* SSB mutants (16). Whether or not VirE2 protein has any enzymatic activities remains to be tested. The *virE* locus also has the additional capacity to encode a 7-kDa polypeptide at its 5′ end (124), and whether or not this polypeptide interacts with VirE2 to provide an additional activity remains to be determined. Clones containing only *virE2* sequences in *E. coli* express a polypeptide with the expected molecular weight for VirE2 protein, and cell extracts from this heterologous host exhibit strong binding to ss DNA; thus, it is unlikely that the 5′ *virE* product is required for ss DNA binding.

The assignment of SSB function to the *virE* locus may help to explain the unusual phenotype of *virE* mutants. First, *virE* mutants are avirulent only on some plant hosts (35, 99, 128). Perhaps in the absence of *vir*-specific SSB, the *Agrobacterium*-chromosome-specified SSB may be used to protect the T-strand from degradation in some but not all plant hosts. Second, *virE* mutants can be complemented for virulence if they are coinoculated on plants with a wild-type *A. tumefaciens* strain (78; D. Corbin and S. Stachel, unpublished results). These data suggested that the *virE* protein might function extracellularly outside the bacterial cell, potentially as part of a T-DNA/protein complex; that *virE* specifies a DNA-binding protein supports these earlier hypotheses. The VirE2 protein may have additional roles; for example, that *virE* mutants are not blocked in T-strand synthesis (102, 115) does not necessarily mean that SSB is not required for this reaction. Since most SSBs have a high affinity for ss DNA without regard to its sequence composition, chromosomal SSB could provide the necessary function in the absence of the *virE* product. Potentially, *A. tumefaciens* has evolved to contain a T-DNA-transfer-specific SSB which has an added function in the plant cell, for example, to facilitate integration of the T-DNA into the plant cell genome (17a, 35).

3. Other potential candidates for the structure of the T-DNA transfer intermediate molecule

Two other DNAs have been proposed to be intermediates in T-DNA transfer, a double-stranded (ds) T-DNA circle (3, 62, 71, 127) and a ds linear T-DNA (55, 61; Z. Koukolikova-Nicola and B. Hohn, unpublished results). Both types of DNAs were observed primarily following transfection of *E. coli* with bacteriophage lambda containing *vir*-induced DNA. This method is sensitive and was specifically designed to select only free ds T-DNA molecules. The particular Ti plasmid used for these studies contained a T-DNA region of 50 kbp and a *cos* site that promotes packaging of DNA into lambda virions. The linear ss T-DNA form is more abundant than the circular form (Koukolikova-Nicola and Hohn, unpublished data). The circular form has been suggested to result from a low frequency of recombination between border-sequence direct repeats instigated by their nicking (101). Hybridization tests indicated that the ds T-DNA forms are not (101) or are only barely (Koukolikova-Nicola and Hohn, unpublished data) detectable. However, the frequency of recovery of T-DNA molecules by the packaging method suggested that they are at least as abundant as T-strands. These conflicting data do not allow a simple resolution of whether or not ds linear T-DNA molecules may indeed be true T-DNA intermediate molecules. Further unresolved results are that *vir*-induced ds cleavages at the T-DNA borders are observed only when total DNA is digested with restriction enzymes bracketing the T-DNA borders (54, 114); without enzymatic treatment ds breaks are not observed. Conceivably, the T-DNA region from which a free T-DNA copy is being made is a multistranded and topologically complex structure which becomes altered in an unexpected fashion following subjection to lambda packaging or enzymatic treatment.

Recent studies further characterize the formation, albeit rare, of T-DNA circles in *vir*-induced *A. tumefaciens* (109). Two types of circles containing recombinant T-DNA were observed, a major class representing precise site-specific recombination between both T-DNA borders and a minor class representing recombination events either utilizing only one T-DNA border and other Ti plasmid sequences or involving Ti plasmid sequences other than the T-DNA borders. The observed T-DNA homologous circles are proposed to result from ss nicks at the T-DNA borders which unwind sufficiently to promote recombination over short (less than 20 bp) stretches of DNA with local homology to border contiguous sequences.

The ss T-strand is the most plausible T-DNA intermediate molecule. T-DNA circles or ds breaks at the T-DNA borders to produce a linear ds T-DNA would lead to a loss of the T-DNA region on the Ti plasmid. It seems unlikely that the T-DNA transfer process would have evolved to be suicidal. Furthermore, the polarity of the T-strands and the fact that border nicks occur predominantly on the bottom strand of the T-DNA region fit well with the genetic data on the polarity of the T-DNA borders. For the discussion which follows, it is assumed that the T-DNA transfer intermediate is the T-strand molecule.

4. Model for the mechanism of T-DNA transfer

Just as the activation of *vir* gene expression is based on an evolutionarily conserved mechanism for signal transduction, the T-DNA transfer process might be related to systems for DNA transfer between bacterial cells, such as virus infection or conjugation. While the T-strand may be packaged into a virus-like particle potentially coated with VirE2, the facts that T-strands are not produced in abundant quantities and that T-DNA transfer requires close physical contact between *A. tumefaciens* and the plant cell suggest that the T-DNA transfer process is not analogous to viral infection per se. In bacterial conjugation exemplified by the F factor (52, 122, 123; see below), DNA is transferred as a linear single strand from the donor to the recipient cell. This requires cell-cell contact, occurs through a specialized channel, and entails nicking within a specific origin of transfer sequence. Mobilization of the nicked strand in a 5′-to-3′ direction is accomplished by unwinding the donor strand by displacement concomitant to replacement strand synthesis on the template donor DNA strand (123).

The aspects of T-DNA transfer which may correspond to bacterial conjugation can be summarized as follows. Border nicks are analogous to the nick at the origin of conjugative DNA transfer. The T-strand is analogous to the single strand transferred to the recipient bacterium, and the AS-induced internal T-region structures are analogous to replacement-strand-synthesis intermediates in conjugation. Furthermore, bacterial conjugation requires direct contact between donor and recipient cells, and the same requirement holds for the T-DNA transfer process. Thus, it has been proposed that T-DNA is bacterial conjugation applied to plant cells (101, 104).

Strong support for this model derives from the recent demonstration that the origin of transfer of the conjugative plasmid pRSF1010 can substitute for the T-DNA borders in directing DNA transfer to plant cells from *A. tumefaciens* (11). This hybrid transfer system also requires an intact Ti-plasmid *vir* region and a region of pRSF1010 encoding polypeptides involved in plasmid mobilization. The *oriT* of pRSF1010 and its cognate mobilization proteins presumably generate a conjugative DNA transfer intermediate which is then mobilized to plant cells by using *vir*-specific transfer machinery. Thus, in order to understand T-DNA transfer better, we also have to become familiar with fundamental principles of bacterial conjugation. (Note that since no chapter in this volume covers bacterial conjugation, the following section is especially long, and serves partly as a review of this subject.)

D. Comparison of T-DNA Transfer to Bacterial Conjugation

Conjugation is a process whereby DNA is transferred from a donor to a recipient bacterium by a mechanism involving specialized cell-to-cell contact. The functions which mediate this process are encoded by conjugative plasmids, of which there are a wide variety in both gram-negative and gram-positive bacteria. While the ultimate consequence, i.e., DNA transfer, is the same in all the different systems, the detailed mechanisms vary. For example, plasmids from one incompatibility group or family do not complement transfer-deficient mutants in another group. The best-studied system is that of the F (fertility) conjugative plasmid. Only the most salient features of F conjugation are mentioned here as a focus for comparison with the *Agrobacterium* T-DNA transfer process. The reader is directed to several excellent recent reviews (52, 122, 123) for the

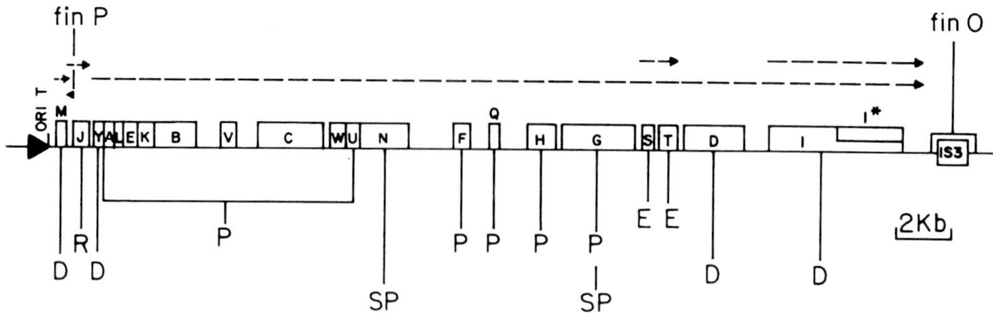

Figure 6. Genetic and functional map of the F transfer region. Boxes indicate the placement and relative size of the different *tra* genes. Dashed arrows indicate *tra* transcripts. The solid arrow at the left indicates the origin of conjugal DNA transfer, which occurs away from the *tra* genes. The functional assignments of the *tra* genes are shown below: P, pilus biosynthesis and assembly; D, conjugal donor DNA metabolism; SP, stabilization of mating pair; E, exclusion of nonproductive (donor-donor) mating; and R, regulation of conjugation process, which is also mediated by *finO* and *finP* (shown above). Redrawn from Ippen-Ihler and Minkley (52).

original citations for the bulk of the work described below; only references not cited in these reviews are given in the present chapter. Although F conjugation has been studied for more than 40 years, it is still far from understood. Its complexity serves also as a hint that bacterial-plant conjugation may be even more intricate.

The F plasmid is 100 kbp in size, and 33.3 kbp of contiguous sequences is devoted to plasmid fertility. The fertility region contains at least 25 transfer (*tra*) genes whose products are required for five different functions in conjugation: (i) conjugal DNA metabolism; (ii) biosynthesis (and assembly) of F pilus; (iii) mating pair stabilization; (iv) surface exclusion to prevent mating between two F-containing cells; and (v) regulation of the expression of transfer functions. Figure 6 presents a physical, genetic, and functional map of the F transfer region, and Table 2 summarizes the general characteristics of F-mediated conjugation.

1. Conjugal DNA Metabolism

Generation of F conjugal DNA. The key site on F at which conjugative DNA processing is initiated is the origin of transfer (*oriT*). This site is asymmetrical on the F plasmid and is oriented so that the transfer region is transferred last (arrow, Fig. 6). *oriT* is recognized by the *traI*/*traY* endonuclease (110), which introduces a nick on the transferred strand of F. The 5′ end of this nick provides the leading terminus for the linear transfer of F out of the donor cell. The TraI protein also has helicase and ATPase activities essential for unwinding the transferred strand (110). The TraY protein is located in the cell envelope, and its association with TraI at this site may help to move the transferred strand through the donor cell membrane. It has also been proposed, although not yet proven, that the endonuclease may also act as part of a protein-DNA transfer complex which "pilots" the F strand into the recipient cell.

Table 2. Characteristics of F-mediated bacterial conjugation

Steps in conjugation	Characteristics
Processing of F DNA	1. Specific *oriT* site which is preferentially nicked on one DNA strand; requires a site-specific endonuclease
	2. 5′-to-3′ directional transfer of single-stranded donor DNA; requires helicase(s), an energy source, and probably single-stranded-DNA-binding protein
	3. Replacement-strand DNA synthesis; may require specific primase, but otherwise probably utilizes cellular enzymes for DNA synthesis and repair
Intercellular interactions	1. Mating signal; structure unknown, but must exist since donor cells initiate DNA transfer events in the presence of recipient cells even without close stable mating pair formation
	2. Elaboration of extended pilus structures
	3. Formation of a stable mating pair; probably requires association between membrane components in donor and recipient cells
	4. Formation of a specialized channel for DNA transfer
	5. Exclusion of sibling mating

The nicking and directional 5′-to-3′ transfer of F DNA is potentially one of the best-understood aspects of bacterial conjugation and seems most similar to the T-DNA transfer process; i.e., border nicks and T-strands correspond to nicks at *oriT* and the F strand. Furthermore, genetic studies suggest that T-DNA transfer occurs in a polar right-to-left direction, and T-strands are proposed to be generated in a 5′ (right)-to-3′ (left) direction dictated by border nicks on the lower strand of the T-DNA border repeats. Thus, the genetic and physical data suggest that T-DNA transfer occurs in a 5′-to-3′ direction analogous to bacterial conjugation.

Whereas there is only one origin of transfer in F which results in the transfer of the entire F plasmid, T-DNA transfer mobilizes only the T-DNA region, not the rest of the Ti plasmid. This latter result is a consequence of the presence of a left copy of the T-DNA border sequence which, when nicked, terminates the transfer of the internal T-DNA region. The left border in nopaline Ti plasmids can also function to transfer DNA to the left, i.e., to the outside, of the T-DNA region. Normally such transfer is not noticed using wild-type two-border Ti plasmids since no genes are present to the left of the T-DNA which confer a phenotype in transformed plant cells. In vitro constructs where selectable marker genes are placed to the left of the T-DNA region do, however, demonstrate that transfer can occur to the left of the T-DNA region (59). More direct evidence that multiple T-DNA borders can produce multiple transferable T-DNA copies derives from naturally occurring octopine Ti plasmids containing three adjacent T-DNAs bracketed by four T-DNA border sequences. DNA homologous to all three T-DNAs has been observed in transformed plant cells, and T-strands corresponding to all (six) border-to-border combinations of nicked Ti plasmids are observed in *vir*-induced *A. tumefaciens* (Fig. 5). Thus, directly oriented copies of the 25-bp border repeats provide *A. tumefaciens* with multiple "*oriT*" genes. Although the question has not been directly addressed, F-derived plasmid constructs carrying multiple copies of directly oriented *oriT* genes might efficiently transfer DNA segments bracketed by *oriT*.

Note that reversing the orientation of the right T-DNA border blocks transfer of the adjacent (leftward) T-DNA region only when the T-DNA is contained on a large plasmid such as Ti. The T-DNA transfer on smaller plasmids, up to 50 kbp in size, is unaffected by the orientation of the borders (56, 88). The inefficiency of reverse right-border constructs to direct T-DNA transfer from the Ti plasmid is reminiscent of polarized chromosomal transfer in *E. coli* Hfr strains directed by an integrated F factor. That is, when the border is reversed, T-strand synthesis is directed away from the T-DNA (tumor-forming genes), which is reached only if the entire 200 kbp of the circular Ti plasmid is traversed.

Protein products which mediate conjugative donor-strand DNA processing. Besides the *traI*/*traY* endonuclease, two other F *tra* genes, *D*, and *M*, are required for processing of conjugal F DNA. The exact function of the *traD* product is unknown, but its location at the inner membrane and its nonspecific binding to ss and ds DNAs are consistent with the hypothesis that it forms part of a pore for DNA export. TraM is also an inner membrane protein that recently has been shown to bind specifically to F DNA in the region of *oriT*. It has been suggested that TraM responds (somehow) to the signal that a mating pair has formed and activates donor-strand unwinding and replacement-strand synthesis (see below). The TraM protein might potentially aid in the production of a stretch of ss DNA in the region of *oriT* and provide a suitable substrate for the initiating of unwinding by the *traI* and cellular helicases.

Thus, only four F-specific protein products (TraD, -I, -M, -Y) are required to modify and interact with F DNA during donor-strand transfer. Only two *vir* proteins, VirD1 and VirD2, have been proven to be necessary for the production of T-DNA transfer intermediate molecules. The *virD* locus is 4.6 kbp in size, and DNA sequencing predicts that it encodes four polypeptides of 16, 47, 21, and 76 kDa, respectively (54, 73, 133). The 16-kDa (VirD1) and 47-kDa (VirD2) proteins encoded by the 5′ end of the locus together provide the border endonuclease (54, 102, 130; G. DeVos and P. Zambryski, manuscript in preparation). By analogy to the model proposed for F DNA transfer, these proteins may be part of a covalent protein-DNA complex which "pilots" the T-strand during its transfer; in support of this hypothesis, the VirD2 polypeptide copurifies with T-strands under conditions which should denature protein from DNA (E. Howard, G. DeVos, B. Winsor, and P. Zambryski, unpublished results). The functions of the proteins encoded by the 3′ half of the *virD* locus are unknown. Since the entire *virD* locus is required for plant cell transformation, the putative VirD3 and VirD4 products may act at later stages.

Following nicking, the rate-limiting step for T-DNA transfer may be unwinding of the T-strand from duplex DNA. Several different helicases may be needed to displace the transferred DNA effectively during F transfer, one helicase to initiate unwinding by providing a localized region of gapped nonduplex DNA and a second helicase to unwind the substrate DNA processively and rapidly. Both types of helicases likely have a high energy requirement which may be supplied by an endogenous (i.e., with the helicase itself) or exogenous ATPase activity. Other gyrase or topoisomerase functions may also be invoked. T-strand production also may require such additional DNA enzymes.

In the F system, DNA transfer and conjugative DNA synthesis are initiated in response to an as yet unidentified signal generated by the formation of a stable mating pair. Since all the *tra* genes are constitutively expressed (see below), this signal is not necessary to stimulate nicking at *oriT*. Nicking and religation at *oriT* may occur continuously in donor cells. In contrast, in the *Agrobacterium* system, nicking at the T-DNA borders is under tight control, which is dependent on plant-induced *vir* gene expression. In further contrast, free, transferable T-strand molecules are generated (in the presence of plant wound factors but) in the absence of recipient plant cells. The latter result implies that the mechanisms responsible for unwinding of the T-strand and its replacement synthesis on the Ti plasmid are not likely to occur exactly as in the F system.

One further DNA-associated protein, SSB, is proposed to be an essential component of conjugative DNA transfer. SSB may act at several different steps to (i) maintain the nicked site in a nonduplex state to enhance helicase activity, (ii) promote correct initiation and elongation during replacement-strand DNA synthesis, and (iii) protect and stabilize the donor DNA strand during its transit to the recipient cell. While many conjugative plasmids carry their own SSBs, the proof of their role awaits experiments using an *E. coli* strain conditionally defective for SSB. In the case of F, the *ssb* locus is located outside the *tra* region in the segment of DNA which is transferred early into the recipient cell. This location may be advantageous in permitting rapid synthesis of new SSB in the recipient cell and thereby provide better protection of the newly arrived ss DNA. The requirement for plasmid-specified SSBs may not be absolute, since in their absence the chromosomally encoded SSB might provide much of the necessary function. *A. tumefaciens* encodes its own *vir*-regulated, and hence T-DNA-transfer-specific, SSB in the *virE* locus. As indicated above, this protein is not absolutely essential for plant cell transformation; *virE* mutants are defective for T-DNA transfer only on certain plant hosts.

In summary, several types of protein products are necessary for the production of a conjugative DNA (or T-DNA) molecule. The first, the origin of transfer endonuclease, is sequence specific and thus an integral part of each specific conjugative system. The second, helicase, may be generalized or system specific if it is part of the endonuclease itself or if it interacts with the endonuclease at the nick site. The third, SSB, may be provided by the conjugative plasmid or the host genome. A fourth type of function is envisaged to transport the DNA through the donor and recipient cell membranes, and would interact with DNA or a protein-DNA complex within the hydrophobic membrane environment. The *traD* and *traM* loci of F are candidates to provide such functions, although activities have not been demonstrated biochemically. A final class of DNA-associated proteins required for conjugal DNA transfer is likely to include those which participate in more general steps of DNA metabolism, such as DNA synthesis and repair; these would be host specified.

Conjugative DNA synthesis. Transfer of a single strand of F DNA is associated with synthesis of a replacement strand in the donor cell and of a complementary strand in the recipient cell. Experiments using temperature-sensitive *E. coli* mutants have shown that DNA polymerase holoenzyme is primarily responsible for conjugative DNA synthesis. Donor-cell replacement-strand synthesis presumably utilizes the 3′ OH at the *oriT* nick site as the priming site; potentially an RNA molecule synthesized de novo by RNA polymerase in the recipient cell acts as the primer for replacement-strand DNA synthesis there. Many conjugative plasmids (although not F) transfer proteins with primase function along with DNA during conjugation. This transmission seems to be highly specific, since there is no general transfer of donor cell proteins (14).

In *A. tumefaciens*, replacement synthesis of the lower strand of the T-DNA region presumably uses the 3′ OH of the nicked right T-DNA border as an initiation (priming) site. No *vir* proteins have been implicated in this step; host enzymes may be involved by analogy to F conjugal DNA synthesis. Second-strand synthesis of the T-DNA, before or after its integration into the plant cell genome, may involve plant cell enzymes. Alternatively, the T-DNA–protein transfer complex may carry proteins which stimulate copying of the T-strand in the plant cell.

Note that in the F system, DNA transfer depends on unwinding the donor strand induced by the formation of a stable mating pair. Thus, replacement-strand synthesis is normally coupled to DNA transfer. In *A. tumefaciens* the T-strand can be produced in the absence of recipient plant cells. This latter result may reflect a genuine difference in the mechanisms of DNA transfer between bacterial conjugation and *A. tumefaciens*-plant cell transformation. It is also possible that the addition of the plant signal molecule AS to *A. tumefaciens* in laboratory experiments is a way of artificially "fooling" *A. tumefaciens*, since in nature the presence of AS indicates that susceptible plant cells are nearby.

2. Initiation of donor-recipient interaction mediated by F pili

Bacteria carrying F have one to three F pili that establish the physical contact between donor and recipient cells needed for conjugative DNA transfer. Pili are composed of a single-subunit protein, pilin, arranged as repeating units elongated along four intertwined helices. Pilin is the 13.2-kDa polypeptide product of the *traA* locus and is processed to 7.0 kDa in the mature pilin subunits. F pili are hollow cylindrical filaments 8 nm in diameter with a central axial hole of 2 nm, and they can extend up to 2 μm from the cell surface. While F pili are relatively simple in composition, more than 13 kbp of the 33-kbp coding capacity of the F *tra* region is devoted to their synthesis and assembly. The TraQ product cleaves the pilin precursor. *traL*, *-E*, *-K*, *-B*, *-V*, *-C*, *-W*, *-U*, *-F*, *-H*, and *-G* gene products erect the pilus structure, although the exact roles of each of the proteins have not yet been determined. Some may constitute an integral part of the pilus, for example, as a transmembrane or basal body anchor. Others may transport the pilin subunits through the membranes during pilus assembly. Still others may modify pilus proteins by glucosylation, phosphorylation, or acetylation.

A. tumefaciens seems not to carry obvious pili; *vir*-induced and noninduced cells do not differ in appearance under the electron microscope (G. Kuldau and P. Zambryski, unpublished observations). The chromosomal loci *chvA*, *chvB* (28, 29, 84; Zorreguieta et al., in preparation), and *pscA* (106) may provide the pilus-analogous function which allows *A. tumefaciens* to form an effective contact with recipient plant cells. These genes may have a general role which is essential in mediating bacterial-plant cell attachment, since related genes are found in other soil bacteria. For example, the *ndvA*, *ndvB*, and *exoC* loci of *Rhizobium meliloti* (31, 72), required to initiate nitrogen-fixing-nodule development, are highly homologous to the *chvA*, *chvB*, and *pscA* loci of *A. tumefaciens*. Whereas *R. meliloti* actually invades the interior of the plant, *A. tumefaciens* only transfers a DNA segment from the outside to the inside of individual plant cells. The fact that chromosomally encoded products are involved in two different bacterial-plant interactions suggests that they only mediate a nonspecific aspect of the initial contact process.

Elongated conjugative pili are not a formal requirement for conjugal DNA transfer, since some systems, exemplified by gram-positive bacteria such as those of the genus *Streptococcus*, transfer DNA without the use of pili (reviewed in reference 18). In fact, there may be additional parallels between T-DNA transfer and the transfer of DNA between streptococci. For example, it is striking that streptococci transfer drug resistance factors between donor and recipient replicons present in different bacterial cells by a mechanism termed conjugative transposition. Although *A. tumefaciens* may or may not utilize proteinaceous surface protrusions for T-DNA transfer, some sort of specialized pore structure that interacts with the plant cell membrane and facilitates T-DNA transfer seems likely. The *virB* locus may provide the structural components of a DNA transfer channel. Three *virB*-specific proteins of 25, 33, and 80 kDa have been localized specifically to the cell envelope fraction of *vir*-induced *A. tumefaciens* (32) and each is abundant, suggesting a structural rather than an enzymatic role in the transformation process.

F pili are used to find an appropriate recipient cell; once a stable pair (see below) has been established, the cells are brought into direct contact, apparently by retraction of the F pili and disassembly of F pilin subunits. Actual DNA transfer may then proceed through a transmembrane pore in the hypothesized basal structure of the pilus. In this model, DNA transfer itself does not require an extended pilus. In the case of the *A. tumefaciens* T-DNA transfer process, initial contact between *A. tumefaciens* and the recipient plant cell might be accomplished by polysaccharide components on the bacterial and plant cell surfaces. By analogy to the F plasmid conjugation system, T-DNA transfer may only proceed when a close contact between bacterial and plant cell membranes is formed; specific *vir* (*B?*) polypeptides may bring about this close association.

3. Stabilization of the mating pair

Mutation in *traN* or the C-terminal region of *traG* results in mutants which make F pili but are defective in both aggregation with recipient cells and DNA transfer. These mutants are not defective in any steps of processing donor DNA for transfer, nor presumably receipt of the mating signal from recipient cells, since they are not blocked for replacement synthesis of the donor DNA strand. *traG* and *traN* products may interact directly with the major outer membrane protein OmpA of the recipient cell, since *ompA* mutants are also defective in the formation of stable mating pairs. The existence of functions which aid in the formation of stable mating pairs during bacterial conjugation serves to point out the types of (as yet unidentified) functions which may participate also in the transformation of plant cells by *A. tumefaciens*.

4. Prevention of nonproductive mating

The products of the surface exclusion genes *traA* and *traT* prevent sibling mating. The TraA protein is located in the inner membrane and may block DNA transfer per se. The TraT product is a lipoprotein located in the outer membrane. It has been proposed that the TraT protein competively binds the pilus tip. Electron micrographs of purified TraT reveal it to be an oligomeric, donut-shaped molecule with an outer diameter of 18.5 nm and a central region appropriate for binding the 8-nm-diameter pilus filament. Whereas there has been no systematic study to suggest or exclude the possibility of *Agrobacterium-Agrobacterium* T-DNA transfer, a priori it seems there should be a mechanism which favors transferring the T-DNA molecule to plant cells. For transfer of Ti plasmids between bacterial cells, *A. tumefaciens* has evolved a separate conjugative operon located outside of the T-DNA and *vir* regions on the Ti plasmid (25, 48).

5. Regulation of F transfer

While the ancestral form of F was repressed for transfer, modern F has an IS3 element inserted in one (*finO*) of the two loci (*finO*, *finP*) which regulate the transcription of F transfer proteins. As a consequence, the F transfer genes are constitutively expressed. Repression of transfer is probably more frequent and favored as a regulatory strategy among conjugal plasmids; an inducible transfer system permits rapid spread of DNA without continuous risk of epidemic spread of donor-specific phage. F transfer is rare within homogeneous populations because the F-encoded TraS and TraT proteins on the cell surface interfere with mating. Regulation of F transfer gene expression can be supplied in *trans* by compatible *finO*$^+$ F-like plasmids; control is exerted at the level of transcription. While the T-DNA transfer system is not normally observed (as yet) in the absence of plant cell factors, there is no reason to suspect that such expression would be harmful to the cell. Furthermore, the "normal" bacterium-bacterium conjugative system is constitutive in many commonly used strains of *A. tumefaciens* (48). The only requirement for constitutive expression is that there is no loss of the genes essential to the transfer process. In the F system, actual DNA transport out of the donor cell is most likely designed to be dependent on the recognition of the appropriate recipient cell. In this regard, the conservative production of a T-DNA transfer molecule by excision from the T-DNA region on the Ti plasmid as invoked in some models would result in the loss of the T-DNA region; such a mechanism would likely be selected against, since any mutation which resulted in constitutive *vir* gene expression would result in the loss of the T-DNA region from the Ti plasmid.

In summary, out of the eight different properties of F-specific bacterial conjugation listed in Table 2, the only property which may not be shared with the T-DNA transfer system is the elaboration of an extended pilus structure. Future studies will reveal how well the analogy between *Agrobacterium*-plant cell transformation and bacterial conjugation holds.

E. Integration of the T-DNA into the Plant Genome

While some of the early reactions that lead to T-DNA transfer are partially understood, little is known about later events of T-DNA integration into the plant chromosomes. Studies have focused on the structure of the stably integrated T-DNA. Although these analyses do not lead directly to the formulation of detailed mechanistic models for the integration process, they do provide a foundation for future model building.

1. T-DNA copy number and chromosomal location

Hybridization analyses reveal that single copies are frequent, but on average three T-DNA copies are found integrated into the genomes of various plant species, including tobacco (22, 67, 96, 107), petunia (57, 60, 116), tomato (17), sunflower (111), morning glory (95), and *Crepis capilaris* (4); occasionally, up to 20 to 50 copies have been detected (9). Multiple copies can be either linked in tandem arrays or unlinked at different chromosomal locations. In the most extensive study (K. Feldman and M. Christianson, manuscript in preparation), 161 individual transformants of *Arabidopsis thaliana* were analyzed genetically: 55% of the transformants segregated for one, 20% for two unlinked, 6% for two linked, 5% for three unlinked, and 1% for four T-DNA segments. The remaining 12% did not segregate in a simple Mendelian fashion. In a separate study, isozymes and restriction fragment length polymorphism (RFLP) markers in 12 chromosomes were used to analyze 10 independent transformed lines of tomato (17). Nine out of 10 of the T-DNA inserts mapped to different chromosomal locations; the 10th insert could not be mapped. Furthermore, in situ hybridization of four transformed lines of *C. capillaris* indicated that the T-DNA element was inserted into four

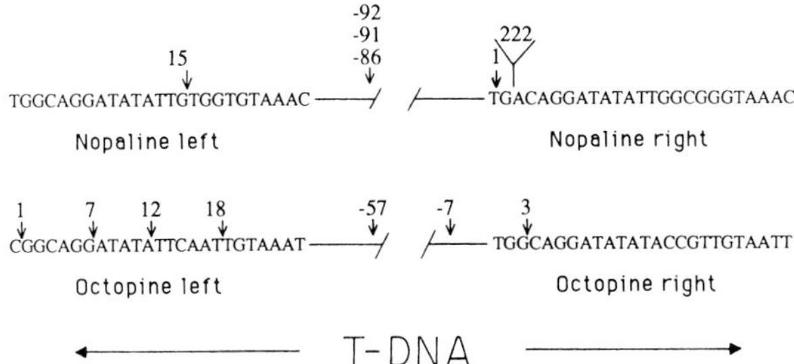

Figure 7. Endpoints of the T-DNA in transformed plant cell DNA. Numbers and arrows indicate the positions of the ends of T-DNAs in transformed plant cell DNA relative to the 25-bp T-DNA border repeats. The nopaline endpoints (−92, −91, −86, 1, 2, 2, 2) are from Zambryski et al. (131), and endpoint 15 is from Yadav et al. (126). Octopine endpoint 1 is from Simpson et al. (93), 7 and 18 are from Kwok et al. (65), and 12, −57, −7, and 3 are from Holsters et al. (49).

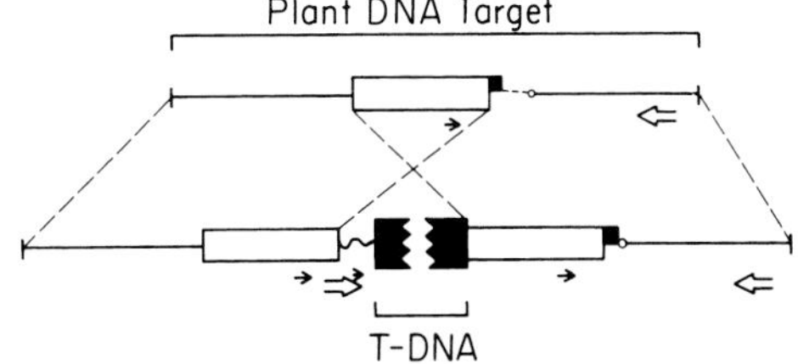

Figure 8. Rearrangements of target plant DNA resulting from T-DNA insertion. Top: Unoccupied target plant DNA; bottom: the left and right junctions of the target plant DNA following insertion of the T-DNA element. Large solid (interrupted) box, T-DNA. Open boxes, 158-bp duplications of target plant DNA at both the right and left borders following T-DNA insertion. Additional rearrangements on the right include a short deletion of target sequences (dashed line) and a C-to-T transition (dot). Additional rearrangements on the left include "filler" DNA (wavy line); arrows indicate homology of filler DNA to other DNA sequences. Adapted from Gheysen et al. (40).

out of the six possible chromosomes (4). These results, coupled with the ability of *A. tumefaciens* to transform a wide range of dicotyledonous plant species, indicate that T-DNA insertion is not dependent on a specific target DNA sequence. This issue is of both fundamental interest and practical importance in the use of *A. tumefaciens* as a generalized vector for gene transfer to plants.

Tandem arrangements of the T-DNA can be composed of either direct or indirect repeats (e.g., 60, 67, 96), and both inverted and direct repeats may be present in a single T-DNA array. Both single-copy and repeated T-DNA structures are equally capable of expressing internal T-DNA sequences. Variability in expression is most often due to insertion into different chromosomal sites, i.e., position-effect variation, or to insufficient physiological standardization of the conditions used for plant growth prior to analysis (57). There is, however, some evidence that multiple T-DNA copies, especially inverted copies, are unstable (57, 60); thus, to minimize variability in the expression of T-DNA of interest, it would be best to utilize transformants with single (or few directly repeated) T-DNAs.

2. Nucleotide sequence analyses of T-DNA insertion events

Nucleotide sequencing of cloned T-DNA-plant junctions reveals that the plant sequences immediately adjacent to the T-DNA borders are unrelated (48, 65, 93, 95, 126, 131); the only shared feature between these plant target sequences is that they are enriched for As and Ts. In terms of the T-DNA, these studies reveal that junction points at the right T-DNA end are close to or within the right 25-bp border repeat sequence; the junction points at the left end are spread over 100 bp internal to and including the left 25-bp border sequence (Fig. 7). Thus, T-DNA insertion occurs more precisely on the right than on the left, which suggests that T-DNA integration, like the generation of the transferable T-strand copy, is directed by the right T-DNA border.

To better understand how T-DNA integration occurs, one needs to analyze plant DNA target sites prior to and after T-DNA insertion. Only one study, however (40), has compared the nucleotide sequences of the plant DNA portions of a single T-DNA insert with the original unoccupied plant DNA target site. Interestingly, the data show that plant DNA can undergo rearrangements as a result of T-DNA insertion (Fig. 8). The most striking alteration of target sequences is the generation of a direct repeat of 158 bp at the right and left T-DNA junctions. Besides this duplication event, there are short deletion and insertion events at the end of the 158-bp segment. Comparable studies of additional T-DNA insertion events will help provide a better understanding of the frequency and extent of the rearrangements during integration.

3. Potential mechanism of T-DNA integration

There is no evidence for multimeric, free, transferable T-DNA copies in *A. tumefaciens* just prior to plant cell transformation, i.e., following activation of *vir* gene expression. Presumably the tandem arrays of T-DNA inserts often observed in transformed plant cell DNA are produced during or after T-DNA is transferred to the plant cell. Duplication of T-DNA copies as well as of plant target DNA sequences (Fig. 8) may reflect common processes. The variety of rearrangements observed suggests that T-DNA insertion is a multistep process involving several different types of recombination, replication, and repair activities most likely mediated by host-plant-encoded enzymes.

There are two general mechanisms for DNA insertion into the genome of a eucaryotic cell. First the DNA element itself encodes some or most of the enzymes involved in integration; this first class (I) is exemplified by transposons such as the corn element *Ac* (e.g., 89; N. V. Fedoroff, this volume) and retroviruses (H. Varmus and P. Brown, this volume). A second class (II) of insertion is mediated by host enzymes; such integrations are observed upon introduction of DNA through microinjection, electroporation, or transfection and upon infection with some viruses. Class I events are relatively efficient and relatively undisturbing; usually a short segment (up to 10 bp) of target sequence is duplicated at the ends of the insertion element, and further rearrangements are rare. In contrast, class II events can cause major disruptions of DNA contiguity, such as during the translocation of cellular oncogenes (38). Also, deletion of target DNA is typical, and the extent may vary from a few (26) to several thousand (27) base pairs. At a particular polyomavirus-host DNA joint, a 37-bp filler sequence is an inverted perfect duplication of a single-copy host sequence found 650 bp away from the junction site (27). Similarly, there is a 33-bp filler sequence at the left of the T-DNA junction in the plant target analyzed (40) (Fig. 8) which is homologous to an inverted host sequence 200 bp from the T-DNA–plant DNA junction. That filler DNA sequences are found near insertion sites in both plant and animal cell DNAs implies that the (repair?) mechanism responsible for their generation may occur throughout eucaryotic cells in general.

Thus, T-DNA integration may resemble class II events in the variety of rearrangements which occur both to the T-DNA element and to the target plant sequences. However, it differs from class II events in its efficiency. Only microinjection of DNA directly into the nucleus shows a comparable transformation efficiency (12). In this regard, it is possible that the T-DNA is directed preferentially into the nucleus. Note also that in the integration event depicted in Fig. 8 the right T-DNA–plant junction contains a direct transition from T-DNA to plant DNA sequences; in contrast, the left T-DNA junction contains scrambled "filler" sequences. The precision of the right junction compared to the left may be explained if T-DNA integration initiates with the right T-DNA end, consistent with the observed functional and physical polarity of the T-DNA border sequences. Polar T-DNA transfer or integration could be mediated by proteins linked to the right end of the T-DNA during its transit; such T-DNA transfer proteins may also function to target the T-DNA–protein complex efficiently to the plant cell nucleus.

The efficiency of the T-DNA transfer process also may be due in part to the nature of the proposed T-DNA transfer intermediate molecule, the T-strand. While it is unknown whether the T-strand is altered (e.g., into a duplex form) prior to transfer or integration, its single strandedness does not conflict with a role in transfer and integration. For example, most models for general recombination involve invasion of target sequences by single-stranded donor DNA (30, 121). Linear duplex DNAs are thought to be inactive as recombination intermediates until rendered single stranded at their ends by cellular exonucleases (7, 63, 70).

While it is still too early to formulate a firm model for T-DNA insertion into plant DNA, the following steps are consistent with the available information. First, the T-strand is transferred to the plant cells as a DNA-protein complex. Second, following localization of the complex to the cell nucleus, a protein at the 5′ (right) end of the T-strand interacts with a nicked sequence in plant DNA. Third, local torsional strain on the plant DNA results in the production of a second nick on the opposite strand of the target sequence at a variable distance from the first nick. Fourth, each end of the T-strand is ligated to plant DNA and its homologous strand is copied by cellular enzymes. Fifth, repair and replication of the staggered nick in the plant target result in the production of a repeated sequence (of variable length), plus additional sequence rearrangements ("filler" DNA, deletion, etc.) at the ends of the inserted T-DNA element. Insertion of the large T-DNA may trigger errors in replication and repair at the insertion site and sometimes lead to the formation of direct and inverse repeats of T-DNA.

IV. SUMMARY AND PROSPECTS

There are at least seven steps in the transfer of the T-DNA molecule from *A. tumefaciens* to the plant cell: (i) recognition of a susceptible plant cell, (ii) induction of *vir* gene expression, (iii) production of a transferable T-DNA copy, (iv) transfer of the T-DNA complex to the bacterial membrane, (v) transfer of the T-DNA complex through the bacterial membrane and the plant cell membrane, (vi) transfer of the T-DNA complex through the plant cytoplasm and through the nuclear membrane, and finally (vii) integration of the T-DNA into the plant nuclear genome. It is clear that we are part of the way toward understanding steps (i) through (iii); the major unsolved questions in these early stages of T-DNA transfer are related to defining the protein products which mediate the production of the T-DNA transfer complex. To sort out steps (iv) and (v), and potentially step (vi), will require that we become more knowledgeable about membrane structure and function. Step (vii), the end result of the coordinated action of 20 or more bacterial gene products, seems less tractable. The identification of the *Agrobacterium*-specified components of the transfer complex should, it is hoped, elucidate these final steps, perhaps involving bacterial protein products which localize the T-DNA complex to the nucleus or insert the T-DNA molecule into plant DNA. Several plant products probably are also involved in the last steps of T-DNA transfer. The wounding of plant cells stimulates DNA replication and cell proliferation, and the attendant changes in recombination or DNA repair enzyme activities may encourage T-DNA integration.

While the early steps of the T-DNA transfer process seem to use well-conserved bacterial processes, plant pathways normally related to plant growth and development may be recruited for the later steps. In this regard, the *Agrobacterium*-plant cell interaction provides an avenue into the exploration of fundamental plant processes: events at the cell surface and the internal network of membranes, and the replication and rearrangement of DNA sequences and chromatin structure in the nucleus.

Acknowledgments. I thank my co-workers for obtaining the results which made this review possible; their names are given in the by-lines of the articles of which I am coauthor. I especially thank Scott Stachel for his determination to perform a variety of analyses to help solve many questions related to the biology of the *Agrobacterium*-plant cell interaction. I am grateful to my colleagues for sending preprints and sharing their unpublished observations. I thank Marc Van Montagu and Jeff Schell of the University of Gent, Belgium, for support of my research during the many years of my stay in their lab.

The work in my laboratory is currently supported by grant DMB-8617772 from the National Science Foundation.

LITERATURE CITED

1. **Akiyoshi, D. E., H. Klee, R. M. Amasino, E. W. Nester, and M. P. Gordon.** 1984. T-DNA of *Agrobacterium tumefaciens* encodes an enzyme of cytokinin biosynthesis. *Proc. Natl. Acad. Sci. USA* **81:**5994–5998.
2. **Albright, L. M., M. F. Yanofsky, B. Leroux, D. Ma, and E. W. Nester.** 1987. Processing of the T-DNA of *Agrobacterium tumefaciens* generates border nicks and linear, single-stranded T-DNA. *J. Bacteriol.* **169:**1046–1055.
3. **Alt-Moerbe, J., B. Rak, and J. Schroeder.** 1986. A 3.6 kbp segment from the *vir* region of Ti plasmids contains genes responsible for border sequence-directed production of T-region circles in *E. coli*. *EMBO J.* **5:**1129–1135.
4. **Ambros, P. F., A. J. M. Matzke, and M. A. Matzke.** 1986. Localization of *Agrobacterium rhizogenes* T-DNA in plant chromosomes by *in situ* hybridization. *EMBO J.* **5:** 2073–2077.
5. **An, G., B. D. Watson, S. Stachel, M. P.Gordon, and E. W. Nester.** 1985. New cloning vehicles for transformation of higher plants. *EMBO J.* **4:**277–284.
6. **Ashby, A. M., M. D. Watson, and C. H. Shaw.** 1987. A Ti plasmid determined function is responsible for chemotaxis of *A. tumefaciens* towards the plant wound compound acetosyringone. *FEMS Microbiol. Lett.* **41:**189–192.
7. **Ayares, D., L. Chekuri, K. Y. Song, and R. Kucherlapati.** 1986. Sequence homology requirements for intermolecular recombination in mammalian cells. *Proc. Natl. Acad. Sci. USA* **83:**5199–5203.
8. **Barker, R. F., K. B. Idler, D. V. Thompson, and J. D. Kemp.** 1983. Nucleotide sequence of the T-DNA region from the *A. tumefaciens* octopine Ti plasmid pTi15955. *Plant Mol. Biol.* **2:**335–350.
9. **Barton, K. A., A. N. Binns, A. J. M. Matzke, and M. D. Chilton.** 1983. Regeneration of intact tobacco plants containing full length copies of genetically engineered T-DNA and transmission of T-DNA to R1 progeny. *Cell* **32:** 1033–1043.
10. **Bevan, M.** 1984. Binary *Agrobacterium* vectors for plant transformation. *Nucleic Acids Res.* **12:**8711–8721.
11. **Buchanan-Wollaston, V., J. E. Passiatore, and F. Cannon.** 1987. The *mob* and *oriT* mobilization functions of a bacterial plasmid promote its transfer to plants. *Nature* (London) **328:**172–175.
12. **Capecchi, M. R.** 1980. High efficiency transformation by direct microinjection of DNA into cultured mammalian cells. *Cell* **22:**479–488.
13. **Chase, J. W., and K. R. Williams.** 1986. Single stranded DNA binding proteins required for DNA replication. *Annu. Rev. Biochem.* **55:**103–136.
14. **Chatfield, L. K., and B. M. Wilkins.** 1984. Conjugative transfer of IncI1 plasmid DNA primase. *Mol. Gen. Genet.* **197:**461–466.
15. **Chilton, M.-D., M. H. Drummond, D. J. Merlo, D. Sciaky, A. L. Montoya, M. D. Gordon, and E. W. Nester.** 1977. Stable incorporation of plasmid DNA into higher plant cells: the molecular basis of crown gall tumorigenesis. *Cell* **11:** 263–271.
16. **Christie, P. J., J. E. Ward, S. C. Winans, and E. W. Nester.** 1988. The *Agrobacterium tumefaciens virE2* gene product is a single-stranded DNA-binding protein that associates with T-DNA. *J. Bacteriol.* **170:**2659–2667.
17. **Chyi, Y. S., R. A. Jorgensen, D. Goldstein, S. D. Tanksley, and F. Loaiza-Figueroa.** 1986. Locations and stability of *Agrobacterium*-mediated T-DNA insertions in the *Lycopersicon* genome. *Mol. Gen. Genet.* **204:**64–69.

17a. **Citovsky, V., G. DeVos, and P. Zambryski.** 1988. A novel single stranded DNA binding protein, encoded by the *virE* locus, is produced following activation of the *A. tumefaciens* T-DNA transfer process. *Science* **240:**501–504.

18. **Clewell, D. G., and C. Gawron-Burke.** 1986. Conjugative

transposons and the dissemination of antibiotic resistance in *Streptococci*. *Annu. Rev. Microbiol.* **40**:635–659.

19. **Darvill, A., and P. Albersheim.** 1984. Phytoalexins and their elicitors—a defense against microbial infection in plants. *Annu. Rev. Plant Physiol.* **35**:243–275.
20. **Das, A.** 1988. The *A. tumefaciens vir*E operon encodes a single stranded DNA binding protein. *Proc. Natl. Acad. Sci. USA* **85**:2909–2913.
21. **Das, A., S. E. Stachel, P. Ebert, P. Allenza, A. Montoya, and E. Nester.** 1986. Promoters of *A. tumefaciens* Ti plasmid virulence genes. *Nucleic Acids Res.* **14**:1355–1364.
22. **DeBeuckeleer, M., M. Lemmers, G. De Vos, L. Willmitzer, M. Van Montagu, and J. Schell.** 1981. Further insight on the transferred DNA of octopine crown gall. *Mol. Gen. Genet.* **183**:283–288.
23. **Deblaere, R., B. Bytebier, H. De Greve, F. Deboeck, J. Schell, M. Van Montagu, and J. Leemans.** 1985. Efficient octopine Ti plasmid-derived vectors for *Agrobacterium*-mediated gene transfer to plants. *Nucleic Acids Res.* **13**:4777–4788.
24. **De Framond, A., K. A. Barton, and M.-D. Chilton.** 1983. Mini-Ti: a new vector strategy for plant genetic engineering. *Bio/Technology* **1**:262–269.
25. **DeGreve, H., H. Decraemer, J. Seurinck, M. Van Montagu, and J. Schell.** 1981. The functional organization of the octopine *A. tumefaciens* plasmid pTiB6S3. *Plasmid* **6**:235–248.
26. **Dejean, A., L. Bougueleret, K. H. Grzeschik, and P. Tiollais.** 1986. Hepatitis B virus DNA integration in a sequence homologous to v-*erb*-A and steroid receptor genes in a hepatocellular carcinoma. *Nature* (London) **322**:70–72.
27. **Della Valle, G., R. G. Fenton, and C. Basilico.** 1981. Polyoma large T antigen regulates the integration of viral DNA sequences into the genome of transformed cells. *Cell* **23**:347–355.
28. **Douglas, C. J., W. Halperin, and E. W. Nester.** 1982. *Agrobacterium tumefaciens* mutants affected in attachment to plant cells. *J. Bacteriol.* **152**:1265–1275.
29. **Douglas, C. J., R. J. Staneloni, R. A. Rubin, and E. W. Nester.** 1985. Identification and genetic analysis of an *Agrobacterium tumefaciens* chromosomal virulence region. *J. Bacteriol.* **161**:850–860.
30. **Dressler, D., and H. Potter.** 1982. Molecular mechanisms in genetic recombination. *Annu. Rev. Biochem.* **51**:727–761.
31. **Dylan, T., L. Ielpi, S. Stanfield, L. Kashyap, C. Douglas, M. Yanofsky, E. Nester, D. R. Helinski, and G. Ditta.** 1986. *Rhizobium meliloti* genes required for nodule development are related to chromosomal virulence genes in *A. tumefaciens*. *Proc. Natl. Acad. Sci. USA* **83**:4403–4407.
32. **Engstrom, P., P. Zambryski, M. Van Montagu, and S. E. Stachel.** 1987. Characterization of *Agrobacterium tumefaciens* virulence proteins induced by the plant factor acetosyringone. *J. Mol. Biol.* **197**:635–645.
33. **Firmin, J. L., K. E. Wilson, L. Rossen, and A. W. B. Johnston.** 1986. Flavonoid activation of nodulation genes in Rhizobium reversed by other compounds present in plants. *Nature* (London) **423**:90–92.
34. **Fraley, T. R., S. G. Rogers, and R. B. Horsch.** 1986. Genetic transformation in higher plants. *Crit. Rev. Plant Sci.* **4**: 41–46.
35. **Gardener, R. C., and V. Knauf.** 1986. Transfer of *Agrobacterium* DNA to plants requires a T-DNA border but not the *vir*E locus. *Science* **231**:725–727.
36. **Garfinkel, D. J., and E. W. Nester.** 1980. *Agrobacterium tumefaciens* mutants affected in crown gall tumorigenesis and octopine catabolism. *J. Bacteriol.* **144**:732–743.
37. **Geilan, J., M. De Beuckeleer, J. Seurinck, F. DeBoeck, H. De Greve, M. Lemmers, M. Van Montagu, and J. Schell.** 1984. The complete nucleotide sequence of the TL-DNA of the *A. tumefaciens* plasmid pTiAch5. *EMBO J.* **3**:835–846.
38. **Gerondakis, S., S. Cory, and J. M. Adams.** 1984. Translocation of the *myc* cellular oncogene to the immunoglobulin heavy chain locus in murine plasmacytomas is an imprecise reciprocal exchange. *Cell* **36**:973–982.
39. **Gheysen, G., P. Dhaese, M. Van Montagu, and J. Schell.** 1985. DNA flux across genetic barriers: the crown gall phenomenon, p. 11–49. *In* B. Hohm and E. Dennis (ed.), *Genetic Flux in Plants*. Springer, Vienna.
40. **Gheysen, G., M. Van Montagu, and P. Zambryski.** 1987. Integration of *A. tumefaciens* T-DNA involves rearrangements of target plant DNA sequences. *Proc. Natl. Acad. Sci. USA* **84**:6169–6173.
41. **Gietl, C., Z. Koukolikova-Nicola, and B. Hohn.** 1987. Mobilization of T-DNA from *Agrobacterium* to plant cells involves a protein that binds single stranded DNA. *Proc. Natl. Acad. Sci. USA* **84**:9006–9010.
42. **Herrera-Estrella, L., M. DeBlock, P. Zambryski, M. Van Montagu, and J. Schell.** 1985. *Agrobacterium* as a vector system for the introduction of genes into plants, p. 61–94. *In* J. H. Dodds (ed.), *Plant Genetic Engineering*. Cambridge University Press, New York.
43. **Hille, J., I. Klasen, and R. B. Schilperoort.** 1982. Construction and application of R prime plasmids, carrying different segments of an octopine Ti plasmid from *A. tumefaciens* for complementation of *vir* genes. *Plasmid* **7**:107–118.
44. **Hirooka, T., and C. I. Kado.** 1986. Location of the right boundary of the virulence region on *Agrobacterium tumefaciens* plasmid pTiC58 and a host-specifying gene next to the boundary. *J. Bacteriol.* **168**:237–243.
45. **Hirooka, T., P. M. Rogowsky, and C. I. Kado.** 1987. Characterization of the *vir*E locus of *Agrobacterium tumefaciens* plasmid pTiC58. *J. Bacteriol.* **169**:1529–1536.
46. **Hoekema, A., P. R. Hirsch, P. J. J. Hooykaas, and R. A. Schilperoort.** 1983. A binary plant vector strategy based on separation of *vir*- and T-region of the *A. tumefaciens* Ti plasmid. *Nature* (London) **310**:115–120.
47. **Hoekema, A., P. W. Roelvink, P. J. J. Hooykaas, and R. A. Schilperoort.** 1984. Delivery of T-DNA from the *A. tumefaciens* chromosome into plant cells. *EMBO J.* **3**:2485–2490.
48. **Holsters, M., B. Silva, F. Van Vliet, C. Genetello, M. De Block, P. Dhaese, A. Depicker, D. Inze, G. Engler, R. Villarroel, M. Van Montagu, and J. Schell.** 1980. The functional organization of the nopaline *A. tumefaciens* plasmid pTiC58. *Plasmid* **3**:212–230.
49. **Holsters, M., R. Villarroel, J. Gielen, J. Seurinck, H. De Greve, M. Van Montagu, and J. Schell.** 1983. An analysis of the boundaries of the octopine TL-DNA in tumors induced by *Agrobacterium tumefaciens*. *Mol. Gen. Genet.* **190**:35–41.
50. **Inze, D., A. Follin, M. Van Lijsebettens, C. Simoens, C. Genetello, M. Van Montagu, and J. Schell.** 1984. Genetic analysis of the individual T-DNA genes of *Agrobacterium tumefaciens*; further evidence that two genes are involved in indole-3-acetic acid synthesis. *Mol. Gen. Genet.* **194**: 265–274.
51. **Inze, D., A. Follin, H. Van Onckelen, P. Rudelsheim, J. Schell, and M. Van Montagu.** 1987. Functional analysis of the T-DNA *onc* genes, p. 181–196. *In* J. E. Fox and M. Jacobs (ed.), *Molecular Biology of Plant Growth Control*. Alan R. Liss, Inc., New York.

52. **Ippen-Ihler, K. A., and E. G. Minkley.** 1986. The conjugation system of F, the fertility factor of *Escherichia coli. Annu. Rev. Genet.* **20:**593–624.

53. **Janssens, A., C. Genetello, M. Van Montagu, and P. Zambryski.** 1986. Plant cells induce transcription of the *Agrobacterium tumefaciens* nopaline pTiC58 virulence region. *Plant Sci.* **47:**185–193.

54. **Jayaswal, R. K., K. Veluthambi, S. B. Gelvin, and J. L. Slightom.** 1987. Double-stranded T-DNA cleavage and the generation of single stranded T-DNA molecules in *Escherichia coli* by a *virD*-encoded border-specific endonuclease from *Agrobacterium tumefaciens. J. Bacteriol.* **169:**5035–5045.

55. **Jen, G., and M. D. Chilton.** 1986. The right border region of pTiT37 T-DNA is intrinsically more active than the left border region in promoting T-DNA transformation. *Proc. Natl. Acad. Sci. USA* **83:**3895–3899.

56. **Jin, S., T. Komari, M. P. Gordon, and E. W. Nester.** 1987. Genes responsible for the supervirulence phenotype of *Agrobacterium tumefaciens* A281. *J. Bacteriol.* **169:**4417–4425.

57. **Jones, J. D. G., D. E. Gilbert, K. L. Grady, and R. A. Jorgensen.** 1987. T-DNA structure and gene expression in petunia plants transformed by *A. tumefaciens* C58 derivatives. *Mol. Gen. Genet.* **207:**478–485.

58. **Joos, H., D. Inze, A. Caplan, M. Sormann, M. Van Montagu, and J. Schell.** 1983. Genetic analysis of T-DNA transcripts in nopaline crown galls. *Cell* **32:**1057–1067.

59. **Joos, H., B. Timmerman, M. Van Montagu, and J. Schell.** 1983. Genetic analysis of transfer and stabilization of *Agrobacterium* DNA in plant cells. *EMBO J.* **2:**2151–2160.

60. **Jorgensen, R. A., C. Snyder, and J. D. G. Jones.** 1987. T-DNA is organized predominantly in inverted repeat structures in plants transformed with *A. tumefaciens* C58 derivatives. *Mol. Gen. Genet.* **207:**471–477.

61. **Koukolikova-Nicola, Z., L. Albright, and B. Hohn.** 1987. The mechanism of T-DNA transfer from *Agrobacterium tumefaciens* to the plant cell, p. 109–148. *In* T. Hohn and J. Schell (ed.), *Plant DNA Infectious Agents.* Springer-Verlag, New York.

62. **Koukolikova-Nicola, Z., R. D. Shillito, B. Hohn, K. Wang, M. Van Montagu, and P. Zambryski.** 1985. Involvement of circular intermediates in the transfer of T-DNA from *Agrobacterium tumefaciens* to plant cells. *Nature* (London) **313:**191–196.

63. **Kucherlapati, R. S., E. M. Eves, K. Y. Song, B. S. Morse, and O. Smithies.** 1984. Homologous recombination between plasmids in mammalian cells can be enhanced by pretreatment of input DNA. *Proc. Natl. Acad. Sci. USA* **81:**3153–3157.

64. **Kustu, S., K. Sei, and J. Keener.** 1986. Nitrogen regulation in enteric bacteria. *Symp. Soc. Gen. Microbiol.* **39:**139–154.

65. **Kwok, W. W., E. W. Nester, and M. P. Gordon.** 1985. Unusual plasmid DNA organization in an octopine crown gall tumor. *Nucleic Acids Res.* **13:**459–471.

66. **Leemans, J., R. Deblaere, L. Willmitzer, H. De Greve, J. P. Hernalsteens, M. Van Montagu, and J. Schell.** 1982. Genetic identification of functions of TL-transcripts in octopine crown galls. *EMBO J.* **1:**147–152.

67. **Lemmers, M., M. DeBeuckeleer, M. Holsters, P. Zambryski, A. Depicker, J. P. Hernalsteens, M. Van Montagu, and J. Schell.** 1980. Internal organization, boundaries and integration of the Ti plasmid DNA in nopaline crown gall tumors. *J. Mol. Biol.* **144:**353–376.

68. **Leroux, B., M. F. Yanofsky, S. C. Winans, J. E. Ward, S. F. Ziegler, and E. W. Nester.** 1987. Characterization of the *virA* locus of *Agrobacterium tumefaciens*: a transcriptional regulator and host range determinant. *EMBO J.* **6:**849–856.

69. **Lichtenstein, C. P., and S. L. Fuller.** 1987. Vectors for the genetic engineering of plants, p. 103–183. *In Genetic Engineering*, vol. 6. Academic Press, Inc., New York.

70. **Lin, F. W., K. Sperle, and N. Sternberg.** 1984. Model for homologous recombination during transfer of DNA into mouse L cells: role for DNA ends in the recombination process. *Mol. Cell. Biol.* **4:**1020–1034.

71. **Machida, Y., S. Usami, A. Yamamoto, Y. Niwa, and I. Takebe.** 1986. Plant inducible recombination between the 25 bp border sequences of T-DNA in *A. tumefaciens. Mol. Gen. Genet.* **204:**374–382.

72. **Marks, J. R., T. J. Lynch, J. E. Karlinsey, and M. F. Thomashow.** 1987. *Agrobacterium tumefaciens* virulence locus *pscA* is related to the *Rhizobium meliloti exoC* locus. *J. Bacteriol.* **169:**5835–5837.

73. **Melchers, L. S., D. V. Thompson, K. B. Idler, R. A. Schilperoort, and P. J. J. Hooykaas.** 1986. Nucleotide sequence of the virulence gene *virG* of *Agrobacterium tumefaciens* octopine Ti plasmid: significant homology between *virG* and the regulatory genes *ompR*, *phoB*, and *dye* of *E. coli. Nucleic Acids Res.* **24:**9933–9942.

74. **Morris, R. O.** 1986. Genes specifying auxin and cytokinin biosynthesis in phytopathogens. *Annu. Rev. Plant Physiol.* **37:**509–538.

75. **Ninfa, A., and B. Magasanik.** 1986. Covalent modification of the glnG product, NR1, by the glnL product, NRII, regulates transcription of the glnALG operon in *E. coli. Proc. Natl. Acad. Sci. USA* **83:**5909–5913.

76. **Nixon, B. T., C. W. Ronson, and F. M. Ausubel.** 1986. Two-component regulatory systems responsive to environmental stimuli share strongly conserved domains with the nitrogen assimilation regulatory genes ntrB and ntrC. *Proc. Natl. Acad. Sci. USA* **83:**7850–7854.

77. **Ooms, G., P. M. Klapwijk, J. A. Poulis, and R. A. Schilperoort.** 1980. Characterization of Tn*904* insertions in octopine Ti plasmid mutants of *Agrobacterium tumefaciens. J. Bacteriol.* **144:**82–91.

78. **Otten, L., H. DeGreve, J. Leemans, R. Hain, P. Hooykass, and J. Schell.** 1984. Restoration of virulence of *vir* region mutants of *A. tumefaciens* strain B6S3 by coinfection with normal and mutant *Agrobacterium* strains. *Mol. Gen. Genet.* **195:**159–163.

79. **Peralta, E. G., R. Hellmiss, and R. Ream.** 1986. *overdrive*, a T-DNA transmission enhancer on the *A. tumefaciens* tumor-inducing plasmid. *EMBO J.* **5:**1137–1142.

80. **Peralta, E. G., and L. W. Ream.** 1985. T-DNA border sequences required for crown gall tumorigenesis. *Proc. Natl. Acad. Sci. USA* **82:**5112–5116.

81. **Peters, K. N., J. W. Frost, and S. R. Long.** 1986. A plant flavone, luteolin, induces expression of *Rhizobium meliloti* nodulation genes. *Science* **233:**977–980.

82. **Powell, B. S., G. K. Powell, R. O. Morris, P. M. Rogowsky, and C. I. Kado.** 1987. Nucleotide sequence of the virG locus of the *Agrobacterium tumefaciens* plasmid pTiC58. *Mol. Microbiol.* **1:**309–316.

83. **Porter, S. G., M. F. Yanofsky, and E. W. Nester.** 1987. Molecular characterization of the *virD* operon from *A. tumefaciens. Nucleic Acids Res.* **15:**7503–7517.

84. **Puvanesarajah, V., F. M. Schell, G. Stacey, C. J. Douglas, E. W. Nester.** 1985. Role for 2-linked β-D-glucan in the virulence of *Agrobacterium tumefaciens. J. Bacteriol.* **164:**102–106.

85. **Redmond, J. W., M. Bartley, M. A. Djordjevic, R. W. Innes, P. L. Kuempel, and B. Rolfe.** 1986. Flavones induce expression of nodulation genes in *Rhizobium*. *Nature* (London) 323:632–635.
86. **Rogers, S. G., and H. Klee.** 1987. Pathways to plant genetic manipulation employing *Agrobacterium*, p. 179–203. *In* T. Hohn and J. Schell (ed.), *Plant DNA Infectious Agents*. Springer-Verlag, New York.
87. **Ronson, C. W., B. T. Nixon, and F. M. Ausubel.** 1987. Conserved domains in bacterial regulatory proteins that respond to environmental stimuli. *Cell* 5:579–581.
88. **Rubin, R. A.** 1986. Genetic studies on the role of octopine T-DNA border regions in crown gall tumor formation. *Mol. Gen. Genet.* 202:312–320.
89. **Saedler, H., and P. Nevers.** 1985. Transposition in plants: a molecular model. *EMBO J.* **4**:585–590.
90. **Schroder, G., S. Waffenschmidt, E. W. Weiler, and J. Schrolder.** 1984. The T-region of Ti plasmids encodes for an enzyme synthesizing indole-3-acetic acid. *Eur. J. Biochem.* **138**:387–391.
91. **Shaw, C. H., M. D. Watson, G. H. Carter, and C. H. Shaw.** 1984. The right hand copy of the nopaline Ti plasmid 25 bp repeat is required for tumor formation. *Nucleic Acids Res.* **12**:6031–6041.
92. **Simoens, C., T. Alliotte, R. Mendel, A. Muller, M. Van Lijsebettens, J. Schell, M. Van Montagu, and D. Inze.** 1986. A binary vector for transferring genomic libraries to plants. *Nucleic Acids Res.* **14**:8073–8090.
93. **Simpson, R. B., P. J. O'Hara, W. Kwok, A. L. Montoya, C. Lichtenstein, M. P. Gordon, and E. W. Nester.** 1982. DNA from the A6S/2 crown gall tumor contains scrambled Ti plasmid sequences near its junctions with the plant DNA. *Cell* **29**:1005–1014.
94. **Simpson, R. B., A. Spielman, L. Margossian, and T. D. McKnight.** 1986. A disarmed binary vector from *A. tumefaciens* functions in *A. rhizogenes*. *Plant Mol. Biol.* **6**: 403–415.
95. **Slightom, J. L., L. Jouanin, F. Leach, R. F. Drong, and D. Tepfer.** 1985. Isolation and identification of TL-T-DNA/ plant junctions in *Convolvulus arvenis* transformed by *Agrobacterium rhizogenes* strain A4. *EMBO J.* **4**:3069–3077.
96. **Spielman, A., and R. B. Simpson.** 1986. T-DNA structure in transgenic tobacco plants with multiple independent integration sites. *Mol. Gen. Genet.* **205**:34–41.
97. **Stachel, S. E., G. An, C. F. Flores, and E. W. Nester.** 1985. A Tn3 transposon for the random generation of beta-galactosidase gene fusions: application to the analysis of gene expression in *Agrobacterium*. *EMBO J.* **4**:891–898.
98. **Stachel, S. E., E. Messens, M. Van Montagu, and P. Zambryski.** 1985. Identification of the signal molecules produces by wounded plant cells which activate the T-DNA transfer process in *Agrobacterium tumefaciens*. *Nature* (London) 318:624–629.
99. **Stachel, S. E., and E. W. Nester.** 1986. The genetic and transcriptional organization of the *vir* region of the A6 Ti plasmid of *Agrobacterium*. *EMBO J.* **5**:1445–1454.
100. **Stachel, S. E., E. W. Nester, and P. Zambryski.** 1986. A plant cell factor induces *Agrobacterium tumefaciens vir* gene expression. *Proc. Natl. Acad. Sci. USA* **83**:379–383.
101. **Stachel, S. E., B. Timmerman, and P. Zambryski.** 1986. Generation of single-strand T-DNA molecules during the initial stages of T-DNA transfer from *Agrobacterium tumefaciens* to plant cells. *Nature* (London) 322:706–712.
102. **Stachel, S. E., B. Timmerman, and P. Zambryski.** 1987. Activation of *Agrobacterium tumefaciens vir* gene expression generates multiple single-stranded T-strand molecules from the pTiA6 T-region: requirement for 5′ *vir*D products. *EMBO J.* **6**:857–863.
103. **Stachel, S. E., and P. Zambryski.** 1986. *vir* A and *vir*G control the plant-induced activation of the T-DNA transfer process of *Agrobacterium tumefaciens*. *Cell* **46**:325–333.
104. **Stachel, S. E., and P. C. Zambryski.** 1986. *Agrobacterium tumefaciens* and the susceptible plant cell: a novel adaptation of extracellular recognition and DNA conjugation. *Cell* **47**:155–157.
105. **Tempe, J., and A. Petit.** 1982. Opine utilization by *Agrobacterium*, p. 451–459. *In* G. Kahl and J. Schell (ed.), *Molecular Biology of Plant Tumors*. Academic Press, Inc., New York.
106. **Thomashow, M. F., J. E. Karlinsey, J. R. Marks, and R. E. Hurlbert.** 1987. Identification of a new virulence locus in *Agrobacterium tumefaciens* that affects polysaccharide composition and plant-cell attachment. *J. Bacteriol.* **169**:3209–3216.
107. **Thomashow, M. F., R. Nutter, A. Montoya, M. P. Gordon, and E. W. Nester.** 1980. Integration and organization of Ti plasmid sequences in crown gall tumors. *Cell* **19**:729–739.
108. **Thomashow, L. S., S. Reeves, and M. F. Thomashow.** 1984. Crown gall oncogenesis: evidence that a T-DNA gene from the *Agrobacterium* Ti plasmid pTiA6 encodes an enzyme that catalyzes synthesis of indole acetic acid. *Proc. Natl. Acad. Sci. USA* **81**:5071–5075.
109. **Timmerman, B., M. Van Montagu, and P. Zambryski.** 1988. Site-specific recombination in *Agrobacterium*: vir-induced excision of the T-DNA from the Ti plasmid. *J. Mol. Biol.* **203**:373–384.
110. **Traxler, B. A., and E. G. Minkley.** 1987. Revised genetic map of the distal end of the F transfer operon: implications for DNA helicase I, nicking at *ori*T, and conjugal DNA transport. *J. Bacteriol.* **169**:3251–3259.
111. **Ursic, D., J. L. Slightom, and J. D. Kemp.** 1983. *A. tumefaciens* T-DNA integrates into multiple sites of the sunflower crown gall genome. *Mol. Gen. Genet.* **190**:494–503.
112. **Van Haaren, M. J. J., N. J. A. Sedee, R. A. Schilperoort, and P. J. J. Hooykaas.** 1987. Overdrive is a T-region transfer enhancer which stimulates T-strand production in *A. tumefaciens*. *Nucleic Acids Res.* **15**:8983–8997.
113. **Velton, J., and J. Schell.** 1985. Selection-expression plasmid vectors for use in genetic transformation of higher plants. *Nucleic Acids Res.* **13**:6981–6998.
114. **Veluthambi, K., R. K. Jayaswal, and S. B. Gelvin.** 1987. Virulence genes A, G, and D mediate the double stranded border cleavage of T-DNA from the *Agrobacterium* Ti plasmid. *Proc. Natl. Acad. Sci. USA* **84**:1881–1885.
115. **Veluthambi, K., W. Ream, and S. B. Gelvin.** 1988. Virulence genes, borders, and overdrive generate single-stranded T-DNA molecules from the A6 Ti plasmid of *Agrobacterium tumefaciens*. *J. Bacteriol.* **170**:1523–1532.
116. **Wallroth, M., A. G. M. Gerats, S. G. Rogers, R. T. Fraley, and R. B. Horsch.** 1986. Chromosomal localization of foreign genes in Petunia hybrida. *Mol. Gen. Genet.* **202**: 6–15.
117. **Wang, K., C. Genetello, M. Van Montagu, and P. Zambryski.** 1987. Sequence context of the T-DNA border repeat element determines its relative activity during T-DNA transfer to plant cells. *Mol. Gen. Genet.* **210**:338–346.
118. **Wang, K., L. Herrera-Estrella, M. Van Montagu, and P. Zambryski.** 1984. Right 25 bp terminus sequences of the nopaline T-DNA is essential for and determines direction of DNA transfer from *Agrobacterium* to the plant genome. *Cell* **38**:35–41.

119. **Wang, K., S. E. Stachel, B. Timmerman, M. Van Montagu, and P. Zambryski.** 1987. Site-specific nick occurs within the 25 bp transfer promoting border sequence following induction of *vir* gene expression in *Agrobacterium tumefaciens*. *Science* **235:**587–591.
120. **Ward, J. E., D. Akiyoski, D. Regier, A. Datta, M. P. Gordon, and E. W. Nester.** 1988. Characterization of the virB operon from an *Agrobacterium tumefaciens* Ti plasmid. *J. Biol. Chem.* **263:**5804–5814
121. **Whitehouse, H. L. K. (ed.).** 1982. *Genetic Recombination*. John Wiley and Sons, Inc., New York.
122. **Willetts, N., and R. Skurray.** 1987. Structure and function of the F factor and mechanism of conjugation, p. 1110–1133. *In* F. C. Neidhardt, J. L. Ingraham, K. B. Low, B. Magasanik, M. Schaechter, and H. E. Umbarger (ed.), *Escherichia coli and Salmonella typhimurium: Cellular and Molecular Biology*, vol. 2. American Society for Microbiology, Washington, D.C.
123. **Willetts, N., and B. Wilkins.** 1984. Processing of plasmid DNA during bacterial conjugation. *Microbiol. Rev.* **48:** 24–41.
124. **Winans, S. C., P. Allenza, S. E. Stachel, K. E. McBride, and E. W. Nester.** 1987. Characterization of the *vir*E operon of the *Agrobacterium* Ti plasmid pTiA6. *Nucleic Acids Res.* **15:**825–837.
125. **Winans, S. C., P. R. Ebert, S. E. Stachel, M. P. Gordon, and E. W. Nester.** 1986. A gene essential for *Agrobacterium* virulence is homologous to a family of positive regulatory loci. *Proc. Natl. Acad. Sci. USA* **83:**8278–8282.
126. **Yadav, N. S., J. Vanderleyden, D. R. Bennett, W. M. Barnes, and M.-D. Chilton.** 1982. Short direct repeats flank the T-DNA on a nopaline Ti plasmid. *Proc. Natl. Acad. Sci. USA* **79:**6322–6326.
127. **Yamamoto, A., M. Iwahaski, M. F. Yanovsky, E. W. Nester, I. Takebe, and Y. Machida.** 1987. The proximal region in the *virD* locus of *Agrobacterium tumefaciens* is necessary for the plant inducible circularization of T-DNA. *Mol. Gen. Genet.* **206:**174–177.
128. **Yanofsky, M., B. Lowe, A. Montoya, R. Rubin, W. Krul, M. Gordon, and E. W. Nester.** 1985. Molecular and genetic analysis of factors controlling host range in *Agrobacterium tumefaciens*. *Mol. Gen. Genet.* **201:**237–246.
129. **Yanofsky, M. F., and E. W. Nester.** 1986. Molecular characterization of a host-range determining locus from *Agrobacterium tumefaciens*. *J. Bacteriol.* **168:**244–250.
130. **Yanofsky, M. F., S. G. Porter, C. Young, L. M. Albright, M. P. Gordon, and E. W. Nester.** 1986. The *vir*D operon of *Agrobacterium tumefaciens* encodes a site-specific endonuclease. *Cell* **47:**471–477.
131. **Zambryski, P., A. Depicker, K. Kruger, and H. Goodman.** 1982. Tumor induction by *Agrobacterium tumefaciens*: analysis of the boundaries of T-DNA. *J. Mol. Appl. Genet.* **1:**361–370.
132. **Zambryski, P., L. Herrera-Estrella, M. De Block, M. Van Montagu, and J. Schell.** 1984. The use of the Ti plasmid of *Agrobacterium* to study the transfer and expression of foreign DNA in plant cells: new vectors and methods, p. 253–278. *In* J. Setlow and A. Hollaender (ed.), *Genetic Engineering, Principles and Methods*, vol. 6. Plenum Publishing Corp., New York.
133. **Zambryski, P., M. Holsters, K. Kruger, A. Depicker, J. Schell, M. Van Montagu, and H. M. Goodman.** 1980. Tumor DNA structure in plant cells transformed by *A. tumefaciens*. *Science* **209:**1385–1391.
134. **Zambryski, P., H. Joos, C. Genetello, J. Leemans, M. Van Montagu, and J. Schell.** 1983. Ti plasmid vector for the introduction of DNA into plant cells without alteration of their normal regeneration capacity. *EMBO J.* **2:**2143–2150.

Chapter 13

Transposable Elements in *Saccharomyces cerevisiae*

JEF D. BOEKE

Jef D. Boeke ■ Department of Molecular Biology and Genetics, Johns Hopkins University School of Medicine, 75 North Wolfe Street, Baltimore, Maryland 21205.

I. INTRODUCTION

The yeast *Saccharomyces cerevisiae* contains several transposons or transposonlike sequences. These reside in both nuclear and mitochondrial DNA, although no single yeast transposon sequence has been found in both compartments of the cell. The basic features of the *S. cerevisiae* transposons covered in

Table 1. Properties of *S. cerevisiae* transposable elements

Name	Terminal repeat type	Target site duplication	Copy number	Location
Ty1	δ, direct repeat	5	30	Nucleus
Ty2	δ, direct repeat	5	10	Nucleus
Ty3	σ, direct repeat	5	2–4	Nucleus
δ		5	±100	Nucleus
σ		5	20–30	Nucleus
τ		5	15–25	Nucleus
Ω			0–1	Mitochondrion

this book, Ty1, Ty2, Ty3, δ, σ, τ, and Ω, are summarized in Table 1. Of these elements, only Ω is found within the mitochondrial DNA; it is discussed elsewhere (B. Dujon and L. Belcour, this volume). The transposons seem to vary in their mode of transposition, deduced in most cases from their structural properties. Most yeast transposons so far discovered are called **retrotransposons** because their mode of transposition resembles retroviral replication (H. Varmus and P. Brown, this volume) in many ways. Ty1 elements have been shown directly to transpose via a reverse transcription process; Ty2 and Ty3 show considerable structural similarity to Ty1 and thus probably transpose in a similar way. Retroviral proviruses are flanked by long terminal repeat (LTR) sequences. Similarly, δ elements form the LTRs of Ty1 and Ty2 elements; σ elements form the LTRs of the Ty3 element; τ elements may represent the LTR of an as yet undiscovered retrotransposon. In contrast, the mitochondrial element Ω apparently transposes directly through a DNA form, probably via a homology-dependent mechanism very reminiscent of yeast mating-type switching (A. Klar, this volume). Finally, coding sequences with homology to reverse transcriptase and RNase H have been identified within certain fungal mitochondrial introns (127; R. Doolittle, personal communication); the relationships of these open reading frames to transposable elements (if any) are unknown. Further information of Ty elements may be found in previous reviews (8, 10, 13, 75, 125, 152, 170, 184, 190).

Curiously, no transposons of the *Drosophila* P element and maize *Ac*/*Ds* element types (W. R. Engels, this volume; N. V. Fedoroff, this volume), characterized by short, inverted terminal repeats and the fact that they are located in the nucleus, have been found in *S. cerevisiae*, although among some more complex, larger eucaryotes this class of transposable element may be at least as widely distributed as the retrotransposons (76, 77). A third, less common class of eucaryotic transposons, typified by the *Drosophila* F, G, and I elements (D. J. Finnegan, this volume) and mammalian L1/LINE elements (C. A. Hutchison III, S. C. Hardies, D. D. Loeb, W. R. Shehee, and M. H. Edgell, this volume) and characterized by a reverse-transcriptase-like open reading frame but no LTR sequences, is also unknown in yeasts. Perhaps the constraints of a small nuclear genome are somehow more amenable to exploitation by retrotransposons than by the other classes of eucaryotic transposons.

This chapter covers the nuclear retrotransposons (and their associated LTR sequences) of *S. cerevisiae*. Because most is known about the Ty1 and δ elements, aspects of these prototypic elements will be emphasized. Where data exist on the other transposable elements of *S. cerevisiae*, they will be compared to the prototype elements. The chapter includes the transposition process itself, the regulation of the transposition process, the effects of yeast transposons on adjacent genes, and the distribution of yeast transposons both within the genome and among different strains and species of yeasts. The relationship between these transposable elements and the genes of the host yeast will be examined in a consideration of the *SPT* genes, a set of host genes whose products apparently interact intimately with the Ty elements and their LTR sequences, especially the δ elements. Finally, the effects of yeast transposons, which are families of dispersed repeat sequences, on genomic stability will be considered.

II. NOMENCLATURE

There is no generally accepted nomenclature for yeast transposon names, genes, or the gene products they encode. In this section, I propose the following conventions, in accord with standard nomenclatural rules for yeast genetics (164, 188) and previously proposed guidelines for procaryotic transposons (24) and largely consistent with protein nomenclatural conventions currently in use for retroviruses, for referring to yeast transposons, their open reading frames, and their proteins.

A. Element Names

Yeast transposons will be written Ty1, Ty2, Ty3, etc., except for Ω, which is already in common usage. Ty1 through Ty3 are currently accounted for. Note that neither Ty nor the number is italicized and that the second letter is written in lower case as in bacterial Tn elements, *Caenorhabditis* Tc elements, etc. It is at this point impractical to give each isolate of a Ty element a new number, particularly since they often change in sequence during transposition (14, 16) and because so many isolates have already been numbered in a nonsystematic way; hence, it is proposed to give new numbers in this series only to new families of elements which clearly differ in a substantive way from existing Ty1, Ty2, and Ty3 families. Previous isolates with names such as Ty917 and Ty912 should thus be written Ty2-917 and Ty1-912 to indicate the family to which they belong and to avoid confusion. Elements which resemble LTRs of retrotransposons are given Greek letter designations (δ, σ, and τ being accounted for).

B. Gene Names

The wild-type genes, being generally dominant (11), will be written in capital italics (e.g., *TYA*, *TYB*). Mutant alleles, whether generated in vitro or naturally occurring, will be written in lower case if recessive and upper case if dominant (e.g., *tyb-2098*, a naturally occurring mutation in *TYB*). When possible, the allele number will refer to the base mutated in the Ty element, using as reference points the published Ty1 (Ty1-912) sequence (38) and Ty2 (Ty2-117) sequence (189).

Usually, an author will refer only to one class of Ty element, so that the class of element (1, 2, or 3) can be inferred from the context. It may be desirable (e.g., in comparative studies) to distinguish the *TYA* and *TYB* open reading frames from different classes of Ty elements as follows: *TYA1*, *TYA2*, and *TYA3*; *TYB1*, *TYB2*, and *TYB3*.

C. Protein Names

The proteins encoded by these genes will be prefixed with the letter p and the estimated molecular weight of the protein obtained from sodium dodecyl sulfate-polyacrylamide gel electrophoresis analysis, followed by a hyphen and the gene from which it derives written in capital roman letters (e.g., p58-TYA). Although there will inevitably be some initial variations in these molecular weight estimations due to such factors as differences in electrophoresis conditions, the molecular weights should eventually converge to a mutually acceptable figure. This convention is in accord with that used by retrovirologists. Note that alternative systems of protein nomenclature have been proposed (123, 131). In the nomenclature suggested by the Kingsmans' group (123), Ty-specified proteins are named p1, p2, p3, etc., in the order of their discovery. The advantage of their system is that differences in molecular weight estimations will not affect the name of the protein. However, it is hoped that the system proposed above will be more informative to readers in that the protein name indicates the reading frame and its approximate molecular weight. It is hoped that these proposals, which incorporate some aspects of all of the currently used conventions, will be generally adopted.

III. STRUCTURE OF *S. CEREVISIAE* TRANSPOSONS

A. General DNA Structure

The restriction maps of the "complete" *S. cerevisiae* transposons Ty1, Ty2, and Ty3 are diagrammed in Fig. 1. Like true retroviruses, Ty elements are flanked by two identical or nearly identical LTR sequences (Table 2) (the LTR is called δ for Ty1 and Ty2 and σ for Ty3). The LTR sequences can be divided into three domains, U3, R, and U5, by analogy to retroviral sequences (Fig. 2; see also section on RNA structure, below). The region between the two LTR sequences, which contains the open reading frames, is often referred to as ϵ (epsilon) in the case of Ty1 and Ty2; it has no special name in the case of Ty3.

The δ, σ, and τ sequences are compared to each other and to the Moloney murine leukemia virus (MoMLV) LTR in Fig. 3. The retrotransposons are remarkably similar to each other in structure; the LTR-like sequences δ, σ, and τ also share certain structural similarities which suggest that they are functionally related as well. These similarities include their total size of 334 to 371 base pairs (bp); the presence of short terminal inverted repeats conforming to the consensus TGTTG. . .CAACA (δ sequences conform only partially to this); and certain blocks of short sequence identities scattered throughout these elements with relatively similar spacing (86). Finally, intact Ty1, Ty2, Ty3, δ, σ, and τ elements are nearly always flanked by 5-bp duplications of host sequence (30, 39, 48, 69, 84).

Table 2. LTR sequences in Ty elements

Element	Length (bp)	Identical?	References
Ty1-912	334	Yes	38
Ty1-ADH2-2[c]	334	Yes	191
Ty1-ADH2-8[c]	334	Yes	191
Ty1-ADH2-7[c]	337	Yes	191
Ty1-ADH2-3[c]	332	No	191
Ty1-ADH2-6[c]	332	No	191
Ty1-p109	331	Yes	93
Ty1-H3	334	No	11
Ty1-B10	338	Yes	84
Ty1-D15	338	Yes	84
Ty2-117	332	Yes	189
Ty2-917	332/333	No	69, 149, 152
Ty3-pSBS12	340	Yes	39

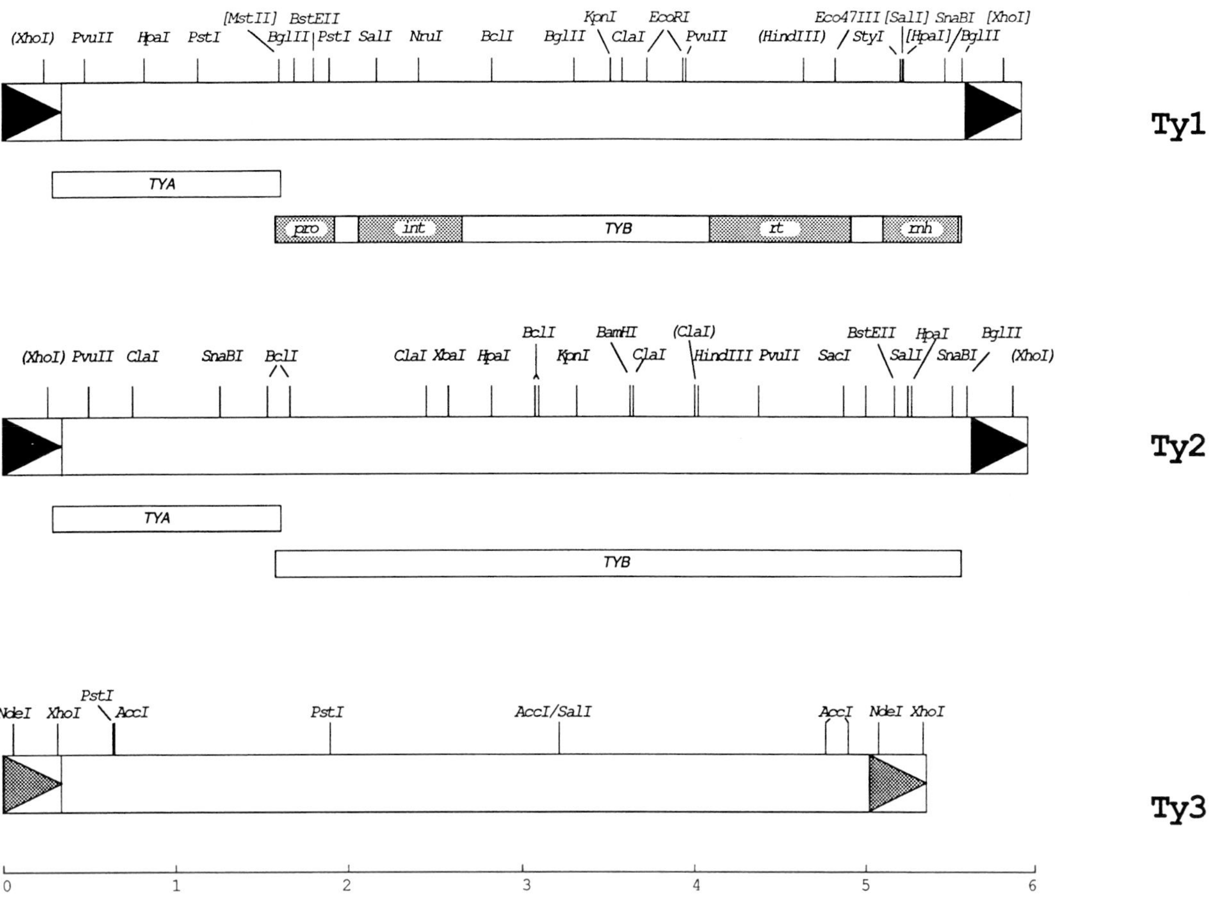

Figure 1. Restriction maps of Ty1, Ty2, and Ty3. Selected restriction sites are shown. Polymorphic restriction sites are indicated in parentheses or square brackets. Sites in parentheses are found in the prototypic elements shown but not in certain other Ty elements; sites in square brackets are found in certain other Ty elements but not in the prototypic elements shown. The prototypic elements shown are Ty1-H3 (Ty1) (11) and Ty2-117 (Ty2) (189). Boxed black triangles, δ elements; boxed shaded triangles, σ elements; triangles point in the direction of transcription of the Ty elements. The *TYA* and *TYB* open reading frames are indicated for Ty1 and Ty2; sequence analysis for Ty3 is not yet complete. The domains of amino acid sequence similarity to retroviral *pol* genes are indicated by shaded regions in the Ty1 diagram and labeled as follows: pro, protease; int, integrase; rt, reverse transcriptase, rnh, RNase H. These domains are found at similar positions in Ty2.

B. Open Reading Frames

The open reading frames of Ty1 and Ty2 show considerable similarity in amino acid sequence to each other (189) and to certain regions of vertebrate retroviruses as well as to retrotransposons from a variety of sources (38, 130) (R. F. Doolittle, D. F. Feng, M. S. Johnson, and M. A. McClure, *Q. Rev. Biol.*, in press). An alignment of the *TYB* open reading frame with the *pol* regions of MoMLV and the *Drosophila* retrotransposons *copia* and *17.6* is presented in Fig. 4.

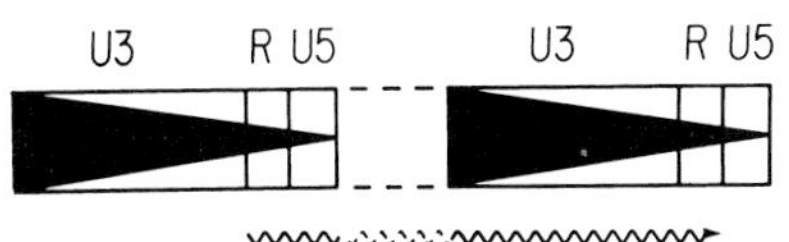

Figure 2. U3, R, and U5 domains of LTR sequences. Diagram indicates the definitions of the domains of LTR sequences in general. Boxed triangles, LTR sequences; wavy line, RNA.

The Ty1 and Ty2 families have been shown to differ mainly within the *TYA* and *TYB* open reading frames; the ends are relatively similar (82, 108). Comparison of the amino acid sequences indicates that within *TYB*, the protease, integrase, reverse transcriptase, and RNase H regions are relatively well conserved, but a region of unknown function (sometimes referred to as the ? domain, although it is not certain whether this is truly a separate functional domain of the protein [130]) is considerably divergent (189).

Although some *Drosophila* retrotransposons have a third open reading frame which may correspond to an *env* gene (96, 122, 159), there is no such gene in Ty1 or Ty2.

```
(A)     10         20         30         40         50
TGTTGGAATA GAAATCAACT ATCATCTACT AACTAGTATT TACATTACTA

        60         70         80         90        100
GTATATTATC ATATACGGTG TTAGAAGATG ACGCAAATGA TGAGAAATAG
                                            UUUUUUUUUUU

       110        120        130        140        150
TCATCTAAAT TAGTGGAAGC TGAAACGCAA GGATTGATAA TGTAATAGGA
UUUUUUUUUUUUUUUUUUUUUUUUUUUUUUU
                RRRRRRRRRRRRRRR

       160    (G) 170        180        190        200
TCAATGAATA TAAACATATA AAATGATGAT AATAATATTT ATAGAATTGT
     TTTTTTTTTTTTTTTTTTTTTTT

       210        220        230        240        250
GTAGAATTGC AGATTCCATT TTGAGGATTC CTATATCCTC GAGGAGAACT
                                            *

       260        270        280        290        300
TCTAGTATAT TCTGTATACC TAATATTATA GCCTTTATCA ACAATGGAAT
                                               +++
                                             AAAAAA
       310        320        330
CCCAACAATT ATCTCAACAT TCACCCAATT CTCA
```

```
(B)     10         20         30         40         50
TGTTGTATCT CAAAATGAGA TATGTCAGTA TGACAATACG TCATCCTAAA

        60         70         80         90        100
CGTTCATAAA ACACATATGA AACAACCTTA TAACAAAACG AACAACATGA
     aaaaa aaa   aaaa aaaa

       110        120        130        140        150
GATAAAACCC GTCCTTCCCT AGCTGAACTA CCAAAAGTAT AAATGCCTGA
                                           TTTTTTT

       160        170        180        190        200
ACAATTAGTT TAGATCCGAG ATTCCGCGCT TCCACCACTT AATATGATTC

       210        220        230        240        250
ATATTTTATA TAATATATAA GATAAGTAAC ATTCCGTGAA TTAATCTGAT
       TTTTTTTTTTTTT    *

       260        270        280        290        300
AAACTGTTTT GACAACTGGT TACTTCCCTA AGACCGTTTA TATTAGGATT

       310        320        330        340
GTCAAGACAC TCCGGTATTA CTCGAGCCCG TAATACAACA
```

```
(C)     10         20         30         40         50
TGTTGGAACG AGAGTAATTA ATAGTGACAT GAGTTGCTAT GGTAACAATT

        60         70         80         90        100
CAATGCTTAC ATCGTATATT AATGTACAAC TCGTATACGT TTAAGTGTGA

       110        120        130        140        150
TTGCGCCTAT TGCAGAAGGA ATGTTAAACG AGAAGCTCAG ACAATACTGA

       160        170        180        190        200
AGCTGTGTTA AAGACCTATT AGTTGAACAT GTTATGGTAG GTACATATAT

       210        220        230        240        250
GAGGAATATG AGTCGTCACA TCAATGTATA GTAACTACCG GAATCACTAT

       260        270        280        290        300
TATATTGGTC ATGATTAATA TGACCAATCG GCGTGTGTTT TATATACCTC

       310        320        330        340        350
TCTTATTTAG TATAAGAAGA TCAGTAATTA TTTCTTCATT AATACTAATT

       360        370
TTTAACCTCT AATTATCAAC A
```

```
(D)     10         20         30         40         50
TGAAAGACCC CACCTGTAGG TTTGGCAAGC TAGCTTAAGT AACGCCATTT

        60         70         80         90        100
TGCAAGGCAT GGAAAAATAC ATAACTGAGA ATAGAGAAGT TCAGATCAAG

       110        120        130        140        150
GTCAGGAACA GATGGAACAG CTGAATATGG GCCAAACAGG ATATCTGTGG

       160        170        180        190        200
TAAGCAGTTC CTGCCCCGGC TCAGGGCCAA GAACAGATGG TCCCCAGATG

       210        220        230        240        250
CGGTCCAGCC CTCAGCAGTT TCTAGAGAAC CATCAGATGT TTCCAGGGTG

       260        270        280        290        300
CCCCAAGGAC CTGAAATGAC CCTGTGCCTT ATTTGAACTA ACCAATCAGT

       310        320        330        340        350
TCGCTTCTCG CTTCTGTTCG CGCGCTTCTG CTCCCCGAGC TCAATAAAAG

       360        370        380        390        400
AGCCCACAAC CCCTCACTCG GGGCGCCAGT CCTCCGATTG ACTGAGTCGC
                       *

       410        420        430        440        450
CCGGGTACCC GTGTATCCAA TAAACCCTCT TGCAGTTGCA TCCGACTTGT
                    AA AAAAAAAAAA AAAAAAAAA

       460        470        480        490        500
GGTCTCGCTG TTCCTTGGGA GGGTCTCCTC TGAGTGATTG ACTACCCGTC

       510
AGCGGGGGTC TTTCA
```

Figure 3. LTR sequences of Ty elements and MoMLV. The sequences of selected δ, σ, and τ sequences are shown. Note their similar sizes and that all begin and end with TG. . .CA sequences. Although the MoMLV LTR and other mammalian retrovirus LTRs are considerably longer than the *S. cerevisiae* LTRs, avian retrovirus LTRs are often about the size of the *S. cerevisiae* LTRs (182). (A) δ sequence. Underlined sequences are sequences of 10 bp or more which are identical in a large collection of sequenced δ elements (190). The UUU symbols underline a region which may function as a UAS (114); the TTT symbols underline a putative TATA region (Kapakos et al., submitted); the RRR symbols underline the δ copy of the reverse complement sequence reported by Warmington et al. (189); the (G) is above a nucleotide substitution which apparently inactivates the δ promoter (Durbin, Ph.D. Thesis). The asterisk underlies a major site of initiation of transcription; AAA, nucleotide positions at which pol(A) are added (61); +++, nucleotides specifying AUG of the *TYA* open reading frame. Direction of transcription is from left to right. Sequence shown is Ty1-912 δ (38), which is identical to the 3′ δ of Ty1-H3 (11). (B) σ sequence. Symbols: aaa underlies sequences responsible for α-factor inducibility of σ transcription; TTT underlies putative TATA boxes; asterisk underlies a major transcript start site. Sequence shown is from pFD2 (48); the reverse complement of the original sequence is given in order to orient this LTR such that transcription would proceed from left to right (39). (C) τ sequence. The direction of transcription (if any) of τ elements is unknown, hence the order of the sequence is as originally published. Sequence shown is from pFG26 (30). (D) MoMLV LTR sequence. Sequence shown is 5′ LTR from BALB/Mo mouse provirus (183). Asterisk indicates the start site of transcription; AAA, site of polyadenylation.

```
                                                                   * *
Ty1-912          TISTTFTLGQELTESTVNHTNHSDDEL    PGHLLL   DSGASRTLIR SAHHIHSASSN
Ty2-117            TTNLVLGQQQKESKPTHTIDSNDEL    PDHLLI   DSGRSQTLVR  AHYLHHATPN
COPIA            QVQTATSHGIAFMVKEVNNTSVMDN       CGFVL   DSGASDHLINDESLYTDSVEVV
17.6            TGRKFSATSLGKPQYITIKYKENN         LKCLI   DTGSTVNMTSKNIF  D LPIQ
Mo-MLV      TLDDQGGQGQDPPPEPR ITLKVGGQP          VTFLV   DTGAQHSVLTQN  PGP LSDK

Ty1-912          PDINVVDAQKRNIP     INAIGDLQFHFQDNTKTSIKVLH    TPNIAYDLLSLNELAAVDITACF
Ty2-117          SEINIVDAQKQDIP     INAIGNLHFNFQNGTGTSIKALH    TPNIAYDLLSLSELANQNITACF
COPIA            PPLKIAVAKQGEFI     YATKRGIVRLRNDHEITLEDVLF    CKEAAGNLMSVKRLQEAGMSIEFD
17.6             NTSTFIHTSNGPLIVN   KSIIIPSKILFPTTNEFLLHPF      SENYDLLLGRKLLAEAKATISY
Mo-MLV           SA  WVQGATGG       KRYRWTTDRKVHLATGKVTHSFL   HVPDCPYPLLGRDLLTKLKAQIHF
```

PRO

```
                   *    *                               *  *                             *
Ty1-912   STRKYPYFI HRMLAHANAQTIRYSLKNNTITYFNESDVDWSSAIDYQ CPDCLIGKSTKHRHIKGSRLKYQNSYEPFQYLHTDIFGPVHNLPNSAPSYFISFTD
Ty2-117   SVNKYPYPLIHRMLGHANFRSIQKSLKKNAVTYLKESDIEWSNACTYQ CPDCLIGKSTKHRHVKGSRLKYQESYEPFQYLHTDIFGPVHHLPKSAPSYFISFTD
COPIA     AKHKNNFRLWHERFGHISDGKL LEIKRKNM  FSDQSLLNNLELSCEICEPCLNGKQARL PFKQLKDKTHIK RPLFVVHSDVCGPITPVTLDDKNYFVIFVD
17.6      AEFKELILTAHEKLLHPGIQKTTKLFG   ETYYFPNSQLLIQNII NECSICNLAKTEHR NFDMPTKTTPKEPHCREKFMIDI   YS    SEGKHY   VS
Mo-MLV     TFELDEFLHQ LTHLSFSKMKALLERSHSPYYMLNRDRTLKNIT ETCKACAQVNASKS AVKQGTRVRGHR  PGTHWEIDF    TEIKPGLYGYKY LLVF

                                                     *                                       *
Ty1-912   ETTKFRWVYPLHDRREDSILDVFTTILAFIKNQFQASVLVIQMDRGSEYTNRTLHKFLEKNGITPCYTTTADSRAHGVAERLNRTLLDDCRTQLQCSGLPN
Ty2-117   EKTRFQWVYPLHDRREESILNVFTSILAFIKNQFNARVLVIQMDRGSEYTNKTLHKFFTNRGITACYTTTADSRAHGVAERLNRTLLNDCRTLLHCSGLPN
COPIA     QFTHYCVTYLIKYKSD  VFSMFQDFVAKSEAHFNLKVVYLYIDNGREYLSNEMRQFCVKKGISYHLTVPHTPQLNGVSERMIRTITEKARTMVSGAKLDK
17.6      CIDIYSKFATLEEIKTKDWIECKNALM RIFNQLG KPKLLKADRDGAFSSLALKRWLESEEVELQLNT  TKTGVADIERLHKTINEKIRIIKTSDDEE
Mo-MLV     IDTFSGWIEAFPTKKETAKVVTKKLLEEIFPRFG MPQVLGTDNGPAFVSKVTQTVADLLGIDWKLHCAYRPQSSGQVERMNRTI  KETLTKLTLATGS
```

INT

```
Ty1-912   MKTWDTDEYYDRKEIDPKRVINSMFIFNKKRDGTHKRRF   VARG DIQHPDTYDSGMQSNTVHHYALMTSLSLALDNNYYITQLD  ISSAYLYAD    IKEELYIRPPPHLG
Ty2-117   MNTWDTNKYYDRNDIDPKKVINSMFIFNKKRDGTHKARF   VARG DIQHPDTYDSDMQSNTVHHYALMTSLSIALDNDYYITQLD  ISSAYLYAD    IKEELYIRPPPHLG
COPIA              NKNIVDSRWVFSVKYN ELGNPIRYKARL   VARGFTQKYQIDYEETFAP VARISSFRFILSLVIQYNLKVHQMD  VKTAFLNGT    LKEEIYMRLPQGISC
17.6         YSKYSYPQAYEQE VESQIQD MLNQGIIRTSNSPY   NSPIWVVPK KQ DASGKQKFRIVIDYRKLNEITVGDRHPIPNMD  EILGKLGRC    NYFTTIDLAKGFHQI
Mo-MLV       IKQYPMSQEARL GIKPHIQR LLDQGILVPCQSPW   NTPLLPVK     KPGTNDYRPVQDLREVNKRVEDIHPTVPN PYNLLSGLPPSH    QWYTVLDLKDAFFCL

                                                                          **
Ty1-912   MNDKLIRLKKSLYELKQ       SGANWYETIKSYLIQQCGMEEVRGWSCVF       KNSQVTICLFVDDMVLFS     KNLNSNKR IIEKLKM       QY DTKIINLGES
Ty2-117   LNDKLLRLRKSLVGLKQ       SGANWYETIKSYLINCCDMQEVRGWSCVF       KNSQVTICLFVDDMILFS     KDLNANKK IITTLKK       QY DTKIINLGES
COPIA     NSDNVCKLNKAIYGLKQ       AARCWFEVFEQAL KECEFVNSSVDRCIYILDKGNINENIYVLLYVDDVVIATGDM  TRMNNFKRYLMEKFRM       TD LENIKHFIGI
17.6      EMDPESVSKTA FSTKH       GHYEY LRMPFGLKNAPATFQRCMNDILR       PLLNKHCLVYLDDIIVFSTSL DEHLQSLGLVFEKLAKA       NLKLQLDKCEF L
Mo-MLV    RLHPTSQPLFA FEWRDP EMGISGQLTW TRLPQGFKNSPTLFDEALHRDLADF  RIQHPDLILLQYVDDLLLAATSE LDCQQGTRALLQTLGNL       GYRASAKKA QIC

Ty1-912   DEEIQYDILGL EIKYQRGKYMKLGME  NSLTEKIP KLNVPLNPK    GRKLSAPGQPG      LYIDQDELEIDEDQYKEK     VHEMQKLIGLASYVGYKF RFD
Ty2-117   DNEIQYDILGL EIKYQRSKYMKLGME  KSLTEKLP KLNVPLNPK    GKKLRAPGQPG      HYIDQDELEIDEDEYKEK     VHEMQKLIGLASYVGYKF RFD
COPIA     RIEMQEDKIYLSQSAYVKKILSKFNMENCNAVSTPLPSKINYELLNS    DEDCNTPCRSL      IGCLMYIMLCTRPDLTTA     VNILSRYSSKNNSELWQ  NLK
17.6      KQETTFLGHVLTPDGILPNPEKIEAIQKYPIP TKPKEIKAFLGLTG    YYRKFIPNFAD      IAKPMTKCLKKNMKIDTT     NPEYDSAFKKLKYLISEDPILKVPD
Mo-MLV    QKQVKYLGYLLKEGQRWLTEARKETVMGQPTPKTPRQLREFL GTAGFCRLWI PGFAEMAAP    L    Y PLTKTGTLFN     WGPDQQKAYQEIKQALLTAPALGLPD
```

RT

```
                        *                                     *                            *
Ty1-912   PDNKLVAISDASY GNQPYYKSQIGNIF LLNGKVIGGKSTKASLTCTSTTEAEIHAVSESI  PLL           NKKPIIKGLLTDSRST
Ty2-117   PDNKLVAISDASY GNQPYYKSQIGNIF LLNGKVIGGKSTKASLTCTSTTEAEIHAVSEAI  PLLNNLSHLVQELNKKPIIKGLLTDSRST
COPIA     FENKIIGYVDSDWAGSEIDRKSTTGYLFKMFDFNLICWNTKRQNSVAASSTEAEYMALFEAVREALW           KLENPIK IYEDNQGC
17.6      FTKKFTLTTDASDVALGAVLS   QDGHPLSYISR   TLNEHEINYSTIEKELLAIVWATK TFR           HYLLGRHFEISSDHQPL
Mo-MLV    PDADHTWYTDGSSLLQE GQR   KAGAAVTTETEV  IWAKALDAGTSAQRAELIALTQA   LK           MAEG KKLNVYTDSRYA

                                                                        *
Ty1-912   IS             IIKSTNEEKFRNRFFGTKAMRLRDEVSGNNLYVYYIETK        KNIADVMTKPLPIKRFK
Ty2-117   IS             IIKSTNEEKFRNRFFGTKAMRLRDEVSGNNLYVYYIETK        KNIADVMTKPLPIKTFK
COPIA     IS             IANNPSCHK RAKHIDIKYHFAREQVQNNVICLEYIPTE        NQLADIFTKPLPAARFVE
17.6      SW             LYRMK DPNSKL   TR  WRVKL SEFDFDIKYIKGKE        NCVADALSRIKLEETYLSEQTQ
Mo-MLV    FATAHIHGEIYRRRGLLTSEGKEIKNKDEILALLKALFLPKRL  SIIHCPGHQKGHSAEARGNRMADQAARKAAITETPD
```

RNH

Figure 4. Amino acid sequence alignments. Alignments of the Ty1-912 and Ty2-117 sequences to the two *Drosophila* retrotransposons *17.6* and *copia* and to the retrovirus MoMLV are shown for the *TYB* reading frame. These sequence alignments (except for the Ty2 alignment) were kindly provided by R. F. Doolittle (Doolittle et al., in press). Abbreviations are as in Fig. 1. The asterisks indicate residues which were identical in the sequences compared by Doolittle et al. (in press): protease (PRO): 14 retrovirus/retrotransposon sequences, four pepsins; integrase (INT): 12 retrovirus/retrotransposon sequences; reverse transcriptase (RT): 15 retrovirus/retrotransposon/DNA-genome virus sequences; RNase H (RNH): 15 retrovirus/retrotransposon/DNA-genome virus sequences and *Escherichia coli* RNase H. The Ty1-912 DNA sequence coordinates corresponding to the termini of each homology domain are as follows: protease, 1576 to 1917; integrase, 2059 to 2673; reverse transcriptase, 4084 to 4953; RNase H, 5095 to 5535.

C. Heterogeneity of Structure within Families of *S. cerevisiae* Transposons

Typical *S. cerevisiae* strains contain about 25 to 30 Ty1 elements and about 10 Ty2 elements (23; unpublished data). Within these families, a certain degree of structural variability is found. Among the variations in structure that are known within the Ty1 family are deletions, insertions, inversions, and base substitutions of various types. The Ty2 family contains some base substitutions, but, on the whole, Ty2 seems to be structurally more uniform than the Ty1 family, perhaps because of its lower copy number.

1. Deletions

Although large deletions occasionally occur during the transposition process, most genomic Ty elements appear to be full length or nearly full length. Among a collection of approximately 50 spontaneous Ty1 insertions into a target plasmid bearing a promoterless *his3* gene, *his3Δ4*, one (H4) bears a deletion of approximately 2 kilobases (kb). Also within this collection, approximately 10% of the Ty1 elements had small deletions within the ϵ region very near to the 3′ end (J. D. Boeke and D. J. Garfinkel, unpublished data). The significance of the small deletions is unknown, but it is likely that they would render the Ty element unable to produce functional *TYB* proteins. More recently, among a collection of seven Ty1 elements which transposed into bacteriophage λ in vitro (see below), one element was found to have sustained a 3-kb deletion (S. Pelger, D. Eichinger and J. D. Boeke, unpublished data). In none of these cases has the detailed structure of such deleted Ty elements been determined, but it is likely that the fusion junctions occur at short direct repeats within the element (70, 137).

The solo LTR sequences δ, σ, and τ are often found in truncated form in the vicinity of tRNA genes (48, 92, 134, 180). The significance of this is not obvious, but it may reflect a high frequency of rearrangements in these regions of the genome; δ elements which interrupt other copies of δ, σ, and τ (as though by previous transposition events) are also common.

2. Insertions

Insertion variants of Ty1 have also been reported; the first of these is Ty1-61, a Ty from chromosome III (108), which contains a 1.2-kb insertion near the 5′ end of the element. (A recent publication disputes this claim, however [188a].) A few Ty1 elements bearing much smaller insertions were also found in the collection of Ty1 elements mentioned above. In some of these, restriction analysis suggests that these represent partial duplications of Ty sequence. One partially sequenced Ty3 element may have an internal duplication within the coding region (S. Sandmeyer, personal communication). Again, the detailed structure of these variants awaits definitive sequence analysis, but it seems likely that they represent defective elements.

3. An inversion

Errede et al. (67) reported the structure of Ty-1-CYC7-H2, a Ty1 element which activated the transcription of the *CYC1* gene and on which a great deal of molecular genetic research has been done. The 5′ δ of this element has a most unusual structure: although the U5 region of this δ is normal, its U3 region and part of its R region are missing, and in their place is an inverted copy of a portion of R and all of U5. The generation of this inversion is attributed to an error in reverse transcription of Ty RNA. The element is presumed to be transcriptionally and hence transpositionally nonfunctional, because much of its 5′ δ is missing.

4. Point mutations

The most common type of heterogeneity among different copies of Ty elements is point mutation of the base substitution type. For example, comparison of the sequences of three Ty1 elements shows that they tend to be remarkably similar in length. Most of these nucleotide changes are probably innocuous in that they are third-position changes in codons. However, there are 16 coding differences between Ty1-912 (38) and Ty1-H3 (11). That these point mutations can have profound consequences is certain, as recent analysis of another Ty1 element, Ty1-173 (11), has shown. This element, unlike the functional Ty1 element Ty1-H3, cannot give rise to high-frequency transposition when fused to the inducible *GAL1* promoter (see also Section IV below). Analysis of hybrids between Ty1-173 and Ty1-H3 sequences showed that the sequences responsible for preventing transposition reside between a *Pst*I site at position 1897 and a *Sal*I site at position 2174. Sequence analysis showed that a point mutation in nucleotide 2098 results in a Leu → Ile substitution in *TYB* in Ty1-173. When this mutation alone was introduced into wild-type Ty1-H3 sequences, Ty1-H3 was no longer able to give rise to high-frequency transposition when fused to the *GAL1* promoter (11).

Comparison of another sequenced Ty1 element, Ty1-p109 (93), to the Ty1-912 and Ty1-H3 sequences reveals several apparent frameshift mutations. These mutations almost certainly destroy the function of this element, because two of the blocks of conserved amino acids among retroviruses and retrotransposons (blocks M3 and II [38, 142]) are partially out of frame in this Ty1 element. However, these frameshifts might be attributable to sequencing errors, as they occur near the ends of sequencing runs, where gel resolution is not expected to be as good.

5. LTR sequences (δ, σ, and τ)

The most variable yeast transposon sequences are the δ sequences. As these elements appear not to encode any protein products (except for the first few codons of *TYA* in the U5 region of δ), they are presumably more able to drift in sequence than the ε region. However, certain signals are present in these LTR sequences which are presumably essential for their functions as segments of intact retrotransposons. These include the upstream activation sequence (UAS), TATA, and initiation sites required for the initiation of transcription, sites required for the polyadenylation and termination of transcription, terminal sequences such as the conserved TG. . .CA terminal dinucleotides, which probably play a direct role in the transposition (integration) of the reverse transcripts into the DNA, and possibly packaging signals and translational signals for the Ty RNA. It has also been proposed that a reverse complement sequence, one copy of which is found in the U3 region and the other in a region within ε near the 5′ LTR, functions as "alignment" signals which fold the RNA so as to ensure appropriate juxtaposition of sequences in the Ty RNA (189). This might be important for some of the strand transfer reactions involved in the reverse transcription process (10, 87). Thus, it is instructive to examine which regions of LTR sequences are conserved both within and among the δ, σ, and τ families (86, 190).

Those δ elements not associated with Ty1 and Ty2 elements (i.e., "solo" δ elements) have greater sequence divergence than those associated with an intact retrotransposon, presumably because they no longer "need" many of the above functions (83, 156). They apparently decay in the genome by the accumulation of point mutations and may represent "experimental DNA" with which the cell might eventually evolve new functions (75). Some δ sequences have been isolated which have 70% or less sequence identity with Ty-associated δ sequences (156).

Among 12 Ty elements studied for which both δ sequences have been determined, 8 have identical and 4 have nonidentical pairs of δ sequences, these differing at three to six nucleotides (Table 2).

Solo σ and τ elements appear to be somewhat more conserved in structure than solo δ elements. Perhaps this is because they serve a function of some kind or, in the case of τ, because of a relatively low copy number. Alternatively, these sequences may not have existed in the yeast genome as long as δ sequences have.

D. RNA Structure

Because *S. cerevisiae* transposons are members of multigene families, studies on Ty RNA must be interpreted with caution. Certain families, such as the Ty1 family, show considerable structural variability; therefore, the transcription patterns of these elements may not be uniform. Ty1 and Ty2 RNAs are among the most abundant polyadenylated RNAs in *S. cerevisiae*; they apparently comprise 5 to 10% of *S. cerevisiae* polyadenylated RNA in haploid cells during log-phase growth (however, this estimate was not quantitative and has not yet been reproduced in other laboratories) (59).

The structure of the major species of Ty1 mRNA has been determined and is very indicative of the relationship between Ty elements and retroviruses (59, 61). Ty2 mRNA probably has the same general structure as Ty1 mRNA, since Ty1 and Ty2 elements have extremely similar δ termini. The major species of Ty RNA found in normal *S. cerevisiae* cells is a 5.6-kb molecule which initiates about 240 nucleotides into the 5′ LTR. The 3′ end of this transcript, which is polyadenylated, was determined by cDNA cloning experiments to vary between positions 291 and 297 nucleotides into the 3′ LTR. Hence, the RNA molecule is terminally repetitious for about 50 nucleotides; this structure allows domains of the Ty LTR to be defined in a manner similar to that used in describing retroviruses, namely, the region of the LTR unique to the 5′ end of Ty RNA is called U5, the region unique to the 3′ end is called U3, and the repeated region of the transcript is called R (Fig. 2). Another species of Ty RNA is reportedly several hundred nucleotides shorter at the 3′ end, suggesting an inefficient internal termination site; however, the data on this point are rather weak and the experiments have not been reproduced in other laboratories. A transcript similar to the major Ty RNA but approximately 800 nucleotides shorter at the 5′ end is the predominant product of Ty transcription seen in yeast cells bearing a mutation in the *SPT3* (196; K. Durbin, Ph.D. thesis, Cornell University, Ithaca, N.Y., 1985), *SPT7*, or *SPT8* (195) genes. In these *spt* mutants (discussed in more detail in Section V.3), full-length Ty RNA is virtually undetectable. The 5′-truncated transcript noted above is much less abundant than is the normal Ty RNA. A similar-sized transcript is often seen in wild-type cells, but it is not known whether this shorter molecule is identical in structure to the species observed in the *spt* mutants. Many minor species of Ty RNA can be seen in Northern (RNA) blot analysis; however, these tend to be nonreproducible and to vary with strain background. A fairly abundant 2.2-kb Ty RNA species has been observed by several groups (126, 172). Its detailed structure is uninvestigated.

Ty3, perhaps because it is a member of a much smaller multigene family, produces a more homogeneous transcription pattern. Interestingly, Ty3 and σ transcripts are induced in *S. cerevisiae* cells of **a** mating type by the mating pheromone α factor (and in α cells by **a** factor) (39, 180) (see Section V.A.7). When Northern blots are probed with an internal (non-σ) probe, only 5.2-kb (full-length) and 3.1-kb species are detected. Whereas the 5.2-kb transcript is inducible by pheromone, the 3.1-kb transcript is constitutive. The ends of the 5.2-kb transcript have been determined by S1 mapping experiments; this RNA initiates at positions 222, 226, and 234 nucleotides in the 5′ LTR (with most molecules initiating at position 222). The 3′ ends vary between position 265 and the end of the LTR. Hence, the 5.2-kb Ty3 RNA is also terminally repetitious. The structure of the 3.1-kb transcript is not known, although it does hybridize to an LTR probe, suggesting that it is either 5′ or 3′ coterminal with the 5.2-kb transcript (or both) (39).

Another abundant α-factor-inducible transcript, which is 650 bp long and hybridizes to σ (the LTR of Ty3) probes, has been reported (180). This transcript initiates within the σ element and reads into adjacent cellular sequences, away from an adjacent $tRNA^{Trp}$ gene. The function of this transcript, if any, is unknown. A 450-nucleotide, pheromone-induced transcript also hybridized to a σ probe but was not structurally characterized.

IV. MECHANISM OF Ty ELEMENT TRANSPOSITION

The mechanism of Ty element transposition was inferred early on from the structural analysis of Ty elements and their transcripts and comparison of these to the structures of retroviral proviruses and genomic RNA molecules, respectively. Because these entities are so similar, it was proposed that the mechanism of Ty1 transposition might be a replicative one, in which a Ty1 transcript would be reverse transcribed; following this, the reverse transcript (cDNA) would integrate into a new site. Later work on sequence analysis of Ty1 elements supported this model: Ty1 elements were found to contain an open reading frame (*TYB*) which had statistically significant, albeit weak, amino acid sequence similarities to retroviral *pol* genes (38, 189). Furthermore, the sequence analysis together with studies of various gene fusions to the *TYB* reading frame established that *TYB* protein is synthesized as a *TYA-TYB* readthrough precursor protein (38, 124, 193). This readthrough requires a frameshift event; a similar frameshift event is known to occur in Rous sarcoma virus (and other retroviruses) between the *gag* and *pol* genes (98, 99).

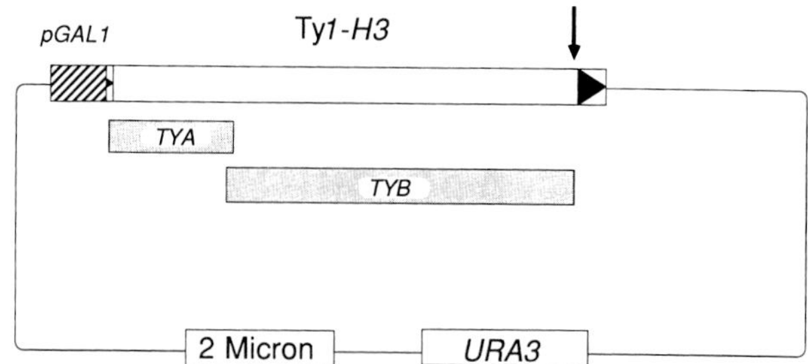

Figure 5. Structure of pGTy1-H3 plasmids. A controllable *S. cerevisiae* promoter (the *GAL1* promoter; hatched box) has been fused with Ty1-H3 in such a way that the resulting transcript initiates at the same nucleotide as does a native Ty1 element transcript. The Ty1 promoter sequence (U3 region) has been deleted from the 5′ δ element in the process. The position of the open reading frames (*TYA* and *TYB*) is indicated by shaded boxes. The heavy arrow indicates a site (a *Bgl*II site; see Fig. 2) at which foreign DNAs have been inserted without disrupting transposition (13, 14, 16, 18, 100, 198). The boxes labeled *URA3* and 2 Micron represent segments of *S. cerevisiae* DNA containing a selectable marker used to maintain the plasmid and a segment of the 2μm circle plasmid conferring the ability to replicate in high copy number, respectively. The simple line represents pBR322 sequences.

A functional correlation between retroviruses and Ty1 elements could be made when it was shown that Ty1 elements transpose through an RNA intermediate (14) and that Ty1 encodes a reverse transcriptase activity (85, 126). These findings were made possible by the refinement of sensitive assays for Ty1 element transposition (see Section V.D for details about transposition assays), as well as the construction of plasmids which contained Ty elements fused to strong promoters. One of these promoter fusions, called pGTy1-H3, consists of the yeast *GAL1* promoter fused to a Ty1 element called Ty1-H3. A diagram of the structure of pGTy1-H3 plasmids is given in Fig. 5. This Ty element is functional for transposition by itself, as can be readily demonstrated in marking experiments in which various pieces of non-Ty DNA can act as passengers in the transposition process. Moreover, induction of transposition in cells carrying pGTy1-H3 plasmids results in elevated levels of transposition of the chromosomal copies of Ty1 elements, suggesting that one or more components expressed from pGTy1-H3 is normally lacking or limiting in the cell (14, 18).

Direct evidence for an RNA intermediate in Ty element transposition was provided by an intron marking experiment using a *GAL1*/Ty1-H3 fusion plasmid. A 600-bp fragment of the *S. cerevisiae RP51* gene, which contains an intron of 400 bp as well as flanking exon sequences, was introduced into a site in the Ty which could tolerate passenger DNA. Transposition was induced with galactose, and progeny transposons (identified by their *RP51*-exon sequences), were studied. Most of these progeny transposons had precisely lost the intron sequences during the transposition process, as determined by fine-structure restriction mapping, Southern analysis, and sequence analysis (14).

Cells containing pGTy1-H3 and similar plasmids have been found by several groups to produce viruslike particles (Ty-VLPs) which morphologically resemble certain intracellular retroviral particles. Moreover, these particles contain the above-mentioned reverse transcriptase activity, Ty RNA, a primer for Ty RNA reverse transcription (85, 126), double-stranded Ty DNA (57a, 198), and several Ty-encoded proteins (3, 85, 131, 200). Thus, it seems extremely likely that these particles are essential intermediates in the transposition process. Recently obtained evidence strongly supports this notion (see below). These observations led to the development of the model for Ty element transposition shown in Fig. 6, consisting of several steps, each of which will be considered separately below. The steps in the process are: transcription of Ty elements, packaging of Ty RNA (particle formation), reverse transcription of Ty RNA, transport of Ty DNA to the nucleus, and integration of the reverse transcripts into target DNA. It is likely that the mechanisms of Ty1, Ty2, and Ty3 transposition are similar overall; many of the experiments done with Ty1 have also been carried out on Ty2 (Boeke and Garfinkel, unpublished data), but not yet on Ty3.

The observation of Ty-VLPs in cells undergoing Ty transposition raised the intriguing question of the possible infectivity of the VLPs as well as questions about the transposition process itself. Could horizontal transmission be a factor in the wide distribution of Ty elements among *S. cerevisiae* strains? There is no evidence for extracellular forms of these particles, and experiments designed to detect intercellular (i.e., infectious) transposition events have all been negative (85; H. Xu and J. D. Boeke, unpublished data). It is interesting, however, that recent work on yeast killer virus, which is not infectious in mitotic cultures (173), shows that purified killer particles can "infect" both yeast spheroplasts and yeast mating mixtures (60). Similar experiments have so far failed to provide evidence for such infectivity of Ty-VLPs (S. Youngren, H. Levin, J. D. Boeke, and D. J. Garfinkel, unpublished data).

A. Transcription of Ty Elements

Transcription of *S. cerevisiae* Ty elements appears to be similar to transcription of other cellular genes. RNA polymerase II is apparently the transcribing enzyme, because Ty transcription is sensitive to α-amanitin (103) and Ty RNA synthesis decreases rapidly in an *rpb1* temperature-sensitive mutant (a mutation in an RNA polymerase II subunit) at the nonpermissive temperature (136). Ty transcripts are presumably polyadenylated, because Ty RNA binds to oligo(dT)-cellulose.

The proper transcription of Ty elements is dependent upon a number of host gene products which have been identified genetically. These are a subset of the *SPT* genes, which were originally identified as suppressors of Ty insertions. The effect on transcription of one class of these mutations, consisting of the *spt3*, *spt7*, and *spt8* genes, has already been mentioned. Mutations in these three genes prevent transcriptional initiation from occurring at the normal position in the 5′ LTR (δ) of Ty1 (196; Durbin, Ph.D. thesis) and Ty2 (F. Winston, personal communication) elements and from solo δ elements (165, 196; Durbin, Ph.D. thesis) (the effect on σ and Ty3 transcription is uninvestigated). In Ty1 and Ty2, a small amount of transcription from an internal promoter does occur in these *spt* mutants; the net result is a low level of Ty transcripts in the cell, most or all of which are 800 nucleotides shorter than normal at the 5′ end (196; Durbin, Ph.D. thesis). (See Section V.2 for further discussion of the relationship of *SPT* genes and Ty element transcription.)

The transcription of Ty3 elements and certain σ elements can be induced by exposure of **a** cells to the mating pheromone α factor, as will be discussed in detail in Section V.A.7.

B. Particle Formation

S. cerevisiae cells undergoing high-frequency transposition of Ty1 and Ty2 elements (i.e., those containing an induced *GAL1*/Ty plasmid) contain large amounts of viruslike particles (Ty-VLPs) (86;

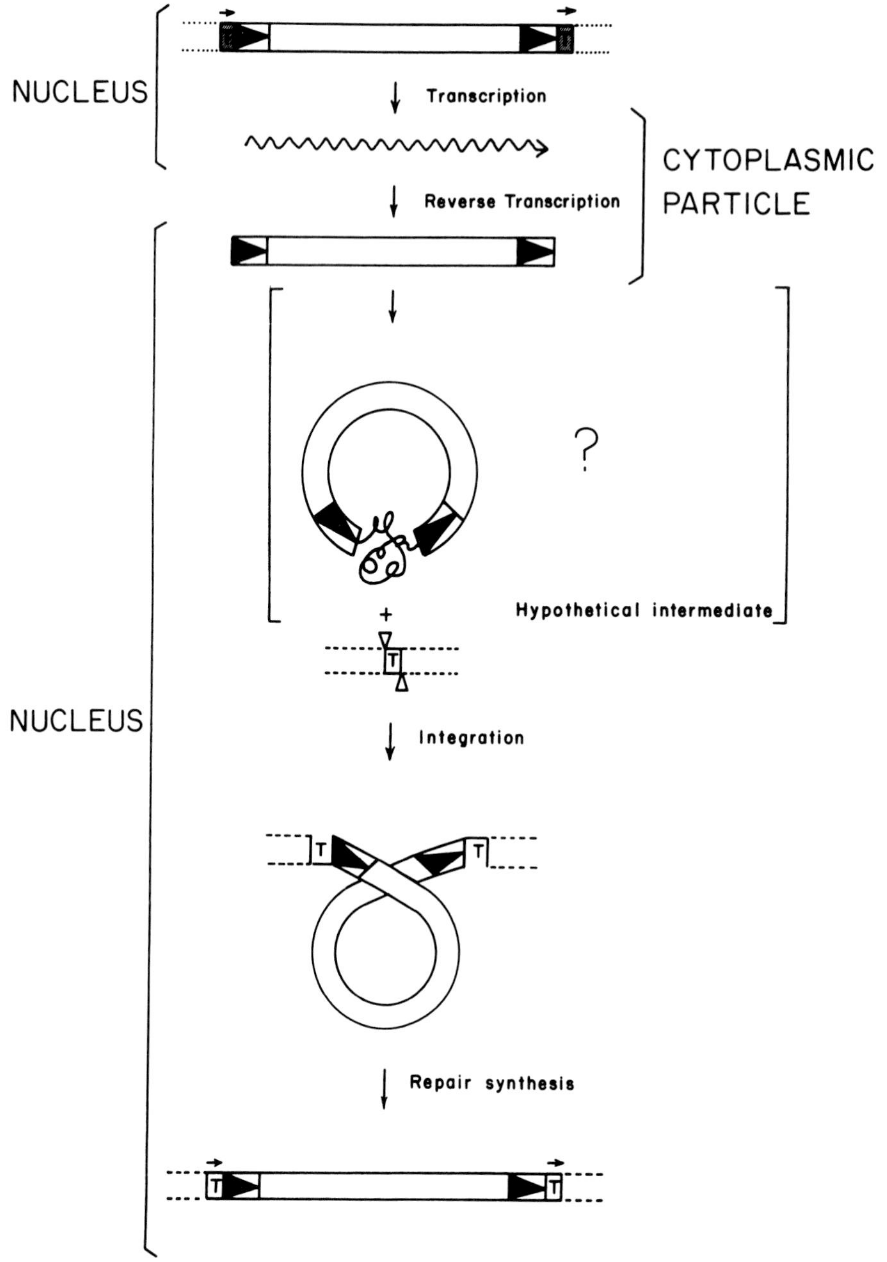

Figure 6. Model for Ty transposition mechanism. Boxed triangles represent LTR sequences; open box represents the internal sequence of a Ty element; dotted and dashed lines represent parental and progeny flanking host DNAs, respectively. The parental Ty element (top line) is flanked by a 5-bp host sequence repeat (shaded boxes labeled T). The wavy line represents the major Ty transcript, produced in the nucleus. The Ty RNA is packaged into a cytoplasmic particle, the Ty-VLP. Reverse transcription of this RNA gives rise to predominantly linear Ty DNA molecules. These may form a complex with integrase protein (squiggly line between LTRs), as indicated in the diagram of the hypothetical intermediate. A new target site in the nuclear DNA (open box labeled T) is cleaved so as to produce a 5-bp "staggered cut," presumably by integrase (cut sites are indicated by open triangles, shown arbitrarily as leaving a 5′ overhang). The free end of the target DNA is joined to the ends of the Ty DNA during the integration reaction. Repair synthesis fills in the terminal gaps to complete the transposition process.

D. Eichinger, S. Pelger, and J. D. Boeke, unpublished data). Because these particles contain Ty RNA molecules, Ty-encoded reverse transcriptase, as well as capsid proteins specified by the *TYA* gene, it is very likely that Ty-VLPs are intermediates in the transposition process. Certain mutant Ty elements, which direct the synthesis of morphologically aberrant particles, are unable to transpose (see below). Moreover, it has recently been shown that a particle that copurifies with Ty-VLPs contains double-stranded Ty

DNA (mostly in the form of linear molecules) and all of the macromolecular components required for transposition in vitro into a naked bacteriophage λ target (Eichinger and Boeke, unpublished data) (see Section IV.E).

Although little is known about the details of the formation of Ty-VLPs, at least two steps in the process can be imagined. These are the packaging of Ty RNA into particles and proteolytic cleavage of the p58-TYA and p190-TYA/TYB precursor proteins (primary translation products) into end products.

1. Packaging of Ty RNA and particle assembly

Ty RNA can be found inside Ty-VLPs. Cellular mRNAs are packaged inefficiently if at all. When *GAL1/lacZ* RNA was overexpressed in cells producing Ty-VLPs, none of the abundant *lacZ* mRNA was detected within the Ty1-VLPs (200). Similar results were obtained with the less abundant *URA3* transcripts (85). These results strongly indicate that Ty RNA is packaged into VLPs in a specific manner. Efforts to localize a specific "packaging signal" in Ty RNA have been inconclusive. In retroviruses, two separate regions of the genomic RNA molecule have been implicated in packaging. In every vertebrate retrovirus studied, a region between the 5′ LTR and the beginning of the *gag* reading frame (the "leader" region) has proven to be essential for packaging (107, 121). Because Ty1 and Ty2 differ from the vertebrate retroviruses (and, for that matter, from all other sequenced retrotransposons) in that the *TYA* coding frame (the equivalent of *gag*) initiates **within** the 5′ LTR sequence, there is no corresponding "leader" region in Ty1 and Ty2. An additional packaging signal is found near the 3′ LTR (168).

Mini-Ty elements have been constructed which lack large portions of internal sequence. These can transpose at high efficiency when they are accompanied by a "helper" Ty which overproduces *TYA* and *TYB* gene products (H. Xu and J. D. Boeke, manuscript in preparation). Mini-Ty RNA is packaged normally; from this type of analysis one can say what regions of the genome are **not** essential for packaging (Fig. 7). Much of the internal region of Ty1 (positions 818 to 5562) can be deleted from mini-Ty elements without affecting their transposition frequency. However, deletions removing a region (base pairs 478 to 818) within *TYA* result in a drastic drop in transposition frequency in those mini-Tys. The primary cause of this drop is probably the removal of enhancer sequences (Section V.A.4), which also lie within this segment, as transcripts of this particular mini-Ty are much less abundant, which would explain the decrease of transposition of this mini-Ty.

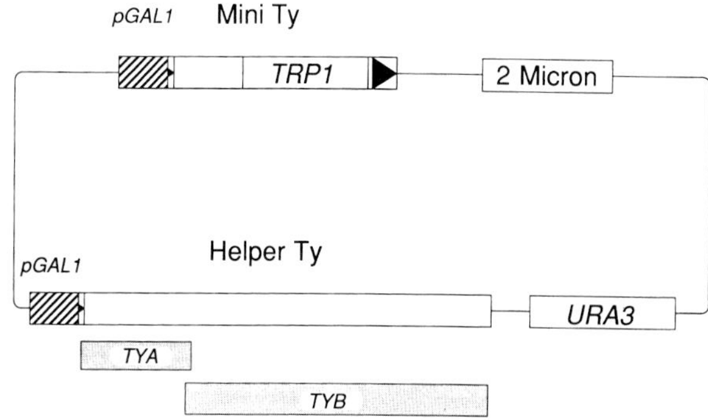

Figure 7. Mini-Ty plasmid structure. Mini-Ty (upper box) contains the selectable marker *TRP1* and all elements needed in *cis* for transposition. The mini-Ty contains only R and U3 regions at its 5′ end and an intact δ at its 3′ end (just like the canonical pGTy1-H3 [helper-independent] plasmids). Helper Ty (lower box) contains intact open reading frames but lacks the 3′ δ sequence and hence is itself unable to transpose. Transcription of both *GAL1*/Ty fusions starts at the point where *GAL1* and Ty sequences are fused. *URA3*, selectable marker for maintaining plasmid; 2 Micron, 2-μm sequence required for high-copy-number replication of the plasmid. Note that both mini-Ty and helper Ty are driven by the yeast *GAL1* promoter (hatched box).

The packaging of Ty RNA is one part of a complex assembly process in which Ty RNA, primer tRNA, capsid (*TYA*) proteins, and *TYB* proteins must be brought together in the appropriate configurations. A simple model outlining a possible assembly mechanism is given in Fig. 8. Because unprocessed p58-TYA proteins are able to assemble into a form of Ty-VLP in the apparent absence of *TYB* proteins (3, 131), it seems likely that this form of the Ty-VLP represents a precursor (pre-Ty-VLP) to the mature Ty-VLP. It may be that *TYB* proteins gain access to pre-Ty-VLPs by virtue of the fact that they are synthesized as p190-TYA/TYB precursor molecules; the amino-terminal part of this protein might function as a p58-TYA protein molecule and thus ensure packaging of *TYB*-derived protein sequences in the particle. A prediction of this model (which is undoubtedly a gross oversimplification of the actual process of particle assembly) is that the C-terminal ends of p58-TYA would be oriented toward the inside of the particle, ensuring that the *TYB*-derived portion of the p190-TYA/TYB molecule would end up inside the particle. It is interesting to note in this regard that variants of p58-TYA, molecules with different C-terminal extensions derived from foreign genes, can still assemble into recognizable VLP-like structures (2, 120). In this model, proteolytic processing events catalyzed by the *TYB*-encoded protease subsequent to the formation of a pre-Ty-VLP would then give rise to mature Ty-VLPs.

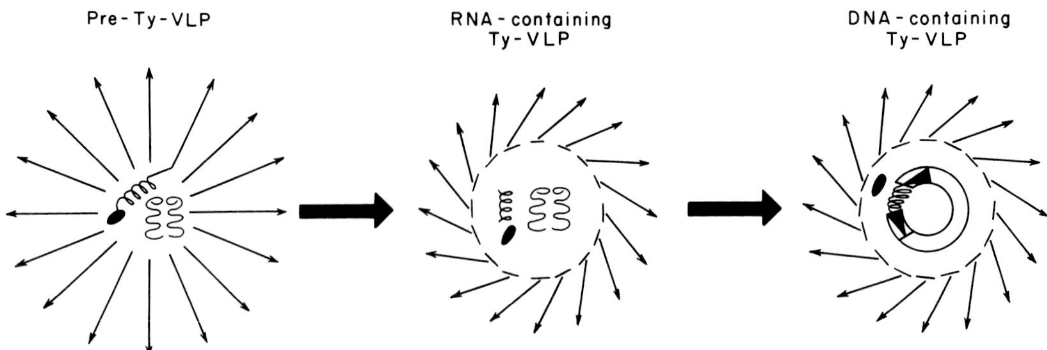

Figure 8. Particle structure model. A hypothetical model for the sequence of events in particle assembly is presented. Note that this model includes many details for which there is little or no direct experimental evidence. For example, it is not known whether there are discrete populations of RNA-containing and DNA-containing Ty-VLPs; few of the details of the proteolytic processing in Ty-VLPs are known; there is no direct evidence for interaction of Ty integrase with the termini; there is no physical evidence for dimers of Ty RNA inside Ty-VLPs; and the number of copies of the various Ty-encoded proteins per VLP is unknown. The arrowheads represent the N termini of p58-TYA (short arrows in pre-Ty-VLPs) and the N termini of p190-TYA/TYB (long arrow with attachments). The helical region of p190-TYA/TYB represents protease/integrase; the filled oblong represents reverse transcriptase/RNase H (as pointed out in the text, it is unknown whether these are actually synthesized as discrete entities). The shorter arrows in the mature VLPs represent a processed form of p58-TYA; the dashes at their bases represents a segment of p58-TYA which has been processed off. The two attached wavy lines symbolize a pair of Ty transcripts; the open box represents the resultant Ty reverse transcript.

2. Protein processing

There is ample evidence that Ty1-encoded proteins from both *TYA* and *TYB* are processed proteolytically; moreover, a Ty-encoded protease in the *TYB* open reading frame appears to be responsible for many or all of these cleavage events. Early studies on cells containing high-copy-number plasmids bearing fusions between the *PGK* (phosphoglycerokinase) promoter and Ty1 elements provided suggestive evidence that TYA protein existed in at least two forms, originally called p1 and p2 (50, 123). In subsequent work, these proteins have also been referred to, respectively, as pro-TYA and TYA (131) and as p58-TYA and p54-TYA (200). Furthermore, *TYB* proteins are also apparently processed. Sequence analysis of several Ty1 and Ty2 elements indicates that the *TYB* open reading frame would encode a protein of about 150 kilodaltons (kDa) (171), but a *TYB*-encoded protein of about 90 kDa, called p90-TYB (85), has been reported. This protein requires functional Ty-encoded protease for biosynthesis (200).

Proteolytic processing is not essential for early events in Ty-VLP formation; a pre-Ty-VLP can be seen which is morphologically distinguishable from normal Ty-VLPs in cells overproducing the *TYA* open reading frame alone. Yet another form of aberrant particle is seen in certain linker insertion mutants which destroy transposition and processing of both *TYA* and *TYB* protein processing but not reverse transcriptase activity (13, 200). These latter particles appear "cracked" in the electron microscope, as though they have some sort of maturation defect. The cracked particles may have a defect in appropriate packaging of nucleic acids, as they are less capable of carrying out an endogenous reverse transcriptase reaction than are wild-type particles (200).

Protease mutants of Ty1 also produce aberrant *TYB* proteins. Various antisera directed against synthetic peptides corresponding to the *TYB* open reading frame as well as *trpE/TYB* fusion proteins made in *E. coli* recognize a 90-kDa protein in cells overproducing a functional Ty1 element, but cells overproducing protease mutant Ty1 elements make a protein of about 190 kDa, probably equivalent to the full-length *TYA/TYB* fusion protein/primary translation product, and little or no p90-TYB (200).

C. Reverse Transcription

Ty RNA is apparently reverse transcribed in Ty-VLPs, which are found predominantly in the cytoplasm (85, 126). The reverse transcriptase is Ty encoded and cosediments with Ty-VLPs and Ty RNA (75, 126). This activity can be detected by standard assays for reverse transcriptase, utilizing homopolymer primer-template complexes (90).

1. Priming events

The reverse transcriptase is functionally associated with Ty RNA and some form of primer molecule, because an endogenous reaction, free of exoge-

Ty element	3' end of internal segment (putative polypurine stretch) : U3 region of LTR	U5 region of LTR : 5' end of internal segment (tRNA priming site)
Ty*1-H3*	TAGATCTATTACATTATGGGTGGTA:TGTTGGAATAGAAA	TCACCCAATTCTCA:TGGTAGCGCCTGTGCTTCGGTT
Ty*2-117*	TAGATCTATTACATTATGGGTGGTA:TGTTGGAATAAAAA	CCACGTTCTCTTCA:TGGTAGCGCCTATGCTTCGGTT
Ty*3-pSBS12*	AAGACAACCCTGAGAGAGAGGAAGA:TGTTGTATCTCAAA	CCCGTAATACAACA:CCTGGTAGCGTTAAAGGTTACT

Figure 9. LTR-adjacent (internal to δ and σ elements) sequences in Ty1, Ty2, and Ty3. Sequences just inside the 5′ and 3′ LTR sequences are indicated. The putative "polypurine stretch" [(+)-strand primer binding site] is underlined, as is the homology to the 3′ end of initiator methionine-tRNA (31, 56, 167).

nously added primer or template, can be carried out with preparations of purified Ty-VLPs (85, 126). The accepted model for retroviral replication posits a (−) strong-stop DNA as the initial product of reverse transcription; this is a molecule which in vertebrate retroviruses is always primed by a cellular tRNA (10, 87). A DNA species corresponding to (−) strong-stop DNA with a 75-nucleotide, base-sensitive, RNase-sensitive component at its 5′ end is one of the products of this reaction. This RNA, presumed to be the primer, has many properties which suggest that it is initiator methionine-tRNA (A. Bystrom, J. D. Boeke, and G. R. Fink, unpublished data). Ty1, Ty2, and Ty3 all share 8 (Ty3) to 10 (Ty1 and Ty2) nucleotides of perfect sequence complementarity to the 3′ end of this tRNA molecule (Fig. 9). This homology is found exactly at the boundary between the LTR sequence and the internal part of Ty1 and Ty2 elements (38, 56, 189); in Ty3 elements there are two extra bases between the end of the LTR and the beginning of the tRNA homology, reminiscent of the retroviral sequence arrangement (39). Thus, it may be that these three transposable elements compete for the same tRNA primer. *S. cerevisiae* has four to seven genes for this tRNA, depending on the strain (31, 186) (A. Bystrom and G. R. Fink, manuscript in preparation); the coding sequences of these tRNA genes are identical (31) (Bystrom and Fink, in preparation).

Presumably, the priming tRNA is packaged into virions together with Ty RNA or with reverse transcriptase. After the (−) strong-stop molecule is synthesized using the tRNA as a primer, it must be transferred to the 3′ end of the RNA template in a manner similar to that described for retroviruses, at which point a nearly complete (−) strand could be synthesized. Degradation of the RNA template by an RNase H activity would produce a single-stranded Ty DNA molecule. Priming by an as yet unidentified entity would result in a (+) strong-stop molecule (which itself has not yet been identified for any Ty element); a similar transfer or displacement of the (+) strong-stop DNA would allow a completely double-stranded DNA molecule to be synthesized by further polymerization by reverse transcriptase. A purine-rich sequence which may function as a primer binding site for (+) strong-stop DNA is very evident in Ty3, but less so in Ty1 and Ty2, just inside the 3′ LTR (Fig. 9). (+) strong-stop molecules have been identified for several *Drosophila* retrotransposons; the nature of the primer for these is uncertain (5).

2. Recombination during reverse transcription

There is evidence that during or immediately after reverse transcription, recombination or recombinationlike events similar to those that occur during retroviral replication can take place during Ty1 element transposition. Transposition of marked Ty elements is often (at least 25% of the time) accompanied by loss of restriction sites in the parental pGTy1-H3 molecule (or acquisition of sites absent from it) (14, 16). These sequence polymorphisms all correspond to known sequence polymorphisms present in other, genomic copies of Ty elements. These sequence changes do not occur when marked Ty elements transpose from a pGTy1-H3 plasmid in *spt3* mutant cells, in which genomic Ty element transcription is greatly reduced (16). Probably more than one Ty RNA molecule is packaged into Ty-VLPs (this has not yet been examined directly), and during reverse transcription of the (−) strand, hybrid reverse transcripts are formed (Fig. 10).

During transposition of Ty elements marked with two tandem copies of a marker segment (a bacterial *neo* gene), very high frequency (80 to 100%) deletion between the *neo* sequences was observed, even when these were interrupted by a different sequence (198). This suggests that these recombination events can occur at a very high frequency intramolecularly. Duplication of marker sequences is only very rarely observed, whereas deletion can occur at frequencies as high as 100% (198; Xu and Boeke, unpublished data).

3. Structure of reverse transcripts

The detailed structure of the completed Ty1 reverse transcripts is under investigation. At least three types of structures have been seen in cells

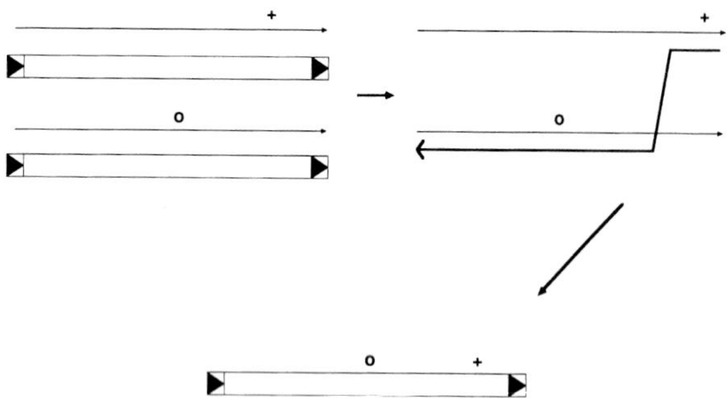

Figure 10. Recombination during reverse transcription. A simple copy-choice model (42) is presented which may explain observed homologous recombinationlike phenomena observed during transposition of marked Ty1 elements (12, 16). The thin arrows represent Ty transcripts; the heavy arrow represents a reverse transcript. The o and + signs represent genetic markers on the donor Ty elements. Terminal sequences have been omitted for clarity.

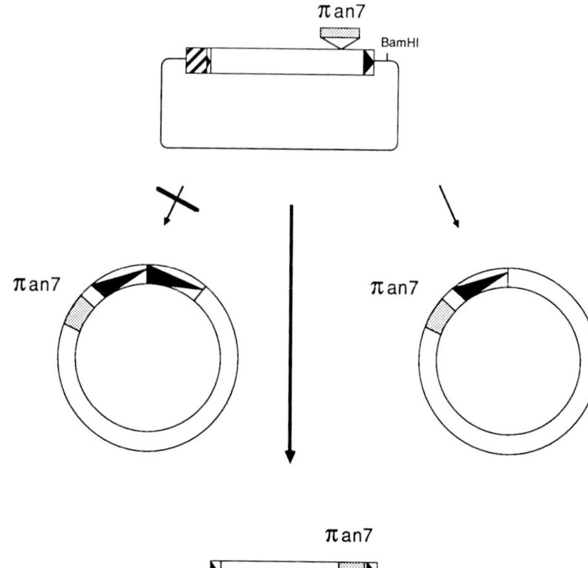

Figure 11. Expected forms of Ty reverse transcripts. Three forms of unintegrated retroviral DNA have been reported: linear, one-LTR circle, and two-LTR circle; the equivalent molecules have been sought for Ty as shown here. A pGTy1-H3 plasmid marker with plasmid πan7 (163, 187) (see text) was introduced into cells, and attempts were made to recover circular reverse transcripts by cutting total DNA with an enzyme (*Bam*HI) which cuts pGTy1-H3/πan7 but not the circular reverse transcripts. Although many one-LTR circles have been recovered by transformation into *E. coli* and selection for πan7 sequences, no two-LTR circles have been found. Direct gel analysis of reverse transcripts in Ty-VLPs has indicated that most reverse transcripts are linear (57a).

infected with retroviruses, namely, linear molecules, one-LTR circles, and two-LTR circles (185). Similar DNA molecules corresponding to various retrotransposons have been isolated from *Drosophila* cells (4, 79, 80, 95, 129). Until recently, two-LTR circles were widely believed to be the active species in retroviral integration based on experiments done on spleen necrosis virus (139). However, experiments designed to isolate two-LTR Ty DNA circles have been unsuccessful, although molecules with one LTR were readily isolated (Fig. 11). Circular forms of Ty DNA have been found by other groups in total *S. cerevisiae* DNA, but the detailed structure of these has not been determined (7, 135). Moreover, the most abundant form of DNA found within Ty-VLPs appears to be a full-length duplex linear reverse transcript; circular forms were not readily detected (57a). Recent results with MoMLV preintegration complexes suggest that the amount of circular DNA in them is insufficient to account for the amount of in vitro integration observed, but that the amount of linear DNA is sufficient. None of these results preclude the existence of a transient, covalently closed circular intermediate or of a protein-linked circular intermediate.

D. Transport to the Nucleus

Since reverse transcripts are found within Ty-VLPs which are predominantly if not entirely located within the cytoplasm (Fig. 12), this genetic information must somehow be transported into the nucleus, where Ty DNA transposition intermediates can find and insert into their targets. The fact that a Ty DNA transposition intermediate which is capable of integrating into naked bacteriophage λ DNA cosediments with Ty-VLPs (see Section IV.E) suggests that in vivo a particle of similar size enters into the nucleus before carrying out a transposition event. This is surprising, because the diameter of Ty-VLPs as measured in thin-section electron micrographs is about 60 nm (measured by independent investigators using different fixation procedures). The largest particles previously shown to be capable of entering the nucleus (presumably via nuclear pores) are ferritin particles (linked to simian virus 40 [SV40] nuclear localization peptide) of about 10 nm; in similar experiments immunoglobulin M molecules, estimated to be 25 to 40 nm in size and linked to the same SV40 peptide, were unable to enter the nucleus (111). These experiments were carried out with animal cells, which probably have nuclei and nuclear pores of the same size or larger. Naturally, it is possible that the particle or nucleoprotein complex which actually enters the nucleus is flexible enough to squeeze through a small hole, or it might just perch at the nuclear envelope and "inject" its DNA through the pore (or other entry site) into those target DNA sequences which happen to reside near the entry site at that moment.

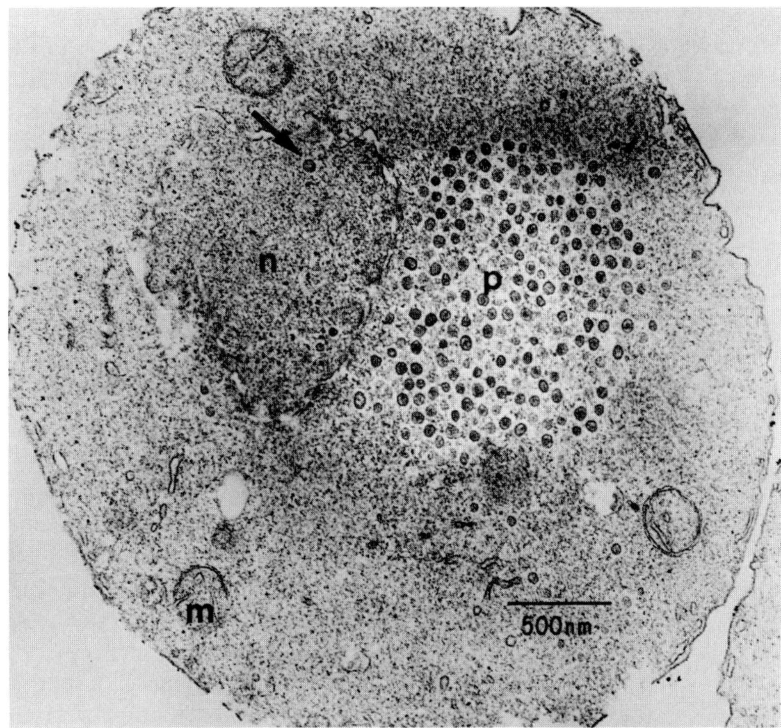

Figure 12. Photograph of Ty-VLPs. Ty-VLPs can readily be seen in thin sections of *S. cerevisiae* cells containing plasmids which overproduce Ty gene products. Note that these are predominantly cytoplasmic, but occasionally a VLP-like entity may be seen within the nucleus (arrow). The significance of the nuclear particles is uncertain. Abbreviations: n, nucleus; m, mitochondrion; p, particles.

E. Integration of Reverse Transcripts

Little is known about the mechanics of integration of the reverse transcripts found in the VLPs, but recently a system was developed which allows the study of this process (57a). This cell-free system is very similar to that developed by Brown et al. (21) to study MoMLV integration. A *GAL1*/Ty1-H3 plasmid is marked with the *Escherichia coli* miniplasmid πan7, an 885-bp plasmid carrying the *E. coli supF* tRNA suppressor gene (163, 187). This plasmid, like other pGTy1-H3 plasmids, directs the synthesis of Ty-VLPs and allows for the transposition of the Ty1-H3πan7 construct into target DNA when the host cells are grown in inducing medium.

The cell-free system for the study of Ty transposition is based on the following procedure. Ty-VLPs purified from the above cells are mixed with concatemers of bacteriophage λ (strain λ · gtWES · λB) DNA (112). This λ strain cannot plaque on wild-type (*sup*^{0}) strains of *E. coli*, because they carry amber mutations in the essential *W*, *E*, and *S* genes. Insertion of a Ty element into the λ DNA (which can be tolerated because λ · gtWES is considerably shorter than wild-type λ-phage DNA) results in a viable phage carrying its own suppressor tRNA gene. Such viable phage can be recovered at a relatively high efficiency and contain DNA of the expected structure, namely an intact Ty element flanked by 5-bp duplications of λ DNA in a nonessential region of the phage genome (57a).

Thus, the Ty-VLPs contain both a substrate (Ty DNA) and one or more protein components required for the reaction. The reaction requires only Ty-VLPs and a divalent cation in order to occur; deoxynucleotide triphosphates and high-energy cofactors do not stimulate the reaction. Ty-VLPs purified over two density gradients are fully competent to carry out the reaction, suggesting that soluble factors are not required and that host factors required for the transposition reaction are limited to those copurifying with the particle (57a).

V. REGULATION OF Ty ELEMENT TRANSPOSITION

Transposition, which can be deleterious to the host cell if allowed to get out of hand, must be tightly regulated by the cell. All transposons studied in sufficient detail are found to regulate their transposition. Tn*10*, one of the best-studied transposable elements, shows many layers and types of host cell regulation (N. Kleckner, this volume). Several types of regulation have been observed or proposed for Ty elements, although at this point the relative importance of these different types of regulation is unknown. Regulation can be imagined both at the level of synthesis of gene products (which includes Ty RNA, the genetic material for transposition) and at postsynthetic levels.

A. Transcriptional Regulation

1. UAS sequences

Most retroviruses contain regulatory sequences 5′ to the transcription start site in the 5′ LTR sequence. Recently, two blocks of sequence important for transcription of the Ty2 element Ty2-917 were proposed on the basis of deletion experiments (114). One of these corresponds to a TATA box region, and the other may represent a UAS of some sort. It may be difficult to establish definitively whether there is a UAS sequence within the δ sequence itself by such deletion experiments because of the unknown effects of variations in flanking sequences. The particular DNA sequences residing upstream of the 5′ δ can have a profound effect on the ability of the δ promoter to direct transcription (165). Experiments in which a Ty UAS sequence has been transplanted into another promoter region to assess its function (91) have not yet been carried out. If such constructs could be shown to be regulated by *SPT3*, this result would help establish the existence of a UAS important in Ty element transcriptional regulation.

2. *MAT* regulation

The first reported regulation of Ty element RNA levels was determined by the constitution of the *MAT* locus (mating type) (59, 61). The level of Ty RNA is severalfold higher (the exact amount varies with strain background) in *MAT* **a** or α cells than in *MAT* **a**/α diploids. This is not due to an effect of ploidy, because **a**/**a** and α/α diploids show high Ty transcript levels as well. This **a**1/α2 (132) downregulation requires the products of the *STE7*, *STE11*, and *STE12* genes, which are also responsible for the regulation of other haploid-specific genes encoding **a**- and α-factor pheromone, pheromone receptors, and other factors involved in the mating process (26, 42a, 64, 72–74).

MAT regulation of the transcription of Ty1 and Ty2 elements and genes adjacent to them requires sequences internal to the Ty element; solo δ sequences do not activate adjacent gene transcription in a *MAT*-regulated manner, unlike Ty1 and Ty2 elements at exactly the same positions (37, 149, 151). Originally, it seemed that one sequence or set of sequences might be responsible for both *MAT* regulation of Ty transcription itself and that of the adjacent gene. There is currently considerable disagreement as to exactly which sequences in Ty1 and Ty2 elements are necessary for *MAT* regulation. Deletion analysis placed the sequences required for *MAT* regulation of Ty2-917 between nucleotides 820 and 854 (J. G. Kapakos, J. J. Clare, and P. J. Farabaugh, submitted for publication). Other transcriptional studies examined *MAT* regulation of genes adjacent to Ty1 elements. One group (66) concluded that sequences from nucleotides 668 to 799 were needed for *MAT* regulation of *CYC7-H2*, a gene adjacent to a Ty1 element. Another group concluded that the sequences conferring *MAT* regulation on the adjacent *PGK* gene (in an artificial construct) lie between nucleotides 384 and 478 of Ty1-15 (147). Unfortunately, it may be that neither of these results is general. The Ty1 element adjacent to *CYC7-H2* has a 5′ δ with an aberrant structure (summarized above, Section III.C.3) and is probably not transcriptionally active. In the studies of Ty1-15, the investigators inserted fragments containing only a portion of the 5′ δ sequence (lacking the U3 region) upstream of *PGK*. Presumably in this case also, the Ty element activating the adjacent gene is not transcribed. Clearly, this type of study also needs repeating, using Ty elements with normal, intact 5′ δ sequences.

3. *SPT* and *ROC/TYE* gene regulation

A rather large number of genes (at least seven) seem to be required for the proper transcription of Ty elements, suggesting that these genes play some role in regulating the proper expression of Ty elements. The number of host genes which play a role in activating transcription from δ promoters suggests that Ty element transcriptional initiation is tightly regulated by the host. Yet, Ty mRNA is one of the most abundant mRNAs in the cell. Although many *SPT* and *ROC/TYE* genes have been identified, it is not clear exactly how these genes might regulate the frequency or timing of Ty transposition events. See Table 3 for a summary of the phenotypes of *spt*, *roc*, and *tye* mutants.

***SPT* genes.** The *SPT* genes were originally identified as suppressors of Ty insertions. That is, the *SPT* genes were initially identified using Ty or δ insertion mutations in the regulatory regions of reporter genes such as *HIS4* and *LYS2*. Phenotypic revertants in

Table 3. *SPT* and *ROC* genes and their possible roles

Gene	Location	Phenotypes	References
SPT1		Suppresses *his4-912* and *his4-917*	149
SPT2	V(R)	Suppresses *his4-912*δ, *his4-917*	148, 166, 194
SPT3	IV(R)	Suppresses δ element, *lys2-61*, and *his4-917* insertions; reduces Ty transcription; sporulation defect; mating defect	166, 194, 196
SPT4	VII(R)	Suppresses δ element insertions; some alleles are methylmethane sulfonate sensitive	194
SPT5		Suppresses δ element insertions	194
SPT6	VII(R) (=*SSN20*)	Suppresses δ element insertions	41
SPT7	II(R)	Suppresses δ element insertions; reduces Ty transcription	195
SPT8	XII(R)	Suppresses δ element insertions; reduces Ty transcription	195
SPT9		Suppresses δ element insertions and *his4-917*; one allele is temperature-sensitive lethal	F. Winston, personal communication
SPT10		Suppresses *his4-917*δ insertion	Winston, personal communication
SPT11/*SPT12*	IV(R) (=*HTA1*/*HTB1*)}	Suppress δ element insertions; some alleles suppress *his4-917*	40
SPT13, *SPT14*		Suppress *his4-917* and *lys2-61* but not *his4-912* and δ insertions; *spt13* mutants have sporulation defect; *spt13* gene disruption is viable; one *spt14* allele is somewhat temperature sensitive; *spt14* gene disruption is inviable	71
SPT15		Suppresses *his4-917*δ; reverse phenotype of *lys2-173R2*	Winston, personal communication
SPT21	XIII }	Suppresses whole Ty elements within 5′ end of structural genes	Natsoulis and Boeke, unpublished data
SPT22	X }	Suppresses whole Ty elements within 5′ end of structural genes	Natsoulis and Boeke, unpublished data
ROC1	II }	Reduces overproduction of transcript of gene adjacent to Ty element	55; E. Jacobs, personal communication
ROC2	}		
TYE1, *-2*, *-3*, *-4*		Same as above; *TYE1* and *TYE2* have mating defects; they are not allelic to *SPT3*, *SPT7*, or *SPT8*	37; Winston, personal communication

which the silenced gene was reexpressed often arose due to change at one of several unlinked loci. Many if not all of these genes affect the transcriptional regulation either of Ty elements themselves or of the reporter genes adjacent to Ty elements or solo δ elements. Some of these genes are the same as genes that have been identified in other systems and which are thought to be important in general transcription; for example, *SPT6* is the same as *SSN20*, a gene important in the transcriptional regulation of the *SUC2* gene (needed for sucrose metabolism) (41, 133), and *SPT11* and *SPT12* are histone H2 genes (40).

As mentioned above (Section III.D), the three *SPT* genes, *SPT3*, *SPT7*, and *SPT8*, are essential for initiating the transcription of Ty elements. In *spt3*, *spt7*, and *spt8* mutants, Ty transcription is distinctly aberrant. Very little (<5%) Ty transcript of normal length is seen. The site of action of the *SPT3* gene product may be the U3 region of the δ sequence, as *GAL1*/Ty fusions are expressed more or less normally in *spt3* mutant cells as judged by Northern blot analysis (Durbin, Ph.D. thesis; K. Chapman and J. D. Boeke, unpublished data). This and the fact that *SPT3* acts on solo δ promoters suggests that the *SPT3* gene product may be (or may regulate) a transcription factor required for initiation at δ promoters. Not surprisingly, transposition of Ty1 elements occurs at greatly reduced frequency in *spt3* mutant cells unless a *GAL1*/Ty fusion is introduced into the cells (16). Recently, *SPT3* mutants have also been shown to have decreased levels of mRNA for α factor and **a** factor (94). Transcription of other genes appears to be normal.

***ROC/TYE* genes.** Additional host mutations affecting the expression of Ty elements define the *ROC* and *TYE* genes. In contrast to the *SPT* genes, *ROC* and *TYE* genes were identified by mutations that reverse the phenotype of a Ty insertion that had previously activated the expression of adjacent genes (37, 55). *ROC* refers to ROAM mutation (see Section VII.B.1) control; *TYE* signifies Ty-mediated expression. A *roc1* mutation seems to affect the transcription of Ty elements in general as well as that of the adjacent DNA; recent data suggest that mutations in this gene decrease Ty1 transcription to a great extent but do not affect Ty2 transcription (I. LaLoux, F. DuBois, and E. Jacobs, personal communication).

4. Enhancer and silencer sequences

Ty elements contain, in addition to "conventional" UAS sequences, an important block of internal regulatory sequences downstream from the transcription start site, including sequences showing homology to SV40 and "core" enhancer sequences (65, 153). There is disagreement among different groups concerning the precise location of these sequences in various Ty elements, but some of this may be attributable to the fact that different Ty elements as well as different reporter genes are studied by different groups. Sequences significantly similar to the SV40 and core enhancer sequences have been found within both Ty1 (65) and Ty2 (153) sequences. These sequences lie within a block of sequence which is important in regulating the transcription of Ty itself as well as that of the host DNA adjacent to the 5′ δ. Although these sequences resemble mammalian enhancers in that they apparently function downstream of the initiation site for Ty transcription, there is disagreement as to whether they work in both orientations. Errede et al. (66) reported two sequences in Ty1 which are important for enhancing gene expression. One of these (coordinates 240 to 478) does not show *MAT* regulation and can be inverted without significantly affecting adjacent gene expression, but its deletion abolished adjacent gene expression (66). A second segment, extending from coordinates 478 to 1223, which encompasses the "enhancer" homology regions as well as certain homologies to **a**1/α2 control regions, was required for full stimulation of adjacent gene expression and for *MAT* regulation (43). Rathjen et al. (147) report a Ty1 sequence (coordinates 384 to 478) which shows *MAT* regulation and cannot be inverted. In both these studies, intact 5′ δ elements were not present in the Ty element adjacent to the gene whose expression was monitored, raising the question of the generality of these findings.

Sequences important for the enhancement of adjacent gene expression in Ty2 elements were initially identified by the experiments of Roeder et al. (153), who found that a single base change just adjacent to the enhancer core homology sequence of Ty2-917 was responsible for the conversion of *his4-917* (a mutation caused by Ty2-917, which does not stimulate the transcription of the adjacent *HIS4* gene) to *his4-917(467)*, a His$^+$ derivative. Recent deletion experiments (Kapakos et al., submitted) confirm that this block of sequence is important and in addition identify a second region within Ty2-917 which stimulates adjacent gene transcription. These experiments on adjacent gene transcription may not be generalizable, because the element used (Ty2-917) normally shuts off expression of the adjacent gene and because, again, an intact element was not used. In fact, when the intact Ty2-917 element was used, no stimulation of adjacent gene expression was seen, suggesting that active transcription of this element might decrease transcription of the adjacent gene.

The effect of the *SPT3* gene on the ability of Ty1 elements to enhance adjacent gene function has also been investigated. In most cases, when the Ty element is inserted within 200 bp of the reporter gene (*his3Δ4*), the reporter gene is expressed in both *SPT3* and *spt3* backgrounds. However, when Ty elements are inserted at distances significantly greater than 200 bp from the reporter gene, its activation is observed only in the *spt3* strain background. Thus, the Ty enhancer seems to be able to work at a distance of up to 200 bp away from *his3Δ4* in wild-type backgrounds but up to 700 bp in *spt3* backgrounds in which the δ promoter is inactive. This result suggests that (i) the state of transcription of the Ty element can profoundly influence the extent of expression of the adjacent gene (*his3Δ4*) and (ii) the enhancer can work over much longer distances if the δ promoter is silenced somehow (16). Ty activation of adjacent genes at long distances has also been reported at the *DUR1/2* locus (29, 44; G. Chisholm and T. Cooper, XII International Conference on Yeast Genetics and Molecular Biology, 1984, abstr. no. J9, p. 203).

5. Induction of Ty transcription by UV light and DNA-damaging agents

Rolfe et al. (154) recently reported that Ty element transcription can be increased as much as 10-fold by UV light irradiation. Similar results have been obtained in another laboratory, as well as preliminary evidence that Ty transposition also is stimulated by UV irradiation (122a). It is interesting to note that in the δ promoter region there are several sequences closely related to the sequence ATGATT found repeated several times upstream of another UV-induced gene, the *CDC9* (DNA ligase) gene of *S. cerevisiae* (9). Preliminary studies of γ-irradiation- and chemically induced mutagenesis suggested that such treatments may also stimulate Ty transposition (128; C. Stuewer, W. Vogel, and U. Hagen, *Yeast* 2:s371).

6. Carbon source and Ty transcription

The carbon source used by *S. cerevisiae* influences the transcription of Ty elements. In glycerol-grown cultures, there appear to be fewer Ty transcripts than in glucose-grown cultures (172). It was also reported (172) that *MAT* regulation of Ty transcription and of adjacent *ADH2* expression is much more evident in glycerol-grown cultures.

7. σ Element transcription: pheromone inducibility

Recently, the σ element has been shown to respond transcriptionally to treatment with the yeast mating pheromone α factor. Van Arsdell et al. (180) demonstrated the existence of several α-factor-inducible transcripts hybridizing to a σ-specific probe. Two of these were particularly intense, a 650-nucleotide species, which was studied in detail, and a transcript of 5.2 kb, which is apparently the mRNA of the Ty3 transposon. The 650-nucleotide species arises from a segment of DNA previously isolated on the basis of its encoding a hormone-responsive transcript (169). This region of DNA contains two truncated δ elements, one intact δ element, one intact and one truncated σ element, and a tryptophan-tRNA gene. The α-factor-inducible transcript initiates in the intact σ and reads away from the tRNA and through (in order) the remainder of the intact σ, the truncated σ, a truncated δ, and an intact δ sequence, presumably terminating in flanking single-copy sequence. It is curious that the length of this sequence suggests that it does not terminate in the intact δ, even though it is in the appropriate orientation to terminate the transcript (180). A similar LTR-transcriptional readthrough phenomenon has been observed with the *Drosophila* mutation w^{a}, which was induced by a *copia* element (201). Perhaps certain LTR sequences have defective termination or polyadenylation signals.

It is not clear whether all σ elements respond to α factor in the same way; the pattern of transcripts hybridizing to σ probes suggests that not all transcripts reading out of σ elements are equally abundant. This may simply reflect differences in the stability of the RNAs produced.

The σ element alone appears to be responsible for the α-factor-responsive regulation, because a σ element/*SUC2* fusion construct directs the α-factor-dependent synthesis of invertase (the product of the *SUC2* gene) (180). The specific sequence probably responsible for responding to pheromone induction in the σ element as well as in other genes responding to α-factor stimulation has the consensus sequence ATGAAACA (22, 178, 181). Deletion of these sequences from the σ element results in a loss of pheromone induction (S. van Arsdell and J. Thorner, personal communication). The pheromone-inducible nature of the σ element is reminiscent of the hormone-inducible enhancer sequences in mouse mammary tumor virus.

B. Posttranscriptional Regulation

Ty1 and Ty2 gene expression shows certain other features which may present possible targets of regulation of the transposition process. These include the mode of expression of the important *TYB* protein (ribosomal frameshifting), the phosphorylation of Ty-encoded proteins, which has been demonstrated for *TYA* proteins (123), and the proteolytic processing of p58-TYA and p190-TYA/TYB primary translation products into smaller proteins, as well as general translational control. Thus far, however, no direct demonstrations of regulation of the transposition process through such mechanisms have been made.

C. Regulation by Accumulation of Defective Copies

Because many Ty elements bear sequence heterogeneities which suggest that they encode nonfunctional gene products (see above), it seems possible that many or most Ty elements in the genome are defective. On the other hand, sequence analysis has indicated the presence of intact open reading frames in all fully sequenced Ty elements, and preliminary experiments suggest that as many as half of the Ty1 and Ty2 elements which are fused to a *GAL1* promoter are able to induce high levels of transposition, although this is based on a rather small number of elements tested (45a; unpublished data). That there are defective elements with normal open reading frame structure is clear from analysis of Ty1-173, a Ty1 element bearing a specific Leu → Ile mutation in *TYB*. This mutation, *tyb-2098*, renders Ty1-173 completely unable to induce transposition when fused to the *GAL1* promoter. Hybridization experiments with oligonucleotide probes suggest that seven copies of Ty1 in the genome carry this lesion (11). This mutation appears to prevent proper proteolytic processing of *TYB* proteins (the *TYB* protein p90-TYB is produced at very low levels) and lies within a putative metal-binding domain (104) in the integrase region.

Ty element transcription probably varies in strength from copy to copy of the element, because flanking sequences can affect the potency of the LTR promoter (165). Thus, it may be that the precise level of transposition in a cell is dictated by the relative strength of the LTR promoters lying upstream of the functional copies of Ty elements. Promoters of various strengths might easily be "shuffled" via the general homologous recombination system of *S. cerevisiae* with functional or nonfunctional Ty elements, thereby modulating the overall level of functional Ty gene products present in the cell (11, 12). Such "promoter-shuffling" regulation would explain the fact that although Ty element transposition, Ty-VLP formation, and p90-TYB levels increase 20- to 100-fold in the presence of *GAL*/Ty plasmids, the total amount of Ty transcription increases by a much smaller amount (12). This type of regulation may account for the basal level of transposition in a strain on an evolutionary time scale; other mechanisms probably exist to regulate it on a shorter time scale.

D. Transposition Rates In Vivo

Two assay systems which allow the accurate measurement of the rate of transposition of native Ty1 elements (i.e., in the absence of Ty-overproducing plasmids) have been described. In both cases, the assays make use of the mechanism by which Ty transpositions give rise to the expression of a silent gene. In the first system, developed by Paquin and Williamson (140, 141), Ty transposition into two loci, *ADH2* and *ADH4*, is measured as the frequency of antimycin A-resistant growth in the presence of glucose in Δ*adh1* strains. Cells lacking *ADH1* activity are unable to grow fermentatively; respiration is inhibited by adding antimycin A to the plates. This leads to a strong selection for fermentative growth; insertion of Ty1 or Ty2 elements at *ADH2* allows *ADH2* to be expressed in the presence of glucose (where it is normally repressed). The ability of Ty insertions to allow growth by inserting at *ADH4* is unclear but acts presumably by activating a cryptic *ADH* function which is normally repressed. An advantage of this system is that under appropriate conditions (of low temperature), essentially 100% of the antimycin A-resistant colonies are derived from Ty transposition events; a disadvantage is that the Ty elements are more difficult to analyze and clone out, as they are in the chromosomes at two different loci. In the second system, originally described by Scherer et al. (162), a promoterless (nonexpressed) *HIS3* gene, *his3*Δ*4*, is introduced via a centromeric (± single copy) plasmid into an *S. cerevisiae* strain which is completely deleted for the *HIS3* locus on the chromosome. Ty transpositions into the bacteriophage λ DNA upstream of the *HIS3* coding sequences activate the transcription of the promoter-deleted *HIS3* sequences, giving rise to a His$^+$ phenotype. An advantage of this system is that plasmids bearing the Ty insertion can be readily recovered from *S. cerevisiae* and studied in great detail; the disadvantage of this system is that even under optimal conditions, a small fraction of the His$^+$ revertants obtained derive from events unrelated to Ty element transposition, so that further physical characterization of the plasmids must be performed in order to establish the "Ty fraction," that is, the fraction of revertants actually caused by Ty insertion (14, 16).

Because these assays rely on the insertion of Ty elements into a relatively small promoter region, and because the Ty insertions recovered are nearly all in one orientation, the target size for transposition events which can be recovered is extremely small, on the order of a few hundred base pairs. Under optimal conditions, the rate of transposition (summed for all Ty elements capable of activating expression of the target gene) was estimated to be about 10^{-7} per cell division for the *ADH2* and *ADH4* loci (combined) (141) and 10^{-7} per cell division at *his3*Δ*4* (16). A frequency of Ty1 insertion at the *SUP4-o* locus of 9×10^{-8} has recently been estimated by Giroux et al. (88). At *his3*Δ*4*, the effective target size is known to be about 200 bp (14, 16). Assuming that this 200 bp forms as good a target as the rest of the genome (which may not be true!) and that essentially the entire genome is disruptible (89), a Ty1 transposition event would occur in about 1% of all cell divisions. Assuming that all 30 or so copies of Ty1 elements transpose and activate *his3*Δ*4* equally well (many, if not all of them, can activate *his3*Δ*4*; the fraction capable of transposing is uncertain), this would yield a mean rate of transposition per Ty1 element of between 10^{-4} and 10^{-3} per generation. These calculations are of course very crude and subject to all of the caveats mentioned, but they provide a ballpark figure for the Ty1 transposition rate. It is unlikely that the rate is more than 10-fold higher than the range calculated above, but it may well be 10-fold lower. Naturally, a small fraction of these insertions will be lethal (in a haploid strain) (89), and an unknown fraction of these insertions will put their host at a selective disadvantage and so will tend to disappear from the population by selective pressure; therefore, it is difficult to estimate from these numbers the accumulation rate of Ty elements in *S. cerevisiae* cells.

In both of the transposition assay systems mentioned, Ty1 insertions outnumber Ty2 insertions by a considerable amount (about 25:1 in the *his3Δ4* system) (Boeke and Garfinkel, unpublished data; C. Paquin and V. Williamson, personal communication), although the copy number of Ty1 elements in the genome is only two- to threefold higher than the Ty2 copy number in most strains. Those Ty2 transpositions recovered show as strong a His^+ phenotype as do the Ty1 insertions. It is difficult to know whether this difference in numbers actually reflects a lower transposition frequency of Ty2 elements or simply that only a small fraction of Ty2 elements is capable of activating *his3Δ4* and the *ADH* genes. Variants of Ty2 elements with different strengths of activation of the adjacent *HIS4* gene have been described (151, 153).

1. Temperature effects

Low temperatures stimulate transposition in various organisms; *S. cerevisiae* Ty1 elements are very temperature sensitive for transposition. Paquin and Williamson originally demonstrated this for transpositions at *ADH2* (140) and confirmed it at *ADH4* (141). Transposition at *his3Δ4* follows the same pattern (16). The Ty1 transposition rate is relatively high at low temperatures (124) but decreases by 20-fold at 30°C and is difficult to detect at 37°C. Curiously, overexpression of a Ty element (with *GAL*/Ty plasmids) partially overcomes this temperature sensitivity in vivo, allowing high-frequency transposition at 30 but not 37°C (Xu and Boeke, unpublished data). It is likely that the temperature sensitivity of Ty1 transposition is due to the inherent thermolability of the reverse transcriptase, which shows a temperature sensitivity profile in vitro reminiscent of the pattern of transposition temperature sensitivity (85).

2. Carbon source and *MAT* effects

Paquin and Williamson (141) investigated the effects of carbon source and constitution of the *MAT* locus on Ty1 transposition, using the *ADH2/4* system. Slightly lower transposition frequencies were observed on glycerol than on glucose. These differences parallel the reported effects of carbon source on Ty transcription (172); the effect on transposition is not as great as the effect on transcription. Similarly, although *MAT* **a**/α diploids transposed at a somewhat lower frequency than isogenic **a**/**a** and α/α diploids, this difference was not as great as expected, given the relative abundance of Ty RNAs in these strains.

VI. TRANSPOSON-ENCODED GENE PRODUCTS

A. The *TYA/gag* Gene

1. Structure

The *TYA* open reading frame of both Ty1 and Ty2 encodes a protein of about 50 kDa. Perhaps because of its very high proline content (nearly 9%), the primary translation product of the gene migrates anomalously on gels as a ca. 58-kDa species (p58-TYA), although the molecular weight predicted from the DNA sequence is closer to 50 kDa. A proline-rich region (>20% proline) is found near the amino terminus. Clare and Farabaugh (38) pointed out a putative helix-turn-helix motif in the C-terminal region of the protein which may represent a nucleic acid-binding domain. This C-terminal region is also extremely basic (20% basic residues). This putative nucleic acid-binding domain lies in a region of *TYA* which is in a position similar to that of a putative metal-binding domain found in all other retroviruses and retrotransposons compared (45; Doolittle et al., in press). An amino acid sequence showing some similarity to a region of the *tnpA* gene from Tn*3* has also been reported (93). Antibodies directed against Ty-VLPs (3), synthetic peptides predicted by the *TYA* sequence (13, 200), and purified p58-TYA (131) all cross-react with p58-TYA, p54-TYA, and smaller *TYA* proteins. The p58-TYA protein is the most abundant *TYA* protein in cells overexpressing *TYA* alone or in protease mutants of plasmids overproducing the whole Ty element (3, 131, 200). Clearly, p58-TYA is processed into more than one smaller protein by the *TYB*-encoded protease; a major product of the proteolysis reaction is p54-TYA, which is perhaps the most abundant *TYA* protein in the cell under normal conditions or when an intact Ty1 element is overexpressed. When *TYA* proteins in Ty-VLPs are examined, however, different patterns of proteins are seen by different groups. In all cases, several smaller species of *TYA* protein are observed. Two groups working with plasmid constructs overproducing Ty1 elements Ty1-15 and Ty1-9C found that p54-TYA (also called p2 or TYA protein) is the major capsid protein, but a third group working on Ty1-H3 found that two smaller proteins, p41-TYA and p38-TYA, were the dominant epitopes in Ty1-

VLPs. These differences cannot be ascribed to migration differences due to different gel electrophoresis conditions because these workers found p58-TYA and p54-TYA on the same gels to be the dominant *TYA* protein species in total cell protein extracts (200). The significance of these results is uncertain but may be related to differences in the methods used to purify the Ty-VLPs.

TYA proteins are apparently phosphorylated (123). The site(s) and extent of phosphorylation of TYA proteins and the state of phosphorylation of *TYA* proteins in the Ty-VLPs have not been determined, nor has the protein kinase responsible been identified. Other modifications of *TYA* proteins have not been described; the absence of N-terminal glycine makes it unlikely that *TYA* proteins are myristylated in the manner of many other *gag* proteins. Since the role of the myristic acid group in retroviruses is probably to interact with membrane or *env* protein or both (which Ty-VLPs apparently lack), there is probably no "need" for it in Ty.

2. Function

Recent studies provide strong evidence that the primary function of the *TYA* proteins is in the formation of the capsid or shell of the VLP (which normally also contains Ty RNA and reverse transcriptase and possibly other molecules such as integrase and protease). The best evidence that *TYA* specifies the capsid proteins is the fact that overproduction of *TYA* alone results in the accumulation of pre-Ty-VLPs in the cell (3, 131). The pre-Ty-VLPs probably contain Ty RNA (S. Youngren and D. J. Garfinkel, personal communication). Since packaging of Ty RNA in normal Ty-VLPs is specific, at least one of the *TYA* proteins must carry out specific binding of Ty RNA and perhaps also of priming tRNA, although in retroviruses the priming tRNA is thought to be bound to reverse transcriptase (145). Another possible function of the Ty-VLPs is in the delivery of reverse transcripts to target DNA in the nucleus. Perhaps Ty-VLPs also sequester the reverse transcriptase from host mRNAs, reducing the likelihood of generating potentially deleterious pseudogenes (75, 118).

B. The *TYB/pol* Gene

The *TYB* gene encodes proteins which carry out several functions essential to the transposition process. Domains of the *TYB* open reading frame have been identified by their weak but significant amino acid sequence similarity to their retroviral counterparts. The function of the reverse transcriptase or RNA-dependent DNA polymerase domain of *pol* genes is well known; equally important functions are carried out by the protease, integrase, and RNase H as well as a long region of *TYB* which lies between the integrase and the reverse transcriptase homology regions and is **not** similar to any retroviral sequence. In-frame linker insertion mutations in all of these domains can abolish transposition, suggesting that they are all essential in the transposition process (J. D. Boeke and G. Monokian, unpublished data).

1. Mode of expression (frameshifting)

One of the most interesting aspects of the *TYB* open reading frame is its mode of expression. Apparently, all retroviruses and retrotransposons are faced with a similar dilemma in gene expression; more moles of capsid (*gag*) proteins are needed than are moles of *pol* protein, yet the same mRNA (equivalent to genomic RNA) is used to express both. Essentially all retroviruses and retrotransposons use various forms of "attenuation" (the term is used very loosely here to indicate a downmodulation of expression) to achieve this goal (reviewed in reference 10); in almost all cases this "attenuation" is translational. In the case of Ty elements, there is now considerable evidence that *TYB* proteins are made initially as a *TYA/TYB* fusion protein, p190-TYA/TYB (38, 124, 193). This appears to be carried out by a specific ribosomal frameshifting event; recent evidence suggests that only 14 bp of sequence (13/14 conserved between Ty1 and Ty2) from the region where *TYA* and *TYB* overlap is sufficient to effect this frameshifting event within the heterologous *HIS4* gene (38a). This readthrough event occurs at about a 20% frequency; it cannot be due to RNA splicing or RNA editing, as the RNA from this region has been sequenced directly and shows no evidence of variability in structure within the region of the overlap between *TYA* and *TYB* (38a). The detailed structure of p190-TYA/TYB in the overlap region is unknown. It is suspected that a specific host tRNA gene can act as a frameshift-suppressor tRNA at a specific region within this 14-nucleotide sequence (38a). The detailed mechanism of Ty element frameshifting is likely to differ from that used by Rous sarcoma virus, mouse mammary tumor virus, and human immunodeficiency virus (97–99), which have −1 frameshifts rather than the +1 frameshift found in Ty1 and Ty2. Also, retroviral −1 frameshifting probably requires a stem-loop structure 3′ to the apparent site of frameshifting which seems not to be required for Ty frameshifting (97, 98).

2. *TYB* proteins

Only one protein deriving from the *TYB* open reading frame has been identified immunologically, p90-TYB. This protein bears epitopes from the integrase homology region as well as the region between the integrase and reverse transcriptase domains (57a, 85, 200), but it is not clear which other segments of *TYB* are included in it. A most unexpected property of this protein is that it is virtually undetectable in normal yeast cells by immunoblotting, but the protein is abundant in cells bearing an induced pGTy1-H3 plasmid.

3. Protease domain

The N-terminal sequences of the *TYB* open reading frame, like the *pol* genes of most retroviruses and retrotransposons, contain significant amino acid similarities to acid proteases such as pepsin (130, 174; Doolittle et al., in press). Recent studies have shown directly that this domain of *TYB* is responsible for the proteolytic processing both of *TYB* proteins themselves (200) as well as of *TYA* proteins (3, 13, 131, 200). This protease function is also essential for the transposition process (200). The molecular weight of the protease molecule is unknown; it is not even known whether it exists by itself or as part of a polyfunctional protein.

4. Integrase domain

The integrase protein is thought to carry out the actual transposition process itself, that is, the integration of the completed reverse transcript into the host target sequences. A consequence of integrase action must be the duplication of a short stretch of target DNA; the *S. cerevisiae* integrases all make 5-bp duplications. The Ty1 integrase protein has recently been identified as p90-TYB on the basis of immunological studies. p90-TYB is contained within Ty-VLPs, which are competent to carry out transposition in a cell-free system. Linker insertion mutations (J. D. Boeke and G. Monokian, manuscript in preparation) and a spontaneous point mutation (11) in the integrase domain are able to greatly reduce Ty element transposition. The sequence organization of Ty1 and Ty2 (and the *Drosophila* retrotransposon *copia*) differs from that of all other sequenced retroviruses and retrotransposons in that the integrase domain is found 5′ to the reverse transcriptase and RNase H coding regions, suggesting that these elements form a separate subfamily of retrotransposons. An interesting structural feature of integrases is a possible metal-binding domain (zinc finger?) found near the N-terminal portion of the homologous region (104). The Ty1-173 mutation discussed above lies within this portion of integrase.

5. Region of unknown function

The region of *TYB* between the integrase and reverse transcriptase domains (Fig. 2) has no known function; it may simply be a very variable region of the reverse transcriptase or of integrase. This domain is essential for Ty transposition, because numerous linker insertion mutations in it inactivate Ty1 element transposition (Boeke and Monokian, unpublished data).

6. Reverse transcriptase (RNA-dependent DNA polymerase) domain

Gene disruptions and frameshift mutations affecting the region of *TYB* showing similarities in sequence to retroviral reverse transcriptases inactivate the reverse transcriptase activity associated with Ty1-VLPs (3, 85); in-frame linker insertions in the region abolish transposition (Boeke and Monokian, unpublished data). Other mutations, such as large in-frame deletions of other portions of *TYB*, destroy transposition ability but leave the reverse transcriptase activity unaffected (200). These experiments indicate that the "RT" domain indeed encodes the reverse transcriptase enzyme found in Ty-VLPs.

7. RNase H domain

The RNase H domain of retroviruses and retrotransposons was recently identified by sequence comparison studies (104). As in most other retrotransposons and retroviruses, the Ty RNase H is downstream of the reverse transcriptase. Whether it exists as a separate protein product is unknown; in retroviruses it is "tethered" to the reverse transcriptase via a short stretch of amino acids.

C. Reverse Transcriptase and RNase H Activities Found in Yeast Cells

An RNase H activity isolated from yeast cells was described several years ago (105). Its relationship to Ty elements, if any, is unclear, but recent work on this interesting enzyme has shown that it has an associated reverse transcriptase activity (106). Although the authors claim that this enzyme is not Ty encoded, some of the enzymatic properties are very similar to those of Ty-encoded reverse transcriptase (such as its low-temperature optimum).

VII. EFFECTS OF *S. CEREVISIAE* TRANSPOSONS ON EXPRESSION OF NEIGHBORING GENES

A. Transpositions into Coding Regions

Transpositions of Ty1 elements into the coding regions of *S. cerevisiae* genes generally result in a completely null phenotype. Such insertions include the mutation *ura3-52*, which is widely used as a convenient nonreverting allele (155). Several dozen additional mutations in the *URA3* structural gene have similar properties: their reversion rate is well below 10^{-8}, suggesting that precise excision is generally undetectable in the case of Ty1 elements (G. Natsoulis and J. D. Boeke, unpublished data). Insertion mutations well within the *LYS2* structural gene behave in exactly the same way (Natsoulis and Boeke, unpublished data). Exceptions to these cases include the Ty1 insertions *lys2-173* (166), *ura3-109*, and *ura3-153* (G. Natsoulis, W. Thomas, M.-C. Roghmann, F. Winston, and J. D. Boeke, submitted for publication). These insertions within coding regions do revert, but never by precise excision. These three mutations are all near the 5′ end of the target gene, suggesting that they can revert via events which give rise to a functional or partially functional fusion protein.

A Ty insertion at the *URA2* locus has also been isolated which lies within the coding region; this element was selected because it activated the expression of the C-terminal activity of this gene, which encodes a bifunctional protein (6, 68).

B. Transpositions into Noncoding Regions

Ty element transpositions into noncoding regions are far more variable and complex with regard to phenotype, as they disrupt regulatory rather than coding sequences. When Ty elements inactivate a target gene, the resultant mutations invariably revert, often in many interesting ways, including mutations in the unlinked suppressor (*SPT*) genes mentioned above. Ty insertions in noncoding regions can be conveniently classified as either activating transpositions (i.e., stimulating the expression of a previously silent target gene) or inactivating transpositions (preventing the expression of a gene).

1. Activating transpositions

Activating transpositions have been observed when Ty elements have transposed next to a multitude of different *S. cerevisiae* genes which were normally not expressed for a variety of reasons (summarized in Table 4). Such activations can occur either in the chromosome or on episomal plasmids. With few exceptions, these activating Ty insertion mutations have the following general properties. (i) The Ty element is oriented such that it is pointing away from the target gene; that is, it forms a divergent transcription unit with the target gene. (ii) The adjacent gene shows some degree of *MAT* transcriptional regulation not unlike that of Ty itself (Section V.A.2). Usually, the Ty element lies within 200 bp of the target gene, although exceptions to this have already been mentioned (Section V.A.4). This class of activating Ty insertion mutations has been referred to as ROAM mutations (regulated overproducing alleles under mating signals [63]). Apparently, this class of insertion mutations responds to enhancerlike sequences within the Ty element which are under the influence of the *MAT* locus.

Four interesting exceptions to these generalities have been described. A very rare class of activating insertion in which the Ty element is in the same transcriptional orientation as that of the target gene has been observed in two systems. Among the collection of Ty elements isolated as His$^+$ revertants of *his3Δ4*, a very small fraction of the Ty element insertions (<5%) do not form divergent transcription units with the *HIS3* sequences; that is, they are transcribed in the same direction as *his3Δ4*. None of these His$^+$ revertants show *MAT* regulation, whereas the Ty insertions which form divergent transcription units do show *MAT* regulation (unpublished data). Similarly, among His$^+$ revertants of the Ty1 insertion *his4-912* (see below), which is transcribed in the same direction as *HIS4*, one case of a His$^+$ gene convertant which contained a new Ty1 sequence replacing Ty1-912 was found; this His$^+$ convertant did not show *MAT* regulation (151). The mechanism of activation of these Ty elements which are transcribed in the same direction as the target gene thus does not appear to use a *MAT*-regulated enhancer; perhaps their mechanism of activation is via the fortuitous formation of a promoter element by the appropriate juxtaposition of specific Ty and target sequences. Another interesting case is posed by the *cyc1-Δ512* mutation. This deletion near the 3′ end of *CYC1* apparently removes the transcriptional termination sequences, resulting in a Cyc$^-$ phenotype. Among the revertants of this mutation was a Ty1 element which transposed into the region 3′ of *cyc1-Δ512*; the Ty is in the same transcriptional orientation as is *CYC1*. Apparently, the *CYC1* transcript can now terminate within a 5′ δ element of the Ty1, restoring function to the gene. This is the only case known where a Ty element 3′ to a target gene affects its function (110, 202).

Table 4. Genes activated by Ty insertion

Gene	Location	Phenotype	*MAT* regulated?	Transcription of Ty[a]	References
ADH2	5′ end	*ADH2* is expressed in glucose-containing media	Yes	D	32–35, 192
CYC7	5′ end	*CYC7* is overexpressed 20- to 40-fold	Yes	D	63, 64, 158
DUR1/2	5′ end	*DUR1/2* is expressed constitutively	Yes	D	29, 44, 55, 113[b]
CAR1	5′ end	Constitutive, high-level expression of *CAR1*	Yes	D	49, 54, 55, 102
CAR2	5′ end	Constitutive, high-level expression of *CAR2*	Yes	D	47, 55, 101
his3-4[c]	5′ end	Promoterless *HIS3* gene is expressed	Yes	D	14, 16, 162
his3-4[d]	5′ end	Promoterless *HIS3* gene is expressed	No	T	16; and unpublished data
URA2	Structural gene	Activation of C-terminal activity of bifunctional protein			6, 68
PHO5	5′ end		Yes	D	175, 176
cyc1-512	3′ end	Terminatorless *CYC1* gene is expressed	No	T	110, 202
neo	5′ end	Expression of normally silent bacterial gene in yeast	Yes	D	179
CYR1	5′ end	Decreased expression of *CYR1* results in heat shock resistance	Yes	D	Iida, personal communication
HMLα	?	Activation of expression of silent cassette	?	?	—[e]

[a] With respect to target gene; D, divergent transcription; T, transcribed in tandem (same orientation as target gene).
[b] Also G. Chisholm and T. Cooper, XII International Conference on Yeast Genetics and Molecular Biology, 1984, abstr. no. J9, p. 203.
[c] Major class.
[d] Minor class.
[e] K. Weinstock, D. J. Garfinkel, and J. Strathern, personal communication.

Recently, Ty1 transpositions which activated the *HMLα* silent cassette were isolated as α-factor-resistant mutants (M. Mastrangelo, K. Weinstock, D. J. Garfinkel, and J. Strathern, personal communication). The detailed mechanism of this activation is unknown.

2. Inactivating transpositions

Ty insertions which inactivate gene expression by insertion into a promoter region have also been isolated. Ty and δ insertions at the *HIS4* (25, 69, 149, 150), *LYS2* (57, 166) (Natsoulis et al., submitted), *URA3* (Natsoulis et al., submitted), *CAN1* (C. Wilke, S. Heidler, and S. Liebman, personal communication), and *CYR1* (=*HSR1*) (H. Iida, personal communication) loci have been isolated. In general, such mutations can be caused by Ty elements in either orientation, although there is a bias for Ty elements which inactivate gene expression to be oriented in the same direction as the target gene (Natsoulis and Boeke, unpublished data); both Ty1 and Ty2 elements can inactivate target genes in this way. The main mechanism of gene inactivation appears to be by the separation of essential elements of the promoter of the target gene from each other. For example, the well-studied *his4-912* element is inserted between the *HIS4* UAS and the TATA box, whereas the *his4-917* Ty insertion lies between the TATA box and the initiation site for *HIS4* transcription (149). One series of inactivating Ty insertions in the *LYS2* gene promoter lies between the TATA box and the initiation site; a rarer class lies between the UAS and the TATA box (57, 81, 166; Natsoulis and Boeke, unpublished data). In all of these cases, the cause of the inactivation of the gene appears to be the loss of promoter function of the gene (i.e., no transcription).

Mutations of this type can revert in many interesting ways. For example, the *his4-912* mutation, which is His$^-$ at all temperatures, reverts to a His^{cs} (cold-sensitive) phenotype when the Ty element undergoes δ-δ recombination, leaving behind a solo δ element called *his4-912*δ (150). Although the solo δ element acts as a promoter, the transcript it produces, which reads through into *HIS4*, is probably nonfunctional because it contains out-of-frame AUG codons, which would reduce the ability of the *HIS4* open reading frame to be translated. Since *his4-912*δ is

His$^-$ at low temperature, further reversion experiments were carried out on it. Many of the first *spt* (previously called *spm*) mutations were isolated in this way (149, 194). For example, *spt3* mutations shut off the δ promoter, restoring *HIS4* transcription (196, 197). A point mutation within the δ in a sequence likely to serve as the TATA box for Ty transcription was also isolated (Durbin, Ph.D. thesis) (Fig. 3). Mutations in the same *SPT* genes isolated by reversion of *his4-912*δ could also be isolated as suppressors of Ty and δ insertions in *LYS2* (166). Revertants of other Ty insertions have arisen by a multitude of complex homology-dependent genomic rearrangements involving the Ty element in question (see Section IX below).

Recently, insertions at the *CYR1* locus have been obtained by a novel procedure. *S. cerevisiae* strains stably resistant to heat shock (also called *HSR1* for heat shock resistant) were isolated. Two independent mutants of this type contain Ty insertions near the 5′ end of the *CYR1* gene (Iida, personal communication).

VIII. DISTRIBUTION OF TRANSPOSONS IN THE YEAST GENOME

A. Variability within Species

Within different *S. cerevisiae* strains, Ty1, Ty2, δ, σ, and τ elements are widely distributed. In fact, it is difficult to find strains lacking Ty elements, although strains bearing a low number of copies have been isolated from the wild. Wild strains tend to have lower Ty copy numbers than do laboratory strains. A strain of *Saccharomyces uvarum* (considered by modern taxonomists to be part of the *S. cerevisiae* species group) containing only one Ty element has been found. This element is apparently not closely related to either Ty1 or Ty2 (M. Ciriacy and P. Philippsen, personal communication).

B. Variability among Yeast Species

The Ty1, Ty2, σ, and τ elements, for the most part, are undetectable by Southern hybridization techniques in genera of yeast other than *Saccharomyces* (such as *Candida*, *Pichia*, and *Hansenula*) (144, 146; G. Natsoulis, C. Kurtzman, and J. D. Boeke, unpublished data). Even within the genus *Saccharomyces*, the distribution of these elements seems to be limited to what modern-day taxonomists consider to be *S. cerevisiae* (143, 144, 146), except that a small amount of hybridization of Ty1 probe is seen in the rather distantly related species *Saccharomyces kluyveri* (Natsoulis et al., unpublished data; Weinstock et al., personal communication). Certain *Saccharomyces* "species" now considered part of the *S. cerevisiae* species group, such as *Saccharomyces norbensis*, appear to contain Ty2 elements but no Ty1 elements, although the converse situation has not been found (75; Natsoulis et al., unpublished data).

C. Position of Transposons in the *S. cerevisiae* Genome

Yeast Ty1 and Ty2 elements are distributed more or less randomly among *S. cerevisiae* chromosomes (75, 109). The Ty3 elements have been tentatively localized to chromosome pair VII + XV and chromosome IX in strain AB972 and to chromosomes XII, VII, XI, and IX in strain YP148 (160). From the limited general mapping of Ty elements which has been done (109), it is uncertain whether some domains of chromosomes are much richer in Ty elements than others. There is considerable evidence that Ty, δ, σ, and τ elements are all very abundant in the vicinity of tRNA genes (20, 30, 48, 58, 83, 160, 161, 171). This is particularly true of σ elements, where 15 of 17 elements analyzed from the same strain were at positions −15 to −19 from different tRNA genes, −17 being the most prevalent site of insertion. This truly remarkable target specificity of σ and Ty3 elements is interesting because tRNA transcription initiation is thought to occur near these very sites (62). It has been suggested that the Ty3 transposition complex may recognize tRNA transcription factors as part of its transposition process (160).

The general preference of these transposons for tRNA-coding regions is unlikely to be attributable to a general lack of nondisruptible sequences elsewhere in the genome; it has recently been demonstrated that about 90% of randomly generated gene disruptions in *S. cerevisiae* can be tolerated with no apparent phenotype (89). Thus, tRNA gene regions may be particularly good targets for retrotransposon insertion, or, alternatively, the transposon sequences may actually be selectively advantageous there. Direct evidence for the latter point has been difficult to obtain. Sigma elements have no gross effect on the expression of the adjacent tRNA gene (161). Certain tRNA genes have been isolated both with and without adjacent Ty elements from different strains. Effects of the Ty1 elements on *S. cerevisiae* tRNA gene transcription in frog oocytes have been measured; the presence of a Ty element increases the transcription severalfold in this rather heterologous system (92, 134). Interestingly, Ty element insertions turned up as a major cause of mutations knocking out the tRNA locus *SUP4-o* (88), suggesting that tRNA genes are good targets for Ty insertions. Although these results suggest that tRNA-containing regions of the genome are preferred targets for transposition, it is formally possible that transposable elements are more difficult to remove from these parts of the chromosomes (by homologous recombination events) than from protein-coding regions of the chromosome.

D. Specificity of Insertion

Specificity of *S. cerevisiae* transposon insertion has only been studied systematically for the Ty1 element. The first data on this were obtained from examination of spontaneous mutations in the *LYS2* gene (mutations in this gene are easily obtained because they give rise to an α-amino-adipate-resistant phenotype [28]). A small fraction of these were shown by Southern blotting experiments to consist of Ty1 insertions at *LYS2* (57, 166). Of 10 such insertions isolated by two groups, 9 fell fairly near the 5′ end of the *LYS2* gene. Subsequent isolation of a much larger number (about 40) of insertions at the *LYS2* locus confirms this general pattern: there are two hot spots for Ty1 insertion, one just upstream of the coding region and one inside it near the 5′ end. Transpositions into the ~3 kb of remaining coding sequences occur also, but at a much lower frequency (Natsoulis et al., submitted).

The most systematic study done so far has been at the *S. cerevisiae URA3* gene, where 82 Ty insertion sites have been sequenced. *ura3* mutants can also be selected directly using the drug 5-fluoroorotic acid (15, 17). This drug seems to select only for essentially complete knockout of *URA3* gene function, so insertions in the 5′ noncoding region are only very rarely recovered (2/82 insertions). This is in contrast to the α-amino-adipate selection, where mutants with *LYS2* function reduced by only a small amount can survive (some are Lys$^+$ phenotypically). Of the mutations within the coding region of *URA3*, however, there is again a hot spot near the 5′ end; 12/82 insertions are found in a single-nucleotide hot spot near the 5′ end, and about 50% of the insertions are found in the first 20% of the gene (Natsoulis et al., submitted). Similar results have recently been obtained in a study of insertions into the *CAN1* (arginine permease) gene, where a positive selection is also available (Wilke et al., personal communication). Thus, it seems that in general the 5′ regions are preferred targets for Ty1 transposition. This may reflect the open chromatin structure found near the 5′ ends of genes; a bulky, particulate transposition intermediate may have easier access to open chromatin.

Recently, Giroux et al. (88) have shown that five of six Ty1 transpositions into *SUP4-o* occur within the anticodon. This remarkable specificity suggests that this region of the tRNA gene is an exceptionally good target. Interestingly, this region falls between essential elements of the internal promoter sequence of the tRNA gene.

It has recently been shown that Ty1 elements can insert into the transcriptionally "silenced" *HML*α locus, showing that nontranscribed regions can act as targets for transcription (Mastrangelo et al., personal communication).

IX. *S. CEREVISIAE* TRANSPOSONS AND GENOME STABILITY: PARTICIPATION IN HOMOLOGOUS RECOMBINATION EVENTS

Like many transposable elements, *S. cerevisiae* transposons are responsible for a considerable variety of genomic rearrangements aside from that of primary transposition, which has already been considered in detail above. Many of these rearrangements are probably unrelated to the transposition process itself and are the consequence of the action of cellular homologous recombination systems upon families of repeated sequences which are widely scattered around the *S. cerevisiae* genome. The genomic rearrangements which have been reported include recombination between two δ elements, which usually results in deletion of an intervening ε segment, gene conversion between pairs or even trios of Ty elements, deletions and inversions between pairs of appropriately oriented Ty elements, and chromosomal translocations occurring at Ty elements. Moreover, DNA rearrangements involving solo δ elements are frequently observed by loss of a tRNA gene *SUP4-o*, which lies in a region containing several δ elements (156); these probably play a very important role in reshaping the *S. cerevisiae* genome because solo δ elements are so abundant. All of the events discussed below involve Ty1, Ty2, or δ elements; no such events have been studied for σ, τ, or the recently found Ty3 elements. Ty3 elements may not be significant contributors to genomic instability because of their low copy number (usually two to four per genome [160]); however, σ elements are numerous enough to be a significant factor in genome rearrangement.

A. LTR-LTR (δ-δ) Recombination

Perhaps the most frequently reported homologous recombination event involving Ty elements (as well as other retrotransposons and retroviruses) is LTR-LTR recombination. In the most commonly

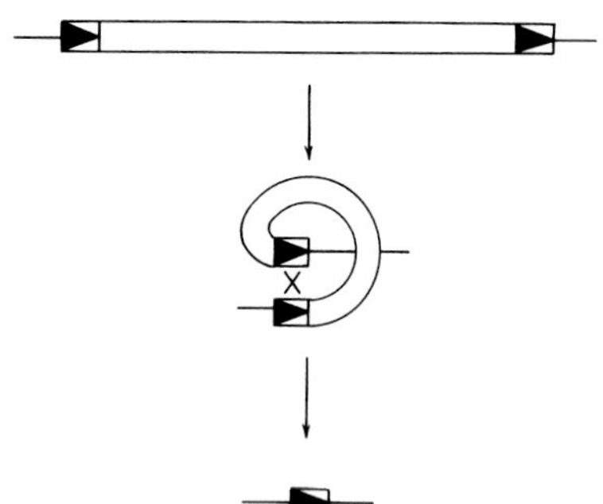

Figure 13. δ-δ recombination. δ elements flanking a Ty ε sequence (or other sequence) can recombine with each other as shown, resulting in deletion of the intervening material (the one-LTR circular molecule produced would presumably be lost by dilution, as it cannot replicate). Similar deletion events could occur by gene conversion events in which a single δ is the donor and a complete Ty element the recipient or by unequal crossing over (19, 36).

observed version of this event, the central ϵ sequences of a Ty element are lost as a consequence of a single-crossover homologous recombination event involving the corresponding LTR sequences which are in direct orientation (Fig. 13). LTRs which do not flank Ty elements also recombine with one another frequently; depending on their orientation relative to one another, the intervening material can be either deleted (direct orientation) or inverted (inverse orientation). Perhaps one reason for the frequent discovery of these phenomena is that deletion of an ϵ segment has no effect on viability, but often influences the expression of genes adjacent to the Ty element. The frequency of LTR-LTR recombination has been estimated to be about 10^{-5} at the Ty insertion *his4-912* marked with the *S. cerevisiae URA3* gene (194), which has identical LTRs but only about 1×10^{-7} to 3×10^{-7} at the *SUP4* locus. The δ elements at *SUP4* are closer together than δ elements in an intact Ty and have only 70% sequence identity (156). Both of these events are dependent on the function of the *RAD52* gene. Assuming that all Ty elements in the genome have roughly similar deletion frequencies, this would suggest that in haploid strains, a Ty element is lost during normal mitotic growth about once per 3,000 cell divisions. Note that this rate is considerably lower than the predicted rate of transposon copy number increase due to transposition. This may help to explain the increase in Ty element copy number in the genomes of laboratory strains, which are nearly always grown as haploids. Because deletion of Ty elements will have unpredictable effects on the phenotype, these deletion events may of course lead to a selective advantage or disadvantage. In nature, where *S. cerevisiae* cells normally exist as diploids, another important and presumably efficient mechanism for Ty element copy number reduction exists in the form of meiotic or mitotic homozygosis of segments of DNA which are heterozygous for a Ty element. Curiously, preliminary studies of mitotic and meiotic recombination events involving heterozygous Ty elements suggest that gene conversion often preferentially removes the uninterrupted copy rather than the Ty element-containing copy (A. Vincent and T. Petes, *Yeast* **2**:s408). Ty transposition is also expected to be lower in **a**/α diploids because of the lower abundance of Ty RNA (141). If Ty element copy numbers are to remain roughly constant in populations of *S. cerevisiae* cells, the overall loss rate must be balanced by events which increase the net number of Ty elements. Two such events have been observed: new primary transpositions and replacement of solo LTRs by intact Ty elements, another homologous recombination event which can occur frequently (36). δ-δ recombination is very frequently observed during *S. cerevisiae* transformation (179; D. J. Garfinkel and G. R. Fink, unpublished data).

Solo LTR (δ) elements have recently been isolated by two groups as the causative agent of mutations in a tRNA gene (88) and the *URA3* gene (Natsoulis et al., submitted). In the latter case, the mutations were recovered from cells containing an induced *GAL*/Ty plasmid. Of about 80 transpositions into *URA3*, two were solo δ elements, whereas the rest were Ty1 elements. Two possible explanations for these results are that δ elements can transpose independently of the intact Ty element or, more likely, that Ty1 transposition is associated with stimulation of homologous recombination in the vicinity of the insertion (these two models are compared in Fig. 14). Ty transposition in diploid *S. cerevisiae* can give rise to homozygous copies of the same Ty insertion at the *LYS2* locus at a high frequency, supporting the latter possibility (Y. Kassir, F. Dietrich, and G. R. Fink, personal communication).

B. Gene Conversion between Ty Elements

Homologous recombination events in mitotically growing *S. cerevisiae* consist largely of gene conversion events rather than reciprocal recombination events. Gene conversion events between different Ty elements are quite common and are readily observed in mitotically growing cells when one has the means of detecting them. Roeder and Fink (151) used the Ty insertion mutation *his4-917* to this end; the insertion mutation causes a His$^-$ phenotype, which reverts to His$^+$ as a consequence of gene conversion events involving certain other Ty elements as donors. Kupiec and Petes (110a, 110b) have recently com-

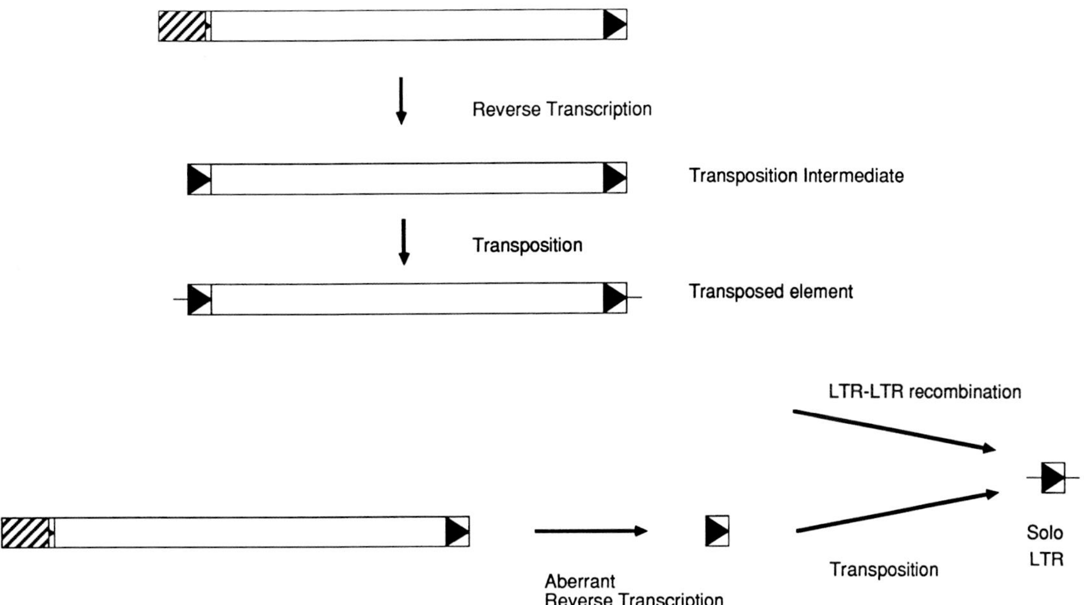

Figure 14. Two models for the generation of solo δ insertions during Ty transposition. In the first (more likely) model (above), a normal transposition event is followed by a homologous recombination event. The frequency of the latter may be stimulated in the vicinity of the transposition event because the transposition event left behind DNA in the form perhaps of nicks or gaps in the DNA. Alternatively (below), a solo δ sequence might be formed within the particle at some frequency [by annealing of (−) and (+) strong-stop DNAs and filling in?], complete with all terminal sequences necessary for integration. These might then be capable of integrating directly. Hatched box, *GAL1* promoter; boxed triangles, LTR sequences.

pleted detailed studies of a variety of Ty gene conversion events occurring mitotically and meiotically.

C. Deletions

Just as deletions and inversions can occur between LTR sequences located near each other, deletions and inversions between intact Ty elements have also been reported. Two well-characterized deletion events of this type result from recombination between a pair of Ty1 elements on chromosome X and a pair of Ty2 elements on chromosome III. In the first case, the 13-kb deleted region is called *DEL1* and encompasses the three genes *CYC1*, *OSM1*, and *RAD7*. High-frequency deletion (10^{-6} to 10^{-5}) is only observed in certain *S. cerevisiae* strains which happen to bear both of these Ty1 elements. These events are reported to be *RAD52* independent and are unaffected by mutations in a variety of other *RAD* genes (53, 117). In the second case, the deletion occurs between Ty2-117, near *LEU2*, and Ty2-917, a spontaneous Ty insertion adjacent to the *HIS4* locus originally isolated by virtue of its His$^-$ phenotype. The latter deletion is remarkable in that it removes 25 to 40 kb of DNA, presumably including many genes, but is haploviable (116, 178). This deletion mutation has been isolated repeatedly mitotically and meiotically. In at least one of these Ty2-117/Ty2-917 deletions, the single Ty element remaining at the deletion boundary was not a Ty2 element but a Ty1 element, suggesting that at least three Ty elements participated in the recombination event(s) that gave rise to the deletion (151).

A similar deletion event on chromosome III may be responsible for the curious multiple phenotypes of independently isolated *pet18* mutants. *pet18* mutations always result in three phenotypes: temperature sensitivity, loss of mitochondrial function (Pet$^-$), and loss of killer plasmid (Mak$^-$). It has been shown that this is a deletion mutation which removes several genes; the deletion lies in a region which contains many Ty and δ elements (177; I. Chiu and G. R. Fink, personal communication).

Deletions between an intact Ty element and solo δ elements also occur quite frequently in the *DEL1* region mentioned above in strains bearing only one intact Ty element. A variety of different structures remain after the deletion event occurs (52). Finally, certain *S. cerevisiae* strains in which δ-mediated inversion formation at *SUP4* occurs at extremely high frequency (perhaps 1 in 20 cell divisions) have been found. This instability seems to be localized to the *SUP4* region; it does not segregate as a single gene, but can be inherited meiotically (146a). The *SUP4* region contains many δ elements (the two regions flank a unique, essential tRNA gene). The

SUP4 region undergoes frequent δ-δ recombination, resulting in loss of the nonessential *SUP4* locus and flanking DNA (156, 157).

D. Inversions

Although inversion events between relatively closely spaced δ elements have been reported (156), those occurring between intact Ty elements seem to be less common. An inversion associated with reversion of the *his4-912* mutation was reported by Chaleff and Fink (25) but has not been characterized at the molecular level. One breakpoint in the inversion mapped to *HIS4*, the other to *HML*.

E. Translocations

Translocations between *his4-912* and chromosomes I and XII have also been observed by classical genetic techniques, but molecular analysis of these events has not been carried out (25).

F. What Limits Homologous Recombination between Ty Elements?

The great variety of observed homologous recombination events discussed above opens the possibility of rampant destruction in the genome mediated (passively) by Ty elements. However, the *S. cerevisiae* genome seems remarkably stable, even during meiosis, when recombination frequencies increase by order of magnitude. Preliminary results suggest that gene conversion events between Ty elements may occur at much lower frequency than conversion events between heteroalleles of the *HIS3* genes placed on different chromosomes (M. Kupiec and T. Petes, *Yeast* **2**:s202). What then limits the frequencies of Ty-Ty recombination events to tolerable levels? Several ideas have been proposed.

1. The heterogeneity model

Ty elements are known to contain heterologies; an important question is whether these are sufficiently prevalent to block or discourage homologous recombination events. In the heterogeneity model, it is argued that the heterology between the many copies of Ty1 and Ty2 elements in the genome limits the ability of these elements to engage in or complete homologous recombination events. Clear predictions of this model are that Ty elements must contain fairly regularly spaced heterogeneities which are sufficient to block recombination and that introduction of a number of identical copies of Ty elements into the same genome would destabilize it, particularly meiotically.

Ty1 and Ty2 elements differ by two rather major blocks of nonhomology (108), and these may well prevent recombination between the two classes to a large extent, since Ty1/Ty2 recombinants are only very rarely detected (M. Kupiec, personal communication). Ty1 elements, however, are rather similar to each other in sequence. Three Ty1 elements have now been sequenced in their entirety; stretches of perfect nucleotide sequence identity as long as 1,315 bp exist between Ty1-912 and Ty1-H3, as well as several stretches of hundreds of bases of perfect identity. Similarly, very long stretches of sequence identity can be found between Ty1-p109 and the other two Ty1 elements sequenced. Finally, two sequenced Ty2 elements, Ty2-117 and Ty2-917, are 99% identical in sequence throughout their length (189; P. Farabaugh, personal communication). Thus, this theory is not supported by sequence data for Ty1 and Ty2 elements.

S. cerevisiae strains have been constructed which carry double or more than the normal copy number of Ty1 elements; these show essentially no meiotic defect when crossed by wild-type strains or even by each other. Since these strains were constructed by induction of a *GAL*/Ty plasmid carrying a single Ty element, it is likely that many of its progeny are identical or nearly so, yet in preliminary studies no increase in meiotic lethality is observed (the mitotic homologous recombination frequencies in these strains have not been assessed [D. J. Eichinger and J. D. Boeke, unpublished data]). These observations argue strongly against the "heterogeneity" model.

2. Cold-spot hypothesis

It may be that Ty elements have accumulated in the genome in regions which happen to be rather infrequently involved in homologous recombination events. This might occur either because Ty elements have some innate propensity to seek out and insert into these regions or because they have simply accumulated there, because Ty elements accumulating in "hot" regions would tend to be lost over time. In the extreme view of this model, insertions in hot regions are selected against during meiosis. Such variations in recombination frequencies across a chromosome certainly exist, although systematic correlation with the position of Ty elements has not been carried out. There are, however, examples of cases, such as the *LEU2* region, which is relatively hot for homologous recombination but contains abundant Ty and δ elements (115). Comprehensive studies of the map locations of Ty elements have not been carried out, so that a rigorous testing of this model will be difficult. It is unlikely that mitotic recombination or meiotic gene conversion mechanisms select against Ty insertions in heterozygotes; if anything, Ty insertions are preferentially inherited (A. Vincent and T. Petes, *Yeast* 2:s408).

3. Recombination suppressor model

Suppression of recombination between Ty elements and δ elements may require a special active mechanism such as a protein which binds to Ty elements or δ sequences to inhibit their ability to participate in recombination reactions. A search for mutations which affect the frequency of δ-mediated rearrangements has turned up a gene, termed *EDR* (enhanced δ recombination), which may play such a role. In *edr* mutants, particularly null alleles, δ-mediated rearrangements and gene conversion events between Ty elements are both elevated (157; J. Wallis and R. Rothstein, personal communication). Further analysis of the *EDR* gene is likely to elucidate the mechanism by which recombination events involving Ty and δ elements are limited. The experiments mentioned above in which Ty element copy number was artificially increased in the genome by repeated induction of high-frequency transposition are most consistent with the "recombination suppressor" model.

X. WHAT ARE YEAST TRANSPOSONS GOOD FOR?

Yeast transposons are definitely useful tools for those of us interested in manipulating yeasts, but there remains the question of exactly what these transposons are doing for the yeast cell. Their wide distribution within *S. cerevisiae* and the wide distribution of retrotransposons in general suggest either that they are incredibly successful parasites or that, alternatively, retrotransposons may serve the host cell in some way. It seems likely that if the former was once true, the host cell may have evolved in such a way that the latter is now also true.

A. Uses of Ty Elements as Tools

Ty elements have been exploited in at least three ways by molecular biologists for practical purposes: (i) pGTy vectors have been used to build strains bearing many copies of Ty elements carrying a gene of interest; (ii) pGTy vectors have been used in transposon tagging experiments, where genes which are difficult to clone by other methods are "tagged" by a marked Ty transposition and then cloned by selection for the marker; and (iii) *TYA* protein fusions have been used to make hybrid Ty-VLPs bearing epitopes of a gene of interest which can easily be purified in quantity.

1. Disseminating integrated copies of a gene of interest by transposition

The pGTy1-H3 plasmids which have been useful for studies of the mechanism of transposition have recently been marked with entire genes rather than short segments. So far, the yeast *TRP1* (18) and *HIS3* (85a) genes and the bacterial *neo* (18) and *galK* (100) genes have been inserted into pGTy1-H3 vector derivatives and successfully transposed to multiple sites in the genome by long-term growth on inducing (galactose) medium. In the case of the *galK* gene, the coding sequence was fused to strong yeast promoters prior to transposition; importantly, good expression of *galK* was observed (100). In all of the above cases, gene expression of the marker gene imbedded within the transposed Ty element was observed. Typically, the mean number of marked transposon insertions per genome is 2 to 3 under ideal conditions and, surprisingly, is fairly independent of insertion length, with insertions of between 40 and 1,000 bp in length (larger insertions give rise to somewhat less transposition). The resultant strains are generally healthy; by screening about 20 random colonies, strains bearing as many as 7 to 10 copies of the amplified gene can usually be found. Thus, this would seem to be a useful strategy because many integrated copies of the gene are easily obtained; moreover, these should be much more stable than a multicopy plasmid for overproducing the gene product of interest.

2. Transposon tagging

Derivatives of pGTy1-H3 marked with *neo* (Futcher, personal communication) and *HIS3* (85a) have been successfully used to tag genes in *S. cerevisiae*; these can be easily cloned by virtue of the fact that the marker genes in the Ty element can be selected for in *E. coli*. The πan7-marked derivative of pGTy1-H3 mentioned above (57a) should be useful for this purpose also, since it already carries both a replicon and a selectable marker (*supF*) usable in *E. coli* (85a). In contrast to *S. cerevisiae*, much of the transposon tagging in higher eucaryotes has been with unmarked elements; tagging with these engineered Ty elements is equivalent in power to the use of the naturally drug-resistant Tn elements of bacteria.

3. Hybrid Ty-VLPs

The Kingsmans' group has recently reported success with a novel method for producing Ty-VLPs bearing epitopes of a gene of interest by simply overproducing *TYA* fusion proteins with foreign material fused to the C-terminal end of *TYA*; both interferon and human immunodeficiency virus *env* peptides have been incorporated into Ty-VLPs in this way (2, 120). These will certainly be useful in the production of antibodies for research purposes, much as *E. coli trpE* and *lacZ* fusion proteins are. The Ty-VLPs may also be useful as an alternative to recombinant vaccinia virus particles for eventual vaccine production (2, 119). A potential advantage of these recombinant *TYA*-fusion proteins over certain other types of fusion protein systems is their ease of purification due to their particulate nature.

B. Are Yeast Transposons Useful to Yeasts?

There has been considerable debate about whether transposons in general are simply "selfish DNA" (46, 51, 138) or actually carry out useful work for the cell. There is probably some truth to both assertions; transposons are certainly genome parasites able to harm their hosts, but under appropriate conditions they can confer a selective advantage upon their host, for example, when it is adapting to a new environment (1, 27). The notion that introns derive from ancient transposons is another example of this idea; an originally deleterious aspect of a genome parasite (gene disruption) is recruited by the host as a regulatory mechanism (78).

There is direct evidence that Ty element hybridization patterns change when *S. cerevisiae* cells are grown, for example, in chemostats (1). The consistent occurrence of yeast transposon sequences in the vicinity of tRNA genes suggests the possibility of a functional role of yeast transposons there (20, 30, 48, 58, 83, 160, 161, 171). It is also notable that the *MAT*-regulated activation of adjacent genes provided by Ty elements may be particularly advantageous to laboratory yeast strains, which are usually continually maintained in a haploid **a** or α "mater" state mitotically by virtue of the fact that they are heterothallic. Interestingly, in the *S. uvarum* strain with only one Ty element (mentioned above), it was possible to delete this element without any obvious phenotypic change observed in the cells (M. Ciriacy, personal communication). However, this strain may be unusual (after all, it had only one Ty element to begin with) and it may contain members of a transposon family undetectable with Ty1 probes. Also, this strain still had a relatively normal number of δ elements in its genome. Answering the questions about the potential usefulness of yeast transposons definitively may require the ability to introduce experimentally functional *S. cerevisiae* transposons into the genome of a distantly related organism, such as *Saccharomyces pombe*, or, alternatively, to remove all of the copies of the transposon in question from a more representative *S. cerevisiae* strain. The latter should not be difficult to do with Ty3.

A further potentially useful feature of yeast transposon sequences (which is deleterious if allowed to occur at too high a frequency) is their ability to act as substrates for homology-dependent genome rearrangements. These are probably useful to the organism in the evolutionary time scale in adapting to a new environment and eventually in speciation. Thus, again, it seems that aspects of a parasite which may originally have been deleterious have over evolutionary time been made use of by the host.

Acknowledgments. I gratefully acknowledge the generous communication of unpublished manuscripts and data by my colleagues. I thank S. Edmunds, D. J. Garfinkel, R. Rothstein, S. Sandmeyer, and F. Winston for helpful comments on the manuscript and J. Olsen for expert assistance in preparing the manuscript.

My research was supported by grants from the National Institutes of Health and from the Searle Scholars Foundation/Chicago Community Trust.

LITERATURE CITED

1. **Adams, J., and P. W. Oeller.** 1986. Structure of evolving populations of *Saccharomyces cerevisiae*: adaptive changes are frequently associated with sequence alterations involving mobile elements belonging to the Ty family. *Proc. Natl. Acad. Sci. USA* **83:**7124–7127.
2. **Adams, S. E., K. M. Dawson, K. Gull, S. M. Kingsman, and A. J. Kingsman.** 1987. The expression of hybrid HIV:Ty virus-like particles in yeast. *Nature* (London) **329:**68–70.
3. **Adams, S. E., J. Mellor, K. Gull, R. B. Sim, M. F. Tuite, S. M. Kingsman, and A. J. Kingsman.** 1987. The functions and relationships of Ty-VLP proteins in yeast reflect those of mammalian proteins. *Cell* **49:**111–119.
4. **Arkhipova, I. R., T. V. Gorelova, Y. V. Ilyin, and N. G. Schuppe.** 1984. Reverse transcription of Drosophila mobile dispersed genetic element RNAs: detection of intermediate forms. *Nucleic Acids Res.* **12:**7533–7548.
5. **Arkhipova, I. R., A. M. Mazo, V. A. Cherkasova, T. V. Gorelova, N. G. Schuppe, and Y. V. Ilyin.** 1986. The steps of reverse transcription of Drosophila mobile dispersed genetic elements and U3-R-U5 structure of their LTRs. *Cell* **44:** 555–563.
6. **Bach, M.** 1984. Ty1-promoted gene expression of aspartate transcarbamylase in the yeast Saccharomyces cerevisiae. *Mol. Gen. Genet.* **194:**395–401.
7. **Ballario, P., P. Filetici, N. Junakovic, and F. Pedone.** 1983.

Ty1 extrachromosomal circular copies in Saccharomyces cerevisiae. *FEBS Lett.* **155**:225–229.

8. **Baltimore, D.** 1985. Retroviruses and retrotransposons: the role of reverse transcription in shaping the eukaryotic genome. *Cell* **40**:481–482.
9. **Barker, D. G., J. H. M. White, and L. H. Johnston.** 1985. The nucleotide sequence of the DNA ligase gene (CDC9) from Saccharomyces cerevisiae: a gene which is cell-cycle regulated and induced in response to DNA damage. *Nucleic Acids Res.* **13**:8323–8337.
10. **Boeke, J. D.** 1987. Retrotransposons, p. 59–103. *In* E. Domingo, J. Holland, and P. Ahlquist (ed.), *RNA Genetics*, vol. II. *Retroviruses, Viroids and RNA Recombination.* CRC Press, Inc., Boca Raton, Fla.
11. **Boeke, J. D., D. Eichinger, D. Castrillon, and G. R. Fink.** 1988. The *Saccharomyces cerevisiae* genome contains functional and nonfunctional copies of transposon Ty1. *Mol. Cell. Biol.* **8**:1432–1442.
12. **Boeke, J. D., D. Eichinger, and G. R. Fink.** 1987. Regulation of yeast Ty element transposition, p. 169–182. *In* M. E. Lambert, J. F. McDonald, and I. B. Weinstein (ed.), *Eukaryotic Transposable Elements as Mutagenic Agents.* Cold Spring Harbor Laboratory, Cold Spring Harbor, N.Y.
13. **Boeke, J. D., and D. J. Garfinkel.** 1987. Yeast Ty elements as retroviruses, p. 15–39. *In* M. J. Leibowitz and Y. Koltin (ed.), *Viruses of Fungi and Lower Eukaryotes.* Marcel Dekker Publications, Inc., New York.
14. **Boeke, J. D., D. J. Garfinkel, C. A. Styles, and G. R. Fink.** 1985. Ty elements transpose through an RNA intermediate. *Cell* **40**:491–500.
15. **Boeke, J. D., F. Lacroute, and G. R. Fink.** 1984. A positive selection for mutants lacking orotidine 5′-phosphate decarboxylase activity in yeast: 5-fluoro-orotic acid resistance. *Mol. Gen. Genet.* **197**:345–346.
16. **Boeke, J. D., C. A. Styles, and G. R. Fink.** 1986. *Saccharomyces cerevisiae SPT3* gene is required for transposition and transpositional recombination of chromosomal Ty elements. *Mol. Cell. Biol.* **6**:3575–3581.
17. **Boeke, J. D., J. Trueheart, G. Natsoulis, and G. R. Fink.** 1987. 5-Fluoro-orotic acid as a selective agent in yeast molecular genetics. *Methods Enzymol.* **154**:164–174.
18. **Boeke, J. D., H. Xu, and G. R. Fink.** 1988. A general method for the chromosomal amplification of genes in yeast. *Science* **239**:280–282.
19. **Breilmann, D., J. Gafner, and M. Ciriacy.** 1985. Gene conversion and reciprocal exchange in a Ty-mediated translocation in yeast. *Curr. Genet.* **9**:553–560.
20. **Brodeur, G. M., S. B. Sandmeyer, and M. V. Olson.** 1983. Consistent association between sigma elements and tRNA genes in yeast. *Proc. Natl. Acad. Sci. USA* **80**:3292–3296.
21. **Brown, P. O., B. Bowerman, H. E. Varmus, and J. M. Bishop.** 1987. Correct integration of retroviral DNA in vitro. *Cell* **49**:347–356.
22. **Bulawa, C. E., M. Slater, E. Cahib, J. Au-Young, A. Sburlati, L. Adair, and P. W. Robbins.** 1986. The Saccharomyces cerevisiae structural gene for chitin synthetase (CHS1) is not required for chitin synthesis in vivo. *Cell* **46**:213–225.
23. **Cameron, R. R., E. Y. Loh, and R. W. Davis.** 1979. Evidence for transposition of dispersed repetitive DNA families in yeast. *Cell* **16**:739–751.
24. **Campbell, A., D. E. Berg, D. Botstein, E. M. Lederberg, R. P. Novick, P. Starlinger, and W. Szybalski.** 1979. Nomenclature of transposable elements in prokaryotes. *Gene* **5**: 197–206.
25. **Chaleff, D. T., and G. R. Fink.** 1980. Genetic events associated with an insertion mutation in yeast. *Cell* **21**:227–237.
26. **Chaleff, D. T., and K. Tachell.** 1985. Molecular cloning and characterization of the *STE7* and *STE11* genes of *Saccharomyces cerevisiae. Mol. Cell. Biol.* **5**:1878–1886.
27. **Chao, L., C. Vargas, B. B. Spear, and E. C. Cox.** 1983. Transposable elements as mutator genes in evolution. *Nature* (London) **303**:633–635.
28. **Chattoo, B. B., F. Sherman, D. A. Azubalis, T. A. Fjellstedt, D. Mehvert, and M. Ogur.** 1979. Selection of lys2 mutants of the yeast Saccharomyces cerevisiae by the utilization of alpha-amino-adipate. *Genetics* **93**:51–65.
29. **Chisholm, G., and T. Cooper.** 1984. *cis*-Dominant mutations which dramatically enhance *DUR1, 2* gene expression without affecting its normal regulation. *Mol. Cell. Biol.* **4**:947–955.
30. **Chisholm, G. E., F. S. Genbauffe, and T. G. Cooper.** 1984. Tau, a repeated DNA sequence in yeast. *Proc. Natl. Acad. Sci. USA* **81**:2965–2969.
31. **Cigan, M., and T. F. Donahue.** 1986. The methionine initiator tRNA genes of yeast. *Gene* **41**:343–348.
32. **Ciriacy, M.** 1975. Genetics of alcohol dehydrogenase in Saccharomyces cerevisiae. *Mol. Gen. Genet.* **138**:157–164.
33. **Ciriacy, M.** 1975. Genetics of alcohol dehydrogenase in Saccharomyces cerevisiae. I. Isolation and genetic analysis of adh mutants. *Mutat. Res.* **29**:315–326.
34. **Ciriacy, M.** 1976. Cis-dominant regulatory mutations affecting the formation of glucose-repressible alcohol dehydrogenase (ADHII) in Saccharomyces cerevisiae. *Mol. Gen. Genet.* **145**:327–333.
35. **Ciriacy, M.** 1979. Isolation and characterization of further cis- and trans-acting regulatory elements involved in the synthesis of glucose-repressible alcohol dehydrogenase (ADHII) in Saccharomyces cerevisiae. *Mol. Gen. Genet.* **176**:427–431.
36. **Ciriacy, M., and D. Breilman.** 1982. Delta sequences mediate DNA rearrangements in Saccharomyces cerevisiae. *Curr. Genet.* **6**:55–61.
37. **Ciriacy, M., and V. M. Williamson.** 1981. Analysis of mutations affecting Ty-mediated gene expression in Saccharomyces cerevisiae. *Mol. Gen. Genet.* **182**:159–163.
38. **Clare, J., and P. Farabaugh.** 1985. Nucleotide sequence of a yeast Ty element: evidence for an unusual mechanism of gene expression. *Proc. Natl. Acad. Sci. USA* **82**:2829–2833.

38a. **Clare, J. J., M. Belcourt, and P. J. Farabaugh.** 1988. Efficient translational frameshifting occurs within a conserved sequence of the overlap between the two genes of a yeast Ty1 transposon. *Proc. Natl. Acad. Sci USA* **85**:6816–6820.

39. **Clark, D. J., V. W. Bilanchone, L. J. Haywood, S. L. Dildine, and S. B. Sandmeyer.** 1988. A yeast sigma composite element, Ty3, has properties of a retrotransposon. *J. Biol. Chem.* **263**:1413–1423.
40. **Clark-Adams, C. D., D. Norris, M. A. Osley, J. S. Fassler, and F. Winston.** 1988. Changes in histone gene dosage alter transcription in yeast. *Genes Dev.* **2**:150–159.
41. **Clark-Adams, C. D., and F. Winston.** 1987. The *SPT6* gene is essential for growth and is required for delta-mediated transcription in *Saccharomyces cerevisiae. Mol. Cell. Biol.* **7**:679–686.
42. **Coffin, J. M.** 1979. Structure, replication, and recombination of retroviral genomes: some unifying hypotheses. *J. Gen. Virol.* **42**:1–26.

42a. **Company, M., C. Adler, and B. Errede.** 1988. Identification of a Ty1 regulatory sequence responsive to *STE7* and *STE12*. *Mol. Cell. Biol.* **8**:2545–2554.

43. **Company, M., and B. Errede.** 1987. Cell-type-dependent gene activation of yeast transposon Ty1 involves multiple regulatory determinants. *Mol. Cell. Biol.* 7:3205–3211.

44. **Cooper, T. G., and G. Chisolm.** 1984. Position-dependent, Ty-mediated enhancement of DUR1,2 gene expression, p. 289–303. *In* M. Simon and I. Herskowitz (ed.), *Genome Rearrangement.* Alan R. Liss, Inc., New York.

45. **Covey, S. N.** 1986. Amino acid sequence homology in gag region of reverse transcribing elements and the coat protein of cauliflower mosaic virus. *Nucleic Acids Res.* **14**:623–633.

45a. **Curcio, M. J., N. J. Sanders, and D. J. Garfinkel.** 1988. Transpositional competence and transcription of endogenous Ty elements in *Saccharomyces cerevisiae*: implications for regulation of transposition. *Mol. Cell. Biol.* **8**:3571–3581.

46. **Dawkins, R.** 1976. *The Selfish Gene.* Oxford University Press, Oxford.

47. **Degols, G., J.-C. Jauniaux, and J.-M. Wiame.** 1987. Molecular characterization of transposable-element-associated mutations that lead to constitutive L-ornithine aminotransferase expression in Saccharomyces cerevisiae. *Eur. J. Biochem.* **165**:289–296.

48. **DelRey, F. J., T. F. Donahue, and G. R. Fink.** 1982. Sigma, a repetitive element found adjacent to tRNA genes of yeast. *Proc. Natl. Acad. Sci. USA* **79**:4138–4142.

49. **Deschamps, J., E. Dubois, and J.-M. Wiame.** 1979. L-Ornithine transaminase synthesis in Saccharomyces cerevisiae: regulation by inducer exclusion. *Mol. Gen. Genet.* **174**:225–232.

50. **Dobson, M. J., J. Mellor, A. M. Fulton, N. A. Roberts, B. A. Bowen, S. M. Kingsman, and A. J. Kingsman.** 1984. The identification and high level expression of a protein encoded by the yeast Ty element. *EMBO J.* **3**:1115–1119.

51. **Doolittle, W. F., and C. Sapienza.** 1980. Selfish genes, the phenotype paradigm and genome evolution. *Nature* (London) **284**:601–603.

52. **Downs, K. M., G. Brennan, and S. W. Liebman.** 1987. Deletions extending from a single Ty1 element in *Saccharomyces cerevisiae. Mol. Cell. Biol.* **5**:3451–3457.

53. **Downs, K. M., A. E. Szwast, and S. W. Liebman.** 1983. Analysis of the effect of radiation repair mutations on the DEL1 mutator region of Saccharomyces cerevisiae. *Curr. Genet.* **7**:57–61.

54. **Dubois, E., D. Hiernaux, M. Grenson, and J.-M. Wiame.** 1978. Specific induction of catabolism and its relation to repression of biosynthesis in arginine metabolism of Saccharomyces cerevisiae. *J. Mol. Biol.* **122**:383–406.

55. **Dubois, E., E. Jacobs, and J.-C. Jauniaux.** 1982. Expression of the ROAM mutations in Saccharomyces cerevisiae: involvement of trans-acting regulatory elements and relation with the Ty1 transcription. *EMBO J.* **1**:1133–1139.

56. **Eibel, H., J. Gafner, A. Stotz, and P. Philippsen.** 1981. Characterization of the yeast mobile genetic element Ty1. *Cold Spring Harbor Symp. Quant. Biol.* **45**:609–617.

57. **Eibel, H., and P. Philippsen.** 1984. Preferential integration of yeast transposable element Ty1 into a promoter region. *Nature* (London) **307**:386–388.

57a. **Eichinger, D. J., and J. D. Boeke.** 1988. The DNA intermediate in yeast Ty1 element transposition copurifies with virus-like particles: cell-free Ty1 transposition. *Cell* **54**: 955–966.

58. **Eigel, A., and H. Feldmann.** 1982. Ty1 and delta elements occur adjacent to several tRNA genes in yeast. *EMBO J.* **1**:1245–1250.

59. **Elder, R. T., E. Y. Loh, and R. W. Davis.** 1983. RNA from the yeast transposable element Ty1 has both ends in the direct repeats, a structure similar to retrovirus RNA. *Proc. Natl. Acad. Sci. USA* **80**:2432–2436.

60. **Elder, R. T., T. P. St. John, D. T. Stinchcomb, and R. W. Davis.** 1980. Studies on the transposable element Ty1 of yeast. I. RNA homologous to Ty1. *Cold Spring Harbor Symp. Quant. Biol.* **45**:581–584.

61. **El-Sherbeini, M., and K. Bostian.** 1987. Viruses in fungi: infection of yeast with the K1 and K2 killer viruses. *Proc. Natl. Acad. Sci. USA* **84**:4293–4297.

62. **Engelke, D. R., P. Gegenheimer, and J. Abelson.** 1985. Nucleolytic processing of a tRNA-Arg-tRNA-Asp dimeric precursor by a homologous component from Saccharomyces cerevisiae. *J. Biol. Chem.* **260**:1271–1279.

63. **Errede, B., T. S. Cardillo, F. Sherman, E. Dubois, J. Deschamps, and J. M. Wiame.** 1980. Mating signals control expression of mutations resulting from insertion of a transposable repetitive element adjacent to diverse yeast genes. *Cell* **22**:427–436.

64. **Errede, B., T. S. Cardillo, G. Wever, and F. Sherman.** 1980. Studies on transposable elements in yeast. I. ROAM mutations causing increased expression of yeast genes: their activation by signals directed toward conjugation functions and their formation by insertion of Ty1 repetitive elements. *Cold Spring Harbor Symp. Quant. Biol.* **45**:593–602.

65. **Errede, B., M. Company, J. D. Ferchak, C. A. Hutchison, and W. S. Yarnell.** 1985. Activation regions in a yeast transposon have homology to mating type control sequences and to mammalian enhancers. *Proc. Natl. Acad. Sci. USA* **82**:5423–5427.

66. **Errede, B., M. Company, and C. A. Hutchison III.** 1987. Ty1 sequence with enhancer and mating-type-dependent regulatory activities. *Mol. Cell. Biol.* **7**:258–265.

67. **Errede, B., M. Company, and R. Swanstrom.** 1986. An anomalous Ty1 structure attributed to an error in reverse transcription. *Mol. Cell. Biol.* **6**:1334–1338.

68. **Exinger, F., and F. Lacroute.** 1979. Genetic evidence for the creation of a reinitiation site by mutation inside the yeast URA2 gene. *Mol. Gen. Genet.* **173**:109–113.

69. **Farabaugh, P. J., and G. R. Fink.** 1980. Insertion of the eukaryotic transposable element Ty1 creates a 5 base pair duplication. *Nature* (London) **286**:352–356.

70. **Farabaugh, P. J., U. Schmeissner, M. Hofer, and J. H. Miller.** 1978. Genetic studies of the lac repressor. VII. On the molecular nature of spontaneous hotspots in the lacI gene of Escherichia coli. *J. Mol. Biol.* **126**:847–863.

71. **Fassler, J. S., and F. Winston.** 1988. Isolation and analysis of a novel class of suppressor of Ty insertion mutations in Saccharomyces cerevisiae. *Genetics* **118**:203–212.

72. **Fields, S., D. T. Chaleff, and G. F. Sprague.** 1988. Yeast *STE7*, *STE11*, and *STE12* genes are required for expression of cell-type-specific genes. *Mol. Cell. Biol.* **8**:551–556.

73. **Fields, S., and I. Herskowitz.** 1985. The yeast STE12 product is required for expression of two sets of cell-type-specific genes. *Cell* **42**:923–930.

74. **Fields, S., and I. Herskowitz.** 1987. Regulation by the yeast mating-type locus of *STE12*, a gene required for cell-type-specific expression. *Mol. Cell. Biol.* **7**:3818–3821.

75. **Fink, G. R., J. D. Boeke, and D. J. Garfinkel.** 1986. The mechanism and consequences of retrotransposition. *Trends Gen.* **2**:118–123.

76. **Finnegan, D. J., and D. H. Fawcett.** 1986. Transposable elements in Drosophila melanogaster. *Oxford Surv. Eukaryotic Genes* **3**(Suppl.):1–12.

77. **Finnegan, D. J., and D. H. Fawcett.** 1986. Transposable

elements in Drosophila melanogaster. *Oxford Surv. Eukaryotic Genes* **3**:1–62.

78. **Flavell, A.** 1985. Introns continue to amaze. *Nature* (London) **316**:574–575.

79. **Flavell, A. J., and D. Ish-Horowicz.** 1981. Extrachromosomal circular copies of the eukaryotic transposable element copia in cultured Drosophila cells. *Nature* (London) **292**: 591–595.

80. **Flavell, A. J., and D. Ish-Horowicz.** 1983. The origin of extrachromosomal circular copia elements. *Cell* **34**: 415–419.

81. **Fleig, U. N., R. D. Pridmore, and P. Philippsen.** 1986. Construction of LYS2 cartridges for use in genetic manipulations of Saccharomyces cerevisiae. *Gene* **46**:237–245.

82. **Fulton, A. M., J. Mellor, M. J. Dobson, J. Chester, J. R. Warmington, K. J. Indge, S. G. Oliver, P. de la Paz, W. Wilson, A. J. Kingsman, and S. M. Kingsman.** 1985. Variants within the yeast Ty sequence family encode a class of structurally conserved proteins. *Nucleic Acids Res.* **13**: 4097–4112.

83. **Gafner, J., E. M. DeRobertis, and P. Philippsen.** 1983. Delta sequences in the 5′ non-coding region of yeast tRNA genes. *EMBO J.* **2**:583–591.

84. **Gafner, J., and P. Philippsen.** 1980. The yeast transposon Ty1 generates duplications of target DNA on insertion. *Nature* (London) **286**:414–418.

85. **Garfinkel, D. J., J. D. Boeke, and G. R. Fink.** 1985. Ty element transposition: reverse transcriptase and virus-like particles. *Cell* **42**:507–517.

85a. **Garfinkel, D. J., M. F. Mastrangelo, N. J. Sanders, B. K. Shafer, and J. N. Strathern.** 1988. Transposon tagging using Ty elements in yeast. *Genetics* **120**:95–108.

86. **Genbauffe, F. S., G. E. Chisholm, and T. G. Cooper.** 1984. Tau, sigma, and delta. *J. Biol. Chem.* **259**:10518–10525.

87. **Gilboa, E., S. W. Mitra, S. Goff, and D. Baltimore.** 1979. A detailed model of reverse transcription and a test of crucial aspects. *Cell* **18**:93–100.

88. **Giroux, C. N., J. R. A. Mix, M. K. Pierce, S. E. Kohalmi, and B. A. Kunz.** 1988. DNA sequence analysis of spontaneous mutations in the *SUP4-o* gene of *Saccharomyces cerevisiae*. *Mol. Cell. Biol.* **8**:978–981.

89. **Goebl, M. G., and T. D. Petes.** 1986. Most of the yeast genomic sequences are not essential for cell growth and division. *Cell* **46**:983–992.

90. **Goff, S., P. Traktman, and D. Baltimore.** 1981. Isolation and properties of Moloney murine leukemia virus mutants: use of a rapid assay for release of virion reverse transcriptase. *J. Virol.* **38**:239–248.

91. **Guarente, L., R. R. Yocum, and P. Gifford.** 1982. A GAL10-CYC1 hybrid yeast promoter identifies the GAL4 regulatory region as an upstream site. *Proc. Natl. Acad. Sci. USA* **79**:7410–7414.

92. **Hauber, J., P. Nelbock, U. Pilz, and H. Feldmann.** 1986. Enhancer-like stimulation of yeast tRNA gene expression by a defined region of the Ty element micro-injected into Xenopus oocytes. *Biol. Chem. Hoppe-Seyler* **367**: 1141–1146.

93. **Hauber, J., P. Nelbock-Hochstetter, and H. Feldmann.** 1985. Nucleotide sequence and characteristics of a Ty element from yeast. *Nucleic Acids Res.* **13**:2745–2758.

94. **Hirschorn, J. N., and F. Winston.** 1988. *SPT3* is required for normal levels of a-factor and α-factor in *Saccharomyces cerevisiae*. *Mol. Cell. Biol.* **8**:822–827.

95. **Ilyin, Y. V., N. G. Schuppe, N. V. Lyubomirskaya, T. V. Gorelova, and I. R. Arkhopova.** 1984. Circular copies of mobile dispersed genetic elements in cultured Drosophila melanogaster cells. *Nucleic Acids Res.* **12**:7517–7531.

96. **Inouye, S., S. Yuki, and K. Saigo.** 1986. Complete nucleotide sequence and genome organization of a Drosophila transposable genetic element, 297. *Eur. J. Biochem.* **154**: 417–425.

97. **Jacks, T., M. D. Power, F. R. Masiarz, P. A. Luciw, P. J. Barr, and H. E. Varmus.** 1988. Characterization of ribosomal frameshifting in HIV-1 gag-pol expression. *Nature* (London) **331**:280–283.

98. **Jacks, T., K. Townsley, H. E. Varmus, and J. Majors.** 1987. Two efficient ribosomal frameshifting events are required for synthesis of mouse mammary tumor virus gag-related polyproteins. *Proc. Natl. Acad. Sci. USA* **84**:4298–4302.

99. **Jacks, T., and H. E. Varmus.** 1985. Expression of the Rous sarcoma virus pol gene by ribosomal frameshifting. *Science* **230**:1237–1242.

100. **Jacobs, E., M. Dewerchin, and J. D. Boeke.** 1988. Retrovirus-like vectors for Saccharomyces cerevisiae: integration of foreign genes controlled by efficient promoters into yeast chromosomal DNA. *Gene* **67**:259–269.

101. **Jauniaux, J. C., E. Dubois, M. Crabeel, and J. M. Wiame.** 1981. DNA and RNA analysis of arginase regulatory mutants in Saccharomyces cerevisiae. *Arch. Int. Physiol. Biochim.* **89**:B111–B112.

102. **Jauniaux, J. C., E. Dubois, S. Vissers, M. Crabeel, and J. M. Wiame.** 1982. Molecular cloning, DNA structure, and RNA analysis of the arginase gene in Saccharomyces cerevisiae. A study of cis-dominant regulatory mutants. *EMBO J.* **1**: 1125–1132.

103. **Jerome, J. F., and J. A. Jaehning.** 1986. mRNA transcription in nuclei isolated from *Saccharomyces cerevisiae*. *Mol. Cell. Biol.* **6**:1633–1639.

104. **Johnson, M. S., M. A. McClure, D.-F. Feng, J. Gray, and R. F. Doolittle.** 1986. Computer analysis of retroviral pol genes: assignment of enzymatic functions to specific sequences and homologies with nonviral enzymes. *Proc. Natl. Acad. Sci. USA* **83**:7648–7652.

105. **Karwan, R., H. Blutsch, and U. Wintersberger.** 1983. Physical association of a DNA polymerase stimulating activity with a ribonuclease H purified from yeast. *Biochemistry* **22**:5500–5507.

106. **Karwan, R., C. Kuhne, and U. Wintersberger.** 1986. Ribonuclease H(70) from Saccharomyces cerevisiae possesses cryptic reverse transcriptase activity. *Proc. Natl. Acad. Sci. USA* **83**:5919–5923.

107. **Katz, R. A., R. W. Terry, and A. M. Skalka.** 1986. A conserved *cis*-acting sequence in the 5′ leader of avian sarcoma virus RNA is required for packaging. *J. Virol.* **59**:163–167.

108. **Kingsman, A. J., R. L. Gimlich, L. Clarke, A. C. Chinault, and J. Carbon.** 1981. Sequence variation in dispersed repetitive sequences in Saccharomyces cerevisiae. *J. Mol. Biol.* **145**:619–632.

109. **Klein, H. L., and T. D. Petes.** 1984. Genetic mapping of Ty elements in *Saccharomyces cerevisiae*. *Mol. Cell. Biol.* **4**: 329–339.

110. **Kotval, J., K. S. Zaret, S. Consaul, and F. Sherman.** 1983. Revertants of a transcription termination mutant of yeast contain diverse genetic alterations. *Genetics* **103**:367–388.

110a. **Kupiec, M., and T. D. Petes.** 1988. Allelic and ectopic recombination between Ty elements in yeast. *Genetics* **119**: 549–559.

110b. **Kupiec, M., and T. D. Petes.** 1988. Meiotic recombination between repeated transposable elements in *Saccharomyces cerevisiae. Mol. Cell. Biol.* 8:2942–2954.

111. **Lanford, R. E., P. Kanda, and R. C. Kennedy.** 1986. Induction of nuclear transport with a synthetic peptide homologous to the SV40 T antigen transport signal. *Cell* **46:** 575–582.

112. **Leder, P., D. Tiemeier, and L. Enquist.** 1969. EK2 derivatives of bacteriophage lambda useful in the cloning of DNA from higher organisms: the lambda gtWES system. *Science* **196:**175–177.

113. **Lemoine, Y., E. Dubois, and J.-M. Wiame.** 1978. The regulation of urea amidolyase of Saccharomyces cerevisiae. Mating type influence on a constitutivity mutation acting in cis. *Mol. Gen. Genet.* **166:**251–258.

114. **Liao, X.-B., J. J. Clare, and P. J. Farabaugh.** 1987. The UAS site of a Ty2 element of yeast is necessary but not sufficient to promote maximal transcription of the element. *Proc. Natl. Acad. Sci. USA* **84:**8520–8524.

115. **Lichten, M., R. H. Borts, and J. E. Haber.** 1987. Meiotic gene conversion and crossing over between dispersed homologous sequences occurs frequently in Saccharomyces cerevisiae. *Genetics* **115:**233–246.

116. **Liebman, S., P. Shalit, and S. Picologlou.** 1981. Ty elements are involved in the formation of deletions in DEL1 strains of Saccharomyces cerevisiae. *Cell* **26:**401–409.

117. **Liebman, S. W., and K. M. Downs.** 1980. The RAD52 gene is not required for the function of the DEL1 mutator gene in Saccharomyces cerevisiae. *Mol. Gen. Genet.* **179:**703–705.

118. **Linial, M.** 1987. Creation of a processed pseudogene by retroviral infection. *Cell* **49:**93–102.

119. **Mackett, M., G. L. Smith, and B. Moss.** 1982. Vaccinia virus: a selectable eukaryotic cloning and expression vector. *Proc. Natl. Acad. Sci. USA* **79:**7415–7419.

120. **Malim, M. H., S. E. Adams, K. Gull, A. J. Kingsman, and S. M. Kingsman.** 1987. The production of hybrid Ty:IFN virus-like particles in yeast. *Nucleic Acids Res.* **15:** 7571–7580.

121. **Mann, R., R. C. Mulligan, and D. Baltimore.** 1983. Construction of a retrovirus packaging mutant and its use to produce helper-free defective retrovirus. *Cell* **33:**153–159.

122. **Marlor, R. L., S. M. Parkhurst, and V. G. Corces.** 1986. The *Drosophila melanogaster* gypsy transposable element encodes putative gene products homologous to retroviral proteins. *Mol. Cell. Biol.* **6:**1129–1134.

122a. **McEntee, K., and V. Bradshaw.** 1988. Effects of DNA damage on transcription and transposition of Ty retrotransposons of yeast. *In* M. E. Lambert, J. F. McDonald, and I. B. Weinstein (ed.), *Eukaryotic Transposable Elements of Mutagenic Agents.* Cold Spring Harbor Laboratory, Cold Spring Harbor, N.Y.

123. **Mellor, J., A. M. Fulton, M. J. Dobson, N. A. Roberts, W. Wilson, A. J. Kingsman, and S. M. Kingsman.** 1985. The Ty transposon of Saccharomyces cerevisiae determines the synthesis of at least three proteins. *Nucleic Acids Res.* **13:** 6249–6263.

124. **Mellor, J., A. M. Fulton, M. J. Dobson, W. Wilson, S. M. Kingsman, and A. J. Kingsman.** 1985. A retrovirus-like strategy for the expression of a fusion protein encoded by yeast transposon Ty1. *Nature* (London) **313:**243–246.

125. **Mellor, J., A. J. Kingsman, and S. M. Kingsman.** 1986. Ty, an endogenous retrovirus of yeast? *Yeast* **2:**145–152.

126. **Mellor, J., M. H. Malim, K. Gull, M. F. Tuite, S. McCready, T. Dibbayawan, S. M. Kingsman, and A. J. Kingsman.** 1985. Reverse transcriptase activity and Ty RNA are associated with virus-like particles in yeast. *Nature* (London) **318:** 583–586.

127. **Michel, F., and B. F. Lang.** 1985. Mitochondrial class II introns encode proteins related to the reverse transcriptase of retroviruses. *Nature* (London) **316:**641–643.

128. **Morawetz, C.** 1987. Effect of irradiation and mutagenic chemicals on the generation of ADH2-constitutive mutants in yeast. Significance for the inducibility of Ty transposition. *Mutat. Res.* **177:**53–60.

129. **Mossie, K. G., M. W. Young, and H. E. Varmus.** 1985. Extrachromosomal DNA forms of copia-like transposable elements, F elements and middle repetitive DNA sequences in Drosophila melanogaster. *J. Mol. Biol.* **181:**31–43.

130. **Mount, S. M., and G. M. Rubin.** 1985. Complete nucleotide sequence of the *Drosophila* transposable element copia: homology between copia and retroviral proteins. *Mol. Cell. Biol.* **5:**1630–1638.

131. **Muller, F., K.-H. Bruhl, K. Freidel, K. V. Kowallik, and M. Ciriacy.** 1987. Processing of Ty1 proteins and formation of Ty1 virus-like particles in Saccharomyces cerevisiae. *Mol. Gen. Genet.* **207:**421–429.

132. **Nasmyth, K. A.** 1982. Molecular genetics of yeast mating type. *Annu. Rev. Genet.* **16:**439–500.

133. **Neigeborn, L., J. L. Celenza, and M. Carlson.** 1987. *SSN20* is an essential gene with mutant alleles that suppress defects in *SUC2* transcription in *Saccharomyces cerevisiae. Mol. Cell. Biol.* **7:**672–678.

134. **Nelbock, P., R. Stucka, and H. Feldmann.** 1985. Different patterns of transposable elements in the vicinity of tRNA genes in yeast: a possible clue to transcriptional modulation. *Biol. Chem. Hoppe-Seyler* **366:**1041–1051.

135. **Newlon, C. S., R. J. Devenish, and L. R. Lipschitz.** 1982. Mapping autonomously replicating segments on a circular derivative of chromosome III, p. 52–68. *In Proceedings of the Berkeley Workshop on Recent Advances in Yeast Molecular Biology.* Lawrence Berkeley Laboratory, Berkeley, Calif.

136. **Nonet, M., C. Scafe, J. Sexton, and R. Young.** 1987. Eucaryotic RNA polymerase conditional mutant that rapidly ceases mRNA synthesis. *Mol. Cell. Biol.* **7:**1602–1611.

137. **Omer, C. A., K. Pogue-Geile, R. Guntaka, K. A. Staskus, and A. J. Faras.** 1983. Involvement of directly repeated sequences in the generation of deletions of the avian sarcoma virus *src* gene. *J. Virol.* **47:**380–382.

138. **Orgel, L. E., and R. H. C. Crick.** 1980. Selfish DNA: the ultimate parasite. *Nature* (London) **284:**604–607.

139. **Panganiban, A. T., and H. M. Temin.** 1984. Circles with two tandem LTRs are precursors to integrated retrovirus DNA. *Cell* **36:**673–679.

140. **Paquin, C. E., and V. M. Williamson.** 1984. Temperature effects on the rate of Ty transposition. *Science* **226:**53–55.

141. **Paquin, C. E., and V. M. Williamson.** 1986. Ty insertions at two loci account for most of the spontaneous antimycin A resistance mutations during growth at 15°C of *Saccharomyces cerevisiae* strains lacking *ADH1. Mol. Cell. Biol.* **6:** 70–79.

142. **Patarca, R., and W. A. Haseltine.** 1984. Sequence similarity among retroviruses—erratum. *Nature* (London) **309:**728.

143. **Pedersen, M. B.** 1985. DNA sequence polymorphisms in the genus Saccharomyces. II. Analysis of the genes RDN1, HIS4, LEU2 and Ty transposable elements in Carlsberg, Tuborg, and 22 Bavarian brewing strains. *Carlsberg Res. Commun.* **50:**263–272.

144. **Pedersen, M. B.** 1986. DNA sequence polymorphisms in the genus Saccharomyces. III. Restriction endonuclease fragment

patterns of chromosomal regions in brewing and other yeast strains. *Carlsberg Res. Commun.* **51**:163–183.

145. **Peters, G. G., and J. Hu.** 1980. Reverse transcriptase as the major determinant for selective packaging of tRNAs into avian sarcoma virus particles. *J. Virol.* **36**:692–700.

146. **Philippsen, P., H. Eibel, J. Gafner, and A. Stotz.** 1983. Ty elements and the stability of the yeast genome, p. 189–200. *In* M. Korhola and E. Vaisanen (ed.), *Gene Expression in Yeast. Proceedings of the Alko Yeast Symposium Helsinki.* Foundation for Biotechnical and Industrial Fermentation Research, Helsinki.

146a. **Picologlou, S., M. E. Dicig, P. Kovarik, and S. W. Liebman.** 1988. The same configuration of Ty elements promotes different types and frequencies of rearrangements in different yeast strains. *Mol. Gen. Genet.* **211**:272–281.

147. **Rathjen, P. D., A. J. Kingsman, and S. M. Kingsman.** 1987. The yeast ROAM mutation-identification of the sequences mediating host gene activation and cell-type control in the yeast retrotransposon, Ty. *Nucleic Acids Res.* **15**: 7309–7324.

148. **Roeder, G. S., C. Beard, M. Smith, and S. Keranen.** 1985. Isolation and characterization of the *SPT2* gene, a negative regulator of Ty-controlled yeast gene expression. *Mol. Cell. Biol.* **5**:1543–1553.

149. **Roeder, G. S., P. J. Farabaugh, D. T. Chaleff, and G. R. Fink.** 1980. The origins of gene instability in yeast. *Science* **209**: 1375–1380.

150. **Roeder, G. S., and G. R. Fink.** 1980. DNA rearrangements associated with a transposable element in yeast. *Cell* **21**: 239–249.

151. **Roeder, G. S., and G. R. Fink.** 1982. Movement of yeast transposable elements by gene conversion. *Proc. Natl. Acad. Sci. USA* **79**:5621–5625.

152. **Roeder, G. S., and G. R. Fink.** 1983. Transposable elements in yeast, p. 299–326. *In* J. A. Shapiro (ed.), *Mobile Genetic Elements.* Academic Press, Inc., New York.

153. **Roeder, G. S., A. B. Rose, and R. E. Perlman.** 1985. Transposable element sequences involved in the enhancement of yeast gene expression. *Proc. Natl. Acad. Sci. USA* **82**:5428–5432.

154. **Rolfe, M., A. Spanos, and G. Banks.** 1986. Induction of yeast Ty element transcription by ultraviolet light. *Nature* (London) **319**:339–340.

155. **Rose, M., and F. Winston.** 1984. Identification of a Ty insertion within the coding sequence of the S. cerevisiae URA3 gene. *Mol. Gen. Genet.* **193**:557–560.

156. **Rothstein, R., C. Helms, and N. Rosenberg.** 1987. Concerted deletions and inversions are caused by mitotic recombination between delta sequences in *Saccharomyces cerevisiae. Mol. Cell. Biol.* **7**:1198–1207.

157. **Rothstein, R. J.** 1984. Double-strand break repair, gene conversion and post-division segregation. *Cold Spring Harbor Symp. Quant. Biol.* **49**:629–637.

158. **Rothstein, R. J., and F. Sherman.** 1980. Dependence on mating type of the overproduction of iso-cytochrome C in the yeast Saccharomyces cerevisiae. *Genetics* **94**:891–898.

159. **Saigo, K., W. Kugimiya, Y. Matsuo, S. Inouye, K. Yoshioka, and S. Yuki.** 1984. Identification of the coding sequence for a reverse transcriptase-like enzyme in a transposable genetic element in Drosophila melanogaster. *Nature* (London) **312**: 659–661.

160. **Sandmeyer, S. B., V. W. Bilanchone, D. J. Clark, P. Morcos, G. F. Carle, and G. M. Brodeur.** 1988. Sigma elements are position-specific for many different yeast tRNA genes. *Nucleic Acids Res.* **16**:1499–1515.

161. **Sandmeyer, S. B., and M. V. Olson.** 1982. Insertion of a repetitive element at the same position in the 5′-flanking regions of two dissimilar yeast tRNA genes. *Proc. Natl. Acad. Sci. USA* **79**:7674–7678.

162. **Scherer, S., C. Mann, and R. W. Davis.** 1982. Reversion of a promoter deletion in yeast. *Nature* (London) **298**:815–819.

163. **Seed, B.** 1983. Purification of genomic sequences from bacteriophage libraries by recombination and selection in vivo. *Nucleic Acids Res.* **11**:2427–2445.

164. **Sherman, F., and C. W. Lawrence.** 1974. Saccharomyces, p. 359–393. *In* R. C. King (ed.), *Handbook of Genetics*, vol. I. Plenum Publishing Corp., New York.

165. **Silverman, S. J., and G. R. Fink.** 1984. Effects of Ty insertions on *HIS4* transcription in *Saccharomyces cerevisiae. Mol. Cell. Biol.* **4**:1246–1251.

166. **Simchen, G., F. Winston, C. A. Styles, and G. R. Fink.** 1984. Ty-mediated gene expression of the LYS2 and HIS4 genes of Saccharomyces cerevisiae is controlled by the same SPT genes. *Proc. Natl. Acad. Sci. USA* **81**:2431–2434.

167. **Simsek, M., and U. L. RajBhandary.** 1972. The primary structure of yeast initiator methionine tRNA. *Biochem. Biophys. Res. Commun.* **49**:508–515.

168. **Sorge, J., W. Ricci, and S. H. Hughes.** 1983. *cis*-Acting RNA packaging locus in the 115-nucleotide direct repeat of Rous sarcoma virus. *J. Virol.* **48**:667–675.

169. **Stetler, G. L., and J. Thorner.** 1984. Molecular cloning of hormone-responsive genes from the yeast Saccharomyces cerevisiae. *Proc. Natl. Acad. Sci. USA* **80**:1144–1148.

170. **Stucka, R., J. Hauber, and H. Feldmann.** 1986. Conserved and non-conserved features among the yeast Ty elements. *Curr. Genet.* **11**:193–200.

171. **Stucka, R., J. Hauber, and H. Feldmann.** 1987. One member of the tRNA (Glu) gene family in yeast codes for a minor GAGtRNA(Glu) species and is associated with several short transposable elements. *Curr. Genet.* **12**:323–328.

172. **Taguchi, A. K. W., M. Ciriacy, and E. T. Young.** 1984. Carbon source dependence of transposable element-associated gene activation in *Saccharomyces cerevisiae. Mol. Cell. Biol.* **4**:61–68.

173. **Tipper, D. J., and K. A. Bostian.** 1984. Double-stranded ribonucleic acid killer systems in yeast. *Microbiol. Rev.* **48**:125–156.

174. **Toh, H., M. Ono, K. Saigo, and T. Miyata.** 1985. Retroviral protease-like sequence in the yeast transposon Ty1. *Nature* (London) **315**:691.

175. **Toh-e, A., Y. Kaneko, J. Akimaru, and Y. Oshima.** 1983. An insertion mutation associated with constitutive expression of repressible acid phosphatase in Saccharomyces cerevisiae. *Mol. Gen. Genet.* **191**:339–346.

176. **Toh-e, A., and Y. Oshima.** 1984. Regulation of acid phosphatase synthesis in Saccharomyces cerevisiae, p. 396–399. *In* T. Hasegawa (ed.), *Proceedings of the First Intersectional Congress of Microbiological Societies.* Science Council of Japan, Tokyo.

177. **Toh-e, A., and Y. Sahashi.** 1985. The PET18 locus of Saccharomyces cerevisiae: a complex locus containing many genes. *Yeast* **1**:159–171.

178. **Trueheart J., J. D. Boeke, and G. R. Fink.** 1987. Two genes required for cell fusion during yeast conjugation: evidence for a pheromone-induced surface protein. *Mol. Cell. Biol.* **7**:2316–2328.

179. **Tschumper, G., and J. Carbon.** 1986. High frequency excision of Ty elements during transformation of yeast. *Nucleic Acids Res.* **14**:2989–3001.

180. **Van Arsdell, S. W., G. L. Stetler, and J. Thorner.** 1987. The

yeast repeated element sigma contains a hormone-inducible promoter. *Mol. Cell. Biol.* 7:749–759.

181. **Van Arsdell, S. W., and J. Thorner.** 1987. Hormonal regulation of gene expression in yeast, p. 325–332. *In* D. K. Granner, M. G. Rosenfeld, and S. Chang (ed.), *Transcriptional Control Mechanisms.* Alan R. Liss, Inc., New York.

182. **Van Beveren, C., J. Coffin, and S. Hughes.** 1985. Nucleotide sequences complemented with functional and structural analysis, p. 567–1148. *In* R. Weiss, N. Teich, H. Varmus, and J. Coffin (ed.), *RNA Tumor Viruses. Molecular Biology of Tumor Viruses.* Cold Spring Harbor Laboratory, Cold Spring Harbor, N.Y.

183. **Van Beveren, C., J. G. Goddard, A. Berns, and I. M. Verma.** 1980. Structure of Moloney murine leukemia viral DNA: nucleotide sequence of the 5′ long terminal repeat and adjacent cellular sequences. *Proc. Natl. Acad. Sci. USA* 77:3307–3311.

184. **Varmus, H. E.** 1985. Reverse transcriptase rides again. *Nature* (London) **314**:583–584.

185. **Varmus, H. E., and T. Swanstrom.** 1984. *RNA Tumor Viruses.* Cold Spring Harbor Laboratory, Cold Spring Harbor, N.Y.

186. **Venegas, A., E. Gonzalez, P. Bull, and P. Valenzuela.** 1982. Isolation and structure of a yeast initiator tRNAmet gene. *Nucleic Acids Res.* **10**:1093–1096.

187. **Villarreal, L. P., and N. J. Soo.** 1985. Comparison of the transient late region expression of SV40 DNA and SV40-based shuttle vectors: development of a new shuttle vector that is efficiently expressed. *J. Mol. Appl. Genet.* **3**:62–71.

188. **Von Borstel, R. C.** 1969. Yeast genetics supplement. *Microb. Genet. Bull.* (Yeast Genetics Suppl.) **31**:26–28.

188a. **Warmington, J. R., R. P. Green, C. S. Newlon, and S. G. Oliver.** 1987. Polymorphisms on the right arm of chromosome III associated with Ty transposition and recombination events. *Nucleic Acids Res.* **15**:8963–8982.

189. **Warmington, J. R., R. B. Waring, C. S. Newlon, K. J. Indge, and S. G. Oliver.** 1985. Nucleotide sequence characterization of Ty 1-17, a class II transposon from yeast. *Nucleic Acids Res.* **13**:6679–6693.

190. **Williamson, V. M.** 1983. Transposable elements in yeast. *Int. Rev. Cytol.* **83**:1–25.

191. **Williamson, V. M., D. Cox, E. T. Young, D. W. Russell, and M. Smith.** 1983. Characterization of transposable element-associated mutations that alter yeast alcohol dehydrogenase II expression. *Mol. Cell. Biol.* **3**:20–31.

192. **Williamson, V. M., E. T. Young, and M. Ciriacy.** 1981. Transposable elements associated with constitutive expression of yeast alcohol dehydrogenase II. *Cell* **23**:605–614.

193. **Wilson, W., M. H. Malim, J. Mellor, A. J. Kingsman, and S. M. Kingsman.** 1986. Expression strategies of the yeast retrotransposon Ty: a short sequence directs ribosomal frameshifting. *Nucleic Acids Res.* **14**:7001–7016.

194. **Winston, F., D. T. Chaleff, B. Valent, and G. R. Fink.** 1984. Mutations affecting Ty-mediated expression of the HIS4 gene of Saccharomyces cerevisiae. *Genetics* **107**:179–197.

195. **Winston, F., C. Dollard, E. A. Malone, J. Clare, J. G. Kapakos, P. Farabaugh, and P. L. Mineheart.** 1987. Three genes required for trans-activation of Ty element transcription in yeast. *Genetics* **115**:649–656.

196. **Winston, F., K. J. Durbin, and G. R. Fink.** 1984. The SPT3 gene is required for normal transcription of Ty elements in S. cerevisiae. *Cell* **39**:675–682.

197. **Winston, F., and P. L. Minehart.** 1986. Analysis of the yeast SPT3 gene and identification of its product, a positive regulator of Ty transcription. *Nucleic Acids Res.* **14**: 6885–6900.

198. **Xu, H., and J. D. Boeke.** 1987. High frequency deletion between homologous sequences during retrotransposition of Ty elements in Saccharomyces cerevisiae. *Proc. Natl. Acad. Sci. USA* **84**:8553–8557.

199. **Yoshinaka, Y., I. Katoh, T. D. Copeland, and S. Oroszlan.** 1985. Murine leukemia virus protease is encoded by the gag-pol gene and is synthesized through suppression of an ember termination codon. *Proc. Natl. Acad. Sci. USA* **82**: 1618–1622.

200. **Youngren, S. D., J. D. Boeke, N. J. Sanders, and D. J. Garfinkel.** 1987. Functional organization of the retrotransposon Ty from *Saccharomyces cerevisiae*: Ty protease is required for transposition. *Mol. Cell. Biol.* **8**:1421–1431.

201. **Zachar, Z., D. Davison, D. Garza, and P. M. Bingham.** 1985. A detailed developmental and structural study of the transcriptional effects of insertion of the copia transposon into the white locus of Drosophila melanogaster. *Genetics* **111**:495–515.

202. **Zaret, K. S., and F. Sherman.** 1984. Mutationally altered 3′ ends of yeast CYC1 mRNA affect transcript stability and translational efficiency. *J. Mol. Biol.* **176**:107–135.

Chapter 14

Maize Transposable Elements

NINA V. FEDOROFF

Nina V. Fedoroff ■ Department of Embryology, Carnegie Institution of Washington, 115 West University Parkway, Baltimore, Maryland 21210.

I. INTRODUCTION

Transposable genetic elements were discovered in maize (*Zea mays*) by Barbara McClintock. Although their ability to move, affect gene expression, and cause chromosomal rearrangements was already well understood through genetic experiments in maize by the 1950s, it was another decade before they were discovered in bacteria, two decades before their ubiquity became apparent, and three decades before the profundity of McClintock's discovery was recognized with the award of a Nobel Prize. The clarity and completeness of her genetic analyses continue to astonish students of maize transposable elements.

Molecular studies on maize transposable elements began less than 10 years ago. Today, virtually every aspect of their structure, function, and regulation has been illuminated by the results of molecular analyses, often using mutants isolated and characterized decades earlier. Much still remains to be discovered, and the genetic treasury is far from exhausted. Moreover, the pace of understanding has accelerated with the demonstration that certain maize elements can be introduced into and function in readily manipulable dicot transformation systems (2, 3). The study of maize transposable elements now extends from the corn field to in vitro mutagenesis of cloned DNA fragments. The structure and expression of the several maize elements are already relatively well characterized, and progress is being made in understanding the manner in which expression of the elements is regulated. Perhaps the central remaining mystery is how to make sense of transposable elements in our concepts of development and evolution.

The most striking property of maize transposable elements is their ability to sense developmental parameters. Elements transpose and promote genetic rearrangements at characteristic times and frequencies in the development of the plant. There are indications that the elements can be differentially expressed within the organism and that they can be preprogrammed to undergo changes in expression at specified intervals, in certain parts of the plant, or in different areas of a single tissue. McClintock originally gave them the designation **controlling elements** because she understood that such genetic elements were separate from a gene's structure, yet could impose a novel developmental pattern of expression on the gene (80, 81). A contemporary view might be that transposable elements subvert genes, superimposing their own regulatory mechanisms on gene expression. But the deeper truth is that both are subject to the same developmental game plan, and an understanding of how transposable elements in particular are developmentally programmed will tell us something of how the plant does it in general.

II. MAIZE AS A GENETIC SYSTEM

The distinction of maize as a higher eucaryotic genetic system lies in its special reproductive structures and characteristics (17, 59, 123, 167). Each individual organism produces both male and female gametes. The plant is ordinarily self-fertile, but the male and female reproductive structures are physically separate, and controlled fertilization, either self-pollination or cross-pollination, is readily accomplished. The maize ear, which is the female reproductive organ, bears hundreds of **ovaries**, each of which develops into a kernel upon fertilization (Fig. 1). As in other plants, seed development commences with a double fertilization event. Two sperm cells are introduced into the **embryo sac** from each fertilizing

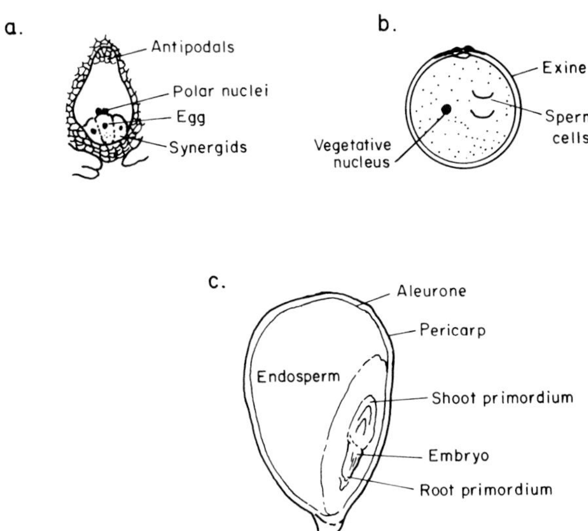

Figure 1. Structure of the maize gametophytes and kernel. (a) Embryo sac; (b) pollen grain; (c) kernel.

pollen grain (Fig. 1b). These contain sister haploid nuclei and are, in most cases, genetically identical. One sperm cell fuses with the egg cell to form the diploid zygote that develops into the embryo (Fig. 1c). The second sperm cell fuses with the large haploid central cell of the embryo sac. The central cell contains the two polar nuclei, which in turn are sister nuclei of the haploid egg nucleus. Fusion of a sperm nucleus with the two nuclei of the central cell produces the triploid primary endosperm nucleus of the central cell, which develops into the nutritive **endosperm** of the kernel (Fig. 1c). The surface layer or layers of the endosperm differentiate into the **aleurone** late in development (Fig. 1c). The ovary wall, which is maternal tissue, develops into the tough, protective **pericarp** layer. As a consequence of double fertilization, the embryo and endosperm receive the same genetic contributions from the male and female parents. Thus, except for a difference in ploidy and genetic changes attending mitotic divisions, the embryo and endosperm are genetically identical.

The maize endosperm is a bulky, readily visualized tissue. Many genes are expressed in the endosperm, and among these are a significant number in which null mutations are not lethal to plant development. These include genes in the pathways leading to starch, storage protein, and anthocyanin pigment biosynthesis (reviewed in reference 17). Such genes have been essential and richly revealing in the genetic study of transposable elements. The fact that the genetic constitution of the endosperm parallels that of the embryo has meant that rare mutations observed in genes expressed in the endosperm are most often carried in the embryo of the same kernel. Because the endosperm belongs genetically to the next generation, the use of endosperm markers permits facile analysis of hundreds of progeny for each cross performed. The dispensability of certain genes has made possible the analysis of mutations of all types. Genes in the anthocyanin pigment biosynthetic pathway have been especially useful because they permit detection of quite subtle changes in levels of gene expression.

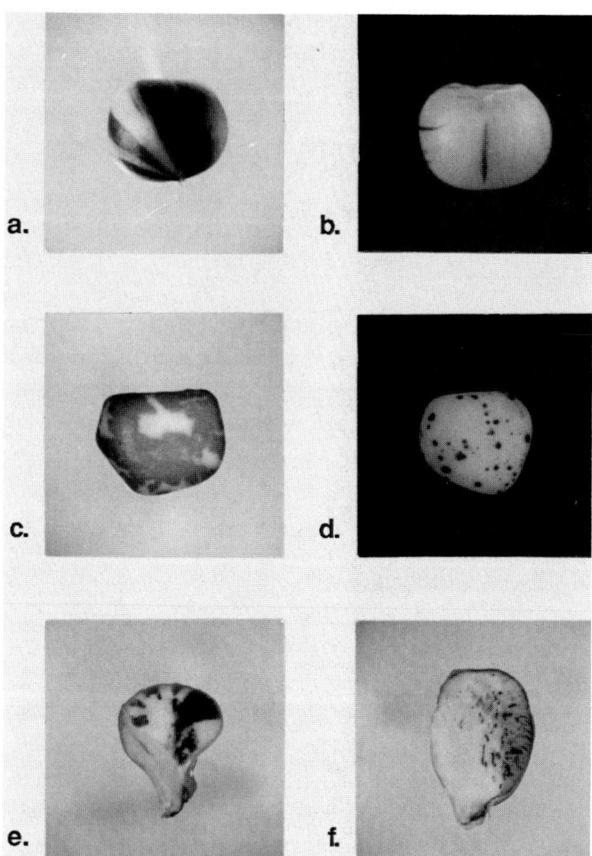

Figure 2. Developmental differences in variegation patterns. (a and b) Variegated expression of the *p* locus in pericarp tissue associated with insertion of the *Mp* element. (c and d) Variegated expression of the *a* locus in aleurone tissue associated with insertion of elements belonging to the *Spm* family. (e and f) Variegated expression of the *waxy* locus in endosperm tissue associated with insertion of elements belonging to the *Ac-Ds* family. In each pair, the left and right kernels show the variegation patterns associated with excision of the resident element early and late, respectively, in the development of the tissue.

III. TRANSPOSABLE ELEMENTS AND MAIZE DEVELOPMENT

The properties of maize transposable elements have facilitated the study of plant development. Insertion mutations are commonly unstable, and they revert during both plant and kernel development. Insertion mutations reduce or prevent gene expression, giving null or intermediate phenotypes. Reversion occurs by excision of the element, generally restoring gene function and the original phenotype. Hence plants with transposable element mutations can be **variegated**, exhibiting clones of revertant cells on a mutant background. The size and shape of a given clone of revertant cells are determined by the developmental timing of excision and by the pattern of cell division in the tissue within which it occurred (96). The developmental timing of excision is determined by the element and can be altered by mutations within the element, as will be discussed in detail in subsequent sections. Differences in variegation pattern caused by changes in the developmental time of excision are illustrated in successive panels of Fig. 2. The shape of a revertant sector is determined by the pattern of cell division and differs dramatically from tissue to tissue, as illustrated for endosperm, aleurone, and pericarp tissues of the kernels in Fig. 2.

McClintock's early understanding that transposable element insertions impose new and highly specific patterns of gene expression and that such patterns can change during development in a programmed way led her to designate the insertions **controlling elements.** She also understood that a single maize element could control the pattern of expression of several genes with related insertions (80, 81). This hierarchic relationship among the elements and genes with insertion mutations further supported the concept that the elements were central developmental control mechanisms. The insights into element structure and function gained from molecular analysis permit a new perspective on the relationship between transposable elements and development, as will become apparent in the ensuing discussion.

IV. MAIZE TRANSPOSABLE ELEMENT FAMILIES

Maize transposable elements were initially subdivided into groups on the basis of their ability to interact genetically. The groups have been designated **families** and comprise structurally related elements. The initial genetic grouping was based on the existence of two types of insertion mutations, one of which was inherently unstable and the other of which was unstable only in the presence of a transposable element elsewhere in the genome. These two types of mutations were called **autonomous** and **nonautonomous** (72), and the elements were subsequently given the corresponding designations (42). The autonomous element in each element family is a single, structurally conserved element (4, 7, 40, 41, 64, 111, 112). The nonautonomous elements of a family comprise a structurally heterogeneous group of elements, some of which are mutant, transposition-defective elements. Other nonautonomous elements are less similar or virtually unrelated in structure to the autonomous element, except for their termini.

Four maize element families have been studied in some detail genetically. These are the *Activator-Dissociation* (*Modulator*), *Suppressor-mutator* (*Enhancer-Inhibitor*), *Mutator*, and *Dotted* element families (reviewed in references 31, 42–44, 105, and 106). Several other families have been identified genetically (45, 47, 121, 140, 141, 146) or through the cloning of insertions (23, 57, 58). The present review will concentrate on the first three of the well-studied element families because they are the most thoroughly characterized at the molecular level.

The *Activator-Dissociation* (*Ac-Ds*) element family was identified and studied by McClintock (68–71, 74). It was through the study of elements belonging to the *Ac-Ds* family that McClintock first demonstrated transposition (70). Unstable mutations caused by an element of the same family were studied earlier by Emerson (33–35). These studies were continued by Brink and his students, who designated the element *Modulator* (*Mp*) and showed it to be genetically interchangeable with *Ac* (5, 12). The *Suppressor-mutator* (*Spm*) element family was identified and studied by McClintock (74–76, 78). Mutations caused by the same element family were identified by Rhoades and Dempsey (128) and studied by Peterson (114–117), who called the element *Enhancer* (*En*) and later showed it to be genetically interchangeable with the *Spm* element. The *Dotted* (*Dt*) element family was identified by Rhoades (124) and studied by several investigators, including Rhoades (125–127), Neuffer (103, 107, 108), and Doerschug (24–27).

The *Mutator* (*Mu*) element family, although less well understood than the *Ac-Ds* and *Spm* families, is included in the present review because there is evidence that its transposition mechanism is quite different from that of the other elements. *Mu* elements have been isolated from unstable insertion mutations that arise at high frequency in maize strains having the *Mutator* trait, first identified and described by Robertson (132) as an anomalously high frequency of spontaneous mutations.

V. THE *Ac-Ds* ELEMENT FAMILY

A. The Genetic Characteristics of *Ac* and *Ds* Elements

McClintock (67, 68) first identified the *Ds* element as a specific site of **chromosome breakage** or **dissociation.** Chromosome breakage at *Ds* did not occur spontaneously, but required the presence of a second genetic element, designated the *Activator*, for its ability to activate chromosome breakage at *Ds* (69). In McClintock's subsequent investigations, it became apparent that these genetic elements, both originally called **loci**, had unusual properties. Both could occasionally either disappear or change their map positions, and in 1948 McClintock first formally reported that the *Ds* locus could transpose (70).

Among the earliest strains with transposed *Ds* elements, McClintock (70, 71, 75) identified one with a new mutation at the *c* locus on the short arm of chromosome 9. It was through the detailed genetic analysis of this mutation that McClintock understood the relationship between unstable mutations and transposition. She established that *Ds*, identified by its ability to break the chromosome, had transposed to the *c* locus in the mutant strain. The new mutation was stable in the absence of an *Ac* element, but reverted both somatically and in the germ line when an *Ac* was present. Germinal revertants of the *c* gene mutation no longer showed chromosome breakage at the locus, thus revealing that the instability of the mutation was attributable to the excision of the *Ds* element.

In subsequent studies on a variety of mutations, McClintock found that the *Ac* element could also cause insertion mutations (74, 81, 91). These differed from *Ds* mutations by their inherent instability. That is, an *Ac* insertion could spontaneously excise, while a *Ds* insertion could excise only in the presence of an *Ac* element. She reported that not all *Ds* elements broke chromosomes and that chromosome breakage was rarely associated with an *Ac* insertion. Furthermore, she reported that an *Ac* mutation could change directly into one that had the genetic properties of a *Ds* mutation, suggesting that at least some *Ds* elements might be mutant *Ac*'s (79, 81, 88). Moreover, an *Ac* element could undergo reversible inactivation. When inactive, the *Ac* element behaved genetically as a *Ds* element, but differed from a true *Ds* element by its ability to return to an active form (90, 92). The various genetic properties of the *Ac* and *Ds* elements are summarized diagrammatically in Fig. 3.

b. Transposition

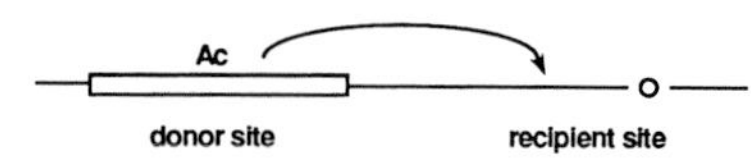

c. Mutation

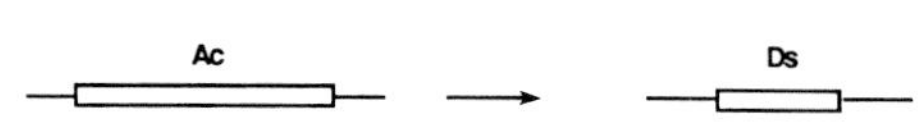

d. Inactivation

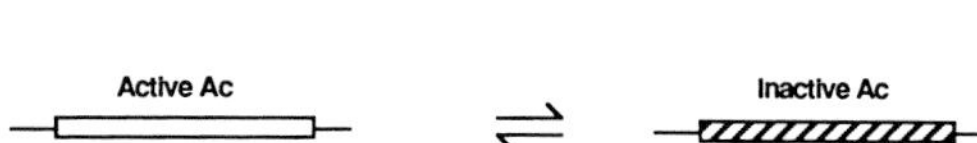

Figure 3. Genetic properties of the *Ac-Ds* element family. Elements belonging to the *Ac-Ds* family can break chromosomes, transpose, mutate, and become reversibly inactivated.

B. Structure of *Ac* and *Ds* Elements

1. The *Ac* element is a unique small transposable element

The first *Ac* element was isolated from the *waxy* (*wx*) locus of maize, which encodes an abundant starch-biosynthetic enzyme (41, 101, 157). Several additional elements have since been isolated, including the genetically interchangeable *Mp* element, and all are similar or identical, suggesting that the transpositionally competent member of the family is structurally unique (7, 15, 37). The *Ac* element is 4.6 kilobases (kb) in length and has short, 11-base-pair (bp) imperfect inverted terminal repetitions (IRs; 100, 122; Fig. 4). A single transcription unit has been identified for the element (62). It extends over most of the element's length and encodes a 3.5-kb mRNA comprising 5 exons and encoding a 2,421-nucleotide open reading frame (ORF). Several sites of transcription initiation have been identified between 302 and 357 bp from the left or 5′ end of the element (62).

2. *Ds* elements are structurally heterogeneous

Ds elements have proved surprisingly varied in structure, ranging from *Ac* elements with deletions to sequences that are virtually unrelated to that of the *Ac* element, except for the terminal IRs (4, 29, 30, 32, 41, 99, 160). In addition to the several *Ds* elements shown in Fig. 4, whose origin from an *Ac* element has been documented at both genetic and molecular levels, a 2-kb element that consists of a 1-kb sequence from each end of the *Ac* element has been recovered from both the *wx* and *shrunken* (*sh*) loci (*Ds6* in Fig. 4). This element is also found in the double *Ds* element described below and may correspond to the original *Ds* element identified by McClintock (69). Curiously, this 2-kb *Ds* element has perfect 11-bp terminal IRs, unlike either of the *Ac* elements that have been sequenced and found to have terminal IRs that differ from each other by a single base pair (Fig. 4; 32). Although the genetic history of the *Ds6* element does not permit the definitive conclusion that it originated by a deletion within an *Ac* element, its structure indicates that it did.

Several loci have yielded a short element, designated *Ds1*, that is structurally unrelated to the *Ac* element, except for the identity of its terminal IRs (142, 160, 170). The *Ds1* elements comprise a small

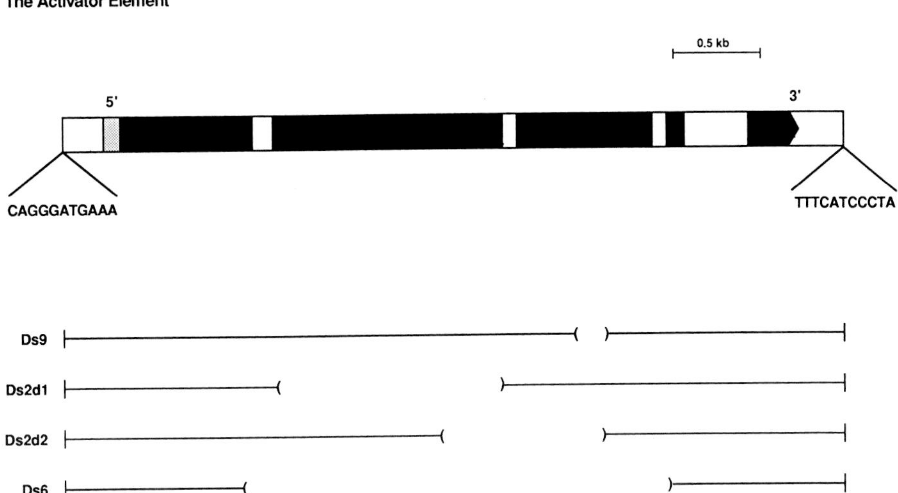

Figure 4. Structure of the *Ac* element and several *Ds* elements. The *Ac* element is 4.6 kb in length and has 11-bp imperfect terminal IRs. The 5′ and 3′ ends of the *Ac* transcription unit are indicated, and the left and right ends of the element are correspondingly designated the 5′ and 3′ ends. The stippled box represents the region of the element within which transcription start sites map. Filled boxes correspond to exons, and the open boxes between them represent introns. The interrupted lines below the *Ac* element represent the structure of the several *Ds* elements. The line corresponds to the part of the *Ac* sequence, and the gap represents the *Ac* sequence missing from the *Ds* element.

family that is quite homogeneous in both length and sequence (110). A longer *Ds* element, named *Ds2*, has been cloned from the *adh* and *bz2* loci. It has *Ac* IRs, as well as additional *Ac*-homologous sequences interspersed with nonhomologous sequences (99, 163).

C. Transposition and Chromosome Breakage

1. *Ac* transposition is nonreplicative

Genetic studies on transposition were carried out primarily by Brink and by Greenblatt, using an insertion allele of the *p* locus with an *Mp* element (5, 12, 48–55). These studies have been reviewed in detail elsewhere (42, 43), and only the conclusions are summarized here (Fig. 5). The element undergoes nonreplicative transposition either during or after chromosome replication, giving rise to dissimilar daughter chromatids, one of which retains the element at the donor site and the other of which does not. The element commonly inserts in an unreplicated recipient site, but it can also insert after replication of the recipient site. When the element inserts into an unreplicated site, both the donor and nondonor daughter chromatids bear a transposed *Ac* element. This inference, initially made from genetic analyses, has recently been confirmed by analyzing

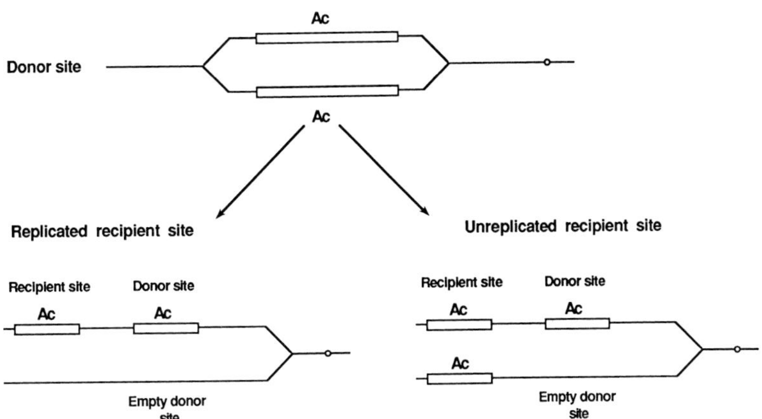

Figure 5. *Ac* transposition mechanism. As described in the text, the *Ac* donor and recipient sites are commonly on the same chromosome. *Ac* transposes from the donor site on one daughter chromatid to either an unreplicated or a replicated recipient site.

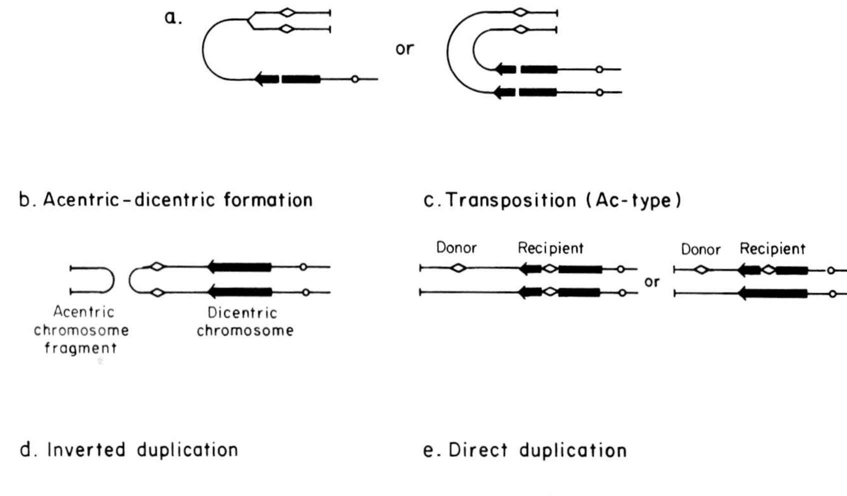

Figure 6. Chromosome rearrangements associated with *Ds*. (a) Partially or completely replicated chromosome bearing a *Ds* element (diamond), cleaved at both ends of the element as well as at a recipient site elsewhere on the same chromosome, represented by the filled arrow. (b through e) Different types of chromosomal rearrangements that result from rejoining of the replicated or partially replicated sister chromatids in different ways: (b) daughter chromatids are joined to each other at the site of *Ds* insertion, giving a U-shaped acentric chromosome fragment and a dicentric chromosome; (c) the *Ds* element from one daughter chromatid moves to the recipient site, either before or after replication, to give a transposed *Ds* element by the *Ac* type of transposition mechanism shown in Fig. 5. (d) The *Ds* element is transposed as in diagram c, but the chromosome fragment extending from the donor to the recipient site is then inserted into the same chromatid, generating an inverted duplication extending from the donor site (marked by a vertical bar) to the recipient site. It is expected that a deleted chromosome will also be produced in such a rearrangement event, although such extensively deleted chromosomes have not been recovered. (e) Cleaved ends of the chromatids are rejoined in an alternative way to generate a direct duplication of the chromosome fragment extending from donor to recipient site. The chromosomal rearrangements depicted in diagrams d and c are not the only ones that can be obtained, but correspond to the two *Ds*-associated chromosome 9 rearrangements analyzed by McClintock (73).

the lengths of restriction fragments bearing the newly transposed element in progeny receiving the donor and nondonor daughter cells of a single transposition event (15).

The recipient site is frequently on the same chromosome (61%), and of these, 40% are within 4 map units of the donor site (53). Moreover, those recipient sites that map closest to the donor site are exclusively on the distal side of the donor site, while recipient sites at a greater distance from the donor site are located distal and proximal to the donor site equally frequently. Unlike most genetically silent sequences with homology to the element, newly transposed elements are located in undermethylated regions of the genome (15).

2. Certain *Ds* elements break chromosomes

The capacity of a *Ds* element to promote chromosome breakage and to transpose was initially documented by McClintock using a multiply marked chromosome 9 (see references 42 and 74 for illustrative kernels). The initial chromosomal cleavage event occurs during or after chromosome replication (Fig. 6a). The ends of the cleaved chromosome fragments either fail to separate or are ligated to each other, yielding a U-shaped acentric fragment and a dicentric chromosome (Fig. 6b). The acentric fragment is subsequently lost, permitting expression of recessive markers borne on the homolog. The dicentric chromosome breaks at mitosis, generating fragment chromosomes lacking telomeres, which again replicate to produce dicentric chromosomes, perpetuating what McClintock designated the **breakage-fusion-bridge cycle** (65). Since marker loss is initiated by cleavage and chromatid fusion at the *Ds* element, the pattern of subsequent marker loss depends on the insertion site. The position at which McClintock originally identified *Ds* was proximal to the *waxy* (*wx*) locus, and it was not until the *Ds* element transposed to more distal sites on chromosome 9 that it became apparent that the first chromosome break at *Ds* is followed by additional marker loss attributable to the breakage-fusion-bridge cycle (69–73). McClintock used the changes in the pattern of marker loss that accompany transposition of the element to identify transposed *Ds* elements (73).

3. Transposition of *Ds* can be accompanied by chromosomal rearrangements

Although *Ds* elements can undergo simple transposition events of the type described above for *Ac* (Fig. 6c), transposition of the chromosome-breaking *Ds* element that McClintock (73) first identified and studied was frequently accompanied by chromosomal rearrangements. Indeed, she noted that a large proportion of kernels that showed phenotypic evidence of a transposed *Ds* element failed to germinate, suggesting that lethal rearrangements frequently accompany movement of the chromosome-breaking *Ds* element.

Two of the intrachromosomal rearrangements that McClintock analyzed in detail are represented schematically in Fig. 6d and e (73). In these rearranged chromosomes, *Ds* transposition was accompanied by the duplication of the entire chromosome segment between the donor and recipient sites, either in inverted (Fig. 6d) or direct (Fig. 6e) order. These observations, taken together with those on transposition and acentric-dicentric formation (depicted diagrammatically in Fig. 6b and c), indicate that transposition is accompanied by chromosome cleavage at both donor and recipient sites. Because duplications extend from donor to recipient site and include the element itself, transposition of the chromosome-breaking *Ds* element appears to differ from *Ac* transposition by the common occurrence of breaks at the donor site on both daughter chromatids, rather than on only one of them.

In transposition, chromosome ends are rejoined in such a way that chromosomal continuity is restored and the element is removed to the recipient site (Fig. 6b). The most straightforward interpretation of the rearrangements and chromosome aberrations that are associated with the chromosome-breaking *Ds* element is that the sister chromatids cleaved at physically juxtaposed donor and recipient sites can be rejoined in a variety of ways. The resulting daughter chromatids may be joined as a dicentric chromosome and a U-shaped acentric fragment (Fig. 7b) or have direct or inverted duplications extending from donor to recipient site (Fig. 7d and e).

4. A "nontransposing" *Ds* element

McClintock identified several *Ds* mutations at the *sh* and *bz* loci which had the unusual genetic characteristic that the mutation reverted without apparent loss of the resident *Ds* element, as judged by the continued association of a *Ds* chromosome breakage site with the locus (76–79, 81). Several of these have been analyzed in detail at the molecular level and have provided insight into the structure of chromosome-breaking *Ds* elements and *Ds*-associated rearrangements (21, 168).

The chromosome-breaking *Ds* element isolated from the *sh-m6233* allele of the *shrunken* (*sh*) locus is a double element in which one 2-kb *Ds* element, resembling the *Ds6* isolated from the *wx* locus (Fig. 4), is inserted into the center of itself in the reverse orientation (Fig. 7a and b; 32, 168). Another allele of the same locus, which McClintock designated *sh-m5933*, proved to have a more complex structure (21). It contains an extensive insertion of about 30 kb within the coding sequence at the locus (Fig. 7c). The insertion is flanked on one side by the double *Ds* element depicted in Fig. 7b and on the other by a deleted double *Ds* element lacking half of one of its constituent elements (Fig. 7c). The large insertion itself comprises an inverted duplication that extends for at least 8 kb from the flanking *Ds* elements (arrows, Fig. 7c). Moreover, there is a duplication of part of the insertion, its deleted *Ds* element, and the 5′-terminal end of the *sh* gene nearby on the same chromosome (Fig. 7c).

The ability of the *sh-m5933* mutation to revert without concomitant loss of *Ds*-mediated chromosome breakage at the locus was explained by the discovery that *Sh* revertants lack the insertion within the gene, but retain the duplicated segment and its attendant *Ds* element (Fig. 7d; 21). Evidence that chromosome breakage in the revertants was mediated by the *Ds* element within the duplicated segment came from the demonstration that one revertant with a reduced frequency of chromosome breakage had lost a complete 2-kb *Ds* element and retained only half of one of the original *Ds* elements in the duplicated segment of the rearranged *sh* locus (Fig. 7e).

Although the *sh-m5933* rearrangement appears to be more complex than other *Ds*-mediated rearrangements, it may have arisen by similar chromosome breakage and rejoining events. The chromosome in which the mutant allele arose carried a *Ds* element just distal to the *sh* locus. Figure 8 depicts a scheme of chromosome breakage at the *Ds* donor and *sh* recipient sites, replication, and rejoining that accounts for the structure of the *sh-m5933* allele. To explain the complexity of the insertion within the *sh* locus, it is only necessary to postulate that chromosome breakage occurred in the replicated flanking sequences adjacent to the element at the donor site (Fig. 8b), generating a small duplication that accompanied the duplicated *Ds* element to the recipient site (Fig. 8c). It is then necessary to postulate that religation of the daughter chromatids gave rise to the larger duplication that includes part of the rearranged *sh* locus (Fig. 8d).

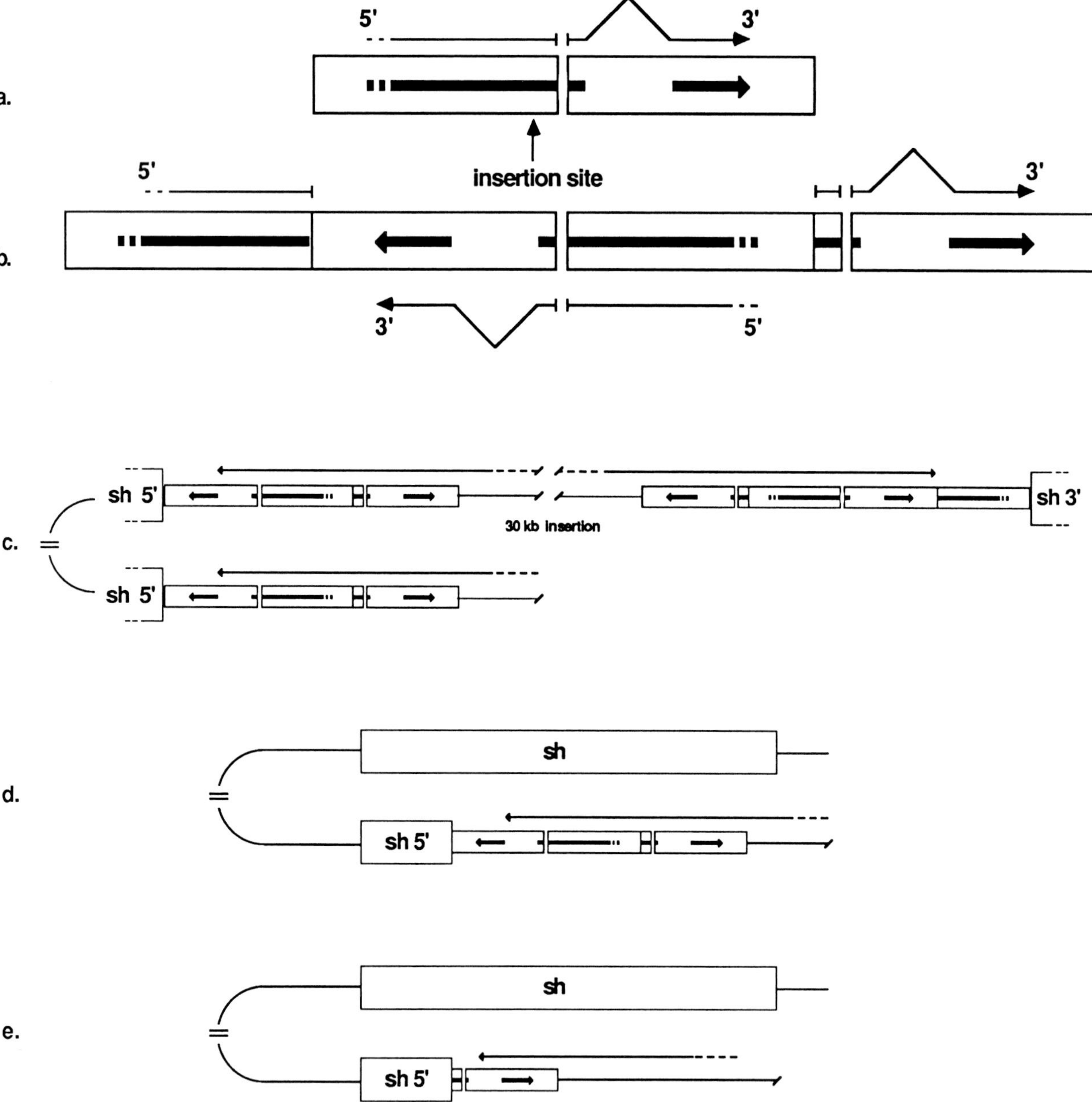

Figure 7. Structure of the double *Ds* element and the *sh-m5933* allele. (a) Structure of the 2-kb *Ds-6* element (Fig. 4), showing the remaining portions of the *Ac* transcription unit. (b) Structure of the double *Ds* element. The 2-kb *Ds* element depicted in part a, inserted in inverse orientation at the site indicated by the arrow in part a, generates the double *Ds* element recovered from the *sh-m6233* allele. (c) Structure of the *sh-m5933* allele. The *sh* locus of the *sh-m5933* allele is interrupted by a 30-kb insertion, much of which comprises a long inverted duplication (arrows) flanked by *Ds* elements. A complete double *Ds* element is at the end of the insertion adjacent to the 3′ end of the gene, while the *Ds* structure adjacent to the 5′ end of the gene comprises 1 1/2 *Ds* elements, as indicated in the diagram. Located on the same chromosome is a duplication which includes the 5′ end of the gene and part of the insertion, as shown. Its orientation with respect to the *sh* locus is not known. (d) Structure of *Sh* revertants of the *sh-m5933* allele. In most revertants, the large insertion in the *sh* gene has been excised, but the partial duplication of the *sh* gene and its adjacent *Ds*-containing insertions remain. (e) Structure of an exceptional *Sh* revertant that exhibits late, infrequent chromosome breakage. The structure resembles that of the revertant chromosome depicted in part d, except that the complete 2-kb *Ds* element is no longer present in the duplication.

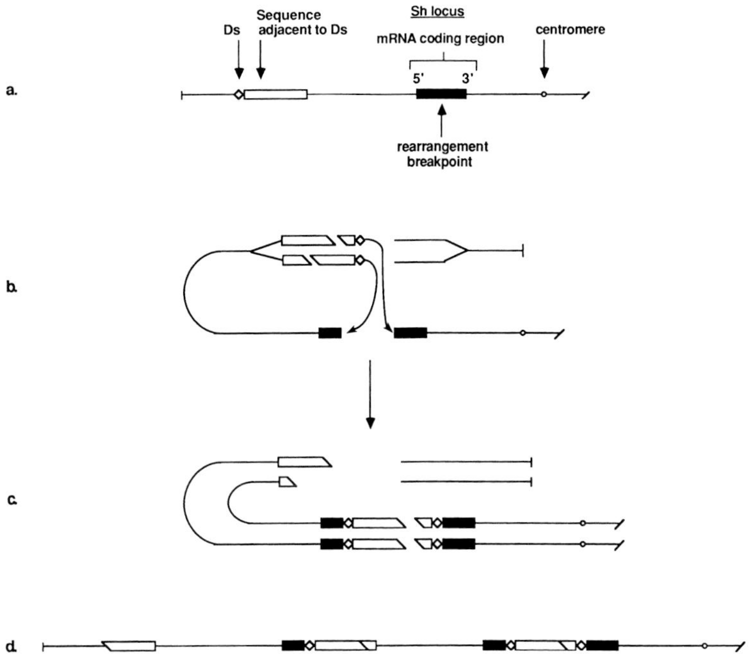

Figure 8. Origin of the *sh-m5933* rearrangement. (a) Structure of an unrearranged chromosome with *Ds* at a site distal to the *sh* locus. (b) Cleavage of a partially replicated chromosome at and adjacent to the *Ds* element, followed by transfer of the replicated *Ds*-containing chromosome fragments to an unreplicated recipient site. (c) Replication of the rearranged recipient site. (d) Rejoining of the rearranged chromosomes to yield the rearranged, partially duplicated chromosome of the *sh-m5933* allele.

D. The Molecular Mechanism of Transposition

1. The chromatid selectivity of transposition is disrupted by the double *Ds* element

Although nothing is known about the molecular mechanism of *Ac* transposition, important inferences can be drawn from the types of rearrangements that have been observed in strains with chromosome-breaking *Ds* elements. *Ac* transposition generally involves only one of the two replicated daughter chromatids, which implies an asymmetry in the recognition of parental strands by the enzymes involved in the chromosome breakage and rejoining events. The rearrangements that accompany movement of the double *Ds* element necessitate cleavage of both replicated daughter chromatids at the donor site and can result in the movement of both copies of the element. This suggests that the structure of the double *Ds* element interferes with the inherent strand asymmetry of the transposition process. Since the hallmark of the double *Ds* element is the inverted duplication of the element, it follows that it is the presence of some element sequence in both direct and inverted orientations that disrupts the asymmetry. The implication is that nonsymmetrical sequences within the element are involved in determination of the chromosome cleavage pattern at the donor site.

The observation that movement of the double *Ds* commonly involves larger chromosome rearrangements that result from a failure to cleave the donor site at the element's terminus suggests that the double structure interferes with correct cleavage of the element at one of its termini. That is, in some cases the rearrangements appear to have been initiated by a cut at an element end and resolved by a second cut outside the element. This indicates that the structure of the double *Ds* interferes with both the strand selectivity of the initial cleavage event and the location of the second cleavage event that defines the length of DNA to be transposed.

2. Could asymmetric cleavage sites determine strand selectivity?

Methylation of element sequences is implicated in the control of both transposition and expression of transposable elements in bacteria and in maize (14, 16, 131, 151). Although there is no detailed information about the role of methylation in transposition of maize elements, it is known that transposition of Tn*10* is more frequent from hemimethylated DNA than from fully methylated DNA (131). Moreover, the donor duplex is much more active when methylated in one strand than in the other. This implies that the ability to cleave daughter chromatids differently requires an asymmetric hemimethylated recognition site. An analogous asymmetry of a methylatable cleavage site can explain the tendency of the double *Ds* element to produce chromosomal aberrations. That is, for example, the active cleavage site might be a sequence such as AAC*GTT/TTGCAA, while its hemimethylated complement AACGTT/TTGC*AA is cleaved much less readily. This would explain the observation that only one of the two daughter chromatids normally participates in a transposition event (Fig. 5). Because of its internal inverted duplication, active cleavage sites would be present on both daughter chromatids carrying the double *Ds* element. Moreover, if unique asymmetric recognition sites at each element end determined the order and polarity of the initial and terminal cleavages of transposition (perhaps relative to the overall polarity of DNA replication), then the structure of the double *Ds* element might result in transposition-initiating cleavage events that do not terminate correctly at an element end, thereby leading to a chromosomal rearrangement.

3. Sequences other than termini are necessary for transposition

The first hints that there are sequences inside the element's IRs that are critical to the excision process have come from experiments on *Ac* and *Ds* elements introduced into tobacco cells by *Agrobacterium tumefaciens* Ti plasmid-mediated transformation (3). Since the *Ac* element promotes its own transposition in tobacco cells, such experiments permit the analysis by in vitro mutagenesis of sequences involved in transposition. In preliminary studies, it has been found that the terminal 11-bp IRs do not carry sufficient information for excision of a sequence by a *trans*-acting *Ac* element (unpublished data). Moreover, deletion of the sequence between 0.05 and 0.18 kb from the element's 5′ end and slightly upstream of its site of transcription initiation immobilizes it (B. Baker and G. Coupeland, personal communication). It appears, therefore, that internal element sequences are required for transposition. These sequences may contribute to determining the chromatid specificity of excision, as well.

4. Empty donor sites retain the target site duplication

Like most other transposable elements, the *Ac* and *Ds* elements generate a short duplication of target sequences upon insertion (32, 122, 160). The target duplication is 8 bp, and part or all of the duplication remains when the element excises (3, 122, 138, 164, 168). As first became evident from the analysis of several revertants of a *Ds1* mutation in the *adh* locus (138), the structure of the former insertion site is variable (Table 1). The most common features of such sites, termed empty donor sites, are a deletion of several central nucleotides and the transversion of one or two central nucleotides.

Although they are not yet distinguishable experimentally, two proposals have been advanced to explain these common structural features of empty donor sites in plants. Saedler and Nevers (139) have postulated that single-strand nicks are introduced at each end of the target duplication during excision, exposing the ends to nuclease attack as well as to DNA repair enzymes. They attribute the central transversions to strand switching by repair enzymes. In their work on transposable elements in *Antirrhinum majus*, Coen et al. (18) have found terminal homologies to maize elements, as well as similar heterogenities in empty donor site sequences. They have advanced the alternative proposal that complementary strands at the donor site are transiently religated in hairpin structures, whose resolution and repair generate both deletions and short inversions.

5. *Ac* encodes a transposase

There is evidence that an *Ac*-encoded gene product is involved in excision and transposition. The *Ac* and *Ds* elements isolated from the *Ac wx-m9* allele of the *waxy* locus and its immediate *Ds* derivative, the *wx-m9* allele, are virtually identical elements at the same insertion site. The *Ds9* element differs from the *Ac* element by a 194-bp deletion within the transcribed region (Fig. 4), indicating that interruption of the corresponding genetic function had converted the autonomously transposable *Ac* element to a transposition-defective *Ds* element (41). The same transposition-defective phenotype is associated with deletions at other points within the transcription unit (Fig. 4; 4, 28). Because an *Ac* element can mobilize a *Ds* element, it follows that an element-encoded gene product is required for transposition. This conclusion is supported by the observation that the *Ac* element promotes its own transposition, as well as that of *Ds* elements, when introduced into tobacco cells on an *Agrobacterium* Ti plasmid (2, 3).

E. Regulation of *Ac* Transposition

The frequency and timing of *Ac* transposition, as well as expression of the element, are subject to genetic control. These aspects of *Ac* function are as yet understood only poorly or not at all at the molecular level. But as with other aspects of *Ac*

Table 1. Empty donor site sequences

Element	Genotype	Allele	Sequence
Ac-Ds family[a]	Wild type	*Adh*	—GGGACTGA—
	Mutant	*adh1-FM335*	—GGGACTGA *Ds1* GGGACTGA—
	Revertant, maize	*adh1-RV1*	—GGGACTGTCGGACTGA—
		adh1-RV2	—GGGACTGTCCGACTGA—
		adh1-RV3	—GGGACTGTC ACTGA—
		adh1-RV4	—GGGACTG . . GGACTGA—
	Wild type	*Wx*	—CATGGAGA—
	Mutant	*Ac wx-m9*	—CATGGAGA *Ac9* CATGGAGA—
	Revertant, maize	*wx9-r1*	—CATGGAGA . . TGGAGA—
	Revertant, tobacco	*T-wx1*	—CATGGAGA . . TGGAGA—
	Revertant, tobacco	*T-wx2*	—CATGGAGTGATGGAGA—
	Revertant, *Arabidopsis*		—CATGGA . TGATGGAGA—
	Wild type	*Sh*	—CTTGTCCC—
	Mutant	*sh-m6233*	—CTTGTCCC *Ds* CTTGTCCC—
	Revertant	*sh-m6233 r1*	—CTTGTC . . CTTGTCCC—
Spm family[b]	Wild type	*Wx*	—GTT—
	Mutant	*wx-m8*	—GTT*Spm-8*GTT—
	Revertant	*wx*$^{+}$*-1*	—GTT ATT—
	Revertant	*wx*$^{+}$*-2*	—GTC GTT—
	Wild type	*Wx*	—TCAAGTT CAAC—
	Somatic excisions		—TCAAGTT GTTCAAC—
			—TCAAGT . GTTCAAC—
			—TCAAG . . GTTCAAC—
			—TCAA . . . GTTCAAC—
			—TCAAGTT ATTCAAC—
			—TCAAGTACGTTCAAC
			—TCAA . . . . C—

[a] References: *adh*, 138, 160; *wx*, 3, 122, 164; *sh*, 168.
[b] References: *wx*, 139, 154.

function, the genetic observations define and identify the element properties that seek molecular explanation.

1. Transposition is developmentally delayed as *Ac* copy number increases

The developmental timing of *Ac* transposition becomes progressively later as the number of *Ac* copies per genome increases (12, 70). This manifests itself as a reduction in the frequency of large revertant sectors produced by excision events early in development of the tissue (see Fig. 2 and reference 42 for illustrative kernels). The reduction is observed whether the reverting mutation is caused by the *Ac* element itself or is measured by the *trans*-activation of excision of a *Ds* insertion mutation or *Ds*-mediated chromosome breakage (70). This seemingly simple phenomenon is surprisingly subtle and complex.

To begin with, each different tissue displays the same type of delay in the developmental timing of excision, regardless of the precise number of elapsed cell divisions. Thus, for example, both the kernel endosperm and the plant grown from it subsequently will display progressively smaller revertant sectors as *Ac* copies accumulate (70). Even within a tissue, such as the endosperm, the excision events can reflect not the overall pattern of development, but rather the destiny of each sublineage of cells. The overall pattern of endosperm divisions is cambial, with late cell divisions confined to the periphery, while the internal cells increase in size and become polyploid (123). Yet a fine pattern of revertant clones can be observed throughout the endosperm with an *Ac* or a *Ds* mutation at the *wx* locus and two or three copies of the *Ac* element in the endosperm (Fig. 2f; 96, 123, 147). This suggests that the timing of excision is related to the physiological state of the cells (147) and the number of remaining divisions in a given cell lineage.

2. Additional genetic factors can affect the frequency and developmental timing of *Ac* transposition

The results of McClintock's early genetic analyses suggest that there are additional heritable factors that affect the developmental timing of *Ac* transposition. She identified elements which underwent either a transposition or a functional change during either the first or second nuclear division in endosperm development in over 90% of the kernels examined (70, 74). She also noted heritable differences among *Ac* elements in the frequency and developmental timing of *Ac*-activated excision of *Ds* elements or *Ds*-mediated chromosome breakage. She designated these "states" of the element, signifying that they were heritable but relatively labile (70). More recently, Rhoades and Dempsey (129, 130) and Schwartz (148) have reported the identification of *Ac* elements whose activity can be distinguished from that of the standard element by the timing and frequency of transposition, as well as the ability to *trans*-activate excision of *Ds* elements or stimulate chromosome breakage at *Ds*.

3. Germinal and somatic excision events may be controlled independently

Although the molecular bases of the developmental control mechanisms of *Ac* are not yet understood, the results of molecular studies have contributed some insights. Two *Ac* elements have been isolated from different alleles of the *wx* locus that show marked differences in the developmental patterns of element excision. These are the *Ac wx-m9* and *Ac wx-m7* alleles. The *Ac* elements have been sequenced and found to be identical (100, 122). The *Ac* element of the *Ac wx-m9* allele shows a typical delay in the developmental timing of transposition with increasing element dosage, while that of the *Ac wx-m7* allele was believed not to, excising early in endosperm development even when three copies of the element were present.

Recently, Schwarz (149) has provided evidence that the lack of a dosage effect of the *Ac wx-m7* allele may be more apparent than real. He reported that although the resident *Ac* element often excises from the *wx* locus premeiotically or during the first or second endosperm divisions, the remaining endosperm tissue exhibits the usual developmental delay in element excision with increasing *Ac* copy number. He further suggested that the *Ac* element tends to reinsert into the locus in those cell lineages in which the element excised early and that reinsertion also shows the normal response to *Ac* copy number. Although these different behaviors of the *Ac* element are associated with different insertion sites within the locus, the location does not appear to determine the element's behavior. That is, the early-reversion property of the *Ac wx-m7* element can transpose away from the locus with the element, and *Ac wx-m9* derivatives can be isolated that show behavior typical of the *Ac wx-m7* allele. Moreover, control of these properties by separate factors in the genetic background is unlikely, since each allele exhibits its characteristic behavior in heterozygotes (D. Schwartz, personal communication).

If neither the sequence nor the location of the element is responsible for such differences in element behavior and they cannot be attributed to extra-element genetic factors, then it is possible that they are attributable to reversible DNA modifications that affect the timing of element expression. Perhaps the property of the *Ac wx-m7* allele that merits most emphasis is the lack of correspondence between the early, often germinal, excision events and the later pattern of somatic excision (149). Although in many *Ac* insertion alleles the frequency of germinal events decreases with the dosage-dependent delay in the developmental timing of excision, this does not appear to be an absolute correlation. In the *Ac wx-m7* allele, early excision events leading to germinal reversion or large somatic revertant sectors coexist with a late somatic excision pattern. These observations suggest that early and late expression of the element or early and late excision may be controlled independently.

4. *Ac* probably encodes a negative regulator of transposition

The simplest interpretation of the *Ac* dosage effect is that the element encodes a repressor of element expression that accumulates with increasing numbers of *Ac* elements. Several strains with *Ds* elements that arose from *Ac* elements by short internal deletions have been analyzed genetically for the ability of the mutant element to contribute to the dosage effect. The small deletions identified for the *wx-m9 Ds* element and a *Ds* derivative of the *Ac bz-m2* allele (Fig. 4) suffice to render the element unable to contribute to the dosage effect (28). If the element's major transcript is its only transcript, then the corresponding protein is responsible for both the dosage effect and the element's *trans*-acting transposase activity.

An additional indication that the element's major transcript encodes the relevant gene product comes from the observation that there is an increase in the abundance of the transcript with increasing element dosage (62). It has been proposed that increasing element numbers are directly responsible for the negative dosage effect by competing for a limiting supply of a critical cellular DNA-binding factor (149, 152). This explanation appears unlikely in view of the fact that the virtually full-length *Ds* elements of the *wx-m9* and *bz-m2* alleles do not contribute to the

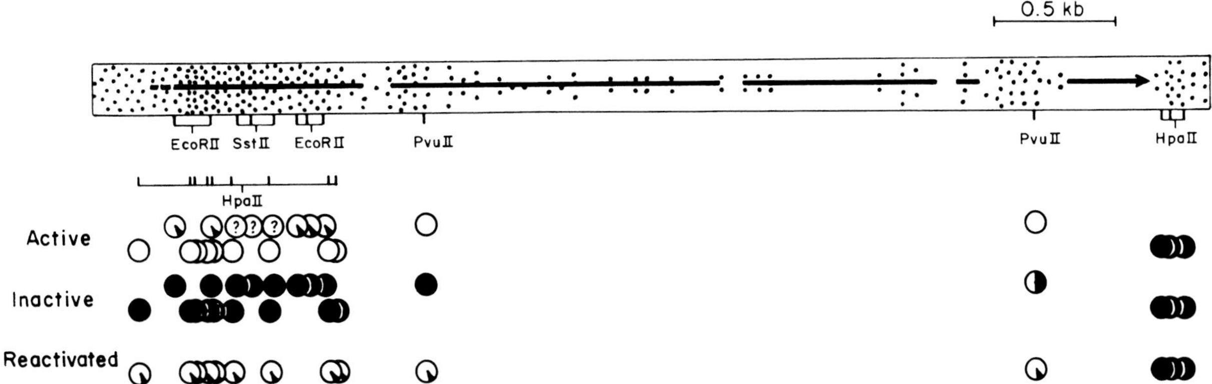

Figure 9. Methylation of several restriction sites in active and inactive *Ac* elements. The box represents the *Ac* element, and the dots represent the number of methylation-sensitive C residues (CGs and CNGs) in each 0.05-kb interval along the element's length. The arrow within the element represents the transcript, with the arrowhead corresponding to its 3′ end. The locations of *Eco*RII, *Sst*II, *Pvu*II, and *Hpa*II restriction sites are indicated below the diagram, and their methylation state in active, inactive, and reactivated *Ac* elements is represented by the circles below a given restriction site and aligned with it vertically. Filled, empty, and partially filled circles represent methylated, unmethylated, and partially methylated sites, as inferred from sensitivity to cleavage with the indicated enzyme. Open circles with question marks correspond to sites whose methylation state could not be determined.

dosage effect (28), nor does a reversibly inactivated, but genetically intact *Ac* element (90, 92).

The inverse relationship between the timing of somatic excision and the abundance of the element-encoded transcript indicates that the element is not regulated at the transcriptional level. Rather, it appears that high concentrations of the element-encoded gene product interfere with excision of the resident element. It is impossible to distinguish on the basis of existing evidence whether the inhibition is effected at the level of the element, possibly by reversible sequence modifications, or by interfering with the function of the element's *trans*-acting transposition function. Nonetheless, because the same intraelement deletions inactivate both the element-encoded transposition function and its apparent negative regulatory function, it is likely that the same element-encoded protein participates in both.

5. The *Ac* element can undergo reversible inactivation

The *Ac* element can be inactivated by a reversible genetic mechanism (90, 92). An inactive element can neither *trans*-activate excision of a *Ds* element, promote *Ds*-mediated chromosome breakage, nor contribute to the *Ac* dosage effect (90, 92). The inactive state of the element is readily distinguishable from a mutation to a *Ds* element by the ability of the element to return to an active state (91, 92).

Initial insight into the structural differences between genetically active and inactive *Ac* elements came from the observation that the single genetically active element in an *Ac*-containing strain could be distinguished structurally from the several other *Ac*-homologous sequences in the genome with restriction enzymes whose ability to cleave DNA was sensitive to methylation of C residues within the cleavage site (41). Subsequent studies have provided evidence that reversible inactivation of *Ac* is correlated with methylation of the element (16, 151).

Reactivation of a genetically inactive element is correlated with hypomethylation of sites near the element's 5′ end (16, 151), a region which is characterized by a high concentration of methylatable CG dinucleotides and CNG trinucleotides (Fig. 9). The results of recent studies suggest that only some methylation differences are correlated with the state of the element's activity, while others are subject to different methylation patterns when propagated through the pollen and egg parents (150). Schwartz and Dennis (151) also noted that reactivation of an inactive element may be a multistep process genetically, because kernels with somatic sectors containing reactivated *Ac* elements occur in sectors on ears of plants that commence development with an inactive element.

F. Effect of *Ac* and *Ds* Elements on Gene Structure and Function

1. *Ds* elements can be spliced from gene transcripts

Most, but not all, of the *Ac* and *Ds* mutations that have been studied have a null phenotype. One of the few exceptions is the *Ac wx-m9* allele and its *Ds* derivative, *wx-m9* (89). Both were initially distinctive because they gave a phenotype intermediate between the wild-type *Wx* and null *wx* phenotypes (89).

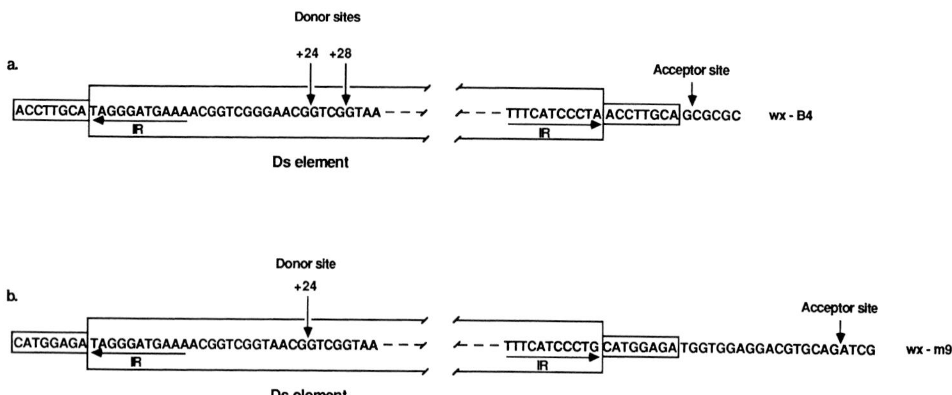

Figure 10. Splicing of the *Ds* sequence from *wx* gene transcripts. The interrupted box represents the *Ds* element inserted in the *wx* gene in the *wx-B4* and *wx-m9* alleles. The small boxes flanking the *Ds* element enclose the 8-bp target site duplication. The splice donor sites are located 24 and 28 bp from the 3′ end of the element as it is depicted in Fig. 4, and the acceptor sites lie just outside of the element in the *wx* gene sequence at the *Ds* insertion site.

Endosperm starch granule proteins from plants carrying the *wx-m9* allele contain small amounts of several electrophoretically distinguishable *Wx* proteins (152). RNA from such endosperm tissue contains large amounts of an mRNA that is homologous to the locus and is the same size as the gene's normal transcript (169).

Early indications that a *Ds* element could be spliced out of a transcript came from analyses of the transcripts made from an *adh* allele containing a *Ds1* insertion (110). Recently, cDNA clones were prepared from the mRNA isolated from endosperm tissue of plants carrying the *wx-m9* allele (169). The results of these studies show that the *wx* locus transcripts include aberrantly spliced mRNAs in which almost all of the inserted element is removed from the transcript. Analysis of similarly normal-length transcripts from a phenotypically null *Ds* mutant (*wxB4*) of the same locus by cDNA cloning has revealed a similar splicing pattern that results in the removal of most of the inserted element from the transcript (169). These observations have permitted the identification of several donor splice sites at the 3′ end of the *Ac* element. As illustrated in Fig. 10, donor sites at the end of the element are spliced to acceptor sites just outside of the element. Selection of the acceptor sites appears to be guided by homology of the element's terminus to the plant consensus splice acceptor sequence (169). Because there are three different splice donor sites at the element's 3′ end, different splicing patterns can and do occur in a population of transcripts, some of which restore the correct reading frame to mRNAs processed from a gene with an insertion in an exon. Wessler et al. (169) speculate that the splicing mechanism may have evolved to ensure some compensatory gene function in insertion mutants.

2. Protein structure is altered in stable alleles containing empty donor sites

As described in an earlier section, *Ac* elements leave molecular traces of insertion and excision events (Table 1). These are various imperfect versions of the target sequence duplication generated on insertion (122, 138, 168). Such imprecise excision events undoubtedly account for much of the variety noted many years ago in the phenotypes of stable derivatives that arise by excision of an element inserted within a gene (75, 84, 88). The *Ac* insertion site in the *Ac wx-m9* allele, for example, is in the 10th exon of the *wx* gene, which is expressed at a low level in the mutant because the element transcript is spliced out of the precursor (169). Somatic and germinal excisions of the resident element give phenotypes ranging from null to wild type (148). However, kernels with wild-type sectors or *Wx* germinal revertants produce a *Wx* protein with a slightly reduced electrophoretic mobility (152), and a cloned *Wx* revertant allele was found to have retained 6 bp of the original 8-bp duplication at the former *Ac* insertion site (122). Its *Wx* phenotype is probably attributable to the restoration of reading frame by the deletion of 2 bp of the original duplication during excision. Stable alleles with different phenotypes have not been analyzed at the molecular level, but probably reflect the effect on the *Wx* protein's primary structure of different types of empty donor site sequences.

VI. THE *Suppressor-mutator* ELEMENT FAMILY

The *Suppressor-mutator* (*Spm*) element family, like the *Ac-Ds* element family, comprises both auton-

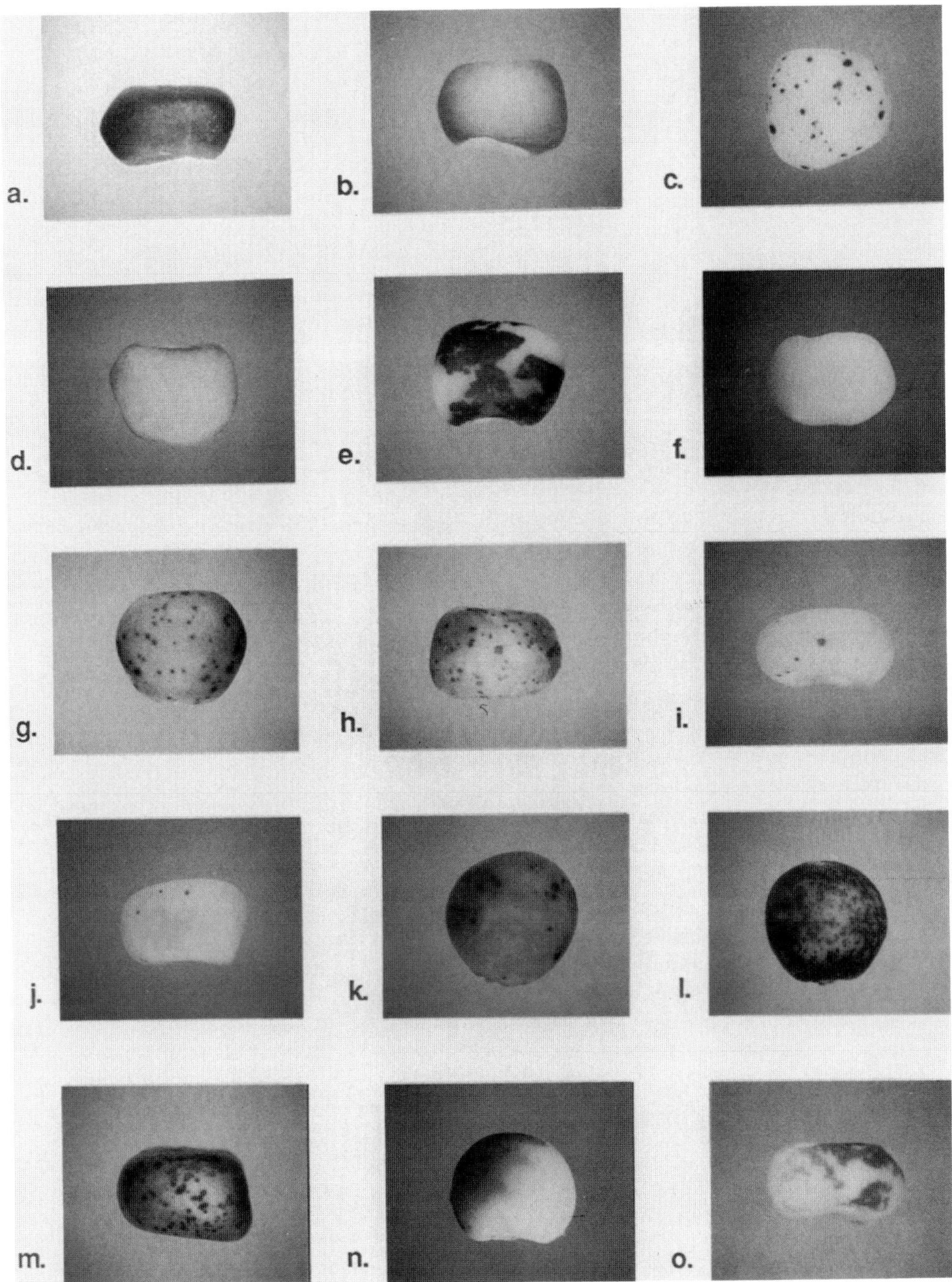

Figure 11. Kernels exhibiting the phenotypes associated with *Spm* insertion mutations. Kernels shown in the figure carry the wild-type *A* allele of the *a* locus (panel a) or derivatives of the *a-m1* (b through e) or *a-m2* (f through o) *Spm* insertion alleles of the locus. (a) *A*; (b) *a-m1-5719A1*, no *Spm-s*; (c) *a-m1-5719A1*, with *Spm-s*; (d) *a-m1-6078, no Spm-s*; (e) *a-m1-6078*, with *Spm-s*; (f) *a-m2-7995*, no *Spm-s*; (g) *a-m2-7995*, with *Spm-s*; (h) *a-m2-7977B*, with *Spm-s*; (i) *a-m2-8004*, with *Spm-s*; (j) *a-m2-8167B*, with *Spm-s*; (k) *a-m2-8011*, no *Spm-s*; (l) *a-m2-8011*, with *Spm-s*; (m) *a-m2-7991A1*; (n) *a-m2-7991A1* (*Spm-i* returning to active state); (o) *a-m2-7995*, preset pattern. (See color plates, this volume.)

omous and nonautonomous elements. McClintock designated the autonomous element of the family *Spm* (78), but did not give the nonautonomous element a separate name. Peterson (114) designated the autonomous element *Enhancer* (*En*) and the nonautonomous element *Inhibitor* (*I*). The nonautonomous elements have also been designated **receptor** elements (117) or **defective *Spm*** (***dSpm***) elements (4), as they will be called in the present text.

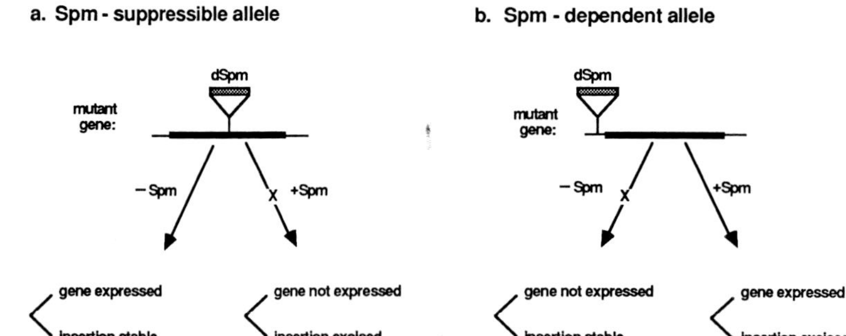

Figure 12. Diagrammatic representation of the interactions between the *Spm* element and genes with *dSpm* insertions.

A. Genetic Definition of *Spm* and *dSpm* Elements

The genetic properties of *Spm* and *dSpm* insertions are remarkably varied and have provided much insight into transposable element function. The element was named for a common type of *dSpm* insertion allele which shows a reduced but not a null phenotype in the absence of an autonomous element (Fig. 11a and b). Introduction of an autonomous *Spm* element into the same genome inhibits or suppresses expression of the gene with the *dSpm* insertion mutation and promotes reversion of the mutation, giving the characteristic variegated phenotype (Fig. 11c). The autonomous element was originally named for its ability to suppress expression of the gene in this type of mutant allele, as well as to promote mutation to the wild-type allele of the affected locus (78). Such alleles have been designated ***Spm-suppressible*** (Fig. 12a; 64). The corresponding functions of the element are the **suppressor** and **mutator (transposase)** functions (78).

Although most insertions of an *Spm* element disrupt gene expression, McClintock identified and studied a different type of *Spm* insertion allele at the *a* locus, in which gene expression was not completely interrupted by the insertion (75, 87). This mutant, designated the original *a-m2* allele, had an autonomous *Spm* element inserted at the locus in such a way that the element and gene were coexpressed (87). McClintock selected a number of spontaneous derivatives of this allele, in some of which the resident element had become transposition defective by virtue of a mutation in the element (88). Such *dSpm* derivatives of the original *a-m2* allele give a null phenotype in the absence of a *trans*-acting autonomous *Spm* element (Fig. 11f), but most express the *a* gene in the presence of an autonomous *Spm* element, which also promotes excision of the *dSpm* element from the locus (Fig. 11g through j). Such alleles have been designated ***Spm-dependent*** (64). Figure 12 summarizes diagrammatically the relationship between the mutant gene and the *trans*-acting *Spm* element for *Spm-dependent* and *Spm-suppressible* alleles.

Several genetically distinguishable forms of the autonomous *Spm* element have been recognized and given separate names. McClintock (79, 82, 83, 86, 89) referred to the standard autonomous element as an *Spm-s* to distinguish it from a type of derivative, designated a *weak Spm* (*Spm-w*), that *trans*-activates excision at a reduced frequency and later in development than an *Spm-s*. As described in subsequent sections, the *Spm* element can undergo reversible inactivation (83). The inactive form of the element has been designated *Spm-i*. Elements that alternate frequently between active and inactive phases have been called *cycling Spm* (*Spm-c*) elements, and very stably inactive elements have been designated *cryptic Spm*'s (*Spm-cr*) (38, 39, 42). McClintock (82) also identified an element, termed *Modifier*, that enhances *Spm*-activated excision, but only in the presence of another *Spm* element.

B. Structure of *Spm* (*En*) and *dSpm* Elements

1. *Spm* and *En* are almost identical small transposable elements

An *En* element cloned from the *wx* gene (109, 112) and an *Spm-s* element cloned from the *a* gene (4) have been sequenced and are virtually identical (64, 111). The 8.3-kb *Spm* and *En* elements differ by 4 nucleotides in length and at 6 additional single nucleotides. The element encodes a single major transcript which commences 0.2 kb from the element's 5′ end and terminates 0.4 kb from its 3′ end (111). The primary transcript is processed to yield a 2.4-kb mRNA with 11 exons (Fig. 13). The mRNA has an 0.4-kb untranslated leader, followed by a 4.4-kb intron. The transcript appears to encode a protein of 621 amino acids, commencing at an AUG codon in the second exon (111). There are two large, virtually

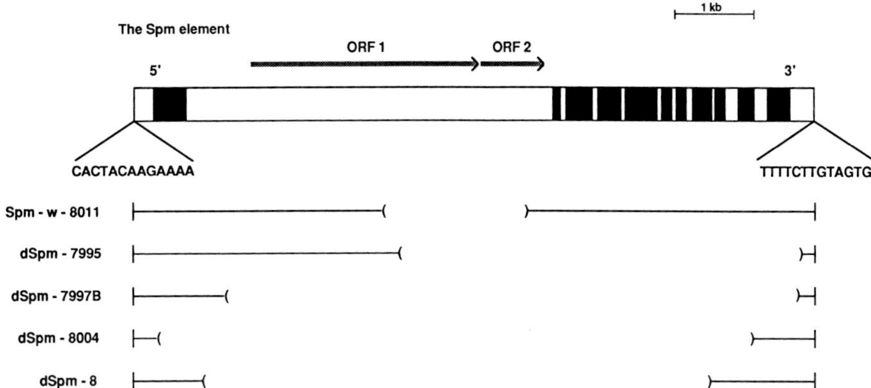

Figure 13. Structure of the *Spm* element. The *Spm* element is represented by the open box. The 5′ and 3′ ends of the transcription unit are indicated, and the left and right ends of the element are correspondingly designated the 5′ and 3′ ends. The filled areas between the ends of the transcription unit correspond to exons, and the empty areas correspond to introns. The two large ORFs within the element's major intron are represented by arrows over the element. Below the *Spm* diagram are interrupted lines representing the various *dSpm* elements whose designations are shown at the far left. The interruption in each line represents the internal sequence deleted from each *dSpm* element.

contiguous ORFs, 2,714 and 761 nucleotides long within the first intron (64, 111).

The *Spm* element's ends have striking structural features, depicted diagrammatically in Fig. 14. Its termini are 13-bp perfect IRs. Commencing within a few nucleotides of each terminal IR are the **subterminal repetitive regions**, which extend to a distance of 180 bp from the element's 5′ end and 299 bp from its 3′ end (46, 64, 155). The subterminal repetitive regions consist of several copies of a short sequence, repeated in both orientations (Fig. 14). Comparison of the repetitive elements yields the 12-bp consensus sequence CCGACACTCTTA, to which the various copies are 75 to 90% homologous (64, 155). As noted by Schwarz-Sommer et al. (155), the arrangement of the direct and inverted repeats is such that the element's ends can be drawn as an extended, interrupted stem-loop structure. In addition to the subterminal repetitive regions, the *Spm* element, like the *Ac* element, has a strikingly GC-rich sequence near its 5′ end that is also internally repetitive (38, 39, 64, 111).

2. *dSpm* elements are deleted *Spm* elements

By contrast to what was found in the *Ac-Ds* family, all of the *dSpm* elements that have so far been cloned and sequenced are closely related in structure to the *Spm* element. Almost all of them are deletion derivatives, and the sequences missing from several of those that have been analyzed are indicated in Fig. 13, together with the element designation.

C. Genetic Functions Encoded by the *Spm* Element

1. The mutator function of *Spm* is its transposase

The *Spm* element encodes a transposition function that promotes its own excision and transposi-

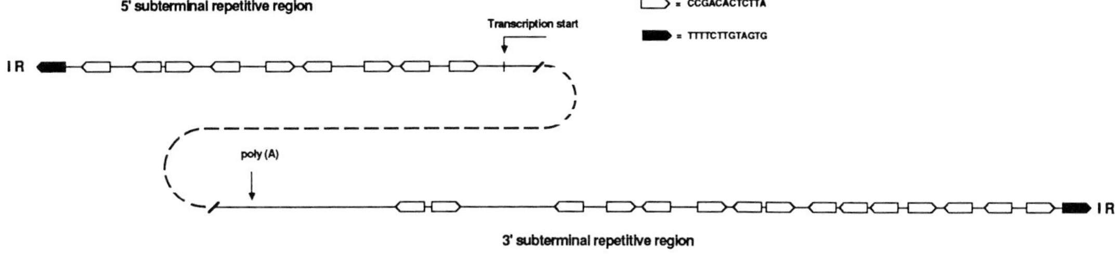

Figure 14. Structure of *Spm* ends. The repetitive structure of the ends of the *Spm* element is represented diagrammatically: top line, 5′ end; bottom line, 3′ end. The transcription start and poly(A) addition sites are indicated. Filled arrows represent the 13-bp inverted terminal IRs; open arrows represent the location and orientation of sequences with marked homology to the 12-bp consensus sequence shown in the diagram. The regions comprising the closely spaced repeats represented by open arrows are designated the subterminal repetitive regions.

tion, as well as that of defective elements (83, 87, 91). This function was originally designated the mutator function and is called the **transposase** in the present discussion. Information about the element sequence encoding the transposase has been obtained from studies on two types of internally mutant elements, *Spm-w* and *dSpm* elements. *Spm-w* elements comprise a relatively common type of mutant *Spm* element and show a reduction relative to the *Spm-s* element in their ability to promote their own transposition, as well as that of a *dSpm* element located elsewhere in the genome (compare Fig. 11k and l; 82, 83, 86, 89). The *Spm-w-8011* element has been analyzed at the molecular level and shown to have a 1.6-kb deletion within its 4.2-kb intron (Fig. 13; 64). The deletion eliminates parts of each ORF within the intron, implying that neither need be intact for expression of the transposase.

Maize plants carrying the *Spm-w* element contain less of the *Spm* transcript than plants carrying an *Spm-s* element (64). A deletion that overlaps those in the *Spm-w* element at the left (5′) end, but extends to the right (3′) end of the element, renders it transposition defective (*dSpm-7995*; Fig. 11g and 13). The most straightforward interpretation of these observations is that the element's transposase is encoded by its major transcript (64).

2. The suppressor function of *Spm* regulates gene expression

Perhaps the most striking property of the *Spm* transposable element family is its ability to control gene expression. Genes can be under either the positive or the negative control of the *Spm* element (Fig. 12; 64, 91). Moreover, mutations within the element can alter both the basal level of expression of a gene with an *Spm* insertion mutation and the pattern of gene expression resulting from excision of the element during plant development (79, 88–92, 116, 120). As molecular information about the mutant genes and *Spm* elements accumulates, it becomes increasingly evident that the properties of *Spm* insertion mutations provide insight into different aspects of transposable element function and regulation. In the ensuing discussion, the effects of *Spm* mutations on gene expression are analyzed from this perspective. The thesis is developed that insertion of the transposable element brings the gene under the control of the element's own regulatory mechanisms. Although the variety of interactions among *Spm* elements and between the elements and genes with insertion mutations has prompted proposals that there are two or more different element-encoded regulatory mechanisms (105), the hypothesis presented here is that the diverse effects are different manifestations of a single element-encoded regulatory function, initially designated the suppressor function. Moreover, it is proposed to be a positive regulatory function required both for expression of the element and for maintenance of the element in a genetically active state (38, 39, 64).

3. *Spm* insertions can affect gene expression in opposite ways

In an *Spm-suppressible* allele (Fig. 12), the *dSpm* insertion does not by itself interrupt gene expression completely, although most such alleles show reduced gene expression relative to a wild-type allele of the locus (Fig. 11b and d; 61, 92). Gene expression is inhibited completely, however, when a *trans*-acting *Spm* element is present together with the *Spm-suppressible* allele (Fig. 11c and e). By contrast, the gene of an *Spm-dependent* allele is expressed only in the presence of the *trans*-acting *Spm* element (Fig. 11f through j; 88). Thus the *trans*-acting *Spm* element provides a gene product that can either activate or inhibit expression of a gene containing a *dSpm* insertion. This implies that an element-encoded protein interacts with the element sequence inserted at the locus to alter the gene's ability to be expressed.

4. Sequences at element ends mediate *Spm* control of gene expression

Spm and *dSpm* alleles of the *wx*, *bz*, and *a* loci have been cloned and analyzed, as have wild-type alleles of each (37, 40, 64, 109, 144, 153, 155–157, 161). The structural information available from several studies is presented in diagrammatic form in Fig. 15. The various derivatives whose structures are shown in the figure arose from four independent insertion mutations designated the *bz-m13* allele of the *bz* locus (102), the *a-m1* and *a-m2* alleles of the *a* locus (75), and the *wx-m8* allele of the *wx* locus (87). The *bz-m13* and *a-m1* alleles are *Spm-suppressible*, while the *a-m2* alleles are *Spm-dependent*. The *wx-m8* allele has a null phenotype in both the presence and absence of an *Spm* element. The *dSpm* elements recovered from the loci range in length from 0.9 to 8.3 kb. Almost all represent deleted derivatives of the complete element. The *bz-m13*, *wx-m8*, and *a-m1-6078* alleles arose by insertion of the same 2.2-kb *dSpm* element.

Elements ends contain the sequences that mediate the *Spm* element's activation or inhibition of gene

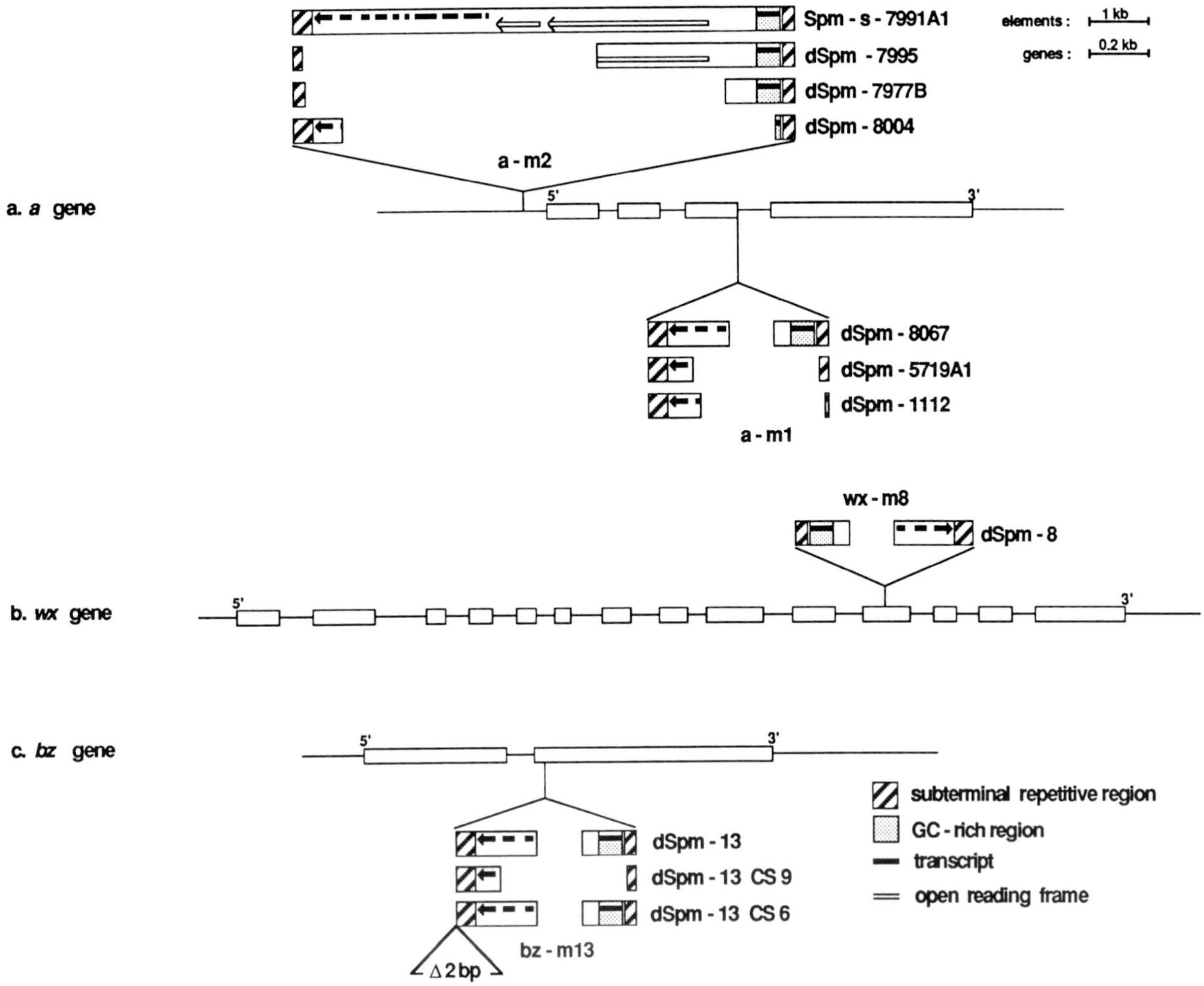

Figure 15. *Spm* insertions at the *a*, *wx*, and *bz* loci. The structure of the transcription unit is shown for the *a*, *wx*, and *bz* genes in parts a, b, and c; open blocks represent the exons, and the lines between them correspond to introns of the respective genes. The location and structure of several *Spm* insertions relative to each of the transcription units are represented above and below the corresponding gene diagram for the *a-m1*, *a-m2*, *wx-m8*, and *bz-m13* alleles. The top diagram in each set of insertions corresponds to the element inserted in the original mutant allele bearing the corresponding designation. This was a complete *Spm* element, in the case of the *a-m2* alleles, and was the 2.2-kb element designated *dSpm-8* in Fig. 13 for the *wx-m8*, *a-m8*, *a-m1*, and *bz-m13* alleles. The subterminal repetitive regions of *Spm* are represented by the shaded boxes at element ends, and its GC-rich region is represented by the stippled box. The exon structure of the transcript is represented by the broken, filled arrows, while the ORFs are represented by open arrows. The portion of these structural features retained in each deleted derivative is indicated. The gap in the 2.2-kb *dSpm-8* element corresponds to a 6.1-kb deletion. The diagrams on the lower lines of each set represent the structure of elements recovered from derivatives showing altered element functions, as discussed in the text. All those shown in the present diagram have further internal deletions in the inserted element, and the portion of the original element retained in the derivative is shown.

expression. This conclusion follows from the observation that alleles with short, extensively deleted elements retain the overall properties of the original mutant allele (64, 91, 143, 161). As will be discussed subsequently, some structural changes within the element affect its response to the *trans*-acting element in subtle ways. Nonetheless, all of the alleles with deleted elements depicted in Fig. 15a and c retain the basic *Spm-suppressible* or *Spm-dependent* character of the original mutation.

5. The location of the *dSpm* insertion determines how *Spm* controls the gene

In both the *Spm-suppressible* and *Spm-dependent* alleles depicted in Fig. 15a and c, the orientation of the element's transcription unit is opposite to that of the gene. Since the difference between *Spm-suppressible* and *Spm-dependent* alleles is determined neither by the precise element sequence inserted nor by the orientation of the insertion, it appears likely that it is determined by the location of the insertion relative to the transcription unit of the gene. In *Spm-suppressible* alleles of both loci, the element is inserted within an exon of the gene, while the insertion site is 99 bp upstream of the transcription start site of the *a* gene in the *Spm-dependent a-m2* alleles (156). Thus the inserted element behaves as an *Spm*-dependent enhancer sequence for the *a* gene of the *a-m2* alleles, but inhibits gene expression in the presence of *Spm* when inserted within the coding sequence.

6. The effect of mutations on the *Spm* suppressor function

Despite its 1.6-kb internal deletion, the *Spm-w-8011* element (Fig. 13) retains the ability to activate expression of *Spm-dependent* alleles and inhibit expression of *Spm-suppressible* alleles (87, 94). McClintock concluded from the original genetic observations that the suppressor function of *Spm* was responsible for its ability to activate expression of *Spm-dependent* alleles, as well as its ability to inhibit expression of *Spm-suppressible* alleles. This conclusion is supported by the observation that intraelement deletions that render the element transposition defective simultaneously eliminate its ability to affect expression of genes with *dSpm* insertions (64).

7. Is there a single element-encoded gene product?

Although the ability of the *Spm* element to *trans*-activate transposition is functionally separable from its ability to influence expression of genes with *dSpm* insertions, there is no evidence that the functions can be separated by mutations or other genetic changes that affect element expression. Mutations to the *Spm-w* phenotype reduce transposition frequency (Fig. 11k), but clearly do not disrupt the transposase- or suppressor-coding sequences (64, 83, 84). All mutations isolated to date that render the element transposition defective simultaneously eliminate its ability to control expression of genes with *dSpm* insertions (64, 91).

McClintock reported that *Spm* elements can undergo reversible changes in genetic activity (83–85, 87, 88, 95). An inactive element can neither transpose, *trans*-activate transposition, nor exert either positive or negative control over expression of genes with *dSpm* insertions. It is functionally distinguishable from a *dSpm* element by its ability to return to an active form, either spontaneously or in the presence of another *Spm* element (83). Some *Spm* isolates undergo multiple changes in activity during development; these have been designated *cycling Spm* (*Spm-c*) elements (see reference 42 for illustrative kernel; 84, 85). Inactivation and reactivation of the element simultaneously affect both its transposase function and its ability to control expression of genes with *dSpm* insertion mutations (64, 83, 88). These observations raise the possibility that the element's transposition and its suppressor functions are associated with or require a common element-encoded gene product (64).

8. *dSpm* elements can be spliced out of a gene transcript

Further insight into the molecular basis of the ability of *Spm* to influence gene expression requires an understanding of the mechanism that permits genes with *dSpm* insertions to be expressed. Transcripts produced by genes with the same *dSpm* element inserted in opposite orientations are markedly different (46, 60). The same 2.2-kb *dSpm* element is responsible for the *Spm-suppressible bz-m13* allele of the *bz* locus (Fig. 15c) and the *wx-m8* allele of the *wx* locus (Fig. 15b). Unlike the *bz-m13* allele, the *wx-m8* allele exhibits a null phenotype in both the presence and absence of an *Spm* element. Both insertions are within exons of the respective genes, but the elements are inserted in opposite orientations. In the *wx-m8* allele, the orientation of the element's transcription unit is the same as that of the gene. The 2.2-kb *dSpm* element responsible for both mutations contains the element's transcription initiation and termination sequences.

Analysis of *wx* locus transcripts made from the *wx-m8* allele in the absence of an autonomous *Spm* indicates that transcription begins at the gene's normal transcription start site, but terminates prematurely within the element, probably at the element's termination site (46). By contrast, normal-length transcripts of the mutant *bz* locus of the *bz-m13* allele are relatively abundant (60). cDNA copies of the mutant gene have been cloned and sequenced, revealing that the element sequence is spliced from the primary transcript. The splice joins a donor site at an exon/intron boundary in the gene to an acceptor site in the element's 5′ terminus (Fig. 16; 60). The splice removes all but 2 nucleotides of the element's sequence, as well as 38 nucleotides of the gene, but restores the original reading frame. The element-mediated splicing reaction alters the structure of the protein-coding sequence. This appears to account for the observation that the flavonoid glucosyltransferase encoded by the mutant allele has a reduced thermal stability (60).

Although analogous studies on transcripts of the *Spm-suppressible a-m1* alleles of the *a* locus have not been done, the orientation of the insertion relative to

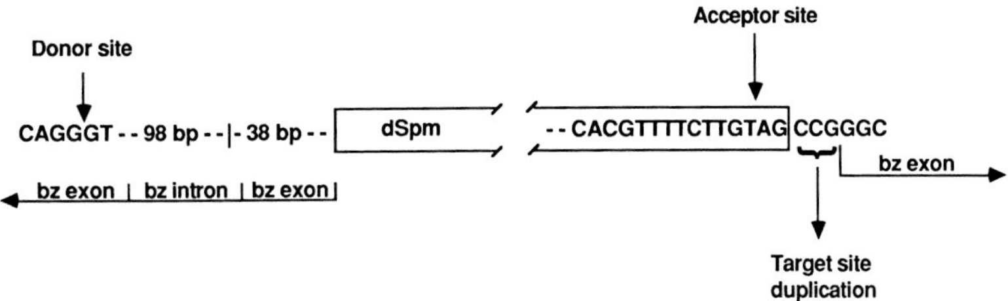

Figure 16. Splicing of a *dSpm* element from a *bz* gene transcript. The 0.9-kb *dSpm* element inserted at the *bz* locus in the *bz-m13* CS9 allele is eliminated from the *bz* gene transcript by splicing of the normal splice donor site, at the 5′ end of the gene's only intron, to an acceptor sequence 2 bp from the end of the inserted *dSpm* element, represented by the box.

the transcription unit is the same as it is in the *bz-m13* mutation (Fig. 15a; 161). Thus gene expression in *Spm-suppressible* alleles may generally be attributable to splicing of the element sequences from the mutant gene's transcript. In some of her initial studies on derivatives of the original *a-m1* allele, McClintock noted that both the basal level of gene expression and the pattern of somatic reversion could be altered by the same mutation (Fig. 11b and d; 79, 91). The original allele arose by insertion of a 2.2-kb *dSpm*. Derivatives, first designated states of the locus, often arise by further deletions within the element (Fig. 15a; 161). The 0.9-kb *dSpm* element in the *a-m1-5719A1* allele (Fig. 11b and c), for example, was derived from the 2.2-kb *dSpm* element in the *a-m1-6078* allele (Fig. 11d and e) by an internal deletion that removes the sequence surrounding the element's transcription start site and extends into the subterminal repetitive region upstream of it (Fig. 15a; 161). The *a-m1-5719A1* allele shows a higher level of *a* gene expression than does the *a-m1-6078* allele (Fig. 11b and d). Such changes may affect either the amount or structure of the transcript.

9. The *dSpm* element of *Spm*-suppressible alleles interferes with gene expression

Because an *Spm* element effects only the expression of loci with *dSpm* insertions, its ability to inhibit gene expression must be mediated by an interaction between an element-encoded gene product and the element sequences at the locus. Although the *wx-m8* allele of the *wx* locus does not exhibit the suppressible phenotype, the element at the locus is the same as that inserted in the *Spm-suppressible bz-m13* allele. Analysis of the *wx* locus transcripts made from this allele suggests that the aberrant transcript that terminates within the element made in the absence of a *trans*-acting *Spm* either is not synthesized or does not accumulate in the presence of the element (46). Thus an interaction between the element-encoded gene product and the element sequence may act at the transcriptional level either directly to block transcription or indirectly to prevent initiation of transcription. Alternatively, the interaction may be at the RNA level, serving to destabilize the transcript.

As noted earlier, the element sequences that mediate suppression of gene expression are near element ends. All deletion derivatives of the *bz-m13* and *a-m1* alleles continue to exhibit suppression of gene expression in the presence of an *Spm* element (143, 161). Since the element ends are internally repetitive, it may be that there are binding sequences for the element's suppressor function at both element ends.

10. The *dSpm* element of the *Spm*-dependent *a-m2* alleles behaves as an enhancer sequence

By contrast to what is observed with *Spm-suppressible* alleles, the *dSpm* insertion in the *a-m2* alleles exhibits the genetic behavior of an *Spm*-dependent enhancer sequence (64, 88). The structure of several different *dSpm* derivatives is represented diagrammatically in Fig. 13 and 15, and the corresponding phenotypes are shown in Fig. 11f through j. Element ends suffice to mediate the positive effect of *Spm* on gene expression (64). The capacity of the various deleted elements to mediate activation of *a* gene expression is not affected by deletion of internal element sequences extending to within 200 bp of the element's 3′ end. However, deletion of sequences between 0.3 and 1.1 kb from its 5′ end, but not including its transcription initiation site, markedly reduces its ability to activate expression of the adjacent gene (*dSpm-8004* element; Fig. 11i and 15a).

The element sequences that are required to activate expression of the *a* gene are probably located at the element's 5′ end. Similar levels of gene expression are observed with elements ranging in length from 8.3 to 1.3 kb. In alleles with a full-length element, the element's 3′ end and the gene's normal upstream sequences are located at a distance of more than 8 kb from the gene's transcription start site. It is therefore likely that the sequences that mediate activation of the adjacent *a* gene are near the transcription initiation site of *Spm* at its left end.

The element's transcription start site and the 0.2-kb subterminal repetitive regions retained in the *dSpm-8004* element suffice to activate expression of the gene (Fig. 11i and 15; 64). However, sequences downstream from the element's transcription start site, including the GC-rich first exon and part of its first intron, are required for maximal expression of the adjacent *a* gene (Fig. 11 and 15). Hence sequences at the element's transcription start site and within its transcription unit may well serve as *Spm*-dependent promoter and enhancer elements for the adjacent *a* gene.

11. *Spm*-dependent activation of the *a* gene reveals an *Spm*-encoded positive regulatory mechanism

The ability of an *Spm* element to activate expression of the *a* gene in the *a-m2* alleles precisely parallels its ability to reactivate a complete but inactive *Spm* element. McClintock (83, 95) initially reported that an inactive *Spm* element is active in the presence of an active element, but segregates as an inactive element in progeny. Similar observations have been made with an *Spm-i* derivative of an *a-m2-7991A1* allele with a complete *Spm* inserted at the locus (38, 39, 64). As illustrated in Fig. 11m and n, inactivation of the resident *Spm-s* element of the *a-m2-7991A1* allele is accompanied by loss of *a* gene expression. Both element activity, as judged by excision of the element, and *a* gene expression are inactivated and reactivated simultaneously in this allele (Fig. 11n; 64, 88). Both the *Spm* element and the *a* gene of the *a-m2-7991A1* allele can be reactivated by an *Spm-w* element elsewhere in the genome (38, 39).

Assuming that element control of *a* gene expression is mediated by the same protein-element sequence interactions, it follows that the molecular studies on the *a-m2* alleles identify the sequences responsible for the element's ability to regulate its own expression (64). These include sequences upstream of the transcription start site, as well as an enhancerlike sequence within the transcription unit, located between the transcription start site and 1.1 kb from the element's 5′ end (38).

Taken together, the foregoing observations suggest that the *Spm* element's ability to both activate and inhibit expression of genes with *dSpm* insertion mutations reflects the operation of an element-encoded positive regulatory mechanism. Both activation of *Spm-dependent* alleles and inhibition of *Spm-suppressible* alleles are mediated by an interaction between an element-encoded gene product and sequences at element ends. The difference in the consequences of such interactions for expression of the mutant gene appears to be attributable to differences in the location of the element with respect to the gene. Thus the element function initially termed the suppressor appears to be its positive regulatory gene product.

12. Coding sequences for the *Spm* transposase and positive regulator may be the same or overlap

It is not yet clear whether the element's positive regulatory function is associated with or requires the same gene product as its transposition function. It is a persistent observation that the element's ability to affect expression of *dSpm* alleles is genetically inseparable from its ability to promote excision. Perhaps the most direct molecular indication for a close association of the activities involved in regulation and transposition comes from the analysis of intraelement deletions (64, 144, 153, 161).

McClintock (79) reported that mutations affecting the timing and frequency of *dSpm* excision occur only in the presence of an autonomous *Spm* element. Many such mutant elements arose by intraelement deletions (Fig. 15). The implication is that an element-encoded gene product is involved in the origin of the deletions. The endpoints of intraelement deletions are commonly located in or at one end of a sequence with homology to the consensus repeat of the subterminal repetitive region (Fig. 14; 64, 144, 161). This implies that the element-encoded protein responsible for deletion formation recognizes the subterminal repeat sequence. The subterminal repetitive region comprises almost all of the 0.2-kb sequence upstream of the element's transcription initiation site (Fig. 14), and the sequences that mediate *Spm*-dependent gene expression in the *a-m2* alleles are also within 275 bp of the element's 5′ end (64). These observations suggest the possibility that a single element-encoded polypeptide recognizes and cleaves element ends in the normal transposition process and binds to internal sequences to promote transcription of the element (64).

D. Regulation of *Spm* Expression

1. *Spm* expression is subject to negative control

As initially reported by McClintock, the *Spm* can undergo a reversible change in its ability to be expressed (83–85, 87, 88, 95). An inactive element (*Spm-i*) is distinguishable from a defective element in two ways. It is able to return to an active state spontaneously, and it can be transiently reactivated by an active element (95). McClintock's initial inference that *Spm* elements interact was based on the phenotypes observed with an *Spm-suppressible* allele in the presence of *Spm* elements that differed markedly in the duration of active and inactive phases (83, 95). These studies revealed the inactivation and reactivation processes to be complex genetic phenomena.

Inactive elements differ widely in the stability of the inactive state, as well as the developmental pattern of element reactivation (85). Some derivatives, designated *cycling Spm* (*Spm-c*) elements (42), undergo inactivation and reactivation so frequently that the plant is mosaic for element expression. Using an *Spm-suppressible* allele of the *a2* locus, McClintock (95) found that an *Spm-c* element of this type exhibits a dosage effect. As the number of copies of the element increases, the area of the kernel exhibiting the phenotype associated with the inactive element decreases (see reference 42 for illustrative kernels). McClintock interpreted this to mean that the elements undergo inactivation independently, with the result that the probability that all of the resident elements are simultaneously inactive decreases with increasing element copy number.

By contrast to the *Spm-c* element, other inactive elements (*Spm-i*) remain inactive throughout much of the plant's life cycle, returning to an active state infrequently and either just premeiotically, to give occasional kernels exhibiting element activity, or postmeiotically, to give kernel sectors with an active *Spm* (Fig. 11n). Despite the relative genetic stability of its inactive state, an *Spm-i* element contributes to the above-described dosage effect in the same way as an *Spm-c* element does (95). The *Spm-i* element segregates as an inactive element in progeny, indicating that its state of inactivation is not heritably altered. McClintock concluded from these observations that the *Spm-c* element, which is initially active during plant development, activates the *Spm-i* element when both are present in the same plant (95).

McClintock (82) also identified a transposable element, which she designated *Modifier*, whose properties are similar to that of an *Spm-i* element. The *Modifier* element enhances the ability of either an *Spm-w* or an *Spm-s* to promote excision of a *dSpm* element, but it has no independent effect on a *dSpm* element. That is, it exhibits the properties of an active element, but only in the presence of another active element. A *Modifier* element differs from an *Spm-i* element in its inability to return to a stably active form. The *Modifier* therefore resembles an *Spm* element which has a mutation that affects its ability to maintain itself in a active state.

Finally, *Spm* elements can exist in a stably inactive state, designated *cryptic* (*Spm-cr*), which is

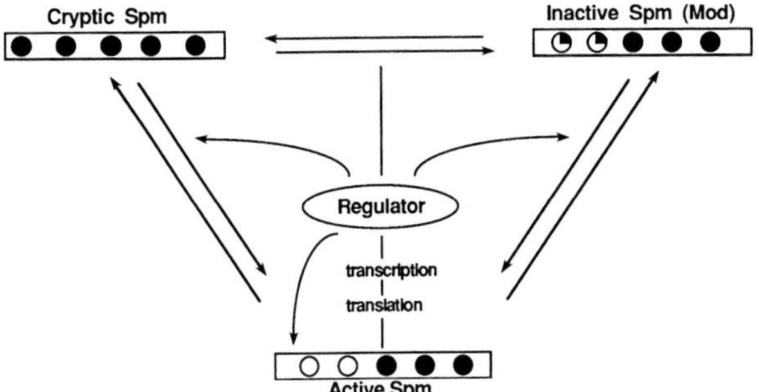

Figure 17. Diagrammatic summary of the regulatory interactions among *Spm* elements. The putative regulatory gene product is termed the *Regulator*. Active, inactive, and cryptic elements are represented as differing in the methylation pattern of the GC-rich regulatory region (see Fig. 18), depicted by the two leftmost circles. The methylation state is represented by filled circles for fully methylated, open circles for unmethylated, and partially filled circles for partially methylated. The element-encoded regulatory gene product acts both to activate expression of an inactive element (or a *Modifier* element, designated Mod) and to maintain expression of the active element. The same gene product is believed to promote the heritable conversion of cryptic and inactive elements to active elements.

distinguishable from the foregoing states by its stability and its response to a *trans*-activating element (36, 38, 39). Unlike an *Spm-i* element, an *Spm-cr* is not invariably, but only occasionally active in the presence of an active element. Nonetheless, an *Spm-cr* shows evidence of interacting with an active element. An active element enhances the frequency with which an *Spm-cr* element is heritably activated (36). Although the *Spm-i* element tends to return to the inactive state when it segregates away from the active element, passage through the heterozygote increases the probability that it will remain active (unpublished data).

Overall, the *Spm* element exists in three genetically distinguishable states of activity (Fig. 17). The fully active and the cryptic states are quite stable. Both the inactivation of an active element and the reactivation of a cryptic one are rare events. As discussed below, reactivation of a cryptic element can be increased by a variety of physical and genetic manipulations. By contrast, the class of elements designated inactive is quite unstable. Elements of this class exhibit the intriguing genetic property of existing in a variety of distinguishable, relatively heritable states that differ in the developmental timing and frequency of element reactivation. As a class, the *Spm-i* forms are distinguishable genetically from the cryptic form by their full activity in the presence of an active element. The genetic and developmental characteristics of each type of element, and what is known about the genetic and molecular mechanisms responsible for the different states of element activity, are considered in the following several sections.

2. *Spm* elements exhibit heritable developmental patterns of inactivation and reactivation

Spm-c and *Spm-i* elements differ in the developmental timing of element inactivation and reactivation. The *Spm-c* type of element is active during most of development, undergoing late inactivation events and very occasional reactivation events (95; see illustrative kernels in reference 42). The *Spm-i* elements remain inactive during most of the developmental cycle, undergoing occasional late reactivation events. In general, a given *Spm-i* derivative has a characteristic frequency and timing of element reactivation. These are heritable but fairly labile properties of the *Spm-i* derivative. Moreover, *Spm-i* elements can exhibit different states of activity in different parts of the same plant and even in different areas of a single tissue. For example, a plant with an *Spm-i* element commonly exhibits extensive reactivation of the element in tiller ears while maintaining the element's inactive state in both ears produced on the main stalk (unpublished data).

Striking nonclonal patterns of element activity and inactivity are exhibited by two types of *En* alleles studied by Peterson (118). In the *En-crown* and *En-flow* alleles, the *En* element is active only in the top or crown of the kernel and in the lower portion of the kernel, respectively. Moreover, it is a common observation that kernels with a reactivated *Spm-i* element are clustered at the base and the tip of an ear. Since the ear's developmental pattern is such that development of the base and tip is somewhat retarded relative to that of the central portion of the ear, it may be that such differences are determined, once again, by developmental timing cues.

3. Cryptic and inactive elements can be heritably reactivated

McClintock's studies on transposable elements grew from genetic experiments in which cryptic transposable elements were activated (65–68). She carried out crosses in which a broken chromosome 9 was contributed by each parent, with the result that the broken homologs not only fused to create dicentrics, but also interacted with other chromosomes to produce a variety of chromosomal rearrangements (65, 66, 97). Among the progeny of these plants, McClintock observed numerous unstable mutations. Later, both McClintock and Doerschug showed that the same genetic configuration could stimulate the activation of the *Dotted* element, representing a family of elements not activated in the initial experiments in McClintock's cultures (27, 73, 75).

Other investigators have shown that agents that directly induce chromosome damage, such as UV, X-ray, and gamma irradiation, also activate cryptic elements (8, 11, 104, 166). Recent studies on maize tissue culture document the progressive accumulation of cells with damaged and rearranged chromosomes (63). Plants regenerated from cultured cells exhibit mutations of all types and frequently contain newly activated transposable elements (113). The results of such experiments have established a clear correlation between chromosome damage and activation of cryptic transposable elements, although the connections are not necessarily causal.

Active *Spm* elements promote the genetic activation of both cryptic and inactive elements. The genetic distinction between a cryptic and an inactive element resides in its response to an active *Spm* element. An *Spm-i* element can be reactivated in all progeny kernels by introducing a *trans*-acting *Spm* element, while only a small fraction, commonly no more than 0.1 to 1%, of kernels with an *Spm-cr* element exhibit activation of the element (36, 38, 39). Transactivated elements tend to segregate as inactive elements in progeny (36, 95; unpublished data). However, some of the *trans*-activated elements remain heritably active, whether they originate from an *Spm-cr* or an *Spm-i* (36; unpublished observations).

4. *Spm* can heritably alter element and gene expression

The genetic interactions among active, inactive, and cryptic *Spm* elements indicate that the element's positive regulatory mechanism has both an immediate and a heritable effect on element expression. Not only can an active element transiently activate an inactive or a cryptic one, but it also promotes the conversion of both to a heritably active form, as described above. Given the parallel between *Spm* control of its own expression and expression of the *a* gene of the *a-m2* alleles, these observations may illuminate a puzzling property of certain *a-m2* alleles. This property, termed **presetting** by McClintock (89, 90, 92, 93), is exhibited by the *a-m2-7995* and *a-m2-7977B* alleles (Fig. 11g and h; Fig. 15a).

Both alleles have a colorless phenotype when propagated continuously in the absence of an *Spm* element, indicating that the *a* gene is not expressed. In the presence of an *Spm* element, both exhibit an intermediate level of *a* gene expression as well as deeply pigmented revertant sectors (Fig. 11g and h). However, when either allele is exposed to an *Spm* element during plant development, it retains the ability to express the *a* gene for some time after the loss or meiotic segregation of the element away from the locus. An example of the preset expression of the *a* gene is shown in Fig. 11o. Kernels showing preset expression commonly exhibit pigmented sectors, as if the kernel commenced development able to express the *a* gene but the gene progressively lost the capacity to be expressed during development.

McClintock used the term presetting to designate the early participation of the *Spm* element in predisposing the gene to later expression. Preset expression does not require the continuous presence of the element throughout development. Preset expression of the *a* gene occurs even when the element is lost or inactivated during development, giving rise to a large plant or ear sector lacking element activity. The level of gene expression is the same within a kernel, but differs from kernel to kernel, implying that the event or setting that determines it occurs at meiosis. Preset patterns occur only when a plant carrying an allele capable of being preset commences development with an active *Spm*, but they are generally not heritable, implying that the setting can be erased at or before a subsequent meiosis (90, 91). Occasional kernels on ears grown from plants exhibiting preset expression of the *a* gene show a similar pattern of *a* gene expression; most show the colorless phenotype characteristic of the alleles in the absence of an *Spm* element.

The significance of presetting is twofold. First, it shows that the *a* gene of the *a-m2* alleles can be expressed in the absence of the *Spm* element under certain circumstances. This suggests that the inactivity of the *a* gene is not simply a function of the *dSpm* insertion, but involves an additional inactivating mechanism. Second, it implies that an *Spm*-encoded gene product not only *trans*-activates expression of the gene, but can interfere with the inactivating mechanism in a way that is heritable through a number of cell divisions after the element is segregated away from the mutant gene, either mitotically or meiotically. Presetting of the *a* gene by an *Spm* element parallels the *Spm* element's ability to promote the heritable activation of cryptic and inactive *Spm* elements and may be a consequence of the same molecular mechanisms (64).

Presetting is not unique to the *a-m2* alleles. McClintock (93) studied presetting using *dSpm* insertions at the *c2* locus. The results of these studies strengthen the parallels between presetting of a gene with a *dSpm* insertion and the mechanism that controls the activity of the element. McClintock (93) observed, for example, that a genetic event or setting that permits expression of the gene in sporophytic tissue can be erased at meiosis, so that kernels exhibit no sign of gene expression, nor do plants grown from the kernels. Studies on elements undergoing inactivation and reactivation show a parallel tendency to be reset at meiosis, giving progeny with a new characteristic pattern of element expression. Moreover, preset expression of the gene can be markedly different in ears on the main stalk and tillers of a given plant, as can the state of element activity.

5. *Spm* inactivation is accompanied by methylation of element sequences

The foregoing summary of genetic studies on *Spm* inactivation reveals the operation of a unique kind of genetic mechanism. Its characteristics are that the element can exist in a large number of relatively heritable states between the two extremes of the fully active state and completely silent, cryptic state. Each state is characterized by the developmental timing and frequency of the transition from the active to the inactive state, or vice versa. A given state is relatively but not completely heritable and can be altered or reset during plant development and at meiosis.

The results of recent studies show that element inactivation is correlated with the methylation of sequences near the transcription start site (39). The element's first intron is extremely GC rich relative to the remainder of the element (Fig. 18). Not surprisingly, the CG dinucleotide and CNG trinucleotide sequences that are methylated in plant DNA are common in this region (56, 64, 111).

The results of studies on the *a-m2-8167B* allele, which contains a full-length *Spm* element, indicate that methylation of sequences in the GC-rich region of the element shown in Fig. 18 can have the same phenotypic consequences as deletion of the sequence (39). As is evident in Fig. 11i and j, the *a-m2-8167B* allele exhibits a phenotype very similar to that of the *a-m2-8004* allele. It shows a low level of *Spm*-dependent *a* gene expression, as well as a low excision frequency. The element is structurally indistinguishable from an *Spm* element (64), but differs from it by its inaccessibility to digestion by methylation-sensitive restriction enzymes within the GC-rich region (39). These results are summarized diagrammatically in Fig. 18.

Evidence that methylation of sequences within the GC-rich region is correlated with the element's state of genetic activity has been obtained from studies on plants with an *Spm-i* element at the *a* locus (39). Although these studies are as yet preliminary, the results show a clear correlation between increased methylation of the GC-rich region and decreased element expression. Outside of the GC-rich region, the element appears to be extensively methylated regardless of whether it is genetically active or inactive (Fig. 18), although adjacent sites in the *a* locus are not methylated (39).

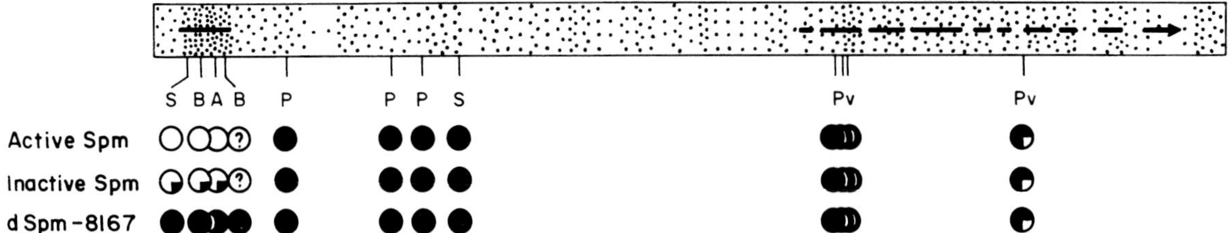

Figure 18. Methylation of several restriction sites in active and inactive *Spm* elements. The box represents the *Spm* element, and the dots represent the number of methylatable C residues (CGs and CNGs) in each 0.1-kb interval along its length. The broken arrow within the element represents the exon structure of its transcription unit. The locations of cleavage sites for the methylation-sensitive enzymes *Sal*I (S), *Bgl*I (B), *Ava*I (A), *Pst* (P), and *Pvu*II (Pv) are indicated below the diagram. The filled, open, and partially filled circles correspond to methylated, unmethylated, and partially methylated sites, as judged by susceptibility to cleavage by each enzyme. The circles vertically aligned with each site show the methylation state of each site in active and inactive elements, as well as in the *dSpm-8167 B* element. Open circles with a question mark represent sites whose methylation state is not known.

The hypothesis that the *Spm* element is inactivated by methylation within the GC-rich region is attractive because it is compatible with the unusual inheritance pattern of inactive elements. Differences in the developmental timing and frequency of reactivation among different types of inactive elements might reflect the extent and precise pattern of methylation within the GC-rich region, for example. Both the heritability and the reversibility of a given state of inactivity are compatible with the known mechanisms of inheritance of methylation patterns. In Fig. 17, the cryptic state is depicted as completely methylated, while the inactive state is depicted as only partially methylated.

6. Genetic basis of *Spm* inactivation

The genetic basis of the *Spm* inactivation mechanism is not understood. Although the ability to activate an *Spm-i* is a property of an active element, there is no comparable information about the genetic factors required for element inactivation. The ability to inactivate the resident element shows some heritability (unpublished data). Moreover, among progeny of a plant in which the resident element becomes inactive, some are capable of silencing active elements introduced by an appropriate genetic cross. This ability is correlated with a high stability of the inactive state of the resident *Spm* element. Whether this is attributable to genes unrelated to *Spm* elements or with the presence of defective and inactive *Spm* elements in a given genome is not readily apparent.

7. *Spm* autoregulates and reprograms element expression

The observations that an active *Spm* can activate an inactive one imply that *Spm* encodes a positive regulator of its own expression (Fig. 17). The parallels between *Spm* control of *a* gene expression in the *Spm-dependent a-m2* alleles and its control of element expression suggest that both reflect the operation of the same element-encoded regulatory mechanism. Sequences that mediate *Spm*-dependent *a* gene expression include sequences upstream of the element's transcription initiation site, as well as a downstream enhancerlike sequence that includes the element's GC-rich untranslated first exon (64). This, in turn, suggests that the *Spm* regulator interacts with sequences at and near the transcription start site to maintain transcriptional activity of its promoter. Active and inactive elements differ in the extent of methylation of the GC-rich enhancer.

The resemblance between presetting of *a* gene expression by *Spm* and its ability to promote the heritable reactivation of inactive and cryptic elements offers some insight into the responsible element sequences. Presetting is observed with *dSpm* elements in the *a-m2* series that retain at least the GC-rich region near the element's 5′ end (*dSpm-7995* and *dSpm-7977B*), but not with elements that either lack the GC-rich region (*dSpm-8004*) or exhibit extensive methylation of the GC-rich region (*dSpm-8167B*). Thus the region of the element within which activity-associated changes in methylation pattern have been observed is the same region that mediates preset expression of the *a* gene. Analysis of the presetting phenomenon indicates that the *Spm* element exerts its influence on the gene with a *dSpm* insertion early in plant development, possibly by promoting the reversal of a previously programmed genetic setting. *Spm*-independent molecular events later in development or at meiosis impose a new setting for later expression of the gene, and these settings can in turn be erased at or after meiosis. What this means in molecular terms is not yet known, but the reversibility of such genetic settings and the necessity for the element's GC-rich region for presetting are consistent with the involvement of methylation events.

Presetting reveals the operation of a system that can "reprogram" gene expression in development in the same way that the activity of inactive *Spm* elements can be reprogrammed. Assuming that both involve the same molecular events, then presetting contributes the insight that the *Spm* element acts early in development to promote later reprogramming of the element's subsequent state of activity.

If the changes in element expression are effected by changes in the methylation pattern of the element's GC-rich region, then the study of presetting and *Spm-i* elements implicates a dynamic genetic system capable of altering preestablished methylation patterns during development, as well as at meiosis. Moreover, the element's ability to promote directed heritable changes in element expression suggests that its regulatory gene product participates in the process either directly or indirectly. One possibility is that the interaction of the element-encoded regulator with element sequences to promote activation of element expression generally interferes with the maintenance of the preexisting methylation pattern during replication, either directly or indirectly, by virtue of increased transcription of the element. Other possibilities are that an element-encoded gene product induces or activates an enzymatic pathway involved in demethylation and de novo methylation of DNA sequences, or that it promotes specific changes in the methylation patterns by interacting with element sequences.

E. Excision and Transposition of the *Spm* Element

1. *Spm* probably transposes by a nonreplicative mechanism

Although studies of *Spm* transposition have not been as extensive as those of *Ac* transposition, the available information suggests that they transpose by similar nonreplicative mechanisms. The *Spm* element does not exhibit the marked dosage-dependent difference in the timing and frequency of excision that facilitated studies on *Ac* transposition (91). However, there is a relationship between element copy number and variegation pattern in the special case of the *Spm-c* element (95). Twinned sectors revealing differences in *Spm* copy number appear on kernels carrying an *Spm-c* element and an appropriate *dSpm* allele (95). The phenotypes of the twinned sectors suggest that an element lost from one daughter cell at a mitotic division is gained by the other daughter cell. The existence of twinned sectors implies that a transposition event has removed the element from only one of the two sister chromatids at a mitotic division and that the reinserted element subsequently segregated at mitosis with the chromosome bearing the nontransposed element, as has been documented for the *Ac* element (Fig. 5).

McClintock analyzed plants grown from kernels in which the *Spm* element had excised from the *a* locus of the original *a-m2* allele, reporting that although 40% of the stable mutants still had one or more *Spm* elements in the genome, less than 10% retained the *Spm* element at or near the donor site (88). Peterson (119) reported transposition of the element to a recipient site on the same chromosome as the donor site in about 25% of progeny in which the element had transposed.

2. Empty donor sites are heterogeneous in sequence

The *Spm* element generates a 3-bp duplication on insertion and excises by the same type of imperfect mechanism as do other plant transposable elements (154, 155). Several empty donor sites have been sequenced and are shown in Table 1. Although it has not been reinvestigated directly, McClintock's early observation that excision of the *Spm* element from the *a-m2* alleles commonly gave stable mutants with null, pale, or mottled phenotypes is probably attributable to the small variations in the structure of the empty donor site associated with different excision events (64, 87, 88).

3. *Spm* elements have *cis* determinants of transposition frequency

McClintock reported that mutant alleles with *dSpm* insertions frequently undergo what she termed "changes of state" (79, 83, 84, 94). Each change is a heritable mutation that alters the timing and frequency of excision of the resident *dSpm* element in the presence of a *trans*-acting *Spm* (Fig. 11c and e; also see references 42 and 91 for additional illustrative kernels). Several such derivatives have been cloned and sequenced (64, 144, 153, 161). Their structure provides insight into element sequences involved in the *cis* determination of excision (and presumably transposition) frequency.

Sequences that influence the ability of a *dSpm* element to excise are located near and at element ends. The same 2.2-kb *dSpm* element has been cloned from the *wx*, *bz*, and *a* loci, from all of which it excises early in development and at a high frequency (91, 143). The element comprises 0.8 kb from the 5′ end and 1.3 kb from the 3′ end of the *Spm* sequence and constitutes a simple deletion derivative of the original element (46, 144, 153).

A number of derivatives of the 2.2-kb *dSpm* element showing reduced excision frequency have been cloned from both the *a* and the *bz* loci (79, 91, 144, 153, 161), as have several such derivatives of the *Spm* element in the *a-m2* alleles (64). Almost all of the mutations that reduce transposition frequency are caused by intraelement deletions, most of which remove sequences at one end or the other of the element. One of the altered elements recovered from the *bz* locus has a terminal 2-bp deletion (Fig. 15; 144). The deletion reduces but does not eliminate transposition, indicating that although the element termini are important in excision, some sequence changes are tolerated.

The suggestion that the subterminal repetitive regions are *cis* determinants of excision frequency was made by Schwarz-Sommer et al. (153) on the basis of the observation that an element with a deletion extending to within 0.2 kb of the element's 5′ end showed a reduced excision frequency (*dSpm-5719A1*; Fig. 11c and 15). However, a deletion derivative of the original *a-m2* allele that retains all of the subterminal repetitive regions, but still exhibits a low excision frequency, was subsequently identified (*dSpm-8004* in Fig. 15a; 64). Sequence comparisons of different *dSpm* elements have permitted the localization of the 5′ *cis* determinant of transposition frequency outside of the subterminal repetitive region in the interval between 275 and 860 bp from the element's 5′ end. It should be noted that the element sequence that serves as an enhancer of *a* gene expression in the *a-m2* alleles is located in the same region of the *Spm* element (64).

The subterminal repetitive regions of the element may also be or contain sequences directly involved in excision and transposition. Elements with deletions extending into the subterminal repetitive region at the 3′ end of the element exhibit a reduced excision frequency (64, 143, 144). However, the possibility of additional important sequences located just outside of the 3′ subterminal repetitive region has not been eliminated.

4. Methylation of element sequences reduces excision frequency

Although almost all of the *dSpm* elements that exhibit a reduced excision frequency have sustained deletions, two exceptional elements have been described that appear to have no structural alteration relative to the parent element (39, 64, 143; V. Raboy, J. Schiefelbein, and O. Nelson, personal communication). In both cases, the element has been found to be methylated at sites within the GC-rich region (Fig. 18). The reduction in excision frequency is genetically reversible (unpublished data).

In summary, analysis of mutations that reduce transposition frequency has revealed an unexpected complexity in the factors that determine the frequency. Not only do the element's terminal IRs and subterminal repetitive regions appear to be involved, but so does a region within the element's transcription unit. Moreover, similar reductions in transposition frequency result from either deletion or extensive methylation of certain internal element sequences. It is a striking and perhaps significant observation that methylation of the same region of the element reduces the element's ability both to be expressed and to be excised by the transposase supplied by a *trans*-acting element. Since the methylated sequence is far from the terminal IRs involved in transposition, it appears unlikely that methylation directly interferes with excision and transposition. These observations suggest that there may be a requirement for element transcription for its transposition or that common binding protein may be involved in both element-specific transcription and transposition.

VII. THE *Mutator* SYSTEM

The *Mutator* (*Mu*) system of maize was initially identified by virtue of its ability to increase the spontaneous mutation rate by a factor of 50 or more (132). Many of the mutations that arose in plants exhibiting the *Mutator* trait were unstable, suggesting the involvement of transposable elements. In subsequent molecular studies, a family of related transposable elements has been identified using the insertions recovered from unstable mutant alleles identified from plants with *Mutator* activity (8, 109, 158). Although the overall properties of the system are much less well understood than those of the *Ac-Ds* and *Spm* transposable element families, there are both similarities and differences of considerable interest.

A fundamental property of the *Mu* system is its non-Mendelian inheritance pattern. This is manifested in two ways. First, 90% of the progeny obtained from a cross between a *Mu* parent and a non-*Mu* parent exhibit the mutator trait (132). Second, the trait is reversibly lost in about 10% of the outcross progeny or by inbreeding of plants with *Mutator* activity (133, 134). Considerable progress has been made recently in understanding the genetic and molecular bases of these odd inheritance patterns.

As will be detailed below, several elements with common structural features have been recovered

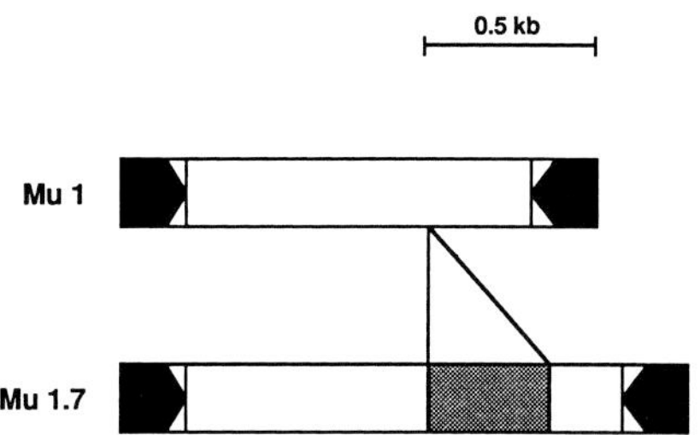

Figure 19. Structure of the *Mu1* and *Mu1.7* elements. The open boxes represent the elements, and the filled portions represent the extended terminal IRs. *Mu1* is 1,367 bp long and *Mu1.7* is 1,745 bp long. The IRs are 0.2 kb in length. The *Mu1* and *Mu1.7* elements are similar in sequence; *Mu1.7* is longer than *Mu1* by a 385-bp internal sequence, as indicated.

from *Mu* plants. Although it has not been possible as yet to demonstrate that any one of them is an autonomous element analogous to the *Ac* and *Spm* elements, all appear to be similarly affected by the genetic mechanisms that are responsible for the manifestations of the *Mutator* trait and are therefore members of a transposable element family.

A. Structure of *Mu* Elements

The insertion sequence that has been recovered most frequently from the unstable mutant alleles arising in *Mu* stocks has been designated *Mu1* (Fig. 19; 6, 10, 158). It is 1.4 kb in length and has the structural features of a small transposable element. It has 0.2-kb IRs that are 95% homologous and is flanked by 9-bp direct repeats. Although there are several short ORFs in the element's sequence (6), their significance is unclear since it is not known whether *Mu1* can transpose autonomously. The *Mu1* element, and a closely related but slightly longer element designated *Mu1.7* (Fig. 19), are present in multiple copies (10 to 100) in the genomes of *Mu* plants, but in few copies in non-*Mu* plants (10, 13). The *Mu1* element is more abundant than the *Mu1.7* element by a factor of 2 to 30 in different mutator stocks (13). The results of sequence comparisons show the elements to be similar and suggest that the 1.4-kb element was derived from either the 1.7-kb element or a longer one by an internal deletion (162). Several additional elements with ends that are similar to those of *Mu1* and *Mu1.7* have recently been recovered from unstable mutant alleles of genes for which probes exist, as well as by direct cloning of sequences homologous to the short elements (V. L. Chandler, personal communication). In these elements, the *Mu* IRs flank unrelated internal sequences. The significance of the various members of the *Mu* element family is not known, since a relationship among elements like that defined for autonomous and nonautonomous members of the *Ac-Ds* and *Spm* element families has not been demonstrated.

B. Transposition of *Mu* Elements

The observation that a large fraction of the progeny plants obtained upon outcrossing of *Mutator* plants exhibit the trait suggested the involvement of multiple genetic elements (132). The further observation that many mutations are unstable and attributable to *Mu* insertions suggests that these elements are responsible for the *Mutator* trait (9, 132, 134, 135, 165). Furthermore, the inheritance pattern of the *Mu* elements parallels that of the *Mutator* trait (1).

When plants containing 10 to 50 copies of *Mu* elements are outcrossed, progeny plants are found to have approximately the same number of copies (1). Moreover, it is generally observed that preexisting insertion sites are partitioned as would be expected for Mendelian traits, while the additional elements represent new insertion sites. These observations suggest that the *Mu* transposition mechanism is replicative and differs from the nonreplicative mechanism deduced for the *Ac* and *Spm* elements. A physical basis for replicative transposition of the *Mu* element family was provided by the discovery that the elements can exist in a circular, nonchromosomal form. Both *Mu1* and *Mu1.7* elements exist as free, supercoiled circles (159).

C. Modification of *Mu* Elements

The identification of unstable *Mu* insertions in genes in the anthocyanin pigment pathway has facilitated studies on the loss of instability observed in some outcrosses and during inbreeding of stocks exhibiting the *Mutator* trait. The loss of instability of a *Mu* insertion mutation is often, but not always, associated with increased levels of methylation of *Mu* element sequences throughout the plant's genome (9, 14). The *Mu* elements in plants exhibiting instability of the insertion mutation are hypomethylated with respect to other genomic sequences, such as the histone genes (166). It has been estimated that essentially all of the methylatable C residues are modified in histone genes, while only about half of the methylatable C residues in the *Mu* sequences of a *Mutator* stock are modified. Loss of instability of a *Mu* insertion mutation is associated with complete methylation of element sequences (166).

The stabilization of *Mu* insertion mutations due to methylation can be reversed either by appropriate crosses to mutator lines or by gamma-irradiation (9, 135, 137, 165, 166). Reacquisition of the unstable phenotype is associated with a decrease in the extent of methylation of *Mu* sequences throughout the genome.

D. Relationship between the *Mutator* Trait and *Mu* Element Instability

Recent experiments have addressed the relationship between the *Mutator* trait, as measured by a high germinal mutation frequency, and the somatic instability of a *Mu* insertion mutation at the *a* locus (136). Although the losses of somatic and germinal instability commonly occur together, this is not invariably the case. A high germinal mutation rate has been observed in some lines that have lost somatic instability of the *Mu* insertion allele, and some lines exhibit a high somatic instability together with a low germinal mutation rate.

Perhaps the clearest association between the genetic and molecular observations is that between element methylation and loss of somatic instability of a *Mu* insertion mutation (165). In view of the observations made in the *Spm* element system, that element sequence modifications can affect both an element's ability to be expressed and its ability to be excised by a *trans*-acting element, it is difficult to interpret the lack of complete correlation between somatic instability of a *Mu* insertion mutation and the mutator trait. In general, it appears likely that *Mu* elements are responsible for the high germinal mutation rate that defines the *Mutator* trait. However, until the element or gene responsible for *trans*-activating transposition in the *Mu* system is identified, this conclusion remains tentative.

VIII. USING TRANSPOSABLE ELEMENTS

A. Isolation of Maize Genes

Maize transposable elements belonging to all three of the families discussed here have been used in cloning maize genes. The first gene isolated using a transposable element was the *bz* locus (37), and its facile isolation demonstrated the feasibility of using elements to clone a gene whose product had not been identified, but which was marked by the insertion of a cloned and characterized transposable element. The efficiency of the initial experiments in transposon tagging was attributable to the structural identity and low copy number of the *Ac* element. Several strategies have since been devised to extend the utility for gene isolation of transposable elements whose genomic redundancy is higher than that of the *Ac* element.

The first major advance, described by O'Reilly et al. (109), was the use of two different elements to identify a set of genomic DNA fragments with homology to each element from different maize strains carrying corresponding element insertions at the *a* locus. The cloned fragments were subsequently cross-hybridized to identify those with homology to each other. This permitted the use of both *Spm* and *Mu*, the use of each of which alone is complicated by its high genomic redundancy. An additional improvement in using transposable elements for gene isolation developed from the observation that most sequences homologous to *Spm* are methylated in maize genomic DNA, while some sites within active elements are not (19). This has permitted the use of methylation-sensitive restriction endonucleases in the identification and isolation of genomic DNA fragments corresponding to a gene with an active element (19, 145). Although the high redundancy of the *Mu* element reduces its utility in gene isolation, its usefulness has recently been improved by combining the use of the element to identify *Mu*-containing genomic DNA fragments cloned from a strain with an insertion mutation at the *bz2* locus with the further screening of the cloned fragments for homology to mRNAs present in strains with a wild-type allele of the locus, but not in strains from which the gene was deleted (98).

B. Transposon Tagging of Maize Genes

The recent rapid progress in isolating maize genes with transposable element probes has been possible because of the availability of genetically well-characterized insertion mutations at many loci. The mutant alleles were identified and studied over many decades by two generations of classical geneticists. The future ability to use transposable elements to identify and clone genes involved in maize development and traits of agronomic value, such as disease resistance, depends on the efficiency with which new insertion mutations can be produced and identified genetically. Knowledge of how a given transposable element moves can substantially enhance recovery of transposable element mutations in a gene of interest. It has been reported for both *Mu* and *Spm* elements that the frequency of heritable transposition events is substantially higher through the male germ line than through the female (20, 102, 134). The careful studies of Greenblatt (53) revealed that *Ac* transposes most frequently to nearby sites on the same chromosome on one side of the donor site. Major differences have been noted in the overall frequency of germinal mutations for the different transposable element system studied to date (reviewed in reference 20). These may reflect differences in transposition mechanism as well as target site specificity.

C. Maize Transposable Elements Are Mobile in Other Plants

The observation that the maize *Ac* element transposes in tobacco cells when introduced on an *Agrobacterium* Ti plasmid has opened the possibility of using the well-characterized maize elements to tag genes in plants other than maize (3). The *Ac* element excises from the donor site on the T-DNA in up to 70% of transformed tobacco cells, implying not only that the element is efficiently expressed and its transcript is properly processed in a dicot, but that the dicot can provide any additional factors that are required for transposition of the maize element (2). The *Ac* element is also mobile in carrot and *Arabidopsis thaliana*, two other commonly used laboratory plants (164). These observations open the possibility of analyzing transposable elements by in vitro mutagenesis techniques.

Several different procedures have been used to introduce *Ac* into plant cells. These include *Agrobacterium*-mediated introduction of an *Ac*-containing nontumorigenic T-DNA by cocultivation of regenerating protoplasts with the bacteria, as well cotransformation of intact plants with the T-DNAs of an *Ac*-containing Ti plasmid and an *Agrobacterium rhizogenes* Ri plasmid (2, 3, 164). In addition, the *Ac* element has been introduced into tobacco cells by DNA transformation of protoplasts (22).

The success of transposon tagging in plants into which maize transposable elements can be introduced will be enhanced by further advances in the control of transposition frequency, as well as refinements in the techniques for distinguishing between insertion mutations and other genetic changes that are often associated with the passage of plants through tissue culture.

IX. TRANSPOSABLE ELEMENTS AND EVOLUTION

Molecular studies on mutations caused by maize transposable elements have revealed that they contribute to genetic change at both fine and coarse structural levels. The insertions themselves can either abolish or, in essence, reprogram gene expression in development, and a chromosomal site that has experienced an insertion rarely returns to its previous sequence when the element transposes away. The mutations that result from the insertion and excision of transposable elements contribute to sequence diversity at the relatively fine resolution of a few base pairs (Table 1). More extensive structural rearrangements of all types have also been associated with active transposable elements (21, 73, 97). Moreover, maize elements have the capacity to act in *trans* to promote excision and transposition of defective elements, as well as intraelement deletions and chromosomal rearrangements with endpoints at homologous insertions. Thus a single active element can make extensive contributions to genetic change throughout the genome.

An important recent insight into transposable element functions is that maize elements can undergo reversible, methylation-associated changes that affect both their ability to express *trans*-acting functions and their ability to move in response to transposase produced by *trans*-acting elements. Equally important is the realization that active elements can inactivate themselves by intraelement deletions. Thus, the mutagenic effects of transposable elements are kept in check by both reversible and irreversible mechanisms, some and perhaps even all of which are element encoded. There is increasing evidence that elements interact with each other to influence both immediate expression levels and the heritability of expression (64). And there is ample evidence that inactive elements can be activated by a variety of agents and manipulations that perturb genome structure. Thus, transposable elements increasingly emerge as a dynamic, environmentally responsive system of genetic change and genome reorganization.

LITERATURE CITED

1. **Alleman, M., and M. Freeling.** 1986. The *Mu* transposable elements of maize: evidence for transposition and copy number regulation during development. *Genetics* **112:** 107–119.
2. **Baker, B., G. Coupland, N. Fedoroff, P. Starlinger, and J. Schell.** 1987. Phenotypic assay for excision of the maize controlling element *Ac* in tobacco. *EMBO J.* **6:**1547–1554.
3. **Baker, B., J. Schell, H. Lörz, and N. V. Fedoroff.** 1986. Transposition of the maize controlling element *Activator* in tobacco. *Proc. Natl. Acad. Sci. USA* **83:**4844–4848.
4. **Banks, J., J. Kingsbury, V. Raboy, J. W. Schiefelbein, O. Nelson, Jr., and N. Fedoroff.** 1985. The *Ac* and *Spm* controlling element families in maize. *Cold Spring Harbor Symp. Quant. Biol.* **50:**307–311.
5. **Barclay, P. C., and R. A. Brink.** 1954. The relation between Modulator and Activator in maize. *Proc. Natl. Acad. Sci. USA* **40:**1118–1126.
6. **Barker, R. F., D. V. Thompson, D. R. Talbot, J. Swanson, and J. L. Bennetzen.** 1984. Nucleotide sequence of the maize transposable element *Mu1*. *Nucleic Acids Res.* **12:** 5955–5967.
7. **Behrens, U., N. Fedoroff, A. Laird, M. Müller-Neumann, P. Starlinger, and J. Yoder.** 1984. Cloning of *Zea mays* con-

trolling element *Ac* from the *wx-m7* allele. *Mol. Gen. Genet.* **194**:346–347.

8. **Bennetzen, J. L.** 1984. Transposable element *Mu1* is found in multiple copies only in Robertson's mutator maize lines. *J. Mol. Appl. Genet.* **2**:519–524.
9. **Bennetzen J. L.** 1985. The regulation of *Mutator* function and *Mu1* transposition. *UCLA Symp. Mol. Cell. Biol.* **35**: 343–354.
10. **Bennetzen, J. L., J. Swanson, W. C. Taylor, and M. Freeling.** 1984. DNA insertion in the first intron of maize *Adh1* affects message levels: cloning of progenitor and mutant *Adh1* alleles. *Proc. Natl. Acad. Sci. USA* **81**:4125–4128.
11. **Bianchi, A., F. Salamini, and F. Restaino.** 1969. Concomitant occurrence of different controlling elements. *Maize Genet. Coop. Newsl.* **43**:91.
12. **Brink, R. A., and R. A. Nilan.** 1952. The relation between light variegated and medium variegated pericarp in maize. *Genetics* **37**:519–544.
13. **Chandler, V. L., C. J. Rivin, and V. Walbot.** 1986. Stable non-*Mutator* lines of maize have elements homologous to the *Mu1* transposable element. *Genetics* **114**:1007–1021.
14. **Chandler, V. L., and V. Walbot.** 1986. DNA modification of a maize transposable element correlates with loss of activity. *Proc. Natl. Acad. Sci. USA* **83**:1767–1771.
15. **Chen, J., I. M. Greenblatt, and S. L. Dellaporta.** 1987. Transposition of *Ac* from the *P* locus of maize into unreplicated chromosomal sites. *Genetics* **117**:109–116.
16. **Chomet, P. S., S. Wessler, and S. L. Dellaporta.** 1987. Inactivation of the maize transposable element *Activator* (*Ac*) is associated with DNA modification. *EMBO J.* **6**: 295–302.
17. **Coe, E. H., Jr., and M. G. Neuffer.** 1977. The genetics of corn, p. 111–223. *In* G. F. Sprague (ed.), *Corn and Corn Improvement.* American Society for Agronomy, Madison, Wis.
18. **Coen, E. S., R. Carpenter, and C. Martin.** 1986. Transposable elements generate novel spatial patterns of gene expression in *Antirrhinum majus. Cell* **47**:285–296.
19. **Cone, K. C., F. A. Burr, and B. Burr.** 1986. Molecular analysis of the maize anthocyanin regulatory locus C1. *Proc. Natl. Acad. Sci. USA* **83**:9631–9635.
20. **Cone, K. C., R. J. Schmidt, B. Burr, and F. A. Burr.** 1988. Advantages and limitations of using *Spm* as a transposon tag, p. 149–160. *In* O. E. Nelson (ed.), *Plant Transposable Elements.* Plenum Publishing Corp., New York.
21. **Courage-Tebbe, U., H.-P. Döring, N. Fedoroff, and P. Starlinger.** 1983. The controlling element *Ds* at the *Shrunken* locus in *Zea mays*: structure of the unstable *sh-m5933* allele and several revertants. *Cell* **34**:383–393.
22. **Czernilofsky, A. P., R. Hain, B. Baker, and V. Wirtz.** 1986. Studies on the structure and functional organization of foreign DNA integrated into the genome of *Nicotiana tobacum. DNA* **5**:473–482.
23. **Dellaporta, S. L., P. S. Chomet, J. P. Mottinger, J. A. Wood, and S. M. Yu.** 1984. Endogenous transposable elements associated with virus infection in maize. *Cold Spring Harbor Symp. Quant. Biol.* **49**:321–328.
24. **Doerschug, E.** 1968. Transpositions of Dt_1. *Maize Genet. Coop. Newsl.* **42**:22.
25. **Doerschug, E.** 1968. Stability of Dt_1. *Maize Genet. Coop. Newsl.* **42**:24–26.
26. **Doerschug, E.** 1968. Activation cycles of Dt^{TB}. *Maize Genet. Coop. Newsl.* **42**:26–28.
27. **Doerschug, E. B.** 1973. Studies of Dotted, a regulatory element in maize. *Theor. Appl. Genet.* **43**:182–189.
28. **Dooner, H., J. English, E. Ralston, and E. Weck.** 1986. A single genetic unit specifies two transposition functions in the maize element *Activator. Science* **234**:210–211.
29. **Dooner, H. K., E. Weck, S. Adams, E. Ralston, and M. Favreau.** 1985. A molecular genetic analysis of insertions in the *bronze* locus in maize. *Mol. Gen. Genet.* **200**:240–246.
30. **Doring, H.-P., M. Freeling, S. Hake, M. A. Johns, and R. Kunze.** 1984. A *Ds* mutation of the *Adh1* gene in *Zea mays* L. *Mol. Gen. Genet.* **193**:199–204.
31. **Doring, H.-P., and P. Starlinger.** 1986. Molecular genetics of transposable elements in plants. *Annu. Rev. Genet.* **20**: 175–200.
32. **Doring, H.-P., E. Tillmann, and P. Starlinger.** 1984. DNA sequence of the maize transposable element *Dissociation. Nature* (London) **307**:127–130.
33. **Emerson, R. A.** 1914. The inheritance of recurring somatic variation in variegated ears of maize. *Am. Nat.* **48**:87–115.
34. **Emerson, R. A.** 1917. Genetical studies of variegated pericarp in maize. *Genetics* **2**:1–35.
35. **Emerson, R. A.** 1929. The frequency of somatic mutation in variegated pericarp of maize. *Genetics* **14**:488–511.
36. **Fedoroff, N.** 1986. Activation of *Spm* and *Modifier* elements. *Maize Genet. Coop. Newsl.* **60**:18–20.
37. **Fedoroff, N., D. Furtek, and O. Nelson, Jr.** 1984. Cloning of the *Bronze* locus in maize by a simple and generalizable procedure using the transposable controlling element *Ac. Proc. Natl. Acad. Sci. USA* **81**:3825–3829.
38. **Fedoroff, N., P. Masson, and J. Banks.** 1987. Regulation of the maize *Suppressor-mutator* element, p. 63–70. *In* M. E. Lambert, J. F. McDonald, and I. B. Weinstein (ed.), *Eukaryotic Transposable Elements as Mutagenic Agents.* Cold Spring Harbor Laboratory, Cold Spring Harbor, N.Y.
39. **Fedoroff, N., P. Masson, J. Banks, and J. Kingsbury.** 1988. Positive and negative regulation of the *Suppressor-mutator* element, p. 1–16. *In* O. E. Nelson (ed.), *Plant Transposable Elements.* Plenum Publishing Corp., New York.
40. **Fedoroff, N., M. Shure, S. Kelly, M. Johns, D. Furtek, J. Schiefelbein, and O. Nelson, Jr.** 1984. Isolation of *Spm* controlling elements from maize. *Cold Spring Harbor Symp. Quant. Biol.* **49**:339–345.
41. **Fedoroff, N., S. Wessler, and M. Shure.** 1983. Isolation of the transposable maize controlling elements *Ac* and *Ds. Cell* **35**:243–251.
42. **Fedoroff, N. V.** 1983. Controlling elements in maize, p. 1–63. *In* J. Shapiro (ed.), *Mobile Genetic Elements.* Academic Press, Inc., New York.
43. **Fincham, J. R. S., and G. R. K. Sastry.** 1974. Controlling elements in maize. *Annu. Rev. Genet.* **8**:15–50.
44. **Freeling, M.** 1984. Plant transposable elements and insertion sequences. *Annu. Rev. Plant Physiol.* **35**:271–298.
45. **Friedemann, P., and P. A. Peterson.** 1982. The *Uq* controlling element system in maize. *Mol. Gen. Genet.* **187**:19–29.
46. **Gierl, A., Z. Schwarz-Sommer, and H. Saedler.** 1985. Molecular interactions between the components of the En-I transposable element system of *Zea mays. EMBO J.* **4**: 579–583.
47. **Gonella, J. A., and P. A. Peterson.** 1977. Controlling elements in a tribal maize from Columbia: Fcu, a two unit system. *Genetics* **85**:629–645.
48. **Greenblatt, I. M.** 1966. Transposition and replication of modulator in maize. *Genetics* **53**:361–369.
49. **Greenblatt, I. M.** 1968. The mechanism of modulator transposition in maize. *Genetics* **58**:585–597.
50. **Greenblatt, I. M.** 1974. Proximal-distal polarity of Modula-

tor transpositions upon leaving the P locus. *Maize Genet. Coop. Newsl.* **48**:188–189.

51. **Greenblatt, I. M.** 1974. Modulator: a modifier of crossing over. *Maize Genet. Coop. Newsl.* **48**:189–191.
52. **Greenblatt, I. M.** 1974. Movement of Modulator in maize: a test of an hypothesis. *Genetics* 77:671–678.
53. **Greenblatt, I. M.** 1984. A chromosomal replication pattern deduced from pericarp phenotypes resulting from movements of the transposable element, *modulator*, in maize. *Genetics* **108**:471–485.
54. **Greenblatt, I. M., and R. A. Brink.** 1962. Twin mutations in medium variegated pericarp maize. *Genetics* **47**:489–501.
55. **Greenblatt, I. M., and R. A. Brink.** 1963. Transposition of Modulator in maize into divided and undivided chromosome segments. *Nature* (London) **197**:412–413.
56. **Gruenbaum, Y., T. Naveh-Many, H. Cedar, and A. Razin.** 1981. Sequence specificity of methylation in higher plant DNA. *Nature* (London) **292**:860–862.
57. **Hehl, R., N. S. Shepherd, and H. Saedler.** 1985. DNA sequence homology among members of the *Cin1* repetitive DNA family in maize and teosinte. *Maydica* **30**:199–207.
58. **Johns, M. A., J. Mottinger, and M. Freeling.** 1985. A low copy number, *copia*-like transposon in maize. *EMBO J.* **4**:1093–1102.
59. **Kiesselbach, T. A.** 1951. A half-century of corn research. *Am. Sci.* **39**:629–655.
60. **Kim, H.-Y., J. W. Schiefelbein, V. Raboy, D. B. Furtek, and O. E. Nelson, Jr.** 1987. RNA splicing permits expression of a maize gene with a defective *Suppressor-mutator* transposable element in an exon. *Proc. Natl. Acad. Sci. USA* **84**: 5863–5867.
61. **Klein, A. S., and O. E. Nelson, Jr.** 1983. Biochemical consequences of the insertion of a *Suppressor-mutator* (*Spm*) receptor at the bronze-1 locus in maize. *Proc. Natl. Acad. Sci. USA* 80:7591–7595.
62. **Kunze, R., U. Stochaj, J. Laufs, and P. Starlinger.** 1987. Transcription of transposable element *Activator* (*Ac*) of *Zea mays* L. *EMBO J.* **6**:1555–1563.
63. **Lee, M., and R. L. Phillips.** 1987. Genomic rearrangements in maize induced by tissue culture. *Genome* **29**:122–128.
64. **Masson, P., R. Surosky, J. Kingsbury, and N. V. Fedoroff.** 1987. Genetic and molecular analysis of the *Spm-dependent a-m2* alleles of the maize *a* locus. *Genetics* **177**:117–137.
65. **McClintock, B.** 1942. The fusion of broken ends of chromosomes following nuclear fusion. *Proc. Natl. Acad. Sci. USA* **28**:458–463.
66. **McClintock, B.** 1942. Maize genetics. *Carnegie Inst. Wash. Year Book* **41**:181–186.
67. **McClintock, B.** 1945. Cytogenetic studies of maize and Neurospora. *Carnegie Inst. Wash. Year Book* **44**:108–112.
68. **McClintock, B.** 1946. Maize genetics. *Carnegie Inst. Wash. Year Book* **45**:176–186.
69. **McClintock, B.** 1947. Cytogenetic studies of maize and Neurospora. *Carnegie Inst. Wash. Year Book* **46**:146–152.
70. **McClintock, B.** 1948. Mutable loci in maize. *Carnegie Inst. Wash. Year Book* **47**:155–169.
71. **McClintock, B.** 1949. Mutable loci in maize. *Carnegie Inst. Wash. Year Book* **48**:142–154.
72. **McClintock, B.** 1950. The origin and behavior of mutable loci in maize. *Proc. Natl. Acad. Sci. USA* **36**:344–355.
73. **McClintock, B.** 1950. Mutable loci in maize. *Carnegie Inst. Wash. Year Book* **49**:157–167.
74. **McClintock, B.** 1951. Chromosome organization and genic expression. *Cold Spring Harbor Symp. Quant. Biol.* **16**: 13–47.
75. **McClintock, B.** 1951. Mutable loci in maize. *Carnegie Inst. Wash. Year Book* **50**:174–181.
76. **McClintock, B.** 1952. Mutable loci in maize. *Carnegie Inst. Wash. Year Book* **51**:212–219.
77. **McClintock, B.** 1953. Mutation in maize. *Carnegie Inst. Wash. Year Book* **52**:227–237.
78. **McClintock, B.** 1954. Mutations in maize and chromosomal aberrations in Neurospora. *Carnegie Inst. Wash. Year Book* **53**:254–260.
79. **McClintock, B.** 1955. Controlled mutation in maize. *Carnegie Inst. Wash. Year Book* **54**:245–255.
80. **McClintock, B.** 1956. Intranuclear systems controlling gene action and mutation. *Brookhaven Symp. Biol.* **8**:58–74.
81. **McClintock, B.** 1956. Controlling elements and the gene. *Cold Spring Harbor Symp. Quant. Biol.* **21**:197–216.
82. **McClintock, B.** 1956. Mutation in maize. *Carnegie Inst. Wash. Year Book* **55**:323–332.
83. **McClintock, B.** 1957. Genetic and cytological studies of maize. *Carnegie Inst. Wash. Year Book* **56**:393–401.
84. **McClintock, B.** 1958. The suppressor-mutator system of control of gene action in maize. *Carnegie Inst. Wash. Year Book* **57**:415–429.
85. **McClintock, B.** 1959. Genetic and cytological studies of maize. *Carnegie Inst. Wash. Year Book* **58**:452–456.
86. **McClintock, B.** 1961. Some parallels between gene control systems in maize and bacteria. *Am. Nat.* **95**:265–277.
87. **McClintock, B.** 1961. Further studies of the suppressor-mutator system of control of gene action in maize. *Carnegie Inst. Wash. Year Book* **60**:469–476.
88. **McClintock, B.** 1962. Topographical relations between elements of control systems in maize. *Carnegie Inst. Wash. Year Book* **61**:448–461.
89. **McClintock, B.** 1963. Further studies of gene-control systems in maize. *Carnegie Inst. Wash. Year Book* **62**:486–493.
90. **McClintock, B.** 1964. Aspects of gene regulation in maize. *Carnegie Inst. Wash. Year Book* **63**:592–602.
91. **McClintock, B.** 1965. The control of gene action in maize. *Brookhaven Symp. Biol.* **18**:162–184.
92. **McClintock, B.** 1965. Components of action of the regulators Spm and Ac. *Carnegie Inst. Wash. Year Book* **64**: 527–534.
93. **McClintock, B.** 1967. Regulation of pattern of gene expression by controlling elements in maize. *Carnegie Inst. Wash. Year Book* **65**:568–578.
94. **McClintock, B.** 1968. The states of a gene locus in maize. *Carnegie Inst. Wash. Year Book* **66**:20–28.
95. **McClintock, B.** 1971. The contribution of one component of a control system to versatility of gene expression. *Carnegie Inst. Wash. Year Book* **70**:5–17.
96. **McClintock, B.** 1978. Development of the maize endosperm as revealed by clones, p. 217–237. *In* S. Subtelny and I. M. Sussex (ed.), *The Clonal Basis of Development.* Academic Press, Inc., New York.
97. **McClintock, B.** 1978. Mechanisms that rapidly reorganize the genome. *Stadler Symp.* **10**:25–47.
98. **McLaughlin, M., and V. Walbot.** 1987. Cloning of a mutable *bz2* allele of maize by transposon tagging and differential hybridization. *Genetics* **117**:771–776.
99. **Merckelbach, A., H.-P. Döring, and P. Starlinger.** 1986. The aberrant *Ds* element in the *Adh1-2F11::Ds* allele. *Maydica* **31**:109–122.
100. **Müller-Neumann, M., J. I. Yoder, and P. Starlinger.** 1984. The DNA sequence of the transposable element *Ac* of *Zea mays* L. *Mol. Gen. Genet.* **198**:19–24.
101. **Nelson, O. E., and H. W. Rines.** 1962. The enzymatic

deficiency in the *waxy* mutant of maize. *Biochem. Biophys. Res. Commun.* **9**:297–300.

102. **Nelson, O. E., Jr., and A. S. Klein.** 1984. The characterization of an *Spm*-controlled *bronze*-mutable allele in maize. *Genetics* **106**:769–779.
103. **Neuffer, M. G.** 1965. Crossing over in heterozygotes carrying different mutable alleles at the A1 locus in maize. *Genetics* **52**:521–528.
104. **Neuffer, M. G.** 1966. Stability of the suppressor element in two mutator systems at the A1 locus in maize. *Genetics* **53**:541–549.
105. **Nevers, P., and H. Saedler.** 1977. Transposable genetic elements as agents of gene instability and chromosome rearrangements. *Nature* (London) **268**:109–115.
106. **Nevers, P., N. Shepherd, and H. Saedler.** 1985. Plant transposable elements. *Adv. Bot. Res.* **12**:102–203.
107. **Nuffer, M. G.** 1955. Dosage effect of multiple Dt loci on mutation of *a* in the maize endosperm. *Science* **121**: 399–400.
108. **Nuffer, M. G.** 1961. Mutation studies at the A1 locus in maize. I. A mutable allele controlled by Dt. *Genetics* **46**: 625–640.
109. **O'Reilly, C., N. S. Shepherd, A. Pereira, Z. Schwarz-Sommer, I. Bertram, D. S. Robertson, P. A. Peterson, and H. Saedler.** 1985. Molecular cloning of the *a1* locus of *Zea mays* using the transposable elements *En* and *Mu1*. *EMBO J.* **4**:877–882.
110. **Peacock, W. J., E. S. Dennis, W. L. Gerlach, M. M. Sachs, and D. Schwartz.** 1984. Insertion and excision of *Ds* controlling elements in maize. *Cold Spring Harbor Symp. Quant. Biol.* **49**:347–354.
111. **Pereira, A., H. Cuypers, A. Gierl, Z. Schwarz-Sommer, and H. Saedler.** 1986. Molecular analysis of the En/Spm transposable element system of *Zea mays*. *EMBO J.* **5**:835–841.
112. **Pereira, A., Z. Schwarz-Sommer, A. Gierl, I. Bertram, P. A. Peterson, and H. Saedler.** 1985. Genetic and molecular analysis of the Enhancer (En) transposable element system of *Zea mays*. *EMBO J.* **4**:17–23.
113. **Peschke, V. M., R. L. Phillips, and B. G. Gengenbach.** 1987. Discovery of transposable element activity among progeny of tissue culture-derived maize plants. *Science* **238**:804–807.
114. **Peterson, P. A.** 1953. A mutable pale green locus in maize. *Genetics* **38**:682–683.
115. **Peterson, P. A.** 1960. The pale green mutable system in maize. *Genetics* **45**:115–133.
116. **Peterson, P. A.** 1961. Mutable *a1* of the En system in maize. *Genetics* **46**:759–771.
117. **Peterson, P. A.** 1965. A relationship between the *Spm* and *En* control systems in maize. *Am. Nat.* **99**:391–398.
118. **Peterson, P. A.** 1966. Phase variation of regulatory elements in maize. *Genetics* **54**:249–266.
119. **Peterson, P. A.** 1970. The *En* mutable system in maize. III. Transposition associated with mutational events. *Theor. Appl. Genet.* **40**:367–377.
120. **Peterson, P. A.** 1970. Controlling elements and mutable loci in maize: their relationship to bacterial episomes. *Genetica* **41**:33–56.
121. **Peterson, P. A.** 1980. Instability among the components of a regulatory element transposon in maize. *Cold Spring Harbor Symp. Quant. Biol.* **45**:447–455.
122. **Pohlman, R. F., N. V. Fedoroff, and J. Messing.** 1984. The nucleotide sequence of the maize controlling element *Activator*. *Cell* **37**:635–643.
123. **Randolph, L. F.** 1936. Developmental morphology of the caryopsis in maize. *J. Agric. Res.* **53**:881–916.
124. **Rhoades, M. M.** 1936. The effect of varying gene dosage on aleurone colour in maize. *J. Genet.* **33**:347–354.
125. **Rhoades, M. M.** 1938. Effect of the Dt gene on the mutability of the a_1 allele in maize. *Genetics* **23**:377–397.
126. **Rhoades, M. M.** 1941. The genetic control of mutability in maize. *Cold Spring Harbor Symp. Quant. Biol.* **9**:138–144.
127. **Rhoades, M. M.** 1945. On the genetic control of mutability in maize. *Proc. Natl. Acad. Sci. USA* **31**:91–95.
128. **Rhoades, M. M., and E. Dempsey.** 1950. New mutable loci. *Maize Gen. Coop. Newsl.* **24**:50.
129. **Rhoades, M. M., and E. Dempsey.** 1982. The induction of mutable systems in plants with the high-loss mechanism. *Maize Genet. Coop. Newsl.* **56**:21–26.
130. **Rhoades, M. M., and E. Dempsey.** 1987. Do strong *Ac2* alleles represent duplications or triplications of a single *Ac* element? *Maize Genet. Coop. Newsl.* **61**:24–25.
131. **Roberts, D., B. C. Hoopes, W. R. McClure, and N. Kleckner.** 1985. IS10 transposition is regulated by DNA adenine methylation. *Cell* **43**:117–130.
132. **Robertson, D. S.** 1978. Characterization of a mutator system in maize. *Mutat. Res.* **51**:21–28.
133. **Robertson, D. S.** 1983. A possible dose-dependent inactivation of mutator (*Mu*) in maize. *Mol. Gen. Genet.* **191**:86–90.
134. **Robertson, D. S.** 1985. Genetic studies on the loss of *Mu* mutator activity in maize. *Genetics* **113**:765–773.
135. **Robertson, D. S.** 1985. Differential activity of the maize mutator *Mu* at different loci and in different cell lineages. *Mol. Gen. Genet.* **200**:9–13.
136. **Robertson, D. S., D. W. Morris, P. H. Stinard, and B. A. Roth.** 1988. Germline and somatic Mutator activity: are they functionally related?, p. 17–42. *In* O. E. Nelson (ed.), *Plant Transposable Elements.* Plenum Publishing Corp., New York.
137. **Robertson, D. S., P. S. Stinard, J. G. Wheeler, and D. W. Morris.** 1985. Genetic and molecular studies on germinal and somatic mutability in *Mutator*-induced aleurone mutants of maize, p. 317–331. *In* M. Freeling (ed.), *Plant Genetics.* Alan R. Liss, Inc., New York.
138. **Sachs, M. M., W. J. Peacock, E. S. Dennis, and W. L. Gerlach.** 1983. Maize *Ac/Ds* controlling elements—a molecular viewpoint. *Maydica* **28**:289–303.
139. **Saedler, H., and P. Nevers.** 1985. Transposition in plants: a molecular model. *EMBO J.* **4**:585–590.
140. **Salamini, F.** 1980. Controlling elements at the *opaque*-2 locus of maize: their involvement in the origin of spontaneous mutation. *Cold Spring Harbor. Symp. Quant. Biol.* **45**:467–476.
141. **Salamini, F.** 1980. Genetic instability at the *opaque*-2 locus of maize. *Mol. Gen. Genet.* **179**:497–507.
142. **Schiefelbein, J. W., D. B. Furtek, V. Raboy, J. A. Banks, N. V. Fedoroff, and O. E. Nelson, Jr.** 1985. Exploiting transposable elements to study the expression of a maize gene, p. 445–459. *In* M. Freeling (ed.), *Plant Genetics.* Alan R. Liss, Inc., New York.
143. **Schiefelbein, J. W., V. Raboy, N. V. Fedoroff, and O. E. Nelson, Jr.** 1985. Deletions within a defective *Suppressor-mutator* element in maize affect the frequency and developmental timing of its excision from the *bronze* locus. *Proc. Natl. Acad. Sci. USA* **82**:4783–4787.
144. **Scheifelbein, J. W., V. Raboy, H.-Y. Kim, and O. E. Nelson, Jr.** 1988. Molecular characterization of *Suppressor-mutator* (*Spm*)-induced mutations at the *bronze*-1 locus in maize: the *bz-m13* alleles, p. 261–278. *In* O. E. Nelson (ed.), *Plant Transposable Elements.* Plenum Publishing Corp., New York.

145. **Schmidt, R. J., F. A. Burr, and B. Burr.** 1987. Transposon tagging and molecular analysis of the maize regulatory locus *opaque-2*. *Science* **238**:960–963.
146. **Schnable, P. S., and P. A. Peterson.** 1986. Distribution of genetically active Cy transposable elements among diverse maize lines. *Maydica* **31**:59–81.
147. **Schwartz, D.** 1984. Analysis of the *Ac* transposable element dosage effect in maize. *Mol. Gen. Genet.* **196**:81–84.
148. **Schwartz, D.** 1985. Differential activity of transposed *wx-m9 Ac* derivatives on various *Ds* elements, p. 391–403. *In* M. Freeling (ed.), *Plant Genetics*. Alan Liss, Inc., New York.
149. **Schwartz, D.** 1986. Analysis of the autonomous *wx-m7* transposable element mutant of maize. *Maydica* **31**: 123–129.
150. **Schwartz, D.** 1988. Comparison of methylation of the male and female-derived *wx-m9 Ds-cy* allele in endosperm and sporophyte, p. 351–354. *In* O. E. Nelson (ed.), *Plant Transposable Elements*. Plenum Publishing Corp., New York.
151. **Schwartz, D., and E. Dennis.** 1986. Transposase activity of the *Ac* controlling element in maize is regulated by its degree of methylation. *Mol. Gen. Genet.* **205**:476–482.
152. **Schwartz, D., and C. S. Echt.** 1982. The effect of *Ac* dosage on the production of multiple forms of the *Wx* protein by the *wx-m-9* controlling element mutation in maize. *Mol. Gen. Genet.* **187**:410–413.
153. **Schwarz-Sommer, Z., A. Gierl, R. Berndtgen, and H. Saedler.** 1985. Sequence comparison of 'states' of *a1-m1* suggests a model of Spm (En) action. *EMBO J.* **4**:2439–2443.
154. **Schwarz-Sommer, Z., A. Gierl, H. Cuypers, P. A. Peterson, and H. Saedler.** 1985. Plant transposable elements generate the DNA sequence diversity needed in evolution. *EMBO J.* **4**:591–597.
155. **Schwarz-Sommer, Z., A. Gierl, R. B. Klösgen, U. Wienand, P. A. Peterson, and H. Saedler.** 1984. The Spm (En) transposable element controls the excision of a 2-kb DNA insert at the *wxm8* allele of *Zea mays*. *EMBO J.* **3**:1021–1028.
156. **Schwarz-Sommer, Z., N. Shepherd, E. Tacke, A. Gierl, W. Rohde, L. Leclercq, M. Mattes, R. Berndtgen, P. A. Peterson, and H. Saedler.** 1987. Influence of transposable elements on the structure and function of the *A1* gene of *Zea mays*. *EMBO J.* **6**:287–294.
157. **Shure, M., S. Wessler, and N. Fedoroff.** 1983. Molecular identification and isolation of the *Waxy* locus in maize. *Cell* **35**:235–242.
158. **Strommer, J. N., S. Hake, J. Bennetzen, W. C. Taylor, and M. Freeling.** 1982. Regulatory mutants of the maize *Adh1* gene caused by DNA insertions. *Nature* (London) **300**: 542–544.
159. **Sundaresan, V., and M. Freeling.** 1987. An extrachromosal form of the *Mu* transposons of maize. *Proc. Natl. Acad. Sci. USA* **84**:4924–4928.
160. **Sutton, W. D., W. L. Gerlach, D. Schwartz, and W. J. Peacock.** 1984. Molecular analysis of *Ds* controlling element mutations at the *Adh1* locus of maize. *Science* **223**: 1265–1268.
161. **Tacke, E., Z. Schwarz-Sommer, P. A. Peterson, and H. Saedler.** 1986. Molecular analysis of states of the *A* locus of *Zea mays*. *Maydica* **31**:83–91.
162. **Taylor, L. P., and V. Walbot.** 1987. Isolation and characterization of a 1.7 kb transposable element from a mutator line of maize. *Genetics* **117**:297–307.
163. **Theres, K., and P. Starlinger.** 1986. Molecular cloning of *bz2-m*. *Maize Genet. Coop. Newsl.* **60**:40.
164. **Van Sluys, M. A., J. Tempé, and N. Fedoroff.** 1987. Studies on the introduction and motility of the maize *Activator* element in *Arabidopsis thaliana* and *Daucus carota*. *EMBO J.* **6**:3881–3889.
165. **Walbot, V.** 1986. Inheritance of Mutator activity in *Zea mays* as assayed by somatic instability of the *bz2-mu1* allele. *Genetics* **114**:1293–1312.
166. **Walbot, V., A. B. Britt, K. Luehrsen, M. McLaughlin, and C. Warren.** 1988. Regulation of mutator activities in maize, p. 121–136. *In* O. E. Nelson (ed.), *Plant Transposable Elements*. Plenum Publishing Corp., New York.
167. **Weatherwax, P.** 1923. *The Story of the Maize Plant*. University of Chicago Press, Chicago.
168. **Weck, E., U. Courage, H.-P. Döring, N. Fedoroff, and P. Starlinger.** 1984. Analysis of *sh-m6233*, a mutation induced by the transposable element *Ds* in the sucrose synthase gene of *Zea mays*. *EMBO J.* **3**:1713–1716.
169. **Wessler, S. R., G. Baran, and M. Varagona.** 1987. The maize transposable element *Ds* is spliced from RNA. *Science* **237**: 916–918.
170. **Wessler, S. R., G. Baran, M. Varagona, and S. L. Dellaporta.** 1986. Excision of the *Ds* produces *waxy* proteins with a range of enzymatic activities. *EMBO J.* **5**:2427–2432.

Chapter 15

Consequences and Mechanisms of Transposition in *Antirrhinum majus*

ENRICO S. COEN, TIM P. ROBBINS, JORGE ALMEIDA, ANDREW HUDSON, and ROSEMARY CARPENTER

I. INTRODUCTION

Genetic instability in *Antirrhinum majus* (snapdragon) has received the attention of scientists for over a century. Only within the past 5 years, however, has it been possible to analyze the molecular basis of this instability. This analysis has led to the characterization of three transposable elements in *A. majus* and to a greater understanding of the mechanisms by which they excise from one position on the chromosome and integrate at another. An important consequence of this transposition is that it can change the quantity and spatial pattern of gene expression, and this has allowed transposable elements to be used for probing gene regulation. The possibility of using transposable elements of *A. majus* to isolate and analyze a variety of genes involved in plant development now exists and should lead to a further understanding of the mechanisms of transposition and of the ways in which transposition produces genetic and phenotypic change. In this review we consider our current knowledge of the properties of transposable elements and their relation to genetic variation in *A. majus*.

A general notion among many biologists of the 19th century was that changes which occurred during somatic development could be inherited to some degree. This has not proved to be the case in animals, where the early separation of germ line and soma ensures that purely somatic events cannot be passed on to offspring. However, in the case of plants the separation of soma and germ line is not laid down early in development so that sometimes somatic events can be inherited. For this reason, 19th-century biologists were particularly interested in the phenomenon of bud variation, in which a part or sector of a plant showing an altered phenotype could give rise to progeny, either by vegetative propagation or occasionally through seed, which inherited the new character (21). Such events are normally rare but can occur with high frequency in certain types of variegated plants. As an example of this phenomenon, Darwin (21) cited a variety of *A. majus* which had white flowers with red spots and stripes. *A. majus*

Enrico S. Coen, Tim P. Robbins, Jorge Almeida, Andrew Hudson, and Rosemary Carpenter ■ Agricultural and Food Research Council, Institute of Plant Science Research, John Innes Institute, Colney Lane, Norwich NR4 7UH, United Kingdom.

normally produces an inflorescence comprising about 10 to 20 flowers on the upper half of the main stem. Lateral branches produced from nodes in the lower part of the main stem also terminate in an inflorescence. An inflorescence similar to that mentioned by Darwin had already been described and illustrated in 1838 in *Paxton's Magazine of Botany* (1) (Fig. 1). Darwin observed occasional mosaic plants of this variety in which some lateral branches had variegated flowers (as in Fig. 2b) but other branches on the same plant had full red flowers (Fig. 2a).

The first systematic study of this phenomenon was made in the 1890s by De Vries (24) on the variegated line of *A. majus*. He took self-fertilized seed from the occasional lateral branches which had full red flowers and showed that a high proportion of progeny (71%) had full red flowers throughout the plant, the rest of the progeny having variegated flowers. Selfed seed taken from variegated flowers gave mainly variegated progeny, but a small proportion of plants (2 to 30%) had full red flowers throughout. The greater the size and frequency of red spots on the parental flowers, the more full red progeny were produced. De Vries concluded that variegated plants showed heritable reversion to full red, through bud variation and through seed. However, the observation that full red progeny from variegated plants frequently gave rise to about 25% variegated and 75% full red progeny was interpreted to indicate that full red plants could themselves revert to variegated forms. De Vries was unaware at that time of Mendel's results, so he was unable to distinguish between the phenomenon of reversion from variegated to full red and chromosome segregation which subsequently occurred in the progeny of full red plants (see below).

A full interpretation of De Vries' results was given in 1914 by Emerson (27), who was studying the related phenomenon of variegated ears in maize. Emerson suggested that the variegation in both maize and *A. majus* was due to an unstable recessive allele giving no color, which could change in a cell to a stable dominant allele conferring full red color. This change could occur during the somatic development of a flower to give red spots or sectors; the earlier the change occurred during flower development, the larger the sector of color. If the change occurred very early in the ontogeny of the plant, a whole lateral branch having full red flowers could be produced. Thus Emerson was the first to understand clearly that the phenomenon of bud variation in *A. majus* was simply an early manifestation of the same process that gave rise to spots within a flower. Emerson further explained the appearance of full red progeny by proposing that the change to the stable dominant allele could occur in tissue giving rise to gametes. These full red progeny would generally be heterozygous for the dominant allele and the unstable recessive and would hence give progeny segregating about three full red to one variegated.

These pioneering studies on instability were pursued further by Baur and his colleagues (3, 79). They chose *A. majus* as a convenient species for genetic analysis because it exhibited abundant genetic variation, was easy to grow, produced over 200 seeds per capsule, and had large flowers which were easy to self-fertilize or emasculate and cross-pollinate. *A. majus* can go through three generations per year and can be vegetatively propagated. Their analysis showed that in *A. majus* unstable alleles also existed for genes controlling other aspects of plant development such as flower or plant morphology. In these cases, instability could not be observed through a variegated phenotype since genotypic changes in individual cells had no easily observed phenotypic effect. Rather, alleles of these genes were considered unstable if they frequently gave rise to revertant progeny, presumably by reversion in tissue giving rise to gametes. Baur further showed that unstable alleles could give rise to derivative alleles conferring novel phenotypes. For example, the unstable allele conferring variegated flowers described by De Vries, termed *pallida*^{recurrens}, was shown to give rise to numerous *pallida* alleles which conferred different stable intensities of flower pigmentation (3, 30).

In 1949 McClintock (55) showed in maize that the change of unstable recessive alleles to the dominant form was due to the movement of a short segment of chromosome (a transposable element). The recessive allele gives no color to the plant because it carries a transposable element inserted in a gene required for pigment synthesis. Excision of the element and its transposition to a new site, in either somatic or germinal tissue, regenerated a functional allele and hence full red color. McClintock's proof of transposition focused on certain unstable alleles of maize. Although instability in *A. majus* was known to share many similarities with that of maize (27, 39), not until the advent of molecular analysis was it possible to show unequivocally that transposable elements also caused flower color variegation in *A. majus* (9, 10, 89).

II. ISOLATION AND STRUCTURE OF TRANSPOSABLE ELEMENTS IN *A. MAJUS*

Plant transposable elements have been most readily studied genetically when inserted in genes required for the synthesis of anthocyanin pigment. In

Figure 1. Inflorescence of an *A. majus* variety (called *caryophylloides*) showing variegated flowers. Illustration taken from *Paxton's Magazine of Botany* (1) by kind permission of the John Innes Foundation.

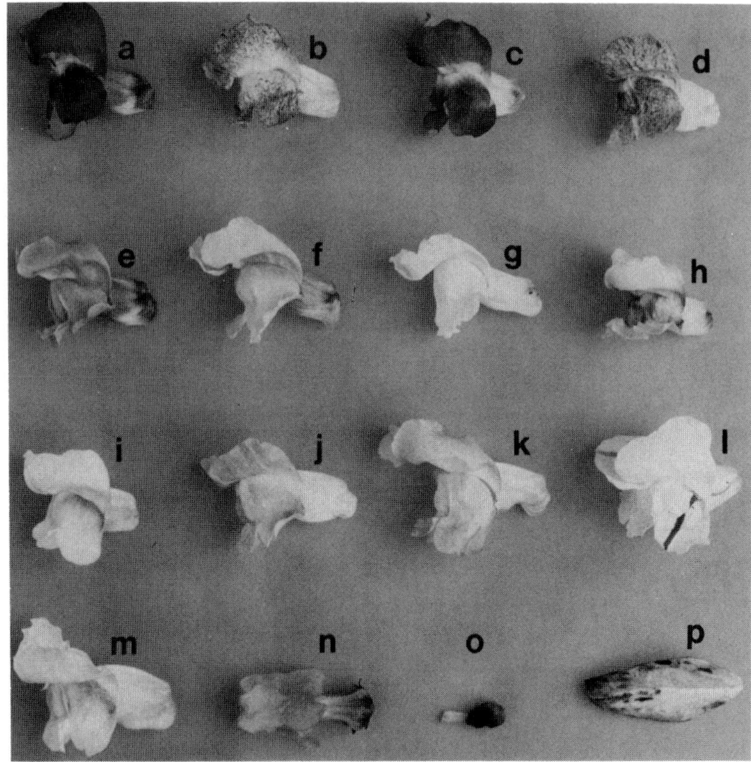

Figure 2. Various mutations in *A. majus*: (a) full red wild type; (b) *pal*^{rec}-2 (the variegated flower studied by Darwin and De Vries); (c) *niv*^{rec}-98; (d) *niv*^{rec}-53, which also carries the *delila* mutation preventing pigmentation of the corolla tube; (e) *pal*-518; (f) *pal*-33; (g) *pal*-15; (h) *pal*-41; (i) *niv*-525; (j) offspring from a cross between a line containing *niv*^{rec}-53 and a line carrying *niv*-44 (both parents carry the *delila* mutation, which blocks pigmentation of the corolla tube); (k) *pal*-42; (l) *niv*-540; (m) *niv*-543; (n) *cyloidea*^{radialis}, which also contains the mutations *eosinea* (pink) and *sulfurea* (yellow), giving an overall apricot color; (o) *deficiens*^{globifera}; (p) variegated leaf mutant. (See color plates, this volume.)

A. majus, two genes in the pigment pathway have been most intensively studied. The *nivea* (*niv*) gene encodes the enzyme chalcone synthase, which catalyzes an early step in the pathway (76, 78); the *pallida* (*pal*) gene encodes dihydroflavonol 4-reductase, catalyzing a later step (78; unpublished results) (Fig. 3). Mutations in these loci can be distinguished phenotypically, since null alleles of the *niv* locus give albino flowers whereas null alleles of *pal* give ivory flowers with a yellow area of pigmentation around the lower lip of the corolla. This yellow pigment (aurone) is synthesized from an early intermediate in the pathway, whose synthesis is blocked by *niv* mutations but is not affected by mutations at *pal* or other loci such as *incolorata* which act at later steps in the pathway.

The *niv* locus was cloned from genomic DNA of *A. majus*, using a chalcone synthase cDNA clone from parsley as a probe (89). This then allowed the isolation of the unstable *nivea*^{recurrens}-53 (*niv*^{rec}-53)

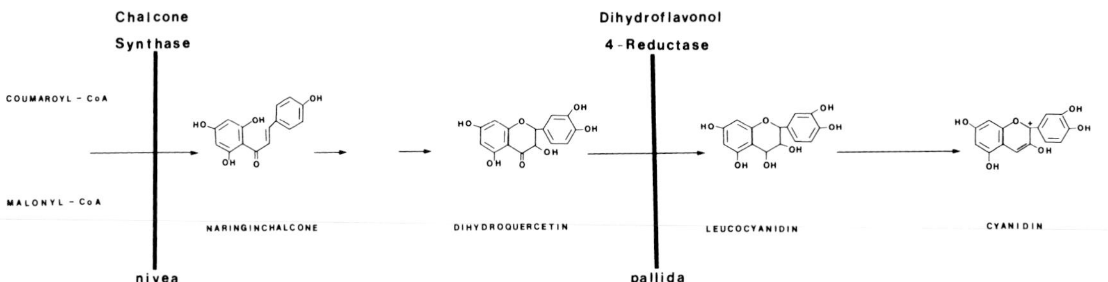

Figure 3. Pathway to anthocyanin synthesis in *A. majus*, showing the enzymatic steps affected by *pal* and *niv* mutations. The several steps between naringinchalcone and dihydroquercetin, which include the step affected by *incolorata* mutations, are not shown separately.

Table 1. Transposable elements in *A. majus*

Allele	Flower phenotype	First documented	Element	Element size (kb)	Element position (relative to gene)	Length of inverted repeat (bp)	Length of target duplication (bp)	References
niv^{rec}-53	Red sites on white background	1936	Tam1	15	17 bp upstream of TATA box	13	3	9, 10, 51
niv-44	Stable white	1955	Tam2	5	First exon/intron boundary	14	3	83
niv^{rec}-98	Red sites on pale red background	1979	Tam3	3.5	29 bp upstream of TATA box	12	8	74
pal^{rec}-2	Red sites on ivory background	1838	Tam3	3.5	41 bp upstream of TATA box	12	5	1, 19, 53

allele, which confers a variegated (so-called *recurrens*) phenotype (Fig. 2d). This allele carries a 15-kilobase (kb) transposable element, Tam1 (Tam is an abbreviation of transposon *A. majus*), inserted in the promoter region of the *niv* gene (9, 10). Analysis of two further *niv* alleles, *niv*-44, which conferred a stable null phenotype, and niv^{rec}-98, which gave a variegated phenotype on a pale red background, enabled two further elements, Tam2 and Tam3, to be isolated (74, 83). The main features and origins of these elements are summarized in Table 1. The Tam elements are generally present in 10 to 30 copies in the genome, as judged by the number of bands seen in Southern hybridization of genomic DNA with Tam probes, although the number of bands observed can depend on the region of the element used as a probe so it is not possible to assign a precise copy number. It is unlikely that these three transposable elements include all active elements in the genome of *A. majus*, since only three independent *niv* mutations have so far been analyzed.

The isolation of Tam3 from the *niv* locus allowed the *pal* locus to be cloned by transposon tagging (see Section V), since the pal^{rec}-2 allele has Tam3 inserted in the *pal* promoter (19, 53) (Table 1). The precise origin of the pal^{rec}-2 allele is unknown, but it can be directly traced back to De Vries' studies and probably corresponds to the variegated plant illustrated in 1838 (Fig. 1) since this shows red sectors on flowers with yellow pigmentation around the lower lip.

In common with many transposons isolated from other species, all of the Tam elements have inverted repeats at their termini (12 to 14 base pairs [bp]) and are flanked by direct duplications (3 to 8 bp) of host sequence (the target site). The elements may be divided into two families. Tam1 and Tam2 have homologous termini (49), and both produce target duplications of 3 bp. The ends of these elements are also homologous to 12 out of 13 bp of the termini of the *Spm* and *En* elements of *Zea mays* (10, 83) and to the terminal 5 bp of the Tgm1 element of *Glycine max* (86). All of these elements also produce 3-bp target duplications, suggesting that they may belong to the same family of transposable elements (10).

Tam3 is quite distinct from Tam1 and Tam2 but has 7 out of 11 terminal nucleotides homologous with the termini of the *Ac* element of *Z. mays* (74). *Ac* produces 8-bp target duplications, which is the same length as the Tam3 target duplications described for three different Tam3 insertions at the *niv* locus (74; A. Hudson, R. Carpenter, and E. Coen, manuscript in preparation). However, unlike other plant transposable elements described, Tam3 can be flanked by various lengths of target site duplication since a 5-bp target duplication flanks Tam3 in pal^{rec}-2 (19). It is not clear whether this variation in target length reflects different properties of the Tam3 copies at the *niv* and *pal* loci or an intrinsic variability shared by all Tam3 copies.

III. EXCISION OF TRANSPOSABLE ELEMENTS

Transposition of many elements in plants is believed to occur via excision of a copy of the transposable element from a donor site and its integration at a recipient site. The best evidence for this comes from studies on the *Ac* element of maize, where the fate of an excised *Ac* copy can be followed by virtue of the effect of different *Ac* doses on transposition of a nonautonomous *Ds* element (33, 35). These studies showed that excision of *Ac* is usually, and perhaps always, associated with its integration at a new site. Such an analysis has not

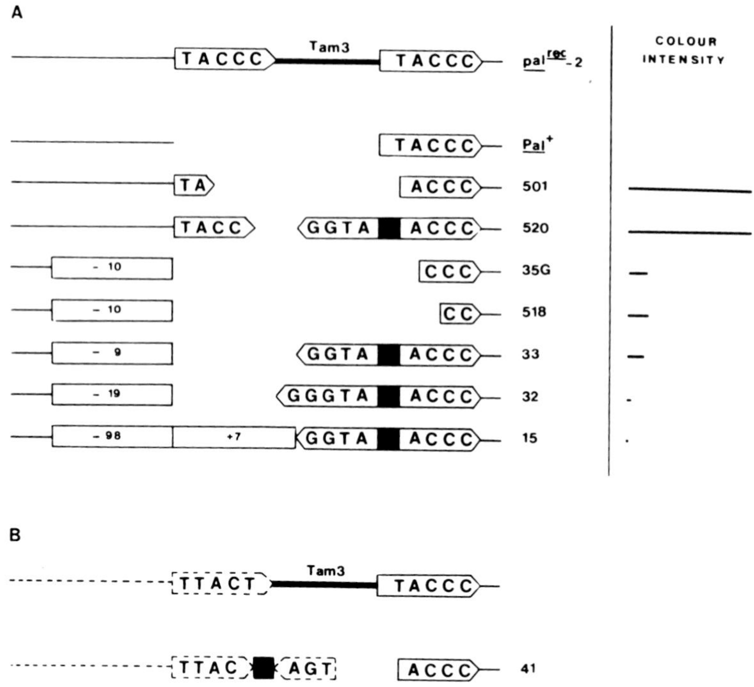

Figure 4. Structures of *pal* alleles. (A) The pal^{rec}-2 allele is shown at the top. Tam3 is indicated by a thick line (not drawn to scale) and flanking sequences by thin lines. The target duplication is indicated in boxed arrows. Only the sequence derived from the target duplication is shown. The length of the solid bars on the right indicates the relative intensity of flower pigmentation, as measured by anthocyanin content. No value is given for the Pal^+ wild-type allele since it is maintained in a very different genetic background from the other alleles. The Pal^+ wild-type allele is shown below pal^{rec}-2, and transcription of this allele initiates about 70 bp to the right of the target site. Various alleles derived from pal^{rec}-2 are also illustrated. Gaps are shown in the sequences at the site of Tam3 excision to align the left and right flanking sequences. The numbers inside the open boxes indicate the sizes in base pairs of deletions (−) or insertions (+). The solid squares represent a single nucleotide missing at the center of symmetry of inverted duplications. The phenotypes conferred by the alleles are described in the text and illustrated in Fig. 2. The data are taken from reference 19 and Almeida et al. (in preparation). (B) The allele shown above *pal*-41 is its hypothetical progenitor and is thought to have been derived from pal^{rec}-2 by an inversion which replaced the sequence flanking the left of Tam3 by a new sequence (dotted line; see text). Both alleles illustrated have been aligned with the sequences flanking the right of Tam3 shown in panel A. Data are from Robbins et al. (in press).

been possible in *A. majus*, where the fate of single copies of Tam elements cannot be followed genetically. However, the analysis of several mutant alleles produced by reintegration of Tam elements suggests that these elements also transpose by an excision/integration mechanism (see Section IV). Although an interaction between donor and recipient sites probably takes place during the transposition process, it is convenient to discuss first the excision step separately. One reason for this is that excision can have a multitude of genetic and phenotypic consequences which are independent of the fate of the excised transposon. In this section we shall examine excision, considering in turn its phenotypic consequences, mechanisms, and control.

A. Phenotypic Consequences of Excision

A detailed understanding of the consequences of excision has been arrived at by the analysis of stable alleles that have arisen in the progeny of lines carrying unstable pal^{rec} or niv^{rec} alleles. Usually, the progenies of such lines resemble their parents in phenotype, but a variable proportion carry alleles conferring stable phenotypes. Most of these stable alleles confer a full red color on the flowers (revertants) and have been shown to be due to excision of transposable elements in germinal tissue (10, 41, 43, 54, 74). Less frequently, a number of stable mutants conferring reduced intensities of pigmentation ranging from almost full red to very pale, or determining coloration confined to certain areas of the flower, have been identified (11, 12, 30, 37, 73). These stable alleles have also been shown to result from transposable element excision (18, 19, 73).

The structures around excision sites in many of the stable alleles derived from unstable pal^{rec} and niv^{rec} alleles are represented in Fig. 4 and 5, respectively. In 21 excision sites so far sequenced, including

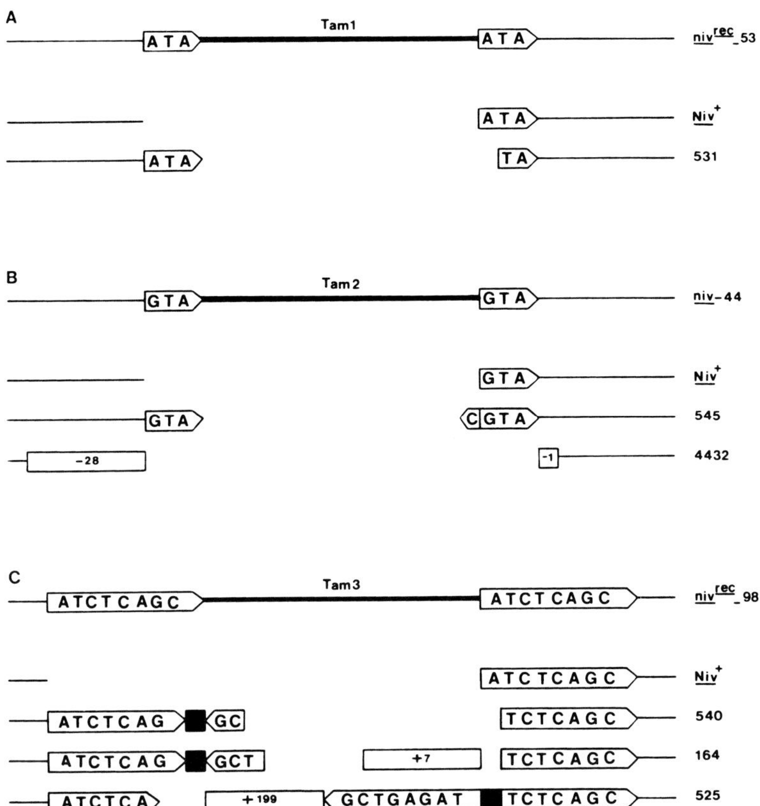

Figure 5. Structures of *niv* alleles. The meaning of the symbols is the same as in Fig. 4. The unstable alleles carrying Tam1, Tam2, and Tam3, which are located in different positions in the *niv* locus, are represented in panels A, B, and C, respectively. In each case Niv^+ is also shown. All the other alleles were derived from excisions of the transposable elements. Two other revertants from *niv*-44, the sequence of which is identical to that of *niv*-545, have been reported (41). Data are from references 10, 18, 43, and 49 and Hudson et al. (in preparation). The sequences of an additional four Tam1 excision sites have recently been published (73) but are not shown.

8 full red revertants, it has been found that the wild-type sequence around the excision site has not been precisely restored following excision. Since selection of full red phenotypes carries no bias towards finding mutated sequences, imprecise restoration of excision sites appears to be a general feature of Tam element excision in *A. majus*.

The phenotypic effects of the various mutated sequences depend on the location of the transposable element in the progenitor allele. Alleles pal^{rec}-2, niv^{rec}-98, and niv^{rec}-53 carry transposable elements in the regulatory regions of the *pal* or *niv* gene, respectively 41, 29, and 17 bp upstream of the TATA box (10, 19, 74, 75). In all full red revertants from these lines that have been examined, the sequences flanking the target duplication were conserved relative to wild type but the target duplications were modified, with the result that the revertants carry additional DNA sequences, ranging from 1 bp (*pal*-501) to 16 bp (*niv*-164). As assayed phenotypically, such insertions do not disturb the normal *cis*-regulatory functions of these regions. Tam2 is inserted in the first exon/intron junction in *niv*-44 (83). The three independent Tam2 excision events that gave rise to full red revertants all resulted in 4-bp insertions which disrupt the wild-type splice donor sequence (41, 43) (*niv*-545; Fig. 5B). However, the newly created sequence still fits the consensus for splice donor sites and results in a full red phenotype.

One of the best-characterized series of alleles which alter the intensity and distribution of pigment are those derived from pal^{rec}-2 (see Fig. 2 and 4). One of these, *pal*-518, confers reduced pigmentation correlating with decreased levels of *pal* transcript (14%) relative to that in full red flowers (Fig. 2c). Originally this mutation was considered to have only a quantitative effect on gene expression (19). However, close inspection of the flowers shows that, in addition to reducing the level of *pal* expression, this allele also confers an altered distribution of color since pigmentation in the flower lobes is much more reduced, compared to wild type, than pigmentation in the flower tube (compare Fig. 2a and e). This altered distribution correlates with a change in the distribution of *pal* transcript in flowers (J. Almeida, R. Carpenter, T. Robbins, C. Martin, and E. Coen, manuscript in preparation). In *pal*-518, 3 bp of the 5-bp Tam3 target sequence and 10 bp to its left are deleted relative to wild type (Fig. 4A). The differences in transcript level and distribution between *pal*-518 and alleles conferring full red color, such as *pal*-501 or *pal*-520, are very probably not due to differences in genetic background since the lines that carry these alleles have descended by self-pollination from a common pal^{rec}-2 progenitor which was inbred for many generations. Therefore, the sequence missing from *pal*-518 is necessary in *cis* for the wild-type level and distribution of *pal* transcription.

Other alleles which confer a phenotype similar to that conferred by *pal*-518 but of lower intensity, *pal*-35G, *pal*-33, and *pal*-32, carry deletions of 9, 10, and 19 bp to the left of the Tam3 excision site (Fig. 4A). As with *pal*-518, all these alleles confer less pigmentation in the flower lobes than in the tube. In the case of *pal*-33 and *pal*-32, excision of Tam3 has also generated an inverted duplication of 4 or 5 bp of the right copy of the target duplication, similar to the one carried by *pal*-520 which confers full red color (Fig. 4A). The *pal*-32 and *pal*-33 alleles were not derived in the same genetic background as *pal*-518, and crosses between the lines indicate that the lower intensity of pigmentation in *pal*-33 relative to *pal*-518 (see Fig. 2e and f) is a consequence of differences in unlinked genetic factors (Almeida et al., in preparation). However, the much lower intensity of pigmentation conferred by *pal*-32 as compared with *pal*-518 and *pal*-33 is not due to the effects of genetic background and is probably due to the larger 19-bp deletion it carries.

The remaining alleles shown in Fig. 4 also determine reduced levels of *pal* transcription and altered patterns of flower color. The *pal*-15 ($pal^{tubocolorata}$) allele causes pigmentation to be restricted mainly to the base of the flower tube (Fig. 2g). In *pal*-15, a short segment of 7 bp, located 125 bp upstream of the excision site, is directly duplicated adjacent to the target. The *pal*-41 allele shows the most radical alteration of the *pal* promoter: the entire region 5′ to the Tam3 excision site has been replaced by a foreign sequence normally located about 5 map units from the *pal* locus, which has been brought to its new position by an inversion (Fig. 4B; T. P. Robbins, R. Carpenter, and E. S. Coen, *EMBO J.*, in press).

One explanation for the generation of specific patterns of pigmentation by these sequence alterations is that the wild-type *pal* promoter is composed of a set of sequences that respond to diverse regulatory signals, spatially arranged in the flower. Novel spatial patterns would be produced by mutations that change the interpretation of these signals by modifying the affinity of the *pal* promoter for different regulatory molecules.

Such a model can be invoked to explain the generation of patterns in which the corolla tube is preferentially pigmented. The lobes of full red flowers are more intensely pigmented than the tubes. Therefore, an overall reduction in *pal* transcription would be expected to give flowers which still have darker lobes than tubes. However, alleles carrying deletions extending to the left of the target give flowers with tubes more intensely pigmented than lobes to varying degrees (e.g., *pal*-518, *pal*-35G, *pal*-33, and *pal*-32). This indicates that these deletions reduce *pal* expression in the lobes by a disproportionately large amount. The region deleted in these alleles might therefore be involved in interactions with transcription factors that specifically enhance *pal* expression in the lobes. Evidence for differential regulation of *pal* transcription in tubes and lobes comes from studies on the interaction between *pal* and the unlinked recessive gene *delila* (*del*). Plants which are homozygous for *del* and carry a wild-type Pal^{+} allele have flowers with full red lobes and unpigmented tubes. This has been shown to correlate with the absence of *pal* transcript in the tubes and normal levels in the lobes (Almeida et al., in preparation), confirming that there are lobe-specific *pal* transcription factors.

In *pal*-15, a deletion of 98 bp to the left of Tam3 results in lack of pigmentation in the whole corolla except for a ring at the base of the tube (Fig. 2g). This ring of pigment is formed at a very early stage during flower development in both *pal* mutants and wild type, suggesting that the base of the tube might represent an area where *pal* expression is subjected to different control mechanisms, presumably independently of sequences within approximately 100 bp to the left of the Tam3 insertion site. It appears, therefore, from the analysis of this set of alleles, that separable *cis*-acting regulatory elements might confer spatial-specific expression on *pal*. The involvement of complex sets of *cis*-acting sequences in conferring cell type-specific expression has also been postulated as an explanation for many of the alleles of the *yellow* gene which confer different patterns of pigmentation in *Drosophila melanogaster* (15, 31).

The pattern produced by *pal*-41 is quite distinct from that of the tube-specific alleles (Fig. 2h) and requires a different explanation. It is possible that this pattern partly reflects an interaction with signals different from those that normally interact with the *pal* promoter, since in this allele the entire region to the left of the Tam3 excision site has been replaced by a foreign sequence. For example, in *D. melanogaster* an inversion between the *Antp* gene and a nearby gene results in the exchange of *cis*-acting sequences, with the result that the *Antp* gene is brought under the control of the promoter of the other gene (68). However, the foreign sequence brought next to the *pal* gene in *pal*-41 does not promote transcription in flowers when in its original position (Robbins et al., in press). This indicates that the pattern observed in *pal*-41 may reflect a position effect of the adjacent foreign chromatin rather than the introduction of a new promoter. Such effects have been noted for particular P element-mediated transformants of *D. melanogaster*; the insertion of the *white* gene at different positions in the genome can result in novel distributions of the red eye pigment (52).

Many stable alleles of the *niv* locus that confer reduced intensities or altered spatial patterns of flower color have also been derived from niv^{rec}-53, *niv*-44, and niv^{rec}-98 (12, 37, 49, 73). In alleles derived from niv^{rec}-53, deletions in the *niv* promoter caused by the excision of Tam1 have been shown to result in reduction of pigmentation and *niv* transcript levels (73). Unlike the mutations which cause changes in the rate and pattern of transcription initiation, *niv*-4432, which was derived by excision of Tam2 from the first exon/intron junction, carries a deletion that removes 32 bp of the first exon, resulting in a null phenotype (49) (Fig. 5B).

The majority of mutations in genes encoding enzymes involved in pigment biosynthesis, such as *niv* and *pal* described above, reduce the overall quantity of gene expression. These mutations are recessive to wild type since a single dose of the wild-type allele encodes sufficient gene product to confer the full red phenotype. However, several alleles derived from niv^{rec}-98 have been described which are semidominant to wild type (12). One of these, *niv*-525, has been analyzed extensively (18). When homozygous, *niv*-525 gives very pale flowers with pigment concentrated in the tubes and around the lower lip of the lobes (Fig. 2i). The flowers of Niv^+/*niv*-525 heterozygotes show the same spatial distribution of color but of a slightly darker intensity. Analysis of chalcone synthase protein and mRNA from flowers of *niv*-525 and Niv^+ homozygotes and Niv^+/*niv*-525 heterozygotes has shown that there are no detectable qualitative changes in the products of *niv* expression. Homozygous *niv*-525 flowers show a sharp decrease in chalcone synthase RNA levels (1 to 2% of wild type). In the absence of an interaction between the two alleles, a transcript level of about 50% would be expected in Niv^+/*niv*-525 heterozygotes. However, the amount of *niv* RNA produced by the heterozygous genotype is significantly below such expectation (12.5% of wild type), indicating that *niv*-525 inhibits Niv^+ expression in *trans*. *niv*-525 carries an inverted duplication of 207 bp with its axis of symmetry at the site of Tam3 excision. The duplicated region includes the 8-bp target site and extends a further 199 bp, spanning part of the promoter, the entire untranslated leader, and part of the first exon (Fig. 5C).

One possible model for the semidominance of *niv*-525 is that it produces a short antisense RNA molecule. A region of 1.2 kb upstream of the TATA box, which is required for normal *niv* expression (47), remains intact in *niv*-525 and could drive transcription from a TATA box-like sequence located within the inverted duplication near the junction of the inverted leader and coding region. This would produce an antisense RNA transcript about 40 bp of which could hybridize with the leader of the wild-type *niv* transcript, resulting in its degradation in the nucleus (18). One problem with this model is that previous studies using DNA transformation in plants indicate that inhibition by antisense RNA requires a 100-fold excess of DNA expressing antisense over DNA expressing sense RNA or that the antisense RNA is driven by a stronger promoter than the sense RNA (26, 64). However, these experiments involved the synthesis of antisense RNA from random locations in the nucleus, whereas the *niv*-525 allele is at a chromosome position homologous to the position of the inhibited Niv^+ allele. A strong effect of chromosome position on the ability of a dominant allele at the *white* locus, w^{DZL}, to act in *trans* to reduce expression of a w^+ allele has been described in *D. melanogaster* (8). Inhibition of w^+ only results if the w^{DZL} allele is at a homologous chromosome position, and this may reflect chromosome pairing between homologs during gene expression. Such effects of chromosome position, termed transvection in *Drosophila* studies, might be explained by invoking inhibitory labile RNA species (e.g., antisense RNA) which require close proximity between the interacting genes to maintain a high local concentration (45).

B. Mechanisms of Excision

Models of transposable element excision in plants are based primarily on the sequences of excision sites. Figures 4 and 5 show many of the excision sites characterized in *A. majus*. These studies show that the wild-type sequence is not precisely restored on excision. Several excision sites contain rearrangements confined to the target duplication (*pal*-501, *pal*-520, *niv*-531, *niv*-540, *niv*-545). Some have inverted duplications of different extents of the target (e.g., 4 bp in *pal*-520, 1 bp in *niv*-545). Similar structures have also been observed at excision sites of

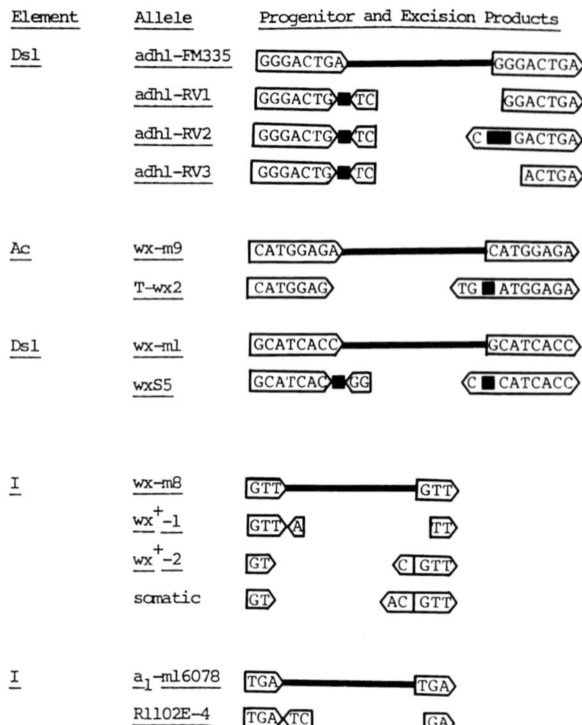

Figure 6. Inverted duplications produced by excision of transposable elements in maize. The meaning of the symbols is as in Fig. 4. Data are from references 2, 65, 70, 71, and 87.

the *Ac/Ds* or *I/dSpm* elements of maize (Fig. 6). A third group have alterations extending beyond the target sequence, which include deletions (*pal*-518, *pal*-35G, *pal*-33, *pal*-32, *pal*-15, *niv*-4432) and long inverted duplications (*niv*-525). Finally, two cases have been analyzed in which short sequences normally located adjacent to or 125 bp upstream of the target are directly duplicated near the target sequence (*niv*-164 and *pal*-15). A feature common to all these events is that, unlike other transposon-induced mutations such as excisions of *Drosophila* P elements (82), no trace of the transposon is left at the target site after excision. This indicates that excision of plant transposable elements usually involves removal of the entire element and that the variability in excision sites results from imprecise restoration of the wild-type sequence after the element has been removed.

To account for these sequence alterations, we imagine two basic steps in excision, (i) removal of the transposable element and (ii) restoration of the excision site. It is possible that the two steps are tightly coupled and occur in a single protein-DNA complex possibly involving an interaction between the inverted repeats at the ends of the transposon.

In the current models of plant transposable element excision (18, 19, 59, 65), cuts introduced at specific positions result in complete removal of the transposable element. The models differ, however, in the positions of the cuts. The model proposed by Saedler and Nevers (66) assumes that cuts staggered by the length of the target sequence are introduced at both copies of the target duplication. Such cuts are postulated to explain the duplication of the target site on integration and could therefore be carried out by the same enzyme. In another model, removal results from single-strand cuts at the nucleotides flanking the transposon which, for some elements, can be staggered by 1 bp (e.g., Tam3/*Ac*) and for others (e.g., Tam2/*Spm*) can generate blunt ends (18, 19).

In the model of Saedler and Nevers, the initial staggered cuts generate termini that become available to exonuclease degradation and/or DNA polymerase activity which can fill gaps from the 3′ recessed ends, using the chromosomal strand as a template. Different combinations of the exonuclease and polymerase activities can explain the multitude of excision products, including deletions extending beyond the excision site or partially degraded or replicated copies of the target. To explain inverted duplications of the target, the model proposes that the element is not released immediately after the initial cuts, so that DNA polymerase can switch template from the chromosomal strand to the single-stranded tail that remains attached to the transposon. This model does not easily explain the generation of inverted duplications extending beyond the target duplication, as is the case in *niv*-525 (Fig. 5C). The initial cut at the removal step would have to occur 199 bp to the right of Tam3. The resulting long single-stranded tail attached to the transposon would then be copied after DNA polymerase strand switching while being kept protected from exonuclease degradation.

Because of the detailed structure of the inverted duplications at excision sites, we prefer an alternative model (18, 19). As illustrated in Fig. 7 for *pal*-520, single-strand cuts staggered by 1 bp are introduced at nucleotides flanking the transposon (step A), producing two DNA molecules each having an overhang of one nucleotide. The free ends are ligated (step B) to form two hairpin structures, which are resolved by further nicking (step C). A similar enzymatic reaction can be carried out by the bacteriophage ϕX174 A* protein, which can nick double-stranded DNA and self-ligate the ends to form hairpins (84). It is not possible to determine from the sequence of excision sites whether the 1-bp overhang produced in step A is 3′ (as illustrated in Fig. 7) or 5′. This is because identical products would arise if step A generated a 5′ overhang followed by resolution cuts on the strand of the hairpin opposite to that illustrated in Fig. 7C. After resolution, ends are ligated (step D) and repli-

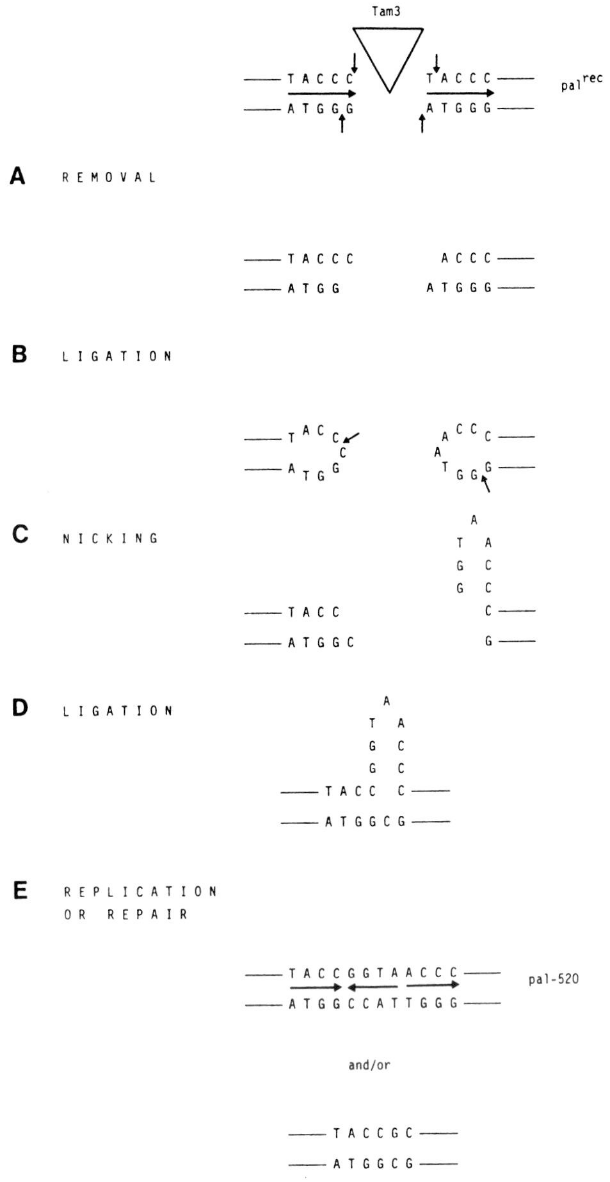

Figure 7. General model for the excision of plant transposable elements. The horizontal arrows indicate sequences of the target duplication and their orientation, and the smaller arrows represent endonucleolytic cuts. The particular sequence of events shown describes the generation of allele *pal*-520.

cation of the heteroduplex gives two reciprocal products (step E). Alternatively, the heteroduplex intermediate might be corrected by gene conversion. In the example illustrated, the resolving cuts (step C) are introduced at positions within the target duplication. Resolution in this way would lead to the small sequence changes confined to the target which are observed in most excisions which give full red revertants. However, a number of the structures represented in Fig. 4 and 5 would require cuts beyond the target duplication. Thus, for instance, in *pal*-33 the cut in the right arm would have been identical to the one that generated *pal*-520, but the cut in the left arm would have occurred 9 bp to the left of the target. In the case of *pal*-520, the hypothetical reciprocal product arising from replication is also a rearrangement of the target. In *pal*-33, however, the reciprocal of the deletion would be a longer inverted duplication, equivalent to that found in *niv*-525.

An attractive feature of this model is that it can account for the formation of acentric/dicentric chromosomes resulting from the activity of *Ds* elements in maize (19; see N. V. Fedoroff, this volume). The structures of acentric/dicentric chromosomes and inverted duplications produced by the activity of *Ds* elements in maize have led to the suggestion that hairpin DNA molecules may be involved (29). McClintock (55) has suggested that excision and the formation of dicentric chromosomes are alternative events associated with *Ds* elements. These observations may be explained if the hairpins, produced during *Ds* excision, sometimes fail to be resolved at one or both ends before mitosis. Replication of the hairpin molecule in the next cell cycle would result in dicentric or acentric chromosomes. The proportion of attempted excisions in which resolution fails to occur would depend on the element structure, which may interfere with correct resolution (see Fedoroff, this volume). Thus, as observed by McClintock (55), a *Ds* element showing high frequency of excision (correct resolution) would produce fewer chromosome breaks (failed resolution).

A regular feature of Tam3 excision events is that 1 bp is missing at the axis of symmetry in all eight inverted duplications sequenced (Fig. 4 and 5). The proposed staggering of initial cuts by 1 bp accounts for this. Excisions of maize *Ac/Ds* elements also produce inverted duplications, with 1 bp missing at the axis of symmetry, in six of seven cases published (Fig. 6). One exception, *adh1-RV2*, in which two nucleotides are missing, could reflect occasional errors in the removal cuts. The similarity between *Ac/Ds* excision products and those produced by Tam3 suggests that these elements excise following 1-bp staggered cuts. This provides a further example of a property shared between Tam3 and *Ac* and supports the notion that they belong to the same family of transposable elements.

The sequence of three independent Tam2 excision sites (see *niv*-545, Fig. 5B), shows a 1-bp perfect palindrome without loss of a nucleotide at the axis of symmetry. This suggests that the initial cuts were unstaggered and generated two blunt-ended DNA molecules, the termini of which were at the ends of Tam2. Palindromes resulting from excisions of *I/dSpm* elements belonging to the *En/Spm* family of maize also have no nucleotides missing from the axis of symmetry in all four cases described (Fig. 6). These observations suggest that the Tam2 and *Spm/En* elements all excise following the production of blunt ends. These elements have homologous termini and produce 3-bp target duplications, indicating that the production of blunt ends during excision is a further property shared by this family of elements.

To account for the structures of the two types of inverted duplications, the Saedler and Nevers model would have to assume that on template switching, DNA polymerase regularly misses the first nucleotide of the single-stranded tail attached to the ends of Tam3 and *Ac*, but not that of Tam2 or *En/Spm*.

Finally, the short sequence located between the target duplications in *niv*-164 has been explained by "stuttering" of DNA polymerase which might have copied this sequence from a region adjacent to the target site (74). This explanation was originally described in terms of the Saedler and Nevers excision model but is equally consistent with the hairpin model since the substrate proposed for the reaction could have arisen after steps A or C in Fig. 7. However, it is not easy to explain the structure of *pal*-15 by polymerase stuttering, since in this case the short duplicated sequence is not found adjacent to the target but 125 bp away.

C. Control of Excision

When Tam elements are located in regions of pigment genes which tolerate small sequence changes, such as promoter regions, most excisions restore gene expression. This allows the rate of somatic excision to be estimated by the frequency of pigmented sites on the flower and the rate of germinal excision to be estimated by the frequency of revertant progeny. If the rate of somatic excision is high, it can also be estimated from the Southern hybridizations of genomic DNA with the host gene as a probe. The relative intensity of restriction fragments corresponding to the host gene carrying or lacking the transposable element indicates the proportion of cells in which the element has excised. In general, all of these estimates of excision rates correlate with each other (12, 39, 53) so that different estimation methods have given the same conclusions concerning the control of excision rates. If an element is located in a region of the host gene which is less tolerant of sequence changes, many excisions will not restore gene function, so the frequency of phenotypic reversion will be an underestimate of the true rate of excision. However, the rate of phenotypic reversion is still useful to study relative excision rates, since it should be a constant proportion of the excision rate.

The excision rate of the transposable elements of *A. majus* can be influenced by both environmental and genetic factors. Tam3 excision from both the *pal*^{rec}-2 and *niv*^{rec}-98 alleles is approximately 1,000 times more frequent at 15°C than at 25°C (12, 39). The rates of Tam1 excision from the *niv*^{rec}-53 allele, and of Tam2 from *niv*-44, also increase at lower temperatures, but only about 10-fold (37; A. Hudson, R. Carpenter, and E. S. Coen, unpublished results). This suggests that the degree of temperature sensitivity is an intrinsic property of the element. This may reflect altered levels or activities of transposition proteins or repressors of transposition. Alternatively, temperature may have an effect indirectly by changing the rate of chromosome replication and cell division. Analogous temperature sensitivity in transposition has been described for the Ty element in *Saccharomyces cerevisiae* (58) and Tn*3* in *Escherichia coli* (50). In all these cases the higher rate of transposition occurs at the lower temperature.

Genetic factors influencing the rate of excision are of two types: activators and repressors. The Tam2 element has been shown to be activated to transpose in certain genetic backgrounds (41, 43). Originally Tam2 was isolated from the *niv*-44 allele, which is stable, indicating that Tam2 did not excise from this allele (Table 1). However, when the line carrying *niv*-44 was crossed to either of two other stable *niv* mutant lines, spots of pigment were sometimes observed on the flowers of the F_1 progeny. Germinal revertants conferring full red pigmentation derived from these plants were found to have the Tam2 target duplication (Fig. 5B) but no Tam2 element, indicating that Tam2 had been activated to excise. In both F_1 hybrids which showed Tam2 activation, the other homolog contained a stable derivative allele of *niv*^{rec}-53::Tam1. The factor activating Tam2 was mapped close to the stable *niv* allele (43), which in one case was found to have a copy of Tam1 which had lost 5 bp from one terminal repeat (41) (*niv*-46; see Fig. 8C). These results strongly suggested that Tam2 was activated to excise by Tam1 resident at the *niv* locus. Since several copies of Tam1 are seen in Southern hybridizations of both the *niv*-44 line and the stable lines derived from *niv*^{rec}-53, the results also indicated that only the copy of Tam1 at the *niv* locus was able to activate transposition. Considering the homology of the termini of Tam1 and Tam2, it seems possible that Tam1 at the *niv* locus produces a transposase which acts on the ends of Tam2, as well as Tam1, to cause excision. Tam2, and perhaps many of the other Tam1 copies in the genome, may be defective elements which require a functional copy of Tam1 to be present in order to transpose.

The rates of Tam1 and Tam2 excision, based on frequencies of pigmented sites on flowers, are also affected by repressors carried in the genetic background of the line in which the *niv*-44 allele is maintained (37, 43, 49). The inheritance of Tam1 and Tam2 repression shows some strikingly non-Mendelian features. Repression increases over several generations of inbreeding and can also increase during the lifetime of an individual. Furthermore, apparently isogenic plants can show different levels of repression, and sometimes reciprocal differences are observed in genetic crosses. The non-Mendelian behavior may reflect cytoplasmic transmission of a repressor which stimulates its own production, as has been proposed for P cytotype in *Drosophila* sp. (see W. R. Engels, this volume), or a progressive modification of the elements, perhaps by methylation, as has been found for the *Mu* and *Ac* elements of maize (6, 13, 17; Fedoroff, this volume). The repressors might act at the ends of Tam1 and Tam2, making them a poorer substrate for transposition proteins, or may reduce the production of transposition proteins from Tam1.

When Tam1 is active in the *niv*^{rec}-53 allele, the red sites produced by excision of the element are seen against a white background, which indicates that Tam1 completely blocks chalcone synthase expression unless it excises. Repression of Tam1 excision often results in expression from the *niv*^{rec}-53 allele, resulting in a diffuse pigmentation which may be granular or streaky in appearance (39) (Fig. 2j). This expression does not arise because of any rearrangement at the locus (49). One possible explanation of the correlation between repression and inhibition is that the transposition protein complex, assembled on the element ends, physically hinders the access of transcription factors to the *niv* promoter, thus preventing background expression (see Section IV.A). Repressors that prevented the formation of this transposition complex would allow some *niv* gene expression.

Variation in the size and frequency of red spots in different lines carrying the *pal*^{rec} allele had already been noted by De Vries (24), who also showed that lines with higher frequencies of large red sectors gave rise to higher proportions of revertant offspring. A major cause of this variation in reversion frequency is due to polymorphism for a semidominant repressor allele, *Stabilizer* (*St*). Plants homozygous for *St* have a Tam3 excision frequency 1,000 times lower than plants homozygous for the *st* allele, while *St*/*st* heterozygotes have an intermediate frequency (12, 40). The *St* allele only affects Tam3 excision frequency and does not affect Tam1. The specific interaction between Tam3 and *St* suggests that *St* might be a Tam3-related element which encodes a repressor of transposition, although Southern hybridization with a Tam3 probe has not so far revealed a Tam3 copy which segregates with *St*.

Unlike the *Ac* element of maize, no strong effect of Tam element dosage on excision frequency or timing has been observed by comparing plants carrying two doses of an unstable allele (homozygotes) with plants carrying one dose (heterozygotes). However, the interpretation of dosage effects in *A. majus* is difficult because it is usually not possible to define the number of autonomous elements in a genotype and because various activators and repressors may also be present.

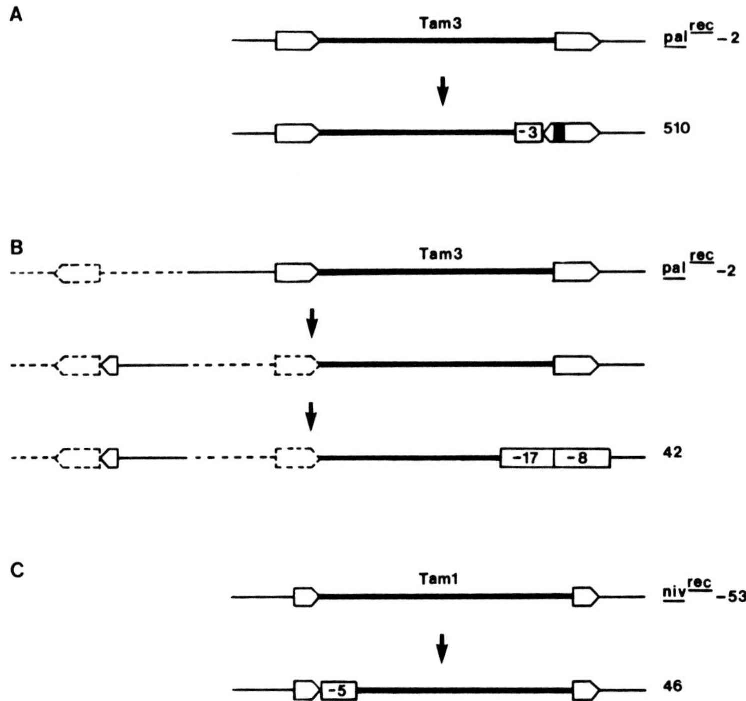

Figure 8. Structure of unstable alleles and their derivatives. The meaning of symbols is the same as in Fig. 4, although the sequence of target sites is not shown in detail. (A) Structure of *pal*-510 and its progenitor, pal^{rec}-2. (B) Structure of *pal*-42 and its progenitors. The broken boxed arrow indicates a sequence normally located about 5 map units away from the *pal* locus as illustrated for pal^{rec}-2. In the intermediate, illustrated immediately below pal^{rec}-2, an inversion has brought the broken boxed sequence next to the *pal* locus. The *pal*-42 allele was derived from this intermediate by a deletion at the right end of Tam3. (C) Structure of *niv*-46 and its progenitor niv^{rec}-53. Data taken from reference 41; Robbins et al., in press; and Hudson et al., in preparation.

IV. INTEGRATION OF TRANSPOSABLE ELEMENTS

A. Phenotypic Effects of Integration

The phenotypic consequences of various transposable elements can be considered systematically by grouping the insertions according to the element's position within the locus: promoter, exon, or intron. Promoter insertions represent the most interesting class because they can interfere with normal control of transcription. The pal^{rec}-2 allele contains Tam3 close to the start of *pal* transcription (Table 1). This results in abolition of anthocyanin pigment synthesis throughout the flower except in the revertant sectors caused by Tam3 excision (Fig. 2b). The simplest interpretation of the inhibitory effect of the Tam3 insertion is that it separates sequences in the promoter region that need to be near each other for effective gene transcription. However, the analysis of alleles derived from pal^{rec}-2 indicates that this view of Tam3 as a passive DNA spacer is an oversimplification.

The *pal*-510 allele was derived from pal^{rec}-2 and displays a much reduced level of somatic instability against a light background pigmentation. *pal*-510 carries a 3-bp deletion at the downstream end of Tam3 and a small inverted duplication of the flanking sequence, with the loss of one nucleotide at the axis of symmetry (Fig. 8A). The pale background pigmentation observed in the *pal*-510 allele indicates that Tam3 no longer fully inhibits *pal* gene expression. No readthrough transcript from Tam3 into the *pal* gene could be detected in *pal*-510, suggesting that the 3-bp deletion relieves inhibition by intact Tam3 rather than allowing gene activation from within the element (Hudson et al., in preparation).

A similar phenomenon is observed with a structurally more complex allele derived from pal^{rec}-2. The *pal*-42 allele is thought to have arisen in two distinct steps (Fig. 8B). The first was a large-scale inversion resulting from two double-strand breaks, one at the upstream end of Tam3 and the other at a position approximately 5 map units from the *pal* locus (to the left in Fig. 8B; Robbins et al., in press). A short sequence, present once at the left breakpoint in the progenitor pal^{rec}-2, has been duplicated in the process and flanks the inversion (Fig. 8B, broken boxed arrows). The second step giving rise to the *pal*-42 allele was a deletion of 25 bp at the downstream border of Tam3, including 17 bp at the end of the element (Fig. 8B). This deletion at the end of Tam3 reduces the frequency of somatic excision and permits some degree of background expression not observed with the progenitor *pal*^{rec}-2 (Fig. 2k).

The Tam1 insertion in the promoter of the *niv*^{rec}-53 allele (Table 1) inhibits *niv* gene expression. This allele confers a phenotype similar to that conferred by *pal*^{rec}-2, i.e., abolition of all background pigmentation (Fig. 2d). In the derived allele *niv*-46, a deletion of the five terminal nucleotides at the upstream end of Tam1 results in a loss of instability and a very pale background pigmentation, indicating a low level of *niv* gene activity (41) (Fig. 8C).

The observation that three independent small alterations at the termini of elements inserted in promoter regions (Fig. 8; *pal*-510, *pal*-42, and *niv*-46) affect both transposition and adjacent gene activity suggests a model for the mechanism of inhibition by the intact element. The termini of transposable elements are believed to be the sites at which transposition proteins bind, and the reduced excision frequency probably reflects a reduced ability of transposase or an associated protein to act at the deleted termini. In the model, binding of transposition proteins to intact Tam elements in gene regulatory regions excludes transcription proteins from these regions. In the case of transposition-defective Tam elements, decreased binding of transposition proteins permits greater access by the transcription proteins. A similar model has been proposed to explain the phenotypic effects of the maize *En*/*Spm* element (69), although in this case transcription elongation rather than initiation is affected (see Fedoroff, this volume). One difficulty with this model is how binding at the upstream end of Tam1, 15 kb from the start of transcription, could influence gene activity unless the termini of Tam1 are held together in a stable complex during gene expression throughout flower development. A further difficulty involves *niv*^{rec}-98, which contains a Tam3 insertion in the promoter of the *niv* gene in a position comparable to that of *pal*^{rec}-2 (Table 1). However, in contrast to *pal*^{rec}-2, the *niv*^{rec}-98 allele confers a significant background level of expression (Fig. 2c). These differences might reflect dissimilar promoter structures of the *niv* and *pal* genes or the opposite orientations of Tam3 with respect to transcription in these two alleles.

Perhaps the most complex phenotype resulting from a transposable element insertion and rearrangement in the promoter is *niv*-543 (Fig. 2m). This allele was derived from *niv*^{rec}-98, and preliminary analysis suggests that it contains an intact Tam3 element flanked by an inverted duplication of about 200 bp similar to *niv*-525 (Section III.A; E. Coen and R. Carpenter, unpublished results). *niv*-543 demonstrates somatic instability (Fig. 2m) in addition to being semidominant, as documented earlier for the stable *niv*-525 allele.

Insertions within transcribed regions can affect mRNA elongation, termination, and processing. The null phenotype due to *niv*-44 Tam2 insertion may reflect altered RNA processing. Tam2 disrupts the normal splice donor site at the first exon/intron boundary (Table 1), although the upstream *niv*/Tam2 border (GTGTACAC) might still act as a splice donor site since it matches 5 of 8 nucleotides of the normal *Niv*$^{+}$ donor site (GTGTAAGA). Tam2 may also contain additional splice donor/acceptor sites which result in aberrant splicing, as is known for maize elements (48, 88). Alternatively, Tam2 may contain a termination signal as has been found for *Spm* of maize (32).

Perhaps the simplest result of Tam insertion to understand is in the case of *niv*-540. This allele was derived from *niv*^{rec}-98 and contains Tam3 inserted in the last exon of the *niv* gene, with the result that all background pigmentation is abolished (Fig. 2l).

B. Mechanisms of Integration

The relationship between excision and integration of transposable elements in plants has been most intensively studied in maize for the *Ac* element (14, 33–35). Starting with a plant carrying one copy of an active *Ac* at a single locus, it was possible to monitor the fate of the excised *Ac* element by virtue of its dosage effect on the excision frequency of *Ac* and *Ds* elements (see Fedoroff, this volume). By studying large twin sectors, all the products of a transposition event could be recovered after mitotic division. The results showed that excision is generally, perhaps always, followed by integration at a new site, suggesting that excision and integration represent two steps in a conservative transposition event. However, the mechanism which linked the two steps was not clear from these studies since it was not possible to study intermediates in the transposition pathway. One way to investigate these intermediates is to analyze the products of aberrant transpositions. Detailed analysis of several of these aberrant events in *A. majus* indicates that Tam3 transposes in a manner similar to *Ac*, and suggests a model of transposition which requires a physical association of donor and recipient sites and involves the sequential transfer of the element termini.

We propose that, during transposition, the donor and recipient sites associate (Fig. 9A). A double-strand break is first made at one end of the element, and the flanking DNA is self-ligated to give a hairpin (step a) as in excision (Fig. 7; Section III.B). In addition, a staggered double-strand break in the target DNA results in overhanging ends of a length corresponding to the target duplication characteristic for the element (3 bases for Tam1 and Tam2, 5 or 8 bases for Tam3; see Table 1). The released end of the element then ligates to one of the overhanging ends at the target site such that donor and recipient sequences are temporarily linked via the element (Fig. 9, step b). A second cycle of breakage and ligation for the other end of the element results in the transfer of the entire element to the recipient site (step c). Transposition is completed by the ligation of the two hairpin structures at the donor site, resulting in the imprecise restoration of the original sequence as documented in Section III.B (Fig. 9, step d). DNA repair synthesis at the recipient site would produce the characteristic target site duplication flanking the insertion (broken arrows, step d).

The evidence which leads us to propose the sequential transfer of the element termini comes from aberrant excision events in *A. majus* in which the element remains at a locus but one end of the element, and sometimes also flanking DNA, has undergone sequence changes. This is illustrated by *pal*-510, in which Tam3 has lost 3 bp from one end and is flanked by a short inverted duplication of the target lacking a single nucleotide at the axis of symmetry (Fig. 8A). This would reflect an aborted excision event in which only one Tam3 end is cleaved, digested by nuclease, and religated to the flanking sequence. The inverted duplication of the flanking sequence with loss of a nucleotide would reflect the formation of a hairpin intermediate as postulated for Tam3 excision (Section III.B). The absence of change at the other Tam3 end supports the view that transposition is a sequential process involving first one end and then the other. More extensive inverted duplications adjacent to one Tam3 end (54; *niv*-543, Section IV.A) may have arisen similarly. As with normal excision (Section III.B), reciprocal cases in which a flanking sequence was deleted rather than duplicated would also be expected. This is illustrated by *pal*-42, which has a 17-bp deletion at one end of Tam3 with an 8-bp deletion of flanking DNA (Fig. 8B). *niv*-46 has a 5-bp deletion at one end of Tam1 and does not have any change in the flanking sequence, reflecting a more precise religation to the flanking sequence (41) (Fig. 8C). Events involving cleavage at only one end could also explain the ability at one half of a *Ds* element to cause chromosome breaks in maize (20, 25). In this case only one *Ds* end is present, so that its cleavage would produce one hairpin without a partner. Failure to resolve this hairpin by nicking and religation to the half-*Ds* would give a chromosome break and produce a dicentric or acentric chromosome (see Section III.B).

The evidence for the association of donor and recipient sites during transposition comes primarily from the structure of the large inversion of *pal*-42 (Fig. 8B). Inversions associated with eucaryotic transposable elements are often explained by recombination between pairs of elements present at both ends of the breakpoints (22, 62, 68). In the case of *pal*-42, Tam3 is only found at one end of the inversion and there is no evidence for the presence of Tam3 at the other breakpoint in either the progenitor *pal*^{rec}-2 or the derivatives *pal*-42 and *pal*-41. The other breakpoint does, however, have a short sequence of 5 bp which has been duplicated in the rearrangement and flanks both ends of the inversion in *pal*-42 (Fig. 8B). Tam3 integration can also produce a 5-bp duplication of host DNA, suggesting that *pal*-42 arose by an attempted integration of Tam3 at the inversion breakpoint. According to this view, the junctions at the inversion breakpoints in *pal*-42 have resulted from aberrant ligation of the cleaved ends from the *pal* Tam3 donor site to the ends of the recipient site, supporting the notion that the two sites are physically associated during transposition (Fig. 9B) (Robbins et al., in press).

The result of the aberrant transposition shown in Fig. 9 is an inversion flanked by the duplicated target site (broken arrows). The alternative ligation of the transferred Tam3 end to the end of the recipient site opposite to the one diagrammed in Fig. 9B (step b) would result in a deletion rather than an inversion adjacent to an intact element. Larger deletions, some of which are lethal when homozygous, have been found adjacent to apparently intact copies of Tam3 at the *niv* locus (54).

The model presented above bears certain similarities to a model proposed for transposition in procaryotes which also requires a physical association of donor and recipient sites (72). This model was put forward to explain transposition by a replicative pathway in which excision does not occur. However, the model has subsequently been modified to include procaryotic elements which transpose by a conservative mechanism, such as Tn*5* (7) and Tn*10* (5), by introducing additional single-strand cuts which release the element from the donor site before replication of the element can occur (for review, see reference 23). Such a unified model for replicative and conservative transposition could be extended to include plant transposable elements by invoking a mechanism for resealing the donor site (e.g., by resolving and ligating two hairpin structures). In

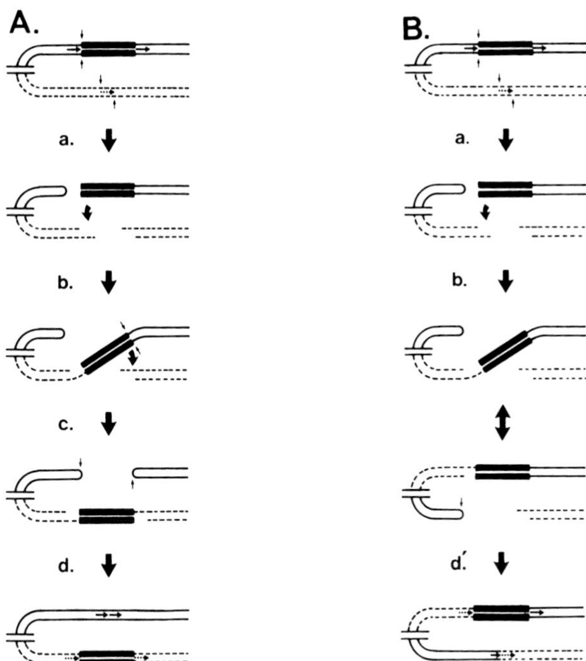

Figure 9. Model for both normal and aberrant transposition in *A. majus*. The model shows the particular case of intrachromosomal transposition of an element (thick solid lines) from a donor site (thin solid lines) to a recipient site (thin broken lines) which are separated by an unspecified distance. Each line represents a single strand of the DNA duplex. (A) Normal transposition. The element is initially flanked by a direct duplication of donor site sequence (solid arrows between DNA strands) and ends up flanked by a direct duplication of recipient site sequence (broken arrows between DNA strands). The small arrows perpendicular to the DNA strands indicate sites of single-strand cleavage. The model for normal transposition comprises the following steps: (a) double-strand break at one end of element, staggered double-strand break at recipient site; (b) first end of element ligates to recipient site; (c) double-strand break at second end of element, second end of element ligates to recipient site; (d) resealing of donor site by hairpin-hairpin ligation, DNA repair synthesis to give target site duplication. (B) Model for aberrant transposition giving rise to rearrangement. Model comprises the same initial steps (a and b), but the transfer of the second end of the element (step c in model A) fails to occur. The resulting structure has been redrawn to make clearer the consequences of the subsequent ligation (step d') of a donor site hairpin to the remaining free end at the recipient site. DNA repair synthesis completes the target site duplication, which flanks a rearrangement rather than the element as in model A.

procaryotes, intramolecular transposition by the replicative pathway is known to generate either deletions flanking an element or inversions flanked by a replicated element (36). If replicative transposition occurred occasionally for Tam3, this model could explain the origin of the *pal*-42 inversion, although it is necessary to assume a subsequent excision of Tam3 at one inversion breakpoint. Replicative transposition could also explain a rearrangement flanked by two copies of Tam3 derived from *niv*^{rec}-98 (54).

A common feature of these models for the origin of *pal*-42 is the requirement for an association of donor and recipient sites during the transposition process. This rules out models of transposition in which an excised element diffuses freely in the nucleoplasm. Transposition should also occur preferentially between those loci that are most likely to become associated in the interphase nucleus. This might explain why two-thirds of transpositions of *Ac* in maize are to linked positions (85) and almost half are within 5 map units (34). The most likely site to become associated with a donor site might be closely linked due to the spatial arrangement of chromatin in the nucleus. In this respect, transposition would be expected to be under physical constraints similar to other processes requiring association between sequences at different chromosome positions. For example, nonallelic gene conversion also occurs more frequently between sequences on the same chromosome (46) than between sequences on different chromosomes (67).

The mechanism of transposition described here is mainly based on studies of the Tam3 element of *A. majus* and of *Ac* in maize. It is likely that the transposable elements Tam1/Tam2 of *A. majus* and *En/Spm* of maize transpose by a similar mechanism. The structure of their excision sites is similar to those found for Tam3 and *Ac*, although the palindromes produced show a consistent difference of one nucleotide at the axis of symmetry (see Section III.B). Furthermore, genetic studies show that excision of *En/Spm* in maize is correlated with integration at new sites, often on the same chromosome (56, 60), consistent with the conservative mechanism of transposition described here.

The phenomena associated with Tam elements can be compared with those documented for elements from other eucaryotes that are thought to transpose by an excision/integration-type mechanism. In *D. melanogaster* there is evidence that at least two types of element may transpose by such a mechanism, the P elements (52) and the large TE element (16, 44). However, the excision of both types of element differs significantly from that of Tam elements, since part of the element remains at the excision site (see Section III.B above and Engels, this volume). P elements are known to be associated with chromosome rearrangements, predominantly simple inversions, that can often be explained in terms of breaks at element termini (28). However, some inversions do not have P elements at both breakpoints, either in the progenitor or after rearrangement, and may arise by the same mechanism as the *pal*-42 inversion. The large TE element shows a marked preference for transposition to linked positions on the same chromosome (44), and the same may be true

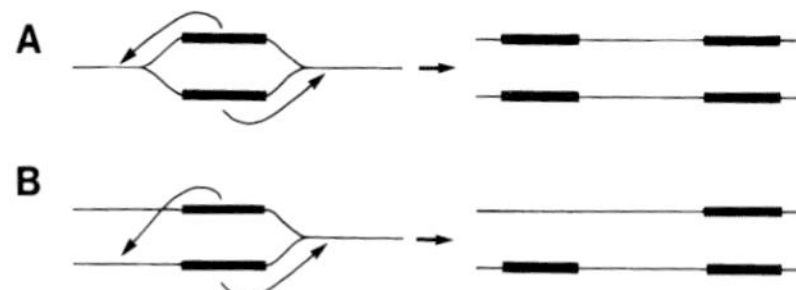

Figure 10. Models to explain the structure of the *niv*-540 allele. Each line represents a double strand of DNA, and the thick solid lines represent Tam3. Both models propose that after the passage of a replication fork the two copies of Tam3 produced both transpose to new sites. In model A, both copies of Tam3 transpose into unreplicated DNA. In model B, one copy of Tam3 transposes into unreplicated DNA while the other copy transposes into replicated DNA of the sister chromatid. The resulting chromatids after completion of replication are shown to the right in each case. In all cases the original site occupied by the element has been vacated.

for up to a third of P element transpositions (52). This suggests that these elements might also transpose by a mechanism requiring association of donor and recipient sites, although direct evidence for this is still lacking.

If Tam transposition does not cause the replication of Tam, how is the Tam copy number maintained or increased against selective pressure? We imagine that elements may preferentially transpose during S phase from replicated to unreplicated sites, thereby increasing the copy number in one sister chromatid. Evidence of transposition following chromosome replication in *A. majus* is provided by *niv*-540, an allele that is derived by the loss of Tam3 from its original position in the promoter and that carries two new Tam3 copies, one approximately 3 kb upstream and the other a comparable distance downstream of the excision site (Hudson et al., in preparation). A simple explanation for the origin of *niv*-540 is that the new Tam3 copies resulted from reintegration of the excised Tam3, consistent with an excision/integration mechanism which favors closely linked sites. The observation of two new Tam3 copies suggests that transposition occurred after replication of the donor site, into either replicated or unreplicated sites (Fig. 10). This interpretation is compatible with studies in maize which show that *Ac* transposition occurs after replication of the donor site (14, 35). If both target sites were unreplicated, as shown in Fig. 9A, it is necessary to propose an origin of replication in or close to Tam3. This suggests the interesting possibility that Tam3 may carry its own origin of replication. If this origin ensured early replication in the cell cycle, it would raise the probability that the element transposed into an unreplicated site and hence increased in copy number. It is interesting that the procaryotic element Tn*10* also transposes after chromosome replication (61), and perhaps this is a general feature of elements that transpose predominantly by a conservative pathway.

C. Factors Affecting Integration

De novo integration at a locus is a rare event since a single gene represents a relatively small target in the genome. However, it has been possible to monitor the frequency of Tam3 integration by using a line of *A. majus*, TR.75, which has full red wild-type flowers and was originally derived from a revertant of *pal*^{rec}-2. When the TR.75 line is crossed to a plant carrying a null allele of a gene required for pigment synthesis, the F_1 heterozygote shows white or pale sites of pigmentation on a full red background. These sites have been proposed to be caused by integration of a transposable element (later shown to be Tam3) into the wild-type allele of the heterozygote, thus blocking its function (38). Integration presumably also occurs in the TR.75 parent line but is not easily observed since both alleles of an anthocyanin gene would have to be inactivated to inhibit pigmentation. The frequency of white sites shows the same response to temperature and to the *Stabilizer* allele that Tam3 excision does (Section III.C), which provides strong evidence that excision and integration are two parts of the same process of conservative transposition.

As discussed in the previous section, *Ac* preferentially integrates at positions closely linked to the excision site, and in two cases Tam3 has been shown to integrate very close (3 kb away) to the excision site. The frequency of integration at a locus may therefore be much greater if a transposable element is located nearby. The *niv*-540 allele has a copy of Tam3 located about 3 kb upstream from the *niv* gene and a second Tam3 copy located in the third exon (Section IV.A). A phenotype similar to that of the TR.75 line is conditioned by a derivative of the *niv*-540 allele in which the copy of Tam3 in the exon has excised but the upstream copy has been retained. When heterozygous with a *niv* null allele, this allele gives full red flowers with a high frequency of white sites (Hudson et al., in preparation). The white sites may be a result of transposition of the remaining copy of Tam3 from its upstream position into the coding region, inactivating the *niv* gene. This interpretation is supported by the finding that plants in which the upstream Tam3 copy has transposed away from *niv* have full red flowers without white sites. The high frequency of white sites observed in the presence of the upstream Tam3 suggests that short distance transpositions of this type are relatively common, consistent with the model of transposition which involves association between donor and recipient sites (see previous section).

V. TRANSPOSABLE ELEMENTS AS GENETIC TOOLS

The twin properties of transposable elements, excision and integration, allow them to be used as powerful tools for the study of genes in *A. majus*. The integration of an element into a gene can result in a mutant phenotype and may allow the gene to be isolated by using the transposable element as a molecular tag. The property of excision allows proof that the correct gene has been isolated by tagging and also provides a means of generating large numbers of mutations at specific positions in the gene in an isogenic background. These properties may be used to study any gene in which mutations cause an easily scored phenotypic effect. To illustrate the application of these principles in *A. majus*, we first consider the isolation and analysis of the *pallida* locus.

As described in Section IV.C, the TR.75 line was thought to contain a transposable element showing a high rate of integration in somatic tissue and having properties similar to those of the element responsible for the *pal*^{rec} mutation. If this interpretation were correct, it should have been possible to recover germinal integrations of this element. For this reason, the TR.75 line was grown at 15°C, to enhance transposition of the element, and crossed to lines homozygous for null alleles of the *niv*, *inc*, or *pal* loci. If the transposable element were to integrate into the wild-type allele of one of these loci in the germinal tissue of the TR.75 parent, a new unstable phenotype should be observed in the F_1 heterozygote. Out of about 10,000 plants screened, 1 showed a new unstable phenotype. This plant had been derived from the cross between TR.75 and a plant homozygous for null alleles of both the *niv* and *inc* loci. Genetic analysis showed that this phenotype was due to a new allele of the *niv* locus, called *niv*^{rec}-98 (38). The availability of a probe for the *niv* locus enabled this allele to be cloned; it was found to carry a 3.5-kb transposable element, Tam3, inserted in the *niv* promoter (74).

Assuming that the transposable element most active in the TR.75 line is of the same type as the one responsible for the *pal*^{rec} mutation, the Tam3 element trapped by the *niv* locus should be homologous to the element causing the *pal*^{rec} phenotype. This was tested by probing *Eco*RI-digested DNA (*Eco*RI does not cut within Tam3) from plants homozygous for *pal*^{rec} or *Pal*$^{+}$ alleles with Tam3. Among a total of about 15 hybridizing bands, 1 of 7.8 kb was consistently seen in DNA from *pal*^{rec} plants but not from *Pal*$^{+}$ plants. This band was cloned and shown to contain a 3.5-kb copy of Tam3, flanked on each side by about 2 kb of single-copy DNA. Proof that this corresponded to the *pal* locus was obtained by using the flanking sequence to probe *Eco*RI-digested DNA from *pal*^{rec} and full red revertant plants. All the *pal*^{rec} plants gave a band of 7.8 kb, whereas four independent revertants had bands of 4.3 kb due to excision of Tam3 (53).

As described in previous sections, the availability of a clone of the *pal* locus has enabled many alleles resulting from Tam3 excision or integration to be studied in detail. In several cases the alterations at the *pal* locus were produced in an isogenic background by self-pollination, so that subtle quantitative changes in gene expression could be studied. It should now be possible to extend this type of analysis to other loci in *A. majus* which control various aspects of plant development.

As mentioned in the introduction, unstable mutations in genes involved in diverse aspects of *A. majus* development were extensively described in the earlier part of this century (79). For example, *deficiens*^{globifera} is a recessive allele that results in a homoeotic change: sepals grow in place of petals and stigmatic tissue in place of stamens (Fig. 2o). This allele can revert gametically to wild type. Furthermore, small areas of petal tissue can sometimes be seen on *deficiens*^{globifera} flowers, indicating somatic reversion early in flower development. Similarly, a recessive allele, *cycloidea*^{radialis} (Fig. 2n), which produces flowers with radial symmetry (actinomorphic), regularly reverts to the wild-type allele, giving flowers with the normal bilateral symmetry (zygomorphic). The *cycloidea*^{radialis} mutation is also homoeotic, since all the petals in the symmetrical flower resemble the lower middle petal of the wild-type flower. It is tempting to speculate that these unstable mutations are due to the insertion of transposable elements into genes involved in flower morphogenesis. However, it is unlikely that they are due to integration of an element for which a probe is available, since the mutations were not produced in a manner designed to introduce one specific type of element. The isolation and analysis of these genes therefore requires new unstable mutations to be produced by using lines which are known to carry active copies of a specific transposable element.

In most cases the insertion of a transposable element into a locus will produce a recessive allele. Two approaches may be adopted to screen for these types of mutation. One approach involves crossing a line carrying highly active transposable elements with a line homozygous for a recessive allele at the desired locus. As described for the Tam3 integration at *niv*, large numbers of F_1 plants (10^4 to 10^5) need to be screened for any exceptional progeny which show the mutant phenotype. Such progeny would, optimally, be heterozygous for the recessive allele and a new unstable allele of the locus produced by integration of the transposable element. One problem with this approach is discriminating genetically between the new unstable allele and the original recessive allele of the locus used in the cross. This is not a problem for cell autonomous characters such as flower pigmentation, since the new unstable allele can be identified by the variegated phenotype it confers. However, for genes involved in plant morphogenesis, distinguishing the unstable allele may be difficult unless genetic markers tightly linked to the locus are used in the parents of the original cross. A second disadvantage of this approach is that only one locus is screened for, so that tagging many loci would require large-scale screening for each locus independently.

The second approach to generating new insertions, which does not suffer from these disadvantages, involves self-pollination of large numbers of plants. We have pursued this approach in *A. majus* by growing the unstable TR.75 line, or lines closely related to it, at 15°C to enhance Tam3 transposition specifically. Large numbers of progeny from these plants were grown (6,500) and self-pollinated, and an average of three seeds per plant were grown to screen for any recessive mutations which may have been produced by Tam3 integration in the grandparents grown at 15°C. Several mutations, affecting the morphology or pigmentation of flowers, leaves, or the whole plant, have been recovered, although it is not yet known how many are due to Tam3 insertion (R. Carpenter and E. Coen, unpublished results). Some of the pigmentation mutants have a variegated phenotype, consistent with their being induced by insertion of transposable elements. For example, a variegated leaf mutant has sectors of dark green pigmentation on a yellow background (Fig. 2p). The number of dark green spots increases dramatically when the plant is grown at 15°C, as expected if the mutation is due to Tam3 integration. It should now be possible to obtain germinal revertants of some of these mutants. Molecular and genetic analysis of such revertants should establish how many of the mutations are due to Tam3 integration and, hence, whether the genes involved can be isolated by using the Tam3 probe.

Once genes have been isolated by this method, it should be possible to exploit the mutagenic property of Tam3 to produce allelic series similar to those described for the *pal* locus. The range of phenotypes obtained would clearly depend on the particular site of Tam3 insertion in the gene. This seems feasible since allelic series at many loci which alter morphology have already been described in *A. majus* (79). For example, several alleles of the *deficiens* locus confer phenotypes intermediate between *deficiens*^{globifera} and wild type (3, 42) (Fig. 11). Furthermore, since this approach to gene isolation involves only self-pollination, it should be possible to produce allelic series in isogenic backgrounds to allow comparisons between alleles conferring subtle differences in phenotype.

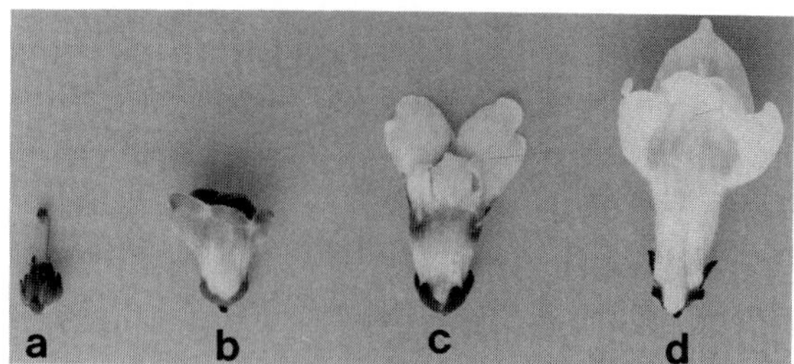

Figure 11. Allelic series at the *deficiens* locus. Flowers shown are homozygous for (a) *deficiens*^{globifera}; (b) *deficiens*^{nicotianoides}; (c) *deficiens*^{chlorantha}; (d) *Deficiens*$^{+}$.

VI. EVOLUTIONARY IMPLICATIONS

The ability of transposable elements to generate a high degree of genetic diversity suggests that they may have important consequences for evolution in natural populations. However, there is little information on the genetic and molecular basis of natural variation, so assessing the evolutionary importance of transposable elements has to be based on indirect evidence.

Although plant transposable elements have only been characterized in detail in two species, *A. majus* and *Z. mays*, it is likely that they are ubiquitous. Unstable alleles with features typical of mutations caused by transposable elements are known in over 30 plant species (57). Many stable alleles may also be due to transposable elements, since alleles (e.g., *niv*-44) have been described which carry inactive elements. Cloned transposable elements from *A. majus* have been used to probe DNA from natural populations of *A. majus* or DNA from other species representative of the *Antirrhinum* genus. All eight *Antirrhinum* species investigated contain copies of all three Tam elements, and in the case of Tam3, the overall internal organization and length of the element was conserved (E. Coen, S. Mackay, and R. Carpenter, unpublished results). A copy of Tam3 from the New World species *Antirrhinum nuttalianum* was cloned and showed a restriction map similar to that of Tam3 from the European *A. majus*, indicating that Tam3 may have been conserved since the continents separated. No Tam3 sequences were detected by Southern hybridization with species outside the *Antirrhinum* genus with the exception of *Misopates orontium*, which is phylogenetically so close to *Antirrhinum* that it has often been classified as *Antirrhinum orontium* (63). These observations do not rule out the possibility that many of the nonhybridizing species contain Tam3-related DNA, since the sequences may have diverged to such a degree that they are no longer detectable by these methods. This interpretation is supported by the observation that transposable elements from *Z. mays* show sequence homology at their ends with those of *A. majus*, although not sufficient to result in DNA hybridization (10, 74, 83). This suggests the possibility that transposable elements in monocotyledonous (*Z. mays*) and dicotyledonous (*A. majus*) plants share a common progenitor which existed over 100 million years ago (77), but that during the evolution of each lineage they have diverged and acquired their individual characteristics. This requires a mechanism which allows transposable elements to maintain themselves in the genome for long periods of time and which also ensures that similarity is maintained between different copies of an element within a species even though substantial divergence may occur between families of elements in different species.

As described earlier, the Tam elements of *A. majus* studied to date probably transpose by an excision/integration mechanism rather than by replicative transposition. If excision and integration simply involved the movement of an element from one position to another on the chromosome, without any increase in copy number, elements would eventually be lost from sexual populations since deleterious mutations caused by integration in genes would result in overall selection against the transposable element. However, as described in Section IV.B, transposition can occur after DNA replication and into unreplicated sites, resulting in a net increase in copy number. This strategy for replication is particularly effective if, as suggested, the element carries an origin of replication which acts early in the cell cycle and hence continually ensures that most transpositions will be into unreplicated DNA. Such an origin would be favored in evolution since elements which lost the origin would be at a replicative disadvantage compared with those that did not.

Given the ability of Tam3 to increase in copy number, there must also be constraints which prevent excessive amplification, since there are generally 15 to 30 copies of Tam3 in the genome (judging by Southern blots). Selection against harmful mutations induced by integration may be one constraint, although the ability of certain genes to repress transposition of Tam elements (see Section IV.C) indicates genetic constraints on copy number. These repressors may themselves be related to the elements they affect and hence be part of an autoregulatory system in which the elements control their own copy number. The current copy number of Tam3 in a population may reflect an equilibrium in which the rates of gain and loss of elements are similar. Such an equilibrium would result in a low rate of turnover of elements in the population, ensuring that different copies of the same family of elements retain some sequence similarity while divergence may occur between families in different species.

What might the consequences of the continual presence of transposable elements in the genome be for the phenotype of the organism? One of the most important properties of transposable elements is their ability to generate genetic variation. This has several consequences for natural populations. Particular regions of the genome which have recently carried transposable elements may be very polymorphic because of numerous different imprecise excisions or rearrangements. When these polymorphic regions occur near promoters or other regulatory regions, they will give rise to quantitative variation which could be the basis of much of the variation described as polygenic in natural populations. For example, an allelic series of the *rosea* locus is responsible for much of the color variation in natural populations of *A. majus* (4). Perhaps this allelic series has arisen from imprecise excisions of a transposable element from the *rosea* locus promoter. Promoter mutations may also change the spatial pattern of gene expression by changing the way in which the promoter interacts with spatially distributed transcription factors. Changes in regulatory regions are likely to be responsible for producing diversity in spatial patterns in loci such as *yellow* in *D. melanogaster* (15), *succinea* in the ladybird *Harmonia axyridis* (81), and the *R* locus of *Z. mays* (80). The ability of elements to generate numerous alleles in regulatory regions may therefore be important in the evolution of new patterns of gene expression and hence in the production of phenotypic novelty.

Acknowledgments. We would like to thank Brian Harrison and Joachim Bollmann for critical comments on the manuscript and Andrew Davis and Peter Scott for photography.

LITERATURE CITED

1. **Anonymous.** 1838. *Paxton's Magazine of Botany* 5:55–56.
2. **Baker, B., J. Schell, H. Lörz, and N. Fedoroff.** 1986. Transposition of the maize controlling element "activator" in maize. *Proc. Natl. Acad. Sci. USA* 83:4844–4848.
3. **Baur, E.** 1924. Untersuchungen über das Wesen, die Entstehung und Vererbung von Rassenunterschieden bei *Antirrhinum majus*. *Bibl. Genet.* 4:1–70.

4. **Baur, E.** 1933. Artumgrenzung und Artbildung in der Gattung *Antirrhinum*, Sektion Antirrhinastrum. *Z. Indukt. Abstamm. Vererbungsl.* **63**:256–302.
5. **Bender, J., and N. Kleckner.** 1986. Genetic evidence that Tn10 transposes by a nonreplicative mechanism. *Cell* **45**: 801–815.
6. **Bennetzen, J. L.** 1987. Covalent DNA modification and the regulation of *Mutator* element transposition in maize. *Mol. Gen. Genet.* **208**:45–51.
7. **Berg, D. E.** 1983. Structural requirement for IS*50*-mediated gene transposition. *Proc. Natl. Acad. Sci. USA* **80**:792–796.
8. **Bingham, P. M., and Z. Zachar.** 1985. Evidence that two mutations, w^{DZL} and z^1, affecting synapsis-dependent genetic behavior of *white* are transcriptional regulatory mutations. *Cell* **40**:819–825.
9. **Bonas, U., H. Sommer, B. J. Harrison, and H. Saedler.** 1984. The transposable element Tam1 of *Antirrhinum majus* is 17 kb long. *Mol. Gen. Genet.* **194**:138–143.
10. **Bonas, U., H. Sommer, and H. Saedler.** 1984. The 17kb Tam1 element of *Antirrhinum majus* induces a 3bp duplication upon integration into the chalcone synthase gene. *EMBO J.* **3**: 1015–1019.
11. **Carpenter, R., E. S. Coen, A. D. Hudson, and C. R. Martin.** 1984. Transposable genetic elements and genetic instability in *Antirrhinum. John Innes Inst. Annu. Rep.* **73**:52–64.
12. **Carpenter, R., C. Martin, and E. S. Coen.** 1987. Comparison of genetic behaviour of the transposable element Tam3 at two unlinked pigment loci in *Antirrhinum majus. Mol. Gen. Genet.* **207**:82–89.
13. **Chandler, V. I., and V. Walbot.** 1986. DNA modification of a maize transposable element correlates with loss of activity. *Proc. Natl. Acad. Sci. USA* **83**:1767–1771.
14. **Chen, J., I. M. Greenblatt, and S. L. Dellaporta.** 1987. Transposition of Ac from the *P* locus of maize into unreplicated chromosomal sites. *Genetics* **117**:109–116.
15. **Chia, W., G. Howes, M. Martin, Y. Meng, K. Moses, and S. Tsubota.** 1986. Molecular analysis of the *yellow* locus of *Drosophila. EMBO J.* **5**:3597–3605.
16. **Chia, W., S. McGill, R. Karp, D. Gubb, and M. Ashburner.** 1985. Spontaneous excision of a large composite transposable element of *Drosophila melanogaster. Nature* (London) **316**: 81–83.
17. **Chomet, P. S., S. Wessler, and S. L. Dellaporta.** 1987. Inactivation of the maize transposable element *Activator* (*Ac*) is associated with its DNA modification. *EMBO J.* **6**: 295–302.
18. **Coen, E. S., and R. Carpenter.** 1988. A semi-dominant allele, *niv-525*, acts in *trans* to inhibit the expression of its wild-type homologue in *Antirrhinum majus. EMBO J.* **7**:877–883.
19. **Coen, E. S., R. Carpenter, and C. Martin.** 1986. Transposable elements generate novel spatial patterns of gene expression in *Antirrhinum majus. Cell* **47**:285–296.
20. **Courage-Tebbe, U., H. P. Döring, N. Fedoroff, and P. Starlinger.** 1983. The controlling element *Ds* at the *Shrunken* locus in Zea mays: structure of the unstable *sh-m5933* allele and several revertants. *Cell* **34**:383–393.
21. **Darwin, C. R.** 1868. *The Variation of Animals and Plants under Domestication.* John Murray, London.
22. **Davis, P. S., M. W. Shen, and B. H. Judd.** 1987. Asymmetrical pairings of transposons in and proximal to the white locus of *Drosophila* account for four classes of regularly occurring exchange products. *Proc. Natl. Acad. Sci. USA* **84**:174–178.
23. **Derbyshire, K. M., and N. D. F. Grindley.** 1986. Replicative and conservative transposition in bacteria. *Cell* **47**:325–327.
24. **de Vries, H.** 1910. *The Mutation Theory 2.* Open Court Publishing Co., Peru, Ill. (German original published in 1903, Verlag von Veit u. Co., Leipzig.)
25. **Döring, H.-P., R. Garber, B. Nelsen, and E. Tillmann.** 1985. Transposable element Ds and its chromosomal rearrangements, p. 355–367. *In* M. Freeling (ed.), *Plant Genetics.* Alan R. Liss, Inc., New York.
26. **Ecker, J. R., and R. W. Davis.** 1986. Inhibition of gene expression in plant cells by expression of anti-sense RNA. *Proc. Natl. Acad. Sci. USA* **83**:5372–5376.
27. **Emerson, R. A.** 1914. The inheritance of a recurring somatic variation in variegated ears of maize. *Am. Nat.* **48**:87–115.
28. **Engels, W. R., and C. R. Preston.** 1984. Formation of chromosome rearrangements by P factors in *Drosophila. Genetics* **107**:657–678.
29. **Fedoroff, N.** 1983. Controlling elements in maize, p. 1–63. *In* J. Shapiro (ed.), *Mobile Genetic Elements.* Academic Press, Inc., New York.
30. **Fincham, J. R. S., and B. Harrison.** 1967. Instability at the *Pal* locus in *Antirrhinum majus.* II. Multiple alleles produced by mutation of one original unstable allele. *Heredity* **22**: 211–227.
31. **Geyer, P. K., and V. G. Corces.** 1987. Separate regulatory elements are responsible for the complex pattern of tissue-specific and developmental transcription of the *yellow* locus in *Drosophila melanogaster. Genes Dev.* **1**:996–1004.
32. **Gierl, A., Z. Schwarz-Sommer, and H. Saedler.** 1985. Molecular interactions between the components of the En-I transposable element system of *Zea mays. EMBO J.* **4**:579–583.
33. **Greenblatt, I. M.** 1974. Movement of Modulator in maize: a test of an hypothesis. *Genetics* **77**:671–678.
34. **Greenblatt, I. M.** 1984. A chromosome replication pattern deduced from pericarp phenotypes resulting from movements of the transposable element Modulator, in maize. *Genetics* **108**:471–485.
35. **Greenblatt, I. M., and R. A. Brink.** 1962. Twin mutations in medium variegated pericarp maize. *Genetics* **47**:489–501.
36. **Grindley, N. D. F., and R. R. Reed.** 1985. Transpositional recombination in prokaryotes. *Annu. Rev. Biochem.* **54**: 863–896.
37. **Harrison, B. J., and R. Carpenter.** 1973. A comparison of the instabilities at the *nivea* and *pallida* loci in *Antirrhinum majus. Heredity* **31**:309–323.
38. **Harrison, B. J., and R. Carpenter.** 1979. Resurgence of genetic instability in *Antirrhinum majus. Mutat. Res.* **63**:47–66.
39. **Harrison, B. J., and J. R. S. Fincham.** 1964. Instability at the *Pal* locus in *Antirrhinum majus.* 1. Effects of environment on frequencies of somatic and germinal mutation. *Heredity* **19**: 237–258.
40. **Harrison, B. J., and J. R. S. Fincham.** 1968. Instability at the *Pal* locus in *Antirrhinum majus.* 3. A gene controlling mutation frequency. *Heredity* **23**:67–72.
41. **Hehl, R., H. Sommer, and H. Saedler.** 1987. Interaction between the Tam1 and Tam2 transposable elements of *Antirrhinum majus. Mol. Gen. Genet.* **207**:47–53.
42. **Hertwig, P.** 1926. Ein neuer Fall von multiplem Allelomorphismus bei Antirrhinum. *Z. Indukt. Abstamm. Vererbungsl.* **41**:42–47.
43. **Hudson, A., R. Carpenter, and E. S. Coen.** 1987. De novo activation of the transposable element Tam2 of *Antirrhinum majus. Mol. Gen. Genet.* **207**:54–59.
44. **Ising, G., and K. Block.** 1980. Derivation-dependent distribution of insertion sites for a *Drosophila* transposon. *Cold Spring Harbor Symp. Quant. Biol.* **65**:527–544.
45. **Jack, J. W., and B. H. Judd.** 1979. Allelic pairing and gene regulation: a model for the *zeste-white* interaction in *Dro-*

sophila melanogaster. Proc. Natl. Acad. Sci. USA **76:** 1368–1372.

46. **Jackson, J. A., and G. R. Fink.** 1981. Gene conversion between duplicated genetic elements in yeast. *Nature* (London) **292:**306–310.
47. **Kaulen, H., J. Schell, and F. Kreuzaler.** 1986. Light induced expression of the chimeric chalcone synthase-NPTII gene in tobacco cells. *EMBO J.* **5:**1–8.
48. **Kim, H.-Y., J. W. Schiefelbein, V. Raboy, D. B. Furtek, and O. E. Nelson, Jr.** 1987. RNA splicing permits expression of a maize gene with a defective Suppressor-mutator transposable element insertion in an exon. *Proc. Natl. Acad. Sci. USA* **84:**5863–5867.
49. **Krebbers, E., R. Hehl, R. Piotrowiak, W.-E. Lonnig, H. Sommer, and H. Saedler.** 1987. Molecular analysis of paramutant plants of *Antirrhinum majus* and the involvement of transposable elements. *Mol. Gen. Genet.* **209:**499–507.
50. **Kretschmer, P. J., and S. N. Cohen.** 1979. Effect of temperature on translocation frequency of the Tn*3* element. *J. Bacteriol.* **139:**515–519.
51. **Kuckuck, H.** 1936. Uber vier neue Serien multipler Allele bei *Antirrhinum majus. Z. Indukt. Abstamm. Vererbungsl.* **71:** 429–440.
52. **Levis, R., T. Hazelrigg, and G. M. Rubin.** 1985. Effects of genomic position on the expression of transduced copies of the *white* gene of *Drosophila. Science* **229:**558–561.
53. **Martin, C. R., R. Carpenter, H. Sommer, H. Saedler, and E. S. Coen.** 1985. Molecular analysis of instability in flower pigmentation of *Antirrhinum majus*, following isolation of the *pallida* locus by transposon tagging. *EMBO J.* **4:**1625–1630.
54. **Martin, C. R., S. MacKay, and R. Carpenter.** 1988. Large scale chromosomal restructuring is induced by the transposable element Tam3 at the *nivea* locus of *Antirrhinum majus. Genetics* **119:**171–184.
55. **McClintock, B.** 1949. Mutable loci in maize. *Carnegie Inst. Wash. Year Book* **48:**142–154.
56. **McClintock, B.** 1962. Topographical relations between elements of control systems in maize. *Carnegie Inst. Wash. Year Book* **61:**448–461.
57. **Nevers, P., N. Shepherd, and H. Saedler.** 1986. Plant transposable elements. *Adv. Bot. Res.* **12:**103–203.
58. **Paquin, C. E., and V. M. Williamson.** 1986. Ty insertions at two loci account for most spontaneous antimycin A resistance mutations during growth at 15°C of *Saccharomyces cerevisiae* strains lacking ADH1. *Mol. Cell. Biol.* **6:**70–79.
59. **Peacock, W. J., E. S. Dennis, W. L. Gerlach, M. M. Sachs, and D. Schwartz.** 1984. Insertion and excision of Ds controlling elements in maize. *Cold Spring Harbor Symp. Quant. Biol.* **49:**347–354.
60. **Peterson, P. A.** 1970. The *En* mutable system in maize. *Theor. Appl. Genet.* **40:**367–377.
61. **Roberts, D., B. C. Hoopes, W. R. McClure, and N. Kleckner.** 1985. IS10 transposition is regulated by DNA adenine methylation. *Cell* **43:**117–130.
62. **Roeder, G. S., and G. R. Fink.** 1983. Transposable elements in yeast, p. 299–328. *In* J. Shapiro (ed.), *Mobile Genetic Elements.* Academic Press, Inc., New York.
63. **Rothmaler, W.** 1956. Taxonomische Monographie der Gattung *Antirrhinum. Feddes Repert Specierum Nov. Regni Veg. Beih.* **136:**1–124.
64. **Rothstein, S. J., J. DiMaio, M. Strand, and D. Rice.** 1987. Stable and heritable inhibition of the expression of nopaline synthase in tobacco expressing antisense RNA. *Proc. Natl. Acad. Sci. USA* **84:**8439–8443.
65. **Sachs, M. M., W. J. Peacock, E. S. Dennis, and W. L. Gerlach.** 1983. Maize Ac/Ds controlling elements: a molecular viewpoint. *Maydica* **28:**289–302.
66. **Saedler, H., and P. Nevers.** 1985. Transposition in plants: a molecular model. *EMBO J.* **4:**585–590.
67. **Scherer, S., and R. W. Davis.** 1980. Recombination of dispersed repeated DNA sequences in yeast. *Science* **209:** 1380–1383.
68. **Schneuwly, S., A. Kuroiwa, and W. J. Gehring.** 1987. Molecular analysis of the dominant homoeotic *Antennapedia* phenotype. *EMBO J.* **6:**201–206.
69. **Schwartz-Sommer, Z., A. Gierl, R. Berntgen, and H. Saedler.** 1985. Sequence comparison of "states" of *a1-m-1* suggests a model for Spm (En) action. *EMBO J.* **4:**2439–2443.
70. **Schwartz-Sommer, Z., A. Gierl, H. Cuypers, P. A. Peterson, and H. Saedler.** 1985. Plant transposable elements generate the DNA sequence diversity needed in evolution. *EMBO J.* **4:**591–597.
71. **Schwartz-Sommer, Z., N. Shepherd, E. Tacke, A. Gierl, W. Rhode, L. Leclerq, M. Mattes, R. Berndtgen, P. A. Peterson, and H. Saedler.** 1987. Influence of transposable elements on the structure and function of the *A1* gene of *Zea mays. EMBO J.* **6:**287–294.
72. **Shapiro, J. A.** 1979. Molecular model for the transposition and replication of bacteriophage Mu and other transposable elements. *Proc. Natl. Acad. Sci. USA* **76:**1933–1937.
73. **Sommer, H., U. Bonas, and H. Saedler.** 1988. Transposon-induced alterations in the promoter region affect transcription of the chalcone synthase gene of *Antirrhinum majus. Mol. Gen. Genet.* **211:**49–55.
74. **Sommer, H., R. Carpenter, B. J. Harrison, and H. Saedler.** 1985. The transposable element Tam3 of *Antirrhinum majus* generates a novel type of sequence alteration upon excision. *Mol. Gen. Genet.* **199:**225–231.
75. **Sommer, H., and H. Saedler.** 1986. Structure of the chalcone synthase gene of *Antirrhinum majus. Mol. Gen. Genet.* **202:** 429–434.
76. **Spiribille, R., and G. Forkmann.** 1982. Genetic control of chalcone synthase activity in flowers of *Antirrhinum majus. Phytochemistry* **21:**2231–2234.
77. **Stewart, W. N.** 1983. *Paleobotany and the Evolution of Plants.* Cambridge University Press, Cambridge.
78. **Strickland, R. G., and B. J. Harrison.** 1974. Precursors and genetic control of pigmentation. 1. Induced biosynthesis of pelargonidin, cyanidin and delphinidin in *Antirrhinum majus. Heredity* **33:**108–112.
79. **Stubbe, H.** 1966. *Genetik and Zytologie von Antirrhinum L. sect. Antirrhinum.* VEB Gustav Fischer, Jena, German Democratic Republic.
80. **Styles, E. D., O. Ceska, and K. T. Seah.** 1973. Developmental differences in action of R and B alleles in maize. *Can. J. Genet. Cytol.* **15:**59–72.
81. **Tan, C. C.** 1945. Mosaic dominance in the inheritance of color patterns in the lady-bird beetle, *Harmonia axyridis. Genetics* **31:**195–210.
82. **Tsubota, S., and P. Schedl.** 1986. Hybrid dysgenesis-induced revertants of insertions at the 5′ end of the *rudimentary* gene in *Drosophila melanogaster*: transposon-induced control mutations. *Genetics* **114:**165–182.
83. **Upadhyaya, K. C., H. Sommer, E. Krebbers, and H. Saedler.** 1985. The paramutagenic line *niv*-44 has a 5kb insert, Tam2, in the chalcone synthase gene of *Antirrhinum majus. Mol. Gen. Genet.* **199:**201–207.
84. **van der Ende, A., S. A. Langeveld, R. Teertstra, G. A. van

sophila melanogaster. *Proc. Natl. Acad. Sci. USA* **76:** 1368–1372.

46. **Jackson, J. A., and G. R. Fink.** 1981. Gene conversion between duplicated genetic elements in yeast. *Nature* (London) **292:**306–310.
47. **Kaulen, H., J. Schell, and F. Kreuzaler.** 1986. Light induced expression of the chimeric chalcone synthase-NPTII gene in tobacco cells. *EMBO J.* **5:**1–8.
48. **Kim, H.-Y., J. W. Schiefelbein, V. Raboy, D. B. Furtek, and O. E. Nelson, Jr.** 1987. RNA splicing permits expression of a maize gene with a defective Suppressor-mutator transposable element insertion in an exon. *Proc. Natl. Acad. Sci. USA* **84:**5863–5867.
49. **Krebbers, E., R. Hehl, R. Piotrowiak, W.-E. Lonnig, H. Sommer, and H. Saedler.** 1987. Molecular analysis of paramutant plants of *Antirrhinum majus* and the involvement of transposable elements. *Mol. Gen. Genet.* **209:**499–507.
50. **Kretschmer, P. J., and S. N. Cohen.** 1979. Effect of temperature on translocation frequency of the Tn3 element. *J. Bacteriol.* **139:**515–519.
51. **Kuckuck, H.** 1936. Uber vier neue Serien multipler Allele bei *Antirrhinum majus*. *Z. Indukt. Abstamm. Vererbungsl.* **71:** 429–440.
52. **Levis, R., T. Hazelrigg, and G. M. Rubin.** 1985. Effects of genomic position on the expression of transduced copies of the *white* gene of *Drosophila*. *Science* **229:**558–561.
53. **Martin, C. R., R. Carpenter, H. Sommer, H. Saedler, and E. S. Coen.** 1985. Molecular analysis of instability in flower pigmentation of *Antirrhinum majus*, following isolation of the *pallida* locus by transposon tagging. *EMBO J.* **4:**1625–1630.
54. **Martin, C. R., S. MacKay, and R. Carpenter.** 1988. Large scale chromosomal restructuring is induced by the transposable element Tam3 at the *nivea* locus of *Antirrhinum majus*. *Genetics* **119:**171–184.
55. **McClintock, B.** 1949. Mutable loci in maize. *Carnegie Inst. Wash. Year Book* **48:**142–154.
56. **McClintock, B.** 1962. Topographical relations between elements of control systems in maize. *Carnegie Inst. Wash. Year Book* **61:**448–461.
57. **Nevers, P., N. Shepherd, and H. Saedler.** 1986. Plant transposable elements. *Adv. Bot. Res.* **12:**103–203.
58. **Paquin, C. E., and V. M. Williamson.** 1986. Ty insertions at two loci account for most spontaneous antimycin A resistance mutations during growth at 15°C of *Saccharomyces cerevisiae* strains lacking ADH1. *Mol. Cell. Biol.* **6:**70–79.
59. **Peacock, W. J., E. S. Dennis, W. L. Gerlach, M. M. Sachs, and D. Schwartz.** 1984. Insertion and excision of Ds controlling elements in maize. *Cold Spring Harbor Symp. Quant. Biol.* **49:**347–354.
60. **Peterson, P. A.** 1970. The *En* mutable system in maize. *Theor. Appl. Genet.* **40:**367–377.
61. **Roberts, D., B. C. Hoopes, W. R. McClure, and N. Kleckner.** 1985. IS10 transposition is regulated by DNA adenine methylation. *Cell* **43:**117–130.
62. **Roeder, G. S., and G. R. Fink.** 1983. Transposable elements in yeast, p. 299–328. *In* J. Shapiro (ed.), *Mobile Genetic Elements*. Academic Press, Inc., New York.
63. **Rothmaler, W.** 1956. Taxonomische Monographie der Gattung *Antirrhinum*. *Feddes Repert Specierum Nov. Regni Veg. Beih.* **136:**1–124.
64. **Rothstein, S. J., J. DiMaio, M. Strand, and D. Rice.** 1987. Stable and heritable inhibition of the expression of nopaline synthase in tobacco expressing antisense RNA. *Proc. Natl. Acad. Sci. USA* **84:**8439–8443.
65. **Sachs, M. M., W. J. Peacock, E. S. Dennis, and W. L. Gerlach.** 1983. Maize Ac/Ds controlling elements: a molecular viewpoint. *Maydica* **28:**289–302.
66. **Saedler, H., and P. Nevers.** 1985. Transposition in plants: a molecular model. *EMBO J.* **4:**585–590.
67. **Scherer, S., and R. W. Davis.** 1980. Recombination of dispersed repeated DNA sequences in yeast. *Science* **209:** 1380–1383.
68. **Schneuwly, S., A. Kuroiwa, and W. J. Gehring.** 1987. Molecular analysis of the dominant homoeotic *Antennapedia* phenotype. *EMBO J.* **6:**201–206.
69. **Schwartz-Sommer, Z., A. Gierl, R. Berntgen, and H. Saedler.** 1985. Sequence comparison of "states" of *a1-m-1* suggests a model for Spm (En) action. *EMBO J.* **4:**2439–2443.
70. **Schwartz-Sommer, Z., A. Gierl, H. Cuypers, P. A. Peterson, and H. Saedler.** 1985. Plant transposable elements generate the DNA sequence diversity needed in evolution. *EMBO J.* **4:**591–597.
71. **Schwartz-Sommer, Z., N. Shepherd, E. Tacke, A. Gierl, W. Rhode, L. Leclerq, M. Mattes, R. Berndtgen, P. A. Peterson, and H. Saedler.** 1987. Influence of transposable elements on the structure and function of the *A1* gene of *Zea mays*. *EMBO J.* **6:**287–294.
72. **Shapiro, J. A.** 1979. Molecular model for the transposition and replication of bacteriophage Mu and other transposable elements. *Proc. Natl. Acad. Sci. USA* **76:**1933–1937.
73. **Sommer, H., U. Bonas, and H. Saedler.** 1988. Transposon-induced alterations in the promoter region affect transcription of the chalcone synthase gene of *Antirrhinum majus*. *Mol. Gen. Genet.* **211:**49–55.
74. **Sommer, H., R. Carpenter, B. J. Harrison, and H. Saedler.** 1985. The transposable element Tam3 of *Antirrhinum majus* generates a novel type of sequence alteration upon excision. *Mol. Gen. Genet.* **199:**225–231.
75. **Sommer, H., and H. Saedler.** 1986. Structure of the chalcone synthase gene of *Antirrhinum majus*. *Mol. Gen. Genet.* **202:** 429–434.
76. **Spiribille, R., and G. Forkmann.** 1982. Genetic control of chalcone synthase activity in flowers of *Antirrhinum majus*. *Phytochemistry* **21:**2231–2234.
77. **Stewart, W. N.** 1983. *Paleobotany and the Evolution of Plants*. Cambridge University Press, Cambridge.
78. **Strickland, R. G., and B. J. Harrison.** 1974. Precursors and genetic control of pigmentation. 1. Induced biosynthesis of pelargonidin, cyanidin and delphinidin in *Antirrhinum majus*. *Heredity* **33:**108–112.
79. **Stubbe, H.** 1966. *Genetik and Zytologie von Antirrhinum L. sect. Antirrhinum*. VEB Gustav Fischer, Jena, German Democratic Republic.
80. **Styles, E. D., O. Ceska, and K. T. Seah.** 1973. Developmental differences in action of R and B alleles in maize. *Can. J. Genet. Cytol.* **15:**59–72.
81. **Tan, C. C.** 1945. Mosaic dominance in the inheritance of color patterns in the lady-bird beetle, *Harmonia axyridis*. *Genetics* **31:**195–210.
82. **Tsubota, S., and P. Schedl.** 1986. Hybrid dysgenesis-induced revertants of insertions at the 5′ end of the *rudimentary* gene in *Drosophila melanogaster*: transposon-induced control mutations. *Genetics* **114:**165–182.
83. **Upadhyaya, K. C., H. Sommer, E. Krebbers, and H. Saedler.** 1985. The paramutagenic line *niv*-44 has a 5kb insert, Tam2, in the chalcone synthase gene of *Antirrhinum majus*. *Mol. Gen. Genet.* **199:**201–207.
84. **van der Ende, A., S. A. Langeveld, R. Teertstra, G. A. van**

Arkel, and P. J. Weisbeek. 1981. Enzymatic properties of the bacteriophage ϕX174 A* protein on superhelical ϕX174 DNA: a model for the termination of the rolling circle DNA replication. *Nucleic Acids Res.* **9**:2037–2053.

85. **Van Schaik, N. W., and R. A. Brink.** 1959. Transpositions of Modulator, a component of the variegated pericarp allele in maize. *Genetics* **44**:725–738.

86. **Vodkin, L. O., R. R. Rhodes, and R. B. Goldberg.** 1983. A cA lectin gene insertion has a structural feature of a transposable element. *Cell* **34**:1027–1031.

87. **Wessler, S., G. Baran, M. Varagona, and S. L. Dellaporta.** 1986. Excision of *Ds* produces *waxy* proteins with a range of enzymatic activities. *EMBO J.* **5**:2427–2432.

88. **Wessler, S. R., G. Baran, and M. Varagona.** 1987. The maize transposable element Ds is spliced from RNA. *Science* **237**: 916–918.

89. **Wienand, U., H. Sommer, Z. Schwarz, N. Shepherd, H. Saedler, F. Kreuzaler, H. Ragg, E. Fautz, K. Hahlbrock, B. J. Harrison, and P. Peterson.** 1982. A general method to identify plant structural genes among genomic DNA DNA clones using transposable element induced mutations. *Mol. Gen. Genet.* **187**:195–201.

Chapter 16

P Elements in *Drosophila melanogaster*

WILLIAM R. ENGELS

William R. Engels ■ Laboratory of Genetics, University of Wisconsin, Madison, Wisconsin 53706.

I. INTRODUCTION

There has been an explosion of knowledge about P elements in the 5 years since my last general review of the field (59). The new information comes from many sources, but much of it was stimulated by Rubin and Spradling's (174, 209) discovery that P elements could be used for germ line transformation by the injection of DNA into embryos. As a result, many investigators working on unrelated questions made use of this procedure, generating information on P elements in the process. Even more importantly, the transformation technique could be applied directly to the study of P elements themselves. First the P element transposase and more recently its regulatory functions have been successfully studied by means of in vitro-modified P elements carrying visible marker genes.

In this review, I will concentrate on work done within the past 5 years. However, some of the older studies, which were covered only lightly in the previous review, have taken on increased importance in

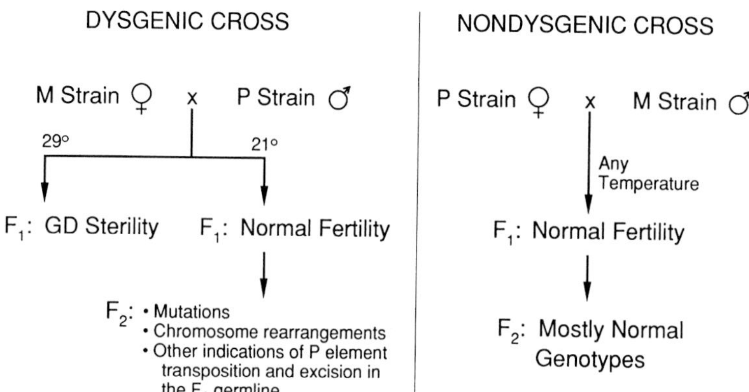

Figure 1. P-M hybrid dysgenesis. The cross on the left yields "gonadal dysgenic" (GD) sterile progeny in both sexes if the higher rearing temperature is used. The frequency of sterility depends on the particular P and M strains. Other than GD sterility, there are no visible abnormalities in the F_1 hybrids, but high frequencies of genetic aberrations are seen in the next generation. The reciprocal cross, shown at right, yields adults with no GD sterility. Genetic aberrations are increased in the next generation, but not as much as in the progeny of dysgenic hybrids.

view of recent developments. These will also be discussed in some detail. I begin with a brief outline of the early work on P elements. Additional background information is available from previous reviews (25, 34, 59, 154, 193, 205a).

II. BACKGROUND

A. Initial Observations of Hybrid Dysgenesis

The study of P elements was started by population geneticists sampling chromosomes from natural populations. In 1960, Hiraizumi and Crow (94) performed a series of crosses between wild-caught *Drosophila melanogaster* males and females from a genetically marked laboratory stock. They were interested in the variation in fitness effects of chromosomes in natural populations. What they found was *SD*, the *segregation distorter* gene, which occupied their research and that of several other laboratories for many years. When Hiraizumi returned to the fitness problem in the early 1970s, using a similar crossing scheme, something unexpected turned up once again. This time it was male recombination (93), a trait not normally seen in this species. His finding led eventually to the discovery of P elements.

Hiraizumi's observation was the first of several loosely connected enigmata being pursued more or less independently by Picard, Berg, the Kidwells, Sved, Green, and others (13, 84, 95, 113, 114, 116, 160, 171, 203, 214, 215, 226). The abnormal traits included high mutability, segregation distortion, and sterility, in addition to male recombination. These traits were usually seen in crosses involving strains recently taken from wild populations. With a 1977 report by Kidwell et al. (117), these observations coalesced into a unified (but still enigmatic) syndrome called "hybrid dysgenesis." Kidwell et al. found that most strains could be classified into one of two categories, called M and P for maternal and paternal, such that the F_1 hybrids from M female × P male crosses were frequently sterile, especially when reared at elevated temperatures. When these hybrids were fertile, high frequencies of mutations, chromosome rearrangements, and male recombination could be detected by examining their offspring (43, 55, 65, 91, 95, 115, 174, 203, 216, 234–236). Thus, the germ cells of the M × P hybrids were genetically unstable. This syndrome was not seen within either class of strains. When the reciprocal cross was performed, the traits could still be detected, but only at much reduced levels (Fig. 1). This reciprocal cross difference became a hallmark of hybrid dysgenesis.

B. Strain Types

Strains recently taken from natural populations in North America typically fell into the P classification, whereas established laboratory stocks were mostly M strains. It soon became clear, however, that the P-M classification system was an oversimplification. Further study showed that it was necessary to postulate at least two independent dysgenesis systems to explain all the data (104). The second system, called I-R, was responsible for some of the early reports (160, 171) and is reviewed elsewhere in this volume (D. J. Finnegan, Chapter 18, this volume). Even within the P-M system, classification was complicated by the discovery of strain types, such as Q and M′, that did not fit the original categories (15, 35, 67, 104, 190, 195, 197). These types will be discussed in more detail elsewhere in this review. The word "strain" itself loses its meaning when applied to offspring from experimental crossing schemes or to populations that are heterogeneous or not in equilibrium. The P-M classification system is still useful, but only as a first approximation.

C. P Factors as the Basis of Hybrid Dysgenesis

Further work showed that P-M hybrid dysgenesis was caused by multiple factors mapping to many locations in the P strain genomes (54). These "P factors" acted independently, and their genomic distribution varied among P strains. Attempts to map these factors in more detail by meiotic recombination yielded ambiguous results, presumably because of the multiplicity and mobility of the elements (201, 202, 236). However, the finding of chromosome breakage hot spots (14, 68, 163, 194, 231) made it possible to localize them with considerable precision. Approximately 85% of the breakpoints in dysgenesis-induced chromosome rearrangements were found to occur at a small number of cytological points (68). The positions of such points varied among chromosomes derived from unrelated P strains, and they did not occur on M chromosomes, leading to the suggestion (68) that they were the sites of P factors. These observations provided a strong indication that P factors were actually a family of transposable elements present in P strains but not in M strains. The M female × P male cross somehow mobilized these elements, resulting in hybrid dysgenesis.

Another line of evidence supporting this interpretation was the instability of some dysgenesis-induced mutations (53, 82). One such allele at the *singed* bristle locus was called sn^w (53, 57). It was especially unstable, mutating to a more extreme form (sn^e) and reverting toward wild type [$sn^{(+)}$: nearly wild type] at frequencies exceeding 50% in some cases, but only in dysgenic crosses (Fig. 2). The hypothesis that sn^w was actually a P factor insertion mutation was reinforced by the finding that sn^w coincided with the appearance of a new chromosome breakage hot spot at cytological position 7D1-2, the site of the *singed* locus (68).

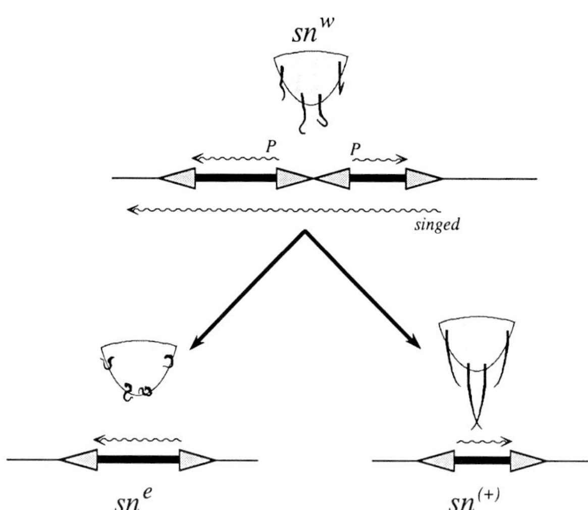

Figure 2. The sn^w allele and its derivatives. The scutellum and four major bristles are shown for each phenotype. The sn^w allele mutates at high frequencies in two other forms, sn^e and $sn^{(+)}$, in the germ line of dysgenic flies (53). Its insertion is composed of two P elements of lengths 1.15 and 0.95 kb, oriented in opposite ways. Deletions of one or the other of these elements gives rise to the corresponding derivative allele (167). Wavy lines indicate transcription. In the case of transcription of the *singed* gene, the direction is known (O'Hare, personal communication) but the exact boundaries of the transcripts have not been determined.

D. Isolation of P Element DNA

The P factor hypothesis was confirmed when a transposable element insertion was found in a dysgenesis-induced allele at the *white* eyes locus (173), which had been cloned previously (16). Homologous elements were found at the sites of chromosome breakage hot spots and dysgenesis-induced mutations (15), thus confirming the identity of these elements as P factors. Predictably, most P strains were found to carry multiple copies of these elements, typically 30 to 50, and most M strains had none. An important exception was the M strain known as Muller-5 Birmingham (hereafter simply Birmingham), which was seen to carry many P elements scattered throughout the genome (15, 64, 190, 197). The elements in Birmingham were apparently inactive. Strains of this kind, which behave in genetic crosses as M strains yet have P element homology in their DNA, have come to be called M′ strains.

O'Hare and Rubin (155) distinguished two main structural classes of P elements. One class, called "complete" P elements, was 2,907 base pairs (bp) in length and highly conserved. The other class was heterogeneous, consisting of various shorter versions of the complete element. These are sometimes called "defective" P elements. In a sample of 12 elements cloned from the genome of one P strain, called π_2, there were 8 defective and 4 complete P elements. The existence of internally deleted P elements provided an explanation for inactive P elements, such as those in the Birmingham strain, and also for nonautonomous P elements as described below (Section IV.A).

E. Two Kinds of Regulation

1. Germ line specificity

Several lines of evidence indicated that P-M dysgenesis was restricted to the germ line. First, the sterility (called GD sterility, for "gonadal dysgenesis") observed when dysgenic hybrids were reared above 25°C resulted from a failure of germ line development (65, 182). All somatically derived tissues appeared intact, but the gonads themselves were greatly reduced or absent in both larvae and adults. Another indication was the lack of somatic recombination (144, 220), which should have been readily detectable if it occurred at frequencies comparable to recombination in the male germ line. Finally, there was no evidence for elevated somatic mosaicism of the sn^{w} allele, despite the finding that essentially 100% of the germ lines were mosaic in dysgenic flies (53). These observations were generally interpreted as evidence for a tissue-specific regulatory function.

2. Cytotype

Another type of regulation was evident from the lack of dysgenesis within the P strains themselves and in the progeny of P strain females mated to M strain males. Genetic analysis showed that this regulation was due to multiple factors on all the chromosomes of the P strains, presumably the P factors themselves (54, 105). Dysgenesis was suppressed by a cellular state called the "P cytotype," which was established only in the presence of P chromosomes. The alternative state, which permitted hybrid dysgenesis, was called the "M cytotype."

The determination of cytotype in a given individual occurred by an unusual mode of inheritance involving both chromosomal and maternal transmission. Flies with mixed P and M ancestry usually had either P or M cytotype, rather than an intermediate state (54, 105). The actual proportion of P cytotype progeny from a given cross was difficult to predict and depended on the particular strains. Maternal age and rearing temperature also seemed to have an effect (168, 170), but part of it might have been due to selection within the germ line such that cells with greater dysgenic activity were less likely to survive. A variety of genetic techniques were employed to derive the following qualitative generalizations (54, 57, 105, 217, 233, 237): (i) increasing the number of P factors in the genome increases the frequency of the P cytotype, provided the maternal cytotype is held constant; (ii) the P cytotype is more frequent in the progeny of P cytotype mothers than M cytotype mothers, provided the genomic contributions are held constant; (iii) in the absence of P factors, the cytotype is always M. Thus, cytotype is determined jointly by an individual's own genotype and its mother's cytotype. Experiments in which both contributions are variable can lead to unpredictable results (85, 86, 188). Rules i and iii account for P and M strains having the corresponding cytotype. Rule ii accounts for the reciprocal cross effect. The mechanism underlying these rules of inheritance is not known, but some possibilities will be discussed elsewhere in this review (Section V.B).

F. MR: an Alternative Terminology

Hiraizumi and co-workers first observed male recombination in flies heterozygous for "extracted" second chromosomes (93, 95, 203). That is, one homolog of chromosome 2 was recently derived from a natural population, but most or all of the rest of the genome came from laboratory (presumably M) stocks. Hiraizumi, Green, Slatko, and others therefore assumed that these effects were due to a single "primary" site on chromosome 2 (202, 204). Second chromosomes of this kind were called MR, which originally meant "male recombination," but it has been reinterpreted in various writings as "mitotic recombination" or "mutation/recombination." It now seems likely that most or all of the effects attributed to MR chromosomes were actually the result of mobilization of P elements by P-M hybrid dysgenesis (85). The difference is primarily one of terminology.

To minimize confusion, I will stay with the P-M terminology for the remainder of this article, even when discussing the works of authors who use MR. One reason for preferring P-M is that it does not place special emphasis on a particular chromosome or a particular site; P elements can be anywhere in the genome. Another reason is that cytotype and the reciprocal cross effect are easier to discuss in P-M terms. Workers in the MR camp generally did not accept Kidwell's notion of P and M strain types. Consequently, when they observed reciprocal cross differences, they attributed it to the MR chromosomes being active only when they were inherited through the male line (85, 86). It is now clear, however, that P elements can be highly active when they are inherited from a female, provided she has the M cytotype (54, 64, 166, 190, 197). Finally, the use of MR terminology is confusing when the P-M and I-R dysgenesis systems are discussed together. Note that strains with MR chromosomes are neither M in the P-M system nor R in the I-R system.

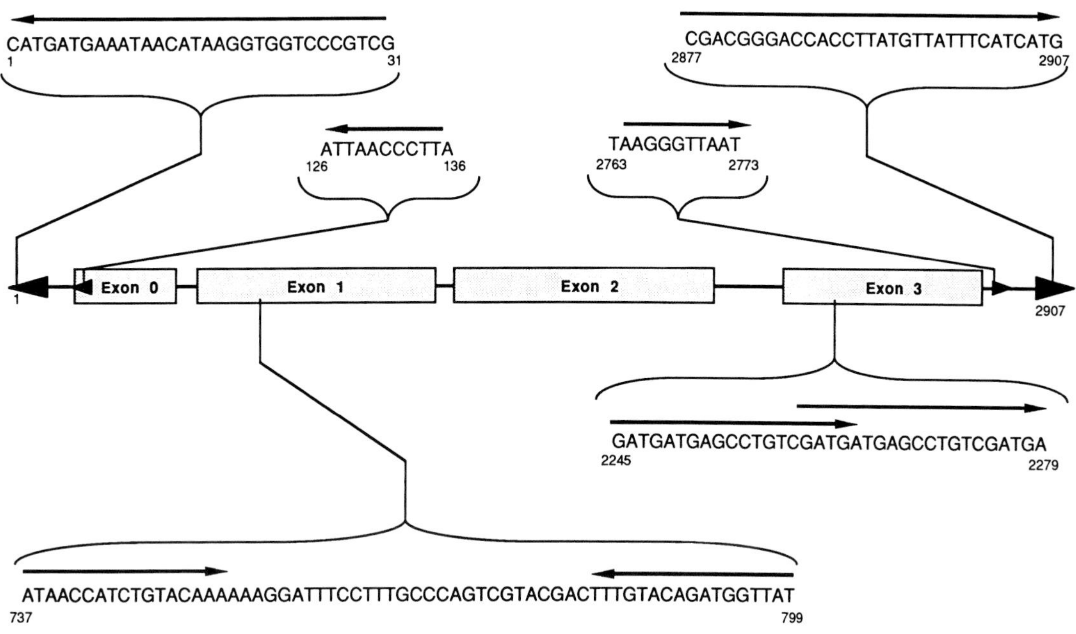

Figure 3. The complete P element and its repeat structures. The sequence was obtained by O'Hare and Rubin (155).

III. STRUCTURE OF P ELEMENTS

A. Repeats

The sequence of a complete P element (155) revealed several interesting repeat structures (Fig. 3). There are perfect inverted repeats of 31 bp at the termini and an 11-bp subterminal inverted repeat starting about 125 bp from each end. The importance of these repeats for mobility will be discussed below. There are direct repeats of 8 bp of genomic DNA on either side of the insert, presumably formed by a staggered cut at the time of insertion. Internally, there is an inverted 17-bp repeat separated by 29 nonrepeated bp. There is also a 20-bp overlapping direct repeat. No functional significance has been found for these two structures, both of which lie within coding regions for the transposase gene, as will be described below. Homologous exchange events within these repeats would maintain the reading frame, but would result in substitution of 11 amino acids in the case of the inverted repeat and the loss of 5 amino acids in the direct repeat. However, such events would be difficult to detect, and none has been found to date.

B. Defective P Elements

Sequencing revealed that the shorter P elements in π_2 were derivable from the complete sequence by internal deletions (155). The sizes and locations of these deletions were highly variable, but they did not overlap the terminal repeats. A large proportion, but not all, of the breakpoints occurred within short (2- to 6-bp) direct repeats. Most cases were simple deletions, but more complex changes were also seen. These included a 3-bp direct duplication and a 7-bp direct triplication of the region immediately adjacent to one of the breakpoints.

These deleted elements are highly heterogeneous. Many of them are probably unique within a given genome. However, Black et al. (17) found one particular deleted element which had a high copy number in many unrelated strains. This element, dubbed "KP," is 1,154 bp in length, with a simple deletion from position 808 to 2560. KP elements might account for half or more of all P-homologous sequences in some natural populations.

C. Other Sequence Variations

Aside from their internal deletions, P elements are remarkably homogeneous within *D. melanogaster*. O'Hare and Rubin (155) found two polymorphic sites (positions 32 and 33) among the elements they examined from π_2, but Sakoyama et al. (176) found no differences from the canonical sequence among 2,109 bp sequenced from an element in a strain unrelated to π_2. The KP element has a substitution in position 32, the same as one of the polymorphisms found in π_2, but no other base changes (17). Other studies employing restriction maps rather than sequencing also show wide variety in deletions, but near homogeneity otherwise (98, 176). Current data do not permit an accurate estimate of nucleotide

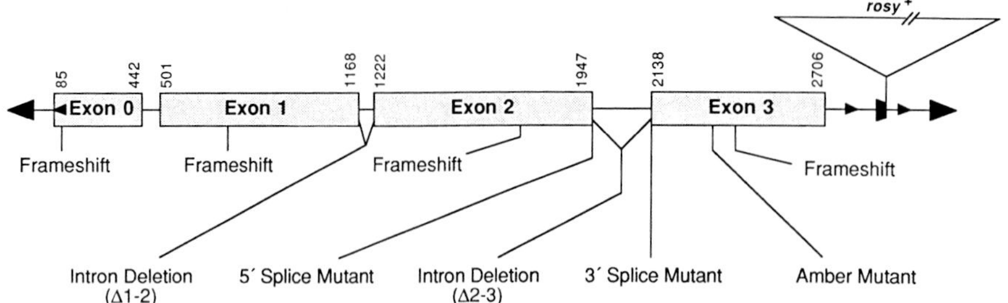

Figure 4. The Pc[*ry*] element and its mutant derivatives. Pc[*ry*] was constructed by Karess and Rubin (102), and the exon boundaries were found by Laski et al. (126). Frameshift mutations were produced by filling in restriction sites for *Alu*I, *Xho*I, *Eco*RI, and *Sal*I (left to right) (102). The splice mutants are synonymous base substitutions (126). Intron deletions were made at the boundaries determined from cDNA in the case of Δ1-2 and from general splice consensus sequences for Δ2-3 (126).

diversity, but it seems likely that P elements are less diverse than many other parts of the *D. melanogaster* genome. Implications of this homogeneity for the recent evolutionary history of P elements are discussed below (Section X).

IV. EXCISION AND TRANSPOSITION

A. Autonomous versus Nonautonomous P Elements

Just as P elements exist in two structural classes, complete and defective, there are also two functional categories: autonomous and nonautonomous. Autonomous P elements are defined as those that can transpose and excise on their own. Nonautonomous elements can also transpose and excise, but only in the presence of at least one autonomous P element. Some authors have used the terms "P factor" and "P element" to make this distinction (15, 155, 209). However, I will use "P element" as a general term for any member of the P family and use "autonomous" to distinguish self-sufficient elements.

The existence of nonautonomous P elements was first indicated by the sn^w allele, which became stable when separated from other P elements in the genome (60). Its hypermutability was restored when it was reunited with any chromosome from a P strain. Spradling and Rubin (209) made use of this property as an assay for autonomous P elements. They injected a plasmid carrying a copy of the complete (2,907-bp) P element into sn^w (M background) embryos. Transpositions of the complete element from its plasmid to the genome were recovered by a screen for sn^w hypermutability. This experiment demonstrated the feasibility of P element transformation, described in more detail below (Section XI), and it simultaneously confirmed the autonomy of the 2,907-bp P element. Rubin and Spradling further showed that autonomous P elements could be used as "helpers" in transformation; that is, a complete P element, coinjected with a marked nonautonomous element, could catalyze the transposition of the marked element (174).

The ability to mobilize nonautonomous P elements defines an element-encoded gene function generally referred to as **transposase**. The P element sequence revealed four long open reading frames (ORFs), denoted ORF0, -1, -2, and -3, all on the same strand (155). These were considered prime candidates for encoding the transposase.

B. The Transposase Gene

As we will see below, harnessing the P element transposase for use in germ line transformation has allowed the genetic dissection of many *Drosophila* genes. In fact, the transposase gene itself has been particularly well characterized by these methods.

1. Genetic analysis by in vitro mutagenesis

Karess and Rubin (102) set out to analyze the transposase gene by means of site-directed in vitro mutagenesis of an autonomous P element. Germ line transformation was used to place each modified element in the genome, where it was tested for sn^w hypermutability to indicate transposase production. The authors' first task was to create an autonomous P element whose presence could be monitored irrespective of its transposase function. They used the wild-type allele of *rosy*, an eye color gene, to construct the element called Pc[*ry*] (Fig. 4). This element, which was easily detected in a ry^- genetic background, was seen to be autonomous by virtue of its ability to transpose from an injected plasmid to chromosomal sites without a helper plasmid and to promote sn^w hypermutability once integrated into the genome.

Karess and Rubin then produced four mutated version of Pc[*ry*], each with a frameshift mutation at a restriction site in one of the four ORFs (see Fig. 4). Each element was then returned to the genome with the aid of a helper element. None of them mobilized sn^w, nor was any of the six pairwise combinations complementary. It was therefore concluded that the transposase was encoded by a single gene composed of four exons spanning most of the P element's length.

2. The transposase message

Analysis of the transposase mRNA was complicated by the presence of many irrelevant RNA fragments found in ordinary P strains (102). These fragments can be produced by defective P elements and also by genomic promoters with P element insertions downstream. The solution was again provided by the use of strains transformed with Pc[*ry*]. Such strains have only one or a small number of P elements and produce a much simpler set of RNA fragments. With this approach, Karess and Rubin (102) and Laski et al. (126) determined that transposase came from a 2.5-kb message starting at or near nucleotide position 85 and terminating after a polyadenylation signal at position 2696. The two sets of splice sites connecting the first three ORFs were also determined from the Pc[*ry*] transcripts (Fig. 4). The third splice shown in Fig. 4 was not present in these transcripts and was determined by a different method which will be discussed below when I take up the subject of P element regulation (Section V.A). To confirm these findings, Laski et al. showed that the two intron deletion mutations shown in Fig. 4 did produce transposase, whereas the *amber* mutant and the splice junction mutants did not (126).

3. The transposase protein

From this analysis and from the assumption that translation starts at position 153, the site of the initial ATG, Laski et al. predicted that the transposase was an 86,854-dalton protein. Such a protein would have a short region near the ORF1-ORF2 boundary with a secondary structure similar to that of sequence-specific DNA-binding proteins (126). Rio et al. (165) then used immunochemical techniques to identify a protein of the predicted size in cultured cells transformed with another modified autonomous P element (*hs*πΔ2-3). They also detected transposase activity in these cells by means of an excision assay in which a precise P element excision, or a near-precise excision leaving a multiple of 3 bp, restored the function of a plasmid-borne β-galactosidase gene.

4. Some questions remain

These experiments provided a clear picture of the P element transposase gene. However, there were also some unexplained findings. In addition to the 2.5-kb transcript, there was a less abundant 3.0-kb transcript which appeared to start at the same position but terminated outside the P element. No function is known for this transcript. Another unexpected finding was an alternative splice site for ORF1-ORF2 (position 1156 instead of 1168) which was used in a small proportion of the cDNAs from cultured cells. The resulting message encodes an expected polypeptide very similar to the transposase, but lacking four amino acids in the region suspected for playing a role in DNA binding. The alternatively spliced product has no detectable transposase activity (126), but it is not known whether it plays some other role. Finally, Satta et al. (178) noted that the ORF2-ORF3 intron contains a 9-base sequence that matches the resolvase binding sites of the procaryotic transposon Tn3. A somewhat weaker match also occurred in the P element's 31-base terminal repeats. The functional significance of these sequences has yet to be established.

C. Characteristics of P Element Excision

1. Types of excisions

P elements can undergo three kinds of changes that are generally referred to as excisions. First, there are internal excisions, which leave behind some part of the terminal regions of the element. These range from the removal of only a few nucleotide pairs to losses of nearly the entire element. Events of this kind are presumed to be the origin of defective P elements. The second type of event removes both the P element and a stretch of genomic DNA flanking it. In some cases these events remove sequences on both sides of the element. Finally, precise excisions have been observed. All of the P element sequence plus one copy of the 8-bp genomic duplication is lost, thus restoring the original sequence. The possibility that all of these events can be explained by a single mechanism will be discussed below.

2. Sources of excision data and frequency estimates

Data are plentiful, and they come from a variety of detection methods. Since each detection method selects for a different spectrum of molecular events, the observed excision rates depend critically on which screening procedure was used. Furthermore, there is considerable site-to-site variability within each method, so that the typical rates given in the list below should not be taken as applicable to every case.

Reversion of P element insertion mutations. When a P element inserts in a gene, the resulting allele is often unstable in the germ line of dysgenic flies. It can revert to wild type or remutate to other allelic forms. Starting with the unstable *singed* alleles of Golubovsky et al. (81, 82, 241), many cases have been observed at various loci. Molecular studies of these events have been conducted for several loci, including *white* (130, 155, 173), *RpII215* (184, 224), *rudimentary* (223), *Sex-lethal* (177), *singed* (167), and *yellow* (29). The plasmid excision assay of a P element in *Escherichia coli* β-galactosidase also falls into this category (165). The events recovered in these studies were primarily internal excisions, but in some cases chromosome rearrangements and even secondary P element insertions were found (177, 223; W. Eggleston and W. Engels, unpublished data). Typical frequencies are in the range of 1% per generation for dysgenic flies. In cases where the original insertion is within a coding region, this screening method selects for precise excisions plus certain nearly precise excisions which do not alter the reading frame. The reversion frequencies are greatly reduced in such cases (173, 228).

Disappearance of in situ hybridization sites. P element excisions can also be detected as the loss of in situ hybridization sites. An advantage of this method is that it allows detection of events that do not result in any phenotypic change. However, the resolution limits are such that this technique fails to detect internal excisions that leave behind large (>0.5-kb) P element fragments, or excisions of P elements that happen to lie close (<50 kb) to other P elements. In one large-scale study (12; W. Benz, personal communication), it was estimated that excision events detectable in this way occur at the rate of approximately 1.8% (193/~11,000) per element each dysgenic generation. In another study (50), using a different genetic background, the estimate was 0.4% (9/2,420).

Loss or change of expression of genes carried within P elements. Another way to measure excision is to screen for the loss of a visible marker gene carried by a P element. Such marked P elements can be constructed in vitro and placed in the genome by transformation (174). P elements carrying wild-type alleles of the eye color genes *white* and *rosy* have been used in this way (37, 129, 166). Figure 5 shows an example of the sort of mating used to recover this kind of excision and to estimate the rate. The ob-

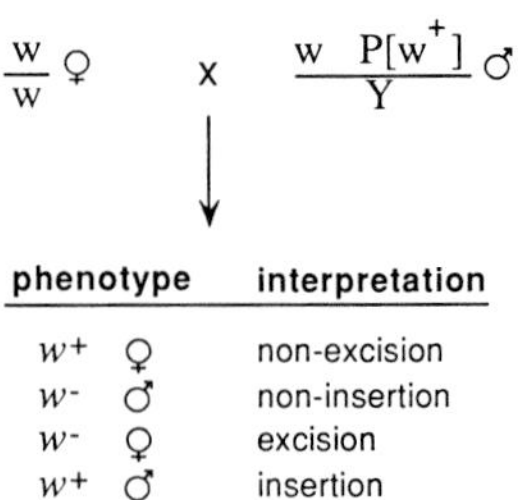

Figure 5. Method for detecting excisions and transpositions of a marked P element. In this example, a P element, P[w^+], is marked with the wild-type allele of the *white* (*w*) gene. Other visible markers, such as *rosy* (*ry*), would work equally well. The male has a mutant allele of *white* and also the P[w^+] element on the X chromosome, giving him wild-type eyes. He is presumed to be of the M cytotype and have a transposase source somewhere in the genome. The females are homozygous for *white* and therefore have white eyes. Since the P[w^+] element is on the X chromosome, normal segregation would result in only wild-type daughters and *white*-eyed sons. Transposition of the P[w^+] element to an autosomal site can produce wild-type sons, and complete or partial excision events would yield *white*-eyed daughters or daughters with less than wild-type eye pigmentation.

served rates seem to depend on several variables, such as the source of transposase, the developmental temperature of the male, and especially the genomic position of the insert (166), but they are typically on the order of 5%. In the majority of cases, the loss of gene function is not accompanied by complete loss of the P element, suggesting that most of the events are internal excisions (37). Even very small internal deletions can be detected by this method provided they remove a necessary component of the marker gene. This sensitivity probably accounts for the relatively high observed excision rates, but the large size of marked P elements relative to naturally occurring ones might also play a role. Significantly, some of the internal excisions had both breakpoints within the marker gene, indicating that the site of the breakage need not involve P element sequences (37).

Loss or change of expression of genes flanking P element insertions. Another method involves selecting for mutations of a gene near a P element insertion. The result is usually an excision which removes some genomic material flanking the element. It has been used at the loci *Beadex* (53, 143), *rudimentary* (223), and *Sex-lethal* (177), with frequencies on the order of 0.1%. It should be noted that these loci are all on the X chromosome. All of the mutations at the *Beadex* and *rudimentary* loci and some of those at *Sex-lethal* came from crossing procedures in which the mutated chromosome can only be recovered if it is viable and fertile in males. This means that any large deletion taking out an essential locus would be eliminated in these screens. The true rate at which P elements create flanking deletions might therefore be

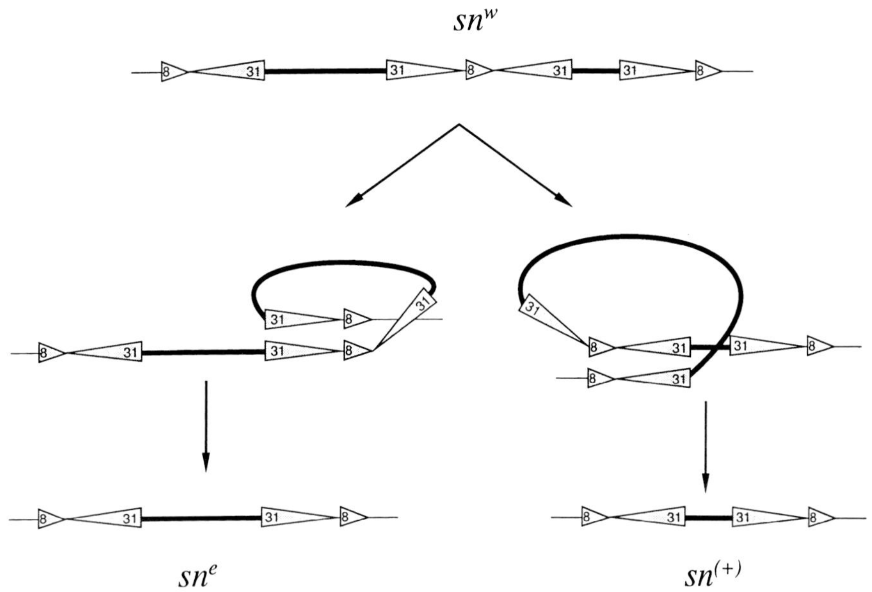

Figure 6. Alignment of direct repeats in sn^w. The top diagram shows the structure of sn^w suggested by Roiha et al. (167). In the middle diagram, the two sets of 39-bp direct repeats are shown juxtaposed to indicate how deletions within each repeat can lead to the sn^e and $sn^{(+)}$ structures. The final step could be any event which removes the looped portion of the sequence. Reciprocal recombination is not necessarily implied. In fact, the lack of any reported cases of P elements inverting their orientation suggests that the recombinational events are not reciprocal.

considerably greater than 0.1%. Most of the deletions were less than 7 kb; a few were longer, but the exact length was undetermined. The longer ones occurred at the *Sex-lethal* locus, where some of the crossing procedures permitted recovery of recessive lethals (177).

3. Double P elements have enhanced excision rates

The hypermutable *singed* allele, sn^w, discussed previously (see Fig. 2), is unusual for its extreme instability. Ordinary M × P dysgenic crosses yield $sn^w \rightarrow sn^{(+)}$ and $sn^w \rightarrow sn^e$ mutation rates totaling 20 to 60% (53, 57). When the element P[ry^+ Δ2-3](99B) is used as a transposase source (discussed in Section XI.A; see references 64 and 166), the mutability can exceed 90%. The sn^w allele has recently been shown to consist of a double P element insertion (167). Two defective P elements are inserted in reverse orientation into the same 8-bp target sequence (Fig. 2 and 6). A mutation results when one of the elements excises, yielding either sn^e or $sn^{(+)}$, depending on which element has excised. These excisions appear to be precise at the level of Southern blots, with one of the two elements entirely missing and the other remaining intact. At least one $sn^{(+)}$ mutation was shown by sequencing to be a precise excision of the 1.15-kb element (167).

Thus, the P elements in sn^w are unusual in two respects. First, their excision rates are much greater than that of a typical P element. Second, the proportion of precise or nearly precise excision is greatly enhanced. It now appears that these may be general properties of double P elements. W. Eggleston (personal communication) has examined 69 double inserts, most within 30 bp of each other, at the *singed* and *yellow* loci. In some cases the two elements were in reverse orientation with respect to one another, as was the case of sn^w, whereas others were directly oriented. The majority of them had rates of instability in the range of 10 to 30%, as detected by various phenotypic changes. Moreover, the proportion of these changes that appeared to be precise excisions by Southern blots was usually 20 to 100%, which is much higher than what has been reported for single P elements. As will be discussed below (Section IV.D), the explanation for this behavior may lie in the directly repeated sequences formed by double P elements.

4. General features of P element excision

We can draw several general conclusions from the various studies cited above. First, P element excision is a transposase-dependent process. All excision events recovered by the above detection methods occurred in the germ line of dysgenic flies. Other methods have been used to find excisions in nondysgenic flies (162), but such events are extremely rare. Therefore, the M cytotype and at least one autonomous P element somewhere in the genome are both necessary for high frequencies of excision to occur. Second, the most common type of event is the internal excision, as opposed to precise excisions or flanking deletions. There are, however, three apparent exceptions to this rule: (i) in double P elements such as *sn*^{w}, the predominant event appears to be precise excision of one of the two elements; (ii) it appears that P elements with a wild-type homolog undergo precise excisions at elevated rates (J. Sved, W. Eggleston, and W. Engels, unpublished data); (iii) the unusual allele *sn*^{cm} is reported to be an exception (89). The third generalization is that all of the observed excision events occurred in situ. That is, they represent the loss of material from an element that existed in the genome the previous generation, rather than the creation of a new, internally deleted element through transposition. Furthermore, when transposition of a doubly marked P element is selected by screening for only one of the two markers, the unselected marker is almost always recovered (127). This observation indicates that transposition is not normally accompanied by internal excisions.

One of the early suggestions was that P element excisions occur by faulty DNA synthesis as new P elements are created by a replicative transposase process (155). This model provides an explanation for the involvement of transposase. However, this model appears to be ruled out by the last generalization listed above, since it requires that the excisions appear at sites of new P element insertions, as opposed to happening in situ. One way to explain the requirement of transposase in the production of in situ excisions is to postulate that loss of P element sequences occurs at the donor site, rather than the recipient site, in a transposition event. Alternatively, one can hypothesize that excisions are incidental effects of transposase activity and occur independently of particular transposition events. Such a model is proposed below.

D. A Single Mechanism of P Element Excision

1. Do P elements excise precisely?

There is no question that precise excisions of P elements can occur. In fact, the first sequenced P element excision was found to be precise (155). However, further work has shown that the majority of excisions that appear to be precise on the level of Southern blotting are actually internal excisions leaving behind small (<40-bp) fragments of the element (165, 184, 223). In most cases both breakpoints are located asymmetrically within the 31-bp terminal repeats. Table 1 summarizes data from several reports of P element excisions from the *white*, *RpII215*,

Table 1. Precise versus nearly precise excisions

Repeat	No. of excisions	
	Precise	Nearly precise
Perfect 8-base repeat[a]	7	13
Mismatched 8-base repeat[b]	0	10

[a] Data were compiled from studies of the *white* (155), *rudimentary* (223), and *RpII215* (224) loci and from the tissue culture excision assay of Rio et al. (165).

[b] These tissue-culture excision events were recovered from a plasmid in which the 8 bases flanking the P element (inserted within a β-galactosidase gene) contained two base mismatches (165). The flanking DNA sequence was also different from that used in the other excision assays.

rudimentary, and β-galactosidase genes. It is clear that the nearly precise excisions are much more frequent than true precise excisions. Furthermore, the detection methods in all of these studies tend to favor precise excisions over nearly precise ones. For example, in the cases of *white* and β-galactosidase, where the P insertion is in the coding region, all precise excisions would be recovered as phenotypic reversions, whereas no more than one-third of the imprecise ones (the non-frameshifts) would be recovered. This means that the true ratio of imprecise to precise excisions is probably much greater than suggested by Table 1. Moreover, as discussed below, there is a tendency for breakpoints to occur at direct repeats. As a result, the 8-bp direct repeat generated by the insertion probably makes precise excisions slightly more frequent than excisions involving other breakpoints. Therefore, I propose that precise P element excisions do not occur by any specific mechanism, such as "reversal" of the insertion process (155) or a conservative mode of transposition. Instead, they may be merely special cases of the same mechanism that produces internal deletions.

2. The importance of direct repeats

Several lines of evidence support the idea that excision breakpoints tend to occur between copies of short direct repeats, deleting the intervening base pairs. First, as noted above, there was a strong tendency for the breakpoints in defective P elements sequenced by O'Hare and Rubin (155) to lie at the sites of direct repeats of 2 to 6 bp. The breakpoint of the KP element (14) is also at a 2-bp direct repeat. Second, Rio et al. (165) point out that a large proportion of the nearly precise deletions they observed had breakpoints in a direct repeat of 6 out of 7 bp contained within the 31-bp reverse repeat. Similarly, four of the five nearly precise excisions of Tsubota and Schedl (223) at the *rudimentary* locus occurred at 2- or 3-bp direct repeats, including one case in which a breakpoint was in the flanking DNA.

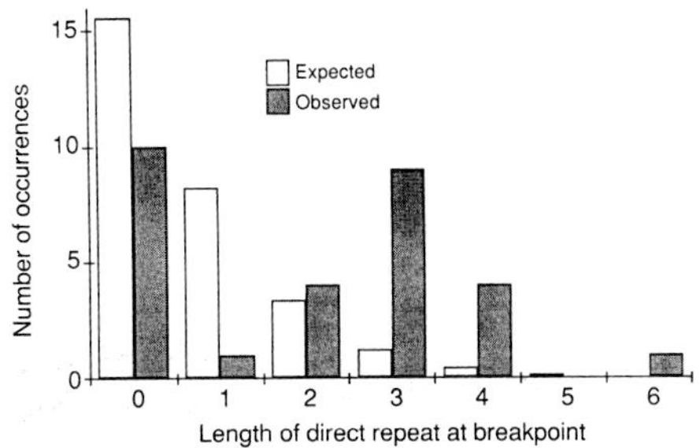

Figure 7. Observed and expected direct repeats at deletion breakpoints. The sequences of 29 defective P elements and partial deletion events from six published sources (17, 29, 153, 155, 165, 223) were compared with the complete P element sequence (155) to determine the number of directly repeated base pairs at the new junction. Precise excisions were not included, nor were deletions generating duplications or new base pairs. The expected numbers were generated by computer from 10^6 trials. Each trial consisted of selecting two points at random within the complete P element sequence, constructing the corresponding deletion, and counting the maximum number of directly repeated base pairs touching the breakpoint.

Figure 7 shows a comparison between the observed numbers of direct repeats at the breakpoints and the expected numbers if all breakpoints were equally likely. There is a clear excess of deletions connecting directly repeated sequences, over the expected numbers. Finally, Table 1 suggests that the frequency of precise excisions relative to nearly precise ones is significantly decreased for a P element insert in which the 8-bp flanking repeat is interrupted by two mismatches.

3. Direct repeats might explain the unusual behavior of sn^w and other double P inserts

The exceptional behavior of double P elements such as sn^w can also be explained by postulating a common excision mechanism that favors breakpoints in direct repeats. Note that sn^w is unusual in two ways. First, excisions of either of the two P elements occur at abnormally high rates, many times that of most single P elements. Second, the great majority of these excisions appear to be precise or nearly so. To explain this behavior, note that the double P element configuration proposed for sn^w (167) results in two sets of 39-bp direct repeats formed by the combination of the 31-bp inverted repeats and the 8-bp duplication (Fig. 6). As pointed out by Roiha et al. (167), pairing these repeats as shown in Fig. 6 and deleting the intervening region would result in the exact structures reported for sn^e and $sn^{(+)}$. The relatively high frequencies of these events compared with excisions of single elements or the simultaneous loss of both sn^w elements could then be explained by the greater length of the direct repeat in sn^w (39 bp) as compared with that of single P elements (8 bp).

The recent finding by Eggleston (personal communication), that most double P element insertions, regardless of whether the two elements are in direct or reverse orientation, display hypermutability and an excessive proportion of precise or nearly precise excisions, is in good agreement with this model. In at least some of these cases, the two elements were separated by more than 8 bp. In such cases the length of the direct repeat is expected to be 31 rather than 39 bp, but this is still a relatively long direct repeat compared to single P elements. In cases in which the two elements are in direct orientation, the length of the direct repeat will be greater still, depending on what internal deletions are present in the two elements. More detailed analyses of these double elements and their derivatives will indicate whether models similar to that in Fig. 6 are applicable.

4. A possible mechanism for P element deletions

On the basis of the generalizations just discussed, I suggest that all excision events can be explained by a single mechanism. The necessary features of such a mechanism are: (i) transposase function is involved; (ii) it causes deletions within preexisting P elements; and (iii) it favors breakpoints in direct repeats. Several mechanisms with these characteristics are conceivable. For example, DNA repair systems might preferentially link fragments with short regions of homology. Therefore, if transposase is assumed to produce random (i.e., uniformly distributed) breaks in the donor P element and occasionally in flanking DNA, then all known types of excisions could be explained.

Another possibility is that deletions occur by faulty DNA replication. To account for the excision of preexisting P elements, these events are presumed to occur when the chromosome replicates as part of the normal cell cycle, as opposed to special DNA synthesis during transposition. To explain the involvement of transposase, assume that one of the effects of transposase activity is to change the spatial arrangement of the P element DNA. Specifically, the terminal sequences are brought into close proximity. A likely result of this abnormal configuration is a high frequency of errors in DNA replication, especially when short direct repeats are juxtaposed. The relatively high proportion of nearly precise excisions is then explained by the close proximity of the P element termini. Other events, such as internal excisions in which neither break is near a terminus, and deletions of flanking DNA, might result from intermediate structures prior to the juxtaposition of the termini or from a more general disruption of the normal spatial arrangement at the time of replication.

An advantage of this mechanism over the breakage-repair model is that it provides a better explanation of the duplication and triplication events involving short regions immediately adjacent to the breakpoint. As noted above, two such structures were observed in defective P elements sequenced by O'Hare and Rubin (155), who discussed replication fork "slippage" as a possible mechanism. For further discussion of transposable element excision mechanisms, see the chapters by M. Meuth and S. D. Ehrlich (this volume).

E. Transposition

1. Frequency of transposition events

I will use the terms "transposition" and "insertion" frequency (or rate) synonymously and define them as the expected number of new elements produced by a given P element per organismal generation. In the following discussion, I will attempt to convert data gleaned from various sources into estimates of this quantity.

These transposition rates are sensitive to several variables, the most obvious being transposase and cytotype. Without transposase no transposition occurs. Transposition is also greatly reduced, but not entirely eliminated, in the P cytotype (50, 162). Another variable that is likely to be important is element size. The data of Spradling (207) indicate that smaller P elements transform embryos more readily than larger ones, suggesting that their transposition rates are similarly affected. There is also some indication that transposition rates increase with developmental temperature (166). In addition, sequences placed within in vitro-modified P elements also have a large effect in some cases (207; C. Berg and A. Spradling, personal communication). Even with all of the above variables held constant, there are still substantial differences among P elements (166) which, as will be discussed below, are probably due to genomic position effects.

Several independent estimates of the insertion frequency have been obtained by counting the number of new in situ hybridization sites to appear on the X chromosome per generation in the progeny of dysgenic males. This number should then be multiplied by 9 to correct for transpositions to the autosomes, which will be about 8/9 of the events in males, and divided by the number of P elements which could serve as the donor site in the genome of the dysgenic parent. The resulting estimates, which can be considered as an average over all available donor sites, are remarkably consistent. The data of Bingham et al. (15) yield a transposition rate estimate of 0.25, based upon 18 insertions onto 22 chromosomes and assuming 30 potential donor sites. Two similar experiments done on much larger scales yielded almost identical estimates. Benz and Engels (12) scored 878 insertion events in 1,168 chromosome-generations. With an estimated 25 to 30 donor sites per generation, these data are consistent with a transposition rate of 0.25. The second study (50), which included 193 insertion events and employed a more accurate estimate of the number of donor sites, also yielded an estimate of 0.25 insertions per element-generation. Thus, the value of 0.25 per element can be taken as a fairly sound estimate of the transposition rate in ordinary M × P hybrids. The critical importance of this estimate will be clear below when transposition mechanisms are discussed.

A much higher transposition rate was obtained by Robertson et al. (166) when the defective P elements on the second chromosome of strain Birmingham were mobilized by a transposase-producing P element called P[ry^+ Δ2-3](99B). A total of 100 new insertions were seen on 29 X chromosomes with 17 potential donor sites. The resulting transposition rate, 1.82, is more than sevenfold greater than that obtained by ordinary M × P crosses described above. Several factors are likely to contribute to this difference. (i) Birmingham elements are all thought to be defective and might be smaller, on the average, than the P elements in typical P strains. The smaller size would imply a greater transposition rate. (ii) The transposase activity of P[ry^+ Δ2-3](99B) might be greater than that of P strains. (iii) Neither Birmingham nor the P[ry^+ Δ2-3](99B) strain is thought to carry any P elements capable of making the repressor of P element transposition thought to be responsible for P cytotype (see Section V.B).

Transposition frequencies of marked P elements can be measured by any of several variations of the cross in Fig. 5. For the example shown, the estimate would be (9/4) × (number of w^+ males)/(total males). The factor of 9/4 is needed because only four of the nine chromosome arms on which an insertion might occur are included in a Y-bearing sperm. One advantage of this approach is that it allows estimates of the mobility of individual P elements as opposed to the average of a large number of elements. Data of this kind come from Levis et al. (129) and Robertson et al. (166), plus unpublished data from my laboratory. The resulting transposition rates are highly variable, with most of them falling in between 0.01 and 0.20, with an average near 0.03. Much of the variability can be explained by element size, with the larger elements being less mobile (compare the larger P[w^+] elements with the smaller P[ry^+] elements in Table 3 of reference 166), and by differences in transposase source, with P[ry^+ Δ2-3](99B) generally causing more transposition than M × P crosses. Residual variability not accounted for by the above variables probably reflects differences in the genomic positions of the elements, as will be discussed below under the topic of genomic effects (Section IX.A). On the whole, transposition rates for marked P elements were significantly lower than those measured using in situ hybridization for naturally occurring P elements. Again, the difference is probably explained by the much smaller size of the naturally occurring P elements. Some of the difference might also be explained by internal excisions which eliminate expression of the marker gene, thereby rendering their subsequent transposition undetectable.

2. Target sites for insertion

Casual inspection of the distribution of P element insertion sites plotted on the scale of the whole genome or whole chromosomes (12, 15, 129, 166, 169, 207, 210; W. Eanes, personal communication; unpublished data from my laboratory) suggests an approximately uniform distribution with a few "hot" and "cold" regions and other minor irregularities. However, close inspection of the data reveals that the choice of target sites is severely influenced by unknown factors on all levels of genomic organization. Moreover, the in situ hybridization data of W. Benz and W. Eanes (personal communication) strongly indicate strain-specific effects in target site preference. Some of the factors that are probably important in determining insertion sites are as follows.

First, there may be a strong preference for euchromatic sites over heterochromatic sites, as indicated by the following lines of evidence. (i) No labeling in the chromocenter appears when polytene chromosomes from P strains are hybridized in situ with P element probes (e.g., references 15 and 50). This observation contrasts sharply with the heavy chromocentric labeling seen for most other transposable elements (50, 71). (ii) A sample of 746 dysgenesis-induced chromosome rearrangements with one selected breakpoint at cytological position 17C and one or more unselected breakpoints was examined by Engels and Preston (69). Approximately 85% of the breakpoints were at P element sites, but only 3 of the 845 unselected breakpoints were heterochromatic. (iii) Pooled data from published (129, 207, 210) and unpublished sources show the cytological locations of many P element insertions which were detected by expression of an element-internal marker. These insertions were obtained by transformation with marked P elements or by mobilization of previously transformed P elements. Almost all of the sites were visible by in situ hybridization, and therefore euchromatic. Individually, each of the above arguments has weaknesses: (i) in situ hybridization is likely to miss some chromocentric P elements because of underreplication of heterochromatin in polytene chromosomes; (ii) differences in chromatin structure might make chromosome breakage less likely at heterochromatic P element sites; (iii) marked P elements inserted into heterochromatin might sometimes fail to express their marker gene and therefore go undetected. However, when all the arguments are taken as a whole, I believe they make a good case for preferential insertion into the euchromatic portion of the genome. This is not to say that all P element insertions are euchromatic—at least two cases of Y-chromosomal P insertions have been detected—but rather that euchromatic sites are heavily favored.

Insertional site preferences also exist on a much finer scale. One report (187) based on electron microscopic examination of in situ hybridization suggested that interband sites were favored, but such data are difficult to interpret. Other studies have shown drastic differences in P element insertion rates when different genetic loci were compared. At one extreme is the *singed* locus, which mutates by P element insertions at frequencies between 0.1 and 1% (53, 84, 166, 196). In contrast, the *Adh* gene failed to acquire any P element insertions in a screen of approximately 10^7 gametes from dysgenic hybrids (L. Craymer and E. Meyerowitz, personal communication). This variability is also evident over relatively short distances. Simmons and Lim (194) measured the mutation rates of 17 loci in a short segment of the X chromosome called the *zeste-white* region. They found widely disparate rates which were uncorrelated with the target sizes of the loci estimated previously by chemical or radiation mutagenesis. Further evidence for locus-specific insertion rates comes from lethal mutation data (47) and Kidwell's (111) compilation of dysgenesis-induced mutability studies from many laboratories.

There is probably a tendency for P elements to integrate in or near other P elements. Formation of double elements such as sn^w (Fig. 2) would be highly unlikely otherwise. Recent evidence suggests that double elements are formed more frequently than would be expected if there were no preference for insertion at P element sites (W. Eggleston, personal communication; K. O'Hare, personal communication).

Finally, hot spots for P element insertion have been pinpointed by molecular methods. One was found in the *white* locus, where three of four P element insertions occurred at a single nucleotide site (155). One of the inserts was in the opposite orientation relative to the other two, suggesting that target site specificity is based upon the symmetrical portions of the P element's sequence. The *singed* gene has at least two hot spots separated by fewer than 100 bp (167). These sites contained 8 of 10 P element insertions. A hot spot in the *Notch* locus was indicated by Southern blot experiments (103), but only two insertions were sequenced and they did not coincide in position.

What is the basis of the insertional specificity of P elements? One important factor is probably the 8-bp target sequence which is duplicated upon insertion. O'Hare and Rubin (155) examined the sequences flanking 18 P element insertions and found a consensus sequence: GGCCAGAC. The hot spot sites at *singed* and *Notch* match this consensus in at least 4 of 8 bp, and the *white* hot spot matches it perfectly. However, there are several non-hot spot sites at both *singed* and *Notch* that match the consensus better than the hot spots, indicating that the preference for this consensus is quite weak. Much of the target site preference must therefore be determined by features not immediately apparent from inspection of the flanking DNA sequence. One suggestion (103, 222) is that hot spots have a tendency to lie near the transcription start sites of genes, especially genes with germ line activity, although there is probably some ascertainment bias favoring hot spots with obvious phenotypic effects. On the whole, when one looks at individual P element insertion sites that have been analyzed at various loci (28, 29, 52, 96a, 177, 184, 222, 223), as opposed to just hot spots, those at or near transcription start sites do seem to outnumber those in the amino acid coding region and other areas. Further work will be needed to confirm and explain the relationship between transcription sites and P element insertion sites.

3. The *cis* requirements for P element transposition

As mentioned above, P elements with large internal deletions, or with even larger additions, such as w^+ and ry^+ genes, can still transpose readily provided there is a suitable transposase source. They cannot jump, however, without the proper termini (102). To determine the precise requirements for mobility, Mullins et al. (M. Mullins, D. Rio, and G. Rubin, personal communication) devised a method to test whether any given sequence can serve as a right-hand (i.e., 3′) terminus. They began with a construct (Fig. 8a) which contains three terminal sequences plus two distinguishable marker genes, which I will call X and Y. The idea was to use germ line transformation to test whether the middle sequence could function as a P element end by allowing recovery of X without Y. Mullins et al. injected a series of P elements derived from this construct (Fig.

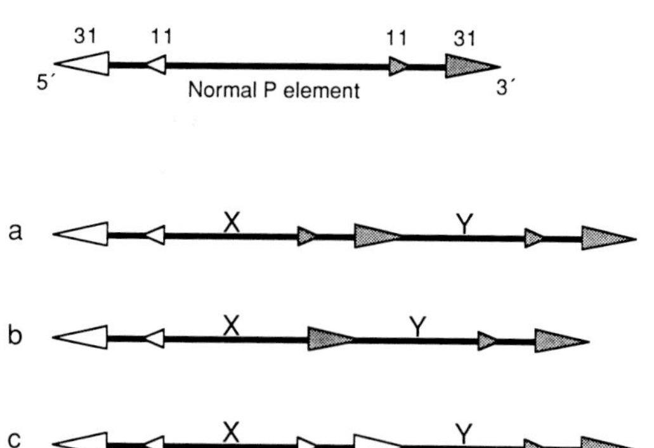

Figure 8. Constructs used to determine the *cis* requirements for transposition. The experiments (see text) were performed by Mullins et al. (personal communication).

8b and c) and selected for the incorporation of the X marker. By testing for the presence of Y in these inserted elements, they arrived at the following conclusions. (i) When two functional 3′ termini are present, the terminus most proximal to the 5′ end is used preferentially. Thus, only marker X is recovered from constructs such as that in Fig. 8a, in which both 3′ ends are functional. When the proximal 3′ end is not functional (Fig. 8b), then the distal one is used and both markers are recovered. (ii) Both the 31-bp inverted repeat and sequences in the region of the 11-bp inverted repeat (Fig. 3) are required for a functional 3′ end. (iii) Some, but not all, of the sequences between the 31-bp repeat and the 11-bp repeat are important for transposition. (iv) An inverted copy of the 5′ end of the P element cannot substitute for the 3′ terminus, indicating that the *cis* requirements for transposition are asymmetrical. (v) The 8-bp direct duplication of flanking DNA is not required for transposition. (Note, however, that these repeats do seem to be important for precise P element excision; see above and Table 1.)

4. Mechanism of transposition

No satisfactory model for P element transposition has yet been proposed. The best we can do at the present is list several types of models that appear to be particularly unsatisfactory in light of current data. In the following discussion, I will assume that there is a single mechanism that can account for all P element transposition. Under this assumption, models that fail to explain any aspect of the data can be discarded. It is possible, of course, that there is actually more than one mechanism of P element transposition, in which case some of the hypotheses rejected in the following discussion are actually viable. However, in the absence of any specific evidence for multiple mechanisms, the assumption of a single model provides the best working hypothesis.

First, conservative transposition models, i.e., those requiring excision of an element from one site and reinsertion elsewhere without replication, seem unlikely. I say this despite two reports to the contrary. The first is that of Steller and Pirrotta (212), who argued in favor of conservative transposition on the basis of a high frequency of new inserts found in flies selected for reversions of a P element insertion mutation. These reversions were not analyzed, but were assumed to be precise excisions. Subsequent studies showed this assumption to be incorrect (V. Pirrotta, personal communication). In addition, the frequency of insertions in individuals that were not selected for a reversion was not examined, thereby leaving open the possibility that the insertions were already present in the stock. Such insertions might have occurred in their stock in a previous generation and have reached a high frequency in the population by the time the reversions were selected. Therefore, it is not necessary to postulate conservative transposition to explain this result. Hawley et al. (89) also propose conservative transposition to explain their finding of a stock from which they recovered both transpositions and excisions of a particular type of defective P element. However, this interpretation must be considered tentative for several reasons. (i) There was no direct correspondence between particular insertion and excision events, which were recovered in different gametes. (ii) The phenotypic reversions were not shown to be precise excisions of the P element. However, six of them appeared precise at the level of Southern blotting. (iii) The authors' assumption that the donor site for the observed insertions was a nearby P element of the same size might be incorrect.

The primary difficulty for conservative models is that precise excision events are much less frequent than transposition. The best data for precise excision frequencies are those of Tsubota and Schedl (223), who sequenced all their apparent excision events of a P element inserted in the *rudimentary* gene. The resulting estimate is 3/7,793 = 0.0004 precise excision per generation from dysgenic crosses. We can also use data from the *white* and *RpII215* loci if we assume that all complete phenotypic reversions are precise excisions of the inserted P element. The resulting estimates, however, must be considered upper limits since only one reversion at each locus has been sequenced and shown to be a precise excision (155, 184). For the *white* locus, combining data from two sources (173, 228) gives an estimate of 0.001. For the *RpII215* locus, it is 0.007, obtained by multiplying the reversion rate estimated by Voelker et al. (224) by the proportion of their reversions that appear to be fully wild type. In all three sets of data, the inserts were defective P elements, and the reversions were obtained using M × P crosses. Therefore, the comparable estimates of transposition frequency are those obtained by in situ hybridization following a dysgenic cross. Three such experiments discussed above all yielded data consistent with approximately 0.25 transposition event per element per generation (12, 15, 50). We can conclude, therefore, that an average P element produces new insertions much more frequently than it undergoes precise excision. The fold difference can be estimated to lie in the range of 36 to 625, with the higher value coming from the most appropriate data.

These estimates make it difficult, but not yet impossible, to sustain a conservative transposition model. Such a model would require that the great majority of precise excision events are not recovered or are not recognized for some reason. One such reason could be that chromosomes with precise excisions are usually lost. However, the rate of dysgenesis-induced chromosome loss appears to be too low for this explanation (Benz, personal communication). Alternatively, we could postulate the transposition is essentially conservative, but a small amount of DNA synthesis still occurs, resulting in part of the P element sequence left at the donor site. Such events would not be recognized as precise excisions. However, even if we include the nearly precise excision events (Table 1), the insertion rate still seems to exceed the excision rate, but the difference is no longer overwhelming. It is possible that enough new synthesis occurs at the donor site to replace most or all of the original element. However, it is not clear that such a model, which is currently being tested in my laboratory, can properly be called "conservative."

Another argument against conservative transposition is the lack of correlation between excision and insertion events. The model for excision proposed above illustrates one way in which excisions might occur independently of transposition, despite the requirement for transposase in both processes. A more direct line of evidence was recently provided by D. Coen and co-workers (personal communication), who used a scheme similar to that in Fig. 5 to look for correlations between excision and insertion events within the germ line of individual males. The background genotype was held constant. Despite large sample sizes, no such correlation was detected. Finally, P elements have been found to replicate autonomously in *Saccharomyces cerevisiae* (42), suggesting the presence of an origin of replication which might be involved in replicative transposition.

A second class of unsatisfactory models are those that require a cointegrate structure, as in the case of bacterial transposon Tn*3* (D. Sherratt, this volume). According to such models, there should be a high frequency of chromosome rearrangements due to occasional failure in the resolution of cointegrates. Indeed, chromosome rearrangements are frequently produced by hybrid dysgenesis, but they are of the kind expected from random breakage and rejoining of chromosomes, rather than successive two-point rearrangements as predicted if cointegrate structures were the cause (69). Furthermore, new P element sites, as detected by in situ hybridization, are not consistently generated at the breakpoints, as would be expected if these rearrangements were equivalent to cointegrate structures (69).

Lastly, it appears unlikely that P elements transpose by means of an RNA intermediate and reverse transcription. The presence of introns and the absence of long terminal repeats in P elements (155) make it hard to see how such a mechanism could reproduce the entire element. Moreover, no full-length transcripts of P elements have been detected (102, 126), even in flies where transposition should be frequent.

In conclusion, much has been learned about how P elements transpose, but there is still no single mechanism to provide a satisfactory explanation for all the data.

V. REGULATION

A. Germ Line Specificity

The germ line is the first tissue to form in the *Drosophila* embryo. It starts as a small cluster of pole cells at the posterior end of the syncytium and retains its identity throughout development. One of the more remarkable features of hybrid dysgenesis is that its effects are strictly limited to this cell lineage. P elements are almost completely stable in somatic tissues despite their enormous rates of transposition and excision in the germ line (see above). Moreover, they retain their high germ line activity during most of development, as indicated by the lengthy temperature-sensitive period for some dysgenic traits (65, 182) and the wide range of cluster sizes for mutations and other events among the progeny of dysgenic flies.

This tissue specificity comes about by differential splicing of the transposase message. The 2-3 intron, which must be removed before functional transposase can be produced, is properly spliced out in germ cells but not in somatic cells. This mechanism was discovered by Laski et al. (126), who were investigating the P element transcripts. It was not possible to study germ line RNA directly, since the germ cells compose only a minute fraction of the organism at any stage of development. These cells cannot be isolated on a large enough scale for RNA studies. Therefore, Laski et al. concentrated on somatic RNA. They found that replacing the P element's normal promoter with a heat shock promoter greatly increased transcript abundance, but there was still no evidence for somatic transposition or excision. Furthermore, the RNA they analyzed always retained the 2-3 intron, whose presence was predicted from previous work (102, 155). They reasoned that this splice may be limited to the germ line and therefore not detected by their methods. To test this possibility, they constructed a modified version of the Pc[*ry*] element called P[ry^+ Δ2-3], as shown in Fig. 4. This element lacked the 2-3 intron, whose boundaries were conjectured on the basis of known splice site consensus sequences. When they placed P[ry^+ Δ2-3] into the genome, they found high levels of somatic P element activity indicated by mobilization of nonautonomous P elements (sn^w and P[w^+]) resulting in somatic mosaics (see panel C of the frontispiece of this volume). They concluded that the absence of hybrid dysgenesis in somatic cells was entirely due to the failure of the 2-3 splice to occur somatically.

B. Cytotype

1. The repressor model of cytotype

As discussed above (see Section II), cytotype is the regulation mechanism that accounts for the reciprocal cross difference in hybrid dysgenesis. The P cytotype is the state in which transposition and excision are suppressed both in the germ line and in somatic cells. One way of explaining cytotype is to assume that at least some P elements produce a gene product that serves as a repressor of P element activity. Motivation for this idea comes from the observed increase in P cytotype frequency whenever P strain-derived chromosomes are added, provided the maternal cytotype is held constant (54, 105), and from the disappearance of P cytotype when all P elements are removed (217). However, some additional complexity must be added to this model to account for the maternal component of cytotype determination. For example, we could assume that the repressor exhibits positive regulation of its own production (155). In that case, transmission of P elements through the male would break the positive feedback cycle, since *Drosophila* sperm are unlikely to carry any repressor whereas eggs probably would. Alternatively, some P elements themselves might exist extrachromosomally with limited self-replication plus the ability to produce the repressor. Either mechanism could result in the mixture of chromosomal and maternal inheritance observed for cytotype.

Another constraint on this model is imposed by the finding that most of the P element's length is used to encode the transposase (see Fig. 4). This means that the hypothetical repressor must be extremely small, or else overlap the transposase. The idea of transposase and repressor sharing part of their sequences is attractive because it parallels the regulatory systems of some procaryotic transposons (D. E. Berg, this volume). In addition, it suggests some possible mechanisms of repression. For example, a modified transposase molecule might prevent P element mobility by competing with the real transposase for binding sites on P element termini, especially if its modification caused it to bind more effectively than the transposase. Alternatively, a defective transposase molecule might act by "poisoning" a multimeric transposase protein. Models of this kind are consistent with the data of Steller and Pirrotta (212), who studied a specially modified P element with a heat shock promoter causing it to greatly overproduce its transposase transcript. In the P cytotype, this element's ability to cause sn^{w} mutability was decreased by about half, but it was not completely eliminated. One interpretation is that the increased abundance of transposase partially overcame the repressor. Another possibility is that repressor acts through transcriptional regulation and that the heat shock promoter caused some attenuation of this effect.

2. The ORF2-ORF3 splice is not required for P cytotype

If an element-encoded repressor is involved in the P cytotype, it probably does not utilize the ORF2-ORF3 splice. This conclusion comes from the finding (64) that the somatic P element activity brought about by a P[ry^{+} Δ2-3] element is fully suppressed by the P cytotype. This means that the P[ry^{+} Δ2-3] element responds to the P cytotype even in somatic cells, where, as discussed above, the ORF2-ORF3 splice is not made. Further evidence for this point is provided by reports of two strains, both derived from Japanese populations, which appear to have the P cytotype but lack complete P elements (88a, 153, 176). One of these, designated WY113, was first reported to have no P element homology (98) and then later described as having a large P element with a deletion in ORF3 (153). (No explanation for this difference was suggested.) Several other P elements were also present in WY113 and might also have played a role in its P cytotype.

3. Evidence from transformed M strains

Additional support for the repressor model comes from the study of M strains which had been given autonomous P elements through transformation. Such elements can spread throughout the genome in the course of several generations to reach large copy numbers (2, 36, 162a, 181). Several such lines acquired the P cytotype, thus confirming the hypothesis that P elements can bring about the P cytotype. However, the switch from M to P cytotype came only after many generations, usually more than 15, during which time the P element copy number remained comparable to that of typical P strains. One line with multiple P elements failed to switch to the P cytotype after 4 years (36). These findings indicate that the P cytotype requires something more than merely a large number of complete P elements. I suggest below that part of this "something" is the presence of P elements at particular genomic positions, and that at least some defective P elements can contribute to the P cytotype, depending on their site of integration.

4. Evidence from cytotype-dependent mutations

Further support for the repressor model comes from the finding of cytotype-dependent mutations. These are P element insertion mutations which place some aspect of gene expression under cytotype regulation. We have recovered three independent *singed* mutations of this kind that have a more extreme bristle phenotype in the M cytotype than in the P cytotype (H. Robertson and W. Engels, unpublished data). Some *singed* mutations have a pleiotropic effect on female fertility and egg morphology. This effect can also be cytotype dependent, as indicated by five independently derived alleles which cause more extreme female sterility and more abnormal egg morphology in the P cytotype than in the M cytotype. The sn^{w} allele, described above for its hypermutability (Fig. 2 and 6), is one of these five. In addition, Williams et al. (227a, 227b) have recovered a P element insertion in the *vestigial* wings locus which is nearly wild type in the P cytotype but has extreme *vestigial* wings in the M cytotype. Finally, Coen and co-workers (personal communication) found a case in which a *white* gene carried within a P element (denoted P[w^{+}]) yielded wild-type eye color in the M cytotype and a brown color in the P cytotype. Note that the cytotype effect on the *singed* sterility alleles and the P[w^{+}] insert was in the opposite direction from that seen with the *vestigial* and *singed* bristle alleles. Such complex regulatory behavior is analogous to the case of the *Spm* system in maize (N. V.

dichotomize the species into maternally and paternally contributing strains, the former being primarily laboratory stocks and the latter predominating in nature. In both systems, crosses between males from a paternally contributing stock and females from a maternally contributing stock result in a syndrome of aberrant germ line traits, but the details of these traits differ between the two systems (104). Both systems are regulated by mechanisms whose transmission involves a combination of maternal and chromosomal components; "reactivity" in the I-R system corresponds to the M cytotype in the P-M system. Finally, each system is based upon a family of transposable elements whose autonomous subclass is present only in the paternally contributing strains.

2. Functional independence

As the two systems were studied in more detail, the similarity began to break down. The P and I transposable elements are quite dissimilar in structure and transposition mechanism (Finnegan, Chapter 18, this volume). Kidwell (104) showed that appropriate interstrain crosses could produce progeny which were dysgenic in either the P-M or the I-R system without dysgenesis in the other system. Studies using sn^{w} hypermutability showed specifically that I-R dysgenesis did not mobilize P elements (60), and a large-scale in situ hybridization experiment (50) showed that P-M dysgenesis did not mobilize I elements. Thus, all indications were that the two systems were closely analogous, but neither homologous nor interdependent.

3. The missing RP strain category

The last vestige of possible nonindependence between the two systems lies in their strain distributions. Surveys of many wild and laboratory populations (104, 106) failed to find any strains with P factors but not I factors ("RP" strains). The other three possible combinations (IP, IM, RM) were common. Did the missing category mean that P elements are not able to spread in the absence of I factors? To answer this question, several workers attempted to synthesize RP strains by genetic crosses (120) or by direct transformation of an RM strain with cloned P elements (2). Both efforts were successful, yielding strains which tested as RP. Therefore, the absence of naturally occurring RP strains probably reflects the histories of I and P elements in the species, as discussed below, rather than any functional dependence.

B. Are Other Transposable Elements Mobilized by P-M Hybrid Dysgenesis?

1. Initial reports claimed massive cross-mobilization

According to several reports, P-M hybrid dysgenesis mobilizes not only P elements but most other families of transposable elements as well (76, 77, 79, 131), particularly elements of the retroviruslike structure reviewed elsewhere in this volume (P. M. Bingham and Z. Zachar). Given the many structural and functional differences between P elements and retroviruslike elements, such a finding seems unlikely a priori. In addition, the majority (10/13) of P-M dysgenesis-induced lethal mutations were found to correspond to new P element insertions visible by in situ hybridization (47). Those lethals without visible in situ sites might still possess P elements with insufficient homology with the probe to show up. If P-M dysgenesis were mobilizing other transposable elements to a significant degree, then many of the lethals should have been caused by insertions of non-P elements. The same conclusion can be drawn from dysgenesis-induced visible mutations (e.g., references 28, 29, 96a, 103, 166, 167, 173, 177, 222, 224), although rare exceptions have been reported (33b, 173). To add to the doubts, there were problems in the design of the cross-mobilization experiments, making it impossible to distinguish between new transposition events and preexisting variability in the stocks. I shall discuss these problems in more detail shortly.

2. Follow-up studies indicated no cross-mobilization

Two experiments avoided the ambiguity of preexisting variability, and both resulted in strong evidence against mobilization of non-P elements by hybrid dysgenesis. The first experiment was by Woodruff et al. (228), who measured the reversion frequency of mutations known to be caused by insertion of one of several transposable elements, including P, *Foldback*, *copia*, *gypsy* (*mdg-4*), *B104* (*roo*), *F*, and *hobo*. Large samples of progeny were scored from both dysgenic and nondysgenic parents. The P element mutations reverted much more frequently in the dysgenic conditions than in the nondysgenic conditions, as expected. However, the non-P element mutations reverted at about the same rate in both sets of crosses. Woodruff et al. pointed out that their conclusion is limited to those events that alter the mutations' phenotype, usually excisions. Their experiment was not designed to detect an increase in insertion rates.

mechanism could result in the mixture of chromosomal and maternal inheritance observed for cytotype.

Another constraint on this model is imposed by the finding that most of the P element's length is used to encode the transposase (see Fig. 4). This means that the hypothetical repressor must be extremely small, or else overlap the transposase. The idea of transposase and repressor sharing part of their sequences is attractive because it parallels the regulatory systems of some procaryotic transposons (D. E. Berg, this volume). In addition, it suggests some possible mechanisms of repression. For example, a modified transposase molecule might prevent P element mobility by competing with the real transposase for binding sites on P element termini, especially if its modification caused it to bind more effectively than the transposase. Alternatively, a defective transposase molecule might act by "poisoning" a multimeric transposase protein. Models of this kind are consistent with the data of Steller and Pirrotta (212), who studied a specially modified P element with a heat shock promoter causing it to greatly overproduce its transposase transcript. In the P cytotype, this element's ability to cause sn^{w} mutability was decreased by about half, but it was not completely eliminated. One interpretation is that the increased abundance of transposase partially overcame the repressor. Another possibility is that repressor acts through transcriptional regulation and that the heat shock promoter caused some attenuation of this effect.

2. The ORF2-ORF3 splice is not required for P cytotype

If an element-encoded repressor is involved in the P cytotype, it probably does not utilize the ORF2-ORF3 splice. This conclusion comes from the finding (64) that the somatic P element activity brought about by a P[ry^{+} Δ2-3] element is fully suppressed by the P cytotype. This means that the P[ry^{+} Δ2-3] element responds to the P cytotype even in somatic cells, where, as discussed above, the ORF2-ORF3 splice is not made. Further evidence for this point is provided by reports of two strains, both derived from Japanese populations, which appear to have the P cytotype but lack complete P elements (88a, 153, 176). One of these, designated WY113, was first reported to have no P element homology (98) and then later described as having a large P element with a deletion in ORF3 (153). (No explanation for this difference was suggested.) Several other P elements were also present in WY113 and might also have played a role in its P cytotype.

3. Evidence from transformed M strains

Additional support for the repressor model comes from the study of M strains which had been given autonomous P elements through transformation. Such elements can spread throughout the genome in the course of several generations to reach large copy numbers (2, 36, 162a, 181). Several such lines acquired the P cytotype, thus confirming the hypothesis that P elements can bring about the P cytotype. However, the switch from M to P cytotype came only after many generations, usually more than 15, during which time the P element copy number remained comparable to that of typical P strains. One line with multiple P elements failed to switch to the P cytotype after 4 years (36). These findings indicate that the P cytotype requires something more than merely a large number of complete P elements. I suggest below that part of this "something" is the presence of P elements at particular genomic positions, and that at least some defective P elements can contribute to the P cytotype, depending on their site of integration.

4. Evidence from cytotype-dependent mutations

Further support for the repressor model comes from the finding of cytotype-dependent mutations. These are P element insertion mutations which place some aspect of gene expression under cytotype regulation. We have recovered three independent *singed* mutations of this kind that have a more extreme bristle phenotype in the M cytotype than in the P cytotype (H. Robertson and W. Engels, unpublished data). Some *singed* mutations have a pleiotropic effect on female fertility and egg morphology. This effect can also be cytotype dependent, as indicated by five independently derived alleles which cause more extreme female sterility and more abnormal egg morphology in the P cytotype than in the M cytotype. The sn^{w} allele, described above for its hypermutability (Fig. 2 and 6), is one of these five. In addition, Williams et al. (227a, 227b) have recovered a P element insertion in the *vestigial* wings locus which is nearly wild type in the P cytotype but has extreme *vestigial* wings in the M cytotype. Finally, Coen and co-workers (personal communication) found a case in which a *white* gene carried within a P element (denoted P[w^{+}]) yielded wild-type eye color in the M cytotype and a brown color in the P cytotype. Note that the cytotype effect on the *singed* sterility alleles and the P[w^{+}] insert was in the opposite direction from that seen with the *vestigial* and *singed* bristle alleles. Such complex regulatory behavior is analogous to the case of the *Spm* system in maize (N. V.

Fedoroff, this volume). Northern (RNA) blots by K. O'Hare (personal communication) showed that sn^w females produced the normal *singed* transcripts in the M cytotype, but the transcripts associated with female fertility effects were missing in the P cytotype. Therefore, cytotype dependence works at the level of transcription, at least for the female fertility effects at *singed*. The simplest way to explain cytotype-dependent mutations is to assume that a repressor molecule binds to the P element, thereby altering the adjacent gene's expression.

5. At least two repressor-making P elements have been found

The Pc[*ry*] element and its mutational derivatives proved their usefulness once again when they were applied to the problem of P element repressor. The use of strains transformed with these elements, as opposed to ordinary P strains, avoids the inherent ambiguities that result when multiple, heterogeneous P elements are present.

Two of the elements shown in Fig. 4 were found to mimic the P cytotype by the following criteria (Robertson and Engels, unpublished data). First, they altered expression of cytotype-dependent alleles in the same way as P cytotype. Second, they suppressed somatic mosaicism in sn^w; P[ry^+ Δ2-3] flies. Third, they partially suppressed sn^w hypermutability in the germ line. One of the elements is designated P[ry^+ *Sal*I](89D) to indicate that it carries a frameshift mutation in ORF3 (at the *Sal*I restriction site) and is integrated at cytological position 89D. The second element, P[ry^+ 1949G](96B), is one of three integrations of the 3′ splice junction mutant (Fig. 4). Its effects on dysgenesis and cytotype-dependent alleles were qualitatively similar to those of P[ry^+ *Sal*I](89D), but somewhat less pronounced. Another potential repressor-making P element has been reported (153), but it has not yet been checked for its effect on cytotype-dependent alleles or tested in isolation from other potential repressor-making P elements.

Many other insertions of the same or similar elements in different genomic sites failed to yield significant evidence of repressor. Therefore, it seems that the genomic position at which these elements integrate is critical for repressor production. Also, note that both repressor makers carry mutations that would prevent the proper expression of information in exon 3. This finding agrees well with the previous conclusion (see above) that the ORF2-ORF3 splice is not essential for the P cytotype.

6. Do elements P[ry^+ *Sal*I](89D) and P[ry^+ 1949G](96B) encode P cytotype?

The connection between cytotype and the effects of the two repressor-making P elements is not firmly established. These elements clearly mimic the P cytotype in the tests described above, but they fail to resemble cytotype in at least one respect. There was no detectable maternal component in the inheritance of the suppression effects of P[ry^+ *Sal*I](89D), as would be expected if this element were producing the P cytotype (Robertson and Engels, unpublished data). Further work will be needed to determine whether the repressor encoded by these elements is responsible for the P cytotype.

7. Molecular nature of the repressor

The gene product responsible for the suppression effect by these elements has not yet been determined. If one assumes that transcription initiation, termination, and splicing all occur for repressor just as they do for transposase, then the repressor-making elements P[ry^+ *Sal*I](89D) and P[ry^+ 1949G](96B), as well as the putative repressor maker of Nitasaka et al. (153), were all expected to produce various truncated forms of the transposase molecule in which most or all of ORF3 is missing. Indeed, Rio et al. (165) suggested that the 66-kilodalton protein produced in the absence of the ORF2-ORF3 splice might act as a repressor. One possibility is that truncation of the transposase at any of several points near the ORF2-ORF3 boundary results in a molecule with some suppression capability. However, this model fails to explain the very pronounced position effects, which suggest that the difference between transposase and repressor may be more complicated than simple truncation. Better understanding of the nature of the repressor will require a detailed analysis of the gene products of these repressor-making elements plus further studies of the position effects.

C. Other Kinds of Regulation

1. Suppression of P element activity in M′ strains

There are other mechanisms besides the P cytotype that can have a suppressing effect on P element activity. This conclusion came from experiments involving a class of strains called M′. As discussed previously, such strains have the M cytotype, as indicated by hybrid dysgenesis when crossed to P strain males, but their chromosomes harbor many P elements. Genetic (64, 197) and molecular (17) tests showed that Birmingham, the first M′ strain to be studied (15), is devoid of autonomous P elements. However, it soon became clear that other M′ strains did have low levels of transposase activity, revealing that autonomous P elements were present (100).

What was keeping these elements relatively inactive and allowing M′ strains to be stable over long periods of time? The mechanism was clearly different from the P cytotype. First, it did not show the pronounced bimodality associated with cytotype determination. When a genotype includes both P and M chromosomes, the cytotype is almost always either M or P, not intermediate (54). However, various combinations of M and M′ chromosomes result in all levels of dysgenesis when crossed to P strain males (20, 108, 190, 197). Furthermore, this suppression was fully transmissible through the male line (108, 190, 197), as opposed to the partially maternal inheritance of cytotype. (One exception to the above rule was reported by Simmons et al. [197], who found what they interpret as a "weak" P cytotype in one subline of Birmingham. Another subline, studied in my laboratory [unpublished data], showed no such effect.)

2. Some of the suppression can be explained by titration of transposase

Simmons et al. (190, 197) pointed out that the nonautonomous P elements in M′ chromosomes might partially suppress sn^{w} hypermutability by competing for transposase. They reasoned that the P element termini probably serve as substrate for transposase at some point in the transposition process. Therefore, the addition of many nonautonomous elements, each with a set of P element termini to bind transposase but no gene to produce it, might leave less transposase to act upon the P elements at sn^{w}. Simmons et al. showed that each of the Birmingham chromosomes had a suppressing effect on sn^{w} mutability, as would be expected from their model, since each Birmingham chromosome contains many defective P elements. Furthermore, they found that Birmingham chromosomes did not suppress GD sterility; in fact, they enhanced it. This result is also expected from the titration model, since the action of transposase on nonautonomous P elements probably contributes to GD sterility (see below), and the titration of transposase by such elements would not detract from the overall effect. Further support for this idea came from microinjection studies in which defective P elements injected into dysgenic embryos were shown to reduce sn^{w} mutability by about 30% (W. Eggleston, M. Fortini, and W. Engels, submitted for publication). A nearly complete P element with only its terminal sequences missing did not have this effect, indicating that the P element ends were needed for the titration effect.

3. Some M′ chromosomes suppress GD sterility

Kidwell (108) found that chromosomes from another M′ strain, called Sexi, could partially suppress GD sterility. As mentioned above, the titration effect can account for suppression of sn^{w} mutability, but it cannot suppress GD sterility. Black et al. (17) suggested that KP elements, a specific class of defective P elements, might be responsible for this suppression. KP elements are unusual for their high copy number in *Sexi* and other M′ strains, but they are not found in Birmingham. The idea of some functional role for KP elements seems plausible a priori, since it would help explain their high copy in many M′ strains, where natural selection would be expected to favor anything that diminishes GD sterility. Although Black et al. claimed to have genetic data supporting a regulatory role for KP elements, further evidence is clearly needed. To provide such evidence, an experiment would need the following features, which were not incorporated in the work reported (17): (i) the effects of KP elements must be separated from the possible effects of other P elements on the same chromosomes; (ii) appropriate genetic markers and balancers should be used so that the genotypes of the tested individuals can be determined; (iii) the possibility of selective differences between chromosomes should be taken into account in interpreting any long-term changes in stocks; (iv) if the differences in dysgenesis levels are subtle, then the experiments must be designed to allow statistical analysis. In any case, the final answer is likely to be complicated, since two of the strains used by Black et al. (Gomel and Kibris) appear from their data to possess many KP elements, yet exhibit no suppression of GD sterility. Furthermore, I. Boussy (personal communication) found that the level of regulation in Australian M′ lines is uncorrelated with the number of KP elements.

VI. INTERACTIONS WITH OTHER TRANSPOSABLE ELEMENTS

A. The P-M and I-R Systems Are Independent

1. Similarities between the two systems

The I-R system of hybrid dysgenesis is reviewed elsewhere in this volume (Finnegan, Chapter 18, this volume). On the basis of pure phenomenology, the P-M and I-R systems look remarkably similar. Both dichotomize the species into maternally and paternally contributing strains, the former being primarily laboratory stocks and the latter predominating in nature. In both systems, crosses between males from a paternally contributing stock and females from a maternally contributing stock result in a syndrome of aberrant germ line traits, but the details of these traits differ between the two systems (104). Both systems are regulated by mechanisms whose transmission involves a combination of maternal and chromosomal components; "reactivity" in the I-R system corresponds to the M cytotype in the P-M system. Finally, each system is based upon a family of transposable elements whose autonomous subclass is present only in the paternally contributing strains.

2. Functional independence

As the two systems were studied in more detail, the similarity began to break down. The P and I transposable elements are quite dissimilar in structure and transposition mechanism (Finnegan, Chapter 18, this volume). Kidwell (104) showed that appropriate interstrain crosses could produce progeny which were dysgenic in either the P-M or the I-R system without dysgenesis in the other system. Studies using sn^{w} hypermutability showed specifically that I-R dysgenesis did not mobilize P elements (60), and a large-scale in situ hybridization experiment (50) showed that P-M dysgenesis did not mobilize I elements. Thus, all indications were that the two systems were closely analogous, but neither homologous nor interdependent.

3. The missing RP strain category

The last vestige of possible nonindependence between the two systems lies in their strain distributions. Surveys of many wild and laboratory populations (104, 106) failed to find any strains with P factors but not I factors ("RP" strains). The other three possible combinations (IP, IM, RM) were common. Did the missing category mean that P elements are not able to spread in the absence of I factors? To answer this question, several workers attempted to synthesize RP strains by genetic crosses (120) or by direct transformation of an RM strain with cloned P elements (2). Both efforts were successful, yielding strains which tested as RP. Therefore, the absence of naturally occurring RP strains probably reflects the histories of I and P elements in the species, as discussed below, rather than any functional dependence.

B. Are Other Transposable Elements Mobilized by P-M Hybrid Dysgenesis?

1. Initial reports claimed massive cross-mobilization

According to several reports, P-M hybrid dysgenesis mobilizes not only P elements but most other families of transposable elements as well (76, 77, 79, 131), particularly elements of the retroviruslike structure reviewed elsewhere in this volume (P. M. Bingham and Z. Zachar). Given the many structural and functional differences between P elements and retroviruslike elements, such a finding seems unlikely a priori. In addition, the majority (10/13) of P-M dysgenesis-induced lethal mutations were found to correspond to new P element insertions visible by in situ hybridization (47). Those lethals without visible in situ sites might still possess P elements with insufficient homology with the probe to show up. If P-M dysgenesis were mobilizing other transposable elements to a significant degree, then many of the lethals should have been caused by insertions of non-P elements. The same conclusion can be drawn from dysgenesis-induced visible mutations (e.g., references 28, 29, 96a, 103, 166, 167, 173, 177, 222, 224), although rare exceptions have been reported (33b, 173). To add to the doubts, there were problems in the design of the cross-mobilization experiments, making it impossible to distinguish between new transposition events and preexisting variability in the stocks. I shall discuss these problems in more detail shortly.

2. Follow-up studies indicated no cross-mobilization

Two experiments avoided the ambiguity of preexisting variability, and both resulted in strong evidence against mobilization of non-P elements by hybrid dysgenesis. The first experiment was by Woodruff et al. (228), who measured the reversion frequency of mutations known to be caused by insertion of one of several transposable elements, including P, *Foldback*, *copia*, *gypsy* (*mdg-4*), *B104* (*roo*), *F*, and *hobo*. Large samples of progeny were scored from both dysgenic and nondysgenic parents. The P element mutations reverted much more frequently in the dysgenic conditions than in the nondysgenic conditions, as expected. However, the non-P element mutations reverted at about the same rate in both sets of crosses. Woodruff et al. pointed out that their conclusion is limited to those events that alter the mutations' phenotype, usually excisions. Their experiment was not designed to detect an increase in insertion rates.

The second large-scale experiment, however, could detect both insertions and excisions. Eggleston et al. (50) used in situ hybridization to monitor the activity of 19 transposable element families, including those listed above plus *flea*, *springer*, *opus*, *412*, *297*, *I factor*, *G*, *Doc*, *Delta88*, and three other retroviruslike elements. (See reference 71 for further descriptions of these elements.) All tested chromosomes were derived from a single X chromosome at the initiation of the experiment. This chromosome was marked with sn^{w} to allow continuous monitoring of P-M dysgenesis. A backcrossing scheme utilizing attached-X females was employed to follow the lineage of all tested chromosomes. The results showed that all transposable elements except P had only trivial background levels of mobility in both dysgenic and nondysgenic lines. Therefore, the earlier claims that other elements are mobilized by P-M dysgenic crosses are probably incorrect.

3. Possible explanations for the discrepancy

How can we reconcile the initial reports of extensive cross-mobilization between P elements and other mobile element families with the unambiguously negative results from subsequent studies? In the case of the report of Lewis and Brookfield (131), the most likely explanation is that standing variability in their base populations was misinterpreted as new insertion events. Their procedure was to cross an M strain (Canton S) and a P strain (Harwich) in both ways as small mass matings to initiate lines which were analyzed 14 generations later by in situ hybridization. In the meantime, the Harwich and Canton S chromosomes segregated and recombined freely. No genetic markers were used. Differences in hybridization sites at the end of the experiment might reflect new transposition events, as the authors assume, but they might also reflect polymorphisms in the original Canton S or Harwich stocks. In fact, the authors noted "many" such polymorphisms in their data, but argued that such false insertions and excisions should occur equally in the dysgenic and nondysgenic lines and therefore could be ignored. However, the two sets of crosses were not symmetrical. The dysgenic lines are expected to carry a much greater proportion of the Canton S genome than the nondysgenic lines. This difference results from the segregation distortion effect against P chromosomes in dysgenic flies (93) and also from the sex chromosome difference between reciprocal crosses. Therefore, if we assume that the non-P elements in Canton S were more polymorphic than those in Harwich, the data can be explained without resorting to cross-mobilization.

Gerasimova and colleagues (76–79, 147) reported cross-mobilization in the form of transpositional "bursts." They collected flies with unusual phenotypes from a stock called "ct^{MR2}" and found chromosomes in which the in situ hybridization sites for several transposable element families were drastically different from those of the canonical ct^{MR2} chromosome. The authors assumed that these differences were due to P-M hybrid dysgenesis, since the ct^{MR2} strain came from a P-M dysgenic cross. However, there were no nondysgenic controls, leaving open the interpretation that the bursts were unrelated to P-M dysgenesis. In addition, we must also consider the possibility that most or all of these observations were not real transposition events. The primary problem is that these chromosomes were collected in a way which makes it impossible to know their lineage with certainty. Recessive alleles and foreign chromosomes can exist within a stock such as ct^{MR2} for many generations without detection. By the time the foreign chromosome is detected, perhaps by phenotypic differences in a hemizygous male, it might have recombined one or more times with a homolog in the ct^{MR2} genome to yield a chromosome with mixed ancestry. The uncertainty as to the origin of these chromosomes is accentuated by inconsistencies in the data. Two distinct origins were reported for each of the alleles cm^{MR5} and w^{MR5}, with different "parental" chromosomes in the latter case (77). The g^{MR1} allele also has two distinct origins reported in reference 77 and a third described in reference 79. In most cases, these apparent multiple occurrences can be resolved by assuming that these alleles preexisted in the base stock and were rerecovered several times.

Gerasimova et al. also assert that their ct^{MR2} stock has no variability for transposable element sites except what is associated with gross phenotypic changes. They provide no clear documentation for this claim; in fact, some of the figures in reference 79 show differences in *mdg-1* and *copia* sites which appear to contradict it. In addition, they report that during transpositional bursts, *copia*-like elements have a strong tendency to reinsert into their own long terminal repeat sequences, which had been left behind from partial excision events in previous generations (78). Again, however, without knowing the lineages of these chromosomes, it is impossible to be certain that these "reinsertions" are not merely the rerecovery of the original insertion, which may have remained in the stock.

At least some of the phenotypic changes recovered by Gerasimova et al. were undoubtedly the result of real transposable element activity. In their mutations of the ct^{MR2} allele itself, the authors observed partial excisions of the inserted *mdg-4* element, also known as *gypsy*, or insertions of another transposable element (147). However, given the many uncertainties in their data, the conclusion that P-M hybrid dysgenesis causes transpositional bursts of many transposable element families, or even that such bursts occur at all, seems premature. If these bursts are real, it would be a discovery of considerable interest and importance. Despite this importance, indeed because of it, the finding must be documented with the utmost genetic rigor. The powerful genetic techniques available to *D. melanogaster* researchers make such a rigorous approach highly feasible.

VII. P ELEMENT-INDUCED CHANGES IN THE *DROSOPHILA* GENOME

A. Male Recombination

The first trait identified as part of the hybrid dysgenesis syndrome still remains mysterious. Although its molecular basis is unknown, the genetic features of male recombination were determined from several studies (55, 95, 97, 200, 216, 229, 230, 236). The conclusions from these experiments are as follows. (i) Male recombination occurs in the germ line of dysgenic flies, usually premeiotically. (ii) Its frequency is dependent upon which P strain is used, but it is usually on the order of 1% per chromosome arm. This is much less frequent than female (meiotic) recombination. (iii) The distribution of recombination sites also depends upon the P strain and clearly differs from the meiotic recombination pattern. In euchromatic regions, the male recombination map distances are closer to the physical length than are meiotic maps. Nevertheless, very little male recombination occurs in the heterochromatin. (iv) The events underlying male recombination probably also occur in dysgenic females, as indicated by altered distribution of crossovers. (v) The "interchromosomal effect," in which suppression of meiotic recombination by chromosome rearrangements in one area of the genome increases recombination in others, does not occur for male recombination. (vi) There is little or no positive interference.

To explain male recombination, we might suppose that it occurs by a mechanism similar to that of radiation-induced chromosome rearrangements. According to this view, chromosome breakage brought about by P element mobilization leads to nonhomologous rejoining of fragments. If the breakpoints are on opposite homologs, the result can be male recombination. A simple test of this model was conducted by Sved (216). He reasoned that if male recombination were frequently between nonhomologous sites, then many of the recombinant chromosomes would have duplications or deficiencies. The latter should usually be lethal as homozygotes. However, when he compared the recombinant chromosomes with nonrecombinants, he found approximately the same frequency of recessive lethals. This experiment was subsequently repeated by two other groups (97, 200), who arrived at the same conclusion. The pooled data from all three experiments are: 28/198 lethals among the recombinant chromosomes and 20/179 among the nonrecombinants. Therefore, male recombination seems to occur by a very precise mechanism resulting in homologous exchange. This is quite unlike the chaotic events leading to chromosome rearrangements. It is not known whether male recombination events tend to occur at the sites of P elements, as do chromosome rearrangements (see Section VII.C).

B. Mutations

1. Types of P-induced mutational events

Like most transposable elements, P elements cause mutations. They do it by at least three kinds of events: excisions, chromosome rearrangements, and insertions. Excisions, as discussed above, sometimes remove genomic sequences flanking the element, resulting in mutations of nearby genes (37, 143, 177). Chromosome rearrangements, primarily inversions, also tend to mutate genes close to existing P element sites, which is where breakpoints usually occur (68, 69, 163, 194). Insertion mutations, however, do not require a preexisting P element close to the affected gene. The donor site can be anywhere in the genome, or even on an extrachromosomal plasmid put in by injection.

2. Average rate of insertional mutations

As discussed previously, there is tremendous variation in the rates at which different loci acquire P element insertions. In ordinary M × P dysgenic crosses, the per-locus rates range from being too small to measure to very high frequencies, approaching 0.01 per generation. We can calculate the mutation rate of an average locus by using the estimate of 0.25 as the expected number of new insertions produced by each P element donor site per generation (see Section IV.E). Assuming (i) there are 25 potential donor elements in an M/P dysgenic hybrid, (ii) that the recipient gene's effective target size is 5 kb, and (iii) that all insertions are in the estimated 150,000 kb of euchromatic DNA, then the mutation rate would be approximately 0.0002 per gene per generation.

Another approach to estimating this quantity is to use the copious data from P element mutagenesis screens conducted for various loci in many laboratories. Kidwell compiled a list of such studies from a wide variety of sources (111). She pointed out that these data come from many diverse experimental techniques, and few of the experiments were designed with mutation rate estimates in mind. Therefore, a numerical average of the results would not be meaningful. However, a casual inspection of the list indicates that the estimate of 0.0002 computed above is in the right range for most loci.

3. Recessive lethal mutations

Another mutation rate of interest is the frequency of lethal mutations. Simmons et al. (192, 198) estimated the lethal rate per generation at 0.03 for a P strain X chromosome in dysgenic males. This value, however, includes not only insertion mutations, but also mutations due to excisions and rearrangements, since the mutated chromosome already carried many P element sites at which such events could occur. To estimate the lethal rate due to insertions alone, it is necessary to use an M strain chromosome as the target. By this method, the same authors estimated that the lethal rate just from insertion events is approximately 0.008. They interpreted this difference to mean that the majority of lethal mutations on a P chromosome result from excisions, rearrangements, and other changes occurring at P element sites.

The lethal mutation rate estimate of 0.008 seems too low relative to the "average" mutation rate of 0.0002 discussed above. Together, these estimates imply that there are only about 40 essential genes on the X chromosome, whereas the actual number is probably 20 times that. To resolve this inconsistency, recall that P element activity occurs throughout development. This means that many recessive lethal mutations on the X chromosome of males, especially those occurring early in development, will kill the germ cell. Such events will not be represented in these data. As a result, the lethal rates indicated above are probably underestimates of the true frequency of recessive lethal events. This bias could, in principle, be avoided by measuring lethal frequencies on the X chromosome in females or on the autosomes of either sex, provided the M and P chromosomes can be distinguished.

What fraction of P element insertions result in recessive lethal mutations? Spradling and Rubin (210) estimate this value at 30%, but as pointed out by Simmons et al. (198), their estimate is probably inflated by the presence of lethal mutations already on the chromosome prior to the P element insertion. A better estimate would be the ratio of the lethal mutation rate to the average number of new P element in situ hybridization sites on the X chromosome. That procedure yields an estimate of about 1% lethal events per insertion (198). However, consideration of the cell selection bias just discussed implies that the true value is somewhat greater than the 1% estimate. More recent data (33a) are consistent with 10% lethal mutations per insert.

4. Do P elements interact with other mutagens?

Many studies have been aimed at detecting synergism between P-M hybrid dysgenesis and conventional mutagens, i.e., radiation (139, 206) and chemicals (19, 45, 70). Some of these studies suggested that the combination of hybrid dysgenesis and a mutagen yielded a mutation rate slightly greater than the sum of the separate effects from hybrid dysgenesis and the mutagen. If these interactions are real (see below), they might be explained by the "two-hit" nature of chromosome rearrangements or possibly by inhibition of repair mechanisms by hybrid dysgenesis.

The statistical significance of these putative interactions is difficult to assess from the available data. The premeiotic nature of dysgenesis-induced mutations results in clusters of mutants among the progeny of individual crosses. This process makes the resulting mutation rate estimates notoriously unstable. Standard parametric statistical procedures used in the above studies and many others (e.g., reference 228) are not reliable. The practice used in most of the above studies, of discarding all but one member of each mutant cluster, also introduces bias and alters the definition of mutation rate (56). (Note that the nonmutants from dysgenic crosses can also be considered as clustered. This clustering is less obvious than that of the mutants, but it is equally real. Therefore, to be consistent, those who would disregard all but one mutant from each cross must also disregard all but one nonmutant.) A valid assessment of questions such as whether mutation rates are synergistic requires experimental designs in which data are compartmentalized into many independent units which can be analyzed by nonparametric methods (33, 56).

5. Effects of P element insertions on gene expression

P element insertion mutations have now been studied at many loci, including *w*, *y*, *r*, *Sxl*, *RpII215*, *Hsp28*, *Notch*, and *sn* (28, 29, 52, 96a, 103, 130, 155, 167, 177, 222–224). Some insertions are in coding regions (29, 96a, 130, 155), but most of them are in either the 5′ untranslated leader sequences (29,

52, 96a, 103) or the control regions 5′ to the transcript (177, 222–224).

The alleles with P elements inserted within the amino acid coding regions behave as amorphs. The others, however, include a full range of phenotypes. The nature of their regulatory effects covers a remarkably broad spectrum, defying any attempt at simplifying generalizations. Some of them increase the abundance of the transcript, and others decrease it (96a, 222, 223). They can alter the developmental timing (29, 52, 223) or the tissue specificity (29, 222) of expression. Some result in complex interallelic complementation patterns (177) or change the response to heat shock (52). Several cases (described above) result in cytotype-dependent expression, and at least one insertion allele (described below) responds to an extragenic suppressor mutation. This wealth of effects is partly due to differences in the position of the element within the gene, but much of its also depends on the structure of the inserted element. The latter is seen by the wide variety of new phenotypes that can result from internal changes in the inserted P element (29, 37, 177, 223, 224). There is at least some indication that P elements whose transcription is in the same direction as that of the mutated gene cause a more severe effect than inserts at the same position but with the reverse orientation (29, 167). This trend suggests that the P element's own termination and splice sites might interfere with a gene's expression. Some of the effects P elements have on gene expression might come about from the chromatin structure at the element's termini. Nuclease-hypersensitive sites appear at P element ends, but the effect does not extend into flanking DNA (51, 52). The dazzling variety of effects P elements can have on gene expression probably reflects an equally impressive array of mechanisms.

6. Insertional reversions

Note from Fig. 2 that the phenotype of the double P element (*sn*^{w}) is less severe than that of the single insert, *sn*^{e}. In addition, Salz et al. (177) found two cases of reversion of a P-induced *Sex-lethal* allele resulting from the insertion of a second P element. These observations suggest that the addition of a P element can actually cause partial reversion of some insertion mutations. Eggleston (personal communication) has recently tested this idea by collecting full or partial reversions of P element insertion mutations at the *singed* and *yellow* loci. Among 145 such reversions analyzed by Southern blots, 15 to 40%, depending on the allele, were actually new insertions near the original P element. In all cases the new insert was 5′ of the coding region. Restriction maps were consistent with these new inserts being P elements, as was confirmed by sequencing in 2 cases and hybridization to a P element probe in 12 others. The mechanism by which these insertions revert the phenotype is not yet known.

C. Chromosome Rearrangements

It is evident in many chapters in the present volume and similar reviews that the production of chromosome rearrangements is one of the hallmarks of transposable element activity. P elements are no exception. Several investigators (14, 68, 69, 86, 163, 194, 231, 239) have observed rearrangements large enough to be visible on the level of salivary gland chromosome cytology, and others have studied them by molecular methods (167, 177, 223). The frequency of rearrangements on the X chromosome alone has been estimated at 10% per generation in the germ line of dysgenic males (14, 69), making rearrangements one of the most common effects of P element mobilization.

In one large-scale study, a collection of 746 rearrangements was obtained by selecting mutations at the *hdp-b* locus, which was near a P element insert on the X chromosome (68, 69). Thus one of the breakpoints of each rearrangement was considered **selected**, and the remaining breakpoints were unselected. These rearrangements were analyzed by cytology, genetics, and, in some cases, in situ hybridization, leading to the following generalizations. (i) Approximately 85% of the unselected breakpoints occurred at the sites of P elements to within a few hundred base pairs. (ii) The remaining 15%, called **sporadics**, were distributed uniformly about the chromosome. (iii) All but one of the unselected breakpoints were on the X chromosome, suggesting a preference for intrachromosomal rearrangements. Part of this preference might reflect a preponderance of X chromosomal P element sites or selection against X-autosome translocations which often cause sterility in males. (iv) Most of the events were simple inversions, but there were also many complex rearrangements with three, four, or five breakpoints. They occurred in the relative proportions expected if the number of breakpoints were determined by a Poisson process. (v) Approximately one-fourth (14/58) of the breakpoints at the site of a P element resulted in loss of the element, as defined by the disappearance of an in situ hybridization site. In one extreme case, a three-point rearrangement, in which each breakpoint was at a P element site, lost all three elements. (vi) One-third (6/18) of the sporadic breakpoints analyzed by in situ hybridization resulted in a gain of a P element site. This fraction becomes slightly smaller (6/24) if we include three rearrangements analyzed in another study (87). (vii) Most inversions that retained P elements at both breakpoints reverted readily in dysgenic flies to restore the normal chromosome sequence and return the *hdp-b* gene to wild type. (viii) These restored chromosomes could undergo further rearrangements selected by the *hdp-b* method. (ix) Some of the rearrangements included large duplications, indicating that the event occurred after DNA synthesis in the cell cycle. (x) The overall distribution of rearrangement types was of the kind expected if multiple chromosome breakage was followed by random rejoining of fragments. It did not fit the expected distribution if the rearrangements occurred by a sequence of two-break events, as required if the primary mechanism is homologous recombination between P element sites or the formation of cointegrate structures. These two mechanisms were also ruled out by, respectively, the frequent loss of P element sites at the breakpoints and the failure of most sporadic breakpoints to result in new P element insertions.

Molecular descriptions of three of the rearrangements were obtained by Roiha et al. (167) and by O'Hare (personal communication). In one case, the rearranged chromosome is most easily described as the sequence that would result from a homologous exchange between the larger of the two sn^w elements (Fig. 2) and the complete P element near *hdp-b*, which lies in the opposite orientation. There was also a concomitant loss of the smaller P element at sn^w. The other two rearrangements were more complex, involving the loss of the terminal repeat of one of the P elements and the synthesis of part of a new terminus of another P element.

At this early stage in the study of P-induced chromosome rearrangements, we can only conclude that the process is a chaotic one, resulting in many kinds of sequence changes. The model suggested above to explain P element excisions was one of frequent aberrant changes occurring as a by-product of the presence of transposase. The same may prove to be true for chromosome rearrangements.

D. Effects of P Element Activity on Fitness and Other Quantitative Traits

1. Fitness

The greatly elevated rates of mutations and chromosome rearrangements in dysgenic germ lines give rise to a measurable loss of fitness or its components in subsequent generations when the chromosomes are made homozygous (73, 116, 137, 149, 240). Fitzpatrick and Sved (73) found that a single generation of dysgenesis reduced the homozygous fitness of second chromosomes by 10 to 20% relative to control chromosomes that had been through a nondysgenic P × M cross. Mackay (137) found a much greater effect relative to M chromosome controls which had been through no P × M or M × P crosses. However, she saw the same fitness reduction in the nondysgenic P × M controls. A possible explanation is that the balancer P strain used in these experiments was not homogeneous for the P cytotype. Therefore, the fitness decrement might have accumulated during several generations of dysgenesis for both sets of crosses.

There is a large body of work (reviewed in reference 191) aimed at measuring the effects on fitness of spontaneous mutations in *Drosophila*. Although most of the conclusions from this work are probably valid, some of them must be reevaluated in view of the strong effect hybrid dysgenesis-induced mutations can have on fitness. In particular, studies such as those of Mukai and Yamaguchi (150), in which chromosomes from a natural population were crossed to laboratory stocks as a first step toward measuring their fitness effects, might actually have measured the effects of new mutations produced in the laboratory by P element activity. Some earlier attempts of this kind (148, 157) are probably unaffected by hybrid dysgenesis, since the lethal mutation rates measured at the same time were similar to what is seen in the absence of P elements.

2. Other quantitative traits

There is some uncertainty about how much impact P-induced mutations have on quantitative traits. Mackay (135, 136) found that several generations of dysgenesis had a surprisingly large effect on the response to artificial selection for abdominal bristle numbers. However, R. Phillis (submitted for publication) found that dysgenesis had no detectable effect on selection for geotaxis. A third experiment, by Torkamanzehi et al. (221a), was essentially a replicate of Mackay's procedure, but did not produce the same pronounced effect from dysgenesis. If anything, there was a greater response in the nondysgenic controls.

One possible reason for these conflicting results is that most of the effect seen by Mackay might have been due to specific P element sites on the chromosomes of the P strain she used (Harwich). Since the data of Simmons et al. (198) indicate that the majority of dysgenesis-induced lethal mutations on P chromosomes occur at preexisting P elements sites, we can postulate that the same is true of polygenic mutations. Independent evidence (73) shows that

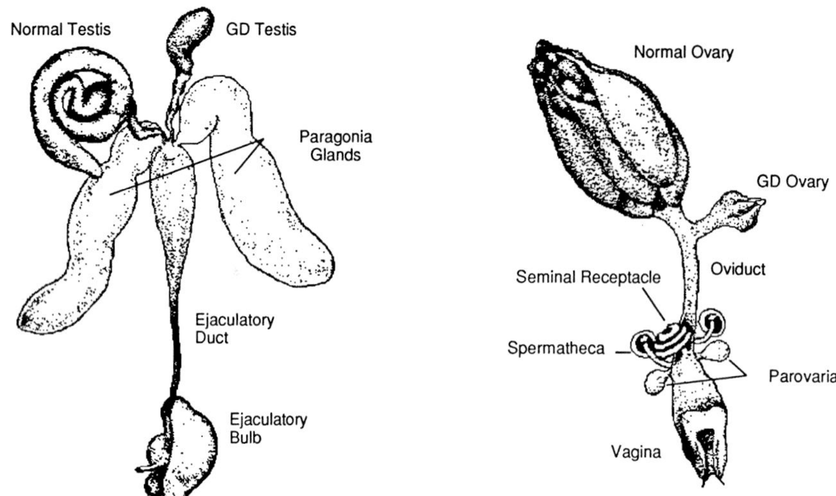

Figure 9. Morphology of GD sterility. The adult male (left) and female (right) reproductive systems of *Drosophila*. The drawings are modified from reference 159 to show the effects of GD sterility. In each case, the left gonad is normal and the right one is that of a GD sterile adult (65). GD sterility can result in both gonads being rudimentary or in unilateral cases resembling these drawings.

chromosomes of the Harwich stock are highly variable in their response to P element mutagenesis, indicating polymorphisms within the stock. It is possible, therefore, that the particular line in Mackay's experiment had one or more P elements near bristle-affecting loci, but the line used by Torkamanzehi et al. did not. Mackay also found a very pronounced effect of dysgenesis on the abdominal bristle variability in homozygotes for an M-derived chromosome that had been exposed to dysgenesis (138). Since this chromosome had no P elements prior to the experiment, such effects cannot be explained by existing variability in the Harwich strain. However, these chromosomes came from the same experiment mentioned above (137), in which the number of generations of dysgenesis might have been much larger than the author assumed. Further work will be needed to resolve this issue.

VIII. CELL-LETHAL AND DOMINANT LETHAL EFFECTS OF P ELEMENT ACTIVITY

A. Some Dysgenic Traits Can Be Explained by Cell Lethality

The remaining abnormalities associated with hybrid dysgenesis are segregation ratio distortion, embryo lethality, GD sterility, and pupal lethality. This last trait requires a Δ2-3 element as a source of somatic transposase. I believe all of these traits can be explained by postulating a cytotoxic effect of P element mobilization. This effect could come about through a high frequency of dominant lethal mutations caused by chromosome breakage.

1. Segregation ratio distortion and embryo lethality

While scoring the progeny of dysgenic males, Hiraizumi noticed that P chromosomes were recovered less frequently than their M homologs (93). This distortion can be interpreted as the result of a high frequency of dominant lethal mutations occurring preferentially on the P chromosomes and being expressed in the F_2 embryos. Recall from the above discussion that the majority of recessive lethal mutations were the result of events at the sites of P elements. If the same were true of dominant lethals, then segregation distortion in the direction Hiraizumi observed is the expected outcome. The finding of embryonic lethality in the progeny of dysgenic males and females (48, 49, 107) agrees well with this interpretation. In some cases, the dominant lethal mutation might prevent formation of a functional gamete, thus explaining Matthews's (142) observation of degenerate spermatids in the testes of dysgenic males.

2. Gonadal dysgenic (GD) sterility

The gonads of dysgenic males and females frequently fail to develop properly if the rearing temperature exceeds 27°C (1, 65, 117, 182). The resulting adults are sterile and possess agametic gonads. The somatically derived tissues appear normal, but only rudiments of the germinal tissues are present (Fig. 9). Pole cell transplantation experiments have shown that this effect, called GD sterility, is a cell-autonomus trait (151). The frequency of GD sterility is well correlated with the amount of P transposase activity as measured by sn^w mutability, but there are some outliers (60, 88a, 123, 189). GD sterility is often used as a convenient assay for hybrid dysgenesis, but low levels of transposase activity are more readily detected by sn^w mutability (100).

The early suggestion (65) that GD sterility results from cell death in the early developing germ line has been confirmed recently by cytological observations (152). The first pycnotic cells appeared in embryos after 6 to 7 h of development at 29°C, which agrees well with the onset of the temperature-sensitive period (see below). Cell death was most frequent during the larval period in which germ cells undergo exponential growth, but some dying cells were also observed prior to germ cell division.

The underlying mechanism of GD sterility might be dominant lethal chromosome breakage similar to the events postulated above to explain segregation ratio distortion, but occurring early enough to kill the germ cells prior to adult gonadal development. This model is in good agreement with the observation that most cell death occurs during the period of rapid cell division, when the germ cells are expected to be most sensitive to chromosome breakage. In addition, it explains why males are somewhat less sensitive to GD sterility than females. Male gonads have many more germ cells during larval development than female gonads. Therefore, it is more likely that at least some of them will survive to produce an adult gonad.

3. Pupal lethality in the presence of a Δ2-3 element

The discovery that transposase can be expressed in somatic tissues by in vitro-modified P elements with the 2-3 intron deleted (126) made it possible to study the effects of P-M hybrid dysgenesis in somatic cells (64). One such element, called P[ry^+ Δ2-3](99B), was found to be especially useful in these studies because of its high transposase activity levels and its inability to undergo normal levels of transposition and excision (166). When this element was combined with chromosomes from the M′ strain Birmingham, the result was death of the organism at the pupal stage (64). Engels et al. (64) postulated that this effect was analogous to GD sterility, except that it affected the imaginal disks and histoblasts as well as the germ line. Cell death by chromosome breakage is not expected to occur in larval tissues, which are primarily polytene and nondividing. The imaginal disk and histoblast cells, however, do undergo cell division during larval development, but these cells are not essential for larval survival. Thus, the effects of dominant cell-lethal events during larval development are delayed until pupation, as has been seen following X-irradiation (225). This interpretation is in good agreement with the cytological studies by R. Phillis (unpublished data), who found frequent occurrences of chromosome bridges and fragments in dividing neuroblast cells from larvae of the kind destined to die as pupae.

B. Temperature Sensitivity of Cytotoxic Effects

1. The range of temperature effects

GD sterility, embryo lethality, and somatic pupal lethality are all highly temperature dependent. In each case, the cell-lethal effects increase monotonically with temperature. For embryo lethality and GD sterility, the severity of the effect is approximately linear with temperature between 21 and 29°C, with very little effect below 21°C (107, 118). Pupal lethality occurs at much lower temperatures and is not entirely abolished even at the lowest possible developmental temperature (64). It increases sharply above 16°C with a slope that is dependent upon the genotype. In the presence of P[ry^+ Δ2-3](99B) and the complete Birmingham genome, pupal lethality approaches 100% even at 19°C. Paradoxically, segregation distortion does not display a monotonic relationship with developmental temperature. Instead, the effect increases between 19 and 25°C and then decreases at 29°C (55). One possibility is that at 29°C most cell lineages in which dominant lethal events can occur have already been eliminated prior to meiosis.

2. Developmental timing of temperature sensitivity

The temperature-sensitive periods (TSPs) of these traits agree well with the above model of cell-lethal mutations. For GD sterility, the TSP begins in late embryos at about the time of the onset of rapid cell division and gradually falls off during the second larval instar (65, 118). The decrease in sensitivity at the later larval stages can be explained by the very large number of germ cells that are present at that stage. This makes it unlikely that a shift to a higher temperature will kill all the germ cells. The TSP for embryo lethality extends through all stages of larval and adult life of the parents (107), as would be expected if it reflects the production of mutations in the germ line. Somatic pupal lethality has a TSP beginning in the early larval stages and extending to about the time of pupation (64). This period corresponds to the time during which larval tissues are resistant to chromosome damage (as indicated by radiation studies [225]) but the imaginal disks and histoblasts are sensitive.

3. Why is temperature important?

One explanation for the importance of temperature is that transposase production might be enhanced at higher temperatures. Indeed, other dysgenic traits such as *sn*^{w} hypermutability do show a temperature effect (166). However, the traits associated with cell lethality seem to be much more sensitive to high temperature than the other dysgenic traits. At 21°C, for example, there is virtually no GD sterility, but *sn*^{w} is still highly mutable. Therefore, we can speculate that there is a temperature-dependent step needed to prevent cell lethality that is not required for other traits. For example, there might be a temperature-sensitive DNA repair process that is needed to prevent dysgenesis-induced chromosome breaks from becoming dominant lethal mutations. In good agreement with this interpretation is the finding (205; see Section IX.B) that some mutations in the DNA repair system greatly increase segregation ratio distortion and embryo lethality, but have little or no effect on male recombination or insertional mutagenesis.

C. Importance of "Target" P Elements

Perhaps the strongest argument that the cell-lethal traits are caused by chromosome breakage, as opposed to a more direct effect of transposase on the dividing cells as previously thought (59), is the finding that transposase itself is insufficient for these effects. This fact came to light in experiments by Simmons et al. (197), using a chromosome called *T-5* which had one or more strong transposase-making P elements but very few other P elements, and by Engels et al. (64), who worked with the stable transposase-making element, P[*ry*$^{+}$ Δ2-3](99B), in a strain with no other P elements. In both cases it was found that the transposase-making elements caused no GD sterility by themselves, even at 29°C. However, high frequencies of GD sterility resulted when one of these transposase-making strains was crossed (in either direction) with the Birmingham strain. This strain, as discussed previously, has approximately 60 defective P element sites on its chromosomes, but no complete P elements. Any one of the major Birmingham chromosomes was shown to cause GD sterility in combination with a transposase-making P element.

The same pattern was seen for somatic pupal lethality. Neither the P[*ry*$^{+}$ Δ2-3](99B) element nor the Birmingham genome was able to cause pupal lethality by itself, but when the P[*ry*$^{+}$ Δ2-3](99B) element was combined with any one of the major Birmingham chromosomes, the result was 100% pupal lethality at 29°C (64).

These observations were interpreted to mean that cell-lethal effects result from the interaction of transposase with chromosomally integrated ("target") P elements on the Birmingham chromosomes. Such interactions were hypothesized to cause dominant lethal and cell-lethal chromosome breaks.

IX. EFFECTS OF THE *DROSOPHILA* GENOME ON P ELEMENTS

A. *cis* Effects

1. Flanking genomic sequences affect P element expression

The instability of autonomous P elements makes it difficult to measure the transposase production of individual elements. However, there are two lines of indirect evidence involving nonautonomous P elements that suggest that P element expression is affected by the genomic site of insertion. First, it is clear from several studies that genes carried by in vitro-modified P elements are expressed in different ways, depending on their genomic locations (28, 90, 127, 129, 210). When two genes are carried within a P element, their expression levels are positively correlated ($r \approx 0.5$) over insertion sites, indicating that position effects are not entirely gene specific (127). Furthermore, a β-galactosidase gene driven by the P element promoter showed strong position effects (157a). This effect will be discussed in more detail below (Section XI.B) under the topic of germ line transformation. Second, there is the pronounced position effect evident in the determination of P element repressor function (see Section V.B).

2. Stability of P elements can be determined by their genomic insertion site

Some P elements have very low frequencies of transposition and excision, even in the presence of high levels of transposase function, which can be assayed by mobility of nonautonomous P elements elsewhere in the genome. In the case of P[*ry*$^{+}$ Δ2-3](99B) mentioned previously (166), it is not known whether this stability is determined by intrinsic factors, such as a small internal deletion, or by its genomic position. In another case, however, there is evidence that the genomic position is responsible. A P element marked with the wild-type allele of the *white* eye gene (P[w^+]), inserted at cytological position 17B, was found to have abnormally low levels of transposition and excision, as measured by crosses similar to that shown in Fig. 5. However, when a rare transposition to a new position (22A) was finally recovered, the new site and many other sites subsequently derived from it displayed normal transposition and excision levels (C. Preston and W. Engels, unpublished data). These findings are interpreted as indicating that the stability of the original P[w^+] site at 17B was due to its position. A change internal to the element could still be responsible for the altered stability, but such a change must be reversible, thus ruling out internal deletions.

In addition to these cases of nearly stable elements, there is also considerable quantitative variability in insertion and excision rates among different insertions of the same marked P element in the presence of a constant transposase source (166; R. Phillis, personal communication). This variability is most easily explained in terms of *cis* effects from flanking DNA, which might, for example, affect accessibility to transposase.

B. Effects of Mutations in DNA Repair

Many loci involved with DNA repair in *D. melanogaster* have been identified in screens for meiotic abnormalities (*mei* mutants) or hypersensitivity to mutagens (*mus* mutants). Some of these, such as *mei-41* and *mus-101*, are thought to function in postreplication repair. Others, such as *mei-9*, are considered to be involved with excision repair (75).

Slatko et al. (205) tested the effects of some of these loci on P-M dysgenic traits. Specifically, they looked at recessive lethal mutation induction on an M-derived target chromosome (and thus, presumably, P element insertion mutations), male recombination, segregation ratio distortion, and embryo lethality, all in the presence of mutations at *mei-41*, *mei-9*, *mus-101*, and the combination of *mei-41* and *mei-9*. There was no discernible effect on male recombination or insertional mutagenesis by any of these genotypes. However, *mei-41* and *mus-101* had pronounced effects on segregation distortion and embryo lethality, increasing the severity of both traits. The *mei-41* mutation was particularly effective, reducing the recovery of the P-derived autosome to less than 10%, as opposed to about 40% in a wild-type background. Embryo lethality was increased from about 20% to nearly 50%. These effects were not seen in the P cytotype, indicating that they were specifically due to hybrid dysgenesis. Therefore, it seems that defects in the postreplication repair system enhance at least some of the dysgenic traits attributable to cell-lethal and dominant lethal events. However, neither postreplication repair nor excision repair defects appear to affect male recombination or insertion frequencies.

An apparently contradictory result was reported by Eeken and Sobels (44), who studied forward mutations at several visible loci. One of their loci was *singed*, where dysgenesis-induced mutations are very likely to be P element insertions (166, 167). They found increases in *singed* mutation rates in the presence of *mei-9*, *mei-41*, and the double mutation. However, even their highest *singed* mutation rate, which occurred in the *mei-9 mei-41* double mutant, was less than half that observed by Simmons et al. (196) in an experiment where there were no repair-deficient mutations present. This discrepancy suggests that the mutability at individual loci might be strongly dependent upon the particular strains used. (The *mei-41*, *mei-9*, etc., stocks presumably had different genetic backgrounds). The rate of lethal mutations, which represents an average over many loci, might be more reliable for studies of this kind. In addition, the statistical problems discussed above in relation to the interactions between P elements and conventional mutagens also apply to both sets of studies on repair mutations. Therefore, conflicting conclusions based on statistically "significant" findings are not unexpected.

Eeken and Sobels also looked at reversions of one of their P-induced *singed* mutations (46). Their data indicated a slightly lowered reversion rate in the presence of repair mutations, especially *mei-41*. The authors suggested that this decreased rate might be explained by some of the *singed* revertants being eliminated by their association with dominant lethal mutations, which are made more frequently by *mei-41*. However, a similar experiment utilizing sn^w gave a different result (unpublished data). In the presence of *mei-41*, recovery of the sn^w chromosome was greatly reduced, presumably due to extreme segregation distortion as expected from the results of Slatko et al. (205). However, those derivatives of sn^w that were recovered included apparently normal proportions of sn^e and $sn^{(+)}$.

C. Suppression of Phenotypic Effects by *su(s)*

The phenotypes of some insertion mutations in *D. melanogaster* are subject to modification by alleles of "suppressor" loci. Many such cases are reviewed elsewhere in this volume (Bingham and Zachar). Until recently, it was thought that this behavior was limited to retroviruslike transposable elements. However, M. Simmons (personal communication) has recently found that the sn^{w} bristle phenotype is strongly affected by mutations at *suppressor-of-sable* [*su(s)*]. Searles and Voelker (186) had previously shown that *su(s)* suppresses (i.e., restores to wild type) certain insertion alleles of the retroviruslike element *412*. Simmons found that the bristles of sn^{w}, but not those of its sn^{e} derivatives, appeared wild type in the presence of at least some alleles of *su(s)*. It is not yet known whether *su(s)* affects other P insertion mutations, or whether P element mobility is altered in the presence of *su(s)* mutations.

D. Involvement of Other *Drosophila* Genes

P elements are relatively simple, yet their behavior is complex. It seems likely, therefore, that they make use of host functions for their mobility. Recently, D. Rio (personal communication) has found an activity (presumably a protein) in *Drosophila* tissue culture cells that binds specifically to the P element termini. It binds to the outermost 16 bp of each 31-bp repeat, irrespective of the flanking DNA sequence. The source of this activity is apparently a genomic locus, since it came from P element-free cells. Its specificity for the termini suggests a role in transposition, but no functional tests have been conducted. The inability of P elements to transpose normally in species outside the genus *Drosophila* (see Section X.D) also suggests that they require gene products that are available only in *Drosophila* species. The identification and analysis of host genes involved in P element activity are likely to be among the important areas of research in the next few years.

X. POPULATION AND EVOLUTIONARY BIOLOGY OF P ELEMENTS

A. Population Dynamics

1. Data from mixed populations

One approach to describing the population dynamics of P elements is to set up laboratory populations with a mixture of P and M or P and M′ chromosomes and monitor the changes in cytotype, GD sterility potential, etc., over time. The general conclusion from studies of this kind (8, 17, 83, 92, 117a, 119, 121) is that changes can occur in either the P or M direction, with the changes toward P predominating. It is not clear what role, if any, P element transposition and excision play in these changes. The classical forces of population genetics, drift, and natural selection, acting on ordinary loci linked to the P chromosomes, might account for some or even all of the observed changes.

2. Studies of transformed lines

Some of these uncertainties can be avoided through the use of inbred lines which have acquired complete P elements by transformation. This approach minimizes the effects of selective differences at linked loci. It also makes it easier to detect P element changes by molecular tests, since the starting population is relatively simple. In studies by Anxolabéhère et al. (2) and Daniels et al. (36), the initial populations were polymorphic for a small but undetermined number of P elements. After many generations, some of them became P strains, with the P cytotype developing only very gradually over 20 to 30 generations. Others became M′ strains, and still others remained in a dysgenic condition even after 80 generations. Molecular data, which were limited to Southern gels on mass-extracted DNA from the populations, indicated that new P element sites and internally deleted P elements were generated during the course of the experiment, especially in those lines evolving toward P strains.

To follow the course of events more closely, a third experiment by Preston and Engels (162a) was conducted in which the initial population had a single complete P element site. Lines were maintained by full sib matings and tested at each generation by in situ hybridization to determine the number of P elements and their chromosomal positions. Several lines lost their transposase activity, presumably by complete or partial excision of the P elements. Others, however, retained a single copy for a variable number of generations, usually 5 to 15, and then rapidly increased their copy number to 20 or more over only 2 to 5 generations. At this point, the lines became extremely difficult to maintain, presumably because of deleterious P-induced mutations being made homozygous by the close inbreeding. Four such lines were lost, but a fifth one survived after cessation of brother-sister matings. This line went on to evolve into a P strain with very slow development of the P cytotype, similar to the cases reported previously (2, 36).

These experiments leave many uncertainties about the population dynamics of P elements. Data from the mass population experiments are too crude to provide a complete picture. In addition, the demographic parameters of the experimental populations, which may be crucial in determining the outcome, were largely unmonitored. On the other hand, the sib-mating experiments, utilizing in situ hybridization at each generation, provide more detailed data, but they are too far removed from natural population conditions to be generally applicable. Much work remains to be done in this area. Theoretical studies, reviewed elsewhere in this volume (J. W. Ajioka and D. L. Hartl), might provide some guidance for future approaches.

3. Why is the P cytotype so slow to develop?

One generalization from these experiments is that the P cytotype can appear following the introduction of P elements, but it takes many generations. This finding seems at odds with earlier results (54, 105, 217) indicating that P strain chromosomes can change the cytotype from M to P in only one or a few generations. One possible explanation is that most P elements, including complete ones, do not produce the hypothetical repressor responsible for P cytotype, and many generations are usually required before a repressor maker appears. As discussed above, the ability to produce repressor is highly dependent upon an element's genomic position, and it might also require an internal deletion. Therefore, it may take many generations in which P element positions are randomly reshuffled and internal excisions are generated before a repressor-making P element is produced. Once such elements appear, natural selection would be expected to bring them to a high copy number. Thus, natural P strains, which are the product of a long history of this kind of selection, are expected to have more repressor makers than the "synthetic" P strains produced by these experiments.

B. Distribution of P Elements in *D. melanogaster* Natural Populations

1. Sources of data

D. melanogaster is a cosmopolitan species. The task of describing its P element distribution is therefore a formidable one; fortunately, many investigators have contributed to the work. The data consist of intensive studies of particular populations (66, 98, 156, 169, 176, 188, 221, 232) as well as surveys of major geographical areas (3, 5, 6, 9, 17, 23, 23a, 24, 106, 112). Some studies employed genetic assays consisting of crosses to standard reference P and M strains (3–6, 8, 17, 23, 24, 66, 98, 106, 112, 156, 188). The presence or absence of dysgenic traits in the offspring provided information on both transposase and cytotype or other regulation. In other studies, direct data on the numbers and kinds of P elements were obtained by in situ hybridization, Southern gels, and squash blots (3, 9, 17, 23a, 98, 169, 176, 221).

2. Some broad patterns

The following generalizations can be derived from these studies. (i) There is very little hybrid dysgenesis in natural populations. Some populations are stable because of suppression such as the P cytotype, some produce only low levels of transposase, and some are stable for both of these reasons (4–6, 9, 23, 24, 66, 98, 106, 112, 156, 188, 232). The last kind of population is called a Q strain (67, 88a, 103, 195) and is characterized by the absence of dysgenic traits following crosses to either P males or M females. (ii) All natural populations that have been tested have at least some genomic P elements (9, 17, 23a, 98, 153, 169, 176, 221). Therefore, wild strains with the M cytotype are actually M′ strains. (iii) P element sites are highly polymorphic within populations (169). (iv) Q strains can be found in almost all geographical regions. Some areas have a mixture of P and Q strains, and others have primarily M′ and Q (4, 5, 9, 23, 24, 66, 98, 106, 112, 156, 188, 232). Figure 10 shows how these types are distributed (5). (v) In Europe and central Asia, the P and Q lines generally have more P element copies than the M′ strains. The frequency of Q populations gradually declines from west to east (9). (vi) In eastern Australia, the predominant strain type changes rapidly from P to Q to M′ as one moves from north to south. Unlike the European and central Asian strains, the Australian M′ strains do not have fewer P elements than the P and Q strains. However, they probably have fewer complete P elements (23, 23a, 24). (vii) KP elements are common in all populations in Europe and the Soviet Union (17). They are also found in eastern Australia, increasing in frequency from north to south (Boussy, personal communication). However, they are rare or nonexistent in North American populations.

C. P Elements in Other Species

Sequences homologous to P elements are common in many species in the subgenus *Sophophora* and probably elsewhere in the genus *Drosophila* (7, 10, 39, 41, 124, 125, 211). They are less common outside of *Drosophila*, but some homology has been detected in other diptera, including some species outside the Drosophilidae family (10). In many cases, the distribution of P elements does not follow phylogenetic groupings. Interestingly, P elements do not occur naturally in *D. simulans* or *D. mauritiana*, the species most closely related to *D. melanogaster* (27), but there are P-like elements in most species of the much more distantly related *D. willistoni* group (39, 41, 124, 125, 211). Other species within the *D.*

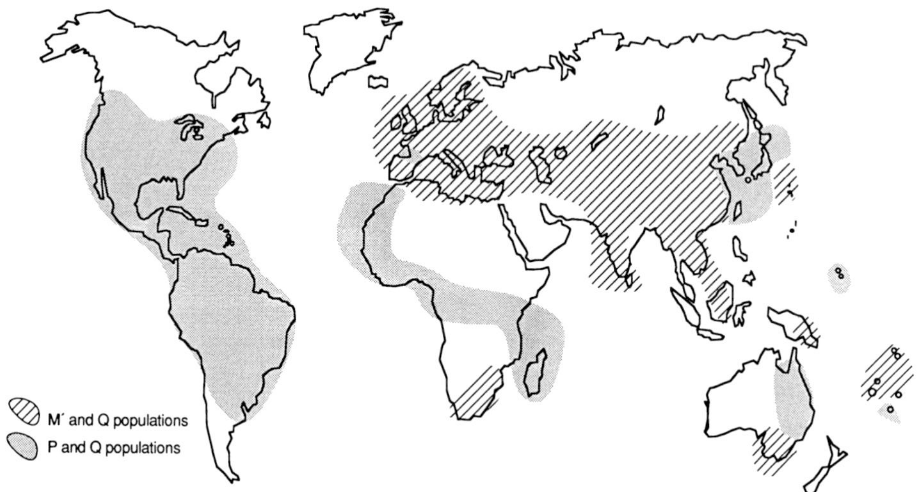

Figure 10. Global distribution of strain types. Unshaded regions indicate areas where no samples have been tested. The data come from references 4, 5, 9, and 112.

melanogaster group do have P-hybridizing sequences, but these are less closely homologous to the canonical P sequence than are the *D. willistoni* group elements (39, 125). The distribution of P-like elements in the genomes of at least some of the other species tends to differ from that in *D. melanogaster* by a greater homogeneity of chromosomal positions between unrelated strains and by the frequent occurrence of elements in the heterochromatin (39, 125).

Two elements from *D. nebulosa*, a member of the *D. willistoni* species group, have been sequenced (124). They were selected among several cloned P-like elements for having restriction maps that looked most like that of the complete P element and were therefore the best candidates for functional transposable elements. However, the sequences showed that there were many base changes which would destroy any open reading frames in the transposase gene and probably knock out the *cis* functions as well. Overall, there were 5 to 6% mismatched bases and many small additions or deletions of base pairs between the *D. nebulosa* elements and the standard P element sequence from *D. melanogaster*. Lansman et al. (124) reported that these elements failed to function either as transposase sources or as transposable elements when injected into *D. melanogaster*, but no sample sizes were given. Interestingly, the two *D. nebulosa* elements were more closely related to the standard P element sequence than they were to each other, suggesting that they both diverged from a common ancestor similar to the *D. melanogaster* P element. One complete P element has also been sequenced from *D. willistoni* (K. Peterson and M. Kidwell, personal communication). Unlike the *D. nebulosa* elements, this sequence did match the *D. melanogaster* P element very closely, and the element was transpositionally active when injected into an M strain.

D. P Elements Transplanted into Other Species

1. Other species in the family Drosophilidae

D. hawaiiensis is a distant relative of *D. melanogaster* with no closely homologous P-like elements. It differs from *D. melanogaster* in many ways: it is much larger, endemic to the Hawaiian Islands, and adapted to high altitudes and low temperatures. Yet Brennan et al. (26) found integrated P elements in 2 of 24 *D. hawaiiensis* recipients of injected complete P elements. Furthermore, these elements continued to transpose in subsequent generations, even producing chromosome rearrangements with P element in situ hybridization sites at the breakpoints. Similar behavior was seen when P elements were put into the much more closely related species, *D. simulans* (40, 180, 181). Again, the elements were able to transpose and undergo partial excision within the *D. simulans* lines. Scavarda and Hartl (181) reported no GD sterility of the kind shown in Fig. 9. However, in another study, Daniels et al. (S. Daniels, A. Chovnick, and M. Kidwell, personal communication) found that GD sterility was clearly present in *D. simulans* lines which had acquired P elements. Some lines obtained by these investigators eventually developed the P cytotype, as indicated by reciprocal cross effects for both GD sterility and male recombination.

O'Brochta and Handler (153a) have tested several drosophilid species, using a modification of the excision assay developed by Rio et al. (165) described above. Instead of putting the *hs*π Δ2-3 helper element and the β-galactosidase fusion DNA into cultured cells, they injected them directly into embryos. Excision events were detected in *D. simulans*, *D. willistoni*, *D. melanica*, *D. grimshawi*, *Chymomyza proncemis*, and *Zaprionus tuberculatus* species, but the frequencies were 2- to 10-fold less than in *D. melanogaster*. There was also preliminary evidence for germ line specificity in *D. simulans*, as indicated by a lack of excisions when the helper element carried a normal 2-3 intron.

The general conclusion from these studies is that P elements can transpose in *Drosophila* species where they do not occur naturally. At least in transformed lines of *D. simulans*, they also display cytotype regulation and some of the effects of hybrid dysgenesis. This conclusion raises the possibility that if such transformed strains were released into natural populations, P elements would spread through the species. Such an "experiment" would be irreversible, and its consequences for the species are unpredictable. Therefore, researchers working with transformed lines of species other than *D. melanogaster* are urged to take every precaution to avoid accidental release.

2. Species outside of the family *Drosophilidae*

To date, P element transformation has not been successful outside the genus *Drosophila*. Miller et al. (146) injected over 2,000 mosquito embryos and recovered only one integration event which had clearly occurred by a mechanism other than P element transposition. The embryo excision assay, when applied to other dipteran species (153a), indicated that P activity is either extremely low or nonexistent in the families Tephritidae (fruit flies) and Sphaeroceridae (dung flies). Some excision events, albeit at a very low frequency, were recovered from shore fly (Ephydridae) embryos, which are much closer to *Drosophila* spp. Surprisingly, a similar assay in mammalian cells did yield a low frequency of excisions in some tests (164). The transposase source in this case was a chromosomally integrated gene driven by a heat shock promoter. Rio et al. also tested a P element cDNA construct in the yeast *S. cerevisiae*. They found evidence for DNA breakage, but no indication of transposition or excision. All of these experiments were performed with transposase sources in which at least the 2-3 intron had been removed, and with the transposase gene driven by a promoter which is expected to function in the tested species. The most likely explanation for the lack of transposition or excision activity, therefore, is that these processes require additional functions that are specific to *Drosophila* species.

E. The Origin of P Elements

1. History of P and M strains

One of the intriguing questions in the study of P elements is why they are completely absent from old laboratory stocks but ubiquitous in nature. This question has been discussed recently from several points of view (63, 109, 219). The simplest explanation seemed at first to be that the P elements had been lost from older experimental stocks as a result of laboratory conditions (58). However, the large numbers of defective P elements in most natural populations make this appear unlikely. Once all complete P elements are lost, there is no apparent way of ridding the genome of any defective P elements that are fixed in the strain. Even the highly active elements at sn^{w} have no measurable excision rate in the absence of transposase (60, 64, 197). Therefore, the invasion of a population by P elements is probably irreversible.

An alternative hypothesis is that P elements originated in *D. melanogaster* only recently, spreading rapidly throughout natural populations. If this happened within the past 30 years, it would explain why older laboratory strains do not have P elements (104, 106). According to this view, "populations are often invaded by (presumably parasitic) elements of which the P element is an example" (15). However, this model is only reasonable if the rate at which new transposable elements enter the genome is extraordinarily high. If this rate were less than one new element per 1,000 years, then standard statistical reasoning requires rejection (at the 5% significance level) of the notion that P elements originated in *D. melanogaster* during the past 50 years. In fact, the true rate is probably several orders of magnitude less than 1/1,000. This conclusion comes from noting that the total quantity of moderately repetitive DNA in the genome allows for only about 30 to 60 typical families of transposable elements, most of which are estimated to have been present in the genome for more than 4×10^7 years, according to their species distributions (141, 172, 208). Therefore, it is highly unlikely that a new element would have entered the genome within the time span required by this model. Combining this reasoning with the finding that complete I factors (Finnegan, Chapter 18, this volume) are also missing from many laboratory stocks, but are common in nature, makes it very clear that another explanation is needed.

To get around these problems, several authors (59, 106) have suggested that the difference between P and M strains is related to the drastic changes in the demography of *D. melanogaster* due to human activity during the current century. The geographical elements in drosophilids and nondrosophilids. *Proc. Natl. Acad. Sci. USA* **85**:6052–6056.

154. **O'Hare, K.** 1985. The mechanism and control of P element transposition in *Drosophila melanogaster*. *Trends Genet.* **1**:250–254.

155. **O'Hare, K., and G. M. Rubin.** 1983. Structure of P transposable elements and their sites of insertion and excision in the *Drosophila melanogaster* genome. *Cell* **34**:25–35.

156. **Ohishi, K., E. Takanashi, and S. I. Chigusa.** 1982. Hybrid dysgenesis in natural populations of *Drosophila melanogaster* in Japan. I. Complete absence of the P factor in an island population. *Jpn. J. Genet.* **57**:423–428.

157. **Ohnishi, O.** 1977. Spontaneous and ethyl methanesulfonate-induced mutations controlling viability in *Drosophila melanogaster*. II. Homozygous effect of polygenic mutations. *Genetics* **87**:529–545.

157a. **O'Kane, C. J., and W. J. Gehring.** 1987. Detection *in situ* of genomic regulatory elements in Drosophila. *Proc. Natl. Acad. Sci. USA* **84**:9123–9127.

158. **Orr-Weaver, T. L., and A. C. Spradling.** 1986. Drosophila chorion gene amplification requires an upstream region regulating s18 transcription. *Mol. Cell. Biol.* **6**:4624–4633.

159. **Patterson, J. T.** 1943. The Drosophilidae of the Southwest, p. 7–216. *In* J. T. Patterson (ed.), *Studies in the Genetics of Drosophila*, vol. 3. University of Texas Publication.

160. **Picard, G., and P. L'Héritier.** 1971. A maternally inherited factor inducing sterility in *Drosophila melanogaster*. *Drosophila Inform. Serv.* **46**:54.

161. **Pirrotta, V., H. Steller, and M. P. Bozzetti.** 1985. Multiple upstream regulatory elements control the expression of the Drosophila *white* gene. *EMBO J.* **4**:3501–3508.

162. **Preston, C. R., and W. R. Engels.** 1984. Movement of P elements within a P strain. *Drosophila Inform. Serv.* **60**: 169–170.

162a. **Preston, C. R., and W. R. Engels.** 1989. Spread of P transposable elements in inbred lines of *Drosophila melanogaster*. *Prog. Nucleic Acid Res. Mol. Biol.*, in press.

163. **Raymond, J. D., T. R. Laverty, and M. J. Simmons.** 1986. Cytogenetic analysis of lethal X chromosomes derived from dysgenic hybrids of *Drosophila melanogaster*. *Drosophila Inform. Serv.* **63**:111–114.

164. **Rio, D. C., G. Barnes, F. A. Laski, J. Rine, and G. M. Rubin.** 1988. Evidence for Drosophila P element transposase activity in mammalian cells and yeast. *J. Mol. Biol.* **200**:411–415.

165. **Rio, D. C., F. A. Laski, and G. M. Rubin.** 1986. Identification and immunochemical analysis of biologically active Drosophila P element transposase. *Cell* **44**:21–32.

166. **Robertson, H. M., C. R. Preston, R. W. Phillis, D. Johnson-Schlitz, W. K. Benz, and W. R. Engels.** 1988. A stable genomic source of P element transposase in *Drosophila melanogaster*. *Genetics* **118**:461–470.

167. **Roiha, H., G. M. Rubin, and K. O'Hare.** 1988. P element insertions and rearrangements at the *singed* locus of *Drosophila melanogaster*. *Genetics* **119**:75–83.

168. **Ronsseray, S.** 1986. P-M system of hybrid dysgenesis in *Drosophila melanogaster*: thermic modifications of the cytotype can be detected for several generations. *Mol. Gen. Genet.* **205**:23–27.

169. **Ronsseray, S., and D. Anxolabéhère.** 1986. Chromosomal distribution of P and I transposable elements in a natural population of *Drosophila melanogaster*. *Chromosoma* **94**: 433–440.

170. **Ronsseray, S., D. Anxolabéhère, and G. Périquet.** 1984. Hybrid dysgenesis in *Drosophila melanogaster*: influence of temperature on cytotype determinations in the P-M system. *Mol. Gen. Genet.* **196**:17–23.

171. **Rosenfeld, A., A. Carpenter, and L. Sandler.** 1971. A non-chromosomal factor causing female sterility in *D. melanogaster*. *Drosophila Inform. Serv.* **47**:85.

172. **Rubin, G. M.** 1983. Dispersed repetitive DNAs in Drosophila, p. 329–361. *In* J. A. Shapiro (ed.), *Mobile Genetic Elements*. Academic Press, Inc. (London), Ltd., London.

173. **Rubin, G. M., M. G. Kidwell, and P. M. Bingham.** 1982. The molecular basis of P-M hybrid dysgenesis: the nature of induced mutations. *Cell* **29**:987–994.

174. **Rubin, G. M., and A. C. Spradling.** 1982. Genetic transformation of Drosophila with transposable element vectors. *Science* **218**:348–353.

175. **Rubin, G. M., and A. C. Spradling.** 1983. Vectors of P element mediated gene transfer in Drosophila. *Nucleic Acids Res.* **11**:6341–6351.

176. **Sakoyama, Y., T. Todo, S. I. Chigusa, T. Honjo, and S. Kondo.** 1985. Structures of defective P transposable elements prevalent in natural Q and Q-derived M strains of *Drosophila melanogaster*. *Proc. Natl. Acad. Sci. USA* **82**:6236–6239.

177. **Salz, H. K., T. W. Cline, and P. Schedl.** 1987. Functional changes associated with structural alterations induced by mobilization of a P element inserted in the Sex-lethal gene of Drosophila. *Genetics* **117**:221–231.

178. **Satta, Y., T. Gojobori, T. Maruyama, and S. I. Chigusa.** 1985. Tn3 resolvase-like sequence in P transposable element of *Drosophila melanogaster*. *Jpn. J. Genet.* **60**:261–266.

179. **Satta, Y., T. Gojobori, T. Maruyama, K. Saigo, and S. I. Chigusa.** 1985. Homology between P transposable element of *Drosophila melanogaster* and bacterial transposase gene of Tn3. *Jpn. J. Genet.* **60**:499–503.

180. **Scavarda, N. J., and D. L. Hartl.** 1984. Interspecific DNA transformation in Drosophila. *Proc. Natl. Acad. Sci. USA* **81**:7615–7619.

181. **Scavarda, N. J., and D. L. Hartl.** 1987. Germ line abnormalities in *Drosophila simulans* transfected with the transposable P element. *J. Genet.* **66**:1–15.

182. **Schaefer, R. E., M. G. Kidwell, and A. Fausto-Sterling.** 1979. Hybrid dysgenesis in *Drosophila melanogaster*: morphological and cytological studies of ovarian dysgenesis. *Genetics* **92**:1141–1152.

183. **Scholnick, S., B. Morgan, and J. Hirsh.** 1983. The cloned *dopa decarboxylase* gene is developmentally regulated when reintegrated into the Drosophila genome. *Cell* **34**:37–45.

184. **Searles, L. L., A. L. Greenleaf, W. E. Kemp, and R. A. Voelker.** 1986. Sites of P element insertion and structures of P element deletions in the 5′ region of *Drosophila melanogaster RpII215*. *Mol. Cell. Biol.* **6**:3312–3319.

185. **Searles, L. L., R. S. Jokerst, P. M. Bingham, R. A. Voelker, and A. L. Greenleaf.** 1982. Molecular cloning of sequences from a Drosophila RNA polymerase II locus by P element transposon tagging. *Cell* **31**:585–592.

186. **Searles, L. L., and R. A. Voelker.** 1986. Molecular characterization of the Drosophila *vermillion* locus and its suppressible alleles. *Proc. Natl. Acad. Sci. USA* **83**:404–408.

187. **Semeshin, V. F., E. S. Belyaeva, I. F. Zhimulev, J. T. Lis, G. Richards, and M. Bourouis.** 1986. Electron microscopical analysis of Drosophila polytene chromosomes. IV. Mapping of morphological structures appearing as a result of transformation of DNA sequences into chromosomes. *Chromosoma* **93**:461–468.

188. **Simmons, G.** 1986. Gonadal dysgenesis determinants in a

cells, they injected them directly into embryos. Excision events were detected in *D. simulans*, *D. willistoni*, *D. melanica*, *D. grimshawi*, *Chymomyza proncemis*, and *Zaprionus tuberculatus* species, but the frequencies were 2- to 10-fold less than in *D. melanogaster*. There was also preliminary evidence for germ line specificity in *D. simulans*, as indicated by a lack of excisions when the helper element carried a normal 2-3 intron.

The general conclusion from these studies is that P elements can transpose in *Drosophila* species where they do not occur naturally. At least in transformed lines of *D. simulans*, they also display cytotype regulation and some of the effects of hybrid dysgenesis. This conclusion raises the possibility that if such transformed strains were released into natural populations, P elements would spread through the species. Such an "experiment" would be irreversible, and its consequences for the species are unpredictable. Therefore, researchers working with transformed lines of species other than *D. melanogaster* are urged to take every precaution to avoid accidental release.

2. Species outside of the family *Drosophilidae*

To date, P element transformation has not been successful outside the genus *Drosophila*. Miller et al. (146) injected over 2,000 mosquito embryos and recovered only one integration event which had clearly occurred by a mechanism other than P element transposition. The embryo excision assay, when applied to other dipteran species (153a), indicated that P activity is either extremely low or nonexistent in the families Tephritidae (fruit flies) and Sphaeroceridae (dung flies). Some excision events, albeit at a very low frequency, were recovered from shore fly (Ephydridae) embryos, which are much closer to *Drosophila* spp. Surprisingly, a similar assay in mammalian cells did yield a low frequency of excisions in some tests (164). The transposase source in this case was a chromosomally integrated gene driven by a heat shock promoter. Rio et al. also tested a P element cDNA construct in the yeast *S. cerevisiae*. They found evidence for DNA breakage, but no indication of transposition or excision. All of these experiments were performed with transposase sources in which at least the 2-3 intron had been removed, and with the transposase gene driven by a promoter which is expected to function in the tested species. The most likely explanation for the lack of transposition or excision activity, therefore, is that these processes require additional functions that are specific to *Drosophila* species.

E. The Origin of P Elements

1. History of P and M strains

One of the intriguing questions in the study of P elements is why they are completely absent from old laboratory stocks but ubiquitous in nature. This question has been discussed recently from several points of view (63, 109, 219). The simplest explanation seemed at first to be that the P elements had been lost from older experimental stocks as a result of laboratory conditions (58). However, the large numbers of defective P elements in most natural populations make this appear unlikely. Once all complete P elements are lost, there is no apparent way of ridding the genome of any defective P elements that are fixed in the strain. Even the highly active elements at sn^w have no measurable excision rate in the absence of transposase (60, 64, 197). Therefore, the invasion of a population by P elements is probably irreversible.

An alternative hypothesis is that P elements originated in *D. melanogaster* only recently, spreading rapidly throughout natural populations. If this happened within the past 30 years, it would explain why older laboratory strains do not have P elements (104, 106). According to this view, "populations are often invaded by (presumably parasitic) elements of which the P element is an example" (15). However, this model is only reasonable if the rate at which new transposable elements enter the genome is extraordinarily high. If this rate were less than one new element per 1,000 years, then standard statistical reasoning requires rejection (at the 5% significance level) of the notion that P elements originated in *D. melanogaster* during the past 50 years. In fact, the true rate is probably several orders of magnitude less than 1/1,000. This conclusion comes from noting that the total quantity of moderately repetitive DNA in the genome allows for only about 30 to 60 typical families of transposable elements, most of which are estimated to have been present in the genome for more than 4×10^7 years, according to their species distributions (141, 172, 208). Therefore, it is highly unlikely that a new element would have entered the genome within the time span required by this model. Combining this reasoning with the finding that complete I factors (Finnegan, Chapter 18, this volume) are also missing from many laboratory stocks, but are common in nature, makes it very clear that another explanation is needed.

To get around these problems, several authors (59, 106) have suggested that the difference between P and M strains is related to the drastic changes in the demography of *D. melanogaster* due to human activity during the current century. The geographical range of the species, its effective population number, and the migration rate are all thought to have increased enormously in recent years. Accordingly, the spread of P elements and complete I factors is a once-only occurrence rather than part of a normal evolutionary process. There are two ways such population changes could have brought about the invasion of P elements. First, P elements might have been present in isolated subpopulations for a very long time, but probably not prior to the divergence from *D. simulans*. The recent demographic changes then increased migration sufficiently to allow them to spread to the rest of the species. Other transposable elements, such as complete I factors, might have existed in different subpopulations and also spread when the species became homogeneous. Alternatively, the broadening of its geographical range might have brought *D. melanogaster* into contact with other organisms from which it acquired P elements through horizontal transfer. The close sequence homology between P elements in *D. melanogaster* and those in nembers of the *D. willistoni* group argues strongly for a recent horizontal transfer. Surveys of laboratory strains taken from nature at various times and places suggest P elements spread in North and South America prior to becoming common in other parts of the world (5, 106).

2. Where did P elements come from?

If P elements first entered the *D. melanogaster* genome since the divergence from *D. simulans* (approximately 1 million years ago), and perhaps as recently as 50 years ago, then much of their evolution must have occurred in another species. On the basis of preliminary evidence discussed above, it appears that P elements are better adapted to function in *D. melanogaster* and its close relatives than in less closely related species. This suggests that they came from one of the closely related species, or else that they have been in *D. melanogaster* long enough for some additional adaptation to occur. The remarkably close sequence match between P elements in *D. willistoni* and *D. melanogaster* (Peterson and Kidwell, personal communication) makes the former species a good candidate for the original source of horizontal transfer. Further studies will also be needed to determine the range of species in which P elements are fully functional.

3. Horizontal transfer of P elements in nature

The notion that P elements entered the species within the past 1 million years implies that there is some mechanism of interspecific horizontal transfer that can occur in nature. No such mechanism has yet been identified, but one suggestion is that insect viruses might serve as vectors (145). An investigation of this possibility is in progress in my laboratory.

The observation that P elements can spread rapidly in laboratory populations of *D. simulans* after transformation by microinjection raises the question of why these elements have not already invaded *D. simulans* and many other *Drosophila* species. This question is especially puzzling since *D. simulans* and *D. melanogaster* share habitats worldwide. It might mean that interspecific horizontal transfer by any mechanism is an exceedingly rare event in the wild, which happened to occur in *D. melanogaster* but not *D. simulans*.

XI. P ELEMENTS AS TOOLS FOR THE STUDY OF *DROSOPHILA* GENETICS

P elements are now a standard part of the *Drosophila* geneticist's toolkit. Their high degree of mobility in the M cytotype and near stability in the P cytotype make them almost ideal for use in molecular studies of *Drosophila* genes. P element insertion mutations have been used to clone genes by a process called "transposon tagging" and also to generate new alleles by secondary mutagenesis. Perhaps even more important is their use as germ line transformation vectors. Although other methods are under development (R. K. Blackman and W. M. Gelbart, this volume), P elements currently provide the only practical method for putting DNA into the *Drosophila* genome. Detailed protocols for these procedures are available from several sources (61, 110, 166, 207). The following is a general overview of these methods, with emphasis on the most recent developments.

A. P Element Insertion Mutations

1. Use in transposon tagging

The idea of using P element insertions to clone genes was suggested by Bingham et al. (16) and first applied by Searles et al. (185). Since then it has been used successfully to isolate DNA from many *Drosophila* genes. It is primarily useful for loci where the gene product is not known but an efficient mutation screen is available. Knowledge of the approximate map position of the gene is also helpful for applying this technique. The first step is obtaining a P element insertion mutation at the gene of interest. Methods for getting and verifying such mutations are discussed below. A library from the mutant strain is then constructed, using a restriction enzyme such as

*Bam*HI which does not cut more than once within the P element sequence. Cloned fragments which hybridize to the P sequence are then expected to contain genomic DNA flanking the P insertion, possibly including the gene of interest. If the mutant strain has more than one P element, then the cloned fragments must be screened by in situ hybridization to M strain chromosomes to determine which ones carry flanking DNA from the desired location. It is therefore advantageous to obtain mutant strains with few or no extraneous P elements. Finally, the selected clones are used to probe an M strain library to recover the wild-type DNA of the gene.

2. Use in generating new variability

P element insertions in or near a gene of interest can also be mobilized to recover new alleles of the gene through imprecise excision, chromosome rearrangements, and other changes. This method has been used several times (37, 177, 223, 224; Eggleston, personal communication) to obtain variants that would have been hard to come by otherwise. The procedure is to screen for new phenotype variants in the progeny of dysgenic flies carrying the insertion. This method is most useful when the insertion mutation itself has a phenotype that is intermediate between the wild type and the null allele, thus allowing recovery of changes in either direction. The resulting new alleles can be internal deletions of the P element, deletions of flanking sequences, or secondary insertions.

3. Obtaining P insertions from M × P dysgenic crosses

The first method used to obtain P insertion mutations was that of screening the progeny of dysgenic hybrids. In a typical F_1 screen, dysgenic males are crossed to females with appropriate genetic markers or chromosome constructions to allow detection of the mutation. It is preferable for these females to have the P cytotype so that the newly recovered mutation will be stable (62). It is also desirable to design the cross so that the new mutation will occur on an M-derived chromosome. That way the number of extraneous P elements linked to the target gene is minimized. Unfortunately, the goal of minimizing the number of extraneous P elements in the genome and that of recovering the mutation in the P cytotype are mutually exclusive when ordinary M × P crosses are used.

As discussed above, a typical mutation rate from crosses of this kind is 0.0002, but there is wide variation between loci. As Kidwell's (111) list of mutagenesis experiments from many laboratories shows, mutations at some genes can be recovered easily with relatively small sample sizes, whereas other genes have such a low mutation rate that the method is simply not practical. There is currently no way for a researcher to know in advance whether his gene will be one of the refractory ones.

4. Alternative transposase sources

Some of the shortcomings of using M × P crosses to generate insertions can be overcome by the use of stable transposase-making P elements. These are special P elements which produce high levels of transposase activity, but which, for one reason or another, are not mobile themselves. One such element, P[ry^+ Δ2-3](99B), was mentioned above (166). This element produces higher levels of sn^w mutability by itself than the entire genome of any P strain yet tested. At the same time, its own transposition and excision rates are negligible. It makes transposase in both germ line and somatic tissues as a result of the removal of its 2-3 intron (126), and it carries a $rosy^+$ gene as a marker. Another such element is called *Jumpstarter* (33a). This element is unmarked and is less stable than P[ry^+ Δ2-3](99B), making it more difficult to maintain. However, its transposase is limited to the germ line, which is advantageous in some cases. A marked element with an *hsp70* promoter is also available (239a). Finally, M. Mullins and G. Rubin (personal communication) are developing a chromosomally integrated P element with one of its termini deleted to make it stable and with its transposase gene driven by an *hsp70* promoter. The resulting element might prove to be highly useful for procedures requiring a genomic transposase source.

5. Alternative ammunition elements

The advantage of using stable transposase-making elements of the kind just described is that they can be combined with nonautonomous P elements, which are mobilized to produce insertion mutations. Nonautonomous elements used in this way are called "ammunition" elements. Crosses can be designed so that the transposase maker is present for only a single generation and the resulting insertion mutation is recovered in a transposase-free background. Therefore, there is no need to recover the mutation in the P cytotype, and the number of extraneous P elements in the mutant genome can be minimized.

One strategy is to use many small but highly mobile nonautonomous P elements as ammunition.

Robertson et al. (166) combined chromosome 2 of the M′ strain Birmingham with P[ry^+ Δ2-3](99B) on chromosome 3 to screen for new insertions on the X chromosome. The Birmingham second chromosome, known as *Birm2*, has 17 nonautonomous P elements but no autonomous ones. There was an average of two new X-chromosomal inserts per generation. This is approximately twice the insertion rate seen with M × P crosses employing P strains such as π_2 and Harwich. In a mutagenesis screen of 10,000 gametes, this rate will yield an average of one new insertion per kilobase of euchromatic DNA.

Alternatively, one can use "smart" ammunition, such as pUChsneo, developed by Steller and Pirrotta (212). This element has the necessary resistance gene and replication origin to form a clonable plasmid upon circularization, and it can be selected in *Drosophila* by virtue of its neomycin resistance gene. Once such an element is inserted into the gene of interest, cloning the gene is relatively easy. Obtaining the insertion mutation, however, requires considerably more effort if this kind of element is used. One reason is that only one or a small number of pUChsneo elements are usually available in a given genome. In addition, elements of this kind are larger than the *Birm2* elements, which, as discussed above, is likely to mean a lower transposition rate. As a result, it is usually not practical to use pUChsneo to screen for mutations at a specific locus. Instead, the usefulness of pUChsneo and similar elements lies in the construction of large "libraries" of stocks, each of which has an insertion at a different random site. Such a library was recently made by Cooley et al. (33a), who used their *Jumpstarter* element to mobilize pUChsneo and selected for transposition of the neomycin resistance gene. They produced 1,300 lines, each with a single insert of the pUChsneo element in a transposase-free background. Many of these lines had visible or lethal mutations, usually caused by the insertion. A similar procedure has employed an enhanced transposase source (239a).

In summary, nonautonomous P elements make the best ammunition for mutagenesis. A chromosome like *Birm2* with many small P elements is most effective for specific-locus mutagenesis, but in vitro-modified elements can also be useful for some purposes.

6. Confirming that the mutation is a P insertion

Mutations produced by the above methods are usually, but not always, P element insertions. One way of confirming the presence of a P element is to show by in situ hybridization that a new P element site is present at the cytological position of the target gene. Further evidence can be obtained by showing that the mutation tends to revert at high frequencies in dysgenic flies.

7. Detection of regulatory elements

Recently O'Kane and Gehring (157a) have used the expression of the *lacZ* gene in a P element vector to detect developmentally regulated genes. Some insertions of the vector resulted in tissue- or stage-specific expression of *lacZ*, thus indicating a genomic position subject to such regulation.

B. Germ Line Transformation

The spread of P elements through *D. melanogaster* may have been rapid in evolutionary terms, but it was slow compared to their invasion of *Drosophila* technology, which began in 1982. That was when Rubin and Spradling (174, 209) found that P elements could be used to transfer cloned genes into *Drosophila* embryos. The procedure was simple. Plasmid DNA bearing a P element, into which the gene of interest had been inserted, was injected into young M cytotype embryos prior to the cellularization of the germ line. Transposase was supplied either by P elements in the genome or by coinjected "helper" elements. The plasmid-borne P element then transposed to one of the chromosomes, apparently by the normal transposition mechanism, as indicated by the lack of plasmid flanking sequences at the chromosomal insertion sites. Transformed individuals could then be recovered in the next generation. A detailed and updated protocol is provided by Spradling (207).

This method has been used successfully to study gene regulation, amplification, chromosome puffing, population variability, evolutionary divergence, and other topics (11, 21, 22, 30, 32, 38, 72, 74, 80, 88, 90, 96, 101, 127, 129, 140, 158, 161, 175, 183, 199, 212, 218, 227). There are, however, two weaknesses in the method. First, the effort required to obtain each independent transformant is too great to permit cloning of genes by transformation of random fragments followed by complementation tests. Second, P element insertions occur at more or less random genomic sites, which results in the expression of the transformed gene being subject to unpredictable euchromatic position effects. These effects can be overcome to some extent by the inclusion of "buffer" sequences, especially on the 5′ side of the gene, but there is still some residual variability due to genomic position (207).

1. Alternative transposase sources

The first transformation experiments were performed with transposase supplied by autonomous P elements, which were either endogenous or coinjected. These autonomous elements were often recovered along with the transduced gene, thus causing instability in subsequent generations. To overcome this problem, Karess and Rubin (102) introduced the "wings clipped" helper element, called pπ25.7wc. This was a P element with its transposase gene intact, but with part of its 3′ terminal repeat removed, thus preventing it from transposing. The pπ25.7wc element could then be coinjected with the marked vector element to yield stable transformed lines. Unfortunately, some autonomous elements still got into the genome as a result of recombination between the helper and the vector plasmids prior to integration (102). The risk of such recombination can be minimized by injecting the helper and vector in a ratio favoring the vector, usually 20:1, and by using vectors with no more P homology than necessary. Another nonautonomous but transposase-making helper element, called phsπ, in which the transposase gene is driven by an *hsp70* promoter, has also been successful (88).

As an alternative to a coinjected helper element, it is possible to use a stable genomic transposase source, such as P[ry^+ Δ2-3](99B). One advantage of this method is that it does not require that the recipient germ cell take up both the helper and the vector, nor does the injected DNA have to be expressed before integration can take place. A dominant marker on the P[ry^+ Δ2-3](99B) chromosome makes it possible to select only those transformed progeny in which the transposase-making element has already been eliminated by segregation. This method has been used to obtain high frequencies of transformation for two marked vector elements (166).

2. Alternative vectors

The transformation process can be facilitated by the use of specially constructed P element vectors containing polylinker sequences into which the gene of interest (the "passenger" gene) can be cloned. These vectors also carry a selectable marker to be used for identifying transformants. Such a marker is especially important when the passenger gene has no readily scored phenotype, or when its ability to produce the phenotype is uncertain. Vectors are available with *rosy*$^+$ (175), *white*$^+$ (122; Pirrotta, personal communication), *ADH* (J. Posakony, cited in reference 101), and neomycin resistance (212). The *white*$^+$ vectors are particularly useful due to their high transformation frequency and their intermediate level of expression in most integration sites. The latter property helps in selecting homozygotes.

A cosmid vector is available for transducing very large segments (88). This technique is especially valuable for studying large genes or for determining the extent of a gene. Sequences greater than 40 kb have been transformed in this way, but the frequency of transformation is significantly reduced as compared with smaller vectors (88). In general, transformation frequencies tend to decrease with increasing size of the vector (207). In addition, some DNA sequences seem to have an adverse effect on the rate of transformation when they are placed within a P element, but the reason for this effect is not understood (207).

3. The problem of position effects

In most cases, transformed genes retain their normal qualitative expression, including developmental and tissue specificity, provided enough of the surrounding sequences are included. There are, however, quantitative position effects which cannot be avoided (28, 90, 127, 129, 210). These effects are especially troublesome to experimenters interested in relatively subtle differences in gene expression. Laurie-Ahlberg and Stam (127) have shown that it is possible to deal with this problem through careful statistical analysis of a large number of insertion sites. The required number of inserts depends upon the magnitude of the difference being measured, the degree of variability between insertion sites, and the error variance of the measurements. They estimated the sample sizes required to detect differences in expression levels for two alleles of *Adh* (designated Fast and Slow) and one of *rosy* (xanthine dehydrogenase). The required sample size can be inconveniently large for differences in expression level less than 15% (Fig. 11). Note that the much larger sample sizes for ADH-Slow relative to ADH-Fast are probably due to the intrinsically lower level of expression of ADH-Slow, which results in a greater relative measurement error.

Fortunately, it is no longer necessary to obtain large samples by repeated use of the transformation procedure. Once the first insert has been obtained, it can be used to generate an array of new sites by further transposition. The stable genomic transposase sources discussed above are particularly useful for this process. For example, Robertson et al. (166) used the P[ry^+ Δ2-3](99B) element to obtain a large number of secondary transpositions of various *rosy*$^+$ and *white*$^+$ elements. As before, the crosses can be

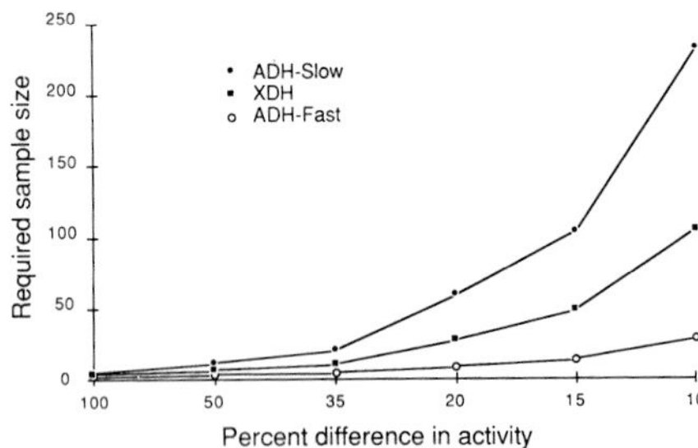

Figure 11. Sample sizes for transformation experiments. For each enzyme studied by Laurie-Ahlberg and Stam (127), the ordinate indicates the number of transformants required for an 80% probability of detecting the indicated activity difference at the 95% significance level, using a standard one-tailed *t*-test.

designed so that each new insertion is recovered in a transposase-free background.

C. Utility of P Elements for Work with Other Species

As discussed previously, several studies have been aimed at extending the use of P elements to other species (31, 146, 153a, 164). In one case, the presence of P element sequences increased the transfection frequency in mammalian cells (31), but the effect was apparently due to an enhancement of expression rather than integration; no true P element transposition was involved. In fact, the general conclusion from these studies is that P elements are not capable of transposition outside the genus *Drosophila.* Fusing the P element's transposase gene with a different promoter and removing its introns—particularly the one with tissue-specific splicing—were shown to be insufficient to promote transposition in other species. It is possible that other modifications to the P element sequence may yet result in transposition in heterologous systems. However, researchers are reminded to proceed with caution in this direction, keeping in mind the demonstrated ability of P elements to spread rapidly through populations, often with deleterious consequences.

XII. COMPARISON WITH OTHER TRANSPOSABLE ELEMENTS

A. A Class of Structurally Similar Elements

As pointed out by Streck et al. (213), the P family is one member of a group of transposable elements which share certain structural features. These elements include *Ac* and *Ds* in maize (Fedoroff, this volume), *hobo* in *D. melanogaster* (Blackman and Gelbart, this volume), Tam3 in *Antirrhinum majus* (E. S. Coen, T. P. Robbins, J. Almeida, A. Hudson, and R. Carpenter, this volume), and *1723* in *Xenopus laevis* (D. Carroll, D. S. Knutzon, and J. E. Garrett, this volume). All of these elements terminate in short inverted repeats which share some sequence similarity (Fig. 12). Furthermore, they all duplicate 8 bp of the target sequence upon insertion. Both P and *hobo* show a strong preference for target octamers ending in the dinucleotide AC. There is no obvious homology between these elements' internal sequences, but both P and *hobo* have interior repeat structures (compare Fig. 3 of this article with Fig. 2 of Streck et al. [213]). The functional or evolutionary significance of these similarities is not known.

B. Similarities in Transposition and Regulation

As discussed above, most of the known transposition mechanisms do not seem to apply to P elements. However, there are hints of similarities. Satta et al. (178, 179) pointed out that there is weak but significant amino acid homology between P element sequences and the Tn*3* (Sherratt, this volume) resolvase and transposase. In addition, the likelihood that the P element repressor shares common sequences with its transposase is similar to the situation described for Tn*5* (Berg, this volume). In plant elements, there is abundant evidence for the involvement of methylation in the regulation of transposable elements (Fedoroff, this volume). Although there is no evidence that such a mechanism could be operating in *Drosophila* spp., it might help explain the unusual mode of inheritance by cytotype and also reactivity in the I-R system.

C. Other Cases of Hybrid Dysgenesis

As discussed previously, there is a second system of hybrid dysgenesis in *D. melanogaster* called the I-R system. It resembles the P-M system in the

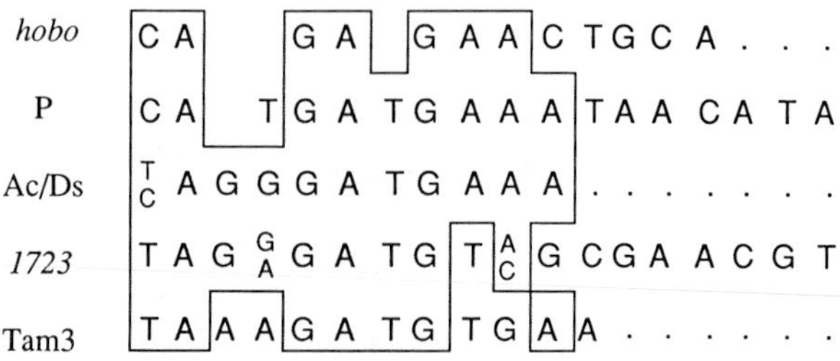

Figure 12. Terminal repeats of several transposable elements. Modified from Streck et al. (213). Both nucleotides are shown for elements in which the two ends can vary. The left end of each displayed sequence is the element's terminus. Short gaps in the P and *hobo* sequences were introduced to improve the alignment. All of the elements produce 8-bp duplications of the target sequence.

distribution of complete elements and in some aspects of its regulation (Finnegan, Chapter 18, this volume). However, there is no clear mechanistic basis for these similarities.

More recently, the *hobo* element has been reported to show high transpositional activity in some strains (18, 238; Blackman and Gelbart,. this volume). These elements produce chromosome rearrangements with many similarities to the rearrangements produced by P elements. The *hobo* element also forms the basis for another strain classification system: strains with many euchromatic copies of *hobo* are called H for *hobo*-containing, and the others are called E for empty (213). These designations will probably become less clear cut as more strains are studied in detail. The analogy between *hobo* and the P-M system of hybrid dysgenesis is quite weak. First, there are no obvious reciprocal differences in the two kinds of E × H hybrids with respect to their *hobo* activation. Furthermore, it is not yet clear that such crosses are even necessary to activate *hobo*. In a long series of experiments, Lim and colleagues (99, 128, 132–134) have characterized a system of high mutability and chromosome rearrangements called *Uc* for "unstable chromosome." Recent evidence from cloning the sites of instability and in situ hybridization (J. Lim, *Proc. Natl. Acad. Sci. USA*, in press) has indicated that *hobo* is invariably present at the sites of instability. Reversion of *Uc*-induced lethal mutations coincided with the disappearance of the corresponding *hobo* site. Since *Uc* has been unstable in inbred stocks for more than 10 years, it seems clear that outcrossing is not necessary to activate *hobo*. It is also clear that *hobo* is not activated by P-M dysgenic crosses (50). Further work will be required to determine whether the behaviors of *hobo* and P are truly analogous.

XIII. CONCLUSIONS

In the 17 years since Hiraizumi isolated his first "MR" chromosome, P elements have revealed much about their own workings and about general questions of the genetics, biology, and evolution of transposable elements. By serving as mutagens, transposon tags, and transformation vehicles, they have opened new approaches to many other questions in the wider area of *Drosophila* genetics. I believe we are still very far from exhausting what we have to learn from the study of P elements. However, if that day ever comes, it will only be necessary for Hiraizumi or someone else to sample yet again from a natural population of *Drosophila* and discover what else is waiting for us.

Acknowledgments. I am grateful to the many people who contributed comments and suggestions, especially Chris Preston, Randy Phillis, Jim Crow, W. Eggleston, W. Benz, I. Boussy, H. Robertson, M. Simmons, J. Lim, and C. Denniston. However, these people are not responsible for the speculations and any possible misinterpretations presented here. I also thank the various researchers who sent me preprints and granted permission to discuss their unpublished data.

This work was supported by Public Health Service grants GM 30948 and GM 35099 from the National Institutes of Health and by U.S. Department of Agriculture grant 87-CRCR-1-2517.

LITERATURE CITED

1. **Angus, D. S., and J. A. Raisbeck.** 1979. A transmissible factor involved in hybrid sterility in *Drosophila melanogaster*. *Genetica* 50:81–87.
2. **Anxolabéhère, D., H. Benes, D. Nouaud, and G. Périquet.** 1987. Evolutionary steps and transposable elements in *Drosophila melanogaster*: the missing RP type obtained by genetic transformation. *Evolution* **41**:846–853.
3. **Anxolabéhère, D., L. Charles-Palabost, A. Fleuriet, and G. Périquet.** 1988. Temporal surveys of French populations of *Drosophila melanogaster*: P-M system, enzymatic polymorphism and infection by the sigma virus. *Heredity* **61**: 121–131.
4. **Anxolabéhère, D., H. Kai, D. Nouaud, G. Périquet, and S. Ronsseray.** 1984. The geographical distribution of P-M hybrid dysgenesis in *Drosophila melanogaster*. *Genet. Sel. Evol.* **16**:15–26.
5. **Anxolabéhère, D., M. G. Kidwell, and G. Périquet.** 1988. Molecular characteristics of diverse populations are consistent with a recent invasion of *Drosophila melanogaster* by mobile P elements. *Mol. Biol. Evol.* **5**:252–269.
6. **Anxolabéhère, D., D. Nouaud, and G. Périquet.** 1982. Cytotype polymorphism of the P-M system in two wild populations of *Drosophila melanogaster*. *Proc. Natl. Acad. Sci. USA* **79**:7801–7803.
7. **Anxolabéhère, D., D. Nouaud, and G. Périquet.** 1985. Séquences homologues a l'élément P chez des espèces de Drosophila du groupe obscura et chez *Scaptomyza pallida* (Drosophilidae). *Genet. Sel. Evol.* **17**:579–584.
8. **Anxolabéhère, D., D. Nouaud, G. Périquet, and S. Ronsseray.** 1986. Evolution des potentialités dysgénésiques du système P-M dans des populations expérimentales mixtes P, Q, M et M' de *Drosophila melanogaster*. *Genetica* **69**: 81–95.
9. **Anxolabéhère, D., D. Nouaud, G. Périquet, and P. Tchen.** 1985. P element distribution in Eurasian populations of *Drosophila melanogaster*: a genetic and molecular analysis. *Proc. Natl. Acad. Sci. USA* **82**:5418–5422.
10. **Anxolabéhère, D., and G. Périquet.** 1987. P-homologous sequences in diptera are not restricted to the Drosophilidae family. *Genet. Iber.*, in press.
11. **Bargiello, T. A., F. R. Jackson, and M. W. Young.** 1984. Restoration of circadian behavioural rhythms by gene transfer in Drosophila. *Nature* (London) **312**:752–754.
12. **Benz, W. K., and W. R. Engels.** 1984. Evidence for replica-

tive transposition of P elements in *Drosophila melanogaster*. *Genetics* **107**:s10.

13. **Berg, R. L.** 1974. A simultaneous mutability rise at the *singed* locus in two out of three Drosophila melanogaster populations studied in 1973. *Drosophila Inform. Serv.* **51**: 100–102.
14. **Berg, R. L., W. R. Engels, and R. A. Kreber.** 1980. Site-specific X-chromosome rearrangements from hybrid dysgenesis in *Drosophila melanogaster*. *Science* **210**:427–429.
15. **Bingham, P. M., M. G. Kidwell, and G. M. Rubin.** 1982. The molecular basis of P-M hybrid dysgenesis: the role of the P element, a P strain-specific transposon family. *Cell* **29**: 995–1004.
16. **Bingham, P. M., R. Levis, and G. M. Rubin.** 1981. The cloning of the DNA sequences from the *white* locus of *Drosophila melanogaster* using a novel and general method. *Cell* **25**:693–704.
17. **Black, D. M., M. S. Jackson, M. G. Kidwell, and G. A. Dover.** 1987. KP elements repress P-induced hybrid dysgenesis in D. melanogaster. *EMBO J.* **6**:4125–4135.
18. **Blackman, R., R. Grimaila, M. Koehler, and W. Gelbart.** 1987. Mobilization of *hobo* elements residing with the *decapentaplegic* gene complex: suggestion of a new hybrid dysgenesis system in *Drosophila melanogaster*. *Cell* **49**: 497–505.
19. **Blount, J. L., R. C. Woodruff, and S. J. Hudson.** 1985. Interaction between mobile DNA-element-induced lethal mutations and chemical mutagens in the hybrid dysgenic system of *Drosophila melanogaster*. *Mutat. Res.* **149**:33–40.
20. **Boedigheimer, M., and M. J. Simmons.** 1987. Studies on the ability of an M′ strain to suppress P element-induced hybrid dysgenesis. *Drosophila Inform. Serv.* **66**:26–28.
21. **Bonner, J. J., C. Parks, J. Parker-Thornberg, M. A. Mortin, and H. R. B. Pelham.** 1984. The use of promoter fusions in Drosophila genetics: isolation of mutations affecting the heat shock response. *Cell* **37**:979–991.
22. **Bourouis, M., and G. Richards.** 1985. Remote regulatory sequences of the Drosophila glue gene *sgs3* as revealed by P-element transformation. *Cell* **40**:349–357.
23. **Boussy, I. A.** 1987. A latitudinal cline in P-M gonadal dysgenesis potential in Australian *Drosophila melanogaster* populations. *Genet. Res.* **49**:11–18.

23a. **Boussy, I. A., M. J. Healy, J. G. Oakeshott, and M. G. Kidwell.** 1988. Molecular analysis of the P-M gonadal dysgenesis cline in eastern Australian *Drosophila melanogaster*. *Genetics* **119**:889–902.

24. **Boussy, I. A., and M. G. Kidwell.** 1987. The P-M hybrid dysgenesis cline in eastern Australian *Drosophila melanogaster*: discrete P, Q and M regions are nearly contiguous. *Genetics* **115**:737–745.
25. **Bregliano, J. C., and M. G. Kidwell.** 1983. Hybrid dysgenesis determinants, p. 363–410. *In* J. A. Shapiro (ed.), *Mobile Genetic Elements*. Academic Press, Inc. (London) Ltd., London.
26. **Brennan, M. D., R. G. Rowan, and W. J. Dickinson.** 1984. Introduction of a functional P element into the germ line of *Drosophila hawaiiensis*. *Cell* **38**:147–151.
27. **Brookfield, J. F. Y., E. Montgomery, and C. Langley.** 1984. Apparent absence of transposable elements related to the P elements of D. melanogaster in other species of Drosophila. *Nature* (London) **310**:330–332.
28. **Chang, D.-Y., B. Wisely, S.-M. Huang, and R. A. Voelker.** 1986. Molecular cloning of *suppressor of sable*, a *Drosophila melanogaster* transposon-mediated suppressor. *Mol. Cell. Biol.* **6**:1520–1528.
29. **Chia, W., G. Howes, M. Martin, Y. Meng, K. Moses, and S. Tsubota.** 1986. Molecular analysis of the *yellow* locus in Drosophila. *EMBO J.* **13**:3597–3605.
30. **Clark, S. H., and A. Chovnick.** 1986. Studies of normal and position-affected expression of *rosy* region genes in *Drosophila melanogaster*. *Genetics* **114**:819–840.
31. **Clough, D. W., H. M. Lepinske, R. L. Davidson, and R. V. Storti.** 1985. *Drosophila* P element-enhanced transfection in mammalian cells. *Mol. Cell. Biol.* **5**:898–901.
32. **Cohen, R. S., and M. Messelson.** 1985. Separate regulatory elements of the heat-inducible and ovarian expression of the Drosophila *hsp26* gene. *Cell* **43**:737–746.
33. **Conover, W. J.** 1980. *Practical Nonparametric Statistics*, 2nd ed. John Wiley & Sons, New York.

33a. **Cooley, L., R. Kelley, and A. Spradling.** 1988. Insertional mutagenesis of the Drosophila genome with single P elements. *Science* **239**:1121–1128.

33b. **Coté, B., W. Bender, D. Curtis, and A. Chovnick.** 1986. Molecular mapping of the *rosy* locus in *Drosophila melanogaster*. *Genetics* **112**:769–783.

34. **Crow, J. F.** 1984. The P-factor: a transposable element in Drosophila, p. 257–273. *In* E. H. Y. Chu and W. M. Generoso (ed.), *Mutation, Cancer and Malformation*. Plenum Publishing Corp., New York.
35. **Daniels, S. B., I. A. Boussy, A. Tukey, M. Carrillo, and M. G. Kidwell.** 1987. Variability among "true M" lines in P-M gonadal dysgenesis potential. *Drosophila Inform. Serv.* **66**:37–39.
36. **Daniels, S. B., S. H. Clark, M. G. Kidwell, and A. Chovnick.** 1987. Genetic transformation of *Drosophila melanogaster* with an autonomous P element: phenotypic and molecular analyses of long-established transformed lines. *Genetics* **115**: 711–723.
37. **Daniels, S. B., M. McCarron, C. Love, S. H. Clark, and A. Chovnick.** 1985. Dysgenesis induced instability of *rosy* locus transformation in *Drosophila melanogaster*: analysis of excision events and the selective recovery of control element deletions. *Genetics* **109**:95–117.
38. **Daniels, S. B., M. McCarron, C. Love, S. H. Clark, and A. Chovnick.** 1986. The underlying bases of gene expression differences in stable transformants of the *rosy* locus in *Drosophila melanogaster*. *Genetics* **113**:265–285.
39. **Daniels, S. B., and L. D. Strausbaugh.** 1986. The distribution of P element sequences in Drosophila: the *willistoni* and *saltans* species groups. *J. Mol. Evol.* **23**:138–148.
40. **Daniels, S. B., L. D. Strausbaugh, and R. A. Armstrong.** 1985. Molecular analysis of P element behavior in *Drosophila simulans* transformants. *Mol. Gen. Genet.* **200**:258–265.
41. **Daniels, S. B., L. D. Strausbaugh, L. Ehrman, and R. Armstrong.** 1984. Sequences homologous to P elements occur in *Drosophila paulistorum*. *Proc. Natl. Acad. Sci. USA* **81**:6794–6797.
42. **Danilevskaya, O. N., E. V. Kurenova, B. A. Leibovitch, A. Y. Shevelev, I. A. Bass, and R. B. Khesin.** 1984. Telomeres and P element of *Drosophila melanogaster* contain sequences that replicate autonomously in *Saccharomyces cerevisiae*. *Mol. Gen. Genet.* **197**:342–344.
43. **Eanes, W. F., C. Wesley, J. Hey, D. Houle, and J. W. Ajioka.** 1988. The finess consequences of P-element insertion in *Drosophila melanogaster*. *Genet. Res.* **52**:17–26.
44. **Eeken, J. C. J., and F. H. Sobels.** 1981. Modification of MR mutator activity in repair-deficient strains. *Mutat. Res.* **83**: 191–200.
45. **Eeken, J. C. J., and F. H. Sobels.** 1983. The effect of two

chemical mutagens ENU and MMS on MR-mediated reversion of an insertion-sequence mutation in *Drosophila melanogaster*. *Mutat. Res.* **110:**297–310.

46. **Eeken, J. C. J., and F. H. Sobels.** 1983. The influence of deficiencies in DNA-repair on MR-mediated reversion of an insertion-sequence mutation in *Drosophila melanogaster*. *Mutat. Res.* **110:**287–295.
47. **Eeken, J. C. J., F. H. Sobels, V. Hyland, and A. P. Schalet.** 1985. Distribution of MR-induced sex-linked recessive lethal mutations in *Drosophila melanogaster*. *Mutat. Res.* **150:** 261–275.
48. **Eggleston, P.** 1984. Hybrid dysgenesis in *Drosophila melanogaster*: the frequency and distribution of male recombination events. *Heredity* **52:**189–202.
49. **Eggleston, P.** 1984. Correlation in the induction and response of SF and GD sterility. *Drosophila Inform. Serv.* **60:**99–101.
50. **Eggleston, W. B., D. M. Johnson-Schlitz, and W. R. Engels.** 1988. P-M hybrid dysgenesis does not mobilize other transposable element families in D. melanogaster. *Nature* (London) **331:**368–370.
51. **Eisenberg, J. C., and S. C. R. Elgin.** 1986. Chromatic structure of a P element transduced *Hsp28* gene in *Drosophila melanogaster*. *Mol. Cell. Biol.* **6:**4126–4129.
52. **Eisenberg, J. C., and S. C. R. Elgin.** 1987. $Hsp28^{stl}$: a P-element insertion mutation that alters the expression of a heat shock gene in *Drosophila melanogaster*. *Genetics* **115:** 333–340.
53. **Engels, W. R.** 1979. Extrachromosomal control of mutability in *Drosophila melanogaster*. *Proc. Natl. Acad. Sci. USA* **76:**4011–4015.
54. **Engels, W. R.** 1979. Hybrid dysgenesis in *Drosophila melanogaster*: rules of inheritance of female sterility. *Genet. Res.* **33:**219–236.
55. **Engels, W. R.** 1979. Germline aberrations associated with a case of hybrid dysgenesis in *Drosophila melanogaster* males. *Genet. Res.* **33:**137–146.
56. **Engels, W. R.** 1979. The estimation of mutation rates when premeiotic events are involved. *Environ. Mutagen.* **1:**37–43.
57. **Engels, W. R.** 1981. Germline hypermutability in Drosophila and its relation to hybrid dysgenesis and cytotype. *Genetics* **98:**565–587.
58. **Engels, W. R.** 1981. Hybrid dysgenesis in Drosophila and the stochastic loss hypothesis. *Cold Spring Harbor Symp. Quant. Biol.* **45:**561–565.
59. **Engels, W. R.** 1983. The P family of transposable elements in Drosophila. *Annu. Rev. Genet.* **17:**315–344.
60. **Engels, W. R.** 1984. A trans-acting product needed for P factor transposition in Drosophila. *Science* **226:**1194–1196.
61. **Engels, W. R.** 1985. Guidelines for P element transposon tagging. *Drosophila Inform. Serv.* **61:**1–1.
62. **Engels, W. R.** 1985. A set of P-cytotype balancer chromosomes. *Drosophila Inform. Serv.* **61:**70–71.
63. **Engels, W. R.** 1986. On the evolution and population genetics of hybrid-dysgenesis-causing transposable elements in Drosophila. *Philos. Trans. R. Soc. Lond. Ser.* **312:**205–215.
64. **Engels, W. R., W. K. Benz, C. R. Preston, P. L. Graham, R. W. Phillis, and H. M. Robertson.** 1987. Somatic effects of P element activity in *Drosophila melanogaster*: pupal lethality. *Genetics* **117:**745–757.
65. **Engels, W. R., and C. R. Preston.** 1979. Hybrid dysgenesis in *Drosophila melanogaster*: the biology of male and female sterility. *Genetics* **92:**161–175.
66. **Engels, W. R., and C. R. Preston.** 1980. Components of hybrid dysgenesis in a wild population of *Drosophila melanogaster*. *Genetics* **95:**111–128.
67. **Engels, W. R., and C. R. Preston.** 1981. Characteristics of a "neutral" strain in the P-M system of hybrid dysgenesis. *Drosophila Inform. Serv.* **56:**35–37.
68. **Engels, W. R., and C. R. Preston.** 1981. Identifying P factors in Drosophila by means of chromosome breakage hotspots. *Cell* **26:**421–428.
69. **Engels, W. R., and C. R. Preston.** 1984. Formation of chromosome rearrangements by P factors in Drosophila. *Genetics* **107:**657–678.
70. **Fahmy, M. J., and O. G. Fahmy.** 1985. The MR P-type transposable elements and the genetic activities of mutagens and carcinogens in *Drosophila melanogaster*. I. N,N-Dimethylnitrosamine (DMN). *Mutat. Res.* **152:**169–185.
71. **Finnegan, D. J., and D. H. Fawcett.** 1986. Transposable elements in *Drosophila melanogaster*. *Oxford Surv. Eukaryotic Genes* **3:**1–62.
72. **Fischer, J. A., and T. Maniatis.** 1986. Regulatory elements involved in Drosophila *Adh* gene expression are conserved in divergent species and separate elements mediate expression in different tissues. *EMBO J.* **5:**1275–1289.
73. **Fitzpatrick, B., and J. A. Sved.** 1986. High levels of fitness modifiers induced by hybrid dysgenesis in *Drosophila melanogaster*. *Genet. Res.* **48:**89–94.
74. **Garabedian, M. J., B. M. Shepherd, and P. C. Wensink.** 1986. A tissue-specific transcription enhancer from the Drosophila *yolk protein 1* gene. *Cell* **45:**859–867.
75. **Generoso, W., M. Shelby, and F. de Serres (ed.).** 1980. *DNA Repair and Mutagenesis in Eukaryotes*. Plenum Publishing Corp., New York.
76. **Gerasimova, T. I., Y. V. Ilyin, L. J. Mizrokhi, L. V. Semjonova, and G. P. Georgiev.** 1984. Mobilization of the transposable element *mdg4* by hybrid dysgenesis generates a family of unstable *cut* mutations in *Drosophila melanogaster*. *Mol. Gen. Genet.* **193:**488–492.
77. **Gerasimova, T. I., L. V. Matyunina, Y. V. Ilyin, and G. P. Georgiev.** 1984. Simultaneous transposition of different mobile elements: relation to multiple mutagenesis in *Drosophila melanogaster*. *Mol. Gen. Genet.* **194:**517–522.
78. **Gerasimova, T. I., L. V. Matyunina, L. J. Mizrokhi, and G. P. Georgiev.** 1985. Successive transposition explosions in *Drosophila melanogaster* and reverse transpositions of mobile dispersed genetic elements. *EMBO J.* **4:**3773–3779.
79. **Gerasimova, T. I., L. J. Mizrokhi, and G. P. Georgiev.** 1984. Transposition bursts in genetically unstable *Drosophila melanogaster*. *Nature* (London) **309:**714–716.
80. **Goldberg, D., J. Posakony, and T. Maniatis.** 1983. Correct developmental expression of a cloned *alcohol dehydrogenase* gene transduced into the Drosophila germ line. *Cell* **34:** 59–73.
81. **Golubovsky, M. D.** 1978. Two types of instability of *singed* alleles isolated from populations of *Drosophila melanogaster* during mutation outburst in 1973. *Drosophila Inform. Serv.* **53:**171.
82. **Golubovsky, M. D., Y. N. Ivanov, and M. M. Green.** 1977. Genetic instability in *Drosophila melanogaster*: putative multiple insertion mutants at the *singed* bristle locus. *Proc. Natl. Acad. Sci. USA* **74:**2973–2975.
83. **Good, A. G., and D. A. Hickey.** 1987. Hybrid dysgenesis in *Drosophila melanogaster*: the elimination of P elements through backcrossing to an M-type strain. *Genome* **29:** 195–200.
84. **Green, M. M.** 1977. Genetic instability in *Drosophila melanogaster*-

nogaster: de novo induction of putative insertion mutations. *Proc. Natl. Acad. Sci. USA* 74:3490–3493.

85. **Green, M. M.** 1984. Genetic instability in *Drosophila melanogaster*: on the identity of the MR and P-M mutator systems. *Biol. Zentralbl.* **103**:1–8.

86. **Green, M. M.** 1986. Genetic instability in *Drosophila melanogaster*: the genetics of an MR element that makes complete P insertion mutations. *Proc. Natl. Acad. Sci. USA* **83**:1036–1040.

87. **Green, M. M., M.-T. Yamamoto, and G. L. G. Miklos.** 1987. Genetic instability in *Drosophila melanogaster*: cytogenetic analysis of MR-induced X-chromosome deficiencies. *Proc. Natl. Acad. Sci. USA* **84**:4533–4537.

88. **Haenlin, M., H. Steller, V. Pirrotta, and E. Mohier.** 1985. A 43 kilobase cosmid P transposon rescues the $fs(1)^{K10}$ morphogenetic locus and three adjacent Drosophila developmental mutants. *Cell* **40**:827–837.

88a. **Hagiwara, N., E. Nakamura, E. T. Matsuura, and S. I. Chigusa.** 1987. Hybrid dysgenesis in natural populations of *Drosophila melanogaster* in Japan. II. Strains which cannot induce P-M dysgenesis may completely suppress functional P element activity. *Genet. Res.* **50**:105–111.

89. **Hawley, R. S., R. A. Steuber, C. H. Marcus, R. Sohn, D. M. Baronas, M. L. Cameron, A. E. Zitron, and J. W. Chase.** 1988. Molecular analysis of an unstable P element insertion at the *singed* locus of *Drosophila melanogaster*: evidence for intracistronic transposition of a P element. *Genetics* **119**: 85–94.

90. **Hazelrigg, T., R. Levis, and G. M. Rubin.** 1984. Transformation of *white* locus DNA in Drosophila: dosage compensation, *zeste* interaction, and position effects. *Cell* **36**: 469–481.

91. **Henderson, S. A., R. C. Woodruff, and J. N. Thompson, Jr.** 1978. Spontaneous chromosome breakage at male meiosis associated with male recombination in *Drosophila melanogaster*. *Genetics* **88**:93–107.

92. **Hihara, F., N. Hisamatsu, and T. Hirota.** 1985. Hybrid dysgenesis in *Drosophila melanogaster*: type conversions observed in the recently established isofemale lines and hybrid lines originated from MxP crosses. *Jpn. J. Genet.* **60**:199–214.

93. **Hiraizumi, Y.** 1971. Spontaneous recombination in *Drosophila melanogaster* males. *Proc. Natl. Acad. Sci. USA* **68**: 268–270.

94. **Hiraizumi, Y., and J. F. Crow.** 1960. Heterozygous effects on viability, fertility, rate of development, and longevity of Drosophila chromosomes that are lethal when homozygous. *Genetics* **45**:1071–1083.

95. **Hiraizumi, Y., B. Slatko, C. Langley, and A. Nill.** 1973. Recombination in *Drosophila melanogaster* male. *Genetics* **73**:439–444.

96. **Hiromi, Y., A. Kuroiwa, and W. J. Gehring.** 1985. Control elements of the Drosophila segmentation gene *fushi tarazu*. *Cell* **43**:603–613.

96a. **Howes, G., M. O'Connor, and W. Chia.** 1988. On the specificity and effects on transcription of P element insertions at the *yellow* locus of *Drosophila melanogaster*. *Nucleic Acids Res.* **16**:3039–3052.

97. **Isackson, D. R., T. K. Johnson, and R. E. Denell.** 1981. Hybrid dysgenesis in Drosophila: the mechanism of T-007-induced male recombination. *Mol. Gen. Genet.* **184**: 539–543.

98. **Iwano, M., K. Hattori, K. Saigo, Y. Matsuo, T. Mukai, and T. Yamazaki.** 1984. Cloning of P elements from P and Q strains in natural populations of *Drosophila melanogaster* in Japan. *Jpn. J. Genet.* **59**:403–409.

99. **Johnson-Schlitz, D., and J. K. Lim.** 1987. Cytogenetics of *Notch* mutations arising the unstable X chromosome *Uc* of *Drosophila melanogaster*. *Genetics* **115**:701–709.

100. **Jongeward, G., M. J. Simmons, and E. Heath.** 1987. The instability of a P element insertion mutation is affected by chromosomes derived paternally from a pseudo-M strain of *D. melanogaster*. *Drosophila Inform. Serv.* **66**:77–80.

101. **Kamb, A., L. E. Iverson, and M. A. Tanouye.** 1987. Molecular characterization of *Shaker*, a Drosophila gene that encodes a potassium channel. *Cell* **50**:405–413.

102. **Karess, R. E., and G. M. Rubin.** 1984. Analysis of P transposable element functions in Drosophila. *Cell* **38**:135–146.

103. **Kelley, M. R., S. Kidd, R. L. Berg, and M. W. Young.** 1987. Restriction of P-element insertions at the *Notch* locus of *Drosophila melanogaster*. *Mol. Cell. Biol.* **7**:1545–1548.

104. **Kidwell, M. G.** 1979. Hybrid dysgenesis in *Drosophila melanogaster*: the relationship between the P-M and I-R interaction systems. *Genet. Res.* **33**:105–117.

105. **Kidwell, M. G.** 1981. Hybrid dysgenesis in *Drosophila melanogaster*: the genetics of cytotype determination in a neutral strain. *Genetics* **98**:275–290.

106. **Kidwell, M. G.** 1983. Evolution of hybrid dysgenesis determinants in *Drosophila melanogaster*. *Proc. Natl. Acad. Sci. USA* **80**:1655–1659.

107. **Kidwell, M. G.** 1984. Hybrid dysgenesis in *Drosophila melanogaster*: partial sterility associated with embryonic lethality in the P-M system. *Genet. Res.* **44**:11–28.

108. **Kidwell, M. G.** 1985. Hybrid dysgenesis in *Drosophila melanogaster*: nature and inheritance of P element regulation. *Genetics* **111**:337–350.

109. **Kidwell, M. G.** 1986. Molecular and phenotypic aspects of hybrid dysgenesis systems, p. 169–198. *In* S. Karlin and E. Nevo (ed.), *Evolutionary Processes and Theory*. Academic Press, Inc., New York.

110. **Kidwell, M. G.** 1986. P-M mutagenesis, p. 59–82. *In* D. B. Roberts (ed.), *Drosophila: a Practical Approach*. IRL Press, Oxford.

111. **Kidwell, M. G.** 1987. A survey of success rates using P element mutagenesis in *Drosophila melanogaster*. *Drosophila Inform. Serv.* **66**:81–86.

112. **Kidwell, M. G., T. Frydryk, and J. B. Novy.** 1983. The hybrid dysgenesis potential of *Drosophila melanogaster* strains of diverse temporal and geographical natural origins. *Drosophila Inform. Serv.* **51**:97–100.

113. **Kidwell, M. G., and J. F. Kidwell.** 1975. Cytoplasm-chromosome interactions in *Drosophila melanogaster*. *Nature* (London) **253**:755–756.

114. **Kidwell, M. G., and J. F. Kidwell.** 1976. Selection for male recombination in *Drosophila melanogaster*. *Genetics* **84**: 333–351.

115. **Kidwell, M. G., J. F. Kidwell, and P. T. Ives.** 1977. Spontaneous, non-reciprocal mutation and sterility in strain crosses of *Drosophila melanogaster*. *Mutat. Res.* **42**:89–98.

116. **Kidwell, M. G., J. F. Kidwell, and M. Nei.** 1973. A case of high rate of spontaneous mutation affecting viability in *Drosophila melanogaster*. *Genetics* **75**:133–153.

117. **Kidwell, M. G., J. F. Kidwell, and J. A. Sved.** 1977. Hybrid dysgenesis in *Drosophila melanogaster*: a syndrome of aberrant traits including mutation, sterility, and male recombination. *Genetics* **86**:813–833.

117a. **Kidwell, M. G., K. Kimura, and D. M. Black.** 1988. Evolution of hybrid dysgenesis potential following P element contamination in *Drosophila melanogaster. Genetics* **119:** 815–828.

118. **Kidwell, M. G., and J. B. Novy.** 1979. Hybrid dysgenesis in *Drosophila melanogaster*: sterility resulting from gonadal dysgenesis in the P-M system. *Genetics* **92:**1127–1140.
119. **Kidwell, M. G., J. B. Novy, and S. M. Feeley.** 1981. Rapid unidirectional change of hybrid dysgenesis potential in Drosophila. *J. Hered.* 72:32–38.
120. **Kidwell, M. G., and H. M. Sang.** 1986. Hybrid dysgenesis in *Drosophila melanogaster*: synthesis of RP strains by chromosomal contamination. *Genet. Res.* **47:**181–185.
121. **Kiyasu, P. K., and M. G. Kidwell.** 1984. Hybrid dysgenesis in *Drosophila melanogaster*: the evolution of mixed P and M populations maintained at high temperature. *Genet. Res.* **44:**251–259.
122. **Klemenz, R., U. Weber, and W. J. Gehring.** 1987. The *white* gene as a marker in a new P element vector for gene transfer in Drosophila. *Nucleic Acids Res.* **15:**3947–3959.
123. **Kocur, G. J., E. A. Drier, and M. J. Simmons.** 1986. Sterility and hypermutability in the P-M system of hybrid dysgenesis in *Drosophila melanogaster. Genetics* **114:**1147–1163.
124. **Lansman, R. A., R. O. Shade, T. A. Grigliatti, and H. W. Brock.** 1987. Evolution of P transposable elements: sequences of *Drosophila nebulosa* P elements. *Proc. Natl. Acad. Sci. USA* **84:**6491–6495.
125. **Lansman, R. A., S. N. Stacey, T. A. Grigliatti, and H. W. Brock.** 1985. Sequences homologous to the P mobile element of *Drosophila melanogaster* are widely distributed in the subgenus Sophophora. *Nature* (London) **318:**561–563.
126. **Laski, F. A., D. C. Rio, and G. M. Rubin.** 1986. Tissue specificity of Drosophila P element transposition is regulated at the level of mRNA splicing. *Cell* **44:**7–19.
127. **Laurie-Ahlberg, C. C., and L. F. Stam.** 1987. Use of P element-mediated transformation to identify the molecular basis of naturally occurring variants affecting *Adh* expression in *Drosophila melanogaster. Genetics* **115:**129–140.
128. **Laverty, T. R., and J. K. Lim.** 1982. Site-specific instability in *Drosophila melanogaster*: evidence for a transposition of destabilizing element. *Genetics* **101:**461–476.
129. **Levis, R., T. Hazelrigg, and G. M. Rubin.** 1985. Effects of genomic position on the expression of transduced copies of the *white* gene of Drosophila. *Science* **229:**558–561.
130. **Levis, R., K. O'Hare, and G. Rubin.** 1984. Effects of transposable element insertions on RNA encoded by the *white* gene of Drosophila. *Cell* **38:**471–481.
131. **Lewis, A. P., and J. F. Y. Brookfield.** 1987. Movement of *Drosophila melanogaster* transposable elements other than P elements in a P-M dysgenic cross. *Mol. Gen. Genet.* **208:** 506–510.
132. **Lim, J. K.** 1979. Site-specific instability in *Drosophila melanogaster*: the origin of the mutation and cytogenetic evidence for site specificity. *Genetics* **93:**681–701.
133. **Lim, J. K.** 1981. Site-specific intrachromosomal rearrangements in *Drosophila melanogaster*, cytogenetic evidence for transposable elements. *Cold Spring Harbor Symp. Quant. Biol.* **45:**553–560.
134. **Lim, J. K., M. J. Simmons, J. D. Raymond, N. M. Cox, R. F. Doll, and T. P. Culbert.** 1983. Homologue destabilization by a putative transposable element in *Drosophila melanogaster. Proc. Natl. Acad. Sci. USA* **80:**6624–6627.
135. **Mackay, T. F. C.** 1984. Jumping genes meet abdominal bristles: hybrid dysgenesis-induced quantitative variation in *Drosophila melanogaster. Genet. Res.* **44:**231–237.
136. **Mackay, T. F. C.** 1985. Transposable element-induced response to artificial selection in *Drosophila melanogaster. Genetics* **111:**351–374.
137. **Mackay, T. F. C.** 1986. Transposable element-induced fitness mutations in *Drosophila melanogaster. Genet. Res.* **48:**77–87.
138. **Mackay, T. F. C.** 1987. Transposable element-induced polygenic mutations in *Drosophila melanogaster. Genet. Res.* **49:**225–233.
139. **Margulies, L., D. I. Briscoe, and S. S. Wallace.** 1986. The relationship between radiation-induced and transposon-induced genetic damage in Drosophila oogenesis. *Mutat. Res.* **162:**55–68.
140. **Marsh, J. L., P. D. L. Gibbs, and P. M. Timmons.** 1985. Developmental control of transduced *dopa decarboxylase* genes in D. melanogaster. *Mol. Gen. Genet.* **198:**393–403.
141. **Martin, G., D. Wiernasz, and P. Schedl.** 1983. Evolution of Drosophila repetitive-dispersed DNA. *J. Mol. Evol.* **19:** 203–213.
142. **Matthews, K. A.** 1981. Developmental stages of genome elimination resulting in transmission ratio distortion of the T-007 male recombination (MR) chromosome of *Drosophila melanogaster. Genetics* **97:**95–111.
143. **Mattox, W. W., and N. Davidson.** 1984. Isolation and characterization of the Beadex locus of *Drosophila melanogaster*: a putative *cis*-acting negative regulatory element for the *heldup-a* gene. *Mol. Cell. Biol.* **4:**1343–1353.
144. **McElwain, M. C.** 1986. The absence of somatic effects of P-M hybrid dysgenesis in *Drosophila melanogaster. Genetics* **113:**897–918.
145. **Miller, D. W., and L. K. Miller.** 1982. A virus mutant with an insertion of a *copia*-like element. *Nature* (London) **299:** 562–564.
146. **Miller, L. H., R. K. Sakai, P. Romans, R. W. Gwadz, P. Kantoff, and H. G. Coon.** 1987. Stable integration and expression of a bacterial gene in the mosquito *Anopheles gambiae. Science* 237:779–781.
147. **Mizrokhi, L. J., L. A. Obolendova, A. F. Priimägi, Y. V. Ilyin, T. I. Gerasimova, and G. P. Georgiev.** 1985. The nature of unstable insertion mutations and reversions in the locus *cut* of D. melanogaster: molecular mechanism for transposon memory. *EMBO J.* **4:3781–3787.**
148. **Mukai, T.** 1964. The genetic structure of natural populations of *Drosophila melanogaster*. I. Spontaneous mutations rate of polygenes controlling viability. *Genetics* **50:**1–19.
149. **Mukai, T., M. Baba, M. Akiyama, N. Uowaki, S. Kusakabe, and F. Tajima.** 1985. Rapid change in mutation rate in a local population of *Drosophila melanogaster. Proc. Natl. Acad. Sci. USA* **82:**7671–7675.
150. **Mukai, T., and O. Yamaguchi.** 1974. The genetic structure of natural populations of *Drosophila melanogaster.* XI. Genetic variability in a local population. *Genetics* **76:**339–336.
151. **Niki, Y.** 1986. Germline autonomous sterility of P-M dysgenic hybrids and their application to germline transfers in *Drosophila melanogaster. Dev. Biol.* **113:**255–258.
152. **Niki, Y., and S. I. Chigusa.** 1986. Developmental analysis of the gonadal sterility of P-M hybrid dysgenesis in *Drosophila melanogaster. Jpn. J. Genet.* **61:**147–156.
153. **Nitasaka, E., T. Mukai, and T. Yamazaki.** 1987. Repressor of P elements in *Drosophila melanogaster*: cytotype determination by a defective P element with only open reading frames 0 through 2. *Proc. Natl. Acad. Sci. USA* **84:**7605–7608.

153a. **O'Brochta, D. A., and A. M. Handler.** 1988. Mobility of P elements in drosophilids and nondrosophilids. *Proc. Natl. Acad. Sci. USA* **85**:6052–6056.

154. **O'Hare, K.** 1985. The mechanism and control of P element transposition in *Drosophila melanogaster*. *Trends Genet.* **1**:250–254.

155. **O'Hare, K., and G. M. Rubin.** 1983. Structure of P transposable elements and their sites of insertion and excision in the *Drosophila melanogaster* genome. *Cell* **34**:25–35.

156. **Ohishi, K., E. Takanashi, and S. I. Chigusa.** 1982. Hybrid dysgenesis in natural populations of *Drosophila melanogaster* in Japan. I. Complete absence of the P factor in an island population. *Jpn. J. Genet.* **57**:423–428.

157. **Ohnishi, O.** 1977. Spontaneous and ethyl methanesulfonate-induced mutations controlling viability in *Drosophila melanogaster*. II. Homozygous effect of polygenic mutations. *Genetics* **87**:529–545.

157a. **O'Kane, C. J., and W. J. Gehring.** 1987. Detection *in situ* of genomic regulatory elements in Drosophila. *Proc. Natl. Acad. Sci. USA* **84**:9123–9127.

158. **Orr-Weaver, T. L., and A. C. Spradling.** 1986. Drosophila chorion gene amplification requires an upstream region regulating s18 transcription. *Mol. Cell. Biol.* **6**:4624–4633.

159. **Patterson, J. T.** 1943. The Drosophilidae of the Southwest, p. 7–216. *In* J. T. Patterson (ed.), *Studies in the Genetics of Drosophila*, vol. 3. University of Texas Publication.

160. **Picard, G., and P. L'Héritier.** 1971. A maternally inherited factor inducing sterility in *Drosophila melanogaster*. *Drosophila Inform. Serv.* **46**:54.

161. **Pirrotta, V., H. Steller, and M. P. Bozzetti.** 1985. Multiple upstream regulatory elements control the expression of the Drosophila *white* gene. *EMBO J.* **4**:3501–3508.

162. **Preston, C. R., and W. R. Engels.** 1984. Movement of P elements within a P strain. *Drosophila Inform. Serv.* **60**: 169–170.

162a. **Preston, C. R., and W. R. Engels.** 1989. Spread of P transposable elements in inbred lines of *Drosophila melanogaster*. *Prog. Nucleic Acid Res. Mol. Biol.*, in press.

163. **Raymond, J. D., T. R. Laverty, and M. J. Simmons.** 1986. Cytogenetic analysis of lethal X chromosomes derived from dysgenic hybrids of *Drosophila melanogaster*. *Drosophila Inform. Serv.* **63**:111–114.

164. **Rio, D. C., G. Barnes, F. A. Laski, J. Rine, and G. M. Rubin.** 1988. Evidence for Drosophila P element transposase activity in mammalian cells and yeast. *J. Mol. Biol.* **200**:411–415.

165. **Rio, D. C., F. A. Laski, and G. M. Rubin.** 1986. Identification and immunochemical analysis of biologically active Drosophila P element transposase. *Cell* **44**:21–32.

166. **Robertson, H. M., C. R. Preston, R. W. Phillis, D. Johnson-Schlitz, W. K. Benz, and W. R. Engels.** 1988. A stable genomic source of P element transposase in *Drosophila melanogaster*. *Genetics* **118**:461–470.

167. **Roiha, H., G. M. Rubin, and K. O'Hare.** 1988. P element insertions and rearrangements at the *singed* locus of *Drosophila melanogaster*. *Genetics* **119**:75–83.

168. **Ronsseray, S.** 1986. P-M system of hybrid dysgenesis in *Drosophila melanogaster*: thermic modifications of the cytotype can be detected for several generations. *Mol. Gen. Genet.* **205**:23–27.

169. **Ronsseray, S., and D. Anxolabéhère.** 1986. Chromosomal distribution of P and I transposable elements in a natural population of *Drosophila melanogaster*. *Chromosoma* **94**: 433–440.

170. **Ronsseray, S., D. Anxolabéhère, and G. Périquet.** 1984. Hybrid dysgenesis in *Drosophila melanogaster*: influence of temperature on cytotype determinations in the P-M system. *Mol. Gen. Genet.* **196**:17–23.

171. **Rosenfeld, A., A. Carpenter, and L. Sandler.** 1971. A non-chromosomal factor causing female sterility in *D. melanogaster*. *Drosophila Inform. Serv.* **47**:85.

172. **Rubin, G. M.** 1983. Dispersed repetitive DNAs in Drosophila, p. 329–361. *In* J. A. Shapiro (ed.), *Mobile Genetic Elements*. Academic Press, Inc. (London), Ltd., London.

173. **Rubin, G. M., M. G. Kidwell, and P. M. Bingham.** 1982. The molecular basis of P-M hybrid dysgenesis: the nature of induced mutations. *Cell* **29**:987–994.

174. **Rubin, G. M., and A. C. Spradling.** 1982. Genetic transformation of Drosophila with transposable element vectors. *Science* **218**:348–353.

175. **Rubin, G. M., and A. C. Spradling.** 1983. Vectors of P element mediated gene transfer in Drosophila. *Nucleic Acids Res.* **11**:6341–6351.

176. **Sakoyama, Y., T. Todo, S. I. Chigusa, T. Honjo, and S. Kondo.** 1985. Structures of defective P transposable elements prevalent in natural Q and Q-derived M strains of *Drosophila melanogaster*. *Proc. Natl. Acad. Sci. USA* **82**:6236–6239.

177. **Salz, H. K., T. W. Cline, and P. Schedl.** 1987. Functional changes associated with structural alterations induced by mobilization of a P element inserted in the Sex-lethal gene of Drosophila. *Genetics* **117**:221–231.

178. **Satta, Y., T. Gojobori, T. Maruyama, and S. I. Chigusa.** 1985. Tn3 resolvase-like sequence in P transposable element of *Drosophila melanogaster*. *Jpn. J. Genet.* **60**:261–266.

179. **Satta, Y., T. Gojobori, T. Maruyama, K. Saigo, and S. I. Chigusa.** 1985. Homology between P transposable element of *Drosophila melanogaster* and bacterial transposase gene of Tn3. *Jpn. J. Genet.* **60**:499–503.

180. **Scavarda, N. J., and D. L. Hartl.** 1984. Interspecific DNA transformation in Drosophila. *Proc. Natl. Acad. Sci. USA* **81**:7615–7619.

181. **Scavarda, N. J., and D. L. Hartl.** 1987. Germ line abnormalities in *Drosophila simulans* transfected with the transposable P element. *J. Genet.* **66**:1–15.

182. **Schaefer, R. E., M. G. Kidwell, and A. Fausto-Sterling.** 1979. Hybrid dysgenesis in *Drosophila melanogaster*: morphological and cytological studies of ovarian dysgenesis. *Genetics* **92**:1141–1152.

183. **Scholnick, S., B. Morgan, and J. Hirsh.** 1983. The cloned *dopa decarboxylase* gene is developmentally regulated when reintegrated into the Drosophila genome. *Cell* **34**:37–45.

184. **Searles, L. L., A. L. Greenleaf, W. E. Kemp, and R. A. Voelker.** 1986. Sites of P element insertion and structures of P element deletions in the 5′ region of *Drosophila melanogaster RpII215*. *Mol. Cell. Biol.* **6**:3312–3319.

185. **Searles, L. L., R. S. Jokerst, P. M. Bingham, R. A. Voelker, and A. L. Greenleaf.** 1982. Molecular cloning of sequences from a Drosophila RNA polymerase II locus by P element transposon tagging. *Cell* **31**:585–592.

186. **Searles, L. L., and R. A. Voelker.** 1986. Molecular characterization of the Drosophila *vermillion* locus and its suppressible alleles. *Proc. Natl. Acad. Sci. USA* **83**:404–408.

187. **Semeshin, V. F., E. S. Belyaeva, I. F. Zhimulev, J. T. Lis, G. Richards, and M. Bourouis.** 1986. Electron microscopical analysis of Drosophila polytene chromosomes. IV. Mapping of morphological structures appearing as a result of transformation of DNA sequences into chromosomes. *Chromosoma* **93**:461–468.

188. **Simmons, G.** 1986. Gonadal dysgenesis determinants in a

natural population of *Drosophila melanogaster*. *Genetics* 114:897–918.

189. **Simmons, G.** 1987. Sterility-mutability correlation. *Genet. Res.* 50:73–76.

190. **Simmons, M. J., and L. M. Bucholz.** 1985. Transposase titration in *Drosophila melanogaster*: a model of cytotype in the P-M system of hybrid dysgenesis. *Proc. Natl Acad. Sci. USA* **82**:8119–8123.

191. **Simmons, M. J., and J. F. Crow.** 1977. Mutations affecting fitness in Drosophila populations. *Annu. Rev. Genet.* **11**: 49–78.

192. **Simmons, M. J., N. A. Johnson, T. M. Fahey, S. M. Nellett, and J. D. Raymond.** 1980. High mutability in male hybrids of *Drosophila melanogaster*. *Genetics* **96**:479–490.

193. **Simmons, M. J., and R. E. Karess.** 1985. Molecular and population biology of hybrid dysgenesis. *Drosophila Inform. Serv.* **61**:2–7.

194. **Simmons, M. J., and J. K. Lim.** 1980. Site specificity of mutations arising in dysgenic hybrids of *Drosophila melanogaster*. *Proc. Natl. Acad. Sci. USA* 77:6042–6046.

195. **Simmons, M. J., J. D. Raymond, M. J. Boedigheimer, and J. R. Zunt.** 1987. The influence of nonautonomous P elements on hybrid dysgenesis in *Drosophila melanogaster*. *Genetics* **117**:671–685.

196. **Simmons, M. J., J. D. Raymond, T. Culbert, and T. Laverty.** 1984. Analysis of dysgenesis-induced lethal mutations on the X chromosome of a Q strain of *Drosophila melanogaster*. *Genetics* **107**:49–63.

197. **Simmons, M. J., J. D. Raymond, N. Johnson, and T. Fahey.** 1984. A comparison of mutation rates for specific loci and chromosome regions in dysgenic hybrid males of *Drosophila melanogaster*. *Genetics* **106**:85–94.

198. **Simmons, M. J., J. D. Raymond, T. R. Laverty, R. F. Doll, N. C. Raymond, G. J. Kocur, and E. A. Drier.** 1985. Chromosomal effects on mutability in the P-M system of hybrid dysgenesis of *Drosophila melanogaster*. *Genetics* **111**:869–884.

199. **Simon, J. A., C. A. Sutton, R. B. Lobell, R. L. Glaser, and J. T. Lis.** 1985. Determinants of heat shock-induced chromosome puffing. *Cell* **40**:805–817.

200. **Sinclair, D. A. R., and T. A. Grigliatti.** 1985. Investigation of the nature of P-induced male recombination in *Drosophila melanogaster*. *Genetics* **110**:257–279.

201. **Slatko, B. E.** 1978. Evidence for newly induced genetic activity responsible for male recombination induction in *Drosophila melanogaster*. *Genetics* **90**:105–124.

202. **Slatko, B. E., and M. M. Green.** 1980. Genetic instability in *Drosophila melanogaster*: mapping the mutator activity of an MR strain. *Biol. Zentralbl.* **99**:149–155.

203. **Slatko, B. E., and Y. Hiraizumi.** 1973. Mutation induction in the male recombination strains of *Drosophila melanogaster*. *Genetics* 75:643–649.

204. **Slatko, B. E., and Y. Hiraizumi.** 1975. Elements causing male crossing over in *Drosophila melanogaster*. *Genetics* 81:313–324.

205. **Slatko, B. E., J. M. Mason, and R. C. Woodruff.** 1984. The DNA transposition system of hybrid dysgenesis in *Drosophila melanogaster* can function despite defects in host DNA repair. *Genet. Res.* **43**:159–171.

205a. **Snyder, M., and W. F. Doolittle.** 1988. P elements in *Drosophila*: selection at many levels. *Trends Genet.* **4**: 147–149.

206. **Sobels, F. H., and J. C. J. Eeken.** 1981. Influence of the MR (mutator) factor on X-ray-induced genetic damage. *Mutat. Res.* **83**:201–206.

207. **Spradling, A. C.** 1986. P element-mediated transformation, p. 175–197. *In* D. B. Roberts (ed.), *Drosophila: a Practical Approach*. IRL Press, Oxford.

208. **Spradling, A. C., and G. M. Rubin.** 1981. Drosophila genome organization: conserved and dynamic aspects. *Annu. Rev. Genet.* **15**:219–264.

209. **Spradling, A. C., and G. M. Rubin.** 1982. Transposition of cloned P elements into Drosophila germ line chromosomes. *Science* **218**:341–347.

210. **Spradling, A. C., and G. M. Rubin.** 1983. The effect of chromosomal position on the expression of the Drosophila *xanthine dehydrogenase* gene. *Cell* **34**:47–57.

211. **Stacey, S. N., R. A. Lansman, H. W. Brock, and T. A. Grigliatti.** 1986. Distribution and conservation of mobile elements in the genus Drosophila. *Mol. Biol Evol.* **3**:522–534.

212. **Steller, H., and V. Pirrotta.** 1986. P transposons controlled by the heat shock promoter. *Mol. Cell. Biol.* **6**:1640–1649.

213. **Streck, R. D., J. E. MacGaffey, and S. K. Beckendorf.** 1986. The structure of *hobo* transposable elements and their insertion sites. *EMBO J.* **5**:3615–3623.

214. **Sved, J. A.** 1973. Short term heritable changes affecting viability in *Drosophila melanogaster*. *Nature* (London) **241**: 453–454.

215. **Sved, J. A.** 1976. Hybrid dysgenesis in *Drosophila melanogaster*: a possible explanation in terms of spatial organization of chromosomes. *Aust. J. Biol. Sci.* **29**:375–388.

216. **Sved, J. A.** 1978. Male recombination in dysgenic hybrids of *Drosophila melanogaster*: chromosome breakage or mitotic crossing over? *Aust. J. Biol. Sci.* **31**:303–309.

217. **Sved, J. A.** 1987. Hybrid dysgenesis in *Drosophila melanogaster*: evidence from sterility and Southern hybridization that P cytotype is not maintained in the absence of chromosomal P factors. *Genetics* **115**:121–127.

218. **Tansey, T., M. D. Mikus, M. Dumoulin, and R. Storti.** 1987. Transformation and rescue of a flightless Drosophila tropomyosin mutant. *EMBO J.* **6**:1375–1385.

219. **Temin, H. M., and W. R. Engels.** 1983. Movable genetic elements and evolution, p. 173–202. *In* J. W. Pollard (ed.), *Evolutionary Theory, Paths Into the Future*. J. Wiley & Sons, Inc., New York.

220. **Thompson, J. N., Jr., R. C. Woodruff, and G. B. Schaeffer.** 1978. An assay of somatic recombination lines of *Drosophila melanogaster*. *Genetics* **49**:77–80.

221. **Todo, T., Y. Sakoyama, S. I. Chigusa, A. Fukunaga, T. Honjo, and S. Kondo.** 1984. Polymorphism in distribution and structure of P elements in natural populations of *Drosophila melanogaster* in and around Japan. *Jpn. J. Genet.* 59:441–451.

221a. **Torkamanzehi, A., C. Moran, and F. W. Nicholas.** 1988. P element-induced mutation and quantitative variation in *Drosophila melanogaster*: lack of enhanced response to selection in lines derived from dysgenic crosses. *Genet. Res.* 51:231–238.

222. **Tsubota, S., M. Ashburner, and P. Schedl.** 1985. P-element-induced control mutations at the *r* gene of *Drosophila melanogaster*. *Mol. Cell. Biol.* **5**:2567–2574.

223. **Tsubota, S., and P. Schedl.** 1986. Hybrid dysgenesis-induced revertants of insertions at that 5′ end of the *rudimentary* gene in *Drosophila melanogaster*: transposon-induced control mutations. *Genetics* **114**:165–182.

224. **Voelker, R. A., A. L. Greenleaf, H. Gyurkovics, G. B. Wisely, S. Huang, and L. L. Searles.** 1984. Frequent imprecise excision among reversions of a P element-caused lethal mutation in Drosophila. *Genetics* **107**:279–294.

225. **Waddington, C. H.** 1942. Some developmental effects of X-rays in Drosophila. *J. Exp. Biol.* **19:**101–117.

226. **Waddle, R. R., and I. I. Oster.** 1974. Autosomal recombination in males of *Drosophila melanogaster* caused by a transmissible factor. *J. Genet.* **61:**177–183.

227. **Wakimoto, B., L. Kalfayan, and A. C. Spradling.** 1986. Developmentally regulated expression of Drosophila chorion genes introduced at diverse chromosomal positions. *J. Mol. Biol.* **187:**33–45.

227a. **Williams, J. A., S. S. Pappu, and J. B. Bell.** 1988. Molecular analysis of hybrid dysgenesis-induced derivatives of a P-element allele at the *vg* locus. *Mol. Cell. Biol.* **8:**1489–1497.

227b. **Williams, J. A., S. S. Pappu, and J. B. Bell.** 1988. Suppressible P element alleles of the *vestigial* locus in *Drosophila melanogaster*. *Mol. Gen. Genet.* **212:**370–374.

228. **Woodruff, R. C., J. L. Blount, and J. N. Thompson, Jr.** 1987. Hybrid dysgenesis in D. melanogaster is not a general release mechanism for DNA transpositions. *Science* **237:**1206–1207.

229. **Woodruff, R. C., B. Slatko, and J. N. Thompson, Jr.** 1978. Lack of an interchromosomal effect associated with spontaneous recombination in males of *Drosophila melanogaster*. *Ohio J. Sci.* **78:**310–317.

230. **Woodruff, R. C., and J. N. Thompson, Jr.** 1977. An analysis of spontaneous recombination in *Drosophila melanogaster* males: isolation and characterization of male recombination lines. *Heredity* **38:**291–307.

231. **Yamaguchi, O., and T. Mukai.** 1974. Variation of spontaneous occurrence rates of chromosomal aberrations in the second chromosomes of *Drosophila melanogaster*. *Genetics* **78:**1209–1221.

232. **Yamamoto, A., F. Hihara, and T. K. Watanabe.** 1984. Hybrid dysgenesis in *Drosophila melanogaster*: predominance of Q factor in Japanese populations and its change in the laboratory. *Genetica* **63:**71–77.

233. **Yannopoulos, G.** 1978. Progressive resistance against the male recombination factor 31.1 MRF acquired by *Drosophila melanogaster*. *Experientia* **34:**1000–1001.

234. **Yannopoulos, G.** 1978. Studies on male recombination in a Southern Greek *Drosophila melanogaster* population: (c) chromosomal abnormalities at male meiosis; (d) cytoplasmic factor responsible for the reciprocal cross effect. *Genet. Res.* **31:**187–196.

235. **Yannopoulos, G.** 1978. Studies on the sterility induced by the male recombination factor 31.1 MRF in *Drosophila melanogaster*. *Genet. Res.* **32:**239–247.

236. **Yannopoulos, G.** 1979. Ability of the male recombination factor 31.1 MRF to be transposed to another chromosome in *Drosophila melanogaster*. *Mol. Gen. Genet.* **176:**247–253.

237. **Yannopoulos, G., and M. Pelecanos.** 1977. Studies on male recombination in a Southern Greek *Drosophila melanogaster* population: (a) effect of temperature; (b) suppression of male recombination in reciprocal crosses. *Genet. Res.* **29:**231–238.

238. **Yannopoulos, G., N. Stamatis, M. Monastirioti, P. Hatzopoulos, and C. Louis.** 1987. *hobo* is responsible for the induction of hybrid dysgenesis by strains of *Drosophila melanogaster* bearing the male recombination factor 23.5MRF. *Cell* **49:**487–495.

239. **Yannopoulos, G., A. Zacharopoulou, and N. Stamatis.** 1982. Unstable chromosome rearrangements associated with male recombination in *Drosophila melanogaster*. *Mutat. Res.* **96:**41–51.

239a. **Yoshihara, M., E. Takasu-Ishikawa, Y. Sakai, Y. Hotta, and H. Okamoto.** 1988. Single P element insertional mutagenesis in *Drosophila melanogaster*. *Proc. Jpn. Acad. Sci.* **64:**172–176.

240. **Yukuhiro, K., K. Harada, and T. Mukai.** 1985. Viability mutations induced by the P elements in *Drosophila melanogaster*. *Jpn J. Genet.* **50:**531–537.

241. **Yurchenko, N. N., I. K. Zakharov, and M. D. Golubovsky.** 1984. Unstable alleles of the *singed* locus in *Drosophila melanogaster* with reference to a transposon marked with a visible mutation. *Mol. Gen. Genet.* **194:**279–285.

Chapter 17

Retrotransposons and the FB Transposon from *Drosophila melanogaster*

PAUL M. BINGHAM and ZUZANA ZACHAR

I. OBJECTIVES OF CHAPTER

Transposable elements of eucaryotes are a large polyphyletic group. This probably reflects the fact that these elements are genomic parasites and that the gene pools of sexual populations constitute very rich niches for such parasites. Here we review first the current understanding of retrotransposons of *Drosophila melanogaster*. Second, we review studies of a second group of transposons called the foldback or FB elements. In each case, we focus on the following issues: (i) mechanism of transposition, (ii) regulation of transposition, (iii) evolutionary origin, (iv) contemporary interactions with the host genome, and (v) mutagenic effects on the host genome and their potential evolutionary implications.

II. *DROSOPHILA* RETROTRANSPOSONS

A. Overview

Ten independent retrotransposon families have been extensively characterized in *D. melanogaster* (Fig. 1, Table 1). (These transposons were originally called "*copia*-like" elements for historical reasons; however, the more descriptive generic term retrotransposon is preferable; see below.) Each family is

Paul M. Bingham and Zuzana Zachar ■ Department of Biochemistry, State University of New York, Stony Brook, New York 11794.

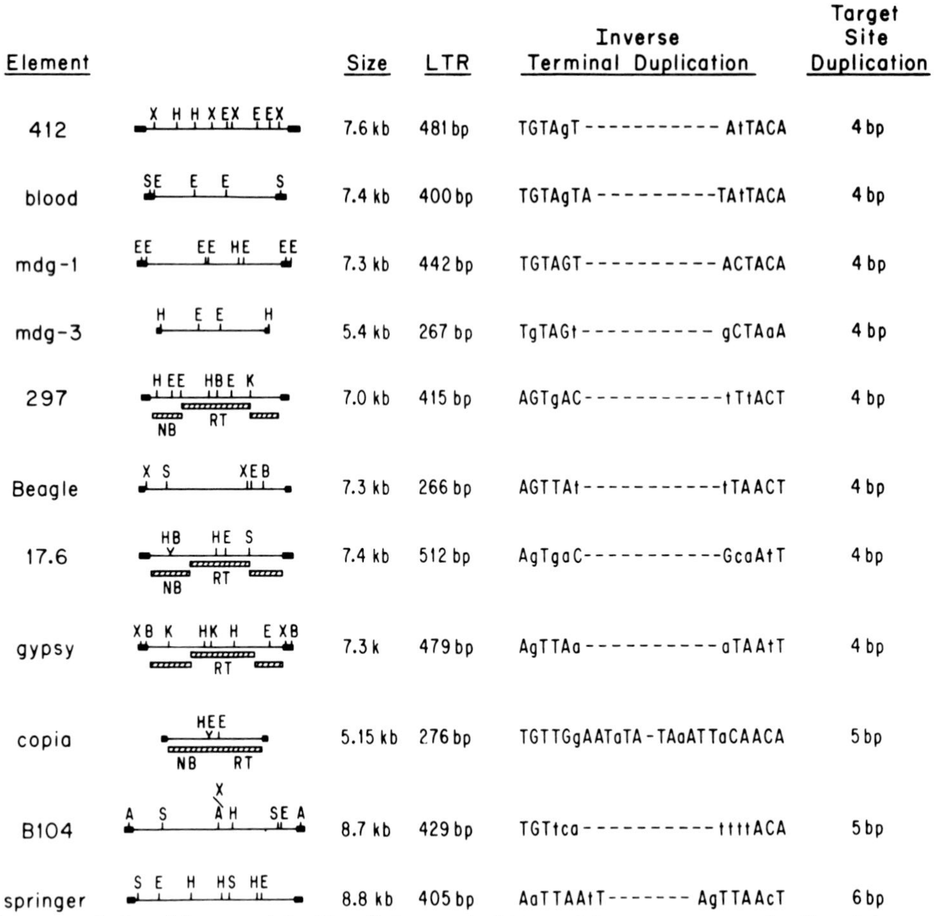

Element		Size	LTR	Inverse Terminal Duplication	Target Site Duplication
412	X H H XEX EEX	7.6 kb	481 bp	TGTAgT----------- AtTACA	4 bp
blood	SE E E S	7.4 kb	400 bp	TGTAgTA---------TAtTACA	4 bp
mdg-1	EE EE HE EE	7.3 kb	442 bp	TGTAGT---------- ACTACA	4 bp
mdg-3	H E E H	5.4 kb	267 bp	TgTAGt---------- gCTAaA	4 bp
297	H EE HBE K; NB RT	7.0 kb	415 bp	AGTgAC-----------tTtACT	4 bp
Beagle	X S XEB	7.3 kb	266 bp	AGTTAt-----------tTAACT	4 bp
17.6	HB HE S; NB RT	7.4 kb	512 bp	AgTgaC-----------GcaAtT	4 bp
gypsy	XB K HK H E XB; RT	7.3 k	479 bp	AgTTAa----------aTAAtT	4 bp
copia	HEE; NB RT	5.15 kb	276 bp	TGTTGgAATaTA-TAaATTaCAACA	5 bp
B104	A S X AH SE A	8.7 kb	429 bp	TGTtca----------ttttACA	5 bp
springer	S E H HS HE	8.8 kb	405 bp	AaTTAAtT------- AgTTAAcT	6 bp

Figure 1. Summary of selected features of the 10 well-characterized *Drosophila* retrotransposon families. Column 1 consists of the element family name followed immediately by a map of the cutting sites for selected restrictions enzymes. (Restriction enzymes: A, *Ava*I; B, *Bgl*II; E, *Eco*RI; H, *Hin*dIII; K, *Kpn*I; S, *Sal*I; X, *Xho*I.) The sizes of characterized copies of the element are given in column 2 and the sizes of LTRs in column 3. The sequences at the termini of characterized inserted copies of the element are given in column 4, and the number of bases of host sequence duplicated at insertion points for the elements is given in column 5. The approximate number of copies of each element per haploid genome in available fly populations is as follows: *412*, 40; *blood*, 10; *mdg-1*, 25; *mdg-3*, 15; *297*, 30; *Beagle*, 50; *17.6*, 40; *gypsy*, 10; *copia*, 60; *B104*, 80; and *springer*, 6. These copy number figures are quite similar in different populations for most of these elements (98; also see references in the text).

represented by ca. 10 to 80 copies per haploid genome. These copies are generally distributed individually and largely randomly throughout the host genome.

Evidence indicates that most dispersed, repeated *Drosophila* DNA sequences are transposons (e.g., 98). Dispersed repeated DNA sequences make up roughly 10% (1.6×10^7 base pairs [bp]) of the *Drosophila* genome (62). Well-characterized retrotransposon families (Fig. 1) make up a total of ca. 2.8×10^6 bp or ca. 18% of the presumed transposon complement in *D. melanogaster*. Moreover, because new retrotransposon families continue to be discovered (see below), these elements may actually constitute the majority of transposons in *D. melanogaster*.

B. General Features of Retrotransposon Structure and Transposition

Retrotransposons transpose by means of an RNA intermediate by a mechanism equivalent to that of retroviruses. Indeed, much of our current picture of retrotransposition is derived from extrapolation from the extensive studies of vertebrate retroviruses. Selected features of this life cycle are diagrammed in Fig. 2 and described below. The reader is also referred to the chapters in this volume by H. Varmus and P. Brown (retroviruses) and J. D. Boeke (yeast retrotransposons). *Drosophila* retrotransposons are ca. 5 to 9 kilobases (kb) in size. Their termini consist of direct duplications (200 to 500 bases in size) referred to as long terminal repeats (LTRs; Fig. 2).

Table 1. Selected structural features of *Drosophila* retrotransposons[a]

Retrotransposon family	Presumed tRNA primer-binding site	Presumed second (plus)-strand primer-binding site
Arg-tRNA[b]	TGGCGACCGTGACA	
mdg-1	]TGGCGACCGTGACAAAGG	AAAAGGAGGGAGA[
blood	]TGGCGACCGTGAGGCAGG	AAAAAAGAGGGGAGG[
412	]TGGCGACCGTGACAGTCG	AAAAGGAGGGAGA[
RSV[c]	]TTTGGTGACCCCGACGTGAT	AGGGAGGGGGAAA[
B104 (*roo*)	]TTTGGTCAATCG	
MSV[d]	]TTTGGGGGCTCG	
Lys-tRNA[b]	T GGCGCCCAAC	
mdg-4 (*gypsy*)	T] GGCGCCCAACCAAAATCT	GAGGGGGGAGT[
Beagle	T] GGCG	GGAG[
17.6	T] GGCGCAGTCGATGTGAT	AAGGGAAGGGA[
297	T] GGCGCAGTCGGTAGGAT	AAGGGAAGGGG[
Ser-tRNA[b]	T GGCGCAGTCGGTAGGAT	
mdg-3	]ATCTCAGAAGTGGGATTAA	AGGACGGCCGAGT[
copia	]GGTTATGGGCCCAGTCCAT	GAGGGGGCG[
tRNA[Met e]	GGTTATGGGCCCAG	

[a] Column 1 is a list of various *Drosophila* retrotransposons and a list of sequences for comparison to parts of those transposons. Column 2 shows the sequences of presumed *Drosophila* retrotransposon tRNA primer-binding sites and the sequences of sites or tRNAs to which they are to be compared. Column 3 shows the sequence of the presumed second (plus)-strand primer site (see Fig. 2 and text) for the *Drosophila* elements. The end of the 5′ LTR is indicated by] and the beginning of the 3′ LTR by [. References to these sequence results are given in the text.
[b] Complements of the 5′ portions of the indicated *Drosophila* tRNA.
[c] Rous sarcoma virus.
[d] Murine sarcoma virus.
[e] Complement of a portion of a *Drosophila* $tRNA^{Met}$.

The retrotransposition cycle is initiated when an RNA copy of a preexisting inserted transposon is made (Fig. 2). This transcript begins at the 5′ boundary of a subsegment of the 5′ LTR designated R and proceeds through the U5 portion of the 5′ LTR, the large central, nonrepetitive portion of the transposon, the U3 portion of the 3′ LTR, and finally the 3′ boundary of the R portion of the 3′ LTR, where it is terminated (polyadenylated) (Fig. 2). This RNA transcript is copied into double-stranded DNA by the enzyme reverse transcriptase, which is encoded by the transposon. Reverse transcription apparently occurs largely or entirely in a particle consisting of structural and nucleic acid-binding protein(s) encoded by the transposon and enzymatic activities, including reverse transcriptase and the RNA transcript itself. Two steps in this rather complex process are especially relevant to the discussion below. First, synthesis of the minus (first) DNA strand is primed by a 3′ hydroxyl group provided by a host tRNA that base pairs to the retrotransposon transcript immediately 3′ to the 5′ LTR (designated tRNA PBS in Fig. 2). The first DNA made is copied from the 5′ end of the RNA template (consisting of the U5 and R segments) (leftward in Fig. 2) and accumulates as the "first-strand strong-stop" species, apparently reflecting a pause in reverse transcription. First-strand synthesis is eventually continued, and synthesis of the plus (second) DNA strand proceeds from an initiation or primer site immediately 5′ to the 3′ LTR (rightward in Fig. 2). This second-strand synthesis is templated by the first DNA strand and proceeds initially to the 5′ end of this template. As a result, a short DNA segment consisting of the U3, R, and U5 segments plus a small portion of the tRNA primer accumulates as the "second-strand strong-stop" species, apparently also reflecting a pause before the completion of second-strand synthesis.

The end product of reverse transcription is a complete, double-stranded, linear copy of the transposon containing full-length terminal duplications (LTRs) (Varmus and Brown, this volume). This copy can be circularized by joining its ends (Fig. 2). The circular form of the transposon apparently can function as a substrate for integration at new sites in the host genome (Varmus and Brown, this volume; but also see Boeke, this volume). By analogy with retroviruses, integration is presumed to require specific small (4 to 12 bases) inverted repeats at the ends of the LTRs and specific transposon-encoded enzymatic functions (Varmus and Brown, this volume). Moreover, retrotransposon integration is associated with the duplication of four to six bases of host sequence at the insertion site. Such target site duplications presumably arise by staggered cleavage of the two strands of the target site. All insertions of members of

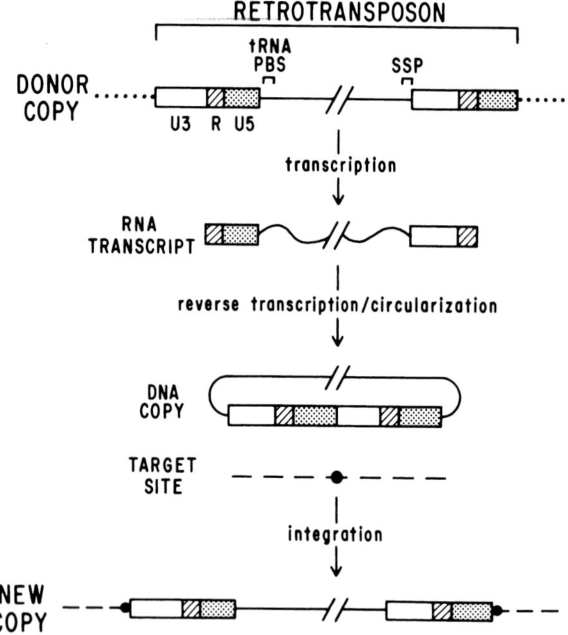

Figure 2. The retrotransposon life cycle. Duplicative transposition is achieved by the series of steps diagrammed. A previously existing (donor) copy is transcribed beginning in one of two identical LTRs (boxed portions of the diagram) and ending in the other. (Typical LTRs are 200 to 500 bases in length and the central, nonrepetitive portion of the transposon, symbolized by the solid line, is typically 4.5 to 9 kb in size.) This RNA transcript is used to template the synthesis of a double-stranded DNA copy by the transposon-coded reverse transcriptase through a complex series of steps described in the text. [The sites of the tRNA primer-binding site (tRNA PBS) and the second (plus)-strand primer site (SSP) used in this process are indicated, as are the U3, R, and U5 subsegments of the LTR.] This process results in the regeneration of two complete copies of the LTR sequence. This double-stranded DNA copy is cyclized by joining its ends and is integrated at a new site in the host genome. (Host sequences at the donor site are symbolized by a dotted line and at the site of the new copy by a dashed line.) *Drosophila* retrotransposons (like most transposons) duplicate small segments (typically four to seven bases; symbolized by the dot on the target site and new copy diagrams) of host material at the site of insertion (see reference 35 for a review of this issue).

the same retrotransposon family are associated with the same size target site duplication, presumably reflecting the cleavage specificity of the transposon-encoded integrase activity.

C. *Drosophila* Retrotransposons

Drosophila retrotransposons are recognized by virtue of structural and functional similarities to well-characterized vertebrate retroviruses. The complete DNA sequences of representative *copia* (34, 70), *17.6* (81), *297* (48), *412* (99), and *gypsy* (63) *Drosophila* retrotransposons have been determined. In addition, DNA sequences of the terminal portions of the following retrotransposon families have been determined: *mdg-1* (56), *B104* (82; also called *roo*), *Beagle* (90), *mdg-3* (69), and *blood* (7). Selected results of these studies are summarized in Fig. 1 and Table 1.

Similarities between these transposons and retroviruses are as follows. First, the ends of *Drosophila* retrotransposons consist of direct terminal repetitions similar in size and sequence to retroviral LTRs (Fig. 1). Retroviral LTRs contain characteristic sequences responsible for the initiation of transcription and polyadenylation defining the boundaries between the U3, R, and U5 portions of the LTRs (Fig. 2; also see Varmus and Brown, and Boeke, this volume). The 5′ and 3′ termini of the *copia* (34) and *B104* (82) retrotransposon transcripts have been determined and indicate that the U3, R, and U5 segments of *B104* are 380, 17, and 32 bp long, respectively; the U3, R, and U5 segments of *copia* are 140, 84, and 52 bp long, respectively. The positions of the polyadenylated termini of *B104* and *copia* transcripts reflect specific polyadenylation signals. The analysis of presumed first- and second-strand strong-stop species from the *mdg-1*, *mdg-2*, and *gypsy* (*mdg-4*) elements (2) indicates that these elements also have the expected R-U5 . . . U3-R structure (where the dots indicate the large central, nonrepetitive portion of the transposon). Detailed comparisons of sequences of the LTRs of *17.6*, *297*, *412*, *gypsy*, and *B104* with those of retroviruses indicate that these *Drosophila* elements also contain transcription start sites and polyadenylation signals in the expected positions (55, 63, 82, 96).

Second, all sequenced *Drosophila* retrotransposons (with the possible exception of *17.6*) have short, inverted terminal repeats similar to those at the ends of retroviral LTRs (Fig. 1). Moreover, most vertebrate retroviral LTRs contain the sequence TG at one terminus and the sequence CA at the other. TG and CA are also at the termini of a subset of *Drosophila* retrotransposons consisting of *412*, *blood*, *mdg-1*, *B104*, and *copia* (see Fig. 1 for details).

Third, each sequenced *Drosophila* retrotransposon (with the possible exception of *mdg-3*) contains a purine-rich stretch immediately 5′ to the 3′ LTR (on the plus strand; Table 1). This structure is present in all retroviruses and constitutes the primer site for second (plus)-strand synthesis (SSP in Fig. 2) (95; Varmus and Brown, this volume).

Fourth, most sequenced *Drosophila* retrotransposons contain a sequence related to the tRNA primer-binding sites of vertebrate retroviruses (Table 1). This sequence begins with bases TGG (complementary to CCA at the 3′ end of tRNAs) and is immediately 3′ to the 5′ LTR (Fig. 2, Table 1). This region of *blood*, *412*, and *mdg-1* retrotransposons is in each case very similar in sequence to the Rous sarcoma virus tRNA primer-binding site (7) (Table 1) and closely matches a *Drosophila* arginine-tRNA (7, 99) (Table 1). The presumed tRNA primer-binding site for the *B104* transposon is very similar to the corresponding site of Moloney sarcoma virus (82) (Table 1). *B104* also has the two additional nucleotides between the LTR sequence and the tRNA primer-binding site characteristic of most retroviruses (93, 95; Varmus and Brown, this volume). In the case of retroviruses, these two nucleotides are present in the initial DNA copy of the retroviral RNA but are eliminated during integration. This seems to be also true for *B104*. The absence of these extra nucleotides in most *Drosophila* retrotransposons suggests that no nucleotides are eliminated during the integration of the final double-stranded DNA product of their reverse transcription (Fig. 2; see below). The presumed tRNA primer-binding sites for *297*, *17.6*, *Beagle*, and *gypsy* extend one base into the 5′ LTR (46) (Table 1). The *297* primer-binding site has extensive homology to a *Drosophila* serine-tRNA (46), and the *gypsy* primer-binding site has extensive homology to a lysine-tRNA species (3, 63) (Table 1). Although the primer-binding sites for these elements are probably correctly identified, their positions initially seemed paradoxical: the entire LTR must be within the DNA segment primed by the tRNA (see above; Varmus and Brown, this volume). A simple resolution of this paradox is to assume that the terminal A of the tRNA is removed to produce the primer that actually is used. This is attractive as a strategy for dealing with charged tRNAs which occupy the primer sites but which, without processing (removal of the amino acid), would probably be unable to prime DNA synthesis. *copia* and *mdg-3* represent a third distinct subgroup of *Drosophila* retrotransposons in not having a conventional tRNA primer-binding site (Table 1). This seeming inconsistency was recently resolved in the case of *copia* by the demonstration that an internal portion of the *Drosophila* initiator $tRNA^{Met}$ is very likely used as a primer (54). Apparently, either a degradation product of $tRNA^{Met}$ initially binds to *copia* RNA or the entire tRNA binds and is cleaved during or before initiation of first (minus)-strand synthesis. It is conceivable that *mdg-3* first-strand priming occurs by an analogous strategy.

Fifth, the large blocks of coding sequence in the *17.6* (81), *copia* (34, 70), *gypsy* (63), *412* (99), and *297* (48) retrotransposons show extensive similarities to corresponding vertebrate retroviral sequences (Fig. 1). Specifically, retroviruses encode reverse transcriptase activities in the central portion, called the *pol* gene, of the plus strand. The plus strand of each of these five *Drosophila* retrotransposons also contains coding sequence with significant similarity to *pol* genes. Moreover, the 5′ portion of retroviral plus-strand sequences encodes the *gag* polyprotein whose components complex with the viral RNA, reverse transcriptase, and other components to make the core particle of the virus. Parts of the *gag* polyprotein can bind nucleic acids, and limited but probably significant similarities to these parts of *gag* are seen near the 5′ end of the plus strands of *copia*, *17.6*, and *297* (Fig. 1). Further, the 3′ portion of retroviral *pol* genes encodes an endonuclease implicated in integration into the host chromosome (the last step in Fig. 2) (33; Varmus and Brown, this volume). The 3′ part of the presumed *pol* genes of *17.6*, *297*, *412*, and *gypsy* appears to encode a similar endonuclease domain. *copia* also seems to encode such an endonuclease, but from a segment 5′ instead of 3′ to the presumed reverse transcriptase-coding sequence. (The yeast Ty retrotransposon is similar to *copia* in organization in this respect [Boeke, this volume], suggesting a relatively recent common ancestor to the Ty and *copia* elements [70].) Lastly, part of a plus-strand retroviral open reading frame between *gag* and *pol* encodes a protease activity that cleaves retroviral polyproteins into functional components (Varmus and Brown, this volume). A similar amino acid sequence is encoded by the corresponding portions of *17.6*, *297*, *gypsy*, and *copia*.

Sixth, extrachromosomal DNA copies of retroviruses are generated during the viral life cycle (Varmus and Brown, this volume) (Fig. 2). These include circular forms with one LTR, circular forms with two LTRs, and linear double-stranded DNAs. Extrachromosomal circular forms have been detected in *Drosophila* tissue culture cells for each of the following retrotransposons: *copia*, *412*, *gypsy*, *297*, and *mdg-1* (38, 69, 88). Most circular forms of *412* (88) and *mdg-1* and *mdg-3* (45) contain only a single LTR rather than the two LTRs expected of possible transposition intermediates (Fig. 2), whereas most circular forms of *copia* have the expected two-LTR structure (37). Extrachromosomal linear forms of *copia* have also been found (36, 37). The free *copia* linear forms have the same ends as integrated copies of *copia* and thus also correspond in structure to predicted transposition intermediates (54). Consistent with their proposed origin as reverse transcripts of *copia* mRNAs, synthesis of these linear DNAs continues when the major cellular DNA polymerase is inhibited with aphidicholine (36).

Seventh, retrovirus infection entails entry into the host cell of characteristic particles containing the viral RNA, parts of *gag* polyprotein, reverse transcriptase, and other proteins. *copia* RNA is found in particles (termed viruslike particles) resembling these intracellular viral particles in structure and apparently also containing reverse transcriptase activity (89).

Eighth, reverse transcription of retroviral RNAs results in characteristic "strong-stop" DNA species (see above; Varmus and Brown, this volume). The expected strong-stop species are observed in *Drosophila* tissue culture cells for the *mdg-1*, *mdg-3*, and *gypsy* (*mdg-4*) retrotransposons (2). Reverse transcription is also expected to produce characteristic larger intermediates consisting of single-stranded DNA or RNA-DNA heteroduplexes approximately coextensive with the full-length retrotransposon transcript. Molecules apparently of this form derived from *mdg-1* and *mdg-3* have been reported (1).

In summary, an extensive array of observations provides compelling evidence that *Drosophila copia*-like elements are retrotransposons, transposing by a process that is the intracellular equivalent of the life cycle of vertebrate retroviruses.

D. Origin and Relationships of *Drosophila* Retrotransposons

The origin of retrotransposons and retroviruses is currently largely conjectural. However, several observations in various *Drosophila* species have important implications.

Several drosophilid species have been characterized sufficiently that their evolutionary relationship to one another and to *D. melanogaster* is relatively clear (reference 58 and references therein). This, in turn, has allowed retrospective study of the presumptive origin of *Drosophila* retrotransposons (16, 29, 64). Sequences with homology to *copia* and *297* in Southern blots are present in the closely related species *D. simulans* and *D. mauritiana*, but are missing from the more distant species *D. yakuba* and *D. erecta*. In contrast, *412* is more widely represented; it is present in all closely related species tested (*D. simulans*, *D. mauritiana*, *D. yakuba*, and *D. erecta*) as well as in the more distantly related *D. pseudoobscura*.

These findings suggest that retrotransposons persist in the gene pools of drosophilid species for moderately long periods of time. *412* was apparently present in the ancient population ancestral to both the *D. melanogaster* and *D. pseudoobscura* species groups and persists in contemporary descendant species. *copia* and *297*, on the other hand, apparently invaded a more recent population ancestral to *D. melanogaster*, *D. mauritiana*, and *D. simulans* and persist in their contemporary descendants.

Additional inferences are less clear. While these data are simply interpreted to indicate that *copia* and *297* invaded the gene pool of the population immediately ancestral to *D. melanogaster*, *D. simulans*, and *D. mauritiana*, alternative interpretations exist. For example, retrotransposon persistence in a gene pool requires ongoing transposition which is potentially quite mutagenic to the element itself (e.g., 13). The rate of accumulation of base change in retrotransposons might be both substantially greater than that in the host genome and different for different retrotransposons. Thus, since studies of the presence of element families reviewed above depend on hybridization assays, hypothetical, diverged copies of *copia* and *297* in species more distantly related to *D. melanogaster* could have evaded detection.

In summary, evidence suggests that retrotransposons can persist in the gene pools of drosophilid species for at least moderately long periods of time and persist in at least some cases for very long periods of time. Given the relatively small number of retrotransposon families in *D. melanogaster* (Fig. 1; text above), it follows that the probability of a new retrotransposon arising in or entering (from unknown external sources) the gene pool of a drosophilid species is probably extremely low. Note that this is in contrast to the P element transposon, which has apparently invaded the *D. melanogaster* gene pool during the last several decades (W. R. Engels, this volume).

Information about evolutionary relationships also comes from detailed structural studies of retrotransposons reviewed above (Fig. 1, Table 1). The *297* and *17.6* retrotransposon families show extensive similarities in their protein-coding sequences (48). The *297*, *17.6*, *Beagle*, and *gypsy* (*mdg-4*) retrotransposon families all seem to be evolutionarily related, based on similarities between tRNA primer-binding sites, inverse terminal duplications, target site recognition (see below), and target site duplications (Fig. 1, Table 1). The same structural criteria indicate that *412*, *blood*, and *mdg-1* are evolutionarily related. Lastly, *copia*, *springer*, *B104*, and *mdg-3* are not obviously closely related to either of the two groups above or to one another.

In overview, studies of retrotransposons in drosophilid species contribute to the view that retrotransposons are ancient and further suggest that descendants of a single retrotransposon family can diverge into separate element families that can coexist in the same gene pool.

E. Regulation of Retrotransposon Transcription and Transposition

In the context of *Drosophila* transposons as germ line parasites, the question arises as to whether they are obligatory germ line parasites, having no infectious, viral form that allows efficient horizontal transmission. Though this issue remains to be resolved, the evidence suggests that most *Drosophila* transposons do not produce infectious particles. First, the *copia* transposon does not have the capacity to encode the equivalent of the retroviral *env* protein (95; Varmus and Brown, this volume) (Fig. 1). Second, while *297*, *17.6*, *gypsy*, and *412* have coding capacity in the 3′ portions of their plus strands sufficient for an *env*-like product (Fig. 1), the two elements (*297* and *17.6*) whose sequence in the region has been examined in detail do not appear to encode an *env*-like protein (48).

In view of these considerations, it is surprising that all known *Drosophila* retrotransposons are transcribed at relatively high levels in somatic tissues. These transcripts mostly correspond in structure to the full-length RNA copy of the generalized retrotransposon diagrammed in Fig. 2. (Minor classes of transcripts apparently resulting from processing of the full-length transcript are also seen in some cases.) Moreover, somatic retrotransposon transcription is developmentally programmed in patterns idiosyncratic to individual transposon families (74, 84). Further, all retrotransposons so far examined show extensive amplification in copy number (typically at least severalfold) in *Drosophila* tissue culture cell lines (reference 2 and references therein). Thus, transposition seems to occur in cultured cells, which are presumably equivalent to somatic cells. In view of these considerations, a simple interpretation is that *Drosophila* retrotransposons transpose promiscuously, i.e., in both the germ line and the soma. Note that this would be in contrast to the P element, which restricts its transposition to the germ line (10, 57; Engels, this volume).

The germ line copy number of some *Drosophila* retrotransposons (98) is remarkably constant (varying by less than twofold), indicating that copy number is controlled in some way in at least some cases. There is no direct information, however, as to how this is achieved. While autogenous regulation analogous to that of the P element (Engels, this volume) may exist, it is also conceivable that the host ultimately mediates copy number control. For example, it has been pointed out (68) that recombination between copies of a transposon at different sites (27, 40) could help maintain a low copy number. Such recombination events will frequently produce inviable gametes. (For example, they can generate reciprocal translocations, which often lead to dicentric and acentric chromosomes.) The probability of such recombination increases as an exponential function of copy number, rendering this process potentially very effective in controlling copy number. Moreover, these events might require elements of the *Drosophila* meiotic recombination machinery, thus occurring only in the germ line. The absence of meiotic recombination may account for the apparent loss of copy number control seen in the continuously cultivated somatic cells represented by tissue culture cell lines (see above).

F. Retrotransposons as Complex Insertional Mutators

A significant fraction of retrotransposon copies are in genetically active euchromatin but have no discernible effect on the host phenotype. For example, a typical wild-type laboratory strain has about 20 copies of *copia* transposon distributed among the euchromatic chromosome arms as assessed by in situ hybridization of salivary gland polytene chromosomes (9, 98). Further, genetically silent insertions in and around known genetic loci are commonly observed (e.g., 4, 51, 100). Nevertheless, retrotransposon insertion is responsible for a very large fraction of known spontaneous mutations in *Drosophila* (4, 100).

Drosophila retrotransposons are *polII* transcription units with associated transcript start site and terminus formation signals (see above). Moreover, these transcription units typically show developmentally programmed, somatic expression, indicating the presence of specialized promoter or enhancer elements (see above). Thus, retrotransposons can cause mutations both by simple insertional gene inactivation and by more complex effects on the expression of target host genes. Cases of inactivation of genetic regulatory elements by insertion include the following: *B104* insertion into an enhancerlike element causing the w^{sp1} allele at the *white* locus (reference 28 and references therein), the *Beagle* insertion into the TATA box of a larval cuticle protein gene (90), the *297* insertion into the TATA box of a histone gene (47), and the *gypsy* insertion in a *cis*-acting negative regulatory element, creating a *Beadex* allele at the *heldup* locus (65). Relatively few retrotransposon insertions in exon sequences of host genes are known, possibly as a result of experimental bias against full null alleles (see reference 100 for a discussion) or the tendency of at least some retrotransposons to insert in relatively AT-rich sequences (see below). The best-characterized case of exon insertion is the $w^{bd81b11}$ allele due to *copia* in the *white* locus (10, 71, 103). There is also at least one exon insertion of *B104* and at least one of *copia* among six characterized spontaneous *rosy* mutant alleles (22, 26). Other cases which could represent exon insertions are *B104* at *glued* (91) and *copia* and *gypsy* insertions in the Hw^{Ua} and Hw^{1} alleles, respectively (17).

A second common cause of mutation inactivation of host genes by retrotransposon insertion is transcription termination (transcript terminus formation). These effects are documented for the leaky mutant alleles at the *white* locus *white-apricot* (w^{a}), *white-buff* (w^{bf}), and *white-blood* (w^{bl}) due to insertions of *copia*, *B104*, and *blood*, respectively (7, 9, 11, 60, 71, 76, 100, 103). In each of these cases, the transposon is in a *white* intron and allows processing of that intron to occur in full-length transcripts. However, the inserted transposon also contains one or more polyadenylation signals, and primary transcripts of these alleles are thus of two types. Most are polyadenylated in transposon sequences, producing a truncated transcript encoding a truncated (inactive) polypeptide. The minority transcript is full-length, properly spliced, structurally wild-type, mature mRNA.

The w^{a} allele has been analyzed in detail and thus is of special interest as an example of this phenomenon (71, 76; see reference 103 for additional details). w^{a} results from the insertion of *copia* into the second *white* intron in the same transcriptional orientation as *white*. The majority of w^{a} primary transcripts are polyadenylated in (or very near) the 3′ *copia* LTR and are thus truncated. About 5% of w^{a} primary transcripts are spliced using the normal *white* splicing signals, thus generating wild-type *white* mRNA. (One or both of these alternative RNA processing events is influenced by a class of second-site suppressor loci discussed below.)

The w^{bl} (*white-blood*) allele results from the insertion of the *blood* element into the second *white* intron, which causes temperature-dependent effects on terminus formation or on the splicing of *white* transcripts (7).

Polyadenylated terminus formation for host gene transcripts in inserted retrotransposon sequences is also apparent in several other cases. The *copia* insertions in the $w^{hd81b11}$ (103) *white* mutant allele and the Hw^{Ua} (17) *Hairy-wing* mutant allele are in the opposite transcriptional orientation to *white* and *Hairy-wing*, respectively, and produce polyadenylated terminus formation early in transposon sequences (i.e., at a point near the 3′ end of the transposon transcription unit). The *B104* insertion at *glued* is in the same transcriptional orientation as the *glued* transcription unit and produces polyadenylated terminus formation of *glued* transcripts early in the transposon (presumably in the 5′ LTR) (91). The insertion of a retrotransposon in the *Drosophila* tropomyosin gene has a complex effect on the processing of primary transcripts of the allele (52). The precise mechanistic basis of their effect is unclear; however, developmentally regulated transcript terminus formation is a possibility. Given this sample of analyzed cases, it seems likely that termination effects are involved in producing mutant phenotypes in many retrotransposon insertion alleles.

A third class of effects have been seen for retrotransposon insertions in vertebrates (P. Soriano, T. Gridley, and R. Jaenisch, this volume) and *Saccharomyces cerevisiae* (reference 97 and references therein; Boeke, this volume). These result when transposon-borne enhancer or promoter elements influence the behavior of nearby host transcription start sites. Three possible cases of such effects are documented in *D. melanogaster*. The y^{2} (*yellow-two*) mutant allele at the *yellow* locus (18, 73) results from the insertion of the *gypsy* (*mdg-4*) retrotransposon about 800 bases 5′ to the *yellow* transcription unit. The *gypsy* and *yellow* transcription units are oriented in opposite directions, thus juxtaposing their 5′ regulatory regions. (This arrangement is commonly seen in analogous cases in *S. cerevisiae* and vertebrates.) The evidence that *gypsy* regulatory sequences are influencing expression of the *yellow* promoter in y^{2} is the capacity of a *gypsy*-specific suppressor locus [*su(Hw)*; see below for additional details] to influence phenotypic expression of y^{2} even though the *gypsy* insertion is well outside the *yellow* transcription unit (73). The $Antp^{NS}$ allele at the *Antennapedia* locus apparently results from an intron insertion of the *B104* retrotransposon (85). $Antp^{NS}$ is a dominant, "gain-of-function" mutant allele whose phenotype results from *Antennapedia* expression in tissues where it is normally silent. This suggests that *B104* is activating illicit expression of the *Antennapedia* promoter (85). The Hw^{1} allele results from *gypsy* insertion into the T5 transcription unit of the *achaete-scute* gene complex (17). Hw^{1} is a dominant, gain-of-function mutant allele and the Hw^{1} gypsy insertion leads to substantial overproduction of T5 transcripts (17). This T5 hyperexpression occurs at the same developmental stages (late larval and pupal) at which *gypsy* is heavily expressed, suggesting that the T5 promoter is being stimulated by *gypsy*-internal regulatory elements.

In addition to those reviewed above, there are a number of cases in which the basis for the effects of retrotransposon insertion on expression of the host remains to be established. Among these are *gypsy* insertions at several loci, including *bithorax* (4), *forked* (47), and others (67), and *412* insertions at *vermillion* (86).

Table 2. Sequences of retrotransposon insertion sites[a]

Retrotransposon family	Insertion site sequence[b]	Locus of insertion site	References
17.6	AAAGATATATATTTTTT	Histone	55
	TTATACATATACATATG	Unselected	47
	GAGTATATATATATTTA	Unselected	47
	TAATATATATATATAAT	Unselected	47
	ATACATATATACATATA	Unselected	47
	TCATGTATATGTATGTA	Unselected	47
Beagle	GATGATATATAAAC	Larval cuticle protein	90
297	ATTGCTATATAGTAGG	Histone	47
gypsy	TACATA	*scute*^{1}	39
	TACATA	*bithorax*^{3}	39
	TACATA	*bithorax*^{34e}	39
	TATATA	*forked*^{1}	63
	TATACA	Unselected	3
	TACACA	Unselected	3
412	TAAGGGCTTGTATCGC	Unselected	96
	CGGCATCTGGCCCAAC	Unselected	96
mdg-1	TGGAGTCGATAAAAAA	Unselected	56
	ATAT	Unselected	
blood	TCAATTGGTAATTGGA	*white*^{bl}	7
copia	TCTTAATAAAGGGTCCAA	*white*^{a}	71, 103
	AAGATGGCAACCATCTGC	*white*^{81b11}	71
	CCCTCATCATCCGTCCAA	Unselected	32
	GACTGAGTATTATTACCA	Unselected	32
	CTCTCGGCCAGAAATACT	Unselected	32
springer	TATATA	Tropomyosin	52
B104	TACGGCACCTTAGC	Unselected	82
	AGCATATTG	Unselected	82

[a] Host sequences immediately surrounding insertions of various *Drosophila* retrotransposons.
[b] Target site duplication is underlined.

G. Insertion Site Selection by *Drosophila* Retrotransposons

The details of insertion site selection by *Drosophila* transposons are poorly understood, with the possible exception of the P element (Engels, this volume). This problem is particularly severe as a majority of well-studied transposon insertions were isolated on the basis of specific mutant phenotypes. Thus, the role of experimental selection is generally impossible to assess. In spite of these limitations, several observations have informative implications.

First, the retrotransposons *17.6*, *297*, and *Beagle* appear to have a strong predisposition to insert in AT-rich sequences (47) (Table 2); in the case of *17.6*, six of six characterized insertions are associated with a target site duplication of the sequence ATAT (47). Presumably in part as a result of this preference for AT-rich insertion sites, several insertions of this group of elements are found to occur in TATA box promoter elements (47, 90). It is noteworthy that the *Drosophila* P element (Engels, this volume) tends to insert in or very near transcription start sites. This has been interpreted as suggesting a role for the structure of the target site chromatin complex in determining target site preference (Engels, this volume). Insertion of *17.6*, *Beagle*, and *297* in TATA promoter elements (see above) is consistent with a similar interpretation.

Second, the *gypsy* element appears to have a strong preference for insertion in the sequence TAYAYA (Y indicates a pyrimidine, underlining indicates the target site duplication) (39, 63).

Third, none of the remaining characterized *Drosophila* retrotransposons show an obvious preference for specific target site sequences (Table 2). However, all but *copia* and *B104* seem to have a preference for relatively AT-rich sequences (Table 2). We also note that two apparently independent *copia* insertions into the *white* gene occur at or very near the identical location (79). This suggests that *copia* insertion has quite specific insertion site requirements in spite of the heterogeneity of sequences at the actual insertion site.

Finally, individual transposon copies causing deleterious mutations by insertion are eliminated from the host gene pool by the same selective forces eliminating the mutation. As a result, there may be some selection for transposons that tend to insert in genetically silent regions, and strong insertion site preferences may thus exist. It is noteworthy that protein-coding sequences in *D. melanogaster* have a lower AT content than noncoding sequences. Thus, the apparent tendency for at least some *Drosophila* transposons to insert in AT-rich sequences may represent an artifact leading to rare insertion into coding sequences with concomitantly low probability of producing a deleterious mutation by insertion.

H. Host Regulatory Loci Whose Information Is Parasitized by Retrotransposons

The developmentally programmed, somatic transcription of *Drosophila* retrotransposons implies that these elements parasitize host regulatory information. Retrotransposons are thus attractive for the analysis of host regulatory machinery.

In practice, retrotransposon transcription units offer particular advantages for such studies. As noted above, retrotransposon insertion causes a large fraction of spontaneous mutations in *D. melanogaster* (4, 67, 73, 100). In a subset of these mutations, expression of the host gene and that of the inserted retrotransposon are coupled. These "reporter" alleles allow genetic screens for host genes that regulate retrotransposon expression (67, 72, 73, 103) (see discussion of w^a and y^2 above). Among host genes so identified are allele-specific suppressors and enhancers that interact quite specifically with individual retrotransposon families (67, 72, 73, 103). (Note that these *Drosophila* loci are phenomenologically similar to the *Spt* loci in *Saccharomyces cerevisiae* [Boeke, this volume].)

The strict allele specificity of these suppressors and enhancers argues that they do not influence phenotypes of the modified alleles for generalized or trivial reasons. Rather, this specificity indicates that these suppressors and enhancers are likely to be involved in processes idiosyncratic to individual retrotransposons, which may include developmental programming of expression. Several allele-specific suppressor loci are currently being analyzed in detail in several laboratories.

The *suppressor-of-white-apricot* [*su*(w^a)] locus suppresses (makes more nearly wild type) the phenotype of the w^a *copia* insertion allele at *white* (see above; see reference 103 and references therein). Detailed studies of the *su*(w^a) locus (21, 102, 104) indicate that *su*(w^a) encodes a relatively large polypeptide (ca. 105 kilodaltons) containing an unusual domain rich in arginine and serine amino acids. A similar domain is found in the protamine class of nucleic acid-binding proteins (21). Production of this polypeptide is apparently autogenously regulated at the level of splicing. Specifically, the *su*(w^a) protein appears to block removal of the first of seven introns from its primary transcript, thus preventing formation of the mature mRNA coding for it (102; see reference 8 for a review). *su*(w^a) autoregulation at the level of RNA processing is congruent with the observation that its *trans* effects on the w^a allele apparently involve processing of the w^a primary transcript (polyadenylation or splicing) (103).

Recessive mutant alleles at the *suppressor-of-Hairy-wing* [*su(Hw)*] locus suppress *gypsy* insertion alleles at *bithorax*, *yellow*, *Hairy-wing*, *scute*, *diminutive*, *cut*, *lozenge*, *forked*, *Beadex*, and *hairy* loci (see reference 74 for a review). *su(Hw)* has recently been cloned by V. Corces and colleagues (personal communication). Recessive mutant alleles at the *suppressor-of-sable* locus [*su(s)*] suppress *412* insertion alleles at *purple*, *speck*, and *vermillion* loci (20, 86). *su(s)* has recently been cloned (20). Recessive mutant alleles at the *suppressor-of-forked* [*su(f)*] locus suppress *gypsy* insertions at the *forked* and *lozenge* loci, also suppressed by mutant alleles at *su(Hw)*, and enhance (make more mutant) the w^a *copia* insertion allele at *white* (see reference 74 for a recent review). *su(f)* has recently been cloned by K. O'Hare and colleagues (personal communication).

The partial overlap between the allele specificities of *su*(w^a) and *su(f)* on one hand (both interact with the w^a allele [42]) and *su(f)* and *su(Hw)* on the other (suppress overlapping subsets of *gypsy* insertion alleles; see reference 74 for additional details) suggests that allele-specific suppressor loci represent components of the same or related circuitry. The function of this circuitry in the host is obscure; however, the genetic analysis of *su(f)* indicates that it is an essential gene, suggesting important host function (31, 80). Vertebrate retroviruses are sensitive to and presumably monitor the host cell cycle (see reference 95 and references therein). Thus, one attractive speculation for the role of allele-specific suppressor loci is in cell cycle regulation.

I. Retrotransposons, Genetic Instability, and Chromosome Rearrangement

Retrotransposon insertion does not result in the high level of genetic instability that is associated with some other classes of transposons such as FB (see below) and P (Engels, this volume). Indeed, many of the relatively stable mutations originally used by Morgan and his colleagues in establishing Mendelian-Morganian classical genetics have been shown to result from retrotransposon insertion (4, 67, 73,

100). As a result of this relative stability, only a few cases of secondary mutation associated with retrotransposon-insertion alleles have been examined.

First, "LTR excisions" are thought to occur by homologous recombination between the 5′ and 3′ LTRs of a complete element, resulting in a single LTR in place of the original complete element. LTR excisions are known to occur at Ty insertions in *S. cerevisiae* (Boeke, this volume) and in various vertebrate retroviral insertions (87, 94, 97). LTR excisions in *D. melanogaster* spp. have been found as partial revertants of the w^a *copia* insertion allele at the *white* locus (19, 103) and probably occur at other loci as well (18). While the rates of LTR excision have not been carefully measured, they are clearly very low under most circumstances, of the order of 1 per 10^5 gametes or lower (e.g., 103).

Second, nonhomologously placed retrotransposon copies interact apparently by recombination to duplicate, delete, or invert the intervening host segment. (Analogous events are well documented for the *S. cerevisiae* Ty element; Boeke, this volume.) Two examples have been characterized in *D. melanogaster* (27, 41). In both studies, small, local deletions and tandem duplications involving portions of the *white* locus were shown to result from apparently homologous recombination between two copies of the same retrotransposon at different locations and occurred at frequencies of about 1 in 10^5 to 10^4 gametes (27, 41). A case probably involving a similar mechanism is the small inversion associated with the $Antp^{73b}$ mutation (53, 83). It is noteworthy that all three of these events involve transposons relatively near one another (separated by only several hundred kilobases). Whether this represents a bias in the occurrence of these mutations or selective elimination of rearrangements involving larger segments of the host genome (presumably much more deleterious, on average) remains to be clarified.

In summary, secondary mutational events engendered by preexisting retrotransposons appear to result from homologous recombination involving transposon sequences. Moreover, retrotransposition involves the donor transposon only as a transcription template (see above; Fig. 2). Retrotransposons probably do not encode functions that mediate such recombination events. Further, such events occur at rates one or more orders of magnitude lower than those associated with elements like FB (below) and P (Engels, this volume). Thus, the results are most simply interpreted as indicating that integrated retrotransposons participate in the rearrangement of contiguous host sequences only passively, representing dispersed homologous sequences which are occasionally acted on by host recombination/repair machinery to produce inadvertent rearrangement.

III. THE FB TRANSPOSON FAMILY

A. Overview

Drosophila FB transposons constitute a single, heterogeneous family. The properties of FB elements indicate that they transpose by a mechanism quite unrelated to that of retrotransposons (see below). In this section, we review (i) the structure of FB transposons, (ii) the inferences that can be drawn about the mechanism and control of FB transposition, (iii) the ability of FB elements to cause rearrangement of host chromosomes, including transposition of large host chromosomal segments, and (iv) the capacity of FB-mediated chromosomal rearrangement to create unexpectedly complex new patterns of host gene expression.

B. Structure of FB Transposons

The termini of FB transposons are large inverse repetitions (92) (Fig. 3). These inversely repeated terminal segments have a complex, internally repetitious structure (77, 78, 92), which is best illustrated by the fully sequenced *FB4* element (77). Within the ca. 220 bases at each end of the *FB4* transposon, a 10-base sequence is found repeated several times. Proceeding inward, one next encounters a ca. 375-base segment containing several imperfect copies of a 20-base repeating sequence which subsumes the more distal 10-base repeat. Proceeding further inward, one next encounters a ca. 500-base sequence consisting entirely of a 31-base tandemly repeated sequence subsuming the more distal 20-base repeat.

The terminal repeats of FB elements vary from several hundred base pairs to several kilobases (59, 77, 78, 92), largely because of variation in the number of copies of the 31-bp repeats (59, 77, 78, 92). Otherwise, the structures of the inversely symmetric ends of FB are strongly (though not precisely) conserved.

The central portions of FB elements (sequences between the inverse repeats) are heterogeneous (15, 59, 77, 78, 92) (Fig. 3). Many elements have little or no additional sequence information in their central portions; such FBs consist largely or entirely of two copies (often of different lengths) of the inverse repeats (59, 92) (Fig. 3). However, several FBs have been analyzed that carry several kilobases of additional sequence in their central portions (Fig. 3). In one case, *FB4*, there is ca. 1.7 kb of additional DNA that is not normally associated with an FB element and probably represents a second transposon (called

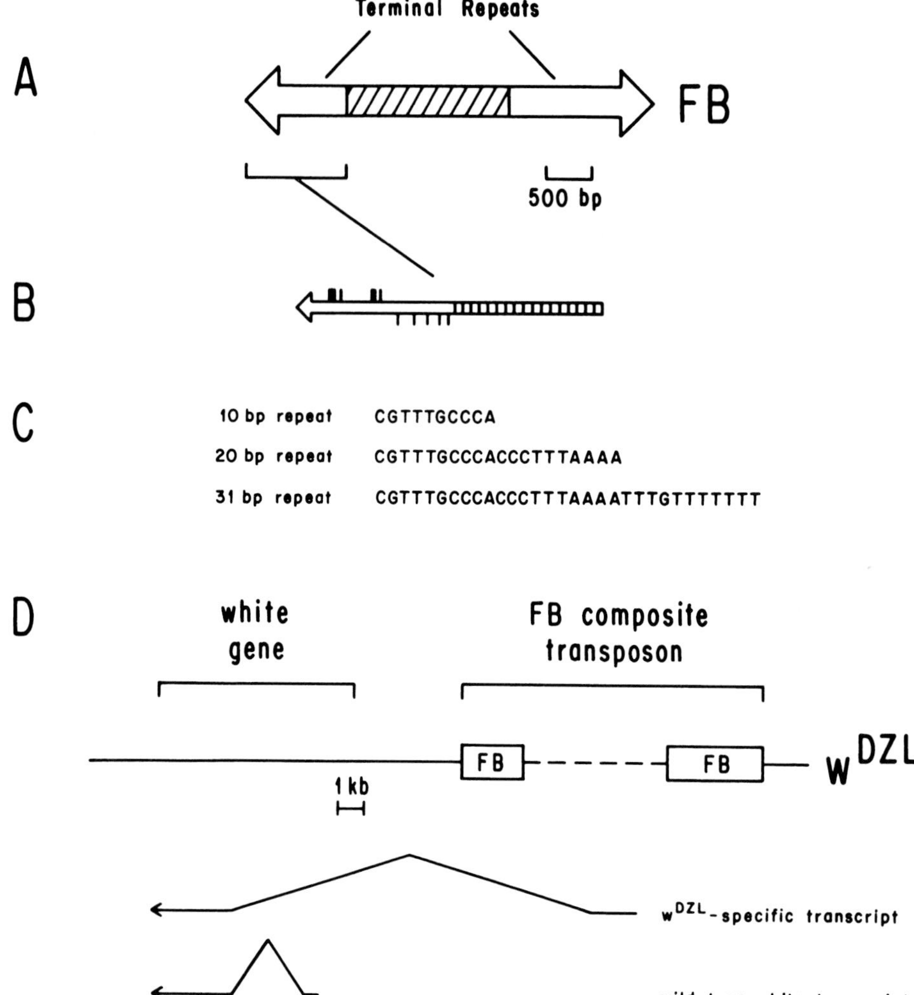

Figure 3. Structures of an FB and an FB composite transposon. (A) Structure of a generalized FB transposon. Its ends are two inversely repeated sequences (open arrows) bounding a central, nonrepetitive segment (hatched portion). The terminal repeat segments are internally repetitive as diagrammed in the expanded view in panel B. This internal repetition consists of several copies of a 10-base repeat (marks above the arrow) near the end of the repeat, which is first expanded to a 20-base repeat (marks below the arrow) and then a 31-base repeat (marks inside the arrow) as one approaches the central portion of the transposon. (C) Sequences of the 10-, 20-, and 31-bp repeats. The sequences given are typical, though individual copies sometimes differ slightly from one another (77). The overall length of an FB terminal repeat segment is highly variable, resulting (largely) from variation in the number of 31-base repeat segments. The central, nonrepetitive segment is also highly variable in size (from ca. 4.5 kb to less than 100 bases [77; also see text]). (D) Structure of the w^{DZL} mutant allele at the *white* locus. This allele results from the insertion of a composite FB transposon next to *white* (reference 12 and references therein). This composite consists of a central segment (dashed line) of *Drosophila* DNA (the unique segment of the transposon) normally present in one copy per haploid genome on another chromosome. This unique segment is bounded by two members of the FB transposon family. This entire composite transposon was inserted at *white* in a single event. A normally silent promoter in the unique segment is activated in specific tissues when juxtaposed to *white*, leading to the production of a mature transcript containing both unique-segment and *white* sequences as diagrammed. Also diagrammed is the wild-type *white* transcript. For both transcripts horizontal lines represent exon sequences and diagonal lines introns. (Additional introns in the extreme 3′ portions of both transcripts are omitted for simplicity.)

HB) as evidenced by its structure (terminal inverse repetitions) and its status as a dispersed repeated sequence (15, 78). Thus, *FB4* has a "transposon-within-a-transposon" structure. Given that HB and FB elements can apparently exist separately, their association in *FB4* is presumably fortuitous. In a second case, in three independently isolated FB elements, each contained a central segment of ca. 4 kb between the two inverse repeats (59, 75; N. Smyth-Templeton and S. S. Potter, personal communication). This 4-kb central segment is represented several times per haploid genome, always in association with the terminal, inversely repeated FB sequences (15, 59, 75, 77, 78, 92, 100). Sequence analysis of the 4-kb central segment of one of these FB elements reveals three long open reading frames, suggesting substantial protein-coding capability (Smyth-Templeton and Potter, personal communication).

In summary, the FB transposon family has an unusually complex organization. As a result, though substantial effort has been invested in its characterization, several important basic issues of structure and function remain to be resolved. However, at this writing, the following scenario is reasonable and, arguably, likely. The original or parental FB element was structurally similar to the 4-kb central-segment elements. The inversely repeated terminal sequences are presumably involved in transposition. The 4-kb central segment may code for one or more polypeptide products necessary to allow or regulate FB transposition. FB transposition (like P element transposition; Engels, this volume) might be highly error prone, leading to the frequent production of deletion variants of FB elements which constitute the majority of FB elements in the FB family. Such deleted elements would lack element-coded peptides, and their transposition would be dependent on FB elements having intact peptide-coding sequences. Only those deleted elements capable of being mobilized in *trans* by a functional FB element would persist in the gene pool. Note that this scenario is quite analogous to the functional, structural, and historical organization of the better understood P element transposon family (Engels, this volume).

C. Mechanism of FB Transposition

Though the detailed mechanism of FB transposition remains obscure, several general comments can be made. First, FB transposition is very unlikely to involve RNA intermediates; FB structure is unlike that known or expected of transposons using RNA intermediates (see above; see also D. J. Finnegan, this volume, Chapter 18). In contrast, the presence of inversely repeated terminal segments is reminiscent of the P element transposon (Engels, this volume) and of bacterial insertion sequences and drug resistance transposons (N. Kleckner, this volume).

Second, FB transposition is likely to involve DNA strand scission events at the boundaries of the donor transposon copy and/or transposon-internal recombination events. This is an inference from the high frequency of FB-mediated chromosome rearrangement events involving, at least in part, previously existing FB copies (see below). Specifically, such high-frequency rearrangement events suggest participation of transposon activities (for example, transposases or resolvases). In one such scenario, a resolvaselike activity would produce host chromosome rearrangements as a result of inappropriate recombination or failure of recombination between transposon copies. In a second scenario, rearrangements could result from the host repair of inadvertent chromosome breaks produced by a transposaselike activity acting on the termini of preexisting FB elements.

Third, FB insertion is associated with a target site duplication of nine bases. No obvious site specificity for insertion emerges from comparing the three available cases (71, 92).

D. Control of FB Transposition

Control of FB transposition, like mechanism, is poorly understood. Two related observations make up the total of available information. First, mutable FB insertion alleles at the *white* locus (see below), w^{c} (M. M. Green, personal communication) and w^{DZL} (unpublished results), initially grew more stable during passage in laboratory culture (involving extensive de facto inbreeding) in the months immediately after their isolation. While both alleles are still comparatively quite unstable, mutation rates of both have declined by roughly an order of magnitude. This establishment of a more stable condition during inbreeding is reminiscent of P element transposition (Engels, this volume), in turn suggesting some sort of active regulation of FB transposition.

Second, suggestive evidence exists that high mutability can be restored to w^{DZL} stocks partially stabilized in long-term culture by outcrossing to other laboratory strains (unpublished results). These results are reminiscent of the outcross effects involved in the activation of the P (Engels, this volume) and I (Finnegan, this volume, Chapter 18) elements; however, the studies of FB elements have been complicated by unidentified factors leading to large variation in the magnitude of the effect and must be regarded as no more than suggestive.

In summary, limited, suggestive evidence exists that FB transposition is actively regulated. However, this regulation does not appear to be as complete as the control of the P element (Engels, this volume), for FB retains substantial capacity to promote rearrangement under all known conditions.

E. FB-Mediated Simple Chromosome Rearrangement

Our understanding of FB-mediated chromosome rearrangement derives from the analysis of "genetic instabilities" that have proven on subsequent molecular analysis to correlate with the presence of an FB element. Such unstable alleles present themselves by producing new mutant derivatives spontaneously at high frequencies. (The elevated spontaneous mutant frequencies are of the order of one new derivative per 100 to 1,000 chromosomes under optimal conditions. Most of these new variants are FB mediated; see below.)

Evidence indicates that deletions, inversions, and reciprocal translocations with one or both breakpoints at sites of preexisting FB insertions arise very frequently. This evidence comes largely from two groups of studies. First, various classes of spontaneous derivatives of the unstable w^c allele have been characterized genetically and cytogenetically (43, 44). Some of these proved to be deletions beginning at or near the locus of the w^c mutation (*white*). Subsequent molecular analysis showed w^c to result from the insertion of an FB copy into the *white* locus (23, 59; see above). (The w^c FB insertion actually occurred in one of two copies of a tandemly duplicated *white* segment present in the w^i allele parental to w^c [23, 59].) Molecular studies further showed that a series of spontaneous derivatives of w^c apparently resulted from recombination between two previously existing FB copies (25). (Reversions of w^c to wild type or to the parental w^i allele also occur at high rates; however, the presence of tandemly duplicated copies of the site into which the FB element is inserted in this allele renders the interpretation of these excisions as "simple" transposon-mediated events ambiguous [24].)

Second, the unstable w^{DZL} allele produces deletions, inversions, and reciprocal translocations with one breakpoint at or near the locus of the w^{DZL} mutation (*white*) within the resolution of cytogenetic analysis (5, 6). Subsequent molecular analysis showed that the w^{DZL} allele results from the insertion of a composite FB transposon immediately to the right of the *white* locus (12, 59, 100) and that breakpoints for several w^{DZL}-generated rearrangements mapped to the site of insertion of this transposon (100; unpublished observations). Moreover, the very frequent reversion of w^{DZL} by the deletion of the central portion of the composite w^{DZL} transposon (see below and Fig. 3 for details) is interpretable as the deletion of segments bounded by two FB elements through recombination between the elements. (A subsequent study [66] showed that a highly selected subset of deletions from the w^{DZL} strain have breakpoints at sites other than that of the w^{DZL} transposon insertion. While the significance of these "anomalous" breaks remains to be defined, it seems likely that they result from a minority deletion-generating process not directly related to the w^{DZL} transposon.)

In summary, evidence shows that FB elements can participate in generating rearrangements of host chromosomal sequences. Rearrangements involving individual FB elements occur with frequencies of about 1 per 1,000 chromosomes. This high frequency suggests that FB genetic functions are involved in catalyzing or stimulating these rearrangements.

F. FB Elements Form Complex, Composite Transposons

In addition to the chromosome rearrangements described above, FB elements are able to promote an additional, especially interesting class of chromosome rearrangements. These are insertional translocations resulting from the transposition of FB composite transposons. Such composite transposons consist of a segment of normally unique *Drosophila* DNA flanked by two FB elements (Fig. 3). In a general sense, these composites have structures analogous to the composite bacterial drug resistance transposons (Kleckner, this volume).

Two cases of this have been characterized in extensive molecular detail. First is the related family of FB composite transposons designated TEs (49, 50). TEs arose in a particular strain and in most cases carry the closely linked *white* and *roughest* genes on chromosomal segments of several hundred kilobases. Progeny carrying such transposed *white* and *roughest* genes are observed at frequencies of ca. 1 per 10,000. Several have been characterized at the molecular level (40, 70), and the results have several implications. (i) TEs consist of host chromosomal DNA segments mobilized by FB elements flanking the segment. (ii) The size and endpoints of transposable chromosomal segments are heterogeneous. This suggests that different, preexisting FBs are used for different individual transposition events. This, in turn, suggests that the strain producing the TEs was unusual not simply by virtue of the presence of a particular FB element or pair of FB elements around *white* and *roughest*, but rather, in part, by virtue of some characteristic (as yet unknown) that allows FB elements to be unusually active. (ii) The FB elements at the termini of the characterized TEs are themselves heterogeneous. Represented are both the 4-kb central-segment FB element and deleted FB elements (40; see above).

Second is the w^{DZL} transposon. This transposon consists of a segment of ca. 6.5 kb of normally unique *Drosophila* DNA (referred to below as the unique segment) flanked by two FB elements (59, 100) (Fig. 3). The FB elements at the termini of the w^{DZL} transposon are about 2 and 4 kb in size, respectively, and lack most of the 4-kb central segment (59; see above for discussion). The w^{DZL} allele arose in an available inbred stock (5). This parental w^+ allele has no additional sequences at the point of insertion of the w^{DZL} transposon, indicating that the w^{DZL} composite transposon inserted into *white* in a single event (61, 100). Thus, the composite w^{DZL} transposon can apparently transpose as a unit to new sites. Each FB element of this composite transposon can participate in generating deletions of host sequences flanking the transposon (100), and the two FB elements participate in deleting the unique segment of the transposon at very high frequency through a process interpretable as recombination between FB elements (59, 100).

Composite transposons are attractive candidates for central players in the evolution of genomes of host organisms. The insertion of a composite transposon juxtaposes two unrelated host segments, potentially creating new combinations of genetic elements. The w^{DZL} allele represents the clearest example of this phenomenon (12, 101) (Fig. 3). The unique segment of the w^{DZL} transposon carries a promoter that is silent throughout late larval, pupal, and adult life in its original location on the second chromosome. However, when juxtaposed to *white*, this promoter is activated in a specific tissue (apparently eye pigment cells). This activation probably requires *white* locus genetic elements and is associated with the failure to activate efficiently the normal *white* promoter in this same tissue. This activation of the transposon unique-segment promoter results in a primary transcript beginning in the transposon and extending out of the transposon and through the *white* transcription unit (Fig. 3). A 5′ splice site in the unique segment interacts with the 3′ splice site at the beginning of the second *white* exon to produce a composite mature transcript in which the first exon is derived from transposon sequences and the remaining exons are the second through last *white* exons (Fig. 3). In other tissues in w^{DZL} individuals, the transposon promoter is not activated and the normal *white* promoter functions as in w^+ individuals.

These results have two important, related implications. First, regulatory genetic elements can interact over long distances and in unexpectedly complex ways. This allows novel patterns of gene expression to be produced by single mutational events as in the w^{DZL} case. Second, the efficient formation of the composite w^{DZL} transcript involving both transposon unique-segment and *white* sequences demonstrates that the presence of the FB transposon in an intron does not prevent splicing. It follows that the insertion of composite FB transposons into introns of transcription units will sometimes result in the introduction of one or more exons from transposon sequences into the mature transcript produced from the start site of the transcription unit undergoing the insertion.

In view of its capacity to make such potentially evolutionarily significant mutations, the capacity of FB elements to subsequently "disappear" from such mutations is noteworthy. Specifically, existing FB copies apparently undergo high-frequency, transposon-internal deletion events (25, 78). At least some of these events inactivate the FB element for production of additional genetic instability (25). Thus, selectively advantageous alleles resulting from FB-mediated events can become stable alleles on a geological time scale.

G. Implications of the Possible Rarity of FB-Like Transposons

One of the most conspicuous attributes of FB elements is their conspicuousness. All well-characterized, highly unstable alleles in *D. melanogaster* result either from the P element (Engels, this volume) or FB elements (see above). This suggests that there may be a relatively small number of transposons in *D. melanogaster* of the group of which FB elements are members. This is in contrast to retrotransposons, which are represented by a large number of different specific elements (see above).

The presence of the P element in a population has a substantial potential cost in fertility and, presumably, in mutational load as a result (apparently) of its propensity to make chromosome breaks (Engels, this volume). The same seems likely to be true of FB elements. Perhaps smaller numbers of transposons capable of such dramatic short-term effects on host fitness can persist in a population than is true for transposons that have less drastic effects such as retrotransposons.

The FB transposon is present in all species closely related to *D. melanogaster* that have been examined: *D. simulans*, *D. mauritiana*, *D. yakuba*, and *D. teissieri* (16). This indicates that the FB transposon is likely to have been present in the gene pools of populations ancestral to this group of species. These considerations suggest a unique opportunity to assess the evolutionary role of transposon-mediated chromosome rearrangement. If such rearrangements arise frequently in natural populations, they should be responsible for some of the selectively significant genetic changes distinguishing different members of the *D. melanogaster* species group. While this is difficult to test directly at this juncture, a corollary of this prediction is probably testable. Specifically, many of the presumably more common neutral chromosome rearrangement changes that become fixed in individual species by stochastic processes should prove to be FB-mediated events if such events contribute substantially to the evolutionarily significant behavior of these species. It will likely be of substantial value to test this prediction.

Acknowledgments. Work from the authors' laboratory was supported by Public Health Service grant GM32003 from the National Institutes of Health to P.M.B.

LITERATURE CITED

1. **Arkhipova, I. R., T. V. Gorelova, Y. V. Ilyin, and N. G. Schuppe.** 1984. Reverse transcription of Drosophila mobile dispersed genetic element RNAs: detection of intermediate forms. *Nucleic Acids Res.* **12**:7533–7548.
2. **Arkhipova, I. R., A. M. Mazo, V. A. Cherkasova, T. V. Gorelova, N. G. Schuppe, and Y. V. Ilyin.** 1986. The steps of reverse transcription of Drosophila mobile dispersed genetic elements and U3-R-U5 structure of their LTRs. *Cell* **44**: 555–563.
3. **Bayev, A. A., N. V. Lyubomirsdaya, E. B. Dzhumagaliev, E. V. Ananiev, I. G. Amiantova, and Y. V. Ilyin.** 1984. Structural organization of transposable element *mdg-4* from Drosophila melanogaster and a nucleotide sequence of its long terminal repeats. *Nucleic Acids Res.* **12**:3707–3723.
4. **Bender, W., M. Akam, F. Korch, P. A. Beachy, M. Peifer, P. Spierer, E. B. Lewis, and D. Hogness.** 1983. Molecular genetics of the bithorax complex of Drosophila melanogaster. *Science* **221**:23–29.
5. **Bingham, P. M.** 1980. The regulation of white locus expression: a dominant mutant allele at the white locus of Drosophila melanogaster. *Genetics* **95**:341–353.
6. **Bingham, P. M.** 1981. A novel dominant mutant allele at the white locus of Drosophila melanogaster is mutable. *Cold Spring Harbor Symp. Quant. Biol.* **45**:519–525.
7. **Bingham, P. M., and C. H. Chapman.** 1986. Evidence that *white-blood* is a novel type of temperature-sensitive mutation resulting from temperature-dependent effects of a transposon insertion on formation of *white* transcript. *EMBO J.* **5**:3343–3351.
8. **Bingham, P. M., T.-B. Chou, I. Mims, and Z. Zachar.** 1988. On/off regulation of gene expression at the level of splicing. *Trends Genet.* **4**:134–138.
9. **Bingham, P. M., and B. H. Judd.** 1981. A copy of the *copia* transposon is very tightly linked to the w^a allele at the *white* locus of D. melanogaster. *Cell* **25**:705–711.
10. **Bingham, P. M., M. G. Kidwell, and G. M. Rubin.** 1982. The molecular basis of P-M hybrid dysgenesis: the role of the P element, a P-strain specific transposon family. *Cell* **29**: 995–1004.
11. **Bingham, P. M., R. Levis, and G. M. Rubin.** 1981. Cloning of DNA sequences from the *white* locus of D. melanogaster by a novel and general method. *Cell* **25**:693–704.
12. **Bingham, P. M., and Z. Zachar.** 1985. Evidence that two mutations, w^{DZL} and z^1, affecting synapsis-dependent genetic behavior of *white* are transcriptional regulatory mutations. *Cell* **40**:819–825.
13. **Boeke, J. D., D. J. Garfinkel, C. A. Styles, and G. R. Fink.** 1985. Ty elements transpose through an RNA intermediate. *Cell* **40**:491–500.
14. **Boggs, R. T., P. Gregor, S. Idriss, J. M. Belote, and M. McKeown.** 1987. Regulation of sexual differentiation in D. melanogaster via alternative splicing of RNA from the transformer gene. *Cell* **50**:739–747.
15. **Brierley, H. L., and S. S. Potter.** 1985. Distinct characteristics of loop sequences of two Drosophila foldback elements. *Nucleic Acids Res.* **13**:485–500.
16. **Brookfield, J. F. Y., E. Montgomery, and C. H. Langley.** 1984. Apparent absence of transposable elements related to the P elements of D. melanogaster in other species of Drosophila. *Nature* (London) **310**:330–332.
17. **Campuzano, S., L. Balcells, R. Villares, L. Carramolino, L. Garcia-Alonso, and J. Modelell.** 1986. Excess function hairy-wing mutations caused by gypsy and copia transposable elements inserted within structural genes of the achaete-scute locus of Drosophila. *Cell* **44**:303–312.
18. **Campuzano, S., L. Carramolino, C. V. Cabrera, M. Ruiz-Gomez, R. Villares, A. Boronat, and J. Modollel.** 1985. Molecular genetics of the *achaete-scute* gene complex of D. melanogaster. *Cell* **40**:327–388.
19. **Carbonara, B. D., and W. J. Gehring.** 1985. Excision of copia element in a revertant of the white-apricot mutation of Drosophila melanogaster leaves behind one long terminal repeat. *Mol. Gen. Genet.* **199**:1–6.
20. **Chang, D.-Y., B. Wisely, S.-M. Huang, and R. A. Voelker.** 1986. Molecular cloning of suppressor of sable, a *Drosophila melanogaster* transposon-mediated suppressor. *Mol. Cell. Biol.* **6**:1520–1528.
21. **Chou, T.-B., Z. Zachar, and P. M. Bingham.** 1987. Developmental expression of a regulatory gene is programmed at the level of splicing. *EMBO J.* **7**:4095–4104.
22. **Clark, S. H., C. McCarron, C. Love, and A. Chovnick.** 1986. On the identification of the *rosy* locus DNA in Drosophila melanogaster; intragenic recombination mapping of mutations associated with insertions and deletions. *Genetics* **112**:755–767.
23. **Collins, M., and G. M. Rubin.** 1982. Structure of the Drosophila mutable allele, *white-crimson* and its *white-ivory* and wild type derivatives. *Cell* **30**:71–79.
24. **Collins, M., and G. M. Rubin.** 1983. High-frequency precise excision of the Drosophila foldback transposable element. *Nature* (London) **303**:259–260.
25. **Collins, M., and G. M. Rubin.** 1984. Structure of chromosomal rearrangements induced by the FB transposable element in Drosophila. *Nature* (London) **308**:323–327.
26. **Cote, B., W. Bender, D. Curtis, and A. Chovnick.** 1986. Molecular mapping of the *rosy* locus in Drosophila melanogaster. *Genetics* **112**:769–783.
27. **Davis, P. S., M. W. Shen, and B. H. Judd.** 1987. Asymmetrical pairings of transposons in and proximal to the white locus of Drosophila account for four classes of regularly occurring exchange products. *Proc. Natl. Acad. Sci. USA* **84**:174–178.
28. **Davison, D., C. H. Chapman, C. Wedeen, and P. M. Bingham.** 1985. Genetic and physical studies of a portion of the *white* locus participating in transcriptional regulation and in synapsis-dependent interactions in Drosophila adult tissues. *Genetics* **110**:479–494.
29. **Dowsett, A. P.** 1983. Closely related species of Drosophila can contain different libraries of middle repetitive DNA sequences. *Chromosoma* **88**:104–108.
30. **Dowsett, A. P., and M. W. Young.** 1982. Differing levels of dispersed repetitive DNA among closely related species of Drosophila. *Proc. Natl. Acad. Sci. USA* **79**:4570–4574.
31. **Dudick, M. E., T. R. F. Wright, and L. L. Brothers.** 1974. The developmental genetics of the temperature-sensitive lethal allele of the suppressor of forked, $1(1)su(f)^{ts67g}$, in Drosophila melanogaster. *Genetics* **76**:487–510.
32. **Dunsmuir, P., W. J. Brorein, M. A. Simon, and G. M. Rubin.** 1980. Insertion of the Drosophila transposable element *copia* generates a 5 base pair duplication. *Cell* **21**:575–579.
33. **Duyk, S., J. Leis, M. Longiaru, and A. M. Skalka.** 1983.

Selective cleavage in the avian retroviral long terminal repeat sequence by the endonuclease associated with the alphabeta form of avian reverse transcriptase. *Proc. Natl. Acad. Sci. USA* **80**:6745–6749.

34. **Emori, Y., T. Shiba, S. Inouye, S. Yuki, and K. Saigo.** 1985. The nucleotide sequences of *copia* and *copia*-related RNA in Drosophila virus-like particles. *Nature* (London) **315**: 773–776.
35. **Finnegan, D. J., and D. H. Fawcett.** 1986. Transposable elements in Drosophila melanogaster. *Oxf. Surv. Eukaryotic Genes* **3**:1–62.
36. **Flavell, A. J.** 1984. Role of reverse transcription in the generation of extrachromosomal *copia* mobile genetic elements. *Nature* (London) **310**:514–516.
37. **Flavell, A. J., and C. Brierley.** 1986. The termini of extrachromosomal linear copia elements. *Nucleic Acids Res.* **14**:3659–3669.
38. **Flavell, A. J., and D. Ish-Horowicz.** 1981. Extrachromosomal circular copies of the eukaryotic transposable element *copia* in cultured Drosophila cells. *Nature* (London) **292**: 591–594.
39. **Freund, R., and M. Meselson.** 1984. Long terminal repeat nucleotide sequence and specific insertion of the gypsy transposon. *Proc. Natl. Acad. Sci. USA* **81**:4462–4464.
40. **Goldberg, M. J., R. Paro, and W. J. Gehring.** 1982. Molecular cloning of the *white* locus of Drosophila melanogaster using a large transposable element. *EMBO J.* **1**:93–98.
41. **Goldberg, M. L., J.-Y. Sheen, W. J. Gehring, and M. M. Green.** 1983. Unequal crossing-over associated with asymmetrical synapsis between nomadic elements in the Drosophila melanogaster genome. *Proc. Natl. Acad. Sci. USA* **80**:5017–5021.
42. **Green, M. M.** 1959. Spatial and functional properties of pseudoalleles at the *white* locus in Drosophila melanogaster. *Heredity* **13**:302–315.
43. **Green, M. M.** 1967. The genetics of a mutable gene at the white locus of Drosophila melanogaster. *Genetics* **56**: 467–482.
44. **Green, M. M., and G. Lefevre.** 1972. The cytogenetics of mutator gene induced X-linked lethals in Drosophila melanogaster. *Mutat. Res.* **16**:59–64.
45. **Ilyin, Y. V., N. G. Schuppe, N. V. Lyumbomirskya, T. V. Gorelova, and I. R. Archipova.** 1984. Circular copies of mobile dispersed genetics element in cultured Drosophila melanogaster cells. *Nucleic Acids Res.* **12**:7517–7531.
46. **Inouye, S., K. Saigo, K. Yamada, and Y. Kuchino.** 1986. Identification and nucleotide sequence determination of a potential primer tRNA for reverse transcription of a Drosophila retrotransposon, 297. *Nucleic Acids Res.* **14**: 3031–3043.
47. **Inouye, S., S. Yuki, and K. Saigo.** 1984. Sequence-specific insertion of the Drosophila transposable element *17.6*. *Nature* (London) **310**:332–333.
48. **Inouye, S., S. Yuki, and K. Saigo.** 1986. Complete nucleotide sequence and genome organization of a Drosophila transposable genetic element, *297*. *Eur. J. Biochem.* **154**: 417–425.
49. **Ising, G., and K. Block.** 1981. Derivation-dependent distribution of insertion sites for a Drosophila transposon. *Cold Spring Harbor Symp. Quant. Biol.* **45**:527–549.
50. **Ising, G., and C. Ramel.** 1976. The behavior of a transposable element in Drosophila melanogaster, p. 947–954. *In* M. Ashburner and E. Novitski (ed.), *The Genetics and Biology of Drosophila*, vol. 1b. Academic Press, Inc. (London), Ltd., London.
51. **Karess, R. E., and G. M. Rubin.** 1982. A small tandem duplication is responsible for the unstable *white-ivory* mutation in Drosophila. *Cell* **30**:63–69.
52. **Karlik, C. C., and E. A. Fyrberg.** 1985. An insertion within a variably spliced Drosophila tropomyosin gene blocks accumulation of only one encoded isoform. *Cell* **41**:57–66.
53. **Kaufman, T. C., R. Lewis, and B. Wakimoto.** 1983. Cytogenetic analysis of chromosome 3 in Drosophila melanogaster: the homeotic gene complex in polytene chromosome interval 84A-B. *Genetics* **94**:115–133.
54. **Kikuchi, Y., Y. Ando, and T. Shiba.** 1986. Unusual priming mechanism of RNA-directed DNA synthesis in *copia* retrovirus-like particles of Drosophila. *Nature* (London) **323**: 824–826.
55. **Kugimiya, W., H. Ikenaga, and K. Saigo.** 1983. Close relationship between the long terminal repeats of avian leukosis-sarcoma virus and *copia*-like movable genetic elements of Drosophila. *Proc. Natl. Acad. Sci. USA* **80**:3193–3197.
56. **Kulguskin, V. V., Y. V. Ilyin, and G. P. Georgiev.** 1981. Mobile dispersed genetic element MDG1 of Drosophila melanogaster: nucleotide sequence of long terminal repeats. *Nucleic Acids Res.* **9**:3451–3463.
57. **Laski, F. A., D. C. Rio, and G. M. Rubin.** 1986. Tissue specificity of Drosophila P element transposition is regulated at the level of mRNA splicing. *Cell* **44**:7–19.
58. **Lemeuneir, F., and M. Ashburner.** 1976. Relationships within the melanogaster species subgroup of the genus Drosophila (Sophophora). II. Phylogenetic relationship between six species based upon polytene chromosome banding sequences. *Proc. R. Soc. London* **193**:275–294.
59. **Levis, R., M. Collins, and G. M. Rubin.** 1982. FB elements are the common basis for the instability of the w^{DZL} and w^c Drosophila mutations. *Cell* **30**:551–565.
60. **Levis, R., K. O'Hare, and G. M. Rubin.** 1984. Effects of transposable element insertions on RNA encoded by the *white* gene of Drosophila. *Cell* **36**:471–481.
61. **Levis, R., and G. M. Rubin.** 1982. The unstable w^{DZL} mutation of Drosophila is caused by a 13 kb insertion that is imprecisely excised in revertants. *Cell* **30**:543–550.
62. **Manning, J. E., C. W. Schmid, and N. Davidson.** 1975. Interspersion of repetitive and nonrepetitive DNA sequences in the Drosophila melanogaster genome. *Cell* **4**:141–155.
63. **Marlor, R. L., S. M. Parkhurst, and V. G. Corces.** 1986. The Drosophila melanogaster gypsy transposable element encodes putative gene products homologous to retroviral proteins. *Mol. Cell. Biol.* **6**:1129–1134.
64. **Martin, G., D. Wiernasz, and P. Schedl.** 1983. Evolution of Drosophila repetitive-dispersed DNA. *J. Mol. Evol.* **19**: 203–213.
65. **Mattox, W. M., and N. Davidson.** 1984. Isolation and characterization of the *Beadex* locus of Drosophila melanogaster: a putative *cis*-acting negative regulatory element for the *heldup-a* gene. *Mol. Cell. Biol.* **4**:1343–1353.
66. **McGinnis, W., and S. K. Beckendorf.** 1983. Association of a Drosophila transposable element of the roo family with chromosomal deletion breakpoints. *Nucleic Acids Res.* **11**: 737–751.
67. **Modelell, J., W. Bender, and M. Meselson.** 1983. Drosophila melanogaster mutations suppressible by the suppressor of Hairy-wing are insertions of a 7.3kb mobile element. *Proc. Natl. Acad. Sci. USA* **80**:1678–1682.
68. **Montgomery, E., B. Charlesworth, and C. H. Langley.** 1987. A test for the role of natural selection in the stabilization of

transposable element copy number in a population of Drosophila melanogaster. *Genet. Res.* **49:**31–41.

69. **Mossie, K. B., M. W. Young, and H. E. Varmus.** 1985. Extrachromosomal DNA forms of *copia*-like transposable elements, F elements and middle repetitive DNA sequences in Drosophila melanogaster. *J. Mol. Biol.* **182:**31–42.
70. **Mount, S. M., and G. M. Rubin.** 1985. Complete nucleotide sequence of the Drosophila transposable element copia: homology between copia and retroviral proteins. *Mol. Cell. Biol.* **5:**1630–1638.
71. **O'Hare, K., C. Murphy, R. Levis, and G. M. Rubin.** 1984. DNA sequence of the *white* locus in Drosophila melanogaster. *J. Mol. Biol.* **180:**437–455.
72. **Parkhurst, S. M., and V. G. Corces.** 1985. *forked*, *gypsys* and suppressors in Drosophila. *Cell* **41:**429–437.
73. **Parkhurst, S. M., and V. G. Corces.** 1986. Interactions among the *gypsy* transposable element and the *yellow* and *suppressor-of-Hairy-wing* loci in Drosophila melanogaster. *Mol. Cell. Biol.* **6:**46–53.
74. **Parkhurst, S. M., and V. G. Corces.** 1986. Retroviral elements and suppressor genes in Drosophila. *BioEssays* **5:** 52–57.
75. **Paro, R., M. L. Goldberg, and W. J. Gehring.** 1983. Molecular analysis of large transposable elements carrying the *white* locus of Drosophila melanogaster. *EMBO J.* **2:** 853–860.
76. **Pirrotta, V., and C. Brockl.** 1984. Transcription of the Drosophila *white* locus and some of its mutants. *EMBO J.* **3:**563–568.
77. **Potter, S. S.** 1982. DNA sequence of a foldback transposable element in Drosophila. *Nature* (London) **297:**201–204.
78. **Potter, S. S.** 1982. DNA sequence analysis of a Drosophila foldback transposable element rearrangement. *Mol. Gen. Genet.* **188:**107–110.
79. **Rubin, G. M., M. G. Kidwell, and P. M. Bingham.** 1982. The molecular basis of PM hybrid dysgenesis: the nature of induced mutation. *Cell* **29:**987–994.
80. **Russel, M. A.** 1974. Pattern formation in the imaginal discs of a temperature-sensitive cell-lethal mutant of Drosophila melanogaster. *Dev. Biol.* **40:**24–39.
81. **Saigo, K., W. Kugimiya, Y. Matsuo, S. Inouye, K. Yoshioka, and S. Yuki.** 1984. Identification of the coding sequence for a reverse transcriptase-like enzyme in a transposable genetic element in Drosophila melanogaster. *Nature* (London) **312:** 659–661.
82. **Scherer, G., C. Tschudi, J. Perera, H. Delius, and V. Pirrotta.** 1982. *B104*, a new dispersed repeated gene family in Drosophila melanogaster and its analogies with retroviruses. *J. Mol. Biol.* **157:**435–452.
83. **Schneuwly, S., A. Kuroiwa, and W. J. Gehring.** 1987. Molecular analysis of the dominant homeotic Antennapedia phenotype. *EMBO J.* **6:**201–206.
84. **Schwartz, H. E., T. J. Lockett, and M. W. Young.** 1982. Analysis of transcripts from two families of nomadic DNA. *J. Mol. Biol.* **157:**49–58.
85. **Scott, M. P., A. J. Weiner, T. F. Hazelrigg, B. A. Polisky, V. Pirrotta, F. Scalenghe, and T. C. Kaufman.** 1983. The molecular organization of the Antennapedia locus of Drosophila. *Cell* **35:**763–776.
86. **Searles, L. L., and R. A. Voelker.** 1986. Molecular characterization of the Drosophila vermillion locus and its suppressible alleles. *Proc. Natl. Acad. Sci. USA* **83:**404–408.
87. **Seperack, P. K., M. C. Strobel, D. J. Corrow, N. A. Jenkins, and N. G. Copeland.** 1987. Somatic and germline reverse mutation rates of the retrovirus-induced dilute coat-color mutation of DBA mice. *Proc. Natl. Acad. Sci. USA* **85:** 189–192.
88. **Shepard, B. M., and D. J. Finnegan.** 1984. Structure of circular copies of the *412* transposable element present in Drosophila melanogaster tissue culture cells and isolation of a free *412* long terminal repeat. *J. Mol. Biol.* **180:**21–40.
89. **Shiba, T., and K. Saigo.** 1983. Retrovirus-like particles containing RNA homologous to the transposable element *copia* in Drosophila melanogaster. *Nature* (London) **302:** 119–124.
90. **Snyder, M. P., D. Kimbrell, M. Hunkapiller, R. Hill, J. Fristrom, and N. Davidson.** 1982. A transposable element that splits the promoter inactivates a Drosophila cuticle protein gene. *Proc. Natl. Acad. Sci. USA* **79:**7430–7434.
91. **Swaroop, A., M. E. Paco-Larson, and A. Garen.** 1985. Molecular genetics of a transposon-induced dominant mutation in the Drosophila *glued* locus. *Proc. Natl. Acad. Sci. USA* **82:**1751–1755.
92. **Truett, M. A., R. S. Jones, and S. S. Potter.** 1981. Unusual structure of the FB family of transposable elements in Drosophila. *Cell* **24:**753–763.
93. **Varmus, H. E.** 1983. Retroviruses, p. 411–505. *In* J. A. Shapiro (ed.), *Mobile Genetic Elements.* Academic Press, Inc., New York.
94. **Varmus, H. E., H. Quintrell, and S. Ortiz.** 1981. Retroviruses as mutagens: insertion and excision of a nontransforming provirus alter expression of a resident transforming provirus. *Cell* **25:**23–36.
95. **Varmus, H. E., and R. Swanstrom.** 1985. Supplement: replication of retroviruses, p. 75–134. *In* R. Weiss, N. Teich, H. Varmus, and J. Coffin (ed.), *RNA Tumor Viruses.* Cold Spring Harbor Laboratory, Cold Spring Harbor, N.Y.
96. **Will, B. M., A. A. Bayev, and D. J. Finnegan.** 1981. Nucleotide sequence of terminal repeats of 412 transposable elements of Drosophila melanogaster. *J. Mol. Biol.* **153:** 897–915.
97. **Winston, F., T. D. Schaleff, B. Valent, and G. R. Fink.** 1984. Mutations affecting Ty mediated expression of the HIS4 gene of Saccharomyces cerevisiae. *Genetics* **107:**179–197.
98. **Young, M. W.** 1979. Middle repetitive DNA: a fluid component of the Drosophila genome. *Proc. Natl. Acad. Sci. USA* **76:**6274–6278.
99. **Yuki, I., S. Inouye, S. Ishimaru, and K. Saigo.** 1986. Nucleotide sequence characterization of a Drosophila retrotransposon, *412*. *Eur. J. Biochem.* **158:**403–410.
100. **Zachar, Z., and P. M. Bingham.** 1982. Regulation of *white* locus expression: the structure of mutant alleles at the *white* locus of Drosophila melanogaster. *Cell* **30:**529–541.
101. **Zachar, Z., C. H. Chapman, and P. M. Bingham.** 1985. On the molecular basis of transvection effects and the regulation of transcription. *Cold Spring Harbor Symp. Quant. Biol.* **50:**337–346.
102. **Zachar, Z., T.-B. Chou, and P. M. Bingham.** 1987. Evidence that a regulatory gene autoregulates splicing of its transcript. *EMBO J.* **7:**4105–4111.
103. **Zachar, Z., D. Davison, D. Garza, and P. M. Bingham.** 1985. A detailed developmental and structural study of the transcriptional effects of insertion of the *copia* transposon into the white locus of Drosophila melanogaster. *Genetics* **111:**495–515.
104. **Zachar, Z., D. Garza, T.-B. Chou, J. Goland, and P. M. Bingham.** 1987. Molecular cloning and genetic analysis of the suppressor-of-white-apricot locus from Drosophila melanogaster. *Mol. Cell. Biol.* **7:**2498–2505.

Chapter 18

The I Factor and I-R Hybrid Dysgenesis in *Drosophila melanogaster*

D. J. FINNEGAN

I. INTRODUCTION

Hybrid dysgenesis is a particularly interesting system for studying DNA transposition because of its obvious biological consequences and the ease with which it can be manipulated experimentally. In this review I shall concentrate on recent molecular studies related to I-R dysgenesis. This system was discovered and genetically characterized by L'Héritier and his colleagues in France. These early studies have been reviewed in detail by Bregliano et al. (4), Bregliano and Kidwell (3), and Kidwell (26). In this review I have tried to observe the following convention when referring to I factors and I elements. By "I factors" I mean genetic determinants that stimulate I-R dysgenesis and 5.4-kilobase (kb) DNA sequences that have been shown to have this activity. By "I elements" I mean segments of DNA that contain all, or part, of the sequence of an I factor. I elements include both complete and incomplete I factors.

I-R hybrid dysgenesis is the appearance of a set of unusual characteristics in the progeny produced by crossing particular strains of *Drosophila melanogaster*. These characteristics are reduced fertility, X chromosome nondisjunction, and increased frequencies of mutations. There may well be nondisjunction of autosomes as well, but this has not been assayed. Strains of *D. melanogaster* can be classified as being one or other of two types—I, or inducer, and R, or reactive—with respect to this phenomenon, and I-R dysgenesis is manifest by female progeny of crosses between R-strain females and I-strain males. The progenies of all other crosses are apparently normal (42, 44). The female progeny from an I-R dysgenic cross are known as SF females (*stérilité femelle*), while the female progeny from the reciprocal nondysgenic cross are known as RSF females (Fig. 1). The fertility of SF females can vary from cross to cross, as there is quantitative variation between the responses produced by different reactive and inducer strains (8). Some strains do not give a significant dysgenic response when crossed either as males with a reactive strain or as females with an inducer strain (28). These

D. J. Finnegan ■ Department of Molecular Biology, University of Edinburgh, King's Buildings, Mayfield Road, Edinburgh EH9 3JR, Scotland.

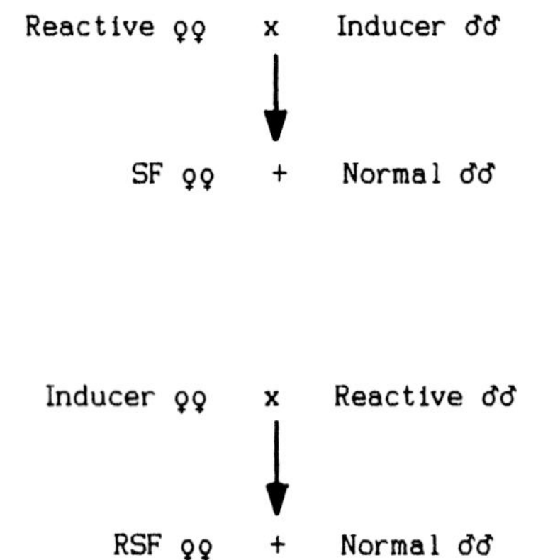

Figure 1. Nature of the progeny produced by reciprocal crosses between inducer and reactive strains of *D. melanogaster*.

strains are said to be neutral (45) and are interpreted as being very weakly reactive.

The sterility associated with I-R dysgenesis is quite distinctive and can be distinguished easily from the GD sterility associated with P-M dysgenesis. SF females have ovaries that appear normal, and they lay normal numbers of fertilized eggs, but these eggs stop developing during early cleavage divisions prior to blastoderm formation (30, 31). In contrast, females exhibiting GD sterility have malformed ovaries and lay reduced numbers of eggs. The proportion of eggs that are laid by SF females and that hatch can vary from 0% to normal levels, and this value, called "hatchability," is used as a measure of SF sterility. This depends on the strength of the inducer and reactive strains that were crossed to produce the SF females, but is independent of the males with which the SF females are mated. The proportion of eggs that hatch increases with the age of the SF females and can reach normal levels 15 to 20 days after eclosion (6).

II. GENETIC PROPERTIES OF I FACTORS

The genetic determinants that control the inducer state are known as I factors. These are present on the chromosomes of inducer strains and are inherited stably in a Mendelian fashion from heterozygous males produced by crossing reactive females and inducer males (42). I factors behave quite differently in SF females, where they are unstable and can transfer from one chromosome to another. Picard (42) used strains carrying marked chromosomes with crossover suppressors to show that, in SF females, I factors can transpose from chromosomes of the inducer parent to homologous chromosomes from the reactive parent. He was also able to show that I factors could transpose to nonhomologous chromosomes. The frequency with which chromosomes from a reactive strain acquire I factors by this process, known as chromosome contamination, can be as high as 100% per generation, indicating that I factors can transpose very efficiently in SF females. They also transpose to new sites on the chromosomes of the inducer parent, but the frequency with which this occurs is more difficult to measure. Chromosomes from the reactive fathers of RSF females acquire I factors in a similar way, but the frequency with which this occurs is at least fivefold less than in isogenic SF females (42). The frequency of I-factor transposition has not been measured precisely, as neither the number of potential donor elements nor the number of transposed copies has been determined for any dysgenic cross. The frequency of chromosome contamination could underestimate the number of transposition events, as only functional elements are scored in such an experiment, and chromosomes getting more than one copy would be scored as single events. Nevertheless, the frequency of I-factor transposition in SF females is of the same order as that of P elements in a P-M dysgenic cross (W. Engels, this volume).

The number of genetically active I factors may vary from one inducer strain to another. Some strains are heterozygous for chromosomes carrying I factors (i^+ chromosomes) and chromosomes that do not (i^0 chromosomes). The i^0 chromosomes are stable in an inducer strain unless they acquire an I factor by recombination. This has made it possible to map I factors genetically and to show that they have different chromosome locations in different strains (41).

The levels of sterility induced in SF females descended from a particular reactive strain vary according to the source of the I factors present, but there is no strong additive effect when more than one I factor is present (37). Some differences in inducer activity are intrinsic to the I factors themselves (36). Others appear to be due to position effects, as the inducer activity associated with some elements increases after transposition (39).

III. MUTATIONS RESULTING FROM I-FACTOR ACTIVITY

Mutations occur at unusually high frequencies in the germ line of SF females and are recovered in flies of the next generation. They probably result from events at or just before meiosis, as they appear in isolated individuals or as very small clusters (44). A variety of mutations have been recovered and studied on X chromosomes. They may be associated with recessive or dominant phenotypes and may behave as point mutations or chromosome rearrangements.

The frequencies of these mutations are correlated with the levels of SF sterility and X chromosome nondisjunction, suggesting that these aspects of the dysgenic phenotype have a common origin but vary from locus to locus. The *white* and *cut* genes can mutate with a frequency of about 10^{-4}, while the *singed* and *forked* genes are less mutable (44). This contrasts with P-M dysgenesis, in which *singed* is highly mutable, and suggests that I factors and P factors stimulate mutations differentially. Proust and Prudhommeau (48) have studied sex-linked lethal mutations arising in SF and RSF females. The frequency of lethals from females showing high levels of SF sterility was about 5%, while the frequency in the corresponding RSF females was at least 20-fold less than this. In one experiment the frequency of lethals fell from 6 to 1% as the SF females aged, consistent with there being a causal relationship between lethal mutation frequency and age.

Proust and Prudhommeau (48) have analyzed 58 lethal mutations in detail. These mutations arose on chromosomes of both the reactive and inducer parents, and 15 were associated with visible chromosome rearrangements. They were not distributed at random, but were concentrated at two hot spots, one near *cut* and the other in the *yellow* region.

IV. MOLECULAR STRUCTURE OF I FACTORS

The genetic studies discussed so far suggested that at least some mutations produced during I-R dysgenesis might be due to insertion of I factors into the genes in question. This has been confirmed by the molecular analysis of I-R-induced mutations. Bucheton et al. (9) and Sang et al. (50) determined the structure of the *white* locus on chromosomes carrying eight *white* mutations, w^{IR1} through w^{IR8}, that arose in SF females. Two of them are due to deletions with one endpoint in the *white* gene. The other six are associated with insertions of indistinguishable 5.4-kb elements. The element associated with one of these mutations, w^{IR1}, was presumed to be a functional I factor since this mutation cannot be separated by recombination from I-factor activity (38).

The insertions associated with mutations w^{IR1} and w^{IR3} have been cloned (Fig. 2), as have the ends of the elements associated with the other four *white* mutations (9, 17). Each element is flanked by a target site duplication, but these vary in length from one element to another (Fig. 3). In some cases it is not possible to determine the lengths of these duplications precisely, as they are related to the sequence of one end of the I factor. Three of the elements, those associated with mutations w^{IR1}, w^{IR3}, and w^{IR4}, have

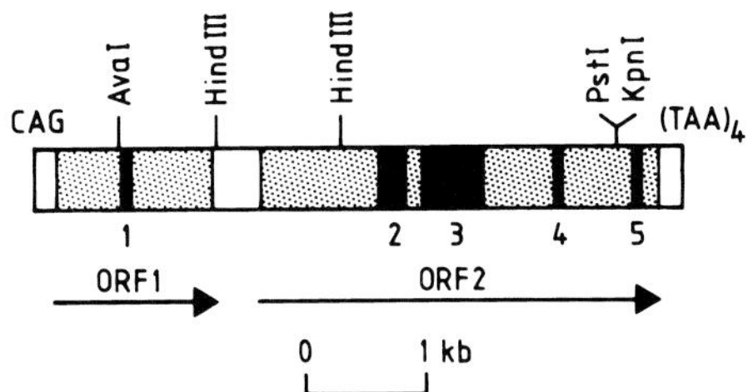

Figure 2. Map of a complete I factor. This map is based on the sequence of the I factor associated with the w^{IR1} mutation. There are no sites for the restriction enzymes *Bam*HI, *Eco*RI, *Sac*I, *Sal*I, *Sma*I, and *Xho*I in this element. The two largest open reading frames, ORF1 and ORF2, are stippled. Five regions of amino acid sequence that may be associated with particular functions are indicated by the solid shading. These are (1) a possible nucleic acid-binding sequence; (2) a sequence that is highly conserved in elements structurally similar to the I factor; (3) a possible reverse transcriptase; (4) a possible RNase H; (5) a possible metal-binding "finger."

inserted at the same site but have different target site duplications.

The base sequence of the putative I factor associated with the w^{IR1} mutation has been determined completely (17). It is 5,371 base pairs (bp) long and is flanked by a target site duplication of 12 bp. It has no terminal repeats, but has four copies of the sequence TAA at the 3′ end of one strand (Fig. 2). This sequence organization is a general property of I factors and has been found at the ends of the five other I elements inserted within the *white* gene and one associated with a mutation of the *Bithorax* complex, bx^{F31} (35). The ends of these elements are highly conserved, and the only differences that have been found are in the third base at the left-hand end, which is either G or T, and in the number of TAA repeats at the right-hand end (Fig. 3). All the coding capacity of the I factor appears to be in one strand. This contains two long open reading frames, ORF1 of 1,278 bp and ORF2 of 3,258 bp (Fig. 2). These are separated by 471 bp including another open reading

MUTATION	LEFT-HAND END	RIGHT-HAND END	TARGET SITE DUPLICATION
w^{IR1}	CATTACC	$TCA(TAA)_4$	TAATATGCAAAT
w^{IR3}	CAGTACC	$TCA(TAA)_5$	TTAATATGCAAAAT
w^{IR4}	CAGTACC	$TCA(TAA)_5$	TTAATATGCAAAT
w^{IR2}	CATTACC	$TCA(TAA)_7$	TTTACTGCAGAG
w^{IR5}	CAGTACC	$TCA(TAA)_6$	TCCGAAATAACT
w^{IR6}	CAGTACC	$TCA(TAA)_6$	TAACAACCAG
bx^{F31}	CAGTACC	$TCA(TAA)_6$	TAAAAGGCCGAAA

Figure 3. Base sequences found at the ends of I factors. The first column indicates the mutations concerned. The second and third columns show sequences at the left- and right-hand ends of the I factors associated with them. The last column shows the target site duplications next to each insertion. We cannot be sure of the precise length of the duplications associated with w^{IR1}, w^{IR6}, and bx^{F31}. In each case, one copy of the TAA repeat could be part either of the target site duplication or of the I factor.

```
I Factor   NFIFTDGSKI    98aa  GNELADQAAK
RNAase H   veIFTDGScl   118aa  cdELAraAAm
RSV        ptvFTDaSss   108aa  adsqAtfqAy
```

Figure 5. The region of ORF2 of the I factor that may encode an RNase H. The second line gives the amino acid sequence from two regions of *E. coli* RNase H (24). The first and third lines show regions of ORF2 of the I factor (17) and the *pol* gene of Rous sarcoma virus (RSV; 51) that match the *E. coli* sequences. The I-factor sequence and sequences that match it exactly are shown in capital letters. Sequences that differ from that of the I factor are shown in lowercase.

in development, possibly in late embryos (Engels, this volume), whereas I factors probably transpose at, or shortly before, meiosis.

VI. GENOMIC DISTRIBUTION OF I-FACTOR SEQUENCES

I-factor sequences have been found in the genomic DNA of all *D. melanogaster* strains that have been tested so far, whether inducer or reactive. Southern transfer experiments have been used to probe DNAs from five inducer and five reactive strains, with restriction fragments covering the whole of the I factor (9). The patterns of hybridization of the reactive-strain DNAs were remarkably similar to each other, even though these strains represented three independent isolates from the wild. The I elements revealed by this experiment must be nonfunctional since they are in reactive strains and must transpose infrequently, if at all. The pattern of hybridization seen with DNA from inducer strains included most of the bands common to reactive strains, together with several additional bands, many of which differed from strain to strain. This was particularly obvious when probing with end fragments, as is expected with transposable elements.

I-factor probes hybridize to sequences in and around the chromocenter of salivary gland chromosomes from reactive strains and to the chromocenter and euchromatic sites on the arms of inducer chromosomes (9). The number and location of euchromatic sites vary from one strain to another, suggesting that the elements that they contain include active I factors. Inactive and nontransposing I elements are probably concentrated in centromeric regions, although some could be at euchromatic sites in inducer strains. I elements may transpose at a low level in inducer strains, as the number and location of euchromatic sites can vary between individuals within a particular strain or population (32; Biemont and Gautier, *Heredity*, in press; A. Pélisson, personal communication).

The DNAs of reactive and inducer strains can be distinguished easily in Southern transfer experiments designed to detect large internal fragments from complete I factors. Prominent bands of hybridization can only be seen with DNA from inducer strains (9, 11). The intensity of these bands corresponds to about 10 copies per haploid genome for inducer strains and only about 1 copy, or less, for reactive strains. This sets an upper limit to the number of complete I factors in these strains.

VII. STRUCTURES OF INCOMPLETE I ELEMENTS

Crozatier and co-workers (14a) have analyzed the structures of 14 I elements cloned at random from reactive and inducer strains. None of the I elements from the reactive strain is complete, as they all contain large internal deletions or other rearrangements and have lost one or both ends. Each element is adjacent to moderately repeated sequences, as expected if they come from pericentromeric regions. The ends of three of these elements have been sequenced. Two have lost the 3′ end, and the other has lost the 5′ end. These terminal deletions probably took place after the elements had inserted in centromeric DNA, as none of them is flanked by a target site duplication.

Ten I elements from an inducer strain have been studied in detail. Their structures were compared with that of a complete I factor and were found to be of three types. Two of the 10 appear to be complete, 2 are truncated at their 5′ ends, and 6 have internal and terminal deletions like the I elements found in the reactive strain. The two complete I elements and one of the 5′-truncated elements are flanked by nonrepeated DNA, and in situ hybridization with these sequences indicates that they have come from sites on the chromosome arms. The remaining elements are flanked by repeated sequences and presumably came from pericentromeric sites.

The ends of three of these elements, one complete, one 5′ truncated, and one with several rearrangements, have been sequenced. The ends of the complete element are identical to those of the w^{IR1} I factor and are flanked by a probable target site duplication. The 5′-truncated element is also flanked by a potential target site duplication, and its 3′ end is equivalent to that of the w^{IR1} I factor. These elements may have inserted in the genome recently, as Southern hybridization experiments indicate that they are not present in all inducer strains. The third I element differs from a functional I factor in many places. It has an intact 3′ end, but has lost its 5′ end, has one large internal deletion, has several base substitutions and a small deletion in the region sequenced, and is not flanked by a target site duplication. This element is surrounded by sequences that are repeated in the genome and presumably belongs to the class of I elements that are present in both inducer and reactive strains and that correspond to the inactive, nontransposing I elements identified in Southern transfer experiments by Bucheton et al. (9).

The I elements that are flanked by repeated sequences differ from the w^{IR1} I factor by gross rearrangements and many base substitutions and small additions or deletions, and those that have an intact 3′ end terminate with the sequence TAA $(TAAA)_n$ or a more complex A+T-rich sequence. The overall divergence in the regions that have been sequenced is about 7%. This contrasts with less than 1% divergence for I elements that have transposed to sites on the chromosome arms (14a). The observation that the only rearrangements associated with these elements are 5′-terminal deletions suggests that more complex rearrangements are not the results of transposition. These are probably the results of spontaneous mutational events that have occurred over long periods of time in elements that have accumulated in centromeric heterochromatin.

VIII. SPECTRUM OF MUTATIONS INDUCED BY I-R DYSGENESIS

Not all mutations induced by I-R hybrid dysgenesis are due to insertions of 5.4-kb I factors. There is no evidence that the two deletions of the *white* gene that arose in SF females were stimulated by I-R dysgenesis, as there are no I-factor sequences associated with either of them (17), but other types of rearrangements have been found among I-R-induced mutations of the *yellow* gene (18a; I. Busseau, A. Pélisson, and A. Bucheton, personal communication). Three of these mutations are associated with insertions, but only one, y^{IR8}, contains a 5.4-kb element like those described above. The other two mutations, y^{IR3} and y^{IR4}, have insertions of incomplete I elements that are truncated by different amounts at their 5′ ends. The 3′ ends of these elements are like those of complete I factors, except that one of them has the terminal sequence $TAA(TAAA)_3$ rather than simply a run of TAA repeats. Both elements are flanked by target site duplications, suggesting that they were not deleted after integration. One cannot say whether they are exact copies of deleted donor elements or whether they were truncated during transposition, but in either case they indicate that the sequence at the 5′ end of complete elements is not required for integration.

The two remaining mutations are associated with large rearrangements, both of which have I-factor sequences at at least one breakpoint. One of them, y^{IR5}, has a cytologically visible inversion with an apparently complete I factor at the breakpoint within *yellow* (Fig. 6a). The other end of the inversion has not been studied at the molecular level. There is no cytologically visible rearrangement on chromosomes carrying the other mutation, y^{IR6}, but it is also associated with an inversion. This starts within *yellow* and extends proximally for an unknown distance. There is a complete I factor at the *yellow* breakpoint and a 5′-truncated I element at the other (Fig. 6b). The organization of sequences surrounding these elements suggests that this rearrangement is the result of two events, (i) insertion of a truncated I element at *yellow* followed by (ii) recombination between this element and a complete I factor proximal to it, inverting the DNA between them (Fig. 6c). The recombination step could have been stimulated by I-R dysgenesis directly or could have taken place after transposition and independently of it.

Mutations induced by I-R dysgenesis are stable, and there is no evidence for excision of I elements as none of these mutations has been found to revert in SF females. Tests for reversion of the w^{IR1} mutation during I-R dysgenesis generated other mutant *white* alleles but no wild-type revertants. The w^{IR1} I factor is inserted in an intron of the *white* gene, and flies carrying this allele have red-brown eyes, as compared to the red wild-type color. Flies carrying dysgenesis-induced derivatives of w^{IR1} have either a paler eye color or an extreme white phenotype. These alleles appeared about eight times more frequently on w^{IR1} chromosomes than on either w^+ chromosomes in the same SF females or the w^{IR1} chromosome in nondysgenic flies. Some of the w^{IR1} derivatives have been analyzed at the molecular level (18a; M. Lynch, J. Sved, H. Sang, and D. J. Finnegan, unpublished data). Several are associated with deletion of sequences to one side of the I factor. One, A30, has a 2.4-kb deletion that starts immediately to the right of the I factor. Both ends of the I factor are intact, but the number of TAA repeats at the 3′ end has increased from four to seven, a structure that could not have been generated by a simple deletion (Fig. 7a). This mutation could have been formed in a similar fashion to y^{IR6}. In this case, a second I element would have inserted 2.4 kb to the right of the w^{IR1} element and then have recombined with it. The intervening DNA would have been deleted if the two elements were in the same orientation, and the 3′ end of the I element left at *white* would be that of the new insertion rather than of w^{IR1} (Fig. 7b). The presence of I-factor sequences at the breakpoints of these

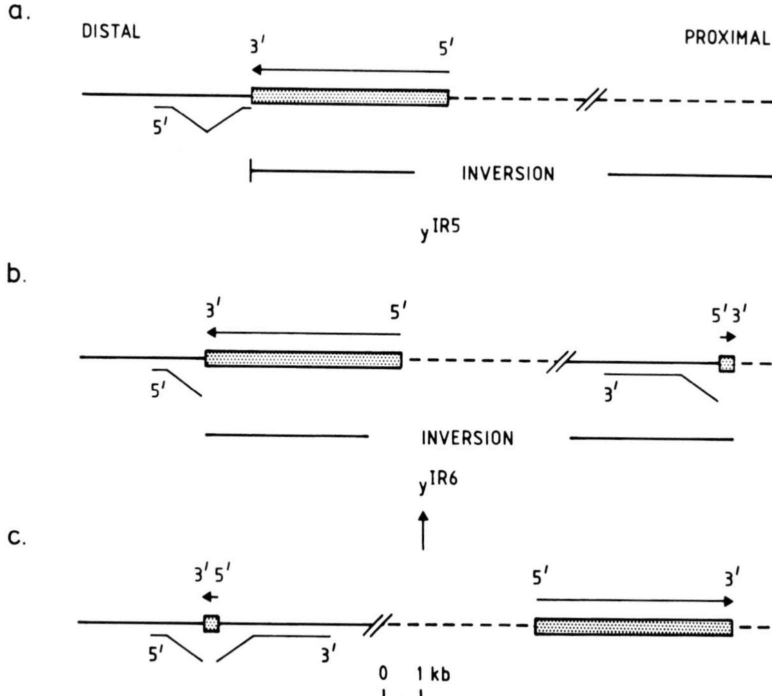

Figure 6. Inversions associated with I-R-induced mutations. (a) Map of the inversion associated with the y^{IR5} mutation. The solid line indicates chromosomal DNA from the *yellow* locus. The dashed line indicates chromosomal DNA that is normally proximal to *yellow*. The stippled boxes indicate I elements. The horizontal line immediately below the map indicates the two exons of the *yellow* gene, with the intervening intron indicated by the dip (13). The I element at the distal breakpoint appears to be complete. The proximal breakpoint has not been analyzed. (b) Map of the inversion associated with the y^{IR6} mutation. The I element at the distal breakpoint appears to be complete, while that at the proximal breakpoint is truncated at the 5′ end. The length of the inversion is not known. (c) Hypothetical intermediate in the formation of the y^{IR6} inversion, showing the truncated I element inserted into the *yellow* gene prior to recombination with a complete I factor lying proximal to it. A more detailed explanation is given in the text.

rearrangements in *white* and *yellow* is probably typical of I-R-induced mutations, as J. Proust (personal communication) has found that I-factor probes hybridize to at least one breakpoint in 35 of 36 such rearrangements on the X chromosome.

IX. REGULATION OF I-R DYSGENESIS

The phenomenon of I-R hybrid dysgenesis can be explained by saying that I factors are held in check in inducer strains by regulatory molecules not present in reactive strains. Inhibition of I-factor activity would be relieved in SF females because I factors from inducer males enter a cellular environment in which they are no longer regulated. Expression of I factors might be greater in RSF females than in inducer strains because of a slight dilution of regulatory molecules in the nuclei of zygotes formed by union of inducer oocytes and reactive sperm. The most likely sources of these regulatory molecules are the I factors themselves, possibly the potential nucleic acid-binding domains of ORF1 or ORF2 or both (Fig. 2).

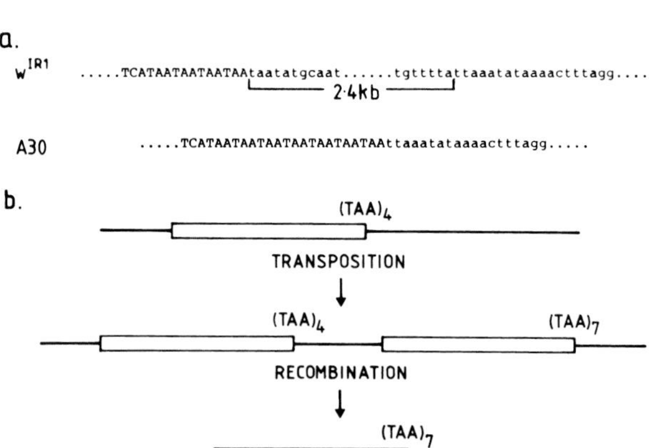

Figure 7. Structure and possible origin of the A30 mutation. (a) The first line shows the base sequence at the right-hand end of the w^{IR1} I factor and of the *white* gene 2.4 kb to the right. The second line shows the sequence of the corresponding region of A30. The I-factor sequence is in capital letters, and the *white* sequence is shown in lowercase. (b) A possible two-step process for production of A30 from w^{IR1}. The first step is transposition of an I element into the *white* gene 2.4 kb to the right of the w^{IR1} I factor. This is shown as a complete 5.4-kb I element, but could equally well be a 5′-truncated element. The second step is recombination between these two I elements, deleting the intervening 2.4 kb and leaving behind a recombinant element.

The effects of I-R dysgenesis do not disappear in one generation. The daughters of SF females have reduced fertility due to SF sterility, indicating that I-factor expression can continue for at least one generation (43). The fertility of the daughters of SF females is lower when their mothers are mated with reactive-strain males and higher if they are mated with inducer males. These observations could be explained if I-factor expression is inhibited by a subset of I elements and more than one of these elements are required for complete inhibition. They need not be complete I factors, but must be able to transpose or be produced by transposition. They would then control their own copy number by blocking transposition. The number of regulatory elements in SF females would be about half that in the corresponding inducer strain, bringing it below the threshold required to prevent dysgenesis. Transposition in SF females would increase the number of regulatory elements, but might not raise it above the threshold in one generation. The number of generations required to reach the threshold would increase if SF females were mated to reactive males, as this would reduce the number of regulatory elements in the following generation.

Pélisson and Bregliano (40) followed the behavior of a strain of flies established by intercrossing the progeny of SF females. The genotype of the parental reactive strain was reconstructed in the daughters of the SF females, the chromosomes of this dysgenic line differing from the original reactive chromosomes only by I elements acquired through chromosome contamination. The fertility of about 1,000 females was tested at each generation, and 60 to 120 of the least fertile individuals were used to continue the line. The proportion of females showing SF sterility dropped from 80 to 20% over the first four generations and then remained constant. The number of potentially functional I factors present at various generations was measured by in situ hybridization, scoring the number of euchromatic sites of I sequences, and by Southern transfer experiments designed to reveal fragments internal to I factors. The reactive strain used to start the experiment had no I elements at euchromatic sites, but they quickly accumulated there, reaching an average of about six sites per genome within the first three generations. They remained fairly constant thereafter, although there was considerable variation in the number of euchromatic elements among individuals within a generation. The results of the Southern transfer experiments also indicated that the number of I elements increased early in the experiment and then reached a plateau (40).

At first sight these results suggest that SF sterility can be maintained in the absence of transposition, perhaps because fertility is more sensitive to low levels of I-factor expression. This need not be the case, as transposition may have taken place at each generation, the number of I elements remaining fairly constant because selection of the least fertile females at each generation also selected for low numbers of I elements. This was tested by measuring the number of euchromatic I elements in the progeny of the 60 most fertile and 60 least fertile females at generation 15. The average number of I elements from the fertile females, 10.7, was higher than that from the infertile females, 8.7, although the difference was barely significant. The number of euchromatic I elements also increased when the line was continued for eight generations without selection, but again the difference was slight (40). Unfortunately the techniques used in this experiment do not distinguish between functional and nonfunctional elements, as it is the number of functional elements that is likely to be critical for regulating dysgenesis. Nevertheless, these results are consistent with the idea that SF sterility and transposition continue until the number of I elements present reaches a threshold above which a strain will become inducer.

X. WHAT IS REACTIVITY?

Reactivity is a cellular environment that permits I-factor expression as recognized by reduced fertility, transposition of I factors, and increased mutation and nondisjunction. Not all reactive strains exhibit the same level of reactivity, and strong- and weak-reactive strains can be distinguished in crosses with standard inducer males. Some reactive stocks are heterogeneous and contain females giving different responses when mated with inducer males. Lines containing only strong- or weak-reactive females can be derived from such stocks by selection (5, 45).

The female progeny of crosses between strong- and weak-reactive strains show a level of reactivity similar to that of their mothers, although some females with intermediate levels of reactivity are found (5, 10). This suggests that reactivity levels are maternally inherited, but can be modified by the nuclear genome. The ability of individual chromosomes to modify reactivity has been tested by crossing females from a strong-reactive strain with males heterozygous for strong- and weak-reactive chromosomes that were marked genetically (43). Females with each of the eight possible genotypes with respect to chromosomes X, *2*, and *3* were mated with inducer males, and the fertility of their female progeny was measured. All females were infertile, as expected since they were descended from a strong-reactive strain through the maternal line, but their fertility varied according to the number of weak-reactive chromosomes carried by their mothers. All chromosomes from the weak-reactive strain could modify reactivity, indicating that several genetic factors are involved. This was confirmed in an equivalent experiment in which various numbers of chromosomes from a strong-reactive strain were introduced into the genome of females descended from a weak-reactive strain through the maternal line.

The level of reactivity of these strains was ultimately determined by their nuclear genotypes, as flies descended from strong-reactive mothers, but having chromosomes X, *2*, and *3* from a weak-reactive strain, eventually reached a low level of reactivity. This took about 10 generations, suggesting that reactivity is controlled by both nuclear and cytoplasmic factors and that the cytoplasmic factors can be perpetuated for several generations.

The defective I elements present in the genomes of both reactive and inducer strains are likely candidates for chromosomal factors modifying reactivity. If there are molecules that regulate I-factor expression directly in inducer strains, then a high level of reactivity could simply be the complete absence of such molecules. Intermediate levels of reactivity might then be due to low levels of regulation mediated by a subset of I elements located on all chromosomes. These could be elements that have their regulatory sequences intact but are otherwise defective, or I factors subject to position effects. Strong- and weak-reactive strains could have different numbers of these particular I elements, or their expression might be modified by their chromosomal environment. Variations in reactivity are not due to large quantitative differences in the number of I elements, as no differences between I-factor sequences could be detected in two pairs of strong- and weak-reactive strains, each descended from a heterogeneous founder stock (9). This latter study was performed using Southern transfer experiments, so small differences in I-factor structure might not have been noticed.

An alternative explanation for variations in reactivity is that reactive strains express something that is required for hybrid dysgenesis in addition to functional I factors and that expression of genes coding for these factors is inhibited in inducer strains. The chromosomal determinants responsible for various levels of reactivity might then be different alleles of these genes. There is no evidence to support this model, however.

The most likely site for a cytoplasmically inherited determinant would seem to be the mitochondrial genome, although a comparison of mitochondrial DNAs from inducer and reactive strains by heteroduplexing and restriction mapping detected no differences between them (47). A more interesting comparison might be between mitochondrial DNAs from strong- and weak-reactive strains. The number of repeated sequences in mitochondrial DNA in eggs laid by *Drosophila mauritiana* females changes with maternal age (55), and similar changes in *D. melanogaster* might explain changes in the fertility of SF females with age and the difference between strong- and weak-reactive lines.

Reactivity is also affected by nongenetic factors such as temperature and aging. The hatchability of eggs laid by SF females reared at 20°C increases two- to threefold after the temperature is raised to 25 or 29°C, suggesting that there is a temperature-sensitive period late in oogenesis (46). If these females are returned to 20°C, then the hatchability of their eggs falls again and may go below its original level (6).

Although these changes in SF sterility could be regarded as effects of aging and temperature on the interaction between I-factor expression and oogenesis, they are cumulative and can be passed from one generation to the next (6). In studies of two lines from a strong-reactive strain, one maintained with females 2 to 4 days old and the other with females 45 to 55 days old, the "young" line continued to be strongly reactive whereas the "old" line became weakly reactive after about three generations (7). This effect was reversible, and the old line gave a strong-reactive response within about five generations if there was a switch to using young females as parents. K. Luning (*Hereditas*, in press) has studied changes in reactivity due to aging and has concluded that these must be due to variations in the number or activity of several cytoplasmic elements.

The effects of heat treatment are also cumulative; lines propagated from eggs laid a few days after placing females at 29°C show reduced levels of reactivity. This effect falls off if flies are again maintained at 20°C, but 10 generations or so may elapse before reactivity returns to its original level. A change in temperature during oogenesis may be more important than the temperature itself, as flies held continuously at 29°C maintain the level of reactivity shown at 20°C (7).

As yet there is no satisfactory explanation for the effects of temperature and aging on the inheritance of reactivity. Presumably they modulate the expression of chromosomally or extrachromosomally inherited genetic determinants, possibly incomplete I elements, but the molecular basis for this is not known.

XI. MECHANISM OF I-FACTOR TRANSPOSITION

There is no direct evidence to indicate the mechanism by which I elements transpose, but there are several clues. The absence of terminal repeats, the 5′-truncated elements that have recently transposed, and the reverse transcriptase-like sequence of ORF2 all suggest that transposition involves reverse transcription of an RNA intermediate. This RNA must include the entire sequence of a donor element, since the ends of complete I factors are well conserved and the transposed copy of the w^{IR3} I factor differs from the donor element only by one TAA repeat.

The structures of recently transposed incomplete I elements are consistent with this RNA being synthesized from left to right in Fig. 2 and then reverse transcribed in the opposite direction. Truncated elements would result from premature termination of reverse transcription. The promoter for this RNA must lie within the I factor itself, since the w^{IR3} factor can transpose from a 6.2-kb restriction fragment that does not include a known promoter (Pritchard et al., in press). This internal promoter is unlikely to be recognized by RNA polymerase III, as the first 200 bp of the I factor do not match the *boxA* and *boxB* sequences characteristic of PolIII promoters (20) and both strands contain oligo(T) regions that are known to serve as PolIII terminators in *Xenopus* sp. (2). We do not know whether the truncated elements associated with y^{IR3} and y^{IR4} are exact copies of deleted donor elements or whether they were truncated during transposition. The latter seems more likely, as elements that are deleted at their 5′ ends should have lost the promoter required for synthesis of a transposition intermediate. No full-length I-factor transcripts have been detected, but these might be produced only in a subset of female germ cells during dysgenesis. This RNA could be reverse transcribed and integrated directly, or it could be packaged into particles having the product of ORF1 as the major structural protein and including the reverse transcriptase encoded by ORF2.

Nothing is known about the priming of reverse transcription or the mechanism by which I elements integrate. The 5′-truncated elements that are flanked by target site duplications indicate that the left-hand end of complete I factors is not required for integration. Neither ORF1 nor ORF2 has a sequence similar to the endonuclease domains of retroviral *pol* genes, and I elements may integrate at staggered single-strand breaks that occur fortuitously in chromosomal DNA. This would explain the variable lengths of the target site duplications flanking I elements. The I factors associated with the w^{IR1}, w^{IR3}, and w^{IR4} mutations have inserted at the same site, but this may indicate that this sequence is frequently nicked, even in the absence of I-R dysgenesis, rather than that an I-factor-encoded endonuclease is involved in integration.

A family of transposable elements, the Cin4 elements, with a DNA sequence organization similar to that of I elements, has been found in the genome of *Z. mays* (53). Like I elements, they have no terminal repeats, have an A-rich region at the 3′ end of one strand, and encode a putative reverse transcriptase. An elegant model has been proposed for their transposition in which the 3′ ends of Cin4 transcripts associate with the 5′ ends of chromosomal DNA in regions with staggered single-strand breaks, the 3′ end of the opposite DNA strand acting as the primer for reverse transcription (53). The same could be true of I elements. The products of ORF2 or ORF1 or both could bind the UAA repeats at the 3′ ends of transposition intermediates and to the 5′ end of a staggered break at the target site (Fig. 8b). Many of the target site duplications determined so far contain TAA, in some cases as the first three bases. This may reflect a preference of the ORF2 or ORF1 products for $^{U}_{T}$AA and might contribute to variation in the number of TAA repeats at the 3′ ends of I elements. This cannot be the only mechanism for gaining TAA repeats, as the transposed copy of the w^{IR3} I factor has an extra copy even though TAA is not the first three bases of the w^{IR3} target site duplication (Fig. 3).

In the Cin4 model, integration is favored by base-pairing between sequences within Cin4 RNA and the target site, and reverse transcription terminates at these points of association to give truncated elements. This is unlikely to be the case for I elements. There is no relationship between the target site duplications adjacent to 5′-truncated I elements and the I-factor sequence next to their breakpoints, and in any case, complete I elements frequently transpose without deletion. A possible mechanism for I-factor transposition is shown in Fig. 8.

The defective I elements that are found embedded in repeated sequences in both inducer and reactive strains probably do not transpose. No elements of this type have been found associated with I-R-induced mutations or at other euchromatic sites, and they are not flanked by target site duplications. The elements that have been sequenced so far would not be expected to code for a functional reverse transcriptase because of changes in the region encoding ORF2 (14a), and those that lack the 5′ end would be unable to synthesize a transposition intermediate. Transcription of these elements may also be inhibited because of their chromosomal location. If the products of ORF1 and ORF2 bind tightly to the RNA

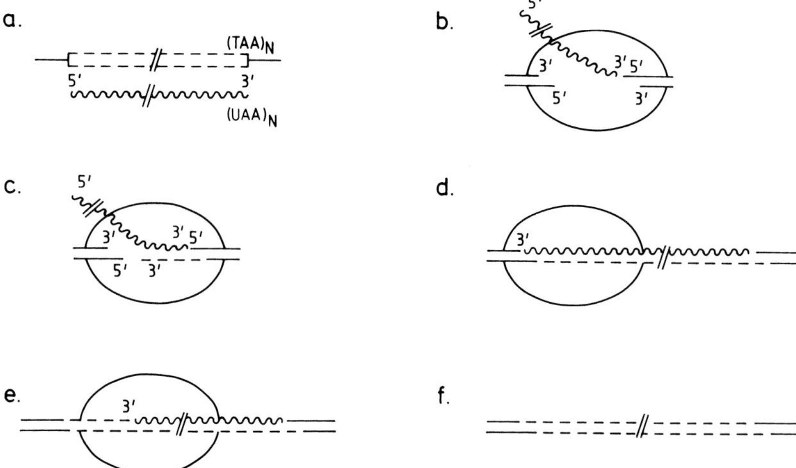

Figure 8. A possible mechanism for I-factor transposition. (a) A full-length I-factor RNA is transcribed using an internal promoter. (b) An I-factor-encoded protein, the product of ORF2 or ORF1 or both, binds the 3′ end of this RNA and a site on the chromosome with staggered single-strand breaks. The 3′ end of the RNA is associated with the 5′ end of one DNA strand. The 5′ end is protruding in this figure, but could equally well be recessed. (c) The reverse transcriptase activity of the ORF2 product uses the 3′ end of the opposite strand of chromosomal DNA to prime synthesis of the first I-factor strand. (d) Reverse transcription continues, and the newly synthesized DNA strand is ligated to the 5′ end of chromosomal DNA at the opposite of the single-strand break. This is shown occurring after reverse transcriptase has copied the complete I-factor RNA, but could happen prematurely to give a truncated element. (e) The ORF2 product now synthesizes the second I-factor strand, using its RNase H activity to degrade RNA template for the first strand as it goes. (f) The second strand is completed and is ligated to chromosomal DNA. This completes synthesis and integration of a new I factor and creates a target site duplication from the sequence between the staggered single-strand breaks that were present at the site of insertion. I-factor RNA is indicated by wavy lines, I-factor DNA is shown by broken lines, and chromosomal DNA is indicated by unbroken lines. The oval represents a complex of the products of ORF1 and ORF2.

from which they have been translated, and in so doing block further translation, then transposition of defective elements using proteins supplied in *trans* might be infrequent or nonexistent, even if they were transcribed.

There may well be host factors required for transposition, but these have yet to be analyzed. One intriguing possibility is that the products of genes required for meiotic recombination might be involved in transposition, since both are confined to females. One role they could play would be to generate the staggered single-strand breaks in chromosomal DNA that serve as targets for transposition. This could be investigated by observing the effect on I-R dysgenesis of mutations that reduce meiotic recombination.

The molecular basis of SF sterility is completely unknown. Perhaps the simplest explanation would be that it is a direct consequence of the transposition of I elements. There are abnormalities in mitosis in embryos laid by SF females, and most of the affected embryos die after two or three nuclear divisions with fragmented chromosomes and abnormal spindle fibers (30, 31). These might result if many single- or double-stranded breaks were left in chromosomes as a result of incomplete transposition events.

XII. PHYLOGENETIC DISTRIBUTION OF I-FACTOR SEQUENCES

Sequences hybridizing to I-factor probes have been detected in all but 1 of the 21 members of the *D. melanogaster* species group tested so far, and in four species outside it (11, 56). This observation suggests that I elements are an old component of the genomes of the *D. melanogaster* species group. The elements most like the 5.4-kb I factor of *D. melanogaster* are found in *Drosophila simulans*, *D. mauritiana*, and *Drosophila sechellia*, and DNAs from these species give a hybridization pattern similar to that of an inducer strain in a Southern transfer experiment (11). About two-thirds of the long I elements in *D. simulans* contain an additional *Hin*dIII site within the 2.3-kb *Hin*dIII-*Pst*I fragment used to distinguish inducer and reactive strains of *D. melanogaster*, and this site is present in all of the long I elements of *D. mauritiana* and *D. sechellia*. Very similar patterns of hybridization were found with DNAs from 19 different *D. simulans* strains, many recently established from the wild, and none resembled that of a reactive strain.

The distribution of I-factor sequences on the salivary gland chromosomes of *D. simulans* and *D.*

mauritiana is also like that of inducer strains in that they are concentrated in the chromocenter with several additional sites on the chromosome arms. The actual positions of these sites of hybridization were different in chromosomes from two strains of *D. simulans*, suggesting that I elements can transpose in this species (54).

The genome of *Drosophila teissieri* also contains several copies of conserved I elements, but their structure is quite unlike that of *D. melanogaster* elements as few of the internal restriction fragments of complete I factors can be detected in digests of *D. teissieri* DNA or in cloned *D. teissieri* elements (11, 54). These elements are located in the chromocenter and on the chromosome arms, and some may be transposable, because the positions of these sites were different when chromosomes of two strains were compared.

XIII. EVOLUTION OF I-R DYSGENESIS IN *D. MELANOGASTER*

Soon after they realized that strains of *D. melanogaster* are of two types with respect to I-R dysgenesis, Picard et al. (45) noticed that reactive strains were only found among old laboratory stocks, while inducer flies could be found in the wild. This was confirmed when about 250 strains established between 1920 and 1980 were tested (28). Reactive strains are only common among the oldest stocks, although one neutral strain was established between 1970 and 1980.

Two explanations have been put forward to account for this relationship between the age of a stock and its inducer-reactive status (27). One suggestion is that *D. melanogaster* is naturally inducer and that reactive flies appear as a result of laboratory culture (8). This is equivalent to the "stochastic loss" model proposed to account for the existence of M strains in the P-M system of hybrid dysgenesis (16). The alternative proposal is that all *D. melanogaster* strains were reactive until the 1930s. At this time I factors appeared in one or more populations, from which they spread during the next 50 years, so that today all populations are inducer. These I factors might have entered *D. melanogaster* horizontally from another species or have arisen de novo from I elements which must have been present in all reactive strains (9, 11). These I elements would presumably be the relics of active I factors that had been in the genome in former times. The idea of horizontal spread of I factors is equivalent to the "recent invasion" hypothesis for P factors (1, 25). These possibilities are not mutually exclusive, and there is no direct experimental evidence to support either. No one has demonstrated either an inducer strain becoming reactive or an I factor appearing spontaneously, although Luning (in press) has preliminary evidence for the latter.

Acknowledgments. I am grateful to my colleagues in Edinburgh and Clermont-Ferrand, with whom I have had many stimulating discussions regarding the molecular basis of I-R dysgenesis, and to all those who have allowed me to present their unpublished data. I am also indebted to S. Lake, M. Lynch, M. Pritchard, and J. Prosser for their comments on the manuscript, to A. Wilson for artwork, and to G. Brown for photography.

Some of the experiments described in this review were carried out with financial support of project grants from the Medical Research Council.

LITERATURE CITED

1. **Bingham, P. M., M. G. Kidwell, and G. M. Rubin.** 1982. The molecular basis of P-M hybrid dysgenesis: the role of the P element, a P-strain specific transposon family. *Cell* **29:** 995–1004.
2. **Bogenhagen, D. F., and D. D. Brown.** 1981. Nucleotide sequences in Xenopus 5S DNA required for transcription termination. *Cell* **24:**261–270.
3. **Bregliano, J. C., and M. G. Kidwell.** 1983. Hybrid dysgenesis determinants, p. 363–410. *In* J. Shapiro (ed.), *Mobile Genetic Elements*. Academic Press, Inc., New York.
4. **Bregliano, J. C., G. Picard, A. Bucheton, A. Pelisson, J. M. Lavigem, and P. L'Heritier.** 1981. Hybrid dysgenesis in *Drosophila melanogaster*. *Science* **207:**606–611.
5. **Bucheton, A.** 1973. Contribution à l'étude de la sterilité femelle non mendelienne chez *Drosophila melanogaster*. Transmission héréditaire des degrés d'efficacité du facteur "reacteur." *C. R. Acad. Sci. Paris Ser. D* **276:**641–644.
6. **Bucheton, A.** 1978. Non-mendelian female sterility in *Drosophila melanogaster*: influence of aging and thermic treatments. *Heredity* **41:**357–369.
7. **Bucheton, A.** 1979. Non-mendelian female sterility in *Drosophila melanogaster*: influence of aging and thermic treatments. III. Cumulative effects induced by these factors. *Genetics* **93:**131–142.
8. **Bucheton, A., J. M. Lavige, G. Picard, and P. L'Heritier.** 1976. Non-mendelian female sterility in *Drosophila melanogaster*: quantitative variations in the efficiency of inducer and reactive strains. *Heredity* **36:**305–314.
9. **Bucheton, A., R. Paro, H. M. Sang, A. Pélisson, and D. J. Finnegan.** 1984. The molecular basis of I-R hybrid dysgenesis: identification, cloning and properties of the I factor. *Cell* **38:**153–163.
10. **Bucheton, A., and G. Picard.** 1978. Non-mendelian female sterility in *Drosophila melanogaster*: hereditary transmission of reactivity levels. *Heredity* **40:**207–223.
11. **Bucheton, A., M. Simonelig, C. Vaury, and M. Crozatier.** 1986. Sequences similar to the I transposable element involved in I-R hybrid dysgenesis in *D. melanogaster* occur in other *Drosophila* species. *Nature* (London) **322:**650–652.
12. **Burke, W. D., C. C. Calalang, and T. H. Eickbush.** 1987. The

site-specific ribosomal insertion element type II of *Bombyx mori* (R2Bm) contains the coding sequence for a reverse transcriptase-like enzyme. *Mol. Cell. Biol.* 7:2221–2230.

13. **Chia, W., G. Howe, M. Martin, Y. B. Meng, K. Moses, and S. Tsubota.** 1986. Molecular analysis of the *yellow* locus of *Drosophila. EMBO J.* 5:3597–3605.

14. **Covey, S. N.** 1986. Amino acid sequence homology in *gag* region of reverse transcribing elements and the coat protein gene of cauliflower mosaic virus. *Nucleic Acids Res.* **14:** 623–633.

14a. **Crozatier, M., C. Vaury, I. Busseau, A. Pélisson, and A. Bucheton.** 1988. Structure and genomic organization of I elements involved in I-R hybrid dysgenesis in *Drosophila melanogaster. Nucleic Acids Res.* **16:**9199–9213.

15. **Di Nocera, P. P., M. E. Digan, and I. Dawid.** 1983. A family of oligo-adenylated transposable sequences in *Drosophila melanogaster. J. Mol. Biol.* **168:**715–727.

16. **Engels, W.** 1981. Hybrid dysgenesis in Drosophila and the stochastic loss hypothesis. *Cold Spring Harbor Symp. Quant. Biol.* **45:**561–565.

17. **Fawcett, D. H., C. K. Lister, E. Kellett, and D. J. Finnegan.** 1986. Transposable elements controlling I-R hybrid dysgenesis in D. melanogaster are similar to mammalian LINEs. *Cell* **47:**1007–1015.

18. **Finnegan, D. J.** 1988. I-factors in *Drosophila melanogaster* and similar elements in other eukaryotes, p. 271–285. *In* A. J. Kingsman, S. Kingsman, and K. Chater (ed.), *Transposition.* Cambridge University Press, Cambridge.

18a. **Finnegan, D. J., I. Busseau, M. Lynch, D. H. Fawcett, C. K. Lister, A. Pélisson, H. M. Sang, and A. Bucheton.** 1988. The structure of mutations induced by I-R hybrid dysgenesis in *Drosophila melanogaster*, p. 209–215. *In* M. Lambert, J. McDonald, and I. B. Weinstein (ed.), *Banbury Report 30: Eukaryotic Transposable Elements as Mutagenic Agents.* Cold Spring Harbor Laboratory, Cold Spring Harbor, N.Y.

19. **Galibert, F., E. Mandart, F. Fitoussi, P. Tiollais, and P. Charnay.** 1979. Nucleotide sequence of the hepatitis B virus genome (subtype ayw) cloned in *E. coli. Nature* (London) **281:**646–650.

20. **Galli, G., H. Hofstetter, and M. L. Birnsteil.** 1981. Two conserved sequence blocks within eukaryotic tRNA genes are major promoter elements. *Nature* (London) **294:**626–631.

21. **Hattori, M., S. Kuhara, O. Takenaka, and Y. Sakaki.** 1986. L1 family of repetitive DNA sequences in primates may be derived from a sequence encoding a reverse transcriptase-related protein. *Nature* (London) **321:**625–628.

22. **Johnson, M. S., M. A. McClure, D. F. Feng, T. Gray, and R. F. Doolittle.** 1986. Computer analysis of retroviral *pol* genes: assignment of enzymatic functions to specific sequence and homologies with nonviral enzymes. *Proc. Natl. Acad. Sci. USA* **83:**7648–7652.

23. **Kamer, G., and P. Argos.** 1984. Primary structure comparison of RNA-dependent polymerases from plant, animal and bacterial viruses. *Nucleic Acids Res.* **12:**7269–7282.

24. **Kanaga, S., and R. J. Crouch.** 1983. DNA sequence of the gene coding for *Escherichia coli* ribonuclease H. *J. Biol. Chem.* **258:**1276–1281.

25. **Kidwell, M. G.** 1979. Hybrid dysgenesis in *Drosophila melanogaster*: the relationship between the *P*-M and *I*-R interaction systems. *Genet. Res.* **33:**205–217.

26. **Kidwell, M. G.** 1983. Intraspecific hybrid sterility, p. 125–153. *In* M. Ashburner, H. L. Carson, and J. N. Thompson (ed.), *Genetics and Biology of Drosophila*, vol. 3c. Academic Press, Inc. (London) Ltd., London.

27. **Kidwell, M. G.** 1983. Evolution of hybrid dysgenesis determinants in *Drosophila melanogaster. Proc. Natl. Acad. Sci. USA* **80:**1655–1659.

28. **Kidwell, M. G., T. Frydryk, and J. B. Novy.** 1983. The hybrid dysgenesis potential of *Drosophila melanogaster* strains of diverse temporal and geographical natural origins. *Drosophila Information Service* **59:**63–69.

29. **Kimmel, B. E., O. K. Moiyoi, and J. R. Young.** 1987. Ingi, a 5.2-kilobase dispersed sequence element from *Trypanosoma brucei* that carries half of a smaller mobile element at either end and has homology with mammalian LINEs. *Mol. Cell. Biol.* **7:**1465–1475.

30. **Lavige, J. M.** 1986. I-R system of hybrid dysgenesis in *Drosophila melanogaster*: further data on the arrest of development of the embryos from SF females. *Biol. Cell.* **56:** 207–216.

31. **Lavige, J. M., and P. Lecher.** 1982. Mitoses anormales dans les embryos à développement bloqué dans le système I-R de dysgenesie hybride chez *Drosophila melanogaster. Biol. Cell.* **44:**9–14.

32. **Leigh Brown, A. J., and J. E. Moss.** 1987. Transposition of the I element and *copia* in a natural population of *Drosophila melanogaster. Genet. Res.* **49:**121–128.

33. **Loeb, D. D., R. W. Padgett, S. C. Hardies, W. R. Shehee, M. B. Comer, M. H. Edgell, and C. A. Hutchison.** 1986. The sequence of a large L1Md element reveals a tandemly repeated 5′ end and several features found in retrotransposons. *Mol. Cell. Biol.* **6:**168–182.

34. **Miller, J., A. D. McLachlan, and A. Klug.** 1985. Repetitive zinc-binding domains in the protein transcription factor IIIA from *Xenopus* oocytes. *EMBO J.* **4:**1609–1614.

35. **Peifer, M., and W. Bender.** 1986. The anterobithorax and bithorax mutations of the bithorax complex. *EMBO J.* **5:** 2293–2303.

36. **Pélisson, A.** 1978. Non-mendelian female sterility in *Drosophila melanogaster*: variations of chromosomal contamination when caused by chromosomes of various inducer efficiencies. *Genet. Res.* **32:**113–122.

37. **Pélisson, A.** 1979. The I-R system of hybrid dysgenesis in *Drosophila melanogaster*: influence of SF sterility of their inducer and reactive paternal chromosomes. *Heredity* **43:** 423–428.

38. **Pélisson, A.** 1981. The I-R system of hybrid dysgenesis in *Drosophila melanogaster*: are I factor insertions responsible for the mutator effect of the I-R interaction? *Mol. Gen. Genet.* **183:**123–129.

39. **Pélisson, A., and J. C. Bregliano.** 1981. The I-R system of hybrid dysgenesis in *Drosophila melanogaster*: construction and characterisation of a non-inducer stock. *Biol. Cell.* **40:** 159–164.

40. **Pélisson, A., and J. C. Bregliano.** 1987. Evidence for rapid limitation of the I element copy number in a genome submitted to several generations of I-R hybrid dysgenesis in *Drosophila melanogaster. Mol. Gen. Genet.* **207:**306–313.

41. **Pélisson, A., and G. Picard.** 1979. Non-mendelian female sterility in *Drosophila melanogaster*: I-factor mapping on inducer chromosomes. *Genetica* **50:**141–148.

42. **Picard, G.** 1976. Non-Mendelian female sterility in *Drosophila melanogaster*: hereditary transmission of I factor. *Genetics* **83:**107–123.

43. **Picard, G.** 1978. Non mendelian female sterility in *Drosophila melanogaster*: sterility in the daughter progeny of SF and RSF females. *Biol. Cell.* **31:**235–244.

44. **Picard, G., J. C. Bregliano, A. Bucheton, J. M. Lavige, A. Pélisson, and M. G. Kidwell.** 1978. Non-mendelian female

sterility and hybrid dysgenesis in *Drosophila melanogaster*. *Genet. Res.* 32:275–287.

45. **Picard, G., A. Bucheton, J. M. Lavige, and A. Fleuriet.** 1972. Contribution à l'étude d'un phenomène de sterilité à determinisme non mendelien chez *Drosophila melanogaster*. *C. R. Acad. Sci. Paris Ser. D* **275:**933–936.
46. **Picard, G., J. M. Lavige, A. Bucheton, and J. M. Bregliano.** 1977. Non-mendelian female sterility: physiological pattern of embryo lethality. *Biol. Cell.* **29:**89–98.
47. **Picard, G., and D. Wolstenholme.** 1980. SF sterility in *Drosophila melanogaster*: a comparative study of mitochondrial DNA molecules from flies of inducer and reactive strains. *Biol. Cell.* **38:**157–162.
48. **Proust, J., and C. Prudhommeau.** 1982. Hybrid dysgenesis in *Drosophila melanogaster*. I. Further evidence for, and characterisation of, the mutator effect of the inducer-reactive interaction. *Mutat. Res.* **95:**225–235.
49. **Saigo, K., W. Kugimya, Y. Matsuo, S. Inouye, K. Yoshioka, and S. Yuki.** 1984. Identification of the coding sequence for a reverse transcriptase-like enzyme in a transposable genetic element in *Drosophila melanogaster*. *Nature* (London) **312:** 659–661.
50. **Sang, H. M., A. Pélisson, A. Bucheton, and D. J. Finnegan.** 1984. Molecular lesions associated with *white* gene mutations induced by I-R hybrid dysgenesis in *Drosophila melanogaster*. *EMBO J.* **3:**3079–3085.
51. **Schwartz, D. E., R. Tizard, and W. Gilbert.** 1983. Nucleotide sequence of Rous sarcoma virus. *Cell* **32:**853–869.
52. **Schwartzberg, P., J. Colicelli, and S. P. Goff.** 1984. Construction and analysis of deletion mutants in the *pol* gene of Moloney murine leukemia virus: a new viral function required for productive infection. *Cell* **37:**1043–1052.
53. **Schwarz-Sommer, Z., L. Leclercq, and H. Saedler.** 1987. Cin4, an insert altering the structure of the *A1* gene in *Zea mays*, exhibits properties of nonviral retrotransposons. *EMBO J.* **6:**3873–3880.
54. **Simonelig, M., C. Bazin, A. Pélisson, and A. Bucheton.** 1987. Transposable and non-transposable elements homologous to the I factor involved in I-R hybrid dysgenesis in *Drosophila melanogaster* coexist in various Drosophila species. *Proc. Natl. Acad. Sci. USA* **84:**1141–1145.
55. **Solignac, M., J. Genermont, M. Monnerot, and J.-C. Mounolou.** 1987. Drosophila mitochondrial genetics: evolution of heteroplasmy through germ line cell divisions. *Genetics* **117:**687–696.
56. **Stacy, S. N., R. A. Lansman, H. W. Brock, and T. Grigliatti.** 1986. Distribution and conservation of mobile elements in the genus *Drosophila*. *Mol. Biol. Evol.* **6:**522–534.
57. **Temin, H. M., and S. Mizutani.** 1970. RNA-dependent DNA polymerase in virions of Rous sarcoma virus. *Nature* (London) **226:**1211–1213.
58. **Toh, H., H. Hayashida, and T. Miyata.** 1983. Sequence homology between retroviral reverse transcriptase and putative polymerases of hepatitis B virus and cauliflower mosaic virus. *Nature* (London) **305:**827–829.
59. **Verma, I. M.** 1975. Studies on reverse transcription of RNA tumor viruses. I. Localization of thermolabile DNA polymerase and RNase H activities on one polypeptide. *J. Virol.* **15:**121–126.

Chapter 19

F and Related Elements in *Drosophila melanogaster*

D. J. FINNEGAN

I. INTRODUCTION

The genome of *Drosophila melanogaster* contains at least five families of transposable elements, in addition to I elements (D. J. Finnegan, this volume), that are structurally related to mammalian L1 elements and other retroposons (12). The common features of these elements are that they have no terminal repeats and they have a run of A's, preceded by one or more copies of the polyadenylation signal AATAAA, at the 3′ end of one strand. These elements are called F (7) (Fig. 1a), G (6) (Fig. 1b), *Doc* (1) (Fig. 1c), *jockey* (9) (Fig. 1d), and D (11). Other sequences that may be of this class are the type II insertions present in a proportion of 28S rRNA genes. These elements have a poly(A) sequence at the 3′ end of one strand, but are not flanked by a target site duplication and seem to be present only in rDNA (13).

II. F AND G ELEMENTS

The first F element to be discovered, 101F, was found within a nonnucleolar copy of one of the intervening sequences, the type I sequences, found within many 28S rRNA genes in *D. melanogaster* (3). The base sequence of 101F has been compared with those of three other F elements that were cloned because they hybridize to 101F (7). Each element is flanked by a target site duplication that varies in length from 8 to 13 base pairs. The right-hand (3′) ends of these elements are conserved and terminate with a short poly(A) sequence, but their left-hand ends are variable. A fifth F element, Fw, is associated with the w^{1+A} mutation, a derivative of w^{1} (8). This F element is similar to the others, but has a large deletion of the left-hand end (5, 7). There are about 50 to 100 F elements in the genomes of the *D. melanogaster* strains that have been tested. About half are on the chromosome arms, and half are in centromeric DNA. The distributions of these elements are different on chromosomes from different strains, indicating that they are transposable (3, 10). The longest F elements cloned so far are about 4.7 kilobases, but longer elements may exist.

The Fw element is 3,542 base pairs long and contains two long open reading frames (5). The amino acid sequence of the second of these, ORF2, contains a region similar to the polymerase domains of retroviral reverse transcriptases. Of 14 amino acid residues that are invariant in viral reverse transcriptases, 10 are found in ORF2, and there is considerable similarity between the amino acid sequence encoded by ORF2 and those of the corresponding open reading frames of related elements (5; see Fig. 4 of Finnegan, this volume). The first open reading frame, ORF1, starts at the beginning of Fw and extends for 366 base pairs, but may well be longer in F elements that are not truncated at their left-hand ends. The polypeptide encoded by ORF1 may interact with RNA or DNA since it contains one complete and two incomplete copies of the motif $CX_2CX_4HX_4C$ found in nucleic acid-binding seg-

D. J. Finnegan ■ Department of Molecular Biology, University of Edinburgh, King's Buildings, Mayfield Road, Edinburgh EH9 3JR, Scotland.

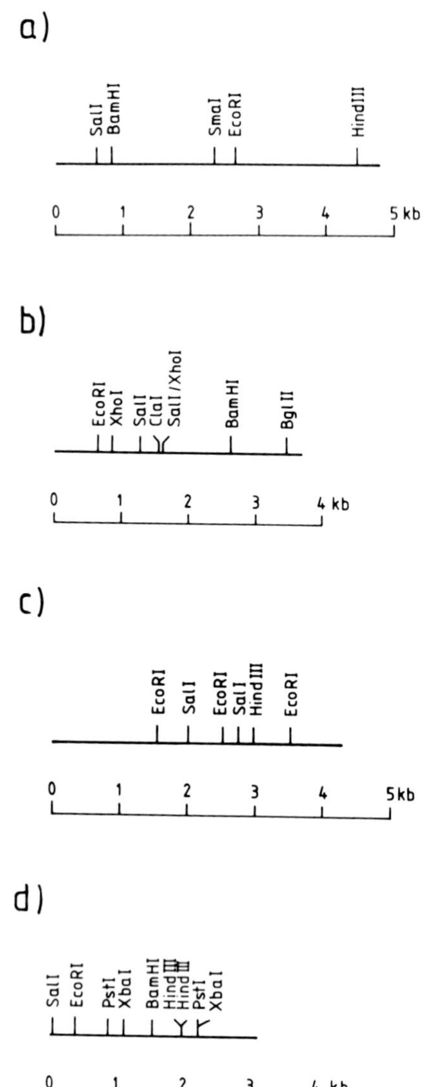

Figure 1. Restriction maps of F, G, *Doc*, and *jockey* elements. These maps are oriented so that their poly(A) sequences are to the right. (a) Consensus map of an F element (6). There are no sites for enzymes *Cla*I, *Pvu*I, and *Xho*I. (b) Map of the G1 element (4). There are no sites for enzyme *Sma*I. (c) Map of the *Doc* element associated with the bx^3 mutation (1). (d) Map of the *jockey* element inserted in an *mdg4/gypsy* element (Y. Ilyin and G. Georgiev, personal communication). There are no sites for enzymes *Bgl*II, *Kpn*I, and *Xho*I, and the *Sal*I site is present in only a few copies of *jockey*. kb, Kilobases.

ments of retroviral *gag* polypeptides and the coat protein of cauliflower mosaic virus (2).

A curious feature of F elements is that the DNA sequence at their extreme 3′ termini is very similar to that of *suffix* elements of *D. melanogaster* (15, 16). These are repeated about 300 to 400 times in the genome, often at the 3′ ends of genes where they are oriented so that their transcribed strand starts with a poly(T) sequence (15). *suffix* elements are not flanked by target site duplications, so either they have moved around the genome by a mechanism that does not generate target site duplications, or such duplications have been obscured by mutation. The sequences of *suffix* elements themselves are well conserved, as five genomic copies have been found to differ from a consensus sequence by only about 1% (15).

The structure of F elements and the fact that they encode a reverse transcriptase-like polypeptide suggest that they transpose by reverse transcription of a polyadenylated RNA intermediate. This RNA probably comes from a few "master" elements that are transcribed either because they are adjacent to suitable promoters or because they contain an internal promoter. Transposed and deleted elements are probably unable to transpose again. There is, as yet, no direct evidence for RNA transposition intermediates. Dawid et al. (3) found low levels of F transcripts in poly(A)$^-$ RNA from embryos, but could find none in poly(A)$^+$ RNA.

The first G elements to be studied were found inserted within F elements (6). They resemble F elements in having a polyadenylation signal and poly(A) tract at the 3′ end of one strand, and one G element has been shown to be flanked by a target site duplication. These elements are mostly inserted in repeated sequences in centromeric DNA and are fairly stable as judged by the results of Southern transfer experiments in which genomic DNAs from different *D. melanogaster* strains are used. When they transpose, they probably do so via RNA intermediates, as G elements also encode a reverse transcriptase-like polypeptide (4).

III. *Doc*, *jockey*, AND D ELEMENTS

The original *Doc* element was found in the Bithorax region of chromosomes carrying the bx^3 mutation but not on other third chromosomes, although it is not responsible for the bx^3 mutant phenotype (1). Similar elements are present at both break points of an inversion associated with the $Antp^{73b}$ mutation (14). They lie in opposite orientation to each other, and the $Antp^{73b}$ inversion probably arose by recombination between them. They have no terminal repeats, but there is a poly(A) region at the 3′ end of the same strand in each. The sequences adjacent to these elements indicate that they would have been flanked by target site duplications before the recombination event that created the $Antp^{73b}$ inversion. Another *Doc* element is inserted in the *white* gene of chromosomes carrying the w^1 mutation (K. O'Hare, personal communication).

There are about 50 *jockey* elements in the *D. melanogaster* genome, but only 1 has been studied in detail. This was found inserted in a copy of the *copia*-like element *mdg4/gypsy* associated with a mutation of the *cut* gene (9). It is flanked by a target site duplication, has no terminal repeats, and has a poly(A) sequence at the 3′ end of one strand (Y. Ilyin, personal communication). The results of Southern transfer experiments indicate that *jockey* elements are well conserved but that their chromosome distributions are variable. D elements are the least well characterized elements of this type. The only copy analyzed so far is a 380-base-pair sequence inserted within the *dunce* locus (11). It has no terminal repeats, has a poly(A) sequence at the 3′ end of one strand, and is flanked by a target site duplication. There are 50 to 100 D elements in the genome, but there is no indication as to whether they are all as short as this copy.

IV. CONCLUSIONS

These five families of transposable elements presumably share a common evolutionary history, although there is no significant DNA sequence homology between them. Only one or two elements of this type have been found in any other eucaryotic genome. In mammals this is probably because L1 elements have multiplied to such high copy numbers. Mammalian genomes may contain a similar spectrum of these elements, most of which have escaped attention because of their low copy number.

Acknowledgments. I am grateful to all those who have allowed me to mention their unpublished data.

LITERATURE CITED

1. **Bender, W., M. Akam, F. Korch, P. A. Beachy, M. Peifer, P. Spierer, E. B. Lewis, and D. S. Hogness.** 1983. Molecular genetics of the bithorax complex of *Drosophila melanogaster*. *Science* **221**:23–29.
2. **Covey, S. N.** 1986. Amino acid sequence homology in *gag* region of reverse transcribing elements and the coat protein gene of cauliflower mosaic virus. *Nucleic Acids Res.* **14**: 623–633.
3. **Dawid, I. B., E. O. Olong, P. P. Di Nocera, and M. L. Pardue.** 1981. Ribosomal insertion-like elements in Drosophila melanogaster are interpersed with mobile sequences. *Cell* **25**: 399–408.
4. **Di Nocera, P. P.** 1988. Close relationship between non-viral retroposons in *Drosophila melanogaster*. *Nucleic Acids Res.* **16**:4041–4052.
5. **Di Nocera, P. P., and G. Casari.** 1987. Related polypeptides are encoded by Drosophila F elements, I factors and mammalian L1 sequences. *Proc. Natl. Acad. Sci. USA* **84**:5843–5847.
6. **Di Nocera, P. P., and I. B. Dawid.** 1983. Interdigitated arrangement of two oligo(A)-terminated DNA sequences in *Drosophila*. *Nucleic Acids Res.* **11**:5475–5482.
7. **Di Nocera, P. P., M. E. Digan, and I. Dawid.** 1983. A family of oligo-adenylated transposable sequences in *Drosophila melanogaster*. *J. Mol. Biol.* **168**:715–727.
8. **Karess, R., and G. M. Rubin.** 1982. A small tandem duplication is responsible for the unstable *white*-ivory mutation in Drosophila. *Cell* **30**:63–69.
9. **Mizrokhi, L. J., L. A. Obolenkova, A. F. Priimagi, Y. V. Ilyin, T. I. Gerasimova, and G. P. Georgiev.** 1985. The nature of unstable insertion mutations and reversions in the locus *cut* of *Drosophila melanogaster*: molecular mechanism of transposition memory. *EMBO J.* **4**:3781–3787.
10. **Pardue, M. L., and I. B. Dawid.** 1981. Chromosomal locations of two DNA segments that flank ribosomal insertion-like sequences in *Drosophila*: flanking sequences are mobile elements. *Chromosoma* **83**:29–43.
11. **Pittler, S. J., and R. L. Davis.** 1987. A new family of the poly-deoxyadenylated class of Drosophila transposable elements identified by a representative member at the *dunce* locus. *Mol. Gen. Genet.* **208**:325–328.
12. **Rogers, J. E.** 1985. The origin and evolution of retroposons. *Int. Rev. Cytol.* **93**:188–280.
13. **Roiha, H., J. R. Miller, L. C. Woods, and D. M. Glover.** 1981. Arrangements and rearrangements of sequences flanking the two types of rDNA insertions in *D. melanogaster*. *Nature* (London) **290**:749–753.
14. **Schneuwly, S., A. Kuroiwa, and W. J. Gehring.** 1987. Molecular analysis of the dominant homeotic *Antennapedia* phenotype. *EMBO J.* **6**:201–206.
15. **Tchurikov, N. A., A. K. Ebralidz, and G. P. Georgiev.** 1986. The suffix sequence is involved in processing the 3′ ends of different mRNAs in *Drosophila melanogaster*. *EMBO J.* **5**:2341–2347.
16. **Tchurikov, N. A., A. K. Naumova, E. S. Zelentsova, and G. P. Georgiev.** 1982. A cloned unique gene of Drosophila melanogaster contains a repetitive 3′ exon whose sequence is present at the 3′ ends of many different mRNAs. *Cell* **28**:365–373.

Chapter 20

The Transposable Element *hobo* of *Drosophila melanogaster*

RONALD K. BLACKMAN and WILLIAM M. GELBART

I. INTRODUCTION

The *hobo* element is a recent addition to the collection of mobile elements known in *Drosophila melanogaster*. As will be described below, the physical structure, genomic distribution, and genetic behavior of *hobo* are reminiscent of the properties of the P-element system (W. R. Engels, this volume), although notable differences between the two exist. The I-element system (D. J. Finnegan, this volume) also has genetic analogies with P, although its properties are distinct from those of *hobo* (2) and will not be discussed further.

The extent of the similarity between P and *hobo* is unclear at this time. In part, this uncertainty reflects historical differences in the early studies of the two systems. The P element was identified as a molecular entity only after years of study on the phenomenology of P-M hybrid dysgenesis, in which hypermutability was only one of several traits exhibited by the progeny of certain outcrosses (Engels, this volume). In contrast, the *hobo* element was cloned and characterized about 5 years ago (18), but was only implicated in genetic instabilities within the past 2 years. Indeed, since the instabilities associated with *hobo* have only been characterized in a few systems preselected for their hypermutability, it is unclear whether these initial observations are even representative of the responses of "typical" *hobo*-containing strains. Thus, there are several characteristics of the P-element system, including the phenomena of hybrid dysgenesis and cytotype, for which comparable information for *hobo* is limited. It is the purpose of this review to outline the current information on the molecular and genetic properties of *hobo*, as gleaned from the initial descriptions of the system, and to compare these properties with those of P.

II. MOLECULAR ANALYSIS

A. Distribution of *hobo* Elements

Two classes of strains have been defined with regard to *hobo* elements (25). H (or *hobo*-containing) strains contain 3.0-kilobase (kb) elements (complete elements, see below) and, typically, numerous smaller derivatives of the element. E (or empty) strains lack

Ronald K. Blackman and William M. Gelbart ■ Department of Cellular and Developmental Biology, Harvard University, 16 Divinity Avenue, Cambridge, Massachusetts 02138-2097.

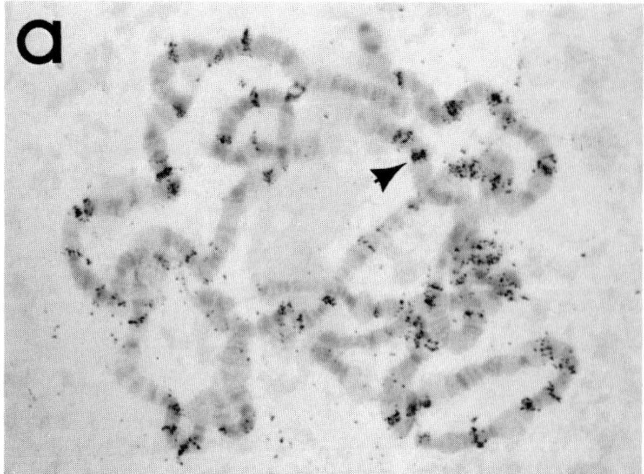

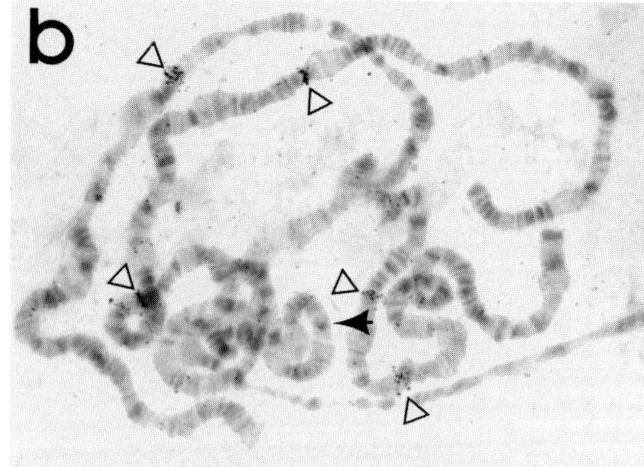

Figure 1. Genomic distribution of *hobo* elements. Salivary gland polytene chromosomes were analyzed by hybridization in situ with tritium-labeled *hobo* probes. (a) Chromosomes of the OR^S strain, the strain from which $dpp^{d\text{-}blk}$ is derived (see text and reference 2), hybridized to plasmid pRG2.6X (2), a probe that locates all *hobo* elements. Approximately 75 euchromatic sites are labeled. (b) OR^S chromosomes hybridized to plasmid pRG0.9R (2), a probe specific for full-length *hobo* elements in this strain. Hybridization signal, noted by open triangles, is found at five sites. Note that no labeling of the chromocenter is apparent in either panel. The location of the *decapentaplegic* gene, 22F1,2, is indicated by a filled arrowhead.

all such elements. These strain designations are based solely on molecular criteria.

In most strains examined, the number of 3.0-kb elements in the genome is low, about 2 to 10 copies (2, 25; unpublished results). Two exceptions are the related dpp^{d1} and dpp^{d1R+} strains, which have 30 to 50 copies of 3-kb elements (2). However, further analysis revealed microheterogeneity among these elements, suggesting that not all are fully functional (unpublished results). The low copy number of complete *hobo* elements suggests that numerous intact elements may be deleterious, perhaps because they promote inordinately high levels of instability. Alternatively, full-length elements may be refractory to transposition.

The smaller ("defective") *hobo* elements result from internal deletions within full-length copies (25), equivalent to those found in defective P elements (19). While there may be 30 to 75 defective *hobo* elements per genome, they usually constitute only a few size classes (2, 25). Different H strains contain different size classes of defective elements. The homogeneity of defective elements within a given strain is in contrast to the heterogeneous size distribution typically found for P elements within a strain (19). The spread of identical defective elements throughout the genome suggests that (i) the rate of *hobo* element transposition is greater than its rate of internal deletion (25) and (ii) there is a strong preference for the amplification of defective elements rather than complete elements (see above).

Hybridization in situ to H-strain polytene chromosomes (Fig. 1) shows that the *hobo* elements are located in euchromatic sites throughout the genome. No labeling of the chromocenter is observed. P elements show a similar distribution (1).

Both H and E strains contain other DNA fragments which weakly hybridize to *hobo* element probes when assayed by genomic Southern blots (2, 25). These uncharacterized sequences probably lack at least one of the termini of the *hobo* element and, in fact, may be functionally unrelated to *hobo*. These sequences cannot be localized by in situ hybridization to polytene chromosomes, suggesting that they reside in underreplicated heterochromatic sequences (Fig. 1; J. Lim, personal communication).

hobo appears to be more frequently represented in long-established laboratory strains than is P (unpublished results) and has been observed in *D. simulans* and *D. mauritiana* (25), sibling species of *D. melanogaster*. With regard to the P-M and H-E systems, we have encountered strains that are PH, MH, and ME by molecular criteria; our initial surveys have not uncovered PE strains (unpublished results).

B. *hobo* Element Sequence Organization

The ubiquitous presence of the 3.0-kb size class of *hobo* elements in H strains suggested (25) that these are the "complete," fully functional *hobo* elements, i.e., elements capable of transposing themselves and mediating the transposition of other *hobo* elements. One 3.0-kb element, called $hobo_{108}$, has

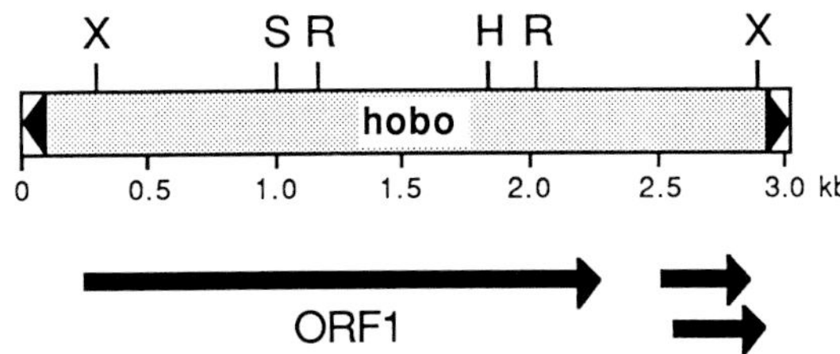

Figure 2. Structure of $hobo_{108}$, a putative complete *hobo* element sequenced by Streck et al. (25). The 12-bp inverted terminal repeats are depicted by black arrowheads. The sequence of the left end repeat is CAGAGAACTGCA. Below the element, the approximate extents of three open reading frames are shown. Sites of cleavage for *Xho*I (X), *Sal*I (S), *Eco*RI (R), and *Hin*dIII (H) are noted.

been sequenced (25) (Fig. 2). Although $hobo_{108}$ has not been tested for function, we have not found restriction site differences between it and another element that permits germ line transformation (see below).

Sequence comparisons of $hobo_{108}$ and P reveal some shared structural features but no obvious sequence similarity (19, 25). The $hobo_{108}$ sequence contains 3,016 base pairs (bp), including 12-bp inverted terminal repeats (Fig. 2). The P element is 2,907 bp long with 31-bp inverted terminal repeats (19). These are the only two known elements of *D. melanogaster* with short, precisely duplicated, inverted repeats (6). Both P and *hobo* produce 8-bp duplications of target sites upon insertion into the genome (19, 25). Some site specificity for *hobo* insertion sites has been observed: all sequenced duplicated target sites end in AC or begin with TTTC (25).

At present, the actual transcriptional and translational products of *hobo* remain unknown, but some inferences can be made from the $hobo_{108}$ DNA sequence. The *hobo* element contains a 2-kb open reading frame (ORF1, Fig. 2). If no internal mRNA splicing occurs, ORF1 is capable of encoding a 73-kilodalton protein. The codon usage for ORF1 is not typical of most *D. melanogaster* genes although it is similar to that of the P element (25). Other smaller open reading frames are also present. There is no amino acid sequence similarity between the P-element polypeptide (9, 11) and the predicted *hobo*-encoded protein. It should be noted that no direct proof exists that $hobo_{108}$ is a functional full-length *hobo* element. It will be important to obtain sequence information on "full-length" elements which have been shown, probably by means of germ line transformation experiments (see below), to be sufficient in *trans* to mobilize other *hobo* elements.

III. GENETIC PROPERTIES OF *hobo* MOBILIZATION

hobo has recently been implicated as an agent of genetic instability in four strains of *D. melanogaster*. However, since much of this research was undertaken without knowledge of the involvement of *hobo*, alternate forms of nomenclature, based on the history of the experimental system, have led to a confusing array of strain names. Two of the strains, dpp^{d1} (unpublished results) and $dpp^{d\text{-}blk}$ (2), were first examined because of the occurrence of spontaneous mutations of the *decapentaplegic* (*dpp*) gene. Each allele contains an independent *hobo* insertion within *dpp* which is capable of participating in numerous additional rearrangements. Instability involving other *hobo* elements in the genome is also seen in these strains. In a third experimental system, *hobo* insertions cause hypermutability and chromosomal rearrangements in the X chromosome of what is termed the unstable chromosome (*Uc*) strain (Lim, personal communication). Finally, *hobo* insertions are associated with chromosomal aberrations produced by the actions of the so-called male recombination factor *23.5MRF* (7, 27). We will refer to this as the *23.5MRF* strain. In this section, we summarize the generalizations that have emerged from these initial reports regarding the genetic behavior of *hobo* and the mutations it causes. While the rates of hypermutability observed in these strains may not be representative of all *hobo*-containing strains, the types of chromosomal instability they produce should be.

A. Conditions Promoting Germ Line Instability

In the P-M system, crosses of P-strain males to M-strain females promote the mobilization of P elements in the germ lines of their progeny, producing hypermutability and other events reflecting chromosomal damage, a syndrome termed hybrid dysgenesis (10; Engels, this volume). The reciprocal cross, M males to P females, is much less effective at promoting instability. Crosses of H and E strains also promote germ line hypermutability in their progeny. In one study, the $dpp^{d\text{-}blk}$ strain, containing a *hobo* element within the *dpp* gene, was used. Genetic assays for further disruption of the gene identified new *dpp* mutations in the germ lines of 1 to 5% of the progeny of H × E crosses (2). Both the H male × E female and the E male × H female crosses gave nearly the same frequency of mosaic germ lines. However, the H male × E female crosses generally produced clusters of two or more mutant flies, whereas the reciprocal crosses typically led to events generating single mutant individuals. Thus, mutability in the H male × E female cross is three- to fourfold higher than that of the reciprocal cross if the rates of recovery of mutant individuals are compared. By comparison, even when cluster size is taken into account, the difference in mutability of the reciprocal crosses in the P-M system is at least three- to fourfold higher than that observed in the H-E system (23).

hobo element instability has also been observed in established stocks (ostensibly H × H crosses) of the four *hobo* strains (7, 8, 12–15, 27–30; unpublished results), as well as in some H × H outcrosses (unpublished results). These results suggest that these particular H strains do not repress *hobo* mobilization. Whether or not this is a property of all H strains remains to be explored.

In the case of the P-M system, the only truly stable crosses are M × M, i.e., crosses between strains lacking full-length P elements. Studies of a strain containing numerous internally deleted *hobo* elements (including one in *dpp*) but lacking complete elements indicate that the H-E system behaves analogously (unpublished results). Crosses of this strain to E mates failed to produce new *dpp* mutations, showing that the *hobo* element in *dpp* was at least 30- to 100-fold less active in this new strain. This suggests that full-length *hobo* elements provide an activity essential for the mobilization of *hobo* elements (presumably a transposase).

B. Somatic Instability?

Normal P elements are immobile in somatic cells (17). In contrast, several observations indicate that *hobo* may be somatically active. In both the *Uc* and *23.5MRF* strains, chromosomal aberrations were found in single nuclei or sectors of nuclei in salivary gland polytene chromosome squashes (13, 14, 29). The breakpoints of these chromosomal rearrangements correlate well with known sites of *hobo* elements (27; Lim, personal communication), suggesting that these rearrangements were *hobo* mediated. In a third system, somatic reversion of the dpp^{d1} phenotype has been observed in homozygous individuals (W. M. Gelbart, unpublished results). Because the dpp^{d1} mutation results from an inversion between two *hobo* elements (unpublished results), these reversion events probably arise by the somatic reinversion of this chromosomal segment. However, all of these observations were anecdotal in nature, and the possibility that *hobo* is active in somatic cells needs further testing.

C. Gonadal Dysgenesis

With the $dpp^{d\text{-}blk}$, *Uc*, and *23.5MRF* strains, specific crosses led to atrophy of gonadal tissues similar to gonadal dysgenesis produced by the P-M system (2, 13, 26, 27). The $dpp^{d\text{-}blk}$ strain is devoid of P elements by molecular criteria, so in this case the possible involvement of P-M hybrid dysgenesis is ruled out (2). By genetic analysis, the *Uc* strain is also an M strain (Lim, personal communication). The *23.5MRF* strain contains P elements, confounding the analysis of this system (27). Because most crosses involving *hobo* elements do not promote gonadal dysgenesis, we cannot be certain that infertility results directly from *hobo* mobilization. However, the fact that three systems displaying *hobo* activity exhibited the uncommon syndrome of gonadal dysgenesis suggests that *hobo* may be involved.

D. *hobo*-Mediated Hybrid Dysgenesis?

P-M hybrid dysgenesis is a genetically defined syndrome associated with chromosomal instability, temperature-dependent infertility, male recombination, transmission ratio distortion, and chromosome nondisjunction (Engels, this volume). At present, *hobo* has been shown to promote genetic instability and possibly gonadal dysgenesis as well. Testing of the other features has not been attempted. Nonreciprocity is a second aspect of P-M hybrid dysgenesis; P × M crosses are more dysgenic than M × P crosses. But the more stable M × P cross still induces significant chromosomal instability (23), and even P × P crosses are active, although much less than either hybrid cross (20). As discussed above, *hobo* mobilization has been observed in H × E and E × H crosses at nearly equal rates and also in H × H crosses. Thus, for these strains, outcrosses are not necessary for *hobo* mobilization. Additional analysis will be needed to determine the genetic, cellular, and environmental factors which can modulate *hobo* activity in these and other strains. Only then can a full evaluation of any contribution of *hobo* to hybrid dysgenesis be determined.

E. Properties of *hobo*-Mediated Mutations

In each of the four *hobo* systems, cytologically visible chromosomal rearrangements are routinely observed. They account for 40% of the *dpp* mutations recovered in screens with the $dpp^{d\text{-}blk}$ chromosome (2) and 30% of the lethal mutations recovered in the *Uc* system (13, 14). The rearrangements usually occur between sites of preexisting *hobo* elements and, in most cases, the elements remain at the breakpoints (2, 27; Lim, personal communication; unpublished results). It is also common to find rearrangements between nearby *hobo* elements. A striking example of this is found for a particular derivative of the *Uc* chromosome which produces *Notch* mutations at a rate of 3.0% per generation (8). All of these are small deletions removing the material between two closely situated *hobo* elements (Lim, personal communication).

hobo mobilization also promotes submicroscopic lesions, including deletions as small as 1.1 kb, small inversions, and new *hobo* insertions (2, 7, 27; Lim, personal communication; unpublished results). Molecular studies (2; Lim, personal communication; unpublished results) have shown that deletion junctions usually retain a *hobo* element.

In the four systems, essentially every *hobo*-mediated rearrangement is confined to a single chromosome arm even though *hobo* elements are distributed throughout the genome (2, 13, 14, 28–30). Even when multiple inversions are present, nearly all are paracentric. A similar absence of interchromosomal translocations has been observed in the P-M system. Because X-chromosome rearrangements were being surveyed, the authors interpreted their findings in terms of the male sterility of X-autosome translocations (3, 5). The results with *hobo* call for a reexamination of whether P-M also induces only intrachromosomal events. The preponderance of *hobo*-mediated paracentric rearrangements might reflect certain aspects of chromosomal architecture within the nuclei of germ line cells. In the salivary gland polytene nucleus, each chromosome arm lies within in its own domain (16). If such domains exist in germ line nuclei, recombinant intermediates or diffusible signals promoting mobilization may be spatially restricted, constraining rearrangements to a single chromosome arm.

IV. GERM LINE TRANSFORMATION

P elements have been used to great advantage as transformation vectors to mediate the insertion of microinjected DNA into the germ line cells of *Drosophila* embryos (21, 24). Sequences placed within the P element are integrated into the recipient chromosomes when the P-specific transposase is also present in the cell. Transposase activity can be supplied by the embryos themselves (if produced in a dysgenic cross) or by a coinjected plasmid carrying an autonomous P element (4, 21, 24). We review here our recent results showing that the *hobo* element can be used to mediate germ line transformation in a similar fashion (R. K. Blackman, M. M. D. Koehler, R.

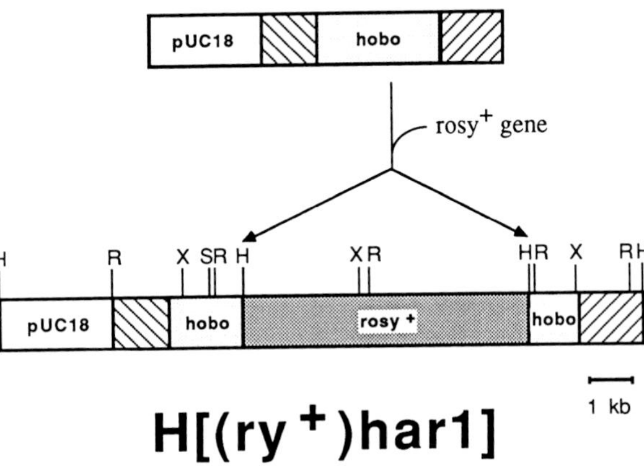

Figure 3. Structure of the H[(ry^+)har1] transposon. A 3-kb *hobo* element and additional flanking sequences (diagonally striped regions) were subcloned into the pUC18 plasmid vector (top line). This plasmid was partially digested with *Hin*dIII, and a 7.3-kb *Hin*dIII fragment containing the $rosy^+$ eye marker gene (22) was inserted into the *hobo* element, producing plasmid H[(ry^+)har1]. Restriction sites of the transposon are abbreviated as in Fig. 2.

Grimaila, and W. M. Gelbart, submitted for publication).

First, we constructed a plasmid, called H[(ry^+) har1] (abbreviated H[har1]), which contained the $rosy^+$ (ry^+) eye color marker gene inserted within a subcloned *hobo* (Fig. 3). This marker gene allowed a phenotypic assay for the integration of the injected DNA. The inserted ry^+ gene disrupted the major *hobo* open reading frame (ORF1), presumably preventing the production of functional transposase by the vector. The first injections used ry^- embryos from an H × E cross, so that endogenous full-length elements (and, therefore, transposase) were present. The *hobo* element marked with ry^+ was efficiently integrated into the germ line (28% of fertile G_0 individuals carried the transposon in their germ lines).

A second set of injection experiments utilized a cloned 3.0-kb *hobo* element to provide the *hobo* products needed in *trans* for the germ line integration of the H[har1] sequences. When this 3.0-kb element and H[har1] were coinjected into embryos from a ry^- E strain, ry^+ transformants were recovered from 25% of the fertile G_0 individuals. In contrast, comparable injections of the H[har1] construct alone failed to produce any germ line transformants in recipients of the same ry^- E strain.

Transformants from both injection series were analyzed molecularly and genetically. Southern blot analysis showed that integration occurred near, if not at, the termini of the *hobo* element without internal rearrangement of the *hobo* or ry^+ sequences. None of the sequences in the H[har1] plasmid that flank *hobo* were integrated. Only a single copy of the transposon was inserted at a given site, and the target sites were widely distributed in the genome.

The properties of *hobo* element transformation are identical to those found for P-element transformation. We speculate that these common properties arise because both elements possess inverted terminal repeats and not because the two elements can share transposase or other regulatory molecules. By use of the P and *hobo* transformation systems, an examination of the independence of the two elements can be made in defined genetic backgrounds and with the controlled introduction of P or *hobo* transposase.

The availability of two transformation systems will also allow the shuttling of modified P and *hobo* elements into the genomes, using the other transposable element. This shuttling system will be useful in the identification of *cis* sequences necessary for *hobo* mobilization. *hobo* elements, altered by in vitro mutagenesis, can be introduced into the genome by P-mediated transformation and assayed for function independent of their own ability to integrate. Finally, *hobo*-mediated transformation may allow the transformation of strains not amenable to P transformation, including strains of the P cytotype and other species in which P is inactive.

V. CONCLUSIONS

Much of the current genetic information on *hobo* element mobilization was documented before the involvement of *hobo* was realized. Still, the analysis is sufficiently complete that the parallels of the *hobo* and P systems are striking. It is clear that there are differences between the systems, however. The study of these may reveal much about the control of transposable element mobilization.

The advent of a germ line transformation system based on the *hobo* element should provide important avenues for the study of the mobilization of this element and should augment the P-element tools available for the in vitro manipulation of *Drosophila* genes.

Acknowledgments. We are extremely grateful to Johng Lim for his permission to cite his results prior to publication, to the editors for their many excellent suggestions, and to Macy Koehler for her many contributions.

This work was funded by a Public Health Service grant to W.M.G. from the National Institutes of Health and by a National Research Service Award fellowship to R.K.B.

LITERATURE CITED

1. **Bingham, P. M., M. G. Kidwell, and G. M. Rubin.** 1982. The molecular basis of P-M hybrid dysgenesis: the role of the *P* element, a P-strain-specific transposon family. *Cell* **29:** 995–1004.
2. **Blackman, R. K., R. Grimaila, M. M. D. Koehler, and W. M. Gelbart.** 1987. Mobilization of *hobo* elements residing within the decapentaplegic gene complex: suggestion of a new hybrid dysgenesis system in *Drosophila melanogaster*. *Cell* **49:** 497–505.
3. **Engels, W. R.** 1979. Germ line aberrations associated with a case of hybrid dysgenesis in *Drosophila melanogaster* males. *Genet. Res.* **33:**137–146.
4. **Engels, W. R.** 1984. A *trans*-acting product needed for *P* factor transposition in *Drosophila*. *Science* **226:**1194–1196.
5. **Engels, W. R., and C. R. Preston.** 1984. Formation of chromosome rearrangements by *P* factors in *Drosophila*. *Genetics* **107:**657–678.
6. **Finnegan, D. J., and D. H. Fawcett.** 1986. Transposable elements in *Drosophila melanogaster*, p. 1–62. *In* N. Maclean (ed.), *Oxford Surveys on Eucaryotic Genes*, vol. 3. Oxford University Press, Oxford.
7. **Hatzopoulos, P., M. Monastirioti, G. Yannopoulos, and C. Louis.** 1987. The instability of the TE-like mutation *Dp(2;2)GYL* of *Drosophila melanogaster* is intimately associated with the *hobo* element. *EMBO J.* **6:**3091–3096.
8. **Johnson-Schlitz, D., and J. K. Lim.** 1987. Cytogenetics of *Notch* mutations arising in the unstable X chromosome *Uc* of *Drosophila melanogaster*. *Genetics* **115:**701–709.
9. **Karess, R. E., and G. M. Rubin.** 1984. Analysis of *P* transposable element functions in *Drosophila*. *Cell* **38:**135–146.
10. **Kidwell, M. G., J. F. Kidwell, and J. A. Sved.** 1977. Hybrid dysgenesis in *Drosophila melanogaster*: a syndrome of aberrant traits including mutation, sterility and male recombination. *Genetics* **86:**813–833.
11. **Laski, F. A., D. C. Rio, and G. M. Rubin.** 1986. Tissue specificity of *Drosophila P* element transposition is regulated at the level of mRNA splicing. *Cell* **44:**7–19.
12. **Laverty, T. R., and J. K. Lim.** 1982. Site-specific instability in *Drosophila melanogaster*: evidence for transposition of destabilizing element. *Genetics* **101:**461–476.
13. **Lim, J. K.** 1979. Site-specific instability in *Drosophila melanogaster*: the origin of the mutation and cytogenetic evidence for site specificity. *Genetics* **93:**681–701.
14. **Lim, J. K.** 1981. Site-specific intrachromosomal rearrangements in *Drosophila melanogaster*: cytogenetic evidence for transposable elements. *Cold Spring Harbor Symp. Quant. Biol.* **45:**553–560.
15. **Lim, J. K., M. J. Simmons, J. D. Raymond, N. M. Cox, R. F. Doll, and T. P. Culbert.** 1983. Homologue destabilization by a putative transposable element in *Drosophila melanogaster*. *Proc. Natl. Acad. Sci. USA* **80:**6624–6627.
16. **Mathog, D., M. Hochstrasse, Y. Gruenbaum, H. Saumweber, and J. Sedat.** 1984. Characteristic folding pattern of polytene chromosomes in *Drosophila* salivary gland nuclei. *Nature* (London) **308:**414–421.
17. **McElwain, M. C.** 1986. The absence of somatic effects of P-M hybrid dysgenesis in *Drosophila melanogaster*. *Genetics* **113:** 897–918.
18. **McGinnis, W., A. W. Shermoen, and S. K. Beckendorf.** 1983. A transposable element inserted just 5′ to a *Drosophila* glue protein gene alters gene expression and chromatin structure. *Cell* **34:**75–84.
19. **O'Hare, K., and G. M. Rubin.** 1983. Structures of *P* trans-

posable elements and their sites of insertion and excision in the *Drosophila melanogaster* genome. *Cell* **34**:25–35.

20. **Preston, C. R., and W. R. Engels.** 1984. Movement of *P* elements within a P strain. *Drosophila Inf. Serv.* **60**:169–170.
21. **Rubin, G. M., and A. C. Spradling.** 1982. Genetic transformation of *Drosophila* with transposable element vectors. *Science* **218**:348–353.
22. **Rubin, G. M., and A. C. Spradling.** 1983. Vectors for *P* element-mediated gene transfer in *Drosophila. Nucleic Acids Res.* **11**:6341–6351.
23. **Simmons, M. J., N. A. Johnson, T. M. Fahey, S. M. Nellett, and J. D. Raymond.** 1980. High mutability in male hybrids of *Drosophila melanogaster. Genetics* **96**:479–490.
24. **Spradling, A. C., and G. M. Rubin.** 1982. Transposition of cloned *P* elements into *Drosophila* germ line chromosomes. *Science* **218**:341–347.
25. **Streck, R. D., J. E. MacGaffey, and S. K. Beckendorf.** 1986. The structure of *hobo* transposable elements and their insertion sites. *EMBO J.* **5**:3615–3623.
26. **Yannopoulos, G., N. Stamatis, and J. C. J. Eeken.** 1986. Differences in the cytotype and hybrid dysgenesis inducer ability of different P-strains of *Drosophila melanogaster. Experientia* **42**:1283–1285.
27. **Yannopoulos, G., N. Stamatis, M. Monastirioti, P. Hatzopoulos, and C. Louis.** 1987. *hobo* is responsible for the induction of hybrid dysgenesis by strains of *Drosophila melanogaster* bearing the male recombination factor *23.5MRF. Cell* **49**: 487–495.
28. **Yannopoulos, G., N. Stamatis, A. Zacharopoulou, and M. Pelecanos.** 1981. Differences in the induction of specific deletions and duplications by two male recombination factors isolated from the same *Drosophila* natural population. *Mutat. Res.* **83**:383–393.
29. **Yannopoulos, G., N. Stamatis, A. Zacharopoulou, and M. Pelecanos.** 1983. Site specific breaks induced by the male recombination factor 23.5 MRF in *Drosophila melanogaster. Mutat. Res.* **108**:185–202.
30. **Yannopoulos, G., A. Zacharopoulou, and N. Stamatis.** 1982. Unstable chromosome rearrangements associated with male recombination in *Drosophila melanogaster. Mutat. Res.* **96**: 41–51.

Chapter 21

Transposable Element *mariner* in *Drosophila* Species

DANIEL L. HARTL

I. INTRODUCTION

This chapter is a brief review of the transposable element *mariner* which focuses on the somewhat unusual characteristics of expression of the element, particularly its ability to be expressed in somatic cells and its distribution among species of *Drosophila*. Although the *mariner* element occurs in species closely related to *Drosophila melanogaster*, such as *Drosophila mauritiana* and *Drosophila simulans*, it does not, insofar as is known, occur in *D. melanogaster* itself.

My group's interest in transposable elements in species closely related to *D. melanogaster* was stimulated in part by reports (7, 8) that the sibling species *D. simulans* and *D. mauritiana* contain only about 14% as much middle repetitive DNA as *D. melanogaster*. *D. simulans* and *D. mauritiana* appear to have diverged from a common ancestor with *D. melanogaster* only about 4 million years ago (3, 6). Since a significant fraction of middle repetitive DNA in *Drosophila* spp. is presumed to be associated with transposable elements (16), it seemed as if the complements of transposable elements in the genomes of these species were changing more rapidly than expected.

At about the same time, there was much discussion of hybrid dysgenesis induced by the P element and the I element in certain strains of *D. melanogaster* (W. R. Engels, this volume; D. J. Finnegan, this volume, Chapter 18), and consideration was given to a possible role of such phenomena in the origin of postmating reproductive isolation during speciation (1). If this was true, then high degrees of transposable element activity might be expected in the germ line of species hybrids. Any transposable elements whose transposition was activated by interspecific hybridization might be expected to be identified as the agents responsible for specific mutations occurring in F_2 hybrids of interspecific crosses in which one of the sexes in the F_1 hybrid was fertile. *D. simulans* and *D. mauritiana* appeared ideal for such experiments because the females of the interspecific crosses are fertile (12). Any new transposable elements recovered from the experiments might shed light on general questions concerning the evolution of transposable elements and their distribution among species. Moreover, the experiments used to advantage the well-developed classical and molecular genetics of *D. melanogaster*.

II. DISCOVERY OF *mariner*

The initial genetic hybridization experiments between *D. simulans* and *D. mauritiana* were carried out by David Haymer and Laurel Mapes (unpublished results). Details of the experiments are of little

Daniel L. Hartl ■ Department of Genetics, Washington University School of Medicine, St. Louis, Missouri 63110-1095.

any kind. However, the control series of females from intraspecific crosses within the Cambridge wild-type strain of *D. mauritiana* gave one exceptional male with peach-colored rather than the normal brick-red eyes. The mutation proved to be X linked, and its failure to complement the known *D. mauritiana white* allele w^1 quickly showed that the new mutation was an allele of *white*, which was accordingly designated *white-peach* (w^{pch}).

III. SOMATIC AND GERMINAL MUTATION

During the initial and subsequent crosses with w^{pch}, it was observed that the allele was somewhat unstable in somatic cells (9, 10). The instability is observed phenotypically as mosaic eye color, with one or more pigmented patches on an otherwise white-peach background. The pigmented patches are usually unilateral and range in size from single ommatidia up to pigmented areas covering about 20 to 30% of the eye. In typical *D. mauritiana* strains containing w^{pch}, flies with somatic mosaicism occur at a frequency of approximately 4×10^{-3} (9). This typical form of somatic mosaicism is not genetically transmitted, because the infrequent flies that exhibit somatic mosaicism give rise to nonmosaic progeny. There is also a more dramatic form of somatic mosaicism in which the mosaicism is genetically transmitted, as described below.

In addition to somatic instability, the w^{pch} allele is unstable in germ cells (9, 10). Exceptional phenotypes resulting from germ line mutations are primarily wild type or bleached-white. Germ line mutation rates differ substantially between the sexes. In the male germ line, the overall rate of mutation of w^{pch} to wild type and bleached-white is approximately 2×10^{-3} to 4×10^{-3}; it is lower in females by a factor of at least 2 (9, 10). By contrast, the rates of somatic mutation giving rise to eye-color mosaics are approximately equal in the two sexes (9). Germ line mutations of w^{pch} appear to occur primarily during gametogenesis or in gametes themselves. This is inferred from the finding that the exceptional progeny from germ line mutations do not occur in clusters (groups of exceptional siblings), as would be expected if the mutations occurred premeiotically (9, 10).

Not all white-eyed exceptional progeny derived from white-peach flies have null alleles of the *white* gene. Complementation tests revealed one class of white-eyed flies that were actually double mutants in which the X chromosome contained both w^{pch} and a second mutation originally designated plum (*pm*), located on the X chromosome approximately 66 map

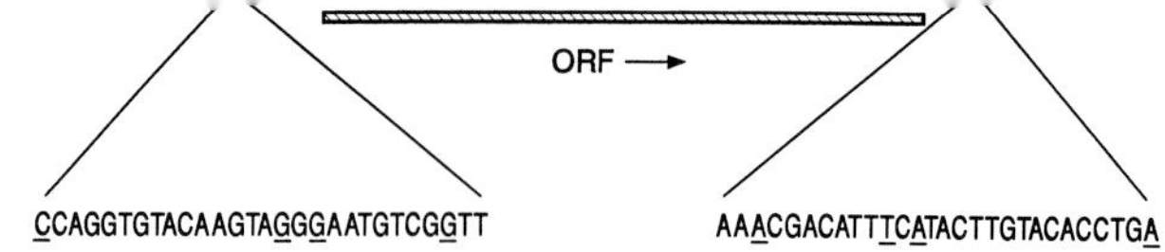

Figure 1. Sequence organization of the *mariner* element inserted in the w^{pch} allele. The 28-bp inverted repeats contain four mismatches (underlined). Putative promoter elements CCACC, CCATA, and CATA occur at nucleotides 98, 118, and 143, respectively, and a ribosome-binding sequence TGCAGTCAAC occurs at position 162. An ATG at nucleotide 172 initiates an open reading frame (ORF) of 1,038 nucleotides terminating in the codons GAA (Glu) and TAA (ochre terminator). The overlapping AATAA in these codons may serve as a polyadenylation signal. Restriction site abbreviations are as follows: RS, *Rsa*I; SS, *Ssp*I; S, *Sal*I; X, *Xho*II; N, *Nhe*I.

units from w^{pch} (10). Subsequent crosses showed that the *pm* mutations were alleles of the *D. melanogaster* gene *garnet* (unpublished results), and the gene symbol *pm* has therefore been superseded by *g*. At least in some strains of *D. mauritiana*, the *garnet* locus appears to mutate readily when in the presence of w^{pch} (10).

IV. CLONING AND CHARACTERIZATION

Somatic and germinal instability of w^{pch} suggested that the mutation might result from insertion of a transposable element. Preliminary Southern blots using DNA probes from the *D. melanogaster white* gene (2) revealed a novel DNA sequence of approximately 1.3 kilobases inserted just 5′ of the first exon, and which is lost when the w^{pch} allele reverts to wild type (9, 11). Construction of bacteriophage λ libraries and screening with appropriate probes from the *white* gene yielded phage containing the novel inserted DNA flanked with sequences from *white* (11). Southern blots of genomic DNA of *D. mauritiana* using probes from the w^{pch} insert showed that the inserted DNA was present in multiple copies in the genome of *D. mauritiana*, and different strains contained the DNA sequences at different sites (11). These experiments identified the inserted DNA in w^{pch} as a putative transposable element, which was designated *mariner*.

The genetic organization of the *mariner* element present in the w^{pch} allele is outlined in Fig. 1. See reference 11 for the complete nucleotide sequence. As shown in Fig. 1, *mariner* has an inverted-repeat type of sequence organization containing 28-base-pair (bp) inverted repeats with four mismatches. The total length of the element is 1,286 bp, making it considerably smaller than either the P element (W. R. Engels, this volume) or the *hobo* element (R. K.

Blackman and W. M. Gelbart, this volume). Indeed, the *mariner* element is one of the smallest eucaryotic transposable elements known, comparable in size to bacterial insertion sequences.

One unusual feature of *mariner* is the presence of a single uninterrupted reading frame of 1,035 nucleotides flanked by the inverted repeats. This reading frame, if transcribed and translated, would code for a polypeptide of 345 amino acids. Codon usage is typical of that in other *Drosophila* genes. The open reading frame is preceded by acceptable *Drosophila* promoter elements, and the putative AUG translational initiator codon is preceded by a sequence with an excellent match to other translational start sequences in *Drosophila* spp. At the other end of the open reading frame, the sequence GAA TAA may do double duty in coding for the final amino acid (GAA, Glu) and termination (TAA, ochre), as well as providing a poly(A) addition signal. Although the nucleotide sequence of *mariner* suggests the presence of a single functional open reading frame, appropriate cDNA clones have yet to be sequenced for confirmation. Present data cannot critically exclude the possibility that the element may contain one or more very small introns of lengths that are exact multiples of three nucleotides.

The *mariner* element inserted in w^{pch} is not of small size merely because it is a deletion derivative of a larger element found in the same genome. In studies of *mariner* elements present in the genome of the sibling species *D. mauritiana*, *D. simulans*, *D. yakuba*, and *D. teissieri*, including seven elements sequenced in their entirely (see below), no elements of length greater than 1,286 bp have been identified (K. Maruyama and D. L. Hartl, unpublished results). It is therefore presumed that 1,286 bp represents the full length of the element. All seven *mariner* elements sequenced so far contain the 28-bp inverted repeats, and the number of mismatches ranges from two to four. All of the full-length elements so far sequenced also conserve the single long open reading frame.

Although *mariner* elements longer than 1,286 bp have not been identified, smaller elements containing internal deletions are sometimes found. In most genomes, the deletion derivatives are not prevalent. However, in the genome of at least one strain of *D. teissieri*, half or more of the elements appear to contain deletions (Maruyama and Hartl, unpublished results).

Little is known about the insertional specificity of *mariner*. In the w^{pch} allele, the element is inserted into the untranslated leader sequence in the following context: 5′-TGGCGTA/TAAACCG-3′, where the slash marks the insertion of the *mariner* sequence (11). The TA immediately flanking the point of insertion is a putative 2-bp duplication. The flanking TA duplication is also characteristic of other *mariner* insertions (Maruyama and Hartl, unpublished results).

V. DISTRIBUTION AMONG SPECIES

The *D. melanogaster* species group is divided into 10 named species subgroups, including the one designated the *D. melanogaster* species subgroup (13, 14). The *D. melanogaster* species subgroup consists of eight currently recognized sibling species that fall into the following clusters based on degree of relatedness: (a) *D. melanogaster*, *D. simulans*, *D. mauritiana*, and *D. sechellia*; (b) *D. yakuba* and *D. teissieri*; and (c) *D. erecta* and *D. orena* (13). With the exceptions of *D. melanogaster* and *D. simulans*, which are cosmopolitan in distribution, all of these species are African or island endemics (15).

The *mariner* element is widespread among members of the *D. melanogaster* species subgroup, and its copy number in the genome varies widely. Among species in this subgroup, the element is most abundant in *D. mauritiana*, where it occurs in 20 to 30 copies. Naturally occurring strains of *D. simulans* usually contain two copies, and those of *D. sechellia* contain one or two copies. Somewhat greater variation in copy number occurs among laboratory strains of *D. simulans*, in which the copy number varies from 0 to about 10 (Maruyama and Hartl, unpublished results). These estimates of copy number and those below are based on Southern hybridizations of genomic DNA and do not necessarily represent the number of functional copies. Strains of *D. melanogaster* containing *mariner* sequences have not been found.

Similar variation in copy number also occurs in the other species of the *D. melanogaster* species subgroup. The genome of *D. yakuba* contains approximately 4 copies of *mariner*, that of *D. teissieri* about 10, and that of *D. erecta* none. Unlike other species in the group, *D. teissieri* contains many copies of *mariner* that are deletion derivatives of the full-length element missing the large *Xho*II-*Nhe*I restriction fragment (Fig. 1). The significance of this observation is unclear. It must be emphasized that only limited numbers of strains of the species in this branch of the phylogeny are available for investigation. Two strains of *D. yakuba* and one of each of *D. teissieri* and *D. erecta* have been studied (Maruyama and Hartl, unpublished results); the one extant strain of *D. orena* contains no *mariner* elements.

The *mariner* element also occurs in *Drosophila* species outside the immediate *D. melanogaster* subgroup (Maruyama and Hartl, unpublished results). Among other species within the *D. melanogaster* complex, *mariner* has been found in the *D. montium*

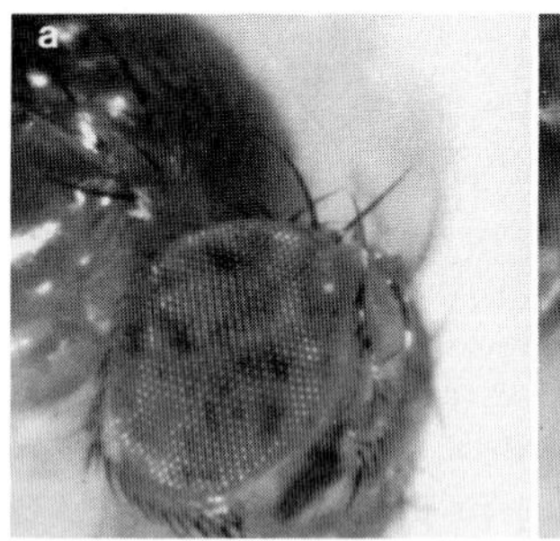

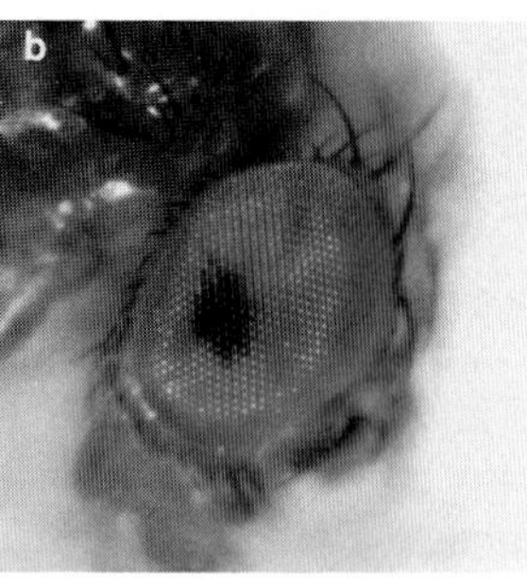

Figure 2. Somatic mosaicism resulting from the *mariner* mosaic element (*Mos*). Both offspring are from a mother with genotype w^{pch}/w^{pch}; *Mos*/+. (a) Typical somatic mosaic resulting from the *Mos* factor, genotype w^{pch}/w^{pch}; *Mos*/*Mos*; (b) maternal effect mosaic, genotype w^{pch}/w^{pch}; +/+. For the colors of the background and pigmented spots, see color plates, this volume.

subgroup (seven species examined), but not thus far in the *D. takahashii* (five species examined) or *D. ananassae* (one species examined) subgroups. To this point, the *mariner* element has not been identified in *Drosophila* species groups other than the *D. melanogaster* complex, but the number of species examined is inadequate to delineate the true distribution of the element among species.

VI. INHERITED SOMATIC MOSAICISM

Among transposable elements so far studied in *Drosophila* spp., the *mariner* element is unique in that it shows high rates of mutation in somatic cells. Also unprecedented is the finding of certain strains in which the somatic mutability is itself an inherited trait (Fig. 2). The initial example of inherited mosaicism arose in a series of routine crosses involving w^{pch} *D. mauritiana* parents, in which several offspring with eye color mosaicism were observed in the same vial. Subsequent breeding tests showed that the eye color mosaicism was an inherited trait, and a mosaic stock was established.

Due to the paucity of genetic markers in *D. mauritiana*, genetic analysis of the mosaicism was carried out in *D. simulans* strains into which both the w^{pch} allele and the inherited mosaicism had been introduced by several generations of backcrossing (5). Crosses of the *D. simulans* strains confirmed the conclusion that the somatic mosaicism was inherited as a simple Mendelian dominant and localized the factor to the third chromosome (5). Southern blots of genomic digests of DNA from nonmosaic and mosaic w^{pch} flies demonstrated that at the molecular level the mechanism of the somatic mosaicism was the excision of the *mariner* element from w^{pch} in somatic cells. The mosaic strains were also observed to have a high rate of mutation of w^{pch} in the germ line (5).

By analogy with other transposable element systems, it seemed reasonable to suppose that the *Mos* (mosaic) factor was actually a copy of *mariner* located in the third chromosome that results in high rates of excision of *mariner* elements. This model was supported by four findings (13a):

(i) *mariner* elements in addition to the one located in w^{pch} are also somatically excised at a higher rate.

(ii) The *Mos* factor also undergoes a high rate of germ line mutation (approximately 3×10^{-3} per generation), resulting in the occurrence of nonmosaic flies in homozygous *Mos* stocks.

(iii) The *Mos* factor can itself transpose, giving rise to stocks in which the mosaic factor is located in the X chromosome or in chromosome 2.

(iv) Homozygous *Mos* stocks can be shown to have a characteristic restriction fragment containing *mariner* that is deleted among nonmosaic revertants arising in the stock.

Item (iv) has allowed identification of the genomic restriction fragment containing the putative *mariner* element responsible for the mosaicism. The element has been cloned, and its characterization is in progress. Direct confirmation that the mosaicism results from the cloned element has been obtained from transformation experiments in which the element and some of its flanking sequences have been reintroduced into the genome. Flies of genotype w^{pch} that have been transformed with the cloned *Mos* DNA show inherited somatic mosaicism (M. Medhora and D. L. Hartl, unpublished results).

VII. DYSGENESIS

Here I use the term **dysgenesis** in its widest sense to mean the creation of generally deleterious genetic effects, in this case mutations. In the case of the transposable element P, dysgenesis is associated with hybridization between particular types of strains, hence the term **hybrid dysgenesis** (Engels, this volume). However, with respect to the *mariner* element, dramatic differences between reciprocal crosses have not been observed for somatic or germ line events.

On the other hand, an effect of interspecific crossing has been observed with *mariner*. Haymer and Marsh (9) studied w^{pch}-containing female hybrids of *D. mauritiana* and *D. simulans*. In these crosses, the frequency of reversion of w^{pch} in the germ line of the hybrid females was significantly greater than in nonhybrids. The frequency of spontaneous nonheritable somatic mosaicism was also increased to a level of approximately 1% in the offspring of the hybrid females (9). Rates of spontaneous somatic mosaicism of up to 10% in interspecific crosses in which w^{pch} is segregating have been observed by J. Coyne (unpublished results), including somatic mosaics among the sterile hybrid females between *D. mauritiana* and *D. melanogaster*. The mechanisms involved in the somatic destabilization of w^{pch} in interspecific crosses have not been investigated.

Phenotypically wild-type strains of *D. mauritiana* that are homozygous for the inherited mosaicism factor *Mos* undergo a type of dysgenesis resulting in very high forward mutation rates (D. Garza, G. Bryan, and D. L. Hartl, unpublished results). In one series of experiments, the overall rate of new visible X-linked mutations was approximately 5×10^{-4} per X chromosome per generation. Mutations were recovered in a large number of X-linked genes (not all of them identified), including five new alleles of *white* and two of *yellow*. The new *white* and *yellow* mutations have all been shown to result from the insertion of DNA sequences of approximately 1.3 kilobases (the size of *mariner*) in or near the gene. In two of the *white* alleles, the insertion has been shown to be a *mariner* element inserted within the same *Bam*HI restriction fragment as occurs in the w^{pch} allele (not necessarily at exactly the same position). The insertions in the other *white* alleles occur elsewhere in the gene. Interestingly, in the presence of the *Mos* factor, but not in its absence, flies containing any of the five *white* alleles are eye color mosaics and those containing either of the *yellow* alleles are body color mosaics. The occurrence of body color mosaicism in the case of the *yellow* mutations demonstrates genetically that the effects of *Mos* are not restricted to the *mariner* element inserted in the *white* gene.

VIII. MATERNAL EFFECTS IN MOSAICISM

The *mariner* element is also associated with a unique type of maternal effect in which factors transmitted through the egg result in high rates of somatic excision in the progeny (4). The maternal effect was discovered in the course of the backcrosses in which the heritable mosaic factor *Mos* was transferred from *D. mauritiana* into *D. simulans*. When mosaic females of *D. simulans* that are heterozygous for the *Mos* factor (genotype *Mos*/+) are crossed with *D. simulans* nonmosaic w^{pch} males, two distinct classes of mosaic offspring result, which occur in approximately equal frequency. In one class, the mosaicism resembles that in the mother in that frequent reversion events occur at different stages during eye development, resulting in multiple patches of pigmented tissue. These offspring transmit the mosaicism to their offspring and are genotypically *Mos*/+. In the other class of mosaic offspring, the mosaicism is manifested as a small number of reversion events occurring relatively early in eye development, resulting in a small number of relatively large pigmented patches. The mosaics in this class do not transmit the mosaicism to their offspring and are genotypically +/+. The reciprocal cross of w^{pch} *D. simulans Mos*/+ males with +/+ females yields mosaic and nonmosaic offspring in the expected 1:1 ratio.

The availability of a *D. simulans* attached X chromosome has permitted genetic imprinting to be excluded in the maternal mosaicism (4). With genetic imprinting, the *mariner* elements in the w^{pch} alleles in heterozygous *Mos*/+ mothers might be conditioned to undergo excision in the progeny, even in the absence of the *Mos* factor. However, when attached X females heterozygous for *Mos* are crossed with w^{pch} males, the +/+ male offspring still manifest the maternal effect type of mosaicism, even though they receive their w^{pch} allele from their fathers.

IX. SUMMARY AND SIGNIFICANCE

Current views of transposable elements in *Drosophila* spp. have been strongly influenced by the characteristics of a small number of elements isolated from *D. melanogaster* that have been exceptionally well studied. Although *mariner* has been discovered only recently and is far from completely understood, it differs from other known *Drosophila* transposable elements in a number of important particulars. First, the element is exceptionally small for a eucaryotic transposable element (1,286 bp), and it appears to contain a single, long, uninterrupted reading frame (11). Copies of the element that contain internal deletions do occur (11; Maruyama and Hartl, unpublished results), but in most genomes they appear to be less common than is the case with the P element (Engels, this volume) or the *hobo* element (Blackman and Gelbart, this volume).

The *mariner* element gives unmistakable evidence for somatic expression, which is detected phenotypically by somatic mosaicism resulting from the excision of *mariner* during development. Excision of *mariner* from the *white-peach* allele results in pigmented sectors on an otherwise peach-colored eye, and excision of *mariner* from insertion sites near or in the *yellow* gene results in mosaic body color.

Eye color mosaics occurring in *white-peach* flies have been the most extensively studied. In *D. mauritiana*, somatic mutations resulting in somatic mosaicism occur at approximately the same frequency as germinal mutations, roughly 10^{-3} (9). This type of spontaneous mosaicism is not inherited, and its frequency may be increased dramatically in certain interspecific crosses (Coyne, unpublished results).

A dramatic inherited form of somatic mosaicism resulting from somatic excision of *mariner* also occurs. This type of mosaicism can be traced to the presence of a particular *mariner* element (*Mos*) at a location other than near the *white* gene, because w^{pch} mosaic strains that have undergone reversion and are no longer mosaic are missing the copy of *mariner* identified with *Mos*. In the original mosaic strain, the *Mos* element was present in the third chromosome, but it has been shown to be capable of transposition to the second chromosome or to the X chromosome. The molecular basis of the change from normal *mariner* to *Mos* factor has not been determined.

A novel type of maternal transmission has been observed in strains of *D. simulans* into which the w^{pch} allele and the *Mos* factor have been introduced. The non-*Mos*-bearing progeny of females that are heterozygous for *Mos* are themselves somatic mosaics, but they do not transmit the mosaicism to their progeny. This phenomenon does not involve genetic imprinting of the w^{pch} alleles in the mother, but rather the maternal transmission of substances that promote *mariner* excision (4).

The presence of the *Mos* factor in the genome leads to a type of dysgenesis detected as a high rate of mutation resulting from new *mariner* insertions (Garza et al., unpublished results). However, the dysgenic effects of the *Mos* factor do not appear to depend on hybridization or to differ in reciprocal crosses.

The several characteristics by which *mariner* can be distinguished from other transposable elements in *Drosophila* suggest that it may be unrealistic to try to categorize transposable elements into a few simple classes, for example, elements with inverted repeats contrasted with retroviruslike elements. Rather, the unique quirks and features of every *Drosophila* transposable element so far analyzed suggest that the elements represent a diverse assemblage of highly evolved DNA sequences that differ among themselves in many particulars, and the set of criteria by which the elements may be most usefully classified has yet to be determined.

Acknowledgments. I am indebted to several excellent students who have contributed to the *mariner* story by their observations and experiments, including Glenn Bryan, Danny Garza, David Haymer, James Jacobson, Ann MacPeek, Kyoko Maruyama, Meetha Medhora, and Nancy Scavarda. Glenn Bryan, Meetha Medhora, and Kyoko Maruyama also made important contributions to the manuscript.

This work was supported by National Institutes of Health grant GM33741.

LITERATURE CITED

1. **Bingham, P. M., M. G. Kidwell, and G. M. Rubin.** 1982. The molecular basis of P-M hybrid dysgenesis: the role of the P element, a P-strain-specific transposon family. *Cell* **29:** 995–1004.
2. **Bingham, P. M., R. Levis, and G. M. Rubin.** 1981. Cloning of DNA sequences from the *white* locus of *D. melanogaster* by a novel and general method. *Cell* **25:**693–704.
3. **Bodmer, M., and M. Ashburner.** 1984. Conservation and change in the DNA sequences coding for alcohol dehydrogenase in sibling species of *Drosophila. Nature* **309:**425–430.
4. **Bryan, G. J., and D. L. Hartl.** 1988. Maternally inherited transposon excision in *Drosophila simulans. Science* **240:** 215–217.
5. **Bryan, G. J., J. W. Jacobson, and D. L. Hartl.** 1987. Heritable somatic excision of a *Drosophila* transposon. *Science* **235:** 1636–1638.
6. **Coyne, J.** 1985. Genetic studies of three sibling species of *Drosophila* with relationship to theories of speciation. *Genet. Res.* **46:**169–192.
7. **Dowsett, A. P.** 1983. Closely related species of *Drosophila* can contain different libraries of middle repetitive DNA sequences. *Chromosoma* **88:**104–108.
8. **Dowsett, A. P., and M. W. Young.** 1982. Differing levels of dispersed repetitive DNA among closely related species of *Drosophila. Proc. Natl. Acad. Sci. USA* **79:**4570–4574.
9. **Haymer, D. S., and J. L. Marsh.** 1986. Germ line and somatic instability of a *white* mutation in *Drosophila mauritiana* due to a transposable element. *Dev. Genet.* **6:**281–291.
10. **Jacobson, J. W., and D. L. Hartl.** 1985. Coupled instability of two X-linked genes in *Drosophila mauritiana*: germinal and somatic instability. *Genetics* **111:**57–65.
11. **Jacobson, J. W., M. M. Medhora, and D. L. Hartl.** 1986. Molecular structure of a somatically unstable transposable element in *Drosophila. Proc. Natl. Acad. Sci. USA* **83:** 8684–8688.
12. **Lachaise, D., J. R. David, F. Lemeunier, L. Tsacas, and M. Ashburner.** 1986. The reproductive relationships of *Drosophila sechellia* with *D. mauritiana, D. simulans*, and *D. melanogaster* from the afrotropical region. *Evolution* **40:** 262–271.
13. **Lemeunier, F., and M. Ashburner.** 1984. Relationships within the *melanogaster* species subgroup of the genus Drosophila (Sophophora). *Chromosoma* **89:**343–351.

13a. **Medhora, M. M., A. H. MacPeek, and D. L. Hartl.** 1988. Excision of the *Drosophila* transposable element *mariner*: identification and characterization of the *Mos* factor. *EMBO J.* **7:**2185–2189.

14. **Okada, T.** 1981. Oriental species, including New Guinea, p. 261–289. *In* M. Ashburner, H. L. Carson, and J. N. Thompson, Jr. (ed.), *The Genetics and Biology of Drosophila*, vol. 3a. Academic Press, Inc., New York.
15. **Tsacas, L., D. Lachaise, and J. R. David.** 1981. Composition and biogeography of the afrotropical Drosophilid fauna, p. 197–259. *In* M. Ashburner, H. L. Carson, and J. N. Thompson, Jr. (ed.), *The Genetics and Biology of Drosophila*, vol. 3a. Academic Press, Inc., New York.
16. **Young, M. W.** 1979. Middle repetitive DNA: a fluid component of the *Drosophila* genome. *Proc. Natl. Acad. Sci. USA* **76:**6274–6278.

Chapter 22

Mobile Elements in *Caenorhabditis elegans* and Other Nematodes

D. G. MOERMAN and R. H. WATERSTON

I. INTRODUCTION

Transposable elements have recently been described in several species of nematodes: *Caenorhabditis elegans*, *Caenorhabditis briggsae*, *Ascaris lumbricoides*, and *Panagrellus redivivus*. Because of the intense interest in *C. elegans* as an experimental organism for developmental genetic studies and the availability of sophisticated genetics (103), most is known about transposons in this species. This review focuses principally on Tc1 (Tc = transposon) of *C. elegans* (26), the best understood element in nematodes. Other elements in *C. elegans* and also elements in other species of nematodes will be briefly surveyed. The interested reader should also see two recent related reviews (20, 43).

The genome of *C. elegans* is 8×10^7 base pairs (bp) in extent, the smallest known for any metazoan (93). There are six chromosomes per haploid set (73), and about 83% of *C. elegans* DNA behaves as

D. G. Moerman ■ Department of Zoology, University of British Columbia, Vancouver, British Columbia, Canada V6T 2A9. **R. H. Waterston** ■ Department of Genetics, Washington University, St. Louis, Missouri 63110.

single-copy sequence in renaturation experiments (93). The repeated sequences are of several types, including functional genes (44, 81, 99), inverted repeat or "foldback" sequences (23, 93), and short repeated sequences of a few hundred nucleotides (23). The global arrangement of these short repeats is of the "short-period-interspersion" or "*Xenopus*" pattern (14, 23).

Some of the repetitive sequences consist of transposable elements, and at least five distinct families of elements have been identified in *C. elegans*, Tc1 through Tc5. The sequence of one Tc1 element has been determined (85) and shows that Tc1 resembles bacterial insertion sequence elements with terminal inverted repeats and a central open reading frame (45). The complete sequences for any members of the other transposon families have not been determined, but the data suggest that Tc2, Tc3, and Tc5 are also insertion sequence-like in structure and that Tc4 is foldbacklike in structure (78). No "retrotransposon-like" (2) elements have been identified in *C. elegans*, although such elements have been described in *A. lumbricoides* (1) and *P. redivivus* (55).

II. Tc1: GENERAL FEATURES

A. Identification and Structure of Tc1

The most thoroughly studied nematode transposon, Tc1, was initially identified as the agent responsible for extensive restriction fragment length polymorphisms between the two common laboratory strains of *C. elegans*, variety Bristol, N2 (English strain [8]), and variety Bergerac, BO (French strain [73]). Southern blot analyses using unique-sequence probes identified several restriction fragments that were 1.6 kilobases (kb) longer in BO DNA than in N2 DNA (21). Comparison of the cloned fragments from each strain revealed that this 1.6-kb length difference was due to the insertion of a dispersed repeated sequence present in N2 at about 30 copies and in BO at over 300 copies per haploid genome (26, 54). A survey of several wild strains revealed that this sequence is ubiquitous in *C. elegans*. Most strains are "low copy number" like N2, but a few "high-copy-number" strains similar to BO were also identified (26, 54). The initial element seemed to be structurally similar to insertion sequence elements, and because of its structure and its repeated nature, the sequence was assumed to be a transposable element and was named Tc1 (85).

The first Tc1 sequenced from BO, contained in plasmid pCe(Be)T1, is 1,610 bp in length, contains 54-bp perfect terminal inverted repeats, and has two open reading frames (ORFs) on the same DNA strand (85) (Fig. 1). The larger ORF could encode a basic polypeptide of 273 amino acids. Both TATA and CAAT box sequences for transcription initiation are present 5′ of this putative coding domain. Like Ty elements in *Saccharomyces cerevisiae* (31), Tc1 has an enhancer core sequence (D. G. Moerman, unpublished results). A second ORF, contained within the first but out of frame with it, could potentially encode a 112-amino-acid polypeptide (Fig. 1). There are, however, no appropriate TATA or CAAT sequences near this second ORF.

A striking feature of the Tc1 family is the generally homogeneous size of its members, even in strains with over 300 copies. For example, *Hae*III restriction digest of pCe(Be)T1 yields five fragments from the element, two of approximately 0.1 kb and three of approximately 0.5 kb, which together span most of the element. Probing genomic *Hae*III digests of N2 and BO DNA with a Tc1 probe reveals that most hybridizing material is restricted to bands of these sizes (26, 54). Similar results have been obtained with *Eco*RV, which cuts within the inverted repeats of Tc1 to yield a 1.6-kb fragment. On Southern blots of N2 and BO genomic DNA cut with *Eco*RV, the majority of the hybridizing signal is at 1.6 kb, but a few discrete minor bands below and above the size can also be detected (24, 70). This conservation of length is in contrast to the length heterogeneity of P elements of *Drosophila* sp. (74) and *Ac*/*Ds* elements of maize (34, 77), with which Tc1 shares other structural features (detailed below). How this length is maintained, and the relationship, if any, between length maintenance and the transposition process, are unknown.

Despite their length homogeneity, there is microheterogeneity among the Tc1 elements of the BO strain. Microheterogeneity has primarily been detected by comparing restriction digests of isolated elements with the canonical element pCe(Be)T1. Approximately 10% of the BO Tc1 elements contain a *Hin*dIII site not present in pCe(Be)T1, and N2 harbors at least one copy of this variant Tc1 (83). Similarly, a different subset of BO Tc1 elements contains an *Eco*RI site (17). A *Tha*I Tc1 polymorphism has also been found among BO Tc1 elements (I. Mori and D. G. Moerman, unpublished results). The minor *Tha*I variant is present in about 10% of the Tc1 population, and this variant again makes up a different subset of the Tc1 population than either the *Hin*dIII or *Eco*RI variants.

One other Tc1 element, *st137*::Tc1, has been entirely sequenced (76) (Fig. 2). It was isolated after a recent transposition in the BO strain. There are six nucleotide alterations in the large ORF of this ele-

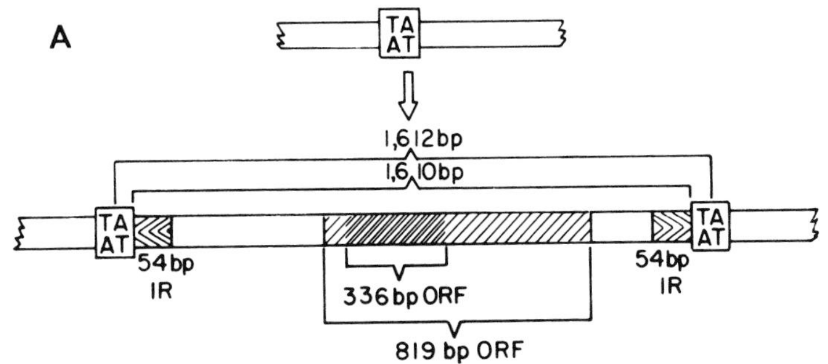

B

```
   1 5'CAGTGCTGGCCAAAAAGATATCCACTTTTGGTTTTTTGTGTGTAACTTTTTTCTCAAGCATCCATTTGACTTGAATTTTT
                 inverted repeat
  81   CCGTGTGCATAAAGCGAAATGTTACGCAAATTTGCGGACCAAACATTACATGATTATCGATTTTTTCTGAATTTTATTTC
 161   AATTTTTTGATTTTTTCGTTTTTCCAATTTTCATTATTTTTTTTGAATTATCAATAAAACGCACTCTGTTTGTTGCACTG
 241   GATTTGTTTGGTTGATAAATTATTTTTAAGGTATGGTAAAATCTGTTGGGTGTAAAAATCTTTCCTTGGACGTCAAGAAA
 321   GCCATTGTAGCTGGCTTCGAACAAGGAATACCCACGAAAAGCTCGCGCTGCAAATTCAACGTTCTCCGTCGACTATTTGG
 401   AAAGTAATCAAGAAGTACCAAACTGAGGTGAGTTCGAAAAATATTATTTTTTAATAATAAATGTTTAGAAATCCGTCGCT
                         CAAT                                  TATA
 481   TTGAGAATCTCGCCCGGCAGGCCTCGAGTGACAACCCATAGGATGGATCGCAACATCCTCCGATCAGCAAGAGAAGATCC
                                                 Met1
 561   GCATAGGACCGCCACGGATATTCAAATGATTATAAGTTCTCCAAATGAACCTGTACCAAGTAAACGAACTGTTCGTCGAC
                                                   Met2
 641   GTTTACAGCAAGCAGGACTACACGGACGAAAGCCAGTCAAGAAACCGTTCATCAGTAAGAAAAATCGCATGGCTCGAGTT
 721   GCGTGGGCAAAAGCGCATCTTCGTTGGGGACGTCAGGAATGGGCTAAACACATCTGGTCTGACGAAAGCAAGTTCAATTT
 801   GTTCGGGAGTGATGGAAATTCCTGGGTACGTCGTCCTGTTGGCTCTAGGTACTCTCCAAAGTATCAATGCCCAACCGTTA
 881   AGCATGGAGGTGGGAGCGTCATGGTGTGGGGGTGCTTCACCAGCACTTCCATGGGCCCACTAAGGAGAATCCAAAGCATT
                                                                 End2
 961   ATGGATCGTTTTCAATACGAAAACATCTTTGAAACTACAATGCGACCCTGGGCACTTCAAAATGTGGGCCGTGGCTTCGT
1041   GTTTCAGCAGGATAACGATCCTAAGCATACTTCTCTTCATGTGCGTTCATGGTTTCAACGTCGTCATGTGCATTTGCTCG
1121   ATTGGCCAAGTCAGTCTCCGGACTTGAATCCAATAGAGCATTTGTGGGAAGAGTTGGAAAGACGTCTTGGAGGTATTCGG
1201   GCTTCAAATGCAGATGCCAAATTCAACCAGTTGGAAAACGCTTGGAAAGCTATCCCCATGTCAGTTATTCACAAGCTGAT
1281   CGACTCGATGCCACGTCGTTGTCAAGCTGTTATTGATGCAAACGGATACGCGACAAAGTATTAAGCATAATTATGTTGTT
                                                                  End1
1361   TTTAAATCCAATTGCTCATATTCCGGTACTTTAATTGTCATTTCCTTGCAACCTCGGTTTTTTCAATATTTCTAGTTTTT
1441   CGATTTTTTTGAATTTTTCTGAAGTTTTTTCAAAATCTGTTGAACATTTTTGATGAATATTGTGTTTTTAGATTTTGTGA
1521   ACACTGTGGTGAAGTTTCAAAACAAAATAACCACTTAGAAAAAAGTTACACACAAAAAACCAAAAGTGGATATCTTTTTG
                                  polyA                                inverted repeat
1601   GCCAGCACTG3'  1610 bp
```

Figure 1. (A) Major features of Tc1 and its insertion target site, based on nucleotide sequencing (85) and showing the terminal inverted repeats (IR) and the two longest open reading frames (ORF). (Adapted from reference 43.) (B) The 1,610-bp sequence of Tc1. Only the sequence of the plus strand (same sense as the putative mRNAs) is presented. The 54-bp perfect inverted terminal repeats are underlined. Possible transcriptional control signals consisting of a TATA box, a CAAT box, and a possible poly(A) addition site are indicated and underlined. The initiation and termination codons of the two hypothetical reading frames are marked. (Adapted from reference 85.)

ment relative to pCe(Be)T1, and one of these generates an in-phase amber stop codon that would truncate this putative polypeptide by 100 amino acids.

There is also microheterogeneity among the Tc1 elements in N2. A *Hin*dIII polymorphism was mentioned above. In addition, restriction analysis of 17 cloned Tc1 elements from N2 (of 30 total) shows that 1 has a 55-nucleotide insert, 2 have 700-bp deletions, and 2 have at least single-site polymorphisms (L. Harris, personal communication). The possible role of these sequence variants in Tc1 regulation is unknown.

B. Tc1 Transposition

Definitive evidence that Tc1 can transpose in the germ line came from genetic studies of spontaneous mutants. In a survey of nine *C. elegans* strains,

```
position     1  2          28 29 30 31        175        188        274
             MetAsp        AsnGluProVal       Gln        Ser        END
Ce(Be)T1     ATGGAT------AATGAACCTGTA-------CAG-------TCA-------TAA
                 *            *  *    *        *          *
st137::Tc1   ATGAAT------AAAGGACCTTTA-------TAG-------TCC-------TAA
             MetAsn        LysGlyProLeu       END        Ser        END
```

Figure 2. Comparison of the long ORF between Ce(Be)T1 and *St137*::Tc1. A total of six nucleotides (asterisk) differ between the two variants, leading to five alterations in the encoded polypeptide. Most critically, at position 175 the Gln codon of Ce(Be)T1 is changed to an amber stop codon in *st137*::Tc1. (Adapted from reference 76.)

including the N2 and BO strains described earlier, only the BO strain was found to possess mutator activity (68). The muscle gene *unc-22* was used as an indicator of mutability (8). Mutations in this gene when homozygous disrupt organization of the myofilament lattice of the body wall musculature (101; D. G. Moerman, Ph.D. thesis, Simon Fraser University, Burnaby, British Columbia, Canada, 1980) and result in slow-moving animals. In addition, homozygosity for *unc-22* mutations results in an easily detectable continuous fine twitch of the body wall muscles, a phenotype not associated routinely with any of the other *unc* genes. Heterozygotes can be distinguished from either wild-type or homozygous mutant animals because under normal culture conditions they resemble the wild type, but in the presence of an acetylcholine agonist, such as a solution of 1% nicotine alkaloid, they twitch rapidly; in contrast, wild-type animals are rigidly paralyzed (65). The nicotine assay for *unc-22* mutants is sensitive and allows mutations to be detected even at frequencies as low as 10^{-7} to 10^{-8} per gamete. Revertants are also easily detected by their improved movement.

The forward spontaneous mutation frequency for *unc-22* in eight of the nine strains was less than or equal to 10^{-6} per gamete, but for the BO strain it was 10^{-4} per gamete. The mutant alleles reverted at high frequency, and both the forward mutation and reversion frequencies were sensitive to alterations of the genetic background (68). These are all characteristics associated with previously identified transposable element systems.

When this mutator activity was found, the *unc-22* gene had not been isolated, and so it was not possible to determine directly which transposable element system was responsible for this activity. Studies of transposable elements in other systems had led to the idea of "transposon tagging," or using these elements to recover DNA from mutationally defined genes (6, 33, 62, 64, 91). The *unc-22* gene was cloned using this approach, with the assumption that Tc1 was the causative element. With the gene in hand, mutator activity in the BO strain was shown to represent Tc1 transposition: spontaneous unstable *unc-22* mutants have Tc1 inserted within *unc-22*, and *unc-22* revertants result from the excision of Tc1 from the gene (67). The cloning of *unc-22* and the tagging and cloning of a second gene, *lin-12* (38), confirmed the usefulness of transposon tagging in *C. elegans*, and this protocol is now widely used in *C. elegans* to isolate genes.

An alternative approach (17) exploited the previously cloned *unc-54* gene (56) to provide the first direct evidence that the BO mutator activity is associated with Tc1 transposition. The *unc-54* gene encodes a myosin heavy chain (30, 59, 58), and a selective system equal in sensitivity to the one described for *unc-22* is available for isolating *unc-54* mutants (75). Mutations in *unc-54* result in thin, paralyzed animals. A mutation in another muscle-affecting gene, *unc-105(n490)*, results in animals with hypercontracted muscle, which therefore appear dumpy and uncoordinated. Mutations in *unc-54* are epistatic to *unc-105(n490)*, and the thin, paralyzed, doubly mutant animals can be readily observed even on a crowded plate of dumpy, uncoordinated *n490* animals (75).

Several spontaneous *unc-54* mutations in N2 and in N2/BO hybrid strains were isolated using this genetic screen (17, 18). The frequency for spontaneous forward mutations in *unc-54* was only slightly higher in the mixed N2/BO hybrid strain than in the N2 strain, but while none of the *unc-54* mutations in N2 were the result of Tc1 insertion (18), two-thirds of the *unc-54* mutations isolated in the N2/BO hybrid strain resulted from Tc1 insertion (17). Like *unc-22*::Tc1 mutations, the *unc-54*::Tc1 alleles were unstable.

C. Target Site for Tc1 Insertion

Like many elements in procaryotes and eucaryotes (5, 29, 36, 40, 46, 47, 50, 74), Tc1 exhibits insertion site specificity. A comparison of two Bergerac strain Tc1 sites, Ce(Be)T1 and 2001, to their corresponding "empty" sites in the Bristol strain revealed that Tc1 inserted at a TA dinucleotide (85, 86). A further 14 Tc1 insertion sites have been sequenced, 13 of which are the result of de novo transpositions, and in each case the insertion has taken place at a TA dinucleotide (19, 69, 83). In all 16 examples, Tc1 is flanked on both sides by a TA

A.

Insertion site	5′ Sequence 3′			Reference
pCe(Be)T1	GCA	CATATATTTG	AAA	(85,86)
pCe2001	ACA	TATTTATGTA	CTT	(26,86)
stP1	AGA	GATATAGGTT	T--	(69)
Tc1(Hin)	TCA	GTCATAACTA	ACG	(83)
unc-22(st136::Tc1)	AAG	GATGTACATT	GAA	(69)
unc-22(st137::Tc1)	CTT	GATGTACCAG	GAA	"
unc-22(st139::Tc1)	ACC	AATGTACCAT	CTT	"
unc-22(st140::Tc1)	CCA	GTGGTAGCTT	CTC	"
unc-22(st141::Tc1)	AAC	GACATATCCC	AAA	"
unc-22(st185::Tc1)	CGG	CATCTATGTC	GCC	"
lin-12(e1979::Tc1)	GCA	TGTATATGTA	AAC	"
unc-15(r408::Tc1)	GAT	GAGATATGTG	TGT	"
unc-54(r323::Tc1)	GCT	CATCTACGTA	AGT	(19)
unc-54(r322::Tc1)	GAA	GAGATACCAA	GAG	"
unc-54(r327::Tc1)	CGC	AAGTTATGGA	GGA	"
unc-54(r360::Tc1)	GAG	ATAATACCGA	GTG	"

B.

Position Relative to Insertion Site

		−8	−7	−6	−5	−4	−3	−2	−1	+1	+2	+3	+4	+5	+6	+7	+8
Number of Occurrences Among 16 DNA Strands	G	6	3	3	8	1	4	4	0	0	2	7	2	3	6	3	3
	A	5	5	7	3	12	1	8	0	16	1	1	3	7	5	6	5
	T	1	1	3	2	3	9	2	16	0	7	1	10	4	2	4	4
	C	4	7	3	3	0	2	2	0	0	6	7	1	2	3	2	3
Consensus:					G	A	G/T	A/G	T	A	T/C	G/C	T				

Figure 3. Consensus target site for the insertion of Tc1. (A) DNA sequences of 16 Tc1 insertion sites. In each case, Tc1 inserted into the central TA dinucleotide. For each site, the strand shown is the one that maximizes the score of the derived consensus site. (B) The matrix derived from the strands shown in panel A. The consensus sequence (below) was derived from the matrix by accepting as significant any base or combination of bases whose probability of occurrence by chance is less than 5%. Underlined values indicate significant bases. (See references 19 and 69 for details; adapted from reference 19.)

dinucleotide, suggesting a 2-bp duplication of the target site like that seen for many other transposable elements (29, 39, 40, 74). Such duplications are generally thought to reflect staggered cutting at the target site prior to insertion (39). Alternatively, the palindromic structure of the TA dinucleotide introduces ambiguity into the definition of the termini of Tc1, i.e., Tc1 might be 1,612 bp instead of 1,610 bp long and contain T at one end and A at the other. If this latter possibility is true, then Tc1 would insert between T and A following flush rather than staggered cleavage of the target site, which would be unusual but not without precedent for an element with an insertion sequence-like structure, as detailed below.

Early hints that Tc1 exhibits additional insertion site specificity (19, 69) (Fig. 3) came from three observations. First, the two Tc1 flanking regions initially sequenced are similar (85, 86). Second, *unc-22* is a preferred mutational target relative to most genes (68). Third, Tc1 insertions within *unc-54* are not distributed randomly (17). The sequence of seven Tc1 insertions at six sites in *unc-22*, plus three Tc1 insertions in other genes and two previously sequenced sites (85, 86), suggested a 9-bp consensus sequence for Tc1 insertion (69). Additional sequencing of four sites within *unc-54* (19) suggested two modifications and yielded the consensus shown in Fig. 3.

The only absolute requirements for Tc1 insertion are the −1 T and the +1 A, conserved among all 16 sequenced sites. There is obvious ambiguity at 4 sites with T or G at the −3 site, a pyrimidine at the +2 site, and G or C at the +3 site. Despite this degeneracy, sites within *unc-54* that are hit repeatedly when Tc1 inserts into *unc-54* give the best match in the entire *unc-54* coding sequence when compared to the consensus matrix derived from sites in other genes (19, 69), demonstrating the importance of the target sequence.

The target sequence has interesting features: it is AT rich, is palindromic, and contains alternating purines and pyrimidines (69). The AT bias is close to that of the *C. elegans* genome, which is 64% AT (93). Many of the sequenced sites are even more strongly palindromic than the consensus sequence. The *Escherichia coli* transposon Tn*10* also recognizes a palindromic sequence, and Halling and Kleckner (40) have proposed that Tn*10* transposase may be conceptually analogous to type II restriction enzymes, which cleave about the middle of a palindromic sequence. Perhaps a Tc1 transposase also acts in this manner. Unlike Tn*10* or other elements that recognize specific target sequences, Tc1 does not duplicate the entire recognition sequence. As a result, when Tc1 inserts into the middle of the recognition sequence, it destroys the consensus site.

The alternation of purines and pyrimidines occurs at seven or eight positions from −4 to +4 in the target motif (69). This also represents a preference, not an absolute requirement, since only some sites show a perfect alteration throughout the target region. The observation of an alternating purine/pyrimidine tract is intriguing. Although this tract is relatively short, it may be able to adopt the left-handed or Z DNA conformation (reviewed in reference 79). In *Ustilago* sp., a protein involved in recombination only binds to Z DNA (51), and in *Drosophila* sp., two retrotransposons prefer sites consisting of alternating purines and pyrimidines (36, 47).

Although Tc1 transposition prefers specific target sequences for integration, parameters other than the primary sequence may influence the frequency of integration. Tc1 insertion into *unc-22* is about 100-fold more frequent than into *unc-54* (11, 17; discussed in reference 19). The different sizes of the genes (30 versus 7 kb), and a postulated larger number of target sites in *unc-22* (69), account for only a 10- to 20-fold difference in mutation frequencies of *unc-22* and *unc-54* (69). A greater preference for sites in *unc-22* because they perhaps match better the optimal target might account for some of the residual fivefold difference, but there may be other factors influencing Tc1 insertion beyond the specific target sequence.

D. Tc1 Excision

1. Somatic excision

Tc1 elements undergo somatic excision from genomic sites in both N2 and BO strains at high frequency. This was originally seen in Southern blots of restriction-digested BO DNA hybridized with unique-sequence probes flanking Tc1 insertion sites (26). Two bands, one containing Tc1 and one that is 1.6 kb smaller, representing the empty target site, are found with most, if not all, Tc1 sites (17, 26, 42, 66, 67). Empty sites typically accumulate to about 1 to 5% of the filled sites in populations of animals of mixed ages (for example, see reference 19), but under certain conditions, such as prolonged starvation, and for certain sites, the fraction of empty sites can approach 50% (24; D. G. Moerman, G. M. Benian, and R. H. Waterston, unpublished results).

The majority of the excision observable by Southern analysis occurs in somatic cells (17, 25, 66). In early experiments, the inheritance of anonymous dimorphic restriction fragments was studied by establishing multiple subclones and analyzing their DNAs. The Tc1 element-containing fragments were shown to be more stably inherited than would be expected if the high-frequency excision occurred in the germ line (25). In addition, analysis of different developmental stages showed that the fraction of empty sites is low in embryos, accumulates during larval development, and then decreases as more germ line cells are added in the adult (25), an expected result if excision occurs primarily in somatic tissue. Later, excision was demonstrated to continue in starved animals in the absence of growth (22, 24) (see below, Fig. 5).

Somatic excision has been verified directly by observing mosaic animals among populations of Tc1-induced mutants of *unc-54* and *unc-22* (17, 66). The *unc-54* myosin heavy chain and the *unc-22* proteins are present only in the cells in which the genes are transcribed and only in muscle cells. Mosaic animals among the progeny of *unc-54*::Tc1 mutants could be detected as uncoordinated egg-laying-competent animals because of the following facts. *unc-54* mutants are normally defective in egg laying in addition to exhibiting paralysis, since the *unc-54* myosin heavy-chain function is required in egg-laying muscles as well as in body wall muscles (30). Fourteen of the 95 body wall muscle cells and the 16 egg-laying muscle cells derive from a common stem cell (94), so that animals which had regained the ability to lay eggs might also have some normal body wall muscle cells. Indeed, when the body wall muscles of these egg-laying-competent animals were examined using polarized light microscopy, a few cells exhibited a near-wild-type pattern of birefringence against the background of mutant cells (17). These few cells presumably had lost the Tc1 from the *unc-54* gene and were able to express functional myosin heavy chain.

Mosaic animals among the progeny of *unc-22*::Tc1 mutants were demonstrated directly using an antibody directed against the *unc-22* protein (Fig. 4). Rare muscle cells containing the *unc-22* protein were found side by side with cells that did not contain the antigen (66). Antigen-positive cells were surprisingly difficult to find, since on Southern blots the excision band observed is about 1 to 5% of the insertion band in intensity. This may be explained by findings that many excision events are imprecise (see below), and

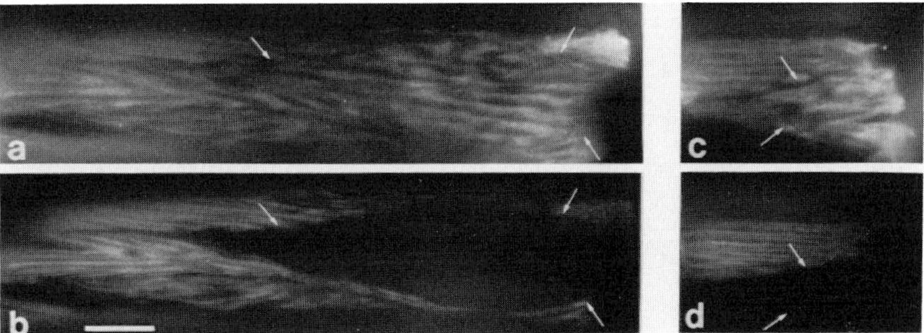

Figure 4. Somatic excision of Tc1 from *unc-22*. Animals homozygous for the mutation *unc-22(st136*::Tc1) were stained with either an antimyosin monoclonal antibody (DM 28.2; a, c) or a polyclonal antibody to the *unc-22* protein (R11-3; b, d), and the muscle was viewed by fluorescence microscopy. Most muscle cells stained only with the antimyosin antibody, but as shown in the paired photos (a and b; c and d), rare cells stained with both antimyosin and anti-*unc-22* antibodies. The boundaries of one spindle-shaped muscle cell are indicated by the small arrows. Bar, 10 μm. (From reference 66.)

only a fraction yield an in-frame product detectable by the antibody.

At those sites examined, somatic excision occurs at a similar rate per unit time. For example, arresting the development of first-stage larvae (L1) by starvation does not alter the rate of excision (22, 24) (Fig. 5). In fact, in cultures of starved, nongrowing L1 which exhibited no cell division, excision continued for over 2 weeks of culture and approached 10% for some sites. Excision apparently does not require

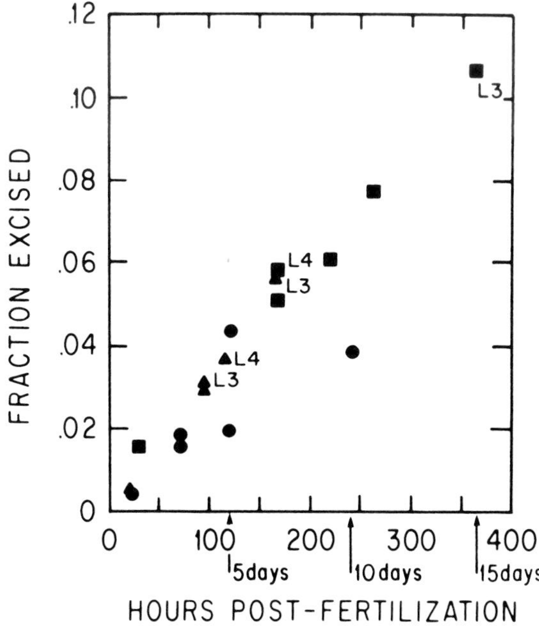

Figure 5. Somatic excision of Tc1 from three anonymous sites with time. First-stage larvae were growth arrested by starvation, and the excision of Tc1 from each site (Tc1.1 [●], Tc1.2 [▲], and Tc1.3 [■]) was measured by Southern hybridization analysis. Unlabeled points are for first-stage larvae arrested at 16°C for the times shown. L3 indicates worms fed and harvested as third-stage larvae at the times shown; L4 indicates worms similarly harvested as fourth-stage larvae. (From reference 22.)

chromosomal DNA replication, in contrast to maize, where transposition is correlated with DNA replication (32, 37).

2. Germ line excision

Although high-frequency excision appears limited to somatic cells, Tc1 is unstable in the BO germ line; germ line reversion of *unc-22*::Tc1 alleles occurs at a frequency of about 2×10^{-3} (68); reversion of *unc-54*::Tc1 alleles occurs at a frequency of 10^{-5} to less than 10^{-7}, depending upon their location within the gene (17, 19). Clusters of revertants among progeny are occasionally observed with *unc-22*::Tc1 alleles, suggesting that some germ line reversion is premeiotic in origin (Moerman, unpublished results). For inserts in *unc-22* and *unc-54*, the requirement for restoration of function places constraints on the types of excision events that can be recovered as revertants, and therefore the observed frequencies should be viewed as a lower limit for germ line excision. Mutations in genes that yield higher germ line reversion frequencies have been recovered. For example, Tc1 alleles of *unc-52* and *unc-15* revert at frequencies approaching 10^{-2} (Moerman, unpublished results), and a *dpy-19* spontaneous allele has a reversion frequency of 10^{-2} (M. Finney and H. R. Horvitz, unpublished results). These latter germ line reversion frequencies approach those estimated for somatic reversion rates using Southern blot analysis. Thus, the discrepancies between somatic and germ line excision may not be as great as has been inferred (19, 22, 24).

3. Sequencing somatic and germ line excision sites

Sequencing of cloned sites after Tc1 somatic and germ line excision revealed that most Tc1 excision

```
                                  A. Somatic Excision
Excision events at anonymous sites:
        Site Tc1.1
                 N2 site                         5'-AAATATAT-3'
                 Filled BO site              AAATAcag...Tc1...ctgTATAT
          Empty site:
                 Clone name
                 #1                                AAATATAT
                 #2                                AAATATAT
                 #3                                AAATATAT
                 #4                                AAATATAT
                 #5                                AAATATAT
                 #6                                AAATATAT
        Site Tc1.2
                 N2 site                         5'-ATTTATGT-3'
                 Filled BO site              ATTTAcag...Tc1...ctgTATAT
          Empty site:
                 Clone name
                 #1                              ATTTAtgTATGT
                 #2                              ATTT----ATGT
                 #3                              ATTT----ATGT
                 #4                              ATTTAtgTATGT
                 #5                              ATTTAtgTATGT

unc-54 excision events:
        Site 1.
                 unc-54(+)                       5'-ATCTACGT-3'
                 unc-54(r323::Tc1)           ATCTAcag...Tc1...ctgTACGT
                 Clone name
                 TR#29                          ATCTA---gTACGT
                 TR#30                          ATCTAca--TACGT
                 TR#31                          ATCTA--tgTACGT
                 TR#32                          ATCTA--tgTACGT
                 TR#33                          ATCTA------CGT
        Site 2.
                 unc-54(+)                       5'-AGATACCA-3'
                 unc-54(r322::Tc1)           AGATAcag...Tc1...ctgTACCA
                 Clone name
                 TR#25                          AGATA--tgTACCA
                 TR#26                          AGATA--tgTACCA
                 TR#27                          AGATA--tgTACCA
                 TR#28                          AGATAca--TACCA
```

Figure 6. DNA sequence comparisons of Tc1 excision sites. The DNA sequences of excision events are shown relative to wild type and the Tc1 insertion mutant from which they were derived. (A) Somatic excision clones from two anonymous sites (88) and from two sites in *unc-54* (19). (B) Germ line excision events from the same two *unc-54* sites as in panel A (19) and from a single *unc-22* site (49; Kiff et al., unpublished data). The duplicated TA dinucleotide target sequence is underlined. Nucleotides of the Tc1 element are given in lower case. Extra nucleotides resulting from imprecise excision are aligned where possible with the insertion mutant from which they were derived. In those cases where the extra nucleotides appear to represent duplication of flanking sequences, both the extra nucleotides and the sequence from which they might derive are doubly underlined and are bold. Sequence identities possibly involved in deficiency formation are also doubly underlined and in bold. (· · ·) Sequences omitted from the figure; (---) gaps introduced to permit alignment of flanking sequences.

events are imprecise (19, 24, 49, 88) (Fig. 6), as is common in other transposon systems (15, 89, 90, 96, 102). The precision of somatic excision from two anonymous sites in the BO strain was examined (88) (Fig. 6A). Of five clones recovered from one site, two showed precise excision and three were imprecise. Each of the latter had the same 4-bp insertion. At a second site, six of six excisions of Tc1 were precise. Somatic excision of Tc1 from two sites within *unc-54* has also been examined (19) (Fig. 6B). Only one of nine sequenced clones restored the wild-type sequence; seven others had 4-bp inserts and one had a 3-bp insert. In both the anonymous sites and the *unc-54* sites, the inserts contained an extra TA plus one or two nucleotides from one or the other end of Tc1.

Germ line Tc1 excision events have been analyzed from sites in *unc-22* and *unc-54*. Among 51 revertants from a single *unc-22*::Tc1 insertion, 47 were phenotypically wild type and four were partial revertants. The empty Tc1 sites of four randomly selected full revertants and the four partial revertants were cloned and sequenced (49) (Fig. 6B). Only one of the four seemingly full revertants had regained the ancestral wild-type sequence. Two other full revertants had duplications of either 15 or 21 bp 5′ of the insertion site, and the fourth had a 2-bp substitution whose sequence suggested that DNA cleavages for excision had been shifted 2 bp at each end of the element. Of the four partial revertants, one was due to small duplication of DNA just 5′ of the insertion site, but the remaining three had large deletions, 1 to

```
                         B. Germline reversion
unc-54 excision events:
   Site 1.
          unc-54(+)                         5'-ATCTACGT-3'
          unc-54(r323::Tc1)              ATCTAcag...Tc1...ctgTACGT
        and unc-54(r328::Tc1)
   WT revertants
          unc-54(r851)                      ATCTA-tgTACGT
          unc-54(r852)                      ATCTA-tgTACGT
          unc-54(r853)                      ATCTA-tgTACGT
          unc-54(r854)                      ATCTA-tgTACGT
          unc-54(r855)                      ATCTA-tgTACGT
   Partial revertants
          unc-54(r837)                      ATCT-ctgTACGT
          unc-54(r661)                      AT-----gTACGT

   Site 2.
          unc-54(+)                         5'-AGATACCA-3'
          unc-54(r322::Tc1)              AGATAcag...Tc1...ctgTACCA
   WT revertants
          unc-54(r847)                      AGATAca---ACCA
          unc-54(r848)                      AGATAcatgTACCA
          unc-54(r849)                      AGATAcatgTACCA
          unc-54(r850)                      AGATAcatgTACCA

unc-22 excision events:

   unc-22(+)           TGGTTCTCCAATTTTGGGAT------------------------ATGTCGTTGAACGT
   unc-22(st192::Tc1)  TGGTTCTCCAATTTTGGGATAcag.....Tc1........ctgTATGTCGTTGAACGT
WT revertants
   unc-22(st534)       TGGTTCTCCAATTTTGGGAT------------------------ATGTCGTTGAACGT
   unc-22(st532)       TGGTTCTCCAATTTTGG------------------------tgTATGTCGTTGAACGT
   unc-22(st527)       TGGTTCTCCAATTTTGGGATAcaATTTTGGGA---------tgTATGTCGTTGAACGT
   unc-22(st533)       TGGTTCTCCAATTTTGGGATAcTTCTCCAATTTTGGGA---tgTATGTCGTTGAACGT
Partial revertants
 Associated with substitutions:
   unc-22(st530)       TGGTTCTCCAATTTTGGGATTTTGGG------------------ATGTCGTTGAACGT
 Associated with deletions:
   unc-22(+)           TGGTTCTCCAATTTTGGGATA...972 bp....CAATTTGGATGCC
   unc-22(st531)       TGGTTCTCCAATTTTGG----984 bp deficiency----ATGCC

   unc-22(+)           ACCACTTGAAGTTTC...1246 bp...TATGTCGTTGAACGTTTTGAGAAG
   unc-22(st529)       ACCACTTGAAGTTTC-------1263 bp deficiency-----TGAGAAG

   unc-22(+)           TGGTTCTCCAATTTTGGGATA.....1969 bp......CCAAGTATCGAGGTTCCAAATC
   unc-22(st528)       TGGTTCTCCAATTTTG---1974 bp deficiency--TTCCAAATCGAGGTTCCAAATC
```

Figure 6. *Continued.*

2 kb in size. One endpoint of each deletion was within a few bases of the insertion site and all three were in frame, yielding partially functional, although much shorter, *unc-22* polypeptides (49; J. E. Kiff, D. G. Moerman, and R. H. Waterston, unpublished data).

Of 11 germ line excision events from two sites in *unc-54*, 9 were phenotypically wild type and 2 were partial revertants (Fig. 6B). All were examined in detail; none represented precise excision (19). Three of four full revertants of the first *unc-54* mutant had an identical 6-bp insert, and the fourth had a 3-bp insert. The 6-bp insert consisted of a TA dinucleotide plus 2 bp identical to each terminus of Tc1. These insertions maintain the translational reading frame. The five full revertants of the other *unc-54* mutant all had the same 4-bp insert. This insert contains an extra TA dinucleotide plus 2 bp from the 3′ terminus of Tc1 and is identical to the 4-bp insert of seven of the nine somatic excision events from the two *unc-54*::Tc1 mutants (this includes two inserts which have the complementary 4 bp at the 5′ end of the sites) and to the 4-bp insert of three of the five events from one anonymous site. The functionality of these alleles despite the 4-bp insert is ascribed to a shift in the 5′ splice site and a fortuitous restoration of the reading frame. The two partial revertants of the second mutant contain 4- and 2-bp inserts, respectively.

The majority of Tc1 excision events are thus imprecise in somatic and germ line tissue, even in cases where Tc1 is located within exons and a wild-type phenotype was selected. The similarity of a large fraction of somatic and germ line events for the *unc-54* mutants has led to the speculation that perhaps these events are each mediated by the same factor, e.g., the putative Tc1 transposase (19). Both *Ac* in maize (32) and the P element in *Drosophila* sp.

(53) can produce transposases which function in both germ line and somatic tissues.

The excision events observed from the different sites show distinctive features. Those in *unc-54* would seem to involve cuts within Tc1 such that a piece of Tc1 is left behind at the site. In addition, for the much more frequent excision from the *unc-22* site, flanking DNA is often either duplicated or deleted. In contrast, one anonymous site yielded entirely precise excision. These differences from site to site suggest that flanking sequences influence the excision process.

E. Extrachromosomal Tc1 DNA

Extrachromosomal forms of Tc1 have been detected in the Bergerac strain (84, 87), just as extrachromosomal forms of other transposable elements have been detected in other organisms, including Tn*10* in *E. coli* (71) and Mu in maize (95). A 1.6-kb linear molecule is the predominant form, but supercoiled and relaxed circles are also present. The abundance of these extrachromosomal species is approximately 0.1 to 1.0 copy per cell (87).

The linear form is very similar or identical in sequence to the integrated form. *Eco*RV, which cuts 20 bp from each end of Tc1, slightly reduces the size of the linear DNA. Restriction enzyme digestion of the circular forms suggests that they are also intact Tc1 elements. Cleavage of the circles by *Eco*RV indicates that they contain at least one inverted repeat, and size measurements suggest they contain two. To determine how the inverted repeats are joined, the circular forms were digested with either *Rsa*I or *Pst*I (87). If Tc1 upon forming a circle includes a T and an A at its ends, it will contain an *Rsa*I site (CTGTACAG). However, if the circular Tc1 lacks the TA, then a *Pst*I site (CTGCAG) should be generated. The circles were cleaved by *Rsa*I and not cleaved by *Pst*I, indicating that they include the T and A.

This result has implications for the mechanism of transposition. As noted earlier, Tc1 always inserts at a TA dinucleotide, and elements are flanked by TA dinucleotides at each end. If the circles constitute transposition intermediates, the result implies that Tc1 makes flush, not staggered, cuts, and would indicate that Tc1 does not generate a target site duplication. Only one example of an element with similar structure has been found which inserts without duplication: the bacterial transposon Tn*554* (72). On the other hand, if the circles represent dead-end excision products, no inferences can be drawn from them about the transposition process.

It has been suggested that these extrachromosomal Tc1 copies are the products of high-frequency somatic excision (84, 87). However, extrachromosomal copies of Tc1 do not increase in starved larvae even though Tc1 continues to excise (24, 87). For example, in first-stage BO larvae starved for 242 h, the amount of excision at each site is on average about 10%, corresponding to an average of more than 60 excision events per diploid cell. There is still less than one extrachromosomal Tc1 copy per cell in these starved animals. Possibly these extrachromosomal copies are unstable; perhaps they reinsert (87); or perhaps they are a side product. With regard to the latter possibility, it should be recalled that many excision products appear to be imprecise, in contrast to the homogeneity of the circles as judged by *Rsa*I digestion.

The correlation of excision frequency with transposition frequency and the discovery of linear and circular extrachromosomal Tc1 have led to models in which Tc1 transposition is conservative, involving a cut-and-paste mechanism. In these models, Tc1 excision represents an early step in the transposition pathway. The excised linear molecule would become circularized, find an appropriate target, and reinsert. Reinsertion of a circular molecule via site-specific recombination is certainly possible (10), and interestingly the DNA sequence at the point of Tc1 circularization bears homology to the Tc1 insertion site consensus (19). Many steps in this pathway are analogous to the conservative mechanism of bacteriophage λ insertion and to that postulated for Tn*5* (4) and Tn*10* (3). Although this conservative pathway for Tc1 transposition is attractive, direct evidence for such a mechanism is lacking and the role, if any, of the circular form in the pathway is uncertain.

F. Effects of Tc1 Insertion on Gene Expression

Depending on its site of insertion within a gene, Tc1 can produce a range of phenotypic effects, from a very mild mutant phenotype to a phenotype similar to that of a bona fide null allele. At all four *unc-54* insertion sites, Tc1 causes a phenotype similar to amber and deficiency alleles (17). In *lin-12*, Tc1 alleles are also similar to existing null alleles (38). By contrast, none of the Tc1 inserts in the 12 different sites of *unc-22* result in a phenotype as severe as amber or deficiency alleles (66, 68). In *unc-15*, both *unc-15*::Tc1 weak alleles (S. Rioux, H. Kagawa, R. P. Anderson, and R. H. Waterston, unpublished results) and null-like alleles (D. G. Moerman and M. Fedor, unpublished results) have been isolated. Clearly, the type of gene and the relative position of Tc1 within or near a gene influence the type of mutant phenotype produced, a fact that should be considered by those attempting transposon tagging of their favorite gene.

The effects of Tc1 on the message of its host gene are diverse. In the most thoroughly studied example, six Tc1 alleles of *unc-22* were examined on Northern (RNA) blots, and five of the six yielded messages larger than the N2 wild-type product (66). In these five cases, Tc1 did not appear to interrupt transcription of *unc-22*, but instead was simply incorporated into the *unc-22* message. This result was independent of Tc1 orientation. Results similar to these were observed for a *tra-2*::Tc1 allele (P. Okkema and J. Kimble, unpublished results), and in maize, a *Ds* element inserted into the alcohol dehydrogenase locus is also transcribed as part of the message (16). A perhaps more expected observation was found with a *lin-12*::Tc1 allele, which has a truncated message (38; I. Greenwald, personal communication).

The sixth *unc-22*::Tc1 allele examined and three *unc-15*::Tc1 alleles, all mild in phenotype, have messages that are neither truncated nor elongated (66; S. Rioux, unpublished results). Message levels may be reduced in these mutants, but careful quantitative measurements have not been done. In each case, the Tc1 insert is at or near the 5′ end of the gene, but the transcriptional start site has not been determined for either gene. Therefore, it is not clear whether Tc1 is 5′ of the gene and downregulating its transcription, or is in an intron and interfering with processing of its transcript.

In general, the effects of Tc1 on message correlate with effects on phenotype: a *lin-12*::Tc1 null allele has a truncated message; an extremely mild *unc-22*::Tc1 and three mild *unc-15*::Tc1 alleles have normal-length messages. The intermediate *unc-22*::Tc1 alleles that have Tc1 incorporated into messages longer than wild-type messages are more puzzling. Perhaps these alleles produce a truncated polypeptide that retains enough function to alleviate partially the severity of the phenotype. For some *unc-22*::Tc1 alleles, a truncated polypeptide has been found (66; D. G. Moerman, R. Barstead, L. Shriefer, and R. H. Waterston, unpublished results).

III. REGULATION OF Tc1 TRANSPOSITION

A. Correlation of Germ Line Excision and Insertion

The recovery of unstable *unc-22*::Tc1 alleles provided the first opportunity to study the activation of Tc1 transposition in the germ line. Both the induction frequency of *unc-22*::Tc1 mutations and their stability are sensitive to genetic background (68). In the BO strain, the forward spontaneous mutation frequency at *unc-22* is approximately 10^{-4}, as compared to 10^{-6} to 10^{-7} in N2. BO-derived *unc-22* Tc1 mutations revert at a frequency of about 10^{-3} in a BO genetic background, but when placed in a primarily N2 genetic background, they revert at a frequency of less than 10^{-6}. These same mutations when reintroduced into a BO/N2 mixed background once again become unstable (68). Strains in which the *unc-22* alleles become unstable also exhibit elevated forward mutation rates at the *unc-22* locus. These observations suggest that excision and transposition of some Tc1 elements are regulated by *trans*-acting factors; otherwise, unstable *unc-22* alleles could not be stabilized by altering the genetic background. Further studies have confirmed that Tc1 insertion and excision activities are correlated (70, 100; also see reference 11; details are given below). Thus, germ line excision and insertion may be regulated in the same manner and, as suggested earlier, may reflect different stages of the transposition process.

B. Effects of Tc1 Copy Number

Tc1 copy number per se has little or no influence on Tc1 transposition frequencies. Of nine strains examined for their ability to generate spontaneous unstable *unc-22* alleles, BO is uniquely active (68), yet three other strains, DH424, FR1, and BL1, are also high-copy-number strains for Tc1 (26, 54). The BO, FR1, and BL1 strains are all Bergerac strains isolated by either V. Nigon or J. Brun of the Lyon nematode laboratory in France (20, 68). BO is therefore unique even among Bergerac strains. In another study, spontaneous *unc-54* mutations were isolated in N2, DH424, and BO, and of these three strains, only BO yielded Tc1-generated mutations (17). In a third study (70, 100), a spontaneous *unc-22*::Tc1 mutation, initially isolated in BO and then stabilized in an N2 background, was crossed back again to BO. From this cross, several isogenic mixed N2/BO lines containing the *unc-22*::Tc1 mutation were established. Measurements of Tc1 copy number and reversion frequency at the *unc-22* site (Fig. 7) indicate that Tc1 copy number is not the major variable regulating Tc1 transposition. Reversion frequencies among the various lines ranged from 10^{-3} to less than 10^{-7} and were not correlated with Tc1 copy number. In the extreme, more recently constructed strains have less than 60 copies of Tc1, rather than the more than 300 copies typical of BO, but these strains have activities comparable to BO (70).

C. Mutators: Factors Regulating Tc1 Activity

The identity, behavior, and structure of factors regulating Tc1 activity (termed *mut*, for mutators)

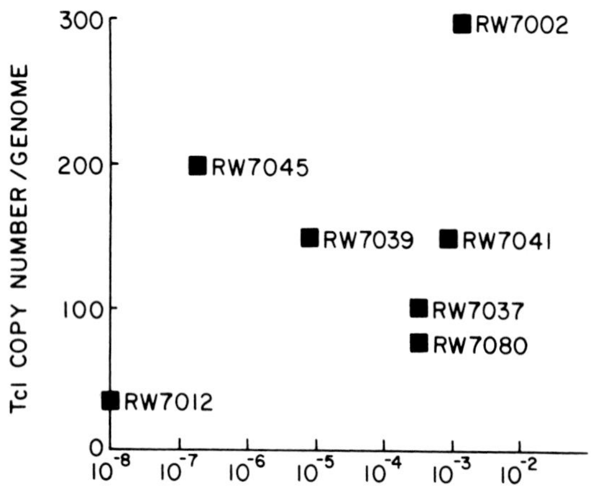

Figure 7. Copies of Tc1 versus mutator activity in N2/BO hybrids. After a single outcross of the BO strain with an N2-derived strain, seven lines were established after 10 generations of selfing. The copy number of Tc1 in each of the strains was estimated by dot blot hybridization, and the mutator activity was estimated by measuring the reversion of an *unc-22*::Tc1 mutation. Little correlation was observed between reversion frequency and copy number. (From reference 100.)

are currently under intensive investigation. Early genetic experiments failed to identify a single major factor responsible for Tc1 activity after replacing single BO chromosomes by their N2 homologs (68). In addition, unstable *unc-22* mutations isolated in BO were stabilized only after repeated outcrosses to N2. These studies indicated that the factors in BO are polygenic and are present in multiple sites in the genome.

Further experiments with BO/N2 hybrids suggest that there may be relatively few *mut* loci in BO. In the comparison of Tc1 copy number and Tc1 reversion frequency (above), several mixed BO/N2 lines were established, each of which could be monitored for activity because they each contained an *unc-22*::Tc1 allele. Three of 11 strains had severely reduced mutator activity (70). The measurement of Tc1 copy number in these strains suggests that there was a mild selection against the BO chromosomes, but, even so, the large null-activity class would not be expected if mutator activity arises from many different sites scattered about the BO genome. This result combined with those described in the previous paragraph suggests that there may be a small number of mutators spread over a few chromosomes. One of the active BO/N2 hybrid lines was further characterized: the principal *mut* activity in this strain, called *mut-4*, was mapped to a 1-map-unit interval near *unc-37* on chromosome I (70).

The *mut-4* activity derived from the BO strain itself appears to be mobile. After crossing a *mut-4*, *unc-22*::Tc1-bearing strain, RW7037, to N2 male animals, 19 *unc-22*::Tc1 isogenic lines were established from the F_2 progeny. Fifteen of the 19 strains had lost the mutator activity, but 1 of the remaining 4 strains had mutator activity at least equal to that of the parental RW7037 strain. Genetic mapping of the mutator activity revealed that this strain contained two sites of mutator activity, the original *mut-4* plus a second, designated *mut-5*, located near *dpy-10* on chromosome II (70). No activity was associated with chromosome II in the parental *mut-4* strain. This observation led to the following specific test for mutator mobility.

If the mutator itself is capable of transposition, it could insert into the *unc-22* gene. Such an allele could not be stabilized by placement into an otherwise inactive background because it is inherently active. Accordingly, spontaneous *unc-22* mutations were isolated in a *mut-4* strain that was also mutant at the closely linked *dpy-5* locus, the *dpy-5 mut-4* chromosome was replaced by the N2 homolog, and then each *unc-22* mutant strain was examined for germ line stability. The elimination of *mut-4* on chromosome I should yield stable *unc-22* lines, as that was the principal source of activity in the parent strain, unless the mutator activity had transposed to a new site. Of 26 tested mutations, 24 were stable after outcrossing, but 2 remained unstable and had reversion frequencies comparable to those of typical BO-derived *unc-22*::Tc1 mutations. The new activity in one of these strains, designated *mut-6*, has been positioned near *dpy-13* on chromosome IV (70) but outside of *unc-22*. Although this experiment did not capture the mutator in *unc-22*, it may have demonstrated the capacity of the mutator for transposition. Although transposition is the most likely explanation for the occurrence of *mut-5* and *mut-6* mutators, other explanations exist. For instance, the mutators might have arisen de novo or may have resulted from inter- or intrachromosomal gene conversion events involving inactive elements, like the mechanism generating Ty variability in *S. cerevisiae* (82). The high frequency of occurrence of these mutators makes it unlikely that they result from simple spontaneous mutations or from gene conversion, which generally occurs at a much lower frequency in other eucaryotes than it does in *S. cerevisiae*.

Based on analogy to the *Ac/Ds* system of maize (32) and the P system (28) and *mariner* element of *Drosophila* sp. (63), one interpretation of these observations on mutators is that there are autonomous and nonautonomous Tc1 elements and the latter can be mobilized by the former. The *unc-22*::Tc1 alleles that can be stabilized by altering the genetic background are clearly nonautonomous elements, but

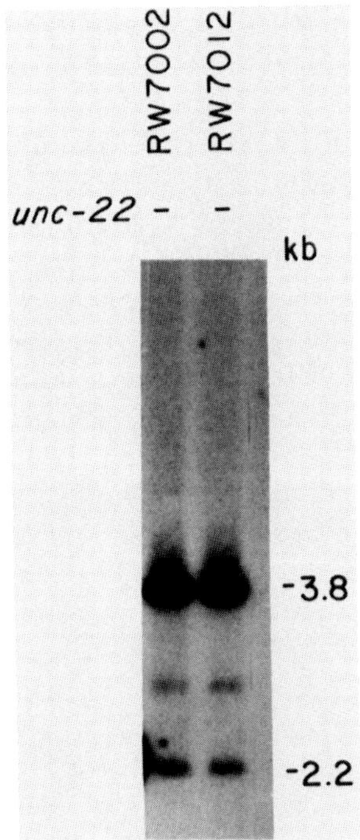

Figure 8. Southern hybridization analysis of somatic excision in strains with and without germ line Tc1 mutator activity (68). DNAs from strains RW7002 and RW7012 were hybridized with a unique probe spanning the *unc-22*::Tc1 insertion site. The *unc-22(st136*::Tc1) mutation in the RW7012 strain reverts in the germ line at a frequency of less than 10^{-7}, whereas reversion of the same mutation in the RW7002 strain is about 10^{-3}. Despite this difference in germ line reversion, little if any difference exists between the two strains in the relative intensities of the somatic excision band at 2.2 kb and the insertion band at 3.8 kb. (From reference 67.)

whether the mutators are indeed the autonomous Tc1 elements remains unknown. In this regard, the microheterogeneity observed among Tc1 isolates takes on added significance, since such variability may underlie the differences between inactive and autonomous Tc1 elements.

D. Tissue-Specific Regulation of Tc1 Activity

Excision of Tc1 occurs in germ line and somatic tissue, and although these events appear superficially similar, there is a major difference: germ line excision of Tc1 is more sensitive to genetic background than is somatic excision. This differential control of Tc1 movement was first demonstrated in studies of *unc-22* (67, 68). Germ line excision of Tc1 from *unc-22* differs 1,000-fold between BO (10^{-3} excision frequency) and N2 (less than 10^{-6} excision frequency). In contrast, somatic excision varies little, if at all, for the same *unc-22*::Tc1 alleles in BO and N2 backgrounds (67) as estimated by the ratio of excision to insertion bands on Southern blots (Fig. 8). This result has since been confirmed for several other Tc1 insertion mutations, with not more than a fivefold reduction in somatic excision of N2 relative to BO (19, 22, 42).

Dramatic confirmation of the idea that Tc1 activity is regulated in a tissue-specific manner came from mutants in which Tc1 transposition and excision are elevated in the germ line but unaffected in the soma (11). By screening 1,500 ethyl methanesulfonate (EMS)-mutagenized, cloned N2/BO hybrid animals for elevated reversion frequencies of an *unc-54*::Tc1 mutant, eight strains were isolated with reversion frequencies that were 4- to 100-fold higher than the parental strain. The mutations were presumably all EMS induced, since none was isolated spontaneously from 1,095 control animals. Although the selection scheme was based on Tc1 reversion, transposition of the element was also increased, and the increases in reversion and transposition frequencies correlated in the various strains. This is in agreement with results obtained during the mapping of the BO mutators and suggests that insertion and excision are regulated by the same factors.

Two of the eight EMS-induced mutations have been mapped. The *mut-3* locus is near *unc-54* at the right end of chromosome I, and *mut-2* is more centrally located on chromosome I (M. Finney, unpublished results). The *mut-2* locus is most intriguing because it elevates Tc1 transposition frequency 20-fold above the activity reported for BO (17, 68). A strain, TR679, carrying *mut-2* has become especially popular for transposon tagging and has been dubbed "high hopper." However, the strain is sickly, has few viable progeny, and has an increased frequency of X chromosome nondisjunction, perhaps as a consequence of its high mutator activity (11, 43).

The relationship of these EMS-induced alleles of *mut* loci (e.g., *mut-2*) to the mutators endogenous to the BO strain (e.g., *mut-4*) is unknown. The endogenous mutators are possibly transposons, but it is not clear whether the EMS-induced mutators are Tc1 elements, related factors, or even host regulatory genes, whose products interact with transposons. In this regard, it should be noted that the BO mutator activity has only been shown to activate Tc1 in more than 25 mutations examined, whereas in a comparable number of mutations, *mut-2* has been shown to cause transposition of both Tc3 and Tc4 and perhaps other transposons.

Control of transposition can be achieved in a variety of ways including the following: "genomic shock" can activate transposons in maize (61); copy number of *Ac* can influence the time and frequency of

Ds transposition in maize (59); copy number can also affect Tn*5* transposition in bacteria (48); a host gene, that for the Dam methylase, controls Tn*10* and Tn*5* transposition in part by regulating the synthesis of their respective transposases (80); germ line transposition of P is affected by "cytotype" in *Drosophila* sp. (27); and the differential regulation of P excision in germ and soma in *Drosophila* sp. is controlled by tissue-specific splicing (53). Somatic excision of Tc1 is constitutive, while germ line excision of Tc1 is regulated. Models involving repressors, activators, transposase, etc., can be built but cannot yet be readily distinguished. In this regard, the isolation and characterization of a bona fide mutator would be most helpful.

E. A Spontaneous De Novo Tc1 Mutator Activator

The mutator loci described so far either are endogenous to BO or were isolated in BO derivatives after EMS mutagenesis. However, there is one example of a mutator occurring spontaneously in a non-mutator strain. While using a CB30 [*sma-1(e30)*] strain as a genetic marker to map Tc1 polymorphisms, it was noted that the strain had about 50 extra copies of Tc1 relative to N2 (C. Trent, unpublished results). This observation was puzzling, since CB30 was derived from N2 after EMS mutagenesis and therefore, like N2, should have had 30 copies of Tc1 (8). More surprising, the strain has mutator activity. Spontaneous unstable *unc-22* alleles have been isolated, and these have Tc1 inserted in the locus (C. Trent and C. Link, unpublished results). The earliest frozen version of CB30 (frozen in 1969) has a Tc1 pattern on a Southern blot identical to that of N2 and does not have mutator activity (J. Hodgkin, unpublished results). In contrast, the Cambridge, England, working stock of *sma-1(e30)*, designated CB4000, is similar to the Trent CB30 strain. It contains more than 100 copies of Tc1, grows poorly, has mutator activity, and has slightly increased X chromosome nondisjunction (Hodgkin, unpublished results). This working strain has been propagated continuously on plates for more than 15 years. The best estimate is that Tc1 activation occurred between 1969 and 1975 (J. Hodgkin, personal communication).

These observations on CB30 (CB4000) are clearly exciting and puzzling. They give us a glimpse at the possible origin of mutator strains, including BO, and study of the strain may reveal how a quiescent strain can be transformed into a mutator strain.

IV. MOLECULAR APPROACHES TO IDENTIFYING MUTATORS

Identification of the genes and proteins controlling Tc1 transposition should provide insight into the mechanism of transposition and might provide the experimentalist with additional ways to use Tc1. The ability to locate specific mutators by genetic mapping raises the possibility of cloning the segments containing the activity. The development of the physical map of the nematode genome (12) will provide a means for recovering DNA for any mapped locus. If the BO mutators are indeed autonomous Tc1 elements, analysis of DNAs from sibling strains either with or without mutator activity should reveal that element consistently associated with mutator activity. Strains containing *mut-5* in a very low copy number background (<50 Tc1 hybridizing fragments) have been constructed and are being used in mapping experiments (70). Once candidate cloned DNAs have been recovered, they can be tested by reintroduction via transformation for their ability to activate excision (35).

The protein product(s) of Tc1 can be produced in *E. coli* from cloned sequences either alone or as part of a fusion protein. Antibodies produced against the protein(s) are being developed and could provide direct information about the tissues of expression in strains differing in activity. Alternatively, the *E. coli*-produced protein might be active, and in vitro assays for its activity are being developed. Studies of the protein(s) would be greatly aided by the identification of a bona fide autonomous element.

V. OTHER TRANSPOSABLE ELEMENTS IN *C. ELEGANS*

Five other repetitive element families that have characteristics of transposons have been found in *C. elegans*, and four of these have been shown to transpose.

A. Tc2

Tc2 elements were fortuitously discovered during a search for abnormally sized Tc1 elements as possible sources of mutator activity (24). A fusion between Tc1 and Tc2 had generated a Tc1-hybridizing *Eco*RV fragment larger than the standard Tc1 1.6-kb fragments, but this fusion element segregates from mutator activity (A. Levitt and S. W. Emmons, unpublished results; I. Mori, D. G. Moerman, and R. H. Waterston, unpublished results).

The copy number of Tc2 varies among strains; N2 and other strains that are low in Tc1 copy number have about five copies of Tc2, while BO and other strains that are high in Tc1 copy number have about 30 copies of Tc2. Most copies are not associated with Tc1 sequences and are about 2.1 kb in extent. They are more heterogeneous in size than are Tc1 sequences.

Tc2 appears to be mobile. From examining Tc2 in a series of strains derived from N2/BO hybrids, elements were found in positions that did not correlate with Tc2 positions in either parental strain (Levitt and Emmons, unpublished results). In fact, the number of polymorphisms observed among these strains suggests a substantial transposition activity for Tc2, but no insertion of Tc2 into known genes has been identified.

B. Tc3, Tc4, and Tc5

Analysis of mutations arising in the TR679 *mut-2* strain (11; J. Collins and P. Anderson, unpublished results; Finney and Horvitz, unpublished results) has revealed three other non-cross-hybridizing elements capable of transposition, designated Tc3, Tc4, and Tc5. In general, the copy number of these elements is similar in N2, BO, and other non-*mut-2* strains but is higher in *mut-2*-containing strains, providing further evidence of their activation by *mut-2* and reinforcing the notion that the endogenous BO mutators and *mut-2* may have different specificities.

In the most extensive analysis of *mut-2* activities, 60 spontaneous *unc-22* mutations were recovered in a *mut-2* background and analyzed using the cloned *unc-22* gene (Collins and Anderson, unpublished results). Of 11 mutations associated with novel insertions, 3 were due to the insertion of an element designated Tc3 and another 2 were due to the insertion of Tc5. Like Tc1, both Tc3 and Tc5 excise at high frequency from the germ line in *mut-2* strains, but neither Tc3 nor Tc5 exhibits somatic excision detectable by Southern analysis. Tc3 is 2.5 kb in length and exists in about 15 copies in non-*mut-2* strains. Cloned copies of the element share several conserved restriction sites. Tc5 inserts into *unc-22* are 3.2 kb long, and restriction analysis of the two independent inserts suggests that they are very similar. Only four to six copies of the element are present in non-*mut-2* strains.

Another element, designated Tc4 and found mobilized in the TR679 *mut-2* strain, was initially identified as a 1.8-kb insert in the *unc-86* gene (Finney and Horvitz, unpublished results). Later, *ced-4*::Tc4 alleles were also identified, and these alleles show germ line instability (J. Yuan and H. R. Horvitz, unpublished results). Non-*mut-2* strains, like N2 and BO, have about 20 copies of Tc4, and there are some polymorphisms detectable between strains. Preliminary structural studies of Tc4 suggest that it may be a foldback element: it has symmetrical restriction sites and is unstable when cloned in bacterial plasmids (J. Yuan, M. Finney, and H. R. Horvitz, unpublished results). Like Tc1 (66), Tc4 may lack transcriptional stop signals, because *ced-4*::Tc4 alleles yield messages that are longer than wild-type messages which contain Tc4 hybridizing material (Yuan and Horvitz, unpublished results).

A hint as to how this apparent coactivation of different element families by *mut-2* might be mediated comes from comparing the partial sequence of a Tc3 insert to the sequence of Tc1 (J. Collins, B. Forbes, A. Rushforth, and P. Anderson, unpublished results). The Tc3 element, like Tc1, inserts at a TA dinucleotide, although the flanking sequences do not match the consensus sequence for Tc1 insertion. The terminal 8 bp of Tc3 matches with eight of the nine terminal bases of Tc1. The terminal inverted repeats of Tc3, in contrast to the 54-nucleotide inverted repeats of Tc1, extend to at least 90 bases. The two Tc3 termini match perfectly for 70 bp, are mismatched by a 1-bp deletion, and then are matched perfectly for another 20 bases (J. Collins, personal communication). Beyond the first 9 bp, the inverted repeats of Tc1 and Tc3 are not similar. However, this initial similarity may be enough to account for coactivation by a permissive transposase.

C. Foldback Sequences

The first foldback sequence identified in *C. elegans* (as yet unnamed) was initially recognized as a Bristol-Bergerac polymorphism (21). The sequence is 1.7 kb in length, and electron microscopy reveals that it has 0.85-kb inverted repeats with no significant separation between the repeats (D. Dreyfus and S. W. Emmons, unpublished results). It is present at about 20 copies per genome in several strains. Some differences have been noted in the position of the element among various wild isolates, but the variability is low (about 10% overall). The candidate element is distinct from Tc4 (Dreyfus and Emmons, unpublished results) but has at least one feature in common with both Tc1 and Tc3: at the ends of the foldback element there are nine consecutive nucleotides which are identical to the termini of the inverted repeats of Tc1.

VI. TRANSPOSABLE ELEMENTS IN OTHER NEMATODE SPECIES

A. *Caenorhabditis briggsae*

Tc1-like sequences are present in *C. briggsae*, another free-living nematode related to *C. elegans* (41). About 30 bands that cross-hybridize are found in the Indian strain G16 (A. Fodor, unpublished results) with low to moderate stringency. Several Tc1 hybridizing clones were isolated from a G16 genomic library, and partial characterization of these clones revealed that, unlike Tc1 in *C. elegans*, this repetitive element family is quite heterogeneous. Polymorphisms have been detected for these repeated sequences in three *C. briggsae* strains, G16, Z, and BO, suggesting that the Tc1-like sequences in *C. briggsae* are mobile. This family of elements has been designated TCb ("transposon *C. briggsae*"; also known as Barney).

One element, called TCb1, has been partially sequenced (41) and found to have a large ORF similar in size and sequence to the large ORF found in Tc1. These two ORF sequences are 71% identical at the DNA level and are 75% identical at the amino acid level. The second ORF detected in Tc1 is not present in TCb1. Outside of the ORF, sequence similarity drops to about 40%, except in the inverted repeat, where the sequences are 68% identical (36 of 53 bp). Unfortunately, the true extent of the inverted repeat in TCb1 is unknown because only one end of the element is sequenced. This conservation of the ORF and inverted repeats across species supports the idea that these elements are related, and that the ORF and inverted repeats are the functionally relevant portions of the element.

The presence of Tc1 hybridizing sequences in another species of nematode should not come as a surprise. Clearly, Tc1 is distributed widely within *C. elegans*, and P hybridizing sequences are widespread within the genus *Drosophila* (13, 52, 92). However, Tc1-like sequences are not limited to nematodes. A search of the EMBL sequence data bank revealed that Tc1 and TCb1 elements have significant similarity to the HB family of transposable elements of *Drosophila melanogaster* (41). HB elements are 1,654 bp in length, have 30-bp terminal inverted repeats, and are present at about 20 copies per genome in *D. melanogaster* (9). Allowing for three small deletions in HB, the overall ORF amino acid similarity between Tc1 and HB is about 30%. The regions around the HB ORF have no similarity to Tc1, but, like TCb1, the inverted repeats have some similarity to the inverted repeats of Tc1. Whether such a widespread distribution of Tc1-like sequences reflects horizontal transmission or is a reflection of an early evolutionary origin of Tc1-like sequences is a matter for speculation. Many nematodes parasitize insects, including the Drosophilidae, and screening of such parasitic nematodes might give further evidence on the likelihood of horizontal transmission.

B. *Panagrellus redivivus*

In studies of *P. redivivus*, a gonochoristic (male-female), free-living nematode, a strain, called C15, was found in which spontaneous twitcher mutants, phenotypically similar to *unc-22* mutants, occur at a frequency of 2×10^{-4} (55). This frequency is comparable to that reported for spontaneous *unc-22* mutants in the *C. elegans* strain BO and is at least two orders of magnitude higher than the N2 spontaneous frequency (68). Tc1 is not present in *P. redivivus* (55), but the *unc-22*-like sequence of *P. redivivus* was isolated using the cloned *unc-22* gene from *C. elegans* (67), and in turn the wild-type clones were used to isolate *unc-22*-like sequences from two mutants. One of these mutations was a deficiency, but the other did have a large insert within *unc-22* (55).

The insert is 4.8 kb in size, and in five independent wild isolates of *P. redivivus* examined, 10 to 50 copies of the element were found; the C15 strain contains 20 copies. Copies of the element are dispersed and several polymorphisms were observed between strains (55). Structurally the element is probably retrotransposonlike, since it has direct terminal repeats of at least 170 bp. This is the first transposon described in *Panagrellus* sp. and it has been designated PAT-1 ("*Panagrellus* transposon"). The overall structures of PAT-1 elements in the C15 strain are similar, but not identical. A relatively abundant 900-bp transcript from PAT-1 has also been observed (55). Neither this transcript nor homologous DNA sequences to PAT-1 have been found in *C. elegans*.

The observation that spontaneous twitcher mutants in *C. elegans* and *P. redivivus* are both due to transposon insertion and the ease of recovering twitcher mutants using acetylcholine agonists suggest that *unc-22* homologs may be useful as general "transposon traps" for isolating transposons in other nematode species (55).

C. *Ascaris lumbricoides*

In the large parasitic nematode *A. lumbricoides* var. *suum*, a retroviruslike element was identified in studies of satellite-DNA-containing clones (1). A study of transposable elements was initiated as part of a larger project investigating chromatin diminution in *Ascaris* sp. Because mobile elements cause chromosome breaks (59) and genetic rearrangements (60), there has been speculation that they may have a role in chromatin diminution, noted originally in *Ascaris* sp. (7) and later observed in other nematodes, copepods, insects, and ciliates (97; M.-C. Yao, this volume).

This element is 7.5 kb long and has 256-bp terminal direct repeats that are similar in overall structure to retroviral long terminal repeats (98; H. Varmus and P. Brown, this volume). Each direct repeat is bounded by TG. . .CA, and within each direct repeat the first and last 12 nucleotides are inverted repeats. Other similarities to a retrovirus include a 19-bp stretch similar to a *Drosophila* $tRNA^{Arg}$ (104) at a position comparable to a retroviral tRNA primer-binding site and a 5-bp oligopurine tract (GGGAG) analogous to a primer-binding site for second-strand synthesis upstream of the right direct repeat. A common feature of transposable elements is duplication of a target site upon insertion, and each of the four elements examined was flanked by a 5-bp duplication (1).

This retroviruslike element is designated Tas ("transposon *Ascaris*"). Tas is present at about 50 copies per genome and is dispersed. Most Tas elements are intact, but a few solo long terminal repeats were detected. Several Tas polymorphisms were identified among five individual animals, suggesting that the element is indeed mobile and the elements are believed to be transcriptionally active (1).

Two restriction site variants of Tas are known, Tas-1 and Tas-2. The Tas-1 group accounts for about two-thirds and Tas-2 for the remainder of the genomic copies of Tas. During chromatin diminution, the variants are differentially eliminated: one-quarter of Tas-1 variants are eliminated, a proportion which corresponds to the overall chromatin loss during the diminution process (97). In contrast, all Tas-2 elements are eliminated from the soma. Whether Tas-1, Tas-2, or other transposons have a role in chromatin diminution is unknown. The function of chromatin diminution itself is a mystery but presumably involves some selective advantage.

Acknowledgments. We thank Ikue Mori for helpful discussions and comments on the manuscript and the many members of the *C. elegans* community for sharing unpublished data.

Preparation of this review was supported in part by a Natural Sciences and Engineering Research Council (Canada) grant to D.G.M. and by a U.S. Public Health Service grant to R.H.W.

LITERATURE CITED

1. **Aeby, P., A. Spicher, Y. De Chastonay, F. Muller, and H. Tobler.** 1986. Structure and genomic organization of proretrovirus-like elements partially eliminated from the somatic genome of *Ascaris lumbricoides*. *EMBO J.* **5**:3353–3360.
2. **Baltimore, D.** 1985. Retroviruses and retrotransposons: the role of reverse transcription in shaping the eukaryotic genome. *Cell* **40**:481–482.
3. **Bender, J., and N. Kleckner.** 1986. Genetic evidence that Tn10 transposes by a nonreplicative mechanism. *Cell* **45**: 801–815.
4. **Berg, D. E.** 1983. Structural requirement for IS50-mediated gene transposition. *Proc. Natl. Acad. Sci. USA* **80**:792–796.
5. **Berg, D. E., J. Lodge, C. Sasakawa, D. K. Nag, S. H. Phadnis, K. Weston-Hafer, and G. F. Carle.** 1984. Transposon Tn5: specific sequence recognition and conservative transposition. *Cold Spring Harbor Symp. Quant. Biol.* **49**:215–226.
6. **Bingham, P. M., R. Levis, and G. M. Rubin.** 1981. Cloning of DNA sequences from the *white* locus of *D. melanogaster* by a novel and general method. *Cell* **25**:693–704.
7. **Boveri, T.** 1887. Uber Differenzierung der Zellkerne während der Furchung des Eies von *Ascaris megalocephala*. *Anat. Anz.* **2**:688–693.
8. **Brenner, S.** 1974. The genetics of *Caenorhabditis elegans*. *Genetics* **77**:71–94.
9. **Brierley, H. L., and S. S. Potter.** 1985. Distinct characteristics of loop sequences of two *Drosophila* foldback transposable elements. *Nucleic Acids Res.* **13**:485–500.
10. **Campbell, A.** 1983. Bacteriophage lambda, p. 66–103. *In* J. A. Shapiro (ed.), *Mobile Genetic Elements*. Academic Press, Inc., New York.
11. **Collins, J., B. Saari, and P. Anderson.** 1987. Activation of a transposable element in the germ line but not the soma of *Caenorhabditis elegans*. *Nature* (London) **328**:726–728.
12. **Coulson, A., J. E. Sulston, S. Brenner, and J. Karn.** 1986. Toward a physical map of the genome of the nematode *C. elegans*. *Proc. Natl. Acad. Sci. USA* **81**:6784–6788.
13. **Daniels, S. B., L. D. Strausbaugh, L. Erman, and R. Armstrong.** 1984. Sequences homologous to P elements occur in *Drosophila paulistorum*. *Proc. Natl. Acad. Sci. USA* **81**: 6794–6797.
14. **Davidson, E. H., B. R. Hough, C. S. Amenson, and R. J. Britten.** 1973. General interspersion of repetitive and nonrepetitive sequence elements in the DNA of Xenopus. *J. Mol. Biol.* **77**:1–23.
15. **Dooner, H. K., and O. E. Nelson, Jr.** 1979. Heterogeneous glucosyltransferase in purple derivatives from a controlling element-suppressed bronze mutant in maize. *Proc. Natl. Acad. Sci. USA* **76**:2369–2371.
16. **Doring, H. P., M. Freeling, S. Haje, M. A. Johns, R. Kunze, A. Merckelbach, F. Salamini, and P. Starlinger.** 1984. A *Ds*-mutation of the *Adh1* gene in *Zea mays* L. *Mol. Gen. Genet.* **193**:199–204.
17. **Eide, D., and P. Anderson.** 1985. Transposition of Tc1 in the nematode *C. elegans*. *Proc. Natl. Acad. Sci. USA* **82**: 1756–1760.
18. **Eide, D., and P. Anderson.** 1985. The gene structures of spontaneous mutations affecting a *Caenorhabditis elegans* myosin heavy chain gene. *Genetics* **109**:67–79.
19. **Eide, D., and P. Anderson.** 1988. Insertion and excision of the *Caenorhabditis elegans* transposable element Tc1. *Mol. Cell. Biol.* **8**:737–746.
20. **Emmons, S. W.** 1988. The genome, p. 47–79. *In* W. B.

Wood (ed.), *The Nematode Caenorhabditis elegans*. Cold Spring Harbor Laboratory, Cold Spring Harbor, N.Y.

21. **Emmons, S. W., M. R. Klass, and D. Hirsh.** 1979. Analysis of the constancy of DNA sequences during the development and evolution of the nematode *Caenorhabditis elegans*. *Proc. Natl. Acad. Sci. USA* **76:**1333–1337.
22. **Emmons, S. W., S. Roberts, and K. S. Ruan.** 1986. Evidence in a nematode for regulation of transposon excision by tissue-specific factors. *Mol. Gen. Genet.* **202:**410–415.
23. **Emmons, S. W., B. Rosenzweig, and D. Hirsh.** 1980. Arrangement of repeated sequences in the DNA of the nematode *Caenorhabditis elegans*. *J. Mol. Biol.* **144:**481–500.
24. **Emmons, S. W., K. S. Ruan, A. Levitt, and L. Yesner.** 1985. Regulation of Tc1 transposable elements in *Caenorhabditis elegans*. *Cold Spring Harbor Symp. Quant. Biol.* **50:** 313–319.
25. **Emmons, S. W., and L. Yesner.** 1984. High-frequency excision of transposable element Tc1 in the nematode *Caenorhabditis elegans* is limited to somatic cells. *Cell* **36:** 599–605.
26. **Emmons, S. W., L. Yesner, K. S. Ruan, and D. Katzenberg.** 1983. Evidence for a transposon in *Caenorhabditis elegans*. *Cell* **32:**55–65.
27. **Engels, W. R.** 1979. Hybrid dysgenesis in Drosophila melanogaster: rules of inheritance of female sterility. *Genet. Res.* **33:**119–136.
28. **Engels, W. R.** 1983. The P family of transposable elements in *Drosophila*. *Annu. Rev. Genet.* **17:**315–344.
29. **Engler, J. A., and M. P. van Bree.** 1981. The nucleotide sequence and protein-coding capability of the transposable element IS5. *Gene* **14:**155–163.
30. **Epstein, H. F., R. H. Waterston, and S. Brenner.** 1984. A mutant affecting the heavy chain of myosin. *J. Mol. Biol.* **90:**291–300.
31. **Errede, B., M. Company, J. D. Ferchak, C. A. Hutchison III, and W. S. Yarnell.** 1985. Activation regions in a yeast transposon have homology to mating type control sequences and to mammalian enhancers. *Proc. Natl. Acad. Sci. USA* **82:**5423–5427.
32. **Fedoroff, N.** 1983. Controlling elements in maize, p. 1–63. *In* J. A. Shapiro (ed.), *Mobile Genetic Elements*. Academic Press, Inc., New York.
33. **Fedoroff, N., D. Furtek, and O. Nelson, Jr.** 1984. Cloning of the bronze locus in maize by a simple and generalizable procedure using the transposable controlling element Activator (Ac). *Proc. Natl. Acad. Sci. USA* **81:**3825–3829.
34. **Fedoroff, N., S. Wessler, and M. Shure.** 1983. Isolation of the transposable maize controlling elements Ac and Ds. *Cell* **35:**235–242.
35. **Fire, A.** 1986. Integrative transformation of *C. elegans*. *EMBO J.* **5:**2673–2680.
36. **Freund, R., and M. Meselson.** 1984. Long terminal repeat nucleotide sequence and specific insertion of the *gypsy* transposon. *Proc. Natl. Acad. Sci. USA* **81:**4462–4464.
37. **Greenblatt, I. M., and R. A. Brink.** 1962. Twin mutations in medium variegated pericarp maize. *Genetics* **47:**489–501.
38. **Greenwald, I.** 1985. *lin-12*, a nematode homeotic gene, is homologous to a set of mammalian proteins that includes growth factor. *Cell* **43:**583–590.
39. **Grindley, N. D. F.** 1978. IS1 insertion generates duplication of a nine base pair sequence at its target site. *Cell* **13:** 419–426.
40. **Halling, S., and N. Kleckner.** 1982. A symmetrical six-base-pair target site sequence determines Tn10 insertion specificity. *Cell* **28:**155–163.
41. **Harris, L. J., D. L. Baillie, and A. M. Rose.** 1988. Sequence identity between an inverted repeat family of transposable elements in *Drosophila* and *Caenorhabditis*. *Nucleic Acids Res.*, in press.
42. **Harris, L. J., and A. M. Rose.** 1986. Somatic excision of transposable element Tc1 from the Bristol genome of *Caenorhabditis elegans*. *Mol. Cell. Biol.* **6:**1782–1786.
43. **Herman, R., and J. Shaw.** 1987. The transposable genetic element Tc1 in the nematode *Caenorhabditis elegans*. *Trends Genet.* **3:**222–225.
44. **Hirsh, D., G. N. Cox, J. M. Kramer, D. Stinchcomb, and R. Jeffries.** 1985. Structure and expression of the collagen genes of *C. elegans*. *Ann. N.Y. Acad. Sci.* **460:**163–171.
45. **Iida, S., J. Meyer, and W. Arber.** 1983. Prokaryotic IS elements, p. 59–221. *In* J. A. Shapiro (ed.), *Mobile Genetic Elements*. Academic Press, Inc., New York.
46. **Ikenaga, H., and K. Saigo.** 1982. Insertion of a movable genetic element, 297, into the T-A-T-A box for the H3 histone gene of *Drosophila melanogaster*. *Proc. Natl. Acad. Sci. USA* **79:**4143–4147.
47. **Inouye, S., S. Yuki, and K. Saigo.** 1984. Sequence-specific insertion of the Drosophila transposable genetic element 17.6. *Nature* (London) **310:**332–333.
48. **Johnson, R., and W. Reznikoff.** 1984. Copy number control of Tn5 transposition. *Genetics* **107:**9–18.
49. **Kiff, J. E., D. G. Moerman, L. A. Schriefer, and R. H. Waterston.** 1988. Transposon-induced deletions in *unc-22* of *Caenorhabditis elegans* associated with almost normal gene activity. *Nature* (London) **331:**631–633.
50. **Klaer, R., S. Kuhn, E. Tillman, H. J. Fritz, and P. Starlinger.** 1981. The sequence of IS4. *Mol. Gen. Genet.* **181:**169–175.
51. **Kmiec, E. B., K. J. Angelides, and W. K. Holloman.** 1985. Left-handed DNA and the synaptic pairing reaction promoted by *Ustilago* rec1 protein. *Cell* **40:**139–145.
52. **Lansman, R. A., S. N. Stacey, T. A. Grigliatti, and H. W. Brock.** 1985. Sequences homologous to the P mobile element of *Drosophila melanogaster* are widely distributed in the subgenus *Sophophora*. *Nature* (London) **318:**561–563.
53. **Laski, F., D. Rio, and G. M. Rubin.** 1986. Tissue specificity of Drosophila P element transposition is regulated at the level of mRNA splicing. *Cell* **44:**7–19.
54. **Liao, L. W., B. Rosenzweig, and D. Hirsh.** 1983. Analysis of a transposable element in *Caenorhabditis elegans*. *Proc. Natl. Acad. Sci. USA* **80:**3585–3589.
55. **Link, C., J. Graf-Witsel, and W. B. Wood.** 1987. Isolation and characterization of a nematode transposable element from *Panagrellus redivivus*. *Proc. Natl. Acad. Sci. USA* **84:**5325–5329.
56. **MacLeod, A. R., J. Karn, and S. Brenner.** 1981. Molecular analysis of the *unc-54* myosin heavy chain of *Caenorhabditis elegans*. *Nature* (London) **291:**386–390.
57. **MacLeod, A. R., R. H. Waterston, and S. Brenner.** 1977. An internal deletion mutant of a myosin heavy chain in *C. elegans*. *Proc. Natl. Acad. Sci. USA* **74:**5336–5340.
58. **MacLeod, A. R., R. H. Waterston, R. M. Fishpool, and S. Brenner.** 1977. Identification of the structural gene for a myosin heavy chain in *C. elegans*. *J. Mol. Biol.* **114:** 133–140.
59. **McClintock, B.** 1951. Chromosome organization and genic expression. *Cold Spring Harbor Symp. Quant. Biol.* **16:** 13–47.
60. **McClintock, B.** 1956. Controlling elements and the gene. *Cold Spring Harbor Symp. Quant. Biol.* **21:**197–216.
61. **McClintock, B.** 1984. The significance of responses of the genome to challenge. *Science* **226:**792–801.

62. **Mattox, W. W., and N. Davidson.** 1984. Isolation and characterization of the *Beadex* locus of *Drosophila melanogaster*: a putative *cis*-acting negative regulatory element for the *heldup-a* gene. *Mol. Cell. Biol.* **4:**1343–1353.
63. **Medhora, M. M., A. H. MacPeek, and D. L. Hartl.** 1988. Excision of the *Drosophila* transposable element *mariner*: identification and characterization of the *Mos* factor. *EMBO J.* 7:2185–2189.
64. **Modolell, J., W. Bender, and M. Meselson.** 1983. *Drosophila melanogaster* mutations suppressible by the suppressor of hairy wing are insertions of a 7.3 kilobase mobile element. *Proc. Natl. Acad. Sci. USA* **80:**1678–1682.
65. **Moerman, D. G., and D. L. Baillie.** 1979. Genetic organization in *Caenorhabditis elegans*: fine structure analysis of the *unc-22* gene. *Genetics* **91:**95–103.
66. **Moerman, D. G., G. M. Benian, R. J. Barstead, L. Schriefer, and R. H. Waterston.** 1988. Identification and intracellular localization of the *unc-22* gene product of *Caenorhabditis elegans*. *Genes Dev.* **2:**93–105.
67. **Moerman, D. G., G. M. Benian, and R. H. Waterston.** 1986. Molecular cloning of the muscle gene *unc-22* in *Caenorhabditis elegans* by Tc1 transposon tagging. *Proc. Natl. Acad. Sci. USA* **83:**2579–2583.
68. **Moerman, D. G., and R. H. Waterston.** 1984. Spontaneous unstable *unc-22 IV* mutations in *C. elegans* var. Bergerac. *Genetics* **108:**859–877.
69. **Mori, I., G. M. Benian, D. G. Moerman, and R. H. Waterston.** 1988. The transposon Tc1 of *C. elegans* recognizes specific target sequences for integration. *Proc. Natl. Acad. Sci. USA* **85:**861–864.
70. **Mori, I., D. G. Moerman, and R. H. Waterston.** 1988. Analysis of a mutator activity necessary for germline transposition and excision of Tc1 transposable elements in *Caenorhabditis elegans*. *Genetics* **120:**397–407.
71. **Morisato, D., and N. Kleckner.** 1984. Transposase promotes double strand breaks and single strand joints at Tn10 termini *in vitro*. *Cell* **39:**181–190.
72. **Murphy, E., and S. Lofdahl.** 1984. Transposition of Tn554 does not generate a target duplication. *Nature* (London) **307:**292–294.
73. **Nigon, V.** 1949. Les modalities do la reproduction et le determinism du sex chez quelques Nematodes libres. *Ann. Sci. Nat. Zool.* **11:**1–132.
74. **O'Hare, K., and G. M. Rubin.** 1983. Structures of P transposable elements and their sites of insertion and excision in the *Drosophila melanogaster* genome. *Cell* **34:**25–35.
75. **Park, E.-C., and H. R. Horviz.** 1986. *C. elegans unc-105* mutations affect muscle and are suppressed by other mutations that affect muscle. *Genetics* **113:**853–867.
76. **Plasterk, R. H. A.** 1987. Differences between Tc1 elements from the *C. elegans* strain Bergerac. *Nucleic Acids Res.* **15:**10050.
77. **Pohlman, R. F., N. V. Fedoroff, and J. Messing.** 1984. The nucleotide sequence of the maize controlling element Activator. *Cell* **37:**635–643.
78. **Potter, S. S., M. Truett, M. Phillips, and A. Maher.** 1980. Eukaryotic transposable genetic elements with inverted terminal repeats. *Cell* **20:**639–647.
79. **Rich, A., A. Nordheim, and A. H. J. Wang.** 1984. The chemistry and biology of left-handed Z-DNA. *Annu. Rev. Biochem.* **53:**791–846.
80. **Roberts, D., B. C. Hoopes, W. R. McClure, and N. Kleckner.** 1985. IS10 transposition is regulated by DNA adenine methylation. *Cell* **43:**117–130.
81. **Roberts, S. B., M. Sanicola, S. W. Emmons, and G. Childs.** 1987. Molecular characterization of the histone gene family of *C. elegans*. *J. Mol. Biol.* **196:**27–38.
82. **Roeder, G. S., and G. R. Fink.** 1983. Transposable elements in yeast, p. 300–328. *In* J. A. Shapiro (ed.), *Mobile Genetic Elements*. Academic Press, Inc., New York.
83. **Rose, A. M., L. J. Harris, N. R. Mawji, and W. R. Morris.** 1985. Tc1(Hin): a form of the transposable element Tc1 in *Caenorhabditis elegans*. *Can. J. Biochem. Cell Biol.* **63:** 752–756.
84. **Rose, A. M., and T. P. Snutch.** 1984. Isolation of the closed circular form of the transposable element Tc1 of *Caenorhabditis elegans*. *Nature* (London) **311:**485–486.
85. **Rosenzweig, B., L. Liao, and D. Hirsh.** 1983. Sequence of the *C. elegans* transposable element Tc1. *Nucleic Acids Res.* **11:**4201–4209.
86. **Rosenzweig, B., L. Liao, and D. Hirsh.** 1983. Target sequences for the *C. elegans* transposable element Tc1. *Nucleic Acids Res.* **11:**7137–7140.
87. **Ruan, K. S., and S. W. Emmons.** 1984. Extrachromosomal copies of transposon Tc1 in the nematode *Caenorhabditis elegans*. *Proc. Natl. Acad. Sci. USA* **81:**4018–4022.
88. **Ruan, K. S., and S. W. Emmons.** 1987. Precise and imprecise somatic excision of the transposon Tc1 in the nematode *C. elegans*. *Nucleic Acids Res.* **15:**6875–6881.
89. **Schwartz-Sommer, Z., A. Gierl, H. Cuypers, P. A. Peterson, and H. Saedler.** 1985. Plant transposable elements generate the DNA sequence diversity needed in evolution. *EMBO J.* **4:**591–597.
90. **Searles, L., A. Greenleaf, W. Kemp, and R. Voelker.** 1986. Sites of P element insertion and structures of P element deletions in the 5′ region of *Drosophila melanogaster RpII215*. *Mol. Cell. Biol.* **6:**3312–3319.
91. **Searles, L. L., R. S. Jokerst, P. M. Bingham, R. A. Voelker, and A. L. Greenleaf.** 1982. Molecular cloning of sequences from a Drosophila RNA polymerase II locus by P element transposon tagging. *Cell* **31:**585–592.
92. **Stacey, S. N., R. A. Lansman, H. W. Brock, and T. A. Grigliatti.** 1986. Distribution and conservation of mobile elements in the genus *Drosophila*. *Mol. Biol. Evol.* **3:** 522–534.
93. **Sulston, J. E., and S. Brenner.** 1974. The DNA of *C. elegans*. *Genetics* **77:**95–104.
94. **Sulston, J. E., and H. R. Horvitz.** 1977. Post-embryonic cell lineages of the nematode, *Caenorhabditis elegans*. *Dev. Biol.* **56:**110–156.
95. **Sundareson, V., and M. Freeling.** 1987. An extrachromosomal form of the Mu transposons of maize. *Proc. Natl. Acad. Sci. USA* **84:**4924–4928.
96. **Sutton, W., W. Gerlach, D. Schwartz, and W. Peacock.** 1984. Molecular analysis of *Ds* controlling element mutations at the *Adh1* locus of maize. *Science* **223:**1265–1268.
97. **Tobler, H.** 1986. The differentiation of germ and somatic cell lines in nematodes, p. 1–69. *In* W. Hennig (ed.), *Germline-Soma Differentiation, Results and Problems in Cell Differentiation*, vol. 13. Springer-Verlag KG, Berlin.
98. **Varmus, H. E.** 1983. Retroviruses, p. 411–503. *In* J. A. Shapiro (ed.), *Mobile Genetic Elements*. Academic Press, Inc., New York.
99. **Ward, S., D. Burke, J. E. Sulston, A. R. Coulson, D. G. Albertson, D. Ammons, M. Klass, and E. Hogan.** 1988. Genomic organizations of major sperm protein genes and pseudogenes in the nematode *Caenorhabditis elegans*. *J. Mol. Biol.* **199:**1–13.

100. **Waterston, R. H., D. G. Moerman, G. M. Benian, R. J. Barstead, I. Mori, and R. Francis.** 1986. Muscle genes and proteins in *Caenorhabditis elegans*. *UCLA Symp. Mol. Cell. Biol.* **29:**605–617.
101. **Waterston, R. H., J. N. Thomson, and S. Brenner.** 1980. Mutants with altered muscle structure in *Caenorhabditis elegans*. *Dev. Biol.* 77:271–302.
102. **Wessler, S. R., G. Baran, M. Varagona, and S. L. Dellaporta.** 1986. Excision of Ds produces waxy proteins with a range of enzymatic activities. *EMBO J.* **5:**2427–2432.
103. **Wood, W. B. (ed.).** 1988. *The Nematode Caenorhabditis elegans*. Cold Spring Harbor Laboratory, Cold Spring Harbor, N.Y.
104. **Yen, P. H., and N. Davidson.** 1980. The gross anatomy of a tRNA cluster at region 42A of the *D. melanogaster* chromosome. *Cell* **22:**137–148.

Chapter 23

Mobile Genetic Elements in the Cellular Slime Mold *Dictyostelium discoideum*

RICHARD A. FIRTEL

I. INTRODUCTION

The cellular slime mold *Dictyostelium discoideum* has become an important system in the attempt to understand the molecular mechanisms controlling cellular differentiation. Chemotaxis, signal transduction, and cell-type-specific gene expression are all integral parts of the tightly controlled life cycle. *D. discoideum* grows as vegetative amoebae; however, upon depletion of the food source, single cells initiate multicellular differentiation that culminates in the formation of a fruiting body composed of predominantly two cell types, ~20% stalk cells and 80% spores (19, 20). Within 6 h of the onset of starvation, individual amoebae start to aggregate into a multicellular organism consisting of ~10^5 cells. The aggregation process is mediated chemotactically by cyclic AMP (cAMP) and is initiated as a consequence of cAMP pulses emitted by cells which later become the aggregation center. The cAMP binds to the cell-surface cAMP receptor on surrounding cells, and adenylate cyclase becomes activated via G (guanine-nucleotide-binding) proteins, resulting in the production and release of cAMP into the surrounding medium. The cAMP activates receptors on adjacent cells, thus relaying the signal outward from the aggregation center. A desensitization or adaptation response then ensues, and receptors are not capable of further activation of intracellular processes until the ligand concentration is reduced by an extracellular phosphodiesterase. The receptors will then deadapt and again become responsive to ligand. In vivo, this pulsatile or oscillating relay system is initiated once every 6 to 7 min. The cell responds to the cAMP signal by a cascade of events which includes an intracellular signal transduction pathway as well as an overall movement of cells up the cAMP concentration gradient (see references 12, 17, and 18 for details). During this process, a number of genes are induced as a consequence of activation of the cell-surface receptor by cAMP (10, 17, 22). Aggregation continues until a loose mound of cells is formed. The cAMP signal continues to oscillate within the aggregate, and differentiation of two different cell types, the prestalk and prespore cells, is induced. These cells are precursors to the mature stalk cells and spores of the fruiting body. Induction of cell-type-specific expression of new classes of genes also involves cAMP interacting with the cell-surface receptor (10, 22, 32, 35, 36). Recent analyses of these processes have centered on the interaction between the signal transduction systems, including the receptor and associated G proteins, and the activation of gene expression. Biochemical and molecular analyses of the signal transduction pathways in vitro combined with identification of the *cis*-acting sequences and *trans*-acting factors that control expression of cell-type-specific genes have led to a better understanding of the developmental pathway in this organism (10, 12, 18, 22, 35).

Richard A. Firtel ■ Department of Biology, Center for Molecular Genetics, M-034, University of California, San Diego, La Jolla, California 92093.

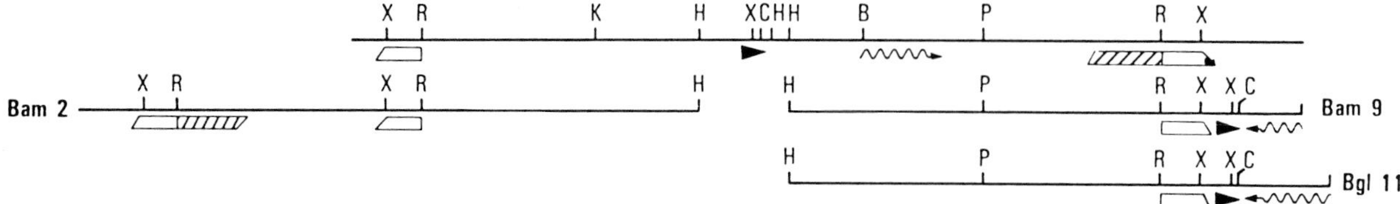

Figure 1. Restriction map and structure of Tdd-1. Top line shows a composite structure of Tdd-1 elements. The line represents a stretch of chromosomal DNA. The ends of the Tdd-1 element are indicated by the positions of the open boxes (see below). There is substantial restriction site polymorphism among individual repeats, indicating a degree of nucleotide sequence divergence. *Bam*2, *Bam*9, and *Bgl*11 represent three independent isolates of partial Tdd-1 elements obtained from a *Hin*dIII/*Bam*HI and a *Hin*dIII/*Bgl*II restriction digest library of *Dictyostelium* DNA cloned into the *Hin*dIII and *Bam*HI restriction sites of pXF3 (28). The open box under the map indicates the inverted repeats of Tdd-1. This box is pointed at one end, showing that the LTRs are inverted repeats (see text). The blackened tip of the right-hand repeat indicates that it extends beyond the left-hand repeat. The filled-in arrowhead indicates a 110-bp imperfect repeat encompassing the *Xba* and *Cla* sites of Tdd-1, which is found outside of the repeat insertion site in cloned *Bam*9 and *Bgl*11. The wavy line represents a segment between the *Bgl*II and *Pvu*II sites within Tdd-1 that is also found, as an inverted repeat, in the flanking regions outside the Tdd-1 element in clones *Bam*9 and *Bgl*11 and other clones, again suggesting a preferential insertion of Tdd-1 into Tdd-1 related sequences. The right end of the Tdd-1 sequence (box containing slashes) contains the internal complementary repeat (ICR) and is also found in the flanking region of the repeat in clone *Bam*2 (Fig. 2). The 4.5-kb transcript extends from within the left LTR through the 4.1-kb central region into the right LTR. The 1.4-kb heat shock transcript extends from within the right LTR into the element. The majority of the elements contain the conserved sites as determined from the mapping of cloned elements and from genomic mapping. As such, they are included in the map of the upper element, which is a "consensus" Tdd-1 element. Restriction site symbols: X, *Xba*I; R, *Eco*RI; K, *Kpn*I; C, *Cla*I; B, *Bgl*II; P, *Pvu*II.

As part of the approach to understanding cellular differentiation in *D. discoideum*, many inducible genes have been isolated initially as cDNA clones. Analyses of genomic sequences corresponding to developmentally regulated and cell-type-specific genes led unexpectedly to the identification of three transposable elements. The first, Tdd-1 (DIRS-1), appears to be a retrotransposon containing two transcription units of opposite polarity (4, 6, 28, 39) (Fig. 1). One is a major developmentally regulated 4.5-kilobase (kb) transcript believed to encode proteins involved in the retrotransposition of this element, and the other, 1.4 kb in size, is a heat-shock-inducible RNA (3–6, 8, 28, 29, 37–39). This second transcript presumably does not encode a functional protein, since it does not contain an extended open reading frame (ORF) (4).

The other two elements, Tdd-2 and Tdd-3, are considered mobile genetic elements because they are present at different chromosomal locations in different laboratory strains and wild-type isolates of *D. discoideum*, but do not appear to be retrotransposons (26). Although the exact size of either element has not been exactly determined, they appear to be ~4 to 5 kb in length. Tdd-2 and Tdd-3 are both present in ~20 to 30 copies per haploid genome, including partial elements, a copy number that is considerably lower than that seen for Tdd-1 (26). Unlike Tdd-1, they lack long terminal repeats (LTRs) and are not known to be transcribed during either vegetative growth or multicellular differentiation. An unusual aspect of Tdd-2 and Tdd-3 is that both contain a short (22-base-pair [bp]) homologous sequence (see below) and structural analysis suggests that one transposon preferentially inserts into the other (26). In this chapter, I review our understanding of these three mobile genetic elements. In addition, I also discuss the recent analysis of a *D. discoideum* strain that mutates at a high frequency, producing cells with new phenotypes, a process reminiscent of switching in *Candida albicans* (D. R. Soll, this volume).

II. ANALYSIS OF Tdd-1

Tdd-1, or DIRS-1, was independently isolated by two laboratories studying different molecular processes. DIRS-1 (*Dictyostelium* interspersed repeat sequence) was identified by the Lodish laboratory as a cloned restriction fragment which they believed to be an interspersed repeat sequence associated with a number of different developmentally regulated mRNAs. Their first report (39) showed that the cloned genomic restriction fragment hybridized to a heterogeneously sized population of RNAs that were expressed at low levels during vegetative growth and induced during multicellular development. In addition, the element hybridized to many different restriction fragments of genomic DNA. Independently, Rosen et al. (28) identified the transposon Tdd-1 while examining heat-shock-inducible genes in *D. discoideum* and found that it specifies a 1.4-kb heat shock mRNA and a major 4.5-kb, developmentally regulated transcript of opposite polarity. In addition, the element contains sequences complementary to a large number (>50) of smaller transcripts, most likely derived from partial or truncated elements. Tdd-1 emerges as an interesting, complex element with an unusual structure and overlapping pattern of gene expression.

Tdd-1 is 4.8 kb in length with 313-bp inverted repeats (Fig. 1 and 2). Interestingly, the LTRs have several nucleotide sequence mismatches between the left and right LTRs that are conserved between different elements (4, 28, 38) (Fig. 2). Moreover, the right terminal repeat extends 36 bp past this common 313-bp sequence, while the left LTR probably extends 1 bp past this sequence, as determined by sequence analysis of left and right LTRs from several Tdd-1 elements. Thus, in contrast to LTRs in other retrotransposons, the two ends are of different lengths and are inverted rather than direct. Also, the last 34 bp of the left LTR and a portion of the right LTR are present within the core region of Tdd-1 just inside the right LTR (4, 28) (Fig. 1). The repeat sequence has been designated the internal complementary repeat (ICR) (4).

There are ~50 copies of the intact Tdd-1 element per haploid genome and ~50 to 100 copies of partial or truncated elements. Examination of the hybridization pattern of Tdd-1 sequences in genomic restriction digests of three closely related laboratory stocks all derived from the axenic strain Ax-3, as well as different wild-type isolates of *D. discoideum*, indicates that Tdd-1 is present in all strains and is mobile. Another interesting aspect of Tdd-1 concerns its sites of insertion within the genome. Analysis of ~20 genomic clones containing whole or partial Tdd-1 elements indicates that the flanking sequences surrounding Tdd-1 consist of partial Tdd-1 elements. The interpretation is that Tdd-1 preferentially inserts into other Tdd-1 sequences (3, 28, 29) (Fig. 1), although it is possible that insertion of Tdd-1 causes rearrangements that result in extensive duplications of part of the Tdd-1 element.

Complete Tdd-1 elements specify two major transcripts, a 4.5-kb, developmentally regulated RNA that is initiated within the left LTR and extends into the right LTR, and a 1.4-kb RNA under separate regulation. The 4.5-kb RNA is expressed at relatively low levels within vegetative, growing cells and at ~20-fold higher levels in the multicellular stages (6, 28, 37). Analysis of transcripts complementary to Tdd-1 shows a heterogeneously sized population of >50 RNAs that probably are initiated from the same site at the 4.5-kb transcript within the left LTR. These are believed to represent transcripts from partial elements, which explains the initial suggestion (39) of an interspersed repeat sequence complementary to a large number of developmentally regulated messages.

The 1.4-kb transcript is induced ~20-fold upon heat shock treatment of vegetative or developing cells expressed from the complementary strand (leftward from the right LTR). The segment transcribed during heat shock is entirely contained within the segment that specifies the 4.5-kb transcript. DNA sequence analysis reveals a sequence that is homologous to the highly conserved heat shock regulatory element found upstream of heat shock genes from a number of eucaryotic organisms, including *Saccharomyces cerevisiae* (5). When the right-hand portion of Tdd-1, including the right LTR, is transformed into *S. cerevisiae* cells, it is also inducible by heat shock (5, 8, 28, 29, 38).

The major 4.5-kb transcript has three ORFs (4). ORF-1 is ~1 kb in length and is the only one initiating with a methionine. ORF-2 is ~2,100 nucleotides in length and overlaps ~170 nucleotides with ORF-1. ORF-3 is ~1 kb in length and is completely contained within ORF-2 and overlaps ~57 nucleotides with ORF-1. ORF-3 contains a 200-amino-acid sequence that bears homology to reverse transcriptase from retroviruses as well as other eucaryotic transposons, such as Ty1 from *S. cerevisiae* and *copia* from *Drosophila* (2, 4, 13, 14, 16, 24). Analysis of the proteins expressed by adjacent and slightly overlapping ORFs in retrotransposons suggests that, at least in some cases, a shift in the reading frame occurs during translation so that the protein is encoded by two reading frames (7, 23). Since ORF-3 has homology to retrovirus reverse transcriptase, it is presumed that the ORF-3 product serves in an equivalent manner for Tdd-1, although the presence of reverse transcriptase in *D. discoideum* has not yet been documented. Since ORF-1 alone initiates with a methionine, it is presumed that this represents the start of translation of the 4.5-kb mRNA. Except for the suggestion based upon the ORF-3 reverse transcriptase homology, there is no evidence that any of the ORFs actually specify proteins in vivo.

From the structure of the element and the amino acid sequence homology, it has been proposed that Tdd-1 is a retrotransposon (4). This implies that the 4.5-kb mRNA is copied by a reverse transcriptase, resulting in DNA copies that insert into new sites (H. Varmus and P. Brown, this volume). It is particularly interesting that this RNA is preferentially transcribed during the developmental stage of *D. discoideum.* It is not known if Tdd-1 is also transcribed during the formation of macrocysts, the probable sexual cells, whether Tdd-1 encodes proteins that are essential during *Dictyostelium* differentiation, or whether transcription during this time of development actu-

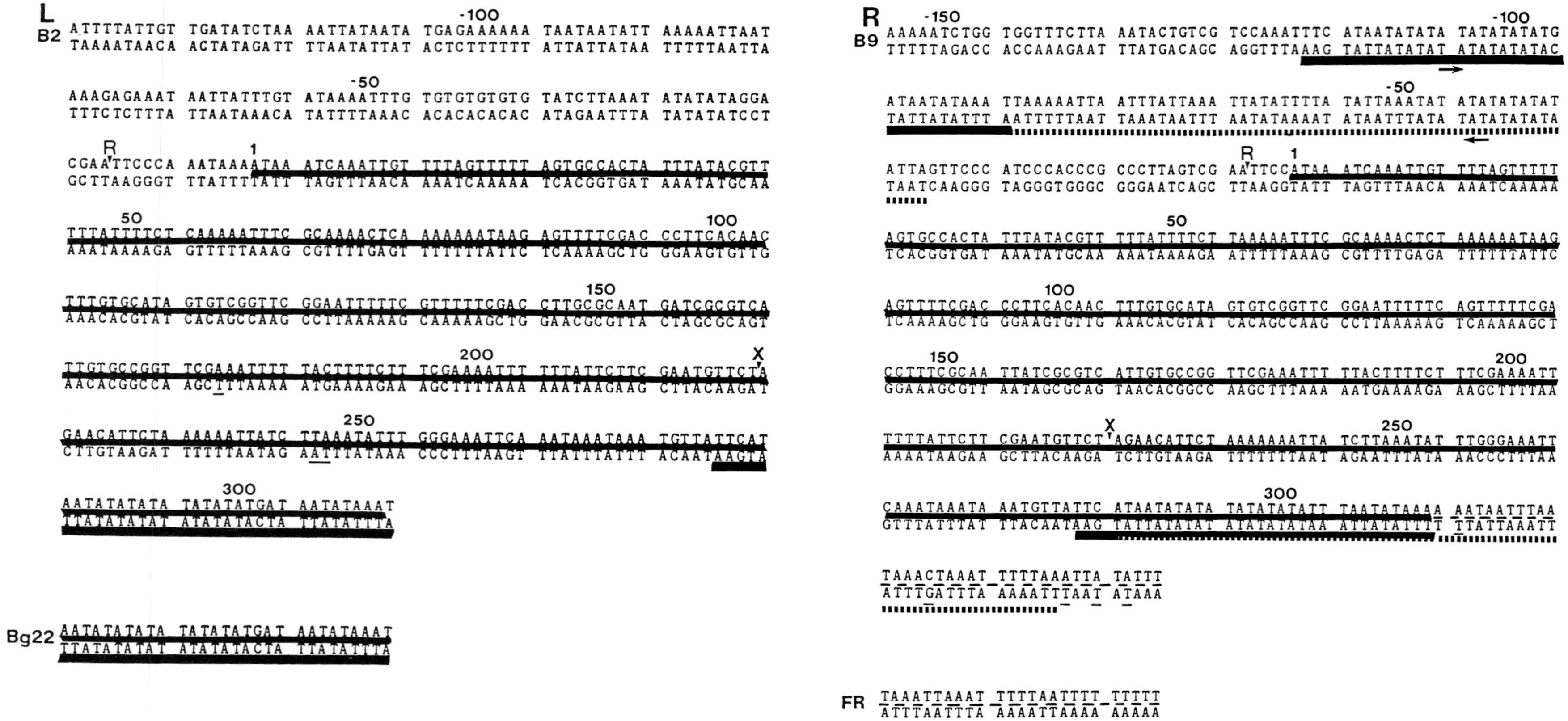

L

```
                                         -100
B2 ATTTTATTGT TGATATCTAA AATTATAATA TGAGAAAAAA TAATAATATT AAAAATTAAT
   TAAAATAACA ACTATAGATT TTAATATTAT ACTCTTTTTT ATTATTATAA TTTTTAATTA

                         -50
   AAAGAGAAAT AATTATTTGT ATAAAATTTG TGTGTGTGTG TATCTTAAAT ATATATAGGA
   TTTCTCTTTA TTAATAAACA TATTTTAAAC ACACACACAC ATAGAATTTA TATATATCCT

       R           1
   CGAATTCCCA AATAAAATAA ATCAAATTGT TTTAGTTTTT AGTGCCACTA TTTATACGTT
   GCTTAAGGGT TTATTTTATT TAGTTTAACA AAATCAAAAA TCACGGTGAT AAATATGCAA

       50                                                  100
   TTTATTTTCT CAAAAATTTC GCAAAACTCA AAAAAATAAG AGTTTTCGAC CCTTCACAAC
   AAATAAAAGA GTTTTTAAAG CGTTTTGAGT TTTTTTATTC TCAAAAGCTG GGAAGTGTTG

                                              150
   TTTGTGCATA GTGTCGGTTC GGAATTTTTC GTTTTTCGAC CTTGCGCAAT GATCGCGTCA
   AAACACGTAT CACAGCCAAG CCTTAAAAAG CAAAAAGCTG GAACGCGTTA CTAGCGCAGT

                                   200                            X
   TTGTGCCGGT TCGAAATTTT TACTTTTCTT TCGAAAATTT TTTATTCTTC GAATGTTCTA
   AACACGGCCA AGCTTTAAAA ATGAAAAGAA AGCTTTTAAA AAATAAGAAG CTTACAAGAT

                         250
   GAACATTCTA AAAAATTATC TTAAATATTT GGGAAATTCA AATAAATAAA TGTTATTCAT
   CTTGTAAGAT TTTTTAATAG AATTTATAAA CCCTTTAAGT TTATTTATTT ACAATAAGTA

              300
   AATATATATA TATATATGAT AATATAAAT
   TTATATATAT ATATATACTA TTATATTTA

Bg22 AATATATATA TATATATGAT AATATAAAT
     TTATATATAT ATATATACTA TTATATTTA
```

R

```
   -150                                                         -100
B9 AAAAATCTGG TGGTTTCTTA AATACTGTCG TCCAAATTTC ATAATATATA TATATATATG
   TTTTTAGACC ACCAAAGAAT TTATGACAGC AGGTTTAAAG TATTATATAT ATATATATAC

                                                     -50
   ATAATATAAA TTAAAAATTA ATTTATTAAA TTATATTTTA TATTAAATAT ATATATATAT
   TATTATATTT AATTTTTAAT TAAATAATTT AATATAAAAT ATAATTTATA TATATATATA

                                    R   1
   ATTAGTTCCC ATCCCACCCG CCCTTAGTCG AATTCCATAA ATCAAATTGT TTTAGTTTTT
   TAATCAAGGG TAGGGTGGGC GGGAATCAGC TTAAGGTATT TAGTTTAACA AAATCAAAAA

                         50
   AGTGCCACTA TTTATACGTT TTTATTTTCT TAAAAATTTC GCAAAACTCT AAAAAATAAG
   TCACGGTGAT AAATATGCAA AAATAAAAGA ATTTTTAAAG CGTTTTGAGA TTTTTTATTC

              100
   AGTTTTCGAC CCTTCACAAC TTTGTGCATA GTGTCGGTTC GGAATTTTTC AGTTTTTCGA
   TCAAAAGCTG GGAAGTGTTG AAACACGTAT CACAGCCAAG CCTTAAAAAG TCAAAAAGCT

       150                                                     200
   CCTTTCGCAA TTATCGCGTC ATTGTGCCGG TTCGAAATTT TTACTTTTCT TTCGAAAATT
   GGAAAGCGTT AATAGCGCAG TAACACGGCC AAGCTTTAAA AATGAAAAGA AAGCTTTTAA

                        X                         250
   TTTTATTCTT CGAATGTTCT AGAACATTCT AAAAAAATTA TCTTAAATAT TTGGGAAATT
   AAAATAAGAA GCTTACAAGA TCTTGTAAGA TTTTTTTAAT AGAATTTATA AACCCTTTAA

                                   300
   CAAATAAATA AATGTTATTC ATAATATATA TATATATATT TAATATAAAA AATAATTTAA
   GTTTATTTAT TTACAATAAG TATTATATAT ATATATATAA ATTATATTTT TTATTAAATT

   TAAACTAAAT TTTTAAATTA TATTT
   ATTTGATTTA AAAATTTAAT ATAAA

FR TAAATTAAAT TTTTAATTTT TTTTT
   ATTTAATTTA AAAATTAAAA AAAAA
```

Figure 2. DNA sequence of the inverted repeats of Tdd-1. The LTR sequence is numbered from inside the element moving toward the ends. The first nucleotide is designated 1. Nucleotides inside the LTR within the Tdd-1 element are given negative numbers. This unusual numbering system was used since the left and right elements are of different lengths (see text). (**L**) The sequence of a left terminal repeat from clone *Bam2* in Fig. 1, labeled B2 in this figure. The sequence starts internal to the *Eco*RI site (R) and extends past the *Xba*I site (X) to the point at which the sequences of the two left clones diverge. The thinner solid line between the sequences of the two strands indicates the inverted terminal repeat (bases 1 to 312; see **R** panel). The bottom line presents the sequence of a second left LTR clone (labeled Bg22) around the point of divergence, which is used to define the left end of Tdd-1. The underlined nucleotides indicate positions within the last 140 bp where the two sequences differ. There are several insertions, deletions, and substitutions in the sequence close to the *Eco*RI site (not shown). The heavy black line underlying both strands at the end of the terminal repeat denotes a 34-bp sequence which is duplicated internally to the right-hand *Eco*RI site (see **R** panel). (**R**) The sequence of a right terminal repeat of clone *Bam9* (Fig. 1), labeled B9 in this figure. The sequence starts internal to the *Eco*RI site and extends past the *Xba*I site to the point at which the sequences of the two right-end clones (*Bam9*, *Bgl*II) diverge. The thinner solid line between the sequences of the two strands indicates the inverted terminal repeat (bases 1 to 313; see **L** panel). The bottom line presents the sequence of a second right LTR (labeled FR) around the point of divergence, which is used to define the right end of Tdd-1. The underlined nucleotides indicate positions within the last 100 bp where the two sequences differ. The heavy black underline indicates a 34-bp sequence found at the ends of the terminal repeats as well as internal to the right-hand *Eco*RI site. However, note that this 34-bp region from the left and right LTRs can be distinguished by conserved GA residues at positions 302 and 303 in the left LTR that is a conserved, single-T residue at position 304 in the right LTR. The point of divergence between the sequences of B9 and FR, which defines the right end of Tdd-1, is found 36 bp beyond the end of the right terminal repeat. This region is indicated by a thin dashed line between the two lines of sequence (bases 314 to 349). The ICR (see text) consists of the last 34 bp of the left LTR (heavy underline, extending from nucleotide −119 in the **R** sequence through nucleotide −86) and the region from nucleotides −87 through −33 in the **R** sequence (dashed underline) that is an inverted repeat of most of the end of the right LTR (nucleotides 286 through 340, also dashed underline), although it is inverted relative to the end of the right LTR. The polarities of both of these sections of the ICR are indicated by arrows. An arrowhead on the **R** and **L** panels indicates the end of the element on the right and left.

ally is causally related to the mobility of the element in vivo.

The 1.4-kb heat shock RNA does not have an extended ORF and thus presumably is not translated (4, 28). It is possible that heat-shock-induced RNA has no function in vivo. The accumulated 1.4-kb heat shock transcript is not very abundant, suggesting that this promoter is not strong. Alternatively, one might speculate that the 1.4-kb heat shock RNA might regulate the expression of the 4.5-kb RNA posttranscriptionally under conditions of stress. It is conceivable that it is deleterious to express proteins encoded by Tdd-1 under such physiological conditions and that the 1.4-kb heat-shock-induced transcript, an antisense transcript of the 4.5-kb RNA, could modulate the expression of the Tdd-1 gene products.

The left and right LTRs of Tdd-1 are not the same length and contain nucleotide changes between the LTRs that are conserved among different Tdd-1 elements (28, 38). These differences are difficult to reconcile with models for replication of this element based on the retrovirus mechanism. An alternative model (4) involves hybridization to and priming off of the ICR sequence within the 4.1-kb internal region. The ICR contains nucleotides at specific sites that are identical to the end of the right LTR and different in the left LTR (4, 28). By this model, it would be possible to replicate the 4.5-kb RNA and maintain the unique structural features of the two LTRs.

III. STRUCTURES OF Tdd-2 AND Tdd-3

Elements Tdd-2 and Tdd-3 were identified in the upstream flanking regions of two tightly linked members of the discoidin I multigene family (26, 27, 31). Discoidin I is a developmentally regulated lectin whose function may be analogous to that of fibronectin in higher eucaryotes (15, 34). Discoidin I is expressed during early development and appears to be laid out as a blanket on which the cells migrate during aggregation (11, 21, 25, 26, 30, 31). Several functional studies, including antisense mutagenesis, by which discoidin I expression was blocked by expressing an RNA complementary to endogenous discoidin I mRNA, indicates that discoidin I has an important function during the aggregation phase of development (1, 9, 15, 34). In the wild-type isolate NC-4, discoidin I is encoded by a three-member gene family. Two genes, DiscI-β and -γ, are separated by <1 kb and are in the same orientation (Fig. 3). A pseudogene (DiscI-ψ) is ~5 kb downstream from the β and γ genes (left in Fig. 3). Finally, the DiscI-α gene is more than 20 kb from the DiscI-β and -γ genes (25–27, 31).

Initial analysis of the regions surrounding DiscI-β and -γ indicated that they contain sequences that are moderately repeated within the *Dictyostelium* genome. DNA blot analysis of a number of laboratory stocks (see below) and separate wild-type isolates showed substantial restriction site polymorphism surrounding these genes that apparently resulted from the insertion or deletion or both of mobile elements in different strains (27, 31). Further analysis of laboratory stocks of the axenic strain Ax-3, originally derived from NC-4, has helped to elucidate the nature of these repeated elements (26, 27). One Ax-3 derivative, Ax-3L, contained a duplicated region of the DiscI-β and -γ gene region, but this duplication lacked the Tdd-2 element present in NC-4. The Tdd-2 element is still present at the original locus. In contrast, strain Ax-3K derived from Ax-3L lacks this duplication, but has instead a new repeat element, Tdd-3, upstream from the β and γ genes (Fig. 3). The two laboratory stocks Ax-3L and Ax-3K were derived from a parental Ax-3 line and had been maintained in different laboratories for ~5 years. This indicated that the region is unstable and suggested that the instability is linked to these repeat elements. The DiscI-β/γ region in NC-4 has a single mobile repeat element, Tdd-2, present ~5 kb upstream from the DiscI-β and -γ genes. The duplicated β/γ region in Ax-3L lacks this element and an additional repeat sequence lying downstream from the β/γ cluster. The data suggest that the duplication of the DiscI-β/γ region in Ax-3L resulted from an unequal crossing over between two Tdd-2 sequences. This results in complementary duplication and deletion of the DiscI-β/γ region. Examination of Ax-3K showed that another element, Tdd-3, had inserted into Tdd-2 near the end proximal to the DiscI-β/γ genes (Fig. 4 and discussion below). These rearrangements are described in Fig. 3 and 5.

The possibility that Tdd-3 preferentially inserts into Tdd-2 was further analyzed. A clone (pDdB14) containing a Tdd-2 element that is not linked to the DiscI-β and -γ genes was isolated from a genomic library of strain Ax-3L using a Tdd-2 probe (Fig. 4A). This Tdd-2 element contained an insertion of Tdd-3 near the site Tdd-3 occupied in the DiscI-β/γ-linked Tdd-2 element of strain Ax-3K. Single-copy flanking sequences were used to investigate the structure of that particular region in the wild-type, ancestral strain NC-4. The locus contained a Tdd-2 element but lacked a Tdd-3 insertion.

These results suggest that Tdd-3 inserts preferentially within Tdd-2 elements. While Tdd-2 and Tdd-3 show little overall sequence homology at stan-

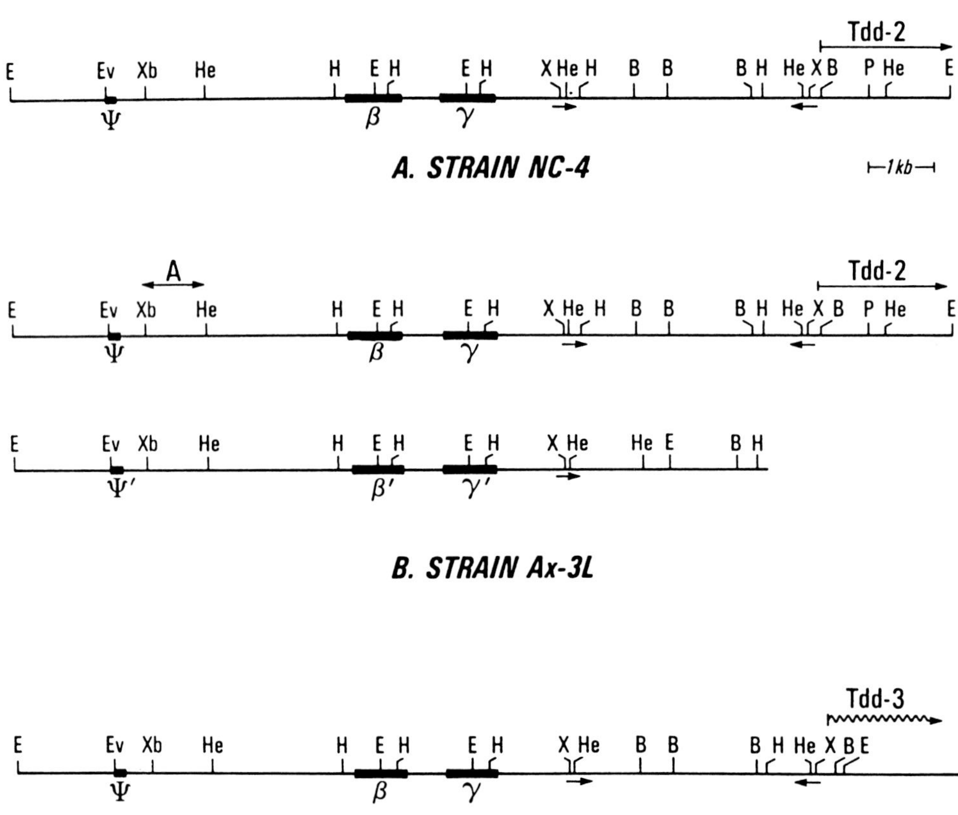

Figure 3. Linked discoidin I genes of NC-4 and derivative strains. Shown are the restriction maps of the regions containing the linked β and γ genes. These regions also contain a 139-bp pseudogene fragment ψ (25). Also shown are the short, inverted repeats upstream from the DiscI-γ gene (short horizontal arrows) (25) and the locations of the Tdd-2 and Tdd-3 mobile elements. To emphasize similarities, the regions that are homologous among the strains have been drawn with identical sites, based on the restriction sites of strain Ax-3L. However, not all of these sites have been checked on genomic DNA blots in strains NC-4 and Ax-3K. For strain Ax-3L, there are two regions shown. The bottom strand is the duplicated region that has lost the Tdd-2 element. It is not known if the duplication is adjacent to the original sequence or if it has been transposed or translocated to a different region of the genome. The map was constructed from genomic clones isolated from strains Ax-3L and Ax-3K and from mapping of the genomic organization of these loci using DNA blots. The small arrows under the region containing *Xho*I and *Hae*III restriction sites indicate an inverted repeat. Restriction site symbols: B, *Bgl*II; E, *Eco*RI; Ev, *Eco*RV; H, *Hin*dIII; He, *Hae*III; P, *Pst*I; X, *Xho*I; and Xb, *Xba*I. (Figure is taken from reference 26.)

→

Figure 4. Analysis of Tdd-3 insertion into Tdd-2. (A) Restriction map of clone pDdB14. A *Bam*HI library of Ax-3L DNA in λ BF101 was screened with the *Pst*I-*Eco*RI fragment of the Tdd-2 element in pDdDiscI-CD (Fig. 3). The restriction map of the 12-kb *Bam*HI insert of pDdB14 is shown. The locations of the various regions of repeat elements Tdd-2 and Tdd-3 are shown above. The dotted arrow indicates that the exact location of the right end of the Tdd-2 element is not known. Restriction site symbols: Bh, *Bam*HI; X, *Xho*I; Xb, *Xba*I; P, *Pst*I; Ev, *Eco*RV; H, *Hin*dIII. B′, B″, and B‴ designate specific *Bgl*II sites referred to in panel B. (B) Nucleotide sequences of the Tdd-3 insertion site in pDdB14. The orientation of the sequence is the same as that of the map in panel A. The upper region of the figure shows the sequences leftward from the B′ *Bgl*II site. The lower portion of the sequence shows the sequence rightward from the *Xho*I site 4.2 kb away. The restriction sites X, B′, B″, and B‴ refer to sites in the map of panel A. The solid lines underneath the sequences show regions of Tdd-2 DNA. The wavy lines show regions of Tdd-3 DNA. The boxes show the 22-bp region of homology near the left termini of Tdd-2 and Tdd-3. The horizontal arrows over the sequences show the duplication of Tdd-2 DNA presumably caused by insertion of the Tdd-3 element. Arrows are drawn over the 9-bp duplication. Since the exact structure of the Tdd-3 left end is not known, it is also possible that the duplication extends to the A residue 1 bp farther leftward. The A residue rightward is not part of the duplication. (C) Nucleotide sequences around the Tdd-3 insertion into Tdd-2 in the DiscI-β/γ region. The nucleotide sequences rightward from the far *Xho*I site upstream from the DiscI-γ gene in strains Ax-3L and Ax-3K are shown. The orientation of the sequences is the same as that of the map in Fig. 3. Regions of homology between this region of the two strains are designated by asterisks above the sequence. Upper line, Sequence from the *Xho*I site of Ax-3L. Lower line, Sequence from the *Xho*I site of Ax-3K. Other symbols are as in the legend to panel B.

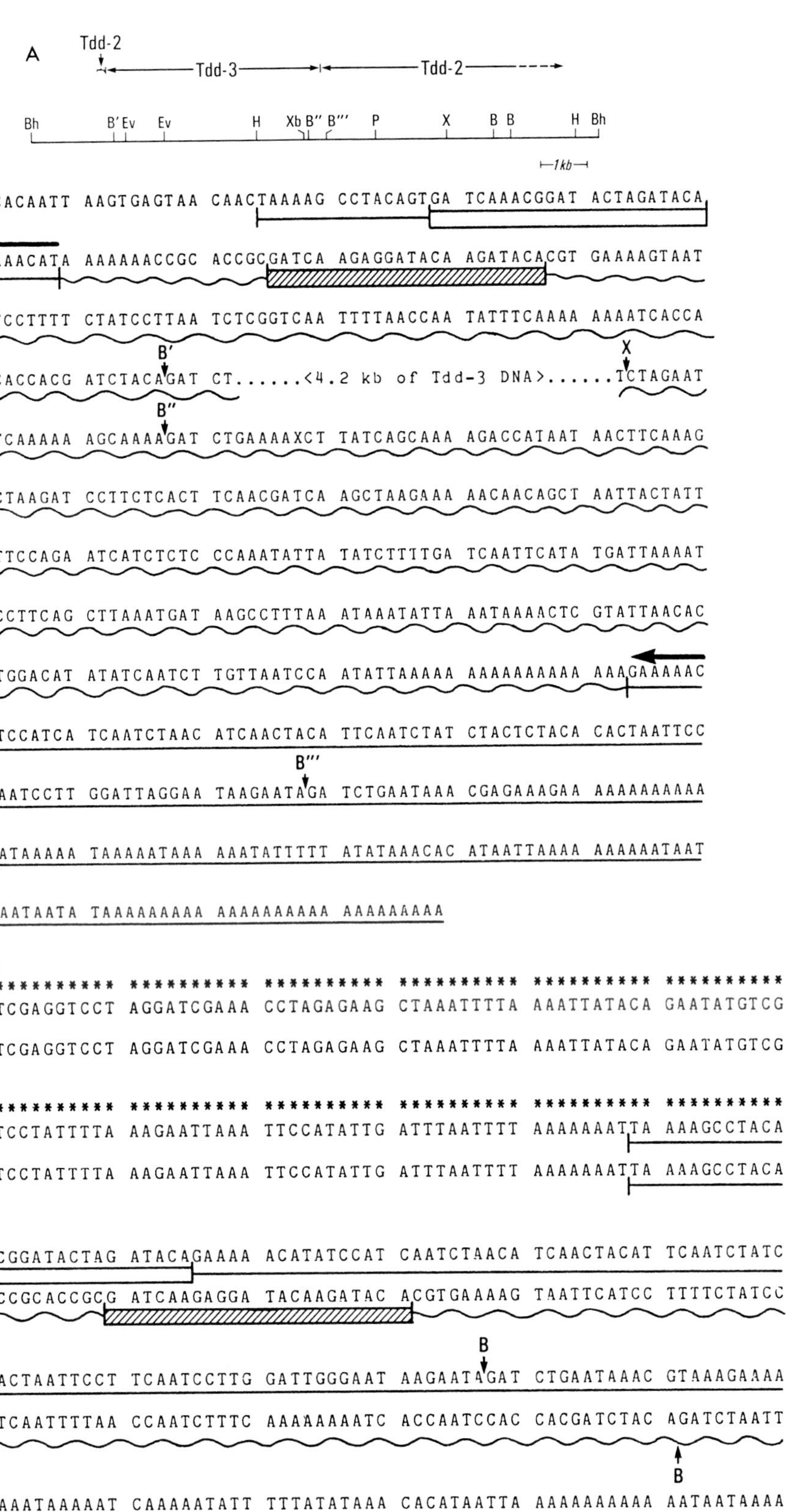
A
Tdd-2
Tdd-3
Tdd-2
Bh B′ Ev Ev H Xb B″ B‴ P X B B H Bh
1kb
B
CACCACAATT AAGTGAGTAA CAACTAAAAG CCTACAGTGA TCAAACGGAT ACTAGATACA
GAAAAACATA AAAAAACCGC ACCGCGATCA AGAGGATACA AGATACACGT GAAAAGTAAT
TCATCCTTTT CTATCCTTAA TCTCGGTCAA TTTTAACCAA TATTTCAAAA AAAATCACCA
B′
X
ATCCACCACG ATCTACAGAT CT......<4.2 kb of Tdd-3 DNA>......TCTAGAAT
B″
ACATCAAAAA AGCAAAAGAT CTGAAAAXCT TATCAGCAAA AGACCATAAT AACTTCAAAG
ATCCTAAGAT CCTTCTCACT TCAACGATCA AGCTAAGAAA AACAACAGCT AATTACTATT
GTATTCCAGA ATCATCTCTC CCAAATATTA TATCTTTTGA TCAATTCATA TGATTAAAAT
AACCCTTCAG CTTAAATGAT AAGCCTTTAA ATAAATATTA AATAAAACTC GTATTAACAC
AGATGGACAT ATATCAATCT TGTTAATCCA ATATTAAAAA AAAAAAAAAA AAAGAAAAAC
ATATCCATCA TCAATCTAAC ATCAACTACA TTCAATCTAT CTACTCTACA CACTAATTCC
B‴
TTCAATCCTT GGATTAGGAA TAAGAATAGA TCTGAATAAA CGAGAAAGAA AAAAAAAAAA
AAAATAAAAA TAAAAATAAA AAATATTTTT ATATAAACAC ATAATTAAAA AAAAAATAAT
AAAAATAATA TAAAAAAAAA AAAAAAAAAA AAAAAAAAA
C
X
[Ax3L]..CTCGAGGTCCT AGGATCGAAA CCTAGAGAAG CTAAATTTTA AAATTATACA GAATATGTCG
[Ax3K]..CTCGAGGTCCT AGGATCGAAA CCTAGAGAAG CTAAATTTTA AAATTATACA GAATATGTCG
CTAGTTCGAT TCCTATTTTA AAGAATTAAA TTCCATATTG ATTTAATTTT AAAAAAATTA AAAGCCTACA
CTAGTTCGAT TCCTATTTTA AAGAATTAAA TTCCATATTG ATTTAATTTT AAAAAAATTA AAAGCCTACA
GTTGATCAAA CGGATACTAG ATACAGAAAA ACATATCCAT CAATCTAACA TCAACTACAT TCAATCTATC
GTTAAAAAAA CCGCACCGCG ATCAAGAGGA TACAAGATAC ACGTGAAAAG TAATTCATCC TTTTCTATCC
B
TACTCTACAC ACTAATTCCT TCAATCCTTG GATTGGGAAT AAGAATAGAT CTGAATAAAC GTAAAGAAAA
TTAATCTCGG TCAATTTTAA CCAATCTTTC AAAAAAAATC ACCAATCCAC CACGATCTAC AGATCTAATT
B
AAAAAAATAA AAATAAAAAT CAAAAATATT TTTATATAAA CACATAATTA AAAAAAAAAA AATAATAAAA
TACCAAATCG CCGATTCAAA GGAATCCCAT CGAGATGAAT TC....... ....
E
ATAATATAAA AAAAAAAAAA AAAAAAAAAA A.....[Ax3L]
....[Ax3K]

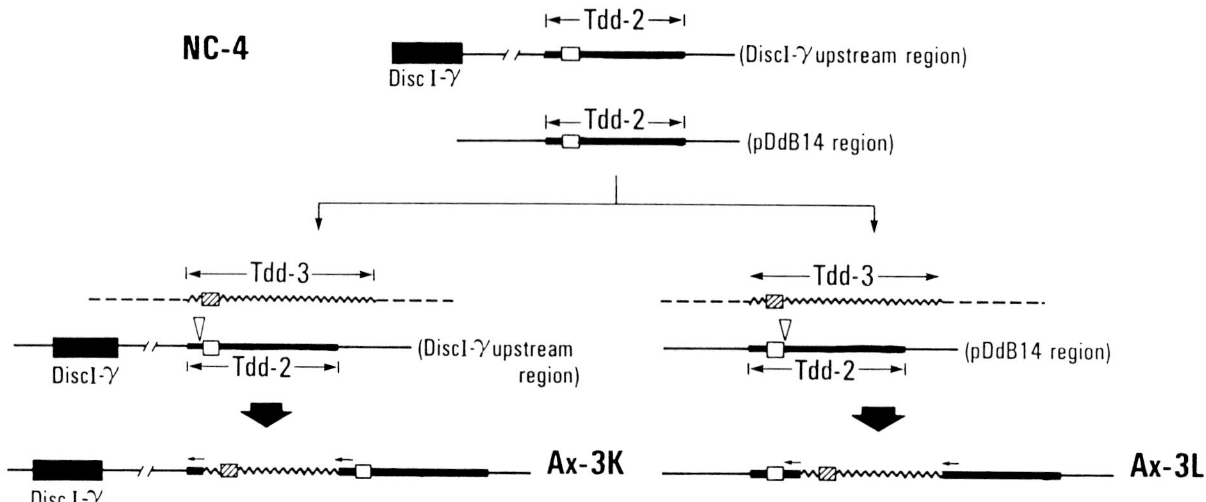

Figure 5. Transposition of the Tdd-3 element into two copies of the Tdd-2 element at different genetic loci. A model to explain the observed differences between Tdd-2 and Tdd-3 at the pDdB14 and DiscI-β/γ regions is shown. Thick solid lines represent the Tdd-2 element, wavy lines represent the Tdd-3 element, and the open and hatched boxes represent the 22-bp homology regions. The thinner solid lines represent unique DNA sequences, and thus the regions surrounding the Tdd-2 element in the DiscI-γ and pDdB14 loci are not the same. (The figure is not drawn to scale.) Strain NC-4 has one copy of the Tdd-2 element in the pDdB14 region. The Tdd-3 element preferentially inserts at the site of the 22-bp homology. In strain Ax-3L, the Tdd-3 element has inserted in the DdB14 locus at the right side of the 22-bp homology, splitting the Tdd-2 element at the site shown and causing a short direct duplication of the target site DNA. In strain Ax-3K, a Tdd-3 element has inserted into the DiscI-γ upstream region, this time to the left of the homology. It is assumed that this results in short, direct duplication at this target site also. The mechanism of insertion is unknown. See text and Fig. 3 and 4 for details. (Figure is taken from reference 26.)

dard stringencies of hybridization, there is a 22-bp sequence homology near one end of each element (Fig. 4B and C). Figure 4C shows the sequence around the Tdd-3 element found in Tdd-2 in the DiscI-β/γ locus in strain Ax-3K. As can be seen, the Tdd-3 element also inserts near this 22-bp homologous region. Of interest is the fact that in these two examples, the 22-bp regions of the two Tdd-2 elements are identical and the two regions in Tdd-3 are identical. While the number of cases studied is limited, the data suggest that this sequence may be involved in the apparently preferential insertion of Tdd-3 into Tdd-2 elements. The insertion event results in a 9- to 10-bp duplication of the target site DNA (26). The uncertainty of the extent of the duplication (9 or 10 bp) is a consequence of the fact that the insertion target site is extremely AT-rich.

Structural analysis of Tdd-2 and Tdd-3 indicates that the cloned elements lack terminal direct or inverted repeats. However, both ends of the Tdd-3 element in pDdB14 terminate in poly(dA) sequences of 17 and 18 nucleotides, respectively. Again, because of the very AT-rich nature of *Dictyostelium* genomic DNA, it is not clear if these sequences represent a short terminal repeat.

Conserved internal restriction fragments spanning most of the of Tdd-2 and Tdd-3 elements have been used to hybridize to RNA isolated from cells at different developmental stages. There is no evidence for transcription of either element. Several hundred base pairs has been sequenced from the ends of both elements and no extended ORFs have been found.

IV. EVIDENCE FOR OTHER MOBILE GENETIC ELEMENTS IN *D. DISCOIDEUM*

A number of genetic experiments suggest the presence of additional mobile elements. Soll and co-workers have been examining mutants that alter the relative length of the interphase between starvation and the onset of aggregation (33). One such mutant, FM-1, behaves as though a mutation occurs in a single locus and is unstable and reverts to wild-type or other phenotypes at frequencies of 3×10^{-2}, as compared with a switching rate of 5×10^{-3} in wild-type cells as determined by the formation of sectors with different phenotypes within colonies (33). A derivative of FM-1, FM-1(S1), shows a four- to sixfold higher frequency of switching between different phenotypes when grown on bacteria on nutrient agar. Analyses of these variants show sectors with a variety of phenotypes arising from single clones. These phenomena appear similar to switching observed in *C. albicans*, which may also be associated with a mobile genetic element (33; Soll, this volume).

A

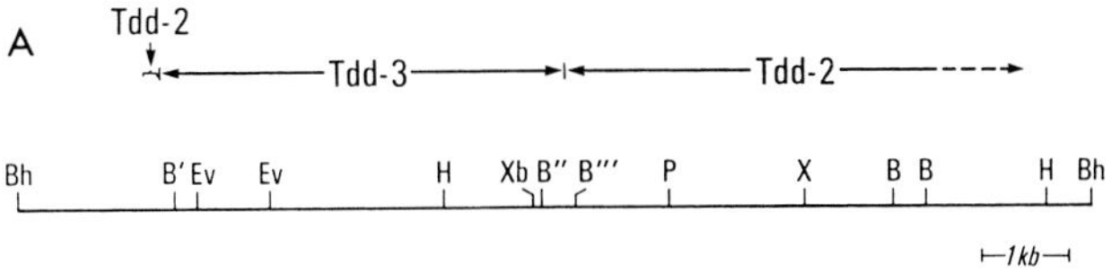

Tdd-2

Tdd-3 — Tdd-2

Bh B′ Ev Ev H Xb B″ B‴ P X B B H Bh

1kb

B

```
CACCACAATT AAGTGAGTAA CAACTAAAAG CCTACAGTGA TCAAACGGAT ACTAGATACA
GAAAAACATA AAAAAACCGC ACCGCGATCA AGAGGATACA AGATACACGT GAAAAGTAAT
TCATCCTTTT CTATCCTTAA TCTCGGTCAA TTTTAACCAA TATTTCAAAA AAAATCACCA
                  B'                                          X
ATCCACCACG ATCTACAGAT CT......<4.2 kb of Tdd-3 DNA>......TCTAGAAT
                  B"
ACATCAAAAA AGCAAAAGAT CTGAAAAXCT TATCAGCAAA AGACCATAAT AACTTCAAAG
ATCCTAAGAT CCTTCTCACT TCAACGATCA AGCTAAGAAA AACAACAGCT AATTACTATT
GTATTCCAGA ATCATCTCTC CCAAATATTA TATCTTTTGA TCAATTCATA TGATTAAAAT
AACCCTTCAG CTTAAATGAT AAGCCTTTAA ATAAATATTA AATAAAACTC GTATTAACAC
AGATGGACAT ATATCAATCT TGTTAATCCA ATATTAAAAA AAAAAAAAAA AAAGAAAAAC
ATATCCATCA TCAATCTAAC ATCAACTACA TTCAATCTAT CTACTCTACA CACTAATTCC
                              B'''
TTCAATCCTT GGATTAGGAA TAAGAATAGA TCTGAATAAA CGAGAAAGAA AAAAAAAAAA
AAAATAAAAA TAAAAATAAA AAATATTTTT ATATAAACAC ATAATTAAAA AAAAAATAAT
AAAAATAATA TAAAAAAAAA AAAAAAAAAA AAAAAAAAA
```

C

```
       X
       ***********  ********** ********** ********** ********** **********
[Ax3L]..CTCGAGGTCCT AGGATCGAAA CCTAGAGAAG CTAAATTTTA AAATTATACA GAATATGTCG

[Ax3K]..CTCGAGGTCCT AGGATCGAAA CCTAGAGAAG CTAAATTTTA AAATTATACA GAATATGTCG

********** ********** ********** ********** ********** ********** **********
CTAGTTCGAT TCCTATTTTA AAGAATTAAA TTCCATATTG ATTTAATTTT AAAAAAATTA AAAGCCTACA

CTAGTTCGAT TCCTATTTTA AAGAATTAAA TTCCATATTG ATTTAATTTT AAAAAAATTA AAAGCCTACA

***
GTTGATCAAA CGGATACTAG ATACAGAAAA ACATATCCAT CAATCTAACA TCAACTACAT TCAATCTATC

GTTAAAAAAA CCGCACCGCG ATCAAGAGGA TACAAGATAC ACGTGAAAAG TAATTCATCC TTTTCTATCC

                                           B
TACTCTACAC ACTAATTCCT TCAATCCTTG GATTGGGAAT AAGAATAGAT CTGAATAAAC GTAAAGAAAA

TTAATCTCGG TCAATTTTAA CCAATCTTTC AAAAAAAATC ACCAATCCAC CACGATCTAC AGATCTAATT
                                                                 B

AAAAAAATAA AAATAAAAAT CAAAAATATT TTTATATAAA CACATAATTA AAAAAAAAAA AATAATAAAA

TACCAAATCG CCGATTCAAA GGAATCCCAT CGAGATGAAT TC........ .......... ..........
                                          E

ATAATATAAA AAAAAAAAAA AAAAAAAAAA A.....[Ax3L]

.......... .......... .......... ......[Ax3K]
```

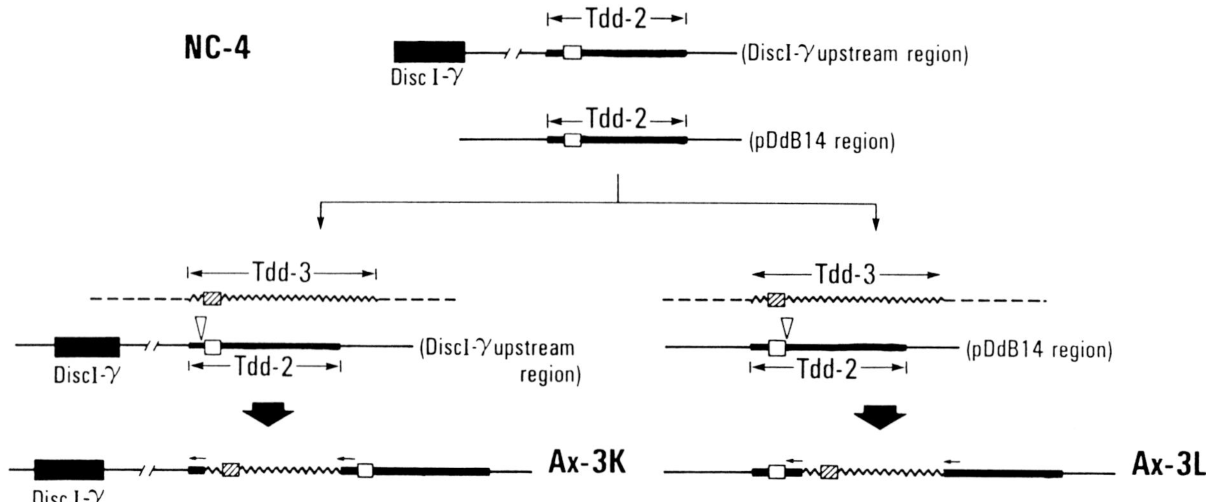

Figure 5. Transposition of the Tdd-3 element into two copies of the Tdd-2 element at different genetic loci. A model to explain the observed differences between Tdd-2 and Tdd-3 at the pDdB14 and DiscI-β/γ regions is shown. Thick solid lines represent the Tdd-2 element, wavy lines represent the Tdd-3 element, and the open and hatched boxes represent the 22-bp homology regions. The thinner solid lines represent unique DNA sequences, and thus the regions surrounding the Tdd-2 element in the DiscI-γ and pDdB14 loci are not the same. (The figure is not drawn to scale.) Strain NC-4 has one copy of the Tdd-2 element in the pDdB14 region. The Tdd-3 element preferentially inserts at the site of the 22-bp homology. In strain Ax-3L, the Tdd-3 element has inserted in the DdB14 locus at the right side of the 22-bp homology, splitting the Tdd-2 element at the site shown and causing a short direct duplication of the target site DNA. In strain Ax-3K, a Tdd-3 element has inserted into the DiscI-γ upstream region, this time to the left of the homology. It is assumed that this results in short, direct duplication at this target site also. The mechanism of insertion is unknown. See text and Fig. 3 and 4 for details. (Figure is taken from reference 26.)

dard stringencies of hybridization, there is a 22-bp sequence homology near one end of each element (Fig. 4B and C). Figure 4C shows the sequence around the Tdd-3 element found in Tdd-2 in the DiscI-β/γ locus in strain Ax-3K. As can be seen, the Tdd-3 element also inserts near this 22-bp homologous region. Of interest is the fact that in these two examples, the 22-bp regions of the two Tdd-2 elements are identical and the two regions in Tdd-3 are identical. While the number of cases studied is limited, the data suggest that this sequence may be involved in the apparently preferential insertion of Tdd-3 into Tdd-2 elements. The insertion event results in a 9- to 10-bp duplication of the target site DNA (26). The uncertainty of the extent of the duplication (9 or 10 bp) is a consequence of the fact that the insertion target site is extremely AT-rich.

Structural analysis of Tdd-2 and Tdd-3 indicates that the cloned elements lack terminal direct or inverted repeats. However, both ends of the Tdd-3 element in pDdB14 terminate in poly(dA) sequences of 17 and 18 nucleotides, respectively. Again, because of the very AT-rich nature of *Dictyostelium* genomic DNA, it is not clear if these sequences represent a short terminal repeat.

Conserved internal restriction fragments spanning most of the of Tdd-2 and Tdd-3 elements have been used to hybridize to RNA isolated from cells at different developmental stages. There is no evidence for transcription of either element. Several hundred base pairs has been sequenced from the ends of both elements and no extended ORFs have been found.

IV. EVIDENCE FOR OTHER MOBILE GENETIC ELEMENTS IN *D. DISCOIDEUM*

A number of genetic experiments suggest the presence of additional mobile elements. Soll and co-workers have been examining mutants that alter the relative length of the interphase between starvation and the onset of aggregation (33). One such mutant, FM-1, behaves as though a mutation occurs in a single locus and is unstable and reverts to wild-type or other phenotypes at frequencies of 3×10^{-2}, as compared with a switching rate of 5×10^{-3} in wild-type cells as determined by the formation of sectors with different phenotypes within colonies (33). A derivative of FM-1, FM-1(S1), shows a four- to sixfold higher frequency of switching between different phenotypes when grown on bacteria on nutrient agar. Analyses of these variants show sectors with a variety of phenotypes arising from single clones. These phenomena appear similar to switching observed in *C. albicans*, which may also be associated with a mobile genetic element (33; Soll, this volume).

Subsequent molecular analysis should clarify whether these instabilities are due to mobile DNA.

V. SUMMARY

The three mobile genetic elements from *D. discoideum* are of particular interest because of their structural and transcriptional properties. There does not appear to be a direct relationship between any of these elements and known phenotypic mutations affecting growth or development. This may be related to the apparent preferential insertion of Tdd-1 into partial Tdd-1 sequences and of Tdd-3 into Tdd-2. Thus, while the elements are quite mobile within the genome, mutagenesis by insertion into genes is an unlikely event. Of potential interest as an aid to studying development in this system is the identification of high-frequency mutating strains by Soll and collaborators (33). Molecular characterization of this system could lead to a genetically functional system for the isolation of mutants in which the genes of interest are tagged by a mobile element, possibly one not described in this review. The use of such mutants would allow rapid cloning of the relevant gene, aid in its analysis, and greatly increase our ability to dissect the regulatory pathways that regulate cellular differentiation in *D. discoideum*.

Acknowledgments. I thank A. Hjorth for helpful suggestions and D. Soll for communicating unpublished results.

The preparation of this manuscript was supported by grants from the U.S. Public Health Service.

LITERATURE CITED

1. **Alexander, S., T. M. Shinnick, and R. A. Lerner.** 1983. Mutants of *Dictyostelium discoideum* blocked in expression of all members of the developmentally regulated discoidin multigene family. *Cell* **34:**467–475.
2. **Boeke, J. D., D. J. Garfinkel, C. A. Styles, and G. R. Fink.** 1985. Ty elements transpose through an RNA intermediate. *Cell* **40:**491–500.
3. **Cappello, J., S. M. Cohen, and H. F. Lodish.** 1984. *Dictyostelium* transposable element DIRS-1 preferentially inserts into DIRS-1 sequences. *Mol. Cell. Biol.* **4:**2207–2213.
4. **Cappello, J., K. Handelsman, and H. F. Lodish.** 1985. Sequence of *Dictyostelium* DIRS-1: an apparent retrotransposon with inverted terminal repeats and an internal circle junction sequence. *Cell* **43:**105–115.
5. **Cappello, J., C. Zuker, and H. F. Lodish.** 1984. Repetitive *Dictyostelium* heat-shock promoter functions in *Saccharomyces cerevisiae*. *Mol. Cell. Biol.* **4:**591–598.
6. **Chung, S., C. Zuker, and H. F. Lodish.** 1983. A repetitive and apparently transposable DNA sequence in *Dictyostelium discoideum* associated with developmentally regulated RNAs. *Nucleic Acids Res.* **11:**4835–4906.
7. **Clare, J., and P. Farabaugh.** 1985. Nucleotide sequence of a yeast Ty1 element: evidence for an unusual mechanism of gene expression. *Proc. Natl. Acad. Sci. USA* **82:**2829–2833.
8. **Cohen, S. M., J. Cappello, and H. F. Lodish.** 1984. Transcription of *Dictyostelium discoideum* transposable element DIRS-1. *Mol. Cell. Biol.* **4:**2332–2340.
9. **Crowley, T. E., W. Nellen, R. H. Gomer, and R. A. Firtel.** 1985. Phenocopy of discoidin I-minus mutants by antisense transformation in *Dictyostelium*. *Cell* **43:**633–641.
10. **Datta, S. D., S. K. O. Mann, A. Hjorth, R. Gomer, P. Howard, D. Armstrong, C. Reymond, C. Silan, and R. A. Firtel.** 1987. cAMP-regulated gene expression during *Dictyostelium* development is mediated by the cell-surface cAMP receptor, p. 33–61. *In* W. F. Loomis (ed.), *Genetic Regulation of Development. 45th Symposium for the Society of Developmental Biology*. Alan R. Liss, Inc., New York.
11. **Devine, J. M., A. S. Tsang, and J. G. Williams.** 1982. Differential expression of the members of the discoidin I multigene family during growth and development of *Dictyostelium discoideum*. *Cell* **28:**793–800.
12. **Devreotes, P. N.** 1982. Chemotaxis, p. 117–168. *In* W. F. Loomis (ed.), *The Development of Dictyostelium discoideum*. Academic Press, Inc., New York.
13. **Emori, Y., T. Shiba, S. Kanaya, S. Inouye, S. Yuki, and K. Saigo.** 1985. The nucleotide sequence of copia and copia-related RNA in *Drosophila* virus-like particles. *Nature* (London) **315:**773–776.
14. **Flavell, A. J.** 1984. Role reverse transcription in the generation of extrachromosomal copia mobile genetic elements. *Nature* (London) **310:**514–516.
15. **Gabius, H.-J., W. R. Springer, and S. H. Barondes.** 1985. Receptor for the cell binding site of discoidin I. *Cell* **42:**449–456.
16. **Garfinkel, D. J., J. D. Boeke, and G. R. Fink.** 1985. Ty element transposition: reverse transcriptase and virus-like particles. *Cell* **42:**507–517.
17. **Gerisch, G.** 1987. Cyclic AMP and other signals controlling cell development and differentiation in *Dictyostelium*. *Annu. Rev. Biochem.* **56:**853–879.
18. **Janssens, P. M. W., and P. J. M. Van Haastert.** 1987. Molecular basis of transmembrane signal transduction in *Dictyostelium discoideum*. *Microbiol. Rev.* **51:**396–418.
19. **Loomis, W. F., Jr. (ed.).** 1975. *Dictyostelium discoideum, a Developmental System*. Academic Press, Inc., New York.
20. **Loomis, W. F. Jr. (ed.).** 1982. *The Development of Dictyostelium discoideum*. Academic Press, Inc., New York.
21. **Ma, G., and R. A. Firtel.** 1978. Regulation of the synthesis of two carbohydrate binding proteins in *Dictyostelium discoideum*. *J. Biol. Chem.* **253:**2924–2932.
22. **Mann, S. K. O., S. Datta, P. Howard, A. Hjorth, C. Reymond, C. Silan, and R. A. Firtel.** 1987. Cyclic AMP regulation of gene expression during *Dictyostelium* development. *UCLA Symp. Mol. Cell. Biol. New Ser.* **51:**308–328.
23. **Mellor, J., S. M. Fulton, M. J. Dobson, W. Wilson, S. M. Kingsman, and A. J. Kingsman.** 1985. A retrovirus-like strategy for expression of a fusion protein encoded by yeast transposon Ty1. *Nature* (London) **313:**243–246.
24. **Mount, S. M., and G. M. Rubin.** 1985. Complete nucleotide sequence of the *Drosophila* transposable element copia: homology between copia and retroviral proteins. *Mol. Cell. Biol.* **5:**1630–1638.
25. **Poole, S. J., and R. A. Firtel.** 1984. Conserved structural features are found upstream from the three coordinately regulated discoidin I genes of *Dictyostelium discoideum*. *J. Mol. Biol.* **172:**203–220.

26. **Poole, S. J., and R. A. Firtel.** 1984. Genomic instability and mobile genetic elements in regions surrounding two discoidin I genes of *Dictyostelium discoideum*. *Mol. Cell. Biol.* **4:** 671–680.
27. **Poole, S., R. A. Firtel, E. Lamar, and W. Rowekamp.** 1981. Sequence and expression of the discoidin I gene family in *Dictyostelium discoideum*. *J. Mol. Biol.* **153:**273–289.
28. **Rosen, E., A. Siversten, and R. A. Firtel.** 1983. An unusual transposon encoding heat shock inducible and developmentally regulated transcripts in *Dictyostelium*. *Cell* **35:**253–251.
29. **Rosen, E., A. Sivertsen, R. A. Firtel, S. Wheeler, and W. F. Loomis.** 1984. Heat shock genes of *Dictyostelium*, p. 257–278. *In* B. G. Atkinson and D. B. Walden (ed.), *Changes in Gene Expression in Response to Environmental Stress*. Academic Press, Inc., New York.
30. **Rosen, S. D., J. A. Kafka, D. L. Simpson, and S. H. Barondes.** 1973. Developmentally regulated, carbohydrate-binding protein in *Dictyostelium discoideum*. *Proc. Natl. Acad. Sci. USA* **70:**2554–2557.
31. **Rowekamp, W., S. Poole, and R. A. Firtel.** 1980. Analysis of multigene family coding the developmentally regulated carbohydrate-binding protein Discoidin-I in *D. discoideum*. *Cell* **20:**495–505.
32. **Schaap, P.** 1986. Regulation of size and pattern in the cellular slime molds. *Differentiation* **33:**1–16.
33. **Soll, D. R., L. Mitchell, B. Kraft, S. Alexander, R. Finney, and B. Varnum-Finney.** 1987. Characterization of a timing mutant of *Dictyostelium discoideum* which exhibits "high frequency switching." *Dev. Biol.* **120:**25–37.
34. **Springer, W. R., D. N. W. Cooper, and S. H. Barondes.** 1984. Discoidin I is implicated in cell-substratum attachment and ordered cell migration of *Dictyostelium discoideum* and resembles fibronectin. *Cell* **39:**557–564.
35. **Williams, J. G., C. Ceccarelli, S. McRobbie, H. Mahbubani, R. R. Kay, A. Early, M. Barks, and K. A. Jermyn.** 1987. Direct induction of *Dictyostelium* prestalk gene expression by DIF provides evidence that DIF is a morphogen. *Cell* **49:**185–192.
36. **Williams, J. G., C. J. Pears, K. A. Jermyn, D. M. Driscoll, H. Mahbubani, and R. R. Kay.** 1986. The control of gene expression during cellular differentiation of *Dictyostelium discoideum*, p. 277–298. *In* I. Booth and C. Higgins (ed.), *Regulation of Gene Expression*. Cambridge University Press, Cambridge.
37. **Zuker, C., J. Cappello, R. L. Chisholm, and H. F. Lodish.** 1983. A repetitive *Dictyostelium* gene family that is induced during differentiation and by heat shock. *Cell* **34:**997–1005.
38. **Zuker, C., J. Cappello, H. F. Lodish, P. George, and S. Chung.** 1984. *Dictyostelium* transposable element DIRS-1 has 350-base-pair inverted terminal repeats that contain a heat shock promoter. *Proc. Natl. Acad. Sci. USA* **81:**2660–2664.
39. **Zuker, C., and H. F. Lodish.** 1981. Repetitive DNA sequences cotranscribed with developmentally regulated *Dictyostelium discoideum* mRNAs. *Proc. Natl. Acad. Sci. USA* **78:** 5386–5390.

Chapter 24

Transposable Elements in *Xenopus* Species

DANA CARROLL, DEBORAH S. KNUTZON, and JAMES E. GARRETT

I. INTRODUCTION

Essentially all species carry mobile DNA elements in their genomes, and *Xenopus* spp. are no exception. Anyone who thinks it particularly appropriate, however, to find jumping genes in a frog has not witnessed the inelegant flounderings of this normally aquatic amphibian when stranded on dry land.

This review mentions candidate mobile elements of several families, but emphasizes the Tx1 and Tx2 elements, which have been studied in our laboratory. These are the best characterized of the *Xenopus* elements, and their properties broaden the scope of identified features of transposable elements. In addition, they blur the classification of existing elements by exhibiting characteristics of two previously distinct classes. They have the well-defined ends and inverted terminal repeats of elements thought to move via a DNA intermediate (e.g., P elements of *Drosophila* spp., *Ac* elements of maize), while the coding sequences of the Tx elements are very similar to those of families speculated to undergo RNA-mediated transposition (e.g., mammalian long interspersed repetitive elements, *Drosophila* I factors). Resolution of this apparent conflict presents an experimental challenge and promises to advance significantly our general understanding of mobile DNAs.

II. SHORT, DISPERSED REPEATS

Global aspects of DNA sequence organization were characterized first in *Xenopus laevis*, and its genome has thus become a standard among vertebrates (6). About 15% of its DNA consists of short, repetitive sequences that are interspersed with longer, nonrepetitive segments. The repeats must have been dispersed by some mechanism, and it is conceivable that a large fraction of them will ultimately prove to be mobile by some definition. Of the few that have been examined at the sequence level, some have characteristics similar to those of genuine transposable elements. Examples for which mobility has been suggested are described briefly here, and some of their properties are listed in Table 1.

The Vi elements (34) have inverted, terminal repeats (Table 2) and apparently make target duplications (3 base pairs [bp]). The sequenced copies, all from the vitellogenin loci, are highly variable in length and in sequence between these ends. Polymorphisms involving Vi elements—one in the vitellogenin

Dana Carroll and Deborah S. Knutzon ■ Department of Biochemistry, University of Utah School of Medicine, Salt Lake City, Utah 84132. **James E. Garrett** ■ Institute for Advanced Biomedical Research, Oregon Health Sciences University, Portland, Oregon 97201.

Table 1. Catalog of putative *Xenopus* transposable elements

Element	Size (bp)	Copy no.	% of genome	Internal repeats[a] (bp)	References
Tx1					
c	15,000	100	0.05	393	4, 15; unpublished data
d	8,000	650	0.2	393	4, 15; unpublished data
Tx2					
c	15,000	50	0.025	400	4, 15; unpublished data
(d)	(8,000)	(?)	(?)	(?)	4, 15; unpublished data
1723	8,000	8,500	2.4	185	23
Vi	112–469	7,500	0.1		34
JH12	?	44,000	0.2+		29
REM1	500+	25,000	0.4+		22
X132/X11	?	100,000	0.2+	77–79	37, 38

[a] The existence of internal tandem repeats that vary in number within an element family is a feature of several of the families listed here, but is not generally found among mobile DNAs.

B2 gene among contemporary animals, the other a more ancient difference between duplicated B1 and B2 genes—are not simple insertion/deletions, so the evidence for transposition is equivocal. A suggestion of function for the Vi elements, perhaps in the regulation of gene expression, emerges from the observation of related sequences at similar locations upstream of three estrogen-inducible genes: *X. laevis* vitellogenin B2, chicken vitellogenin, and chicken apoVLDLII (42).

Support for a transposable lifestyle is even less compelling for several other families of short, dispersed repeats. Members of the JH12 family (29) are present at different locations in the duplicated *X. laevis* globin gene clusters I and II (W. Knochel, personal communication). The sequence structure of individual JH12 family members has not been extensively investigated. The REM1 element (17, 22) is characterized by inverted terminal repeats, but again characterization is at an early stage. The two available sequences match for only about 130 bp, and their terminal inverted repeats are quite different. Members of the X132/X11 family (38, 39) are found as dispersed clusters of tandem repeats of a 77- to 79-bp unit. Since the few characterized examples terminate at different points within the repeating unit, it is difficult to say what the junctions with flanking DNA look like.

Table 2. Terminal sequences of *Xenopus* elements[a]

Element	Inverted terminal repeat	Match	Target duplication (bp)	References
Tx1	T L 5′-CCCTTTAAGTGCCAGCAGA. . . * * * R 5′-CCTTTTAAATGCCA - CAGA. . . C	16/19	4	15
1723	G L 5′-TAGGGATGTAGCGAACGT. . . * * R 5′-TAGAGATGTCGCGAACGT. . . G G	16/18	8	23, 28
Vi	T A AC L 5′-GGGGCACATTTAC. . . R 5′-GGGGCACATTTAC. . . T A G G	13/13	3	34
REM1-Vtg	L 5′-TTGAAGGGATTCTGTCATG. . . * * R 5′-TTAAAGGGATACTGTCATG. . .	17/19	?	17

[a] In illustrating the inverted terminal repeats, the 5′ strands at the left (L) and right (R) ends of the elements are compared. Asterisks indicate mismatches within otherwise identical sequences. Variations on the most common sequences are shown above and below the continuous sequences. There is some uncertainty whether the REM1-Vtg ends are truly the common ends of the family, since only this one example has been sequenced to the apparent ends.

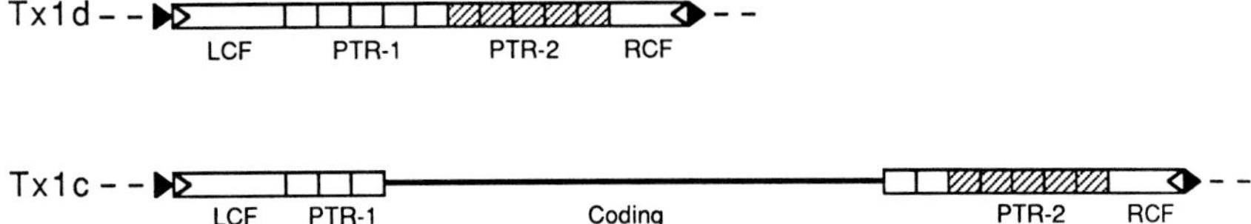

Figure 1. Schematic diagrams of the versions of Tx1 elements. Both complete (Tx1c) and deleted (Tx1d) versions are flanked by target duplications (▶) and contain inverted terminal repeats (▷), two types of 400-bp internal tandem repeats (PTR-1 and PTR-2), and regions known as left and right common flanks (LCF, RCF). In addition, Tx1c includes a 7-kb central region containing apparent protein-coding sequences. (Sources: 4, 15, and unpublished results.)

These elements do not suggest any very satisfying generalizations. It would be interesting to investigate whether the short dispersed sequences are internal repeats or perhaps defective versions of larger elements of the type exemplified by Tx1, Tx2, and 1723 (Table 1, and see below).

III. 1723 ELEMENTS

The first *Xenopus* repetitive elements shown to have structural characteristics very reminiscent of transposable elements were called 1723 (23). Typical members of this family are long (about 8 kb), dispersed repeats, of high copy number (Table 1). Like DNA-based mobile elements, they have discrete ends marked by short, inverted terminal repeats (16/18-bp match; Table 2); they are flanked consistently by an 8-bp apparent target duplication; and there is some suggestion of a low degree of target specificity. Perhaps significantly, there is also a longer inverted repeat (about 180 bp) recessed roughly 200 bp from each end. Like the Tx elements described in detail below, 1723 family members show a variable number of internal tandem repeats (the unit is about 185 bp). Although the sequence of the 1723 inverted terminal repeat is not obviously related to that of other *Xenopus* elements, its outermost 8 bp is identical to that of members of the maize *Ac/Ds* family, and lesser matches to several other elements have been noted (41).

The most direct evidence for 1723 mobility is the existence of an insertion polymorphism at the *X. laevis* proenkephalin locus *A1*, due to the presence or absence of a family member (28). Although this element is considerably smaller than full-length 1723 elements, it has the canonical inverted terminal repeats and is flanked by an 8-bp target duplication. A study of transcripts complementary to 1723 revealed abundant, but heterogeneous and largely nuclear, RNA copies of both strands (24). This likely represents readthrough from neighboring transcription units and not intentional gene copies destined for translation.

IV. Tx ELEMENTS

A. Structure of Tx1 Elements

The structures of the two major versions of the Tx1 element found in the *X. laevis* genome (4, 15) are illustrated in Fig. 1, and general features of the family are listed in Table 1. Previous characterization focused on the more abundant Tx1d version, which we now believe to be deleted, and possibly defective, compared to the complete Tx1c. We have since discovered that a substantial subset of all Tx1 elements have the central region of Tx1c, which contains the major (and perhaps only) protein-coding sequences of the family. Listed below are some of the properties of the Tx1 elements (4, 15; unpublished results).

(i) There are about 750 members of the Tx1 family, broadly dispersed around the *X. laevis* genome; 100 of these are Tx1c and the remainder are Tx1d.

(ii) Much of the Tx1d sequence consists of paired clusters of tandemly repeated sequences. Each of the repeating units, PTR-1 and PTR-2, is 393 ± 2 bp in length. The PTR-1 and PTR-2 consensus sequences are 63% identical to each other. The number of PTR repeats is variable among elements, the most common number being seven of each type.

(iii) The outside ends of the elements consist of inverted terminal repeats in which 16 of 19 bp at the two ends match (Table 2). A curious feature is that one position, 3 bp from the outside end, is variable, but the left and right ends of any given element are always mismatched at that position. That is, when C is found at position 3 of the left inverted repeat, the right inverted repeat has T at the corresponding site, and vice versa.

(iv) An apparent target duplication of 4 bp flanks each element; in all five cases examined the duplication is the same: 5′-TTAA-3′. Because this sequence is its own inverse complement, it could also be construed as part of the terminal inverted repeat. One additional T on the left side of all elements is conserved, suggesting a target site specificity for

TTTAA. Beyond that on both sides, the sequences flanking independently cloned elements bear no relation to one another.

(v) Between the termini and the internal repeat clusters are sequences called left and right common flanks (LCF and RCF, respectively). The LCF (1,400 bp) is made up largely of diverged copies of PTR-1 repeats; only the 400 to 500 bp nearest the left end is not obviously related to PTR-1. In contrast, the RCF is 998 bp long and has a sharp border with PTR-2 sequences.

(vi) The RCF contains several distinctive sequence features. Phased clusters of A · T base pairs in one region cause a DNA bend, which is manifested as low electrophoretic mobilities of fragments containing this segment. Such bends have been found at replication origins (46), in gene regulatory sequences (2), and in a number of other contexts (e.g., 26), but it is not possible to draw a general conclusion concerning their biological relevance. The RCF also contains long stretches of alternating purines and pyrimidines, largely based on $(GT)_n$. While these could potentially adopt the left-handed Z form (32), no conclusions regarding function are possible. A number of direct and inverted repeats in the terminal 85 bp of RCF can be drawn in alternative stem-loop structures, with stems of high predicted stabilities but of unknown significance.

(vii) The only structural difference between Tx1c and Tx1d is the absence in Tx1d of a 7-kb sequence that is bracketed by PTR-1 repeats in Tx1c. This sequence is present in the *X. laevis* genome only in Tx1c elements, never in isolation, so it is believed to be an integral part of the elements. It contains long, convincing open reading frames (ORFs) (discussed below).

(viii) Tx1 sequences are present in the other *Xenopus* species tested, *Xenopus borealis*, *Xenopus mulleri*, and *Xenopus tropicalis*, but not in other genera, including those phylogenetically closest to *Xenopus*, *Hymenochirus* and *Pipa* (8). The families in *X. borealis* and *X. mulleri* contain clusters of both PTR-1 and PTR-2 repeats, with copy numbers similar to those in *X. laevis*, but little additional structural analysis has been performed.

B. Evidence for Mobility

The above structural features are highly suggestive of a mobile lifestyle for Tx1. Discrete elements are widely distributed in the genome. Like many eucaryotic and procaryotic transposable elements, Tx1 has short, inverted terminal repeats, is flanked by apparent target duplications, and shows target specificity.

The most compelling evidence for transposition of Tx1, however, is the existence of a genomic site that is polymorphic among individual frog chromosomes for the presence or absence of a Tx1d element (15). Both the unoccupied and the occupied sites have been sequenced. They have precisely the relationship expected for pre- and postinsertion sequences: they are identical except for the presence of Tx1d and the duplication of TTAA, which is present in one copy at the unoccupied site. Other weak arguments support the notion that the unoccupied site is preinsertion rather than postexcision (15). In any case, the Tx1 element must have jumped into or out of this site at least once in the history of the species. Similar examination of other occupied sites among unrelated animals suggests that the rate of transposition is low (15).

C. Tx2 Elements

In the course of characterizing the Tx1 family, we discovered a related family of elements, which we call Tx2. We have examined only "complete" versions (i.e., ones carrying the central coding region) in the Tx2 family, and this characterization is incomplete. Our present knowledge is summarized in Fig. 2, by comparison with Tx1c.

The coding region of Tx2c is similar enough to cross-hybridize with probes from the corresponding area of Tx1c, but the classes have distinct restriction maps. There is evidence for internal, 400-bp tandem repeats in Tx2 elements, and we have sequences of a few examples from the right end of the coding region which match PTR-1 and PTR-2 at about the 70% level. The left end of Tx2c aligns well with the LCF of Tx1, and again the sequences are about 70% identical. The 19-bp Tx2 sequence corresponding to the left inverted terminal repeat of Tx1 differs from it at five positions. The right end of Tx2 has not yet been sequenced.

The LCF and PTR regions have no apparent coding capacity (15). Assuming they are under little or no constraint, their divergence allows us to calculate that the minimum time of evolution of the Tx1 and Tx2 families from a common ancestor is about 30 to 40 million years (44).

In the coding regions of both Tx1c and Tx2c, there is a long, methionine-initiated open reading frame (ORF2) of 1,304 amino acids on the right side of both sequences and a shorter one (ORF1) of about 780 amino acids on the left side (Fig. 2). Detailed comparisons between these ORFs in Tx1c and Tx2c indicate that they are under selective pressure. As shown in Fig. 2, the nucleotide sequences in both

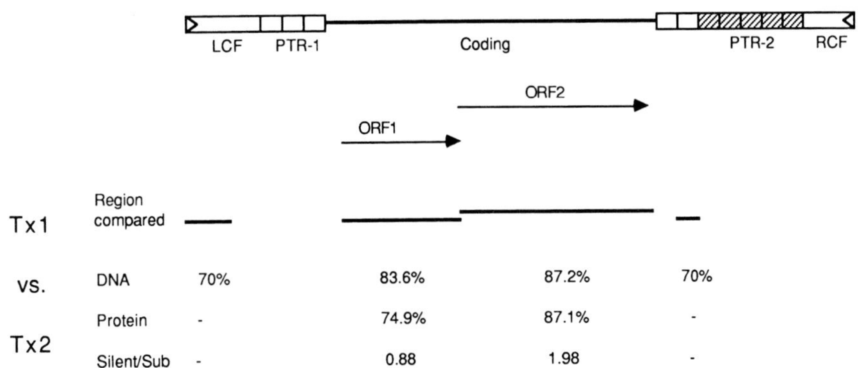

Figure 2. Schematic comparison of Tx1c and Tx2c sequences. The various regions of Tx1c are designated as in Fig. 1. These are compared with corresponding portions of Tx2c. The major open reading frames of both elements are designated by arrows. For sequenced regions of the two elements, the percentages of positions at which they match are given below the diagrams. The lowest line shows the ratio of nucleotide sequence changes that are silent with respect to predicted protein sequence to those that result in amino acid substitutions. (Source: J. E. Garrett, D. S. Knutzon, R. Sharma, and D. Carroll, unpublished results.)

ORFs are better conserved than in the noncoding regions. The predicted amino acid sequences are also well conserved, particularly in ORF2. When conservative amino acid substitutions are included, the similarity between Tx1 and Tx2 products rises to 84.1% in ORF1 and to 94.2% in ORF2. Furthermore, a large proportion of the nucleotide sequence differences are silent with respect to amino acid sequence, particularly in ORF2. Thus, we believe that these ORFs encode genuine products of the elements.

D. Comparison with Other Mobile Elements

The predicted amino acid sequences of ORF2 in Tx1c and Tx2c are not only closely related to each other, but show a remarkable degree of similarity to the long ORF product of mammalian LINEs (37). The comparison has been done most thoroughly with the mouse element L1Md (Fig. 3). The segments of greatest similarity include those that have been identified as regions of weak match between LINEs and viral reverse transcriptases (21, 27). In the 325-amino-acid portion compared in Fig. 3, the Tx1 and L1Md products are identical at 98 positions (30%); the similarity rises to 52% when conservative substitutions are included. In these same regions, Tx1c and Tx2c ORFs match the large predicted protein products of the *Drosophila* I factor (12), the *Drosophila* F element (9), and putative mobile elements from silkmoth (3, 45), trypanosomes (25), and maize (35), although not as well as they match the LINE ORF. The similarities among the LINE, I factor, F element, and Tx polypeptides are all greater than those between any of them and the genuine reverse transcriptases.

In addition, the predicted Tx ORF1 product shares with I factor ORF1 a copy of a sequence found in retroviral *gag* proteins and thought to be involved in nucleic acid binding (12). This segment conforms to the amino acid sequence motif $CX_2CX_4HX_4C$.

While the above matches to the Tx elements are encouraging, they also create confusion regarding the classification of the *Xenopus* elements. Eucaryotic mobile DNAs are usually classified by structural similarities and on the basis of predicted transposition mechanisms (14, 16). Elements having short, inverted terminal repeats, e.g., P elements of *Drosophila* (31), *Ac* elements of maize (10), and Tc1 elements of nematodes (11), are generally thought to move via a DNA intermediate. Excision is seen with these elements and may be connected mechanistically with transposition to new sites. Elements with long, direct terminal repeats, e.g., Ty elements of yeast (13), *copia*-like elements of *Drosophila* (14), and mammalian IAP elements (14), are retroviruslike in many respects and are supposed to move via an RNA intermediate. This has been demonstrated for Ty elements (1), and plausible particle-associated RNA intermediates have been identified in other cases (36).

It has been suggested that LINEs, I factors, and F elements represent a second class of RNA-based mobile elements (9, 12, 27). While they have neither short nor long terminal repeats, they possess some properties of RNA-derived processed pseudogenes (43) (variable apparent target duplications, A-rich 3′ ends, truncated copies) and have the above-mentioned weak matches to reverse transcriptases. The conundrum is that Tx1 and Tx2 elements have ends very like those of the DNA-based elements but coding sequence similarity with putative RNA-based

```
Tx2                 K     N       H               V            H       N  L  T  M
Tx1             ERLET PITLDELSQA LRLMPHNKSP GLDGLTIEFF QFFWDTLGPD FHRVLTEAFK KGELPLSCRR
                ..*.. **.  *.    .  .*  *** * ** . **. * * . * *   *..        * ** *
L1Md            DHLNS PISPKEIEAV INSLPTKKSP GPDGFSAEFY QTFKEDLIPI LHKLFHKIEV EGTLPNSFYE
                . ..   *.  *. .  ...*    .* * .  *                ". ***  *      .*...
I Fact          QTIEE NITYLELSSA LQTL-KGCAP GLNRISYGMI KNSSHTTKNR ITKLFNEIFN SH-IPQAYKT

Tx2                                                                V                     V
Tx1        AVLSLLPK-KG DLRLIKNWRP VSLLSTDYKI VAKAISLRLK SVLAEVIHPD QSYTVPGRTI FDNVFLIRDL
           * ..*.**    *   * *.** .**.  * ** . * .. *.    .  .**** *   .**    . *.    ..
L1Md       ATITLIPKPQK DPTKIENFRP ISLMNIDAKI LNKILANRIQ EHIKAIIHPD QVGFIPGMQG WFNIRKSINV
           . *  * **.  * **   .** ***    *** *.**.* *.    .     .  .  *     *         .*
I Fact     SLIIPILKPNT DKTKTSSYRP ISLNCCIAKI LDKIIAKRLW WLVTYNNLIN DKQFGFK-KG KSTSDCLLYV
                . *           .*  .     + .  .
RTase          IRKASGS        YRL L----HDLRA VNA

Tx2             KA                           L               S    L  T
Tx1        LHFARRTGLS LA-FLSLDQEK AFDRVDHQYL IGTLQAYSF-G PQFVGYLKTM YASAECLVKI NWSLTAPLAF
           .*.  .        .*** **  ***...* .. *  *.  .  * * ..  .*.. *. .   .*. *     ...
L1Md       IHYINKLKDK NHMIISLDAEK AFDKIQHPFM IKVLERSGIQG P-YLNMIKAI YSKPVANIKV NGEKLEAIPL
              * * * .    ...** .  ***..    . *  *.     * *  .  **    *     ..*       ..**
I Fact     DYLITKSKMH TS-LVTLDFSR AFDRVGVHSI IQQLQEWKT-G PKIIKYIKNF MSNRKITVRV GPHTSSPLPL
                          . **     *  .
RTase                    LMVLDLKD CFFSIPL

Tx2                        A                     R
Tx1        KRGVRQGCPL SGQLYSLAIE PFLCLLRKRL TGLVLKEPDM RVVLSAYADD VILVAQDLVD LERAQ-ECQEV
           * * ****** *. *. . .*      .*        .  .    * .* ***  .*.   *  .   *   .  .
L1Md       KSGTRQGCPL SPYLFNIVLE VLARAIRQQK EIKGIQIGKE EVKISLLADD MIVYISDPKN STRELLNLINS
             *  ** .. *  ** *  .  *.  *   * ***               ***   . *   ** .  .
I Fact     FNGIPQGSTI SVILFLIAFN KLSNIISLHK EIK------- ---FNAYADD FFLIINFNKN TNTN
                ** .  *  .  . .       .+                     + **   +
RTase        VLPQGMTC SPTICQLVVG QVLEPLRLK                MLHYMDD LLL

Tx2                    T         P H   S  S         T V
Tx1        YAAASSARIN WSKSSGLLEG SLKV-DFLPPA FRDISWESKI IKYLGVYLSA
           ..   . .**   ** . * .   *  .          *  .   ****** *.
L1Md       FGEVVGYKIN SNKSMAFLYT KNKQAEKEIRE TTPFSIVTNN IKYLGVTLTK
                *  .    *                          **   .* **.**
I Fact          GASLS LSKCQ                        VT-S LKILGITLNN
                .  +    *                               . +**  *
RTase           GFTIS PDKVQ                          PG VQYLGYKL
```

Figure 3. Comparison of predicted amino acid sequences in the long ORFs of Tx1, Tx2, mouse long interspersed repetitive element (L1Md), and *Drosophila* I factor (I Fact) with each other and with viral reverse transcriptase (RTase) (the sequence shown is from Rous sarcoma virus). The sequence of the Tx2 product is identical to that of Tx1 except where indicated. In other comparisons, an asterisk denotes identity between adjacent sequences and a dot denotes a conservative substitution, based on the groupings (K,R,H), (D,E,N,Q), (A,G,P,S,T), (L,I,V,M), (F,W,Y), (C). In the comparisons with RTase, asterisks and dots are awarded if all sequences match with these criteria; plus sign indicates that the Tx elements and one of L1Md or I factor are identical to RTase. (Sources: Tx1, Tx2, unpublished results; L1Md, reference 27; I factor and RTase, reference 12.)

elements. The issue cannot be resolved at present. If the LINE-like elements do move through an RNA intermediate, it will have to be determined how they regenerate or capture a promoter to permit subsequent transpositions. Alternatively, it may be that the similarity to reverse transcriptases, which is weak in any case, is misleading and the sequence conservation reflects properties of DNA-binding or DNA-synthesizing proteins more generally.

V. POSSIBLE UTILITY OF *XENOPUS* MOBILE ELEMENTS

Why would anyone choose to study transposable elements in relatively obscure organisms like *Xenopus*? Various considerations (1- to 2-year generation time, lack of convenient genetic markers, etc.) make it impractical to monitor the movement of naturally occurring elements in defined frog lineages.

Nonetheless, the frogs are well suited to three types of study involving mobile elements.

First, because early development is easily observed and conveniently manipulated with *Xenopus* (e.g., 18, 30), developmental patterns of expression, and possibly of mobilization, of various elements can be investigated. Second, the elements might be marshalled as transformation vectors, as has been done so successfully with P elements in *Drosophila* (33, 40). As developmentally regulated genes of *Xenopus* are being isolated (e.g., 7), this would provide a way to reintroduce modified versions of these genes into the genome for studies of regulatory sequences.

Finally, we are particularly excited about the prospect of using the oocyte injection procedure to elucidate details of transposition mechanisms. *Xenopus* oocytes have tremendous capacity to process injected molecules in biologically relevant fashions (19, 20). For example, DNA molecules injected in the oocyte nucleus can be assembled into normal-looking chromatin (19), transcribed from genuine promoters (20), and recombined through homologous sequences (5). We are optimistic that providing engineered components of a *Xenopus* transposable element to the milieu of an oocyte nucleus will allow jumps to occur on an experimentally exploitable time scale. The components could then be manipulated to dissect portions essential for mobility, specificity, etc. Given the amphimorphic nature of the Tx1 and Tx2 elements, such studies could be most illuminating.

Acknowledgments. We are grateful to B. K. Kay, W. Knochel, G. Spohr, and W. Wahli for responding to our request for updated information, in some cases with permission to quote unpublished results. The work on the Tx elements was initiated by Brenda Lam, and contributions to the analysis of their nucleotide sequences were made by Rajesh Sharma and Charles Hussey.

Work in our laboratory has been supported by a research grant from the National Institutes of Health.

LITERATURE CITED

1. **Boeke, J. D., D. J. Garfinkel, C. A. Styles, and G. R. Fink.** 1985. Ty elements transpose through an RNA intermediate. *Cell* **40**:491–500.
2. **Bossi, L., and D. M. Smith.** 1984. Conformational change in the DNA associated with an unusual promoter mutation in a tRNA operon of *Salmonella*. *Cell* **39**:643–652.
3. **Burke, W. D., C. C. Calalang, and T. H. Eickbush.** 1987. The site-specific ribosomal insertion element type II of *Bombyx mori* (R2Bm) contains the coding sequence for a reverse transcriptase-like enzyme. *Mol. Cell. Biol.* **7**:2221–2230.
4. **Carroll, D., J. E. Garrett, and B. S. Lam.** 1984. Isolated clusters of paired tandemly repeated sequences in the *Xenopus laevis* genome. *Mol. Cell. Biol.* **4**:254–259.
5. **Carroll, D., S. H. Wright, R. K. Wolff, E. Grzesiuk, and E. B. Maryon.** 1986. Efficient homologous recombination of linear DNA substrates after injection into *Xenopus laevis* oocytes. *Mol. Cell. Biol.* **6**:2053–2061.
6. **Davidson, E. H., B. R. Hough, C. S. Amenson, and R. J. Britten.** 1973. General interspersion of repetitive with nonrepetitive sequence elements in the DNA of *Xenopus*. *J. Mol. Biol.* **77**:1–23.
7. **Dawid, I. B., and T. D. Sargent.** 1986. Molecular embryology in amphibians: new approaches to old questions. *Trends Genet.* **2**:47–50.
8. **Deuchar, E. M.** 1975. *Xenopus: The South African Clawed Frog*. John Wiley & Sons, Inc., New York.
9. **Di Nocera, P. P., and G. Casari.** 1987. Related polypeptides are encoded by *Drosophila* F elements, I factors, and mammalian L1 sequences. *Proc. Natl. Acad. Sci. USA* **84**: 5843–5847.
10. **Doring, H.-P., and P. Starlinger.** 1984. Barbara McClintock's controlling elements: now at the DNA level. *Cell* **39**:253–259.
11. **Emmons, S. W., and L. Yesner.** 1984. High-frequency excision of transposable element Tc1 in the nematode *Caenorhabditis elegans* is limited to somatic cells. *Cell* **36**:599–605.
12. **Fawcett, D. H., C. K. Lister, E. Kellett, and D. J. Finnegan.** 1986. Transposable elements controlling I-R hybrid dysgenesis in *D. melanogaster* are similar to mammalian LINEs. *Cell* **47**:1007–1015.
13. **Fink, G. R., J. D. Boeke, and D. J. Garfinkel.** 1986. The mechanism and consequences of retrotransposition. *Trends Genet.* **2**:118–122.
14. **Finnegan, D. J.** 1985. Transposable elements in eukaryotes. *Int. Rev. Cytol.* **93**:281–326.
15. **Garrett, J. E., and D. Carroll.** 1986. Tx1: a transposable element from *Xenopus laevis* with some unusual properties. *Mol. Cell. Biol.* **6**:933–941.
16. **Georgiev, G. P.** 1984. Mobile genetic elements in animal cells and their biological significance. *Eur. J. Biochem.* **145**: 203–220.
17. **Gerber-Huber, S., D. Nardelli, J.-A. Haefliger, D. N. Cooper, F. Givel, J.-E. Germond, J. Engel, N. M. Green, and W. Wahli.** 1987. Precursor-product relationship between vitellogenin and the yolk proteins as derived from the complete sequence of a *Xenopus* vitellogenin gene. *Nucleic Acids Res.* **15**: 4737–4760.
18. **Gerhart, J. C.** 1980. Mechanisms regulating pattern formation in the amphibian egg and early embryo, p. 133–316. *In* R. F. Goldberger (ed.), *Biological Regulation and Development*, vol. 2. Plenum Publishing Corp., New York.
19. **Gurdon, J. B., and D. A. Melton.** 1981. Gene transfer in amphibian eggs and oocytes. *Annu. Rev. Genet.* **15**:189–218.
20. **Gurdon, J. B., and M. P. Wickens.** 1983. The use of *Xenopus* oocytes for the expression of cloned genes. *Methods Enzymol.* **101**:370–386.
21. **Hattori, M., S. Kuhara, O. Takenaka, and Y. Sakaki.** 1986. L1 family of repetitive DNA sequences in primates may be derived from a sequence encoding a reverse transcriptase-related protein. *Nature* (London) **321**:625–628.
22. **Hummel, S., W. Meyerhof, E. Korge, and W. Knochel.** 1984. Characterization of highly and moderately repetitive 500 bp EcoRI fragments from *Xenopus laevis* DNA. *Nucleic Acids Res.* **12**:4921–4938.
23. **Kay, B. K., and I. B. Dawid.** 1983. The 1723 element: a long, homogeneous, highly repeated DNA unit interspersed in the genome of *Xenopus laevis*. *J. Mol. Biol.* **170**:583–596.
24. **Kay, B. K., M. Jamrich, and I. B. Dawid.** 1984. Transcription

of a long, interspersed, highly repeated DNA element in *Xenopus laevis*. *Dev. Biol.* **105**:518–525.

25. **Kimmel, B. E., O. K. Ole-Moiyoi, and J. R. Young.** 1987. Ingi, a 5.2-kb dispersed sequence element from *Trypanosoma brucei* that carries half of a smaller mobile element at either end and has homology with mammalian LINEs. *Mol. Cell. Biol.* 7:1465–1475.

26. **Koo, H.-S., H.-M. Wu, and D. M. Crothers.** 1986. DNA bending at adenine-thymine tracts. *Nature* (London) **320:** 501–506.

27. **Loeb, D. D., R. W. Padgett, S. C. Hardies, W. R. Shehee, M. B. Comer, M. H. Edgell, and C. A. Hutchison III.** 1986. The sequence of a large L1Md element reveals a tandemly repeated 5′ end and several features found in retrotransposons. *Mol. Cell. Biol.* **6**:168–182.

28. **Martens, G. J. M., and E. Herbert.** 1984. Polymorphism and absence of leu-enkephalin sequences in proenkephalin genes in *Xenopus laevis*. *Nature* (London) **310**:251–254.

29. **Meyerhof, W., E. Korge, and W. Knochel.** 1987. Characterization of repetitive DNA transcripts isolated from a *Xenopus laevis* gastrula-stage cDNA clone bank. *Roux's Arch. Dev. Biol.* **196**:22–29.

30. **Newport, J., and M. Kirschner.** 1982. A major developmental transition in early *Xenopus* embryos. I. Characterization and timing of cellular changes at the midblastula stage. *Cell* **30**:675–686.

31. **O'Hare, K., and G. M. Rubin.** 1983. Structures of P transposable elements and their sites of insertion and excision in the *Drosophila melanogaster* genome. *Cell* **34**:25–35.

32. **Rich, A., A. Nordheim, and A. Wang.** 1984. The chemistry and biology of left-handed Z DNA. *Annu. Rev. Biochem.* **53**:791–846.

33. **Rubin, G. M., and A. Spradling.** 1982. Genetic transformation of *Drosophila* with transposable element vectors. *Science* **218**:348–353.

34. **Schubiger, J.-L., J.-E. Germond, B. ten Heggeler, and W. Wahli.** 1985. The Vi element. A transposon-like repeated DNA sequence interspersed in the vitellogenin locus of *Xenopus laevis*. *J. Mol. Biol.* **186**:491–503.

35. **Schwarz-Sommer, Z., L. Leclercq, E. Gobel, and H. Saedler.** 1987. Cin4, an insert altering the structure of the A1 gene in *Zea mays*, exhibits properties of nonviral retrotransposons. *EMBO J.* **6**:3873–3880.

36. **Shiba, T., and K. Saigo.** 1983. Retrovirus-like particles containing RNA homologous to the transposable element *copia* in *Drosophila melanogaster*. *Nature* (London) **302**:119–124.

37. **Singer, M. F., and J. Skowronski.** 1985. Making sense out of LINES: long interspersed repeat sequences in mammalian genomes. *Trends Biochem. Sci.* **10**:119–122.

38. **Spohr, G., W. Reith, and I. Sures.** 1981. Organization and sequence analysis of a cluster of repetitive DNA elements from *Xenopus laevis*. *J. Mol. Biol.* **151**:573–592.

39. **Spohr, G., C. Reymond, W. Reith, I. Sures, and M. Crippa.** 1983. Structural analysis of repetitive sequence elements transcribed in early development of *Xenopus laevis*. *Mol. Biol. Rep.* **9**:33–38.

40. **Spradling, A., and G. M. Rubin.** 1982. Transposition of cloned P elements into *Drosophila* germ line chromosomes. *Science* **218**:341–347.

41. **Streck, R. D., J. E. McGaffey, and S. K. Beckendorf.** 1986. The structure of hobo transposable elements and their insertion sites. *EMBO J.* **5**:3615–3623.

42. **Van het Schip, F. R. Strijker, J. Samallo, M. Gruber, and G. Ab.** 1986. Conserved sequence motifs upstream from the coordinately expressed vitellogenin and apoVLDLII genes of chicken. *Nucleic Acids Res.* **14**:8669–8680.

43. **Vanin, E. F.** 1985. Processed pseudogenes: characteristics and evolution. *Annu. Rev. Genet.* **19**:253–272.

44. **Wilson, A. C., H. Ochman, and E. M. Prager.** 1987. Molecular time scale for evolution. *Trends Genet.* **3**:241–247.

45. **Xiong, Y., and T. H. Eickbush.** 1988. The site-specific ribosomal DNA insertion element R1Bm belongs to a class of non-long-terminal-repeat retrotransposons. *Mol. Cell. Biol.* **8**:114–123.

46. **Zahn, K., and F. R. Blattner.** 1985. Sequence-induced DNA curvature at the bacteriophage λ origin of replication. *Nature* (London) **317**:451–453.

Chapter 25

TU Elements and Puppy Sequences

BARBARA HOFFMAN-LIEBERMANN, DAN LIEBERMANN, and STANLEY N. COHEN

I. INTRODUCTION

Since the original discovery of transposable controlling elements in maize four decades ago (117), many mobile genetic elements have been identified in procaryotes and eucaryotes, as detailed throughout this book. These elements move from one genetic locus to another without significant sequence homology; this involves an "illegitimate" or nonhomologous "recombination" mechanism that generally duplicates a short sequence of recipient DNA at the site of insertion. In procaryotes, depending on the element, transposition involves either replication of the element without its excision from the donor site or excision

Barbara Hoffman-Liebermann and Dan Liebermann ■ Department of Biochemistry and Biophysics, University of Pennsylvania School of Medicine, Philadelphia, Pennsylvania 19104. **Stanley N. Cohen** ■ Department of Genetics, Stanford University School of Medicine, Stanford, California 94305.

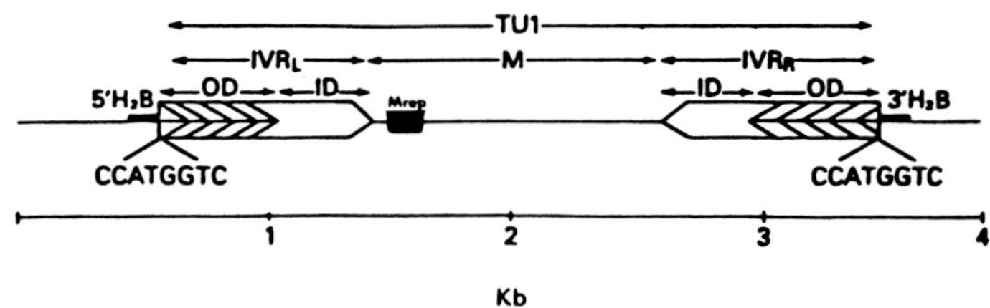

Figure 1. General structure of TU1. The large open arrows represent the terminal long IVRs with their OD and ID demarcated. The middle segment (M) and the Mrep lie between the IVRs. The internally repetitive nature of the OD is represented by the arrowheads. The thick solid line represents flanking H2B sequences, and the 8-bp direct repeats of "host" histone H2B-coding sequence flanking TU1 are indicated.

and then reinsertion of the element into a recipient site (6, 38, 99). In eucaryotes, RNA-mediated as well as DNA-mediated mechanisms have been implicated in transposition. RNA-mediated transposition, which involves transcription of DNA into RNA, reverse transcription back into cDNA, and then insertion of this cDNA into another site, has been shown for retroviruses (162) and retroposons (166) in eucaryotic species ranging from yeasts to humans. In contrast, DNA-mediated transposition, probably involving a DNA excision/reinsertion mechanism, has been shown for only a few transposable elements in plants (40), fruit flies (128), and nematodes (52).

This chapter summarizes the properties of TU elements, a heterogeneous family of modularly structured transposons discovered in sea urchins, which are echinoderms, close relatives of chordates, as evident from evolutionary (160) and early embryonic developmental studies (88). The structure and distribution of these elements suggest that they undergo DNA-mediated transposition and may also have a role in the generation of genetic diversity. Observations indicating that sequences homologous to TU element inverted repeats are present in many eucaryotic species, including humans, are reviewed. Among these are "Puppy" elements, whose structural and conformational characteristics suggest biological functions related to gene regulation, recombination, and replication, which also are functions associated with transposition. Finally, our notions concerning TU-Puppy sequence interrelationships and role(s) in the evolution of eucaryotic genomes are summarized.

II. DISCOVERY OF TU ELEMENTS IN THE SEA URCHIN *STRONGYLOCENTROTUS PURPURATUS* AND STRUCTURAL CHARACTERISTICS OF TU1

The transposable element TU1 was discovered during studies of sea urchin histone genes termed orphons (24, 110) because they are dispersed and not in the tandem arrays seen with most histone genes (81, 95, 96, 116). One variant clone, λjw110, contained an H2B histone-coding region that was split by the insertion of nonhistone DNA. As sea urchin histone genes do not contain introns, it seemed possible that the DNA segment interrupting the H2B sequences might be a transposable element.

An 8-base-pair (bp) stretch of the H2B-coding sequence was duplicated as direct repeats at each end of the insert (Fig. 1), further suggesting that the inserted sequence, termed TU1, is a transposable element. The sequence of this interrupted orphon H2B gene was found to start 7 bp downstream from the cap site of the authentic H2B gene. Moreover, a poly(A) stretch was adjacent to the 3′ terminus of the gene and a 6-bp sequence was repeated adjacent to the 5′ end of the gene and just beyond the poly(A) stretch. Thus, the sequence into which TU1 appears to have transposed may itself correspond to a histone retroposon pseudogene.

Electron microscope analysis of snapback structures containing both TU1 and the interrupted H2B gene revealed that TU1 has terminal inverted repeats (IVRs) 840 bp long and a middle sequence ~1.2 kb long (Fig. 1 and 3A). The IVRs of TU1 contain two distinct sequence domains: an outer domain (OD) of 512 bp, composed of a series of 15-bp direct repeats having the consensus sequence GACA$^{\mathrm{A}}_{\mathrm{T}}$TTGCTC CGCC, elongated in some instances by the addition of GAAGC; and an inner domain (ID) that consists of a different nonrepeating nucleotide sequence, portions of which are AT rich (Fig. 2). Another distinctive feature of TU1 is a 5-bp segment of nonhomology termed the "keyhole" just 3 bp from the ends of each IVR (Fig. 2c); this allows the arbitrary designation of left and right IVRs containing the keyhole sequences 5′-TCCAT-3′ and 5′-AACCC-3′, respectively. With this exception, the two IVR sequences are identical for at least the first 100 bp (110). In contrast, the more interior regions exhibit some sequence divergence, which becomes more prominent in the IVR-ID (Fig. 2d), where mismatches are clus-

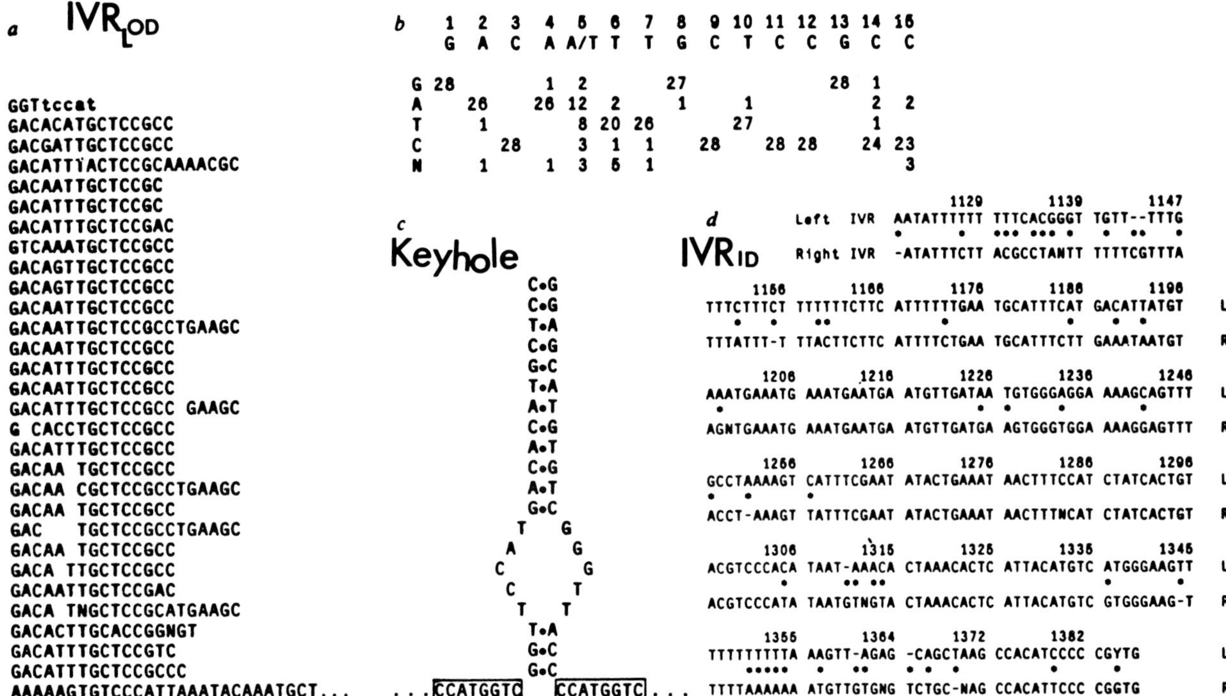

Figure 2. Nucleotide sequences of selected regions of TU1. (a) The sequence of the left inverted repeat outer domain (IVR_L-OD) is arranged to display its internally repetitious structure. The IVRs are identical for at least the first 99 bases, except for the 5-bp nonhomology at bases 4–8 (shown in lowercase letters). The five-base segment is used to identify the IVR_L (left) sequence, 5′-TCCAT, and IVR_R (right) sequence, 5′-AACCC. (b) The canonical sequence of the 15-bp tandem repeats that comprise the IVR-OD domain which was derived by scoring the base assignment at every position as shown. (c) The secondary structure formed by one strand of TU1 in a snapback configuration. The 8-bp repeat of the H2B-coding sequence is boxed and the five-base keyhole loopout structure is shown. (d) Analogous sequences of the left and right inverted repeat inner domains (IVR-ID) were aligned (110). The left IVR sequences represent those starting from position 1120 of the entire sequenced fragment (110). Any ambiguity in the sequence is represented by N.

tered at the ends (83). The segment termed Mrep, which is located at the boundary between the left IVR and the middle segment (Fig. 1), is also found unassociated with the TU1 IVR-OD or middle segments at many genomic locations (Fig. 4 below). Sequence analysis reveals that Mrep is ~320 bp long and contains a short polypurine/polypyrimidine DNA sequence element ("Puppy"; see below) $(TC)_{11}TT(TC)_5$; its termini consist of almost perfect 51-bp inverted repeats whose inner borders are flanked by 7- to 8-bp direct repeats (28). Further analysis of Mrep-related sequences in *S. purpuratus* may establish whether Mrep is a transposon that has integrated into TU.

III. TU ELEMENTS IN *S. PURPURATUS*: A HETEROGENEOUS FAMILY OF MODULARLY STRUCTURED TRANSPOSONS

Several cloned TU elements and the heterogeneous TU element population in *S. purpuratus* were studied using two approaches (83). The frequency of physical linkage of various TU domains in the genome was ascertained by hybridizing each of multiple plaque lifts from a λ phage library of sea urchin genomic DNA to a TU-derived domain-specific probe and determining the frequencies of phage plaques which hybridize to different combinations of probes. Additional information was obtained from Southern blots, using total genomic DNA. These studies indicated that there are about 400 TU elements per haploid *S. purpuratus* genome and that they fall into two major classes (83).

One class, comprising ~40% of TU elements, contains elements that are quite uniform in structure and closely resemble TU1. Two of four cloned TU1-like elements, TU3 and TU4 (Fig. 3), contain insertions of DNA segments that are not in TU1 (TU3IS and TU4IS; Fig. 3). Both of these insertions are themselves repetitive sequences, found at other loci which vary from one sea urchin individual to another. Most of these are not associated with the IVR-OD segment of TU elements. Thus, the additional sequences of TU3 and TU4 may themselves be

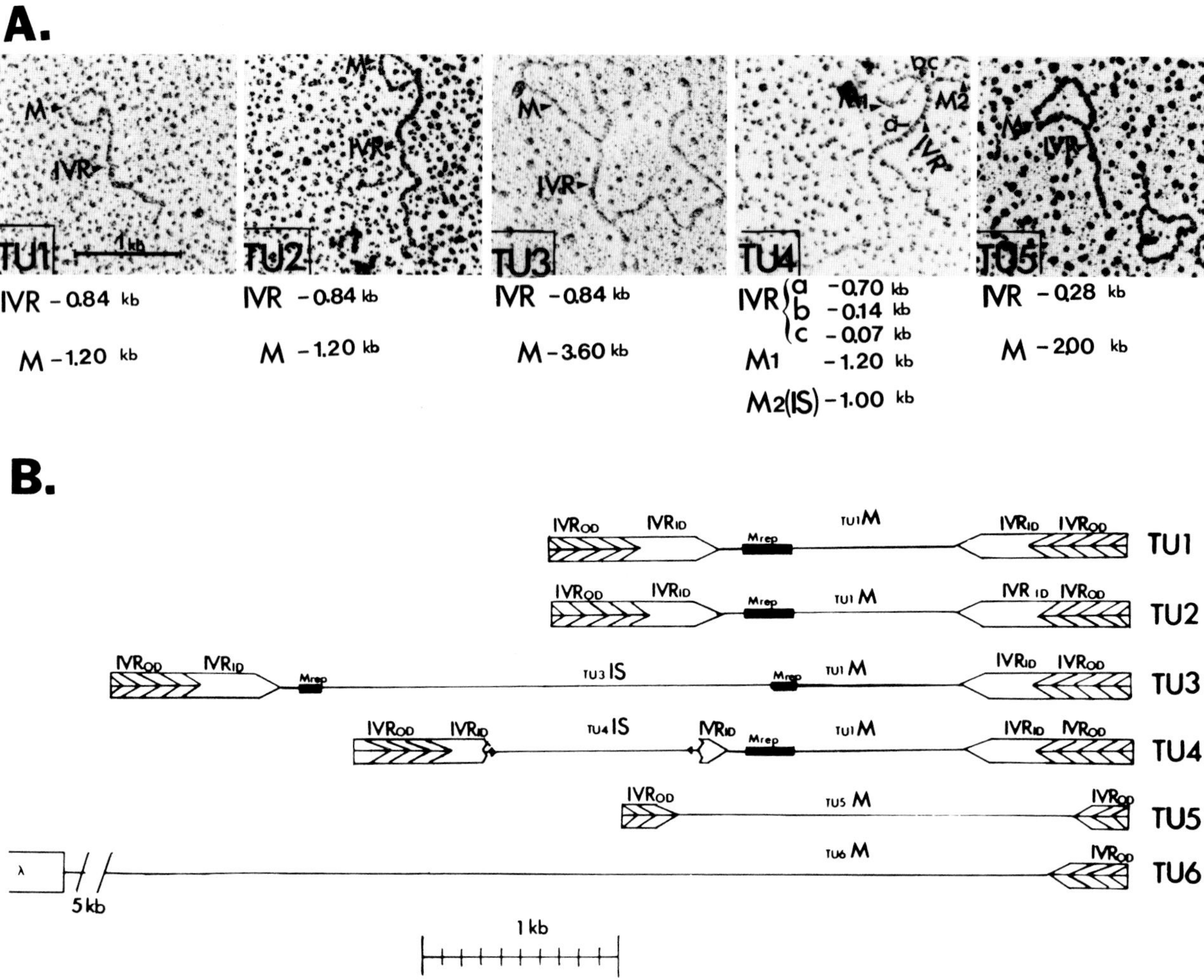

Figure 3. (A) Electron micrographs of TU element snapback structures. Lengths of the IVR and M segments of each element are indicated. (B) Schematic summary diagram of the structures of cloned TU elements (83). Note that four of the elements contain TU1-like middles. TU1 and TU2 have the same structure; however, although they are similar, they are not identical at the nucleotide level. TU3 and TU4 contain, in addition, insertion sequences designated TU3IS and TU4IS, respectively. Also note that TU5 and TU6 contain different nonhomologous middle segments and their IVRs are composed of IVR-ODs only.

members of yet other uncharacterized families of transposons. Further studies have corroborated these notions for TU4IS, which was found to be a member of a family of mobile elements designated Tsp transposons (28, 29). Since two of the six cloned members of the TU element family studied contain additional inserted sequences, TU elements may serve as hot spots for transposition of other transposable elements, as has been observed for DIRS-1 transposons in *Dictyostelium* sp. (21), *Ds* elements in maize (41), and certain transposons in procaryotes (100).

TU elements of the second class, comprising ~45% of TU elements, are very heterogeneous. They have IVRs that vary in size and share homology with only the IVR-OD segment of TU1. The cloned members of this class, TU5 and TU6, contain middle sequences having no homology to the TU1 middle region or to each other (Fig. 3 and 4). TU5 includes IVRs of 280 bp and a middle segment of ~2 kb. The TU6 clone consists of a 400-bp region with homology to the TU1 IVR-OD segment and a middle segment at least 9 kb long. Since only one IVR-OD has been identified in TU6, the possibility that this element consists of only a single IVR analogous to a bacterial IS element (86), single σ sequences of a yeast Ty1 element (140), or a solo retroviral LTR (162) cannot be excluded. However, the absence of obvious inverted repeats at both ends of the IVR-OD sequence of TU6 and of short direct repeats in the adjacent cellular DNA suggests that the entire TU6 element may consist of two IVRs and a middle segment longer than 9 kb. The middle segments of all cloned TU elements contain sequence elements that are members of families of moderately to highly

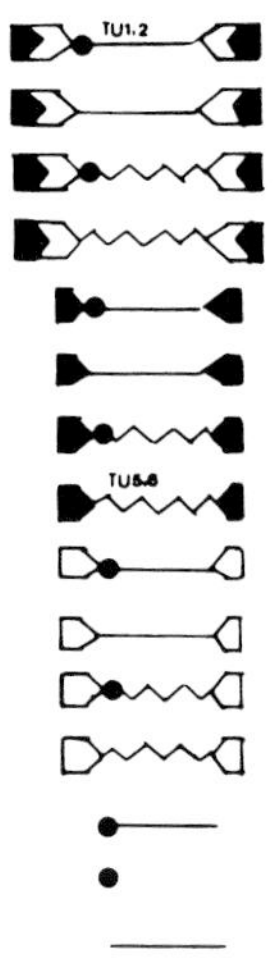

Linkage Group	Possible Structure	Number of Plaques: PU	Dr	Fr	Pi
1. IVR_{OD}—IVR_{ID}—M_{REP}-M		132	600	—	—
2. IVR_{OD}—IVR_{ID}———M		—	—	—	—
3. IVR_{OD}—IVR_{ID}—M_{REP} —		9	16	8	—
4. IVR_{OD}—IVR_{ID} ———		24	19	56	—
5. IVR_{OD}———M_{REP}-M		—	—	46	—
6. IVR_{OD}————M		—	—	—	—
7. IVR_{OD}———M_{REP} —		50	25	45	—
8. IVR_{OD} ————		140	280	460	300
9. IVR_{ID}—M_{REP}—M		6	8	—	—
10. IVR_{ID}———M		—	—	—	—
11. IVR_{ID}—M_{REP} —		—	—	—	—
12. IVR_{ID}————		21	16	12	—
13. M_{REP}-M		—	—	22	—
14. M_{REP} —		6500	9500	9000	100
15. ——M		—	—	—	

Figure 4. Linkage analysis for IVR-OD, IVR-ID, TU1 M, and TU1 Mrep sequences in different sea urchin species. The linkage analysis was carried out with plaque lifts from one genome equivalent (for *S. purpuratus*) or one half-genome equivalent (for other species) of the respective λ phage genomic libraries, using IVR-OD, IVR-ID, TU1 M, and TU1 Mrep hybridization probes (83). Indicated are the number of plaques per haploid genome in each linkage group. Schematic diagrams of possible element structures are shown. Further support for the putative structures shown was obtained from restriction enzyme analysis of cloned DNA (114). Straight lines, TU1 M; wavy lines, any M other than TU1. Also indicated are cloned elements in which the deduced structure is as depicted (Fig. 3B). PU, *S. purpuratus*; Dr, *S. drobachiensis*; Fr, *S. franciscanus*; Pi, *L. pictus*.

repetitive sequences found at many loci, and may be members of still other unknown families of mobile genetic elements that have transposed into TU elements.

In addition to containing two major classes of TU elements, the *S. purpuratus* genome includes IVR-ID sequences not associated with IVR-OD sequences; a subset of these sequences are linked to segments homologous to the middle region of TU1 (Fig. 4). These observations indicate the existence of still other types of TU-related elements that may have served as building blocks in the generation of the TU element family, as further discussed in Section V.

In conclusion, the studies reviewed thus far indicate that TU elements are modular structures composed of IVRs plus other sequence domains that are themselves members of dispersed repetitive sequence families, some of which may correspond to other, presently uncharacterized transposon families. Modularity analogous to that found for the TU elements is seen with certain transposable elements of procaryotes (for example, see reference 100), elements of the foldback family of transposons in *Drosophila* spp. (see Section V), and the *Ac/Ds* family of transposons in maize (40).

IV. UBIQUITY AND DIVERSITY OF TU ELEMENTS AMONG DIFFERENT SEA URCHIN SPECIES

The genomes of sea urchins contain multiple families of dispersed repetitive sequence elements that differ even among closely related species (2, 15, 16, 36, 91, 139). The structural heterogeneity of the TU family in *S. purpuratus* and the modularity of TU elements prompted a search for TU-related sequences in the DNAs of other sea urchin species (109).

Homology to TU element sequences was tested in *Strongylocentrotus drobachiensis*, a species closely related to *S. purpuratus* and believed to have diverged from a common ancestor ~7 million years ago; *Strongylocentrotus franciscanus*, a less-related species thought to have diverged from a common ancestor ~25 million years ago; and the distantly related sea urchin species *Lytechinus pictus*, of the order Temnopleuroida rather than Echinoida (family Strongylocentrotidae), to which the other three species belong. Divergence from an ancestor common to both orders probably occurred ~200 million years ago (44). TU domain-specific probes derived from characterized TU elements of *S. purpuratus* were used for interspecies hybridization in Southern transfer and linkage analysis experiments.

Sequences having homology to TU segments abundant in the *S. purpuratus* genome, including IVR-ID, the middle segments of TU1, TU5, and TU6, and TU4IS, are also abundant in *S. drobachiensis* and *S. franciscanus*, but are absent from the genome of distantly related *L. pictus*. TU3IS, scarce even in *S. purpuratus*, was found only in *S. drobachiensis*. In contrast, multiple sequences highly homologous to the IVR-OD segment are present in the genomes of each of these four species. This evolutionary conservation was not seen for other TU element domains or for other sea urchin repetitive sequences, including a member of the csp2108 sequence family that is present in a variety of other sea urchin species (144).

Analysis of λ phage recombinant genomic clones containing IVR-OD hybridizing sequences of the four sea urchin species indicated that two segments homologous to the IVR-OD regions are present within a distance of less than 15 kb. Moreover, linkage analysis indicated that IVR-OD regions are in close physical proximity to homologs of other TU element domains in species closely related to *S. purpuratus* (Fig. 4). Thus, the IVR-OD homologs in other sea urchin species are also organized into TU-element-like structures, i.e., two IVR-OD segments bracketing a middle region. Data obtained from linkage analysis experiments indicated that the frequency of TU elements having a particular middle segment varies

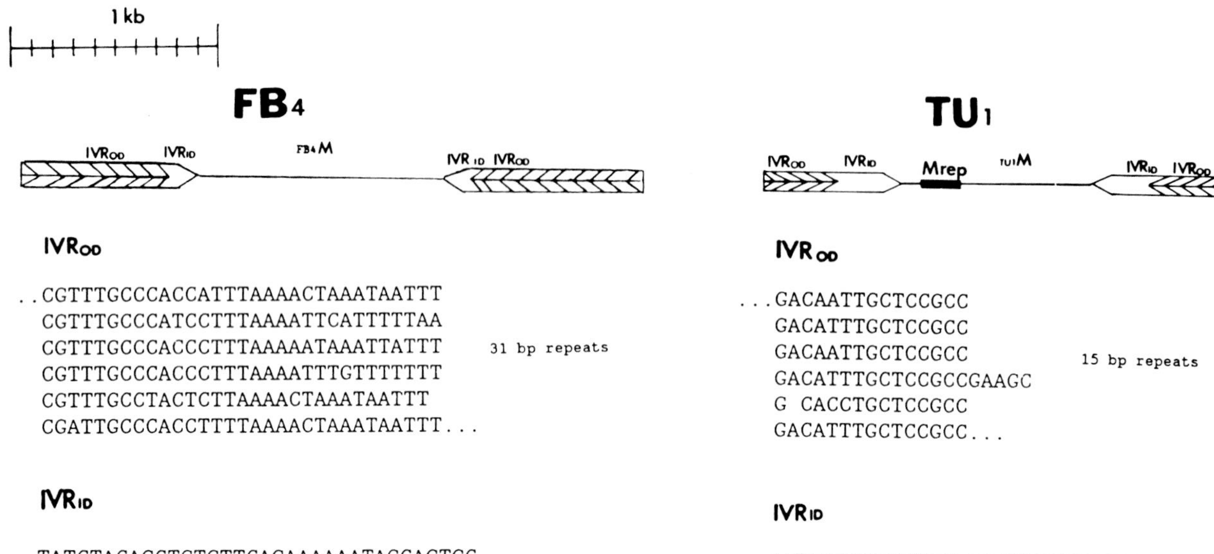

Figure 5. *Drosophila* FB transposons versus sea urchin TU elements. Displayed are diagrams comparing FB4 (131) to TU1 (110) (the IVR-ID is not drawn to scale; see below), as well as selected sequences derived from their IVR-OD and IVR-ID domains. The 31-bp tandem repeats of FB4 make up the inner regions of the FB4 IVR-OD, and shorter versions of this repeat are found at the IVR-OD termini. The 15-bp tandem repeats of TU1, elongated somewhat on occasion, comprise the entire TU1 IVR-OD, except for the 5-bp keyhole loopout structure at the IVR termini (Fig. 2). The FB4 IVR-ID is only 33 bp in length. The TU1 IVR-ID is much longer, ~330 bp in length. Only the terminal 33 bp, located at the TU1 IVR-ID/IVR-OD junction, is shown. Few FB elements harbor the FB4 middle (M) segment (14), whereas elements with the TU1 M segment are abundant in sea urchin genomes.

substantially among different sea urchin species (109).

Additional linkage analysis experiments (109) (Fig. 4) indicated that TU1-like elements, composed of IVRs with ODs and IDs flanking a TU1 middle segment that contains an Mrep domain, are abundant only in *S. purpuratus* and its close relative *S. drobachiensis*. Surprisingly, in *S. franciscanus*, TU elements with a TU1-like middle segment including the Mrep contain IVRs having homology to OD but not ID sequences. Moreover, *S. franciscanus* contains additional TU1-middle homologs linked to Mrep, but not to IVR-OD or IVR-ID homologs (Fig. 4, linkage group 13). The TU1-middle homolog differs in size from those observed in *S. purpuratus* and *S. drobachiensis*, whether or not linked to IVR-OD homologs. In the genomes of these three species of the Strongylocentrotidae family, TU IVR-ID sequences not linked to other known TU element domains were observed as well. This raises the possibility of yet another uncharacterized family of TU-related transposons with IVR-ID segments as their termini.

In conclusion, different types of TU elements predominate in different echinoderm species, supporting the notion that the TU family is heterogeneous, consisting of modularly structured transposons whose component parts reflect the genetic diversity seen even among closely related sea urchin species (3, 15, 16, 36, 42, 43, 155). These mobile genetic elements are widespread, present in species that probably diverged ~200 million years ago. The outer domains of their inverted repeat sequences have been remarkably conserved.

V. TU-LIKE TRANSPOSONS MAY BE UBIQUITOUS AMONG EUCARYOTES

The TU elements have features in common with the foldback (FB) elements of *Drosophila* sp. (130, 131, 158). Both have long terminal inverted repeats composed of short tandem repeats that share some sequence homology (Fig. 5 and 7). Particularly striking are the structural similarities between FB4 and TU1; both elements include IVRs that have an OD and a nonrepeating ID flanking a middle segment (Fig. 5). However, most TU elements contain middle segments (simple foldbacks lacking a middle segment are rare or nonexistent; Fig. 3 [82, 108]), whereas most FB elements of *Drosophila* sp. do not have middle segments (13). In one study (130, 131, 158), only 1 of 10 FB elements (i.e., FB4) contained a middle segment. While this segment is only rarely (1 of 23) associated with FB IVR sequences, it is common in the HB family of transposable elements,

A.

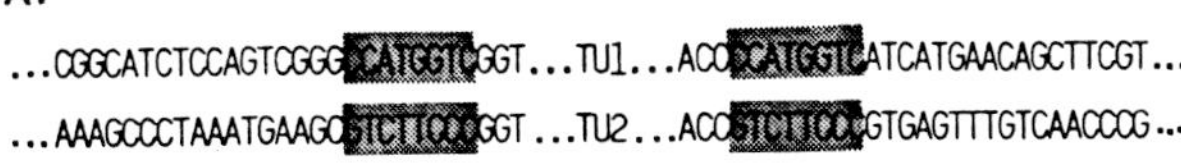

B.

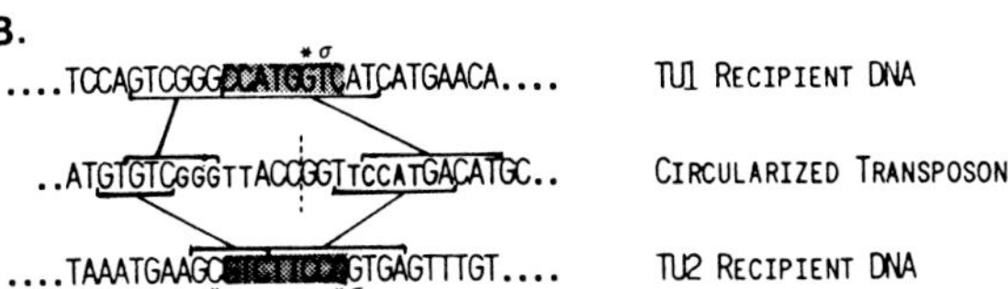

Figure 6. (A) The 8-bp direct repeats at the TU element junction with non-TU DNA (83, 110). The 8-bp duplication of the sequence flanking each TU element is shaded. (B) Homologies between the IVR termini and sequences at and near the insertion site. The 5-bp nonhomology forming the keyhole loop at the IVR termini is printed in smaller type. The 8-bp sequence insertion site duplication is shaded. The regions of homology are indicated in the figure: *, mismatch; σ, loopout. TU1 recipient DNA is the sequence of genomic DNA (histone H2B gene) at the site of TU1 insertion. Circularized transposon is the sequence of the termini of TU1 and TU2 when circularized (conceptually). TU2 recipient DNA is the sequence of genomic DNA at the site of TU2 insertion.

which have the short IVR-ID sequences of FB4 as their termini. Thus, FB4 could be a composite transposon formed by insertion of an HB element within an FB element. This is reminiscent of the TU1 middle and IVR-ID sequences that are not linked to TU IVR-OD sequences (Fig. 4). Another *Drosophila* foldback element with a middle segment, FBwc (30, 31, 107, 108), is always found to be flanked by FB IVRs (107, 129). The recently described Tel1/TBE transposons in ciliated protozoa (23, 82) contain IVR-ODs composed of C_4A_4 telomeric repeats and may represent yet another family in this apparently ubiquitous class of transposable elements.

VI. EVIDENCE FOR MOBILITY OF TU ELEMENTS AND SPECULATIONS AS TO THEIR MODE OF TRANSPOSITION

The findings that TU1 inserted within an H2B gene is flanked by 8-bp direct repeats of H2B-coding sequences and that TU2 is flanked by another unique 8-bp direct repeat (Fig. 6) are consistent with the idea that TU elements have transposed, but do not indicate that these elements are currently active transposons. Several lines of evidence suggest that this may be the case. First, no H2B genes in a second individual of the same species contain TU1 (110), indicating variation in the location of TU1. Second, a highly polymorphic banding pattern of restriction fragments hybridizing to IVR-OD or TU1-middle probes was observed with Southern blots of *S. purpuratus* genomic DNA from five different individuals collected at the same oceanographic location. This latter observation is consistent with variability in TU element integration sites among different *S. purpuratus* individuals, although the possibility that part of this polymorphism is due to restriction site polymorphism or size variation among TU1-like elements cannot be excluded. The development of inbred sea urchin strains capable of propagation under laboratory conditions (E. Davidson, personal communication) should permit conclusive tests of the idea that TU elements in sea urchins undergo transposition.

The structural similarity of TU elements to certain procaryotic transposons (100, 101) and to the *Drosophila* FB elements suggests that TU element transposition is DNA mediated without an RNA intermediate (38, 70). By extension of this analogy, TU elements may transpose via an excision/reinsertion transposition mechanism, as probably is the case for the *Drosophila* FBs (30–32, 94, 107, 108). The high frequency of precise excisions of eucaryotic transposons such as the *Drosophila* FBs (31) and P elements (128) or the Tc transposons in nematodes (52) suggests the involvement of a transposase to catalyze this event. This frequency contrasts with the rare precise excision seen for procaryotic transposons such as Tn*10* and Tn*5*; the latter occurs independently of transposase functions and seems to involve the terminal inverted repeats in a mechanism similar to spontaneous deletion formation (1, 48, 55).

The IVR-ODs of TU elements, although consisting of tandemly repeated sequences, contain short open reading frames of up to 60 codons that potentially might encode a transposase. In procaryotes, IS*10* and IS*50*, which comprise the inverted termini of Tn*10* and Tn*5*, respectively, encode transposases (70, 99). In this regard, we note that developmentally regulated RNA transcripts with homology to the TU IVR-OD are made at early stages of sea urchin embryogenesis (unpublished data). Alternatively, the TU1 middle, which also has open reading frames (unpublished results), may encode for a transposase, as does the internal segment of Tn*3* (25, 64, 65, 76), the *Drosophila* P element (103, 128, 138), and possibly the *Ac* elements in maize (40, 56). Consistent with this view, developmentally regulated transcripts containing sequences homologous to the TU1 middle segment are also made (unpublished data). It has also been suggested that cellular genes can encode transposition functions (14, 31, 129).

The terminal sequences of IVRs of different TU elements are highly conserved within one element as well as between elements. Analogous conservation of terminal sequences has been observed for many other

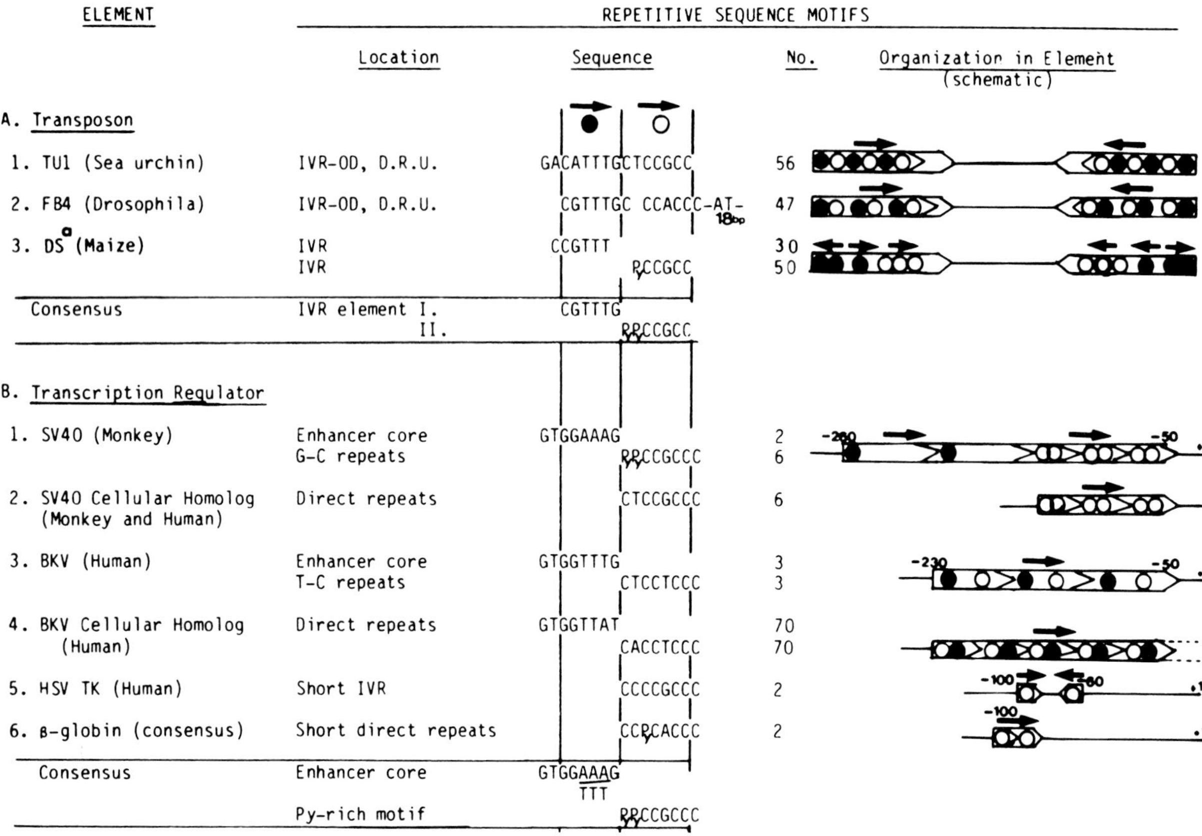

Figure 7. Repetitive sequence motifs common to transposon IVRs and transcription regulators. Arrows in the schematic diagram indicate the 5′-to-3′ orientation of sequence motifs. For transposons, the number of repetitive motifs includes both IVRs. In the case of the FB, simian virus 40, BK virus (BKV), and BKV cellular homologs, one or both of the common sequence motifs are contained within larger repeating units (31, 72, 68, and 21 bp in length, respectively). References for elements: FB (131), DS (41), simian virus 40 (5, 47, 54, 60, 63, 74, 75, 97, 122, 165), simian virus 40 cellular homolog (33, 46, 118, 135, 143), BKV (146), BKV cellular homolog (141), herpes simplex virus thymidine kinase (47, 119–121), and β-globin (39). D.R.U., Direct repeating unit. Note that the DS segment shown ([a]) is part of a double DS that originates from an inverted insertion of one DS into another one (41).

transposons (147), including the IVRs of *Drosophila* foldback elements (130, 158), suggesting that such terminal repeats are important for transposition. The short homologies between the IVR-OD sequences of TU1 and TU2 termini and sequences within the insertion site region (Fig. 6) are consistent with the notion that TU element termini might serve as a substrate for a putative transposase. These regions of homology include the 5-bp segment forming the keyhole loop. Analogous homologies between sequences near the transposon ends and sequences at the insertion site have been found for some procaryotic transposons (59, 62, 159).

VII. SEQUENCES HOMOLOGOUS TO THE TU IVR-OD IN DIVERSE EUCARYOTIC SPECIES INCLUDING HUMANS

Sequences homologous to the IVR-OD were found in DNA from *Saccharomyces cerevisiae* and a variety of multicellular eucaryotes, including maize, fruit fly, *Aplesia*, frog (*Xenopus* sp.), chicken, mouse, monkey, and human, but not in procaryotic DNA (109). A polymorphic pattern of hybridization of the IVR-OD probe was observed with DNA obtained from six unrelated humans (109).

The IVR-OD segments of TU elements and of FB IVRs contain direct repeat motifs that resemble each other and also resemble motifs in the IVRs of the double *Ds* element of maize (41) (Fig. 7). In each case, multiple copies are tandemly repeated within the IVR segment of the transposon; moreover, two consensus sequences, one of which is highly pyrimidine rich in one strand and purine rich in the other, have been identified (Fig. 7). Further analysis indicated that these TU1, FB4, and *Ds* consensus sequences are similar to sequences found in repeat units in certain cellular and viral transcription regulatory regions (Fig. 7). The broad phylogenetic distribution of sequences homologous to the TU IVR-OD se-

A.

```
Hut 2

TTTTTTTCTTCTTCTCC
        TTCTCCTCCTCCTCC
        TTCTCCTCCTCC
        TTCTCCTCCTCCTCC
        TTCTCCTCCTCCTCC
        TTCTCC
        TTCTCCTCCTCC
        TCTTCCTCCTCC
        TTCTCCTCCTCCTCC
        TTCTCC
        TTCTCCTCCTCC
        TCTTCCTCCTCC
        TTCTCCTCCTCCTCCTCCTCCTCCTCCTCCTCCTCC
        TGCTCCTCCTCCTCC
        TGCTCCTCCTCCTCC
        TGCTCCTCCTCCTCC
        TGCTCCTCCTCCTCC
        TGCTCCTTC
        TGCTCC
        TGCTCCTCC
        TGCTCC
        TTCTCCTCCTTCTCCTTCTCCTTCTCCTTCGCTTTTTT

          TTC(TCC)n
          TGC(TCC)n
```

```
Hut 17

GGAGCATCTCC
     ATCTCCTCC
     ATCTCCTCCTCCTCCTCC
     ATCTCCTCCTCCTCCTCC
     ATCTCCTCCTCC
     ATCTCCTCCTCCTCCTCCTCCTCC
     ATCTCCTCCTCCTCC
     ATCTCCTCCTCCTCCTCCTCCTCCTCCTCC
     ATCTCCTCCTCCTCCTCCTCC
     ATCTCCTCCTCCTCCTCC
     ATCTCCTCCTCCTCCTCCTCC
     ATCTCCTCCTCCTCCTGCTCCCCCCCCC

       ATC(TCC)n
```

B.

```
Hut2       ...TCCTCCTGCTCCTCC...
               *   ****** **
IVROD      ...GACATTTGCTCCGCC...
               *** * **** **
Hut17      ...TCCATCTCCTCCTCC...
```

Figure 8. (A) Nucleotide sequences of Hut2 and Hut17. The sequences are arranged to display their internally repetitious structures. The six Ts flanking the Hut2 element and the inverted repeats flanking Hut17 are underlined with arrows. The predominant variant triplets in each sequence are indicated below the sequence. (B) Homologies between tandem repeats of Hut2 and Hut17 and the tandem repeat of the TU IVR-OD.

quence and the similarity of the IVR-OD repeating motif to transcriptional regulatory regions suggested that these conserved sequences are possibly important in transposition, recombination, replication, and/or the regulation of transcription. In several instances, more than one of these processes have been shown to involve common sequence motifs recognized by common protein factors (9, 14, 92, 163).

VIII. PUPPY ELEMENTS: A CLASS OF SEQUENCES IN HUMANS AND OTHER EUCARYOTES THAT RESEMBLE TU INVERTED REPEAT SEQUENCES

A. Puppys with Homology to TU Inverted Repeat Sequences

Screening of a human genomic library with a TU IVR-OD probe revealed several classes of sequences that have homology to this structural domain of the TU transposons. Several cloned fragments containing one abundant IVR-OD-related class were studied further (84). Nucleotide sequence analysis of the IVR-OD-hybridizing regions of two clones, Hut2 and Hut17, revealed purine/pyrimidine asymmetry (pPu/pPy or "Puppy" sequences) on the two DNA strands (Fig. 8), as in the TU IVR-OD region. In both cases, the pyrimidine-rich strand consists of the trimer TCC tandemly repeated many times. Different variations in this trimer repeat distinguish the Puppy sequences of Hut2 from Hut17. The Hut17 sequence is 221 bp long and is flanked by 6-bp inverted repeats, one of which includes two of the trimers that make up the Puppy sequence; adjacent to the 6-bp right IVR of Hut17 is a stretch of seven Cs in the pyrimidine-rich strand. Hut2 is 313 bp long and is flanked on both sides by six Ts in the pyrimidine-rich strand. For both λ clones, Hut2 and Hut17 are the only elements with homology to the TU IVR-OD.

B. A Variety of Puppys Exist in the Human Genome as Well as in Other Eucaryotic Genomes

The extent of variability of Puppy sequences in the human genome was examined by hybridization of a Hut2 probe to multiple plaque lifts from a human genomic library at stringencies that detect 70, 80, or 90% sequence homology. To detect other human Puppy sequences and to further distinguish among those detected with the Hut2 probe, a probe from a repeated chicken DNA sequence with the motif TC TCC tandemly repeated 32 times (called 21H [45]) was also used in hybridizations to the human genomic library. Both experiments yielded an estimate of at least 46,000 Puppy elements per haploid human genome (84).

These studies (84) showed that human Puppy sequence elements can be grouped according to the extent of their homology with specific DNA probes. Those that hybridize at high stringency (90% homology) with Hut2 make up one family and those that hybridize with the 21H probe under the same conditions comprise another family. Assuming that the sequences hybridizing with one or the other of these probes only at lower stringency belong to still other families, we infer that there is a superfamily of Puppy elements that includes at least the various families identified by hybridization at different stringencies with Puppy-containing probes. Given the previously identified stretches of purine/pyrimidine asymmetry (Table 1), and assuming that most of these exist and are repetitive in the human genome, a conservative extrapolation is that at least 100,000 Puppy elements are present per haploid human genome. Interestingly, 9 of the 10 human Puppy sequence elements cloned and analyzed by us were found to be in close physical proximity to *Alu* family elements (SINEs [short interspersed repetitive elements]; unpublished data), as has been found also for some members of the LINE (long interspersed repetitive elements) family of repeats (124).

Additional studies revealed Puppy families in other eucaryotic species, including yeasts and maize, with some species-specific differences in their distribution (84). The species conservation of the human Hut2 and chicken 21H Puppy families, together with observations indicating that other segments of purine/pyrimidine asymmetry occur in DNA from many higher organisms (7) (Table 1), support the notion that the superfamily of Puppy sequence elements exists in all eucaryotes.

Analysis of a variety of DNA sequences composed of tandemly repeated units (<100 bp; Table 1) expands the concept of "Puppyness" even further. Two distinguishable classes of Puppy elements seem to exist. One class includes Puppys whose tandem direct repeating units display extreme purine/pyrimidine asymmetry between individual strands of duplex DNA (i.e., Table 1, sequences 6 and 22). The second class includes Puppys whose tandem repeating units consist of alternating domains of purine or pyrimidine asymmetry on each strand of the duplex DNA, such as Puppy sequences 1 and 12 in Table 1, the minisatellite G 33.1 (87) and the TU IVR-OD (Fig. 2a), respectively. This expansion suggests that Puppys comprise the majority of "simple sequence DNA" (most notably including TG repeats [23]) that punctuate the genomes of most or all eucaryotic species.

C. Puppys Can Alter DNA Conformation In Vitro as Well as In Vivo

Short tracts of purine/pyrimidine asymmetry (10 to 42 bp long) found near the 5′ ends of certain genes are S1 sensitive in protein-free supercoiled DNA (53, 57, 85, 112, 127, 142, 145). For the α- and β-globin genes, S1-sensitive sites were found 5′ to the coding sequence following treatment of isolated nuclei with S1 nuclease. This in vivo S1 sensitivity was shown to be developmentally regulated (102), correlating with globin gene expression. S1-sensitive sites have also been identified in a tract of GA sequences in cloned DNAs of histone gene spacer regions (68, 80). Such S1 sensitivity probably reflects an altered DNA conformation, perhaps left-handed helices or triplexes (105), but not single-stranded DNA (20, 133). Proteins may be involved in generating, stabilizing, or recognizing some of these S1-sensitive regions in vivo (49, 50).

Both the Hut2 and Hut17 Puppy elements were S1 sensitive in vitro on supercoiled plasmids (84). Hut2 sequences can form S1 nuclease-sensitive structures in vivo in stable transformants of mouse Ltk$^-$ cells transfected with a plasmid containing the Puppy Hut2 element inserted upstream from the herpes simplex virus thymidine kinase gene promoter. Endogenous Hut2 Puppy sequences in HeLa cells are also S1 sensitive, indicating that the endogenous Hut2 Puppy sequence can assume an altered DNA conformation in its original chromosomal location. In the case of the stably transfected L cells, not all of the integrated exogenously derived Puppy tracts were S1 sensitive in vivo. Thus, while Puppy nucleotide sequences impart the potential for altering DNA conformation, other factors may be involved in determining whether this conformation actually is assumed by a particular Puppy sequence. Hut2 Puppy tracts showing S1 sensitivity in vitro or in vivo are cleaved at multiple sites that span the Puppy tract, indicating that the conformation of the entire 300-bp region is altered (84). That other Puppy sequence elements also adopt altered conformations in situ is

Table 1. pPu/pPy (Puppy) sequence elements in various animals[a]

Sequence	Animal (reference[s])
1. (TCCTGTGGCCCCAGATGTGGGTGCCCCTPyCACGCGGGGGTTCTTTCTGCCCTCCCCACCCTGCA)$_{14}$	Human (87)
2. (AAAGGGTGGGCAGGAAGTGGAG(TG)$_3$CCTGCTTCCCTTCCCTGTCTTGTCCTGGAAACTC)$_{26}$	Human (87)
3. (GGGGACAGTCCTGCCTGCCCACTCCAGCCACCTCCTACTTC)$_5$	Human (87)
4. (GCCCTTCCTCCGGAGCCCTCCTCCAGCCCTTCCTCCA)$_{18}$	Human (87)
5. (CCCACCACCAGCCCAGGGCCTCACTGCTTCCTG)$_3$	Human (87)
6. (CCCPyPyCTCCACCCPyTTCCTGCCCACCTCCCGG)$_6$	Human (87)
7. (AGCTGCCCCTCCTGCTGCTCCAGGTGCTCC)$_{36}$	Human (78)
8. (TGCTCCCGCATCTTCTCCTCC)$_{46}$	Human (78)
9. (TAGCTTGCCCCTGCTCCTTC)$_{15}$	Rat (73)
10. (ACCCGCGTCCCCTCCTGC)$_{55}$	Human (78)
11. (CCTCCCCCTCCTGCCCPu)$_{14}$	Human (87)
12. (GACAATTGCTCCGCC)$_{17–28}$	Sea urchin (110)
13. (TCCACCTGCCCACCTC)$_{29}$	Human (87)
14. (CCCCTCCCCACTGT)$_{12–39}$	Human (132)
15. (CCCCACACCCCTGT)$_{26–209}$	Human (4)
16. (AGCTCTCACCTCCC)$_9$	Mouse (123)
17. (PyCCTTTCPyCTTTT)$_5$	Human (69)
18. (TCTCCC)$_{12}$	Mouse (66, 137)
19. (CCCTGC)$_{17}$	Human (78)
20. (TCTCC)$_{32–36}$	Mouse (5, 37), chicken (45, 115)
21. (CTTCC)$_{27}$	Chicken (115)
22. (TTTCC)$_{55}$	Chicken (115)
23. (TCCC)$_8$	Chicken (57)
24. (TTTC)$_{11–43}$	Human (125), mouse (66)
25. (CCCG)$_{35–52}$	Human (132)
26. (TCC)$_{10–416}$	Human, mouse (27, 78), rat (133)
27. (TC)$_{27–150}$	Mouse (5, 67, 68, 137), sea urchin (153)
28. pPu/pPy without apparent direct repeating unit	Human (26, 113, 114), chicken (77), fruit fly (151)

[a] In all cases the sequence of the pyrimidine-rich strand is shown. Purine domains, when present, are underlined. The subscript next to the sequences indicates the number of direct repeating units in the region of purine/pyrimidine asymmetry. All the direct repeating units of Puppy 8, contained within 100 bp of the human beta-interferon promoter, are not contiguous.

suggested by the finding that monoclonal antibodies against triplex DNA (a non-B, S1-sensitive conformation believed to be adopted by Puppys [105]) bind to eucaryotic chromosomes (104).

D. Possible Biological Functions for Puppy Sequences in Eucaryotic Genomes

Several independent observations suggest that Puppy elements may carry out specific biological functions. (i) A superfamily of Puppy elements is probably present in DNAs of all eucaryotes as a major component of simple-sequence DNA implicated in gene regulation (72, 81, 164, 167), gene conversion and recombination (18, 19, 80, 87, 95, 149, 157), and replication (8). (ii) Sequences homologous to the Epstein-Barr virus IR3 region, which consists of stretches of purine/pyrimidine asymmetry (78), are found on every human chromosome except Y (79), which contains few functional genes (51, 101). (iii) At least some Puppy elements assume non-B DNA structural conformations. Various biological roles have been suggested for non-B DNA, including regulation of gene expression, recombination, and replication (111, 136). The sequence of a particular Puppy tract, its length, and its chromosomal location may all be relevant to its putative biological role. The structural complexity of the Puppy superfamily may reflect functional complexity, consistent with a role(s) in the complex temporal and tissue-specific regulation of gene expression during cell development.

Upstream promoter elements for constitutive transcription in yeasts are Puppy sequences (152). The prevalence of asymmetric purine/pyrimidine stretches at 5′ ends of genes and the presence of Puppy tracts in enhancers (61, 67, 69) are consistent with a possible role for short Puppy sequence motifs in transcriptional regulation. Longer Puppy tracts, including Puppy sequence elements that vary in size from more than 50 bp up to several kilobases in length, have been observed frequently several kilobases upstream from transcription start sites (115), within introns (66, 73, 115, 132, 134) and exons (78) of genes, as well as at 3′ regions of genes (151, 153).

Recently, it was shown that highly recurring sequence elements 6 to 19 bp in length, the majority of which display extreme purine/pyrimidine asymmetry, are abundant over entire regions of a variety of genetic loci (including gene exons and introns) and are often homologous to transcription regulatory sequence motifs (10). Such observations suggest a possible role for some short Puppy sequence motifs related to DNA organization and the regulation of gene expression; one can speculate that these elements mediate interaction with the nuclear matrix, recently shown to be a dynamic protein scaffold upon which many cellular processes, including DNA replication and transcription, occur (see reference 9 and references therein).

Transcripts hybridizing with different Puppys are present at different abundances in different tissues and in different stages of development (A. B. Troutt, D. Liebermann, B. Hoffmann-Liebermann, S. N. Cohen, and L. H. Kedes, submitted for publication). Previous findings of shared repetitive sequence elements on sets of developmentally regulated mRNAs in various animal species (11, 22, 36, 98, 168), inhibition of translation of a number of muscle-specific mRNAs following interaction with a small RNA molecule (126), and identification of regions of cross homology among cDNAs of myosin light chain, myosin heavy chain, and a chicken 7S RNA (150) may all reflect the importance of transcribed repetitive sequences. Finally, it is also conceivable that the function for some transcribed Puppys may be found in the polypeptides for which they encode, as suggested for the opa-transcribed repeats in *D. melanogaster* (168). Since the regulated Hut2-Puppy transcripts contain preferentially the purine strand of the element (Troutt et al., submitted), the Hut2-Puppy would predominantly encode the triplets for Arg, Glu, Gly, and Lys.

IX. TU ELEMENTS AND PUPPY SEQUENCES: INTERRELATIONSHIPS AND ROLE IN EUCARYOTIC GENE EVOLUTION

A. TU Element-Puppy Sequences: Structural and Evolutionary Relationships

What is the evolutionary relationship of the Puppy sequences to the TU element transposons? Do they represent building blocks from which segments of TU element domains have evolved? Are they remnants of sequences once organized into domains of a widely prevalent group of transposable elements? Unfortunately, there are currently no definitive answers to such questions. It is not known whether even some of the Puppy elements represent inverted repeat termini of TU-like transposons. The λ phage clones containing Hut2 and Hut17, as well as genomic clones of some other not yet fully characterized Puppy sequences, contain only one Puppy element within ~15 kb of genomic DNA. The sizes of middle segments of TU elements vary extensively (Fig. 3). Truncated elements containing only one Puppy sequence, equivalent to one IVR, in genomic inserts of clones may have been chosen inadvertently for initial analysis, and Puppys engaged in TU-like superstructures may await discovery.

In several instances, Puppy sequences were found within introns of genes (66, 73, 115, 132, 134). Although these Puppys may constitute relics or "footprints" of imprecise excisions of TU-like transposons, it is possible that some or all Puppys have never been components of TU-like elements. Alternatively, transposons containing Puppy sequences as terminal inverted repeats may represent only a portion of Puppys in the genome; they may reflect a biological role of Puppy sequences which have led, in the course of evolution, to their assembly and successful utilization as inverted repeat termini of the TU class of transposons. Additional experiments are needed to test if and to what extent inverted repeats of Puppy elements can mobilize DNA segments between them.

B. A Possible Role for TU and Puppy Sequences in the Generation of Genetic Diversity and Evolution of Eucaryotic Genomes

As noted above, dispersed repetitive sequence elements of eucaryotes have long been implicated in a range of biological functions, including the regulation of gene expression (11, 12, 15–17, 22, 34–36, 71, 93, 98, 126, 148, 154, 161, 164, 167, 168), recombination (18, 19, 80, 87, 93, 95, 106, 149, 157), and replication (8, 89, 90, 93). The bulk of the dispersed repetitive sequence element population in various organisms consists of mobile genetic elements (58, 147, 166), which are believed to play a role in genome evolution (3, 15, 16, 36, 42, 43, 155). The widespread occurrence of "cryptic simplicity" in both coding and noncoding eucaryotic DNA (156) has led to the suggestion that ubiquitous slippagelike mechanisms, most obviously involving dispersed repetitive simple-sequence DNA, are a major source of genetic variation in all regions of genomes. Moreover, highly recurrent sequence elements, many corresponding to short Puppy tracts, are abundant over entire regions of genetic loci and are often homologous to regulatory sequences and protein-binding sites (10).

It is not difficult to envision how transposition/imprecise excision events and DNA slippagelike mechanisms combined with gene conversion and unequal crossing-over events of TU-Puppy sequence elements could generate genetic diversity affecting the molecular and phenotypic gestalt of organisms both during the programmed development of an individual and on an evolutionary time scale.

X. SUMMARY, CONCLUSIONS, AND FUTURE PROSPECTS

TU elements delineate a novel class of transposons that may be ubiquitous among animal species, including humans. The modular structure and long terminal inverted repeats of this type of eucaryotic transposable element resemble certain bacterial transposons and suggest a DNA-mediated mechanism for transposition, so far documented in eucaryotes for some transposable elements in plants, fruit flies, and nematodes. In addition to terminal inverted repeats, TU elements contain other sequence domains that themselves may be members of different families of dispersed repetitive sequences, suggesting that TU elements may play a role in the generation of genetic diversity.

The OD of the TU IVR termini is composed of short tandem repeats containing polypurine/polypyrimidine-rich reiterated sequence motifs similar to motifs in the IVRs of certain transposons in *Drosophila* sp. and maize and also in transcription control regions of a variety of viral and cellular genes. Finally, sequences homologous to TU IVR-OD sequences are present in the genome of a variety of eucaryotic species, including human. These observations suggest biological functions for this type of sequence that have been conserved phylogenetically.

Human homologs of the TU IVR-OD sequence are members of a superfamily of polypurine/polypyrimidine (Puppy) sequence elements that seem to comprise the majority of eucaryotic simple sequence DNA. Puppy sequences can alter DNA conformation, assayed by S1 sensitivity, in vivo and in vitro.

A great deal remains to be learned about TU element structures, their transcriptional and translational products, and their mechanism of transposition. Additional studies also are needed to better understand TU-Puppy sequence interrelationships and the possible functions of the intriguing Puppy sequences in diverse eucaryotic organisms.

Acknowledgments. The studies reviewed here were carried out with the support of grants from the National Institutes of Health and the American Cancer Society.

This work was performed with the collaboration of L. Kedes.

LITERATURE CITED

1. **Albertini, A. M., M. Hofer, M. P. Calos, and J. H. Miller.** 1982. On the formation of spontaneous deletions: the importance of short sequence homologies in the generation of large deletions. *Cell* **29:**319–328.

2. **Anderson, D. M., R. H. Scheller, J. W. Posakony, L. B. McAllister, S. G. Trabert, C. Beoll, R. J. Britten, and E. H. Davidson.** 1981. Repetitive sequences of the sea urchin genome: distribution of members of specific repetitive families. *J. Mol. Biol.* **145**:5–28.
3. **Baltimore, D.** 1985. Retroviruses and retroposons: the role of reverse transcription in shaping the eucaryotic genome. *Cell* **40**:481–482.
4. **Bell, G. I., M. J. Selby, and W. J. Rutter.** 1982. The highly polymorphic region near the human insulin gene is composed of simple tandemly repeating sequences. *Nature* (London) **295**:31–35.
5. **Bennett, K. L., R. E. Hill, D. F. Pietras, M. Woodworth-Gutai, C. Kane Haas, J. M. Houston, J. K. Heath, and N. D. Hastie.** 1984. Most highly repeated dispersed DNA families in the mouse genome. *Mol. Cell. Biol.* **4**:1561–1571.
6. **Berg, D. E.** 1977. Insertion and excision of the transposable kanamycin resistance determinant Tn5, p. 205–212. *In* A. I. Bukhari, J. A. Shapiro, and S. C. Adhya (ed.), *DNA Insertion Elements, Plasmids, and Episomes.* Cold Spring Harbor Laboratory, Cold Spring Harbor, N.Y.
7. **Birnboim, H. C., R. R. Sederoff, and M. C. Paterson.** 1979. Distribution of polypyrimidine · polypurine segments in DNA from diverse organisms. *Eur. J. Biochem.* **98**:301–307.
8. **Blackburn, E. H., and J. W. Szostak.** 1984. The molecular structure of centromeres and telomeres. *Annu. Rev. Biochem.* **53**:163–194.
9. **Blackwell, T. K., M. W. Moore, G. D. Yancopoulos, H. Suh, S. Lutzker, E. Selsing, and F. W. Alt.** 1987. Recombination between immunoglobulin variable region gene segments is enhanced by transcription. *Nature* (London) **324**:585–589.
10. **Bodnar, J. W., and D. C. Ward.** 1987. Highly recurring sequence elements identified in eucaryotic DNAs by computer analysis are often homologous to regulatory sequences or protein binding sites. *Nucleic Acids Res.* **15**:1835–1851.
11. **Bozzoni, I., A. Togoni, P. Pierandrei-Amaldi, E. Beccaru, M. Buonigiorno-Nardelli, and F. Amaldi.** 1982. Isolation and structural analysis of ribosomal protein genes in *Xenopus laevis*: homology between sequences present in the gene and in several different messenger RNAs. *J. Mol. Biol.* **161**: 353–371.
12. **Brickell, D. M., D. S. Latchman, D. Murphy, K. Willison, and P. W. J. Rigby.** 1983. Activation of a Qa/Tla class I major histocompatibility antigen gene is a general feature of oncogenesis in the mouse. *Nature* (London) **306**:756–760.
13. **Brierley, H. L., and S. S. Potter.** 1985. Distinct characteristics of loop sequences of two *Drosophila* foldback transposable elements. *Nucleic Acids Res.* **13**:485–500.
14. **Brill, S. J., S. Dinardo, K. Voelkel-Meiman, and R. Sternglanz.** 1987. Need for DNA topoisomerase activity as a swivel for DNA replication for transcription of ribosomal RNA. *Nature* (London) **326**:414–416.
15. **Britten, R. J.** 1981. DNA sequence organization and repeat sequences. *Chromosomes Today* **7**:9–23.
16. **Britten, R. J.** 1982. Genomic alterations in evolution, p. 41–64. *In* J. T. Bonner (ed.), *Evolution and Development.* Springer-Verlag KG, Berlin.
17. **Britten, R. J., and E. H. Davidson.** 1969. Gene regulation for higher cells: a theory. *Science* **165**:349–357.
18. **Bullock, P., J. Miller, and M. Botchan.** 1986. Effects of poly d(pGpT) · d(pApC) and poly d(pCpG) · d(pGpC) repeats on homologous recombination in somatic cells. *Mol. Cell. Biol.* **6**:3948–3953.
19. **Campbell, D. A., M. D. Van Bree, and J. C. Boothroyd.** 1984. The 5′ limit of transposition and upstream barren region of a trypanosome VSG gene: tandem 76 base-pair repeats (TAA)90. *Nucleic Acids Res.* **12**:2759–2774.
20. **Cantor, C. R., and A. Efstratiadis.** 1984. Possible structures of homopurine-homopyrimidine S1-hypersensitive sites. *Nucleic Acids Res.* **12**:8059–8072.
21. **Cappello, J. S., M. Cohen, and H. F. Lodish.** 1984. *Dictyostelium* transposable element DIRS-1 preferentially inserts into DIRS-1 sequences. *Mol. Cell. Biol.* **4**:2207–2213.
22. **Carpenter, C. D., A. M. Bruskin, P. E. Hardin, M. J. Keast, J. Anstrom, A. L. Tyner, B. P. Brandhorst, and W. H. Klein.** 1984. Novel proteins belonging to the troponin c superfamily are encoded by a set of mRNAs in sea urchin embryos. *Cell* **36**:663–671.
23. **Cherry, J. M., and E. H. Blackburn.** 1985. The internally located telomeric sequences in the germ-line chromosomes of Tetrahymena are at the ends of transposon-like elements. *Cell* **43**:747–758.
24. **Childs, G., R. Maxson, R. Cohn, and L. H. Kedes.** 1981. Orphons: dispersed genetic elements derived from tandem repetitive genes in eucaryotes. *Cell* **23**:651–663.
25. **Chou, J., P. G. Lemaux, M. Casadaban, and S. N. Cohen.** 1979. Transposition protein of Tn3: identification and characterization of an essential repressor controlled gene product. *Nature* (London) **282**:801–806.
26. **Christophe, D., B. Cabrer, A. Bacolla, H. Targovnik, V. Pohl, and G. Vassart.** 1985. An unusually long poly(purine)-poly(pyrimidine) sequence is located upstream from the human thyroglobulin gene. *Nucleic Acids Res.* **13**: 5127–5143.
27. **Cohen, J. B., K. Effron, G. Rechavi, Y. Ben-Neriah, R. Zakut, and D. Givol.** 1982. Simple DNA sequences in homologous flanking regions near immunoglobulin V_H genes: a role in gene interaction? *Nucleic Acids Res.* **11**: 3353–3369.
28. **Cohen, J. B., B. Hoffman-Liebermann, and L. H. Kedes.** Structure and unusual characteristics of a new family of transposable elements in the sea urchin *Strongylocentrotus purpuratus. Mol. Cell. Biol.* **5**:2804–2813.
29. **Cohen, J. B., D. Liebermann, and L. H. Kedes.** 1985. Tsp transposons: a heterogeneous family of mobile sequences in the genome of the sea urchin *Stronglyocentrotus purpuratus. Mol. Cell. Biol.* **5**:2814–2825.
30. **Collins, M., and G. M. Rubin.** 1982. Structure of the *Drosophila* mutable allele, white-crimson and its white-ivory and wild-type derivatives. *Cell* **30**:71–79.
31. **Collins, M., and G. M. Rubin.** 1983. High frequency precise excision of the *Drosophila* foldback transposable element. *Nature* (London) **303**:259–260.
32. **Collins, M., and G. M. Rubin.** 1984. Structure of chromosomal rearrangements induced by the FB transposable element in *Drosophila. Nature* (London) **308**:323–327.
33. **Conrad, S. E., and M. R. Botchan.** 1982. Isolation and characterization of human DNA fragments with nucleotide sequence homologies with the simian virus 40 regulatory region. *Mol. Cell. Biol.* **2**:949–965.
34. **Davidson, E. H., and R. J. Britten.** 1979. Regulation of gene expression: possible role of repetitive sequences. *Science* **204**:1052–1059.
35. **Davidson, E. H., H. T. Jacobs, and R. J. Britten.** 1983. Very short repeats and coordinate induction of genes. *Nature* (London) **301**:468–470.
36. **Davidson, E. H., and J. W. Posakony.** 1982. Repetitive sequence transcripts in development. *Nature* (London) **297**: 633–635.
37. **Davis, M. M., S. K. Kim, and L. E. Hood.** 1980. DNA-

sequences mediating class switching in α-immunoglobulins. *Science* **209**:1360–1365.

38. **Derbyshire, K. M., and D. F. Grindley.** 1986. Replicative and conservative transposition in bacteria. *Cell* **47**:325–327.
39. **Dierks, P., A. Van Ooyen, M. D. Cochran, C. Dobkin, J. Reiser, and C. Weissman.** 1983. Three regions upstream from the cap site are required for efficient and accurate transcription of the rabbit β-globin gene in mouse 3T6 cells. *Cell* **32**:695–706.
40. **Doring, H. P., and P. Starlinger.** 1986. Molecular genetics of transposable elements in plants. *Annu. Rev. Genet.* **20**: 175–200.
41. **Doring, H. P., E. Tillman, and P. Starlinger.** 1983. DNA sequence of the maize transposable element dissociation. *Nature* (London) **307**:127–130.
42. **Douglas, H. E., and J. W. Valentine.** 1984. "Hopeful monsters," transposons, and Metazoan radiation. *Proc. Natl. Acad. Sci. USA* **81**:5482–5483.
43. **Dover, G.** 1982. Molecular drive: a cohesive mode of species evolution. *Nature* (London) **299**:111–116.
44. **Durham, J. W.** 1986. Classification, p. 270–295. *In* R. C. Moore (ed.), *Treatise on Invertebrate Paleontology (U) Echinodermata 3 (1)*. Geological Society of America and University of Kansas Press, New York.
45. **Dybvig, K., C. D. Clark, G. Aliperti, and M. J. Schlesinger.** 1983. A chicken repetitive DNA sequence that is highly sensitive to single strand specific endonucleases. *Nucleic Acids Res.* **11**:8495–8508.
46. **Dynan, W. S., J. D. Safer, W. S. Lee, and R. Tjian.** 1985. Transcription factor Sp1 recognizes promoter sequences from the monkey genome that are similar to the simian virus 40 promoter. *Proc. Natl. Acad. Sci. USA* **82**:4915–4919.
47. **Dynan, W. S., and R. Tjian.** 1985. Control of eucaryotic messenger RNA synthesis by sequence specific DNA binding proteins. *Nature* (London) **316**:774–777.
48. **Egner, C., and D. E. Berg.** 1981. Excision of transposon Tn5 is dependent on the inverted repeats but not on the transposase function of Tn5. *Proc. Natl. Acad. Sci. USA* **78**: 459–463.
49. **Emerson, B. M., and G. Felsenfeld.** 1984. Specific factor conferring nuclease hypersensitivity at the 5′ end of the chicken adult β-globin gene. *Proc. Natl. Acad. Sci. USA* **81**:95–99.
50. **Emerson, B. M., C. D. Lewis, and G. Felsenfeld.** 1985. Interaction of specific nuclear factors with the nuclease-hypersensitive region of the chicken adult β-globin gene: nature of the binding domain. *Cell* **41**:21–30.
51. **Emery, A. E. H., and D. L. Rimoin (ed.).** 1983. *Principles and Practice of Medical Genetics*, vol. 1, p. 58–59. Churchill-Livingstone, Ltd., Edinburgh.
52. **Emmons, S. W., and L. Yesner.** 1984. High frequency excision of transposable element Tc1 in the nematode *Caenorhabditis elegans* is limited to somatic cells. *Cell* **36**: 599–605.
53. **Evans, T., E. Schon, G. Gora-Maslak, J. Patterson, and A. Efstradiadis.** 1984. S1-hypersensitive sites in eukaryotic promoter regions. *Nucleic Acids Res.* **12**:8043–8058.
54. **Everett, R. D., D. Baty, and P. Chambon.** 1983. The repeated GC-rich motifs upstream from the TATA box are important elements of the SV40 early promoter. *Nucleic Acids Res.* **11**:2447–2464.
55. **Farabaugh, P. J., V. Schmeisser, M. Hoffer, and J. H. Miller.** 1978. Genetic studies on the lac repressor. VII. On the molecular nature of the spontaneous hot spots in the lac I gene of *E. coli*. *J. Mol. Biol.* **126**:847–857.
56. **Fedoroff, N., S. Wessler, and M. Shure.** 1983. Isolation of the transposable maize controlling elements Ac and Ds. *Cell* **35**:235–242.
57. **Finer, M. H., E. J. Fodor, H. Boedtker, and P. Doty.** 1984. Endonuclease S1-sensitive site in chicken pro-α 2(I) collagen 5′ flanking gene region. *Proc. Natl. Acad. Sci. USA* **81**: 1659–1663.
58. **Finnegan, D. J.** 1985. Transposable elements in eukaryotes. *Int. Rev. Cytol.* **93**:281–326.
59. **Foster, T. J., V. Lundblad, S. Hanley-Way, S. M. Halling, and N. Kleckner.** 1981. Three Tn10-associated excision events: relationships to transposition and role of direct and inverted repeats. *Cell* **23**:215–227.
60. **Fromm, M., and P. Berg.** 1982. Deletion mapping of DNA regions required for SV40 early region promoter function in vivo. *J. Mol. Appl. Genet.* **1**:457–481.
61. **Fujita, T., H. Shibuya, H. Hotta, K. Yamanishi, and T. Tanguchi.** 1987. Interferon-β gene regulation: tandemly repeated sequences of a synthetic 6bp oligomer function as a virus-inducible enhancer. *Cell* **77**:357–367.
62. **Galas, D. J., M. P. Calos, and J. H. Miller.** 1980. Sequence analysis of Tn9 insertions in the lacZ gene. *J. Mol. Biol.* **144**:19–41.
63. **Gidoni, D., S. W. Dynan, and R. Tjian.** 1984. Multiple specific contacts between a mammalian transcription factor and its cognate promoter. *Nature* (London) **312**:409–413.
64. **Gill, R., F. Heffron, G. Dougan, and S. Falkow.** 1978. Analysis of sequences transposed by complementation of two classes of transposition-deficient mutants of Tn3. *J. Bacteriol.* **136**:742–756.
65. **Gill, R., F. Heffron, and S. Falkow.** 1979. Identification of the protein encoded by the transposable element Tn3 which is required for its transposition. *Nature* (London) **282**: 797–801.
66. **Gilliam, A. C., A. Shen, J. E. Richards, F. R. Blattner, J. F. Mushinski, and P. W. Tucker.** 1984. Illegitimate recombination generates a class switch from C_μ to C_δ in an IgD-secreting plasmacytoma. *Proc. Natl. Acad. Sci. USA* **81**: 4164–4168.
67. **Gillies, S. D., V. Folsom, and S. Tonegawa.** 1984. Cell type specific enhancer element associated with a mouse MHC gene, E_B. *Nature* (London) **310**:594–597.
68. **Glikin, G. C., G. Gargiulo, L. Rena-Descalzi, and A. Worcel.** 1983. *Escherichia coli* single-strand binding protein stabilizes specific denatured sites in superhelical DNA. *Nature* (London) **303**:770–774.
69. **Goodbourn, S., K. Zinn, and T. Maniatis.** 1985. Human β-interferon gene expression is regulated by an inducible enhancer element. *Cell* **41**:509–520.
70. **Grindley, N. D., and R. R. Reed.** 1985. Transpositional recombination in procaryotes. *Annu. Rev. Biochem.* **54**: 836–896.
71. **Hamada, H., M. G. Petrino, and T. Kakunaga.** 1982. A novel repeated element with Z-DNA-forming potential is widely found in evolutionary diverse eukaryotic genomes. *Proc. Natl. Acad. Sci. USA* **79**:6465–6469.
72. **Hamada, H., M. Seidman, B. H. Howard, and C. M. Gorman.** 1984. Enhanced gene expression by the poly(dT-dG) · poly(dc-dA) sequence. *Mol. Cell. Biol.* **4**:2622–2630.
73. **Harris, S. E., P. Mansson, D. B. Tully, and B. Burkhart.** 1983. Seminal vesicle secretion IV gene: allelic difference due to a series of 20 base-pair direct repeats within an intron. *Proc. Natl. Acad. Sci. USA* **80**:6460–6464.
74. **Hartzell, S. W., B. J. Byrne, and K. N. Subramanian.** 1984.

Mapping of the late promoter of simian virus 40. *Proc. Natl. Acad. Sci. USA* **81**:23–27.

75. **Hartzell, S. W., J. Yamaguchi, and K. N. Subramanian.** 1983. SV40 deletion mutants lacking the 21bp repeated sequence are viable but have noncomplementable deficiencies. *Nucleic Acids Res.* **11**:1601–1616.

76. **Heffron, F., B. J. McCarthy, H. Ohtsubo, and E. Ohtsubo.** 1979. DNA sequence analysis of the transposon Tn3: three genes and three sites involved in transposition in Tn3. *Cell* **18**:1153–1163.

77. **Helig, R., R. Muraskowsky, and J. Mandel.** 1982. The ovalbumin gene family. The 5′ end region of the X and Y genes. *J. Mol. Biol.* **156**:1–19.

78. **Heller, M., E. Flemington, E. Kieff, and P. Deininger.** 1985. Repeat arrays in cellular DNA related to the Epstein-Barr virus IR3 repeat. *Mol. Cell. Biol.* **5**:457–465.

79. **Heller, M., A. Henderson, and E. Kieff.** 1982. Repeat array in Epstein-Barr virus DNA is related to cell DNA sequences interspersed on human chromosomes. *Proc. Natl. Acad. Sci. USA* **79**:5916–5920.

80. **Hentschel, C. C.** 1982. Homocopolymer sequences in the spacer of a sea urchin histone gene repeat are sensitive to S1 nuclease. *Nature* (London) **295**:714–716.

81. **Hentschel, C., and M. Brinstiel.** 1981. The organization and expression of histone gene families. *Cell* **25**:301–313.

82. **Herrick, G., S. Cortintour, D. Dawson, D. Ang, R. Sheets, A. Lee, and K. Williams.** 1985. Mobile elements bounded by C4A4 telomeric repeats in *Oxytricha fallax*. *Cell* **43**: 759–768.

83. **Hoffman-Liebermann, B., D. Liebermann, L. H. Kedes, and S. N. Cohen.** 1985. TU elements: a heterogeneous family of modularly structured eukaryotic transposons. *Mol. Cell. Biol.* **5**:991–1001.

84. **Hoffman-Liebermann, B., D. Liebermann, A. Troutt, L. H. Kedes, and S. N. Cohen.** 1986. Human homologues of TU transposon sequences: polypurine/polypyrimidine sequence elements that can alter DNA conformation in vitro and in vivo. *Mol. Cell. Biol.* **6**:3632–3642.

85. **Htun, H., E. Lund, and J. E. Dahlberg.** 1984. Human U1 RNA genes contain an unusually sensitive nuclease S1 cleavage site within the conserved 3′ flanking region. *Proc. Natl. Acad. Sci. USA* **81**:7288–7292.

86. **Iida, S., J. Meyer, and W. Arber.** 1983. Prokaryotic IS elements, p. 159–221. *In* J. Shapiro (ed.), *Mobile Genetic Elements*. Academic Press, Inc., New York.

87. **Jeffreys, A. J., V. Wilson, and S. L. Thein.** 1985. Hypervariable "mini satellite" regions in human DNA. *Nature* (London) **314**:67–73.

88. **Jeffries, R. P. S.** 1986. *The Ancestry of the Vertebrates*. British Museum of Natural History, London.

89. **Jelinek, W. R., T. P. Toomey, L. Leinwand, C. H. Duncan, P. A. Biro, P. V. Choudary, S. M. Weissman, C. M. Rubin, C. M. Houck, P. L. Deininger, and C. W. Schmid.** 1980. Ubiquitous, interspersed repeated sequences in mammalian genomes. *Proc. Natl. Acad. Sci. USA* **77**:1398–1402.

90. **Johnson, E. M., and W. R. Jelinek.** 1986. Replication of a plasmid bearing a human Alu-family repeat in monkey cos-7 cells. *Proc. Natl. Acad. Sci. USA* **83**:4660–4664.

91. **Johnson, S. A., E. H. Davidson, and R. J. Britten.** 1984. Insertion of a short repetitive sequence (D881) in a sea urchin gene: a typical interspersed repeat. *J. Mol. Evol.* **20**:195–201.

92. **Jones, K. A., J. T. Kadonga, P. J. Rosenfeld, T. J. Kelley, and R. Tjian.** 1987. A cellular DNA-binding protein that activates eukaryotic transcription and DNA replication. *Cell* **48**:79–87.

93. **Kao, F.** 1985. Human genome structure. *Int. Rev. Cytol.* **96**:51–88.

94. **Karess, R. E., and G. M. Rubin.** 1982. A small tandem duplication is responsible for the unstable white-ivory mutation in *Drosophila*. *Cell* **30**:63–69.

95. **Kedes, L. H.** 1979. Histone genes and histone messengers. *Annu. Rev. Biochem.* **48**:837–870.

96. **Kedes, L. H., R. H. Cohn, J. C. Lowry, A. C. Y. Chang, and S. N. Cohen.** 1975. The organization of sea urchin histone genes. *Cell* **6**:359–369.

97. **Khoury, G., and P. Gruss.** 1983. Enhancer elements. *Cell* **33**:313–314.

98. **Kimmel, A. R., and R. A. Firtel.** 1985. Sequence organization and developmental expression of an interspersed repetitive element and associated single copy DNA sequences in *Dictyostelium discoideum*. *Mol. Cell. Biol.* **5**:2123–2130.

99. **Kleckner, N.** 1981. Transposable elements in prokaryotes. *Annu. Rev. Genet.* **15**:341–404.

100. **Kopecko, D. J., J. Brevet, and S. N. Cohen.** 1976. Involvement of multiple translocating DNA segments and recombinational hotspots in the structural evolution of bacterial plasmids. *J. Mol. Biol.* **108**:333–360.

101. **Kuo, M. T., and W. Plunkett.** 1985. Nick-translation of metaphase chromosomes: *in vitro* labeling of nuclease-hypersensitive regions in chromosomes. *Proc. Natl. Acad. Sci. USA* **82**:854–858.

102. **Larsen, A., and H. Weintraub.** 1982. An altered DNA conformation detected by S1 nuclease occurs at specific regions in active chick globin chromatin. *Cell* **29**:609–622.

103. **Laski, F. A., D. L. Rio, and G. M. Rubin.** 1986. Tissue specificity of *Drosophila* P element transposition is regulated at the level of mRNA splicing. *Cell* **44**:17–19.

104. **Lee, J. S., G. D. Burkholder, L. J. P. Latimer, B. L. Houg, and R. P. Braun.** 1987. A monoclonal antibody to eucaryotic chromosomes. *Nucleic Acids Res.* **15**:1047–1061.

105. **Lee, J. S., M. L. Woodsworth, L. J. P. Latimer, and A. R. Morgan.** 1984. Poly(pyrimidine) · poly(purine) synthetic DNA containing 5-methylcytosine form stable triplexes at neutral pH. *Nucleic Acids Res.* **12**:6603–6614.

106. **Lehrman, M. A., W. J. Schneider, T. C. Sudhof, M. S. Brown, J. L. Goldstein, and D. W. Russell.** 1985. Mutation in LDL receptor: Alu-Alu recombination deletes exons encoding transmembrane and cytoplasmic domains. *Science* **227**:140–146.

107. **Levis, R., M. Collins, and G. M. Rubin.** 1982. FB elements are the common basis for the instability of the W^{DZL} and W^{C} *Drosophila* mutations. *Cell* **30**:551–565.

108. **Levis, R., and G. M. Rubin.** 1982. The unstable W^{DZL} mutation of *Drosophila* is caused by a 13 kilobase insertion that is imprecisely excised in phenotypic revertants. *Cell* **30**:543–550.

109. **Liebermann, D., B. Hoffman-Liebermann, A. Troutt, L. H. Kedes, and S. N. Cohen.** 1986. Sequences from the sea urchin TU transposons are conserved among multiple eucaryotic species, including humans. *Mol. Cell. Biol.* **6**: 218–226.

110. **Liebermann, D., B. Hoffman-Liebermann, J. Weinthal, G. Childs, R. Maxson, A. Mauron, S. N. Cohen, and L. H. Kedes.** 1983. An unusual transposon with long terminal inverted repeats in the sea urchin *Strongylocentrotus purpuratus*. *Nature* (London) **306**:342–347.

111. **Lilly, D.** 1986. Bent molecules—how and why? *Nature* (London) **320**:487.

112. **Mace, H. A. F., H. R. B. Pelham, and A. A. Travers.** 1983. Association of an S1 nuclease-sensitive structure with short direct repeats 5′ of *Drosophila* heat shock genes. *Nature* (London) **304**:555–557.
113. **Maeda, N., J. B. Bliska, and O. Smithies.** 1983. Recombination and balanced chromosome polymorphism suggested by DNA sequences 5′ to the human δ-globin gene. *Proc. Natl. Acad. Sci. USA* **80**:5012–5016.
114. **Maeda, N., F. Yang, D. R. Barnett, B. H. Bowman, and O. Smithies.** 1984. Duplication within the haptoglobin Hp^2 gene. *Nature* (London) **309**:131–135.
115. **Maroteaux, L., R. Heilig, D. Dupret, and J. L. Mandel.** 1983. Repetitive satellite-like sequences are present within or upstream from 3 avian protein-coding genes. *Nucleic Acids Res.* **11**:1227–1243.
116. **Maxson, R., T. Mohun, R. Whn, and L. H. Kedes.** 1983. Expression and organization of histone genes. *Annu. Rev. Genet.* **17**:239–277.
117. **McClintock, B.** 1956. Controlling elements and the gene. *Cold Spring Harbor Symp. Quant. Biol.* **21**:197–216.
118. **McCutchan, T. F., and M. F. Singer.** 1981. DNA sequences similar to those around the simian virus 40 origin of replication are present in the monkey genome. *Proc. Natl. Acad. Sci. USA* **78**:95–99.
119. **McKnight, S. L.** 1982. Functional relationships between transcriptional control signals of the thymidine kinase gene of herpes simplex virus. *Cell* **31**:355–365.
120. **McKnight, S. L., and R. Kingsbury.** 1982. Transcriptional control signals of a eukaryotic protein-coding gene. *Science* **217**:316–324.
121. **McKnight, S. L., R. C. Kingsbury, A. Spence, and M. Smith.** 1984. The distal transcription signals of the herpesvirus tk gene share a common hexanucleotide control sequence. *Cell* **37**:253–262.
122. **Mishoe, H., J. N. Brady, M. Radonovich, and N. P. Salzman.** 1984. Simian virus 40 guanine-cytosine-rich sequences function as independent transcriptional control elements in vitro. *Mol. Cell. Biol.* **4**:2911–2920.
123. **Miyakake, S., T. Yokota, F. Lee, and K. Arai.** 1985. Structure of the chromosomal gene for murine interleukin 3. *Proc. Natl. Acad. Sci. USA* **82**:316–320.
124. **Miyake, T., K. Migita, and Y. Sakaki.** 1983. Some kpn1 family members are associated with the Alu family in the human genome. *Nucleic Acids Res.* **11**:6837–6846.
125. **Moss, M., and D. Gallwitz.** 1983. Structure of two human β-actin-related processed genes one of which is located next to a simple repetitive sequence. *EMBO J.* **2**:757–761.
126. **Mroczkowski, B., T. L. McCarthy, J. D. Zezza, P. W. Bragg, and J. M. Heywood.** 1984. Small RNAs involved in gene expression of muscle-specific proteins. *Exp. Biol. Med.* **9**: 277–283.
127. **Nickol, J. M., and G. Felsenfeld.** 1983. DNA conformation at the 5′ end of the chicken adult β-globin gene. *Cell* **35**:467–477.
128. **O'Hare, K., and G. M. Rubin.** 1983. Structures of P transposable elements and their sites of insertion and excision in the *Drosophila melanogaster* genome. *Cell* **34**:25–35.
129. **Paro, R., M. L. Goldberg, and W. J. Gehring.** 1983. Molecular analysis of large transposable elements carrying the white locus of *Drosophila melanogaster*. *EMBO J.* **2**: 853–860.
130. **Potter, S., M. Truett, M. Phillips, and A. Maher.** 1980. Eukaryotic transposable genetic elements with inverted terminal repeats. *Cell* **20**:639–647.
131. **Potter, S. S.** 1982. DNA sequence of a foldback transposable element in *Drosophila*. *Nature* (London) **297**:201–204.
132. **Proudfoot, N. J., A. Gil, and T. Maniatis.** 1982. The structure of the human zeta-globin gene and a closely linked nearly identical pseudogene. *Cell* **31**:553–563.
133. **Pulleyblank, D. E., D. B. Haniford, and A. R. Morgan.** 1985. A structural basis for S1 nuclease sensitivity and double-stranded DNA. *Cell* **42**:271–280.
134. **Qasba, P. K., and S. K. Safaya.** 1984. Similarity of the nucleotide sequences of rat α-lactalbumin and chicken lysozyme genes. *Nature* (London) **308**:377–380.
135. **Queen, C., S. T. Lord, T. F. McCutchan, and M. F. Singer.** 1981. Three segments from the monkey genome that hybridize to simian virus 40 have common structural elements. *Mol. Cell. Biol.* **1**:1061–1068.
136. **Rich, A., A. Nordheim, and A. H. J. Wang.** 1984. The chemistry and biology of left-handed Z-DNA. *Annu. Rev. Biochem.* **53**:791–846.
137. **Richards, J. E., A. C. Gilliam, A. Shen, P. W. Tucker, and F. R. Blattner.** 1983. Unusual sequences in the murine immunoglobulin μ-δ heavy-chain region. *Nature* (London) **306**:483–487.
138. **Rio, D. L., F. A. Laski, and G. M. Rubin.** 1986. Identification and immunochemical analysis of biologically active *Drosophila* P element transposase. *Cell* **44**:21–32.
139. **Roberts, J. W., J. W. Grula, J. W. Posakony, R. Hudspeth, E. H. Davidson, and R. J. Britten.** 1983. Comparison of sea urchin and human mtDNA: evolutionary rearrangement. *Proc. Natl. Acad. Sci. USA* **80**:4614–4618.
140. **Roeder, G. S., and G. R. Fink.** 1983. Transposable elements in yeast, p. 299–328. *In* J. Shapiro (ed.), *Mobile Genetic Elements*. Academic Press, Inc., New York.
141. **Rosenthal, N., M. Kress, P. Gruss, and G. Khoury.** 1983. BK virus enhancer element and a human cellular homolog. *Science* **222**:749–755.
142. **Ruiz-Carrillo, A.** 1984. The histone gene is flanked by S1 hypersensitive structures. *Nucleic Acids Res.* **12**:6473–6492.
143. **Saffer, J. D., and M. F. Singer.** 1984. Transcription from SV40-like monkey and DNA sequences. *Nucleic Acids Res.* **12**:4769–4788.
144. **Scheller, R. H., D. M. Anderson, J. W. Posakony, L. B. McAllister, J. R. Britten, and E. H. Davidson.** 1981. Repetitive sequences of the sea urchin genome II: subfamily structure and evolutionary conservation. *J. Mol. Biol.* **149**: 15–39.
145. **Schon, E., T. Evans, J. Welsh, and A. Efstratiadis.** 1983. Conformation of promoter DNA: fine mapping of S1-hypersensitive sites. *Cell* **35**:837–848.
146. **Seif, I., G. Khoury, and R. Dhar.** 1979. The genome of human papovavirus BKV. *Cell* **18**:963–977.
147. **Shapiro, J. A. (ed.).** 1983. *Mobile Genetic Elements*. Academic Press., Inc., New York.
148. **Sharp, P. A.** 1987. Transplicing: variation on a familiar theme? *Cell* **50**:147–148.
149. **Shen, S. H., J. L. Slighton, and O. Smithies.** 1981. The history of the human fetal globin gene duplication. *Cell* **26**:191–203.
150. **Siddigui, M. A. Q., P. Khandekar, M. Krauskopf, C. Mendola, A. M. Zarraga, and C. Saidopet.** 1984. Control of muscle gene expression: cardiac 7S RNA is homologous to 3′ noncoding myosin and repetitive DNA sequences. *Exp. Biol. Med.* **9**:269–276.
151. **Southgate, R., A. Ayme, and R. Voellmy.** 1983. Nucleotide sequence analysis of the *Drosophila* small heat shock gene cluster at locus 67B. *J. Mol. Biol.* **165**:35–57.

152. **Stuhl, K.** 1985. Naturally occurring poly (dA-dT) sequences are upstream promoter elements for constitutive transcription in yeast. *Proc. Natl. Acad. Sci. USA* **82:**8419–8423.
153. **Sures, I., J. Lowry, and L. H. Kedes.** 1978. The DNA sequence of sea urchin (*S. purpuratus*) H2A, H2B and H3 histone coding and spacer regions. *Cell* **15:**1033–1044.
154. **Sutcliffe, J. G., R. J. Milner, J. M. Gottesfeld, and R. A. Lerner.** 1984. Identifier sequences are transcribed specifically in brain. *Nature* (London) **308:**237–241.
155. **Syvanen, M.** 1984. The evolutionary implications of mobile genetic elements. *Annu. Rev. Genet.* **18:**271–293.
156. **Tautz, D., T. Martin, and G. A. Dover.** 1986. Cryptic simplicity in DNA is a major source of genetic variation. *Nature* (London) **322:**652–656.
157. **Treco, D., and N. Arenheim.** 1986. The evolutionary conserved repetitive sequence d(TG · AC)n promotes reciprocal exchange and generates unusual recombinant tetrads during yeast meiosis. *Mol. Cell. Biol.* **6:**3934–3947.
158. **Truett, M. A., R. S. Jones, and S. S. Potter.** 1981. Unusual structure of the FB family of transposable elements in *Drosophila*. *Cell* **24:**753–762.
159. **Tu, C. D., and S. N. Cohen.** 1980. Translocation specificity of the Tn3 element: characterization of sites of multiple insertions. *Cell* **19:**151–160.
160. **Valentine, J. W.** 1978. The evolution of multicellular plants and animals. *Sci. Am.* **239:**140–158.
161. **VanderPloeg, L. H. T.** 1986. Discontinuous transcription and splicing in trypanosomes. *Cell* **47:**479–480.
162. **Varmus, H. E.** 1983. Retroviruses, p. 411–503. *In* J. A. Shapiro (ed.), *Mobile Genetic Elements.* Academic Press, Inc., New York.
163. **Voelkel-Meiman, K., R. L. Keid, and G. S. Roeder.** 1987. Recombination-stimulating sequences in yeast ribosomal DNA correspond to sequences regulating transcription by RNA polymerase. I. *Cell* **48:**1071–1079.
164. **Wang, A. J. H., G. J. Quigly, F. J. Koplak, J. L. Crawford, J. H. Van Boom, G. VanderMarel, and A. Rich.** 1979. Molecular structure of a left handed double helical DNA fragment at atomic resolution. *Nature* (London) **282:** 680–686.
165. **Weiher, H., M. Konig, and P. Gruss.** 1983. Multiple point mutations affecting the simian virus 40 enhancer. *Science* **219:**626–631.
166. **Weiner, A. M., P. L. Deininger, and A. Efstratiatis.** 1986. Nonviral retroposons in genes, pseudogenes, and transposable elements generated by the reverse flow of genetic information. *Annu. Rev. Biochem.* **55:**631–661.
167. **Weintraub, H., and M. Groudine.** 1976. Chromosomal subunits in active genes have an altered conformation. *Science* **193:**848–856.
168. **Wharton, K. A., B. S. Yedvobrick, K. G. Finnerty, and S. Artavanis-Tsakonos.** 1985. Opa: a novel family of transcribed repeats shared by the notch locus and other developmentally regulated loci in *D. melanogaster. Cell* **40:** 55–62.

Chapter 26

LINEs and Related Retroposons: Long Interspersed Repeated Sequences in the Eucaryotic Genome

CLYDE A. HUTCHISON III, STEPHEN C. HARDIES, DANIEL D. LOEB, W. RONALD SHEHEE, and MARSHALL H. EDGELL

Clyde A. Hutchison III, Daniel D. Loeb, W. Ronald Shehee, and Marshall H. Edgell ■ Department of Microbiology and Immunology, University of North Carolina at Chapel Hill, Campus Box 7290, FLOB, Chapel Hill, North Carolina 27514. **Stephen C. Hardies** ■ Department of Biochemistry, University of Texas Health Science Center at San Antonio, 7703 Floyd Curl Drive, San Antonio, Texas 78284-7760.

I. INTRODUCTION

Interspersed repetitive sequences, originally detected as rapidly reannealing components of genomic DNA, are ubiquitous in the genomes of higher eucaryotes as diverse as flies, sea urchins, frogs, birds, rodents, and humans (6, 16, 27, 41, 66, 100). Hypotheses concerning their roles range from those in which they are the main agents of regulation (7, 15) to those in which they are simply neutral hitchhikers without effect on host phenotype (23, 80). Several recent reviews cover material complementary to that presented here (28, 32, 35, 90, 91, 95, 108, 113).

It has long been hoped that studies of the structure and organization of interspersed repetitive families would suggest possible functions for such sequences. With the advent of restriction enzymes, a number of specific families of repetitive DNA were characterized. Many were originally named for the characteristic discrete bands seen above the continuum of single-copy fragments in a restriction digest of total genomic DNA (e.g., the *Kpn* and *Alu* repeats). Such discrete bands reflect the presence of restriction sites at specific locations within the majority of members of a repeat family. Structural investigations have revealed two basic types of interspersed repetitive families, termed LINEs ("long interspersed sequences" or "long interspersed nuclear elements") and SINEs ("short . . ." etc.) by Singer (107).

A. SINEs

SINEs are typically less than 500 base pairs (bp) in length and occur in approximately 10^5 copies per genome. Prominent SINEs include the *Alu* family of primates and the B1 family of rodents. This class of repetitive element is described in detail elsewhere in this volume (see the chapter by P. L. Deininger). At the time of Singer's proposed classification, the structure of several SINEs was known at the nucleotide sequence level.

B. LINEs

In contrast, the structure of LINEs was initially much less well understood. These were described as elements greater than 5 kilobases (kb) in length and present at copy numbers of greater than 10^4 per mammalian genome. Examples cited by Singer (107) were the *Kpn*I family of primates (1, 42, 65, 99) and MIF-1 of rodents (8, 12, 50, 75). Analysis of these elements during the past 5 years has led to a more complete and unified picture of their structure. Full-length LINEs elements were shown to be about 7 kb in the mouse genome (31, 120) and about 6.5 kb in the human genome (43). Homology between mouse LINEs (then called MIF-1 or *Bam*HI) and a human LINEs element (*Kpn*I family) indicated that these were both representatives of a single family of elements (11, 110; F. H. Burton, C. F. Voliva, M. H. Edgell, and C. A. Hutchison III, *DNA* **2**:82, 1983), which is widespread in mammals (11). This family was termed L1, for LINE-1 (120). A two-letter designation is appended to specify the source of an L1 element. Thus, the mouse L1 is called L1Md (LINE-1 from *Mus domesticus*), and human L1 is called L1Hs (LINE-1 from *Homo sapiens*). This notation is useful because the L1 elements within a species are more similar in sequence than are those from different species (Table 1, below). Recently, certain repetitive elements from insects (9, 20, 22, 34, 84), protozoans (58), and higher plants (102) have been shown to be similar structurally to mammalian L1.

C. Structure of LINE-1 (L1)

DNA sequence analysis of full-length L1 elements, described in detail below, has revealed a simple genetic structure with interesting similarities to and differences from retroviruses and related transposable elements such as *copia* and Ty. The consensus L1 structure has a poly(A) tail at one end and is flanked by short direct repeats (generally 5 to 15 bp). L1 does not have long terminal repeats (LTRs). These features suggest that L1 elements are produced by reverse transcription of polyadenylated mRNAs and insertion of the reverse transcript copies into the genome. Rogers (90) coined the term **retroposon** to describe such elements. **Retrotransposon**, in contrast, refers to elements which have LTRs and thus are more similar to retroviruses. L1 contains two large open reading frames (ORF-1 [about 375 codons] and ORF-2 [about 1,300 codons]), totaling to about 5 kb in length (64). ORF-2 contains amino acid sequence motifs characteristic of reverse transcriptases (48, 64). The hypothesis that ORF-2 encodes an enzyme which functions in the transposition of L1 underlies much of current L1 research.

D. Are There Other LINEs Families?

In common usage, the term LINEs has become synonymous with L1, although it was originally coined to include all elements of a particular size (>5 kb) and copy number (>10^4). There is now need to ask whether L1 is the only LINEs family as originally defined. In the mouse genome, most of the major bands of repetitive DNAs in restriction digests of total genomic DNA have been shown by Southern blotting experiments to be homologous to L1 (copy number of about 10^5) (11). This indicates that L1 is the predominant family of LINEs in the mouse genome. These experiments do not rule out the possibility of other elements, 5- to 10-fold less abundant, which would still fit the original definition of LINEs. Certain mouse retrotransposons, such as the intracisternal A particle, are not considered LINEs since they are found at too low a copy number. In the human genome, an element named THE 1 (transposonlike human element 1) (83) does come close to satisfying the original definition of LINEs (about 10^4 copies), but it is only 2.3 kb long.

In light of present knowledge, we think it is useful to redefine LINEs as **active retroposons** which, like L1, encode proteins likely to mediate their own transposition. This evolved definition of LINEs excludes **passive retroposons**, which do not encode such functions, even though these apparently result from a process involving reverse transcription. Passive retroposons include SINEs and pseudogene copies of sequences such as globin genes. If new LINEs families are found which do not share obvious sequence homology with L1, then we propose they be designated L2, L3, etc.

E. Unanswered Questions

Current research on L1 addresses fundamental questions of LINEs transposition, evolution, and effects on host cells.

What functions are encoded in the open reading frames of L1 elements? When and where are they expressed in the host organism? Are they useful to the host organism?

Do the 5′ ends of L1 contain promoters, enhancers, or other regulatory elements? Can L1 regulate expression of nearby host genes? If so, how?

Do the many copies of L1 elements reflect proliferation from a very few ancestral elements? If so, how has this replication and transposition occurred? How many copies and which copies of L1 are functional in this proliferation? How do rearrangements in L1 sequences such as loss of the 5′ end occur? Can L1 transposition be harnessed as a tool for insertional mutagenesis in mammals?

What role does L1 play in overall genome organization and evolution? Does recombination between L1 elements (and other retroposons) at different loci keep the genome in a state of flux and thereby speed evolution (104)?

II. STRUCTURE OF L1 ELEMENTS

Canonical structures for full-length L1 elements have been determined in two animals, mice (64, 105) and humans (47, 48, 111). An inherent difficulty in determining such canonical structures stems from the

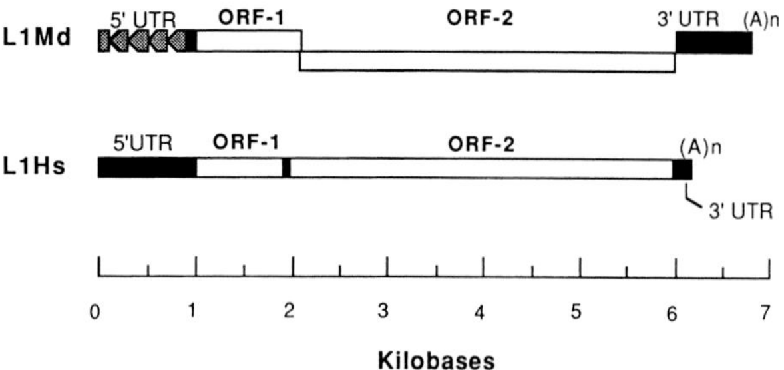

Figure 1. Canonical mouse and human L1 structures. Potential protein-coding regions are indicated by open boxes. The 208-bp repeats of L1Md are represented by the shaded areas. Other noncoding sequences are indicated by black boxes.

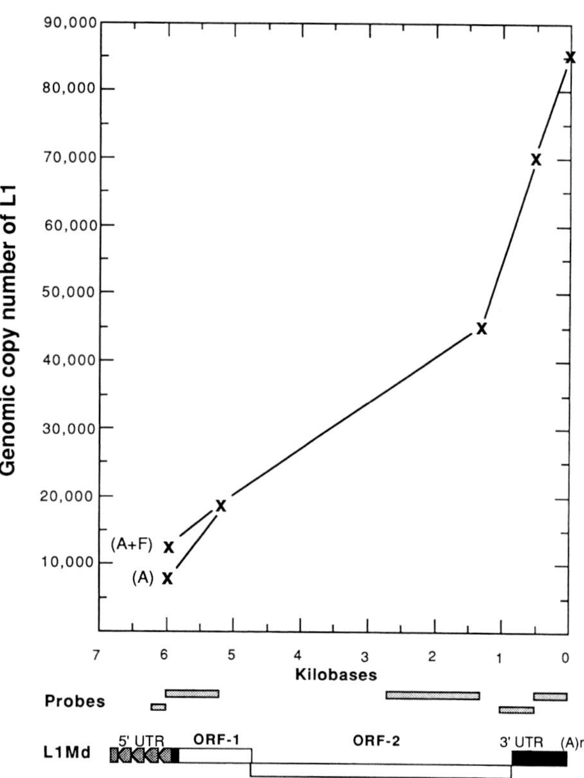

Figure 2. Genomic copy number of L1Md. The vertical axis represents the haploid copy number and the horizontal axis represents the position of the L1Md probe used in the copy number measurement. The 3′ end of L1 is at the right (28; Comer, unpublished results).

need to sequence multiple L1 isolates because individual elements frequently harbor mutations which are not representative of the L1 family of that species. Despite this difficulty, the structures of mouse and human L1 are quite well known, and L1 elements of several other species are partially characterized. A comparison of these different L1 sequences reveals a consensus mammalian L1 structure. Important features of mouse and human L1 elements (Fig. 1) distinguish LINEs from other types of mobile genetic elements such as retroviruses, retrotransposons, and procaryotic transposons. These are: (i) 10^4 to 10^5 copies per haploid genome, (ii) lengths of 6 to 7 kb (for complete elements), (iii) frequent deletion of a variable-length segment corresponding to the 5′ end of the L1 element (5′ truncation phenomenon), (iv) an adenine-rich 3′ terminus, (v) short direct repeats (SDRs; usually <20 bp) flanking the element, and (vi) open reading frames (ORFs) within the element.

In addition to these common features, the L1 elements of each animal species also have unique structural features.

A. Mouse L1 (L1Md)

1. Common features of L1Md

One end or terminus of individual L1Md elements (designated the 3′ end; Fig. 1) contains a polyadenylation signal (AATAAA) followed by a variable-length (up to 50 bp) adenine-rich tail (31, 39, 120). This is one of the features which suggested that individual L1Md elements are generated via an mRNA intermediate that is reinserted into new genomic locations after reverse transcription. The end containing the A tail is referred to as the 3′ end by analogy with the structure of polyadenylated mRNA.

The presence of many copies of this element in the genome was one of the first features determined. Probing a Southern blot of mouse genomic DNA with an L1Md probe results in an intense smear, with several discrete bands (when restriction enzymes that cut at two or more sites within the element are used). These L1-specific bands are also commonly visible in an ethidium bromide-stained gel of the same digest (12). The smear of L1-hybridizing DNA results from fragments extending from a restriction site within L1 to a site in flanking chromosomal DNA, and reflects the many truncated elements as well as elements with mutations in restriction sites. The apparent copy number (per haploid genome) can range from 10^4 to 10^5, depending on the segment of L1Md used as the probe (3, 4, 39; M. B. Comer, unpublished results). The largest L1Md members have been estimated to be about 7 kb long from genomic restriction mapping (11, 31, 120) and sequencing (64, 105).

Most members of the L1Md family are not full length, but are truncated at apparently random distances from a common 3′ end (31, 120), a phenomenon referred to as 5′ truncation. As a consequence, the 3′ end is more highly repeated (about 10^5 copies) than the 5′ end (about 10^4 copies [3, 4, 39; Comer, unpublished results]) (Fig. 2).

Most L1Md elements are flanked by small direct

repeats of chromosomal origin, usually less than 20 bp; these repeats are taken as evidence of L1 mobility (121, 123). As in the case of target site duplications associated with classical transposons, they are thought to result from the formation of staggered nicks, followed by gap filling during insertion. They are found flanking both full-length elements and truncated elements, indicating that the truncated elements are inserted as such, rather than arising by deletion from full-length preexisting elements.

Examination of the nucleotide sequences of two large L1Md elements and the 5′ half of a third element reveals over 5 kb of open reading frame, organized into two frames of 1,137 bp (ORF-1) and 3,900 bp (ORF-2) that overlap by 14 bp (64, 105). ORF-2 has several domains with amino acid sequence homology to reverse transcriptases from viruses and retrotransposons of diverse organisms (64; see D. J. Finnegan, this volume, Chapter 18, Fig. 4) and also homology to the metal-binding domains of nucleic acid-binding proteins, such as those of retroviruses and TFIIIA (29).

2. Unique features of L1Md

The 5′ end of each full-length L1Md element consists of one of two alternative 200-bp tandemly repeating motifs (A type [208 bp] and F type [206 bp]; 31, 64, 81). These motifs may be unique to L1Md; they are not found in the L1 elements of rat or human (14, 47). The A-type and F-type tandem repeats are not sequence homologous, and only one of these types is found at the end of any given element (31, 64, 81; Comer, unpublished results). Three elements with different numbers of A-type repeats ($4\frac{2}{3}$, $2\frac{2}{3}$, and $1\frac{2}{3}$) have been sequenced (105). Interestingly, the 5′-most repeat in each array is approximately two-thirds of a copy, although the endpoints are not at the same site in each isolate (105). The 5′ element in some F-type arrays is also two-thirds of a copy, but others end within a few nucleotides of a complete copy (81).

A structural polymorphism also exists in ORF-1 of L1Md. Individual elements contain different numbers of a 42-base sequence. Two elements examined contain two copies, and a third element contains three copies (105). Although the significance of this polymorphism is not understood, we note that it is a multiple of 3 bp within an open reading frame and thus preserves the reading frame. Rat L1 elements (see below) also contain a polymorphism of a small tandem repeat in an equivalent region (14, 114).

B. Human L1 (L1Hs)

Like L1Md, the L1Hs elements are present at a high copy number (107,000 copies per haploid genome; 51) and exhibit 5′ truncation. Unit-length members are about 6 kb long (1, 47). The 3′ end is adenine rich, and short direct repeats flank most elements. L1Hs also has two open reading frames, of 1,122 and 3,852 bp (111) (similar to the 1,137- and 3,900-bp ORFs of L1Md). The ORFs of L1Hs, in contrast to those of L1Md, are in the same reading frame. The 3′ end of ORF-1 and the 5′ end of ORF-2 are defined by two TAA termination codons which are 33 bp apart. ORF-2 of L1Hs, like that of L1Md, exhibits amino acid homology to reverse transcriptase (48) and to nucleic acid-binding proteins (29), as does ORF-2 of mouse L1.

The L1Hs family contains a polymorphism for a 132-bp insert 5′ to ORF-1 (47, 111). About half of the elements not truncated before this region contain this additional 132 bp.

C. L1 Elements in Other Species

L1Rn of the rat *Rattus norvegicus* is the third-most-studied L1 family. Several truncated or partial elements and one full-length element have been sequenced (e.g., 14, 15, 22). L1Rn also exhibits the common features of this family: high copy number (125), unit length between 6 and 7 kb (14), 5′ truncation (97), a 3′ adenine-rich terminus (14, 97, 114), short direct repeats flanking the element (14, 97, 114), and large open reading frames (14, 114), but with some notable exceptions. 5′ Truncation is reported to be less frequent in rat than in mouse or human elements, although different experiments were performed with each animal (14). Although open reading frames are present, they remain poorly defined because of apparent chain-terminating and frameshift mutations in the two examples sequenced (14, 114).

L1 elements from primates (slow loris and African green monkey), rabbit, dog, and cat have also been sequenced. The features seen in these species are generally consistent with those in mouse and human L1, and the following features merit comment. Rabbit L1 (L1Oc) has a 3′ untranslated region (the sequence between the conserved terminator of ORF-2 and the 3′ terminus) of about 900 bp (18), in contrast to about 200 bp for L1Hs (47, 111) and about 700 bp for L1Md (64, 105) and L1Rn of rat (14, 97). The L1 element of the prosimian slow loris also has two large ORFs (48). The green monkey (63), dog (55), and cat (29) L1 families contain some elements which have internal rearrangements (inver-

Table 1. Sequence divergence between LINE-1 of mouse and those of other mammals[a]

Animal	LINE-1 element	Divergence from L1Md (%)		Evolutionary separation from *M. domesticus* (Myr)
		DNA	Protein	
Mouse (*Mus domesticus*)	L1Md	4	9	
Mouse (*Mus platythrix*)	L1Mc	9	11	12
Rat (*Rattus norvegicus*)	L1Rn	15	17	20
Rabbit (*Oryctolagus cuniculus*)	L1Oc	37	51	65–80
Human (*Homo sapiens*)	L1Hs	33	40	65–80

[a] DNA sequence divergence of L1Md within *M. domesticus* was calculated by averaging the pairwise divergences between 12 sequences from the BALB/c mouse (70) for a region of 300 nucleotides preceding the highly conserved termination codon at the 3′ end of ORF-2. These same sequences were used to calculate amino acid sequence divergence within *M. domesticus*. Padding characters were added to one of the sequences to correct for two single-base deletions before translation. The DNA sequence divergence between *M. domesticus* and *M. platythrix* was calculated by averaging the pairwise divergences between the 12 *M. domesticus* sequences and 10 *M. platythrix* sequences. The amino acid sequence divergence was calculated by comparing the translation of the *M. domesticus* consensus sequence with the *M. platythrix* consensus. Each of the other values was obtained by comparing the sequence of the mouse element L1MdA2 (64) with a single sequence each from a rat element (LINE 3 [14]), a rabbit element (18), and a human element (TβG41 [47]). These sequences were aligned with L1MdA2 by the addition of padding characters, where necessary, before translation.

sions or deletions) compared to the consensus structures for these species.

D. Consensus Full-Length Mammalian L1 Structure

A generalized or consensus full-length mammalian L1 element is 6 to 7 kb long and has the following features. (i) At the 5′ end is a segment that does not encode a protein, which is 800 bp long in the human element and varies in size in the mouse element (from at least 0.3 to 1.3 kb or more) due to a variable number of copies of 200-bp tandem repeats. The extreme 5′ end of this domain can have an unusually high G+C content (e.g., mouse, human, and rat L1) or a repeating motif (mouse L1). The possibility of regulatory sequences in this region remains to be tested. (ii) A central region contains two ORFs of about 1 and 4 kb in at least the mouse and human L1 elements, which may be separated by a small distance or overlap by a few nucleotides. (iii) A 3′ non-protein-coding region of variable length (ranging from 200 bp in human to 900 bp in rabbit) is followed by an adenine-rich 3′ terminus. (iv) There are many partial elements, truncated at their 5′ ends. (v) Short direct repeats flank both complete and truncated L1 elements.

III. HOW WIDESPREAD ARE LINEs?

A. LINE-1 (L1) Occurs throughout Mammalia

Hybridization studies show that L1 is present throughout marsupial and placental mammalian orders (11). Southern blots of restriction digests of total genomic DNAs from diverse animals probed with DNA from different parts of L1Md showed that every mammal tested, including marsupials, contained L1Md homologous sequences. The extent of homology decreased with increasing phylogenetic distance from the mouse. These results are consistent with a model according to which current L1 elements are descended from a single family present in the ancestor to all mammals. They are not easily explained on the basis of an alternative model in which L1 has spread more recently by horizontal transmission between species.

Estimates of the extent of DNA sequence divergence among L1Md elements and between L1Md and L1 of several other mammals are presented in Table 1. Although this is a useful illustration of the degree of relatedness of L1 elements, there are two important qualifications to these numerical values:

(i) The extent of sequence divergence is not constant along the length of the element. The 5′ end of the element, including the N-terminal half of ORF-1, and the 3′ end beyond the ORF-2 termination codon have diverged most rapidly. The C-terminal half of ORF-1, and ORF-2, have diverged more slowly. Amino acid sequence motifs within ORF-2 which are characteristic of reverse transcriptases are the most highly conserved parts of the sequence. For uniformity, the values in Table 1 are calculated for 300 nucleotides preceding the highly conserved termination codon at the 3′ end of ORF-2.

(ii) The amount of sequence data and the degree of intactness of the L1s from which the data are derived vary for the different species studied. In some cases, sequences of full-length elements with intact ORFs are available, and in other cases, only obvi-

Table 2. Properties of L1 and related nonmammalian retroposons[a]

Host	Element	Copies	Location	SDR (bp)	LTRs?	Size (kb)	3′ End	ORF-1 (a.a.)	ORF-2 (a.a.)	5′ Truncation?
Mammals	L1	~10^5	Dispersed	5–15	No	6–7	$(A\ rich)_{8-50}$	379	1,300	Yes (majority)
Drosophila sp.	I	10 (inducer), ≤1 (reactive)	Dispersed	10–14	No	5.4	$(TAA)_{4-7}$	426	1,086	Yes
Drosophila sp.	F	60–80	Dispersed	8–13	No	4.7	$(A)_{12-30}$	≥122	859	Yes
Drosophila sp.	G	20	Tandem arrays with rDNA	9[b]	No	~4	$(A)_{19}$[b]	NS	NS	NR
Bombyx mori	R2	25	28S rDNA-specific site	None	No	4.2	GAAAA	None	1,151	NR
Zea mays	Cin4	50–100	Dispersed	3–16	No	≥7	$(A)_{6-11}$	≥931	1,163	Yes (majority)
Trypanosoma brucei	Ingi	200	Dispersed	4[c]	No	5.2	$(A)_{7-14}$	386[d]	1,120[d]	NR (not majority)

[a] Abbreviations: SDR, short direct repeat; LTR, long terminal repeat; a.a., amino acids; NS, not sequenced; NR, none reported.
[b] Several junctions were sequenced, but these may represent a single integration event (22).
[c] Only three cases sequenced (58); this may vary.
[d] This is a hypothetical size inferred from pseudogene sequences (58).

ously defective elements were analyzed (see footnote to Table 1).

The data also demonstrate a phenomenon termed concerted evolution. Although there is about 30% divergence between L1Md and L1Hs, the DNA sequence divergence among L1Md elements is only 4%.

Sequences of the L1 elements from mammals more distant from mouse and human (e.g., the marsupials koala and wallaby) have not been determined. However, the conditions that showed specific hybridization with L1Md probes indicate that the most conserved parts of the element in marsupials must be at least 70% homologous to L1Md at the DNA sequence level (11).

B. Nonmammalian Retroposons

Hybridization studies (11) did not detect L1-specific sequences in birds (duck) or invertebrates (sea urchin). However, the possibility of equivalent LINEs too divergent from the mouse L1 probes used to be detected is not excluded. This is emphasized by findings of repetitive elements with structural similarities to mammalian L1 in invertebrates such as *Drosophila* spp. (20, 22, 34, 84), the protozoan *Trypanosoma brucei* (58), the silkworm *Bombyx mori* (9), and the higher plant *Zea mays* (102). These elements are distantly related to each other and to L1 and in most cases do not share obvious sequence homology at either the DNA or amino acid sequence level. Their relatedness is indicated by (i) conserved features of sequence organization (Table 2) and (ii) patches of sequence homology which contain amino acid sequence motifs that are also common to well-characterized reverse transcriptases (Finnegan, this volume, Chapter 18, Fig. 4).

1. L1-like *Drosophila* elements

The D (84), F (20, 21), and G (22) elements and the I factor (34, 35) of *Drosophila* spp. all have structural properties of retroposons. They have one A-rich end, lack LTRs, and are flanked by short direct repeats. Sequence analysis of F and I reveals long ORFs with homology to reverse transcriptases and to DNA-binding proteins (Finnegan, this volume, Chapter 18). Both F and I exhibit 5′ truncation equivalent to that in mammalian LINEs. The I factor was first found as a transposable element causing a type of hybrid dysgenesis (Finnegan, this volume, Chapter 18). F elements are also considered transposable, since they are located at different sites on chromosomes of different *Drosophila* sp. strains and are responsible for known mutations (Finnegan, this volume, Chapter 19).

The one D element sequenced (84) is only 0.38 kb in length and contains no ORFs longer than 24 codons. It is therefore more similar to mammalian SINEs than to LINEs.

The G element (22) is about 4 kb in length and occurs at about 20 copies per haploid genome. The majority of G elements are found in tandem arrays of units which consist of a G element plus a segment of the nontranscribed spacer of rDNA that is less than 1 kb in length. However, a tandem array of a unit consisting of a G element plus a segment of F element sequence about 1 kb in length has also been found. No complete G element sequence has been reported.

2. Ingi, a protozoan L1-like element

An element with structural similarities to LINES found in *Trypanosoma brucei*, and named ingi, is 5.2 kb long, contains one A-rich end, and is flanked by short direct repeats (58). It occurs at about 200 copies per haploid genome. One ingi element (ingi-3) has been completely sequenced (58). It appears to be a pseudogene since it contains termination codons interrupting an open reading frame, some of which are absent from other partially sequenced elements (58). The ingi ORF exhibits a sequence similarity to ORF-2 of mammalian LINEs that is highly significant statistically and encodes amino acid sequence motifs found in reverse transcriptases (58). These are shown aligned with L1Md and other retroposon and retrotransposon sequences in Fig. 4 of Chapter 18 of this volume.

The 5′ 253 bp and the 3′ 253 bp of the ingi-3 sequence are homologous to the two halves of a previously described short repetitive element of *T. brucei* named RIME, which contains a 160-codon ORF (46). The RIME element could have arisen by deletion of the central 4.7 kb of the 5.2-kb ingi element, or ingi could have arisen by insertion into RIME. These short RIME elements are somewhat less abundant than ingi elements (58).

3. Ribosomal insertion elements in insects

A fraction of the 28S rRNA genes (>10%) in several insect species are interrupted by non-rDNA elements about 5 kb in length (Fig. 3). These elements have been classified into two groups, type I (or R1) and type II (or R2), based on the sequences at their junctions with rDNA (17, 87, 93).

Type I elements are found at a specific site about two-thirds of the way from the 5′ end of the 28S rRNA gene. They are flanked by a 14-bp duplication of rDNA sequence (17, 87, 93).

Type II elements are located 75 bp upstream from the type I insertion site. The *Drosophila* type II elements (17, 87) have poly(A) tails 13 to 22 nucleotides in length at one end. Those of *B. mori* end in GAAAA (9). Many of the *Drosophila* elements are truncated at the 5′ end (17, 87), while such truncated *Bombyx* elements have not been found. Type II elements are not flanked by short direct repeats. Type II elements occur at about 25 copies per *B. mori* genome. About 20 copies are in 28S rDNA (out of about 240 rDNA units in total) and 5 are at undetermined locations elsewhere in the genome (9). The sequence of one complete *B. mori* type II element (R2Bm) showed that it is 4.2 kb long and contains a 1,151-codon ORF encoding amino acid sequence motifs that are conserved in reverse transcriptases (Finnegan, this volume, Chapter 18, Fig. 4). The predicted amino acid sequence shows a significant similarity to ORF-2 of mammalian L1 (9).

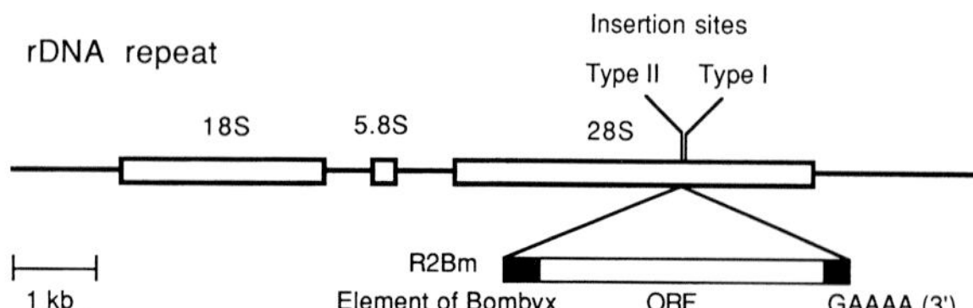

Figure 3. Insertion sites for ribosomal insertion sequences of insects. Transcription of the rDNA is from left to right. Open boxes indicate the positions of the mature 18S, 5.8S, and 28S rRNA sequences within the rDNA repeat. The R2Bm type II insertion sequence of *B. mori* is drawn to scale. Untranslatable 5′ and 3′ regions are filled and the ORF is open. Transcription of the R2Bm ORF is from left to right.

4. The Cin4 element of *Zea mays*

The Cin4 element of *Z. mays* is an interspersed repetitive sequence with a copy number of 50 to 100 per diploid genome (102). Individual Cin4 elements terminate at one end with 6 to 11 A residues and are flanked by short direct repeats of 3 to 16 bp. The largest element studied is about 7 kb in length, and six other mapped elements are truncated at their 5′ ends to different extents. Overlapping DNA sequences derived from the analysis of seven Cin4 clones were combined to reveal two ORFs, 2,793 bp (ORF-1) and 3,489 bp (ORF-2) in length. ORF-2 also contains amino acid sequence motifs that are characteristic of reverse transcriptases (Finnegan, this volume, Chapter 18, Fig. 4).

C. Is THE 1 Another Family of LINEs?

THE 1 (transposonlike human element 1) is a 2.3-kb element terminated at both ends by 350-bp LTRs each of which begins with the sequence 5′-TG. . . and ends with . . .CA-3′ (83). The internal region of the element is present at 10,000 copies in the human genome, generally associated with LTRs, whereas the isolated LTRs are present at 30,000 copies per genome. The element is flanked by 5-bp direct repeats. One representative of the THE 1 family has been sequenced (83). In spite of its structural similarities to retroviruses and other retrotransposons, no sequence similarity between THE 1 and other elements has been discerned. The longest ORF found is only 129 codons long, and thus the one sequence known may be that of a defective element.

In copy number, THE 1 conforms to the original definition of LINEs. In size, it is intermediate between

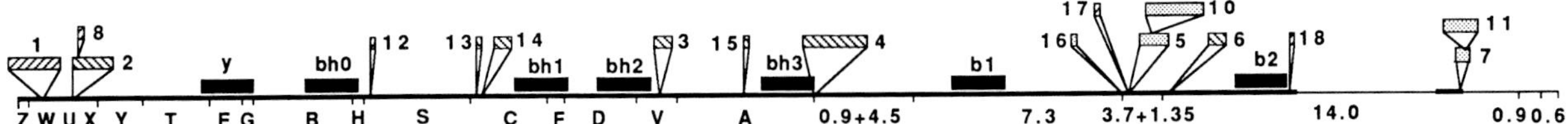

Figure 4. Location of L1Md elements in the BALB/c mouse β-globin locus. The region depicted is 65 kb in length. Solid boxes represent β-globin genes and pseudogenes. Boxes hatched in either direction represent the two orientations of the L1 elements (all are 5′ truncated) (leftward hatching, 5′ to 3′; rightward hatching, 3′ to 5′). Stippled boxes represent rearranged L1 elements. The numbers 1 to 18 indicate L1 elements L1Md-1 through L1Md-18. The letters and numbers below the line refer to the *Eco*RI restriction fragments from the locus (54); downstrokes indicate *Eco*RI sites. Heavy line, Region analyzed; light line, region not analyzed.

SINEs and previously described LINEs, and thus illustrates how arbitrary these definitions can be. It seems most reasonable to classify THE 1 with the retrotransposons because it is terminated by LTRs, although there is no direct evidence that its movement involves the reverse transcription of an RNA intermediate. As noted above, we use the term LINEs to mean retroposons which appear to encode reverse transcriptase involved in their own transposition.

IV. GENOMIC DISTRIBUTION OF L1

Many experiments have shown that L1 elements are highly dispersed in mammalian genomes. (i) Southern blots of genomic DNAs with L1-specific probes result in heterogeneous smears of L1-containing fragments (11, 12, 30, 50). The size distribution of L1 fragments seen depends on the restriction endonucleases used for DNA cleavage and on the L1 segment used as the probe. Such distributions indicate that L1 sequences are dispersed, not clustered in tandem repetition. (ii) In situ hybridization to metaphase chromosome squashes demonstrates L1-related sequences at many locations (4, 67). (iii) L1-containing DNAs are found in the main band rather than in satellite DNA bands in buoyant density gradients, although the distribution in the main band has been reported to be nonuniform (50, 115). (iv) About 1/10 (African green monkey [42]) to 1/3 (mouse [4, 120; Comer, unpublished results]) of the clones in recombinant DNA libraries with average insert sizes of 15 kb contain L1-related sequences. It has been estimated that L1Md elements constitute about 10% of mouse genomic DNA, based upon probing duplicate plaque lifts of such libraries with probes spanning L1Md to count elements and determine the size distribution of truncated elements (Fig. 2).

Additional information comes from the fine structure mapping and sequencing of cloned DNAs. L1 elements have been found within genes and pseudogenes for β-globins (1, 10, 18, 120), immunoglobulins (26, 39, 40, 123), insulin (114), c-*myc* (55), MHC (62), and factor VIII (56) and in the introns of the mouse kallikrein (71) and mouse and human MHC genes (57, 61). The association of L1 with processed pseudogenes (inactive gene copies resulting from insertion of reverse transcripts of mRNA) is reported to be more frequent than expected at random (47, 97). They are also associated with (both inserted into and adjacent to) other repetitive sequences including other L1 elements (121), SINEs of the primate *Alu* family (36), and alpha satellite (42, 85, 86, 118). Finally, searches of sequence data bases often reveal previously undiscovered L1 elements in the sequences upstream or downstream of the gene being studied (47, 92). For example, we found 17 L1 elements in a 65-kb segment of the mouse β-globin complex (Fig. 4). Of the 58 kb sequenced, 13 kb is L1Md. Some of these L1 elements differ by up to 40% from the L1Md consensus sequence and were not detected by Southern hybridization. Thus, even the value of 10% may underestimate the true fraction of the mouse genome represented by L1 elements.

V. MOBILITY OF L1

Although there is no experimental system for detecting L1 transposition in the laboratory, a variety of evidence indicates that L1 elements do transpose. The best evidence comes from findings of polymorphisms for the presence or absence of L1 at specific loci: the mouse β-globin complex (10, 11a, 106), the rat *Igh* and *Mlvi-2* loci (26), the rat insulin locus (60), the dog and human c-*myc* genes (55; B. Morse and S. Astrin, personal communication), and the human gene for blood clotting factor VIII (56). Figure 5 shows nine cases in which both the DNA flanking an L1 element and the uninterrupted target site have been sequenced. In eight cases, the L1 element is flanked by SDRs of a 7- to 15-bp sequence which occurs only once in the empty site. This shows that the SDRs flanking L1 elements are actually target site duplications.

Polymorphisms for the presence of L1 elements at particular sites could be due to L1 insertion or to

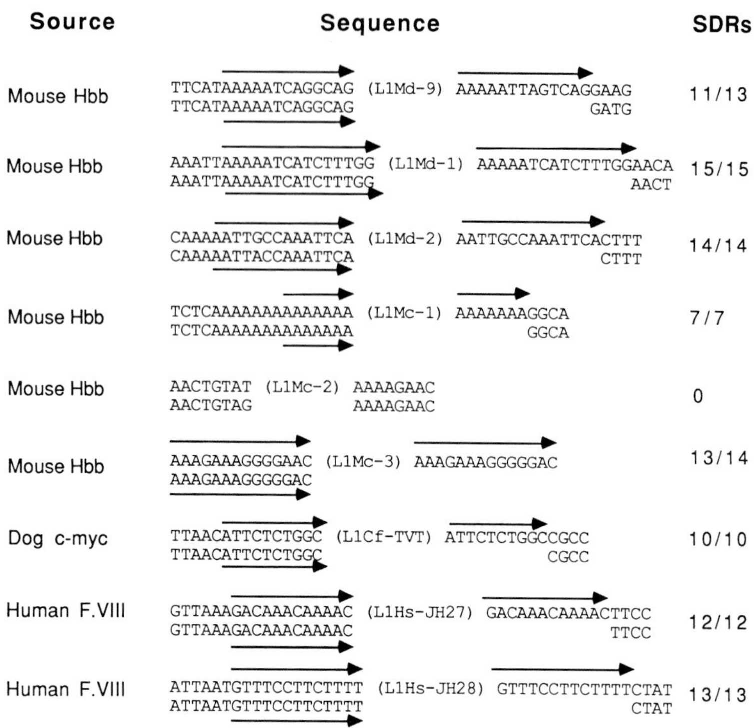

Figure 5. L1 flanking sequences and uninterrupted target sites. The species and genetic locus for each L1 element are listed under Source. References: For the first mouse *Hbb* example, reference 10; for the remaining mouse *Hbb* examples, Casavant et al. (submitted); for dog c-*myc*, reference 55; for human factor VIII, reference 56. Each L1 element with its flanking sequences is shown with the A-rich 3′ end on the right. The sequence of the uninterrupted target site is indicated below each element. A gap is introduced into the target site at the point of L1 insertion for alignment. Arrows indicate the SDRs flanking each L1 and the region of the target which is duplicated. The number of matches and the lengths of the SDRs are indicated (matches/length) for each L1 element in the final column.

precise excision of the element plus one SDR. There is no direct evidence for L1 excision, although it could conceivably occur by homologous recombination (deletion) between the SDRs. The existence of SDRs indicates that L1 can transpose, independent of whether or not it undergoes excision.

One study compared the presence or absence of two particular L1 elements originally identified in inbred mice (one in the β-globin complex and one in an intron of the albumin gene) in a number of different wild mice (2). These specific L1 insertions were found only in some *M. domesticus* and *Mus musculus* individuals and not in any more distantly related mice. This indicates that they are due to a recent insertion and are not relics of an ancestral chromosome from which the elements had been excised in some species.

A. L1 Transposition Is Rapid Enough to Explain Concerted Evolution

Members of the L1 family within a species are more similar to each other than are L1 elements from different species (70) (Table 1). This phenomenon, called concerted evolution, is quite different from that found with independently evolving genes. For example, the mouse and human embryonic β-like globin genes are much more similar to each other than either is to the adult β-globin gene from the same species. Two different explanations for concerted evolution of L1 have been proposed (44, 70). (i) Most L1 sequences within a species may have derived from recent proliferation of a small number (one to three) of sequences known as **molecular drivers.** (ii) Alternatively, preexisting L1 elements may have undergone gene conversion from one or a few master copies (24).

The rapid transposition model predicts that most L1 elements occupy unique species-specific sites. The gene conversion model predicts that most sites of L1 insertion in a particular species would also be occupied by L1 in closely related species. To distinguish between the two models, the locations of L1 elements in the β-globin loci of two mouse species were compared (11a). Of 10 cases where the site of an L1 insertion was examined in a divergent chromosome, the site was vacant in 9 cases. The homologous loci compared were *M. domesticus* versus *Mus caroli* (6 million years [Myr] divergence; 4/4), *M. domesticus* BALB/c (*Hbb*^{d}) versus C57BL (*Hbb*^{s}) (4 Myr divergence; 3/3), and two chromosomes from a noninbred *M. caroli* (1 Myr divergence; 2/3). These results are consistent with a rate of transposition high enough to account for concerted evolution of L1 sequences in the mouse genome without invoking gene conversion.

It is not clear whether L1 movement occurs at a constant rate throughout evolution or in short bursts. The establishment of stable allelic differences with respect to L1 location implies L1 mobility in germ line tissue, independent of whether L1 transposition also occurs in somatic cells. The data described above do not give a measure of the total rate of transposition within the germ line of an individual, since only those transposition events which have become genetically fixed in the population were observed. Nor do the data give a measure of the frequency of somatic L1 transpositions.

B. Historically Recent L1 Transposition Events

Several specific L1 transposition events have been detected which have occurred within identified individuals or their parents.

In one study, three male patients with hemophilia were found to have L1 insertions that inactivate the X-linked clotting factor VIII gene (56; H. Kazazian, personal communication). Two insertions were at different sites in coding sequence (56) and a third was in an intron (Kazazian, personal communication). They were not present in either copy of the maternal factor VIII gene (nor in the paternal copy), indicating that L1 moved either in the maternal gametes or in the patients themselves. The possibility that these insertions are present in the germ lines of the affected individuals has not been tested.

In another study, mammary tumor cells were found to have an L1 insertion in an intron of one copy of the c-*myc* gene (Morse and Astrin, personal communication). Other cell types tested from the same 47-year-old woman did not have the same insertion, indicating that L1 moved during her lifetime. Only one end of this L1 insertion has been sequenced, and the existence of SDRs flanking the element has not been established, and so this case has not yet been shown to result from a normal L1 transposition event. It is not known whether this L1 insertion is involved in tumorigenesis.

These above results demonstrate that L1 transposition is occurring in the current mammalian genome. They argue against alternative models in which current L1 inserts would reflect much earlier transposition activity. These are also clear cases of allelic differences for the presence or absence of L1 which result from L1 insertion, and are not simply due to the excision of preexisting elements.

VI. EXPRESSION OF L1

The movement of an L1 element is thought to require its proper transcription as a discrete mRNA, followed by translation of that mRNA. This section reviews evidence and speculation concerning L1 transcription and translation and the regulation of these processes.

A. Implications of the L1 DNA Sequence for Its Expression

Two different repetitive motifs (A and F) are present at the 5′ ends of full-length L1Md elements (31, 64, 81, 124). It has been proposed that A and F monomer sequences are used to regulate tissue-specific L1 transcription, either as a specific enhancer or as a promoter (64). The 5′ ends of A-type elements tend to be at about the two-thirds point, about 78 bp from the 5′ end of the 208-bp monomer unit (105). Just 5′ to this endpoint is a 5/6-bp match with a consensus promoter sequence for several housekeeping genes (CCGCCC) (72, 74, 88, 127), suggesting that this endpoint corresponds to a region of transcription initiation (64).

Characterized full-length L1Hs elements, in contrast, do not have a repetitive 5′ motif, and instead end at a specific sequence (43). Consequently, the transcription of L1Hs and L1Md may be regulated differently. Primer extension studies of L1Hs full-length RNAs from the germ line-like Ntera2D1 cells indicate that transcription is initiated at the same position that defines the 5′ end of full-length genomic elements (109, 111, 113). This implies that the signals for the initiation of L1Hs transcription in Ntera2D1 cells either lie within the transcribed sequence or extend beyond the 5′ end of L1Hs defined by sequence analyses. It is possible, for example, that a small number of master L1Hs elements contain promoters at their 5′ ends and that such master elements have not been sequenced.

Expression of L1-encoded proteins may also be controlled at the level of translation. ORF-1 and ORF-2 of mouse L1 overlap by 14 bp such that a +1 or −2 frameshift would generate an ORF-1/ORF-2 fusion protein. Similar short out-of-frame overlaps are found with the *tya* and *tyb* genes of *Saccharomyces cerevisiae* Ty elements (13, 49) and the *gag* and *pol* genes of several (but not all) retroviruses. The ORF arrangements are similar to L1Md, except that the retroviral ORFs would be fused by a +2 or −1 frameshift (rather than a +1 or −2 frameshift) (101). Such overlaps can be used to control the relative amounts of the two proteins produced; synthesis of the downstream protein requires a translational frameshifting event (13, 52, 73; H. Varmus and P. Brown, this volume). L1Hs appears to have two nonoverlapping ORFs separated by termination codons (109, 111), and L1Hs may thus rely on UAA suppression to regulate the relative levels of ORF-1 and ORF-2 proteins.

L1 ORFs may also be expressed as fusion products due either to insertion into an active gene or to subsequent illegitimate recombination. For example, an L1 element was involved in a secondary rearrangement of an active immunoglobulin gene in an Abelson murine leukemia virus-transformed pre-B-cell line (126). L1Md contains a sequence similar to recognition sequences involved in the recombination of immunoglobulin gene segments just downstream of the site of this recombination (126). This secondary rearrangement fused the immunoglobulin reading frame with that of the L1 element. It is unclear whether this L1-immunoglobulin gene fusion is expressed, but as a general principle, rearrangements of this sort must provide a way in which L1-encoded proteins or fragments can be expressed, often in a tissue-specific manner.

B. Studies of L1 Transcription

L1 transcription is required for its proliferation, according to current models of L1 transposition. Typically, when Northern (RNA) blots of total cellular RNA are hybridized with L1 probes, a smear of hybridizing RNA representing heterogeneously sized polymerase II transcripts is seen. Similar size distributions are seen with RNAs from a variety of mouse (31, 53, 115), human (98, 103, 112, 116), and rat (114, 125) cell lines and tissues. The intensity of the smear is variable, indicating that some cells produce more L1 RNA than others. There are several possible sources of the L1 RNA size heterogeneity. L1 elements which are truncated to various degrees may be located within the transcription units of other genes. Examples include (i) L1Md in the first intron of a kallikrein gene (71) (here the L1 sequence is expected to be spliced out of the primary transcript of that gene), and (ii) L1Hs just downstream of the promoter of an integrated human leukemia virus (79). The heterogeneously sized L1 RNAs observed on Northern blots may also be degradation products of longer RNAs. These longer RNAs may be fortuitously transcribed from exogeneous promoters, or they may be full-length RNAs initiated from L1 promoters.

Results detailed in Section V indicate that L1 can transpose in the germ line and suggest that functional L1 transcripts be sought in germ line cells. Human Ntera2D1 and mouse F9 are teratocarcinoma cell lines which are considered equivalent to germ line cells. Full-length discrete L1 transcripts have been found in cytoplasmic poly(A)-selected RNA from Ntera2D1 cells (112) and in cytoplasmic RNA from F9 cells (S. Schichman, unpublished results). These transcripts consist specifically of the strand which encodes ORF-1 and ORF-2. A cDNA library was made from the discrete 6.5-kb transcript of Ntera2D1 cells (113). DNA sequence analysis of these cDNAs indicates the presence of three subsets of transcripts (113). All cDNAs had significant ORFs, but each contained termination codons that would have blocked translation into polypeptides of the expected length. Evolutionary analysis of L1Md sequences (see below) indicates that elements capable of transposition contain intact ORFs. We believe it is unlikely that L1Hs and L1Md differ in this respect, and therefore conclude that the majority of the full-length transcripts observed in Ntera2D1 cells are probably not from functional (transposition-competent) elements.

Two discrete mRNA L1 transcripts (both sense strand; 8 and 18 kb) have been found in poly(A)-containing cytoplasmic RNA from several murine lymphoid cell lines (25). The sequences of these RNA species are not known, so their homogeneity is not known, nor is it clear whether they come from functional elements.

Several discrete cytoplasmic transcripts containing L1Hs sequences but shorter than 5 kb have been described (59, 98). These transcripts probably do not reflect expression that would lead to transposition, since they are too short to contain the entire L1 coding capacity.

C. L1 Protein Products?

L1-encoded proteins have not been detected in vivo, and the only evidence that L1 elements encode active or functional protein products comes from the finding of highly conserved ORFs in L1Md (64, 105, 109, 111) and L1Hs (47, 48, 96). The ORFs from these elements have been compared by analyzing the ratio of codon changes that result in amino acid replacement to those that are silent (R/S analysis). When there is no selection for protein function, as in pseudogenes, replacement changes are about three times more common than silent changes. When a sequence is under selection for protein function, replacement and silent changes are about equally frequent. R/S analysis of mouse and human L1 ORFs indicates that the 3′ half of ORF-1 and the entire ORF-2 have evolved under selection for protein function (64). Analysis of evolutionary trees of mouse genomic L1 sequences (see Section VII) indicates that only L1 sequences under selection for protein function proliferate (44), in accord with the model that L1-encoded *cis*-acting proteins mediate L1 proliferation and transposition.

Two basic strategies are being used in attempts to detect and analyze the functions of L1-encoded proteins. Since the putative ORF-2 product has homology to several reverse transcriptases, reverse transcriptase activity has been sought after expression of this ORF in bacteria and *S. cerevisiae*, a strategy used successfully with cauliflower mosaic virus (117), Rous sarcoma virus (94), and human immunodeficiency virus (33). However, no activity has been detected in a standard reverse transcriptase assay (unpublished results).

Another strategy which has often been successful for detecting predicted protein products involves making specific antisera from fusion proteins (122) or specific peptides (19). This strategy has been used with L1Md and L1Hs elements, and polyclonal antisera have been obtained which react specifically with translation products of L1 sequences made in *Escherichia coli* (unpublished results). However, as yet antibodies have not revealed significant levels of L1-encoded proteins in mammalian cells.

VII. MOLECULAR EVOLUTION OF THE LINE-1 FAMILY

The members of the L1 family are ultimately derived from a sequence which is active both in encoding a protein or proteins and in generating a large number of copies which are dispersed in the genome. Observations that most copies are defective due to 5′ truncation or point mutations in the coding sequence and are apparently incapable of further movement have made finding an active L1 sequence a difficult needle-in-the-haystack problem. In contrast, the abundance of L1 DNA sequence data provides an unusual opportunity for a detailed analysis of the molecular evolution of the element. This section reviews how evolutionary analysis has been used to complement the biochemical characterization of the L1 family.

A. Concerted Evolution

The marked homogeneity of L1 elements within a species relative to their divergence between species means that L1 sequences must evolve in concert. The

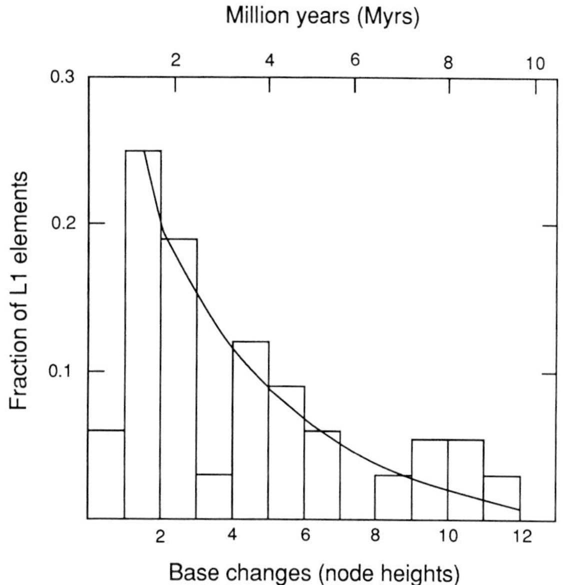

Figure 6. Age distribution of L1Md elements. The age of each element was estimated from the number of base changes in the element since the last node. The rate used to convert base changes to age was 4.1×10^{-9} changes per site per year (70). The curve represents an approximation of the distribution by an exponential function with a half-life of 2 Myr. (Figure redrawn from data in reference 44.)

effect can be observed by restriction mapping (8, 42) or sequencing studies (70). Dover (24) estimated that, in the mouse genome, the spread of diagnostic variants occurred within 1 to 2 Myr, 100-fold faster than would be expected due to random genetic exchanges. The effect of concerted evolution on the distribution of L1 sequences is strikingly portrayed in the work of Martin et al. (70). An evolutionary tree was constructed from the sequences of 30 L1 elements and the age of each element was determined from its branch length. A plot of the resulting age distribution (Fig. 6) shows that half of the sequences currently present in the mouse genome have been derived within the last 3 Myr.

Dover, noting the fast rate of concerted evolution in this family (then called MIF-1), used it along with other examples to illustrate the theory of "molecular drive" (24). Molecular drive, broadly defined, is an evolutionary trend driven not by natural selection, but by a nonrandom molecular genetic mechanism. During concerted evolution of repetitive sequences, molecular drive occurs if there are preferred sequence donors in the homogenizing process (24, 77, 78), thus (i) increasing the rate of homogenization over that attained by random exchanges and (ii) driving the sequences of the family toward that of the preferred donors.

Duplicative movement has been indicated as the

major homogenizing process for mouse L1 (reviewed in Section V), although gene conversion may also play a role (44). The appearance of homogenization is achieved because the older and more highly diverged copies are either deleted or diluted. Major tenets of the theory of molecular drive apply to the duplicative movement model as follows.

(i) The process of homogenization is accelerated by the preferential use of a few specific templates for generating progeny L1 elements. Analysis of evolutionary trees representing the ancestry of 30 L1 elements from rodents showed that only one to three lineages in each species of mouse ever give rise to new branches (44). These special lineages correspond to L1 sequences that are preferentially making new copies, so they were named molecular drivers. A proposed mechanistic basis for the preference is that the special sequences may be transcribed in the germ line and encode a functional reverse transcriptase that mediates copying of the mRNA into DNA and thus the movement to new sites (64, 69, 96).

(ii) Homogenization of the family toward the sequence(s) of the molecular driver(s) is directly seen on the evolutionary trees. Each younger generation of branches originates from a molecular driver and the corresponding sequences are less divergent than are those of the preceding generation. Thus, L1 sequences that are defective due to 5′ truncation often still have long open reading frames.

(iii) The prediction (24) that most sequence variants (caused by L1 insertion at some particular site) will be abundant in the population should be as valid for duplicative movement as for gene conversion models. This is because the spread of variant chromosomes through the population by sexual reproduction in diploid organisms is expected to be much more rapid than either duplicative movement or gene conversion. The extent to which the average L1 insertion has spread has not been experimentally tested; a more extensive theoretical treatment, however, has appeared (76).

Unlike gene conversion, duplicative insertions increase the total number of sequences in the genome. The rate of insertion observed in rodents could not have been sustained since the time of the mammalian radiation without a balancing rate of loss. This has led to the speculation that elements are deleted by recombination between the short direct repeats (44). However, other possibilities exist. (i) Natural selection against chromosomes with too many L1 insertions could limit the spread of new insertions into the population and inhibit their transmission into future generations. (ii) The current rate of 25,000 L1 insertions per Myr in mice may not be typical of the L1 family in other species and at much earlier times. Thus, a species may not be required to tolerate a high rate of insertion for a long time.

B. Concerted Evolution in Primates

A precedent for a reduced rate of movement may be found in the primate L1 sequences. L1 sequences are considerably more divergent in primates than in mice, suggesting a lower rate of homogenization in primates. Most, if not all, sequenced primate L1s are defective due to frameshifts and rearrangements; however, open reading frames appear in consensus sequences (96, 108, 109). This suggests that concerted evolution and molecular drive are operating, although at a much lower rate. Elements that are rearranged beyond typical 5′ truncation or are missing structural features, such as the small flanking direct repeats (113), are more prevalent in primates than in rodents. Some of these defects may represent a normal accumulation of mutations over a long period since the initial insertion event. For instance, it would be difficult to recognize small direct repeats after about 20 Myr of divergence. Similarly, one can expect deletions, fusions of independently inserted repeats, and insertions of newer repeats into older ones. However, in addition, some L1 elements have internal inversions apparently formed by the movement mechanism itself (see Section VIII.E, below).

C. Reading Frames

Long open reading frames are observed in rodent L1 elements (64, 69) and in primate consensus sequences (96, 108, 109). Evidence for natural selection on the encoded amino acid sequence has been obtained by comparing the frequencies of amino acid replacement and synonymous mutations. This has implied the expression of a functional protein from L1 elements in primates, rabbit, and rodents (18, 44, 64, 69). Selection has been found in multiple molecular lineages within individual rodent species (44), suggesting the presence of more than one functional L1 gene per species. These analyses indicated that about half of all possible one-base changes that would have caused amino acid replacements were selected against. This is similar to the extent of selection operating on globin genes.

D. The Number and Kinds of Rodent L1 Elements

1. Defective elements

Because of 5′ truncation, most rodent L1 elements contain only fragments of the reading frame and accordingly evolve as typical pseudogenes (44).

Supporting evidence includes the observations that (i) their lineages on an evolutionary tree contain no branches, indicating that these L1 copies have not undergone further duplicative movements, and (ii) the mutations in these lineages are randomly distributed between replacement and synonymous sites. The observed mutations deviate slightly from this expectation, which is discussed in Section VII.D.3, below. Most, if not all, L1 repeats that have been cloned and sequenced are defective elements.

Sequences from candidates for active elements can be examined for characteristics of defective elements. These include obvious defects such as nonsense codons and frameshifts, and also a random distribution of replacement and synonymous mutations indicating the absence of selective pressure on the encoded amino acid sequence. Both full-length genomic L1 sequences (105, 113) and cDNA clones (113; G. Raschella, unpublished results) have been detected that have these properties, thus eliminating them from consideration as putative active elements. However, not all defective elements can be detected in this way. In Section VII.E, below, we discuss how sequence analysis can be used to identify definitively sequences that should be active.

2. Molecular drivers

An analysis of evolutionary trees based on L1Md sequences (44) revealed duplicative events in only one to three lineages per species, indicating that the number of fully active elements may be very small. These lineages represent the molecular drivers discussed in Section VII.A, above. Tabulation of replacement and synonymous mutations on each branch of the tree showed that the molecular drivers are under selection for the encoded protein(s). Thus, it was concluded that elements that are active in retroposition are required in order to express active proteins. These sequences have not been cloned, but are reconstructed ancestors of present-day defective L1s. It is unclear if the multiple molecular drivers represent functional L1 elements at separate loci or are alleles of a single functional L1 element. At the extreme, the data are compatible with a single functional L1 locus per haploid genome.

3. Other active L1 elements

Finally, some sequences have undergone natural selection for the encoded protein but have not become molecular drivers (44). They were revealed by the ratios of replacement and synonymous mutations on the pseudogene branches in L1 evolutionary trees. Each branch, regardless of its length, deviated from the ratio expected for pseudogenes in a way suggesting that it was under selection for the first million years of its existence. This can be conceptualized in terms of the molecular driver passing its sequence to other functional L1 elements which only last about 1 Myr, during which time they produce a small number of defective copies and then become quiescent or are lost. These other active sequences share the selective pressure and copy-making potential of molecular drivers; but, in contrast to molecular drivers, they do not succeed in surviving long term or in populating a large fraction of the subsequent population of repeats with their progeny. The number of nondriver active L1 sequences could be as small as one or as large as the number of full-length elements in the genome (on the order of thousands).

These active nondriver sequences are the key to understanding the evolutionary dynamics of the L1 family. Unfortunately, numerous possibilities fit the data. (i) They could represent fully active elements with the potential to be molecular drivers, but which lost out to the sequence now recognized as the molecular driver in selfish competition to expand the number of their progeny. (ii) They could be alleles of the molecular driver that never spread to a large portion of animals in the population. (ii) They could be members of a family of active genes which includes the molecular driver, and whose individual identity was periodically erased by gene conversion from the molecular driver.

With this background, note that our conclusion that duplicative transposition drives the concerted evolution of L1 sequences applies specifically to the production of the defective L1 copies. We do not know either the intensity or mechanism of homogenization among active L1 sequences.

E. Construction of Active Sequences

Sequence analysis can supplement the strategy of testing candidates for active L1 elements for the capability to produce a protein. The latter test is not foolproof, since there may be defective L1 elements that are transcribed, translated, and produce a protein with partial enzymatic activity, but which is prevented from fulfilling its full biological function by some point mutation.

To circumvent the problem of identifying and cloning a fully active L1 element from among the approximately 100,000 defective elements, one can use sequence analysis to define a recent ancestor with certified properties of a molecular driver and then rebuild the sequence using recombinant DNA techniques. The key to this approach is in the definition of a molecular driver. (i) An ancestral sequence is defined to be a molecular driver because a number of its progeny L1 elements have been cloned and sequenced. Therefore, its duplicative capacity cannot be in doubt. (ii) Selective pressure on the ancestral sequence can be evaluated by measuring the frequencies of replacement and synonymous mutations, thus providing assurance that its encoded protein is fully active. Such a sequence has been defined (105), and its coding region has been reconstructed from restriction fragments of present-day L1Md sequences (unpublished results). This construct can now be used as a source of the encoded proteins.

F. Rate of Accumulation of Point Mutations

The accumulation of point mutations within defective rodent L1 elements has occurred at the rate expected of pseudogenes (44). L1 molecular drivers show an acceleration of the rate of mutations at synonymous sites, coupled with a suppression of mutations at replacement sites (44). Apparently, the sequences are subjected to twice as many mutations per unit time as are pseudogenes, and then half of them are eliminated due to deleterious effects on the encoded protein. The source of the accelerated mutation rate is not known, but could come from error-prone reverse transcription if the molecular driver is itself moving around the genome.

G. Subfamilies of L1

Subfamilies of rodent L1s have been defined by analyses of restriction patterns (8) and DNA sequences from various subregions (44, 64, 81, 124; Raschella, unpublished results). In general, the subfamily structures in various regions have not been correlated with each other. Interpretation is complicated by the existence of the following two different kinds of subfamilies in which all of the members share common mutations. (i) Clades are subfamilies neither of which is ancestral to the other. The number of clades corresponds to the number of molecular drivers. (ii) One subfamily can be ancestral to the other. A new subfamily of this kind can be defined for every mutation in a molecular driver. Construction of an evolutionary tree is required in order to distinguish between these two kinds of subfamilies. A complete description of the subfamilies could, in principle, define the extent to which recombination or gene conversion affects the molecular drivers. Subgroups have also been described for primate L1s (42, 47, 56, 89, 113).

H. Summary

The evolutionary modeling has assisted the characterization of L1 on the following issues. (i) A rate of movement of L1 sequences in rodents was projected (44) and subsequently verified (see Section V). (ii) An assay based on counting replacement and synonymous changes was provided to determine whether a collection of cDNAs was enriched for functional sequences (see Section VII.D). (iii) The finding that those branches of the evolutionary tree which give rise to new L1 elements are under selection for the protein sequences encoded by the ORFs supports the view that these proteins are required for L1 transposition and suggests that they are *cis* acting. And (iv) an ancestral sequence was defined that is projected to be active, both in terms of engaging in and in encoding factors needed for transposition (see Section VII.E).

VIII. MECHANISMS OF L1 PROPAGATION IN THE GENOME

We use the term propagation to refer to the mechanism which has produced a population of L1 elements that are dispersed throughout the genome and are apparently evolving in concert. L1 propagation can be explained by (i) the processes of duplicative transposition and (ii) processes whereby L1 elements are lost from the genome. This section will discuss molecular models of L1 propagation.

Two problems make identification and characterization of the molecular intermediates associated with L1 transposition difficult, as follows. (i) While the rate of dispersal of L1 elements within the genome is rapid in evolutionary terms, it is not clear whether L1 transposition events take place in animals or in cell lines within the time frame of an experiment. (ii) The only candidates for transposition intermediates which have been detected are L1 transcripts. They constitute a mixed population of molecules, most of which are probably not competent to transpose because their ORFs are interrupted by termination codons (discussed in Section VI above).

Current models of L1 propagation therefore draw heavily on sequences of individual L1 elements. Model building based on the evolutionary analysis of sequences is further complicated, since it is not clear whether L1 is simply a selfish transposable element or whether selection may operate at the level of some L1-encoded function needed by the host (28). A number of features have been found in L1 elements which need to be fit into any L1 duplicative transposition model.

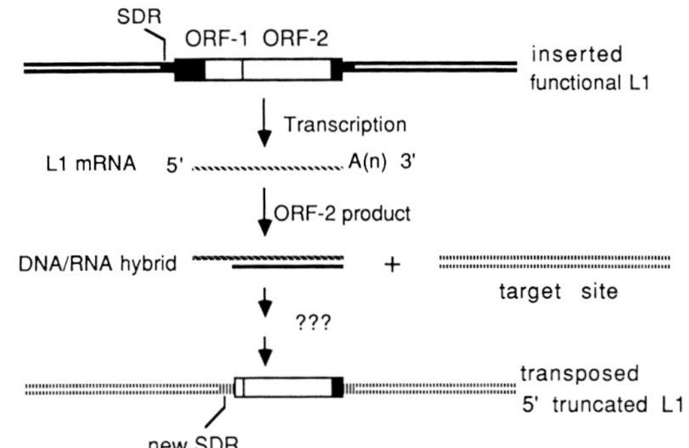

Figure 7. L1 transposition according to the retroposon model. Protein-coding sequence is represented by open boxes and noncoding sequence by solid boxes.

A. L1 Sequence Features Suggest a Model for Transposition

The retroposon model of L1 movement (Fig. 7), in which cDNA copies of L1 mRNA are inserted throughout the genome (31, 120), is based on several features of the L1 structure described above, as follows. (i) L1 elements are flanked by short direct repeats, presumed to result from staggered cuts in target DNA during insertion (31, 120). (ii) An A-rich sequence at one end, preceded by the polyadenylation signal AATAAA (31, 39, 120, 123) suggests that L1 elements are cDNA copies of a poly(A)$^+$ mRNA. (iii) The conserved ORF-2 (4 kb) sequence could encode a protein with homology to reverse transcriptases (48, 82, 119). This suggests that an L1-encoded reverse transcriptase makes cDNA copies of L1 mRNA, which are then inserted into the genome. Because the parental L1 element is not lost, this process results in duplicative transposition (Fig. 7).

5′ Truncation flows naturally from the model, assuming that reverse transcription is often incomplete in vivo, as it is in vitro.

B. L1 Integration Sites

L1 elements are bounded by SDRs (121), which implies that the genomic DNA into which the element inserts is broken via staggered, single-stranded cleavages and that the gaps formed as the element is joined to these ends are filled in by a DNA polymerase (38, 45). SDRs vary in length, generally from 5 to 15 bases, and in sequence, indicating that neither the location of the breaks nor the spacing between them is highly specific. An analysis of rat L1 elements suggested an insertional preference for A-rich regions (14, 37). The reported bias is small, however, and is not obvious for L1Md elements in the mouse β-globin locus (121) or in primates. The reported A-rich insertion site preference may reflect a preference for a particular chromatin structure or a species-specific feature. The clustering of L1 sequences within the Giemsa-dark bands of human chromosomes has been reported and has been attributed to insertional preference for a particular chromatin state (68).

It is not known whether L1 encodes a nuclease which cleaves the target DNA during insertion. It seems quite possible that target DNA nicking is due to other processes. Target DNA might initially simply be nicked and held in place by the complementary strand and hence be resealable. Alternatively, the ends might be separated and hence not easily resealed. That is, the insertional template might be either nicked or broken DNA. If it is broken DNA, produced by host cell factors, the insertion of an L1 element to join the broken ends would constitute a repair event. Such a process would presumably be useful in the survival of the host genome and could serve as a focus of selection.

C. L1 Integration Mechanisms

The molecular details of L1 integration are likely to differ from those of other transposable elements and are unknown. The primer(s) for L1 cDNA synthesis has not been identified. It has been proposed that a 3′ end of the DNA nicked or broken at the insertion site serves as primer for the 3′ end of L1 and other retroposon RNAs (90, 102, 121). Specifically, we suggest that following the synthesis of the ORF-1 and ORF-2 products on cytoplasmic polysomes, the L1 proteins remain associated with the L1 mRNA molecule as a nucleoprotein complex that moves to the nucleus. There is a sequence with homology to zinc-finger DNA-binding proteins in ORF-2, and we suggest that this allows the complex to bind DNA and find a nick or break which provides a 3′ end from which the first strand of cDNA synthesis is primed.

It has been proposed that to complete this process, the other end of the broken target DNA associates with the 5′ end of retroposon RNA either by ligation (90) or by base pairing (102, 121). The 3′ end of the first-strand cDNA is ligated to the 5′ terminus of this broken end, and integration is completed by repair synthesis that produces the second strand of the retroposon cDNA, accompanied by degradation of the RNA template by RNase H (102). A possible integration model of this sort has been described for the *Drosophila* I factor (Finnegan, this volume, see Fig. 8 of Chapter 18).

Since retrotransposons use tRNA molecules to prime cDNA synthesis, the alternate possibility that L1 cDNA synthesis is also primed by a small cellular

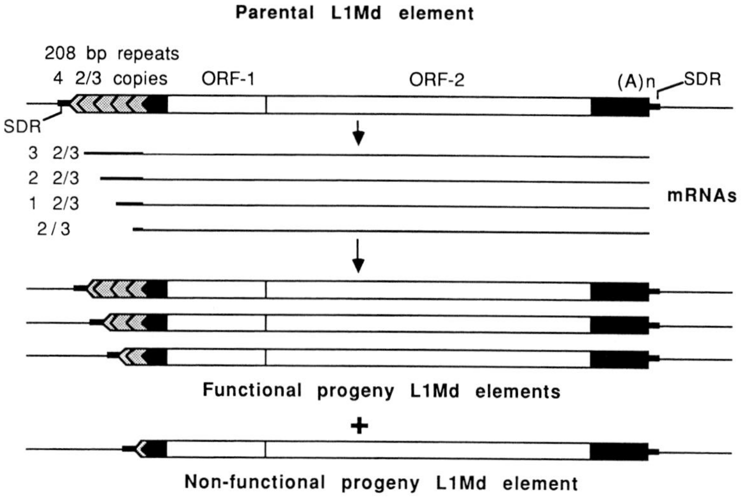

Figure 8. Promoter arrays and L1 transposition. Open boxed regions indicate protein-coding sequence. Chevrons indicate 208-bp repeats, which are postulated to contain a promoter element. Other noncoding regions are in black. The portion of the mRNAs transcribed from the 208-bp repeats is indicated by a thicker line.

RNA must also be considered (Varmus and Brown, this volume). Recently, Bertling et al. (W. Bertling, C. A. Hutchison III, and M. H. Edgell, unpublished results) have found a tRNA-like sequence within the 208-bp repeat at the 5′ end of L1Md, oriented so that the L1 transcript would carry the complement of the tRNA-like sequence near its 5′ end. Similar structures were found in the same orientation in the L1 of every species from which sequence is available and in the *Drosophila* I and the trypanosome ingi elements. We do not know whether the similarity of these sequences to tRNAs is statistically or functionally significant.

D. Maintenance of L1 Regulatory Sequences

New copies of a selfish transposable element should be able to initiate further rounds of transposition. Assuming that L1 mRNAs are Pol II transcripts, the direct insertion of a cDNA copy of a transcript into the genome would not be expected to give rise to a new transcriptional unit because Pol II promoters are normally external (5′) to the transcript. LTRs containing Pol II promoters present in retrotransposons solve this problem for these elements (Varmus and Brown, this volume; J. D. Boeke, this volume), but none of the "full-length" L1 elements has been found to carry such LTRs. On the other hand, L1Md elements carry a 5′ structural motif which appears to provide an alternative solution to this problem of how a Pol II promoter might transpose with the L1Md element. The 5′ ends of L1Md elements contain a tandem array of a 208-bp unit containing a putative Pol II initiation sequence (28, 64). Each repeat contains enough DNA for one nucleosome, suggesting that an array of promoters in a specific phase with respect to chromatin structure is involved. An element carrying an array of $N + 2/3$ copies of the 208-bp repeat would be expected to produce N different transcript species (Fig. 8). Transcription initiation from any but the 3′-most initiation sequence in this region would therefore give rise to a transcript carrying a complete Pol II transcription initiation sequence. Each newly transposed element would contain fewer copies of the promoter than its parent (Fig. 8), but unequal crossing over between L1 elements on homologous chromosomes could lead to expansion of this promoter array. Although this model is appealing for the transposition of functional L1Md elements, it cannot be general, since no equivalent sequences have been found associated with L1 elements in primates or in the rat. The proliferation of these other elements may depend on fortuitous insertion near promoters that are active in the germ line.

E. L1 Elements with 5′ Inversions

In addition to full-length and simple 5′-truncated elements, other rearranged L1 structures have been described in primates (56, 63, 85) and rodents (40, 71). Recently, D. Loeb (unpublished results) has analyzed a series of L1 elements in the mouse β-globin locus. In addition to 5′-truncated elements, five examples of another type of L1 element were found that formally can be derived from the truncated elements by inversion of the 5′ end or by an inversion plus a small deletion (Fig. 9).

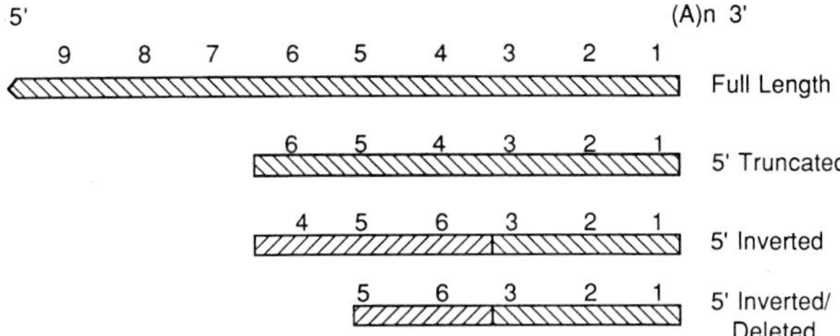

Figure 9. Variant L1 structures. The three common types of noncanonical L1 structures are compared to the canonical or full-length element. The numbers are an arbitrary indication of position along the element. The direction of hatching indicates the orientation of each element or segment thereof.

All four classes of L1 elements (Fig. 9), including those with inversions, are flanked by short direct repeats. It therefore appears that the inversions arose prior to or during insertion, not by rearrangement of previously inserted elements. Each strand of an internally inverted L1 element contains sequence from both strands of the canonical full-length structure. We propose that all of the variant L1 structures are due to aberrant cDNA synthesis similar to that often seen in vitro (Fig. 10). If, following synthesis of the first cDNA strand, RNase H activity degrades the L1 RNA template, the 3′ end of the cDNA strand may sometimes fold back on itself and prime second-strand synthesis from diverse locations, as happens in vitro. This results in a double-stranded cDNA with a single-stranded loop at one end (Fig. 10), which can be integrated only after cleavage of the loop. Complete digestion of the loop results in a 5′-truncated L1 element shorter than the original first cDNA strand. A single nick at one end of the loop would result in an L1 with a 5′ inversion. Partial digestion results

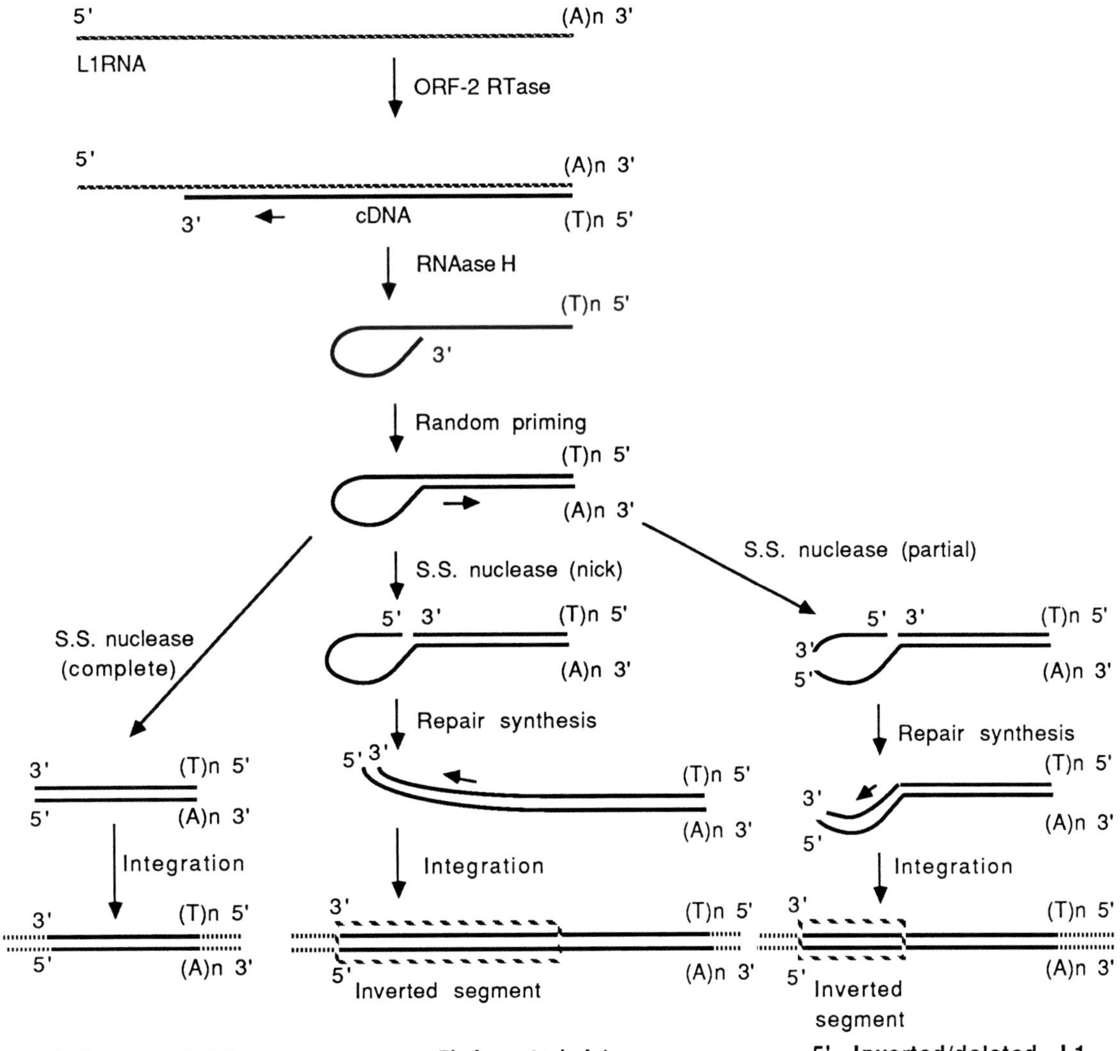

Figure 10. A model for 5′ inversion of L1. A possible mechanism to generate 5′-truncated, 5′-inverted, and 5′-inverted/deleted L1 elements. L1 DNA strands are indicated by solid lines. RNA and flanking DNA are indicated by broken lines.

in L1 elements with a 5′ inversion plus a deletion (Fig. 10).

Although this model for generating variant L1 elements is in accord with the general retroposon model of L1 transposition, it does not bear on the integration mechanism itself. The variant double-stranded structures might be synthesized either free in the cell (as depicted in Fig. 10) or attached to one end of the broken DNA at the integration site (as discussed in Section VIII.C above).

IX. BIOLOGICAL SIGNIFICANCE OF LINEs

L1 elements seem to be present in every mammal, and related retroposons are found in most if not all higher eucaryotes. Their structural and functional properties raise questions concerning the relationship between L1 and its host.

A. Are LINEs Selfish?

Sequence analyses have shown that L1 ORFs are under selection for protein function. This implies that L1-encoded protein is advantageous, either for the host animal or for the proliferation of L1 itself. In the former case, the L1 family might provide a reverse transcriptase which is used by the host animal and also litters the genome with its own processed pseudogenes. In the latter case, retroposons might behave only as selfish DNAs that proliferate independently of any beneficial effects on the host (23, 80). In this case, many active elements would vie to populate the genome with their progeny. Those that happen to avoid an inactivating mutation would proliferate. Evolutionary analysis indicates that a small number (one to three) of L1 sequences (the molecular drivers) have generated those progeny elements that can undergo further rounds of transposition (44). This does not fit the simple model of a large population of selfish elements.

The situation might actually be intermediate between the extreme models presented above: a fundamentally selfish retroposon might also have some beneficial effects on host gene expression which would confer a selective advantage on the host.

B. Effects of L1 on Host Gene Expression

A transposable element might reasonably be expected to affect host gene expression by (i) disruption of a gene by insertion within the gene, (ii) insertion in the vicinity of a gene, thereby interfering with the normal regulation of the gene, and (iii) insertion that places the host gene under the control of regulatory sequences within the transposable element. The first two effects would be expected for any sequence which can transpose at random throughout the genome.

Insertion of L1 into the coding region of the human factor VIII gene is an example of direct gene disruption (56). In the cases of L1 insertion within or near genes, the consequences for host gene expression are unclear. An L1 is inserted upstream from exon 1 of the c-*myc* gene in canine transmissible venereal tumor, but is absent at this site in normal tissues (55). There is no direct evidence to support L1 involvement in the observed elevated c-*myc* RNA levels or in tumor formation. Another case involves the insertion of L1 within the c-*myc* gene in a human breast carcinoma (Morse and Astrin, personal communication). In addition to the rearranged c-*myc*, the normal c-*myc* gene is amplified about 20-fold in the tumor cells, and there is no direct evidence to support the involvement of L1 in tumorigenesis.

Although inconclusive, these findings are provocative, and suggest that L1 transposition may be a significant source of both germ line and somatic mutation in mammals.

C. Other Potential Consequences of L1

In addition to effects on host gene expression, other possible effects of L1 on the host can be envisaged.

1. Repair of broken DNA

If L1 does not itself direct target DNA cleavage, it may act to repair breaks caused by other means. The short direct repeats flanking L1 insertions presumably represent repaired ends resulting from staggered breaks on the two strands of the target DNA. These ends are of variable length and of no apparent sequence specificity, consistent with the lack of an L1 integrase. A molecular model has been proposed for such L1 insertion at preexisting breaks in DNA (28, 121).

2. Recombination between L1 elements

Homology-based unequal crossovers between L1 sequences (which make up about 10% of the genome) may be a significant source of translocations and other major genome rearrangements.

3. Chromosome pairing

The distribution of repetitive sequences along chromosomes might provide a pattern for the recognition of homologous chromosomes in the initial stage of pairing at meiosis (68). This might entail a bivalent protein which binds specifically to L1 sequence and to another molecule of the same protein on the homologous chromosome. This would cause strong pairing only when the array of irregularly spaced L1 elements in two homologous chromosomes was aligned in register, thereby favoring legitimate homologous recombination events relative to unequal crossing over.

4. Speciation

The idea that L1 might be involved in speciation is suggested by the finding that the *Drosophila* hybrid dysgenesis factor I is an L1-like retroposon. If L1 transposition is induced in crosses between species in which the L1 sequences are sufficiently diverged, then the resulting hybrid dysgenesis (W. R. Engels, this volume; Finnegan, this volume, Chapter 18) could provide a barrier to cross-species genetic exchange. We are not aware, however, of any evidence for L1-induced dysgenesis in interspecies hybrids. Another model for the involvement of L1 in speciation is based on the chromosome pairing scheme just outlined (Section IX.C.3). In this case, major differences in the pattern of the locations and the sequence of L1 on homologous chromosomes could prevent proper pairing and result in the sterility of hybrids.

L1 may also affect the survival of species due to the genetic load caused by L1-induced mutations. The fitness of host species would therefore be affected by mutations altering the rate of L1 transposition.

5. Production of processed pseudogenes

If certain L1 elements do encode a reverse transcriptase which functions in transposition, then L1 might cause transposition of other, unrelated sequences. It might mediate the transposition of passive retroposons, which do not encode reverse transcriptase, including SINEs and processed pseudogene copies of cellular genes such as globin genes. This would require that the L1-encoded transposition functions operate in *trans*, at least occasionally.

X. FUTURE PROSPECTS

The prospect of reaching into the genome and extracting the L1 elements which represent molecular drivers from the 80,000 copies of L1 remains remote. There are, however, alternative strategies for exploring L1 transposition. Transposition of the *S. cerevisiae* retrotransposon Ty was greatly accelerated by placing its transcription under the control of a strong inducible promoter (5). The exciting possibility of using this approach for L1Md is enhanced because the likely sequence of a functional ancestral element has been deduced (105) and reconstructed by in vitro cloning (unpublished results). Success with this approach could also be important technically in designing a transposon tagging system in mice. If L1 transposition could be turned on by an inducible promoter, then a mutagenesis system comparable to hybrid dysgenesis might result. An advantage of such a system over gene tagging by microinjection of DNA or retrovirus infection of embryos would be its convenience, since conventional mating experiments could be used to produce mice with tagged genes.

Working out the population genetics of the L1 family in mammals by using only the sequence "fingerprints" left by the selective processes of evolution leaves a number of ambiguities in the models. However, the use of L1-like retroposons in *Drosophila* spp. should permit new analyses of the molecular biology of retroposons, much as the Ty element of *S. cerevisiae* has proven a valuable model for the study of retrotransposons.

The L1 family and related active retroposons represent a major class of genetic elements among higher eucaryotes. Their possible roles in the genetic variability, population dynamics, and perhaps even formation or extinction of species remain unclear. Recent advances in our understanding of the structure and function of such elements now permit the formulation of testable theories concerning their biological consequences.

Acknowledgments. This review was written while C.A.H. was on sabbatical leave at the Medical Research Council Laboratory of Molecular Biology, Cambridge, England. We thank the Medical Research Council for facilities, and B. G. Barrell, in whose laboratory this chapter was assembled. We especially thank the many people who made their unpublished results available to us.

We thank the Kenan Foundation for financial support. Work in the authors' laboratories on LINES is funded by Public Health Service grant GM21313 from the National Institutes of Health (to C.A.H. and M.H.E.).

LITERATURE CITED

1. **Adams, J. W., R. E. Kaufman, P. J. Kretschmer, M. Harrison, and A. W. Nienhuis.** 1980. A family of long reiterated DNA sequences, one copy of which is next to the human beta-globin gene. *Nucleic Acids Res.* 8:6113–6128.
2. **Bellis, M., V. Jubier-Maurin, B. Dod, F. Vahlerbergne, A.**

Laurent, C. Senglat, F. Bonhomme, and G. Roizes. 1987. Distribution of two recently inserted long interspersed elements of the L1 repeat family at the Alb and βh3 loci in wild mice. *J. Mol. Evol.* **4**:351–363.

3. **Bennett, K. L., and N. D. Hastie.** 1984. Looking for relationships between the most repeated dispersed DNA sequences in the mouse: small R elements are found associated consistently with long MIF repeats. *EMBO J.* **3**:467–472.
4. **Bennett, K. L., R. E. Hill, D. F. Pietras, M. Woodworth-Gutai, C. Kane-Haas, J. M. Houston, J. K. Heath, and N. D. Hastie.** 1984. Most highly repeated dispersed DNA families in the mouse genome. *Mol. Cell. Biol.* **4**:1561–1571.
5. **Boeke, J. D., D. J. Garfinkel, C. A. Styles, and G. R. Fink.** 1985. Ty elements transpose through an RNA intermediate. *Cell* **40**:491–500.
6. **Bonner, J., W. T. Garrard, J. Gottesfeld, D. S. Holmes, J. S. Sevall, and M. Wilkes.** 1973. Functional organization of the mammalian genome. *Cold Spring Harbor Symp. Quant. Biol.* **38**:303–310.
7. **Britten, R. J., and E. H. Davidson.** 1969. Gene regulation for higher cells: a theory. *Science* **165**:349–357.
8. **Brown, S. D. M., and G. Dover.** 1981. Organization and evolutionary progress of a dispersed repetitive family of sequences in widely separated rodent genomes. *J. Mol. Biol.* **150**:441–466.
9. **Burke, W. D., C. C. Calalang, and T. H. Eickbush.** 1987. The site-specific ribosomal insertion element type II of *Bombyx mori* (R2Bm) contains the coding sequence for a reverse transcriptase-like enzyme. *Mol. Cell. Biol.* **7**:2221–2230.
10. **Burton, F. H., D. D. Loeb, S. F. Chao, C. A. Hutchison III, and M. H. Edgell.** 1985. Transposition of a long member of the L1 major interspersed DNA family into the mouse β-globin gene locus. *Nucleic Acids Res.* **13**:5071–5084.
11. **Burton, F. H., D. D. Loeb, C. F. Voliva, S. L. Martin, M. H. Edgell, and C. A. Hutchison III.** 1986. Conservation throughout mammalia and extensive protein-encoding capacity of the highly repeated DNA long interspersed sequence one. *J. Mol. Biol.* **187**:291–304.

11a. **Casavant, N. C., S. C. Hardies, F. D. Funk, M. B. Comer, M. H. Edgell, and C. A. Hutchison III.** 1988. Extensive movement of LINES ONE sequences in β-globin loci of *Mus caroli* and *Mus domesticus*. *Mol. Cell. Biol.* **8**:4669–4674.

12. **Cheng, S.-M., and C. L. Schildkraut.** 1980. A family of moderately repetitive sequences in mouse DNA. *Nucleic Acids Res.* **8**:4075–4090.
13. **Clare, J., and P. Farabaugh.** 1985. Nucleotide sequence of a yeast Ty element: evidence for an unusual mechanism of gene expression. *Proc. Natl. Acad. Sci. USA* **82**:2829–2933.
14. **D'Ambrosio, E., S. D. Waitzkin, F. R. Witney, A. Salemme, and A. V. Furano.** 1986. Structure of the highly repeated, long interspersed DNA family (LINE or L1Rn) of the rat. *Mol. Cell. Biol.* **6**:411–424.
15. **Davidson, E. H., and R. J. Britten.** 1979. Regulation of gene expression: possible role of repetitive sequences. *Science* **204**:1052–1059.
16. **Davidson, E. H., B. R. Hough, C. S. Amenson, and R. J. Britten.** 1973. General interspersion of repetitive with nonrepetitive sequence elements in the DNA of *Xenopus*. *J. Mol. Biol.* 77:1–23.
17. **Dawid, I. B., and M. L. Rebbert.** 1981. Nucleotide sequence at the boundaries between gene and insertion regions in the rDNA of *Drosophila melanogaster*. *Nucleic Acids Res.* **9**: 5011–5020.
18. **Demers, G. W., K. Brech, and R. C. Hardison.** 1986. Long interspersed L1 repeats in rabbit DNA are homologous to L1 repeats of rodents and primates in an open-reading frame region. *Mol. Biol. Evol.* **3**:179–190.
19. **Dewey, R. E., D. H. Timothy, and C. S. Levings III.** 1987. A mitochondrial protein associated with cytoplasmic male sterility in the T cytoplasm of maize. *Proc. Natl. Acad. Sci. USA* **84**:5374–5378.
20. **Di Nocera, P. P., and G. Casari.** 1987. Related polypeptides are encoded by *Drosophila* F elements, I factors, and mammalian L1 sequences. *Proc. Natl. Acad. Sci. USA* **84**: 5843–5847.
21. **Di Nocera, P. P., M. E. Digan, and I. B. Dawid.** 1983. A family of oligo-adenylate-terminated transposable sequences in *Drosophila melanogaster*. *J. Mol. Biol.* **168**:715–727.
22. **Di Nocera, P. P., F. Graziani, and G. Lavorgna.** 1986. Genomic and structural organization of *Drosophila melanogaster* G elements. *Nucleic Acids Res.* **14**:675–691.
23. **Doolittle, W. F., and C. Sapienza.** 1980. Selfish genes, the phenotype paradigm and genome evolution. *Nature* (London) **284**:601–603.
24. **Dover, G.** 1982. Molecular drive: a cohesive mode of species evolution. *Nature* (London) **299**:111–117.
25. **Dudley, J. P.** 1987. Discrete high molecular weight RNA transcribed from the long interspersed repetitive element L1Md. *Nucleic Acids Res.* **15**:2581–2592.
26. **Economou-Pachnis, A., M. A. Lohse, A. V. Furano, and P. N. Tsichlis.** 1985. Insertion of long interspersed repeated elements at the *Igh* (immunoglobulin heavy chain) and *Mlvi-2* (Moloney leukemia virus integration 2) loci of rats. *Proc. Natl. Acad. Sci. USA* **82**:2857–2861.
27. **Eden, F. C., J. P. Hendrick, and S. S. Gottlieb.** 1978. Homology of single copy and repeated sequences in chicken, duck, Japanese quail, and ostrich DNA. *Biochemistry* **17**: 5113–5121.
28. **Edgell, M. H., S. C. Hardies, D. D. Loeb, W. R. Shehee, R. W. Padgett, F. H. Burton, M. B. Comer, N. C. Casavant, F. D. Funk, and C. A. Hutchison III.** 1987. The L1 family in mice, p. 107–129. *In* G. Stamatoyannopolos and A. W. Nienhuis (ed.), *Developmental Control of Globin Gene Expression*. Alan R. Liss, Inc., New York.
29. **Fanning, T., and M. F. Singer.** 1987. The LINE-1 DNA sequences in four mammalian orders predict proteins that conserve homologies to retrovirus proteins. *Nucleic Acids Res.* **15**:2251–2260.
30. **Fanning, T. G.** 1982. Characterization of a highly repetitive family of DNA sequences in the mouse. *Nucleic Acids Res.* **10**:5003–5013.
31. **Fanning, T. G.** 1983. Size and structure of the highly repetitive BAM HI element in mice. *Nucleic Acids Res.* **11**:5073–5091.
32. **Fanning, T. G., and M. F. Singer.** 1987. LINE-1: a mammalian transposable element. *Biochim. Biophys. Acta* **910**: 203–212.
33. **Farmerie, W. G., D. D. Loeb, N. C. Casavant, C. A. Hutchison III, M. H. Edgell, and R. Swanstrom.** 1987. Expression and processing of the AIDS virus reverse transcriptase in *E. coli*. *Science* **236**:305–308.
34. **Fawcett, D. H., C. K. Lister, E. Kellett, and D. J. Finnegan.** 1986. Transposable elements controlling I-R hybrid dysgenesis in D. melanogaster are similar to mammalian LINEs. *Cell* **47**:1007–1015.
35. **Finnegan, D. J.** 1988. I factors in *Drosophila melanogaster* and similar elements in other eukaryotes, p. 271–285. *In* A. J. Kingsman, S. Kingsman, and K. Chater (ed.), *Transposition*. Cambridge University Press, Cambridge.
36. **Fujita, A., M. Hattori, O. Takenaka, and Y. Sakaki.** 1987.

The L1 family (KpnI family) sequence near the 3′ end of the human beta-globin gene may have been derived from an active L1 sequence. *Nucleic Acids Res.* **15**:4007–4020.

37. **Furano, A. V., C. C. Somerville, P. N. Tsichlis, and E. D'Ambrosio.** 1986. Target sites for the transposition of rat long interspersed repeated DNA elements (LINEs) are not random. *Nucleic Acids Res.* **14**:3717–3727.
38. **Galas, D. J., and M. Chandler.** 1981. On the molecular models of transposition. *Proc. Natl. Acad. Sci. USA* **78**: 4858–4862.
39. **Gebhard, W., T. Meitinger, J. Höchtl, and H. G. Zachau.** 1982. A new family of interspersed repetitive DNA sequences in the mouse genome. *J. Mol. Biol.* **157**:453–471.
40. **Gebhard, W., and H. G. Zachau.** 1983. Organization of the R family and other interspersed repetitive DNA sequences in the mouse genome. *J. Mol. Biol.* **170**:255–270.
41. **Graham, D. E., B. R. Neufeld, E. H. Davidson, and R. J. Britten.** 1974. Interspersion of repetitive and non-repetitive DNA sequences in the sea urchin genome. *Cell* **1**:127–137.
42. **Grimaldi, G., and M. F. Singer.** 1983. Members of the KpnI family of long interspersed repeated sequences join and interrupt alpha-satellite in the monkey genome. *Nucleic Acids Res.* **11**:321–338.
43. **Grimaldi, G., J. Skowronski, and M. F. Singer.** 1984. Defining the beginning and end of *Kpn*I family elements. *EMBO J.* **3**:1753–1759.
44. **Hardies, S. C., S. L. Martin, C. F. Voliva, C. A. Hutchison III, and M. H. Edgell.** 1986. An analysis of replacement and synonymous changes in the rodent L1 repeat family. *Mol. Biol. Evol.* **3**:109–125.
45. **Harshey, R. M., and A. I. Bukhari.** 1981. A mechanism of DNA transposition. *Proc. Natl. Acad. Sci. USA* **78**: 1090–1094.
46. **Hasan, G., M. J. Turner, and J. S. Cordingley.** 1984. Complete nucleotide sequence of an unusual mobile element from *Trypanosoma brucei. Cell* **37**:333–341.
47. **Hattori, M., S. Hidaka, and Y. Sakaki.** 1985. Sequence analysis of a Kpn I family member near the 3′ end of human beta-globin gene. *Nucleic Acids Res.* **13**:7813–7827.
48. **Hattori, M., S. Kuhara, O. Takenaka, and Y. Sakaki.** 1986. L1 family of repetitive DNA sequences in primates may be derived from a sequence encoding a reverse transcriptase-related protein. *Nature* (London) **321**:625–628.
49. **Hauber, J., P. Nelböck-Hochstetter, and H. Feldmann.** 1985. Nucleotide sequence and characteristics of a Ty element from yeast. *Nucleic Acids Res.* **13**:2745–2758.
50. **Heller, R., and N. Arnheim.** 1980. Structure and organization of the highly repeated and interspersed 1.3 kb Eco RI-Bgl II sequence family in mice. *Nucleic Acids Res.* **8**: 5031–5042.
51. **Hwu, H. R., J. W. Roberts, E. H. Davidson, and R. J. Britten.** 1986. Insertion and/or deletion of many repeated DNA sequences in human and higher ape evolution. *Proc. Natl. Acad. Sci. USA* **83**:3875–3879.
52. **Jacks, T., and H. Varmus.** 1985. Expression of the Rous sarcoma virus *pol* gene by ribosomal frameshifting. *Science* **230**:1237–1242.
53. **Jackson, M., D. Heller, and L. Leinwand.** 1985. Transcriptional measurements of mouse repeated DNA sequences. *Nucleic Acids Res.* **13**:3389–3403.
54. **Jahn, C. L., C. A. Hutchison III, S. J. Phillips, S. Weaver, N. L. Haigwood, C. F. Voliva, and M. H. Edgell.** 1980. DNA sequence organization of the β-globin complex in the BALB/c mouse. *Cell* **21**:159–168.
55. **Katzir, N., G. Rechavi, J. B. Cohen, T. Unger, F. Simoni, S. Segal, D. Cohen, and D. Givol.** 1985. "Retroposon" insertion into the cellular oncogene c-*myc* in canine transmissible venereal tumor. *Proc. Natl. Acad. Sci. USA* **82**:1054–1058.
56. **Kazazian, H. H., Jr., C. Wong, H. Youssoufian, A. F. Scott, D. G. Phillips, and S. E. Antonarakis.** 1988. Haemophilia A resulting from *de novo* insertion of *L1* sequences represents a novel mechanism for mutation in man. *Nature* (London) **332**:164–166.
57. **Kelly, A., and J. Trowsdale.** 1985. Complete nucleotide HLA-DPβ sequence of a functional HLA-DP beta gene and the region between the DPβ 1 and DPα 1 genes: comparison of the 5′ ends of HLA class II genes. *Nucleic Acids Res.* **13**:1607–1621.
58. **Kimmel, B. E., O. K. Ole-Moiyoi, and J. R. Young.** 1987. Ingi, a 5.2-kb dispersed sequence element from *Trypanosoma brucei* that carries half of a smaller mobile element at either end and has homology with mammalian LINEs. *Mol. Cell. Biol.* **7**:1465–1475.
59. **Kole, L. B., S. R. Haynes, and W. R. Jelinek.** 1983. Discrete and heterogeneous high molecular weight RNAs complementary to a long dispersed repeat family (a possible transposon) of human DNA. *J. Mol. Biol.* **165**:257–286.
60. **Lakshmikumaran, M. S., E. D'Ambrosio, L. A. Laimans, D. T. Lin, and A. V. Furano.** 1985. Long interspersed repeated DNA (LINE) causes polymorphism at the rat insulin 1 locus. *Mol. Cell. Biol.* **5**:2197–2203.
61. **Larhammar, D., U. Hammerling, M. Denaro, T. Lund, R. Flavell, L. Rask, and P. A. Peterson.** 1983. Structure of the murine immune response I-A beta locus: sequence of the I-A beta gene and an adjacent beta-chain second domain exon. *Cell* **34**:179–188.
62. **Larhammar, D., B. Servenius, L. Rask, and P. A. Peterson.** 1985. Characterization of an HLA DRβ pseudogene. *Proc. Natl. Acad. Sci. USA* **82**:1475–1479.
63. **Lerman, M. I., R. E. Thayer, and M. F. Singer.** 1983. Kpn I family of long interspersed repeated DNA sequences in primates: polymorphism of family members and evidence for transcription. *Proc. Natl. Acad. Sci. USA* **80**:3966–3970.
64. **Loeb, D. D., R. W. Padgett, S. C. Hardies, W. R. Shehee, M. B. Comer, M. H. Edgell, and C. A. Hutchison III.** 1986. The sequence of a large L1Md element reveals a tandemly repeated 5′ end and several features found in retrotransposons. *Mol. Cell. Biol.* **6**:168–182.
65. **Maio, J. J., F. L. Brown, W. G. McKenna, and P. R. Musich.** 1981. Toward a molecular paleontology of primate genomes. *Chromosoma* **83**:127–144.
66. **Manning, J. E., C. W. Schmid, and N. Davidson.** 1975. Interspersion of repetitive and nonrepetitive DNA sequences in the *Drosophila melanogaster* genome. *Cell* **4**:141–155.
67. **Manuelidis, L.** 1982. Repeated DNA sequences and nuclear structure, p. 263–285. *In* G. A. Dover and R. B. Flavell (ed.), *Genome Evolution and Phenotypic Variation.* Academic Press, Inc. (London), Ltd., London.
68. **Manuelidis, L., and D. C. Ward.** 1984. Chromosomal and nuclear distribution of the HindIII 1.9 kb human DNA repeat segment. *Chromosoma* **91**:28–38.
69. **Martin, S. L., C. F. Voliva, F. H. Burton, M. H. Edgell, and C. A. Hutchison III.** 1984. A large interspersed repeat found in mouse DNA contains a long open reading frame that evolves as if it encodes a protein. *Proc. Natl. Acad. Sci. USA* **81**:2308–2312.
70. **Martin, S. L., C. F. Voliva, S. C. Hardies, M. H. Edgell, and C. A. Hutchison III.** 1985. Tempo and mode of concerted evolution in the L1 repeat family of mice. *Mol. Biol. Evol.* **2**:127–140.

71. **Mason, A. J., B. A. Evans, D. R. Cox, J. Shine, and R. I. Richards.** 1983. Structure of mouse kallikrein gene family suggests a role in specific processing of biologically active peptides. *Nature* (London) **303**:300–307.

72. **McKnight, S. L., and R. Kingsbury.** 1982. Transcriptional control signals of a eucaryotic protein-coding gene. *Science* **217**:316–324.

73. **Mellor, J., S. M. Fulton, M. J. Dobson, W. Wilson, S. M. Kingsman, and A. J. Kingsman.** 1985. A retrovirus-like strategy for expression of a fusion protein encoded by yeast transposon Ty1. *Nature* (London) **313**:243–246.

74. **Melton, D. W., D. S. Konecki, J. Brennand, and C. T. Caskey.** 1984. Structure, expression, and mutation of the hypoxanthine phosphoribosyltransferase gene. *Proc. Natl. Acad. Sci. USA* **81**:2147–2151.

75. **Meunier-Rotival, M., P. Soriano, G. Cuny, F. Strauss, and G. Bernardi.** 1982. Sequence organization and genomic distribution of the major family of interspersed repeats of mouse DNA. *Proc. Natl. Acad. Sci. USA* **79**:355–359.

76. **Ohta, T.** 1986. Population genetics of an expanding family of mobile genetic elements. *Genetics* **113**:145–159.

77. **Ohta, T., and G. A. Dover.** 1983. Population genetics of multigene families that are dispersed into two or more chromosomes. *Proc. Natl. Acad. Sci. USA* **80**:4079–4083.

78. **Ohta, T., and G. A. Dover.** 1984. The cohesive population genetics of molecular drive. *Genetics* **108**:501–521.

79. **Okamato, T., M. S. Reitz, Jr., M. F. Clarke, L. L. Jagodzinski, and F. Wong-Staal.** 1986. Activation of a novel *Kpn* I transcript by downstream integration of a human T-lymphotropic virus type 1 provirus. *J. Biol. Chem.* **261**:4615–4619.

80. **Orgel, L. E., and F. H. C. Crick.** 1980. Selfish DNA: the ultimate parasite. *Nature* (London) **284**:604–607.

81. **Padgett, R. W., C. A. Hutchison III, and M. H. Edgell.** 1988. The F-type 5′ motif of mouse L1 elements: a major class of L1 termini similar to the A-type in organization but unrelated in sequence. *Nucleic Acids Res.* **16**:739–749.

82. **Patarca, R., and W. A. Haseltine.** 1984. Sequence similarity among retroviruses. *Nature* (London) **309**:728.

83. **Paulson, K. E., N. Deka, C. W. Schmid, R. Misra, C. W. Schindler, M. G. Rush, L. Kadyk, and L. Leinwand.** 1985. A transposon-like element in human DNA. *Nature* (London) **316**:359–361.

84. **Pittler, S. J., and R. L. Davis.** 1987. A new family of the poly-deoxyadenylated class of *Drosophila* transposable elements identified by a representative member at the *dunce* locus. *Mol. Gen. Genet.* **208**:325–328.

85. **Potter, S. S.** 1984. Rearranged sequences of a human *Kpn* I element. *Proc. Natl. Acad. Sci. USA* **81**:1012–1016.

86. **Potter, S. S., and R. S. Jones.** 1983. Unusual domains of human alphoid satellite DNA with contiguous non-satellite sequences: sequence analysis of a junction region. *Nucleic Acids Res.* **11**:3137–3153.

87. **Rae, P. M. M., B. D. Kohorn, and R. P. Wade.** 1980. The 10 kb *Drosophila virilis* 28S rDNA intervening sequence is flanked by a direct repeat of 14 base pairs of coding sequence. *Nucleic Acids Res.* **8**:3491–3504.

88. **Reynolds, G. A., S. K. Basu, T. F. Osborne, D. J. Chin, G. Gil, M. S. Brown, J. L. Golstein, and K. L. Luskey.** 1984. HMG CoA reductase: a negatively regulated gene with unusual promoter and 5′ untranslated regions. *Cell* **38**: 275–285.

89. **Rogan, P. K., J. Pan, and S. M. Weissman.** 1987. L1 repeat elements in the human epsilon-G-gamma-globin gene intergenic region: sequence analysis and concerted evolution within this family. *Mol. Biol. Evol.* **4**:327–342.

90. **Rogers, J.** 1985. The origin and evolution of retroposons. *Int. Rev. Cytol.* **93**:187–279.

91. **Rogers, J.** 1986. The origin of retroposons. *Nature* (London) **319**:725.

92. **Rogers, J. H.** 1985. Long interspersed sequences in mammalian DNA. Properties of newly identified specimens. *Biochim. Biophys. Acta* **824**:113–120.

93. **Roiha, H., J. R. Miller, L. C. Woods, and D. M. Glover.** 1981. Arrangements and rearrangements of sequences flanking the two types of rDNA insertion in *D. melanogaster. Nature* (London) **290**:749–753.

94. **Roth, M., N. Tanese, and S. P. Goff.** 1985. Purification and characterization of murine retroviral reverse transcriptase expressed in *Escherichia coli. J. Biol. Chem.* **260**: 9326–9335.

95. **Sakaki, Y.** 1987. RNA-mediated (?) dispersion of the L1 family of long interspersed repetitive DNA in mammalian genomes. *Mol. Biol. Med.* **4**:193–197.

96. **Sakaki, Y., M. Hattori, A. Fujita, K. Yoshioka, S. Kuhara, and O. Takenaka.** 1986. The LINE-1 family of primates may encode a reverse transcriptase-like protein. *Cold Spring Harbor Symp. Quant. Biol.* **51**:465–469.

97. **Scarpulla, R. C.** 1985. Association of a truncated cytochrome *c* processed pseudogene with a similarly truncated member from a long interspersed repeat family of rat. *Nucleic Acids Res.* **13**:763–775.

98. **Schmeckpeper, B. J., A. F. Scott, and K. D. Smith.** 1984. Transcripts homologous to a long repeated DNA element in the human genome. *J. Biol. Chem.* **259**:1218–1225.

99. **Schmeckpeper, B. J., H. F. Willard, and K. D. Smith.** 1981. Isolation and characterization of cloned human DNA fragments carrying reiterated sequences common to both autosomes and the X chromosome. *Nucleic Acids Res.* **9**: 1853–1872.

100. **Schmid, C. W., and P. L. Deininger.** 1975. Sequence organization of the human genome. *Cell* **6**:345–358.

101. **Schwartz, D. E., R. Tizard, and W. Gilbert.** 1983. Nucleotide sequence of Rous sarcoma virus. *Cell* **32**:853–869.

102. **Schwarz-Sommer, Z., L. Leclercq, E. Gobel, and H. Saedler.** 1987. Cin4, an insert altering the structure of the *A1* gene in *Zea mays*, exhibits properties of noviral retrotransposons. *EMBO J.* **6**:3873–3880.

103. **Shafit-Zagardo, B., F. L. Brown, P. J. Zavodny, and J. J. Maio.** 1983. Transcription of the *Kpn* I families of long interspersed DNAs in human cells. *Nature* (London) **304**: 277–280.

104. **Sharp, P. A.** 1983. Conversion of RNA to DNA in mammals: *Alu*-like elements and pseudogenes. *Nature* (London) **301**: 471–472.

105. **Shehee, W. R., S.-F. Chao, D. D. Loeb, M. B. Comer, C. A. Hutchison III, and M. H. Edgell.** 1987. Determination of a functional ancestral sequence and definition of the 5′ end of A-type mouse L1 elements. *J. Mol. Biol.* **196**:757–767.

106. **Shyman, S., and S. Weaver.** 1985. Chromosomal rearrangements associated with LINE elements in the mouse genome. *Nucleic Acids Res.* **13**:5085–5093.

107. **Singer, M. F.** 1982. SINEs and LINEs: highly repeated short and long interspersed sequences in mammalian genomes. *Cell* **28**:433–434.

108. **Singer, M. F., and J. Skowronski.** 1985. Making sense out of LINES: long interspersed repeat sequences in mammalian genomes. *Trends Biochem. Sci.* **10**:119–122.

109. **Singer, M. F., J. Skowronski, T. G. Fanning, and S. Mongkolsuk.** 1988. The functional potential of the human LINE-1 family of interspersed repeats. *In Eukaryotic Transposable*

Elements as Mutagenic Agents: Banbury Report 30. Cold Spring Harbor Laboratory, Cold Spring Harbor, N.Y., in press.

110. **Singer, M. F., R. E. Thayer, G. Grimaldi, M. I. Lerman, and T. G. Fanning.** 1983. Homology between the *Kpn*I primate and *Bam*HI (MIF-1) rodent families of long interspersed, repeated sequences. *Nucleic Acids Res.* **11:**5739–5745.

111. **Skowronski, J., T. G. Fanning, and M. F. Singer.** 1988. Unit-length Line-1 transcripts in human teratocarcinoma cells. *Mol. Cell Biol.* **8:**1385–1397.

112. **Skowronski, J., and M. F. Singer.** 1985. Expression of a cytoplasmic LINE-1 transcript is regulated in a human teratocarcinoma cell line. *Proc. Natl. Acad. Sci. USA* **82:** 6050–6054.

113. **Skowronski, J., and M. F. Singer.** 1986. The abundant LINE-1 family of repeated DNA sequences in mammals: genes and pseudogenes. *Cold Spring Harbor Symp. Quant. Biol.* **51:**457–464.

114. **Soares, M. B., E. Schon, and A. Efstratiadas.** 1985. Rat LINE1: the origin and evolution of a family of long interspersed middle repetitive DNA elements. *J. Mol. Evol.* **22:**117–133.

115. **Soriano, P., M. Meunier-Rotival, and G. Bernardi.** 1983. The distribution of interspersed repeats is nonuniform and conserved in the mouse and human genomes. *Proc. Natl. Acad. Sci. USA* **80:**1816–1820.

116. **Sun, L., K. E. Paulson, C. W. Schmid, L. Kadyk, and L. Leinwand.** 1984. Non-*Alu* family interspersed repeats in human DNA and their transcriptional activity. *Nucleic Acids Res.* **12:**2669–2690.

117. **Takatsuji, H., H. Hirochika, T. Fukushi, and J.-E. Ikeda.** 1986. Expression of cauliflower mosaic virus reverse transcriptase in yeast. *Nature* (London) **319:**240–243.

118. **Thayer, R. E., and M. F. Singer.** 1983. Interruption of an alpha-satellite array by a short member of the KpnI family of interspersed, highly repeated monkey DNA sequences. *Mol. Cell. Biol.* **3:**967–973.

119. **Toh, H., H. Hayashida, and T. Miyata.** 1983. Sequence homology between retroviral reverse transcriptase and putative polypolymerases of hepatitis B virus and cauliflower mosaic virus. *Nature* (London) **305:**827–829.

120. **Voliva, C. F., C. L. Jahn, M. B. Comer, C. A. Hutchison III, and M. H. Edgell.** 1983. The L1Md long interspersed repeat family in the mouse: almost all examples are truncated at one end. *Nucleic Acids Res.* **11:**8847–8859.

121. **Voliva, C. F., S. L. Martin, C. A. Hutchison III, and M. H. Edgell.** 1984. Dispersal process associated with the L1 family of interspersed repetitive DNA sequences. *J. Mol. Biol.* **178:**795–813.

122. **Weinstock, G. M., C. apRhys, M. L. Berman, B. Hampar, D. Jackson, T. J. Silhavy, J. Weisemann, and M. Zweig.** 1983. Open reading frame expression vectors: a general method for antigen production in *E. coli* using protein fusions to B-galactosidase. *Proc. Natl. Acad. Sci. USA* **80:**4432–4436.

123. **Wilson, R., and U. Storb.** 1983. Association of two different repetitive DNA elements near immunoglobulin light chain genes. *Nucleic Acids Res.* **11:**1803–1817.

124. **Wincker, P., V. Jubier-Maurin, and G. Roizes.** 1987. Unrelated sequences at the 5′ end of mouse LINE-1 repeated elements define two distinct sub-families. *Nucleic Acids Res.* **15:**8593–8606.

125. **Witney, F. R., and A. V. Furano.** 1984. Highly repeated DNA families in the rat. *J. Biol. Chem.* **259:**10481–10492.

126. **Yancopoulos, G. D., R. A. DePinho, K. A. Zimmerman, S. G. Lutzker, N. Rosenberg, and F. W. Alt.** 1986. Secondary genomic rearrangement events in pre-B cells: V_HDJ_H replacement by LINE-1 sequence and directed class switching. *EMBO J.* **5:**3259–3266.

127. **Yang, J. K., J. N. Masters, and G. Attardi.** 1984. Human dihydrofolate reductase gene organization: extensive conservation of the (GC)-rich 5′ noncoding sequence and strong intron size divergence from homologous mammalian cells. *J. Mol. Biol.* **176:**169–187.

Chapter 27

SINEs: Short Interspersed Repeated DNA Elements in Higher Eucaryotes

PRESCOTT L. DEININGER

I. OVERVIEW

A. Introduction

There are a number of different classes of repeated DNA sequences found interspersed throughout the genomes of mammals and other higher eucaryotes. Some are thought to represent mobile genetic elements, whereas others, such as the variable-number tandem repeats (66), show no evidence of genetic mobility. Some repetitive elements are thought to be mobile via RNA intermediates. Retrotransposons, such as *copia* (64) in *Drosophila* spp. and THE1 in humans (74), have structures similar to those of

Prescott L. Deininger ■ Department of Biochemistry and Molecular Biology, Louisiana State University Medical Center, 1901 Perdido Street, New Orleans, Louisiana 70112.

Table 1. A collection of representative SINE families

Family	Species distribution	Copy number	% of genome	Divergence from consensus (%)	References
Alu	Primates	500,000	5	14–16	19, 78
Alu subA	Primates	75,000	0.7	6–8	96, 113
Alu type IIa	Galago	200,000 total	2	6	13
Alu type IIb	Galago			9	14
Monomer	Galago	200,000	0.8	12	15
B1 (rodent type II)	Rodents	80,000	0.3	8	49
B2a (rodent type IIa)	Rodents	80,000	0.7	10	80
B2b (rodent type IIb)	Mouse			4	80
Identifier (ID)	Rat	130,000	0.5	4	85, 98
	Mouse	12,000	0.05	ND[a]	
	Hamster	2,500	0.01	ND	
Rabbit C	Rabbit	170,000	1.7	?	37
Artiodactyl C-BCS	Goat and cow	300,000	3	20	97, 110
art2	Goat	100,000	1.8	ND	23
Salmon SINE	Salmon	ND	?	2	59

[a] ND, Not determined.

retroviruses (112). Other elements, termed retroposons (112), use RNA intermediates in their duplication but do not have a retroviruslike structure. In this group are the processed pseudogenes (108), which represent RNA sequences from mRNAs or other small RNAs which have been reverse transcribed and inserted back into the genome (a process termed retroposition [79]). These are present at only very low copy numbers. The majority of the mammalian interspersed repeated DNAs, however, fall into two high-copy-number retroposon families of sequences referred to as short and long interspersed elements (SINEs and LINEs, respectively) (94). The SINEs were initially differentiated from the LINEs purely based on length (94), with the SINEs typically ranging from 75 to as much as 500 base pairs (bp) in length. In contrast, the LINEs can extend up to 7 kilobases in length. A more meaningful definition of SINEs and LINEs, which will be used in this review, is that the SINEs contain a promoter for RNA polymerase III which is involved in their transcription, while the LINEs are transcribed by RNA polymerase II. SINEs generally lack open reading frames and do not code for specific enzymes ("transposases") which mediate the insertion process, but rather employ cellular mechanisms for retroposition. LINEs do have coding capacity and may well code for enzymes involved in their retroposition (83). Both the LINEs and the SINEs represent very efficient retroposition events compared to the infrequent retropositions associated with processed pseudogenes (reviewed in references 108 and 112). All retroposons appear to share similar mechanisms of retroposition, although the mechanisms probably differ in specific details.

The general properties of the SINEs and their evolution have been discussed in a number of excellent reviews (18, 80, 89, 90, 112). Many SINE families consist of more than 100,000 individual family members (Table 1) that are all of about the same length (except for the A-rich region at the 3′ end; see below) and exhibit 70 to >98% sequence homology. Characteristics of major SINE families are presented in Table 1. A given family is often represented by a consensus sequence, determined by sequencing a number of family members and aligning them to find the most common nucleotide at each position.

The "generic" SINE sequence contains an internal RNA polymerase III promoter, an A-rich 3′ end (on the strand corresponding to the transcript), and flanking direct repeats (Fig. 1). The A-rich 3′ end is quite variable in length and exact sequence, and constitutes the region of most heterogeneity among members of any given family (Fig. 2). The 3′ A-rich end regions vary from less than 8 to longer than 50 bp and are often mixed with base pairs containing bases other than dA. In fact, simple-sequence repeats of the form $(XA_y)_n$, where X represents any other base (see reference 80 for review), are often found in this region. Other 3′-end patterns are also found (Fig. 2), including other simple repeating sequences. Some artiodactyl (bovine and goat) repeat families lack either A richness or even a simple tandem repeat structure (97, 110) (Fig. 2).

The direct repeats that flank the SINEs (Fig. 1 and 2) are not part of the repeated DNA family member itself, but derive by duplication of target sequences at the site of integration. These direct repeats vary in size from a few base pairs to ≥30 bp in length and are generally A rich (14, 80, 90) (Fig. 2). In addition, as many as one-third of the members

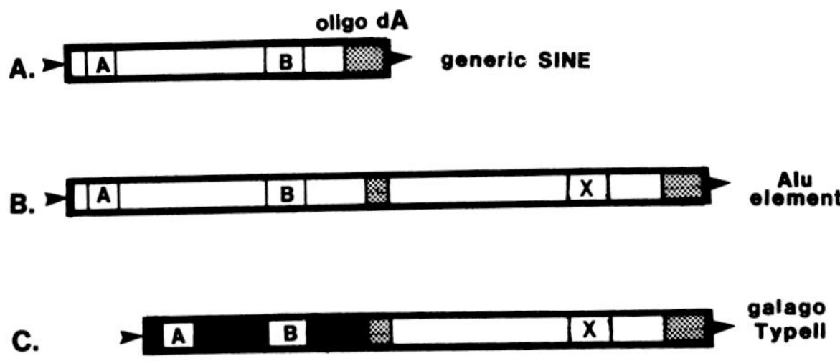

Figure 1. (A) Features of the generic SINE. The typical SINE sequence contains two major features, an A-rich 3′ end which is quite variable in length and exact sequence among different SINE family members (shaded region) and an internal RNA polymerase III promoter. The promoter has a typical bipartite structure as represented by the A and B boxes. This entire repeated sequence is typically flanked by short direct repeats (indicated by arrowheads) formed during its integration into the genome (see Fig. 2). (B) The *Alu* family. The human *Alu* family has all of the features of the generic SINE, but also has a dimerized structure. The right half of the sequence is clearly related to the left half, but has 31 extra bases (marked with an X) and does not contain an active RNA polymerase III promoter. There is a short A-rich region between the two halves of the dimer, and typically a much longer, more variable A-rich region at the 3′ end of the repeat. (C) The galago type II family. In the prosimian galago, there is a different form of composite *Alu* family in addition to the normal form seen in panel B. The new form is a fusion of a different SINE family member (a Monomer family member; blackened region), which contains a different RNA polymerase III promoter, with the right half of the *Alu* family member. This helps to illustrate the capacity of some SINEs (e.g., Monomer family members) to be able to fuse with and carry other sequences to new locations.

of some families are not flanked by obvious direct repeats.

B. The Human *Alu* Family

The *Alu* family is the most abundant interspersed repeated DNA family in primates (16) and was the first to be well defined. A number of authors refer to other SINE families as "*Alu*-like" families. The *Alu* family was originally defined as a fraction of renatured, repetitive duplex DNA containing a distinctive *Alu*1 cleavage site (41). The family is made up of about 500,000 copies (78) of an approximately 300-bp-long sequence and thus makes up about 5% of the human genome. There is about one *Alu* element every 5,000 bp on the average, distributed fairly randomly in the human genome. The sequences of individual *Alu* elements vary broadly from the consensus, with an average variation of 14% (19, 96, 113). In addition to a well-defined RNA polymerase III promoter (32, 73, 76) and the 3′ A-rich region, the *Alu* family also has a dimerlike structure (19) (Fig. 1). The consensus sequences for the two dimer halves are matched at about 68% of positions, and the right half has an extra 31 bases not found in the left half. The RNA polymerase III promoter is in the left half, with no apparent promoter function in the right half, probably due to divergence from the consensus promoter sequence (32) (Fig. 3). Thus, the promoter in the left half directs the transcription of the entire element.

A related repeated sequence is found in the genome of a prosimian (*Galago crassicaudatus*), the type II family. The right half of this SINE is almost identical to the right half of the *Alu* family, but is entirely different in its left half. The type II family probably resulted from a fusion of two different SINE elements (13) (Fig. 1). Here again, the left half contains the RNA polymerase III promoter needed for transcription. There is no evidence that the right half of the sequence of either family has an important role, suggesting that they may be carried along solely by virtue of their linkage to the RNA polymerase III promoter-containing sequence.

```
Alu family.

tkA         AAAAGAATGTTGG.......................TAGAAGAACAAAAACAAAACAAAAAGAATGGTGC
tkD     ATAAATAACTGGTTTTC.......................ATAAAAAAATAAATAAATAACTGGTTTTC
tubA          AAAGACGAGGA.......................AAAAAAAAAAAAAAAAAAAGAGGA
tubJ        AAATAATAACAAAT.......................AAAA (TAA)8 TAAATAATAACAATT
Mlvi-2              GAAAATGT.......................AAAAAAAAAAAAAAAAAAAAAAAAAAAAGAAAATGT
gor       AAAGAGATGCAATCT.......................AAAAAAAAAAAAAAAAAAAAAAAAAAGAGATGCAATCTC

Salmon SINE

Sm3                  TTATAAT........136bp..........ATATATTATTATATTATTATTATATATTATAAT
Sm2                  ATACGGT........136bp..........A (TA)3 (TTATTAT)2 (ATT)6 (AT)4 ATACAGT

Goat C-BCS

5'BVC      TTCTATTGTTAGTA........200bp.......CTCACTCACTCACTCGACTTCTATTGTCAGTA
EVIVS       TAAGGCTGTTCAC........200bp.......TTGCATGCATGCACAGTAAGGCTGCCCAT
BAIVS5'           CTAACCT........200bp.......TTTCACTTTACTAACCT
```

Figure 2. The sequences of SINE direct repeats and 3′ A-rich regions. Sequences are presented for several members from each of three SINE families. The *Alu* family members are discussed further in reference 21, the salmon SINEs in reference 59, and the goat C-BCS family in reference 97. The dotted lines represent the approximately 260-bp *Alu* family or the distance marked in the other families. The flanking direct repeats are underlined. The 3′ A-rich regions are shown, with large repeating patterns marked in parentheses.

```
                         5'        A-box                       B-box

tRNA POL III con         7 bp...TGGCNNAGTNGG...17-60bases....GGTTCGANNCC....

Met tRNA                 .......TGGCGCAGcGGa....31 bp........GGaTCGAAACC....

galago Type II           14 bp..TaGCaCAGTGGt....32 bp........GGTTCGAACCC....

Alu consensus left       3 bp...gGGCGCgGTGGc....59 bp........aGTTCGAGACC....
Alu consensus right      .......gGGCGTgGTGGc....58 bp........GaggCGgAGgt....

B1 consensus             3 bp...gGGCATgGTGGt....49 bp........aGTTCGAGGCC....

7SL RNA gene             3 bp...gGGCGCaGTGGc....59 bp........aGTTCtgGGCt....
```

Figure 3. The RNA polymerase III promoter. The consensus sequence for the two portions of the A and B boxes of the tRNA RNA polymerase III promoter is presented, with N at positions where any base is functional and with typical spacings from the 5′ end of the transcript and between the A and B boxes presented. In the promoters for the SINE consensus sequences, the Met-tRNA, and the 7SL RNA gene, uppercase letters indicate agreement with the tRNA consensus and lowercase letters indicate disagreement. Note the poor match of the *Alu* right half to the promoter sequence.

II. ORIGIN AND EVOLUTION OF THE SINEs

A. Ancestral Sources of SINEs: Small Nontranslatable RNAs

Since the SINEs contain an RNA polymerase III promoter, it is thought that they derive from RNA polymerase III-transcribed genes. Surprisingly, the evidence indicates that they were derived from such genes quite recently in evolutionary time. The first datum in this regard is the similarity of the human *Alu* family to the 7SL RNA gene (55, 102). 7SL RNA is an important part of the signal recognition particle complex, which is involved in transmembrane protein transport (109). The *Alu* family consensus and 7SL RNA genes share about 90% sequence similarity, although the 7SL RNA gene has about 150 bp in its middle that is not found in the *Alu* family. The B1 family, found in mouse and other rodents, is similarly related to the 7SL RNA gene. The B1 sequence, like the *Alu* sequence, has a deletion relative to the 7SL RNA gene sequence, but in addition contains a small duplication. Unlike the *Alu* family, the B1 family does not have a dimeric structure (like the generic SINE; Fig. 1A).

Although originally thought to be species-specific variants of the same repeat family, *Alu* and B1 families may have arisen independently from 7SL RNA genes (112). An alternative view is that the *Alu* or B1 family was an ancestor of the 7SL RNA gene. I consider this incorrect for several reasons. (i) 7SL RNA has a known function, unlike *Alu*. (ii) 7SL RNA genes are found in all metazoan organisms (101, 103), whereas the *Alu* family is more restricted in species distribution (18). (iii) The *Alu* family members are diverging much more rapidly than is the 7SL RNA gene.

The next important finding concerning the origin of SINEs was the discovery that the Monomer family in the prosimian galago was related to a methionine-tRNA gene (15) (Fig. 4). The two sequences exhibit nearly 70% sequence identity and are almost exactly colinear. Since the Monomer family is not present in other species, not even in other primates, it must have been derived from the tRNA gene relatively recently. Analysis of SINE families from other mammalian species also showed each family to be derived from different tRNA genes (15, 35, 52, 80, 84). Thus, it does not appear that the different SINE families once had a common origin and diverged in a species-specific manner. Instead, it appears that the

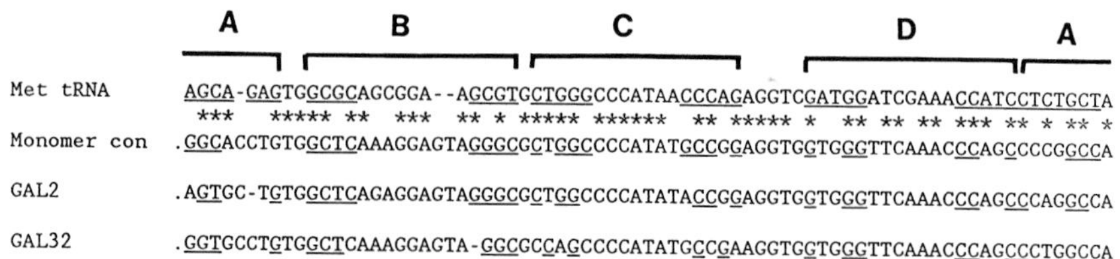

Figure 4. The galago Monomer family and its similarity to Met-tRNA. The human Met-tRNA gene sequence is shown, with the various base-paired stems marked and the regions which base pair underlined. The regions marked A represent the aminoacyl stem, B represents the dihydrouridine stem and loop, C is the anticodon stem and loop, and D is the pseudouridine stem and loop. The equivalent stem regions are underlined for both the Monomer consensus and two individual Monomer family members, *GAL2* and *GAL32*. Although there may be alternative ways to fold these sequences, note the general decrease in base-pairing capability. The positions of sequence identity between the Met-tRNA gene and the Monomer consensus are marked with asterisks.

SINE families arose independently, each from a different RNA polymerase III-transcribed gene (18).

B. How SINEs Arise

SINEs are not just 7SL or tRNA pseudogenes. The consensus sequences of individual SINEs do not represent fully functional tRNA genes (15, 84) (Fig. 3). If SINEs were generated directly as pseudogenes from a functional tRNA gene (or 7SL RNA gene), their consensus would coincide much more closely with the sequence of the parental tRNA gene. Something must occur to make a specific gene much more effective at duplicating itself for it to become a SINE. Requirements for the formation of a new SINE family are likely to include (i) an RNA polymerase III promoter which is functional in the germ line, (ii) an A-rich stretch at the 3′ end (despite some possible exceptions, considered below), (iii) a terminator for RNA polymerase III transcription downstream from the 3′ end of the repeat, and (iv) a gene sequence which is nonessential and altered early in its evolution so that, as it is amplified, it will not interfere with the unaltered progenitor form of the gene.

I propose that the relevant changes must start on an RNA polymerase III-transcribed gene that is nonessential. This is likely to be one copy of a multicopy gene family that is free to evolve to a new function or become a functionless pseudogene. This gene or pseudogene would then accumulate mutations over a fairly long period of time. Most SINE families appear to be about 30% divergent from their ancestral, functional genes. At a neutral evolution rate, this much divergence would require about 60 million years. This divergence probably accumulates prior to rapid amplification as a repeated DNA family, because once a great deal of amplification occurs, the consensus sequence is likely to become relatively fixed.

The progenitor gene must also have obtained a 3′ A-rich region. There are three models for this event. (i) The RNA polymerase III terminator could have been deleted from the gene, bringing a naturally A-rich region into position. (ii) The progenitor gene may have been duplicated by retroposition and inserted into an A-rich region of the genome. (iii) The transcript from the progenitor gene may have been aberrantly polyadenylated prior to its integration into the genome via retroposition (112). The first model seems unlikely for many SINE families, as many of the A-rich regions (e.g., in *Alu*, B1, rat ID, and galago Monomer) coincide quite closely with the mature 3′ end of the RNA specified by the parent gene for that SINE family. Because the SINE sequences do not contain a terminator for RNA polymerase III transcription (24, 32), termination must be signaled by sequences downstream of the repeat. Termination of RNA polymerase III transcription appears to require a string of at least four T residues (6), and this would occur quite frequently by chance alone.

The last requirement of SINEs is that the promoter must almost certainly allow strong germ line expression, because only retroposition in the germ line can result in heritable sequence amplification.

C. Appearance of *Alu* Sequences in the Primate Globin Region

Sequence data for orthologous genes (the same genes in different species) allow an analysis of the evolution of individual SINE family members found within those gene regions and indicate whether an individual SINE has been in that location since the divergence of the two species in question. For instance, by analyzing the human and chimpanzee α-globin gene regions, which contain seven *Alu* family members (87), it was found that (i) no *Alu* family members had integrated or been deleted from that region and (ii) the orthologous *Alu* family members appeared to be accumulating mutations at the neutral rate characteristic of introns and intergenic regions (approximately 0.5% per million years). These data suggest that in the last 5 million years, since the divergence of human and chimp, there has been relatively little movement of the *Alu* elements and that they are under no apparent selective pressure. Similar conclusions emerge from studies of the β-globin cluster, where seven *Alu* elements are found in common between human and orangutan (46). Further analysis of two such *Alu* family members showed that they had predated the New World monkey-human divergence (57a). In contrast, the α-globin gene region in galago contains only five *Alu* family members, and none are at the same sites as the *Alu* elements in the human α-globin gene region (86). Thus, most of the *Alu* family members appear to have integrated in the last 55 million years (the time of human-galago divergence), but with relatively little recent amplification. This is consistent with the failure to find *Alu* elements outside of the primates, indicating that *Alu* probably did not exist as a repeated DNA family 65 million years ago, when primate and rodent lineages diverged (18). Several examples of *Alu* family members which have been inserted recently in evolution (25, 100) suggest, however, that some amplification is still occurring.

These data from sequence analysis contrast sharply with bulk, solution hybridization experiments that suggest that as much as two-thirds of the

Alu family members may have integrated since the divergence of the great apes (42). Although it is possible that the sequencing studies presented above are from nonrepresentative regions of the genome, it is more likely that there is a significant degree of error in the hybridization data.

D. SINE Subfamilies and Species Specificities

Distinct subfamilies of sequences can be found within some of the SINE families (Table 1). A mouse-specific subfamily of the rodent B2 family varies distinctly in sequence from the majority of B2 family members in mouse and in other rodent genomes (80). There are two major subfamilies of the galago type II repeat (14), and there are at least two (96) and probably three (113) major subfamilies in the human *Alu* family. Each subfamily is distinguished by small point mutations, insertions, and deletions that are common only within the subfamily. For the *Alu* family, a subfamily consisting of about 15% of the family members varies from its subfamily consensus by only 9% (96), whereas the typical *Alu* family member differs from the *Alu* consensus by 14%. This observation and the species specificity of the mouse type II subfamily suggest that some subfamilies may form after appreciable amplification of a SINE family has occurred (96, 113). In addition to subfamilies, some of the SINE families differ significantly between related species (e.g., the primate *Alu* family, which differs in small but significant ways between galago and human [16, 96]).

How can such subfamilies have amplified rapidly enough to make up a significant portion of the SINE family members? It has been proposed that massive gene conversion events homogenize repeated DNA sequences within a species, or alternatively that old SINEs could be gradually deleted as new SINEs are integrated (these models are reviewed in reference 16). Although dispersed *Alu* family members may undergo very low levels of gene conversion (57a), studies of orthologous globin regions suggest that both gene conversion and removal of SINE sequences are too rare to account for the observed patterns of divergence. A subfamily could arise early in the amplification of a SINE family. If one member undergoes some sequence variation and then significant amplification of each family member occurs, the proportion that fit the modified subfamily will roughly correspond to the number of SINE family members present when the one was modified. Thus, variation of sequences in a highly amplified family is less likely to contribute to a change in the consensus sequence or to major subfamilies, especially if most SINE copies are likely to undergo further amplification.

Very different patterns of evolution and subfamily formation are expected if only a small subset of family members is capable of active retroposition. There is in fact growing evidence that only one or a very few SINE family members contribute significantly to amplification. For example, the three most recently inserted *Alu* family members in the primate genome are all very similar in sequence, suggesting that they were derived from a single *Alu* family member (21). If there is generally only one or a few "master template" SINEs, changes in these master elements would have a large effect upon progeny sequences, and relatively large subfamilies could form quite late in the amplification process.

E. Are There Ancient SINE Families?

Each SINE family is present in only a limited range of species (i.e., primate or rodent specific; Table 1). Because each SINE family appears to be derived independently from a different RNA polymerase III-transcribed gene (reviewed in references 18 and 112), SINEs could not have existed as repeated DNA families at the divergence of the major mammalian orders 65 to 80 million years ago. However, the divergence from the parental RNA polymerase III-transcribed genes may have begun before then. Of the SINEs, the *Alu* family seems to be the oldest (with the possible exception of some of the artiodactyl repeats; Table 1), because it has more internal heterogeneity (mismatching) than other families and is present through the oldest groupings of species. The human *Alu* family members show about 75% sequence identity with each other or with other primate *Alu* family members, on the average. If an older family had 10 to 15% more mismatching, it would not have been detected by traditional hybridization techniques, but enough sequencing studies of large genomic regions have been carried out that computer searches should have detected any such sequences that were as highly amplified as other SINE families. Thus, ancient SINE families would have to be both low in copy number and highly mismatched to escape detection in any species in which large amounts of genome have been sequenced. There has apparently been a burst of SINE activity since the mammalian radiation.

F. Nonmammalian SINEs

Because most SINE amplifications occurred after the mammalian radiation, is it possible that it is largely a mammalian phenomenon? Only a few of the interspersed repeated sequences found in the nonmammalian species studied to date have the features characteristic of mammalian SINEs (i.e., RNA polymerase III promoter, A-rich region, and flanking direct repeats). In contrast, the majority of interspersed repeats in mammals are SINEs.

Several SINEs in nonmammalian species (newt [28], tortoise [28], and salmon [59]) appear to be derived from tRNA genes and thus could be considered SINE-like. Although they share about the same degree of similarity to tRNA genes as do the mammalian SINEs, in general they lack other features expected of the SINEs. For instance, the newt and tortoise sequences do not show the typical 3′ A-rich region or a recognizable flanking direct repeat, suggesting that their mechanism of interspersion differs from that of mammalian retroposition. In contrast, the salmon sequences do have direct repeats, and the family members analyzed have 3′ regions consisting of alternating A's and T's, which would be unusual but not unprecedented in mammalian SINEs (Fig. 2). This salmon family does appear to be an authentic SINE sequence on the basis of its RNA polymerase III transcription, tRNA origin, 3′ region, and flanking direct repeats. Analysis of additional family members will be required to assess which features are general to the family and thus reflect fundamental variation in the SINE retroposition mechanism. This family appears to have amplified very recently (less than 5 million years ago; Table 1).

G. Can SINEs Take Over the Genome?

There is no direct evidence that individual SINE family members undergo excision, although homologous recombination between the flanking direct repeats has been proposed (80, 90). If they are not frequently removed and retroposition continues to make extra copies, they could increase until limited by natural selection. The human *Alu* family appears to be both the oldest and the highest-copy-number SINE family (Table 1). Eventually, however, the *Alu* family members might accumulate sequence variation until they no longer represent members of a recognizable repeated DNA family. In this extreme case, the ultimate copy number would be controlled by the rate of sequence amplification balanced by the accumulation of mutations and the random elimination of sequence members and perhaps natural selection.

Theoretically, as the copy number of a SINE family increases, the number of actively retroposing sequences could also increase. Thus, SINE amplification rates might increase rapidly. However, data on the orthologous *Alu* elements in the globin gene regions indicate that they have recently been amplifying at a slower rate than previously. Most of the *Alu* element amplification must have occurred between 15 and 55 million years ago. Similar bodies of data do not exist for other SINE families, but because most of them are younger families (18), they are more likely to be still actively amplifying, as suggested by examples of recently inserted family members causing specific genomic polymorphisms (3, 72, 91).

There are a number of possible explanations for why a SINE family might lose its ability to amplify. (i) Only a very few of the SINE members may be capable of "active" retroposition (see Section III, below). Any change that affects the ability of one of these master copies to retropose will significantly affect the spread of that family. (ii) A change might occur in one or more species that affects the retroposition process. These include epidemics of retroviral infection or LINE activation to supply reverse transcriptase, specific inhibitors of SINE family transcription, or other, as yet unknown factors associated with the retroposition process.

III. MECHANISM OF SINE RETROPOSITION

The basic model for retroposition of SINEs involves RNA polymerase III transcription of the SINE gene, reverse transcription of the RNA, and integration into a new genomic site (Fig. 5). This model shares many features with models invoked for other retroposons (e.g., LINEs and processed pseudogenes). Most steps are based on structural and evolutionary data, and critical tests await the ability to reproduce part or all of the retroposition process in culture. Such an assay appears to have been developed using a human lung carcinoma cell line with the *Escherichia coli* Xgpt gene introduced as a selectable marker of mutation events (55a). A single *Alu* retroposition event was detected out of 126 mutants analyzed, suggesting that it may now be possible to test some of the proposals discussed below.

A. Transcription by RNA Polymerase III (In Vitro)

The first, and possibly rate-limiting, step is transcription of a nuclear SINE (DNA) sequence by RNA polymerase III. The SINEs all contain an internal RNA polymerase III promoter as determined by a match to the consensus promoter sequence (Fig. 3) and confirmed in many cases by in vitro transcription using RNA polymerase III. In the cases tested in vitro, the 5′ end of the SINE transcript has coincided exactly with the left end of the repeated DNA sequence (32, 43). These results helped spark the original proposal that the *Alu* family was spread

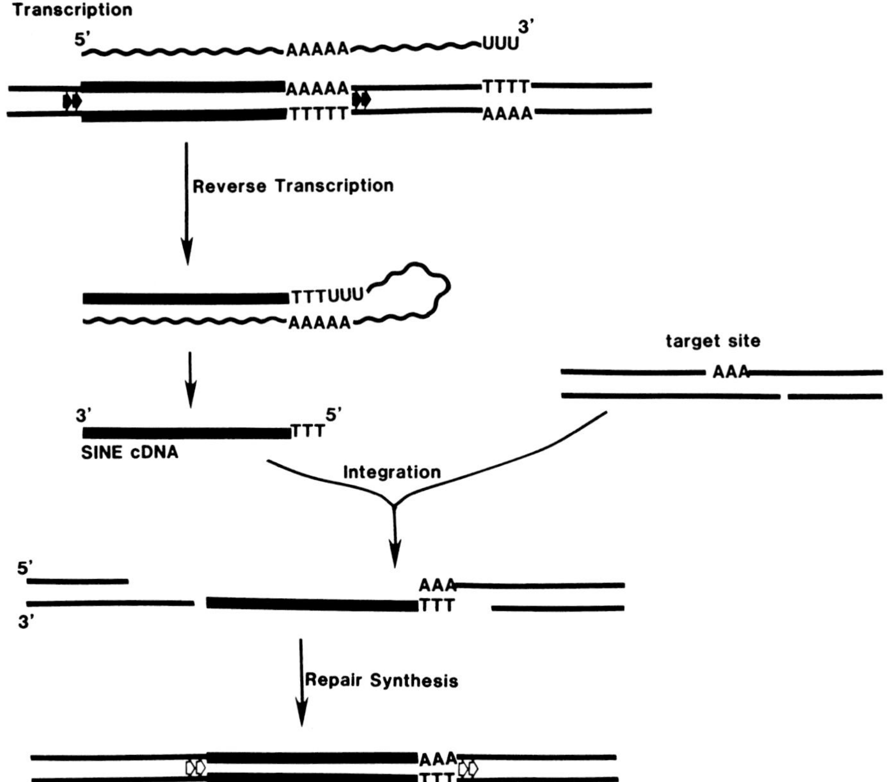

Figure 5. A proposed mechanism for SINE retroposition. The first step that must occur is transcription of the repeated DNA sequence. The repeat is represented by a heavy line, its flanking sequence by thinner lines, and the transcript by a wavy line. Transcription initiates at the beginning of the repeat, adjacent to the flanking direct repeat (double solid arrows), continues through the entire repeat, and terminates in flanking sequence. This transcript is suggested to be capable of self-priming reverse transcription by priming with its terminal U residues on the 3′ A-rich region of the repeat transcript. Removal of the RNA will then leave a single-stranded cDNA copy of the entire repeat with no flanking sequences. This cDNA must then integrate into a genomic site with staggered nicks. It is hypothesized that an A richness at the nick site may interact with the T-rich cDNA end to stabilize the interaction. Repair synthesis at the junctions will then result in formation of a newly integrated repeated DNA family member with a different flanking direct repeat (double hollow arrows). Many of these steps are hypothetical and a number of alternatives for some of them are presented in the text.

through the genome via RNA-mediated transposition (43, 106).

Although I will not review all available information on SINE RNA polymerase III promoters, several points are particularly relevant. First, because the promoter is within the gene and therefore its transcript, the promoter will be duplicated during the retroposition process. This means that, in principle, each new copy could potentially be transcribed and form more copies. This alone could contribute to the high rate of transposition of the SINEs relative to other retroposons (such as processed pseudogenes), although recent data (discussed below) call this into question. Second, because the SINEs appear to be derived from tRNA or 7SL RNA genes, we might expect the promoters of these genes to be very similar to SINE promoters. Lastly, there may be quite a variation in the promoter strength of members within the same family, reflecting sequence variation among individual family members (92). It is found that, although the majority of the SINEs from several families studied produce a transcript in vitro, most have relatively weak promoters. A few do have unusually strong promoters, however, and, in general, the SINE family members whose consensus promoters correlate best with the consensus of that family are most efficiently transcribed in vitro (G. R. Daniels and P. L. Deininger, unpublished results). Since the consensus is our best approximation to the sequence of the most actively retroposing sequence of a family, this outcome provides a correlation between promoter strength and retroposition rate.

B. Transcription by RNA Polymerase III (In Vivo)

Although in vitro data are of some value in understanding promoters in SINEs, it is the in vivo transcription in the germ line that is important to

SINE proliferation. There are very few data concerning germ line SINE transcription, so I will review the general data on in vivo expression of SINEs in other tissues because of its implications for germ line SINE transcription, and also for insight into the possible function(s) or impact(s) of SINEs.

Although most of the individual SINE family members are transcribed by RNA polymerase III in vitro, different results are often observed in vivo. Careful study of HeLa and certain other human cells did not reveal any significant levels of RNA that were attributable to specific RNA polymerase III transcription of *Alu* family DNA (27, 75) (one must differentiate *Alu*-specific RNA transcripts from the larger transcripts of other genes that fortuitously contain *Alu* family members in introns or 3′ noncoding regions). Some SINEs, such as the rodent B2 family (33), do seem to be generally transcribed. Others, like the rat ID family, are apparently only transcribed in specific tissues, such as nervous system (99) and testes (60), as well as a wide variety of tissue culture lines. The B2 family has also been shown to be transcribed to a higher extent in transformed cell lines (11, 26, 33, 65, 95). Very recently, a specific *Alu*-related transcript has also been found in monkey and human brains (111), but not in several other tissues tested. Thus, there appears to be a great deal of regulation of expression of these repeated DNA families.

SINE families are capable of producing in vivo transcripts, and this transcription can be regulated in a tissue-specific manner. It is important to know whether all or just a few of the SINE family members are transcribed. Sequence analysis showed that the brain-specific *Alu* transcripts were homogeneous, thus indicating that one or a few identical *Alu* family members are transcribed (111). A similar situation has been proposed for the rat ID family (17), based on cloning of random cDNAs for the nervous system-specific BC1 transcript. The clones showed the ID-specific sequences at their 5′ ends followed by a long A-rich region, and also contained a 3′ stretch of sequence unrelated to the ID family consensus. Also, the ID portions of 9 of 10 cDNA clones were identical in sequence, whereas individual genomic ID elements typically vary at multiple positions. This suggests that just one or a few nearly identical genes are responsible for the majority of the BC1 transcription. A similar cDNA cloning study suggests that different genes or subsets of genes are coding for the related BC1 and BC2 transcripts (61). These studies also showed more heterogeneity in the transcripts, suggesting that more than a single member is capable of transcription, but that these may be a very small subset of the ID family members.

What can explain the high degree of regulation and specificity of transcription of these SINE families? The *Alu* family is derived from a 7SL RNA gene, and 7SL transcription requires a specific 37-bp upstream activating sequence in addition to its internal promoter (105). 7SL pseudogenes (made via retroposition) are generally not transcribed in vivo because the upstream sequences are not duplicated during retroposition (104). Similarly, the *Alu* promoter may be dependent on such upstream sequences, so that almost all *Alu* retropositions result in an inactive element. Transcription of many tRNA-based SINEs is probably regulated similarly because some tRNA genes are also controlled by flanking sequences in a tissue-specific manner (48, 115).

In this view, only rarely will a SINE element be inserted near appropriate activating sequences. Alternatively, a key step in forming an active SINE may be mutations which make its promoter independent of flanking sequences. This view is supported by findings of SINE family members, including *Alu*, which have been introduced back into cells and been found to be transcriptionally active (116; V. K. Slagel and P. L. Deininger, unpublished results). However, these cases involve transient transfections, where the transcription was assayed within a few days of transfection. We have found that, using a galago SINE family member which also expresses under these conditions, the same SINE in long-term transformants is almost always transcriptionally silent (Slagel and Deininger, unpublished results). This suggests that there is a regulatory aspect of established chromatin structure or environment that is lacking in the transient experiments. The integrated environment of the introduced SINEs assayed in the long-term experiment probably mimics the typical environment of newly integrated SINE family members better than does the transient assay, and therefore supports the concept that the internal promoter is not sufficient by itself in vivo.

C. Termination and Processing of Transcripts

In vitro transcription studies have shown that most SINEs do not contain a signal for termination of RNA polymerase III transcription (24, 32). Instead, the transcription begins at the left end of the SINE, proceeds through the entire repeat, and terminates in the flanking sequence (Fig. 5). The consensus RNA polymerase III terminator simply involves four or more T's in a row (6) and occurs fairly readily by chance. Each SINE family member will be integrated into a different location and thus potentially could specify a transcript with a unique 3′ end. On the other hand, most of the RNA polymerase III-transcribed RNAs from the SINEs appear to be fairly discrete in size. Three different major RNA species,

BC1, BC2, and T3, are transcribed from the rat ID family (60) and one from the primate *Alu* family (111). As discussed above, some or all of these transcripts may be the result of transcription from a single family member, thus producing a discrete RNA species. Other data suggest that many heterogeneous transcripts might be produced by RNA polymerase III and then processed to a homogeneous size. This is supported by analysis of RNAs from cloned DNAs carrying *Alu*, B2, or rat ID family members injected into *Xenopus* oocytes (1, 35). A nuclear transcript is made that is processed to a smaller, stable transcript, which in the case of the rat ID family is comparable in size to the in vivo T3 transcript from testes (60). The biological significance of results of *Xenopus* oocyte injections is not known. This transient system may also be allowing transcription that would not occur if the SINE were integrated into the chromosome, and it is not clear that similar processing happens in mammalian cells. Another observation of these studies is that ID transcripts are more stable than rodent type II transcripts from family members in the same recombinant DNA clone (35), indicating that RNA stability may also be important in the overall expression levels of individual SINEs.

Most discrete in vivo SINE transcripts also appear to be polyadenylated, as judged by selection on oligo(dT)-cellulose. In some cases, polyadenylation might occur after processing, and this could also affect RNA stability. However, the structure of cloned BC1 cDNAs shows that the poly(A) is actually derived from A-rich regions coded within the BC1 gene (17).

D. Reverse Transcription

In any successful retroposition event, a SINE RNA molecule must be transcribed into DNA (either single or double stranded). This probably requires a primer and certainly requires a polymerase capable of reverse transcriptase activity. Just how this occurs is one of the major mysteries in the proposed mechanism of SINE proliferation (Fig. 5).

The transcripts of SINE family members may be capable of self-priming their own reverse transcription (43). Since the transcripts routinely contain a long A-rich region, and the RNA polymerase III transcript should have three or four U's at the 3′ end (6, 17), the three U's may fold back on the A-rich region and prime reverse transcription, resulting in a DNA copy of the SINE with none of the flanking region (as illustrated in Fig. 5). A number of U3 snRNA pseudogenes appear to have been primed by the folding back of the U3 small nuclear RNA (snRNA) on itself (4) and thus provide a precedent for self-priming. There is no direct evidence for the self-priming of SINEs, however, as they, unlike U3 snRNA, cannot form a thermodynamically stable structure. SINE reverse transcription could also be primed by an intermolecular interaction (106), for example, using the 3′ end of one transcript to prime reverse transcription from the A-rich region of another. Evidence that priming can be intermolecular comes from processed pseudogenes that must be primed on their poly(A) tails, as they lack the self-complementary sequences that would be necessary for any such intramolecular priming event. Self-priming of reverse transcription could be another explanation for the very high efficiency of SINE retroposition relative to the formation of processed pseudogenes.

It has been proposed that reverse transcription may be primed either directly by the genomic DNA at the target site of integration (63) or by a simple-sequence DNA synthesized de novo at the target site by a telomerelike addition mechanism (80). The latter possibility was suggested to allow priming on 3′ regions which are simple-sequence repeats, similar to telomeric sequences. These models do not seem likely because most processed pseudogenes lack introns, and thus they appear to be made from cytoplasmic RNA (112), suggesting that their retroposition process is initiated in the cytoplasm.

The final question regarding reverse transcription concerns the source and availability of the reverse transcriptase. Most mammalian genomes are, or have been, exposed to numerous retroviruses which could supply reverse transcriptase, either during the infection or from endogenous retroviral sequences (5, 58, 77, 93). Although many endogenous retroviruses are defective, lacking functional reverse transcriptase genes, others are not defective and can be rescued from the genome under the proper circumstances. In addition, LINE elements (L1) code for a reverse transcriptase-like protein (83) that similarly might participate in SINE retroposition. Small RNA molecules can be packaged into retroviral particles and be reverse transcribed (56). Packaging should facilitate reverse transcription and may be another explanation for the extremely high efficiency of SINE retroposition relative to longer transcripts that will package less effectively (e.g., in processed pseudogene formation). Packaging may also foster an "infection-like" process in which RNAs made in somatic cells enter the germ line, thereby helping to explain processed pseudogenes of genes not normally expressed in the germ line (108).

Very different patterns of retroposition might be expected from a constant endogenous source of reverse transcriptase than from viral infection as the source. In the case of infection, retroposition might occur at a higher rate in species which have high levels of infection. Although the endogenous retroviruses in primates (77, 93) provide evidence for infection of the primate germ line by retroviruses as much as 25 million years ago, there is little evidence for massive infection by retroviruses of humans in recent times. Will the new influx of retroviruses (such as human immunodeficiency virus) set off a new round of active SINE retroposition?

With very few exceptions, the SINE family members are full length. That is, the reverse transcriptase must have proceeded to, or within a few bases of, the 5′ end of the RNA transcript. This is in contrast to the much longer LINEs, which are very heterogeneous at their 5′ ends. Complete SINE reverse transcription may simply be due to the short length of SINEs. The observation that the tRNA-derived SINEs have significantly less ability to form RNA secondary structures than do their parental tRNAs (15) (Fig. 4) suggests that elimination of the possible interference with reverse transcription that such structures could cause may be an important step in the formation of the SINE progenitor. It is also possible that reverse transcription is not always complete, but that full-length reverse transcripts integrate preferentially.

E. Integration into the Genome

In principle, either single- or double-stranded SINE nucleic acid (RNA, DNA, or RNA:DNA hybrids) may interact with the chromatin during the integration process. I believe that SINE reverse transcription is probably primed in the cytoplasm (112); accordingly, either a DNA molecule (which could be single or double stranded) or a DNA:RNA hybrid is a likely candidate for interaction at the site of integration.

We know that the new target site must be nicked to allow the entry of the new sequences. Nicking of the genome on both strands followed by repair synthesis makes the flanking direct repeats at the integration site (Fig. 5). Analyses of the direct repeats have shown that they are in general very A rich and vary widely in length. This length variation suggests that SINEs do not use specific integration enzymes, as is the case for retroviruses and transposons, but instead take advantage of nicks generated by other, nonspecific cellular enzymes.

One group of enzymes proposed for the nicking activity are the topoisomerases, enzymes that relax the genome during replication and transcription (107). It is proposed that the transient covalent intermediate formed between the topoisomerase and the genomic DNA may serve as an activated group to facilitate ligation to the 3′ end of the SINE cDNA (Fig. 5). At least in the case of one SINE family member, the 3′ end and direct repeats are preferential sites for nicking activity by DNA topoisomerase I in vitro (76a). Although topoisomerase I is generally thought to be nonspecific in its nicking activity, hot spots for DNA cleavage have been reported (9). These sites are A rich and at least partially resemble the 3′ terminus and direct repeats of SINEs. It is still not clear that topoisomerase I will show specificity for all SINE elements, because the 3′ A-rich region is the most variable region of a SINE family. Alternatively, the activities of DNA replication and repair or random nicking from chemicals or radiation might also be involved. Since integration requires some synthesis of new DNA, it may occur preferentially during DNA replication or repair. In support of this idea, retroposition of a human *Alu* family member in cultured cells was only observed after UV irradiation (55a).

Not only are the integration sites of SINEs A rich, but the A richness is predominantly at the left end of the direct repeat (closest to the 3′ end of the inserted element) (14, 80), as had been recognized for processed pseudogenes (108). The A richness of the integration sites has several interesting ramifications. First, it shows that integration is not random. Target sequence selection reflects either preferential nicking activity or the selection of specific sequences during the integration process (i.e., via specific sequence interactions between the SINE sequence and the target site). Second, since the 3′ ends of the SINE families are generally A rich, when they integrate into a new site they generally make that site even more A rich. Thus, the 3′ ends of SINEs are improved integration sites for more SINE copies, resulting in a tendency to form perfect tandem dimers (reviewed in references 80 and 112). This extends to other retroposons: SINEs and LINEs have been found adjacent to each other. In several cases, it appears that the integration of one element adjacent to another fused those two elements so that they could then retropose as a single unit (e.g., the *Alu* family [80] and the galago type II family [13]). Third, the presence of the A richness at one end of the direct repeat and its similarity to the A-rich 3′ end of the SINE indicate the possibility of some pairing between these sequences during integration. For example, a T-rich 3′ overhang in the target could prime reverse transcription directly on the RNA (63) (although direct RNA priming seems unlikely for retroposons, such as processed pseudogenes, that almost always involve cytoplasmic RNA species [112]) or the T-rich region of the first cDNA strand could pair directly to the 5′ A-rich overhang (14) (Fig. 5).

It has been suggested that the critical step in integration involves the ligation of the 3′ end of the cDNA to the integration site (Fig. 5), perhaps using a topoisomerase-activated intermediate (107). The interaction at the A-rich end would then merely stabilize the structure during the repair synthesis. Such stabilization has been proposed to explain the almost complete lack of deletions at the A-rich (3′) end of SINE family members (112), in contrast to snRNA pseudogenes, which lack an A-rich end and are often truncated at the 3′ end (107).

There are also limited sequence similarities between the 5′ ends of some SINE family members and their adjacent direct repeats (90, 107). This suggests an alternative possibility, that base pairing may also stabilize the interaction of the 3′ end of the cDNA with the target (5′ end of the SINE; Fig. 5) and help prime repair DNA synthesis across the SINE sequence during the integration. Not all SINE family members show such sequence similarities, and thus they may insert by other mechanisms. Finally, integration may involve other recombination or strand displacement mechanisms involving a double-stranded cDNA instead of the single-stranded molecule shown in Fig. 5.

Besides a preference for A-rich integration regions, it is possible that other factors aid integration into specific regions of the genome. Electron microscopic studies of renatured repeated sequences indicate a broad, near-random distribution in the human genome, with some clustering (20), which is confirmed by DNA sequencing studies as well as in situ hybridization to chromosome spreads (47). *Alu* elements are found in or around most genes, but several genes, including human genes for β-tubulin (53), thymidine kinase (29), tissue plasminogen activator (31), and prothrombin (30), have an exceptionally high abundance of SINEs in their introns. This may reflect a more open chromatin structure associated with such genes in the germ line (96) or other factors, such as the presence of preferential sites for nicking of the genome.

IV. SPECULATION ON THE POSSIBLE FUNCTIONS OF SINES

It has often been assumed that any sequence as prevalent as SINEs (Table 1) must serve an important cellular function. Postulated functions include origins of DNA replication (33, 45), signals for RNA processing (33, 45) and stability (12, 34), control of transcription (8), and organization of chromatin structure. A number of other effects on genome structure and evolution are also possible. For example, SINEs may facilitate recombination (54), act as limits to gene conversions (39), and move unrelated DNA segments throughout the genome either via retroposition of sequences adjacent to SINEs (117) or by facilitating recombination. These differ from the previously mentioned functions in the sense that the cell or organism could live without them. They may just affect the long-term adaptability of the species.

The relative species specificity and recent evolutionary origin (18) of the different SINEs and the variation in SINE copy number from one species to another make it difficult to envision how any given family could have any critical function. SINEs may have no function but to be "selfish," only involved in their own amplification (22, 68).

A. How Many SINE Family Members Are Important to the Cell?

Even if specific SINE elements carry out important functions, it does not seem likely that all family members are crucial to that function. Certain SINE elements have deletions that eliminate much of the sequence. Additionally, because *Alu* family members are all in different positions in the galago versus human α-globin regions (86), the exact location of any given SINE cannot be important. A given SINE family can differ greatly in copy number among related species (e.g., 100,000 ID members in rat, but only 2,500 members in Chinese hamster [85]; Table 1). In addition, evolutionary analysis of the rate of change in individual *Alu* family members suggests that they are mutating at a neutral rate (46, 57a, 87). Thus, there seems to be no evolutionary pressure to conserve the sequence of entire families members, indicating that most individual SINE family members are likely to be expendable.

On the other hand, perhaps only a limited number of SINE members are functional, and the majority represent retroposed copies which are essentially pseudogenes. In this case, studies of the most frequent elements would generally relate only to the nonessential copies. In favor of this possibility are the observations that most of the ID- and *Alu*-specific transcripts are from a single or small group of identical genes (17, 111) and that the most recently inserted *Alu* family members are all very closely related (21). The sequence that gave rise to these new *Alu* inserts is not closely related to the sequence responsible for the brain-specific *Alu* transcript in primates, however. This suggests that either SINE transcripts may form which are not capable of retroposition or that elements transcribed in brain are not necessarily transcribed in germ line and therefore cannot contribute to SINE amplification. The evidence that the different SINE families are clearly derivatives of tRNA and 7SL RNA genes is consistent with the possibility that none of these elements is necessarily functional. The significant changes in *Alu* family consensus sequence throughout evolution (21) suggest only moderate selection pressure on the amplifying sequences. However, it is also possible that the single expressing gene was derived from a different RNA polymerase III-transcribed gene but has adapted into a functional gene. A suggestion that this may occur comes from the observation that the rodent 4.5S gene is closely related to the 7SL gene and may have been derived from it, but, unlike the situation revealed by the data on SINEs, this RNA gene appears to be under selective pressure (36, 38). For that matter, one cannot rule out the possibility that in various ways the other SINE family members may be gradually adapting to represent important functional elements in some of the ways that are discussed below.

B. SINEs as Tissue-Specific Regulators of Transcription

This general proposal is one of the oldest explanations of the presence of repeated DNA sequences in eucaryotic genomes (8) and was originally based purely on studies of copy number and distribution of repeated DNA sequences. The model held that if it was necessary for a number of genes to be coordinately regulated, i.e., at a specific state of differentiation, this might be achieved by means of common DNA regulatory elements near each of these genes. This model has been upheld in the sense that there are certainly common sequence motifs regulating a number of genes. However, none of the regulatory sequences have proven to be retroposons.

The rat ID family of SINEs has been postulated to act as nervous system-specific regulators of gene transcription (61, 98, 99). This was based on the observed neural specificity of RNA polymerase III transcription of ID-related sequences (99), along with an observation that genes which include ID elements within their introns were preferentially transcribed into heterogeneous nuclear RNA (hnRNA) in brain tissue (99). This latter observation is uncertain, because other groups found no appreciable difference in the in vivo abundance of hnRNA transcripts with ID elements in brain relative to other tissues (34, 57, 71, 85) and because ID-containing genes were not preferentially transcribed in brain (34). It is now difficult to conceive of a mechanism by which a mobile, high-copy-number element could be responsible for tissue-specific gene regulation. A global role in transcription regulation therefore appears unlikely for SINEs.

Sequence analysis had also led to a suggestion that the B2 SINE may decrease the stability of some RNA transcripts by contributing A+T-rich destabilizing sequences to the mRNA (12). The presence of rat ID family members in transcripts apparently alters mRNA stabilities in normal versus transformed cells (34). Thus, rather than influencing gene expression through transcription rates, SINEs could have the same effect by selectively altering mRNA stabilities. These observations suggest that specific SINE elements might be adapted to functional roles in specific instances. This possibility as well as proposals that SINEs could affect RNA splicing, chromatin organization, or other functions are still speculative in nature.

C. Why Are SINE Transcripts Regulated?

There is a great deal of cell-type-specific variation of expression of several of the SINEs. The high synthesis of the rat ID-related, T3 transcript in testes (60) is consistent with the need for germ line expression for SINE amplification. Remaining transcriptionally inactive in most tissues other than germ line will minimize the potential damage caused by transcription or retroposition of SINEs in somatic tissues. Certain other transposing elements, such as *Drosophila* P elements (51), have developed elaborate strategies to limit their transposition to the germ line.

The regulation of RNA polymerase III-specific transcripts from SINE families in a somatic tissue-specific manner (e.g., neural cells [60, 111]) or in transformed cells (26, 33, 65, 95) is more difficult to explain in terms of the needs of the retroposons themselves. These observations would be consistent with the SINEs (at least the limited number of SINEs and SINE-related sequences that are transcribed) carrying out a tissue-specific function.

The regulation of SINE transcription can also be explained without involving a function. These transcripts could reflect fortuitous integration in a chromosomal site that can activate a nearby SINE promoter. Or the SINEs might be derived from a gene that is somewhat tissue specific in its expression (as are some tRNA and 5S genes), and this specificity may be at least partially due to sequences internal to the SINE sequence and therefore spread during retroposition. Lastly, it is possible that the factors for germ line expression of SINEs needed for their amplification are also present in some other cells. For instance, the transformation of cells may cause a dedifferentiated state that resembles the germ line for

RNA polymerase III transcription. There might also be a specific trait that is common to the germ line and neural tissue, such as more promiscuous RNA polymerase III transcription or a higher stability for RNA polymerase III-transcribed RNAs.

D. Possible Impact of SINEs on the Host Cell

Although data on possible functions for SINEs are still incomplete and quite controversial, it is likely that SINEs have a major impact on their genomes. The most obvious is their mutagenic potential due to the disruption of sequences at the site of their integration. SINE integrations into exons and into regulatory or other functional regions would often cause null alleles and might be selected against even in heterozygous diploids. SINE integrations into introns and intergenic regions are more likely to be neutral. The tendency for SINEs to integrate into A-rich regions (14, 80) should favor integration into such nonselected regions and thereby minimize the potential negative impacts of their amplification.

Even in introns and intergenic regions, SINEs may have an impact on genome function. For instance, multiple SINE family members in introns could allow inter- and intramolecular base pairing in hnRNA molecules. Such interactions might inhibit or alter the processing of hnRNAs. However, a number of genes have very high numbers of SINEs in their introns without obviously impairing the processing of their transcripts: for example, there are 13 *Alu* family members in the introns of the human thymidine kinase gene (29) and yet there is no evidence for alternate or abnormal processing.

SINEs might affect the expression of nearby RNA polymerase II-transcribed genes. Data from transient transfections show that a rat ID sequence can repress or activate transcription directed by a nearby RNA polymerase II promoter (62). Whether this same effect is seen in chromosomally located SINEs is not clear. For instance, if the effect requires the RNA polymerase III promoter to be active and most SINE family members are not transcriptionally active, then we would expect little or no effect.

SINEs have also been found in the 3′ noncoding exons of numerous genes without deleterious effects. The B2 family elements have an apparently functional poly(A) addition signal which can be used as the signal for polyadenylation of transcripts terminated with a B2 element (50, 82). Several *Alu* elements have also been found in the 3′ ends of protein-coding regions within genes (7, 10), resulting in a portion of the *Alu* sequence being translated and the *Alu* element supplying the translation terminator. Thus, the insertion of these sequences in new sites could result in mRNA truncation, altered polyadenylation, and modified protein structures.

E. SINEs and Recombination Processes

The presence of so many short, homologous sequences throughout mammalian genomes has implications for recombination. It is surprising that mammalian chromosomes are relatively stable, given the possibilities for homologous recombination that these sequences present. Data on SINE involvement in recombination come from analyses of naturally occurring mutants, primarily of the globin and low-density lipoprotein (LDL) receptor genes. The LDL receptor gene has numerous *Alu* family members in its introns, 3′ noncoding region, and flanking regions (114). Five naturally occurring insertion/deletion mutants of this gene which produce defective receptors have been cloned and analyzed, four of which involve *Alu-Alu* recombination (40, 54). These events led to either deletions or duplications, and thus to mutant (defective) LDL receptor genes. Surprisingly, only two of the four *Alu-Alu* recombinations appear to be homologous. A fifth LDL receptor mutant involves a deletion between a single *Alu* family member and an unrelated sequence (54). Studies of the globin region reveal a number of mutant genes (e.g., those responsible for thalassemias) that also involve deletions between an *Alu* family sequence and some unrelated sequence (44, 67, 69, 70). None involve *Alu-Alu* recombination, but the globin region has a relatively low density of *Alu* family members. A sex chromosome rearrangement has also involved an *Alu-Alu* homologous recombination (81), and a duplication of the entire growth hormone gene early in human evolution may have occurred via *Alu-Alu* recombination (2). The number of recombinations in which *Alu* elements were involved is much higher than expected for random recombinations, especially for the LDL receptor gene. However, it would appear that the *Alu* family, and possibly other SINEs, may be almost as heavily involved in illegitimate as in homologous recombination events.

F. How Might SINE Elements Be Involved in Illegitimate Recombination?

It is possible that the presence of an RNA polymerase III promoter may allow *Alu* or other SINEs to be more accessible to recombination processes because of factors that interact with them or because of the transcription process itself.

A recent observation provides another highly speculative possibility. The 3′ A-rich region of one *Alu* family member has been shown to be a hot spot for topoisomerase I interaction in vitro (76a). Topoisomerase I is thought to be fairly nonspecific, with only a few hot spots reported (9). Interestingly, those hot spots were also A rich and had some of the characteristics of the 3′ ends of the SINEs. Whether this single observation on an *Alu* family member will prove general is not clear. However, if topoisomerase I (or any other DNA-nicking enzyme) interacts preferentially at these sequences, the resulting nicks might foster strand displacement during both homologous and illegitimate recombination.

G. How Might SINE Elements Limit Gene Conversion?

Studies of the globin gene regions of several species provide evidence that SINEs can help to limit the extent of gene conversion events (39, 88). The globin genes constitute a multigene family whose members began to evolve separately after duplication. However, periodically one of the duplicate genes has undergone gene conversion from another copy, either partially or completely. Increasing divergence of the duplicate genes lessens the probability of gene conversion. A SINE integration in one copy of a duplicate gene results in a discontinuity that apparently blocks (at least on some occasions) the progression of gene conversion events in both the human α-globin (39) and goat β-globin (88) genes. By limiting the extent of gene conversion, SINE sequences may spread gene diversification and the evolution of new functions.

H. Can SINEs Mobilize Other Sequences?

The formation of several composite SINE families (*Alu* and galago type II) in which new sequences are fused to a SINE and become part of its functionally retroposing unit (80, 112) indicates that SINEs could mobilize other sequences. Although these examples involve fusions with other portions of SINE sequences, the formal possibility exists for any sequence between a SINE family member and a long A-rich region that does not also contain an RNA polymerase III terminator to become part of a retroposing unit. There is one example of a sequence that is located between two artiodactyl SINEs that retroposed one time as a unit with them, resulting in the duplication of this sequence (117). This sort of retroposon "sandwich" may not be common, but it does supply the very powerful possibility of moving otherwise single-copy sequences to new locations. This could contribute to altered gene expressions and to phenomena such as exon shuffling. There is no evidence that this is occurring on a broad scale, however.

In summary, the SINEs represent one of the most successful of all of the mobile genetic elements, as judged by their high copy number. Their mobility via retroposition is very similar to that of other retroposons, such as processed pseudogenes, but is much more efficient, for as-yet-undefined reasons. The recent origin of SINE families suggests that the whole process of retroposition may have only started relatively recently. These and other data make it difficult to envision a cellular function for the SINE elements, although it is equally clear that their amplification and widespread presence in mammalian genomes must have had, and is probably continuing to have, major effects on genomic structure and function.

LITERATURE CITED

1. **Adeniyi-Jones, S., and M. Zasloff.** 1985. Transcription, processing and nuclear transport of a B1 Alu RNA species complementary to an intron of the murine alpha-fetoprotein gene. *Nature* (London) **317:**81–84.
2. **Barsh, G. S., P. H. Seeburg, and R. E. Gelinas.** 1983. The human growth hormone gene family: structure and evolution of the chromosomal locus. *Nucleic Acids Res.* **11:** 3939–3958.
3. **Barta, A., R. I. Richards, J. D. Baxter, and J. Shine.** 1981. Primary structure and evolution of rat growth hormone gene. *Proc. Natl. Acad. Sci. USA* **78:**4867–4871.
4. **Bernstein, L., S. M. Mount, and A. M. Weiner.** 1983. Pseudogenes for human small nuclear RNA U3 appear to arise by integration of self-primed reverse transcripts of the RNA into new chromosomal sites. *Cell* **32:**461–472.
5. **Bishop, J. M.** 1978. Retroviruses. *Annu. Rev. Biochem.* **47:**35–88.
6. **Bogenhagen, D. F., S. Sakonju, and D. D. Brown.** 1980. A control region in the center of the 5S RNA gene directs specific initiation of transcription. II. The 3′ border of the region. *Cell* **19:**27–35.
7. **Brann, M. R., and L. V. Cohen.** 1987. Diurnal expression of transducin mRNA and translocation of transducin in rods of rat retina. *Science* **235:**585–587.
8. **Britten, R. J., and E. H. Davidson.** 1969. Gene regulation for higher cells: a theory. *Science* **165:**349.
9. **Busk, H., B. Thomsen, B. J. Bonven, E. Kjeldsen, O. F. Nielsen, and O. Westergaard.** 1987. Preferential relaxation of supercoiled DNA containing a hexadecameric recognition sequence for topoisomerase I. *Nature* (London) **327:** 638–640.
10. **Caras, I. W., M. A. Davitz, L. Rhee, G. Weddell, D. W. Martin, Jr., and V. Nussenzweig.** 1987. Cloning of decay-accelerating factor suggests novel use of splicing to generate two proteins. *Nature* (London) **325:**545–549.
11. **Carey, M. F., K. Singh, M. Botchan, and N. R. Cozzarelli.** 1986. Induction of specific transcription by RNA polymerase III in transformed cells. *Mol. Cell. Biol.* **6:**3068–3076.
12. **Clemens, M. J.** 1987. A potential role for RNA transcribed from B2 repeats in the regulation of mRNA stability. *Cell* **49:**157–158.
13. **Daniels, G. R., and P. L. Deininger.** 1983. A second major

class of Alu family repeated DNA sequences in a primate genome. *Nucleic Acids Res.* **11:**7595–7610.

14. **Daniels, G. R., and P. L. Deininger.** 1985. Integration site preferences of the Alu family and similar repetitive DNA sequences. *Nucleic Acids Res.* **13:**8939–8954.
15. **Daniels, G. R., and P. L. Deininger.** 1985. Repeat sequence families derived from mammalian tRNA genes. *Nature* (London) **317:**819–822.
16. **Daniels, G. R., M. Fox, D. Lowensteiner, C. Schmid, and P. L. Deininger.** 1983. Species specific homogeneity of the primate Alu family of repeated DNA sequence. *Nucleic Acids Res.* **11:**7579–7593.
17. **DeChiara, T. M., and J. Brosius.** 1987. Neural BC1 RNA: cDNA clones reveal nonrepetitive sequence content. *Proc. Natl. Acad. Sci. USA* **84:**2624–2628.
18. **Deininger, P. L., and G. R. Daniels.** 1986. The recent evolution of mammalian repetitive DNA elements. *Trends Genet.* **2:**76–80.
19. **Deininger, P. L., D. J. Jolly, C. M. Rubin, T. Friedmann, and C. W. Schmid.** 1981. Base sequence studies of 300 nucleotide renatured repeated human DNA clones. *J. Mol. Biol.* **151:** 17–33.
20. **Deininger, P. L., and C. W. Schmid.** 1976. An electron microscope study of the DNA sequence organization of the human genome. *J. Mol. Biol.* **106:**773–790.
21. **Deininger, P. L., and V. K. Slagel.** 1988. Recently amplified *Alu* family members share a common parental *Alu* sequence. *Mol. Cell. Biol.* **8:**4566–4569.
22. **Doolittle, W. F., and C. Sapienza.** 1980. Selfish genes, the phenotype paradigm and genome evolution. *Nature* (London) **284:**601–603.
23. **Duncan, C. H.** 1987. Novel Alu-type repeat in artiodactyls. *Nucleic Acids Res.* **15:**1340.
24. **Duncan, C. H., P. A. Biro, P. V. Choudary, J. T. Elder, R. R. C. Wang, B. G. Forget, J. K. deRiel, and S. M. Weissman.** 1979. RNA polymerase III transcription units are interspersed among human non-α-globin genes. *Proc. Natl. Acad. Sci. USA* **76:**5095–5099.
25. **Economou-Pachnis, A., and P. N. Tsichlis.** 1985. Insertion of an Alu SINE in the human homologue of the Mlvi-2 locus. *Nucleic Acids Res.* **13:**8379–8387.
26. **Edwards, D. R., C. L. J. Parfett, and D. T. Denhardt.** 1985. Transcriptional regulation of two serum-induced RNAs in mouse fibroblasts: equivalence of one species to B2 repetitive elements. *Mol. Cell. Biol.* **5:**3280–3288.
27. **Elder, J. T., J. Pan, C. H. Duncan, and S. M. Weissman.** 1981. Transcriptional analysis of interspersed repetitive polymerase III transcription units in human DNA. *Nucleic Acids Res.* **9:**1171–1189.
28. **Endoh, H., and N. Okada.** 1986. Total DNA transcription *in vitro*: a procedure to detect highly repetitive and transcribable sequences with tRNA-like structures. *Proc. Natl. Acad. Sci. USA* **83:**251–255.
29. **Flemington, E., H. D. Bradshaw, Jr., V. Traina-Dorge, V. Slagel, and P. L. Deininger.** 1987. Sequence, structure and promoter characterization of the human thymidine kinase gene. *Gene* **52:**267–277.
30. **Friezner Degen, S. J., and E. W. Davie.** 1987. The nucleotide sequence of the gene for human prothrombin. *Biochemistry* **26:**6165–6177.
31. **Friezner Degen, S. J., B. Rajput, and E. Reich.** 1986. The human tissue plasminogen activator gene. *J. Biol. Chem.* **261:**6972–6985.
32. **Fuhrman, S. A., P. L. Deininger, P. LaPorte, T. Friedmann, and E. P. Geiduschek.** 1981. Analysis of transcription of the human Alu family ubiquitous repeating element by eukaryotic RNA polymerase III. *Nucleic Acids Res.* **9:**6439–6456.
33. **Georgiev, G. P., D. A Kramerov, A. P. Ryskov, K. G. Skryabin, and E. M. Lukanidin.** 1983. Dispersed repetitive sequences in eukaryotic genomes and their possible biological significance. *Cold Spring Harbor Symp. Quant. Biol.* **47:** 1109–1121.
34. **Glaichenhaus, N., and F. Cuzin.** 1987. A role for ID repetitive sequences in growth and transformation-dependent regulation of gene expression in rat fibroblasts. *Cell* **50:** 1081–1089.
35. **Gutierrez-Hartmann, A., and J. D. Baxter.** 1986. Stable accumulation of a rat truncated repeat transcript in Xenopus oocytes. *Proc. Natl. Acad. Sci. USA* **83:**3106–3110.
36. **Harada, F., N. Kato, and H. Hoshino.** 1979. Series of 4.5S RNAs associated with poly(A)-containing RNAs of rodent cells. *Nucleic Acids Res.* **7:**909–917.
37. **Hardison, R. C., and R. Printz.** 1985. Variability with the rabbit C repeats and sequences shared with other SINEs. *Nucleic Acids Res.* **13:**1073–1088.
38. **Haynes, S. R., T. P. Toomey, L. Leinwand, and W. R. Jelinek.** 1981. The chinese hamster Alu-equivalent sequence: a conserved highly repetitious interspersed DNA sequence in mammals has a structure suggestive of a transposable element. *Mol. Cell. Biol.* **1:**573–583.
39. **Hess, J. F., G. M. Fox, C. Schmid, and C.-K. J. Shen.** 1983. Molecular evolution of the human adult α-like globin gene region: insertion and deletion of Alu family repeats and non-Alu DNA sequences. *Proc. Natl. Acad. Sci. USA* **80:** 5970–5974.
40. **Horsthemke, B., U. Beisiegel, A. Dunning, J. R. Havinga, R. Williamson, and S. Humphries.** 1987. Unequal crossing-over between two alu-repetitive DNA sequences in the low-density-lipoprotein-receptor gene. *Eur. J. Biochem.* **164:** 77–81.
41. **Houck, C. M., F. P. Rinehart, and C. W. Schmid.** 1979. A ubiquitous family of repeated DNA sequences in the human genome. *J. Mol. Biol.* **132:**289–306.
42. **Hwu, H. R., J. W. Roberts, E. H. Davidson, and R. J. Britten.** 1986. Insertion and/or deletion of many repeated DNA sequences in human and higher ape evolution. *Proc. Natl. Acad. Sci. USA* **83:**3875–3879.
43. **Jagadeeswaran, P., B. G. Forget, and S. M. Weissman.** 1981. Short, interspersed repetitive DNA elements in eukaryotes: transposable DNA elements generated by reverse transcription of RNA pol III transcripts? *Cell* **26:**141–142.
44. **Jagadeeswaran, P., D. Tuan, B. G. Forget, and S. M. Weissman.** 1982. A gene deletion ending in the midpoint of a repetitive DNA sequence in one form of hereditary persistence of fetal globin. *Nature* (London) **296:**469–470.
45. **Jelinek, W. R., T. P. Toomey, L. Leinwand, C. Duncan, P. A. Biro, P. V. Choudary, S. M. Weissman, C. M. Rubin, C. M. Houck, P. L. Deininger, and C. W. Schmid.** 1980. Ubiquitous interspersed repeated sequences in mammalian genomes. *Proc. Natl. Acad. Sci. USA* **77:**1398–1402.
46. **Koop, B. F., M. M. Miyamoto, J. E. Embury, M. Goodman, J. Czelusniak, and J. L. Slightom.** 1986. Nucleotide sequence and evolution of the orangutan epsilon globin gene region and surrounding Alu repeats. *J. Mol. Evol.* **24:**94–102.
47. **Korenberg, J. R., and M. C. Rykowski.** 1988. Human genome organization: Alu, Lines, and the molecular structure of metaphase chromosome bands. *Cell* **53:**391–400.
48. **Korn, L. J.** 1982. Transcription of Xenopus 5S ribosomal RNA genes. *Nature* (London) **295:**101–105.
49. **Krayev, A. S., D. A. Kramerov, K. G. Skryabin, A. P.**

Ryskov, A. A. Bayev, and G. P. Georgiev. 1980. The nucleotide sequence of the ubiquitous repetitive DNA sequence B1 complementary to the most abundant class of mouse foldback RNA. *Nucleic Acids Res.* **8:**1201–1215.

50. **Kress, M., Y. Barra, J. G. Seidman, G. Khoury, and G. Jay.** 1984. Functional insertion of an Alu type 2 (B2 SINE) repetitive sequence in murine class I genes. *Science* **226:** 974–977.

51. **Laski, F. A., D. C. Rio, and G. M. Rubin.** 1986. The tissue specificity of Drosophila P element transposition is regulated at the level of mRNA splicing. *Cell* **44:**7–19.

52. **Lawrence, C. B., D. P. McDonnell, and W. J. Ramsey.** 1985. Analysis of repetitive sequence elements containing tRNA-like sequences. *Nucleic Acids Res.* **13:**4239–4252.

53. **Lee, M. G.-S., L. Loomis, and N. J. Cowan.** 1984. Sequence of an expressed human β-tubulin gene containing ten Alu family members. *Nucleic Acids Res.* **12:**5823–5836.

54. **Lehrman, M. A., J. L. Goldstein, D. W. Russell, and M. S. Brown.** 1987. Duplication of seven exons in LDL receptor gene caused by Alu-Alu recombination in a subject with familial hypercholesterolemia. *Cell* **48:**827–835.

55. **Li, W.-Y., R. Reddy, D. Henning, P. Epstein, and H. Busch.** 1982. Nucleotide sequence of 7S RNA (homology to Alu DNA and LA 4.5S RNA). *J. Biol. Chem.* **257:**5136–5142.

55a. **Lin, C., J. Goldthwaite, and D. Samols.** 1988. Identification of *Alu* transposition in human lung carcinoma cells. *Cell* **54:**153–159.

56. **Linial, M., E. Medeiros, and W. S. Hayward.** 1978. An avian oncovirus mutant (SE 21Q1b) deficient in genomic RNA: biological and biochemical characterization. *Cell* **15:** 1371–1381.

57. **Lone, Y. C., M. P. Simon, A. Kahn, and J. Marie.** 1986. Sequences complementary to the brain-specific "identifier" sequences exist in L-type pyruvate kinase mRNA (a liver-specific messenger) and in transcripts especially abundant in muscle. *J. Biol. Chem.* **261:**1499–1502.

57a. **Maeda, N., C.-I. Wu, J. Bliska, and J. Reneke.** 1988. Molecular evolution of intergenic DNA in higher primates: pattern of DNA changes, molecular clock, and evolution of repetitive sequences. *Mol. Biol. Evol.* **5:**1–20.

58. **Martin, M. A., T. Bryan, S. Rasheed, and A. S. Khan.** 1981. Identification and cloning of endogenous retroviral sequences present in human DNA. *Proc. Natl. Acad. Sci. USA* **78:**4892–4896.

59. **Matsumoto, K., K. Murakami, and N. Okada.** 1986. Gene for lysine tRNA1 may be a progenitor of the highly repetitive and transcribable sequences present in the salmon genome. *Proc. Natl. Acad. Sci. USA* **83:**3156–3160.

60. **McKinnon, R. D., P. Danielson, M. A. D. Brow, F. E. Bloom, and J. G. Sutcliffe.** 1987. Expression of small cytoplasmic transcripts of the rat identifier element in vivo and in cultured cells. *Mol. Cell. Biol.* **7:**2148–2154.

61. **McKinnon, R. D., P. Danielson, M. A. Brow, M. Godbout, J. B. Watson, and J. G. Sutcliffe.** 1987. The neuronal identifier sequence as a positive regulatory element for neuronal gene expression, p. 78–89. *In* S. Easter, K. Barald, and B. Carlson (ed.), *From Message to Mind: Directions in Developmental Neurobiology.* Sinauer Associates, Sunderland, Mass.

62. **McKinnon, R. D., T. M. Shinnick, and J. G. Sutcliffe.** 1986. The neuronal identifier element is a cis-acting positive regulator of gene expression. *Proc. Natl. Acad. Sci. USA* **83:** 3751–3755.

63. **Moos, M., and D. Gallwitz.** 1983. Structure of two human β-actin related processed genes one of which is located next to a simple repetitive sequence. *EMBO J.* **2:**757–761.

64. **Mount, S. M., and G. M. Rubin.** 1985. Complete nucleotide sequence of the *Drosophila* transposable element *copia*: homology between *copia* and retroviral proteins. *Mol. Cell. Biol.* **5:**1630–1638.

65. **Murphy, D., P. M. Brickell, D. S. Latchman, K. Willison, and P. W. J. Rigby.** 1983. Transcripts regulated during normal embryonic development and oncogenic transformation share a repetitive element. *Cell* **35:**865–871.

66. **Nakamura, Y., M. Leppert, P. O'Connell, R. Wolff, T. Holm, M. Culver, C. Martin, E. Fujimoto, M. Hoff, E. Kumlin, and R. White.** 1987. Variable number of tandem repeat (VNTR) markers for human gene mapping. *Science* **235:**1616–1622.

67. **Nichols, R. D., N. Fischel-Ghodsian, and D. R. Higgs.** 1987. Recombination at the human α-globin gene cluster: sequence features and topological constraints. *Cell* **49:**369–378.

68. **Orgel, L. E., and F. H. C. Crick.** 1980. Selfish DNA: the ultimate parasite. *Nature* (London) **284:**604–607.

69. **Orkin, S. H., and A. Michelson.** 1980. Partial deletion of the α globin structural gene in human alpha-thalassaemia. *Nature* (London) **286:**538–540.

70. **Ottolenghi, S., and B. Giglioni.** 1982. The deletion in a type of δ^0-β^0 thalassaemia begins in an inverted Alu I repeat. *Nature* (London) **300:**770–771.

71. **Owens, G. P., N. Chaudhari, and W. E. Hahn.** 1985. Brain "identifier sequence" is not restricted to brain: similar abundance in nuclear RNA of other organs. *Science* **229:** 1263–1265.

72. **Page, C. S., S. Smith, and H. M. Goodman.** 1981. DNA sequence of the rat growth hormone gene: location of the 5′ terminus of the growth hormone mRNA and identification of an internal transposon-like element. *Nucleic Acids Res.* **9:**2087–2104.

73. **Paolella, G., M. A. Lucero, M. H. Murphy, and F. E. Baralle.** 1983. The Alu family repeat promoter has a tRNA-like bipartate structure. *EMBO J.* **2:**691–696.

74. **Paulson, K. E., N. Deka, C. Schmid, R. Misra, C. Schindler, M. Rush, L. Kadyk, and L. Leinwand.** 1985. A transposon-like element in human DNA. *Nature* (London) **316:** 359–363.

75. **Paulson, K. E., and C. W. Schmid.** 1986. Transcriptional inactivity of Alu repeats in HeLa cells. *Nucleic Acids Res.* **14:**6145–6158.

76. **Perez-Stable, C., T. M. Ayres, and C.-K. J. Shen.** 1984. Distinctive sequence organization and functional programming of an Alu repeat promoter. *Proc. Natl. Acad. Sci. USA* **81:**5291–5295.

76a. **Perez-Stable, C., C. Shen, and C.-K. Shen.** 1988. Enrichment and depletion of HeLa topoisomerase I recognition sites among specific types of DNA elements. *Nucleic Acids Res.* **16:**7975–7993.

77. **Rabin, H., C. V. Benton, M. A. Tainsky, N. R. Rice, and R. V. Gilden.** 1979. Isolation and characterization of an endogenous type C virus of rhesus monkeys. *Science* **204:** 841–842.

78. **Rinehart, F. P., T. G. Ritch, P. L. Deininger, and C. W. Schmid.** 1981. Renaturation rate studies of a single family of interspersed repeated DNA sequences in human deoxyribonucleic acid. *Biochemistry* **20:**3003–3010.

79. **Rogers, J.** 1983. Retroposons defined. *Nature* (London) **301:**460.

80. **Rogers, J.** 1985. The origin and evolution of retroposons. *Int. Rev. Cytol.* **93:**187–279.

81. **Rowyer, F., M.-C. Simmler, D. C. Page, and J. Weissenbach.** 1987. A sex chromosome rearrangement in a human XX male caused by Alu-Alu recombination. *Cell* **51**:417–425.

82. **Ryskov, A. P., P. L. Ivanov, D. A. Kramerov, and G. P. Georgiev.** 1983. Mouse ubiquitous B2 repeat in polysomal and cytoplasmic poly(A)$^+$ RNAs: unidirectional orientation and 3′-end localization. *Nucleic Acids Res.* **18**:6541–6559.

83. **Sakaki, Y., M. Hattori, A. Fujita, K. Yoshioka, S. Kuhara, and O. Takenaka.** 1986. The LINE-1 family of primates may encode a reverse transcriptase-like protein. *Cold Spring Harbor Symp. Quant. Biol.* **51**:465–469.

84. **Sakamoto, K., and N. Okada.** 1985. Rodent type 2 Alu family, rat identifier sequence, rabbit C family, and bovine or goat 73-bp repeat may have evolved from tRNA genes. *J. Mol. Evol.* **22**:134–140.

85. **Sapienza, C., and B. St.-Jacques.** 1986. "Brain-specific" transcription and evolution of the identifier sequence. *Nature* (London) **319**:418–420.

86. **Sawada, I., and C. W. Schmid.** 1986. Primate evolution of the α-globin gene cluster and its Alu-like repeats. *J. Mol. Biol.* **192**:693–709.

87. **Sawada, I., C. Willard, C.-K. J. Shen, B. Chapman, A. C. Wilson, and C. W. Schmid.** 1985. Evolution of the Alu family repeats since the divergence of human and chimpanzee. *J. Mol. Evol.* **22**:316–322.

88. **Schimenti, J. C., and C. H. Duncan.** 1984. Ruminant globin gene structures suggest an evolutionary role for Alu-type repeats. *Nucleic Acids Res.* **12**:1641–1655.

89. **Schmid, C. W., and W. R. Jelinek.** 1982. The Alu family of dispersed repetitive sequences. *Science* **216**:1065–1070.

90. **Schmid, C. W., and C.-K. J. Shen.** 1986. The evolution of interspersed repetitive DNA sequences in mammals and other vertebrate, p. 323–358. *In* R. J. MacIntyre (ed.), *Molecular Evolutionary Genetics*. Plenum Publishing Corp., New York.

91. **Schuler, L. A., J. L. Weber, and J. Gorski.** 1983. Polymorphism near the rat prolactin gene caused by insertion of an Alu-like element. *Nature* (London) **305**:159–160.

92. **Shen, C.-K. J., and T. Maniatis.** 1982. The organization, structure and *in vitro* transcription of Alu family RNA polymerase III transcription units in the human α-like globin gene cluster. *J. Mol. Appl. Genet.* **1**:343–360.

93. **Sherwin, S. A., and G. J. Todaro.** 1979. A new endogenous primate type C virus isolated from the Old World monkey *Colobus polykomos*. *Proc. Natl. Acad. Sci. USA* **76**: 5041–5045.

94. **Singer, M.** 1982. SINEs and LINEs: highly repeated short and long interspersed sequences in mammalian genomes. *Cell* **28**:433–434.

95. **Singh, K., M. Carey, S. Saragosti, and M. Botchan.** 1985. Expression of enhanced levels of small RNA polymerase III transcripts encoded by the B2 repeats in simian virus 40-transformed cells. *Nature* (London) **314**:553–556.

96. **Slagel, V., E. Flemington, V. Traina-Dorge, H. Bradshaw, and P. Deininger.** 1987. Clustering and subfamily relationships of the Alu family in the human genome. *Mol. Biol. Evol.* **4**:19–29.

97. **Spence, S. E., R. M. Young, K. J. Garner, and J. B. Lingrel.** 1985. Localization and characterization of members of a family of repetitive sequences in the goat β globin locus. *Nucleic Acids Res.* **13**:2171–2186.

98. **Sutcliffe, J. G., R. J. Milner, F. E. Bloom, and R. A. Lerner.** 1982. Common 82-nucleotide sequence unique to brain RNA. *Proc. Natl. Acad. Sci. USA* **79**:4942–4946.

99. **Sutcliffe, J. G., R. J. Milner, J. M. Gottesfeld, and R. A. Lerner.** 1984. Identifier sequences are transcribed specifically in brain. *Nature* (London) **308**:237–241.

100. **Trabuchet, G., Y. Chebloune, P. Savatier, J. Lachuer, C. Faure, G. Verdier, and V. M. Nigon.** 1987. Recent insertion of an Alu sequence in the β-globin gene cluster of the gorilla. *J. Mol. Evol.* **25**:288–291.

101. **Ullu, E., V. Esposito, and M. Melli.** 1982. Evolutionary conservation of the human 7S RNA sequence. *J. Mol. Biol.* **161**:195–201.

102. **Ullu, E., S. Murphy, and M. Melli.** 1982. Human 7S RNA consists of a 140 nucleotide middle repetitive sequence inserted in an Alu sequence. *Cell* **29**:195–202.

103. **Ullu, E., and C. Tschudi.** 1984. Alu sequences are processed 7SL RNA genes. *Nature* (London) **312**:171–172.

104. **Ullu, E., and A. M. Weiner.** 1984. Human genes and pseudogenes for the 7SL RNA component of signal recognition particle. *EMBO J.* **3**:3303–3310.

105. **Ullu, E., and A. M. Weiner.** 1985. Upstream sequences modulate the internal promoter of the human 7SL RNA gene. *Nature* (London) **318**:371–374.

106. **VanArsdell, S. W., R. A. Denison, L. B. Bernstein, A. M. Weiner, T. Manser, and R. F. Gesteland.** 1981. Direct repeats flank three small nuclear RNA pseudogenes in the human genome. *Cell* **26**:11–17.

107. **VanArsdell, S. W., and A. M. Weiner.** 1984. Pseudogenes for human U2 small nuclear RNA do not have a fixed site of 3′ truncation. *Nucleic Acids Res.* **12**:1463–1471.

108. **Vanin, E.** 1984. Processed pseudogenes: characteristics and evolution. *Biochim. Biophys. Acta* **782**:231–241.

109. **Walter, P., and G. Blobel.** 1982. Signal recognition particle contains a 7S RNA essential for protein translocation across the endoplasmic reticulum. *Nature* (London) **299**:691–698.

110. **Watanabe, Y., T. Tsukadam, M. Notake, S. Nakanishi, and S. Numa.** 1982. Structural analysis of repetitive DNA sequences in the bovine corticotropin-β-lipotropin precursor gene region. *Nucleic Acids Res.* **10**:1459–1469.

111. **Watson, J. B., and J. G. Sutcliffe.** 1987. Primate brain-specific cytoplasmic transcript of the *Alu* repeat family. *Mol. Cell. Biol.* **7**:3324–3327.

112. **Weiner, A. M., P. L. Deininger, and A. Efstratiadis.** 1986. Nonviral retroposons: genes, pseudogenes, and transposable elements generated by the reverse flow of genetic information. *Annu. Rev. Biochem.* **55**:631–661.

113. **Willard, C., H. T. Nguyen, and C. W. Schmid.** 1987. Existence of at least three distinct Alu subfamilies. *J. Mol. Evol.* **26**:180–186.

114. **Yamamoto, T., C. G. Davis, M. S. Brown, W. J. Schneider, M. L. Casey, J. L. Goldstein, and D. W. Russell.** 1984. The human LDL receptor: a cysteine rich protein with multiple Alu sequences in its mRNA. *Cell* **39**:27–38.

115. **Young, L. S., N. Takahashi, and K. U. Sprague.** 1986. Upstream sequences confer distinctive transcriptional properties on genes encoding silkgland-specific tRNA Ala. *Proc. Natl. Acad. Sci. USA* **83**:374–378.

116. **Young, P. R., R. W. Scott, D. H. Hamer, and S. M. Tilghman.** 1982. Construction and expression *in vivo* of an internally deleted mouse alpha fetoprotein gene: presence of a transcribed Alu-like repeat within the first intervening sequences. *Nucleic Acids Res.* **10**:3099–3116.

117. **Zelnick, C. R., D. J. Burks, and C. H. Duncan.** 1987. Mobilization of genomic sequences by flanking Alu-type repetitive DNA. *Nucleic Acids Res.* **15**:10437–10453.

Chapter 28

Bacterial DNA Inversion Systems

ANNA C. GLASGOW, KELLY T. HUGHES, and MELVIN I. SIMON

I. INTRODUCTION

Phase variation may be defined as an alteration in the expression of a phenotype involving either the on-off switching of the phenotype or the expression of any one of a number of alternate phenotypes or serotypes (antigenic variation). There are numerous examples of phenotypic behavior in procaryotes that suggest that some form of phase variation is operating (Table 1). In some cases, little is known about the precise mechanism that drives these changes; however, phase variation in many bacteria has been shown to involve targeted, or programmed, DNA rearrangements that alter gene expression. Spontaneous mutation affecting gene expression occurs at a very low frequency, usually less than 1 mutational event per 10^6 cells per generation. However, programmed DNA rearrangements controlling gene expression, which have been characterized in bacterial (and eucaryotic) systems (for review, see reference 10), occur at frequencies ranging from 1 in 10^2 to 10^5 per generation. Often these programmed DNA rearrangements serve to preadapt the organism to changes in its environment (see below). They may also be developmentally regulated, such as the site-specific deletion in the nitrogen fixation genes of *Anabaena* that occurs during heterocyst differentiation (45, 95). Programmed or developmental rearrangements can occur by a variety

Anna C. Glasgow, Kelly T. Hughes, and Melvin I. Simon ■ Department of Biology 147-75, California Institute of Technology, Pasadena, California 91125.

Table 1. Examples of phase variations in procaryotes

Observation	Organism	Mechanism	References
Flagellar variation	*Salmonella typhimurium*	Site-specific inversion of DNA segment	This chapter
Fimbrial variation	*Escherichia coli*	Site-specific inversion of DNA segment	This chapter
Pilin variation	*Moraxella bovis*	Site-specific inversion of DNA segment	This chapter
Pilin variation[a]	*Neisseria gonorrhoeae*	Duplicative transposition of genes (gene conversion)	50, 143, 144, 161, 162
Opacity protein variation[a]	*Neisseria gonorrhoeae*	Duplication/deletion of nucleotides in leader peptide sequence	155, 156
Flagellar variation	*Campylobacter coli*	DNA rearrangement	49, 52
Pigment and flagellar variation	*Serratia marcescens*	?	16, 114
Pigment variation	*Rhodopseudomonas spheroides*	?	46
Pigment variation	*Streptomyces reticuli*	Deletion of gene by recombination between reiterated sequences	142
Pigment variation	*Myxococcus xanthus*	DNA rearrangement	17, 175
Bioluminescence variation	*Vibrio harveyi*	?	83, 150
Vi surface antigen variation	*Citrobacter freundii*	Precise/insertion/deletion of IS*1*	7, 11a, 153
Extracellular polysaccharide variation	*Pseudomonas atlantica*	Precise insertion/deletion of DNA sequence	8a
Serotype variation[b]	*Borrelia hermsii*	Duplicative transposition of genes (site-specific recombination?)	99, 119, 120
Agglutinogen variation	*Bordetella pertussis*	?	127
Tail fiber (host range) variation	Bacteriophages Mu and P1	Site-specific inversion of DNA segment	This chapter
Variation in alkaline phosphatase synthesis	*Escherichia coli* (*phoR* mutant)	Physiological switch?	171–173

[a] For reviews, see reference 55 and J. Swanson and J. M. Koomey, this volume.
[b] For review, see A. Barbour, this volume.

of mechanisms; e.g., they may involve insertion, duplicative transposition, deletion, or inversion of DNA sequences (Table 1).

One form of DNA rearrangement which is involved in fimbrial phase variation in *Escherichia coli* and flagellar phase variation in *Salmonella typhimurium*, as well as phase variation in other bacteria and host-range variation in several bacteriophages (see below), is the inversion of DNA. In various systems, the invertible DNA segment may either contain a promoter which is inverted relative to a stationary structural gene (as in *E. coli* and *S. typhimurium* phase variation) or contain structural genes which are switched relative to a stationary promoter (as in host-range variation in bacteriophages Mu and P1). The simplicity of this mechanism for control of gene expression has stimulated extensive characterization of inversion in several bacterial and phage systems. This review will focus on site-specific DNA inversion systems and homologous recombination resulting in inversions in bacteria.

II. CONTROL OF GENE EXPRESSION BY INVERSION

A. Type 1 Fimbriae Variation in *E. coli*

The type 1 fimbriae (pili) of *E. coli* mediate the mannose-sensitive attachment of the bacteria to epithelial cells, erythrocytes, and leukocytes (8, 107, 113, 134), thus playing an important role in the virulence of *E. coli* (31, 176; for reviews see references 32 and 152). Fimbrial expression undergoes an on-off switch at the rate of approximately 1 change per 1,000 bacteria per generation (11, 159, 160). In mapping and complementation studies with nonfimbriate mutants, Brinton and co-workers found that

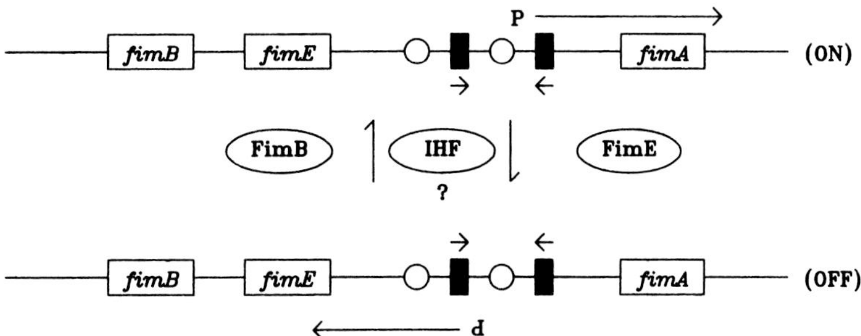

Figure 1. Inversion control of type 1 fimbrial expression. Inversion of a DNA segment containing an active promoter (P) controls the expression of *fimA*, which encodes the major structural protein of the type 1 fimbriae of *E. coli*. The invertible segment is bounded by 9-bp inverted repeats (solid boxes) of the sequence TTGGGGCCA. The protein FimB directs inversion of the DNA segment to the ON orientation, whereas FimE directs the switching to the OFF orientation. The role of IHF in this inversion is uncertain (?); the putative binding sites for IHF are indicated by open circles (27, 33).

there were at least three *pil* genes, located at 98 min on the *E. coli* chromosome, which were required for type 1 fimbriae (pili) formation (12, 160). In vivo operon fusions of the *lac* genes to *fim* (*pil*) have been used to probe the regulation of expression of the type 1 fimbriae (30, 111). Orndorff et al. (111) and Freitag et al. (37) demonstrated with *lacZ* operon fusions to *fimA* that regulation of expression of the type 1 fimbriae involved a tightly linked *cis*-acting element and *trans*-active factors. The close linkage of the switch-controlling element to *fimA*, which encodes the 17-kilodalton pilin protein (86, 110), was determined by P1 transduction of chromosomal *fimA-lac* fusions (111). The *cis* nature of the controlling element was indicated by the independent switching phenotype of a specialized transducing λ bacteriophage containing the *fimA-lac* operon fusion ($Lac^+ \leftrightarrow Lac^-$) upon transduction of a fim^+ strain ($Fim^+ \leftrightarrow Fim^-$) (37). Specialized λ phages which were defective for alternate expression of *fimA-lacZ* in Δ*fim* strains could be complemented for switching in fim^+ strains, indicating that a *trans*-acting factor was also involved in phase transition. Abraham et al. (2) found that the *cis*-acting element controlling fimbrial expression was an invertible DNA segment containing the promoter for *fimA* (*pilA*). Southern blot analysis of DNA from Fim^+ (ON) and Fim^- (OFF) clonal variants of *E. coli* revealed that a DNA rearrangement was associated with the phase variation. The DNA sequence of the ON and OFF variants of λ specialized transducing phages containing the *fimA-lac* fusion confirmed that inversion of a 314-base-pair (bp) DNA segment, flanked by 9-bp inverted repeats, altered the orientation of a promoter sequence immediately upstream of the *fimA* gene (Fig. 1).

B. Flagellar Phase Variation in *S. typhimurium*

One of the first identified DNA rearrangements controlling gene expression in bacteria was the inversion of the H segment of *S. typhimurium* which controls phase variation of the flagellar antigen. In 1922, Andrewes discovered that *S. typhimurium* alternately exhibited two serotypes which corresponded to its flagellar antigens (4). This phase transition from one flagellar antigen to the other occurred at frequencies ranging from 10^{-5} to 10^{-3} per bacterial division (157) and presumably enabled part of the bacterial population to elude the immune response of the host. Lederberg and Iino (96) found that the control of the expression of the two unlinked flagellin genes, *H1* and *H2*, cotransduced with the *H2* locus independent of whether the donor strain in the transduction was expressing *H1* or *H2*. Therefore, the phase of *H2* expression (ON or OFF) or the "state" of the *H2* gene in the donor determined the expression of *H1* in the recipient. They suggested that the heritable state of the *H2* gene was due to a genetic modification similar to the mechanism of action of controlling elements in maize described by McClintock (96, 98; N. V. Fedoroff, this volume). In retrospect, after detailed genetic and molecular characterization of the *Salmonella* phase variation system (described below), this comparison still seems quite valid. Silverman and Simon (147) have suggested that the controlling element of the *Salmonella* phase variation system may have evolved from the interaction of transposable elements with the regulatory region of an ancestral flagellin gene.

Genetic studies undertaken to define the controlling factors involved in flagellar phase variation in *S. typhimurium* revealed a *trans*-acting factor (Rh1) which represses the expression of the *H1* gene and is cotranscribed with *H2* (38, 115, 158). Therefore, coexpression of *H2* and the adjacent *rH1* gene, encoding the repressor of *H1*, results in synthesis of only the *H2* flagellin. Alternatively, when the "controlling element" turns off expression of *H2* and *rH1*, the *H1* flagellin is synthesized (149) (Fig. 2). The nature of the controlling element was discovered

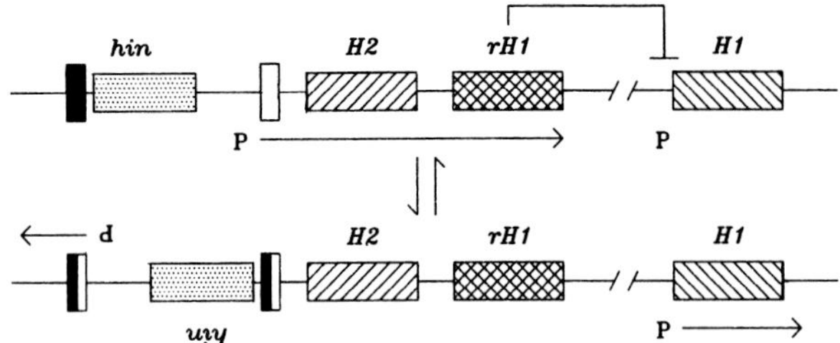

Figure 2. Inversion-controlled expression of the flagellin genes of *S. typhimurium*. Inversion of a 996-bp DNA segment between two short inverted repeats (solid/open boxes) switches the orientation of a promoter (P) within the invertible segment. In one orientation (ON) the *H2* flagellin gene and *rH1* are expressed from this promoter. The product of *rH1* negatively regulates transcription from the *H1* flagellin gene promoter (P). In the other orientation (OFF), the promoter is uncoupled from the *H2* operon, *rH1* is not expressed, and the repression of *H1* expression is relieved.

when genetic and physical analyses of the cloned *H2* region revealed an invertible DNA segment upstream of *H2* which contained a promoter capable of activating transcription of genes fused to the invertible segment (148, 179, 180). Deletions extending into the invertible segment eliminated inversion of the DNA and phase variation (148). Mutational and DNA-sequence analyses of the inversion region (146, 181) demonstrated that the orientation of the *H2* promoter (and therefore the expression of *H2* and *rH1*) is controlled by inversion of the 996-bp DNA segment which contains the *H2* promoter and encodes a protein required for its inversion (Hin). Two *cis*-acting sites, the 26-bp inverted repeats which bound the inversion region, were also found to be essential for this DNA rearrangement, which controls flagellin gene expression.

C. Pilin Phase Variation in *Moraxella bovis*

Another phase variation system in bacteria that appears to use an invertible DNA segment as a controlling element is found in *Moraxella bovis*. The switch between expression of the alpha- and beta-pilin proteins of *M. bovis* involves the inversion of an approximately 2-kilobase (kb) region of DNA encoding the carboxy-terminal variable region of each protein (96a, 97). The control of pilin expression by DNA inversion was indicated by genomic Southern hybridization analysis of strains expressing alpha or beta pilin. Interestingly, the DNA sequence of the inversion region within the beta-pilin gene exhibits 58% homology with the left inverted repeat of the invertible H segment of *S. typhimurium* (Fig. 3), suggesting that these control elements may be related.

D. Bacteriophage Host-Range Variation and Related Inversion Systems

Regulation of gene expression by inversion is not limited to phase variation in bacteria; bacteriophages Mu and P1 (and related phages) have homologous invertible DNA regions which control their host range (19, 61, 167). The advantage conferred by the low-frequency switching between host-range deter-

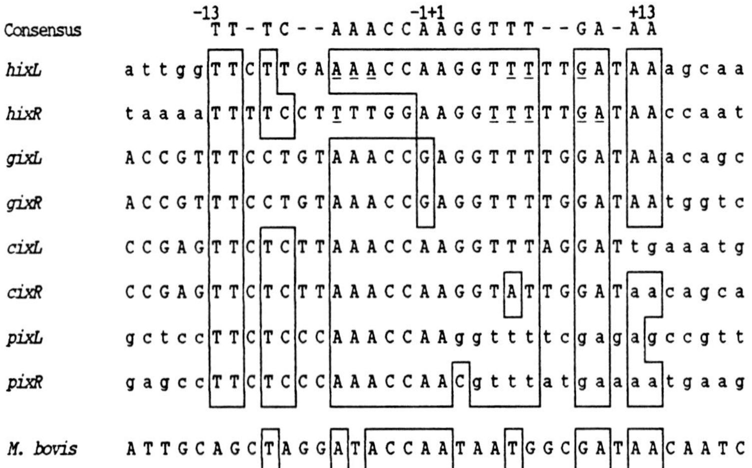

Figure 3. Recombination sites of the Hin-related inversion systems. The 26-bp consensus sequence was determined by comparison of the sequence surrounding the recombination sites of the Hin, Gin, Cin, and Pin systems, which are aligned here for maximum homology (58, 121, 147). Bases conserved between the recombination sites are boxed. Underlined bases in the *hixL* and *hixR* sequences are in base pairs which are contacted by Hin, as indicated by methylation interference assays (Glasgow et al., manuscript in preparation). The uppercase letters represent bases within the inverted repeats of each system; the lowercase letters indicate the bases outside of the inverted repeats in the ON orientation. Below the well-characterized recombination sites the sequence of one of the proposed recombination sites for pilin gene inversion in *M. bovis* is shown (aligned for maximum homology with *hixL*) (96a, 97).

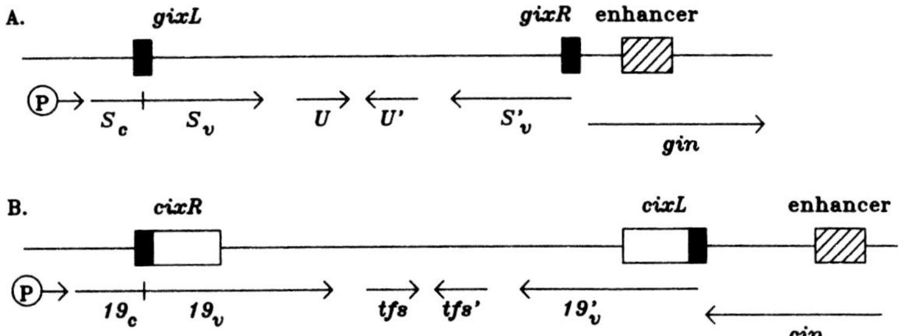

Figure 4. Invertible regions of bacteriophages Mu and P1. (A) The G(+) orientation of the invertible DNA segment of bacteriophage Mu. The expression of the *S*, *U*, *S'*, and *U'* genes is controlled by inversion of the G segment. In the G(+) orientation (shown), the variable carboxy-terminal portion of the *S* gene (S_v) is coupled to its constant amino-terminal portion (S_c). The expression of both *S* and *U* from the external promoter (P) results in the formation of tail fibers which recognize LPS receptors on *E. coli* K-12. In the G(−) orientation, *S'* and *U'* are expressed, resulting in the formation of tail fibers which recognize different LPS receptors on a variety of bacterial hosts, including *E. coli* C, *C. freundii*, and *S. sonnei* (168). Inversion of the 3,000-bp segment is mediated by the product of the *gin* gene located immediately adjacent to the invertible region. The recombination sites (*gixL* and *gixR*) are within the 34-bp inverted repeats (solid/open boxes) bounding the G segment. The enhancer sequence (hatched box) located near the 5' end of the *gin* gene stimulates the inversion reaction. (B) The C segment of bacteriophage P1, which has a 3-kb internal DNA segment which is homologous to the G segment of Mu and is flanked by two 0.62-kb inverted repeats (solid/open boxes) containing the *cixL* and *cixR* recombination sites (solid boxes). The genes encoded within the invertible DNA segment, *19*, *19'*, *tfs*, and *tfs* (tail fiber specificity), share extensive homology with the corresponding *S*, *S'*, *U*, and *U'* genes of Mu and have analogous functions (79). Note that the orientation of the *cin* gene with respect to the invertible segment is inverted relative to that of the *gin* gene; it has been proposed that this occurred by site-specific recombination between a pseudo-recombination site within the promoter region of *cin* and the *cixR* recombination site (58).

minants for Mu and P1 is similar to that of flagellar variation for *S. typhimurium*; i.e., the programmed DNA rearrangement preadapts part of the phage population to changes in the environment by expanding its host range, thereby increasing the chances of survival.

Heteroduplex mapping of the DNA of bacteriophage Mu revealed a 3-kb inversion loop (G segment) in the phage genome (23, 63). A related phage, D108, also exhibits G-segment inversion (43). DNA isolated from phage produced by induction of a Mu lysogen contained both orientations of the G segment, while DNA from phage resulting from infection of *E. coli* K-12 had primarily one orientation, identified as G(+) (24). Similar to the inversion control of flagellar variation in *S. typhimurium*, it was found that the inversion of the G segment controlled alternation of the host range of Mu (76, 168). In the G(+) orientation, the phage could adsorb to the lipopolysaccharide (LPS) receptors of *E. coli* K-12; in the G(−) orientation, the phage could no longer infect *E. coli* K-12 but now was capable of infecting other enteric bacteria, including *E. coli* C, *Citrobacter freundii*, *Shigella sonnei*, *Serratia marcescens*, and *Enterobacter cloacae* (76, 168). Initial genetic characterization of the G segment mapped the essential genes *S* and *U* to this inversion region (62). Further genetic analysis of the cloned inversion region demonstrated that two alternate sets of genes *S*-*U* [G(+)] and *S'*-*U'* [G(−)] are expressed, depending on the orientation of the G segment (44, 168). Genes *S* and *S'* share the coding region for the constant amino-terminal segment; however, the carboxy-terminal region varies, depending on the orientation of the G segment (Fig. 4). The *S*, *U*, *S'*, and *U'* genes have been shown to be required for the formation of the tail fibers of the Mu phage particle (48; F. Grundy, Ph.D. thesis, University of Wisconsin, Madison, 1984) which mediate the interaction of phage with different LPS receptors on various bacterial hosts. Therefore, in the G(+) orientation, expression of *S* and *U* results in tail fibers which mediate adsorption to one type of LPS receptor, and in the G(−) orientation, expression of *S'* and *U'* produces tail fibers capable of adsorbing to a different type of LPS receptor, thereby varying the host range of Mu.

Bacteriophage P1, which is unrelated to phage Mu, has an invertible DNA region called the C segment. The 4.2-kb C segment contains two large (0.62 kb) inverted repeats flanking a 3-kb internal segment which is homologous to the G segment of Mu (19). Inversion of the C segment also controls the phage host specificity (58, 65), with the C-segment containing genes homologous in DNA sequence and function to *S*, *U*, *S'*, and *U'* (19, 166, 167, 170). A P1-related phage, P7, also contains an invertible DNA segment controlling its host range which is homologous to the C segment of P1 (18).

Mutational analysis has shown that Mu and P1 encode *trans*-acting factors which are essential for the inversion of the G and C segments, respectively (68, 78). The genes encoding these recombinase functions, *gin* (Mu) and *cin* (P1), are located outside of the inversion region (Fig. 4), unlike the *hin* gene of *S. typhimurium*, which is located within the invertible H segment (Fig. 2). Complementation studies have shown that the Gin, Cin, and Hin recombinases are functionally similar, i.e., each recombinase is capable of mediating inversion of the G, C, or H segments (68, 77, 93). The G and C segments also share DNA sequence homology with the H segment within the inverted repeats which bound the inversion regions (58, 77); these inverted repeats are the sites at which the recombinational crossover event occurs (see below).

Three other systems that involve site-specific inversion and are probably related to those of bacteriophage Mu and P1 (and *S. typhimurium*) have been identified, but their function is unknown. The *e*14 element of *E. coli*, which may be a defective prophage, contains an invertible DNA region called the P segment (122). Inversion of the P segment appears to control the alternate expression of two proteins whose coding sequences overlap the inversion region, as for those of the *S* and *S′* genes of the G segment of Mu. However, there is no sequence homology between the tail fiber genes of Mu and these genes of unknown function in the P segment (122). The homology between the *e*14 inversion system and those of Mu and P1 again resides in the recombinase responsible for inversion of the P segment (Pin) and the inverted repeats on either side of the inversion region (122). Pin can complement the inversion function of Mu and *S. typhimurium* (34, 169). The phage P1-related plasmid p15B has a complex, multiple-invertible segment which shows little homology to the C segment encoding the host-range functions of P1; however, it does show DNA sequence similarity within the inverted repeats at the crossover sites. Also, its recombinase Cin(P15) can functionally replace Cin(P1) (67; W. Arber, J. Meyer, R. Hiestand-Nauer, H. Sandmeir, and S. Iida, personal communication). Several of the related staphylococcal penicillinase plasmids contain an invertible DNA segment (*inv*) which is located near the β-lactamase gene (145). The *inv* region in plasmid pI524 of *Staphylococcus aureus* is 2.2 kb, flanked by 0.65-kb inverted repeats (104). Site-specific recombination between the inverted repeats of *inv* in the closely related plasmid pI9789*blaI.blaZ* requires a protein encoded by a gene on the plasmid (S.-J. Rowland and K. G. H. Dyke, personal communication). The predicted amino acid sequence of this protein (Bin) shows ~39% amino acid identity with the recombinases of the inversion systems of *S. typhimurium*, bacteriophages Mu and P1, and the *e*14 element of *E. coli* (Fig. 5). It has been proposed that Bin is related to the Hin family of recombinases and mediates the inversion of the *inv* region of the staphylococcal plasmids (131a).

Another site-specific inversion system is the IncI1 plasmid pR64 "shufflon" system, which contains four DNA segments flanked by inverted repeats (91). The multiple DNA rearrangements exhibited by this "clustered inversion region" are similar to those of the complex p15b system (see above); however, the recombinase of the shufflon system, encoded by the *rci* gene, shows homology to the integrases of the lambdoid phages rather than to the recombinases of the Cin-related inversion systems (A. Kuko, A. Kusukawa, and T. Komano, personal communication). Therefore, the shufflon is probably unrelated to Cin(P15) and its homologous systems.

III. MOLECULAR MECHANISMS OF SITE-SPECIFIC DNA INVERSION

A. FimB/FimE-Mediated Inversion in *E. coli*

1. Role of *fimB* and *fimE* in the inversion reaction

The inversion-controlled expression of type 1 fimbriae of *E. coli* resembles phase variation of the flagellar antigens of *S. typhimurium* in that the inversion of a promoter-containing DNA segment controls expression of the pilin structural gene *fimA*. However, there is no *trans* complementation of *fim* inversion by the *hin* gene product (37) and no sequence homology between the 9-bp inverted repeats of the *fim*-controlling element and the larger inverted repeats of the H segment of *S. typhimurium* (2) (see legends of Fig. 1 and 3). The *trans*-acting factors involved in this phase transition, while not yet completely defined, appear to effect a unique control of inversion. Two regulatory genes, *fimB* and *fimE*, located upstream of *fimA* each controls unidirectional inversion of the promoter-containing DNA segment (87). In the absence of both *fimB* and *fimE*, the invertible element is locked in either the ON or OFF orientation as seen with switch-defective λ specialized phages (*fimA-lacZ* fusions) which remained Lac^+ or Lac^- upon transduction of a Δ*fim* strain (37). However, when only *fimB* is inactivated, phase variation goes in one direction, to Fim^- (OFF), but not from Fim^- to Fim^+ (ON) (88).

The *fimB* and *fimE* genes have been cloned, separately and together, on plasmids which were compatible with a plasmid containing the rest of the genes required for fimbriation, including the invertible DNA segment with the promoter for *fimA* in the

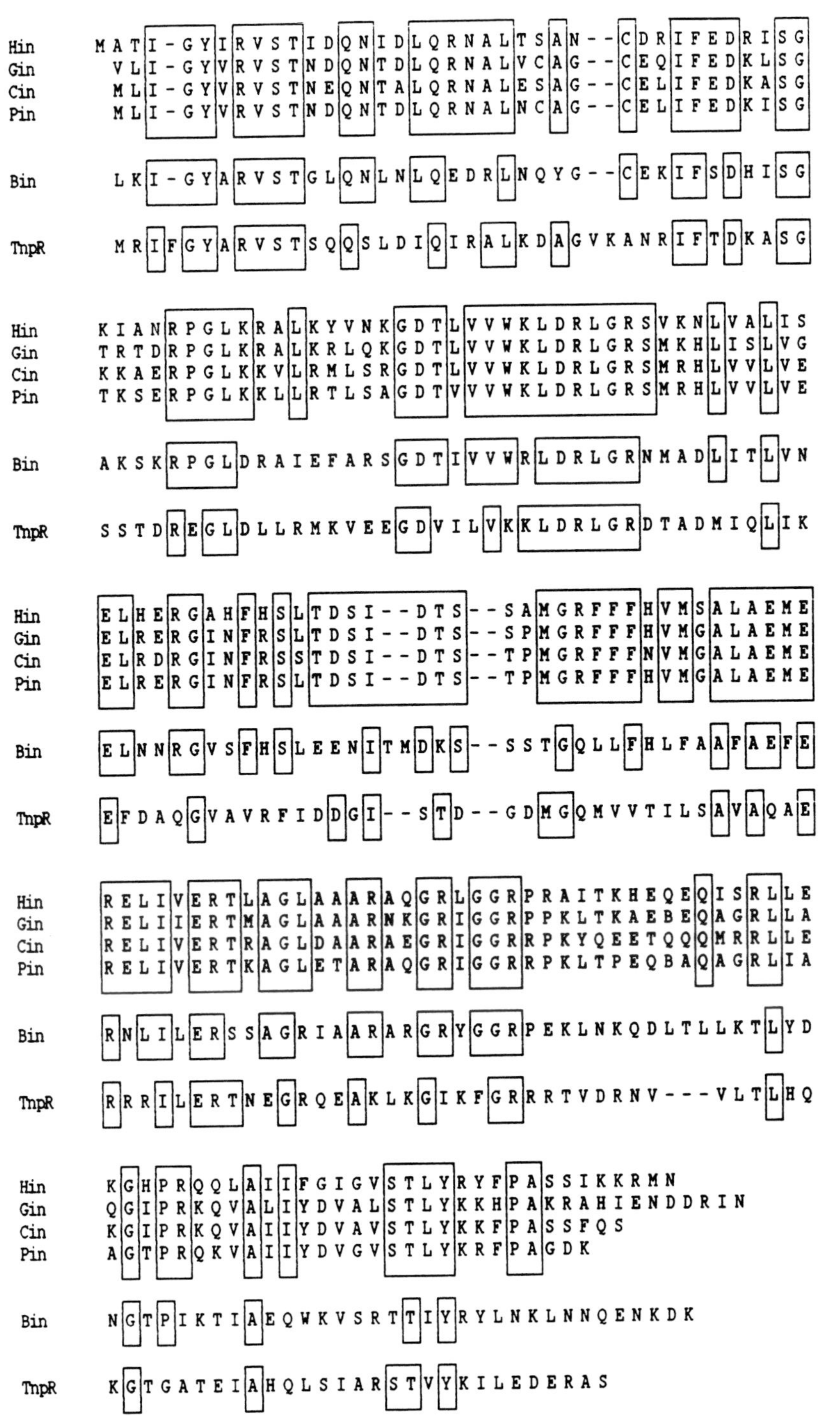

Figure 5. Comparison of the Hin-related recombinases and Bin of *S. aureus* and TnpR of γδ. The amino acid sequences of Hin (181), Gin (116), Cin (58), Pin (116), Bin (131a), and TnpR (57) are aligned with some gaps for maximum homology. Amino acid residues shared by Hin, Gin, Cin, and Pin are boxed. Identical residues between Bin or TnpR and the homologous invertases are boxed separately.

OFF orientation (87). The ability of the *fimB* and *fimE* gene products to complement in *trans* the inversion of the promoter-containing segment was measured by fimbriation of transformed Δ*fim* strains. It was found that the $fimB^+$ $fimE^+$ and the $fimB^+$ plasmids complemented the Fim OFF plasmids for the phase switch to the ON mode. The $fimE^+$ plasmid, which could not effect the complementation of switching to the ON mode, could complement inversion to the OFF orientation of the Fim^+ (ON) variant plasmid. These phase transitions correlated with the physical inversion of the DNA segment

```
FIM B   H M L R H S C G F A L A N M G - I D T R L I Q D Y L G H R N - I R H T V W Y T A
FIM E   H M L R H A C G Y E L A E R G - A D T R L I Q D Y L G H R N - I R H T V R Y T A
P2      H A L R H S F A T H F M I N G - G S I I T L Q R I L G H T R - I E Q T M V Y A H
186     H V L R H T F A S H F M M N G - G N I L V L Q R V L G H T D - I K M T M R Y A H
P22     H D L R H T W A S W L V Q A G - V P I S V L Q E M G G W E S - I E M V R R Y A H
P1      H S A R V G A A R D M A R A G - V S I P E I M Q A G G W T N - V N I V M N Y I R
λ       H E L R S L S A - R L Y E K Q - I S D K F A Q H L L G H K S - D T M A S Q Y R -
φ80     H D M R R T I A T N L S E L G - C P P H V I E K L L G H Q M - V G V M A H Y N L
P4      H G F R T M A R G A L G E S G L W S D D A I E R Q L S H S E R N N V R A A Y I H
```

Figure 6. Comparison of the C-terminal regions of the integrase proteins of the lamboid phages and the FimB and FimE proteins of *E. coli*. The amino acid sequences for FimB (residues 140 to 178) and FimE (residues 135 to 173) are aligned for maximum homology with the carboxy-terminal homologous regions of the integrase proteins (starting at residue 306 for λ integrase) (5, 27, 33). Conserved residues (exact matches only) between one or both Fim proteins and at least two of the integrases are boxed (33).

containing the *fimA* promoter, as determined by restriction analysis of the plasmid. In addition, in the presence of excess FimE, an intact *fim* gene cluster (on the chromosome or on a plasmid) exhibits phase switching predominantly (80%) to the OFF orientation, and the opposite is observed when FimB is in excess. The presence of both FimB and FimE in excess did not affect the normal phase control system. Based on these results, it has been proposed that the *fimB* gene product mediates inversion of the 314-bp DNA segment to the ON promoter orientation, resulting in expression of *fimA*, and that the *fimE* gene product mediates inversion to the OFF promoter orientation, resulting in the failure to express *fimA* (87) (Fig. 1). Therefore, the control of fimbrial phase variation in *E. coli* must be tightly linked to the control of FimB and FimE activity or expression. Other host proteins that have been shown to interact with or affect the Fim inversion system are described below.

2. Other host proteins involved in FimB/FimE-directed inversion

Another function encoded within the *fim* gene cluster had been proposed to affect type 1 fimbrial phase variation, since strains carrying a mutation in this locus were hyperfimbriated (Hyp); i.e., they exhibited an increase in the number and length of fimbriae per cell (109). This mutant phenotype of *hyp* is similar to that of *fimE* (increased fimbriation). *hyp* was thought to correlate with *fimE* because it maps to approximately the same position in the Fim operon as *fimE* and encodes a protein the same size as the *fimE* gene product (23 kilodaltons) (87, 108, 109). However, *hyp* and *fimE* probably are different genes whose products control the expression of *fimA* by different mechanisms, since insertion mutations in the *hyp* gene do not affect phase variation exhibited by a chromosomal *fimA-lacZ* fusion and appear to affect the level of transcription of the *fimA* gene (twofold increase) only when the phase switch is in the ON orientation (111). As discussed earlier, *fimE* mutations affect expression of *fimA* by controlling inversion of the promoter-containing DNA segment to the OFF orientation.

A gene unlinked to the *fim* gene cluster, *flu* (fluffing), also has been reported to affect fimbriation (25). However, the Flu⁻ phenotype was quite undefined, affecting several surface properties which influenced aggregation (fluffing) of cells cultured in liquid media.

Whether FimB and FimE are the actual recombinases responsible for the inversion reaction controlling fimbrial phase variation or if they are simply accessory proteins controlling the direction of the recombination reaction (and another gene such as *flu* encodes the recombinase) has not been experimentally determined. However, the amino acid sequences of the FimB and FimE proteins, based on their DNA sequences, show 48% identity with each other and significant homology with the integrases (Int) found in the lambdoid phages (87). There is approximately 50% amino acid identity between FimE and FimB and the carboxy-terminal region of the Int recombinases (Fig. 6). This portion of the Int proteins is proposed to contain the active sites for the recombinase and topoisomerase activities of Int (5, 27, 33; J. F. Thompson and A. Landy, this volume). The homology between FimB/FimE and phage integrases supports the proposal that FimB and FimE are site-specific recombinases mediating the inversion reaction and prompted examination of the role of the integration host factor (IHF) in fimbrial phase variation. The binding of IHF to its recognition sequences in the λ phage attachment site is required for normal integration of λ into the host chromosome (41, 164, 165; Thompson and Landy, this volume). IHF is composed of two subunits which are encoded by the *himA* and *himD/hip* genes (35, 102, 105). Transduction of an *E. coli* strain containing a *fimA-lacZ* operon fusion with *himA* or *himD* deletion/insertion mutations resulted in severely reduced rates of phase variation, as measured by β-galactosidase activity (27, 33). Southern blot analysis of the transductants revealed that the invertible DNA segment was "locked" in the ON or OFF orientation. Providing *himA* or *himD* gene product from a plasmid restored phase variation, indicating that IHF either is required for the actual inversion reaction or regulates the expression of the factor required for the recombination process, e.g., *fimB* or *fimE*.

Regulatory roles for IHF other than in integrative recombination of λ include transcriptional and translational regulation of several bacterial and bacteriophage genes (for review, see reference 28). In fact, transcription of the *fimA-lacZ* fusion is reduced approximately sevenfold in a *himA* or *himD* mutant when the phase switch is locked in the ON position as compared to transcription in the fim^+ strain. This may indicate a possible regulatory role for IHF in the transcription of *fimA* other than in the inversion of the *fimA* promoter (27).

The location of IHF consensus binding sequences (21) near the left 9-bp inverted repeat of the invertible DNA segment and immediately upstream of the *fimA* promoter (Fig. 2) lends some support to the proposed roles of IHF in the DNA rearrangement resulting in fimbrial phase variation and in the expression of *fimA* (27, 33). However, binding of IHF to these sites has not been demonstrated. Although the molecular mechanism of fimbrial phase variation is still being characterized, it appears to be a unique inversion system in bacteria.

B. Hin-Mediated Inversion and Related Site-Specific Recombination Systems

Comparison of the inversion systems of *S. typhimurium*, bacteriophages Mu and P1, and the defective prophage *e*14 of *E. coli* has revealed 60 to 70% identity in amino acid sequences among the recombinases that mediate the site-specific inversion reactions (Fig. 5) and significant homology in the DNA sites with which the recombinases interact (Fig. 3) (for reviews, see references 90 and 121). The similarity among these systems is most likely to extend to the site-specific recombination reaction itself, and therefore it is useful to discuss the work on the molecular mechanism of inversion for all of these systems together, noting similarities and significant differences among the systems.

The identification of the components of the inversion reaction has been facilitated by the development of in vitro recombination systems for Hin-, Gin-, and Cin-mediated inversion of a plasmid substrate (51, 67, 69, 72, 101, 118). The in vitro reaction conditions are simply buffer, NaCl, and Mg^{2+}; no energy cofactor is required. The DNA substrate and protein requirements are more reflective of the complexity of the inversion reaction.

1. DNA substrate requirements for inversion

The Hin-related recombinases exhibit strong orientation dependence, utilizing efficiently only recombination sites in an inverted configuration on supercoiled DNA substrates in vitro (69, 101, 118). In vivo a very low level of recombination is seen between directly repeated recombination sites and between recombination sites located on independent replicons (67, 117, 137).

Comparison of the inverted repeat sequences of the related inversion systems reveals a conserved 26-bp sequence exhibiting dyad symmetry; it consists of two imperfect 12-bp inverted repeats separated by a 2-bp core (116, 121, 147) (Fig. 3). Nuclease and chemical protection studies have shown that the recombinase-binding site overlaps this 26-bp consensus sequence in the Hin, Gin, and Cin inversion systems (15, 51, 74; A. Glasgow, M. Bruist, and M. Simon, manuscript in preparation). To determine whether the 26-bp region within the *hixL* recombination site (Hin system) was sufficient for site-specific inversion, a recombination substrate was constructed which contained two synthetically derived 26-bp *hixL* sequences in inverted repeat configuration (72). The rate of Hin-mediated recombination supported by this substrate was less than 5% of the recombination rate for a substrate containing the wild-type inversion region. However, a substrate containing a synthetically derived *hixL* site and the wild-type *hixL* site, which included approximately 200 bp of adjacent sequence from the invertible H segment, gave the wild-type rate of recombination. Analysis of deletion, substitution, and insertion mutations within the region adjacent to the wild-type *hixL* site identified a 60-bp sequence (Fig. 7) which enhanced recombination approximately 20-fold. Similar enhancer sequences have been found in the Gin (75, 141) and Cin (64) systems. These results indicated that two inverted 26-bp recombination sites are sufficient for the reaction, but recombination is significantly stimulated by the presence of a recombinational enhancer sequence.

The recombination sites. The extent of the consensus 26-bp sequence that is required for recombi-

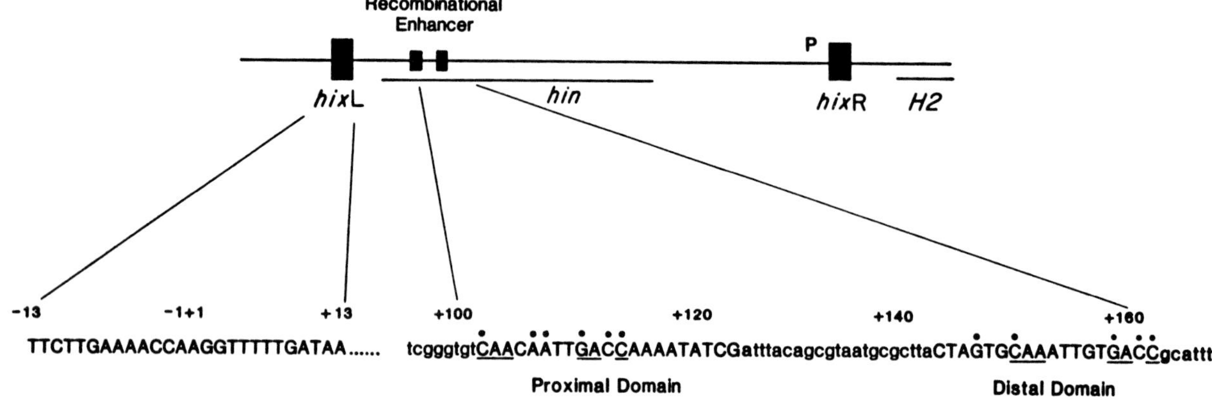

Figure 7. *cis*-acting sites required for Hin-mediated inversion. The approximate positions of the recombination sites *hixL* and *hixR* and the recombinational enhancer are shown for the ON orientation of the H segment, in which the promoter (P) is in the proper orientation for expression of the *H2* flagellin gene. The sequences for *hixL* and the enhancer are given. The proximal and distal domains are indicated by uppercase letters, and the proposed consensus sequence within each domain is underlined. Dots indicate bases which are in base pairs contacted by Fis as determined by methylation interference assays (13).

nase recognition and recombinational activity has been tested in both the Hin (64a, 72; Glasgow et al., manuscript in preparation) and Gin (74) systems. Using mutant recombination sites derived from synthetic oligonucleotides, it has been determined that the outer 2 bp of the consensus 26-bp sequence are not essential; removal of these base pairs does, however, result in a reduction in recombinase binding affinity and in the rate of inversion. It was also established that both 12-bp inverted repeats within the 26-bp recombination site are required for efficient inversion frequencies, even though Hin and Gin will specifically recognize and bind to single consensus half sites (−1 through +13 bp from the center of dyad symmetry) of *hixL* and *gixL*, respectively. Essentially no binding of Hin is seen with the isolated half site of *hixR* that deviates significantly from the consensus sequence (Fig. 3). Increasing the spacing between the 12-bp inverted repeats of *hixL* from 2 bp to 3 or 5 bp only slightly reduces the binding affinity of Hin for the recombination site but eliminates recombination (see below).

The base pairs that are involved in specific protein-DNA interactions between Hin and its recombination sites *hixL* and *hixR* have been determined in vitro by chemical interference assays (Glasgow et al., manuscript in preparation) and in vivo by analysis of site mutations affecting the binding of Hin to the recombination site (64a). The results of these studies indicate that the critical contacts for Hin binding to the recombination sites are in adjacent major and minor grooves within the 12-bp inverted repeats of *hixL* and *hixR* (Fig. 3).

Although the sequence of the central 2 bp which separates the inverted repeats of the recombination site is apparently not important for the binding of the recombinase, complementarity between the recombination sites at this core sequence is required for the recombination reaction. Johnson and Simon (72) demonstrated in vitro that no recombination is seen with a substrate containing a wild-type *hixL* site and a mutant site in which the central 2 bp had been changed from AA to AT, while efficient recombination is obtained with a substrate containing two AT mutant sites. Iida and Hiestand-Nauer (66) have also noted that efficient recombination occurs in vivo only between Cin recombination sites (*cix*) which are homologous within the 2-bp core; but, significantly, the low-frequency recombination that does occur between recombination sites with only one common base pair at the 2-bp core results in localized conversion of the mismatched residue. These results suggest that a 2-bp staggered cut is made by Cin, followed by reciprocal strand exchange, at the central 2-bp sequence. A 2-bp staggered cut flanking the central AA of the *hixL* site has been observed in vitro with Hin-cleaved DNA fragments (R. Johnson, unpublished results cited in reference 73). Mismatches at this core sequence presumably would interfere with base pairing of the exchanging strands prior to ligation by the recombinase. It is also possible that homology at the core sequence is required for proper alignment of the recombination sites prior to cleavage and strand exchange.

The recombinational enhancer. In addition to the two recombination sites, a third sequence corresponding to the enhancer function is required for efficient recombination. The enhancer sequences of the Hin, Gin, and Cin inversion systems are located near the 5′ end of the homologous recombinase genes (Fig. 4 and 6). Although the size of the enhancer region varies from 60 bp in the Hin system (13, 72) to 100 (74, 75) or 170 bp (141) in the Gin system, these recombinational enhancers are functionally interchangeable (at least for the Hin and Cin systems) (64). In each of these recombination systems it has been demonstrated that the inversion enhancer functions in *cis* in a distance- and orientation-independent manner (64, 72, 75). There is one significant exception to the distance-independent action of the Hin recombinational enhancer: the enhancer cannot function when it is too close (within 30 bp) to a *hix* recombination site. This inability to function is not due to steric interference between proteins bound to the recombination site and the enhancer sequence, since nuclease protection studies show that the recombination sites and the enhancer are occupied by the appropriate proteins (13) (see below). Furthermore, the addition of a second enhancer sequence at a greater distance from the recombination site (120 bp) restores enhancer activity (R. Johnson and M. Simon, unpublished results cited in reference 73), suggesting a mechanistic requirement for a minimum distance between the recombination site and the enhancer sequence (see below).

In the presence of a host factor, called factor II or Fis (factor for inversion stimulation), this enhancer sequence can stimulate the initial rate of recombination more than 150-fold (70). Fis is required for the stimulation of inversion by the recombinational enhancer both in the in vitro assays for inversion (51, 70, 75, 89) and in vivo (68a). Nuclease and chemical protection studies and DNA gel retardation assays have shown that Fis binds specifically to sequences within the enhancer regions of the Hin (13), Gin (74, 89), and Cin (51) systems.

The sequences required for Fis binding and enhancer activity have been characterized by methylation interference assays and mutational analysis of the Hin recombinational enhancer (13). The enhancer region extends from +103 to +163 from the center of the *hixL* site and consists of two domains, called the proximal and distal domains (relative to *hixL*) (Fig. 7). Mutations within either domain eliminate enhancer activity. DNA gel retardation assays with the isolated proximal and distal domains and the full enhancer region show that Fis interacts with each domain independently, binding with a fivefold-higher affinity to the distal domain than to the proximal domain. Comparison of the sequences of the proximal and distal domains and of the enhancer regions of the homologous inversion systems revealed a possible consensus sequence for Fis binding: CAPu-a--TGA-C (13). This sequence is within the boundaries of the proximal and distal domains, determined by deletion and substitution analysis, and spans all but one of the bases contacted by Fis in each domain, as resolved by methylation protection and interference experiments (Fig. 7). This sequence is also found within the regions protected by Fis from DNase I cleavage in the enhancers of the Gin and Cin systems (13, 51, 74).

The spacing between the proximal and distal domains, measured from end to end of the directly repeated consensus sequence, is 48 bp or approximately 4.5 turns of the DNA helix. This spacing is conserved in the Gin and Cin recombinational enhancers. The disposition of the contacts made by Fis at its binding sites in the Hin recombinational enhancer (from methylation binding interference experiments) indicates that Fis binds primarily to one face of the DNA helix. The spacing of approximately 4.5 turns of the helix between the Fis binding sites places the Fis molecules at each binding site on nearly opposite sides of the DNA helix. The importance of the spatial relationship of the Fis binding sites was addressed by Johnson et al. (71) in experiments measuring enhancer activity of and Fis binding to insertion mutations that altered the spacing between the proximal and distal domains of the Hin recombinational enhancer. It was found that insertion of nonintegral turns of the DNA helix (+2, 5, 14, 18 bp, etc.) substantially reduced (≤20% of wild-type activity) or abolished enhancer activity, while insertion of approximately one complete turn of the DNA helix (+10 bp) restored enhancer activity to ~70 to 80% of wild-type activity. Insertion of more than one turn of the DNA helix (+21, 32, and 43 bp) only partially restored enhancer activity. The requirement for the proper helical orientation of the Fis binding sites for enhancer activity might suggest that cooperative interactions between Fis molecules facilitate binding to the recombinational enhancer, analogous to cooperative binding of the λ repressor to operator sites correctly positioned on the DNA helix (60). However, DNA-binding assays (gel retardation assays and DNase I cleavage protection studies) indicated that the binding affinity of Fis for the proximal and distal domains was not significantly altered for any of the insertion mutants (71). These results were consistent with the binding studies described earlier which showed that Fis bound independently to the proximal and distal domains. However, since the binding assays were performed with linear DNA, it still is possible that the helical dependence of enhancer activity is due to cooperative interactions that only occur on superhelical DNA.

The recombinational enhancer appears to have another unusual structural feature that may play a role in the function of the enhancer. Johnson et al. (71) reported that DNA fragments containing the enhancer sequence migrate aberrantly on polyacrylamide gels. It was also observed that the electrophoretic mobility of Fis-enhancer complexes on polyacrylamide gels was altered depending on the spatial orientation of the enhancer domains. These results are consistent with an altered DNA structure, such as a bend or kink, being associated with the Fis binding domains. It has been reported (135, 182) that a DNA fragment containing two bends (or kinks) exhibits altered relative mobilities depending on whether the bends are oriented toward or away from one another (changing the end-to-end distance of the DNA fragment). Thus, changing the spacing between bent sequences by a nonintegral number of helical turns would shift the mobility of a DNA fragment relative to its mobility when integral turns of the DNA helix are inserted between the bends, as seen with the insertion mutants of the enhancer DNA. The results of circular permutation assays (174) have indicated that both sequence-directed and protein-induced bends are associated with the enhancer DNA (A. Glasgow and M. Simon, unpublished data). Such altered DNA structures have been found associated with sites within other recombination systems, including the λ phage attachment site (128), the *res* site of the γδ resolvase recombination system (54), and the ends of the IS*1* insertion element (123), as well as with regions near replication origins of bacteriophage λ (177), plasmid pSC101 (154), and SV40 (132). It has been suggested for most of these systems that the role of the altered DNA is to promote the formation of complex nucleoprotein structures (29).

2. Recombinases and host factors required for inversion

The recombinases. The *trans*-acting Hin, Gin, Cin, and Pin functions mediate the site-specific inversion reaction at the recombination sites of the homologous inversion systems (68, 78, 122, 146). The Hin and Gin recombinases, which have been purified to near homogeneity, are capable of carrying out the recombination reaction alone, albeit at very low rates of recombination (70, 81, 100). These recombinases also have site-specific topoisomerase activity, which has the same substrate and protein requirements as the recombination reaction (69, 74). In experiments using conditions that lead to possible recombination intermediates, it was shown that Hin could catalyze the double-stranded cleavage of DNA at the central 2 bp of the recombination site, leaving Hin covalently associated with the 5′ end of the DNA cut site (R. Johnson and M. Simon, unpublished results reported in reference 73). Unlike the topoisomerase activity, this double-strand cleavage does not require the presence of Fis or the recombinational enhancer.

The recombinases (Hin, Gin, and Cin) are very hydrophobic, 21-kilodalton proteins which form insoluble, inactive aggregates at low ionic strength (51, 70, 100). This aggregation has made it difficult to overproduce and purify the recombinases in an active form. The active binding species of the recombinase, determined for Gin isolated from glycerol gradients, is a monomer (100). DNase I "footprinting" studies have shown that Hin, Gin, and Cin bind specifically to each inverted repeat of the palindromic 26-bp consensus site within the recombination sites (51, 69, 74; Glasgow et al., manuscript in preparation). There do not appear to be cooperative interactions between the recombinase and Fis, bound to their respective sites, on linear DNA fragments (13, 51). The possible interactions in the course of the inversion reaction between the recombination sites and the enhancer site, independently bound by the recombinase and Fis, respectively, are discussed below.

In these inversion systems the recombinase is required in stoichiometric rather than catalytic amounts. This is also true for other site-specific recombinases, including the Int protein of λ (105), the Cre recombinase of phage P1 (3), and the γδ resolvase (124). This may reflect the participation of the recombinase in the formation of a complex nucleoprotein structure required for recombination (29).

The homologous DNA invertases are also related to the resolvases of the Tn*3* class of transposons, although they are unable to complement one another (116, 151; D. Sherratt, this volume). Like the DNA invertases, the resolvases show orientation dependence, only catalyzing site-specific deletion between directly repeated *res* sites (124). The amino acid sequence homology between Hin and the γδ resolvase (approximately 30%) suggested that Hin protein structure may be similar to that of the resolvase (116, 151) (Fig. 5). The resolvase consists of two functional domains, a carboxy-terminal DNA-binding domain (43 amino acids) and an amino-terminal domain (140 amino acids) that provides the catalytic and dimerization functions (1, 53, 106). Bruist et al. (14) synthesized two peptides corresponding to the last 52 and 31 amino acids of Hin which are homologous to the 43-amino-acid DNA-binding domain of resolvase. DNase I cleavage protection studies demonstrated that this 52-mer specifically bound to the *hix* half sites, while the 31-mer which contained the helix-turn-helix DNA-binding motif shared by many DNA-binding proteins (112) could not bind to the *hix* sites. Further studies have shown that the 52-mer defines the complete DNA-binding domain required for site-specific binding to the *hix* sites; peptides corresponding to less than the last 51 amino acids of Hin (e.g., 50 and 49 amino acids) do not bind to the *hix* sites (J. Sluka, S. Horvath, A. Glasgow, M. Simon, and P. Dervan, manuscript in preparation). These results suggest that regions of the DNA-binding domain of Hin other than the helix-turn-helix motif may be involved in sequence recognition.

Host factors Fis and HU. A host factor requirement for efficient site-specific inversion has been demonstrated in vitro for the Hin, Gin, and Cin inversion systems. Two host factors purified from crude extracts (derived from *E. coli*) were required for maximum rates of Hin-mediated recombination in vitro; they were the histonelike protein HU and a previously uncharacterized protein called factor II (also found in *S. typhimurium*) (70, 72). A single host factor was shown to be required for efficient Gin-mediated inversion, called Fis (factor for inversion stimulation) (75, 89) or GHF (Gin host factor) (81), and for Cin-mediated inversion (also called Fis) (51). Fis can substitute for factor II in Hin-mediated inversion, and the two are now assumed to be the same protein (the designation for factor II has been changed to Fis) (13). Other site-specific recombination systems, such as the λ integration/excision system and the inversion system of *E. coli* controlling type 1 fimbrial phase variation, require the host protein IHF for efficient recombination (27, 33, 85, 101; for review, see reference 28). However, strains carrying mutations in the genes encoding the two subunits of IHF still have full inversion stimulatory activity in the Hin and Gin systems (70, 89).

The gene encoding Fis (*fis*) has recently been cloned, sequenced, and mapped between *aroE* and *fabE* at 72 min on the *E. coli* chromosome (68a). The isolation of stable mutant strains which have null mutations in the *fis* gene indicates that Fis is not essential for cell growth. However, the *fis* mutant strains supported only very low levels of Hin-mediated inversion.

Fis has been purified from *E. coli* extracts to near homogeneity; it is a 12-kilodalton basic protein that exists in solution as a dimer (70, 89). Fis binds specifically to the sequences within the recombinational enhancer and is required for enhancer-mediated stimulation of the site-specific inversion reactions. Fis can also stimulate recombination at a low level in the absence of enhancer DNA (70, 75), perhaps utilizing secondary (nonessential) Fis binding sites that have been identified in DNase I footprints of the enhancer region of the homologous inversion systems (13, 51, 74).

Fis has recently been found to be an additional component of the site-specific excisive recombination pathway of bacteriophage λ (163). Fis binds specifically to a sequence within the λ attachment site, *attP*, which overlaps one of the two binding sites for Xis, and facilitates the binding of Xis to its adjacent site. The cooperative interactions between Fis and Xis stimulate the excision reaction at low Xis concentrations in vitro; Fis does not affect integrative recombination. It was demonstrated, using in vivo dimethyl sulfate protection studies (163) and Hin inversion assays (68a), that the concentration of Fis apparently drops drastically in stationary-phase cells, thus linking excisive recombination of λ, and perhaps inversion-controlled phase and host-range variation, to the growth phase of the host cells.

It is interesting that the binding of Fis to the λ *attP* site induces a bend in the DNA (163), as was seen with Fis binding to the recombinational enhancer of the Hin system. Since the Fis binding sites in these two recombination systems share no apparent homology (13), it is possible that Fis recognizes a structural feature of the DNA (bend or kink) rather than, or in addition to, a specific DNA recognition sequence.

Fis shares some properties and functional analogies with IHF, which also acts in λ site-specific recombination. Like Fis, IHF is a basic, heat-stable protein which can locally alter the DNA structure upon binding (40, 123, 126a, 154). Both proteins are thought to be involved in promoting the formation of nucleoprotein structures which facilitate synapsis of the DNA strands required for recombination (29, 73). However, the amino acid sequence of Fis (98 amino acids), derived from its DNA sequence, exhibits no homology with that of IHF or similar DNA-binding proteins such as HU (68a). Fis does share ~49% amino acid identity with the carboxy-terminal DNA-binding region of NtrC (NR1), which interacts with the transcriptional enhancer associated with the RpoN (σ^{54})-dependent promoters of the Ntr (nitrogen assimilation) and Nif (nitrogen fixation) operons (68a, 126).

As mentioned earlier, HU is a second host factor that stimulates Hin-mediated inversion. HU is a small, heat-stable protein that consists of two subunits of similar amino acid sequence (94, 131). It is found in relative abundance (30,000 to 50,000 dimers per cell) in most bacterial species. Analogous to the eucaryotic histones, HU is capable of condensing DNA, wrapping the DNA into a nucleosomelike structure (130; for review, see reference 28). HU has been shown to function in reactions in *E. coli* which require complex protein-DNA structures. HU stimulates the initiation of replication from the *oriC* locus in an in vitro assay in which it is part of a prepriming complex formed in conjunction with the products of *dnaA*, *dnaB*, and *dnaC* (26, 39). In vitro assays for two other site-specific recombination reactions besides Hin inversion have shown an essential or stimulatory role for HU. In the transposition of bacteriophage Mu, HU and the Mu A and B functions are the only proteins required for the initial strand transfer reaction in transposition (22). HU and IHF have been shown to stimulate the in vitro transposition of Tn*10* (103).

In the Hin-mediated inversion reaction there is a tenfold stimulation of recombination in the presence of HU with a wild-type DNA substrate (70). HU stimulation is dependent on the distance between the recombinational enhancer and the recombination sites. When the enhancer is 350 bp or more from the closest recombination site, the stimulation of recombination by HU drops to less than twofold. Therefore, the greatest effect of HU on the inversion reaction is seen when the enhancer is relatively close to one of the recombination sites. The optimal concentration of HU dimers for the inversion reaction (40 to 50 dimers per substrate DNA molecule) is only approximately 20% of the amount required to cover the DNA molecule. This suggests the possibility that locally high concentrations of HU perform a specific role in the recombination reaction. It has been proposed that HU facilitates the formation of a synaptic complex between the recombinational enhancer and the recombination site (see below), which is a role consistent with the dependence of HU stimulation on the relative distance between the enhancer and recombination site.

The fact that the absence of HU has little effect on the Gin- and Cin-mediated inversion reactions (51, 89) may be explained at least in part by the nature of the DNA substrates. The recombinational enhancer in the Cin inversion system is located over 400 bp from the nearest recombination site, and therefore, by extrapolation from the Hin system, the presence of HU would have less than a twofold effect on the inversion rate. In the wild-type substrate for Gin-mediated inversion, the enhancer is less than 100 bp from the right recombination site; if the wild-type substrate was used in the assays for the effect of HU on inversion, it is surprising that no stimulation of recombination by HU was seen. This discrepancy has not been resolved.

C. Topology of the DNA Inversion

In order to understand the mechanisms of synapsis and strand exchange in the inversion reaction, the topological changes that result from and occur during the recombination reaction have been determined for Gin-mediated inversion (74, 80, 82). When uniform topoisomers of the inversion DNA substrate are reacted with Gin and the products are separated by gel electrophoresis, a change in linking number of +4 is seen (Fig. 8). Successive rounds of recombination change the linking number in increments of +4 (+8 and +12 for the second and third rounds of recombination). These results are consistent with strand exchange occurring by a double-strand break followed by rotation and religation. The γδ resolvase-mediated recombination also results in a +4 change in linking number, indicating that strand exchange occurs by a similar mechanism (9). The nature of cleavage of the recombination site is also similar for resolvase and the Hin invertase; both give a 2-bp 3′ overhang with the recombinase bound to the 5′-phosphoryl end (125; M. Bruist, R. Johnson, and M. Simon, unpublished results reported in reference 73). Inversion mediated by the Cre protein of phage P1 between inverted repeats of the *loxP* sites, by the λ integrative/excisive recombination proteins using inversely repeated *attP*, *attB* or *attL*, *attR* sites, and by the FLP function in the inversion system of the *Saccharomyces cerevisiae* 2μm plasmid results in a change in linking number of +2. A single-strand-break mechanism involving the formation of Holliday structures and branch migration for recombination has been proposed for recombination mediated by these recombinases (for reviews, see references 80 and 133).

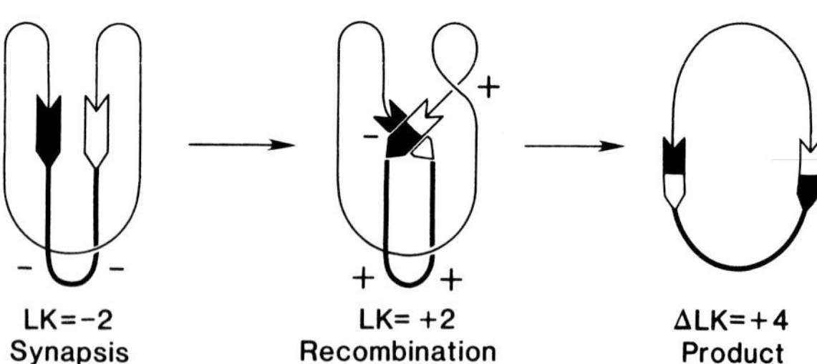

Figure 8. Topology of site-specific inversion. The model for site-specific DNA inversion as presented by Kanaar and van de Putte (80) and Bruist et al. (13) is illustrated. Synapsis between two inverted recombination sites (solid and open arrows) traps two negative (−) interdomainal supercoil nodes (linking number LK = −2). The domains are distinguished by thick and thin lines. In recombination, the recombinase (not shown) bound to each inverted repeat makes a double-strand cut and then strand crossover (depicted by the split arrows) occurs, creating a negative interdomainal node. The plectonemic wrapping of the recombination sites in the negatively supercoiled substrate favors right-handed rotation (180°) of the strands, which introduces a positive (+) intradomainal node. Also, the sign of the interdomainal nodes is changed to + when the relative orientation of the two domains is inverted in the recombination reaction. Therefore, the change in linking number from synapsis (LK = −2) to the recombination product (LK = +2) is +4. For a detailed explanation, see references 20, 80, and 82.

The site-specific relaxation of DNA (loss of four negative supercoils) by Gin has the same DNA substrate and protein requirements as the recombination reaction. Examination of the topology of the DNA inversion products by gel electrophoresis revealed unknotted circles (Fig. 8). This result indicates that synapsis between the recombination sites would trap a defined number of DNA supercoils between the two sites. Simple random collision of recombination sites traps a variable number of supercoils between the sites, resulting in complex, knotted recombination products. Inversion mediated by the λ recombinases Int and Xis, by the 2μm plasmid FLP protein, and by Cre of phage P1 (to some extent) gives such knotted recombination products, indicating that random collision of the recombination sites leads to productive synapsis. Like Gin, resolvase gives simple reaction products, singly linked catenanes, indicating that the productive synaptic complex between directly repeated *res* sites also has a defined topology. Several models have been proposed for this type of ordered site synapsis which involve diffusion-driven "tracking" of the recombinase along the DNA between recombination sites, or "slithering" of DNA segments past each other until protein-bound recombination sites are aligned for synapsis; these models are discussed in detail in recent reviews (42, 80). A "two-step synapsis" model which is consistent with the topological specificity of resolvase-mediated recombination has been proposed by Boocock et al. (9). The recombination sites (*res*) of the resolvase system have three subsites, only one of which recombines (47). In this model the *res* sites find each other by random collision, but only collisions resulting in wrapping of properly oriented accessory subsites (step 1), which then directs the alignment of the crossover sites (step 2), lead to strand exchange. The conformational constraints within a circular DNA molecule on the plectonemic wrapping of the resolvase-bound accessory subsites of *res* provide a "topological filter" that selects the precise topology of the productive synaptic complex. Although the resolvase specifically catalyzes excision and the Gin-related recombinases catalyze inversion, these systems share many topological parameters as well as sequence homology between their recombinases (80, 82), suggesting that similar mechanisms for synapsis and strand exchange may be involved in their recombination pathways.

D. The Role of the Recombinational Enhancer in Site-Specific Inversion

The inversion reaction mediated by Hin, Gin, or Cin is unique among the recombinase systems in that it requires an enhancer sequence which has many properties analogous to those of the transcriptional enhancers (84). A model for the role of the recombinational enhancer in the inversion reaction has been proposed for the Hin inversion system (13, 71, 73) (Fig. 9). Hin and Fis interact independently with their respective sites on the DNA molecule, with Fis bound to each domain of the enhancer on opposite sides of the DNA helix. Subsequent protein-protein interactions between Fis and Hin loop out the intervening DNA between the enhancer and the recombination sites and direct the Hin-bound recombination sites into the proper topological alignment for strand exchange to occur. The recombinational enhancer therefore acts to assemble Hin and the recombination sites into a productive synaptic complex. This role for the enhancer in the inversion reaction is analogous to that suggested for the *res* accessory subsites in resolvase-mediated recombination (see above). The proposal that the recombinational enhancer is part of a specific intermediate in the strand exchange reaction is supported by the fact that the enhancer sequence and Fis are required for both the topoisomerase and recombination activity of Hin (and Gin) which involves strand exchange, but the binding and cleavage of the recombination sites by Hin does not require the enhancer sequence.

This "looping synapsis" model incorporates the definitive features, both functional and structural, of the enhancer. The enhancer can function at variable distances from the recombination sites and in either orientation. The only distance requirement is that there be sufficient DNA between the enhancer and recombination sites for the DNA to loop around so that the required Fis-Hin interactions for synapsis of the recombination sites could occur. This is consistent with the fact that the enhancer does not function when less than 30 bp from the recombination site. The distance-dependent stimulation of recombination by HU suggests that HU may stimulate recombination by binding to the intervening sequences between the enhancer and *hix* sites, facilitating or stabilizing DNA bending in the formation of the synaptic complex.

The role of the enhancer as a topological filter directing the proper configuration of the recombination sites is supported by the dependence of enhancer function on the proper spatial relationship between the two enhancer domains. Rotation of the Fis binding sites (proximal and distal domains) by half a turn of the DNA helix would alter the topology of the synaptic complex determined by the Fis-Hin interactions; such alterations in the enhancer structure have been shown to abolish enhancer function (71). A recombination intermediate containing the specific

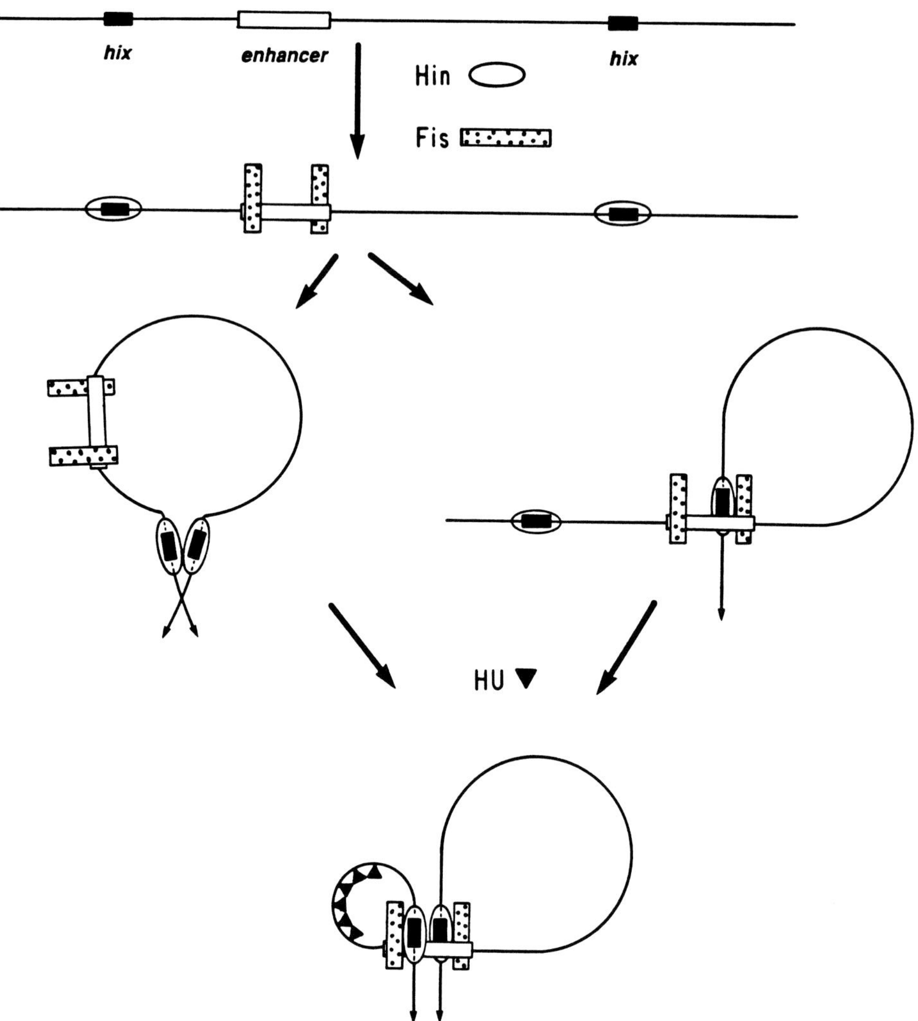

Figure 9. Looping-synapsis model for site-specific DNA inversion. Hin and Fis bind independently to their respective sites. These proteins are shown as single entities, although Fis probably binds as a dimer to each domain of the enhancer and multimeric forms of Hin may bind to both halves of each recombination site. Interaction between Fis, bound to different sides of the DNA helix, and Hin, bound to each recombination site, results in the formation of a productive synaptic complex (see text).

DNA crossover nodes defined by the Fis-enhancer complex as shown in Fig. 9 should give the topological recombination products observed for Gin-mediated inversion. Hin-mediated double-strand cleavage, clockwise rotation of the strands, and religation would result in an unknotted product relaxed by exactly four supercoils (Fig. 8).

IV. CHROMOSOMAL INVERSIONS

Thus far, this chapter has focused on site-specific recombination involving homologies of less than 30 bp. Inversions have also been shown to occur by homologous recombination of the bacterial chromosome between larger regions of homology. Programmed rearrangements resulting from inversion of large chromosomal regions by homologous recombination have not been found. Studies of chromosome rearrangements in bacteria suggest that there may be important mechanisms that militate against such inversions.

Among the types of chromosomal rearrangements found in bacteria, the transposition, deletion, and duplication events are common, while inversion rearrangements are rare (56, 129). A priori, one might expect inversions between chromosome homologies to be as common as duplications and deletions. The rarity of occurrence of inversions in bacteria remains a mystery, but recent work has provided some clues about chromosome inversions and the maintenance of the general structure of the bacterial chromosome. Only a few examples of chromosomal inversion mutations have been reported among tens of thousands of mutants characterized in *E. coli* and *S. typhimurium*. The most notable have occurred since the divergence of *E. coli* and *S. typhimurium* from a common ancestor. While the DNA sequence between *E. coli* and *S. typhimurium* has diverged greatly, the gene order of known markers on the chromosome remains essentially identical (6, 136). The exception is a large region flanking the replication terminus, whose gene order is inverted between the two organisms. It is not known what selective mechanism is involved in the conservation of gene order.

The first screens set up to select directly for chromosome inversion mutations did not yield such mutants, suggesting that there are constraints on the ability to form such a chromosomal rearrangement or that it is not viable (92, 178; E. B. Konrad, Ph.D. thesis, Harvard University, Cambridge, Mass., 1969). Since these initial studies, inversion mutants have been isolated and have been found to be viable (59, 138, 139; M. B. Schmid, Ph.D. thesis, University of Utah, Salt Lake City, 1981). Recently two groups have independently devised four selection schemes to allow for the direct selection of inversion mutants. The major advantage of these schemes is that they allow selection of inversion mutants with predetermined endpoints which can be distributed throughout the chromosome (36; A. Segall, Ph.D. thesis, University of Utah, Salt Lake City, 1987; M. Mahan, A. Segall, and J. Roth, personal communication). The results of these selections suggest that the ability to generate large chromosomal inversions depends on the region of the chromosome one is trying to invert, and not on whether the replication origin or terminus is included within the inverted region as suggested by earlier results.

Despite the rare occurrence of inversion mutants in early procaryotic genetic studies, recent experiments have demonstrated that chromosomal inversions of probably any region of the chromosome can occur and are viable. There are mechanistic constraints on the ability to isolate inversions by intrachromosomal (reciprocal) recombination, which can be bypassed using a two-fragment (nonreciprocal) transduction cross (Segall, Ph.D. thesis). These constraints may have retarded evolution of large programmed inversion systems using homologous recombination. It is still not clear what the genetic selection pressures are that maintain the order of genes on *E. coli* and *S. typhimurium* chromosomes. Inversions between ribosomal cistrons in *E. coli* were characterized for their stability compared to isogenic strains with the normal gene order (C. Hill and J. Gray, personal communication). It was found that cells without the inversions would outgrow the inversion mutants, suggesting a selective advantage for the wild-type gene order. Obvious disruptions resulting from large inversions are a transient increase or decrease in gene dosage when a gene is placed closer to or further from the replication origin (140). Inversions covering the origin or terminus would alter the symmetry between the origin and terminus and would probably lead to an increase in cell generation time. Thus, even with these recent advances in understanding the nature of inversions and the structure of the bacterial chromosome, the mechanism controlling both remains a mystery.

V. CONCLUSIONS

Programmed DNA rearrangements controlling gene expression have been identified in many different organisms (for review, see reference 10). Many of these rearrangements may be developmentally programmed or simply controlled by the relative availability of the proteins required for the rearrangements. The function of the rearrangements is as varied as the organisms in which they occur. The programmed inversions which have thus far been found to occur in bacteria primarily control the expression of surface structural proteins. The clonal variation resulting from low-frequency DNA inversions preadapts a portion of the population to changes in the environment, such as specific antibodies raised by an infected host (antigenic variation) or the appearance of a new bacterial host with different phage receptors (phage host-range variation).

The inversion systems discussed in this chapter appear to have some evolutionary relationship to each other and to other site-specific recombination systems (for reviews, see references 79, 116, 121, 133, and 147). There are homologies between the recombinases and their recognition sites, as well as shared host factors. The mechanism of recombination also appears to be shared between some of these systems. However, it is the unique features of these inversion systems—the recombinational enhancer of the Hin-related inversion systems and the unidirectional inversion functions FimB and FimE of the *E. coli* phase variation inversion system—which promise to provide further insight into the recombination process.

In contrast to spontaneous mutational DNA rearrangements, which include inversion, deletion, and transposition of DNA sequences, and are for the most part deleterious, programmed DNA rearrangements have survived and are presumably of some selective advantage. It is not always apparent that the individual organism is aided by the programmed rearrangement; however, in many cases, e.g., antigenic variation, it is clear that there is survival value for the clonal population. There also appear to be constraints on spontaneous rearrangements that require homologous recombination. Thus, chromosomal inversions mediated by RecA are peculiarly rare in bacteria. This mechanism may contribute to chromosomal stability over evolutionary time scales. The identification of the selective mechanisms controlling Rec-mediated chromosomal inversion should help in the understanding of the structure of the bacterial chromosome and of the DNA rearrangements involved in the evolution of related bacterial strains.

LITERATURE CITED

1. **Abdel-Meguid, S. S., N. D. F. Grindley, N. S. Templeton, and T. A. Steitz.** 1984. Cleavage of the site-specific recombination protein γδ resolvase: the smaller of the two fragments binds DNA specifically. *Proc. Natl. Acad. Sci. USA* **81**:2001–2005.
2. **Abraham, J. M., C. S. Freitag, J. R. Clements, and B. I. Eisenstein.** 1985. An invertible element of DNA controls phase variation of type I fimbriae of *Escherichia coli*. *Proc. Natl. Acad. Sci. USA* **82**:5724–5727.
3. **Abremski, K., and R. Hoess.** 1984. Bacteriophage P1 site-specific recombination. Purification and properties of the cre recombinase protein. *J. Biol. Chem.* **259**:1509–1514.
4. **Andrewes, F. W.** 1922. Studies in group-agglutination. I. The *Salmonella* group and its antigenic structure. *J. Pathol. Bacteriol.* **25**:1509–1514.
5. **Argos, P., A. Landy, K. Abremski, J. B. Egan, E. Haggard-Ljungqvist, R. H. Hoess, M. L. Kahn, B. Kalionis, S. V. L. Narayana, L. S. Pierson III, N. Sternberg, and J. Leong.** 1986. The integrase family of site-specific recombinases: regional similarities and global diversity. *EMBO J.* **5**:433–440.
6. **Bachman, B. J.** 1983. Linkage map of *Escherichia coli* K-12, edition 7. *Microbiol. Rev.* **47**:180–230.
7. **Baron, L. S., D. J. Kopecko, S. M. McCowen, N. J. Snellings, E. M. Johnson, W. C. Reid, and C. A. Life.** 1982. Genetic and molecular studies of the regulation of atypical citrate utilization and variable Vi antigen expression in enteric bacteria, p. 175–194. *In* A. Hollander (ed.), *Genetic Engineering of Microorganisms for Chemicals*. Academic Press, Inc., New York.
8. **Bar-Shavit, Z., I. Ofek, R. Goldman, D. Mirelman, and N. Sharon.** 1977. Mannose residues on phagocytes as receptors for the attachment of *Escherichia coli* and *Salmonella typhi*. *Biochem. Biophys. Res. Commun.* **78**:455–460.

8a. **Bartlett, D. H., M. E. Wright, and M. Silverman.** 1988. Variable expression of polysaccharide in the marine bacterium *Pseudomonas atlantica* is controlled by genome rearrangement. *Proc. Natl. Acad. Sci. USA* **85**:3923–3927.

9. **Boocock, M. R., J. L. Brown, and D. J. Sherratt.** 1987. Topological specificity in Tn3 resolvase catalysis, p. 703–718. *In* T. J. Kelley and R. McMacken (ed.), *DNA Replication and Recombination*. Alan R. Liss, Inc., New York.
10. **Borst, P., and D. R. Greaves.** 1987. Programmed gene rearrangements altering gene expression. *Science* **235**: 658–667.
11. **Brinton, C. C., Jr.** 1959. Nonflagellar appendages of bacteria. *Nature* (London) **183**:782–786.
12. **Brinton, C. C., P. Gemski, S. Falkow, and L. S. Baron.** 1961. Location of the piliation factor on the chromosome of *Escherichia coli*. *Biochem. Biophys. Res. Commun.* **5**: 293–298.
13. **Bruist, M. F., A. C. Glasgow, R. C. Johnson, and M. I. Simon.** 1987. Fis binding to the recombinational enhancer of the Hin DNA inversion system. *Genes Dev.* **1**:762–772.
14. **Bruist, M. F., S. J. Horvath, L. E. Hood, T. A. Steitz, and M. I. Simon.** 1987. Synthesis of site-specific DNA binding peptide. *Science* **235**:777–780.
15. **Bruist, M. F., R. C. Johnson, M. B. Glaccum, and M. I. Simon.** 1984. Characterization of Hin-dependent DNA inversion and binding *in vitro*, p. 63–75. *In* M. I. Simon and I. Herskowitz (ed.), *Genome Rearrangement*. Alan R. Liss, Inc., New York.
16. **Bunting, M. I.** 1946. The inheritance of color in bacteria with special reference to *Serratia marcescens*. *Cold Spring Harbor Symp. Quant. Biol.* **11**:25–32.
17. **Burchard, R. P., A. C. Burchard, and J. H. Parish.** 1977. Pigmentation phenotype instability in *Myxococcus xanthus*. *Can. J. Microbiol.* **23**:1657–1662.
18. **Chow, L. T., T. R. Broker, R. Kahmann, and D. Kamp.** 1978. Comparison of the G DNA inversion in bacteriophage Mu, P1 and P7, p. 55–56. *In* D. Schlessinger (ed.), *Microbiology—1978*. American Society for Microbiology, Washington, D.C.
19. **Chow, L. T., and A. I. Bukhari.** 1976. The invertible DNA segments of coliphages Mu and P1 are identical. *Virology* **74**:242–248.
20. **Cozzarelli, N. R., M. A. Krasnow, S. P. Gerrard, and J. H. White.** 1984. A topological treatment of recombination and topoisomerases. *Cold Spring Harbor Symp. Quant. Biol.* **49**:383–400.
21. **Craig, N., and H. Nash.** 1984. *E. coli* integration host factor binds to specific sites in DNA. *Cell* **39**:707–716.
22. **Craigie, R., D. Arndt-Jovin, and K. Mizuuchi.** 1985. A defined system for the DNA strand-transfer reaction at the initiation of bacteriophage Mu transposition: protein and DNA substrate requirements. *Proc. Natl. Acad. Sci. USA* **82**:7570–7574.
23. **Daniell, E., J. Abelson, J. S. Kim, and N. Davidson.** 1973. Heteroduplex structures of bacteriophage Mu DNA. *Virology* **51**:237–239.
24. **Daniell, E., W. Boram, and J. Abelson.** 1973. Genetic mapping of the inversion loop in bacteriophage Mu DNA. *Proc. Natl. Acad. Sci. USA* **70**:2153–2156.
25. **Diderichsen, B.** 1980. *flu*, a metastable gene controlling surface properties of *Escherichia coli*. *J. Bacteriol.* **141**: 858–867.
26. **Dixon, N., and A. Kornberg.** 1984. Protein HU in the enzymatic replication of the chromosomal origin of *Escherichia coli*. *Proc. Natl. Acad. Sci. USA* **81**:424–428.
27. **Dorman, C. J., and C. F. Higgins.** 1987. Fimbrial phase variation in *Escherichia coli*: dependence on integration host factor and homologies with other site-specific recombinases. *J. Bacteriol.* **169**:3840–3843.
28. **Drlica, K., and J. Rouviere-Yaniv.** 1987. Histone-like proteins of bacteria. *Microbiol. Rev.* **51**:301–319.
29. **Echols, H.** 1986. Multiple DNA-protein interactions govern-

ing high-precision DNA transactions. *Science* **233:** 1050–1056.

30. **Eisenstein, B. I.** 1981. Phase variation of type 1 fimbriae in *Escherichia coli* is under transcriptional control. *Science* **214:**337–339.
31. **Eisenstein, B. I.** 1982. Genetic control of type 1 fimbriae in *Escherichia coli*, p. 308–311. *In* D. Schlessinger (ed.), *Microbiology—1982.* American Society for Microbiology, Washington, D.C.
32. **Eisenstein, B. I.** 1987. Pathogenic mechanisms of *Legionella pneumophila* and *Escherichia coli. ASM News* **53:**621–622.
33. **Eisenstein, B. I., D. Sweet, V. Vaughn, and D. I. Friedman.** 1987. Integration host factor is required for the DNA inversion that controls phase variation in *Escherichia coli. Proc. Natl. Acad. Sci. USA* **84:**6506–6510.
34. **Enomoto, M., K. Oosawa, and H. Momuta.** 1983. Mapping of the *pin* locus coding for a site-specific recombinase that causes flagellar-phase variation in *Escherichia coli* K-12. *J. Bacteriol.* **156:**663–669.
35. **Flamm, E. L., and R. A. Weisberg.** 1985. Primary structure of the *hip* gene of *Escherichia coli* and of its product, the beta subunit of integration host factor. *J. Mol. Biol.* **183:**117–128.
36. **Francois, V., J. Louarn, J. Patte, and J. Louarn.** 1987. A system for *in vivo* selection of genomic rearrangements with predetermined endpoints in *Escherichia coli* using modified Tn*10* transposons. *Gene* **56:**99–108.
37. **Freitag, C. S., J. M. Abraham, J. R. Clements, and B. I. Eisenstein.** 1985. Genetic analysis of the phase variation control of expression of type 1 fimbriae in *Escherichia coli. J. Bacteriol.* **162:**668–675.
38. **Fujita, H., S. Yamaguchi, and T. Iino.** 1973. Studies on H-O variants in *Salmonella* in relation to phase variation. *J. Gen. Microbiol.* **76:**127–134.
39. **Funnell, B. E., T. A. Baker, and A. Kornberg.** 1986. Complete enzymatic replication of plasmids containing the origin of the *Escherichia coli* chromosome. *J. Biol. Chem.* **261:** 5616–5624.
40. **Gamas, P., M. G. Chandler, P. Prentki, and D. J. Galas.** 1987. *Escherichia coli* integration host factor binds specifically to the ends of the insertion sequence IS*1* and to its major insertion hot-spot in pBR322. *J. Mol. Biol.* **195:** 261–272.
41. **Gardner, J. F., and H. A. Nash.** 1986. Role of *Escherichia coli* IHF protein in lambda site-specific recombination: a mutation analysis of binding sites. *J. Mol. Biol.* **191:** 181–189.
42. **Gellert, M., and A. Nash.** 1987. Communication between segments of DNA during site-specific recombination. *Nature* (London) **325:**401–404.
43. **Gill, G. S., R. C. Hull, and R. Curtiss III.** 1981. Mutator bacteriophage D108 and its DNA: an electron microscopic characterization. *J. Virol.* **37:**420–430.
44. **Giphart-Gassler, M., R. H. A. Plasterk, and P. van de Putte.** 1982. G inversion in bacteriophage Mu: a novel way of gene splicing. *Nature* (London) **297:**339–342.
45. **Golden, J. W., S. J. Robinson, and R. Haselkorn.** 1985. Rearrangement of nitrogen fixation genes during heterocyst differentiation in the cyanobacterium *Anabaena. Nature* (London) **314:**419–423.
46. **Griffiths, M., and R. Stanier.** 1956. Some mutational changes in the photosynthetic pigment system in *Rhodopseudomonas spheroides. J. Gen. Microbiol.* **14:**698–706.
47. **Grindley, N. D. F., M. R. Lauth, R. G. Wells, R. J. Wityk, J. J. Salvo, and R. R. Read.** 1982. Transposon-mediated site-specific recombination: identification of three binding sites for resolvase at the *res* sites of $\gamma\delta$ and Tn3. *Cell* **30:**19–27.
48. **Grundy, F. J., and M. M. Howe.** 1984. Involvement of the invertible G segment in bacteriophage Mu tail fiber biosynthesis. *Virology* **134:**296–317.
49. **Guerry, P., S. M. Logan, and T. J. Trust.** 1988. Genomic rearrangements associated with antigenic variation in *Campylobacter coli. J. Bacteriol.* **170:**316–319.
50. **Haas, R., and T. F. Meyer.** 1986. The repertoire of silent pilus genes in *Neisseria gonorrhoeae*: evidence for gene conversion. *Cell* **44:**107–115.
51. **Haffter, P., and T. A. Bickle.** 1987. Purification and DNA binding properties of Fis and Cin, two proteins required for the bacteriophage P1 site-specific recombination systems, *Cin. J. Mol. Biol.* **198:**579–587.
52. **Harris, L. A., S. M. Logan, P. Guerry, and T. J. Trust.** 1987. Antigenic variation of *Campylobacter* flagella. *J. Bacteriol.* **169:**5066–5071.
53. **Hatfull, G. F., and N. D. F. Grindley.** 1986. Analysis of $\gamma\delta$ resolvase mutants *in vitro*: evidence for an interaction between serine-10 of resolvase and site 1 of *res. Proc. Natl. Acad. Sci. USA* **83:**5429–5433.
54. **Hatfull, G. F., S. M. Noble, and N. D. Grindley.** 1987. The $\gamma\delta$ resolvase induces an unusual DNA structure at the recombinational crossover point. *Cell* **49:**103–110.
55. **Heckels, J. E.** 1986. Gonococcal antigenic variation and pathogenesis, p. 77–94. *In* T. H. Birkbeck and C. W. Penn (ed.), *Antigenic Variation in Infectious Diseases.* Special Publications of the Society for General Microbiology, vol. 19. IRL Press, Oxford.
56. **Heffron, F.** 1983. Tn*3* and its relatives, p. 223–260. *In* J. A. Shapiro (ed.), *Mobile Genetic Elements.* Academic Press, Inc., Orlando, Fla.
57. **Heffron, F., J. McCarthy, M. Ohtsubo, and E. Ohtsubo.** 1979. DNA sequence analysis of the transposon Tn3: three genes and three sites involved in transposition of Tn3. *Cell* **18:**1153–1163.
58. **Hiestand-Nauer, R., and S. Iida.** 1983. Sequence of the site-specific recombinase gene *cin* and of its substrates serving in the inversion of the C segment of bacteriophage P1. *EMBO J.* **2:**1733–1740.
59. **Hill, C. W., and W. Harnish.** 1981. Inversions between ribosomal RNA genes of *Escherichia coli. Proc. Natl. Acad. Sci. USA* **78:**7069–7072.
60. **Hochschild, A., and M. Ptashne.** 1986. Homologous interactions of λ repressor and λCro with the λ operator. *Cell* **44:**925–933.
61. **Howe, M. M.** 1980. The invertible G segment of phage Mu. *Cell* **21:**605–606.
62. **Howe, M. M., J. W. Schumm, and A. L. Taylor.** 1979. The *S* and *U* genes of bacteriophage Mu are located in the invertible G segment of Mu DNA. *Virology* **92:**108–124.
63. **Hsu, M. T., and N. Davidson.** 1974. Electron microscope heteroduplex study of the heterogeneity of Mu phage and prophage DNA. *Virology* **58:**229–239.
64. **Huber, H. E., S. Iida, W. Arber, and T. A. Bickle.** 1985. Site-specific DNA inversion is enhanced by a DNA sequence in *cis. Proc. Natl. Acad. Sci. USA* **82:**3776–3780.

64a. **Hughes, K. T., P. Youderian, and M. I. Simon.** 1988. Phase variation in *Salmonella*: analysis of Hin recombinase and *hix* recombination site interaction in vivo. *Genes Dev.* **2:**937–948.

65. **Iida, S.** 1984. Bacteriophage P1 carries two related sets of

genes determining its host range in the invertible C segment of its genome. *Virology* **134**:421–434.

66. **Iida, S., and R. Hiestand-Nauer.** 1986. Localized conversion at the crossover sequences in the site-specific DNA inversion system of bacteriophage. *Cell* **45**:71–79.

67. **Iida, S., H. Huber, R. Hiestand-Nauer, J. Meyer, T. A. Bickle, and W. Arber.** 1984. The bacteriophage P1 site-specific recombinase cin: recombination events and DNA recognition sequences. *Cold Spring Harbor Symp. Quant. Biol.* **49**:769–777.

68. **Iida, S., J. Meyer, M. E. Kennedy, and W. Arber.** 1982. A site-specific conservative recombination system carried by bacteriophage P1. Mapping the recombinase gene *cin* and the crossover sites *cix* for the inversion of the C segment. *EMBO J.* **1**:1445–1453.

68a. **Johnson, R. C., C. A. Ball, D. Pfeffer, and M. I. Simon.** 1988. Isolation of the gene encoding the Hin recombinational enhancer binding protein. *Proc. Natl. Acad. Sci. USA* **85**: 3484–3488.

69. **Johnson, R. C., M. B. Bruist, M. B. Glaccam, and M. I. Simon.** 1984. *In vitro* analysis of Hin-mediated site-specific recombination. *Cold Spring Harbor Symp. Quant. Biol.* **49**:751–760.

70. **Johnson, R. C., M. F. Bruist, and M. I. Simon.** 1986. Host protein requirements for *in vitro* site-specific DNA inversion. *Cell* **46**:531–539.

71. **Johnson, R. C., A. C. Glasgow, and M. I. Simon.** 1987. Spatial relationship of the Fis binding sites for Hin recombinational enhancer activity. *Nature* (London) **329**:462–465.

72. **Johnson, R. C., and M. I. Simon.** 1985. Hin-mediated site-specific recombination requires two 26 bp recombination sites and a 60 bp recombinational enhancer. *Cell* **41**:781–791.

73. **Johnson, R. C., and M. I. Simon.** 1987. Enhancers of site-specific recombination in bacteria. *Trends Genet.* **3**: 262–267.

74. **Kahmann, R., G. Mertens, A. Klippel, B. Brauer, F. Rudt, and C. Koch.** 1987. The mechanism of G inversion, p. 681–690. *In* T. J. Kelley and R. McMacken (ed.), *DNA Replication and Recombination.* Alan R. Liss, Inc., New York.

75. **Kahmann, R., F. Rudt, C. Koch, and G. Mertens.** 1985. G inversion in bacteriophage Mu DNA is stimulated by a site within the invertase gene and a host factor. *Cell* **41**:771–780.

76. **Kamp, D.** 1981. Invertible DNA: the G segment of bacteriophage Mu, p. 73–75. *In* D. Schlessinger (ed.), *Microbiology—1981.* American Society for Microbiology, Washington, D.C.

77. **Kamp, D., and R. Kahmann.** 1981. The relationship of two invertible segments in bacteriophage Mu and *Salmonella typhimurium. Mol. Gen. Genet.* **184**:564–566.

78. **Kamp, D., R. Kahmann, D. Zipser, T. R. Broker, and L. T. Chow.** 1978. Inversion of the G DNA segment of phage Mu controls phage infectivity. *Nature* (London) **271**:577–580.

79. **Kamp, D., E. Kardas, W. Ritthaler, R. Sandulache, R. Schmucker, and B. Stern.** 1984. Comparative analysis of invertible DNA in phage genomes. *Cold Spring Harbor Symp. Quant. Biol.* **49**:301–311.

80. **Kanaar, R., and P. van de Putte.** 1987. Topological aspects of site-specific DNA inversions. *Bioessays* **7**:195–200.

81. **Kanaar, R., P. van de Putte, and N. R. Cozzarelli.** 1986. Inversion of the G segment of phage Mu *in vitro* is stimulated by a host factor. *Biochim. Biophys. Acta* **866**:170–177.

82. **Kanaar, R., P. van de Putte, and N. Cozzarelli.** 1988. Gin-mediated DNA inversion: product structure and the mechanism of strand exchange. *Proc. Natl. Acad. Sci. USA* **85**:752–756.

83. **Keynan, A., and J. W. Hastings.** 1961. Bioluminescence in *V. harveyi. Woods Hole Biol. Bull.* **121**:375–378.

84. **Khoury, G., and P. Gruss.** 1983. Enhancer elements. *Cell* **33**:313–314.

85. **Kikuchi, A., and H. A. Nash.** 1978. The bacteriophage lambda *int* gene product. *J. Biol. Chem.* **253**:7149–7157.

86. **Klemm, P.** 1984. The *fimA* gene encoding the type-1 fimbrial subunit of *Escherichia coli*: nucleotide sequence and primary structure of the protein. *Eur. J. Biochem.* **143**:395–399.

87. **Klemm, P.** 1986. Two regulatory *fim* genes, *fimB* and *fimE* control the phase variation of type I fimbriae in *Escherichia coli. EMBO J.* **5**:1389–1393.

88. **Klemm, P., B. J. Jørgensen, I. van Die, H. de Ree, and H. Bergmans.** 1985. The *fim* genes responsible for synthesis of type 1 fimbriae in *Escherichia coli*, cloning and genetic organization. *Mol. Gen. Genet.* **199**:410–414.

89. **Koch, C., and R. K. Kahmann.** 1986. Purification and properties of the *Escherichia coli* host factor required for inversion of the G segment in bacteriophage Mu. *J. Biol. Chem.* **261**:15673–15678.

90. **Koch, C., G. Mertens, F. Rudt, R. Kahmann, R. Kanaar, R. H. A. Plasterk, P. van de Putte, R. Sandulache, and D. Kamp.** 1987. The invertible G segment, p. 75–91. *In* N. Symonds, A. Toussaint, P. van de Putte, and M. M. Howe (ed.), *Phage Mu.* Cold Spring Harbor Laboratory, Cold Spring Harbor, N.Y.

91. **Komano, T., A. Kubo, and T. Nisioka.** 1987. Shufflon: multi-inversion of four contiguous DNA segments of plasmid R64 creates seven different open reading frames. *Nucleic Acids Res.* **15**:1165–1172.

92. **Konrad, E. B.** 1977. Method for the isolation of *Escherichia coli* mutants with enhanced recombination between chromosomal duplications. *J. Bacteriol.* **130**:167–172.

93. **Kutsukake, K., and T. Iino.** 1980. Inversions of specific DNA segments in flagellar phase variation of *Salmonella* and inversion systems of bacteriophages P1 and Mu. *Proc. Natl. Acad. Sci. USA* **77**:7238–7341.

94. **Laine, B., D. Kmiecik, P. Sazutiere, G. Biserte, and M. Cohen-Solal.** 1980. Complete amino-acid sequences of DNA-bindings proteins HU-1 and HU-2 from *Escherichia coli. Eur. J. Biochem.* **103**:447–461.

95. **Lammers, P. J., J. W. Golden, and R. Haselkorn.** 1986. Identification and sequence of a gene required for a developmentally regulated DNA excision in *Anabaena. Cell* **44**: 905–911.

96. **Lederberg, J., and T. Iino.** 1956. Phase variation in *Salmonella. Genetics* **41**:743–757.

96a. **Marrs, C. F., W. W. Reuhl, G. K. Schoolnik, and S. Falkow.** 1988. Pilin gene phase variation of *Moraxella bovis* is caused by an inversion of the pilin gene. *J. Bacteriol.* **170**:3032–3039.

97. **Marrs, C. F., G. Schoolnik, J. M. Koomey, J. Hardy, J. Rothbard, and S. Falkow.** 1985. Cloning and sequencing of a *Moraxella bovis* pilin gene. *J. Bacteriol.* **163**:132–139.

98. **McClintock, B.** 1956. Controlling elements and the gene. *Cold Spring Harbor Symp. Quant. Biol.* **21**:197–216.

99. **Meier, J. T., M. I. Simon, and A. Barbour.** 1985. Antigenic variation is associated with DNA rearrangements in a relapsing fever borrelia. *Cell* **41**:403–409.

100. **Mertens, G., H. Fuss, and R. Kahmann.** 1986. Purification and properties of the DNA invertase Gin encoded by bacteriophage Mu. *J. Biol. Chem.* **261**:15668–15672.

101. **Mertens, G., A. Hoffmann, H. Blocker, R. Frank, and R.**

Kahmann. 1984. *Gin*-mediated site-specific recombination in bacteriophage Mu DNA: overproduction of the protein and inversion *in vitro*. *EMBO J.* 3:2415–2421.

102. **Miller, H. I., and D. I. Friedman.** 1980. An *E. coli* gene product required for lambda site-specific recombination. *Cell* 20:711–719.

103. **Morisato, D., and N. Kleckner.** 1987. Tn10 transposition and circle formation in vitro. *Cell* 51:101–111.

104. **Murphy, E., and R. P. Novick.** 1979. Physical mapping of *Staphylococcus aureus* penicillinase plasmid pI 524: characterization of an invertible region. *Mol. Gen. Genet.* 175: 19–30.

105. **Nash, H. A., and C. Robertson.** 1981. Purification and properties of the *Escherichia coli* protein factor required for lambda integrative recombination. *J. Biol. Chem.* 256:9246–9253.

106. **Newman, B. J., and N. D. F. Grindley.** 1984. Mutants of the γδ resolvase: a genetic analysis of the recombination function. *Cell* 38:463–469.

107. **Ofek, I., D. Mirelman, and N. Sharon.** 1977. Adherence of *Escherichia coli* to human mucosal cells mediated by mannose receptors. *Nature* (London) 265:623–625.

108. **Orndorff, P. E., and S. Falkow.** 1984. Organization and expression of genes responsible for type 1 piliation in *Escherichia coli*. *J. Bacteriol.* 159:736–744.

109. **Orndorff, P. E., and S. Falkow.** 1984. Identification and characterization of a gene product that regulates type 1 piliation in *Escherichia coli*. *J. Bacteriol.* 160:61–66.

110. **Orndorff, P. E., and S. Falkow.** 1985. Nucleotide sequence of *pilA*, the gene encoding the structural component of type 1 pili in *Escherichia coli*. *J. Bacteriol.* 162:454–457.

111. **Orndorff, P. E., P. A. Spears, D. Schauer, and S. Falkow.** 1985. Two modes of control of *pilA*, the gene encoding type 1 pilin in *Escherichia coli*. *J. Bacteriol.* 164:321–330.

111a. **Ou, J. T., L. S. Baron, F. A. Rubin, and D. J. Kopecko.** 1988. Specific insertion and deletion of insertion sequence *1*-like DNA element causes the reversible expression of the virulence capsular antigen Vi of *Citrobacter freundii* in *Escherichia coli*. *Proc. Natl. Acad. Sci. USA* 85:4402–4405.

112. **Pabo, C. O., and R. T. Sauer.** 1984. Protein-DNA recognition. *Annu. Rev. Biochem.* 53:293–321.

113. **Parry, S. H., and D. M. Rooke.** 1985. Adhesions and colonization factors of *Escherichia coli*, p. 79–155. *In* M. Sussman (ed.), *The Virulence of Escherichia coli: Reviews and Methods*. Special Publications of the Society of General Microbiology, vol. 13. Academic Press (London), Ltd., London.

114. **Paruchuri, D. K., and R. M. Harshey.** 1987. Flagellar variation in *Serratia marcescens* is associated with color variation. *J. Bacteriol.* 169:61–65.

115. **Pearce, U. B., and B. A. D. Stocker.** 1967. Phase variation of flagellar antigens in *Salmonella*: abortive transduction studies. *J. Gen. Microbiol.* 49:335–347.

116. **Plasterk, R. H. A., A. Brinkman, and P. van de Putte.** 1983. DNA inversions in the chromosome of *E. coli* and in bacteriophage Mu: relationship to other site-specific recombination systems. *Proc. Natl. Acad. Sci. USA* 80:5355–5358.

117. **Plasterk, R. H. A., T. A. M. Ilmer, and P. van de Putte.** 1983. Site-specific recombination by Gin of bacteriophage Mu: inversions and deletions. *Virology* 127:24–36.

118. **Plasterk, R. H. A., R. Kanaar, and P. van de Putte.** 1984. A genetic switch *in vitro*: DNA inversion by Gin protein of phage Mu. *Proc. Natl. Acad. Sci. USA* 81:2689–2692.

119. **Plasterk, R. H. A., M. I. Simon, and A. G. Barbour.** 1985. Transposition of structural genes to an expression sequence on a linear plasmid causes antigenic variation in the bacterium *Borrelia hermsii*. *Nature* (London) 318:257–263.

120. **Plasterk, R. H. A., M. I. Simon, and A. G. Barbour.** 1986. Molecular basis for antigenic variation in a relapsing fever Borrelia, p. 127–146. *In* T. H. Birkbeck and C. W. Penn (ed.), *Antigenic Variation in Infectious Diseases*. Special Publications of the Society for General Microbiology, vol. 19. IRL Press, Oxford.

121. **Plasterk, R. H. A., and P. van de Putte.** 1984. Genetic switches by DNA inversions in prokaryotes. *Biochim. Biophys. Acta* 782:111–119.

122. **Plasterk, R. H. A., and P. van de Putte.** 1985. The invertible P-DNA segment in the chromosome of *Escherichia coli*. *EMBO J.* 4:237–242.

123. **Prentki, P., M. Chandler, and D. J. Galas.** 1987. *Escherichia coli* integration host factor bends the DNA at the ends of IS1 and in an insertion hotspot with multiple IHF binding sites. *EMBO J.* 6:2479–2487.

124. **Reed, R. R.** 1981. Transposon-mediated site-specific recombination: a defined *in vitro* system. *Cell* 25:713–719.

125. **Reed, R. R., and N. D. F. Grindley.** 1981. Transposon-mediated site-specific recombination *in vitro*: DNA cleavage and protein-DNA linkage at the recombination site. *Cell* 25:721–728.

126. **Reitzer, L. J., and B. Magasanik.** 1986. Transcription of *glnA* in *E. coli* is stimulated by activator bound to sites far from the promoter. *Cell* 45:785–792.

126a. **Robertson, C. A., and H. A. Nash.** 1988. Bending of the bacteriophage λ attachment site by *Escherichia coli* Integration Host Factor. *J. Biol. Chem.* 263:3554–3557.

127. **Robinson, A., C. J. Duggleby, A. R. Gorringe, and I. Livey.** 1986. Antigenic variation in *Bordetella pertussis*, p. 147–163. *In* T. H. Birkbeck and C. W. Penn (ed.), *Antigenic Variation in Infectious Diseases*. Special Publications of the Society for General Microbiology, vol. 19. IRL Press, Oxford.

128. **Ross, W., and A. Landy.** 1982. Anomalous electrophoretic mobility of restriction fragments containing the att region. *J. Mol. Biol.* 156:523–529.

129. **Roth, J. R., and M. B. Schmid.** 1981. Arrangement and rearrangement of the bacterial chromosome. *Staedler Symp.* 13:53–70.

130. **Rouviere-Yaniv, J., J. Germond, and M. Yaniv.** 1979. *E. coli* DNA binding protein HU forms nucleosome-like structure with circular double-stranded DNA. *Cell* 17:265–274.

131. **Rouviere-Yaniv, J., and F. Gros.** 1975. Characterization of a novel, low molecular weight DNA-binding protein from *Escherichia coli*. *Proc. Natl. Acad. Sci. USA* 72:3428–3432.

131a. **Rowland, S.-J., and K. G. H. Dyke.** 1988. A DNA invertase from *Staphylococcus aureus* is a member of the Hin family of site-specific recombinases. *FEMS Microbiol. Lett.* 50: 253–258.

132. **Ryder, K., S. Silver, A. L. Deluca, E. Fanning, and P. Tegtmeyer.** 1986. An altered DNA conformation in origin region I is a determinant for the binding of SV40 large T antigen. *Cell* 44:719–725.

133. **Sadowski, P.** 1986. Site-specific recombinases: changing partners and doing the twist. *J. Bacteriol.* 165:341–347.

134. **Salit, I. E., and E. C. Gotschlich.** 1977. Hemagglutination by purified type 1 *Escherichia coli* pili. *J. Exp. Med.* 146: 1169–1181.

135. **Salvo, J. J., and N. D. F. Grindley.** 1987. Helical phasing between DNA bends and the determination of bend direction. *Nucleic Acids Res.* 15:9771–9779.

136. **Sanderson, K. E., and J. R. Roth.** 1983. Linkage map of

Salmonella typhimurium, edition VI. *Microbiol. Rev.* **47:** 410–453.

137. **Scott, T. N., and M. I. Simon.** 1982. Genetic analysis of the mechanism of the *Salmonella* phase variation site-specific recombination system. *Mol. Gen. Genet.* **188:**313–318.
138. **Schmid, M. B., and J. R. Roth.** 1983. Selection and endpoint distribution of bacterial inversion mutations. *Genetics* **105:** 537–557.
139. **Schmid, M. B., and J. R. Roth.** 1983. Genetic methods for analysis and manipulation of inversion mutations in bacteria. *Genetics* **105:**517–537.
140. **Schmid, M. B., and J. R. Roth.** 1987. Gene location affects expression level in *Salmonella typhimurium*. *J. Bacteriol.* **169:**2872–2875.
141. **Schmucker, R., W. Ritthaler, B. Stern, and D. Kamp.** 1986. DNA inversion in bacteriophage Mu: characterization of the inversion site. *J. Gen. Virol.* **67:**1123–1133.
142. **Schrempf, H.** 1983. Reiterated sequences within the genome of *Streptomyces*, p. 131–142. *In* K. Chater, C. A. Cullis, D. A. Hopwood, A. W. B. Johnston, and H. W. Woolhouse (ed.), *Proceedings of the Fifth John Innes Symposium*. Croom Helm, London.
143. **Segal, E., E. Billyard, M. So, S. Storzbach, and T. F. Meyer.** 1985. Role of chromosomal rearrangement in *N. gonorrhoeae* pilus variation. *Cell* **40:**293–300.
144. **Segal, E., P. Hagblom, H. S. Seifert, and M. So.** 1986. Antigenic variation of gonococcal pilus involves assembly of separated silent gene segments. *Proc. Natl. Acad. Sci. USA* **83:**2177–2181.
145. **Shalita, Z., E. Murphy, and R. P. Novick.** 1980. Penicillinase plasmids of *Staphylococcus aureus*: structural and evolutionary relationships. *Plasmid* **3:**291–311.
146. **Silverman, M., and M. I. Simon.** 1980. Phase variation: genetic analysis of switching mutants. *Cell* **19:**845–854.
147. **Silverman, M., and M. Simon.** 1983. Phase variation and related systems, p. 537–557. *In* J. Shapiro (ed.), *Mobile Genetic Elements*. Academic Press, Inc., New York.
148. **Silverman, M., J. Zieg, M. Hilmen, and M. Simon.** 1979. Phase variation in *Salmonella*: genetic analysis of a recombinational switch. *Proc. Natl. Acad. Sci. USA* **76:**391–395.
149. **Silverman, M., J. Zieg, and M. Simon.** 1979. Flagellar-phase variation: isolation of the *rh1* gene. *J. Bacteriol.* **137:** 517–523.
150. **Simon, M. I., and M. Silverman.** 1983. Recombinational regulation of gene expression in bacteria, p. 211–227. *In* J. Beckwith, J. Davies, and J. A. Gallant (ed.), *Gene Function in Prokaryotes*. Cold Spring Harbor Laboratory, Cold Spring Harbor, N.Y.
151. **Simon, M., J. Zieg, M. Silverman, G. Mandel, and R. Doolittle.** 1980. Phase variation: evolution of a controlling element. *Science* **209:**1370–1374.
152. **Smyth, C. J.** 1986. Fimbrial variation in *Escherichia coli*, p. 95–126. *In* T. H. Birkbeck and C. W. Penn (ed.), *Antigenic Variation in Infectious Diseases*. Special Publications of the Society for General Microbiology, vol. 19. IRL Press, Oxford.
153. **Snellings, N. J., E. M. Johnson, D. J. Kopecko, H. H. Collins, and L. S. Baron.** 1981. Genetic regulation of variable Vi antigen expression in a strain of *Citrobacter freundii*. *J. Bacteriol.* **145:**1010–1017.
154. **Stenzel, T. T., P. Patel, and D. Bastia.** 1987. The integration host factor of *Escherichia coli* binds to bent DNA at the origin of replication of the plasmid pSC101. *Cell* **49:**709–711.
155. **Stern, A., M. Brown, P. Nickel, and T. F. Meyer.** 1986. Opacity genes in Neisseria gonorrhoeae: control of phase and antigenic variation. *Cell* **47:**61–71.
156. **Stern, A., P. Nickel, T. F. Meyer, and M. So.** 1984. Opacity determinants of Neisseria gonorrhoeae: gene expression and chromosomal linkage to the gonococcal pilus gene. *Cell* **37:**447–456.
157. **Stocker, B. A. D.** 1949. Measurement of the rate of mutation of flagellar antigenic phase in *Salmonella typhimurium*. *J. Hyg.* **47:**398–413.
158. **Suzuki, H., and T. Iino.** 1975. Absence of messenger ribonucleic acid specific for flagellin in non-flagellate mutants of *Salmonella*. *J. Mol. Biol.* **95:**549–546.
159. **Swaney, L. M., Y.-P. Liu, K. Ippen-Ihler, and C. C. Brinton, Jr.** 1977. Genetic complementation analysis of *Escherichia coli* type 1 somatic pilus mutants. *J. Bacteriol.* **130:**506–511.
160. **Swaney, L. M., Y.-P. Liu, C.-M. To, C.-C. To, K. Ippen-Ihler, and C. C. Brinton, Jr.** 1977. Isolation and characterization of *Escherichia coli* phase variants and mutants deficient in type 1 pilus production. *J. Bacteriol.* **130:**495–505.
161. **Swanson, J., S. Bergström, O. Barrera, K. Robbins, and D. Corwin.** 1985. Pilus$^-$ gonococcal variants. Evidence for multiple forms of piliation control. *J. Exp. Med.* **162:** 729–744.
162. **Swanson, J., S. Bergstrom, K. Robbins, O. Barrera, D. Corwin, and J. M. Koomey.** 1986. Gene conversion involving the pilin structural gene correlates with pilus$^+$ pilus$^-$ changes in *Neisseria gonorrhoeae*. *Cell* **47:**267–276.
163. **Thompson, J. F., L. M. de Vargas, C. Foch, R. Kahmann, and A. Landy.** 1987. Cellular factors couple recombination with growth phase: characterization of a new component in the λ site-specific recombination pathway. *Cell* **50:**901–908.
164. **Thompson, J. F., D. Waechter-Brulla, R. I. Gumport, J. F. Gardner, L. Moitoso de Vargas, and A. Landy.** 1986. Mutations in an integration host factor-binding site: effect on lambda site-specific recombination and regulatory implications. *J. Bacteriol.* **168:**1343–1351.
165. **Thompson, J. F., L. Moitoso de Vargas, S. E. Skinner, and A. Landy.** 1987. Protein-protein interactions in a higher-order structure direct lambda site-specific recombination. *J. Mol. Biol.* **195:**481–493.
166. **Tominaga, A., K. Nakamura, and M. Enomoto.** 1986. Isolation of P1 *cin*C(+) and P1 *cin*C(−) mutants and detection of their polypeptides involved in host specificity. *Jpn. J. Genet.* **61:**1–13.
167. **Toussaint, A., N. Lefebvre, J. R. Scott, J. A. Cowan, F. de Bruijn, and A. I. Bukhari.** 1978. Relationships between temperate phages Mu and P1. *Virology* **89:**146–163.
168. **Van de Putte, P., S. Cramer, and M. Giphart-Gassler.** 1980. Invertible DNA determines host specificity of bacteriophage Mu. *Nature* (London) **286:**218–222.
169. **Van de Putte, P., R. Plasterk, and A. Kuijpers.** 1984. A Mu *gin* complementing function and an invertible DNA region in *Escherichia coli* K-12 are situated on the genetic element *e*14. *J. Bacteriol.* **158:**517–522.
170. **Walker, J. T., and D. H. Walker, Jr.** 1981. Structural proteins of coliphage P1, p. 69–77. *In* M. S. DuBow (ed.), *Progress in Clinical Biological Research*. Alan R. Liss, Inc., New York.
171. **Wanner, B. L.** 1980. Novel regulatory mutants of the phosphate regulon in *Escherichia coli* K12. *J. Mol. Biol.* **191:**39–58.
172. **Wanner, B. L.** 1987. Control of *phoR*-dependent bacterial alkaline phosphatase clonal variation by the *phoM* region. *J. Bacteriol.* **169:**900–903.

173. **Wanner, B. L., M. R. Wilmes, and D. C. Young.** 1988. Control of bacterial alkaline phosphatase synthesis and variation in an *Escherichia coli* K-12 *phoR* mutant by adenyl cyclase, the cyclic AMP receptor protein, and the *phoM* operon. *J. Bacteriol.* **170:**1092–1102.

174. **Wu, H.-M., and D. M. Crothers.** 1984. The locus of sequence-directed and protein-induced DNA bending. *Nature* (London) **308:**503–513.

175. **Yee, T., and M. Inouye.** 1984. Two-dimensional S1 nuclease heteroduplex mapping: detection of rearrangements in bacterial genomes. *Proc. Natl. Acad. Sci. USA* **891:**2723–2727.

176. **Zafriri, D., Y. Oron, B. I. Eisenstein, and I. Ofek.** 1987. Growth advantage and enhanced toxicity of *Escherichia coli* adherent to tissue culture cells due to restricted diffusion of products secreted by the cells. *J. Clin. Invest.* **79:**1210–1216.

177. **Zahn, K., and F. Blattner.** 1987. Direct evidence for DNA bending at the lambda replication origin. *Science* **236:** 416–422.

178. **Zieg, J., and S. R. Kushner.** 1977. Analysis of genetic recombination between two partially deleted lactose operons of *Escherichia coli* K-12. *J. Bacteriol.* **131:**123–132.

179. **Zieg, J., M. Silverman, M. Hilmen, and M. Simon.** 1977. Recombinational switch for gene expression. *Science* **196:** 170–172.

180. **Zieg, J., M. Silverman, M. Hilmen, and M. Simon.** 1978. The mechanism of phase variation, p. 411–423. *In* J. H. Miller and W. S. Reznikoff (ed.), *The Operon*. Cold Spring Harbor Laboratory, Cold Spring Harbor, N.Y.

181. **Zieg, J., and M. I. Simon.** 1980. Analysis of the nucleotide sequence of an invertible controlling element. *Proc. Natl. Acad. Sci. USA* **77:**4196–4200.

182. **Zinkel, S. S., and D. M. Crothers.** 1987. DNA bend direction by phase sensitive detection. *Nature* (London) **328:**178–181.

Chapter 29

DNA Inversion in the 2μm Plasmid of *Saccharomyces cerevisiae*

MICHAEL M. COX

I. THE 2μm PLASMID: HISTORICAL OVERVIEW

The isolation of closed circular DNA from the yeast *Saccharomyces cerevisiae* yields a population of DNA molecules with limited size heterogeneity. The major component is approximately 2 μm in circumference (10, 22, 28, 60). The minor species are approximately 3, 4, and 6 μm long (10, 22). The 2-μm circle, referred to as the 2μm plasmid, is present at 60 to 100 copies per cell (11). Restriction analysis showed that there are two major species, related by a large inversion of approximately 40% of the molecule (29). Denaturation and renaturation of nicked or cleaved 2-μm DNA produced DNAs with a dumbbell structure in which two single-stranded loops were separated by a double-stranded stem, indicating the presence of long inverted repeats. The inversion could be explained by recombination between these repeats (23). The 4- and 6-μm DNAs are dimers and trimers of the 2-μm circles, and these could also be accounted for by recombination (55). Recombination between the repeats occurs efficiently in vivo during plasmid propagation (6) and is mediated by a plasmid-encoded site-specific recombination system (7, 8, 17).

The sequence of the 2-μm circle (24) revealed important details of its structure (Fig. 1). The plasmid consists of 6,318 base pairs (bp), with perfect inverted repeats of 599 bp separating unique regions of 2,774 and 2,346 bp. Within the 599-bp repeats are two sets of shorter repeats, a pair of 16-bp inverted repeats and three repeats of a 13-bp sequence, concentrated within 122 bp. The 13-bp repeats are binding sites for the plasmid-encoded site-specific recombinase (FLP, see below). Another set of short repeats, a tandem array containing 5.5 copies of a 62- to 63-bp sequence, is found in the large unique region designated REP3. The 2-μm circle contains four significant open reading frames, designated A, B, C, and D (or FLP, REP1, REP2, and RAF) in order of decreasing size. These encode proteins with molecu-

Michael M. Cox ■ Department of Biochemistry, College of Agricultural and Life Sciences, University of Wisconsin-Madison, Madison, Wisconsin 53706.

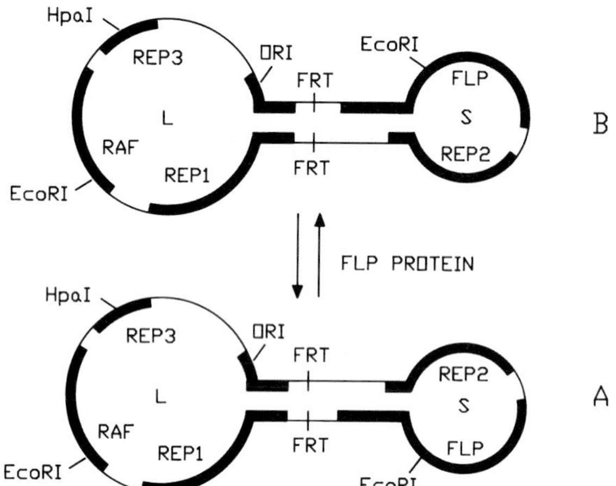

Figure 1. The *S. cerevisiae* 2μm plasmid. Major genetic loci described in the text are designated by filled boxes. The FRT site (Fig. 3) spans the *Xba*I cleavage sites. Interconversion between the A and B forms involves site-specific recombination at the FRT site. The 599-bp inverted repeats are presented as horizontal lines. FLP, REP1, REP2, and RAF correspond, respectively, to open reading frames A, B, C, and D (see text). REP3, also called STB, is a DNA site recognized by REP1/REP2.

lar weights of 48,619, 43,231, 33,196, and 21,289, respectively (24). The origin of replication (ORI) is about 75 bp long and overlaps the end of one of the 599-bp repeats. The plasmid confers no known phenotype on its host other than a slight decrease in the growth rate (14). It may be regarded simply as an intracellular parasite (14).

Replication of the 2-μm circle is under stringent cell cycle control during normal growth (68). Each molecule replicates just once early in S phase. The origin of replication is structurally related to the chromosomal *ars* sequences (45, 66). Replication proceeds bidirectionally and depends completely on host functions (41). Plasmid-encoded proteins function in plasmid segregation during cell division and in the regulation of copy number as outlined below.

II. STABLE MAINTENANCE OF THE 2μm PLASMID: ROLE OF PLASMID GENES

The plasmid possesses two systems that ensure its own stable maintenance. The first facilitates equal partitioning of the plasmid population at mitosis. This system consists of two *trans*-acting factors, the REP1 and REP2 proteins (open reading frames B and C, respectively), and a *cis*-acting element, designated REP3 or STB (34, 37, 65), that appears to be the site of action of REP1 and REP2 proteins (35, 46). REP3 is a centromerelike locus consisting of a series of 62- to 63-bp repeats (described above). The REP1 protein is highly homologous in its carboxy one-half with vimentin, tubulin, and myosin heavy chain (65). It is found in the nucleus and cofractionates with an *S. cerevisiae* nuclear skeletal structure. These and other results led to a model in which the REP proteins form a microtubulelike structure that connects the 2μm plasmid to some nuclear structure that partitions equally between daughters at mitosis (65). The REP system can stabilize plasmids with replication origins other than the 2μm ORI (37).

Copy number is also controlled by a second system, designated FLP, in addition to the REP system. The FLP system is responsible for copy number amplification (54, 64). FLP protein is the only 2μm gene product required for amplification (54). This protein promotes site-specific recombination at two sites designated FRT within the 599-bp inverted repeats. This recombination inverts the two unique sequence regions of the plasmid, and, as described below, this inversion often leads to plasmid amplification.

REP1 and REP2 also regulate copy number by acting in concert to reduce FLP gene transcription (47, 54). The RAF protein, in contrast, increases FLP expression by relieving the inhibition caused by REP1 and REP2 (47). RAF expression is itself decreased by the REP1 and REP2 proteins (47). Many of the details about how this complicated regulatory circuit maintains and fine tunes plasmid copy number remain to be worked out.

III. THE ROLE OF DNA INVERSION IN COPY NUMBER AMPLIFICATION

The role of the FLP DNA inversion system in causing 2μm plasmid amplification has only recently been elucidated. Its function was initially obscured by the observation that deletion of the FLP gene did not reduce the high plasmid copy number (8, 15, 18). These FLP^- plasmids, however, are much less stable than the FLP^+ plasmid (14, 15), and the high copy number was probably due to the strong selection for cells that contained plasmids in this study (15).

A role for the FLP system in copy number amplification was first suggested by Futcher (13) in an elegant explanation of how cell cycle control of replication could be circumvented by site-specific recombination (Fig. 2). Initiation of bidirectional replication at the origin results in two replication forks traveling in opposite directions. In the absence of recombination, these forks meet on the other side of the plasmid, creating two daughter molecules. The 599-bp repeat nearest the origin is replicated before the second 599-bp repeat. If recombination involving these repeats occurs between the replication of the

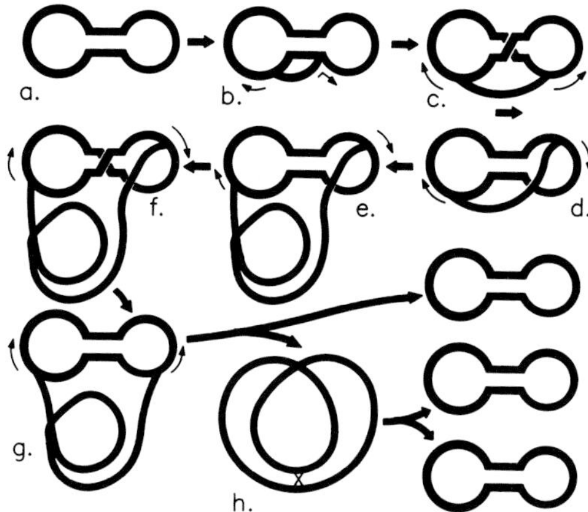

Figure 2. Futcher's (13) model for amplification of 2-μm circle copy number. Inversion following replication of one 599-bp repeat (step c) results in a double rolling circle (d). This results in multiple copies of the 2μm plasmid even though replication is initiated only once. (Adapted from reference 64.)

first and the second copies, the plasmid is converted to a "double rolling circle" (Fig. 2). In this, the two replication forks move in the same direction and produce a multimeric form of the 2μm plasmid. This multimer can be resolved to monomers by FLP-mediated recombination between directly repeated 599-bp repeats. It is not clear if homologous recombination can substitute for the FLP system in the resolution step. All key features of this model have now been substantiated experimentally, with a clear link established between copy number amplification and the FLP system (47, 54, 64). Overexpression of the FLP protein in *S. cerevisiae* results in abnormally high plasmid copy numbers and causes increased cell size and lethality (54). Similar effects are observed with strains of mutants in the chromosomal NIB genes (30, 31). This suggests that the NIB product, along with REP1 and REP2 proteins, regulates FLP gene expression.

IV. THE FLP SITE-SPECIFIC RECOMBINATION SYSTEM

A. Fundamental Properties

The FLP system carries out both intramolecular and intermolecular recombination in vivo (6, 12, 55). No evidence has been found for the involvement of any protein besides the plasmid-encoded FLP protein. The FLP gene has been cloned and expressed in *Escherichia coli* (12, 63). The FLP system functions well in vivo in *E. coli*, carrying out intra- and intermolecular recombination at the specific FRT sites (12). Since no *S. cerevisiae* gene other than FLP was transferred to bacteria in these experiments, it was clear that FLP protein is the only *S. cerevisiae* protein required for recombination.

In vitro recombination systems that utilize extracts from *E. coli* cells expressing FLP protein have been developed in several laboratories (44, 51, 63). Recombination in vitro requires only FLP protein, a substrate, buffer (pH of ~6 to 9), and appropriate concentrations of salt. No divalent cations or high-energy cofactors are needed. The substrate DNA can be relaxed or supercoiled, although the rate of the reaction is increased with supercoiled DNA (see below). The system promotes intramolecular (inversions, deletions) or intermolecular recombination given appropriate substrates, and all results obtained in vitro are consistent with observations made in vivo.

In vitro systems have been developed for several other site-specific recombination systems, primarily from procaryotic sources (56; see J. F. Thompson and A. Landy; H. Varmus and P. Brown; D. Sherratt; and A. C. Glasgow, K. T. Hughes, and M. I. Simon, all this volume). The FLP system ranks among the simplest of these. Unlike bacteriophage λ integration or procaryotic DNA inversion, the FLP system requires no proteins other than the recombinase itself. Unlike the procaryotic invertases and the resolvases from the Tn*3* class of transposable elements, the FLP system will carry out both inversions and deletions (depending on the orientation of the FRT sites). The FRT recombination site employed by the FLP system is among the smallest (about 30 bp; see below). The only known procaryotic system that appears to be analogous to FLP is the Cre-Lox system of bacteriophage P1 (25–27). The added complexity of most of these other recombination systems probably reflects regulation of the rate and direction of the reaction imposed on the fundamental chemistry of site-specific recombination.

B. The Recombination Site: Primary Structure

The recombination site used by the FLP system has been designated FRT (42), and its sequence is given in Fig. 3. It consists of three 13-bp repeats. The first and second are inverted and separated by an 8-bp spacer. The third is a direct repeat, in tandem to the second. Deletion analysis (3, 58) demonstrated

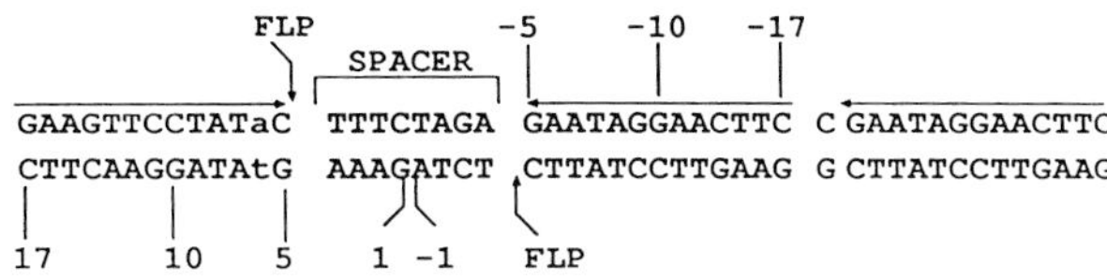

Figure 3. The FRT site. Position numbers are as in reference 9.

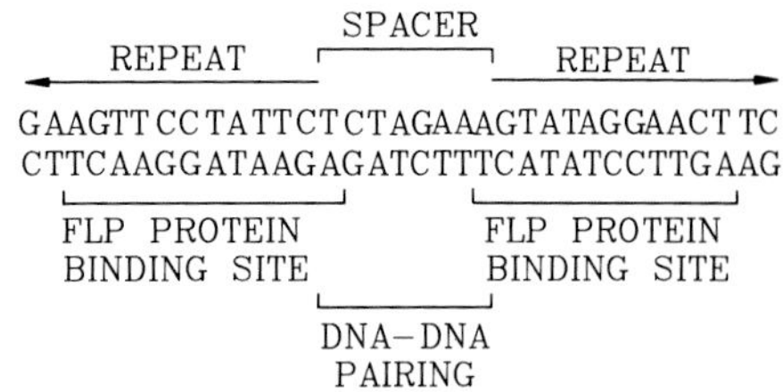

Figure 4. Summary of structure and function within the minimal FRT site.

that the third 13-bp repeat was not necessary for recombination in vitro, although this repeat appears to increase the efficiency of intermolecular recombination slightly in vivo (33). Further deletions from the outer ends of the first and second 13-bp repeats produce little or no effect in vitro until more than 3 bp is removed. As additional base pairs are removed, a gradual reduction in FRT site function is observed, and deletion of 8 bp inactivates the site. The minimal sequence for a fully functional FRT site therefore consists of 28 to 30 bp, including the spacer and 10 bp of each adjacent repeat (3, 58).

The FLP protein cleaves the FRT site at the boundaries of the spacer and adjacent repeats (3, 58), making 8-bp staggered cuts and leaving overhangs with 5′ hydroxyl termini. The protein forms a transient covalent intermediate with the DNA via a 3′ phosphotyrosyl linkage (21).

The functional organization of the minimal FRT site is summarized in Fig. 4. The inverted repeats are binding sites for FLP protein. Use of the spacer is homology dependent, but not strictly sequence dependent. This implies that DNA-DNA pairing occurs between spacer sequences of two reacting FRT sites during recombination. These functions are described below.

1. The FLP-binding sites: protein-DNA interactions

FLP protein-purine contacts, defined by methylation protection experiments, occur within two 12-bp regions, the base pair at either side of the spacer and the proximal 11 bp in the adjacent repeat (9). FLP protein-phosphate contacts are clustered near the point at which the DNA is cleaved (9).

This work is complemented by the extensive mutational analyses of the FLP binding site (2, 42, 51, 59a). The effects of base pair substitutions in the FLP-binding site on recombination in vitro are summarized in Fig. 5. The effects at different positions fall into three classes: either (i) all three changes have small effects, (ii) all three changes have large effects, or (iii) one change has a large effect, while the others have small effects (59a). Interestingly, eight positions fall into class (i), with 31 of 37 mutations tested having effects of 10-fold or less. Only one position (11) falls into class (ii). Most interesting are positions 5 to 7, which fall into class (iii). One change at each position, in each case of transversion, has a large deleterious effect on recombination. The other two changes have either small or no detectable effects. These patterns may provide important clues as details of the interaction between FLP protein and this site are elucidated.

Other patterns observed in these studies suggest functional interactions between FLP-binding sites

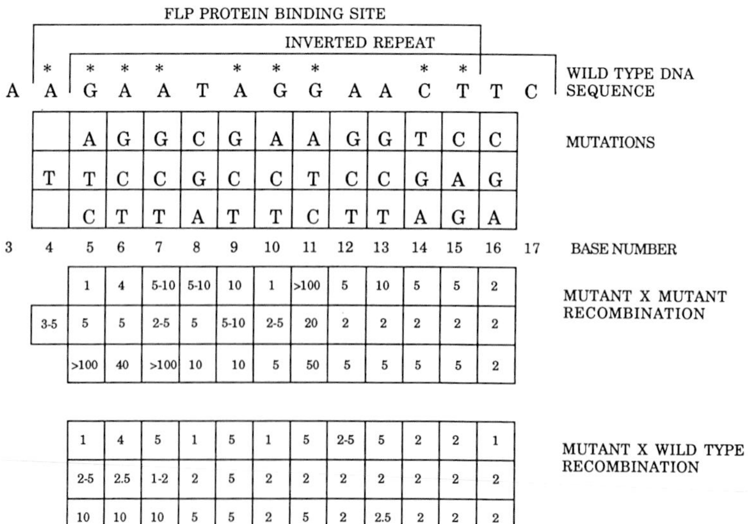

Figure 5. Effects of mutations in the FLP protein-DNA-binding site. Asterisks denote purine contacts identified in methylation protection studies. Numbers denote the factor by which the FLP protein concentration must be raised to produce a minimal reaction relative to a normal recombination site; i.e., higher numbers correspond to less reactive sites.

both within a single FRT and between different FRT sites. An interaction within a single FRT is indicated by the greater than additive effect of mutations at certain positions when present in both rather than in only one of the FLP-binding sites within an FRT site. An interaction between different FRT sites is indicated by the improvement in the reaction efficiency of mutant sites (with single mutations in one FLP-binding site) observed generally when the second FRT site participating in a recombination event is wild type (Fig. 5) (51, 59a).

2. The spacer: homology and directionality

Initial work on the spacer focused on its two most obvious features: size and sequence. These studies at first indicated that the size was important but the sequence was not.

The unaltered spacer contains a cleavage site for the restriction enzyme *Xba*I. Destruction of this *Xba*I site (removal of 4 bp) was found to eliminate inversion in *S. cerevisiae* (7). Similar results were obtained in vivo in *E. coli* (12) and in vitro in *S. cerevisiae* (44), providing an important experimental link with the normal system in *S. cerevisiae.* The importance of spacer size was suggested by the fact that the *Xba*I site mutations altered the size of the spacer (adding or subtracting 4 bp at this *Xba*I site had the same effect [8, 12, 44]). Subsequent tests showed that FRT function was impaired when the spacer size was increased or decreased by 1 bp and inactivated by the addition of 2 bp to the spacer (58).

The spacer sequence, in contrast, initially appeared to be unimportant provided that the spacers of the two sites undergoing reaction were homologous. This was first suggested by the increase in reaction efficiency observed when FRT sites with spacers altered by ±1 bp were reacted with identical mutant sites rather than with a wild-type site (58). Base substitutions at each of the central six positions in the spacer were subsequently found to have no effect on the reaction as long as reactants were homologous (1, 59).

The homology requirement implies that the spacers of two reacting FRT sites are paired at some point in the reaction. This might reflect the formation of a four-strand intermediate prior to DNA cleavage, as suggested for bacteriophage λ integration (38), or events subsequent to DNA cleavage. Similar homology requirements have been noted in other site-specific recombination systems (5, 27, 36, 38, 48). The role of homology has been most thoroughly analyzed in the λ integration reaction. Here, it has been demonstrated that the cleavage events are ordered. DNA cleavage and reunion first occur on one side of the crossover region, creating a "Holliday structure" intermediate. This intermediate is resolved by DNA cleavage and reunion in the complementary strands 7 bp away on the other side of the crossover region (40, 49). Homology in the crossover region (spacer) is required for the branch migration that occurs before the second set of cleavage and reunion events. A possible additional requirement for homology at an earlier step in the reaction (i.e., a four-stranded intermediate) has not been precluded by experimental tests. The evidence argues against an alternative model in which all four cleavage events are simultaneous and homology is required to reanneal the resulting cohesive ends before religation (48). The recombination events mediated by the Cre-Lox system (26) and the FLP system (43a; M. Jayaram, K. L. Crain, R. L. Parsons, and R. M. Harshey, *Proc. Natl. Acad. Sci. USA*, in press) also proceed via Holliday intermediates. In both cases, the intermediates have been isolated and characterized, and they are resolved by the respective recombinase in vitro.

The spacer homology requirement serves an additional special function in the FLP system, the alignment of the two FRT sites in the same orientation during recombination. The spacer is the only asymmetric segment of the FRT site, if the third 13-bp repeat is removed, and is sufficient to define the directionality of the intramolecular reaction (inversion versus deletion). An FRT site with a symmetrical spacer (and two FLP-binding sites) is still functional in vitro, but directionality is abolished (59). A similar observation has been made in the Cre-Lox system (27). Directionality is not restored by the third 13-bp FLP-binding site if the spacer is symmetrical (X. Qian and M. Cox, unpublished results).

The sequence flexibility in the spacer, combined with the homology requirement, makes it possible to construct numerous different functional FRT sites that do not cross-react, and therefore to set up two discrete reactions in the same tube. Abolishing directionality by using a symmetrical spacer has the effect of increasing the number of possible products generated in a given experiment. This ability to establish two distinct reactions in the same tube has proven useful experimentally in the study of reactant turnover (15a) and the internal comparison of several effects of FLP-binding-site mutations in a single experiment (59a).

The general interpretation of studies of simple substitution mutations was that the spacer sequence was not important if homology was retained. However, further analysis has shown that the spacer sequence can affect FRT site function and has led to a reevaluation of spacer function, as described in the next section.

C. The Recombination Site: Higher-Order Structure

While generating mutant FRT sites, several that differed from wild type at several rather than just one position were found that nearly inactivated the site even in homologous reactions. These results do not appear to reflect undetected DNA-protein contacts in the spacer region, because many multiple-base-pair changes (up to five in a single site) had no effect regardless of which positions were altered. No individual base pair substitution in the central 6 bp of the 8-bp spacer had a large effect on recombination. A comparison of many different spacer sequences (62a) indicates that there are at least two features of the general sequence structure of the FRT site that are essential for its function. The first is the set of two nearly symmetrical pyrimidine tracts that begin on each strand in the spacer and continue in opposite directions through the first 9 bp of the adjacent repeats. The second is the predominance of A · T base pairs in the spacer (six of eight). Disruption of the polypyrimidine tract or an increase in spacer G+C content decreases the recombination efficiency.

Significantly, the spacer alterations that inactivate the FRT site do **not** have a deleterious effect on in vitro binding by FLP protein and therefore provide a clear distinction between protein-DNA recognition and site function. A recombination step subsequent to protein binding and prior to product dissociation is affected. The effects of G+C content, and the additional observation that the deleterious effects are moderated if the DNA is supercoiled, suggest that the affected step involves strand separation within the spacer (62a).

These results are relevant to the study of DNA structure-function relationships. Specific DNA sequences take up a variety of non-B DNA structures where the biological functions are still obscure in many cases. In the most extreme case, the FRT site could simply represent a highly degenerate recognition sequence that localizes recombination to that site. The results of the spacer mutations described above, however, demonstrate that the higher-order sequence structure of the FRT site (as opposed to the primary sequence itself) is essential not only for recognition but for site function. The FRT site is hypersensitive to cleavage by S1 nuclease in the absence of FLP protein (S. Umlauf and M. Cox, unpublished results). The FRT site therefore appears to adopt a non-B DNA structure. The effects of A+T content in the spacer and the fact that DNA supercoiling enhances recombination efficiency (62a) indicate that this structure facilitates DNA unwinding during recombination in a manner as yet undefined at the molecular level. Enhanced sensitivity of polypyrimidine tracts to S1 nuclease digestion has been observed in several other studies (39, 53). Polypyrimidine tracts are not standard features of recombination sites used by site-specific systems. The sequences found at these sites, however, do exhibit other motifs (Thompson and Landy; Sherratt; and Glasgow et al., all this volume). Features represented include poly(A) tracts (lambda integration, bacterial DNA invertases) and alternating purine-pyrimidine tracts (Cre-Lox system [27]). Most of these recombination sites do feature A+T-rich regions. The results obtained for the FLP system indicate a need for further study of the functional significance of such DNA sequence arrangements.

D. The FLP Protein

The FLP protein has recently been obtained from extracts of *E. coli* cells harboring appropriate recombinant DNA plasmids and has been purified to near homogeneity by using site-specific DNA chromatography (16, 43). The pure protein is free of detectable nuclease activity and is stable for many months at −70°C. Stability under reaction conditions at 30°C is greatly enhanced by the addition of bovine serum albumin and glycerol. When substrates with two FRT sites are employed, reactions are almost entirely intramolecular. Intermolecular recombination is facilitated by the addition of polyethylene glycol (16).

A general strategy for isolating FLP$^-$ mutations has been described that uses a selection based on loss of antibiotic resistance in bacteria by FLP-mediated deletion unless the cell is FLP$^-$ (19). Site-directed mutagenesis has also been used (52). Mutations that change Tyr-343 to Ser or Phe inactivate the protein but do not decrease its binding to the FRT site. This provides evidence that this is the tyrosine involved in the covalent link to the DNA (52). Mutations at His-305 cause an accumulation of DNA cleavage products that do not complete the recombination process (R. Parsons, P. Prasad, R. M. Harshey, and M. Jayaram, submitted for publication). These early results emphasize the importance of mutational analysis of the FLP gene in dissecting this site-specific recombination reaction pathway.

E. The Recombination Reaction

Site-specific recombination reactions provide an opportunity to examine important principles of protein-DNA interactions. The recombination proteins must locate specific binding sites, juxtapose two sites, cleave a DNA within the two sites, and then religate the resulting ends to new partners with high fidelity. Each of these recombinases must act both as a specific endonuclease and as a DNA ligase. Because of its relative simplicity, the FLP system seems ideal for probing the chemical details of such processes. Until recently, however, many properties of the system were elucidated with partially purified preparations, and, unless noted, the information presented below was obtained with FLP protein fractions that were approximately 5% pure (1, 44).

A DNA-binding assay based on gel retardation has been used to resolve three binding complexes, which apparently correspond to the binding of one, two, or three FLP monomers, respectively, to one, two, or three 13-bp repeats (1, 51, 59a). If an FRT site with a strong deleterious mutation in one FLP-binding site is employed, FLP protein binds and cleaves only the DNA adjacent to the unaltered FLP-binding site (59a). The efficiency of intramolecular recombination decreases as the distance between FRT sites increases, suggesting that tracking may be involved in site juxtaposition (20). This relationship was abolished, however, if KCl was omitted from the reaction, suggesting that under at least some conditions, site juxtaposition may involve random collisions (57). The cleavage of the DNA and formation of a covalent adduct has been described above. As with other site-specific recombination systems and topoisomerases, this covalent adduct preserves the high-energy bond and obviates the need for high-energy cofactors in the reaction.

A common observation with site-specific recombination systems is that levels of recombinase that are at least equimolar with the concentration of available recombination sites are generally required to produce an optimum reaction in vitro (56). This suggests that these proteins do not turn over, and therefore are not true enzymes. Similar observations were made with the early in vitro experiments with FLP protein. However, recent results with purified FLP protein using conditions that enhance its stability indicate that this protein turns over slowly when relaxed DNA substrates are used (15a). The turnover rate is increased 1.5- to 3-fold if supercoiled substrates are used. The apparent turnover number is low, ~0.12 min^{-1} with linear substrates under one set of conditions, and the slow step in the process has not yet been identified.

V. THE EFFECTS OF OTHER PROTEINS ON FLP PROTEIN-PROMOTED RECOMBINATION

Purification of FLP protein was hampered to some degree by variation from one fraction to the next in the relative efficiencies of inter- and intramolecular recombination and by increases and decreases in overall activity (20, 44). Increasing levels of many FLP-protein-containing fractions led first to a general enhancement of intermolecular recombination, while inhibition of the reaction was observed when higher concentrations of the same fraction were used. These effects have been traced to *E. coli* proteins that were present as prominent contaminants at early stages of the FLP protein purification (43). These proteins do not include factors known to play a role in other procaryotic site-specific recombination systems, such as IHF, HU, or FIS (Glasgow et al., this volume; Thompson and Landy, this volume). At least four *E. coli* proteins can inhibit FLP protein action, and each of these has recently been purified to homogeneity (R. Bruckner and M. Cox, unpublished results). One (M_r 27,000) is the H protein (32), a histonelike protein that cross-reacts immunologically with histone H2A. The in vitro properties of the other three proteins (M_r 26,000, 24,000, and 16,000) suggest that they also are DNA-binding proteins. The M_r 26,000 protein is a common contaminant of H-protein preparations (32). N-terminal sequence analysis of each of these proteins has shown, surprisingly, that all four are ribosomal proteins (Bruckner and Cox, unpublished results). H protein is ribosomal protein S3, and the other proteins are S4, S5, and L2. Interestingly, histones produce similar effects on FLP protein-promoted recombination (Bruckner and Cox, unpublished results). The significance of this result is not clear, although these effects almost certainly do not reflect specific interactions with the FLP system. It is worth pointing out, however, that the 2μm plasmid is packaged as chromatin in *S. cerevisiae* and it is conceivable that histones indirectly help regulate or modulate this recombination system in some manner in vivo.

VI. BIOLOGICAL RELEVANCE: COPY NUMBER AMPLIFICATION AND GENE AMPLIFICATION

The mechanism employed by the 2μm plasmid to amplify its copy number may be widespread among eucaryotic plasmids. A number of plasmids with similar overall sequence arrangement have been found in several *Zygosaccharomyces* species, and each of these also possesses a site-specific inversion system (4, 62). These plasmids exhibit little or no homology to the yeast 2-μm circle, except for limited homology at the amino acid level within the FLP genes. If these plasmids evolved independently, their common structural features may have general adaptive significance.

Chromosomal gene amplification can occur in

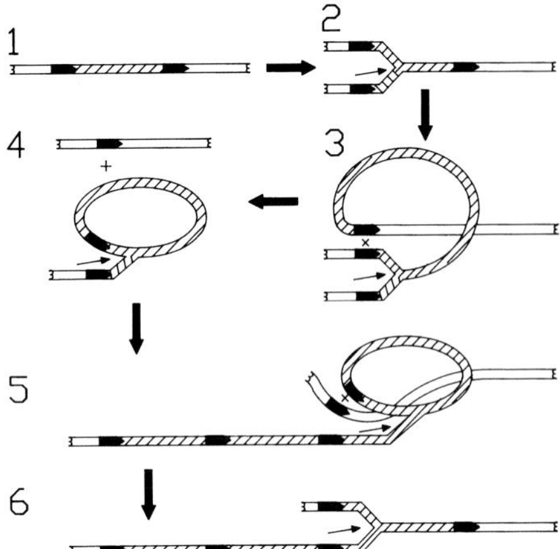

Figure 6. A model for gene amplification. A recombination event (homologous, site specific, illegitimate) occurring between one site in front of a replication fork and one site behind the replication fork will result in the formation of a rolling circle (4). This will result in amplification of the cross-hatched DNA sequences in one of the daughter DNA molecules. This amplified sequence will be present in a tandem array. Shaded sequences represent potential recombination sites. Small arrows denote the movement of the replication fork.

many species as part of a normal developmental process (e.g., rRNA genes, chorion), cell transformation (e.g., oncogenes such as *myc*, *sis*, *ras*), or under strong selection pressure (for a review, see reference 61), and may involve similar recombination events. Mechanisms previously proposed to explain amplification include (i) unequal crossing over following replication or (ii) a cycle of excision of a DNA sequence, its extrachromosomal replication, and then reinsertion into the chromosome as amplified DNA. These models do not satisfactorily explain the inverted duplication structure found in many amplified DNAs. The model for 2μm copy number amplification (13) (Fig. 2) has suggested alternative mechanisms to explain such chromosomal gene amplification events (50). In this model, illegitimate recombination immediately following replication would produce an extrachromosomal circular DNA molecule with the DNA to be amplified in inverted repeats. The second step would involve recombination events to produce a double-rolling-circle-replicating species (as in Fig. 2). After amplification, this DNA would be reinserted into the chromosome.

A more straightforward model (Fig. 6) to explain amplified DNA sequences in tandem arrays again uses features of the Futcher proposal (67; U. Hornemann, personal communication). Its key feature is a recombination event (illegitimate, site specific, or homologous) between two sites, one in front of and one behind the replication fork. This produces a single rolling circle and tandem repeats of the DNA within the circle. A similar recombination event restores the original replication fork, with DNA sequences now amplified on one of the daughter DNA molecules.

In conclusion, the properties of the FLP system outlined above recommend it for eucaryotic site-specific genetic recombination, as a model system for dissection of the basic chemistry of site-specific recombination reactions in general, and as a vehicle to test ideas about the biological functions of diverse DNA rearrangements.

LITERATURE CITED

1. **Andrews, B. J., L. G. Beatty, and P. D. Sadowski.** 1987. Isolation of intermediates in the binding of the FLP recombinase of the yeast plasmid 2-micron circle to its target sequence. *J. Mol. Biol.* **193:**345–358.
2. **Andrews, B. J., M. McLeod, J. R. Broach, and P. D. Sadowski.** 1986. Interaction of the FLP recombinase of the yeast 2-μm plasmid with mutated target sequences. *Mol. Cell. Biol.* **6:**2482–2489.
3. **Andrews, B. J., G. A. Proteau, L. G. Beatty, and P. D. Sadowski.** 1985. The FLP recombinase of the 2-micron circle DNA of yeast: interaction with its target sequences. *Cell* **40:**795–803.
4. **Araki, H., H. Tatsumi, T. Sakurai, A. Jearnpipatkul, K. Ushio, T. Muta, and Y. Oshima.** 1985. Molecular and functional organization of yeast plasmid pSR1. *J. Mol. Biol.* **182:**191–203.
5. **Bauer, C. E., J. F. Gardner, and R. I. Gumport.** 1985. Extent of sequence homology required for bacteriophage λ site-specific recombination. *J. Mol. Biol.* **181:**187–197.
6. **Beggs, J. D.** 1978. Transformation of yeast by a replicating hybrid plasmid. *Nature* (London) **275:**104–108.
7. **Broach, J. R., V. R. Guarascio, and M. Jayaram.** 1982. Recombination within the yeast plasmid 2 micron circle is site-specific. *Cell* **29:**227–234.
8. **Broach, J. R., and J. B. Hicks.** 1980. Replication and recombination functions associated with the yeast plasmid 2 micron circle. *Cell* **21:**501–508.
9. **Bruckner, R. C., and M. M. Cox.** 1986. Specific contacts between the FLP protein of the yeast 2 micron plasmid and its recombination site. *J. Biol. Chem.* **261:**11798–11807.
10. **Clark-Walker, G. D.** 1973. Size distribution of circular DNA from petite-mutant yeast lacking ρ DNA. *Eur. J. Biochem.* **32:**263–267.
11. **Clark-Walker, G. D., and G. L. G. Miklos.** 1974. Localization and quantification of circular DNA in yeast. *Eur. J. Biochem.* **41:**359–365.
12. **Cox, M. M.** 1983. The FLP protein of the yeast 2 micron plasmid: expression of a eukaryotic genetic recombination system in *Escherichia coli*. *Proc. Natl. Acad. Sci. USA* **80:** 4223–4227.
13. **Futcher, A. B.** 1986. Copy-number amplification of the 2 micron circle plasmid of *Saccharomyces cerevisiae*. *J. Theor. Biol.* **119:**197–204.
14. **Futcher, A. B., and B. S. Cox.** 1983. Maintenance of the 2μm circle plasmid in populations of *Saccharomyces cerevisiae*. *J. Bacteriol.* **154:**612–622.
15. **Futcher, A. B., and B. S. Cox.** 1984. Copy number and the

stability of 2μm circle-based artificial plasmids of *Saccharomyces cerevisiae*. *J. Bacteriol.* **157**:283–290.

15a. **Gates, C. A., and M. M. Cox.** 1988. FLP recombinase is an enzyme. *Proc. Natl. Acad. Sci. USA* **85**:4628–4632.

16. **Gates, C. A., L. Meyer-Leon, J. M. Attwood, E. A. Wood, and M. M. Cox.** 1987. Purification of FLP recombinase using sequence-specific DNA affinity chromatography. *UCLA Symp. Mol. Cell. Biol.* **69**:197–206.

17. **Gerbaud, C., P. Fournier, H. Blanc, M. Aigle, H. Heslot, and M. Guerineau.** 1979. High frequency of yeast transformation by plasmids carrying part or entire 2-μm yeast plasmid. *Gene* **5**:233–253.

18. **Gerbaud, C., and M. Guerineau.** 1980. 2 micron plasmid copy number in different yeast strains and repartition of endogenous and 2 micron chimeric plasmids in transformed strains. *Curr. Genet.* **1**:219–228.

19. **Govind, N. S., and M. Jayaram.** 1987. Rapid localization and characterization of random mutations within the 2μ circle site-specific recombinase: a general strategy for analysis of protein function. *Gene* **51**:31–41.

20. **Gronastajski, R. M., and P. D. Sadowski.** 1985. The FLP protein of the 2 micron plasmid of yeast. Inter- and intramolecular reactions. *J. Biol. Chem.* **260**:12320–12327.

21. **Gronastajski, R. M., and P. D. Sadowski.** 1985. The FLP recombinase of the yeast 2 micron plasmid attaches covalently to DNA via a phosphotyrosyl linkage. *Mol. Cell. Biol.* **5**: 3274–3279.

22. **Guerineau, M., C. Grandchamp, J. Paoletti, and P. Slonimski.** 1971. Characterization of a new class of circular DNA molecules in yeast. *Biochem. Biophys. Res. Commun.* **42**: 550–557.

23. **Guerineau, M., C. Grandchamp, and P. Slonimski.** 1976. Circular DNA of a yeast episome with 2 inverted repeats: structural analysis by a restriction enzyme and electron microscopy. *Proc. Natl. Acad. Sci. USA* **73**:3030–3034.

24. **Hartley, J. L., and J. E. Donelson.** 1980. Nucleotide sequence of the yeast plasmid. *Nature* (London) **286**:860–865.

25. **Hoess, R. H., and K. Abremski.** 1985. Mechanism of strand cleavage and exchange in the cre-lox site-specific recombination system. *J. Mol. Biol.* **181**:351–362.

26. **Hoess, R., A. Wierzbicki, and K. Abremski.** 1987. Isolation and characterization of intermediates in site-specific recombination. *Proc. Natl. Acad. Sci. USA* **84**:6840–6844.

27. **Hoess, R. H., A. Wierzbicki, and K. Abremski.** 1986. The role of the *loxP* spacer in P1 site-specific recombination. *Nucleic Acids Res.* **14**:2287–2300.

28. **Hollenberg, C. P., P. Borst, and E. J. F. van Bruggen.** 1970. Mitochondrial DNA. V. A 25-μ closed circular duplex DNA in wild-type yeast mitochondria. Structure and genetic complexity. *Biochim. Biophys. Acta* **209**:1–15.

29. **Hollenberg, G. P., A. Degelmann, B. Kustermann-Kuhn, and H. D. Royer.** 1976. Characterization of 2 μm DNA of *Saccharomyces cerevisiae* by restriction fragment analysis and integration in an *Escherichia coli* plasmid. *Proc. Natl. Acad. Sci. USA* **73**:2072–2076.

30. **Holm, C.** 1982. Clonal lethality caused by the yeast plasmid 2 μ DNA. *Cell* **29**:585–594.

31. **Holm, C.** 1982. Sensitivity to the yeast plasmid 2 μ DNA is conferred by the nuclear allele *nib*1. *Mol. Cell. Biol.* **2**: 985–992.

32. **Hübscher, V., H. Lutz, and A. Kornberg.** 1980. Novel histone H2A-like protein of *Escherichia coli*. *Proc. Natl. Acad. Sci. USA* **77**:5097–5101.

33. **Jarayam, M.** 1985. Two-micrometer circle site-specific recombination: the minimal substrate and possible role of flanking sequences. *Proc. Natl. Acad. Sci. USA* **82**:5875–5879.

34. **Jayaram, M., Y.-Y. Li, and J. R. Broach.** 1983. The yeast plasmid 2 micron circle encodes components required for its high copy propagation. *Cell* **34**:95–104.

35. **Jayaram, M., A. Sutton, and J. R. Broach.** 1985. Properties of REP3: a *cis*-acting locus required for stable propagation of the yeast plasmid 2 micron circle. *Mol. Cell. Biol.* **5**:2466–2475.

36. **Johnson, R. C., and M. I. Simon.** 1985. Hin-mediated site-specific recombination requires two 26 bp recombination sites and a 60 bp recombinational enhancer. *Cell* **41**:781–791.

37. **Kikuchi, Y.** 1983. Yeast plasmid requires a *cis*-acting locus and two plasmid proteins for its stable maintenance. *Cell* **35**:487–493.

38. **Kikuchi, Y., and H. A. Nash.** 1979. Nicking-closing activity associated with bacteriophage λ int gene product. *Proc. Natl. Acad. Sci. USA* **76**:3760–3764.

39. **Kilpatrick, M. W., A. Torri, D. S. Kang, J. A. Engler, and R. D. Wells.** 1986. Unusual structures in the adenovirus genome. *J. Biol. Chem.* **261**:11350–11354.

40. **Kitts, P. A., and H. A. Nash.** 1987. Homology-dependent interactions in phage λ site-specific recombination. *Nature* (London) **329**:346–348.

41. **Livingston, D. A., and D. A. Kupfer.** 1977. Control of *Saccharomyces cerevisiae* 2 micron DNA replication by cell division cycle genes that control nuclear DNA replication. *J. Mol. Biol.* **116**:249–260.

42. **McLeod, M., S. Craft, and J. R. Broach.** 1986. Identification of the crossover site during FLP-mediated recombination in the yeast plasmid 2 micron circle. *Mol. Cell. Biol.* **6**: 3357–3367.

43. **Meyer-Leon, L. C. A. Gates, J. M. Attwood, E. A. Wood, and M. M. Cox.** 1987. Purification of the FLP site-specific recombinase by affinity chromatography and re-examination of basic properties of the system. *Nucleic Acids Res.* **15**: 6469–6488.

43a. **Meyer-Leon, L., L.-C. Huang, S. W. Umlauf, M. M. Cox, and R. B. Inman.** 1988. Holliday intermediates and reaction by-products in FLP protein-promoted site-specific recombination. *Mol. Cell. Biol.* **8**:3784–3796.

44. **Meyer-Leon, L., J. F. Senecoff, R. C. Bruckner, and M. M. Cox.** 1984. Site-specific recombination promoted by the FLP protein of the yeast 2 micron plasmid in vitro. *Cold Spring Harbor Symp. Quant. Biol.* **49**:797–804.

45. **Murray, A. W., and J. W. Szostak.** 1983. Pedigree analysis of plasmid segregation in yeast. *Cell* **34**:961–970.

46. **Murray, J. A. H., and G. Cesarini.** 1986. Functional analysis of the yeast plasmid partition locus STB. *EMBO J.* **5**: 3391–3399.

47. **Murray, J. A. H., M. Scrapa, N. Rossi, and G. Cesarini.** 1987. Antagonistic controls regulate copy number of the yeast 2 μ plasmid. *EMBO J.* **6**:4205–4212.

48. **Nash, H. A., C. E. Bauer, and J. F. Gardner.** 1987. Role of homology in site-specific recombination of bacteriophage λ: evidence against joining of cohesive ends. *Proc. Natl. Acad. Sci. USA* **84**:4049–4053.

49. **Nunes-Düby, S. E., L. Matsumoto, and A. Landy.** 1987. Site-specific recombination intermediates trapped with suicide substrates. *Cell* **50**:779–788.

50. **Passananti, C., B. Cavies, M. Ford, and M. Fried.** 1987. Structure of an inverted duplication formed as a first step in a gene amplification event: implications for a model of gene amplification. *EMBO J.* **6**:1697–1703.

51. **Prasad, P. V., D. Horensky, L.-J. Young, and M. Jayaram.** 1986. Substrate recognition by the 2μm circle site-specific

recombinase: effect of mutations within the symmetry elements of the minimal substrate. *Mol. Cell. Biol.* 5:4329–4334.

52. **Prasad, P. V., L.-J. Young, and M. Jayaram.** 1987. Mutations in the 2-μm circle site-specific recombinase that abolish recombination without affecting substrate recognition. *Proc. Natl. Acad. Sci. USA* **84:**2189–2193.

53. **Pulleyblank, D. E., D. B. Haniford, and A. R. Morgan.** 1985. A structural basis for S1 nuclease sensitivity of double-stranded DNA. *Cell* **42:**271–280.

54. **Reynolds, A. E., A. W. Murray, and J. W. Szostak.** 1987. Roles of the 2μm gene products in stable maintenance of the 2 μm plasmid of *Saccharomyces cerevisiae. Mol. Cell. Biol.* 7:3566–3573.

55. **Royer, H., and C. P. Hollenberg.** 1977. *Saccharomyces cerevisiae* 2-μm DNA: an analysis of the monomer and its multimers by electron microscopy. *Mol. Gen. Genet.* **150:** 271–284.

56. **Sadowski, P.** 1986. Site-specific recombinases: changing partners and doing the twist. *J. Bacteriol.* **125:**341–347.

57. **Sadowksi, P. D., L. G. Beatty, D. Cleary, and S. Ollerhead.** 1987. Mechanisms of action of the FLP recombinase of the 2 micron plasmid of yeast, p. 691–701. *In* R. McMacken and T. Kelly (ed.), *DNA Replication and Recombination.* Alan R. Liss, Inc., New York.

58. **Senecoff, J. F., R. C. Bruckner, and M. M. Cox.** 1985. The FLP recombinase of the yeast 2-micron plasmid: characterization of its recombination site. *Proc. Natl. Acad. Sci. USA* **82:**7270–7274.

59. **Senecoff, J. F., and M. M. Cox.** 1986. Directionality in FLP protein-promoted site-specific recombination is mediated by DNA-DNA pairing. *J. Biol. Chem.* **261:**7380–7386.

59a. **Senecoff, J. F., P. J. Rossmeissl, and M. M. Cox.** 1988. DNA recognition by the FLP recombinase of the yeast 2 μ plasmid: a mutational analysis of the FLP binding site. *J. Mol. Biol.* **201:**405–421.

60. **Sinclair, J. H., B. J. Stevens, P. Sanghavi, and M. Rabinowitz.** 1967. Mitochondrial-satellite and circular DNA filaments in yeast. *Science* **156:**1234–1237.

61. **Stark, G. R., and G. M. Wahl.** 1984. Gene amplification. *Annu. Rev. Biochem.* **57:**447–491.

62. **Toh-e, A., I. Utatsu, A. Utsunomiya, S. Sakamoto, and T. Imura.** 1987. Two micron DNA-like plasmids from non-*Saccharomyces* yeasts, p. 425–437. *In* R. B. Wickner, A. Hinnebusch, A. M. Lambowitz, I. C. Gunsalus, and A. Hollaender (ed.), *Extrachromosomal Elements in Lower Eukaryotes.* Plenum Publishing Corp., New York.

62a. **Umlauf, S. W., and M. M. Cox.** 1988. The functional significance of DNA sequence structure in a site-specific genetic recombination reaction. *EMBO J.* **7:**1845–1852.

63. **Vetter, D., B. J. Andrews, L. Roberts-Beatty, and P. D. Sadowski.** 1983. Site-specific recombination of yeast 2 micron DNA in vitro. *Proc. Natl. Acad. Sci. USA* **80:**7284–7288.

64. **Volkert, F. C., and J. R. Broach.** 1986. Site-specific recombination promotes plasmid amplification in yeast. *Cell* **46:** 541–550.

65. **Volkert, F. C., L. C. Wu, P. A. Fisher, and J. R. Broach.** 1987. Survival strategies of the yeast plasmid 2 micron circle, p. 375–396. *In* R. B. Wickner, A. Hinnebusch, A. M. Lambowitz, I. C. Gunsalus, and A. Hollaender (ed.), *Extrachromosomal Elements in Lower Eukaryotes.* Plenum Publishing Corp., New York.

66. **Williamson, D. H.** 1985. The yeast ARS element, six years on: a progress report. *Yeast* **1:**1–14.

67. **Young, M., and J. Cullum.** 1987. A plausible mechanism for large-scale chromosomal DNA amplification in streptomycetes. *FEBS Lett.* **212:**10–14.

68. **Zakian, V. A., B. J. Brewer, and W. L. Fangman.** 1979. Replication of each copy of the yeast 2 micron DNA plasmid occurs during the S phase. *Cell* **17:**923–934.

Chapter 30

The Interconversion of Yeast Mating Type: *Saccharomyces cerevisiae* and *Schizosaccharomyces pombe*

AMAR J. S. KLAR

I. INTRODUCTION

The central issue in developmental biology is to determine how progeny of different cell types are generated by cells apparently containing the same genome. The main theme now emerging is that such changes are a result of altered and judicious gene expression such that specific sets of cell-type-specific functions are expressed at particular stages of development. But the key question to answer is: what causes these variations in genome utilization in a programmed fashion during the development of an organism? Among the model systems used to address such fundamental questions at the single-cell level, yeasts have been particularly valuable. This review will focus on the program of mating type interconversion which operates at the single-cell level of the distantly related yeasts *Saccharomyces cerevisiae* and *Schizosaccharomyces pombe*. The information gained with these simple systems is interesting and it is expected that this knowledge will help us understand strategies used for the development of multicellular organisms. Furthermore, comparison of the molecular details may be useful in determining how these

Amar J. S. Klar ■ Developmental Genetics Laboratory, Building 539, National Cancer Institute Frederick Facility, P.O. Box B, Frederick, Maryland 21701.

systems have evolved. Given the space limitation, I am forced to limit the scope of this review to key studies on mating type interconversion. For a more comprehensive picture, the reader is directed to several excellent reviews (24, 30, 41, 44, 73, 87, 92).

II. YEASTS SWITCH MATING TYPES IN A PROGRAMMED FASHION: CELL LINEAGE STUDIES

The most attractive feature of both yeasts is the observed asymmetry of cell division such that only one of two sister cells produces progeny with a changed cell type (Fig. 1). Such a production of nonequivalent sisters is the hallmark of eucaryotic differentiation. The two sexual cell types of *S. cerevisiae* are designated **a** and α, which are correspondingly conferred by the *MAT***a** and *MAT*α alleles of the mating type locus. In the so-called homothallic strains, the cells of each mating type interchange to the other type efficiently and regularly. This phenomenon was discovered by Winge (130, 131), who found that haploid spore cells isolated from diploid cells after meiosis and sporulation grew to establish diploid clones which repeat this process. The *MAT***a**/*MAT*α diploid cells cannot mate, but can go through meiosis and sporulation to form four haploid spores. Lindegren and Lindegren (80) isolated natural heterothallic isolates of *S. cerevisiae* that grew into haploid clones containing stable mating type, either **a** or α. The opposite mating type cells can mate to establish nonmating *MAT***a**/*MAT*α diploid cultures. The only difference between homothallic and heterothallic strains is their ability to efficiently switch mating type. The ability to switch is due to a gene called *HO* (for homothallism). Homothallic strains contain a functional *HO* gene, while heterothallic strains contain the inactive allele *ho* (104).

It was known earlier (D. C. Hawthorne, Proc. Int. Congr. Genet **1**:133, 1963, abstr. no. 88) that two zygotes can result from matings between the first four progeny of a homothallic *S. cerevisiae* spore. The systematic cell lineage studies were conducted by Hicks and Herskowitz (42) and Strathern and Herskowitz (115). They employed an interesting property of yeast cells: **a** cells stop their growth in G1 phase of the cell cycle in the presence of a pheromone called α factor. The factor is secreted by α cells and causes **a** cells to make a characteristic irregular "shmoo"-shaped cell (12). The α cells are not affected by the treatment. Similarly, **a** cells secrete a pheromone which arrests only α cells. Thus, switching of individual homothallic cells from one mating type to the other can be followed microscopically without actually mating them. Three rules for switching were thus determined:

(i) The pairs rule. Whenever a switch was found, both progeny of a cell were always switched.

(ii) The mother-daughter asymmetry rule. *S. cerevisiae* cells divide by asymmetric cell division (39). The new cell, called the bud, starts out as a small protrusion from the spherical mother cell which pinches off from the mother at cell division. Only the mother cell, dubbed the experienced cell, produces switched progeny. The newly born daughter cell does not until it has gone through a division, thereby graduating to the status of a mother cell.

(iii) The preference rule. Nearly 73% of mother cells switch, and this switch is primarily or exclusively to the opposite mating type. Thus, a mechanism exists to regularly alternate mating type.

S. pombe is also homothallic; a clone of a single spore contains cells of both mating types such that about 90% of cells form zygotes which undergo meiosis and sporulation (16, 78). This is also due to

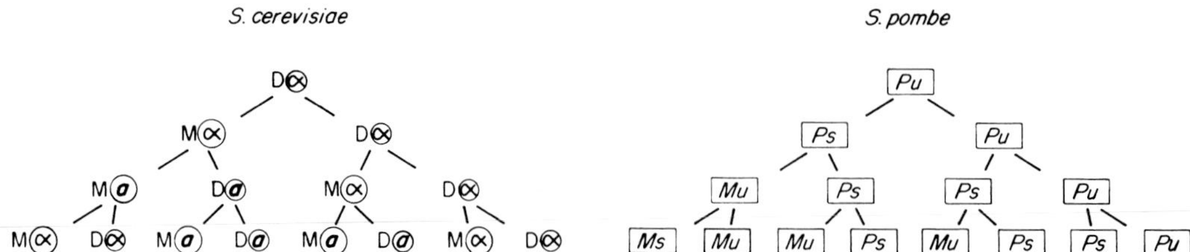

Figure 1. Idealized pattern of switching cell type in cell lineages of *S. cerevisiae* and *S. pombe*. The just-born daughter cells (D) of the budding yeast *S. cerevisiae* are slightly smaller than their mother cells (M), and this is reflected in the figure. The cells of the fission yeast *S. pombe* are rod shaped and divide to produce nearly equal-sized daughter cells. *Pu* and *Mu* are unswitchable cells and *Ps* and *Ms* will produce switched progeny. See text for details. (Data from references 22, 42, 85, and 115, and unpublished observations.)

mating type switching, but in a pattern that is fundamentally different from that of *S. cerevisiae*. The cells of this yeast are rod shaped and divide by fission to produce essentially equal-sized daughter cells. The two cell types, called *P* (for plus) and *M* (for minus), are conferred by the alternate alleles of the *mat1* locus, *mat1-P* and *mat1-M*, respectively. Only recently has the pattern of switching in an *S. pombe* cell lineage been worked out (85). Two aspects of *S. pombe* cell biology made it difficult to determine the pattern of switching at the single-cell level. First, unlike *S. cerevisiae*, the *S. pombe* cells mate only after nutritional starvation, especially for nitrogen (28), making it cumbersome to grow and test mating type on two different media. Second, no diffusible factor (analogous to the α factor discussed above) was known which could be used to determine cell type without actually mating cells to tester cells of known mating type. Such factors have been defined recently (79, 132) but are not useful for cell lineage studies because of a considerable insensitivity of cells to these factors.

Miyata and Miyata (85) defined the pattern shown in Fig. 1 by letting the progeny of a single cell grow on nitrogen-limited starvation media. When the four grandchildren cells, generated by two consecutive divisions from an original cell, were watched microscopically, it was found that in 72 to 94% of cell divisions only two cells mated to form one zygote. The balance presumably failed to switch. The single zygote proceeded to form an ascus containing four meiotic products. In other words, in this so-called Miyata rule, only one of the four progeny from a single cell switches mating type. The remaining cells will later switch following the same rule of one-in-four related cells.

Thus, the pattern of switching is quite different in the two yeasts. The cell fate is determined by cell-autonomous signals, which we wish to understand. Although the pattern of switching is quite different, the overall strategy of change is quite analogous: switches occur by genome rearrangement (see below). Despite many differences in molecular details, both systems can be treated equivalently in that a cell division produces two nonequivalent daughter cells, only one of which is competent to produce switched progeny. The main difference is that in *S. cerevisiae* both daughters of the mother cell are switched, while in *S. pombe* only one daughter of a pair of sister cells is switched (Fig. 1).

The major recent excitement in this field has involved the attempt to understand molecular details of the decision-making events which generate nonequivalent pairs of sister cells. I will first discuss the relevant studies of *S. cerevisiae* and follow it by those with *S. pombe*.

III. STUDIES OF *S. CEREVISIAE*

A. The *S. cerevisiae MAT* Switches by DNA Transposition

1. Discovery of *HO HM*a (*HML*α) and *HM*α (*HMR*a) loci

Figure 2 summarizes a vast amount of work showing that *MAT*a-to-*MAT*α interconversion occurs by a transposition-substitution event in which a copy of either *HML*α or *HMR*a is transmitted to the *MAT* locus, resulting in a mating type switch. *HML* usually contains unexpressed *MAT*α information, while *HMR* usually contains unexpressed *MAT*a information. The main conclusions were derived from genetic studies and subsequently confirmed by molecular approaches and are summarized below.

Takahashi (121) first proposed the existence of three loci, now designated *HO*, *HML*, and *HMR*, which are required for switching in both directions (a⇌α). Takano and Oshima (123) described a naturally occurring diploid which after meiosis segregated two **a**:two homothallic (**a**/α) spores in each tetrad. That is, two **a** spore clones did not switch. This analysis defined the locus called *HM*a, and the "mutation" was designated *hm*a, in contrast to the wild type, designated *HM*a. Thus, *HM*a is required for switching **a** to α. Likewise, Santa Maria and Vidal (106) described a strain which produced two α:two homothallic (**a**/α) segregants in each ascus. The locus defined by this analysis was designated *HM*α, and the mutation was designated *hm*α. Thus, *HM*α is required for switching α to **a**.

Thus, *HO HM*a *HM*α cells switched between **a** and α, *HO HM*a *hm*α only switched **a** to α, and *HO hm*a *HM*α only switched α to **a**. By crossing strains of the latter two types, Naumov and Tolstorukov (96) constructed strains of genotype *HO hm*a *hm*α and must have been surprised to find that they were not heterothallic. Instead, they switched in both directions, similar to strains of genotype *HO HM*a *HM*α. As a way out of this interesting paradox, they suggested that the *hm*a allele may be functionally equivalent to *HM*α, and the *hm*α may be functionally equivalent to *HM*a. Klar and Fogel (59) provided the first evidence that *hm*a was not simply a recessive defect in *HM*a, because the diploid *MAT*α/*MAT*α colony sectors generated by mitotic crossing over of a strain of genotype *HM*a/*hm*a *MAT*α/*MAT*a *hm*α/*hm*α *HO*/*ho* could give rise to **a**/**a** or **a**/α diploid cells.

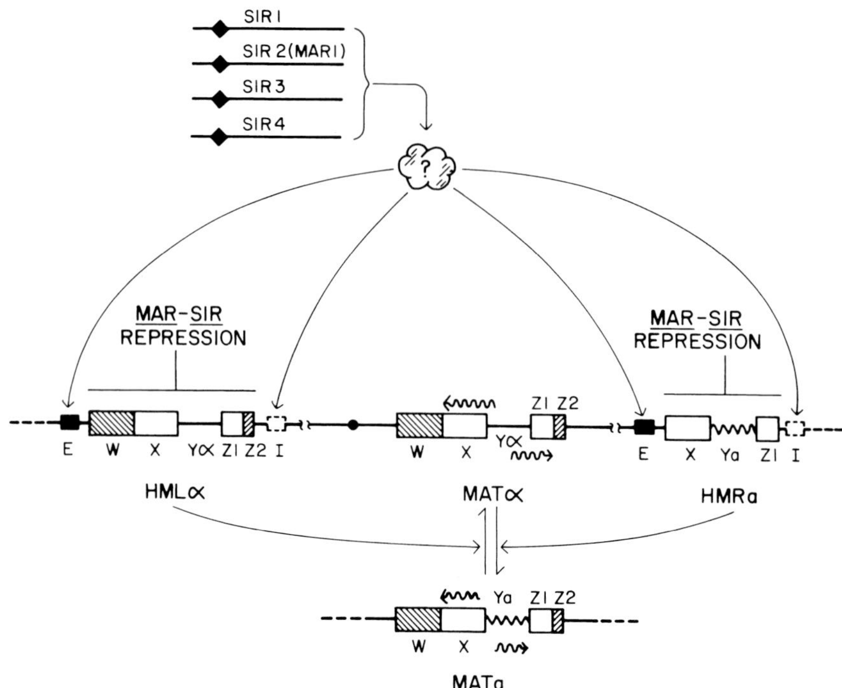

Figure 2. Arrangement of *MAT*, *HML*, and *HMR* elements in chromosome III of *S. cerevisiae*. The interchange of *MAT* alleles occurs by a gene conversion event which replaced the allele-specific *MAT* Y region (Ya is 642 bp, Yα is 747 bp in extent) with that derived from *HML* or *HMR*. The *MAT*-encoded transcripts are indicated by wavy arrows. The *HML* and *HMR* cassettes are kept unexpressed by the *trans*-acting *MAR/SIR* genes and the indicated *cis*-acting E and I sites (for review of the silent gene regulation, see reference 92). The *HML* is about 180 kb to the left of *MAT*, while *HMR* is about 120 kb to the right of *MAT* (37, 66, 118). The solid dot indicates centromere. The boxes W, X, Z1, and Z2 (723, 704, 239, and 88 bp in extent, respectively) represent regions of homology shared by the mating type loci (3, 46, 94, 95, 119). (Figure derived from reference 73).

Note that this strain did not have *HM*α, yet it switched from α to **a**. Thus, *hm***a** must have performed the function of *HM*α, i.e., *hm***a** can promote switching of α to **a**. Subsequent demonstration of codominance of *HM***a** and *hm***a** and of *HM*α and *hm*α came from studies with the diploids constructed either by protoplast fusion or as segregants of tetraploid strains (2, 38). Thus, each *HM* locus can exist in either of two alternate allelic forms. The two loci were renamed to reflect the new understanding. Since the *HM***a** locus maps to the left arm of chromosome III (66, 76), it was named *HML*α (*L* for the left arm of chromosome III; α because it can switch mating type to α) and its alternate allele *hm***a** (which can switch mating type to **a**) was renamed *HML***a**. Similarly, since *HM*α maps to the right arm of chromosome III (37, 66), it was renamed *HMR***a** (*R* for the right arm of chromosome III) and *hm*α was renamed *HMR*α. In conclusion, the *HML***a** or *HMR***a** allele is required for switching from *MAT*α to *MAT***a**, whereas *HMR*α or *HML*α is required for switching from *MAT***a** to *MAT*α. *MAT* maps in the middle of chromosome III, and *HML* is about 180 kilobases (kb) to the left and *HMR* is about 120 kb to the right of *MAT* (37, 66, 118).

2. Genetic tests support the transposition model

Two kinds of models have been considered for mating type switching. Hawthorne (Abstr. Proc. 11th Int. Congr. Genet. 1963) suggested that the *HO* gene "mutates" the *MAT* locus in a directed fashion between the alternative *Mat***a** and *MAT*α alleles. It was imagined that some DNA modification event allowed expression of either *MAT***a** or *MAT*α at the same locus (D. C. Hawthorne, quoted in reference 49). In a specific model of this kind, the flip-flop model, it was presumed that **a** and α determinants shared a regulatory site, say a promoter region, which could be inverted by an intrachromosomal recombination event resulting in alternate expression of *MAT* alleles (11; J. B. Hicks, Ph.D. thesis, University of Oregon, Eugene, 1975). Oshima and Takano (98; see also reference 36) originally proposed the transposition-insertion model, which was based on DNA transposition. They proposed that *HML* and *HMR* code for specific kinds of allele-specific transposable "controlling elements" which attach to the *MAT* locus and thus differentiate it into *MAT***a** or *MAT*α. This proposal was motivated by the work of McClintock (83), who discovered transposable elements which can control the activity of a nearby gene in maize. It was proposed that the *HO* gene controls the removal and insertion of that element. This bold proposal was originally not based on any specific new data, but supporting evidence came from the interesting finding that a naturally occurring allele called *MATα-inc* (*inc* for inconvertible) switched inefficiently (1 in 1,000 cells) to *MAT***a**. That new *MAT***a** allele, when switched, generated the wild-type and highly switchable *MATα* allele. The interpretation by Takano et al. (122) of their result was that the *MATα-inc* controlling element was defective in some function, reducing its ability to be displaced from *MAT*. Once removed, the efficiently switchable *MATα* allele would be derived from the wild-type *HMLα* determinant.

The remarkable property of healing *MAT* mutations by switching was systematically explored by D. C. Hawthorne (quoted in reference 43) and by Hicks and Herskowitz (43). D. C. Hawthorne found that an ochre mutation of *MATα* (*matα-ochre*) could be switched to *MAT***a** and then to "cured" wild-type *MATα*. Hicks and Herskowitz (43) were pleased to find that sterile spores with a *matα1-2* or *matα1-5 HO* genotype could efficiently generate α mating cells within as few as three generations after spore germination. That is, the *matα* mutation was changed to wild-type *MATα*. Subsequently, similar healing by switching was observed with several *mat***a** and other *matα* mutations (62, 114). These observations lead to a specific modification of the controlling element model, the cassette model by Hicks et al. (45). This model proposed that *HML* and *HMR* carry unexpressed α and **a** "cassettes" which could be activated by transposing their copies to the *MAT* locus, resulting in a switch. It was suggested that perhaps the *HM* loci were unexpressed because they lacked promoter elements. Both genetic and physical studies have verified the key features of the transposition model and have discovered an interesting regulatory system which regulates *HML* and *HMR* cassettes. I will use the controlling element and the cassette designations interchangeably throughout this chapter.

Two important genetic predictions of the transposition models were that at least a component of the *MAT* locus should be present at the *HM* loci, and more importantly, that strains bearing mutations in the *HM* loci should transmit those mutations to *MAT* by switching. Both these predictions were shown to hold true by characterizing a spontaneous mutation in the *MAR1* (mating type regulator) locus (61). Because haploid strains with the genotype *HMLα MATα* (or *MAT***a**) *HMR***a** *mar1 ho* exhibit sterility and undergo aberrant sporulation, two phenotypes partly similar to those of *MAT***a**/*MATα* diploids, it was suggested that both **a** and α mating type information is expressed in *mar1* mutants. Thus, the model proposed was that the *MAR1* locus acts as a negative regulator of the hypothesized *MAT* information resident at both *HM* loci. Important evidence supporting this hypothesis was the observation that *HML***a** *MAT***a** *HMR***a** *mar1 ho* strains are not sterile and mate as efficiently as wild-type **a** cells and *HMLα MATα HMRα mar1 ho* mate as αs (61). In other words, the sterility of *mar1* strains can be suppressed by alternate alleles of *HM* loci, and sterility is not dependent on a functional *HO* locus. Sterility can be suppressed also by mutations in *MAT* and *HM* loci. For example, *HMLα mat***a** *hmr***a** *mar1 ho* cells are phenotypically α and the *HML***a** *mat***a** *hmr***a** *mar1 ho* are functional **a** cells. A cross between these strains generated 2α:2**a** products in each ascus, and interestingly, mating type mapped to *HML*. Thus complete *MAT* information must exist at *HML*, which can be expressed in *mar1* mutants (61, 66).

Other mutations with a phenotype equivalent to that of *mar1* have been isolated. The mutation called *ssp1* was previously selected as a suppressor of the sterile phenotype of *matα*$^-$ mutation (Hicks, Ph.D. thesis); the role of *SSP1* as a negative transcriptional regulator of *HM* loci was proposed later, independent of the observations with the *mar1* mutant (103). Based particularly on findings with the *mar1* mutation, another mutation, *cmt1*, which also conferred sterility on haploid cells (32), was interpreted to define another locus, now called *MAR2* or *SIR3* (silent information regulator) (101). A systematic study to characterize such negative regulators employed strains of genotype *HMLα mat***a** *HMRα ho*. Cells of such a strain exhibit **a** mating, but a mutation allowing expression of either *HMLα* or *HMRα* or of both will exhibit an α phenotype. Four such negative regulators, designated *SIR1* (*SSP1*), *MAR1* (*SIR2*), *MAR2* (*CMT1*, *SIR3*), and *SIR4*, have been described (50, 51, 101) (Fig. 2).

The second genetic prediction of the transposition model was verified by demonstrating that mutations from the *HM* loci could be transmitted to *MAT* by mating type interconversion. The first such experiment employed a *mar1* mutant strain. Cells of genotype *HMLα MATα HMR***a** *mar1 ho* are sterile because of the simultaneous expression of **a** and α cassettes from *MAT* and *HM* loci. A scheme to select mutagenized cells which mate as α haploids generated mutations at the *HMR***a** locus, the *hmr***a***-1* mutation (61). After incorporating the *HO* gene by genetic crosses, it was found that a spore of genotype *HMLα MAT***a** *hmr***a***-1 HO* changed first to *MATα* and then, during the next switching cycle, only switched to mutant *mat***a** (60). This strain was a promiscuous mater, since it kept on switching continuously between *mat*a and *MAT*α alleles. Such an experiment was also conducted by employing an ethyl methanesulfonate-induced mutation at *HML*α, called *hml*α*-66*. Cells of this strain alternated between **a** and sterile (*mat*α) cell types (77). These so-called "wounding" experiment studies were also done with strains containing defined mutations at *HML* and *HMR*. A particularly satisfying observation was that cells carrying the *hmr***a**-nonsense or *hml***a**-nonsense mutations gave rise only to *mat***a**-nonsense alleles (55, 63).

Another interesting and independent line of reasoning forming the basis of the cassette model suggested a molecular explanation for recessive lethal genetic rearrangements involving chromosome III (40). These were demonstrated to result from fusions between *MAT* and *HMR* cassettes, which were associated with a large deletion of the interval between these loci (the Hawthorne deletion), and between *MAT* and *HML*, generating a circular chromosome (the Strathern circle). Southern (109) analysis of DNA showed that these events generate recombinant cassettes which activate the *HM* locus information (35, 55, 63, 119). Notably, large circular DNA molecules (Strathern circles) were in fact observed by electron microscopy, supporting the transposition model (118). This analysis demonstrated that all three cassettes are situated as directly oriented elements on chromosome III as drawn in Fig. 2.

3. Molecular confirmation of the transposition model

By employing the then recently established DNA-mediated transformation procedures (8, 47), two groups isolated *MAT* clones by complementation of the mutant *mat* loci of heterothallic strains. In one study, mutant *mat***a** and in another, mutant *mat*α*-ochre* was used as the recipient for DNA transformation. It was found that a plasmid-borne clone of *HML*α or *MAT*α can confer an α phenotype to the cell. When those clones were used as molecular probes, three *Hin*dIII restriction fragments in the total *S. cerevisiae* genome were found to have homology to the probe. The *MAT*α-containing fragment was found to be approximately 100 bp larger than that containing the *MAT***a** allele. Analogous size differences were found in the alternate α and **a** versions of *HML* and *HMR*, thus confirming genetic conclusions and establishing the transposition model (46, 94, 95, 119). As shown in Fig. 2, the allele-specific Y regions are flanked by homology boxes W, X, Z1, and Z2 (3, 46, 94, 95, 119).

B. The Mechanism of *MAT* Interconversion

1. A double-stranded break initiates *MAT* gene conversion

To visualize intermediates in the switching process, it was necessary to identify conditions under which a population of continuously switching cells could be grown. Most homothallic strains have the genotype *HO HML*α *HMR***a**. In these strains, haploid **a** cells efficiently switch to α, and α cells efficiently switch to **a**. Subsequently, sibling cells of opposite mating type mate to form **a**/α cells, in which switching stops (40, 98). In contrast, *HO HML***a** *MAT***a** *HMR***a** cells appear to be stable clones of **a** cells but, in reality, are continuously switching from **a** to **a**. The existence of these homologous switches can be established by employing genetically marked *HM* donors (52, 64). Since no **a**/α diploids are formed, the interconversion of mating type continues. Thus, these strains can be maintained as a population of cells undergoing homologous switching continuously. Similarly, *HO HML*α *MAT*α *HMR*α cells can be maintained as a population undergoing homologous α-to-α switches.

When DNA from heterothallic (*ho*) *S. cerevisiae* was subjected to restriction endonuclease cleavage and Southern analysis, DNA fragments corresponding to *HML*, *MAT*, and *HMR* were identified (46, 94). However, when DNA from the exponentially growing continuously switching cells discussed above was examined, two additional, smaller fragments homologous to the *MAT* probe were detected (117). The combined size of the additional fragments was equal to the fragment containing intact *MAT*. Thus, the *MAT* locus was being cut in vivo. These fragments were present in about 2% of the molar equivalence of a single-copy gene in DNA isolated from an exponentially growing culture. The position of the cut is at the function of the allele-specific Y region and the homology box Z1 (Fig. 2).

In order to determine whether the cut at *MAT* is a step that was dependent on the presence of donor *HM* loci, strains deleted for both *HM* loci were analyzed. The cut was found even in the absence of the donor loci, indicating that the cut is probably an initiation step and not a step in resolving the recombination intermediate (70, 117). Even more gratifying was the observation that in donor-deleted strains, the mother cells grew to produce a pair of inviable cells while the daughters never produced lethal progeny. When the daughter cell became an experienced cell after one cell cycle, it produced lethal progeny. In this so-called pedigree of death, the same rules of switching as described for wild-type strains were observed. Instead of switched cells, however, inviable progeny were generated. This result was interpreted to mean that each and every cell switches by a double-stranded break at *MAT*. Such a break is deadly in donor-deleted cells, presumably because the break cannot be healed in the absence of switching (70).

In exponentially growing cells, only about 2% of *MAT* DNA is cut. Yet the pedigree-of-death observation suggested that each switching-competent cell experiences the double-stranded break. The observed low level of break, therefore, must be because the break is short lived and is quickly repaired by a switching event. It was previously observed that radiation-sensitive *rad52* strains die in the presence of the *HO* gene, providing circumstantial evidence that switching may be initiated by a double-stranded break (81, 126), as *rad52* mutants are presumed to be defective in double-strand-break repair (100, 120).

The *HO* strains contain an activity which can cleave *MAT* and *HM* DNA in vitro at the Y/Z junction. The cleavage in vitro occurs three bases into Z1, with four-base 3′ extensions with a cleavage within the A/T A/T C/G T/A sequence terminating in 3′ hydroxyl groups (76). A sequence of more than 16 nucleotides is required for efficient cleavage (97). Primer extension studies have shown that the nature of the cut made in vivo is the same as those found in vitro (D. Raveh and J. N. Strathern, personal communication). Further, the *HO* gene itself codes for this "Y/Z endonuclease" activity, since transformants of *Escherichia coli* containing the cloned *HO* gene express this activity (75). Based on the *HO* DNA sequence, it is predicted that the *HO* protein has multiple potential zinc-binding fingers which are thought to be protein motifs commonly shared by proteins which interact with DNA (104).

Several mutants isolated either as survivors from the pedigree of death or from *HO rad52* strains confirmed the requirement of a Y/Z junction for switching (70, 128). Both single base substitutions and small deletions can block the *MAT* cleavage, resulting in inconvertible *MAT* loci. It is generally accepted that this double-stranded break initiates the gene conversion event in a mechanism analogous to that suggested for other loci (100, 120). Several models incorporating this feature have been proposed (30, 113, 117).

Recent studies by two groups (14, 113) have followed the fate of the cut ends by physical means. A construct in which the *HO* gene is controlled by the *GAL* promoter was used to induce the endonuclease gene with galactose. When *MAT*a cells containing *Gal-HO* were blocked in G1 by α-factor treatment and then induced by galactose, a large proportion of cells accumulated a double-stranded break at *MAT* within 1 h of induction. Surprisingly, completion of the switches to *MAT*α, as assayed by Southern analysis, took another hour after the break was generated. The switches occurred both when the cells were kept blocked in G1 and when they were allowed to enter S period. The fact that cells can switch in G1 implies that resolution of the gene conversion intermediate can occur in the absence of normal DNA replication; apparently repair synthesis can accomplish the job. The longevity of cut ends seems to be donor dependent; in donor-deleted strains, the ends were short lived, perhaps because of degradation. Also, primer extension analysis has shown that the cut ends are not protected by any mechanism (Raveh and Strathern, personal communication). In standard strains with a normally controlled *HO* gene, the *MAT* switching occurs very fast, as indicated by the fact that the double-stranded break is short lived. It is not clear why the break resulting from *GAL-HO* induction is processed so inefficiently. Possibly the cells are deficient in a function required for efficient utilization of the break. By this model, the standard *HO*-containing strains must have that rate-limiting function already induced in both mother and daughter cells. Clearly, following the fate of cut chromosomes by physical means promises to be very useful approach in understanding molecular details of *MAT* switching.

Recombination required for mating type switching is initiated by the double-stranded break at *MAT* in mitotically growing **a** and α cells but not in **a**/α cells. This is because the *HO* gene is shut off by the **a**/α genotype (53). Most recombination studies in *S. cerevisiae*, however, have been conducted with meiotic cells because they experience a high rate of recombination (27). It was not known how meiotic recombination was initiated. To determine whether the double-stranded break at *MAT* could initiate meiotic *MAT* gene conversion, the cut was introduced in cells undergoing meiosis by employing a galactose-inducible *GAL-HO* construct. In such diploid cells, where *MAT*-to-*MAT* transfer was followed, *MAT* converted in about 14% of asci (74). Also, these events were associated with recombination of flanking markers, a feature that is a hallmark of most of meiotic gene conversion events (27). Furthermore, when the Y/Z cut site was placed at other loci, the *HO*-promoted conversion was also found in both meiotic and mitotic cell division (74, 97). Thus, one can conclude that the double-stranded break found at *MAT* can act as a general and efficient initiator event in gene conversion both in mitosis and in meiosis.

2. How much DNA is transferred?

It is generally believed that most recombination in yeasts occurs between homologous sequences. It is presumed that shared sequences at W, X, Z1, and Z2 regions (Fig. 2) help align the cassettes during the gene conversion process, which mostly transfers heterologous allele-specific Ya or Yα regions. During a switch to the opposite allele, the entire Y region has to be transmitted as such, since the transfer of DNA has to start and terminate in homology boxes. To determine whether a part of the Y region can ever be transmitted, in my laboratory we tested **a**-to-**a** switching in two strains, one containing *mat***a***1-1*, *hmr***a***1-2* and the other *mat***a***1-2*, *hmr***a***1-1* mutant loci combinations. The **a***1-1* and **a***1-2* are two different mutations in Y**a** (61). Nearly 7% of switches generated *MAT***a**$^+$ recombinant loci. This result led to the suggestion that in homologous switches, a heteroduplex at the recipient *MAT* locus is generated and that mismatch repair of such a structure generated the *MAT***a**$^+$ allele. Thus, it was concluded that the entire Y region need not always transpose as a complete (discrete) cassette during a homologous controlling element replacement and that a heteroduplex is generated at *MAT* by transferring a single strand from the donor (67).

Further support for the lack of unique initiation and resolution sites and the presence of a heteroduplex at *MAT* came from studies employing cassettes containing sequence polymorphisms in Z1 and X regions. A single base deletion nine bases to the left of the X/Y boundary was transmitted in 99% of the switches (124). Systematic study of several linker insertion mutations creating restriction site polymorphisms throughout the X and Z1 regions of *MAT*α demonstrated that switches to *MAT***a** are sometimes but not always accompanied by the loss of the restriction site (113). Two interesting points are worth noting. There was a gradient of coconversion such that sites close to Y are transferred at higher frequencies than those away from it. Second, a single pair of sister switched cells may contain nonidentical alleles, one retaining the polymorphism and the other lacking it. Similarly, interaction between *MAT*α and *hmr*α::*LEU2* initiated by inducing *GAL-HO* may generate a pair of sister cells, one containing *LEU2* at *MAT* and the other lacking it (B. Connolly and J. E. Haber, personal communication). In another relevant study, *MAT* has been shown to switch using a donor that was deleted for about 100 bp immediately adjacent to the *HO* endonuclease cleavage site (127). All these results suggest that after the initial break at the Y/Z junction, some exonuclease degradation occurs to expose homologous sequences; these then interact with the donor locus, which ultimately transfers a single strand to the recipient and thus generates a heteroduplex at *MAT*. Mismatch repair should then generate the switched allele. In heterologous switching the large heterology must always be corrected in favor of the donated information. Since the mutations at *MAT* can only rarely be transmitted to *HM* loci, one can assume that the corresponding heteroduplex at the *HM* loci is not formed, or else, if it is formed, it must always be corrected to maintain the original content of the *HM* loci. Alternatively, particularly during switches to the opposite allele, degradation of both strands at the *MAT* Y region occurs. The gapped *MAT* locus then interacts with the donor *HM* locus, resulting in a gene conversion event by the double-stranded-break repair model (100, 120). Both types of models, correction of the presumed heteroduplex at *MAT* and the gap repair, have been entertained (30, 33, 67, 113, 117).

Although *MAT* switching is presumed to occur by the process of gene conversion, one important feature of the lack of association of this conversion with recombination of flanking markers is particularly noteworthy and is not yet explained. In yeasts, most conversion events are associated with the recombination of flanking markers (27). Yeast mating type switching is an intrachromosomal event (33), a feature which may restrict the way in which the presumed Holliday junctions (48) are resolved so as to preclude recombination of flanking markers. Such recombination would generate large deletions and thereby cause lethality. In contrast, when *HO*-promoted gene conversion of *MAT* is achieved in diploid strains where donor and recipient *MAT* loci are situated in homologous chromosomes (interchromosomal event), these events are associated with recombination of flanking markers both in mitosis and in meiosis (69, 74). Why the structure is resolved differently in intra- and interchromosomal events remains an interesting issue open to further experimentation.

3. The mother-daughter switching asymmetry

From the foregoing discussion, it is clear that *MAT* switching most probably occurs in the G1 period and then the newly switched allele is replicated and segregated to both daughter cells of a mother cell. The rules of switching (Fig. 1) are dictated by the pattern of *HO* gene expression. Several recent studies have established that *HO* transcription is highly regulated by combinatorial controls (Fig. 3). At least 1,400 bp of sequence upstream of the translation site is required for proper regulation of *HO* (89). Three kinds of controls are operative: (i) *HO* is expressed in the mother but not in the

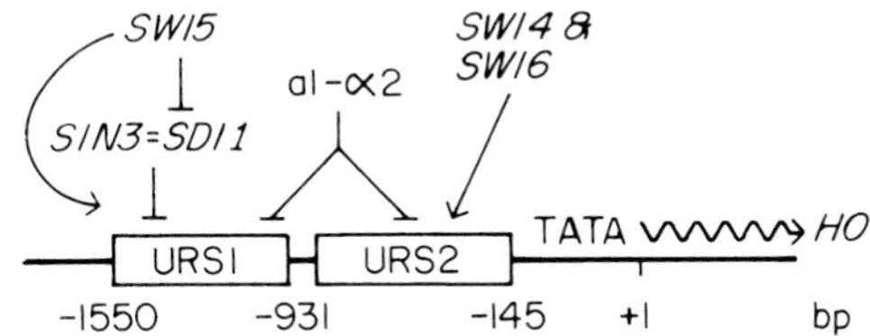

Figure 3. Summary of *HO* gene controls. Arrows indicate formally positive regulatory control and blocked lines indicate formally negative regulatory control. The mother-daughter controls (e.g., *SWI5*) act through the URS1 region, cell cycle control (*SWI4* + *SWI6*) operates through the URS2 region, and the cell type **a**/α control operates by interacting with both URS1 and URS2 elements. The wavy arrow indicates the *HO* transcript. (Figure modified from reference 111.)

bud cell, the mother-daughter control; (ii) *HO* is expressed in **a** and α but not in **a**/α cells, the cell type control; and (iii) *HO* is expressed only late in G1, the cell cycle control.

The region from −1,000 to −1,400 bp, called URS1, is required for the mother-cell-specific expression, while the region from −150 to −900 bp, called URS2, confers cell cycle control. These elements individually can confer specific controls on heterologous test promoters (84, 89). The **a**/α control (116) is thought to be exerted through the 10 copies of a repeated sequence motif present in both URS1 and URS2 elements (89, 104). This sequence is also found upstream of other genes which are similarly repressed by the **a**/α control (25).

In addition to the *cis*-acting sites defined by deletion analysis of the *HO* promoter, at least six *trans*-acting functions encoded by *SWI* (switch) genes are required for proper *HO* regulation. The *swi* mutants fail to express *HO* and therefore fail to switch mating type (10, 31, 110, 111). The specific roles of each of these genes are not fully understood and are being investigated. However, several interesting features have come to light.

Perhaps the most puzzling and fascinating feature is the way in which expression of *HO* is restricted to mother cells (88). It is clear that the control is exerted through URSI and the *SWI5* gene. The key result is that the replacement of the URS1 with the galactose-inducible *GAL* promoter makes *HO* expression galactose inducible, independent of *SWI5*, and relieves switching from the mother-daughter control (52, 90). Mating type switching in this construct is still regulated by cell-cycle and cell-type controls and all other *SWI* genes are still required for switching. When *GAL-HO* is expressed in G1 of both mothers and daughters, then both mothers and daughters switch their mating types. This suggests that daughters must be missing only the URS1 activation mechanism (possibly *SWI5* alone). In other words, in standard strains, all other *SWI* genes must be operative in daughter cells even though *HO* is not transcribed (90).

Both genetic and molecular studies have been undertaken to determine whether the *SWI5* activator operates by directly interacting with the *HO* promoter or by antagonizing a negative regulator of *HO*. An *HO*–β-galactosidase fusion is not expressed in *swi5* mutants because of the lack of *SWI5* gene function. Mutagenized cells of such a strain were screened for the expression of β-galactosidase by looking for blue colonies, using chromogenic X-Gal indicator plates (93, 111). Extragenic suppressors of *swi1* and *swi4* mutations were isolated similarly. These studies identified at least six *SIN* (switch independent) or *SDI* (*swi*-dependent interconversion) loci which formally act as negative control elements for *HO* expression. Several of them suppress mutations in more than one *swi* gene. Among these, only *SIN3* (which is *SDI1*) has been further characterized. It is now believed that at least one function of the *SWI5* product is to antagonize the *SIN3* negative control. Nasmyth et al. (91) suggest that this may occur by directly binding the *SWI5* protein to the *HO* promoter, thus precluding the *SIN3* binding. The *SWI5* gene contains three tandem copies of zinc-finger binding motifs. Furthermore, this binding activity was found in extracts of cells overexpressing *SWI5* and shown to bind at positions at −1,260 and −1,320 bp in URS1 (112). The puzzling observation is that when this *SWI5*-binding site is deleted, *HO* transcription is still *SWI5* dependent. It is suggested that perhaps *SW15* operates by interacting at some other, as-yet-unidentified sites in the *HO* promoter. An activity in addition to *SW15* binds to the URSI region at −1,300 and −1,320 bp and is present in *SIN3* but not in *sin3* mutant cells; this activity is not affected by the *SWI5* genotype (112).

The URS2 region contains 10 copies of an octanucleotide sequence (CACGAAAA) that confers cell cycle regulation on test promoters (10). At least *SWI4* and *SWI6*, and possibly other *SWI* genes, are required for this control. Recent work from Herskowitz's group (B. Andrews, W. Kruger, and I. Herskowitz, personal communication) suggests that URS2 is a "latent" upstream activating sequence which is kept silent by *SWI1* and *SIN1* proteins by a negative regulation. They showed that the upstream activating sequence activity requires *SWI4* protein. Their view is that URS1 activity relieves the repression of URS2. Furthermore, they suggest that the only thing *SWI5* protein does is to antagonize *SIN3* function. Their unpublished evidence is that the *HO* expression level by a *swi5 sin3* double mutant is the same as that of a wild-type strain.

Although the interplay of negative and positive controls operative at the *HO* gene is not fully understood, one thing is clear: there is a combinatorial control which senses many aspects of the *S. cerevisiae* cell cycle. These studies suggest that *SWI5* and *SIN3/SDI1* functions play a key role in determining the mother-daughter asymmetry. It is suggested that the *SWI5* product is somehow asymmetrically distributed to the mother but not to the daughter cell. It is now known that *SWI5* is expressed in the latter part of both mother and daughter cell cycles, long after the time of *HO* transcription (91). Thus, the *SWI5* product expressed late in the cell cycle will be used to elicit *HO* expression in the next cell cycle.

Several models have been considered for asymmetric distribution of the *SWI5* protein to mother and daughter cells. The asymmetric partitioning could be the consequence of an inherently asymmetric cell division: at the time of cell separation, the daughter cell is smaller in size than the mother cell (39). Consequently, there may be a greater decay of *SWI5* in daughter cells, since they have a longer G1 period than do mother cells (93, 111). Because wild-type cells change their phenotype nearly every generation, many cell-type-specific proteins of the parent cell must be inactivated, and newly changed cell-type-specific proteins must be synthesized within the time span of one generation. Therefore, an efficient inactivation (degradation?) mechanism must be operative. This may account for the differential activity of *SWI5* gene product in mother and daughter cells.

In another obvious model, the *SWI5* protein may be sequestered in such a way that only the mother cell inherits it. The *SWI5* protein may contain domains that ensure such partitioning. One subtle but interesting observation which leads to the suggestion that *SWI5* might be distributed to mother cells as an assembled transcriptional complex on the *HO* promoter is worth considering in this context. Stationary-phase mother cells blocked at G1 are capable of transcribing *HO* following inoculation into fresh medium. Interestingly, a deletion of the *HO* upstream regulatory region between −788 and −1,160 bp eliminates this transcription following G1 arrest without affecting *HO* activity in subsequent cycles. It was suggested that *SWI5* may be inherited by mother cells only as a stable transcriptional complex whose stability is reduced in the case of the deleted promoter (91). In this model, the chromatid containing the competent *HO* segregates to the mother cell at cell division. An interesting prediction of such a model is that in those strains where *HO* is inverted in the chromosome, the daughters should switch, while the mothers should not. This model has recently been ruled out by studies with a strain containing an inverted *HO* gene placed near *CDC36* on chromosome IV (57). In this strain, only the pattern of switching found in wild-type strains was observed. Furthermore, when *HO* was placed in chromosome III at *LEU2*, the pattern of switching was again not altered, suggesting that mother-daughter control is not specific to chromosome IV where *HO* is normally located.

In summary, *HO* is produced in a narrow time window late in G1 only in mother cells, where all the negative controls are inoperative, and thus its transcription is allowed. At the time the cassette model was proposed, its proponents knew that only mother cells can switch. Although this interesting finding was not ignored, the simplest version of the cassette model failed to explain it. Perhaps a lesson is to be learned here. When building models, one should not worry about explaining every finding. Some seemingly conflicting observations may actually reflect some of the fine details of the process, as it turned out for the mother-daughter control exerted by regulating *HO* expression and the control of directionality of switching (see below). Nonetheless, why *S. cerevisiae* has evolved a mechanism such that daughter cells should not switch remains an interesting unanswered question.

C. Directionality of Switching

Mother cells of genotype *HO HMLα HMR***a** switch to the opposite mating type in about 73% of cell divisions, which is well over the 50% rate expected should the choice of donor cassette be random (42, 115). Thus, based on their mating phenotype, cells seem to switch primarily (or only) to the opposite allele. From the foregoing discussion, it is not intuitively evident what feature of this system should dictate the donor choice mechanism once the cut has been made at *MAT*.

A hint came from the observation that strains containing opposite alleles of *HM* loci (*HML***a** *HMRα* rather than the usual *HMLα HMR***a**) switch to the opposite *MAT* allele less than 6% of the time (52, 64, 102). To explain this feature, my colleagues and I entertained the hypothesis that **a** cells preferentially choose *HML* (the cassette on the left arm) as the donor while α cells mostly choose *HMR* (the cassette on the right arm). By this logic, *HMLα HMR***a** strains will preferentially switch to opposite alleles, while those with *HML***a** *HMRα* will undergo mostly unscorable futile homologous switches with no phenotypic effects. By employing strains containing genetically marked cassettes, it was demonstrated that *HO HML***a** *MATα HMRα′* (α′ is a genetically marked α cassette) strains efficiently switch *MATα* to

*MAT*α′ and likewise the *HO HML*a′ *MAT*a *HMR*α strains switch *MAT*a to *MAT*a′. Thus, α cells preferentially choose *HMR* as the donor regardless of the genetic content of the latter, and a cells likewise mostly choose *HML* as the donor. The observed preference is on the order of 10 to 1. Furthermore, the preference is not absolute, in the sense that when the preferred donor locus is deleted, the remaining donor is used efficiently. Based on these results, we imagine that the cell type dictates which donor to choose, and that the choice is mediated by competition between two donors. In a cells *HML* wins, and in α cells *HMR* wins, in donating information to *MAT* (64).

Perhaps the particular locations of the *HM* loci in the chromosome make them differentially accessible to the *MAT* locus, and that accessibility is influenced by the cell type. Providing a precedence for such nonrandom chromosome organization in the nucleus are the cytological observations made with *Drosophila* nuclei (82). However, the *HM* loci can be used even when placed on other chromosomes, although with a lesser efficiency (34). As shown in Fig. 2, both *MAT* and *HML* have W and Z2 sequences, while *HMR* does not. This difference in the donors does not dictate preference, since when the *HMR* element is replaced by a 5-kb *HML* sequence, a cells still predominantly choose the left copy of *HML* as donor. Inversion of the *HML* cassette also did not affect the donor choice mechanism (unpublished observations). Clearly, the mechanism of donor choice is a very interesting feature that remains to be worked out. Although considerable attention is being given to it, progress seems slow because a good idea to solve the problem is lacking.

Although beyond the scope of this review, it is important to realize that *MAT* and *HM* loci essentially have the same sequence, yet *MAT* is expressed and can undergo interchange, while *HM* loci fail to do either (3, 72). Both *MAT* alleles encode two transcripts which are divergently transcribed from within the Y region outward (71, 95). Both *trans*-acting, formally negative regulatory *MAR/SIR* functions (discussed above), a positive regulatory *SUM1* (suppressor of *mar*; 65), and *cis*-acting sites flanking *HM* loci called E and I (1, 26) are required to keep them unexpressed (reviewed in reference 92). The position-effect control for transposition is due to the lack of an in vivo cut at the Y/Z boundary in *HM* loci (117). Interestingly, in *mar* mutant strains where *HM* loci are expressed, they can be cleaved by the *HO* endonuclease and consequently can switch efficiently by following the rules of switching discussed earlier (70, 72). Clearly, then, the mechanism that keeps the *HM* loci unexpressed also plays an important role in blocking their switching by making them inaccessible to *HO* endonuclease (70). In summary, *S. cerevisiae* activates the information at the *HM* loci by transposing a copy of the information away from locations of negative control to *MAT*, where it can be expressed and switched efficiently.

IV. STUDIES OF *S. POMBE*

A. The *S. pombe mat1* Interconversion Occurs via DNA Transposition

Although both budding yeasts and fission yeasts are single-celled ascomycete fungi, evolutionarily they are not closely related (105). The term "yeast" is a mycological term indicating a single-celled fungus and it is not meant to imply a close evolutionary relationship. *S. pombe* is a fairly typical eucaryote and primarily lives in a haploid state. The genome is the same size as that of *S. cerevisiae*, but it contains only 3 chromosomes, compared to 17 of *S. cerevisiae*.

Haploid cells of *P* and *M* mating types divide by fission on rich medium, but as the medium becomes depleted, especially of nitrogen, the cells cease division and begin conjugation with those of the opposite type. The zygotic cell immediately enters a round of meiotic division, resulting in the formation of a four-spored ascus (28). Thus, unlike *S. cerevisiae*, the *S. pombe* diploid phase is transient. However, from the mating mix, diploid zygotic cells which are not yet committed to meiosis can be selected by plating on selective media and then maintained as diploids by growing on rich medium. The diploid cells can be induced to undergo meiosis and sporulation to produces asci (28). Unlike *S. cerevisiae*, *mat1* interconversion in *S. pombe* continues in diploid cells (4, 20, 22; and unpublished results). Heterozygosity at the *mat1* locus is required for completion of meiosis and sporulation (reviewed in reference 24).

The *mat1* locus is flanked by centromere proximal *his7* and distal marker *his2* on chromosome II (24). The *P* or *M* cell type is controlled by a complex locus composed of three components, *mat1*, *mat2*, and *mat3* (Fig. 4). The cell type is determined by the particular allele present at *mat1*.

Elegant genetic experiments first established the complexity of mating type determination. Interconversion of mating type was originally proposed to be mediated by inverting a regulatory element at *mat1*, the flip-flop model (17). Genetic analysis of two mutations which generated meiosis-deficient *mat1* alleles, however, suggested the transposition model (19, 23). Those mutations were used to define *mat2* and *mat3* loci, which map genetically at 1 centimor-

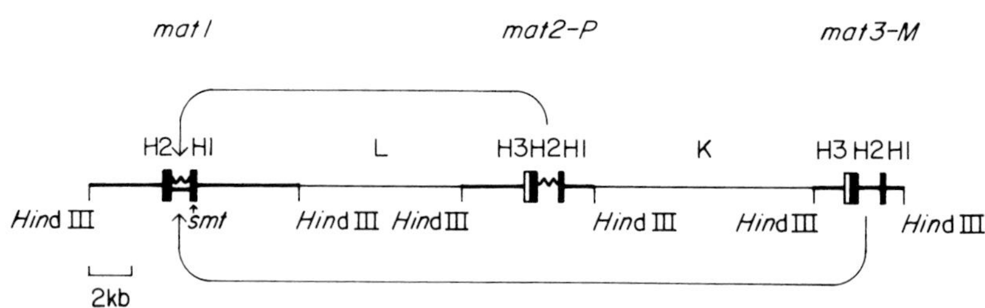

Figure 4. The *mat* region on chromosome II. Thick lines show the 10.4-, 6.3-, and 4.2-kb *Hin*dIII fragments respectively containing *mat1*, *mat2-P*, and *mat3-M* cassettes. The centromere is located about 65 cM to the left of *mat1*. The cassette indicated by the zigzag line contains the *P* allele, and that with the smooth line contains the *M* allele. The homology boxes H1 (59 bp) and H2 (135 bp) presumably allow interaction between the *mat1* recipient and the *mat2* or *mat3* donor loci required for the gene conversion-mediated transposition event. Arrows indicate unidirectional transfer of information from *mat2* and *mat3* to *mat1*. The interval between *mat1* and *mat2* is about 15 kb and is called the L region; the 15-kb *mat2-mat3* interval is called the K region. The *smt* marks the site of the in vivo double-stranded cut in *mat1*. The 57-bp H3 region is present at *mat2* and *mat3* and not at *mat1*. (Figure derived from data presented in references 4, 5, 19, 23, and 54.)

gan (cM) distal to *mat1*. A critical experiment with a strain of genotype *mat1-P mat2-pm* showed a change from *mat1-P* to *mat1-M* and then to *mat1-pm*, strongly arguing for the transposition model (23). In this model, *M* cells contained the *mat1-M* allele at the expression locus *mat1*. The *P* phenotype was hypothesized to result from the insertion of the *P* element (derived from *mat2*) causing disruption of the *mat1-M* allele. Following the exit of the *P* element, the *mat1-M* allele is restored. Likewise, another mutation, B406, now known to reside at *mat3*, was shown to be transmitted to *mat1*, suggesting the existence of the *mat3* locus (19). Whether mutation B406 maps in *mat2* or *mat3* could not be determined in initial tests because genetic recombination between *mat2* and *mat3* has not been observed (19). In other words, genetically *mat2* and *mat3* were inseparable.

Molecular studies with cloned pieces of DNA have confirmed and extended the transposition model and have identified *mat1*, *mat2*, and *mat3* as physically separate loci (Fig. 4). Plasmid DNA-mediated transformation (6) was used to isolate a DNA sequence that conferred sporulation competence to h^{-S}/h^{-S} diploids (7). This sequence was expected to contain a mating type *P* activity. (The h^{-S} strains [see below] maintain stable *M* type because they contain a *P* donor deletion [5].) Using the clone with *P* activity as the probe, it was found that *S. pombe* has three restriction fragments with various degrees of sequence homology to the *P* probe. The *mat1* locus was found to exist in two forms, one containing the *P* allele and the other containing the *M* allele. The *P* information was also found in the *mat2*-containing fragment, while the *M* information was found at a physically separate locus called *mat3* (4, 5).

Further Southern analysis (109) of spontaneous recessive lethal mutations which result from rearrangements of the *mat1* region allowed determination of the order of these loci in the chromosome. One such rearrangement, called h^{+L}, with a stable *P* phenotype, and another, h^{-L}, with a stable *M* phenotype, established the order of all three *mat* elements. The h^{+L} results from a fusion between *mat1* and *mat2* (*mat1:2-P*), resulting in a deletion of the L sequence (Fig. 4). The L sequence is 15 kb in length and is essential for viability. The h^{-L} is a larger deletion and is a fusion between *mat1* and *mat3*; it contains the stable *M* allele. This analysis placed *mat1* in the centromere proximal side, followed by the *mat2*, 15 kb away, and then *mat3* at 30 kb away from *mat1*. The region between *mat1* and *mat2* is termed the L region, and the segment between *mat2* and *mat3* is called the K region (5).

The DNA sequence determination of all three *mat* loci showed that the *P*-specific region is 1,113 bp long (present in both *mat1-P* and *mat2-P*), while the nonhomologous *M*-specific region is 1,127 bp long (present in both *mat1-M* and *mat3-M*) (54). A unique *Bam*HI site is present in the *P* cassette, and a unique *Eco*RI site is present in the *M* cassette (4). Transposition is presumably facilitated by the flanking homologous regions. The proximal 135-bp H2 region and the distal 59-bp H1 sequences are shared by all three loci and are much smaller than the analogous homology sequences of *S. cerevisiae* (Fig. 2). As suggested by the evolutionary divergence between *S. pombe* and *S. cerevisiae* (105), their mating type cassettes do not show any significant sequence homology (3, 54).

Two open reading frames and two transcripts have been defined for both the *P* and *M* alleles of *mat1*. The mRNAs are divergently transcribed from within the cassette to the outside from both strands. One of the two transcripts from each *mat1* allele is required for mating, while both transcripts of each allele are required for sporulation. Under starvation conditions, the *mat1* transcripts are induced (54), and this feature of gene regulation may explain the restriction of cells to mate and undergo meiosis under growth-limiting conditions.

The promoter regions of *mat1*, *mat2*, and *mat3* are identical in DNA sequence; therefore, some position-effect control must operate to keep the *mat2* and *mat3* loci unexpressed. The findings of many *mat* cassette rearrangements of type h^{+L}, h^{-L} (discussed above) and h^{+N}, h^{-U} (discussed below) and that a 1,450-bp *mat1* fragment which contains only 14 bp to the left of H2 and 140 bp to the right of H1 is transcriptionally active (58) suggest that there is no positive regulatory element next to *mat1*; rather, there may be silencer elements next to *mat2* and *mat3*. The most likely candidate for a silencer region is the 57-bp H3 sequence present on the left of *mat2* and *mat3* but not present at *mat1* (54) (Fig. 4).

The *mat2* and *mat3* donor loci in *S. pombe* are separated by a 15-kb interval called K (Fig. 4). Using the value of 4 kb/cM (86), one would expect these markers to be separated by about 4 cM. However, no meiotic recombination in this interval was found among 17,000 meiotic products selected for having undergone recombination for the flanking *his2* and *his7* markers (19). Therefore, genetically, *mat2* and *mat3* are separated by less than 0.001 cM. Essentially these loci segregate as a single locus. In standard strains, *his2* is located 2 cM distal to *mat1* (24). If the *mat2*, *mat3*, and K regions are deleted, then this interval, rather than shrinking in genetic distance, increases to between 12 to 16 cM (68). To explain this paradoxical result, it is suggested that the cold spot of the K region (19) inhibits recombination in flanking intervals (68). The *mat2* and *mat3* interval is clearly unusual, apparently inhibiting classical recombination, yet available for *mat1* switching. The organization making this region unavailable for classical recombination may also serve to keep the *mat2* and *mat3* cassettes transcriptionally inactive. Also, the switches occur primarily (or only) to the opposite mating type. In addition, such a hypothesized higher-order chromatin structure may prevent isomerization of the recombination intermediate and thus block recombination of flanking markers during the intrachromosomal gene conversion event assumed to be required for *mat* interconversion. All these properties may be a consequence of the precise location and the chromatin organization of *mat* cassettes in the nucleus, a theme also suggested for the *S. cerevisiae* *MAT* system (see Section III).

One particularly useful simple feature of *S. pombe* employed in genetic studies of mating type is the fact that sporulating cells accumulate a starchlike compound. Thus, clones capable of switching can sporulate and therefore can be stained black with iodine vapors (28). Hence, slow-switching, or heterothallic, variants can be easily spotted on a plate by a visual screen following iodine staining. Heterothallic strains of class h^{+N} (*P* phenotype), h^{-U} (unstable *M*) or h^{-S} (stable *M*) are generated spontaneously at a frequency of about 10^{-5} to 10^{-3} (for a review, see reference 24). The homothallic colony, designated h^{90}, is stained homogeneously black, while the h^{+N} and h^{-U} strains produce streaks of stain within the colony. The h^{-S} fails to change to any other type and thus fails to stain. It is suggested that h^{+N} and h^{-U} apparently result from errors in the resolution step in which both *mat2* and *mat3* and the region between the cassettes, the K region, are transmitted to the *mat1* region (Fig. 4). This is an extensive gene conversion event in which about 17 kb of DNA is transmitted to *mat1*. The centromere proximal cassette present at *mat1*, *P* in h^{+N} and *M* in h^{-U}, is transcribed, resulting in the observed phenotypes (5). At similar very low rates, each of these rearrangements can switch back to h^{90}. h^{-S} is a fusion of *mat2* and *mat3*, resulting in a deletion of the K region and the loss of *P* information and thus producing a stable *M* phenotype of the h^{-S} cells. The deletion of the 15-kb K region otherwise is not deleterious (5).

In addition to the above rearrangements of the *mat* region, two classes of mutants with a reduced frequency of switching have been described, those which map at *mat1* and those mapping elsewhere. The ones mapping at *mat1* presumably define a *cis*-acting site(s), while unlinked mutants define *trans*-acting functions required for interconversion.

B. The Double-Stranded Break at *mat1* Initiates Mitotic and Meiotic Gene Conversions

Southern blot analysis of *S. pombe* DNA revealed that 20 to 25% of the *mat1* DNA is cut in vivo in homothallic cells (4, 21). By analogy with the *S. cerevisiae* system (see Section III), my colleagues and I proposed that this cut initiates recombination required for *mat1* switching. Although the nature of the cut has not been fully defined, we presume that the site is near the boundary of the allele-specific and the H1 regions. By analyzing DNA isolated from strains blocked at various points in the cell cycle or from a synchronized population of cells, it was shown that a similar amount of the cut is present throughout the cell cycle (4; O. Nielson and R. Egel, personal communication).

Ten complementation groups have been defined genetically as *trans*-acting functions required for switching (21, 29, 107). The *swi1* (switch), *swi3*, and *swi7* genes are required for the formation of the double-stranded cut at *mat1*, as is evidenced by a reduced level of cut DNA in strains harboring mutations in these genes. Mutations in *swi4*, *swi8*, *swi9*, and *swi10* result in a high frequency of rearrangement of the mating type locus generating the h^{+N} and h^{-U} genotypes, indicative of resolution errors of recombination intermediates. Mutations in *swi2*, *swi5*, and *swi6* do not affect levels of the break at *mat1*, but cause defects in switching (21). It appears that mutants of the last class are deficient in utilization of the double-stranded break for switching and that cells can seal the break without switching. Clearly, several of the *swi* genes are involved in functions related to general recombination, since several *swi* mutants cause sensitivity to UV light or γ rays (108). *swi5* mutants have been shown to be defective in recombination at the *ade6* locus (108) and at the *mat1* flanking regions (unpublished observations). Some *swi* mutant combinations, such as *swi1 swi7* mutants and *swi3 swi7* mutants, are lethal, suggesting a role for these genes in other cellular functions. Interestingly, double and triple *swi* mutants show a cumulative reduction in the frequency of *mat1* interconversion only if the genes belong to different groups discussed above (108). This result may mean that gene products of the group *swi1*, *swi3*, and *swi7* form a complex to catalyze a cleavage at *mat1*. The same argument may be proposed for genes of the other two groups.

Several *cis*-acting mutations, called *smt-s*, all of which reduce the rate of switching because they reduce the level of double-stranded breaks, have been isolated (18). One of these mutations, C13-P11, which maps about 0.1 cM distal to *mat1* (18), has been characterized and shown to be a 27-bp deletion. Interestingly, the mutation affects the *mat1* cleavage at a site approximately 50 bp away from the deletion (R. Cafferkey and A. J. S. Klar, unpublished observations). Although a site about 50 bp away from the cut site may be essential for efficient cleavage, an equally likely possibility is that it diminishes cleavage in another way. For example, the deletion may place the cut site in a chromatin structure that is inaccessible to DNA cleavage or for the "imprinting" that is described below.

A quite intriguing observation is that the double-stranded break at *mat1* persists for the whole length of the cell cycle. Although not rigorously demonstrated, it seems likely that the break is made in one cell cycle and healed one generation later by switching (58). Equally intriguing, the break can also be healed without switching, since strains with both donor *mat* loci deleted contain a normal level of break, yet are viable as haploids (68). Even in *swi5* mutants, which are defective in the utilization of the double-stranded break for recombination, the cut ends must be rejoined, with or without switching, to restore the integrity of the chromosome (68). Clearly, some mechanism must exist to protect the broken ends from degradation in vivo.

The persistence of cut DNA at *mat1* for the entire length of the cell cycle clearly implicates a mechanism for long-term protection of these ends. Supportive of this reasoning, both 5′ ends of the cut molecule are found to be protected in vitro from digestion by λ exonuclease, which is specific for free 5′ DNA ends, and one 3′ end is protected from degradation by exonuclease III, an activity specific for free 3′ ends. The other 3′ end is, however, accessible to exonuclease III digestion (unpublished observations). Perhaps specific proteins attached to DNA ends are not fully removed by proteinase K treatment (during the DNA preparation), thus protecting the ends. These same proteins would protect the cut ends in vivo and may form components of the healing mechanism which can operate with or without switching.

The possibility that the cut at *mat1*, which persists for the entire cell cycle in donor-deleted strains, can initiate meiotic gene conversion at *mat1* was tested by crossing *mat1-P* and *mat1-M* strains each deleted for donor *mat2* and *mat3* loci. Although 80% of asci contained nonconvertant 2*P*:2*M* tetrads, 10% contained 3*P*:1*M* and another 10% contained 1*P*:3*M* convertant tetrads. This gene conversion was associated with recombination of flanking markers. None of the asci in *S. pombe* produced a 4:0 pattern (68). A similar experiment with *S. cerevisiae* only generated 4:0 conversion events (74). Thus, in *S. pombe*, only one of two sister chromatids experiences meiotic gene conversion, whereas in *S. cerevisiae*, both chromatids are converted. This meiotic result and the mitotic pattern of switching (Fig. 1) suggest that gene conversion at the mating type locus occurs at different phases of the cell cycle in these yeasts. In *S. pombe*, it is formally a G2 event and in *S. cerevisiae*, at least in mitotic conversion, it is a G1 event.

C. Developmental Asymmetry among Daughter Cells Is a Consequence of Inheriting Nonequivalent Parental DNA Chains

Why some cells switch and others do not (Fig. 1) is a fascinating question. It would seem that the switching potential must be asymmetrically segregated to daughter cells either via the cytoplasm or via the DNA template. In the first model, essential components, such as those encoded by *swi* genes, are asymmetrically segregated, differentially stabilized, or unequally expressed in the daughter cells. This is essentially the same model as proposed for the *HO* regulation in *S. cerevisiae* (see above). In the second model, some semiheritable chromosomal modification, termed "chromosomal imprinting" (after Crouse [15]), occurs on some chromosomes but not on others; the cell inheriting the modified chromosome is postulated to be competent to produce switched progeny. Analysis of the switching pattern of mitotically dividing diploid cells showed that asymmetric segregation of competence for switching is separately determined for the two homologous chromosomes. This observation indicated that the switching potential is not cellular, but segregates with the template (20). Whatever the mechanism, generation of the *mat1* cut must be the rate-limiting step for regulating the frequency of switching. An independent test of the imprinting hypothesis was provided by the finding that meiotic crosses involving strains deleted for *mat2* and *mat3* generate a high rate (20%) of *mat1* conversions in which both 3*P*:1*M* and 1*P*:3*M* aberrant asci in equal proportion are produced (68). The likely candidates to catalyze the epigenetic imprinting event are the gene functions involved in generating the cut at *mat1*, e.g., those of *swi1*, *swi3*, and *swi7* (21). The cross between a *mat1-M swi3*$^-$ mutant and a *mat1-P swi3*$^+$ strain generated aberrant tetrads primarily with 3*M*:1*P* segregants in which only *mat1-P* allele converted. Thus, the *mat1* allele "remembered" its prior history of growth with respect to the *swi3* genotype. (In *S. pombe*, immediately after mating, cells undergo meiosis and sporulation [28].) Similarly, crosses involving a *swi1*$^-$ or a *swi7*$^-$ parent generate meiotic *mat1* conversions in which only the allele contributed by the *swi*$^+$ parent acts as the recipient for conversion (unpublished observations). These results show (i) that the competence for meiotic gene conversion at *mat1* segregates in *cis* with the *mat1* locus, (ii) that the *swi*, *swi3*, and *swi7* functions confer that competence, and (iii) that the presence of these functions in the zygotic cells (as it was provided by the wild-type parent) fails to confer the gene conversion potential to the *mat1* allele that was previously replicated in a respective *swi1*$^-$, *swi3*$^-$ or *swi7*$^-$ background. In other words, chromosomally imprinted functions are catalyzed by these *swi* genes at least a generation before a gene conversion at *mat1* can be realized. Thus, these results fit well with the classical definition of imprinting by Crouse (15).

That some chromosomes are potentiated to switch while others are not may be the consequence of the presence of rate-limiting functions in yeasts which are required to imprint or cleave DNA. Results presented in below argue that imprinting is in fact a site- and strand-specific chromosomal marking within *mat1* which makes *mat1* competent for in vivo cleavage and that switching functions are not present in rate-limiting amounts.

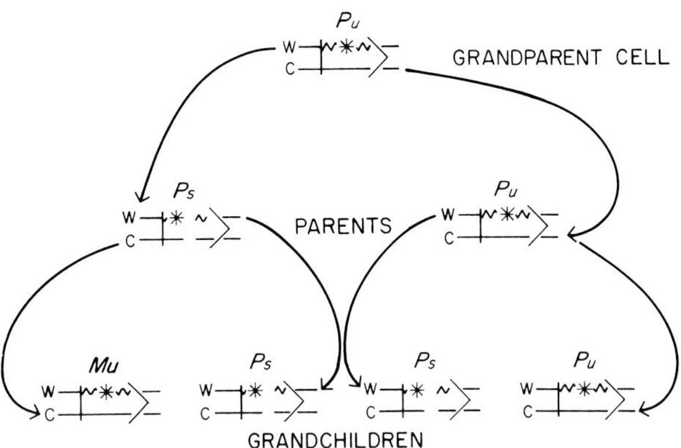

Figure 5. Strand segregation model. Details of the model are presented in the text. W and C indicate the Watson and Crick strands of DNA, respectively; *P* and *M* denote *mat1* alleles; star indicates the strand- and sequence-specific imprinting at *mat1*. A gap in the chromosome indicates the double-stranded DNA break. *Pu* and *Mu* indicate unswitchable *mat1* loci, and *Ps* and *Ms* denote switchable cassettes. Arrows indicate the segregation of specific strands to progeny cells. (Figure modified from reference 58.)

In particular, one wants to address the question of why only one of four descendants of a given chromosome (cell) switches. The decision for a given switch must be made in the grandparent cell two generations before the switched cell is produced (Fig. 1). Apparently, two consecutive asymmetric cell divisions are required to produce a switch of only one grandchild cell. One may imagine a molecular model (58) in which the Watson and Crick strands of a given DNA molecule are not equivalent in their ability to acquire the developmental potential for switching. Specifically, specific *swi* gene functions are hypothesized to catalyze an imprinting event, possibly a sequence-specific DNA modification, such that only one specific chain (say, the Watson strand) is modified at the *mat1* locus (Fig. 5). Perhaps the site of modification is not palindromic; thus, only one of the two strands can be modified, in a fashion analogous to the modification by the *Hin*fIII type III restriction-modification enzyme (99). Replication of the hemimodified chromosome generates two sister chromatids, and the one possessing the old modified Watson chain is cleaved in vivo in G2. Following cell division, the one daughter cell that inherited the broken chromosome produces a switched and an unswitched progeny. Although the specific details of the imprinting event are not known, the key idea, the segregation of DNA strands conferring developmental asymmetry to daughter cells, is testable.

This strand segregation model is based on the

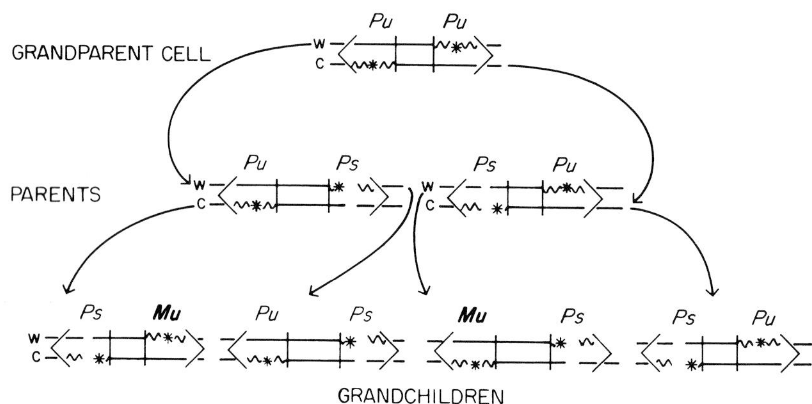

Figure 6. Tests of the strand segregation model employing inverted tandem duplication of *mat1* cassettes. Wide arrows indicate cassette orientation. The genomic duplication was created by inserting a *mat1* cassette to the left of the standard *mat1* cassette as described (58). The two predictions of the model are that both cassettes should never be cleaved simultaneously in the same chromosome and that two cousins in four related grandchildren should switch. Designations are given in legend to Fig. 5. (Figure modified from reference 58.)

assumption (and, in effect, it tests the assumption) that both progeny of a given parental cell are identical in cytoplasmic constituents. It postulates that the developmental asymmetry of sister cells reflects differences in developmental potential imprinted on individual DNA strands. According to this model, a given cell division in any biological system could generate two nonequivalent daughters because one inherits the old Watson and the other inherits the old Crick strand. Similar overall strategies with differences in mechanistic details have been proposed to explain cellular differentiation in metazoans as well as for *S. cerevisiae* and *S. pombe* yeast cell-type determination (13, 20, 49, 56, 129).

A molecular and a genetic test of this model have been designed, based on determining the efficiency of cleavage and the switching pattern of cells containing tandem inverted *mat1* duplications (Fig. 6). The additional cassette was inserted next to the resident *mat1* cassette by DNA-mediated transformation (58). The strand segregation model predicts that one cell from each half of the pedigree should switch a single (but different) cassette. In other words, two cousins in four related cells should switch, each using a different *mat1* locus as the target for switching. Another molecular prediction is that the two *mat1* cassettes in the same chromosome should never be cleaved together, since both cassettes should never switch in the same cell. In contrast, the cytoplasmic segregation model predicts that only one in four cells should switch and that both cassettes in the same chromosome may be cleaved in the same generation.

The switching competence of each cassette was determined molecularly by assaying for the presence or absence of the double-stranded cut. This design exploited the persistence of the cut for the entire length of the cell cycle. With the inverted *mat1* duplication, it was found that the two *mat1* sites on the same chromosome were never cut simultaneously even though both cassettes were individually efficiently cut (58). Thus, a given cell acquires the competence to switch one or the other, but never both cassettes, during the same cell cycle. This result is consistent with the strand segregation model and rules out any cytoplasmic segregation model. Based on this result, it is argued that the switched cell always inherits a particular chromosome which does not contain strands of the original grandparental chromosome. In other words, if the Watson strand is imprinted, the chromosome inheriting the Crick strand receives the switched allele (Fig. 6). Also, this analysis showed that the sequences required for imprinting and cutting are contained within the 1,450-bp *mat1*-containing *Ssp*I fragment that was used to construct the tandem duplication (58).

Testing the key genetic prediction that two (cousins) should switch in four related cells in a typical pedigree analysis of individual cells is nearly impossible because identifying the mating type of a cell requires its successful mating to the cell of the opposite type and forming a zygotic cell. This makes it impossible to determine the phenotype of both a given cell and its progeny in consecutive generations. However, an alternate test is provided by the fact that diploids heterozygous for *mat1-P* and *mat1-M* alleles can undergo meiosis and sporulation, while those homozygous at *mat1* cannot (28). This provides a single-cell assay for determining the contents of *mat1*. Egel and Eie (22) recently used this assay and found that in a diploid of genotype h^{90}/h^{-U}, the wild-type h^{90} chromosome switches from *mat1-M* to *mat1-P* by following Miyata's rule of switching one-

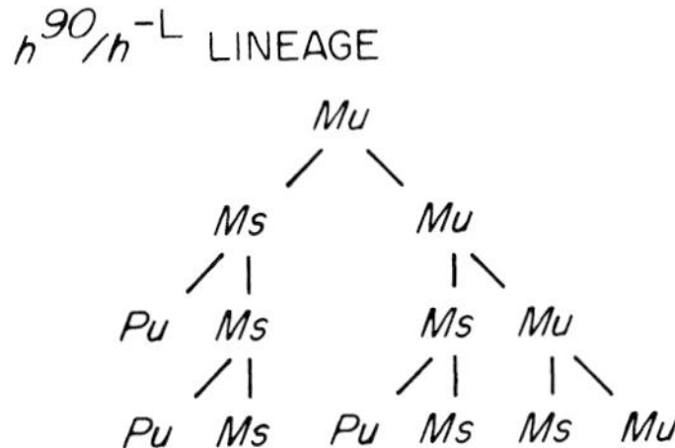

Figure 7. The pattern of switching of the *mat1-M* allele in the h^{90} chromosome to *mat1-P* in h^{90}/h^{-L} diploid cells as determined by assaying for the ability of single cells to sporulate. Only the deduced genotype of the *mat1* allele in the h^{90} chromosome is indicated in a particular lineage (unpublished observations). The cell marked as *Pu* stopped growing because it sporulated.

in-four related cells. The h^{-U} allele employed maintains a stable *M* allele which switches only rarely (see Section IV.A, above). A switch of *mat1-M*/h^{-U} to *mat1-P*/h^{-U} makes the cell capable of sporulation. By this assay, it was possible to demonstrate that h^{90}/h^{-L} and h^{90}/h^{+L} (the h^{+L} and h^{-L} alleles do not switch [5]) diploids also switch their *mat1* on the h^{90} chromosome by following the rule of one-in-four (unpublished observations) (Fig. 7). Both studies show that the sister of the recently switched cell is capable of producing switched progeny in the ensuing cell division; presumably, it inherits the broken chromosome. The switched cell, therefore, must inherit the healed chromosome and could only switch after two generations; otherwise, the one-in-four rule will be violated. Employing this assay of the sporulation competence of each cell, it was found that diploids containing the *mat1* inverted tandem duplication/h^{+L} (or h^{-L}) did indeed generate pedigrees where each member of a pair of sister cells generated a single switched and a single unchanged progeny (unpublished observations). In other words, the predicted violation of Miyata's rule is produced: two cousins in four related cells are found to switch in strains with a *mat1* inverted duplication (Fig. 6). That is, the grandparent cell produced developmentally equivalent daughters. Thus, both single-cell assays and the determination of *mat1* DNA cleavage in exponentially growing cultures establish the strand segregation model and argue strongly against the asymmetric segregation of some cytoplasmic factor(s).

V. CONCLUDING REMARKS

The main issue of generating developmentally nonequivalent daughter cells in two different unicellular organisms is addressed here. Two different yeasts are shown to employ fundamentally different mechanisms. In *S. cerevisiae*, mating type switching is controlled cytoplasmically, that is, via the unequal distribution of potential for *HO*-endonuclease gene expression to progeny cells. In *S. pombe*, the cytoplasmic components of daughter cells are equivalent, and instead cell lineage is controlled by imprinting of the substrate *mat1* DNA. The critical events seem to be quite different; since these are evolutionarily distantly related organisms, it is not surprising that details of the mating type interconversion process are indeed very different. Although the end products of mating type interconversion of both *S. pombe* and *S. cerevisiae* are known, many details of DNA transposition remain to be worked out. It will be useful to have an in vitro switching system available for both yeasts. At least for the *S. pombe* system, the availability of mutants with defects in known steps of recombination will be useful to realize this goal.

In the context of cell lineage studies, one should consider the pattern of switching of another distantly related methylotrophic budding yeast, *Pichia pinus*. This yeast also produces a mixed population of two mating types, also termed **a** and α. The interconversion is repressed during mitotic growth on rich media. During growth-limiting conditions, switching occurs in the late few rounds of mitotic growth. In contrast with patterns of both *S. cerevisiae* and *S. pombe*, switching in *P. pinus* is a random process, determined independently of the position of the cell in its pedigree (125). Perhaps controls related evolutionarily to those operative in *S. cerevisiae* and *S. pombe* have been modified to achieve the *P. pinus* pattern, with the added proviso that the entire process is repressed by growth in rich medium.

The term homothallism for fungi was coined by Blakeslee (9) to indicate sexual dimorphism; in particular, it indicates the capability to self-fertilize between siblings. It is a very widespread phenomenon among fungi. It is an interesting question to ask why such a system of crossing has evolved, considering that homothallism primarily potentiates genetic inbreeding. Perhaps such a system provides an advantage by providing a ready access to gametes of both types, which in fungi after the act of meiosis invariably produce meiotic products better fit to weather adverse conditions.

Acknowledgments. I am immensely thankful to friends in the business who communicated their unpublished work to me and which is quoted as personal communication in the text; to D. Berg, M. Howe, B. Futcher, C. Greider, T. Peterson, and R. Cafferkey for critically reading the manuscript; to S. Arana for preparation of the manuscript; and to M. Ockler, D. Greene, and J. Duffy for preparing the figures. This review was completed while the author was at Cold Spring Harbor Laboratory, Cold Spring Harbor, N.Y.

This work was supported by Public Health Service grant GM25678 (National Institutes of Health) and National Science Foundation grant DCB-8611960 to the author.

LITERATURE CITED

1. **Abraham, J., K. A. Nasmyth, J. N. Strathern, A. J. S. Klar, and J. B. Hicks.** 1984. Regulation of mating-type information in yeast. Negative control requiring sequences both 5′ and 3′ to the regulated region. *J. Mol. Biol.* **176**:307–331.
2. **Arima, K., and I. Takano.** 1979. Evidence for co-dominance of the homothallism genes, *HMα/hmα* and *HMa/hma*, in *Saccharomyces* yeasts. *Genetics* **93**:1–12.
3. **Astell, C. R., L. Ahlstrom-Jonasson, M. Smith, K. Tatchell, K. A. Nasmyth, and B. D. Hall.** 1981. The sequence of the DNAs coding for the mating-type loci of *Saccharomyces cerevisiae*. *Cell* **27**:15–23.
4. **Beach, D. H.** 1983. Cell type switching by DNA transposition in fission yeast. *Nature* (London) **305**:682–688.
5. **Beach, D. H., and A. J. S. Klar.** 1984. Rearrangements of the transposable mating-type cassettes of fission yeast. *EMBO J.* **3**:603–610.
6. **Beach, D. H., and P. Nurse.** 1981. High frequency transformation of the fission yeast. *Nature* (London) **290**:140–142.
7. **Beach, D. H., P. Nurse, and R. Egel.** 1982. Molecular rearrangement of mating-type genes in fission yeast. *Nature* (London) **296**:682–683.
8. **Beggs, J. D.** 1978. Transformation of yeast by a replicating hybrid plasmid. *Nature* (London) **275**:104–109.
9. **Blakeslee, A. F.** 1904. Zygospore formation, a sexual process. *Science* **19**:864–866.
10. **Breeden, L., and K. Nasmyth.** 1987. Cell cycle regulation of *HO*: *cis-* and *trans*-acting regulators. *Cell* **48**:389–397.
11. **Brown, S. W.** 1976. A cross-over shunt model for alternate potentiation of yeast mating type allele. *J. Genet.* **62**:81–91.
12. **Bucking-Throm, E., W. Duntze, and L. H. Hartwell.** 1973. Reversible arrest of haploid cells at the initiation of DNA synthesis by a diffusible sex factor. *Exp. Cell Res.* **76**:99–110.
13. **Cairn, J.** 1975. Mutation selection and the natural history of cancer. *Nature* (London) **255**:197–200.
14. **Connolly, B., C. I. White, and J. E. Haber.** 1988. Physical monitoring of mating type switching in *Saccharomyces cerevisiae*. *Mol. Cell. Biol.* **8**:2342–2349.
15. **Crouse, H. V.** 1960. The controlling element in sex chromosome behavior in *Sciara*. *Genetics* **45**:1429–1443.
16. **Egel, R.** 1977. Frequency of mating-type switching in homothallic fission yeast. *Nature* (London) **266**:172–174.
17. **Egel, R.** 1977. "Flip-flop" control and transposition of mating-type genes in fission yeast, p. 447–455. *In* A. J. Bukhari, J. Shapiro, and S. Adhya (ed.), *DNA Insertion Elements, Plasmids and Episomes*. Cold Spring Harbor Laboratory, Cold Spring Harbor, N.Y.
18. **Egel, R.** 1980. Mating-type switching and mitotic crossing-over at the mating-type locus in fission yeast. *Cold Spring Harbor Symp. Quant. Biol.* **45**:1003–1007.
19. **Egel, R.** 1984. Two tightly linked silent cassettes in the mating-type of *Schizosaccharomyces pombe*. *Curr. Genet.* **8**:199–203.
20. **Egel, R.** 1984. The pedigree pattern of mating type switching in *Schizosaccharomyces pombe*. *Curr. Genet.* **8**:205–210.
21. **Egel, R., D. H. Beach, and A. J. S. Klar.** 1984. Genes required for initiation and resolution steps of mating type switching in fission yeast. *Proc. Natl. Acad. Sci. USA* **81**:3481–3485.
22. **Egel, R., and B. Eie.** 1987. Cell lineage asymmetry in *Schizosaccharomyces pombe*: unilateral transmission of a high-frequency state of mating-type switching in diploid pedigrees. *Curr. Genet.* **12**:429–433.
23. **Egel, R., and H. Gutz.** 1981. Gene activation by copy transposition in mating-type switching of a homothallic fission yeast. *Curr. Genet.* **3**:5–12.
24. **Egel, R., J. Kohli, P. Thuriaux, and K. Wolf.** 1980. Genetics of the fission yeast *Schizosaccharomyces pombe*. *Annu. Rev. Genet.* **14**:77–108.
25. **Errede, B., M. Company, J. D. Ferchak, C. A. Hutchison III, and W. S. Yarnell.** 1985. Activation regions in a yeast transposon have homology to mating type control sequences and to mammalian enhancers. *Proc. Natl. Acad. Sci. USA* **82**:5423–5427.
26. **Feldman, J. B., J. B. Hicks, and J. R. Broach.** 1984. Identification of sites required for repression of a silent mating type locus in yeast. *J. Mol. Biol.* **178**:815–834.
27. **Fogel, S., R. Mortimer, and K. Lusnak.** 1981. Mechanisms of meiotic gene conversion or "Wanderings on a foreign strand," p. 289–339. *In* J. N. Strathern, E. W. Jones, and J. R. Broach (ed.), *The Molecular Biology of the Yeast Saccharomyces. Life Cycle and Inheritance.* Cold Spring Harbor Laboratory, Cold Spring Harbor, N.Y.
28. **Gutz, H., H. Heslot, U. Leupold, and N. Loprieno.** 1974. *Schizosaccharomyces pombe*, p. 395–446. *In* R. C. King (ed.), *Handbook of Genetics*, vol. 1. Plenum Publishing Corp., New York.
29. **Gutz, H., and H. Schmidt.** 1985. Switching genes in *Schizosaccharomyces pombe*. *Curr. Genet.* **9**:325–331.
30. **Haber, J. E.** 1983. Mating-type genes of *Saccharomyces cerevisiae*, p. 559–619. *In* J. Shapiro (ed.), *Mobile Genetic Elements*. Academic Press, Inc., New York.
31. **Haber, J. E., and B. Garvik.** 1977. A new gene affecting the efficiency of mating type interconversions in homothallic strains of *Saccharomyces cerevisiae*. *Genetics* **87**:33–50.
32. **Haber, J. E., and J. P. George.** 1979. A mutation that permits the expression of normally silent copies of mating type information in *Saccharomyces cerevisiae*. *Genetics* **93**:13–35.
33. **Haber, J. E., D. T. Rogers, and J. McCusker.** 1980. Homothallic conversions of yeast mating type genes occur by intrachromosomal recombination. *Cell* **22**:277–289.
34. **Haber, J. E., L. Rowe, and D. T. Rogers.** 1981. Transposition of yeast mating type genes from two translocations of the left arm of chromosome III. *Mol. Cell. Biol.* **1**:1106–1119.
35. **Haber, J. E., B. Weiffenbach, D. T. Rogers, J. McCusker, and L. B. Rowe.** 1980. Chromosomal rearrangements accompanying yeast mating type switching: evidence for a gene-conversion model. *Cold Spring Harbor Symp. Quant. Biol.* **45**:991–1002.
36. **Harashima, S., Y. Nogi, and Y. Oshima.** 1974. The genetic system controlling homothallism in *Saccharomyces* yeasts. *Genetics* **77**:639–650.
37. **Harashima, S., and Y. Oshima.** 1976. Mapping of the homothallic genes, *HMα* and *HMa*, in *Saccharomyces* yeasts. *Genetics* **84**:437–451.
38. **Harashima, S., and Y. Oshima.** 1980. Functional equiva-

lence and co-dominance of homothallic genes, *Hmα/hmα* and *HMα/hma*, in *Saccharomyces* yeasts. *Genetics* **95:** 819–831.

39. **Hartwell, L. H., and M. W. Unger.** 1977. Unequal division in *Saccharomyces cerevisiae* and its implication for the control of cell division. *J. Cell Biol.* **75:**422–435.
40. **Hawthorne, D. C.** 1983. A deletion in yeast and its bearing on the structure of the mating type locus. *Genetics* **48:** 1727–1729.
41. **Herskowitz, I., and Y. Oshima.** 1981. Control of cell type in *Saccharomyces cerevisiae*: mating type and mating-type interconversion, p. 181–209. *In* J. N. Strathern, E. W. Jones, and J. R. Broach (ed.), *Molecular Biology of the Yeast Saccharomyces: Life Cycle and Inheritance.* Cold Spring Harbor Laboratory, Cold Spring Harbor, N.Y.
42. **Hicks, J. B., and I. Herskowitz.** 1976. Inter-conversion of yeast mating types. I. Direct observations of the action of the homothallism (*HO*) gene. *Genetics* **83:**245–258.
43. **Hicks, J. B., and I. Herskowitz.** 1977. Inter-conversion of yeast mating types. II. Restoration of mating ability to sterile mutants in homothallic and heterothallic strains. *Genetics* **85:**373–393.
44. **Hicks, J. B., J. M. Ivy, J. N. Strathern, and A. J. S. Klar.** 1985. Yeast sex: how and why, p. 97–111. *In* H. O. Halvorson (ed.), *The Origin and Evolution of Sex.* Academic Press, Inc., New York.
45. **Hicks, J. B., J. N. Strathern, and I. Herskowitz.** 1977. The cassette model of mating type interconversion, p. 457–462. *In* A. Bukhari, S. Shapiro, and S. Adhya (ed.), *DNA Insertion Elements, Plasmids and Episomes.* Cold Spring Harbor Laboratory, Cold Spring Harbor, N.Y.
46. **Hicks, J. B., J. N. Strathern, and A. J. S. Klar.** 1979. Transposable mating type genes in *Saccharomyces cerevisiae. Nature* (London) **282:**478–483.
47. **Hinnen, A., J. B. Hicks, and G. R. Fink.** 1978. Transformation of yeast. *Proc. Natl. Acad. Sci. USA* **75:**1929–1933.
48. **Holliday, R.** 1964. A mechanism for gene conversion in fungi. *Genet. Res.* **5:**282–304.
49. **Holliday, R., and J. E. Pugh.** 1975. DNA modification mechanisms and gene activity during development. *Science* **187:**226–232.
50. **Ivy, J. M., J. B. Hicks, and A. J. S. Klar.** 1985. Map positions of yeast genes *SIR1*, *SIR3* and *SIR4*. *Genetics* **111:**735–744.
51. **Ivy, J. M., A. J. S. Klar, and J. B. Hicks.** 1986. Cloning and characterization of four *SIR* genes of *Saccharomyces cerevisiae. Mol. Cell. Biol.* **6:**688–702.
52. **Jensen, R., and I. Herskowitz.** 1984. Directionality and regulation of cassette substitution in yeast. *Cold Spring Harbor Symp. Quant. Biol.* **49:**97–104.
53. **Jensen, R., F. Sprague, and I. Herskowitz.** 1983. Regulation of yeast mating type interconversion: feedback control of *HO* gene expression by the mating-type locus. *Proc. Natl. Acad. Sci. USA* **80:**3035–3039.
54. **Kelly, M., J. Burke, M. Smith, A. Klar, and D. Beach.** 1988. Four transcriptionally regulated mating type genes control sexual differentiation in the fission yeast. *EMBO J.* **7:** 1537–1547.
55. **Klar, A. J. S.** 1980. Interconversion of yeast mating type by transposable genes. *Genetics* **95:**631–648.
56. **Klar, A. J. S.** 1987. Determination of the yeast cell lineage (minireview). *Cell* **49:**433–435.
57. **Klar, A. J. S.** 1987. The mother-daughter mating type switching asymmetry of budding yeast is not conferred by the segregation of parental *HO* gene DNA strands. *Genes Dev.* **1:**1059–1064.
58. **Klar, A. J. S.** 1987. Differentiated parental DNA strands confer developmental asymmetry on daughter cells in fission yeast. *Nature* (London) **326:**466–470.
59. **Klar, A. J. S., and S. Fogel.** 1976. The action of homothallism genes in *Saccharomyces* diploids during vegetative growth and the equivalence of *hma* and *HMα* loci. *Genetics* **85:** 407–416.
60. **Klar, A. J. S., and S. Fogel.** 1979. Activation of mating type genes by transposition in *Saccharomyces cerevisiae. Proc. Natl. Acad. Sci. USA* **76:**4539–4543.
61. **Klar, A. J. S., S. Fogel, and K. Macleod.** 1979. *MAR1*—A regulator of *HMα* and *HMa* loci in *Saccharomyces cerevisiae. Genetics* **93:**37–50.
62. **Klar, A. J. S., S. Fogel, and D. N. Radin.** 1979. Switching of a mating-type a mutant allele in budding yeast *Saccharomyces cerevisiae. Genetics* **92:**759–776.
63. **Klar, A. J. S., J. B. Hicks, and J. N. Strathern.** 1981. Irregular transpositions of mating-type genes in yeast. *Cold Spring Harbor Symp. Quant. Biol.* **45:**983–990.
64. **Klar, A. J. S., J. B. Hicks, and J. N. Strathern.** 1982. Directionality of yeast mating-type interconversion. *Cell* **28:**551–561.
65. **Klar, A. J. S., S. N. Kakar, J. M. Ivy, J. B. Hicks, G. P. Livi, and L. M. Miglio.** 1985. *SUM1*—an apparent positive regulator of the cryptic mating-type loci in *Saccharomyces cerevisiae. Genetics* **111:**745–758.
66. **Klar, A. J. S., J. McIndoo, J. B. Hicks, and J. N. Strathern.** 1980. Precise mapping of the homothallism genes, *HML* and *HMR*, in yeast *Saccharomyces cerevisiae. Genetics* **96:** 315–320.
67. **Klar, A. J. S., J. McIndoo, J. N. Strathern, and J. B. Hicks.** 1980. Evidence for a physical interaction between the transposed and the substituted sequences during mating-type gene transposition in yeast. *Cell* **22:**291–298.
68. **Klar, A. J. S., and L. M. Miglio.** 1986. Initiation of meiotic recombination by double-stranded DNA breaks in *Schizosaccharomyces pombe. Cell* **46:**725–731.
69. **Klar, A. J. S., and J. N. Strathern.** 1984. Resolution of recombination intermediates generated during yeast mating type switching. *Nature* (London) **310:**744–748.
70. **Klar, A. J. S., J. N. Strathern, and J. A. Abraham.** 1984. Involvement of double-strand chromosomal breaks for mating-type switching in *Saccharomyces cerevisiae. Cold Spring Harbor Symp. Quant. Biol.* **49:**77–88.
71. **Klar, A. J. S., J. N. Strathern, J. R. Broach, and J. B. Hicks.** 1981. Regulation of transcription in expressed and unexpressed mating-type cassettes of yeast. *Nature* (London) **289:**239–244.
72. **Klar, A. J. S., J. N. Strathern, and J. B. Hicks.** 1981. A position-effect control for gene transposition: state of expression of yeast mating-type genes affects their ability to switch. *Cell* **25:**517–524.
73. **Klar, A. J. S., J. N. Strathern, and J. B. Hicks.** 1984. Developmental pathways in yeast, p. 151–195. *In* R. Losick and L. Shapiro (ed.), *Microbial Development.* Cold Spring Harbor Laboratory, Cold Spring Harbor, N.Y.
74. **Kolodkin, A. L., A. J. S. Klar, and F. W. Stahl.** 1986. Double-strand breaks can initiate meiotic recombination in *S. cerevisiae. Cell* **46:**733–740.
75. **Kostriken, R., and F. Heffron.** 1984. The product of the *HO* gene is a nuclease: purification and characterization of the enzyme. *Cold Spring Harbor Symp. Quant. Biol.* **49:**89–96.
76. **Kostriken, R., J. N. Strathern, A. J. S. Klar, J. B. Hicks, and F. Heffron.** 1983. A site-specific endonuclease essential for

mating-type switching in *Saccharomyces cerevisiae. Cell* **35:**167–174.

77. **Kushner, P. J., L. C. Blair, and I. Herskowitz.** 1979. Control of cell types by mobile genes—a test. *Proc. Natl. Acad. Sci. USA* **76:**5264–5268.
78. **Leupold, U.** 1950. Die Vererbung von Homothallie und Heterothallie bei *Schizosaccharomyces pombe. C. R. Lab. Carlsberg Ser. Physiol.* **24:**381–480.
79. **Leupold, U.** 1987. Sex appeal in fission yeast. *Curr. Genet.* **12:**543–545.
80. **Lindegren, D. C., and G. Lindegren.** 1943. A new method for hybridizing yeast. *Proc. Natl. Acad. Sci. USA* **29:**306–308.
81. **Malone, R., and R. E. Esposito.** 1980. The *RAD52* gene is required for homothallic interconversion of mating-types and spontaneous recombination in yeast. *Proc. Natl. Acad. Sci. USA* **77:**503–507.
82. **Mathog, D., M. Hochstrasser, Y. Gruenbaum, H. Saumweber, and J. Sadat.** 1984. Characteristic folding pattern of polytene chromosomes in *Drosophila* salivary gland nuclei. *Nature* (London) **308:**414–421.
83. **McClintock, B.** 1957. Controlling elements and the gene. *Cold Spring Harbor Symp. Quant. Biol.* **21:**197–216.
84. **Miller, A. M., V. L. MacKay, and K. A. Nasmyth.** 1985. Identification and comparison of two sequence elements that confer cell-type specific transcription in yeast. *Nature* (London) **314:**598–603.
85. **Miyata, H., and M. Miyata.** 1981. Modes of conjugation in homothallic cells of *Schizosaccharomyces pombe. J. Gen. Appl. Microbiol.* **27:**365–371.
86. **Nakaseko, Y., Y. Adachi, S. Funahashi, O. Niwa, and M. Yanagida.** 1986. Chromosome walking shows a highly homologous repetitive sequence present in all the centromere regions of fission yeast. *EMBO J.* **5:**1011–1021.
87. **Nasmyth, K. A.** 1982. Molecular genetics of yeast mating type. *Annu. Rev. Genet.* **16:**439–500.
88. **Nasmyth, K. A.** 1983. Molecular analysis of a cell lineage. *Nature* (London) **302:**670–676.
89. **Nasmyth, K. A.** 1985. At least 1400 base pairs of 5′-flanking DNA is required for the correct expression of the *HO* gene in yeast. *Cell* **42:**213–223.
90. **Nasmyth, K. A.** 1987. The determination of mother-cell-specific mating-type-switching in yeast by a specific regulator of *HO* transcription. *EMBO J.* **6:**243–248.
91. **Nasmyth, K. A., A. Seddon, and G. Ammerer.** 1987. Cell cycle regulation of *SWI5* is required for mother-cell-specific *HO* transcription in yeast. *Cell* **49:**549–559.
92. **Nasmyth, K. A., and D. Shore.** 1987. Transcriptional regulation in the yeast life cycle. *Science* **237:**1162–1170.
93. **Nasmyth, K. A., D. Stillman, and P. Kipling.** 1987. Both positive and negative regulators of *HO* are required for mother-cell-specific mating-type switching in yeast. *Cell* **48:** 579–587.
94. **Nasmyth, K. A., and K. Tatchell.** 1980. The structure of transposable yeast mating-type loci. *Cell* **19:**753–764.
95. **Nasmyth, K. A., K. Tatchell, B. D. Hall, C. Astell, and M. Smith.** 1980. Physical analysis of mating type loci in *Saccharomyces cerevisiae. Cold Spring Harbor Symp. Quant. Biol.* **45:**961–981.
96. **Naumov, G. I., and I. I. Tolstorukov.** 1973. Comparative genetics of yeast. X. Reidentification of mutators of mating-types in *Saccharomyces. Genetika* **9:**82–91.
97. **Nickoloff, J. A., E. Chen, and F. Heffron.** 1986. A 24-bp DNA sequence from the *MAT* locus stimulates intergenic recombination in yeast. *Proc. Natl. Acad. Sci. USA* **83:** 7831–7835.
98. **Oshima, Y., and I. Takano.** 1971. Mating types in *Saccharomyces*: their convertibility and homothallism. *Genetics* **67:**327–335.
99. **Piekarowicz, A., T. A. Bickle, J. C. W. Shepart, and K. Ineichen.** 1981. The DNA sequence recognized by the *Hin*fIII restriction endonuclease. *J. Mol. Biol.* **146:**167–172.
100. **Resnick, M. A., and P. Martin.** 1976. The repair of double-strand breaks in the nuclear DNA of *Saccharomyces cerevisiae* and its genetic control. *Mol. Gen. Genet.* **143:**119–129.
101. **Rine, J., and I. Herskowitz.** 1987. Four genes responsible for a position effect on expression from *HML* and *HMR* in *Saccharomyces cerevisiae. Genetics* **116:**9–22.
102. **Rine, J., R. Jensen, D. Hagen, L. Blair, and I. Herskowitz.** 1980. Pattern of switching and fate of the replaced cassette in yeast mating-type interconversion. *Cold Spring Harbor Symp. Quant. Biol.* **45:**951–960.
103. **Rine, J., J. N. Strathern, J. B. Hicks, and I. Herskowitz.** 1979. A suppressor of mating-type locus mutations in *Saccharomyces cerevisiae*: evidence for and identification of cryptic mating-type loci. *Genetics* **93:**877–901.
104. **Russell, D. W., R. Jensen, M. Zoller, J. Burke, B. Errede, M. Smith, and I. Herskowitz.** 1986. Structure of the *Saccharomyces cerevisiae HO* gene and analysis of its upstream regulatory region. *Mol. Cell. Biol.* **6:**4281–4294.
105. **Russell, P., and P. Nurse.** 1986. *Schizosaccharomyces pombe* and *Saccharomyces cerevisiae*: a look at yeasts divided (minireview). *Cell* **45:**781–782.
106. **Santa Maria, J., and D. Vidal.** 1970. Segregacion anormal del "mating type" en *Saccharomyces.* Inst. Nac. Invest. Agron. Conf. **30:**1–8.
107. **Schmidt, H.** 1987. Strains of *Saccharomyces cerevisiae* with a disrupted *swi1* gene still show mating type switching. *Mol. Gen. Genet.* **210:**486–489.
108. **Schmidt, H., P. Kapitza, and H. Gutz.** 1987. Switching genes in *Schizosaccharomyces pombe*: their influence on cell viability and recombination. *Curr. Genet.* **11:**303–308.
109. **Southern, E. M.** 1975. Detection of specific sequences among DNA fragments separated by gel electrophoresis. *J. Mol. Biol.* **98:**503–505.
110. **Stern, M., R. Jensen, and I. Herskowitz.** 1984. Five *SWI* genes are required for expression of the *HO* gene in yeast. *J. Mol. Biol.* **178:**853–868.
111. **Sternberg, P. W., M. J. Stern, I. Clark, and I. Herskowitz.** 1987. Activation of the yeast *HO* gene by release from multiple negative controls. *Cell* **48:**567–577.
112. **Stillman, D. J., A. T. Bankier, A. Seddon, E. G. Groenhout, and K. A. Nasmyth.** 1988. Characterization of a transcription factor involved in mother cell specific transcription of the yeast *HO* gene. *EMBO J.* **7:**485–494.
113. **Strathern, J. N.** 1988. Control and execution of homothallic switching in *Saccharomyces cerevisiae*, p. 445–464. *In* R. Kucherlapati and G. R. Smith (ed.), *Genetic Recombination.* American Society for Microbiology, Washington, D.C.
114. **Strathern, J. N., L. C. Blair, and I. Herskowitz.** 1979. Healing of *mat* mutations and control of mating type interconversion by the mating type locus in *Saccharomyces cerevisiae. Proc. Natl. Acad. Sci. USA* **76:**3425–3429.
115. **Strathern, J. N., and I. Herskowitz.** 1979. Asymmetry and directionality in production of new cell types during clonal growth: the switching pattern of homothallic yeast. *Cell* **17:**371–381.
116. **Strathern, J. N., J. B. Hicks, and I. Herskowitz.** 1981. Control of cell type in yeast by the mating-type locus. The α1-α2 hypothesis. *J. Mol. Biol.* **147:**357–372.
117. **Strathern, J. N., A. J. S. Klar, J. B. Hicks, J. A. Abraham,**

J. M. Ivy, K. A. Nasmyth, and C. McGill. 1982. Homothallic switching of yeast mating type cassettes is initiated by a double-stranded cut in the *MAT* locus. *Cell* **31**:183–192.

118. **Strathern, J. N., C. S. Newlon, I. Herskowitz, and J. B. Hicks.** 1979. Isolation of a circular derivative of yeast chromosome III: implications for the mechanism of mating type interconversion. *Cell* **18**:309–319.
119. **Strathern, J. N., E. Spatola, C. McGill, and J. B. Hicks.** 1980. Structure and organization of transposable mating type cassettes in *Saccharomyces* yeasts. *Proc. Natl. Acad. Sci. USA* 77:2839–2843.
120. **Szostak, J. W., T. L. Orr-Weaver, R. J. Rothstein, and F. W. Stahl.** 1983. The double-strand-break repair model for recombination. *Cell* **33**:25–35.
121. **Takahashi, T.** 1958. Complementary genes controlling homothallism in *Saccharomyces. Genetics* **43**:705–714.
122. **Takano, I., T. Kusumi, and Y. Oshima.** 1973. An α mating-type allele insensitive to the mutagenic action of the homothallism genes system in *Saccharomyces cerevisiae. Mol. Gen. Genet.* **126**:19–23.
123. **Takano, I., and Y. Oshima.** 1970. Mutational nature of an allele specific conversion of the mating type by the homothallism gene *HOα* in *Saccharomyces. Genetics* **65**:421–427.
124. **Tanaka, K., T. Oshima, H. Araki, S. Harashima, and Y. Oshima.** 1984. Mating type control in *Saccharomyces cerevisiae*: a frameshift mutation at the common DNA sequence, X, of the *HMLα* locus. *Mol. Cell. Biol.* **4**:203–211.
125. **Tolstorukov, I. I., and S. V. Benevolensky.** 1980. Study of mechanisms of mating and self-diploidization in haploid yeast *Pichia pinus*. II. Mutations in the mating type locus. *Genetika* **16**:1335–1341.
126. **Weiffenbach, B., and J. E. Haber.** 1981. Homothallic mating type switching generates lethal breaks in *rad52* strains of *Saccharomyces cerevisiae. Mol. Cell. Biol.* **6**:522–534.
127. **Weiffenbach, B., and J. E. Haber.** 1985. Homothallic switching of *Saccharomyces cerevisiae* mating type genes by using a donor containing a larger internal deletion. *Mol. Cell. Biol.* **5**:2154–2158.
128. **Weiffenbach, B., D. T. Rogers, J. E. Haber, M. Zoller, D. W. Russell, and M. Smith.** 1983. Deletion and single base pair changes in the yeast mating type locus that prevent homothallic mating type conversions. *Proc. Natl. Acad. Sci. USA* **80**:3401–3405.
129. **Williamson, D. H., and D. J. Fennell.** 1981. Non-random assortment of sister chromatids in yeast mitosis, p. 90–102. *In* D. Von Wettstein, J. Friis, M. Kielland-Brant, and A. Stenderup (ed.), *Molecular Genetics in Yeast.* Munksgaard, Copenhagen.
130. **Winge, O.** 1935. On haplophase and dilophase in some *Saccharomyces. C. R. Trav. Lab. Carlsberg Ser. Physiol.* 27:77.
131. **Winge, O., and C. Roberts.** 1949. A gene for diploidization of yeast. *C. R. Trav. Lab. Carlsberg Ser. Physiol.* **24**: 341–346.
132. **Yasuhisa, F., Y. Kaziro, and M. Yamamoto.** 1986. Mating pheromone-like diffusible factor released by *Schizosaccharomyces pombe. EMBO J.* **5**:1991–1993.

Chapter 31

Immunoglobulin Heavy-Chain Class Switching

STUART G. LUTZKER and FREDERICK W. ALT

Stuart G. Lutzker and Frederick W. Alt ■ Howard Hughes Medical Institute and Departments of Biochemistry and Microbiology, The College of Physicians and Surgeons of Columbia University, 630 W. 168th Street, New York, New York 10032.

I. INTRODUCTION

Antibodies or immunoglobulins are specialized proteins that protect the body from invading pathogens and their toxins. B-lineage cells initially express antibodies on their surface as receptors for foreign antigens and, upon stimulation with antigen, differentiate into cells that secrete antibodies into the serum, where they take part in various types of immune reactions. An antibody is composed of two identical heavy (H)-chain proteins and two identical light (L)-chain proteins (Fig. 1). The amino-terminal portions of these two proteins (designated the variable or V region) interact to form the antigen-binding sites of the antibody molecule. The DNA sequences that encode both the H- and L-chain variable regions are assembled from component gene segments during B-lymphocyte development. The combinatorial assortment of these gene segments helps give rise to the vast number of different variable regions necessary to recognize all foreign antigens. The remaining portion of the H- and L-chain proteins is termed the constant (C) region and is encoded by a constant-region gene segment. The heavy-chain locus in mice and humans contains a number of constant-region (C_H) genes located downstream from the rearranged V_H genes. The C_H genes encode different heavy-chain proteins termed isotypes. The constant region of the heavy-chain protein determines effector functions of the antibody molecule. Effector functions include properties such as complement fixation, movement from the bloodstream across the placenta or into interstitial fluids and secretions, and targeting of absorption onto the surface of other hematopoietic cells where interaction with antigen triggers release of substances such as histamine.

Normally, only the C_H gene located immediately 3′ to the productively rearranged V_H gene gives rise to heavy-chain proteins. In newly formed B cells, the C_μ gene is 3′ to the V_H gene and gives rise to μ heavy chains. B-lineage cells terminally differentiated to secrete high levels of antibody frequently produce heavy-chain proteins of other isotypes. This isotype switch results from a DNA recombination event termed class switching that juxtaposes the assembled V_H gene to a new C_H gene other than C_μ. Given the different array of effector functions particular for each heavy-chain isotype, the question of how class switching is regulated is of central importance in understanding how the immune system efficiently reacts to various types of antigens.

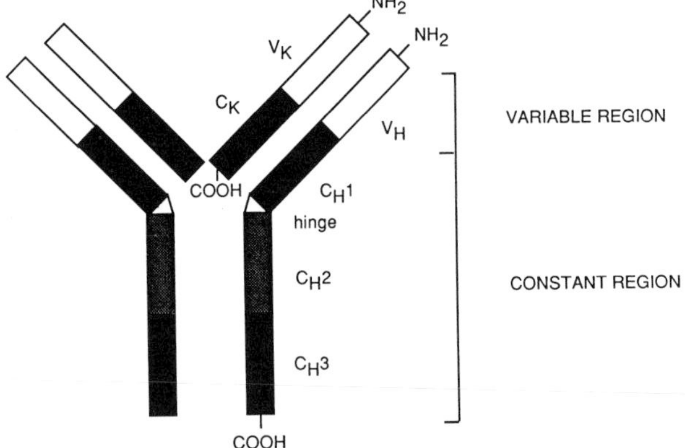

Figure 1. Structure of an immunoglobulin molecule with two identical γ heavy chains and two identical κ light chains. The V_H, C_H, V_κ, and C_κ homology domains are shown as boxes. The hinge region separating the C_H1 and C_H2 domains is depicted as a triangle. The four proteins are linked together via disulfide bonds.

II. VARIABLE-REGION GENES

A. Organization of Variable-Region Gene Segments

A complete heavy-chain variable-region gene is assembled from three distinct groups of germ line gene sequences, the variable (V_H), diversity (D), and joining (J_H) segments (Fig. 2) (for review, see references 114 and 144). The J_H segments (four in mice, six in humans) lie several kilobases upstream from the C_μ constant-region gene. In mice, 12 known D segments lie 1 to 80 kilobases (kb) upstream from the J_H segments; the organization and extent of the human D-segment locus has not been elucidated in detail. There are hundreds of V_H segments in the mouse and human genomes. The large-scale organization of the V_H locus has been studied in more detail in the human than the mouse genome (12). This region covers approximately 2,500 kb of DNA and begins approximately 100 kb 5′ to the constant-region locus. Light-chain variable regions are assem-

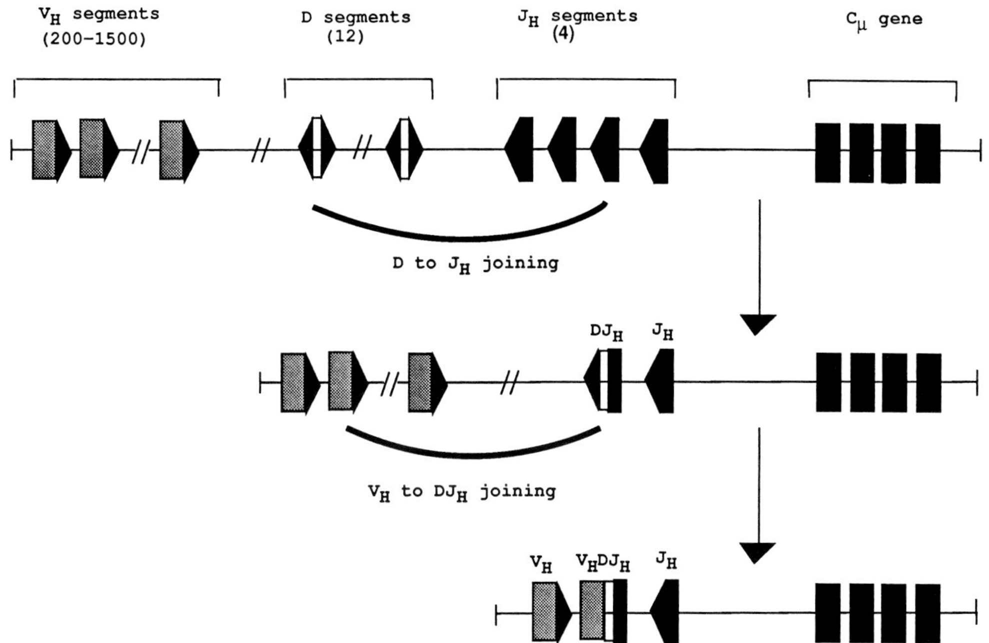

Figure 2. Organization of the mouse heavy-chain variable-region locus and V-D-J joining. Boxes and triangles represent gene segments and heptamer-nonamer recognition elements, respectively. The recognition elements flanking the V_H and J_H segments contain 23-bp spacers; recognition elements flanking the D segments contain 12-bp spacers. The two recombination events resulting in a fully assembled heavy-chain variable-region gene are shown.

bled from V_L and J_L gene segments and do not contain D segments (144). The general organization of the light-chain variable-region locus is similar to that of the heavy chain (144).

B. Regulated Assembly of Variable-Region Genes

All variable-region gene segments, including those which are assembled to form antigenic receptors expressed on the surface of T cells, are flanked by a highly conserved heptamer and a characteristic nonamer sequence which are separated by either a 12- or a 23-base-pair (bp) spacer (for review, see reference 53). The length of the spacer region is apparently crucial for directing the recombination process, because joining occurs only between gene segments which are flanked, respectively, by recognition sequences containing 12- or 23-bp spacers (40, 120). The assembly of heavy- and light-chain variable regions in developing B cells occurs in the fetal liver and the adult bone marrow. Heavy-chain variable-region gene assembly and expression generally precedes that of light chains during B-lymphocyte differentiation (for review, see reference 156). The assembly of a complete heavy-chain variable-region gene from the component gene segments follows an ordered process (8) (Fig. 2). The first step involves D-to-J_H recombination, which generally occurs on both chromosomes; subsequently, V_H segments are appended to the DJ_H intermediate. Expression of the assembled heavy-chain variable-region gene with the downstream C_μ gene segments leads to the production of μ heavy-chain protein, which is found in the cytoplasm of the earliest identified immunoglobulin-containing B-lineage cells, pre-B cells. Upon subsequent assembly to light-chain variable-region genes, pre-B cells give rise to the next B-lineage stage, B lymphocytes, which express both heavy- and light-chain proteins together on their cell surface (sIg^+).

Assembly of variable regions is mediated by a recombination system (referred to as VDJ recombinase) which cuts at the border of the recognition sequences and the coding regions of the involved gene segments (3; for review, see reference 5). Religation occurs between the coding ends to generate a coding joint, and between the ends containing the nonamer-heptamer sequences to create a "reciprocal" joint. Depending on the orientation of the recombining gene segments in the chromosome, this joining mechanism either excises the region of intervening DNA as a free circle or inverts this region without excision (for review, see reference 4). The actual joining of coding segments is imprecise, with the addition and loss of nucleotides at the coding joint. This imprecision generates novel coding sequences in these junction regions and is a mechanism for generating antibody diversity (144). However, another frequent result of this imprecise joining process is that an assembled V_HDJ_H rearrangement often occurs with the V and J segments in different translational reading frames, leading to non- or incompletely translatable heavy-chain transcripts (referred to as nonproductive rearrangements; for review, see reference 4). Production of a heavy-chain protein from a productive rearrangement has been postulated to signal an end to heavy-chain variable-region assembly and to be involved in initiating light-chain variable-region assembly (117; for review, see reference 156).

There are two types of light-chain proteins, termed κ and λ; in both the mouse and human genomes, the genes encoding these light chains are located on separate chromosomes. Light-chain variable-region assembly is also ordered; V_κ-to-J_κ rearrangements usually occur before V_λ-to-J_λ rearrangements (for review, see reference 13).

The ordered (regulated) assembly of immunoglobulin variable-region genes apparently has evolved to ensure that the antibody molecules expressed on the surface of a given B cell are of only one antigen specificity. Interaction of these surface receptors with cognate antigen stimulates resting B cells to proliferate and differentiate into plasma cells which secrete large amounts of antibody; the productive rearrangement of one heavy-chain and one light-chain variable region ensures that the antibody produced is specific for the stimulating antigen. This clonal specificity is the result of allelic exclusion (22, 102) and may be an intrinsic requirement for an immune system based upon clonal selection.

C. An "Accessibility" Model for Control of Variable-Region Rearrangement

The tissue- and stage-specific rearrangement of immunoglobulin and T-cell receptor (TCR) variable-region gene segments requires that the recombination enzymes present in B and T cells be able to distinguish the various gene segments. However, the recognition sequences which flank the component gene segments are quite similar, with no consistent differences that could account for the ordered patterns of rearrangement (53, 144). The fact that the recognition sequences flanking these diverse gene segments are so conserved suggested that they are joined by a common variable-region recombinase. In support of this hypothesis is the finding that exogenous light-chain or TCR variable-region gene segments recombine when introduced on a plasmid into Abelson murine leukemia virus (A-MuLV)-transformed pre-B cells which do not rearrange the corresponding endogenous gene segments (14, 157). These exogenous light-chain and TCR variable-region gene segments, unlike their endogenous counterparts, appear to have a more open chromatin configuration and are believed to be more "accessible" for variable-region recombination. The stage- and tissue-specific control of variable-region gene assembly therefore appears to be regulated by modulating the accessibility of the flanking recognition sequences to a common variable-region recombinase (155).

Although the exact definition of accessibility is not known in molecular terms, studies of variable-region joining in A-MuLV-transformed pre-B cells indicate that accessibility of germ line gene segments correlates with their transcription (14, 155). For example, germ line V_H segments are transcribed in A-MuLV-transformed pre-B cells which are undergoing V_H-to-DJ_H joining but not in cells that contain productive V_HDJ_H (155). Furthermore, recent experiments in transgenic mice indicate that *cis*-acting DNA elements associated with unrearranged TCR V gene segments (perhaps the tissue-specific promoter) promote TCR V-to-DJ rearrangement in T cells but not in B cells (P. Ferrier, F. W. Alt, et al., submitted for publication).

III. CONSTANT-REGION GENES

A. Organization of the Murine C_H Locus

The entire murine C_H locus (chromosome *12*) has been linked on overlapping cloned DNA segments (38). The organization of the eight C_H genes in BALB/c mice is 5′ (variable-region genes)-C_μ-C_δ-$C_\gamma3$-$C_\gamma1$-$C_\gamma2b$-$C_\gamma2a$-C_ε-C_α-3′ (126) (Fig. 3). When paired with light-chain proteins, the eight different heavy-chain proteins encoded by these C_H genes give rise to eight classes of antibodies termed immunoglobulin M (IgM), IgD, IgG3, IgG1, IgG2b, IgG2a, IgE, and IgA. The various C_H genes are thought to have evolved from a primordial C_H gene most similar to C_μ that contained four C_H exons encoding distinct protein domains (75, 118). Gene duplication followed by functional divergence could have given rise to the other C_H genes. C_ε resembles C_μ in that it has four C_H exons (58, 77), while the various C_γ genes and C_α each contain three C_H exons with a short hinge exon separating the C_H1 and C_H2 exons (121, 146, 147) (Fig. 1). The hinge region of the heavy chain imparts segmental flexibility to the two antigen-binding sites of the antibody molecule, allowing for optimal antibody binding to two closely spaced antigenic determinants. C_δ contains only two C_H exons, which are separated by a hinge exon (145). Sequence analyses revealed extensive segmental homology between pairs of C_γ genes (86). For example, based upon nucleotide differences in introns and at

Mouse constant region genes:

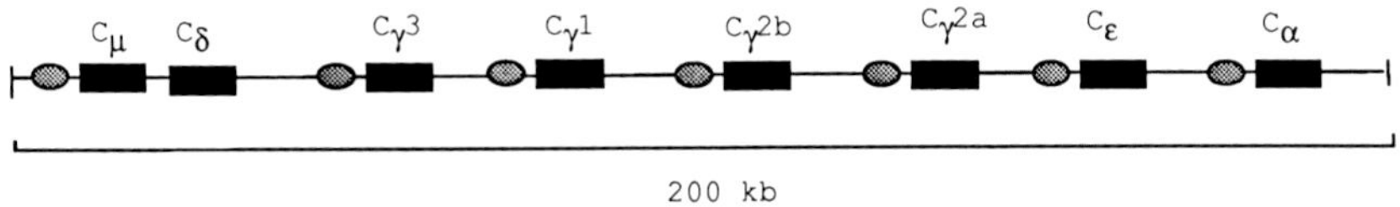

Human constant region genes:

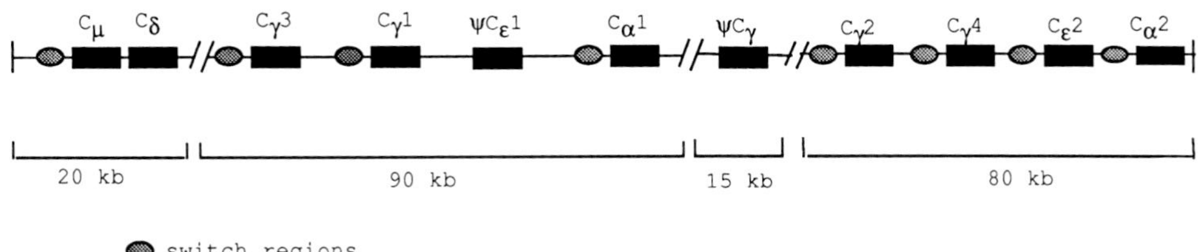

Figure 3. Organization of mouse and human constant-region genes. Boxes represent C_H genes and ovals represent switch regions. The approximate size of each gene cluster is also noted.

silent positions (third base pair) in exons, the mouse γ2b and γ2a genes are more homologous in the C_H2 region and flanking intervening sequences than in other parts of the gene (51). This sharing of genetic information between C_H genes may have arisen through gene conversion or genetic recombination at intervening sequences (exon shuffling). The ongoing evolutionary changes associated with the murine C_H locus are apparent in the case of Japanese wild mice, which have undergone a duplication of the $C_\gamma2a$ gene (125).

B. Organization of the Human C_H Locus

There are nine functional human C_H genes and two pseudo (nonfunctional) C_H genes located on chromosome *14* (73); an additional processed pseudogene ($C_\varepsilon3$) is found on chromosome 9 (10). The organization of the human C_H genes is considerably different from that of the BALB/c mouse. A C_γ-C_γ-C_ε-C_α multigene unit is duplicated downstream of the C_μ and C_δ genes (46) with a pseudo C_γ gene separating the two units (11). The organization of the human C_H locus is as follows: 5′ (variable-region genes)-C_μ-C_δ-$C_\gamma3$-$C_\gamma1$-$\psi C_\varepsilon2$-$C_\alpha1$-ψC_γ-$C_\gamma2$-$C_\gamma4$-$C_\varepsilon1$-$C_\alpha2$-3′ (46, 116) (Fig. 3). Although the functional human C_γ genes are not clustered as they are in the mouse, the four human γ isotype proteins show greater amino acid homology (>90% in nonhinge regions) than do those of the mouse (<80%) (51). This finding suggests that the human C_γ genes may have diverged more recently than their murine counterparts. Segmental homologies similar to those between mouse C_γ genes are also found among the human C_γ genes, indicating that the human C_γ genes also have coevolved (141).

C. Expression of Membrane-Bound and Secreted Heavy-Chain Proteins

Each heavy-chain protein can be expressed as membrane-bound or secreted forms, which differ in their carboxy-terminal sequences (6, 41, 118) (Fig. 4). In the secreted form, the carboxy-terminal amino acids are encoded by the most 3′ C_H exon (C_H4 for C_μ, C_H3 for the C_γ genes, etc.). The transmembrane and short intracytoplasmic portions of the membrane-bound form are encoded by two separate exons (except for C_α, in which they are encoded by a single exon) located downstream of the main C_H exons. These membrane exons are spliced to the 3′-most C_H exon, thereby removing the last few nucleotides that encode the secreted tail. The distinctive mRNA sequences that encode the membrane-bound and secreted forms are generated by differential utilization of cleavage/polyadenylation sites located 5′ and 3′ of the membrane exons and alternate RNA splicing, as outlined in Fig. 4 (69). The intracytoplasmic portions are of various lengths (2 amino acids for C_μ and C_δ genes, 27 amino acids for C_γ genes) and may interact with cellular proteins to transduce the signal generated by antibody-antigen interaction. The transmembrane segment of the heavy-chain protein consists of 26 amino acids, which generally are conserved between C_H regions and are perhaps anchored by a common membrane

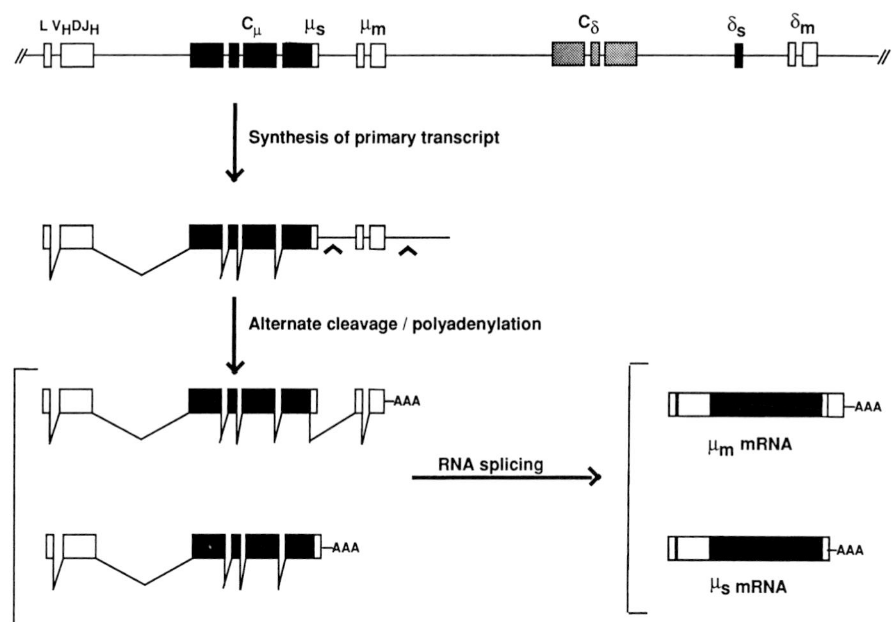

Figure 4. Synthesis of mRNA encoding μ membrane-bound and secreted heavy-chain protein. Exons are denoted as boxes and introns as lines. The two mRNAs differ at their 3′ ends as a result of alternate RNA cleavage and splicing as depicted.

protein (154). Resting B cells synthesize relatively more heavy-chain mRNA encoding the membrane-bound form than that encoding the secreted form. After differentiation into plasma cells, the predominant heavy-chain mRNA produced encodes the secreted form (6, 104, 118).

D. Expression of IgD

Most newly formed B cells express IgM, or both IgM and IgD, on their surface (103). The simultaneous expression of μ and δ heavy-chain proteins results from the production of primary heavy-chain transcripts which continue through the C_μ locus and terminate downstream of the C_δ gene (74, 80) (Fig. 5). Some of these primary transcripts undergo cleavage and polyadenylation immediately downstream of the C_μ exons and give rise to mRNA encoding μ heavy-chain proteins. The production of mRNA encoding δ heavy-chain protein results from the splicing of the primary transcript so that the variable-region exon is joined to the C_δ exons (Fig. 5; for review, see reference 15). It has been hypothesized that isotypes encoded by C_H genes located downstream of C_δ could also be produced by a similar transcriptional mechanism (see below).

IV. HEAVY-CHAIN CLASS SWITCHING

A. DNA Recombination and Isotype Expression

DNA rearrangment within the heavy-chain locus, referred to as class switch recombination, is the common mechanism associated with the high-level expression of non-IgM/IgD antibodies. The deletion of intervening C_H genes first was demonstrated in mouse myeloma cell lines (30, 33, 52, 63, 109, 159), which derive from tumor cells that represent the Ig-secreting plasma-cell stage of B-cell development. For example, in an IgG3-producing myeloma line, the C_H genes 5′ to the expressed $C_\gamma3$ gene (C_μ and C_δ) are deleted and the $C_\gamma3$ gene is brought into a position just downstream from the V_HDJ_H locus (Fig. 6). Subsequent analyses of genomic DNA from various other types of B-lineage tumor cell lines and nontransformed B cells (113) yielded results consistent with this general type of class switch recombination mechanism.

B. Switch Regions

The sites of class switch recombination are clustered in a region of DNA upstream from individual C_H genes. These regions are composed of highly repetitive sequences termed switch (S) regions (35, 62, 80). In the mouse genome, S regions range in size from 1 kb (S_ε) to 10 kb ($S_\gamma1$) and are located a few kilobases upstream from each C_H gene except C_δ (Fig. 3). Switch region repeats share two common pentameric sequences, GAGCT and TGGGG (92, 133). The 3′ portion of S_μ is made up of tandemly repeated blocks of these two sequences (91). The 5′ portion of S_μ contains the repeat sequence YAGGTTG (Y = pyrimidine), which also is found in other switch regions (82). The repeats associated with the

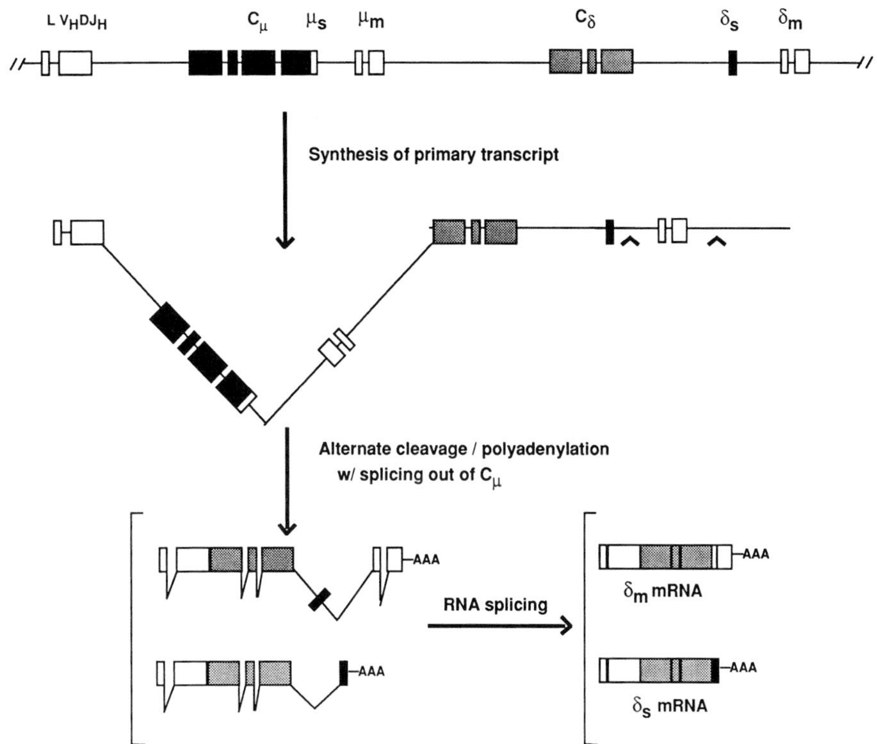

Figure 5. Synthesis of mRNA encoding δ membrane-bound and secreted heavy-chain protein. Exons are depicted as boxes and introns as lines. Transcription proceeds past the C_μ exons and terminates downstream of the C_δ exons. Splicing out of the intervening RNA separating the variable-region and C_δ genes leads to the production of δ, rather than μ, mRNA as depicted.

other switch regions demonstrate a higher level of organization than does S_μ. The γ switch regions are composed of repeat groups of 245 or 295 bp, each composed of five or six repeat units of 49 bp each (63, 88, 91, 92, 133, 139). The overall homology between the 49-bp repeats within a given switch region ranges from 45 to 82%, indicating a lack of strict sequence conservation. The tandem repeats 5′

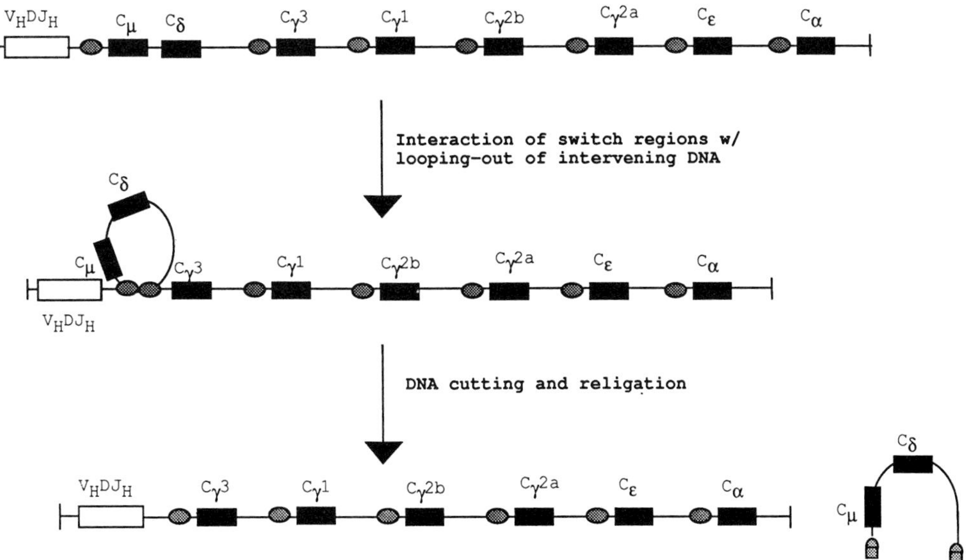

Figure 6. Recombination/deletion model for class switch recombination. The murine heavy-chain locus is depicted as in Fig. 3. A primary μ-to-γ3 class switch is shown, resulting in deletion from the chromosome of the intervening DNA as a linear DNA segment. This DNA is normally lost from the cell (see text for details).

of C_ε are 40 bp long (92), while those upstream of C_α are 80 bp long (35, 94). Based upon the prevalence of the two pentameric sequences mentioned above, S_ε and S_α show the highest degree of homology with S_μ (92, 133); the homology of S_γ regions to S_μ is much less, with the homology decreasing with the position along the chromosome ($S_\gamma3 > S_\gamma1 > S_\gamma2b > S_\gamma2a$). Mouse and human S_μ sequences are composed of the same pentameric sequences, and the homology between S_μ and the other human S regions follows the same pattern as that seen in the mouse sequences (S_ε, $S_\alpha > S_\gamma3 > S_\gamma1 > S_\gamma2 > S_\gamma4$), indicating a general conservation of switch sequences (110, 115, 140).

C. DNA Sequences at Recombination Sites

DNA sequences at the sites of class switch recombination have been determined in a number of transformed B-lineage cells. Comparison of these sequences indicates that no common sequence motif is found near all recombination sites (e.g., reference 92). Thus, switch recombination differs from the site-specific recombination which is directed by the heptamer-nonamer sequences flanking variable-region gene segments. In this regard, class switch recombination occurs within large DNA regions between the various C_H genes, the retained portion of which becomes introns; thus, the exact site of DNA breakage and religation does not affect the reading frame of the processed heavy-chain transcript and the points of recombination need not be precise. However, the sites of class switch recombination do not appear to be randomly distributed over a given switch region. For example, recombination sites within the γ switch regions tend to cluster in parts of the 49-bp repeats which bear the strongest homology with the S_μ-like pentamers GAGCT and GGGGT (105). Many S_μ recombination sites also occur in the 3′ region of S_μ that is composed primarily of these pentamers (reviewed in reference 66). Thus, although there appear to be no strict sequence requirements for class switch recombination, it is possible that the pentamers are recognized preferentially by the class switch recombinase. In many B-lineage cell lines, sites of S_μ rearrangement occur in a 5′ region of S_μ that is composed almost entirely of the repeated sequence YAGGTTGG, suggesting that this sequence also may be preferentially recognized by the class switch enzymes (82).

D. Switch Region Deletions and Secondary Switches

Most of what is known about class switching comes from studies that compare the sites of class switch recombination events in particular B-lineage cell lines (usually transformed cells) to the corresponding germ line DNA sequences. One potential problem with this type of analysis is that internal deletions occur within switch regions. If such deletions occurred after a class switch recombination event, they could obscure the actual site of class switch recombination. Switch region deletions frequently occur in the S_μ region of transformed cell lines (7, 36, 97) and normal B cells (55); similar deletions have been characterized within the S regions upstream of C_γ genes as well (105, 158). It seems likely that such deletions are mediated by class switch recombinase; a number of cell lines with known class switch recombination activity accumulate such deletions during growth in culture (for review, see reference 4). Deletions also occur in switch region sequences that have been molecularly cloned into bacteriophages or plasmids when propagated in bacteria; however, these events seem to be mediated by general homologous recombination mechanisms (64, 81). Another potential difficulty in analyzing switch recombination sites in cells lines is that B cells can also undergo successive class switch events, such as μ to γ2b to ε (48, 92, 94, 105). Thus, if a primary switch (μ to γ2b) resulted in deletion of most of the 3′ portion of S_μ (which contains the pentameric sequences), the secondary switch (γ2b to ε) might then be forced to utilize the YAGGTTG sequences in the 5′ portion of S_μ (66).

E. Recombination of Transfected Switch Regions

An approach to delineating the sequences required for class switching that may avoid some of the potential problems encountered in examining endogenous events would be to introduce class switch recombination substrates that contain different switch sequences into B-lineage cells that are known to have class switch activity (see below), and subsequently to compare the frequency and mechanism of recombination. This general approach has been very rewarding in the study of VDJ recombination (for review, see reference 5), and such a system recently has been utilized to examine the frequency of switch recombination in B-lineage cells versus non-B-lineage cells (97) (see below and Fig. 7 for more details). In the latter studies, the S_μ-to-$S_\gamma2b$ recombinations within the construct utilized the 3′-GAGCT, GGGGT-rich region of S_μ and not the 5′-YAGGT TGG-rich region, further supporting the notion that the pentameric sequences are preferentially recognized by the class switch enzymes. Additional class switch recombination substrates could be constructed to examine how the overall size and compo-

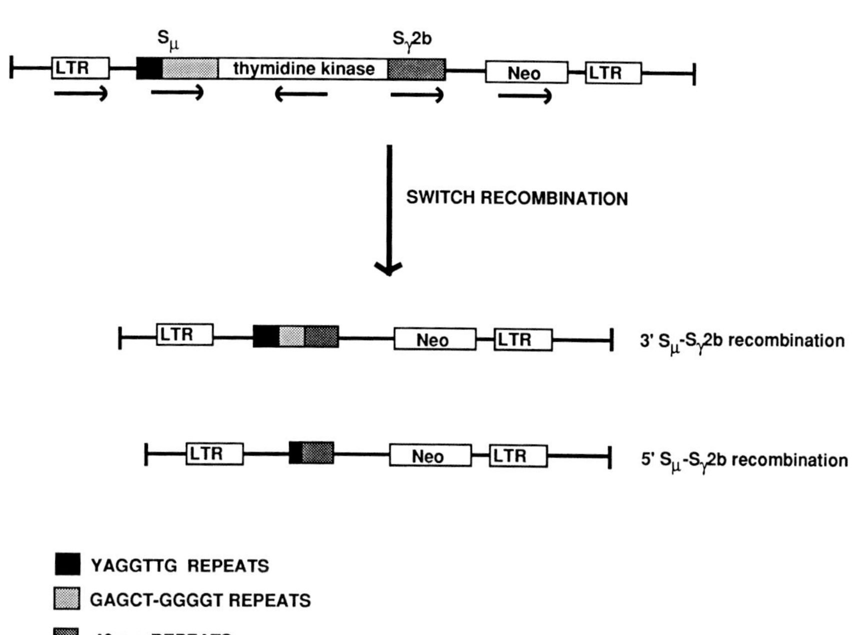

Figure 7. Recombination of exogenous switch sequences within a retroviral construct. Relevant DNA segments are depicted as boxes. The construct was introduced into cell lines deficient in thymidine kinase activity (TK^-). The thymidine kinase gene is expressed from a herpes simplex virus promoter. Switch regions are transcribed from promoter/enhancer elements located within the viral long terminal repeat (LTR) sequences. Cells can be selected for either expression or loss of the thymidine kinase gene by growth in selective media. Two possible switch region-mediated recombinations are depicted.

sition of switch regions influence the frequency of switch recombination, as well as to test other aspects of the recombination mechanism.

V. MECHANISM OF CLASS SWITCHING

A. Intrachromosomal Switching

Switch region recombination results in the juxtaposition of the assembled V_H gene to a new C_H gene and usually occurs intrachromosomally (150, 159). The steps involved in this process are outlined in Fig. 6. The initial event is depicted as the interaction between the two intact switch regions, with the excision of the intervening DNA; however, it is also possible that one switch region is cleaved initially, and the free end generated then interacts with the other intact switch region, leading to the ultimate class switch recombination. As illustrated (Fig. 6), recombinations between two switch regions would result in a deletion of the intervening C_H genes from the chromosome. Although intervening C_H genes are usually lost from cells after a class switch event, they occasionally are found to be retained elsewhere in the genome (e.g., reference 36). The intervening DNA segments between joined variable-region genes are probably deleted as a circle (47, 96). The existence of recombination events within class switch regions leading to inversions supports the possibility that such sequences could also be deleted as a circle (59, 158). However, in the one case where the deleted product of a class switch event was analyzed, it appeared to have been deleted as a linear sequence (37).

B. Interchromosomal Switching

The possibility of switching between chromosomes or sister chromatids was initially invoked to explain the presence of a short stretch of S_α sequences between the S_μ and $S_\gamma 1$ segments in a myeloma cell line (35, 94). However, such recombination events are probably not frequent, as they would result in the accumulation of cells with duplicated C_H genes on the same chromosome. Such duplications do not correlate with the loss of C_μ genes in populations of B cells stimulated to undergo switching (113), nor have they been detected in subclones of cell lines which actively undergo class switching in culture (150). Thus, although interchromosomal switching probably does occur at some frequency (72), the intrachromosomal deletion mechanism appears to account for the majority of class switch events.

C. A Distinct Mechanism Promotes Switch Region Recombination in B-Lineage Cells

The highly repetitive nature of S-region sequences and the lack of specific recombination sequences at switch sites led to the suggestion that class switching may be mediated through a general homologous recombination mechanism rather than a switch region-specific recombination mechanism. It was noted that switch sequences resemble Chi sequences (70), which stimulate generalized recombination in bacteria (128). Furthermore, switch regions recombine when propagated in bacteria or when incubated in a bacterial extract (64), indicating that these sequences are recombinogenic in procaryotes. In these studies, a preference was seen for recombination sites close to TGAG and TGGG sequences, sites which are similar to those seen in mouse myelomas. Homologous recombination between switch sequences also appears to occur in the germ cells of mice since the length of unrearranged switch regions varies extensively in inbred and wild mouse strains (81).

To test whether switch regions recombine more efficiently in B-lymphoid cells than in nonlymphoid cells, the switch recombination constructs mentioned above (97) (Fig. 7) were introduced into A-MuLV-transformed pre-B cells which spontaneously undergo class switching in culture and into transformed mouse fibroblasts (L cells). The construct contained a selectable marker gene (that for thymidine kinase) between the S_{μ} and $S_{\gamma}2b$ sequences. Recombination between the S_{μ} and $S_{\gamma}2b$ segments results in deletion of the thymidine kinase gene and allows the growth of cells harboring the rearranged construct in bromodeoxyuridine. The frequency of bromodeoxyuridine-resistant clones in the pre-B populations ranged from 1×10^{-5} to 5×10^{-5} cells per generation, while in the fibroblast line the frequency was 4×10^{-8} cells per generation. Furthermore, the majority of the bromodeoxyuridine-resistant L-cell subclones inactivated the thymidine kinase gene by a nondeletional mechanism, indicating that the actual rate of S-region recombination in that line is much lower than measured. Although more cell lines must be analyzed to reach a firm conclusion, the much higher class switch recombination frequency in pre-B cells compared to L cells suggests that generalized recombination is not responsible for S-region recombination in B-lineage cells or that this system is somehow modified in B-lineage cells so that switch repeats are recognized more efficiently.

VI. DIRECTED CLASS SWITCHING

A. Class Switching May Be Regulated to Promote Isotype Expression

The majority of circulating B cells in vivo express IgM, or both IgM and IgD, on their surface. In a primary immune response, the reaction of these B cells with the appropriate antigen triggers their proliferation and differentiation into plasma cells which secrete IgM or other isotypes. The isotypes produced in an immune response to a particular antigen often are not random. Murine antibody responses to carbohydrates and soluble proteins generally are restricted to IgG3 (and IgG2b) and IgG1, respectively (101, 119). Infection of mice with DNA and RNA viruses elicits predominantly an IgG2a immune response (34), while parasitic infections elicit an IgE response (61). The ability of different antigens to elicit different isotype responses could result from several, not mutually exclusive, mechanisms. For example, the production of non-IgM antibodies requires class switch recombination which could potentially occur either randomly (30) or through a directed (nonrandom) process (18). If specific switch regions are targeted for recombination, then class switching could be regulated to promote the expression of particular isotypes. Isotype expression could also be regulated by the selective stimulation of a subpopulation of B cells whose C_H genes have already switched, either randomly or by a directed process, to a particular isotype (31). Therefore, the question of whether class switching occurs by random or directed mechanisms is pivotal in elucidating mechanisms that control isotype expression in vivo.

B. Directed Switching in Murine B Cells

Evidence for directed switching comes from analyses of C_H rearrangements in mouse B cells. In one set of experiments (112), spleen cells from C57BL/6 × BALB/c F_1 mice were stimulated to produce IgG1 (see below) and cell populations were isolated that expressed $\gamma 1$ heavy chains encoded by the C57BL/6 allele. Because the two IgH loci in these mice exhibit restriction fragment length polymorphisms of the C_{μ}, $S_{\gamma}1$, and $S_{\gamma}2b$ regions, the extent of rearrangement at each of the two alleles could be compared by Southern blotting. The IgG1-expressing cells showed complete loss of the C_{μ}-hybridizing restriction fragment from the nonproductive allele as well as the productive allele, indicating that both alleles had undergone class switch rearrangement. In addition, the same extent of $S_\gamma 1$ rearrangement was detected on both alleles, while the $S_\gamma 2b$ regions were unrearranged. These data suggest that both alleles had switched to $\gamma 1$. Furthermore, analysis of 21 IgG1-producing hybridoma cell lines which contained nonproductive allele switch recombinations indicated that 17 had switched to $\gamma 1$ on the nonproductive allele as well (152). Hybridoma cell lines expressing other γ isotypes demonstrate a similar pattern of switching on the productive and nonproductive alleles (54). In addition, aberrant class switch events resulting in *c-myc* proto-oncogene translocations frequently involve the same switch region on the nonproductive allele as is rearranged on the productive allele (see below and reference 32). Rearrangements to the same switch region on allelic chromosomes would not be expected if switching were a random process and strongly suggest that specific switch regions can be targeted for recombination.

C. Directed Switching in Human B Cells

C_H rearrangements in human B-cell tumors appear to be more random (17, 151). In one study, only two of seven human B-cell tumor lines which had undergone class switching on both alleles had switched to the same C_H gene; these results are more consistent with a stochastic switching model (17). However, the limited sample size may have influenced the pattern observed in these studies. Also, some of the tumor lines may have undergone multiple class switch rearrangements in vivo prior to transformation; this is less likely to have happened in the mouse IgG1-producing hybridomas discussed above because splenic B cells were stimulated to produce IgG1 and immediately transformed. However, given the fact that the human IgH locus is organizationally more complex than that of the mouse, with duplication and interspersion of C_γ, C_ε, and C_α genes, it is possible that the apparently different results from the murine and human analyses may reflect different mechanisms that evolved in different species to promote expression of C_H genes.

D. Class Switching in B-Lineage Cell Lines

The possibility of directed switching also has been suggested by the switching patterns observed in certain cultured murine B-lineage cell lines. The I.29 B-cell lymphoma undergoes switching from μ to α when stimulated with the B-cell mitogen bacterial lipopolysaccharide (LPS) in vitro (135, 136), while numerous independently isolated A-MuLV-transformed pre-B cells spontaneously switch from μ to $\gamma 2b$ (2, 7, 18, 19, 36, 78, 137, 158). Molecular analyses confirmed that these isotype switches resulted from the recombination of switch regions and thus represented bona fide class switch events. I.29 B cells also switch less frequently to ε and $\gamma 2a$, and one A-MuLV-transformed pre-B cell line spontaneously switches to both $\gamma 2b$ and $\gamma 3$ (137). However, the majority of the switch events in these cell lines appear to involve a specific switch region. These data demonstrate that directed switching can occur in murine B-lineage cells, at least in vitro. The predisposition of I.29 B cells and A-MuLV-transformed pre-B cells to switch to restricted isotypes contrasts with the finding that the progeny of a single IgM-producing murine B cell, when transferred into an irradiated host mouse, are capable of switching to multiple isotypes (48, 49). However, recent studies demonstrated that normal B cells are also committed to switch to specific isotypes in response to certain stimuli (131, 132) and suggest that directed switching does in fact occur in vivo (see below for further discussion).

E. An Accessibility Mechanism for Directed Class Switch Recombination

The likelihood of directed switching in murine B cells raises the question of how recombination between specific S regions is accomplished. One suggestion is that separate switch recombinase enzymes exist to promote each recombination event (59). According to this model, A-MuLV-transformed pre-B cells would express switch recombinases which only recombine S_μ and $S_\gamma 2b$ sequences (59); I.29 B cells would express a switch recombinase specific for S_μ and S_α sequences. Such recombinases would have to detect differences between the switch regions to function efficiently. There is no evidence, direct or indirect, to support or exclude the existence of such a recombination system. Another suggested mechanism is that a single switch recombinase mediates all switch region recombinations and that specificity is accomplished by controlling the ability of the recombinase to interact with specific switch regions, reminiscent of the accessibility model originally proposed to explain the tissue-specific and stage-specific assembly of immunoglobulin and TCR variable-region genes (155). As mentioned above, accessibility for variable-region gene rearrangement appears to correlate with germ line transcription of gene segments prior to rearrangement. Notably, the germ line C_α

gene is transcribed in I.29 B cells that switch to α (136), and the germ line $C_\gamma 2b$ gene is transcribed in A-MuLV-transformed pre-B cells that switch to γ2b (78, 158); correspondingly, germ line C_H genes in these cells that do not undergo frequent switch events are transcriptionally silent.

The germ line γ2b transcripts in A-MuLV-transformed pre-B cells (and normal B cells; see below) initiate in a region 2 kb upstream of $S_\gamma 2b$ and are polyadenylated at the normal $C_\gamma 2b$ polyadenylation sites (78); therefore, transcription occurs through the $S_\gamma 2b$ region and not through other switch regions. This primary transcript is processed so that an exon derived from sequences 5′ of $S_\gamma 2b$ (termed $I_\gamma 2b$) is spliced directly onto the $C_H 1$ exon of $C_\gamma 2b$. Although these transcripts are polyadenylated and accumulate in the cytoplasm, nucleotide sequence analyses and other experiments indicated that this transcript is probably not translatable (7, 78). Recent experiments indicate that germ line ε, γ3, and α transcripts of similar structure are present in B-lineage cells that actively switch to the corresponding C_H genes (134; P. Rothman, S. Lutzker, W. Cook, R. Coffman, and F. W. Alt, *J. Exp. Med.*, in press). In the case of germ line γ3 transcripts, the transcription initiation region also is positioned upstream from the $S_\gamma 3$ region and a polyadenylated germ line γ3 transcript analogous to the germ line γ2b transcript is produced and accumulates in the cytoplasm (P. Rothman, S. Lutzker, and F. W. Alt, manuscript in preparation). Evidence for transcription through the unrearranged γ1 region in cells that switch to this isotype has also been observed (P. Rothman and F. W. Alt, unpublished data); however, distinct germ line γ1 transcripts have not been observed to accumulate in the cytoplasm of such cells. Together these findings support an accessibility model for the control of class switch recombination; the potential physiological relevance of such a mechanism will be discussed in more detail below.

F. Dual-Isotype Expression and Directed Switching in Memory B Cells

A few percent of splenic B cells in mice express both IgM and a second class of antibody other than IgD on their cell surface (103). It has been suggested that some of these cells express both heavy-chain isotypes by alternate RNA processing of a long primary transcript (100) in a manner analogous to that employed for simultaneous IgM and IgD production (see above). In support of this notion, purified surface $IgM^+/IgG1^+$ B cells contain a primary transcript hybridizing to both a C_μ and a $C_\gamma 1$ probe, but not to a C_α probe, and purified surface IgM^+/IgA^+ B cells contain transcripts that hybridize to all three C_H probes (100). In addition, some transformed B cells are capable of expressing dual isotypes on their surface without the apparent deletion of C_H genes (24, 71). It should be noted, however, that some of the dual-isotype-expression studies remain controversial; it has been suggested that some dual-isotype-producing cells have undergone recent class switching and still translate μ mRNA present in their cytoplasm, or have passively absorbed other isotypes onto the cell surface via Fc receptors (65).

Dual-isotype-expressing B cells may represent so-called memory B cells which have interacted with antigen but have not differentiated to the plasma-cell stage (95). These long-lived B cells give rise to a rapid and massive secondary immune response upon subsequent encounter with antigen. The major class of antibody produced during the primary immune response to antigen is IgM, with lesser amounts of IgG and IgA, whereas IgG and IgA dominate the secondary response to antigen. It has been hypothesized that the transcriptional events associated with the production of long transcripts in dual-isotype-producing (memory) B cells may play a role in directing class switch recombination to the switch region upstream of the C_H gene encoding the second isotype (100). If this is true, then accessibility may not simply correlate with transcription through switch sequences per se, because multiple intervening switch regions would also be transcriptionally active in these cells. Recently, however, a B-cell lymphoma that expresses both surface IgM and surface IgG2b without apparent C_H gene deletion has been shown to express germ line γ2b transcripts (60; W. Strober, personal communication; S. Lutzker and F. W. Alt, unpublished data). Thus, mechanisms may exist in B cells to permit both dual isotype expression (perhaps via long transcripts) and targeting of specific switch regions for recombination by an "accessibility" mechanism similar to that described above for A-MuLV-transformed pre-B cells and I.29 B cells. In this context, the recent observations of *trans*-RNA splicing in trypanosomes (76, 89, 138) raises the interesting possibility that dual-isotype expression also occurs via a similar mechanism that would join the variable-region exon of a primary μ heavy-chain transcript to the first C_H exon of a primary germ line C_H-region transcript (rather than alternate splicing of a long transcript). This would provide a mechanism to direct switching to the second expressed isotype in the same cell, perhaps after expression of a class switch recombinase. Although there is no evidence for this mechanism, it could be easily tested and is appealing in its simplicity.

VII. REGULATION OF CLASS SWITCHING AND ISOTYPE EXPRESSION

A. Lipopolysaccharide Stimulates Class Switching in Splenic B Cells

Bacterial LPS, a glycolipid isolated from the cell walls of gram-negative bacteria, belongs to a class of polysaccharide antigens termed thymus-independent because it induces antibody secretion in the absence of thymus-derived (T) lymphocytes or other accessory cells (68, 85; for review, see reference 23). The antibodies secreted in this response are nonspecific, since LPS stimulates polyclonal B-cell proliferation and maturation to the antibody-secreting stage (124); the use of this agent avoids the cumbersome process of isolating B cells that respond to a specific antigen. In mice, LPS (and T-independent antigens in general) stimulates the production of IgM, IgG3, and IgG2b antibodies in splenic B-cell cultures. The IgG secretors appear to arise from cells which originally expressed IgM on their surface, because administration of anti-μ antibodies to the LPS cultures inhibits both the IgM and the IgG responses (9, 67). Similarly, purified surface IgM^+ B cells elicit the same IgG response as do unfractionated cells, while surface IgG^+ cells are not stimulated by LPS to secrete antibody (124). Thus, LPS appears to directly induce γ3 and γ2b class switching in murine splenic B cells rather than selectively stimulating the proliferation or differentiation of IgG2b- and IgG3-producing cells in the population. LPS-stimulated splenic B cells undergo class switching at the frequency approaching 1 in 10 cells per generation (111), a much higher frequency than that observed in transformed cell lines (97). IgG3- and IgG2b-secreting cells appear with identical kinetics in LPS-treated cultures, indicating that the IgG2b-producing cells do not arise from cells that first switch to IgG3 production (123).

B. T Cells and Regulation of Specific Isotype Expression

Although LPS induces switching in splenic B cells independent of other cell types, T lymphocytes have a profound influence on isotype expression both in vivo (142) and in vitro (84). This property is evident in vivo in athymic mice, which have decreased levels of IgG1 and IgG2a in their serum (108) and impaired immune responses to certain antigens (153). T cells have been functionally divided into helper or suppressor T cells (T_H or T_S) based upon their ability to either enhance or suppress B-cell-mediated immune responses. Most of the antigenic responses in vivo require T_H cells that have become activated by the interaction of their surface T-cell receptors with both antigen and accessory proteins (major histocompatibility proteins) on the surfaces of B cells or other antigen-presenting cells (for review, see reference 83). Activated T_H cells secrete soluble substances (lymphokines) which act upon B cells and other hematopoietic cells.

C. Interleukin-4

The effect of T_H cells on isotype production has been studied in vitro with polyclonally activated B cells. For example, the addition of certain activated T_H cells (84) or supernatant fluids which contain their secreted lymphokines (26, 57) to LPS-stimulated B cells results in increased production of IgG1 and IgE with a concomitant decrease in production of IgG3 and IgG2b. The IgG1, IgE-enhancing/IgG3, IgG2b-suppressing activity in these supernatant fluids is mediated by a single protein termed interleukin-4 (IL-4; 27, 93, 127, 148). IL-4 has pleiotropic effects upon B-lineage cells as well as upon other cell types (for review, see reference 99). In terms of its effect on isotype expression, IL-4 together with LPS stimulates surface $IgM^+/IgG1^-$ B cells to produce IgG1 and IgE (28, 57); this is consistent with IL-4 inducing γ1 and ε class switching rather than selecting for IgG1- and IgE-producing cells. Similarly, statistical analysis of limiting-dilution B-cell cultures indicates that IL-4 increases the frequency of B cells which give rise to IgG1 and IgE producers, rather than the number of IgG1- and IgE-producing cells that arise from a single IgG1 and IgE producer (28, 57). Although IL-4, by itself, does not induce IgG1 production, pretreatment of B cells with IL-4 results in increased IgG1 production when the IL-4 is removed and LPS is added (131). This result suggests that IL-4 can commit B cells to undergo γ1 switching in response to an antigenic stimulus.

IL-4 appears to be a more potent stimulus for IgG1 production than for IgE production in LPS-treated B-cell cultures, since low IL-4 concentrations which result in significant IgG1 production do not result in detectable levels of IgE (129). Although the levels of IL-4 required for IgE production in vitro are higher than the levels present in supernatant fluids of cultured T cells, it is possible that the direct interaction of T cells with B cells could result in very high local concentrations of IL-4 (106). Evidence that IL-4 does in fact play a crucial role in IgE production in vivo comes from studies of mice infected with the larvae of the helminth *Nippostrongylus brasiliensis*, which have serum IgE levels 100-fold above normal. The administration of monoclonal anti-IL-4 antibodies at the time of infection diminishes the increase in serum IgE to 5- to 10-fold above normal (43). Additional in vivo data also indicate that lower concentrations of IL-4 are required for IgG1 production than for IgE production (44)

D. Other Lymphokines

Analysis of a panel of T_H cell lines indicates that IL-4 is produced by a distinct subset of T_H cells designated T_H2 (25, 87). T_H2 cells, unlike the other major subset of T_H cells, T_H1, also produce interleukin-5 (IL-5) in response to the same stimuli that elicit IL-4 secretion. IL-5 by itself does not induce IgG1 or IgE production in LPS-stimulated B-cell cultures, nor does it affect the levels of either IgG1 or IgE produced in response to high concentrations of IL-4. At low levels of IL-4, however, IL-5 is able to substantially enhance both IgG1 and IgE production (28). The mechanism by which IL-5 enhances the IL-4-induced expression of IgG1 and IgE is unclear. IL-5 may enhance IgG1 and IgE production in IL-4/LPS-treated B cells by stimulating cells that have already undergone γ1 and ε class switching. Along these same lines, IL-5 (without IL-4) has recently been shown to enhance IgA production in LPS-treated B cells by stimulating proliferation of surface IgA^+ cells (16, 29, 50, 90).

The other major subset of T_H cells, T_H1 cells, produce interleukin-2 (IL-2) and gamma interferon (IFN-γ) (87). In LPS-stimulated B-cell cultures, IL-2 has no direct effect on the production of particular isotypes. IFN-γ, however, stimulates the production of IgG2a and inhibits IgG3 and IgG2b production (130). IFN-γ stimulates IgG2a production in cultures of surface IgM^+ B cells and increases the frequency of cells giving rise to IgG2a-producing clones (132). IFN-γ also appears to commit B cells to undergo γ2a class switching, since pretreatment of B cells with IFN-γ results in IgG2a production when cells are later cultured with LPS alone (132). Thus, the effect of IFN-γ on IgG2a production is very similar to that of IL-4 on IgG1 production.

E. Mitogens and Lymphokines Alter Accessibility of Switch Regions

Certain A-MuLV-transformed pre-B cells and freshly isolated splenic B cells do not express germ line γ2b transcripts at detectable levels; correspondingly, neither of these cell types switches to γ2b when cultured in the absence of LPS. Treatment of A-MuLV-transformed pre-B cells and normal spleen cells with LPS rapidly induces the expression of germ line γ2b transcripts. This induction is followed by γ2b switching, consistent with a cause-and-effect relationship (79). LPS-induced γ2b class switching in splenic B cells appears to occur by the same directed switching mechanism implicated for spontaneous γ2b class switching in A-MuLV-transformed pre-B cells. Similarly, germ line γ3 transcripts also are induced by LPS in splenic B cells prior to γ3 switching (Rothman et al., in preparation). Stimulation with IL-4 inhibits the induction of germ line γ2b and germ line γ3 transcripts by LPS and, correspondingly, inhibits γ2b and γ3 switching in LPS-treated B-cell cultures (79; Rothman et al., in preparation). In addition, IL-4 together with LPS induces germ line ε transcripts in splenic B cells prior to ε class switching (Rothman et al., in press). Together, these results support the proposal that transcriptional/recombinational control elements are associated with the number of C_H genes and that the activity of these elements is regulated by mitogens and T cells (via their production of lymphokines) to directly induce class switching by altering the accessibility of switch regions for recombination.

LPS treatment also induces germ line γ3 transcripts as well as germ line γ2b transcripts in certain independent A-MuLV-transformed pre-B cells; consistent with an accessibility model, the rate of both γ2b and γ3 class switching is increased in these LPS-treated cells (79; Rothman et al., in preparation). LPS induces germ line γ2b transcripts in all A-MuLV-transformed pre-B cells analyzed; germ line γ3 transcripts, however, are induced only in a subset of these cell lines, suggesting that expression of these two germ line transcripts is not strictly coregulated, at least at the pre-B-cell stage. In addition, A-MuLV-transformed pre-B cell lines have been derived which undergo ε switching in response to LPS and IL-4 (Rothman et al., in press). Significantly, IL-4 together with LPS induces germ line ε transcripts in these cells prior to switching. Thus, A-MuLV-transformed pre-B cells provide a model system for examining many aspects of mitogen- and lymphokine-stimulated class switching.

VIII. A MODEL FOR ISOTYPE REGULATION IN VIVO

The classes of antibodies produced in an immune response appear to be determined by the nature of the inducing antigen or perhaps the route of antigen into the host. How this process is regulated and its physiological significance are still important areas for exploration. Below, we outline a (necessarily vague) model for the control of isotype switching that attempts to incorporate some of the recent cellular and molecular findings.

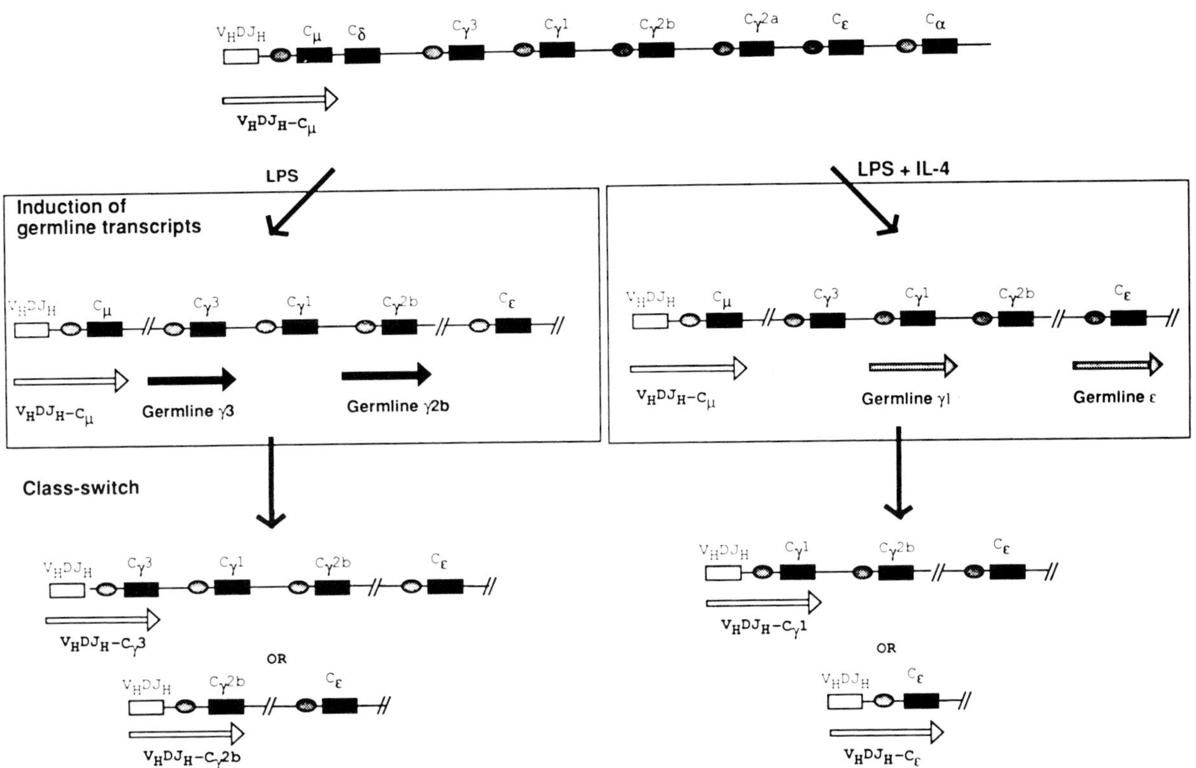

Figure 8. Model for LPS- and IL-4-directed class switching. Surface IgM^+ B cells appear to be committed to switch to γ3 and γ2b; this commitment is manifested by the production of germ line γ2b and germ line γ3 transcripts following polyclonal activation with LPS. IL-4 alters this commitment, resulting in the production of germ line γ1 and ε transcripts upon stimulation with LPS. The LPS induction of germ line C_H-region transcripts prior to class switch recombination is proposed to be related to mechanisms which target the respective switch regions for class switch recombination via an accessibility mechanism (see text for details).

A-MuLV-transformed pre-B cells generally appear programmed to switch to γ2b or γ3, either spontaneously or after LPS treatment. This suggests a predisposition of normal pre-B cells to switch to these isotypes. Likewise, other transformed cell lines representative of early B-lineage cells differentiate in vitro into mature B cells that express IgG2b antibodies (60, 71a, 98), indicating that this predisposition does not result from A-MuLV transformation. In addition, the first surface IgG^+ B cells detected in neonatal liver cultures synthesize IgG2b or IgG3 antibodies (20). Together, these lines of evidence suggest that a commitment to switch to γ2b and γ3 may appear as part of a normal developmental pathway and persist after pre-B cells become functional B cells (Fig. 8, left). This precommitment is manifested by the expression of germ line γ2b and germ line γ3 transcripts in polyclonally (LPS) activated spleen cells. As outlined above, transcription through germ line switch regions may reflect processes that target them in the context of an accessibility mechanism for switch recombination by a general class switch recombinase (Fig. 8, left). Polyclonal activation (in the absence of added lymphokines) thus provides a non-isotype-specific signal for the transcription of these germ line C_H genes, resulting in enhanced switching to IgG2b and IgG3 production. Consistent with such a model, most antigens that directly activate B cells without the help of T cells (i.e., T-independent antigens) also induce the production of IgG2b and IgG3 antibodies, presumably through the same mechanism. Thus, it is possible that the interaction of B cells with antigen or polyclonal activators activates switching via a predetermined pathway resulting in γ2b and γ3 switching (Fig. 8, left).

Alterations of this precommitment appear to be effected by interaction of B cells with T cells or their secreted lymphokines. Thus, antigenic (or polyclonal) stimulation in the presence of T_H2 cells (or exogenous IL-4) results in switching to γ1 (and ε) rather than γ2b and γ3; this change in commitment is again reflected by the apparent change in the commitment of B cells to transcribe new sets of germ line C_H genes after LPS (antigen) treatment (Fig. 8, right). In the context of regulation through accessibility, the altered transcription patterns would reflect class switch targeting of the different set of genes (Fig. 8, right).

Cells pretreated with IL-4 switch to γ1 upon subsequent stimulation with LPS alone, indicating that this altered switch commitment is not transient. In this regard, it is possible that the I.29 lymphoma line, which can be induced by LPS to switch to α or ε, may have been exposed to IL-4 or a similarly acting lymphokine before its transformation.

As suggested above, B cells stimulated by antigens and T cells in vivo which do not undergo switching and terminal differentiation into plasma cells (i.e., memory B cells) may retain the altered switch commitment and upon restimulation with antigen transcribe germ line C_H genes other than γ2b and γ3, followed by switching to these isotypes. Also, in the context of this model, the constitutive expression of germ line C_H transcripts in some cell lines (e.g., A-MuLV-transformed pre-B cells) may relate to some aspect of their transformed phenotype. For example, the fact that these cells are rapidly proliferating in the absence of LPS stimulation suggests that proliferation (or activation) may cause or be required for expression of germ line C_H-region transcripts. Correspondingly, the constitutive expression of germ line γ2b transcripts in A-MuLV-transformed pre-B cells can be reduced by inhibiting proliferation by growth at high cell density (79).

IX. CLASS SWITCHING AND B-CELL TUMORS

Aberrant class switching has been implicated in the development of certain B-cell tumors containing translocations of the c-*myc* proto-oncogene (for review, see reference 32). These translocations frequently occur within switch regions of the nonproductive allele and favor the switch region upstream of the C_H gene expressed on the productive allele. Thus, in IgM-producing human Burkitt lymphomas, IgA-producing mouse plasmacytomas, and IgE-producing rat immunocytomas, the c-*myc* oncogene is frequently translocated into the S_μ, S_α, and S_ε region, respectively, of the nonproductive allele (143; for review, see reference 32). This pattern of switching resembles that seen in hybridomas, as discussed above, and strongly suggests that the chromosomal translocation was influenced by the mechanisms that direct the class switch process. Sequences reminiscent of switch sequences frequently are found near the site of DNA breakage within the c-*myc* locus (21, 39). However, it seems likely that these translocations often may result from a recombination event that resolves an initial class switch-specific scission by illegitimate recombination of the liberated ends into the c-*myc* locus; such a mechanism has been proposed for recombination events between variable-region gene elements and the c-*myc* locus in other B-lineage tumors (32). Chromosomal breakpoints within the c-*myc* region are scattered, occurring mostly within the intron between exon 1 and exon 2, or within exon 1. Some translocation breakpoints, however, occur 5′ of exon 1. In all cases, the c-*myc* gene is joined to the switch region in a head-to-head fashion; however, the nature of the events which result in the activation of the c-*myc* gene is unclear. Studies involving transgenic mice indicate that c-*myc* tumorigenicity does not result merely from the loss of exon 1 or 5′ flanking regions and also show that the IgH locus is critical for activation (1).

X. PROSPECTS

As summarized in this chapter, the expression of different isotypes in humans and in mice results from the duplication and divergence of C_H genes and the evolution of a DNA recombination mechanism which recognizes and joins switch regions. Recent findings support the notion that this process is directed by the accessibility of specific switch regions to the recombinase; differential accessibility of switch regions may be regulated by antigens (mitogens) and T cells (through the production of lymphokines) to promote isotype-specific class switching. Although much progress has been made in understanding the regulation of isotype expression, many intriguing questions remain to be answered.

One area for further research involves the nature of the "enzymes" which mediate class switch recombination. Class switch recombinase does not appear to be expressed in fibroblasts (97), but still may be expressed in cell types other that those of the B lineage. Analysis of recombination sites suggests that the pentamers GAGCT and TGGGG may be preferentially recognized by switch recombinase enzymes, but other sequences may also be recognized. In addition, the large size of switch regions suggests that repetition of switch sequences increases the efficiency of the switch recombinase to interact with the substrate DNA. Although it is possible that class switch recombination and variable-region recombination share some of the same enzymes, this is not necessarily the case. The recognition and recombination requirements of these two recombination systems are quite different and at least one genetic defect (severe combined immunodeficiency) that impairs VDJ recombination (122) apparently does not affect class switching (T. K. Blackwell and F. W. Alt, unpublished data).

The relationships between accessibility and transcription remains to be determined, both for variable-region joining and class switching. Studies involving variable-region recombination in transgenic mice suggest that sequences required for accessibility are also required for transcription, namely an active enhancer and possibly the promoter (Ferrier et al., submitted). It is possible that accessibility is mediated by some of the factors which interact with these sequences and allow RNA polymerase to initiate transcription. The ability of RNA polymerase to unwind transcribed regions may contribute to their accessibility for recombination (45). In *Saccharomyces cerevisiae*, for example, a "hot spot" for DNA recombination has been shown to contain an enhancer element and an RNA polymerase promoter (149). This hot spot enhances recombination when the sequences undergoing recombination are transcribed; positioning of a transcription termination sequence between the promoter and the recombination sequences reduces recombination markedly. Similarly, transcription of two adjacent *S. cerevisiae* gene copies significantly enhances the frequency of crossing over between them (B. Thomas and R. Rothstein, personal communication). Transcribed genes also tend to be found at discrete sites within the nucleus (56); switch recombinase activity may be restricted to those regions of the nucleus. The possibility also remains that the germ line C_H region transcripts themselves may have a direct role in promoting the recombination process.

Many of the questions regarding accessibility could potentially be addressed through class switch constructs introduced into cell lines that undergo switching or into the germ line of transgenic mice. For example, the position and orientation of transcriptional control elements could be changed to determine whether transcription through a switch region is actually required for class switch recombination. In addition, promoters for different RNA polymerases could be compared for their ability to stimulate recombination; a specific DNA configuration or other elements required for accessibility may only be associated with a subset of these promoters.

Many questions remain concerning the ability of mitogens and lymphokines to induce or inhibit the expression of germ line C_H-region transcripts. LPS induces both germ line γ2b and germ line γ3 transcripts in normal spleen cells. The transcriptional control sequences that respond to LPS induction have yet to be determined. Expression of these two germ line transcripts does not appear to be strictly coregulated in A-MuLV-transformed pre-B cells, suggesting that different transcription factors may be involved. IL-4 inhibits both germ line γ2b and germ line γ3 transcripts in normal B cells by an unknown mechanism. Inhibition could take place at the DNA level by inducing or modifying LPS-induced transcription factors which in turn prevent the expression of germ line transcripts, or in the cytoplasm by modifying the transmission of the LPS signal received at the cell surface. IL-4 also induces germ line ε transcripts in spleen cells and in certain A-MuLV-transformed pre-B cells, but only in the presence of LPS. This suggests that a common transcription factor may be induced by LPS which is either modified by IL-4 treatment or is required, along with an IL-4-induced factor, for germ line ε transcription. Identification of the regulatory sequences required for the expression of germ line ε transcripts should be helpful in distinguishing between these two possibilities.

The characterization and cloning of T-cell-derived lymphokines has shed much light on the role T cells play in isotype regulation. Studies of T_H cell lines suggest that these cells can be divided into two categories, those secreting IFN-γ (T_H1) and those secreting IL-4 and IL-5 (T_H2). Nontransformed T_H cells also appear to segregate into two groups based upon their ability to secrete IFN-γ or IL-4 (107). Whether these two cell types develop as distinct cell lineages in vivo or are able to convert from the production of IFN-γ to the production of IL-4 and IL-5 remains to be determined. An important question pertains to how T_H cells function in vivo to promote an isotype-specific response against an invading pathogen. Transformed T_H lines produce varying levels of IL-4, IL-5, and IFN-γ that, based upon results from LPS-treated B cells, could potentially result in different patterns of isotype production in vivo. In addition, IL-4 and IFN-γ antagonize the effects of one another on B cells in vitro (26, 130) as well as in vivo (42). Thus, infection with a parasite that elicits an IgE response must presumably activate T_H2 cells rather than T_H1 cells. Whether there is, in fact, differential activation of T_H1 and T_H2 cells in vivo and the mechanism for this activation remain to be determined. This knowledge could aid in the treatment of disease states associated with deficiencies or the overproduction of specific classes of antibodies.

LITERATURE CITED

1. **Adams, J., A. Harris, C. Pinkert, L. Corcoran, W. Alexander, S. Cory, R. Palmiter, and R. Brinster.** 1985. The c-myc oncogene driven by immunoglobulin enhancer induces lymphoid malignancy in transgenic mice. *Nature* (London) **318:** 533–538.
2. **Akira, S., H. Sugiyama, N. Yoshida, H. Kikutani, Y. Yamamura, and T. Kishimoto.** 1983. Isotype switching in murine pre-B cell lines. *Cell* **34:**545–556.

3. **Alt, F., and D. Baltimore.** 1982. Joining of immunoglobulin H chain gene segments, implications from a chromosome with evidence of three D-J_H fusions. *Proc. Natl. Acad. Sci. USA* **79**:4118–4122.
4. **Alt, F., T. K. Blackwell, R. DePinho, M. Reth, and G. Yancopoulos.** 1986. Regulation of genomic rearrangement events during lymphocyte differentiation. *Immunol. Rev.* **89**:1–30.
5. **Alt, F., T. K. Blackwell, and G. Yancopoulos.** 1987. Development of the primary antibody repertoire. *Science* **238**: 1079–1087.
6. **Alt, F., A. Bothwell, M. Knapp, E. Siden, E. Mather, M. Koshland, and D. Baltimore.** 1980. Synthesis of secreted and membrane-bound immunoglobulin mu heavy chains is directed by mRNAs that differ at their 3′ ends. *Cell* **29**: 293–301.
7. **Alt, F., N. Rosenberg, R. Casanova, E. Thomas, and D. Baltimore.** 1982. Immunoglobulin heavy chain expression and class switching in a murine leukaemia cell line. *Nature* (London) **296**:325–331.
8. **Alt, F., G. Yancopoulos, T. K. Blackwell, C. Wood, E. Thomas, M. Boss, B. Coffman, N. Rosenberg, S. Tonegawa, and D. Baltimore.** 1984. Ordered rearrangement of immunoglobulin heavy chain variable region segments. *EMBO J.* **3**:1209–1212.
9. **Anderson, J., A. Coutinho, and F. Melcher.** 1978. Stimulation of murine B lymphocytes to IgG synthesis and secretion by the mitogens lipopolysaccharide and lipoprotein and its inhibition by anti-immunoglobulin antibodies. *Eur. J. Biochem.* **8**:336–343.
10. **Battey, J., E. Max, W. McBride, D. Swan, and P. Leder.** 1982. A processed human immunoglobulin ε gene has moved to chromosome 9. *Proc. Natl. Acad. Sci. USA* **79**: 5956–5960.
11. **Bech-Hensen, N., P. Linsley, and D. Cox.** 1983. Restriction fragment length polymorphisms associated with immunoglobulin C_γ genes reveal linkage disequilibrium and genomic organization. *Proc. Natl. Acad. Sci. USA* **80**:6952–6956.
12. **Berman, J., S. Mellis, R. Pollock, C. Smith, H. Suh, B. Heinke, C. Kowal, U. Surti, L. Chess, C. Cantor, and F. Alt.** 1988. Content and organization of the human Ig V_H locus: definition of three new V_H families and linkage to the Ig C_H locus. *EMBO J.* **7**:727–738.
13. **Blackwell, T. K., and F. Alt.** 1988. Immunoglobulin genes, p. 1–60. *In* D. Hames and D. Glover (ed.), *Frontiers in Molecular Immunology*. IRL Press, Oxford, United Kingdom.
14. **Blackwell, T. K., M. Moore, G. Yancopoulos, H. Suh, S. Lutzker, E. Selsing, and F. Alt.** 1986. Recombination between immunoglobulin variable region segments is enhanced by transcription. *Nature* (London) **324**:585–589.
15. **Blattner, F., and P. Tucker.** 1984. The molecular biology of IgD. *Nature* (London) **307**:417–422.
16. **Bond, M., B. Shrader, T. Mossman, and R. Coffman.** 1987. A mouse T cell product that preferentially enhances IgA production. II. Physiochemical characterization. *J. Immunol.* **139**:3691–3696.
17. **Borzillo, G., M. Cooper, M. Kubagawa, A. Landay, and P. Burrows.** 1987. Isotype switching in human B lymphocyte malignancies occurs by DNA deletion: evidence for nonspecific switch recombination. *J. Immunol.* **139**:1326–1335.
18. **Burrows, P., G. Beck, and M. Wabl.** 1981. Expression of μ and γ immunoglobulin heavy chains in different cells of a cloned mouse lymphoid line. *Proc. Natl. Acad. Sci. USA* **78**:564–568.
19. **Burrows, P., G. Beck-Engeser, and M. Wabl.** 1983. Immunoglobulin heavy-chain class switching in a pre-B cell line is accompanied by DNA rearrangement. *Nature* (London) **306**:243–246.
20. **Calvert, J., M. Kim, W. Gathings, and M. Cooper.** 1983. Differentiation of B lineage cells from liver of neonatal mice: generation of immunoglobulin isotype diversity in vitro. *J. Immunol.* **131**:1693–1697.
21. **Care, A., L. Cianetti, A. Giampaolo, N. Sposi, V. Zappavigna, F. Mavillo, G. Alimena, S. Amadori, F. Mandelli, and C. Peschle.** 1986. Translocation of c-myc into the immunoglobulin heavy-chain locus in human acute B-cell leukemia. A molecular analysis. *EMBO J.* **5**:905–911.
22. **Cebra, J., J. Colberg, and S. Dry.** 1966. Rabbit lymphoid cells differentiated with respect to α-, γ- and μ-heavy polypeptide chains and to allotypic markers AA1 and AA2. *J. Exp. Med.* **123**:547–557.
23. **Cebra, J., J. Komisar, and P. Schweitzer.** 1984. C_H isotype switching during normal B lymphocyte development. *Annu. Rev. Immunol.* **2**:493–548.
24. **Chen, Y.-W., C. Word, V. Dev, J. Uhr, E. Vitetta, and P. Tucker.** 1986. Double isotype production by a neoplastic B cell line. II. Allelically excluded production of μ and γ1 heavy chains without C_H gene rearrangement. *J. Exp. Med.* **164**:562–579.
25. **Cherwinski, H., J. Schumacher, K. Brown, and T. Mosmann.** 1987. Two types of helper T cell clones. III. Further differences in lymphokine synthesis between T_H1 and T_H2 clones revealed by RNA hybridization, monospecific bioassays, and monoclonal antibodies. *J. Exp. Med.* **166**: 1229–1244.
26. **Coffman, R., and J. Carty.** 1986. A T cell activity that enhances polyclonal IgE production and its inhibition by interferon-γ. *J. Immunol.* **136**:949–954.
27. **Coffman, R., J. Ohara, M. Bond, J. Carty, Z. Zlotnik, and W. Paul.** 1986. B cell stimulatory factor-1 enhances the IgE response of lipopolysaccharide-activated B cells. *J. Immunol.* **136**:4538–4541.
28. **Coffman, R., B. Seymour, D. Lebman, D. Hiraki, J. Christiansen, B. Shrader, H. Cherwinski, H. Savelkoul, F. Finkelman, M. Bond, and T. Mosmann.** 1988. The role of helper T cell products in mouse B cell differentiation and isotype regulation. *Immunol. Rev.* **102**:5–28.
29. **Coffman, R., B. Shrader, J. Carty, T. Mossman, and M. Bond.** 1987. A mouse T cell product that preferentially enhances IgA production. I. Biological characterization. *J. Immunol.* **139**:3685–3690.
30. **Coleclough, C., D. Dooper, and R. Perry.** 1980. Rearrangement of immunoglobulin heavy chain genes during B-lymphocyte development as revealed by studies of mouse plasmacytoma cells. *Proc. Natl. Acad. Sci. USA* **77**:1422–1426.
31. **Cooper, M., A. Lawton, and P. Kincade.** 1972. A two-stage model for development of antibody-producing cells. *Clin. Exp. Immunol.* **11**:143–999.
32. **Cory, S.** 1986. Activation of cellular oncogenes in hemopoietic cells by chromosome translocation. *Adv. Cancer Res.* **47**:189–234.
33. **Cory, S., and J. Adams.** 1980. Deletions are associated with somatic rearrangement of immunoglobulin heavy chain genes. *Cell* **19**:37–46.
34. **Coutelier, J.-P., J. Van der Logt, F. Heessen, G. Warnier, and J. Van Snick.** 1987. IgG2a restriction of murine antibodies elicited by viral infections. *J. Exp. Med.* **165**:64–69.
35. **Davis, M., S. Kim, and L. Hood.** 1980. DNA sequences mediating class switches in α-immunoglobulins. *Science* **209**: 1360–1365.

36. **DePinho, R., K. Kruger, N. Andrews, S. Lutzker, D. Baltimore, and F. Alt.** 1984. Molecular basis of heavy chain class switching and switch region deletion in an Abelson virus transformed cell line. *Mol. Cell. Biol.* **4**:2905–2910.

37. **DePinho, R., G. Yancopoulos, T. K. Blackwell, M. Reth, K. Kruger, S. Lutzker, and F. Alt.** 1986. Regulation of the assembly and expression of immunoglobulin genes: variable region assembly and heavy chain class switching, p. 1–17. *In* M. Ferrarini and B. Pernis (ed.), *The Molecular Basis of B-Cell Differentiation and Function*. Plenum Publishing Corp., New York.

38. **D'Eustachio, P., D. Pravtcheva, K. Marcu, and F. Ruddle.** 1980. Chromosomal location of the structural gene cluster encoding murine immunoglobulin heavy chains. *J. Exp. Med.* **151**:1545–1550.

39. **Dunnick, W., B. Shell, and C. Dery.** 1983. DNA sequences near the site of reciprocal recombination between a c-myc oncogene and an immunoglobulin switch region. *Proc. Natl. Acad. Sci. USA* **80**:7269–7293.

40. **Early, P., H. Huang, M. Davis, K. Calame, and L. Hood.** 1980. An immunoglobulin heavy chain variable region gene is generated from three segments of DNA, V_H, D and J_H. *Cell* **19**:981–992.

41. **Early, P., J. Rogers, M. Davis, K. Calame, M. Bond, R. Wall, and L. Hood.** 1980. Two mRNAs can be produced from a single immunoglobulin μ gene by alternate RNA processing pathways. *Cell* **20**:313–319.

42. **Finkelman, F., I. Katona, T. Mossman, and R. Coffman.** 1988. IFN-γ regulates the isotypes of Ig secreted during in vivo humoral immune responses. *J. Immunol.* **140**: 1022–1027.

43. **Finkelman, F., I. Katona, J. Urban, C. Snapper, J. Ohara, and W. Paul.** 1986. Suppression of in vivo polyclonal IgE response by monoclonal antibody to the lymphokine BSF-1. *Proc. Natl. Acad. Sci. USA* **83**:9675–9678.

44. **Finkelman, F., C. Snapper, J. Mountz, and I. Katona.** 1987. Polyclonal activation of the murine immune system by a goat antibody to mouse IgD. *J. Immunol.* **138**:2826–2830.

45. **Fisher, M.** 1982. DNA unwinding in transcription and recombination. *Nature* (London) **299**:105–106.

46. **Flanagan, J., and T. Rabbitts.** 1982. Arrangement of human immunoglobulin heavy chain constant region genes implies evolutionary duplication of a segment containing γ, ε, and α genes. *Nature* (London) **300**:709–713.

47. **Fujimoto, S., and H. Yamagishi.** 1987. Isolation of an excision product of T-cell receptor α-chain gene rearrangement. *Nature* (London) **327**:242–243.

48. **Gearhart, P., J. Hurwitz, and J. Cebra.** 1980. Successive switching of antibody isotypes expressed within the lines of a B-cell clone. *Proc. Natl. Acad. Sci. USA* **77**:5424–5428.

49. **Gearhart, P., N. Sigal, and N. Klinman.** 1975. Production of antibodies of identical idiotype but diverse immunoglobulin classes by cells derived from a single stimulated B cell. *Proc. Natl. Acad. Sci. USA* **72**:1707–1711.

50. **Harriman, G., D. Kunimoto, J. Elliot, V. Paetkau, and W. Strober.** 1988. The role of IL-5 in IgA B cell differentiation. *J. Immunol.* **140**:3033–3039.

51. **Hayashida, H., T. Miyata, Y. Yamawaki-Kataoka, T. Honjo, J. Wels, and F. Blattner.** 1984. Concerted evolution of the mouse immunoglobulin gamma chains. *EMBO J.* **3**:2047–2053.

52. **Honjo, T., and T. Kataoka.** 1978. Organization of the immunoglobulin heavy chain genes and allelic deletion model. *Proc. Natl. Acad. Sci. USA* **75**:2140–2144.

53. **Hood, L., M. Kronenberg, and T. Hunkapiller.** 1985. T cell antigen receptor and the immunoglobulin supergene family. *Cell* **40**:225–231.

54. **Hummel, M., J. Berry, and W. Dunnick.** 1987. Switch region content of hybridomas: two spleen IgH loci tend to rearrange to the same isotype. *J. Immunol.* **138**:3539–3548.

55. **Hurwitz, J., and J. Cebra.** 1982. Rearrangement between the immunoglobulin heavy chain gene J_H and C_μ regions accompanying normal B lymphocyte differentiation in vitro. *Nature* (London) **299**:742–744.

56. **Hutchinson, N., and H. Weintraub.** 1985. Localization of DNAse I-sensitive sequences to specific regions of interphase nuclei. *Cell* **43**:471–482.

57. **Isakson, P., E. Pure, E. Vitetta, and P. Kramer.** 1982. T cell-derived B cell differentiation factor(s). Effect on the isotype switch of murine B cells. *J. Exp. Med.* **155**:734–748.

58. **Ishida, N., S. Ueda, H. Hayashida, T. Miyata, and T. Honjo.** 1982. The nucleotide sequence of the mouse immunoglobulin epsilon gene: comparison with human epsilon gene sequence. *EMBO J.* **1**:1117–1123.

59. **Jack, H.-M., M. McDowell, C. Steinberg, and M. Wabl.** 1988. Looping out and deletion mechanism for the immunoglobulin heavy-chain class switch. *Proc. Natl. Acad. Sci. USA* **85**:1581–1585.

60. **Jacobs, D., M. Sneller, J. Misplon, L. Edison, D. Kunimoto, and W. Strober.** 1986. T cell induced expression of membrane IgG by 70Z/3 B cells. *J. Immunol.* **137**:55–60.

61. **Jarrett, E., and H. Miller.** 1982. Production and activities of IgE in helminth infection. *Prog. Allergy* **31**:178–233.

62. **Kataoka, T., T. Kawakami, N. Takahasi, and T. Honjo.** 1980. Rearrangement of immunoglobulin γ1-chain gene and mechanism for heavy-chain class switch. *Proc. Natl. Acad. Sci. USA* **77**:919–923.

63. **Kataoka, T., T. Miyata, and T. Honjo.** 1981. Repetitive sequences in class switch recombination regions of immunoglobulin heavy chain genes. *Cell* **23**:357–368.

64. **Kataoka, T., S.-I. Takeda, and T. Honjo.** 1983. *Escherichia coli* extract-catalyzed recombination in switch regions of mouse immunoglobulin. *Proc. Natl. Acad. Sci. USA* **80**: 2666–2670.

65. **Katona, I., J. Urban, and F. Finkelman.** 1985. B cells that simultaneously express surface IgM and IgE in *Nippostrongylus brasilienis*-infected SJA/9 mice do not provide a model for isotype switching without gene deletion. *Proc. Natl. Acad. Sci. USA* **82**:511–515.

66. **Katzenberg, D., and B. Birshtein.** 1988. Sites of class switch recombination in IgG2b and IgG2a producing hybridomas. *J. Immunol.* **140**:3219–3227.

67. **Kearney, J., M. Cooper, and A. Lawton.** 1976. B lymphocyte differentiation induced by lipopolysaccharide. III. Suppression of B cell maturation by anti-mouse immunoglobulin antibodies. *J. Immunol.* **116**:1664–1668.

68. **Kearney, J., and A. Lawton.** 1975. B lymphocyte differentiation induced by lipopolysaccharide. I. Generation of cells synthesizing four major immunoglobulin classes. *J. Immunol.* **115**:671–676.

69. **Kemp, D., G. Morahan, A. Cowman, and A. Harris.** 1983. Production of RNA for secreted immunoglobulin u chains does not require transcription termination 5′ to the μ_m exons. *Nature* (London) **301**:84–86.

70. **Kenter, A., and B. Birshtein.** 1980. Chi, a promoter of generalized recombination of λ phage, is present in immunoglobulin genes. *Nature* (London) **293**:402–404.

71. **Kinashi, T., T. Godal, Y. Noma, N. Ling, Y. Yaoita, and T. Honjo.** 1987. Human neoplastic B cells express more than

two isotypes of immunoglobulins without deletion of heavy chain constant region genes. *Genes Dev.* **1**:465–470.

71a. **Kinashi, T., K. Inaba, T. Tsubata, K. Tashiro, R. Palacios, and T. Honjo.** 1988. Differentiation of an interleukin-3 dependent precursor B cell clone into immunoglobulin-producing cell in vitro. *Proc. Natl. Acad. Sci. USA* **85**: 4473–4477.

72. **Kipps, T., and L. Herzenberg.** 1986. Homologous chromosome recombination generating immunoglobulin allotype and isotype switch variants. *EMBO J.* **5**:263–268.

73. **Kirsch, I., C. Morton, K. Nakahara, and P. Leder.** 1982. Human immunoglobulin heavy chain genes map to a region of translocation in malignant B lymphocytes. *Science* **216**: 301–303.

74. **Knapp, M., C.-P. Liu, N. Newell, R. Ward, P. Tucker, S. Strober, and F. Blattner.** 1982. Simultaneous expression of immunoglobulin μ and δ heavy chains by a cloned B cell lymphoma: a single copy of the V_H gene is shared by two adjacent C_H genes. *Proc. Natl. Acad. Sci. USA* **79**: 2996–3000.

75. **Kuwakami, T., N. Takahashi, and T. Honjo.** 1980. Complete nucleotide sequence of mouse immunoglobulin μ gene and comparison with other immunoglobulin heavy chain genes. *Nucleic Acids Res.* **8**:3933–3945.

76. **Laird, P., J. Zomerdijk, D. De Korte, and P. Borst.** 1987. In vitro labelling of intermediates in the discontinuous synthesis of mRNA's in *Trypanosoma bruceii. EMBO J.* **6**:1055–1062.

77. **Liu, F.-T., K. Alband, J. Sutchliffe, and D. Katz.** 1982. Cloning and nucleotide sequence of mouse immunoglobulin ε chain cDNA. *Proc. Natl. Acad. Sci. USA* **79**:7852–7856.

78. **Lutzker, S., and F. Alt.** 1988. Structure and expression of germline immunoglobulin γ2b transcripts. *Mol. Cell. Biol.* **8**:1849–1852.

79. **Lutzker, S., P. Rothman, R. Pollock, R. Coffman, and F. Alt.** 1988. Mitogen- and IL-4-regulated expression of germline Ig γ2b transcripts: evidence for directed heavy chain class switching. *Cell* **53**:177–184.

80. **Maki, R., W. Roeder, A. Trauhecker, C. Sidman, M. Wabl, L. Rascke, and S. Tonewaga.** 1981. The role of DNA rearrangement and alternate RNA processing in the expression of immunoglobulin δ genes. *Cell* **24**:353–365.

81. **Marcu, K., J. Banerji, N. Penncavage, R. Lang, and N. Arnheim.** 1980. 5′ flanking region of immunoglobulin heavy chain constant region genes displays length heterogeneity in germlines of inbred mouse strains. *Cell* **22**:187–196.

82. **Marcu, K., R. Lang, L. Stanton, and L. Harris.** 1982. A model for the molecular requirements of immunoglobulin heavy chain class switching. *Nature* (London) **298**:87–89.

83. **Marrack, P., and J. Kappler.** 1987. The T cell receptor. *Science* **238**:1073–1079.

84. **Martinez-Alonso, C., A. Coutinho, and A. Augustin.** 1980. Immunoglobulin C gene expression. I. The commitment to IgG subclass of secreting cells is determined by the quality of the non-specific stimuli. *Eur. J. Immunol.* **10**:698–702.

85. **Melchers, F., and J. Anderson.** 1973. Synthesis, surface deposition and secretion of immunoglobulin M in bone-marrow derived lymphocytes before and after mitogen stimulation. *Transplant. Rev.* **14**:76–130.

86. **Miyata, T., T. Yasunaga, Y. Yamawaki-Kataoka, M. Obata, and T. Honjo.** 1980. Nucleotide sequence divergence of mouse immunoglobulin γ1 and γ2b chain genes and the hypothesis of intervening-sequence mediated domain transfer. *Proc. Natl. Acad. Sci. USA* **77**:2143–2147.

87. **Mossman, T., H. Cherwinski, M. Bond, M. Giedlin, and R. Coffman.** 1986. Two types of murine helper T cell clones. I. Definition according to profiles of lymphokine activities and secreted proteins. *J. Immunol.* **136**:2438–2357.

88. **Mowatt, M., and W. Dunnick.** 1986. DNA sequence of the murine γ1 switch segment reveals novel structural elements. *J. Immunol.* **136**:2674–2683.

89. **Murphy, W., K. Watkins, and N. Agabian.** 1987. Identification of a novel Y branch structure as an intermediate in trypanosome mRNA splicing: evidence for trans-splicing. *Cell* **47**:517–525.

90. **Murray, P., D. McKenzie, S. Swain, and M. Kagnoff.** 1987. Interleukin 5 and interleukin 4 produced by Peyer's patch T cells selectively enhance immunoglobulin A expression. *J. Immunol.* **139**:2669–2674.

91. **Nikaido, T., S. Nakai, and T. Honjo.** 1981. Switch region of immunoglobulin C_μ gene is composed of simple tandem repetitive sequences. *Nature* (London) **292**:845–848.

92. **Nikaido, T., Y. Yamakami-Kataoka, and T. Honjo.** 1982. Nucleic acid sequences of switch regions of immunoglobulin C_ϵ and C_γ genes and their comparison. *J. Biol. Chem.* **257**:7322–7329.

93. **Noma, Y., P. Sideras, T. Naito, S. Bergstedt-Lindquist, C. Azuma, E. Severinson, T. Tanabe, T. Kinashi, F. Matsuda, Y. Yaoita, and T. Honjo.** 1986. Cloning of a cDNA encoding the murine IgG1 induction factor by a novel strategy using SP6 promoter. *Nature* (London) **319**:640–645.

94. **Obata, M., T. Kataoka, S. Nakai, H. Yamagishi, N. Takahashi, Y. Yamawaki-Kataoka, T. Nikaido, A. Shimizu, and T. Honjo.** 1981. Structure of a rearranged γ1 chain gene and its implications to immunoglobulin class switch mechanism. *Proc. Natl. Acad. Sci. USA* **78**:2437–2441.

95. **Okamura, K., M. Julius, T. Tsu, and L. Herzenberg.** 1976. Demonstration that IgG memory is carried by IgG-bearing cells. *Eur. J. Immunol.* **6**:467–472.

96. **Okazaki, K., D. Davis, and H. Sakano.** 1987. T cell receptor β gene sequences in the circular DNA of thymocyte nuclei: direct evidence for intramolecular DNA deletion in V-D-J joining. *Cell* **49**:477–485.

97. **Ott, D., F. Alt, and K. Marcu.** 1987. Immunoglobulin heavy chain switch recombination within a retroviral vector in murine pre-B cells. *EMBO J.* **6**:577–584.

98. **Palacios, R., H. Karasuyama, and A. Rolink.** 1987. Ly1+ PRO-B lymphocyte clones. Phenotype, growth requirements and differentiation in vitro and in vivo. *EMBO J.* **6**: 3687–3693.

99. **Paul, W., and J. Ohara.** 1987. B-cell stimulatory factor-1/interleukin-4. *Annu. Rev. Immunol.* **5**:429–459.

100. **Perlmutter, A., and W. Gilbert.** 1984. Antibodies of the secondary response can be expressed without switch recombination in normal B cells. *Proc. Natl. Acad. Sci. USA* **81**:7189–7193.

101. **Perlmutter, R., D. Hansburg, D. Briles, R. Nicolotti, and J. Davie.** 1978. Subclass restriction of murine anticarbohydrate antibodies. *J. Immunol.* **121**:566–572.

102. **Pernis, B., G. Chiappino, A. Kelis, and P. Gell.** 1985. Cellular localization of immunoglobulin with different allotype specificities in rabbit lymphoid tissue. *J. Exp. Med.* **122**:853–875.

103. **Pernis, B., L. Forni, and A. Luzzati.** 1977. Synthesis of multiple immunological classes by single lymphocytes. *Cold Spring Harbor Symp. Quant. Biol.* **41**:175–183.

104. **Perry, R., D. Kelley, C. Coleclough, and J. Kearney.** 1981. Organization and expression of immunoglobulin genes in fetal liver hybridomas. *Proc. Natl. Acad. Sci. USA* **78**: 247–251.

105. **Petrini, J., B. Shell, M. Hummel, and W. Dunnick.** 1987. The immunoglobulin heavy chain switch: structural features of the γ1 recombinant switch regions. *J. Immunol.* **138:** 1940–1946.

106. **Poo, W.-J., L. Conrad, and C. Janeway, Jr.** 1988. Receptor-directed focusing of lymphokine release by helper T cells. *Nature* (London) **332:**178–180.

107. **Powers, G., A. Abbas, and R. Miller.** 1988. Frequencies of IL-2 and IL-4 secreting T cells in naive and antigen-stimulated lymphocyte populations. *J. Immunol.* **140:**3352–3357.

108. **Pritchard, H., J. Riddaway, and H. Micklem.** 1973. Immune responses in congenitally thymus-less mice. II. Quantitative studies of serum immunoglobulin, the antibody response to sheep erythrocytes, and the effect of thymus allografting. *Clin. Exp. Immunol.* **13:**125–138.

109. **Rabbits, T., A. Forster, W. Dunnick, and D. Bentley.** 1980. The role of gene deletion in the immunoglobulin heavy chain switch. *Nature* (London) **283:**351–356.

110. **Rabbits, T., A. Forster, and C. Milstein.** 1981. Human immunoglobulin heavy chain genes: evolutionary comparison of C_μ, C_δ and C_γ genes and associated switch regions. *Nucleic Acids Res.* **9:**4509–4524.

111. **Radbruch, A., C. Bruger, S. Klein, and W. Muller.** 1986. Control of immunoglobulin class switch recombination. *Immunol. Rev.* **89:**69–83.

112. **Radbruch, A., W. Muller, and K. Rajewsky.** 1986. Class switch recombination is IgG1 specific on the active and inactive IgH loci of IgG1-secreting B cell blasts. *Proc. Natl. Acad. Sci. USA* **83:**3954–3957.

113. **Radbruch, A., and F. Sablitzky.** 1983. Deletion of C_μ gene in mouse B lymphocytes upon stimulation with LPS. *EMBO J.* **2:**1929–1935.

114. **Rathbun, G., J. Berman, G. Yancopoulos, and F. Alt.** 1988. Organization and expression of the mammalian heavy chain variable region locus in immunoglobulin genes. *In* F. Alt, T. Honjo, and T. Rabbitts (ed.), *Immunoglobulin Genes.* Academic Press, Inc., New York, in press.

115. **Ravetch, J., I. Kirsch, and P. Leder.** 1980. Evolutionary approach to the question of immunoglobulin heavy chain switching: evidence from cloned human and mouse genes. *Proc. Natl. Acad. Sci. USA* **77:**6734–6738.

116. **Ravetch, J., U. Siebenlist, S. Korsmeyer, T. Waldmann, and P. Leder.** 1981. Structure of the human immunoglobulin μ locus: characterization of embryonic and rearranged J and D genes. *Cell* **27:**583–591.

117. **Reth, M., E. Petrac, P. Wiese, L. Lobel, and F. Alt.** 1987. Activation of V_κ rearrangement in pre-B cells following expression of membrane-bound immunoglobulin heavy chains. *EMBO J.* **6:**3299–3305.

118. **Rogers, J., P. Early, C. Carter, K. Calame, M. Bond, L. Hood, and R. Wall.** 1980. Two mRNAs with different 3′ ends encode membrane-bound and secreted forms of immunoglobulin μ chain. *Cell* **20:**303–312.

119. **Rosenberg, Y., and J. Chiller.** 1979. Ability of antigen-specific helper cells to effect a class-restricted increase in total Ig-secreting cells in spleens after immunization with antigen. *J. Exp. Med.* **150:**517–530.

120. **Sakano, H., R. Maki, Y. Kurosawa, W. Roeder, and S. Tonegawa.** 1980. Two types of somatic recombination are necessary for the generation of complete immunoglobulin heavy chain genes. *Nature* (London) **286:**676–683.

121. **Sakano, H., J. Rogers, K. Huppi, C. Brack, A. Traunecker, R. Maki, R. Wall, and S. Tonegawa.** 1979. Domains and the hinge region of immunoglobulin heavy chain are encoded in separate DNA segments. *Nature* (London) **277:**627–633.

122. **Schuler, W., L. Weiler, A. Schuler, R. Phillips, N. Rosenberg, T. Mak, J. Kearney, R. Perry, and M. Bosma.** 1986. Rearrangement of antigen receptor genes is defective in mice with severe combined immunodeficiency. *Cell* **46:**963–972.

123. **Severinson, E., S. Bergstedt-Lindqvist, W. Van der Loo, and C. Fernandez.** 1982. Characterization of the IgG response induced by polyclonal B cell activation. *Immunol. Rev.* **67:**73–85.

124. **Severinson-Gronowicz, E., C. Doss, F. Assise, E. Vitetta, R. Coffman, and S. Strober.** 1979. Surface immunoglobulin isotypes on cells responding to lipopolysaccharide by IgM and IgG secrection. *J. Immunol.* **123:**2049–2056.

125. **Shimizu, A., Y. Hamaguchi, Y. Taoita, K. Mariwaki, K. Kondo, and T. Honjo.** 1982. Japanese wild mouse, *Mus musculus molossinus*, has duplicated immunoglobulin G2a genes. *Nature* (London) **298:**82–84.

126. **Shimizu, A., N. Takahashi, Y. Yaoita, and T. Honjo.** 1982. Organization of the constant region gene family of the mouse immunoglobulin heavy chain. *Cell* **29:**499–506.

127. **Sideras, P., S. Bergstedt-Lindqvist, and E. Severinson.** 1985. Partial biochemical characterization of IgG1-inducing factor. *Eur. J. Immunol.* **15:**593–598.

128. **Smith, G., S. Kunes, D. Schultz, A. Taylor, and K. Triman.** 1981. Structure of chi hotspots of generalized recombination. *Cell* **24:**429–436.

129. **Snapper, C., F. Finkelman, and W. Paul.** 1988. Differential regulation of IgG1 and IgE synthesis by interleukin-4. *J. Exp. Med.* **167:**183–196.

130. **Snapper, C., and W. Paul.** 1987. Interferon-γ and B cell stimulatory factor-1 reciprocally regulate Ig isotype production. *Science* **236:**944–947.

131. **Snapper, C., and W. Paul.** 1987. B cell stimulatory factor-1 (interleukin-4) prepared resting murine B cells to secrete IgG1 upon subsequent stimulation with bacterial lipopolysaccharide. *J. Immunol.* **139:**10–17.

132. **Snapper, C., C. Peschel, and W. Paul.** 1988. IFN-γ stimulates IgG2a secretion by murine B cells stimulated with bacterial lipopolysaccharide. *J. Immunol.* **140:**2121–2127.

133. **Stanton, L., and K. Marcu.** 1982. Nucleotide sequences and properties of the murine γ3 immunoglobulin heavy chain gene switch region: implications for successive C_γ gene switching. *Nucleic Acids Res.* **10:**5993–6006.

134. **Stavnezer, J., G. Radcliffe, and E. Severinson.** 1987. Specificity of immunoglobulin heavy chain switching by cultured I.29 B lymphoma cells. *Nobel Symp.*, in press.

135. **Stavnezer, J., S. Sirlin, and J. Abbott.** 1985. Induction of immunoglobulin isotype switching in cultured I.29 B lymphoma cells. *J. Exp. Med.* **161:**577–601.

136. **Stavnezer-Nordgren, J., and S. Sirlin.** 1986. Specificity of immunoglobulin heavy chain switch correlates with activity of germline heavy chain genes prior to switching. *EMBO J.* **5:**95–102.

137. **Sugiyama, H., T. Maeda, S. Akira, and S. Kishimoto.** 1986. Class switch from μ to γ3 or γ2b production at pre-B cell stage. *J. Immunol.* **136:**3092–3097.

138. **Sutton, R., and J. Boothroyd.** 1986. Evidence for trans-splicing in trypanosomes. *Cell* **47:**527–535.

139. **Szurek, P., J. Petrini, and W. Dunnick.** 1985. Complete nucleotide sequence of the murine γ3 switch region and analysis of switch recombination sites in two γ3-expressing hybridomas. *J. Immunol.* **135:**620–626.

140. **Takahashi, N., S. Nikai, and T. Honjo.** 1980. Cloning of human immunoglobulin μ gene and comparison with mouse μ gene. *Nucleic Acids Res.* **8:**5983–5991.

141. **Takahashi, N., S. Ueda, M. Obata, T. Nikaido, S. Nakai,**

and T. Honjo. 1982. Structure of human immunoglobulin gamma genes: implications for evolution of a gene family. *Cell* **29:**671–679.

142. **Taylor, R., and H. Wortis.** 1968. Thymus dependence of antibody response. *Nature* (London) **220:**927–928.

143. **Tian, S.-S., and C. Faust.** 1987. Rearrangement of rat immunoglobulin E heavy chain and c-myc in the B cell immunocytoma IR162. *Mol. Cell. Biol.* **7:**2614–2619.

144. **Tonegawa, S.** 1983. Somatic generation of antibody diversity. *Nature* (London) **302:**575–581.

145. **Tucker, P., C.-P. Liu, J. Mushinski, and F. Blattner.** 1980. Mouse immunoglobulin D: messenger RNA and genomic DNA sequences. *Science* **209:**1353–1360.

146. **Tucker, P., K. Marcu, N. Newell, J. Richards, and F. Blattner.** 1979. Structure of the constant and 3′ untranslated regions of the murine γ2b heavy chain messenger RNA. *Science* **206:**1303–1306.

147. **Tucker, P., J. Slightom, and F. Blattner.** 1981. Mouse IgA heavy chain gene sequence: implications for evolution of immunoglobulin hinge exons. *Proc. Natl. Acad. Sci. USA* **78:**7684–7688.

148. **Vitetta, E., J. Ohara, C. Myers, J. Layton, P. Krammer, and W. Paul.** 1985. Serological, biochemical, and functional identity of B cell stimulatory factor-1 and B cell differentiation factor for IgG1. *J. Exp. Med.* **162:**1726–1731.

149. **Voelkel-Meiman, K., R. Keil, and G. Roeder.** 1987. Recombination stimulating sequences in yeast ribosomal DNA correspond to sequences regulating transcription by RNA polymerase I. *Cell* **48:**1071–1079.

150. **Wabl, M., J. Meyer, G. Beck-Engeser, M. Tenkhoff, and P. Burrows.** 1985. Critical test of a sister chromatid exchange model for immunoglobulin heavy-chain class switch. *Nature* (London) **313:**687–689.

151. **Webb, C., M. Cooper, P. Burrows, and J. Griffin.** 1985. Immunoglobulin gene rearrangements and deletions in human Epstein-Barr virus transformed cell lines producing IgG and IgA subclasses. *Proc. Natl. Acad. Sci. USA* **82:** 5495–5499.

152. **Winter, E., U. Krawinkel, and A. Radbruch.** 1987. Directed Ig class switch recombination in activated B cells. *EMBO J.* **6:**1663–1671.

153. **Wortis, H.** 1971. Immunological responses of "nude" mice. *Clin. Exp. Immunol.* **8:**305–320.

154. **Yamawaki-Kataoka, Y., S. Nakai, T. Miyata, and T. Honjo.** 1982. Nucleotide sequences of gene segments encoding membrane domains of immunoglobulin γ chains. *Proc. Natl. Acad. Sci. USA* **79:**2623–2627.

155. **Yancopoulos, G., and F. Alt.** 1985. Developmentally controlled and tissue specific expression of unrearranged V_H gene segments. *Cell* **40:**271–281.

156. **Yancopoulos, G., and F. Alt.** 1986. The regulation of variable region gene assembly. *Annu. Rev. Immunol.* **4:** 339–368.

157. **Yancopoulos, G., T. K. Blackwell, H. Suh, L. Hood, and F. Alt.** 1986. Introduced T cell receptor variable region gene segments recombine in pre-B cells: evidence that B and T cells use a common recombinase. *Cell* **44:**251–259.

158. **Yancopoulos, G., R. DePinho, K. Zimmerman, S. Lutzker, N. Rosenberg, and F. Alt.** 1986. Secondary rearrangement events in pre-B cells: V_HDJ_H replacement by a LINE-1 sequence and directed class switching. *EMBO J.* **5:** 3259–3266.

159. **Yaoita, Y., and T. Honjo.** 1980. Deletion of immunoglobulin heavy chain genes from expressed allelic chromosomes. *Nature* (London) **286:**850–853.

Chapter 32

Site-Specific Chromosome Breakage and DNA Deletion in Ciliates

MENG-CHAO YAO

I. INTRODUCTION

Ciliated protozoa offer some of the most extreme examples of developmental genome reorganization among eucaryotes. Gene amplification, DNA elimination, chromosome fragmentation, new telomere formation, and sequence rearrangements are all known to occur. These processes result in somatic genomes drastically altered in both sequence content and overall organization. Recent studies have revealed that two distinctive mechanisms, site-specific chromosome breakage and DNA deletion, are each intimately involved in this phenomenon. Chromosome breakage occurs at hundreds to thousands of specific sites during ciliate nuclear differentiation, producing DNA fragments ranging from a few kilobases to several thousand kilobases in size which are maintained stably in the subsequent vegetative life. This process is associated with the formation of new telomeres at the broken ends, elimination of sequences around the breakage sites, and amplification of some of the resulting fragments. Internal DNA deletion occurs even more frequently, to remove thousands to tens of thousands of specific DNA segments ranging from about 14 base pairs (bp) to more than 4.5 kilobases (kb). Both single-copy and

Meng-Chao Yao ■ Division of Basic Sciences, Fred Hutchinson Cancer Research Center, 1124 Columbia Street, Seattle, Washington 98104.

repetitive sequences are deleted in a process which can be very highly sequence specific. In some cases it joins previously separated open reading frames. The molecular details of these processes will be the main focus of this chapter.

Studies of genome reorganization have been made in two distinctive groups of ciliates: the holotrichous ciliates, which includes such well-studied members as *Tetrahymena* and *Paramecium* spp., and the more complex hypotrichous ciliates, which include *Oxytricha*, *Stylonychia*, and *Euplotes* spp., among others. Early studies revealed a major difference between the features of genome reorganization in these two groups of organisms. This is now believed to be in the extents of reorganization, and not the nature of the reorganization mechanisms. Studies in both groups will be described and compared throughout this chapter to provide a better understanding of these processes.

II. LIFE CYCLES OF CILIATES

I will give a brief summary of ciliate life cycles, with emphasis on features relevant to genome reorganization. Most ciliates contain two types of nuclei in each cell. These two nuclei differ drastically in structure and activity. This nuclear dimorphism is perhaps the most important feature for consideration here. The well-studied *Tetrahymena thermophila* best illustrates this phenomenon. There is one macronucleus and one micronucleus in each vegetatively growing cell, readily distinguishable by many criteria. The following are most relevant. First, the macronucleus is large, containing approximately 45 times the haploid amount of DNA (45C) just after division. The micronucleus is small and is normally diploid (99). Second, the macronucleus divides amitotically, without detectable condensation of chromosomes or the formation of a spindle apparatus, whereas the micronucleus divides by typical mitosis. Third, the macronucleus is active in transcription, and the micronucleus is essentially inert (reviewed in reference 35), with the possible exception of one identified sequence (89). This suggests that the cell phenotype is dictated by the macronucleus and not the micronucleus. In support of this argument is the observation that genetic heterokaryons always express the phenotypes specified by their macronuclear and not their micronuclear alleles (70, 71). Fourth, the macro- and the micronucleus are not synchronous in DNA synthesis or division, even though they share the same cytoplasm. Other differences, such as histone composition (reviewed in reference 36) and DNA methylation patterns (37, 41, 42), have also been described. In general these differences are found in other holotrichs and hypotrichs as well, although their details may differ.

T. thermophila can grow indefinitely by binary fission, and can also be induced to mate by nutritional starvation. In the mating process cells of different mating types form a pair, and a series of events follow to eventually replace the macro- and micronuclei of the mating pair. First is meiosis, which results in four haploid nuclei, one of which divides mitotically to produce two gametic nuclei, with the other three degenerate. This event is followed by reciprocal fertilization, in which one of the two gametic nuclei in each cell migrates to the opposite mating cell and there fuses with the stationary gametic nucleus to form a zygotic nucleus. The zygotic nucleus then goes through two rounds of mitosis to form four genetically identical nuclei. Two of these develop gradually into macronuclear anlagen; the other two remain as micronuclei. At about this time the pair of cells separate, and the old macronucleus, which up to this point had not changed significantly in morphology, gradually becomes pycnotic and eventually disappears. The cell then resorbs one of the micronuclei and goes through cell division, division of the remaining micronucleus, and separation of the two new macronuclei. The resulting cells each contain one micronucleus and one macronucleus and resume vegetative life (reviewed in reference 77). Genetically the two zygotic nuclei in the mating partners are identical, and the four resulting cells share identical micronuclei, although variation in macronuclear development can lead to each cell having a different macronucleus. Most other ciliate species studied have a similar sexual cycle, but not all stocks used in the laboratory are capable of successful mating.

Based on its role during mating, the micronucleus can be regarded as the equivalent of a germ line nucleus, maintaining the genetic continuity of the species during sexual reproduction. The macronucleus, on the other hand, resembles a somatic cell nucleus of a metazoan, as it is lost following each sexual cycle. Thus, although a single-cell organism, the ciliate displays true germ line and soma differentiation. An important consequence is that irreversible changes in macronuclear DNA are limited to the following vegetative life, and not transmitted sexually to the future generations. The DNA alterations in ciliates are thus comparable to those in mammalian lymphocytes, but differ from those in other unicellular species such as yeasts and trypanosomes, in which programmed alterations of DNA are often reversible.

It is important to stress that even though the holotrichs and the hypotrichs both exhibit nuclear dimorphism, they are morphologically distinct and evolutionarily distant, and are often placed taxonomically in different classes. Their differences are also evident in comparing the micronuclear and the macronuclear genomes. Typically the hypotrich genome is more extensively reorganized. It is readily detectable by cytological examination, which allowed the initial discovery of this phenomenon (5). The holotrichs, in contrast, are easier to handle and better studied both biochemically and genetically. Consequently, although they exhibit less extensive reorganizational processes, it has often been easier to obtain unambiguous answers with them.

III. CHROMOSOME BREAKAGE

A. Sizes of Macronuclear DNA

The first evidence of chromosome breakage in ciliates was derived from direct size measurements of macronuclear DNA of the hypotrich *Stylonychia mytilus*. Initially this entailed sedimentation through sucrose gradients and electron microscope visualization. Both types of studies suggested that the macronuclear DNA was about 1.5×10^6 daltons, which is several orders of magnitude smaller than the expected sizes of micronuclear chromosomal DNA (83). That these small, linear DNAs were not products of random fragmentation during isolation was demonstrated later using the method of agarose gel electrophoresis and Southern hybridization. The macronuclear DNAs of *S. mytilus* and other hypotrichs, including *Oxytricha nova*, *Oxytricha fallax*, *Stylonychia pustulata*, and *Euplotes aediculatus*, were well resolved by agarose gel electrophoresis into distinct bands ranging in size from less than 1 to more than 10 kb (69, 91, 93). This pattern suggested that these macronuclear DNAs are gene sized, that is, each molecule contains a single gene. Support for this idea in *Oxytricha* sp. has come from cloning of DNA molecules, including genes of known function, such as actin, α-tubulin, and rRNA (30, 43, 50, 51, 88, 93, 92). In all cases, each molecule appears to contain a single gene or to specify a single transcript. This interesting observation may have significant consequences in gene regulation, but is not general for all ciliates, since holotrichs have much larger macronuclear DNA. Regardless of its generality and implications, the existence of gene-sized DNA illustrates the extremes to which the breakage process can go in certain ciliate species.

The DNAs of holotrich macronuclei are much larger. In *Tetrahymena* and *Paramecium* spp., except for the ribosomal RNA genes, the macronuclear DNA is too large to be measured accurately by traditional methods. This fact delayed the realization that extensive chromosome breakage is also common in holotrichs. Sucrose gradient sedimentation studies first suggested that the average size of *Tetrahymena* and *Paramecium* macronuclear DNA is 630 and 300 kb, respectively, which are less than 1/10 the sizes expected for micronuclear chromosomal DNA (81). Although those values turned out to be correct, the evidence was not conclusive due to technical limitations. The first conclusive evidence for fragmented macronuclear DNA in holotrichs was derived from studies of *Glaucoma chattoni* (a close relative of *Tetrahymena* sp.), in which some macronuclear DNAs were smaller than 20 kb in size and were readily detectable by agarose gel electrophoresis and Southern hybridization using a *Tetrahymena* telomeric DNA sequence as a probe (55). More recently, *Tetrahymena* and *Paramecium* macronuclear DNAs have also been analyzed using pulse field gel electrophoresis (20, 86). They are resolved into distinctive bands, each less than 10% the size of micronuclear chromosomal DNA, and unique sequences hybridize only to single bands (4, 25).

Thus, in both hypotrichs and holotrichs, the macronuclear DNA is drastically reduced in size compared to micronuclear DNA. The simplest interpretation is that the macronucleus undergoes chromosome breakage during nuclear differentiation. Based on the number of kinds of molecules present in the macronucleus, one estimates that there are between 50 and 200 breakage sites in *T. thermophila* and at least 20,000 such sites in *O. nova*. The reproducible banding and hybridization patterns in pulse field gels suggest site specificities for the breakage. In *T. thermophila*, essentially the same patterns are found in different inbreeding lines; thus, the same breakage sites are used in different lineages or genotypes. This conclusion was confirmed by studies of individual breakage sites.

B. Free Ends of the Macronuclear DNA

The free ends of the macronuclear DNA of ciliates contain special structures. This was first determined using the rRNA genes of *Tetrahymena* sp. (12). There are approximately 9,000 copies of this 21-kb linear molecule in each macronucleus of *T. thermophila* (33, 111). The molecule is a giant palindrome containing two sets of the 17S, 26S, and 5.8S ribosomal RNA genes (53). At the very ends of this molecule are tandem repeats of the hexanucleotide 5′-CCCCAA, with the 5′ end facing toward the end of the molecule. The repeat number is variable, ranging from 20 to 70, as estimated from restriction fragment sizes. This heterogeneity is present even in recently established clonal populations. The exact structure at the very end of the molecule is not known, although there are suggestions that it may contain a 12-nucleotide protruding 3′ end (12, 44; reviewed in reference 11).

The structure of other macronuclear DNA ends in *Tetrahymena* sp. has also been determined. Southern hybridization studies have shown that tandem repeats of C_4A_2 exist widely in the genome (105, 106), and essentially all of these sequences are sensitive to exonuclease Bal31 digestion (112). These results indicated that the C_4A_2 repeats are within 1 kb of the free ends of the macronuclear DNA, whose average size is around 800 kb. Five of these ends have been cloned and sequenced. Each contains a long stretch of direct tandem repeats of the hexanucleotide, with the 5′ end of the CA strand at one end of the insert. These repeats are also heterogeneous in their length, averaging between 0.6 and 1 kb (112, 117). Southern hybridization revealed that C_4A_2 repeats are present in a number of other holotrichs, and Bal31 analysis suggested that they are located near ends of macronuclear DNA (M.-C. Yao and C.-H. Yao, unpublished results). Direct cloning and sequencing has also revealed this feature in the rRNA genes of *G. chattoni* (23). Cloning and sequencing of the telomeric DNA adjacent to the surface antigen A gene of *Paramecium tetraurelia* and the related G gene of *Paramecium primaurelia* showed that the sequences consist of C_4A_2 repeats, with occasional C_3A_3 repeats (8, 32).

The telomeric DNA in macronuclei of hypotrichs is somewhat different. Direct sequencing studies of total macronuclear DNA indicated that in *O. nova* most macronuclear DNA contained repeats of the octanucleotide 5′-CCCCAAAA at their ends (61, 68, 80). There are at least 2.5 copies in the CA strand and at least 4.5 copies in the GT strand, with the two terminal copies of the GT octanucleotide forming a single-stranded protruding 3′ end. This structure is shared by most hypotrichs studied and is found in all cloned macronuclear DNA. Unlike holotrichs, the hypotrichs exhibit little length heterogeneity in the telomeric repeats in clonal populations.

Similar structures, consisting essentially of long stretches of simple sequences rich in A and C, are present as chromosomal DNA termini in diverse eucaryotes (reviewed in references 11, 13, and 22). These structures are believed to function as telomeres important for the stability and propagation of linear DNA ends. Thus, the addition of these telomeric sequences to the broken ends of ciliate chromosomes partly explains their stabilities in vegetative cells.

C. Sequence Signals for Chromosome Breakage

In *T. thermophila*, the specificity of chromosome breakage sites is likely determined by a conserved nucleotide sequence. The first suggestion was derived from studies of the rRNA genes. In the macronucleus, these genes reside in a 21-kb linear palindromic molecule present in roughly 9,000 copies per genome. In contrast, the micronucleus contains only a single copy of an rDNA sequence in the chromosome (108). Thus, the macronuclear rDNA must be derived from this micronuclear copy through an excision and amplification process. Further investigation suggests that this process involves breakage of chromosomal DNA at or near the two ends of the rDNA sequence. Sequences which flank both ends of the micronuclear rDNA are present in the macronucleus but are not amplified. They are part of larger molecules, each containing a free end near the original junction point with the rDNA (101, 114) (Fig. 1). Thus, the formation of rDNA in *T. thermophila* involves breakage of chromosomes at specific sites, formation of palindromic sequences, and selective amplification (for further review of rDNA amplification in *T. thermophila* see references 102 and 104).

Comparison of sequences near the two ends of the micronuclear rDNA revealed a 20-bp homology in the regions immediately flanking the rDNA (Fig. 1) (114). Two identical and one degenerate (with 16/20-bp homology) copies of this sequence are present at the 5′ end, and one copy with a single mismatch is present at the 3′ end of the rDNA. The 5′-end copies are direct repeats spaced 25 bp apart and are in the orientation opposite that of the 3′-end copy. These sequences are located 14 and 19 bp away from the 5′ and 3′ ends of the rDNA, respectively. The flanking regions share no other sequence homology.

A 15-bp portion of this 20-bp sequence is present widely in the genome, and is tightly associated with other chromosome breakage sites in *T. thermophila* (113). Two lines of evidence lead to this conclusion. First, a synthetic oligonucleotide containing the 20-bp sequence was found to hybridize with a large number of similar sequences in the micronuclear genome, but not in the macronuclear genome, suggesting that these sequences are involved in genome reorganization. Seven of these micronuclear sequences were cloned and used as hybridization probes to study the structures of the corresponding regions in the macronucleus. In six cases, the macronuclear DNA fragments detected were sensitive to pretreatment with Bal31, indicating that their locations are telomeric. Thus, chromosome breakage must have occurred at or near the 20-bp sequences in these regions to generate the free ends. These clones

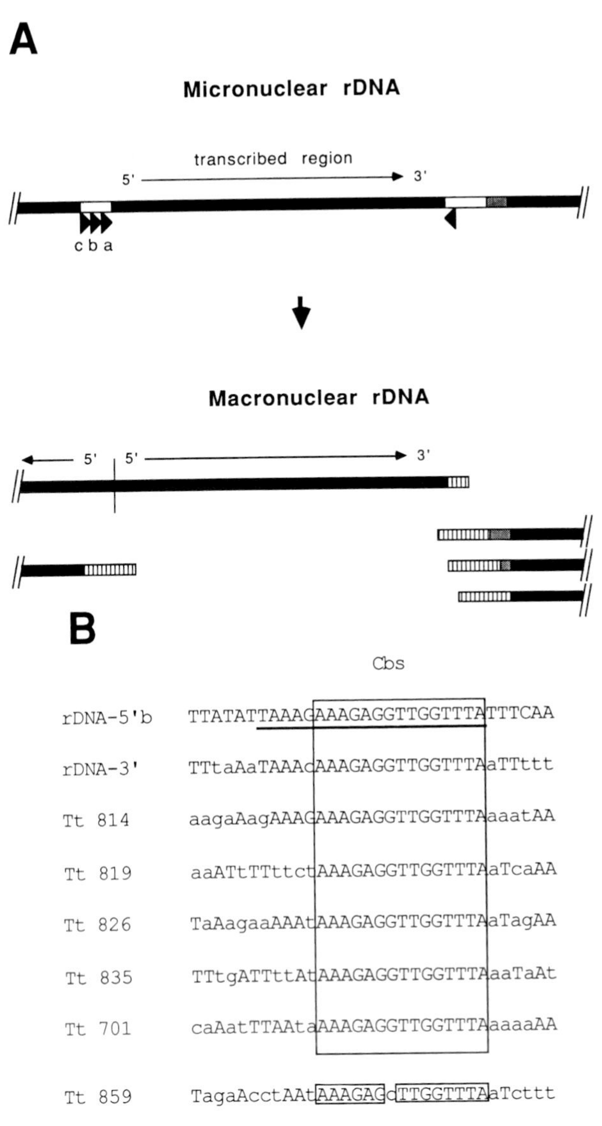

Figure 1. *Tetrahymena* rDNA and chromosome breakage site sequences. (A) The structures of rDNA in the micronucleus and the macronucleus of *T. thermophila*. The solid bars represent the macronucleus-destined sequences and open bars the germ line-limited sequences. The shaded area in the right flanking region represents a sequence which is eliminated only part of the time. Three products of this variable elimination are shown. The striped bars represent telomeric sequences of the macronuclear DNA. The solid triangles represent the conserved sequence Cbs. This figure is not drawn to scale. (B) Cbs (or similar sequences) from eight different regions of the genome, including the two ends of the rDNA. rDNA-5′b contains the second copy of Cbs at the 5′ end of the rDNA as shown in part A. It differs from rDNA-5′a (which contains the first copy of Cbs, not shown) only at the last four positions shown. The 20-bp sequence shared between these and the 3′-end sequence (rDNA-3′) is underlined. rDNA-5′c contains the third, degenerate 5′-end copy. The 15-bp sequence Cbs, which is enclosed in the box, is shared between these and the other five breakage site sequences shown. Tt 859 contains a mismatch within this 15-bp region and is not located at a breakage site. Nucleotides identical to those in rDNA 5′b are shown in upper-case letters. (Part B is modified from Fig. 8 of reference 113.)

were also subjected to sequence analysis. Six were found to contain the same 15-bp portion of the 20-bp sequence. The seventh contains a mismatch in this 15-bp region. Interestingly, this is the same clone which does not contain a breakage site based on Bal31 analysis. Thus, all six clones which contain the shared 15-bp sequence also contain a breakage site. These results strongly suggest an important role for the 15-bp sequence, which is referred to as Cbs (chromosome breakage sequence).

The second approach addresses the issue of whether most breakage sites contain Cbs. Two macronuclear DNA termini, including the C_4A_2 repeats and the adjacent internal sequences, were cloned (117). The internal single-copy sequences were then used as probes to screen for the corresponding micronuclear DNA clones. These clones should contain a breakage site. Subsequent sequencing studies revealed the presence of a copy of Cbs in each of these two clones. Probably as the result of a rare coincidence, the two macronuclear DNA ends used for this study are derived from the two opposite sides of the same breakage junction. This fact became apparent when the two micronuclear DNA clones isolated were found to contain an identical overlapped region, which included sequences from both of the macronuclear DNAs (113). Thus, at least one copy of Cbs is found at this breakage site, as well as the breakage sites located at the two ends of the rDNA. These results provide support for the argument that most breakage sites in *T. thermophila* contain at least one copy of Cbs.

The results in *T. thermophila* raised the interesting possibility that sequences similar to Cbs might also be present in other ciliates. An oligonucleotide probe containing Cbs was found to hybridize well with *Paramecium* DNA (Yao and Yao, unpublished results), but it is not known whether such sequences are associated with breakage sites. In hypotrichs, however, it is clear that Cbs is not present in all, if any, breakage sites. In 11 cases in which micronuclear sequences from regions corresponding to macronuclear telomeres in *O. nova* and *O. fallax* were analyzed, no sequence with close homology to Cbs was found (47, 58, 59, 62). Thus, the conserved sequence in *T. thermophila* is not present in most breakage sites in these hypotrichs. The same studies also failed to reveal any other shared sequence. However, it is still possible that conserved sequences are located further away from the macronucleus-destined sequences, and not within the regions analyzed. In most cases only a portion of the eliminated sequence around a breakage site was determined. Also, breakage point heterogeneity could confuse the issue of where to look for Cbs or other conserved sequences. Since hypotrichs contain about 100-fold more breakage sites than does *T. thermophila*, there could be more than one type of breakage site, each containing a different shared sequence. In this case a larger number of breakage sites would have to be analyzed before finding common sequences.

D. DNA Elimination at Sites of Breakage

The organization of DNA at several breakage sites has now been studied in detail. In both hypotrichs and holotrichs the breakage is associated with loss of DNA. The first example was provided by studies of the 3′ end of the rDNA in *T. thermophila*. At this breakage site approximately 4 kb of DNA are eliminated from the macronucleus (101) (Fig. 1). Elimination of DNA was also shown in seven other sites studied in *T. thermophila*, including the 5′ end of the rDNA (113, 114). In these cases the eliminated regions are much shorter, probably no more than a few hundred base pairs in length. In one case the eliminated region is only 54 bp (113). This is the only case in holotrichs in which both free ends generated from a breakage site have been sequenced. One copy of Cbs is found within this region 14 bp from one end of the deleted segment. No other recognizable feature is present in the remaining sequences, except that they are very A+T rich. These sequence studies were done on DNA from a single line. Hybridization studies using other lines suggested that the exact boundaries of the eliminated region may vary slightly. Sufficient sequence information is now available for the seven different regions containing Cbs. Except for the 5′ end of the rDNA, they all appear to contain only one copy of Cbs flanked on both sides by A+T-rich sequences (average of 87% A+T).

Except in one case, the breakage sites studied in *O. nova* also appear to contain some eliminated sequences. In three cases where conclusive evidence is available, the sizes of the eliminated regions range from 90 to approximately 380 bp. Since more than 90% of micronuclear DNA does not appear in the macronucleus (67, 83), the eliminated regions at some breakage sites must be much longer than 380 bp. It has been suggested that most macronucleus-destined sequences exist in clusters (14, 62), with relatively short eliminated regions located between members of a cluster, and long eliminated regions between clusters. Whether this is true remains to be determined. The one exceptional case is interesting. Not only is there no apparent sequence elimination, but the two adjacent macronuclear DNA termini appear to overlap by 5 bp (62). Unfortunately, one of the sequences used for this determination derives from a different allele, and therefore the result needs to be treated with caution.

E. Telomeric DNA Addition

Essentially all macronuclear DNAs in the ciliates studied contain special sequences at their termini. How are these sequences formed during development? Comparisons between macronuclear and micronuclear DNA indicated that the repeated C_4A_2 or C_4A_4 sequences are not templated in the micronuclear genome (56, 57) and must therefore be added during development. This is one of the first suggestions that nontemplated DNA synthesis occurs in eucaryotes. For *T. thermophila*, sequence information is available from three separate sites (113, 114) in which no repeated C_4A_2 is found in micronuclear DNA at positions corresponding to where it appears in the macronulcear DNA. Hybridization and sequencing also rule out the presence of tandemly repeated C_4A_2 in six other breakage sites in *T. thermophila* (113). The same conclusion is also reached through similar sequencing studies on a number of telomere addition sites in *O. nova* and *O. fallax* (47, 58, 59).

No obvious sequence similarity has been found among sites of telomeric DNA addition, suggesting the absence of common sequence signals for this process. In fact, for some sites the exact points of telomere addition are variable among independently formed macronuclei, or even among different DNA molecules in one macronucleus. The first case reported concerns the surface antigen G gene of *P. primaurelia* (8). This gene is located near the macronuclear telomere. Four independently cloned DNAs from a homozygous strain contained this gene with C_4A_2 repeats starting at different points within 0.8 kb of its 3′ flanking region (8). A similar phenomenon was also observed for the surface antigen A gene in *P. tetraurelia* (31, 32) and also in *Tetrahymena* sp. In a homozygous *Tetrahymena* strain, up to four different telomeric addition points were detected in the region flanking the rDNA 3′ end, with spacings as much as 2 kb (Yao and Yao, unpublished results). Variations in other telomere addition sites in *Tetrahymena* sp. appear to be smaller and not readily detectable by restriction mapping. Large-scale variations have also been documented in hypotrichs. In *O. fallax* both ends of one macronuclear DNA were found to occur at two points, resulting in three different molecules which coexist in the same macronucleus (47). This kind of variation is detected at one end of a macronuclear DNA molecule of *O. nova* (58). Small-scale variation has also been detected in at least one case (47).

The addition sites of telomeric DNA are not always variable. One clear case is the rDNA 3′ end of *T. thermophila*, which happens to be the first example of a telomere addition site sequenced in ciliates (23, 56, 57). The micronuclear DNA in this region contains one copy of C_4A_2. Comparisons with one macronuclear DNA sequence indicated that the addition point of C_4A_2 repeats in this cell line is 5 bp upstream of the single C_4A_2. It was suggested that the presence of a single C_4A_2 might be important for the addition of repeated C_4A_2 to the ends (56). However, subsequent studies failed to locate this sequence at or near the telomere addition points in at least three other sites. Nor has any copy of C_4A_4 been found near telomere addition sites in hypotrichs (46, 47, 58, 59, 62). Thus, the presence of these sequences is not essential for new telomere formation. Nonetheless, one cannot rule out a possible role for the single C_4A_2 at the rDNA 3′ end. Reinvestigation of this problem in several strains of *T. thermophila* revealed that in all but one case the point of telomere addition is identical and occurs precisely at the single copy of C_4A_2. The macronuclear DNA end sequence previously determined (57) happens to be a rare exception. This conclusion is based on sequence studies of three independent lines and oligonucleotide hybridization of several additional lines (M.-C. Yao and Y. Rao, unpublished observations). Thus, in contrast to other sites, the point of telomere addition at the 3′ ends of rDNA is nearly constant and occurs at the single copy of C_4A_2 which is unique to the rDNA. These observations suggest that telomere addition might occur at some preferred sequences, such as C_4A_2, when present and at variable points when they are absent.

An activity which may be responsible for telomeric DNA elongation has recently been identified in *T. thermophila* (38). This activity, which contains both RNAs and proteins as active components (39), is capable of adding G_4T_2 repeats in vitro, without templates, to the 3′ ends of several single-stranded oligonucleotides. These primer oligonucleotides include repeated G_4T_2 and G_4T_4 but not C_4A_2 or sequences with no resemblance to known telomeres. It is suggested that in vivo the added G_4T_2 strand could serve as the template for the synthesis of the C_4A_2 strand through the normal mode of replication. These striking results provide a nice explanation for how telomeric DNA can be maintained in growing eucaryotic cells. The activity has been extracted from both vegetatively growing and conjugating cells, and appears to be more abundant in the latter. It is possible that this activity is responsible for the formation of new telomeres following chromosome breakage during conjugation, as well as for the maintenance of preexisting telomeres in vegetative cells. One difficulty for this interpretation is that, except for one site, there is no specific preexisting sequence at breakage sites for telomere addition. In fact, most of the sites in *T. thermophila* are highly rich in A+T and do not resemble the oligonucleotide sequences used for the in vitro reaction. Thus, a cofactor or a different activity is needed to explain the formation of new telomeres during conjugation.

F. Developmental Schedule of Chromosome Breakage

The developmental schedules for chromosome breakage have been determined in both hypotrichs and holotrichs by cytological and molecular analysis. Macronuclear development in hypotrichs includes a series of distinctive cytological events, which offered some of the earliest evidence for genome reorganization in ciliates (5, 72, 73; reviewed in reference 60). Soon after the second postzygotic division, the macronuclear anlage begins to increase in size and DNA content to approximately 60C in *S. mytilus*. At this stage, polytene chromosomes with discernible banding patterns are seen. Soon afterward, vesicular partitions are formed, which appear to transect polytene chromosomes between bands. The resulting vesicles contain chromatin bodies which probably are derived from individual bands. At about this time, the nuclear DNA content decreases dramatically by approximately 30-fold. This decrease is also seen in the DNA of individual vesicles. Several rounds of DNA replication then follow to bring the nuclear DNA content to approximately 100C, a level comparable to that of a mature macronucleus. It is likely that chromosome breakage occurs sometime after polytene chromosome formation and before the second wave of DNA replication. This would put the breakage event around the vesicle stage. In fact, it has been suggested that the vesicular partitions are related to the mechanism for chromosome breakage. For instance, they might be responsible for breaking chromosomes in between clusters of macronuclear-destined sequences.

Molecular details of chromosome breakage have been determined in conjugating cells of *T. thermophila* and *Euplotes crassus*, from which synchronously developing macronuclei can be purified in reasonable quantities (1, 85). Southern hybridization using rDNA and C_4A_2 repeats as probes showed that chromosome breakage occurs in *T. thermophila* between 12 and 14 h after conjugation begins. These studies failed to reveal any structure which could be interpreted as an intermediate product of breakage without DNA elimination or telomere addition. Thus, telomere addition must be coupled with, or immediately follow, the breakage event in this organism (M.-C. Yao, D. Allis, and C.-H. Yao, unpublished results). Studies in *E. crassus* showed that breakage occurs between 38 and 44 h after conjugation begins, which includes the vesicle stage. Interestingly, molecules of intermediate sizes were found, suggesting a multistep process for chromosome breakage. In addition, the telomeric DNA was longer by 140 bp when it first appeared. It abruptly decreases to the expected length when the macronucleus matures (85). Whether the longer telomere has a particular function during development is not clear.

G. Possible Mechanisms for Chromosome Breakage

The first event in chromosome breakage is probably a double-stranded DNA cleavage. The presence of the conserved sequence Cbs offers a simple explanation for the site specificity of this event in *T. thermophila* and related ciliates. One could imagine that specific proteins (or other macromolecules) which interact with these sequences (Fig. 2) are produced during conjugation and targeted to the macronuclear anlagen. These interactions lead to double-stranded DNA cleavages at or near the binding sites. The initial cleavage points have been hard to determine due to sequence elimination at these sites. It is tempting to suggest that similar mechanisms are also responsible for achieving the site specificity in other ciliates, although in hypotrichs different conserved sequences are probably utilized.

The cleavage reaction is coupled with a reaction that removes DNA from the broken ends, probably involving an exonuclease (Fig. 2). This is followed by the addition of telomeric sequences, which protect the broken ends from further degradation. One could view the exonuclease and telomere addition reactions as two competing processes. Delayed addition of telomeric DNA leads to elimination of DNA around the breakage sites. If neither reaction has a strong sequence specificity, variable telomeric DNA addition points can be expected, and indeed were observed at a number of sites. The one exceptional case, in which a preexisting copy of C_4A_2 is present and used as the predominant site for telomeric addition, can be taken to indicate the preference of the telomere addition activity for this sequence. To carry this idea one step further, one might argue that the activity of the exonuclease is influenced by local chromatin structures. Thus, in some breakage sites the exonuclease reactions proceed for a longer distance than in others before telomere additions occur, creating a longer eliminated region. The 3′ flanking region of *Tetrahymena* rDNA might represent one such site.

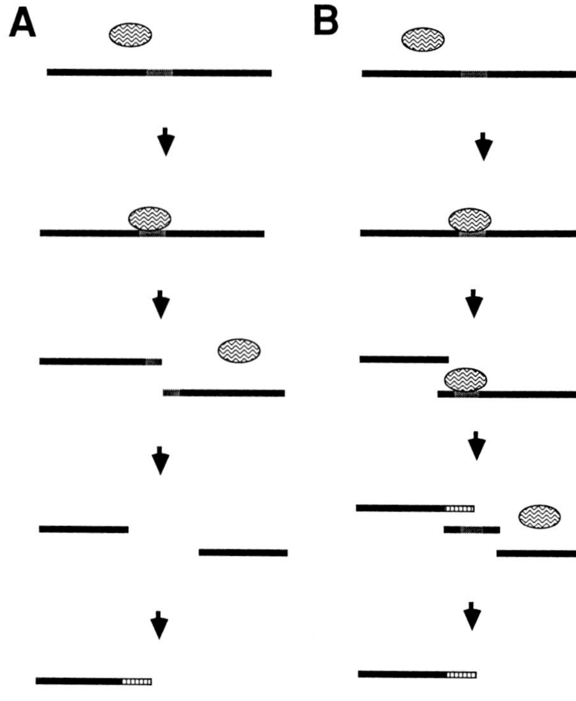

Figure 2. Possible steps involved in chromosome breakage. Two possible schemes are shown to illustrate the steps involved in chromosome breakage. The black bars represent chromosomal DNA at breakage sites and the shaded bars represent Cbs or other conserved breakage site sequences. The striped bars represent the telomeric DNA, which are added to the free ends after breakage occurs. The round object represents an unknown protein which recognizes and binds to the conserved sequence. (A) Scheme calling for an exonuclease activity to eliminate the sequences at the breakage site. (B) Scheme relying on multiple endonuclease cleavages to explain sequence elimination. For details see text.

An exonuclease activity does not offer the only explanation for sequence elimination at breakage sites. The endonucleases might cleave at more than one point around Cbs, thereby eliminating DNA between these points (Fig. 2). The endonucleases could be specific for Cbs, or other, nonspecific endonucleases could work in conjunction with Cbs-recognizing molecules. Such processes could also lead to the observed elimination of variable amounts of DNA.

H. Analysis of Chromosome Breakage through a Transformation System

Further studies of chromosome breakage should profit from two types of approach: analysis of the reaction in vitro, and mutational dissection of *cis*-acting sequences. The former is still at an explorative stage; the recent development of an efficient transformation system (96) has made the latter feasible.

Successful transformation of *T. thermophila* was

first achieved through microinjection into mature and developing macronuclei (96). The donor molecule was a mutant macronuclear rDNA which confers resistance to the antibiotic paromomycin (18, 87). This molecule also differs from the host rDNA in its restriction map and relative strength of replication (or segregation) (66, 79). Under optimal conditions approximately one of two injected cells is phenotypically transformed, and most macronuclear rDNAs in the transformants are of the injected type (96).

To use this system for the study of chromosome breakage, the micronuclear rDNA was cloned from a strain (SL 062) which carried the appropriate resistance and restriction site markers. The cloned rDNA contains 4 kb of the 5′ flanking sequence and 0.5 kb of the 3′ flanking sequence, which includes all copies of Cbs known at these two breakage sites. Upon injection into developing macronuclei of conjugating cells, some of these cloned sequences were properly processed and amplified, such that approximately 4% of the injected cells became drug resistant. The macronuclear rDNA of the transformed cells was indistinguishable from normal mature rDNA and carried the markers of the injected type. This process did not seem to involve conversion of the host rDNA, since a truncated gene containing the selectable genetic markers failed to transform at comparable rates. It is likely that the injected copies are processed directly without reintegration. In this sense it is interesting that both linear and supercoil forms of DNA transform equally well. Thus, all sequence information necessary for chromosome breakage, telomere addition, palindrome formation, and selective amplification appears to be present in the injected molecules. Further analysis using deletion mutants has limited the requirement for the 5′ flanking sequence to less than 200 bp (Yao and Yao, unpublished results). This approach should lead to a better understanding of the *cis*-acting sequences needed for programmed chromosome breakage in *T. thermophila*.

IV. DNA DELETION

A. Occurrence

Chromosome breakage has been known in ciliates for more than 15 years, but evidence for site-specific deletion was not available until 1984, when two cases were reported in *T. thermophila* (19, 107) and one case was found in *O. nova* (59). It is now clear that this process occurs widely in these and other ciliates.

The finding of DNA deletion in *T. thermophila*

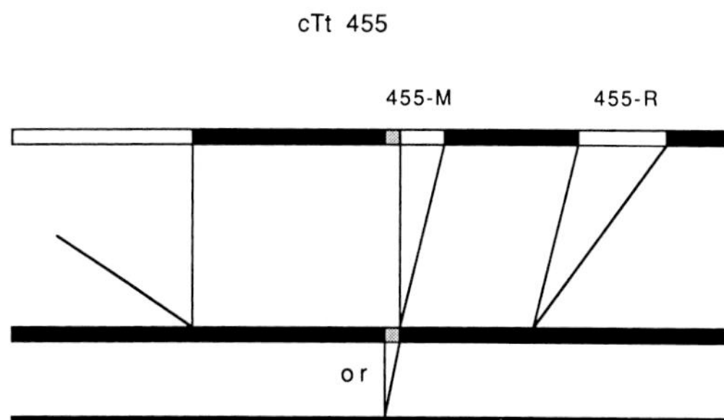

Figure 3. An example of DNA deletion in *T. thermophila*. The micronuclear and the macronuclear DNA from a region of *T. thermophila* (represented by the micronuclear DNA clone cTt 455) is shown to illustrate the phenomenon of DNA deletion. The solid bars represent sequences which are retained in the macronucleus and the open bars represent the deleted regions. The shaded area is deleted only part of the time. Both products of these alternate processes are shown.

derived from an effort to understanding DNA elimination. In this organism the macronuclear genome contains about 85% of the sequence complexity of the micronuclear genome (110). The germ line-limited DNA, which is composed of both single-copy and moderately repetitive sequences (110), has been analyzed through cloning (2, 16, 49, 52, 98, 103, 109, 116) and is found to be distributed throughout the genome (52, 103). One such clone (referred to as cTt 455), which contains both eliminated regions and the flanking macronucleus-destined regions, has been isolated and analyzed in detail. Comparison between this and the corresponding macronuclear sequence revealed the nature of the elimination. Three segments of DNA, with sizes of 0.9, 1.1, and more than 2 kb, are missing from the macronuclear genome. The remaining DNA is joined together in one contiguous piece (107) (Fig. 3). Thus, DNA elimination in this case is the result of internal deletion. It should be mentioned that although *T. thermophila* is diploid, genetic manipulation can be used to construct whole-genome homozygotes, which were used here to avoid confusion due to heterozygosity. In addition, single-copy sequences were always used to assess the identities of the macro- and the micronuclear sequences without complications due to repetitive sequences.

Studies of the single α-tubulin gene provided the other example of DNA deletion in *T. thermophila* (19). Southern hybridization studies using a heterologous probe suggested that two segments of DNA, each approximately 1 kb in size, were deleted from the macronucleus. The segments were within 3 and 6 kb of the 5′ and 3′ ends of the gene, respectively.

Conclusive evidence for DNA deletion was also obtained later in a third case, in which a 4.5-kb piece of DNA, which contains a cluster of C_4A_2 repeats, is deleted from the macronucleus (116). The significance of the internal C_4A_2 repeats in the micronucleus will be addressed in a later section.

If internal deletions of DNA, each a few kilobases in size, are largely responsible for elimination of 15% of the genome, there should be approximately one eliminated region per 10 to 100 kb if they are evenly distributed. This prediction was tested directly and found to be true. Twenty random macronuclear DNA clones, each approximately 15 kb long, were used as probes for Southern hybridization with micronuclear and macronuclear genomes. Of these clones, 17 hybridized to one or a few fragments and 9 of these revealed clear differences between the two genomes. The differences are believed to result from internal DNA deletion. They are not due to chromosome breakage, since none of the hybridizing macronuclear DNA restriction fragments contains a preexisting free end, although other types of rearrangements have not been ruled out. These results suggest that in *T. thermophila* there are approximately 6,800 segments of germ line-limited DNA distributed widely in the genome. If the segments are on the average 2 kb long, they could account for most of the germ line-limited sequences (107). This is in good agreement with studies of other germ line-limited sequences, in which undefined rearrangements are often observed (2, 16, 49, 98).

Studies in hypotrichs are more complicated, due partly to the lack of inbred lines in most species studied and partly to the uncertainty of micronuclear ploidy in some species. Thus, to be certain of the genetic identity of a given DNA, it is often necessary to sequence a number of recombinant DNA clones from both the macronucleus and the micronucleus. Only DNAs with identical sequences except in the regions of rearrangement are considered to be different developmental forms of the same DNA segment.

When the first matching pair of macro- and micronuclear DNAs was obtained by this method from *O. nova*, internal DNA deletions were found (59). These are referred to as IESs (internal eliminated sequences). There are three IESs in this 0.8-kb region: two are 49 bp and the third is 32 bp in size. This macronuclear DNA contains an open reading frame within a transcribed region. One IES is located within the coding region, and one lies in the transcribed noncoding region. This is the first case known in ciliates in which elimination of DNA leads to the formation of an open reading frame. IESs have also been found in three other cases. Two additional cases of DNA deletion were later found in *O. nova*. One of them contains five IESs in a 1-kb region as determined by direct sequence comparison, and the other contains at least five IESs within a 2.3-kb region as determined through restriction mapping (84). One example of DNA deletion involving three IESs was observed within a 2-kb region in *O. fallax*. One of these IESs also appears to interrupt a coding region (46). Based on this analysis, it is concluded that IESs are present in most macronucleus-destined regions of DNA in *Oxytricha* sp. (84). Since there are approximately 20,000 different macronuclear DNA molecules, there could be as many as 60,000 IESs in the genome.

B. Sequence Structure of Deleted DNA

The sequence organization of deleted DNAs is better characterized in hypotrichs partly due to their smaller size and higher abundance. Sequences for nine IESs and their immediate flanking DNAs have been reported (46, 84). The deleted regions range in size from 14 to 548 bp. The most distinctive common feature is the presence of a pair of short direct repeats at their ends. One copy of this repeat is retained in the macronucleus. The lengths of these repeats vary, ranging from 2 to 6 bp, as do their sequences. Other features, such as short inverted repeats, are also present (84), but their locations relative to the ends of IESs are variable. Comparison among IESs revealed very little homology, except that they are rich in A+T (between 75 and 100%). Their small sizes also put severe limits on their coding potentials (84).

Two deleted regions in *T. thermophila* have been completely sequenced. They are neighboring regions 2 kb apart, referred to here as 455-R and 455-M, and are regions covered by the DNA clone cTt 455 mentioned above (107). Deletion 455-R is a 1,092-bp single-copy sequence (7). Deletion 455-M occurs in two sizes, either 908 or 588 bp. This sequence hybridizes to several dozen related sequences in the genome and probably contains a member of a repeated DNA family. The issue of alternate DNA deletion will be addressed further in a later section. Analysis of the nucleotide sequence revealed that, as in hypotrichs, the deleted regions contained a pair of short direct repeats at their ends. The repeats in 455-R are 6 bp in length (7). Their sequence is different from those in 455-M, which are 8 and 5 bp in length for the longer and shorter deletions, respectively (7a; C. F. Austerberry, Ph.D. thesis, Washington University, St. Louis, Mo., 1987) (Fig. 3). The deleted sequences are rich in A+T (between 76 and 85%). Known coding regions in *T. thermophila* contain less than 60% A+T (48). It is unlikely that these two deleted regions code for proteins. However, such a possibility is hard to rule out, especially since in this and some other ciliates only UGA is used as a termination codon (21, 40, 48, 82).

The two alternate forms of deletion 455-M share the same right ends but differ in their left ends. Comparisons between these two left ends revealed an interesting common feature consisting of a stretch of homopurine sequence located 30 to 40 bp away from the deleted regions. These two sequences are 22 to 24 bp in length and include an identical 10-bp segment. Interestingly, a similar structure is also present near the common right end. All three sequences have the same orientation relative to the adjacent deletions (7a; Austerberry, Ph.D. thesis). These features probably have an important role in the formation of the 455-M deletion. However, the same feature is not found in deletion 455-R, and thus it cannot be needed for all deletion events in *T. thermophila*. Since more than 6,000 deletion are made during macronuclear development, it is quite possible that more than one type of signal is used in deletion and that 455-M and 455-R are members of different deletion classes.

C. DNA Deletion during Macronuclear Development

The process of deletion has been characterized by Southern hybridization in synchronously developing nuclei of *T. thermophila* (6). The results show that in most cells both deletions 455-R and 455-M occur between 12 and 14 h after conjugation begins. During these 2 h neither the cell nor the nucleus divides. Microfluorometry analysis shows that the macronuclear anlagen contain 4C of DNA at 12 h and that 70% of them have increased to 8C by 14 h. Thus, deletion in *T. thermophila* occurs at a precise developmental stage and is accompanied by less than one round of DNA replication. All starting materials for deletion are immediately replaced by the final products without nuclear division. These observations put stringent constraints on possible mechanisms for DNA deletion. This Southern blotting study also showed that the deleted DNAs were eliminated as soon as deletion occurs, in good agreement with an in situ hybridization study which had shown elimination of germ line-limited DNA from developing macronuclei at 16 h (115).

D. Variability of DNA Deletion

Within the resolution afforded by agarose gel electrophoresis, DNA deletions in ciliates appear to be invariable in occurrence and size, with a few notable exceptions. Deletion 455-R in *T. thermophila* serves as a well-characterized example of invariable deletion. In all strains analyzed, which includes more than 40 caryonidal lines (clones containing descendants of individual macronuclear anlage), the 455-R deletion always occurs and the products appear to be identical in size (6, 7, 107). Even in a mating population in which millions of nuclei are formed independently, the deletion products are homogeneous in size (6). The sensitivity of such methods indicates that no variant form can contribute more than 5% of the deletions. This is probably typical of DNA deletions in this and other ciliates, although extensive analysis of other regions has not been reported.

Variable deletion has also been shown to occur in *T. thermophila*. The best-studied example is deletion 455-M, in which the deletion is 0.6 or 0.9 kb in length. These two forms are equally abundant in a mating population (6). Investigations of clones originating from single macronuclear anlagen (caryonidal lines) indicate that both forms of deletion occur within single macronuclei (6, 7a; Austerberry, Ph.D. thesis). Since each macronuclear anlage contains at least four copies of the genome during deletion (6) and both forms of deletion are detected at the same stage, it is clear that they are alternate events occurring in the same macronucleus. Examination of over 40 caryonidal lines revealed both forms in most lines, although in some only one is found (7a; Austerberry, Ph.D. thesis). Alternate deletion has also been reported in another region of the *Tetrahymena* genome. In this case each caryonidal line contains one of several forms of the same DNA, presumably as the result of alternate deletion (49). Other variation in DNA elimination or rearrangement in *T. thermophila* has also been found (54, 98).

The precision of deletion in *T. thermophila* has been analyzed using the simple approach of Southern blot hybridization with oligonucleotide probes. Under appropriate conditions, this method detects mismatches of a single nucleotide and permits the precision of deletion to be analyzed without direct cloning and sequencing. Oligonucleotides (24-mers) matching the sequences surrounding the 455-R and 455-M deletion junctions were prepared and used to hybridize with the macronuclear DNA from many caryonidal lines under stringent conditions. For deletion 455-R, which shows no variation at the gross level, the probe junction sequence is present in 71 of 72 lines analyzed (7; Austerberry, Ph.D. thesis; C. Austerberry and M.-C. Yao, unpublished observations). Thus, this deletion is highly specific at the sequence level. This observation provides another indication that DNA deletion in ciliates is strictly regulated.

One of the two alternate forms of deletion 455-M also appears to be highly specific. The junction sequence derived from one 0.6-kb deletion product hybridized well with all 19 lines which contained this form of deletion. In contrast, the 0.9-kb deletion junction sequence from one line hybridized only with 26 of the 35 lines known to contain the 0.9-kb deletion product. The junction sequences in the other lines must be different. One of these other products has been cloned and sequenced and found to differ from the first in having 3 bp more of DNA (a deletion shorter by 3 bp). By oligonucleotide hybridization, this second junction sequence was found in 7 of 20 lines analyzed. Three other lines failed, however, to hybridize with either probe (Austerberry, Ph.D. thesis; C. Austerberry and M.-C. Yao, in preparation). Thus, at least three junction sequences can be created as the result of the 0.9-kb deletion.

Although variation of deletion has not been extensively investigated in hypotrichs, large size variations do not occur in the regions of the limited number of cell lines analyzed. Whether small variations, on the order of a few nucleotides, are also absent is not known.

E. Transposable Elements and DNA Deletion

Although no transposable element has been found in ciliates, two recent reports (24, 45) have shown that DNA sequences resembling transposable elements are present in the micronuclei of *T. thermophila* and *O. fallax*. These sequences have two peculiar properties. First, they are eliminated from the macronucleus and thus may be related to DNA deletion, and second, they contain C_4A_2 or C_4A_4 repeats at their ends, suggesting a possible link with telomeres in this process.

Approximately 200 small clusters of tandemly repeated C_4A_2 sequences have been found in the *Tetrahymena* micronuclear genome (103, 105, 106, 116). Unlike the repeats in the macronucleus, most of these sequences are not located at or near chromosome ends (112). They seem not to be the precursors of the macronuclear telomeres, although this possibility is hard to rule out entirely. Twenty-two of these clusters have been isolated by cloning and studied. In most cases the sequences flanking the C_4A_2 repeats are eliminated during macronuclear development (105, 116), suggesting that the micronuclear C_4A_2 repeats are also eliminated. Elimination is clearly the result of internal deletion in the case of a 4.5-kb segment of DNA including a cluster of C_4A_2 repeats (116). It appears that the internal C_4A_2 repeats of the *Tetrahymena* micronucleus are typical of the many moderately repeated DNA families which are eliminated from the macronucleus as the result of DNA deletion (103, 109).

Analysis of six clusters of C_4A_2 repeats (24) revealed that 50 to 70% of the repeating hexameric units are degenerate. This degeneracy provides further evidence that these C_4A_2 repeats are not the developmental precursors of macronuclear telomeres, which are perfect repeats of C_4A_2. In addition, a conserved 30-bp sequence was found adjacent to the 3′ end of the C_4A_2 strand of each cluster of repeats. This sequence is identical in three cases and slightly different in the other three. In one clone, two clusters of C_4A_2 repeats are present, arranged in opposite orientations, with the C_4A_2 repeats immediately flanking the 30-bp sequences. Genomic Southern hybridization studies suggest that most clusters of C_4A_2 repeats are adjacent to a copy of the 30-bp sequence. Based on these observations, it has been suggested that the C_4A_2 repeats and the 30-bp sequences are the termini of a transposable element, Tel-1 (24). This possibility is strengthened by a parallel, but more detailed study in *O. fallax*.

Two major classes of tandem repeats of C_4A_4 have been found in the micronucleus of *O. fallax* (28, 29). Elements of the first class are located in long stretches of DNA (approximately 4 kb) which are resistant to digestion by most restriction enzymes, but sensitive to the exonuclease Bal31, and are thus believed to be part of the telomeric sequence of micronuclear chromosomes (28). Elements of the second class are much shorter, present in approximately 2,000 copies scattered throughout the genome, and generally not at chromosome ends. Approximately 100 copies of them are slightly more than 20 bp long, but the rest are shorter and are believed to have the same 17-bp sequence: 5′-$C_1A_4C_4A_4C_4$ (45). This sequence is remarkably similar but not identical to that of the macronuclear telomeres of this organism. Many copies of this sequence were found to be linked to members of a repeated sequence family referred to as TBE1, present in approximately 2,000 copies in the genome. The two that have been cloned and analyzed contain features typical of a transposable element. They are 3.8 and 4.0 kb in size and are bounded at both ends by inverted repeats that are 77 or 78 bp in length, of which the final 17 bp is $C_1A_4C_4A_4C_4$. Immediately outside these sequences are 3-bp direct repeats (45). This structure is reminiscent of the terminal inverted repeats and target site duplications of many transposons. That these 3.8- and 4.0-kb elements might in fact be transposons is further supported by the presence of a second copy of this chromosomal DNA segment lacking the transposonlike element but containing one copy of the 3-bp direct repeat. This empty site is flanked by DNA which is similar but not identical to the DNA flanking the putative transposable elements. It is not clear whether the two DNAs are different alleles at the same locus or members of a repetitive sequence family. Although there is no other direct evidence for transposition, these features suggest strongly that TBE1 is, or once was, a transposable element (45).

Essentially all copies of TBE1 are absent from the macronucleus (45). It is tempting to suggest that elimination of these sequences occurs through precise excision, in a manner similar to the deletion of IESs. This explanation could provide a link between DNA deletion and transposition. Unfortunately, no data exist to support this notion.

In both groups of ciliates, the association of internal C_4A_2 or C_4A_4 repeats with putative transposable elements raises the interesting possibility that some transposable elements in ciliates acquire telomeric DNA during transposition. Perhaps DNA molecules with free ends are generated as transposition intermediates and these ends serve as sites of telomeric sequence addition (24, 45). Both the repeat lengths and polarities of these sequences agree with known features of macronuclear telomeric sequences of the respective organisms. One difficulty with this explanation is that the transposition presumably occurs in the micronucleus, whereas telomere addition of the kind proposed occurs primarily in developing macronuclei. It is possible that similar telomere addition processes also occur in the micronuclei or that DNA molecules are able to pass from one nucleus to the other. No data exist to support either possibility.

The relationship between transposition and DNA deletion is also unclear. In both organisms, the putative transposable elements are eliminated from the macronucleus, and in the only case known in *T. thermophila* the C_4A_2 repeats are eliminated through internal deletion. Thus, there may exist a link between these two processes. Nonetheless, the simple notion that elimination is through precise removal of the transposable element cannot be correct. In all cases studied in *T. thermophila*, the internal C_4A_2 repeats are several kilobases away from the two ends of the eliminated regions and thus do not define the ends of these regions (24, 103, 116). Transposable elements might simply be present within germ line-limited sequences because they are more likely to be tolerated there.

F. Possible Mechanisms for DNA Deletion

Perhaps the most striking features of DNA deletion in ciliates are its frequency and precision. There is no other group of organisms, procaryotic or eucaryotic, in which DNA deletion is known to occur precisely at thousands of sites during differentiation. The molecular mechanism responsible for such a process must be highly regulated. The sequence studies conducted so far have failed to reveal any striking features sufficient to explain this kind of regulation and offer no clear indication that this process is related to other well-described systems. There remains the possibility that DNA deletion in ciliates occurs through a molecular mechanism not yet characterized in other organisms.

Excision of certain transposable elements provides some of the rare examples in eucaryotes of regulated DNA deletion, although in most cases excision is neither frequent nor precise (27, 75). It is unlikely that DNA deletion in ciliates occurs by the same mechanisms, since, with the possible exception of the short direct repeats, none of the sequence features associated with typical transposable elements are found associated with the deleted DNA. A second example of regulated DNA deletion in eucaryotes is that associated with the rearrangement of immunoglobulin and T-cell receptor genes in mammals. These deletions are also imprecise at the nucleotide level (reviewed in references 3 and 63). Again there is little resemblance between the sequence arrangements near deletion breakpoints in these specialized vertebrate tissues and those found in ciliates.

Well-regulated, precise deletions occur in procaryotes. The best-studied example is that of lambda prophage excision. The ends of the excised region are defined by the attachment sites which contain a 15-bp core sequence as direct repeats, although additional sequences are also required for the process to occur (65; reviewed in reference 97 and J. F. Thompson and A. Landy, this volume). The direct repeats flanking the deleted DNA in ciliates are much shorter, 2 to 8 bp, and are probably insufficient to support a similar mechanism. Another example is found in *Anabaena*, in which specific segments of DNA are deleted from the genes involved in nitrogen fixation (34, 64; reviewed by R. Haselkorn, this volume). As in ciliates, DNA deletion in this organism occurs synchronously in most differentiating cells and appears to be highly specific. Direct repeats 11 bp in size are present at the ends of one deleted region, and a gene in this region encodes a *trans*-acting factor needed for deletion. Despite superficial similarities, it does not seem that excision in *Anabaena* and DNA deletion in ciliates occur by the same mechanism.

Two characteristics of DNA deletion in ciliates set stringent restrictions on their possible mechanisms. First, the smallest deletion known is only 14 bp in length, thus setting a very restrictive upper limit for the length of *cis*-acting sequences required within one deleted region. Since this is probably insufficient for the sequence specificity seen, some, and perhaps most, of the *cis*-acting sequences are likely located outside the deleted region. In addition, the small size argues against mechanisms requiring alignments of the terminal repeats through the formation of a double-stranded DNA loop, which would be difficult to form over such a short distance. Second, deletion in *T. thermophila* results in the formation of final products and the disappearance of all starting materials without nuclear division. Thus, mechanisms such as unequal crossing over which require nuclear division for the dilution or removal of the starting materials or by-products are not likely to be responsible for this process.

The following provides one example of a possible mechanism of DNA deletion in ciliates. Because terminal direct repeats can be as short as 2 bp, they are insufficient by themselves to account for the sequence specificity observed. Additional sequences outside the deleted region are likely to provide the requisite specificity. The homopurine stretches flanking deletion 455-M of *T. thermophila* might represent one such sequence. If so, one could envision a *trans*-acting factor whose expression is developmentally regulated and which binds or recognizes these sequences (Fig. 4). This interaction leads to cleavage of DNA at the neighboring terminal repeats by either the same or other factors. Such a process could occur near both ends of a segment to be deleted, producing two similar protein-DNA complexes, which then interact with each other, leading to ligation of the two broken ends. The deleted region, presumably with unprotected ends, is probably degraded soon afterward. Comparisons among different deletion regions have not revealed universal sequence features which might serve as recognition sequences. It is possible that there are more than a few classes of deletion, each having a set of unique recognition sequences and *trans*-acting factors (7, 7a; Austerberry, Ph.D. thesis). In this way, the precision of deletion is dependent upon recognition of both the flanking sequences and the terminal repeats. The minor variations at the deletion junctions in *T. thermophila* may be due to occasional failure or variation in recognition of the terminal repeats. On the other hand, the alternative 0.6- and 0.9-kb deletions in 455-M may result from alternate uses of the duplicated flanking sequences. Since the same right end can be joined to two different left ends, it becomes necessary to explain how two adjacent deletion regions are kept separate. The presence of multiple classes of deletions proposed above could provide a simple solution.

The recent development of a transformation system in *T. thermophila* has provided a new approach to understanding DNA deletion in ciliates.

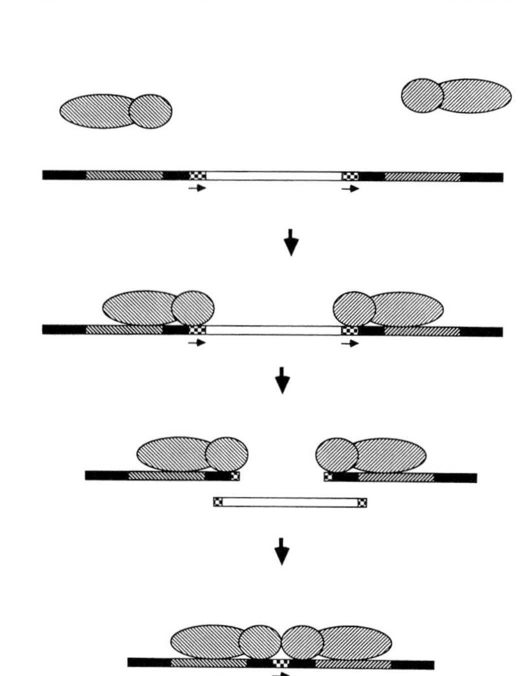

Figure 4. A possible mechanism for DNA deletion. This scheme describes one possible mechanism for DNA deletion in ciliates. The checkered bars with arrowheads indicate the direct repeats at the ends of the deleted region. The striped bars represent postulated recognition sequences which flank the deleted region. The scheme proposes that some unknown *trans*-acting factors bind to these sequences and cleave DNA at the direct repeats, which eventually leads to ligation of the free ends. The deleted region is degraded. See text for details.

The approach utilizes the same system developed for the analysis of chromosome breakage described earlier (96). A segment of micronuclear DNA containing the segment that is deleted in 455-M was inserted into the 3′ noncoding region of the micronuclear rDNA and injected into the developing macronuclear anlagen. Approximately 4% of the injected clones fully processed the injected micronuclear rDNA and expressed the expected drug-resistant phenotype. Insertion of the extra sequence did not appear to affect transformation. When the structure of the inserted DNA was analyzed among the transformed cells, most were found to have gone through proper deletion. Both the 0.9- and 0.6-kb normal deletion products were found, and in 7 of the 10 deletion products examined, the process seemed to be precise as judged by oligonucleotide hybridization (R. Godiska and M.-C. Yao, unpublished results). This preliminary result indicates that deletion can occur outside of its proper chromosomal environment and that it does not require more than the approximately 2 kb of sequence present in the insert to act in *cis*. Through this type of approach, it is likely that the *cis*-acting sequences for deletions in *T. thermophila* will soon be defined.

G. Relationship between DNA Deletion and Chromosome Breakage

Since both DNA deletion and chromosome breakage occur in the same genome in approximately the same developmental stage, it is possible that these two processes share some common steps. For instance, chromosome breakage can be considered as an abortive deletion process in which cleavage occurs without rejoining of the free ends (discussed in references 11 and 47). Although this is an attractive idea, it is not supported by the data. In most sites of either deletion or breakage, the events occur at frequencies of near 100% and in no cases have deletion and breakage been found to share the same site. Thus, deletion and breakage seem not to be alternate processes using the same sequence signals.

The strongest evidence against a common mechanism for deletion and breakage is provided by Cbs. This 15-bp sequence is present in most breakage sites in *T. thermophila*, but has not been found in other regions of the genome, including the two deletion sites studied. This property clearly distinguishes breakage from deletion sites and suggests different mechanisms for these two processes.

None of these arguments contradict the idea of a common evolutionary origin for breakage and deletion. For instance, Cbs might be derived from the recognition sequence for an ancient set of deletion proteins and have evolved to carry out just the cleavage function.

V. BIOLOGICAL SIGNIFICANCE

A. Chromosome Breakage

There is no obvious a priori reason why chromosome breakage must occur in ciliates. Arguments have been made that the gene-sized DNA produced through breakage in hypotrichs is advantageous in gene regulation. However, it does not explain the situation in holotrichs, where the macronuclear DNA is usually more than a few hundred kilobases in size.

In some ciliates, such as *T. thermophila*, a phenomenon called phenotypic assortment occurs during vegetative growth. In this process, a genetically heterozygous strain gives rise to subclones which are phenotypically pure for either allele (reviewed in reference 17) or for differentially marked DNA molecules (7). Assortment is probably a consequence of the macronuclear amitotic division mechanism. One possible advantage of such a process is the generation of somatic diversity among individuals derived from a common ancestor. By phenotypic assortment, a cell which is heterozygous at numerous loci will give rise to an extraordinary number of somatic descendants expressing the phenotype of one allele, the other allele, or both alleles at each locus. Survival of any one of these phenotypes ensures the survival of the germ line. Thus, this property may provide a strong selective advantage for species living in diverse and changing natural environments.

Chromosome breakage is particularly significant in the process of assortment. First, it disconnects most DNA segments from centromeres and thus ensures amitotic division. Second, by extensive fragmentation of the chromosomes, it significantly decreases the size of linkage groups, and thus greatly increases the number of combinations of pure types that can be generated through assortment.

Despite these features, phenotypic assortment has not been detected in some ciliates. For example, in *P. tetraurelia* the macronuclear ploidy level is high and the somatic life span is short; even though its macronucleus undergoes amitosis, phenotypic assortment has not been detected, probably because there is too little time for it to become effective. Despite such exceptions, phenotypic assortment represents one of the most appealing roles for chromosome breakage in most ciliates.

The amplification of rDNA in *T. thermophila* is coupled with chromosome breakage and thus may also be considered as fostering this process. rDNA amplification illustrates graphically how chromosome breakage can provide a way of maintaining different copy numbers for specific sequences in the somatic nucleus. At one extreme is the amplification of rDNA and at the other is the elimination of specific sequences. Although DNA elimination in *T. thermophila* occurs mainly through internal deletion, some sequences eliminated from around breakage sites are lost by other mechanisms: for example, loss of DNA bounded by two breakage sites and/or exonucleolytic erosion at transient ends. Because the known types of deletion can only account for a small proportion of the total sequences eliminated in hypotrichs, most DNA, which is outside of clusters of macronucleus-destined sequences, may be eliminated through chromosome breakage. Between the extremes of rDNA amplification and sequence elimination may be additional processes that achieve more modest levels of copy number variation. Although in the *Tetrahymena* macronucleus most sequences are present in roughly equal numbers, some variations have been reported in hypotrichs (90).

The occurrence of chromosome breakage raises

additional interesting questions. For instance, how are different DNA molecules maintained and transmitted in the somatic nucleus? Since there is no apparent process for equal segregation of sister chromatids, another mechanism may exist to prevent serious genic imbalance from occurring (81). Completely random assortment of DNA molecules into daughter nuclei predicts severe genic loss or imbalance and the formation of inviable subclones. The rates of formation of inviable subclones predicted by such models have not been observed (81). Is each type of molecule individually regulated in terms of replication or segregation or both, and if so, how? This problem is particularly challenging for hypotrichs, whose macronuclei each contains more than 20,000 different molecules which are so short that very little extra DNA seems to be available to serve as *cis*-acting sequences. Recent studies on the control of *Tetrahymena* rDNA replication may begin to shed light on this interesting and potentially important question (66, 96).

B. DNA Deletion

One outcome of DNA deletion is the elimination of germ line-limited DNA. In *T. thermophila* it makes up about 15% of the genome (109, 110). The significance of germ line-limited DNA is not clear in ciliates or other eucaryotes. It might contain information which is important only in the germ line, such as meiosis, nuclear differentiation, or micronucleus-specific regulations of transcription or replication. Recent evidence suggests that a germ line-limited sequence is transcribed in starved *Tetrahymena* sp. cells (89). This would be the first indication of expression of a germ line-limited sequence.

Deletion also leads to joining of previously separated sequences, and in two cases in hypotrichs it creates complete coding sequences, equivalent in its consequences to RNA splicing (46, 59). Thus, deletion can also serve as a mechanism for differential gene activation. Because introns are found in some *Tetrahymena* macronuclear genes (26, 74, 100), DNA deletion can be viewed as complementary to and not a replacement of RNA splicing in ciliates.

Another outcome of DNA deletion is the generation of somatic diversity. In *T. thermophila*, more than one type of product can be generated from a given type of deletion. The differences in the end points of a particular type of deletion can be a few or several hundred base pairs, depending on the mode of variation. As a consequence, different caryonidal clones with identical germ line nuclei can differ in their somatic macronuclear genomes. This somatic diversification is further enhanced by phenotypic assortment, through which alternate deletion products in a single macronuclear anlage are distributed unequally to daughter cells (6). Given the large number of potential deletion sites for such variations, a huge amount of somatic heterogeneity is generated. If just a few of them affect the functioning of the genome, this will have significant influence on the survival of the species. One possible example derives from studies of mating-type determination in *T. thermophila*. The genetics of this determination is rather unusual. Different caryonidal clones with identical germ line genomes can have stable but different mating types (reviewed in references 76 and 77). As many as six different mating types can be generated from one completely homozygous germ line genome. One model invokes variable DNA rearrangements for generating the somatic diversity (76, 78). Techniques are now at hand to test this model directly.

C. Concluding Remarks

Recent studies on ciliates have provided detailed information regarding how sequences can be massively reorganized in a developmentally programmed fashion. The changes that regularly occur during formation of each macronucleus include reductions in chromosomal sequence content and size and alterations in the linear alignment of DNA segments. The scale of these changes is unknown in other organisms, and the precision is among the highest known. Given the simplicity of some ciliates for molecular analysis, these systems offer excellent opportunities for understanding general principles of genome stability and reorganization. The knowledge obtained in these simple systems should have important implications for all eucaryotes.

It is already clear that the types of processes described here are probably not limited to ciliates. Historically, the most familiar case of genome reorganization is the phenomenon of chromatin diminution in the nematode *Ascaris megalocephala*, first observed a century ago (15; reviewed in reference 95). Much of the *Ascaris* genome is eliminated from the somatic cell lineage during early development. Coupled with this process, a single pair of chromosomes is broken into more than 40 fragments visible under a light microscope. The molecular nature of the breakage junctions is unknown, and the possible occurrence of internal DNA deletion remains to be tested. At the current superficial level, chromosome diminution in *A. megalocephala* appears remarkably similar to DNA rearrangement in ciliates and may share the same basic mechanisms. DNA deletion seems to occur in the crustacean *Cyclops turcifer* and *Cyclops divulsus*: in the somatic progenitor cells of early cleavage embryos, blocks of heterochromatin are removed from interstitial regions of chromosomes, and the resulting chromosomes, although smaller by the amount of material removed, are still intact and unchanged in number (9). The molecular details of this internal DNA deletion process in *Cyclops* sp. are largely unknown, but large, circular DNA molecules have been isolated from the appropriate developmental stages, and these may represent the deleted DNA (10).

Processes similar to the ones described here probably occur widely among eucaryotes but are readily detected only when they are extensive. Perhaps ciliates provide such an extreme case simply because they contain two types of nuclei in one cell. The germ line nucleus is released of all responsibility for housekeeping functions, and the somatic nucleus need not deal with the sexual process. This clear separation of roles may have encouraged the evolution of mechanisms for drastic and irreversible genome reorganization.

Ciliates thus provide unique opportunities for examining different roles of chromosomal DNA separated in different nuclei and for understanding how the different stabilities of such DNAs can be developmentally regulated.

Acknowledgments. I wish to thank Dan Gottschling, Rosemary Sweeney, and Ron Godiska for critical reading of this manuscript and the large number of colleagues who shared with me the results of their recent work.

Part of this work was supported by Public Health Service grant GM26210 and Research Career Development Award HD00547 from the National Institutes of Health.

LITERATURE CITED

1. **Allis, C. D., and D. K. Dennison.** 1982. Identification and purification of young macronuclear anlagen from conjugating cells of *Tetrahymena thermophila*. *Dev. Biol.* **93:** 519–533.
2. **Allitto, B. A., and K. M. Karrer.** 1986. A family of DNA sequences is reproducibly rearranged in the somatic nucleus of Tetrahymena. *Nucleic Acids Res.* **14:**8007–8025.
3. **Alt, F. W., T. K. Blackwell, R. A. Depinho, M. G. Reth, and G. D. Yancopoulos.** 1986. Regulation of genome rearrangement events during lymphocyte differentiation. *Immunol. Rev.* **89:**5–30.
4. **Altshuler, M. I., and M.-C. Yao.** 1985. Macronuclear DNA of *Tetrahymena thermophila* exists as defined subchromosomal-sized molecules. *Nucleic Acids Res.* **13:**5817–5831.
5. **Ammermann, D.** 1971. Morphology and development of the macronuclei of the ciliates *Stylonychia mytilus* and *Euplotes aediculatus*. *Chromosoma* **33:**209–238.
6. **Austerberry, C. F., C. D. Allis, and M.-C. Yao.** 1984. Specific DNA rearrangements in synchronously developing nuclei of Tetrahymena. *Proc. Natl. Acad. Sci. USA* **81:**7383–7387.
7. **Austerberry, C. F., and M.-C. Yao.** 1987. Nucleotide sequence structure and consistency of a developmentally regulated DNA deletion in *Tetrahymena thermophila*. *Mol. Cell. Biol.* **7:**435–443.

7a. **Austerberry, C. F., and M.-C. Yao.** 1988. Sequence structures of two developmentally regulated, alternative DNA deletion junctions in *Tetrahymena thermophila*. *Mol. Cell. Biol.* **8:**3947–3950.

8. **Baroin, A., A. Prat, and F. Caron.** 1987. Telomeric site position heterogeneity in macronuclear DNA of *Paramecium primaurelia*. *Nucleic Acids Res.* **15:**1717–1728.
9. **Beerman, S.** 1977. The diminution of heterochromatic chromosomal segments in Cyclops (Crustacea, Copepoda). *Chromosoma* **60:**297–344.
10. **Beerman, S.** 1984. Circular and linear structures in chromatin diminution of Cyclops. *Chromosoma* **89:**321–328.
11. **Blackburn, E. H.** 1986. Telomeres, p. 155–178. *In* J. G. Gall (ed.), *The Molecular Biology of Ciliated Protozoa*. Academic Press, Orlando, Fla.
12. **Blackburn, E. H., and J. G. Gall.** 1978. A tandemly repeated sequence at the termini of the extrachromosomal ribosomal RNA genes in Tetrahymena. *J. Mol. Biol.* **120:**33–35.
13. **Blackburn, E. H., and J. W. Szostak.** 1984. The molecular structure of centromeres and telomeres. *Annu. Rev. Biochem.* **53:**163–194.
14. **Boswell, R. E., C. L. Jahn, A. F. Greslin, and D. M. Prescott.** 1983. Organization of gene and non-gene sequences in micronuclear DNA of *Oxytricha nova*. *Nucleic Acids Res.* **11:**3651–3663.
15. **Boveri, T.** 1887. Uber Differenzierung der Zellkerne wahrend der Furchung des Eies von Ascaris megalocephala. *Anat. Anz.* **2:**688–693.
16. **Brunk, C. F., S. G. S. Tsao, C. H. Diamond, P. S. Ohashi, N. N. G. Tsao, and R. E. Pearlman.** 1982. Reorganization of unique and repetitive sequences during nuclear development in *Tetrahymena thermophila*. *Can. J. Biochem.* **60:**847–853.
17. **Bruns, P. J.** 1986. Genetic organization of Tetrahymena, p. 27–44. *In* J. G. Gall (ed.), *The Molecular Biology of Ciliated Protozoa*. Academic Press, Inc., Orlando, Fla.
18. **Bruns, P. J., A. L. Katzen, L. Martin, and E. H. Blackburn.** 1985. A drug resistant mutation maps in the ribosomal DNA of Tetrahymena. *Proc. Natl. Acad. Sci. USA* **82:**2844–2846.
19. **Callahan, R. C., G. Shalke, and M. A. Gorovsky.** 1984. Developmental rearrangements associated with a single type of expressed alpha-tubulin gene in Tetrahymena. *Cell* **36:** 441–445.
20. **Carle, G. F., and M. V. Olson.** 1984. Separation of chromosomal DNA molecules from yeast by orthogonal-field-alternation gel electrophoresis. *Nucleic Acids Res.* **12:** 5647–5664.
21. **Caron, F., and E. Meyer.** 1985. Does *Paramecium primaurelia* use a different genetic code in its macronucleus? *Nature* (London) **314:**185–188.
22. **Cech, T. R.** 1988. G-strings at chromosome ends. *Nature* (London) **332:**777–778.
23. **Challoner, P. B., and E. H. Blackburn.** 1986. Conservation of sequences adjacent to the telomeric C_4A_2 repeats of ciliate macronuclear ribosomal RNA gene molecules. *Nucleic Acids Res.* **14:**6299–6311.
24. **Cherry, J. M., and E. H. Blackburn.** 1985. The internally located telomeric sequences in the germline chromosomes of Tetrahymena are at the conserved ends of transposon-like elements. *Cell* **43:**747–758.

25. **Conover, R. K., and C. F. Brunk.** 1986. Macronuclear DNA molecules of *Tetrahymena thermophila. Mol. Cell. Biol.* 6:900–905.
26. **Cupples, C. G., and R. E. Pearlman.** 1986. Isolation and characterization of the actin gene from *Tetrahymena thermophila. Proc. Natl. Acad. Sci. USA* **83:**5160–5164.
27. **Daniels, S. B., M. McCarron, C. Love, and A. Chovnick.** 1985. Dysgenesis-induced instability of rosy locus transformation in *Drosophila melanogaster*: analysis of excision events and the selective recovery of control element deletions. *Genetics* **109:**95–117.
28. **Dawson, D., and G. Herrick.** 1984. Telomeric properties of C_4A_4-homologous sequences in micronuclear DNA of *Oxytricha fallax. Cell* **36:**171–177.
29. **Dawson, D., and G. Herrick.** 1984. Rare internal C_4A_4 repeats in the micronuclear genome of *Oxytricha fallax. Mol. Cell. Biol.* **4:**2661–2667.
30. **Elsevier, S. M., H. J. Lipps, and G. Steinbruck.** 1978. Histone genes in macronuclear DNA of the ciliate *Stylonychia mytilus. Chromosoma* **69:**291–306.
31. **Epstein, L. M., and J. D. Forney.** 1984. Mendelian and non-Mendelian mutations affecting surface antigen expression in *Paramecium tetraurelia. Mol. Cell. Biol.* **4:**1583–1590.
32. **Forney, J. D., and E. H. Blackburn.** 1988. Developmentally controlled telomere addition in wild type and mutant paramecia. *Mol. Cell. Biol.* **8:**251–258.
33. **Gall, J. G.** 1974. Free ribosomal RNA genes in the macronucleus of Tetrahymena. *Proc. Natl. Acad. Sci. USA* **71:** 3078–3081.
34. **Golden, J. W., S. J. Robinson, and R. Hasselkorn.** 1985. Rearrangement of nitrogen fixation genes during heterocyst differentiation in the cyanobacterium Anabaena. *Nature* (London) **314:**419–423.
35. **Gorovsky, M. A.** 1973. Macro- and micronuclei of Tetrahymena pyriformis: a model system for studying the structure and the function of eukaryotic nuclei. *J. Protozool.* **20:** 19–25.
36. **Gorovsky, M. A.** 1986. Ciliate chromatin and histones, p. 227–261. *In* J. G. Gall (ed.), *The Molecular Biology of Ciliated Protozoa.* Academic Press, Inc., Orlando, Fla.
37. **Gorovsky, M. A., S. Hattman, and G. L. Pleger.** 1973. (6N) methyl adenine in the nuclear DNA of a eukaryote, *Tetrahymena pyriformis. J. Cell Biol.* **56:**697–701.
38. **Greider, C. W., and E. H. Blackburn.** 1985. Identification of a specific telomere terminal transferase activity in Tetrahymena extracts. *Cell* **43:**405–413.
39. **Greider, C. W., and E. H. Blackburn.** 1987. The telomere terminal transferase of Tetrahymena is a ribonucleoprotein enzyme with two kinds of primer specificity. *Cell* **51:** 887–898.
40. **Hanyu, N., Y. Kuchino, and S. Nishimura.** 1986. Dramatic events in ciliate evolution: alteration of UAA and UAG termination codons due to anticodon mutations in two Tetrahymena tRNAs. *EMBO J.* **5:**1307–1311.
41. **Harrison, G. S., and K. M. Karrer.** 1985. DNA synthesis, methylation and degradation during conjugation in *Tetrahymena thermophila. Nucleic Acids Res.* **13:**73–87.
42. **Harrison, G. S., R. C. Findly, and K. M. Karrer.** 1986. Site-specific methylation of adenine in the nuclear genome of a eukaryote, *Tetrahymena* sp. *Mol. Cell. Biol.* **6:**2364–2370.
43. **Helftenbein, E.** 1985. Nucleotide sequence of a macronuclear DNA molecule coding for α-tubulin from the ciliate Stylonychia lemnae. Special codon usage: TAA is not a translation termination codon. *Nucleic Acids Res.* **13:** 415–433.
44. **Henderson, E., C. C. Hardin, S. K. Walk, I. Tinoco, Jr., and E. H. Blackburn.** 1987. Telomeric DNA oligonucleotides form novel intramolecular structures containing guanine-guanine base pairs. *Cell* **51:**899–908.
45. **Herrick, G., S. Cartinhour, D. Dawson, D. Ang, R. Sheets, A. Lee, and K. Williams.** 1985. Mobile elements bounded by C_4A_4 telomeric repeats in *Oxytricha fallax. Cell* **43:** 759–768.
46. **Herrick, G., S. W. Cartinhour, K. R. Williams, and K. P. Kotter.** 1987. Multiple sequence versions of the *Oxytricha fallax* 81-MAC alternate processing family. *J. Protozool.* **34:**429–434.
47. **Herrick, G., D. Hunter, K. Williams, and K. Kotter.** 1987. Alternative processing during development of a macronuclear chromosome family in Oxytricha. *Genes Dev.* **1:** 1047–1058.
48. **Horowitz, S., and M. A. Gorovsky.** 1985. An unusual genetic code in nuclear genes of Tetrahymena. *Proc. Natl. Acad. Sci. USA* **82:**2452–2455.
49. **Howard, E. A., and E. H. Blackburn.** 1985. Reproducible and variable genomic rearrangements occur in the developing somatic nucleus of the ciliate *Tetrahymena thermophila. Mol. Cell. Biol.* **5:**2039–2050.
50. **Kaine, B. P., and B. B. Spear.** 1980. Putative actin genes in the macronucleus of *Oxytricha fallax. Proc. Natl. Acad. Sci. USA* **77:**5336–5340.
51. **Kaine, B. P., and B. B. Spear.** 1982. Nucleotide sequence of a macronuclear gene for actin in *Oxytricha fallax. Nature* (London) **295:**430–432.
52. **Karrer, K. M.** 1983. Germ line specific DNA sequences are present on all five micronuclear chromosomes in *Tetrahymena thermophila. Mol. Cell. Biol.* **3:**1909–1919.
53. **Karrer, K. M., and J. G. Gall.** 1976. The macronuclear ribosomal DNA of *Tetrahymena pyriformis* is a palindrome. *J. Mol. Biol.* **104:**421–453.
54. **Karrer, K. M., S. Stein-Graves, and B. A. Allitto.** 1984. Micronucleus-specific DNA sequences in an amicronucleate mutant of Tetrahymena. *Dev. Biol.* **105:**121–129.
55. **Katzen, A. L., G. M. Cann, and E. H. Blackburn.** 1981. Sequence specific fragmentation of macronuclear DNA in a holotrichous ciliate. *Cell* **24:**313–320.
56. **King, B. O., and M.-C. Yao.** 1982. Tandemly repeated hexanucleotide at Tetrahymena rDNA end is generated from a single copy during development. *Cell* **31:**177–182.
57. **Kiss, G. B., A. A. Amin, and R. E. Pearlman.** 1981. Two separate regions of the extrachromosomal rDNA of *Tetrahymena thermophila* enable autonomous replication of plasmids in yeast. *Mol. Cell. Biol.* **1:**535–543.
58. **Klobutcher, L. A., M. E. Huff, and G. E. Gonye.** 1988. Alternate use of chromosome fragmentation sites in the ciliated protozoan *Oxytricha nova. Nucleic Acids Res.* **16:** 251–264.
59. **Klobutcher, L. A., C. L. Jahn, and D. M. Prescott.** 1984. Internal sequences are eliminated from genes during macronuclear development in the ciliated protozoan *Oxytricha nova. Cell* **36:**1045–1055.
60. **Klobutcher, L. A., and D. M. Prescott.** 1986. The special case of the hypotrichs, p. 111–154. *In* J. G. Gall (ed.), *The Molecular Biology of Ciliated Protozoa.* Academic Press, Inc., Orlando, Fla.
61. **Klobutcher, L. A., M. T. Swanton, P. Donini, and D. M. Prescott.** 1981. All gene-sized molecules in four species of

hypotrichs have the same terminal sequence and an unusual 3′ terminus. *Proc. Natl. Acad. Sci. USA* **78**:3015–3019.

62. **Klobutcher, L. A., A. M. Vialonis-Walsh, K. Cahill, and R. M. Ribas-Aparicio.** 1986. Gene-sized macronuclear DNA molecules are clustered in micronuclear chromosomes of the ciliate *Oxytricha nova*. *Mol. Cell. Biol.* **6**:3606–3613.
63. **Kronenberg, M., G. Siu, L. Hood, and N. Shastri.** 1986. The molecular genetics of the T-cell antigen receptor and T-cell antigen recognition. *Annu. Rev. Immunol.* **4**:529–591.
64. **Lammers, P. J., J. W. Golden, and B. Haselkorn.** 1986. Identification and sequence of a gene required for developmentally regulated DNA excision in Anabaena. *Cell* **44**: 905–911.
65. **Landy, A., and W. Ross.** 1977. Viral integration and excision: structure of the lambda att sites. *Science* **197**: 1147–1160.
66. **Larson, D. D., E. H. Blackburn, P. C. Yeager, and E. Orias.** 1986. Control of rDNA replication in Tetrahymena involves a cis-acting upstream repeat of a promoter element. *Cell* **47**:229–240.
67. **Lauth, M. R., B. B. Spear, J. Heumann, and D. M. Prescott.** 1976. DNA of ciliated protozoa: DNA sequence diminution during macronuclear development of Oxytricha. *Cell* **7**: 67–74.
68. **Lawn, R. M.** 1977. Gene-sized DNA molecules of the Oxytricha macronucleus have the same terminal sequence. *Proc. Natl. Acad. Sci. USA* **74**:4325–4328.
69. **Lawn, R. M., J. M. Heumann, G. Herrick, and D. M. Prescott.** 1978. The gene-sized DNA molecules of Oxytricha. *Cold Spring Harbor Symp. Quant. Biol.* **42**:483–492.
70. **Mayo, K. A., and E. Orias.** 1981. Further evidence for lack of gene expression in the Tetrahymena micronucleus. *Genetics* **98**:747–762.
71. **Mayo, K. A., and E. Orias.** 1985. Lack of expression of micronuclear genes determining two different enzymatic activities in *Tetrahymena thermophila*. *Differentiation* **28**: 217–224.
72. **Meyer, G. F., and H. J. Lipps.** 1980. Chromatin elimination in the hypotrichous ciliate *Stylonychia mytilus*. *Chromosoma* 77:285–297.
73. **Meyer, G. F., and H. J. Lipps.** 1981. The formation of polytene chromosomes during the macronuclear development of the hypotrichous ciliate *Stylonychia mytilus*. *Chromosoma* **82**:309–314.
74. **Nielsen, H., P. H. Andreasen, H. Dreisig, K. Kristiansen, and J. Engberg.** 1986. An intron in a ribosomal protein gene from Tetrahymena. *EMBO J.* **5**:2711–2717.
75. **O'Hare, K., and G. M. Rubin.** 1983. Structures of P transposable elements and their sites of insertion and excision in the *Drosophila melanogaster* genome. *Cell* **34**:25–35.
76. **Orias, E.** 1981. Probable somatic DNA rearrangements in mating type determination in *Tetrahymena thermophila*: a review and a model. *Dev. Genet.* **2**:185–202.
77. **Orias, E.** 1986. Ciliate conjugation, p. 45–84. *In* J. G. Gall (ed.), *The Molecular Biology of Ciliated Protozoa*. Academic Press, Inc., Orlando, Fla.
78. **Orias, E., and M. P. Baum.** 1984. Mating type differentiation in *Tetrahymena thermophila*: strong influence of delayed refeeding of conjugating pairs. *Dev. Genet.* **4**:145–158.
79. **Pan, W.-C., E. Orias, M. Flacks, and E. H. Blackburn.** 1982. Allele-specific, selective amplification of a ribosomal RNA gene in *Tetrahymena thermophila*. *Cell* **28**:595–604.
80. **Pluta, A. F., B. P. Kain, and B. B. Spear.** 1982. The terminal organization of macronuclear DNA in *Oxytricha fallax*. *Nucleic Acids Res.* **10**:8145–8154.
81. **Preer, J. R., Jr., and L. B. Preer.** 1979. The size of macronuclear DNA and its relationship to models for maintaining genic balance. *J. Protozool.* **26**:14–18.
82. **Preer, J., Jr., L. B. Preer, B. Rudman, and A. Barnett.** 1985. A deviation from the universal code: the gene for surface protein 51A in Paramecium. *Nature* (London) **314**:188–190.
83. **Prescott, D. M., K. G. Murti, and C. J. Bostock.** 1973. Genetic apparatus of *Stylonychia* sp. *Nature* (London) **242**: 597–600.
84. **Ribas-Aparicio, R. M., J. J. Sparkowski, A. E. Proulx, J. D. Mitchell, and L. A. Klobutcher.** 1987. Nucleic acid splicing events occur frequently during macronuclear development in the protozoan *Oxytricha nova* and involve the elimination of unique DNA. *Genes Dev.* **1**:323–336.
85. **Roth, M. R., and D. M. Prescott.** 1985. DNA intermediates and telomere addition during genome reorganization in Euplotes crassus. *Cell* **41**:411–417.
86. **Schwartz, D. C., and C. R. Cantor.** 1984. Separation of yeast chromosome-sized DNAs by pulsed field gradient gel electrophoresis. *Cell* **37**:67–75.
87. **Spangler, E. A., and E. H. Blackburn.** 1985. The nucleotide sequence of the 17S ribosomal RNA gene of *Tetrahymena thermophila* and the identification of point mutations resulting in resistance to the antibiotics paromomycin and hygromycin. *J. Biol. Chem.* **260**:6334–6340.
88. **Spear, B. B.** 1980. Isolation and mapping of the rRNA genes in the macronucleus of *Oxytricha fallax*. *Chromosoma* 77:193–202.
89. **Stein-Gavens, S., J. M. Wells, and K. M. Karrer.** 1987. A germline specific DNA sequence is transcribed in Tetrahymena. *Dev. Biol.* **120**:259–269.
90. **Steinbruck, G.** 1983. Over-amplification of genes in macronuclei of hypotrichous ciliates. *Chromosoma* **88**:156–163.
91. **Steinbruck, G., I. Haas, K. Hellmer, and D. Ammermann.** 1981. Characterization of the macronuclear DNA in five species of ciliates. *Chromosoma* 83:199–208.
92. **Swanton, M. T., A. F. Greslin, and D. M. Prescott.** 1980. Arrangement of coding and non-coding sequences in the DNA molecules coding for rRNAs in *Oxytricha* sp. *Chromosoma* 77:203–215.
93. **Swanton, M. T., J. M. Heumann, and D. M. Prescott.** 1980. Gene-sized DNA molecules of the macronuclei in three species of hypotrichs: size distribution and absence of nicks. *Chromosoma* 77:217–227.
94. **Swanton, M. T., R. M. McCarroll, and B. B. Spear.** 1982. The organization of macronuclear rDNA molecules in four hypotrichous ciliated protozoans. *Chromosoma* **85**:1–9.
95. **Tobler, H.** 1986. The differentiation of germinal and somatic cell lines in nematodes, p. 1–70. *In* W. Hennig (ed.), *Germ Line-Soma Differentiation*. Springer-Verlag, New York.
96. **Tondravi, M. M., and M.-C. Yao.** 1986. Transformation of *Tetrahymena thermophila* by microinjection of ribosomal RNA genes. *Proc. Natl. Acad. Sci. USA* **83**:4369–4373.
97. **Weisberg, R., and R. Landy.** 1983. Site specific recombination in phage lambda, p. 211–250. *In* R. W. Hendrix, J. W. Roberts, F. W. Stahl, and R. A. Weisberg (ed.), *Lambda II*. Cold Spring Harbor Laboratory, Cold Spring Harbor, N.Y.
98. **White, T. C., and S. L. Allen.** 1986. Alternative processing of sequences during macronuclear development in *Tetrahymena thermophila*. *J. Protozool.* **33**:30–38.
99. **Woodard, J., E. S. Kaneshiro, and M. A. Gorovsky.** 1972. Cytochemical studies on the problem of macronuclear subnuclei in Tetrahymena. *Genetics* **70**:251–260.
100. **Wu, M., C. D. Allis, R. Richman, R. G. Cook, and M. A. Gorovsky.** 1986. An intervening sequence in an unusual

histone H1 gene of *Tetrahymena thermophila. Proc. Natl. Acad. Sci. USA* **83**:8674–8678.

101. **Yao, M.-C.** 1981. Ribosomal RNA gene amplification in Tetrahymena may be associated with chromosome breakage and DNA elimination. *Cell* **24**:765–774.
102. **Yao, M.-C.** 1982. Amplification of ribosomal RNA gene in Tetrahymena, p. 127–153. *In* H. Busch and L. Rothblum (ed.), *The Cell Nucleus*, vol. 12. Academic Press, Inc., New York.
103. **Yao, M.-C.** 1982. Elimination of specific DNA sequences from the somatic nucleus of the ciliate Tetrahymena. *J. Cell Biol.* **92**:783–789.
104. **Yao, M.-C.** 1986. Amplification of ribosomal RNA genes, p. 179–201. *In* J. G. Gall (ed.), *The Molecular Biology of Ciliated Protozoa.* Academic Press, Inc., Orlando, Fla.
105. **Yao, M.-C., E. H. Blackburn, and J. G. Gall.** 1978. Amplification of the rRNA genes in Tetrahymena. *Cold Spring Harbor Symp. Quant. Biol.* **43**:1293–1296.
106. **Yao, M.-C., E. Blackburn, and J. G. Gall.** 1981. Tandemly repeated C-C-C-C-A-A hexanucleotide of Tetrahymena rDNA is present elsewhere in the genome and may be related to the alteration of the somatic gene. *J. Cell Biol.* **90**: 515–520.
107. **Yao, M.-C., J. Choi, S. Yokoyama, C. Austerberry, and C.-H. Yao.** 1984. DNA elimination in Tetrahymena: a developmental process involving extensive breakage and rejoining of DNA at defined sites. *Cell* **36**:433–440.
108. **Yao, M.-C., and J. G. Gall.** 1977. A single integrated gene for ribosomal RNA in an eucaryote, *Tetrahymena pyriformis. Cell* **12**:121–132.
109. **Yao, M.-C., and J. G. Gall.** 1979. Alteration of the Tetrahymena genome during nuclear differentiation. *J. Protozool.* **26**:10–13.
110. **Yao, M.-C., and M. A. Gorovsky.** 1974. Comparison of the sequences of the macro- and micronuclear DNA of *Tetrahymena pyriformis. Chromosoma* **48**:1–18.
111. **Yao, M.-C., A. R. Kimmel, and M. A. Gorovsky.** 1974. A small number of cistrons for ribosomal RNA in the germinal nucleus of a eukaryote, *Tetrahymena pyriformis. Proc. Natl. Acad. Sci. USA* **71**:3082–3086.
112. **Yao, M.-C., and C.-H. Yao.** 1981. Repeated hexanucleotide CCCCAA is present near free ends of the macronuclear DNA of Tetrahymena. *Proc. Natl. Acad. Sci. USA* **78**:7436–7439.
113. **Yao, M.-C., K. Zheng, and C.-H. Yao.** 1987. A conserved nucleotide sequence at the site of developmentally regulated chromosomal breakage in Tetrahymena. *Cell* **48**:779–788.
114. **Yao, M.-C., S.-G. Zhu, and C.-H. Yao.** 1985. Gene amplification in *Tetrahymena thermophila*: formation of extrachromosomal palindromic genes coding for rRNA. *Mol. Cell. Biol.* **5**:1260–1267.
115. **Yokoyama, R. W., and M.-C. Yao.** 1982. Elimination of DNA sequences during macronuclear differentiation in *Tetrahymena thermophila*, as detected by in situ hybridization. *Chromosoma* **85**:11–22.
116. **Yokoyama, R., and M.-C. Yao.** 1984. Internal micronuclear DNA regions which include sequences homologous to macronuclear telomeres are deleted during development in Tetrahymena. *Nucleic Acids Res.* **12**:6103–6116.
117. **Yokoyama, R., and M.-C. Yao.** 1986. Sequence characterization of Tetrahymena macronuclear DNA ends. *Nucleic Acid Res.* **14**:2109–2122.

Chapter 33

Excision of Elements Interrupting Nitrogen Fixation Operons in Cyanobacteria

ROBERT HASELKORN

I. INTRODUCTION

Nitrogen fixation, the conversion of atmospheric nitrogen to ammonia, is a unique property of procaryotes. Bacteria capable of fixing nitrogen include facultative enterics such as *Klebsiella pneumoniae*, free-living photosynthetic bacteria such as *Rhodobacter capsulatus*, the soil aerobes *Azotobacter* spp., and root-nodulating *Rhizobium* spp. A few species of unicellular and many species of filamentous cyanobacteria also fix nitrogen. In every one of the organisms mentioned, the synthesis of the nitrogenase enzyme complex is repressed when ammonia or another suitable source of reduced nitrogen is available in the environment. In addition, with the exception of *Azotobacter* spp., in all of these organisms both the synthesis and activity of nitrogenase are compromised by oxygen. Since cyanobacteria evolve oxygen in the light due to the activity of photosystem II, they require special cellular mechanisms to provide a local anaerobic environment for nitrogen fixation.

The regulation of nitrogen fixation (*nif*) gene expression by combined nitrogen and by oxygen is easier to justify than to explain. Nitrogen fixation is an expensive process in terms of ATP and reductant, and the enzyme complex is rather inefficient (19). It is to the advantage of a cell to utilize other nitrogen sources before derepressing the synthesis of nitrogenase. In *K. pneumoniae*, the *nif* operons have unique promoters requiring transcriptional activation and a modified form of RNA polymerase. The activating protein is in turn controlled at the levels of both synthesis and function by a cascade of factors that can be traced back to an enzyme controlled by the intracellular ratio of glutamine to α-ketoglutarate, that is, the ratio of fixed nitrogen to carbon (10).

The nitrogenase enzyme and components of the electron transport chain to nitrogenase are very low potential redox proteins; in air they are irreversibly inactivated in a few seconds. Thus, most bacteria do not express the *nif* genes under aerobic conditions. The control of *nif* gene transcription by oxygen occurs at several levels in *K. pneumoniae*. One involves the inactivation of the *nif*-gene-activating protein and a second involves a change in DNA supercoiling, which is required for transcription of the regulatory *nif* genes. The molecular details of the regulation of *nif* gene expression and of the *nif* gene rearrangements in *Anabaena* described below are not known. However, both nitrogen source depletion and anaerobiosis are required for *nif* gene expression and rearrangement.

Anabaena is a representative of many filamen-

Robert Haselkorn ■ Department of Molecular Genetics and Cell Biology, University of Chicago, 920 East 58th Street, Chicago, Illinois 60637.

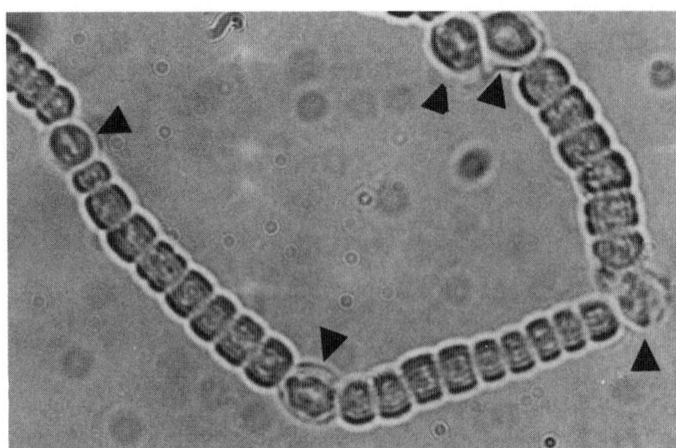

Figure 1. Differentiated filament of *Anabaena* sp. strain 7120. The phase contrast micrograph shows the characteristic spacing of thick-walled heterocysts (arrows) seen in cultures growing on N_2 as nitrogen source.

tous cyanobacterial genera that grow in chains of 50 to 100 uniform vegetative cells when the environment contains reduced nitrogen (usually nitrate or ammonia). When the external supply of reduced nitrogen is exhausted, cells specialized for nitrogen fixation differentiate at regular intervals along each filament (Fig. 1). These cells, called heterocysts, provide the anaerobic environment necessary for nitrogen fixation (11, 23). The differentiation requires roughly one cell generation to complete (approximately 24 to 30 h). During that time, a new cell envelope consisting of an outer polysaccharide layer and an inner glycolipid layer is deposited outside the preexisting cell wall, proteins involved in oxygen evolution and CO_2 fixation are destroyed by proteases, and enzymes for nitrogen fixation and the production of ATP and reductant are induced.

Intercellular communication is critical for this arrangement to work. Vegetative cells continue to fix CO_2; carbohydrate transported from them to the heterocyst provides the substrate for the manufacture of ATP and reductant. The ultimate product of nitrogen fixation in the heterocyst is glutamine, which is exported to neighboring vegetative cells. In spite of the extensive transport of small molecules between the two cell types, gene expression in each cell is radically different. Many genes, including those that encode the proteins of the nitrogenase complex, are expressed exclusively in heterocysts, while others are turned off selectively in heterocysts. Heterocysts do not divide. In a culture fixing nitrogen, one vegetative cell midway between two heterocysts will differentiate in roughly the time required for each nondifferentiating vegetative cell to divide once. This relationship between division and differentiation maintains the heterocyst spacing. The interval between heterocysts is roughly 10 cells in the *Anabaena* strain we use.

To study the transcriptional regulation of heterocyst differentiation, we have cloned and sequenced a number of genes that are expressed differentially in vegetative cells and heterocysts (1, 6, 18, 20, 22). These studies suggested a consensus sequence for promoters of genes expressed in vegetative cells but not of genes expressed only in heterocysts. During experiments designed to isolate high-molecular-weight RNA from heterocysts (in order to screen recombinant DNA libraries for additional heterocyst-specific genes and thereby increase the number of promoter sequences), Golden et al. discovered that their procedure for preparing RNA also yielded high-molecular-weight heterocyst DNA. Cleavage of this DNA with certain restriction enzymes yielded *nif*-homologous fragments that differed in size from those in vegetative-cell DNA (9). These size differences were subsequently shown to reflect two large-scale rearrangements of *nif* DNA during heterocyst differentiation (8, 9).

II. *nif* GENE ORGANIZATION IN *K. PNEUMONIAE*

To understand the significance of these rearrangements, it will be useful to review the organization of the *nif* genes of *K. pneumoniae*, the best-studied diazotroph. This organization is shown in Fig. 2. There are eight operons containing at least 19 genes within 24 kilobases (kb) of DNA (19). The genes *H*, *D*, and *K* encode the polypeptides of the nitrogenase complex itself. The *F* and *J* genes encode a flavodoxin and an oxidoreductase, respectively, required for electron transport from pyruvate to nitrogenase. Genes *Q*, *B*, *V*, *N*, and *E* appear to be involved in the synthesis of the iron-sulfur-molybdenum cofactor that comprises the N_2-binding active site of nitrogenase. The *M* gene is required for "maturation" of the product of gene *H*. The *S* and *U* genes are likewise required for maturation of the gene *K* and *D* products, although the chemical nature of maturation remains unknown. The *X* and *Y* genes have no known function. The *A* and *L* genes encode regulatory proteins needed to turn transcription of the other *nif* operons on and off, respectively (11). These proteins mediate signals that depend on the nitrogen and oxygen status of the cell.

III. *nif* GENE ORGANIZATION IN *ANABAENA* VEGETATIVE CELLS

The organization of the *nif* genes in *Anabaena* vegetative-cell DNA is also shown in Fig. 2. The first

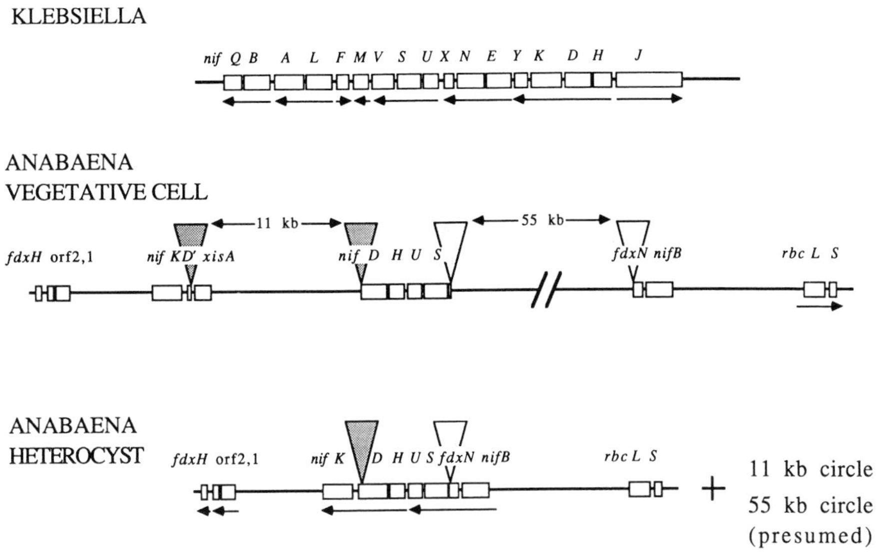

Figure 2. Comparison of the organization of the *nif* genes of *K. pneumoniae* (above), *Anabaena* vegetative cells (middle), and *Anabaena* heterocysts (below). Horizontal arrows indicate transcription units. Shaded triangles show the exact location of the recombination sites for the *nifD* and *nifS* excisions. The excised 11-kb circle contains the *xisA* gene. To the left of the *nifK* gene is several kilobases of uncharacterized DNA followed by two open reading frames labeled orf2,1 (M. Basche, unpublished results) and a gene (*fdxH*) encoding a heterocyst-specific [2Fe-2S] ferredoxin (3). All three are transcribed only under conditions of nitrogenase induction. The ferredoxin is distinct from the [4Fe-4S] ferredoxinlike protein encoded in the *nifB fdxN nifS nifU* operon. The *rbcLS* operon encodes the large and small subunits of ribulose bisphosphate carboxylase; it is transcribed only in vegetative cells, indicated by the horizontal arrow. It is likely that other *nif* genes are located between orf1 and *nifK*.

genes to be mapped were *nifK*, *D*, *H*, and *S* (16, 20). Cloned fragments of *Klebsiella nif* DNA were used to identify and to isolate homologous fragments of *Anabaena* vegetative-cell DNA from recombinant DNA libraries (16). Physical mapping of the *Anabaena* vegetative-cell *nif* DNA showed that it differed in organization from the *Klebsiella* gene in several ways (20). First, although the *Anabaena nifH* and *nifD* genes are adjacent, as in *K. pneumoniae*, *nifK* is about 11 kb from *nifD* (Fig. 2). Second, *nifS* is to the right of *nifH* in *Anabaena*, not to the left as in *K. pneumoniae*. In all, a region of over 30 kb of *Anabaena* vegetative-cell DNA was found to contain only four *nif* genes. Subsequent work by D. Rice showed that regions to the left of *nifK* were transcribed during conditions of nitrogenase induction, indicating that *nif* genes not detected using *Klebsiella* probes might be present there. On the other hand, the DNAs between *nifK* and *nifD* and to the right of *nifS* were not transcribed, and therefore were not likely to contain *nif* genes (20).

The organization of the *nif* genes in vegetative cells and the fact that heterocysts are terminally differentiated in *Anabaena* prompted early attempts to compare the physical maps of *nif* region DNAs from heterocysts and vegetative cells, but those experiments were inconclusive due to the very low molecular weight of DNA obtained from heterocysts. Subsequently, the new procedure for preparing high-molecular-weight DNA from purified heterocysts provided material suitable for restriction and Southern hybridization analysis (9). These studies indicated that DNA in the region of the *nif* genes was rearranged during heterocyst differentiation.

Two rearrangements were observed. Each corresponds to the excision of specific DNA segments from vegetative-cell DNA by site-specific recombination within short, directly repeated sequences. The final organization of the *nif* genes in heterocyst DNA is also shown in Fig. 2.

IV. EXCISION OF AN 11-kb ELEMENT INTERRUPTING *nifD*

The first excision event characterized in detail removes an 11-kb element that interrupts the *nifD* reading frame 44 codons from its 3′ terminus. The rearrangement is easily seen in *Hin*dIII digests of total *Anabaena* DNA. Fragments of 2.9 and 0.7 kb in vegetative-cell DNA are replaced by new fragments of 2.1 and 1.8 kb in heterocyst DNA. The time course of appearance of the heterocyst-specific fragments during differentiation was determined by preparing

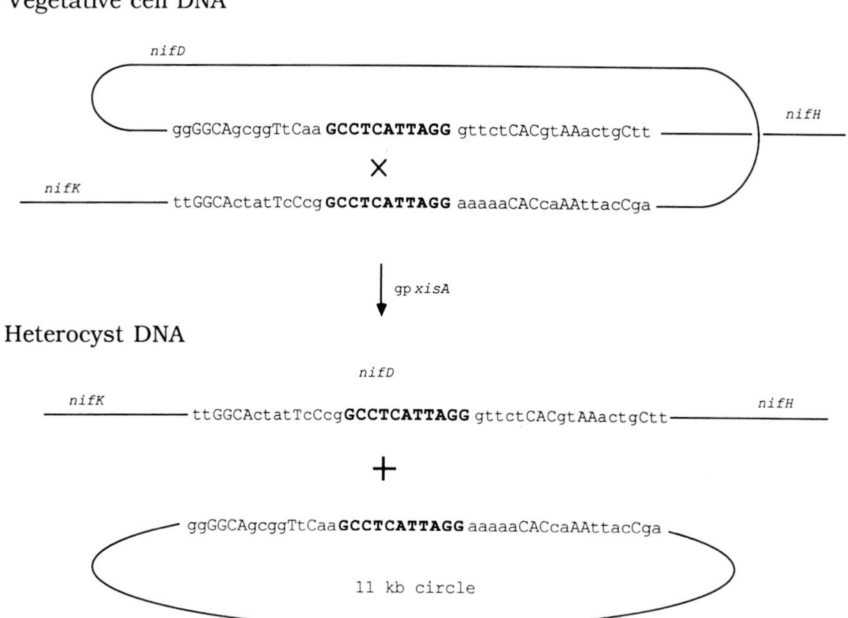

Figure 3. Proposed mechanism of *nifD* rearrangement. The two 11-bp directly repeated sequences that flank the 11-kb element in vegetative-cell DNA form a synapse and recombine within the repeats. The reaction products are two unlinked DNA molecules, each retaining one copy of the 11-bp repeat. Nucleotides in the regions flanking the 11-bp repeat are shown in uppercase letters if they are the same in both sites and in lowercase letters if they differ. Boldface letters indicate the region within which the recombination event must occur.

DNA from vegetative cells, from a differentiating culture at 6-h intervals, and from purified heterocysts after 42 h in N-free medium (9). The *Hin*dIII fragments characteristic of vegetative-cell DNA, identified by *nifD* gene probes, remain unchanged in the 90% of cells that do not become heterocysts. In the other 10% of cells that differentiate into heterocysts, the 2.9- and 0.7-kb *Hin*dIII bands disappear. They are replaced by the new bands derived from the excised element and the deleted chromosome. The rearrangement is first detected in the sample taken at 18 h, roughly the time at which the first mature heterocysts appear.

An 11-kb circular molecule was detected in the differentiating cells by electrophoresis and Southern blotting of these DNA preparations without prior restriction enzyme cutting (9). The circular molecule appeared at the same time as the new *Hin*dIII bands. These studies showed that excision occurs late in heterocyst development, after synthesis of the complex envelope is complete and the internal milieu is made anaerobic (11).

The two products of the *nifD* excision, an 11-kb excised segment and a 6-kb *Eco*RI fragment from the deleted chromosome, were obtained from recombinant libraries of heterocyst DNA. Sequence analysis of the breakpoints showed that the *nifD* rearrangement was the result of a site-specific recombination between direct repeats of 11 base pairs (bp) separated by 11 kb in the vegetative-cell chromosome (9). This recombination results in excision of the 11-kb element as a circular molecule and the fusion of previously distant chromosomal regions (Fig. 3).

The rearrangement has two effects on *nif* gene expression. One is the formation of a *nifHDK* operon. Northern analysis with *nifH*, *D*, and *K* probes identified a common 4.7-kb RNA containing the transcript of all three genes. *nifHDK* mRNA is detectable only after the rearrangement has occurred. Two other transcripts of the *nifHDK* operon are also seen. A 1.1-kb transcript contains only *nifH* sequences, and a 2.8-kb transcript contains *nifH* and *nifD* sequences. In principle both of these transcripts could be made prior to excision of the 11-kb element, but the conditions needed for their appearance (nitrogen depletion, anaerobiosis) are the same as those needed for excision. It has therefore not been possible to determine whether their synthesis requires rearrangement.

The second effect of the *nifD* rearrangement is the replacement of 27 amino acids encoded by the apparent vegetative-cell gene with 43 amino acids encoded by DNA sequences originally 11 kb away. There is good homology between the predicted car-

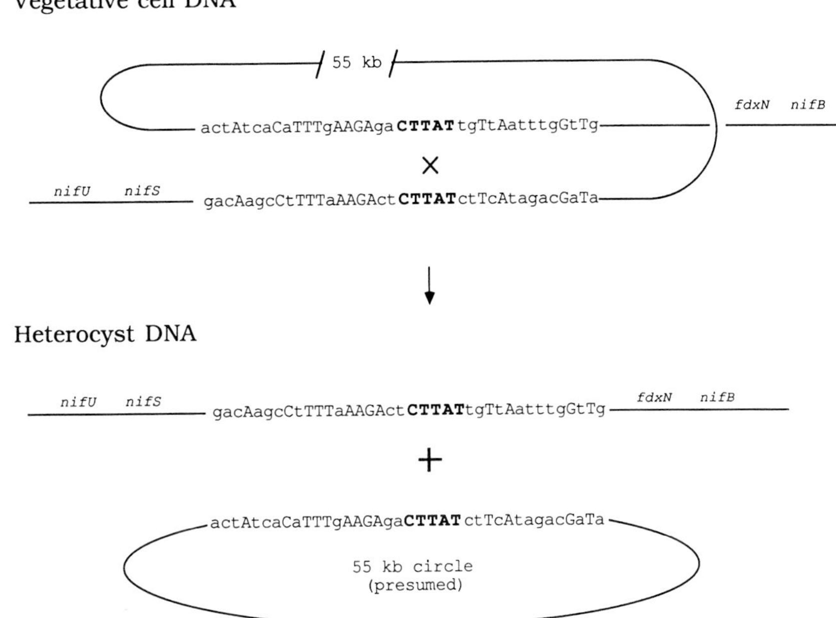

Figure 4. Proposed mechanism of the *nifS*/*B* rearrangement. Upper- and lowercase letters and boldface letters have the same meaning as in Fig. 3. The excision product is presumed to be a 55-kb circle because the joint fragment can be isolated from heterocysts, but the intact circular molecule has not been seen. The putative recombinase for this event has not been detected.

boxy-terminal regions of the protein encoded by the rearranged cyanobacterial *nifD* gene and the *nifD* genes from two species of *Rhizobium* (28/43), whereas the putative product of the nonrearranged *nifD* gene does not show significant homology (1/27). Western blots of total *Anabaena* proteins from heterocysts or from cultures induced anaerobically to make nitrogenase show only one form of the *nifD* protein, indicating that the unrearranged *nifD* gene is not expressed (R. Kranz and J. W. Golden, unpublished results).

V. EXCISION OF THE 55-kb ELEMENT INTERRUPTING THE *nifB fdxN nifS nifU* OPERON

The second rearrangement has one DNA breakpoint close to the 5′ end of the *nifS* gene (Fig. 2). The DNA breakpoints involved in this rearrangement were cloned and all four were sequenced (8). The data show that this rearrangement, like *nifD* excision, involves site-specific recombination between repeated DNA sequences, but of 5 bp, in contrast to 11 bp at *nifD*. The repeated sequences involved in the *nifD* and *nifS* rearrangements are not homologous to each other, and presumably the excision reactions are catalyzed by different enzymes (Fig. 3 and 4). Both the *nifD* 11-bp direct repeats and the *nifS* 5-bp repeats are also flanked by regions of partial homology. Aside from these general structural features, the breakpoints bear no resemblance to the target sequences of known bacterial recombinases, such as the integrase of bacteriophage lambda (4) or the inverting enzyme responsible for *Salmonella* flagellar phase variation (21).

The topology of the *nifS* rearrangement was determined by isolating a set of overlapping cosmid clones from a library of vegetative-cell DNA fragments and mapping them with six restriction enzymes. These studies showed that the two sequences that are the sites of recombination are 55 kb apart, in directly repeated orientation (J. W. Golden, C. Carrasco, M. E. Mulligan, G. Schneider, and R. Haselkorn, in press). Thus, the *nifS* rearrangement is also due to the excision of a (presumably circular) molecule (Fig. 4), although its large size has precluded its detection as an intact circle in heterocyst DNA.

DNA sequence analysis and S1 transcript mapping showed that the 55-kb element interrupts an operon that is expressed only after rearrangement (8). The operon contains four open reading frames (Fig. 2). The proteins encoded by the two open reading frames closest to *nifH* are similar to those of *Klebsiella nifS* and *nifU* (2; M. E. Mulligan, unpublished results), which are required for an as yet unknown step in nitrogenase maturation. The presumptive product of the third open reading frame (*fdxN*) is closely related to that of *psaC*, a chloroplast-encoded iron-sulfur protein that forms part of the photosystem I reaction center (12), and less closely related to bacterial ferredoxins. It probably also contains two [4Fe-4S] centers (17a). Although *fdxN* is transcribed in heterocysts, its role in electron transfer to nitrogenase is uncertain because the heterocyst-specific [2Fe-2S] ferredoxin that donates electrons to *Anabaena* nitrogenase in vitro (3) is encoded by another gene, *fdxH* (Fig. 2; H. Böhme and R. Haselkorn, *Mol. Gen. Genet.*, in press). Next to *fdxN*, closest to the promoter of the operon, is an open reading frame similar to the *Klebsiella nifB* gene, which specifies a protein required for the synthesis of the molybdenum-iron cofactor of nitrogenase.

VI. THE *nifD* SITE-SPECIFIC RECOMBINASE GENE *xisA*

The gene whose product (excisase) is believed responsible for the excision of the *nifD* 11-kb element has been identified and sequenced (15). A large-scale plasmid preparation of pAn207 from *Escherichia coli*, which consists of the vector pBR322 joined to the 17-kb *Eco*RI fragment of *Anabaena* DNA containing the 11-kb element (20), was found to contain two plasmids, one of the expected 21-kb size and a second that is 11 kb smaller. Restriction and Southern analysis of these plasmids showed that the smaller plasmid resulted from the excision of the *nifD* element. Sequence analysis of the rearranged plasmid confirmed precise deletion of the 11-kb element from the cloned *Anabaena* DNA by recombination within the 11-bp repeated sequence (15).

This excision occurred rarely (less than 1%) in *E. coli*. To monitor the excision process more effectively, the 11-kb element was marked with a beta-galactosidase gene by the in vivo insertion of a mini-Mu-*lac* transposon (5). This mini-Mu-*lac* tagging procedure also provided a method of insertional mutagenesis to identify genes required for the excision. In one strain, mini-Mu-*lac* was inserted near the middle of the 11-kb element (insert MX25). This strain produced blue (Lac^+) colonies with white (Lac^-) sectors and an occasional (0.3%) pure white colony on indicator medium. Plasmid DNA analysis showed that Lac^- cells had lost the marked element by excision and failure to replicate.

A different mini-Mu-*lac* insertion, MX32, that mapped inside the element near the *nifK*-proximal 11-bp repeat failed to give rise to white colonies (<0.01%). It could, however, be complemented in *trans* with a compatible plasmid containing the corresponding parental (Xis^+) DNA fragment (15). This indicated that the MX32 insertion disrupted the excisase gene *xisA* necessary for deletion of the *nifD* element in *E. coli*. The *xisA* region was sequenced and found to contain an open reading frame of 1,062 bp that codes for a 41.6-kilodalton polypeptide. The *xisA* gene has been cloned downstream of a bacteriophage T7 promoter in a plasmid expression vector (W. J. Buikema, unpublished results). Cells carrying this plasmid together with the test plasmid MX25 produced nearly 100% white colonies after induction of T7 RNA polymerase synthesis, and thus provide suitable starting material for purification of the excisase.

The identification of an excisase leads to a hypothesis for the observation that excision is restricted to differentiating heterocysts. Excision can occur only in those cells in which the *xisA* gene is transcribed and the gene product is active. In vegetative cells the *xisA* gene could be repressed or its product could be inactivated or both. My colleagues have attempted to detect transcripts of the *xisA* gene in *Anabaena* under a variety of conditions, but the transcripts are too rare to be seen even by a sensitive S1 nuclease protection assay (J. W. Golden, unpublished results). Earlier experiments by D. Rice failed to detect transcripts from any part of the 11-kb element (19). The same is true for the left end of the 55-kb element (19).

VII. CONCLUDING REMARKS

The differentiation of heterocysts poses a number of simple questions of developmental importance: How are cells selected for development? How is induction of the genes that are specifically expressed in heterocysts confined to those cells? What is the signal for that induction? To these questions we now add new ones: How is rearrangement of the *nif* DNA confined to heterocysts? What maintains the nonrearranged sequences in vegetative cells?

Our data show that each interrupting element is found in a single location in *Anabaena* sp. strain 7120. The restriction fragments derived from excised elements, as well as the 11-kb circle, appear to persist as long as the heterocyst lives. Since heterocysts are terminally differentiated cells, we cannot determine whether either circle is capable of reintegrating at a new location. The arrangement of the *nifHDK* genes has been surveyed in over a dozen strains of cyanobacteria (7, 13, 14, 17). The results of these surveys are that, in general, unicellular cyanobacteria, filamentous nonheterocystous cyanobacteria, and the cyanobacterial symbionts in leaf cavities of the fern *Azolla* sp. have the *Klebsiella* organization; that is, the *nifHDK* genes are contiguous. However, in nearly all the heterocyst-forming *Anabaena* and *Nostoc* spp., including strains isolated as symbionts on the hornwort *Anthoceros* sp., the *nifHD* genes are separated from *nifK*. Detailed analysis of two strains, using *xisA* as well as *nif* gene probes, indicated that the *nifD* element is in the same location as in *Anabaena* sp. strain 7120 (17). The *nifS* element has not been surveyed extensively, but in one strain examined, *Anabaena* sp. strain 29413, it is absent.

Since strains can be found in nature in which the *nifD* element, or at least a part of it, is in the same place as in *Anabaena* sp. strain 7120, yet the distribution of *Eco*RI, *Hin*dIII, and other restriction sites in the *nifHDK* region differs from that of *Anabaena* sp. strain 7120, it must be concluded that the *nifD* element entered the *nifD* gene long ago and has persisted there for many generations (17). Previously, my colleagues and I drew attention to the formal similarity of the *nifD* element with bacteriophage lambda (4, 9). In this view, the element would correspond to the chromosome of a virus that integrated in the *nif* region. In order for the filament to survive a period of nitrogen starvation, the interrupting element has to be excised during heterocyst differentiation, which would be accomplished by induction of the excisase gene or by allowing the excisase to function.

Our study of cyanobacteria began in 1963 with the first report of a cyanophage and shifted in 1972 to the mechanisms of heterocyst differentiation and nitrogen fixation. It is satisfying now to appreciate that we may have been studying a lysogen (actually a double lysogen) in which induction is critical for cellular differentiation and in which the "viral" genes confer an advantage on the host in nature.

Acknowledgments. I am grateful to M. E. Mulligan, W. J. Buikema, J. W. Golden, H. Böhme, and J. Meeks for unpublished data cited in this chapter. All of the unpublished results cited were obtained in Chicago except for the topology of the *nifS* excision, elucidated by J. W. Golden at Texas A & M University.

Work in Chicago was supported by grants from the National Institutes of Health and the United States Department of Agriculture Science and Educational Administration through the Competitive Grants Office.

LITERATURE CITED

1. **Belknap, W. R., and R. Haselkorn.** 1983. Cloning and light regulation of expression of the phycocyanin operon of the cyanobacterium *Anabaena*. *EMBO J.* **6:**871–884.
2. **Beynon, J., A. Ally, M. Cannon, F. Cannon, M. Jacobson, V. Cash, and D. Dean.** 1987. Comparative organization of nitrogen fixation-specific genes from *Azotobacter vinelandii* and *Klebsiella pneumoniae*: DNA sequence of the *nifUSV* genes. *J. Bacteriol.* **169:**4024–4029.
3. **Böhme, H., and B. Schrautemeier.** 1987. Comparative characterization of ferredoxins from heterocysts and vegetative cells of *Anabaena variabilis*. *Biochim. Biophys. Acta* **891:**1–7.
4. **Campbell, A.** 1983. Bacteriophage lambda, p. 65–103. *In* J. A. Shapiro (ed.), *Mobile Genetic Elements*. Academic Press, Inc., New York.
5. **Castilho, B. A., P. Olfson, and M. J. Casadaban.** 1984. Plasmid insertion mutagenesis and *lac* gene fusion with mini-Mu bacteriophage transposons. *J. Bacteriol.* **158:**488–495.
6. **Curtis, S. E., and R. Haselkorn.** 1984. Organization and transcription of genes coding for 32 kd thylakoid membrane proteins in *Anabaena*. *Plant Mol. Biol.* **3:**249–258.
7. **Franche, C., and G. Cohen-Bazire.** 1987. Evolutionary divergence in the *nif*HDK gene region among nine symbiotic *Anabaena azollae* and between *Anabaena azollae* and some free-living heterocystous cyanobacteria. *Symbiosis* **3:**159–178.
8. **Golden, J. W., M. E. Mulligan, and R. Haselkorn.** 1987. Different recombination site-specificity of two developmentally regulated genome rearrangements. *Nature* (London) **327:**526–529.
9. **Golden, J. W., S. J. Robinson, and R. Haselkorn.** 1985. Rearrangement of nitrogen fixation genes during heterocyst differentiation in the cyanobacterium *Anabaena*. *Nature* (London) **314:**419–423.
10. **Gussin, G. N., C. W. Ronson, and F. M. Ausubel.** 1986. Regulation of nitrogen fixation genes. *Annu. Rev. Genet.* **20:**567–592.
11. **Haselkorn, R.** 1978. Heterocysts. *Annu. Rev. Plant Physiol.* **29:**319–344.
12. **Hayashida, N., T. Matsuboyashi, K. Shinozaki, M. Sugiura, K. Inoue, and T. Hiyama.** 1987. The identification and characterization of the 9 kd polypeptide gene found in tobacco chloroplast DNA. *Curr. Genet.* **12:**247–250.
13. **Kallas, T., T. Coursin, and R. Rippka.** 1985. Different organization of *nif* genes in heterocystous and non-heterocystous cyanobacteria. *Plant Mol. Biol.* **5:**321–329.
14. **Kallas, T., M.-C. Rebiere, R. Rippka, and N. Tandeau de Marsac.** 1983. The structural *nif* genes of *Gloeothece* and *Calothrix* share homology with those of *Anabaena* but the *Gloeothece* genes have a different arrangement. *J. Bacteriol.* **155:**427–431.
15. **Lammers, P. J., J. W. Golden, and R. Haselkorn.** 1986. Identification and sequence of a gene required for a developmentally regulated DNA excision in *Anabaena*. *Cell* **44:** 905–911.
16. **Mazur, B. J., D. Rice, and R. Haselkorn.** 1980. Identification of blue-green algal nitrogen fixation genes by using heterologous DNA hybridization probes. *Proc. Natl. Acad. Sci. USA* **77:**186–190.
17. **Meeks, J. C., C. M. Joseph, and R. Haselkorn.** 1988. Organization of the *nif* genes in cyanobacteria in symbiotic association with *Azolla* and *Anthoceros*. *Arch. Microbiol.* **150:** 61–71.

17a. **Mulligan, M. E., W. J. Buikema, and R. Haselkorn.** 1988. Bacterial-type ferredoxin genes in the nitrogen fixation regions of the cyanobacterium *Anabaena* sp. strain PCC7120 and *Rhizobium meliloti*. *J. Bacteriol.* **170:**4406–4410.

18. **Nierzwicki-Bauer, S. A., S. E. Curtis, and R. Haselkorn.** 1984. Cotranscription of genes encoding the small and large subunits of RuP_2 carboxylase in the cyanobacterium *Anabaena* 7120. *Proc. Natl. Acad. Sci. USA* **81:**5961–5965.

19. **Orme-Johnson, W. H.** 1985. Molecular basis of nitrogen fixation. *Annu. Rev. Biophys. Biophys. Chem.* **14:**419–459.

20. **Rice, D., B. J. Mazur, and R. Haselkorn.** 1982. Isolation and physical mapping of nitrogen fixation genes from the cyanobacterium *Anabaena* 7120. *J. Biol. Chem.* **257:**13157–13163.

21. **Silverman, M., and M. Simon.** 1983. Phase variation and related systems, p. 537–557. *In* J. A. Shapiro (ed.), *Mobile Genetic Elements*. Academic Press, Inc., New York.

22. **Tumer, N. E., S. J. Robinson, and R. Haselkorn.** 1983. Different promoters for the *Anabaena* glutamine synthetase gene during growth using molecular or fixed nitrogen. *Nature* (London) **306:**337–342.

23. **Wolk, C. P.** 1982. Heterocysts, p. 359–386. *In* N. G. Carr and B. A. Whitton (ed.), *The Biology of Cyanobacteria*. Blackwell Scientific Publications, Ltd., Oxford.

Chapter 34

Mechanisms for Variation of Pili and Outer Membrane Protein II in *Neisseria gonorrhoeae*

JOHN SWANSON and J. MICHAEL KOOMEY

John Swanson ■ Laboratory of Microbial Structure and Function, Rocky Mountain Laboratories, National Institute of Allergy and Infectious Diseases, 903 South 4th Street, Hamilton, Montana 59840. J. Michael Koomey ■ Department of Microbiology and Immunology, University of Michigan Medical School, Ann Arbor, Michigan 48109.

I. INTRODUCTION

Neisseria gonorrhoeae is a highly successful pathogen in humans, its sole host. This success stems from the popular mode of transmission of the bacterium and its seeming ability to outwit the defenses of the host. Although gonococci can reside within tissue culture cells in vitro (120, 147) and have been seen within epithelial cells in clinical specimens (30, 40, 147), their in vivo existence appears to be largely extracellular on mucosal surfaces (40). This lifestyle exposes gonococci to both humoral and cellular elements of the immune system of their host, but these defenses seem quite ineffective toward eradicating or preventing infections by these otherwise fragile and relatively fastidious gram-negative bacteria.

Interactions with the epithelial and immune components of the host are likely mediated by one or more surface-exposed constituents of gonococci, capsule, pili, and both protein and nonprotein molecules of outer membranes. The capsule has been difficult to isolate and remains little characterized (46, 51). Lipopolysaccharide and the H.8 lipoprotein (H.8) are major components of the outer membrane, are expressed constitutively, and exhibit modest degrees of variation (3, 16, 48, 118). Several outer membrane proteins, including proteins I and III (P.I, P.III), are also relatively invariant (34, 41, 56, 76). A P.I polypeptide with porin activity (149) is expressed by all gonococci and its structure is uniform for the variants of a given strain, but both major and minor structural variations occur among P.I molecules of different strains (34, 56, 124). P.III is the least variable outer membrane component examined (57, 76, 134) and recent information indicates that the C-terminal domain of P.III shares amino acid sequence with a periplasmic portion of *ompA* in *Escherichia coli* (33).

Roles in gonococcal virulence or pathogenicity have been suggested for all the above-mentioned surface constituents (34, 94, 121). However, the absence of a laboratory animal model for gonorrheal infection has retarded definition of the role(s) that each surface constituent plays in initiation or maintenance of gonorrheal disease, and such roles are reasonably well understood only for pili and outer membrane protein II (P.II).

Pili and P.II are highly variable, in marked contrast to the relatively stable and constitutively expressed components such as P.I and P.III. Pili and P.II occur in numerous structurally and antigenically distinct forms in variants of a given strain, and neither pili nor P.II is expressed constitutively. In the following, we focus mainly on the events that cause variable expression of these two proteins and thereby underlie the impressive variety of gonococcal surfaces. Pili and P.II both exhibit phase and structural variations, but their controlling mechanisms are independent and distinct.

II. BACKGROUND

A. Pili and P.II Influence Pathogenicity

Twenty years ago it was shown by intraurethral challenge of males that gonococci with particular colonial morphologies (types 1 and 2) were consistently pathogenic; others (types 3 and 4) initially displayed virulence but gave rise to permanently nonpathogenic (type 4) variants after additional passages in vitro (61, 62). Subsequent studies showed that cells in colony types 1 and 2 bear pili (pilus$^+$), whereas colony types 3 and 4 contain nonpiliated organisms (pilus$^-$) (54, 133). In retrospect, the pathogenic and nonpathogenic pilus$^-$ gonococci probably corresponded to reverting (to pilus$^+$) and nonreverting pilus$^-$ phenotypes, respectively (131). The P.II constitutions of the virulent and avirulent colony types were not determined, but judging from the descriptions, types 1 to 3 (dark colored) colonies probably contained P.II$^+$ gonococci while type 4 (light colored) colonies consisted of P.II$^-$ organisms; both piliation and P.II$^+$ phenotypes seemed to correlate with pathogenicity in those studies. Those early results are usually summarized as showing that pilus$^+$ gonococci are pathogenic while pilus$^-$ are not. Verification of this comes from recent studies on male volunteers challenged either with pilus$^+$ gonococci or with pilus$^-$ variants (P$^-$n phenotype) that cannot revert to pilus$^+$. The pilus$^+$ and pilus$^-$ inocula were from the same strain and each contained predominately (>99%) P.II$^-$ organisms. Pilus$^+$ gonococci caused gonorrhea in 7 of 11 males, whereas 0 of 9 males challenged with P$^-$n organisms developed infection (J. Boslego, J. M. Koomey, and J. Swanson, unpublished results).

Evidence that P.II contributes to pathogenicity is less direct. Gonococci isolated from naturally infected males consistently display a P.II$^+$ phenotype, whereas isolates from infected females may be P.II$^-$ or P.II$^+$ (50); the variations among isolates from females correlate both with the anatomical sites sampled and the menstrual histories of the infected women (52, 150). Cervical isolates are usually P.II$^+$ except when obtained at or near menstruation. Gonococci isolated from the fallopian tubes of an infected female are almost uniformly P.II$^-$ (25). In the recent human challenges with pilus$^+$ gonococci mentioned above, input pilus$^+$ organisms were mainly (>99%) P.II$^-$, but virtually all gonococci reisolated from the infected men were P.II$^+$, expressing one or two P.IIs; the P.IIs that they expressed most commonly (IIc, IIf) were not those that predominated among in vitro variants of the input organisms (J. Swanson, unpublished results). These clinical microbiological phenomena point to a clear but ill-defined role for P.II in gonococcal pathogenicity.

B. Colony Form Reveals Pilus and P.II Constitutions

Colony size, shape, and edge morphology provide reliable, sensitive clues to the piliation of gonococci propagated on clear solid medium (127, 130, 131). Pilus$^+$ bacteria produce small, domed colonies, in contrast to the large, flat, or low-convex colonies formed by pilus$^-$ organisms. Even minor sequence differences in pilus subunit polypeptide (pilin) can change the edge morphology of a pilus$^+$ colony. The bases for correlations between piliation and colonial morphology are ill defined; they probably relate to intergonococcal congregation/aggregation as influenced by pilin/pilus structure (hydrophobicity? charge?).

P.II expression also influences gonococcal colony morphology (122, 123, 141, 146). P.II variants produce nonopaque (O$^-$) colonies, whereas the production of one or more P.IIs correlates with the formation of opaque colonies. The various P.II moieties often seem to confer distinctive degrees of colonial opacity (O$^+$, O^{++}, etc.) (127).

C. Pilus and P.II Expressions Change at High Frequency In Vitro

Piliation and P.II variants arise consistently during in vitro passage. The frequencies of these changes in vitro have been estimated (75, 131), but are subject to variables, such as propagation history and medium composition, whose selective effects are not yet well understood. Strains with long histories of laboratory passage typically display piliation and P.II transitions among 0.1 to 1% of their progeny. Some reisolates from experimental infections, however, spawn P.II and/or piliation variants in vitro in >10% of their offspring (Swanson, unpublished results). As mentioned above, phase (pilus$^+ \rightleftharpoons$ pilus$^-$) and structural/antigenic pilus changes are not linked to P.II transitions (P.II$^- \rightleftharpoons$ P.IIa$^+ \rightleftharpoons$ P.IIab$^+$, etc.) (130).

III. PILUS VARIATION

A. Pili and Pilin

Gonococcal pili are 8-nm filaments that vary in length and number on the surface of the organism (54, 133), where their anchoring in the cell wall is not defined. They often emanate from the division septum between sister diplococci (unpublished results). Whether pili radiate centrifugally from the surface of gonococci in a fluid milieu or are closely adherent to the cell wall surface of the organism is not clear (121). Individual pilus filaments are homopolymers of polypeptide subunits (pilin) whose calculated molecular weights are in the range of 17,500 (103); in our hands such a pilin displays electrophoretic migration of approximately 21,000 M_r and diverse pilins vary in M_r from 20,000 to 26,000 (71, 130). Individual cells probably elaborate only a single pilin species at a time, although some studies find pilins of different sizes in preparations subjected to isoelectric focusing or sodium dodecyl sulfate-polyacrylamide gel electrophoresis (96, 130). Pilin is the only polypeptide identified in isolated pili.

Different pili exhibit varied tendencies to aggregate laterally (12); some aggregate weakly, while others yield paracrystalline cable-like arrays that can exceed the diameters of gonococci. It is not clear whether pilin charge or hydrophobic character or both account for the aggregation of pili. Pilin polypeptides have several hydrophobic regions, especially their conserved amino-terminal residues 1 to 50 (47, 103) (Fig. 1). Pili have net negative charges at physiological pH values because of their acidic pI values (4.9 to 5.3) (44, 96). There are only five charged residues in the amino-terminal one-fifth of pilin, but they are abundant in the carboxyl-terminal portion of the polypeptide, where they cluster in four relatively discrete amino acid regions (residues 72 to 84, 100 to 118, 127 to 146, and 152 to C terminus). One of these (127 to 146) contains predominantly negatively charged residues (glutamic acid, aspartic acid), while positively charged amino acids (lysine, arginine) predominate in the others (Fig. 1); all four regions are relatively hydrophilic (98).

B. Pilin Polypeptides Contain Conserved and Variable Domains

An early-recognized characteristic of gonococcal pili was their extreme antigenic heterogeneity (12, 13, 69, 87). Pilin polypeptides from various strains and intrastrain variants possess both common and unique regions as defined initially by immunochemical techniques (13, 87, 103, 144). These common and

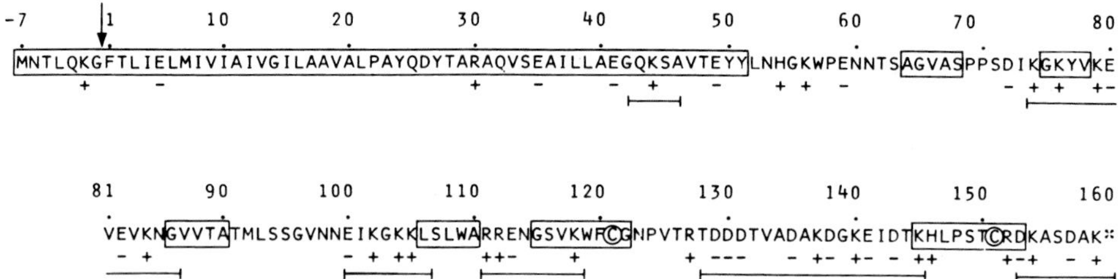

Figure 1. Amino acid sequence of prepilin polypeptide. The deduced sequence for the primary translation product (prepilin) encoded by the single functional pilin gene of strain MS11 is shown. Amino acids are noted in single-letter code, and the numbering of position 1 designates the first residue of mature pilin rather than that of prepilin as in references 6, 132, 135, and 136. The arrow indicates the position of the cleavage to generate mature pilin. Conserved regions are boxed and derived from the sequences shown in Fig. 2. Noted below the sequence are charged amino acids. The major hydrophilic domains of pilin are designated by brackets (⊢⊣) according to data in reference 103. The amino-terminal 51 amino acids of pilin are conserved and are highly hydrophobic. Other conserved stretches include residues 64 to 68, 75 to 78, 85 to 90, 106 to 110, 115 to 122, and 145 to 153. Each hydrophilic domain contains an excess of positively charged amino acids, except for region 128 to 145, where negatively charged residues predominate.

unique antigenic portions of pilins reside, respectively, in CNBr-2 (residues 8 to 92) and CNBr-3 (residues 93 to carboxy terminus), the two major cyanogen bromide cleavage fragments of pilin (103). This early view of pilin polypeptides has been expanded by sequencing several cloned complete pilin genes and pilin mRNAs (6, 38, 79, 85, 135, 136) (Fig. 2). The amino-terminal 51 amino acids of gonococcal pilins sequenced are highly conserved and consist mainly of hydrophobic, uncharged residues. Additional conserved regions include stretches of 9 to 10 amino acids (residues 122 to 132 and 151 to 161) between the two cysteine residues (121 and 151) and four short stretches each of 4 to 6 residues (65 to 68, 75 to 78, 86 to 90, and 106 to 110) (Fig. 1 and 2). The C-terminal two-thirds of pilin polypeptides displays sequence heterogeneity and has been divided into one "hypervariable" and several "semivariable" domains (36, 38). The hypervariable region occurs between the two cysteine residues, and some pilins differ by every amino acid in this entire region. Sequence variation is less marked in the rest of the C-terminal two-thirds of both expressed pilins and several silent pilin genes (Fig. 2).

C. Adherence Function of Pili

Many studies conclude that pilus$^+$ gonococci adhere more avidly to human cells in suspension, tissue culture, or organ culture than do pilus$^-$ organisms (77, 92, 120, 126), and such enhanced adherence is thought to explain the positive influence of pili on pathogenicity. Strain P9 variants that express pili with different-sized subunits exhibit distinctive adherence proclivities for various eucaryotic cells (fresh human buccal cells, Chang conjunctiva, and HEp-2 lines) in vitro (43). The molecular basis for pilus-mediated adherence to eucaryotic cells is not entirely clear, but is probably based on electrostatic forces (148). Blocking the —COOH groups on whole pilus$^+$ gonococci raised the pI of the cells to 8.2 (from 5.3) and greatly enhanced their adherence to WISH tissue culture cells; blocking —NH_2 groups reduced the pI of the cells to 4.0 and markedly reduced their adherence (44). These phenomena are attributed mainly to pili, but individual or cooperative roles for other chemically modified outer membrane components in adherence to tissue culture cells have not been excluded.

Several discrete regions of the pilin molecule appear to participate in pilus-mediated adherence of gonococci to eucaryotic cells. Antisera raised against oligopeptides representing conserved regions 41–50 and 69–84 and variable regions 121–134 and 135–151 *all* inhibited adherence of homologous pilus$^+$ organisms to a human cervical carcinoma cell line (99). Only the antisera against oligopeptides 41–50 and 69–84 interfered with adherence of a heterologous strain; but pilins of many gonococci reisolated from recent experimental infections exhibit sequence changes in their 69–84 regions (135). Monoclonal antibodies with epitopes in other variable portions of pilin inhibited adherence of homologous pili; cross-reactive monoclonals that recognized a conserved epitope bound to all pili, but had no influence on adherence (28, 142, 143). Thus, the participation of common versus variable portions of the pilin molecule in adhesion of pili to epithelial cells is unsettled.

D. Complete and Partial Pilin Genes

One (rarely two) expressed, complete pilin genes and a multitude of silent, partial pilin genes reside in the gonococcal genome (79, 80, 106, 132) and are distributed in strain MS11 (36, 64) (Fig. 3). The complete pilin gene or pilin expression locus contains

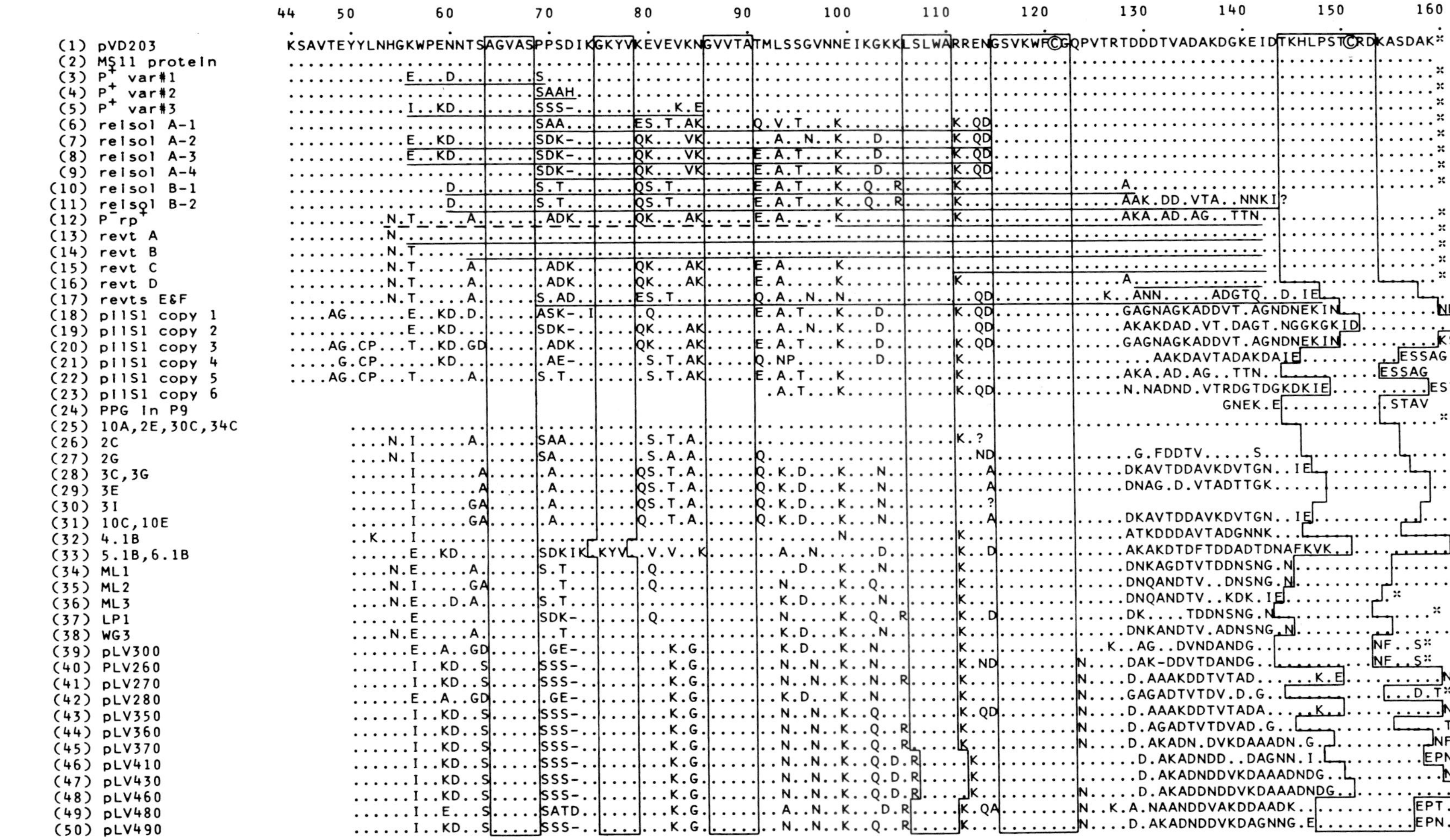

Figure 2. Previously published pilin sequences (1) to (50). One [sequence (2)] was obtained by amino acid sequencing, but the others were deduced from the sequences of cloned complete pilin genes [(1), (39) to (50)], cloned partial pilin genes [(18) to (24)], and pilin mRNAs [(1), (3) to (17), (25) to (38)]. Amino acids are noted by single-letter code and their locations are for mature pilin. Sequences are aligned at cysteine 121. Conserved regions are boxed and are described in Fig. 1 and the text. The amino acid sequence resulting from gene conversion in the variants [(3) to (5) and (12) to (17)] of an MS11 strain that had expressed a pVD203-like [(1)] pilin are underlined. Also included are reisolates from experimental gonorrheal infections [(6) to (11)] initiated by an MS11 strain that expressed pVD203-like pilin [(1)]. Some of these [(8) to (11)] may be products of several sequential gene conversion events (reference 134) (Fig. 5). One variant [(12)] contains a long nucleotide stretch (dashed underline) that had been incorporated prior to the transition that gave rise to this P^-rp^+ phenotype variant; that last change brought in the sequence denoted by the solid underline (reference 135). Sources of these sequences are as follows: (1), reference 6; (2), reference 103: (1), (6) to (11), reference 135; (12) to (17), reference 136; (18) to (23), reference 36; (25) to (38), reference 38; (39) to (50), reference 85. Sequences (3) to (5) are unpublished and represent mRNA sequences of $pilus^+$ morphotype variants of (1) in strain MS11 formed in vitro. Sequence (24) is of a cloned partial pilin gene from strain P9 (88a).

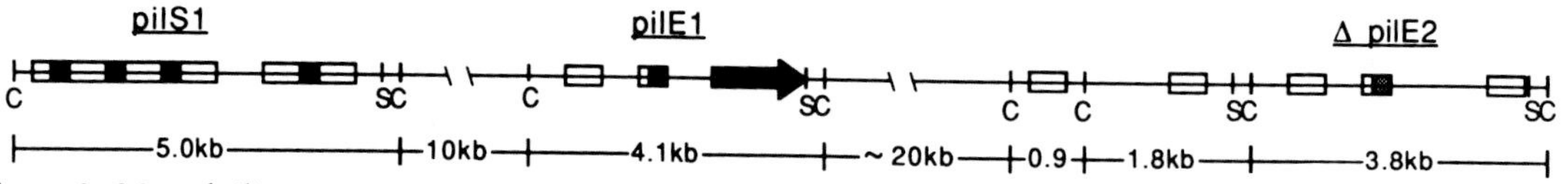

Figure 3. Map of pilin gene sequences in strain MS11 chromosome. Locations for several partial pilin genes (open boxes) and the single complete pilin gene (solid arrow) in the MS11 chromosome are taken from references 64 and 113. Note the relationship of pilin genes to *Cla*I (C) sites and to 65-bp *Cla*I-*Sma*I segments (not to scale). RS1 repeats (solid boxes) lie adjacent to and separate several partial pilin genes. In the strain MS11 depicted here, the 5′ portion, a previously active expression locus *pilE2*, has been deleted, and hence *pilE2* is not expressed.

a single long open reading frame preceded by a strong ribosome-binding site (AGGAG) 7 bases upstream (6, 79). The transcription initiation site, localized through primer extension, has a proper −10 sequence (TATAAT), but no typical −35 region, a feature of many promoters that require positive regulatory factors for transcription (93). The complete pilin gene open reading frame encodes a prepilin polypeptide of 166 to 175 amino acids, variable in length among different strains and intrastrain variants. The primary translation product has a seven-amino-acid amino-terminal portion that likely plays a role in prepilin processing or transport (Fig. 1), and pilin from isolated pili lacks this heptapeptide (47). Prepilins in *Bacteroides*, *Moraxella*, and *Pseudomonas* spp. contain similar, short amino-terminal peptide extensions (29, 74, 101). The open reading frame ends with either a TGA or TAA translation stop signal. TAA codons occur in both the −1 to +1 reading frames (Fig. 4).

An estimated 12 to 20 partial pilin genes are clustered in the chromosome at several silent loci on each side of the complete pilin gene (36, 64) (Fig. 3). One silent locus (*pilS1*) containing six partial pilin genes has been cloned and sequenced (36); they are tandemly arranged and all lack the sequences that correspond to the 5′-terminal one-third of a complete pilin gene. Some *pilS1* partial pilin genes are separated by a common 39- or 40-nucleotide repeat (RS1), while others are separated by intergenic regions consisting of 49-base-pair direct repeats (RS2) and a family of direct and inverted repeats (RS3) possessing a common 6-base-pair sequence (36). The number of codons in the *pilS1* locus corresponding to those in the complete pilin gene varies: one contains codons 31 to 158, several contain codons 44 to 158, and one contains codons 92 to 158. Only one partial pilin gene in *pilS1* contains a translation stop signal analogous to that which terminates the complete pilin gene open reading frame. One highly truncated partial pilin gene (codons 141 to 159 in MS11, codons 122 to 132 in P9) is immediately upstream of the complete pilin gene in strains MS11 and P9 (36, 88a); it is flanked by an RS1 repeat at its 3′ end. Partial and complete pilin genes have identical sequence in six short regions (amino acids 64 to 68, 75 to 78, 87 to 90, 106 to 110, 115 to 122, and 145 to 153) (Fig. 2).

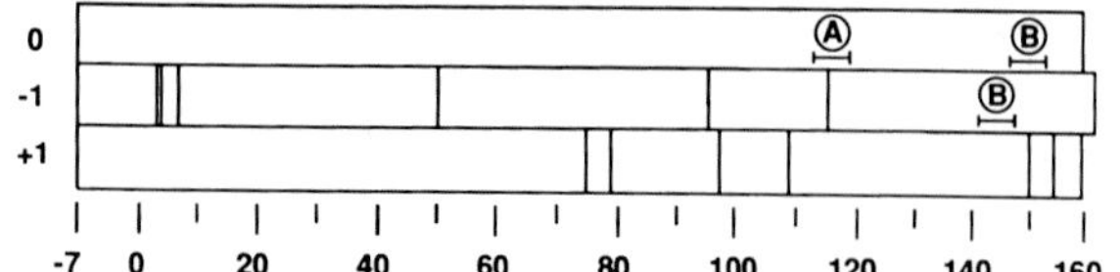

Figure 4. Reading frames of the complete pilin gene. The long open reading frame is represented in the 0 frame. Additions or deletions of single nucleotides at the 5′ end generate premature translation termination codons (vertical lines). Ⓐ and Ⓑ denote conserved amino acid sequences that span cysteine residues 121 (CGQP) and 151 (PSTCR), respectively. Identical five-amino-acid sequences (PSTCR) spanning cysteine 158 are encoded in both 0 and −1 frames Ⓑ. The −1 frame after codon 158 encodes a polypeptide that is two residues longer than the 0 frame. (Figure modified from reference 128.)

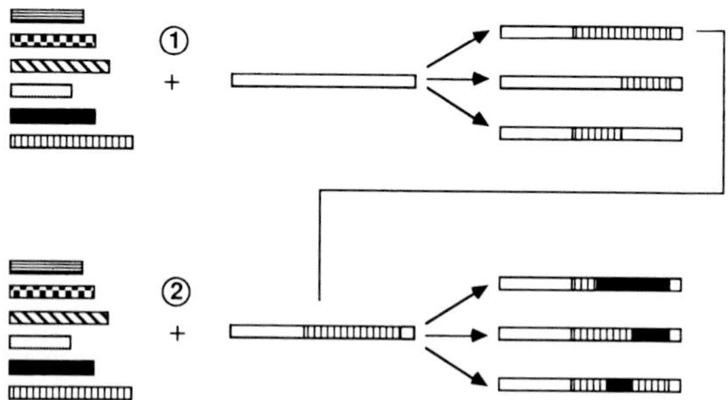

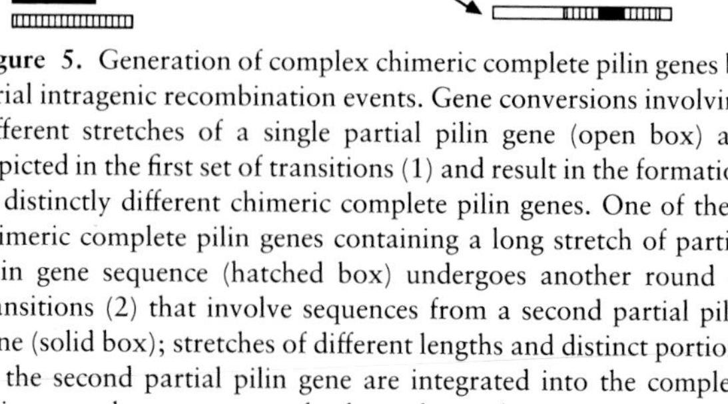

Figure 5. Generation of complex chimeric complete pilin genes by serial intragenic recombination events. Gene conversions involving different stretches of a single partial pilin gene (open box) are depicted in the first set of transitions (1) and result in the formation of distinctly different chimeric complete pilin genes. One of these chimeric complete pilin genes containing a long stretch of partial pilin gene sequence (hatched box) undergoes another round of transitions (2) that involve sequences from a second partial pilin gene (solid box); stretches of different lengths and distinct portions of the second partial pilin gene are integrated into the complete pilin gene, but none completely replaces the sequence inserted through the first gene conversion. The resulting complete pilin genes are quite heterogeneous due to the variable contributions of partial pilin gene-derived sequence. (Figure modified from reference 128.)

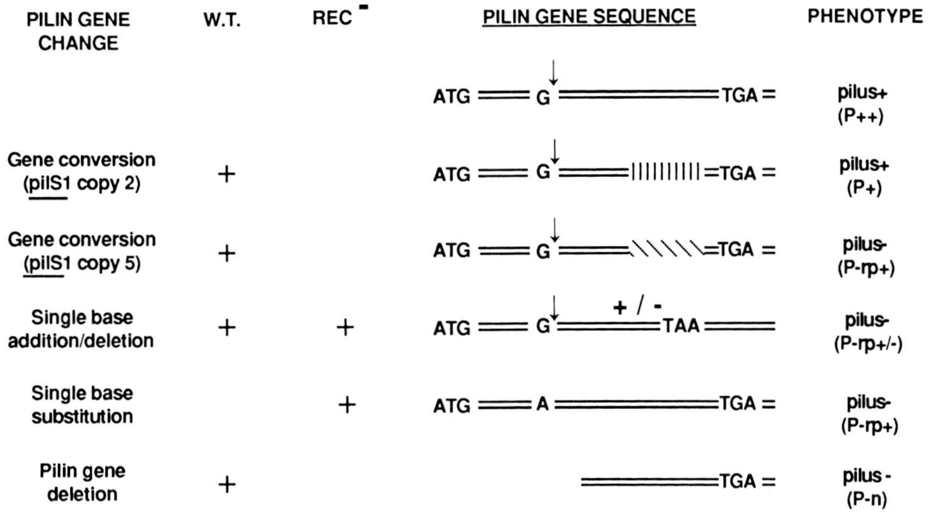

PILIN GENE CHANGE	W.T.	REC$^-$	PILIN GENE SEQUENCE	PHENOTYPE
			ATG ═ G↓ ═══ TGA ═	pilus+ (P++)
Gene conversion (pilS1 copy 2)	+		ATG ═ G↓ ═ ||||||||| ═ TGA ═	pilus+ (P+)
Gene conversion (pilS1 copy 5)	+		ATG ═ G↓ ═ \\\\\\\\\\\\ ═ TGA ═	pilus- (P-rp+)
Single base addition/deletion	+	+	ATG ═ G↓ ═ +/- ═ TAA ═	pilus- (P-rp+/-)
Single base substitution		+	ATG ═ A ═══ TGA ═	pilus- (P-rp+)
Pilin gene deletion	+		═══ TGA ═	pilus - (P-n)

Figure 6. Summary of observed mutations in the complete pilin gene and their corresponding pilus phenotypes. The complete pilin gene sequence for pilus$^+$ gonococci (P^{++} morphotype) is depicted (top line) to include its translation initiation codon (ATG), proteolytic cleavage site for processing prepilin to pilin (arrow), and translation termination codon (TGA). Pilin gene changes were observed in gonococci of wild type (W.T.) or Δ*recA* (REC$^-$) or both. Gene conversion with *pilS1* copy 2 introduces a new sequence (vertical bars) into the complete pilin gene of wild-type organisms that retain pilus$^+$ status but express an antigenically changed pilin, as mirrored by their altered colony morphotype (P$^+$). Gene conversion with *pilS1* copy 5 inserts a missense sequence (diagonal bars) into the complete pilin gene and results in a pilus$^-$ phenotype that can revert to pilus$^+$ (P$^-$rp$^+$) via a subsequent gene conversion. Single-nucleotide base addition/deletion can produce the reverting pilus$^-$ phenotype in either wild-type or Δ*recA* organisms by producing downstream premature translation termination signals (TAA). The pilus$^-$ phenotype also follows single-base substitution (G to A) in the triplet encoding the last amino acid of the leader peptide to produce a prepilin molecule that is not proteolytically processed to the mature pilin molecule; this mutation was observed in Δ*recA* organisms, but likely occurs also in wild type. Deletion of 5′-terminal sequences of the complete pilin gene in wild-type gonococci produces an incomplete, nonfunctional gene and a pilus$^-$ phenotype incapable of reversion (P$^-$n). (Figure modified from reference 128.)

E. Complete Pilin Gene Sequence Changes Correlate with Pilus Phase and Antigenic Transitions

Different complete pilin genes and pilin mRNAs display striking nucleotide sequence diversity [e.g., sequences (25) to (50) in Fig. 2]. Diversity occurs both among the variants of a single strain and among strains (38, 85). When different strains or multistep variants of a given strain are compared, their differences are not generally recognizable as representing known partial pilin gene sequences. However, in single-step pilus variants that arise in vitro (132, 136) and in some in vivo variants, blocks of nucleotides with sequence identical to a known partial pilin gene are identifiable (38, 135). As will become obvious in the following, our view is that such blocks of partial pilin gene sequence become part of the complete pilin gene by nonreciprocal intragenic recombination between the complete pilin gene and a partial pilin gene. This can result with change in the ability to form pili (pilus$^+$, pilus$^-$; phase transition) or expression of a new pilin (Fig. 5, 6, and 7). This mechanism differs from the suggestion of others that the pilus phase change occurs by deletion and reconstitution of pilin gene 5′ sequences (80, 105).

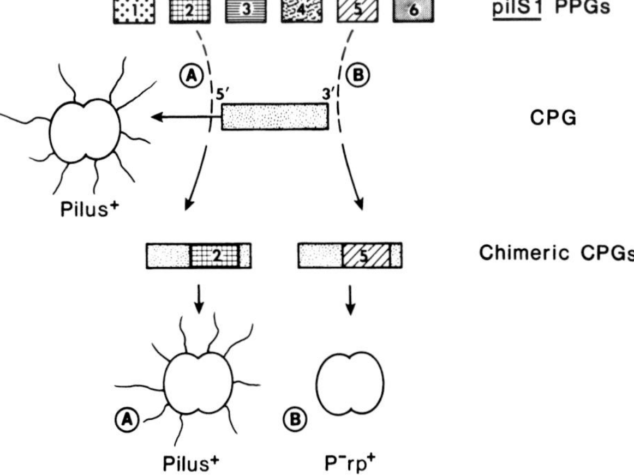

Figure 7. Pilus phase and antigenic transitions. Both proceed by recombinational insertion of partial pilin gene sequence into the complete pilin gene. Two different intragenic recombinations are depicted in pilus$^+$ gonococci in which *pilS1* copy 2 Ⓐ or copy 5 Ⓑ sequences (*pilS1* PPGs) are incorporated into the complete pilin gene (CPG). The resulting chimeric complete genes containing stretches from copy 2 retain pilus$^+$ phenotypes but express different pilin polypeptides. The copy 5-containing complete gene expresses a missense pilin, and pilus$^-$ phenotype results (P$^-$rp$^+$). (Figure modified from reference 128.)

A portion of *pilS1* copy 5 sequence regularly occurs in pilin mRNAs of independently derived pilus$^-$ strains that can revert to pilus$^+$ (designated P^-rp^+). This copy 5 sequence is at least 109 nucleotides long and includes the codons for residues 111 to 141 [sequence (12), Fig. 2]. The copy 5-containing mRNA transcript encodes a pilin that is defective for pilus formation. In pilus$^+$ revertants, part or all of the copy 5-derived sequence in the complete pilin gene is replaced by sequence from another partial pilin gene. Six such revertants contain blocks of new sequences ranging in length from 38 to 254 nucleotides [(13) to (17) in Fig. 2] (136). The junctions between new and preexisting sequences occur in regions that are conserved among all complete and partial pilin genes and thus are difficult to define precisely. Nevertheless, in two of the pilus$^+$ revertants [(15) and (16) in Fig. 2], it is clear that only a portion of parental copy 5 sequence was replaced by incoming partial pilin gene sequence. The copy 5 nucleotide block was entirely replaced in the others [(13), (14), and (17) in Fig. 2].

Of the six sibling revertant complete pilin genes, four contain different lengths of the same partial pilin gene [(13) to (16) in Fig. 2] found by three different labs to be expressed by strain MS11 [(1), (2), and (25) in Fig. 2]; two others contain sequences from a different partial gene [(2) and (17) in Fig. 2]. The basis for the frequent occurrence of this pilin gene sequence in strain MS11 propagated in vitro is not known.

Several gonococcal reisolates from experimentally infected males express pilin mRNAs that contain nucleotide stretches of the partial gene, *pilS1* copy 2 [(7) to (9) in Fig. 2] (127). The copy 2 sequences in turn contain short inserts of partial gene copy 5. These variants probably result from two serial transitions, first involving a relatively large segment of copy 2 and then a smaller segment of copy 5 sequences. Clearly, such serial recombinational changes could generate a highly complex array of chimeric complete pilin genes (Fig. 5).

F. Single-Nucleotide Changes Correlate with Pilus Transitions of Δ*recA* Gonococci

Pilus$^+$ gonococci with a defective *recA* gene exhibit rare pilus$^+$ to pilus$^-$ transitions attributable to single-nucleotide changes in the complete pilin gene sequence (67; unpublished data). Most of these pilus$^-$ variants make mRNAs with a single base deletion from the run of eight cytosines in codons 68 to 70; this causes a frameshift mutation in the complete gene and premature termination of pilin translation (Fig. 4 and 6). Pilus$^+$ revertants of such pilus$^-$ Δ*recA* gonococci have regained an eighth cytosine and thus the correct reading frame. Other pilus$^-$ variants formed in the Δ*recA* strain contained a G-to-A transversion (GGC to AGC) in the codon for the last amino acid of the putative signal heptapeptide (Fig. 6). The corresponding single amino acid change (glycine to serine) blocks prepilin processing and pilin formation. These and several other mRNAs from pilus$^-$ variants formed in Δ*recA* strains did not contain substitutions of large DNA segments like those in pilus transitions of wild-type gonococci; they probably result from classical point mutation, not recombination.

IV. PILUS VARIATION RESULTS FROM INTRAGENIC RECOMBINATION

A. Repair and Recombination

Pilus$^+$ gonococci undergo DNA-mediated transformation of chromosomal markers at frequencies that approach 1% (110). Transformation involves the incorporation of single DNA strands into the preexisting DNA, forming heteroduplex DNA, which may be resolved either by mismatch repair (20) or by the next round of replication. Studies with in vitro-generated mutations have shown that heteroduplexes with mispaired regions as large as 2 to 3 kilobases can be formed (65, 66). Studies of gonococci with a defective *recA* gene support proposals that pilin gene recombination occurs by the formation of equivalent DNA heteroduplexes. Mutations in the *recA* locus reduce or abolish the frequency of transformation for chromosomal markers; they also reduce or abolish the rates of genetic exchange between the partial pilin genes and the complete pilin gene, as scored by the rates of pilus phase variation (67).

The role of the *recA* gene product in pilin gene recombination is somewhat uncertain because *recA* mutations are probably pleiotropic (19). *recA* alone is probably not sufficient to execute the rearrangements, by analogy with other procaryotic systems, and additional gene products involved in DNA metabolism undoubtedly influence pilin gene recombination. These might include gonococcal equivalents of polymerase I, DNA ligase, single-stranded binding proteins, and the *mut* or *uvr* genes of *E. coli* (39, 63, 91, 109, 145). Such activities in gonococci have not yet been characterized.

B. Pilin Gene Conversion

Recombination between partial and complete pilin genes is nonreciprocal; analysis of selected variants showed that partial pilin genes are sequence donors, while the complete pilin gene is the recipient (36, 132, 135). This process has been designated "gene conversion" because of the unequal recovery of pilin alleles following recombination; this usage pertains to end results of the process without implying a specific exchange mechanism (67). There are several potential sources of apparent nonreciprocity, the simplest being that not all recombination products in bacteria are viable, recoverable, or identifiable. Analysis of the basis of variability is complicated since gonococci exist as diplococcal units whose precise chromosomal constitutions and interactions are unknown, and because pilus$^+$ organisms are competent for transformation and undergo autolysis during in vitro cultivation. Asymmetric pilin gene recombination could conceivably occur between elements on the same chromosome, between pilin genes on the two chromosomes of a single cell after their replication but before cell septation, or through the uptake and integration of DNA released from other cells undergoing autolysis. Findings that copy 2 and copy 5 sequences persist at the *pilS1* locus in MS11 bacteria that have undergone many pilin gene conversions show that the integrity of partial pilin genes is maintained. Whatever the method of exchange, P$^-$rp$^+$ convert to pilus$^+$ and then back again to P$^-$rp$^+$ with expression of *pilS1* copy 5 sequences by both P$^-$rp$^+$ organisms (132).

C. Is Transformation Involved in Pilin Gene Conversion?

Transformation of gonococci for chromosomal markers (under saturating conditions) and productive pilin gene conversions occur at similar frequencies, both approaching 1% (110, 131). Two independent studies suggest that pilus phase transitions result from transformation with extracellular DNA. In one study, there was an increase in pilus$^+$ to pilus$^-$ transitions correlated with cessation of active growth and the onset of autolysis (86). However, such results might also have reflected the selective growth advantage of preexisting nonpiliated variants owing to their lack of autoagglutination (67). The second study indicated that pilus$^-$ gonococci could spawn pilus$^+$ revertants after exposure to DNA from pilus$^+$ gonococci (5), but other investigators using the same strains did not obtain the same results (7). These studies are flawed in that only colony morphology was used to judge pilus$^-$ to pilus$^+$ transitions and such reliance on morphology can be unreliable; nor was it clear whether the pilus$^-$ DNA recipients could revert spontaneously to pilus$^+$.

Direct attempts to transform pilus$^+$ gonococci to express serologically different pili were unsuccessful even though it appears that the reagents available at that time were sensitive enough to detect such transformation (87). Other attempts to test this hypothesis by studying the influence of DNase on the rates of pilus phase variation are problematic, since it is difficult to maintain enzymatic activity throughout the period of culture and since DNase (or its contaminants) has different inhibitory effects on piliated and nonpiliated organisms (5, 60).

The idea that transformation is the sole or major basis for pilin gene conversion is also undermined by the observation that reverting pilus$^-$ variants spawn pilus$^+$ revertants via pilin gene conversion at a frequency of 1% despite the fact that they are incompetent for transformation for other chromosomal markers (132; J. M. Koomey, unpublished data). Isolation of well-defined mutants deficient in DNA uptake or autolysis and characterization of their relative rates of pilus phase variation and pilin gene conversion should provide an opportunity to evaluate the role of transformation in these events.

D. Asymmetry in Pilin Gene Conversion

Any mechanism invoked for pilin gene conversion must explain the polarity of the exchange and the curious organization of complete and partial pilin genes. Within clusters of complete and partial pilin genes, the complete pilin gene is situated at the 3′ end of the array. In the few cases analyzed so far, the partial pilin gene source of the complete pilin gene of a variant mapped just upstream of the complete pilin gene, and the sequence of this source gene was not changed (36, 37, 132, 136). Similarly, pilus$^+$ revertants of pilus$^-$ cells also preferentially expressed the sequence of a partial pilin gene that maps immediately upstream of the complete pilin gene (6, 132, 136). In seemingly analogous surface glycoprotein variation of *Trypanosoma equiperdum*, a particular basic gene copy is selectively used; this is thought to reflect the degree of homology between it and the sequence of the current expression locus (97), not its position. Extent of homology alone does not seem to account for the phenomena observed in gonococci. But how the position of a partial gene might influence its use to convert a complete gene is not apparent from current models.

What other factors might influence the asymmetry of pilin gene recombination? What is unique about the complete pilin gene that might make it an acceptor rather than a donor of sequences? Simple transformation models of nonreciprocity do not account for the curious lack of movement of the unique 5′ end of the complete pilin gene into a partial pilin gene locus, an event that should have similar affects on pilin variation.

Studies of homologous recombination and gene conversion in other species indicate that exchange events are often initiated at or near special sites (112) that may increase accessibility of adjacent sequences to strand invasion, or serve as recognition sites for specific enzymes (e.g., Chi sites [108]) or site-specific cleavage (initiating mating-type switching in *Saccharomyces cerevisiae* [68]). If opening the DNA helix makes it more susceptible to strand invasion, then transcription of the complete pilin gene could stimulate recombination, as noted in other systems (9, 26, 45), and this might dictate its being an acceptor rather than a donor of sequence information.

Some insight into mechanisms of pilin gene conversion come from the common structural elements found by sequencing complete and partial genes. All pilin gene clusters have a specific 65-base-pair *Sma*I-*Cla*I fragment at their 3′ border and some loci have intergenic repeat sequences (RS1, RS2, RS3) that might facilitate gene conversion (36) (Fig. 3). All complete genes sequenced have several conserved stretches that also occur in partial genes and may have roles in homologous pairing during recombination (36).

The blocks of conserved sequence in pilin genes might have additional implications for their recombination. Gonococci possess a restriction-modification system involving the enzyme *Ngo*II (18, 138), and all pilin gene copies have *Ngo*II recognition sites (GGCC) within the codons 57 to 58 and 109 to 110. Other *Ngo*II sites are present 3′ to the complete pilin gene and also in codons 39 to 40 of the complete pilin gene and of *pilS1* copy 2. Sporadic cleavage or nicking of these sites by *Ngo*II perhaps when they are undermethylated (e.g., just after replication) might provide single-stranded ends or points of exonuclease attack needed to initiate a recombinational event. It may also be significant that *Hae*III, an isoschizomer of *Ngo*II, can cleave single-stranded DNA, although this activity has not been demonstrated for *Ngo*II. Genes encoding the *Ngo*II restriction-modification system have been cloned in *E. coli* (102) and this should facilitate construction of mutants to study the possible role of *Ngo*II in pilin gene conversion.

V. OUTER MEMBRANE P.II VARIATION

A. P.II Polypeptides

Each gonococcal strain can express a series of highly related species of outer membrane protein II (P.II) (8, 23, 70, 84, 127). When expressed, P.II is a major component of the outer membrane, but P.II synthesis is neither constitutive nor essential. Gonococci can express no, one, or more than one P.II with the amount of each being the same whether or not others are coexpressed. All P.IIs share several properties, including heat modifiability (i.e., retarded migration if solubilized at 100 compared to 37°C), alkaline isoelectric points (pH 9.0 to 10.0) (10, 84), and an association with colony opacity. The various P.II species (IIa, IIb, IIc, etc.) are usually distinguished by their electrophoretic mobilities when solubilized at 100°C; M_r ranges from 29,000 to 36,000. Common and unique regions among them are detectable by peptide mapping, immunochemical probing, and amino acid or DNA sequencing (10, 42, 58, 84, 113, 125). In general, these proteins are exposed on the cell surface and are cleaved into multiple membrane-bound fragments by exogenous proteases (J. Swanson and P. van der Ley, unpublished results). Their surface-exposed portions typically contain unique antigens (8, 129).

B. Variation in P.II Expression

P.II variants constitute 0.1 to 1% of populations for typical gonococcal strains propagated in vitro (75). Diverse strains differ markedly in their frequencies of variants and in the kind they spawn. Recent studies on strain MS11 revealed a preference for expression of P.IIa among in vitro P.II$^+$ variants of several different P.II$^-$ parents (Swanson, unpublished results). P.IIa$^+$ was 2-fold more frequent than P.IIc$^+$, 10-fold more frequent than P.IIb$^+$, P.IId$^+$, P.IIf$^+$, or P.IIi$^+$, and 50- to 100-fold more frequent than P.IId$^+$, P.IIg$^+$, or P.IIh$^+$. The basis for preferred expression of particular P.II species is not known. We are reluctant to ascribe the preferential formation of specific P.II$^+$ variants to multiple copies of the P.IIa of specific P.II genes in strain MS11 because another strain, called JS3, contains three identical copies of the same P.II gene, but its variants express the corresponding polypeptide no more frequently than others specified by single-copy genes (P. van der Ley, personal communication).

A very different spectrum of P.II$^+$ revertants came from recent studies on gonococci reisolated from the urine or semen of males experimentally infected (intraurethrally) with populations that were largely (99.6%) P.II$^-$ (Swanson, unpublished results). All reisolates expressed one of five P.II species (P.IIa, P.IIc, P.IIf, P.IIh, P.IIi). P.IIc$^+$ was most frequent, and was either the dominant or sole P.II expressed by isolates from three subjects early in their infections. Later in infection, a greater variety of P.II species was expressed by the reisolates from each

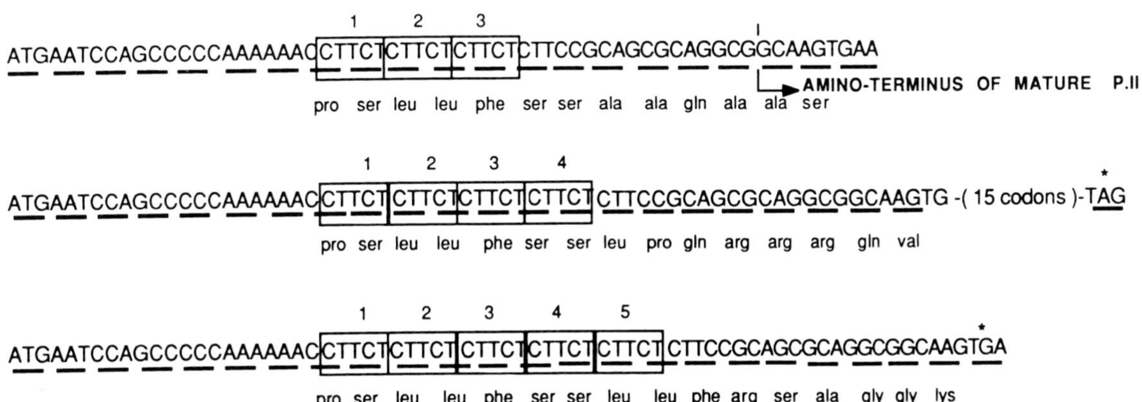

Figure 8. Changes in CTTCT repeats that correlate with on-off switches of P.II genes. An idealized P.II gene is shown with three, four, or five CTTCT repeats (boxed) between its ATG translation initiation codon and the nucleotide triplet that encodes the amino-terminal alanyl residue of the mature P.II. These two codons are in frame only with three copies of the CTTCT repeat; however, leader peptides containing the pentameric core sequence Leu-Leu-Phe-Ser-Ser are encoded by each of the three genes. Several P.II gene transcripts in strain MS11 differ in the location of guanosine residues within the stretch of cytosines immediately upstream of the CTTCT repeats (detected by primer extension reactions carried out with dideoxythymidine or dideoxycytosine to detect the first adenosine or guanosine residues, respectively, upstream of the CTTCT tract).

subject. Most reisolates from semen of one subject expressed P.IIf, whereas most from his urine were P.IIc$^+$. Although relatively few P.IIa$^+$ organisms were isolated from the experimental infections, both the P.IIc$^+$ and the P.IIf$^+$ reisolates gave rise to P.IIac$^+$ and P.IIaf$^+$ variants, respectively, after one or two passages in vitro. Currently, we cannot attribute these phenomena either to selection or to induction of specific mutational events.

Variation is known for gonococci isolated at different times or from different infected sites of the same individual or his or her sexual partners (52, 53, 150). Genital and rectal isolates are predominantly P.II$^+$, whereas those from cervices of menstruating women and from fallopian tubes are predominantly P.II$^-$ (25).

C. P.II Genes

An estimated 7 to 10 P.II genes are scattered in the MS11 chromosome with some located just downstream of pilin gene sequences. Estimates on the number of P.II genes in the gonococcal genome vary according to the strain and techniques used; Southern blotting of DNAs from strains MS11 and JS3 with oligonucleotide probes revealed 9 and 10 hybridizing fragments, respectively (113, 114; Van der Ley, personal communication), while primer extension of mRNAs indicated 7 and 9 different P.II transcripts, respectively, in these two strains (Swanson, unpublished results). Several genes from these two strains have been cloned and sequenced (15, 113, 114); the control regions upstream of the ATG translation initiation codon of each gene are nearly identical. Downstream of ATG but preceding the mature P.II coding region, each gene contains a tract of CTTCT tandem repeats (Fig. 8) varying from 6 to 27 in number (15, 115; Swanson, unpublished results). The variation in the number of these repeats is correlated with the expression of a gene turning on and off, as described below. A hydrophobic leader peptide containing the repetitive sequence Leu-Leu-Phe-Ser-Ser (LLFSS) is encoded by the tract of CTTCT repeats in all three reading frames. That this hydrophobic oligopeptide helps insertion into the membrane is known for a cloned P.II gene with transposon Tn*phoA* (73) inserted into the coding region of the mature protein (B. Belland, personal communication). P.II expressed in *E. coli*, like that in gonococcus, is sensitive to exogenous trypsin in situ, suggesting that in each organism it is transported to and oriented in the outer membranes.

The ATG initiation codon and the triplet encoding the mature protein II amino-terminal residue (alanine) are out of frame in all P.II genes that have been successfully cloned (15, 115) (Fig. 8). Small amounts of P.II are expressed by such cloned genes in *E. coli*, either because of the use of an alternative in-frame initiation codon (ATC, directly downstream from ATG; Fig. 8) or because of ribosomal slippage along the CUUCU repeat-containing message (114). P.II genes whose ATG is in frame with the mature P.II amino-terminal codon are probably "toxic" to *E. coli*. Gonococcal P.I, when expressed by recombinant clones, is probably lethal to *E. coli* (17, 34).

Homology among different cloned P.II genes is variable. Three P.II genes that encode virtually the identical mature P.II have been cloned from a P.II$^-$

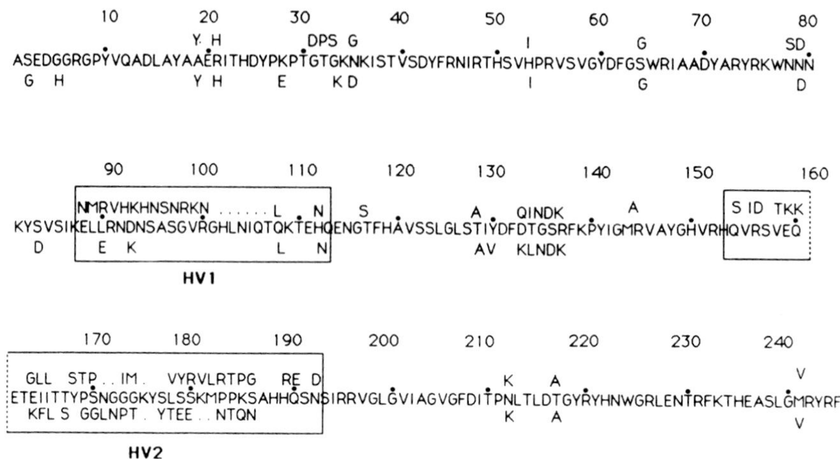

Figure 9. Amino acid sequence comparisons of cloned P.II genes. The amino acid sequences of mature polypeptides encoded by three cloned P.II genes from two different strains are shown. The entire polypeptide sequence for pEX6 from strain JS3 (Van der Ley, personal communication) is shown. Gonococci of that strain have three distinct genes that encode identical P.II moieties. Amino acid differences between this P.II and those encoded by two genes cloned from strain MS11 (V28 and V0 [113]) are noted above and below pEX6, respectively. Their heterologies are concentrated in two hypervariable regions (boxed, HV1, HV2), but many sequence differences occur outside these domains (e.g., 130 to 140).

(predominantly) population in strain JS3. One gene is flanked by unique sequences, while the others are flanked by a different set of nearly identical sequences, suggesting a recent gene duplication (Van der Ley, personal communication). The P.II encoded by the three virtually identical genes is not expressed at correspondingly high frequency by P.II^{+} variants in this strain.

One P.II gene locus was cloned from two strain MS11 variants and displayed many sequence changes (115). Differences were mainly in the 3′ two-thirds of the gene, within two discrete hypervariable regions (HV1, HV2) and the intervening semivariable domain (Fig. 9). The differences may reflect a recombinational insertion of new sequences into the locus (115). The differences between genes from strains MS11 and JS3 are also concentrated in these hypervariable and semivariable regions (Fig. 9) (Van der Ley, personal communication).

D. P.II Expression Switches Correlate with DNA Sequence Changes

Out-of-frame P.II genes are turned on by frameshifts due to gain or loss of CTTCT repeats (78, 115). A transcript appears to be made from each gene, independent of whether a corresponding mature polypeptide is produced (115). mRNA transcripts of switched-off genes are generally truncated (Van der Ley, personal communication), but their 5′ ends can be detected by primer extension or Northern blot hybridization of RNAs from both P.II^{+} and P.II^{-} gonococci. The mRNA transcript of an expressed P.II gene is full length (Van der Ley, personal communication). Of the seven P.II transcripts identified in strain MS11, several can be distinguished by sequence differences in the region just upstream from the CTTCT repeats (J. Swanson and K. Robbins, unpublished results; see legend to Fig. 8).

The increase or decrease in number of CTTCT repeats usually involves just one pentanucleotide, but changes of up to 14 CTTCT repeats can also occur (113, 114; Swanson, unpublished results). Most on-off switches correlate with a change in only one gene. Occasional "silent" CTTCT repeat changes occur in unexpressed P.II genes taking them from one out-of-frame off condition to another.

E. On-Off Changes of P.II Genes Are *recA* Independent

Little is known about the mechanism(s) responsible for changes in number of the CTTCT repeats in P.II genes, but recombination might be involved a priori. However, Δ*recA* gonococci undergo P.II on-off switching that is accompanied by mRNA CTTCT repeat changes at rates similar to wild-type recombination-proficient (*recA*$^{+}$) gonococci. Further, CTTCT repeat changes occur when cloned P.II genes (marked with Tn*phoA* inserts) are grown in *recA E. coli* (Belland, personal communication). These observations show that such events are *recA* independent, equivalent to other illegitimate events involving short repeated sequences (S. D. Ehrlich, this volume).

F. Sequence Exchange among P.II Genes

Blot hybridization with specific oligonucleotide probes shows that P.II gene sequences can be exchanged between the various P.II gene copies (104, 115). This is thought to be a nonreciprocal exchange analogous to that between pilin gene sequences. Consistent, but less conclusive, is the observation of more P.II protein variants than of P.II gene copies in strain MS11. Similar phenomena seem not to occur in Δ*recA* gonococci (Swanson, unpublished results). The frequencies of sequence exchange between P.II genes are estimated to be several orders of magnitude lower than on-off switching by change in CTTCT repeat number.

VI. DISCUSSION: THE NATURE OF PILUS/P.II REGULATION AND VARIATION

Gonococci from human infections display a broader array of variant pilins than during serial passage in vitro (127), and different P.II species are favored in vivo than in vitro (Swanson, unpublished results). Pilus and P.II changes occur early in gonorrheal infection and may reflect the need for expression of particular pilins or P.IIs in diverse mucosal niches for maximal colonization (27, 50, 135). These variations most likely reflect gonococcal responses to millennia of selective pressures by host immune systems.

Pilus phase variation seems to have little or no relevance to gonorrheal infections. Primary gonococcal isolates are invariably piliated (111), whereas pilus$^-$ nonpiliated variants arise during in vitro cultivation, where they have a clear growth advantage. Pilus$^-$ phenotypes result from changes in the pilin structural gene: 5′ deletions (105, 130) and frameshift, nonsense, and missense mutations (6, 67, 132). The rates of pilus$^+$ to pilus$^-$ transitions are at least 30-fold lower than those of pilus$^-$ to pilus$^+$ transition (correction of pilin mutations by intragenic recombination) (67). Pilus$^-$ variants due to 5′ deletions within the complete pilin gene are probably products of aberrant recombinational events between pilin genes. They cannot spawn pilus$^+$ revertants and they are monopathogenic in males. These nonreverting, avirulent pilus$^-$ gonococci probably do not survive long outside of the laboratory environment. Analogous avirulent variants also arise in many other pathogenic microorganisms during their laboratory passage. Although phase variations to the pilus$^-$ condition may not be directly relevant to gonococcal pathogenesis infection, the error-prone nature of recombination and mutation underlying such transitions serves as a model for the types of genetic mechanisms utilized by gonococci to generate diversity. The formation of frameshift and nonsense mutations in pilin genes and the consequent pilus phase transitions constitute a stochastic process, not programmed variation.

The numerous stretches of conserved DNA sequence allow use of different partial pilin genes and various extents of a given partial gene in gene conversion events. This in turn permits expression of an array of chimeric pilins far larger than the basic repertoire of 12 to 20 partial pilin genes. The plasticity of pilin expression is further enhanced by possibilities of crossing over within codon triplets. Two instances of pilin amino acid substitutions due to single base changes in the complete pilin gene have also been documented in Δ*recA* gonococci (67). Both mutations were due to GC-to-AT transitions, attributable to the spontaneous deamination of cytosine to uracil or of 5-methylcytosine to thymine. Classical point mutations therefore probably also contribute to pilin diversification. Although it is not known whether gonococci possess an error-prone repair system analogous to SOS of *E. coli*, gonococci are known to be mutable by mutagens such as those that cause mispairing (14). We have referred to gonococcal pilus variation system as "pilin roulette" to emphasize the great variety of active pilin gene sequences that are possible and the ease of transitions among them (6).

A. Programmed Variation?

The preferential expression of certain P.II proteins suggests the possibility of program or order for the expression of specific family members. Alternatively, the observed pattern might reflect preferred sites for recombination or bacterial growth rate selection. On-off control of P.II expression is attributable to *recA*-independent sequence changes in the number of CTTCT repeats of the corresponding gene, perhaps during DNA replication. The stabilities of the on-off switches in different P.II genes vary widely; nearly half of the clones started by some P.II$^+$ colonies are P.II$^-$, whereas in other P.II$^+$ variants, the P.II$^-$ cells are typically 100-fold less abundant (Swanson, unpublished results). The diversity due to on-off switching of individual P.II genes is enhanced by sequence exchange among them.

B. Control of Analogous Pili in Other Bacteria

Analogous pilus variability is found in other bacterial pathogens. Pili of gonococci belong to a family of type 4 fimbriae (88) found also on *Neisseria meningitidis*, *Moraxella bovis*, *Pseudomonas aeruginosa*, and *Bacteroides nodosus* (29, 47, 55, 74, 100, 119). In each member, the pilins have short leader peptides, a highly conserved amino-terminal stretch of 30 or more amino acids, *N*-methyl phenylalanine at the N terminus of the mature pilin, and a disulfide loop region between two cysteines. Piliation correlates with virulence for all these organisms, with the possible exception of the opportunistic *P. aeruginosa*. Despite the apparent divergence of these disparate pilin genes from a common ancestor, they exhibit several differences. Intrastrain pilus antigenic variation and pilus$^+$ to pilus$^-$ phase transition have been documented only in pathogenic neisseriae and in moraxella. The mechanisms used by *N. meningitidis* are probably identical to those used by *N. gonorrhoeae* (31, 59, 60). In *M. bovis*, in contrast, there are only two pilin types. The change from one to the other results from inversion of a DNA element carrying unique promoter and the 5′ sequences of two truncated, variable, and divergently oriented pilin genes (C. Marrs, personal communication), similar to the inversion that controls tail fiber specificity of bacteriophage Mu (35, 49, 89). There is more pilus phase variation in *M. bovis* than can be ascribed to simple inversion, suggesting that such inversion may be error prone (Marrs, personal communication). Interestingly, the expression of pili in *Neisseria* spp. and *Moraxella* spp. correlates with competence for DNA transformation (11, 31, 110). The lack of both pilus antigenic variation and phase transition in *Bacteroides* and *Pseudomonas* correlates with their having only one pilin gene copy in their genomes (2, 119).

C. Analogies in Borrelia and Trypanosomes

A resemblance to gonococcal surface protein variation is seen in *Borrelia hermsii* and in the surface glycoproteins of trypanosomes (4, 24). Infection by these organisms involves waves of parasitemia by serotype variants whose sequential expressions follow a somewhat predictable but inexact order (4, 21, 81, 82, 97, 116, 139). One obvious distinction between gonococci and the two hemophiles is that variable genes of borrelia and trypanosomes reside on linear chromosomes or plasmids (4, 22, 90, 140). Such linear genomes simplify some of the difficulties in explaining gene conversion on circular chromosomes, since the double-stranded ends of the linear molecules are recombinogenic, and only a single crossover event may be required for productive recombination. These systems are discussed elsewhere (A. Barbour, this volume; J. E. Donelson, this volume).

D. Chicken Immunoglobulin Gene Diversity

Probably the clearest analogy to events in gonococcal pilin variation is the gene conversion events that diversify the variable region of the lambda light-chain locus in birds (95, 137). Similarities between these two diversification processes include the relative genomic organization of the partial variable gene copies (pseudogenes) and the expression locus, the apparent unidirectional character of sequence exchange, and the fact that repeated exchange events with different segments of donor gene copies generate a high level of combinatorial diversity. In both systems, one finds nucleotide sequences in their expression loci that have no counterparts in the partial gene repertoire; this suggests that error-prone or inexact mismatch correction during the gene conversion event may itself be a significant source of diversification. The gonococcal and bird immunoglobulin systems clearly differ from that which generates immunoglobulin diversity in mammals, where alternative joining of multiple families of variable, diversity, and joining elements and subsequent somatic mutations account for diversities (83; S. Lutzker and F. W. Alt, this volume).

VII. CONCLUSIONS

The pilin and P.II gene systems of gonococci are remarkable for the broad range of structurally variant molecules they generate. Variability in both systems is due in large part to recombination between highly related sequences that had probably arisen long ago by gene duplication. In the case of pilin, only one gene can be expressed. It mutates by acquiring a new polynucleotide sequence from a donating partial gene without change in the donor segment. There are more than 12 such partial genes in typical gonococcal chromosomes, each serving as a reservoir of new sequences for the expressed complete gene. Although this process sometimes creates a complete gene that does not express a functional pilin, such pilus$^-$ mutants revert easily to pilus$^+$ by additional recombination events with partial pilin donors, thereby leading to the formation of functional pili. The recombination process that causes pilin diversity appears to be nonreciprocal. The mechanisms of this nonreciprocity, of the control of this process, and of subsequent pilus assembly are unknown. Nonpilin proteins that may participate in processing and translocating pilin molecules to the gonococcus surface have not been identified. The possibility that proteins other than pilins may be the actual adhesion-promoting components of pili as in the enterobacterial pilus system (72) has not been well tested.

Diversity among the P.II outer membrane proteins involves recombinational exchanges among the approximately 7 to 10 gene copies in the multigene P.II family. The on-off switch involves a change in the number of iterated CTTCT repeats at the 5′ end of the P.II gene, thereby changing the reading frame of rest of the gene. It is *recA* independent and may involve slippage during replication or use of an alternative transcription initiation codon. Each P.II gene is turned on or off at frequencies that vary among members of the P.II multigene family. Various members are often coexpressed, and typically produce polypeptides that are structurally and antigenically distinct. The various P.II genes may have resulted from recent gene duplication or from extensive conversion from one or a few common donor genes.

Both pili and P.II strongly affect the surface features and properties of gonococci. The genetic systems that have permitted variation in pilin and P.II molecules provide insight into the evolutionary advantage of variability for gonococcus confronting its human hosts.

Acknowledgments. J.M.K. was supported by Public Health Service grants AI-10615 and AI-19469 from the National Institutes of Health. The studies of J.M.K. were carried out at the Laboratory of Bacteriology and Immunology, The Rockefeller University, New York, N.Y.

We thank the following individuals for contributing unpublished observations: Fred Sparling, Thomas Meyer, Jon Saunders, John Heckels, Milan Blake, Virginia Clark, Peter van der Ley, and Bob Belland. We thank Milan Blake, Emil Gotschlich, Peter van der Ley, and Bob Belland for their helpful conversation and suggestions. We also thank Susan Smaus, Bob Evans, and Gary Hettrick (Rocky Mountain Laboratories) for help in the preparation of the manuscript.

LITERATURE CITED

1. **Abraham, J. M., C. S. Freitag, J. R. Clements, and B. I. Eisenstein.** 1985. An invertible element of DNA controls phase variation of type 1 fimbriae in *Escherichia coli. Proc. Natl. Acad. Sci. USA* **82:**5724–5727.
2. **Anderson, B. J., C. L. Kristo, J. R. Egerton, and J. S. Mattick.** 1986. Variation in the structural subunit and basal protein antigens of *Bacteriodes nodosus* fimbriae. *J. Bacteriol.* **166:** 453–460.
3. **Apicella, M. A., and N. C. Gagliardi.** 1979. Antigenic heterogeneity of the non-serogroup antigen structure of *Neisseria gonorrhoeae* lipopolysaccharide. *Infect. Immun.* **26:**870–874.
4. **Barbour, A. G.** 1988. Molecular mechanisms of antigenic variation in *Borrelia. UCLA Symp. Mol. Cell. Biol. New Ser.* **20:**365–377.
5. **Baron, E. S., and A. K. Saz.** 1978. Genetic transformation of piliation and virulence into *Neisseria gonorrhoeae* T4. *J. Bacteriol.* **133:**972–986.
6. **Bergström, S., K. Robbins, J. M. Koomey, and J. Swanson.** 1986. Piliation control mechanisms in *Neisseria gonorrhoeae. Proc. Natl. Acad. Sci. USA* **83:**3890–3894.
7. **Biswas, G. D., T. Sox, E. Blackman, and P. F. Sparling.** 1977. Factors affecting genetic transformation of *Neisseria gonorrhoeae. J. Bacteriol.* **129:**983–992.
8. **Black, W. J., R. S. Schwalbe, I. Nachamkin, and J. G. Cannon.** 1984. Characterization of *Neisseria gonorrhoeae* protein II phase variation by use of monoclonal antibodies. *Infect. Immun.* **45:**453–457.
9. **Blackwell, T. K., M. W. Moore, G. D. Yancopoulos, H. Suh, S. Lutzker, E. Selsing, and F. W. Alt.** 1986. Recombination between immunoglobulin variable region gene segments is enhanced by transcription. *Nature* (London) **324:**585–589.
10. **Blake, M. S., and E. C. Gotschlich.** 1984. Purification and partial characterization of the opacity-associated proteins of *Neisseria gonorrhoeae. J. Exp. Med.* **159:**452–462.
11. **Bovre, K., and L. O. Froholm.** 1972. Competence in genetic transformation related to colony type and fimbriation in three species of *Moraxella. Acta Pathol. Microbiol. Scand. Sect. B* **80:**649–659.
12. **Brinton, C. C., J. Bryan, J.-A. Dillon, N. Guerina, L. J. Jacobson, A. Labik, S. Lee, A. Levine, S. Lim, J. McMichael, S. Polen, K. Rogers, A. C.-C. To, and S. C.-M. To.** 1978. Uses of pili in gonorrhoea control: role of bacterial pili in disease, purification and properties of gonococcal pili, and progress in the development of a gonococcal pilus vaccine for gonorrhea, p. 155–178. *In* G. F. Brooks, E. C. Gotschlich, K. K. Holmes, W. D. Sawyer, and F. E. Young (ed.), *Immunobiology of Neisseria gonorrhoeae*. American Society for Microbiology, Washington, D.C.
13. **Buchanan, T. M.** 1975. Antigenic heterogeneity of gonococcal pili. *J. Exp. Med.* **141:**1470–1475.
14. **Campbell, L. A., and R. E. Yasbin.** 1984. Mutagenesis of *Neisseria gonorrhoeae*: absence of error-prone repair. *J. Bacteriol.* **160:**288–293.
15. **Cannon, J. G.** 1988. Genetics of protein II of *Neisseria gonorrhoeae. UCLA Symp. Mol. Cell. Biol. New Ser.* **20:** 75–83.
16. **Cannon, J. G., W. Black, I. Nachamkin, and P. Stewart.** 1984. Monoclonal antibody that recognizes an outer membrane antigen common to pathogenic *Neisseria* species but not to most nonpathogenic *Neisseria* species. *Infect. Immun.* **43:**994–999.
17. **Carbonetti, N. H., and P. F. Sparling.** 1987. Molecular cloning and characterization of the structural gene for protein I, the major outer membrane protein of *Neisseria gonorrhoeae. Proc. Natl. Acad. Sci. USA* **84:**9084–9088.
18. **Clanton, D. J., W. S. Riggsby, and R. V. Miller.** 1979. *Ngo*II, a restriction endonuclease from *Neisseria gonorrhoeae. J. Bacteriol.* **137:**1299–1307.
19. **Clark, A. J.** 1973. Recombination-deficient mutants of *E. coli* and other bacteria. *Annu. Rev. Genet.* **7:**67–86.
20. **Claverys, J. P., and S. A. Lacks.** 1986. Heteroduplex deoxyribonucleic acid base mismatch repair. *Microbiol. Rev.* **50:** 133–165.
21. **Coffey, E. M., and W. C. Eveland.** 1967. Experimental relapsing fever initiated by *Borrelia hermsii*. II. Sequential appearance of major serotypes in the rat. *J. Infect. Dis.* **117:**29–34.
22. **De Lange, T., and P. Borst.** 1982. Genomic environment of expression linked extra copies of genes for surface antigens

Chapter 35

DNA Rearrangements and Antigenic Variation in African Trypanosomes

JOHN E. DONELSON

I. INTRODUCTION

African trypanosomes are unicellular protozoans that survive in the bloodstream of their mammalian hosts because their genome undergoes frequent gene rearrangements. These gene rearrangements enable the trypanosomes to evade the immune system of their hosts by periodically switching their surface coat. The way in which this switch is accomplished at the molecular level is now partially understood as the result of much effort by several research groups during the past 10 years.

The trypanosome genome contains as many as 1,000 different genes encoding variant surface glycoproteins (VSGs). Usually only one VSG gene is expressed at any time, and about 10^7 copies of its VSG product completely cover the outer trypanosome membrane. The major function of the VSG is to serve as a barrier to protect other constituents of this outer membrane from the assault of the host immune system. Antibodies are raised against the VSG, but the parasite population manages to escape total destruction because individual parasites occasionally switch spontaneously from the expression of one VSG to another, a process called **antigenic variation.** New antibodies must be raised against the VSG of the

John E. Donelson ■ Department of Biochemistry, University of Iowa, Iowa City, Iowa 52442.

switched parasite and its descendants, enabling the trypanosome population as a whole to stay "one step ahead" of the host immune response.

The trypanosome DNA rearrangements maneuver VSG genes into and out of chromosomal locations called "expression sites" where transcription may occur. These expression sites are always adjacent to chromosomal telomeres. The switch from transcription of one VSG gene to another occurs when a new VSG gene is activated in an expression site. The process is complicated by the fact that several, perhaps many, potential expression sites exist in the genome, yet only one is normally activated at any time.

This extraordinary and seemingly rather chaotic mechanism for sequential VSG expression has spawned many review articles in recent years (16, 18, 20, 21, 35, 39, 57, 128). Most of these articles illustrate both the intricacies required for this parasite-host interaction and the power of molecular biology techniques to decipher its details. Yet many aspects of antigenic variation remain poorly understood.

This chapter will first summarize the relevant biology of trypanosomes and the biochemistry of the VSGs. It will then describe the different DNA rearrangements associated with the switch from one VSG to another. Finally, it will emphasize those features of antigenic variation that are not understood and briefly describe another distinctive feature of trypanosomes called discontinuous transcription.

II. THE BIOLOGY OF AFRICAN TRYPANOSOMES

African trypanosomes spend part of their life cycle in their insect vector, the tsetse fly, and the remainder of the life cycle in the bloodstream of their mammalian host. They can inhabit one of several species of tsetse flies in the genus *Glossina* and usually occur wherever these insects are located. This includes an area of Africa south of the Sahara Desert that is approximately the size of the United States and supports a population of over 200 million people.

African trypanosomes belong to the genus *Trypanosoma* within the order Kinetoplastida. Their species classification is based on morphology, geographical location, hosts they infect and, occasionally, their virulence. *Trypanosoma brucei brucei* has been used for most experimental investigations because it is easy to maintain in laboratory animals and does not survive in human serum, reportedly because it is lysed by the high-density lipoprotein fraction (112). The two subspecies of *T. brucei brucei* that infect humans are *Trypanosoma brucei rhodesiense* (acute infections) and *Trypanosoma brucei gambiense* (chronic infections). Two other important species that have a tremendous impact on the African cattle industry are *Trypanosoma congolense* and *Trypanosoma vivax*. In Africa the species that infect domestic animals are much more prevalent than those that infect humans and have a far greater impact. This has not always been the case. It has been estimated that in the late 1890s two-thirds of the population living near Lake Victoria in Uganda, about 200,000 people, died of the disease. Today the threat of human epidemics on this scale is reduced by constant surveillance and drug treatment. However, the disease continues to be a severe problem for domestic livestock in endemic areas because drug cures do not confer immunity to reinfection.

A. Trypanosome Life Cycle

Unlike many parasites, trypanosomes have a relatively simple life cycle (135). When a tsetse fly bites an infected mammal, trypanosomes can be ingested with the blood meal (Fig. 1). Once in the midgut of the fly, the organisms change from anaerobic to aerobic growth, lose their surface coat containing the VSG, and begin to multiply. After about 3 weeks in the midgut, the parasites move to the salivary glands or proboscis (depending on the trypanosome species) and differentiate into the metacyclic form. Metacyclic trypanosomes reacquire a VSG

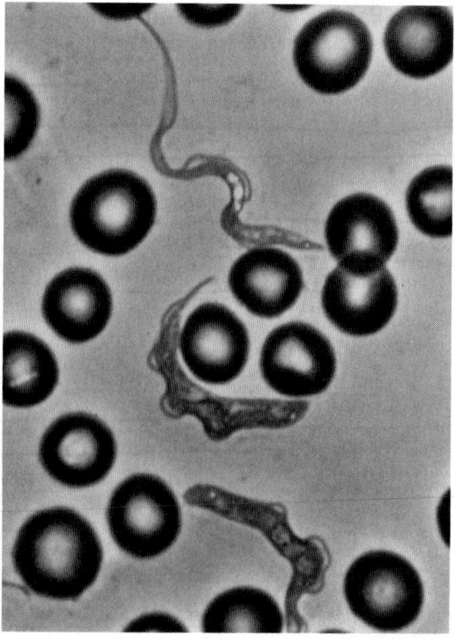

Figure 1. Photograph of long, slender and stumpy forms of bloodstream *Trypanosoma brucei rhodesiense* among erythrocytes.

coat but do not divide. An infected tsetse fly can harbor 20,000 to 50,000 metacyclic trypanosomes, only 1 of which is sufficient to potentially initiate the mammalian infection if transmitted during the fly bite.

Shortly after entering the bloodstream, trypanosomes begin to multiply by binary fission and undergo a variety of additional morphological and metabolic changes (returning to anaerobic glycolysis as the main source of ATP). They continue, however, to express the metacyclic VSGs for about 5 days after the infection (58, 59). Between days 5 and 7 of infection the parasites switch from the expression of metacyclic VSGs to the first bloodstream VSGs, one of which is often the same VSG that was expressed by the trypanosomes ingested by the fly (59). In the bloodstream, trypanosomes divide every 5 to 10 h, depending on the trypanosome isolate and the host species. Bloodstream forms of the parasite persist in the bloodstream and eventually invade the central nervous system. The infected individual ultimately becomes comatose and dies, hence the common name for African trypanosomiasis, sleeping sickness.

Distinct developmental forms of the parasite also occur in the bloodstream. The long, slender forms actively divide and undergo antigenic variation. The short, stumpy forms do not divide but have a more developed mitochondrion and are thought to be the form that infects the insect. The life cycle is completed when a tsetse fly takes up the bloodstream forms while feeding on an infected mammal.

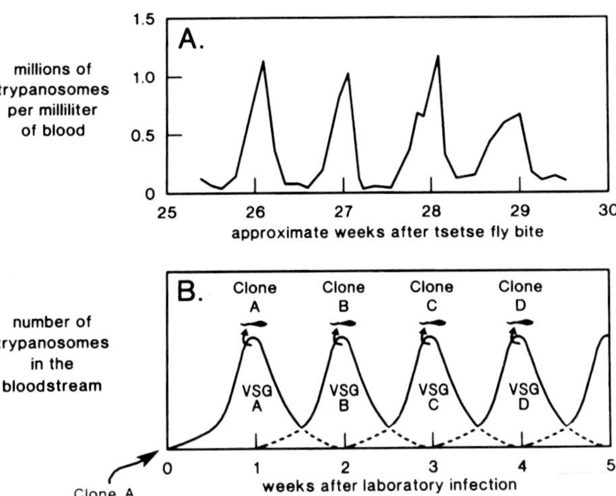

Figure 2. (A) Demonstration of the successive waves of parasitemia during a *Trypanosoma brucei rhodesiense* infection of a human patient (redrawn from reference 113). (B) An example of a carefully controlled infection of *T. brucei rhodesiense* in an immunocompetent laboratory animal. A single trypanosome expressing VSG_A (clone A) is injected into the blood and gives rise to the first peak of parasitemia, in which all of the parasites are expressing VSG_A. The host immune system kills most of the parasites during the next few days, but one or more trypanosomes switches to the expression of VSG_B, which is not recognized by the anti-VSG_A antibodies. These trypanosomes begin the next wave of parasitemia. VSG_B then elicits a second set of antibodies to start the process again. From each peak of parasitemia, individual trypanosomes can be cloned to investigate the VSG structure and the molecular mechanism of antigenic variation. Since the switch to a new VSG occurs spontaneously and individual trypanosomes can have different growth rates, the synchrony of the parasitemia peaks is lost later in the infection.

B. The Phenomenon of Antigenic Variation

During the early investigations of trypanosomiasis at the beginning of this century, it was noticed that the number of trypanosomes in the bloodstream of an infected individual fluctuated dramatically with time. In 1910, Ross and Thomson (113) showed that one peak of parasitemia in the blood of an infected patient was followed a few days later by a low level of parasites and then by another peak about 2 weeks later (Fig. 2A). Demonstrating remarkable insight for the time, they and other investigators speculated that a new wave of parasitemia was composed of trypanosomes that had somehow escaped the action of antibodies directed against the previous parasites. It would be nearly 60 years before the relationship between trypanosomes and the immune system would be better understood.

By the mid-1970s, it had been shown that the trypanosome, including its flagellum, was covered by a single major protein (135). Then, in 1976, Bridgen et al. (23) prepared four cloned trypanosome populations from a single rabbit infected with a cloned strain of *T. brucei brucei* and isolated the different VSGs from these four populations. N-terminal analyses of the four VSGs demonstrated that the first 25 to 30 amino acids of each were remarkably different and not related by potential frameshift changes or point mutations. This provided a more detailed explanation of the antigenic variation observed six decades earlier.

As depicted in Fig. 2B, each peak of parasitemia early in the infection of a laboratory animal by a single trypanosome clone is composed of a population of trypanosomes expressing the same immunologically distinct VSG. About 1 week after the initial infection virtually all of the trypanosomes are expressing VSG_A, the same VSG as in the original organism. By this time the immune system has mounted a response against VSG_A that kills most of the parasites. Before all of the parasites are eliminated, however, one or more switch to the expression of a new, immunologically distinct VSG, VSG_B. The descendants of these switched organisms give rise to the second peak of parasitemia. The immune system must raise new antibodies against this new VSG.

Before all of these parasites are removed, at least one has again switched, to VSG_C, which leads to a third parasitemia, and so on. In one often-cited case a trypanosome infection of an experimental rabbit resulted in the successive expression of over 100 different antigenic types (28).

It has been estimated that laboratory strains of *T. brucei brucei* spontaneously switch VSGs at a rate of 10^{-6} to 10^{-7} per division (77, 94). In a chronically infected large animal this switch rate means that many variants, all derived from the same ancestor, can be in the bloodstream at any time. In the laboratory large numbers of trypanosomes can be obtained from acutely infected animals such as mice or rats. Trypanosomes can also be grown in a variety of culture media, where they mimic the tsetse fly stages without a VSG coat (called procyclic forms), or in a more defined culture medium containing feeder cells or a reducing agent and pyruvate, where they retain the surface coat (7, 63). Immunofluorescent assays and cell sorting techniques have shown that VSG switches can occur in the latter culture conditions, indicating that the immune system of the host does not induce the switch (43, 137). It does, however, select new variants that will give rise to the next parasitemia, since it does not immediately recognize their VSGs.

The order of appearance of new bloodstream variants remains somewhat controversial. In some cases, new VSGs appear more or less at random (25), although some VSGs are detected more frequently than others. In *Trypanosoma equiperdum*, which is transmitted venereally in horses (without an insect vector) a subset of VSGs appears early in the infection, while other VSGs are more likely to occur later in the infection (87). In still other cases, a given VSG has a propensity to follow another (94) or to appear at a specific time after infection (79, 124). It is possible that the genomic location of a VSG gene influences the frequency with which it is expressed and whether it is likely to follow another specific VSG. In a few rare instances, trypanosomes have been detected that are expressing two VSG genes, or at least two VSGs, simultaneously (6, 31, 47). In one of these cases, trypanosomes were trapped in the transition between the expression of a metacyclic VSG and a bloodstream VSG (47).

C. Trypanosome Nomenclature

A standardized nomenclature is now used to identify the different groups of parasites being investigated in the laboratory (142). A brief summary of these terms is as follows. A trypanosome **sample** refers to a trypanosome population collected on a single occasion from a naturally infected insect, animal, or patient. A **primary isolate** is obtained when a sample is introduced into an experimental animal or established in culture. This primary isolate may be a mixture of several species or subspecies that were present in the original sample. Trypanosome **clones** are obtained from an isolate by serial dilution to individual organisms. A clone may be expanded by growing the parasites in immunosuppressed laboratory animals, but the resultant population must be monitored to ensure that a switch to another VSG does not occur during the expansion. Such expanded clones are often cryopreserved as **stabilates** for later use. A trypanosome **serodeme** results when a trypanosome clone is used to establish a chronic infection in an experimental animal. All of the trypanosomes that are derived from this serodeme presumably have the potential to express the same VSGs. Serodemes are named according to the institute at which they were generated, e.g., WRATaR 1 is the Walter Reed Army (Research Institute) Trypanozoon antigen Repertoire 1 serodeme. Trypanosome clones from this serodeme are **variant antigen types** (**VATs**) and are identified by sequential numbers; e.g., WRATat 1.1 and WRATat 1.2 are cloned variant antigen types within the WRATaR 1 serodeme.

III. VARIANT SURFACE GLYCOPROTEINS

The approximately 10^7 VSG molecules on the surface of a trypanosome make up about 5% of the total protein of the organism (126, 127). For a surface, membrane-attached protein this is a remarkably large percentage of the total protein and greatly facilitates its purification (32). Likewise, about 5% of the polyadenylated RNA of the parasite codes for the VSG, which simplifies detection of VSG cDNA in a trypanosome cDNA library by differential first-strand cDNA hybridization or immunoscreening.

The complete or partial nucleotide sequences of the cDNAs for about 20 different VSGs from at least five different trypanosome species have been determined (see reference 39 for a partial summary). These cDNA sequences demonstrate that nascent VSGs contain about 500 amino acids, including an N-terminal signal peptide of 20 to 30 amino acids. In addition, the nucleotide comparisons reveal that the VSGs of most trypanosome species can be placed into two groups based on similarities among the coding regions for the last 50 amino acids of the nascent VSGs, a region that includes a C-terminal hydrophobic tail of 17 or 23 amino acids that is also removed during the maturation of the nascent VSG. In addition, the 80 to 100 nucleotides in the 3′ nontranslated region of the VSG mRNAs display extensive similarities. These sequence similarities at the 3′ ends of the VSG genes often form the downstream boundary of a transposed VSG gene (see below).

The association of the mature VSG with the trypanosome membrane has several interesting features. The protein component of the VSG lies entirely outside the cell membrane and is attached to constituents of the membrane though the alpha-carboxyl of the C-terminal amino acid (serine in one similarity grouping and aspartic acid in the other). An ethanolamine residue covalently links the C terminus of the VSG to several carbohydrate moieties of unknown structure, which in turn are linked to phosphotidylinositol (4, 5, 65, 66). In *T. brucei brucei* the fatty acid groups of this glycolipid are myristic acid ($C_{14:0}$) (49). A similar phosphotidylinositól linkage to surface proteins has recently been identified in unrelated cells, suggesting that it may be a much more common linkage for membrane-associated proteins than was previously appreciated (61, 117).

The conventional VSG purification procedure (32) activates a phospholipase C activity that cleaves the VSG from its lipid attachment. Metal ions inhibit the enzyme, facilitating the preparation of the membrane form of the VSG (29, 30, 62). This enzyme may play a central role in the turnover of the VSG during an antigenic switch.

VSGs associate as dimers of identical molecules, both on the intact organism and in the purified soluble form (3). In addition, a protease-sensitive site is present about two-thirds of the way into all VSGs examined (69). The 0.6-nm-resolution X-ray studies of this N-terminal two-thirds portion of two unrelated VSGs reveal that they have very similar overall structures (52, 91). Their primary sequences are quite different, yet both structures have similar rodlike α-helical bundles composed of several α-helices extending back and forth. Such tertiary structures suggest that the VSGs are arrayed on the trypanosome surface as closely packed dimeric rods in which relatively few amino acid residues are exposed to the outside. This leads to the prediction that changing just a few specific amino acids might be sufficient to generate antigenically distinct VSGs, a prediction consistent with the observation that VSGs with substantial identity throughout their entire primary sequences can be different immunologically (8, 108).

The dimeric nature of VSGs, their generally similar shapes, and their presumed close packing on the surface suggest that the order of appearance of VSGs during antigenic variation might be influenced by the packing compatibility of different VSGs on the surface (79). For example, if new VSG molecules produced after a switch cannot compatibly associate with existing VSG molecules during the transition period, the surface coat could be weakened or disrupted. This might be deleterious to the trypanosome and select against that particular sequential switch. Alternatively, if the new and old VSGs can form stable heterodimers or can associate closely together, the transition of a new surface coat would proceed smoothly for the organism.

IV. CHROMOSOMAL DISTRIBUTION OF VSG GENES

A. Genome Size

One of the ironies in the study of the genetic basis for trypanosome antigenic variation is that it has not used the conventional tools of genetics. For example, no mutants that affect the process of antigenic variation exist and, in fact, no nuclear mutations at all have been characterized. Indeed, very little is known about the genetics of trypanosomes except that DNA exchange between different cells may occur during an insect stage (68, 123). As is the case for other lower eucaryotes, the number of trypanosome chromosomes cannot be determined by cytological staining because they do not condense sufficiently during mitosis. Furthermore, without mutations or genetic crosses, linkage groups cannot be determined. Even the ploidy of the parasite remains somewhat uncertain, although most evidence suggests that bloodstream trypanosomes are diploid in their housekeeping genes while metacyclic parasites are haploid (54, 68, 123, 136, 147). DNA kinetic (C_0t) analysis and colorimetric measurements indicate that the haploid trypanosome genome is about 3.7×10^7 base pairs (bp) in extent, while the nuclear DNA content of bloodstream parasites is twice that (22). This haploid genome size is about 10 times that of *Escherichia coli*, 2 to 3 times that of haploid *Saccharomyces cerevisiae*, and 1/100 that of the human genome.

Curiously, the recent history of trypanosomes may influence their nuclear DNA content. For example, when a mixture of two trypanosome stocks was used to coinfect tsetse flies, "hybrid" trypanosomes emerging from the flies were found to have about 1.5 times as much nuclear DNA as did the parents (139). This increased DNA content was stable for many generations when the parasites were maintained in culture and for at least a limited amount of time during growth in the bloodstream. The hybrid trypanosomes appeared to have twice the number of minichromosomes (see next section) as did the par-

ents, although this increase accounted for only about 20% of the additional DNA. The mechanism by which the trypanosomes acquire, maintain, and eventually lose this additional DNA is not clear, but it may play a role in the assortment and distribution of VSG genes among trypanosome populations in the wild.

B. Chromosome Number

Although traditional approaches have not been useful in determining the number of chromosomes, it is now clear that the trypanosome genome is distributed among a large number of DNA molecules, some of which are quite small by eucaryotic standards. The first evidence for these small chromosomes was the identification on sucrose gradients of a specific VSG gene on an 80-kilobase (kb) "minichromosome" (141). Pulsed field gradient gel electrophoresis (54, 115, 133) resolves trypanosome nuclear DNA into four general size classes (Fig. 3A). There are about 100 minichromosomes of 50 to 100 kb each, 5 to 7 chromosomes of 200 to 700 kb each, several chromosomes in the 2,000-kb range, and some molecules that do not effectively enter the gel, presumably because they either are too large or contain an unusual structure (133). Furthermore, trypanosome clones from the same serodeme sometimes have different sizes and numbers of resolved chromosomes (10). This indicates that trypanosome chromosomes

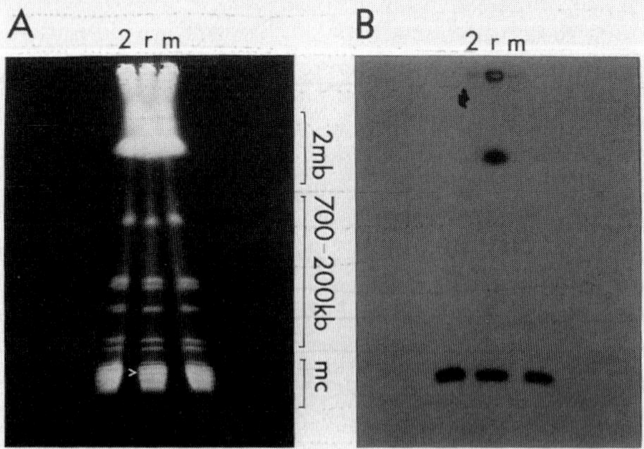

Figure 3. (A) Ethidium bromide stain of a pulsed field gradient agarose gel containing the chromosomes of three closely related *T. brucei brucei* clones (2, r, and m). Minichromosomes of about 100 kb (mc), 200- to 700-kb chromosomes, and 2,000-kb (2-megabase [2mb]) chromosomes can be observed. Chromosomes of still larger size remain in the gel slots or do not resolve. (B) Autoradiogram of a Southern blot of the same gel probed with VSG_r cDNA. The probe hybridizes to the basic copy VSG_r gene on a minichromosome in all three genomes and to an ELC_r gene on a 2-megabase chromosome in the genome of trypanosome clone r. Thus, the expression site is on a different chromosome than is the basic copy gene.

undergo size changes, perhaps by fragmentation and fusion. These frequent chromosomal changes are not due entirely to VSG gene switches, because they also occur in *Trypanosoma cruzi* and *Leishmania major*, which do not undergo antigenic variation (46, 53, 120). Minichromosomes, however, are less prominent in these latter species, suggesting that their presence might correlate with antigenic variation (54).

DNA probes for the VSG C-terminal similarity region hybridize to all chromosome size classes in low-stringency Southern blots (133), indicating that VSG genes are scattered among all chromosomes. Since there are over 100 chromosomes, as many as 200 VSG genes could be linked to telomeres, while the rest are at interior chromosomal locations.

Using similar low-stringency Southern blots of restricted trypanosome DNA, it has been estimated that several hundred to 1,000 VSG genes exist in the genome (134). Since VSG mRNAs consist of about 1,600 nucleotides, the coding regions for all VSG genes may constitute 4% of the genome. As described below, the actual transposition unit in the VSG gene rearrangements is usually 2,500 to 3,500 bp in extent. Thus, as much as 9% of the genome is devoted to encoding the VSGs and accommodating the rearrangements of their genes. Indeed, recombinant DNA cosmid clones have been reported in which several unrelated VSG genes were present (134). To devote nearly 1/10 of the genome to a single family of isogenes may seem excessive, but perhaps a VSG gene repertoire of this size is required for antigenic variation to be effective under natural conditions.

V. REARRANGEMENTS OF VSG GENES

The selection of a specific VSG gene for expression involves transcriptional regulation, since a VSG cDNA hybridizes under high stringency only to RNA from trypanosomes expressing that VSG and not to RNA from trypanosomes producing other VSGs (64, 84). This observation led to the concept that a VSG gene must be at a genomic location called an "expression site" for its transcription to be initiated.

When a VSG cDNA is hybridized under high stringency to restricted genomic DNA in Southern blots, one of two basic results is observed. In some cases an extra copy of the VSG gene is observed in the genomes of the trypanosomes expressing that VSG. This duplicated VSG gene is called the expression-linked copy (ELC) gene (64). The ELC gene is more susceptible to degradation by DNase I in isolated nuclei than is the gene from which it was duplicated, suggesting that it is the gene undergoing transcription (106). Frequently, but not always, the switch to expression of another VSG is accompanied by the loss of this ELC and the appearance of a new ELC. In other cases, there is no duplicated ELC gene for the VSG being expressed. Instead, transcription of a preexisting VSG gene is activated. Inevitably, this activated VSG gene is located near a chromosomal telomere. Attempts to understand the differences between these two basic observations have been the source of much of the interest in trypanosome antigenic variation by molecular biologists.

A. Duplicative Transposition (ELC Formation)

Each of the approximately 1,000 **basic copy** VSG genes contains the complete, or nearly complete, coding region for that VSG. These genes are situated at one of two general chromosomal locations, either at an internal site or near a telomere. The internal genes are never transcribed in their intrachromosomal location. They must be duplicated and translocated to a telomere-linked expression site before their transcription can occur. This duplicative translocation is a **gene conversion**, or a nonreciprocal transfer of nucleotide sequences from one DNA segment to another (50, 122). The transferred region generally replaces the previous segment in the new location and occurs without loss of the duplicated segment.

Telomere-linked basic copy VSG genes, on the other hand, need not undergo gene conversion to be transcribed. They seemingly have the potential to be expressed in situ even though they sometimes do undergo duplication prior to their transcription (24, 78, 97, 98, 109, 146). Since there may be 100 distinct DNA molecules in the trypanosome nucleus, as many as 200 VSG genes could be telomere linked. It is possible that many if not all of these genes are poised for transcription, because, on the basis of Southern blots, it is not essential for them to move from their telomere-linked site to be transcriptionally activated.

Southern blots under high stringency have revealed that many VSG genes occur within families of 2 to 10 different isogenes (40, 100, 103, 145). These isogenes have extensive similarities throughout their entire coding regions rather than just within the C-terminal 50 codons and the 3′ nontranslated regions. This suggests that they evolved by duplications followed by point mutations or other alterations. In some cases these isogenes may arise from ELCs that are not deleted from the genome during the switch to another VSG but "linger" as unexpressed telomere-linked VSG genes, i.e., they remain in the genome to become basic copy genes (24, 78, 144). In other cases, a silent gene conversion of an internal basic copy gene to another internal site may occur, although such an event has not been detected.

Retroposons with inverted sequence boundaries and an internal poly(A) have been fortuitously detected in rRNA genes of trypanosomes and at other genomic locations (1, 60, 72), but there is no evidence that they contribute to the formation of VSG isogenes. The sequences within the boundaries of those retroposons examined contain an open reading frame but do not specify VSGs. Tandem duplication as a mechanism to generate isogenes may also take place, although so far isogenes have been found scattered about the genome rather than adjacent to each other (134). Some evidence exists to suggest that different isogenes may have accumulated point mutations at different rates for unknown reasons (51).

B. Anatomy of Telomere-Linked VSG Genes

The proximity of the ELC and other VSG genes to a telomere was first detected by careful mapping of restriction sites surrounding the gene (19, 40, 129, 143). These initial experiments suggested that virtually every restriction enzyme tested cleaved at a specific site 5 to 20 kb downstream from these VSG genes (usually with no restriction sites in between). The only rational explanation for such a "universal" restriction site was that a double-stranded DNA break, a chromosomal telomere, occurred at that position. This was demonstrated directly using Bal31 exonuclease digestion of the genomic DNA prior to restriction enzyme analysis (36). Bal31 exonuclease progressively degrades linear DNA from the ends, thereby reducing the sizes of restriction fragments that contain a telomere at one end. Bal31 susceptibility has now been used extensively to establish telomere linkage of both expressed and nonexpressed VSG genes (82, 100, 105, 107, 146).

Most of what is known about telomere-linked VSG genes has come from sequence determinations of recombinant DNA clones containing these genes and their flanking regions. Initially, such clones were difficult to obtain because of the lack of flanking restriction sites at which ligation to a cloning vector could be conducted and the tendency of the highly repetitive telomeric DNA to undergo deletions when cloned. Now a few recombinant DNA clones of the ELC or other telomere-linked VSG genes have been acquired by cloning fragments of sheared or Bal31-treated DNA or by identifying rare restriction sites in the flanking "barren" regions (15, 26, 74, 95, 118).

The results of sequence analyses of these cloned telomere-linked VSG genes are summarized in Fig. 4. Downstream from these genes are tandem repeats of

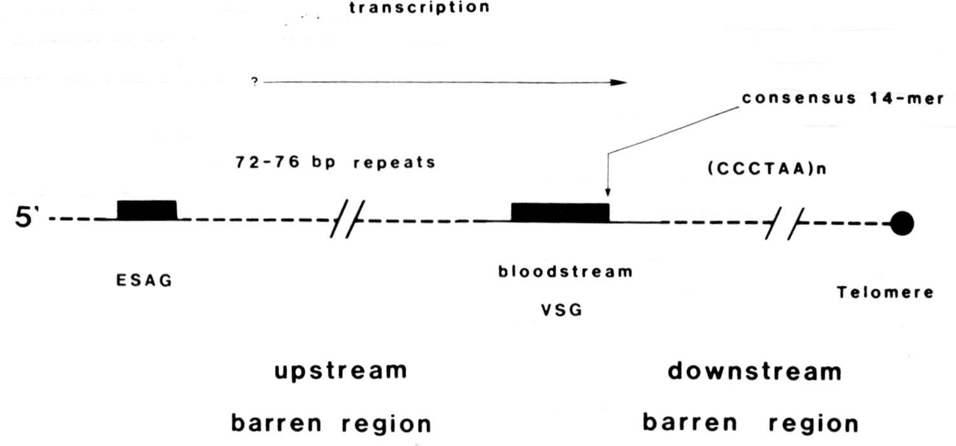

Figure 4. The general structure of telomere-linked basic copy VSG genes and ELC genes. The solid line indicates sequences containing a conventional distribution of restriction sites. The dashed lines indicate "barren regions" of 5 to 40 kb that are made up, at least in part, of the indicated repetitive sequences lacking restriction sites. The large black dot denotes a telomere. The consensus 14-mer refers to a 14-bp sequence that has been found at the ends of all expressed VSG genes. Transcription of the expression-site-associated gene (ESAG) and the adjacent telomere-linked VSG gene is coordinately regulated and occurs in the direction indicated. More than one ESAG-like open translation reading frame may be in front of the upstream barren region.

the hexanucleotide 5′-CCCTAA-3′ that are studded with point changes and interspersed AT-rich segments (15, 131). Although not unambiguously demonstrated, it seems likely that the entire downstream barren region of 5 to 20 kb is composed of these hexameric repeats, AT-rich regions, and related sequences. Bal31 digestion of total genomic DNA preferentially degrades restriction fragments that hybridize to these hexamer repeats, suggesting that most of the repeats occur at telomeres.

The subtelomeric locations of expressed VSG genes permit VSG cDNAs to be used as probes for specific telomeres in Southern blots. Such blots have demonstrated that the distance between a specific VSG gene and its telomere steadily increases with time, and occasionally decreases dramatically due to large deletions (11, 105, 131). Estimates of this growth rate are between 6 and 10 bp, or perhaps one hexamer repeat, per generation. Presumably, periodic deletions of blocks of repeats then reduce the number of repeats, after which telomere growth begins again. This continual change in the number of hexamer repeats adjacent to the telomere may simply be a function of telomere replication (71) and is responsible for the variable sizes of the restriction fragments containing telomere-linked VSG genes.

The downstream (3′) boundary of the ELC transposition unit usually occurs within the coding sequence for the C-terminal hydrophobic tail or the 3′ nontranslated region of the mRNA (12, 38, 90). Thus, the expressed ELC gene for a VSG is often a hybrid in which most of the coding sequence is derived from the basic copy gene and the 3′ end of the gene is provided by the previous VSG gene at that expression site. In some cases (38, 114), the donating basic copy gene does not even possess a termination codon in the correct location, implying that remnants of a previous ELC, or another gene, must contribute the 3′ end of the new ELC. Variation in the exact locations of the transposition boundaries relative to the coding sequence supports the idea that switching from one ELC to another in the same expression site occurs via a gene conversion initiated within common flanking sequences (92, 109, 130). Thus, C-terminal amino acid similarities are required for the folding and deposition of the VSG on the parasite surface, while the corresponding coding sequence similarities participate in the ELC transposition.

The upstream (5′) boundary of the ELC transposition unit typically occurs 1 to 2 kb in front of the initiator methionine codon of the VSG. Since the average VSG coding sequence is about 1.6 kb in extent, this means that the transposition unit consists of between 2.5 and 3.5 kb. In front of this transposition unit is another "barren" region of 5 to 40 kb containing few, if any, restriction sites. In three cases that have been studied, this upstream barren region contains tandem repeats of a 72- to 76-bp sequence (some of which are interrupted by TAA repeats) that are unrelated to the downstream telomeric hexamer repeats (26, 86, 118). A few copies of this 72- to 76-bp repeat are also upstream of many internal basic copy VSG genes (134). These findings suggest that these 5′ flanking repeats serve as homologous upstream "nucleation sites" for the gene conversion event that replaces one ELC with another in an expression site. The conserved coding sequences for the C-terminal 50 amino acids and the 3′ nontranslated region could then serve as the downstream nucleation site for the gene conversion (12, 38, 93). One reservation about this model is that in separate characterizations of two unrelated ELCs, these 72- to 76-bp repeats were not observed at the 5′ transposition boundary (95, 109). In contrast, another repetitive sequence appeared to occur at this boundary. Therefore, additional sequences or factors could also be involved in the gene conversion events. In addition, it should be emphasized that the 5′ barren region is occasionally as large as 40 kb (103) and there is no evidence that this entire region is composed of the same 72- to 76-bp repeats.

Immediately upstream of the 5′ barren regions of expressed telomere-linked VSG genes are often, perhaps always, several other transcribed genes (74). The best studied of these is a family of novel isogenes called the expression-site-associated genes (ESAGs) (33, 34). Transcription of these genes is coordinately regulated with the downstream VSG gene in the expression site. If the VSG gene is transcribed, then the corresponding ESAG is also transcribed. If the VSG gene is not transcribed, then its neighboring ESAG is not expressed. In addition, neither ESAG mRNA nor VSG mRNA is present in procyclic trypanosomes, the culture form of the parasite that does not possess a VSG coat. Southern blots under low stringency indicate that the genome contains 14 to 25 ESAG isogenes. The nucleotide sequences of two different ESAG genes were found to possess 71% identity and encode glycoproteins with 60% amino acid identity. These ESAG proteins do not resemble VSGs in sequence, but do appear to have an N-terminal signal peptide and a hydrophobic component that might act as a membrane anchor. Their presence has been detected directly by immunoprecipitation of ^{35}S-labeled trypanosome extracts, using antisera against a recombinant fusion protein containing an ESAG coding region. In contrast to the VSG, which represents 5% of the total protein, the corresponding ESAG protein makes up probably only about 0.01% of the total protein. The subcellular location of this small amount of ESAG protein and its biological function are unknown.

The number of potential expression sites available to a given VSG gene has been the subject of much debate. The simplest model is one in which the genome contains a unique expression site that a VSG gene must occupy before it can be transcribed (20). Support for this model comes from the demonstration that some VSG genes are repeatedly expressed from a site that has the same restriction pattern (93). However, the situation is much more complex and several lines of evidence indicate that there are at least several different telomeres at which a VSG gene can be transcribed. In the BoTAR serodeme of *T. equiperdum* the BoTat 1 VSG gene appears to be capable of undergoing transcription from three different telomere-linked sites as determined by restriction mapping (88, 89, 111). In the *T. brucei brucei* AnTAR serodeme the AnTat 1.1C VSG gene is transcribed from a site with a different restriction pattern than the site from which previous VSG genes are expressed (109). In addition, Southern blots of pulsed field gradient gels indicate that ELCs of different VSG genes can be located on different-sized chromosomes. For example, Fig. 3B shows a case in which the basic copy VSG gene is on a minichromosome while its corresponding ELC is on a 2,000-kb chromosome. Most expressed VSG genes have been found on large chromosomes (that do not enter a pulsed field gradient gel) or on the midsized 2,000-kb chromosomes, although at least one occurs on a minichromosome (2, 16). The *T. brucei brucei* 221 VSG gene is telomere linked and can be expressed either from its own site on a large chromosome or as an ELC on a 2,000-kb chromosome (133). These examples and others show that the gene conversion leading to an ELC can be a transchromosomal event.  Perhaps the existence of the estimated 14 to 25 ESAGs means that there are a similar number of expression sites (34), but this possibility is difficult to test.

C. Telomere Conversion

The generation of an ELC appears to be essential for an internal basic copy VSG gene to be transcribed (Fig. 5). It is, however, just one of several alternatives available (Fig. 6, left pathway) to a telomere-linked VSG gene for activation.

Sometimes when a telomeric VSG gene is duplicatively transposed to another chromosome, enormous chunks of DNA are transferred, from more than 40 kb upstream of the gene all of the way down, presumably, to the telomere (9, 10). This event appears to be a nonreciprocal duplication of a large telomeric region onto another chromosome (Fig. 6, right pathway). It is equivalent to a gene conversion process in which there is an upstream boundary to the transposition and the telomere itself forms the downstream boundary. As a result, the process has been called **telomere conversion** (19). One scenario for telomere conversion is that it starts out as a simple ELC formation of a telomere-linked basic copy gene in which the upstream nucleation of similar 72- to 76-bp repeats takes place but the downstream nucleation of the C-terminal similarity regions

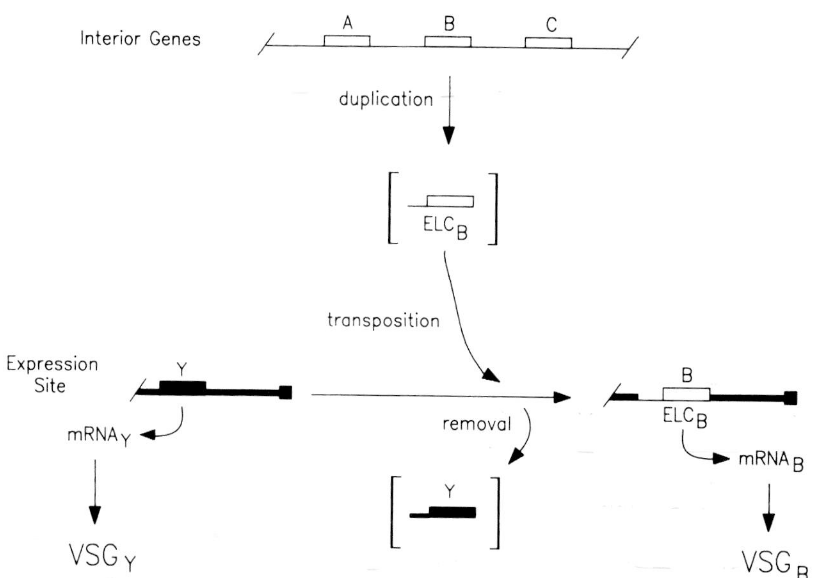

Figure 5. Summary of the activation of VSG genes that are located at interior chromosomal sites. These genes, represented by A, B, and C, can only be transcribed after undergoing a duplicative transposition to form an expression-linked copy (ELC) gene in an expression site linked to a telomere (indicated by the black square). The transposed segment usually includes the VSG coding region and 1 to 2 kb in front of the gene. Often the VSG previously in that expression site, gene Y, is deleted from the genome during the formation of the new ELC. Sometimes, however, the ELC goes to another telomere-linked expression site and gene Y lingers as an unexpressed telomere-linked gene in the old expression site.

does not occur, swinging a new telomere onto the chromosome and activating its linked VSG gene.

D. Telomere Exchange

There is also evidence that an exchange of two telomeres and their linked VSG genes can result in a switch of the VSG gene undergoing transcription. In the example reported (104), restriction mapping indicated that a reciprocal crossover occurred within the upstream (5′) barren regions of two telomere-linked VSG genes (Fig. 6, center pathway), inactivating the expressed VSG gene and activating a previously silent VSG gene. The inactivated gene remained in the genome, presumably now suppressed by the silent telomere environment of one chromosome while the other gene inherited the active environment of the second chromosome. Since the two telomeres and their linked VSG genes appeared to just exchange positions on their respective chromosomes, the process was called **telomere exchange.**

It should be pointed out that there is no evidence that any of the DNA rearrangements is accompanied by an **obligatory** switch in transcription. Many gene rearrangements may result in no change in VSG expression. Others might shut off one VSG gene without activating another, with fatal consequences for the organism. Detection of these silent rearrangements is difficult since there is no way to select for their occurrence. Nevertheless, some fortuitous evidence for their occurrence exists (2, 85, 146).

E. Gene Conversion and VSG Gene Switches

The double-strand break repair model for gene conversion (122) can be used to describe the generation of a new ELC at the molecular level, just as it accounts for experimental data on *S. cerevisiae* mating-type switches (50, 122). One form of this model as applied to ELC formation is illustrated in Fig. 7. An initial double-strand break in the DNA occurs within the 72- to 76-bp repeats upstream of the VSG gene in the telomere-linked expression site. One of the resultant 3′ ends strand invades an equivalent repeat sequence upstream of a basic copy gene and serves as the primer for repair synthesis of one strand. Both strands of the old ELC are degraded (illustrated diagrammatically by the unwinding of this old ELC segment) until the other, newly liberated 3′ end within the C-terminal similarity region can strand invade the corresponding downstream region of the basic copy gene and provide the primer for repair of the complementary strand. Subsequent branch migration followed by cleavage and religation of the two crossover junctions complete the process (see

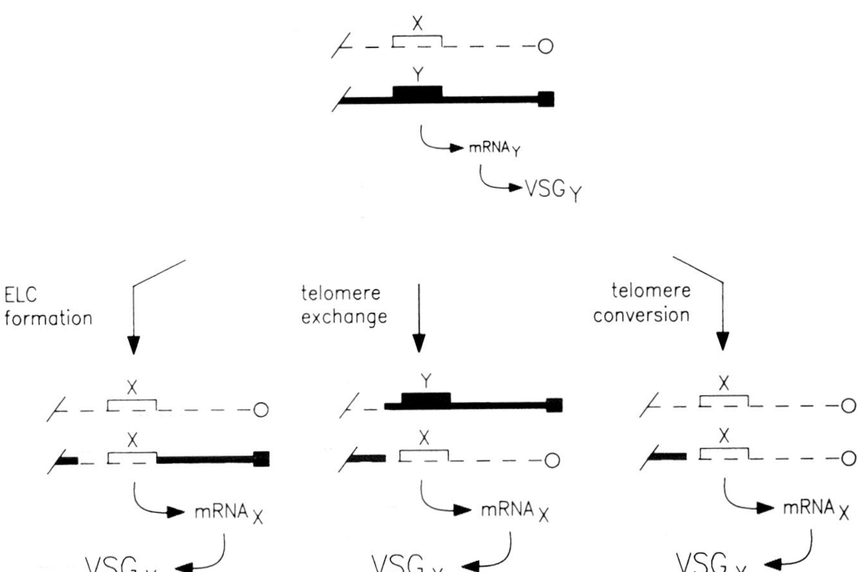

Figure 6. Summary of the DNA rearrangements associated with the expression of VSG genes that are already located near telomeres. At the top of the diagram, gene X is an unexpressed VSG gene near the telomere (open circle) of a chromosome (dashed line). Gene Y is undergoing transcription from an expression site near the telomere (black square) of another chromosome (solid line). The left pathway shows the formation of an ELC of gene X at the expression site on the solid chromosome with the concomitant deletion of gene Y. The center pathway illustrates the activation of gene X, and inactivation of gene Y, by a reciprocal exchange of the telomeric regions, including genes X and Y, between the dashed and solid chromosomes. The right pathway shows the displacement of the telomere, plus gene Y, on the solid chromosome with a duplicated copy of the telomere and flanking region of the dashed chromosome. This telomere conversion event results in the activation of the duplicated gene X on the solid chromosome. In a fourth possible pathway (not shown), gene X is activated in situ on the dashed chromosome, and gene Y on the solid chromosome is correspondingly inactivated, without detectable DNA rearrangements. The signal for the switch in this fourth case is not known.

reference 122 for a more complete discussion of these final steps).

The same model can be used to explain telomere conversion if the donating basic copy gene is itself telomere linked and the initial strand invasion results in duplication of the entire donating region between the upstream repeats and the telomere. The DNA previously at this end of the recipient chromosome, including the previous telomere and its linked VSG gene, is lost. Likewise, the proposed chromosomal crossover event in telomere exchange could be initiated by double-strand cleavage within the upstream repeats of **both** of the telomere-linked genes undergoing the exchange. In this case, no DNA is lost; instead, the initial cleavages facilitate the crossover followed by branch migration and the telomere exchange. Thus, all of the rearrangements of VSG genes described above fit the double-strand cleavage models for gene conversions studied in fungi and other organisms. More complex VSG gene rearrangements involving more than one donating basic copy gene (discussed below) may also be initiated by transient double-stranded cleavages, but additional intermediate steps likely participate in the formation of the final composite VSG gene.

F. In Situ Activation

Finally, and as mentioned earlier, large and easily detected rearrangements are not essential for activation of telomere-linked VSG genes (82, 88, 98, 129, 133, 144). Often, these genes appear to undergo activation in situ. However, failure to detect rearrangements on Southern blots must be interpreted with caution. A potential change might occur far upstream of the VSG gene or involve just a few base pairs. Another possible control mechanism stems from the presence of DNA modifications of unknown structure within or near VSG genes that interfere with restriction enzyme cleavage (13, 102), but a good correlation between VSG expression and nucleotide modification has not been established.

VI. SEGMENTAL VSG GENE CONVERSION

Perhaps the most interesting aspect of VSG gene rearrangements, and one with stimulating implications for the evolution of antigenic variation, is partial, or segmental, gene conversion. ELC formation is usually a complete gene conversion event in

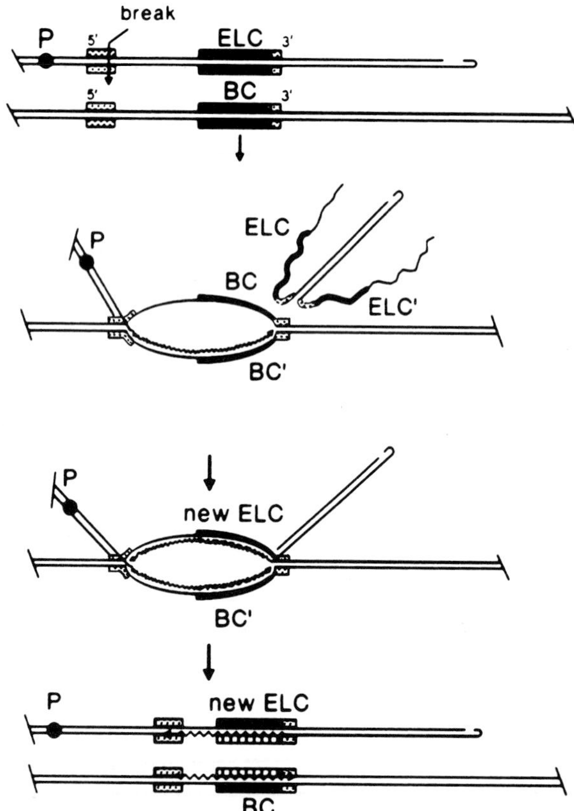

Figure 7. Hypothetical model for a gene conversion event that leads to the formation of a new ELC at a telomere-linked expression site. The model is based on the initial formation of a double-strand break (122) within the 72- to 76-bp repeats upstream of the previous VSG gene at that site. See text for a description of the indicated steps. Unlike the case for *S. cerevisiae* mating-type switches, no evidence exists for the transient presence of this double-strand break upstream of an expression site.

which all of the coding sequence (except sometimes the final few codons) of a basic copy gene is transposed into an expression site. However, as mentioned earlier, the boundaries of the ELC transposition unit are not fixed. Sometimes, a flanking region of a previous ELC remains in place after the new ELC has come in. In most cases, this variability of the insertion/excision sites does not significantly alter the amino acid sequence of the new VSG. Occasionally, however, crossover events occur **within** the coding sequence itself during a VSG gene switch.

One such partial gene conversion event was detected during an examination of the expressed ELC genes from three successively derived trypanosome clones, which, for the purpose of discussion, can be called A, B, and C (108). The ELC genes for VSG_A and VSG_B are closely related and are faithful duplicates of their telomere-linked basic copy isogene equivalents. About 80% of their nucleotide positions are identical. However, the two VSGs, which display 67% identity, are immunologically distinct. The third VSG, VSG_C, is serologically the same as VSG_A, but its ELC is a composite of three VSG genes. The 5′ two-thirds of ELC_C is the same as that of ELC_A, the 3′ one-third is the same as that of ELC_B, and 133 bp in the middle is of unknown origin. This suggests that ELC_B was only partially displaced by an attempt to reexpress ELC_A with perhaps the participation of a third gene that contributed the interior 133-bp segment.

Two other composite ELC genes have been detected in independent isolates of *T. equiperdum*, both of which contain VSG 78 on their surface (87, 114). Each of these two ELCs was also generated by the interaction of three separate basic copy genes. The 3′ ends of both ELCs contained at least 255 bp donated by one of the basic copy genes. The 5′ portions of the two ELCs appeared to be mosaics of the two other, closely related basic copy isogenes. At least one of these 5′ basic copy genes was an incomplete gene that could not code for a complete VSG. The sequence of this pseudogene could only code for an intact VSG after it had been positioned adjacent to a functional 3′ end.

The sequential steps and overall mechanism of the recombinations that led to these composite ELCs are not clear, but must be more complex and with more intermediates than the simple model in Fig. 7. The potential biological significance of these composite genes is less mysterious, however. Since very similar VSGs, such as VSG_A and VSG_B above, can be immunologically distinct, these intermediate isogene recombinations during ELC formation could create new VSGs with antigenic properties different from any previously encoded VSGs. Trypanosomes thus have the potential to produce more VSGs than are simultaneously encoded in their gene pool. This additional capacity for making new VSGs seems to be a natural consequence of the gene rearrangements and expands the potential of trypanosomes for antigenic variation to even further limits.

VII. METACYCLIC VSG GENES

The DNA rearrangements described above involve the VSG genes expressed while the trypanosomes are in the bloodstream. During the trypanosome life cycle, VSGs first appear at the metacyclic stage, the final developmental stage of the parasite in the tsetse fly (83). Each metacyclic trypanosome possesses only one VSG on its surface, but the 20,000 to 50,000 metacyclic trypanosomes in an infected fly usually express collectively a maximum of about 15 different VSGs (48, 58, 59). In a given serodeme, the same metacyclic VSGs are expressed on the same proportion of metacyclic parasites each time trypanosomes are ingested by the fly. After transmission to the bloodstream, these metacyclic VSGs continue to be expressed for about 5 days. Then the trypanosomes switch to the expression of bloodstream VSGs, among the first of which is often the VSG originally ingested by the fly (59).

Since most VSG genes are not expressed at the metacyclic stage, there must be a distinctive regulatory feature to those 15 or so genes that are transcribed at this stage. Several metacyclic VSG cDNAs have been cloned and their corresponding genes examined (44, 80, 82). In virtually all respects the metacyclic VSG genes are similar to bloodstream VSG genes. They are expressed from telomere-linked expression sites, are accompanied by upstream ESAGs, and encode the conserved C-terminal regions and hydrophobic tails. In most cases, the single metacyclic VSG isogene is expressed in situ at the metacyclic stage without detectable DNA rearrangement. This suggests that 15 different expression sites can be activated at the metacyclic stage. The three metacyclic VSG genes of *T. brucei rhodesiense* that have been investigated are not flanked by upstream 72- to 76-bp repeats as determined by Southern blots (80, 82). Since these repeats are likely involved in the initial stages of ELC formation (Fig. 7), perhaps their absence accounts for the lack of rearrangement of these genes.

Metacyclic VSGs can also be expressed in the bloodstream of a chronically infected animal (44, 125). In *T. brucei rhodesiense*, a metacyclic gene and its flanking regions were found to be the same in the genome of bloodstream trypanosomes expressing that VSG as in trypanosomes not expressing that VSG, except for a difference in the number of downstream telomeric repeats (80). In a *T. brucei brucei* serodeme, the VSGs of bloodstream clones AnTat 1.30 and 1.45 appear to be reproducibly present in about 15 and 4% of the metacyclic population, respectively (44). In contrast to the above *T. brucei rhodesiense* result, these genes can be expressed by either ELC formation or in situ activation. However, no characteristic feature of their genomic location or mechanism of expression could be detected that distinguishes them from nonmetacyclic VSG genes. In still another serodeme, tsetse flies were fed on bloodstream trypanosomes expressing a metacyclic VSG, yet the resultant metacyclic population in the flies did not have a significantly increased proportion of organisms expressing that metacyclic VSG (125). These studies collectively suggest that expression of VSGs in the fly and in the bloodstream are independently regulated. The simplest model to account for the results is that metacyclic VSGs are indeed a part of the normal repertoire of VSGs expressed in the bloodstream but possess an additional regulatory feature that activates them at the metacyclic stage.

Thus, we are left with a complex and incomplete picture of the expression sites for VSG genes. The one constant that has stood the test of continued experimentation is that it is **necessary but not sufficient** for a VSG gene to be linked to a telomere if it is to be transcribed. On the basis of the data it may be correct to think of all telomere-linked VSG genes as "on deck" and ready for expression. The factor(s) that activates one, and only one, of these telomere-linked genes to the exclusion of the others remains unknown. Some possibilities, as mentioned above, include DNA modification in a critical regulatory region, other subtle sequence or chromatin changes, or a diffusible, telomere-specific regulatory complex.

VIII. TRANSCRIPTION IN TRYPANOSOMES

Since VSG gene expression is regulated at the level of transcription, a knowledge of the structure of the precursor VSG RNA and the enzymology of its synthesis is essential for understanding the mechanism of antigenic variation. It turns out that formation of **all** mRNA species in trypanosomes, including VSG mRNAs, occurs in a very distinct manner.

A. Discontinuous Transcription

All trypanosome mRNAs have the same 39 nucleotides at their 5′ ends, independent of their coding sequences (17, 101, 110, 132, 138). At the extreme 5′ ends is a cap structure, which is followed by four modified nucleotides, the remaining conserved 35 nucleotides, a normal 5′ nontranslated region, and the coding region (110). These conserved 39 nucleotides have been called the spliced leader or miniexon sequence (99, 132). Their presence is not an exclusive property of antigenic variation, since mRNAs of other Kinetoplastida, such as *Leishmania* spp. and *Crithidia* spp., that do not undergo antigenic variation have spliced leaders. In *T. brucei brucei*, where it has been best studied, the spliced leader is encoded within about 200 copies of a 1.35-kb tandem DNA repeat whose primary transcript of about 140 nucleotides contains the 39 nucleotides and cap structure at its 5′ end (27, 37, 41, 99). The capped 39 nucleotides are donated from this 140-nucleotide transcript to the transcripts of protein-coding genes via a *trans*-splicing mechanism that is only partially understood (76, 96, 121). Since the resultant mRNAs are derived from two independent transcription events, the process is called discontinuous transcription.

The reason for discontinuous transcription is not clear. The spliced leader must play a general role for the cell, since all mRNAs have it. Since the initial 140-nucleotide spliced leader transcript possesses a cap structure, one function is to donate this structure and the other modified nucleotides to the protein-coding mRNAs (81). It seems unlikely that this is the only function, however. Another possible role (16, 56, 74) is based on the observation that some protein-coding genes of trypanosomes are present as 2 to 10 tandem repeats from which a polycistronic precursor RNA containing the gene repeats is synthesized. This polycistronic RNA could be cleaved by *trans* splicing of the leader sequence to generate the 5′ ends of monomer mRNA units while polyadenylation forms the 3′ ends. Other possible, but unsubstantiated, roles for the spliced leader include (i) transport across the nuclear membrane, (ii) stability of the mRNA, (iii) translation initiation by trypanosome ribosomes, or (iv) cytoplasmic compartmentalization.

Stimulating interest even more are recent reports that a few, but not all, mRNAs in other organisms also possess spliced leader-like sequences. In the free-living nematode *Caenorhabditis elegans*, three of the four actin mRNAs contain a 5′-terminal sequence of 22 nucleotides that is encoded within a DNA repeat far removed from the actin genes themselves (75). Since some genes of *C. elegans* have introns, this organism has to accommodate both *cis* and *trans* splicing of RNA. In vaccinia virus, a similar discontinuous synthesis of RNA occurs during viral growth (14, 116). Clearly the mechanism of RNA *trans* splicing and its relationship to the more common *cis* splicing will be an active area of future research.

B. Transcription of VSG Genes

Nuclear run-on experiments indicate that VSG RNA synthesis occurs in the presence of α-amanitin at a concentration that inhibits transcription of genes for tubulin, the spliced leader, and the housekeeping enzymes studied to date (73). Since RNA polymerase I, the enzyme that transcribes rRNA genes, is insensitive to α-amanitin, this result suggests that VSG genes are transcribed either by RNA polymerase I or by an enzyme with similar properties.

Additional support for the possible involvement of RNA polymerase I in VSG transcription comes from a study in which a VSG gene transposition unit with an unusually large upstream region was detected (119). Nuclease S1 protection and primer extension mapped the 5′ end of this 117-VSG transcript to a site about 4,000 nucleotides upstream of the VSG gene. About 60 bp further upstream from this site is a sequence that resembles an RNA polymerase I promoter. Coupled with the α-amanitin insensitivity, this finding suggests that the same polymerase activity may indeed be responsible for the transcription of both rRNA and VSG genes. This in turn indicates that the nucleolus might be the site of VSG gene transcription. If a telomere were confined to the nucleolus with the rRNA genes and RNA polymerase I, then its linked VSG gene would be transcribed. All other telomere-linked VSG genes would be outside the nucleolus and not exposed to the RNA polymerase I activity.

Although very attractive from several standpoints, this model of subnuclear compartmentalization raises as many questions as it answers. What directs a specific telomere to the nucleolus? What excludes all other telomeres from associating with the nucleolus? When a switch occurs, what removes the expressed telomere from the nucleolus and admits another one? Why are most VSG gene transposition units lacking a cotransposed promoter for RNA polymerase I?

In contrast to the above result, most transcripts of VSG genes appear to be initiated much further upstream of the VSG coding region. The best-studied example is the 221 VSG gene of *T. brucei brucei*, in which UV irradiation was used to estimate the size of its primary transcript (70). This procedure is based on the inability of RNA polymerase to traverse pyrimidine dimers induced in the DNA by UV irradiation. Thus, after irradiation of trypanosomes for various times, nuclei were isolated and then incubated with radioactive nucleoside triphosphates, and the labeled RNA was used to probe cloned regions upstream of the telomere-linked VSG gene. This approach indicated that the initial 221 VSG transcription unit is at least 60 kb in length, with the VSG coding region at the extreme 3′ end. This transcriptional unit includes the 5′ repetitive barren region described earlier plus another 40 to 50 kb of upstream sequence. Furthermore, the 60-kb transcription unit is processed down to at least eight (much smaller) stable mRNA species, each of which acquires the spliced leader and two of which encode the VSG and its corresponding ESAG protein (74). All of the resultant mRNAs, except for the VSG mRNA, map upstream of the 5′ barren region. It is not known if these other mRNAs also specify proteins.

It is difficult at this juncture to reconcile these two results: in one case, a VSG gene cotransposes its own promoter to an expression site, while in the other case, a VSG gene utilizes a promoter that is 60 kb upstream. Referring to unpublished data, Johnson et al. (70) say that UV irradiation has been used to detect two transcription units of the VSG gene with the cotransposed promoter, one that is initiated far upstream and one that begins about 4 kb upstream. If this observation is confirmed, it suggests that there are multiple sites for transcription initiation of VSG genes, just as there are multiple ways in which a VSG gene can be maneuvered into an expression site or an expression site can be activated.

IX. QUESTIONS FOR THE FUTURE

During the nearly 10 years since the first report of VSG gene rearrangement (140) a great deal has been learned about this gene family. We know that VSG gene regulation occurs at the transcriptional level, the structures of several VSG genes and their flanking regions have been determined, most of the DNA rearrangements have been shown to be gene conversions, the boundaries of the VSG transcription unit have been identified, and experimental conditions for achieving differential expression of the genes have been found. Nevertheless, the detailed mechanism regulating the mutually exclusive transcription of VSG genes remains a mystery. One reason for this is that mutants affecting antigenic variation have not been found and it is difficult to analyze biological regulation without using genetic approaches. In the case of trypanosomes, the problem is not in generating mutations but in identifying informative ones against a background of other mutations. One can mutagenize trypanosomes in cultures with any of the common mutagens, but a selection technique has not been devised for identifying mutants with a defect in, or an enhancement of, antigenic variation. Thus, molecular biology techniques have to substitute for the tools of genetics.

One recent development that should contribute to the study of antigenic variation in the next few years is the demonstration that electroporation can be used to introduce foreign DNA into trypanosomes (45, 55, 67). In the experiments described to date, the introduced DNA was not stably maintained in the trypanosomes, which means that only transient assays could be performed. Nevertheless, the electroporation approach should permit the design of experiments using reporter genes such as those for chloramphenicol acetyltransferase or luciferase to define more precisely the promoter regions for VSG genes.

From what has been learned about the molecular basis of antigenic variation, it is very unlikely, although perhaps not impossible, that a vaccine against trypanosomiasis will be developed. Likewise, a detailed understanding of the molecular mechanism of VSG gene regulation will probably remain elusive for the near future. There is, however, no shortage of good research projects to undertake. I conclude this chapter with a partial list of unanswered questions about trypanosome antigenic variation, many of which can be addressed experimentally.

Is it indeed essential that the expressed VSG gene be linked to a telomere? Since a basic copy VSG gene and its corresponding ELC are usually about the same intensity on a Southern blot (see Fig. 3B), are these two genes haploid or diploid in a single organism? Why do multiple expression sites exist? What are the functions of the ESAG products and why are they transcribed in *cis* with the expressed VSG gene? Why are VSG genes not transcribed by a conventional RNA polymerase II activity as are the other protein-coding genes in trypanosomes? What survival advantage does discontinuous transcription impart to the parasite? Do specific enzymes catalyze the VSG gene conversions? Why do some trypanosomes of the same serodeme, whose only known difference is the VSG being expressed, have dramatically different growth rates in the bloodstream and in culture? Do the tertiary structures of the individual VSGs influence their sequential order of appearance? These questions and others like them should provide the basis for research on trypanosome antigenic variation for many years.

Acknowledgments. Research performed in my laboratory has been supported by grants from the U.S. Public Health Service and the U.S. Army and an award from the Burroughs-Wellcome Fund.

LITERATURE CITED

1. **Aksoy, S., T. Lalor, J. Martin, L. Van der Ploeg, and F. Richards.** 1987. Multiple copies of a retroposon interrupt spliced leader RNA genes in the African trypanosome, *Trypanosoma gambiense*. *EMBO J.* **6:**3819–3826.
2. **Aline, R., and K. Stuart.** 1985. The two mechanisms for antigenic variation in *Trypanosoma brucei* are independent processes. *Mol. Biochem. Parasitol.* **16:**11–20.
3. **Auffret, C., and M. Turner.** 1981. Variant specific antigens of *Trypanosoma brucei* exist in solution as glycoprotein dimers. *Biochem. J.* **193:**647–650.
4. **Baltz, T., G. Duvillier, C. Giroud, C. Richet, D. Baltz, and P. Degand.** 1983. The variant surface glycoproteins of *Trypanosoma equiperdum*. Identification of a phosphorylated glycopeptide as the cross-reacting determinant. *FEBS Lett.* **158:**174–178.
5. **Baltz, T., C. Giroud, D. Baltz, G. Duvillier, P. Degand, J. Demanille, and R. Pautrizel.** 1982. The variant surface glycoproteins of *Trypanosoma equiperdum* are phosphorylated. *EMBO J.* **1:**1393–1398.
6. **Baltz, T., C. Giroud, D. Baltz, C. Roth, A. Raibaud, and H. Eisen.** 1986. Stable expression of two variant surface glyco-

proteins by cloned *Trypanosoma equiperdum. Nature* (London) **319**:602–604.

7. **Baltz, T., C. Giroud, and J. Crockett.** 1985. Cultivation of a semidefined medium of animal infective forms of *Trypanosoma brucei, T. equiperdum, T. evansi, T. rhodesiense* and *T. gambiense. EMBO J.* **4**:1273–1277.
8. **Barbet, A., W. Davis, and T. McGuire.** 1982. Cross-neutralization of two different trypanosome populations derived from a single organism. *Nature* (London) **300**:453–456.
9. **Bernards, A., T. De Lange, P. Michels, A. Liu, M. Huisman, and P. Borst.** 1984. Two modes of activation of a single surface antigen gene of *Trypanosoma brucei. Cell* **36**: 163–170.
10. **Bernards, A., J. Kooter, P. Michels, R. Moberts, and P. Borst.** 1986. Pulsed field gradient electrophoresis of DNA digested in agarose allows the sizing of the large duplication unit of a surface antigen gene in trypanosomes. *Gene* **42**: 313–322.
11. **Bernards, A., P. Michels, C. Lincke, and P. Borst.** 1983. Growth of chromosome ends in multiplying trypanosomes. *Nature* (London) **303**:592–597.
12. **Bernards, A., L. Van der Ploeg, A. Frasch, P. Borst, J. Boothroyd, S. Coleman, and G. Cross.** 1981. Activation of trypanosome surface glycoprotein genes involves a duplication-transposition leading to an altered 3′ end. *Cell* **27**: 497–505.
13. **Bernards, A., N. van Harten-Loosbroek, and P. Borst.** 1984. Modification of telomeric DNA in *Trypanosoma brucei*: a role in antigenic variation. *Nucleic Acids Res.* **12**: 4153–4170.
14. **Bertholet, C., E. Van Meir, B. ten Heggeler-Bordier, and R. Wittek.** 1987. Vaccine virus produces late mRNAs by discontinuous synthesis. *Cell* **50**:153–162.
15. **Blackburn, E., and P. Challoner.** 1984. Identification of a telomeric DNA sequence in *Trypanosoma brucei. Cell* **36**: 447–457.
16. **Boothrooyd, J.** 1985. Antigenic variation in African trypanosomes. *Annu. Rev. Microbiol.* **39**:473–502.
17. **Boothroyd, J., and G. Cross.** 1982. Transcripts coding for different variant surface glycoproteins in *Trypanosoma brucei* have a short identical exon at their 5′-end. *Gene* **20**: 279–287.
18. **Borst, P.** 1986. Discontinuous transcription and antigenic variation in trypanosomes. *Annu. Rev. Biochem.* **55**: 701–732.
19. **Borst, P., A. Bernards, L. Van der Ploeg, P. Michels, A. Liu, T. De Lange, P. Sloof, G. Veeneman, M. Tromp, and J. Van Boom.** 1983. DNA rearrangements controlling the expression of genes for variant surface antigens in trypanosomes, p. 207–233. *In* K. Chater, D. Cullis, D. Hopwood, A. Johnston, and H. Woolhouse (ed.), *Genetic Rearrangement.* Croom Helm, London.
20. **Borst, P., and G. Cross.** 1982. Molecular basis for trypanosome antigenic variation. *Cell* **29**:291–303.
21. **Borst, P., and D. R. Greaves.** 1987. Programmed gene rearrangements altering gene expression. *Science* **235**: 658–667.
22. **Borst, P., L. Van der Ploeg, J. van Hoek, J. Tas, and T. James.** 1982. On the DNA content and ploidy of trypanosomes. *Mol. Biochem. Parasitol.* **6**:13–23.
23. **Bridgen, P., G. Cross, and J. Bridgen.** 1976. N-terminal amino acid sequences of variant-specific antigens from *Trypanosoma brucei. Nature* (London) **263**:613–614.
24. **Buck, G., S. Longacre, A. Raibaud, U. Hibner, C. Giroud, T. Baltz, D. Baltz, and H. Eisen.** 1984. Stability of expression-linked surface antigen genes in *Trypanosoma equiperdum. Nature* (London) **307**:563–566.
25. **Campbell, G., K. Esser, B. Wellde, and C. Diggs.** 1979. Isolation and characterization of a serodeme of *Trypanosoma rhodesiense. Am. J. Trop. Med. Hyg.* **28**:974–983.
26. **Campbell, D., D. Thornton, and J. Boothroyd.** 1984. The 5′-limit of transposition and upstream barren region of a trypanosome VSG gene. Tandem 76 base-pair repeats flanking $(TAA)_{90}$. *Nature* (London) **12**:2759–2774.
27. **Campbell, D., D. Thornton, and J. Boothroyd.** 1984. Apparent discontinuous transcription of *Trypanosoma brucei* surface antigen genes. *Nature* (London) **311**:350–355.
28. **Capbern, A., C. Biroud, T. Baltz, and P. Maltern.** 1977. *Trypanosoma equiperdum*: étude des variations antigèniques au cours de la trypanosome expérimentale du lapin. *Exp. Parasitol.* **42**:6–13.
29. **Cardosa de Almeida, M., L. Allan, and M. Turner.** 1984. Purification and properties of the membrane form of variant surface glycoproteins (VSGs) from *Trypanosoma brucei. J. Protozool.* **31**:53–59.
30. **Cardosa de Almeida, M., and M. Turner.** 1983. The membrane form of variant surface glycoproteins of *Trypanosoma brucei. Nature* (London) **302**:349–352.
31. **Cornelissen, A., P. Johnson, J. Kooter, L. Van der Ploeg, and P. Borst.** 1985. Two simultaneously active VSG gene transcription units in a single *Trypanosoma brucei* variant. *Cell* **41**:825–832.
32. **Cross, G.** 1975. Identification, purification and properties of clone-specific glycoprotein antigens constituting the surface coat of *Trypanosoma brucei. Parasitology* **71**:393–417.
33. **Cully, D., C. Gibbs, and M. Cross.** 1986. Identification of proteins encoded by variant surface glycoprotein expression site-associated genes in *Trypanosoma brucei. Mol. Biochem. Parasitol.* **21**:189–197.
34. **Cully, D., H. Ip, and G. Cross.** 1985. Coordinate transcription of variant surface glycoprotein genes and an expression site associated gene family in *Trypanosoma brucei. Cell* **42**:173–182.
35. **De Lange, T.** 1986. The molecular biology of antigenic variation in trypanosomes: gene rearrangements and discontinuous transcription. *Int. Rev. Cytol.* **99**:85–117.
36. **De Lange, T., and P. Borst.** 1982. Genomic environment of the expression-linked extra copies of genes for surface antigens of *Trypanosoma brucei* resembles the end of a chromosome. *Nature* (London) **299**:451–453.
37. **De Lange, T., A. Liu, L. Van der Ploeg, P. Borst, M. Tromp, and J. Van Boom.** 1983. Tandem repetition of the 5′ mini-exon of variant surface glycoprotein genes. A multiple promoter for VSG gene transcription? *Cell* **34**:891–900.
38. **Donelson, J., W. Murphy, S. Brentano, A. Rice-Ficht, and G. Cain.** 1983. Comparison of the expression-linked extra copy (ELC) and basic copy (BC) genes of a trypanosome surface antigen. *J. Cell. Biochem.* **23**:1–12.
39. **Donelson, J., and A. Rice-Ficht.** 1985. Molecular biology of trypanosome antigenic variation. *Microbiol. Rev.* **49**: 107–125.
40. **Donelson, J., J. Young, D. Dorfman, and P. Majiwa.** 1982. The ILTat 1.4 surface antigen gene family of *Trypanosoma brucei. Nucleic Acids Res.* **10**:6581–6595.
41. **Dorfman, D., and J. Donelson.** 1984. Characterization of the 1.35 kb DNA repeat until containing the conserved 35 nucleotides at the 5′-termini of VSG mRNAs in *Trypanosoma brucei. Nucleic Acids Res.* **12**:4907–4920.
42. **Dorfman, D., M. Lenardo, L. Reddy, L. Van der Ploeg, and J. Donelson.** 1985. The 5.8S ribosomal RNA gene of *Trypa-*

nosoma brucei: structural and transcriptional studies. *Nucleic Acids Res.* 13:3533–3549.

43. **Doyle, J., H. Hirumi, H. Hirumi, N. Lupton, and G. Cross.** 1980. Antigenic variation in clones of animal-infective *Trypanosoma brucei* derived and maintained in vitro. *Parasitology* 80:359–369.
44. **Dulauw, M.-F., M. Laurent, P. Paindavoine, D. Aerts, E. Pays, D. LeRay, and M. Steinert.** 1987. Characterization of genes coding for two major metacyclic surface antigens in *Trypanosoma brucei*. *Mol. Biochem. Parasitol.* 23:9–17.
45. **Eid, J., and B. Sollner-Webb.** 1987. Efficient introduction of plasmid DNA into *Trypanosoma brucei* and transcription of a transfected chimeric gene. *Proc. Natl. Acad. Sci. USA* 84:7812–7816.
46. **Engman, D., L. Reddy, J. Donelson, and L. Kirchhoff.** 1987. *Trypanosoma cruzi* exhibits inter- and intra-strain heterogeneity in molecular karyotype and chromosomal gene location. *Mol. Biochem. Parasitol.* 22:115–123.
47. **Esser, K., and M. Schoenbechler.** 1985. Expression of two variant surface glycoproteins on individual African trypanosomes during antigenic switching. *Science* 229:190–193.
48. **Esser, K., M. Schoenbechler, and J. Gingerich.** 1982. *Trypanosoma rhodesiense* blood forms express all antigen specificities relevant to protection against metacyclic (insect form) challenge. *J. Immunol.* 129:1715–1718.
49. **Ferguson, M., and G. Cross.** 1984. Myristylation of the membrane form of a *Trypanosoma brucei* variant surface glycoprotein. *J. Biol. Chem.* 259:3011–3015.
50. **Fink, G., and T. Petes.** 1984. Gene conversion in the absence of reciprocal recombination. *Nature* (London) 310:728.
51. **Frasch, A., P. Borst, and J. Van den Burg.** 1982. Rapid evolution of genes coding for variant surface glycoproteins in trypanosomes. *Gene* 17:197–211.
52. **Fregmann, D., P. Metcalf, M. Turner, and D. C. Wiley.** 1984. 6Å-resolution X-ray structure of a variable surface glycoprotein from *Trypanosoma brucei*. *Nature* (London) 311:167–169.
53. **Gibson, W., and P. Borst.** 1986. Size-fractionation of the small chromosomes of Trypanozoon and Nannomonas trypanosomes by pulsed field gradient gel electrophoresis. *Mol. Biochem. Parasitol.* 18:127–140.
54. **Gibson, W., K. Osinga, P. Michels, and P. Borst.** 1985. Trypanosomes of subgenus Trypanozoon are diploid for housekeeping genes. *Mol. Biochem. Parasitol.* 16:231–242.
55. **Gibson, W., T. White, P. Laird, and P. Borst.** 1987. Stable introduction of exogenous DNA into *Trypanosoma brucei*. *EMBO J.* 6:2457–2461.
56. **Gonzalez, A., T. Lerner, M. Huecas, B. Sosa-Pineda, N. Nogueira, and P. Lizardi.** 1985. Apparent generation of a segmented mRNA from two tandem gene families in *Trypansoma cruzi*. *Nucleic Acids Res.* 13:5789–5803.
57. **Hadjuk, S.** 1984. Antigenic variation during the developmental cycle of *Trypanosoma brucei*. *J. Protozool.* 31: 41–42.
58. **Hajduk, S., C. Cameron, J. Barry, and K. Vickerman.** 1981. Antigenic variation in cyclically-transmitted *Trypanosoma brucei*. I. Variable antigen type composition of metacyclic trypanosome populations from the salivary glands of *Glossina morsitans*. *Parasitology* 83:595–607.
59. **Hajduk, S., and K. Vickerman.** 1981. Antigenic variation in cyclically transmitted *Trypanosoma brucei*. Variable antigen type composition of the first parasitaemia in mice bitten by trypanosome-infected *Glossina morsitans*. *Parasitology* 83: 609–621.
60. **Hansan, G., M. Turner, and J. Cordingley.** 1984. Complete nucleotide sequence of an unusual mobile element from *Trypanosoma brucei*. *Cell* 37:333–341.
61. **Hemperly, J., G. Edelman, and B. Cunningham.** 1986. cDNA clones of the neural adhesion molecule (N-CAM) lacking a membrane-spanning region consistent with evidence for membrane attachment via a phosphatidylinositol intermediate. *Proc. Natl. Acad. Sci. USA* 83:9822–9826.
62. **Herald, D., J. Krakow, J. Bangs, G. Hart, and P. Englund.** 1986. A phospholipase C from *Trypanosoma brucei* which selectively cleaves the glycolipid on the variant surface glycoproteins. *J. Biol. Chem.* 261:13813–13819.
63. **Hill, G. C., and H. Hirumi.** 1985. African trypanosomes, p. 193–219. *In* J. B. Jensen (ed.), *In Vitro Cultivation of Protozoan Parasites*. CRC Press, Boca Raton, Fla.
64. **Hoeijmakers, J., A. Frasch, A. Bernards, P. Borst, and G. Cross.** 1980. Novel expression-linked copies of the genes for variant surface antigens in trypanosomes. *Nature* (London) 284:78–80.
65. **Holder, A.** 1983. Carbohydrate is linked through ethanolamine to the C-terminal amino acid of *Trypanosoma brucei* variant surface glycoprotein. *Biochem. J.* 209:261–262.
66. **Holder, A., and G. Cross.** 1981. Glycopeptides from variant surface glycoproteins of *Trypanosoma brucei*. C-terminal location of antigenically cross-reacting carbohydrate moieties. *Mol. Biochem. Parasitol.* 2:135–150.
67. **Hughes, D., and L. Simpson.** 1986. Introduction of plasmid DNA into the trypanosomatid protozoan *Crithidia fasciculata*. *Proc. Natl. Acad. Sci. USA* 83:6058–6062.
68. **Jenni, L., S. Marti, J. Schweizer, B. Betschart, R. Le Page, J. Wells, A. Tait, P. Paindovoine, E. Pays, and M. Steinert.** 1986. Hybrid formation between African trypanosomes during cyclical transmission. *Nature* (London) 322: 173–175.
69. **Johnson, J., and G. Cross.** 1979. Selective cleavage of variant surface glycoproteins from *Trypanosoma brucei*. *Biochem. J.* 178:689–697.
70. **Johnson, P., J. Kooter, and P. Borst.** 1987. Inactivation of transcription by UV irradiation of *T. brucei* provides evidence for a multicistronic transcription unit including a VSG gene. *Cell* 51:273–281.
71. **Karrer, K., and J. Gall.** 1976. The macronuclear ribosomal DNA of tetrahymena priformis is a palindrome. *J. Mol. Biol.* 104:421–454.
72. **Kimmel, B., O. Ole-Moiyoi, and J. R. Young.** 1987. Ingi, a 5.2 kb dispersed sequence element from *Trypanosoma brucei* that carries half of a smaller mobile element at either end and has homology with mammalian LINES. *Mol. Cell. Biol.* 7:1465–1475.
73. **Kooter, J., and P. Borst.** 1984. Alpha-amanitin-insensitive transcription of variant surface glycoprotein genes provides further evidence for discontinuous transcription in trypanosomes. *Nucleic Acids Res.* 12:9457–9472.
74. **Kooter, J., H. van der Spek, R. Wagter, C. d'Oliveria, F. Van der Hoeven, P. Johnson, and P. Borst.** 1987. The anatomy and transcription of a telomeric expression site for variant-specific surface antigens in *T. brucei*. *Cell* 51:261–272.
75. **Krause, M., and D. Hirsh.** 1987. A trans-spliced leader sequence on actin mRNA in *C. elegans*. *Cell* 49:753–761.
76. **Laird, P., J. Zomerdijk, D. de Korte, and P. Borst.** 1987. In vitro labelling of intermediates in the discontinuous synthesis of mRNAs in *Trypanosoma brucei*. *EMBO J.* 6:1055–1062.
77. **Lamont, G., R. Tucker, and G. Cross.** 1986. Analysis of antigen switching rates in *Trypanosoma brucei*. *Parasitology* 92:355–367.
78. **Laurent, M., E. Pays, E. Magnus, N. Van Meirvenne, G.**

Natthijssens, R. Williams, and M. Steinert. 1983. DNA rearrangements linked to expression of a predominant surface antigen gene of trypanosomes. *Nature* (London) **302:** 263–266.

79. **Lee, M. G.-S., and L. H. T. Van der Ploeg.** 1987. Frequent independent duplicative transpositions activate a single VSG gene. *Mol. Cell. Biol.* 7:357–364.

80. **Lenardo, M., K. Esser, A. Moon, L. Van der Ploeg, and J. Donelson.** 1986. Metacyclic VSG genes of *Trypanosoma brucei rhodesiense* are activated *in situ* and their expression is transcriptionally regulated. *Mol. Cell. Biol.* **6:**1991–1997.

81. **Lenardo, M., D. Dorfman, and J. Donelson.** 1985. The spliced leader sequence of *Trypanosoma brucei* has a potential role as a cap donor structure. *Mol. Cell. Biol.* **9:** 2487–2490.

82. **Lenardo, M., A. Rice-Ficht, G. Kelly, K. Esser, and J. Donelson.** 1984. Characterization of the genes specifying two metacyclic variable antigen types in *Trypanosoma brucei rhodesiense. Proc. Natl. Acad. Sci. USA* **81:**6642–6646.

83. **Le Ray, D., J. Barry, and K. Vickerman.** 1978. Antigenic heterogeneity of metacyclic forms of *Trypanosoma brucei. Nature* (London) **273:**300–302.

84. **L'Heureux, M., T. Vervoort, N. Van Meirveene, and M. Steinert.** 1979. Immunological purification and partial characterization of variant-specific surface antigen messenger RNA of *Trypanosoma brucei brucei. Nucleic Acids Res.* **7:**595–609.

85. **Liu, A., P. Michels, A. Bernards, and P. Borst.** 1985. Trypanosome variant surface glycoprotein genes expressed early in infection. *J. Mol. Biol.* **182:**383–396.

86. **Liu, A., L. Van der Ploeg, F. Rijsewijk, and P. Borst.** 1983. The transposition unit of VSG gene 118 of *Trypanosoma brucei.* Presence of repeated elements at its border and absence of promoter-associated sequences. *J. Mol. Biol.* **167:**57–75.

87. **Longacre, S., and H. Eisen.** 1986. Expression of whole and hybrid genes in *Trypanosoma equiperdum. EMBO J.* **5:** 1057–1063.

88. **Longacre, S., U. Hibner, A. Raibaud, H. Eisen, T. Baltz, C. Giroud, and D. Baltz.** 1983. DNA rearrangement and antigenic variation of *Trypanosoma equiperdum.* Multiple expression linked sites in independent isolates of trypanosomes expressing the same antigen. *Mol. Cell. Biol.* **3:** 399–409.

89. **Longacre, S., A. Raibaud, U. Hibner, G. Buck, and H. Eisen.** 1983. DNA rearrangements and antigenic variation in *Trypanosoma equiperdum.* Expression-independent DNA rearrangements in the basic copy of a variant surface glycoprotein gene. *Mol. Cell. Biol.* **3:**410–414.

90. **Majumder, H., J. Boothroyd, and H. Weber.** 1981. Homologous 3′-terminal regions of mRNAs for surface antigens of different antigenic variants of *Trypanosoma brucei. Nucleic Acids Res.* **9:**4745–4753.

91. **Metcalf, P., M. Blum, D. Freymann, M. Turner, and D. Wiley.** 1987. Two variant surface glycoproteins of *Trypanosoma brucei* of different sequence classes have similar 6 Å resolution X-ray structures. *Nature* (London) **325:**84–86.

92. **Michels, P., A. Bernards, L. Van der Ploeg, and P. Borst.** 1982. Characterization of the expression-linked copies of variant surface glycoprotein 118 in two independently isolated clones of *Trypanosoma brucei. Nucleic Acids Res.* **10:**2353–2366.

93. **Michels, P., A. Liu, A. Bernards, P. Sloof, M. Van der Bijl, A. Schinkel, H. Menke, P. Borst, G. Veeneman, M. Tromp, and J. Van Boom.** 1983. Activation of the genes for variant surface glycoproteins 117 and 118 in *Trypanosoma brucei. J. Mol. Biol.* **66:**537–556.

94. **Miller, E., and J. Turner.** 1981. Analysis of antigenic types appearing in first relapse populations of clones in *Trypanosoma brucei. Parasitology* **82:**63–80.

95. **Murphy, W., S. Brentano, A. Rice-Ficht, D. Dorfman, and J. Donelson.** 1984. DNA rearrangements of the variable surface antigen genes of trypanosomes. *J. Protozool.* **31:**65–73.

96. **Murphy, W., K. Watkins, and N. Agabian.** 1987. Identification of a novel Y branch structure as an intermediate in trypanosome mRNA splicing: evidence for trans splicing. *Cell* **47:**517–525.

97. **Myler, P., J. Allison, N. Agabian, and K. Stuart.** 1984. Antigenic variation in African trypanosomes by replacement or activation of alternate telomeres. *Cell* **39:**203–211.

98. **Myler, P., R. Nelson, N. Agabian, and K. Stuart.** 1984. Two mechanisms of expression of a predominant variant antigen gene of *Trypanosoma brucei. Nature* (London) **309:** 282–284.

99. **Nelson, R., M. Parsons, P. Barr, K. Stuart, M. Selkirk, and N. Agabian.** 1983. Sequences homologous to the variant antigen mRNA spliced leader are located in tandem repeats and variable orphons in *Trypanosoma brucei. Cell* **34:** 901–909.

100. **Parsons, M., R. Nelson, G. Newport, M. Milhausen, K. Stuart, and N. Agabian.** 1983. Genomic organization of *Trypanosoma brucei* variant antigen gene families in sequential parasitemias. *Mol. Biochem. Parasitol.* **9:**255–269.

101. **Parsons, M., R. Nelson, J. Watkins, and N. Agabian.** 1984. Trypanosome mRNAs share a common 5′ spliced leader sequence. *Cell* **38:**309–316.

102. **Pays, E., M. Delauw, M. Laurent, and M. Steinert.** 1984. Possible DNA modification in GC dinucleotides of *Trypanosoma brucei* telomeric sequences; relationship with antigenic variation. *Nucleic Acids Res.* **12:**5235–5247.

103. **Pays, E., M. Delauw, S. Van Assel, M. Laurent, T. Vervoort, N. Van Meirvenne, and M. Steinert.** 1983. Modifications of a *Trypanosoma b. brucei* antigen gene repertoire by different DNA recombinational mechanisms. *Cell* **35:**721–731.

104. **Pays, E., M. Gayaux, D. Aerts, N. Van Meirvenne, and M. Steinert.** 1985. Telomeric reciprocal recombination as a possible mechanism for antigenic variation in trypanosomes. *Nature* (London) **316:**562–564.

105. **Pays, E., M. Laurent, K. Delinte, N. Van Meirvenne, and M. Steinert.** 1983. Differential size variations between transcriptionally active and inactive telomeres in *Trypanosoma brucei. Nucleic Acids Res.* **11:**8137–8147.

106. **Pays, E., M. L'Heureux, and M. Steinert.** 1981. The expression-linked copy of surface antigen gene in *Trypanosoma* is probably the one transcribed. *Nature* (London) **292:** 365–367.

107. **Pays, E., M. L'Heureux, and M. Steinert.** 1982. Structure and expression of a *Trypanosoma brucei gambiense* variant specific antigen gene. *Nucleic Acids Res.* **10:**3149–3163.

108. **Pays, E., S. Van Assel, M. Laurent, M. Darville, T. Vervoort, N. Van Meirveene, and M. Steinert.** 1983. Gene conversion as a mechanism for antigenic variation in trypanosomes. *Cell* **34:**371–381.

109. **Pays, E., S. Van Assel, M. Laurent, B. Dero, F. Michiels, P. Kronenberger, G. Matthyssens, N. Van Meirvenne, D. Le Ray, and M. Steinert.** 1983. At least two transposed sequences are associated in the expression site of a surface antigen gene in different trypanosome clones. *Cell* **34:** 359–369.

110. **Perry, K., K. Watkins, and N. Agabian.** 1987. Trypanosome

mRNAs have unusual "cap 4" structures acquired by addition of a spliced leader. *Proc. Natl. Acad. Sci. USA* **84:** 8190–8194.

111. **Raibaud, A., C. Gaillard, S. Longacre, U. Hibner, G. Buck, G. Bernardi, and H. Eisen.** 1983. Genomic environment of variant surface antigen genes of *Trypanosoma equiperdum. Proc. Natl. Acad. Sci. USA* **80:**4306–4310.
112. **Rifkin, M.** 1978. Identification of the trypanocidal factor in normal human serum high density lipoprotein. *Nature* (London) 75:3450–3454.
113. **Ross, R., and D. Thomson.** 1910. A case of sleeping sickness studied by precise enumerative methods. Regular periodic increase in the parasites disclosed. *Proc. R. Soc. London Ser. B* **82:**411–415.
114. **Roth, C. W., S. Longacre, A. Raibaud, T. Baltz, and H. Eisen.** 1986. The use of incomplete genes for construction of a *Trypanosoma equiperdum* variant surface glycoprotein gene. *EMBO J.* **5:**1065–1070.
115. **Schwartz, D., and C. Cantor.** 1984. Separation of yeast chromosome-sized DNA by pulsed field gradient gel electrophoresis. *Cell* **37:**67–75.
116. **Schwer, B., P. Visca, J. Vox, and H. Stunnenberg.** 1987. Discontinuous transcription or RNA processing of vaccine virus late messengers results in a 5′ poly(A) leader. *Cell* **50:**163–169.
117. **Seki, T., T. Moriuchi, H. Chang, R. Denome, and J. Silver.** 1985. Structural organization of the rat thy-1 gene. *Nature* (London) **313:**485–487.
118. **Shah, J., J. Young, B. Kimmel, K. Iams, and R. Williams.** 1987. The 5′ flanking sequence of a *Trypanosoma brucei* variable surface glycoprotein gene. *Mol. Biochem. Parasitol.* **24:**163–174.
119. **Shea, C., M. Lee, and L. Van der Ploeg.** 1987. VSG gene 118 is transcribed from a cotransposed pol-I like promoter. *Cell* **50:**603–612.
120. **Spithill, T., and N. Samaras.** 1985. The molecular karyotype of *Leishmania major* and mapping of α- and β-tubulin gene families to multiple unlinked chromosomal loci. *Nucleic Acids Res.* **13:**4155–4169.
121. **Sutton, R., and J. Boothroyd.** 1986. Evidence for trans splicing in trypanosomes. *Cell* **47:**527–535.
122. **Szostak, J., T. Orr-Weaver, R. Rothstein, and F. Stahl.** 1983. The double-strand-break repair model for recombination. *Cell* **33:**25–35.
123. **Tait, A.** 1980. Evidence for diploidy and mating in trypanosomes. *Nature* (London) **287:**536–538.
124. **Timmers, H. T., T. deLange, J. M. Kooter, and P. Borst.** 1987. Coincident multiple activations of the same surface antigen gene in *Trypanosoma brucei. J. Mol. Biol.* **194:** 81–90.
125. **Turner, C., J. Barry, and K. Vickerman.** 1986. Independent expression of the metacyclic and bloodstream variable antigen repertories in *Trypanosoma brucei rhodesiense. Parasitology* **92:**67–73.
126. **Turner, M.** 1982. Biochemistry of the variant surface glycoproteins of salivarian trypanosomes. *Adv. Parasitol.* **21:** 69–153.
127. **Turner, M.** 1985. The biochemistry of the surface antigens of the African trypanosomes. *Br. Med. Bull.* **41:**137–143.
128. **Van der Ploeg, L.** 1987. Control of variant antigen switching in trypanosomes. *Cell* **51:**159–161.
129. **Van der Ploeg, L., A. Bernards, F. Rijsewijk, and P. Borst.** 1982. Characterization of the DNA duplication-transposition that controls the expression of two genes for variant surface glycoproteins in *Trypanosoma brucei. Nucleic Acids Res.* **10:**593–609.
130. **Van der Ploeg, L., A. Cornelissen, P. Michels, and P. Borst.** 1984. Chromosomes rearrangement in *Trypanosoma brucei. Cell* **39:**213–221.
131. **Van der Ploeg, L., A. Liu, and P. Borst.** 1984. The structure of the growing telomeres of trypanosomes. *Cell* **36:**459–468.
132. **Van der Ploeg, L., A. Liu, P. Michels, T. De Lange, P. Borst, H. Majumder, H. Weber, G. Veeneman, and J. Van Boom.** 1982. RNA splicing is required to make the messenger RNA for a variant surface antigen in trypanosomes. *Nucleic Acids Res.* **10:**3591–3604.
133. **Van der Ploeg, L., D. Schwartz, C. Cantor, and P. Borst.** 1984. Antigenic variation in *Trypanosoma brucei* analyzed by electrophoretic separation of chromosome-sized DNA molecules. *Cell* **37:**77–84.
134. **Van der Ploeg, L., D. Valerio, T. De Lange, A. Bernards, P. Borst, and F. Groveld.** 1982. An analysis of cosmid clones of nuclear DNA from *Trypanosoma brucei* shows that the genes for variant surface glycoproteins are clustered in the genome. *Nucleic Acids Res.* **10:**5905–5923.
135. **Vickerman, K.** 1974. Antigenic variation in African trypanosomes, p. 58–80. *In Parasites in the Immunized Host: Mechanism of Survival. CIBA Foundation Symposium.* Elsevier/North-Holland Publishing Co., Amsterdam.
136. **Vickerman, K.** 1986. Parasitology. Clandestine sex in trypanosomes. *Nature* (London) **322:**113–114.
137. **Vickerman, K., J. Barry, S. Hajduk, and L. Tetley.** 1980. Antigenic variation in trypanosomes, p. 179–190. *In The Host Invader Interplay.* Elsevier/North-Holland/Biomedical Press, Amsterdam.
138. **Walder, J., P. Eder, D. Engman, S. Brentano, R. Walder, D. Knutzon, D. Dorfman, and J. Donelson.** 1986. The 35 nucleotide spliced leader sequence is common to all trypanosome messenger RNAs. *Science* **233:**569–571.
139. **Wells, J., T. Prospero, L. Jenni, and R. LePage.** 1987. DNA contents and molecular karotypes of hybrid *Trypanosoma brucei. Mol. Biochem. Parasitol.* **249:**103–116.
140. **Williams, R., J. Young, and P. Majiwa.** 1979. Genomic rearrangements correlated with antigenic variation in *Trypanosoma brucei. Nature* (London) **282:**847–849.
141. **Williams, R., J. Young, and P. Majiwa.** 1982. Genomic environment of *Trypanosoma brucei* VSG genes: presence of a minichromosome. *Nature* (London) **299:**417–421.
142. **World Health Organization.** 1978. *Bull. WHO* **307:** 467–480.
143. **Young, J., J. Donelson, P. Majiwa, S. Shapiro, and R. Williams.** 1982. Analysis of genomic rearrangements associated with two variable antigen genes of *Trypanosoma brucei. Nucleic Acids Res.* **10:**803–819.
144. **Young, J., E. Miller, R. Williams, and M. Turner.** 1983. Are there two classes of VSG genes in *Trypanosoma brucei*? *Nature* (London) **306:**196–198.
145. **Young, J., J. Shah, G. Matthyssens, and R. Williams.** 1983. Relationship between multiple copies of a *T. brucei* variable surface glycoprotein gene whose expression is not controlled by duplication. *Cell* **32:**1149–1159.
146. **Young, J., M. Turner, and R. Williams.** 1984. The role of duplication in the expression of a variable surface glycoprotein gene of *Trypanosoma brucei. J. Cell Biochem.* **24:** 287–295.
147. **Zampetti-Bosseler, F., J. Schweizer, E. Pays, L. Jenni, and M. Steinert.** 1986. Evidence for haploidy in metacyclic forms of *Trypanosoma brucei. Proc. Natl. Acad. Sci. USA* **83:** 6063–6064.

Chapter 36

Antigenic Variation in Relapsing Fever *Borrelia* Species: Genetic Aspects

ALAN BARBOUR

I. INTRODUCTION

In 1868 a young German physician reported his discovery that epidemic relapsing fever was caused by a spirochete. This event occurred during the formative years of immunology and was not unnoticed. Ehrlich and some other pioneers of this new field were fascinated by the phenomenon of relapses that characterized these spirochetal infections and saw relapsing fever as a useful model for understanding the workings of the immune system (reviewed in reference 16). For this reason and to meet public health needs, relapsing fever was studied by many investigators in several countries. Most of the important features of the biology and immunology of *Borrelia* spp., the etiologic spirochetes of relapsing fever, were, in fact, described at the end of the 19th century and in the first half of this century.

Two breakthroughs of the 20th century, penicillin and DDT (dichlorodiphenyltrichloroethane), effectively reduced the morbidity and mortality of the disease and the magnitude of relapsing fever epidemics. At the same time, the immunologic and microbiologic studies of the disease and its etiologic agent were reaching their technological limits. These forces combined to dramatically reduce scientific interest in relapsing fever after World War II. Relatively little work was done on the relapsing fever borreliae in the succeeding three decades.

With the powerful tools of molecular biology that are available now and the demonstration that a newly discovered member of the genus *Borrelia* causes Lyme disease, a disorder of growing worldwide importance, there was again the impetus to pick up the trail first blazed by scientists of two or more generations ago. Antigenic variation in relapsing fever once more is an intriguing topic for many biologists.

Details of the biology, biochemistry, and immunology of the relapsing fever *Borrelia* spp. will not be covered here. The reader is referred to other reviews for in-depth coverage of these areas and for citations to the works summarized briefly here (1, 4). The subject of the present review, the genetics of these organisms, is also of narrow breadth. But in this case, the restrictions are imposed not by editorial decisions but by the still limited range of our knowledge of this aspect of borreliae.

II. BIOLOGY AND IMMUNOLOGY OF RELAPSING FEVER BORRELIAE

After being bitten by a spirochete-laden tick or louse, a person may experience periods of fever lasting for 3 to 7 days, separated by 4- to 7-day intervals of well-being. During febrile periods large numbers of borreliae can be found circulating in the blood; during afebrile periods very few, if any, or-

Alan Barbour ■ Departments of Microbiology and Medicine, University of Texas Health Science Center, 7703 Floyd Curl Drive, San Antonio, Texas 78284-7758.

ganisms are detectable in the blood. Experimental infections of rodents can be initiated with a single spirochete; the interval between spirochetemic attacks is between 4 and 7 days, as it is in human cases. An infected human or mammal may suffer 10 or more relapses before either dying or recovering completely.

As early as 1918 bacteriologists had noted that spirochetes recovered from the first attack of relapsing fever differed serologically from organisms recovered during relapses. Subsequent studies showed that this antigenic variation was multiphasic rather than simply biphasic. Several antigenically distinct variants or serotypes were eventually identified among the progeny of a single borrelia injected into a mouse; Stoenner and co-workers isolated 25 different serotypes (17). The sequence of serotypes in succeeding relapses was often, but not invariably, the same in different animals. There seems to be a loose program influencing the order of appearance of the various serotypes during the course of experimental infection. The succession of serotypes is not completely predictable, though; serotypes that are more typically seen in late infections occasionally will make up the majority of a first relapse population (6, 17). Of additional importance for models attempting to explain antigen switching was the observation that the hereditary information necessary for expression of a given serotypic determinant was not lost from cells during the course of the infection. That is, a relapse isolate (e.g., serotype Y) recovered from a mouse infected originally with serotype X could yield serotype X again if injected as a single cell into another mouse that had no previous exposure to serotype X.

The generation time of one species, *Borrelia hermsii*, in rats and mice is approximately 6 h. If during the peak of the first relapse there are 10^6 organisms per ml—effectively 10^6 borreliae per mouse—then calculations show that it would take about 5 days for this size of population to have multiplied from a single spirochete. This estimate agrees well with observed intervals between attacks and indicates that variant serotypes are probably present in very low numbers at the time of clearance of the infecting serotype. The frequency of appearance of new serotypes in a population of borreliae growing in a mammalian host or in culture medium has been estimated directly as 10^{-4} to 10^{-3} per cell per generation (17).

III. OUTER MEMBRANE PROTEINS ARE DETERMINANTS OF SEROTYPE SPECIFICITY

In a first step toward understanding the molecular mechanisms of variation, we sought to identify the determinants of serotype specificity. The species under study was *B. hermsii*, a cause of tick-borne relapsing fever in North America (Fig. 1). At the time of these early studies, little was known about the antigenic components of borreliae. Previous immunofluorescence studies had shown that the variable antigens were located on the surface of the cell. Other experiments had indicated that relapse strain-specific antigens were proteinaceous in character.

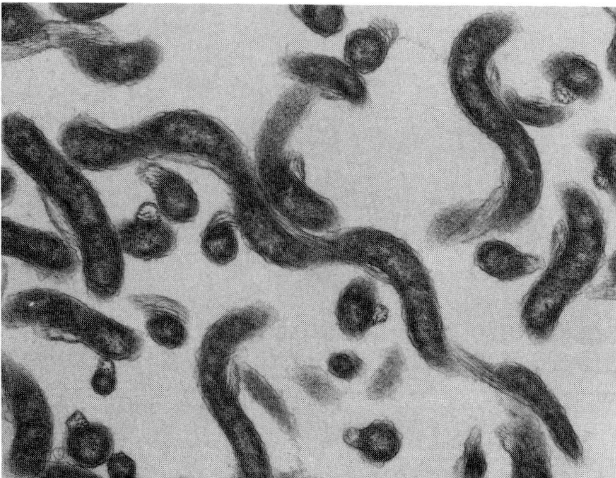

Figure 1. Transmission electron photomicrograph of cells of the relapsing fever agent, *B. hermsii.* A cell shown in a longitudinal section demonstrates the helical shape of the organism and the periplasmic flagella that traverse the cell from one end to another. The diameter of the cell is approximately 250 nm.

Our studies focused on isogenic serotypes 7, 21, and C of Stoenner's original collection. Serotypes 7 and 21 are mouse virulent and are among the more frequent serotypes to be found in mice infected with the HS1 strain of *B. hermsii.* Serotype C predominates in populations of *B. hermsii* which have been passed several times in broth culture medium. Serotype C appears to have lost the ability to mutate to other variants at detectable frequencies.

Monoclonal antibodies specific for each of the three serotypes were produced (6, 7). These monoclonal antibodies were found to react with abundant outer membrane proteins, the variable major proteins (Vmp's), that differed between serotypes in their apparent molecular weights in polyacrylamide gel electrophoresis. Vmp's were isolated by differential detergent solubilization followed by high-performance liquid chromatography. The isolated proteins were then subjected to cyanogen bromide (CNBr) proteolysis, and the resultant peptides were recovered. Using the battery of monoclonal antibodies specific for either serotype 7 or serotype 21, we examined antibody reactivities with whole cells, purified intact Vmp's, and CNBr peptides of Vmp's (2, 7). These studies showed that there was more than one serotype-specific epitope associated with each Vmp and that serotype-specific epitopes were not limited to one portion of a Vmp molecule. There are not lengthy constant or invariable regions in the Vmp's.

One monoclonal antibody that reacted with both serotype 7 and serotype 21 (but not serotype C) was obtained by hyperimmunizing a mouse. Such cross-reactive antibodies had not been found in animals with acute infections, and, thus, it is likely that the epitope for this antibody was not exposed on the surface of the cell. Nevertheless, the antibody indicated that there were conserved regions in the Vmp's. This assumption was confirmed by partial amino acid sequencing of the CNBr fragments of Vmp7 and Vmp21. This analysis revealed short regions of homology as well as lengthier spans of primary structure diversity.

IV. ANTIGENIC VARIATION IS THE CONSEQUENCE OF DNA REARRANGEMENTS

Presented with the above information, it was difficult avoiding the conclusion that the antigenic variation during relapsing fever is the result of a reversible genetic change in the borreliae. The next experimental step was, therefore, examination of this organism at the level of its genome. However, despite the century-old familiarity of microbiologists with borreliae, very little was known about the genetics of this or other genera of spirochetes. One reason for this paucity of information has been the difficulty in growing borreliae in vitro. They grow slowly in broth medium and do not readily form colonies on solid medium. Until recently the guanosine-cytosine contents of the DNA and whether there are bacteriophage or plasmids were unknown (4, 11).

Our entree to the borrelia genome was provided by synthetic oligonucleotide probes designed on the basis of previously determined amino acid sequences (7, 12). One probe was judged likely to be specific for the gene for Vmp7. If Vmp expression was, as we supposed, determined at or before the transcriptional level, then we would expect to find that the Vmp7 message would be detected in serotype 7 cells but not in cells of other serotypes. This was found when RNA extracts of different serotypes were probed with the Vmp7-specific oligonucleotide. Northern (RNA) blots demonstrated that sequences expected to correspond to Vmp mRNA were present exclusively in serotype 7 cells as an mRNA species of 1,100 nucleotides (12). A Vmp21-specific probe, on the other hand, hybridized to mRNA of serotype 21 cells but not to serotype 7 mRNA (14). Thus, different serotypes could be distinguished at the mRNA level.

The Vmp7-specific probe was then used to identify the gene (*vmp-7*) encoding this protein (12). In Southern blots serotype 7 had copies of the *vmp-7* sequences in two different restriction fragments. In contrast, other serotypes had only one of these versions of *vmp-7*. It appeared then that there were expression-linked and silent copies of *vmp-7*. A similar state was also found in serotype 21 (14). That is, *vmp-21* sequences were found in two loci in serotype 21 DNA but in only one of the two restriction fragments in the DNA of other serotypes.

The organization of the Vmp genes themselves was further examined by cloning the different forms of *vmp-7* and *vmp-21* into *Escherichia coli* (12, 14). Whereas some of the cloned genes produced full-length, immunoreactive Vmp7 or Vmp21 protein in *E. coli*, other *vmp* clones did not express the appropriate protein. Whether a *vmp* gene was expressed in either *B. hermsii* or *E. coli* or not appeared to be governed by the regions that flanked it in the *B. hermsii* genome. The expression-linked or active copies of *vmp-7* and *vmp-21* had the same several kilobases of DNA upstream of the gene. On the other hand, the 5′-flanking sequences of the silent forms of *vmp-7* and *vmp-21* were different from one another. The 3′-flanking fragments of the active and silent *vmp* genes were identical for a particular *vmp*. The downstream sequences of different *vmp* genes were not the same.

For the particular antigenic switch examined in most detail, a change in serotype from 7 to 21, it appeared that a silent copy of a *vmp-21* gene had replaced a copy of a *vmp-7* gene at the expression site. The reciprocal product of that switch would be a fusion of the formerly active *vmp-7* with the sequence that had previously been positioned upstream of a silent *vmp-21*. This hypothetical structure was not detectable in the postswitch population of borreliae. Effectively, the old active copy of the *vmp* gene had been destroyed.

These results are consistent with the occurrence of nonreciprocal recombinations involving silent and active loci of different *vmp* genes during antigenic changes. Furthermore, the recombinations appear to be unidirectional; a common sequence is found upstream of active *vmp* genes but not silent genes.

V. THE *vmp* GENES ARE ON LINEAR PLASMIDS

In the course of the studies reviewed above, we noted that what we took to be "chromosomal" DNA of *B. hermsii* comprised three to four closely spaced bands in conventional agarose electrophoresis gels. Pulsed-field gel electrophoresis further resolved the DNA into true chromosomal DNA and discrete extrachromosomal elements (14). The migration of the plasmids in the pulsed-field gels was characteristic of linear DNA, and additional studies showed that the linear plasmids are not circularly permuted (14). The lengths of the plasmids were 23 to 50 kilobases.

Southern blots of the chromosomal DNA and the linear plasmids revealed that the silent and active copies of *vmp-7* and *vmp-21* are arrayed on the plasmids (14). A plasmid of approximately 23 kilobases carries either an expressed *vmp-7* or *vmp-21*.

There are multiple copies of the *vmp*-bearing plasmids (14). The extra copies of *vmp* genes presumably facilitate rapid evolution of the Vmp proteins. Moreover, an extrachromosomal rather than a chromosomal position for the *vmp* genes might save the borrelia from disruption of gene order and essential regions of the genome during *vmp* gene rearrangements.

A linear replicon must protect its ends from exonucleolytic attack. To further characterize the linear plasmid termini, we chose to examine another *Borrelia* species, *B. burgdorferi*, the cause of Lyme disease. The major surface protein genes of this borrelia are also located on a linear plasmid, in this case one of 49 kilobases (3, 10). There appears to be less variability over time in the plasmids of *B. burgdorferi* than those of *B. hermsii*. Thus, structural studies were less likely to be hampered by changes in spirochete populations if *B. burgdorferi* rather than *B. hermsii* was the subject.

The investigations showed that the 49-kilobase linear plasmid of *B. burgdorferi* (and presumably the linear plasmids of *B. hermsii* as well) has covalently closed ends. This terminus architecture is also assumed by members of the poxvirus group, including the vaccinia virus (9, 13). If the borrelial linear plasmids are like the poxviruses in further detail, we would expect to find a palindromic nucleotide sequence conducive to hairpin formation at the ends.

While much remains to be discovered about these novel linear plasmids of *Borrelia* spp., including their method of replication and terminal DNA sequences, it is useful to consider other possible advantages for the borreliae in having their *vmp* genes on linear extrachromosomal replicons. For linear plasmid molecules a single recombination event suffices to complete a switch between two plasmids. If the *vmp* genes were on circular instead of linear plasmids, two recombination events would presumably be necessary: first, for a joint molecule to form, and, second, for this cointegrate to be resolved into two plasmid molecules again.

Southern blots of intact linear plasmids with *vmp*-specific and expression sequence probes provided further evidence that recombination involving these loci is unidirectional. Having seen the input and output of this switch, one would likely conclude that a silent *vmp* gene's translocation from a 25-kilobase-pair plasmid to the 23-kb plasmid containing the expression sequence produces a change in serotype. Movement of the expression sequence on the 23-kilobase-pair plasmid to a different-sized plasmid has not been observed. Nevertheless, occasional transposition of an upstream activating sequence cannot be ruled out, at least until many other serotypes have been examined and the regulation of *vmp* gene expression is better understood.

The finding that only those recombinant plasmids that contained an active copy of a *vmp* gene produced Vmp protein in *E. coli* suggests that control of gene expression is to a large extent determined by the immediate upstream sequence. A *trans*-acting "derepressor" encoded at a distant site is probably not involved. Still, we do not know as yet whether appropriate procaryotic promoter sequences for *vmp* gene transcription might not also be present at the silent site. It is possible, therefore, that the upstream expression sequence fused to a *vmp* gene functions in a way to activate or significantly enhance transcription of the sitting *vmp* gene.

VI. POSSIBLE MECHANISMS OF *vmp* GENE SWITCHING

A molecular model of the switch must successfully account for the following features of antigenic variation in relapsing fever. (i) *vmp* genes are similar to one another but not identical; there may be as much as a 30% difference in nucleotide sequence in some regions of the *vmp-7* and *vmp-21* genes. (ii) In a switch a complete or nearly complete *vmp* gene is replaced at the expression site. The source of the displacing *vmp* gene is a distant site containing a *vmp* gene copy. The intragenic recombination that typifies most pilin gene switches in *Neisseria gonorrhoeae* (J. Swanson and J. M. Koomey, this volume) has not been observed. (iii) The exchange is not reciprocal, and it appears to be unidirectional. (iv) A common upstream expression sequence has been identified, but an analogous downstream site has not been found. (v) The interchange of *vmp* genes takes place between linear plasmids.

Much remains to be learned about the original participants in a switch and the final products. Only one particular switch has been examined in detail, and data on this one event are still scanty. The mouse provides a powerful selection against incomplete or partial antigenic changes, and, thus, those intermediate steps in the process that might have proved useful for analysis have not been available. Espousal of one model rather than another is inappropriate at this time. Nevertheless, consideration of different mechanisms provides a framework for design of future experiments. Different models for the antigenic switch in *B. hermsii* have been discussed previously by Plasterk et al. (14, 15) and Borst and Greaves (8).

One model describes switch events in terms of gene conversion and specifies that linear plasmids exchange genetic material through homologous recombination. The regions of homology that constitute the initial heteroduplex joint are the *vmp* genes themselves. The heteroduplex is resolved by mismatch repair of the strand representing the original active *vmp* gene. Consequently, a *vmp* gene is converted at the expression site without loss of a silent form of the gene. A crossover occurs 5′ to the heteroduplex that forms. The crossover effects translocation of distal regions of the silent gene-bearing plasmid to the expression-linked plasmid. The reciprocal products of this recombination, i.e., the old active gene and its downstream sequence and the upstream region of the translocated silent *vmp* gene, are lost. A variation of this model prescribes that the homologous regions participating in recombination are the 5′ ends of the genes and perhaps a short portion of the adjacent upstream sequence. The heteroduplex extends rightward toward the 3′ end of the *vmp* gene by branch migration.

Recombination presumably initiates with a single-strand or double-strand break in the general area of a silent or active *vmp* gene. If recombinogenic strand disruption began in the Vmp structural genes themselves, one might expect to find bidirectional switches because unless there was differential DNA methylation or the equivalent of chromatin in the borreliae, the putative strand-breaking activity could not distinguish between *vmp* sequences in their silent and active environments.

The evidence so far indicates that translocation is in fact unidirectional. Therefore, unless epigenetic factors further defined the expression state of a *vmp* gene sequence, we must look to the flanking regions for signals to a strand disruption activity. The finding that it is the sequence 5′ to a *vmp* gene that determines whether the gene is active or not suggests that unidirectionality is inherent in the 5′-flanking sequence. For example, an unraveling of DNA in this upstream region might set in motion events resulting in gene conversion. Recombination would initiate in the recipient rather than the donor DNA.

DNA disruption occurring consistently more often in this upstream region implies a degree of site specificity to the switch that is additional to whatever is conferred by *vmp* gene relatedness. The site specificity could be a function of an endonuclease with high affinity for a particular sequence at the expression site or a *cis*-acting, recombination-enhancing sequence similar in function to the chi site of *E. coli.* The mating type switch of *Saccharomyces cerevisiae* is an example of a high-frequency gene conversion process that is initiated by a sequence-specific endonuclease (A. Klar, this volume).

Another factor that could effect site specificity is a transposonlike element upstream of an active *vmp* gene. Complete or partial excision of a transposable element could lead to a double-strand break in the right location for homologous recombination between *vmp* genes to ensue. The hypothetical upstream transposon would not directly affect gene expression, as it would if inserted into or excised from a transcriptional unit. Rather, it would, like one of the maize transposons (N. V. Fedoroff, this volume), bring about change through creation of recombinogenic double-strand breaks in the linear plasmids. This or another mechanism that produces a double-strand break could also account for the nonreciprocality of the switch; in the aftermath of double-stranded breakage the old active *vmp* gene and the 3′ sequences are destroyed.

An illustrative model that incorporates some of the possible events discussed above is shown in Fig. 2. The model depicts a change in serotype of one borrelia cell from 7 to 21. The schematic figure is not drawn to scale and shows only roughly the size and position relationships. It assumes that the silent *vmp* genes in question are on different linear plasmids and that the active *vmp* genes are on a single species of plasmid. Regions of sequence similarity (not identity) constitute the 5′ ends of the *vmp* genes and perhaps the immediate flanking region. Recombination is initiated by a double-strand break in the upstream expression sequence at the junction with the active *vmp* gene. The short sequence that is the site for specific enzymatic cleavage is found only on the expression plasmid. Once the break occurs, a strand from a silent *vmp*-bearing plasmid intercalates at the break site in the region of sequence similarity. From that point rightward the invading *vmp* gene and its distal sequence are copied. If single-strand disruptions occurred at the expression site instead of double-strand cleavages, the invading silent *vmp* gene would be copied and replace the sitting *vmp* gene through gene conversion.

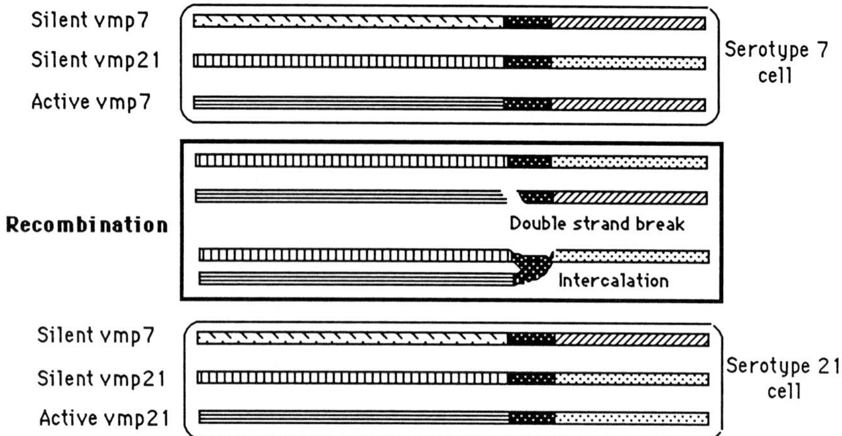

Figure 2. Schematic representation of possible events in a switch of a *B. hermsii* cell from serotype 7 to serotype 21. The cell in its serotype 7 and serotype 21 forms is shown at the top and the bottom. Some of the hypothetical intermediate steps in the recombination are shown in the box in the middle. Different regions of plasmids bearing active or silent *vmp* genes are indicated by the various fill-in patterns constituting the individual idealized plasmids. A region of sequence similarity between different *vmp* genes is shown by white dots on a black background.

It is possible—though, as discussed below, unlikely—that the switch is the consequence of transposition of an element containing an entire *vmp* gene as well as prerequisite terminal repeats. This hypothetical *vmp* gene transposon conceivably also might contain sequences which encode a transposase or recombinase function. There are many examples of transposons bearing genes of selective advantage for a bacterium. Nevertheless, a model specifying that *vmp*-bearing transposons account for reversible antigenic variation of 25 to 30 distinct Vmp proteins soon becomes dizzyingly complicated. In its most pure form the model prescribes that each of the 25 or so *vmp* genes is borne by its own transposable element; during a switch one *vmp* transposon is duplicated, and one of the copies displaces another *vmp* transposon at the expression site. The formidable amount of fine tuning implicit in this model argues against it. (A simpler transposition model assigns the promoter alone to a transposable element; a *vmp* gene is activated when this transposon is inserted upstream of it. However, if this were the case, we would likely have found that both the expressed and silent versions of the *vmp-7* gene were on the same-sized linear plasmid. This was not detected.)

Whatever recombination mechanism is revealed to mediate antigenic variation, there will still remain questions about copy numbers of the plasmids and *vmp* genes. Multiple copies of silent and expressed genes provide conceivable advantages for the borreliae. For one thing, silent *vmp* genes need not be duplicated when they translocate. However, if there are multiple copies of the active genes on individual plasmids, the cell faces the problem of having two types of active *vmp* gene immediately after a switch. A borrelia still placing the old Vmp protein on its surface, even if in a mosaic with the new Vmp, presumably has made little headway in its escape from hostile antibodies. The cell would be well served by a mechanism which either quickly repressed expression of old *vmp* copies or asymmetrically segregated out the active *vmp*-bearing plasmids during the next cell division.

An inability to genetically manipulate spirochetes of the genus *Borrelia* could ultimately limit assessment of the validity of each of the different models. Systems for transformation or transduction would have to be devised. Ideally we would be able to introduce back into borreliae linear plasmids bearing wild-type or mutagenized sequences as well as an antibiotic resistance marker. These techniques are under development. Given the variety of novel biological phenomena that *Borrelia* spp. have already displayed, these efforts to delve further into borrelial genetics seem worthwhile.

Acknowledgments. I thank Paul Barstad, Sven Bergstrom, Cindy Freitag, Claude Garon, Joe Meier, Ron Plasterk, Mel Simon, and Herb Stoenner for their contributions to the studies reviewed here.

LITERATURE CITED

1. **Barbour, A. G.** 1987. Immunobiology of relapsing fever. *Contrib. Microbiol. Immunol.* 8:125–137.
2. **Barbour, A. G., O. Barrera, and R. Judd.** 1983. Structural

analysis of the variable major proteins of *Borrelia hermsii*. *J. Exp. Med.* **158**:2127–2140.

3. **Barbour, A. G., and C. F. Garon.** 1987. Linear plasmids of the bacterium *Borrelia burgdorferi* have covalently-closed ends. *Science* **237**:409–411.
4. **Barbour, A. G., and S. F. Hayes.** 1986. Biology of *Borrelia* species. *Microbiol. Rev.* **50**:381–400.
5. **Barbour, A. G., and H. G. Stoenner.** 1984. Antigenic variation of *Borrelia hermsii*. *UCLA Symp. Mol. Cell. Biol. New Ser.* **20**:123–135.
6. **Barbour, A. G., S. L. Tessier, and H. G. Stoenner.** 1982. Variable major proteins of *Borrelia hermsii*. *J. Exp. Med.* **156**:1312–1324.
7. **Barstad, P. A., J. E. Coligan, M. G. Raum, and A. G. Barbour.** 1985. Variable major proteins of *Borrelia hermsii*: epitope mapping and partial sequence analysis of CNBr peptides. *J. Exp. Med.* **161**:1302–1314.
8. **Borst, P., and D. R. Greaves.** 1987. Programmed gene rearrangements altering gene expression. *Science* **235**:658–667.
9. **DeLange, A. M., M. Reddy, D. Scraba, C. Upton, and G. McFadden.** 1986. Replication and resolution of cloned poxvirus telomeres in vivo generates linear minichromosomes with intact viral hairpin termini. *J. Virol.* **59**:249–259.
10. **Howe, T. R., F. W. LaQuier, and A. G. Barbour.** 1986. Organization of genes encoding two outer membrane proteins of the Lyme disease agent *Borrelia burgdorferi* within a single transcriptional unit. *Infect. Immun.* **54**:207–212.
11. **Hyde, F. W., and R. C. Johnson.** 1984. Genetic relationship of Lyme disease spirochetes to *Borrelia*, *Treponema*, and *Leptospira* spp. *J. Clin. Microbiol.* **20**:151–154.
12. **Meier, J. T., M. I. Simon, A. G. Barbour.** 1985. Antigenic variation is associated with DNA rearrangements in a relapsing fever borrelia. *Cell* **41**:403–409.
13. **Merchlinsky, M., and B. Moss.** 1986. Resolution of linear minichromosomes with hairpin ends from circular plasmids containing vaccinia virus concatemer junctions. *Cell* **45**: 879–884.
14. **Plasterk, R. H. A., M. I. Simon, and A. G. Barbour.** 1985. Transposition of structural genes to an expression sequence on a linear plasmid causes antigenic variation in the bacterium *Borrelia hermsii*. *Nature* (London) **318**:257–263.
15. **Plasterk, R. H. A., M. I. Simon, and A. G. Barbour.** 1986. Molecular basis of antigenic variation in a relapsing fever Borrelia, p. 127–146. *In* T. H. Birkbeck and C. W. Penn (ed.), *Antigenic Variation in Infectious Diseases*. IRL Press, Oxford.
16. **Russell, H.** 1936. Observations on immunity in relapsing fever and trypanosomiasis. *Trans. R. Soc. Trop. Med. Hyg.* **30**: 179–190.
17. **Stoenner, H. G., T. Dodd, and C. Larsen.** 1982. Antigenic variation of *Borrelia hermsii*. *J. Exp. Med.* **156**:1297–1311.

Chapter 37

High-Frequency Switching in *Candida albicans*

DAVID R. SOLL

I. INTRODUCTION

Candida albicans is the major yeast pathogen of humans (2, 19). In healthy individuals, it can live as a commensal in the mouth, gut, and genitalia, but in compromised patients and in a significant number of apparently healthy individuals, it invades host tissue. *C. albicans* is dimorphic (33, 34), growing in a budding yeast form (Fig. 1A) and also in a hyphal form (Fig. 1B). Growth in the hyphal form seems to be important in pathogenicity, since mutants which have lost this ability also exhibit decreased virulence in rat models (31).

C. albicans contains between 35 and 40 fg of DNA per cell in the G1 phase of the cell cycle (24, 33, 35, 41), which is close to the DNA content of diploid *Saccharomyces cerevisiae* (16). Although arguments concerning ploidy still exist (24), a number of tests, including UV survival (20), chemical mutagenesis (20), reassociation kinetics of denatured single-copy DNA (24), and the demonstration of genetic heterozygosity at a number of gene loci (42, 43) indicate that the studied strains of *C. albicans* are diploid. Single-copy sequences compose approximately 85% of the total genome and repetitive sequences approximately 15% (24). Six chromosomes have been separated by orthogonal field, alternating gel electrophoresis (17; C. Bohlman, N. Hancock, and F. Bayliss, Abstr. Annu. Meet. Am. Soc. Microbiol., 1986, K209, p. 228), well below the 12 chromosomes resolved by this technique for *S. cerevisiae* (7), suggesting that the chromosomes may in fact be larger on average in *C. albicans* than in *S. cerevisiae*. There has been no indication of a sexual phase or meiosis in *C. albicans* (40), and the finding of naturally occurring recessive alleles lethal in one strain (43) indicates that haploidization would often be lethal.

It was assumed that the developmental repertoire of *C. albicans* consisted of only the budding (Fig. 1A) and hyphal (Fig. 1B) phenotypes (19, 33). The transition from the bud to the hyphal phenotype can be rapidly induced in well over 90% of a cell population by releasing stationary-phase cells or starved log-phase cells in the budding form (6, 36) into fresh nutrient medium at 37°C, pH 6.7 (32, 34). The reverse transition from the hyphal to the bud phenotype can also be induced in well over 90% of a population of cells growing in the hyphal form by simply lowering the pH from 6.7 to 4.5 (18). These rapid developmental transitions most likely reflect physiological changes, which include differential gene regulation and posttranscriptional changes in the temporal and spatial dynamics of cell cycle events (34), and are not the result of DNA rearrangements. However, it was recently demonstrated that most strains of *C. albicans* and the related species *Candida tropicalis* also possess a second level of phenotypic change in which cells switch heritably but reversibly at frequencies as high as 10^{-2} per cell generation among a limited number of general phenotypes distinguishable by differences in colony morphology and, in some cases, by differences in cellular morphology (1, 25, 29, 30, 37; B. Slutsky, Ph.D. thesis, University of Iowa, Iowa City, 1986). It is likely that

David R. Soll ■ Department of Biology, University of Iowa, Iowa City, Iowa 52242.

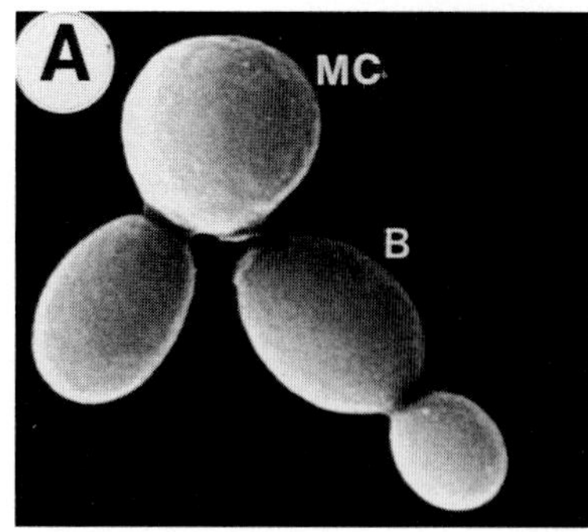

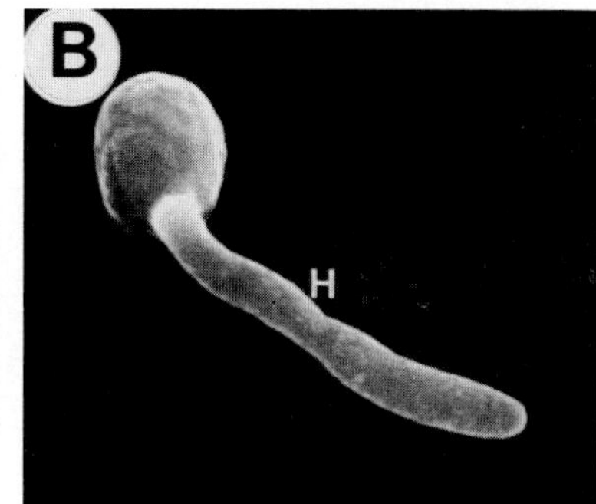

Figure 1. Scanning electron micrographs of (A) a budding cell and (B) hypha-forming cell of *C. albicans* 3153A. Abbreviations: MC, mother cell; B, primary bud; H, hypha.

these switching systems are due to reversible transposition or DNA rearrangement.

II. HIGH-FREQUENCY SWITCHING IN LABORATORY STRAIN 3153A

An early study demonstrated that variant sectors arise frequently in streaks of *C. albicans* on nutrient agar (5). At least 17 variants could be identified by streak macromorphology, several of which were highly unstable during extended culture. Several other reports of variant instability followed this study (e.g., 14, 21), but the full scope of phenotypic switching in *C. albicans* was first realized in a comparison of colony variability in the common laboratory strain 3153A (31). The predominant colony phenotype of this strain is "smooth" (Fig. 2A1), but variant colonies exhibiting abnormal colony morphologies appeared at a frequency of roughly 10^{-4}. These include "star" (Fig. 2A2), "ring" (Fig. 2A3), "irregular wrinkle" (Fig. 2A4), "stipple" (Fig. 2A5), "hat" (Fig. 2A6), and "fuzzy" (Fig. 2A7). These variant morphologies reflected, at least in part, the developmental characteristics of cells in the colony dome (29, 38). In general, there was a correlation between the proportion and length of hyphae in the colony dome and colony wrinkling (Fig. 2D). When cells from individual variant colonies were plated, the majority of the resulting colonies exhibited the variant phenotype. Low doses of UV light (>90% survival) stimulated the formation of variant colonies in wild-type populations almost 100-fold (to a frequency of about 10^{-2}) (29). In the cases of both spontaneous and UV-induced variants, the combined frequencies of switching to other variant phenotypes as well as to the original smooth wild-type phenotype was approximately 10^{-2} (Fig. 2B). An example of a switch from ring to star is presented in Fig. 2C. Although variants can switch to the smooth phenotype ("revertant smooth," Fig. 2A8) at relatively high frequencies (Fig. 2B), most of these clones do not represent true revertants, since they still exhibit a frequency of switching much higher than that of the original smooth wild-type strain (29). In overview, the intriguing characteristics of strain 3153A include (i) low- and high-frequency modes of spontaneous switching of colony phenotype, (ii) a limited number of switch phenotypes, (iii) heritability of the switch phenotypes, (iv) interconversion among variant phenotypes and reversibility at high frequency, and (v) UV stimulation of phenotypic switching.

III. THE "WHITE-OPAQUE TRANSITION": A SECOND HIGH-FREQUENCY SWITCHING SYSTEM

A second system of high-frequency switching of colony phenotypes has been characterized in strain WO-1, isolated from the blood and lungs of a bone marrow transplant patient (30). The majority of cells from this original isolate formed smooth white colonies typical of most *C. albicans* strains (Fig. 3A), but a minority formed wider gray colonies (originally referred to as "opaque" under the lighting employed). Most cells from white colonies again formed white colonies, but between 1 and 10% formed opaque colonies. Cells from opaque colonies in turn usually formed opaque colonies, but a minority again formed white colonies (Fig. 3B). In both white and opaque colonies, sectors of the alternative phenotype appeared frequently after 7 days (Fig. 3C and 3D, respectively), and the appearance of opaque sectors at the edge of white colonies was dramatically stimulated by wrapping petri dishes with Parafilm and incubating them for roughly 12 days (Fig. 3E). Transitions between the white and opaque phenotypes occurred repeatedly with no effect on the subsequent frequency or repertoire of switching phenotypes (1). Strain WO-1 also switched to a wrinkled colony form (Fig. 3F) and two fuzzy colony forms (Fig. 3G and H)

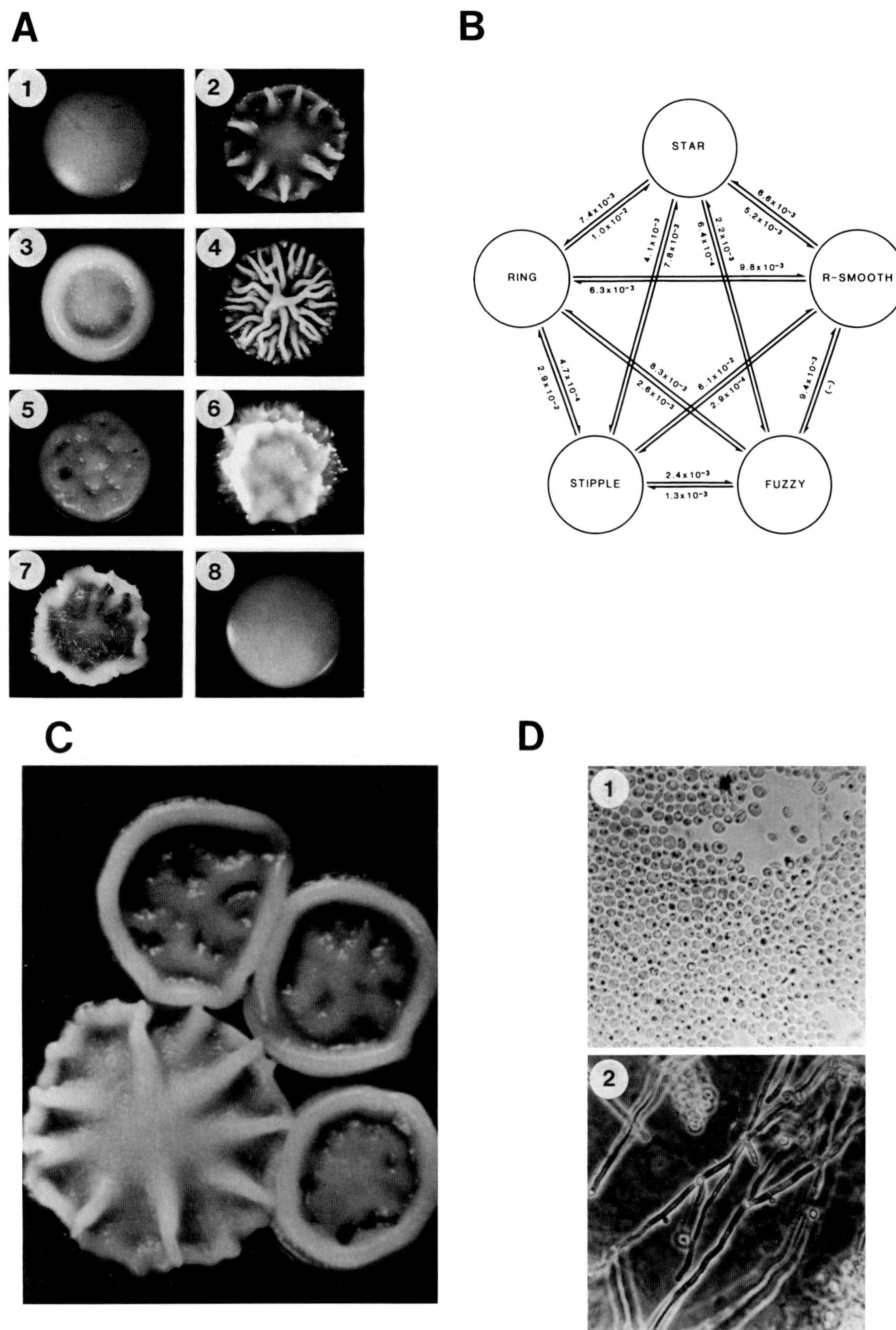

Figure 2. Switching in *C. albicans* 3153A. (A) Switch phenotypes: (1) original smooth, (2) star, (3) ring, (4) irregular wrinkle, (5) stipple, (6) hat, (7) fuzzy, (8) revertant smooth. (B) The frequencies of switching between star, ring, stipple, fuzzy, and revertant smooth (r-smooth). Note that revertant smooth appears similar to original smooth (panel A), but retains a high frequency of spontaneous switching. (C) An example of a switch from ring to star. (D) Cellular phenotypes in the dome of (1) an original smooth colony and (2) a star colony. Note the preponderance of cells in the budding phase in the former and hyphae in the latter colony.

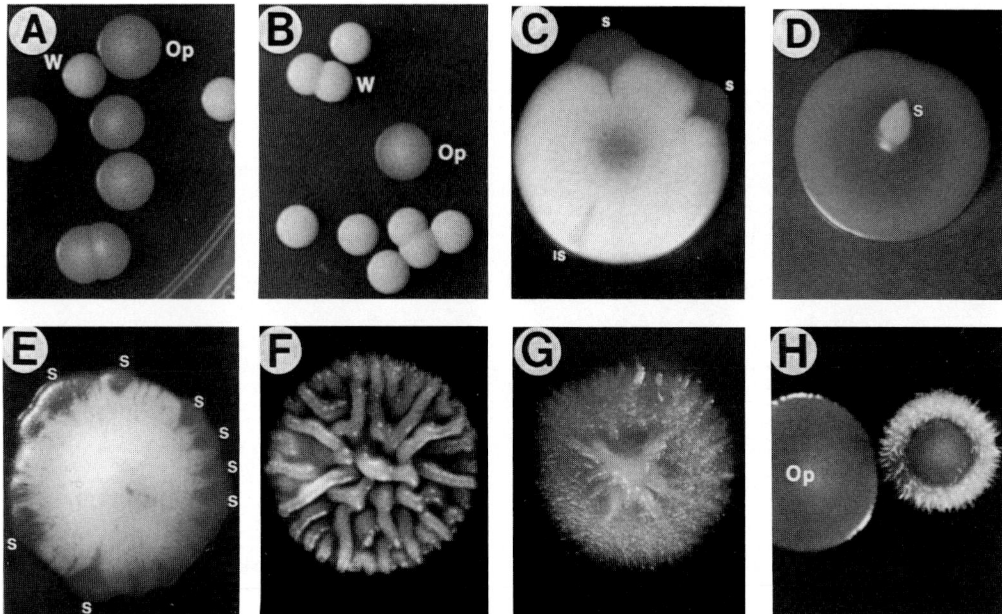

Figure 3. The white-opaque transition in *C. albicans* WO-1. (A) An example of a switch from opaque (Op) to white (W) in cells plated from a single opaque clonal colony. (B) An example of a switch from white (W) to opaque (Op) in cells plated from a single white colony. (C) An example of a white colony with two opaque sectors (S) and a sector of intermediate phenotype (IS) (agar containing phloxine B differentially stains opaque cells [1]). (D) An example of an opaque colony with a central white sector (S) (phloxine B stained). (E) White colony aged on agar containing phloxine B in parafilmed dish exhibiting high-frequency sectoring (S) at colony edge. (F) An example of the irregular wrinkle colony morphology. (G) An example of one type of fuzzy colony. (H) An example of a second type of fuzzy colony morphology, adjacent to an opaque colony (Op). (See color plates, this volume.)

at reduced frequencies. Wrinkled and fuzzy variants also switched back at high frequency to the white and opaque phenotypes (30).

The frequencies of switching between white and opaque phenotypes have been difficult to measure because of differences in generation time between white and opaque cells, enrichment due to differences in spreading dynamics on agar, and apparent differences in switching frequency related to growth conditions (25, 30). The frequencies of white-colony-forming units in opaque colonies and of opaque-colony-forming units in white colonies were recently reported to be about 2×10^{-2} to 3×10^{-2} (1) after 6 days at 25°C (1). In contrast, application of the Luria-Delbruck fluctuation analysis resulted in rates of 5×10^{-4} (opaque to white) and 2×10^{-4} (white to opaque) per cell division (25). These rates of switching can be increased 10- to 100-fold by treating cells with low doses of UV light (>80% survival) (J. A. Anderson, E. Wilson, B. Morrow, and D. R. Soll, unpublished observations). This UV stimulation effect acts via the switching system and cannot be explained by enrichment due to differential sensitivity to UV light, although it has been demonstrated that opaque cells are more sensitive than white cells to UV killing (P. T. Magee, personal communication).

Perhaps the most striking aspect of the white-opaque transition is the unique cellular architecture of the opaque phase (1, 30). In the white phase, budding cells are round (Fig. 4A) and form buds with the same polarity as other strains of *C. albicans* (12, 32, 34). In contrast, opaque cells are bean shaped (Fig. 4B) rather than round, and often one end of the cell is wider than the other. The volume of a stationary-phase white cell is about 33 μm^3, and that of an opaque cell is 114 μm^3 (30). The dry weight of a stationary-phase white cell is on average 21 pg, and that of an opaque cell is 43 pg (30). Therefore, an opaque cell is on average roughly three times larger and two times heavier than a white cell. However, both cell types contain on average one nucleus and similar amounts of DNA (30).

Comparisons of white and opaque budding cells by transmission electron microscopy revealed dramatic differences (30). White budding cells exhibited a single nuclear profile, multiple mitochondrial profiles, a single small vacuole, and compacted ribosome-rich cytoplasm (Fig. 4A). In contrast, opaque cells exhibited an interior dominated by one or more large membrane-bound vesicles containing spaghetti-like material (Fig. 4B) which were originally believed to represent expanded nuclear regions (30). It is now clear that these large vesicles do not represent the cell nucleus (J. Anderson and D. R. Soll, manuscript in

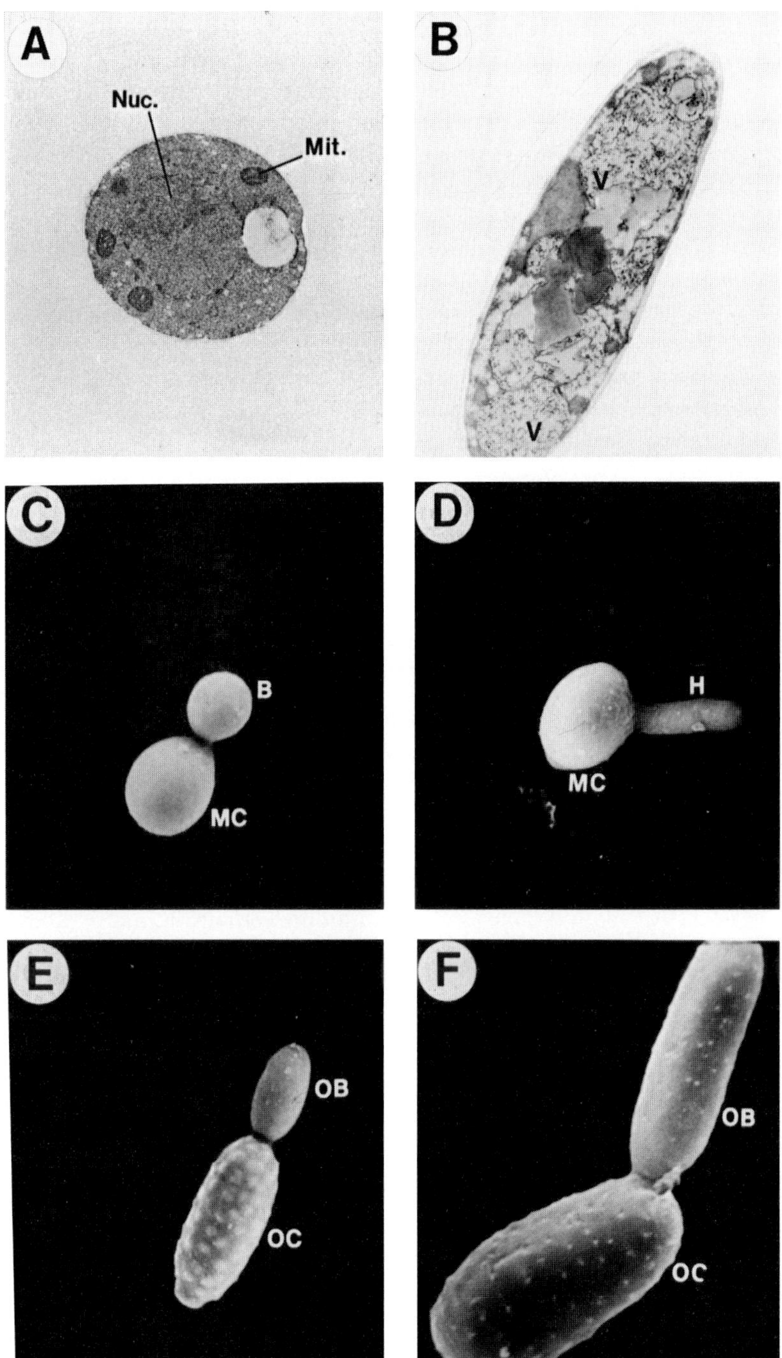

Figure 4. The cellular phenotypes of white and opaque cells in the white-opaque transition in *C. albicans* WO-1. (A) Transmission electron micrograph of a white cell in the budding phenotype which was Glusulase treated to remove the cell wall. (B) Transmission electron micrograph of an opaque cell which was Glusulase treated to remove the cell wall. (C) Scanning electron micrograph of a white cell forming a bud (note the relatively smooth cell wall). (D) Scanning electron micrograph of a white cell forming a hypha (note the relatively smooth cell wall of mother cell and daughter hypha). (E) Scanning electron micrograph of an opaque cell forming an opaque bud (note the "pimples" on the mother cell; pimples will emerge on the daughter bud only after it achieves 50% volume [1]). (F) Scanning electron micrograph of an opaque cell with a maturing bud (note the punctate pattern of dots on the surface of mother cell and daughter bud). Abbreviations: Nuc., nucleus; Mit., mitochondrion; V, large membrane-bound vesicle in opaque cells; MC, white mother cell; B, white bud; H, white hypha; OC, opaque mother cell; OB, opaque bud.

preparation). The nucleus as well as mitochondria and ribosome-filled cytoplasm are pressed by these large vesicles against the cell wall. The function of these vesicles is unknown. The wall of opaque cells also exhibits a unique morphology which can be visualized by scanning electron microscopy (1). The surfaces of white budding cells and white hypha-forming cells are relatively smooth (Fig. 4C and 4D, respectively), whereas the surface of mature opaque cells exhibits an unusual pimpled (Fig. 4E) or punctate (Fig. 4F) surface pattern (Anderson and Soll, in preparation).

White and opaque cells also differ in sugar assimilation patterns (Anderson and Soll, in preparation), temperature sensitivity (25, 30), UV sensitivity (25), and sulfometuron methyl sensitivity and 5-fluoroorotic acid sensitivity (25). More interestingly, opaque cells possess one or more hypha-specific and opaque-specific surface antigens not evident on the surface of white budding cells (1), and do not exhibit the same environmental constraints on the bud-to-hypha transition exhibited by white budding cells (30). Together, these results demonstrate that the reversible transition from the white to the opaque phase results in dramatic and extensive changes in cellular physiology, cellular architecture, and developmental regulation.

IV. DISCUSSION

Besides the switching systems which have been carefully characterized in *C. albicans* strains 3153A and WO-1, a "smooth white-heavy myceliated" transition and a system involving changes in colony diameter have recently been described (37; D. R. Soll, R. Galask, S. Isley, T. V. G. Rao, and D. Stone, submitted for publication). A switching system has also been characterized in *C. albicans* 1001 which includes "smooth," "rough," "wavy," and "fuzzy" colony phenotypes (21), and three distinct switching systems have been described in the related species *C. tropicalis* (38, 39). It appears that in both *C. albicans* and *C. tropicalis*, no strain exhibits more than one switching system defined by the repertoire of colony phenotypes.

We suggest that high-frequency switching plays a role in pathogenesis. Multiple-switch phenotypes of single strains can cohabit the same sites of infection (37) and alternate as the dominant phenotypes through recurrent episodes of vaginal candidiasis (Soll et al., submitted). In addition, several cellular characteristics implicated in *Candida* pathogenesis, including acid protease secretion (8, 23), adhesion to epithelium (27), drug resistance (9), antigenicity (4, 22), and hypha formation (31), are affected by high-frequency switching. Opaque cells in strain WO-1 excrete acid protease at far higher levels than do white cells (T. Ray, C. Payne, and D. R. Soll, manuscript in preparation), exhibit alterations in adhesion to buccal epithelial cells, hydrophobicity, and cohesion (M. J. Kennedy, A. Rogers, L. R. Hanselman, D. R. Soll, and R. J. Yancey, *Mycopathologia*, in press), possess altered surface antigenicity (1), and exhibit changes in the environmental constraints on hypha formation (30). Switch phenotypes of strain 3153A also exhibit changes in the environmental constraints on hypha formation and resistance to 5-fluorocytosine and nystatin (Slutsky, Ph.D. thesis).

The molecular basis of high-frequency switching in *Candida* has not been elucidated, and several possible mechanisms can be considered. The characteristics of switching relevant to molecular models include (i) inheritance of both low- and high-frequency modes of spontaneous switching, (ii) UV stimulation of switching in cells in the low-frequency mode, (iii) multiple-switch phenotypes, and (iv) accurate reversibility or interconversion at high frequency. Several different mechanisms elucidated in other systems are compatible with these observations: (i) movement of silent genes for different switch phenotypes into an expressed locus, as in the *Saccharomyces* mating-type system (11, 13); (ii) transposition of a regulatory sequence to sites near latent genes (3, 10); (iii) an inversion switch as in *Salmonella* flagellar genes (28); (iv) gene duplication; (v) reversible aneuploidy; (vi) metastable physiological switches (without gene rearrangement); and (vii) metastable changes in DNA methylation. In addition, the possibility that the switching mechanism is specified by plasmids or mitochondria rather than chromosomal DNA cannot be excluded.

To test the possibility that a transposon may be involved in *Candida* switching, several species-specific, moderately repetitive genomic sequences have been analyzed for mobility during switching. Two of these sequences, clones Ca3 (also referred to as JH3 [37]) and 27A (27a), which are repeated roughly 20 times throughout the genome, exhibit differences in Southern blot hybridization patterns between strains but are relatively nonmobile within a strain over extended growth periods and between switch phenotypes of the same strain. A third moderately repetitive sequence, clone Ca7 (also referred to as JH7 [37]), can be highly mobile within a strain, but the pattern of transposition does not seem to correlate with switching. Because of the lack of sexuality and an effective genetic system (40), progress has been slow in investigating the molecular basis of phenotypic switching in *C. albicans*. However, the recent development of a transformation system (15) coupled with the increased interest in the molecular biology of this persistent pathogen should speed further analyses.

Acknowledgments. Much of the published and unpublished research reviewed in this chapter was supported by grants from the Public Health Service (grant AI23922 from the National Institutes of Health); the Office of the Vice President of Research and Development, University of Iowa; the Iowa High Technology Council; and the Cecil J. Rusley Memorial Fund.

I am indebted to J. Hicks and associates of Scripps Institute, La Jolla, Calif.; P. Magee and associates of the University of Minnesota; M. Kennedy at Upjohn Corp., Kalamazoo, Mich.; R. Pomes and associates, Universidad Complutense, Madrid, Spain; and S. Sherer and associates of the University of Minnesota for sharing their unpublished results with me. I am also indebted to my collaborators at Iowa, including J. Anderson, G. T. Rao, R. Galask, L. Cundiff, J. Wilson, S. Isley, D. Stone, T. Ray, N. Schroeder, and M. Lohman.

LITERATURE CITED

1. **Anderson, J., and D. R. Soll.** 1987. The unique phenotype of opaque cells in the "white-opaque transition" in *Candida albicans*. *J. Bacteriol.* **169:**5579–5588.
2. **Bodey, G. P., and V. Fainstein (ed.).** 1985. *Candidiasis*. Raven Press, New York.
3. **Boeke, J. D., D. R. Garfinkel, C. A. Styles, and G. R. Fink.** 1985. Ty elements transpose through an RNA intermediate. *Cell* **40:**491–500.
4. **Brawner, D. L., and J. E. Cutler.** 1984. Variability in expression of a cell surface determinant in *Candida albicans* as evidenced by an agglutination monoclonal antibody. *Infect. Immun.* **43:**966–972.
5. **Brown-Thomsen, J.** 1968. Variability in *Candida albicans* (Robin) Berkhout. I. Studies on morphology and biochemical activity. *Hereditas* **60:**355–398.
6. **Buffo, J., M. Herman, and D. R. Soll.** 1984. A characterization of pH-regulated dimorphism in *Candida albicans*. *Mycopathologia* **85:**21–30.
7. **Carle, G. F., and M. V. Olson.** 1985. An electrophoretic karyotype for yeast. *Proc. Natl. Acad. Sci. USA* **82:**3756–3760.
8. **Chattaway, F. W., F. C. Odds, and A. J. E. Barlow.** 1971. An examination of the production of hydrolytic enzymes and toxins by pathogenic strains of *Candida albicans*. *J. Gen. Microbiol.* **67:**255–263.
9. **Defever, K. S., W. L. Whelan, A. L. Rogers, E. S. Beneke, J. M. Veselenak, and D. R. Soll.** 1982. *Candida albicans* resistance to 5-fluorocytosine: frequency of partially resistant strains among clinical isolates. *Antimicrob. Agents Chemother.* **22:**810–815.
10. **Garfinkel, D. J., J. D. Boeke, and G. R. Fink.** 1985. Ty element transposition: reverse transcriptase and virus-like particles. *Cell* **42:**507–517.
11. **Haber, J. E.** 1983. Mating-type genes of *Saccharomyces cerevisiae*, p. 559–619. *In* J. Shapiro (ed.), *Mobile Genetic Elements*. Academic Press, Inc., New York.
12. **Herman, M. A., and D. R. Soll.** 1984. A comparison of volume growth during bud and mycelium formation in *Candida albicans*: a single cell analysis. *J. Gen. Microbiol.* **130:**2219–2228.
13. **Hicks, J. B., J. N. Strathern, and I. Herskowitz.** 1977. The cassette model of mating type interconversion, p. 457–462. *In* A. I. Bukhari, J. A. Shapiro, and S. L. Adhya (ed.), *DNA Insertion, Elements, Plasmids, and Episomes*. Cold Spring Harbor Laboratory, Cold Spring Harbor, N.Y.
14. **Ireland, R., and A. Sarachek.** 1968. A unique minute-rough colonial variant of *Candida albicans*. *Mycopathol. Mycol. Appl.* **35:**346–360.
15. **Kurtz, M. B., M. W. Cortelyou, and D. R. Kirsch.** 1986. Integrative transformation of *Candida albicans*, using a cloned *Candida* ADE2 gene. *Mol. Cell. Biol.* **6:**142–149.
16. **Lauer, G. O., T. M. Roberts, and L. C. Klotz.** 1977. Determination of the nuclear DNA content of *Saccharomyces cerevisiae* and implications of the organization of DNA in yeast chromosomes. *J. Mol. Biol.* **114:**507–526.
17. **Magee, B. B., and P. T. Magee.** 1987. Electrophoretic karyotypes and chromosome numbers in *Candida* species. *J. Gen. Microbiol.* **133:**425–430.
18. **Mitchell, L., and D. R. Soll.** 1979. Commitment to germ tube or bud formation during release from stationary phase in *Candida albicans*. *Exp. Cell Res.* **120:**167–179.
19. **Odds, F. C.** 1978. *Candida and Candidasis*. University Park Press, Baltimore.
20. **Olaiya, A. F., and S. J. Sogin.** 1979. Ploidy determination of *Candida albicans*. *J. Bacteriol.* **140:**1043–1048.
21. **Pomes, R., C. Gil, and C. Nombila.** 1985. Genetic analysis of *Candida albicans* morphological mutants. *J. Gen. Microbiol.* **131:**2107–2113.
22. **Poulain, D., G. Tronchin, B. Lefebvre, and M. O. Husson.** 1982. Antigenic variability between *Candida albicans* blastospores isolated from healthy subjects and patients with *Candida* infection. *Sabouraudia* **20:**173–177.
23. **Remold, H., H. Fasold, and F. Stait.** 1968. Purification and characterization of a proteolytic enzyme from *Candida albicans*. *Biochim. Biophys. Acta* **167:**399–406.
24. **Riggsby, W. S., L. J. Torres-Bauza, J. W. Wills, and T. M. Townes.** 1982. DNA content, kinetic complexity, and the ploidy question in *Candida albicans*. *Mol. Cell. Biol.* **2:**853–862.
25. **Rikkerink, E., B. Magee, and P. Magee.** 1988. Opaque-white: a programmed morphological transition in *Candida albicans*. *J. Bacteriol.* **170:**895–899.
26. **Roeder, G. S., and G. R. Fink.** 1983. Transposable elements in yeast, p. 299–328. *In* J. Shapiro (ed.), *Mobile Genetic Elements*. Academic Press, Inc., New York.
27. **Rotrosen, D., R. A. Calderone, and J. E. Edwards.** 1986. Adherence of *Candida* species to host tissues and plastic surfaces. *Rev. Infect. Dis.* **8:**73–85.

27a. **Scherer, S., and D. A. Stevens.** 1988. A *Candida albicans* dispersed, repeated gene family and its epidemiologic applications. *Proc. Natl. Acad. Sci. USA* **85:**1452–1456.

28. **Silverman, M., and M. Simon.** 1983. Phase variation and related systems, p. 537–557. *In* J. Shapiro (ed.), *Mobile Genetic Elements*. Academic Press, Inc., New York.
29. **Slutsky, B., J. Buffo, and D. R. Soll.** 1985. High frequency switching of colony morphology in *Candida albicans*. *Science* **230:**666–669.
30. **Slutsky, B., M. Staebell, J. Anderson, L. Risen, M. Pfaller, and D. R. Soll.** 1987. "White-opaque transition": a second high-

frequency switching system in *Candida albicans*. *J. Bacteriol.* **169**:189–197.

31. **Sobel, J. D., G. Muller, and H. R. Buchley.** 1984. Critical role of germ tube formation in the pathogenesis of candidal vaginitis. *Infect. Immun.* **44**:576–580.
32. **Soll, D. R.** 1984. The cell cycle and commitment to alternate cell fates in *Candida albicans*, p. 143–162. *In* P. Nurse and E. Streiblova (ed.), *The Microbial Cell Cycle*. CRC Press, Inc., Boca Raton, Fla.
33. **Soll, D. R.** 1985. *Candida albicans*, p. 167–195. *In* P. Szaniszlo (ed.), *Fungal Dimorphism: With Emphasis on Fungi Pathogenic to Humans*. Plenum Publishing Corp., New York.
34. **Soll, D. R.** 1986. The regulation of cellular differentiation in the dimorphic yeast *Candida albicans*, p. 5–11. *In Bioessays*, vol. 5. Cambridge University Press, Cambridge.
35. **Soll, D. R., G. W. Bedell, J. Thiel, and M. Brummel.** 1981. The dependency of nuclear division on volume in the dimorphic yeast *Candida albicans*. *Exp. Cell Res.* **133**:55–62.
36. **Soll, D. R., and M. Herman.** 1983. Growth and the inducibility of mycelium formation in *Candida albicans*: a single cell analysis using a perfusion chamber. *J. Gen. Microbiol.* **129**: 2809–2824.
37. **Soll, D. R., C. J. Langtimm, J. McDowell, J. Hicks, and R. Galask.** 1987. High frequency switching in *Candida* strains isolated from vaginitis patients. *J. Clin. Microbiol.* **85**: 1611–1622.
38. **Soll, D. R., B. Slutsky, S. Mackenzie, C. Langtimm, and M. Staebell.** 1987. Two newly discovered switching systems in *Candida albicans* and their possible roles in oral candidasis, p. 52–59. *In* J. Mackenzie and C. Squier (ed.), *Oral Mucosa Diseases: Biology, Etiology and Therapy*. Laegeforeningens Folarg, Denmark.
39. **Soll, D. R., M. Staebell, C. Langtimm, M. Pfaller, J. Hicks, and T. V. G. Rao.** 1988. Multiple *Candida* strains in the course of a single systemic infection. *J. Clin. Microbiol.* **26**:1448–1459.
40. **Whelan, W. L.** 1987. The genetics of medically important fungi. *Crit. Rev. Microbiol.* **14**:99–170.
41. **Whelan, W., and P. T. Magee.** 1981. Natural heterozygosity in *Candida albicans*. *J. Bacteriol.* **145**:896–903.
42. **Whelan, W. L., R. M. Partridge, and P. Magee.** 1980. Heterozygosity and segregation in *Candida albicans*. *Mol. Gen. Genet.* **180**:107–133.
43. **Whelan, W. L., and D. R. Soll.** 1982. Mitotic recombination in *Candida albicans*: recessive lethal alleles linked to a gene required for methionine biosynthesis. *Mol. Gen. Genet.* **187**: 477–485.

Chapter 38

Illegitimate Recombination in Bacteria

S. DUSKO EHRLICH

I. INTRODUCTION

The genome of an organism can be modified by point mutations, such as base replacements or frameshifts (reviewed in references 68 and 284), or by large DNA rearrangements, such as deletions, insertions, translocations, or duplications (252). The frequency of genome rearrangements relative to point mutations varies. Spontaneous deletions in the *Escherichia coli trp* or *lac* operons are relatively rare, since they occur

S. Dusko Ehrlich ■ Laboratoire de Génétique Microbienne, Institut des Biotechnologies-INRA, Domaine de Vilvert, 78350 Jouy en Josas, France.

with frequencies of 10^{-8} to 10^{-7} and constitute about 10% of all mutations appearing in these operons (77, 81, 236, 302). Deletions are also rare in *Salmonella typhimurium*, occurring with the highest frequency in the *cysC* gene, where they represent 30% of all Cys$^-$ mutations (63). In contrast, deletions and insertions represent a majority (70%) of spontaneous mutations in the *E. coli* bacteriophage P1 (8, 9, 245). Even more frequent are duplications, which occur in *E. coli* with a frequency of about 10^{-4} (6). The highest frequency of spontaneous genome rearrangements has been reported in streptomycetes, where deletions which inactivate the *mel* gene were detected in 1 to 30% of colonies (241). This shows that rearrangements may be a major pathway of genome alteration.

Genome rearrangements are important for evolution. Duplications provide supplementary copies of genes which can accumulate mutations and thus evolve (6, 216). Translocations and deletions alter the environment of a gene and can thus contribute to its integration within novel control circuits. Insertions of foreign material into a genome facilitate horizontal gene transfer and thereby bypass the need for similar functions to evolve repeatedly in different organisms (7, 39). Genome rearrangements are also important medically. They often accompany or cause cancers (38, 55, 304) and hereditary disorders, such as Duchenne and Becker muscular dystrophy (229) or X-linked ichtiosis (303). The occurrence of germ cell mosaicism, detected for both male and female healthy transmitters of Duchenne muscular dystrophy (16, 57), suggests that deletions appear frequently during embryonal development. Finally, genome rearrangements are important in biotechnology, since in vitro-constructed genomes are often found to be unstable (76, 226). The high frequency, the importance, and the ubiquity of genome rearrangements (they have been observed in all organisms in which they have been sought) justify the interest in the mechanisms which underlie them.

Genome rearrangements may result from "legitimate" recombination between long homologous sequences (5, 6), movements of specialized elements (insertion sequences, transposons, phages), and the activities of functions which recombine DNA in a "programmed" way (i.e., DNA inversion [29]), as well as from "illegitimate" recombination between sequences of little or no homology. Mechanisms of legitimate recombination are discussed in several recent reviews (52, 261, 262), while those underlying transposition, phage integration, and programmed rearrangements are presented in detail elsewhere in this volume. Illegitimate recombination is the focus of this chapter. It plays a major role in genome rearrangements, since it concerns all regions of a genome, while the other processes mentioned above concern only particular regions, containing long homologous sequences or the ends of specialized elements.

II. ILLEGITIMATE RECOMBINATION

Illegitimate recombination was originally defined by Franklin as recombination between sequences of little or no homology (80). It now seems appropriate to adopt a somewhat narrower definition, which excludes recombination events resulting from the normal or legitimate activities of transposition or other specialized recombination systems (i.e., DNA inversion, resolution). Franklin speculated that illegitimate recombination may be a consequence of errors of enzymes which break and join DNA or which replicate DNA (80). This proposal, which has remained valid (4), is the framework of this review.

III. DNA BREAK-AND-JOIN ENZYMES

A number of enzymatic systems which break and join DNA mediate illegitimate recombination. These systems could act in two different ways. They could generate recombinants without the intervention of other cellular functions (in this text, "recombinant" refers to the product of illegitimate recombination). Alternatively, they could only trigger illegitimate recombination or generate recombination intermediates, and additional cellular functions could be required to complete the process. The involvement of different DNA breakage and joining systems in illegitimate recombination is reviewed from the perspective of this alternative.

A. Topoisomerases

Topoisomerases (reviewed in references 69 and 87) are enzymes which modulate the superhelicity of DNA. They associate with DNA, break DNA, pass intact DNA strands through the breaks, and finally reseal the breaks. Type 1 and type 2 topoisomerases make single-stranded (ss) and double-stranded (ds) breaks, respectively. *E. coli* topoisomerase I (omega protein, type 1) decreases the superhelicity of DNA, while topoisomerase II (DNA gyrase, type 2) increases it. Topoisomerase I is a 101-kilodalton polypeptide encoded by the *topA* gene. DNA gyrase is composed of two subunits, of 105 and 95 kilodaltons, encoded by the *gyrA* and *gyrB* genes, respectively, and has a tetrameric structure of the A_2B_2 type.

Ikeda et al. showed that DNA gyrase can promote illegitimate recombination. The incubation of plasmid pBR322 and phage lambda DNA in *E. coli* cell extracts led to the formation of lambda-pBR322 recombinants (129). The addition of oxolinic acid, a DNA gyrase inhibitor which acts on the A subunit, stimulated recombination. The stimulation was abolished with gyrase from Nal^r cells, whose A subunit is resistant to oxolinic acid. The addition of purified DNA gyrase to the extracts also stimulated recombination (130). It was proposed that the two A subunits of each tetrameric gyrase molecule bind covalently to DNA and that two tetrameric gyrase molecules, one bound to lambda DNA and the other to pBR322 DNA, associate transiently in an octameric complex. Dissociation of the octamer into tetramers is accompanied by the exchange of one of the A subunits and of DNA strands bound to these subunits. Sealing of the breaks when the gyrase is released generates a recombinant DNA molecule (127).

One prediction of this model is that novel junctions created by gyrase should correspond to known DNA gyrase cleavage sites. These are, unfortunately, rather degenerate. The reported consensus sequences, determined in vitro (198, 199) and in vivo (175), are YRT ↑ GNYNNY and NRT ↑ GRYC$^{T}_{G}$Y, respectively (the arrows indicate the position of cleavage). The consensus sequence of the junctions in pBR322-lambda recombinants generated in cell extracts, NRT ↑ RNNYNY (206), only weakly matches the consensus gyrase cleavage site. Similarly, the junctions in pBR322-lambda recombinants generated in vivo were only weakly homologous to this consensus site (187). Direct tests have shown, however, that three of the five lambda-pBR322 recombination sites are cleaved by DNA gyrase in vitro (128), which supports the above model.

Another prediction of the model is that gyrase alone can carry out the complete recombination process. This was not observed with purified *E. coli* gyrase, since the cell extract was needed along with the enzyme for the reaction to take place (130). A different result was obtained with another type 2 enzyme, the phage T4 topoisomerase, which is also a multiprotein complex (173, 244, 246). Incubation of genetically labeled lambda DNA with the purified enzyme in the absence of cell extract gave rise to recombinants differing from parental genomes by a deletion or a duplication (126). These results support but do not prove conclusively the postulated model. The frequency of recombination (about 10^{-2}) was too low to allow direct visualization of recombinants, and they were detected only after introducing the T4 topoisomerase-treated DNA into *E. coli* cells. The possibility that the topoisomerase did not carry out the complete recombination process but only triggered it by making recombinogenic lesions has therefore not been fully ruled out.

Type I topoisomerases seem to cause genome rearrangements in eucaryotes. Briefly, these enzymes can cleave ssDNA in regions with the potential for intrastrand base pairing (20) and remain bound to the 3′-phosphoryl end. They can then join this end to a 5′-hydroxyl end of ss or dsDNA (19, 107). Short sequences involved in the excision of simian virus 40 provirus from a rat cell line genome (34) resemble topoisomerase recognition sites (73), and several simian virus 40 excision sites coincide with the in vitro topoisomerase cleavage sites (33).

B. Transposases

Transposable elements, discussed in detail elsewhere in this volume, are specialized DNA segments that move (transpose) from one location to another. Their movement is mediated by an element-encoded protein, transposase, acting probably in a complex with host factors. A transposase recognizes the ends of the cognate element, breaks DNA at these ends, and joins the ends to target sequences. Transposases of the Tn*3* family and phage Mu sometimes use sequences other than a transposon end and thus perform illegitimate recombination.

1. Tn*3* family

The transposons of this family originate from gram-negative and gram-positive bacteria (see references 95 and 108 and D. Sherratt, this volume, for reviews). They encode two proteins, a transposase and a resolvase, which act on 34- to 40-base-pair (bp) inverted repeats present at transposon ends and a 120-bp *res* site, respectively. These elements undergo replicative transposition, which can generate deletions and inversions when the target sequence is in the transposon-carrying molecule. In addition to such rearrangements, which result from the "legitimate" transposase activity and are not discussed here, several transposases of the Tn*3* family (Tn*3*, Tn*21*, Tn*1000*, Tn*1721*) were observed to mediate the fusion of donor plasmids containing just one (instead of two) transposon ends with target plasmids (10, 11, 41, 109). This process has been termed "one-ended transposition" (200).

Fusion products were composed of the entire target plasmid and variable amounts of the donor plasmid. The majority of products obtained with Tn*21* or Tn*1721* (>60%) contained more than one entire donor plasmid; the remainder contained only a part of the donor plasmid. Fusions occurred at numerous sites of the target plasmid. One of the novel junctions always contained the transposon end of the donor, while the other junction was variable. Either transposon end (left or right) could mediate fusion in the presence of the cognate transposase. Sequence analysis of four Tn*21* and four Tn*1721*-generated junctions (201) has revealed a short (5 to 6 bp) duplication of the target sequence which flanked the donor material. Sequences at the variable junctions had no obvious homology with the transposon ends. The proposed mechanisms of one-ended transposition are presented below.

2. Bacteriophage Mu

Transposition of phage Mu has been extensively studied (see references 95 and 282 and M. L. Pato, this volume, for reviews). Two phage-encoded proteins (A and B) and at least one host factor (HU) are required for efficient transposition. In addition, about 150 bp at the left and 50 bp at the right phage end are needed (98). The essential features of the ends appear to be the terminal base pair and a series of repeats of a consensus sequence which is the protein A-binding site (35, 54, 97–99). Proteins A and HU interact with the ends to form a nucleoprotein structure ("transpososome") which mediates nicking at the ends and, in the presence of protein B and ATP, the strand transfer from Mu to the target DNA (189, 270). The resulting transposition intermediate can be resolved in two ways. The phage can be replicated, which generates a cointegrate between donor and recipient genomes, or the phage can be transferred without replication from its original location, which gives rise to simple insertions (53).

Plasmids which contain only one end of Mu (either left or right) were found to generate cointegrates, provided that proteins A and B were present (97). As with the Tn*3* family, the cointegrates consisted of the entire target plasmid and more than one entire donor plasmid. One of the junctions was located at the Mu end of the donor DNA, and another junction was variable. Sequence analysis of 12 variable junctions has identified six different junction sites. In five of these, a patchy homology with the protein A binding site and a terminal Mu base pair was found. The remaining site had no homology with the binding site.

3. Mechanisms of one-ended transposition

One-ended transposition can be accounted for in the context of regular transposition, except that the transposase uses one correct end and one deviant sequence, rather than two correct ends. The process could be initiated by cleavage and strand transfer at the correct end and terminated at a deviant sequence (201), in line with asymmetric models of transposition (96). Alternatively, simultaneous cleavage at both the correct and deviant sequences, followed by strand transfer, could occur (97), as hypothesized by the symmetric transposition models (95, 251). The two models postulate that the transposase transfers strands at both junctions. In the case of Tn*21* and Tn*1721*, the hypothesis is supported by the observation that a 5- or 6-bp sequence of target DNA was duplicated in the fusion products as in regular transposition (201).

In all cases, one-ended transposition was less efficient than the regular transposition. The difference was 10^3 times for Mu and Tn*3*, 10^2 times for Tn*1000* and Tn*1721*, and only 10 times for Tn*21*. This may be due to a number of reasons, such as the level of different transposases in the cell, the types of sequences flanking the transposon ends, or the affinity of different transposases for sequences which diverge from the correct transposon ends. Comparison of Tn*3* family transposases, which are all closely related, may allow the features that determine the degree of their substrate specificity to be identified (285).

C. Site-Specific Recombinases

Enzymes which catalyze strand exchanges between specific DNA sequences are called site-specific recombinases. They mediate rearrangements such as the excision and inversion of parts of a genome and the cointegration of two genomes and play a role in gene assembly, gene control, and maintenance of ancillary genetic elements. Various site-specific recombination processes have been detected in procaryotes and lower and higher eucaryotes (see references 29, 136, and 234 and several chapters of the present volume for detailed reviews). The site-specific recombination systems that mediate the integration and excision of lambdoid phages (lambda, ϕ80) and the specific inversion of bacterial DNA segments can cause illegitimate recombination.

1. Lambdoid bacteriophages

Integration of phage lambda into the *E. coli* chromosome and its excision from the chromosome are mediated by a protein complex containing the phage integrase. This complex recombines legitimately specific DNA sequences, named collectively primary attachment sites. Occasionally, it also recombines illegitimately other DNA sequences, and thus mediates phage integration into secondary attachment sites, or deletions of phage and chromosomal sequences. This section focuses on these illegitimate recombination processes.

Integration and excision of phage lambda have been studied extensively in vivo and in vitro (for reviews, see references 208 and 291 and J. F. Thompson and A. Landy, this volume). Integration requires two proteins and two sites. Integrase protein (Int) is encoded by the phage gene *int* (for integration), integration host factor (IHF) by the host genes *hinA* and *hinD*. One of the sites, *attP* (*att* for attachment site), is a 240-bp phage sequence; the other, *attB*, is a 23-bp bacterial sequence (195, 291). Excision requires the same two proteins, and, in addition, the phage-encoded Xis protein (for excision) and the host-encoded Fis protein (factor for inversion stimulation) (279, 291). It also requires two sites which flank the integrated prophage, *attL* and *attR*.

The primary *att* sites have a common 15-bp sequence called the core, usually represented by O, and differ by sequences which flank the core. The *attP* sequences to the left and to the right of the core (called arms) are designated P and P′, respectively (the directions are defined by a conventional representation of the lambda genome, with the head genes at the left end), and the corresponding *attB* sequences are designated B and B′. The central core nucleotide is designated as zero, P extends to −150, P′ to +90, and B and B′ to −11 and +11, respectively. The sequence of *attB* is the following:

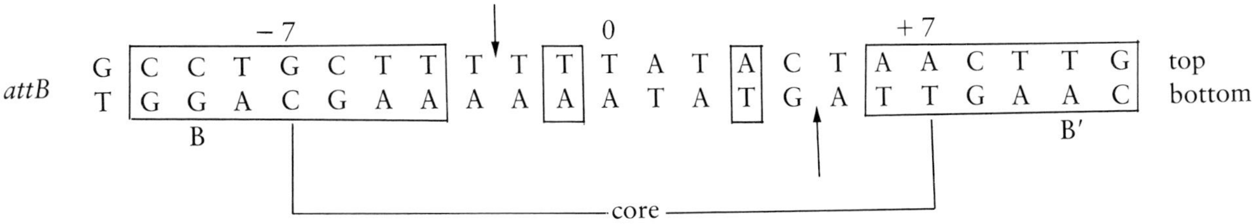

		−7			↓		0						+7		
attB	G	[C C T G C T T]	T T	[T]	T A T	[A]	C T	[A A C T T G]	top
	T	[G G A C G A A]	A A	[A]	A T A	[T]	G A	[T T G A A C]	bottom
		B						↑	B′	

core

(Boxes and arrows represent Int binding and nicking sites, respectively; see below.) Integration and excision reactions can be represented as follows:

$$\text{POP}'\ (attP) + \text{BOB}'\ (attB) \underset{\text{excision}}{\overset{\text{integration}}{\rightleftharpoons}} \text{POB}'\ (attL) + \text{BOP}'\ (attR)$$

The two reactions are catalyzed by a complex nucleoprotein structure (called an intasome [70]) which is formed by proteins interacting with *att*. There are 11 protein-binding sites in *attP*, 7 for Int (2 in P, 2 in O, and 3 in P′), 3 for IHF (2 in P, 1 in P′), 2 for Xis, and 1 for Fis (all in P) (279, 291; Thompson and Landy, this volume). *attB* contains only two binding sites, both for Int. The Int-binding sites are of two types, arm and core. Int bound to the arm sequence $^{C}_{A}$AGT CACTAT bends DNA (IHF and Xis have a similar role), but Int bound to the core sequence CAACTT NNT cleaves and joins DNA and thus mediates strand exchange (291).

Integrative and excisive recombination are not equivalent, as judged by the different protein requirements of the two reactions (Int and IHF for the first, Int, Xis, IHF, and Fis for the second) (71, 279). In addition, both arms (P′ and P) of the *attP* site must be intact for efficient integrative recombination (*attP* × *attB*), while about 60 bp, carrying two Int-, one IHF-, and one Xis-binding sites, can be deleted from the P arm without affecting the efficiency of excisive recombination (*attL* × *attR*) (36). Similarly, a mutation in the P′ arm affects integrative recombination more severely than it does excisive recombination (36).

Strand exchange is initiated by a nick introduced in the top strand between nucleotides −2 and −3 (211). Cleavage generates a free 5′-hydroxyl end and a 3′-phosphate end to which Int remains covalently bound. The energy of the cleaved phosphodiester bond is preserved and is used to join the enzyme-bound DNA strand to an appropriate 5′-hydroxyl nucleotide. Normally, this nucleotide is generated by Int-catalyzed cleavage of another *att* site. The joining exchanges the top strands between the two recombining *att* sites and generates an intermediate of the Holliday type. The intermediate is resolved by nicking the bottom strands between nucleotides +4 and +5 and exchanging them. Since the top strands are exchanged at the position of the left nick (−2/−3), and the bottom strand at the position of the right nick (+4/+5), the resulting recombinant molecule contains material from both parents in the 7-bp region bordered by the nick sites (called the overlap region) and material from one parent to the left of that region and from another parent to the right of that region. Semiconservative replication of the recombinant molecule gives rise to progeny molecules which carry the top- or the bottom-strand information.

Int-mediated recombination between sequences different from primary *att* sites can be considered illegitimate. The frequency of such a process is affected by the homology between the recombining sites. This was first shown by using a phage with a mutant *attP* site. The phage, called *attP-safG* (site affinity, *gal*) was isolated by integration in, followed by excision from, a secondary *attB* site in *gal*. Overlap regions of *attP* and *gal attB* differ at three positions (+1, +2, and +4). The insertion of the wild-type lambda occurred by joining the *attP* and *gal attB* at the right-side nick (between bp +4 and +5), and the excision of the integrated phage involved joining *attR* and *attL* at the left-side nick (between bp −2 and −3). As a consequence, a 7-bp segment from *gal attB* was introduced into the phage attachment site. This sequence was subsequently transferred into the *attB* site by another integration-excision cycle, using the *attP-safG* phage, which yielded a mutant *attB-safG* site (290). The homologous attachment sites (*attP* × *attB*; *attP-safG* × *attB-safG*) recombined efficiently (giving 40 to 80% of lysogenic cells), while nonhomologous sites recombined 100 to 1,000 times less efficiently (290).

Additional *saf* mutants were isolated from *trpC* (*safT*), *malG* (*safM*), and *leu* (*safL*) secondary attachment sites by similar integration-excision procedures (61). One of these, *safT*, differed from the wild-type site in a single position only (C → A at +4). It recombined 4 to 20 times more efficiently in a homologous than in a heterologous cross with the wild-type *att*. Taken together, the results indicate that the homology of the overlap region, but not the exact sequence, affects Int-mediated illegitimate integrative recombination.

The use of in vitro-mutated *att* sites confirmed the importance of the homology in the overlap region (18). Two such sites carried a mutation within that region (T → A or T → G at position −2), while two carried a mutation outside of that region (T → A or T → G at position −3). Efficient in vivo recombination between *attP* and *attB* was observed when the overlap regions were homologous, while a mismatched base within the overlap region decreased the recombination efficiency by 1,000 times (18). Similarly, *att* sites homologous in the overlap region recombined efficiently in vitro, while mismatched sites yielded no detectable recombinants (18, 211). Experiments with heteroduplex *attB* sites, obtained by pairwise annealing of mutant sites, have shown that for efficient recombination it is enough that one of the two DNA strands matches the overlap region of the recombining *attP* site (18, 209).

In addition to the homology in the overlap region, proper spacing of the core Int-binding sites appears necessary for efficient recombination. A mutant site, *attP24*, which contains a stretch of five instead of six T's in the core region, recombined inefficiently with *attB* (291). The homology within the overlap region is preserved in *attP24*, but the two core Int-binding sites are closer together than in the wild-type *attP* site (232). Similarly, *saf* mutants with an additional base in the overlap region (*safM*, *safL*) recombined in a homologous *saf* × *saf* crosses 27 to 45 times less efficiently than did wild-type *att* sites (61). In these mutants, the Int-binding sites in the core are presumably improperly spaced. Interestingly, one of the mutants, *safM*, recombined with the wild-type *att* very efficiently (30 and 80% of *attP* × *attB* in *attP* × *attB safM* and *attP safM* × *attB* crosses, respectively).

Like illegitimate integrative recombination, excisive recombination is affected by homology in the overlap region (291). However, the effect of heterology was less than that observed for integrative recombination. The *safG* mutation decreased the excisive and integrative recombination frequencies on the average by 5.5 and 63 times, respectively (291). Similarly, one of the mutations introduced in vitro in the overlap region (T → A at position −2) decreased excisive and integrative recombination frequencies by 16 and 800 times, respectively (18). These observations indicate that excisive recombination may have less stringent requirements for sequence homology than does integrative recombination. The reasons for this difference are not understood.

Further examples of Int-mediated illegitimate recombination include phage integration in numerous secondary *att* sites, which occurs with an overall frequency 100 to 1,000 times lower than that in the primary *att* site (254, 255, 291). The frequency is even lower for each individual secondary site, since for seven such sites analyzed by sequencing it varied between 10^{-4} and 10^{-10} relative to the primary *attB* site. Illegitimate excisive recombination, which gave rise to deletions in phage lambda ending at the *att* site, was detected by genetic analysis (60, 219, 220) and by sequencing (112). It also generated deletions in a lambda phage carrying a cloned *his3* gene, which all ended in the core region of *attP* (268).

Sequence analysis of secondary *att* sites (221, 231, 278) helps to evaluate whether the entire illegitimate recombination process can be carried out by the Int protein complexes or whether other cellular functions are also involved. Ten of the 15 positions of the *att* core were rather well conserved in 16 sequenced secondary *att* sites. At any of these 10 positions (indicated by an asterisk), the same base occurred 9 to 15 times, which corresponds to 56 to 94% base conservation:

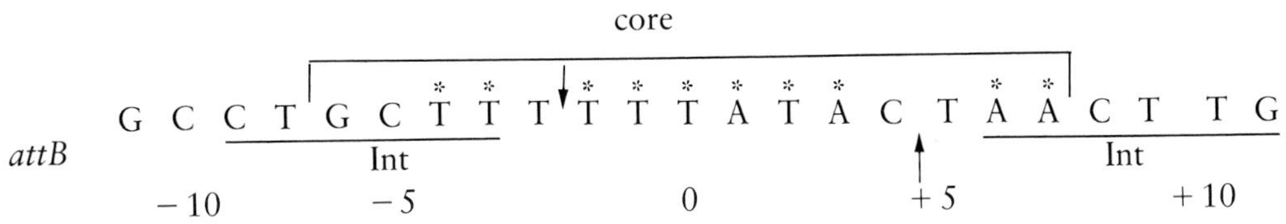

```
                                    core
                 ┌──────────────────────────────────────────┐
                          *  *   ↓*  *  *  *  *  *      *  *
          G C C T G C T T T T T T A T A C T A A C T   T G
attB              ───────────                   ↑   ───────────
                      Int                            Int
        −10          −5             0           +5          +10
```

All the highly conserved bases matched the canonical *att* core bases. Five of these were in the overlap region of the core (between the arrows indicating the positions of the top- and bottom-strand nicks [208]). Mutations at the highly conserved T residue at position −2 (present in 14 of 16 secondary *att* sites) reduced the activity of the site by about 10^3 times (18), while a mutation at the poorly conserved C residue at position +4 (present in 4 of 16 *att* sites) reduced that activity by only about 10 times (61). In contrast to the overlap region, the core Int-binding sites (underlined) were conserved in only three of seven positions. Among the 16 sequences, the average conservation of the 7-bp consensus binding site C$^{AA}_{TG}$CTTNNT (231) was 3.4 bp for the left and 2.7 bp for the right binding sites, while about 1.8 bp (7 × 1/4) would be expected in a random sequence.

Homology in the overlap region of secondary *att* sites could be necessary for Int-mediated strand exchange. Alternatively, it could be necessary for the initiation of recombination by the invasion of the strand nicked by Int at the primary *att* site into a duplex secondary *att* site, without further Int intervention (268). In contrast, homology at Int-binding sites should stimulate the Int-mediated strand exchange without stimulating homology-dependent strand invasion. On the basis of the overall base conservation in the secondary *att* sites, which is obvious in the overlap region but less convincing at the Int-binding sites, it is difficult to exclude homology-mediated strand transfer. In contrast, in two of the secondary *att* sites, the Int-binding sites are conserved better than the overlap region (conservation of 5 of 7 bp in the binding site and of 2 of 7 bp in the overlap region [231]), which suggests that Int-mediated strand exchange may occur at these *att* sites.

Int-mediated strand exchange between the canonical *att* sites is sequential, the top strand being exchanged first (211; see above). Recombination between the mutant *safG* and the wild-type *att* sites gave predominantly recombinants expected from exchange of the top strand only. This suggests that only that strand may have been transferred in such a cross and indicates that the resolution of the resulting intermediate was mediated not by Int (which would transfer the bottom strand, too) but by an unidentified resolvase (290). Transfer of only one strand from the wild-type to a mutant *att* site could possibly also explain the efficient recombination between the canonical *attP* and a deviant bacterial *safM* site having inappropriately spaced Int-binding sites (61). Taken together, the results can be interpreted as follows. (i) Phage lambda Int complex is involved in illegitimate recombination. (ii) The complex can mediate strand exchange between *att* and sequences similar to *att*. (iii) The complex probably does not always carry out the entire process of illegitimate recombination on its own.

It should be noted that the common lambda transducing phages (i.e., lambda *gal* or lambda *bio*) do not result from the activity of the Int protein complex, since their production is not eliminated by mutations in *int* or *xis* (80). Sequence analysis of one of these phages, lambda p*lac5*, has shown that it was generated by recombination between short homologous sequences (133; see also below). Other transducing lambda phages might be generated by DNA gyrase-mediated illegitimate recombination (H. Ikeda, personal communication).

The site-specific recombination system of another lambdoid phage, ϕ80, can also mediate illegitimate recombination (167, 168). This system is similar to that of lambda, since it consists of two structural elements, *attP* and *attB*, and three proteins, Int, Xis, and IHF (166, 168). The *att* sites share a 17-bp core and carry two potential Int-binding sites. In addition, *attP* has an IHF-binding site. Recombination between *attP* and *attB* generates *attL* and *attR*.

Int-mediated illegitimate recombination was demonstrated on a plasmid which carried ϕ80 *xis* and *int* as well as two *att* sites in opposite orientation, *attP* and a secondary *att* site, designated *att*ϕ1.4. The two sites were identical in 11 of 17 core positions and shared a 10-bp stretch of homology interrupted by a single mismatch. Recombination between the sites inverted the segment bordered by the sites, which allowed recovery of the reciprocal recombinants. Three types of plasmids which had undergone inversion were isolated. In all of them, recombination occurred in the region of homology. It entailed either a conversion, which affected *attR* or *attL*, or a 1-bp deletion at a mismatched position. These results can be most simply interpreted by assuming that a double-strand exchange between the *att* sites occurred at a staggered cut, and that the mismatch in the resulting heteroduplex was subsequently repaired (168). Either of the two mechanisms discussed above for lambda (Int- or homology-mediated strand exchange) could have generated the observed product.

2. DNA invertases

Site-specific inversion of DNA sequences in bacteria, reviewed recently by Plasterk and van de Putte (223) and Johnson and Simon (136), is discussed in detail elsewhere in this volume. Briefly, inversion has the role of a genetic switch and controls the expression of genes within or adjacent to the inverted segment. The systems studied include C and G inversions, which control the host range of phages P1 and Mu, respectively, H inversion, which controls phase variation in *S. typhimurium*, and P inversion within the *e*14 element in *E. coli*, which might control an unknown function. Each inversion is catalyzed by a recombinase encoded by a *din* (DNA inversion) gene (122). The recombinases (Cin, Gin, Hin, and Pin, for C, G, H, and P inversion, respectively) are 60 to 70% homologous and act on sequences *cix*, *gix*, *hix*, and *pix*, collectively called *dix* (for DNA inversion cross-over-X) (122), which are also largely homologous (see below). A given recombinase can use a *dix* site from another *din* system, although with a lower efficiency than for a cognate *dix* site (223). Two host factors (Fis and HU) and a *cis*-acting site called enhancer of recombination are also needed for efficient inversion (136).

The *dix* sites are localized in opposite orientation at the borders of invertible segments. They are 26-bp sequences, composed of 12-bp imperfect palindromes which are separated by a central nonpalindromic dinucleotide (*hixR*, the site at the right border of the H segment is an exception, since it is nonpalindromic). Of the 26 bp which forms each *dix* site, 21 bp is conserved in at least six of the eight known *dix* sites. This corresponds to a ≥75% (≥6/8 bp) conservation at these 21 positions. The following *dix* consensus sequence can be deduced from that observation:

```
                                A
T T C T N N N A A A C C - A G G T T T N N G A T A A
              ------->  G  <-------
```

The palindromic part of the sequence is indicated by arrows; the central dinucleotide in *gix* sites is GA; in other *dix* sites, it is AA. Din recombinases are thought to act by introducing a 2-bp staggered cut within the *dix* site at the central dinucleotide, exchanging strands between *dixR* and *dixL* and sealing the cleaved strands.

Din recombinases preferentially catalyze inversions, but can also mediate with a low-frequency cointegration between genomes carrying *dix* sites and deletions within a genome carrying directly repeated *dix* sites (142, 222, 243). Interestingly, Cin enzyme (and presumably other Din recombinases) occasionally mediates recombination not only between *dix* sequences, but also between sequences which are only partially homologous to *dix*, called quasi-*dix*, or *dixQ*, sites (125). Recombination between *cix* and *cixQ* sites, which occurred during the transduction of a pBR322-derived plasmid by a P1 phage and generated a stable cointegrate between the two genomes, was the first such event reported (123). Study of the cointegrate showed that inversions between *cix* and the *cixQ* sites can also occur, although at about a fourfold-lower frequency than between two *cix* sites (125). A weak homology between the *cixQ* site and various *dix* sites was found (122).

Additional *dixQ* sites were isolated by using promoterless genes, which could acquire a promoter upon inversion and were carried on pBR322-related plasmids together with the *cin* gene and one or two *dix* sites (120–122). Inversion of the promoterless *galK* gene by recombination between *cix* and *cixQ* sites revealed four *cixQ* sites (120), while inversion of a promoterless Km^r gene by recombination between a wild-type or a modified *gix* site (*gix** site having AA instead of GA as a central dinucleotide) (120) and *gixQ* sites revealed a number of *gixQ* sites. Five different *gixQ* sites recombined with *gix* and 10 others with *gix**; several of these appeared recurrently. In addition, one pair of *gixQ* sequences recombined to give an inversion. Another such pair was involved in a more complex event (cointegration and inversion).

The consensus *dixQ* sequence can be derived from the analysis of 42 *dixQ* sites:

```
                                    *
          *           *  *  *       A  *           *  *  *                 *  *
 N  N  N  T  T  A  N  A  A  A  N  G  -  A  G  N  T  T  T  N  N  N  N  A  A
                      ------->                    <-------
                                    G
```

At 15 positions of this sequence, the same base is found at least 25 times, which corresponds to ≥59% conservation. Eleven of these 15 positions (indicated by asterisks) match the highly conserved *dix* bases. In addition, a part of the palindrome present in the consensus *dix* sequence is also found in the consensus *dixQ* sequence (indicated by arrows). This observation supports the hypothesis that Cin carries out illegitimate recombination by joining a correct substrate (a *dix* site) to a partly homologous sequence (a *dixQ* site). However, it does not rule out the possibility that certain Cin-dependent recombination events are not generated in this way, since some of the *cixQ* sites did not match the *dix* consensus sequence significantly. For example, only five and six matches of the 21 positions conserved in the *dix* consensus sequence were matched in *gix**Q391 and *gix**Q389, respectively (121). This is equivalent to random-choice expectations (probability of a match of one-fourth for each position, 5.25 matches in 21 bp). In the case of Hin recombinase, sequences which deviated much less from the *dix* consensus sequence (13 or 15 matches at 21 positions) could not be inverted in vitro and were thus a poor substrate for this enzyme (135). It is therefore possible that Cin recombinase sometimes only triggers illegitimate recombination, possibly by breaking DNA at a correct *dix* site (and also at some *dixQ* sites, as shown by recombination between pairs of *dixQ* sites; see above) but does not complete it.

Besides the Din family discussed above, other DNA inversion systems have been described. One such system is carried on the *E. coli* IncI1 plasmid R64 (153). It consists of four DNA segments separated by seven 19-bp repeats. The segments invert independently or in groups by recombination between any two repeats which lie in opposite orientation. The inversion is mediated by a gene function designated *rci* (R cluster inversion). Its role may be to select the production of one of the seven possible proteins encoded in the region. The region has been designated a "shufflon" (154) and has been shown to be present on numerous IncI1 plasmids (153). Another *E. coli* inversion system has recently been described (1). It controls the expression of the *fim* genes, involved in the synthesis of fimbriae. The inverted repeats used by R64 and *fim* systems are not related to each other or to the *dix* sites discussed above. Still another inversion system is carried on the *Staphylococcus aureus* plasmid pI524. A 2.2-kb segment of this plasmid, flanked by 650-bp inverted repeats, inverts independently of the host functions involved in homologous recombination. The system appears capable of mediating cointegration and de letion formation (202). Its biological role is not known. The existence of many different site-specific inversion systems indicates that they may be widespread in nature. Only one of them, Din, has been shown to mediate illegitimate recombination. It would not be surprising if the other systems could mediate it as well.

A remarkable homology exists between DNA invertases and another class of site-specific enzymes, the resolvases encoded by transposons of the Tn*3* family (84, 140, 258). The role of the latter enzymes, discussed in detail elsewhere in this volume, is to mediate recombination between two copies of the same transposon present on a given genome. They act on sites called *res* (95; Sherratt, this volume). These sites are complex, span about 120 bp, and carry three regions which bind the resolvase. Deletions which inactivate any one or two of the three resolvase-binding sequences reduce 5 or 50 times, respectively, the capacity of the shortened *res* site to recombine with the full-size site but do not fully abolish it (292). These observations, together with the fact that they share sequence homology with DNA invertases, suggest that resolvases could potentially mediate illegitimate recombination, using as a substrate sequences other than the *res* site.

D. Origin-Nicking Proteins

Certain phages, such as M13 from *E. coli* (64), and plasmids, such as pC194 from gram-positive bacteria (276, 277), replicate via an ssDNA intermediate. Others, such as conjugative plasmids from gram-negative bacteria, replicate via an ssDNA intermediate during the transfer from one host to another (297). Synthesis of these ss intermediates is in all cases initiated by a replicon-specific protein that nicks the cognate replication origin. The replication proteins of all three classes of replicons can mediate illegitimate recombination.

1. Bacteriophage M13

E. coli ss DNA phages, such as M13 and φX174, replicate by a rolling-circle mechanism (reviewed in references 12 and 13). They encode a protein which initiates replication by nicking a specific site (nick site) within the phage replication origin. The resulting 3′-hydroxyl end is used to prime DNA synthesis, which concomitantly displaces the nicked DNA strand. The protein also terminates phage replication by cleaving the displaced strand at the end of the round of replication and sealing it to form a covalently closed circular ssDNA molecule. This molecule is converted to a ds form by host-encoded enzymes or, alternatively, encapsidated in phage proteins, giving rise to a phage particle.

The involvement of the gene II replication protein of phage M13 (191) in illegitimate recombination (192) was shown by analyzing deletions generated in hybrid plasmids composed of M13mp2, pBR322, and pC194 (a plasmid isolated from *S. aureus* which replicates in *Bacillus subtilis* and *E. coli*) (75, 94, 132, 277). The hybrids were viable in *B. subtilis* but not in *E. coli*. Viable deletion mutants which had lost the M13 sequences were found in *E. coli*. Deletions ended in nearly all of these mutants (23/25) at the M13 replication origin as shown by restriction mapping. In 3 of the 10 plasmids analyzed by sequencing, the deletion endpoints could be assigned unambiguously, since there was no homology between the recombining sequences. One endpoint was constant and corresponded to a nucleotide 5′-phosphate adjacent to the nick site, while the other endpoint was variable. In the remaining seven plasmids, 2- to 6-bp homologies prevented unambiguous localization of the deletion endpoints, but the 5′-phosphate nucleotide adjacent to the nick site was always present within the recombining sequences. If the phage replication protein was inactivated, deletions were 50- to 100-fold less frequent, and none ended at that nucleotide.

Two classes of models could account for the above observations. The gene II protein might initiate replication correctly but terminate synthesis prematurely by cleaving the nascent DNA strand at a sequence resembling the correct termination signal. The resulting circular ss molecule could be converted by host machinery to a ds form as in normal replication. Alternatively, deletion formation could be triggered by nicking the replication origin and completed by other cellular function(s). The first model predicts that sequences at the variable deletion endpoints resemble the correct termination signal, which encompasses 10 bp upstream and 11 to 29 bp downstream of the M13 nick site (67), while the second model does not predict such a resemblance.

To distinguish between the two models, several hundred deletions were analyzed. Three plasmid regions which often recombined with the nick site were found and named hot spots A, B, and C (B. Michel, E. d'Alençon, and S. D. Ehrlich, manuscript in preparation). Up to 40% of deletions ended at a given hot spot in plasmids from which the other two hot spots had been removed. If less than 10% of deletions ended at a hot spot, the hot spot was considered inactive.

Hot spot A corresponded to the pBR322 sequence 1610 to 1700 and contained many different endpoints. No homology with the nick site was observed at any of these. Hot spots B and C corresponded to pC194 sequences 1388 to 1365 and 225 to 248, respectively. Hot spot B was highly homologous (20/25-bp match) with the M13 replication origin, while hot spot C was less homologous (13/25-bp match) with the origin. Unique recombination sites were observed at hot spots B and C.

Hot spots A and C were active only in the host which had a functional *rep* helicase gene. This helicase is needed for the progression of the replication fork initiated at the origin of ssDNA phages (160). Hot spot B was independent of the *rep* gene but required wild-type levels of DNA ligase for its activity (it was inactive in *lig-7* cells, which have thermosensitive ligase, and were kept at the restrictive temperature for 2 h upon receiving the plasmid DNA by transformation). In contrast, hot spots A and C had no such requirement.

In addition to the three "natural" hot spots (A, B, and C), three "artificial" hot spots were generated by inserting in the hybrid plasmid synthetic palindromes of ~30 bp, which lacked any homology with the M13 replication origin. These hot spots recombined with the nick site in the absence of *rep* helicase, did not require the wild-type amounts of ligase, and contained a number of different deletion endpoints. Interestingly, the endpoints were located almost invariably (14/15 cases) at the base or at the side of the palindrome proximal to the origin nick site.

The mechanisms of deletion formation at different hot spots are not understood. However, one can speculate that at least some, and maybe all, hot spots act by arresting the progression of a replication fork. The 3′-hydroxyl nucleotide at the arrested fork could thus be made available for joining to the 5′-phosphate nucleotide at the nick site, possibly by DNA ligase. This speculation is based on the following considerations. (i) Palindromes are known to act as replication pause sites, arresting the fork at positions similar to those observed for deletion endpoints (the base or the side of the palindrome proximal to the site of initiation of replication) (139, 289). (ii) Nonpalindromic pause sites also exist (289) and are recognized by specific polymerases (139). This could explain the observation that the nonpalindromic hot spot A is active only if the M13 replication fork is allowed to progress (in *rep*$^+$ but not *rep* mutant hosts). (iii) Ligase was required for deletion formation at hot spot B. This requirement was not observed at other hot spots, possibly because the mutation used was leaky, and less ligase was needed at these hot spots. Alternatively, at hot spots other than B the deletions may have been generated during the incubation of *lig-7* cells at the permissive temperature necessary to allow cell growth, rather than during the incubation of cells at the restrictive temperature immediately upon transformation. The possible role of DNA ligase in the joining of nonadjacent nucleotides is further discussed in Section IV.F.

Could the hot spots B and C, which are homologous with the M13 replication origin, act as aberrant termination signals for the phage replication protein rather than as replication pause sites? That is unlikely for hot spot B, which required wild-type levels of DNA ligase to be active. This requirement would be superfluous if the joining was mediated by the replication protein. Hot spot C could be a site for aberrant termination by the gene II protein, since it did not require wild-type levels of DNA ligase. However, the homology between this hot spot and the replication origin did not extend throughout the region essential for termination of phage replication, which argues against this interpretation.

Deletions ending at the 3′-hydroxyl nucleotide upstream from the phage nick site rather than the 5′-phosphate nucleotide downstream from that site were also found (116, 237). It was suggested that such deletions resulted from the ligase-catalyzed joining of the 3′-hydroxyl nucleotide neighboring the nick site to a 5′-phosphate nucleotide arising elsewhere in the molecule (192).

Taken together, the results indicate that deletions are most often only triggered but not completed by the M13 replication protein. Other cellular functions complete the formation of deletions triggered by the nicking of the M13 replication origin.

2. Plasmid pC194

A number of ssDNA plasmids which replicate in a similar manner to *E. coli* ssDNA phages have recently been identified in gram-positive bacteria (100, 101, 276, 277). The purified replication protein of one of these, pT181, has been shown to (i) cleave the plasmid replication origin, (ii) cleave ssDNA at sequences resembling the origin nick site, and (iii) seal the origin nick it generates in vitro (148a, 149, 150). The replication protein of another of these ssDNA plasmids, pC194, mediates illegitimate recombination in vivo (193).

Hybrids composed of phage f1 (ssDNA phage related to M13) and plasmids pBR322 and pC194 are stable in *B. subtilis* but not in *E. coli* (see above). When their structure was such that the deletions ending at the f1 phage replication origin (reviewed in the preceding section) could not give rise to viable plasmids in *E. coli*, the less frequent involvement of pC194 replication protein in deletion formation could be analyzed (pC194 is known to replicate in *E. coli*) (56, 94, 100). Stable plasmids were formed in *E. coli* as a result of deletions which in most cases had one end at a single site within the pC194 replication origin. Inactivation of the plasmid replication protein abolished deletion formation at that site. The site at which deletions ended was presumed to be the pC194 nick site (193). The other deletion end was variable.

The consensus sequence derived by analyzing 16 variable deletion ends, $^{C}_{T}TTT^{G}_{A} \uparrow AT$, resembled the sequence flanking the pC194 nick site, CTTTG ↑ AT. This indicates that the pC194 replication protein may generate deletions by correctly initiating and then aberrantly terminating DNA synthesis. Support for this hypothesis comes from the analysis of a plasmid in which nicks would be made in the same strand at the f1 and pC194 replication origins. In a *rep*$^+$ host, which permits the phage replication fork to progress, deletions starting at the nick in pC194 ended in a 1.5-kilobase (kb) region downstream from the f1 origin. In a *rep* mutant host, where the phage replication fork could not progress, deletions ended just 0.2 kb from the f1 origin, either downstream or upstream. This result suggests that ssDNA generated by the phage replication fork in the *rep*$^+$ host was used as a substrate for aberrant termination by the pC194 replication protein.

Correct initiation and aberrant termination of DNA replication by the pC194 *rep* protein can generate deletions which end at a nucleotide immediately downstream from the nick site. Aberrant initiation and correct termination of DNA synthesis would generate deletions ending at a nucleotide immediately upstream from the nick site. Such deletions were observed both in *E. coli* and *Streptococcus pneumoniae* (17, 193), in hybrid plasmids different from the ones described above. Variable endpoints of these deletions matched the established consensus sequence (see above). This implies that the A protein can use a broad set of signals for both initiation and termination of replication and carry out the entire process of illegitimate recombination. In contrast, the M13 replication protein appears to use a much narrower set of substrates, and therefore only triggers illegimate recombination by nicking the replication origin but does not complete it. The reasons for the more stringent specificity of the M13 protein relative to the pC194 protein are not clear.

3. Plasmid F

Many plasmids can be transferred from cell to cell by conjugation (reviewed recently in reference 297). The process was studied most extensively in the *E. coli* plasmid F (296). Transfer is initiated by nicking the transfer origin (*oriT*) by TraYZ endonuclease (encoded by *traY* and *traZ* genes). A 3′-hydroxyl end is thus generated and used as a primer for DNA synthesis, which normally accompanies DNA transfer. DNA helicase I, the product of the *traI* gene, unwinds the dsDNA and generates ssDNA, which is transferred to the recipient cell. The transferred strand is circularized, presumably by TraYZ endonuclease, and converted to a ds form.

Integration of F into the host chromosome generates an Hfr strain (reviewed in reference 179). Occasionally, autonomous replicons containing both F and chromosomal sequences (F′ plasmids) arise by illegitimate recombination during DNA transfer from an Hfr strain (reviewed in references 113 and 178). F′ plasmids can be detected easily in recombination-deficient (*recA*) recipient cells. Up to 98% of such F′ plasmids are unable to transfer conjugally (102, 177, 217) because they lack most of the *tra* operon (Δ*tra* plasmids) (103).

Nine of 10 Δ*tra* F′ plasmids isolated from one Hfr strain had arisen by recombination between a specific sequence in the *oriT* and different chromosomal sequences (104, 117). The recombining *oriT* sequence contained the nick site (R. Deonier, personal communication) (this localization of the nick site differs somewhat from those reported previously in reference 281). One of the chromosomal recombining sequences was determined and found to share 10/12-bp homology with *oriT* (117). It is likely that the TraYZ endonuclease mediates the formation of these Δ*tra* F′ plasmids either by triggering illegitimate recombination, as does the M13 gene II protein, or by carrying it out entirely, as does the pC194 *rep* protein (see above). The transfer-deficient phenotype is due to the loss of different *tra* genes and of the transfer origin.

Not all of the Δ*tra* F′ plasmids form by recombination at the nick site. One of 10 plasmids studied by Horowitz and Deonier (117) and 2 of another set of 4 Tra$^-$ plasmids studied by Guyer et al. (103) were generated differently. The data are not sufficient to indicate the underlying mechanisms. Similarly, Tra$^+$ F′ plasmids did not lose any *tra* genes and thus must have been generated by a process independent of the TraYZ enzyme. They often resulted from classical homologous recombination between directly repeated insertion sequence elements carried on the integrated plasmid and the bacterial chromosome (113). Interestingly, some of Tra$^+$ F′ plasmids underwent spontaneous deletions which generated Δ*tra* F′ plasmids and ended close to (presumably at) *oriT* (37).

Illegitimate recombination was also detected at *oriV1*, which is one of the several vegetative replication origins of plasmid F (212–214). Intermolecular recombination between this origin and plasmid pBR322 sequences yielded F-pBR322 cointegrates. Interestingly, the pBR322 region most often involved in recombination with *oriV1* corresponded to hot spot A, which also recombined frequently with the M13 replication origin in an intramolecular reaction (see above). *oriV1* may thus be a "nicked" replication origin (192). It was also proposed (212) that *oriV1* may be used as a site for the resolution of plasmid dimers into monomers (a *res* site; see Sherratt, this volume).

IV. SHORT HOMOLOGOUS SEQUENCES

Various genome rearrangements can result from recombination between short homologous sequences which are not known to be substrates for specific DNA break-and-join systems. Before different instances and discussing mechanisms of such recombination are described, the meaning of "short" should be defined. Classical homologous recombination in *E. coli* appears to require sequences about 20 bp long, as determined by measuring intermolecular recombination between lambda phage ColE1-type plasmids containing short segments of homology (253, 287). (Somewhat longer homologies, 50 bp, may be needed for phage T4 recombination [260].) Homologies shorter than 20 bp could therefore be tentatively considered to engage solely in illegitimate recombination. The lower limit for significant matching may be defined by considering that in a random sequence the probabilities of finding one, two, or three identical bases at the site of recombination are 1/4, 1/16, and 1/64, respectively. The last value, 0.0156, will be used as statistically significant. "Short" is therefore defined here as being 3 to 20 bp, and the discussion will mostly be confined to recombination events involving segments in this size range.

A. Deletions

Deletions resulting from recombination between short homologous sequences have been detected in different genetic elements (large and small plasmids, phages, and chromosomes) and in different bacteria. They were first described by Miller and co-workers, who studied deletions arising spontaneously in the *E. coli lacI* gene, which was carried on an F′ plasmid (77). About half (7/12) of these deletions were generated by recombination between direct repeats of 5 to 8 bp. The importance of short repeats in deletion formation was demonstrated even more clearly in a subsequent study of deletions ranging from 700 to 1,000 bp (2). Sequence analysis showed that 23 of 24 deletions were formed by recombination between repeats ranging from 4 to 17 bp. The repeats either were strictly homologous or differed by one or two bases. Genetic mapping of >250 deletions superimposed on sequence analysis indicated that most deletions (~60%) involved the longest repeats (17 bp), composed of three matched segments, 6, 3, and 5 bp long, separated by one and two mismatched bases, respectively. One point mutation within the 6-bp segment and another within the 3-bp segment which reduced the overall homology from 14 to 13 of 17 bp decreased the recombination frequency about 10-fold, while another point mutation in the 2-bp mismatch which did not alter the homology did not affect the deletion frequency. In a similar study, 15 of 22 deletions isolated in the *lacI* gene were generated by recombination between short direct repeats of 3 to 9 bp (236). These results show that short homologous repeats play an important role in deletion formation in a large *E. coli* plasmid.

In small *E. coli* plasmids, recombination between short direct repeats was first detected within a derivative of the insertion sequence IS2, named IS2-6, carried on plasmid pSC101 (92). IS2-6, generated by a complex rearrangement (discussed in a later section), carries a 108-bp insert (91). The insert was unstable, and first lost 54 bp by recombination between 9-bp repeats and then the remaining 54 bp by recombination between different 9-bp repeats. Deletions between short repeats also occurred in a pSC101-related plasmid carrying a cloned segment of the *E. coli* chromosome. The segment contained *gal*, *chlD*, and *pgl* genes and insertion sequence IS*1* in the *gal* operon control region. Deletion derivatives which had lost the $chlD^+$ gene and IS*1* were of four types. Two frequent types involved direct repeats of 5 and 9 bp; the remaining two types, which were rare, did not involve direct repeats (264). In another small *E. coli* plasmid, pBR322, deletions which restored the activity of the promoterless Tc^r gene by fusing it to a promoter (226) involved repeats of 7, 8, or 11 bp in three cases, while in the fourth case the deletion arose by crossover within a direct repeat sharing 3 of 4 bp (137). Recombination between pBR322-carried direct repeats matched in 21 of 22 positions was also observed (152).

Several deletions in small plasmids of gram-positive bacteria have also been analyzed. Strong promoters associated with the *S. pneumoniae malM* and *malX* genes (176) or the *S. aureus blaZ* gene (235) could not be stably propagated in *B. subtilis* and were lost by deletion. Five of the nine sequenced *mal* deletion derivatives involved direct repeats of 3 to 13 bp (176), and the one sequenced *blaZ* derivative involved 15/16-bp direct repeats (235). Deletion events fusing a weakly expressed plasmid-carried Cm^r gene (17) to a strong promoter in *S. pneumoniae* involved 6- or 9-bp direct repeats (193). One of the 6-bp repeats contained a nick site within the plasmid replication origin, which indicates that the deletion was mediated by the *rep* protein (see the preceding section). In contrast, the 9-bp repeats are not known to be substrates for a site-specific enzymatic system.

Three spontaneous deletions in *E. coli* bacteriophage T7 involved 7-, 8-, and 10-bp repeats (269). Genetic mapping and restriction analysis indicated that seven other deletions may also have involved repeated sequences (269). A combination of genetic mapping and sequence analysis has indicated that three deletions arising in the rII region of bacteriophage T4 involved direct repeats of 9, 10, and 11 bp (225). Two deletions in the lysozyme gene of T4 similarly involved 8-bp repeats (218).

Recombination between short direct repeats is also a major cause of deletions in the *E. coli* chromosome. A fusion of *lacI* and *lacZ* genes involved a 9/11-bp repeat (30). The formation of phage lambda p*lac5* (133) involved 17/19 bp in lambda and 17/21-bp repeats in *lacY* (256). Four of seven UV-induced deletions of a *trp* operon terminator involved repeats of 3 to 7 bp (300). Finally, two azaquinacrine-induced deletions in the $tRNA^{Tyr}$ $su3^+$ gene involved 11/13-bp repeats (190).

B. Duplications

Duplications can also be generated by recombination between short homologous sequences. Edlund and Normark found that a 10-kb spontaneous duplication in the *ampC* region of the *E. coli* chromosome involved 12-bp repeats (72). A systematic study of duplications forming in an *E. coli* F′ plasmid (295) confirmed this observation. Lac^+ revertants of a strain which carried a promoterless *lacZ* gene resulted from the fusion of the *lacZ* to a nearby promoter. Most of these revertants were unstable in the absence of the Lac^+ selection and carried duplications of 5 to 20 kb, which resulted from the fusion of the *lac* genes to "downstream" (relative to the direction of transcription) promoters. Over 90% of the independent duplications (28 of 30) involved 3- to 8-bp direct repeats, generally adjacent to additional regions of homology. Several repeats were used recurrently, the most frequent (12 of 30 cases) being matched in 14 of 18 bp (295). In another study, a spontaneous duplication in a *Streptomyces lavendulae* plasmid also involved short repeats (5 bp) (207).

C. Cointegrate Formation

In addition to occurring by deletions and duplications, which can be most simply accounted for by intramolecular recombination, intermolecular cointegrate formation can occur by recombination between short homologous sequences. This was shown by analyzing cointegrates between phage lambda and plasmid pBR322 (or a pBR322 derivative) (144). Three of the four isolates resulted from a single recombination event between homologous sequences of 10 or 11 bp. The fourth isolate was formed by two recombination events. Both of these involved homologous sequences, 10 and 13 bp long. Interestingly, several other cointegrates isolated in *recA* cells arose by multiple recombination events between sequences sharing less homology (<5 bp) and are thought to be generated by the host gyrase (187; see Section III.A).

The role of short homologies in cointegrate formation was also documented in *B. subtilis* plasmids (17a). Cointegrates were isolated by selecting for the maintenance of a thermosensitive plasmid (pE194) at the restrictive temperature in cells which harbored a thermoresistant plasmid. Of the 22 cointegrates analyzed, 21 were generated by recombination between 9- to 15-bp homologies, often flanked by additional short regions of homology. Nine different junctions were identified, one of which was used in 12 of the 22 isolates. Interestingly, that particular junction involved a homology of only 11 bp, significantly shorter than the longest straight (15 bp) or patchy (42/49 bp) homology. Similarly, pE194 was found to integrate in the *B. subtilis* chromosome by recombination between sequences homologous in 15 of 16 bp (17b).

Taken together, these results show that recombination between short homologous sequences generates a variety of rearrangements. Given the diversity of these homologies, most rearrangements were probably not formed by sequence-specific DNA break-and-join systems. Possible alternative mechanisms are discussed below, following the description of two related processes, the excision of transposons and the instability of palindromes.

D. Transposon Excision

Most transposons and insertion sequences generate short duplications of the target genome at the site of their insertion (see reference 124 for review). Recombination between the duplications eliminates ("excises") the inserted element and one of the duplications. This process has been studied most thoroughly with Tn*5*, Tn*10*, and phage Mu.

Transposon Tn*5* is 5.7 kb long and contains 1.5-kb terminal inverted repeats, designated IS*50*, which bracket the central region encoding resistance to kanamycin (23; D. E. Berg, this volume). It duplicates a 9-bp sequence of the target upon insertion (237). Precise excision of the transposon restores the original target sequence and can be detected by the reversion of insertion mutations. Studies of mutant elements showed that transposase is not required for excision (23, 74). The excision is therefore not a reversal of transposition (which requires the transposase), and it appears to be carried out exclusively by host functions. The frequency of excision from different sites varied greatly. It ranged from 3×10^{-8} to 2×10^{-4} for Tn*5* carried at different positions within the *lac* gene of an F′ *lac* plasmid. Similarly, larger (50 kb) inserts of Tn*5*-related elements in the Ap^r gene of pBR322 were found to undergo excision (reversion to Ap^r) at frequencies which varied from 1.4×10^{-10} to 2.5×10^{-7} (205). Variation was also observed with perfect palindromes of 22, 32, or 90 bp generated in vitro from these original inserts (59) (palindromes of such length usually do not interfere with plasmid viability; see the following section). Excision frequencies differed by about 100-fold for the same palindrome at different sites. Particularly dramatic was a 3,000-fold difference in excision frequency of the 90-bp palindrome at a pair of adjacent sites which overlapped 8 of 9 bp. These results indicate that different 9-bp repeats recombine with different efficiencies.

The overall Tn*5* excision frequency was significantly reduced, from an average of 10^{-6} to 10^{-9}, with a transposon that had direct rather than inverted repeats of the IS*50* sequences (it was constructed in vitro and transposed with frequencies comparable to those of the wild-type Tn*5*). This suggests that the inverted repeats play a role in transposon excision (23, 74). The effect of the length of the inverted repeats on recombination between the flanking 9-bp repeats was studied with the set of perfect palindromes derived from a 50-kb transposon related to Tn*5* described above (59). Precise excision of the 22-bp palindrome was 3 to 550 times more efficient than that of the entire 50-kb transposon (13 different insertions excising with a frequency of 10^{-10} to 10^{-7} were tested). For all insertions but one, a modest further stimulation (at most 7.5 times) was observed when the length of the palindrome increased to 32 bp. Quite significant stimulation (9 to 314 times) was seen by increasing the palindrome from 32 to 90 bp. The effect was even stronger with the remaining insertion, since the 32- and 90-bp palindromes excised 116 and 18,000 times more frequently than the 22-bp palindrome. These observations were confirmed and extended using several additional palindromes in a comparable size range (18 to 90 bp; K. Weston-Hafer and D. Berg, unpublished results). Similarly, shortening of the inverted repeats from 1,500 to 185 bp decreased the efficiency of Tn*5* excision in *B. subtilis* by about 10 times (220a). Taken together, these results suggest that the inverted repeats stimulate recombination between 9-bp direct repeats and that stimulation increases with the increased length of the inverted repeats.

Excision of transposon Tn*10* resembles that of Tn*5*. Tn*10* is 9.3 kb long, contains 1.4-kb inverted repeats (named IS*10*), and encodes resistance to tetracycline (for review, see reference 147 and N. Kleckner, this volume). It duplicates 9 bp of the target sequence upon insertion (145) and can be excised by recombination between the duplicated sequences. In addition to this process, termed precise excision, Tn*10* can undergo "nearly precise" excision (233). This corresponds to recombination between 23/24-bp direct repeats present within the two IS*10* elements. Nearly precise excision leaves a 50-bp remnant of the transposon flanked by 9-bp direct repeats. This remnant consists of two 23-bp inverted repeats separated by 4 bp. It can also be excised by recombination between the flanking 9-bp direct repeats.

The three types of Tn*10*-related excision events (precise excision, nearly precise excision, and excision of the remnant) do not require transposase (79). Inverted repeats appear to facilitate precise and nearly precise excision, since the frequency of the two events was 10- to 100-fold lower with Tn*10* derivatives having inverted repeats of 400 or 70 bp than with the wild-type transposon. The average excision frequency of the transposon was 10^{-8} and 10^{-6} for precise and nearly precise excision, respectively. More variation from site to site was observed for precise excision of the transposon or of the 50-bp remnant (the range was 30 and 100 times, respectively) than for nearly precise excision (the range was 7 times). Because the recombining direct repeats varied from site to site for precise but not for nearly precise excision, this observation suggests that the nature of the recombining sequences affects the recombination frequency.

Bacteriophage Mu does not contain long terminal inverted repeats and makes only 5-bp direct repeats. Its excision, observed only with pleiotropic mutants which have lost the phage B functions and do not kill their host, differs in two respects from that of transposons Tn*5* and Tn*10*: (i) excision requires the phage transposase (32, 143), and (ii) precise excision, which involves the 5-bp repeats generated by phage insertion, is 10- to 100-fold less frequent than imprecise excision, which involves phage or host DNA sequences or both (32, 143, 204). The excision often involves short homologous sequences (in 11 of 17 cases analyzed) and has been ascribed to replication errors induced by transposase binding to phage ends (204).

E. Palindrome Instability

The existence of natural palindromes and the successful cloning of certain artificial palindromes show that relatively short, head-to-head, inverted repeats can be maintained in bacteria. For instance, a 66-bp uninterrupted palindrome (33-bp inverted repeats) (26, 259), a 48-bp palindrome present in pUC7 (89, 283), and a synthetic 32-bp palindrome (170) could each be propagated on *E. coli* plasmids. Certain longer palindromes, such as a 147-bp palindrome derived from the replication origin of simian virus 40, and two synthetic palindromes of 114 and 140 bp, were also stable when propagated either on high- or low-copy-number *E. coli* plasmids (25, 286). Palindromes in this size range are sometimes unstable: DNA analysis has shown that a 68-bp palindrome derived from pBR322 was unstable at a site in pBR322 and at one of the five sites tested in an R6K-related plasmid (51, 250). The much more sensitive genetic analysis detected instability of 22- to 160-bp palindromes (46, 59). Still longer palindromes are very unstable and affect the viability of the phage or plasmid on which they reside, as indicated by the following observations.

A common experience in early DNA cloning experiments (reported only much later [40, 45]) was that dimeric plasmids composed of two head-to-head monomers ("inverted dimers") were never detected in transformed bacteria. Such DNAs were generated during ligation in vitro (88, 194), which suggested that the long, perfect palindromes are not tolerated in bacteria, a hypothesis confirmed in a systematic study of palindrome instability (44).

Perfect palindromes of 1 or 3 kb were generated

in vitro by deleting a central region and a part of the IS*50* sequences from transposon Tn*5* carried within the Ap^r gene of plasmid pBR322. Two types of transformants were obtained with the resulting molecules. A minority (12 to 35%) harbored plasmids carrying a functional Ap^r gene, restored presumably as a consequence of recombination between the 9-bp repeats flanking the inserted Tn*5*. The remaining transformants contained plasmids with shortened palindromes, generated by deletions affecting the central part of the inverted repeats. The deletions took place by recombination between short direct repeats (4 to 6 bp) as shown by sequence analysis of three plasmids (46). Further recombination between the 9-bp repeats, which flanked the remaining Tn*5* sequences, could restore a functional Ap^r gene in all analyzed plasmids, albeit with vastly different frequencies (10^{-7} to 10^{-2}) (44, 46). Among the possible reasons for the variation was the structure of the remaining inverted repeats (e.g., their length and separation). The instability of perfect palindromes generated by the deletion of a central region of Tn*5* was also documented by Hagan and Warren (105, 106).

Various long palindromes not related to Tn*5* also could not be carried on plasmids, phages, and the bacterial chromosome. Plasmids which would carry inverted repeats made up of a 130-bp sequence derived from pAT153 (a pBR322-related plasmid) could not be constructed, while the corresponding plasmids carrying direct repeats were easily obtained (169). Attempts to construct other palindromes from different pAT153 sequences were also unsuccessful (106). In addition, a 206-bp palindrome, corresponding to the terminal sequence of the minute virus of mice, was unstable when carried on *E. coli* plasmids and underwent deletions by recombination between short, directly repeated sequences (27). Furthermore, an in vitro-constructed lambda phage carrying a 3.2-kb palindrome composed of phage sequences was not viable (163). Similarly, chimeric lambda phages carrying 200- to 500-bp snapback sequences of human origin could not be propagated in recombination-proficient *E. coli* cells (301). A 3.2-kb palindrome inserted into the *E. coli* chromosome was unstable, as judged by the fact that a palindrome-free prophage resulted from lysogenization of *E. coli* by a palindrome-carrying lambda phage (257). Palindrome instability was also detected in bacteria other than *E. coli*, since a palindrome generated in vitro in plasmid pSM19035 underwent deletions in *Streptococcus pyogenes* (21).

Long inverted repeats separated by a central region can be stably maintained in bacteria, as illustrated by the structure of many transposons (146). The length of the central region necessary to stabilize a 1.9-kb palindrome carried on a pBR322-related plasmid was estimated by cloning different DNA segments, ranging from 8 to 876 bp, in its center (286). DNA analysis of 245 resulting plasmids has shown that none had an insert of less than 57 bp, and that only the plasmids which had inserts longer than 150 bp were stable.

It is likely that long palindromes are lethal for any replicon which carries them. The lethality could be due to the formation of hairpin structures by intrastrand base pairing, which possibly interferes with DNA replication (such structures can arrest DNA polymerases [14, 289]). Replicon viability can be restored by (i) deletion of the entire palindrome (44), (ii) deletion of the central part of the palindrome (46), or (iii) insertion of DNA segments in the center of the palindrome (286). In the resulting replicons, hairpin formation is either prevented or probably interfered with. The first two processes, observed in vivo, are due to recombination between short direct repeats.

F. Replication Errors or DNA Breakage?

Various genome rearrangements result from recombination between short homologous sequences. This suggests that they may occur by a common mechanism. Two different classes of mechanisms can be considered, one involving errors of DNA replication and the other involving DNA breakage.

1. Replication errors

The first class of mechanisms is conceptually related to copy-choice DNA recombination (165), which, in its simplest form, can be described as follows. The DNA replication machinery which progresses on a given template may either remain on that template or switch to another, homologous template. If the switching occurs, the newly synthesized DNA is a copy of the first template in the region 5′ from the switch site and a copy of the second template in the region 3′ from that site. The type of rearrangement resulting from the switching would depend on the relative locations of the sequences at which the switching occurred. For example, a deletion would be generated if the replication machinery were displaced by switching to a homologous sequence located on the same genome in the 3′ direction, since the region of the genome flanked by the homologous sequences would not have been replicated. In contrast, a duplication would result if the replication machinery were displaced in the 5′ direction, since the region flanked by the homologous sequences would be replicated twice. Different variants of the basic copy-choice model are discussed below.

The first experimental evidence for template switching was obtained in vitro. Extensive replication of duplex DNA by *E. coli* DNA polymerase I (PolI) was found to generate rapidly renaturing, branched structures (155, 188, 240). These structures indicated that the enzyme replicated one of the two DNA strands while concomitantly displacing the other, and that before the first strand was fully replicated, the enzyme switched to the displaced strand and continued to copy this new template. The branched duplex DNA could result from repeated template switching.

In vitro template switching of PolI on single-stranded (ss) DNA has also been observed (228). The template used was phage M13 carrying a *lacZα* segment in a wrong reading frame. LacZα$^+$ revertants were selected by transformation after DNA replication in vitro. The revertants were obtained at a frequency up to 100-fold higher than that observed if nonreplicated ssDNA was used. They resulted from 50- to 80-bp deletions (in 20% of cases) and frameshift mutations. All deletions occurred between 4- or 5-bp direct repeats (about 60 mutants were sequenced) and were attributed to template switching. Similar deletions formed during in vitro replication of ssDNA templates by the Klenow fragment of PolI (in the presence of 9-aminoacridine [50]) or by eucaryote polymerase β (157). A majority of the PolI-generated frameshift mutations were complex and could most easily be explained by a model involving two template switching events (228). First, a hairpin formed close to the tip of the newly synthesized DNA strand by the annealing of short palindromic sequences, and the polymerase switched from the parental to the newly synthesized strand. Next, a part of the newly synthesized strand was copied and the polymerase switched back to the parental DNA strand. The switching took place at regions of short homology between the newly synthesized and the parental strands.

PolI template switching was also detected in vivo on ColE1-related plasmids which carried two head-to-head, 29-bp *lac* operators (14, 28). In wild-type *E. coli* cells, ColE1 DNA synthesis is initiated by PolI, which extends an RNA primer, and is continued by DNA polymerase III (PolIII) (266; see also reference 58). In cells with a thermosensitive PolIII enzyme incubated at the restrictive temperature, PolI carried out replication from the origin to the site of insertion of the palindromic *lac* operators (about 600 bp). At this location, it switched from the template strand to the displaced strand and proceeded to copy this new template toward the replication origin. It thus generated Θ-like molecules carrying a 600-bp "replication bubble." Heat treatment of such molecules released a 600-bp, double-stranded hairpin molecule and a circular plasmid. Restriction analysis of the hairpin molecules indicated that the site of the turnaround of DNA synthesis coincided with the center of symmetry of the double *lac* operator region.

Template switching was not detected with the same plasmids in wild-type *E. coli* cells, which indicates that PolIII mediates such a process less often than PolI. If, however, these cells were treated with chloramphenicol, template switching was observed. It was suggested that PolIII was exhausted under these conditions more rapidly than PolI, and the elongation was carried out by the latter enzyme (14, 28).

Further support for in vivo template switching was obtained by studying mutations arising within the insertion sequence IS2 (91, 92). This element, inserted in the control region of the galactose operon, turned off expression of the downstream *gal* genes. Gal$^+$ mutations were obtained with a frequency of about 10^{-7}, mostly (>90%) by precise excision of IS2. Two among 107 mutants analyzed, named IS2-6 and IS2-7, were of a different type and corresponded to insertions within IS2 of 108 and 54 bp, respectively. Sequence determination indicated that they might be accounted for by a multistep template switching model. According to this model, the first step consisted of hairpin formation at the tip of the newly synthesized strand, made possible by the presence of 8-bp inverted repeats. The second step corresponded to strand switching and extension of the hairpin, using the newly synthesized strand as template. In the last step, switching back to the original template strand occurred. IS2-7 resulted from such a process. The generation of IS2-6 required an additional cycle of hairpin formation, followed by strand switching. The switching described for IS2-6 and IS2-7 would have taken place on opposite strands of IS2 (90–92). This model has also been proposed for the generation of complex frameshift mutations during in vitro replication of ssDNA by *E. coli* PolI (228; see above).

Another line of evidence which supports template switching in vivo stems from the analysis of frameshift mutations pioneered by Streisinger et al. (267). Mutations in the lysozyme gene of bacteriophage T4 are often due to a deletion or addition of one or two bases in a stretch of repeated bases. They were attributed to the misaligning of two DNA strands during DNA replication. Sequence analysis of spontaneous mutations found in the *E. coli lacI* gene led to a similar conclusion (77, 236). The major mutational hot spot, accounting for about two-thirds of all mutations and consisting of a triple repeat of a tetranucleotide CTGG, was detected in these studies. The mutants acquired an additional repeat, or lost one of the repeats. The former class of mutants reverted frequently to the wild-type sequence by losing the supernumerary repeat. Template switching, displacing the replication machinery forward or backward on the same strand, could easily account for the observed loss or gain of one or several bases.

Various rearrangements observed in vivo (deletions, transposon excision, palindrome instability, cointegrate formation) were suggested to result from polymerase switching between short repeats on an ssDNA template (2, 17a, 23, 74, 77, 79). The template could be generated by DNA unwinding enzymes (helicases) which accompany the replication fork (15, 85, 164, 299). Relatively long regions of ssDNA (1.6 and 6 kb) can be detected in replicating ds phages (lambda and T7, respectively [131, 298]). Replication of ssDNA phages (12, 13, 64) and plasmids (276, 277) also generates ssDNA (genome-size molecules, up to about 10 kb). Conjugal transfer of DNA among bacteria may also lead to the formation of long regions of ssDNA (296). In addition, helicases which participate in DNA recombination (those encoded by *recBCD*, *dda* [148, 230, 272]) or repair (helicase II [111, 156, 184, 215]) rather than in replication, as well as various exonucleases (172), could also generate long ssDNA regions. Relatively large deletions could therefore arise as a consequence of template switching on long, contiguous ssDNA stretches. It can also be envisaged that deletions and duplications may occur by template switching between short stretches of ssDNA separated by long regions of dsDNA (31). In the case of genome fusions (17b, 144), the switching would have to occur between ss regions carried on different DNA molecules.

Transposon excision was shown to occur by template switching on ssDNA. Several lines of evidence indicate that the formation of ssDNA stimulates the excision of Tn*5*. First, this transposon was frequently lost from the *E. coli* ssDNA phage fd (110). Next, in *B. subtilis*, it excised 100 to 1,000 times more frequently from ssDNA plasmids than from dsDNA plasmids or the chromosome (134; L. Jannière and S. D. Ehrlich, manuscript in preparation). Finally, it excised much more frequently (100 to 1,000 times) from F′ plasmids than from the *E. coli* chromosome (74, 115). The excision was stimulated by conjugation and occurred much more frequently in the recipient than in the donor cells (24, 271). Mutations which increased the efficiency of conjugal transfer between "male" cells (*traS* mutations which reduce surface exclusion) stimulated excision, while those that prevented conjugation (*traG*, *T*, *D*, *I*, and *Z* were tested) inhibited excision and abolished the stimulatory effect of the *traS* mutations (271). These results indicate that excision was stimulated because of ssDNA formation during the conjugal transfer of F (24).

Conclusive evidence for the involvement of ssDNA in transposon excision was reported by Brunier et al. (31). A transposon related to Tn*10* (mini-Tn*10-kan* [288]), made up of 78-bp inverted repeats (deriving from IS*10*R) and a 1.5-kb segment encoding Km^r, was used. The transposon was propagated on genomes which could be conditionally activated to produce ssDNA. Derivatives of pBR322 which carry the replication origin of M13 were used, since replication via the pBR322 origin generates no detectable ssDNA (276), while activation of the M13 origin by the phage replication protein (the product of gene II [171, 191]) results in the production of large amounts of ssDNA. Precise transposon excision restored the activity of a genetic marker and could be detected by a simple phenotypic test. Induction of ssDNA synthesis stimulated transposon excision greatly (up to 10^6 times), 100% of cells acquiring the recombinant phenotype within a very short time (less than two bacterial generations in some instances).

Excision by template switching should entail no transfer of material from parental to recombinant molecules. To test this prediction, transposon-carrying dsDNA plasmids were radioactively labeled, the labeling was interrupted, and ssDNA synthesis was induced. Recombination was allowed to proceed for three bacterial generations; plasmid DNA was extracted and analyzed. The proportion of recombinant to parental plasmids was about 2:1 (although all cells had acquired the recombinant phenotype, not all plasmids had recombined). The radioactivity was, however, associated exclusively with the parental DNA (its specific activity was $>10^5$ times higher than that of the recombinant DNA). This shows that transposon excision (precise and nearly precise, occurring at a ratio of 99:1 in the experiment) entailed no transfer of parental material to recombinant molecules and that it therefore occurred by template switching (31).

Further evidence for template switching was obtained by preparing transposon-carrying plasmid DNA in the ss form and introducing it by transformation in cells lacking the M13 replication protein. Only a single round of conversion of ss to dsDNA could take place in these cells, since the resulting ds molecules would continue to replicate in a ds form. Transposon excision was detected during that round of conversion (31).

Template switching between synthetic short direct repeats was also conclusively demonstrated (D. Brunier, B. Peeters, and S. D. Ehrlich, in preparation), using the system developed by Peeters et al. (220a). To generate the repeats, a Cm^r gene carried on plasmids capable of synthesizing ssDNA conditionally in *E. coli* (see above) was cleaved at a unique restriction site. A synthetic oligonucleotide which reproduced the sequence adjacent to the cleaved site and carried in addition a new, unique restriction site was inserted into the cleaved gene. Direct repeats of 9, 18, or 27 bp which flanked a synthetic restriction site were generated in that way. Inverted repeats were generated next to the short direct repeats by inserting the appropriate DNA segments into the synthetic restriction site. They were either short (8 bp, resulting from joining the synthetic and a natural restriction site) or long (300 bp, corresponding to sequences internal to IS*50*), and flanked a 1.5-kb segment encoding resistance to Km. The Cm^r gene was interrupted by insertions, but its integrity could be restored by recombination between the repeats. Recombination frequencies could therefore be determined by measuring the proportion of Cm^r cells in a population of plasmid-carrying Cm^s cells.

Recombination efficiency was almost independent of the length of direct and inverted repeats. It was relatively low in ds plasmids (1 Cm^r cell per 10^4 to 10^5 Cm^s cells was regularly observed in overnight cultures) but was greatly stimulated by the induction of ssDNA synthesis (100% of cells became Cm^r a few generations after induction). No transfer of material from the parental to recombinant molecules occurred during recombination between the 9-bp direct repeats, which shows that the process involved template switching. Interestingly, an abundant transfer was detected with longer direct repeats (18 and 27 bp), indicating that a different mechanism (homologous recombination?) operates on these. Analogous results were obtained with two plasmids containing 9- or 18-bp synthetic direct repeats but lacking completely the adjacent inverted repeats (Brunier et al., in preparation). This shows that the inverted repeats are not required for template switching, but does not rule out the possibility that they may stimulate the switching.

Circumstantial evidence in favor of template switching exists for palindrome instability, which is also mediated by recombination between short homologous sequences. Both nearly precise excision of Tn*10* and palindrome instability occur by recombination between short direct repeats embedded within long inverted repeats (46, 233). The former process was shown to be mediated by template switching (31; see above), and the latter, by analogy, may be mediated in the same way. Further support for this hypothesis comes from the observation that palindromes of 0.53 or 8.4 kb are lost only from replicating and not from nonreplicating phage lambda (162, 257). Template switching can occur only during DNA replication. However, there are certain differences between nearly precise excision and palindrome instability. The first occurs by recombination between relatively long direct repeats (23 of 24 bp), each of which is composed of two inverted 10-bp repeats and a central 4-bp region, while the second appears to be mediated by recombination between direct repeats which are much shorter (4 to 6 bp). Furthermore, the direct repeats which recombine during the first process are equidistant from the ends of the inverted repeats within which they are embedded, and could thus be brought into close proximity by intrastrand annealing of the inverted repeats (see below). In contrast, the direct repeats which were involved in the second process were not equidistant from the ends of the inverted repeats (46, 233). These differences may be important, and it remains to be shown that palindrome instability involves template switching.

Template switching between short direct repeats does not require inverted repeats, but is usually stimulated by their presence (see above). Inverted repeats might stimulate recombination between direct repeats by forming hairpin structures on ssDNA (23, 44, 59, 74, 79, 93). This could facilitate template switching in two ways. First, the recombining direct repeats would be brought into close proximity. Second, replication of ssDNA might be slowed down by hairpin structures (T4 DNA polymerase and *E. coli* polymerase III are temporarily arrested at hairpins in vitro [118, 119, 139, 159]). Separation of the end of the nascent DNA strand from the first copy of the direct repeats could be facilitated by the temporary arrest of DNA synthesis. Its annealing with the second copy of the repeat could be facilitated by the proximity of the two repeats.

An additional role for inverted repeats was suggested by Weston-Hafer and Berg (unpublished), who studied the reversion of mutations generated by the insertion of a Tn*5*-related transposon into the Ap^r gene of pBR322. Direct repeats of 14 and 15 bp, resulting from chance homology between the ends of IS*50* and the Ap^r gene (usually, 9-bp repeats are generated by Tn*5* insertion [237]) were generated in two cases. The repeats differed by 1 bp, corresponding to the terminal base pair of IS*50*. Ap^r revertants obtained from both insertion mutants were of two types, "true" revertants, which had the original gene sequence, and pseudorevertants, which had a base pair of IS*50* origin (the change in sequence did not abolish resistance to ampicillin). The length of the inverted repeats, which varied between 1,542 and 9 bp (intact IS*50* and shorter, in vitro-generated derivatives, respectively), affected the ratio of the two revertant types, a higher proportion of pseudorevertants being found with longer repeats. The formation of pseudorevertants can be explained as follows. The DNA replication machinery could stall within the hairpin structure formed by annealing of the inverted repeats rather than at its base. The inverted repeats, partly separated by DNA synthesis, could reanneal and thus extrude the nascent DNA strand from within the hairpin. The extruded single strand would be partially homologous with the direct repeat following the hairpin and could anneal with it. The region of the extruded strand extending past this homologous stretch would remain single stranded and could be trimmed, which would allow the DNA replication to resume. A sequence internal to the inverted repeat, which would differ from the second direct repeat by 1 bp (the terminal nucleotide of IS*50*), could thus become a part of the nascent DNA strand. This process would generate the observed pseudorevertants. It was suggested that the frequency of extrusion of the nascent strand from the hairpin might depend on the stability of the hairpin and therefore on the length of the inverted repeats (Weston-Hafer and Berg, unpublished). A similar process could be involved in the nearly precise excision of Tn*10* as well as in palindrome instability, which are both mediated by recombination between direct repeats embedded within the inverted repeats. In both processes, the extrusion of the nascent strand from hairpin structures would have to be only partial, and the extruded part would have to anneal with a sequence internal to the inverted repeats.

Besides facilitating template switching by hairpin formation, the inverted repeats were suggested to have other roles, such as stabilizing recombination intermediates or generating entry points for enzymes involved in recombination (79, 182). These roles are not necessarily mutually exclusive.

2. DNA breakage

Recombination between short direct repeats could occur by DNA breakage as well as by replication errors. It was suggested that the process might involve the following steps (49, 59): (i) a ds break is introduced in a DNA molecule between short direct repeats; (ii) ssDNA stretches are generated by exonucleolytic erosion initiated at the break; (iii) short complementary sequences are exposed by erosion anneal; (iv) gaps or ss tails possibly present in the resulting molecule are repaired or trimmed, respectively; (v) nicks formed during repair are sealed by DNA ligase. A deletion eliminating the region between the repeats and one of the repeats could be generated in this way. More complex events, such as inversions, translocations, or duplications, would require the interaction of two DNA molecules.

DNA breakage-mediated recombination between short direct repeats has been demonstrated in plasmids linearized in vitro and introduced by transformation into *E. coli* cells. Such plasmids circularize in vivo and often simultaneously undergo deletion (280). In a systematic study (47–49), linear plasmid DNA with short cohesive ends (pBR322, cleaved with various restriction enzymes and purified by gel electrophoresis) was used to transform competent *E. coli* cells. The transformation efficiency was about 100 times less than that observed with circular DNA. Most of the transformants (~90%) contained plasmids identical to the parent molecule and were presumably generated by in vivo ligation of the short cohesive ends. The remaining transformants harbored modified plasmids, which were most often smaller than the parental plasmid, and resulted from deletions. (More complex rearrangements, such as inversions and duplications, were also observed.) Not surprisingly, blunt-ended linear molecules yielded a higher proportion of deletions (up to 80%), although the transforming efficiency was further reduced by 5 to 10 times (48, 49).

Two types of deletions were found by analyzing plasmids which resulted from *Sal*I-cleaved pBR322. The first was generated by recombination between one short ss end and a complementary sequence present internally in the molecule. The second resulted from recombination between two repeats (20 and 19 bp sharing 16 bp, or 8 bp with one mismatch in the two cases analyzed by sequencing) which were both internal to the molecule (49). Similar events were detected by analyzing two plasmids deriving from pBR322 linearized by *Bam*HI (25, 83). This shows that deletions can result from recombination between short homologous sequences following DNA breakage.

Rearrangements more complex than deletions, which occurred by the joining of DNA molecules with short complementary ss ends, were also observed (42). For that purpose, two different plasmids were used, one carrying the 5′ end of a Cm^r gene on one *Eco*RI segment and the other carrying the 3′ end of that gene on another *Eco*RI segment. End joining of the two *Eco*RI segments could generate a third plasmid, carrying the complete Cm^r gene. The two plasmids were cleaved with *Eco*RI, the segments were introduced by transformation into *E. coli*, and Cm^r clones were obtained. These harbored a plasmid of the expected structure. Similar results were obtained with another pair of *Eco*RI-generating DNA segments, each carrying an antibiotic resistance gene (Ap^r or Km^r), which were joined in vivo to yield a plasmid encoding resistance to the two antibiotics.

The results just presented show that recombination between short homologous sequences can occur in DNA molecules broken in vitro. Chang and Cohen have elegantly shown that such recombination also occurs in DNA molecules broken in vivo (42). They inactivated a Cm^r gene which contained an *Eco*RI site by the insertion of an *Eco*RI segment into that site. The inactive gene was carried on a replicon related to ColE1 (p15) and introduced into *E. coli* by transformation. Cm^r transformants were obtained in cells which had the *Eco*RI restriction/modification system but not in cells which lacked it. The transformants harbored plasmids carrying the intact Cm^r gene, which shows that the *Eco*RI-mediated excision of the inserted segment occurred in vivo. A similar result was obtained upon transformation of *E. coli* with a plasmid constructed such that the excision of one *Eco*RI segment and the inversion of another were required to generate a functional Cm^r gene.

Recombination promoted by *Eco*RI cleavage occurred not only during transformation but also during normal bacterial growth. The propagation of plasmids carrying the inactive Cm^r gene in cells which had the *Eco*RI restriction/modification system resulted in the appearance of derivatives carrying a functional Cm^r gene. Such derivatives did not appear in cells lacking this system. Only one class of recombination events, the joining of cohesive ends, could be detected by the selection procedure used in experiments with DNA which was broken in vivo (42). In view of the results obtained with DNA which was broken in vitro (see above), other events, such as recombination between short homologous sequences distant from the ends, probably also occurred. If so, it is possible that most processes which generate ds breaks in vivo might trigger illegitimate recombination.

It was suggested that enzymes which resolve intermediates in homologous recombination might occasionally introduce recombinogenic ds breaks in DNA (59). Two such enzymes, endonuclease VII of phage T4 and endonuclease I of phage T7, were shown to cleave cruciform structures in vitro (61, 62, 141, 194). The interaction of palindromes, which often neighbor deletion endpoints (93), could generate cruciforms. These might be cleaved in vivo, and the resulting ds breaks could initiate illegitimate recombination. There is no evidence, however, for the role of cruciform-cleaving enzymes in recombination between short homologous sequences. The demonstration that the excision of Tn*10* can occur by template switching (31; see above) suggests that the presumed cleavage of cruciform structures does not play a role in this recombination event.

One can speculate that various DNA break-and-join enzymes discussed in the first part of this chapter may occasionally generate either the ds breaks or the lesions converted into such breaks by some other cellular functions, and thus trigger illegitimate recombination. The protein A of phage Mu, which is needed for transposition (Pato, this volume), may play a similar role. This protein is required for phage excision, which occurs by recombination between short homologous sequences in the vicinity of phage ends (204; see above). Such events are expected to occur in the vicinity of a ds break.

Sealing the breaks is an obligatory step in any recombination process involving DNA breakage. According to the model presented at the beginning of the section, this step would correspond to the sealing of nicks in a fully repaired recombinant molecule. An alternative may be considered. DNA ligase could join nucleotides transiently brought together by the annealing of short ss homologous sequences [see step (iii) of the model, p. 816] and the resulting molecule could then be further processed into a mature recombinant.

Several lines of evidence indicate that ligase can join nucleotides which are juxtaposed only transiently. (i) Cohesive ends of DNA segments generated by restriction enzymes can be joined in vitro at temperatures exceeding those required for their melting; (ii) joining of blunt-ended molecules, well known to be catalyzed by T4 ligase in vitro (247–249) and also detected in vivo with *E. coli* cells (49; D. Petranovic and S. D. Ehrlich, manuscript in preparation), requires transient juxtaposition of nonadjacent nucleotides; and (iii) T4 ligase-catalyzed joining of nucleotides separated by a gap, which was detected in vitro (210), also requires the transient juxtaposition of nonadjacent nucleotides.

To test the role of ligase in recombination between short repeated sequences, the cells which either overproduced the ligase (*lop-11*) or had wild-type ligase levels, or had thermosensitive ligase (*lig-7*) and were kept briefly at the restrictive temperature, were transformed with blunt-ended linear plasmid DNA molecules (Petranovic and Ehrlich, in preparation). The deletions occurred with a frequency of 64, 33, and 4% among transformants obtained in *lop-11*, wild-type, and *lig-7* cells, respectively. Similarly, circularization of linear molecules with cohesive ends was less efficient (by three times) in *lig-7* cells and more efficiently (by three times) in *lop-4* cells than in wild-type cells (47). Interestingly, circular molecules having a nick in each of the two DNA strands transformed the three types of cells with similar efficiency (Petranovic and Ehrlich, in preparation). These results indicate that more ligase is needed for deletion formation than for nick sealing. They suggest that the role of ligase in recombination between short homologous sequences is not simply to seal a nick in a fully repaired molecule, but may be to join nucleotides which were transiently juxtaposed.

G. Genetic Analysis

Transposon excision is the genetically best studied example of recombination between short homologous sequences. A number of mutants which stimulated the precise excision of Tn*10* (named *tex*) were isolated by a papillation test (colonies obtained from ethyl methanesulfonate-mutagenized cells carrying Tn*10* in the *lacZ* gene were screened for increased number of Lac^+ papillae [79]). Many of these could be mapped to previously identified genes, involved either in the methylation-directed mismatch repair of DNA (*dam*, *mutH*, *mutL*, *mutS*, *mutU*) or in recombination (*recBC* genes [181, 182]). Screening of mutations in other genes known to be involved in DNA metabolism has revealed that *ssb* and *mutD* also increase Tn*10* excision frequency (181). Several additional mutations stimulating transposon excision, which did not map to a previously known gene, were isolated using a similar papillation test (*uup* mutants [114]). The excision was affected to different extents by various mutations (114, 181, 182). The highest stimulation, about 100-fold, was observed with *mutU*, *recBCD*, and *uup* mutations; the lowest, about 2-fold, was observed with *mutH* and *mutS*. The same mutations, however, differently affected excision from different sites. This was particularly noticeable for a *mutS* mutation, which stimulated Tn*10* excision 2-fold from a site in *lacZ* and 50-fold from a site in the *hisG* gene. A *mutL* mutation stimulated excision from *lacZ* 11-fold and from *hisG* 50-fold (181).

The template switching model of DNA recombination between short homologous sequences provides a framework for interpreting the effect of various genes known to be involved in methylation-directed mismatch repair. It was suggested that the mismatch repair may involve the following steps (161; see also references 196 and 227 for review): (i) binding of the MutL and MutS proteins at the site of mismatch (shown in vitro for MutS [263]); (ii) loading of DNA helicase II (the product of the *mutU* gene, also known as the *uvrD*, *uvrE*, or *recL* gene [111, 156, 184, 215]) on the DNA, facilitated by the previous MutL and MutS binding; (iii) strand separation by helicase II; (iv) cleavage of the nonmethylated strand by the MutH protein at a GATC sequence on either side of the mismatch (293) (GATC methylation is mediated by the *dam* methylase [86, 185, 186]); and (v) repair of the resulting gap by a DNA polymerase acting in concert with ss binding protein (43, 180).

Replication errors which result in mismatched base pairs or unpaired bases are efficiently repaired by the methylation-directed system. The same system might repair replication errors which generate hairpins and ss loops. MutL and MutS might bind at the base of these structures, and the repair process could continue thereafter in a way similar to that suggested for mismatch bases. An argument against this model is the finding that large ss loops contained in in vitro-formed heteroduplexes between the wild-type lambda phage and a mutant phage carrying IS*1* in the *c*I gene are not repaired by the methylation-directed system (65). However, the system might operate differently on heteroduplex molecules than on disordered structures formed during replication. For example, helicase II could melt out short duplex DNA regions near the nascent end. It would thus eliminate ss loops by allowing the resulting single strand either to reanneal with the correct homologous sequence or to be degraded. In contrast, such melting and repair might be less efficient when the ss loop is far from an end or is not flanked by direct repeats, as in the case of lambda heteroduplexes studied (65).

The efficiency of mismatch repair depends on the type of mismatch (transitions are repaired more efficiently than transversions [66]) and the sequences flanking the mismatch (it increases with increasing G+C content of the sequences neighboring transversion mismatches [138]). The frequency of transposon excision might also depend, at least in part, on the efficiency with which the mismatch repair system acts on structures generated at different sites by template switching. It is interesting, in this context, that the effects of inactivation of the MutS and MutL proteins on the excision of Tn*10* depended on the site of transposon insertion (181; see above).

In general, helicase II deficiency stimulated transposon excision more than did MutL or MutS deficiencies (181; see above). Perhaps helicase loads sites such as the growing end of a DNA strand or ss loops in addition to hypothetical binding sites of MutL and MutS proteins. The greater effect of helicase deficiency would therefore reflect its action on different intermediates in deletion formation.

The effect of mutation in the *mutD* gene (also known as the *dnaQ* gene, encoding the ε subunit of the PolIII holoenzyme) on transposon excision is also in agreement with the template switching model. This mutation inactivates the 3′-to-5′ exonuclease proofreading activity of this replicating enzyme (183, 238, 239). Perhaps the mutant holoenzyme switches templates more frequently than the wild-type holoenzyme because transiently unpaired 3′ ends are not removed. Similarly, *ssb* mutation might affect template switching, since the Ssb protein is known to be involved in DNA replication (reviewed in reference 43).

Certain alternations in the *recBCD* genes, whose products participate in classical recombination between long homologous sequences, also affect recombination between short homologous sequences. The *recBCD* genes encode exonuclease V, which possesses multiple enzymatic activities (3, 275). It acts as an ATP-dependent exonuclease on ds and ssDNA and as an endonuclease on ssDNA. It cleaves DNA endonucleolytically at specific 8-bp Chi sites (224, 273). It also unwinds DNA and generates ss loops (230, 272, 274). Certain *recBCD* mutations, termed *tex*, stimulated the excision of transposons Tn*5* and Tn*10* from the *E. coli* chromosome, while others that are deficient in general recombination (*recB21*, *recC22*) had no effect (182). Stimulation varied from negligible (by 1.3 times) to pronounced (by 140 times), depending on the site of transposon insertion and the particular *texA* mutation. The excision of Tn9, which has long terminal direct repeats (146), was not stimulated, indicating that inverted repeats are necessary. The stimulation was not abolished in TexA Rec$^+$ merodiploids, suggesting that it reflected a modification, not a loss, of RecBCD activity. Recombination between plasmid-carried short direct repeats of 9, 18, and 27 bp which flanked 300-bp inverted repeats was stimulated about 10-fold by another mutation in *recBCD* genes (*recD*) (D. Brunier and S. D. Ehrlich, manuscript in preparation). No stimulation was observed if the long inverted repeats were missing.

An attractive model for the effects of the altered RecBCD protein focuses on its DNA unwinding activity. If the mutant was more effective than the wild type in DNA unwinding, interstrand pairing may have been stimulated. This would result in cruciform structures which could, in turn, facilitate template switching by DNA polymerase.

Other studies indicate that mutations in the *recBC* genes reduce palindrome instability. However, the effect of these mutations is difficult to evaluate, since additional mutations were generally also needed. For example, short, imperfect palindromes of about 150 bp excised from pBR322-related plasmids 10 times less frequently in *recBC sbcB* cells than in the Rec$^+$, *recBC*, or *sbcB* cells (46). Similarly, a plasmid-carried 206-bp palindrome which underwent deletions by recombination between short direct repeats in a Rec$^+$ host could be stably maintained in *recBC sbcB recF* but not in *recBC sbcB* cells (27). Palindromes carried on phage lambda were also more stable in *recBC sbcB* than in Rec$^+$ cells (163). This was shown with in vitro-constructed phages containing a 3.2-kb palindrome, which were used to infect strains carrying mutations in different recombination genes. Phage growth was observed in *recBC sbcB* strains but not in Rec$^+$, *recBC sbcB*, or *recBC sbcA* strains. The 3.2-kb palindrome was not stable, however, even in *recBC sbcB* cells, since the phages grown in these cells carried truncated palindromes of <1 kb. Furthermore, eucaryotic sequences forming secondary structures ("snapback" regions) could be propagated on lambda phages more efficiently in *recBC sbcB* than in Rec$^+$ cells (203, 301).

It is not clear that RecBCD enzyme acts on palindromes directly. First, host permissiveness for palindrome-carrying lambda phages was correlated with inactivation of the nucleolytic activities (by mutation or by the phage Gam function) but not of the recombinational activities of the RecBCD enzyme (294). Second, stimulation of phage recombination by the introduction of a *chi* sequence improved phage viability in Rec$^+$ cells. These observations were interpreted as indicating that the RecBCD enzyme interfered with palindrome maintenance indirectly, by preventing the rolling-circle replication mode of palindrome-carrying phages, which were generally *gam*. Multimeric genomes required for encapsidation (78) therefore could not be generated by replication in Rec$^+$ cells. Also, they could not be generated efficiently by recombination, since the palindromes seemed to reduce the number of phage genomes present, and therefore the amount of substrate available for recombination, in each infected cell (294). Recent experiments indicating that the *sbcC* mutation (present in all currently used *recBC sbcB* strains [174]) is instrumental in palindrome maintenance on phage lambda (D. Leach, personal communication) also call into question the importance of RecBCD protein in palindrome instability.

The involvement of the RecA protein in recombination between short homologous sequences is controversial. The protein, which plays a key role in general recombination (see references 52 and 261 for recent reviews) is not indispensable for deletion formation (2, 92, 137), transposon excision (74, 79), palindrome instability (44, 46, 59), or the formation of cointegrates (144). Its absence did not affect the frequency of some of these processes (transposon excision, palindrome instability, or certain instances of deletion formation [46, 79, 92]), but it did reduce the frequency of others (up to 25-fold for deletion and cointegrate formation [2, 137, 144]). Interestingly, although transposon excision is not affected by the RecA protein, the stimulation of transposon excision by certain mutations (*uvrD*, *texA344*) was abolished in *recA* mutants (181, 182). Taken together, these results indicate that the RecA protein can play a role in recombination between short homologous sequences.

An insight into the role of RecA in recombination mediated by template switching comes from the following observations (Brunier and Ehrlich, in preparation). Synthetic 9-bp direct repeats, bordering either short (8 bp) or long (300 bp) inverted repeats separated by a 1.5-kb central segment and carried on a pBR322-related plasmid, recombined in Rec^+ *E. coli* cells with a frequency of about 10^{-4}. In *recA* cells, the frequency of recombination was the same for the construct with short inverted repeats, but it was reduced 100-fold for the construct with long inverted repeats. These results are most easily explained by supposing that (i) template switching does not require RecA protein (9-bp direct repeats recombined with the same frequency in Rec^+ and *recA* cells provided that they were not adjacent to long inverted repeats), (ii) the structure generated by template switching may be destroyed by an unknown function if a ds region of >8 bp is localized close to the ss loop (presence of long, but not short, inverted repeats leads to a reduction in recombination frequency), and (iii) destruction may be prevented by the RecA protein (recombination in the neighborhood of the long and short inverted repeats is the same in the Rec^+ cells), possibly by its binding to the ssDNA regions and loading onto the adjacent ds regions. A protective role for the RecA protein was envisaged previously to account for the suppression of *uvrD*-mediated stimulation of transposon excision by *recA* mutations (182). The above model would be strengthened if the postulated function that destroys structures generated by template switching (a nuclease? a helicase?) could be identified.

Numerous genes other than *recA* and *recBCD* are involved in general DNA recombination (261). Their role in recombination between short homologous sequences has not been systematically investigated. No effect of *recF* or *recJ* on transposon excision has been found (181).

V. CONCLUSIONS AND PERSPECTIVES

A. Three Categories of Illegitimate Recombination

Many processes, as diverse as the integration-excision of phage lambda and DNA replication, are known to mediate illegitimate recombination. Other processes, not yet identified, may need to be invoked to explain less well-understood cases of illegitimate recombination (151). Possibly all these processes can be classified into three conceptual categories, involving (i) DNA breakage, (ii) DNA replication, and (iii) both breakage and replication. Analogous categories have been proposed in the context of homologous recombination (break-join, copy-choice, and break-copy recombination) (265). Few of the illegitimate recombination processes have been sufficiently well characterized to be unambiguously placed in one of the three categories. It is, however, likely that the process mediated by T4 topoisomerase is of the DNA breakage type (126), while that mediated by template switching between short homologous sequences is of the DNA replication type (31). The majority of illegitimate recombination events at the nicked replication origins require both DNA breakage and DNA replication (192, 193; Michel et al., in preparation). This establishes the existence of the three postulated categories.

B. The "Active End" Hypothesis

An interesting question is whether, notwithstanding their diversity, the various processes that mediate illegitimate recombination have any common features. One such feature seems to be a DNA end. An end, free or bound to a protein, in single- or double-stranded form, is generated by all DNA breakage and DNA replication processes. Unless destroyed by the degradation of the molecule that carries it, one end is eventually joined to another end, either preexisting or generated during the joining process. Prior to joining, an end may be processed either by nuclease degradation or, if it anneals to an appropriate template, by DNA replication. Processing and joining could modify the structure of the DNA within which the end appeared and thus mediate illegitimate recombination. An active end hypothesis can therefore be advanced, stating that all illegitimate processes generate ends in DNA molecules.

A prediction of the active end hypothesis is that any process which generates ends in DNA might initiate illegitimate recombination. Much work would be needed to test this prediction extensively. Two examples may, however, illustrate its validity. Among different processes involved in DNA repair (82, 158), the one which eliminates bulky lesions, such as thymine dimers generated by UV irradiation, has recently been well characterized. A complex formed by UvrA, B, and C proteins nicks the DNA strand which carries the thymine dimer seven bases upstream and four to five bases downstream from the dimer. Another protein, helicase II (*uvrD* gene product), and possibly also DNA polymerase I, releases the complex, together with the lesion-containing oligonucleotide, from the DNA. The resulting gap is repaired by DNA polymerase and DNA ligase (158). The ends (nicks) generated during the process should be able to initiate illegitimate recombination. It has been reported that UV irradiation stimulates deletion formation in *E. coli* (22, 63, 77, 242). Interestingly, among seven UV-induced deletions in a 900-bp region at the end of the *trp* operon (300), five ended at a site localized either 7 bp downstream or 4 bp upstream from a dinucleotide TT. This observation is not statistically significant since the number of deletions analyzed was low (a probability of about 0.1 can be calculated, taking into account the proportion of thymine dimers in the sequence), but, together with the stimulatory effect of UV on deletion formation, it does suggest that the ends generated by the UvrABC complex may initiate illegitimate recombination.

The second example concerns the Tn*10* transposase. This enzyme cleaves the extremities of Tn*10* and therefore generates DNA ends (197). The active end hypothesis predicts that these ends can initiate illegitimate recombination. If recombination took place between the 9-bp repeats which flank the transposon, a molecule should be generated from which the transposon has been precisely excised. Generally, this process cannot be observed, because precise excision of Tn*10* occurs more efficiently by template switching (31), which does not require the transposase. It was, however, observed with a Tn*10* derivative carried at a site in a pBR322-related plasmid from which it excised very infrequently (a frequency of $\sim 10^{-8}$; D. Brunier and S. D. Ehrlich, unpublished results). Induction of high levels of transposase synthesis stimulated by about 100-fold the precise excision of the transposon. This observation conforms to the prediction of the active end hypothesis.

C. Future Prospects

1. Which process and why?

Given that many processes can mediate illegitimate recombination, an intriguing question is which particular process will be most active in a given situation and why. It would not be surprising if the enzymatic systems active on all or most DNA sequences, such as DNA replication machinery or DNA topoisomerases, were more frequently involved in illegitimate recombination than systems with a narrow sequence specificity, such as site-specific recombinases or origin-nicking enzymes. An argument in favor of this speculation is that short homologous sequences between which template switching during replication can occur are very often found at endpoints of different genome rearrangements. However, many rearrangements do not end in short homologous sequences and therefore probably do not result from such erroneous replication. Much more needs to be known about enzymatic systems which mediate illegitimate recombination before the above question can be answered. It will be necessary to characterize known systems more extensively and to identify novel systems. The active end hypothesis could guide the latter endeavor.

2. Control mechanisms

Among many features of illegitimate recombination which should be better known, one is particularly obvious. The efficiency of recombination between homologous, directly repeated sequences of only 9 bp can be so high that within a very few generations (two or three) a majority (>70%) of genomes recombine (31). This high recombination efficiency does not seem to depend on a particular sequence or structure adjacent to the short homologies (Brunier et al., in preparation). Since life as we know it would probably be impossible if illegitimate recombination were generally that frequent, mechanisms to reduce its frequency must exist. In the case of template switching, which mediates the above recombination, these mechanisms might involve either decreasing the polymerase switching efficiency, by limiting the amount of ss template or by using accessory proteins (Ssb protein?), or, alternatively, correcting the structures resulting from switching. In other cases of illegitimate recombination, different correction mechanisms might operate. An understanding of these mechanisms and of the way they respond to various physiological processes and environmental factors will be essential for the comprehension of the stability of genetic material.

Another puzzling feature of illegitimate recombination is the vast difference in frequencies with which different sequences of the same length recombine (10^3- to 10^4-fold) (59). These differences may reflect the specificity of proteins that act on DNA. Knowledge of the parameters involved should contribute to the understanding of fundamental life processes, such as DNA replication and recombination, and allow us to develop tools to control genetic instability, with important consequences for medicine and biotechnology.

Acknowledgments. I would like to thank S. Gruss and M. Young for their critical reading of the manuscript, and F. Haimet for patient preparation of the text.

The work from my laboratory was supported by grants from the Ministère de la Recherche et de l'Enseignement Supérieur (no. 510070) and the Commission des Communautés Européennes (BAP-O141-F).

LITERATURE CITED

1. **Abraham, J. M., C. S. Freitag, J. R. Clements, and B. I. Eisenstein.** 1985. An invertible element of DNA controls phase variation of type 1 fimbriae of *Escherichia coli. Proc. Natl. Acad. Sci. USA* **82**:5724–5727.
2. **Albertini, A. M., M. Hofer, M. P. Calos, and J. M. Miller.** 1982. On the formation of spontaneous deletions: the importance of short sequence homologies in the generation of large deletions. *Cell* **29**:319–328.
3. **Amundsen, S. K., A. F. Taylor, A. M. Chaudhury, and G. R. Smith.** 1986. *recD*: the gene for an essential third subunit of exonuclease V. *Proc. Natl. Acad. Sci. USA* **83**:5558–5562.
4. **Anderson, P.** 1987. Twenty years of illegitimate recombination. *Genetics* **115**:581–584.
5. **Anderson, P., and J. Roth.** 1981. Spontaneous tandem genetic duplications in *Salmonella typhimurium* arise by unequal recombination between rRNA (rrn) cistrons. *Proc. Natl. Acad. Sci. USA* **78**:3113–3117.
6. **Anderson, R. P., and J. R. Roth.** 1977. Tandem genetic duplications in phage and bacteria. *Annu. Rev. Microbiol.* **31**:473–505.
7. **Arber, W.** 1984. Natural mechanisms of microbial evolution, p. 1–14. *In* W. Arber et al. (ed.), *Genetic Manipulation: Impact on Man and Society.* Cambridge University Press, Cambridge.
8. **Arber, W., M. Hümbelin, P. Caspers, H. J. Reif, S. Iida, and J. Meyer.** 1980. Genesis and natural history of IS-mediated transposons. Appendix I: Spontaneous mutations in the *Escherichia coli* prophage P1 and IS-mediated processes. *Cold Spring Harbor Symp. Quant. Biol.* **45**:38–43.
9. **Arber, W., S. Iida, H. Jütte, P. Caspers, J. Meyer, and C. Hänni.** 1978. Rearrangements of genetic material in *Escherichia coli* as observed on the bacteriophage P1 plasmid. *Cold Spring Harbor Symp. Quant. Biol.* **43**:1197–1208.
10. **Arthur, A., E. Nimmo, S. Hettle, and D. Sherratt.** 1984. Transposition and transposition immunity of transposon Tn*3* derivatives having different ends. *EMBO J.* **3**: 1723–1729.
11. **Avila, P., F. de la Cruz, and J. Grinsted.** 1984. Plasmids containing one inverted repeat of Tn*21* can fuse with other plasmids in the presence of Tn*21* transposase. *Mol. Gen. Genet.* **195**:288–293.
12. **Baas, P. D.** 1985. DNA replication of single-stranded *Escherichia coli* phages. *Biochim. Biophys. Acta* **825**:111–139.
13. **Baas, P. D., and H. S. Jansz.** 1988. Single-stranded DNA phage origins. *Curr. Top. Microbiol. Immunol.* **136**:31–70.
14. **Backman, K., M. Betlach, H. W. Boyer, and S. Yanofsky.** 1978. Genetic and physical studies on the replication of ColE1-type plasmids. *Cold Spring Harbor Symp. Quant. Biol.* **43**:69–76.
15. **Baker, T. A., K. Sekimizu, B. E. Funnell, and A. Kornberg.** 1986. Extensive unwinding of the plasmid template during staged enzymatic initiation of DNA replication from the origin of the *Escherichia coli* chromosome. *Cell* **45**:53–64.
16. **Bakker, E., C. Van Broeckhoven, E. J. Bonten, M. J. van de Vooren, H. Veenema, W. Van Hul, G. J. B. Van Ommen, A. Vandenberghe, and P. L. Pearson.** 1987. Germline mosaicism and Duchene muscular dysotrophy mutations. *Nature* (London) **329**:554–556.
17. **Ballester, S., P. Lopez, J. C. Alonso, M. Espinosa, and S. A. Lacks.** 1986. Selective advantage of deletions enhancing chloramphenicol acetyltransferase gene expression in *Streptococcus pneumoniae* plasmids. *Gene* **41**:153–163.

17a. **Bashkirov, V. I., F. K. Khasanov, and A. A. Prozorov.** 1987. Illegitimate recombination in *Bacillus subtilis*: nucleotide sequences in recombinant DNA junctions. *Mol. Gen. Genet.* **210**:578–580.

17b. **Bashkirov, V. I., M. M. Stoilova-Disheva, and A. A. Prozorov.** 1988. Interplasmidic illegitimate recombination in *Bacillus subtilis. Mol. Gen. Genet.* **213**:465–470.

18. **Bauer, C. E., J. F. Gardner, and R. I. Gumport.** 1985. Extent of sequence homology required for bacteriophage lambda site-specific recombination. *J. Mol. Biol.* **181**:187–197.
19. **Been, M. D., and J. J. Champoux.** 1981. DNA breakage and closure by rat liver type 1 topoisomerase: separation of the half-reactions by using a single-stranded DNA substrate. *Proc. Natl. Acad. Sci. USA* **78**:2883–2887.
20. **Been, M. D., and J. J. Champoux.** 1984. Breakage of single-stranded DNA by eukaryotic type I topoisomerase occurs only at regions with the potential for base pairing. *J. Mol. Biol.* **180**:515–531.
21. **Behnke, D., H. Malke, M. Hartman, and F. Walter.** 1979. Post-transformational rearrangement of an *in vitro* reconstructed group-A streptococcal erythromycin resistance plasmid. *Plasmid* **2**:605–616.
22. **Berg, D. E.** 1974. Genetic evidence for two types of gene arrangements in new lambda-*dv* plasmid mutants. *J. Mol. Biol.* **86**:59–68.
23. **Berg, D. E., C. Egner, B. J. Hirschel, L. Johnsrud, R. A. Jorgensen, and T. D. Tlsty.** 1980. Insertion, excision, and inversion of Tn*5*. *Cold Spring Harbor Symp. Quant. Biol.* **45**:115–123.
24. **Berg, D. E., C. Egner, and J. B. Lowe.** 1983. Mechanism of F factor-enhanced excision of transposon Tn*5*. *Gene* **22**:1–7.
25. **Bergsma, D. J., D. M. Olive, S. W. Hartzell, B. J. Byrne, and K. N. Subramanian.** 1982. Cyclization of linear chimeric plasmids *in vivo* by a novel end-to-end joining reaction or by intramolecular recombination: one of the products contains a 147-bp perfect palindrome stable in *Escherichia coli. Gene* **20**:157–167.
26. **Betz, J. L., and J. R. Sadler.** 1981. Variant of a cloned synthetic lactose operator. I. A palindromic dimer lactose operator derived from one strand of the cloned 40-base pair operator. *Gene* **13**:1–12.
27. **Boissy, R., and C. R. Astell.** 1985. An *Escherichia coli recBC sbcB recF* host permits the deletion-resistant propagation of plasmid clones containing the 5′-terminal palindrome of minute virus of mice. *Gene* **35**:179–185.
28. **Bolivar, F., M. C. Betlach, H. L. Heyneker, J. Shine, R. L. Rodriguez, and H. W. Boyer.** 1977. Origin of replication of pBR345 plasmid DNA. *Proc. Natl. Acad. Sci. USA* **74**: 5265–5269.
29. **Borst, P., and D. R. Greaves.** 1987. Programmed gene rearrangements altering gene expression. *Science* **235**: 658–667.
30. **Brake, A. J., A. V. Fowler, I. Zabin, J. Kania, and B.**

Muller-Hill. 1978. Lambda-galactosidase chimeras: primary structure of a *lac* repressor-lambda-galactosidase protein. *Proc. Natl. Acad. Sci. USA* 75:4824–4827.

31. **Brunier, D., B. Michel, and S. D. Ehrlich.** 1988. Copy choice illegitimate DNA recombination. *Cell* **52**:883–892.
32. **Bukhari, A. I.** 1975. Reversal of mutator phage Mu integration. *J. Mol. Biol.* **96**:87–99.
33. **Bullock, P., J. J. Champoux, and M. Botchan.** 1985. Association of crossover points with topoisomerase I cleavage sites: a model for nonhomologous recombination. *Science* **230**:954–958.
34. **Bullock, P., W. Forrester, and M. Botchan.** 1984. DNA sequence studies of simian virus 40 chromosomal excision and integration in rat cells. *J. Mol. Biol.* **174**:55–84.
35. **Burlingame, R. P., M. G. Obukowicz, D. L. Lynn, and M. M. Howe.** 1986. Isolation of point mutations in bacteriophage Mu attachment regions cloned in a lambda : : mini-Mu phage. *Proc. Natl. Acad. Sci. USA* **83**:6012–6016.
36. **Bushman, W., J. F. Thompson, L. Vargas, and A. Landy.** 1985. Control of directionality in lambda site specific recombination. *Science* **230**:906–911.
37. **Buysse, J. M., and S. Palchaudhuri.** 1984. Formation of type II F-primes from unstable Hfrs and their recA-independent conversion of other F-prime types. *Mol. Gen. Genet.* **193**: 543–553.
38. **Cairns, J.** 1981. The origin of human cancers. *Nature* (London) **289**:353–357.
39. **Campbell, A.** 1979. Natural modes of genetic exchange and change, p. 21–29. *In* J. Morgan and W. J. Whelan (ed.), *Recombinant DNA and Genetic Experimentation.* Pergamon Press, Oxford.
40. **Casadaban, M. J., and S. N. Cohen.** 1980. Analysis of gene control signals by DNA fusion and cloning in *Escherichia coli. J. Mol. Biol.* **138**:179–207.
41. **Centola, M. B., M.-M. Tsai, and R. C. Deonier.** 1987. Transposition of Tn*1000*: activity of single or directly repeated termini. *J. Bacteriol.* **169**:5852–5855.
42. **Chang, S., and S. N. Cohen.** 1977. In vivo site-specific genetic recombination promoted by the *Eco*RI restriction endonuclease. *Proc. Natl. Acad. Sci. USA* **74**:4811–4815.
43. **Chase, J. W., and K. R. Williams.** 1986. Single-stranded DNA binding proteins required for DNA replication. *Annu. Rev. Biochem.* **55**:103–136.
44. **Collins, J.** 1980. Instability of palindromic DNA in *Escherichia coli. Cold Spring Harbor Symp. Quant. Biol.* **45**: 409–416.
45. **Collins, J., and B. Hahn.** 1978. Cosmids: a type of plasmid gene-cloning vector that is packageable in vitro in bacteriophage lambda heads. *Proc. Natl. Acad. Sci. USA* **75**: 4242–4246.
46. **Collins, J., G. Volckaert, and P. Nevers.** 1982. Precise and nearly-precise excision of the symmetrical inverted repeats of Tn*5*; common features of *recA*-independent deletion events in *Escherichia coli. Gene* **19**:139–146.
47. **Conley, E. C., and J. R. Saunders.** 1984. Recombination-dependent recircularization of linearized pBR322 plasmid DNA following transformation of *Escherichia coli. Mol. Gen. Genet.* **194**:211–218.
48. **Conley, E. C., V. A. Saunders, V. Jackson, and J. R. Saunders.** 1986. Mechanism of intramolecular recyclization and deletion formation following transformation of *Escherichia coli* with linearized plasmids DNA. *Nucleic Acids Res.* **14**:8919–8932.
49. **Conley, E. C., V. A. Saunders, and J. R. Saunders.** 1986. Deletion and rearrangement of plasmid DNA during transformation of *Escherichia coli* with linear plasmid molecules. *Nucleic Acids Res.* **14**:8905–8917.
50. **Conrad, M., and M. D. Topal.** 1986. Induction of deletion and insertion mutations by 9-aminoacridine. An *in vitro* model. *J. Biol. Chem.* **261**:16226–16232.
51. **Courey, A. J., and J. C. Wang.** 1983. Cruciform formation in a negatively supercoiled DNA may be kinetically forbidden under physiological conditions. *Cell* **33**:817–829.
52. **Cox, M. M., and I. R. Lehman.** 1987. Enzymes of general recombination. *Annu. Rev. Biochem.* **56**:229–262.
53. **Craigie, R., and K. Mizuuchi.** 1985. Mechanism of transposition of bacteriophage Mu: structure of a transposition intermediate. *Cell* **41**:867–876.
54. **Craigie, R., M. Mizuuchi, and K. Mizuuchi.** 1984. Site-specific recognition of the bacteriophage Mu ends by the Mu A protein. *Cell* **39**:387–394.
55. **Croce, C. M.** 1987. Role of chromosome translocations in human neoplasia. *Cell* **49**:155–156.
56. **Dagert, M., I. Jones, A. Goze, S. Romac, B. Niaudet, and S. D. Ehrlich.** 1984. Replication functions of pC194 are necessary for efficient plasmid transduction by M13 phage. *EMBO J.* **3**:81–86.
57. **Darras, B. T., and U. Francke.** 1987. A partial deletion of the muscular dysotrophy gene transmitted twice by an unaffected male. *Nature* (London) **329**:556–558.
58. **Dasgupta, S., H. Masukata, and J.-I. Tomizawa.** 1987. Multiple mechanisms for initiation of ColE1 DNA replication: DNA synthesis in the presence and absence of ribonuclease H. *Cell* **51**:1113–1122.
59. **Dasgupta, U., K. Weston-Hafer, and D. E. Berg.** 1987. Local DNA sequence control of deletion formation in *Escherichia coli* plasmid pBR322. *Genetics* **115**:41–49.
60. **Davis, R. W., and J. S. Parkinson.** 1971. Deletion mutants of bacteriophage lambda. III. Physical structure of attφ. *J. Mol. Biol.* **56**:403–423.
61. **De Massy, B., F. W. Studier, L. Dorgai, E. Appelbaum, and R. A. Weisberg.** 1984. Enzymes and sites of genetic recombination: studies with gene-3 endonuclease of phage T7 and with site-affinity mutants of phage lambda. *Cold Spring Harbor Symp. Quant. Biol.* **49**:715–726.
62. **De Massy, B., R. A. Weisberg, and F. W. Studier.** 1987. Gene 3 endonuclease of bacteriophage T7 resolves conformationally branched structures in double-stranded DNA. *J. Mol. Biol.* **193**:359–376.
63. **Demerec, M.** 1960. Frequency of deletions among spontaneous and induced mutations in *Salmonella. Proc. Natl. Acad. Sci. USA* **46**:1075–1079.
64. **Denhardt, D. T., D. Dressmer, and D. S. Ray (ed.).** 1978. *The Single-Stranded DNA Phages.* Cold Spring Harbor Laboratory, Cold Spring Harbor, N.Y.
65. **Dohet, C., S. Dzidic, R. Wagner, and M. Radman.** 1987. Large non-homology in heteroduplex DNA is processed differently than single base pair mismatches. *Mol. Gen. Genet.* **206**:181–184.
66. **Dohet, C., R. Wagner, and M. Radman.** 1985. Repair of defined single base-pair mismatches in *Escherichia coli. Proc. Natl. Acad. Sci. USA* **82**:503–505.
67. **Dotto, G. P., K. Horiuchi, K. S. Jakes, and N. D. Zinder.** 1983. Signals for the initiation and termination of synthesis of the viral strand of bacteriophage f2. *Cold Spring Harbor Symp. Quant. Biol.* **47**:717–722.
68. **Drake, J. W., and R. H. Baltz.** 1976. The biochemistry of mutagenesis, p. 11–37. *In* J. W. Drake and R. E. Koch (ed.), *Mutagenesis.* Academic Press, Inc., New York.

69. **Drlica, K.** 1984. Biology of bacterial deoxyribonucleic acid topoisomerases. *Microbiol Rev.* **48:**273–289.

70. **Echols, H.** 1986. Multiple DNA-protein interactions governing high-precision DNA transactions. *Science* **233:** 1050–1056.

71. **Echols, H., and G. Guarneros.** 1983. Control of integration and excision, p. 75–92. *In* R. W. Hendrix, J. W. Roberts, F. W. Stahl, and R. A. Weisberg (ed.), *Lambda II.* Cold Spring Harbor Laboratory, Cold Spring Harbor, N.Y.

72. **Edlund, T., and S. Normark.** 1981. Recombination between short DNA homologies causes tandem duplication. *Nature* (London) **292:**269–271.

73. **Edwards, K. A., B. D. Halligan, J. L. Davis, N. L. Nivera, and L. F. Liu.** 1982. Recognition sites of eukaryotic DNA topoisomerase I: DNA nucleotide sequencing analysis of topo I cleavage sites on SV40 DNA. *Nucleic Acids Res.* **10:**2565–2576.

74. **Egner, C., and D. E. Berg.** 1981. Excision of transposon Tn*5* is dependent on the inverted repeats but not on the transposase function of Tn*5*. *Proc. Natl. Acad. Sci. USA* **78:** 459–463.

75. **Ehrlich, S. D.** 1977. Replication and expression of plasmids from *Staphylococcus aureus* in *Bacillus subtilis. Proc. Natl. Acad. Sci. USA* **74:**1680–1682.

76. **Ehrlich, S. D., P. Noirot, M. A. Petit, L. Jannière, B. Michel, and H. te Riele.** 1986. Structural instability of *Bacillus subtilis* plasmids, p. 71–83. *In* J. K. Setlow and A. Hollaender (ed.), *Genetic Engineering*, vol. 8. Plenum Publishing Corp., New York.

77. **Farabaugh, P. J., U. Schmeissner, M. Hofer, and J. H. Miller.** 1978. Genetic studies of the *lac* repressor. VII. On the molecular nature of spontaneous hostspots in the *lac*I gene of *Escherichia coli. J. Mol. Biol.* **126:**847–863.

78. **Feiss, M., and A. Becker.** 1983. DNA packaging and cutting, p. 305–330. *In* R. W. Hendrix, J. W. Roberts, F. W. Stahl, and R. A. Weisberg (ed.), *Lambda II.* Cold Spring Harbor Laboratory, Cold Spring Harbor, N.Y.

79. **Foster, T. J., V. Lundblad, S. Hanley-Way, S. M. Halling, and N. Kleckner.** 1981. Three Tn*10*-associated excision events: relationship to transposition and role of direct and inverted repeats. *Cell* **23:**215–227.

80. **Franklin, N.** 1971. Illegitimate recombination, p. 175–194. *In* A. D. Hershey (ed.), *The Bacteriophage Lambda.* Cold Spring Harbor Laboratory, Cold Spring Harbor, N.Y.

81. **Franklin, N. C.** 1967. Extraordinary recombinational events in *Escherichia coli.* Their independence of the rec+ function. *Genetics* **55:**699–707.

82. **Friedberg, E. C.** 1985. *DNA Repair.* W. H. Freeman and Co., San Francisco.

83. **Garaev, M. M., A. F. Bobkov, A. F. Bobkova, V. N. Kalinin, V. D. Smirnov, Y. E. Khudakov, and T. I. Tikchonenko.** 1982. The site-specific deletion in plasmid pBR322. *Gene* **18:**21–28.

84. **Garnier, T., W. Saurin, and S. T. Cole.** 1987. Molecular characterization of the resolvase genes, *res*, carried by a multicopy plasmid from *Clostridium perfringens*: common evolutionary origin for prokaryotic site-specific recombinases. *Mol. Microbiol.* **1:**371–376.

85. **Geider, K., and H. Hoffmann-Berling.** 1981. Proteins controlling the helical structure of DNA. *Annu. Rev. Biochem.* **50:**233–260.

86. **Geier, G. E., and P. Modrich.** 1979. Recognition sequence of the dam methylase of *Escherichia coli* K12 and mode of cleavage of *Dpn*I endonuclease. *J. Biol. Chem.* **254:** 1408–1413.

87. **Gellert, M.** 1981. DNA topoisomerases. *Annu. Rev. Biochem.* **50:**879.

88. **Gellert, M., K. Mizuuchi, M. H. O'Dea, H. Ohmori, and J. Tomizawa.** 1978. DNA gyrase and DNA supercoiling. *Cold Spring Harbor Symp. Quant. Biol.* **43:**35–40.

89. **Gellert, M., M. H. O'Dea, and K. Mizuuchi.** 1983. Slow cruciform transitions in palindromic DNA. *Proc. Natl. Acad. Sci. USA* **80:**5545–5549.

90. **Ghosal, D., J. Gross, and H. Saedler.** 1978. DNA sequence of IS*2-7* and generation of mini-insertions by replication of IS*2* sequences. *Cold Spring Harbor Symp. Quant. Biol.* **43:** 1193–1197.

91. **Ghosal, D., and H. Saedler.** 1978. DNA sequence of the mini-insertion IS*2-6* and its relation to the sequence of IS*2*. *Nature* (London) **275:**611–617.

92. **Ghosal, D., and H. Saedler.** 1979. IS*2-61* and IS*2-611* arise by illegitimate recombination from IS*2-6*. *Mol. Gen. Genet.* **176:**233–238.

93. **Glickman, B. W., and L. S. Ripley.** 1984. Structural intermediates of deletion mutagenesis: a role for palindromic DNA. *Proc. Natl. Acad. Sci. USA* **81:**512–516.

94. **Goze, A., and S. D. Ehrlich.** 1980. Replication of plasmids from *Staphylococcus aureus* in *Escherichia coli. Proc. Natl. Acad. Sci. USA* **77:**7333–7337.

95. **Grindley, N. D. F., and R. R. Reed.** 1985. Transpositional recombination in prokaryotes. *Annu. Rev. Biochem.* **54:** 863–896.

96. **Grindley, N. D. F., and D. J. Sherratt.** 1978. Sequence analysis at IS*1* insertion sites: models for transposition. *Cold Spring Harbor Symp. Quant. Biol.* **43:**1257–1261.

97. **Groenen, M. A. M., M. Kokke, and P. van de Putte.** 1986. Transposition of mini-Mu containing only one of the ends of bacteriophage Mu. *EMBO J.* **5:**3687–3690.

98. **Groenen, M. A. M., E. Timmers, and P. van de Putte.** 1985. DNA sequences at the ends of the genome of bacteriophage Mu essential for transposition. *Proc. Natl. Acad. Sci. USA* **82:**2087–2091.

99. **Groenen, M. A. M., and P. van de Putte.** 1986. Analysis of the ends of bacteriophage Mu using site-directed mutagenesis. *J. Mol. Biol.* **189:**597–602.

100. **Gros, M. F., H. te Riele, and S. D. Ehrlich.** 1987. Rolling circle replication of single-stranded DNA plasmid pC194. *EMBO J.* **12:**3863–3869.

101. **Gruss, A. D., H. F. Ross, and R. P. Novick.** 1987. Functional analysis of a palindromic sequence required for normal replication of several staphylococcal plasmids. *Proc. Natl. Acad. Sci. USA* **84:**2165–2169.

102. **Guyer, M. S., and A. J. Clark.** 1976. *cis*-dominant, transfer-deficient mutants of the *Escherichia coli* K-12 F sex factor. *J. Bacteriol.* **125:**233–247.

103. **Guyer, M. S., N. Davidson, and A. J. Clark.** 1977. Heteroduplex analysis of tra-delta-F′ plasmids and the mechanisms of their formation. *J. Bacteriol.* **131:**970–980.

104. **Hadley, R. G., and R. C. Deonier.** 1980. Specificity in the formation of delta-tra F′ plasmids. *J. Bacteriol.* **143:** 680–692.

105. **Hagan, C. E., and G. J. Warren.** 1982. Lethality of palindromic DNA and its use in selection of recombinant plasmids. *Gene* **19:**147–151.

106. **Hagan, C. E., and G. J. Warren.** 1983. Viability of palindromic DNA is restored by deletions occurring at low but variable frequency in plasmids of *Escherichia coli. Gene* **24:**317–326.

107. **Halligan, B. D., J. L. Davis, K. A. Edwards, and L. F. Liu.**

1982. Intra- and intermolecular strand transfer by HeLa DNA topoisomerase I. *J. Biol. Chem.* **257**:3995–4000.

108. **Heffron, F.** 1983. Tn3 and its relatives, p. 223–260. *In* J. A. Shapiro (ed.), *Mobile Genetic Elements*. Academic Press, Inc., New York.

109. **Heritage, J., and P. M. Bennett.** 1985. Plasmid fusions mediated by one end of TnA. *J. Gen. Microbiol.* **131**: 1131–1140.

110. **Herrmann, R., K. Neugebauer, H. Zentgraf, and H. Schaller.** 1978. Transposition of a DNA sequence determining kanamycin resistance into the single-stranded genome of bacteriophage fd. *Mol. Gen. Genet.* **159**:171–178.

111. **Hickson, I. D., H. M. Arthur, D. Bramhill, and P. T. Emmerson.** 1983. The *E. coli uvr*D gene product is DNA helicase II. *Mol. Gen. Genet.* **190**:265–270.

112. **Hoess, R. H., and A. Landy.** 1978. Structure of the lambda sites generated by Int-dependent deletions. *Proc. Natl. Acad. Sci. USA* **75**:5437–5441.

113. **Holloway, B., and K. B. Low.** 1987. F-prime and R-prime factors, p. 1145–1153. *In* F. C. Neidhardt, J. L. Ingraham, K. B. Low, B. Magasanik, M. Schaechter, and H. E. Umbarger (ed.), *Escherichia coli and Salmonella typhimurium: Cellular and Molecular Biology*, vol. 2. American Society for Microbiology, Washington, D.C.

114. **Hopkins, J. D., M. Clements, and M. Syvanen.** 1983. New class of mutations in *Escherichia coli* (*uup*) that affect precise excision of insertion elements and bacteriophage Mu growth. *J. Bacteriol.* **153**:384–389.

115. **Hopkins, J. D., M. B. Clements, T.-Y. Liang, R. R. Isberg, and M. Syvanen.** 1980. Recombination genes on the *Escherichia coli* sex factor specific for transposable elements. *Proc. Natl. Acad. Sci. USA* **77**:2814–2818.

116. **Horiuchi, K., J. V. Ravetch, and N. D. Zinder.** 1978. DNA replication of bacteriophage f1 *in vivo*. *Cold Spring Harbor Symp. Quant. Biol.* **43**:389–399.

117. **Horowitz, B., and R. C. Deonier.** 1985. Formation of Delta-tra F′ plasmids: specific recombination at oriT. *J. Mol. Biol.* **186**:267–274.

118. **Huang, C.-C., and J. E. Hearst.** 1980. Pauses at positions of secondary structure during *in vitro* replication of single-stranded fd bacteriophage DNA by T4 DNA polymerase. *Anal. Biochem.* **103**:127–139.

119. **Huang, C.-C., J. E. Hearst, and B. M. Alberts.** 1981. Two types of replication proteins increase the rate at which T4 DNA polymerase traverses the helical regions in a single-stranded DNA template. *J. Biol. Chem.* **256**:4087–4094.

120. **Iida, S., and R. Hiestand-Nauer.** 1986. Localized conversion at the crossover sequences in the site-specific DNA inversion system of bacteriophage P1. *Cell* **45**:71–79.

121. **Iida, S., and R. Hiestand-Nauer.** 1987. Role of the central dinucleotide at the crossover sites for the selection of quasi sites in DNA inversion mediated by the site-specific Cin recombinase of phage P1. *Mol. Gen. Genet.* **208**:464–468.

122. **Iida, S., H. Huber, R. Hiestand-Nauer, J. Meyer, T. A. Bickle, and W. Arber.** 1984. The bacteriophage P1 site-specific recombinase Cin: recombination events and DNA recognition sequences. *Cold Spring Harbor Symp. Quant. Biol.* **49**:769–777.

123. **Iida, S., J. Meyer, and W. Arber.** 1981. Cointegrates between bacteriophage P1 DNA and plasmid pBR322 derivatives suggest molecular mechanisms for P1-mediated transduction of small plasmids. *Mol. Gen. Genet.* **184**:1–10.

124. **Iida, S., J. Meyer, and W. Arber.** 1983. Prokaryotic IS elements, p. 159–221. *In* J. A. Shapiro (ed.), *Mobile Genetic Elements*. Academic Press, Inc., New York.

125. **Iida, S., J. Meyer, K. E. Kennedy, and W. Arber.** 1982. A site-specific, conservative recombination system carried by bacteriophage P1. Mapping the recombinase gene *cin* and the cross-over sites *cix* for the inversion of the C segment. *EMBO J.* **1**:1445–1453.

126. **Ikeda, H.** 1986. Bacteriophage T4 DNA topoisomerase mediates illegitimate recombination *in vitro*. *Proc. Natl. Acad. Sci. USA* **83**:922–926.

127. **Ikeda, H., K. Aoki, and A. Naito.** 1982. Illegitimate recombination mediated *in vitro* by DNA gyrase of *Escherichia coli*: structure of recombinant DNA molecules. *Proc. Natl. Acad. Sci. USA* **79**:3724–3728.

128. **Ikeda, H., I. Kawasaki, and M. Gellert.** 1984. Mechanism of illegitimate recombination: common sites for recombination and cleavage mediated by *E. coli* DNA gyrase. *Mol. Gen. Genet.* **196**:546–549.

129. **Ikeda, H., K. Moriya, and T. Matsumoto.** 1981. In vitro study of illegitimate recombination: involvement of DNA gyrase. *Cold Spring Harbor Symp. Quant. Biol.* **45**: 399–408.

130. **Ikeda, H., and M. Shiozaki.** 1984. Nonhomologous recombination mediated by *Escherichia coli* DNA gyrase: possible involvement of DNA replication. *Cold Spring Harbor Symp. Quant. Biol.* **49**:401–409.

131. **Inman, R. B., and M. Schnös.** 1971. Structure of branch points in replicating DNA: presence of single-stranded connections in lambda DNA branch points. *J. Mol. Biol.* **56**: 319–325.

132. **Iordanescu, S.** 1975. Recombinant plasmid obtained from two different compatible staphylococcal plasmids. *J. Bacteriol.* **124**:597–601.

133. **Ippen, K., J. A. Shapiro, and J. R. Beckwith.** 1971. Transposition of the *lac* region to the *gal* region of the *Escherichia coli* chromosome: isolation of lambda-*lac* transducing bacteriophages. *J. Bacteriol.* **108**:5–9.

134. **Jannière, L., and S. D. Ehrlich.** 1987. Recombination between short repeated sequences is more frequent in plasmids than in the chromosome of *Bacillus subtilis*. *Mol. Gen. Genet.* **210**:116–121.

135. **Johnson, R. C., and M. I. Simon.** 1985. Hin-mediated site-specific recombination requires two 26 bp recombination sites and a 60 bp recombinational enhancer. *Cell* **41**:781–791.

136. **Johnson, R. C., and M. I. Simon.** 1987. Enhancers of site-specific recombination in bacteria. *Trends Genet.* **3**: 262–267.

137. **Jones, I. M., S. B. Primrose, and S. D. Ehrlich.** 1982. Recombination between short direct repeats in a RecA host. *Mol. Gen. Genet.* **188**:486–489.

138. **Jones, M., R. Wagner, and M. Radman.** 1987. Repair of a mismatch is influenced by the base composition of a surrounding nucleotide sequence. *Genetics* **115**:605–610.

139. **Kaguni, L. S., and D. A. Clayton.** 1982. Template-directed pausing in *in vitro* DNA synthesis by DNA polymerase α from *Drosophila melanogaster* embryos. *Proc. Natl. Acad. Sci. USA* **79**:983–987.

140. **Kamp, D., E. Kardas, W. Ritthaler, R. Sandulache, R. Schmucker, and B. Stern.** 1984. Comparative analysis of invertible DNA in phage genomes. *Cold Spring Harbor Symp. Quant. Biol.* **49**:301–311.

141. **Kemper, B., F. Jensch, M. v. Depka-Prondzynski, H.-J. Fritz, U. Borgmeyer, and K. Mizuuchi.** 1985. Resolution of Holliday structures by endonuclease VII as observed in interactions with cruciform DNA. *Cold Spring Harbor Symp. Quant. Biol.* **49**:815–825.

142. **Kennedy, K. E., S. Iida, J. Meyer, M. Stålhammar-Carlemalm, R. Hiestand-Nauer, and W. Arber.** 1983. Genome fusion mediated by the site specific DNA inversion system of bacteriophage P1. *Mol. Gen. Genet.* **189:**413–421.

143. **Khatoon, H., and A. I. Bukhari.** 1981. DNA rearrangements associated with reversion of bacteriophage mu-induced mutations. *Genetics* **98:**1–24.

144. **King, S. R., M. A. Krolewski, S. L. Marvo, P. J. Lipson, K. L. Pogue-Geile, J. H. Chung, and S. R. Jaskunas.** 1982. Nucleotide sequence analysis of *in vivo* recombinants between bacteriophage lambda DNA and pBR322. *Mol. Gen. Genet.* **186:**548–557.

145. **Kleckner, N.** 1979. DNA sequence analysis of Tn*10* insertions: origin and role of 9 bp flanking repetitions during Tn*10* translocation. *Cell* **16:**711–720.

146. **Kleckner, N.** 1981. Transposable elements in prokaryotes. *Annu. Rev. Genet.* **15:**341–404.

147. **Kleckner, N.** 1983. Transposon Tn*10*, p. 261–298. *In* J. A. Shapiro (ed.), *Mobile Genetic Elements.* Academic Press, Inc., New York.

148. **Kodadek, T., and B. M. Alberts.** 1987. Stimulation of protein-directed strand exchange by a DNA helicase. *Nature* (London) **326:**312–314.

148a. **Koepsel, R. R., and S. A. Khan.** 1987. Cleavage of single-stranded DNA by plasmid pT181-encoded RepC protein. *Nucleic Acids Res.* **15:**4085–4097.

149. **Koepsel, R. R., R. W. Murray, W. D. Rosenblum, and S. A. Khan.** 1985. Purification of pT181-encoded repC protein required for the initiation of plasmid replication. *J. Biol. Chem.* **260:**8571–8577.

150. **Koepsel, R. R., R. W. Murray, W. D. Rosenblum, and S. A. Khan.** 1985. The replication initiator protein of plasmid pT181 has sequence-specific endonuclease and topoisomerase-like activities. *Proc. Natl. Acad. Sci. USA* **82:** 6845–6849.

151. **Kokontis, J. M., J. Vaughan, R. G. Harvey, and S. B. Weiss.** 1988. Illegitimate recombination induced by benzo[*a*]pyrene diol epoxide in *Escherichia coli. Proc. Natl. Acad. Sci. USA* **85:**1043–1046.

152. **Kollek, R., W. Oertel, and W. Goebel.** 1980. Site-specific deletion at the replication origin of the antibiotic resistance factor R1. *Mol. Gen. Genet.* **177:**413–419.

153. **Komano, T., S.-R. Kim, and T. Nisioka.** 1987. Distribution of shufflon among IncI plasmids. *J. Bacteriol.* **169:** 5317–5319.

154. **Komano, T., A. Kubo, and T. Nisioka.** 1987. Shufflon: multi-inversion of four contiguous DNA segments of plasmid R64 creates seven different open reading frames. *Nucleic Acids Res.* **15:**1165–1172.

155. **Kornberg, A. (ed.).** 1974. *DNA Synthesis* W. H. Freeman and Co., San Francisco.

156. **Kumura, K., and M. Sekiguchi.** 1984. Identification of the *uvr*D gene product of *Escherichia coli* as DNA helicase II and its production by DNA-damaging agents. *J. Biol. Chem.* **259:**1560–1565.

157. **Kunkel, T. A.** 1985. The mutational specificity of DNA polymerases-α and gamma during *in vitro* DNA synthesis. *J. Biol. Chem.* **260:**12866–12874.

158. **Kushner, S. R.** 1987. DNA repair, p. 1044–1053. *In* F. C. Neidhardt, J. L. Ingraham, K. B. Low, B. Magasanik, M. Schaechter, and H. E. Umbarger (ed.), *Escherichia coli and Salmonella typhimurium: Cellular and Molecular Biology,* vol. 2. American Society for Microbiology, Washington, D.C.

159. **LaDuca, R. J., P. J. Fay, C. Chuang, C. S. McHenry, and R. A. Bambara.** 1983. Site-specific pausing of deoxyribonucleic acid synthesis catalyzed by four forms of *Escherichia coli* DNA polymerase III. *Biochemistry* **22:**5177–5188.

160. **Lane, D. H. E., and D. T. Denhardt.** 1974. The rep mutation. III. Altered structure of the replicating *E. coli* chromosome. *J. Bacteriol.* **120:**805–814.

161. **Längle-Rouault, F., G. Maenhaut-Michel, and M. Radman.** 1987. GATC sequences, DNA nicks and the MutH function in *Escherichia coli* mismatch repair. *EMBO J.* **6:**1121–1127.

162. **Leach, D., and J. Lindsey.** 1986. In vivo loss of supercoiled DNA carrying a palindromic sequence. *Mol. Gen. Genet.* **204:**322–327.

163. **Leach, D. R. F., and F. W. Stahl.** 1983. Viability of lambda phages carrying a perfect palindrome in the absence of recombination nucleases. *Nature* (London) **305:**448–451.

164. **LeBowitz, J. H., and R. McMacken.** 1986. The *Escherichia coli* dnaB replication protein is a DNA helicase. *J. Biol. Chem.* **261:**4738–4748.

165. **Lederberg, J.** 1955. Recombination mechanisms in bacteria. *J. Cell. Comp. Physiol.* **45:**75–107.

166. **Leong, J. M., S. E. Nunes-Düby, C. F. Lesser, P. Youderian, M. M. Susskind, and A. Landy.** 1985. The ϕ80 and P22 attachment sites. *J. Biol. Chem.* **260:**4468–4477.

167. **Leong, J. M., S. Nunes-Düby, A. Oser, C. F. Lesser, P. Youderian, M. M. Susskind, and A. Landy.** 1984. Site-specific recombination systems of phages ϕ80 and P22: binding sites of integration host factor and recombination-induced mutations. *Cold Spring Harbor Symp. Quant. Biol.* **49:**707–714.

168. **Leong, J. M., S. E. Nunes-Düby, and A. Landy.** 1985. Generation of single base-pair deletions, insertions, and substitutions by a site-specific recombination system. *Proc. Natl. Acad. Sci. USA* **82:**6990–6994.

169. **Lilley, D. M. J.** 1981. *In vivo* consequences of plasmid topology. *Nature* (London) **292:**380–382.

170. **Lilley, D. M. J., and A. F. Markham.** 1983. Dynamics of cruciform extrusion in supercoiled DNA: use of a synthetic inverted repeat to study conformational populations. *EMBO J.* **2:**527–533.

171. **Lin, N. S.-C., and D. Pratt.** 1972. Role of bacteriophage M13 gene 2 in viral DNA replication. *J. Mol. Biol.* **72:** 37–49.

172. **Linn, S.** 1982. The deoxyribonucleases of *Escherichia coli,* p. 291–309. *In* S. M. Linn and R. J. Roberts (ed.), *Nucleases.* Cold Spring Harbor Laboratory, Cold Spring Harbor, N.Y.

173. **Liu, L. F., C.-C. Liu, and B. M. Alberts.** 1979. T4 DNA topoisomerase: a new ATP-dependent enzyme essential for initiation of T4 bacteriophage DNA replication. *Nature* (London) **281:**456–461.

174. **Lloyd, R. G., and C. Buckman.** 1985. Identification and genetic analysis of *sbcC* mutations in commonly used *recBC sbcB* strains of *Escherichia coli* K-12. *J. Bacteriol.* **164:** 836–844.

175. **Lockshon, D., and D. R. Morris.** 1985. Sites of reaction of *Escherichia coli* DNA gyrase on pBR322 *in vivo* as revealed by oxolinic acid-induced plasmid linearization. *J. Mol. Biol.* **181:**63–74.

176. **Lopez, P., M. Espinosa, B. Greenberg, and S. A. Lacks.** 1984. Generation of deletions in pneumococcal *mal* genes cloned in *Bacillus subtilis. Proc. Natl. Acad. Sci. USA* **81:**5189–5193.

177. **Low, B.** 1968. Formation of merodiploids in matings with a class of rec-recipient strains of *Escherichia coli* K12. *Genetics* **60:**160–167.

178. **Low, B.** 1972. *Escherichia coli* K-12 F-prime factors, old and new. *Bacteriol. Rev.* **36:**587–607.

179. **Low, K. B.** 1987. Hfr strains of *Escherichia coli* K-12, p. 1134–1137. *In* F. C. Neidhardt, J. L. Ingraham, K. B. Low, B. Magasanik, M. Schaechter, and H. E. Umbarger (ed.), *Escherichia coli and Salmonella typhimurium: Cellular and Molecular Biology*, vol. 2. American Society for Microbiology, Washington, D.C.

180. **Lu, A.-L., K. Welsh, S. Clark, S.-S. Su, and P. Modrich.** 1984. Repair of DNA base-pair mismatches in extracts of *Escherichia coli*. *Cold Spring Harbor Symp. Quant. Biol.* **49:**589–596.

181. **Lundblad, V., and N. Kleckner.** 1984. Mismatch repair mutations of *Escherichia coli* K12 enhance transposon excision. *Genetics* **109:**3–19.

182. **Lundblad, V., A. F. Taylor, G. R. Smith, and N. Kleckner.** 1984. Unusual alleles of *rec*B and *rec*C stimulate excision of inverted repeat transposons Tn*10* and Tn*5*. *Proc. Natl. Acad. Sci. USA* **81:**824–828.

183. **Maki, H., and A. Kornberg.** 1987. Proofreading by DNA polymerase III of *Escherichia coli* depends on cooperative interaction of the polymerase and exonuclease subunits. *Proc. Natl. Acad. Sci. USA* **84:**4389–4392.

184. **Maples, V. F., and S. R. Kushner.** 1982. DNA repair in *Escherichia coli*: identification of the *uvr*D gene product. *Proc. Natl. Acad. Sci. USA* **79:**5616–5620.

185. **Marinus, M. G.** 1987. Methylation of DNA, p. 697–702. *In* F. C. Neidhardt, J. L. Ingraham, K. B. Low, B. Magasanik, M. Schaechter, and H. E. Umbarger (ed.), *Escherichia coli and Salmonella typhimurium: Cellular and Molecular Biology*, vol. 1. American Society for Microbiology, Washington, D.C.

186. **Marinus, M. G., and N. R. Morris.** 1973. Isolation of deoxyribonucleic acid methylase mutants of *Escherichia coli* K-12. *J. Bacteriol.* **114:**1143–1150.

187. **Marvo, S. L., S. R. King, and S. R. Jaskunas.** 1983. Role of short regions of homology in intermolecular illegitimate recombination events. *Proc. Natl. Acad. Sci. USA* **80:** 2452–2456.

188. **Masamune, Y., and C. C. Richardson.** 1971. Strand displacement during deoxyribonucleic acid synthesis at single strand breaks. *J. Biol. Chem.* **246:**1692–2701.

189. **Maxwell, A., R. Craigie, and K. Mizuuchi.** 1987. B protein of bacteriophage Mu is an ATPase that preferentially stimulates intermolecular DNA strand transfer. *Proc. Natl. Acad. Sci. USA* **84:**699–703.

190. **McCorkle, G. M., and S. Altman.** 1982. Large deletion mutants of *Escherichia coli* tRNA1tyr. *J. Mol. Biol.* **155:** 83–103.

191. **Meyer, T. F., K. Geider, C. Kurz, and H. Schaller.** 1979. Cleavage site of bacteriophage fd gene II protein in the origin of viral strand replication. *Nature* (London) **278:**365–367.

192. **Michel, B., and S. D. Ehrlich.** 1986. Illegitimate recombination at the replication origin of bacteriophage M13. *Proc. Natl. Acad. Sci. USA* **83:**3386–3390.

193. **Michel, B., and S. D. Ehrlich.** 1986. Illegitimate recombination occurs between the replication origin of the plasmid pC194 and a progressing replication fork. *EMBO J.* **5:** 3691–3696.

194. **Mizuuchi, K., M. Mizuuchi, and M. Gellert.** 1982. Cruciform structures in palindromic DNA are favored by DNA supercoiling. *J. Mol. Biol.* **156:**229–243.

195. **Mizuuchi, M., and K. Mizuuchi.** 1985. The extent of DNA sequence required for a functional bacterial attachment site of phage lambda. *Nucleic Acids Res.* **13:**1193–1208.

196. **Modrich, P.** 1987. DNA mismatch correction. *Annu. Rev. Biochem.* **56:**435–466.

197. **Morisato, D., and N. Kleckner.** 1984. Transposase promotes double strand breaks and single strand joints at Tn*10* termini *in vivo*. *Cell* **39:**181–190.

198. **Morrison, A., and N. R. Cozzarelli.** 1979. Site-specific cleavage of DNA by *E. coli* DNA gyrase. *Cell* **17:**175–184.

199. **Morrison, A., and N. R. Cozzarelli.** 1981. Contact between DNA gyrase and its binding site on DNA: features of symmetry and asymmetry revealed by protection from nucleases. *Proc. Natl. Acad. Sci. USA* **78:**1416–1420.

200. **Mötsch, S., and R. Schmitt.** 1984. Replicon fusion mediated by a single-ended derivative of transposon Tn*1721*. *Mol. Gen. Genet.* **195:**281–287.

201. **Mötsch, S., R. Schmitt, P. Avila, F. de la Cruz, E. Ward, and J. Grinsted.** 1985. Junction sequences generated by one-ended transposition. *Nucleic Acids Res.* **13:**3335–3342.

202. **Murphy, E., and R. P. Novick.** 1980. Site-specific recombination between plasmids of *Staphylococcus aureus*. *J. Bacteriol.* **141:**316–326.

203. **Nader, W. F., T. D. Edlind, A. Huettermann, and H. W. Sauer.** 1985. Cloning of *Physarum* actin sequences in an exonuclease-deficient bacterial host. *Proc. Natl. Acad. Sci. USA* **82:**2698–2702.

204. **Nag, D. K., and D. E. Berg.** 1987. Specificity of bacteriophage Mu excision. *Mol. Gen. Genet.* **207:**395–401.

205. **Nag, D. K., U. DasGupta, G. Adelt, and D. E. Berg.** 1985. SI50-mediated transposition: specificity and precision. *Gene* **34:**17–26.

206. **Naito, A., S. Naito, and H. Ikeda.** 1984. Homology is not required for recombination mediated by DNA gyrase of *Escherichia coli*. *Mol. Gen. Genet.* **193:**238–243.

207. **Nakano, M. M., H. Ogawara, and T. Sekiya.** 1984. Recombination between short direct repeats in *Streptomyces lavendulae* plasmid DNA. *J. Bacteriol.* **157:**658–660.

208. **Nash, H. A.** 1981. Integration and excision of bacteriophage lambda: the mechanism of conservative site specific recombination. *Annu. Rev. Genet.* **15:**143–167.

209. **Nash, H. A., C. E. Bauer, and J. F. Gardner.** 1987. Role of homology in site-specific recombination of bacteriophage lambda: evidence against joining of cohesive ends. *Proc. Natl. Acad. Sci. USA* **84:**4049–4053.

210. **Nilsson, S. V., and G. Magnusson.** 1982. Sealing of gaps in duplex DNA by T4 DNA ligase. *Nucleic Acids Res.* **10:** 1425–1437.

211. **Nunes-Düby, S. E., L. Matsumoto, and A. Landy.** 1987. Site-specific recombination intermediates trapped with suicide substrates. *Cell* **50:**779–788.

212. **O'Connor, M. B., J. J. Kilbane, and M. H. Malamy.** 1986. Site-specific and illegitimate recombination in the *ori*VI region of the F factor. DNA sequences involved in recombination and resolution. *J. Mol. Biol.* **189:**85–102.

213. **O'Connor, M. B., and M. H. Malamy.** 1984. Role of the F factor *ori*V1 region in *rec*A-independent illegitimate recombination. *J. Mol. Biol.* **175:**263–284.

214. **O'Connor, M. B., and M. H. Malamy.** 1984. Site specific recombination in the *ori*V region of the F factor. *Cold Spring Harbor Symp. Quant. Biol.* **49:**421–434.

215. **Oeda, K., T. Horiuchi, and M. Sejiguchi.** 1982. The *uvr*D gene of *E. coli* encodes a DNA-dependent ATPase. *Nature* (London) **298:**98–100.

216. **Ohno, S.** 1970. *Evolution by Gene Duplication*. Springer-Verlag, New York.

217. **Ou, J. T., and T. F. Anderson.** 1976. F′ plasmids from HfrH and HfrC in recA-*Escherichia coli*. *Genetics* **83:**633–643.

218. **Owen, J. E., D. W. Schultz, A. Taylor, and G. R. Smith.** 1983. Nucleotide sequence of the lysozyme gene of bacteriophage T4. Analysis of mutations involving repeated sequences. *J. Mol. Biol.* **165**:229–248.
219. **Parkinson, J. S.** 1971. Deletion mutants of bacteriophage lambda. II. Genetic properties of att-defective mutants. *J. Mol. Biol.* **56**:385–401.
220. **Parkinson, J. S., and R. J. Huskey.** 1971. Deletion mutants of bacteriophage lambda. I. Isolation and initial characterization. *J. Mol. Biol.* **56**:369–384.
220a. **Peeters, B. P. H., J. H. de Boer, S. Bron, and G. Venema.** 1988. Structural plasmid instability in *Bacillus subtilis*: effect of direct and inverted repeats. *Mol. Gen. Genet.* **212**: 450–458.
221. **Pinkham, J. L., T. Platt, L. W. Enquist, and R. A. Weisberg.** 1980. The secondary attachment site for bacteriophage lambda in the *pro*A/B gene of *Escherichia coli*. *J. Mol. Biol.* **144**:587–592.
222. **Plasterk, R. H. A., A. Brinkman, and P. van de Putte.** 1983. DNA inversions in the chromosome of *Escherichia coli* and in bacteriophage Mu: relationship to other site-specific recombination systems. *Proc. Natl. Acad. Sci. USA* **80**: 5355–5358.
223. **Plasterk, R. H. A., and P. van de Putte.** 1984. Genetic switches by DNA inversions in prokaryotes. *Biochim. Biophys. Acta* **782**:111–119.
224. **Ponticelli, A. S., D. W. Schultz, A. F. Taylor, and G. R. Smith.** 1985. Chi-dependent DNA strand cleavage by RecBC enzyme. *Cell* **41**:145–151.
225. **Pribnow, D., D. C. Sigurdson, L. Gold, B. S. Singer, and C. Napoli.** 1981. I. rII cistrons of bacteriophage T4. DNA sequence around the intercistronic divide and positions of genetic landmarks. *J. Mol. Biol.* **149**:337–376.
226. **Primrose, S. B., and S. D. Ehrlich.** 1981. Isolation of plasmid deletion mutants and study of their instability. *Plasmid* **6**:193–201.
227. **Radman, M., and R. Wagner.** 1986. Mismatch repair in *Escherichia coli*. *Annu. Rev. Genet.* **20**:523–538.
228. **Ripley, L. S., and C. Papanicolaou.** 1988. DNA replication errors: frameshift errors produced by *Escherichia coli* polymerase I, p. 227–235. *In* R. E. Moses and W. C. Summers (ed.), *DNA Replication and Mutagenesis*. American Society for Microbiology, Washington, D.C.
229. **Robertson, M.** 1987. Mapping the disease phenotype. *Nature* (London) **327**:372–373.
230. **Rosamond, J., K. M. Telander, and S. Linn.** 1979. Modulation of the action of the recBC enzyme of *Escherichia coli* K-12 by Ca^{2+}. *J. Biol. Chem.* **254**:8646–8652.
231. **Ross, W., and A. Landy.** 1983. Patterns of lambda Int recognition in the regions of strand exchange. *Cell* **33**: 261–272.
232. **Ross, W., M. Shulman, and A. Landy.** 1982. Biochemical analysis of att-defective mutants of the phage lambda site-specific recombination system. *J. Mol. Biol.* **156**:505–529.
233. **Ross, D. G., J. Swan, and N. Kleckner.** 1979. Nearly precise excision: a new type of DNA alteration associated with the translocatable element Tn*10*. *Cell* **16**:733–738.
234. **Sadowski, P.** 1986. Site-specific recombinases: changing partners and doing the twist. *J. Bacteriol.* **165**:341–347.
235. **Saunders, C. W., B. J. Schmidt, M. S. Mirot, L. D. Thompson, and M. S. Guyer.** 1984. Use of chromosomal integration in the establishment and expression of *blaZ*, a *Staphylococcus aureus* β-lactamase gene, in *Bacillus subtilis*. *J. Bacteriol.* **157**:718–726.
236. **Schaaper, R. M., B. N. Danforth, and B. W. Glickman.** 1986. Mechanisms of spontaneous mutagenesis: an analysis of the spectrum of spontaneous mutation in the *Escherichia coli lacI* gene. *J. Mol. Biol.* **189**:273–284.
237. **Schaller, H.** 1978. The intergenic region and the origins for filamentous phage DNA replication. *Cold Spring Harbor Symp. Quant. Biol.* **43**:401–408.
238. **Scheuermann, R., S. Tam, P. M. J. Burgers, C. Lu, and H. Echols.** 1983. Identification of the ε-subunit of Escherichia coli DNA polymerase III holoenzyme as the *dna*Q gene product: a fidelity subunit for DNA replication. *Proc. Natl. Acad. Sci. USA* **80**:7085–7089.
239. **Scheuermann, R. H., and H. Echols.** 1984. A separate editing exonuclease for DNA replication: the ε subunit of *Escherichia coli* polymerase III holoenzyme. *Proc. Natl. Acad. Sci. USA* **81**:7747–7751.
240. **Schildkraut, C. L., C. C. Richardson, and A. Kornberg.** 1964. Enzymatic synthesis of deoxyribonucleic acid. XVII. Some unusual physical properties of the product primed by native DNA templates. *J. Mol. Biol.* **9**:24–45.
241. **Schrempf, H.** 1985. Genetic instability: amplification, deletion, and rearrangement within *Streptomyces* DNA, p. 436–440. *In* L. Leive, P. Bonventre, J. Morello, S. Schlesinger, S. Simon, and H. Wu (ed.), *Microbiology—1985*. American Society for Microbiology, Washington, D.C.
242. **Schwartz, D. O., and J. R. Beckwith.** 1969. Mutagens which cause deletions in *Escherichia coli*. *Genetics* **61**:371–376.
243. **Scott, T. N., and M. I. Simon.** 1982. Genetic analysis of the mechanism of the Salmonella phase variation site specific recombination system. *Mol. Gen. Genet.* **188**:313–321.
244. **Seasholtz, A. F., and G. R. Greenberg.** 1983. Identification of bacteriophage T4 gene 60 product and a role for this protein in DNA topoisomerase. *J. Biol. Chem.* **258**:1221–1226.
245. **Sengstag, C., and W. Arber.** 1983. IS2 insertion is a major cause of spontaneous mutagenesis of the bacteriophage P1: nonrandom distribution of target sites. *EMBO J.* **2**:67–71.
246. **Setler, G. L., G. J. King, and W. M. Huang.** 1979. T4 DNA-delay proteins, required for specific DNA replication, form a complex that has ATP-dependent DNA topoisomerase activity. *Proc. Natl. Acad. Sci. USA* **76**:3737–3741.
247. **Sgaramella, V.** 1972. Enzymatic oligomerization of bacteriophage P22 DNA and of linear simian virus 40 DNA. *Proc. Natl. Acad. Sci. USA* **69**:3389–3393.
248. **Sgaramella, V., and S. D. Ehrlich.** 1978. Use of the T4 polynucleotide ligase in the joining of flush-ended DNA segments generated by restriction endonucleases. *Eur. J. Biochem.* **86**:531–537.
249. **Sgaramella, V., and H. G. Khorana.** 1972. CXII. Total synthesis of the structural gene for an alanine transfer RNA from yeast. Enzymic joining of the chemically synthesized polydeoxynucleotides to form the DNA duplex representing nucleotide sequence 1 to 20. *J. Mol. Biol.* **72**:427–444.
250. **Shafferman, A., Y. Flashner, and I. Hertman.** 1983. Genetic instability of an artificial palindrome DNA sequence. *J. Biomol. Struct. Dyn.* **1**:729–742.
251. **Shapiro, J. A.** 1979. Molecular model for the transposition and replication of bacteriophage Mu and other transposable elements. *Proc. Natl. Acad. Sci. USA* **76**:1933–1937.
252. **Shapiro, J. A.** 1985. Mechanisms of DNA reorganization in bacteria. *Int. Rev. Cytol.* **93**:25–56.
253. **Shen, P., and H. V. Huang.** 1986. Homologous recombination in *Escherichia coli*: genetic homology and genetic length. *Genetics* **112**:441–457.
254. **Shimada, K., and R. A. Weisberg.** 1972. Prophage lambda at unusual chromosomal locations. I. Location of the secondary

attachment sites and the properties of the lysogens. *J. Mol. Biol.* **63**:483–503.

255. **Shimada, K., R. A. Weisberg, and M. E. Gottesman.** 1975. Prophage lambda at unusual chromosomal locations. III. The components of the secondary attachment sites. *J. Mol. Biol.* **93**:415–429.
256. **Shpakovski, G. V., and Y. A. Berlin.** 1984. Site-specificity of abnormal excision: the mechanism of formation of a specialized transducing bacteriophage lambda-plac5. *Nucleic Acids Res.* **12**:6779–6795.
257. **Shurvinton, C. E., M. M. Stahl, and F. W. Stahl.** 1987. Large palindromes in the lambda phage genome are preserved in a rec+ host by inhibiting lambda DNA replication. *Proc. Natl. Acad. Sci. USA* **84**:1624–1628.
258. **Silverman, M., and M. Simon.** 1983. Phase variation and related systems, p. 537–577. *In* J. A. Shapiro (ed.), *Mobile Genetic Elements.* Academic Press, Inc., New York.
259. **Sinden, R. R., S. S. Broyles, and D. E. Pettijohn.** 1983. Perfect palindromic *lac* operator DNA sequence exists as a stable cruciform structure in supercoiled DNA *in vitro* but not *in vivo. Proc. Natl. Acad. Sci. USA* **80**:1797–1801.
260. **Singer, B. S, L. Gold, P. Gauss, and D. H. Doherty.** 1982. Determination of the amount of homology required for recombination in bacteriophage T4. *Cell* **31**:25–33.
261. **Smith, G. R.** 1987. Mechanism and control of homologous recombination in *Escherichia coli. Annu. Rev. Genet.* **21**: 179–201.
262. **Smith, G. R.** 1988. Homologous recombination in procaryotes. *Microbiol. Rev.* **52**:1–28.
263. **So, S.-S., and P. Modrich.** 1986. *Escherichia coli mutS*-encoded protein binds to mismatch DNA base pairs. *Proc. Natl. Acad. Sci. USA* **83**:5057–5061.
264. **Sommer, H., B. Schumacher, and H. Saedler.** 1981. A new type of IS1-mediated deletion. *Mol. Gen. Genet.* **184**: 300–307.
265. **Stahl, F. W.** 1986. Roles of double strand breaks in generalized genetic recombination. *Prog. Nucleic Acid Res. Mol. Biol.* **33**:169–194.
266. **Staudenbauer, W. L.** 1976. Replication of small plasmids in extracts of *Escherichia coli*: requirement of both DNA polymerases I and III. *Mol. Gen. Genet.* **149**:151–158.
267. **Streisinger, G., Y. Okada, J. Emrich, J. Newton, A. Tsugita, E. Terzaghi, and M. Inouye.** 1966. Frameshift mutations and the genetic code. *Cold Spring Harbor Symp. Quant. Biol.* **33**:77–84.
268. **Struhl, K.** 1981. Deletion, recombination and gene expression involving the bacteriophage lambda attachment site. *J. Mol. Biol.* **152**:517–533.
269. **Studier, F. W., A. H. Rosenberg, M. N. Simon, and J. J. Dunn.** 1979. Genetic and physical mapping in the early region of bacteriophage T7 DNA. *J. Mol. Biol.* **135**: 917–937.
270. **Surette, M. G., S. J. Buch, and G. Chaconas.** 1987. Transposomes: stable protein-DNA complexes involved in the *in vitro* transposition of bacteriophage Mu DNA. *Cell* **49**: 253–262.
271. **Syvanen, M., J. D. Hopkins, T. J. Griffin IV, T.-Y. Liang, K. Ippen-Ihler, and R. Kolodner.** 1986. Stimulation of precise excision and recombination by conjugal proficient F′ plasmids. *Mol. Gen. Genet.* **203**:1–7.
272. **Taylor, A., and G. R. Smith.** 1980. Unwinding and rewinding of DNA by the RecBC enzyme. *Cell* **22**:447–457.
273. **Taylor, A. F., D. W. Schultz, A. S. Ponticelli, and G. R. Smith.** 1985. *Rec*BC enzyme nicking at Chi sites during DNA unwinding: location and orientation-dependence of the cutting. *Cell* **41**:153–163.
274. **Taylor, A. F., and G. R. Smith.** 1985. Substrate specificity of the DNA unwinding activity of the RecBC enzyme of *Escherichia coli. J. Mol. Biol.* **185**:431–443.
275. **Telander-Muskavitch, K. M., and S. Linn.** 1981. *rec*BC-like enzymes: exonuclease V deoxyribonucleases, p. 233–250. *In* P. D. Boyer (ed.), *The Enzymes*, vol. 14. Academic Press, Inc., New York.
276. **te Riele, H., B. Michel, and S. D. Ehrlich.** 1986. Are single-stranded circles intermediates in plasmid DNA replication? *EMBO J.* **5**:631–637.
277. **te Riele, H., B. Michel, and S. D. Ehrlich.** 1986. Single-stranded plasmid DNA in *Bacillus subtilis* and *Staphylococcus aureus. Proc. Natl. Acad. Sci. USA* **83**:2541–2545.
278. **Thomas, M. S., and W. T. Drabble.** 1986. Secondary attachment site for bacteriophage lambda in the *guaB* gene of *Escherichia coli. J. Bacteriol.* **168**:1048–1050.
279. **Thompson, J. F., L. Moitoso de Vargas, C. Koch, R. Kahmann, and A. Landy.** 1987. Cellular factors couple recombination with growth phase: characterization of a new component in the lambda site-specific recombination pathway. *Cell* **50**:901–908.
280. **Thompson, R., and M. Achtman.** 1979. The control region of the F sex factor DNA transfer cistrons: physical mapping by deletion analysis. *Mol. Gen. Genet.* **169**:49–57.
281. **Thompson, R., L. Taylor, K. Kelly, R. Everett, and N. Willetts.** 1984. The F plasmid origin of transfer: DNA sequence of wild-type and mutant origins and location of origin-specific nicks. *EMBO J.* **3**:1175–1180.
282. **Toussaint, A., and A. Résibois.** 1983. Phage Mu: transposition as a life-style, p. 105–158. *In* J. A. Shapiro (ed.), *Mobile Genetic Elements.* Academic Press, Inc., New York.
283. **Vieira, J., and J. Messing.** 1982. The pUC plasmids, an M13mp7-derived system for insertion mutagenesis and sequencing with synthetic universal primers. *Gene* **19**: 259–268.
284. **Walker, G. C.** 1984. Mutagenesis and inducible responses to deoxyribonucleic acid damage in *Escherichia coli. Microbiol. Rev.* **48**:60–93.
285. **Ward, E., and J. Grinsted.** 1987. The nucleotide sequence of the *tnp*A gene of Tn*21*. *Nucleic Acids Res.* **15**:1799–1806.
286. **Warren, G. J., and R. L. Green.** 1985. Comparison of physical and genetic properties of palindromic DNA sequences. *J. Bacteriol.* **161**:1103–1111.
287. **Watt, V. M., C. J. Ingles, M. S. Urdea, and W. J. Rutter.** 1985. Homology requirements for recombination in *Escherichia coli. Proc. Natl. Acad. Sci. USA* **82**:4768–4772.
288. **Way, J. C., M. A. Davis, D. Morisato, D. E. Roberts, and N. Kleckner.** 1984. New Tn*10* derivatives for transposon mutagenesis and for construction of *lacZ* operon fusions by transposition. *Gene* **32**:369–379.
289. **Weaver, D. T., and M. L. de Pamphilis.** 1984. The role of palindromic and non-palindromic sequences in arresting DNA synthesis *in vitro* and *in vivo. J. Mol. Biol.* **180**: 961–986.
290. **Weisberg, R. A., L. W. Enquist, C. Foeller, and A. Landy.** 1983. Role of DNA homology in site-specific recombination. The isolation and characterization of a site affinity mutant of coliphage lambda. *J. Mol. Biol.* **170**:319–342.
291. **Weisberg, R. A., and A. Landy.** 1983. Site-specific recombination in phage lambda, p. 211–250. *In* R. W. Hendrix, J. W. Roberts, F. W. Stahl, and R. A. Weisberg (ed.), *Lambda II.* Cold Spring Harbor Laboratory, Cold Spring Harbor, N.Y.

292. **Wells, R. G., and N. D. F. Grindley.** 1984. Analysis of the gamma-delta *res* site. Sites required for site-specific recombination and gene expression. *J. Mol. Biol.* **179**:667–687.

293. **Welsh, K. M., A.-L. Lu, S. Clark, and P. Modrich.** 1987. Isolation and characterization of the *Escherichia coli mut*H gene product. *J. Biol. Chem* **262**:15624–15629.

294. **Wertman, K. F., A. R. Wyman, and D. Botstein.** 1986. Host/vector interactions which affect the viability of recombinant phage lambda clones. *Gene* **49**:253–262.

295. **Whoriskey, S. K., V.-H. Nghiem, P.-M. Leong, J.-M. Masson, and J. H. Miller.** 1987. Gene amplification in *E. coli*: DNA sequences at the junctures of amplified gene fusions. *Genes Dev.* **1**:227–237.

296. **Willetts, N., and R. Skurray.** 1987. Structure and function of the F factor and mechanism of conjugation, p. 1110–1133. *In* F. C. Neidhardt, J. L. Ingraham, K. B. Low, B. Magasanik, M. Schaechter, and H. E. Umbarger (ed.), *Escherichia coli and Salmonella typhimurium: Cellular and Molecular Biology*, vol. 2. American Society for Microbiology, Washington, D.C.

297. **Willetts, N., and B. Wilkins.** 1984. Processing of plasmid DNA during bacterial conjugation. *Microbiol. Rev.* **48**: 24–41.

298. **Wolfson, J., and D. Dressler.** 1972. Regions of single-stranded DNA in the growing points of replicating bacteriophage T7 chromosomes. *Proc. Natl. Acad. Sci. USA* **69**: 2682–2686.

299. **Wood, E. R., and S. W. Matson.** 1987. Purification and characterization of a new DNA-dependent ATPase with helicase activity from *Escherichia coli. J. Biol. Chem.* **262**: 15269–15276.

300. **Wu, A. M., A. B. Chapman, and T. Platt.** 1980. Deletions of distal sequence affect termination of transcription at the end of the tryptophan operon in *E. coli. Cell* **19**:829–836.

301. **Wyman, A. R., L. B. Wolfe, and D. Botstein.** 1985. Propagation of some human DNA sequences in bacteriophage lambda vectors requires mutant *Escherichia coli* hosts. *Proc. Natl. Acad. Sci. USA* **82**:2880–2884.

302. **Yanofsky, C., and E. S. Lennox.** 1959. Transduction and recombination study of linkage relationships among the genes controlling tryptophan synthesis in *Escherichia coli. Virology* **8**:425–447.

303. **Yen, P. H., E. Allen, B. Marsh, T. Mohandas, N. Wang, R. T. Taggart, and L. J. Shapiro.** 1987. Cloning and expression of steroid sulfatase cDNA and the frequent occurrence of deletions in STS deficiency: implications for X-Y interchange. *Cell* **49**:443–454.

304. **Yunis, J. J.** 1983. The chromosomal basis of human neoplasia. *Science* **221**:227–236.

Chapter 39

Illegitimate Recombination in Mammalian Cells

MARK MEUTH

I. SCOPE

Mammalian cells, like microorganisms, are able to rearrange and recombine DNA sequences having little or no homology. These rearrangements, called illegitimate or nonhomologous recombinations, result in the joining of two dissimilar DNA sequences to produce "simple" alterations such as deletions,

Mark Meuth ■ Imperial Cancer Research Fund, Clare Hall Laboratories, Blanche Lane, South Mimms, Potters Bar, Hertfordshire EN6 3LD, United Kingdom.

insertions, and chromosomal translocations as well as complex structures such as gene amplifications. Illegitimate recombination may involve short regions of sequence homology at the breakpoints of the DNAs that are rejoined, but it does not require the extensive sequence matches of classical homologous recombination. Illegitimate recombination events are widespread. In germ cell lines, they can lead to various inherited disorders; in somatic cells, they can activate oncogenes. Many of these rearrangements of chromosomal genes have now been defined at the DNA sequence level. The objective of this review is to analyze these recent reports, particularly in the context of the systems employed, to determine whether there are common DNA structures, features, or sequences which might indicate the mechanisms of these gene rearrangements.

An alternative approach to analyzing illegitimate recombination in cultured mammalian cells has relied on characterizing the random integration of linear foreign DNAs at sites which have little or no homology with the introduced molecule. The rearrangements resulting from such integrations are similar in many respects to those which occur spontaneously at chromosomal genes. This approach has recently been reviewed (128) and will not be dealt with here.

II. SOME DISTINCTIVE FEATURES OF EUCARYOTIC GENOMES

A. Chromatin Structure

Since illegitimate recombination will be considered in the context of DNA structure and sequence organization, it is important to emphasize distinctions between procaryotic and eucaryotic genomes. Mammalian genomes consist of large amounts of DNA (almost 10^{10} base pairs [bp]), only small portions of which are expressed at any time. This DNA is compartmentalized in the nucleus, where it is packed into protein-DNA complexes (nucleosomes), which in turn are thought to be organized into higher-order looped chromatin structures (85). Specific sites for the attachment of the DNA loop structures to the nuclear scaffold or matrix have been described (105). These sites also appear to function as binding sites for DNA topoisomerase II (30, 31, 45). This DNA organization might influence gene expression, as the loop-nuclear scaffold attachment sites are often associated with regulatory elements for structural genes in the loop (44). Topoisomerase II is strategically poised to influence DNA torsion (supercoiling) within each loop and thus may give regions of DNA bearing active genes a different conformation from regions with inactive ones.

B. Sequence Organization

DNA sequences are also organized in a complex manner. Many well-defined DNA repeat families are highly (though not necessarily uniformly) dispersed throughout mammalian genomes (137). These families are present in varying copy number. Some (such as the mouse *di-2* family [33]) are present in only ~100 copies, while the *Alu* family of short interspersed repetitive elements (SINEs) is present in ~300,000 copies per haploid genome (25, 59, 137; P. L. Deininger, this volume). The families vary in complexity from simple-sequence purine-pyrimidine (dG-dT) repeats, represented about 100,000 times in most mammalian genomes (55), to structures strongly resembling bacterial transposable elements, represented as frequently as 10,000 times in some genomes (123). It has long been suggested that these repeated elements play a role in unequal crossovers, and this will be considered in detail below. Families of tandem repeats are also present in mammalian genomes. These can be functional, coding for essential cellular components (e.g., the genes coding for rRNAs [97]), or structural, such as the α satellite DNAs found at chromosome centromeres (137).

C. Replication

The pattern of DNA replication in complex genomes, which proceeds via initiation from multiple origins in any given chromosome, may also influence illegitimate recombination. These origins of replication apparently function in an ordered manner during a defined period of the cell cycle (the S phase) and must act only once during each S phase. Little is known of the structure of chromosomal origins or how their activity is controlled. Some models of gene amplification envisage the repeated firing of replication origins. Such a programmed series of reinitiations of selected origins appears to cause the normal amplification of chorion genes in *Drosophila* development (142). The overall replication process is generally tightly controlled and highly accurate, as mutations are rare events in mammalian cells, occurring at rates as low as 10^{-8} per cell per generation at various loci (a rate very similar to those found for microorganisms).

III. TOOLS FOR THE ANALYSIS OF ILLEGITIMATE RECOMBINATION

Because of the complexity of mammalian genomes, the analysis of illegitimate recombinational events has depended upon the development of techniques for the molecular analysis of eucaryotic genes and the availability of cloned probes for genomic loci. Consequently, the systems studied reflect the order in which gene probes were isolated.

A. Inherited Disorders

Illegitimate recombination resulting in germ line mutations were first studied in patients suffering from thalassemias or hereditary persistence of fetal hemoglobin (for review, see reference 21). More recently, probes for a number of other important inherited disorders (e.g., familial hypercholesterolemia and Duchenne muscular dystrophy) have become available. Gene rearrangements appear to be a significant source of variation at all these loci, with deletions being the most frequent type.

B. Oncogenes

The molecular cloning of dominant oncogenes provided opportunities for the analysis of somatic rather than germ line mutations. Oncogenes can be generated by point mutations of normal cellular homologs (e.g., the *ras* oncogene) or gene rearrangements (e.g., amplification of the c-*myc* gene; for reviews, see references 10 and 156). Human tumors generally contain numerous chromosomal rearrangements, some of which appear to be specific and diagnostic of the cancer in which they occur (166). Two well-characterized examples of these tumor-specific rearrangements are the Philadelphia chromosome, found in chronic myelogenous leukemia, and the deletion of a defined band of chromosome *13* in retinoblastoma (77). More recently, chromosome walking techniques have been used to clone human genes which, when deleted or altered, lead to defined tumors (e.g., retinoblastoma and osteosarcoma [29, 39]). Molecular analyses of the rearrangements occurring in these tumors will provide further clues to the mechanisms of illegitimate recombination in vivo.

C. Selectable Markers

"Selectable" genetic markers (such as drug resistance [136]) have been widely used to analyze illegitimate recombination in cultured mammalian cells. One advantage of these systems is that mutations are selected in cells growing in defined conditions, making it possible to examine the stresses (chemical or physical) which promote rearrangements; the causes of germ line mutations, in contrast, are much more difficult to characterize. Markers studied in the greatest detail are those giving drug resistance in serial selections by amplification of a specific target gene or in one-step selections by the loss of the enzyme activity that converts the drug to a toxic form. The cloning of amplified genes is obviously facilitated by their elevated copy number and expression. The isolation of the structural genes altered in negative selections has been considerably more difficult and has relied on the cloning of transfected wild-type genes which complement the defect in mutant recipient strains (99). The relative usefulness of various gene targets in the analysis of illegitimate recombination is reflected by another characteristic of eucaryotic genomes, i.e., the large size of many structural genes. Clearly, a small target facilitates the detection, mapping, and analysis of mutations at the molecular level, but may restrict the types of rearrangements observed. The most frequently used target loci vary in size from 4.0 kilobases (kb) for the adenine phosphoribosyltransferase (*aprt*) locus (99) to a less convenient 40 kb for the hypoxanthine-guanine phosphoribosyltransferase (*hprt*) locus (122).

IV. SIMPLE SEQUENCE REARRANGEMENTS IN GERM LINES

A. α Globin

The remarkable variety of naturally occurring mutations in human populations in addition to the early availability of genomic probes led to an extensive analysis of gene rearrangements of the globin gene complexes.

The α-globin locus consists of three functional genes (α1, α2, and the embryonic gene ζ), three pseudogenes, and one gene of unknown function (θ) in a 40-kb segment of the short arm of chromosome *16* (Fig. 1; see reference 21 for review). Most deficiencies in the production of α-globin chains, the α-thalassemias, are relatively large deletions, whereas most β-thalassemias are due to single-base-pair alterations. Twelve naturally occurring deletions in a 170-kb region around the α-globin gene cluster have been carefully characterized (116) (Fig. 1). The deletions range in size from 3.7 to at least 62 kb. The smallest (α-thal-2) results from a crossover between the two virtually identical α-globin genes (~98% match over 1.5 kb [104]). Similarly, the largest deletion, in thalassemia $(\alpha\alpha)^{RA}$, results from a crossover between two *Alu* family members (approximately 80% homologous) that are 62 kb apart (116). The breakpoints in the remaining deletions do not involve such extensive homologies, though at least one breakpoint of each of these deletions is within or near an *Alu* family element. As mentioned previously, *Alu* sequences are distributed throughout the ge-

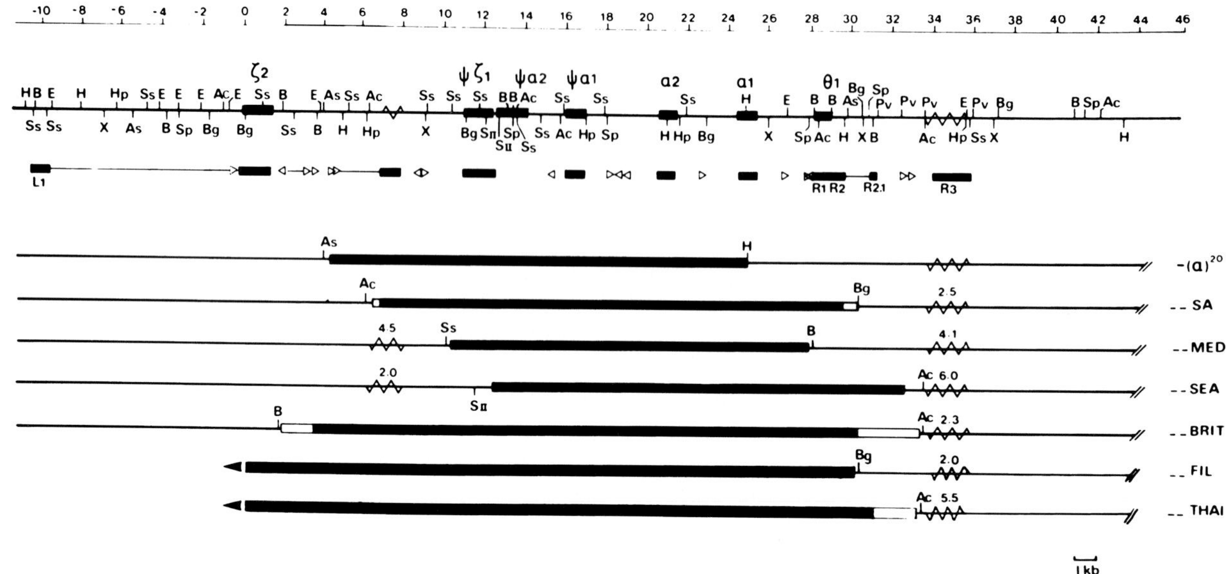

Figure 1. Deletions in the α-globin gene cluster. The cluster of 20- to 30-kb α-globin deletions and the relationship of the breakpoints to *Alu* elements and other features is presented (from reference 116). Two large (>35-kb) α-globin deletions found in Thai and Filipino subjects with α^0-thalassemia are also shown. The upper line shows the distance coordinates in kilobases, with 0 chosen arbitrarily as the ζ2 mRNA cap site. Center: Restriction endonuclease and gene maps. Genes are shown as solid blocks, hypervariable regions as zigzag lines. Enzyme abbreviations: As, *Asp* 718; B, *Bam*HI; Bg, *Bgl*II; E, *Eco*RI; H, *Hin*dIII; Hp, *Hpa*I; N, *Not*I; Pv, *Pvu*II; S, *Sal*I; Ss, *Sst*I; SII, *Sst*II; X, *Xba*I. The next line shows the site and orientation of *Alu* repeat sequences mapped to the region (open triangles), other unidentified repeat families (lines), and unique probes used to map deletions (solid boxes). The deletion cluster is presented at the bottom, with solid black lines representing α-globin sequences deleted and unfilled boxes where endpoints have not been precisely determined by DNA sequencing. The large 62-kb deletion $(\alpha\alpha)^{RA}$ between two direct *Alu* repeats is not indicated on this map, but begins at coordinate +10. (Reprinted with the kind permission of Douglas Higgs.)

nome, though not totally randomly. Because most junctions fall in or near *Alu* sequences (though only one appears to be a simple crossover between two family members), it has been suggested that these elements play a significant role in illegitimate recombination events (116).

In eight of the deletions, the 3′ terminus lies within a 6- to 8-kb segment of DNA (defined as a breakpoint cluster [116]). This region is less than 1 kb upstream of a hypervariable region (67) in the α-globin gene cluster. Another hypervariable region was found near the 5′ termini of several of the deletions, though it is unclear what role (if any) these regions play in the formation of rearrangements. Five of these α-globin deletions also exhibit clustered 5′ breakpoints and are similar in size, much like a set of deletions observed at the β-globin locus (155) (Fig. 1). These mutations might represent deletions of loops of packaged DNA, and thus it will be useful to examine whether these deletion breakpoints are correlated with nuclear scaffolding attachment sites (discussed below).

DNA sequence analyses of the α-globin deletion junctions revealed a number of interesting features, although no single feature was universally present (116). (i) In the α-thalassemia $--^{SEA}$, a GC-rich palindrome capable of forming a stable stem-loop structure was found only 20 bp from the breakpoint. (ii) The breakpoints in the parental DNAs of thalassemia-$(\alpha)^{20.5}$ occur in a region of weak homology between the two (65% over 48 bp), which might aid in their alignment; they occur at two identical trinucleotides separated by over 20 kb and leave just one of the trinucleotides in the mutant allele. (iii) The deletion $--^{MED}$ is more complex. In addition to a deletion of over 20 kb, there is an insertion of 134 bp of unrelated DNA at the novel junction. This insert apparently originated from a region of the α-globin gene cluster 36 kb upstream of the deleted sequences. One and five nucleotides, apparently unrelated to either the donor or recipient DNA, are added at the two novel junctions formed by the insertion. An examination of the sequences of both the donor and the target sites did not reveal any exceptional features (with the possible exception of a nearby *Alu* fragment).

B. β Globin

The human β-globin locus covers 70 kb of DNA on the short arm of chromosome *11* (for review, see reference 21), is larger and more complex than the α-globin locus, and contains more expressed genes. Moreover, these genes, the adult β- and δ-globin genes, the fetal $^{G}\gamma$ and $^{A}\gamma$ globins, and the embryonic ε gene, are arranged in an order reflecting the time of their expression during development. Many β-chain deficiencies have been detected in human populations. Those involving just the β chain are called β-thalassemias, while deletions encompassing both the δ and β genes result in the hereditary persistence of fetal hemoglobin (HPFH) throughout adult life.

Most β-thalassemias result from single-base-pair alterations (21), but over 20 deletions have been identified and many of these have been characterized at the sequence level. Their sizes vary from 619 to greater than 100,000 bp. Several are of similar size (5 or 6 kb) and have 5′ endpoints which fall relatively close in the β-globin gene complex (155). This endpoint clustering suggested that they result from the juxtaposition of distant sequences by their anchorage to the nuclear scaffolding, and that the staggered positions of the 5′ (or 3′) endpoints may indicate a sliding "frame" as sequences move through the anchorage points. Thus, the loop deleted may vary in its 5′ and 3′ endpoints, but be quite constant in size (155). The model is attractive in that it provides a means whereby distant breakpoints are brought into proximity, though, as indicated previously, we do not yet know the relationship of the deletion endpoints to nuclear scaffolding attachment points. There may be other constraints, e.g., viability, since deletions of essential genes would be lethal when homozygous.

No unique deletion "hot spots" have been detected, though one terminus is often within *Alu* (SINE [62, 66, 121, 155]) or L1 (LINE [62]) family repeats. These deletions are not the result of simple homologous recombination, since the other breakpoints lie in single-copy DNA with little homology to the *Alu* family. One particular 3′ terminus lies within a region of the β-globin locus having several interesting structural features (62) (Fig. 2): (i) a portion of an L1 family, (ii) a perfect 160-bp palindrome, and (iii) 41-bp direct repeats which are found elsewhere in the human genome (162). It may be significant that the breakpoint of this deletion is only 3 bp from the midpoint of the palindrome. Deletion termini with similar structures have been detected in some of the selectable systems (64, 110).

The sequences of the novel junctions formed by several of the deletions have been determined. These, in turn, have been used to identify the sequences of the breakpoints in the wild-type β-globin gene (see Fig. 2 for examples). No unique consensus sequence for these breakpoints has emerged, and there are no obvious homologies which might aid the alignment of the DNAs at the crossover points. A deletion causing γδβ-thalassemia can be ascribed to the rejoining of simple blunt end DNA breaks (155) because the endpoints are not similar and no nucleotides were added. Others (e.g., an Indian HPFH [62], an Indian $^{A}\gamma\delta\beta$-thalassemia [68], and another γδβ-thalassemia [155]) involve short direct repeat sequences of 2 to 5 bp. These can be ascribed to simple crossovers between the distant short direct repeat sequences, as one copy of the sequence remains in the mutant allele and no new nucleotides are added. Still others have insertions of new DNA fragments ("orphan" or "filler" DNA) of between 5 and 41 bp (the Indian $^{A}\gamma\delta\beta$ deletion/inversion [68], a $^{G}\gamma^{+}(^{A}\gamma\delta\beta)^{0}$-thalassemia [101], and an Indian β^{0}-thalassemia [143]). The origin of the novel nucleotides introduced into most of these junctions is unknown, although the 41-bp insert in $^{G}\gamma^{+}(^{A}\gamma\delta\beta)^{0}$-thalassemia is homologous to the L1 family repeat

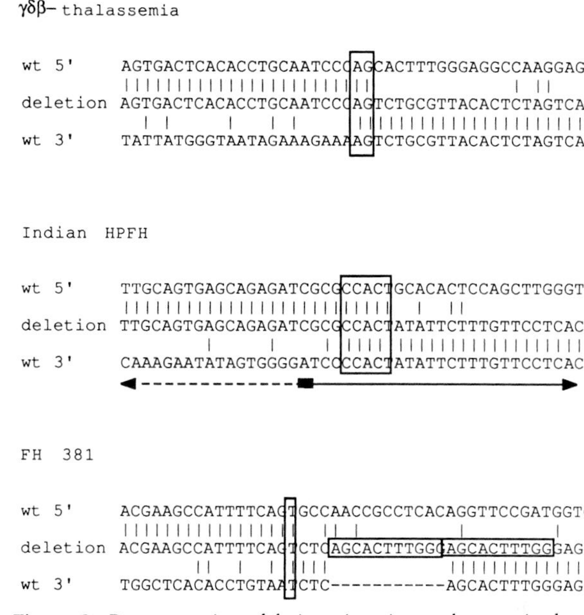

Figure 2. Representative deletion junctions characterized at globin or LDL receptor loci. Junctions are from a γδβ-thalassemia (155), an Indian HPFH (62), and an FH (88). The 5′ breakpoint in the γδβ-thalassemia is in an *Alu* family element 40 kb from the 3′ breakpoint in single-copy DNA. The exchange occurs through a 2-bp overlap (boxed). The breakpoints in the Indian HPFH occur at identical pentanucleotides (though there is little other similarity) separated by 30 kb of DNA. The 5′ breakpoint lies within an *Alu* family repeat, about 40 bp from the poly(A) sequence, while the 3′ breakpoint has several notable features, including a perfect 160-bp palindrome, part of which is denoted here by the broken and solid arrows. A portion of a human L1 (LINE) family repeat was observed 100 bp upstream from this junction. The novel junction formed in FH 381 is more complex, including an 11-bp duplication (boxed) just 3′ of the breakpoint. The 5′ breakpoint is in an exon sequence, while the 3′ breakpoint occurs in an *Alu* family element. Note that the 5′ breakpoint in the γδβ-thalassemia occurs at the same nucleotides which are duplicated in FH 381.

(101). This mutation could be the result of two independent deletions (both terminating within the L1 repeat found in the intervening sequence), rather than one deletion and concomitant insertion (101). The other deletion-plus-insertion mutations can be ascribed to simple nucleotide additions reminiscent of those occurring during immunoglobulin gene rearrangements (S. Lutzker and F. W. Alt, this volume).

Among the mutations reported at the β-globin locus is a complex rearrangement (in an Indian $^{A}\gamma\delta\beta$-thalassemia [68]) consisting of two deletions (of 0.3 and 7.5 kb) and an inversion of an intervening 15.5-kb fragment. Of the two novel junctions formed by this mutation, one contains a new hexanucleotide, while the other involves splicing between two 3-bp direct repeats, leaving one copy of the trinucleotide. Weak homologies that might have aided alignment were detected at the breakpoints which formed the first junction but not at those of the second, with the possible exception of a perfect 12-bp match 240 bp 5′ to the aligned nucleotides.

C. Low-Density Lipoprotein Receptor Locus

Familial hypercholesterolemia (FH) is an autosomal dominant disorder in which the loss of activity of one allele is sufficient for the development of the disease (52). The target gene encodes the low-density-lipoprotein (LDL) receptor, and deficiency of the gene product leads to an impaired ability to transport LDL protein from the plasma into cells, and hence an accumulation of LDL in the blood. The prevalence of the mutant gene in human populations (approximately 1 person in every 500 inherits one copy of the mutant gene) combined with the autosomal dominance of the disorder provides a large amount of material for analysis.

The LDL receptor gene is large, with 18 exons spanning a region of 60 kb on chromosome *19* (90). Five mutations resulting from illegitimate recombination within the gene have been analyzed at the DNA sequence level. The four deletions characterized vary in size from 0.8 to 7.8 kb, but in each case breakpoints are within *Alu* family elements in the introns of the structural gene. Three of the deletions involve exchanges between two *Alu* sequences (63, 89, 90). In two of these, the elements (having about 80% homology in the ~100 bp surrounding the breakpoints) were in the same orientation (63, 89), while in the third they were in opposite orientations (90). The fourth deletion involves an *Alu* element and exon sequences 5 kb away which could form an inverted repeat with the participating *Alu* element (88). This suggested that a stem-loop structure formed by these inverted repeats might have promoted this deletion (as well as the one formed by the *Alu* family members in opposite orientations). The LDL receptor gene is unusually rich in *Alu* family elements and different elements were involved in the formation of each deletion. Three of the *Alu* elements involved mapped to intron 15, perhaps denoting a region highly prone to rearrangement. A fifth mutation at the LDL receptor locus due to unequal exchange between *Alu* family repeats (in this case in exons 2 and 8) is a large (14 kb) tandem duplication (87).

The deletion junctions are neither identical nor ascribable to homologous recombination. One contains an added nucleotide (90), another produces an 11-bp duplication (88) (Fig. 2), and a third includes a 4-bp deletion near the junction (89). The breakpoints lie at different sites in the *Alu* elements, though six of the seven cluster to the small region (~100 bp) between the sequences thought to act as promoter elements for RNA polymerase III, in the left arm of the element. Two of the four β-globin locus deletion breakpoints occurring in *Alu* elements also fall in this region (89) (see Fig. 2 for examples). In contrast, the *Alu-Alu* exchange in the α-thalassemia $(\alpha\alpha)^{RA}$ occurs in the opposite arm of the element (116).

D. Duchenne Muscular Dystrophy Locus

Duchenne muscular dystrophy (DMD) is a prevalent X-linked degenerative disorder of the muscle, confining the victim to a wheelchair by the age of 12 years and proving fatal before the age of 40 years. In a truly dedicated effort, the cDNA for the locus has now been cloned in its entirety, and the analysis of mutant *DMD* genes from affected individuals is well under way (78). The *DMD* gene is the largest yet identified. The cDNA is 14 kb long and the genomic locus is organized into at least 60 exons spanning as much as 2 megabase pairs! In view of the target size, it is not surprising that there is a high rate of new mutation at the locus (representing about one-third of all DMD cases or about 1 in 10,000 live births). At least 50% of these mutations are deletions (26, 38, 78, 84). It is striking that over one-third of these map to just one intron (of unknown proportions) in this massive gene (78). Further work should show whether special structures or sequences cause the mutations within this region or whether the unusual distribution simply reflects a very large intron size.

DMD also affects females, though much less frequently than males. These patients have alterations of the *DMD* locus which result from translocations of the *Xp21* region of the X chromosome (where *DMD* maps) to any one of several autosomes. One such translocation junction, which joins the *DMD* gene to a gene on chromosome *21* encoding the 28S ribosomal subunit, has been cloned and compared with the corresponding parental regions (11). Sequence analysis has revealed deletions of 71 to 72 bp and 16 to 23 bp for the translocated fragment and the chromosome *21* recipient, respectively, accompanied by an insertion of a trinucleotide at the translocation junction. No unusual structural features at the breakpoints were reported, with the exception of some short direct repeats (CGGC) in the ribosomal sequence. This region of the 28S rRNA gene is notable, however, because it is highly variable between individuals (11).

As more probes for human inherited disorders are identified and cloned and the molecular alterations in affected individuals are elucidated, considerably more information will be gained about the nature of illegitimate recombination.

V. TRANSLOCATIONS IN HUMAN AND MOUSE NEOPLASIA

The development of high-resolution banding techniques for metaphase chromosomes allowed the identification of specific chromosomal rearrangements in tumor cells (166). This was closely followed by the recognition that cellular oncogenes or immunoglobulin genes often mapped to the fragments being rearranged (60), thus providing a tool for examining the novel junctions produced. Cloned probes for the known genes were used to obtain novel junction fragments and to analyze their structure at the DNA level. This approach was facilitated by the fact that the rearrangements occurred within the immunoglobulin genes or proto-oncogenes.

Reciprocal translocations resulting in the joining of the first exon or intron of the c-*myc* gene with the immunoglobulin C_H locus were described in mouse plasmacytomas (46, 124). Normal switch mechanisms which allow variable chains to be expressed with different C_H genes were implicated in the translocations by the homologies between certain c-*myc* breakpoints and the normal switch recombination sites, and by the position of the breakpoints in the C_H locus near authentic switch sites. Since the switching region consists of short sequence repeat motifs (118), it is not surprising that several junctions have similar sequences: five of six junctions characterized were 12 bp from the tetranucleotide GAGG (124). There was, however, considerable variation in the detailed structure of the junctions. Deletions (up to 1.6 kb) of participating sequences have been reported in five of the rearrangements, insertions of extraneous nucleotides (as many as 38 bp) in five, and a duplication of 106 bp in one (46, 124). Most of these could be explained by the occurrence of a staggered cut at each of the breakpoints followed by single-strand exonucleolytic digestion or polymerization before ligation to produce deletions or duplications, respectively (46). The staggered cuts apparently can be quite substantial, as in one striking case the deletion was 1.6 kb.

Similarly, immunoglobulin genes and proto-oncogenes are involved in translocations occurring in human B- and T-cell lymphomas and leukemias. In Burkitt lymphoma, the distal end of the long arm of chromosome *8*, the locus of the cellular c-*myc* gene, is translocated to chromosome *14*, *22*, or *2* (168). Cells having this translocation appear to lose normal control of c-*myc* expression (119). More than 60% of non-Burkitt B-cell lymphomas also contain translocations involving *14q32*, the locus of the human immunoglobulin heavy-chain gene, though in these cases c-*myc* is not involved. In some chronic lymphocytic leukemias, characterized by translocation between chromosomes *11* and *14*, breakpoints were found in a 30-bp region of chromosome band *11q13*. The target gene for these translocation breakpoints was consequently designated *bcl-1* (B-cell leukemia/lymphoma-1 [152]). Translocations between chromosomes *14* and *18* are observed in the majority of follicular lymphomas, and similarly the breakpoints on *18q21* clustered in a 100-bp region of a gene called *bcl-2* (150, 151). Like the rearrangements of c-*myc*, the translocations appear to alter the expression of these new oncogenes.

Sequence analyses of the breakpoints in the non-Burkitt lymphomas suggest that the translocations result from the action of the normal V-D-J recombinational mechanism in the pre-B-cell stage of development (24). First, the breakpoints occur in the same sites as would normal immunoglobulin rearrangements (the 5′ region of the J segment); second, extra nucleotides are added at the junctions; and finally, sequences resembling the signal sequences for normal V-D-J joining can be found near the breakpoints on the translocated chromosome *11* and *18* fragments. Consequently, it was proposed that these translocations are rare mutational errors made by the immunoglobulin gene recombinase followed by strong growth selection stemming from oncogene activation.

Chromosomal rearrangements involving c-*myc* have also been reported in T-cell tumors (167). In about 10 to 20% of these leukemias, the *J*α segment of T-cell receptor from chromosome *14* is fused 3′ of the c-*myc* third exon on chromosome *8* (32). Aberrations of the normal process of V-J joining in T-cell receptor rearrangements may account for these events (34).

The best known of these rearrangements is the Philadelphia chromosome, which is present in 96% of patients with chronic myelogenous leukemia (130). This rearrangement is the result of a reciprocal translocation between the most distal ends of chromosomes 9 and 22, resulting in a distinctive and diagnostic marker chromosome. The c-*abl* oncogene, which was mapped to the translocated region of chromosome 9 (60), is transferred in 95% of chronic myelogenous leukemias, resulting in an altered c-*abl* gene product (79). The breakpoints occur within a limited region of chromosome 22 (the breakpoint cluster region, *bcr* [53, 61]), probably because of the stringent selection for the altered c-*abl* product. Of two translocation junctions studied in detail, one appears to be nearly precise, with the addition of just one nucleotide, while the second has a 27-bp insert of no obvious origin. In the case of the first translocation, weak homologies were noted between the chromosome 9 and 22 sequences and a search of data bases revealed homology to the human *Alu* family (61).

In all of these cases, we need to learn what facilitates the exchange between two widely separated chromosomes which do not share any obvious homologies. Are these rare events truly random mutations which are strongly selected phenotypically by their growth advantage? Or are particular factors involved in translocating these specific genes into actively rearranging regions of chromosomes?

VI. OTHER ONCOGENE REARRANGEMENTS/ INSERTIONS

Rearrangements of cellular oncogenes which result in changes in their expression have been reviewed recently (10, 156) and will not be discussed in detail here. However, two, resulting from the insertion of novel sequences, merit special attention, since transposable elements are an important source of genetic flux in many organisms (this volume), but are exceptional in the mammalian systems considered in this review.

The cellular oncogene c-*mos* was activated in a mouse plasmacytoma by the insertion of a 4.7-kb element within the coding region (19). This element contained 335-bp terminal direct repeats and was homologous to the endogenous intracisternal A-particle element. The latter are a group of proretrovirus-like elements present in about 1,000 copies per mouse haploid genome (82). They have a gross structural organization resembling retroviruses but no known extracellular phase. This insertion event was accompanied by a 6-bp target sequence duplication. Two similar cases of insertions of intracisternal A-particle sequences into κ light-chain target genes have also been reported in mouse hybridoma cell lines (58).

Elevated levels of c-*myc* expression in a dog venereal tumor was apparently due to the insertion of a 1.8-kb fragment highly homologous with the primate *Kpn*I (LINE) family (71). The insert was flanked by 10-bp direct repeats of the cellular target site and contained a dA tail, suggesting an mRNA origin. Similar inserts have been observed in inactive pseudogenes (91). More recently, insertions of L1 sequences have been detected in exon 14 of the factor VIII gene in 2 of 240 patients suffering from hemophilia A (73). The insertions of the 3′ portions of the L1 sequence include the poly(A) track and are accompanied by target site duplications. It was proposed that these insertions occurred via an RNA intermediate and preferentially involved exon 14 because the A-rich regions of the exon could base pair with the poly(T) tail of the L1 cDNA.

VII. LOSS OF HETEROZYGOSITY IN TUMORS

The dominant oncogenes (such as *ras* and *myc*; for review, see reference 156) present in various tumors were defined by transfection assays in which nonmalignant cells were transformed by the introduction of the gene. There is evidence for another class of genes in which tumor-predisposing mutations are recessive to the wild-type allele. The best known example of this sort of tumor is retinoblastoma. It was proposed that this cancer developed after two recessive mutations inactivated both alleles of an autosomal locus (77). Individuals predisposed to this tumor inherited a deletion in the chromosome band *13q14* (141), presumably leaving only one copy of the critical (tumor-inhibiting) gene. The molecular analysis of these events became possible when a clone from a chromosome *13*-specific λ phage library was identified which could detect the deletions at band *13q14* in 3 of 37 patients (29). Chromosome walking techniques yielded another 30 kb of surrounding genomic sequences and resulted in the isolation of a cDNA which could detect an mRNA in "normal" cells but not in retinoblastomas (39). Moreover, when this cDNA was used as a probe of genomic DNAs on Southern blots, "deviant" gene patterns were detected, reflecting apparent homozygous deletions, heterozygous deletions, or other rearrangements in at least 30% of the samples originating from tumors, but no alterations were detected in the controls (39). In one report, breakpoints were frequently observed in two particular genomic fragments, suggesting a deletion (or rearrangement) hot spot (41).

Similar events are thought to cause a variety of human tumors: Wilm's tumor, hepatoblastoma, and rhabdomyosarcoma (apparently involving the short arm of chromosome *11* [80]); ductal breast tumors (chromosome *13* [100]); colorectal carcinomas (chromosome *5* [140]); and small-cell lung carcinoma (chromosome *3* [14, 115]), among others. While investigators have understandably concentrated on the clinical and diagnostic applications of these newly developed probes, the mutant genes also constitute a substantial source of data concerning illegitimate recombination processes.

VIII. GENOME REARRANGEMENTS AT SELECTABLE LOCI IN CULTURED MAMMALIAN CELLS

Our knowledge of the mechanisms of illegitimate recombination comes in large part from the examination of mutations selected directly and the subsequent determination of the mutant DNA sequences. This approach is very difficult using cultured mammalian cells because of their slow growth, diploidy, and frequent aneuploidy. Nevertheless, systems for selecting mutations at several well-defined loci have been developed (for reviews, see references 136 and 147), the genomic sequences corresponding to these loci have been cloned, and the structures of mutant DNAs are being analyzed at the sequence level.

The genes encoding enzymes of the purine and pyrimidine salvage pathways are most frequently used as selectable markers. These genes range in size from 4.0 kb (*aprt* [99]) to more than 40 kb (*hprt* [122]). Detailed analysis of the mutations obtained is easiest using small loci such as *aprt*. Rearrangement endpoints, however, are not randomly distributed, and a priori a large locus is more likely to contain a greater variety of rearrangement-provoking structures and provide a larger target for deletions (large genes have a significantly greater proportion of intron sequences which are mostly unaffected by point mutations but remain significant targets for gene rearrangements). Thus, the spectrum of rearrangements at a large gene is potentially more diverse than that at a small gene.

A. *aprt*

The most studied of these loci is the *aprt* gene of cultured Chinese hamster ovary (CHO) cells. The small size of the gene (Fig. 3) allows the detection of deletions and insertions as small as 30 bp and the fine structure mapping of the breakpoints (110, 111). It also facilitates the cloning and sequence analysis of mutant alleles (51, 111). The locus is autosomal, but *aprt* hemizygous strains (the result of large deletions, >20 kb, of one of the two homologs) allow the efficient isolation of new *aprt*$^-$ mutations (1, 12, 110). Consequently, the structure of numerous mutant *aprt* genes has been determined by Southern blotting techniques (15, 54, 102, 110). Most mutations of *aprt* appear to be due to single-base-pair substitutions or rearrangements smaller than 30 bp, because they do not produce any significant changes in gene fragment mobility on Southern blots. About 10% of spontaneous *aprt*$^-$ mutations were mapped to restriction endonuclease sites and were attributable to base pair substitutions or rearrangements of <25 bp (114). Larger alterations (deletions or insertions) were detected in another 8 to 10% of cases. Mutations induced by various chemical and physical agents have also been analyzed. The only treatment found to increase the frequency of rearrangements was γ-irradiation (15, 54), although rearrangements detectable by Southern blotting (>30 bp) still only accounted for a minority (25%) of mutations.

Five deletions at the *aprt* locus (110) (Fig. 3) were similar at the nucleotide level (111), although their size varied from 40 to 2,000 bp. Each occurred between short direct repeat sequences of 2 to 7 bp, leaving one copy of the sequence in the mutant gene (Fig. 4a). Certain tri- and tetranucleotides recurred in these short direct repeats. No other significant homologies were detected in the nucleotides surrounding the breakpoints, and no further alterations were detected near the deletion junction. Deletion termini were not randomly distributed over the *aprt* locus; in fact, 4 of the 10 breakpoints clustered within a 40-bp sequence (Fig. 4b). This region of *aprt* is notable in that (i) it is rich in short direct repeat sequences and inverted repeats which have the potential to form stable stem-loop structures, (ii) it is rich in sequence motifs resembling those in the mouse C_H switch region (118), and (iii) there is a strong homology to a hamster tandem repeat family (G. Phear and M. Meuth, unpublished observations). These features are very similar to the properties of one of the breakpoints of the Indian HPFH mutation at the β-globin locus described earlier (62). In contrast to mutations at the α- and β-globin and the LDL receptor loci, none of the deletions at *aprt* was found within or near *Alu*-equivalent family elements even though the *aprt* gene is flanked on both sides by such repeats. The similarity of the sequences of the *aprt* deletion cluster region to those in the C_H region raises the possibility that some deletions may be caused by

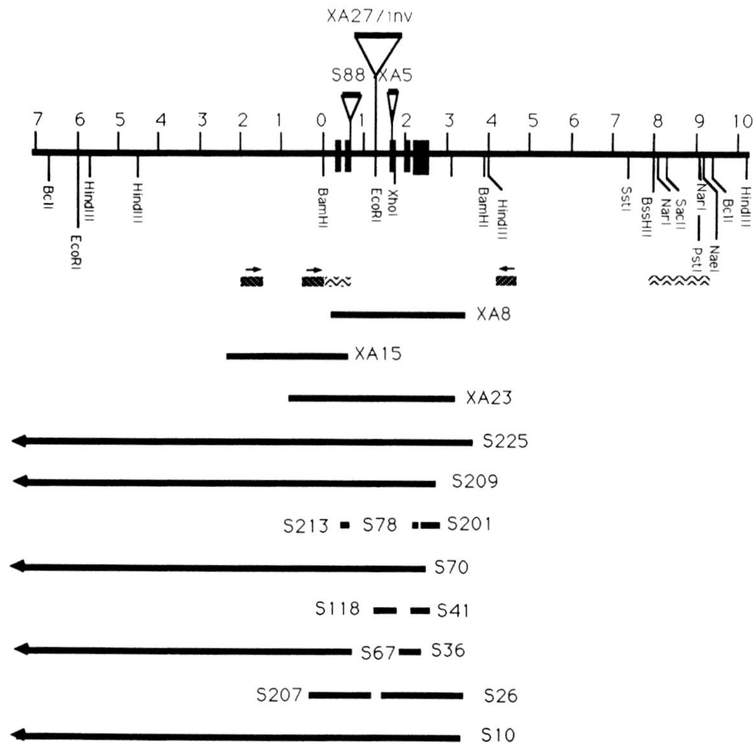

Figure 3. A map of deletions and insertions at the hamster *aprt* locus. The top line shows distance coordinates in kilobase pairs, with the *Bam*HI site just 5′ of the structural gene (exons as filled boxes) arbitrarily chosen as zero. A partial map of restriction endonuclease sites is also included. CHO *Alu*-equivalent repeats are indicated by the striped boxes (orientation indicated by the arrows above the boxes), and CpG ("HTF")-rich islands are indicated by wavy lines. Insertions or inversions (e.g., XA27) are indicated above the map and deletions by the solid lines below the map (one block per deletion). The designation for each mutation is indicated: S, spontaneous mutants; XA, γ-radiation-induced mutants. Arrows on some of the deletions indicate deletions with breakpoints far upstream of the structural gene. Deletions smaller than 40 bp are not presented on this map. (Adapted from reference 103 and C. Miles, G. Sargent, G. Phear, and M. Meuth, unpublished data.)

enzymes which normally rearrange immunoglobulin genes.

Five large deletions extending outside the *Bam*HI fragment bearing the *aprt* structural gene (Fig. 3) were also detected among the spontaneous and γ-radiation-induced mutants (15, 103, 110). These were striking in that all had one breakpoint within the *aprt* gene and the second upstream of *aprt*. None of the breakpoints were downstream of *aprt*. This pattern of deletions suggested that there is an essential gene or structure downstream of *aprt* which would be lethal if deleted in the hemizygous strain. Closer analysis revealed the presence of a DNA region rich in CpG residues (indicative of the presence of an expressed gene [9]) downstream of *aprt* distinct from the CpG island present in the 5′ region of *aprt* (Fig. 3). Other constraints on the pattern of deletions can also be considered. Deletion of an origin of DNA replication could cause cell death during S phase if specific termination signals left large stretches of DNA unreplicated. Alterations of nuclear scaffold attachment sites might prove fatal if the resulting changes in packaging impaired the expression or organization of essential genes. Critical tests of the validity of these ideas await the mapping of such functional sites coupled with a more extensive deletion analysis of representative regions.

The size of deletions of the *aprt* region extending beyond the *Bam*HI fragment have now been determined by pulsed field gel electrophoresis techniques and by probing for the presence of fragments upstream from *aprt*. These analyses indicate that the γ-radiation-induced deletions characterized thus far are relatively small (only about 5 kb maximum in the limited number examined), while the spontaneous deletions may be over 500 kb in size (unpublished observations). Novel junction fragments from eight of these mutant strains have been cloned and the breakpoints determined. The γ-radiation-induced junctions are similar to the spontaneous deletions, in that they are formed by apparently simple crossovers between short direct repeats. Breakpoints of two γ-radiation-induced mutants occurred 2 bp apart, but at a site distant from the region preferred for spontaneous deletions (C. Miles, G. Sargent, G. Phear, and M. Meuth, manuscript in preparation).

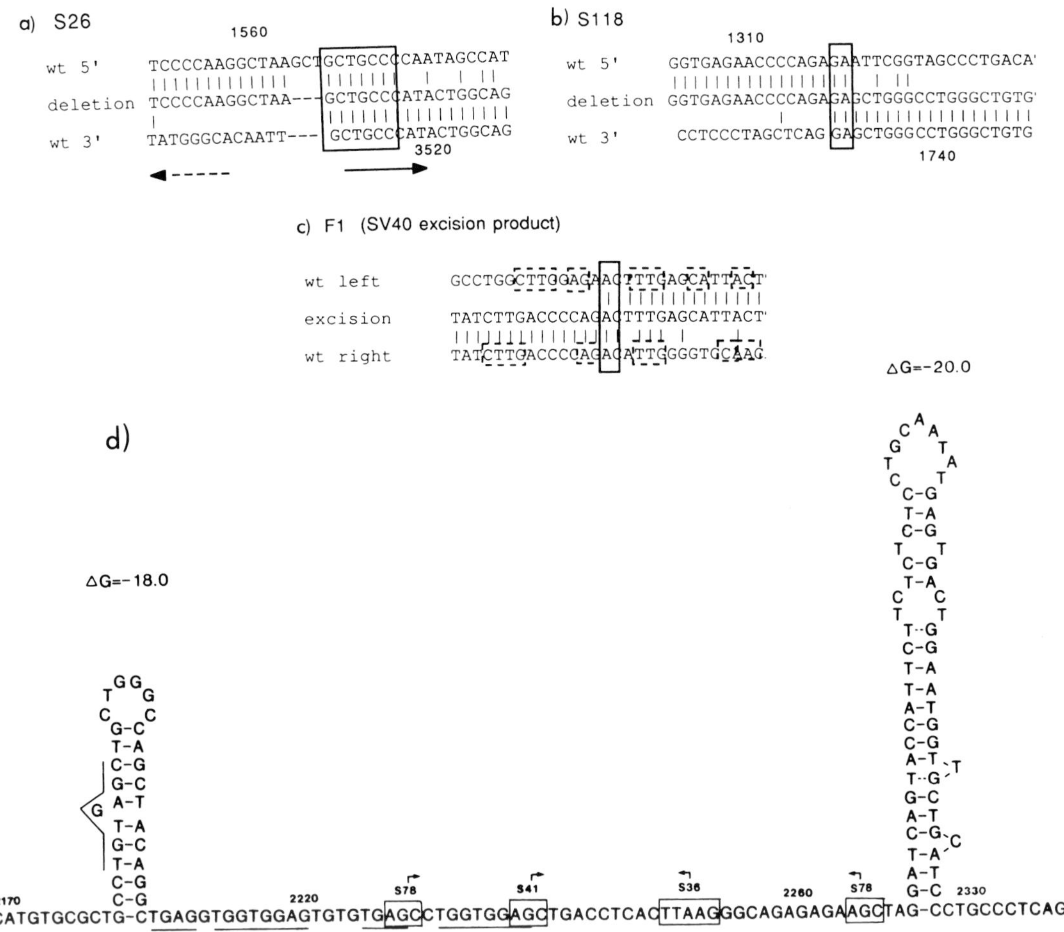

Figure 4. Representative deletion breakpoints and junctions occurring in cultured mammalian cells. (a) S26 and S118 are two junctions from the *aprt* locus (111); F1 is an SV40 excision product (17). The two deletions from *aprt* are relatively small (1,942 and 421 bp, respectively) with breakpoints occurring at short direct repeat sequences (boxed). A dyad symmetry at the 5′ breakpoint in S26 is denoted by arrows. No repetitive family elements were found at or near (within 500 bp) any of these junctions. The "patchy homologies" in F1 which may have aided the formation of heteroduplexes (17) are boxed by dashed lines. (b) A cluster of deletions at the *aprt* locus (111). Breakpoints of the indicated mutant strains falling within this region are boxed. Notable structural features within the region are indicated; short direct repeats are underlined and regions of dyad symmetry are indicated in the form of the projected stem-loop structures (together with the calculated stabilities of the structures; see reference 111 for detailed discussion).

The junction fragments of two very large spontaneous deletions have also been sequenced. One had no particularly striking features (111), while the second contained a complex rearrangement that included amplification of surrounding sequences (112) (to be discussed in detail below).

Insertion mutations are rare at the *aprt* locus, but four have been reported, two spontaneous (110; G. Adair, personal communication) and two radiation induced (15, 54). The two insertions characterized at the DNA sequence level (one spontaneous and one γ-radiation induced; Fig. 5 [15, 103, 113]) bear no resemblance to those induced by transposable elements in bacteria or lower eucaryotes (this volume). The inserts are small (only ~50 to 300 bp); they are accompanied by a deletion of 12 to 13 bp at the target sites, rather than duplication of the surrounding sequences; and they do not contain terminal repeats. The target sites within *aprt* are about 1 kb apart but have some sequence similarities. There is no obvious homology between the target and inserted sequence in the spontaneous mutant, while the inserted fragment in the γ-radiation-induced mutant shows some complementarity with sequences surrounding the target and could form a stable stem-loop structure (15, 113). The target in the γ-radiation-induced insertion is also notable in that it is the site of several other spontaneous mutations, including two small deletions (114) (Fig. 5b).

The sequences inserted are clearly different. The

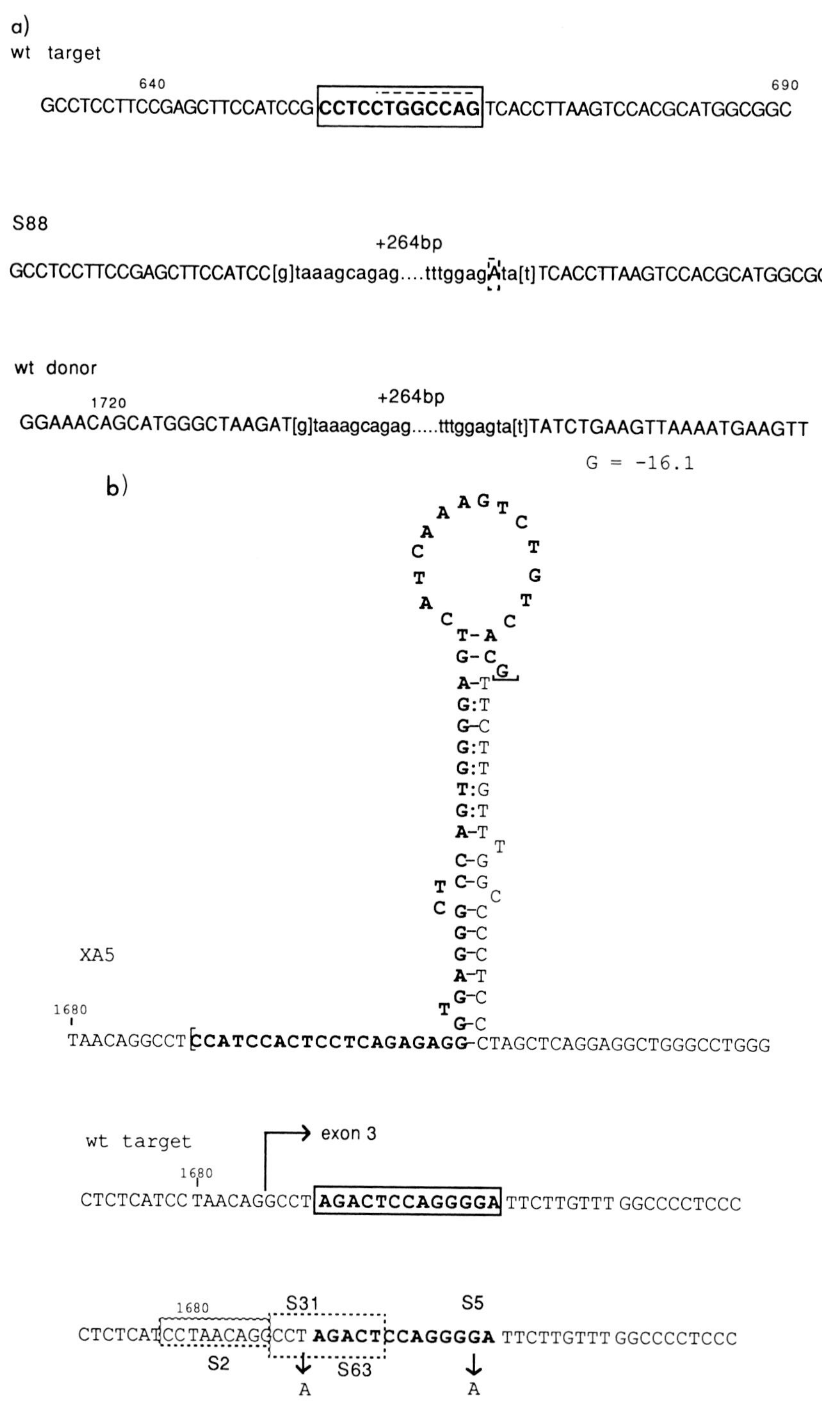

Figure 5. Insertion mutations at the *aprt* locus. (a) The target and donor sequences of the insertion in mutant S88 (103, 113). Nucleotides deleted at the target site are in bold type and boxed (the dashed line above indicates a palindrome). The inserted nucleotides are in lowercase type; the origin of the bracketed nucleotides is ambiguous. The sequence of the inserted nucleotides differed from the sequenced donor region in two places: one, a simple transition (not shown), was due to a simple polymorphism between the two donor alleles, while the origin of the other (a single nucleotide insertion, dashed box) has not been determined. The $[dG\text{-}dT]_{23}$ run (not shown) is 80 bp downstream from the mobilized sequence. (b) The γ-radiation-induced insertion XA5 (15, 103, 113). The sequence of the inserted nucleotides is presented in bold type. The stem-loop structure that the insert could form with adjacent nucleotides is shown (calculated stability indicated), while the nucleotides deleted at the target site are boxed and in bold type. This region of the *aprt* locus is also the site of several spontaneous mutations, including two small deletions (S2 and S63, boxed by the dashed lines) and two base substitutions (S5 and S31). The splice site for exon 3 is also indicated.

γ-radiation-induced mutant is a member of an abundant, highly dispersed, and divergent family, while the insertion in the spontaneous mutant is unique (103, 113; C. Miles and M. Meuth, manuscript in preparation). This repeat family is not related to any of the known dispersed repeats or to families having structures resembling transposable elements. The unique nature of the fragment inserted into the *aprt* gene of the spontaneous mutant allowed the cloning of the donor region and the further analysis of the insertion event. The donor region is largely unique, though the mobilized fragment is flanked on one side by a long, alternating purine-pyrimidine stretch dominated by a 46-bp dG-dT run (Fig. 5a). The donor region in the mutant strain is not altered with respect to sequence or copy number. This suggests that the fragment was duplicated before or during insertion into *aprt*, perhaps as a result of a process initiated by the simple-sequence dG-dT repeat (113). We imagine that the dinucleotide repeat might have produced some distortion at the junction with the mobilized sequence (dG-dT runs are prone to form Z DNA in vitro) and provided a focus for a nonreciprocal exchange with the target site (57, 120, 138, 139).

Several other "small" mutations have been mapped to restriction endonuclease sites in the *aprt* gene (110). A variety of alterations was demonstrated, including small deletions, insertions, or duplications in about one-third of the mutants analyzed (114). The three small deletions analyzed were similar to the larger deletions described above in that they involved short direct repeats, although one was unusual in also containing a novel trinucleotide insert. Two of these small deletions overlapped near the splice junction of exon 3 of *aprt* (Fig. 5b); this was also the target site of the γ-radiation-induced insertion mutation (15). A 9-bp duplication/insertion (114) and a small deletion (also formed between two short 2-bp direct repeats [51]) were found at another site. The small sizes of these rearrangements and the involvement of the short direct repeats make a replication slippage model very attractive (Fig. 6c below).

B. *hprt*

Mutations occurring at the locus coding for the nonessential purine salvage enzyme hypoxanthine-guanine phosphoribosyltransferase have been examined in detail by many groups (for review, see reference 147). At a cellular level, *hprt* offers the same advantages as *aprt*: easy forward and reverse selections, with virtually any type of mutation tolerated. It has the added advantage of being X linked. It is therefore hemizygous in cells from males, and because of the inactive X chromosome, it is functionally hemizygous in cells from females. This makes deficient mutants selectable in virtually any cell line. Unfortunately, it is a very large gene, with nine exons spanning 44 kb (122) in humans. This makes the analysis and mapping of gene rearrangements very difficult and the localization of smaller mutations by simple Southern blotting virtually impossible.

Nevertheless, several collections of spontaneous and induced *hprt*$^-$ mutants of various cultured cell lines have been studied (15, 42, 148, 158). Two isolates among 19 spontaneous and UV-induced *hprt*$^-$ CHO strains had extensive deletions of the structural gene (42). Cytogenetic analysis revealed that these deletions were accompanied by translocations. The frequency of such rearrangements is increased by ionizing radiation (15, 148, 158). Several of these deletions eliminate all coding sequences and are therefore greater than 44 kb in size. Apparently, the strict constraints which apply to *aprt* do not apply to *hprt*. Further studies of deletions at this locus will require DNA probes extending far upstream and downstream from the structural gene.

Mutations occurring in vivo at the *hprt* locus have been examined in human peripheral T cells stimulated to grow in culture (2, 117, 153). In initial studies as many as ~60% of these mutants were gene rearrangements (deletions, etc. [2, 153]). It now appears that the selection conditions employed in the initial studies may have been rather stringent and enriched for mutants having little or no enzyme activity (i.e., those induced by deletions). In more recent studies of larger collections of mutants, the frequency of major rearrangements has indeed been lower (~10% [117]), similar to the frequency of inherited deletions of the *hprt* locus (163). This system is interesting in offering the possibility of comparing mutations induced in vivo in germ line or somatic cells with those isolated from in vitro tissue-cultured cells.

Several recent technical developments should further enhance the attractiveness of the *hprt* locus as a model system. Mutations which escape detection by Southern blotting techniques can be mapped by hybridizing poly(A)-containing mRNA from a mutant strain with an antisense RNA probe, followed by treatment with RNase A to cleave any mismatches in the hybrid (47, 109). Cleavage sites in these hybrids are then precisely mapped by gel electrophoresis and probing with labeled *hprt* sequences. The "polymerase chain reaction" (131, 160) allows the amplification of a specific portion of a target gene using flanking oligonucleotide primers complementary to the two strands. Repeated cycles of polymerization, using a heat-stable DNA polymerase and higher reaction temperatures to increase the specificity of primer annealing, and denaturation result in an exponential increase in the copy number of the specific fragment to such an extent that it can be directly sequenced without further purification or cloning. Undoubtedly, these new approaches will particularly facilitate the study of mutations in large gene targets, although the polymerase chain reaction is applicable to any target with a known sequence.

C. *tk*

Human (13, 96, 165) and hamster (93, 94) DNAs coding for thymidine kinase (TK) have also been cloned. The functional gene is 13 to 16 kb in size and specifies a 1.4-kb mRNA. The gene is autosomal, but a useful human lymphoblastoid strain which is heterozygous ($tk^{+/-}$) for the locus has been reported; in this strain the mutant and wild-type alleles are distinguishable by the loss of a restriction endonuclease site in the mutant allele. *tk* gene fragments from over 200 spontaneous and induced $tk^{-/-}$ mutants produced in this heterozygous strain were examined on Southern blots (161; J. Little, personal communication). The majority of spontaneous and γ-radiation-induced mutations were large deletions that eliminated the entire locus and, in about half of the cases, a closely linked marker (i.e., within the same band on human chromosome *17*) as well. Smaller rearrangements accounted for about 10% of mutations in both collections. Fewer rearrangements were noted among mutations induced by UV light, but surprisingly the DNA alkylating agent ethyl methanesulfonate induced a significant number of deletions or other detectable rearrangements. This finding contrasts with events at hemizygous *aprt* and *hprt* loci, but analyses of ethyl methanesulfonate-induced mutations in a hamster strain having both *aprt* alleles indicate that "small" mutations in one allele are frequently accompanied by the loss of the second to give the enzyme-deficient phenotype (102, 110). None of the *tk* mutant junction sequences has been cloned or sequenced, though clearly many of the mutant genes are amenable to such analyses.

D. *dhfr*

Despite the large size (~25 kb) of the autosomal locus coding for the enzyme dihydrofolate reductase (DHFR), the structure of mutant *dhfr* genes has been examined using a series of 10 genomic probes covering 35 kb of the locus (106, 154). A further 200 kb of surrounding sequence has also been cloned during analyses of the amplification of the locus (see below). Deficient mutants of this essential autosomal gene can be selected using a strain hemizygous for the locus and by adding essential nutrients to the selective medium. In a collection of γ-radiation-induced mutants, several major deletions and inversions were detected (154). Three of the eight deletions eliminated the entire 210-kb region for which probes were available, while four others lost part of the *dhfr* coding region (deletions of 35 to >115 kb). Three mutants had disruptions in the order of *dhfr* sequences indicative of large inversions. This interpretation was confirmed for two of these mutants by alterations in chromosome banding pattern, indicating the participation of massive fragments of DNA. Two UV-induced deletions were also analyzed. One was >95 and the other >210 kb in size. Two breakpoints in γ-radiation-induced mutants fall adjacent to the nuclear matrix attachment sites in the *dhfr* gene, though no rearrangements appeared to alter the position of these sites (70). Deletion junctions are being cloned, but no breakpoint sequences are available.

The patterns of deletions at the *dhfr* and at *aprt* loci differ markedly from those at the α- and β-globin loci with respect to the sizes of the deletions and the distributions of breakpoints. This might reflect the differences in rearrangements which occur at active versus inactive genes or in the structure or packaging of such genes. It should also be pointed out that, with the exception of *tk*, all the selectable systems studied in detail are single-copy hemizygous genes, which in somatic cells precludes unequal crossovers between homologs. The only possible exceptions are unequal crossovers which occur between chromatids just after DNA replication and before segregation during mitosis.

The human major histocompatibility complex offers another potentially useful system for the analysis of illegitimate recombination. Cells expressing an antigen can be selectively killed using antibodies against the antigen and complement, leaving only rare deficient "variants" or mutants (125). Antibodies can be used to select mutations against the antigens on either allele. γ-Radiation-induced mutants which appear to be deletions of several linked genes have been isolated and characterized by two groups (50, 72). These deletions were used in conjunction with Southern blotting techniques to map HLA and other chromosome *6p* sequences (92). Deletion junctions have not been cloned or characterized.

IX. VIRAL DNA EXCISION

Illegitimate recombination resulting in the excision of chromosomally integrated viral molecules were examined by fusing a simian virus 40 (SV40)-

transformed rat cell line with monkey cells permissive for viral replication (17). This resulted in the production of a heterogeneous population of circular molecules containing both viral and cellular sequences. Six recombinant molecules were cloned from this population, and the novel junctions were sequenced. Like the rearrangements described previously, short direct repeat sequences of two to three nucleotides were found at the breakpoints of the excised DNAs (Fig. 4a). "Patchy homologies" may have aided the formation of heteroduplexes between the parental sequences at the crossover point (17), though the statistical significance of these matches relative to those in random DNA sequences is arguable. Inverted repeats with the potential to form stable stem-loop structures were at some but not all breakpoints. The di- and trinucleotides at crossover points were similar to the consensus topoisomerase I (topo I) cleavage site, suggesting that this enzyme was involved in the excisions. An analysis of topo I cleavage sites in the parental molecules revealed such sites at the breakpoints of three excised molecules and one nucleotide away in a fourth (16). Although there is, on the average, one topo I site per 7 bp, it was argued that the association between breakpoints and topo I sites is statistically significant. On the basis of these observations, a model was proposed where topo I played a pivotal role in the excision events. It envisages the enzyme binding at topo I sites and generating single-strand nicks. Breaks on the opposite strand would release DNA-topo I complexes. The enzyme could also participate in the circularization of the excised molecule by rejoining the chromosomal DNA ends (16).

Most eucaryotic cells contain small polydisperse circular DNAs, ranging in size from a few hundred to tens of thousands of base pairs. These molecules may be present at levels of a few to many thousands of copies per cell and represent both repetitive (e.g., *Alu*, L1, or α-like satellite DNA) and unique DNA sequences (e.g., 69, 74, 75, 83, 144). Several of these DNAs have been cloned and sequenced, in many cases along with the regions in the genome from which they originated. These analyses indicate that many of the free circular DNAs are products of illegitimate recombination (excision) between regions of short sequence homologies (83). For example, an L1 repeat sequence cloned from the extrachromosomal DNA of HeLa cells appears to have been excised as a result of a crossover between two 9-bp direct repeats at the sites of the breakpoints (69). The matches may also be more extensive, as in the excision of a class of human tandem repeats strongly resembling alphoid (centromeric) sequences, apparently due to recombination between units of ~170 bp which are 70 to 80% homologous (74, 75).

X. MUTATIONS OF EXOGENOUSLY INTRODUCED GENES

Another approach taken to examine the molecular basis of mutations in mammalian somatic cells is to introduce DNAs which can replicate freely or integrate into the cellular genome as targets for spontaneous or induced mutations. This approach has a number of advantages. The targets can be engineered to guarantee rapid cloning of the genes for molecular analysis and virtually any cell can be used as host, thereby avoiding the arduous task of constructing hemizygous strains or sorting through kilobases of DNA searching for mutations. However, numerous technical difficulties are encountered. The first of these is the high frequency of rearrangement of introduced sequences: in some cases, background mutational frequencies in the target genes are as high as 10^{-2} (18, 126). Frequencies can be lowered by using particular human cell lines (86), but this solution also eliminates the advantage of a broad host range. Exogenously introduced markers which integrate into host chromosomes are often unstable and are deleted entirely at high frequency during selection (149).

Recently several new systems have been described which should overcome many of these problems and thus provide further information about illegitimate recombination mechanisms. A retrovirus-based shuttle vector has been engineered to express the bacterial *gpt* gene as a mutational target in *hprt*-deficient mouse cells (5). A single copy of the vector integrates into the chromosomes of recipient cells and the target gene can be recovered rapidly using the SV40 origin of replication present in the vector. In contrast to the SV40-based vectors, which replicate independently of the host chromosome, the background mutational frequency in this system is low ($\sim 10^{-5}$). From one such mouse strain bearing this target, 43 spontaneously arising mutant genes were recovered and sequenced. Small deletions predominated and half of these were localized to a hot spot consisting of two trinucleotide direct repeats (resulting in the elimination of one of the trinucleotides). The deletions were as large as 1,200 bp (the largest deletion recoverable in the system), though most were smaller than 10 bp. Even though the mutations occurred while the shuttle vector was integrated, the high proportion of small deletions contrasts significantly to the predominance of single-base-pair substitutions observed at the *aprt* locus (51,

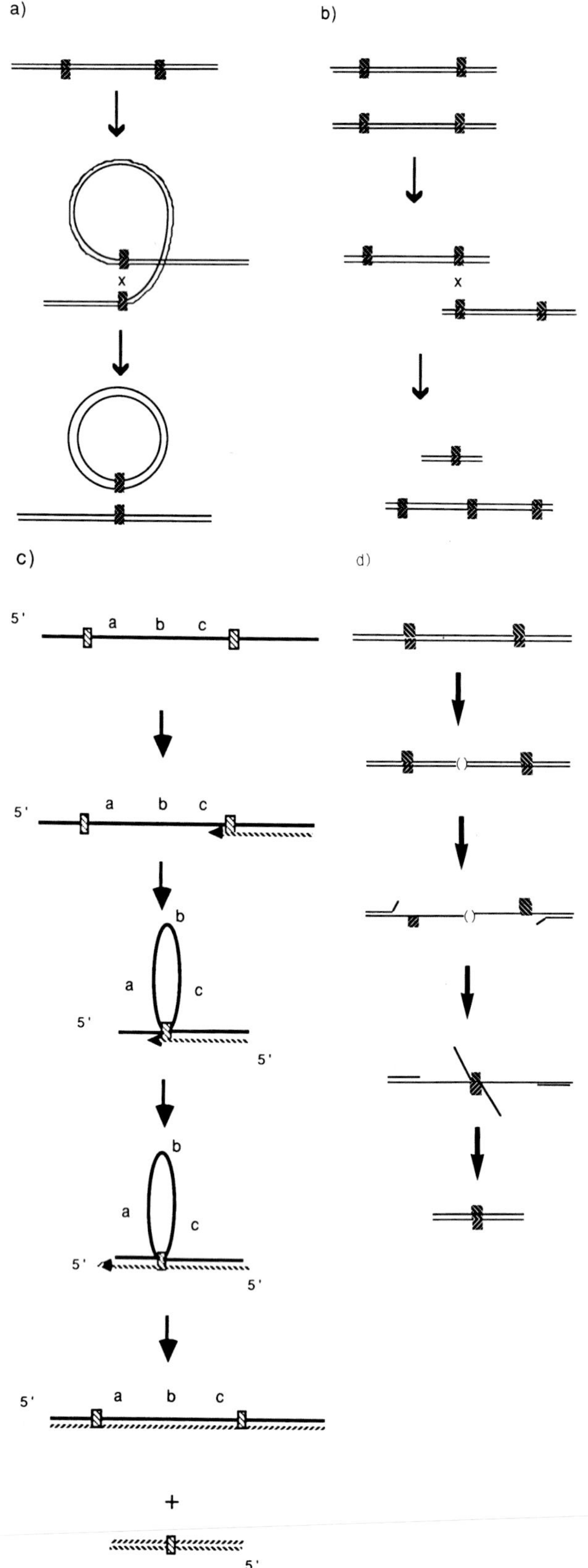

Figure 6. Models for the formation of simple deletions. (a) Intrachromosomal exchange between two short direct repeat sequences (hatched blocks) with the excised molecule most likely joining the pool of polydispersed free circular DNA. (b) Unequal crossover between short direct repeat elements on two unlinked DNA strands to generate deletions or duplications. (c) Deletion formed by slippage of the single-stranded DNA template (solid line here) between two direct repeats (striped blocks) during replication (nascent strand indicated by dashed line). Nucleotides looped out during the first round of replication may be replicated in the next round to give one wild-type molecule and one deleted one. Alternatively, the looped-out strand may be digested by a single-strand endonuclease to give the deletion. It is also possible for the nascent chain (dashed line) to slip after passing the second of the two repeats, creating a duplication in the following round of DNA replication (not illustrated). (d) Deletion formation by exonucleolytic degradation at DNA breaks. Degradation (or melting) of double-stranded molecule proceeds until short complementary sequences (striped boxes) in the single-stranded regions reanneal. Ends would be trimmed and gaps filled to form the novel joint.

114). This difference could reflect some peculiarity of the *gpt* target or the site of integration. The effect of surrounding sequences (or context) could be examined directly by isolating mutants from an independently transformed mouse strain having the target integrated elsewhere in the genome.

A promising extrachromosomally replicating Epstein-Barr virus-based shuttle vector has also been developed utilizing the herpesvirus *tk* gene as a mutational target (27). Spontaneous background mutations of this target are very low (5×10^{-5} or lower), allowing the analysis of weakly mutagenic agents such as γ-radiation. Other target genes could also easily be substituted in this vector.

XI. MODELS FOR SIMPLE REARRANGEMENTS

Although the rearrangements considered thus far are from a variety of sources, their structures are basically very similar. All appear to join two dissimilar DNA molecules to produce a novel junction (Fig. 2 and 4). The rearrangements are simple in the sense that no new DNA fragments larger than 300 bp are introduced and no other alterations are observed. These mutations can be ascribed to intrachromosomal exchanges between short sequence repeats which leave one copy of the sequence at the novel junction (Fig. 6a). Unequal interchromosomal exchanges at repeated sequences (Fig. 6b) have been proposed to explain several of the FH mutations. There is evidence to favor the simple crossover at aligned short direct repeats (Fig. 6a) in that the molecules released by such rearrangements have been recovered in some cases from the free short polydispersed circular DNA molecules found in most mammalian cells. In the cases of SV40 excision (17), T-cell receptor gene rearrangement (40), and tandem repeat sequence excision (74, 75), the liberated molecules have been found in the pool of free circular DNAs. This pool of molecules could also serve as substrates for insertions into target sites, since some DNA families involved in insertion mutations have also been found as free circular molecules (73; Miles and Meuth, in preparation).

It also seems likely that "slipped mispairing" of template (Fig. 6c) or nascent DNA molecules at the replication fork could produce some of the small deletions or duplications observed among the *aprt* mutations (114). It is unlikely that such a slippage mechanism could explain the generation of the larger deletions, as the nucleotide loop would be unstable, but would yield a deletion if the single-stranded region were degraded. An alternative explanation of the deletions is that they arose adjacent to double-strand DNA breaks (possibly induced by chemical or physical damage; Fig. 6d). Nucleotides adjacent to such a break might be digested by exonucleases or even simply melted to allow single strands to reanneal via short complementary sequences. The gaps surrounding these short sequences could then be filled and sealed to generate a novel joint.

A variety of mechanisms for the joining of the ends of the participating DNA molecules could be inferred from the properties of the junctions characterized. "Clean" exchanges between the short direct repeats, perhaps via staggered cuts, dominate the deletions characterized at the *aprt* locus in cultured cells. Germ line mutations show greater variation in the structure of the breakpoints: e.g., blunt end joining of molecules with no similarity, duplications caused by the filling of staggered cuts, and additions of small numbers of nucleotides perhaps by nontemplated synthesis. Many unanswered questions remain. Is the selection of the breakpoints random, or are certain sequences prone to such rearrangements? What are the enzymes involved in their generation? These problems will be considered again below.

XII. COMPLEX REARRANGEMENTS IN AMPLIFIED GENE ARRAYS

Gene amplification is a widespread mechanism by which cells and organisms increase the level of a particular gene product. It occurs in vivo during normal programmed development in some organisms (e.g., 97, 142) and during tumorigenesis (3, 20, 135) and in vitro during serial drug selections in cultured mammalian cells (for reviews, see references 56, 133, 145, and 146). Amplified gene arrays have recently been studied at the nucleotide sequence level by a number of groups, and this review will focus on the properties of these sequences and the relationship of the rearrangements observed to other types of illegitimate recombination.

A. Selection

Gene amplifications giving resistance to cytotoxic drugs have been described for numerous genetic loci in cultured mammalian cells. Generally, cell populations are exposed to gradually increasing concentrations of drug in the medium, with the surviving cells showing elevated levels of the target protein due to an increase in the number of structural gene copies. Low-level gene amplifications have also been obtained in single-step mutants, occurring at rate of 10^{-4} to 10^{-6} (169), which is considerably higher than the mutational rate for most genetic loci (136). The rate of production of novel joints at a particular site (e.g., the *aprt* locus), on the other hand, appears to be very low, 10^{-9} to 10^{-10} mutations per cell per generation, estimated on the basis of a single-step novel joint localized to the *aprt* locus (112). This seeming discrepancy arises because very large tracts of DNA are amplified (from 200 to 10,000 kb), vastly larger than the size of a gene itself, which simply means that it is far more likely that a sequence will be amplified than be involved in novel joint formation. The frequency of gene amplification at some loci is increased by many chemical and physical agents, notably certain DNA-damaging agents and inhibitors of DNA synthesis (76, 127).

B. Structure

The formation of amplified gene arrays is often accompanied by chromosomal rearrangements (35). Amplified DNAs are frequently identifiable cytogenetically by a characteristic banding pattern, termed a homogeneously staining region (7, 8), which may be present at the original site of the structural gene or in other locations (7). Amplified genes can also be maintained extrachromosomally in many cell lines on structures called double minutes (6). Amplified arrays on double minutes tend to be lost in the absence of selection, as they do not contain centromeres and are not partitioned equally to daughter cells (22). The loss of double minutes is often accompanied by an increase in the frequency of homogeneously staining regions (and vice versa), suggesting that double minutes may be precursors of these regions in new chromosomal locations (7, 22).

Since amplified arrays are often detectable by chromosome banding techniques, it is evident that vast amounts of DNA are involved. Even though the target genes undergoing amplification may be quite large (25 to 50 kb), the size of the amplified unit appears to be considerably larger. In the early steps of amplification, the units may be as large as 10,000 kb (49), while much smaller units (e.g., ~240 kb [98]) have been found after multistep selections. It is noteworthy that 10,000 kb represents about 0.1% of the diploid hamster genome (49). Consequently, there would be considerable selection pressure to reduce this unit size in subsequent selections (thus creating further heterogeneous units [4]) to produce the large copy numbers observed (as many as 3,000 [164]) in highly resistant strains. The pattern of novel joints formed during gene amplification is unique to each cell surviving the selection. For two well-studied loci (*dhfr* and *CAD*, encoding the trifunctional protein catalyzing the carbamyl-P synthetase, aspartate transcarbamylase, and dihydroorotase reactions of the pyrimidine pathway) there do not seem to be hot spots prone to rearrangement (4). In contrast, at the adenylate deaminase locus (*AMPD*) nine novel joints were mapped to a 2.6-kb A+T-rich fragment (64). This preferred region for rearrangement is distinctive in having four *Alu*-equivalent repeat elements and many inverted repeat sequences. The joints in these nine cases are not identical (64, 65), and thus appear to be the products of different events.

C. Novel Joints

It is generally accepted that the elucidation of the structures of novel joints formed by the amplified units should contribute significantly to our understanding of the mechanisms of amplification. Several such joints have now been cloned and in some cases sequenced. The first such junction characterized was that of a polyomavirus sequence coding for T antigen joined to a weak enhancer sequence and transfected into a cell strain (Rat-1) which normally exhibits strong growth control (36). These sequences were amplified 20- to 40-fold in transfectants selected by loss of growth control. A more interesting feature of this amplified unit was a very large inverted duplication. Because it was not possible to determine whether this duplication occurred before or after the integration of the introduced sequences, such structures were sought in well-characterized chromosomal gene arrays. A technique to specifically recover large inverted repeats from genomic DNA (37) was used to find inverted duplications of *CAD* or *myc* gene sequences in cell lines or tumors in which these genes had been amplified. Further examples of inverted repeats were subsequently reported for the *aprt* (112), *AMPD* (65), *CAD* (132), and *DHFR* (98) loci, although simple duplications (i.e., direct repeats) were also observed. The finding of inverted repeat structures at 85 to 90% of the novel joints in one amplified gene array at the *dhfr* locus (98) suggested that inverted repeats may be formed during the initial single steps of the process, while other types of joints are produced in subsequent rearrangements.

The fine structure of the inverted repeats isolated from the various amplified arrays was surprisingly similar as well. The repeats are not simple, perfect palindromes; rather, there is a small unduplicated region at each of the joints. Two of these novel joints have been characterized at the nucleotide level and have similar properties. At the *aprt* locus, a large deletion of sequences upstream from the joint (resulting in the deletion of the *aprt* structural gene) is accompanied by a duplication of long stretches (>100 kb) of sequence 672 bp downstream from the joint (112). The duplicated fragment is inverted and rejoined at the deletion breakpoint. The fidelity of the rearranged sequences is preserved, as no other alterations were detected. The structure of the novel joint from the *AMPD* locus is very similar: an unduplicated intervening region of 861 bp separates inverted repeats of >120 kb (65). The region in which the "breakpoints" of the inverted duplication of *AMPD* occurred, however, has some striking properties (64). (i) The sequence is rich in dyad symmetries capable of forming significant stem-loop structures. Both breakpoints of the inverted duplication mapped within such structures. (ii) It contains a "mosaic organization" of *Alu*-equivalent elements (one of which includes a breakpoint). (iii) Numerous other novel joints have been formed in this region during independent amplification events. These include direct repeats ("head-to-tail" joints) as well as inverted repeats ("head-to-head" or "tail-to-tail" joints). In contrast, the "breakpoints" for the inverted duplication at *aprt* did not have such structural properties, nor did they lie near the previously described cluster of deletion termini (Fig. 4b).

D. Cellular Functions

The question of what cellular functions are involved in the production of amplified gene arrays has not been experimentally addressed. Recently, however, cell lines which could provide a vital tool for understanding these functions have been described. Giulotto et al. (48) found that cell strains selected for two single-step amplifications have a significantly higher frequency of subsequent amplification events. These alterations appear dominant in somatic cell hybrids, suggesting that new functions are gained or previously existing ones are present at a higher level (G. Stark, personal communication).

Cells selected by serial selection protocols appear to be unstable in other ways, since mutational rates at independent loci are also increased (28). This increase in the rate of mutation appears to be a result of small lesions of the target genes not detectable by Southern blot analysis (e.g., point mutations [unpublished results]), perhaps suggesting an alteration of replicational fidelity. The investigations are at an early stage, but such cell lines offer a reasonable approach to the analysis of functions participating in gene amplification.

XIII. MODELS FOR GENE AMPLIFICATION

The mechanism most frequently proposed to explain gene amplification involves multiple misfirings at an origin of DNA replication to create an "onion skin" structure in the target DNA region (Fig. 7) (see reference 146 for review). This structure is then resolved by recombination to create a linear array of duplicated genes or, in other cases, extrachromosomal copies. This model has now been challenged in view of the recent findings regarding the structure of the amplified arrays, namely the abundance of the inverted repeat novel joints and the fact that the large size of the units in single-step mutants is far greater than estimates for the size of single replicons (49). A simple mechanism for the generation of such inverted repeats would be a strand switch at a replication fork where the nascent chain replicates around the fork rather than progressing in the normal manner (perhaps during an arrest of DNA replication fork progression; Fig. 8a [23, 112]). The deletion/inverted duplication (as observed for the *aprt* locus) could be formed as a result of the resolution of this structure by a double-strand break at the parental DNA molecule. The unduplicated region could be formed by a loop-out of sequences in advance of the strand switch, leaving one portion of the fork unreplicated (Fig. 8b). In the event at the *AMPD* locus, this loop-out could be stabilized by the complementarity of the sequences extruded, though similar complementarity of the nucleotides "extruded" at *aprt* could only be found 100 bp from the base of the projected loop. While this model could explain the formation of the inverted repeat (head-to-head joint), it does not explain the amplification of the structure. To account for this, Hyrien et al. (65) proposed that the amplification could be produced by a second, similar event at the fork moving in the opposite direction to produce a tail-to-tail joint (Fig. 8c). If a new round of DNA synthesis were initiated

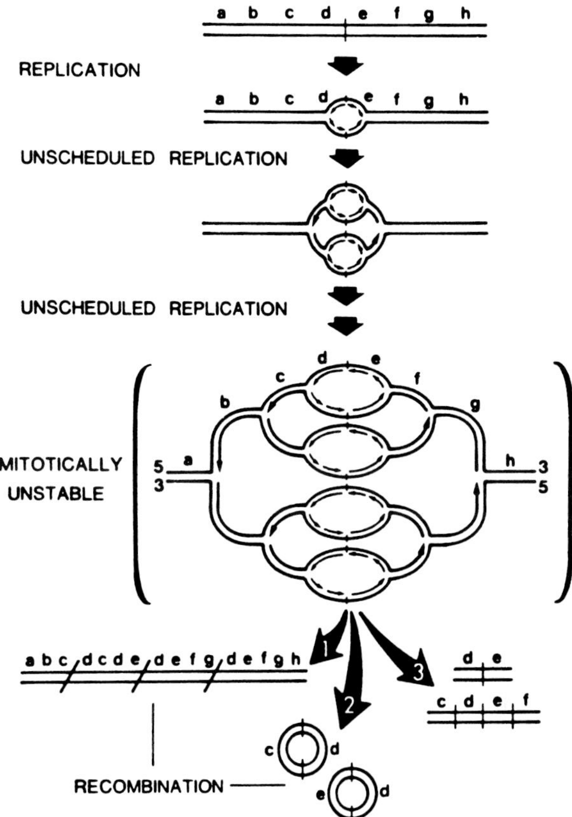

Figure 7. Formation of amplified gene arrays by unscheduled DNA synthesis and recombination (146). Bidirectional replication at an origin generates a bubble that can undergo a further round of replication to produce a nested set of partially replicated duplexes. Note that there are only two continuous DNA molecules. It is possible for linear duplex DNA to become detached from the structure if two replication forks can approach each other very closely (pathway 3). Recombination within the same duplex could generate extrachromosomal circles (pathway 2), while multiple recombinations among different complexes could resolve the structure into an intrachromosomal linear array. (Figure presented with the permission of George Stark.)

at the looped-out structures and the forks progressed along the previously replicated molecules, then two replication forks would pursue each other around this bubble structure, producing replicas of the inverted repeat joint as proposed by Futcher (43) (Fig. 8c). Evidence has been obtained for such a mechanism generating multiple copies of the 2μm circle in *Saccharomyces cerevisiae* (157), though there is little evidence for anything except the original inverted duplication in mammalian cells. Ideally, both head-to-head and tail-to-tail structures should be analyzed in an array unadulterated by subsequent rearrangements. Other structures (e.g., double minutes) could be produced in subsequent steps. The model, though lacking experimental support, is intriguing and could explain some of the features of amplified arrays.

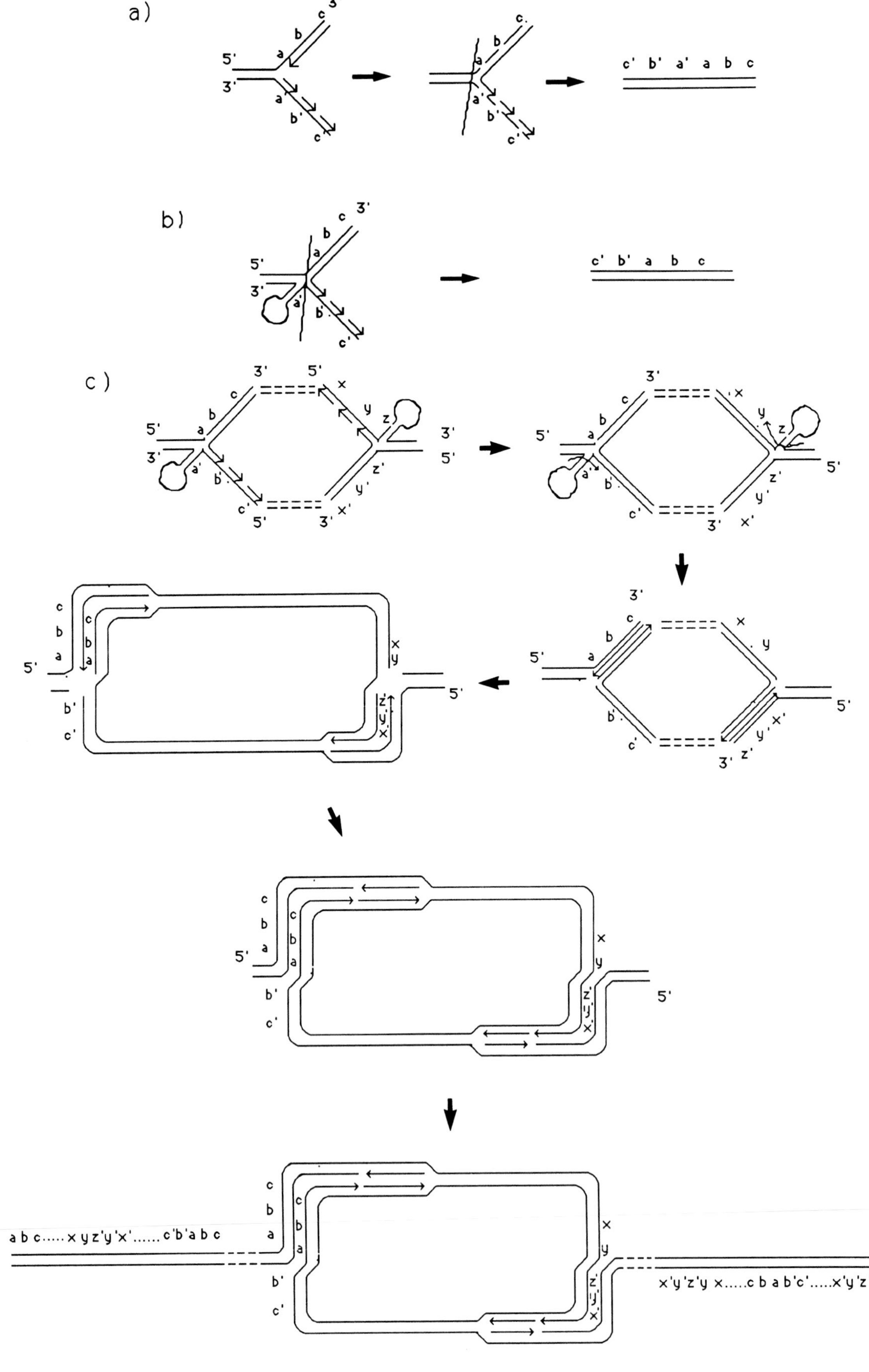
a)
5'
3'
a
b
c
3'
a'
b'
c'
c' b' a' a b c
b)
c' b' a b c
c)
x
y
z
x'
y'
z'
a b c.....x y z'y'x'......c'b'a b c
x'y'z'y x.....c b a b'c'.....x'y'z'

XIV. CONCLUSIONS: ANY ORDER TO THE CHAOS?

Basically I have presented a catalog of illegitimate recombinational events from a variety of cells. It is a diverse collection, and one can justifiably ask whether any uniform explanation will emerge. The obvious rationalization is that multiple mechanisms are involved, as the systems analyzed range from germ lines where the target genes are not expressed, to cultured somatic cells where the selectable loci are by necessity expressed. Surprisingly, similarities emerge in some of the rearrangements from very different sources (Table 1).

At the nucleotide level, certain sequences appear to recur at the breakpoints, particularly in germ cell-derived (inherited) mutations. More specifically, *Alu* family repeat sequences are at or near the breakpoints of most of the characterized deletions responsible for FH (89) and at a significant number of breakpoints of α- (116) and perhaps β-globin (155) deletions. More recently, deletions generated by *Alu*-*Alu* exchanges have been reported involving a Y chromosome locus (*DXYS5*) and a pseudoautosomal gene (129) and at the β-hexosaminidase α-chain gene in a Tay-Sachs patient (108). What the recurrence of *Alu* elements at deletion breakpoints tells us about the mechanism of deletion formation is not clear. Most of the exchanges in both the α- and β-globin loci also involve a non-*Alu* sequence at one of the breakpoints. *Alu* repeats are divergent in sequence (those involved in the exchanges described here are from 70 to 90% homologous) and they are not uniformly distributed around the genome; the LDL receptor gene, for example, is unusually rich in the repeats. Since the majority of the breakpoints within *Alu* sequences lie in the RNA polymerase III promoter region, it has been proposed that transcription or even the binding of transcriptional factors might promote the exchanges (86, 129). The elements could also play a more passive role by aligning DNA sequences packaged in chromatin.

In contrast, *Alu* repeats do not appear to play a significant role in the generation of deletions at the *aprt* locus in cultured hamster cells, although the hot spot for novel joint formation at the *AMPD* locus is rich in these sequences (64). The *aprt* structural gene is flanked both 5′ and 3′ by *Alu* equivalent repeats (Fig. 3), yet only 1 of the 20 or so rearrangements thus far characterized has these elements near (within 500 bp of) the breakpoints. This locus may be unusual, as there are constraints on the generation of deletions; there will, however, be some sort of constraint on the rearrangement of virtually any locus.

Some forms of B- and T-cell leukemias and lymphomas in humans and in mice involve aberrant rearrangements of immunoglobulin loci, in which proto-oncogenes become translocated into an immunoglobulin locus, thus altering the expression or structure of the genes (for summary, see reference 24). The translocations occur in cell lineages actively rearranging immunoglobulin genes, but it is possible that the switching machinery plays a wider role in generating rearrangements in other loci. For example, the region of *aprt* in which there is a cluster of deletion breakpoints has a significant homology with the mouse immunoglobulin heavy-chain switching region. The recurring simple-sequence motifs AGCT and GAGG present in this switch region (118) are found at many of the *aprt* deletion junctions. Very little is known of the enzymes which produce immunoglobulin rearrangements, but these observations suggest that they might be present in many other cell types and promote illegitimate recombination by acting at degenerate sites.

Local sequence features such as short inverted

Figure 8. Models for the formation of amplified gene arrays via inverted repeat structures. (a) Strand switch model for the generation of an inverted repeat. Basically this suggests that the replication complex switches DNA strands, perhaps during an arrest of fork progression, resulting in replication around the fork forming the complex structure illustrated here. The deletion of the sequences adjacent to the inverted repeat as found at *aprt* could be the result of a double-strand break of the parental molecule to resolve the structure. (b) The unduplicated portion, which seems to be a component of the inverted repeats found in amplified arrays thus far, could be the result of a loop-out of sequences in advance of the replication fork on either the leading or lagging strand. The loop-out may be stabilized by inverted repeat sequences at this point as at the *AMPD* locus (65). A similar model was proposed by Berg and co-workers many years ago to account for inverted repeat structures formed in λ*dv* plasmid DNAs (23). (c) A model for the amplification of such inverted repeat units has been proposed by Hyrien et al. (65). The complex structure formed by the strand switch might serve as an initiation point for DNA replication in rare circumstances. In this case it is proposed that the replication fork moves back along the previously replicated strand, thus generating the observed inverted repeats. If a similar series of events occurred at another linked replication fork, even thousands of kilobases away, theoretically two replication forks pursuing each other around the intervening DNA would result (43). This structure could generate the amplified inverted repeat units observed in the arrays. The structure could be resolved in a number of ways to generate a linear array or could undergo a recombination event to be released in the form of a double minute. The crucial event for the production of long tandem arrays by this mechanism is the development of the two replication forks, one chasing the other.

Table 1. Summary of gene rearrangements at various loci in mammalian cells

Locus	Phenotype or selection	Location[a]	Size[b] (kb)	Rearrangements observed	Features at breakpoints[c]	References
α Globin	α-Thalassemia	*16p13*	0.85	Deletions, insertion (small)	SDR, IR, repeat families (A, AA)	116
β Globin	β-Thalassemia	*11p15.5*	1.4	Deletions	SDR, IR	62, 66, 68
	HPFH			Inversion	Repeat families (A, L1)	101, 121, 143, 155
LDL receptor	FH	*19p13.2*	60	Deletions, duplication	SDR, IR, repeat families (A, AA)	63, 87–90
DMD	DMD	*Xp21.2*	~2,000	Deletions		26, 38, 78, 84
				Translocations		11
c-myc	Leukemia, lymphomas	*8q24*	5.2	Translocations	Aberrant V-D-J splicing	24
				Insertions	L1	71
aprt	Drug resistance	Z4, Z7 (CHO) hemizygote characterized		Deletions	SDR, IR	110, 111, 114
				Insertion		113
				Amplification		112
				Inversion[d]		103
hprt	Drug resistance	*Xq26*	40	Deletions		2, 15, 42, 117
				Amplifications		147, 148, 158
tk	Drug resistance	*17q21* heterozygote characterized	16	Deletions		161
dhfr	Drug resistance	Z2 (CHO) hemizygote characterized		Deletions[d], inversions,[d] amplification		106, 154

[a] Chromosomal location of the loci used to analyze illegitimate recombinational events. Sites refer to the human chromosome band unless otherwise indicated. CHO, Chinese hamster ovary.
[b] The sizes presented are the genomic coding regions. The mutations described frequently extend far beyond these boundaries.
[c] A summary of features which have been found at breakpoints of the indicated rearrangements determined by DNA sequencing of junction and wild-type fragments: SDR, short direct repeats (<10 bp) at some breakpoints; IR, inverted repeats (palindromes) found at some breakpoints; repeat families, at least one of the breakpoints in a given deletion fell in a repeat element (A, one breakpoint in an *Alu* repeat; AA, both breakpoints in *Alu* elements; L1, a breakpoint within an L1 repeat).
[d] Mutations induced by some DNA-damaging agent in these selectable systems.

repeats which have the potential to form stable secondary structures can be found at breakpoints in both germ line and somatic cell mutations. A hot spot for novel joint formation in amplified gene arrays at the *AMPD* locus contains a significant inverted repeat as well as short direct repeats. A similar combination of structures can be found at a cluster of deletion breakpoints in the *aprt* locus and at deletion breakpoints of the α- and β-globin loci. Lehrman et al. (88) have proposed models involving inverted repeat structures to account for deletion formation in both FH (one involving *Alu* sequences) and a β-thalassemia. The switching region of the mouse heavy-chain immunoglobulin locus is also rich in inverted repeat structures; the AGCT motif is self-complementary. Despite the presence of structures which could form stable stem-loop structures at deletion breakpoints, there is little evidence that such structures form in vivo. On the other hand, enzymes that bind and cleave stem-loop structures have been purified from bacteria (95, 107) and *S. cerevisiae* (159). It would be very surprising if they did not exist in mammalian cells. It may not even be necessary for stem-loop structures to form, as the inverted repeats may generate deletions by promoting strand switching during DNA replication. The possible roles of other types of sequence features which could interrupt the normal α-helical structure of DNA have not been supported by the rearrangements examined thus far. The only possible exception to this is the simple dG-dT repeat, which has been found near several breakpoints in gene rearrangements (64, 103, 111, 113). It has been suggested that this sequence (which is prone to form Z DNA in model systems) could promote unequal exchange or gene conversion. The significance of these correlations is questionable, however, since dG-dT sequences are fairly uniformly distributed throughout eucaryotic genomes (55) and thus might occasionally turn up at junctions by chance alone.

The breakpoints generated during illegitimate recombination in mammalian somatic cells show a great similarity. Exchanges in most instances occur between short direct repeat sequences (no more than 7 bp in length [111]), possibly via staggered cuts, leaving one copy of the sequence at the junction. There is generally little homology between the two donor sequences which might have aided in their alignment and they are often separated by a great distance. It is not known what brings the termini into contact and facilitates the exchange (maintaining chromosome linearity in the process), though nucleosome packing may play a role. Breakpoints of germ line mutations show much greater variability, suggesting either different inducing agents or formation by different types of events, such as an unequal exchange between the alleles on homologous chromosomes instead of intrachromosomal recombination.

Much new information will soon be available. The potential deletional hot spot in the *DMD* locus is being characterized. Several groups are studying rearrangements at the *aprt* and other selectable loci. New techniques for the characterization of large DNA molecules (134) can now be used to analyze the factors involved in the formation of very large deletions. Although the enzymes which produce the rearrangements are not known, proteins which bind specific DNA structures and sequences have been identified, and recombination enzymes are being purified using assays based on the recombination in vitro of genetically complementing plasmids (81). Illegitimate recombination constitutes a significant form of genetic variation in mammalian cells and is directly involved in tumorigenesis. Given the intrinsic interest in this process, its importance, and the tools now at hand, we can expect a thorough understanding of illegitimate recombination in the near future.

Acknowledgments. I am grateful to Michelle Debatisse and Olivier Hyrien for communicating manuscripts and models before publication. George Stark and Douglas Higgs are thanked for granting permission to use certain figures in this review.

LITERATURE CITED

1. **Adair, G. M., R. L. Stallings, R. S. Nairn, and M. J. Siciliano.** 1983. High-frequency structural gene deletion as the basis for functional hemizygosity of the adenine phosphoribosyltransferase locus in Chinese hamster ovary cells. *Proc. Natl. Acad. Sci. USA* **80:**5961–5964.
2. **Albertini, R. J., J. P. O'Neill, J. A. Nicklas, N. H. Heintz, and P. C. Kelleher.** 1985. Alterations of the *hprt* gene in human *in vivo*-derived 6-thioguanine-resistant T lymphocytes. *Nature* (London) **316:**369–371.
3. **Alitalo, K., M. Schwab, C. C. Lin, H. E. Varmus, and J. M. Bishop.** 1983. Homogeneously staining chromosomal regions contain amplified copies of an abundantly expressed cellular oncogene (c-*myc*) in malignant neuroendocrine cells from a human colon carcinoma. *Proc. Natl. Acad. Sci. USA* **80:**1707–1711.
4. **Ardeshir, F., E. Giulotto, J. Zieg, O. Brinson, W. S. L. Liao, and G. R. Stark.** 1983. Structure of amplified DNA in different Syrian hamster cell lines resistant to *N*-(phosphonacetyl)-L-aspartate. *Mol. Cell. Biol.* **3:**2076–2088.
5. **Ashman, C. R., and R. L. Davidson.** 1987. Sequence analysis of spontaneous mutations in a shuttle vector gene integrated into mammalian chromosomal DNA. *Proc. Natl. Acad. Sci. USA* **84:**3354–3358.
6. **Balaban-Malenbaum, G., and F. Gilbert.** 1977. Double minute chromosomes and the homogeneously staining regions in chromosomes of a human neuroblastoma cell line. *Science* **198:**739–740.

7. **Biedler, J. L.** 1982. Evidence for transient or prolonged extrachromosomal existence of amplified DNA sequences in antifolate-resistant, vincristine-resistant, and human neuroblastoma cells, p. 39–46. *In* R. T. Schimke (ed.), *Gene Amplification*. Cold Spring Harbor Laboratory, Cold Spring Harbor, N.Y.
8. **Biedler, J. L., and B. A. Spengler.** 1976. Metaphase chromosome anomaly: association with drug resistance and cell-specific products. *Science* **191:**185–187.
9. **Bird, A. P.** 1986. CpG-rich islands and the function of DNA methylation. *Nature* (London) **321:**209–213.
10. **Bishop, J. M.** 1987. The molecular genetics of cancer. *Science* **235:**305–311.
11. **Bodrug, S. E., P. N. Ray, I. L. Gonzalez, R. D. Schmickel, J. E. Sylvester, and R. G. Worton.** 1987. Molecular analysis of a constitutional x-autosome translocation in a female with muscular dystrophy. *Science* **237:**1620–1624.
12. **Bradley, W. E. C., and D. Letovanec.** 1982. High-frequency nonrandom mutational event at the adenine phosphoribosyltransferase (*aprt*) locus of sib-selected CHO variants heterozygous for *aprt*. *Somat. Cell Genet.* **8:**51–66.
13. **Bradshaw, H. D., Jr.** 1983. Molecular cloning and cell cycle-specific regulation of a functional human thymidine kinase gene. *Proc. Natl. Acad. Sci. USA* **80:**5588–5591.
14. **Braugh, H., B. Johnson, J. Hovis, T. Yano, A. Gazdar, O. S. Pettengill, S. Graziano, G. D. Sorenson, B. J. Poiesz, J. Minna, M. Linehan, and B. Zbar.** 1987. Molecular analysis of the short arm of chromosome 3 in small-cell and non-small-cell carcinoma of the lung. *N. Engl. J. Med.* **317:** 1109–1113.
15. **Breimer, L. H., J. Nalbantoglu, and M. Meuth.** 1986. Structure and sequence of mutations induced by ionizing radiation at selectable loci in Chinese hamster ovary cells. *J. Mol. Biol.* **192:**669–674.
16. **Bullock, P., J. J. Champoux, and M. Botchan.** 1985. Association of crossover points with topoisomerase I cleavage sites: a model for nonhomologous recombination. *Science* **230:**954–958.
17. **Bullock, P., W. Forrester, and M. Botchan.** 1984. DNA sequence studies of simian virus 40 chromosomal excision and integration in rat cells. *J. Mol. Biol.* **174:**55–84.
18. **Calos, M. P., J. S. Lebkowski, and M. R. Botchan.** 1983. High mutation frequency in DNA transfected into mammalian cells. *Proc. Natl. Acad. Sci. USA* **80:**3015–3019.
19. **Cananni, E., O. Dreazen, A. Klar, G. Rechavi, D. Ryan, J. B. Cohen, and D. Givol.** 1983. Activation of the c-*mos* oncogene in a mouse plasmacytoma by insertion of an endogenous intracisternal A-particle genome. *Proc. Natl. Acad. Sci. USA* **80:**7118–7122.
20. **Carozzi, N., F. Marashi, M. Plumb, S. Zimmerman, A. Zimmerman, L. S. Coles, J. R. E. Wells, G. Stein, and J. Stein.** 1984. Amplification of N-*myc* in untreated human neuroblastomas correlates with advanced disease stage. *Science* **224:**1117–1121.
21. **Collins, F. S., and S. M. Weissman.** 1984. The molecular genetics of human hemoglobin. *Nucleic Acids Res.* **31:** 315–436.
22. **Cowell, J. K.** 1982. Double minutes and homogeneously staining regions: gene amplification in mammalian cells. *Annu. Rev. Genet.* **16:**21–59.
23. **Chow, L. T., N. Davidson, and D. Berg.** 1974. Electron microscope study of the structures of *dv* DNAs. *J. Mol. Biol.* **86:**69–89.
24. **Croce, C. M.** 1987. Role of chromosome translocations in human neoplasia. *Cell* **49:**155–156.
25. **Deininger, P. L., D. J. Dolly, C. M. Rubin, T. Friedmann, and C. W. Schmid.** 1981. Base sequence studies of 300 nucleotide renatured repeated human DNA clones. *J. Mol. Biol.* **151:**17–33.
26. **Den Dunnen, J. T., E. Bakker, E. G. Klein Breteler, P. L. Pearson, and G. J. B. van Ommen.** 1987. Direct detection of more than 50% of the Duchenne muscular dystrophy mutations by field inversion gels. *Nature* (London) **329:**640–642.
27. **Drinkwater, N. R., and D. K. Klinedinst.** 1986. Chemically induced mutagenesis in a shuttle vector with a low background mutant frequency. *Proc. Natl. Acad. Sci. USA* **83:** 3402–3406.
28. **Drobetsky, E., and M. Meuth.** 1983. Increased mutational rates in Chinese hamster ovary cells serially selected for drug resistance. *Mol. Cell. Biol.* **3:**1882–1885.
29. **Dryja, T. P., J. M. Rapport, J. M. Joyce, and R. A. Petersen.** 1986. Molecular detection of deletions involving band q14 of chromosome 13 in retinoblastomas. *Proc. Natl. Acad. Sci. USA* **83:**7391–7394.
30. **Earnshaw, W. C., and M. M. S. Heck.** 1985. Localization of topoisomerase II in mitotic chromosomes. *J. Cell Biol.* **100:**1716–1725.
31. **Earnshaw, W. C., B. Halligan, C. A. Cooke, M. M. S. Heck, and L. F. Liu.** 1985. Topoisomerase II is a structural component of mitotic chromosome scaffolds. *J. Cell Biol.* **100:** 1706–1715.
32. **Erikson, J., L. Finger, L. Sun, A. Ar-Rushdi, K. Nishikura, J. Minowada, J. Finan, B. S. Emanuel, P. C. Nowell, and C. M. Croce.** 1986. Deregulation of *c-myc* by translocation of the α-locus of the T-cell receptor in T-cell leukemias. *Science* **232:**884–886.
33. **Feagin, J. E., D. R. Setzer, and R. T. Schmike.** 1983. A family of repeated DNA sequences, one of which resides in the second intervening sequence of the mouse dihydrofolate reductase gene. *J. Biol. Chem.* **258:**2480–2487.
34. **Finger, L. R., R. C. Harvey, R. C. A. Moore, L. C. Showe, and C. M. Croce.** 1986. A common mechanism of chromosomal translocation in T- and B-cell neoplasia. *Science* **234:**982–985.
35. **Flintoff, W. F., E. Livingston, C. Duff, and R. G. Worton.** 1984. Moderate-level gene amplification in methotrexate-resistant Chinese hamster ovary cells is accompanied by chromosomal translocations at or near the site of the amplified DHFR gene. *Mol. Cell. Biol.* **4:**69–76.
36. **Ford, M., B. Davies, M. Griffiths, J. Wilson, and M. Fried.** 1985. Isolation of a gene enhancer within an amplified inverted duplication after "expression selection." *Proc. Natl. Acad. Sci. USA* **82:**3370–3374.
37. **Ford, M., and M. Fried.** 1986. Large inverted duplications are associated with gene amplification. *Cell* **45:**425–430.
38. **Forrest, S. M., G. S. Cross, A. Speer, D. Gardner-Medwin, J. Burn, and K. E. Davies.** 1987. Preferential deletion of exons in Duchenne and Becker muscular dystrophies. *Nature* (London) **329:**638–640.
39. **Friend, S. H., R. Bernards, S. Rogelj, R. A. Weinberg, J. M. Rapaport, D. M. Albert, and T. P. Dryja.** 1986. A human DNA segment with properties of the gene that predisposes to retinoblastoma and osteosarcoma. *Nature* (London) **323:** 643–646.
40. **Fujimoto, S., and H. Yamagishi.** 1987. Isolation of an excision product of T-cell receptor α-chain gene rearrangements. *Nature* (London) **227:**242–244.
41. **Fung, Y.-K. T., A. L. Murphree, A. T'Ang, J. Qian, S. H. Hinrichs, and W. F. Benedict.** 1987. Structural evidence for

the authenticity of the human retinoblastoma gene. *Science* 236:1657–1661.

42. **Fuscoe, J. C., R. G. Fenwick, D. H. Ledbetter, and C. T. Caskey.** 1983. Deletion and amplification of the HGPRT locus in Chinese hamster cells. *Mol. Cell. Biol.* **3:** 1086–1096.
43. **Futcher, A. B.** 1986. Copy number amplification of the 2μm circle plasmid of *Saccharomyces cerevisiae*. *J. Theor. Biol.* **119:**197–204.
44. **Gasser, S. M., and U. K. Laemmli.** 1986. Cohabitation of scaffold binding regions with upstream enhancer elements of three developmentally regulated genes of D. melanogaster. *Cell* **46:**521–530.
45. **Gasser, S. M., T. Laroche, J. Falquet, E. Boy de la Tour, and U. K. Laemmli.** 1986. Metaphase chromosome structure involvement of topoisomerase II. *J. Mol. Biol.* **188:**613–629.
46. **Gerondakis, S., S. Cory, and J. M. Adams.** 1984. Translocation of the myc cellular oncogene to the immunoglobulin heavy chain locus in murine plasmacytomas is an imprecise reciprocal exchange. *Cell* **36:**973–982.
47. **Gibbs, R. A., and C. T. Caskey.** 1987. Identification and localization of mutations at the Lesch-Nyhan locus by ribonuclease A cleavage. *Science* **236:**303–305.
48. **Giulotto, E., C. Knights, and G. R. Stark.** 1987. Hamster cells with increased rates of DNA amplification, a new phenotype. *Cell* **48:**837–845.
49. **Giulotto, E., I. Saito, and G. R. Stark.** 1986. Structure of DNA formed in the first step of CAD gene amplification. *EMBO J.* **5:**2115–2121.
50. **Gladstone, P., L. Fueresz, and D. Pious.** 1982. Gene dosage and gene expression in the *HLA* region: evidence from deletion variants. *Proc. Natl. Acad. Sci. USA* **79:**1235–1239.
51. **Glickman, B. W., E. A. Drobetsky, and A. J. Grosovsky.** 1987. A study of the specificity of spontaneous and UV-induced mutation at the endogenous *aprt* gene of Chinese hamster ovary cells, p. 167–182. *In* M. M. Moore, D. M. DeMorini, F. de Serres, and K. R. Tindall (ed.), *Mammalian Cell Mutagenesis. Banbury Report 28.* Cold Spring Harbor Laboratory, Cold Spring Harbor, N.Y.
52. **Goldstein, J. L., M. S. Brown, R. G. W. Anderson, D. W. Russell, and W. J. Schneider.** 1983. Receptor-mediated endocytosis: concepts emerging from the LDL receptor system. *Annu. Rev. Cell Biol.* **1:**1–39.
53. **Groffen, J., J. R. Stephenson, N. Heisterkamp, A. de Klein, C. R. Bartram, and G. Grosveld.** 1984. Philadelphia chromosomal breakpoints are clustered within a limited region, bcr, on chromosome 22. *Cell* **36:**93–99.
54. **Grosovsky, A. J., E. A. Drobetsky, P. J. deJong, and B. W. Glickman.** 1986. Southern analysis of genomic alterations in gamma-ray-induced *aprt*-hamster cell mutants. *Genetics* **113:** 405–415.
55. **Hamada, H., M. G. Petrino, and T. Kakunaga.** 1982. A novel repeated element with z-DNA forming potential is widely found in evolutionary diverse eukaryotic genomes. *Proc. Natl. Acad. Sci. USA* **79:**6465–6469.
56. **Hamlin, J. L., J. D. Milbrandt, N. H. Heintz, and J. C. Azizkhan.** 1984. DNA sequence amplification in mammalian cells. *Int. Rev. Cytol.* **90:**31–82.
57. **Haniford, D. B., and D. E. Pulleybank.** 1983. Facile transition of poly[d(TG) · d(CA)] into a left handed helix in physiological conditions. *Nature* (London) **302:**632–634.
58. **Hawley, R. G., M. J. Shulman, and N. Hozumi.** 1984. Transposition of two different intracisternal A particle elements into an immunoglobulin kappa-chain gene. *Mol. Cell. Biol.* **4:**2565–2572.
59. **Haynes, S. R., T. P. Toomey, L. Leinwand, and W. Jelinek.** 1981. The Chinese hamster *Alu*-equivalent sequence: a conserved, highly repetitious, interspersed deoxyribonucleic acid sequence has a structure suggestive of a transposable element. *Mol. Cell. Biol.* **1:**573–583.
60. **Heisterkamp, N., J. Groffen, J. R. Stephenson, N. K. Spurr, P. N. Goodfellow, E. Solomon, B. Carrit, and W. Bodmer.** 1982. Chromosomal localization of human cellular homologues of two viral oncogenes. *Nature* (London) **299:** 747–749.
61. **Heisterkamp, N., K. Stam, and J. Groffen.** 1985. Structural organization of the bcr gene and its role in the Ph′ translocation. *Nature* (London) **315:**758–761.
62. **Henthorn, P. S., D. L. Mager, T. H. J. Huisman, and O. Smithies.** 1986. A gene deletion ending within a complex array of repeated sequences 3′ to the human β-globin gene cluster. *Proc. Natl. Acad. Sci. USA* **83:**5194–5198.
63. **Hobbs, H. H., M. S. Brown, J. L. Goldstein, and D. W. Russell.** 1986. Deletion of exon encoding cysteine-rich repeat of LDL receptor alters its binding specificity in a subject with familial hypercholesterolemia. *J. Biol. Chem.* **261:** 13114–13120.
64. **Hyrien, O., M. Debatisse, G. Buttin, and B. R. de Saint Vincent.** 1987. A hotspot for novel amplification joints in a mosaic of Alu-like repeats and palindromic A+T-rich DNA. *EMBO J.* **6:**2401–2408.
65. **Hyrien, O., M. Debatisse, G. Buitin, and B. R. de Saint Vincent.** 1988. The multicopy appearance of a large inverted duplication and the sequences at the inversion joint suggest a new model for gene amplification. *EMBO J.* **7:**407–417.
66. **Jagadeeswaran, P., D. Tuan, B. G. Forget, and S. M. Weissman.** 1982. A gene deletion ending at the midpoint of a repetitive DNA sequence in one form of hereditary persistence of fetal haemoglobin. *Nature* (London) **296:**469–470.
67. **Jeffreys, A. J., V. Wilson, and S. L. Thein.** 1985. Hypervariable minisatellite regions in human DNA. *Nature* (London) **314:**67–73.
68. **Jennings, M. W., R. W. Jones, W. G. Wood, and D. J. Weatherall.** 1985. Analysis of an inversion within the human beta globin gene cluster. *Nucleic Acids Res.* **13:**2897–2906.
69. **Jones, R. S., and S. S. Potter.** 1985. L1 sequences in HeLa extrachromosomal circular DNA: evidence for circularization by homologous recombination. *Proc. Natl. Acad. Sci. USA* **82:**1989–1993.
70. **Kas, E., and L. A. Chasin.** 1987. Anchorage of the Chinese hamster dihydrofolate reductase gene to the nuclear scaffold occurs in an intragenic region. *J. Mol. Biol.* **198:**677–692.
71. **Katzir, N., G. Rechavi, J. B. Cohen, T. Unger, F. Simoni, S. Segal, D. Cohen, and D. Givol.** 1985. "Retroposon" insertion into the cellular oncogene *c-myc* in canine transmissible venereal tumor. *Proc. Natl. Acad. Sci. USA* **82:**1054–1058.
72. **Kavathas, P., F. H. Bach, and R. DeMars.** 1980. Gamma ray-induced loss of expression of HLA and glyoxalase 1 alleles in lymphoblastoid cells. *Proc. Natl. Acad. Sci. USA* **77:**4251–4255.
73. **Kazazian, H. H., C. Wong, H. Youssoufian, A. F. Scott, D. G. Phillips, and S. E. Antonarakis.** 1988. Haemophila A resulting from *de novo* insertion of L1 sequences represents a novel mechanism for mutation in man. *Nature* (London) **332:**164–166.
74. **Kiyama, R., H. Matsui, and M. Oishi.** 1986. A repetitive DNA family (*Sau*3A family) in human chromosomes: extrachromosomal DNA and DNA polymorphism. *Proc. Natl. Acad. Sci. USA* **83:**4665–4669.
75. **Kiyama, R., K. Okumura, H. Matsui, G. A. P. Bruns, N.**

Kanda, and M. Oishi. 1987. Nature of recombination involved in excision and rearrangement of human repetitive DNA. *J. Mol. Biol.* 198:589–598.

76. Kleinberger, T., S. Etkin, and S. Lavi. 1986. Carcinogen-mediated methotrexate resistance and dihydrofolate reductase amplification in Chinese hamster cells. *Mol. Cell. Biol.* 6:1958–1964.

77. Knudson, A. G., Jr. 1985. Hereditary cancer, oncogenes, and antioncogenes 1. *Cancer Res.* 45:1437–1443.

78. Koenig, M., E. P. Hoffman, C. J. Bertelson, A. P. Monaco, C. Feener, and L. M. Kunkel. 1987. Complete cloning of the Duchenne muscular dystrophy (DMD) cDNA and preliminary genomic organization of the dmd gene in normal and affected individuals. *Cell* 50:509–517.

79. Konopka, J. B., S. M. Watanabe, and O. N. Witte. 1984. An alteration of the human c-*abl* protein in K562 leukemia cells unmasks associated tyrosine kinase activity. *Cell* 37:1035–1042.

80. Koufus, A., M. F. Hansen, N. G. Copeland, N. A. Jenkins, B. C. Lampkin, and W. K. Cavenee. 1985. Loss of heterozygosity in three embryonal tumours suggests a common pathogenetic mechanism. *Nature* (London) 316:330–334.

81. Kucherlapati, R., J. Spencer, and P. D. Moore. 1985. Homologous recombination catalyzed by human cell extracts. *Mol. Cell. Biol.* 5:714–720.

82. Kuff, E. L., L. A. Smith, and K. K. Lueders. 1981. Intracisternal A-particle genes in *Mus musculus*: a conserved family of retrovirus-like elements. *Mol. Cell. Biol.* 1:216–227.

83. Kunisada, T., and H. Yamagishi. 1987. Sequence organization of repetitive sequences enriched in small polydisperse circular DNAs from HeLa cells. *J. Mol. Biol.* 198:557–565.

84. Kunkel, L. M., et al. 1986. Analysis of deletions in DNA from patients with Becker and Duchenne muscular dystrophy. *Nature* (London) 322:73–77.

85. Laemmli, U. K., S. M. Cheng, K. W. Adolph, J. R. Paulson, J. A. Brown, and W. R. Baumbach. 1978. Metaphase chromosome structure: the role of nonhistone proteins. *Cold Spring Harbor Symp. Quant. Biol.* 42:351–340.

86. Lebkowski, J. S., S. Clancy, J. H. Miller, and M. P. Calos. 1985. The *lacI* shuttle: rapid analysis of the mutagenic specificity of ultraviolet light in human cells. *Proc. Natl. Acad. Sci. USA* 82:8606–8610.

87. Lehrman, M. A., J. L. Goldstein, D. W. Russell, and M. S. Brown. 1987. Duplication of seven exons in LDL receptor gene caused by Alu-Alu recombination in a subject with familial hypercholesterolemia. *Cell* 48:827–835.

88. Lehrman, M. A., D. W. Russell, J. L. Goldstein, and M. S. Brown. 1986. Exon-Alu recombination deletes 5 kilobases from the low density lipoprotein receptor gene, producing a null phenotype in familial hypercholesterolemia. *Proc. Natl. Acad. Sci. USA* 83:3679–3683.

89. Lehrman, M. A., D. W. Russell, J. L. Goldstein, and M. S. Brown. 1987. Alu-Alu recombination deletes splice acceptor sites and produces secreted low density lipoprotein receptor in a subject with familial hypercholesterolemia. *J. Biol. Chem.* 272:3354–3361.

90. Lehrman, M. A., W. L. Schneider, T. C. Sudhof, M. S. Brown, J. L. Goldstein, and D. W. Russell. 1985. Mutation in LDL receptor: Alu-Alu recombination deletes exons encoding transmembrane and cytoplasmid domains. *Science* 227:140–146.

91. Lemischka, I., and P. A. Sharp. 1982. The sequences of an expressed rat α-tubulin gene and a pseudogene with an inserted repetitive element. *Nature* (London) 300:330–335.

92. Levine, F., H. Erlich, B. Mach, R. Leach, R. White, and D. Pious. 1985. Deletion mapping of HLA and chromosome 6p genes. *Proc. Natl. Acad. Sci. USA* 82:3741–3745.

93. Lewis, J. A. 1986. Structure and expression of the Chinese hamster thymidine kinase gene. *Mol. Cell. Biol.* 6:1998–2010.

94. Lewis, J. A., K. Shimizu, and D. Zipser. 1983. Isolation and preliminary characterization of the Chinese hamster thymidine kinase gene. *Mol. Cell. Biol.* 3:1815–1823.

95. Lilley, D. M., and B. Kemper. 1984. Cruciform-resolvase interactions in supercoiled DNA. *Cell* 36:413–422.

96. Lin, P.-F., S.-Y. Zhao, and F. H. Ruddle. 1983. Genomic cloning and preliminary characterization of the human thymidine kinase gene. *Proc. Natl. Acad. Sci. USA* 80:6528–6532.

97. Long, E. O., and I. B. Dawid. 1980. Repeated genes in eukaryotes. *Annu. Rev. Biochem.* 49:727–764.

98. Looney, J. E., and J. L. Hamlin. 1987. Isolation of the amplified dihydrofolate reductase domain from methotrexate-resistant Chinese hamster ovary cells. *Mol. Cell. Biol.* 7:569–577.

99. Lowy, I., A. Pellicer, J. F. Jackson, G.-K. Sim, S. Silverstein, and R. Axel. 1980. Isolation of transforming DNA: cloning the hamster aprt gene. *Cell* 22:817–823.

100. Lundberg, C., L. Skoog, W. K. Cavenee, and M. Nordenskjold. 1987. Loss of heterozygosity in human ductal breast tumors indicates a recessive mutation on chromosome 13. *Proc. Natl. Acad. Sci. USA* 84:2372–2376.

101. Mager, D. L., P. S. Henthorn, and O. Smithies. 1985. A Chinese $^{G}\gamma$ + $(^{A}\gamma\delta\beta)^{0}$ thalassemia deletion: comparison to other deletions in the human β-globin gene cluster and sequence analysis of the breakpoints. *Nucleic Acids Res.* 13:6559–6575.

102. Meuth, M., and J. E. Arrand. 1982. Alterations of gene structure in ethyl methane sulfonate-induced mutants of mammalian cells. *Mol. Cell. Biol.* 2:1459–1462.

103. Meuth, M., J. Nalbantoglu, G. Phear, and C. Miles. 1987. Molecular basis of genome rearrangements at the hamster *arpt* locus, p. 183–191. *In* M. M. Moore, D. M. DeMorini, F. de Serres, and K. R. Tindall (ed.), *Mammalian Cell Mutagenesis. Banbury Report 28*. Cold Spring Harbor Laboratory, Cold Spring Harbor, N.Y.

104. Michelson, A. M., and S. H. Orkin. 1983. Boundaries of gene conversion within the duplicated human α-globin genes. *J. Biol. Chem.* 258:15245–15254.

105. Mirkovitch, M.-E. Mirault, and U. K. Laemmli. 1984. Organization of the higher-order chromatin loop: specific DNA attachment sites on nuclear scaffold. *Cell* 39:223–232.

106. Mitchell, P. J., G. Urlaub, and L. Chasin. 1986. Spontaneous splicing mutations at the dihydrofolate reductase locus in Chinese hamster ovary cells. *Mol. Cell. Biol.* 6:1926–1935.

107. Mizuuchi, K., B. Kemper, J. Hays, and R. A. Weisberg. 1982. T4 endonuclease VII cleaves Holliday structures. *Cell* 29:357–363.

108. Myerowitz, R., and N. D. Hogikyan. 1987. A deletion involving Alu sequences in the β-hexoxaminidase α-chain gene of French Canadians with Tay-Sachs disease. *J. Biol. Chem.* 262:15396–15399.

109. Myers, R. M., L. S. Lerman, and T. Maniatis. 1985. A general method for saturation mutagenesis of cloned DNA fragments. *Science* 229:242–247.

110. Nalbantoglu, J., O. Goncalves, and M. Meuth. 1983. Structure of mutant alleles at the *aprt* locus of Chinese hamster ovary cells. *J. Mol. Biol.* 167:575–594.

111. Nalbantoglu, J., D. Hartley, G. Phear, G. Tear, and M. Meuth. 1986. Spontaneous deletion formation at the aprt

locus of hamster cells: the presence of short sequence homologies and dyad symmetries at deletion termini. *EMBO J.* 5:1199–1204.

112. **Nalbantoglu, J., and M. Meuth.** 1986. DNA amplification-deletion in a spontaneous mutation of the hamster *aprt* locus: structure and sequence of the novel joint. *Nucleic Acids Res.* **14:**8361–8371.
113. **Nalbantoglu, J., C. Miles, and M. Meuth.** 1988. Insertion of unique and repetitive DNA fragments into the *aprt* locus of hamster cells. *J. Mol. Biol.* **200:**449–459.
114. **Nalbantoglu, J., G. Phear, and M. Meuth.** 1987. DNA sequence analysis of spontaneous mutations at the *aprt* locus of hamster cells. *Mol. Cell. Biol.* **7:**1445–1449.
115. **Naylor, S. L., B. E. Johnson, J. D. Minna, and A. Y. Sakaguchi.** 1987. Loss of heterozygosity of chromosome 3p markers in small-cell cancer. *Nature* (London) **329:** 451–454.
116. **Nicholls, R. D., N. Fischel-Ghodsian, and D. R. Higgs.** 1987. Recombination at the human α-globin gene cluster: sequence features and topological constraints. *Cell* **49:** 369–378.
117. **Nicklas, J. A., T. C. Hunter, L. M. Sullivan, J. K. Berman, J. P. O'Neill, and R. J. Albertini.** 1987. Molecular analysis of *in vivo hprt* mutations in human T-lymphocytes. I. Studies of low frequency 'spontaneous' mutants by Southern blots. *Mutagenesis* **2:**341–347.
118. **Nikaido, T., S. Nakai, and T. Honjo.** 1981. Switch region of immunoglobulin C_μ gene is composed of simple tandem repetitive sequences. *Nature* (London) **292:**845–848.
119. **Nishikura, K., A. Ar-Rushki, J. Erikson, R. Watt, G. Rovera, and C. M. Croce.** 1983. Differential expression of the normal and of the translocated human c-*myc* oncogenes in B cells. *Proc. Natl. Acad. Sci. USA* **80:**4822–4826.
120. **Nordheim, A., and A. Rich.** 1983. The sequence (dC-dA)n(dG-dT)n forms left-handed Z-DNA in negatively supercoiled plasmids. *Proc. Natl. Acad. Sci. USA* **316:** 1821–1825.
121. **Ottolengthi, S., and B. Giglioni.** 1982. The deletion in a type of δ^0-β^0-thalassaemia begins in an inverted Alu repeat. *Nature* (London) **300:**23–30.
122. **Patel, P. I., P. E. Framson, C. T. Caskey, and A. C. Chinault.** 1986. Fine structure of the human hypoxanthine phosphoribosyltransferase gene. *Mol. Cell. Biol.* **6:**393–403.
123. **Paulson, K. E., N. Deka, C. W. Schmid, R. Misra, C. W. Schindler, M. G. L. Kadyk, and L. Leinwand.** 1985. A transposon-like element in human DNA. *Nature* (London) **316:**359–361.
124. **Piccoli, S. P., P. G. Gaimi, and M. D. Cole.** 1984. A conserved sequence at *c-myc* oncogene chromosomal translocation breakpoints in plasmacytomas. *Nature* (London) **310:**327–330.
125. **Pious, D., M. S. Krangel, L. L. Dixon, P. Parham, and J. L. Strominger.** 1982. HLA antigen structural gene mutants selected with an allospecific monoclonal antibody. *Proc. Natl. Acad. Sci. USA* **79:**7832–7836.
126. **Razzaque, A., H. Mizusawa, and M. M. Seidman.** 1983. Rearrangement and mutagenesis of a shuttle vector plasmid after passage in mammalian cells. *Proc. Natl. Acad. Sci. USA* **80:**3010–3014.
127. **Rice, G. C., C. Hoy, and R. T. Schimke.** 1986. Transient hypoxia enhances the frequency of dihydrofolate reductase gene amplification in Chinese hamster ovary cells. *Proc. Natl. Acad. Sci. USA* **83:**5978–5982.
128. **Roth, D., and J. Wilson.** 1988. Illegitimate recombination in mammalian cells, p. 621–653. *In* R. Kucherlapati and R. Smith (ed.), *Genetic Recombination.* American Society for Microbiology, Washington, D.C.
129. **Rouyer, F., M.-C. Simmler, D. C. Page, and J. Weisenbach.** 1987. A sex chromosome rearrangement in a human XX male caused by Alu-Alu recombination. *Cell* **51:**417–425.
130. **Rowley, J. D.** 1973. A new consistent chromosomal abnormality in chronic myelogenous leukaemia identified by quinacrine fluorescence and Giemsa staining. *Nature* (London) **243:**290–293.
131. **Saiki, R. K., S. Scharf, F. Faloona, K. B. Mullis, G. T. Horn, H. A. Erlich, and N. Arnheim.** 1985. Enzymatic amplification of β-globin genomic sequences and restriction site analysis for diagnosis of sickle cell anemia. *Science* **230:** 1350–1354.
132. **Saito, I., and G. R. Stark.** 1986. Charomids: cosmid vectors for efficient cloning and mapping of large or small restriction fragments. *Proc. Natl. Acad. Sci. USA* **83:**8664–8668.
133. **Schimke, R. T.** 1984. Gene amplification in cultured animal cells. *Cell* **37:**705–713.
134. **Schwartz, D. C., and C. R. Cantor.** 1984. Separation of yeast chromosome-sized DNAs by pulsed field gradient gel electrophoresis. *Cell* **37:**67–75.
135. **Shiloh, Y., J. Shipley, G. M. Brodeur, G. Bruns, B. Korf, T. Donlon, R. R. Schreck, R. Seeger, K. Sakai, and S. A. Latt.** 1985. Differential amplification, assembly, and relocation of multiple DNA sequences in human neuroblastomas and neuroblastoma cell lines. *Proc. Natl. Acad. Sci. USA* **82:** 3761–3765.
136. **Siminovitch, L.** 1976. On the nature of hereditable variation in cultured somatic cells. *Cell* **7:**1–11.
137. **Singer, M. F.** 1982. Highly repeated sequences in mammalian genomes. *Int. Rev. Cytol.* **76:**67–112.
138. **Singleton, C. K., M. W. Kirkpatrick, and R. D. Wells.** 1984. S1 nuclease recognises DNA conformational junctions between left-handed helical (dT-dG)n (dC · dA)n and contiguous right handed sequences. *J. Biol. Chem.* **259:**1963–1967.
139. **Singleton, C. K., S. Klysik, S. M. Stirdivant, and R. D. Wells.** 1982. Left-hand Z-DNA is induced by supercoiling in physiological ionic conditions. *Nature* (London) **299:**312–316.
140. **Solomon, E., R. Voss, V. Hall, W. F. Bodmer, J. R. Jass, A. J. Jeffreys, F. C. Lucibello, I. Patel, and S. H. Rider.** 1987. Chromosome 5 allele loss in human colorectal carcinomas. *Nature* (London) **328:**616–619.
141. **Sparkes, R. S., M. C. Sparkes, M. G. Wilson, J. W. Towner, W. Benedict, A. L. Murphree, and J. J. Yunis.** 1980. Regional assignment of genes for human esterase D and retinoblastoma to chromosome band 13q14. *Science* **208:** 1042–1044.
142. **Spradling, A. C., and A. P. Mahowald.** 1980. Amplification of genes for chorion proteins during oogenesis in Drosophila melanogaster. *Proc. Natl. Acad. Sci. USA* **77:**1096–1100.
143. **Spritz, R. A., and S. H. Orkin.** 1982. Duplication followed by deletion accounts for the structure of an Indian deletion β-thalassemia gene. *Nucleic Acids Res.* **10:**8025–8028.
144. **Stanfield, S. W., and D. R. Helinski.** 1984. Cloning and characterization of small circular DNA from Chinese hamster ovary cells. *Mol. Cell. Biol.* **4:**173–180.
145. **Stark, G. R.** 1986. DNA amplification in drug resistant cells and in tumours. *Cancer Surv.* **5:**1–25.
146. **Stark, G. R., and G. M. Wahl.** 1984. Gene amplification. *Annu. Rev. Biochem.* **53:**447–491.
147. **Stout, J. T., and C. T. Caskey.** 1985. HPRT: gene structure, expression and mutation. *Annu. Rev. Genet.* **19:**127–148.
148. **Thacker, J.** 1986. The nature of mutants induced by ionising radiation in cultured hamster cells. III. Molecular character-

ization of HPRT-deficient mutants induced by γ-rays or α-particles showing that the majority have deletions of all or part of the HPRT gene. *Mutat. Res.* **160**:267–275.

149. **Tindall, K. R., L. F. Stankowski, Jr., R. Machanoff, and A. W. Hsie.** 1984. Detection of deletion mutations in pSV2gpt-transformed cells. *Mol. Cell. Biol.* **4**:1411–1415.

150. **Tsujimoto, T., J. Cossman, E. Jaffe, and C. M. Croce.** 1985. Involvement of the bcl-2 gene in human follicular lymphoma. *Science* **228**:1440–1443.

151. **Tsujimoto, Y., L. R. Finger, J. Yunis, P. C. Nowell, and C. M. Croce.** 1984. Cloning of the chromosome breakpoint of neoplastic B cells with the t(14;18) chromosome translocation. *Science* **226**:1097–1099.

152. **Tsujimoto, Y., E. Jaffe, J. Cossman, J. Gorham, P. C. Nowell, and C. M. Croce.** 1985. Clustering of breakpoints on chromosome 11 in human B-cell neoplasms with the t(11;14) chromosome translocation. *Nature* (London) **315**:340–345.

153. **Turner, D. R., A. A. Morley, M. Haliandros, R. Kutlaca, and B. J. Sanderson.** 1985. *In vivo* somatic mutations in human lymphocytes frequently result from major gene alterations. *Nature* (London) **315**:343–345.

154. **Urlaub, G., P. J. Mitchell, E. Kas, L. A. Chasin, V. L. Funanage, T. T. Myoda, and J. Hamlin.** 1986. Effect of gamma rays at the dihydrofolate reductase locus: deletions and inversions. *Somat. Cell. Mol. Genet.* **12**:555–566.

155. **Vanin, E. F., P. S. Henthorn, D. Kioussis, F. Grosveld, and O. Smithies.** 1983. Unexpected relationships between four large deletions in the human β-globin gene cluster. *Cell* **35**:701–709.

156. **Varmus, H. E.** 1984. The molecular genetics of cellular oncogenes. *Annu. Rev. Genet.* **18**:553–612.

157. **Volkert, F. C., and J. R. Broach.** 1986. Site-specific recombination promotes plasmid amplification in yeast. *Cell* **46**: 541–550.

158. **Vrieling, H., J. W. I. M. Simons, F. Arwert, A. T. Matarajan, and A. A. van Zeeland.** 1985. Mutations induced by X-rays at the HPRT locus in cultured Chinese hamster cells are mostly large deletions. *Mutat. Res.* **144**:281–286.

159. **West, S. C., and A. Korner.** 1985. Cleavage of cruciform DNA structures by an activity from Saccharomyces cerevisiae. *Proc. Natl. Acad. Sci. USA* **82**:6445–6449.

160. **Wong, C., C. E. Dowling, R. K. Saiki, R. G. Higuchi, H. A. Erlich, and H. H. Kazazian, Jr.** 1987. Characterization of β-thalassaemia mutations using direct genomic sequencing of amplified single copy DNA. *Nature* (London) **330**: 384–386.

161. **Yandell, D. W., T. P. Dryja, and J. B. Little.** 1986. Somatic mutations at a heterozygous autosomal locus in human cells occur more frequently by allele loss than by intragenic structural alterations. *Somat. Cell. Genet.* **12**:255–263.

162. **Yang, R., B. Fristensky, A. H. Deutch, R. C. Huang, Y. H. Tan, S. A. Narang, and R. Wu.** 1983. The nucleotide sequence of a new human repetitive DNA consists of eight tandem repeats of 66 base pairs. *Gene* **25**:59–66.

163. **Yang, T. P., P. I. Patel, A. C. Chinault, J. T. Stout, L. G. Jackson, B. M. Hildebrand, and C. T. Caskey.** 1984. Molecular evidence for new mutation at the *hprt* locus in Lesch-Nyhan patients. *Nature* (London) **310**:412–414.

164. **Yeung, C.-Y., E. G. Frayne, M. R. Al-Ubaidi, A. G. Hook, D. E. Ingolia, D. A. Wright, and R. E. Kellems.** 1983. Amplification and molecular cloning of murine adenosine deaminase gene sequences. *J. Biol. Chem.* **258**:15179–15185.

165. **Yun-Fai, L., and Y. Wai Kan.** 1984. Direct isolation of the functional human thymidine kinase gene with a cosmid shuttle vector. *Proc. Natl. Acad. Sci. USA* **81**:414–418.

166. **Yunis, J. J.** 1983. The chromosome basis of human neoplasia. *Science* **221**:227–236.

167. **Zech, L., G. Gahrton, L. Hammarstrom, G. Juliusson, H. Mellstedt, K. H. Robert, and C. I. E. Smith.** 1984. Inversion of chromosome 14 marks human T-cell chronic lymphocyte leukaemia. *Nature* (London) **308**:858–1443.

168. **Zech, L., U. Haglund, K. Nilson, and G. Klein.** 1976. Characteristic chromosomal abnormalities in biopsies and lymphoid-cell lines from patients with Burkitt and non-Burkitt lymphomas. *Int. J. Cancer* **17**:47–56.

169. **Zieg, J., C. E. Clayton, F. Ardeshir, E. Giulotto, E. A. Swyryd, and G. R. Stark.** 1983. Properties of single-step mutants of Syrian hamster cell lines resistant to *N*-(phosphonacetyl)-L-aspartate. *Mol. Cell. Biol.* **3**:2089–2098.

Chapter 40

Mitochondrial DNA Instabilities and Rearrangements in Yeasts and Fungi

BERNARD DUJON and LEON BELCOUR

I. INTRODUCTION

Mitochondrial genomes of all eucaryotic organisms contain many fewer genes than do the nuclear genomes. Yet these genes play a fundamental role in the cell; mutations in them confer a severely deleterious, very often lethal, phenotype. The genetic content of mitochondrial genomes was first established by completely sequencing human mitochondrial DNA (mtDNA) (2) and then recognizing that the set of

Bernard Dujon ■ Unité de Génétique Moléculaire des Levures, Département de Biologie Moléculaire, Institut Pasteur, 25 rue du Docteur Roux, F-75724 Paris Cedex 15, France. **Leon Belcour** ■ Centre de Génétique Moléculaire du Centre National de la Recherche Scientifique, F-91190 Gif sur Yvette, France.

mitochondrial genes is essentially the same in diverse animals including vertebrates (3, 19, 31, 124), insects (33), and nematodes (148). In all these groups, the mitochondrial genome is composed of several hundreds of identical small circular DNA molecules (15 to 16 kilobases [kb]) in which the genes are tightly packed. Perhaps the most significant aspect of an animal mtDNA molecule is that it constitutes a unique and symmetrical pair of transcription units (109), hence limiting genomic rearrangements if function is to be preserved. All genes are expressed from one or the other of two polycistronic RNA molecules (one from each DNA strand). These RNAs originate from two strong promoters located within a unique region that also serves as replication origin (28–30, 108). A third promoter, also in this region, allows higher expression of the rRNA genes. Genes for tRNAs are interspersed among other genes, and the maturation of the tRNAs serves as the processing signal for the formation of mature transcripts of the other genes (108, 115). Rearrangements of the gene order are observed between different groups of animals, but they preserve the functional organization with the two major transcripts.

In contrast with the uniformity of animal mtDNA, higher plants contain larger and more complex mitochondrial genomes with extreme variations in the size and structure of the mtDNA molecules (90, 129). The higher complexity is due in small part to additional genes, but also results from intramolecular recombinational events that generate subgenomic circles from a master genomic circle (98, 119). In some cases, sites of genetic recombination between mtDNA molecules have been determined at the nucleotide level (42). In addition to their main mtDNA, many species of plants contain in their mitochondria small plasmids that replicate autonomously. The role of these plasmids remains obscure, although the presence or absence of some of them has been correlated with cytoplasmic male sterility.

Yeasts and other fungi differ from both animals and plants in that they possess mitochondrial genomes of intermediate complexities, and hence their genes are generally not tightly packed. In the normal situation, the fungal mitochondrial genome is made up of a few dozen identical molecules with sizes varying from ca. 15 kb in some species to ca. 100 kb in others. Perhaps the major difference with animal mtDNA lies in the fact that fungal mtDNAs contain multiple transcription and replication units. In the yeast *Saccharomyces cerevisiae*, for example, transcription is initiated from at least 12 major promoters distributed along the mtDNA molecule in such a way that each primary transcript covers only one or a few neighboring genes (32, 55, 94, 117, 127). Most tRNA genes are clustered in one sector of the map (47), and signals for mRNA processing are separate from the tRNA sequences (116). In addition, short G · C-rich clusters are found at a variety of locations, mostly but not exclusively within intergenic regions. Their presence or absence at given locations is strain dependent (43). In *Neurospora crassa*, palindromic G · C-rich clusters serve as processing signals for the maturation of RNA transcripts (151). Such a functional organization with several transcription units allows a higher degree of polymorphic variation and genomic rearrangement than for animal mtDNA. It is perhaps not so surprising, then, that many instabilities and rearrangements have been observed in yeast and fungal mtDNAs.

Also correlated with the looser genome organization characteristic of fungal (and plant) mitochondria is the presence of introns. Mitochondrial introns differ from nuclear introns of RNA polymerase II-transcribed genes in a number of ways. Some differences probably reflect their origin, while others probably relate to their occurrence in a subcellular compartment in which transcription and translation are not separated. The various organelle introns can be divided into two groups (called I and II) based on putative RNA secondary structures with conserved core elements (104). Several of them, in both group I and group II, are catalysts for their own RNA splicing in vitro (ribozymes) (4, 73, 120, 128, 143–145). In addition, many introns contain long open reading frames coding for specific proteins (47). Of particular relevance to this chapter is the fact that one of them has been shown to serve as a transposase (34), while others are structurally related to reverse transcriptases (106).

II. DELETION AND AMPLIFICATION OF mtDNA IN *SACCHAROMYCES CEREVISIAE*

Major alterations of the mtDNA are relatively common phenomena in yeasts and filamentous fungi. Although of similar molecular nature in the two groups, they result in very different phenotypic expression according to the particular physiology of the cell. In the rare species that are facultative aerobes (such as *S. cerevisiae*), cells with altered mtDNAs survive as respiratory-deficient mutants. In such cells, the population of mtDNA molecules is entirely composed of mutant molecules (homoplasmic cells). In the other species (such as filamentous fungi), similar alterations of mtDNA result in the stoppage of growth. In such cases, mutant molecules largely predominate, but very often a small proportion of

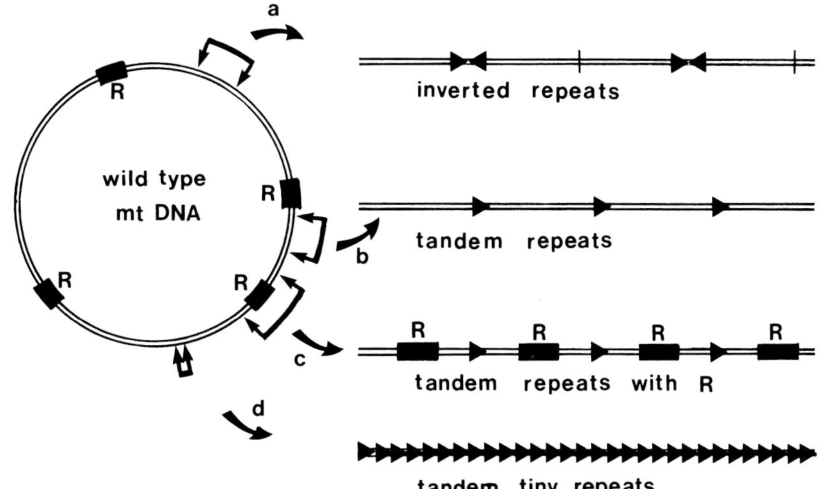

Figure 1. Schematic representation of the deletion and amplification of short segments of mtDNA in [*rho*$^-$] mutants. A short fragment of the wild-type mtDNA is excised by illegitimate recombination between short direct repeats or by saltatory replication (arrows) and amplified as regular tandem or inverted repeats. In event c, the excised segment contains a functional replication origin (R) of the wild-type mtDNA, while in events a, b, and d, they do not. Reiteration of the replication origin along the DNA molecule of [*rho*$^-$], event c, is responsible for an extremely high zygotic suppressiveness. [*rho*$^-$] events a, b, and d show various degrees of low or moderate suppressiveness. In event d, a [*rho*$^-$] mutant is produced by the excision of an extremely short fragment that hence is present in a very high copy number. Solid triangles indicate endpoints of repeats with their orientation. Inverted repeats are separated by a vertical bar.

wild-type molecules can remain (heteroplasmic cells) and serve if growth resumes.

A. The Structure of mtDNA in [*rho*$^-$] Mutants

In the yeast *S. cerevisiae*, respiratory-deficient mutants arise at high frequency (a few percent of new mutants per cell generation). The mutants (called petite or [*rho*$^-$]) result from extensive deletions of the wild-type mtDNA sequences (47). The fragments of mtDNA retained in the mutants range from about one-third of the normal 78-kb mtDNA molecule to just a few dozens of base pairs, and in the extreme case ([*rho*^{0}] mutants) there is no mtDNA left at all. Due to the extent of the deletions and the scattered distribution of essential genes in mtDNA, all [*rho*$^-$] mutants are completely deficient in mitochondrial protein synthesis. Consequently, they are unable to express the mitochondrial genes that remain after deletion. Yet, the fragment preserved is replicated faithfully, and very often transcribed, and the total amount of mtDNA in a [*rho*$^-$] cell resembles that in a wild-type cell. The fragment retained after deletion is amplified many times by regular repetitions, forming long DNA molecules that can reach the size of those in the wild-type cell (47, 96). Thus, the formation of [*rho*$^-$] mutants involves a mechanism of deletion-amplification rather than just a simple deletion. It is unclear whether deletion and amplification are simultaneous. In some [*rho*$^-$] mutants, a multimeric series of small circular molecules is also observed, the unit length of which equals the repeat unit (89, 97). In these cases, the populations of molecules probably result from an equilibrium of integration and excision between the circles, presumably by general homologous recombination. In some [*rho*$^-$] mutants, several multimeric series are present at the same time.

Repetitions along the DNA molecules are generally regular, the fragment retained being repeated either in direct or in inverted orientation (Fig. 1). In the first case (the most frequent among spontaneous [*rho*$^-$] mutants), the repeat unit corresponds to the conserved fragment. In the second case, the repeat unit consists of a palindromic repetition of the conserved fragment. Palindromes may sometimes be imperfect, the fragment conserved being shorter on one side than on the other (132). Sequence analysis carried out to characterize the endpoints of the deletion and amplification indicates that intramolecular recombination between short repeats (a few base pairs) of the wild-type mtDNA molecules is responsible for the formation of most [*rho*$^-$] mutants (14). In some cases, however, endpoints did not reveal obvious repeats. Slippery or abortive replication has been invoked to explain the formation of these mutants, and also for the formation of the [*rho*$^-$] mutants with repeat units so short that they cannot result from intramolecular circularization and excision (21, 58).

[*rho*$^-$] mutations are unusual in that apparently any segment of the wild-type mtDNA can be retained and amplified in a given [*rho*$^-$] mutant, irrespective of whether or not this segment contains a normal replication origin (Fig. 1). Because there is no indication that a single origin would be preserved in all the varied [*rho*$^-$] mutant DNAs, we are forced to conclude that a variety of sequences can function as secondary replication origins if the normal origins are absent. The exact mechanism responsible for the use of such secondary origins remains unclear, but the genetic properties of the various [*rho*$^-$] mutants in crosses to [*rho*$^+$] strains help us to understand part of the phenomenon.

B. Suppressiveness Is the Overall Result of Competition between mtDNA Molecules

"Suppressiveness" is a genetic measure of the efficiency with which the deficient mtDNA molecules of the [*rho*$^-$] mutants can be transmitted to the progeny of crosses in competition with wild-type mtDNA molecules. In such crosses, a fraction of the progeny are [*rho*$^-$] cells, the mtDNA of which is generally indistinguishable from that of their [*rho*$^-$] parent. This fraction varies from near 0 to almost 100%, depending upon the [*rho*$^-$] parent used in the cross. Suppressiveness is a complex parameter that seems to result from a competition between mtDNA molecules for replication and segregation. In fact, the actual origins of replication of wild-type mtDNA were defined and mapped after it was recognized that cloning such a replication origin in the amplified segment of a [*rho*$^-$] mutant causes that mutant to compete more effectively against the wild-type cell (i.e., to be extremely suppressive) (20). Because three or four functional origins (depending upon the strains) are distributed along the wild-type mtDNA molecules, hypersuppressive [*rho*$^-$] mutants can originate from different segments of mtDNA (5, 48). A few additional sequences closely resembling active origins are also present in wild-type mtDNA but are probably inactive, since they contain internal insertions (44, 59). They could not be cloned by selecting hypersuppressive [*rho*$^-$] mutants. Other [*rho*$^-$] mutant DNAs that lack normal replication origins exhibit various degrees of suppressiveness (from near 0 to 70 or 80%), but are never hypersuppressive (more than 90 or 95%). Their range of suppressiveness can also be interpreted as reflecting the efficiency with which various sequences replicate or segregate in the absence of normal replication origins (63, 130). Direct measurement of the rates of DNA synthesis indicates that a replication advantage is important for some [*rho*$^-$] mutants while a segregation advantage is more important for others (27). Particularly remarkable in this respect is the finding of suppressive [*rho*$^-$] mutants that consist of very short segments (70 to 90 base pairs [bp] long) containing only A · T base pairs, implying that even such simple sequences can replicate and segregate efficiently (58).

C. Legitimate and Illegitimate Recombination Involving [*rho*$^-$] mtDNAs

[*rho*$^-$] mtDNAs recombine efficiently with wild-type mtDNAs such that the genes or sequences from their defective genomes can be rescued into functional genomes after a genetic cross (47). A nuclear gene (*PIF1*) that encodes a protein of 857 amino acids with a DNA-binding domain and an N-terminal presequence for targeting to mitochondria is necessary for mtDNA recombination involving some [*rho*$^-$] mutants (61). The *PIF1* protein has homology with excision-repair enzymes, suggesting that it recognizes short, palindromic A · T-rich sequences in these mutants and triggers recombination. Other [*rho*$^-$] mutants, however, do not require the *PIF1* gene product to recombine efficiently with wild-type mtDNA molecules, suggesting that they use a different recombination pathway (60).

[*rho*$^-$] mtDNA molecules are also able to recombine between themselves by legitimate (homology-based) or illegitimate mechanisms (103). Legitimate recombination is probably responsible for the multimeric series of small circular molecules already noted to exist in some [*rho*$^-$] mutants. Illegitimate recombination between two different [*rho*$^-$] mutations obviously creates new genetic arrangements in the resulting mtDNA molecules. Such molecules, however, are still defective, and the novel arrangement needs to be rescued into a [*rho*$^+$] genome by subsequent recombination to become eventually functional. Even though such recombination is possible molecularly, the long-term importance of this mechanism in creating new genome arrangements remains unknown because strains containing the illegitimately recombined [*rho*$^-$] mtDNA molecules are diploids unable to sporulate (because they are respiratory deficient), and hence unable to mate. Triploid crosses have been performed, but did not result in new gene order in the [*rho*$^+$] progeny (F. Michel, Ph.D. dissertation, Université Pierre et Marie Curie, Paris, France, 1982).

The situation is different with nascent [*rho*$^-$] mutants in which several differently deleted mtDNA molecules can be temporarily present together with normal mtDNA molecules. In such cases, intermolecular illegitimate recombination can create new gene orders without the need for subsequent crosses (114).

Such new arrangements, however, often result in very unstable genomes that are difficult to maintain because they tend to produce a high level of new [rho^-] mutants (56, 57).

III. AMPLIFICATION OF mtDNA SEQUENCE IN *ASPERGILLUS* AND *PODOSPORA* SPP.

Amplification of mtDNA sequences, in several aspects similar to that of [rho^-] mutants of *S. cerevisiae*, may be observed in filamentous fungi. "Ragged" mutants of *Aspergillus amstelodami* and senescent cultures of *Podospora anserina* are the best-documented examples. Two main differences between *S. cerevisiae* and filamentous fungi relevant to this chapter should first be recalled. While *S. cerevisiae* is a unicellular organism for which cytoplasmic segregation is accomplished at each cell generation, the filamentous fungi are coenocytic, and this allows continuous cytoplasmic exchanges between cells along a hypha and also between different hyphae via anastomoses. The second major difference, as already pointed out, relates to the energy metabolism: in contrast to *S. cerevisiae*, filamentous fungi are strict aerobes, unable to dispense with mitochondrial functions.

A. Amplification in *Aspergillus amstelodami*

The ragged mutants in *A. amstelodami* are characterized by a slow and erratic growth of the mycelium and by a deficit in cytochromes *b* and aa_3. These traits are transmitted through heterocaryosis independently of nuclear genetic markers (66).

mtDNA preparations from ragged mutants contain, in addition to the intact wild-type mitochondrial genome, a set of subgenomic circular multimeric molecules. Electron microscope, restriction endonuclease, and molecular hybridization studies demonstrated that these molecules are made up of head-to-tail repeated sequences (0.9 to 2.7 kb) that are homologous to single defined segments of the mitochondrial (mt-) chromosome. The size and the origin of the repeated units differ from one mutant to another. Two different regions of the wild-type mt-chromosome (called 1 and 2) may be recovered in amplified molecules. Of seven ragged mutants tested, only one contained mtDNA amplified from region 1, while the other six contained amplifications from region 2. The amplified DNAs in these six mutants all share a common segment of 215 bp (87, 88).

These specific regions have been partially sequenced, and the exact location on the mt-chromosome of the amplified sequences has been determined.

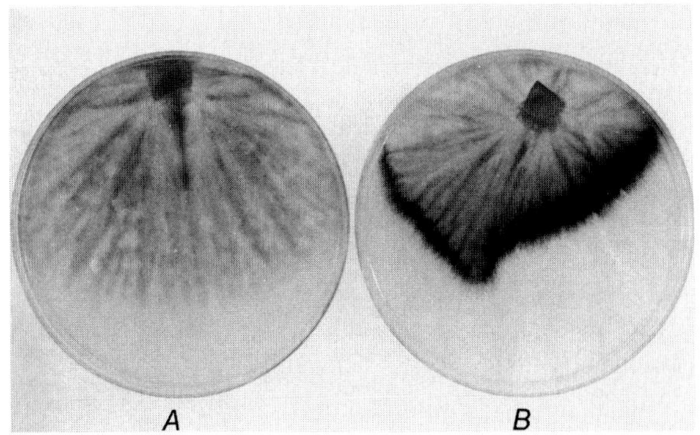

Figure 2. (A) Juvenile and (B) senescent mycelium of *P. anserina* grown on petri dishes. Note the strong pigmentation of cells on the arrested edge of the senescent mycelium.

The results can be summarized as follows. (i) No particular sequences such as direct repeats that suggest a simple mechanism for excision have been detected at the ends of the amplified sequences. (ii) The heads and tails of the amplified sequences may lie either within or outside protein-coding sequences. (iii) The 215-bp segment common to DNAs amplified from region 2 contains a nucleotide sequence that is palindromic, like the putative mitochondrial origins of replication (81, 83).

B. Amplification in Senescent Cultures of *Podospora anserina*

In *P. anserina*, another filamentous ascomycete, indefinite vegetative growth has not been possible, regardless of the wild-type strain or conditions of culture used (122). The irreversible arrest of growth is preceded by a stage called an incubation period in which the mycelium is already committed to death but does not display a deleterious phenotype. Only during the very last few days of growth (continuous growth occurs for several weeks) does the mycelium display defects such as modifications of hyphal morphology, development of a dark pigmentation (Fig. 2), and a strong decrease in cytochromes *b* and aa_3. This phenomenon has been termed senescence, and a mycelium approaching the arrest of growth is said to be in the senescent state (8, 100, 131).

The senescent state is maternally inherited in sexual crosses and is contagious vegetatively. It can also be transmitted from one mycelium to another through anastomoses, independently of migration of nuclei (100, 101, 123).

While DNA isolated from mitochondria of young mycelia always contains randomly broken linear molecules of various sizes and a few 90-kb circular molecules corresponding to the complete wild-type mt-chromosome, DNA isolated from senescent mycelia contains numerous small circular

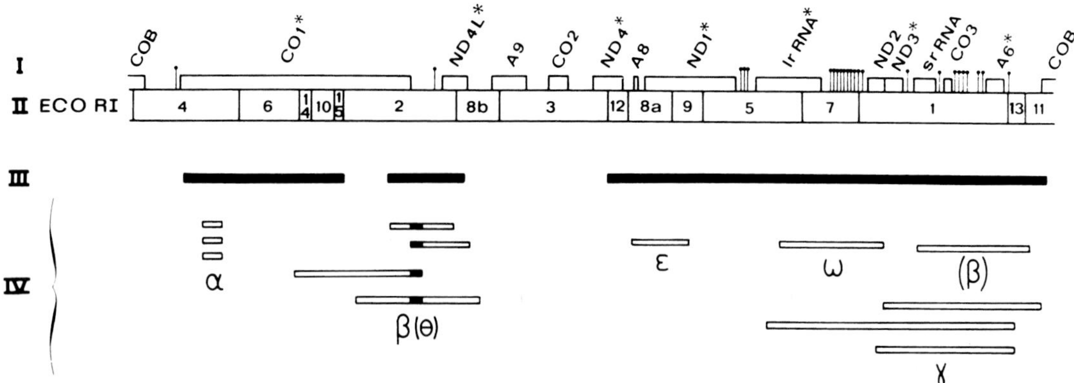

Figure 3. Composite map of the *P. anserina* mt-chromosome. This map is a linear representation of the 94-kb circular chromosome. (Line I) Localization of genes: *COB* for cytochrome *b*; *CO1*, *CO2*, and *CO3*, for subunits 1, 2, and 3 of cytochrome oxidase; *A6*, *A8*, and *A9* for subunits of ATPase; *ND1*, *ND2*, *ND3*, *ND4*, *ND4L*, and *ND5* for subunits of NADH dehydrogenase; *srRNA* and *lrRNA* for small and large rRNAs. Pins indicate positions of known tRNA genes. Starred symbols indicate genes which are known to contain an intron(s). (Line II) *Eco*RI restriction map. (Line III) Thick bars indicating the regions of the chromosome whose sequence has been determined. (Line IV) Schematic localization of senDNAs. The region common to all senDNAs originating from region β is indicated in black. When two different nomenclatures have been used, that of D. J. Cummings is given in parentheses. Data are from references 39, 75, 78, 79, 82, 118, 139, 149, and 150 and from unpublished work of D. J. Cummings and of C. Vierny.

molecules whose sizes make up a multimeric series. These results, deduced by electron microscopy, were confirmed by studies with restriction endonucleases which showed that a head-to-tail, highly reiterated sequence is present in every senescent culture, the size of the repetition unit being in good agreement with the length of the monomeric circular molecule (37, 38, 133). The size and the restriction map of the amplified sequence differ from one senescent culture to another, even in the case of different subcultures derived from the same mycelium (11, 149). Such amplified DNA was found in all the senescent cultures so far studied and was called senDNA.

In all cases studied, Southern hybridization experiments demonstrated that the senDNA monomer is a part of the complete mt-chromosome of young mycelia (11, 74, 82, 149). The location of the senDNA monomers on the restriction and genetic map of the mt-chromosome is shown in Fig. 3.

senDNA α, also called plDNA, is 2,539 bp long and corresponds exactly to the first intron of the *co1* gene (118). This intron belongs to group II (104). Like most fungal mitochondrial introns, it contains a long open reading frame adjacent to and in the same reading frame as the upstream exon. The protein encoded in this intron is closely related to proteins encoded in other class II fungal introns (86, 118) and has scattered but significant homologies with retroviral and retrotransposon reverse transcriptases (106). Amplification of this intron sequence is found in most of the senescent cultures derived from the short-lived *smt*$^-$ and *A* wild-type strains, while it is less often observed in cultures derived from the longer-lived *smt*$^+$ wild-type strain or the *cap*^{r}*1* mitochondrial mutant.

senDNAs derived from other regions have a monomer size varying from 4 to 20 kb. Each of 13 independent senDNAs from the β region contains a common 1,100-bp-long segment (78). The sequence of this segment shows that it comes from the intergenic region downstream of *co1* gene in which intrachromosomal recombination seems to occur frequently (76, 78; C. Vierny, personal communication).

The only property common to all known senDNAs is the presence of at least one intronic open reading frame (12). However, the mitochondrial genome of *P. anserina* is very rich in introns, and it is thus difficult to know the significance of this observation.

In contrast to the situation in ragged mutants of *A. amstelodami*, long mtDNA molecules coexist with the short amplified sequence in senescent cultures of *P. anserina*, but they differ from the wild-type mtDNA. Such molecules also differ from one senescent culture to another and represent several populations. Among them, molecules containing the rRNA-tRNA region of the mt-chromosome are most abundant. Extensive study of these molecules has not been carried out. It is clear, however, that they result from rearrangements of the wild-type mt-chromosome (38, 82, 86).

Mechanisms by which the first molecule of senDNA separates from the mt-chromosome are not fully understood. In some cases, intramolecular illegitimate recombination between short homologous sequences seems most likely. In the case of senDNA ε, for instance, an 11-bp direct repeat sequence is found in the mt-chromosome at the boundaries of the senDNA sequence (39). However, in other cases, no such repeated sequence is found. The involvement of a short consensus sequence was suggested recently for the excision process (139). In the case of senDNA α, for which neither a direct repeat sequence nor the consensus sequence is present, an alternative mechanism based on the intronic nature of the sequence can be proposed. The first molecule of senDNA might originate from splicing at the DNA rather than the RNA level (13, 39, 80, 81, 121). Alternatively, reverse transcription of the excised intronic RNA sequence or of a precursor could account for the formation of the first senDNA molecule, which would be subsequently replicated.

Mechanisms responsible for the amplification of senDNAs are not known either. Autonomous replication of senDNA molecules is the simplest mechanism and has been implied by most authors. However, in filamentous fungi, the origins of replication of the wild-type mt-chromosome have not been localized and thus one cannot decide whether a sequence is able to replicate autonomously in mitochondria (see discussion in reference 125). In the case of senDNA α, amplification through reverse transcription of numerous RNA copies has been invoked. In agreement with this hypothesis, the presence of a reverse transcriptase activity in aged cultures of *P. anserina* has been reported (135).

C. Causal Relationship between Amplification and Rearrangements and the Senescent Phenotype

A strict correlation between the presence of senDNA and the senescent state of the mycelium has been observed during the study of more than 50 independent senescent cultures examined in the laboratory (L. Belcour, unpublished observations). Long-term ethidium bromide treatment of senescent cultures completely reverses the senescent state (perhaps by preferential inhibition of senDNA replication), and ethidium bromide-rejuvenated mycelia recover both a normal life span and a wild-type mtDNA (9). They do not contain any senDNA molecules detectable by DNA-DNA hybridization (77).

Other results suggesting that modifications of mtDNA are responsible for senescence came from the study of mutants selected as escaping senescence. We can summarize the data concerning these mutants as follows. (i) Most mutations allowing a senescent mycelium to resume growth are in mtDNA (10, 76, 78, 80, 81, 141). (ii) These mutations are located in the regions of the mt-chromosome from which senDNAs are excised (13). (iii) About one-third of the mutations are deletions of intron α. However, the boundaries of the deletions do not correspond with that of the intron at the nucleotide level (13, 146).

Although the amplification of a short mtDNA sequence as an extrachromosomal element may be necessary for senescence, it is probably not sufficient, as can be deduced from the following observations. (i) Certain mtDNA sequences in *P. anserina* can be amplified without giving rise to senescence (140). This is the case for some longevity mutants which contain a very short sequence (68 to 368 bp) that is highly amplified as tandem repeats. (ii) The size of the senDNA monomeric unit (more than 2.5 kb) is several times larger than that needed if it contained only an efficient origin of replication. (iii) The existence in senescent cultures of altered mt-chromosomes in addition to amplified senDNA cannot be explained by the autonomous replication of senDNA molecules. This set of data suggests that senDNAs contain information involved in the expression of senescence in addition to a sequence needed for its amplification (78). Since all senDNAs have an intron open reading frame, the overproduction (due to amplification) of the translation product of these sequences might be involved in senescence. As some intron-encoded proteins in yeasts have endonuclease activity (see below) and perhaps reverse transcriptase (see below) or RNA maturase activities as well (6, 7, 26, 70) and as they are normally present at a very low level in wild-type cells, their overproduction might be responsible for the alterations of mtDNA of senescent cultures (12).

The induction of senescence through the transformation of *P. anserina* protoplasts with pBR322 recombinant plasmids containing senDNA α (or a part of it) has been reported (134) but could not be repeated under rigorously controlled conditions (125). An efficient method for the transformation of mitochondria is thus needed to test definitively the proposed causative role of senDNA amplification in senescence.

IV. mtDNA REARRANGEMENT MEDIATED BY PLASMIDS IN FILAMENTOUS FUNGI

A. Plasmids in Filamentous Fungi

We reserve the term plasmid for autonomously replicating elements sharing little or no homology with the resident normal genetic information. Accordingly, neither defective mitochondrial genomes (like those in [*rho*$^-$] mutants of *S. cerevisiae* or

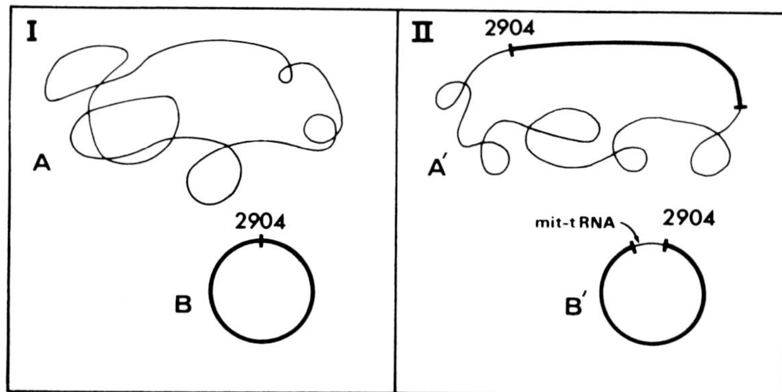

Figure 4. Main modifications of intramitochondrial DNAs in mutants selected by continuous vegetative growth of *N. crassa* and *N. intermedia* containing Mauriceville and Varkud plasmids, respectively. (I) Normal situation: (A) normal mt-chromosome (thin line); (B) mitochondrial plasmid, Mauriceville or Varkud (thick line). Position 2904 corresponds to the transcription start point of the plasmid. (II) Mutant cultures: (A′) mt-chromosome modified by integration of the plasmid; (B′) plasmid DNA modified by integration of a small piece of mtDNA (from data in reference 1).

senDNA molecules in senescent cultures of *P. anserina*) nor linear extrachromosomal rDNA of *Tetrahymena* spp. will be called plasmids (see also reference 84).

Bona fide plasmids have been discovered in several species of filamentous fungi (see references 84 and 110 for reviews). Most are located inside mitochondria; some are circular (the LaBelle, Varkud, and Fidji plasmids of *Neurospora intermedia* [136], the Mauriceville plasmid of *N. crassa* [36], or the Hanalei plasmid of *Neurospora tetrasperma* [113]), while others are linear (plasmids from the phytopathogens *Claviceps purpurea* [138] and *Fusarium solani* [126] and from the commercial mushroom *Agaricus* sp. [107]). Some linear plasmids possess terminal inverted repeats reminiscent of mitochondrial plasmids found in higher plants, such as S1 and S2 in maize (95). Other plasmids have a nuclear or an undetermined cellular location, such as the linear plasmids found in *Ascobolus immersus* (62) and *Morchella conica* (102), or the kalilo element of *N. intermedia* (see below).

Most analyses of the transcriptional activity and information content of these plasmids has come from the Mauriceville and the closely related Varkud plasmids (111, 112). The Mauriceville plasmid, 3,581 bp long, has been completely sequenced. It contains conserved sequence elements resembling group I mitochondrial introns and a long open reading frame that encodes a 710-amino-acid protein. Yet, this protein shows scattered but significant homologies with reverse transcriptases and is thus similar to the product of group II introns. The codon usage in this reading frame is closer to that of intronic mitochondrial reading frames than that of mitochondrial genes. These data led to the suggestion that the Mauriceville and Varkud plasmids are related to the progenitors of present-day mitochondrial introns (112). The plasmid is abundantly transcribed and gives a full-length linear mRNA with discrete 3′ and 5′ ends. No translation product of this reading frame has been detected.

The biological role of fungal plasmids is not known, and it was believed that the presence of these plasmids had no influence on the growth rate or fertility of mycelia because no trait could be associated with the plasmids until recently. However, two instances have now been found in which such plasmids interact with the mitochondrial genome and cause impairment of growth or clonal vegetative death.

B. Integration of a Circular Mitochondrial Plasmid via Reverse Transcription in *Neurospora crassa* and *Neurospora intermedia*

Continuous vegetative growth of strains of *N. crassa* or *N. intermedia* containing the Mauriceville or the Varkud plasmid, respectively, gives rise frequently to "mutant cultures" displaying an abnormal phenotype of growth: slow growth rate at 25°C, stopping followed by restart ("stop-start"), or even clonal death at 37°C. In all cases studied, this growth behavior is correlated with a defect in mitochondrial functions (cytochrome *b* and aa_3 deficiency) and is maternally inherited in sexual crosses (1).

Modifications of both the free plasmid DNA and the mtDNA are observed in growth-impaired mutants, and a reverse transcriptase activity is thought to be responsible for both defects (Fig. 4). The accumulation of altered plasmids is a constant characteristic of these cultures. Most altered plasmids contain a segment from the mtDNA inserted exactly at the transcription start point of the plasmid (position 2904). Deletions in other positions in the plasmid are in some cases also associated with this modification. Determination of the nucleotide sequence near the RNA start point showed that, in the three cases studied, the insertion contained a mitochondrial tRNA sequence. In one instance, the 3′ end of that tRNA contained a C not present in the original mtDNA gene but added posttranscriptionally on the tRNA (as a part of the CCA 3′ tail). This is a strong suggestion that the tRNA sequence from the mtDNA becomes inserted into the plasmid by reverse transcription. mtDNA from most growth variants displayed altered stoichiometry of normal restriction fragments and the appearance of new fragments. This indicated large rearrangements and the existence of several competing populations of mt-chromosomes. In some cases, it has been possible to detect the presence of the plasmid sequence integrated within the mt-chromosome. The inserted sequence corresponds to the plasmid linearized at position 2904, i.e., at the start point of the major plasmid transcript (1).

C. Senescence in *Neurospora intermedia* Mediated by a Linear Plasmid

A phenomenon of senescence in several aspects similar to that described in *P. anserina* occurs in some wild-type strains of *N. intermedia* isolated from the Hawaiian islands. As in *P. anserina*, the progressive deterioration of the vegetative growth ability is accompanied by a progressive decrease in cytochrome *b* and aa_3 content, indicating alterations of mitochondrial functions. Crosses between "senescence-prone" and "stable" strains of *N. intermedia* showed that this alternative property is maternally inherited. A cytoplasmic factor named kalilo is thus responsible for this trait (16, 64).

Comparative studies of mtDNA from juvenile and senescing cultures at various stages of the degenerative process showed that mt-chromosomes containing an insertion of a piece of DNA of foreign origin accumulate progressively in senescing cultures (16). For instance, a restriction fragment, *Eco*RIE, not present in mtDNA of stable *N. intermedia* strains and displaying no homology with it accumulates during senescence to a point where it seems stoichiometric with normal mtDNA restriction fragments. Further tests demonstrated that this fragment is actually integrated in the mt-chromosome in continuity with other mitochondrial DNA sequences.

The free kalilo element is a 9-kb-long, linear plasmid with inverted terminal repeats that integrates in the mtDNA of the senescence-prone strains of *N. intermedia*. The site of integration may be different from one senescent culture to another, but the large rRNA gene seems to be a preferential target. Preferential replication of the mt-chromosome containing the kalilo insert is thought to be responsible for the arrest of growth, since insertions have been observed in indispensable genes (16, 18). However, integration of the kalilo element is not the only modification undergone by mtDNA: multiple integrations, generation of inverted duplications at the ends of the integrated plasmid, and other incompletely characterized rearrangements of the mt-chromosome were also observed (16).

The origin of such a transposable piece of DNA has been an important question. DNA extracted from mitochondrial, nuclear, or cytosolic subcellular fractions from mycelia of senescence-prone and stable strains of *N. intermedia* has been hybridized with probes derived from the integrated kalilo DNA. Unexpectedly, hybridizations revealed a free, 9-kb, linear kalilo plasmid in the nucleus (or in strong association with the nucleus) of senescing strains. This plasmid is absent from strains which do not express senescence (18). It has a restriction map identical to that of the DNA integrated in the mt-chromosome of senescent cultures. It is present in senescence-prone strains of *N. intermedia* at high copy number (about 10 units per nucleus). In summary, senescence results from the integration in mtDNA of a linear plasmid originating from outside the mitochondria followed by overreplication of the mutant mtDNA. The senescent phenotype may result either from the disruption of essential mitochondrial genes or from the expression of the kalilo genes when integrated into mtDNA.

However, the nuclear location of the free kalilo plasmid contradicts the maternal inheritance of the senescence-prone trait. In addition, efficient transmission of the kalilo element can occur, together with transfer of mtDNA, in the absence of any nuclear exchange (15). Thus, the intracellular location of the genetically active kalilo plasmid remains enigmatic. Further studies are needed to elucidate the location and coding ability of kalilo.

D. Conclusion and Speculations on Cytoplasmically Inherited Mycelial Degenerative Syndromes

As we have seen, the instability of the mitochondrial genome is responsible in filamentous fungi for various mycelial degenerative syndromes. In each case, there is an accumulation of defective mt-chromosomes. In some cases, the defective genomes consist of a small part of the normal genome, while in others, they consist of mt-chromosomes in which a plasmid has integrated.

In the case of the ragged mutants of *A. amstelodami*, the amplification of a mitochondrial sequence seems to be by itself the cause of the erratic growth phenotype, probably by competing for replication against the wild-type mtDNA (87). The rationale for this model is that, in filamentous fungi which are coenocytic and obligate aerobes, mutant mtDNA molecules similar to suppressive [*rho*$^-$] DNA of *S. cerevisiae* would: (i) invade normal mycelium using fusion between mitochondria, cytoplasmic continuity between adjacent cells in single hyphae, and anastomoses between hyphae, and (ii) cause the cessation of growth when the concentration of the functional genome falls below a critical threshold (8, 17).

In other cases, several types of incompletely characterized modifications of mtDNA were detected in the impaired mycelia. Suppressiveness has been suggested as a possible explanation for senescence in *P. anserina* (8, 38, 74) and is a favored interpretation for the kalilo-determined senescence of *N. intermedia* (18). However, a toxic effect due to the product encoded by the amplified sequence or its possible recombinogenic properties cannot be excluded (78).

While rearrangements of the mtDNA in *Neurospora* spp. are mediated by genetic elements which are not integral parts of the normal genome (the kalilo, Mauriceville, or Varkud plasmids), in *P. anserina* no foreign nucleic acid is involved in the rearrangements. Strikingly, however, the element most frequently involved in senescence of *P. anserina*, the senDNA α, has a coding capacity for a reverse transcriptase-related protein, as do foreign elements of *Neurospora* spp. While all wild-type strains of *P. anserina* exhibit senescence (100) and contain a sequence homologous to senDNA α (80), *Neurospora* spp. are not known to contain any group II intron open reading frame in their mtDNA and can grow indefinitely unless they contain a plasmid. This raises the possibility that the pathway for senescence could be similar in both *Neurospora* and *Podospora* spp. and involves a reverse transcriptase-like activity.

V. SPLIT MITOCHONDRIAL GENOME IN FILAMENTOUS FUNGI

The plasticity of the mitochondrial genome in fungi is also illustrated by its frequent distribution among several subgenomic molecules. In some mutants of *N. crassa* and *P. anserina*, the mitochondrial genes are on two different molecules instead of on a unique chromosome as in the wild-type strain. Stopper mutants of *N. crassa* are extranuclear mutants characterized by an irregular stop-start phenotype of growth and deficiencies of cytochromes *b* and aa_3 (17). All stopper mutants contain several populations of defective mtDNA molecules whose relative concentrations vary depending on the growing versus the nongrowing stage of the mycelium. During the phase of stopped growth, a circular molecule of about 21 to 24 kb containing the rRNA and most of the tRNA genes predominates (17, 40, 65). However, the size and boundaries of this subgenomic DNA differ somewhat among the different stopper mutants studied.

The stopper mutant E35 also contains a second population of molecules with most mtDNA sequences but is detectable only when mycelium is in the rapidly growing phase. Both types of molecules in the E35 mitochondrial genome lack a 5-kb fragment containing the gene for subunit 5 of NADH dehydrogenase (41). The deletion breakpoints of two independent subgenomic chromosomes of E35 were located in direct repeats that consist almost completely of a nearly pure stretch of nonalternating G · C pairs. Interestingly, in another stopper mutant, the small predominant chromosome is postulated to result from intragenic recombination between two identical Met-tRNA genes (65).

In *P. anserina*, the presence of circular DNA molecules with a contour length several times smaller than that of the wild-type chromosome was detected in mitochondrial DNA from some cytoplasmic mutants (38). More recently, two mitochondrial mutants selected as mycelia having resumed growth from senescent cultures were shown to have a mitochondrial genome composed of at least two subgenomic mt-chromosomes. In both mutants, the breakpoint has occurred in a region of the mt-chromosome shared by all senDNA β's (Fig. 3). An unexpected characteristic of the mtDNA from these mutants is the insertion of stretches of A's and T's a few base pairs long at some of the new junctions (76, 78).

The existence of a split mitochondrial genome in some mutants of filamentous fungi is reminiscent of the normal situation in higher plants in which the mt-chromosome is organized in several physically distinct circular molecules (98, 119).

VI. mtDNA REARRANGEMENT UNDER THE CONTROL OF INTRONS

The mobility of group I and group II introns in organelle genomes is suggested by comparisons between different species of yeasts and fungi or between different strains of the same species. Such comparisons reveal (i) that a given intron may be present or absent at a given site in a gene, and (ii) that closely related introns can be at various locations while different introns can be at similar locations. For example, the intron of the large rRNA gene in *S. cerevisiae* (r1 intron) is more closely related to the second intron of the cytochrome *b* gene in *N. crassa* than to the intron of the large rRNA gene in *N. crassa* (which is inserted at the same site in the rDNA gene) (25). In *Kluyveromyces fragilis*, the large rRNA gene has no intron, but this species possesses an intron very closely related to the r1 intron of *S. cerevisiae* in the gene for subunit 9 of ATPase (50). Similarly, the intron of the cytochrome oxidase subunit I gene in *Schizosaccharomyces pombe* is much more closely related to the same intron of this gene in *Aspergillus nidulans* than are the exons of these two genes. This suggests that the introns were recently inserted at equivalent positions of already divergent genes (85, 147). In fact, cases of sequence relationships between introns that do not follow commonly accepted phylogenetic relationships extend well beyond the mitochondrial genomes. For example, group I introns closely resembling those of filamentous fungi are found in T4 bacteriophage (105). However, despite numerous suggestions as above, the idea that introns are mobile remained speculative since there was no efficient mitochondrial transformation system, and so it was not possible to devise an experiment to test whether a given intron could transpose from one location to another. There is now, however, direct evidence showing that group I introns encode enzymes for their insertion at appropriate sites.

A. The [*omega*] System

The large rRNA gene contains an intron in the segment specifying domain V of the 21S rRNA molecule in many laboratory strains of *S. cerevisiae* (called [*omega*$^+$] strains) but not in all ([*omega*$^-$] strains). The interesting feature of the [*omega*] system is that the simultaneous presence in a zygote of intron-plus and intron-minus copies of the 21S rRNA gene (after mating) results in the presence of an intron-plus allele in almost all progeny. Both genetic and molecular experiments demonstrate that this is due to the insertion of the intron into the previously intronless copies of the gene rather than the selection for intron-plus copies (47, 49, 72). Sequence analysis shows that the insertion of the intron always occurs at the same site (46). This insertion is mediated by a double-strand break within the intronless gene and can be interpreted by the classic double-strand break repair model (137) proposed for yeast nuclear genes (51, 152). Consistent with this model, the two flanking exons are coconverted with the intron insertion over distances of a few hundreds of nucleotides at frequencies that decrease with distance from the site of the double-strand break. Single base substitutions within the 21S rRNA gene near the site of the double-strand break prevent the formation of the break and the subsequent insertion of the intron (51, 52, 152).

B. The r1 Intron Encodes a Double-Strand Endonuclease that Cleaves the Intronless Gene at a Specific Site

Of particular importance for our understanding of the role of mitochondrial introns in genomic rearrangements is the finding that the r1 intron itself encodes the protein that makes the double-strand break in the intronless gene (Fig. 5). This was first suggested by the fact that mutations within the intron open reading frame render the intron unable to insert within the normal intronless gene (72, 99). Expression of the intron-encoded protein in *Escherichia coli* from an engineered reading frame was then used to demonstrate that this protein is itself the double-strand endonuclease, and needs no additional factor to specifically cleave substrate DNA in vivo or in vitro (34). It was further shown, using the enzyme made in *E. coli*, that the endonuclease recognizes the unique sequence at the junction between exons in the intronless gene and generates a 4-bp staggered cut within this sequence at the exact point of intron insertion (35, 50). The endonuclease is unable to recognize the two intron-exon junctions in the intron-plus gene. The intron thus becomes irreversibly inserted and ready to invade new intronless copies of the gene. Interestingly, the recognition sequence for this endonuclease, as determined by random in vitro mutagenesis, extends over more than 18 bp, ensuring a very high specificity for just a single site. Several different mutants of the 18-bp recognition site are still cleaved in vitro, although much less efficiently. Such a minor sequence degeneracy is insufficient to observe mobility during the time course of an experiment because the complexity of the mitochondrial genome is too limited for a secondary site to exist. However, the sequence degeneracy is more than sufficient to predict mobility of that intron on an evolutionary time scale.

The [*omega*] system represents, therefore, the first known case in which an intron encodes the enzyme responsible for its propagation into mitochondrial genomes. This raises evolutionary questions of how [*omega*$^-$] strains have persisted and where the intron originated. The r1 intron is found in only a few isolates in the entire *Saccharomyces* genus, while sequences homologous to this intron are present in most isolates of *Kluyveromyces* spp. This suggests that this intron may have originated in

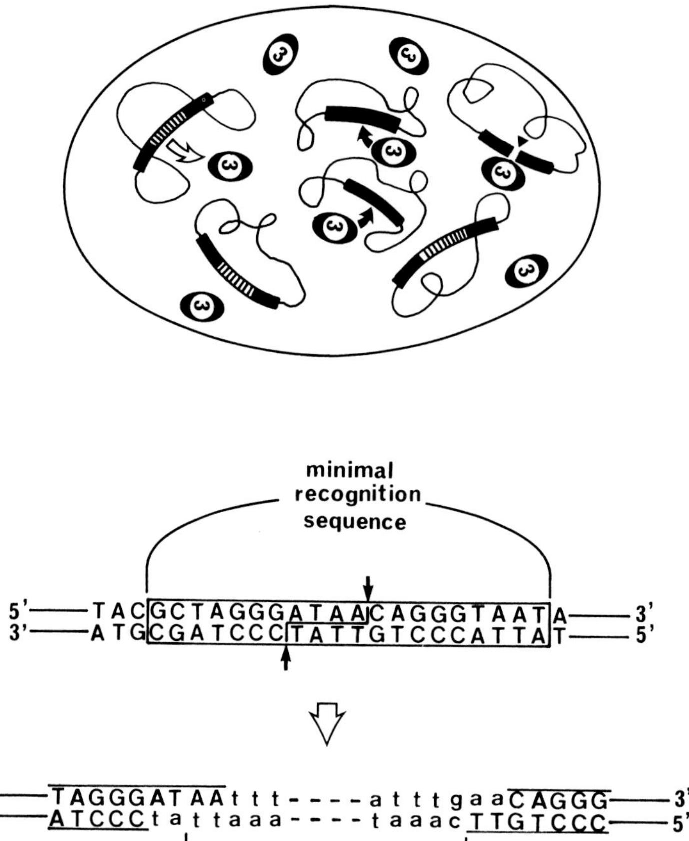

Figure 5. The [*omega*] system: molecular anatomy of a mobile intron. Top: Schematic representation of mtDNA molecules (thin lines) in zygotes issued from a cross between [*omega*$^+$] and [*omega*$^-$] cells. The 21S rRNA gene (solid bar) contains an intron (hatched) on some mtDNA molecules and not on others. The [*omega*] endonuclease (ω) translated from the intron open reading frame (open arrow) recognizes a specific sequence of the intronless genes (solid arrows) and cleaves within this sequence (▼). Bottom: Part of the sequence of the intronless 21S rRNA gene showing the minimal recognition site of the [*omega*] endonuclease (box) and the cuts generated on each strand (arrows). The double-strand break initiates the whole process, during which a copy of the intron becomes stably inserted into the former cut site. Lowercase letters represent the minimal extent of newly synthesized nucleotides. When donor and target regions differ by base substitutions, correction of flanking exon sequences accompanies intron insertion with an efficiency which decreases with distance from the cutting site.

Kluyveromyces spp. and is presently invading *Saccharomyces* spp. (71).

C. The [*omega*]-Like Mobility Is a Property Common to Several (All?) Group I Introns

The [*omega*] system seemed unique for many years, raising questions about the generality of such a phenomenon. However, several other group I introns of mitochondria and chloroplasts from different organisms have recently been found to behave similarly to [*omega*], indicating that this proliferation by induced cleavage and repair from the intron-containing allele is an intrinsic property of many (if not all) group I introns.

The best example comes from the large rRNA gene of the chloroplast of *Chlamydomonas eugametos*. This gene contains six group I introns, one of which (intron 5) behaves like [*omega*] in interspecific crosses with *Chlamydomonas moewusii*, which contains an intronless rRNA gene (91, 92, 142). Like the r1 intron in *S. cerevisiae*, intron 5 of *C. eugametos* contains a long open reading frame (more than 200 codons), the translation product of which has not yet been characterized. However, the product seems a likely candidate for a double-strand endonuclease. If so, it is interesting to note that intron 5 is inserted in the large rRNA gene at a site different from that of the r1 intron of *S. cerevisiae*; hence, the two recognition sequences must be totally different. This implies that endonucleases equivalent to that specified by [*omega*] can differ radically in specificity.

The genus *Chlamydomonas* offers another example of an [*omega*]-like system in the mitochondrial genome (23). An insert of ca. 1 kb in the mitochondrial genome of *Chlamydomonas smithii* that is probably an intron of the cytochrome *b* gene is absent in the interfertile species *Chlamydomonas reinhardtii*. The insert spreads unidirectionally into the *C. reinhardtii* genome in crosses, and partial sequencing of the insert reveals an open reading frame that would encode a protein closely related to the [*omega*] endonuclease (R. Matagne, L. Colleaux, and B. Dujon, unpublished results).

Filamentous fungi offer similar cases of unidirectional spreading of inserts among mitochondrial genomes. Crosses between subspecies of *A. nidulans* differing by six insertions/deletions reveal that three of them are found in all the progeny (53). A similar trait has been described in *Coprinus cinereus* for two different insertions (1.2 kb each) in the region of the cytochrome oxidase subunit I gene (one of them is probably an intron of this gene) (54). Although these systems have not been studied at the molecular level, their present description fits all points of the [*omega*] system.

Finally, the yeast *S. cerevisiae* itself probably harbors more than one mobile intron. Studies of a variety of interfertile species have recently led to the discovery that the fourth intron of the cytochrome oxidase subunit I gene behaves in an [*omega*]-like fashion in crosses with an intronless strain (P. Perlman and co-workers, unpublished results). Interestingly, this intron contains a long open reading frame, the translation product of which was proposed to be a "maturase" involved in RNA splicing (45). The sequence of this protein, however, shares similarities with that of the [*omega*] endonuclease (71), and the protein was recently found to be a sequence-specific endonuclease in vitro that cleaves a sequence different from the [$omega^-$] site (A. Delahodde and C. Jacq, personal communication).

D. Group II Introns May Encode Reverse Transcriptases

Group II introns contain open reading frames differing from those of their group I counterparts by their longer size and different amino acid sequences. Like the group I introns, however, their translation products remain below detectable levels in mitochondria under normal conditions. As already mentioned, amino acid sequence alignment of several group II intron open reading frames revealed significant homology with retroviral reverse transcriptases (106). However, the direct evidence of reverse transcriptase activity is still missing. One problem resides in the difficulty of expressing complete group II intron-encoded proteins in *E. coli*. Only the protein encoded by the group II intron of *S. pombe* has been successfully expressed in native form (50). But even in this case, the efficiency remained very low. Despite this difficulty, the indirect evidence already discussed supports the hypothesis of reverse transcription in mitochondria under the control of group II introns.

VII. CONCLUSION

In preparing this chapter, we deliberately limited ourselves to phenomena directly relevant to DNA instabilities and rearrangements. It is clear, however, that many such phenomena rely on biochemical mechanisms and pathways that also play other fundamental roles in mitochondria. Among those mechanisms, general replication and recombination pathways are obvious candidates. Similarly, some of the intron-encoded proteins, although members of a family of proteins that act at the DNA level, also seem to act as "maturases" at the RNA level.

Whatever the varied causes of mtDNA instabilities, the consequences for the cells or diverse organisms in which they occur are tremendous. In many cases, alterations of mtDNA functions simply result in death. In other cases, severe phenotypes are observed. In humans, for example, deletions in mtDNA are associated with mitochondrial myopathies (69, 93). In these cases, deletions were found in muscle cells, while blood cells remained normal, showing that individuals are mosaics. In addition, muscle cells themselves exhibited heteroplasmy, with both normal and deleted mtDNA molecules present simultaneously. Mitochondrial mosaicism has also been observed in animals and can be transmitted to the progeny, probably because female germ cells are themselves heteroplasmic (22, 67, 68). In plants, where mtDNA rearrangements are common, mtDNA polymorphism has been invoked in cases of cytoplasmic male sterility (129).

Although it is too early to draw general conclusions on the molecular mechanisms involved in the varied cases observed in different organisms, it is possible that findings on yeasts and fungi may help to understand observations made with higher eucaryotes. An example of this situation is perhaps the parallel that can be drawn between the intron-based [*omega*] system in yeast mitochondria and the phenomenon of paramutation described in maize more than 20 years ago (24).

Acknowledgments. This work was supported by grants from Institut National de la Santé et de la Recherche Médicale to B.D. (no. 861007) and to L.B. (no. 861002).

LITERATURE CITED

1. **Akins, R. A., R. L. Kelley, and A. M. Lambowitz.** 1986. Mitochondrial plasmids of *Neurospora*: integration into mitochondrial DNA and evidence for reverse transcription in mitochondria. *Cell* **47**:505–516.
2. **Anderson, S., A. T. Bankier, B. G. Barrel, M. H. L. De Bruijn, A. R. Coulson, J. Drouin, I. C. Eperon, D. P. Nierlich, B. A. Roe, F. Sanger, P. H. Schreier, A. J. H. Smith, R. Staden, and I. G. Young.** 1981. Sequence and organisation of the human mitochondrial genome. *Nature* (London) **290**:457–465.
3. **Anderson, S., M. H. L. De Bruijn, A. R. Coulson, I. C. Eperon, F. Sanger, and I. G. Young.** 1982. Complete sequence of bovine mitochondrial DNA: conserved features of the mammalian mitochondrial genome. *J. Mol. Biol.* **156**: 683–717.
4. **Arnberg, A. C., G. Van der Horst, and H. F. Tabak.** 1986. Formation of lariats and circles in self splicing of the precursor to the large ribosomal RNA of yeast mitochondria. *Cell* **44**:235–242.
5. **Baldacci, G., B. Chérif-Zahar, and G. Bernardi.** 1984. The initiation of DNA replication in the mitochondrial genome of yeast. *EMBO J.* **3**:2115–2120.
6. **Banroques, J., A. Delahodde, and C. Jacq.** 1986. A mitochondrial RNA maturase gene transferred to the yeast nucleus can control mitochondrial mRNA splicing. *Cell* **46**: 837–844.
7. **Banroques, J., J. Perea, and C. Jacq.** 1987. Efficient splicing of two yeast mitochondrial introns controlled by a nuclear encoded maturase. *EMBO J.* **6**:1085–1091.
8. **Belcour, L., and O. Begel.** 1978. Lethal mitochondrial genotypes in *Podospora anserina*: a model for senescence. *Mol. Gen. Genet.* **163**:113–123.
9. **Belcour, L., and O. Begel.** 1980. Life-span and senescence in *Podospora anserina*: effect of mitochondrial genes and functions. *J. Gen. Microbiol.* **119**:505–515.
10. **Belcour, L., O. Begel, A. M. Keller, and C. Vierny.** 1982. Does senescence in *Podospora anserina* result from instability of the mitochondrial genome?, p. 415–421. *In* P. P. Slonimski, P. Borst, and G. Attardi (ed.), *Mitochondrial Genes*. Cold Spring Harbor Laboratory, Cold Spring Harbor, N.Y.
11. **Belcour, L., O. Begel, M. O. Mossé, and C. Vierny.** 1981. Mitochondrial DNA amplification in senescent cultures of *Podospora anserina*: variability between the retained, amplified sequences. *Curr. Genet.* **3**:13–21.
12. **Belcour, L., F. Koll, C. Vierny, A. Sainsard-Chanet, and O. Begel.** 1986. Are proteins encoded in mitochondrial introns involved in the process of senescence in the fungus *Podospora anserina*?, p. 63–71. *In* Y. Courtois, B. Faucheux, B. Forette, D. L. Knook, and J. A. Tréton (ed.), *Modern Trends in Aging Research*. John Libbey Eurotext, London.
13. **Belcour, L., and C. Vierny.** 1986. Variable DNA splicing sites of a mitochondrial intron: relationship to the senescence process in *Podospora*. *EMBO J.* **5**:609–614.
14. **Bernardi, G.** 1982. The origins of replication of the mitochondrial genome of yeast. *Trends Biochem.* **7**:404–408.
15. **Bertrand, H.** 1986. The kalilo senescence factor of *Neurospora intermedia*: a mitochondrial IS-element derived from a nuclear plasmid, p. 93–103. *In* R. B. Wickner, A. Hinnebusch, A. M. Lambowitz, I. C. Gunsalus, and A. Hollaender (ed.), *Extrachromosomal Elements in Lower Eukaryotes*. Plenum Publishing Corp., New York.
16. **Bertrand, H., B. S.-S. Chan, and A. J. F. Griffiths.** 1985. Insertion of a foreign nucleotide sequence into mitochondrial DNA causes senescence in *Neurospora intermedia*. *Cell* **41**:877–884.
17. **Bertrand, H., R. A. Collins, L. L. Stohl, R. R. Goewert, and A. M. Lambowitz.** 1980. Deletion mutants of *Neurospora crassa* mitochondrial DNA and their relationship to the "stop-start" growth phenotype. *Proc. Natl. Acad. Sci. USA* **77**:6032–6036.
18. **Bertrand, H., A. J. F. Griffiths, D. A. Court, and C. K. Cheng.** 1986. An extrachromosomal plasmid is the etiological precursor of kalilo DNA insertion sequence in the mitochondrial chromosome of senescent *Neurospora*. *Cell* **47**:829–837.
19. **Bibb, M. J., R. A. Van Etten, C. T. Wright, M. W. Walberg, and D. A. Clayton.** 1981. Sequence and gene organization of mouse mitochondrial DNA. *Cell* **26**:167–180.
20. **Blanc, H., and B. Dujon.** 1980. Replicator regions of the yeast mitochondrial DNA responsible for suppressiveness. *Proc. Natl. Acad. Sci. USA* **77**:3942–3946.
21. **Bos, J. L., C. F. Van Kreijl, F. H. Ploegaert, J. N. M. Mol, and P. Borst.** 1978. A conserved and unique AT-rich segment in yeast mitochondrial DNA. *Nucleic Acids Res.* **5**: 4563–4578.
22. **Boursot, P., H. Yonekawa, and F. Bonhomme.** 1987. *Mol. Biol. Evol.* **4**:46–55.
23. **Boynton, J. E., E. H. Harris, B. B. Burkhart, P. M. Lamerson, and N. W. Gillham.** 1987. Transmission of mitochondrial and chloroplast genome in crosses of *Chlamydomonas*. *Proc. Natl. Acad. Sci. USA* **84**:2391–2395.
24. **Brink, R. A., E. D. Styles, and J. D. Axtell.** 1968. Paramutation: directed genetic change. *Science* **159**:161–170.
25. **Burke, J. M., C. Breitenberger, J. E. Heckman, B. Dujon, and U. L. RajBhandary.** 1984. Cytochrome b gene of *Neurospora* mitochondria: partial sequence and location of introns at sites different from those in *S. cerevisiae* and *A. nidulans*. *J. Biol. Chem.* **259**:504–511.
26. **Carignani, G., O. Groudinski, D. Frezza, E. Schiavon, E. Bergantino, and P. P. Slonimski.** 1983. An mRNA maturase is encoded by the first intron of the mitochondrial gene for the subunit 1 of cytochrome oxidase in *S. cerevisiae*. *Cell* **35**:733–742.
27. **Chambers, P., and E. Gingold.** 1986. A direct study of the relative synthesis of petite and grande mitochondrial DNA in zygotes from crosses involving suppressive petite mutants of *Saccharomyces cerevisiae*. *Curr. Genet.* **10**:565–571.
28. **Chang, D. D., and D. A. Clayton.** 1984. Precise identification of individual promoters for transcription of each strand of human mitochondrial DNA. *Cell* **36**:635–643.
29. **Chang, D. D., and D. A. Clayton.** 1986. Precise assignment of the light strand promoter of mouse mitochondrial DNA: a functional promoter consists of multiple upstream domains. *Mol. Cell. Biol.* **6**:3253–3261.
30. **Chang, D. D., and D. A. Clayton.** 1986. Precise assignment of the heavy strand promoter of mouse mitochondrial DNA: cognate start sites are not required for transcriptional initiation. *Mol. Cell. Biol.* **6**:3262–3267.
31. **Chomyn, A., P. Mariottini, M. J. W. Cleeter, C. I. Ragan, A. Matsuno-Yagi, Y. Hatefi, R. F. Doolittle, and G. Attardi.** 1985. Six unidentified reading frames of human mitochondrial DNA encode components of the respiratory chain NADH deshydrogenase. *Nature* (London) **314**:592–597.
32. **Christianson, T., and M. Rabinowitz.** 1983. Identification of multiple transcriptional initiation sites on the yeast mitochondrial genome by *in vitro* capping with guanyltransferase. *J. Biol. Chem.* **258**:14025–14033.

33. **Clary, D. O., and D. R. Wolstenholme.** 1985. The mitochondrial DNA molecule of *Drosophila yakuba*: nucleotide sequence, gene organisation and genetic code. *J. Mol. Evol.* **22**:252–271.
34. **Colleaux, L., L. d'Auriol, M. Betermier, G. Cottarel, A. Jacquier, F. Galibert, and B. Dujon.** 1986. Universal code equivalent of a yeast mitochondrial intron reading frame is expressed into *E. coli* as a specific double strand endonuclease. *Cell* **44**:521–533.
35. **Colleaux, L., L. d'Auriol, F. Galibert, and B. Dujon.** 1988. Recognition and cleavage site of the intron encoded *omega* transposase. *Proc. Natl. Acad. Sci. USA* **85**:6022–6026.
36. **Collins, R. A., L. L. Stohl, M. D. Cole, and A. M. Lambowitz.** 1981. Characterization of a novel plasmid DNA found in mitochondria of *N. crassa*. *Cell* **24**:443–452.
37. **Cummings, D. J., L. Belcour, and C. Grandchamp.** 1978. Etude au microscope électronique du DNA mitochondrial de *Podospora anserina* et présence d'une série multimérique de molécules circulaires de DNA dans les cultures sénescentes. *C.R. Acad. Sci.* **187**:157–160.
38. **Cummings, D. J., L. Belcour, and C. Grandchamp.** 1979. Mitochondrial DNA from *Podospora anserina*. II. Properties of mutant DNA and multimeric circular DNA from senescent cultures. *Mol. Gen. Genet.* **171**:239–250.
39. **Cummings, D. J., I. A. MacNeil, J. Domenico, and E. T. Matsuura.** 1985. Excision-amplification of mitochondrial DNA during senescence in *Podospora anserina*. DNA sequence analysis of three unique plasmids. *J. Mol. Biol.* **185**:659–680.
40. **De Vries, H., J. C. de Jonge, P. van't Stant, E. Agsteribbe, and A. Arnberg.** 1981. A "stopper" mutant of *Neurospora crassa* containing two populations of aberrant mitochondrial DNA. *Curr. Genet.* **3**:205–211.
41. **De Vries, H., B. Alzner-DeWeerd, C. A. Breitenberger, D. D. Chang, J. de Jonge, and U. L. RajBhandary.** 1986. The E35 stopper mutant of *Neurospora crassa*: precise localization of deletion endpoints in mitochondrial DNA and evidence that the deleted DNA codes for a subunit of NADH dehydrogenase. *EMBO J.* **5**:779–785.
42. **Dewey, R. E., C. S. Levings III, and D. H. Timothy.** 1986. Novel recombination in the maize mitochondrial genome produces a unique transcriptional unit in the Texas male sterile cytoplasm. *Cell* **44**:439–449.
43. **De Zamaroczy, M., and G. Bernardi.** 1986. The GC clusters of the mitochondrial genome and their evolutionary origin. *Gene* **41**:1–22.
44. **De Zamaroczy, M., G. Faugeron-Fonty, and G. Bernardi.** 1983. Excision sequences in the mitochondrial genome of yeast. *Gene* **21**:193–202.
45. **Dujardin, G., C. Jacq, and P. P. Slonimski.** 1982. Single base substitution in an intron of oxidase gene compensates splicing defects of the cytochrome b gene. *Nature* (London) **298**:628–632.
46. **Dujon, B.** 1980. Sequence of the intron and flanking exons of the mitochondrial 21S rRNA gene of yeast strains having different alleles at the *omega* and *rib1* loci. *Cell* **20**:185–197.
47. **Dujon, B.** 1981. Mitochondrial genetics and functions, p. 505–635. *In* J. Strathern et al. (ed.), *Molecular Biology of the Yeast Saccharomyces*. Cold Spring Harbor Laboratory, Cold Spring Harbor, N.Y.
48. **Dujon, B., and H. Blanc.** 1980. Yeast mitochondria minilysates and their use to screen a collection of hypersuppressive *rho*$^-$ mutants, p. 33–36. *In* A. M. Kroon and C. Saccone (ed.), *The Organization of the Mitochondrial Genome*. Elsevier, Amsterdam.
49. **Dujon, B., M. Bolotin-Fukuhara, D. Coen, J. Deutsch, P. Netter, P. P. Slonimski, and L. Weill.** 1976. Mitochondrial genetics. XI. Mutations at the mitochondrial locus *omega* affecting the recombination of mitochondrial genes in *Saccharomyces cerevisiae*. *Mol. Gen. Genet.* **143**:131–165.
50. **Dujon, B., L. Colleaux, A. Jacquier, F. Michel, and C. Monteilhet.** 1986. Mitochondrial introns as mobile genetic elements: the role of intron encoded proteins, p. 5–27. *In* R. B. Wickner, A. Hinnebush, A. M. Lambowitz, I. C. Gunsalus, and A. Hollaender, (ed.), *Extrachromosomal Elements in Lower Eukaryotes*. Plenum Publishing Corp., New York.
51. **Dujon, B., G. Cottarel, L. Colleaux, M. Betermier, A. Jacquier, L. d'Auriol, and F. Galibert.** 1985. Mechanism of integration of an intron within a mitochondrial gene: a double strand break and the transposase function of an intron encoded protein as revealed by *in vivo* and *in vitro* assays, p. 215–225. *In* E. Quagliarello et al. (ed.), *Achievements and Perspectives of Mitochondrial Research*, vol II. Elsevier/North-Holland, Amsterdam.
52. **Dujon, B., and A. Jacquier.** 1983. Organization of the mitochondrial 21S rRNA gene in *Saccharomyces cerevisiae*: mutants of the peptidyl transferase centre and nature of the omega locus, p. 389–403. *In* R. J. Schweyen et al. (ed.), *Mitochondria 1983*. DeGruyter, Berlin.
53. **Earl, A. J., G. Turner, J. H. Croft, R. B. G. Dales, C. M. Lazarus, H. Lünsdorf, and H. Küntzel.** 1981. High frequency transfer of species specific mitochondrial DNA sequences between members of the *Aspergillaceae*. *Curr. Genet.* **3**: 221–228.
54. **Economou, A., Y. Lees, P. J. Pukkila, M. E. Zolan, and L. A. Casselton.** 1987. Biased inheritance of optional insertions following mitochondrial genome recombination in the basidiomycete fungus *Coprinus cinereus*. *Curr. Genet.* **11**: 513–519.
55. **Edwards, J. C., D. Levens, and M. Rabinowitz.** 1982. Analysis of transcriptional initiation of yeast mitochondrial DNA in a homologous *in vitro* transcription system. *Cell* **31**:337–346.
56. **Evans, R. J., and G. D. Clark-Walker.** 1985. Elevated levels of petite formation in strains of *Saccharomyces cerevisiae* restored to respiratory competence. II. Organization of mitochondrial genomes in strains having high and moderate frequencies of petite mutant formation. *Genetics* **111**: 403–432.
57. **Evans, R. J., K. M. Oakley, and G. D. Clark-Walker.** 1985. Elevated levels of petite formation in strains of *Saccharomyces cerevisiae* restored to respiratory competence. I. Association of both high and moderate frequencies of petite mutant formation with the presence of aberrant mitochondrial DNA. *Genetics* **111**:389–402.
58. **Fangman, W. L., and B. Dujon.** 1984. Yeast mitochondrial genomes consisting of only AT base pairs replicate and exhibit suppressiveness. *Proc. Natl. Acad. Sci. USA* **81**: 7156–7160.
59. **Faugeron-Fonty, G., and C. Goyon.** 1985. Polymorphic variations in the ori sequences from the mitochondrial genomes of different wild-type yeast strains. *Curr. Genet.* **10**:269–282.
60. **Foury, F., and Kolodynski, J.** 1983. *pif* mutation blocks recombination between mitochondrial *rho*$^+$ and *rho*$^-$ genomes having tandemly arrayed repeat units in *Saccharomyces cerevisiae*. *Proc. Natl. Acad. Sci. USA* **80**:5345–5349.
61. **Foury, F., and Lahaye, A.** 1987. Cloning and sequencing of

the *PIF* gene involved in repair and recombination of yeast mitochondrial DNA. *EMBO J.* **6:**1441–1449.

62. **Francou, F.** 1981. Isolation and characterization of a linear DNA molecule in the fungus *Ascobolus immersus. Mol. Gen. Genet.* **184:**440–444.
63. **Gingold, E. B.** 1981. Genetic analysis of the products of a cross involving a suppressive "petite" mutant of *S. cerevisiae. Curr. Genet.* **3:**213–220.
64. **Griffiths, A. J. F., and H. Bertrand.** 1984. Unstable cytoplasms in Hawaiian strains of *Neurospora intermedia. Curr. Genet.* **8:**387–398.
65. **Gross, S. R., T.-S. Hsieh, and P. H. Levine.** 1984. Intramolecular recombination as a source of mitochondrial chromosome heteromorphism in *Neurospora. Cell* **38:**233–239.
66. **Handley, L., and C. E. Caten.** 1973. Vegetative death: a mitochondrial mutation in *Aspergillus amstelodami. Heredity* **31:**136.
67. **Hauswirth, W. W., and P. J. Laipis.** 1982. Mitochondrial DNA polymorphism in a maternal lineage of Holstein cow. *Proc. Natl. Acad. Sci. USA* **79:**4686–4690.
68. **Hauswirth, W. W., M. J. Van De Walle, P. J. Laipis, and P. D. Olivo.** 1987. Heterogenous mitochondrial DNA D-loop sequences in bovine tissue. *Cell* **37:**1001–1007.
69. **Holt, I. J., A. E. Harding, and J. A. Morgan-Hughes.** 1988. Deletions of muscle mitochondrial DNA in patients with mitochondrial myopathies. *Nature* (London) **331:**717–719.
70. **Jacq, C., J. Banroques, A.-M. Becam, P. P. Slonimski, N. Guiso, and A. Danchin.** 1984. Antibodies against a fused 'lacZ-yeast mitochondrial intron' gene product allow identification of the mRNA maturase encoded by the fourth intron of the yeast *cob-box* gene. *EMBO J.* **3:**1567–1572.
71. **Jacquier, A., and B. Dujon.** 1983. The intron of the mitochondrial 21S rRNA gene: distribution in different yeast species and sequence comparison between *Kluyveromyces thermotolerans* and *Saccharomyces cerevisiae. Mol. Gen. Genet.* **192:**487–499.
72. **Jacquier, A., and B. Dujon.** 1985. An intron encoded protein is active in a gene conversion process that spreads an intron into a mitochondrial gene. *Cell* **41:**383–394.
73. **Jacquier, A., and F. Michel.** 1987. Multiple exon-binding sites in class II self-splicing introns. *Cell* **50:**17–29.
74. **Jamet-Vierny, C., O. Begel, and L. Belcour.** 1980. Senescence in *Podospora anserina*: amplification of a mitochondrial DNA sequence. *Cell* **21:**189–194.
75. **Jamet-Vierny, C., O. Begel, and L. Belcour.** 1984. A 20×10^3-base mosaic gene identified on the mitochondrial chromosome of *Podospora anserina. Eur. J. Biochem.* **143:** 389–394.
76. **Koll, F., O. Begel, and L. Belcour.** 1987. Insertion of short poly d(A) d(T) sequences at recombination junctions in mitochondrial DNA of *Podospora. Mol. Gen. Genet.* **209:** 630–632.
77. **Koll, F., O. Begel, A. M. Keller, C. Vierny, and L. Belcour.** 1984. Ethidium bromide rejuvenation of senescent cultures of *Podospora anserina*: loss of senescence-specific DNA and recovery of normal mitochondrial DNA. *Curr. Genet.* **8:** 127–134.
78. **Koll, F., L. Belcour, and C. Vierny.** 1985. A 1100bp sequence of mitochondrial DNA is involved in senescence process in *Podospora*: study of senescent and mutant cultures. *Plasmid* **14:**106–117.
79. **Kück, U., and K. Esser.** 1982. Genetic map of mitochondrial DNA in *Podospora anserina. Curr. Genet.* **5:**143–147.
80. **Kück, U., B. Kappelhoff, and K. Esser.** 1985. Despite mtDNA polymorphism the mobile intron (plDNA) of the *CO1* gene is present in ten different races of *Podospora anserina. Curr. Genet.* **10:**59–67.
81. **Kück, U., H. D. Osiewacz, U. Schmidt, B. Kappelhoff, E. Schulte, U. Stahl, and K. Esser.** 1985. The onset of senescence is affected by DNA rearrangements of a discontinuous mitochondrial gene in *Podospora anserina. Curr. Genet.* **9:**373–382.
82. **Kück, U., U. Stahl, and K. Esser.** 1981. Plasmid-like DNA is part of mitochondrial DNA in *Podospora anserina. Curr. Genet.* **3:**151–156.
83. **Küntzel, H., H. G. Köchel, C. M. Lazarus, and H. Lünsdorf.** 1982. Mitochondrial genes in *Aspergillus*, p. 391–403. *In* P. P. Slonimski, P. Borst, and G. Attardi (ed.), *Mitochondrial Genes.* Cold Spring Harbor Laboratory, Cold Spring Harbor, N.Y.
84. **Lambowitz, A. M., R. A. Akins, R. L. Kelley, S. Pande, and F. E. Nargang.** 1986. Mitochondrial plasmids in *Neurospora* and other filamentous fungi, p. 83–92. *In* R. B. Wickner, A. Hinnebusch, A. M. Lambowitz, I. C. Gunsalus, and A. Hollaender (ed.), *Extrachromosomal Elements in Lower Eukaryotes.* Plenum Publishing Corp., New York.
85. **Lang, B. F.** 1984. The mitochondrial genome of the fission yeast *Schizosaccharomyces pombe*: highly homologous introns are inserted at the same position of the otherwise less conserved *cox1* genes in *Schizosaccharomyces pombe* and *Aspergillus nidulans. EMBO J.* **3:**2129–2136.
86. **Lang, B. F., F. Ahne, and L. Bonen.** 1985. The mitochondrial genome of the fission yeast *Schizosaccharomyces pombe.* The cytochrome b gene has an intron closely related to the first two introns in the *Saccharomyces cerevisiae cox1* gene. *J. Mol. Biol.* **184:**353–366.
87. **Lazarus, C. M., A. J. Earl, G. Turner, and H. Küntzel.** 1980. Amplification of a mitochondrial DNA sequence in the cytoplasmically inherited "ragged" mutant of *Aspergillus amstelodami. Eur. J. Biochem.* **106:**633–641.
88. **Lazarus, C. M., and H. Küntzel.** 1981. Anatomy of amplified mitochondrial DNA in "ragged" mutants of *Aspergillus amstelodami*: excision points within protein genes and a common 215 bp segment containing a possible origin of replication. *Curr. Genet.* **4:**99–107.
89. **Lazowska, J., and P. P. Slonimski.** 1976. Electron microscopy analysis of circular repetitive mitochondrial DNA molecules from genetically characterized *rho*$^-$ mutants of *Saccharomyces cerevisiae. Mol. Gen. Genet.* **146:**61–78.
90. **Leaver, C. J., and M. W. Gray.** 1982. Mitochondrial genome organisation and expression in higher plants. *Annu. Rev. Plant Physiol.* **33:**373–402.
91. **Lemieux, B., M. Turmel, and C. Lemieux.** 1988. Unidirectional gene conversions in the chloroplast of *Chlamydomonas* interspecific hybrids. *Mol. Gen. Genet.* **212:**48–55.
92. **Lemieux, C., and R. W. Lee.** 1987. Non-reciprocal recombination between alleles of the chloroplast 23S rRNA gene in interspecific *Chlamydomonas* crosses. *Proc. Natl. Acad. Sci. USA* **84:**4166–4170.
93. **Lestienne, P., and G. Ponsot.** 1988. Kearns-Sayre syndrome with muscle mitochondrial DNA deletion. *Lancet* **i:**885.
94. **Levens, D., B. Ticho, E. Ackerman, and M. Rabinowitz.** 1981. Transcriptional initiation and 5′ termini of yeast mitochondrial RNA. *J. Biol. Chem.* **256:**5226–5232.
95. **Levings, C. S., and R. R. Sederoff.** 1983. Nucleotide sequence of the S-2 mitochondrial DNA from the S cytoplasm of maize. *Proc. Natl. Acad. Sci. USA* **80:**4055–4059.
96. **Locker, J., A. Lewin, and M. Rabinowitz.** 1979. The structure and organization of mitochondrial DNA from petite yeast. *Plasmid* **2:**155–181.

97. **Locker, J., M. Rabinowitz, and G. S. Getz.** 1974. Electron microscopic and renaturation kinetic analysis of mitochondrial DNA of cytoplasmic petite mutants of *Saccharomyces cerevisiae*. *J. Mol. Biol.* **88:**489–502.

98. **Lonsdale, D. M., T. M. Hodge, and C. M.-R. Fauron.** 1984. The physical map and organization of the mitochondrial genome from the fertile cytoplasm of maize. *Nucleic Acids Res.* **12:**9249–9261.

99. **Macreadie, I. G., R. M. Scott, A. R. Zinn, and R. A. Butow.** 1985. Transposition of an intron in yeast mitochondria requires a protein encoded by that intron. *Cell* **41:**395–402.

100. **Marcou, D.** 1961. Notion de longévité et nature cytoplasmique du déterminant de sénescence chez quelques champignons. *Ann. Sci. Nat. Bot.* **11:**653–764.

101. **Marcou, D., and J. Schecroun.** 1959. La sénescence chez *Podospora anserina* pourrait être due à des particules cytoplasmiques infectantes. *C.R. Acad. Sci.* **248:**280–283.

102. **Meinhardt, F., and K. Esser.** 1984. Linear extrachromosomal DNA in the morel *Morchella conica*. *Curr. Genet.* **8:**15–18.

103. **Michaelis, G., F. Michel, J. Lazowska, and P. P. Slonimski.** 1976. Recombined molecules of mitochondrial DNA obtained from crosses between cytoplasmic petite mutants of *Saccharomyces cerevisiae*: the stoichiometry of parental DNA repeats within the recombined molecules. *Mol. Gen. Genet.* **149:**125–130.

104. **Michel, F., and B. Dujon.** 1983. Conservation of RNA secondary structures in two intron families including mitochondrial-, chloroplast- and nuclear-encoded members. *EMBO J.* **2:**33–38.

105. **Michel, F., and B. Dujon.** 1986. Genetic exchanges between bacteriophage T4 and filamentous fungi. *Cell* **46:**323.

106. **Michel, M., and B. F. Lang.** 1985. Mitochondrial class II introns encode proteins related to the reverse transcriptases of retroviruses. *Nature* (London) **316:**641–643.

107. **Mohan, M., M. J. Meyer, J. B. Anderson, and P. A. Horgen.** 1984. Plasmid-like DNAs in the commercially important mushroom genus *Agaricus*. *Curr. Genet.* **8:**615–619.

108. **Montoya, J., D. Ojala, and G. Attardi.** 1981. Distinctive features of the 5′ terminal sequence of the human mitochondrial mRNAs. *Nature* (London) **290:**465–470.

109. **Murphy, W. I., B. Attardi, C. Tu, and G. Attardi.** 1975. Evidence for complete symmetrical transcription in vivo of mitochondrial DNA in HeLa cells. *J. Mol. Biol.* **99:**809–819.

110. **Nargang, F. E.** 1985. Fungal mitochondrial plasmids. *Exp. Mycol.* **9:**285–293.

111. **Nargang, F. E.** 1986. Conservation of a long open reading frame in two *Neurospora* mitochondrial plasmids. *Mol. Biol. Evol.* **3:**19–28.

112. **Nargang, F. E., J. B. Bell, L. L. Stohl, and A. M. Lambowitz.** 1984. The DNA sequence and genetic organization of a *Neurospora* mitochondrial plasmid suggest a relationship to introns and mobile elements. *Cell* **38:**441–453.

113. **Natvig, D. O., G. May, and J. W. Taylor.** 1984. Distribution and evolutionary significance of mitochondrial plasmids in *Neurospora* sp. *J. Bacteriol.* **159:**288–293.

114. **Oakley, K. M., and G. D. Clark-Walker.** 1978. Abnormal mitochondrial genomes in yeast restored to respiratory competence. *Genetics* **90:**517–530.

115. **Ojala, D., J. Montoya, and G. Attardi.** 1981. tRNA punctuation model of RNA processing in human mitochondria. *Nature* (London) **290:**470–474.

116. **Osinga, K. A., E. DeVries, G. Van der Horst, and H. F. Tabak.** 1984. Processing of a yeast mitochondrial messenger RNA at a conserved dodecamer sequence. *EMBO J.* **3:**829–835.

117. **Osinga, K. A., and H. F. Tabak.** 1982. Initiation of transcription of genes for mitochondrial ribosomal RNA in yeast: comparison of the nucleotide sequence around the 5′ ends of both genes reveals a homologous stretch of nucleotides. *Nucleic Acids Res.* **10:**3617–3626.

118. **Osiewacz, H. D., and K. Esser.** 1984. The mitochondrial plasmid of *Podospora anserina*: a mobile intron of a mitochondrial gene. *Curr. Genet.* **8:**299–305.

119. **Palmer, J. D., and C. R. Shields.** 1984. Tripartite structure of the *Brassica campestris* mitochondrial genome. *Nature* (London) **307:**437–440.

120. **Peebles, C. L., P. S. Perlman, K. L. Mecklenburg, M. L. Petrillo, J. H. Tabor, K. A. Jarrell, and H. L. Cheng.** 1986. A self-splicing RNA excises an intron lariat. *Cell* **44:**213–223.

121. **Picard-Bennoun, M.** 1985. Introns, protein syntheses and aging. *FEBS Lett.* **184:**1–5.

122. **Rizet, G.** 1953. Sur l'impossibilité d'obtenir la multiplication végétative ininterrompue et illimitée de l'ascomycète *Podospora anserina*. *C.R. Acad. Sci.* **237:**838–840.

123. **Rizet, G.** 1957. Les modifications qui conduisent à la sénescence sont-elles de nature cytoplasmique? *C.R. Acad. Sci.* **244:**663–665.

124. **Roe, B. A., D. P. Ma, R. K. Wilson, and J. F. H. Wong.** 1985. The complete nucleotide sequence of the *Xenopus laevis* mitochondrial genome. *J. Biol. Chem.* **260:**9759–9774.

125. **Sainsard-Chanet, A., and O. Begel.** 1986. Transformation of yeast and *Podospora*: innocuity of senescence-specific DNAs. *Mol. Gen. Genet.* **204:**443–451.

126. **Samac, D. A., and S. A. Leong.** 1988. Two linear plasmids in mitochondria of *Fusarium solani* f. sp. *cucurbitae*. *Plasmid* **19:**57–67.

127. **Schinkel, A. H., M. J. A. Groot Koerkamp, G. T. J. Van der Horst, E. P. W. Touw, K. A. Osinga, A. M. Van der Blieck, G. H. Veeneman, J. H. Van Boom, and J. H. Tabak.** 1986. Characterization of the promoter of the large ribosomal RNA gene in yeast mitochondria and separation of mitochondrial RNA polymerase into two different functional components. *EMBO J.* **5:**1041–1047.

128. **Schmelzer, C., and R. J. Schweyen.** 1986. Self-splicing of group II introns in vitro: mapping of the branch point and mutational inhibition of lariat formation. *Cell* **46:**557–565.

129. **Sederoff, R. R.** 1984. Structural variation in mitochondrial DNA. *Adv. Genet.* **22:**1–108.

130. **Sena, E. P., M. Papay, and R. Kuerti.** 1981. Cytoplasmic mixing in *Saccharomyces cerevisiae*: studies on a grande X neutral petite mating and implications for the mechanism of neutrality. *Curr. Genet.* **3:**109–118.

131. **Smith, J. R., and I. Rubenstein.** 1973. The development of senescence in *Podospora anserina*. *J. Gen. Microbiol.* **76:**283–296.

132. **Sor, F., and H. Fukuhara.** 1983. Unequal excision of complementary strands is involved in the generation of palindromic repetitions of *rho*$^-$ mitochondrial DNA in yeast. *Cell* **32:**391–396.

133. **Stahl, U., A. Lemke, P. Tudzynski, U. Kück, and K. Esser.** 1978. Evidence for plasmid like DNA in a filamentous fungus, the ascomycete *Podospora anserina*. *Mol. Gen. Genet.* **162:**341–343.

134. **Stahl, U., P. Tudzynski, U. Kück, and K. Esser.** 1982. Replication and expression of a bacterial-mitochondrial hybrid plasmid in the fungus *Podospora anserina*. *Proc. Natl. Acad. Sci. USA* **79:**3641–3645.

135. **Steinhilber, W., and D. J. Cummings.** 1986. A DNA polymerase activity with characteristics of a reverse transcriptase in *Podospora anserina. Curr. Genet.* **10:**389–392.

136. **Stohl, L. L., R. A. Collins, M. D. Cole, and A. M. Lambowitz.** 1982. Characterization of two new plasmid DNAs found in mitochondria of wild-type *Neurospora intermedia* strains. *Nucleic Acids Res.* **10:**1439–1458.

137. **Szostak, J. W., T. L. Orr-Weaver, R. J. Rothstein, and F. W. Stahl.** 1983. The double strand break repair model for recombination. *Cell* **33:**25–35.

138. **Tudzynski, P., A. Düvell, and B. Oeser.** 1986. Linear plasmids in the phytopathogenic fungus *Claviceps purpurea*, p. 119–127. *In* R. B. Wickner, A. Hinnebusch, A. M. Lambowitz, I. C. Gunsalus, and A. Hollaender (ed.), *Extrachromosomal Elements in Lower Eukaryotes.* Plenum Publishing Corp., New York.

139. **Turker, M. S., J. M. Domenico, and D. J. Cummings.** 1987. Excision-amplification of mitochondrial DNA during senescence in *Podospora anserina.* A potential role for a 11 base-pair consensus sequence in the excision process. *J. Mol. Biol.* **198:**171–185.

140. **Turker, M. S., J. M. Domenico, and D. J. Cummings.** 1987. A novel family of mitochondrial plasmids associated with longevity mutants of *Podospora anserina. J. Biol. Chem.* **262:**2250–2255.

141. **Turker, M. S., J. G. Nelson, and D. J. Cummings.** 1987. A *Podospora anserina* longevity mutant with a temperature sensitive phenotype for senescence. *Mol. Cell. Biol.* **7:** 3199–3204.

142. **Turmel, M., G. Bellemare, and C. Lemieux.** 1987. Physical mapping of differences between the chloroplast DNAs of the interfertile algae *Chlamydomonas eugametos* and *Chlamydomonas moewusii. Curr. Genet.* **11:**543–552.

143. **Van der Horst, G., and H. F. Tabak.** 1985. Self splicing of yeast mitochondrial ribosomal and messenger RNA precursors. *Cell* **40:**759–766.

144. **Van der Horst, G., and H. Tabak.** 1987. New RNA mediated reactions by yeast mitochondrial group I introns. *EMBO J.* **6:**2139–2144.

145. **Van der Veen, R., A. C. Arnberg, G. Van der Horst, L. Bonen, H. F. Tabak, and L. A. Grivell.** 1986. Excised group II introns in yeast mitochondria are lariats and can be formed by self-splicing in vitro. *Cell* **44:**225–234.

146. **Vierny, C., A. M. Keller, O. Begel, and L. Belcour.** 1982. A sequence of mitochondrial DNA is associated with the onset of senescence in a fungus. *Nature* (London) **297:**157–159.

147. **Waring, R. B., T. A. Brown, J. A. Ray, C. Scazzocchio, and R. W. Davies.** 1984. Three variant introns of the same general class in the mitochondrial gene for cytochrome oxidase subunit I in *Aspergillus nidulans. EMBO J.* **3:** 2121–2128.

148. **Wolstenholme, D. R., J. L. McFarlane, R. Okimoto, D. O. Clary, and J. A. Wahleithner.** 1987. Bizarre tRNAs inferred from DNA sequences of mitochondrial genomes of nematode worms. *Proc. Natl. Acad. Sci. USA* **84:**1324–1328.

149. **Wright, R. M., M. A. Horrum, and D. J. Cummings.** 1982. Are mitochondrial structural genes selectively amplified during senescence in *Podospora anserina*? *Cell* **29:**505–515.

150. **Wright, R. M., J. L. Laping, M. A. Horrum, and D. J. Cummings.** 1982. Mitochondrial DNA from *Podospora anserina.* 3. Cloning, physical mapping and localization of ribosomal RNA genes. *Mol. Gen. Genet.* **185:**56–64.

151. **Yin, S., J. Heckman, and U. L. RajBhandary.** 1981. Highly conserved GC rich palindromic DNA sequences flank tRNA genes in *Neurospora crassa* mitochondria. *Cell* **26:**325–335.

152. **Zinn, A. R., and R. A. Butow.** 1985. Non-reciprocal exchange between alleles of the yeast mitochondrial 21S rRNA gene: kinetics and the involvement of a double strand break. *Cell* **40:**887–895.

Chapter 41

Transposable Elements and the Genetic Engineering of Bacteria

CLAIRE M. BERG, DOUGLAS E. BERG, and EDUARDO A. GROISMAN

Claire M. Berg ■ Department of Molecular and Cellular Biology, Box U-131, The University of Connecticut, Storrs, Connecticut 06269-2131. **Douglas E. Berg** ■ Department of Microbiology and Immunology and Department of Genetics, Box 8093, Washington University School of Medicine, 660 S. Euclid, St. Louis, Missouri 63112. **Eduardo A. Groisman** ■ Department of Molecular Biology, Research Institute of Scripps Clinic, La Jolla, California 92037.

I. INTRODUCTION

The first transposable elements to be used for genetic analysis in bacteria were the *Escherichia coli* bacteriophages λ and Mu. Initially, aberrant excision of λ prophages was used to isolate specialized transducing phage carrying bacterial genes that had been adjacent to the integration site (62). Subsequently, genetic manipulations were used to place additional genes near the phage λ primary attachment site (153, 188), to delete the primary attachment site so that the phage integrated at secondary attachment sites (351, 352), and to generate transducing lines directly by induction of pools of secondary-site lysogens (339). This type of in vivo gene cloning was particularly important in the years before in vitro cloning techniques were developed.

The realization that bacteriophage Mu inserts into many locations and can cause mutations when it lysogenizes *E. coli* (373) led to its exploitation to identify and map many genes, starting with those in the *dap* pathway (59). Mu later became the first of the elements engineered to examine patterns of transcriptional and translational regulation and for other sophisticated applications (see reference 24; Table 1). The discovery of antibiotic resistance transposons led to a virtual explosion in the use of transposable elements for in vivo genetic engineering in diverse bacterial species. These elements carried resistance markers that were much easier to select than Mu or λ prophage immunity; they could insert into many sites and transpose in many species, and their generally smaller size facilitated physical mapping and cloning of DNA. An early review by Kleckner et al. (208) was important in focusing attention on the great utility of these elements, which continue to be exploited with ever-increasing sophistication. Transposons are used to identify genes by insertion mutation; to obtain easily selectable or scorable markers in or adjacent to genes of interest; to characterize the organization of operons; to analyze transcriptional and translational regulatory mechanisms, protein conformation, and cellular location; to generate genome rearrangements such as deletions and replicon fusions; to provide regions of homology and useful restriction sites for targeted in vivo and in vitro DNA rearrangements; to introduce specific genes into new hosts or into particular chromosomal sites; to introduce origins of replication or of conjugative DNA transfer; to easily clone large and small contiguous DNA segments in vivo, independent of the natural distribution of restriction sites; to provide mobile primer binding sites that facilitate DNA sequencing; to provide portable, regulatable promoters with which to manipulate patterns of gene expression; and to identify essential genes and sites.

A review of many uses of transposable elements, including lists of known transposon insertion mutations in the genomes of *E. coli* and *Salmonella typhimurium*, was published recently (24). Experimental approaches utilizing particular elements or classes of elements have been detailed elsewhere (8, 58, 97–99, 283, 354, 391, 417, 418).

This chapter deals with uses of wild-type and genetically engineered transposable elements, emphasizing recently developed approaches. Methods for the use of transposons as tools for genetic analysis have been developed primarily in laboratory strains of *E. coli* K-12 and *S. typhimurium* LT2 (reviewed in references 24 and 208), but much of their value comes from their use in other bacterial isolates and species, both gram positive and gram negative. Some of the most commonly used and also the most promising elements are diagrammed in Fig. 1 (p. 889), and a more extensive list is presented in Table 1. Although the gram-negative and gram-positive transposons are treated separately in this table, recent reports of transposition of the gram-positive elements Tn*917* (226), Tn*1545* (87), and Tn*4430* (235) in *E. coli*, and of similarities between some elements from gram-negative and gram-positive bacteria (84, 250, 288; E. Murphy, this volume), show that this distinction is not absolute. Table 2 documents the widespread use of transposable elements in many bacterial species. Their use, in concert with in vitro recombinant DNA technology, has revolutionized the analysis of bacterial species that had not been extensively studied previously.

II. STRUCTURAL DIVERSITY

As described in this volume, bacterial transposable elements range from less than 1 kilobase (kb) (IS*1*) to over 48 kb (bacteriophage λ). The simplest are the insertion sequence (IS) elements, which generally consist only of the genes and sites needed for

Continued on page 893

Table 1. Useful transposable elements

Designation	Marker(s)[a]	Size (kb) and structure	Comments[b] (references)
Gram-negative elements			
Tn*3*[c]	*amp*	5; no IS elements	Duplication: 5 bp. IR: 38 bp. Transposition: Replicative; high frequency to plasmids, low to chromosome; AT-rich regions preferred. Stimulates adjacent deletion formation. Polar in one orientation, stimulates distal transcription in other orientation. Depicted in Fig. 1. (157, 169; Sherratt, this volume)
Tn*3*-Ap,Km	*amp*, *kan*	11.7	Derivative of Tn*3* with replication region and *kan* gene of plasmid R1. (148)
Tn*3*-*trp*	*amp*, *trpB,C,D*		Derivative of Tn*3* with *trpB,C,D* genes. (148)
Tn*3*-HoHo1 (Tn*3*-*lac*)	*amp*, *'lac*	14.25	Derivative of Tn*3*. *lac* operon lacks a functional promoter and Shine-Dalgarno sequences, so both type I and type II fusions are obtained. Transposition defective. (367)
m-Tn*3*-*lac*	*amp*, *'lac*	4.6	Derivative of Tn*3*. Forms type II protein fusions. Resolution via P1 *loxP*; no *res* site. Because of transposition immunity, cannot be used to mutagenize pBR322-derived plasmids. Also derivatives with genes selectable in *S. cerevisiae*. (342)
Tn*2301* (Tn*1*-*oriT*)	*amp*		Derivative of Tn*1* (closely related to Tn*3*) with cloned F-factor conjugation genes. Makes F′-like plasmid insertions and Hfr-like chromosomal insertions. (192)
Tn*951*	*lacZY*	16.6	Duplication: 5 bp. IR: 40 bp, first 38 bp identical to those of Tn*3*. Transposition defective, complemented by Tn*3* (but not Tn*2501*) transposase. Tn*2501* ("minor" transposon) and IS*1* are present in Tn*951*. Resolution of Tn*951* cointegrates by Tn*2501* resolvase. (86)
Tn*2501*	None	6.3	Duplication: 5 bp. IR: 38 bp. Minor transposon of Tn*951*. In Tn*3* family, IRs most closely related to Tn*21*. Tn*2501* promotes its own transposition, but not that of Tn*951*. (263)
Tn*2505* (Tn*951*-*kan*)	*kan*	8.2	Derived from Tn*951* by replacement of 10.5-kb *Bam*HI fragment containing IS*1*, *lac*, and right IR of Tn*2501* with 2.1-kb *kan* fragment from Tn*5*. Transposition defect is complementable by Tn*3* transposase. Is the minor transposon of Tn*2515* (below) in pMBLG2. Can be used to generate deletions by recombination with a copy of Tn*2506* (below) transposed to the same plasmid. (262)
Tn*2515*	*kan*, *tet*		Derived from a plasmid containing Tn*2505* by insertion of a 4.8-kb *Eco*RI fragment containing the Tn*3* transposase and IR. Forms a transposition-proficient major transposon containing the minor transposon Tn*2505*. Distinguishable from Tn*2505* by ability to transpose further and by Tet[r]. (262)
Tn*2506*	*cam*	7.3	Derived in several steps from Tn*2505*. Contains the *cam* gene from pSA. Minor transposon of Tn*2516*. (262)
Tn*2516*	*cam*, *tet*		Major transposon of Tn*2506*. Analogous to Tn*2515*. (262)
Tn*501*	*mer*	8.2	Duplication: 5 bp. IR: 38. In Tn*3* family. Transposase, resolvase, and resistance genes are mercury inducible. Depicted in Fig. 1. (169, 336, 337)
Tn*1721*	*tet*	11.2	Duplication: 5 bp. IR: 38 bp. In Tn*3* family, related to Tn*501*, identical to Tn*1771*. Contains minor transposon, Tn*1722*. *Eco*RI sites 15 bp from end are useful for mapping and for deleting most of transposon. *tet* gene is amplified by homologous recombination involving long flanking direct repeats and is selected by growth with high levels of tetracycline. Depicted in Fig. 1. (75, 337, 387)
Tn*1722*	None	5.6	Minor transposon of Tn*1721* (see Fig. 1). (337, 387)
Tn*4431*	*tet*, *'lux*	15	Derivative of Tn*1721* with promoterless *lux* operon of *Vibrio fischeri*. Contains *lux* in a 7.5-kb *Bam*HI fragment inserted into *Bam*HI site of Tn*1721*. (349)
Tn*1721*-PATE	*tet*, *pate*	22	Derivative of Tn*1721* with polygalacturonate *trans*-eliminase I and II genes from *Klebsiella oxytoca*. (395)

Continued on following page

Table 1—*Continued*

Designation	Marker(s)[a]	Size (kb) and structure	Comments[b] (references)
Tn*1725*	*cam*	8.9	Derivative of Tn*1722* with 3.3-kb *Hin*dIII *cam* fragment of plasmid Sa inserted in *Hin*dIII site. (387)
Tn*1731-bam*	*tet*	7.6	Derivative of Tn*1722* with 2.4-kb *Eco*RI fragment containing *tet* of Tn*1721* from pJOE452 replacing 0.2-kb *Hpa*I fragment of Tn*1722* and with a *Bam*HI site introduced 57 bp from the left end. (387, 388; R. Schmitt, personal communication)
Tn*1732*	*kan*	6.7	Derivative of Tn*1722* with 1.45-kb *Mlu-Pst*I *kan* fragment of Tn*2680* replacing 0.4-kb *Mlu-Pst*I fragment of Tn*1722*. (387)
Tn*1733*	*kan, ble, str*	8.9	Derivative of Tn*1722* with 3.3-kb *Hin*dIII *kan ble str* fragment of Tn*5* into *Hin*dIII site. (387)
Tn*1735Cm*	*cam*, p_{tac}	9.6	Derivative of Tn*1731-bam* with 1.75-kb *Bam*HI-*Hin*dIII p_{tac}-*lacI*^{q} fragment from plasmid pFD115 replacing 3-kb *Bam*HI-*Hin*dIII fragment. High-level, outward transcription inducible by IPTG. Contains 3.2-kb *Hin*dIII *cam* fragment of plasmid SA inserted in *Hin*dIII site. (388)
Tn*1735Km*	*kan*, p_{tac}	9.8	As in Tn*1735Cm*, but with 3.3-kb *Hin*dIII *kan ble str* fragment of Tn*5* inserted in *Hin*dIII site. (388)
Tn*1735Sm*	*str*, p_{tac}	8.4	As in Tn*1735Cm*, but with 2.0-kb *Hin*dIII-*aodA* fragment of pHP45 inserted in *Hin*dIII site. (388)
Tn*1736Tc*	*tet, 'cam*	8.4	Derivative of Tn*1731-bam* with 0.8-kb promotorless *cam* gene of Tn*9* from pCM4 inserted in *Bam*HI site. Forms type I fusions. (388; Schmitt, personal communication)
Tn*1736Km*	*kan, 'cam*	6.8	Derivative of Tn*1731-bam* with 1.45-kb *Bam*HI-*Hin*dIII *kan* fragment of pNeo replacing 3.0-kb *Bam*HI-*Hin*dIII fragment and with promoterless *cam* gene as in Tn*1736Tc*. (388; Schmitt, personal communication)
Tn*1737Km*	*kan, 'lac*	9.8	Derivative of Tn*1731-bam* as in Tn*1736Km* but with 3.8-kb promoterless *lac* fragment from pFD127 in *Bam*HI site. Forms type I fusions. (388; Schmitt, personal communication)
Tn*1737Cm*	*cam, 'lac*	11.1	Derivative of Tn*1731-bam* in several steps with *lac* as in Tn*1737Km* and *cam* from plasmid SA. Forms type I fusions. (388; Schmitt, personal communication)
Tn*1737Sm*	*str, 'lac*	10.6	Derivative of Tn*1731-bam* in several steps with *lac* as in Tn*1737Km* and *str* from pHP45. Forms type I fusions. (388; Schmitt, personal communication)
Gamma-delta (Tn*1000*)	Unknown	6.0	Duplication: 5 bp. IR: 35 bp. Related to Tn*3*. Present on F factor. Transposition: high frequency to plasmid, quite random, but strong preference for AT-rich sequences. Cointegrate formation results in mobilization of chromosome or nonconjugative plasmid. (157, 162, 241; R. Reed, personal communication)
Tn*5*	*kan, ble, str*[d]	5.7; 1.5-kb IS*50* elements inverted	Duplication: 9 kb. IR: 19 bp. Transposition: Conservative; high frequency; quite random but some hot spots found. Transposition to multicopy plasmid is selected by high neomycin resistance. Sometimes stimulates distal transcription in either orientation. *kan* gene encodes *npt-II*. Streptomycin resistance is cryptic in *E. coli*. Depicted in Fig. 1. (34, 36, 41, 43, 256, 297; Berg, this volume)
Tn*5-lac*	*kan, 'lac*	12	Type I operon fusions. *lac* replacing most of IS*50*$_L$. (218)
Tn*5*ORF*lac*	*kan, 'lac*		Type II protein fusions. *lac* replacing most of IS*50*$_L$. (215)
Tn*phoA* (Tn*5phoA*)	*kan, 'phoA*	7.7	Type II protein fusions with *phoA*. *phoA* replacing most of IS*50*L. Fusions used to detect membrane or exported proteins. Depicted in Fig. 1. (253)
Tn*5*seq1	*kan*	3.2	Has outward-facing subterminal T7 and SP6 and transposase gene of IS*50*R). Useful for DNA sequencing and for in vivo and in vitro T7- or SP6-directed transcription. Depicted in Fig. 1. (275)

Continued on following page

Table 1—*Continued*

Designation	Marker(s)[a]	Size (kb) and structure	Comments[b] (references)
Tn*5*seq2	*supF*	0.26	Has one O and one I end of IS*50* and unique subterminal sequences for use as primer binding sites for DNA sequencing. Transposase gene in *cis* on donor plasmid. Used for mutagenesis of λ cloning vectors. Insertions selected by suppression of λ or chromosomal nonsense mutation. Depicted in Fig. 1. (S. Phadnis, H. Huang, and D. E. Berg, in preparation)
Tn*5*tac1	*kan*, p_{tac}	4.6	Has O and I ends and transposase gene of IS*50*R and *lacI*[q] and outward-facing p_{tac}. High-level transcription is inducible by IPTG. Depicted in Fig. 1 (78)
Tn*5*-VB32	*tet*, '*kan*		Type I operon fusions with *kan*. (22)
Tn*5*-131, -132	*tet*	5.7	Tn*10 tet* gene in place of Tn*5* resistance genes. Tn*5*-131 is transposition deficient; Tn*5*-132 is transposition proficient. (37, 324)
Tn*5*-133, -134	*kan*, *tet*	8.4	Contains Tn*10 tet* gene. Tn*5*-133 is transposition deficient; Tn*5*-134 is transposition proficient. (37, 324)
Tn*5*-Tc	*kan*, *tet*		Contains *Sau*3A fragment of RP4 in *Bam*HI site. (356)
Tn*5*-CM	*cam*	3.9	Contains 0.8-kb *Bam*HI promoterless *cam* cartridge from Tn*9* inserted downstream of the *kan* promoter, in place of the Tn*5* central region. (334)
Tn*5*-GM	*gen*	7.5	Contains 4.4-kb *Hin*dIII fragment from pOX38-GM in place of Tn*5* central region. (334)
Tn*5*-TC2	*tet*	5.1	Contains 2.1-kb *Eco*RI-*Pvu*II *tet* fragment from pBR322 in place of Tn*5* central *Bcl*I segment. (334)
Tn*5*-TP	*tmp*	5.2	Contains 2.1-kb *Bam*HI *tmp* fragment from R388 in place of Tn*5* central region. (334)
Tn*5*-*751*	*kan*, *tmp*	9.0	Contains 3.3-kb *Bam*HI *tmp* fragment from R751. (310)
Tn*5*-SM	*str*	5.2	Contains 2.1-kb *Hin*cII fragment with promoterless *str* gene from RSF1010 inserted downstream of *kan* promoter, in place of Tn*5* central region. (334)
Tn*5*-AP	*amp*	5.0	Contains 2.2-kb *Bam*HI fragment from a pSC101::Tn*3* plasmid in place of Tn*5* central region. (334)
Tn*5*-*oriT*	*kan*	6.5	Contains transfer origin of plasmid RK2 inserted as a 760-bp fragment into the *Bam*HI site of Tn*5*. Chromosomal insertions are Hfr-like in presence of helper RK2. Depicted in Fig. 1. (412)
Tn*5*-*mob*	*kan*		Contains transfer origin of plasmid RP4 inserted in *Bam*HI site of Tn*5*. Chromosomal insertions are Hfr-like, and plasmid insertions are F′-like in the presence of helper RP4. (355)
Tn*5*-V	*kan*		Contains replication origin of plasmid pSC101. Useful for in vitro cloning of target genes from other species into *E. coli*, although the active replication origin may limit its usefulness for insertion mutagenesis in the enterics. (137)
Tn*5*-PV	*kan*		Contains transfer and replication origins of plasmid RK2 and T-DNA borders of *Agrobacterium tumefaciens* Ti plasmid. Mediates transfer of target plasmids from *Agrobacterium* sp. to plant cells. Can be complemented for autonomous replication in *trans*. (213)
Tn*5*-*luxAB*	*kan*		Derivative of Tn*5*-PV with promoterless *luxAB* genes replacing most of IS*50*L. Type I fusion probe. (Koncz-Kalman and Schell, personal communication)
Tn*5tox*	*kan*, *tox*	10.5	Has 4.6-kb *Bam*HI *Bacillus thuringiensis* delta endotoxin gene fragment in place of Tn*5* central region. Used for introducing antilepidopteran toxin gene into root-colonizing pseudomonads. Also a 7.3-kb derivative in which the *kan* gene and part of IS*50*R are deleted has been constructed. Transposition defective. (285, 286)
Tn*5*-233	*gen*/*kan*, *str*/*spc*	6.6	Central *Bgl*II fragment of Tn*5* replaced with 3.5-kb *Bam*HI-*Bgl*II *gen*/*kan*-*str*/*spc* fragment of plasmid Sa. Unlike Tn*5*, does not confer resistance to neomycin. Does not suppress transposition of a newly introduced Tn*5*. (106)

Continued on following page

Table 1—*Continued*

Designation	Marker(s)[a]	Size (kb) and structure	Comments[b] (references)
Tn*5*-GmSpSm	*gen, spc, str*	6.8	Central *Bgl*II fragment of Tn*5* replaced with 3.7-kb *Bam*HI fragment of Tn*1696*. (176)
Tn*5*-235	*kan, 'lacZY*	10	*lacZY* inserted into *Bam*HI site in *str*; *lac* transcribed as part of the *kan* operon. (106)
Tn*5-410*	*trpE*	8	Derivative of Tn*5* with *E. coli trpE* gene. (261)
Tn*5.7*	*tmp, str, spc*	7.6	Central *Bgl*II fragment of Tn*5* replaced by 4.6-kb *Sau*3A fragment of Tn*7*. (423)
Tn*5 Bgl*$^+$	*kan, bgl*	11.3	Derivative of Tn*5* with 5.6-kb *Bam*HI insert containing *bgl*$^+$ (cellobiose utilization) gene from *Klebsiella oxytoca*. (395)
Tn*5 Amy*$^+$	*kan, amy*	8.7	Derivative of Tn*5* with 3.0-kb fragment containing *amy*$^+$ (amylase) gene from *Klebsiella oxytoca*. (395)
Tn*7*	*tmp, str, spc*	14	Duplication: 5 bp. IR: About 30 bp, but needs about 105 bp from left end and 75 bp from right end for transposition. Transposes to a single site in the *E. coli* chromosome with a very high frequency and to secondary sites with a lower frequency. Transposes to plasmids. Useful for introducing DNA fragments to a single defined chromosomal site in *E. coli*. Depicted in Fig. 1. (238; Craig, this volume)
Tn*7-lac*	*lacZYA*	11.2	Transposition defective. Also 6.7- to 8.6-kb derivatives with *lac* under the control of different promoters, with different restriction sites or with part of *lacA* deleted, or both. (16; G. F. Barry, personal communication)
Tn*9*	*cam*	2.6; 0.8-kb IS*1* elements as direct repeats	Duplication: 9 (rarely 8) bp. Transposition to many sites, but significant hot spots. Stimulates adjacent deletion formation. One IS element and *cam* can be lost by crossover between IS elements. Usefulness in *E. coli* is compromised by presence of multiple IS*1* elements in genome. Depicted in Fig. 1. (187)
Tn*10*	*tet*	9.3; 1.3-kb IS*10* elements inverted	Duplication: 9 bp. IR: 23. Transposition: To many sites, but hot spots found. Stimulates adjacent deletion and inversion formation. Stimulates low-level distal transcription. Depicted in Fig. 1. (97, 204; Kleckner, this volume)
Tn*10-kan*	*kan*	8.2	Derivative of Tn*10* with Tn*5 kan* gene. Called Kan^r-Tn*10*. (402)
Tn*10-amp*	*amp*	7.1	Inverse Tn*10* transposon with pBR322 *amp* gene in place of *tet*. Transposition reduced. Called *amp* hopper. (130, 402)
Tn*10-cam*	*cam*	9.0	Inverse Tn*10* transposon with Tn*9 cam* gene in place of *tet*. Transposition reduced. Called *cam* hopper or TnHACIO. (130, 402)
Tn*10HH*	*tet*	7.7	Like Tn*10*, but IS*10*L partially deleted; high hopper mutation. Called Tn*10 del*4HH104. Also a 9.4-kb *tet kan* derivative called high hopper *kan-tet*. (402)
Mini-Tn*10*	*tet*	4.0	Transposase gene outside element. Therefore, stable after transposition. Called p_{tac} mini-*tet* (depicted in Fig. 1). Also a 3.5-kb derivative called *del*16*del*17 *tet*, a 2.8-kb *kan* derivative called p_{tac} mini-*kan*, and a 5.4-kb *kan* derivative called *del*16*del*17*kan*^{r}. Can be complemented in *trans* by strongly expressed transposase gene. (402)
Mini-Tn*10*-LK	*kan, 'lac*	4.9	Contains ends of Tn*10*, promoterless *lacZ* gene (forms type II protein fusions), and *kan*. (185)
Mini-Tn*10*-LUK	*kan, 'lac, URA3*	6.1	Derivative of mini-Tn*10*-LK with *S. cerevisiae URA3* gene inserted at unique *Bam*HI site in mini-Tn*10*-LK. (185)
Tn*10-trplac*	*tet, 'lac*	11	Type I operon fusions. Called *trp-lac* fusion hopper. *trp-lac* replacing most of IS*10*L. (402)
m-Tn*10*/*URA3*/*supF*	*URA3, supF*	2.1	Contains *S. cerevisiae URA3* gene and *E. coli supF*. Also a 4.2-kb *URA3 tet* and a 3.2-kb *TRP1 kan* derivative. (364)
Mu	*imm*^{Mu}	37.5	Duplication: 5 bp. IR: 2 bp. Bacteriophage. Transposition: High frequency to chromosome, not recovered in high-copy-number plasmid; quite random, but some hot spots and regional specificity are found. Depicted in Fig. 1. (182, 371, 382; Pato, this volume)

Continued on following page

Table 1—*Continued*

Designation	Marker(s)[a]	Size (kb) and structure	Comments[b] (references)
MupAp1	*amp*	37.5	Derivative of Mu generated by Tn*3* transposition to Mu and spontaneous deletion of nonessential Mu and Tn*3* segments. Plaque forming. G loop locked into (+) orientation. (232)
Mu dI1	*amp*, *'lac*	37	Derivative of Mu with promoterless *lac* genes near the S (right) end of Mu. Type I operon fusions. Useful for studying operon expression. Carries 1.2-kb IS*121* element. Originally called Mu d1. Depicted in Fig. 1. Also derivatives with conditional or complementable transposition defects. (13, 66, 183, 287)
Mu dII301	*amp*, *'lac*	35.6	Derivative of Mu with *lac* genes deleted for promoter and translational start sequences. Type II protein fusions. (65)
Mu dI770-1	*amp*, *kan*, *'lac*		Derivative of Mu dI1 with *kan* gene. Type I fusions. (67)
Mu dII770-301	*amp*, *kan*, *'lac*	35.6	Derivative of Mu dII301 with *kan* gene. Type II fusions. Equivalent to Mu dI770-1. (67)
Mu18-1	*amp*		Mini-Mu derivative of Mu-*amp*, probably generated by inverse transposition of Tn*9*. Retains one IS*1* element which generates further deletions. *amp* inserted into Mu18. Complementable transposition defect. Also a MuA$^+$ transposition-proficient derivative. (311, 381)
Mini-Mu-*kan*	*kan*	16	Mini-Mu derivative of Mu with Tn*903 kan* gene. Complementable transposition defect. (70)
Mini-Mu-*amp*	*amp*	6	Mini-Mu derivative of Mu with Tn*3*-pBR322 hybrid *amp* gene. Also a 10-kb derivative with Tn*3 amp* gene. Complementable transposition defect. (70)
Mini-Mu-*lac*	*lac*	17	Mini-Mu derivative of Mu with functional *lac* cluster. Also a 22-kb derivative with G region that has broader host range. Complementable transposition defect. (70)
Mu dI1678	*amp*, *'lac*		Mini-Mu derivative of Mu dI7701-1 with *kan* and adjacent segment deleted. Type I fusions. (67)
Mu dII1678	*amp*, *'lac*	22.4	Mini-Mu derivative of Mu dII7701-301 with *kan* and adjacent segment deleted. Type II fusions. Equivalent to Mu dI1678. (67)
Mu dI1681	*kan*, *'lac*		Mini-Mu derivative of Mu dI7701-1 with *amp* and adjacent segment deleted. Type I fusions. (67)
Mu dII1681	*kan*, *'lac*	14.2	Mini-Mu derivative of Mu dI7701-301 with *amp* and adjacent segment deleted. Type II fusions. Equivalent to Mu dI1681. (67)
Mu dI1734	*kan*, *'lac*		Mini-Mu derivative of Mu dI1681 with a Mu A and B genes deleted. Type I fusion. Complementable transposition defect. Also called Mu dJ. (67)
Mu dII1734	*kan*, *'lac*	9.7	Mini-Mu derivative of Mu dII1681 with Mu A and B genes deleted. Type II fusion. Complementable transposition defect. Also called Mu dK. Equivalent to Mu dI1734. (67)
Mu d4041	*kan*	7.8	Mini-Mu derivative of Mu dII7701-301 with *amp* and *lac* genes deleted (67). Also a derivative with a cloned mammalian thymidine kinase gene for selection in mammalian cells. (190)
Mu dE	*kan*, *erm*		Derivative of Mu d4041 with *Bam*HI *erm* fragment of pTS19E inserted in *Bam*HI site. *erm* expressed in gram-positive organisms. (225)
Mu dIIPR3	*cam*, *'kan*	4.2	Mini-Mu derivative of Mu with promoterless Tn*903 kan* gene and Tn*9 cam* gene. Forms type II fusions with *kan*. Useful for gene fusions in eucaryotes in which *lac* fusions cannot be used. (307)
Mu dIIPR13	*cam*, *'lac*	8.0	Mini-Mu with promoterless *lacZ* gene and Tn*9 cam* gene. Forms type II fusions. (308)
Mu dPR40	*gen*, *cam*, *'lac*	16	Derivative of Mu dIIPR13 with *gen* gene. (308)
Mu dIIZZ1	*cam*, *'lac*, *LEU2*		Derivative of Mu dIIPR13 with *S. cerevisiae LEU2* gene and 2μm plasmid origin of replication for selection in *S. cerevisiae*. Yeast origin lies in a removable cassette. (92)

Continued on following page

Table 1—*Continued*

Designation	Marker(s)[a]	Size (kb) and structure	Comments[b] (references)
Mu dIIPR46	*gen, cam, 'lac*	17	Derivative of Mu dIIPR40 with *oriT* from pSUP5011. (308)
Mu dIIPR48	*gen, cam, 'lac*	21.7	Derivative of Mu dIIPR46 with *oriRiHRI* from pLJbB11. (308)
Mu d(*lacZ npt-II*)	*'lac, 'kan*		Mini-Mu derivative of Mu with promoterless *lac* and *kan* genes in tandem. Forms type II protein fusions with *lac* and type I operon fusions with *kan* in the same transcript. (231)
Mini-Mu-*tet*	*tet, 'lac*		Mini-Mu derivative of Mu dI1681 with Tn*10 tet* replacing *kan*. (19)
Mini-Mu-*lux*	*kan, 'lux*	15.5	Mini-Mu with *V. fischeri luxAB* (luceriferase) genes. Type I operon fusions; emits light when expressed. (122)
Mini-Mu-*lux*Tc[r]	*tet, 'lux*	18.5	*kan* gene of mini-Mu-*lux* replaced by *tet* gene. (122)
Mini-Mu-*cat*	*cam*		Contains synthetic MuL-end (204 bp) and MuR-end (115 bp) bracketing *cam* gene from pTAPI. Mu A and B genes in *cis* on same plasmid. (291)
Mu d13-1	*kan, 'lac*	10.3	Mini-Mu with transfer origin of RK2. Type I operon fusions. Mu A and B genes deleted. Complementable transposition defect. Useful for both plasmid and chromosomal insertions. (Groisman and Heffron, unpublished data)
Mu d5345	*cam, 'lac*	10.5	Contains *oriT* and *lac* segment for transcriptional fusions. Mu A and B genes deleted. Complementable transposition defect. Useful for both plasmid and chromosomal insertions. (Groisman and Heffron, unpublished data)
m-Mu-*tac*	*kan*, p_{tac}	7	Derivative of Mu dII1681 with *lac* replaced by 270-bp *Bam*HI p_{tac} fragment from pKK223-3. (155)
Mu d-5-3	*kan*, p_{T7}		Contains T7 promoter for regulated expression of both cloned and chromosomal genes. Mu A and B genes deleted. Complementable transposition defect. (Groisman and Heffron, unpublished data)
Mu dII79 (*cos*-mini-Mu)	*amp*		Derivative of Mu dII1681 containing cloned cosmid pHC79. Useful for in vivo cloning and packaging into λ heads. (154)
Mu dII4042	*cam, 'lac*	16.7; plasmid	Derivative of Mu dII1681 with origin of replication from multicopy plasmid P15A. Type II protein fusions. Depicted in Fig. 1. Useful for cloning host genes in vivo and for generalized transduction (Fig. 6). (161, 397)
Mu dII5085	*cam, 'lac*	13.4; plasmid	Derivative of Mu dII4042 with Mu A and B genes deleted. Complementable transposition defect. (159)
Mu d5005	*kan*	7.9; plasmid	Derivative of Mu dII4042 with ColE1-type origin of replication from multicopy plasmid pMB1. *cam*, *ori*-P15A, and *lac* deleted. Useful for cloning large (up to 25 to 30 kb) segments of host genome in vivo. Compatible with P15A-derived plasmids. (159)
Mu dI5086	*kan, 'lac*	14.9; plasmid	Derivative of Mu d5005 with promoterless *lac* operon. Type I operon fusions. (159)
Mu dI5155	*kan, 'lac*	15.6; plasmid	Derivative of Mu dI5086 with origin of transfer of broad-host-range plasmid RK2. Mini-Mu plasmid transferred conjugally in presence of helper RK2. (159)
Mu dI5166	*cam, lac*	15.8; plasmid	Derivative of Mu dI5155 with *cam* in place of *kan*. (159)
Mu dII5117	*spc/str, kan, lac*	21.7; plasmid	Derivative of Mu dII1678 with origin of replication and resistance genes of the low-copy-number plasmid SA. Useful for in vivo cloning of genes that are harmful on high-copy-number plasmids. Compatible with P15A- and pMB1-derived plasmids. (159)
Mu d5294	*kan*, p_{T7}	7.7; plasmid	Derivative of Mu d5005 with outward-facing T7 promoter. Useful for high-level in vitro and in vivo T7-directed transcription of cloned fragment. (Groisman et al., in preparation)
Mu d-P22	*cam*	36.4	Derivative of temperate *Salmonella typhimurium* phage P22 with ends of Mu. Transposes like Mu. Upon induction, DNA replicates in situ in one direction from P22 *pac* site and sequential headfuls are packaged in P22 particles. Two forms,

Continued on following page

Table 1—*Continued*

Designation	Marker(s)[a]	Size (kb) and structure	Comments[b] (references)
			called P and Q, differ in orientation of P22 and consequently in direction of packaging. Useful for amplifying adjacent host DNA, for DNA sequencing, and for transduction. Depicted in Fig. 1 and Fig. 7. (415)
D3112	*imm*^{D3112}	35	Mu-like phage, useful for transposon mutagenesis of *Pseudomonas aeruginosa.* Transposes poorly in *E. coli.* (111, 219)
Mini-D171	*tet*	5.0	Derivative of D3112. Contains *tet* gene from pSC101. Useful for in vivo cloning and generalized transduction. (Darzins and Casadaban, in preparation)
Mini-D165	*kan*	5.1	Derivative of D3112. Contains *kan* gene from Tn*5*. (Darzins and Casadaban, in preparation)
Mini-D214	*tet*	10.9; plasmid	Derivative of D3112. Contains *tet*, *oriT*, *oriV*, and *trfA** from RK2. Useful for in vivo cloning and generalized transduction. (Darzins and Casadaban, in preparation)
Mini-D948	*amp*	11.4; plasmid	Derivative of D3112. Contains *oriT*, *oriV*, and *trfA** from RK2 and the ColE1-type replication origin of pMB1. Useful for in vivo cloning and generalized transduction. (Darzins and Casadaban, in preparation)
Mini-D366	*gen*, *kan*	12	Derivative of D3112. Contains *gen* and *kan* genes, replication origin of plasmid Sa, and transfer origin of plasmid RK2. Useful for in vivo cloning and generalized transduction. (Darzins and Casadaban, in preparation)
Mini-D385.5	*amp*	10.2	Derivative of D3112. Contains *amp* gene and replication origin of *Pseudomonas* sp. plasmid pVSI. Useful for in vivo cloning and generalized transduction. (Darzins and Casadaban, in preparation)
Lambda (λ)	*imm*$^{\lambda}$	49	Transposition due to λ-specific integration. Does not generate duplication of target DNA sequences. Depicted in Fig. 1. Insertions into secondary sites found only when *attB* is deleted. (171, 351; Thompson and Landy, this volume)
λ p*lac*Mu1	*imm*$^{\lambda}$, *'lac*		Derivative of Mu dII301 with lambda genes replacing *amp* and most of the Mu genome. Transposes like Mu and packages like λ. Type II protein fusions. Useful for insertion mutagenesis and in vivo cloning. Depicted in Fig. 1. (55)
λ p*lac*Mu3	*imm*^{21}		Derivative of λ p*lac*Mu1 with *imm*^{21}. (55)
λ p*lac*Mu50	*imm*$^{\lambda}$, *'lac*		Derivative of Mu dI1 with λ genes replacing *amp* and most of the Mu genome. Like λ p*lac*Mu1, but type I operon fusions. (54)
λ p*lac*Mu51	*imm*^{21}, *'lac*		Derivative of λ p*lac*Mu50 with *imm*^{21}. (54)
λ p*lac*Mu9	*kan*, *'lac*		Derivative of λ p*lac*Mu1 with *kan* gene. Type II protein fusions. (54)
λ p*lac*Mu15	*kan*, *'lac*		Derivative of λ p*lac*Mu9 with Mu *A*am1093 substituted for truncated (and partially functional) Mu *A* allele. (E. Bremer, T. J. Silhavy, and G. M. Weinstock, *Gene*, in press)
λ p*lac*Mu53	*kan*, *'lac*		Derivative of λ p*lac*Mu51 with *kan* gene. Type I operon fusions. (54)
λ p*lac*Mu55	*kan*, *'lac*		Derivative of λ p*lac*53 with Mu *A*am1093 substituted for truncated Mu *A* allele. (Bremer et al., in press)
Gram-positive elements			
Tn*916*	*tet*	16.4; no IS elements	Duplication: None. IR: 26 bp. Conjugative transposon (transfer 10^{-8} to 10^{-5} per donor). Related to Tn*918*, Tn*919*, Tn*925*, and Tn*1545*. Inserts into many sites. Transposition can be zygotically induced. Transposition is associated with excision from donor. Circular transposition intermediate transposes. In *E. coli*, precise excision occurs at high frequency and transposition at a much lower frequency. (82–84, 142, 341, 343)

Continued on following page

Table 1—*Continued*

Designation	Marker(s)[a]	Size (kb) and structure	Comments[b] (references)
Tn*1545*	*tet, erm, kan*	25.3	Conjugative transposon related to Tn*916*. More stable in *E. coli* than Tn*916*. Can transpose in *E. coli.* (60, 87)
Tn*917*	*erm*	5.3; no IS elements	Duplication: 5 bp. IR: 38 bp. In Tn*3* family. Related to Tn*551*, Tn*3871*, Tn*4430*, Tn*4451*, and Tn*4452*. Transposition induced by erythromycin. Transposes to chromosome or plasmid. Can transpose in *E. coli.* (347, 379; Murphy, this volume)
Tn*917-lac*	*erm, 'lac*	8.3	Forms type I fusions. (294, 417)
Tn*917-cam*	*erm, 'cam*	6.5	Contains promoterless *cat-86* gene from *Bacillus pumilus.* Forms type I fusions. (417)
Tn*917-lac-cam*	*erm, 'lac, 'cam*	9.6	Contains promoterless *lac* and *cat-86* genes in tandem. Forms type I fusions. (417)
Tn*551*	*erm*	5.3	Duplication: 5 bp. IR: 40 bp. Related to Tn*917*; member of Tn*3* family. Transposes preferentially to the chromosome, with low insertion specificity. (196, 243; Murphy, this volume)
Tn*554*	*erm, spc*	6.7	Duplication: None. IR: None. Transposes by site-specific recombination preferentially to one site (*att554*). No excision detected. Has three transposase genes. (217, 272, 299; Murphy, this volume)

[a] Resistance determinants: *amp*, ampicillin; *ble*, bleomycin; *cam*, chloramphenicol or fusidic acid; *erm*, erythromycin; *gen*, gentamicin; *kan*, kanamycin or neomycin; *mer*, mercury; *spc*, spectinomycin; *str*, streptomycin; *tet*, tetracycline; *tmp*, trimethoprim; *tra*, F-factor transfer genes; *trp*, *E. coli trp* operon; *imm*, phage immunity.

[b] Duplication, Number of base pairs of target DNA duplicated during transposition. Lambda, unlike the other elements described here, inserts by a site-specific mechanism, without duplicating host sequences. The designations of Casadaban and Chou (65) for fusion elements are adopted here. I designates operon fusion elements in which the promoterless reporter gene has a translation start site; II designates gene (protein) fusion elements in which the reporter gene has neither a transcriptional nor a translational start site.

[c] Tn*3* is the most commonly used member of a number of identical or nearly identical ampicillin resistance transposons (e.g., Tn*1*, Tn*2*, Tn*801*, Tn*901*, and Tn*1701*) (Sherratt, this volume).

[d] The *str* gene is cryptic in *E. coli* but can be activated by mutation (257).

Figure 1. Structures of representative transposable elements (not to scale). The vertical filled bars are sites of action of transposition proteins. The hatched bars are the sites of action of resolution proteins (TnpR). Open boxes represent genes involved in transposition and resolution: Mu, A (and B); λ, *int* and *xis*; Tn*5*, *tnp/inh*; Tn*10*, *tnp*; Tn9, A and B; Tn7, A to E; and Tn*3* family (including gamma-delta, Tn*501*, Tn*1721*, and Tn*917*), *tnpA* and *tnpR*. IS*50*L of Tn*5* contains an ochre allele of the *tnp/inh* gene, which is expressed only in ochre-suppressing strains. IS*10*L of Tn*10* contains numerous sequence changes that render its *tnp* gene nonfunctional. Curved arrows indicate the site of action of the repressor of transposition (the Tn*501* repressor may also act at a second site close to *tnpR*). The resistance determinants are: *amp*, ampicillin; *cam*, chloramphenicol; *kan*, kanamycin; *ble*, bleomycin; *str*, streptomycin; *tet*, tetracycline; *dhfr*, dihydrofolate reductase (trimethoprim resistance); *mer*, mercury; *erm*, erythromycin. Truncated *lacZ* and *phoA* reporter genes and also the truncated P22 *sieA* and Tn*1721 tnpA* genes are indicated by a ′ (prime) at the 5′ end. p_{T7}, p_{SP6}, and p_{tac} designate outward-facing promoters from phage T7 and SP6 and a *trp-lac* (*tac*) hybrid promoter, respectively. Other symbols: *rep*, plasmid replication origin; *c*I, R, A, and J, λ genes; *cos*, cohesive end of λ; *erf*, essential recombination function of P22; *pac*, packaging initiation site of P22; O and I, outside and inside IS ends, respectively, of compound transposon; *inh*, inhibitor of Tn*5* transposition; *URA3*, *S. cerevisiae* uracil biosynthetic gene. A minor transposon in Tn*917* is suggested by the DNA sequence (347). The transposase genes and end sequences of most of the elements are not related. The transposase genes and end sequences of members of the Tn*3* family, although related, have diverged sufficiently that they do not interact (although the Tn*3* and gamma-delta resolvases will compensate for each other). For detailed descriptions of these and other elements see the references listed in Table 1, recent books about λ (171) and Mu (371), and the following chapters in this volume: Mu (Pato); Tn*5* (Berg); Tn*10* (Kleckner); Tn9 (Galas and Chandler); Tn7 (Craig); Tn*3*, gamma-delta, Tn*501*, and Tn*1721* (Sherratt); Tn*917* (Murphy); λ (Thompson and Landy).

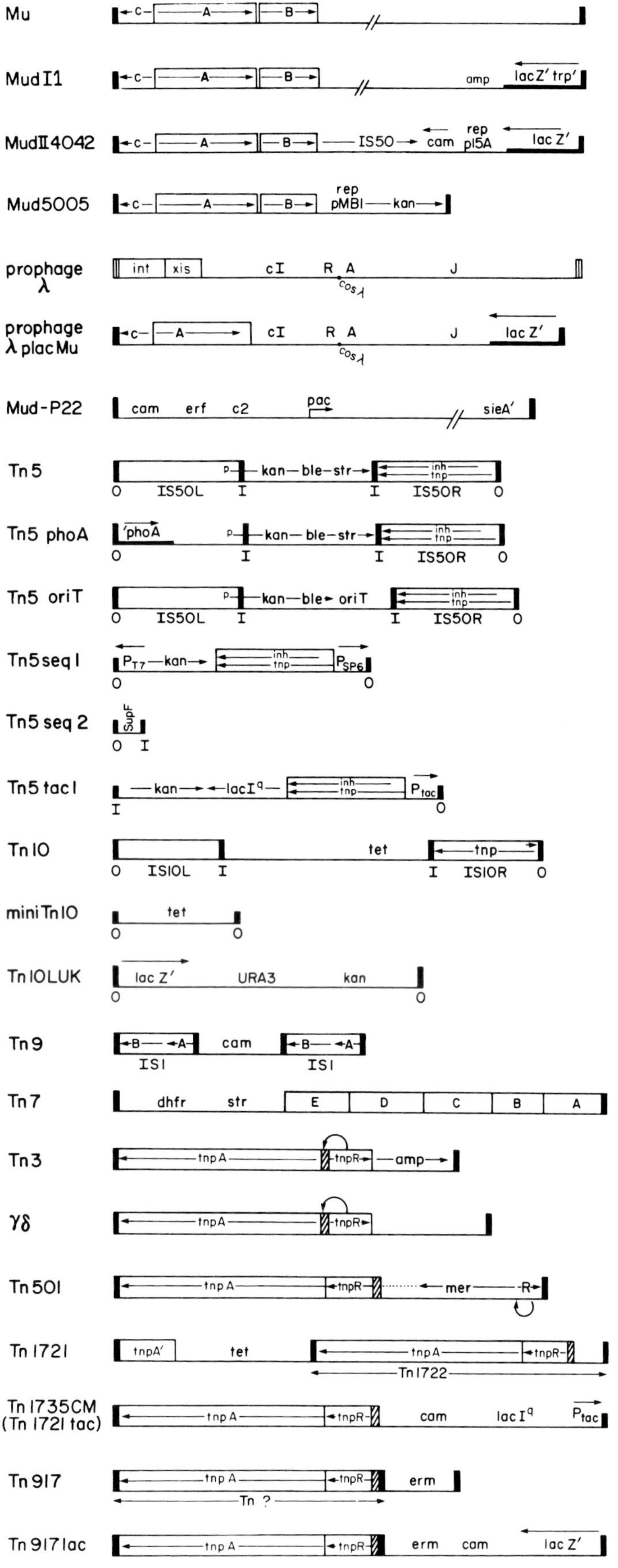
Mu
c
A
B
Mud I1
amp
lacZ′ trp′
MudII4042
IS50
cam
rep
p15A
lac Z′
Mud5005
rep
pMB1
kan
prophage λ
int
xis
cI
R A
J
cos λ
prophage λ placMu
lac Z′
Mud-P22
cam
erf
c2
pac
sieA′
Tn5
p
kan—ble—str
inh
tnp
O
IS50L
I
IS50R
Tn5 phoA
′phoA
Tn5 oriT
kan—ble
oriT
Tn5seq1
P_{T7}
P_{SP6}
Tn5 seq 2
SupF
Tn5 tac1
lacI^q
P_{tac}
Tn10
IS10L
tet
tnp
IS10R
miniTn10
Tn10LUK
lac Z′
URA3
kan
Tn9
IS1
cam
Tn7
dhfr
str
E
D
C
B
A
Tn3
tnpA
tnpR
amp
γδ
Tn501
mer
R
Tn1721
tnpA′
Tn1722
Tn1735CM (Tn1721 tac)
Tn917
erm
Tn ?
Tn917lac

Table 2. Bacterial species in which transposons have been used

Organism	Element	Vector or donor[a]	Reference
Gram-negative species			
Acinetobacter calcoaceticus	Tn*5*	pJB4JI	358
	Tn*10*	pRK2013	118
Acinetobacter sp. strain HO1-N	Tn*5*	pJB4JI	358
Aeromonas salmonicida	Tn*5*	pJB4JI	21
Agrobacterium radiobacter	Tn*5*	pCHR81	271a
A. rhizogenes	Tn*5*	pJB4JI	406
A. tumefaciens	Tn*5*	pJB4JI	140, 179, 301
	Tn*5*	pCHR81	271a
	Tn*3*, Tn*3*-HoHo1	In *E. coli*	109
	Tn*5*-PV, Tn*5*-*luxAB*	pGV748	213; Koncz-Kalman, personal communication
Alcaligenes eutrophus	Tn*5*	pJB4JI	366
	Tn*5*, Tn*5*-*mob*	pSUP5011	221
Azospirillum brasilense	Tn*5*	pJB4JI	117
	Tn*5*	pGS9, pSUP2021, pJB4JI	359, 392
	Tn*5*-*mob*	pSUP5011	392
A. lipoferum	Tn*5*	pGS9, pSUP2021	392
Azotobacter beijerinckii	Tn*5*, Tn*7*, Tn*76*	RP4	290
A. chroococcum	Tn*1*		306
A. vinelandii	Tn*5*-*tet*	pRK2013	S. Phadnis, personal communication
	Tn*5*	pSUP1011	191
	Tn*5*		380
	Tn*3*, Tn*10*	pRK2013	85, 298
Bacteroides fragilis	Tn*4551*	pFD197	360
Bordetella pertussis	Tn*5*	pUW964, pRKTV5	405
	Tn5tac1	pSUP201	B. Cookson and W. Goldman, personal communication
	Tn*7*, Tn*501*	pUW942	404
Bradyrhizobium japonicum	Tn*5*	pSUP1011	323
Bradyrhizobium sp. strain NC92	Tn*5*	pGS9	409
Brucella abortus	Tn*5del22*	pDG4	363
	Tn*5*-*lac*	P1	363
Caulobacter crescentus	Tn*5*-VB32	pJB9JI	22
	Tn*5*	pJB4JI	111
	Tn*7*	RP4	119
	Tn*7*, Tn*5*-*132*	pRK2013	118
Erwinia carotovora	Tn*5*	pJB4JI	Zink et al.[b]
	Tn*5*	pCHR81	271a
	Tn*phoA*	Lambda	175
E. chrysanthemi	Tn*5*	pJB4JI	73, 377
	Mini-Mu	pSUP2021, pULB113	120
E. uredovora	Tn*5*	pJB4JI	108
Escherichia coli K-12	Tn*5*, Tn*10*, Mu d	Various	See reference 24
Klebsiella aerogenes	Tn*5*	P1	304
	Tn*10*, Tn*5*-*mob*, λ, λ *plac*Mu		R. Bender, personal communication
	Mu d1, Mu d5005	Mu	Bender, personal communication
K. pneumoniae	Mu d5005	Mu	104
Legionella pneumophila	Tn*5*	pRK340	200
	Mu dI1681	pRK24.1	266

Continued on following page

Table 2—*Continued*

Organism	Element	Vector or donor[a]	Reference
Methylobacterium organophilum	Tn*5*	In *E. coli*	6
Methylobacterium sp. strain AMI	Tn*5*	R91-5	407
Myxococcus fulvus	Tn*5*	RP4, pME462	Saulnier et al., submitted[c]
M. stipitatus	Tn*5*	RP4, pME462, pCM2019	Saulnier et al., submitted[c]
M. xanthus	Tn*5*	P1	46, 223, 224, 255, 319
	Tn*5-lac*	P1	218, 319
	Tn*5*-V	P1	137
	Tn*5*	RP4, pCM2019	Saulnier et al., submitted[c]
Proteus mirabilis	Mu dII5117	pSC101	160
Pseudomonas aeruginosa	Tn*1*	pME19	280
	Tn*5*	pUW964	365
	Tn*5*, Tn*5-751*	pME305	310
	Tn*501*	Gene replacement	180
	Mini-D3112	D3113	A. Darzins and M. Casadaban, personal communication
P. cepacia	Endogenous IS		236
	Tn*5*	pW12811	300
P. facilis	Tn*5-mob*	pSUP5011	401
P. fluorescens	IS*50*L-*tox*	pSUP1011	284
	Tn*7-lac*	pUC derivatives	16a
P. putida	Mu dI1	R751	357
	Mu dII1681	R388	344
	Tn*5*	pLG221	52
P. solanacearum	Tn*5*	RP4	51
	Tn*5*	pSUP2021	411
	Tn*5*	pW1757	268
P. syringae pv. *phaseolicola*	Tn*5*	NPS3121	240
	Tn*5*	pUW964	292
	Tn*5*	pSUP1011	7
	Tn*5*		S. Patel, personal communication
P. syringae pv. *syringae*	Tn*5*	pSUP1011	7, 269
P. syringae pv. *tabaci*	Tn*5*	pGS9	328
P. syringae pv. *tomato*	Tn*5*	pGS9	91
P. syringae	Tn*903*	pRK2013	230
P. stutzeri (*perfectomarina*)	Tn*5*	pSUP2021	393
Rhizobium fredii	Tn*5*	pJB3JI	376
	Mu d		327
R. leguminosarum	Tn*5*	pJB4JI	44
	Tn*5*-GmSpSm	pXS102	176
R. loti	Tn*5*	pSUP1011	79
R. meliloti	Tn*5*	pJB4JI	112, 128, 259
	Tn*5*	pSUP1011, pSUP2011, pSUP2021	356
	Tn*5*	pRK600	105
	Tn*phoA*	pRK600	241b
	Tn*5*-Tc	pSUP1041	356
	Tn*5-233*	pRK2013	105
	Tn*5-235*	pRK600	105
R. phaseoli	Tn*5*	pJB4JI	44
R. trifolii	Tn*5*	pJB4JI	44, 396
Rhizobium sp., slow growing	Tn*5*	pJB4JI, pSP601	68
Rhizobium sp. strain NGR234	Tn*5*	pACYC184-Mob	74
Rhodobacter capsulatus	Tn*5*	pSUP201	94, 210
	Tn*7*	On R-prime	420
R. sphaeroides	Tn*5*	pSUP202	186
Salmonella typhimurium	Tn*5*, Tn*10*, Mu d	Various	See reference 24

Continued on following page

Table 2—*Continued*

Organism	Element	Vector or donor[a]	Reference
Shigella flexneri	Tn*5*	pCHR81	333
	Tn*10*	F′ (Ts)114*lac*	277
	Mu dII4042	Mu	160
Thiobacillus novellus	Tn*501*	IncP plasmid	96
T. versutus	Tn*1721*	pMD105	96
Thiosphaera pantotropha	Tn*5*	pSUP5011	72
Vibrio anguillarum	Tn3HoHo1	In *E. coli*	378
V. cholerae	Tn*1*	pSJ25	193
	Tn*5*	F	193
	Tn*phoA*	pRK290, pRT733, F′	374, 375
	Tn*5*, Tn*10*	F'_{ts}*lac*	279
	dVcA1	pSJ15	194
V. fischeri	Mini-Mu-*lux*	P1	121
V. harveyi	Tn*5*	P1	18, 19
V. parahaemolyticus	Mini-Mu-*tet*	P1	19
	Mini-Mu-*lux*	P1	20, 122
Xanthobacter sp. strain H4-14	Tn*5*	In *E. coli*	234
Xanthomonas campestris	Tn*5*	pSUP1011	Kingsley and Gabriel[d]
	Tn*4431*	pSa325	348, 349
Yersinia pestis	Mu dI1	In *E. coli*	414
Gram-positive species			
Acholeplasma laidlawii	Tn*916*	pAM120	114
Bacillus anthracis	Tn*916*	*Streptococcus*	189
B. megaterium	Tn*917*	pTV1	50
B. subtilis	Tn*917*	pTV1	329, 389, 419
	Tn*917-lac*	pTV32	242
B. thuringiensis subsp. *israelensis*	Tn*916*	*Streptococcus*	276
Lactobacillus plantarum	Tn*919*	*S. faecalis*	173
Leuconostoc cremoris	Tn*919*	*S. faecalis*	173
Listeria monocytogenes	Tn*916*	*Streptococcus*	199
	Tn*1545*	*L. monocytogenes*	138
	Tn*917*	pTV32	P. Cossart, personal communication
Mycoplasma hominis	Tn*916*	*Streptococcus*	315
M. hyorhinis	Tn*916*	pAM120	113a
M. pulmonis	Tn*916*	pAM120	113a, 114
Staphylococcus aureus	Tn*916*	pAD1, pPD5	195
	Tn*916*	*S. aureus*	413
	Tn*551*	pI258 (Ts)	243
Streptococcus faecalis	Tn*916*		133
S. lactis	Tn*919*	pMG600	174
S. pyogenes	Tn*916*	*S. faecalis*	281
Group B streptococci	Tn*916*	*S. faecalis*	325

[a] In most examples cited, a suicide vector was used to transfer the element. In a few cases, cloned DNA was mutagenized in *E. coli* and then transferred back. Tn*916* and its relatives (Tn*918*, Tn*919*) are conjugative transposons, and in some cases the element was transferred from the donor genome, not on a plasmid.
[b] R. T. Zink, D. A. Feese, and A. K. Chatterjee, *Phytopathology* 72:933, 1982.
[c] P. Saulnier, J. Hanquier, S. Jaqua, H. Reichenbach, and J. F. Guespin-Michel, submitted for publication.
[d] M. T. Kingsley and D. W. Gabriel, Proc. 4th Int. Symp. Molecular Plant-Microbe Interactions, 1988.

transposition. Much more complex are the transposing bacteriophages, such as λ and Mu, that carry genes needed for phage growth and lysogenization as well as integration (transposition). Intermediate in size and complexity are transposons, which usually contain genes for auxiliary traits such as antibiotic or heavy metal resistance as well as genes for transposition. Members of the Tn*916* family also contain information for their own conjugal transfer. Many transposons are composite elements in which the auxiliary genes are bracketed by IS elements in inverted (e.g., Tn*5*, Tn*10*) or direct (e.g., Tn*9*) orientation (Fig. 1). Others, notably members of the Tn*3* family, have only one (Tn*1721*) or no (Tn*3*) smaller transposable elements within them and may have evolved from composite elements by internal deletions (35; D. E. Berg, *in* S. B. Levy and R. Miller, ed., *Gene Transfer in the Environment*, in press). Some elements are found embedded as "minor" transposons within other elements that are related in some cases (Tn*1722* in Tn*1721*, Tn*2501* in Tn*951*), but not in others (IS*1* in Tn*951*) (see Table 1 and Fig. 1).

In most elements, the sequences required in *cis* for transposition (where the transposition proteins act) are present at or near the termini and in many cases are short inverted repeats. For most elements, these sites are in the range of 20 to 40 base pairs (bp) in length (e.g., Tn*3*, Tn*5*, Tn*10*), but in others they are larger and more complex and may not be at the very ends (e.g., λ, Mu, Tn*7*).

III. TRANSPOSITION PRODUCTS

Most elements generate characteristic short direct repeats of DNA sequences where they insert, probably as a result of staggered cuts in target DNAs, joining the element with the overhanging target ends, and filling the gaps. Phage Mu, Tn*7*, and members of the Tn*3* family generate 5-bp repeats; Tn*5*, Tn*9*, and Tn*10* usually generate 9-bp repeats; other less widely used elements generate repeats of from 2 to 20 bp; while Tn*554* and λ insert without duplicating target sequences (see R. L. Charlebois and W. F. Doolittle, D. J. Galas and M. Chandler, Murphy, and M. L. Pato, all in this volume).

Transposition of an element between different replicons (intermolecular transposition) generates either simple insertions (free of donor sequences) or cointegrates (consisting of donor and recipient DNAs joined by direct repeats of the element), depending on the mode of transposition. Conservative ("cut-and-paste") transposition (e.g., Tn*5* and Tn*10*) generates simple insertions (23, 32, 33, 40; see D. E. Berg, this volume, and N. Kleckner, this volume), whereas replicative transposition (e.g., Tn*3* and gamma-delta [Tn*1000*]) generates cointegrates (9, 147, 345; see D. Sherratt, this volume) (Fig. 2). The cointegrates can be resolved by recombination into two smaller DNAs, a genetically unchanged donor molecule and a target molecule, now containing a copy of the element. This recombination is catalyzed by a highly efficient site-specific "resolvase" protein in some cases (e.g., Tn*3* and gamma-delta) and by the less efficient host homologous recombination system in others (e.g., Tn*9* [IS*1*]). Resolution is so efficient for Tn*3* that cointegrates are detected only when mutant elements, lacking resolvase, are used (169; Sherratt, this volume). The free target molecules released by cointegrate resolution are indistinguishable from simple insertion products, while a single reciprocal crossover between homologous insertions in different replicons produces a structure indistinguishable from a cointegrate (Fig. 2, bottom). Some elements, such as Mu and Tn*9*, seem to transpose by both conservative and replicative mechanisms (45, 371; Pato, this volume). For example, the very first insertion of Mu after phage infection is conservative, but transposition during lytic growth is replicative.

In elements such as Tn*10*, Tn*5*, and Tn*9* that have two complete IS elements, the interior IS ends can also participate in transposition, an event referred to as "inverse transposition" (71) or sometimes "inside-out" transposition (130) (Fig. 3). In this case the component IS elements and the DNA between them (the "vector" in direct transposition) move away from the central region of the parental transposon (e.g., deleting the *tet* region of Tn*10*). A segment that may be as large as the entire bacterial chromosome can thus become the central portion of a functional transposon (39, 71, 166, 305, 321). When inverse transposition occurs within one molecule, rearrangement of adjacent DNA also occurs. Depending on the relative orientations of the element and the target site during inverse transposition, the product may retain the two IS elements of the parental transposon, which are now separated by an inverted sequence, or it may consist of two smaller circles, each of which retains one IS element and part of the original replicon (Fig. 4). One daughter circle will generally lack an origin of replication and be lost, while the surviving replicon will contain one IS element and the complementary portion of the parental replicon: it will have a deletion extending from what had been the inside end of that IS element.

Intramolecular replicative transposition also yields different rearrangements, depending upon the relative orientations of the element and target site: two copies of the entire transposon may be separated by portions of the parental replicon (with one portion

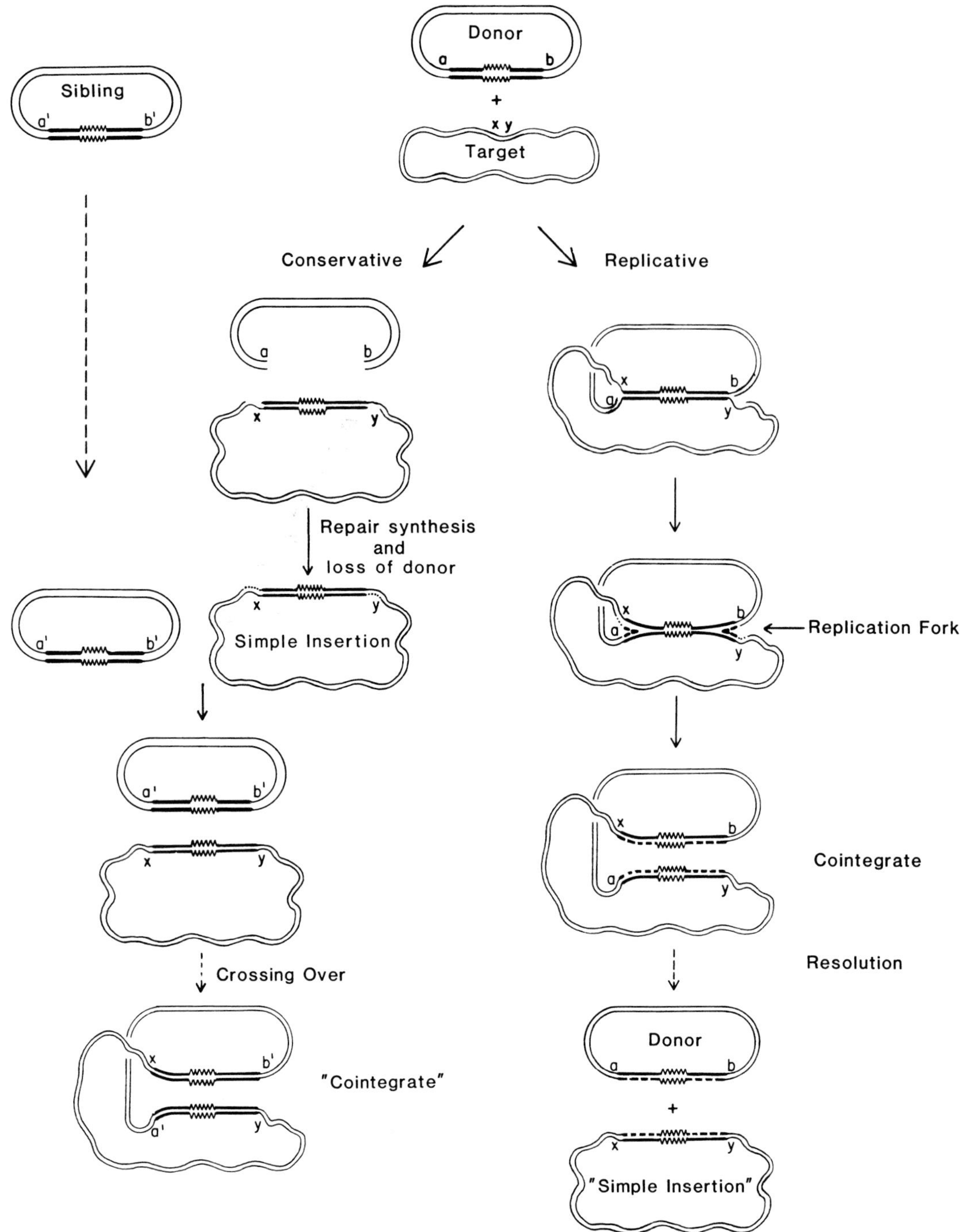

Figure 2. Formation of cointegrates and of simple insertions by both conservative and replicative transposition (modified from reference 24). Transposition is to the x-y target site. The first step in conservative transposition involves excision of the transposable element from the vector (which is degraded) and insertion into a staggered cut in the target. The single-stranded DNA at the site of insertion is repaired to form a characteristic short direct repeat of target sequences bracketing the insert (23, 32, 33). Apparent "cointegrates" could also arise subsequently by a crossover between the transposed element in the target replicon and another element in a sibling of the donor replicon. The first step in replicative transposition involves single-stranded nicks at the element-vector junctions and a staggered cut in the target. Replication of the element after joining of the element and target DNAs results in a cointegrate composed of the two parental replicons bounded by direct repeats of the element (9, 345). An apparent "simple insertion" bracketed by the characteristic short direct repeat of the target sequence could also arise subsequently by a crossover (resolution) between these daughter elements. Resolution is mediated by a site-specific recombinase in some elements (e.g., Tn*3*) or by host-mediated general recombination. In the mobilization of nonconjugative plasmids by conjugative plasmids carrying elements such as gamma-delta and Tn*3*, resolution occurs after transfer into the recipient. Symbols: heavy line, transposable element; a and b, loci adjacent to the element in the donor; x and y, loci bracketing the target site; zigzag, central region of transposon; dotted line, repair synthesis; heavy dashed line, newly replicated strand.

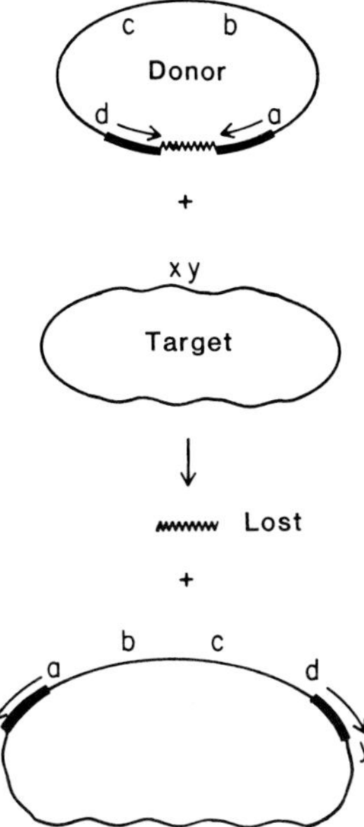

Figure 3. Inverse transposition (modified from reference 24). The inside, rather than the outside, ends of the IS elements of a compound transposon participate in inverse transposition, moving the vector with concomitant destruction of the former interstitial region of the transposon (71). Symbols: Heavy line, IS element; zigzag, central region of transposon.

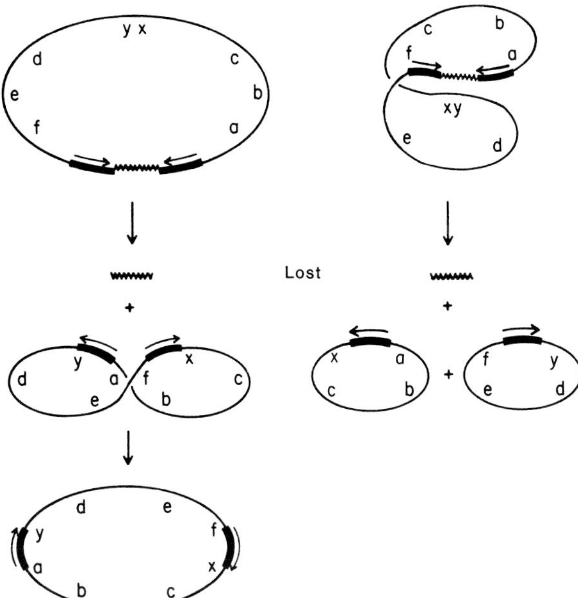

Figure 4. Viable products of intramolecular conservative transposition (modified from reference 24). Transposition is to the x-y target site, with the resulting structures depending upon strand alignment. Only inverse transposition products survive because direct transposition to a vector sequence converts the donor molecule to a linear structure, which is degraded (see Fig. 2). Symbols: Heavy line, IS components of compound transposon; arrows, orientation of the IS components; zigzag, interstitial resistance determinant; a and f, loci adjacent to the element in the donor; x and y, loci bracketing the target site. Left, Transposition in an untwisted molecule results in loss of the central region of the transposon, inversion of one IS component, and inversion of the segment between the two elements. Crossing over between the IS elements would result in two daughters, indistinguishable from those in the right panel. Right, Transposition in a twisted molecule results in loss of the central region of the transposon and the formation of two complementary molecules, each with one IS component. Unless each daughter molecule contains an origin of replication, one will be lost and the other will appear to have suffered a deletion adjacent to the remaining IS element. Many deletions associated with Tn*5*, Tn*9*, and Tn*10* probably arise this way.

inverted), or two daughter circles may be produced, each with a copy of the transposon and a portion of the parental replicon (Fig. 5). Although in this latter case one daughter will be lost during transposition if it lacks an active replication origin, the element is never completely excised from its vector. In contrast, conservative transposition always results in loss of the segment from which transposition took place, which in inverse transposition is the central region of the ancestral transposon (Fig. 4 versus Fig. 5).

Deletions, inversions, and duplications are associated with the movement of certain transposable elements including Mu (382; Pato, this volume), Tn*3* (169, 264), IS*1* (Tn*9*) (4, 61, 309), and IS*10* (Tn*10*) (204, 282, 322; Kleckner, this volume), and infrequently with IS*50* (Tn*5*) (28, 110). Some rearrangements arise following crossing over between insertions at different sites in the same or different replicons. In these cases the transposons function solely as "portable regions of homology" (see references 24 and 208 and Fig. 2, bottom left). Other rearrangements arise as primary products of intramolecular transposition, either replicative (Fig. 5) or conservative (Fig. 4). Intramolecular transposition of Tn*9* and Tn*10* has been used to generate deletions (see Galas and Chandler, this volume, and Kleckner, this volume) and, in the case of Tn*9*, for DNA sequencing (4, 5; see Section XIII).

IV. TARGET SPECIFICITY

Transposable elements vary greatly in target specificity, and their insertion sites often exhibit little or no similarity to element sequences. Some of the most commonly used elements, notably Tn*5*, Tn*10*, and Mu, may have a few to hundreds of target sites in any gene, although, as discussed below, no element has been found to insert completely randomly. Other elements, such as λ, Tn*7*, and Tn*554*, are highly site specific, inserting very efficiently into just a single chromosomal site in a variety of gram-negative (Tn*7*) and gram-positive (Tn*554*) organisms. Yet other elements seem to insert quasi-randomly in some species or replicons, but prefer either particular DNAs or

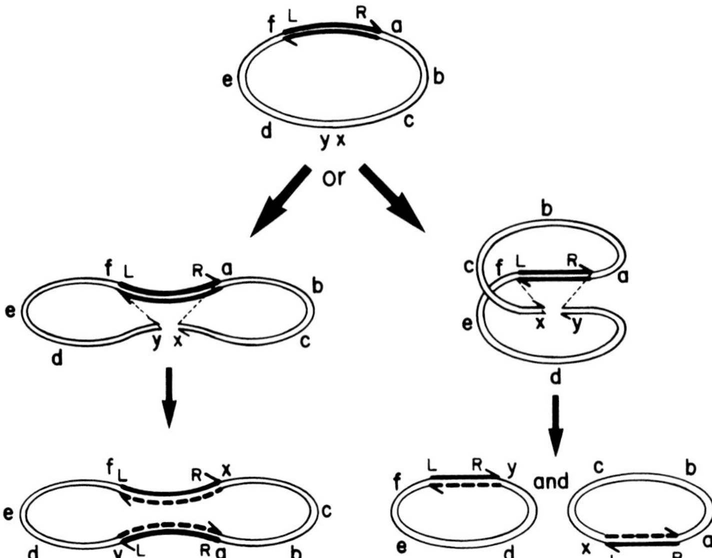

Figure 5. Deletion and inversion products of intramolecular replicative transposition (modified from reference 24). Transposition is to the x-y target site, with the resulting structures depending upon strand alignment. Left, Transposition in an untwisted molecule results in duplication and inversion of the transposon and inversion of the segment between the two daughter elements. Crossing over between the elements would result in reinversion of the segment between them. Right, Transposition in a twisted molecule results in duplication of the transposon and splitting of the molecule into two, each with one daughter element. Unless each daughter molecule contains a replication origin, one will be lost and the other will appear to have suffered a deletion adjacent to the remaining element. Many Tn*3*-, Tn*9* (IS*1*)-, and Mu-associated deletion events can be explained in this way. Symbols: Heavy lines, transposable element (halved arrowheads show strand orientation); heavy dashed line, newly replicated strand; light dashed line, strand attachment; L and R, ends of element.

certain regions of DNA molecules. Especially striking are the preference of Tn*3* and several closely related elements for bacterial plasmids, and their inability to insert into plasmids that already contain a homologous transposon end (termed transposition immunity) (Sherratt, this volume). Members of the Tn*3* family and also Tn*7* and Mu exhibit transposition immunity in plasmids, although structures containing duplicate copies of these elements can be constructed in vitro (3, 94a, 151, 233, 312, 316, 385; see N. L. Craig, this volume; Pato, this volume; Sherratt, this volume). However, Mu transposition immunity seems to be weak in large DNAs because during replicative transposition multiple insertions of Mu occur in the chromosome, and pairs of insertions, that are "trapped" during mini-Mu in vivo cloning (see Section XII), can be as close as 2 kb (161, 381, 399; B. Wang, L. Liu, and C. M. Berg, *Genetics*, in press).

When the high-frequency primary insertion sites used by λ, Tn*7*, or Tn*554* are not present, the insertion frequency is decreased, but secondary sites, which often resemble the primary sites in sequence, are used (see J. F. Thompson and A. Landy, Craig, and Murphy, all this volume). Tn*7* inserts into many sites in plasmids that lack a primary site, but into only one site in the *E. coli* chromosome, and transposition into primary and secondary sites involves different transposon-encoded products (Craig, this volume).

Mu and its derivatives can insert into any region of the bacterial chromosome, and Mu seems to be one of the most random of known elements when insertions in specific regions are mapped, although some hot spots are found (see reference 69 and Pato, this volume). Nevertheless, analyses of mini-Mu generalized transduction and mini-Mu in vivo cloning, discussed below, suggest considerable regional nonrandomness, with both "hot" and "cold" regions, extending over perhaps 20 kb (397, 398; Wang et al., in press). It is not yet clear whether the differences in insertion frequencies observed during replicative transposition are due to actual insertion specificity or to subsequent biases, such as preferential packaging of Mu DNA in certain regions at the end of the lytic cycle. Such a bias may be responsible for the ease with which insertions of mini-Mu, but not of Mu or large derivatives, can be obtained in multicopy plasmids (67; Pato, this volume).

Gamma-delta and Tn*3* insert quite randomly into plasmids, with no significant hot spots (163, 216). The one report of an apparent hot spot for gamma-delta insertion (197) is due to the preferential recovery of plasmids in which a deleterious plasmid-borne gene is inactivated (P. Karlovsky, L. Liu, and C. M. Berg, unpublished data).

Tn*5* will insert into any gene. In many genes, the insertions are quite randomly distributed, while in others significant hot spots are also found (in pBR322, the insertions into *amp* are nearly random, but one-fourth of the insertions into *tet* are at one site [42, 274]). Insertions of Tn*10*, like those of Tn*5*, can be found in many sites. However, hot spots account for a majority of the insertions (over 40% of the insertions in the nine-gene *his* operon of *S. typhimurium* are at one site in *hisG* [209]). Tn*9* (IS*1*) also inserts into many sites, but significant hot regions, rather than hot spots, are found (80% of the insertions into pBR322 are in one 200-bp segment [422]).

Since the insertion specificities of different elements have not been compared under similar conditions, extrapolations must be made from data obtained using different strains under nonuniform conditions. Any of several elements can be used for insertion into a gene or region of interest, but Mu and Tn*5* insert more randomly into the chromosome than do other elements. While Mu exhibits strong regional specificity, it has little or no site specificity within a given region. On the other hand, Tn*5* does not exhibit regional specificity (at least in plasmids and phage), but has occasional insertion hot spots superimposed upon a fairly random background. The patterns of Tn*10* and Tn*9* insertion are more complex, with a relatively few hot spots or preferred regions accounting for most insertions.

For those elements where a number of insertion sites have been characterized, preferred types of sequences, sometimes even consensus sequences, can often be identified. For example, of the elements that make 5-bp target duplications, gamma-delta inserts preferentially into AT-rich sites, with a weak consensus sequence of 5′-ATATT-gamma end (241); Tn*3* inserts preferentially into AT-rich regions that resemble Tn*3* end sequences (386); and Mu inserts preferentially into GC-rich sites (M. Mizuuchi and K. Mizuuchi, personal communication; B. A. Castilho and M. Casadaban, personal communication). Of the elements that make 9-bp duplications, Tn*10* preferentially inserts into sites that are NGCCNGGCN (165), Tn*5* preferentially inserts into sites that have G or C at positions 1 and 9 (41, 241a), and Tn*9* (IS*1*) inserts preferentially into regions of high AT content (422) that have G or C at positions 1 and 9 (264). Thus, transpositional site specificity should be considered in choosing an element when the target DNA is either AT- or GC-rich.

Target recognition often involves more sequence information than that found in the segments that are duplicated at the preferred sites. As discussed above, for those elements that exhibit transposition immunity (Tn*3*, gamma-delta, Tn*7*, Mu) the presence of a transposon end in a plasmid prevents the insertion of additional copies of that element into the same plasmid. In addition, sequence changes of up to 15 bp to one side of a major Tn*5* insertional hot spot strongly decrease insertion into that site (J. K. Lodge and D. E. Berg, unpublished).

V. HOW TO DELIVER TRANSPOSONS

Transposon mutagenesis strategies can be either direct or indirect. In direct mutagenesis the transposon is introduced into the bacterial replicon or species of interest, while in indirect mutagenesis, sometimes referred to as reverse mutagenesis, transposition is carried out in the most convenient host (generally *E. coli*), using protocols to effectively saturate the cloned DNA fragment with transposon insertions. Mutated clones are then returned to the organism of interest, and the phenotypic consequences of the insertion are scored.

Many suicide vectors have been developed that can be used for transposon mutagenesis or delivery. In addition to those listed in Table 2 or cited here, additional vectors are described in depth in several recent books and reviews (134, 317, 410, 417).

A. Direct Transposon Mutagenesis

Two general strategies are employed in direct mutagenesis: either the transposon is introduced into a bacterium, or the transposon's movement to another replicon within the cell is selected, usually by transfer of that replicon to a recipient that is free of the element. Appropriate delivery systems are transduction, conjugation, and transformation. Usually, the transposon is delivered in a vector that cannot replicate in a particular host or under a particular set of conditions, using, for example, plasmids or phages with conditional mutations. Initially, Mu and λ insertions were obtained simply by infecting sensitive cells, selecting lysogens by their immunity to superinfection, and then screening for mutants. Most nonphage transposable elements, in contrast, cannot replicate autonomously, and common strategies for their use involve selection for transposition from one replicon to another and either loss of the donor replicon or transfer of the mutagenized plasmid or phage target molecule to another cell. There are a variety of vectors that are not stably maintained in a chosen recipient strain because of a narrow host range for replication, but broad host range for transfer (223, 224, 407); mutations in the vector (32, 206, 216, 218, 265); mutations in the recipient that make the vector replication deficient (156, 387); plasmid-plasmid incompatibility (131, 304); plasmid-host incompatibility (224); or restriction of vector DNA (25).

A strategy to mutagenize phage DNA involves transposition from the chromosome or a plasmid to a phage DNA target, lytic phage growth, and use of virions to transduce a transposon-free culture (31, 36, 205, 206, 223, 247).

Plasmid DNA can be efficiently mutagenized by selecting for cointegrate-mediated conjugal transfer, for example, by replicative transposition of gamma-delta from the F factor to a nonconjugative plasmid (163) or by conservative transposition of Tn*5* and then recombination between homologous elements (39). In addition, transposition from the chromosome or a low-copy-number plasmid to a multicopy plasmid can be obtained by selecting for high-level antibiotic resistance (42).

Generalized transducing phage can be used to introduce a donor fragment that carries a transposon, but cannot recombine with the chromosome (177, 320). Even if the donor fragment can recombine with the chromosome, transposition, rather than recombination, can be obtained at a useful frequency if the regions of homology bracketing the element are small and the transposition frequency is high (184).

Endogenous transposable elements can insertionally activate cryptic chromosomal genes or inactivate conditional lethal genes (143, 236, 313).

The choice of transposition vector will depend on the particular bacterial species and even the particular strain to be used, and often on the transposition target as well.

1. Phage vectors

Derivatives of phage λ that cannot insert into the chromosome because their attachment site is deleted, and that are also conditionally defective because of nonsense mutations in replication genes, are commonly used as transposon vectors in *E. coli*. Phage λ normally has a narrow host range because its receptor is the product of the *E. coli lamB* gene (340) and it requires specific host gene products for lytic growth (135, 136, 146). Although several members of the family *Enterobacteriaceae* have a membrane protein similar to the *lamB* gene product, most are λ resistant (47, 48, 340). The introduction of a cloned *E. coli lamB* gene into nominally λ-resistant species makes some accessible to λ infection, but unable to support subsequent lytic growth because replication, transcription, or morphogenesis is blocked. Among the organisms which have been rendered λ sensitive by using a cloned *E. coli lamB*$^+$ gene are isolates of *Klebsiella pneumoniae* (107), *S. typhimurium* (107, 167, 168), and *Agrobacterium tumefaciens*, *Pseudomonas aeruginosa*, *Rhizobium meliloti*, and *Azorhizobium sesbaniae* (244).

Phages, such as P1 and P22, that package phage DNAs by a headful mechanism can also be used as vectors. The addition of large DNA segments, such as the 9.3-kb Tn*10* element to P22 or the 12-kb Tn*5*-*lac* element to P1, results in a unit genome that, instead of being terminally redundant, exceeds the packaging capacity of the phage head (70a, 218). In these cases, although the phage head is filled to capacity with DNA, individual particles contain only partial phage genomes. Single infection by such phage can be used to deliver the transposon because the infecting phage DNA cannot circularize and form a stable prophage. However, because the packaged DNAs are permuted, multiplicities of infection greater than one phage per cell permit the formation of complete circular genomes by recombination and successful lytic growth.

Additionally, if the recipient has a strong restriction system that differs from that of the donor strain, the incoming DNA will be restricted, but since the transposon represents a smaller target than the vector and restriction is not instantaneous, transposition can be selected. This approach was used to mutagenize *S. typhimurium* by using P1::Tn*5* grown on *E. coli* (31), but was not efficient enough to mutagenize *E. coli* K-12 with P1::Tn*5* grown on *S. typhimurium* (W. Whalen and C. M. Berg, unpublished data).

Phage P1 can infect, but not lysogenize, many gram-negative species and has been used to deliver Tn*5* and mini-Mu to *Vibrio fischeri* (19, 20) and to deliver Tn*5* and its derivatives to *Myxococcus xanthus* (113, 223, 224), to *Brucella abortus* (363), and, as noted above, to *S. typhimurium*. Mutations that render certain P1-resistant species sensitive to P1 have been obtained (150, 320). In most cases these mutations are uncharacterized, but *galE* mutants of *S. typhimurium* are P1 sensitive because their surface lipopolysaccharides are deficient in galactose residues that normally block phage attachment (289). *galE* mutants of *S. typhimurium* have been extensively used for Mu d1 mutagenesis in which the helper Mu carries host range genes from P1 (90).

2. Plasmid vectors

Any conjugative plasmid that harbors a transposon could, in principle, potentially be used for transposon mutagenesis. Transposition is typically quite rare, however, and suicide plasmids have the advantage that they deliver the element and are then lost. Even rare transposon insertions are thus isolated efficiently in cells that are free of the vector. Mutant F-primes or other plasmids that are conditionally defective in replication or maintenance and are lost at high temperature have been used extensively to obtain transposon insertions simply by selection for the transposon-encoded drug resistance at the restrictive temperature (80, 126, 334).

ColE1-derived plasmids (e.g., pBR322 derivatives, including the pUC series) and P15A-derived plasmids (e.g., pACYC184, pACYC177) cannot replicate in many nonenteric gram-negative species. Derivatives that carry the origin of transfer, but not the replication genes, of the broad-host-range plasmid RK2 (similar or identical to RP4 and RP1) have been used to efficiently deliver Tn*5* and Tn*7* to strains in which they cannot be maintained (356). Analogous suicide vectors could be developed using plasmids that require specific host gene products for their maintenance (e.g., *polA* by ColE1-derived plasmids [203, 412]).

Another plasmid suicide delivery system potentially useful for many gram-negative species consists of a conjugative plasmid with the origin of replication (*oriV*) of R6K, but with the R6K-specific replicase supplied from a gene in the donor chromosome. This plasmid does not replicate after transfer to recipient cells (which lack the replicase gene), and hence rare transposition events can be selected using the transposon-specific resistance trait (265, 374). Since RK2-mediated conjugative transfer has been demonstrated between *E. coli* and several gram-positive species (383), this type of delivery system may have broad usefulness.

Mini-Mu can be delivered using self-transmissible plasmids because entry into nonlysogenic cells often results in zygotic induction of Mu and hence in transposition, often without host cell killing. This approach has been used to mutagenize the gram-negative bacterium *Legionella pneumophila* by a mini-Mu-*lac* reporter element, screening for a change in *lac* phenotype (266).

It is also possible to select for transposition by introducing two plasmids of the same incompatibility group into a strain and selecting for the inheritance of one plasmid and of the transposon from the other (279, 304). This procedure also can be used to select for the replacement of a wild-type allele by a cloned transposon-mutated allele (11, 127, 326).

If a conjugative plasmid carries an element that transposes replicatively, that plasmid can be used for efficient mutagenesis of DNAs cloned in a nonconjugative plasmid, regardless of whether the element itself has a selectable marker. This entails selection for transfer of the resistance determinant of the target nonconjugative plasmid to the recipient as part of a cointegrate (Fig. 1, right) and has been used extensively for mutagenesis by gamma-delta (162, 163). Since gamma-delta is responsible for virtually all F-factor-mediated mobilization of nonconjugative plasmids, selection for transfer of such a plasmid from an F^+ strain results in transfer of a cointegrate that is resolved in the recipient to yield a gamma-delta-mutagenized plasmid. Care must be taken, however, to use a monomeric target plasmid in a *recA* F^+ donor strain because a mutagenized dimeric target plasmid will contain only a single gamma-delta insert and will break down in a *recA*$^+$ recipient to yield nonmutant, as well as mutant, monomeric plasmids. This greatly reduces the mutagenesis efficiency (399; L. Liu and C. M. Berg, unpublished data), and such inadvertent use of dimeric targets is probably responsible for cases in which nonmutant plasmids were recovered following F-mediated mobilization (152, 198). When monomeric target plasmids are used, the absence of a selectable marker in gamma-delta does not affect its use in gene mapping and sequencing (see below) because every transferred plasmid is derived from a cointegrate, and generally every recipient cell will also contain the parental F factor (152). However, for further genetic manipulations, other members of the Tn*3* family which also transpose by cointegrate formation (e.g., derivatives of Tn*3* [342] and Tn*1721* [387, 388]) may be more useful than gamma-delta because their resistance determinants can be selected.

In gram-positive bacteria, where DNA transformation is often efficient (e.g., *Bacillus subtilis*), transformation with purified supercoiled Tn*916* DNA has been used to generate insertions directly (341).

3. Delivery of Mu and Mu-related elements

Phage Mu can infect many enteric bacteria. Its host range is determined by the orientation of the invertible G segment: G(+) phage make tail fibers that allow them to infect *E. coli* K-12, *S. typhimurium*, and *Arizona* spp., and G(−) phage make tail fibers that permit infection of *E. coli* C, *Shigella sonnei*, *Citrobacter freundii*, *Erwinia carotovora*, *Erwinia herbicola*, *Erwinia amylovora*, and *Enterobacter cloacae* (160, 211, 382; Pato, this volume). Other enterics that can be infected by Mu include strains of *Shigella flexneri*, *Serratia marcescens*, and *Klebsiella pneumoniae* (104, 160; R. Harshey, personal communication). The Mu receptors are sugar structures that are part of the lipopolysaccharide (330, 331). Certain bacterial strains are Mu resistant, but can mutate to Mu sensitivity when their resistance is due to blocking of the receptor sites by the presence of additional sugar moieties. Mutants sensitive to phage Mu have been isolated in *S. typhimurium* (123) and *K. pneumoniae* (12, 249) and are attributable to loss of these excess sugars. When the *musA1* mutation of *S. typhimurium* is introduced into *Salmonella enteritidis*, this species becomes Mu sensitive (271). The host range of phage Mu has been extended even further by replacing the Mu G region by the partially related P1 host range genes, yielding Mu hP1, which has been used to deliver Mu derivatives to *S. typhimurium* (90, 252), *Yersinia pestis* (149), and *Klebsiella aerogenes* (E. Groisman and M. Casadaban, unpublished results).

Transposable Mu-like phages have been isolated in several gram-negative genera in addition to *E. coli* (see reference 111). They include D3112 in *P. aeruginosa* (219) and VcA1 of *Vibrio cholerae* (194), thus extending the range of Mu-like mutagenesis systems. An ingenious screen, based upon the expectation that Mu-like phages would carry heterogeneous host DNA fragments at one end of the phage genome, led to a search among *P. aeruginosa* phage for those with a diffuse restriction fragment. This approach led to the discovery of a number of Mu-like isolates (219; see reference 111) and could be extended to screen for Mu-like phage in other species of interest.

A number of deletion derivatives of the *P. aeruginosa* Mu-like phage D3112 that contain a drug resistance marker and other useful traits have been constructed for transposon mutagenesis and in vivo cloning of *P. aeruginosa* and related species (A. Darzins and M. Casadaban, in preparation; see Table 1).

Recently, methods for transient *cis* complementation of mini-Mu elements deleted for the transposition genes have been reported for *E. coli* (67) and for *S. typhimurium* (184) that would be applicable to other Mu-sensitive and P22-sensitive species, respectively. In *E. coli*, a strain dilysogenic for an A^- B^- mini-Mu and an A^+ B^+ Mu helper was induced for transposition and phage growth. Many virions contain mini-Mu packaged along with the left end of a helper phage (carrying the A^+ and B^+ genes), and these yield mini-Mu transpositions in the recipient. In *S. typhimurium*, the mini-Mu elements were placed in the chromosome near an insertion of Mu d1 (66), which is competent for transposition but defective for Mu phage growth. P22-mediated transduction of the mini-Mu marker into another strain yielded primarily single mini-Mu insertion mutants in the recipient due to transposition of the mini-Mu element from transducing DNA fragments that carried mini-Mu plus the transposition genes from the nearby Mu dI1 helper phage. In addition, some generalized transduction of the mini-Mu insertion mutation was obtained. This transposition procedure works because P22 packages 44 kb by a headful procedure, and in this construct the 10-kb mini-Mu element is about 3 to 5 kb from the 37-kb Mu dI1 element.

There are hybrids between Mu and other phage that integrate by means of the Mu transposition system while retaining many of the properties of the other phage. The λ p*lac*Mu derivatives (54, 55; see Fig. 1 and Table 1) are plaque-forming λ phage containing both Mu ends, the Mu A transposition gene, and λ genes needed for lytic growth, but deleted for the λ attachment site. Consequently, λ p*lac*Mu lysogenizes by inserting with the randomness of Mu, and prophage induction results in excision by illegitimate recombination and packaging into the phage head (54, 55, 384).

Mu d-P22 hybrid phage (415) is analogous to λ p*lac*Mu in that insertion is Mu directed, whereas lytic growth and packaging are specified by P22. Thus, it inserts into many sites in the chromosome, where it is "locked in." During lytic P22 growth this element replicates in situ, but does not transpose further. At the end of lytic growth it, like P22, packages sequential headfuls of DNA from a single P22 packaging-initiation (*pac*) site, but since the Mu d-P22 prophage remains in the chromosome, prophage and chromosomal DNA distal to the *pac* site of the inserted element is preferentially packaged (see Fig. 7 and Section XII). These Mu d-P22 phage can also be delivered to *S. typhimurium* as part of a self-transmissible plasmid and recombined into sites of preexisting Mu d prophages (415).

Although one of the major advantages of using transposons is that they function in vivo, efficient Mu transposition in vitro (267) is useful for mutagenesis of plasmid DNA (394).

B. Indirect Transposon Mutagenesis

In indirect or "reverse" mutagenesis, cloned procaryotic or eucaryotic DNA is mutagenized in a suitable host, generally *E. coli*, and then returned to the parental strain by conjugal transfer, transformation, or phage-mediated transduction. This approach requires (i) the use of transposons harboring selectable markers both for *E. coli* (in which the mutagenesis usually takes place) and for the parental species (in which the phenotype is assessed) and (ii) the use of elements able to transpose in *E. coli* but not (or at much reduced frequency) in the other species. When the mutagenized DNA is introduced back into the species of interest, it can integrate via a single crossover in the region of homology on one side of the insert, to yield an unstable duplication, with one mutant and one wild-type allele. A second crossover on the other side of the insert haploidizes the segment, resulting in loss of the vector DNA and replacement of the wild-type by the mutant allele (marker exchange or gene replacement). Tn*5* is frequently used to mutagenize foreign DNA cloned in *E. coli* because its kanamycin resistance marker is readily selected in a broad range of species and because of its quasi-random insertion specificity (see reference 34 and Table 2).

Insertion mutagenesis using plasmid-borne (67) or cosmid-borne (102, 103) mini-Mu elements can be very efficient in *E. coli*. Care must be exercised, however, to avoid zygotically inducing further transposition when the clones are moved to a nonimmune host. Zygotic induction can be avoided by using mini-Mu with a defect in the A or B (transposition) genes, or by using a Mu-lysogenic recipient.

1. Shuttle transposons

"Shuttle" gram-negative transposons, which also contain markers that are selectable in a foreign host, have been engineered for the genetic analysis of DNA cloned in *E. coli*. A mini-Mu element, Mu dE, which carries the gram-positive erythromycin resistance marker has been used to mutagenize *Streptococcus mutans* DNA. Mutagenized plasmids were then returned to *S. mutans* by transformation, selecting for haploid recombinants (225). This element has also been used to mutagenize cloned fragments from *B. subtilis* (cited in reference 225) and *Listeria monocytogenes* (E. Groisman, B. Gicquel-Sanzey, and P. Cossart, unpublished results).

Because many *E. coli* genes are not expressed in *Neisseria gonorrhoeae*, a Tn*3* derivative was constructed with a mutated chloramphenicol resistance gene from Tn*9* that is expressed in *N. gonorrhoeae* to permit genetic analysis of this pathogen (H. S. Seifert, M. Vito, F. Heffron, and M. So, personal communication).

Several transposons have been engineered to harbor both *Saccharomyces cerevisiae* and *E. coli* selectable markers (Table 1, Fig. 1). These include derivatives of Tn*10* with the yeast *URA3* or *TRP1* genes associated with the *E. coli*-selectable *tet*, *kan*, or *supF* genes, and in one case a promoterless *lacZ* gene (185, 364); derivatives of Tn*3* with the yeast *HIS3*, *URA3*, *LEU2*, *TRP1*, or *SUP11* genes associated with *amp* and a promoterless *lacZ* gene (342); a mini-Mu element with *URA3*, *kan*, and the phage T7 promoter for regulated RNA synthesis (E. Groisman and F. Heffron, unpublished results); and a mini-Mu element with *LEU2*, the *S. cerevisiae* 2μm circle origin of replication, *cat*, and a promoterless *lacZ* gene (92). The Tn*3* derivatives utilize the phage P1 *lox/cre* system, rather than Tn*3* *res*, for cointegrate resolution. If *cre* is under the control of a yeast promoter, it will catalyze recombination at *lox* sites in *S. cerevisiae* (335). Therefore, Tn*3* derivatives were constructed that contain two *lox* sites and *cre* under *S. cerevisiae*-specific control. Exchange at *lox* sites eliminates most of the internal sequences of the transposon, leaving only a sequence encoding a 45-amino-acid domain plus a 5-bp duplication of the target DNA. The products of the mutated gene are thus marked with a small epitope against which antibodies are available (M. Hoekstra, D. Burbee, E. Chen, and F. Heffron, unpublished results).

A mini-Mu element that contains the thymidine kinase gene from herpes simplex virus (HSV) has been used to mutagenize and localize nonessential regions of the HSV genome (190, 318). Similarly, Tn*5* has been transposed to cloned HSV fragments in *E. coli* and then recombined in vivo back into the HSV genome, and recombinants have been identified by hybridization of HSV plaques to DNA Tn*5* probe (403). In this way nonessential viral genes can be identified. The most obvious advantage that elements such as mini-Mu and Tn*5* present for studies with herpesviruses (which are over 120 kb in length) is the ease of saturation mutagenesis, without limitations imposed by the distribution of particular restriction sites. Transposons should also facilitate the introduction into vaccinia virus vectors of genes whose protein products would be useful as immunizing antigens.

Shuttle plasmids for the delivery of mutagenized DNA from *E. coli* to a number of other bacterial species include ones for *A. tumefaciens* (143), *Bordetella pertussis* (369), *Campylobacter jejuni* (228), and *R. meliloti* (356).

VI. PHYSIOLOGICAL EFFECTS OF INSERTION MUTATIONS

Transposable elements interrupt target genes and generally cause null-activity, polar mutations because transcription from the normal promoter is terminated within the element. Thus, genes distal to the insertion are not transcribed from the operon promoter (32, 57, 170, 206, 368). Many insertions that completely block transcription from the principal operon promoter nonetheless allow some expression of distal genes because transcription is reinitiated either in or downstream from the element. This can be from an internal promoter in the operon, from within the element if it has an outward-facing promoter (Tn*10* [IS*10*], Tn*3*, gamma-delta, IS2, IS3), or from "hybrid" promoters, composed of both element and target sequences (Tn*5*) (see Berg, Galas and Chandler, Kleckner, and Sherratt, all this volume). Transcription from these secondary promoters is not subject to the normal operon regulation: it is generally low level and is often not detected physiologically, possibly because of the natural transcription termination occurring in untranslated mRNAs (81). Thus, despite such element-associated promoters, transposons can often be used to order genes in operons and to detect significant internal promoters (27, 29, 30, 209). In addition to the natural promoter activity associated with some elements, derivatives of several elements with strong outward-facing regulatable promoters have been constructed to permit the isolation of conditional mutations (see Section X).

Insertion mutations have been found that turn on a nearby cryptic chromosomal gene, but map outside of the gene's promoter or coding region and are not inserted in a regulatory gene. For example, the cryptic *bgl* operon present in the *E. coli* chromosome is activated by insertions of endogenous IS*1* or IS*5* elements upstream of its promoter region (313, 314). Insertions that map outside of structural genes, and in some cases outside of the transcribed region, yet turn off gene expression, have been described for genes cloned in plasmids (241, 295, 370, 399; C. Higgins, personal communication). The distribution of insertions that alter the expression of a plasmid-borne gene is characteristic for each gene-plasmid system studied: in some cases the insertions are clustered within a few hundred base pairs downstream of the gene (241), while in others the insertions may even be kilobases away in the plasmid vector (370; L. Liu, R. Gaffney, and C. M. Berg, unpublished data). Alterations of the plasmid such as inversion of the insert, dimerization, and small deletions drastically affect the expression of these genes without significantly affecting expression of other genes in the plasmid (241, 370).

Many insertions in the chromosome that turn on nearby cryptic genes, as well as insertions in plasmids that turn off adjacent cloned genes, probably affect gene transcription in complementary ways: by inducing changes in the local DNA topology that result in elevated or reduced gene expression, respectively. Although turn-on insertions have been described only for chromosomal genes, and turn-off insertions only for plasmid-borne genes, this probably reflects the relative ease of detection and mapping of each type of change: genes are generally cloned in the active, not inactive, state, and restriction mapping or sequencing, which is readily done with plasmids, is required to determine that an insertion maps outside of the translated portion of a gene.

VII. STABILITY OF INSERTIONS

Two different phenomena are important in the overall stability of transposon insertions: the probability that much or all of the transposon will be excised from its insertion site and the probability that the element will transpose from the original site to new locations.

A. Transposable Element Excision

Mutations induced by most transposons are quite stable. Tn*5*- and Tn*10*-induced mutations can revert by deletion (precise excision) of the element and one copy of the short target sequence duplication, but generally at frequencies of less than 10^{-8} for insertions in the chromosome and less than 10^{-6} for insertions in an F-prime factor. These excisions are effectively spontaneous deletion events, independent of transposition to new sites and of recombination enzymes (116, 132, 274). Significantly higher revertant frequencies may reflect extragenic suppressor mutations. Mutations caused by composite elements whose IS elements are in the direct repeat orientation are also stable, even though the central region and one IS element can be lost by homologous crossing over between the IS components (typically about 1% in *E. coli* [247, 296]), but one IS element bounded by the duplication of target sequences remains at the insertion site. Reversion of insertions of a Tn*5* derivative in which the IS*50* elements are directly repeated is about 100-fold lower than for insertions of Tn*5* wild type, probably because this configuration does not permit the intramolecular pairing that is thought to foster precise excision (116).

Although transposon-induced mutations are generally polar on the expression of distal genes, partial revertants can be obtained by selecting for distal gene function (e.g., $LacY^+$ in *lacZ* insertion mutants). Most revertants of Tn*5* or Tn*10* insertion mutants selected on the basis of distal gene function do not regain target gene function and, hence, are not true revertants. Some can subsequently revert to restore target gene function, while others cannot, depending on whether the deletion causing the partially revertant phenotype extended into the target gene (nonrevertible) or was entirely within the element (generally revertible) (32, 132).

Mutations due to insertion of Mu A^+ B^+ phage do not normally revert (less than 10^{-10}). However, mutations induced by Mu B^- derivatives revert at low frequencies in *recA*$^+$ strains when the Mu A^+ gene is expressed (10^{-6} to 10^{-5}) (57, 88, 202, 273, 373). It has been proposed that this excision is due to abortive transposition (57) or to slippage of nascent strands during DNA synthesis, which is fostered by Mu A protein binding the two Mu ends (273). λ-induced mutations are stable unless the phage *int* and *xis* genes are expressed. When they are, excision is nearly 100% efficient from the primary site, but may occur at 1% or lower frequency from secondary sites (352). The revertibility of Tn*3*-induced mutations has not been reported, but many Tn*3* insertion mutations in F-prime factors are nonrevertible because they are associated with adjacent deletions (264). However, precise excision of Tn*4451*, a member of the Tn*3* family, has recently been reported (1).

For Tn*5* and Tn*10* the reversion frequency depends upon the precise insertion site, whether the insert is in an F plasmid or the chromosome (where it is more stable), and upon the strain background. Any of these parameters may cause differences of 100-fold or more in reversion frequency (116, 274). Very high reversion frequencies are found, for example, with Tn*5* insertions in M13-related phage, probably because the single-strandedness of M13 fosters replication errors, especially when copying palindromic DNA (172). Similarly, the higher reversion of Tn*5*- and Tn*10*-induced mutations in an F plasmid, relative to the chromosome, is probably related to the conjugal transfer that occurs in an F$^+$ culture (38, 372) and which involves single-stranded intermediates. Certain mutant alleles of *recB*, *recC*, and DNA repair genes also significantly increase the deletion of transposable elements as monitored by reversion frequencies (245, 246). Deletions of Tn*5* and Tn*10* may result from errors during DNA replication or occur by DNA breakage (38, 56, 77, 95, 116, 132). Finally, the gram-positive element Tn916 is excised precisely at a very high frequency by a transposon-encoded function in high-copy-number plasmids in *E. coli* (84, 142).

B. Secondary Transposition Events

The transposases of Tn*5*, Tn*9*, Tn*10*, and Mu act preferentially on adjacent sites in *cis*. Thus, their activities are somewhat reduced if the transposase gene is at a distant site in the vector molecule and are drastically reduced if the gene is in a different replicon. Because elements that lack the transposase gene do not move to new sites after transposition (although they can be excised at normal frequencies), it is often advantageous to use complementable transposition-defective derivatives for mutant isolation. The most useful constructs contain the transposition genes in the vector near the element (see Table 2). A somewhat more cumbersome, but still very useful, procedure for stabilizing insertion mutations involves the replacement or inactivation of transposition genes by recombination between the resident transposition-proficient element and a transposition-deficient element introduced on a vector that cannot be maintained in the cell. This has been done for Tn*5* (28, 116, 284), Tn*10*, and Mu (13, 110a).

VIII. MAPPING STRATEGIES

The first uses of transposable elements in chromosome mapping were inadvertent, involving the isolation of Hfr and F-prime strains. In many Hfrs the inserted F factor is found bracketed by a pair of IS2, IS*3*, or gamma-delta elements. In these cases the Hfr chromosome is a cointegrate molecule (178, 346) that arose either by a reciprocal crossover between identical transposable elements in the chromosome and the F factor, or by replicative transposition (see Fig. 2). Similarly, many F-prime factors carrying chromosomal genes appear to have arisen following a reciprocal crossover between homologous IS elements or by intramolecular transposition (Fig. 4 and 5) in Hfr strains. In some cases both IS elements were originally located in chromosomal sequences, while in others one element was located in the F factor and the other in chromosomal DNA (178). In recent years, the inadvertent use of unmarked elements has been largely supplanted by the deliberate use of elements that contain selectable markers. This use of resistance transposons has revolutionized genetic analysis in many species.

A. Large-Scale Mapping

1. Providing a selectable marker

Transposons have been especially valuable in gene mapping because they can be used to place easily selectable markers near the mutation to be mapped or at known sites in the chromosome. Mapping kits utilizing transposons have been constructed in *E. coli* (T. Baker, M. Singer, and C. Gross, personal communication; see reference 400), *S. typhimurium* (222), and *Caulobacter crescentus* (15). For *E. coli*, preliminary mapping can be done using a set of Hfr strains, each with a Tn*10* insertion near its origin of transfer. More detailed mapping can then be done by using a set of F$^-$ strains with Tn*10* or Tn*10*-*kan* at intervals of 1 min or less, so any gene can be located by P1-mediated cotransduction (M. Singer and C. Gross, personal communication). The *S. typhimurium* mapping set contains a comparable collection of mini-Tn*10* insertions for locating genes by P22-mediated cotransduction. Since the strategy generally employed involves selection for inheritance of the drug resistance determinant and scoring for loss of the mutant phenotype, these kits will be particularly useful for mapping genes when the wild-type allele can be screened for, but not selected directly.

2. Oriented chromosome transfer

Transposons provide sites for oriented chromosome transfer in several ways.

Recombination between homologous elements. When the orientation of one copy of the element in a conjugative plasmid, such as the F factor, is known and a second copy of the element is present in the chromosome, a single crossover between these homologous segments inserts F into the chromosome and thus permits efficient oriented transfer of the chromosome from the site of insertion. This strategy has been exploited for mapping in *E. coli* with Mu (421), in *V. cholerae* with Tn*1*, an element nearly identical to Tn*3* (193), and in *S. typhimurium* with Tn*10* (80). This strategy would probably not work as well with elements such as wild-type Tn*5*, in which the central region is small relative to the inverted repeats, because occasional pairing between the "wrong" IS components would give mobilization in the opposite orientation. Mobilization can also be used to determine the direction of transcription of chromosomal genes by using a transposon with a reporter gene such as Mu d-*lac*, Tn*5*-*lac*, or Tn*10*-*lac* (which do not have significant inverted repeats).

Providing an origin of transfer. Tn*1*, Tn*5*, and mini-Mu derivatives have been constructed that contain the transfer origin from a conjugative plasmid (see Table 1). With these elements, the insertion site becomes the origin of Hfr-like oriented transfer, when the insertion is in the chromosome, or F-prime-like oriented transfer, when the insertion is in a plasmid.

In vivo cloning and providing an origin of transfer for the cloned fragment. Mini-Mu elements for in vivo cloning (see Section XII) have been constructed that contain an origin of transfer (158) and can mobilize the chromosome by homologous recombination between the cloned fragment and the corresponding chromosomal segment.

Generation of F-primes and R-primes. Replicative transposition of a mini-Mu element in an F^+ or R^+ strain generates F-prime or R-prime factors that carry segments of the bacterial chromosome bracketed by mini-Mu (mini-Mu-duction) (124, 125). These conjugal plasmids also mobilize the chromosome by crossing over when a homologous segment is in the chromosome. Mini-Mu-duction to RP4 has been used for chromosome mapping of *E. carotovora* (338). However, this plasmid can mediate the transfer of genetic markers from the recipient back to the donor (260).

3. Physical analyses

The availability of physical maps of *E. coli* (361) and of *C. crescentus* (119a), obtained after cleavage by "rare cutter" restriction enzymes and separation of the fragments by pulsed-field gel electrophoresis, allows preliminary localization of transposon insertions by (i) Southern hybridization analysis of the mutated genome using a probe specific for the transposon (362), (ii) the disruption of one fragment when the transposon contains a site cleaved by such rare cutter enzymes, or (iii) the increase in size of one fragment. Wild-type Tn*5* has proven useful in this regard (258, 362) because there is a *Not*I site in each IS*50* element (10).

B. Localized Mutagenesis and Uses of Linked Insertions

Transposons are valuable as agents for localized mutagenesis because they provide selectable markers that can then be used for genetic mapping and for in vitro or in vivo cloning. They also contain restriction sites that are convenient for in vitro mapping and cloning, and their ends can be used as primer binding sites for DNA sequencing.

The object in localized mutagenesis is to obtain insertions in or closely linked to a particular gene. This can be accomplished by mutagenizing a cloned fragment and selecting for recombination into the chromosome and loss of the vector, or by generalized transduction from a randomly mutagenized population, selecting for inheritance of the linked marker. Phage P1 (capacity, 91 kb) has routinely been used as the transducing phage for localized mutagenesis of *E. coli*, yielding insertions scattered within about 2 min of the selected gene (89). If insertions in a larger region are needed, a transducing derivative of phage T4 (capacity, 166 kb) (408, 416) might be used, while if insertions within a few kilobases of the selected gene are needed, mini-Mu elements (that can carry up to 31 kb of bacterial DNA, depending on the element; see Fig. 1 and 6) are useful (89; Wang et al., in press). The set of λ phage containing overlapping segments of the *E. coli* genome (212) should also be useful in this regard. In *S. typhimurium*, phage P22 (capacity, 44 kb) is generally used for localized mutagenesis (97). Given the variety of elements with reporter genes now available (see Table 1), localized

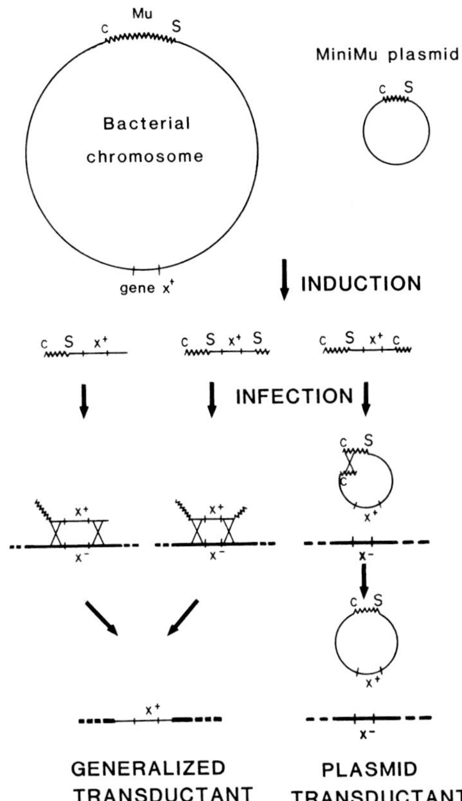

Figure 6. In vivo cloning and generalized transduction by Mu–mini-Mu plasmid double lysogens (161, 397) (from reference 26). Lytic growth of Mu (and the related *P. aeruginosa* phage, D3112) entails repeated cycles of concomitant transposition and replication, resulting in single phage copies inserted at many sites in the host chromosome, often within a few kilobases of each other. Packaging is initiated near the left (c) end of the phage, and a headful of DNA (about 39 kb) is incorporated into each phage particle. The deletion derivatives, called mini-Mu (and mini-D3112), can package up to about 31 kb of adjacent DNA, depending upon the size of the mini-transposon (see Table 1). Consequently, some phage heads will contain at least a portion of a second element. If these two elements are in the same orientation, they can recombine in the recipient to give a viable plasmid that contains the chromosomal DNA linked to the mini-Mu replicon element, analogous to plasmids obtained using conventional in vitro recombinant DNA techniques. Symbols: Light line, host DNA; heavy line, recipient DNA; zigzag, Mu DNA; c, phage left end; S, phage right end. Left, Fate of bacterial DNA in phage that package one copy of mini-Mu plus the physically linked adjacent DNA, which may be entirely bacterial or may also contain the S end of Mu or mini-Mu. A double crossover with the homologous segment in the recipient gives rise to a transductant, free of donor phage sequences. Right, Fate of bacterial DNA in phage that package one copy of mini-Mu plus adjacent DNA, including at least a portion of the c end of mini-Mu or the Mu helper phage. Crossing over in the recipient between the homologous Mu sequences gives rise to a circular molecule containing one copy of mini-Mu plus the host DNA that had been between the two insertions. This molecule can form a plasmid if the mini-Mu element (or the cloned DNA) contains an origin of replication. Subsequent recombination between the cloned fragment and the recipient chromosome can result in gene replacement (not shown).

mutagenesis is also applicable to the analysis of selected regions based on transcription, translation, or transport of the protein product, without screening for a mutant phenotype.

In situations where the products of two or more genes can perform the same function, traditional mapping and cloning can be difficult because different genes will confer the same phenotype (398). However, the use of an insertion linked to one of these genes can facilitate both chromosome mapping and cloning (Wang et al., in press).

C. Detailed Mapping: Operon Organization and Determination of the Direction of Transcription

Because transposons interrupt transcription, they can be used to determine the order of genes in an operon. Genes upstream of an insertion are generally expressed and regulated normally, the gene containing the insertion will have no activity, and downstream genes will have no, or low, unregulated expression (due to low-level, unregulated promoters in the operon or in the transposable element, or to hybrid promoters generated by insertion).

Transposons with promoterless reporter genes or with regulated promoters add another dimension to the analysis of genetic organization. To orient a chromosomal gene, a conjugative plasmid containing the same element can be used for oriented mobilization in conjugation, as described above. To orient a plasmid-borne gene, restriction mapping is used, and if the fragment is large enough, a comparison of this map with a genome restriction map (119a, 212) permits unambiguous mapping and orientation of the cloned fragment. Insertions of engineered transposons, such as Tn*5*tac1 (78), Tn*1721*-*tac* (Tn*1735*) (388), or mini-Mu-*tac* (155), that contain the strong regulated p_{tac} (*trp-lac*) promoter facing outward near one end, are polar when the inducer, IPTG (isopropyl-β-D-thiogalactopyranoside), is absent. If the insertion is in the correct orientation, high-level expression of downstream genes is elicited by addition of IPTG. Tn*5*tac1 and mini-Mu-*tac* are useful for mutagenizing either chromosomal or plasmid-borne genes, while Tn*1721*-*tac* is most efficient as a mutagen for plasmid-borne genes because Tn*1721* only rarely inserts into the chromosome.

Transposons have been constructed with an outward-facing phage T7 promoter for strong in vivo or in vitro transcription from that promoter. A Tn*5* derivative, Tn*5*seq1, which has the T7 promoter at one end and the phage SP6 promoter at the other end, is useful for mutagenizing either plasmid or chromosomal DNA (275), while a mini-Mu plasmid element with a phage T7 promoter has been constructed for direct in vivo cloning and analysis of the orientation of genes in the cloned fragment (E. Groisman, N. Pagratis, and M. Casadaban, manuscript in preparation).

D. Fine-Structure Mapping

Transposon insertions are often used in conjunction with other types of mutations for saturation mutagenesis of chromosomal genes, for several reasons: certain elements insert repeatedly at the same site; large insertions may perturb fine-structure transductional mapping; and point mutations that change just one or a few amino acids are generally more useful than transposon insertions for identifying functional domains of proteins. In addition, intramolecular transposition can yield nested sets of deletions that extend in a single direction from the insertion site (4, 5, 28, 61, 207, 282).

A great value of transposons is in fine-structure analysis of cloned genes, where the insertions can be mapped physically. Transposons have been used to identify the boundaries, the operon structure, and, as described above, the direction of transcription of cloned genes.

IX. REPORTER ELEMENTS AS PROBES FOR GENE EXPRESSION

The use of gene fusions has facilitated analyses of gene regulation, especially for genes whose products are difficult to assay. Fusions have also been instrumental in the discovery of genes that respond to particular environmental signals, such as DNA damage, changes in osmolarity, heat shock, phosphate starvation, anaerobiosis, and thiol reagents (see reference 278). The identification of genes for cell surface and exported proteins (that are likely to be important for pathogenicity) and the study of the cellular localization of proteins have been greatly aided by the ability to easily generate a variety of hybrid proteins (see references 254 and 353).

Early studies were aimed at isolating fusions to the *lac* operon and required a series of genetic manipulations to bring the *lac* operon near the gene of interest (17). A simpler approach used recombination between a Mu prophage in the chromosome and the Mu segment in a λ-*lac*-Mu hybrid phage to obtain deletions that fused the prophage *lac* region to the gene or operon of interest (63). While the quasi-random insertion specificity of Mu made this an improvement over earlier protocols, generating the *lac* fusions still involved many steps.

A novel Mu derivative described in 1979 ushered in a new era in the use of transposable elements. This element, called Mu d1, had been engineered to sensitively monitor transcriptional regulation in any region of the genome: it harbors a promoterless *lac* reporter gene close to one Mu end. The activities and regulation of specific promoters can thus be monitored directly after insertion by the easily scored Lac phenotype (66). Mu d1 is fully competent for transposition, and it can also undergo secondary transposition in a lysogen, although it needs a complementing helper phage for other stages of lytic growth. Derivatives that generate stable Mu B^- insertions (13) or that only transpose in suppressor backgrounds (183) have been developed. The principle of making operon or gene fusions in one step, established by Mu d1, has guided subsequent constructions of diverse reporter transposons (see Table 1).

Transposons with reporter genes have been engineered to generate two types of fusions: "type I" fusions, such as those generated by Mu d1 (now sometimes called Mu dI1), monitor only transcription, because the reporter gene lacks a promoter but contains its own translation initiation signals; "type II" fusions monitor translation as well as transcription, because the reporter gene lacks sequences needed to initiate both transcription and translation (65). Transposons that form type I (transcriptional or operon) fusions need only insert into the target gene in the correct orientation for reporter gene expression, while transposons that form type II (translational or protein) fusions need to insert in both the correct orientation and the correct reading frame for expression. The reporter protein made by a type I fusion is constant in length and amino acid sequence because the fusion occurs upstream of the reporter translation initiation site. In contrast, hybrid proteins formed in type II fusions have an amino terminus of variable length encoded by the target gene and a carboxy-terminal region (reporter) that is constant for any given transposon.

The *lacZ* gene was used as a probe in Mu d1 because its product (β-galactosidase) is easily detected on indicator plates using chromogenic substrates and is easy to assay with precision. *lacZ* has also been incorporated as a type 1 transcription probe into mini-Mu, λ-Mu hybrids, Tn*3*, Tn*5*, Tn*10*, Tn*917*, and Tn*1721* (see Table 1). Type I transcription probe transposons have also been made with the *lux* (luciferase) genes from *V. fischeri* as the reporter for measuring light emission in mini-Mu (122), Tn*1721* (349), and Tn*5* (Z. Koncz-Kalman and J. Schell, personal communication). Other type I transposons contain the *cam* gene from Tn9 in Tn*1721* (388), the *cam* gene from *Bacillus pumilus* in a Tn*917* derivative that allows the generation of type I fusions simultaneously to both *lacZ* and *cam* (417), and the *kan* (*npt-II*) gene from Tn*5* in pHLH1, a mini-Mu element (231). Type II *lac* translation probe transposons have been constructed in Mu (65), mini-Mu (67), and Tn*5* (215; T. Tomcsanyi, K. W. Dodson, and D. E. Berg, unpublished data). Other genes that have been used as translation probes in transposons include the *kan* gene from Tn*903* in mini-Mu (307), the *kan* gene from Tn*5* in Tn*5* (22), and the *phoA* gene from *E. coli* in Tn*5* (253). In addition, one Tn*3*-*lac* probe element, Tn*3*-HoHo1, forms both types of fusions (367). While most gram-negative elements carry *lacZ* and *lacY*, the gram-positive transposon Tn*917* has been engineered to carry only the *lacZ* segment of the *lac* operon because expression of the *lacY* (permease) gene was found to be a lethal event in *B. subtilis* (417).

The *phoA* gene product, alkaline phosphatase, is only active in the periplasm or when excreted, not when sequestered in the cytoplasm. Tn*phoA*, which has a type II *phoA* reporter gene, was designed to use this property. It detects genes that encode exported or membrane proteins, identifies structural features required for protein localization, and identifies protein domains that are outside the cell membrane (253, 254). The *lac* and *phoA* genes are useful, and complementary, reporter genes for type II fusions because the amino acid sequences at their N termini can be replaced without loss of function and because they encode proteins that can only function if excreted (*phoA*) or if present in the cytoplasm.

X. PROMOTER ELEMENTS THAT REGULATE ADJACENT GENE EXPRESSION

Elements have been constructed in vitro that can promote transcription in a regulated manner. These include derivatives of the transposons Tn*1721*, Tn*5*, and mini-Mu that contain the high-level *tac* (*trp-lac*) promoter (78, 155, 388) and of Tn*5* and a mini-Mu plasmid element that contain a phage T7 promoter (for high-level transcription) (275; Groisman et al., in preparation). In addition, the Tn*5* derivative, Tn*5*seq1, that has the T7 promoter at one end also contains a phage SP6 promoter at the other end so that expression of sequences to the left and that to the right of the element are under different transcriptional control (275). When transcripts are made from these elements, adjacent genes will be turned on if "sense" RNA is made and sometimes turned off if "antisense" RNA is made. The Tn*5* derivatives are useful for mutagenizing any target DNA, whereas the Tn*1721* derivatives mutagenize only plasmid DNA efficiently and the mini-Mu plasmid derivative is designed for in vivo cloning and transcriptional studies of the cloned fragment. The elements carrying the *tac* promoter should be especially useful for making novel, easily analyzed conditional mutations and, thus, for identifying genes in many species that are essential or difficult to clone (78).

A Tn*3* derivative with a yeast "silencer" segment that turns off adjacent transcription in eucaryotes (53) has been used to mutagenize *S. cerevisiae* DNA cloned on a shuttle vector in *E. coli* and, when recombined back into the yeast genome, to regulate the expression of chromosomal genes in *S. cerevisiae* (C. Davis, F. Heffron, and C. Hutchinson, unpublished results).

XI. IN VITRO CLONING

Transposons can facilitate in vitro gene cloning by providing resistance markers and arrays of known restriction sites that can be inserted into or next to the gene of interest. Detailed restriction maps of useful natural transposons can be obtained from DNA sequence banks, from other chapters in this volume, and from the following references: Tn*3* (169); Tn*5* (10); Tn*10* (402); gamma-delta (162); Mu (181); λ (93); Tn*917* (347). Restriction maps of the genetically engineered elements can be obtained from the references cited in Table 1.

In cases where a transposon inserted in a gene is used to clone the DNA containing that gene, the wild-type allele can be recovered in the plasmid or phage clone by homologous recombination with the chromosomal allele. In addition, for elements that can undergo precise excision (especially Tn*916* in multicopy plasmids in *E. coli* [142]), the wild-type allele can also be obtained directly from the insertion mutant plasmid by selecting for reversion.

Transposons engineered to contain a plasmid origin of replication that is inactive in the mutagenized strain but functional in another strain can be used for direct in vitro cloning of segments bracketing the insertion. For example, the Tn*5* derivative, Tn*5*-V, containing the plasmid pSC101 origin of replication was used to mutagenize *M. xanthus* (in which pSC101 does not replicate). These insertions were then cloned directly by ligation of cleaved genomic DNAs and transformation of *E. coli* (137). Another derivative, Tn*5*-PV, has the origins of replication and of transfer, but not the replication genes of plasmid RK2 (213). Consequently, insertions iso-

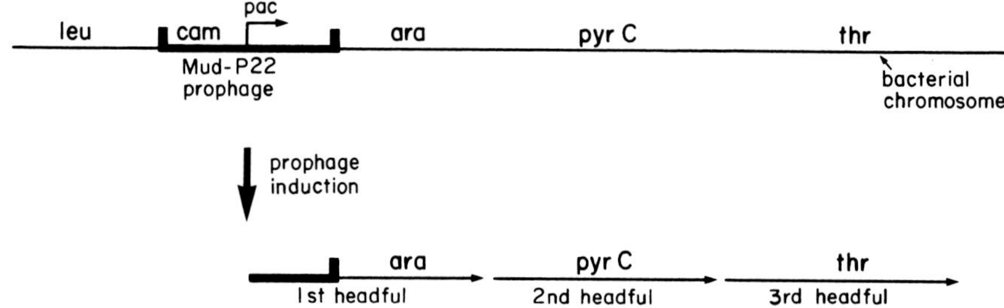

Figure 7. DNA packaging by "locked-in" Mu d-P22 prophage (modified from reference 415). Packaging by the headful mechanism is initiated at the *pac* site in P22 and proceeds from left to right, as drawn. Unlike headful packaging by Mu, several sequential segments are packaged. The first headful contains part of the Mu d-P22 genome and adjacent target DNA, while subsequent headfuls contain only bacterial DNA. Symbols: Heavy line, Mu d-P22 prophage; light line, target DNA; *cam*, chloramphenicol resistance; *pac*, P22 *pac* site; *leu*, *ara*, *pyrC*, and *thr*, bacterial genes.

lated in RK2-free strains can be used to clone adjacent segments into strains carrying the genes needed for RK2 replication. Analogous constructs could easily be made with other plasmid or single-stranded phage replication origins. Generally, the endonuclease(s) used for cloning insertion mutations is one that does not cleave the transposon, but if DNA to only one side of the insertion is sought, an enzyme that cuts the transposon could be used. The major restriction in using transposons that carry an origin of replication for in vitro cloning is that the plasmid origin must not impair cell growth when the transposon is inserted into the chromosome.

XII. IN VIVO CLONING

As noted above, in vivo cloning was accomplished initially by forming F-prime factors and by forming lambdoid specialized transducing phages from prophages inserted at dispersed secondary sites. The limitation imposed by the relatively small number of λ attachment sites had been overcome by using Mu DNA homology to direct the integration of a λ pMu phage into a Mu prophage inserted near the gene of interest (64, 214, 248). More recently, this approach has been simplified by the use of λ-Mu hybrid phage (Fig. 1) that contain Mu ends and insert with the randomness of Mu, but replicate and package DNA as λ phage (54, 55, 384).

Another approach using Mu d-P22 hybrid phage (Fig. 1) has been used for isolating specific chromosomal fragments in *S. typhimurium*. These hybrid elements contain the Mu ends and insert with the randomness of Mu when complemented for Mu A and Mu B functions. Upon prophage induction they replicate under P22 control in situ (without transposing) and package DNA into P22 heads. This involves a headful mechanism starting from the "*pac*" site in the P22 prophage. Chromosomal segments adjacent to the inserted phage are packaged unidirectionally by the headful mechanism, and so 2 to 3 min of adjacent DNA is recoverable in high yield from any single lysogen (415; see Fig. 7). The lysates are well suited for transduction and localized mutagenesis, and the amplified DNAs from phage particles are useful for long-range physical mapping and sequencing from phage particles.

An alternative in vivo cloning strategy is based on the replicative transposition of Mu during lytic growth. Multiple cycles of transposition occur, generating many subgenomic circular DNAs, each with a copy of Mu, and also cointegrates in which a Mu-containing circular DNA is joined to another DNA molecule by duplicate copies of Mu (Fig. 2 and 6). One consequence of multiple transposition cycles is the frequent Mu-mediated translocation of chromosomal genes to a new replicon, with the translocated genes bracketed by copies of Mu or mini-Mu (mini-Mu-duction). In this way, chromosomal genes have been cloned in a conjugative plasmid and their presence has been selected by transfer (124, 125). This approach has been extended by incorporating mini-Mu elements into the broad-host-range plasmid RP4, which has allowed in vivo cloning in several members of the family *Enterobacteriaceae* (390; Groisman and Casadaban, unpublished results). Such plasmids, like F-prime factors and λ transducing phages, have then been used as enriched sources of chromosomal genes for subsequent in vitro subcloning (100, 101, 229). In *P. aeruginosa*, R-prime factors have been successfully generated using the plasmid R68.45, an RP4 derivative that contains a highly transposable IS*8* element (164).

The most generally useful approach to in vivo cloning involves engineered derivatives of Mu (159, 161) and the related *P. aeruginosa* phage D3112 (A. Darzins and M. Casadaban, manuscript in preparation) which contain a plasmid origin of replication and a selectable resistance determinant and are small enough to package up to 31 kb of adjacent DNA into Mu or D3112 phage heads by the headful mechanism. As diagrammed in Fig. 6, many of the progeny phage from a single cycle of lytic growth form plasmids that contain segments ranging from about 2 to 31 kb and come from any region of the chromosome. Some of these engineered mini-Mu or mini-D3112 elements contain a promoterless *lac* reporter gene, a plasmid origin of transfer, a *tac* promoter, a phage T7 promoter, or a combination of these (see Table 1). Derivatives are now available that carry a variety of antibiotic resistance determinants and plasmid replication origins from different incompatibility groups with different inherent copy number (159, 161; A. Darzins and M. Casadaban, personal communication; Table 1). The mini-Mu system has been used to clone genes in several gram-negative species including *E. coli*, *S. typhimurium*, *S. flexneri*, *Proteus mirabilis* (160), *E. herbicola* (48), and *K. pneumoniae* (48, 104). Additional mini-Mu derivatives contain the *cos* site from phage λ, so that circular molecules of the appropriate size can be packaged in vivo into λ phage heads (154, 158). This is useful in permitting the cloning of larger DNA segments because λ packages between 38 and 53 kb in contrast to the 39 kb of Mu.

Appropriate mini-Mu elements have also been used to clone Tn*10* insertions and tightly linked selected genes of interest (isolated by mini-Mu-mediated localized mutagenesis; see Section VIII). This is particularly useful when the gene is difficult to clone directly (Wang et al., in press). It should be noted, however, that one of the most commonly used mini-Mu plasmid elements, Mu dII4042, contains most of one IS*50* element (161). Thus, Tn*5*-containing DNAs cloned by this element are unstable, because homologous recombination generates specific inversion and deletion products in *recA*$^+$ cells (N. Vartak, L. Liu, and C. M. Berg, unpublished data).

In vivo cloning has at least two advantages over traditional in vitro methods: the ease with which large or small fragments carrying the same gene can be obtained, which is especially important when cloning DNA segments whose restriction map is not known, and the colinearity of the cloned piece with the parental chromosome (since chromosomal DNA is not fragmented during cloning, there is no risk of cocloning two noncontiguous DNA segments). The disadvantage associated with the lack of restriction sites at the insert-vector junctures can be mitigated by placing known restriction sites near the ends of the Mu vector.

XIII. DNA SEQUENCING

Chain-termination DNA sequencing entails copying DNA fragments in the presence of chain-terminating dideoxyribonucleotides, starting from an oligonucleotide primer specific for DNA just 5′ (upstream) of the sequence of interest or cloning site (332). Generally only a few hundred base pairs of DNA sequence is determined from any one primer site. More remote parts of longer DNAs have been sequenced using extensive subcloning, various in vitro or in vivo deletion strategies, or downstream primer oligonucleotides.

We and others have exploited transposable elements to develop simplified chain-termination DNA-sequencing strategies. Since transposable elements insert in vivo into many sites, their ends can provide portable sequences for priming DNA synthesis. Strategies have been developed to use the two transposon termini to sequence in both directions from the site of insertion (bidirectional sequencing), or to use only one terminus, often in a single-stranded phage vector, to sequence in one direction (unidirectional sequencing). The only special requirement for bidirectional sequencing is that the primers be complementary to different unique sequences near each end of the element. This can be accomplished using any of a number of natural or genetically engineered elements (see Table 1). Gamma-delta (Tn*1000*), a component of the *E. coli* F factor that has inverted repeats of only 35 bp, is a convenient, relatively random insertion mutagen for DNAs cloned in plasmids (163) and has been used for bidirectional sequencing of plasmid DNA (241).

Tn*5*seq1 and Tn*5*tac1, which, unlike wild-type Tn*5*, have unique subterminal sequences, have been used for bidirectional sequencing of total genomic, as well as plasmid, DNA (Fig. 1; 78, 275). Another Tn*5*-based element, Tn*5*seq2, which is only 264 bp in length, has been designed specifically for the bidirectional sequencing of DNAs cloned in phage λ (Fig. 1; S. H. Phadnis, H. V. Huang, and D. E. Berg, manuscript in preparation). In this case, insertions into the λ phage target are selected by a *supF* (tRNA suppressor) gene in the transposon, using hosts in which only *supF*-bearing λ phage will make plaques.

The first use of a transposon for unidirectional sequencing involved the composite transposon Tn*9* inserted into a plasmid upstream of the contraselectable *galKT*$^+$ genes and a cloned fragment in a *galE* strain (*galE galKT*$^+$ strains are Gal^s). Most Gal^r derivatives arise by intramolecular transposition (Fig. 4 and 5) into or beyond *galKT*, producing deletions that extended from the internal end of one IS*1* component for variable distances into the cloned DNA. In this way a nested set of deletions can be obtained for sequencing one strand (4, 5, 293). Sequences from both strands can be obtained by using complementary plasmids containing the cloned fragment in opposite orientations. The distinct nonrandomness of Tn*9* transposition (422), which may be related to the distribution of integration host factor (IHF) binding sites (302), makes it difficult, however, to obtain a complete set of nested deletions for most target DNAs.

A different strategy for unidirectional sequencing was employed with mini-Mu: insertions were isolated in an M13-hybrid plasmid whose DNA was then packaged as single strands into phage particles. Although only one strand could be sequenced in a single plasmid, a need to first determine the mini-Mu orientation was avoided by use of a mixture of primers, one specific to each end (2).

Although elements with long inverted repeats cannot be used for bidirectional sequencing, they can be used for unidirectional sequencing by using an M13 vector, by separating restriction fragments, or by subcloning the end fragments in vitro. An in vivo method for obtaining appropriate Tn*5* subclones has recently been developed using the mini-Mu in vivo cloning element, Mu dII4042, that contains most of IS*50* from Tn*5* (161). When Tn*5* is cloned using Mu dII4042, the direct repeats of IS*50* elements recombine, yielding plasmids with just a single copy of IS*50* that are suitable for unidirectional sequencing (Vartak et al., unpublished data). Since insertions of wild-type Tn*5* have already been isolated in a variety of bacterial species (24, 34), Mu dII4042 cloning could be used to facilitate sequencing the insertion sites of specific Tn*5* insertions.

Genes that are expressed in plasmids are often located by transposon mutagenesis and restriction analysis. It is possible to use insertions for "shotgun" sequencing without first determining the position of the insertion (2). However, some caution should be exercised if the insertions used for sequencing are chosen by a null phenotype, because with certain (topology-sensitive) promoters insertions that cause a mutant phenotype may map outside of the structural gene (see Section VI).

The choice of a particular strategy and element to be used for DNA sequencing will depend upon whether the gene has already been cloned and subjected to transposon mutagenesis. If not, the base composition of the target DNA and the expected frequency and randomness of insertion of different elements should be considered (see Section IV).

XIV. MIXED BLESSINGS: SOME PROPERTIES OF SPECIFIC ELEMENTS, RESISTANCE DETERMINANTS, AND OTHER TRANSPOSON-BORNE GENES

The special properties of transposable elements have been extraordinarily valuable in genetic analysis and manipulation, although certain characteristics limit the usefulness of individual elements under some conditions. Many properties of transposable elements have been alluded to above and in Table 1, but here we discuss advantages and drawbacks more explicitly. The resistance determinants are discussed under the element in which they were first found, although many have been cloned into other elements or into plasmids.

A. Tn*3* Family

Tn*3* and related elements, like gamma-delta (Tn*1000*), have been particularly useful for mutagenizing plasmid DNAs. The presence of the terminus and ampicillin resistance determinant of Tn*3* in most members of the pBR322 plasmid lineage, including the pUC plasmid series (14), precludes the use of Tn*3* for insertion mutagenesis of these plasmids. Transposition immunity is so specific, however, that these plasmids can be used as targets for gamma-delta and Tn*1721*, two other members of the Tn*3* family, and a plasmid carrying one of these elements can be further mutagenized by either of the others. A pBR322-related plasmid from which the Tn*3* end and *amp* sequences have been removed has been constructed for cloning genes to be mutagenized by Tn*3* (342).

Tn*3* transposes much better at 30°C than at higher temperatures (216), while with gamma-delta the temperature optimum is 37°C (385).

Gamma-delta is found in the *E. coli* F factor plasmid; it is responsible for virtually all F-mediated mobilization of nonconjugative plasmids from F^+ cells by the formation of F-target plasmid cointegrates that are resolved in the recipient cell after transfer. Consequently, gamma-delta is often used to locate cloned genes in pBR322-derived nonconjugative plasmids (163). *recA* strains carrying monomeric target plasmids should be used because dimeric and oligomeric forms of some plasmids predominate in *recA*$^+$ cells and mutagenized dimeric plasmids contain gamma-delta in only one component. Mutagenized dimeric plasmids do not exhibit a mutant phenotype, and they break down in *recA*$^+$ cells to give nonmutant, as well as mutant, plasmids (Liu and Berg, unpublished data).

The absence of a selectable marker in gamma-delta does not significantly affect its usefulness as a mutagen because the coinheritance of the F factor precludes selecting for a transposon-borne determinant. The lack of a selectable marker does, however, limit further manipulations, such as selection for transduction, subcloning, or marker exchange involving the insertion. Although gamma-delta insertions are frequent and quite random in *E. coli*, the bias toward AT-rich sequences (241) suggests that gamma-delta would not be as good a mutagen for GC-rich sequences.

B. Tn*5*

Tn*5* transposes in many gram-negative species with very little insertion specificity. Its central portion contains an operon with three antibiotic resistance genes, *kan* (*neo*), *ble*, and *str* (256). The *kan* gene was the first resistance determinant discovered in Tn*5* and is used extensively in other engineered transposons and in plasmids because it can be selected in animal and plant cells as well as in procaryotes. It encodes resistance to the aminoglycosides kanamycin, neomycin, and G418 (Berg, this volume). In *E. coli*, kanamycin is usually used because it is highly toxic to sensitive cells at concentrations as low as 10 μg/ml, yet a single chromosomal copy of Tn*5* confers resistance to greater than 50 μg/ml. In contrast, neomycin at concentrations of 10 to 20 μg/ml is only partially inhibitory; as the concentration is increased to 50 μg/ml, sensitive cells are killed more efficiently, but the nominally resistant cells also show reduced viability (D. Berg, unpublished data). Thus, for the isolation of chromosomal insertions kanamycin has become the antibiotic of choice. Transposition of Tn*5* from the chromosome to a resident multicopy plasmid can be selected in medium containing 250 μg of neomycin per ml (42).

Tn*5* does not normally confer streptomycin resistance in *E. coli*, but mutations both on the chromosome and in Tn*5* that permit the *str* gene to be expressed in *E. coli* have been isolated (257, 303). Recently, clinical isolates of multiply antibiotic-resistant *E. coli* were shown to carry a Tn*5* element that confers a streptomycin resistance phenotype. The *str* (*rps*) gene in these isolates is identical to that of some laboratory mutants (144). The *str* gene is expressed to various extents in other bacterial species (see Berg, this volume).

The third gene in Tn*5* confers resistance to bleomycin (145, 256) and to the related antibiotic phleomycin, which can be used as a selectable marker in *S. cerevisiae* (141).

Transposition of wild-type Tn*5* is stimulated 5- to 10-fold in *dam* cells (defective in adenine methylation) because there are two adenine methylation sites (5′-GATC) in the transposase gene promoter. The IS*50* inside ends also contain adenine methylation sites, and the transposition of a derivative with two inside ends is stimulated about 300-fold, making it even more active than wild-type Tn*5* in *dam* cells (412a; K. Dodson and D. Berg, *Gene*, in press).

When Tn*5*-induced mutations are transduced by P1 or P22 from one strain to another, often more than half of the Kan^r transductants have Tn*5* transposed to a new site, usually in the phage genome (25). Consequently, when selection for kanamycin resistance is used to transduce or clone a particular chromosomal segment containing Tn*5*, the product should be tested to ensure that Tn*5* has not moved.

Although wild-type Tn*5* cannot readily be used for DNA sequencing because it contains long inverted repeats, a number of suitable asymmetric derivatives have been isolated, including one of just 264 bp for use with DNAs cloned in phage λ (see Table 1). When Tn*5* insertions are cloned in vivo using the mini-Mu plasmid element Mu dII4042, clones containing just one IS*50* element and the adjoining host DNA are generated by crossing over with an IS*50* fragment present in mini-Mu (see Table 1 and Sections XII and XIII).

C. Tn*7*

Tn7 is good for mutagenizing large plasmids. Because there is only one preferred site for Tn7 insertion in the chromosome of *E. coli* and related genera, DNA can be cloned in Tn7 and introduced into that site to assess the effect of cloned DNAs in single copy in a uniform genetic milieu (16a). This specificity precludes the use of Tn7 as a chromosomal mutagen unless the attachment site is deleted or the Tn7 gene required for insertion into the primary site is mutant (see Craig, this volume).

D. Tn*10*

In contrast to the situation with most other resistance genes, the level of resistance to tetracycline conferred by Tn*10* decreases as the *tet* gene copy number increases (this is not true for the related *tet* gene of pBR322) (76, 129). When Tn*10* is inserted in the *E. coli* chromosome, it confers resistance to 20 μg of tetracycline per ml, whereas on a multicopy plasmid it confers resistance to only 5 μg or less of tetracycline per ml, depending upon the strain and variant of Tn*10* used. This multicopy effect appears to be due to the harmful effects of an excess of the membrane-bound Tet protein (237, 270). Tn*10* mutants that confer resistance to higher levels of tetracycline when plasmid borne can be readily isolated (270), and most are due to promoter mutations that reduce the amount of Tet protein synthesized (K. Bertrand, personal communication). Despite the lethality of wild-type Tn*10* on multicopy plasmids in medium with 20 μg of tetracycline per ml, Tn*10* can be cloned in *E. coli* in the absence of tetracycline, or in the presence of low tetracycline levels (2.5 to 5 μg/ml) in appropriate strains (26, 239; E. Groisman, unpublished data). To avoid background growth by Tet^s cells, plates containing 2.5 μg of tetracycline per ml must be fresh and incubated in the dark to retard tetracycline decay (Groisman, unpublished data). Cases of Tn*10* cloning using higher levels of tetracycline are probably due to spontaneous mutations in the *tet* gene (26; Bertrand, personal communication).

Tet^s derivatives of strains carrying Tn*10* in the chromosome can be isolated by penicillin enrichment (129) or by plating on special medium that kills cells containing wild-type Tn*10* (Bochner medium) (49, 251). Among the Tet^s survivors are cells that have undergone a deletion of the *tet* gene and either a deletion or inversion of adjacent DNA as a result of intramolecular inverse transposition (Fig. 4). This ability to select directly for the loss of the *tet* gene can be used to obtain nested deletions extending from the insertion site. It should be remembered, however, that because these deletions are generated by inverse transposition, they retain one copy of IS*10* (Fig. 4). The Tet^r survivors on Bochner medium are also mutant because, in addition to permitting the cell to survive on this medium, the mutated *tetA* allele can be cloned on a multicopy plasmid in the presence of 20 μg of tetracycline per ml (Wang et al., in press).

Because it is possible to select for either the presence or the absence of a functional *tet* gene, Tn*10* can be used as a selective marker in crosses and in localized mutagenesis whether it is in the donor or the recipient (227).

Tn*10*, like Tn*5*, is not stable upon transduction by either P1 or P22 (25, 205).

A number of host- and transposon-encoded functions affect Tn*10* transposition frequency (see Kleckner, this volume), and a mutation in IS*10* that increases the transposition frequency 10- to 100-fold, called "high hooper," has been incorporated into several useful derivatives (402).

E. Mu and Its Derivatives

Wild-type Mu was widely used as an insertion mutagen until more versatile derivatives were engineered (even though it lacks a selectable drug resistance marker) because Mu transposes at a very high frequency and with relatively low specificity. A great variety of derivatives with and without selectable markers are now available, including ones that can be used as probes for transcription and translation, as cloning vehicles, and as mobile sources of transcription promoters (Table 1). While Mu and the larger of its derivatives are very useful for mutagenizing chromosomal DNA, insertions into multicopy plasmids are not recovered. Small mini-Mu elements are valuable for the mutagenesis of both plasmid and chromosomal DNA (67).

Mu is a bacteriophage, as well as a transposable element, and at the end of the lytic cycle the DNAs of the many integrated prophages are packaged into virions by a headful mechanism from the left prophage end. Appropriate deletion derivatives of Mu can thus be used for in vivo cloning and also for efficient generalized transduction (Fig. 6).

Among the drawbacks of Mu are a somewhat inefficient *trans* complementation for transposition, especially where several transposition events are required, such as in mini-Mu-duction and in vivo cloning. The presence of both A and B transposition genes in some elements can lead to secondary transposition events, particularly at temperatures higher than 30°C (because most Mu and mini-Mu derivatives used harbor a temperature-sensitive repressor) or after transfer to a nonlysogenic host (zygotic induction).

F. Tn*916*

The conjugative transposon Tn*916*, and the related Tn*1545*, are becoming the transposons of choice for many gram-positive bacteria because they can be introduced by self-mediated conjugal transfer into species that are not amenable to transformation or that lack well-characterized transducing bacteriophages. Tn*916* is particularly well suited for the analysis of cloned fragments containing insertions because it shows a high frequency of precise excision in *E. coli*, leading to the easy recovery of the wild-type allele.

The drawbacks of these elements include their relatively large size (16.4 kb for Tn*916* and 25.3 kb for Tn*1545*), which usually involves the use of cosmids for cloning the insertions. These elements are still being characterized genetically, and smaller derivatives with useful features have not yet been constructed.

G. Tn*917*

The Tn*3*-related transposon Tn*917*, originally from *S. faecalis* (347, 379), and engineered promoter-probe derivatives such as Tn*917-lac* (see Table 1) are well suited for insertion mutagenesis in diverse gram-positive species. Promoter-probe derivatives are especially valuable for analyzing complex developmental pathways such as sporulation in *B. subtilis* (50, 329, 389, 417–419).

XV. HOW TO CHOOSE THE RIGHT TRANSPOSON FOR A PARTICULAR EXPERIMENT

In the past few years, many transposon and vector delivery systems have been described. The variety is so great that often any of several different elements seem appropriate. Several features should be kept in mind when choosing among transposons and vectors, because there is no such a thing as "the" universally ideal transposon. The following is a summary of considerations that we have found useful.

1. Which organism is being used? Will the transposon mutagenesis take place on a cloned fragment in *E. coli* or in another species? Has the ability of the transposon to jump in that species been tested?

2. Will a plasmid or the chromosome be mutagenized? Some transposons are promiscuous, but others show a strong preference for one type of target replicon. Also, the transposon delivery system may depend upon the target.

3. What is the purpose of the experiment? Is it transposon mutagenesis? In vivo cloning? Generation of gene fusions? A preliminary step for DNA sequencing? Is one insertion in a region sufficient or is it important to get many well-distributed insertions?

4. What drug resistance markers can be used? Many strains are naturally resistant to antibiotics or antimicrobial agents active on laboratory strains of *E. coli*, *Staphylococcus aureus*, etc., and others may already contain a drug-resistant plasmid or transposon. Derivatives of Tn*3*, Tn*5*, Tn*10*, and Mu with a wide variety of drug resistance markers have been constructed.

5. If the target strain is a pathogen, what resistance markers do we not want to introduce into strains that might escape?

6. What is the G+C content of the organism? For example, Tn*3* and gamma-delta preferentially transpose to AT-rich regions and might be good choices for the mutagenesis of cloned DNAs from *P. mirabilis* (39% G+C) or *S. cerevisiae* (38% G+C) DNA, but not *S. marcescens* (which is 61% G+C).

7. Is a shuttle vector needed?

8. Should a system with transposons lacking the transposase be used, exploiting complementation from a transposase gene that is either *cis* in the vector or in *trans*? This will vary with the transposon used and the long-range experimental goals. However, some transposases are very strongly *cis* acting (e.g., Tn*5* and Tn*10*), and for others *trans* complementation is sometimes inefficient (e.g., Mu). Transposase-proficient transposons can be converted to stable derivatives by recombination with defective elements or by use of conditional mutant (nonsense-suppressible) transposase gene alleles.

9. Once the insertion is generated, will there be need to restore the original phenotype by precise excision of the transposon? If so, Tn*916* or perhaps Tn*5*, Tn*10*, or X mutants of Mu would be good choices, but Tn*3* would probably not be. For plasmid mutagenesis, keep in mind that the frequency of precise excision of elements with inverted repeats (Tn*5* and Tn*10*) is stimulated by several orders of magnitude in plasmids that replicate via single-stranded intermediates.

10. Will gene fusions to host promoters, translation initiation signals, or other regulatory sequences be generated? Here the choice of reporter elements is wide and elements are available for obtaining both transcriptional (operon) fusions and translational (hybrid protein) fusions. Many engineered transposons combine several useful features (cassettes) in one element to facilitate genetic analysis. For example, elements used for insertion mutagenesis and gene fusion, like mini-Mu replicons or λ-Mu-*lac*, can be used for both genetic mapping and in vivo cloning.

11. Will there be need to generate mutations with conditional phenotypes, dependent on expression of nearby genes? Here elements with strong, controllable, outward-facing promoters, such as *tac*, will be useful.

XVI. CONCLUDING REMARKS

In recent years transposable elements have been increasingly exploited as tools for mutational analysis and in vivo genetic engineering, initially just in *E. coli* but subsequently in a diverse array of gram-negative and gram-positive microorganisms and eucaryotes. Among their major uses are insertion mutagenesis and the introduction of linked selectable markers; analysis of transcriptional and translational regulation and protein localization; the introduction of engineered genes into organisms; and DNA sequencing.

Acknowledgments. We thank our colleagues for valuable discussions and for providing unpublished results and ideas.

Our work was supported by grants from the

United States Public Health Service (AI-19919 [C.M.B.] and GM-37138 [D.E.B.]), from the National Science Foundation (DMB-8802310 [C.M.B.] and DMB-8608193 [D.E.B.]), and from the Lucille P. Markey Charitable Trust (D.E.B.). E.A.G. is a Fellow of The Jane Coffin Childs Memorial Fund for Medical Research. His work was supported by the Philippe Foundation, by the Association pour le Developpement de l'Institut Pasteur, and by grants from the United States Public Health Service (GM-29067 to M. Casadaban) and from the National Science Foundation (DCB-8602478 to F. Heffron).

LITERATURE CITED

1. **Abraham, L. J., and J. I. Rood.** 1988. The *Clostridium perfringens* chloramphenicol resistance transposon Tn*4451* excises precisely in *Escherichia coli. Plasmid* **19:**164–168.
2. **Adachi, T., M. Mizuuchi, E. A. Robinson, E. Appella, M. H. O'Dea, M. Gellert, and K. Mizuuchi.** 1987. DNA sequence of the *E. coli gyrB* gene: application of a new sequencing strategy. *Nucleic Acids Res.* **15:**771–784.
3. **Adzuma, K., and K. Mizuuchi.** 1988. Target immunity of Mu transposition reflects a differential distribution of Mu B protein. *Cell* **53:**257–266.
4. **Ahmed, A.** 1984. Use of transposon-promoted deletions in DNA sequence analysis. *J. Mol. Biol.* **178:**941–948.
5. **Ahmed, A.** 1987. Use of transposon-promoted deletions in DNA sequence analysis. *Methods Enzymol.* **155:**177–204.
6. **Allen, L. N., and R. S. Hanson.** 1985. Construction of broad-host-range cosmid cloning vectors: identification of genes necessary for growth of *Methylobacterium organophilum* on methanol. *J. Bacteriol.* **161:**955–962.
7. **Anderson, D. M., and D. Mills.** 1985. The use of transposon mutagenesis in the isolation of nutritional and virulence mutants in two pathovars of *Pseudomonas syringae. Phytopathology* **75:**104–108.
8. **Arber, W.** 1983. A beginner's guide to lambda biology, p. 381–394. *In* R. W. Hendrix, J. W. Roberts, F. W. Stahl, and R. A. Weisberg (ed.), *Lambda II.* Cold Spring Harbor Laboratory, Cold Spring Harbor, N.Y.
9. **Arthur, A., and D. Sherratt.** 1979. Dissection of the transposition process: a transposon-encoded site-specific recombination system. *Mol. Gen. Genet.* **175:**267–274.
10. **Auerswald, E. A., G. Ludwig, and H. Schaller.** 1980. Structural analysis of Tn*5*. *Cold Spring Harbor Symp. Quant. Biol.* **45:**107–113.
11. **Avtges, P., R. G. Kranz, and R. Haselkorn.** 1985. Isolation and organization of genes for nitrogen fixation in *Rhodopseudomonas capsulata. Mol. Gen. Genet.* **201:**363–369.
12. **Bachhuber, M., W. J. Brill, and M. M. Howe.** 1976. Use of bacteriophage Mu to isolate deletions in the *his-nif* region of *Klebsiella pneumoniae. J. Bacteriol.* **128:**749–753.
13. **Baker, T. A., M. M. Howe, and C. A. Gross.** 1983. Mu dX, a derivative of Mu d1 (*lac* Ap^r) which makes stable *lacZ* fusions at high temperature. *J. Bacteriol.* **156:**970–974.
14. **Balbas, P., X. Soberon, E. Merino, M. Zurita, H. Lomeli, F. Valle, N. Flores, and F. Bolivar.** 1986. Plasmid vector pBR322 and its special-purpose derivatives—a review. *Gene* **50:**3–40.
15. **Barrett, J. T., R. H. Croft, D. M. Ferber, C. J. Gerardot, P. V. Schoenlein, and B. Ely.** 1982. Genetic mapping with Tn*5*-derived auxotrophs of *Caulobacter crescentus. J. Bacteriol.* **151:**888–898.
16. **Barry, G. F.** 1986. Permanent insertion of foreign genes into the chromosomes of soil bacteria. *Bio/Technology* **4:**446–449.

16a. **Barry, G. F.** 1988. A broad-host-range shuttle system for gene insertion into the chromosome of Gram-negative bacteria. *Gene* **71:**75–83.

17. **Beckwith, J. R., E. R. Signer, and W. Epstein.** 1966. Transposition of the *lac* region of *E. coli. Cold Spring Harbor Symp. Quant. Biol.* **31:**393–401.
18. **Belas, R., A. Mileham, D. Cohn, M. Hilmen, M. Simon, and M. Silverman.** 1982. Bacterial bioluminescence: isolation and expression of the luciferase genes from *Vibrio harveyi. Science* **218:**791–793.
19. **Belas, R., A. Mileham, M. Simon, and M. Silverman.** 1984. Transposon mutagenesis of marine *Vibrio* spp. *J. Bacteriol.* **158:**890–896.
20. **Belas, R., M. Simon, and M. Silverman.** 1986. Regulation of lateral flagella gene transcription in *Vibrio parahaemolyticus. J. Bacteriol.* **167:**210–218.
21. **Belland, R. J., and T. J. Trust.** 1985. Synthesis, export, and assembly of *Aeromonas salmonicida* A-layer analyzed by transposon mutagenesis. *J. Bacteriol.* **163:**877–881.
22. **Bellofatto, V., L. Shapiro, and D. A. Hodgson.** 1984. Generation of a Tn*5* promoter probe and its use in the study of gene expression in *Caulobacter crescentus. Proc. Natl. Acad. Sci. USA* **81:**1035–1039.
23. **Bender, J., and N. Kleckner.** 1986. Genetic evidence that Tn*10* transposes by a nonreplicative mechanism. *Cell* **45:**801–815.
24. **Berg, C. M., and D. E. Berg.** 1987. Uses of transposable elements and maps of known insertions, p. 1071–1109. *In* F. C. Neidhardt, J. L. Ingraham, K. B. Low, B. Magasanik, M. Schaechter, and H. E. Umbarger (ed.), *Escherichia coli and Salmonella typhimurium: Cellular and Molecular Biology.* American Society for Microbiology, Washington, D.C.
25. **Berg, C. M., C. A. Grullon, A. Wang, W. A. Whalen, and D. E. Berg.** 1983. Transductional instability of Tn*5*-induced mutations: generalized and specialized transduction of Tn*5* by bacteriophage P1. *Genetics* **105:**259–263.
26. **Berg, C. M., L. Liu, B. Wang, and M.-D. Wang.** 1988. Rapid identification of bacterial genes that are lethal when cloned on multicopy plasmids. *J. Bacteriol.* **170:**468–470.
27. **Berg, C. M., and K. J. Shaw.** 1981. Organization and regulation of the *ilvGEDA* operon in *Salmonella typhimurium* LT2. *J. Bacteriol.* **145:**984–989.
28. **Berg, C. M., K. J. Shaw, and D. E. Berg.** 1980. The *ilvG* gene is expressed in *Escherichia coli* K-12. *Gene* **12:**165–170.
29. **Berg, C. M., K. J. Shaw, L. Sarokin, and D. E. Berg.** 1981. Probing the organization and regulation of bacterial operons with transposons, p. 1121–1123. *In* D. Schlessinger (ed.), *Microbiology—1981.* American Society for Microbiology, Washington, D.C.
30. **Berg, C. M., K. J. Shaw, J. Vender, and M. Borucka-Mankiewicz.** 1979. Physiological characterization of polar Tn*5*-induced isoleucine-valine auxotrophs in *Escherichia coli* K12: evidence for an internal promoter in the *ilv0GEDA* operon. *Genetics* **93:**309–319.
31. **Berg, C. M., W. A. Whalen, and L. B. Archambault.** 1983. Role of alanine-valine transaminase in *Salmonella typhimurium* and analysis of an *avtA*::Tn*5* mutant. *J. Bacteriol.* **155:**1009–1014.
32. **Berg, D. E.** 1977. Insertion and excision of the transposable kanamycin resistance determinant Tn*5*, p. 205–212. *In* A. I.

Bukhari, J. A. Shapiro, and S. L. Adhya (ed.), *DNA Insertion Elements, Plasmids, and Episomes*. Cold Spring Harbor Laboratory, Cold Spring Harbor, N.Y.

33. **Berg, D. E.** 1983. Structural requirement for IS*50*-mediated gene transcription. *Proc. Natl. Acad. Sci. USA* **80:**792–796.
34. **Berg, D. E., and C. M. Berg.** 1983. The prokaryotic transposable element Tn*5*. *Bio/Technology* **1:**417–435.
35. **Berg, D. E., C. M. Berg, and C. Sasakawa.** 1984. The bacterial transposon Tn*5*: evolutionary inferences. *Mol. Biol. Evol.* **1:**411–422.
36. **Berg, D. E., J. Davies, B. Allet, and J.-D. Rochaix.** 1975. Transposition of R factor genes to bacteriophage lambda. *Proc. Natl. Acad. Sci. USA* **72:**3628–3632.
37. **Berg, D. E., C. Egner, B. J. Hirschel, J. Howard, L. Johnsrud, R. A. Jorgensen, and T. D. Tlsty.** 1980. Insertion, excision, and inversion of Tn*5*. *Cold Spring Harbor Symp. Quant. Biol.* **45:**115–123.
38. **Berg, D. E., C. Egner, and J. B. Lowe.** 1983. Mechanism of F factor-enhanced excision of transposon Tn*5*. *Gene* **22:**1–7.
39. **Berg, D. E., L. Johnsrud, L. McDivitt, R. Ramabhadran, and B. J. Hirschel.** 1982. Inverted repeats of Tn*5* are transposable elements. *Proc. Natl. Acad. Sci. USA* **79:**2632–2635.
40. **Berg, D. E., T. Kazic, S. H. Phadnis, K. W. Dodson, and J. K. Lodge.** 1988. Mechanism and regulation of transposition, p. 107–129. *In* A. Kingsman, S. Kingsman, and K. Chater (ed.), *Transposition*. Cambridge University Press, Cambridge.
41. **Berg, D. E., J. Lodge, C. Sasakawa, D. K. Nag, S. H. Phadnis, K. Weston-Hafer, and G. F. Carle.** 1984. Transposon Tn*5*: specific sequence recognition and conservative transposition. *Cold Spring Harbor Symp. Quant. Biol.* **49:** 215–226.
42. **Berg, D. E., M. Schmandt, and J. B. Lowe.** 1983. Specificity of transposon Tn*5* insertion. *Genetics* **105:**813–828.
43. **Berg, D. E., A. Weiss, and L. Crossland.** 1980. Polarity of Tn*5* insertion mutations in *Escherichia coli*. *J. Bacteriol.* **142:**439–446.
44. **Beringer, J. E., J. L. Beynon, A. V. Buchanan-Wollaston, and A. W. B. Johnston.** 1978. Transfer of the drug-resistance transposon Tn*5* to *Rhizobium*. *Nature* (London) **276:**633–634.
45. **Biel, S. W., and D. E. Berg.** 1984. Mechanism of IS*1* transposition in *E. coli*: choice between simple insertion and cointegration. *Genetics* **108:**319–330.
46. **Blackhart, B. D., and D. R. Zusman.** 1985. Cloning and complementation analysis of the "frizzy" genes of *Myxococcus xanthus*. *Mol. Gen. Genet.* **198:**243–254.
47. **Bloch, M.-A., and C. Desaymard.** 1985. Antigenic polymorphism of the LamB protein among members of the family *Enterobacteriaceae*. *J. Bacteriol.* **163:**106–110.
48. **Bloch, M. A., and O. Raibaud.** 1986. Caracterisation préliminaire d'un nouveau système permettant l'assimilation du maltose par *Escherichia coli*. *Ann. Inst. Pasteur Microbiol. B* **137:**145–153.
49. **Bochner, B. R., H.-C. Huang, G. L. Schieven, and B. N. Ames.** 1980. Positive selection for loss of tetracycline resistance. *J. Bacteriol.* **143:**926–933.
50. **Bohall, N. A., Jr., and P. S. Vary.** 1986. Transposition of Tn*917* in *Bacillus megaterium*. *J. Bacteriol.* **167:**716–718.
51. **Boucher, C., B. Message, D. Debieu, and C. Zischek.** 1981. Use of P-1 incompatibility group plasmids to introduce transposons into *Pseudomonas solanacearum*. *Phytopathology* **71:**639–642.
52. **Boulnois, G. J., J. M. Varley, G. S. Sharpe, and F. C. H. Franklin.** 1985. Transposon donor plasmids, based on CoIIb-P9, for use in *Pseudomonas putida* and a variety of other Gram negative bacteria. *Mol. Gen. Genet.* **200:**65–67.
53. **Brand, A. H., L. Breeden, J. Abraham, R. Sternglanz, and K. Nasmyth.** 1985. Characterization of a "silencer" in yeast: a DNA sequence with properties opposite to those of a transcriptional enhancer. *Cell* **41:**41–48.
54. **Bremer, E., T. J. Silhavy, and G. M. Weinstock.** 1985. Transposable λ p*lac*-Mu bacteriophages for creating *lacZ* operon fusions and kanamycin resistance insertions in *Escherichia coli*. *J. Bacteriol.* **162:**1092–1099.
55. **Bremer, E., T. J. Silhavy, J. M. Weisemann, and G. M. Weinstock.** 1984. λ p*lac*Mu: a transposable derivative of bacteriophage lambda for creating *lacZ* protein fusions in a single step. *J. Bacteriol.* **158:**1084–1093.
56. **Brunier, D., B. Michel, and S. D. Ehrlich.** 1988. Copy choice illegitimate DNA recombination. *Cell* **52:**883–892.
57. **Bukhari, A. I.** 1975. Reversal of mutator phage Mu integration. *J. Mol. Biol.* **96:**87–99.
58. **Bukhari, A. I., and E. Ljungquist.** 1977. Bacteriophage Mu: methods for cultivation and use, p. 749–756. *In* A. I. Bukhari, J. A. Shapiro, and S. L. Adhya (ed.), *DNA Insertion Elements, Plasmids, and Episomes*. Cold Spring Harbor Laboratory, Cold Spring Harbor, N.Y.
59. **Bukhari, A. I., and A. L. Taylor.** 1971. Genetic analysis of diaminopimelic acid- and lysine-requiring mutants of *Escherichia coli*. *J. Bacteriol.* **105:**844–854.
60. **Caillaud, F., and P. Courvalin.** 1987. Nucleotide sequence of the ends of the conjugative shuttle transposon Tn*1545*. *Mol. Gen. Genet.* **209:**110–115.
61. **Calos, M. P., and J. H. Miller.** 1980. Molecular consequences of deletion formation mediated by the transposon Tn*9*. *Nature* (London) **285:**38–41.
62. **Campbell, A.** 1971. Genetic structure, p. 13–44. *In* A. D. Hershey (ed.), *The Bacteriophage Lambda*. Cold Spring Harbor Laboratory, Cold Spring Harbor, N.Y.
63. **Casadaban, M. J.** 1975. Fusion of the *Escherichia coli lac* genes to the *ara* promoter: a general technique using bacteriophage Mu-1 insertions. *Proc. Natl. Acad. Sci. USA* **72:** 809–813.
64. **Casabadan, M. J.** 1976. Transposition and fusion of the *lac* genes to selected promoters in *Escherichia coli* using bacteriophage lambda and Mu. *J. Mol. Biol.* **104:**541–555.
65. **Casadaban, M. J., and J. Chou.** 1984. *In vivo* formation of hybrid protein β-galactosidase gene fusions in one step with a new transposable Mu-*lac* transducing phage. *Proc. Natl. Acad. Sci. USA* **81:**535–539.
66. **Casadaban, M. J., and S. N. Cohen.** 1979. Lactose genes fused to exogenous promoters in one step using a Mu-*lac* bacteriophage: *in vivo* probe for transcriptional control sequences. *Proc. Natl. Acad. Sci. USA* **76:**4530–4533.
67. **Castilho, B. A., P. Olfson, and M. J. Casadaban.** 1984. Plasmid insertion mutagenesis and *lac* gene fusion with mini-Mu bacteriophage transposons. *J. Bacteriol.* **158:** 488–495.
68. **Cen, Y., G. L. Bender, M. J. Trinick, N. A. Morrison, K. F. Scott, P. M. Gresshoff, J. Shine, and B. G. Rolfe.** 1982. Transposon mutagenesis in rhizobia which can nodulate both legumes and the nonlegume *Parasponia*. *Appl. Environ. Microbiol.* **43:**233–236.
69. **Chaconas, G.** 1987. Transposition of Mu DNA in vivo, p. 137–157. *In* N. Symonds, A. Toussaint, P. van de Putte, and M. M. Howe (ed.), *Phage Mu*. Cold Spring Harbor Laboratory, Cold Spring Harbor, N.Y.
70. **Chaconas, G., F. J. deBruijn, M. Casadaban, J. R. Lupski, T. J. Kwoh, R. M. Harshey, M. S. DuBow, and A. I.**

Bukhari. 1981. *In vitro* and *in vivo* manipulations of bacteriophage Mu DNA: cloning of Mu ends and construction of mini-Mu's carrying selectable markers. *Gene* **13:**37–46.

70a. **Chan, R. K., D. Botstein, T. Watanabe, and Y. Ogata.** 1972. Specialized transduction of tetracycline resistance by phage P22 in *Salmonella typhimurium*. II. Properties of a high-frequency transducing lysate. *Virology* **50:**883–898.

71. **Chandler, M., E. Roulet, L. Silver, E. Boy de la Tour, and L. Caro.** 1979. Tn*10* mediated integration of the plasmid R100.1 into the bacterial chromosome: inverse transposition. *Mol. Gen. Genet.* **173:**23–30.

72. **Chandra, T. S., and C. G. Friedrich.** 1986. Tn*5*-induced mutations affecting sulfur-oxidizing ability (*sox*) of *Thiosphaera pantotropha*. *J. Bacteriol.* **166:**446–452.

73. **Chatterjee, A. K., K. K. Thurn, and D. A. Feese.** 1983. Tn*5*-induced mutations in the enterobacterial phytopathogen *Erwinia chrysanthemi*. *Appl. Environ. Microbiol.* **45:** 644–650.

74. **Chen, H., M. Batley, J. W. Redmond, and B. G. Rolfe.** 1985. Alteration of the effective nodulation properties of a fast-growing, broad-host-range *Rhizobium* due to changes in exopolysaccharide synthesis. *J. Plant Physiol.* **120:**331–349.

75. **Choi, C.-L., J. Grinsted, J. Altenbuchner, R. Schmitt, and M. S. Richmond.** 1981. Transposons Tn*501* and Tn*1721* are closely related. *Cold Spring Harbor Symp. Quant. Biol.* **45:**64–66.

76. **Chopra, I., S. W. Shales, J. M. Ward, and L. J. Wallace.** 1981. Reduced expression of Tn*10*-mediated tetracycline-resistance in *Escherichia coli* containing more than one copy of the transposon. *J. Gen. Microbiol.* **126:**45–54.

77. **Chow, L. T., N. Davidson, and D. Berg.** 1974. Electron microscope study of the structures of lambda-dv DNAs. *J. Mol. Biol.* **86:**69–89.

78. **Chow, W.-Y., and D. E. Berg.** 1988. Tn*5*tac1, a derivative of transposon Tn*5* that generates conditional mutations. *Proc. Natl. Acad. Sci. USA* **85:**6468–6472.

79. **Chua, K.-Y., C. E. Pankhurst, P. E. MacDonald, D. H. Hopcroft, B. D. W. Jarvis, and D. B. Scott.** 1985. Isolation and characterization of transposon Tn*5*-induced symbiotic mutants of *Rhizobium loti*. *J. Bacteriol.* **162:**335–343.

80. **Chumley, F. G., R. Menzel, and J. R. Roth.** 1979. Hfr formation directed by Tn*10*. *Genetics* **91:**639–655.

81. **Ciampi, M. S., M. B. Schmid, and J. R. Roth.** 1982. Transposon Tn*10* provides a promoter for transcription of adjacent sequences. *Proc. Natl. Acad. Sci. USA* **79:** 5016–5020.

82. **Clewell, D. B., S. E. Flannagan, Y. Ike, J. M. Jones, and C. Gawron-Burke.** 1988. Sequence analysis of termini of conjugative transposon Tn*916*. *J. Bacteriol.* **170:**3046–3052.

83. **Clewell, D. B., and C. Gawron-Burke.** 1986. Conjugative transposons and the dissemination of antibiotic resistance in streptococci. *Annu. Rev. Microbiol.* **40:**635–659.

84. **Clewell, D. B., E. Senghas, J. M. Jones, S. E. Flannagan, M. Yamamoto, and C. Gawron-Burke.** 1988. Transposition in *Streptococcus*: structural and genetic properties of the conjugative transposon Tn*916*. *Symp. Soc. Gen. Microbiol.* **43:**43–58.

85. **Contreras, A., and J. Casadesus.** 1987. Tn*10* mutagenesis in *Azotobacter vinelandii*. *Mol. Gen. Genet.* **209:**276–282.

86. **Cornelis, G., H. Sommer, and H. Saedler.** 1981. Transposon Tn*951* (TnLac) is defective and related to Tn*3*. *Mol. Gen. Genet.* **184:**241–248.

87. **Courvalin, P., and C. Carlier.** 1987. Tn*1545*: a conjugative shuttle transposon. *Mol. Gen. Genet.* **206:**259–264.

88. **Couturier, M.** 1976. The integration and excision of bacteriophage Mu-1. *Cell* **7:**155–163.

89. **Cronan, J. E., Jr.** 1983. Use of Mu phages to isolate transposon insertions juxtaposed to given genes of *Escherichia coli*. *Curr. Microbiol.* **9:**245–252.

90. **Csonka, L. N., M. M. Howe, J. L. Ingraham, L. S. Pierson III, and C. L. Turnbough, Jr.** 1981. Infection of *Salmonella typhimurium* with coliphage Mu*d1* (Ap^r *lac*): construction of *pyr::lac* gene fusions. *J. Bacteriol.* **145:**299–305.

91. **Cuppels, D. A.** 1986. Generation and characterization of Tn*5* insertion mutations in *Pseudomonas syringae* pv. *tomato*. *Appl. Environ. Microbiol.* **51:**323–327.

92. **Daignan-Fornier, B., and M. Bolotin-Fukuhara.** 1988. In vivo functional characterization of a yeast nucleotide sequence: construction of a mini-Mu derivative adapted to yeast. *Gene* **62:**45–54.

93. **Daniels, D. A., J. L. Schroeder, W. Szybalski, F. Sanger, and F. R. Blattner.** 1983. A molecular map of coliphage lambda, p. 469–517. *In* R. W. Hendrix, J. W. Roberts, F. W. Stahl, and R. A. Weisberg (ed.), *Lambda II*. Cold Spring Harbor Laboratory, Cold Spring Harbor, N.Y.

94. **Daniels, G. A., G. Drews, and M. H. Saier.** 1988. Properties of a Tn*5* insertion mutant defective in the structural gene (*fruA*) of the fructose-specific phosphotransferase system of *Rhodobacter capsulatus* and cloning of the *fru* regulon. *J. Bacteriol.* **170:**1698–1703.

94a. **Darzins, S., N. Kent, M. Buckwalter, and M. Casadaban.** 1988. Bacteriophage Mu site required for transposition immunity. *Proc. Natl. Acad. Sci. USA* **85:**6826–6830.

95. **DasGupta, U., K. Weston-Hafer, and D. E. Berg.** 1987. Local DNA sequence control of deletion formation in *Escherichia coli* plasmid pBR322. *Genetics* **115:**41–49.

96. **Davidson, M. S., P. Roy, and A. O. Summers.** 1985. Transpositional mutagenesis of *Thiobacillus novellus* and *Thiobacillus versutus*. *Appl. Environ. Microbiol.* **49:**1436–1441.

97. **Davis, R. W., D. Botstein, and J. R. Roth.** 1980. *A Manual for Genetic Engineering: Advanced Bacterial Genetics*. Cold Spring Harbor Laboratory, Cold Spring Harbor, N.Y.

98. **de Bruijn, F. J.** 1987. Transposon Tn*5* mutagenesis to map genes. *Methods Enzymol.* **154:**175–196.

99. **de Bruijn, F. J., and J. R. Lupski.** 1984. The use of transposon Tn*5* mutagenesis in the rapid generation of correlated physical and genetic maps of DNA segments cloned into multicopy plasmids—a review. *Gene* **27:**131–149.

100. **Deeley, M. C., and C. Yanofsky.** 1981. Nucleotide sequence of the structural gene for tryptophanase of *Escherichia coli* K-12. *J. Bacteriol.* **147:**787–796.

101. **de Mendoza, D., D. Clark, and J. E. Cronan, Jr.** 1981. One-step gene amplification by Mu-mediated transposition of *E. coli* genes to a multicopy plasmid. *Gene* **15:**27–32.

102. **de Mendoza, D., H. C. Gramajo, and A. L. Rosa.** 1986. In vivo cloning of DNA into multicopy cosmids by mini-Mu-cosduction. *Mol. Gen. Genet.* **205:**546–549.

103. **de Mendoza, D., and A. L. Rosa.** 1985. Cloning of mini-Mu bacteriophage in cosmids: in vivo packaging into phage lambda heads. *Gene* **39:**55–59.

104. **d'Enfert, C., A. Ryter, and A. P. Pugsley.** 1987. Cloning and expression in *Escherichia coli* of the *Klebsiella pneumoniae* genes for production of surface localization and secretion of the lipoprotein pullulanase. *EMBO J.* **6:**3531–3538.

105. **de Vos, G. F., T. M. Finan, E. R. Signer, and G. C. Walker.** 1984. Host-dependent transposon Tn*5*-mediated streptomycin resistance. *J. Bacteriol.* **159:**395–399.

106. **de Vos, G. F., G. C. Walker, and E. R. Signer.** 1986. Genetic

manipulations in *Rhizobium meliloti* utilizing two new transposon Tn*5* derivatives. *Mol. Gen. Genet.* **204:**485–491.

107. **DeVries, G. E., C. K. Raymond, and R. A. Ludwig.** 1984. Extension of bacteriophage λ host range: selection, cloning, and characterization of a constitutive λ receptor gene. *Proc. Natl. Acad. Sci. USA* **81:**6080–6084.

108. **Doten, R. C., and R. P. Mortlock.** 1985. Characterization of xylitol-utilizing mutants of *Erwinia uredovora*. *J. Bacteriol.* **161:**529–533.

109. **Douglas, C. J., R. J. Staneloni, R. A. Rubin, and E. W. Nester.** 1985. Identification and genetic analysis of an *Agrobacterium tumefaciens* chromosomal virulence region. *J. Bacteriol.* **161:**850–860.

110. **Driver, R. P., and R. P. Lawther.** 1985. Restriction endonuclease analysis of the *ilvGEDA* operon of members of the family *Enterobacteriaceae*. *J. Bacteriol.* **162:**1317–1319.

110a. **Druger-Liotta, J., V. J. Prange, D. G. Overdier, and L. N. Csonka.** 1986. Selection of mutations that alter the osmotic control of transcription of the *Salmonella typhimurium proU* operon. *J. Bacteriol.* **169:**2449–2459.

111. **Dubow, M. S.** 1987. Transposable Mu-like phages, p. 201–213. *In* N. Symonds, A. Toussaint, P. van de Putte, and M. M. Howe (ed.), *Phage Mu*. Cold Spring Harbor Laboratory, Cold Spring Harbor, N.Y.

112. **Duncan, M. J.** 1981. Properties of Tn*5*-induced carbohydrate mutants in *Rhizobium meliloti*. *J. Gen. Microbiol.* **122:**61–67.

113. **Dworkin, M., and D. Kaiser.** 1985. Cell interactions in myxobacterial growth and development. *Science* **230:**18–24.

113a. **Dybvig, K., and J. Alderete.** 1988. Transformation of *Mycoplasma pulmonis* and *Mycoplasma hyorinis*: transposition of Tn916 and formation of cointegrate structures. *Plasmid* **20:**33–41.

114. **Dybvig, K., and G. H. Cassell.** 1987. Transposition of gram-positive transposon Tn*916* in *Acholeplasma laidlawii* and *Mycoplasma pulmonis*. *Science* **235:**1392–1394.

115. **Egelhoff, T. T., R. F. Fisher, T. W. Jacobs, J. T. Mulligan, and S. R. Long.** 1985. Nucleotide sequence of *Rhizobium meliloti* 1021 nodulation genes: *nodD* is read divergently from *nodABC*. *DNA* **4:**241–248.

116. **Egner, C., and D. Berg.** 1981. Excision of transposon Tn*5* is dependent on the inverted repeats but not the transposase function of Tn*5*. *Proc. Natl. Acad. Sci. USA* **78:**459–463.

117. **Elmerich, C., and C. Franche.** 1982. *Azospirillum* genetics: plasmids, bacteriophages and chromosome mobilization. *Experientia* **42**(Suppl.)**:**9–17.

118. **Ely, B.** 1985. Vectors for transposon mutagenesis of nonenteric bacteria. *Mol. Gen. Genet.* **200:**302–304.

119. **Ely, B., and R. H. Croft.** 1982. Transposon mutagenesis in *Caulobacter crescentus*. *J. Bacteriol.* **149:**620–625.

119a. **Ely, B., and C. J. Gerardot.** 1988. Use of pulse-field-gradient gel electrophoresis to construct a physical map of the *Caulobacter crescentus* genome. **68:**323–333.

120. **Enard, C., A. Diolez, and D. Expert.** 1988. Systemic virulence of *Erwinia chrysanthemi* 3937 requires a functional iron assimilation system. *J. Bacteriol.* **170:**2419–2426.

121. **Engebrecht, J., K. Nealson, and M. Silverman.** 1983. Bacterial bioluminescence: isolation and genetic analysis of functions from *Vibrio fischeri*. *Cell* **32:**773–781.

122. **Engebrecht, J., M. Simon, and M. Silverman.** 1985. Measuring gene expression with light. *Science* **227:**1345–1347.

123. **Faelen, M., M. Mergeay, J. Gerits, A. Toussaint, and N. LeFebvre.** 1981. Genetic mapping of a mutation conferring sensitivity to bacteriophage Mu in *Salmonella typhimurium* LT2. *J. Bacteriol.* **146:**914–919.

124. **Faelen, M., A. Toussaint, and A. Resibois.** 1979. Mini-Muduction: a new mode of gene transfer mediated by mini-Mu. *Mol. Gen. Genet.* **176:**191–197.

125. **Faelen, M., A. Toussaint, M. Van Montagu, S. Van den Elsacker, G. Engler, and J. Schell.** 1977. *In vivo* genetic engineering: the Mu-mediated transposition of chromosomal DNA segments onto transmissible plasmids, p. 521–530. *In* A. I. Bukhari, J. A. Shapiro, and S. Adhya (ed.), *DNA Insertion Elements, Plasmids, and Episomes*. Cold Spring Harbor Laboratory, Cold Spring Harbor, N.Y.

126. **Fields, P. I., R. V. Swanson, C. G. Haidaris, and F. Heffron.** 1986. Mutants of *Salmonella typhimurium* that cannot survive within the macrophage are avirulent. *Proc. Natl. Acad. Sci. USA* **83:**5189–5193.

127. **Flynn, J. L., and D. E. Ohman.** 1988. Use of a gene replacement cosmid vector for cloning alginate conversion genes from mucoid and nonmucoid *Pseudomonas aeruginosa* strains: *algS* controls expression of *algT*. *J. Bacteriol.* **170:**3228–3236.

128. **Forrai, T., E. Vincze, Z. Banfalvi, G. G. Kiss, G. S. Randhawa, and A. Kondorosi.** 1983. Localization of symbiotic mutations in *Rhizobium meliloti*. *J. Bacteriol.* **153:**635–643.

129. **Foster, T. J.** 1975. Tetracycline-sensitive mutants of the F-like R factors R100 and R100-1. *Mol. Gen. Genet.* **137:**85–88.

130. **Foster, T. J., M. A. Davis, D. E. Roberts, K. Takeshita, and N. Kleckner.** 1981. Genetic organization of transposon Tn*10*. *Cell* **23:**201–213.

131. **Foster, T. J., T. G. B. Howe, and K. M. V. Richmond.** 1975. Translocation of the tetracycline resistance determinant from R100-1 to the *Escherichia coli* K-12 chromosome. *J. Bacteriol.* **124:**1153–1158.

132. **Foster, T. J., V. Lundblad, S. Hanley-Way, and N. Kleckner.** 1981. Three Tn*10* associated excision events: relationship to transposition and role of direct and inverted repeats. *Cell* **23:**215–227.

133. **Franke, A. E., and D. B. Clewell.** 1981. Evidence for a chromosome-borne resistance transposon (Tn*916*) in *Streptococcus faecalis* that is capable of "conjugal" transfer in the absence of a conjugative plasmid. *J. Bacteriol.* **145:**494–502.

134. **Franklin, F. C. H.** 1985. Broad host range cloning vectors for Gram-negative bacteria, p. 165–184. *In* D. M. Glover (ed.), *DNA Cloning: a Practical Approach*, vol. 1. IRL Press, Oxford.

135. **Friedman, D. I., and M. Gottesman.** 1983. Lytic mode of lambda development, p. 21–51. *In* R. W. Hendrix, J. W. Roberts, F. W. Stahl, and R. A. Weisberg (ed.), *Lambda II*. Cold Spring Harbor Laboratory, Cold Spring Harbor, N.Y.

136. **Furth, M. E., and S. H. Wickner.** 1983. Lambda DNA replication, p. 145–173. *In* R. W. Hendrix, J. W. Roberts, F. W. Stahl, and R. A. Weisberg (ed.), *Lambda II*. Cold Spring Harbor Laboratory, Cold Spring Harbor, N.Y.

137. **Furuichi, T., M. Inouye, and S. Inouye.** 1985. Novel one-step cloning vector with a transposable element: application to the *Myxococcus xanthus* genome. *J. Bacteriol.* **164:**270–275.

138. **Gaillard, J. L., P. Berche, and P. Sansonetti.** 1986. Transposon mutagenesis as a tool to study the role of hemolysin in the virulence of *Listeria monocytogenes*. *Infect. Immun.* **52:**50–55.

139. **Galas, D. J., M. P. Calos, and J. H. Miller.** 1980. Sequence analysis of Tn9 insertions in the *lacZ* gene. *J. Mol. Biol.* **144:**19–41.

140. **Garfinkel, D. J., and E. W. Nester.** 1980. *Agrobacterium*

tumefaciens mutants affected in crown gall tumorigenesis and octopine catabolism. *J. Bacteriol.* **144:**732–743.

141. **Gatignol, A., M. Baron, and G. Tiraby.** 1987. Phleomycin resistance encoded by the *ble* gene from transposon Tn*5* as a dominant selectable marker in *Saccharomyces cerevisiae. Mol. Gen. Genet.* **207:**342–348.
142. **Gawron-Burke, C., and D. B. Clewell.** 1984. Regeneration of insertionally inactivated streptococcal DNA fragments after excision of transposon Tn*916* in *Escherichia coli*: strategy for targeting and cloning of genes from gram-positive bacteria. *J. Bacteriol.* **159:**214–221.
143. **Gay, P., D. Le Coq, M. Steinmetz, T. Berkelman, and C. I. Kado.** 1985. Positive selection procedure for entrapment of insertion sequence elements in gram-negative bacteria. *J. Bacteriol.* **164:**918–921.
144. **Genilloud, O., J. Blazquez, P. Mazodier, and F. Moreno.** 1988. A clinical isolate of transposon Tn*5* expressing streptomycin resistance in *Escherichia coli. J. Bacteriol.* **170:** 1275–1278.
145. **Genilloud, O., M. C. Garrido, and F. Moreno.** 1984. The transposon Tn*5* carries a bleomycin-resistance determinant. *Gene* **32:**225–233.
146. **Georgopoulos, C., K. Tilly, and S. Casjens.** 1983. Lambdoid phage head assembly, p. 279–304. *In* R. W. Hendrix, J. W. Roberts, F. W. Stahl, and R. A. Weisberg (ed.), *Lambda II.* Cold Spring Harbor Laboratory, Cold Spring Harbor, N.Y.
147. **Gill, R., R. Heffron, G. Dougan, and S. Salkow.** 1978. Analysis of sequences transposed by complementation of two classes of transposition-deficient mutants of transposition element Tn*3*. *J. Bacteriol.* **136:**742–756.
148. **Goebel, W., W. Lindenmaier, F. Pfeifer, H. Schrempf, and B. Schelle.** 1977. Transposition and insertion of intact, deleted and enlarged ampicillin transposon Tn*3* from mini-R1 (Rsc) plasmids into transfer factors. *Mol. Gen. Genet.* **157:** 119–129.
149. **Goguen, J. D., J. Yother, and S. C. Straley.** 1984. Genetic analysis of the low-calcium response in *Yersinia pestis* Mu d1(Ap *lac*) insertion mutants. *J. Bacteriol.* **160:**842–848.
150. **Goldberg, R. B., R. A. Bender, and S. L. Streicher.** 1974. Direct selection for P1-sensitive mutants of enteric bacteria. *J. Bacteriol.* **118:**810–814.
151. **Goto, N., A. Mochizuki, Y. Inagaki, S. Horiuchi, T. Tanaka, and R. Nakaya.** 1987. Identification of the DNA sequence required for transposition immunity of the γδ sequence. *J. Bacteriol.* **169:**4388–4390.
152. **Goto, N., A. Shoji, S. Horiuchi, and R. Nakaya.** 1984. Conduction of nonconjugative plasmids by F′ *lac* is not necessarily associated with transposition of the γδ sequence. *J. Bacteriol.* **159:**590–596.
153. **Gotesman, S., and J. R. Beckwith.** 1969. Directed transposition of the arabinose operon: a technique for the isolation of specialized transducing bacteriophages for any *Escherichia coli* gene. *J. Mol. Biol.* **44:**117–127.
154. **Gramajo, H. C., and D. de Mendoza.** 1987. A *cos*-mini-Mu vector for *in vivo* DNA cloning. *Gene* **51:**85–90.
155. **Gramajo, H. C., A. M. Viale, and D. de Mendoza.** 1988. Expression of cloned genes by in vivo insertion of *tac* promoter using a mini-Mu bacteriophage. *Gene* **65:** 305–314.
156. **Grindley, N. D. F., and W. S. Kelley.** 1976. Effects of different alleles of the *E. coli* K-12 *polA* gene on the replication of nontransferring plasmids. *Mol. Gen. Genet.* **143:**311–318.
157. **Grindley, N. D. F., and R. R. Reed.** 1985. Transpositional recombination in prokaryotes. *Annu. Rev. Biochem.* **54:** 863–896.
158. **Groisman, E. A., and M. J. Casadaban.** 1986. *In vivo* DNA cloning with a mini-Mu replicon cosmid and a helper lambda phage. *Gene* **51:**77–84.
159. **Groisman, E. A., and M. J. Casadaban.** 1986. Mini-Mu bacteriophage with plasmid replicons for in vivo cloning and *lac* gene fusing. *J. Bacteriol.* **168:**357–364.
160. **Groisman, E. A., and M. J. Casadaban.** 1987. Cloning of genes from members of the family *Enterobacteriaceae* with mini-Mu bacteriophage containing plasmid replicons. *J. Bacteriol.* **169:**687–693.
161. **Groisman, E. A., B. A. Castilho, and M. J. Casadaban.** 1984. *In vivo* DNA cloning and adjacent gene fusing with a mini-Mu-*lac* bacteriophage containing a plasmid replicon. *Proc. Natl. Acad. Sci. USA* **81:**1480–1483.
162. **Guyer, M. S.** 1978. The γδ sequence of F is an insertion sequence. *J. Mol. Biol.* **126:**347–365.
163. **Guyer, M. S.** 1983. Uses of the transposon γδ in the analysis of cloned genes. *Methods Enzymol.* **101:**362–363.
164. **Haas, D., and B. W. Holloway.** 1978. Chromosome mobilization by the R plasmid R68.45: a tool in *Pseudomonas* genetics. *Mol. Gen. Genet.* **158:**229–237.
165. **Halling, S. M., and N. Kleckner.** 1982. A symmetrical six-base pair target site sequence determines Tn*10* insertion specificity. Cell **28:**155–163.
166. **Harayama, S., T. Oguchi, and T. Iino.** 1984. The *E. coli* K-12 chromosome flanked by two IS*10* sequences transposes. *Mol. Gen. Genet.* **197:**62–66.
167. **Harkki, A., H. Karkku, and E. T. Palva.** 1987. Use of λ vehicles to isolate *ompC-lacZ* gene fusions in *Salmonella typhimurium* LT2. *Mol. Gen. Genet.* **209:**607–611.
168. **Harkki, A., and E. T. Palva.** 1984. Application of phage lambda technology to *Salmonella typhimurium*: construction of a lambda-sensitive *Salmonella* strain. *Mol. Gen. Genet.* **195:**256–259.
169. **Heffron, F.** 1983. Tn*3* and its relatives, p. 223–260. *In* J. A. Shapiro (ed.), *Mobile Genetic Elements.* Academic Press, Inc., New York.
170. **Heffron, F., C. Rubens, and S. Falkow.** 1975. Translocation of a plasmid DNA sequence which mediates ampicillin resistance: molecular nature and specificity of insertion. *Proc. Natl. Acad. Sci. USA* **72:**3623–3627.
171. **Hendrix, R. W., J. W. Roberts, F. W. Stahl, and R. A. Weisberg (ed.).** 1983. *Lambda II.* Cold Spring Harbor Laboratory, Cold Spring Harbor, N.Y.
172. **Herrmann, R., K. Neugebauer, H. Zentgraf, and H. Schaller.** 1978. Transposition of a DNA sequence determining kanamycin resistance into the single stranded genome of bacteriophage fd. *Mol. Gen. Genet.* **159:**171–178.
173. **Hill, C., C. Daly, and G. F. Fitzgerald.** 1985. Conjugative transfer of the transposon Tn*919* to lactic acid bacteria. *FEMS Microbiol. Lett.* **30:**115–119.
174. **Hill, C., C. Daly, and G. F. Fitzgerald.** 1987. Development of high-frequency delivery system for transposon Tn*919* in lactic streptococci: random insertion in *Streptococcus lactis* subsp. *diacetylactis* 18–16. *Appl. Environ. Microbiol.* **53:** 74–78.
175. **Hinton, J. C. D., and G. P. C. Salmond.** 1987. Use of Tn*phoA* to enrich for extracellular enzyme mutants of *Erwinia carotovora* subspecies *carotovora. Mol. Microbiol.* **1:**381–386.
176. **Hirsch, P. R., C. L. Wang, and M. J. Woodward.** 1986. Construction of a Tn*5* derivative determining resistance to

gentamicin and spectinomycin using a fragment cloned from R1033. *Gene* **48**:203–209.

177. **Holley, E. A., and J. W. Foster.** 1982. Bacteriophage P22 as a vector for Mu mutagenesis in *Salmonella typhimurium*: isolation of *nad-lac* and *pnc-lac* gene fusions. *J. Bacteriol.* **152**:959–962.

178. **Holloway, B., and K. B. Low.** 1987. F-prime and R-prime factors, p. 1145–1153. *In* F. C. Neidhardt, J. L. Ingraham, K. B. Low, B. Magasanik, M. Schaechter, and H. E. Umbarger (ed.), *Escherichia coli and Salmonella typhimurium: Cellular and Molecular Biology*. American Society for Microbiology, Washington, D.C.

179. **Hooykaas, P. J. J., R. Peerbolte, A. J. G. Regensburg-Tuink, P. de Vries, and R. A. Schilperoort.** 1983. A chromosomal linkage map of *Agrobacterium tumefaciens* and a comparison with the maps of *Rhizobium* spp. *Mol. Gen. Genet.* **188**:12–17.

180. **Horn, J. M., and D. E. Ohman.** 1988. Transcriptional and translational analyses of *recA* mutant alleles in *Pseudomonas aeruginosa*. *J. Bacteriol.* **170**:1637–1650.

181. **Howe, M. M.** 1987. Genetic and physical maps, p. 271–273. *In* N. Symonds, A. Toussaint, P. van de Putte, and M. M. Howe (ed.), *Phage Mu*. Cold Spring Harbor Laboratory, Cold Spring Harbor, N.Y.

182. **Howe, M. M., and E. G. Bade.** 1975. Molecular biology of bacteriophage Mu. *Science* **190**:624–632.

183. **Hughes, K. T., and J. R. Roth.** 1984. Conditionally transposition-defective derivative of Mu*d*I(Amp, Lac). *J. Bacteriol.* **159**:130–137.

184. **Hughes, K. T., and J. R. Roth.** 1988. Transitory *cis* complementation: a method for providing transposition functions to defective transposons. *Genetics* **119**:9–12.

185. **Huisman, O., W. Raymond, K.-U. Froehlich, P. Errada, N. Kleckner, D. Botstein, and M. A. Hoyt.** 1987. A Tn*10-lacZ-kanR-URA3* gene fusion transposon for insertion mutagenesis and fusion analysis of yeast and bacterial genes. *Genetics* **116**:191–199.

186. **Hunter, C. N.** 1988. Transposon Tn*5* mutagenesis of genes encoding reaction centre and light-harvesting LH1 polypeptides of *Rhodobacter sphaeroides*. *J. Gen. Microbiol.* **134**:1481–1489.

187. **Iida, S., J. Meyer, and W. Arber.** 1983. Prokaryotic IS elements, p. 159–221. *In* J. A. Shapiro (ed.), *Mobile Genetic Elements*. Academic Press, Inc., New York.

188. **Ippen, K., J. A. Shapiro, and J. R. Beckwith.** 1971. Transposition of the *lac* region to the *gal* region of the *Escherichia coli* chromosome: isolation of lambda *lac* transducing phages. *J. Bacteriol.* **108**:5–9.

189. **Ivins, B. E., S. L. Welkos, G. B. Knudson, and D. J. LeBlanc.** 1988. Transposon Tn*916* mutagenesis in *Bacillus anthracis*. *Infect. Immun.* **56**:176–181.

190. **Jenkins, F. J., M. J. Casadaban, and B. Roizman.** 1985. Application of the miniMu-phage target-sequence specific insertional mutagenesis of the herpes simplex genome. *Proc. Natl. Acad. Sci. USA* **82**:4773–4777.

191. **Joerger, R. D., R. Premakumar, and P. E. Bishop.** 1986. Tn*5*-induced mutants of *Azotobacter vinelandii* affected in nitrogen fixation under Mo-deficient and Mo-sufficient conditions. *J. Bacteriol.* **168**:673–682.

192. **Johnson, D. A., and N. S. Willetts.** 1980. Tn*2301*, a transposon construct carrying the entire transfer region of the F plasmid. *J. Bacteriol.* **143**:1171–1178.

193. **Johnson, S. R., and W. R. Romig.** 1979. Transposon-facilitated recombination in *Vibrio cholerae*. *Mol. Gen. Genet.* **170**:93–101.

194. **Johnson, S. R., and W. R. Romig.** 1981. *Vibrio cholerae* conjugative plasmid pSJ15 contains transposable prophage dVcA1. *J. Bacteriol.* **146**:632–638.

195. **Jones, J. M., S. C. Yost, and P. A. Pattee.** 1987. Transfer of the conjugal tetracycline resistance transposon Tn*916* from *Streptococcus faecalis* to *Staphylococcus aureus* and identification of some insertion sites in the staphylococcal chromosome. *J. Bacteriol.* **169**:2121–2131.

196. **Kahn, S. A., and R. P. Novick.** 1980. Terminal nucleotide sequences of Tn*551*, a transposon specifying erythromycin resistance in *Staphylococcus aureus*: homology with Tn*3*. *Plasmid* **4**:148–154.

197. **Karlovsky, P., and M. Vaskova.** 1986. Hot spot for Tn*1000* insertions in cloned repressor gene of the L phage. *Plasmid* **16**:219–221.

198. **Karlovsky, P., and M. Vaskova.** 1987. Tn*1000* insertional mutagenesis of cloned repressor gene of the phage L: plasmid oligomerization in the presence of F′*lac*. *Folia Microbiol.* **32**:185–195.

199. **Kathariou, S., P. Metz, H. Hof, and W. Goebel.** 1987. Tn*916*-induced mutations in the hemolysin determinant affecting virulence of *Listeria monocytogenes*. *J. Bacteriol.* **169**:1291–1297.

200. **Keen, M. G., E. D. Street, and P. S. Hoffman.** 1985. Broad-host-range plasmid pRK340 delivers Tn*5* into the *Legionella pneumophila* chromosome. *J. Bacteriol.* **162**:1332–1335.

201. **Kennedy, C., and A. Toukdarian.** 1987. Genetics of azotobacters: applications to nitrogen fixation and related aspects of metabolism. *Annu. Rev. Microbiol.* **41**:227–258.

202. **Khatoon, H., and A. I. Bukhari.** 1981. DNA rearrangements associated with reversion of bacteriophage Mu-induced mutations. *Genetics* **98**:1–24.

203. **Kingsbury, D. T., and D. R. Helinski.** 1973. Temperature-sensitive mutants for the replication of plasmids in *Escherichia coli*: requirement for deoxyribonucleic acid polymerase I in the replication of the plasmid ColE1. *J. Bacteriol.* **114**:1116–1124.

204. **Kleckner, N.** 1983. Transposon Tn*10*, p. 261–298. *In* J. A. Shapiro (ed.), *Mobile Genetic Elements*. Academic Press, Inc., New York.

205. **Kleckner, N., D. F. Barker, D. G. Ross, and D. Botstein.** 1978. Properties of the translocatable tetracycline-resistance element Tn*10* in *Escherichia coli* and bacteriophage lambda. *Gene* **90**:427–461.

206. **Kleckner, N., R. K. Chan, B.-K. Tye, and D. Botstein.** 1975. Mutagenesis by insertion of a drug-resistance element carrying an inverted repetition. *J. Mol. Biol.* **97**:561–575.

207. **Kleckner, N., K. Reichardt, and D. Botstein.** 1979. Inversions and deletions of the *Salmonella* chromosome generated by the translocatable tetracycline-resistance element Tn*10*. *J. Mol. Biol.* **127**:89–115.

208. **Kleckner, N., J. Roth, and D. Botstein.** 1977. Genetic engineering *in vivo* using translocatable drug-resistance elements: new methods in bacterial genetics. *J. Mol. Biol.* **116**:125–159.

209. **Kleckner, N., D. A. Steele, K. Reichardt, and D. Botstein.** 1979. Specificity of insertion by the translocatable tetracycline-resistance element Tn*10*. *Genetics* **92**:1023–1040.

210. **Klipp, W., B. Masepohl, and A. Puhler.** 1988. Identification and mapping of nitrogen fixation genes of *Rhodobacter capsulatus*: duplication of a *nifA-nifB* region. *J. Bacteriol.* **170**:693–699.

211. **Koch, C., G. Mertens, F. Rudt, R. Kahmann, R. Kanaar, R. H. A. Plasterk, P. van de Putte, R. Sandulache, and D.**

Kamp. 1987. The invertible G segment, p. 75–91. *In* N. Symonds, A. Toussaint, P. van de Putte, and M. M. Howe (ed.), *Phage Mu*. Cold Spring Harbor Laboratory, Cold Spring Harbor, N.Y.

212. **Kohara, Y., K. Akiyama, and K. Isono.** 1987. The physical map of the whole *E. coli* chromosome: application of a new strategy for rapid analysis and sorting of a large genomic library. *Cell* **50:**495–508.

213. **Koncz, C., Z. Koncz-Kalman, and J. Schell.** 1987. Transposon Tn*5* mediated gene transfer into plants. *Mol. Gen. Genet.* **207:**99–105.

214. **Kondoh, H., B. R. Paul, and M. M. Howe.** 1980. Use of λ pMU bacteriophages to isolate λ specialized transducing bacteriophages carrying genes for bacterial chemotaxis. *J. Virol.* **35:**619–628.

215. **Krebs, M. P., and W. S. Reznikoff.** 1988. Use of a Tn*5* derivative that creates *lacZ* translational fusions to obtain a transposition mutant. *Gene* **63:**277–285.

216. **Kretschmer, P. J., and S. N. Cohen.** 1977. Selected translocation of plasmid genes: frequency and regional specificity of translocation of the Tn*3* element. *J. Bacteriol.* **130:**888–899.

217. **Krolewski, J. J., E. Murphy, R. P. Novick, and M. G. Rush.** 1981. Site-specificity of the chromosomal insertion of *Staphylococcus aureus* transposon Tn*554*. *J. Mol. Biol.* **152:** 19–33.

218. **Kroos, L., and D. Kaiser.** 1984. Construction of Tn*5lac*, a transposon that fuses *lacZ* expression to exogenous promoters, and its introduction into *Myxococcus xanthus*. *Proc. Natl. Acad. Sci. USA* **81:**5816–5820.

219. **Krylov, V. N., V. G. Bogush, and J. Shapiro.** 1980. Bacteriophages of *Pseudomonas aeruginosa*, the DNA structure of which is similar to the structure of phage Mu1 DNA. I. General description, localization of endonuclease-sensitive sites in DNA, and structure of homoduplexes of phage D3112. *Soviet Genet.* **16:**528–536.

220. **Kuhn, I., F. H. Stephenson, H. W. Boyer, and P. J. Greene.** 1986. Positive-selection vectors utilizing lethality of the *Eco*RI endonuclease. *Gene* **44:**253–263.

221. **Kuhn, M., D. Jendrossek, C. Frund, A. Steinbuchel, and H. G. Schlegel.** 1988. Cloning of the *Alcaligenes eutrophus* alcohol dehydrogenase gene. *J. Bacteriol.* **170:**685–692.

222. **Kukral, A. M., K. L. Strauch, R. A. Maurer, and C. G. Miller.** 1987. Genetic analysis in *Salmonella typhimurium* with a small collection of randomly spaced insertions of transposon Tn*10*Δ*16*Δ*17*. *J. Bacteriol.* **169:**1787–1793.

223. **Kuner, J. M., L. M. Avery, D. E. Berg, and A. D. Kaiser.** 1981. Uses of transposon Tn*5* in the genetic analysis of *Myxococcus xanthus*, p. 128–131. *In* D. Schlessinger (ed.), *Microbiology—1981*. American Society for Microbiology, Washington, D.C.

224. **Kuner, J. M., and D. Kaiser.** 1981. Introduction of transposon Tn*5* into *Myxococcus* for analysis of developmental and other nonselectable mutants. *Proc. Natl. Acad. Sci. USA* **78:**425–429.

225. **Kuramitsu, H. K.** 1987. Utilization of a mini-Mu transposon to construct defined mutants in *Streptococcus mutans*. *Mol. Microbiol.* **1:**229–231.

226. **Kuramitsu, H. K., and M. J. Casadaban.** 1986. Transposition of the gram-positive transposon Tn*917* in *Escherichia coli*. *J. Bacteriol.* **167:**711–712.

227. **Kwan, H. S., and K. K. Wong.** 1986. A general method for isolation of Mu*d*1-8(Ap^r *lac*) operon fusions in *Salmonella typhimurium* LT2 from Tn*10* insertion strains: *chlC*:: Mud1–8. *Mol. Gen. Genet.* **205:**221–224.

228. **Labigne-Roussel, A., P. Courcoux, and L. Tompkins.** 1988. Gene disruption and replacement as a feasible approach for mutagenesis of *Campylobacter jejuni*. *J. Bacteriol.* **170:** 1704–1708.

229. **Laird, A. J., D. W. Ribbons, G. C. Woodrow, and I. G. Young.** 1980. Bacteriophage Mu-mediated gene transposition and *in vitro* cloning of the enterochelin gene cluster of *Escherichia coli*. *Gene* **11:**347–357.

230. **Lam, B. S., G. A. Strobel, L. A. Harrison, and S. T. Lam.** 1987. Transposon mutagenesis and tagging of fluorescent *Pseudomonas*: antimycotic production is necessary for control of Dutch elm disease. *Proc. Natl. Acad. Sci. USA* **84:**6447–6451.

231. **Lång, H., T. Teeri, S. Kurkela, E. Bremer, and E. T. Palva.** 1987. A plasmid vector for simultaneous generation of *lacZ* protein fusions and *npt*-II operon fusions in vivo. *FEMS Microbiol. Lett.* **48:**305–310.

232. **Leach, D., and N. Symonds.** 1979. The isolation and characterisation of a plaque-forming derivative of bacteriophage Mu carrying a fragment of Tn*3* conferring ampicillin resistance. *Mol. Gen. Genet.* **172:**179–184.

233. **Lee, C.-H., A. Bhagwat, and F. Heffron.** 1983. Identification of a transposon Tn*3* sequence required for transposition immunity. *Proc. Natl. Acad. Sci. USA* **80:**6765–6769.

234. **Lehmicke, L. G., and M. E. Lidstrom.** 1985. Organization of genes necessary for growth of the hydrogen-methanol autotroph *Xanthobacter* sp. strain H4-14 on hydrogen and carbon dioxide. *J. Bacteriol.* **162:**1244–1249.

235. **Lereclus, D., J. Mahillon, G. Menou, and M.-M. Lecadet.** 1986. Identification of Tn*4430*, a transposon of *Bacillus thuringiensis* functional in *Escherichia coli*. *Mol. Gen. Genet.* **204:**52–57.

236. **Lessie, T. G., and T. Gaffney.** 1986. Catabolic potential of *Pseudomonas cepacia*, p. 439–481. *In* I. C. Gunsalus, J. R. Sokatch, and L. N. Ornston (ed.), *The Bacteria*, vol. X. Academic Press, Inc., New York.

237. **Levy, S. B.** 1984. Resistance to the tetracyclines, p. 291–240. *In* L. E. Bryan (ed.), *Antimicrobial Drug Resistance*. Academic Press, Inc., New York.

238. **Lichtenstein, C., and S. Brenner.** 1982. Unique insertion site of Tn*7* in the *E. coli* chromosome. *Nature* (London) **297:** 601–603.

239. **Lin, R.-J., M. Capage, and C. W. Hill.** 1984. A repetitive DNA sequence, *rhs*, responsible for duplications within the *Escherichia coli* K-12 chromosome. *J. Mol. Biol.* **177:**1–18.

240. **Lindgren, P. B., R. C. Peet, and N. J. Panopoulos.** 1986. Gene cluster of *Pseudomonas syringae* pv. "*phaseolicola*" controls pathogenicity of bean plants and hypersensitivity on nonhost plants. *J. Bacteriol.* **168:**512–522.

241. **Liu, L., W. Whalen, A. Das, and C. M. Berg.** 1987. Rapid sequencing of cloned DNA using a transposon for bidirectional priming: sequence of the *Escherichia coli* K-12 *avtA* gene. *Nucleic Acids Res.* **15:**9461–9469.

241a. **Lodge, J. K., K. Weston-Hafer, and D. E. Berg.** 1988. Transposon Tn*5* target specificity: preference for insertions at G/C pairs. *Genetics* **120:**645–650.

241b. **Long, A., S. McCune, and G. C. Walker.** 1988. Symbiotic loci of *Rhizobium meliloti* identified by random Tn*phoA* mutagenesis. *J. Bacteriol.* **170:**4257–4265.

242. **Love, P. E., M. J. Lyle, and R. E. Yasbin.** 1985. DNA-damage-inducible (*din*) loci are transcriptionally activated in competent *Bacillus subtilis*. *Proc. Natl. Acad. Sci. USA* **82:**6201–6205.

243. **Luchanski, J. B., and P. A. Pattee.** 1984. Isolation of transposon Tn*551* insertions near chromosomal markers of interest in *Staphylococcus aureus*. *J. Bacteriol.* **159:**894–899.

244. **Ludwig, R. A.** 1987. Gene tandem-mediated selection of coliphage-receptive *Agrobacterium*, *Pseudomonas*, and *Rhizobium* strains. *Proc. Natl. Acad. Sci. USA* **84**:3334–3348.

245. **Lundblad, V., and N. Kleckner.** 1985. Mismatch repair mutations of *Escherichia coli* K12 enhance transposon excision. *Genetics* **109**:3–19.

246. **Lundblad, V., A. F. Taylor, G. R. Smith, and N. Kleckner.** 1984. Unusual alleles of *recB* and *recC* stimulate excision of inverted repeat transposons Tn*10* and Tn*5*. *Proc. Natl. Acad. Sci. USA* **81**:824–828.

247. **MacHattie, L. A., and J. A. Shapiro.** 1978. Chromosomal integration of phage λ by means of a DNA insertion element. *Proc. Natl. Acad. Sci. USA* **75**:1490–1494.

248. **MacNeil, D., M. M. Howe, and W. J. Brill.** 1980. Isolation and characterization of lambda specialized transducing bacteriophages carrying *Klebsiella pneumoniae nif* genes. *J. Bacteriol.* **141**:1264–1271.

249. **MacNeil, T., W. J. Brill, and M. M. Howe.** 1978. Bacteriophage Mu-induced deletions in a plasmid containing the *nif* (N_2 fixation) genes of *Klebsiella pneumoniae*. *J. Bacteriol.* **134**:821–829.

250. **Mahillon, J., and D. Lereclus.** 1988. Structural and functional analysis of Tn*4430*: identification of an integrase-like protein involved in the co-integrate-resolution process. *EMBO J.* **7**:1515–1526.

251. **Maloy, S. R., and W. D. Nunn.** 1981. Selection for loss of tetracycline resistance by *Escherichia coli*. *J. Bacteriol.* **145**: 1110–1112.

252. **Maloy, S. R., and J. R. Roth.** 1983. Regulation of proline utilization in *Salmonella typhimurium*: characterization of *put*::Mu d(Ap *lac*) operon fusions. *J. Bacteriol.* **154**:561–568.

253. **Manoil, C., and J. Beckwith.** 1985. Tn*phoA*: a transposon probe for protein export signals. *Proc. Natl. Acad. Sci. USA* **82**:8129–8133.

254. **Manoil, C., and J. Beckwith.** 1986. A genetic approach to analyzing membrane protein topology. *Science* **233**:1403–1408.

255. **Martinez-Laborda, A., M. Elias, R. Ruiz-Vazquez, and F. J. Murillo.** 1986. Insertions of Tn*5* linked to mutations affecting carotenoid synthesis in *Myxococcus xanthus*. *Mol. Gen. Genet.* **205**:107–114.

256. **Mazodier, P., P. Cossart, E. Giraud, and F. Gasser.** 1985. Completion of the nucleotide sequence of the central region of Tn*5* confirms the presence of three resistance genes. *Nucleic Acids Res.* **13**:195–205.

257. **Mazodier, P., O. Genilloud, E. Giraud, and F. Gasser.** 1986. Expression of Tn*5*-encoded streptomycin resistance in *E. coli*. *Mol. Gen. Genet.* **204**:404–409.

258. **McClelland, M., R. Jones, Y. Patel, and M. Nelson.** 1987. Restriction endonucleases for pulsed field mapping of bacterial genomes. *Nucleic Acids Res.* **15**:5985–6005.

259. **Meade, H. M., S. R. Long, G. B. Ruvkun, S. E. Brown, and F. M. Ausubel.** 1982. Physical and genetic characterization of symbiotic and auxotrophic mutants of *Rhizobium meliloti* induced by transposon Tn*5* mutagenesis. *J. Bacteriol.* **149**: 114–122.

260. **Mergeay, M., P. Lejeune, A. Sadouk, J. Gerits, and L. Fabay.** 1987. Shuttle transfer (or retrotransfer) of chromosomal markers mediated by the plasmid pULB113. *Mol. Gen. Genet.* **209**:61–70.

261. **Meyer, R., G. Boch, and J. Shapiro.** 1979. Transposition of DNA inserted into deletions of the Tn*5* kanamycin resistance element. *Mol. Gen. Genet.* **171**:7–13.

262. **Michiels, T., and G. Cornelis.** 1986. Tn*951* derivatives designed for high-frequency plasmid-specific transposition and deletion mutagenesis. *Gene* **43**:175–181.

263. **Michiels, T., G. Cornelis, K. Ellis, and J. Grinsted.** 1987. Tn*2501*, a component of the lactose transposon Tn*951*, is an example of a new category of class II transposable elements. *J. Bacteriol.* **169**:624–631.

264. **Miller, J. H., M. P. Calos, D. Galas, M. Hofer, D. E. Buchel, and B. Muller-Hill.** 1980. Genetic analysis of transpositions in the *lac* region of *Escherichia coli*. *J. Mol. Biol.* **144**:1–18.

265. **Miller, V. L., and J. J. Mekalanos.** 1988. A novel suicide vector and its use in construction of insertion mutations: osmoregulation of outer membrane proteins and virulence determinants in *Vibrio cholerae* requires *toxR*. *J. Bacteriol.* **170**:2575–2583.

266. **Mintz, C. S., and H. A. Shuman.** 1987. Transposition of bacteriophage Mu in the legionnaires disease bacterium. *Proc. Natl. Acad. Sci. USA* **84**:4645–4649.

267. **Mizuuchi, K.** 1983. *In vitro* transposition of bacteriophage Mu: a biochemical approach to a novel replication reaction. *Cell* **35**:785–794.

268. **Morales, V. M., and L. Sequeira.** 1985. Suicide vector for transposon mutagenesis in *Pseudomonas solanacearum*. *J. Bacteriol.* **163**:1263–1264.

269. **Morgan, M. K., and A. K. Chatterjee.** 1985. Isolation and characterization of Tn*5* insertion mutants of *Pseudomonas syringae* pv. *syringae* altered in the production of the peptide phytotoxin syringotoxin. *J. Bacteriol.* **164**:14–18.

270. **Moyed, H. S., T. T. Nguyen, and K. P. Bertrand.** 1983. Multicopy Tn*10 tet* plasmids confer sensitivity to induction of *tet* gene expression. *J. Bacteriol.* **155**:549–556.

271. **Muller, K.-H., T. J. Trust, and W. W. Kay.** 1988. Unmasking of bacteriophage Mu lipopolysaccharide receptors in *Salmonella enteritidis* confers sensitivity to Mu and permits Mu mutagenesis. *J. Bacteriol.* **170**:1076–1081.

271a.**Murata, N., H. Murata, H. Fujii, and C. Sasakawa.** 1988. Behavior of R388*rep*(Ts)::Tn*5*, a thermosensitive Tn*5* vector, and its derivatives in *Erwinia carotovora* and some species of *Rhizobiaceae*. *Curr. Microbiol.* **17**:293–298.

272. **Murphy, E., and S. Lofdahl.** 1984. Transposition of Tn*554* does not generate a target duplication. *Nature* (London) **307**:292–294.

273. **Nag, D. K., and D. E. Berg.** 1987. Specificity of bacteriophage Mu excision. *Mol. Gen. Genet.* **207**:395–401.

274. **Nag, D. K., U. DasGupta, G. Adelt, and D. E. Berg.** 1985. IS*50*-mediated inverse transposition: specificity and precision. *Gene* **34**:17–26.

275. **Nag, D. K., H. V. Huang, and D. E. Berg.** 1988. Bidirectional chain termination nucleotide sequencing: transposon Tn*5*-seq1 as a mobile source of primer sites. *Gene* **64**: 135–145.

276. **Naglich, J. G., and R. E. Andrews, Jr.** 1988. Introduction of the *Streptococcus faecalis* transposon Tn*916* into *Bacillus thuringiensis* subsp. *israelensis*. *Plasmid* **19**:84–93.

277. **Nassif, X., M.-C. Mazert, J. Mounier, and P. J. Sansonetti.** 1987. Evaluation with an *iuc*::Tn*10* mutant of the role of aerobactin production in the virulence of *Shigella flexneri*. *Infect. Immun.* **55**:1963–1969.

278. **Neidhardt, F. C., J. L. Ingraham, K. B. Low, B. Magasanik, M. Schaechter, and H. E. Umbarger (ed.).** 1987. *Escherichia coli and Salmonella typhimurium: Cellular and Molecular Biology*. American Society for Microbiology, Washington, D.C.

279. **Newland, J. W., B. A. Green, and R. K. Holmes.** 1984. Transposon-mediated mutagenesis and recombination in *Vibrio cholerae*. *Infect. Immun.* **45**:428–432.

280. **Nicas, T. I., and B. H. Iglewski.** 1984. Isolation and characterization of transposon-induced mutants of *Pseudomonas aeruginosa* deficient in production of exoenzyme S. *Infect. Immun.* **45:**470–474.

281. **Nida, K., and P. P. Cleary.** 1983. Insertional inactivation of streptolysin S expression in *Streptococcus pyogenes. J. Bacteriol.* **155:**1156–1161.

282. **Noel, K. D., and G. F.-L. Ames.** 1978. Evidence for a common mechanism for the insertion of the Tn*10* transposon and for the generation of Tn*10*-stimulated deletions. *Mol. Gen. Genet.* **166:**217–223.

283. **Noti, J. D., M. N. Jagadish, and A. A. Szalay.** 1987. Site-directed Tn*5* and transplacement mutagenesis: methods to identify symbiotic nitrogen fixation genes in slow-growing *Rhizobium. Methods Enzymol.* **154:**197–217.

284. **Obukowicz, M. G., F. J. Perlak, S. L. Bolten, K. Kusano-Kretzmer, E. J. Mayer, and L. S. Watrud.** 1987. IS*50L* as a non-self transposable vector used to integrate the *Bacillus thuringiensis* delta-endotoxin gene into the chromosome of root-colonizing pseudomonads. *Gene* **51:**91–96.

285. **Obukowicz, M. G., F. J. Perlak, K. Kusano-Kretzmer, E. J. Mayer, S. L. Bolten, and L. S. Watrud.** 1986. Tn*5*-mediated integration of the delta-endotoxin gene from *Bacillus thuringiensis* into the chromosome of root-colonizing pseudomonads. *J. Bacteriol.* **168:**982–989.

286. **Obukowicz, M. G., F. J. Perlak, K. Kusano-Kretzmer, E. J. Mayer, and L. S. Watrud.** 1986. Integration of the delta-endotoxin gene of *Bacillus thuringiensis* into the chromosome of root-colonizing strains of pseudomonads using Tn*5*. *Gene* **45:**327–331.

287. **O'Connor, M. B., and M. H. Malamy.** 1983. A new insertion sequence, IS*121*, is found on the Mu dI1(Ap *lac*) bacteriophage and the *Escherichia coli* K-12 chromosome. *J. Bacteriol.* **156:**669–679.

288. **Olson, E. R., and S.-T. Chung.** 1988. Transposon Tn*4556* of *Streptomyces fradiae*: nucleotide sequence of the ends and the target sites. *J. Bacteriol.* **170:**1955–1957.

289. **Ornellas, E. P., and B. A. D. Stocker.** 1974. Relation of lipopolysaccharide character to P1 sensitivity in *Salmonella typhimurium. Virology* **60:**491–502.

290. **Owen, D. J., and A. C. Ward.** 1985. Transfer of transposable drug-resistance elements Tn*5*, Tn*7*, and Tn*76* to *Azotobacter beijerinckii*: use of plasmid RP4::Tn76 as a suicide vector. *Plasmid* **14:**162–166.

291. **Patterson, T. A., D. L. Court, G. Dubuc, J. J. Michniewicz, J. Goodchild, A. I. Bukhari, and S. A. Narang.** 1986. Transposition studies of mini-Mu plasmids constructed from the chemically synthesized ends of bacteriophage Mu. *Gene* **50:**101–109.

292. **Peet, R. C., P. B. Lindgren, D. K. Willis, and N. J. Panopoulos.** 1986. Identification and cloning of genes involved in phaseolotoxin production by *Pseudomonas syringae* pv. "*phaseolicola.*" *J. Bacteriol.* **166:**1096–1105.

293. **Peng, Z.-G., and R. Wu.** 1987. A new and simple rapid method for sequencing DNA. *Methods Enzymol.* **155:** 214–231.

294. **Perkins, J. B., and P. J. Youngman.** 1986. Construction and properties of Tn*917-lac*, a transposon derivative that mediates transcriptional gene fusions in *Bacillus subtilis. Proc. Natl. Acad. Sci. USA* **83:**140–144.

295. **Pfeifer, F., M. Betlach, R. Martienssen, J. Friedman, and H. W. Boyer.** 1983. Transposable elements of *Halobacterium halobium. Mol. Gen. Genet.* **191:**182–188.

296. **Phadnis, S. H., and D. E. Berg.** 1985. *recA*-independent recombination between repeated IS*50* elements is not caused by an IS*50*-encoded function. *J. Bacteriol.* **161:**928–932.

297. **Phadnis, S. H., and D. E. Berg.** 1987. Identification of base pairs in the IS*50* O end needed for IS*50* and Tn*5* transposition. *Proc. Natl. Acad. Sci. USA* **84:**9118–9122.

298. **Phadnis, S. H., and H. K. Das.** 1987. Use of the plasmid pRK 2013 as a vehicle for transposition in *Azotobacter vinelandii. J. Biosci.* **12:**131–135.

299. **Phillips, S., and R. Novick.** 1979. Tn*554*: a repressible site-specific transposon in *Staphylococcus aureus. Nature* (London) **278:**476–478.

300. **Pidcock, K. A., and T. L. Stull.** 1988. Tn*5* transposon mutagenesis of *Pseudomonas cepacia* by an IncW suicide vector. *J. Infect. Dis.* **157:**1098–1099.

301. **Pischl, D. L., and S. K. Farrand.** 1983. Transposon-facilitated chromosome mobilization in *Agrobacterium tumefaciens. J. Bacteriol.* **153:**1451–1460.

302. **Prentki, P., M. Chandler, and D. J. Galas.** 1987. *Escherichia coli* integration host factor bends the DNA at the ends of IS*1* and in an insertion hotspot with multiple IHF binding sites. *EMBO J.* **6:**2479–2487.

303. **Putnoky, D., G. B. Kiss, I. Ott, and A. Kondorosi.** 1983. Tn*5* carries a streptomycin resistance determinant downstream from the kanamycin resistance gene. *Mol. Gen. Genet.* **191:**282–294.

304. **Quinto, M., and R. A. Bender.** 1984. Use of bacteriophage P1 as a vector for Tn*5* insertion mutagenesis. *Appl. Environ. Microbiol.* **47:**436–438.

305. **Raleigh, E. A., and N. Kleckner.** 1984. Multiple IS*10* rearrangements in *Escherichia coli. J. Mol. Biol.* **173:**437–461.

306. **Ramos, J. L., and R. L. Robson.** 1985. Isolation and properties of mutants of *Azotobacter chroococcum* defective in aerobic nitrogen fixation. *J. Gen. Microbiol.* **131:** 1449–1458.

307. **Ratet, P., and F. Richaud.** 1986. Construction and uses of a new transposable element whose insertion is able to produce gene fusions with the neomycin-phosphotransferase-coding region of Tn*903*. *Gene* **42:**185–192.

308. **Ratet, P., J. Schell, and F. J. de Bruijn.** 1988. Mini-Mu*lac* transposons with broad host-range origins of conjugal transfer and replication designed for gene regulation studies in *Rhizobiaceae. Gene* **63:**41–52.

309. **Reif, H.-J., and H. Saedler.** 1977. Chromosomal rearrangements in the *gal* region of *E. coli* K12 after integration of IS*1*, p. 81–91. *In* A. I. Bukhari, J. A. Shapiro, and S. L. Adhya (ed.), *DNA Insertion Elements, Plasmids, and Episomes.* Cold Spring Harbor Laboratory, Cold Spring Harbor, N.Y.

310. **Rella, M., A. Mercenier, and D. Haas.** 1985. Transposon insertion mutagenesis of *Pseudomonas aeruginosa* with a Tn*5* derivative: application to physical mapping of the *arc* gene cluster. *Gene* **33:**293–303.

311. **Resibois, A., A. Toussaint, F. van Gijsegem, and M. Faelen.** 1981. Physical characterization of mini-Mu and mini-D108. *Gene* **14:**103–113.

312. **Reyes, O., A. Beyou, C. Mignotte-Vieux, and F. Richaud.** 1987. Mini-Mu transduction: *cis*-inhibition of the insertion of Mud transposons. *Plasmid* **18:**183–192.

313. **Reynolds, A. E., J. Felton, and A. Wright.** 1981. Insertion of DNA activates the cryptic *bgl* operon in *E. coli* K-12. *Nature* (London) **293:**625–629.

314. **Reynolds, A. E., S. Mahadevan, J. Felton, and A. Wright.** 1985. Activation of the cryptic *bgl* operon: insertion sequences, point mutations, and changes in superhelicity affect promoter strength. *UCLA Symp. Mol. Cell. Biol. New Ser.* **20:**265–277.

315. **Roberts, M. C., and G. E. Kenny.** 1987. Conjugal transfer of transposon Tn*916* from *Streptococcus faecalis* to *Mycoplasma hominis*. *J. Bacteriol.* **169:**3836–3839.

316. **Robinson, M. K., P. M. Bennett, and M. H. Richmond.** 1977. Inhibition of Tn*A* translocation by Tn*A*. *J. Bacteriol.* **129:**407–414.

317. **Rodriguez, R. L., and D. T. Denhardt (ed.).** 1988. *Vectors: a Survey of Molecular Cloning Vectors and Their Uses.* Butterworths, Boston.

318. **Roizman, B., and F. J. Jenkins.** 1985. Genetic engineering of novel genomes of large DNA virus. *Science* **229:**1208–1214.

319. **Romeo, J. M., and D. R. Zusman.** 1987. Cloning of the gene for myxobacterial hemagglutinin and isolation and analysis of structural gene mutations. *J. Bacteriol.* **169:**3801–3808.

320. **Rosenfeld, S. A., and J. E. Brenchley.** 1980. Bacteriophage P1 as a vehicle for Mu mutagenesis of *Salmonella typhimurium*. *J. Bacteriol.* **144:**848–851.

321. **Rosner, J. L., and M. S. Guyer.** 1980. Transposition of IS*1*-BI0-IS*1* from a bacteriophage λ derivative carrying the IS*1*-*cat*-IS*1* transposon (Tn9). *Mol. Gen. Genet.* **178:**111–120.

322. **Ross, D. G., J. Swan, and N. Kleckner.** 1979. Physical structures of Tn*10*-promoted deletions and inversions: role of 1400 bp inverted repetitions. *Cell* **16:**721–732.

323. **Rostas, K., P. R. Sista, J. Stanley, and D. P. S. Verma.** 1984. Transposon mutagenesis of *Rhizobium japonicum*. *Mol. Gen. Genet.* **197:**230–235.

324. **Rothstein, S. J., R. A. Jorgensen, J. C.-P. Yin, Z. Yong-Di, R. C. Johnson, and W. S. Reznikoff.** 1980. Genetic organization of Tn*5*. *Cold Spring Harbor Symp. Quant. Biol.* **45:**99–105.

325. **Rubens, C. E., M. R. Wessels, L. M. Heggen, and D. L. Kasper.** 1987. Transposon mutagenesis of type III group B *Streptococcus*: correlation of capsule expression with virulence. *Proc. Natl. Acad. Sci. USA* **84:**7208–7212.

326. **Ruvkun, G. B., and F. M. Ausubel.** 1981. A general method for site-directed mutagenesis in prokaryotes. *Nature* (London) **289:**85–88.

327. **Sadowsky, M. J., E. R. Olson, V. E. Foster, R. M. Kosslak, and D. P. S. Verma.** 1988. Two host-inducible genes of *Rhizobium fredii* and characterization of the inducing compounds. *J. Bacteriol.* **170:**171–178.

328. **Salch, Y. P., and P. D. Shaw.** 1988. Isolation and characterization of pathogenicity genes of *Pseudomonas syringae* pv. *tabaci*. *J. Bacteriol.* **170:**2584–2591.

329. **Sandman, K., R. Losick, and P. Youngman.** 1987. Genetic analysis of *Bacillus subtilis spo* mutations generated by Tn*917*-mediated insertional mutagenesis. *Genetics* **117:** 603–617.

330. **Sandulache, R., P. Prehm, D. Expert, A. Toussaint, and D. Kamp.** 1985. The cell wall receptor for bacteriophage Mu G(−) in *Erwinia* and *Escherichia coli* C. *FEMS Microbiol. Lett.* **28:**307–310.

331. **Sandulache, R., P. Prehm, and D. Kamp.** 1984. Cell wall receptor for bacteriophage Mu G(+). *J. Bacteriol.* **160:** 299–303.

332. **Sanger, F., S. Nicklen, and A. R. Carlson.** 1977. DNA sequencing with chain-terminating inhibitors. *Proc. Natl. Acad. Sci. USA* **74:**5463–5467.

333. **Sasakawa, C., K. Kamata, T. Sakai, S. Makino, M. Yamada, N. Okada, and M. Yoshikawa.** 1988. Virulence-associated genetic regions comprising 31 kilobases of the 230-kilobase plasmid in *Shigella flexneri* 2a. *J. Bacteriol.* **170:**2480–2484.

334. **Sasakawa, C., and M. Yoshikawa.** 1987. A series of Tn*5* variants with various drug-resistance markers and suicide vector for transposon mutagenesis. *Gene* **56:**283–288.

335. **Sauer, B.** 1987. Functional expression of the *cre-lox* site-specific recombination system in the yeast *Saccharomyces cerevisiae*. *Mol. Cell. Biol.* **7:**2087–2096.

336. **Schmitt, R., J. Altenbuchner, and J. Grinsted.** 1982. Complementation of transposition function encoded by transposons Tn*501* (Hg^R) and Tn*1721* (Tet^R), p. 359–370. *In* S. B. Levy, R. C. Clowes, and E. L. Koenig (ed.), *Molecular Biology, Pathogenicity and Ecology of Bacterial Plasmids.* Plenum Publishing Corp., New York.

337. **Schmitt, R., J. Altenbuchner, K. Wiebauer, W. Arnold, A. Puhler, and F. Schoffl.** 1981. Basis of transposition and gene amplification by Tn*1721* and related tetracycline-resistance transposons. *Cold Spring Harbor Symp. Quant. Biol.* **45:** 59–65.

338. **Schoonejans, E., and A. Toussaint.** 1983. Utilization of plasmid pULB113 (RP4::mini-Mu) to construct a linkage map of *Erwinia carotovora* subsp. *chrysanthemi*. *J. Bacteriol.* **154:**1489–1492.

339. **Schrenk, W. J., and R. A. Weisberg.** 1975. A simple method for making new transducing lines of coliphage lambda. *Mol. Gen. Genet.* **137:**101–107.

340. **Schwartz, M., and L. Le Minor.** 1975. Occurrence of the bacteriophage lambda receptor in some *Enterobacteriaceae*. *J. Virol.* **15:**679–685.

341. **Scott, J. R., P. A. Kirchman, and M. G. Caparon.** 1988. An intermediate in transposition of the conjugative transposon Tn*916*. *Proc. Natl. Acad. Sci. USA* **85:**4809–4813.

342. **Seifert, H. S., E. Y. Chen, M. So, and F. Heffron.** 1986. Shuttle mutagenesis: a method of transposon mutagenesis for *Saccharomyces cerevisiae*. *Proc. Natl. Acad. Sci. USA* **83:**735–739.

343. **Senghas, E., J. M. Jones, M. Yamamoto, C. Gawron-Burke, and D. B. Clewell.** 1988. Genetic organization of the bacterial conjugative transposon Tn*916*. *J. Bacteriol.* **170:**245–249.

344. **Shapiro, J., and P. M. Brinkley.** 1984. Programming of DNA rearrangements involving Mu prophages. *Cold Spring Harbor Symp. Quant. Biol.* **49:**313–320.

345. **Shapiro, J. A.** 1979. Molecular model for the transposition and replication of bacteriophage Mu and other transposable elements. *Proc. Natl. Acad. Sci. USA* **76:**1933–1937.

346. **Sharp, P. A., S. N. Cohen, and N. Davidson.** 1973. Electron microscope heteroduplex studies of sequence relations among plasmids of *Escherichia coli*. II. Structure of drug resistance (R) and F factors. *J. Mol. Biol.* **75:**235–255.

347. **Shaw, J. H., and D. B. Clewell.** 1985. Complete nucleotide sequence of macrolide-lincosamide-streptogramin B-resistance transposon Tn*917* in *Streptococcus faecalis*. *J. Bacteriol.* **164:**782–796.

348. **Shaw, J. J., and C. I. Kado.** 1987. Direct analysis of the invasiveness of *Xanthomonas campestris* mutants generated by Tn*4431*, a transposon containing a promoterless luciferase cassette for monitoring gene expression, p. 57–60. *In* D. P. S. Verma and N. Brisson (ed.), *Molecular Genetics of Plant-Microbe Interactions.* Martinus Nijhoff, Dordrecht, The Netherlands.

349. **Shaw, J. J., L. G. Settles, and C. I. Kado.** 1988. Transposon Tn*4431* mutagenesis of *Xanthomonas campestris* pv. *campestris*: characterization of a nonpathogenic mutant and cloning of a locus for pathogenicity. *Mol. Plant-Microbe Interact.* **1:**39–45.

350. **Sherratt, D., A. Arthur, and M. Burke.** 1980. Transposon-specified, site-specific recombination systems. *Cold Spring Harbor Symp. Quant. Biol.* **45:**275–281.

351. **Shimada, K., R. A. Weisberg, and M. E. Gottesman.** 1972.

Prophage lambda at unusual chromosomal locations. I. Location of the secondary attachment sites and the properties of the lysogens. *J. Mol. Biol.* **63**:483–503.

352. **Shimada, K., R. A. Weisberg, and M. E. Gottesman.** 1973. Prophage λ at unusual chromosomal locations. II. Mutations induced by phage in *E. coli* K-12. *J. Mol. Biol.* **80**:297–314.

353. **Silhavy, T. J., and J. R. Beckwith.** 1985. Uses of *lac* fusions for the study of biological problems. *Microbiol. Rev.* **49**: 398–418.

354. **Silhavy, T. J., M. L. Berman, and L. W. Enquist.** 1984. *Experiments with Gene Fusions.* Cold Spring Harbor Laboratory, Cold Spring Harbor, N.Y.

355. **Simon, R.** 1984. High frequency mobilization of gram-negative bacterial replicons by the *in vitro* constructed Tn*5*-Mob transposon. *Mol. Gen. Genet.* **196**:413–420.

356. **Simon, R., U. Priefer, and A. Puhler.** 1983. A broad host range mobilization system for *in vivo* genetic engineering: transposon mutagenesis in Gram negative bacteria. *Bio/Technology* **1**:784–791.

357. **Simon, V., and W. Shumann.** 1987. In vivo formation of gene fusions in *Pseudomonas putida* and construction of versatile broad-host-range vectors for direct subcloning of Mu d1 and Mu d2 fusions. *Appl. Environ. Microbiol.* **53**:1649–1654.

358. **Singer, J. T., and W. R. Finnerty.** 1984. Insertional specificity of transposon Tn*5* in *Acinetobacter* sp. *J. Bacteriol.* **157**: 607–611.

359. **Singh, M., and W. Klingmuller.** 1986. Transposon mutagenesis in *Azospirillum brasilense*: isolation of auxotrophic and Nif⁻ mutants and molecular cloning of the mutagenized *nif* DNA. *Mol. Gen. Genet.* **202**:136–142.

360. **Smith, C. J., and H. Spiegel.** 1987. Transposition of Tn*4551* in *Bacteroides fragilis*: identification and properties of a new transposon from *Bacteroides* spp. *J. Bacteriol.* **169**:3450–3457.

361. **Smith, C. L., J. Enonome, A. Shutt, S. Kloo, and C. R. Cantor.** 1987. A physical map of the *E. coli* genome. *Science* **236**:1448–1453.

362. **Smith, C. L., and R. D. Kolodner.** 1988. Mapping of *Escherichia coli* chromosomal Tn*5* and F insertions by pulsed field gel electrophoresis. *Genetics* **119**:227–236.

363. **Smith, L. D., and F. Heffron.** 1987. Transposon Tn*5* mutagenesis of *Brucella abortus*. *Infect. Immun.* **55**:2774–2776.

364. **Snyder, M., S. Elledge, and R. W. Davis.** 1986. Rapid mapping of antigenic coding regions and constructing insertion mutations in yeast genes by mini-Tn*10* "transplason" mutagenesis. *Proc. Natl. Acad. Sci. USA* **83**:730–734.

365. **Sokol, P. A.** 1987. Tn*5* insertion mutants of *Pseudomonas aeruginosa* deficient in surface expression of ferripyochelin-binding protein. *J. Bacteriol.* **169**:3365–3368.

366. **Srivastava, S., M. Urban, and B. Friedrich.** 1982. Mutagenesis of *Alcaligenes eutrophus* by insertion of the drug-resistance transposon Tn*5*. *Arch. Microbiol.* **131**:203–207.

367. **Stachel, S. E., G. An, C. Flores, and E. W. Nester.** 1985. A Tn*3* *lacZ* transposon for the random generation of β-galactosidase gene fusions: application to the analysis of gene expression in *Agrobacterium*. *EMBO J.* **4**:891–898.

368. **Starlinger, P., and H. Saedler.** 1976. IS-elements in microorganisms. *Curr. Top. Microbiol. Immunol.* **75**:111–153.

369. **Stibitz, S., W. Black, and S. Falkow.** 1986. The construction of a cloning vector designed for gene replacement in *Bordetella pertussis*. *Gene* **50**:133–140.

370. **Stokes, H. W., and B. G. Hall.** 1984. Topological repression of gene activity by a transposable element. *Proc. Natl. Acad. Sci. USA* **81**:6115–6119.

371. **Symonds, N., A. Toussaint, P. van de Putte, and M. M. Howe (ed.).** 1987. *Phage Mu.* Cold Spring Harbor Laboratory, Cold Spring Harbor, N.Y.

372. **Syvanen, M., J. D. Hopkins, T. J. Griffin IV, T.-Y. Liang, K. Ippen-Ihler, and R. Kolodner.** 1986. Stimulation of precise excision and recombination by conjugal proficient F′ plasmids. *Mol. Gen. Genet.* **203**:1–7.

373. **Taylor, A. L.** 1963. Bacteriophage-induced mutation in *E. coli*. *Proc. Natl. Acad. Sci. USA* **50**:1043–1051.

374. **Taylor, R. K., C. Manoil, and J. J. Mekalanos.** 1988. Broad host range vectors for delivery of Tn*phoA*: use in genetic analysis of secreted virulence determinants of *Vibrio cholerae*. *J. Bacteriol.* **170**:1455–1487.

375. **Taylor, R. K., V. L. Miller, D. B. Furlong, and J. J. Mekalanos.** 1987. Use of *phoA* gene fusions to identify a pilus colonization factor coordinately regulated with cholera toxin. *Proc. Natl. Acad. Sci. USA* **84**:2833–2837.

376. **Thomas, P. M., L. D. Kuykendall, and J. S. Angle.** 1986. Mobilization of Tn*5* insertions from *Rhizobium fredii* by pJB3JI. *Appl. Environ. Microbiol.* **52**:206–208.

377. **Thurn, K. K., and A. K. Chatterjee.** 1985. Single-site chromosomal Tn*5* insertions affect the export of pectolytic and cellulolytic enzymes in *Erwinia chrysanthemi* EC16. *Appl. Environ. Microbiol.* **50**:894–898.

378. **Tolmasky, M. E., L. A. Actis, and J. H. Crosa.** 1988. Genetic analysis of the iron uptake region of the *Vibrio anguillarum* plasmid pJM1: molecular cloning of genetic determinants encoding a novel *trans* activator of siderophore biosynthesis. *J. Bacteriol.* **170**:1913–1919.

379. **Tomich, P., F. An, and D. B. Clewell.** 1980. Properties of an erythromycin-inducible transposon Tn*917* in *Streptococcus faecalis*. *J. Bacteriol.* **141**:1366–1374.

380. **Toukdarian, A., and C. Kennedy.** 1986. Regulation of nitrogen metabolism in *Azotobacter vinelandii*: isolation of *ntr* and *glnA* genes and construction of *ntr* mutants. *EMBO J.* **5**:399–407.

381. **Toussaint, A.** 1985. Bacteriophage Mu and its use as a genetic tool, p. 117–146. *In* J. Scaife, D. Leach, and A. Galizzi (ed.), *Genetics of Bacteria.* Academic Press, Inc. (London), Ltd., London.

382. **Toussaint, A., and A. Resibois.** 1983. Phage Mu: transposition as a life-style, p. 105–158. *In* J. A. Shapiro (ed.), *Mobile Genetic Elements.* Academic Press, Inc., New York.

383. **Trieu-Cuot, P., C. Carlier, P. Martin, and P. Courvalin.** 1987. Plasmid transfer by conjugation from *Escherichia coli* to gram-positive bacteria. *FEMS Microbiol. Lett.* **48**: 289–294.

384. **Trun, N. J., and T. J. Silhavy.** 1987. Characterization and *in vivo* cloning of *prlC*, a suppressor of signal sequence mutations in *Escherichia coli* K12. *Genetics* **116**:513–521.

385. **Tsai, M.-M., R. Y.-P. Wong, A. T. Hoang, and R. C. Deonier.** 1987. Transposition of Tn*1000*: in vivo properties. *J. Bacteriol.* **169**:5556–5562.

386. **Tu, C. P., and S. N. Cohen.** 1980. Translocation specificity of the Tn*3* element: characterization of sites of multiple insertions. *Cell* **19**:151–160.

387. **Ubben, D., and R. Schmitt.** 1986. Tn*1721* derivatives for transposon mutagenesis, restriction mapping and nucleotide sequence analysis. *Gene* **41**:145–152.

388. **Ubben, D., and R. Schmitt.** 1987. A transposable promoter and transposable promoter probes derived from Tn*1721*. *Gene* **53**:127–134.

389. **Vandeyar, M. A., and S. A. Zahler.** 1986. Chromosomal insertions of Tn*917* in *Bacillus subtilis*. *J. Bacteriol.* **167**: 530–534.

390. **van Gijsegem, F., and A. Toussaint.** 1982. Chromosome transfer and R-prime formation by an RP4::mini-Mu derivative in *Escherichia coli*, *Salmonella typhimurium*, *Klebsiella pneumoniae*, and *Proteus mirabilis*. *Plasmid* **7**:30–44.

391. **van Gijsegem, F., A. Toussaint, and M. Casadaban.** 1987. Mu as a genetic tool, p. 215–250. *In* N. Symonds, A. Toussaint, P. van de Putte, and M. M. Howe (ed.), *Phage Mu*. Cold Spring Harbor Laboratory, Cold Spring Harbor, N.Y.

392. **Vanstockem, M., K. Michiels, J. Vanderleyden, and A. P. van Gool.** 1987. Transposon mutagenesis of *Azospirillum brasilense* and *Azospirillum lipoferum*: physical analysis of Tn*5* and Tn*5*-mob insertion mutants. *Appl. Environ. Microbiol.* **53**:410–415.

393. **Viebrock, A., and W. G. Zumft.** 1987. Physical mapping of transposon Tn*5* insertions defines a gene cluster functional in nitrous oxide respiration by *Pseudomonas stutzeri*. *J. Bacteriol.* **169**:4577–4580.

394. **Waddell, C. S., and N. L. Craig.** 1988. Tn*7* transposition: two transposition pathways directed by five Tn*7*-encoded genes. *Genes Devel.* **2**:137–149.

395. **Walker, M. J., and J. M. Pemberton.** 1988. Construction of transposons encoding genes for β-glucosidase, amylase and polygalacturonate *trans*-eliminase from *Klebsiella oxytoca* and their expression in a range of gram-negative bacteria. *Curr. Microbiol.* **17**:69–75.

396. **Walton, D. A., and B. E. B. Moseley.** 1981. Induced mutagenesis in *Rhizobium trifolii*. *J. Gen. Microbiol.* **124**: 191–195.

397. **Wang, B., L. Liu, E. A. Groisman, M. J. Casadaban, and C. M. Berg.** 1987. High frequency generalized transduction by miniMu plasmid phage. *Genetics* **116**:201–206.

398. **Wang, M.-D., L. Buckley, and C. M. Berg.** 1987. Cloning of genes that suppress an *Escherichia coli* K-12 alanine auxotroph when present in multicopy plasmids. *J. Bacteriol.* **169**:5610–5614.

399. **Wang, M.-D., L. Liu, B. Wang, and C. M. Berg.** 1987. Cloning and characterization of the *Escherichia coli* K-12 alanine-valine transaminase *avtA* gene. *J. Bacteriol.* **169**: 4228–4234.

400. **Wanner, B. L.** 1986. Novel regulatory mutants of the phosphate regulon in *Escherichia coli* K-12. *J. Mol. Biol.* **191**:39–58.

401. **Warrelmann, J., and B. Friedrich.** 1986. Mutants of *Pseudomonas facilis* defective in lithoautotrophy. *J. Gen. Microbiol.* **132**:91–96.

402. **Way, J. C., M. A. Davis, D. Morisato, D. E. Roberts, and N. Kleckner.** 1984. New Tn*10* derivatives for transposon mutagenesis and for construction of *lacZ* operon fusions by transposition. *Gene* **32**:369–379.

403. **Weber, P. C., M. Levine, and J. C. Glorioso.** 1987. Rapid identification of nonessential genes of herpes simplex virus type 1 by Tn*5* mutagenesis. *Science* **236**:576–579.

404. **Weiss, A. A., and S. Falkow.** 1983. Transposon insertion and subsequent donor formation promoted by Tn*50* in *Bordetella pertussis*. *J. Bacteriol.* **153**:304–309.

405. **Weiss, A. A., E. L. Hewlett, G. A. Myers, and S. Falkow.** 1983. Tn*5*-induced mutations affecting virulence factors of *Bordetella pertussis*. *Infect. Immun.* **42**:33–41.

406. **White, F. F., and E. W. Nester.** 1980. Hairy root: plasmid encodes virulence traits in *Agrobacterium rhizogenes*. *J. Bacteriol.* **141**:1134–1141.

407. **Whitta, S., M. I. Sinclair, and B. W. Holloway.** 1985. Transposon mutagenesis in *Methylobacterium* AM1 (*Pseudomonas* AM1). *J. Gen. Microbiol.* **131**:1547–1549.

408. **Wilson, G. G., K. K. Y. Young, G. J. Edlin, and W. Konigsberg.** 1979. High-frequency generalised transduction by bacteriophage T*4*. *Nature* (London) **280**:80–82.

409. **Wilson, K. J., V. Anjaiah, P. T. C. Nambiar, and F. M. Ausubel.** 1987. Isolation and characterization of symbiotic mutants of *Bradyrhizobium* sp. (*Arachis*) strain NC92: mutants with host-specific defects in nodulation and nitrogen fixation. *J. Bacteriol.* **169**:2177–2186.

410. **Wu, R., and L. Grossman (ed.).** 1987. *Methods in Enzymology*, vol. 153. Academic Press, Inc., New York.

411. **Xu, P., S. Leong, and L. Sequeira.** 1988. Molecular cloning of genes that specify virulence in *Pseudomonas solanacearum*. *J. Bacteriol.* **170**:617–622.

412. **Yakobson, E. A., and D. G. Guiney, Jr.** 1984. Conjugal transfer of bacterial chromosomes mediated by the RK2 plasmid transfer origin cloned into transposon Tn*5*. *J. Bacteriol.* **160**:451–453.

412a. **Yin, J. C., M. P. Krebs, and W. S. Reznikoff.** The effect of *dam* methylation on Tn*5* transposition. *J. Mol. Biol.* **199**: 35–45.

413. **Yost, S. C., J. M. Jones, and P. A. Pattee.** 1988. Sequential transposition of Tn*916* among *Staphylococcus aureus* protoplasts. *Plasmid* **19**:13–20.

414. **Yother, J., T. W. Chamness, and J. D. Goguen.** 1986. Temperature-controlled plasmid regulon associated with low calcium response in *Yersinia pestis*. *J. Bacteriol.* **165**: 443–447.

415. **Youderian, P., P. Sugiono, K. L. Brewer, N. P. Higgins, and T. Elliott.** 1988. Packaging specific segments of the Salmonella chromosome with locked-in Mud-P22 prophages. *Genetics* **118**:581–592.

416. **Young, K. K. Y., and G. Edlin.** 1983. Physical and genetical analysis of bacteriophage T4 generalized transduction. *Mol. Gen. Genet.* **192**:241–246.

417. **Youngman, P.** 1987. Plasmid vectors for recovering and exploiting Tn*917* transpositions in Bacillus and other Gram-positive bacteria, p. 79–103. *In* K. G. Hardy (ed.), *Plasmids: a Practical Approach*. IRL Press, Washington, D.C.

418. **Youngman, P., P. Zuber, J. B. Perkins, K. Sandman, M. Igo, and R. Losick.** 1985. New ways to study developmental genes in spore-forming bacteria. *Science* **228**:285–291.

419. **Youngman, P. J., J. B. Perkins, and R. Losick.** 1983. Genetic transposition and insertional mutagenesis in *Bacillus subtilis* with *Streptococcus faecalis* transposon Tn*917*. *Proc. Natl. Acad. Sci. USA* **80**:2305–2309.

420. **Youvan, D. C., J. T. Elder, D. E. Sandlin, K. Zsebo, D. P. Alder, N. J. Panopoulos, B. L. Marrs, and J. E. Hearst.** 1982. R-prime site-directed transposon Tn*7* mutagenesis of the photosynthetic apparatus in *Rhodopseudomonas capsulata*. *J. Mol. Biol.* **162**:17–41.

421. **Zeldis, J. B., A. I. Bukhari, and D. Zipser.** 1973. Orientation of prophage Mu. *Virology* **55**:289–294.

422. **Zerbib, D., P. Gamas, M. Chandler, P. Prentki, S. Bass, and D. Galas.** 1985. Specificity of insertion of IS*1*. *J. Mol. Biol.* **185**:517–524.

423. **Zsebo, K. M., F. Wu, and J. E. Hearst.** 1984. Tn*5.7* construction and physical mapping of pRPS404 containing photosynthetic genes from *Rhodopseudomonas capsulata*. *Plasmid* **11**:182–184.

Chapter 42

Retroviral Tagging in Mammalian Development and Genetics

PHILIPPE SORIANO, THOMAS GRIDLEY, and RUDOLF JAENISCH

I. INTRODUCTION

Retroviruses are RNA viruses which replicate through a DNA intermediate (see review in this volume by H. Varmus and P. Brown). The retroviral DNA integrates with high efficiency into the host genome and, upon integration, forms a structure known as the provirus. Proviral integration has allowed the study of oncogenes, which could be activated by proviral insertion (19, 28, 84, 95), or the identification of new putative oncogenes by frequent proviral insertion in specific loci in tumor cells (14, 16, 49, 64, 100). More recently, the availability of defective retroviruses transducing foreign genes has allowed high-efficiency gene transfer in cells which were not previously transfectable. This may ultimately permit somatic "gene therapy," for instance, by transfer of genes into bone marrow cells, or the study in tissue culture of specific cell types by immortalization with oncogene-containing retroviruses.

Following integration in the host genome, the provirus is maintained as a stable genetic element in the infected cell and in its progeny. Infected cells can be detected by expression of viral genes. These two basic properties, integration as a stable genetic element and expression of the viral genes, have allowed retroviruses to be used as cell lineage markers in vertebrate development. In addition, insertion of viral sequences into the germ line has been successful for chromosomal marking and for the generation of insertional mutants. This review will focus on the use of retroviruses to study these issues in animals. We will describe the retroviruses used in these studies, methods for introducing them into animals, and their use as cell lineage markers and insertion mutagens.

II. RETROVIRUSES AND EMBRYO INFECTION

A. Retroviruses and Retroviral Vectors

Most retroviruses used for cell lineage studies and as insertion mutagens in mice are derivatives of Moloney murine leukemia virus (M-MuLV). M-MuLV is a replication-competent virus which causes thymic leukemia, lymphosarcoma, or hepatosplenomegaly in infected mice (for a review, see reference

Philippe Soriano ■ Howard Hughes Medical Institute, Institute for Molecular Genetics and Department of Cell Biology, Baylor College of Medicine, 1 Baylor Plaza, Houston, Texas 77030. **Thomas Gridley and Rudolf Jaenisch** ■ Whitehead Institute for Biomedical Research, Nine Cambridge Center, Cambridge, Massachusetts 02142, and Department of Biology, Massachusetts Institute of Technology, Cambridge, Massachusetts 02139.

Table 1. Introduction of proviral genomes in animals

Stage of infection	Proviral sequences in:		Expression from:	
	Germ line	Somatic cells	Viral LTR	Internal promoter
Preimplantation	Frequent	All tissues	Rare	Variable
Postimplantation	Rare	All tissues	High	Variable
Postnatal		Hematopoietic stem cells; at site of injection (retina)	Variable	Variable

105). The complete nucleotide sequence of the virus has been determined (85). Detailed studies of the life cycle of this virus (see review in this volume by Varmus and Brown) has allowed the manipulation of M-MuLV for use as a vector to transduce foreign genes in a wide variety of cell types. Defective retroviruses have been constructed and propagated in the absence of helper virus in specific packaging cell lines. The rationale for the construction of such cell lines is that only RNA genomes containing a specific packaging sequence, called ψ, can be packaged in retroviral particles (4, 56). Packaging cell lines can therefore be constructed by transfection of a proviral genome lacking the ψ sequence; the resulting cell line synthesizes all proteins necessary for the production of viral particles, but the RNAs transcribed from the ψ-defective proviral genome cannot be packaged. Such cell lines were initially used for the production of defective retroviruses, which contain the ψ sequence as well as the long terminal repeats (LTRs) and immediately adjacent sequences required for efficient reverse transcription. The first cell line of this type used extensively was the ψ2 cell line. Viruses produced by the ψ2 line are ecotropic; i.e., they can only infect murine cells. Packaging cell lines allowing production of amphotropic retroviruses (10, 60, 61, 90), capable of infecting both mouse cells and cells from other species, have also been produced. Amphotropic retroviruses have so far been shown to be useful for the study of cell lineages in chickens (J. Sanes, personal communication; C. Cepko, personal communication).

It has recently been shown that the introduction of new viral genomes into these cells can lead to the production of replication-competent virus, possibly by recombination (60). The absence of replication-competent virus is critical for studies of cell lineage. To avoid the accumulation of replication-competent virus, new cell lines called CRE and CRIP (O. Danos and R. Mulligan, personal communication) and GP + E86 (56a) have been constructed. In these cell lines, the viral proteins are encoded by two different transfected DNAs. These constructs also lack the packaging site and contain the simian virus 40 (SV40) instead of the M-MuLV polyadenylation sequence. Only very complex and presumably rare recombination events could thus lead to the formation of replication-competent virus in these cells.

The first defective retroviruses contained exogenous genes under the control of the viral promoter in the LTR (8, 56). Efforts to express these genes in transgenic mice derived by embryo infection were hampered, however, by inactivity of the viral promoter (36, 102). This promoter is inactive in preimplantation embryos, as well as in embryonal carcinoma cells (1, 21, 32, 51, 97), and the block to viral expression, once established at the preimplantation stage, is generally maintained at later stages of development (40). Methylation of the proviral genome has been implicated in the maintenance of the inactive state (38). Because of the host block in expression (and thus replication), a nondefective virus could be used as a lineage marker when introduced into preimplantation embryos (93). In contrast, the viral promoter has been shown to be very active in postimplantation embryos (33).

Because of the inactivity of the viral promoter in early embryos, much work has been devoted to the construction of retroviral vectors containing exogenous genes under the control of an internal promoter (31, 76, 91, 98). Transgenic mice derived from embryos infected with such vectors have not always shown expression of the transgene (31, 76), but in at least two instances, tissue-specific or generalized expression has been observed (91, 98). It is likely that both patterns of expression will be used in future studies of cell lineage. To avoid possible interference of the viral promoter with the internal promoter, crippled retroviruses have been produced by transfection of constructs with an enhancerless or promoterless 3′ LTR (17, 27, 91, 111). Upon replication of the retrovirus, this deletion is transferred to the 5′ LTR. Such constructs have been shown to be functional both in transgenic mice and in animals lethally irradiated and injected with bone marrow cells (17, 91).

B. Embryo Infection

Embryos can be infected with retroviruses at many different stages of development (Table 1). Preimplantation (4 to 8 cells) mouse embryos can be exposed to infectious virus for a short time (32) or cultured up to the morula-blastocyst stage (about 32 to 64 cells) on a monolayer of virus-producing cells (35, 37). They can then be reimplanted into the uteri of foster mothers to allow further development. Alternatively, virus can be microinjected into the cavity of the blastocyst in culture (35). Embryos can also be infected following implantation into the uterus from days 7 to 11, when the embryo has reached hundreds of thousands of cells, by direct injection of virus or virus-producing cells through the uterine wall (33, 77, 83, 92, 99). Endogenous proviral copies have also been shown to be acquired in the germ line at a high frequency in AKR (67, 75, 96) and SWR/J-RF/J hybrid (41) mice. The mechanism by which such viruses are acquired has not been elucidated.

About 70% of the animals resulting from preimplantation embryo infection with M-MuLV (titer of 5×10^6 PFU/ml) contain one or several copies of the proviral genome (35, 93). The percentage of infected embryos has been much lower when other viruses have been used, possibly as a result of lower titers (31, 91, 98, 102). In contrast, virus microinjected into postimplantation embryos replicates in cells of all somatic tissues and efficiently spreads throughout the embryo (40). Interestingly, infection of primordial germ cells by this procedure is extremely rare, and only four transgenic mouse strains have been derived in this way from over 1,500 gametes tested (35, 92). Infection of postimplantation embryos with defective retroviruses results in only a fraction of the cells being infected (99), and this approach is therefore suitable for the analysis of cell lineages (77).

III. RETROVIRUSES AS CELL LINEAGE MARKERS

A. Early Development

Preimplantation mouse embryos are easily accessible to experimental manipulation, and cell lineages have been studied by directly labeling cells with dyes or by examining the developmental potential of individual cells in aggregation or injection chimeras (65). Postimplantation mammalian embryos are not easily accessible, and the study of cell lineages has been much more difficult. The direct approach of labeling individual cells with dyes is limited by the rapid dilution of the dye during subsequent cell divisions. The construction of chimeras between embryos carrying different genetic markers, therefore, has been the predominant method for the study of cell lineage during postimplantation development. The distribution of cells of the two different genotypes has been used to determine the number of cells which are set aside to form the embryo or individual somatic tissues, as well as to determine the time of allocation (108). The use of such chimeras for this purpose has, however, a number of inherent problems (58). In most experiments, noncongenic strains were used, and cells of one genotype were selected over others (104), resulting in distortion of cell interactions during development of the chimeric animal. The study of cell lineages late in development is further limited by the fact that genetic heterogeneity is only introduced during chimera formation, at a relatively early stage.

The ideal marker for cell lineage studies would leave the embryo undisturbed, would be innocuous, and would be detectable over many cell divisions. Genetic mosaicism within single developing embryos, introduced by retroviral markers, in contrast to genetic heterogeneity of chimeras derived from several embryos, is well suited for this purpose because retroviruses efficiently infect mouse embryos at different stages of development. In the first study of this type (93), preimplantation mouse embryos were infected with a retrovirus and then reintroduced into the uterus to allow further development. The virus used was replication competent and contained a bacterial suppressor tRNA gene in each LTR (Fig. 1A). The bacterial tRNA gene provided a unique proviral probe for Southern blot analysis and allowed selective cloning of the provirus and flanking sequences (68). As discussed above, the block in viral expression at the preimplantation stage allowed the use of a replication-competent virus as a lineage marker. Quantitative analysis of the proviruses in mosaic animals showed equal molarities of individual proviral bands in all somatic tissues of a given animal. The equal contribution of virus-labeled cells to all tissues indicates that the cells which give rise to the embryo proper must intermingle extensively prior to definitive tissue allocation. Grafts of cells into ectopic sites (2), as well as in vitro explantation experiments (89), have suggested that at the late egg cylinder stage (about 10,000 cells), different segments have acquired different prospective fates. It is presumably at the primitive streak stage, which marks the onset of gastrulation, that the definitive body plan of the fetus is established. Cell mixing, therefore, must already have occurred prior to this stage.

An estimate of the number of cells set aside to

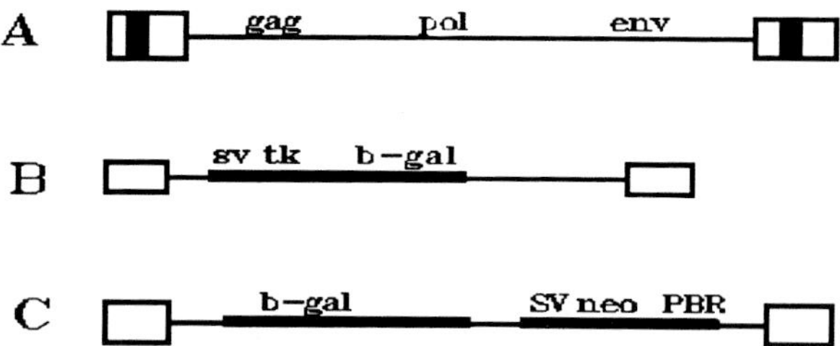

Figure 1. Maps of proviral genomes used for cell lineage studies. (A) Replication-competent retrovirus used for lineage studies following infection of preimplantation embryos (93). *gag*, *pol*, and *env* are the normal viral functions (see Varmus and Brown, this volume). Boxes at each end of the provirus represent LTRs. The black boxes represent a bacterial suppressor gene useful for selective cloning. (B) Replication-deficient retrovirus used for lineage studies in postimplantation embryos (77). A β-galactosidase gene (*b-gal*) is placed under the control of a thymidine kinase gene (*tk*) and the SV40 enhancers (sv). (C) Replication-competent virus used for lineage studies in the retina (66). The β-galactosidase gene (*b-gal*) is under the control of the viral promoter in the LTR, and a *neo* gene is under the control of the SV40 promoter (SV). PBR, Origin of replication of pBR322.

form the somatic tissues of the fetus was provided by the lowest intensities of proviral bands in the mosaic animals. This approach assumes that such founder cells contribute equally to all resulting tissues by equal rates of division and death (72), an idea supported by the equal molarities of the proviruses in all somatic tissues of a given animal. The lowest molarity detected for the 30 proviruses studied was 0.12. This suggests that approximately eight cells (1/0.12 = 8) are allocated to form the embryo. If the proviruses had integrated subsequently to the time of allocation of such cells, the actual number of such founder cells could be lower. A previous estimate, from the genotype distribution in chimeric animals, indicated that a minimum of three cells was set aside to form the embryo proper (62).

The last point of this study concerned the allocation of cells to the germ line. It has been established that in other species the germ line is determined very early. Results obtained with injection chimeras have shown that both somatic and germ cells originate from the primitive ectoderm (20). In addition, the analysis of X-inactivation mosaics suggested that the germ line may be allocated as late as at gastrulation (59). Breeding data of the mosaic animals indicated that for almost 50% of the proviruses, no correlation between contribution to somatic tissues and to the germ line was found. These results suggest that the cells which are destined to form the germ line are allocated prior to or at the time of embryo foundation, when cells are definitively allocated to the somatic lineages.

An important difference between studies in which retroviral mosaics are used and those in which chimeras are used is that an equal extent of mosaicism is observed in virtually all tissues of retroviral mosaics, whereas a wide difference of genotype distribution among different tissues is characteristic for individual aggregation chimeras (73, 106). The genotype variation seen between tissues of a chimeric animal therefore may be a consequence of experimental disturbance inherent in the process of chimera formation and may not reflect cell-cell interaction during development of the normal embryo. Chimeric animals have been used to determine the progenitor cell number for a variety of adult tissues. Estimates of the number of cells allocated to form the Purkinje cells of the cerebellum, the retina, or the facial ganglia range from 8 to 12 (63, 107) and are conspicuously similar to the number which we have estimated to form the embryo proper. Since chimeras are formed during preimplantation development, all such estimates must reduce to the number of original fetal progenitor cells present at the introduction of genetic heterogeneity (108) and may reflect early allocation events at embryo foundation rather than later developmental processes that occur at tissue foundation or during morphogenesis.

Because retroviral labeling in these experiments was performed at the preimplantation stage, the data do not provide information on cell proliferation which occurs subsequent to gastrulation. Nonetheless, the retroviral labeling approach has been criticized as fitting a highly "deterministic" model instead of an apparently more flexible, competing, and mutually exclusive "stochastic" model (78–81). The stochastic model is based on the shape of clone size distribution in the gut, aorta, and retina of chimeric animals (78–81); in these three tissues, however, cell division takes place continuously after gastrulation or after birth, and therefore any pattern of chimeric patches will not reflect early developmental processes. In addition, as the authors of this model admit, the contribution of cell migration and cell mingling to the final patch pattern cannot be reliably evaluated by the statistical analyses used (78). Finally, the use of chimeric animals to deduce cell allocations in development has inherent problems as discussed above. Patch size distribution can be affected by unequal proliferation of the two chimeric components, as is seen, for example, in the heterogeneous distribution of patches along the length of the gut (78). Because the shape and distribution of patch sizes are probably affected by the disturbance of normal developmental processes due to the experimental aggregation of embryos used to generate the chimeras, data obtained by this approach probably do not provide useful information on cellular allocation and proliferation events preceding gastrulation and tissue foundation during normal development.

More recently, methods which allow the identification of singly infected cells have been devised for the study of cell lineages (66, 77, 101). These methods involve the infection of postimplantation embryos (77) or neonatal animals (66, 101) by defective retroviruses encoding a protein which can be easily distinguished histochemically by using specific antibodies. The marker used to date to follow lineage relationships in mammals has been the gene encoding the bacterial β-galactosidase (66, 77). Cells expressing this gene stain blue upon incubation with the chromogenic substrate X-Gal (5-bromo-4-chloro-3-indolyl-β-D-galactopyranoside). The main advantage of this approach is that single cells can be identified. On the other hand, expression of the foreign gene could potentially affect development, and possibly not all cells express the protein sufficiently to allow detection. Experiments with tissue culture cells suggest that limited expression may indeed be a general problem with this marker (54). The method therefore only establishes lineage relationships between those cells expressing β-galactosidase.

The approach of tagging with a β-galactosidase retrovirus has been used at two different stages of mammalian development, at the postimplantation stage (77) and after birth (101). In the study of postimplantation cell lineages, a retrovirus encoding β-galactosidase under the control of an internal SV40 promoter (Fig. 1B) was used to infect embryos by injection through the uterine wall between days 7 and 11 of development (77). Blue-staining cells were always detected in clusters, suggesting that these cells constitute clones. Lineage information was obtained for both visceral yolk sac and the skin. A similar approach was used to investigate cell lineage relationships in the rat retina (66, 101). The retrovirus used contained the β-galactosidase gene expressed from the Moloney LTR and also a neomycin phosphotransferase gene expressed from an SV40 promoter (Fig. 1C). Cells staining blue were detected in small clusters and always showed radial orientation in the retina, with little lateral migration, suggesting clonality. Even toward the end of development, a single retinal progenitor was shown to give rise to a wide variety of cell types, including rods, bipolar cells, amacrine cells, and Muller glial cells. Cell fate may accordingly be determined in large part by environmental factors. Recently, cell lineages in rat cerebral cortex have also been examined, using similar methods (103a).

Tagging with β-galactosidase retroviruses has also been performed in chicken embryos (Cepko, personal communication; Sanes, personal communication). Studies of cell lineages at postgastrulation stages should be facilitated in chicken embryos because they are much more accessible to experimental manipulation than mouse embryos are. For lineage analysis in chickens, amphotropic retroviruses are currently used, although avian retroviruses may ultimately prove more suitable. Preliminary studies have shown clones of cells in the optic tectum and the brain cortex showing radial orientation, again suggesting clonality (Sanes, personal communication; see color plates, this volume).

B. Development of the Hematopoietic System

Bone marrow cells injected into γ-ray-irradiated recipient animals can reconstitute the hematopoietic system, and cell lineages can be followed in the reconstitution process, provided such cells can be tracked in the recipient animal. Once again, genetic labeling is the method of choice in such studies. Because bone marrow cells are difficult to transfect, retroviruses seemed well suited for such studies. Bone marrow cells infected with retroviruses were shown to participate in the reconstitution of the hematopoietic system of the animal (43, 109). Although transduced hematopoietic cells may behave differently upon transplantation than they would otherwise, this approach has yielded important results on the ontogeny of the hematopoietic system (see below). In addition, much effort has been directed to obtain expression of a variety of transduced genes (e.g., adenosine deaminase and hypoxanthine phosphoribosyltransferase genes) in the recipient (3, 9, 27, 29, 45, 55, 57, 110) with, in most cases, little or no success. Long-term and erythroid-specific expression of a human β-globin gene has been detected, however, in one set of experiments (17).

In two reports, injection of bone marrow cells infected with retroviruses into W/W^v mice (15) or CBA mice (45) led to the identification of multipotent precursors capable of generating both myeloid and lymphoid progeny. In these studies, the site of retrovirus integration was used to identify progeny of a single cell. In another report (50) many different classes of stem cells were identified: cells with both myeloid and lymphoid potential which contributed to repopulation of all hematopoietic organs and all anatomical locations, cells with both myeloid and lymphoid potential but which did not contribute to all members of one lineage, cells that gave rise only to the myeloid lineage, and cells that gave rise only to a subset (T or B) of the lymphoid lineage. Secondary transplant experiments also showed that cells which did not contribute extensively to hematopoiesis in a primary recipient contributed much more extensively

Table 2. Retroviruses as insertion mutagens

Mutant	Procedure used	Mutant phenotype	Affected gene
Mov 13	Postimplantation embryo infection	Embryonic death (day 13)	α1(I) collagen
Mov 34	Preimplantation embryo infection	Embryonic death (day 6)	? (Transcripts identified)
HPRT	Infection of ES cells	None observed	HPRT
Dilute	Endogenous retrovirus	Light coat color	? (Transcripts identified)

in a secondary recipient. In addition, periodic sampling of blood from primary recipients revealed a dynamic behavior of stem cell clones, possibly the result of sequential use of different stem cell clones. Such clonal fluctuation has been shown to vary with time following reconstitution (88).

IV. RETROVIRUSES AS CHROMOSOMAL MARKERS

Three transgenic mouse strains have been derived which carry the provirus on the sex chromosomes (25, 92, 98). The most interesting results emerged from analyses of the Mov 15 strain (25), in which the provirus had inserted in the pseudoautosomal region. This region is located at the tip of both the X and the Y chromosomes and can undergo recombination during meiosis. Loci in this region are termed pseudoautosomal because they do not show strict sex linkage. The Mov 15 provirus was shown to recombine with *Tdy*, the primary sex-determining locus which is in the strictly sex-linked portion of the Y chromosome, in 10 to 20% of male meioses. The most striking result was obtained when homozygous Mov 15 males were crossed with wild-type females. Instead of the expected 100% heterozygous progeny, 7% of the offspring had lost the provirus and 7% had acquired a second proviral copy. This behavior of the unique proviral marker indicates unusually frequent unequal recombination in the pseudoautosomal region. Mouse sequences flanking the provirus were cloned and shown to be tandemly repeated and highly polymorphic between different mouse strains. It is possible that the high rate of unequal crossing over is due to the presence of these repeated sequences, as suggested previously (87).

Crosses were performed involving Mov 15 animals and animals mutant in the steroid sulfatase gene (94), which is in the distal part of the pseudoautosomal region and shows no genetic linkage with *Tdy* (44), to further study recombination in the pseudoautosomal region. Recombination between the Mov 15 and steroid sulfatase markers was sevenfold higher in male meiosis than in female meiosis. Similar disparity between male and female meiosis has been seen with human pseudoautosomal markers (74). This is consistent with a historic prediction of Koller and Darlington (46) that pairing and proper disjunction of chromosome pairs should be accompanied by chiasma formation and recombination. In this model, meiotic recombination between the X and the Y chromosomes (male meiosis) can only occur in the pseudoautosomal region, whereas recombination between two X chromosomes (female meiosis) is not constrained. In addition, three double crossovers between the two pseudoautosomal markers and sexual phenotype were detected among 127 offspring analyzed. This is in contrast to observations made in the human genome, where no double crossovers have been detected, and to a previous model suggesting a single obligatory crossover in the pseudoautosomal region to explain the absence of any sex linkage of distal markers (7). One possible explanation for these results is that the pseudoautosomal region is larger in mice than in humans. The availability of additional mouse pseudoautosomal markers will be useful to test this hypothesis.

V. RETROVIRUSES AS INSERTION MUTAGENS

Numerous mutations, both spontaneous and induced by radiation or chemicals, have allowed the mapping of many genes on mouse chromosomes. It has nevertheless been difficult to reach an understanding of the molecular defect which causes a given phenotype. Insertion mutagenesis by the introduction of exogenous DNA into the germ line provides a promising alternative approach for a molecular analysis of mutant phenotypes (reviewed in references 22 and 23). This approach is attractive because the introduced DNA serves both as a mutagen by disrupting or altering the expression of a gene and as a tag for molecular cloning of the affected gene. We review here four cases of mutations generated by retroviruses (Table 2).

Insertion mutants have been produced in transgenic mice both by retroviral infection and by microinjection of DNA into zygotes (23). In general, mutations caused by microinjected DNA have been difficult to analyze because of deletions, duplications, or translocations of both insert and host sequences (reviewed in reference 23). In contrast, DNA blot analyses of several strains derived from retroviral infection have revealed no alterations of host flanking sequences (see reference 92). This permitted a molecular analysis of the induced mutations. Interestingly, retrovirus-induced insertion mutations are produced in mice at a high frequency, possibly as a result of preferential integration of retroviruses near hypersensitive sites often associated with transcribed sequences (11, 71, 103; for discussion, see reference 92).

The first mutation produced by experimental infection of mouse embryos was observed in the Mov 13 strain (34). In this strain, proviral insertion leads to a recessive lethal phenotype. Homozygous embryos die at day 13 of development due to rupture of blood vessels, leading to a general necrosis (52). The provirus had inserted into the first intron of the α1(I) collagen gene (24, 82), interfering with transcription of the gene. Changes in the chromatin conformation (6) and in the methylation pattern (39) of the collagen gene were observed. Nuclear run-on experiments indicated that initiation of transcription of the gene was reduced to 1 to 5% of that of the wild-type allele (26). This mutation has been useful for the study of the role of α1(I) collagen during development (47). Organ rudiments from wild-type, heterozygous, and homozygous day 12 embryos were cultured in vitro and found to develop normally, independently of the genotype of the donor embryo. This result implies either that α1(I) collagen has no role in organ morphogenesis or that it can be replaced by other fibrillar collagens.

In a second embryonic lethal strain, Mov 34, embryos homozygous for the proviral insertion die shortly after implantation in the uterus, before reaching the egg cylinder stage (92). The provirus is inserted toward the 5′ end of an abundantly transcribed gene expressed at similar levels in many different tissues and cell lines. Densitometric analysis suggests that integration of the provirus at the Mov 34 locus decreases the steady state of this transcript. Preliminary analysis of cDNA clones has revealed no sequence homology with any known protein (unpublished data). Thus, in both Mov 13 and Mov 34, proviral insertion has occurred near the 5′ end of a gene and interferes with its expression. Further study will be required to determine whether the Mov 34 mutation, like Mov 13, is associated with changes in methylation and chromatin structure and, thus, whether this constitutes a general mechanism by which retroviruses cause mutant phenotypes.

In at least one instance, a mutation has been generated by the spontaneous insertion of a retrovirus. The original dilute coat color mutation in DBA/2J (*d*) mice is associated with the endogenous Emv 3 provirus (42). Revertants have lost most of the provirus but retain a single LTR (12). Single-copy cellular DNA probes from regions both 5′ and 3′ of the provirus have been isolated (69) and are being used to map a number of radiation-induced mutant alleles in the nearby dilute-short ear complex. Three transcripts from the dilute locus have recently been detected in wild-type mice by the use of these probes; some of these transcripts are missing or altered in dilute mice (N. Jenkins and N. Copeland, personal communication). It has also been thought that another endogenous provirus, Emv 15, may have caused the lethal yellow mutation, since it was associated with this mutation in three inbred strains (13). This has now become unlikely, since other strains lack the provirus but still retain the yellow coat color allele (53, 86). Recently, an endogenous retrovirus has been shown to be the cause of the hairless (*hr*) mutation in mice (98a).

An exciting possibility for the generation of developmental mutants involves the use of embryonic stem (ES) cells (18). ES cells are derived from the inner cell mass of the blastocyst and retain their capacity to colonize somatic and germ line tissues upon chimera formation. Progeny mice therefore can be derived from the cloned ES cells by breeding of the chimeric animals (5). In one study, retroviruses were used to infect ES cells repeatedly (70). The resulting transgenic animals contained many more (about 10) proviral copies than expected from direct embryo infection. The breeding of these mice for homozygosity will probably reveal additional insertion mutations. In another study (48), ES cell clones were infected with retroviruses and then selected for the loss of the purine salvage enzyme hypoxanthine-guanosine phosphoribosyltransferase (HPRT). The HPRT gene is located on the X chromosome in mice as well as in humans. (HPRT deficiency is the primary cause of Lesch-Nyhan syndrome, a human neurological disorder.) Mutant ES cells were then used to form germ line chimeric mice containing the mutant HPRT gene. Interestingly, analysis of the progeny of these chimeras showed that males hemizygous for the mutant HPRT gene are fully viable and do not exhibit any major neurological disorders. Primary fibroblasts derived from hemizygous males exhibited the expected complete deficiency in HPRT activity. This work and an accompanying report (30) represent the first instance of selected insertion mutagenesis in mice and selection for loss of particular gene function. It is probable that much effort will be directed in the next few years in which ES cells will be used for obtaining specific mouse mutants.

VI. PROSPECTS AND CONCLUSIONS

Retroviruses are uniquely suited for both developmental and genetic studies. Because retroviruses can infect mouse embryos at different stages of development, they can be used as direct lineage markers in undisturbed embryos. The formation of genetic mosaics may therefore circumvent many of the limitations of the traditional approaches that relied on chimera formation between different embryos. The use of such retroviral mosaics should be valuable for understanding allocation events during postimplantation development, a period which has previously been difficult to study.

For genetic studies, the use of retroviruses to generate insertion mutations in transgenic mice has the advantage over DNA microinjection that the provirus integrates into the host genome without causing sequence rearrangements. Insertion mutagenesis by retroviruses is attractive, therefore, because once a strain carrying a mutation has been identified, isolation of the mutated gene is unlikely to pose a serious obstacle for the subsequent molecular and biological analysis of the mutant. The identification of additional insertion mutants is likely to give new insights into mammalian embryogenesis. The availability of ES cells which have retained their capacity to colonize the germ line should also allow mutagenesis of specific genes in mice. Retroviral tagging therefore opens new and exciting experimental approaches to problems of mammalian development.

Acknowledgments. We thank Connie Cepko, Neal Copeland, Nancy Jenkins, and Josh Sanes for communicating unpublished data and for discussions.

T.G. was supported by a postdoctoral fellowship (PF-2645) from the American Cancer Society. Work from our laboratory was supported by Public Health Service grants HD-19105 from the National Institutes of Health and PO1-CA38497 from the National Cancer Institute.

LITERATURE CITED

1. **Barklis, E., R. C. Mulligan, and R. Jaenisch.** 1986. Chromosomal position or virus mutation permits retrovirus expression in embryonal carcinoma cells. *Cell* **47:**391–399.
2. **Beddington, R.** 1982. An autoradiographic analysis of tissue potency in different regions of the embryonic ectoderm during gastrulation in the mouse. *J. Embryol. Exp. Morphol.* **69:**265–284.
3. **Belmont, J. W., J. Henkel-Tigges, S. M. W. Chang, K. Wager-Smith, R. E. Kellems, J. E. Dick, M. C. Magli, R. A. Phillips, A. Bernstein, and C. T. Caskey.** 1986. Expression of human adenosine deaminase in murine haematopoietic progenitor cells following retroviral transfer. *Nature* (London) **322:**385–387.
4. **Bender, M. A., T. D. Palmer, R. E. Gelinas, and A. D. Miller.** 1987. Evidence that the packaging signal of Moloney murine leukemia virus extends into the *gag* region. *J. Virol.* **61:** 1639–1646.
5. **Bradley, A., M. Evans, M. H. Kaufman, and E. Robertson.** 1984. Formation of germline chimeras from embryo-derived teratocarcinoma cell lines. *Nature* (London) **309:**255–256.
6. **Breindl, M., K. Harbers, and R. Jaenisch.** 1984. Retrovirus-induced lethal mutation in collagen I gene of mice is associated with altered chromatin structure. *Cell* **38:**9–16.
7. **Burgoyne, P. S.** 1982. Genetic homology and crossing over in the X and Y chromosomes of mammals. *Hum. Genet.* **60:**85–90.
8. **Cepko, C. L., B. E. Roberts, and R. C. Mulligan.** 1984. Construction and applications of a highly transmissible murine retrovirus shuttle vector. *Cell* **37:**1053–1062.
9. **Chang, S. M. W., K. Wager-Smith, T. Y. Tsao, J. Henkel-Tigges, S. Vaishnav, and C. T. Caskey.** 1987. Construction of a defective retrovirus containing the human hypoxanthine phosphoribosyltransferase cDNA and its expression in cultured cells and mouse bone marrow. *Mol. Cell. Biol.* **7:** 854–863.
10. **Cone, R. D., and R. C. Mulligan.** 1984. High efficiency gene transfer into mammalian cells: generation of helper-free recombinant retrovirus with broad mammalian host range. *Proc. Natl. Acad. Sci. USA* **81:**6349–6353.
11. **Conklin, K. F., and M. Groudine.** 1986. Varied interactions between proviruses and adjacent host chromatin. *Mol. Cell. Biol.* **6:**3999–4007.
12. **Copeland, N. G., K. W. Hutchison, and N. A. Jenkins.** 1983. Excision of the DBA ecotropic provirus in dilute coat-color revertants of mice occurs by homologous recombination involving the viral LTRs. *Cell* **33:**379–387.
13. **Copeland, N. G., N. A. Jenkins, and B. K. Lee.** 1983. Association of the lethal yellow (A^y) coat color mutation with an ecotropic murine leukemia virus genome. *Proc. Natl. Acad. Sci. USA* **80:**247–249.
14. **Cuypers, H. T., G. Selten, W. Quint, M. Zijlstra, E. R. Maandag, W. Boelens, P. van Wezenbeek, C. Melief, and A. Berns.** 1984. Murine leukemia virus-induced T-cell lymphomagenesis: integration of proviruses in a distinct chromosomal region. *Cell* **37:**141–150.
15. **Dick, J. E., M. C. Magli, D. Huszar, R. A. Phillips, and A. Bernstein.** 1985. Introduction of a selectable gene into primitive stem cells capable of long-term reconstitution of the hemopoietic system of W/W^v mice. *Cell* **42:**71–79.
16. **Dickson, C., R. Smith, S. Brooke, and G. Peters.** 1984. Tumorigenesis by mouse mammary tumor virus: proviral activation of a cellular gene in the common integration site int-2. *Cell* **37:**529–536.
17. **Dzierzak, E. A., T. Papayannopoulou, and R. C. Mulligan.** 1988. Lineage-specific expression of a human β-globin gene in murine bone marrow transplant recipients reconstituted with retrovirus-transduced stem cells. *Nature* (London) **331:** 35–41.
18. **Evans, M. J., and M. H. Kaufman.** 1981. Establishment in culture of pluripotential cells from mouse embryos. *Nature* (London) **292:**154–156.
19. **Fung, Y. K. T., W. G. Lewis, L. B. Crittenden, and H. J.**

Kung. 1983. Activation of the cellular oncogene c-erbB by LTR insertion: molecular basis for induction of erythroblastosis by avian leukosis virus. *Cell* **33**:357–368.

20. **Gardner, R. L., M. F. Lyon, E. P. Evans, and M. D. Burtenshaw.** 1985. Clonal analysis of X-chromosome inactivation and the origin of the germ line in the mouse embryo. *J. Exp. Embryol. Morphol.* **88**:349–363.
21. **Gautsch, J., and M. Wilson.** 1983. Restriction of Moloney murine leukemia virus growth in teratocarcinoma: involvement of factors other than DNA methylation. *Cold Spring Harbor Conf. Cell Proliferation* **10**:363–378.
22. **Goff, S. P.** 1987. Insertional mutagenesis to isolate genes. *Methods Enzymol.* **151**:489–502.
23. **Gridley, T., P. Soriano, and R. Jaenisch.** 1987. Insertional mutagenesis in mice. *Trends Genet.* **3**:162–166.
24. **Harbers, K., M. Kuehn, H. Delius, and R. Jaenisch.** 1984. Insertion of retrovirus into the first intron of α1(I) collagen gene leads to embryonic lethal mutation in mice. *Proc. Natl. Acad. Sci. USA* **81**:1504–1508.
25. **Harbers, K., P. Soriano, U. Muller, and R. Jaenisch.** 1986. High frequency of unequal recombination in pseudoautosomal region shown by proviral insertion in transgenic mouse. *Nature* (London) **324**:682–685.
26. **Hartung, S., R. Jaenisch, and M. Breindl.** 1986. Retrovirus insertion inactivates mouse α1(I) collagen gene by blocking initiation of transcription. *Nature* (London) **320**:365–367.
27. **Hawley, R. G., L. Covvarubias, T. Hawley, and B. Mintz.** 1987. Handicapped retroviral vectors efficiently transduce foreign genes into hematopoietic cells. *Proc. Natl. Acad. Sci. USA* **84**:2406–2410.
28. **Hayward, W. S., B. G. Neel, and S. M. Astrin.** 1981. Activation of a cellular onc gene by promoter insertion in ALV-induced lymphoid leukosis. *Nature* (London) **290**: 475–480.
29. **Hock, R. A., and A. D. Miller.** 1986. Retrovirus-mediated transfer and expression of drug resistant genes in human haematopoietic progenitor cells. *Nature* (London) **320**: 275–277.
30. **Hooper, M., K. Hardy, A. Handyside, S. Hunter, and M. Monk.** 1987. HPRT-deficient (Lesch-Nyhan) mouse embryos delivered from germline colonization by cultured cells. *Nature* (London) **326**:292–295.
31. **Huszar, D., R. Balling, R. Kothary, M. C. Magli, N. Hozumi, J. Rossant, and A. Bernstein.** 1985. Insertion of a bacterial gene into the mouse germ line using an infectious retrovirus vector. *Proc. Natl. Acad. Sci. USA* **82**:8587–8591.
32. **Jaenisch, R.** 1976. Germ line integration and Mendelian transmission of the exogenous Moloney leukemia virus. *Proc. Natl. Acad. Sci. USA* **73**:1260–1264.
33. **Jaenisch, R.** 1980. Retroviruses and embryogenesis: microinjection of Moloney leukemia virus into midgestation mouse embryos. *Cell* **19**:181–188.
34. **Jaenisch, R., K. Harbers, A. Schnieke, J. Löhler, I. Chumakov, D. Jähner, D. Grotkopp, and E. Hoffman.** 1983. Germline integration of Moloney murine leukemia virus at the Mov13 locus leads to recessive lethal mutation and early embryonic death. *Cell* **32**:209–216.
35. **Jaenisch, R., D. Jähner, P. Nobis, I. Simon, J. Löhler, K. Harbers, and D. Grotkopp.** 1981. Chromosomal position and activation of retroviral genomes inserted into the germ line of mice. *Cell* **24**:519–529.
36. **Jähner, D., K. Haase, R. C. Mulligan, and R. Jaenisch.** 1985. Insertion of the bacterial gpt gene into the germ line of mice by retroviral infection. *Proc. Natl. Acad. Sci. USA* **82**: 6927–6931.
37. **Jähner, D., and R. Jaenisch.** 1980. Integration of Moloney leukemia virus into the germ line of mice: correlation between genotype and virus activation. *Nature* (London) **287**: 456–458.
38. **Jähner, D., and R. Jaenisch.** 1984. DNA methylation in early mammalian development, p. 189–219. *In* A. Razin, A. Cedar, and A. D. Riggs (ed.), *DNA Methylation: Biochemistry and Biological Significance.* Springer-Verlag, New York.
39. **Jähner, D., and R. Jaenisch.** 1985. Retrovirus induced de novo methylation of flanking host sequences correlates with gene inactivity. *Nature* (London) **315**:594–597.
40. **Jähner, D., H. Stuhlmann, C. L. Stewart, K. Harbers, J. Löhler, I. Simon, and R. Jaenisch.** 1982. De novo methylation and expression of retroviral genomes during mouse embryogenesis. *Nature* (London) **298**:623–628.
41. **Jenkins, N. A., and N. G. Copeland.** 1985. High frequency germ-line acquisition of ecotropic MuLV proviruses in SWR/J-RF/J hybrid mice. *Cell* **43**:811–819.
42. **Jenkins, N. A., N. G. Copeland, B. A. Taylor, and B. K. Lee.** 1981. Dilute (d) coat color mutation of DBA/2J mice is associated with the site of integration of an ecotropic MuLV genome. *Nature* (London) **293**:370–374.
43. **Joyner, A., G. Keller, R. A. Phillips, and A. Bernstein.** 1983. Retroviral mediated transfer of a bacterial gene into mouse hematopoietic progenitor cells. *Nature* (London) **305**: 206–208.
44. **Keitges, E., M. Rivest, M. Siniscalco, and S. M. Gartler.** 1985. X-linkage of steroid sulfatase in the mouse is evidence for a functional Y linked allele. *Nature* (London) **315**: 226–227.
45. **Keller, G., C. Paige, E. Gilboa, and E. F. Wagner.** 1985. Expression of a foreign gene in myeloid and lymphoid cells derived from multipotential haematopoietic precursors. *Nature* (London) **318**:149–154.
46. **Koller, P. C., and C. D. Darlington.** 1934. The genetical and mechanical properties of sex chromosomes 1 Rattus norvegicus. *J. Genet.* **29**:159–173.
47. **Kratochwil, K., M. Dziadek, J. Lohler, K. Harbers, and R. Jaenisch.** 1986. Normal epithelial branching morphogenesis in the absence of collagen I. *Dev. Biol.* **117**:596–606.
48. **Kuehn, M. R., A. Bradley, E. J. Robertson, and M. J. Evans.** 1987. A potential animal model for Lesch-Nyhan syndrome through introduction of HPRT mutations in mice. *Nature* (London) **326**:295–298.
49. **Lemay G., and P. Jolicoeur.** 1984. Rearrangement of a DNA sequence homologous to a cell-virus junction fragment in several Moloney murine leukemia virus-induced rat thymomas. *Proc. Natl. Acad. Sci. USA* **81**:38–42.
50. **Lemischka, I. R., D. H. Raulet, and R. C. Mulligan.** 1986. Developmental potential and dynamic behavior of hematopoietic stem cells. *Cell* **45**:917–927.
51. **Linney, E., B. Davis, J. Overhauser, E. Chao, and H. Fan.** 1984. Non function of a Moloney murine leukaemia virus regulatory sequence in F9 embryonal carcinoma cells. *Nature* (London) **308**:470–472.
52. **Löhler, J., R. Timpl, and R. Jaenisch.** 1984. Embryonic lethal mutation in mouse collagen I gene causes rupture of blood vessels and is associated with erythropoietic and mesenchymal cell death. *Cell* **38**:597–607.
53. **Lovett, M., Z. Cheng, E. M. Lamella, T. Yokoi, and C. J. Epstein.** 1987. Molecular markers for the agouti coat color locus of the mouse. *Genetics* **115**:747–754.
54. **MacGregor, G. R., A. E. Mogg, J. F. Burke, and C. T. Caskey.** 1987. Histochemical staining of clonal mammalian

cell lines expressing E. coli β-galactosidase indicates heterogeneous expression of the bacterial gene. *Somatic Cell Mol. Genet.* **13**:253–265.

55. **Magli, M. C., J. E. Dick, D. Huszar, A. Bernstein, and R. A. Phillips.** 1987. Modulation of gene expression in multiple hematopoietic cell lineages following retroviral vector gene transfer. *Proc. Natl. Acad. Sci. USA* **84**:789–793.

56. **Mann, R., R. C. Mulligan, and D. Baltimore.** 1983. Construction of a retrovirus packaging mutant and its use to produce helper-free defective retrovirus. *Cell* **33**:153–159.

56a. **Markowitz, D., S. Goff, and A. Bank.** 1988. A safe packaging line for gene transfer: separating viral genes on two different plasmids. *J. Virol.* **62**:1120–1124.

57. **McIvor, R. S., M. J. Johnson, A. D. Miller, S. Pitts, S. R. Williams, D. Valerio, D. W. Martin, and I. M. Verma.** 1987. Human purine nucleoside phosphorylase and adenosine deaminase: gene transfer into cultured cells and murine hematopoietic stem cells by using recombinant amphotropic retroviruses. *Mol. Cell. Biol.* **7**:838–846.

58. **McLaren, A.** 1976. *In Mammalian Chimeras.* Cambridge University Press, Cambridge.

59. **McMahon, A., M. Fosten, and M. Monk.** 1983. X chromosome inactivation mosaicism in the three germ layers and the germ line of the mouse embryo. *J. Embryol. Exp. Morphol.* **74**:207–220.

60. **Miller, A. D., and C. Buttimore.** 1986. Redesign of retrovirus packaging cell lines to avoid recombination leading to helper virus production. *Mol. Cell. Biol.* **6**:2895–2902.

61. **Miller, A. D., M.-F. Law, and I. M. Verma.** 1985. Generation of helper-free amphotropic retroviruses that transduce a dominant-acting, methotrexate-resistant dihydrofolate reductase gene. *Mol. Cell. Biol.* **5**:431–437.

62. **Mintz, B.** 1974. Gene control in mammalian development. *Annu. Rev. Genet.* **8**:411–470.

63. **Mintz, B., and S. Sanyal.** 1970. Clonal origin of the mouse visual retina mapped for genetically mosaic eyes. *Genetics* **64**(Suppl.):43–44.

64. **Nusse, R., and H. E. Varmus.** 1982. Many tumors induced by the mouse mammary tumor virus contain a provirus integrated in the same region of the host genome. *Cell* **31**:99–109.

65. **Pedersen, R. A.** 1987. Potency, lineage, and allocation in preimplantation mouse embryos, p. 3–33. *In* J. Rossant and R. A. Pedersen (ed.), *Experimental Approaches to Mammalian Embryonic Development.* Cambridge University Press, Cambridge.

66. **Price, J., D. Turner, and C. Cepko.** 1987. Lineage analysis in the vertebrate nervous system by retrovirus-mediated gene transfer. *Proc. Natl. Acad. Sci. USA* **84**:156–160.

67. **Quint, W., H. van der Putten, F. Janssen, and A. Berns.** 1982. Mobility of endogenous ecotropic murine leukemia viral genomes within mouse chromosomal DNA and integration of a mink cell focus-forming virus-type recombinant provirus in the germ line. *J. Virol.* **41**:901–908.

68. **Reik, W., H. Weiher, and R. Jaenisch.** 1985. Replication-competent Moloney murine leukemia virus carrying a bacterial suppressor tRNA gene: selective cloning of proviral and flanking host sequences. *Proc. Natl. Acad. Sci. USA* **82**:1141–1145.

69. **Rinchik, E. M., L. B. Russell, N. G. Copeland, and N. A. Jenkins.** 1986. Molecular genetic analysis of the dilute-short ear (d-se) region of the mouse. *Genetics* **112**:321–342.

70. **Robertson, E., A. Bradley, M. Kuehn, and M. Evans.** 1986. Germ-line transmission of genes introduced into cultured pluripotential cells by retroviral vector. *Nature* (London) **323**:445–448.

71. **Rohdewohld, H., H. Weiher, W. Reik, R. Jaenisch, and M. Breindl.** 1987. Retrovirus integration and chromatin structure: Moloney murine leukemia proviral integration sites map near DNase I-hypersensitive sites. *J. Virol.* **61**:336–343.

72. **Rossant, J.** 1986. Retroviral mosaics: a new approach to cell lineage analysis in the mouse embryo. *Trends Genet.* **2**: 302–303.

73. **Rossant, J., and V. M. Chapman.** 1983. Somatic and germ-line mosaicism in interspecific chimeras between Mus musculus and Mus caroli. *J. Embryol. Exp. Morphol.* **73**: 193–205.

74. **Rouyer, F., M. C. Simmler, C. Johnsson, G. Vergnaud, H. J. Cooke, and J. Weissenbach.** 1986. A gradient of sex linkage in the pseudoautosomal region of the human sex chromosomes. *Nature* (London) **319**:291–295.

75. **Rowe, W. P., and C. A. Kozak.** 1980. Germ-line reinsertions of AKR murine leukemia virus genomes in AKV-1 congenic mice. *Proc. Natl. Acad. Sci. USA* **77**:4871–4874.

76. **Rubenstein, J. L. R., J. F. Nicolas, and F. Jacob.** 1984. Construction of a retrovirus capable of transducing and expressing genes in multipotential embryonic cells. *Proc. Natl. Acad. Sci. USA* **81**:7137–7140.

77. **Sanes, J. R., J. L. R. Rubenstein, and J. F. Nicolas.** 1986. Use of a recombinant retrovirus to study postimplantation cell lineage in mouse embryos. *EMBO J.* **5**:3133–3142.

78. **Schmidt, G. H., D. J. Garbutt, M. M. Wilkinson, and B. A. J. Ponder.** 1985. Clonal analysis of intestinal crypt populations in mouse aggregation chimaeras. *J. Embryol. Exp. Morphol.* **85**:121–130.

79. **Schmidt, G. H., and B. A. J. Ponder.** 1987. From patterns to clones in chimaeric tissues. *Bioessays* **6**:104–108.

80. **Schmidt, G. H., M. M. Wilkinson, and B. A. J. Ponder.** 1986. Non-random spatial arrangement of clone sizes in chimaeric retinal pigment epithelium. *J. Embryol. Exp. Morphol.* **91**: 197–208.

81. **Schmidt, G. H., M. M. Wilkinson, and B. A. J. Ponder.** 1986. Clonal analysis of chimaeric patterns in aortic endothelium. *J. Embryol. Exp. Morphol.* **93**:267–280.

82. **Schnieke, A., K. Harbers, and R. Jaenisch.** 1983. Embryonic lethal mutation in mice induced by retrovirus insertion into the α1(I) collagen gene. *Nature* (London) **304**:315–320.

83. **Sharpe, A. H., R. Jaenisch, and R. M. Ruprecht.** 1987. Retroviruses and mouse embryos: a rapid model for neurovirulence and transplacental antiviral therapy. *Science* **236**: 1671–1674.

84. **Sheng-Ong, G. L. C., M. Potter, J. F. Mushinski, S. Lavu, and E. P. Reddy.** 1984. Activation of the c-myb locus by viral insertional mutagenesis in plasmacytoid lymphosarcomas. *Science* **226**:1077–1080.

85. **Shinnick, T., R. Lerner, and J. Sutcliffe.** 1981. Nucleotide sequence of Moloney murine leukemia virus. *Nature* (London) **293**:543–548.

86. **Siracusa, L. D., L. B. Russell, N. A. Jenkins, and N. G. Copeland.** 1987. Allelic variation within the Emv-15 locus defines genomic sequences closely linked to the agouti locus on mouse chromosome 2. *Genetics* **117**:85–92.

87. **Smith, G. P.** 1976. Evolution of repeated DNA sequences by unequal crossovers. *Science* **191**:528–535.

88. **Snodgrass, R., and G. Keller.** 1987. Clonal fluctuation within the haematopoietic system of mice reconstituted with retrovirus-infected stem cells. *EMBO J.* **6**:3955–3960.

89. **Snow, M. H. L.** 1981. Autonomous development of parts isolated from primitive-streak-stage mouse embryos. Is de-

velopment clonal? *J. Embryol. Exp. Morphol.* **65**(Suppl.): 269–287.

90. **Sorge, J., D. Wright, V. D. Erdman, and A. E. Cutting.** 1984. Amphotropic retrovirus vector system for human cell gene transfer. *Mol. Cell. Biol.* **4**:1730–1737.

91. **Soriano, P., R. D. Cone, R. C. Mulligan, and R. Jaenisch.** 1986. Tissue specific and ectopic expression of genes introduced into transgenic mice by retroviruses. *Science* **234**: 1409–1413.

92. **Soriano, P., T. Gridley, and R. Jaenisch.** 1987. Retroviruses and insertional mutagenesis: proviral integration at the Mov 34 locus leads to early embryonic death. *Genes Dev.* **1**: 366–375.

93. **Soriano, P., and R. Jaenisch.** 1986. Retroviruses as probes for mammalian development: allocation of cells to the somatic and germ cell lineages. *Cell* **46**:19–29.

94. **Soriano, P., E. A. Keitges, E. A. Schorderet, K. Harbers, S. M. Gartler, and R. Jaenisch.** 1987. High rate of recombination and double crossovers in the mouse pseudoautosomal region during male meiosis. *Proc. Natl. Acad. Sci. USA* **84**:7218–7220.

95. **Steffen, D.** 1984. Proviruses are adjacent to c-myc in some murine leukemia virus-induced lymphomas. *Proc. Natl. Acad. Sci. USA* **81**:2097–2101.

96. **Steffen, D. L., B. A. Taylor, and R. A. Weinberg.** 1982. Continuing germ line integration of AKV proviruses during the breeding of AKR mice and derivative recombinant inbred strains. *J. Virol.* **42**:165–175.

97. **Stewart, C., H. Stuhlmann, D. Jahner, and R. Jaenisch.** 1982. De novo methylation, expression, and infectivity of retroviral genomes introduced into embryonal carcinoma cells. *Proc. Natl. Acad. Sci. USA* **79**:4098–4102.

98. **Stewart, C. L., S. Schuetze, M. Vanek, and E. F. Wagner.** 1987. Expression of retroviral vectors in transgenic mice obtained by embryo infection. *EMBO J.* **6**:383–388.

98a. **Stoye, J. P., S. Fenner, G. E. Greenoak, C. Moran, and J. M. Coffin.** 1988. Role of endogenous retroviruses as mutagens: the hairless mutation of mice. *Cell* **54**:383–391.

99. **Stuhlmann, H., R. D. Cone, R. C. Mulligan, and R. Jaenisch.** 1984. Introduction of a selectable gene into different animal tissue by a retrovirus recombinant vector. *Proc. Natl. Acad. Sci. USA* **81**:7151–7155.

100. **Tsichlis, P. N., P. G. Strauss, and L. F. Hu.** 1983. A common region for proviral integration in Mo-MuLV-induced rat thymic lymphomas. *Nature* (London) **302**:445–449.

101. **Turner, D. L., and C. L. Cepko.** 1987. A common progenitor for neurons and glia persists in rat retina late in development. *Nature* (London) **328**:131–136.

102. **van der Putten, H., F. M. Botteri, A. D. Miller, M. G. Rosenfeld, H. Fan, R. M. Evans, and I. M. Verma.** 1985. *Proc. Natl. Acad. Sci. USA* **82**:6148–6152.

103. **Vijaya, S., D. L. Steffen, and H. L. Robinson.** 1986. Acceptor sites for retroviral integrations map near DNase I-hypersensitive sites in chromatin. *J. Virol.* **60**:683–692.

103a. **Walsh, C., and C. L. Cepko.** 1988. Clonally related cortical cells show several migration patterns. *Science* **241**:1342–1345.

104. **Warner, C. M., J. L. Mc Ivor, and T. J. Stephens.** 1977. Chimeric drift in allophenic mice. *Transplantation* **24**: 183–193.

105. **Weiss, R., N. Teich, H. Varmus, and J. Coffin** (ed.). 1984. *RNA Tumor Viruses.* Cold Spring Harbor Laboratory, Cold Spring Harbor, N.Y.

106. **West, J. D., T. B. M. Linke, and M. Dunwald.** 1984. Investigation of variability among mouse aggregation chimaeras and X-chromosome inactivation mosaics. *J. Embryol. Exp. Morphol.* **84**:309–329.

107. **Wetts, R., and K. Herrup.** 1982. Cerebellar Purkinje cells are descended from a small number of progenitors committed during early development: quantitative analysis of Lurcher chimeric mice. *J. Neurosci.* **2**:1494–1498.

108. **Wilkins, A. S.** 1986. *Genetic Analysis of Development.* John Wiley & Sons, Inc., New York.

109. **Williams, D. A., I. R. Lemischka, D. G. Nathan, and R. C. Mulligan.** 1984. Introduction of new genetic material into pluripotent haematopoietic stem cells of the mouse. *Nature* (London) **310**:476–480.

110. **Williams, D. A., S. H. Orkin, and R. C. Mulligan.** 1986. Retrovirus mediated transfer of human adenosine deaminase gene sequences into cells in culture and into murine hematopoietic cells in vivo. *Proc. Natl. Acad. Sci. USA* **83**: 2566–2570.

111. **Yu, S. F., T. von Ruden, P. W. Kantoff, C. Garber, M. Seiberg, U. Rüther, W. F. Anderson, E. F. Wagner, and E. Gilboa.** 1986. Self-inactivating retroviral vectors designed for transfer of whole genes into mammalian cells. *Proc. Natl. Acad. Sci. USA* **83**:3194–3198.

Chapter 43

Population Dynamics of Transposable Elements

JAMES W. AJIOKA and DANIEL L. HARTL

I. INTRODUCTION

Our original assignment was to write about the evolution of transposable elements. We soon realized that this was overly ambitious in view of the tremendous diversity of different types of transposable elements and the different biological characteristics of their host organisms. The diversity among transposable elements is so great that writing a chapter-sized piece on the evolution of transposable elements would be a little like writing one on the evolution of genes: it must consist either of general comments so abstract as to be of little predictive value in particular cases or of a series of brief vignettes amounting to little more than annotated bibliography.

In limiting our purview, we have chosen to focus on the population dynamics of transposable elements, and in particular, elements found in *Drosophila melanogaster* and *Escherichia coli*. The rationale for choosing these species is like that of the proverbial man in the dark looking for his lost wallet under a lamp because that was where the light was shining. We have focused on *D. melanogaster* and *E. coli* because that is where most of the light of data shines, at least most of the data relevant to inferences about population dynamics. By population dynamics we mean the biological and statistical processes that determine the abundance of transposable elements within species. The review contains only a brief discussion of general evolutionary considerations of

James W. Ajioka and Daniel L. Hartl ■ Department of Genetics, Washington University School of Medicine, St. Louis, Missouri 63110-1095.

the sort that emerge from comparisons of DNA sequences. Although substantial DNA sequence information about transposable elements exists, the evolutionary relevance of much of it is speculative or anecdotal, and no general principles of long-term transposable element evolution have emerged that differ from already well-established principles of molecular evolution. However, the population dynamics of transposable elements is unique and of great interest in itself, and much has been learned by analyzing distributions and abundances of these elements in the context of population genetic models. This conjunction is where data and theory overlap and illuminate each other. Comments on wider issues in the evolution of transposable elements are included in the final section on evolutionary implications.

II. TRANSPOSABLE ELEMENTS IN *DROSOPHILA MELANOGASTER*

As far back as 1923, H. J. Muller complained that studies of mutation were merely a dingy basement beneath the imposing building called heredity (85). Quite so, and for good reason. Lacking any understanding of the chemical nature of the gene and any means of experimentally inducing mutations, geneticists focused most of their efforts on gene transmission and expression, leaving mutation to take care of itself. This has all changed, and it is abundantly clear that mutations result from any of a variety of fascinating molecular processes, the understanding of which represents an exciting contemporary challenge to molecular genetics.

On the other hand, in much of evolutionary biology, mutation is still generally treated as a black box from which new variants of any imagined sort can arise. This may be justified when interest is in the characterization and maintenance of genetic variation, but not so when considering the ultimate origins of mutation and their role in long-term evolution (40, 72). The spectacular advances in molecular biology have demonstrated that not all mutational events are equally likely, nor are they capable of producing the same kinds of phenotypes. The discovery of transposable elements and of their association with spontaneous mutation in *D. melanogaster* (P. M. Bingham and Z. Zachar, this volume) has contributed enormously to our understanding of the molecular basis of phenotypic variation. Transposition is the first mutational process which lends itself to experimental genetic analysis at the population level.

As defined by DNA reassociation kinetics, the *D. melanogaster* genome is composed of three fractions of DNA: highly repetitive, moderately repetitive, and unique DNA sequences (77). The moderately repetitive fraction makes up about 17% of the total genomic DNA (111), and roughly three-fourths of this fraction consists of dispersed sequences presumed to be transposable elements. Moreover, the elements examined so far show a high degree of polymorphism among strains in their locations in the genome (5, 37, 91, 104, 106, 111). Studies using randomly cloned DNA suggest that *D. melanogaster* may contain 10 to 100 different families of transposable elements (111). However, any random set of clones is biased against low-copy-number families. When only clones with no cross-homology to others are considered and the results are combined with families of transposable elements identified from insertions into mutant alleles, the resulting estimate of the total number of families of transposable elements in *D. melanogaster* becomes about 30 (J. Hey and W. Eanes, unpublished results).

Many families of transposable elements in *Drosophila* have been well characterized. Different families can exhibit very different structural and biological properties. As a class, retrotransposons (Bingham and Zachar, this volume), exemplified by the *copia* and *gypsy* families, appear to be substantially less mobile than elements responsible for dysgenic phenomena, such as the P element (W. R. Engels, this volume), the I element (D. J. Finnegan, this volume, Chapter 18), *hobo* (R. K. Blackman and W. M. Gelbart, this volume), or *mariner* (D. L. Hartl, this volume). Nevertheless, theoretical models combined with empirical data do provide insight into population dynamics, even though rates of transposition and excision vary dramatically among different classes of transposable elements, and these parameters strongly influence the implications of models. This simply means that theoretical models must be interpreted with caution and applied only to the types of elements for which they are valid. For example, with a few exceptions noted later, theoretical models designed to simulate the population dynamics of *Drosophila* retrotransposons are inappropriate for mammalian long interspersed repeated sequences (LINEs) (C. A. Hutchison III, S. C. Hardies, D. D. Loeb, W. R. Shehee, and M. H. Edgell, this volume) or short interspersed repeated sequences (SINEs) (P. L. Deininger, this volume), because in these cases the sites containing elements are occupied at frequencies at or near fixation in the population.

III. EUCARYOTIC POPULATION MODELS

Since transposition as a mutational process is likely to be an important determinant of genetic variation, assessing the forces which influence the distribution and dynamics of transposable elements in natural populations has become a significant issue in evolutionary biology. Over the past several years, a body of mathematical theory has been developed to evaluate the consequences of such factors as replicative transposition, excision, selection, and stochastic processes. An explicit assumption incorporated into almost all models is the concept suggested by Orgel and Crick (90) and Doolittle and Sapienza (24) that transposable elements should be viewed as selfish DNA. That is, DNA sequences with sufficient rates of transposition can spread through genomes and populations in the absence of favorable selection. At least within the genome of *D. melanogaster*, with a few exceptions (1; B. Charlesworth, unpublished results), the vast majority of sites are occupied at low frequency in the chromosomes in the population, and the absence of any other evidence for favorable selection argues that the selfish DNA concept is reasonable to invoke as a first approximation. Most population models also assume that, while transposition is a random process, it occurs homogeneously in time. That is, the models make no provision for transpositions occurring in discrete bursts punctuating periods of relative quiescence.

Although transposable elements occur in a great variety of organisms, few species are amenable to population genetic study. In higher organisms, *D. melanogaster* has been the primary focus both experimentally and theoretically in the attempt to define the parameters which influence the population dynamics of transposables elements. *D. melanogaster* lends itself to this kind of study because the giant salivary chromosomes serve as a convenient physical map for identifying chromosomal sites of insertion. The technique of in situ hybridization (66; for an example, see Finnegan, this volume, Chapter 18) allows cytogenetic resolution to the level of Bridges' lettered subdivisions (8), giving 115 to 120 independently identifiable "sites" per chromosome arm, although the number of sites per chromosome arm varies because the sections near the chromocenter are often difficult to score. Therefore, much of the following discussion centers around experiments with *D. melanogaster* and the associated theory.

In order to assess the impact of forces which affect the distribution and dynamics of transposable elements in natural populations, most theoretical models begin by incorporating a few essential features of transposition. Using B. Charlesworth's (submitted for publication) notation, one denotes by u the probability per generation that a member of a transposable element family will give rise to a new copy within the germ line of an individual. If this probability is dependent upon the total number n of elements of the family present in the same genome, then the probability of transposition is a function u_n. In the case of autoregulation of transposition, u_n is a strictly decreasing function of n. That is, rates of transposition will be greater with fewer copies of the element. Similarly, a given copy of a transposable element has some probability v of excision per generation. With some transposable elements, such as P elements, transposition is often associated with excision (Engels, this volume), but the relation between transposition and excision is sufficiently obscure with other types of elements, such as retrotransposons (Bingham and Zachar, this volume), that v is often treated as independent of copy number. If u_n is greater than v when n is relatively small, then the expected net increase in copy number in the germ line of an individual equals $n(u_n - v)$. Thus, within a population, the expected change in mean copy number per generation can be approximated by

$$\Delta n = n(u_n - v)$$

This relation implies that, if transposition is regulated, an equilibrium copy number $\hat{n}$ will occur when $u_{\hat{n}} = v$. However, if the rate of transposition exceeds that of excision, the absence of regulation results in the fixation of elements at all possible insertion sites. Although the P element probably autoregulates transposition (Engels, this volume), there is no direct evidence that other *Drosophila* elements, such as the retrotransposon families, autoregulate their transposition (Bingham and Zachar, this volume). In the absence of autoregulation, adverse selection must be invoked to offset an uncontrolled increase in copy number. Consequently, in considering the population dynamics of transposable elements, whether autoregulated or not, it is important to consider the effects on fitness of mutations that result from insertion of the elements.

IV. DETERMINANTS OF POPULATION DYNAMICS

As implied by the discussion above, if replicative transposition of an element is not governed by some autoregulatory mechanism, theory predicts that the element will continue to increase in number until all available sites are occupied. However, the pattern that has emerged from studies of several families of transposable elements in *D. melanogaster* is that many target sites that are occupied at all are occupied at only a low frequency (1, 70, 82, 93).

Molecular studies of several genes in *D. melanogaster* have shown that a substantial fraction of spontaneous mutation results from the insertion of transposable elements (Bingham and Zachar, this volume). Like other mutagenic agents, insertional mutagenesis may be expected to produce mainly deleterious mutations, and therefore the fitness of individuals possessing a transposable element may decrease as the number of copies of the element increases. Charlesworth and Charlesworth (19) investigated conditions under which a balance between replicative transposition and selection results in an equilibrium number of copies of a transposable element. They showed that a nontrivial equilibrium is possible when the relation between w_n, the fitness of an individual containing n copies of the element, and n is

$$d(\ln w_n)/dn = v - u$$

where u and v are again the transposition and excision rates, respectively. Furthermore, the equilibrium is globally stable when the logarithm of fitness declines at a rate greater than linearly as the copy number increases. A linear relation between replicative transposition and selection implies that the equilibrium mean fitness is approximately

$$\bar{w} = \exp[-(u - v)\bar{n}]$$

where $\bar{n}$ is the mean copy number at equilibrium (10, 18). This result implies that, for the many retrotransposon families in which $\bar{n}$ = 30 to 50, and for which $u - v$ is thought to be on the order of 10^{-4} to 10^{-5}, the genetic load is very small, and thus very little selection is necessary to maintain the equilibrium copy number. However, these estimates of the effect of selection on the stabilization of the number of copies of an element depend critically on the poorly known quantities u and v.

A. Effects of Transposable Elements on Fitness

The theoretical result that only weak selection against a transposable element is sufficient to maintain an equilibrium number of copies implies that direct estimation of the selective effects of individual insertion mutations occurring in natural populations might be very difficult. However, several studies exploiting P-M hybrid dysgenesis have provided both direct and indirect estimates of the effects of individual P element insertions as well as the general effects of P-M dysgenic mutations on relative viability. Although the P element mobilizes at an extremely high rate under dysgenic conditions, the effect on fitness of insertion mutations occurring during dysgenesis may nevertheless serve as a working model for other families of transposable elements.

Classical experiments by Mukai's group on the accumulation of mutations affecting viability have indicated an average reduction in viability per homozygous mutation of about 1 to 2% (83, 84). These and other researchers agree that the most frequently detected class of spontaneous mutations affecting viability are those with small detrimental effects when homozygous; these mutations are also usually slightly detrimental when heterozygous, with selection coefficients against heterozygotes of about 0.7% (102). If a substantial fraction of spontaneous mutations in these types of experiments results from transposable elements, then the detrimental effects of individual insertion mutations should be experimentally detectable. These considerations provide the rationale for the studies of P-M hybrid dysgenesis and fitness.

Several studies have attempted to assess the effects on fitness of mutations resulting from P element insertions that occur during P-M hybrid dysgenesis (29, 38, 112, 113). Because the mathematical theory summarized above relies critically upon the rates of transposition and excision, it is worth emphasizing again that rates of transposition and excision during P-M hybrid dysgenesis are several orders of magnitude greater than the values that apply to retrotransposons (Engels, this volume; Bingham and Zachar, this volume). Despite this caveat, the results from studies of P-M dysgenesis may serve as a paradigm for distinguishing the direct effects on fitness of insertions and secondary effects of chromosomal rearrangements.

Two studies indicate that the overall mutagenic effects of P-M hybrid dysgenesis have a substantial impact on fitness. A wild-type second chromosome derived from a Harwich population averaged a 10 to 20% reduction in total fitness relative to nondysgenic control chromosomes when passed through a single generation of P-M hybrid dysgenesis (38). Since the Harwich-derived second chromosome contains P elements, the reduction in fitness is likely to result from both insertion mutations and P element-mediated rearrangements. A similar study of nonlethal mutations in a P element-bearing chromosome after nine consecutive generations of P-M hybrid dysgenesis resulted in an average reduction in homozygous viability of 64% (112). Again, because the original parental chromosome carried P elements, this effect is likely to be due both to insertional mutations and to P element-mediated rearrangements.

In contrast to these studies, the passage of a parental X chromosome free of P elements through one generation of P-M hybrid dysgenesis reduced average viability about 1% per insertion mutation, as estimated by regression of relative viabilities of strains containing independently recovered X chromosomes against the number of P elements newly inserted into the X chromosome in each strain (29). While the average reduction in viability was 1% per insertion mutation, the 95% confidence interval ranged from near 0 to about 2%. This estimate includes only the viability component of fitness, whereas that of the Harwich chromosome (38) includes the potentially important fertility component as well.

The conclusion from these experiments is that P-M dysgenesis substantially reduces viability and other components of fitness. Moreover, the reduction in fitness may be resolved into two distinct components, mutations resulting from P element insertion per se, and the deleterious effects of P element-mediated chromosomal rearrangements. Taken at face value, the 1% average reduction in viability per insertion estimated by Eanes et al. (29), compared with the 64% reduction estimated by Yukuhiro et al. (112), would suggest that secondary rearrangements, the breakpoints of which occur primarily at P element insertion sites occupied prior to dysgenic mobilization (2, 35), cause a large fraction of the mutations that reduce fitness.

Comparisons of lethal mutation rates between chromosomes that contain P elements prior to hybrid dysgenesis and those that lack P elements also support the distinction between fitness effects due to direct mutagenic effects of P elements and those due to secondary chromosomal rearrangements. Simmons et al. (103) estimated that the P element-bearing chromosomes ν_6 and π_2 sustained new recessive lethal mutations at rates of 1.36 and 3.04% per generation, respectively. Similarly, Eanes et al. (29) estimated that the average lethal mutation rate among 100 X chromosomes from a natural population in Lincoln, Massachusetts, following passage through one generation of hybrid dysgenesis, was 2.6%. In contrast, the lethal mutation rate for X chromosomes lacking P elements prior to hybrid dysgenesis was 0.4%.

There are at least two explanations for the difference in lethal mutation rates among X chromosomes initially containing P elements as compared with X chromosomes lacking P elements. First, the P element may not insert into all possible target sites at random. For example, from the hybrid-dysgenesis-induced lethal mutation rate of 0.4%, Eanes et al. (29) estimated that only 76 vital genes in the X chromosome are targets for P element insertion. This estimate is far smaller than Lefevre and Watkins' (69) estimate of the total number of vital X chromosomal genes as 614. The disparity in the estimates argues that P element insertion is nonrandom.

Second, chromosomal rearrangements such as duplications, deletions, and inversions may be more deleterious on the average than simple insertions. The classic study of segmental aneuploidy in *D. melanogaster* (76) showed that, although most small deletions are viable when heterozygous, many reduce average viability or cause sterility in one or both sexes. Furthermore, the deleterious effects of small deletions or duplications (segmental aneuploidy) result primarily from additive effects of genes which individually reduce viability only slightly. Since P element-induced rearrangements may include small deletions and duplications, hybrid dysgenesis is expected to have effects that are more deleterious than would be expected from insertions alone.

Although P-M hybrid dysgenesis is atypical in that an unusually high level of P element mobilization occurs, the differences in fitness effects resulting directly from new insertions and those resulting indirectly from chromosomal rearrangements may be important for affecting the distributions of transposable elements in general. In particular, elements that tend to promote rearrangements may be more strongly selected against than elements that do not.

B. Indirect Estimates of Selection

An ingenious alternative approach to assessing the effects of transposable element insertions on fitness in *D. melanogaster* has exploited the differential exposure to selection of genes in the X chromosome and those in the autosomes. Selection is different because X-linked genes are exposed to selection in the hemizygous state in males. Moreover, because there are fewer X chromosomes than autosomes in a population, the effective population number, which determines the magnitude of random genetic drift for X-linked genes, is three-quarters of that for autosomes. Therefore, the frequency of mutant alleles should be smaller in the X chromosome than in autosomes, although the discrepancy actually depends on the degree of dominance of the mutant alleles. It is reasonable to assume that insertion mutations are not completely dominant in their effects on fitness, and consequently the relative abundance of any particular element should be smaller in the X chromosome than in the autosomes. The lower abundance reflects the average effect of insertions.

Montgomery et al. (81) developed models to test the effects of transposable element insertions on fitness. The null hypothesis assumes that all insertions are selectively neutral, and an alternative assumes that insertions have additive detrimental effects on fitness. Taking into account the relative lengths and effective population numbers of X chromosomes and autosomes, the neutral model predicts that 17% of occupied sites should occur in the X chromosome, whereas the detrimental selection model predicts that 11% of occupied sites should occur in the X chromosome. Data on the chromosomal locations of retrotransposons *B104* (also known as *roo*) and *copia* did not deviate significantly from the neutral model, whereas those of retrotransposon *412* differed significantly in the direction expected for detrimental selection. Recent data on chromosomal occupancies of *copia* in a laboratory population recently derived from the wild (Canary Island) also support selection in exhibiting X linkage of only about 12% of the occupied sites (109). In contrast to the situation with retrotransposons, the chromosomal distribution of sites occupied by P elements in a natural population in Botswana, Africa, is inconsistent with both models in that the occupied sites were relatively more frequent in the X chromosome than in the autosomes (29; but see below).

Differences in chromosomal occupancies of transposable elements suggest that different families of transposable elements product different biological effects, and also that tests of the models may be sensitive to the idiosyncrasies of a given family of elements. For example, in the Botswana population (1), there are hot spots of P element occupation near the tip of the X chromosome. (Hot spots of occupation in natural populations are not necessarily hot spots of insertion, because natural chromosomes have been subjected to selection and the sites that remain occupied may not accurately reflect the a priori distribution of insertions.) In Botswana, one site near the tip is occupied in most X chromosomes examined, thereby adding nearly one element to the mean number of P elements in these X chromosomes. However, if these subtelomeric sites are removed from the data, the remaining distribution does not deviate significantly from the neutral model (29). Secondary mutagenic effects not associated with insertion per se, such as duplications and deletions, may further obscure the differential effects of selection between the X chromosome and the autosomes, because such lesions could result in a greater degree of dominance than simple insertion mutations. As the average dominance per mutation increases, the expected difference between the X chromosome and the autosomes decreases. Chromosomal rearrangements are well documented in the case of P elements, and may occur with retrotransposons as well (see below).

C. Chromosomal Rearrangements

Theoretical considerations suggest that selection against insertion events per se is unlikely to be a major mechanism for regulating copy number, because relatively weak selection can prevent transposable elements from becoming established in the genome unless transposition rates are quite high (18, 19; Charlesworth, submitted). The data regarding P elements reviewed earlier imply that simple insertion mutations are substantially less deleterious on the average than are secondary rearrangements associated with dysgenesis. If this inference extends to other families of transposable elements, then selection may not act effectively against transposable elements until they accumulate to relatively high levels in the population.

Based on these kinds of considerations, Montgomery et al. (81) raise the possibility that the secondary rearrangements that are particularly important for selection result from recombination between homologous elements at different chromosomal sites, a phenomenon that has been observed experimentally (23, 46, 50). Langley et al. (68) begin the modeling of this process by assuming that the fitness of an individual is a function of the proportion of euploid gametes that it produces. Suppose there are j different types of a particular transposable element differing in their tendency to undergo unequal exchange. Let κ_{kl} denote the probability of an unequal exchange occurring between elements of type k and type l that produces an aneuploid gamete. If the aneuploid gamete formed results in dominant lethality and κ_{kl} is small, the fitness of an individual with $\mathbf{n}$ copies of elements, where the vector $\mathbf{n} = (n_1, n_2, \ldots, n_j)$ and n_i is the number of copies of element type i, is approximated by

$$w(\mathbf{n}) = 1 - \Sigma\ \kappa_{kl} n_k n_l$$

where the sum is over all pairs k, l. Thus, the probability of forming an aneuploid gamete is a quadratic function of the copy number, and this is a sufficient selective force to stabilize copy number in the absence of autoregulation.

To test this hypothesis, Langley et al. (68) used two approaches. The first compares chromosomal regions known to have different rates of recombination, and the second contrasts the densities of transposable elements along the X chromosome and the autosomes.

Langley et al. (68) first compared the occurrence of the abundant retrotransposon *B104* at the tip, middle, and proximal regions of the chromosome, adjusting for size and gene density of each subdivision. Since recombination is reduced at both the distal and proximal regions of the chromosome compared to the middle (25, 75), the hypothesis predicts that transposable elements should be relatively more abundant in these regions. If free exchange between any pair of elements is allowed, then the ratio of densities of transposable elements between two regions i and j should be inversely proportional to the ratio of the respective recombination rates, namely,

$$n_i/n_j = L_i c_j / L_j c_i$$

where L_i and L_j are the relative sizes of regions i and j, respectively, and c_i and c_j are the total rates of recombination in regions i and j, respectively. If exchange is limited to elements within a small region, then the expected ratio of densities of elements is given by

$$n_i/n_j = (L_i/L_j)(c_j/c_i)^{1/2}$$

In the comparisons of *B104* densities, the elements tended to be more abundant in the proximal region than expected from random insertion. However, the pairwise ratio of element abundances for the three regions did not reflect the predictions of either exchange model, because elements appeared to be too common in the proximal region and too rare at the tip.

The second approach exploits the differences in X chromosome complements and meiotic recombination between *D. melanogaster* males and females. Since recombination occurs only in females, and females carry two X chromosomes and males only one, two-thirds of the total X chromosomes in the population can undergo recombination in any generation. In contrast, both males and females are diploid for the autosomes, so half of the autosomes in the population can undergo recombination in each generation. Therefore, if unequal exchange between elements during meiosis results in aneuploid gametes, the rate of elimination of X-linked elements should be proportionately greater than for autosomal elements. From this basic feature, Langley et al. (68) tested two models. The first allows recombination between any pair of elements in the genome, and the second restricts recombination to elements in the same region. The first model, like the deleterious insertion model, predicts 11% of the total elements in a given family to be X linked, whereas the second model predicts 15% X linkage and the neutral insertion model predicts 17% X linkage. For the *B104* element, 16% of the insertions were X linked, which is inconsistent with the free exchange model, but very close to predictions of the restricted exchange model. However, the data are also consistent with the neutral insertion model (81).

From this analysis, one cannot be definitive about the importance of unequal exchange between elements in regulating copy number. There are several reasons why the densities of *B104* across the proximal, middle, and tip regions of the X chromosome may not reflect the relative rates of recombination. The assumption that the physical length of a given region reflects the probability of an element inserting into that region may be invalid. In the case of the P element, it is clear that new insertions into the X chromosome are nonrandom (1; Engels, this volume). Since the insertional specificity of retrotransposons such as *B104* is unknown, it is difficult to evaluate this assumption. The interpretation is also obscured if selection against deleterious insertions or excisions is not uniform along the chromosome. The similarity of P element insertion profiles for de novo (sheltered from selection) and naturally occurring (subjected to selection) X chromosomes (1) argues that uniformity in excision and selection may be a valid assumption. On the other hand, the P element may not serve very well as a test for the general models themselves, since its rates of excision are so high as to confound the process of unequal exchange.

Comparison of the densities of transposable elements between the autosomes and the X chromosome is likely to mitigate uncertainties in the above assumptions, because much of the nonuniformity within chromosomes may average out in the comparisons among chromosomes, particularly with abundant elements such as *B104*. In the *B104* case, the data fit the model of restricted exchange quite well and reject the free exchange model. Although the neutral insertion hypothesis could not be rejected, models of unequal exchange and other recombination-dependent processes merit further investigation. As Langley et al. (68) point out, these kinds of mechanisms may also be important with respect to SINE and LINE sequences in mammals, in which each occupied position along the chromosomes appears to be fixed in the population (Hutchison et al., this volume; Deininger, this volume).

D. Rates of Transposition and Excision

If mutations resulting from the insertion of transposable elements are usually deleterious, as the data suggest, then what mechanisms can account for the spread of transposable elements through entire populations? As noted above, theoretical arguments imply that even relatively weak selection is sufficient to stop the spread of transposable elements. At least two possible explanations may circumvent this apparent paradox. First, simple insertion mutations may be less deleterious than other types of mutations mediated by transposable elements, and may be nearly selectively neutral. For example, the fitness effect of P element insertions into the X chromosome is consistent with near neutrality, since the lower 95% confidence limit on the estimated selection coefficient was near zero (29). Second, when transposable elements are first introduced into the genome and the copy number is low, the initial rates of transposition may be high enough to overcome excision of the elements and elimination by selection. The second possibility is likely for the P element, where rates of transposition of up to one new insertion event per chromosome per generation have been observed. Unfortunately, it is difficult to assess the rates of transposition and excision in natural populations, and estimates obtained under laboratory conditions may be unrealistic for many families of elements. Since these parameters are central to theoretical models, it is important to consider the possibility that both environmental conditions and genetic background may have a substantial effect on transposition and on excision in natural populations. Observations of unstable mutant stains (49, 61, 73), coordinate transposition of different elements (44, 71, 80), and transposition "bursts" (43, 45) suggest that conditions may occur in which normally quiescent elements undergo relatively rapid mobilization.

Coordinate transposition of different elements was first suggested during a P-M hybrid dysgenesis screen for mutations at the *white* locus in *D. melanogaster* (94). Two of seven independently derived *white* mutations recovered from a cross of P males (Harwich) by M females (H-40) were due to insertions of the retrotransposon *copia*. In a similar P-M hybrid dysgenesis screen with P males (Harwich) crossed with M females (*cu kar* ry^{+0}), an insertion of *copia* into the *rosy* locus was recovered (22). In a systematic study of this phenomenon, sublines isolated after an initial dysgenic cross of P males (Harwich) by M females (Canton S) were monitored through 14 generations of sib matings (71). Excluding occupied sites that may have been polymorphic in the parental stocks in the F_2 generation, five new insertions each were observed for *copia* and *412* and four new insertions for the F element. No new insertions were found in the nondysgenic control lines. Unfortunately, it is difficult to interpret this and the earlier results with *white* and *rosy* because the experiments were potentially complicated by dysgenic effects resulting from the transposable element *hobo* (6, 110; Blackman and Gelbart, this volume). Although P-M hybrid dysgenesis is not by itself a stimulus for coordinate transposition of unrelated elements (33, 108), there is the formal possibility that outcrosses of the Harwich strain result in mobilization of *copia* and other transposable elements because of interactions between P-M and *hobo* hybrid dysgenesis or because some presently unrecognized regulatory system goes awry.

In another highly mutable genetic system, there seem to be at least two responding elements, *hobo* and the retrotransposon *gypsy*. Lim et al. (73) have demonstrated that an X-linked factor in a strain designated *Uc* is responsible for rates of lethal mutation on the order of 1 to 6% per X chromosome per generation, compared with 0.04% per generation in controls. Despite the fact that the origin of the X-linked mutation in the *Uc* strain is unclear, it seems that *hobo* dysgenesis is at least partly responsible for the genetic instability, as many of the mutations and rearrangement breakpoints are associated with *hobo* elements (Blackman and Gelbart, this volume). However, the X-linked factor, called the *L* factor, also appears to mobilize *gypsy*, which inserts into the *cut* locus, and approximately half the *cut* mutants recovered are also recessive lethals (62). Although *hobo* dysgenesis seems to be part of the *Uc* mutable system, the basis for concomitant movement of *gypsy* elements remains obscure. Nonetheless, the *Uc* mobilization of the retrotransposon *gypsy* is frequent compared with stable lines, where the rate of transposition is below the limit of detection.

Other mutator systems with less dramatic effects have also been shown to involve increased mobilization of retrotransposons, as indicated by the reversion of mutant alleles having insertions of known retrotransposon elements (4, 14, 15). Interestingly, these same indicator alleles also appear to be susceptible to reversion by ionizing radiation (48). One example of such a mutator system results from the presence of a mutator gene designated *mu*. Green (48, 49) has shown that the *yellow* mutant allele y^2, which contains a *gypsy* insertion (14, 15), reverts to y^+ at the rate of approximately 4×10^{-4} in the presence of *mu* and at a rate of 6×10^{-5} after X-ray treatment, but rarely spontaneously (no revertants in 140,000 chromosomes tested [108]).

Although the origin of the mutator *mu* is unclear (49), other genes with mutator activity unlinked to *mu* have been found in natural populations. Ives (61) recovered the partially dominant mutator *hi* from a Florida population screened for the production of spontaneous recessive lethals; *hi* increased the induction of X-linked recessive lethals 5- to 10-fold.

An unstable *Drosophila* strain containing the *cut* allele ct^{MR2} (due to *gypsy* insertion) appears to undergo several transposition events or bursts of transposition in the same germ cell, in which new insertions and excision events have both been observed (43, 44, 80), and the reversion rate of ct^{MR2} was about 1×10^{-3}. Moreover, reverse mutations of *carmine* or *rudimentary*, which are not due to *gypsy*, occurred in 80 to 97% of X chromosomes containing ct^{MR2} revertants (44). New mutations arising in the unstable lines are generally not associated with *gypsy*, and at least one, a *garnet* mutation, appears to be due to a *copia* insertion inasmuch as its reversion is correlated with the loss of a *copia* element at the cytological position of *garnet* (44). Although the rate of mobilization of transposable elements in these cases is not as dramatic as in the *Uc* strain, the observation of simultaneous mobilization of different element raises the possibility that genetic regulatory systems not related to a specific transposable element may alter the population dynamics and distribution of several unrelated families of transposable elements.

Taken together, these studies of mutator systems demonstrate that rates of transposition and excision may be dramatically increased by genetic instability. Since it is not clear at what frequency these or analogous mutator systems occur in natural populations, it is difficult to assess their overall impact on transposable element mobilization. However, because theoretical models depend critically upon the magnitude of transposition and excision rates, the occurrence of mutator systems in natural populations warrants further study.

E. Finite-Population Effects

To evaluate the relative effects of random genetic drift in relation to transposition, excision, and selection on the population dynamics and distribution of transposable elements, Charlesworth and Charlesworth (19) and Langley et al. (67; see also 64) developed models for finite populations. These stochastic models provide a quasiequilibrium probability distribution of occupied sites. Although the ultimate fate of transposable elements in all the stochastic models is their loss from the population, because loss is an absorbing barrier, the rate of loss is so slow that the probability distribution can be regarded as nearly stationary (18; Charlesworth, submitted). With a given set of parameters estimated from experimental data, the models predict the number of sites occupied 1, 2, 3, 4,. . ., times in a sample of chromosomes, which is called the occupancy profile. The expected occupancy profile is compared with the observed result and an appropriate goodness-of-fit test is carried out. With Charlesworth and Charlesworth's (19) notation, the parameter α, defined by

$$\alpha = 4N_e u\tilde{n}/(2m - \tilde{n})$$

includes the effects of drift and transposition for a particular site, where N_e is the effective local population size, u is the rate of transposition, $\tilde{n}$ is the expected mean copy number of elements, approximated by the equilibrium copy number $\bar{n}$ for a large population, and m is the haploid number of possible insertion sites. The parameter β, defined by

$$\beta = 4N_e(s + v)$$

includes the effects of drift, excision, and selection at the site, where s is proportional to the loss in fitness at element copy number $\bar{n} = \tilde{n}$, and v is the rate of excision. Using estimates of α and β derived from data for the retrotransposons *copia*, *297*, and *412* (48), one obtains excellent agreement between observed and expected occupancy profiles (18; Charlesworth, submitted). In addition, the estimates of β are so large (10 to 50) that random genetic drift is relatively weak as compared with selection and excision in determining the copy number of these elements. (For random genetic drift to play a major role, β should be on the order of 1.)

F. Advantageous Mutations

It has often been proposed that transposable elements induce favorable mutations or rearrangements that could be spread by selection and contribute to the maintenance of transposable elements in the population (e.g., 9, 36, 105). However, the low occupancy of individual sites in *D. melanogaster* (1, 70, 82, 93) is inconsistent with the simplest form of the idea, because if some occupied sites confer a selective advantage, then their occupancy frequency should be very high. Thus, as Charlesworth (submitted) has pointed out, the favorable-mutation hypothesis at the level of individual selection does not adequately explain the occupancy data.

An alternative hypothesis is that transposable elements are favored because of beneficial effects they contribute to the entire population, not necessarily to the individuals that contain them. For example, the entire population might benefit in some way from the mutagenic effects of transposable elements. However, hypotheses that invoke group selection are subject to many other problems and reservations (78). All group selection models must explain how elements can spread initially in the face of individual selection against deleterious insertion mutations.

Nevertheless, transposable elements might occasionally cause favorable mutations or chromosomal rearrangements even if these events do not occur with sufficient frequency to have a significant impact on the population dynamics or distribution of the elements. This possibility remains theoretical, since no favorable mutations due to the action of a transposable element have been found in a eucaryote. (However, as discussed below, occasional favorable effects of transposable elements have been demonstrated in procaryotes.) It is interesting in this context to cite two independent insertions near the *G6pd* gene (glucose 6-phosphate dehydrogenase) in *D. melanogaster*, which are found at frequencies of 10 and 30% in two different populations (W. Eanes, unpublished results). Although such high frequencies can also be explained by random genetic drift, which seems an appropriate null hypothesis until proven otherwise, one cannot exclude the possibility that the high occupancies are maintained by selection favoring chromosomes with the insertions.

V. TRANSPOSABLE ELEMENTS IN *ESCHERICHIA COLI*

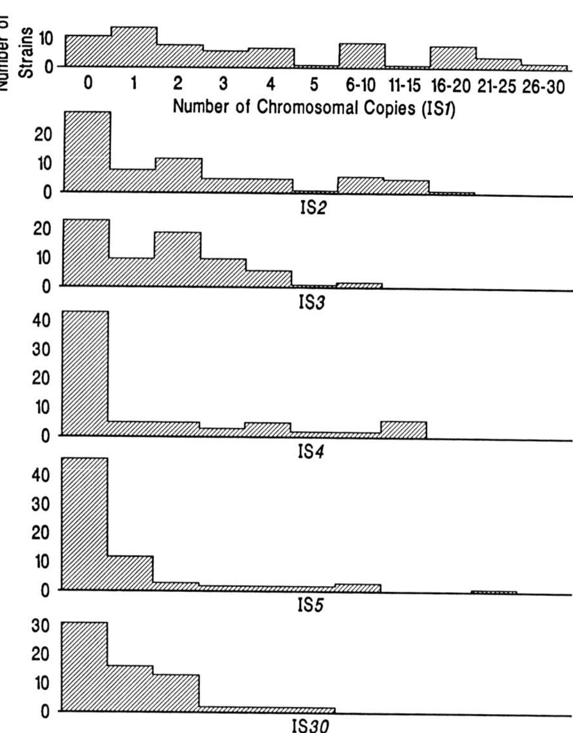

Figure 1. Number of chromosomal copies of insertion sequence elements among 71 strains in the ECOR reference collection of natural isolates of *E. coli*. The distributions differ markedly in the proportion of uninfected strains (IS*1* and IS*5*) and in the extent of the right-hand tail (IS*1* and IS*30*). Data from reference 97.

Most data relevant to the population dynamics of transposable elements in bacteria derive from studies of insertion sequences in *E. coli*. Insertion sequences are transposable DNA sequences usually ranging in size from about 1 to 2 kilobase pairs (D. J. Galas and M. Chandler, this volume). They terminate in short nucleotide sequences that are perfect or nearly perfect inverted repeats, and they generally contain at least one long open reading frame and often one or more shorter ones. Insertion sequences often encode a protein needed for transposition ("transposase") and, at least in some cases, genes for one or more proteins or RNA transcripts that regulate transposition (11, 51, 60, 65).

Natural isolates of *E. coli* are polymorphic for the presence or absence of individual insertion sequences (28, 47, 59, 87). Figure 1 illustrates the distribution of five insertion sequences among a representative sample of 71 natural isolates of *E. coli* (the ECOR sample, described below). The sequences differ in their occurrence among isolates (compare, for example, the number of strains lacking IS*1* versus those lacking IS*5*) and in the largest numbers that occur (e.g., IS*1* versus IS*30*). Sequences related to IS*1* occur in approximately 200 copies in *Shigella dysenteriae* (89), which is very closely related to *E. coli* (107). The reason for the large difference in abundance of IS*1* in the two species is unknown.

Insertion sequences also occur in both large and small plasmids, which enables them to be disseminated among strains as they hitchhike along during plasmid transfer. As discussed below, horizontal transfer of insertion sequences has a marked effect on their population dynamics, which strongly distinguishes their population biology from that of eucaryotic transposable elements.

VI. GENERAL EVOLUTIONARY CONSIDERATIONS

Insertion sequences play at least three important roles in the evolution of bacterial chromosomes and plasmids. First, insertion sequences act as generalized mutagens by producing insertion mutations or chromosome rearrangements. Second, insertion sequences can mobilize the transposition of other DNA sequences, including some that are beneficial under certain circumstances. Third, insertion sequences exhibit characteristics of parasitic or selfish DNA (24, 90), and this also affects their distribution and abundance.

As agents of mutation, insertion sequences inactivate genes directly by inserting into coding regions and disrupting them (7, 60; C. M. Berg, D. E. Berg, and E. A. Groisman, this volume). They also insert outside of coding regions and alter gene expression or activate cryptic genes (20, 21, 42, 63, 92, 96, 100,

114). As with certain eucaryotic transposable elements, secondary chromosomal rearrangements are also of evolutionary significance in bacteria. They can produce chromosome rearrangement by replicative transposition, recombination between elements at different sites in the genome, imprecise excision, and other mechanisms, and they provide recombination sites for the fusion of distinct replicons (41, 79, 86, 95).

Pairs of insertion sequences can transpose as a single unit and mobilize the DNA sandwiched between them, which may include selectively advantageous genes, such as those for resistance to antibiotics or heavy metals. Such composite transposons are well documented (D. E. Berg, this volume; N. Kleckner, this volume), and their ability to incorporate themselves into transmissible plasmids accounts in large part for their ability to spread rapidly when favored by selection, for example, in the spread of antibiotic resistance. As plasmids containing the resistance transposons become widely disseminated through horizontal transfer and natural selection, the insertion sequences that created the transposon, and other elements present by chance in the same plasmids, hitchhike along (11). Some antibiotic resistance transposable elements do not contain terminally repeated insertion sequences, for example, Tn*3* (D. Sherratt, this volume) and Tn*7* (N. L. Craig, this volume). However, the theory is independent of the molecular structure of the elements and should apply to transposable elements of any sort.

Insertion sequences can also be considered as types of parasitic or selfish DNA elements in that they can be maintained in populations solely by their ability to replicate and transpose. Since excision of insertion sequences occurs at a much lower rate than transposition (34, 39), the number of copies of insertion sequences in the genome might be expected to increase without limit unless checked by autoregulation of transposition or natural selection against strains containing the insertion sequences. As with eucaryotic elements, a sufficient selective disadvantage might result merely from the mutagenic and recombinogenic effects of these elements. Possible reductions in fitness they might cause are offset in part by their dissemination by plasmid transfer (12, 24, 54, 90). On the other hand, as discussed below, the presence of insertion sequences can be beneficial to the host under certain conditions, either through the production of beneficial insertion mutations (17) or through physiological mechanisms that do not involve transposition or rearrangement (55).

The ability of insertion sequences to produce advantageous insertion mutations or chromosomal rearrangements and to mobilize selectively advantageous genes is undoubtedly significant in their long-term evolution and dissemination. However, from the standpoint of short-term population dynamics, the parasitic or selfish characteristics may be paramount in determining the distribution and abundance of elements in a particular bacterial species. That is, as regards the general question of the evolutionary effects of insertion sequences, although they may occasionally create favorable mutations or have other beneficial effects, the putative selective advantages, though arguably important in long-term evolution, may not normally occur with sufficient intensity or duration to affect the short-term population dynamics, which is instead dominated by the characteristics of selfish DNA (57, 58, 97, 98).

VII. POPULATION DYNAMICS OF INSERTION SEQUENCES

The hypothesis that the distribution and abundance of insertion sequences are determined primarily by the selfish characteristics of transposition and plasmid-mediated transfer, with runaway accumulation of elements checked by negative selection or autoregulation, has been supported by extensive data analysis and model fitting using data on the distribution and abundance of insertion sequences among natural isolates of *E. coli*. As in the case of models of eucaryotic elements, transposition is assumed to occur homogeneously in time and not in discrete bursts.

A. Distribution and Abundance

Natural isolates of *E. coli* are polymorphic for the presence or absence of insertion sequences (47, 59, 87, 97, 99). For example, among 71 natural isolates studied to determine the copy numbers of each of the insertion sequences IS*1*, IS*2*, IS*3*, IS*4*, IS*5*, and IS*30* (97), the proportion of strains containing no copies of each element in either the chromosome or plasmids was 15% (IS*1*), 39% (IS*2*), 32% (IS*3*), 61% (IS*4*), 65% (IS*5*), and 51% (IS*30*). Histograms of the distributions are given in Fig. 1. *E. coli* also contains other insertion sequences (Galas and Chandler, this volume), but we assume that the six in Fig. 1 are representative.

The 71 strains constitute the ECOR reference set of *E. coli*, a heterogeneous collection of strains containing some urinary tract pathogens and some nonpathogenic isolates from animals and humans from North America and Sweden; they represent three major types of strains, grouped by principal components analysis of electrophoretic enzyme variation (88). Within each type, the isolates are genetically more similar to each other than they are to members of the other types, although the genetic distinctions among the types are by no means absolute. Roughly speaking, genetic differences among the types correspond to those among races in sexual eucaryotic organisms. Judged on differences in enzyme electrophoretic mobility, the ECOR strains are representative of the full range of genetic variation that occurs in *E. coli* (88).

Based on statistical analysis of their distribution and abundance in the ECOR strains, the insertion sequences could be classified according to the apparent strength of autoregulation of transposition. Altogether, the 71 strains contained 1,055 chromosomal sites and 118 plasmid sites occupied by insertion sequences. Relative to insertion sequences IS*1* and IS*5*, the distributions of numbers of copies among strains fit population models that incorporate weak or no autoregulation. For IS*2*, IS*4*, and IS*30*, the distributions fit models with moderate autoregulation. For IS*3*, the number of copies among strains fits a population model with strong autoregulation. Assumptions of the models and methods of analysis are summarized below.

B. Branching Process Models

In population models of sexual eucaryotes such as *Drosophila* spp., sites in the genome that can be occupied by transposable elements can be treated as nearly independent of each other as long as recombination between sites occurs at a greater rate than transposition. Under these assumptions, the numbers of copies of elements are expected to conform to a Poisson distribution, which is reasonably in accord with observations (19, 64, 67).

However, recombination in natural populations of *E. coli* occurs at such a low rate that the assumption of independence of occupied sites is unwarranted. Population models for transposable elements that are valid for *E. coli* must assume nearly clonal reproduction (98). The most general kind of model so far analyzed is one in which the number of copies of a transposable element changes according to a continuous time branching process in which the rate of transposition of the element and the fitness of strains containing it are simple functions of the number of copies (97). The model includes no explicit provision for excision of elements, because the excision process occurs at a much lower rate than transposition (34, 39).

The branching process model stipulates that a host organism can exist in any number n of states ($n = 0, 1, 2, \ldots$), where n is the number of genomic sites in the host that are occupied by the transposable element. Hosts in state 0 (uninfected) become converted to state 1 (one occupied site) by infection, which occurs at the rate u per generation. Individuals in state n (n occupied sites) can undergo single-transposition events to change to state $n + 1$ with a transposition rate $T(n)$ that is a function of n itself. Letting the transposition rate $T(n)$ be a function of n allows for the possibility of autoregulation. The transposition model is a formal one that applies equally to elements that replicate while transposing and to those that undergo cut-and-paste transposition. For examples of the latter, see Berg (this volume) and Kleckner (this volume).

Offsetting the increase in copy number by transposition is the assumption that host cells infected with the element have a reduced fitness, resulting either from a slower growth rate or a higher death rate than uninfected cells. Uninfected cells (state 0) are assumed to reproduce at the net rate R relative to that of infected cells in state n, for which the net rate of reproduction equals $R - D(n)$. Thus, $D(n)$ is the reduction in fitness resulting from n copies of the transposable element. Letting $D(n)$ be a function of n allows for fitness effects that differ according to the number of occupied sites.

With these assumptions, the equilibrium frequency of cells in the population that have n occupied sites is given by (98)

$$\lambda(n) = \mu(n)/L, \qquad L = \sum_{k=0}^{\infty} \mu(k) \qquad (n \geq 0)$$

where

$$\mu(n) = \frac{1}{T(n) + D(n) - u} \prod_{k=1}^{n-1} \frac{T(k)}{T(k) + D(k) - u} \qquad (n \geq 1)$$

where $\mu(0) = 1/u$ is the reciprocal of the rate of infection. The equilibrium exists if $T(n) + D(n) - u$ is bounded from below by a positive number and if the class of uninfected hosts does not become extinct, and provided that the sum L is finite.

C. Fitting the Models to Data

The branching process model contains the functions $T(n)$ and $D(n)$, which can in theory be estimated by fitting the observed distributions of occupied sites to the expected equilibrium distributions. One general approach would be to define the functions as power functions with $T(n) = Tn^{\alpha}$, say, and $D(n) = Dn^{\beta}$, and then estimate α and β. However, this

Table 1. Nine models for fitting data from ECOR strains[a]

Model	Form of $T(n)$	Form of $D(n)$
CC	T	D
HC	T/n	D
CR	T	$D\sqrt{n}$
RC	$T\sqrt{n}$	D
RR	$T\sqrt{n}$	$D\sqrt{n}$
DR	$T\sqrt{n}$	$D\sqrt{n}$
RL	$T\sqrt{n}$	Dn
LL	Tn	Dn
LQ	Tn	Dn^2

[a] The two letters that designate the models refer to the functional forms of the transposition function and the fitness function, respectively, where C denotes constant, H harmonic ($1/n$), R root ($\sqrt{n}$), D inverse root ($1/\sqrt{n}$), L linear (n), and Q quadratic (n^2). Although the models might appear to contain three parameters (T, D, and u), there are really just two because only their ratios matter in the equilibrium distribution.

approach does not work well because the maximum likelihood surface is ill behaved (S. A. Sawyer, unpublished results).

An alternative approach is to define several families of models based on specific assumptions about the functional forms of $T(n)$ and $D(n)$, and then to test the resulting expected equilibrium distributions against the observed distributions to determine which family or families of models fit the best. In practice, the nine models listed in Table 1 were considered for fitting to data from the ECOR strains.

Methods of fitting the models to the ECOR data and assessments of goodness of fit are discussed by Sawyer et al. (97). The overall result is that all insertion sequence distributions can be fit very well by at least one model, and sometimes by several models. For insertion sequences IS2, IS4, and IS*30*, the observed distributions are all amenable to model fitting. The CC model provides an excellent fit, but not to the exclusion of other models, many of which fit almost equally well. There seems little basis to choose among the models CC, CR, RL, and RR, except that the CC model is the simplest in that both the rate of transposition and the effect on fitness are independent of copy number. However, the best models all imply that transposition must be autoregulated, with the rate of transposition per copy being proportional to $1/n$ in the CC and CR models and to $1/\sqrt{n}$ in the RL and RR models.

In the case of IS*1*, all reasonable models can be rejected except for the LL and LQ models. Essentially the same situation holds for IS*5*. The implication for autoregulation is that, when the rate of transposition is given by $T(n) = nT$ (as in the LL and LQ models), then the probability of transposition per copy equals $T(n)/n = T$, a constant, irrespective of the number of elements. This means that transposition of the elements is essentially unregulated (or that autoregulation is relatively weak), which is also suggested by the relatively long tails on the observed distributions of copy number of IS*1* and IS*5*. In the models, the weak autoregulation is offset by relatively strong detrimental effects of the insertion sequences, with fitness decreasing either linearly (LL model) or as the square (LQ model) of the copy number.

The population distribution of insertion sequence IS*3* differs from all the others in that only the HC and DR models give acceptable fits. This is perhaps because the distribution of copy numbers is bimodal, with modes at $n = 0$ and $n = 2$. However, it is not known whether the bimodality is peculiar to the ECOR collection. In any case, the transposition functions $T(n) = T/n$ (HC model) and $T(n) = T/\sqrt{n}$ (DR model) imply that autoregulation of transposition is quite strong, with the rates of transposition per element being proportional to $1/n^2$ or $1/n^{3/2}$, respectively. At the same time, the fitness functions $D(n) = D$ (HC model) and $D(n) = D\sqrt{n}$ (DR model) imply relatively mild effects of IS*3* on fitness.

The general conclusions based on comparisons of the nine models defined above are supported by maximum likelihood calculations performed by Sawyer (unpublished results). The general maximum likelihood model is one in which $T(n) = Tn^{\alpha}$ and $D(n) = Dn^{\beta}$, where α and β are two additional parameters to be estimated. As noted, the maximum likelihood surface is poorly behaved in that β appears to lie on a maximum likelihood line, and point estimates of β cannot be obtained. Point estimates of α are closely in agreement with the following models: IS*1*, $T(n) = Tn$; IS2, $T(n) = T/\sqrt{n}$; IS*3*, $T(n) = T/n^2$; IS4, $T(n) = Tn^{1/4}$; IS*5*, $T(n) = Tn$; and IS*30*, $T(n) = T/\sqrt{n}$. The sampling variances of the estimates of α are rather large, and the general failure of the maximum likelihood approach renders them suspect. However, for whatever they are worth, the point estimates support the previous analysis in implying that autoregulation of transposition for IS*1* and IS*5* is weak, that of IS2, IS4, and IS*30* is moderate, and that of IS*3* is very strong.

D. The Role of Plasmids

Among the ECOR strains, most strains contain one or more large (greater than about 40 kilobases) or small (smaller than about 7.5 kilobases) plasmids. Interestingly, few plasmids are intermediate in size (56, 101). Most strains contain 0 to 4 large plasmids (mean ± standard error = 1.28 ± 0.13) and 0 to 7 small plasmids (mean ± standard error = 2.30 ± 0.30). Only 9/71 strains contained no detectable plasmids of any size. There is no significant correlation in either the presence or the number of large and small plasmids among strains.

Insertion sequences can occupy sites in plasmids as well as in the bacterial chromosome. Approximately 10% of the insertion sequences in the ECOR collection were in plasmids, and those elements that were frequent in the chromosome were also frequent in plasmids. For the five insertion sequences analyzed in the ECOR collection, the frequency of occurrence of insertion sequences in the chromosome or in plasmids appeared to be close to random (57). That is, 2 × 2 contingency tables for presence versus absence of insertion sequences in the chromosome and plasmids are generally nonsignificant.

Plasmids play a key role in the dissemination of insertion sequences and have an important and unexpected consequence for the population dynamics, as discussed below.

E. Correlations among Unrelated Insertion Sequences

Rather to our surprise, unrelated insertion sequences had a tendency to occur together in the ECOR strains. That is, the strains showed highly significant correlations in the presence or absence of unrelated insertion sequences in the chromosome. Highly significant associations were also found among insertion sequences present in the plasmid complements. Such correlations were unanticipated because biological evidence had suggested that insertion sequences were independent in processes of infection and transposition. How can unrelated elements be independent at the mechanistic level and still be correlated at the population level?

Some of the statistical associations can be attributed to stratification in the sample, which occurs in the form of the different types or races of strains that are represented. For example, if two different insertion sequence elements occur preferentially within one type of strain, then this will result in a positive correlation in occurrence across the entire sample. However, the effects of stratification can be minimized by separately calculating the correlations within each of the three types and averaging these values. Averaged across the types of strains, the mean pairwise correlation coefficient in the occurrence of insertion sequences equals 0.171 ± 0.044, which is highly significant.

F. Effects of Plasmid Transmission on Correlations

Actually, the major part of the correlation results from a subtle effect of plasmid-mediated transmission. This correlation comes from the possibility of simultaneous plasmid transmission of two or more unrelated insertion sequences even when the elements occur independently in the plasmid pools. Positive correlations in the presence of unrelated insertion sequences should be expected on theoretical grounds. This has been demonstrated in a branching process model for plasmid-mediated transmission in which there are no interactions affecting transposition or fitness (58). The observed correlations are in good agreement with the expectations of the model (58).

With plasmid-mediated transmission of insertion sequences, the expected Pearson correlation coefficient of the presence or absence of two unrelated insertion sequences is given by (58)

$$\rho = \frac{u_1u_2(s_1 + s_2 - s_3) + u_3s_1s_2}{(\Delta_1\Delta_2)^{1/2}}$$

where

$$\Delta_1 = \frac{s_1}{s_1 - u_1}[u_1u_2(s_1 - s_3) + u_1s_2(s_3 - u_1) + u_3s_2(s_1 - u_1)]$$

and

$$\Delta_2 = \frac{s_2}{s_2 - u_2}[u_1u_2(s_2 - s_3) + u_2s_1(s_3 - u_2) + u_3s_1(s_2 - u_2)]$$

In these equations, u_1, u_2, and u_3 are, respectively, the probabilities of becoming infected with insertion sequence 1 only, insertion sequence 2 only, or both, in a given time interval; and s_1, s_2, and s_3 correspond, respectively, to the selection coefficients against strains containing insertion sequence 1 only, insertion sequence 2 only, or both. In this model, the effects of insertion elements on fitness are assumed to be independent of copy number as long as $n \geq 1$, which corresponds to earlier models in which $D(n) = D$ (a constant). With this assumption, the ratio u/D determines the relative proportions of infected and uninfected strains, and u/T determines the distribution of copy number among infected hosts.

The explicit equation for ρ implies that there will still be a positive correlation among insertion sequences even when the effects of the elements on fitness are strictly additive ($s_3 = s_1 + s_2$). The correlation results from the occurrence of doubly infected plasmids ($u_3 \neq 0$).

With plasmid transmission and selection parameters estimated from the ECOR strain data (58), the expected pairwise correlation between unrelated insertion sequences, averaged across all pairs of insertion sequences and all three types of strains, was

0.142. This compares well with the observed average correlation of 0.171 ± 0.044. Moreover, within each type of strain taken separately (to avoid complications of stratification), the expected correlation is within one standard error of the observed. Therefore, the observed correlations in the presence or absence of unrelated insertion sequences among natural isolates of *E. coli*, far from being mysterious, are merely the consequence of their plasmid-mediated transmission.

G. Beneficial Effects of Transposable Elements

From the standpoint of short-term population dynamics, the ability of insertion sequences and other transposable elements to increase in number by transposition is offset by their elimination by deletion or due to their harmful effects on fitness. Complicating this simple situation are occasional beneficial effects on fitness resulting from altered regulation of genes near insertion sites or from the mobilization of advantageous genes (e.g., antibiotic resistance).

As demonstrated in the discussion above, occasional beneficial effects of insertion sequences need not be invoked to account quantitatively for the population dynamics of insertion sequences. The branching process models, which assume that insertion sequences are detrimental, are sufficient. However, all that can be concluded from this result is that the postulate of occasional favorable effects of insertion sequences is not **necessary** to account for the observed distributions of copy number. Beneficial effects could very well occur, and be very important in the evolution of transposable elements, without necessarily having a detectable effect on the equilibrium distributions of copy number.

One reason for being cautious in discounting possible favorable effects of transposable elements in bacteria is that favorable effects have frequently been demonstrated in the laboratory. Several bacterial transposable elements and temperate bacteriophages (which should, for present purposes, be regarded as highly specialized transposable elements) improve growth rate under the intense competitive conditions of bacterial chemostats in which coisogenic strains undergo competition for limited amounts of an essential resource, such as glucose (reviewed in 12, 27, 53, 54).

Examples of bacteriophages resulting in beneficial selective effects include P1cm, P2-186p, Mu (30), and λ (26, 31, 32, 74). Additional examples of favorable selective effects are provided by the composite transposons Tn*10* (16, 17) and Tn*5* (3, 55) or their component insertion sequences IS*10* and IS*50*, respectively. The selective advantage can be substantial, approximately 1.5% per generation in favor of λ lysogens (31, 74) and approximately 2.5% per generation in favor of strains containing Tn*5* (3, 55).

Although the mechanisms of these selective effects are not well understood, there is some information. For example, the favorable effect of Tn*10* or IS*10* depends on the ability to induce a specific favorable insertion mutation denoted *fit*:IS*10* at approximately 71 min on the *E. coli* genetic map (16, 17). A contrary example is provided by Tn*5* and IS*50*. Although the favorable selective effect requires the transposase or possibly the transposition inhibitor coded by IS*50*R, the effect does not require actual transposition. Strains that have greatly increased in frequency in chemostats due to favorable selection have not in general undergone transposition of IS*50*R (55). Two models are consistent with these observations, namely, a mutational model, in which the effect of IS*50*R occurs through a generalized increase in the mutation rate, and a physiological model, in which the gene products of IS*50*R enhance growth rate in some unknown manner but through effects other than those directly related to transposition or mutation (55).

From the standpoint of the evolution of transposable elements in natural populations, the significance of the favorable selective effects in chemostats is unclear. For example, the effect of IS*50*R in chemostats, while dramatic, is transient, and the selective advantage of IS*50*R-bearing strains disappears after about 50 to 75 generations of competition (3, 55). While the selective advantage of IS*50*R may occur only under certain restricted conditions and be temporary, fluctuating nutrient conditions are likely to be important in natural environments. However, it is not necessary to postulate favorable effects of IS*50* in *E. coli* under natural conditions, because IS*50* does not normally occur in *E. coli*. The original Tn*5* was discovered in a resistance plasmid in *Klebsiella* sp., and among a sample of natural *E. coli* isolates IS*50* does not occur (D. Berg and S. Phadnis, unpublished observations). On the other hand, with regard to the general issue of favorable effects, one must hesitate to dismiss the potential evolutionary significance of observations that apply to all temperate bacteriophage and to all transposable elements so far studied in chemostats.

VIII. EVOLUTIONARY IMPLICATIONS OF TRANSPOSABLE ELEMENTS

As the focus of this review we chose the population dynamics of transposable elements and emphasized the details of how the selfish characteristics of transposable elements allow them to persist in the genome over periods of time that are short relative to the evolutionary time scale. We also concentrated on transposable elements in *D. melanogaster* and *E. coli* because most of the relevant data are available from these organisms.

Considered from a broad evolutionary perspective, the principal implications of transposable elements are similar in both procaryotes and eucaryotes. From the standpoint of short-term population dynamics, the ability of the transposable elements to increase in number by transposition is offset by their elimination by means of deletion or due to their harmful effects on fitness. Complicating this simple situation are occasional beneficial fitness effects resulting from altered regulation of genes near insertion sites, from the mobilization of favored genes (e.g., antibiotic resistance), and perhaps in some cases from the very presence of transposable elements themselves.

On the other hand, the short-term population dynamics of transposable elements are quite different in *D. melanogaster* and *E. coli* because (i) recombination is common in *D. melanogaster* and rare in *E. coli*, and (ii) horizontal transfer of transposable elements is infrequent in *D. melanogaster* and common in *E. coli*. The first difference implies that genomic sites of transposable element insertion can be regarded as nearly independent in *D. melanogaster* but not in *E. coli*, and the second difference implies that horizontal transfer is negligible relative to the short-term population dynamics in *D. melanogaster* but not in *E. coli*. These differences require that population models of transposable elements must be somewhat different in *D. melanogaster* and *E. coli*, but each class of models performs quite well when evaluated in terms of data from its own domain.

In focusing on short-term population dynamics, we have neglected many other important and interesting features of the evolution of transposable elements. The long-term evolutionary implications of transposable elements are discussed in references 12, 13, 54, and 105. It is appropriate here to address some of these issues briefly, in order to put our discussion in a broader context.

The first issue concerns the origin(s) of transposable elements. Do they evolve de novo or are they escapees from normal cellular processes? They are clearly very ancient and probably had multiple origins. Some transposable elements, for example, eucaryotic retrotransposons and certain bacteriophages, are viruses or are clearly related to viruses. Others, for example, the mouse L1 elements, appear to derive from germ line reverse transposition of normal cellular genes (52). Between these relatively clear-cut examples are many transposable elements the origins of which are completely obscure.

A second issue is whether transposable elements began their existence as selfish DNA, in the sense of not increasing the fitness of the host, or whether they persisted initially because of some benefit and later capitalized on their selfish characteristics. In modern bacteria, transposable elements obviously play a key role in mobilizing genes and enabling their spread under suitable selection pressures. Examples of heavy metal tolerance and antibiotic resistance come immediately to mind. However, a logical problem in evaluating the significance of these observations can result from confounding the evolutionary role of transposable elements today, when they are highly evolved, with effects they might have had at the time of origin, when the characteristics that render them so versatile today were still inchoate and undeveloped.

Apart from their ability to mobilize genes, transposable elements act as agents of mutation. In procaryotes, specific favorable mutations can result from the activities of insertion sequences. Examples in *E. coli* include activation of the cryptic *bgl* operon by insertion of IS*1* or IS*5* (92) and creation of a novel *gnd* promoter by a deletion apparently mediated by IS*5* (R. Miller, D. Hartl, and D. Dykhuizen, unpublished results). However, the fact that insertion sequences are not found fixed at any position in the genome of *E. coli* and that many strains are uninfected with any particular insertion sequence element (97) suggest that their favorable effects on mutation or any other process are not large or pervasive. Indeed, all satisfactory population models for insertion sequence elements in *E. coli* must assume a mildly harmful effect of the presence of these elements (58, 97, 98).

On the other hand, the occurrence of favorable effects of certain temperate bacteriophages and insertion sequences in chemostats suggests that, under these and presumably other conditions, the elements may contribute to fitness. Subtle effects such as these might have been sufficient to maintain transposable elements for long periods during their early evolution, and indeed occasional beneficial effects on fitness even in present-day organisms cannot be excluded by the population models. Campbell (12) has emphasized that the long-term persistence of transposable elements might require favorable selection:

> Although an element with a negative effect on the host can be maintained by over-replication, long-term selection in competition with other hosts should ultimately eliminate it. Thus long-term survival may require that the element earn its keep, i.e. whatever cost is entailed in perpetuating the element should at least be balanced by some positive

contribution to the organism's phenotype. Mere possession by the element of a gene conferring an advantage will not suffice. In that case, selection should favor retention of the gene and loss of the rest of the element. However, the elements themselves not only survive, but, especially in the case of the larger phages and conjugative plasmids, have achieved a structural and regulatory complexity indicating a long and successful evolution. Hence their ability to replicate and disseminate themselves must be of some value to the host.

For more discussion of this aspect of transposable element evolution, see reference 54.

A third set of issues in transposable element evolution relates to the evolution of DNA sequences of the elements themselves and their rate of evolution relative to that of the host. Addressing these issues requires defining the extent and types of sequence variation in transposable elements within species. There has been little emphasis on DNA sequence variation in transposable elements within species (but see Hartl, this volume). However, the theory of population genetics is well equipped to deal with intraspecific DNA sequence variation. Inferences can be drawn concerning such phenomena as mutation, recombination, random genetic drift, the occurrence and nature of selection, and horizontal transfer. Even sequence variation that the molecular biologist may regard as dull, such as silent-site variation, can sometimes prove to be a rich store of evolutionary information when analyzed in terms of population genetics. Perhaps the most extensive study of DNA sequence variation of transposable elements within species is that of the L1 family of LINE elements in mice, which provided critical data concerning the mechanism of proliferation of the elements, their concerted evolution, divergence in DNA sequence, and clearance from the genome (52; Hutchison et al., this volume).

Taking DNA sequence variation a step further, sequence comparisons among transposable elements bear on the mechanisms by which the elements themselves evolve and diversify. Extensive DNA sequence comparisons are available for families of related transposable elements in both procaryotes and eucaryotes. Many of these allow the phylogenetic history of the family of elements to be deduced, and they open the way for analysis of rates of molecular evolution, identification of conserved or hypervariable regions, and even the coevolution of transposable elements and their hosts. Each of these stories of long-term evolution belongs in the context of the biology of the transposable elements themselves, and where the information is available, it is reviewed in the appropriate specialized chapters throughout this volume.

Acknowledgments. We are grateful to numerous colleagues who shared their data and conclusions with us prior to publication, and to Brian Charlesworth, Daniel Dykhuizen, and Douglas E. Berg for their helpful suggestions for improving the manuscript.

This work was supported by Public Health Service grants GM33741 and GM30201 from the National Institutes of Health.

LITERATURE CITED

1. **Ajioka, J. W., and W. F. Eanes.** 1988. The accumulation of P-elements on the tip of the X chromosome in populations of *Drosophila melanogaster*. *Genet. Res.*, in press.
2. **Berg, R. E., W. R. Engels, and R. A. Kreber.** 1980. Site-specific X-chromosome rearrangements from hybrid dysgenesis in *Drosophila melanogaster*. *Science* **210**:427–443.
3. **Biel, S. W., and D. L. Hartl.** 1983. Evolution of transposons: natural selection for Tn*5* in *Escherichia coli* K12. *Genetics* **103**:581–592.
4. **Bingham, P. M., and B. H. Judd.** 1981. A copy of the *copia* transposable element is very tightly linked to the w^a allele at the *white* locus of *D. melanogaster*. *Cell* **25**:705–711.
5. **Bingham, P. M., M. G. Kidwell, and G. M. Rubin.** 1982. The molecular basis of P-M hybrid dysgenesis: the role of the P element, a P-strain-specific transposon family. *Cell* **29**: 995–1004.
6. **Blackman, R. K., R. Grimalia, M. M. D. Kohler, and W. M. Gelbart.** 1987. Mobilization of *hobo* elements residing within the decapentaplegic gene complex: suggestions of a new hybrid dysgenesis system in *Drosophila melanogaster*. *Cell* **49**:497–505.
7. **Botstein, D., and D. Shortle.** 1985. Strategies and applications of in vitro mutagenesis. *Science* **229**:1193–1201.
8. **Bridges, C. B.** 1938. A revised map of the salivary gland X-chromosome of *Drosophila melanogaster*. *J. Hered.* **29**: 11–13.
9. **Brookfield, J. Y. F.** 1982. Interspersed repetitive DNA sequences are unlikely to be parasitic. *J. Theor. Biol.* **94**: 281–299.
10. **Brookfield, J. Y. F.** 1986. The population biology of transposable elements. *Phil. Trans. R. Soc. London Sect. B* **312**:217–226.
11. **Calos, M. P., and J. H. Miller.** 1980. Transposable elements. *Cell* **20**:579–595.
12. **Campbell, A.** 1981. Evolutionary significance of accessory DNA elements in bacteria. *Annu. Rev. Microbiol.* **35**:55–83.
13. **Campbell, A.** 1983. Transposons and their evolutionary significance, p. 258–279. *In* M. Nei and R. K. Koehn (ed.), *Evolution of Genes and Proteins*. Sinauer Associates, Sunderland, Mass.
14. **Campuzano, S. L., L. Balcells, R. Villares, L. Garcia-Alonso, and J. Modolell.** 1986. Excess function *Hairy-wing* mutations caused by *gypsy* and *copia* insertions within the structural genes of the *acheate-scute* locus of *Drosophila*. *Cell* **44**:302–312.
15. **Campuzano, S., L. Carramolino, C. V. Cabrera, M. Ruiz-Gomez, R. R. Villares, A. Boronat, and J. Modelell.** 1985. Molecular genetics of the *acheate-scute* gene complex in *D. melanogaster*. *Cell* **40**:327–338.
16. **Chao, L., and S. M. McBroom.** 1985. Evolution of trans-

posable elements: an IS*10* insertion increases fitness in *Escherichia coli. Mol. Biol. Evol.* **2**:359–369.

17. **Chao, L., C. Vargas, B. B. Spear, and E. C. Cox.** 1983. Transposable elements as mutator genes in evolution. *Nature* (London) **303**:633–635.
18. **Charlesworth, B.** 1985. The population genetics of transposable elements, p. 213–232. *In* T. Ohta and K. Aoki (ed.), *Population Genetics and Molecular Evolution*. Academic Press, Inc., New York.
19. **Charlesworth, B., and D. Charlesworth.** 1983. The population dynamics of transposable elements. *Genet. Res.* **42**: 1–27.
20. **Charlier, D., J. Piette, and N. Glansdorf.** 1982. IS*3* can function as a mobile promoter in *E. coli. Nucleic Acids Res.* **10**:5935–5948.
21. **Ciampi, M. S., M. B. Schmid, and J. R. Roth.** 1982. Transposon Tn*10* provides a promoter for transcription of adjacent genes. *Proc. Natl. Acad. Sci. USA* **79**:5016–5020.
22. **Clark, S., M. McCarron, C. Love, and A. Chovnick.** 1986. On the identification of the rosy locus DNA in *Drosophila melanogaster* and intragenic recombination mapping of mutations associated with insertions and deletions. *Genetics* **112**:755–767.
23. **Davis, P. S., M. W. Shen, and B. H. Judd.** 1987. Asymmetrical pairings of transposons in and proximal to the *white* locus of *Drosophila* account for four classes of regularly occurring exchange products. *Proc. Natl. Acad. Sci. USA* **84**:174–178.
24. **Doolittle, W. F., and C. Sapienza.** 1980. Selfish genes, the phenotype paradigm and genome evolution. *Nature* (London) **284**:601–603.
25. **Dubinin, N. P., N. N. Sokolov, and G. G. Tiniakov.** 1937. Crossover between the genes "yellow", "acheate" and "scute". *Drosophila Inform. Serv.* **8**:76.
26. **Dykhuizen, D. E., J. H. Campbell, and B. G. Rolfe.** 1978. The influence of a λ prophage on the growth rate of *Escherichia coli. Microbios* **23**:99–113.
27. **Dykhuizen, D. E., and D. L. Hartl.** 1983. Selection in chemostats. *Microbiol. Rev.* **47**:150–168.
28. **Dykhuizen, D. E., S. A. Sawyer, L. Green, R. D. Miller, and D. L. Hartl.** 1985. Joint distribution of insertion elements IS*4* and IS*5* in natural isolates of *Escherichia coli. Genetics* **111**:219–231.
29. **Eanes, W. F., C. Wesley, J. Hey, D. Houle, and J. W. Ajioka.** 1988. The viability and fitness consequences of P-element insertion in *Drosophila melanogaster. Genet. Res.* **52**:17–26.
30. **Edlin, G., L. Lin, and R. Bitner.** 1977. Reproductive fitness of P1, P2, and Mu lysogens. *J. Virol.* **21**:560–564.
31. **Edlin, G., L. Lin, and R. Kudrna.** 1975. λ lysogens of *E. coli* reproduce more rapidly than nonlysogens. *Nature* (London) **255**:735–737.
32. **Edlin, G., R. C. Tait, and R. L. Rodriguez.** 1984. A bacteriophage λ cohesive ends (*cos*) DNA fragment enhances the fitness of plasmid-containing bacteria growing in energy-limited chemostats. *Bio/Technology* **2**:251–254.
33. **Eggleston, W. B., D. M. Johnson-Schlitz, and W. R. Engels.** 1988. P-M hybrid dysgenesis does not mobilize other transposable element families in *D. melanogaster. Nature* (London) **331**:368–331.
34. **Egner, C., and D. E. Berg.** 1981. Excision of transposon Tn*5* is dependent on the inverted repeats but not on the transposase function of Tn*5*. *Proc. Natl. Acad. Sci. USA* **78**: 459–463.
35. **Engels, W. R., and C. R. Preston.** 1984. Formation of chromosomal rearrangements by *P* factors in *Drosophila. Genetics* **107**:657–678.
36. **Fink, G. R., J. D. Boeke, and D. J. Garfinkel.** 1986. The mechanism and consequences of retrotransposition. *Trends Genet.* **5**:118–123.
37. **Finnegan, D. J., G. M. Rubin, M. W. Young, and D. S. Hogness.** 1978. Repeated gene families in *Drosophila melanogaster. Cold Spring Harbor Symp. Quant. Biol.* **42**:1053–1063.
38. **Fitzpatrick, B. J., and J. A. Sved.** 1986. High levels of fitness modifiers induced by hybrid dysgenesis in *Drosophila melanogaster. Genet. Res.* **48**:89–94.
39. **Foster, T. J., V. Lundblad, S. Hanley-Way, S. M. Halling, and N. Kleckner.** 1981. Three Tn*10*-associated excision events: relationship to transposition and role of direct and inverted repeats. *Cell* **23**:215–227.
40. **Futuyma, D. J.** 1986. *Evolutionary Biology*. Sinauer Associates, Sunderland, Mass.
41. **Gaffney, T. D., and T. G. Lessie.** 1987. Insertion-sequence-dependent rearrangements of *Pseudomonas cepacia* plasmid pTGL1. *J. Bacteriol.* **169**:224–230.
42. **Georgopoulos, C., N. McKitterick, G. Herrick, and H. Eisen.** 1982. An IS*4* transposition causes a 13-bp duplication of phage lambda DNA and results in constitutive expression of the *cI* and *cro* gene products. *Gene* **20**:83–90.
43. **Gerasimova, T. I., L. V. Matyunina, Y. V. Ilyin, and G. P. Georgiev.** 1984. Successive transposition explosions in *Drosophila melanogaster* and reverse transpositions of mobile dispersed genetic elements. *EMBO J.* **4**:3773–3779.
44. **Gerasimova, T. I., L. V. Matyunina, Y. V. Ilyin, and G. P. Georgiev.** 1984. Simultaneous transposition of different mobile elements: relation to multiple mutagenesis in *Drosophila melanogaster. Mol. Gen. Genet.* **194**:517–522.
45. **Gerasimova, T. I., L. J. Mizrokhi, and G. P. Georgiev.** 1984. Transposition bursts in genetically unstable *Drosophila melanogaster. Nature* (London) **309**:714–716.
46. **Goldberg, M. L., J.-Q. Sheen, W. J. Gehring, and M. M. Green.** 1983. Unequal crossing-over associated with asymmetrical synapsis between nomadic elements in the *Drosophila melanogaster* genome. *Proc. Natl. Acad. Sci. USA* **80**:5017–5021.
47. **Green, L., R. D. Miller, D. E. Dykhuizen, and D. L. Hartl.** 1984. Distribution of DNA insertion element IS*5* in natural isolates of *Escherichia coli. Proc. Natl. Acad. Sci. USA* **81**:4500–4504.
48. **Green, M. M.** 1961. Back mutation in *Drosophila melanogaster*. I. X-ray induced back mutation at the *yellow*, *scute*, and *white* loci. *Genetics* **46**:671–682.
49. **Green, M. M.** 1970. The genetics of a mutator gene in *Drosophila melanogaster. Mutat. Res.* **10**:353–363.
50. **Green, M. M.** 1982. Genetic instability in *Drosophila melanogaster*: deletion induction by insertion sequences. *Proc. Natl. Acad. Sci. USA* **79**:5367–5369.
51. **Grindley, N. D. F., and R. R. Reed.** 1985. Transpositional recombination in prokaryotes. *Annu. Rev. Biochem.* **54**: 863–896.
52. **Hardies, S. C., S. L. Martin, C. F. Voliva, C. A. Hutchison III, and M. H. Edgell.** 1986. An analysis of replacement and synonymous changes in the rodent L1 repeat family. *Mol. Biol. Evol.* **3**:109–125.
53. **Hartl, D. L., and D. E. Dykhuizen.** 1984. The population genetics of *Escherichia coli. Annu. Rev. Genet.* **18**:31–68.
54. **Hartl, D. L., D. E. Dykhuizen, and D. E. Berg.** 1984. Accessory DNAs in the bacterial gene pool: playground for coevolution, p. 233–245. *In* D. Evered and G. M. Collins

(ed.), *Ciba Symposium 102: Origins and Development of Adaptations*. Pitman Books, London.

55. **Hartl, D. L., D. E. Dykhuizen, R. D. Miller, L. Green, and J. de Framond.** 1983. Transposable element IS*50* improves growth rate of *E. coli* cells without transposition. *Cell* **35:**503–510.
56. **Hartl, D. L., M. Medhora, L. Green, and D. E. Dykhuizen.** 1986. The evolution of DNA sequences in *Escherichia coli. Phil. Trans. R. Soc. London Ser. B* **312:**191–204.
57. **Hartl, D. L., and S. A. Sawyer.** 1988. Multiple correlations among insertion sequences in the genome of *Escherichia coli*, p. 91–106. *In* A. J. Kingsman, S. M. Kingsman, and K. F. Chater (ed.), *Transposition*. SGM Symposium Series, vol 43. Cambridge University Press, Cambridge.
58. **Hartl, D. L., and S. A. Sawyer.** 1988. Why do unrelated insertion sequences occur together in the genome of *Escherichia coli*? *Genetics* **118:**537–541.
59. **Hu, M., and R. C. Deonier.** 1981. Comparison of IS*1*, IS2, and IS*3* copy number in *Escherichia coli* strains K-12, B and C. *Gene* **16:**161–170.
60. **Iida, S., J. Meyer, and W. Arber.** 1983. Prokaryotic IS elements, p. 159–221. *In* J. A. Shapiro (ed.), *Mobile Genetic Elements*. Academic Press, Inc., New York.
61. **Ives, P. T.** 1945. The genetic structure of American populations of *Drosophila melanogaster*. *Genetics* **30:**167–196.
62. **Jack, J.** 1985. Molecular organization of the *cut* locus of *Drosophila melanogaster*. *Cell* **42:**869–876.
63. **Jaurin, B., and S. Normark.** 1983. Insertion of IS2 creates a novel *ampC* promoter in *Escherichia coli*. *Cell* **32:**809–816.
64. **Kaplan, N. L., and J. Y. F. Brookfield.** 1983. Transposable elements in Mendelian populations. III. Statistical results. *Genetics* **104:**485–495.
65. **Kleckner, N.** 1981. Transposable elements in prokaryotes. *Annu. Rev.. Genet.* **15:**341–404.
66. **Langer-Safer, P. R., M. Levine, and D. Ward.** 1982. Immunological method for mapping genes on *Drosophila* polytene chromosomes. *Proc. Natl. Acad. Sci. USA* **79:**4381–4385.
67. **Langley, C. H., J. Y. F. Brookfield, and N. Kaplan.** 1983. Transposable elements in Mendelian populations. III. A theory. *Genetics* **104:**457–451.
68. **Langley, C. H., E. Montgomery, R. Hudson, N. Kaplan, and B. Charlesworth.** 1988. On the role of unequal exchange in the containment of transposable element copy number. *Genet. Res.*, in press.
69. **Lefevre, G., and W. Watkins.** 1986. The question of the total gene number in *Drosophila melanogaster*. *Genetics* **113:** 869–895.
70. **Leight-Brown, A. J., and J. E. Moss.** 1987. Transposition of the I-element and *copia* in a natural population of *Drosophila melanogaster*. *Genet. Res.* **49:**121–128.
71. **Lewis, A. P., and J. Y. F. Brookfield.** 1987. Movement of *Drosophila melanogaster* transposable elements other than P-elements in a P-M hybrid dysgenic cross. *Mol. Gen. Genet.* **208:**506–510.
72. **Lewontin, R. C.** 1974. *The Genetic Basis of Evolutionary Change*. Columbia University Press, New York.
73. **Lim, J. K., M. J. Simmons, J. D. Raymond, N. M. Cox, R. F. Doll, and T. P. Culbert.** 1983. Homologue destabilization by a putative transposable element in *Drosophila melanogaster*. *Proc. Natl. Acad. Sci. USA* **80:**6624–6627.
74. **Lin, L., R. Bitner, and G. Edlin.** 1977. Increased reproductive fitness of *Escherichia coli* lambda lysogens. *J. Virol.* **21:**554–559.
75. **Lindsley, D. L., and L. Sandler.** 1977. The genetic analysis of meiosis in female *Drosophila*. *Phil. Trans. R. Soc. London Ser. B* **227:**295–312.
76. **Lindsley, D. L., L. Sandler, B. S. Baker, A. T. C. Carpenter, R. E. Denell, J. C. Hall, P. A. Jacobs, G. L. Miklos, B. K. Davis, R. C. Gethmana, R. W. Hardy, A. Hessler, S. M. Miller, H. Nozawa, D. M. Parry, and M. Gould-Somero.** 1972. Segmental aneuploidy and the gross genetic structure of the *Drosophila* genome. *Genetics* **71:**157–184.
77. **Manning, J. E., C. W. Schmid, and N. Davidson.** 1975. Interspersion of repetitive and non-repetitive sequences in the *Drosophila melanogaster* genome. *Cell* **4:**141–155.
78. **Maynard Smith, J.** 1976. Group selection. *Q. Rev. Biol.* **51:**277–283.
79. **Miller, R. D., D. E. Dykhuizen, L. Green, and D. L. Hartl.** 1984. Specific deletion occurring in the directed evolution of 6-phosphogluconate dehydrogenase in *Escherichia coli. Genetics* **108:**765–772.
80. **Mizrokhi, L. J., L. A. Obolenkova, A. F. Primagi, Y. V. Ilyin, T. I. Gerasimova, and G. P. Georgiev.** 1985. The nature of unstable insertion mutations and reversions in the locus *cut* of *D. melanogaster*: molecular mechanism of transposon memory. *EMBO J.* **4:**3781–3787.
81. **Montgomery, E. A., B. Charlesworth, and C. H. Langley.** 1987. A test for the role of natural selection in the stabilization of transposable element copy number in a population of *Drosophila melanogaster*. *Genet. Res.* **49:**31–41.
82. **Montgomery, E. A., and C. H. Langley.** 1983. Transposable elements in Mendelian populations. II. Distribution of three copia-like elements in a natural population of *Drosophila melanogaster*. *Genetics* **104:**473–483.
83. **Mukai, T.** 1964. The genetic structure of natural populations of *Drosophila melanogaster*. I. Spontaneous mutation rate of polygenes controlling viability. *Genetics* **50:**1–19.
84. **Mukai, T., S. I. Chigusa, L. E. Mettler, and J. F. Crow.** 1972. Mutation rate and dominance of genes affecting viability in *Drosophila melanogaster*. *Genetics* **72:**335–355.
85. **Muller, H. J.** 1923. Mutation. *Eugen. Genet. Family* **1:** 106–112.
86. **Nevers, P., and H. Saedler.** 1977. Transposable elements as agents of instability and chromosomal rearrangements. *Nature* (London) **268:**109–115.
87. **Nyman, K., H. Ohtsubo, D. Davidson, and E. Ohtsubo.** 1983. Distribution of insertion element IS*1* in natural isolates of *Escherichia coli*. *Mol. Gen. Genet.* **189:**516–518.
88. **Ochman, H., and R. K. Selander.** 1984. Standard reference strains of *Escherichia coli* from natural populations. *J. Bacteriol.* **157:**690–693.
89. **Ohtsubo, H., K. Nyman, W. Doroszkiewicz, and E. Ohtsubo.** 1981. Multiple copies of iso-insertion sequence IS*1* in *Shigella dysenteriae* chromosomes. *Nature* (London) **292:** 640–643.
90. **Orgel, L. E., and F. H. C. Crick.** 1980. Selfish DNA: the ultimate parasite. *Nature* (London) **284:**604–607.
91. **Potter, S. S., W. J. Brorein, Jr., P. Dunsmuir, and G. M. Rubin.** 1979. Transposition of elements of the *412*, *copia* and *297* dispersed repeated gene families in *Drosophila*. *Cell* **17:**415–427.
92. **Reynolds, A. E., J. Felton, and A. Wright.** 1981. Insertion of DNA activates the cryptic *bgl* operon in *E. coli*. *Nature* (London) **293:**625–629.
93. **Ronsseray, S., and D. Anxolabehere.** 1987. Chromosomal distribution of P and I transposable elements in a natural population of *Drosophila melanogaster*. *Chromosoma* **94:** 433–440.
94. **Rubin, G. M., M. G. Kidwell, and P. M. Bingham.** 1982. The

molecular basis of P-M hybrid dysgenesis: the nature of induced mutations. *Cell* **29:**987–994.

95. **Saedler, H., G. Cornelis, J. Cullum, B. Schumacher, and H. Sommer.** 1980. Is*1*-mediated rearrangements. *Cold Spring Harbor Symp. Quant. Biol.* **45:** 93–98.
96. **Saedler, H., H. J. Reif, S. Hu, and N. Davidson.** 1974. IS2, a genetic element for turn-off and turn-on of gene activity in *E. coli. Mol. Gen. Genet.* **132:**265–289.
97. **Sawyer, S. A., D. E. Dykhuizen, R. F. BuBose, L. Green, T. Mutangadura-Mhlanga, D. F. Wolczyk, and D. L. Hartl.** 1987. Distribution and abundance of insertion sequences among natural isolates of *Escherichia coli. Genetics* **115:** 51–63.
98. **Sawyer, S. A., and D. L. Hartl.** 1986. Distribution of transposable elements in prokaryotes. *Theor. Popul. Biol.* **30:**1–17.
99. **Schoner, B., and R. G. Schoner.** 1981. Distribution of IS*5* in bacteria. *Gene* **16:**347–352.
100. **Scordilis, G. E., H. Ree, and T. G. Lessie.** 1987. Identification of transposable elements which activate gene expression in *Pseudomonas cepacia. J. Bacteriol.* **169:**8–13.
101. **Silver, R. P., W. Aaronson, A. Sutton, and R. Schneerson.** 1980. Comparative analysis of plasmids and some metabolic characteristics of *Escherichia coli* K1 from diseased and healthy individuals. *Infect. Immun.* **29:**200–206.
102. **Simmons, M. J., and J. F. Crow.** 1977. Mutations affecting fitness in *Drosophila* populations. *Annu. Rev. Genet.* **11:** 49–78.
103. **Simmons, M. J., J. D. Raymond, N. A. Johnson, and T. M. Fahey.** 1984. A comparison of mutation rates for specific loci and chromosome regions in dysgenic hybrid males of *Drosophila melanogaster. Genetics* **106:**85–94.
104. **Strobel, E., P. Dunsmuir, and G. M. Rubin.** 1979. Polymorphisms in the chromosomal locations of elements of the *412*, *copia* and *297* dispersed repeated gene families in *Drosophila. Cell* **17:**429–439.
105. **Syvanen, M.** 1984. The evolutionary implications of mobile genetic elements. *Annu. Rev. Genet.* **18:** 271–293.
106. **Truett, M. A., R. S. Jones, and S. S. Potter.** 1981. Unusual structure of the FB family of transposable elements in *Drosophila. Cell* **24:**753–763.
107. **Whittam, T. S., H. Ochman, and R. K. Selander.** 1983. Multilocus genetic structure in natural populations of *Escherichia coli. Proc. Natl. Acad. Sci. USA* **80:**1751–1755.
108. **Woodruff, R. C., J. L. Blount, and J. N. Thompson, Jr.** 1987. Hybrid dysgenesis in *D. melanogaster* is not a general release mechanism for DNA transposition. *Science* **237:**1206–1208.
109. **Yamaguchi, O., T. Yamazaki, K. Saigo, T. Mukai, and A. Robertson.** 1987. Distributions of three transposable elements, *P*, *297*, and *copia*, in natural populations of *Drosophila melanogaster. Jpn. J. Genet.* **62:**205–216.
110. **Yannopoulos, G., N. Stamatis, M. Monastirioti, P. Hatzopoulos, and C. Louis.** 1987. *hobo* is responsible for the induction of hybrid dysgenesis by strains of *Drosophila melanogaster* bearing recombination factor 23.5MRF. *Cell* **49:**487–495.
111. **Young, M. W.** 1979. Middle repetitive DNA: a field component of the *Drosophila* genome. *Proc. Natl. Acad. Sci. USA* **76:**6274–6278.
112. **Yukuhiro, K., K. Harada, and T. Mukai.** 1985. Viability mutations induced by P-elements in *Drosophila melanogaster. Jpn. J. Genet.* **60:**531–537.
113. **Yukuhiro, K., and T. Mukai.** 1986. Increased detrimental load possibly caused by a transposon in a local population of *Drosophila melanogaster. Jpn. J. Genet.* **61:**24–43.
114. **Zafarullah, M., D. Charlier, and N. Glansdorff.** 1981. Insertion of IS*3* can "turn on" a silent gene in *Escherichia coli. J. Bacteriol.* **146:**415–417.

INDEX